Contents—1995 HVAC APPLICATIONS

Contents—1994 REFRIGERATION

1997 ASHRAE® HANDBOOK

FUNDAMENTALS

Inch-Pound Edition

American Society of Heating, Refrigerating and Air-Conditioning Engineers, Inc.

1791 Tullie Circle, N.E., Atlanta, GA 30329

(404) 636-8400 http://www.ashrae.org

DEDICATED

TO THE ADVANCEMENT OF

THE PROFESSION

AND ITS ALLIED INDUSTRIES

Comments, criticisms, and suggestions regarding the subject matter are invited. Any errors or omissions in the data should be brought to the attention of the Editor. If required, an errata sheet will be issued at approximately the same time as the next Handbook. Notice of any significant errors found after that time will be published in the *ASHRAE Journal.*

ISBN 1-883413-44-3

CONTENTS

DUCT AND PIPE DESIGN

GENERAL

INDEX

CONTRIBUTORS

In addition to the Technical Committees, the following individuals contributed significantly to this volume. The appropriate chapter numbers follow each contributor's name.

Roy R. Crawford (1, 6)
Carrier Corporation

Keith E. Herold (1, 19)
University of Maryland at College Park

Anthony M. Jacobi (1)
University of Illinois

Thomas H. Kuehn (1)
University of Minnesota

Horacio Perez-Blanco (1)
Pennsylvania State University

William J. Plzak (1)
The Trane Company

Rick J. Couvillion (2)
University of Arkansas

Arthur E. Bergles (3)
Rensselaer Polytechnic Institute

Michael M. Ohadi (3, 4, 5)
University of Maryland

David M. Pratt (3, 4)
Wright Laboratory

Naim Z. Azer (4)

Albert C. Kent (5)
Southern Illinois University

Russell A. Cooper (7)
Jaffe Holden Scarbrough Acoustics, Inc.

Richard J. Peppin (7)
Larson-Davis

Mark E. Schaffer (7)
McKay Conant Brook

Larry G. Berglund (8)
Tohoku University

John R. Breckenridge (8)

Joseph A. Yoshida (8)
Nihon University, College of Science & Technology

Michael S. Crandall (9)
M.S. Crandall Group, Inc.

Michael J. Hodgson (9)
University of Connecticut Health Center

Albert J. Heber (10)
Purdue University

Gerald L. Riskowski (10)
University of Illinois

Yuanhui Zhang (10)
University of Illinois

Roger C. Brook (11)
Michigan State University

Brian C. Krafthefer (13)
Honeywell, Inc.

Carolyn Margaret Kerr (13)
InAIR Environmental, Ltd.

Richard D. Rivers (13)
Environmental Quality Sciences, Inc.

Kendall K. Brown (14)
University of Alabama in Huntsville

Richard T. Stonier (14)
Solomat

Barry Taylor (14)
National Institute of Standards and Technology

Helmut Feustel (15)
Lawrence Berkeley National Laboratory

Michael Lepage (15)
RWDI

Michael A. Ratcliff (15)
Ratcliff Associates

David J. Wilson (15)
University of Alberta

David L. Grumman (16)
Grumman/Butkus Associates

Thomas A. Butcher (17)
Brookhaven National Laboratory

Douglas W. DeWerth (17)
AGA Research

Richard F. Krajewski (17)
Brookhaven National Laboratory

Earl M. Clark (18)
DuPont Fluorochemicals

Mark O. McLinden (18, 19)
National Institute of Standards and Technology

Eric W. Lemmon (19)
National Institute of Standards and Technology

Rajiv R. Singh (19)
AlliedSignal, Inc.

Anthony M. Deroo (20)
Dow Chemical Company

Lewis G. Harriman III (21, 26)
Mason-Grant Company

Andre O. Desjarlais (22, 23)
Oak Ridge National Laboratory

William P. Goss (22, 24)
University of Massachusetts

Hugo Hens (22)
University of Leuven

Vagn Korsgaard (22)
Hygrowick-International

Anton TenWolde (22, 23)
USDA Forest Service

Harold A. Trethowen (22, 23)
Building Research Association of New Zealand

Jeffrey E. Christian (23, 39)
Oak Ridge National Laboratory

William B. Rose (23)
University of Illinois

Brian A. Rock (25)
University of Kansas

Max H. Sherman (25)
Lawrence Berkeley National Laboratory

Grenville K. Yuill (25)
Pennsylvania State University

Donald G. Colliver (26)
University of Kentucky

Drury B. Crawley (26)
U.S. Department of Energy

Robert J. Morris (26)
Environment Canada

Marc S. Plantico (26)
National Climatic Data Center

James H. Norman (27)
AAA Enterprises

Thomas B. Romine, Jr. (28)
Romine, Romine & Burgess, Inc.

Christopher J. Barry (29)
Libby Owens Ford Company

Stephen C. Carpenter (29)
Enermodal Engineering, Ltd.

Brian P. Crooks (29)
Cardinal IG

Dragan Curcija (29)
Carli, Inc.

William C. duPont (29)
Lawrence Berkeley National Laboratory

Hakim Elmahdy (29)
National Research Council

John F. Hogan (29)
Seattle Department of Construction and Land Use

Joseph H. Klems (29)
Lawrence Berkeley National Laboratory

Michael E. McCabe (29)
National Institute of Standards and Technology

CONTRIBUTORS (*Concluded*)

Ross McCluney (29)
Florida Solar Energy Center

Mary Susan Reilly (29)
Enermodal Engineering, Inc.

David B. Tait (29)
Tait Solar Company

John L. Wright (29)
University of Waterloo

Michael J. Brandemuehl (30)
University of Colorado

Jeff S. Haberl (30)
Texas A&M University

Jan F. Kreider (30)
University of Colorado

Jeffrey D. Spitler (30)
Oklahoma State University

Fred S. Bauman (31)
University of California

Leslie L. Christianson (31)
University of Illinois

Herman F. Behls (32)
Behls and Associates

Patrick J. Brooks (32)
United McGill Corporation

Robert J. Tsal (32)
NETSAL & Associates

Albert W. Black (33)
McClure Engineering Associates

William J. Coad (33)
McClure Engineering Associates

Richard W. Eiden (37)
Richard Eiden Associates

Ravisankar Ganta (37)
Bechtel

John R. Sosoka (37)
PSI Engineers, Inc.

M. Peter Scofield (39)
Building Energy Associates

Adrian N. Tuluca (39)
Steven Winters Associates

ASHRAE HANDBOOK COMMITTEE

Donald M. Eppelheimer, Chair

1997 Fundamentals Volume Subcommittee: **Harold G. Lorsch,** Chair

Larry G. Berglund **Brian C. Krafthefer** **Ramon Pons** **George Reeves**

ASHRAE HANDBOOK STAFF

Robert A. Parsons, Editor **Adele J. Brandstrom,** Associate Editor
Christina D. Tate, Assistant Editor

Scott A. Zeh, Nancy F. Thysell, and **Stefan R. Moore,** Publishing Services

Frank M. Coda, Publisher

W. Stephen Comstock, Director
Communications and Publications

ASHRAE TECHNICAL COMMITTEES AND TASK GROUPS

SECTION 1.0—FUNDAMENTALS AND GENERAL

1.1 Thermodynamics and Psychrometrics
1.2 Instruments and Measurements
1.3 Heat Transfer and Fluid Flow
1.4 Control Theory and Application
1.5 Computer Applications
1.6 Terminology
1.7 Operation and Maintenance Management
1.8 Owning and Operating Costs
1.9 Electrical Systems
1.10 Energy Resources

SECTION 2.0—ENVIRONMENTAL QUALITY

2.1 Physiology and Human Environment
2.2 Plant and Animal Environment
2.3 Gaseous Air Contaminants and Gas Contaminant Removal Equipment
2.4 Particulate Air Contaminants and Particulate Contaminant Removal Equipment
2.6 Sound and Vibration Control
2.7 Seismic Restraint Design
TG Global Climate Change

SECTION 3.0—MATERIALS AND PROCESSES

3.1 Refrigerants and Brines
3.2 Refrigerant System Chemistry
3.3 Refrigerant Contaminant Control
3.4 Lubrication
3.5 Desiccant and Sorption Technology
3.6 Corrosion and Water Treatment
3.8 Refrigerant Containment

SECTION 4.0—LOAD CALCULATIONS AND ENERGY REQUIREMENTS

4.1 Load Calculation Data and Procedures
4.2 Weather Information
4.3 Ventilation Requirements and Infiltration
4.4 Thermal Insulation and Moisture Retarders
4.5 Fenestration
4.6 Building Operation Dynamics
4.7 Energy Calculations
4.9 Building Envelope Systems
4.10 Indoor Environmental Modeling
TG Smart Building Systems

SECTION 5.0—VENTILATION AND AIR DISTRIBUTION

5.1 Fans
5.2 Duct Design
5.3 Room Air Distribution
5.4 Industrial Process Air Cleaning (Air Pollution Control)
5.5 Air-to-Air Energy Recovery
5.6 Control of Fire and Smoke
5.7 Evaporative Cooling
5.8 Industrial Ventilation
5.9 Enclosed Vehicular Facilities
5.10 Kitchen Ventilation

SECTION 6.0—HEATING EQUIPMENT, HEATING AND COOLING SYSTEMS AND APPLICATIONS

6.1 Hydronic and Steam Equipment and Systems
6.2 District Heating and Cooling
6.3 Central Forced Air Heating and Cooling Systems
6.4 In-Space Convection Heating
6.5 Radiant Space Heating and Cooling
6.6 Service Water Heating
6.7 Solar Energy Utilization
6.8 Geothermal Energy Utilization
6.9 Thermal Storage
6.10 Fuels and Combustion

SECTION 7.0—PACKAGED AIR-CONDITIONING AND REFRIGERATION EQUIPMENT

7.1 Residential Refrigerators and Food Freezers
7.4 Unitary Combustion-Engine-Driven Heat Pumps
7.5 Room Air Conditioners and Dehumidifiers
7.6 Unitary Air Conditioners and Heat Pumps

SECTION 8.0—AIR-CONDITIONING AND REFRIGERATION SYSTEM COMPONENTS

8.1 Positive Displacement Compressors
8.2 Centrifugal Machines
8.3 Absorption and Heat Operated Machines
8.4 Air-to-Refrigerant Heat Transfer Equipment
8.5 Liquid-to-Refrigerant Heat Exchangers
8.6 Cooling Towers and Evaporative Condensers
8.7 Humidifying Equipment
8.8 Refrigerant System Controls and Accessories
8.10 Pumps and Hydronic Piping
8.11 Electric Motors—Open and Hermetic

SECTION 9.0—AIR-CONDITIONING SYSTEMS AND APPLICATIONS

9.1 Large Building Air-Conditioning Systems
9.2 Industrial Air Conditioning
9.3 Transportation Air Conditioning
9.4 Applied Heat Pump/Heat Recovery Systems
9.5 Cogeneration Systems
9.6 Systems Energy Utilization
9.7 Testing and Balancing
9.8 Large Building Air-Conditioning Applications
9.9 Building Commissioning
9.10 Laboratory Systems
9.11 Clean Spaces
TG Combustion Gas Turbine Inlet Air Cooling Systems
TG Tall Buildings

SECTION 10.0—REFRIGERATION SYSTEMS

10.1 Custom Engineered Refrigeration Systems
10.2 Automatic Icemaking Plants and Skating Rinks
10.3 Refrigerant Piping, Controls, and Accessories
10.4 Ultra-Low Temperature Systems and Cryogenics
10.5 Refrigerated Distribution and Storage Facilities
10.6 Transport Refrigeration
10.7 Commercial Food and Beverage Cooling, Display and Storage
10.8 Refrigeration Load Calculations
10.9 Refrigeration Application for Foods and Beverages

PREFACE

The Fundamentals Handbook covers basic principles and includes data for the entire technology of the HVAC&R industry. Although design data and information change little over time, research sponsored by ASHRAE and others continues to generate new information for the ASHRAE Handbooks. In addition, the technical committees that prepare the chapters strive not only to provide new information, but also to clarify existing information, delete obsolete material, and reorganize chapters to make the information more understandable and easier to use. In this *1997 ASHRAE Handbook* the following changes and additions are worth noting.

- Chapter 1, Thermodynamics and Refrigeration Cycles, has an improved style of analysis and method of calculating thermodynamic properties. The chapter also discusses zeotropic refrigerant mixtures and includes numerical examples to show how the second law of thermodynamics can be applied to actual refrigeration cycles. The information on absorption refrigeration cycles has been clarified, and example analyses of various absorption cycles are included.
- Chapter 6, Psychrometrics, now includes equations for calculating standard pressure when elevation or temperature is known.
- Chapter 8, Thermal Comfort, includes more information on thermoregulation to help in understanding the physiology underlying comfort and its relation to the thermal environment. New material about the effects of clothing insulation and the perception of draft discomfort is also included.
- Chapter 9, Indoor Environmental Health, has been substantially rewritten. It now includes new information that describes the various health sciences, summarizes diseases associated with the indoor environment, compares pertinent indoor air quality standards, and introduces the principles of industrial hygiene.
- Chapter 14, Measurement and Instruments, includes new sections on carbon dioxide measurement and data logging devices.
- Chapter 15, Airflow Around Buildings, presents simplified methods for estimating the effect of changes in terrain on wind speed profiles. A field-validated model has led to a new procedure for calculating rooftop exhaust stack height. The new model produces stack heights that are about two-thirds as high as the 1993 *ASHRAE Handbook* requirement.
- Chapter 16, Energy Resources, has been moved here from the 1995 *ASHRAE Handbook—Applications*. Basic energy data have been updated.
- Chapter 17, Combustion and Fuels, now includes information on NO_x emissions from uncontrolled fuel-burning equipment. Methods for reducing NO_x are also discussed.
- Chapter 19, Thermophysical Properties of Refrigerants, now includes data for the zeotropic blends R-404A, R-404C, and R-410A, and the azeotropic blend R-507A. Most of the CFC refrigerants have been retained to assist in making comparisons. Revised formulations have been used for most of the hydrocarbon refrigerants and the cryogenic fluids.
- Chapters 22 and 23, Thermal and Moisture Control in Insulated Assemblies, contain more information on moisture transport and control. The effects of moisture on the building and its occupants are discussed in more detail. New recommendations and construction details for moisture control in three types of climates and in attics, roofs, and crawl spaces are included.
- Chapter 25, Ventilation and Infiltration, now has information on nonresidential ventilation, infiltration degree-days, air change effectiveness, and age of air.
- Chapter 26, Climatic Design Information, has been substantially expanded. The chapter includes new heating, cooling, dehumidification, and wind design conditions for 1442 locations.
- Chapter 29, Fenestration, includes new models for calculating heat transfer in glazing cavities. The solar heat gain section has been rewritten. New sections on condensation resistance, complex shading systems, annual energy performance, and durability have been added.
- Chapter 30, Energy Estimating and Modeling Methods, has been substantially rewritten. The chapter provides an overview of the various methods available for estimating energy use. A sample heat balance calculation is included.
- Chapter 32, Duct Design, has additional information on the thermal gravity (stack) effect and on duct system leakage.
- Chapter 33, Pipe Sizing, includes a new section on steam condensate systems.
- Chapter 37, Fundamentals of Control, has been taken from Chapter 42 in the 1995 *ASHRAE Handbook—Applications*. This move divides the information on controls into two topic areas—one covers fundamentals and the other covers the applications of controls.
- Chapter 39, Building Envelopes, is a new chapter that will be moved to the 1999 *ASHRAE Handbook—Applications*. The technical committee completed the chapter in 1996, and the Handbook Committee decided to place it temporarily in this volume.
- Old Chapter 29, Cooling and Freezing Times of Foods, and Chapter 30, Thermal Properties of Foods, are not included in this Handbook. They will be included in the 1998 *ASHRAE Handbook—Refrigeration*.

Each Handbook is published in two editions. One edition contains inch-pound (I-P) units of measurement, and the other contains the International System of Units (SI).

Look for corrections to the 1994, 1995, and 1996 volumes of the Handbook series that have been noted since March 1995 on the Internet at http://www.ashrae.org. Any changes to this volume will be reported in the 1998 *ASHRAE Handbook* and on the Internet.

If you have suggestions on improving a chapter or you would like more information on how you can help revise a chapter, e-mail bparsons@ashrae.org; write to Handbook Editor, ASHRAE, 1791 Tullie Circle, Atlanta, GA 30329; or fax (404) 321-5478.

Robert A. Parsons
ASHRAE Handbook Editor

CHAPTER 1

THERMODYNAMICS AND REFRIGERATION CYCLES

THERMODYNAMICS is the study of energy, its transformations, and its relation to states of matter. This chapter covers the application of thermodynamics to refrigeration cycles. The first part reviews the first and second laws of thermodynamics and presents methods for calculating thermodynamic properties. The second and third parts address compression and absorption refrigeration cycles, the two most common methods of thermal energy transfer.

THERMODYNAMICS

A **thermodynamic system** is a region in space or a quantity of matter bounded by a closed surface. The surroundings include everything external to the system, and the system is separated from the surroundings by the system boundaries. These boundaries can be movable or fixed, real or imaginary.

The concepts that operate in any thermodynamic system are **entropy** and **energy**. Entropy measures the molecular disorder of a system. The more mixed a system, the greater its entropy; conversely, an orderly or unmixed configuration is one of low entropy. Energy has the capacity for producing an effect and can be categorized into either stored or transient forms as described in the following sections.

Stored Energy

Thermal (internal) energy is the energy possessed by a system caused by the motion of the molecules and/or intermolecular forces.

Potential energy is the energy possessed by a system caused by the attractive forces existing between molecules, or the elevation of the system.

$$\mathrm{PE} = mgz \tag{1}$$

where

m = mass
g = local acceleration of gravity
z = elevation above horizontal reference plane

Kinetic energy is the energy possessed by a system caused by the velocity of the molecules and is expressed as

$$\mathrm{KE} = mV^2/2 \tag{2}$$

where V is the velocity of a fluid stream crossing the system boundary.

The preparation of the first and second parts of this chapter is assigned to TC 1.1, Thermodynamics and Psychrometrics. The third part is assigned to TC 8.3, Absorption and Heat Operated Machines.

Chemical energy is energy possessed by the system caused by the arrangement of atoms composing the molecules.

Nuclear (atomic) energy is energy possessed by the system from the cohesive forces holding protons and neutrons together as the atom's nucleus.

Transient Energy

Heat (Q) is the mechanism that transfers energy across the boundary of systems with differing temperatures, always toward the lower temperature.

Work is the mechanism that transfers energy across the boundary of systems with differing pressures (or force of any kind), always toward the lower pressure. If the total effect produced in the system can be reduced to the raising of a weight, then nothing but work has crossed the boundary.

Mechanical or **shaft work** (W) is the energy delivered or absorbed by a mechanism, such as a turbine, air compressor, or internal combustion engine.

Flow work is energy carried into or transmitted across the system boundary because a pumping process occurs somewhere outside the system, causing fluid to enter the system. It can be more easily understood as the work done by the fluid just outside the system on the adjacent fluid entering the system to force or push it into the system. Flow work also occurs as fluid leaves the system.

$$\text{Flow Work (per unit mass)} = pv \tag{3}$$

where p is the pressure and v is the specific volume, or the volume displaced per unit mass.

A **property** of a system is any observable characteristic of the system. The **state** of a system is defined by listing its properties. The most common thermodynamic properties are temperature T, pressure p, and specific volume v or density ρ. Additional thermodynamic properties include entropy, stored forms of energy, and enthalpy.

Frequently, thermodynamic properties combine to form other properties. **Enthalpy** (h), a result of combining properties, is defined as

$$h \equiv u + pv \tag{4}$$

where u is internal energy per unit mass.

Each property in a given state has only one definite value, and any property always has the same value for a given state, regardless of how the substance arrived at that state.

A **process** is a change in state that can be defined as any change in the properties of a system. A process is described by specifying the initial and final equilibrium states, the path (if identifiable), and the interactions that take place across system boundaries during the process.

A **cycle** is a process or a series of processes wherein the initial and final states of the system are identical. Therefore, at the conclusion of a cycle, all the properties have the same value they had at the beginning.

A **pure substance** has a homogeneous and invariable chemical composition. It can exist in more than one phase, but the chemical composition is the same in all phases.

If a substance exists as liquid at the saturation temperature and pressure, it is called **saturated liquid.** If the temperature of the liquid is lower than the saturation temperature for the existing pressure, it is called either a **subcooled liquid** (the temperature is lower than the saturation temperature for the given pressure) or a **compressed liquid** (the pressure is greater than the saturation pressure for the given temperature).

When a substance exists as part liquid and part vapor at the saturation temperature, its quality is defined as the ratio of the mass of vapor to the total mass. Quality has meaning only when the substance is in a saturated state; i.e., at saturation pressure and temperature.

If a substance exists as vapor at the saturation temperature, it is called **saturated vapor**. (Sometimes the term **dry saturated vapor** is used to emphasize that the quality is 100%.) When the vapor is at a temperature greater than the saturation temperature, it is **superheated vapor**. The pressure and temperature of superheated vapor are independent properties, since the temperature can increase while the pressure remains constant. Gases are highly superheated vapors.

FIRST LAW OF THERMODYNAMICS

The first law of thermodynamics is often called the **law of the conservation of energy**. The following form of the first law equation is valid only in the absence of a nuclear or chemical reaction.

Based on the first law or the law of conservation of energy for any system, open or closed, there is an energy balance as

$$\begin{bmatrix}\text{Net Amount of Energy}\\ \text{Added to System}\end{bmatrix} = \begin{bmatrix}\text{Net Increase in Stored}\\ \text{Energy of System}\end{bmatrix}$$

or

Energy In – Energy Out = Increase in Energy in System

Figure 1 illustrates energy flows into and out of a thermodynamic system. For the general case of multiple mass flows in and out of the system, the energy balance can be written

$$\sum m_{in}\left(u + pv + \frac{V^2}{2} + gz\right)_{in} - \sum m_{out}\left(u + pv + \frac{V^2}{2} + gz\right)_{out} + Q - W = \left[m_f\left(u + \frac{V^2}{2} + gz\right)_f - m_i\left(u + \frac{V^2}{2} + gz\right)_i\right]_{system} \quad (5)$$

The steady-flow process is important in engineering applications. Steady flow signifies that all quantities associated with the system do not vary with time. Consequently,

$$\sum_{\substack{\text{all streams}\\ \text{leaving}}} \dot{m}\left(h + \frac{V^2}{2} + gz\right) - \sum_{\substack{\text{all streams}\\ \text{entering}}} \dot{m}\left(h + \frac{V^2}{2} + gz\right) + \dot{Q} - \dot{W} = 0 \quad (6)$$

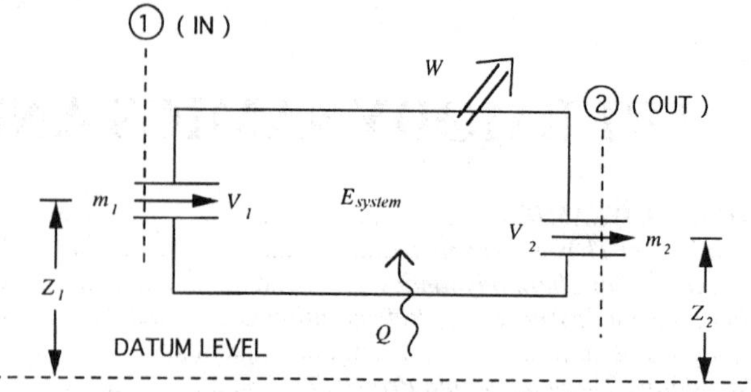

Fig. 1 Energy Flows in General Thermodynamic System

where $h = u + pv$ as described in Equation (4).

A second common application is the closed stationary system for which the first law equation reduces to

$$Q - W = [m(u_f - u_i)]_{system} \quad (7)$$

SECOND LAW OF THERMODYNAMICS

The second law of thermodynamics differentiates and quantifies processes that only proceed in a certain direction (irreversible) from those that are reversible. The second law may be described in several ways. One method uses the concept of entropy flow in an open system and the irreversibility associated with the process. The concept of irreversibility provides added insight into the operation of cycles. For example, the larger the irreversibility in a refrigeration cycle operating with a given refrigeration load between two fixed temperature levels, the larger the amount of work required to operate the cycle. Irreversibilities include pressure drops in lines and heat exchangers, heat transfer between fluids of different temperature, and mechanical friction. Reducing total irreversibility in a cycle improves the cycle performance.

In an open system, the second law of thermodynamics can be described in terms of entropy as

$$dS_{system} = \frac{\delta Q}{T} + \delta m_i s_i - \delta m_e s_e + dI \quad (8)$$

where

dS_{system} = total change within system in time dt during process
$\delta m_i s_i$ = entropy increase caused by mass entering
$\delta m_e s_e$ = entropy decrease caused by mass leaving
$\delta Q/T$ = entropy change caused by reversible heat transfer between system and surroundings
dI = entropy created caused by irreversibilities (always positive)

Equation (8) accounts for all entropy changes in the system. Rearranged, this equation becomes

$$\delta Q = T[(\delta m_e s_e - \delta m_i s_i) + dS_{sys} - dI] \quad (9)$$

In integrated form, if inlet and outlet properties, mass flow, and interactions with the surroundings do not vary with time, the general equation for the second law is

$$(S_f - S_i)_{system} = \int_{rev} \frac{\delta Q}{T} + \sum(ms)_{in} - \sum(ms)_{out} + I \quad (10)$$

In many applications the process can be considered to be operating steadily with no change in time. The change in entropy of the system is therefore zero. The irreversibility rate, which is the rate of entropy production caused by irreversibilities in the process can be determined by rearranging Equation (10)

$$\dot{I} = \sum(\dot{m}s)_{out} - \sum(\dot{m}s)_{in} - \int \frac{\dot{Q}}{T_{surr}} \quad (11)$$

Equation (6) can be used to replace the heat transfer quantity. Note that the absolute temperature of the surroundings with which the system is exchanging heat is used in the last term. If the temperature of the surroundings is equal to the temperature of the system, the heat is transferred reversibly and Equation (11) becomes equal to zero.

Equation (11) is commonly applied to a system with one mass flow in, the same mass flow out, no work, and negligible kinetic or potential energy flows. Combining Equations (6) and (11) yields

$$\dot{I} = \dot{m}\left[(s_{out} - s_{in}) - \frac{h_{out} - h_{in}}{T_{surr}}\right] \quad (12)$$

In a cycle, the reduction of work produced by a power cycle or the increase in work required by a refrigeration cycle is equal to the absolute ambient temperature multiplied by the sum of the irreversibilities in all the processes in the cycle. Thus the difference in the reversible work and the actual work for any refrigeration cycle, theoretical or real, operating under the same conditions becomes

$$\dot{W}_{actual} = \dot{W}_{reversible} + T_0 \sum \dot{I} \quad (13)$$

THERMODYNAMIC ANALYSIS OF REFRIGERATION CYCLES

Refrigeration cycles transfer thermal energy from a region of low temperature T_R to one of higher temperature. Usually the higher temperature heat sink is the ambient air or cooling water. This temperature is designated as T_0, the temperature of the surroundings.

The first and second laws of thermodynamics can be applied to individual components to determine mass and energy balances and the irreversibility of the components. This procedure is illustrated in later sections in this chapter.

Performance of a refrigeration cycle is usually described by a **coefficient of performance**. COP is defined as the benefit of the cycle (amount of heat removed) divided by the required energy input to operate the cycle, or

$$\text{COP} \equiv \frac{\text{Useful refrigerating effect}}{\text{Net energy supplied from external sources}} \quad (14)$$

For a mechanical vapor compression system, the net energy supplied is usually in the form of work, mechanical or electrical, and may include work to the compressor and fans or pumps. Thus

$$\text{COP} = \frac{Q_i}{W_{net}} \quad (15)$$

In an absorption refrigeration cycle, the net energy supplied is usually in the form of heat into the generator and work into the pumps and fans, or

$$\text{COP} = \frac{Q_i}{Q_{gen} + W_{net}} \quad (16)$$

In many cases the work supplied to an absorption system is very small compared to the amount of heat supplied to the generator so the work term is often neglected.

Application of the second law to an entire refrigeration cycle shows that a completely reversible cycle operating under the same conditions has the maximum possible Coefficient of Performance. A measure of the departure of the actual cycle from an ideal reversible cycle is given by the **refrigerating efficiency**:

$$\eta_R = \frac{\text{COP}}{(\text{COP})_{rev}} \quad (17)$$

The Carnot cycle usually serves as the ideal reversible refrigeration cycle. For multistage cycles, each stage is described by a reversible cycle.

EQUATIONS OF STATE

The equation of state of a pure substance is a mathematical relation between pressure, specific volume, and temperature. When the system is in thermodynamic equilibrium

$$f(p, v, T) = 0 \quad (18)$$

The principles of statistical mechanics are used to (1) explore the fundamental properties of matter, (2) predict an equation of state based on the statistical nature of a particulate system, or (3) propose a functional form for an equation of state with unknown parameters that are determined by measuring thermodynamic properties of a substance. A fundamental equation with this basis is the **virial equation**. The virial equation is expressed as an expansion in pressure p or in reciprocal values of volume per unit mass v as

$$\frac{pv}{RT} = 1 + B'p + C'p^2 + D'p^3 + \ldots \quad (19)$$

$$\frac{pv}{RT} = 1 + (B/v) + (C/v^2) + (D/v^3) + \ldots \quad (20)$$

where coefficients B', C', D', etc., and B, C, D, etc., are the virial coefficients. B' and B are second virial coefficients; C' and C are third virial coefficients, etc. The virial coefficients are functions of temperature only, and values of the respective coefficients in Equations (19) and (20) are related. For example, $B' = B/RT$ and $C' = (C - B^2)/(RT)^2$.

The ideal gas constant R is defined as

$$R = \lim_{p \to 0} \frac{(pv)_T}{T_{tp}} \quad (21)$$

where $(pv)_T$ is the product of the pressure and the volume along an isotherm, and T_{tp} is the defined temperature of the triple point of water, which is 491.69°R. The current best value of R is 1545.32 ft·lb$_f$/(lb mole·°R).

The quantity pv/RT is also called the **compressibility factor**, i.e. $Z = pv/RT$ or

$$Z = 1 + (B/v) + (C/v^2) + (D/v^3) + \ldots \quad (22)$$

An advantage of the virial form is that statistical mechanics can be used to predict the lower order coefficients and provide physical significance to the virial coefficients. For example, in Equation (22), the term B/v is a function of interactions between two molecules, C/v^2 between three molecules, etc. Since the lower order interactions are common, the contributions of the higher order terms are successively less. Thermodynamicists use the partition or distribution function to determine virial coefficients; however, experimental values of the second and third coefficients are preferred. For dense fluids, many higher order terms are necessary that can neither be satisfactorily predicted from theory nor determined from experimental measurements. In general, a truncated virial expansion of four terms is valid for densities of less than one-half the value at the critical point. For higher densities, additional terms can be used and determined empirically.

Digital computers allow the use of very complex equations of state in calculating *p-v-T* values, even to high densities. The Benedict-Webb-Rubin (B-W-R) equation of state (Benedict et al. 1940) and the Martin-Hou equation (1955) have had considerable use, but should generally be limited to densities less than the critical value. Strobridge (1962) suggested a modified Benedict-Webb-Rubin relation that gives excellent results at higher densities and can be used for a *p-v-T* surface that extends into the liquid phase.

The B-W-R equation has been used extensively for hydrocarbons (Cooper and Goldfrank 1967):

$$P = (RT/v) + (B_oRT - A_o - C_o/T^2)/v^2 + (bRT - a)/v^3 + (a\alpha)/v^6 + [c(1 + \gamma/v^2)e^{(-\gamma/v^2)}]/v^3T^2 \qquad (23)$$

where the constant coefficients are A_o, B_o, C_o, a, b, c, α, γ.

The Martin-Hou equation, developed for fluorinated hydrocarbon properties, has been used to calculate the thermodynamic property tables in Chapter 17 and in ASHRAE *Thermodynamic Properties of Refrigerants* (Stewart et al. 1986). The Martin-Hou equation is as follows:

$$p = \frac{RT}{v-b} + \frac{A_2 + B_2T + C_2e^{(-kT/T_c)}}{(v-b)^2} + \frac{A_3 + B_3T + C_3e^{(-kT/T_c)}}{(v-b)^3} + \frac{A_4 + B_4T}{(v-b)^4} + \frac{A_5 + B_5T + C_5e^{(-kT/T_c)}}{(v-b)^5} + (A_6 + B_6T)e^{av} \qquad (24)$$

where the constant coefficients are A_i, B_i, C_i, k, b, and α.

Strobridge (1962) suggested an equation of state that was developed for nitrogen properties and used for most cryogenic fluids. This equation combines the B-W-R equation of state with an equation for high density nitrogen suggested by Benedict (1937). These equations have been used successfully for liquid and vapor phases, extending in the liquid phase to the triple-point temperature and the freezing line, and in the vapor phase from 18 to 1800°R, with pressures to 150,000 psi. The equation suggested by Strobridge is accurate within the uncertainty of the measured *p-v-T* data. This equation, as originally reported by Strobridge, is

$$p = RT\rho + \left[Rn_1T + n_2 + \frac{n_3}{T} + \frac{n_4}{T^2} + \frac{n_5}{T^4}\right]\rho^2 + (Rn_6T + n_7)\rho^3 + n_8T\rho^4 + \rho^3\left[\frac{n_9}{T^2} + \frac{n_{10}}{T^3} + \frac{n_{11}}{T^4}\right]\exp(-n_{16}\rho^2) + \rho^5\left[\frac{n_{12}}{T^2} + \frac{n_{13}}{T^3} + \frac{n_{14}}{T^4}\right]\exp(-n_{16}\rho^2) + n_{15}\rho^6 \qquad (25)$$

The 15 coefficients of this equation's linear terms are determined by a least-square fit to experimental data. Hust and Stewart (1966) and Hust and McCarty (1967) give further information on methods and techniques for determining equations of state.

In the absence of experimental data, Van der Waals' principle of corresponding states can predict fluid properties. This principle relates properties of similar substances by suitable reducing factors, i.e., the *p-v-T* surfaces of similar fluids in a given region are assumed to be of similar shape. The critical point can be used to define reducing parameters to scale the surface of one fluid to the dimensions of another. Modifications of this principle, as suggested by Kamerlingh Onnes, a Dutch cryogenic researcher, have been used to improve correspondence at low pressures. The principle of corresponding states provides useful approximations, and numerous modifications have been reported. More complex treatments for predicting property values, which recognize similarity of fluid properties, are by generalized equations of state. These equations ordinarily allow for adjustment of the *p-v-T* surface by introduction of parameters. One example (Hirschfelder et al. 1958) allows for departures from the principle of corresponding states by adding two correlating parameters.

CALCULATING THERMODYNAMIC PROPERTIES

While equations of state provide *p-v-T* relations, a thermodynamic analysis usually requires values for internal energy, enthalpy, and entropy. These properties have been tabulated for many substances, including refrigerants (See Chapters 6, 19, and 36) and can be extracted from such tables by interpolating manually or with a suitable computer program. This approach is appropriate for hand calculations and for relatively simple computer models; however, for many computer simulations, the overhead in memory or input and output required to use tabulated data can make this approach unacceptable. For large thermal system simulations or complex analyses, it may be more efficient to determine internal energy, enthalpy, and entropy using fundamental thermodynamic relations or curves fit to experimental data. Some of these relations are discussed in the following sections. Also, the thermodynamic relations discussed in those sections are the basis for constructing tables of thermodynamic property data. Further information on the topic may be found in references covering system modeling and thermodynamics (Stoecker 1989, Howell and Buckius 1992).

At least two intensive properties must be known to determine the remaining properties. If two known properties are either *p*, *v*, or *T* (these are relatively easy to measure and are commonly used in simulations), the third can be determined throughout the range of interest using an equation of state. Furthermore, if the specific heats at zero pressure are known, specific heat can be accurately determined from spectroscopic measurements using statistical mechanics (NASA 1971). Entropy may be considered a function of *T* and *p*, and from calculus an infinitesimal change in entropy can be written as follows:

$$ds = \left(\frac{\partial s}{\partial T}\right)_p dT + \left(\frac{\partial s}{\partial p}\right)_T dp \qquad (26)$$

Likewise, a change in enthalpy can be written as

$$dh = \left(\frac{\partial h}{\partial T}\right)_p dT + \left(\frac{\partial h}{\partial p}\right)_T dp \qquad (27)$$

Using the relation $Tds = dh - vdp$ and the definition of specific heat at constant pressure, $c_p \equiv (\partial h/\partial T)_p$, Equation (27) can be rearranged to yield

$$ds = \frac{c_p}{T}dT + \left[\left(\frac{\partial h}{\partial p}\right)_T - v\right]\frac{dp}{T} \qquad (28)$$

Equations (26) and (28) combine to yield $(\partial s/\partial T)_p = c_p/T$. Then, using the Maxwell relation $(\partial s/\partial p)_T = -(\partial v/\partial T)_p$, Equation (26) may be rewritten as

$$ds = \frac{c_p}{T}dT - \left(\frac{\partial v}{\partial T}\right)_p dp \tag{29}$$

This is an expression for an exact derivative, so it follows that

$$\left(\frac{\partial c_p}{\partial p}\right)_T = -T\left(\frac{\partial^2 v}{\partial T^2}\right)_p \tag{30}$$

Integrating this expression at a fixed temperature yields

$$c_p = c_{po} - \int_0^p T\left(\frac{\partial^2 v}{\partial T^2}\right) dp_T \tag{31}$$

where c_{p0} is the known zero pressure specific heat, and dp_T is used to indicate that the integration is performed at a fixed temperature. The second partial derivative of specific volume with respect to temperature can be determined from the equation of state. Thus, Equation (31) can be used to determine the specific heat at any pressure.

Using $Tds = dh - vdp$, Equation (29) can be written as

$$dh = c_p dT + \left[v - T\left(\frac{\partial v}{\partial T}\right)_p\right] dp \tag{32}$$

Equations (28) and (32) may be integrated at constant pressure to obtain

$$s(T_1, p_0) = s(T_0, p_0) + \int_{T_0}^{T_1} \frac{c_p}{T} dT_p \tag{33}$$

and

$$h(T_1, p_0) = h(T_0, p_0) + \int_{T_0}^{T_1} c_p dT \tag{34}$$

Integrating the Maxwell relation $(\partial s/\partial p)_T = -(\partial v/\partial T)_p$ gives an equation for entropy changes at a constant temperature as

$$s(T_0, p_1) = s(T_0, p_0) - \int_{p_0}^{p_1} \left(\frac{\partial v}{\partial T}\right)_p dp_T \tag{35}$$

Likewise, integrating Equation (32) along an isotherm yields the following equation for enthalpy changes at a constant temperature

$$h(T_0, p_1) = h(T_0, p_0) + \int_{p_0}^{p_1} \left[v - T\left(\frac{\partial v}{\partial T}\right)_p\right] dp \tag{36}$$

Internal energy can be calculated from $u = h - pv$.

Combinations (or variations) of Equations (33) through (36) can be incorporated directly into computer subroutines to calculate properties with improved accuracy and efficiency. However, these equations are restricted to situations where the equation of state is valid and the properties vary continuously. These restrictions are violated by a change of phase such as evaporation and condensation, which are essential processes in air-conditioning and refrigerating devices. Therefore, the Clapeyron equation is of particular value; for evaporation or condensation it gives

$$\left(\frac{dp}{dT}\right)_{sat} = \frac{s_{fg}}{v_{fg}} = \frac{h_{fg}}{Tv_{fg}} \tag{37}$$

where

s_{fg} = entropy of vaporization
h_{fg} = enthalpy of vaporization
v_{fg} = specific volume difference between vapor and liquid phases

If vapor pressure and liquid (or vapor) density data are known at saturation, and these are relatively easy measurements to obtain, then changes in enthalpy and entropy can be calculated using Equation (37).

Phase Equilibria for MultiComponent Systems

To understand phase equilibria, consider a container full of a liquid made of two components; the more volatile component is designated *i* and the less volatile component *j* (Figure 2A). This mixture is all liquid because the temperature is low—but not so low that a solid appears. Heat added at a constant pressure raises the temperature of the mixture, and a sufficient increase causes vapor to form, as shown in Figure 2B. If heat at constant pressure continues to be added, eventually the temperature will become so high that only vapor remains in the container (Figure 2C). A temperature–concentration (*T-x*) diagram is useful for exploring details of this situation.

Figure 3 is a typical *T-x* diagram valid at a fixed pressure. The case shown in Figure 2A, a container full of liquid mixture with mole fraction $x_{i,0}$ at temperature T_0, is point 0 on the *T-x* diagram. When heat is added, the temperature of the mixture increases. The point at which vapor begins to form is the **bubble point**. Starting at point 0, the first bubble will form at temperature T_1, designated by point 1 on the diagram. The locus of bubble points is the **bubble point curve**, which provides bubble points for various liquid mole fractions x_i.

When the first bubble begins to form, the vapor in the bubble may not have the *i* mole fraction found in the liquid mixture. Rather, the mole fraction of the more volatile species is higher in the vapor than in the liquid. Boiling prefers the more volatile species, and the *T-x* diagram shows this behavior. At T_1, the vapor forming bubbles have an *i* mole fraction of $y_{i,1}$. If heat continues to be added, this preferential boiling will deplete the liquid of species *i* and the temperature required to continue the process will increase. Again, the *T-x* diagram reflects this fact; at point 2 the *i* mole fraction in the liquid is reduced to $x_{i,2}$ and the vapor has a mole fraction of $y_{i,2}$. The temperature required to boil the mixture is increased to

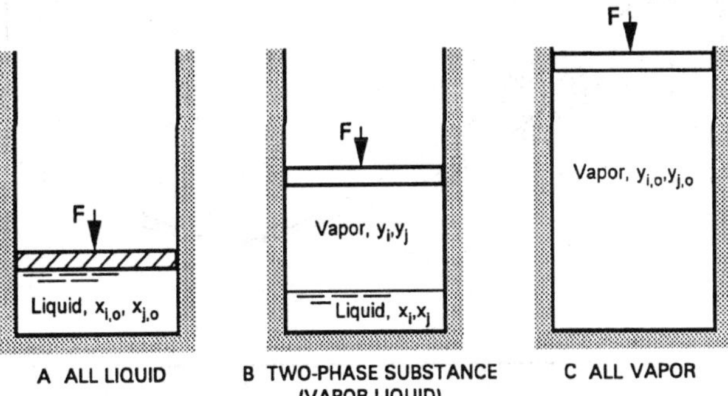

Fig. 2 Mixture of *i* and *j* Components in Constant Pressure Container

T_2. Position 2 on the *T-x* diagram could correspond to the physical situation shown in Figure 2B.

If the constant-pressure heating continues, all the liquid eventually becomes vapor at temperature T_3. At this point the *i* mole fraction in the vapor $y_{i,3}$ equals the starting mole fraction in the all liquid mixture $x_{i,1}$. This equality is required for mass and species conservation. Further addition of heat simply raises the vapor temperature. The final position 4 corresponds to the physical situation shown in Figure 2C.

Starting at position 4 in Figure 3, the removal of heat leads to 3, and further heat removal would cause droplets rich in the less volatile species to form. This point is called the **dew point**, and the locus of dew points is called the **dew-point curve**. The removal of heat will cause the mixture to reverse through points 3, 2, 1, and to starting point 0. Because the composition shifts, the temperature required to boil (or condense) this mixture changes as the process proceeds. This mixture is therefore called **zeotropic**.

Most mixtures have *T-x* diagrams that behave as previously described, but some have a markedly different feature. If the dew point and bubble point curves intersect at any point other than at their ends, the mixture exhibits what is called **azeotropic** behavior at that composition. This case is shown as position a in the *T-x* diagram of Figure 4. If a container of liquid with a mole fraction x_a were boiled, vapor would be formed with an identical mole fraction y_a. The addition of heat at constant pressure would continue with no shift in composition and no temperature glide.

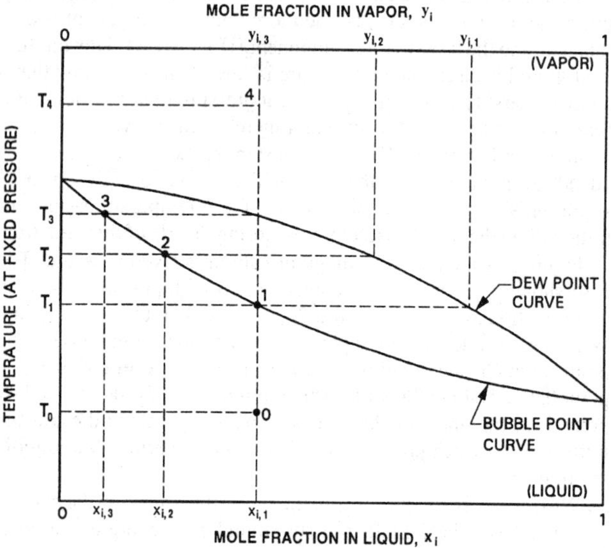

Fig. 3 Temperature-Concentration (*T-x*) Diagram for Zeotropic Mixture

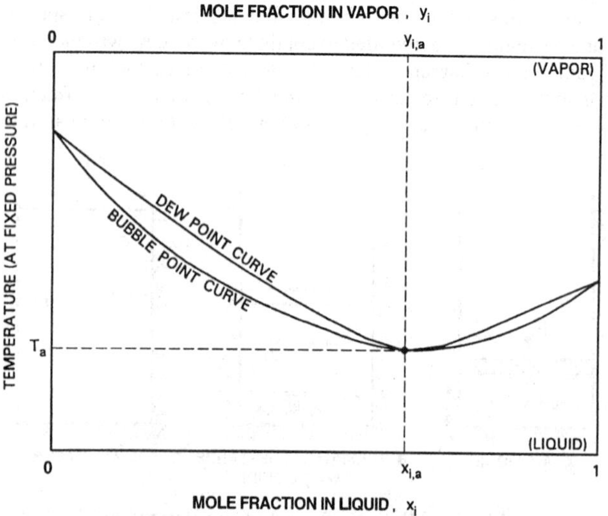

Fig. 4 Azeotropic Behavior Shown on *T-x* Diagram

Perfect azeotropic behavior is uncommon, while near azeotropic behavior is fairly common. The azeotropic composition is pressure dependent, so operating pressures should be considered for their impact on mixture behavior. Azeotropic and near-azeotropic refrigerant mixtures find wide application. The properties of an azeotropic mixture are such that they may be conveniently treated as pure substance properties. Zeotropic mixtures, however, require special treatment, using an equation-of-state approach with appropriate mixing rules or using the fugacities with the standard state method (Tassios 1993). Refrigerant and lubricant blends are a zeotropic mixture and can be treated by these methods (see Thome 1995 and Martz et al. 1996a, b).

COMPRESSION REFRIGERATION CYCLES

CARNOT CYCLE

The Carnot cycle, which is completely reversible, is a perfect model for a refrigeration cycle operating between two fixed temperatures, or between two fluids at different temperatures and each with infinite heat capacity. Reversible cycles have two important properties: (1) no refrigerating cycle may have a coefficient of performance higher than that for a reversible cycle operated between the same temperature limits, and (2) all reversible cycles, when operated between the same temperature limits, have the same coefficient of performance. Proof of both statements may be found in almost any text book on elementary engineering thermodynamics.

Figure 5 shows the Carnot cycle on temperature-entropy coordinates. Heat is withdrawn at the constant temperature T_R from the region to be refrigerated. Heat is rejected at the constant ambient temperature T_0. The cycle is completed by an isentropic expansion and an isentropic compression. The energy transfers are given by

$$Q_0 = T_0(S_2 - S_3)$$

$$Q_i = T_R(S_1 - S_4) = T_R(S_2 - S_3)$$

$$W_{net} = Q_o - Q_i$$

Thus, by Equation (15),

$$\text{COP} = \frac{T_R}{T_0 - T_R} \tag{38}$$

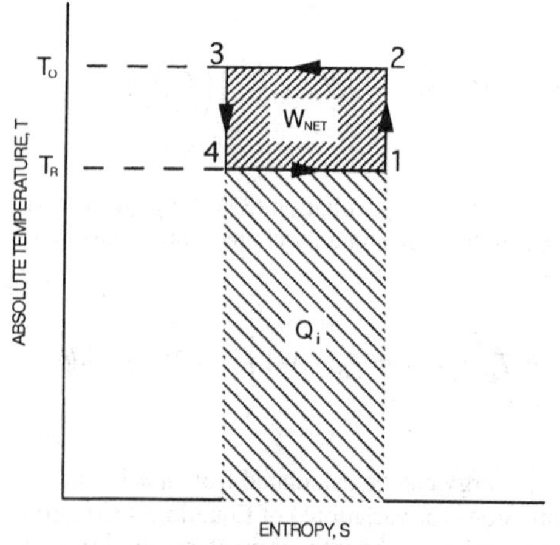

Fig. 5 Carnot Refrigeration Cycle

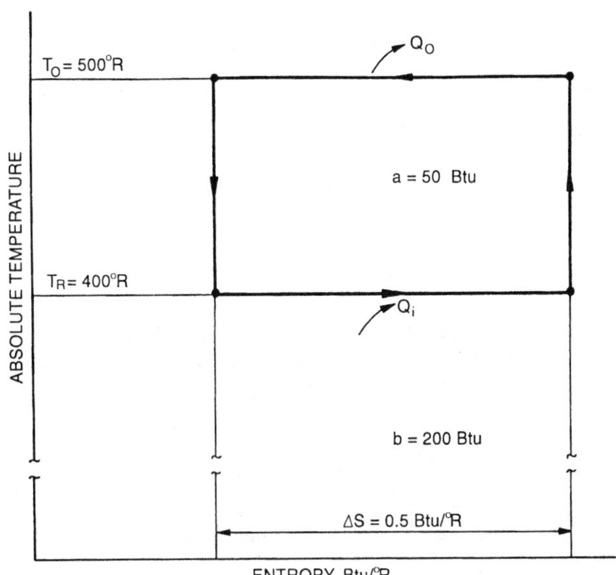

Fig. 6 Temperature-Entropy Diagram for Carnot Refrigeration Cycle of Example 1

Example 1. Determine entropy change, work, and coefficient of performance for the cycle shown in Figure 6. Temperature of the refrigerated space T_R is 400°R and that of the atmosphere T_0 is 500°R. Refrigeration load is 200 Btu.

Solution:

$$\Delta S = S_1 - S_4 = Q_i / T_R = 200/400 = 0.500 \text{ Btu/°R}$$

$$W = \Delta S(T_0 - T_R) = 0.5(500 - 400) = 50 \text{ Btu}$$

$$\text{COP} = Q_i/(Q_o - Q_i) = Q_i/W = 200/50 = 4$$

Flow of energy and its area representation in Figure 6 is:

Energy	Btu	Area
Q_i	200	b
Q_o	250	$a + b$
W	50	a

The net change of entropy of any refrigerant in any cycle is always zero. In Example 1 the change in entropy of the refrigerated space is ΔS_R = −200/400 = −0.5 Btu/°R and that of the atmosphere is ΔS_o = 250/500 = 0.5 Btu/°R. The net change in entropy of the isolated system is $\Delta S_{total} = \Delta S_R + \Delta S_o = 0$.

The Carnot cycle in Figure 7 shows a process in which heat is added and rejected at constant pressure in a two-phase region of a refrigerant. Saturated liquid at state 3 expands isentropically to the low temperature and pressure of the cycle at state d. Heat is added isothermally and isobarically by evaporating the liquid phase refrigerant from state d to state 1. The cold saturated vapor at state 1 is compressed isentropically to the high temperature in the cycle at state b. However the pressure at state b is below the saturation pressure corresponding to the high temperature in the cycle. The compression process is completed by an isothermal compression process from state b to state c. The cycle is completed by an isothermal and isobaric heat rejection or condensing process from state c to state 3.

Applying the energy equation for a mass of refrigerant m yields (all work and heat transfer are positive)

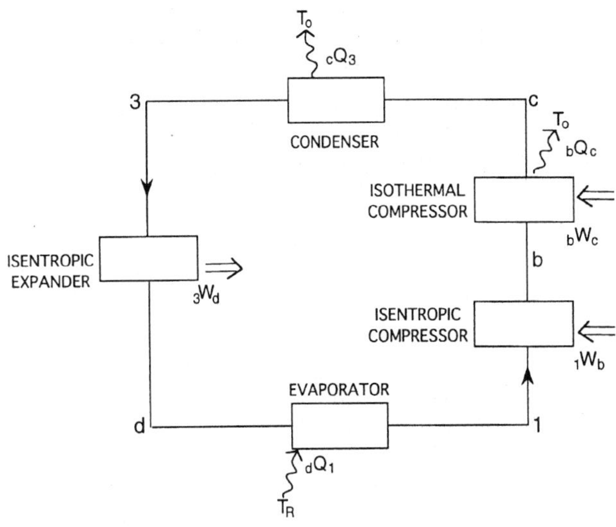

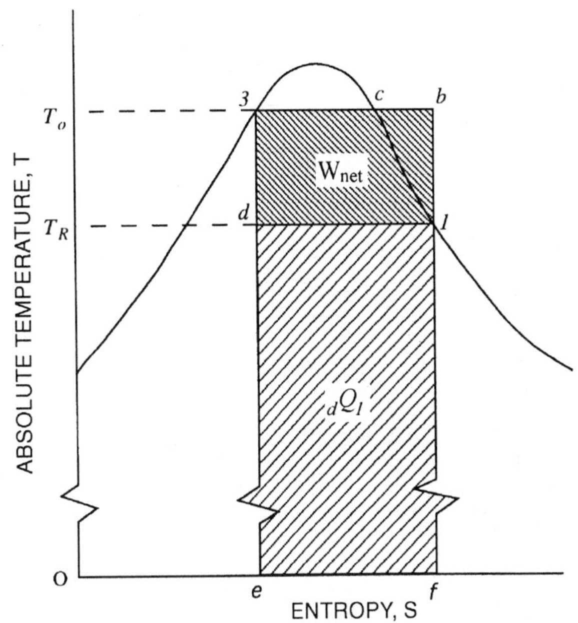

Fig. 7 Carnot Vapor Compression Cycle

$$_3W_d = m(h_3 - h_d)$$

$$_1W_b = m(h_b - h_1)$$

$$_bW_c = T_0(S_b - S_c) - m(h_b - h_c)$$

$$_dQ_1 = m(h_1 - h_d) = \text{Area def1d}$$

The net work for the cycle is

$$W_{net} = {}_1W_b + {}_bW_c - {}_3W_d = \text{Area d1bc3d}$$

and

$$\text{COP} = \frac{_dQ_1}{W_{net}} = \frac{T_R}{T_0 - T_R}$$

THEORETICAL SINGLE-STAGE CYCLE USING A PURE REFRIGERANT OR AZEOTROPIC MIXTURE

A system designed to approach the ideal model shown in Figure 7 is desirable. A pure refrigerant or an azeotropic mixture can be used to maintain constant temperature during the phase changes by maintaining a constant pressure. Because of such concerns as high initial cost and increased maintenance requirements a practical machine has one compressor instead of two and the expander (engine or turbine) is replaced by a simple expansion valve. The valve throttles the refrigerant from high pressure to low pressure. Figure 8 shows the theoretical single-stage cycle used as a model for actual systems.

Applying the energy equation for a mass of refrigerant m yields

$$
\begin{aligned}
{}_4Q_1 &= m(h_1 - h_4) \\
{}_1W_2 &= m(h_2 - h_1) \\
{}_2Q_3 &= m(h_2 - h_3) \\
h_3 &= h_4
\end{aligned}
\tag{39}
$$

The constant enthalpy throttling process assumes no heat transfer or change in potential or kinetic energy through the expansion valve.

The coefficient of performance is

$$\mathrm{COP} = \frac{{}_4Q_1}{{}_1W_2} = \frac{h_1 - h_4}{h_2 - h_1} \tag{40}$$

The theoretical compressor displacement CD (at 100% volumetric efficiency), is

$$\mathrm{CD} = \dot{m}v_3 \tag{41}$$

which is a measure of the physical size or speed of the compressor required to handle the prescribed refrigeration load.

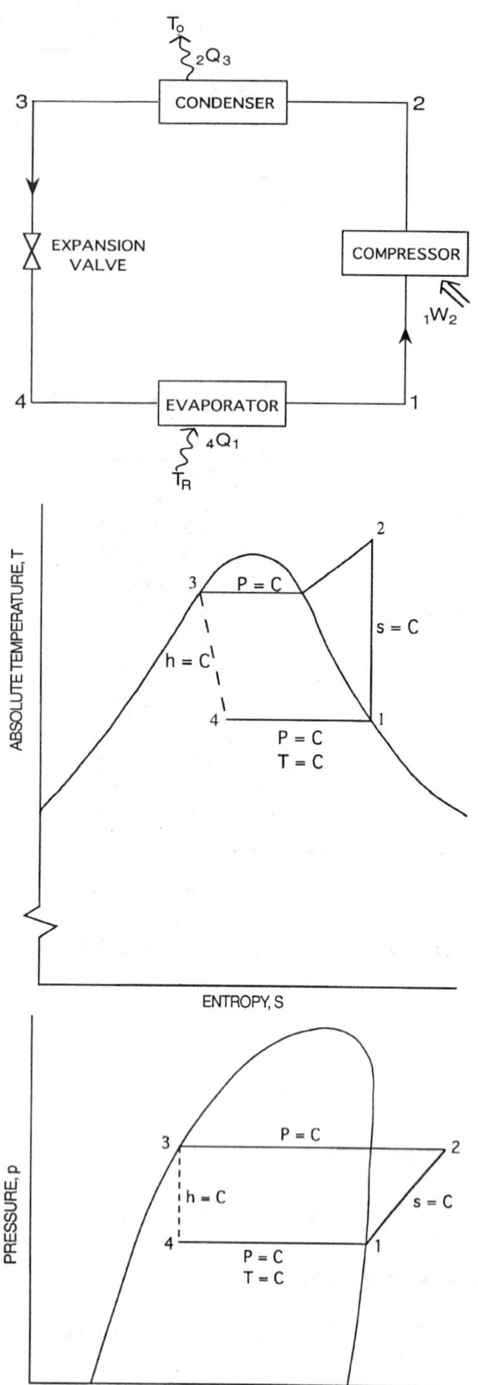

Fig. 8 Theoretical Single-Stage Vapor Compression Refrigeration Cycle

Example 2. A theoretical single-stage cycle using R134a as the refrigerant operates with a condensing temperature of 90°F and an evaporating temperature of 0°F. The system produces 15 tons of refrigeration. Determine (a) the thermodynamic property values at the four main state points of the cycle, (b) the coefficient of performance of the cycle, (c) the cycle refrigerating efficiency, and (d) rate of refrigerant flow.

Solution:

(a) Figure 9 shows a schematic *p-h* diagram for the problem with numerical property data. Saturated vapor and saturated liquid properties for states 1 and 3 are obtained from the saturation table for R134a in Chapter 19. Properties for superheated vapor at state 2 are obtained by linear interpolation of the superheat tables for R134a in Chapter 19. Specific volume and specific entropy values for state 4 are obtained by determining the quality of the liquid-vapor mixture from the enthalpy.

$$x_4 = \frac{h_4 - h_f}{h_g - h_f} = \frac{41.430 - 12.090}{103.015 - 12.090} = 0.3227$$

$$v_4 = v_f + x_4(v_g - v_f) = 0.011187 + 0.3227(2.1587 - 0.011187)$$
$$= 0.7047\ \mathrm{ft^3/lb}$$

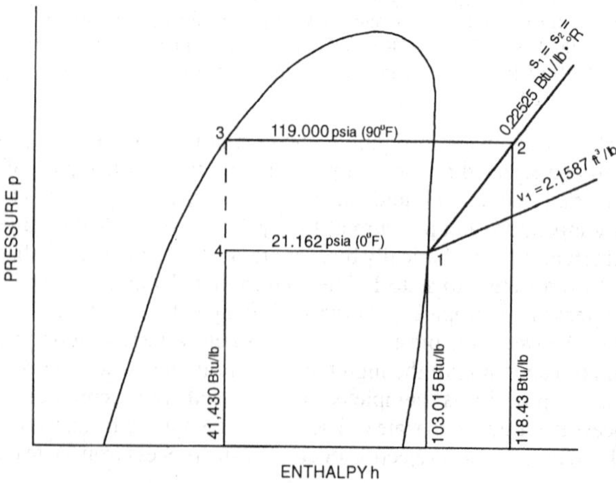

Fig. 9 Schematic *p-h* Diagram for Example 2

Table 1 Thermodynamic Property Data for Example 2

State	t, °F	p, psia	v, ft³/lb	h, Btu/lb	s, Btu/lb °R
1	0	21.162	2.1587	103.015	0.22525
2	103.6	119.00	0.4238	118.43	0.22525
3	90.0	119.00	0.01359	41.430	0.08518
4	0	21.162	0.7047	41.430	0.09127

$$s_4 = s_f + x_4(s_g - s_f) = 0.02744 + 0.3227(0.22525 - 0.02744)$$
$$= 0.09127 \text{Btu/(lb} \cdot {}^\circ\text{R)}$$

The property data are tabulated in Table 1.

(b) By Equation (40)

$$\text{COP} = \frac{103.015 - 41.430}{118.43 - 103.015} = 4.00$$

(c) By Equation (17)

$$\eta_R = \frac{\text{COP}(T_3 - T_1)}{T_1} = \frac{(4.00)(90)}{459.6} = 0.78 \text{ or } 78\%$$

(d) The mass flow of refrigerant is obtained from an energy balance on the evaporator. Thus

$$\dot{m}(h_1 - h_4) = \dot{Q}_i = 15 \text{ tons}$$

$$\text{and } \dot{m} = \frac{(15 \text{ tons})(200 \text{Btu/min ton})}{(103.015 - 41.430)\text{Btu/lb}} = 48.7 \text{ lb/min}$$

The saturation temperatures of the single-stage cycle have a strong influence on the magnitude of the coefficient of performance. This influence may be readily appreciated by an area analysis on a temperature-entropy (*T-s*) diagram. The area under a process line on a *T-s* diagram is directly proportional to the thermal energy added or removed from the working fluid. This observation follows directly from the definition of entropy.

In Figure 10 the area representing Q_o is the total area under the constant pressure curve between states 2 and 3. The area representing the refrigerating capacity Q_i is the area under the constant pressure line connecting states 4 and 1. The net work required W_{net} equals the difference $(Q_o - Q_i)$, which is represented by the shaded area shown on Figure 10.

Because COP = Q_i/W_{net}, the effect on the COP of changes in evaporating temperature and condensing temperature may be observed. For example, a decrease in evaporating temperature T_E significantly increases W_{net} and slightly decreases Q_i. An increase in condensing temperature T_C produces the same results but with less effect on W_{net}. Therefore, for maximum coefficient of performance, the cycle should operate at the lowest possible condensing temperature and at the maximum possible evaporating temperature.

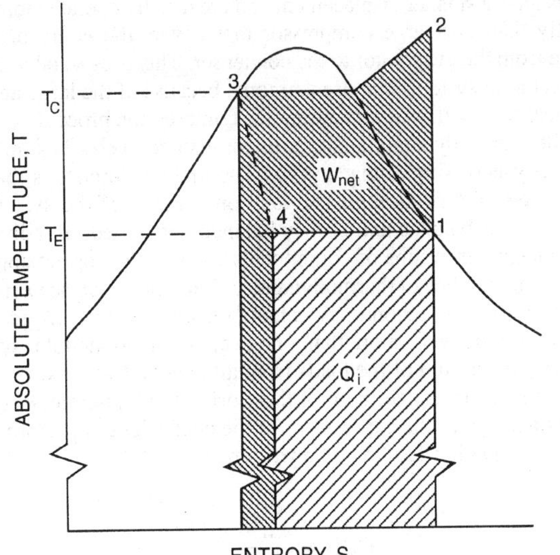

Fig. 10 Areas on *T-s* Diagram Representing Refrigerating Effect and Work Supplied for Theoretical Single-Stage Cycle

LORENZ REFRIGERATION CYCLE

The Carnot refrigeration cycle includes two assumptions which make it impractical. The heat transfer capacity of the two external fluids are assumed to be infinitely large so the external fluid temperatures remain fixed at T_0 and T_R (they become infinitely large thermal reservoirs). The Carnot cycle also has no thermal resistance between the working refrigerant and the external fluids in the two heat exchange processes. As a result, the refrigerant must remain fixed at T_0 in the condenser and at T_R in the evaporator.

The Lorenz cycle eliminates the first restriction in the Carnot cycle and allows the temperature of the two external fluids to vary during the heat exchange. The second assumption of negligible thermal resistance between the working refrigerant and the two external fluids remains. Therefore the refrigerant temperature must change during the two heat exchange processes to equal the changing temperature of the external fluids. This cycle is completely reversible when operating between two fluids, each of which has a finite but constant heat capacity.

Figure 11 is a schematic of a Lorenz cycle. Note that this cycle does not operate between two fixed temperature limits. Heat is added to the refrigerant from state 4 to state 1. This process is assumed to be linear on *T-s* coordinates, which represents a fluid with constant heat capacity. The temperature of the refrigerant is increased in an isentropic compression process from state 1 to state 2. Process 2-3 is a heat rejection process in which the refrigerant temperature decreases linearly with heat transfer. The cycle is concluded with an isentropic expansion process between states 3 and 4.

The heat addition and heat rejection processes are parallel so the entire cycle is drawn as a parallelogram on *T-s* coordinates. A Carnot refrigeration cycle operating between T_0 and T_R would lie between states 1, a, 3, and b. The Lorenz cycle has a smaller refrigerating effect than the Carnot cycle and more work is required. However this cycle is a more practical reference to use than the Carnot cycle when a refrigeration system operates between two single phase fluids such as air or water.

The energy transfers in a Lorenz refrigeration cycle are as follows where ΔT is the temperature change of the refrigerant during the two heat exchange processes.

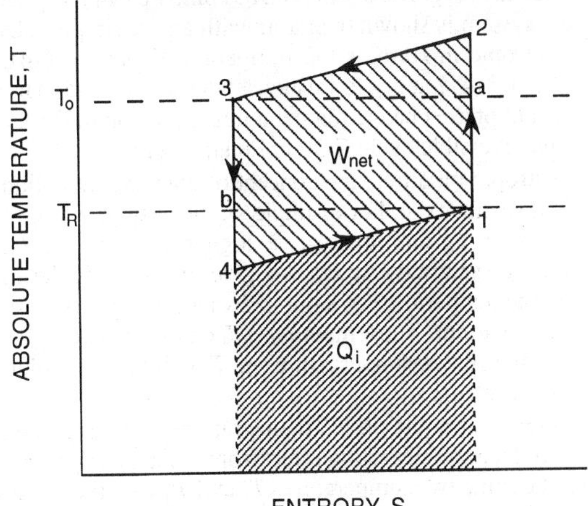

Fig. 11 Processes of Lorenz Refrigeration Cycle

$$Q_0 = (T_0 + \Delta T/2)(S_2 - S_3)$$
$$Q_i = (T_R - \Delta T/2)(S_1 - S_4) = (T_R - \Delta T/2)(S_2 - S_3)$$
$$W_{net} = Q_0 - Q_R$$

Thus by Equation (15),

$$\text{COP} = \frac{T_R - (\Delta T/2)}{T_O - T_R + \Delta T} \quad (42)$$

Example 3. Determine the entropy change, the work required, and the coefficient of performance for the Lorenz cycle shown in Figure 11 when the temperature of the refrigerated space is $T_R = 400°R$, the ambient temperature is $T_0 = 500°R$, the ΔT of the refrigerant is 10°R and the refrigeration load is 200 Btu.

Solution:

$$\Delta S = \int_4^1 \frac{Q_i}{T} = \frac{Q_i}{T_R - (\Delta T/2)} = \frac{200}{395} = 0.5063 \text{ Btu/°R}$$

$$Q_O = [T_O + (\Delta T/2)]\Delta S = (500 + 5)0.5063 = 255.68 \text{ Btu}$$

$$W_{net} = Q_O - Q_R = 255.68 - 200 = 55.68 \text{ Btu}$$

$$\text{COP} = \frac{T_R - (\Delta T/2)}{T_O - T_R + \Delta T} = \frac{400 - (10/2)}{500 - 400 + 10} = \frac{395}{200} = 3.591$$

Note that the entropy change for the Lorenz cycle is larger than for the Carnot cycle at the same temperature levels and the same capacity (see Example 1). That is, the heat rejection is larger and the work requirement is also larger for the Lorenz cycle. This difference is caused by the finite temperature difference between the working fluid in the cycle compared to the bounding temperature reservoirs. However, as discussed previously, the assumption of constant temperature heat reservoirs is not necessarily a good representation of an actual refrigeration system because of the temperature changes that occur in the heat exchangers.

THEORETICAL SINGLE-STAGE CYCLE USING ZEOTROPIC REFRIGERANT MIXTURE

A practical method to approximate the Lorenz refrigeration cycle is to use a fluid mixture as the refrigerant and the four system components shown in Figure 8. When the mixture is not azeotropic and the phase change processes occur at constant pressure, the temperatures change during the evaporation and condensation processes and the theoretical single-stage cycle can be shown on *T-s* coordinates as in Figure 12. This can be compared with Figure 10 in which the system is shown operating with a pure simple substance or an azeotropic mixture as the refrigerant. Equations (14), (15), (39), (40), and (41) apply to this cycle and to conventional cycles with constant phase change temperatures. Equation (42) should be used as the reversible cycle COP in Equation (17).

For zeotropic mixtures, the concept of constant saturation temperatures does not exist. For example, in the evaporator, the refrigerant enters at T_4 and exits at a higher temperature T_1. The temperature of saturated liquid at a given pressure is the **bubble point** and the temperature of saturated vapor at a given pressure is called the **dew point**. The temperature T_3 on Figure 12 is at the bubble point at the condensing pressure and T_1 is at the dew point at the evaporating pressure.

An analysis of areas on a *T-s* diagram representing additional work and reduced refrigerating effect from a Lorenz cycle operating between the same two temperatures T_1 and T_3 with the same value for ΔT can be performed. The cycle matches the Lorenz cycle most closely when counterflow heat exchangers are used for both the condenser and the evaporator.

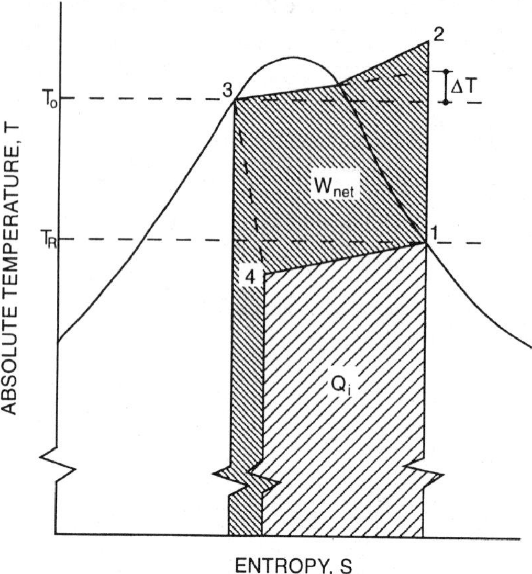

Fig. 12 Areas on *T-s* Diagram Representing Refrigerating Effect and Work Supplied for Theoretical Single-Stage Cycle Using Zeotropic Mixture as Refrigerant

In a cycle that has heat exchangers with finite thermal resistances and finite external fluid capacity rates, Kuehn and Gronseth (1986) showed that a cycle which uses a refrigerant mixture has a higher coefficient of performance than a cycle that uses a simple pure substance as a refrigerant. However, the improvement in COP is usually small. The performance of the cycle that uses a mixture can be improved further by reducing the thermal resistance of the heat exchangers and passing the fluids through them in a counterflow arrangement.

MULTISTAGE VAPOR COMPRESSION REFRIGERATION CYCLES

Multistage vapor compression refrigeration is used when several evaporators are needed at various temperatures such as in a supermarket or when the temperature of the evaporator becomes very low. Low evaporator temperature indicates low evaporator pressure and low refrigerant density into the compressor. Two small compressors in series have a smaller displacement and are usually operate more efficiently than one large compressor that covers the entire pressure range from the evaporator to the condenser. This is especially true in refrigeration systems that use ammonia because of the large amount of superheating that occurs during the compression process.

The thermodynamic analysis of multistage cycles is similar to the analysis of single stage cycles. The main difference is that the mass flow differs through various components of the system. A careful mass balance and energy balance performed on individual components or groups of components ensures the correct application of the first law of thermodynamics. Care must also be exercised when performing second law calculations. Often the refrigerating load is comprised of more than one evaporator, so the total system capacity is the sum of the loads from all evaporators. Likewise the total energy input is the sum of the work into all compressors. For multistage cycles the expression for the coefficient of performance given in Equation 15 should be written as

$$\text{COP} = \frac{\sum Q_i}{W_{net}} \quad (43)$$

When compressors are connected in series, the vapor between stages should be cooled to bring the vapor to saturated conditions

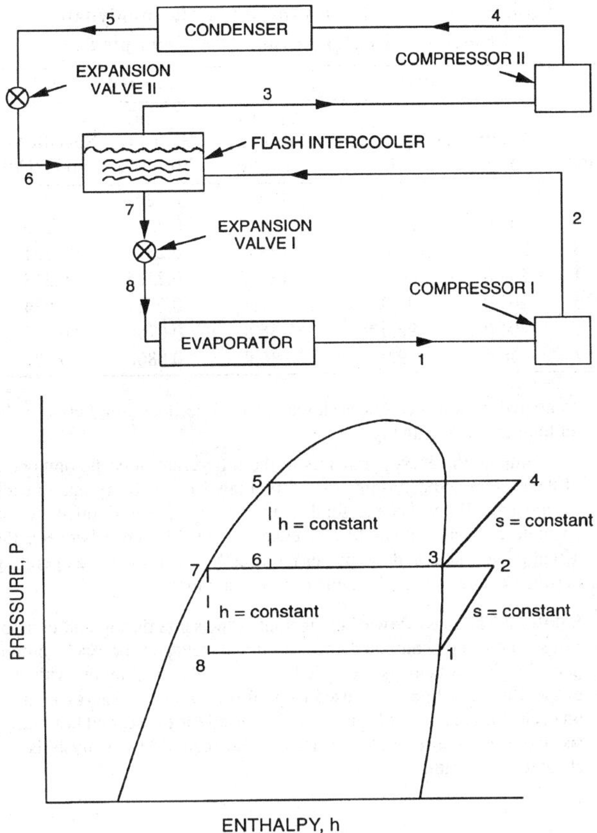

Fig. 13 Schematic and Pressure-Enthalpy Diagram for Dual-Compression, Dual-Expansion Cycle of Example 4

before proceeding to the next stage of compression. Intercooling usually minimizes the displacement of the compressors, reduces the work requirement, and increases the COP of the cycle. If the refrigerant temperature between stages is above ambient, a simple intercooler that removes heat from the refrigerant can be used. If the temperature is below ambient, which is the usual case, the refrigerant itself must be used to cool the vapor. This is accomplished with a flash intercooler. Figure 13 shows a cycle with a flash intercooler installed.

The superheated vapor from compressor I is bubbled through saturated liquid refrigerant at the intermediate pressure of the cycle. Some of this liquid is evaporated when heat is added from the superheated refrigerant. The result is that only saturated vapor at the intermediate pressure is fed to compressor II. A common assumption is to operate the intercooler at about the geometric mean of the evaporating and condensing pressures. This operating point provides the same pressure ratio and nearly equal volumetric efficiencies for the two compressors. Example 4 illustrates the thermodynamic analysis of this cycle.

Example 4. Determine the thermodynamic properties of the eight state points shown in Figure 13, the mass flows, and the COP of this theoretical multistage refrigeration cycle when R-134a is the refrigerant. The saturated evaporator temperature is 0°F, the saturated condensing temperature is 90°F, and the refrigeration load is 15 tons. The saturation temperature of the refrigerant in the intercooler is 40°F, which is nearly at the geometric mean pressure of the cycle.

Solution:

Thermodynamic property data are obtained from the saturation and superheat tables for R-134a in Chapter 19. States 1, 3, 5, and 7 are obtained directly from the saturation table. State 6 is a mixture of liquid and vapor. The quality is calculated by

Table 2 Thermodynamic Property Values for Example 4

State	Temperature, °F	Pressure, psia	Specific Volume, ft³/lb	Specific Enthalpy, Btu/lb	Specific Entropy, Btu/lb·°R
1	0.00	21.162	2.1587	103.015	0.22525
2	48.296	49.724	0.9785	110.49	0.22525
3	40.00	49.724	0.9534	108.705	0.22172
4	95.666	119.000	0.4110	116.47	0.22172
5	90.00	119.000	0.01359	41.430	0.08518
6	40.00	49.724	0.2000	41.430	0.08708
7	40.00	49.724	0.01252	24.694	0.05359
8	0.00	21.162	0.3095	24.694	0.05486

$$x_6 = \frac{h_6 - h_7}{h_3 - h_7} = \frac{41.430 - 24.694}{108.705 - 24.694} = 0.19921$$

Then

$$v_6 = v_7 + x_6(v_3 - v_7) = 0.01252 + 0.19921(0.9534 - 0.012521)$$

$$= 0.2000 \text{ ft}^3/\text{lb}$$

$$s_6 = s_7 + x_6(s_3 - s_7) = 0.05359 + 0.19921(0.22172 - 0.05359)$$

$$= 0.08708 \text{ Btu/lb} \cdot {}^\circ\text{R}$$

Similarly for state 8,

$$x_8 = 0.13862,\ v_8 = 0.3095 \text{ ft}^3/\text{lb},\ s_8 = 0.05486 \text{ Btu/lb} \cdot {}^\circ\text{R}$$

States 2 and 4 are obtained from the superheat tables by linear interpolation. The thermodynamic property data are summarized in Table 2.

The mass flow through the lower circuit of the cycle is determined from an energy balance on the evaporator.

$$\dot{m}_1 = \frac{\dot{Q}_i}{h_1 - h_8} = \frac{15 \text{ tons (200 Btu/min ton)}}{(103.015 - 24.696)\text{Btu/lb}} = 38.30 \text{ lb/min}$$

$$\dot{m}_1 = \dot{m}_2 = \dot{m}_7 = \dot{m}_8$$

For the upper circuit of the cycle,

$$\dot{m}_3 = \dot{m}_4 = \dot{m}_5 = \dot{m}_6$$

Assuming the intercooler has perfect external insulation, an energy balance on it is used to compute $\dot{m}_3$.

$$\dot{m}_6 h_6 + \dot{m}_2 h_2 = \dot{m}_7 h_7 + \dot{m}_3 h_3$$

Rearranging and solving for $\dot{m}_3$

$$\dot{m}_3 = \dot{m}_2 \frac{h_7 - h_2}{h_6 - h_3} = 38.30 \text{ lb/min} \frac{24.694 - 110.49}{41.430 - 108.705} = 48.84 \text{ lb/min}$$

$$\dot{W}_I = \dot{m}_1(h_2 - h_1) = 38.30 \text{ lb/min}(110.49 - 103.015)\text{Btu/lb}$$

$$= 286.3 \text{ Btu/min}$$

$$\dot{W}_{II} = \dot{m}_3(h_4 - h_3) = 48.84 \text{ lb/min}(116.47 - 108.705)\text{Btu/lb}$$

$$= 379.2 \text{ Btu/min}$$

$$\text{COP} = \frac{\dot{Q}_i}{\dot{W}_I + \dot{W}_2} = \frac{15 \text{ tons(200 Btu/min ton)}}{(286.3 + 379.2)\text{Btu/min}} = 4.51$$

Examples 2 and 4 have the same refrigeration load and operate with the same evaporating and condensing temperatures. The two-stage cycle in Example 4 has a higher COP and less work input than the single stage cycle. Also the highest refrigerant temperature leaving the compressor is about 96°F for the two stage cycle versus about 104°F for the single stage cycle. These differences are more pronounced for cycles operating at larger pressure ratios.

ACTUAL REFRIGERATION SYSTEMS

Actual systems operating steadily differ from the ideal cycles considered in the previous sections in many respects. Pressure drops occur everywhere in the system except in the compression process. Heat transfers occur between the refrigerant and its environment in all components. The actual compression process differs substantially from the isentropic compression assumed above. The working fluid is not a pure substance but a mixture of refrigerant and oil. All of these deviations from a theoretical cycle cause irreversibilities within the system. Each irreversibility requires additional power into the compressor. It is useful to understand how these irreversibilities are distributed throughout a real system. Insight is gained that can be useful when design changes are contemplated or operating conditions are modified. Example 5 illustrates how the irreversibilities can be computed in a real system and how they require additional compressor power to overcome. The input data have been rounded off for ease of computation.

Example 5. An air-cooled, direct-expansion, single-stage mechanical vapor-compression refrigerator uses R-22 and operates under steady conditions. A schematic drawing of this system is shown in Figure 14. Pressure drops occur in all piping and heat gains or losses occur as indicated. Power input includes compressor power and the power required to operate both fans. The following performance data are obtained:

Ambient air temperature,	t_O	= 90°F
Refrigerated space temperature,	t_R	= 20°F
Refrigeration load,	$\dot{Q}_E$	= 2 tons
Compressor power input,	$\dot{W}_{comp}$	= 3.0 hp
Condenser fan input,	$\dot{W}_{CF}$	= 0.2 hp
Evaporator fan input,	$\dot{W}_{EF}$	= 0.15 hp

Refrigerant pressures and temperatures are measured at the seven locations shown on Figure 14. Table 3 lists the measured and computed thermodynamic properties of the refrigerant neglecting the dissolved oil. A pressure-enthalpy diagram of this cycle is shown in Figure 15 compared with a theoretical single stage cycle operating between the air temperatures t_R and t_O.

Table 3 Measured and Computed Thermodynamic Properties of Refrigerant 22 for Example 5

	Measured		Computed		
State	Pressure, psia	Temperature, °F	Specific Enthalpy, Btu/lb	Specific Entropy, Btu/lb·°R	Specific Volume, ft³/lb
1	45.0	15.0	106.4	0.2291	1.213
2	44.0	25.0	108.1	0.2330	1.276
3	210.0	180.0	128.8	0.2374	0.331
4	208.0	160.0	124.8	0.2314	0.318
5	205.0	94.0	37.4	0.0761	0.014
6	204.0	92.0	36.8	0.0750	0.014
7	46.5	9.0	36.8	0.0800	0.308

Compute the energy transfers to the refrigerant in each component of the system and determine the second law irreversibility rate in each component. Show that the total irreversibility rate multiplied by the absolute ambient temperature is equal to the difference between the actual power input and the power required by a Carnot cycle operating between t_R and t_O with the same refrigerating load.

Solution: The mass flow of refrigerant is the same through all components, so it is only computed once through the evaporator. Each component in the system is analyzed sequentially beginning with the evaporator. Equation (6) is used to perform a first law energy balance on each component and Equation (13) is used for the second law analysis. Note that the temperature used in the second law analysis is the absolute temperature.

Evaporator:

Energy balance

$$ {}_7\dot{Q}_1 = \dot{m}(h_1 - h_7) = 24{,}000 \text{ Btu/h} $$

$$ \dot{m} = \frac{24000}{(106.4 - 36.8)} = 345 \text{ lb/h} $$

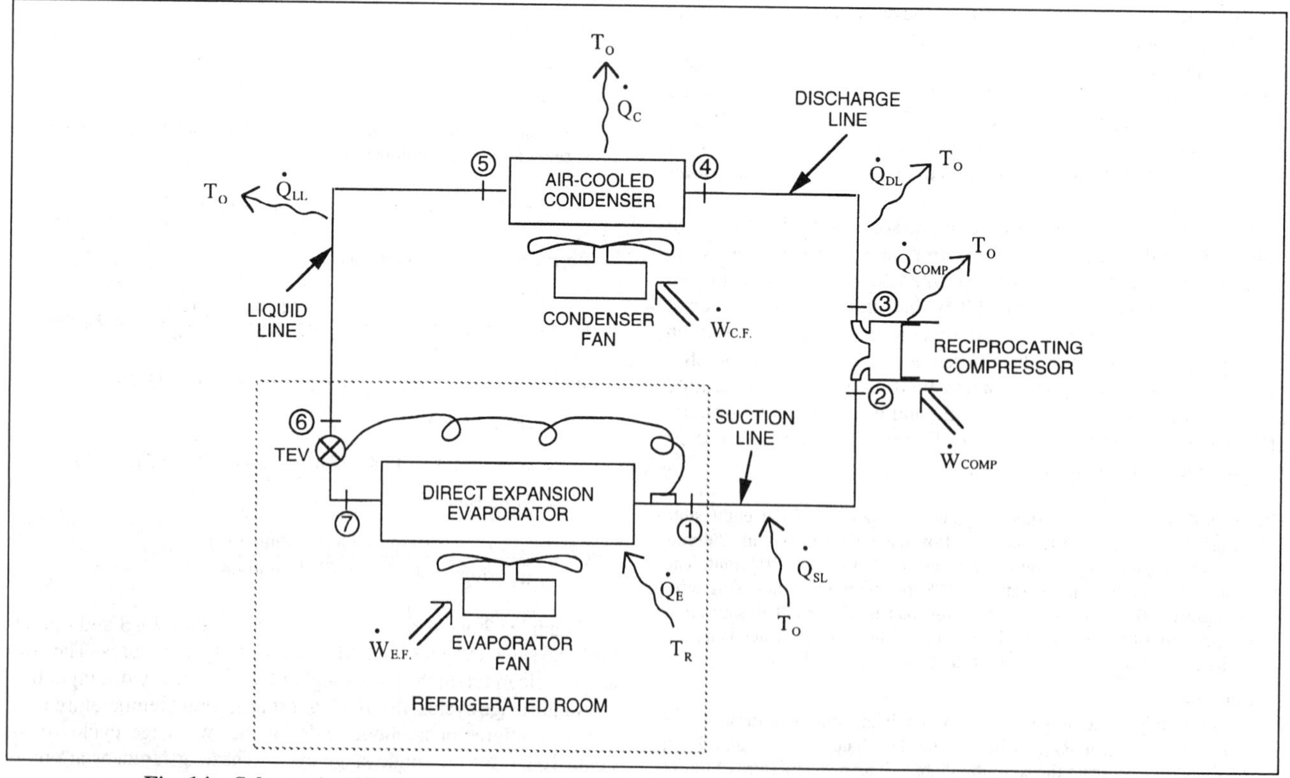

Fig. 14 Schematic of Real, Direct-Expansion, Mechanical Vapor-Compression Refrigeration System

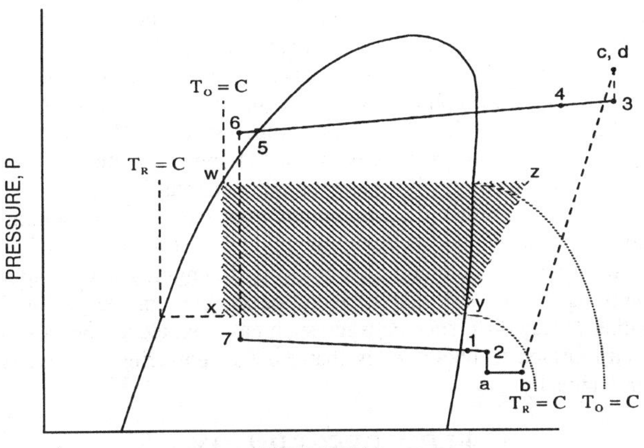

Fig. 15 Pressure-Enthalpy Diagram of Actual System and Theoretical Single-Stage System Operating Between Same Inlet Air Temperatures T_R and T_0.

Second law

$$_7\dot{I}_1 = \dot{m}(s_1 - s_7) - \frac{_7\dot{Q}_1}{T_R}$$

$$= 345(0.2291 - 0.0800) - \frac{24000}{479.67}$$

$$= 1.405 \text{ Btu/h} \cdot {}^\circ\text{R}$$

Suction Line:

Energy balance

$$_1\dot{Q}_2 = \dot{m}(h_2 - h_1)$$

$$= 345(108.1 - 106.4) = 586 \text{ Btu/h}$$

Second law

$$_1\dot{I}_2 = \dot{m}(s_2 - s_1) - \frac{_1\dot{Q}_2}{T_O}$$

$$= 345(0.2330 - 0.2291) - 586/549.67$$

$$= 0.279 \text{ Btu/h} \cdot {}^\circ\text{R}$$

Compressor:

Energy balance

$$_2\dot{Q}_3 = \dot{m}(h_3 - h_2) + {_2\dot{W}_3}$$

$$= 345(128.8 - 108.1) - 3.0(2545)$$

$$= -494 \text{ Btu/h}$$

Second law

$$_2\dot{I}_3 = \dot{m}(s_3 - s_2) - \frac{_2\dot{Q}_3}{T_O}$$

$$= 345(0.2374 - 0.2330) - (-494/549.67)$$

$$= 2.417 \text{ Btu/h} \cdot {}^\circ\text{R}$$

Discharge Line:

Energy balance

$$_3\dot{Q}_4 = \dot{m}(h_4 - h_3)$$

$$= 345(124.8 - 128.8) = -1380 \text{ Btu/h}$$

Second law

$$_3\dot{I}_4 = \dot{m}(s_4 - s_3) - \frac{_3\dot{Q}_4}{T_O}$$

$$= 345(0.2314 - 0.2374) - (-1380/549.67)$$

$$= 0.441 \text{ Btu/h} \cdot {}^\circ\text{R}$$

Table 4 Energy Transfers and Irreversibility Rates for Refrigeration System in Example 5

Component	$\dot{Q}$, Btu/h	$\dot{W}$, Btu/h	$\dot{I}$, Btu/h·°R	$\dot{I}/\dot{I}_{total}$, %
Evaporator	24000	0	1.405	19
Suction line	586	0	0.279	4
Compressor	–494	7635	2.417	32
Discharge line	–1380	0	0.441	6
Condenser	–30153	0	1.278	17
Liquid line	–207	0	≈ 0	≈ 0
Expansion device	0	0	1.725	23
Totals	–7648	7635	7.545	

Condenser:

Energy balance

$$_4\dot{Q}_5 = \dot{m}(h_5 - h_4)$$

$$= 345(37.4 - 124.8) = -30{,}153 \text{ Btu/h}$$

Second law

$$_4\dot{I}_5 = \dot{m}(s_5 - s_4) - \frac{_4\dot{Q}_5}{T_O}$$

$$= 345(0.0761 - 0.2314) - (-30{,}153/549.67)$$

$$= 1.278 \text{ Btu/h} \cdot {}^\circ\text{R}$$

Liquid Line:

Energy balance

$$_5\dot{Q}_6 = \dot{m}(h_6 - h_5)$$

$$= 345(36.8 - 37.4) = -207 \text{ Btu/h}$$

Second law

$$_5\dot{I}_6 = \dot{m}(s_6 - s_5) - \frac{_5\dot{Q}_6}{T_O}$$

$$= 345(0.0750 - 0.0761) - (-207/549.67)$$

$$= 0 \text{ Btu/h} \cdot {}^\circ\text{R}$$

Expansion Device:

Energy balance

$$_6\dot{Q}_7 = \dot{m}(h_7 - h_6) = 0$$

Second law

$$_6\dot{I}_7 = \dot{m}(s_7 - s_6)$$

$$= 345(0.0800 - 0.0750) = 1.725 \text{ Btu/h} \cdot {}^\circ\text{R}$$

These results are summarized in Table 4. For the Carnot cycle

$$COP_{Carnot} = \frac{T_R}{T_o - T_R} = \frac{479.67}{70} = 6.852$$

The Carnot power requirement for the 2 ton load is

$$\dot{W}_{Carnot} = \frac{\dot{Q}_E}{COP_{Carnot}} = \frac{24000}{6.852} = 3502 \text{ Btu/h}$$

The actual power requirement for the compressor is

$$\dot{W}_{comp} = \dot{W}_{Carnot} + \dot{I}_{total}T_o = 3502 + 7.545 \times 549.67 = 7649 \text{Btu/h}$$

This result is within computational error of the measured power input to the compressor of 7635 Btu/h.

The analysis demonstrated in Example 5 can be applied to any actual vapor compression refrigeration system. The only required information for the second law analysis is the refrigerant thermodynamic state points and mass flow rates and the temperatures in which the system is exchanging heat. In this example, the extra compressor power required to overcome the irreversibility in each component is determined. The component with the largest loss is the compressor. This loss is due to motor inefficiency, friction losses, and irreversibilities due to pressure drops, mixing, and heat transfer between the compressor and the surroundings. The unrestrained expansion in the expansion device is the next largest loss. This loss could be reduced by using an expander rather than a throttling process. An expander may be economical on large machines.

All heat transfer irreversibilities on both the refrigerant side and the air side of the condenser and evaporator are included in the analysis. The refrigerant pressure drop is also included. The only items not included are the air-side pressure drop irreversibilities of the two heat exchangers. However these are equal to the fan power requirements as all the fan power is dissipated as heat.

An overall second law analysis, such as in Example 5, shows the designer those components with the most losses, and it helps determine which components should be replaced or redesigned to improve performance. However, this type of analysis does not identify the nature of the losses. A more detailed second law analysis in which the actual processes are analyzed in terms of fluid flow and heat transfer is required to identify the nature of the losses (Liang and Kuehn 1991). A detailed analysis will show that most irreversibilities associated with heat exchangers are due to heat transfer, while pressure drop on the air side causes a very small loss and the refrigerant pressure drop causes a negligible loss. This finding indicates that promoting refrigerant heat transfer at the expense of increasing the pressure drop usually improves performance.

ABSORPTION REFRIGERATION CYCLES

Absorption cycles are primarily heat-operated cycles in which heat is pumped with a minimum of work input. As with vapor compression cycles, absorption cycles can be operated in either a heating or cooling mode. This discussion is restricted to cooling applications because such applications dominate the market. Absorption cooling machines are available in sizes ranging from 3 to 2000 tons of refrigeration. These machines are configured for direct-fired operation as well as for waste heat or heat integration applications.

Figure 16 is a simple schematic of a heat engine and a heat pump together in a single package. Such a combination interacts with the surroundings at three temperature levels, which is typical of an absorption cycle. In the cooling mode, the driving heat must be supplied at the highest temperature in the cycle. The refrigeration effect is provided at the lowest temperature in the cycle. The sum of these heat inputs to the cycle is then rejected at the intermediate temperature. From a thermodynamic standpoint, the refrigeration effect can be accomplished with zero work input. In fact such cycles, called diffusion-absorption cycles, are widely used to refrigerate food in recreational vehicles and hotel rooms. Although such cycles work effectively at the loads required for a food refrigerator (140 Btu/h of cooling), larger machines generally use mechanically driven pumps to circulate the internal fluid. Thus, for most absorption cycles, a small work input of about 1% of the heat input must be supplied as electric power input. Frequently, this work input is ignored when describing the thermal performance of an absorption machine. In practice however, the design, operation, and maintenance associated with the pumps must be considered.

Absorption technology competes with engine-driven vapor compression and desiccant refrigeration systems for the gas-fired market. A key difference between such gas-fired technologies and electric-driven technologies is that the fuel powering the unit is burned locally.

FLOW DESCRIPTION

Key processes in the absorption cycle are the absorption and desorption of refrigerant. The cycle has five main components as shown in Figure 17: the generator (sometimes called desorber), the condenser, the evaporator, the absorber, and the solution heat exchanger. Starting with state point 4 at the generator exit, the stream consists of absorbent-refrigerant solution, which flows to the absorber via the heat exchanger. From points 6 to 1, the solution absorbs refrigerant vapor (10) from the evaporator and rejects heat to the environment. The solution rich in refrigerant (1) flows via the heat exchanger to the generator (3). In the generator thermal energy is added and refrigerant (7) boils off the solution. The refrigerant vapor (7) flows to the condenser, where heat is rejected as the refrigerant condenses. The condensed liquid (8) flows through a flow restrictor to the evaporator. In the evaporator, the heat from the load evaporates the refrigerant, which then flows (10) to the absorber. A portion of the refrigerant leaving the evaporator leaves as liquid spillover (11).

The state points of absorption cycles are usually represented in a Dühring chart (Figure 18). In this chart, refrigerant saturation temperature and its corresponding pressure are plotted versus the solution temperature. The lines of constant solution concentration are straight lines of decreasing slopes for increasing concentrations. In this schematic, the lines represent constant aqueous lithium bromide concentration, with water as the refrigerant. The solution at the exit of the generator (point 4) is cooled to point 5 in the heat exchanger. In the absorber, the solution concentration decreases to that of 1. The solution is then pumped to the generator via a heat exchanger, where its temperature is raised to that of 3. In the generator the solution is reconcentrated to yield 4 again. The refrigerant from the generator condenses at 8 and evaporates at 10 to return to the absorber.

Because absorption machines are thermally activated, large amounts of power input are not required. Hence, where power is expensive or unavailable, and gas, waste, geothermal or solar heat is available, absorption machines provide reliable and quiet cooling.

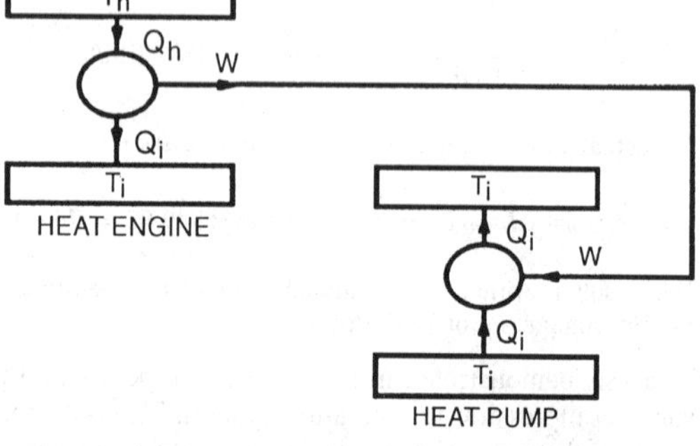

Fig. 16 Absorption Refrigeration Machine as Combination of Heat Engine and Heat Pump

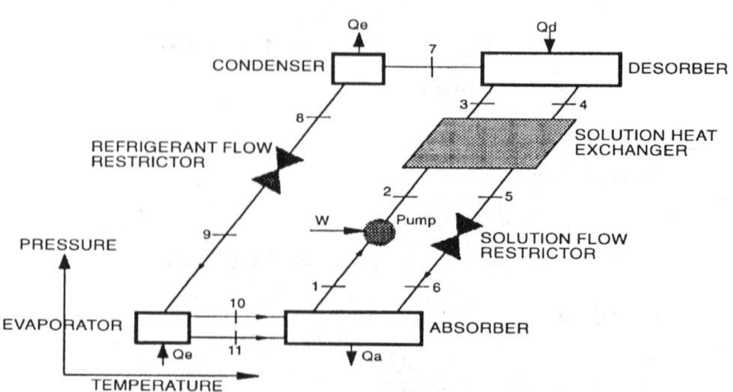

Fig. 17 Single-Effect Lithium Bromide/Water Absorption Cycle

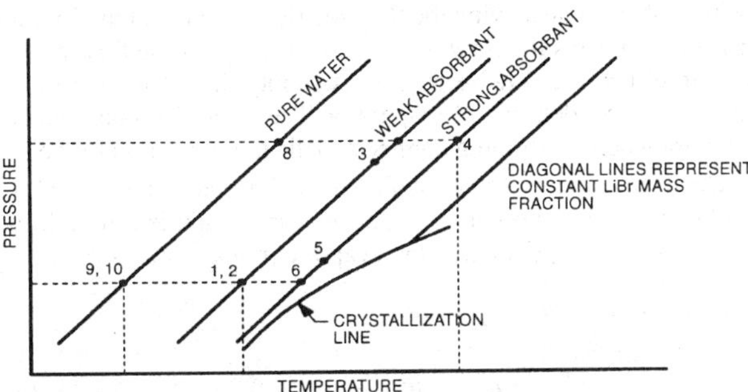

Fig. 18 Single-Effect Lithium Bromide/Water Absorption Cycle Superimposed on Dühring Plot

The usual figure of performance of an absorption cycle is the COP (coefficient of performance), which is defined as the ratio of the evaporator heat to the generator heat. Because it takes about the same amount of heat to boil the refrigerant in both the generator and evaporator, it might be assumed that single effect cycles are capable of a COP of 1. Yet, the best single effect machines reach COPs of only 0.5 to 0.7. The losses responsible for the COP degradation are traced to the following four phenomena:

1. **Circulation loss.** When the cold solution from the absorber (1), is heated in the solution heat exchanger (3), the temperature at 3 is always less than the saturation temperature corresponding to the generator pressure and solution concentration, even for cycles with high heat exchanger effectiveness. Hence, heat must be added to boil the solution, which increases the generator heat input.
2. **Heat of mixing.** Separating the refrigerant from the solution requires about 15% more thermal energy than merely boiling the refrigerant. This additional energy must be supplied to break the intermolecular bonds formed between the refrigerant and absorbent in solution. The heat of mixing also increases the generator heat input.
3. **Expansion loss.** As the refrigerant expands from the condenser to the evaporator, a mixture of liquid and vapor enters the evaporator. Not all of the refrigerant is available as liquid because some vapor was already produced by the expansion process. Thus, the evaporator heat transfer is reduced when vapor forms in the expansion process. This loss can be reduced by subcooling the liquid from the condenser.
4. **Reflux condenser loss.** In the ammonia-water cycle (Figure 21) another loss is introduced due to the volatility of water. In this cycle the refrigerant is ammonia and the absorbent is water. In the generator water vapor evaporates along with the ammonia. However, for proper operation, the water vapor must be removed from the ammonia vapor. The water vapor is separated in a distillation column, which has a reflux coil that condenses some ammonia-water. The heat removed in the reflux coil must be added to the generator, thus decreasing the COP.

In addition, other losses occur during transient operating conditions. For instance, if more refrigerant is produced than can be handled by the evaporator, the refrigerant is directly returned to the absorber via a spillover (point 11 in Figures 17 and 18). Liquid refrigerant returned directly to the absorber is a loss, and machines of recent design are tightly controlled to avoid this loss during transients.

To attain higher COPs, a double effect cycle is used. In this cycle an additional generator and condenser are added to a single effect cycle. The heat input to the high temperature generator is used to drive off refrigerant, which on condensing drives a lower temperature generator to produce yet more refrigerant. In this way the heat input to the higher temperature generator is used twice, and the arrangement is called double effect. Typical COPs of double effect machines range from 1.0 to 1.2.

CHARACTERISTICS OF REFRIGERANT-ABSORBENT PAIRS

Few solutions work as suitable absorbent-refrigerant pairs. The materials that make up the refrigerant-absorbent pair should meet the following requirements to be suitable for absorption refrigeration:

Absence of Solid Phase. The refrigerant-absorbent pair should not form a solid phase over the range of composition and temperature to which it might be subjected. If a solid forms, it presumably would stop flow and cause equipment to shut down.

Volatility Ratio. The refrigerant should be much more volatile than the absorbent so the two can be separated easily. Otherwise, cost and heat requirements can prohibit separation.

Affinity. The absorbent should have a strong affinity for the refrigerant under conditions in which absorption takes place. This affinity (1) causes a negative deviation from Raoult's law and results in an activity coefficient of less than unity for the refrigerant; (2) allows less absorbent to be circulated for the same refrigerating effect so sensible heat losses are less; and (3) requires a smaller liquid heat exchanger to transfer heat from the absorbent to the pressurized refrigerant-absorbent solution. However, calculations by Jacob et al. (1969) indicate that strong affinity has some disadvantages. This affinity is associated with a high heat of dilution; consequently, extra heat is required in the generator to separate the refrigerant from the absorbent.

Pressure. Operating pressures, largely established by physical properties of the refrigerant, should be moderate. High pressures require the use of heavy-walled equipment, and significant electrical power may be required to pump the fluids from the low-pressure side to the high-pressure side. Low pressure (vacuum) requires the use of large volume equipment and special means of reducing pressure drop in refrigerant vapor flow.

Stability. High chemical stability is required because fluids are subjected to severe conditions over many years of service. Instability could cause the undesirable formation of gases, solids, or corrosive substances.

Corrosion. Because absorption fluids can corrode materials used in constructing equipment, corrosion inhibitors are used.

Safety. Fluids must be nontoxic and nonflammable if they are in an occupied dwelling. Industrial process refrigeration is less critical in this respect.

Transport Properties. Viscosity, surface tension, thermal diffusivity, and mass diffusivity are important characteristics of the refrigerant and absorbent pair. For example, a low fluid viscosity promotes heat and mass transfer and reduces pumping power.

Latent Heat. The refrigerant's latent heat should be high so the circulation rate of the refrigerant and absorbent can be kept at a minimum.

Environmental Soundness. The working pairs must be safe, nonflammable, and devoid of lasting environmental effects.

No known refrigerant-absorbent pair meets all requirements listed. However, lithium bromide-water and ammonia-water offer excellent thermodynamic performance and they have little long-term environmental effect. The ammonia-water pair meets most requirements, but its volatility ratio is low, and it requires high operating pressures. Furthermore, ammonia is a Safety Code Group 2 fluid (*ASHRAE Standard* 15), which restricts its use indoors.

Advantages of the water-lithium bromide pair include high safety, high volatility ratio, high affinity, high stability, and high latent heat. However, this pair tends to form solids. Because the refrigerant turns to ice at 32°F, the pair cannot be used for low-temperature refrigeration. Lithium bromide crystallizes at moderate concentrations, especially when it is air cooled, which typically limits the pair to applications where the absorber is water cooled. However, using a combination of salts as the absorbent can reduce this crystallizing tendency enough to permit air cooling (Macriss 1968). Other disadvantages of the water-lithium bromide pair include the low operating pressures it requires and the lithium bromide solution's high viscosity. Proper equipment design can overcome these disadvantages.

Other intriguing refrigerant-absorbent pairs include the following (Macriss and Zawacki 1989):

- Ammonia-salt
- Methylamine-salt
- Alcohol-salt
- Ammonia-organic solvent
- Sulfur dioxide-organic solvent
- Halogenated hydrocarbons-organic solvent
- Water-alkali nitrate
- Water-hydroxide
- Ammonia-water-salt

Several refrigerant-absorbent pairs appear suitable for certain cycles and may solve some of problems associated with the traditional pairs. However, stability, corrosion, and property information on several of them is limited. Also some of the fluids are somewhat hazardous.

EXAMPLE ABSORPTION CYCLE CALCULATIONS

The following examples contrast the performance of absorption machines with similar refrigeration capacities.

Single-Effect Water-Lithium Bromide Cycle

Based on the assumptions and inputs listed in Table 5, the results in Table 6 can be calculated as follows. The property values were obtained from McNeely (1979).

Evaporator Analysis. The state at the evaporator exit is known from the input temperature value and the assumption that the refrigerant is pure water. Thus, the enthalpy and pressure are known at points 10 and 11. The enthalpy at point 9 is determined from a throttling process model applied to the refrigerant flow restrictor. This model requires that the pressure at point 8 must be known. This pressure can be deduced from the given values of the solution mass fraction and the temperature at the saturated state at point 4. Once the pressure at 8 is known, then the temperature and enthalpy are also known. The throttling model yields the result that $h_9 = h_8$. Because the pressure is low in the evaporator, the specific volume of the two-phase state leaving the flow restrictor is quite high. This significantly impacts the flow regime downstream of the flow restriction. In effect, the flow restriction is simply the refrigerant piping between the condenser and evaporator. The expansion takes place at the point where this piping connects to the larger evaporator vessel.

Once the enthalpy values at all ports connected to the evaporator are known, mass and energy balances can be applied to yield the mass flow of refrigerant and the evaporator heat transfer rate.

Mass balance on evaporator

$$\dot{m}_9 = \dot{m}_{10} + \dot{m}_{11} \tag{44}$$

Energy balance on evaporator

$$\dot{Q}_e = \dot{m}_{10}h_{10} + \dot{m}_{11}h_{11} - \dot{m}_9 h_9 \tag{45}$$

The input value for $\dot{Q}_e$ and the enthalpy values at all three states are known. Therefore, Equations (44) and (45) and the input ratio between the flow rates m_{10} and m_{11} represent three equations in three unknowns, and they can be solved to yield the flow rates (as shown in Table 6).

Absorber Analysis. With the known values of m_{10} and m_{11} from the evaporator model, the mass flow around the absorber can be determined by writing mass balances around that device, as follows.

Overall mass balance on absorber

$$\dot{m}_{10} + \dot{m}_{11} + \dot{m}_6 = \dot{m}_1 \tag{46}$$

Lithium bromide mass balance on absorber

$$\dot{m}_6 x_6 = \dot{m}_1 x_1 \tag{47}$$

The mass fractions in Equation (47) are inputs. Thus, Equations (46) and (47) represent two equations in the two unknowns m_1 and m_6 that can be solved to give the values in Table 6.

The heat transfer rate in the absorber can be determined from the enthalpy values at each of the connected state points. The enthalpy at point 1 is determined from the input mass fraction and the assumption that the state is saturated liquid at the same pressure as the evaporator. The enthalpy value at point 6 is determined from a throttling model on the solution flow restrictor

Table 5 Assumptions and Inputs for Single-Effect Lithium Bromide/Water Model (Figure 17)

Assumptions
Steady state refrigerant is pure water
No pressure changes except through the flow restrictors and the pump
States at points 1, 4, 8, and 11 are saturated liquid
State at point 10 is saturated vapor
Flow restrictors are adiabatic
Pump is isentropic
No jacket heat losses

Inputs		
Capacity	$\dot{Q}_e$	500 tons (refrig.)
Evaporator temperature	t_{10}	41.1°F
Desorber solution exit temperature	t_4	209.6°F
Weak solution mass fraction	x_1	59.5% LiBr
Strong solution mass fraction	x_4	64.6% LiBr
Solution heat exchanger exit temp.	t_3	170.3°F
Desorber vapor exit temperature	t_7	200°F
Liquid carryover from evaporator	$\dot{m}_{11}$	2.5% of $\dot{m}_{10}$

Table 6 State Point Data for Single-Effect Lithium Bromide/Water Cycle of Figure 17

Point	h Btu/lb	$\dot{m}$ lb/min	p psi	Q Fraction	t °F	x % LiBr
1	50.6	1299.1	0.127	0	108.3	59.5
2	50.6	1299.1	1.260		108.3	59.5
3	79.3	1299.1	1.260		170.3	59.5
4	106.9	1196.6	1.260	0	209.6	64.6
5	75.7	1196.6	1.260		137.0	64.6
6	75.7	1196.6	0.127	0.003	127.8	64.6
7	1149.4	102.6	1.260		200.0	0
8	77.5	102.6	1.260	0	109.6	0
9	77.5	102.6	0.127	0.064	41.1	0
10	1078.6	100.1	0.127	1.0	41.1	0
11	9.1	2.5	0.127		41.1	0

$COP_c = 0.700$ $\dot{Q}_d = 8.565 \times 10^6$ Btu/h

$\varepsilon = 0.717$ $\dot{Q}_e = 6.000 \times 10^6$ Btu/h

$\dot{Q}_a = 7.969 \times 10^6$ Btu/h $\dot{Q}_{shx} = 2.240 \times 10^6$ Btu/h

$\dot{Q}_c = 6.596 \times 10^6$ Btu/h $\dot{W} = 0.034$ hp

which yields $h_6 = h_5$. The enthalpy at point 5 is determined from an energy balance on the solution heat exchanger as discussed in the next section. Once the enthalpy values at all connected points are known, the energy balance on the absorber can be written as follows.

Energy balance on absorber

$$\dot{Q}_a = \dot{m}_{10}h_{10} + \dot{m}_{11}h_{11} + \dot{m}_6h_6 - \dot{m}_1h_1 \quad (48)$$

Note that the assumption that the solution leaving the absorber is saturated is not accurate. Finite driving potentials in the absorber require that the solution leave subcooled. By assuming saturated liquid, the model would be expected to over predict performance but this effect is beyond the scope of the present treatment.

Solution Heat Exchanger. The solution heat exchanger transfers heat from the high temperature solution stream to the low temperature solution stream. The energy balance is written assuming an adiabatic shell.

Energy balance on solution heat exchanger

$$\dot{m}_2h_2 + \dot{m}_4h_4 = \dot{m}_3h_3 + \dot{m}_5h_5 \quad (49)$$

In this example, the input value specified for the solution temperature at point 3 allows the enthalpy at point 3 to be determined from the known temperature and mass fraction (note that the mass fraction is the same for states 1, 2 and 3). In general, the state at point 3 is subcooled liquid. However, since the pressure has an insignificant effect on the enthalpy of subcooled liquid solution at the conditions of interest, the saturated value at the same temperature and mass fraction is an adequate approximation. The enthalpy at state 4 is known from the input temperature and mass fraction and the assumption that state 4 is saturated liquid. The enthalpy at state 2 is determined from an isentropic pump model as described in the following paragraph. Once three of the enthalpy values are known, Equation (49) can be solved for the remaining enthalpy value at state 5. The temperature at state 5 can also be determined from the enthalpy value (state 5 is also subcooled liquid, in general).

Solution Pump. The minimum solution pump work is obtained from an isentropic pump model as

$$\dot{W} = \dot{m}_1v_1(p_2 - p_1) \quad (50)$$

where it is assumed that the specific volume of the liquid solution does not change appreciably from state 1 to state 2.

Desorber. An energy balance on the desorber determines the heat input required to drive the machine as

$$\dot{Q}_d = \dot{m}_4h_4 + \dot{m}_7h_7 - \dot{m}_3h_3 \quad (51)$$

The enthalpy values at points 3 and 4 have been discussed previously. The enthalpy at point 7 can be determined if the temperature at point 7 is known. In this example, the temperature at point 7 is an input value. In general, point 7 is a superheated pure water vapor state and the enthalpy can be obtained once the pressure and temperature are known.

Condenser. To determine condenser heat, an energy balance on the condenser is

$$\dot{Q}_c = \dot{m}_7(h_7 - h_8) \quad (52)$$

Both of the enthalpy values in Equation (52) have been discussed previously. Thus, the condenser heat can be obtained directly.

Coefficient of Performance. The typical measure of cycle performance is the coefficient of performance defined as

$$COP_c = \dot{Q}_e/\dot{Q}_d \quad (53)$$

The COP value calculated for this example cycle is 0.7, which is typical for a single-effect lithium bromide/water machine.

Double-Effect Absorption

Single-effect absorption technology provides a peak cooling COP of approximately 0.7 and operates with heat input temperatures in the range 190 to 250°F. When higher temperature heat sources are available, absorption technology can be reconfigured to achieve higher COP. One such configuration is a parallel-flow double-effect design as shown schematically in Figures 19 and 20. By adding an additional stage as a topping cycle on a single-effect cycle, the machine becomes a cascade in which the heat rejection from the high temperature stage fires the lower temperature stage. The result, in practice, is cooling COP values in the range of 1.0 to 1.2, depending on the design and application.

Double-effect cycle calculations can be performed in a manner similar to that illustrated for the single-effect cycle. Mass and energy balances of the model shown in Figure 20 were calculated

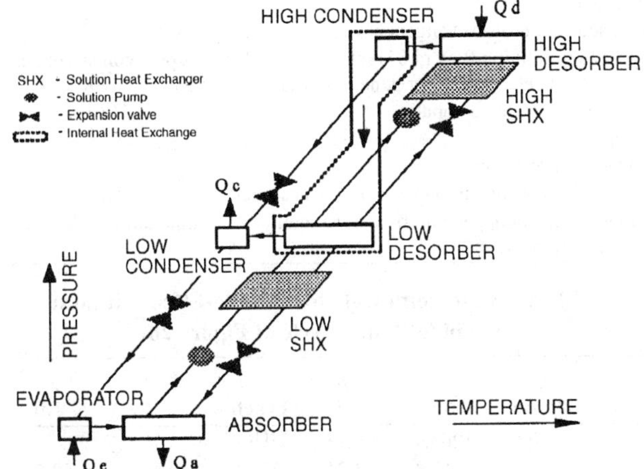

Fig. 19 Double-Effect Lithium Bromide/Water Absorption Cycle

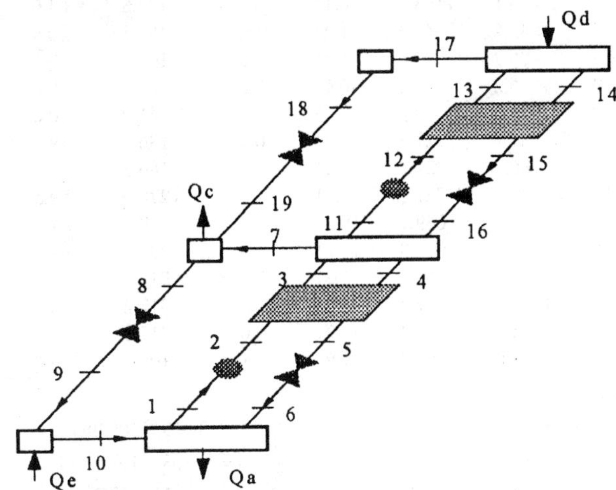

Fig. 20 Double-Effect Lithium Bromide/Water Absorption Cycle with State Points

using the inputs and assumptions listed in Table 7. The results are shown in Table 8. Note that the COP value obtained is high compared to what is obtained in practice. The COP is quite sensitive to several inputs and assumptions. In particular, the effectiveness of the solution heat exchangers and the driving temperature difference between the high temperature condenser and the low temperature generator are two parameters that influence the COP strongly.

Table 7 Inputs and Assumptions for Double-Effect Lithium Bromide/Water Model

Inputs		
Capacity	$\dot{Q}_e$	500 tons refrig.
Evaporator temperature	t_{10}	41.1°F
Desorber solution exit temperature	t_{14}	339.3°F
Condenser/absorber low temperature	$t_1 = t_8$	108.3°F
Solution heat exchanger effectiveness	ε	0.6
Assumptions		
Steady state		
Refrigerant is pure water		
No pressure changes except through the flow restrictors and the pump		
States at points 1, 4, 8, 11, 14 and 18 are saturated liquid		
State at point 10 is saturated vapor		
Temperature difference between high temperature condenser and low temperature generator is 9°F		
Parallel flow		
Both solution heat exchangers have same effectiveness		
Upper loop solution flow rate is selected such that the upper condenser heat exactly matches the lower generator heat requirement		
Flow restrictors are adiabatic		
Pumps are isentropic		
No jacket heat losses		
No liquid carryover from evaporator to absorber		
Vapor leaving both generators is at the equilibrium temperature of the entering solution stream		

Table 8 State Point Data for Double-Effect Lithium Bromide/Water Cycle of Figure 20

No.	h Btu/lb	$\dot{m}$ lb/min	p psia	Q Fraction	t °F	x % LiBr
1	50.6	1263.4	0.13	0.0	108.3	
2	50.6	1263.4	1.21		108.3	59.5
3	78.3	1263.4	1.21		168.1	59.5
4	106.2	1163.7	1.21	0.0	208.0	64.6
5	76.1	1163.7	1.21		137.9	64.6
6	76.1	1163.7	0.13	0.004	127.8	64.6
7	1143.2	42.3	1.21		186.2	0.0
8	76.2	99.8	1.21	0.0	108.3	0.0
9	76.2	99.8	0.13	0.063	41.1	0.0
10	1078.6	99.8	0.13	1.0	41.1	0.0
11	86.7	727.3	1.21	0.0	186.2	59.5
12	86.7	727.3	16.21		186.2	59.5
13	129.4	727.3	16.21		278.0	59.5
14	162.7	669.9	16.21	0.0	339.3	64.6
15	116.4	669.9	16.21		231.6	64.6
16	116.4	669.9	1.21	0.008	210.3	64.6
17	1197.4	57.4	16.21		312.2	0.0
18	185.0	57.4	16.21	0.0	217.0	0.0
19	185.0	57.4	1.21	0.105	108.3	0.0

COP_c = 1.195	$\dot{Q}_e$ = 6.000×10^6 Btu/h
Δt = 9.0°F	$\dot{Q}_{gh}$ = 5.019×10^6 Btu/h
ε = 0.600	$\dot{Q}_{shx1}$ = 2.103×10^6 Btu/h
$\dot{Q}_a$ = 7.936×10^6 Btu/h	$\dot{Q}_{shx2}$ = 1.862×10^6 Btu/h
$\dot{Q}_{cg}$ = 3.488×10^6 Btu/h	$\dot{W}_{p1}$ = 0.032 hp
$\dot{Q}_c$ = 3.085×10^6 Btu/h	$\dot{W}_{p2}$ = 0.258 hp

Ammonia-Water Cycle

An ammonia-water single-stage refrigeration cycle (Figure 21) resembles a lithium bromide-water refrigeration cycle with the exception of two components unique to ammonia/water: (1) a rectifier and (2) a refrigerant heat exchanger. Both components are needed because the vapor pressure of the absorbent (water) is sufficiently high that the water content of the vapor becomes a design issue. The rectifier removes water vapor from the vapor leaving the generator by a fractional distillation process. Distillation occurs in a counterflow mass transfer device such as a packed tower, a bubble tower, or a tower with sieve trays. Reflux, provided by a partial condenser at the top of the column, purifies the rising ammonia vapors coming from the generator. The reflux, after performing its purification function, must be returned to the generator. The generator supplies the rectification heat, which reduces the cycle COP.

Even with a sophisticated rectifier, pure ammonia vapor cannot be obtained at the outlet. A small water fraction, on the order of 0.1% by mass, is enough to cause the refrigerant in the evaporator to exhibit a temperature glide of 40°F (this assumes a single-pass evaporator where the temperature changes along the length of the flow path, such as an in-tube design). The refrigerant heat exchanger uses the energy in the liquid stream coming from the condenser to evaporate the high temperature fraction of the refrigerant. This heat exchanger accounts for the water content in the refrigerant without requiring a blow-down (spillover) system. Inclusion of the refrigerant heat exchanger allows the rectifier to be less effective without any penalty on performance.

In smaller systems, the ammonia vapor purity off the top of the tower (or rectifier/analyzer) is generally less than that in the example. As a result, the tower and condenser pressure are less, but water contamination of the refrigerant is larger. An alternative to providing a refrigerant heat exchanger is to constantly bleed liquid as blow-down (spillover) from the evaporator to the absorber.

In large systems with blow-down (spillover), a vertical liquid leg under the evaporator provides a relatively inactive area and accumulates ammonia that is rich in water. The blow-down (spillover) line taps into this liquid leg. At an evaporator pressure of 75 psia, a pool evaporator containing 10% by mass water increases the evaporator temperature from 41.1 to 45.8°F for a 4.7°F penalty (Jennings and Shannon 1938). However, if allowed to accumulate without blow-down, the water content will exceed 10% eventually and reach a point where the evaporator temperature is too high to produce the desired cooling.

In lithium bromide-water systems, the cooling tower water is fed in series to the absorber and then to the condenser. In ammonia-water systems, the cooling tower water is fed first to the condenser to keep the high-end pressure as low as possible. In both cases, the cooling water can be piped in parallel to improve efficiencies; however, this requires high coolant flow rates and excessively large cooling towers.

In the ammonia-water cycle, the reflux for the rectifier can be created by a separate condenser or by the main condenser. Reflux

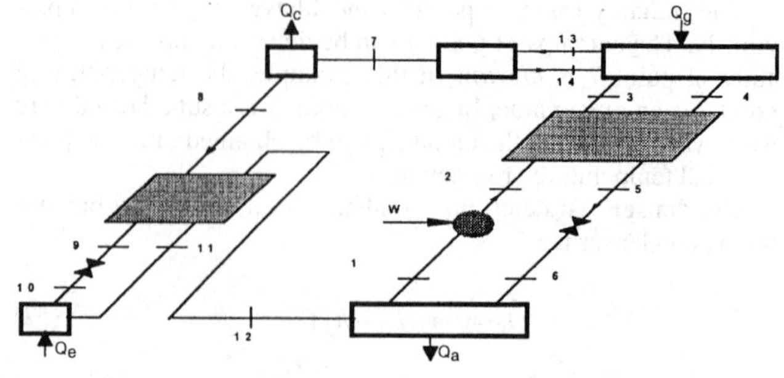

Fig. 21 Single-Effect Ammonia/Water Absorption Cycle

Table 9 Inputs and Assumptions for Single-Effect Ammonia/Water Model of Figure 21

Inputs		
Capacity	$\dot{Q}_e$	500 tons refrig.
High side pressure	p_{high}	211.8 psia
Low side pressure	p_{low}	74.7 psia
Absorber exit temperature	t_1	105°F
Generator exit temperature	t_4	203°F
Rectifier vapor exit temperature	t_7	131°F
Solution heat exchanger eff.	ε_{shx}	0.692
Refrigerant heat exchanger eff.	ε_{rhx}	0.629

Assumptions

Steady state
No pressure changes except through the flow restrictors and the pump
States at points 1, 4, 8, 11, and 14 are saturated liquid
States at point 12 and 13 are saturated vapor
Flow restrictors are adiabatic
Pump is isentropic
No jacket heat losses
No liquid carryover from evaporator to absorber
Vapor leaving the generator is at the equilibrium temperature of the entering solution stream

can flow by gravity or it can be pumped to the top of the tower. Ammonia-water machines do not experience crystallization as lithium bromide-water machines do, so controls can be simpler. Also, the corrosion characteristics of ammonia-water solutions are less severe, although inhibitors are generally used for both systems. Lithium bromide-water systems use combinations of steel, copper, and copper-nickel materials for shells and heat transfer surfaces; but no copper-bearing materials can be used in ammonia-water systems because ammonia rapidly destroys copper.

The mass and energy balances around the cycle shown in Figure 21 were calculated based on the assumptions and input values in Table 9. The conditions were selected to approximately match the conditions chosen for the lithium bromide/water cycle discussed in the previous section. Note that the ammonia/water working fluid allows significantly lower evaporator temperature than that found in this example. The cycle solution is summarized in Table 10 where state point data for all connecting points are given. For most components, the mass and energy balances are performed in an identical fashion as that discussed earlier in relation to the lithium bromide/water example. However, because the ammonia/water cycle has two new components, some additional discussion is needed.

The rectifier model assumes that the rectification process is reversible. This provides a thermodynamic lower bound for the heat transfer required in the rectifier (and thus gives an upper bound on the COP). The key aspect of the reversible model is the assumption that the reflux leaving the rectifier (14) is in equilibrium with the vapor entering the rectifier (13). In a real rectifier, potential differences would be needed to drive the purification process and these potential differences would be evident at the bottom of the column.

Another aspect of the reversible model is the requirement that heat be extracted from the column along the entire length. In a more typical column design, the body of the column would be adiabatic and all the heat transfer would occur in the reflux condenser at the top of the column. These differences between the reversible model and a real rectifier are significant, but the simplified model is used here because it shows the overall trends. Bogart (1981) provides a more detailed discussion of rectifier design.

The refrigerant heat exchanger uses a single phase fluid on the hot side (8–9) and an evaporating mixture on the cold side (11–12). Because the flow rate is the same on both sides, the temperature profiles are not expected to match. However, because the composition of the evaporating fluid changes as it passes through the heat exchanger, the temperature also changes significantly. As a result

Table 10 State Point Data for Single-Effect Ammonia/Water Cycle of Figure 21

No.	h Btu/lb	$\dot{m}$ lb/min	p psia	Q fraction	t °F	x, fraction NH_3
1	−24.55	1408.2	74.7	0.0	105.0	0.50094
2	−24.05	1408.2	211.8		105.5	0.50094
3	38.47	1408.2	211.8		163.0	0.50094
4	83.81	1203.0	211.8	0.0	203.0	0.41612
5	10.61	1203.0	211.8		135.6	0.41612
6	10.61	1203.0	74.7	0.006	132.0	0.41612
7	579.51	205.2	211.8	1.0	131.0	0.99809
8	76.61	205.2	211.8	0.0	100.1	0.99809
9	35.28	205.2	211.8		64.1	0.99809
10	35.28	205.2	74.7	0.049	41.1	0.99809
11	522.55	205.2	74.7	0.953	42.8	0.99809
12	563.88	205.2	74.7	1.0	87.0	0.99809
13	613.91	209.9	211.8	1.0	174.5	0.98708
14	51.72	4.6	211.8	0.0	174.5	0.50094

$COP_c = 0.571$	$\dot{Q}_e = 6.00 \times 10^6$ Btu/h
$\Delta t_{rhx} = 36.00$°F	$\dot{Q}_g = 1.051 \times 10^7$ Btu/h
$\Delta t_{shx} = 30.1$°F	$\dot{Q}_{rhx} = 5.089 \times 10^5$ Btu/h
$\varepsilon_{rhx} = 0.629$	$\dot{Q}_r = 5.805 \times 10^5$ Btu/h
$\varepsilon_{shx} = 0.692$	$\dot{Q}_{shx} = 5.283 \times 10^6$ Btu/h
$\dot{Q}_a = 9.784 \times 10^6$ Btu/h	$\dot{W} = 9.22$ hp
$\dot{Q}_c = 6.192 \times 10^6$ Btu/h	

the temperature profiles match reasonably well. Table 10 shows that the terminal temperature differences are 21.3°F and 13.1°F at the 9–11 and 8–12 ends respectively. The fact that the temperature profiles match well is one of the reasons why this heat exchanger works so effectively.

While operating under nearly identical conditions the overall COP of the ammonia/water cycle is 0.57 as compared to 0.70 for the lithium bromide/water example. The reduction in COP for the ammonia/water cycle can be traced to two major factors: (1) the solution heat exchanger and (2) the rectifier. The solution heat exchanger in the ammonia/water cycle carries more load due to the larger specific heat of the liquid solution as compared to aqueous lithium bromide. Typical specific heat values at the conditions of interest are 1.1 Btu/lb·°F for liquid ammonia/water and 0.45 Btu/lb·°F for aqueous lithium bromide, as computed for the solution heat exchanger examples in Table 6 and Table 10. This difference between the two working fluids implies greater sensitivity to solution heat exchanger design for ammonia/water cycles.

The rectifier heat loss is overwhelmed in this example by the solution heat exchanger loss. An estimate can be obtained by comparing the heat transfer rate in the solution heat exchangers from the two examples. In general, the losses in such a device are approximately proportional to the heat transfer rate and the average temperature difference driving the heat transfer. The fact that the heat transfer duty in the ammonia/water cycle is much greater due to the high specific heat of the liquid is the key to the difference.

NOMENCLATURE

c_p specific heat at constant pressure
COP coefficient of performance
g local acceleration of gravity
h enthalpy, Btu/lb
I irreversibility
$\dot{I}$ irreversibility rate
m mass
$\dot{m}$ mass flow, lb/min
p pressure
Q heat energy, Btu
$\dot{Q}$ rate of heat flow, Btu/h

R	ideal gas constant
s	entropy, Btu/lb·°R
S	total entropy
t	temperature, °F
T	absolute temperature, °R
u	internal energy
W	mechanical or shaft work
$\dot{W}$	rate of work, power
v	specific volume, ft^3/lb
V	velocity of fluid
x	mass fraction (of either lithium bromide or ammonia)
x	vapor quality (fraction)
z	elevation above horizontal reference plane
Z	compressibility factor
ε	heat exchanger effectiveness
η	efficiency

Subscripts

a	absorber
c	condenser or cooling mode
C	condensing conditions
cg	condenser to generator
d	desorber (generator)
e	evaporator
fg	fluid to vapor
gh	high temperature generator
o, 0	reference conditions, usually ambient
p	pump
R	refrigerating or evaporator conditions
rhx	refrigerant heat exchanger
shx	solution heat exchanger

REFERENCES

Benedict, M., G.B. Webb, and L.C. Rubin. 1940. An empirical equation for thermodynamic properties of light hydrocarbons and their mixtures. *Journal of Chemistry and Physics* 4:334.

Benedict, M. 1937. Pressure, volume, temperature properties of nitrogen at high density, I and II. *Journal of American Chemists Society* 59(11): 2224.

Cooper, H.W. and J.C. Goldfrank. 1967. B-W-R Constants and new correlations. *Hydrocarbon Processing* 46(12):141.

Hirschfelder, J.O. et al. 1958. Generalized equation of state for gases and liquids. *Industrial and Engineering Chemistry* 50:375.

Howell, J.R. and R.O. Buckius. 1992. *Fundamentals of Engineering Thermodynamics*, 2nd Ed. McGraw-Hill, New York.

Hust, J.G. and R.D. McCarty. 1967. Curve-fitting techniques and applications to thermodynamics. *Cryogenics* 8:200.

Hust, J.G. and R.B. Stewart. 1966. Thermodynamic property computations for system analysis. *ASHRAE Journal* 2:64.

Jacob, X., L.F. Albright, and W.H. Tucker. 1969. Factors affecting the coefficient of performance for absorption air-conditioning systems. *ASHRAE Transactions* 75(1):103.

Jennings, B.H. and F.P. Shannon. 1938. The thermodynamics of absorption refrigeration. *Refrigerating Engineering* 35(5):338.

Kuehn, T.H. and R.E. Gronseth. 1986. The effect of a nonazeotropic binary refrigerant mixture on the performance of a single stage refrigeration cycle. Proceedings *International Institute of Refrigeration Conference*, Purdue University, p. 119.

Liang, H. and T.H. Kuehn. 1991. Irreversibility analysis of a water to water mechanical compression heat pump. *Energy* 16(6):883.

Macriss, R.A. 1968. Physical properties of modified LiBr solutions. AGA Symposium on Absorption Air-Conditioning Systems, February.

Macriss, R.A. and T.S. Zawacki. 1989. Absorption fluid data survey: 1989 update. Oak Ridge National Laboratories *Report* ORNL/Sub84-47989/4.

Martin, J.J. and Y. Hou. 1955. Development of an equation of state for gases. AICHE *Journal* 1:142.

Martz, W.L., C.M. Burton, and A.M. Jacobi. 1996a. Liquid-vapor equilibria for R-22, R-134a, R-125, and R-32/125 with a polyol ester lubricant: Measurements and departure from ideality. ASHRAE *Transactions* 102(1):367-74.

Martz, W.L., C.M. Burton and A.M. Jacobi. 1996b. Local composition modeling of the thermodynamic properties of refrigerant and oil mixtures. *International Journal of Refrigeration* 19(1):25-33.

McNeely, L.A. 1979. Thermodynamic properties of aqueous solution of lithium bromide. *ASHRAE Transactions* 85(1):413.

NASA. 1971. SP-273. US Government Printing Office, Washington, D.C.

Stewart, R.B., R.T. Jacobsen, and S.G. Penoncello. 1986. *ASHRAE Thermodynamic properties of refrigerants*. ASHRAE, Atlanta, GA.

Strobridge, T.R. 1962. The thermodynamic properties of nitrogen from 64 to 300 K, between 0.1 and 200 atmospheres. National Bureau of Standards Technical Note 129.

Stoecker, W.F. 1989. *Design of thermal systems*, 3rd Ed. McGraw-Hill, New York.

Tassios, D.P. 1993. *Applied chemical engineering thermodynamics*. Springer-Verlag, New York.

Thome, J.R. 1995. Comprehensive thermodynamic approach to modeling refrigerant-lubricant oil mixtures. *International Journal of Heating, Ventilating, Air Conditioning and Refrigeration Research* 1(2):110.

BIBLIOGRAPHY

Alefeld, G. and R. Radermacher. 1994. *Heat conversion systems*. CRC Press, Boca Raton.

Bogart, M. 1981. *Ammonia absorption refrigeration in industrial processes*. Gulf Publishing Co., Houston, TX.

Briggs, S.W. 1971. Concurrent, crosscurrent, and countercurrent absorption in ammonia-water absorption refrigeration. *ASHRAE Transactions* 77(1):171.

Herold, K.E., R. Radermacher, and S.A. Klein. 1996. *Absorption chillers and heat pumps*. CRC Press, Boca Raton.

Jain, P.C. and G.K. Gable. 1971. Equilibrium property data for aqua-ammonia mixture. *ASHRAE Transactions* 77(1):149.

Moran, M.J. and Shapiro, H. 1995. *Fundamentals of engineering thermodymanics*, 3rd Ed. John Wiley and Sons, Inc. New York.

Stoecker, W.F. and L.D. Reed. 1971. Effect of operating temperatures on the coefficient of performance of aqua-ammonia refrigerating systems. *ASHRAE Transactions* 77(1):163.

Van Wylen, C.J. and R.E. Sonntag. 1985. *Fundamentals of classical thermodynamics*, 3rd ed. John Wiley and Sons, Inc., New York.

CHAPTER 2

FLUID FLOW

FLOWING fluids in heating, ventilating, air-conditioning, and refrigeration systems transfer heat or mass (as particles). This chapter introduces the basics of fluid mechanics that are related to HVAC processes, reviews pertinent flow processes, and presents a general discussion of single-phase fluid flow analysis.

FLUID PROPERTIES

Fluids differ from solids in their reaction to shearing. When placed under shear stress, a solid deforms only a finite amount, whereas a fluid deforms continuously for as long as the shear is applied. Both liquids and gases are fluids. Although liquids and gases differ strongly in the nature of molecular actions, their primary mechanical differences are in the degree of compressibility and liquid formation of a free surface (interface).

Fluid motion can usually be described by one of several simplified modes of action or models. The simplest is the ideal-fluid model, which assumes no resistance to shearing; flow analysis is well developed (Baker 1983, Schlichting 1979, Streeter and Wylie 1979), and when properly interpreted, is valid for a wide range of applications. Nevertheless, the effects of viscous action may need to be considered. Most fluids in HVAC applications can be treated as Newtonian, where the deformation is directly proportional to the shearing stress. Turbulence, which complicates fluid behavior, does not depend on the viscous nature of a fluid, but viscosity does tend to influence turbulence.

Density

The density ρ of a fluid is its mass per unit volume. The densities of air and water at standard conditions of 68°F and 14.696 psi (sea level atmospheric pressure) are

$$\rho_{water} = 62.3 \text{ lb}_m/\text{ft}^3$$

$$\rho_{air} = 0.075 \text{ lb}_m/\text{ft}^3$$

Viscosity

Viscosity is the resistance of adjacent fluid layers to shear. For shearing between two parallel plates, each of area A and separated by distance Y, the tangential force F per unit area required to slide one plate with velocity V parallel to the other is proportional to V/Y:

$$F/A = \mu(V/Y)$$

where the proportionality factor μ is the **absolute viscosity** or **dynamic viscosity** of the fluid. The ratio of the tangential force F to area A is the **shearing stress** τ, and V/Y is the **lateral velocity gradient** (Figure 1A). In complex flows, velocity and shear stress may

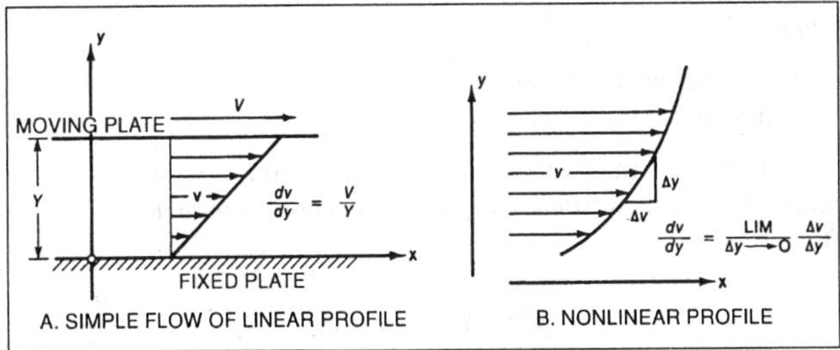

Fig. 1 Velocity Profiles and Gradients in Shear Flows

vary across the flow field; this is expressed by the following differential equation:

$$\tau = \mu \frac{dv}{dy} \tag{1}$$

The velocity gradient associated with viscous shear for a simple case involving flow velocity in the x direction but of varying magnitude in the y direction is illustrated in Figure 1B.

Absolute viscosity μ depends primarily on temperature. For gases (except near the critical point), viscosity increases with the square root of the absolute temperature, as predicted by the kinetic theory. Liquid viscosity decreases with increasing temperature. Viscosities of various fluids are given in Chapter 36.

Absolute viscosity has dimensions of force · time/length^2. At standard conditions, the absolute viscosities of water and air are

$$\mu_{water} = 6.7 \times 10^{-4} \text{ lb}_m/\text{ft} \cdot \text{s} = 2.1 \times 10^{-5} \text{ lb}_f \cdot \text{s}/\text{ft}^2$$

$$\mu_{air} = 1.2 \times 10^{-5} \text{ lb}_m/\text{ft} \cdot \text{s} = 3.7 \times 10^{-7} \text{ lb}_f \cdot \text{s}/\text{ft}^2$$

The centipoise is another common unit of viscosity (1 centipoise = 1 g/(s·m) = 1 mPa·s). At standard conditions, water has a viscosity close to 1.0 centipoise.

In fluid dynamics, **kinematic viscosity** ν is the ratio of absolute viscosity to density:

$$\nu = \mu/\rho$$

The stoke (1 cm^2/s) and centistoke (1 mm^2/s) are common units for kinematic viscosity. At standard conditions, the kinematic viscosities of water and air are

$$\nu_{water} = 1.08 \times 10^{-5} \text{ ft}^2/\text{s}$$

$$\nu_{air} = 1.7 \times 10^{-4} \text{ ft}^2/\text{s}$$

Note that the inch-pound system of units requires the conversion factor $g_c = 32.1740 \text{ lb}_m \cdot \text{ft/s}^2 \cdot \text{lb}_f$ to make some equations containing

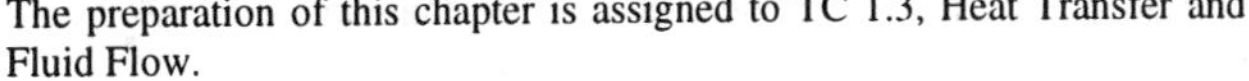

The preparation of this chapter is assigned to TC 1.3, Heat Transfer and Fluid Flow.

lb_f and lb_m dimensionally consistent. The conversion factor g_c is not shown in equations in this chapter but is to be used as needed.

BASIC RELATIONS OF FLUID DYNAMICS

This section considers homogeneous, constant-property, incompressible fluids and introduces fluid dynamic considerations used in most analyses.

Continuity

Conservation of matter applied to fluid flow in a conduit requires that

$$\int \rho v \, dA = \text{constant}$$

where

v = velocity normal to the differential area dA
ρ = fluid density

Both ρ and v may vary over the cross section A of the conduit. If both ρ and v are constant over the cross-sectional area normal to the flow, then

$$\dot{m} = \rho V A = \text{constant} \quad (2a)$$

where $\dot{m}$ is the mass flow rate across the area normal to the flow. When flow is effectively incompressible, ρ = constant; in pipeline and duct flow analyses, the average velocity is then $V = (1/A)\int v dA$. The continuity relation is

$$Q = AV = \text{constant} \quad (2b)$$

where Q is the volumetric flow rate. Except when branches occur, Q is the same at all sections along the conduit.

For the ideal-fluid model, flow patterns around bodies (or in conduit section changes) result from displacement effects. An obstruction in a fluid stream, such as a strut in a flow or a bump on the conduit wall, pushes the flow smoothly out of the way, so that behind the obstruction, the flow becomes uniform again. The effect of fluid inertia (density) appears only in pressure changes.

Pressure Variation Across Flow

Pressure variation in fluid flow is important and can be easily measured. Variation across streamlines involves fluid rotation (vorticity). Lateral pressure variation across streamlines is given by the following relation (Bober and Kenyon 1980, Olson 1980, Robertson 1965):

$$\frac{\partial}{\partial r}\left(\frac{p}{\rho} + gz\right) = \frac{v^2}{r} \quad (3)$$

where

r = radius of curvature of the streamline
z = elevation

This relation explains the pressure difference found between the inside and outside walls of a bend and near other regions of conduit section change. It also states that pressure variation is hydrostatic ($p + \rho gz$ = constant) across any conduit where streamlines are parallel.

Bernoulli Equation and Pressure Variation along Flow

A basic tool of fluid flow analysis is the Bernoulli relation, which involves the principle of energy conservation along a streamline. Generally, the Bernoulli equation is not applicable across streamlines. The first law of thermodynamics can be applied to mechanical flow energies (kinetic and potential) and thermal energies: heat is a form of energy and energy is conserved.

The change in energy content ΔE per unit mass of flowing material is a result from the work W done on the system plus the heat Q absorbed:

$$\Delta E = W + Q$$

Fluid energy is composed of kinetic, potential (due to elevation z), and internal (u) energies. Per unit mass of fluid, the above energy change relation between two sections of the system is

$$\Delta\left(\frac{v^2}{2} + gz + u\right) = E_M - \Delta\left(\frac{p}{\rho}\right) + Q$$

where the work terms are (1) the external work E_M from a fluid machine (E_M is positive for a pump or blower) and (2) the pressure or flow work p/ρ. Rearranging, the energy equation can be written as the *generalized Bernoulli equation*:

$$\Delta\left(\frac{v^2}{2} + gz + \frac{p}{\rho}\right) + \Delta u = E_M + Q \quad (4)$$

The term in parentheses in Equation (4) is the *Bernoulli constant*:

$$\frac{p}{\rho} + \frac{v^2}{2} + gz = \pi \quad (5a)$$

In cases with no viscous action and no work interaction, π is constant; more generally its change (or lack thereof) is considered in applying the Bernoulli equation. The terms making up π are fluid energies (pressure, kinetic, and potential) per mass rate of fluid flow. Alternative forms of this relation are obtained through multiplication by ρ or division by g:

$$p + \frac{\rho v^2}{2} + \rho gz = p + \frac{\rho v^2}{2} + \gamma z = \rho\pi \quad (5b)$$

$$\frac{p}{\rho g} + \frac{v^2}{2g} + z = \frac{p}{\gamma} + \frac{v^2}{2g} + z = \frac{\pi}{g} \quad (5c)$$

where $\gamma = \rho g$ is the specific weight.

The first form involves energies per volume flow rate, or pressures; the second involves energies per mass flow rate, or heads. In gas flow analysis, Equation (5b) is often used with the γz term dropped as negligible. Equation (5a) should be used when density variations occur. For liquid flows, Equation (5c) is commonly used. Identical results are obtained with the three forms if the units are consistent and the fluids are homogeneous.

Many systems of pipes or ducts and pumps or blowers can be considered as one-dimensional flow. The Bernoulli equation is then considered as velocity and pressure vary along the conduit. Analysis is adequate in terms of the section-average velocity V of Equation (2a) or (2b). In the Bernoulli relation [Equations (4) and (5)], v is replaced by V, and variation across streamlines can be ignored; the whole conduit is now taken as one streamline. Two- and three-dimensional details of local flow occurrences are still significant, but their effect is combined and accounted for in factors.

The kinetic energy term of the Bernoulli constant is expressed as $\alpha V^2/2$, where the **kinetic energy factor** ($\alpha > 1$) expresses the ratio of the true kinetic energy of the velocity profile to that of the mean flow velocity.

For laminar flow in a wide rectangular channel, α = 1.54. For a pipe, α = 2.0.

Heat transfer Q may often be ignored. The change of mechanical energy into internal energy Δu may be expressed as E_L. Flow analysis involves the change in the Bernoulli constant ($\Delta\pi = \pi_2 - \pi_1$) between stations 1 and 2 along the conduit, and the Bernoulli equation can be expressed as

$$\left(\frac{p}{\rho} + \alpha\frac{V^2}{2} + gz\right)_1 + E_M = \left(\frac{p}{\rho} + \alpha\frac{V^2}{2} + gz\right)_2 + E_L \quad (6a)$$

or, dividing by g, in head form as

$$\left(\frac{p}{\gamma} + \alpha\frac{V^2}{2g} + z\right)_1 + H_M = \left(\frac{p}{\gamma} + \alpha\frac{V^2}{2g} + z\right)_2 + H_L \quad (6b)$$

The factors E_M and E_L are defined as positive, where $gH_M = E_M$ represents energy added to the conduit flow by pumps or blowers, and $gH_L = E_L$ represents energy dissipated, that is, converted into heat as mechanically nonrecoverable energy. A turbine or fluid motor thus has a negative H_M or E_M. For conduit systems with branches involving inflow or outflow, the total energies must be treated, and analysis is in terms of $\dot{m}\pi$ and not π.

When real-fluid effects of viscosity or turbulence are included, the continuity relation in Equation (2b) is not changed, but V must be evaluated from the integral of the velocity profile, using time-averaged local velocities.

In fluid flow past fixed boundaries, the velocity at the boundary is zero and shear stresses are produced. The equations of motion then become complex and exact solutions are difficult to find, except in simple cases.

Laminar Flow

For steady, fully developed laminar flow in a parallel-walled conduit, the shear stress τ varies linearly with distance y from the centerline. For a wide rectangular channel,

$$\tau = \left(\frac{y}{b}\right)\tau_w = \mu\frac{dv}{dy}$$

where

τ_w = wall shear stress = $b\,(dp/ds)$
$2b$ = wall spacing
s = flow direction

Because the velocity is zero at the wall ($y = b$), the integrated result is

$$v = \left(\frac{b^2 - y^2}{2\mu}\right)\frac{dp}{ds}$$

This is the *Poiseuille-flow parabolic velocity profile* for a wide rectangular channel. The average velocity V is two-thirds the maximum velocity (at $y = 0$), and the longitudinal pressure drop in terms of conduit flow velocity is

$$\frac{dp}{ds} = -\left(\frac{3\mu V}{b^2}\right) \quad (7)$$

The parabolic velocity profile can also be derived for the axisymmetric conduit (pipe) of radius R but with a different constant. The average velocity is then half the maximum, and the pressure drop relation is

$$\frac{dp}{ds} = -\left(\frac{8\mu V}{R^2}\right) \quad (8)$$

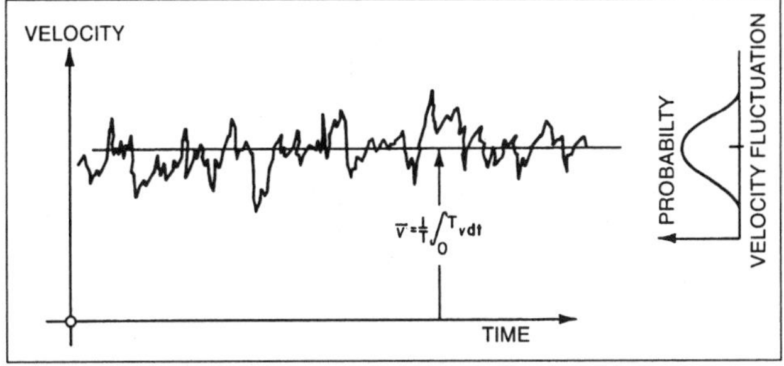

Fig. 2 Velocity Fluctuation at Point in Turbulent Flow

Turbulence

Fluid flows are generally turbulent, involving random perturbations or fluctuations of the flow (velocity and pressure), characterized by an extensive hierarchy of scales or frequencies (Robertson 1963). Flow disturbances that are not random, but have some degree of periodicity, such as the oscillating vortex trail behind bodies, have been erroneously identified as turbulence. Only flows involving random perturbations without any order or periodicity are turbulent; the velocity in such a flow varies with time or locale of measurement (Figure 2).

Turbulence can be quantified by statistical factors. Thus, the velocity most often used in velocity profiles is the temporal average velocity $\bar{v}$, and the strength of the turbulence is characterized by the root-mean-square of the instantaneous variation in velocity about this mean. The effects of turbulence cause the fluid to diffuse momentum, heat, and mass very rapidly across the flow.

The **Reynolds number** Re, a dimensionless quantity, gives the relative ratio of inertial to viscous forces:

$$\text{Re} = VL/\nu$$

where

L = characteristic length
ν = kinematic viscosity

In flow through round pipes and tubes, the characteristic length is the diameter D. Generally, laminar flow in pipes can be expected if the Reynolds number, which is based on the pipe diameter, is less than 2000. Fully turbulent flow exists when $\text{Re}_D > 10{,}000$. Between 2000 and 10,000, the flow is in a transition state and predictions are unreliable. In other geometries, different criteria for the Reynolds number exist.

BASIC FLOW PROCESSES

Wall Friction

At the boundary of real-fluid flow, the relative tangential velocity at the fluid surface is zero. Sometimes in turbulent flow studies, velocity at the wall may appear finite, implying a fluid slip at the wall. However, this is not the case; the difficulty is in velocity measurement (Goldstein 1938). Zero wall velocity leads to a high shear stress near the wall boundary and a slowing down of adjacent fluid layers. A velocity profile develops near a wall, with the velocity increasing from zero at the wall to an exterior value within a finite lateral distance.

Laminar and turbulent flow differ significantly in their velocity profiles. Turbulent flow profiles are flat compared to the more pointed profiles of laminar flow (Figure 3). Near the wall, velocities of the turbulent profile must drop to zero more rapidly than those of the laminar profile, so the shear stress and friction are much greater in the turbulent flow case. Fully developed conduit flow may be characterized by the **pipe factor**, which is the ratio of average to maximum (centerline) velocity. Viscous velocity profiles result in

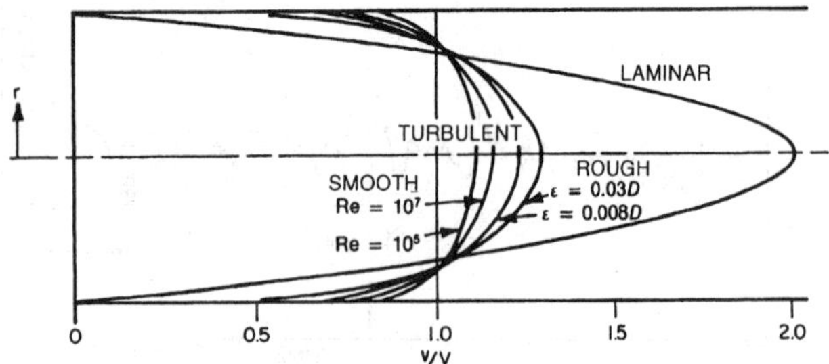

Fig. 3 Velocity Profiles of Flow in Pipes

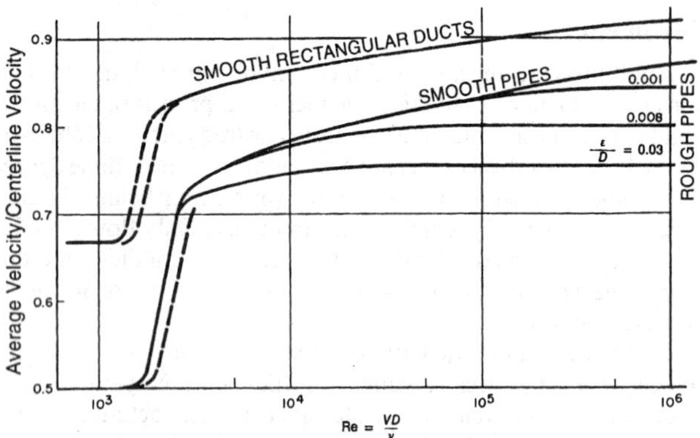

Fig. 4 Pipe Factor for Flow in Conduits

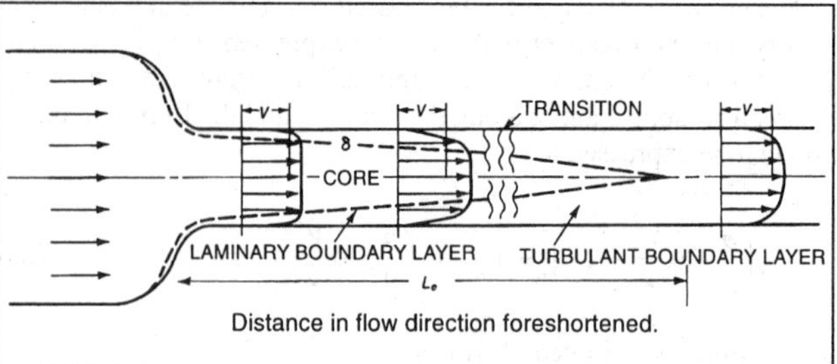

Fig. 5 Flow in Conduit Entrance Region

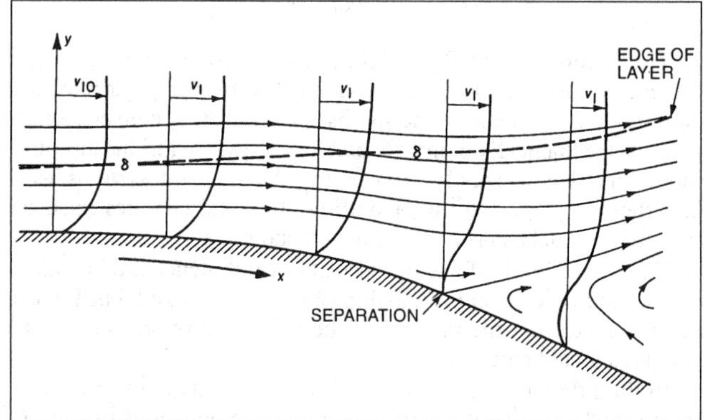

Fig. 6 Boundary Layer Flow to Separation

pipe factors of 0.667 and 0.50 for wide rectangular and axisymmetric conduits. Figure 4 indicates much higher values for rectangular and circular conduits for turbulent flow. Due to the flat velocity profiles, the kinetic energy factor α in Equation (6) ranges from 1.01 to 1.10 for fully developed turbulent pipe flow.

Boundary Layer

In most flows, the friction of a bounding wall on the fluid flow is evidenced by a boundary layer. For flow around bodies, this layer (which is quite thin relative to distances in the flow direction) encompasses all viscous or turbulent actions, causing the velocity in it to vary rapidly from zero at the wall to that of the outer flow at its edge. Boundary layers are generally laminar near the start of their formation but may become turbulent downstream of the transition point (Figure 5). For conduit flows, spacing between adjacent walls is generally small compared with distances in the flow direction. As a result, layers from the walls meet at the centerline to fill the conduit.

A significant boundary-layer occurrence exists in a pipeline or conduit following a well-rounded entrance (Figure 5). Layers grow from the walls until they meet at the center of the pipe. Near the start of the straight conduit, the layer is very thin (and laminar in all probability), so the uniform velocity core outside has a velocity only slightly greater than the average velocity. As the layer grows in thickness, the slower velocity near the wall requires a velocity increase in the uniform core to satisfy continuity. As the flow proceeds, the wall layers grow (and the centerline velocity increases) until they join, after an entrance length L_e. Application of the Bernoulli relation of Equation (5) to the core flow indicates a decrease in pressure along the layer. Ross (1956) shows that although the entrance length L_e is many diameters, the length in which the pressure drop and head loss rates significantly exceed those for fully developed flow is on the order of 10 diameters for turbulent flow in smooth pipes.

In more general boundary-layer flows, as with wall layer development in a diffuser or for the layer developing along the surface of a strut or turning vane, pressure gradient effects can be severe and may even lead to separation. The development of a layer in an adverse-pressure gradient situation (velocity v_1 at edge $y = \delta$ of layer decreasing in flow direction) with separation is shown in Figure 6. Downstream from the separation point, fluid backflows near the wall. Separation is due to frictional velocity (thus local kinetic energy) reduction near the wall. Flow near the wall no longer has energy to move into the higher pressure imposed by the decrease in v_1 at the edge of the layer. The locale of this separation is difficult to predict, especially for the turbulent boundary layer. Analyses verify the experimental observation that a turbulent boundary layer is less subject to separation than a laminar one because of its greater kinetic energy.

Flow Patterns with Separation

In technical applications, flow with separation is common and often accepted if it is too expensive to avoid. Flow separation may be geometric or dynamic. Dynamic separation is shown in Figure 6. Geometric separation (Figures 7 and 8) results when a fluid stream passes over a very sharp corner, as with an orifice; the fluid generally leaves the corner irrespective of how much its velocity has been reduced by friction.

For geometric separation in orifice flow (Figure 7), the outer streamlines separate from the sharp corners and, because of fluid inertia, contract to a section smaller than the orifice opening, the **vena contracta**, with a limiting area of about six-tenths of the orifice opening. After the vena contracta, the fluid stream expands rather slowly through turbulent or laminar interaction with the fluid along its sides. Outside the jet, fluid velocity is small compared with that in the jet and is very disturbed. Strong turbulence vorticity helps spread out the jet, increases the losses, and brings the velocity distribution back to a more uniform profile. Finally, at a considerable distance downstream, the velocity profile returns to the fully developed flow of Figure 3.

Other geometric separations (Figure 8) occur at a sharp entrance to a conduit, at an inclined plate or damper in a conduit, and at a sudden expansion. For these, a vena contracta can be identified; for

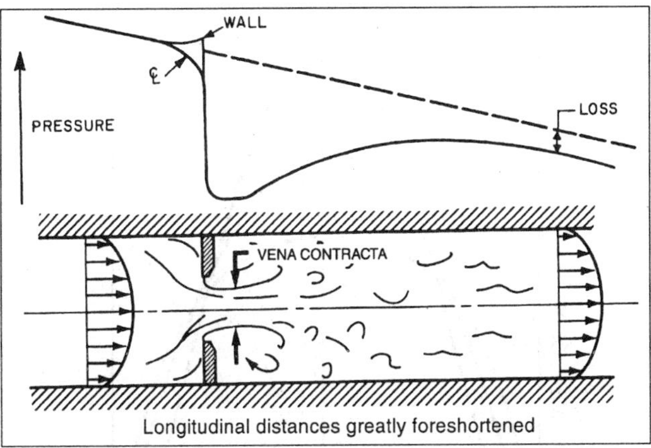

Fig. 7 Geometric Separation, Flow Development, and Loss in Flow Through Orifice

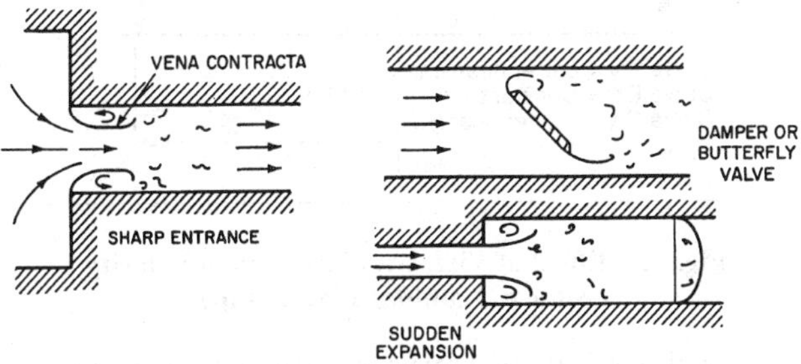

Fig. 8 Examples of Geometric Separation Encountered in Flows in Conduits

sudden expansion, its area is that of the upstream contraction. Ideal-fluid theory, using free streamlines, provides insight and predicts contraction coefficients for valves, orifices, and vanes (Robertson 1965). These geometric flow separations are large loss-producing devices. To expand a flow efficiently or to have an entrance with minimum losses, the device should be designed with gradual contours, a diffuser, or a rounded entrance.

Flow devices with gradual contours are subject to separation that is more difficult to predict, because it involves the dynamics of boundary layer growth under an adverse pressure gradient rather than flow over a sharp corner. In a diffuser, which is used to reduce the loss in expansion, it is possible to expand the fluid some distance at a gentle angle without difficulty (particularly if the boundary layer is turbulent). Eventually, separation may occur (Figure 9), which is frequently asymmetrical because of irregularities. Downstream flow involves flow reversal (backflow) and excess losses exist. Such separation is termed **stall** (Kline 1959). Larger area expansions may use splitters that divide the diffuser into smaller divisions less likely to have separations (Moore and Kline 1958). Another technique for controlling separation is to bleed some low-velocity fluid near the wall (Furuya et al. 1976). Alternatively, Heskested (1965, 1970) shows that suction at the corner of a sudden expansion has a strong positive effect on geometric separation.

Drag Forces on Bodies or Struts

Bodies in moving fluid streams are subjected to appreciable fluid forces or drag. Conventionally expressed in coefficient form, drag forces on bodies can be expressed as

$$D = C_D \rho A V^2/2 \tag{9}$$

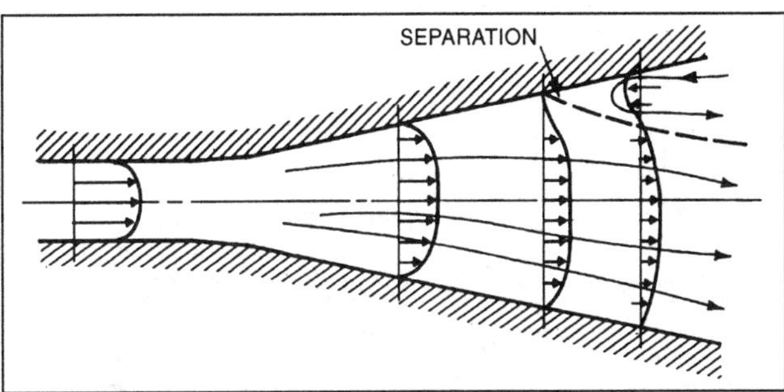

Fig. 9 Separation in Flow in Diffuser

Table 1 Drag Coefficients

Body Shape	$10^3 < \text{Re} < 2 \times 10^5$	$\text{Re} > 3 \times 10^5$
Sphere	0.36 to 0.47	~0.1
Disk	1.12	1.12
Streamlined strut	0.1 to 0.3	< 0.1
Circular cylinder	1.0 to 1.1	0.35
Elongated rectangular strut	1.0 to 1.2	1.0 to 1.2
Square strut	~2.0	~2.0

where A is the projected (normal to flow) area of the body. The **drag coefficient** C_D depends on the body's shape and angularity and on the Reynolds number of the relative flow in terms of the body's characteristic dimension.

For Reynolds numbers of 10^3 to above 10^5, the C_D of most bodies is constant due to flow separation, but above 10^5, the C_D of rounded bodies drops suddenly as the surface boundary layer undergoes transition to turbulence. Typical C_D values are given in Table 1; Hoerner (1965) gives expanded values.

For a strut crossing a conduit, the contribution to the head loss of Equation (6b) is

$$H_L = C_D \left(\frac{A}{A_c}\right)\left(\frac{V^2}{2g}\right) \tag{10}$$

where

A_c = conduit cross-sectional area
A = area of the strut facing the flow

Cavitation

Liquid flow with gas- or vapor-filled pockets can occur if the absolute pressure is reduced to vapor pressure or less. In this case, a cavity or series of cavities forms, because liquids are rarely pure enough to withstand any tensile stressing or pressures less than vapor pressure for any length of time (John and Haberman 1980, Knapp et al. 1970, Robertson and Wislicenus 1969). Robertson and Wislicenus (1969) indicate significant occurrences in various technical fields, chiefly in hydraulic equipment and turbomachines.

Initial evidence of cavitation is the collapse noise of many small bubbles that appear initially as they are carried by the flow into regions of higher pressure. The noise is not deleterious and serves as a warning of the occurrence. As flow velocity further increases or pressure decreases, the severity of cavitation increases. More bubbles appear and may join to form large fixed cavities. The space they occupy becomes large enough to modify the flow pattern and alter performance of the flow device. Collapse of the cavities on or near solid boundaries becomes so frequent that the cumulative impact in time results in damage in the form of cavitational erosion of the surface or excessive vibration. As a result, pumps can lose efficiency or their parts may erode locally. Control valves may be noisy or seriously damaged by cavitation.

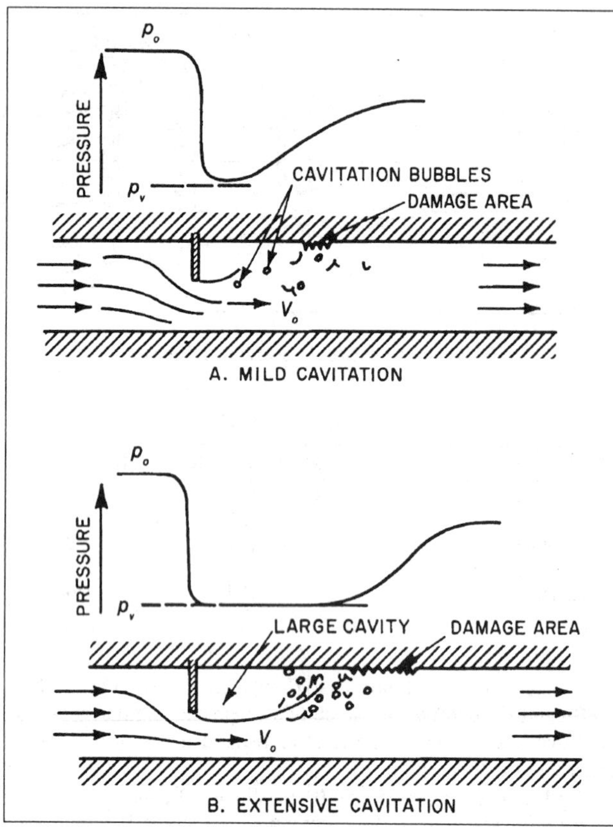

Fig. 10 Cavitation in Flows in Orifice or Valve

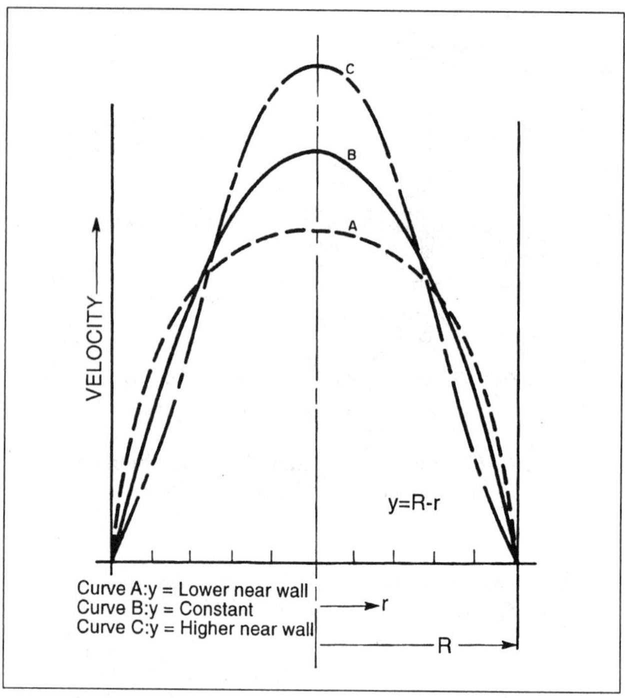

Fig. 11 Effect of Viscosity Variation on Velocity Profile of Laminar Flow in Pipe

Cavitation in orifice and valve flow is indicated in Figure 10. With high upstream pressure and a low flow rate, no cavitation occurs. As pressure is reduced or flow rate increased, the minimum pressure in the flow (in the shear layer leaving the edge of the orifice) eventually approaches vapor pressure. Turbulence in this layer causes fluctuating pressures below the mean (as in vortex cores) and small bubble-like cavities. These are carried downstream into the region of pressure regain where they collapse, either in the fluid or on the wall (Figure 10A). As the pressure is reduced, more vapor- or gas-filled bubbles result and coalesce into larger ones. Eventually, a single large cavity results that collapses further downstream (Figure 10B). The region of wall damage is then as many as 20 diameters downstream from the valve or orifice plate.

Sensitivity of a device to cavitation occurrence is measured by the **cavitation index** or **cavitation number**, which is the ratio of the available pressure above vapor pressure to the dynamic pressure of the reference flow:

$$\sigma = \frac{2(p_o - p_v)}{\rho V_o^2} \tag{11}$$

where p_v is the vapor pressure, and the subscript o refers to appropriate reference conditions. Valve analyses use such an index in the specific form $(H_d - H_v)/(H_u - H_d)$ where u and d refer to upstream and downstream locales and H is the head, or pressure divided by the liquid specific weight. For the globe valve, the discharge coefficient changes at index values below 0.8 (Ball 1957). For the gate valve, the discharge coefficient begins to be reduced for index values below 1.5 at 10% opening and below 2.0 at 40% opening. With flow-metering devices such as orifices, venturis, and flow nozzles, there is little cavitation, because it occurs mostly downstream of the flow regions involved in establishing the metering action.

The detrimental effects of cavitation can be avoided by operating the liquid-flow device at high enough pressures. When this is not possible, the flow must be changed or the device must be built to withstand cavitation effects. Some materials or surface coatings are more resistant to cavitation erosion than others, but none is immune. Surface contours can be designed to delay the advent of cavitation.

Nonisothermal Effects

When appreciable temperature variations exist, the primary fluid properties (density and viscosity) are no longer constant, as usually assumed, but vary across or along the flow. The Bernoulli equation in the form of Equations (5a) through (5c) must be used, because volumetric flow is not constant. With gas flows, the thermodynamic process involved must be considered. In general, this is assessed in applying Equation (5a), written in the form:

$$\int \frac{dp}{\rho} + \frac{V^2}{2} + gz = \pi \tag{12}$$

Effects of viscosity variations also appear. With nonisothermal laminar flow, the parabolic velocity profile (Figure 3) is no longer valid. For gases, viscosity increases as the square root of absolute temperature, and for liquids, it decreases with increasing temperature. This results in opposite effects.

For fully developed pipe flow, the linear variation in shear stress from the wall value τ_w to zero at the centerline is independent of the temperature gradient. In the section on Laminar Flow, τ is defined as $\tau = (y/b)\tau_w$, where y is the distance from the centerline and $2b$ is the wall spacing. For pipe radius $R = D/2$ and distance from the wall $y = R - r$ (see Figure 11), then $\tau = \tau_w(R - y)/R$. Then, solving Equation (1) for the change in velocity gives

$$dv = \left[\frac{\tau_w(R - y)}{R\mu}\right]dy = -\left(\frac{\tau_w}{R\mu}\right) r\,dr \tag{13}$$

When the fluid has a lower viscosity near the wall than at the center (due to external heating of liquid or cooling of gas via heat transfer through the pipe wall), the velocity gradient is steeper near the wall and flatter near the center, so the profile is generally flattened. When liquid is cooled or gas is heated, the velocity profile becomes more pointed for laminar flow (Figure 11). Calculations were made for such flows of gases and liquid metals in pipes (Deissler 1951). Occurrences in turbulent flow are less apparent. If enough heating is applied to gaseous flows, the viscosity increase can cause reversion to laminar flow.

Buoyancy effects and gradual approach of the fluid temperature to equilibrium with that outside the pipe can cause considerable variation in the velocity profile along the conduit. Thus, Colborne and Drobitch (1966) found the pipe factor for upward vertical flow of hot air at a Reynolds number less than 2000 reduced to about 0.6 at 40 diameters from the entrance, then increased to about 0.8 at 210 diameters, and finally decreased to the isothermal value of 0.5 at the end of 320 diameters.

Compressibility

All fluids are compressible to some degree; their density depends on the pressure. Steady liquid flow may ordinarily be treated as incompressible, and incompressible flow analysis is satisfactory for gases and vapors at velocities below about 4000 to 8000 fpm, except in long conduits.

For liquids in pipelines, if flow is suddenly stopped, a severe pressure surge or water hammer is produced that travels along the pipe at the speed of sound in the liquid. This pressure surge alternately compresses and decompresses the liquid. For steady gas flows in long conduits, a decrease in pressure along the conduit can reduce the density of the gas significantly enough to cause the velocity to increase. If the conduit is long enough, velocities approaching the speed of sound are possible at the discharge end, and the Mach number (the ratio of the flow velocity to the speed of sound) must be considered.

Some compressible flows occur without heat gain or loss (adiabatically). If there is no friction (conversion of flow mechanical energy into internal energy), the process is reversible as well. Such a reversible adiabatic process is called isentropic, and follows the relationship

$$p/\rho^k = \text{constant}$$
$$k = c_p/c_v$$

where k, the ratio of specific heats at constant pressure and volume, has a value of 1.4 for air and diatomic gases.

The Bernoulli equation of steady flow, Equation (12), as an integral of the ideal-fluid equation of motion along a streamline, then becomes

$$\int \frac{dp}{\rho} + \frac{V^2}{2} = \text{constant} \tag{14}$$

where, as in most compressible flow analyses, the elevation terms involving z are insignificant and are dropped.

For a frictionless adiabatic process, the pressure term has the form

$$\int_1^2 \frac{dp}{\rho} = \frac{k}{k-1}\left(\frac{p_2}{\rho_2} - \frac{p_1}{\rho_1}\right) \tag{15}$$

Then, between stations 1 and 2 for the isentropic process,

$$\frac{p_1}{\rho_1}\left(\frac{k}{k-1}\right)\left[\left(\frac{p_2}{p_1}\right)^{(k-1)/k} - 1\right] + \frac{V_2^2 - V_1^2}{2} = 0 \tag{16}$$

Equation (16) replaces the Bernoulli equation for compressible flows and may be applied to the stagnation point at the front of a body. With this point as station 2 and the upstream reference flow ahead of the influence of the body as station 1, $V_2 = 0$. Solving Equation (16) for p_2 gives

$$p_s = p_2 = p_1\left[1 + \left(\frac{k-1}{2}\right)\frac{\rho_1 V_1^2}{kp_1}\right]^{k/(k-1)} \tag{17}$$

where p_s is the stagnation pressure.

Because kp/ρ is the square of the acoustic velocity a and the Mach number $M = V/a$, the stagnation pressure relation becomes

$$p_s = p_1\left[1 + \left(\frac{k-1}{2}\right)M_1^2\right]^{k/(k-1)} \tag{18}$$

For Mach numbers less than one,

$$p_s = p_1 + \frac{\rho_1 V_1^2}{2}\left[1 + \frac{M_1}{4} + \left(\frac{2-k}{24}\right)M_1^4 + \ldots\right] \tag{19}$$

When $M = 0$, Equation (19) reduces to the incompressible flow result obtained from Equation (5a). Appreciable differences appear when the Mach number of the approaching flow exceeds 0.2. Thus a pitot tube in air is influenced by compressibility at velocities over 13,000 fpm.

Flows through a converging conduit, as in a flow nozzle, venturi, or orifice meter, also may be considered isentropic. Velocity at the upstream station 1 is negligible. From Equation (16), velocity at the downstream station is

$$V_2 = \sqrt{\frac{2k}{k-1}\left(\frac{p_1}{\rho_1}\right)\left[1 - \left(\frac{p_2}{p_1}\right)^{(k-1)/k}\right]} \tag{20}$$

The mass flow rate is

$$\dot{m} = V_2 A_2 \rho_2 =$$
$$A_2\sqrt{\frac{2k}{k-1}(p_1\rho_1)\left[\left(\frac{p_2}{p_1}\right)^{2/k} - \left(\frac{p_2}{p_1}\right)^{(k+1)/k}\right]} \tag{21}$$

The corresponding incompressible flow relation is

$$\dot{m}_{in} = A_2\rho\sqrt{2\Delta p/\rho} = A_2\sqrt{2\rho(p_1 - p_2)} \tag{22}$$

The compressibility effect is often accounted for in the **expansion factor** Y

$$\dot{m} = Y\dot{m}_{in} = A_2 Y\sqrt{2\rho(p_1 - p_2)} \tag{23}$$

Y is 1.00 for the incompressible case. For air (k = 1.4), a Y value of 0.95 is reached with orifices at p_2/p_1 = 0.83 and with venturis at about 0.90, when these devices are of relatively small diameter (D_2/D_1 less than 0.5).

As p_2/p_1 decreases, the flow rate increases, but more slowly than for the incompressible case because of the nearly linear decrease in Y. However, the downstream velocity reaches the local acoustic value and the discharge levels off at a value fixed by upstream pressure and density at the critical ratio:

$$\left.\frac{p_2}{p_1}\right|_c = \left(\frac{2}{k+1}\right)^{k/(k-1)} = 0.53 \text{ for air} \tag{24}$$

At higher pressure ratios than critical, **choking** (no increase in flow with decrease in downstream pressure) occurs and is used in some flow control devices to avoid flow dependence on downstream conditions.

FLOW ANALYSIS

Fluid flow analysis is used to correlate pressure changes with flow rates and the nature of the conduit. For a given pipeline, either the pressure drop for a certain flow rate, or the flow rate for a certain pressure difference between the ends of the conduit, is needed. Flow analysis ultimately involves comparing a pump or blower to a conduit piping system for evaluating the expected flow rate.

Generalized Bernoulli Equation

Internal energy differences are generally small and usually the only significant effect of heat transfer is to change the density ρ. For gas or vapor flows, use the generalized Bernoulli equation in the pressure-over-density form of Equation (6a), allowing for the thermodynamic process in the pressure-density relation:

$$\int_1^2 \frac{dp}{\rho} + \alpha_1\frac{V_1^2}{2} + E_M = \alpha_2\frac{V_2^2}{2} + E_L \tag{25a}$$

The elevation changes involving z are negligible and are dropped. The pressure form of Equation (5b) is generally unacceptable when appreciable density variations occur, because the volumetric flow rate differs at the two stations. This is particularly serious in friction-loss evaluations where the density usually varies over considerable lengths of conduit (Benedict and Carlucci 1966). When the flow is essentially incompressible, Equation (25a) is satisfactory.

Example 1. Specify the blower to produce an isothermal airflow of 400 cfm through a ducting system (Figure 12). Accounting for intake and fitting losses, the equivalent conduit lengths are 60 and 165 ft and the flow is isothermal. The head at the inlet (station 1) and following the discharge (station 4), where the velocity is zero, are the same. The frictional losses H_L are evaluated as 24.5 ft of air between stations 1 and 2, and 237 ft between stations 3 and 4.

Solution: The head form of the generalized Bernoulli relation is used in place of Equation (25a), which also could be used:

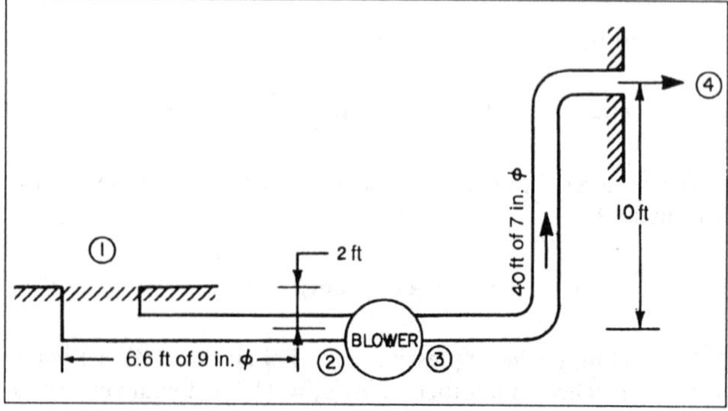

Fig. 12 Blower and Duct System for Example 1

$$(p_1/\rho_1 g) + \alpha_1(V_1^2/2g) + z_1 + H_M$$
$$= (p_2/\rho_2 g) + \alpha_2(V_2^2/2g) + z_2 + H_L \tag{25b}$$

The term $V_1^2/2g$ can be calculated as follows:

$$A_1 = \pi\left(\frac{D}{2}\right)^2 = \pi\left(\frac{9/12}{2}\right)^2 = 0.44 \text{ ft}^2$$

$$V_1 = Q/A_1 = \left(\frac{400 \text{ ft}^3}{\text{min}}\right)\left(\frac{1 \text{ min}}{60 \text{s}}\right)/0.44 \text{ ft}^2 = 15.1 \text{ ft/s}$$

$$V_1^2/2g = (15.1)^2/2(32) = 3.56 \text{ ft}$$

The term $V_2^2/2g$ can be calculated in a similar manner.

In Equation (25b), H_M is evaluated by applying the relation between any two points on opposite sides of the blower. Because conditions at stations 1 and 4 are known, they are used, and the location-specifying subscripts on the right side of Equation (25b) are changed to 4. Note that $p_1 = p_4 = p$, $\rho_1 = \rho_4 = \rho$, and $V_1 = V_4 = 0$. Thus,

$$(p/\rho g) + 0 + 2 + H_M = (p/\rho g) + 0 + 10 + (24.5 + 237)$$

so $H_M = 269.5$ ft of air. For standard air, this corresponds to 3.89 in. of water.

The pressure (or head) difference measured across the blower (between stations 2 and 3), is often taken as the head H_M. It can be obtained by calculating the static pressure at stations 2 and 3. Applying Equation (25b) successively between stations 1 and 2 and between 3 and 4 gives

$$(p_1/\rho g) + 0 + 2 + 0 = (p_2/\rho g) + (1.06 \times 3.56) + 0 + 24.5$$
$$(p_3/\rho g) + (1.03 \times 9.70) + 0 + 0 = (p_4/\rho g) + 0 + 10 + 237$$

where α just ahead of the blower is taken as 1.06, and just after the blower as 1.03; the latter value is uncertain because of possible uneven discharge from the blower. Static pressures p_1 and p_4 may be taken as zero gage. Thus,

$$p_2/\rho g = -26.2 \text{ ft of air}$$
$$p_3/\rho g = 237 \text{ ft of air}$$

The static head across the blower, the difference between these two numbers, is 263.2 ft, which is *not* the H_M calculated after Equation (25b) as 269.5 ft. The apparent discrepancy results from ignoring the velocity heads at stations 2 and 3. Actually, H_M is the change in total head (pressure head plus true kinetic energy) across the machine:

$$H_M = (p_3/\rho g) + \alpha_3(V_3^2/2g) - [(p_2/\rho g) + \alpha_2(V_2^2/2g)]$$
$$= 237 + (1.03 \times 9.70) - [-26.2 + (1.06 \times 3.54)]$$
$$= 247 - (-22.5) = 269.5 \text{ ft of air}$$

The required blower head is the same, no matter how it is evaluated. It is the specific energy added to the system by the machine. Only when the conduit size and velocity profiles on both sides of the machine are the same is E_M or H_M simply found from $\Delta p = p_3 - p_2$.

Conduit Friction

The loss term E_L or H_L of Equation (6a) or (6b) accounts for friction caused by conduit-wall shearing stresses and losses from conduit-section changes. H_L is the loss of energy (ft·lb$_f$) per unit weight (lb$_f$) of flowing fluid.

In real-fluid flow, a frictional shear occurs at bounding walls, gradually influencing the flow further away from the boundary. A lateral velocity profile is produced and flow energy is converted into heat (fluid internal energy), which is generally unrecoverable (a loss). This loss in fully developed conduit flow is evaluated through the **Darcy-Weisbach equation**:

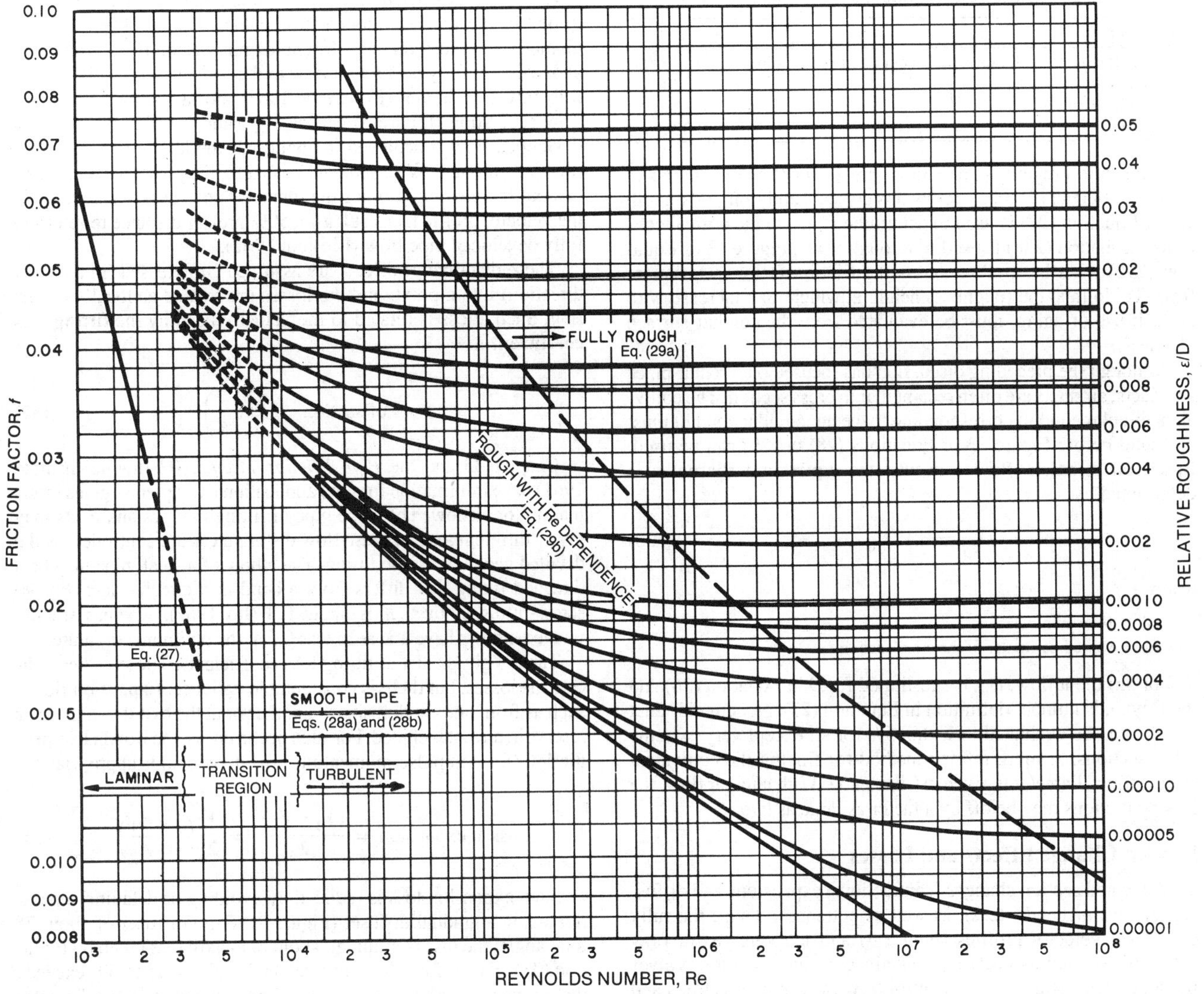

Fig. 13 Relation Between Friction Factor and Reynolds Number
(Moody 1944)

$$(H_L)_f = f\left(\frac{L}{D}\right)\left(\frac{V^2}{2g}\right) \tag{26}$$

where L is the length of conduit of diameter D and f is the **friction factor**. Sometimes a numerically different relation is used with the **Fanning friction factor** (one-quarter of f). The value of f is nearly constant for turbulent flow, varying only from about 0.01 to 0.05.

For fully developed laminar-viscous flow in a pipe, the loss is evaluated from Equation (8) as follows:

$$(H_L)_f = \frac{L}{\rho g}\left(\frac{8\mu V}{R^2}\right) = \frac{32 L \nu V}{D^2 g} = \frac{64}{VD/\nu}\left(\frac{L}{D}\right)\left(\frac{V^2}{2g}\right) \tag{27}$$

where $\mathrm{Re} = VD/\nu$ and $f = 64/\mathrm{Re}$. Thus, for laminar flow, the friction factor varies inversely with the Reynolds number.

With turbulent flow, friction loss depends not only on flow conditions, as characterized by the Reynolds number, but also on the nature of the conduit wall surface. For smooth conduit walls, empirical correlations give

$$f = \frac{0.3164}{\mathrm{Re}^{0.25}} \quad \text{for Re} < 10^5 \tag{28a}$$

$$f = 0.0032 + \frac{0.221}{\mathrm{Re}^{0.237}} \quad \text{for } 10^5 < \mathrm{Re} < 3 \times 10^6 \tag{28b}$$

Generally, f also depends on the wall roughness ε. The mode of variation is complex and best expressed in chart form (Moody 1944) as shown in Figure 13. Inspection indicates that, for high Reynolds numbers and relative roughness, the friction factor becomes independent of the Reynolds number in a *fully-rough flow regime*. Then

$$\frac{1}{\sqrt{f}} = 1.14 + 2\log(D/\varepsilon) \tag{29a}$$

Values of f between the values for smooth tubes and those for the fully-rough regime are represented by **Colebrook's natural roughness function**:

$$\frac{1}{\sqrt{f}} = 1.14 + 2\log(D/\varepsilon) - 2\log\left[1 + \frac{9.3}{\mathrm{Re}(\varepsilon/D)\sqrt{f}}\right] \tag{29b}$$

Table 2 Effective Roughness of Conduit Surfaces

Material	ε, ft
Commercially smooth brass, lead, copper, or plastic pipe	0.000005
Steel and wrought iron	0.00015
Galvanized iron or steel	0.0005
Cast iron	0.00085

A transition region appears in Figure 13 for Reynolds numbers between 2000 and 10,000. Below this critical condition, for smooth walls, Equation (27) is used to determine f; above the critical condition, Equation (28b) is used. For rough walls, Figure 13 or Equation (29b) must be used to assess the friction factor in turbulent flow. To do this, the roughness height ε, which may increase with conduit use or aging, must be evaluated from the conduit surface (Table 2).

Although the preceding discussion has focused on circular pipes and ducts, air ducts are often rectangular in cross section. The equivalent circular conduit corresponding to the noncircular conduit must be found before Figure 13 or Equations (28) or (29) can be used. Based on turbulent flow concepts, the **equivalent diameter** is determined by

$$D_{eq} = 4A/P_w \tag{30}$$

where

A = flow area
P_w = wetted perimeter of the cross section

For turbulent flow, D_{eq} is substituted for D in Equation (26) and the Reynolds number definition in Equation (27). Noncircular duct friction can be evaluated to within 5% for all except very extreme cross sections. A more refined method for finding the equivalent circular duct diameter is given in Chapter 32. With laminar flow, the loss predictions may be off by a factor as large as two.

Section Change Effects and Losses

Valve and section changes (contractions, expansions and diffusers, elbows or bends, tees), as well as entrances, distort the fully developed velocity profiles (Figure 3) and introduce extra flow losses (dissipated as heat) into pipelines or duct systems. Valves produce such extra losses to control flow rate. In contractions and expansions, flow separation as shown in Figures 8 and 9 causes the extra loss. The loss at rounded entrances develops as the flow accelerates to higher velocities. The resulting higher velocity near the wall leads to wall shear stresses greater than those of fully developed flow (Figure 5). In flow around bends, the velocity increases along the inner wall near the start of the bend. This increased velocity creates a secondary motion, which is a double helical vortex pattern of flow downstream from the bend. In all these devices, the disturbance produced locally is converted into turbulence and appears as a loss in the downstream region.

The return of disturbed flow to a fully developed velocity profile is quite slow. Ito (1962) showed that the secondary motion following a bend takes up to 100 diameters of conduit to die out but the pressure gradient settles out after 50 diameters.

With laminar flow following a rounded entrance, the entrance length depends on the Reynolds number:

$$L_e/D \approx 0.06 \text{ Re} \tag{31}$$

At Re = 2000, a length of 120 diameters is needed to establish the parabolic profile. The pressure gradient reaches the developed value of Equation (26) much sooner. The extra head drop is 1.2 velocity head; the change in profile from uniform to parabolic results in a drop of 1.0 velocity head (since $\alpha = 2.0$), and the rest is due to excess friction. With turbulent flow, 80 to 100 diameters following the rounded entrance are needed for the velocity profile to become fully developed, but the friction loss per unit length reaches a value close to that of the fully developed flow value more quickly. After six diameters, the loss rate at a Reynolds number of 10^5 is only 14% above that of fully developed flow in the same length, while at 10^7, it is only 10% higher (Robertson 1963). For a sharp entrance, the flow separation (Figure 8) causes a greater disturbance, but fully developed flow is achieved in about half the length required for a rounded entrance. With sudden expansion, the pressure change settles out in about eight times the diameter change ($D_2 = D_1$), while the velocity profile takes at least a 50% greater distance to return to fully developed pipe flow (Lipstein 1962).

These disturbance effects are assumed compressed (in the flow direction) into a point, and the losses are treated as locally occurring. Such a loss is related to the velocity head by the **fitting loss coefficient** K:

$$\text{Loss of section} = K\left(\frac{V^2}{2g}\right) \tag{32}$$

Chapter 33 and the *Pipe Friction Manual* (Hydraulic Institute 1961) have information for pipe applications. Chapter 32 gives information for airflow. The same type of fitting in pipes and ducts may give a different loss, because flow disturbances are controlled by the detailed geometry of the fitting. The elbow of a small pipe may be a threaded fitting that differs from a bend in a circular duct. For 90 screw-fitting elbows, K is about 0.8 (Ito 1962), whereas smooth flanged elbows have a K as low as 0.2 at the optimum curvature.

Table 3 gives a list of fitting loss coefficients. These values indicate the losses, but there is considerable variance. Expansion flows, such as from one conduit size to another or at the exit into a room or reservoir, are not included. For such occurrences, the **Borda loss prediction** (from impulse-momentum considerations) is appropriate:

$$\text{Loss at expansion} = \frac{(V_1 - V_2)^2}{2g} = \frac{V_1^2}{2g}\left(1 - \frac{A_1}{A_2}\right)^2 \tag{33}$$

Such expansion loss is reduced by avoiding or delaying separation using a gradual diffuser (Figure 9). For a diffuser of about 7° total angle, the loss is minimal, about one-sixth that given by Equation (33). The diffuser loss for total angles above 45 to 60° exceeds that of the sudden expansion, depending somewhat on the diameter ratio of the expansion. Optimum design of diffusers involves many factors; excellent performance can be achieved in short diffusers with splitter vanes or suction. Turning vanes in miter bends produce the least disturbance and loss for elbows; with careful design, the loss coefficient can be reduced to as low as 0.1.

For losses in smooth elbows, Ito (1962) found a Reynolds number effect (K slowly decreasing with increasing Re) and a minimum loss at a bend curvature (bend radius to diameter ratio) of 2.5. At this optimum curvature, a 45° turn had 63%, and a 180° turn approximately 120%, of the loss of a 90° bend. The loss does not vary linearly with the turning angle because secondary motion occurs.

Use of coefficient K presumes its independence of the Reynolds number. Crane Co. (1976) found a variation with the Reynolds number similar to that of the friction factor; Kittridge and Rowley (1957) observed it only with laminar flow. Assuming that K varies with Re similarly to f, it is convenient to represent fitting losses as adding to the effective length of uniform conduit. The **effective length** of a fitting is then

$$L_{eff}/D = K/f_{ref} \tag{34}$$

where f_{ref} is an appropriate reference value of the friction factor. Deissler (1951) uses 0.028, and the air duct values in Chapter 32 are based on an f_{ref} of about 0.02. For rough conduits, appreciable errors can occur if the relative roughness does not correspond to that used when f_{ref} was fixed. It is unlikely that the fitting losses involving separation are affected by pipe roughness. The effective length method for fitting loss evaluation is still useful.

When a conduit contains a number of section changes or fittings, the values of K are added to the fL/D friction loss, or the L_{eff}/D of the fittings are added to the conduit length L/D for evaluating the total loss H_L. This assumes that each fitting loss is fully developed and its disturbance fully smoothed out before the next section change. Such an assumption is frequently wrong, and the total loss can be overestimated. For elbow flows, the total loss of adjacent bends may be over- or underestimated. The secondary flow pattern following a radius elbow is such that when one elbow follows another, perhaps in a different plane, the secondary flow production of the second elbow may reinforce or partially cancel that of the first. Moving the second elbow a few diameters can reduce the total loss (from more than twice the amount) to less than the loss from one elbow. Screens or perforated plates can be used for smoothing velocity profiles (Wile 1947) and flow spreading. Their effectiveness and loss coefficients depend on their amount of open area (Baines and Peterson 1951).

Compressible Conduit Flow

When friction loss is included, as it must be except for a very short conduit, the incompressible flow analysis previously considered applies until the pressure drop exceeds about 10% of the initial pressure. The possibility of sonic velocities at the end of relatively long conduits limits the amount of pressure reduction achieved. For an inlet Mach number of 0.2, the discharge pressure can be reduced to about 0.2 of the initial pressure; for an inflow at $M = 0.5$, the discharge pressure cannot be less than about $0.45p_1$ in the adiabatic case and about $0.6p_1$ in isothermal flow.

Analysis of such conduit flow must treat density change, as evaluated from the continuity relation in Equation (2), with the frictional occurrences evaluated from wall roughness and Reynolds number correlations of incompressible flow (Binder 1944). In evaluating valve and fitting losses, consider the reduction in K caused by compressibility (Benedict and Carlucci 1966). Although the analysis differs significantly, isothermal and adiabatic flows involve essentially the same pressure variation along the conduit, up to the limiting conditions.

Control Valve Characterization

Control valves are characterized by a discharge coefficient C_d. As long as the Reynolds number is greater than 250, the orifice equation holds for liquids:

$$Q = C_d A_o \sqrt{2\Delta P/\rho} \tag{35}$$

where

A_o = area of orifice opening
P = absolute pressure

The discharge coefficient is about 0.63 for sharp-edged configurations and 0.8 to 0.9 for chamfered or rounded configurations. For gas flows at pressure ratios below the choking critical [Equation (24)], the mass rate of flow is

$$\dot{m} = C_d A_o C_1 \left(\frac{P_u}{\sqrt{T_u}}\right) \sqrt{\frac{P_d}{P_u}} \sqrt{1 - \left(\frac{P_d}{P_u}\right)^{(k-1)/k}} \tag{36}$$

where

$C_1 = \sqrt{2k/R(k-1)}$
k = ratio of specific heats at constant pressure and volume
R = gas constant
T = absolute temperature
u, d = subscripts referring to upstream and downstream positions

Incompressible Flow in Systems

Flow devices must be evaluated in terms of their interaction with other elements of the system, for example, the action of valves in modifying flow rate and in matching the flow-producing device (pump or blower) with the system loss. Analysis is via the general Bernoulli equation and the loss evaluations noted previously.

A valve regulates or stops the flow of fluid by throttling. The change in flow is not proportional to the change in area of the valve opening. Figures 14 and 15 indicate the nonlinear action of valves in controlling flow. A gate valve opening controls the flow in a pipeline discharging water from a tank (Figure 14). The fitting loss coefficient K values are those of Table 3; the friction factor f is 0.027. The degree of control also depends on the conduit L/D ratio. For a relatively long conduit, the valve must be nearly closed before its high K value becomes a significant portion of the loss. Figure 15 shows a control damper (essentially a butterfly valve) in a duct discharging air from a plenum held at constant pressure. With a long duct, the damper does not affect the flow rate until it is about one-quarter closed. Duct length has little effect when the damper is more than half closed. The damper closes the duct totally at the 90° position ($K = \infty$).

Flow in a system (pump or blower and conduit with fittings) involves interaction between the characteristics of the flow-producing device (pump or blower) and the loss characteristics of the

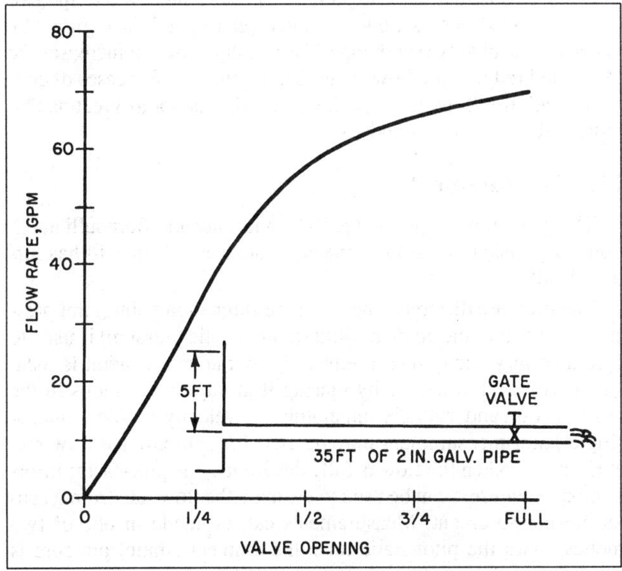

Fig. 14 Valve Action in Pipeline

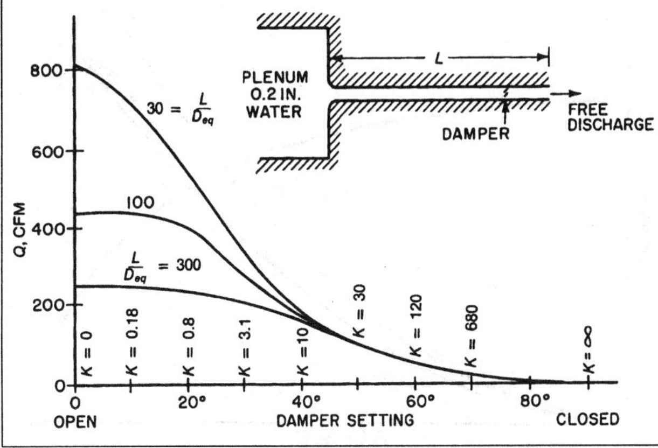

Fig. 15 Effect of Duct Length on Damper Action

Table 3 Fitting Loss Coefficients of Turbulent Flow

Fitting	Geometry	$K = \frac{\Delta p/\rho g}{V^2/2g}$
Entrance	Sharp	0.50
	Well-rounded	0.05
Contraction	Sharp ($D_2/D_1 = 0.5$)	0.38
90° Elbow	Miter	1.3
	Short radius	0.90
	Long radius	0.60
	Miter with turning vanes	0.2
Globe valve	Open	10
Angle valve	Open	5
Gate valve	Open	0.19 to 0.22
	75% open	1.10
	50% open	3.6
	25% open	28.8
Any valve	Closed	∞
Tee	Straight through flow	0.5
	Flow through branch	1.8

pipeline or duct system. Often the devices are centrifugal, in which case the head produced decreases as the flow increases, except for the lowest flow rates. System head required to overcome losses increases roughly as the square of the flow rate. The flow rate of a given system is that for which the two heads match, where the two curves of head versus flow rate intersect (point 1 in Figure 16). When a control valve (or damper) is partially closed, it increases the losses and reduces the flow (point 2 in Figure 16). For cases of constant head, the flow decrease due to valving is not as great as that indicated in Figures 14 and 15.

Flow Measurement

The general principles noted (the continuity and Bernoulli equations) are basic to most fluid-metering devices. Chapter 14 has further details.

The pressure difference between the stagnation point (total pressure) and that in the ambient fluid stream (static pressure) is used to give a point velocity measurement. Flow rate in a conduit is measured with a pitot device by placing it at various locations in the cross section and spatially integrating the velocity profile found. A single point measurement may be used for approximate flow rate evaluation. When the flow is fully developed, the pipe-factor information of Figure 4 can be used to estimate the flow rate from a centerline measurement. Measurements can be made in one of two modes. With the pitot-static tube, the ambient (static) pressure is found from pressure taps along the side of the forward-facing portion of the tube. When this portion is not long and slender, static pressure indication will be low and velocity indication high; as a

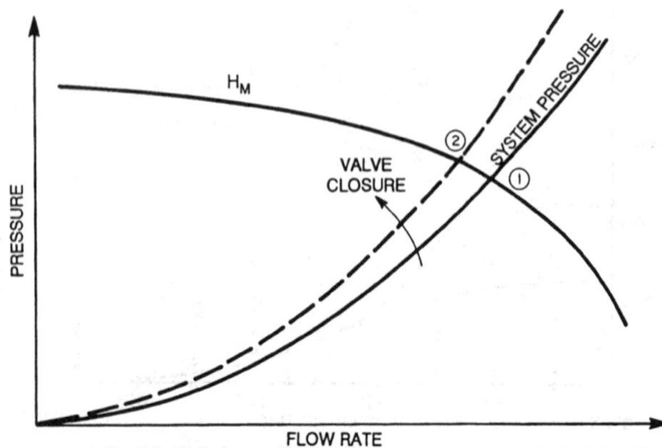

Fig. 16 Matching of Pump or Blower to System Characteristics

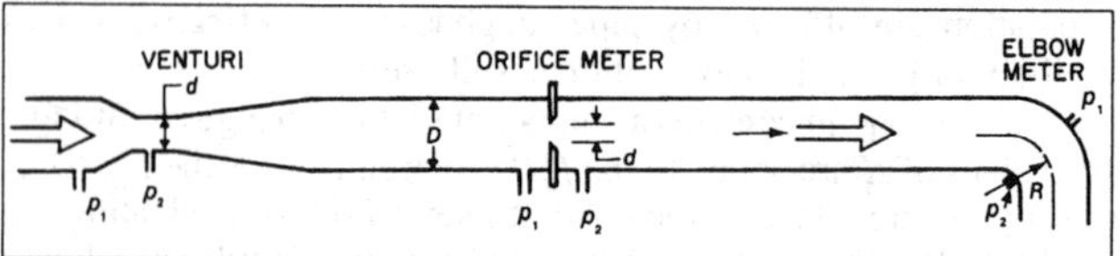

Fig. 17 Differential Pressure Flowmeters

result, a tube coefficient less than unity must be used. For parallel conduit flow, wall piezometers (taps) may take the ambient pressure, and the pitot tube indicates the impact (total pressure).

The venturi meter, flow nozzle, and orifice meter are flow rate metering devices based on the pressure change associated with relatively sudden changes in conduit section area (Figure 17). The elbow meter (also shown in Figure 17) is another differential pressure flowmeter. The flow nozzle is similar to the venturi in action, but does not have the downstream diffuser. For all these, the flow rate is proportional to the square root of the pressure difference resulting from fluid flow. With the area change devices (venturi, flow nozzle, and orifice meter), a theoretical flow rate relation is found by applying the Bernoulli and continuity equations in Equations (6) and (2) between stations 1 and 2:

$$Q_{theor} = \frac{\pi d^2}{4} \sqrt{\frac{2g\Delta h}{1-\beta^4}} \tag{37}$$

where

$\Delta h = h_1 - h_2 = \Delta p/\rho g$ (h = static pressure head)
$\beta = d/D$ = ratio of throat (or orifice) diameter to conduit diameter

The actual flow rate through the device can differ because the approach flow kinetic energy factor α deviates from unity and because of small losses. More significantly, the jet contraction of orifice flow is neglected in deriving Equation (37), to the extent that it can reduce the effective flow area by a factor of 0.6. The effect of all these factors can be combined into the discharge coefficient C_d:

$$Q = C_d Q_{theor} = C_d\left(\frac{\pi d^2}{4}\right)\sqrt{\frac{2g\Delta p}{1-\beta^4}} \tag{38}$$

Sometimes an alternate coefficient is used of the form

$$\frac{C_d}{\sqrt{1-\beta^4}}$$

For compressible fluid metering, the expansion factor Y as described by Equation (23) must be included, and the mass flow rate is

$$\dot{m} = C_d Y \rho Q_{theor} = C_d Y\left(\frac{\pi d^2}{4}\right)\sqrt{\frac{2\rho\Delta p}{1-\beta^4}} \tag{39}$$

Values of Y depend primarily on the pressure ratio p_2/p_1, and also on the metering device and k value of the particular gas.

The general mode of variation in C_d for orifices and venturis is indicated in Figure 18 as a function of Reynolds number and, to a lesser extent, diameter ratio β. For Reynolds numbers less than 10, the coefficient varies as $\sqrt{Re}$.

The elbow meter employs the pressure difference between inside and outside the bend as the metering signal (Murdock et al. 1964). A momentum analysis gives the flow rate as

$$Q_{theor} = \frac{\pi D^2}{4}\sqrt{\frac{R}{2D}(2g\Delta h)} \tag{40}$$

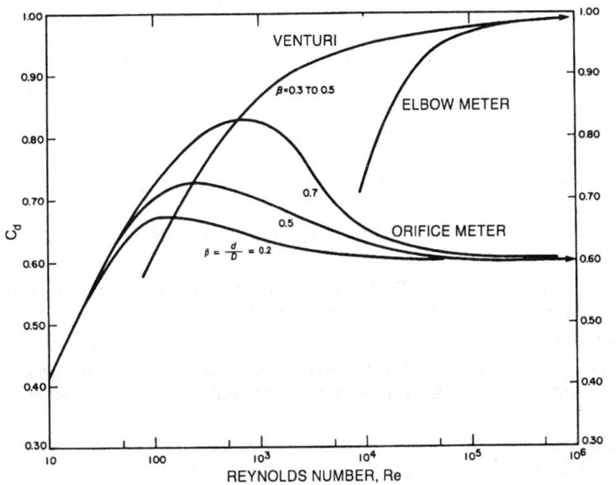

Fig. 18 Flowmeter Coefficients

where R is the radius of curvature of the bend. Again, a discharge coefficient C_d is needed; as in Figure 18, this drops off for the lower Reynolds numbers (below 10^5). These devices are calibrated in pipes with fully developed velocity profiles, so they must be located far enough downstream of sections that modify the approach velocity.

Unsteady Flow

Conduit flows are not always steady. In a compressible fluid, the acoustic velocity is usually high and the conduit length is rather short, so the time of signal travel is negligibly small. Even in the incompressible approximation, system response is not instantaneous. If a pressure difference Δp is applied between the conduit ends, the fluid mass must be accelerated and wall friction overcome, so a finite time passes before the steady flow rate corresponding to the pressure drop is achieved.

The time it takes for an incompressible fluid in a horizontal constant-area conduit of length L to achieve steady flow may be estimated by using the unsteady flow equation of motion with wall friction effects included. On the quasi-steady assumption, friction is given by Equation (26); also by continuity, V is constant along the conduit. The occurrences are characterized by the relation

$$\frac{dV}{d\theta} + \left(\frac{1}{\rho}\right)\frac{dp}{ds} + \frac{fV^2}{2D} = 0 \tag{41}$$

where

θ = time
s = distance in the flow direction

Since a certain Δp is applied over the conduit length L,

$$\frac{dV}{d\theta} = \frac{\Delta p}{\rho L} - \frac{fV^2}{2D} \tag{42}$$

For laminar flow, f is given by Equation (27), and

$$\frac{dV}{d\theta} = \frac{\Delta p}{\rho L} - \frac{32\mu V}{\rho D^2} = A - BV \tag{43}$$

Equation (43) can be rearranged and integrated to yield the time to reach a certain velocity:

$$\theta = \int d\theta = \int \frac{dV}{A - BV} = -\frac{1}{B}\ln(A - BV) \tag{44}$$

and

$$V = \frac{\Delta p}{L}\left(\frac{D^2}{32\mu}\right)\left[1 - \frac{\rho L}{\Delta p}\exp\left(\frac{-32\nu\theta}{D^2}\right)\right] \tag{45a}$$

For long times ($\theta \to \infty$), this indicates steady velocity as

$$V_\infty = \frac{\Delta p}{L}\left(\frac{D^2}{32\mu}\right) = \frac{\Delta p}{L}\left(\frac{R^2}{8\mu}\right) \tag{45b}$$

as by Equation (8). Then, Equation (45a) becomes

$$V = V_\infty\left[1 - \frac{\rho L}{\Delta p}\exp\left(\frac{-f_\infty V_\infty \theta}{2D}\right)\right] \tag{46}$$

where

$$f_\infty = \frac{64\nu}{V_\infty D}$$

The general nature of velocity development for starting-up flow is derived by more complex techniques; however, the temporal variation is as given above. For shutdown flow (steady flow with $\Delta p = 0$ at $\theta > 0$), the flow decays exponentially as $e^{-\theta}$.

Turbulent flow analysis of Equation (41) also must be based on the quasi-steady approximation, with less justification. Daily et al. (1956) indicate that the frictional resistance is slightly greater than the steady-state result for accelerating flows, but appreciably less for decelerating flows. If the friction factor is approximated as constant,

$$\frac{dV}{d\theta} = \frac{\Delta p}{\rho L} - \frac{fV^2}{2D} = A - BV^2$$

and, for the accelerating flow,

$$\theta = \frac{1}{\sqrt{AB}}\tanh^{-1}\left(V\sqrt{\frac{B}{A}}\right)$$

or

$$V = \sqrt{\frac{A}{B}}\tanh(\theta\sqrt{AB})$$

Because the hyperbolic tangent is zero when the independent variable is zero and unity when the variable is infinity, the initial ($V = 0$ at $\theta = 0$) and final conditions are verified. Thus, for long times ($\theta \to \infty$),

$$V_\infty = \sqrt{\frac{A}{B}} = \sqrt{\frac{\Delta p/\rho L}{f_\infty/2D}} = \sqrt{\frac{\Delta p}{\rho L}\left(\frac{2D}{f_\infty}\right)}$$

which is in accord with Equation (26) when f is constant (the flow regime is the fully rough one of Figure 13). The temporal velocity variation is then

$$V = V_\infty \tanh(f_\infty V_\infty \theta/2D) \tag{47}$$

In Figure 19, the turbulent velocity start-up result is compared with the laminar one in Figure 19, where initially the turbulent is steeper but of the same general form, increasing rapidly at the start but reaching V_∞ asymptotically.

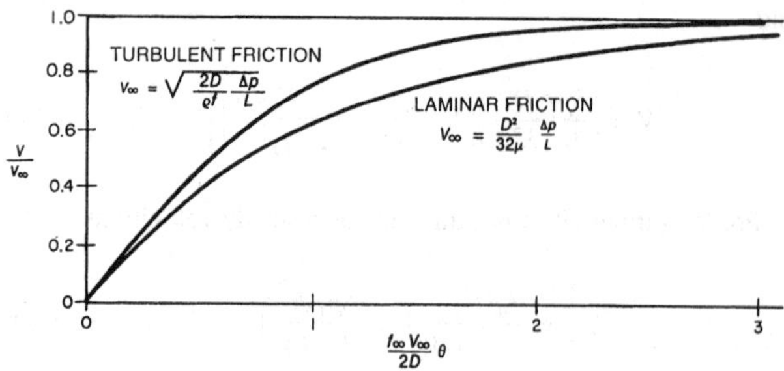

Fig. 19 Temporal Increase in Velocity Following Sudden Application of Pressure

NOISE FROM FLUID FLOW

Noise in flowing fluids results from unsteady flow fields and can be at discrete frequencies or broadly distributed over the audible range. With liquid flow, cavitation results in noise through the collapse of vapor bubbles. The noise in pumps or fittings (such as valves) can be a rattling or sharp hissing sound. It is easily eliminated by raising the system pressure. With severe cavitation, the resulting unsteady flow can produce indirect noise from induced vibration of adjacent parts. See Chapter 43 of the 1995 *ASHRAE Handbook—Applications* for more information on sound control.

The disturbed laminar flow behind cylinders can be an oscillating motion. The shedding frequency f of these vortexes is characterized by a Strouhal number $St = fd/V$ of about 0.21 for a circular cylinder of diameter d, over a considerable range of Reynolds numbers. This oscillating flow can be a powerful noise source, particularly when f is close to the natural frequency of the cylinder or some nearby structural member so that resonance occurs. With cylinders of another shape, such as impeller blades of a pump or blower, the characterizing Strouhal number involves the trailing edge thickness of the member. The strength of the vortex wake, with its resulting vibrations and noise potential, can be reduced by breaking up the flow with downstream splitter plates or boundary-layer trip devices (wires) on the cylinder surface.

Noise produced in pipes and ducts, especially from valves and fittings, is associated with the loss through such elements. The sound pressure of noise in water pipe flow increases linearly with the head loss; the broad-band noise increases, but only in the lower frequency range. Fitting-produced noise levels also increase with fitting loss (even without cavitation) and significantly exceed noise levels of the pipe flow. The relation between noise and loss is not surprising because both involve excessive flow perturbations. A valve's pressure-flow characteristics and structural elasticity may be such that for some operating point it oscillates, perhaps in resonance with part of the piping system, to produce excessive noise. A change in the operating point conditions or details of the valve geometry can result in significant noise reduction.

Pumps and blowers are strong potential noise sources. Turbomachinery noise is associated with blade-flow occurrences. Broadband noise appears from vortex and turbulence interaction with walls and is primarily a function of the operating point of the machine. For blowers, it has a minimum at the peak efficiency point (Groff et al. 1967). Narrow-band noise also appears at the blade-crossing frequency and its harmonics. Such noise can be very annoying because it stands out from the background. To reduce this noise, increase clearances between impeller and housing, and space impeller blades unevenly around the circumference.

REFERENCES

Baines, W.D. and E.G. Peterson. 1951. An investigation of flow through screens. *ASME Transactions* 73:467.

Baker, A.J. 1983. *Finite element computational fluid mechanics*. McGraw-Hill, New York.

Ball, J.W. 1957. Cavitation characteristics of gate valves and globe values used as flow regulators under heads up to about 125 ft. *ASME Transactions* 79:1275.

Benedict, R.P. and N.A. Carlucci. 1966. *Handbook of specific losses in flow systems*. Plenum Press Data Division, New York.

Binder, R.C. 1944. Limiting isothermal flow in pipes. *ASME Transactions* 66:221.

Bober, W. and R.A. Kenyon. 1980. *Fluid mechanics*. John Wiley and Sons, New York.

Colborne, W.G. and A.J. Drobitch. 1966. An experimental study of non-isothermal flow in a vertical circular tube. *ASHRAE Transactions* 72(4):5.

Crane Co. 1976. Flow of fluids. Technical *Paper* No. 410. New York.

Daily, J.W., et al. 1956. Resistance coefficients for accelerated and decelerated flows through smooth tubes and orifices. *ASME Transactions* 78:1071.

Deissler, R.G. 1951. Laminar flow in tubes with heat transfer. National Advisory Technical *Note* 2410, Committee for Aeronautics.

Furuya, Y., T. Sate, and T. Kushida. 1976. The loss of flow in the conical with suction at the entrance. *Bulletin of the Japan Society of Mechanical Engineers* 19:131.

Goldstein, S., ed. 1938. *Modern developments in fluid mechanics*. Oxford University Press, London. Reprinted by Dover Publications, New York.

Groff, G.C., J.R. Schreiner, and C.E. Bullock. 1967. Centrifugal fan sound power level prediction. *ASHRAE Transactions* 73(II): V.4.1.

Heskested, G. 1965. An edge suction effect. *AIAA Journal* 3:1958.

Heskested, G. 1970. Further experiments with suction at a sudden enlargement. *Journal of Basic Engineering, ASME Transactions* 92D:437.

Hoerner, S.F. 1965. *Fluid dynamic drag*, 3rd ed. Published by author, Midland Park, NJ.

Hydraulic Institute. 1961. *Pipe friction manual*. New York.

Ito, H. 1962. Pressure losses in smooth pipe bends. *Journal of Basic Engineering, ASME Transactions* 4(7):43.

John, J.E.A. and W.L. Haberman. 1980. *Introduction to fluid mechanics*, 2nd ed. Prentice Hall, Englewood Cliffs, NJ.

Kittridge, C.P. and D.S. Rowley. 1957. Resistance coefficients for laminar and turbulent flow through one-half inch valves and fittings. *ASME Transactions* 79:759.

Kline, S.J. 1959. On the nature of stall. *Journal of Basic Engineering, ASME Transactions* 81D:305.

Knapp, R.T., J.W. Daily, and F.G. Hammitt. 1970. *Cavitation*. McGraw-Hill, New York.

Lipstein, N.J. 1962. Low velocity sudden expansion pipe flow. *ASHRAE Journal* 4(7):43.

Moody, L.F. 1944. Friction factors for pipe flow. *ASME Transactions* 66:672.

Moore, C.A. and S.J. Kline. 1958. Some effects of vanes and turbulence in two-dimensional wide-angle subsonic diffusers. National Advisory Committee for Aeronautics, Technical *Memo* 4080.

Murdock, J.W., C.J. Foltz, and C. Gregory. 1964. Performance characteristics of elbow flow meters. *Journal of Basic Engineering, ASME Transactions* 86D:498.

Olson, R.M. 1980. *Essentials of engineering fluid mechanics*, 4th ed. Harper and Row, New York.

Robertson, J.M. 1963. A turbulence primer. University of Illinois (Urbana, IL), Engineering Experiment Station *Circular* 79.

Robertson, J.M. 1965. *Hydrodynamics in theory and application*. Prentice-Hall, Englewood Cliffs, NJ.

Robertson, J.M. and G.F. Wislicenus, ed. 1969 (discussion 1970). *Cavitation state of knowledge*. American Society of Mechanical Engineers, New York.

Ross, D. 1956. Turbulent flow in the entrance region of a pipe. *ASME Transactions* 78:915.

Schlichting, H. 1979. *Boundary layer theory*, 7th ed. McGraw-Hill, New York.

Streeter, V.L. and E.B. Wylie. 1979. *Fluid mechanics*, 7th ed. McGraw-Hill, New York.

Wile, D.D. 1947. Air flow measurement in the laboratory. *Refrigerating Engineering*: 515.

CHAPTER 3

HEAT TRANSFER

HEAT is energy in transit due to a temperature difference. The thermal energy is transferred from one region to another by three modes of **heat transfer**: conduction, convection, and radiation. Heat transfer is among a group of energy transport phenomena that includes mass transfer (see Chapter 5), momentum transfer or fluid friction (see Chapter 2), and electrical conduction. Transport phenomena have similar rate equations, in which flux is proportional to a potential difference. In heat transfer by conduction and convection, the potential difference is the temperature difference. Heat, mass, and momentum transfer are often considered together because of their similarities and interrelationship in many common physical processes.

This chapter presents the elementary principles of single-phase heat transfer with emphasis on heating, refrigerating, and air conditioning. Boiling and condensation are discussed in Chapter 4. More specific information on heat transfer to or from buildings or refrigerated spaces can be found in Chapters 24 through 30 of this volume and in Chapter 26 of the 1994 *ASHRAE Handbook—Refrigeration*. Physical properties of substances can be found in Chapters 18, 22, 24, and 36 of this volume and in Chapter 30 of the 1993 *ASHRAE Handbook—Fundamentals*. Heat transfer equipment, including evaporators, condensers, heating and cooling coils, furnaces, and radiators, is covered in the 1996 *ASHRAE Handbook—Systems and Equipment*. For further information on heat transfer, see the section on Bibliography.

HEAT TRANSFER PROCESSES

Thermal Conduction. This is the mechanism of heat transfer whereby energy is transported between parts of a continuum by the transfer of kinetic energy between particles or groups of particles at the atomic level. In gases, conduction is caused by elastic collision of molecules; in liquids and electrically nonconducting solids, it is believed to be caused by longitudinal oscillations of the lattice structure. Thermal conduction in metals occurs, like electrical conduction, through the motion of free electrons. Thermal energy transfer occurs in the direction of decreasing temperature, a consequence of the second law of thermodynamics. In solid opaque bodies, thermal conduction is the significant heat transfer mechanism because no net material flows in the process. With flowing fluids, thermal conduction dominates in the region very close to a solid boundary, where the flow is **laminar** and parallel to the surface and where there is no eddy motion.

Thermal Convection. This form of heat transfer involves energy transfer by fluid movement and molecular conduction (Burmeister 1983, Kays and Crawford 1980). Consider heat transfer to a fluid flowing inside a pipe. If the Reynolds number is large enough, three different flow regions exist. Immediately adjacent to the wall is a **laminar sublayer** where heat transfer occurs by thermal conduction; outside the laminar sublayer is a transition region called the **buffer layer**, where both eddy mixing and conduction effects are significant; beyond the buffer layer and extending to the center of the pipe is the **turbulent region**, where the dominant mechanism of transfer is eddy mixing.

In most equipment, the main body of fluid is in turbulent flow, and the laminar layer exists at the solid walls only. In cases of low-velocity flow in small tubes, or with viscous liquids such as oil (i.e., at low Reynolds numbers), the entire flow may be laminar with no transition or turbulent region.

When fluid currents are produced by external sources (for example, a blower or pump), the solid-to-fluid heat transfer is termed **forced convection**. If the fluid flow is generated internally by nonhomogeneous densities caused by temperature variation, the heat transfer is termed **free convection** or **natural convection**.

Thermal Radiation. In conduction and convection, heat transfer takes place through matter. In thermal radiation, there is a change in energy form from internal energy at the source to electromagnetic energy for transmission, then back to internal energy at the receiver. Whereas conduction and convection are affected primarily by temperature difference and somewhat by temperature level, the heat transferred by radiation increases rapidly as the temperature increases.

Although some generalized heat transfer equations have been mathematically derived from fundamentals, they are usually obtained from correlations of experimental data. Normally, the correlations employ certain dimensionless numbers, shown in Table 1, that are derived from dimensional analysis or analogy.

STEADY-STATE CONDUCTION

For steady-state heat conduction in one dimension, the Fourier law is

$$q = -(kA)\frac{dt}{dx} \tag{1}$$

where

q = heat flow rate, Btu/h
k = thermal conductivity, Btu·ft/h·ft²·°F
A = cross-sectional area normal to flow, ft²
dt/dx = temperature gradient, °F/ft

Equation (1) states that the heat flow rate q in the x direction is directly proportional to the temperature gradient dt/dx and the cross-sectional area A normal to the heat flow. The proportionality factor is the thermal conductivity k. The minus sign indicates that the heat flow is positive in the direction of decreasing temperature. Conductivity values are sometimes given in other units, but consistent units must be used in Equation (1).

The preparation of this chapter is assigned to TC 1.3, Heat Transfer and Fluid Flow.

Table 1 Dimensionless Numbers Commonly Used in Heat Transfer

Name	Symbol	Value[a]	Application
Nusselt number	Nu	hD/k, hL/k, $q''D/\Delta tk$, or $q''L/\Delta tk$	Natural or forced convection, boiling or condensing
Reynolds number	Re	GD/μ or $\rho VL/\mu$	Forced convection
Prandtl number	Pr	$\mu c_p/k$	Natural or forced convection, boiling or condensing
Stanton number	St	h/Gc_p	Forced convection
Grashof number	Gr	$L^3\rho^2\beta g\Delta t/\mu^2$ or $L^3\rho^2 g\Delta t/T\mu^2$	Natural convection (for ideal gases)
Fourier number	Fo	$\alpha\tau/L^2$	Unsteady-state conduction
Peclet number	Pe	GDc_p/k or Re Pr	Forced convection (small Pr)
Graetz number	Gz	GD^2c_p/kL or Re Pr D/L	Laminar convection

[a]A list of the other symbols used in this chapter appears in the section on Symbols.

Equation (1) may be integrated along a path of constant heat flow rate to obtain

$$q = k\left(\frac{A_m}{L_m}\right)\Delta t = \frac{\Delta t}{R} \tag{2}$$

where

A_m = mean cross-sectional area normal to flow, ft²
L_m = mean length of heat flow path, ft
Δt = overall temperature difference, °F
R = thermal resistance, °F·h/Btu

Thermal resistance R is directly proportional to the mean length L_m of the heat flow path and inversely proportional to the conductivity k and the mean cross-sectional area A_m normal to the flow. Equations for thermal resistances of a few common shapes are given in Table 2. Mathematical solutions to many heat conduction problems are addressed by Carslaw and Jaeger (1959). Complicated problems can be solved by graphical or numerical methods such as described by Croft and Lilley (1977), Adams and Rogers (1973), and Patankar (1980).

Analogy to Electrical Conduction. Equation (2) is analogous to Ohm's law for electrical circuits: thermal current (heat flow) in a **thermal circuit** is directly proportional to the thermal potential (temperature difference) and inversely proportional to the thermal resistance. This electrical-thermal analogy can be used for heat conduction in complex shapes that resist solution by exact analytical means. The thermal circuit concept is also useful for problems involving combined conduction, convection, and radiation.

OVERALL HEAT TRANSFER

In most steady-state heat transfer problems, more than one heat transfer mode is involved. The various heat transfer coefficients may be combined into an overall coefficient so that the total heat transfer can be calculated from the terminal temperatures. The solution to this problem is much simpler if the concept of a thermal circuit is employed.

Local Overall Heat Transfer Coefficient—Resistance Method

Consider heat transfer from one fluid to another by a three-step steady-state process: from a warmer fluid to a solid wall, through the wall, then to a colder fluid. An **overall heat transfer coefficient** U based on the difference between the bulk temperatures $t_1 - t_2$ of the two fluids is defined as follows:

$$q = UA(t_1 - t_2) \tag{3}$$

where A is the surface area. Because Equation (3) is a definition of U, the surface area A on which U is based is arbitrary; it should always be specified in referring to U.

Table 2 Solutions for Some Steady-State Thermal Conduction Problems

System	R in Equation $q = \Delta t/R$
Flat wall or curved wall if curvature is small (wall thickness less than 0.1 of inside diameter) 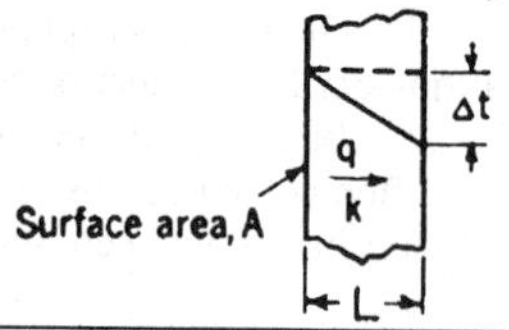	$R = \frac{L}{kA}$
Radial flow through a right circular cylinder 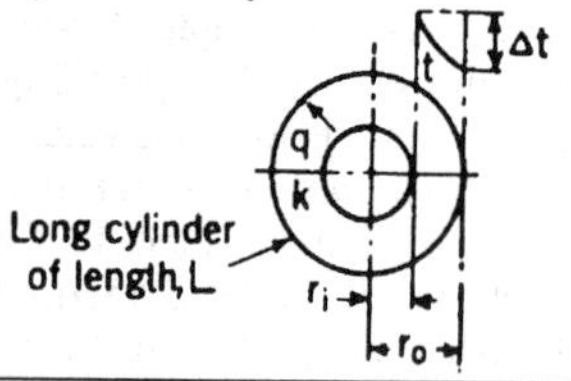	$R = \frac{\ln(r_o/r_i)}{2\pi kL}$
Buried cylinder 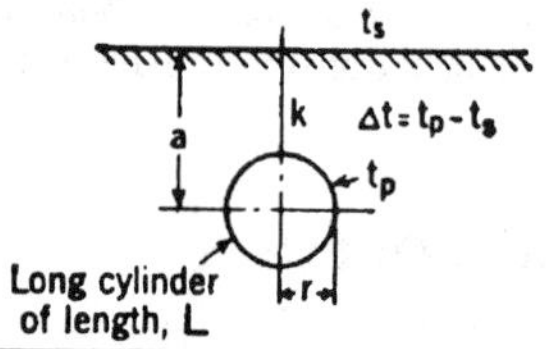	$R = \frac{\ln\left[(a + \sqrt{a^2 - r^2})/r\right]}{2\pi kL} = \frac{\cosh^{-1}(a/r)}{2\pi kL} \quad (L \gg 2r)$
Radial flow in a hollow sphere 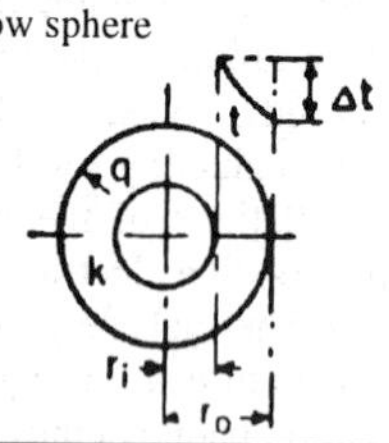	$R = \frac{(1/r_i - 1/r_o)}{4\pi k}$

L, r, a = dimensions, ft
k = thermal conductivity at average material temperature, Btu·ft/h·ft²·°F
A = surface area, ft²

The temperature drops across each part of the heat flow path are

$$t_1 - t_{s1} = qR_1$$
$$t_{s1} - t_{s2} = qR_2$$
$$t_{s2} - t_2 = qR_3$$

where t_{s1}, and t_{s2} are the warm and cold surface temperatures of the wall, respectively, and R_1, R_2, and R_3 are the thermal resistances. Because the same quantity of heat flows through each thermal resistance, these equations combined yield the following:

$$\frac{t_1 - t_2}{q} = \frac{1}{UA} = R_1 + R_2 + R_3 \qquad (4)$$

As shown above, the equations are analogous to those for electrical circuits; for thermal current flowing through n resistances in *series*, the resistances are additive.

$$R_o = R_1 + R_2 + R_3 + \ldots + R_n \qquad (5)$$

Similarly, **conductance** is the reciprocal of resistance, and for heat flow through resistances in *parallel*, the conductances are additive:

$$C = \frac{1}{R_o} = \frac{1}{R_1} + \frac{1}{R_2} + \frac{1}{R_3} + \ldots + \frac{1}{R_n} \qquad (6)$$

For convection, the thermal resistance is inversely proportional to the **convection coefficient** h_c and the applicable surface area:

$$R_c = \frac{1}{h_c A} \qquad (7)$$

The thermal resistance for radiation is written similarly to that for convection:

$$R_r = \frac{1}{h_r A} \qquad (8)$$

The **radiation coefficient** h_r has no physical significance but is useful in computations. It is a function of the temperatures, radiation properties, and geometrical arrangement of the enclosure and the body in question.

Resistance Method Analysis. Analysis by the resistance method can be illustrated by considering heat transfer from air outside to cold water inside an insulated pipe. The temperature gradients and the nature of the resistance analysis are shown in Figure 1.

Because air is sensibly transparent to radiation, some heat transfer occurs by both radiation and convection to the outer insulation surface. The mechanisms act in parallel on the air side. The total transfer then passes through the insulating layer and the pipe wall by thermal conduction, and then by convection and radiation into the cold water stream. (Radiation is not significant on the water side because liquids are sensibly opaque to radiation, although water transmits energy in the visible region.) The contact resistance between the insulation and the pipe wall is assumed negligible.

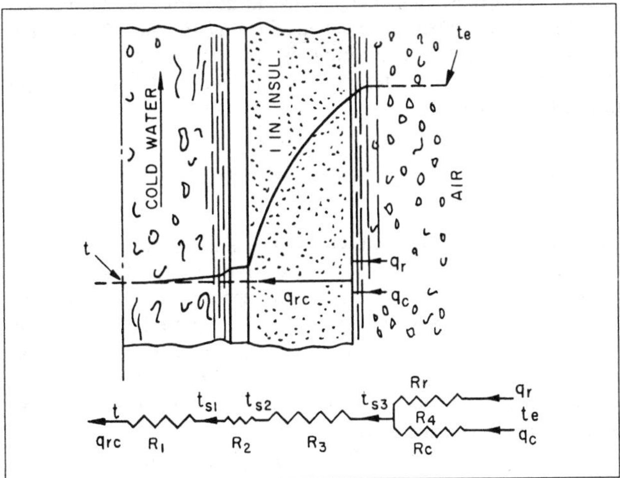

Fig. 1 Thermal Circuit Diagram for Insulated Cold Water Line

The heat transfer rate q_{rc} for a given length L of pipe may be thought of as the sum of the rates q_r and q_c flowing through the parallel resistances R_r and R_c associated with the surface radiation and convection coefficients. The total flow then proceeds through the resistance R_3 offered to thermal conduction by the insulation, through the pipe wall resistance R_2, and into the water stream through the convection resistance R_1. Note the analogy to direct current electricity. A temperature (potential) drop is required to overcome resistances to the flow of thermal current. The total resistance to heat transfer R_o is the sum of the individual resistances:

$$R_o = R_1 + R_2 + R_3 + R_4 \qquad (9)$$

where the resultant parallel resistance R_4 is obtained from

$$\frac{1}{R_4} = \frac{1}{R_r} + \frac{1}{R_c} \qquad (10)$$

If the individual resistances can be evaluated, the total resistance can be obtained from this relation. The heat transfer rate for the length of pipe L can be established by

$$q_{rc} = \frac{t_e - t}{R_o} \qquad (11)$$

For a unit length of the pipe, the heat transfer rate is

$$\frac{q_{rc}}{L} = \frac{t_e - t}{R_o L} \qquad (12)$$

The temperature drop Δt through each individual resistance may then be calculated from the relation:

$$\Delta t_n = R_n q_{rc} \qquad (13)$$

where n = 1, 2, and 3.

Mean Temperature Difference

When heat is exchanged between two fluids flowing through a heat exchanger, the local temperature difference Δt varies along the flow path. Heat transfer may be calculated using

$$q = UA\Delta t_m \qquad (14)$$

where U is the overall coefficient of heat transfer from fluid to fluid, A is an area associated with the coefficient U, and Δt_m is the appropriate mean temperature difference.

For parallel flow or counterflow exchangers and for any exchanger in which one fluid temperature is substantially constant, the mean temperature difference is

$$\Delta t_m = \frac{\Delta t_1 - \Delta t_2}{\ln(\Delta t_1/\Delta t_2)} = \frac{\Delta t_1 - \Delta t_2}{2.3\log(\Delta t_1/\Delta t_2)} \qquad (15)$$

where Δt_1, and Δt_2 are the temperature differences between the fluids at each end of the heat exchanger. Δt_m is called the **logarithmic mean temperature difference**. For the special case of $\Delta t_1 = \Delta t_2$, which leads to an indeterminate form of Equation (15), $\Delta t_m = \Delta t_1 = \Delta t_2$.

Equation (15) for Δt_m is true only if the overall coefficient and the specific heat of the fluids are constant through the heat exchanger, and no heat losses occur (often well-approximated in practice). Parker et al. (1969) give a procedure for cases with variable overall coefficient U.

Calculations using Equation (14) and Δt_m are convenient when terminal temperatures are known. In many cases, however, the temperatures of the fluids leaving the exchanger are not known. To avoid trial-and-error calculations, an alternate method involves the use of three nondimensional parameters, defined as follows:

1. Exchanger Heat Transfer Effectiveness ε

$$\varepsilon = \frac{(t_{hi} - t_{ho})}{(t_{hi} - t_{ci})} \quad \text{when } C_h = C_{min}$$
$$\varepsilon = \frac{(t_{co} - t_{ci})}{(t_{hi} - t_{ci})} \quad \text{when } C_c = C_{min} \tag{16}$$

where

$C_h = (\dot{m}\, c_p)_h$ = hot fluid capacity rate, Btu/h·°F
$C_c = (\dot{m}\, c_p)_c$ = cold fluid capacity rate, Btu/h·°F
C_{min} = smaller of capacity rates C_h and C_c
t_h = terminal temperature of hot fluid, °F. Subscript i indicates entering condition; subscript o indicates leaving condition.
t_c = terminal temperature of cold fluid, °F. Subscripts i and o are the same as for t_h.

2. Number of Exchanger Heat Transfer Units (NTU)

$$\text{NTU} = \frac{A U_{avg}}{C_{min}} = \frac{1}{C_{min}} \int_A U\, dA \tag{17}$$

where A is the area used to define overall coefficient U.

3. Capacity Rate Ratio Z

$$Z = \frac{C_{min}}{C_{max}} \tag{18}$$

Generally, the heat transfer effectiveness can be expressed for a given exchanger as a function of the number of transfer units and the capacity rate ratio:

$$\varepsilon = f\,(\text{NTU, Z, flow arrangement}) \tag{19}$$

The effectiveness is independent of the temperatures in the exchanger. For any exchanger in which the capacity rate ratio Z is zero (where one fluid undergoes a phase change; e.g., in a condenser or evaporator), the effectiveness is

$$\varepsilon = 1 - \exp(-\text{NTU}) \tag{20}$$

Heat transferred can be determined from

$$q = C_h(t_{hi} - t_{ho}) = C_c(t_{co} - t_{ci}) \tag{21}$$

Combining Equations (16) and (21) produces an expression for heat transfer rate in terms of entering fluid temperatures:

$$q = \varepsilon C_{min}(t_{hi} - t_{ci}) \tag{22}$$

The proper mean temperature difference for Equation (14) is then given by

$$\Delta t_m = \frac{(t_{hi} - t_{ci})\varepsilon}{\text{NTU}} \tag{23}$$

The effectiveness for **parallel flow exchangers** is

$$\varepsilon = \frac{1 - \exp[-\text{NTU}(1 + Z)]}{1 + Z} \tag{24}$$

For $Z = 1$,

$$\varepsilon = \frac{1 - \exp(-2\ \text{NTU})}{2} \tag{25}$$

The effectiveness for **counterflow exchangers** is

$$\varepsilon = \frac{1 - \exp[-\text{NTU}(1 - Z)]}{1 - Z \exp[-\text{NTU}(1 - Z)]} \tag{26}$$

$$\varepsilon = \frac{\text{NTU}}{1 + \text{NTU}} \quad \text{for } Z = 1 \tag{27}$$

Incropera and DeWitt (1996) and Kays and London (1984) show the relations of ε, NTU, and Z for other flow arrangements. These authors and Afgan and Schlunder (1974) present graphical representations for convenience.

TRANSIENT HEAT FLOW

Often, the heat transfer and temperature distribution under unsteady-state (varying with time) conditions must be known. Examples are (1) cold storage temperature variations on starting or stopping a refrigeration unit; (2) variation of external air temperature and solar irradiation affecting the heat load of a cold storage room or wall temperatures; (3) the time required to freeze a given material under certain conditions in a storage room; (4) quick freezing of objects by direct immersion in brines; and (5) sudden heating or cooling of fluids and solids from one temperature to a different temperature.

The equations describing transient temperature distribution and heat transfer are presented in this section. Numerical methods are the simplest means of solving these equations because numerical data are easy to obtain. However, with some numerical solutions and off-the-shelf software, the physics that drives the energy transport can be lost. Thus, analytical solution techniques are also included in this section.

The fundamental equation for unsteady-state conduction in solids or fluids in which there is no substantial motion is

$$\frac{\partial t}{\partial \tau} = \alpha \left(\frac{\partial^2 t}{\partial x^2} + \frac{\partial^2 t}{\partial y^2} + \frac{\partial^2 t}{\partial z^2} \right) \tag{28}$$

where thermal diffusivity α is the ratio $k/\rho c_p$; k is thermal conductivity; ρ, density; and c_p, specific heat. If α is large (high conductivity, low density and specific heat, or both), heat will diffuse faster.

One of the most elementary transient heat transfer models predicts the rate of temperature change of a body or material being held at constant volume with uniform temperature, such as a well-stirred reservoir of fluid whose temperature is changing because of a net rate of heat gain or loss:

$$q_{net} = (Mc_v)\frac{dt}{d\tau} \tag{29}$$

where M is the mass of the body, and c_v is its specific heat at constant volume. q_{net} is algebraic, with positive being into the body and negative being out of the body. If the heating occurs at constant pressure, c_v should be replaced by c_p; however, for liquids and solids, c_v and c_p are nearly equal, and c_p can be used with negligible error. The term q_{net} may include heat transfer by conduction, convection, or radiation and is the difference between the heat transfer rates into and out of the body.

From Equations (28) and (29), it is possible to derive expressions for temperature and heat flow variations at different instants and different locations. Most common cases have been solved and

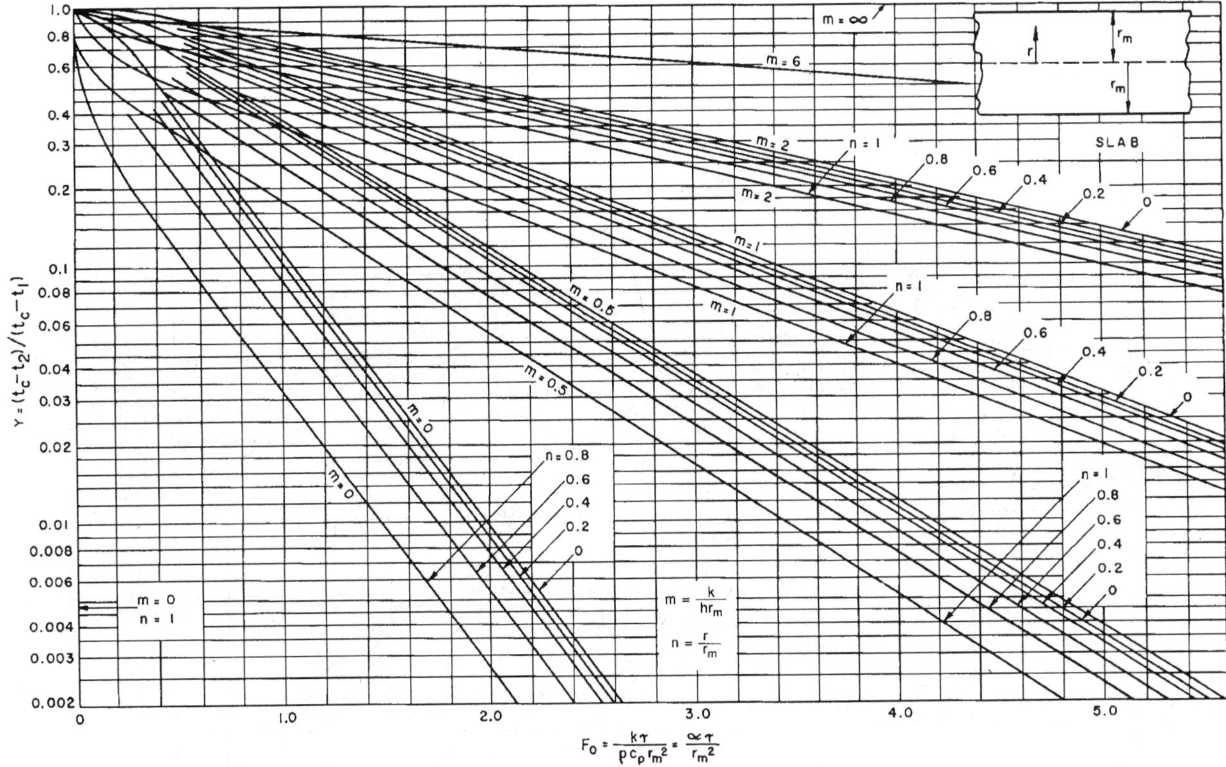

Fig. 2 Transient Temperatures for Infinite Slab

presented in graphical forms (Jakob 1957, Schneider 1964, Myers 1971). In other cases, it is simpler to use numerical methods (Croft and Lilley 1977, Patankar 1980). When convective boundary conditions are required in the solution of Equations (28) and (29), h values based on steady-state correlations are often used. However, this approach may not be valid when rapid transients are involved.

Estimating Cooling Times

Cooling times for materials can be estimated (McAdams 1954) by Gurnie-Lurie charts (Figures 2, 3, and 4), which are graphical solutions for the heating or cooling of infinite slabs, infinite cylinders, and spheres. These charts assume an initial uniform temperature distribution and no change of phase. They apply to a body exposed to a constant temperature fluid with a constant surface convection coefficient of h.

Using Figures 2, 3, and 4, it is possible to estimate both the temperature at any point and the average temperature in a homogeneous mass of material as a function of time in a cooling process. It is possible to estimate cooling times for rectangular-shaped solids, cubes, cylinders, and spheres.

From the point of view of heat transfer, a cylinder insulated on its ends behaves like a cylinder of infinite length, and a rectangular solid insulated so that only two parallel faces allow heat transfer behaves like an infinite slab. A thin slab or a long, thin cylinder may be also considered infinite objects.

Consider a slab having insulated edges being cooled. If the cooling time is the time required for the center of the slab to reach a temperature of t_2, the cooling time can be calculated as follows:

1. Evaluate the temperature ratio $(t_c - t_2)/(t_c - t_1)$.

   where

   t_c = temperature of cooling medium
   t_1 = initial temperature of product
   t_2 = final temperature of product at center

   Note that in Figures 2, 3, and 4, the temperature ratio $(t_c - t_2)/(t_c - t_1)$ is designated as Y to simplify the equations.

2. Determine the radius ratio r/r_m designated as n in Figures 2, 3, and 4.

   where

   r = distance from centerline
   r_m = half thickness of slab

3. Evaluate the resistance ratio k/hr_m designated as m in Figures 2, 3, and 4.

   where

   k = thermal conductivity of material
   h = heat transfer coefficient

4. From Figure 2 for infinite slabs, select the appropriate value of $k\tau/\rho c_p r_m^2$ designated as F_o in Figures 2, 3, and 4.

   where

   τ = time elapsed
   c_p = specific heat
   ρ = density

5. Determine τ from the value of $k\tau/\rho c_p r_m^2$.

Temperature Distribution in Finite Objects

Finite objects can be formed from the intersection of infinite objects. For example, the solid of intersection of an infinite cylinder and an infinite slab is a finite cylinder with a length equal to the thickness of the slab and a radius equal to that of the cylinder (Figure 5). Intersection of three infinite slabs with the same thickness produces a cube; intersection of three dissimilar slabs forms a finite rectangular solid.

The temperature in the finite object can be calculated from the temperature ratio Y of the infinite objects that intersect to form the finite object. The product of the temperature ratios of the infinite

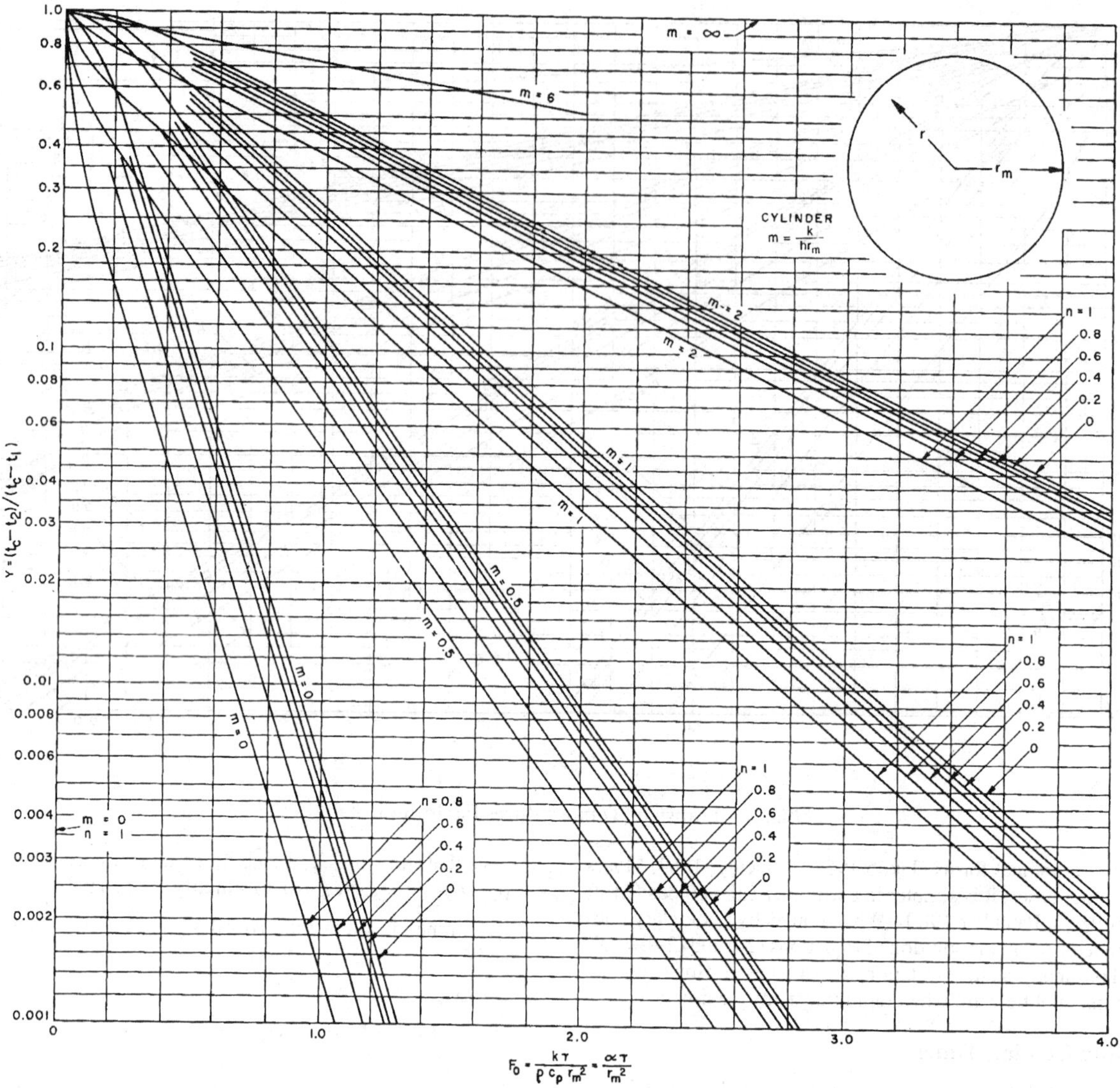

Fig. 3 Transient Temperatures for Infinite Cylinder

objects is the temperature ratio of the finite object; for example, for the finite cylinder of Figure 5,

$$Y_{fc} = Y_{is} Y_{ic} \tag{30}$$

where

Y_{fc} = temperature ratio of finite cylinder
Y_{is} = temperature ratio of infinite slab
Y_{ic} = temperature ratio of infinite cylinder

For a finite rectangular solid,

$$Y_{frs} = (Y_{is})_1 (Y_{is})_2 (Y_{is})_3 \tag{31}$$

where Y_{frs} = temperature ratio of finite rectangular solid, and subscripts 1, 2, and 3 designate three infinite slabs.

Heat Exchanger Transients

Determination of the transient behavior of heat exchangers is becoming increasingly important in evaluating the dynamic behavior of heating and air-conditioning systems. Many studies of the transient behavior of counterflow and parallel flow heat exchangers have been conducted; some are listed in the section on Bibliography.

THERMAL RADIATION

Radiation, one of the basic mechanisms for energy transfer between different temperature regions, is distinguished from conduction and convection in that it does not depend on an intermediate material as a carrier of energy but rather is impeded by the presence of material between the regions. The radiation energy transfer process is the consequence of energy-carrying electromagnetic waves that are emitted by atoms and molecules due to changes in their energy content. The amount and characteristics of radiant energy emitted by a quantity of material depend on the nature of the material, its microscopic arrangement, and its absolute temperature. Although rate of energy emission is independent of the surroundings, the **net** energy transfer rate depends on the temperatures and spatial relationships of the surface and its surroundings.

Blackbody Radiation

The rate of thermal radiant energy emitted by a surface depends on its absolute temperature. A surface is called **black** if it can absorb all incident radiation. The total energy emitted per unit time per unit area of black surface W_b to the hemispherical region above it is given by the **Stefan-Boltzmann law**.

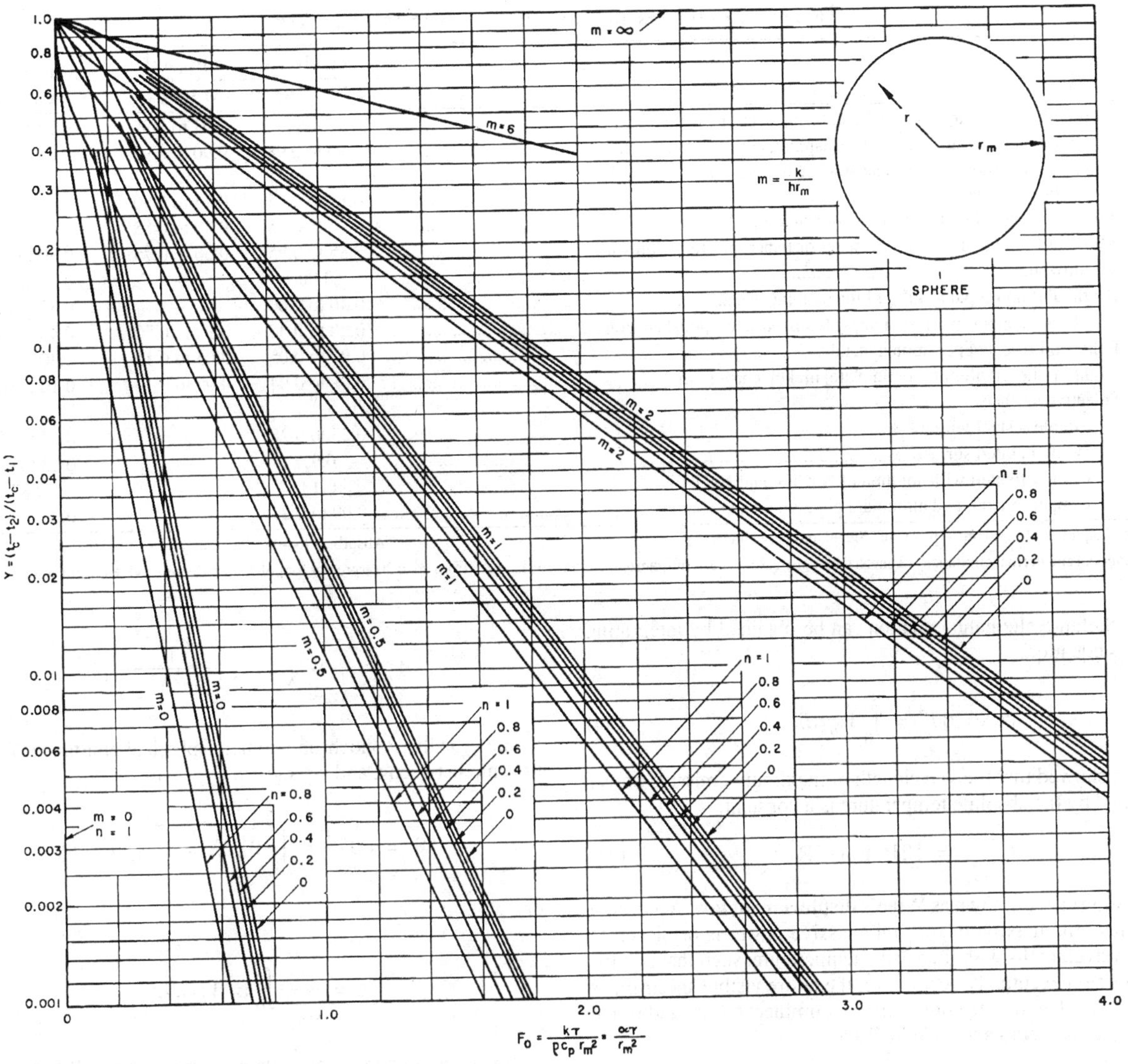

Fig. 4 Transient Temperatures for Spheres

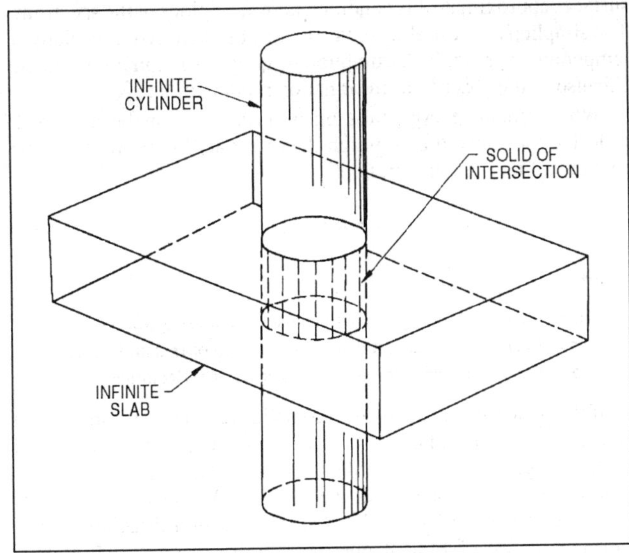

Fig. 5 Finite Cylinder of Intersection from Intersection of Infinite Cylinder and Infinite Slab

$$W_b = \sigma T^4 \tag{32}$$

where W_b is the total rate of energy emission per unit area, and σ is the Stefan-Boltzmann constant (0.1714×10^{-8} Btu/h·ft^2·°R^4).

The heat radiated by a body comprises electromagnetic waves of many different frequencies or wavelengths. Planck showed that the spectral distribution of the energy radiated by a blackbody is

$$W_{b\lambda} = \frac{C_1 \lambda^{-5}}{e^{C_2/\lambda T} - 1} \tag{33}$$

where

$W_{b\lambda}$ = monochromatic emissive power of blackbody, Btu/h·ft^2·μm
λ_1 = wavelength, μm
T = temperature, °R
C_1 = first Planck's law constant = 1.1870×10^8 Btu·μm^4/h·ft^2
C_2 = second Planck's law constant = 2.5896×10^4 μm·°R

$W_{b\lambda}$ is the **monochromatic emissive power**, defined as the energy emitted per unit time per unit surface area at wavelength λ per unit wavelength interval around λ; that is, the energy emitted per unit time per unit surface area in the interval $d\lambda$ is equal to $W_{b\lambda} d\lambda$.

Table 3 Emittances and Absorptances for Some Surfaces[a]

Class	Surfaces	Total Normal Emittance[b] At 50 to 100 °F	Total Normal Emittance[b] At 1000 °F	Absorptance for Solar Radiation
1	A small hole in a large box, sphere, furnace, or enclosure	0.97 to 0.99	0.97 to 0.99	0.97 to 0.99
2	Black nonmetallic surfaces such as asphalt, carbon, slate, paint, paper	0.90 to 0.98	0.90 to 0.98	0.85 to 0.98
3	Red brick and tile, concrete and stone, rusty steel and iron, dark paints (red, brown, green, etc.)	0.85 to 0.95	0.75 to 0.90	0.65 to 0.80
4	Yellow and buff brick and stone, firebrick, fireclay	0.85 to 0.95	0.70 to 0.85	0.50 to 0.70
5	White or light cream brick, tile, paint or paper, plaster, whitewash	0.85 to 0.95	0.60 to 0.75	0.30 to 0.50
6	Window glass	0.90	—	c
7	Bright aluminum paint; gilt or bronze paint	0.40 to 0.60	—	0.30 to 0.50
8	Dull brass, copper, or aluminum; galvanized steel; polished iron	0.20 to 0.30	0.30 to 0.50	0.40 to 0.65
9	Polished brass, copper, monel metal	0.02 to 0.05	0.05 to 0.15	0.30 to 0.50
10	Highly polished aluminum, tin plate, nickel, chromium	0.02 to 0.04	0.05 to 0.10	0.10 to 0.40
11	Selective surfaces			
	Stainless steel wire mesh	0.23 to 0.28	—	0.63 to 0.86
	White painted surface	0.92	—	0.23 to 0.49
	Copper treated with solution of $NaClO_2$ and NaOH	0.13	—	0.87
	Copper, nickel, and aluminum plate with CuO coating	0.09 to 0.21	—	0.08 to 0.93

[a]See also Chapter 36, McAdams (1954), and Siegel and Howell (1981).

[b]Hemispherical and normal emittance are unequal in many cases. The hemispherical emittance may vary from up to 30% greater for polished reflectors to 7% lower for nonconductors.

[c]Absorbs 4 to 40% depending on its transmittance.

The Stefan-Boltzmann equation can be obtained by integrating Planck's equation:

$$W_b = \sigma T^4 = \int_0^{\infty} W_{b\lambda} d\lambda \quad (34)$$

Wien showed that the wavelength of maximum emissive power multiplied by the absolute temperature is a constant:

$$\lambda_{max} T = 5216 \ \mu\text{m} \cdot {}^\circ\text{R} \quad (35)$$

Equation (35) is known as **Wien's displacement law**. According to this law, the maximum spectral emissive power is displaced to shorter wavelengths with increasing temperature, such that significant emission eventually occurs over the entire visible spectrum as shorter wavelengths become more prominent. For additional details, see Incropera and DeWitt (1996).

Actual Radiation

Substances and surfaces diverge variously from the Stefan-Boltzmann and Planck laws. W_b and $W_{b\lambda}$ are the maximum emissive powers at a surface temperature. Actual surfaces emit and absorb less readily and are called **nonblack**. The emissive power of a nonblack surface at temperature T radiating to the hemispherical region above it is written as

$$W = \varepsilon W_b = \varepsilon \sigma T^4 \quad (36)$$

where ε is known as the **hemispherical emittance**. The term emittance conforms to physical and electrical terminology; the suffix "ance" denotes a property of a piece of material as it exists. The ending "ivity" denotes a property of the bulk material independent of geometry or surface condition. Thus, emittance, reflectance, absorptance, and transmittance refer to actual pieces of material. Emissivity, reflectivity, absorptivity, and transmissivity refer to the properties of materials that are optically smooth and thick enough to be opaque.

The emittance is a function of the material, the condition of its surface, and the temperature of the surface. Table 3 lists selected values; Siegel and Howell (1981) and Modest (1993) have more extensive lists.

The monochromatic emissive power of a nonblack surface is similarly written as

$$W_\lambda = \varepsilon_\lambda W_{b\lambda} = \varepsilon_\lambda \left(\frac{C_1 \lambda^{-5}}{e^{C_2/\lambda T} - 1} \right) \quad (37)$$

where ε_λ is the monochromatic hemispherical emittance. The relationship between ε and ε_λ is given by

$$W = \varepsilon \sigma T^4 = \int_0^{\infty} W_\lambda d\lambda = \int_0^{\infty} \varepsilon_\lambda W_{b\lambda} d\lambda$$

or

$$\varepsilon = \frac{1}{\sigma T^4} \int_0^{\infty} \varepsilon_\lambda W_{b\lambda} d\lambda \quad (38)$$

If ε_λ does not depend on λ, then, from Equation (38), $\varepsilon = \varepsilon_\lambda$. Surfaces with this characteristic are called **gray**. Gray surface characteristics are often assumed in calculations. Several classes of surfaces approximate this condition in some regions of the spectrum. The simplicity is desirable, but care must be exercised, especially if temperatures are high. Assumption of grayness is sometimes made because of the absence of information relating ε_λ and λ.

When radiant energy falls on a surface, it can be absorbed, reflected, or transmitted through the material. Therefore, from the first law of thermodynamics,

$$\alpha + \tau + \rho = 1 \quad (39)$$

where

α = fraction of incident radiation absorbed or **absorptance**
τ = fraction of incident radiation transmitted or **transmittance**
ρ = fraction of incident radiation reflected or **reflectance**

If the material is opaque, as most solids are in the infrared, $\tau = 0$ and $\alpha + \rho = 1$. For a black surface, $\alpha = 1$, $\rho = 0$, and $\tau = 0$. Platinum black and gold black are as black as any actual surface and have absorptances of about 98% in the infrared. Any desired degree of blackness can be simulated by a small hole in a large enclosure. Consider a ray of radiant energy entering the opening. It will undergo many internal reflections and be almost completely absorbed before it has a reasonable probability of passing back out of the opening.

Certain flat black paints also exhibit emittances of 98% over a wide range of conditions. They provide a much more durable surface than gold or platinum black and are frequently used on radiation instruments and as standard reference in emittance or reflectance measurements.

Kirchhoff's law relates emittance and absorptance of any opaque surface from thermodynamic considerations; it states that for any surface where the incident radiation is independent of angle or where the surface is diffuse, $\varepsilon_\lambda = \alpha_\lambda$. If the surface is gray, or the incident radiation is from a black surface at the same temperature, then $\varepsilon = \alpha$ as well, but many surfaces are not gray. For most surfaces listed in Table 3, absorptance for solar radiation is different from emittance for low-temperature radiation. This is because the wavelength distributions are different in the two cases, and ε_λ varies with wavelength.

The foregoing discussion relates to total hemispherical radiation from surfaces. Energy distribution over the hemispherical region above the surface also has an important effect on the rate of heat transfer in various geometric arrangements.

Lambert's law states that the emissive power of radiant energy over a hemispherical surface above the emitting surface varies as the cosine of the angle between the normal to the radiating surface and the line joining the radiating surface to the point of the hemispherical surface. This radiation is **diffuse radiation**. The Lambert emissive power variation is equivalent to assuming that radiation from a surface in a direction other than normal occurs as if it came from an equivalent area with the same emissive power (per unit area) as the original surface. The equivalent area is obtained by projecting the original area onto a plane normal to the direction of radiation. Black surfaces obey the Lambert law exactly. The law is approximate for many actual radiation and reflection processes, especially those involving rough surfaces and nonmetallic materials. Most radiation analyses are based on the assumption of gray diffuse radiation and reflection.

In estimating heat transfer rates between surfaces of different geometries, radiation characteristics, and orientations, it is usually assumed that

- All surfaces are gray or black
- Radiation and reflection are diffuse
- Properties are uniform over the surfaces
- Absorptance equals emittance and is independent of the temperature of the source of incident radiation
- The material in the space between the radiating surfaces neither emits nor absorbs radiation

These assumptions greatly simplify problems, although results must be considered approximate.

Angle Factor

The distribution of radiation from a surface among the surfaces it irradiates is indicated by a quantity variously called an interception, a view, a configuration, or an angle factor. In terms of two surfaces i and j, the **angle factor** F_{ij} from surface i to surface j is defined as the fraction of diffuse radiant energy leaving surface i that falls directly on j (i.e., is intercepted by j). The angle factor from j to i is similarly defined, merely by interchanging the roles of i and j. This second angle factor is not, in general, numerically equal to the first. However, the reciprocity relation $F_{ij}A_i = F_{ji}A_j$, where A is the surface area, is always valid. Note that a concave surface may "see itself" ($F_{ii} \neq 0$), and that if n surfaces form an enclosure,

$$\sum_{j=1}^{n} F_{ij} = 1 \tag{40}$$

The angle factor F_{12} between two surfaces is

$$F_{12} = \frac{1}{A_1}\int_{A_1}\int_{A_2} \frac{\cos\phi_1 \cos\phi_2}{\pi r^2}\, dA_1\, dA_2 \tag{41}$$

where dA_1, and dA_2 are elemental areas of the two surfaces, r is the distance between dA_1 and dA_2, and ϕ_1 and ϕ_2 are the angles between the respective normals to dA_1 and dA_2 and the connecting line r. Numerical, graphical, and mechanical techniques can solve this equation (Siegel and Howell 1981, Modest 1993). Numerical values of the angle factor for common geometries are given in Figure 6.

Calculation of Radiant Exchange Between Surfaces Separated by Nonabsorbing Media

A surface radiates energy at a rate independent of its surroundings and absorbs and reflects incident energy at a rate dependent on its surface condition. The net energy exchange per unit area is denoted by q or q_j for unit area A_j. It is the rate of emission of the surface minus the total rate of absorption at the surface from all radiant effects in its surroundings, possibly including the return of some of its own emission by reflection off its surroundings. The rate at which energy must be supplied to the surface by other exchange processes if its temperature is to remain constant is q; therefore, to define q, the total radiant surroundings (in effect, an enclosure) must be specified.

Several methods have been developed to solve certain problems. To calculate the radiation exchange at each surface of an enclosure of n opaque surfaces by simple, general equations convenient for machine calculation, two terms must be defined:

G = irradiation; total radiation incident on surface per unit time and per unit area

J = radiosity; total radiation that leaves surface per unit time and per unit area

The radiosity is the sum of the energy emitted and the energy reflected:

$$J = \varepsilon W_b + \rho G \tag{42}$$

Because the transmittance is zero, the reflectance is

$$\rho = 1 - \alpha = 1 - \varepsilon$$

Thus,

$$J = \varepsilon W_b + (1 - \varepsilon)G \tag{43}$$

The net energy lost by a surface is the difference between the radiosity and the irradiation:

$$q/A = J - G = \varepsilon W_b + (1 - \varepsilon)G - G \tag{44}$$

Substituting for G in terms of J from Equation (43),

$$q = \frac{W_b - J}{(1 - \varepsilon)/\varepsilon A} \tag{45}$$

Consider an enclosure of n isothermal surfaces with areas of $A_1, A_2, \ldots, A_n$, emittances of $\varepsilon_1, \varepsilon_2, \ldots, \varepsilon_n$, and reflectances of $\rho_1, \rho_2, \ldots, \rho_n$, respectively.

The irradiation of surface i is the sum of the radiation incident on it from all n surfaces:

$$G_i A_i = \sum_{j=1}^{n} F_{ji} J_j A_j = \sum_{j=1}^{n} F_{ij} J_j A_i$$

or

$$G_i = \sum_{j=1}^{n} F_{ij} J_j$$

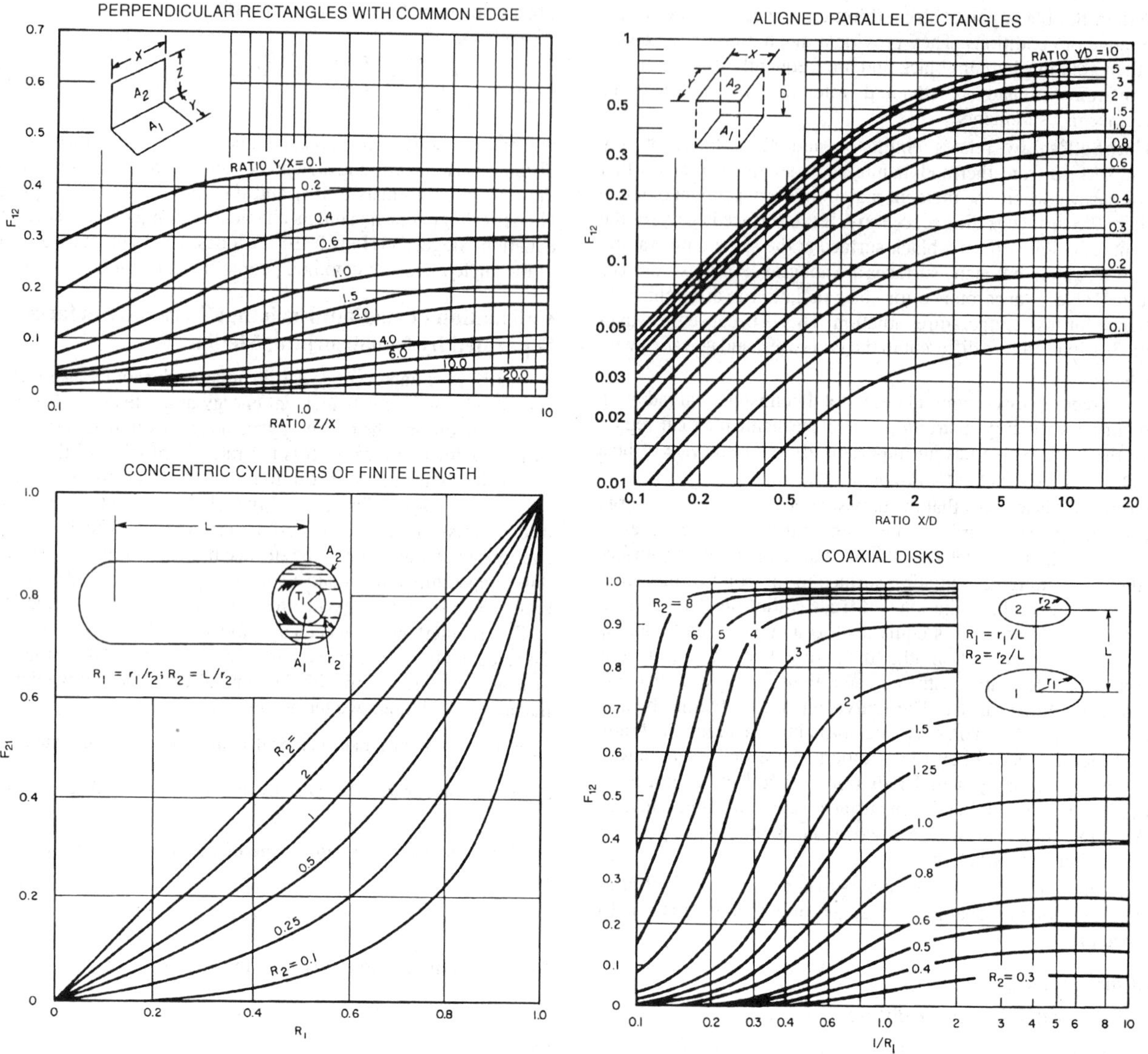

Fig. 6 Radiation Angle Factor for Various Geometries

Substituting in Equation (44) yields the following simultaneous equations when each of the n surfaces is considered:

$$J_i = \varepsilon_i W_{bi} + (1 - \varepsilon_i) \sum_{j=1}^{n} F_{ij} J_j \qquad i = 1, 2, \ldots, n \tag{46}$$

Equation (46) can be solved manually for the unknown Js if the number of surfaces is small. The solution for more complex enclosures requires a computer.

Once the radiosities (Js) are known, the net radiant energy lost by each surface is determined from Equation (45) as

$$q_i = \frac{W_{bi} - J_i}{(1 - \varepsilon_i)/\varepsilon_i A_i}$$

If the surface is black, Equation (45) becomes indeterminate, and an alternate expression must be used, such as

$$q_i = \sum_{j=1}^{n} J_i A_i F_{ij} - J_j A_j F_{ji}$$

or

$$q_i = \sum_{j=1}^{n} F_{ij} A_i (J_i - J_j) \tag{47}$$

since

$$F_{ij} A_i = F_{ji} A_j$$

All diffuse radiation processes are included in the aforementioned enclosure method, and surfaces with special characteristics are assigned consistent properties. An opening is treated as an equivalent surface area A_e with a reflectance of zero. If energy enters the enclosure diffusely through the opening, A_e is assigned an equivalent temperature; otherwise, its temperature is taken as zero. If the loss through the opening is desired, q_2 is found. A window in the enclosure is assigned its actual properties.

A surface in **radiant balance** is one for which radiant emission is balanced by radiant absorption; heat is neither removed from nor supplied to the surface. Reradiating surfaces (insulated surfaces with $q_{net} = 0$), can be treated in Equation (46) as being perfectly reflective (i.e., $\varepsilon = 0$). The equilibrium temperature of such a surface can be found from

$$T_k = \left(\frac{J_k}{\sigma}\right)^{0.25}$$

once Equation (46) has been solved for the radiosities.

Use of angle factors and radiation properties as defined assumes that the surfaces are diffuse radiators—a good assumption for most nonmetals in the infrared region, but a poor assumption for highly polished metals. Subdividing the surfaces and considering the variation of radiation properties with angle of incidence improves the approximation but increases the work required for a solution.

Radiation in Gases

Elementary gases such as oxygen, nitrogen, hydrogen, and helium are essentially transparent to thermal radiation. Their absorption and emission bands are confined mainly to the ultraviolet region of the spectrum. The gaseous vapors of most compounds, however, have absorption bands in the infrared region. Carbon monoxide, carbon dioxide, water vapor, sulfur dioxide, ammonia, acid vapors, and organic vapors absorb and emit significant amounts of energy.

Radiation exchange by opaque solids is considered a surface phenomenon. Radiant energy does, however, penetrate the surface of all materials. The absorption coefficient gives the rate of exponential attenuation of the energy. Metals have large absorption coefficients, and radiant energy penetrates only a few hundred angstroms at most. Absorption coefficients for nonmetals are lower. Radiation may be considered a surface phenomenon unless the material is transparent. Gases have small absorption coefficients, so the path length of radiation through gas becomes very significant.

Beer's law states that the attenuation of radiant energy in a gas is a function of the product $p_g L$ of the partial pressure of the gas and the path length. The monochromatic absorptance of a body of gas of thickness L is then given by

$$\alpha_{\lambda L} = 1 - e^{-\alpha \lambda L} \tag{48}$$

Because absorption occurs in discrete wavelengths, the absorptances must be summed over the spectral region corresponding to the temperature of the blackbody radiation passing through the gas. The monochromatic absorption coefficient α_λ is also a function of temperature and pressure of the gas; therefore, detailed treatment of gas radiation is quite complex.

Estimated emittance for carbon dioxide and water vapor in air at 75°F is a function of concentration and path length (Table 4). The values are for a hemispherically shaped body of gas radiating to an element of area at the center of the hemisphere. Among others, Modest (1993), Siegel and Howell (1981), and Hottel and Sarofim (1967) describe geometrical calculations in their texts on radiation transfer. Generally, at low values of $p_g L$, the mean path length L (or equivalent hemispherical radius for a gas body radiating to its surrounding surfaces) is four times the mean hydraulic radius of the enclosure. A room with a dimensional ratio of 1:1:4 has a mean path length of 0.89 times the shortest dimension when considering radiation to all walls. For a room with a dimensional ratio of 1:2:6, the mean path length for the gas radiating to all surfaces is 1.2 times the shortest dimension. The mean path length for radiation to the 2 by 6 face is 1.18 times the shortest dimension. These values are for cases where the partial pressure of the gas times the mean path length approaches zero ($p_g L \approx 0$). The factor decreases with increasing values of $p_g L$. For average rooms with approximately 8 ft ceilings and relative humidity ranging from 10 to 75% at 75°F, the effective path length for carbon dioxide radiation is about 85% of the ceiling height, or 6.8 ft. The effective path length for water vapor is about 93% of the ceiling height, or 7.4 ft. The effective emittance of the water vapor and carbon dioxide radiating to the walls, ceiling, and floor of a room 16 ft by 48 ft with 8 ft ceilings is in the following tabulation.

Relative Humidity, %	ε_g
10	0.10
50	0.19
75	0.22

The radiation heat transfer from the gas to the walls is then

$$q = \sigma A_w \varepsilon_g (T_g^4 - T_w^4) \tag{49}$$

The examples in Table 4 and the preceding text indicate the importance of gas radiation in environmental heat transfer problems. Gas radiation in large furnaces is the dominant mode of heat transfer, and many additional factors must be considered. Increased pressure broadens the spectral bands, and interaction of different radiating species prohibits simple summation of the emittance factors for the individual species. Departures from blackbody conditions necessitate separate calculations of the emittance and absorptance. McAdams (1954) and Hottel and Sarofim (1967) give more complete treatments of gas radiation.

Table 4 Emittance of CO_2 and Water Vapor in Air at 75 °F

Path Length, ft	CO_2, % by Volume		Relative Humidity, %			
	0.1	0.3	1.0	10	50	100
10	0.03	0.06	0.09	0.06	0.17	0.22
100	0.09	0.12	0.16	0.22	0.39	0.47
1000	0.16	0.19	0.23	0.47	0.64	0.70

NATURAL CONVECTION

Heat transfer involving motion in a fluid due to the difference in density and the action of gravity is called **natural convection** or **free convection**. Heat transfer coefficients for natural convection are generally much lower than those for forced convection, and it is therefore important not to ignore radiation in calculating the total heat loss or gain. Radiant transfer may be of the same order of magnitude as natural convection, even at room temperatures, because wall temperatures in a room can affect human comfort (see Chapter 8).

Natural convection is important in a variety of heating and refrigeration equipment: (1) gravity coils used in high-humidity cold storage rooms and in roof-mounted refrigerant condensers, (2) the evaporator and condenser of household refrigerators, (3) baseboard radiators and convectors for space heating, and (4) cooling panels for air conditioning. Natural convection is also involved in heat loss or gain to equipment casings and interconnecting ducts and pipes.

Consider heat transfer by natural convection between a cold fluid and a hot surface. The fluid in immediate contact with the surface is heated by conduction, becomes lighter, and rises because of the difference in density of the adjacent fluid. The viscosity of the fluid resists this motion. The heat transfer is influenced by (1) gravitational force due to thermal expansion, (2) viscous drag, and (3) thermal diffusion. Gravitational acceleration g, coefficient of thermal expansion β, kinematic viscosity $\nu = \mu/\rho$, and thermal diffusivity $\alpha = k/\rho c_p$ affect natural convection. These variables are included in the dimensionless numbers given in Equation (1) in Table 5. The Nusselt number Nu is a function of the product of the Prandtl number Pr and the Grashof number Gr. These numbers, when combined, depend on the fluid properties, the temperature difference Δt between the surface and the fluid, and the characteristic length L of the surface. The constant c and the exponent n depend on the physical configuration and the nature of flow.

Table 5 Natural Convection Heat Transfer Coefficients

I. General relationships	$\mathrm{Nu} = c(\mathrm{Gr\,Pr})^n$	(1)
	$h = c\dfrac{k}{L}\left(\dfrac{L^3\rho^2\beta g\Delta t}{\mu^2}\right)_f^n\left(\dfrac{\mu c_p}{k}\right)_f^n$	(2)
Characteristic length L		
Vertical plates or pipes	L = height	
Horizontal plates	L = length	
Horizontal pipes	L = diameter	
Spheres	$L = 0.5 \times$ diameter	
Rectangular block, with horizontal length L_h and vertical length L_v	$1/L = (1/L_h) + (1/L_v)$	
II. Planes and pipes		
Horizontal or vertical planes, pipes, rectangular blocks, and spheres (excluding horizontal plates facing downward for heating and upward for cooling)		
(a) Laminar range, when Gr Pr is between 10^4 and 10^8	$\mathrm{Nu} = 0.56(\mathrm{Gr\,Pr})^{0.25}$	(3)
(b) Turbulent range, when Gr Pr is between 10^8 and 10^{12}	$\mathrm{Nu} = 0.13(\mathrm{Gr\,Pr})^{0.33}$	(4)
III. Wires		
For horizontal or vertical wires, use L = diameter, for Gr Pr between 10^{-7} and 1	$\mathrm{Nu} = (\mathrm{Gr\,Pr})^{0.1}$	(5)
IV. With air		
$\mathrm{Gr\,Pr} = 1.6 \times 10^6 L^3\Delta t$ (at 70°F, L in ft, Δt in °F)		
(a) Horizontal cylinders		
Small cylinder, laminar range	$h = 0.27(\Delta t/L)^{0.25}$	(6)
Large cylinder, turbulent range	$h = 0.18(\Delta t)^{0.33}$	(7)
(b) Vertical plates		
Small plates, laminar range	$h = 0.29(\Delta t/L)^{0.25}$	(8)
Large plates, turbulent range	$h = 0.19(\Delta t)^{0.33}$	(9)
(c) Horizontal plates, facing upward when heated or downward when cooled		
Small plates, laminar range	$h = 0.27(\Delta t/L)^{0.25}$	(10)
Large plates, turbulent range	$h = 0.22(\Delta t)^{0.33}$	(11)
(d) Horizontal plates, facing downward when heated or upward when cooled		
Small plates	$h = 0.12(\Delta t/L)^{0.25}$	(12)

Natural convection cannot be represented by a single value of exponent n, but it can be divided into three regions:

1. **Turbulent** natural convection, for which n equals 0.33
2. **Laminar** natural convection, for which n equals 0.25
3. A region that has GrPr less than for laminar natural convection, for which the exponent n gradually diminishes from 0.25 to lower values

Note that for wires, the GrPr is likely to be very small, so that the exponent n is 0.1 [Equation (5) in Table 5].

To calculate the natural-convection heat transfer coefficient, determine GrPr to find whether the boundary layer is laminar or turbulent; then apply the appropriate equation from Table 5. The correct characteristic length indicated in the table must be used. Because the exponent n is 0.33 for a turbulent boundary layer, the characteristic length cancels out in Equation (2) in Table 5, and the heat transfer coefficient is independent of the characteristic length, as seen in Equations (7), (9), and (11) in Table 5. Turbulence occurs when length or temperature difference is large. Because the length of a pipe is generally greater than its diameter, the heat transfer coefficient for vertical pipes is larger than for horizontal pipes.

Convection from horizontal plates facing downward when heated (or upward when cooled) is a special case. Because the hot air is above the colder air, theoretically no convection should occur. Some convection is caused, however, by secondary influences such as temperature differences on the edges of the plate. As an approximation, a coefficient of somewhat less than half the coefficient for a heated horizontal plate facing upward can be used.

Because air is often the heat transport fluid, simplified equations for air are given in Table 5. Other information on natural convection is available in the section on Bibliography under Heat Transfer, General.

Observed differences in the comparison of recent experimental and numerical results with existing correlations for natural convective heat transfer coefficients indicate that caution should be used when applying coefficients for (isolated) vertical plates to vertical surfaces in enclosed spaces (buildings). Bauman et al. (1983) and Altmayer et al. (1983) developed improved correlations for calculating natural convective heat transfer from vertical surfaces in rooms under certain temperature boundary conditions.

Natural convection can affect the heat transfer coefficient in the presence of weak forced convection. As the forced-convection effect (i.e., the Reynolds number) increases, "mixed convection" (superimposed forced-on-free convection) gives way to the pure forced-convection regime. In these cases, other sources describing combined free and forced convection should be consulted, since the heat transfer coefficient in the mixed-convection region is often larger than that calculated based on the natural- or forced-convection calculation alone. Metais and Eckert (1964) summarize natural-, mixed-, and forced-convection regimes for vertical and horizontal tubes. Figure 7 shows the approximate limits for horizontal tubes. Other studies are described by Grigull et al. (1982).

FORCED CONVECTION

Forced air coolers and heaters, forced air- or water-cooled condensers and evaporators, and liquid suction heat exchangers are examples of equipment that transfer heat primarily by forced convection.

When fluid flows over a flat plate, a **boundary layer** forms adjacent to the plate. The velocity of the fluid at the plate surface is zero and increases to its maximum free stream value just past the edge of the boundary layer (Figure 8). Boundary layer formation is important because the temperature change from plate to fluid (thermal resistance) is concentrated here. Where the boundary layer is thick, thermal resistance is great and the heat transfer coefficient is small.

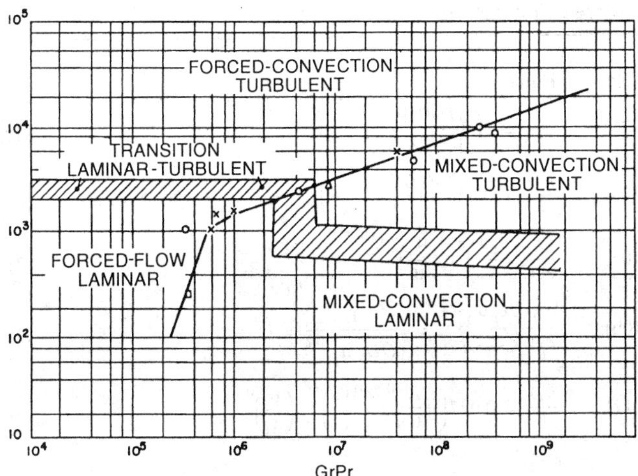

Fig. 7 Regimes of Free, Forced, and Mixed Convection for Flow-Through Horizontal Tubes

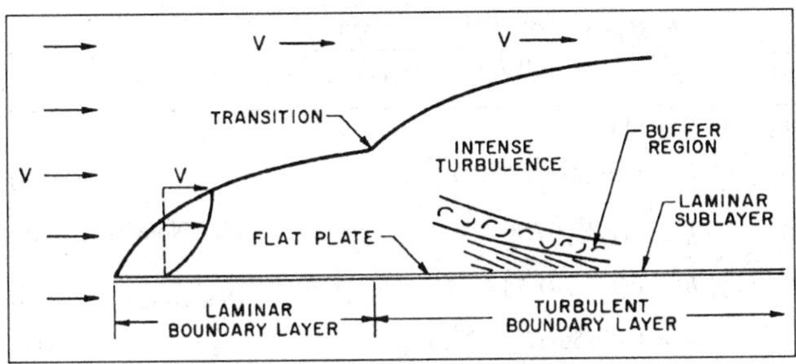

Fig. 8 Boundary Layer Buildup on Flat Plate (Vertical Scale Magnified)

At the leading edge of the plate, boundary layer thickness is theoretically zero and the heat transfer coefficient is infinite. Flow within the boundary layer immediately downstream from the leading edge is laminar and is known as **laminar forced convection**. As flow proceeds along the plate, the laminar boundary layer increases in thickness to a critical value. Then, turbulent eddies develop within the boundary layer, except for a thin **laminar sublayer** adjacent to the plate.

The boundary layer beyond this point is a **turbulent boundary layer**, and the flow is **turbulent forced convection**. The region between the breakdown of the laminar boundary layer and the establishment of the turbulent boundary layer is the **transition region**. Because the turbulent eddies greatly enhance heat transport into the main stream, the heat transfer coefficient begins to increase rapidly through the transition region. For a flat plate with a smooth leading edge, the turbulent boundary layer starts at Reynolds numbers, based on distance from the leading edge, of about 300,000 to 500,000. In blunt-edged plates, it can start at much smaller Reynolds numbers.

For long tubes or channels of small hydraulic diameter, at sufficiently low flow velocity, the laminar boundary layers on each wall grow until they meet. Beyond this point, the velocity distribution does not change, and no transition to turbulent flow takes place. This is called **fully developed laminar flow**. For tubes of large diameter or at higher velocities, trânsition to turbulence takes place and **fully developed turbulent flow** is established (Figure 9). Therefore, the length dimension that determines the critical Reynolds number is the hydraulic diameter of the channel. For smooth circular tubes, flow is laminar for Reynolds numbers below 2100 and turbulent above 10,000.

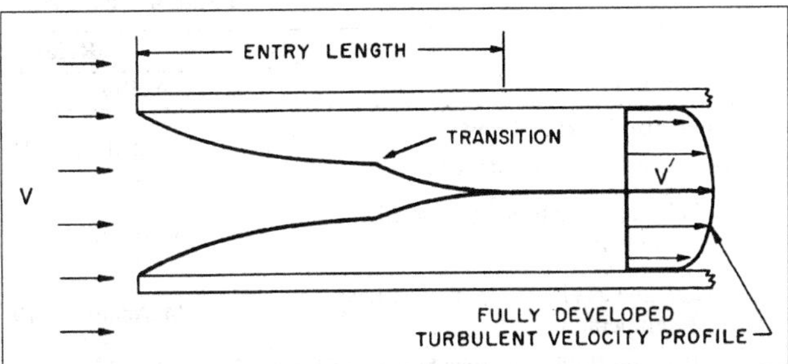

Fig. 9 Boundary Layer Buildup in Entry Length of Tube or Channel

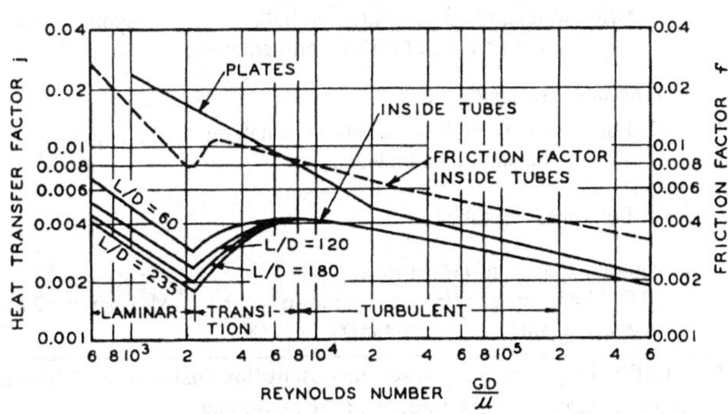

Fig. 10 Typical Dimensionless Representation of Forced-Convection Heat Transfer

Table 6 lists various forced-convection correlations. In the generalized, dimensionless formula of Equation (1) in Table 6, heat transfer is determined by flow conditions and by the fluid properties, as indicated by the Reynolds number and the Prandtl number. This equation can be modified to Equation (4) in Table 6 to get the **heat transfer factor** j. The heat transfer factor is related to the **friction factor** f by the interrelationship of the transport of momentum and heat; it is approximately $f/2$ for turbulent flow in straight ducts. These factors are plotted in Figure 10.

The characteristic length D is the diameter of the tube, outside or inside, or the length of the plane plate. For other shapes, the hydraulic diameter D_h is used:

$$D_h = 2r_h = 4 \times \frac{\text{Cross-sectional area for flow}}{\text{Total wetted perimeter}}$$

This reduces to twice the distance between surfaces for parallel plates or an annulus.

Simplified equations applicable to common fluids under normal operating conditions appear in Equations (8) through (25) of Table 6. Figure 11 gives graphical solutions for water.

Techniques to Augment Forced-Convection Heat Transfer

As discussed by Bergles (1985), techniques applied to augment heat transfer can be classified as passive methods, which require no direct application of external power, or as active schemes, which require external power. Examples of passive techniques include rough surfaces, extended surfaces, displaced promoters, and vortex flow devices. Examples of active techniques include mechanical aids, surface vibration, fluid vibration, and electrostatic fields. The effectiveness of a given augmentation technique depends largely on the mode of heat transfer or the type of heat exchanger to which it is applied.

Table 6 Equations for Forced Convection

Description	Reference: Author	Page	Eq. No.	Equation		
I. Generalized correlations						
(a) Turbulent flow inside tubes	Jakob	491	(23-36)	$\frac{hD}{k} = c\left(\frac{GD}{\mu}\right)^m\left(\frac{\mu c_p}{k}\right)^n$		(1)
(1) Using fluid properties based on bulk temperature t	McAdams	219	(9-10a)	$\frac{hD}{k} = 0.023\left(\frac{GD}{\mu}\right)^{0.8}\left(\frac{\mu c_p}{k}\right)^{0.4}$	(See Note a)	(2)
(2) Same as (1), except μ at surface temperature t_s	McAdams	219	(9-10c)	$\frac{h}{c_p G}\left(\frac{c_p\mu}{k}\right)^{2/3}\left(\frac{\mu_s}{\mu}\right)^{0.14} = \frac{0.023}{(GD/\mu)^{0.2}}$		(3)
(3) Using fluid properties based on film temperature $t_f = 0.5(t_s + t)$, except c_p in Stanton modulus	McAdams	219	(9-10b)	$\frac{h}{c_p G}\left(\frac{c_p\mu}{k}\right)_f^{2/3} = \frac{0.023}{(GD/\mu_f)^{0.2}} = j$		(4)
(4) For viscous fluids (viscosities higher than twice water), using viscosity μ at bulk temperature t and μ_s at surface temperature t_s	Jakob	547	(26-12)	$\frac{hD}{k} = 0.027\left(\frac{GD}{\mu}\right)^{0.8}\left(\frac{\mu c_p}{k}\right)^{1/3}\left(\frac{\mu}{\mu_s}\right)^{0.14}$		(5)
(b) Laminar flow inside tubes						
(1) For large D or high Δt, the effect of natural convection should be included	Jakob	544	(26-5)	$\frac{hD}{k} = 1.86\left[\left(\frac{GD}{\mu}\right)\left(\frac{c_p\mu}{k}\right)\left(\frac{D}{L}\right)\right]^{1/3}\left(\frac{\mu}{\mu_s}\right)^{0.14}$		(6)
(2) For very long tubes				When $\left(\frac{GD}{\mu}\right)\left(\frac{c_p\mu}{k}\right)\left(\frac{D}{L}\right) < 20$, Eq. (6) should not be used		
(c) Annular spaces, turbulent flow All fluid properties at bulk temperature except μ_s at surface temperature t_s	McAdams	242	(9-32c)	$\frac{h}{c_p G}\left(\frac{c_p\mu}{k}\right)^{2/3}\left(\frac{\mu_s}{\mu}\right)^{0.14} = \frac{0.023}{(DeG/\mu)^{0.2}}$		(7)
II. Simplified equations for gases, turbulent flow inside tubes (Units are in lb_m, h, ft, °F, and Btu.)						
(a) Most common gases, turbulent flow (assuming $\mu = 0.0455$ lb_m/ft · h and $\mu c_p/k = 0.78$)	Obtained from Eq. (2)			$h = 0.0144(c_p G^{0.8}/D^{0.2})$		(8)
(b) Air at ordinary temperatures	Obtained from Eq. (2)			$h = c(G^{0.8}/D^{0.2})$	(See Note b)	(9)
(c) Fluorinated hydrocarbon refrigerant gas at ordinary pressures	Obtained from Eq. (2)			$h = c(G^{0.8}/D^{0.2})$	(See Note b)	(10)
(d) Ammonia gas at approximately 150°F, 300 psi	Obtained from Eq. (2)			$h = 0.00756(G^{0.8}/D^{0.2})$		(11)
At 0°F, 24 psi	Obtained from Eq. (2)			$h = 0.00604(G^{0.8}/D^{0.2})$		(12)
III. Simplified equations for liquids, turbulent flow inside tubes (Units are in lb_m, h, ft, °F, and Btu.)						
(a) Water at ordinary temperatures, 40 to 200°F. V is velocity in fps, D is tube ID in inches.	McAdams	228	(9-19)	$h = \frac{150(1 + 0.011t)V^{0.8}}{D^{0.2}}$		(13)
(b) Fluorinated hydrocarbon refrigerant liquid	Obtained from Eq. (2)			$h = c(G^{0.8}/D^{0.2})$	(See Note b)	(14)
(c) Ammonia liquid at approximately 100°F	Obtained from Eq. (2)			$h = 0.0156(G^{0.8}/D^{0.2})$		(15)
(d) Oil heating, approximate equation	Brown and Marco	146	(7-15)	$h = 0.034V/\mu_f^{0.63}$		(16)
(e) Oil cooling, approximate equation	Brown and Marco	146	(7-15)	$h = 0.0255V/\mu_f^{0.63}$		(17)
IV. Simplified equations for air						
(a) Vertical plane surfaces, V of 16 to 100 fps (room temperature)[c]	McAdams	249	(9-42)	$h' = 0.5V^{0.78}$		(18)
(b) Vertical plane surfaces, $V < 16$ fps (room temperature)[c]	McAdams	249	(9-42)	$h' = 0.99 + 0.21V$		(19)
(c) Single cylinder cross flow (film temperature = 200°F) $1000 < GD/\mu_f < 50{,}000$	McAdams	261	(10-3c)	$h = 0.026(G^{0.6}/D^{0.4})$		(20)
(d) Single sphere $17 < GD/\mu_f < 70{,}000$	McAdams	265	(10-6)	$h = 0.37\frac{k_f}{D}\left(\frac{GD}{\mu_f}\right)^{0.6}$		(21)
V. Gases flowing normal to pipes (dimensionless)						
(a) Single cylinder Re from 0.1 to 1000	McAdams	260	(10-3)	$\frac{hD}{k_f} = 0.32 + 0.43\left(\frac{GD}{\mu}\right)^{0.52}$		(22)
Re from 1000 to 50,000	McAdams	260	(10-3)	$\frac{hD}{k_f} = 0.24\left(\frac{GD}{\mu_f}\right)^{0.6}$		(23)
(b) Unbaffled staggered tubes, 10 rows. Approximate equation for turbulent flow[d]	McAdams	272	(10-11a)	$\frac{hD}{k_f} = 0.33\left(\frac{G_{max}D}{\mu_f}\right)^{0.6}\left(\frac{\mu c_p}{k}\right)_f^{1/3}$		(24)
(c) Unbaffled in-line tubes, 10 rows. Approximate equation for turbulent flow[d] ($G_{max}D/\mu_f$) from 2000 to 32,000	McAdams	272	(10-11a)	$\frac{hD}{k_f} = 0.26\left(\frac{G_{max}D}{\mu_f}\right)^{0.6}\left(\frac{\mu c_p}{k}\right)_f^{1/3}$		(25)

[a]McAdams (1954) recommends this equation for heating and cooling. Others recommend exponents of 0.4 for heating and 0.3 for cooling, with a change in constant.

[b]Table 7 in Chapter 2 of the 1981 *ASHRAE Handbook—Fundamentals* lists values for c.

[c]h' is expressed in Btu/h·ft²·°F based on initial temperature difference.

[d]G_{max} is based on *minimum* free area. Coefficients for tube banks depend greatly on geometrical details. These values approximate only.

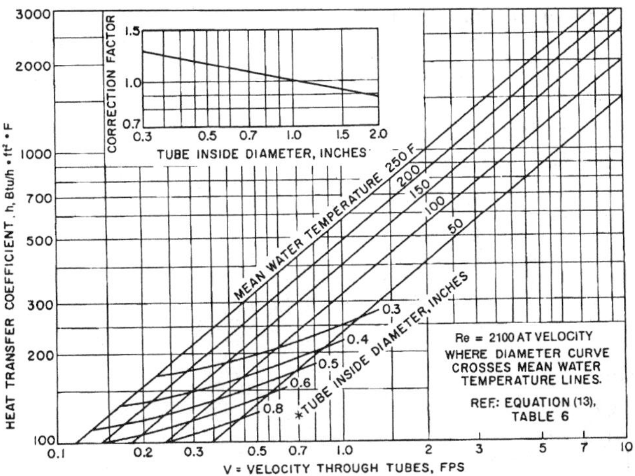

Fig. 11 Heat Transfer Coefficient for Turbulent Flow of Water Inside Tubes

When augmentation is used, the dominant thermal resistances in Equation (9) should be considered; that is, do not invest in reducing an already low thermal resistance (increasing an already high heat transfer coefficient). Additionally, heat exchangers with a large number of heat transfer units (NTU) show relatively small gains in effectiveness with augmentation [see Equations (26) and (27)]. Finally, the increased friction factor that accompanies the heat transfer augmentation must be considered.

Passive Techniques. Several examples of tubes with internal roughness or fins are shown in Figure 12. Rough surfaces of the spiral repeated rib variety are widely used to improve in-tube heat transfer with water, as in flooded chillers. The roughness may be produced by spirally indenting the outer wall, forming the inner wall, or inserting coils. Longitudinal or spiral internal fins in tubes can be produced by extrusion or forming and give a substantial increase in the surface area. The fin efficiency (see the section on Fin Efficiency) can usually be taken as unity. Twisted strips can be inserted as original equipment or as retrofit devices.

The increased friction factor may not require increased heat loss or pumping power if the flow rate can be adjusted or if the length of the heat exchanger can be reduced. Nelson and Bergles (1986) discuss this issue of performance evaluation criteria, especially for HVAC applications.

Of concern in chilled water systems is the fouling that in some cases may seriously reduce the overall heat transfer coefficient U. In general, fouled enhanced tubes perform better than fouled plain tubes, as shown in recent studies of scaling of cooling tower water (Knudsen and Roy 1983) and particulate fouling (Somerscales et al. 1991).

Fire-tube boilers are frequently fitted with **turbulators** to improve the turbulent convective heat transfer coefficient constituting the dominant thermal resistance. Also, due to the high gas temperatures, radiation from the convectively heated insert to the tube wall can represent as much as 50% of the total heat transfer. (Note, however, that the magnitude of the convective contribution decreases as the radiative contribution increases because of the reduced temperature difference.) Two commercial bent-strip inserts, a twisted-strip insert, and a simple bent-tab insert are depicted in Figure 13. Design equations, for convection only, are included in Table 7. Beckermann and Goldschmidt (1986) present procedures to include radiation, and Junkhan et al. (1985, 1988) give friction factor data and performance evaluations.

Several enhanced surfaces for gases are depicted in Figure 14. The offset strip fin is an example of an interrupted fin that is often found in compact plate fin heat exchangers used for heat recovery from exhaust air. Design equations are included in Table 7. These equations are comprehensive in that they apply to laminar and transitional flow as well as to turbulent flow, which is a necessary feature because the small hydraulic diameter of these surfaces drives the Reynolds number down. Data for other surfaces (wavy, spine, louvered, etc.) are given in the section on Bibliography.

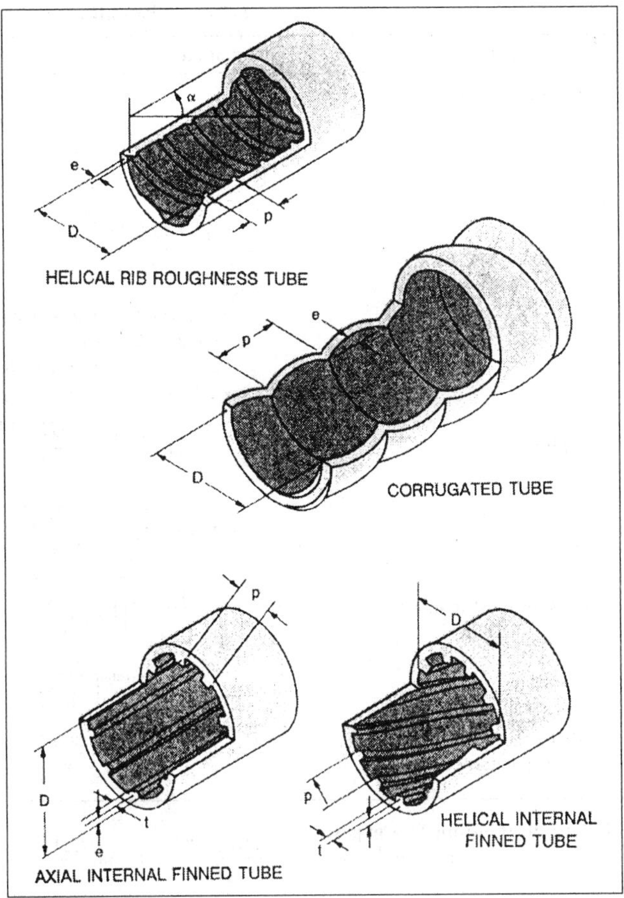

Fig. 12 Typical Tube-Side Enhancements

Active Techniques. Among the various active augmentation techniques, several mechanical aids, including stirring of the fluid by mechanical means, rotation of the heat transfer surface, and use of electrostatic fields have significantly increased the forced convective heat transfer. While mechanical aids are used in appropriate applications (e.g., surface scraping, baking, and drying processes), the electrostatic technique has been demonstrated only on prototype heat exchangers. The electrostatic or **electrohydrodynamic** (EHD) augmentation technique uses electrically induced secondary motions to destabilize the thermal boundary layer near the heat transfer surface, thereby substantially increasing the heat transfer coefficients at the wall. The magnitude and nature of enhancements are a function of (1) electric field parameters such as field potential, field polarity, pulse versus steady discharge, electrode geometry and electrode spacing; (2) flow field parameters such as mass flow rate, temperature, density, and electrical permittivity of the working fluid; and (3) the heat transfer surface type such as smooth, porous, or integrally finned/grooved configurations.

The EHD effect is generally applied by placing wire or plate electrodes parallel and adjacent to the heat transfer surface. Figure 15 presents four electrode configurations for augmentation of forced-convection heat transfer in tube flows. A high-voltage electric field charges the electrode and establishes the electrical body forces required to initiate and sustain augmentation.

Table 7 Equations for Augmented Forced Convection

Description	Equation
I. Turbulent in-tube flow of liquids	
(a) Spiral repeated rib[a]	$\frac{h_a}{h_s} = \left\{\left[1 + 2.64\left(\frac{GD}{\mu}\right)^{0.036}\left(\frac{e}{d}\right)^{0.212}\left(\frac{p}{d}\right)^{-0.21}\left(\frac{\alpha}{90}\right)^{0.29}\left(\frac{c_p\mu}{k}\right)^{-0.024}\right]^7\right\}^{1/7}$
	$\frac{f_a}{f_s} = \left\{1 + \left[29.1\left(\frac{GD}{\mu}\right)^{w}\left(\frac{e}{d}\right)^{x}\left(\frac{p}{d}\right)^{y}\left(\frac{\alpha}{90}\right)^{z}\left(1 + \frac{2.94}{n}\right)\sin\beta\right]^{15/16}\right\}^{16/15}$
	where $w = 0.67 - 0.06(p/d) - 0.49(\alpha/90)$; $x = 0.37 - 0.157(p/d)$; $y = -1.66 \times 10^{-6}(GD/\mu) - 0.33(\alpha/90)$; $z = 4.59 + 4.11 \times 10^{-6}(GD/\mu) - 0.15(p/d)$
(b) Fins[b]	$\frac{hD_h}{k} = 0.023\left(\frac{c_p\mu}{k}\right)^{0.4}\left(\frac{GD_h}{\mu}\right)^{0.8}\left(\frac{A_F}{A_{Fi}}\right)^{0.1}\left(\frac{A_i}{A}\right)^{0.5}(\sec\alpha)^3$
	$f_h = 0.046\left(\frac{GD_h}{\mu}\right)^{-0.2}\left(\frac{A_F}{A_{Fi}}\right)^{0.5}(\sec\alpha)^{0.75}$ Note that in computing the Reynolds number for (b) and (c) there is allowance for the reduced cross-sectional area.
(c) Twisted-strip inserts[c]	$\frac{hD_h}{k} = F\left\{0.023\left[1 + \left(\frac{\pi}{2y}\right)^2\right]^{0.4}\left(\frac{GD_h}{\mu}\right)^{0.8}\left(\frac{c_p\mu}{k}\right)^{0.4} + 0.193\left[\left(\frac{GD_h}{\mu y}\right)^2\frac{D_h}{D_i}\frac{\Delta\rho}{\rho}\left(\frac{c_p\mu}{k}\right)\right]^{1/3}\right\}$
	$f_{h,iso} = 0.127^{-0.406}\left(\frac{GD_h}{\mu}\right)^{-0.2}$
(d) Twisted-strip inserts for an evaporator (cooling)[c]	$\frac{hD_h}{k} = 0.023F\left[1 + \left(\frac{\pi}{2y}\right)^2\right]^{0.4}\left(\frac{GD_h}{\mu}\right)^{0.8}\left(\frac{c_p\mu}{k}\right)^{0.4}$
II. Turbulent in-tube flow of gases	
(a), (b) Bent-strip inserts[d]	$\frac{hD}{k}\left(\frac{T_w}{T_b}\right)^{0.45} = 0.258\left(\frac{GD}{\mu}\right)^{0.6}$
	$\frac{hD}{k}\left(\frac{T_w}{T_b}\right)^{0.45} = 0.208\left(\frac{GD}{\mu}\right)^{0.63}$
(c) Twisted-strip inserts[d]	$\frac{hD}{k}\left(\frac{T_w}{T_b}\right)^{0.45} = 0.122\left(\frac{GD}{\mu}\right)^{0.65}$
(d) Bent-tab inserts[d]	$\frac{hD}{k}\left(\frac{T_w}{T_b}\right)^{0.45} = 0.406\left(\frac{GD}{\mu}\right)^{0.54}$ Note that in computing the Reynolds number there is no allowance for the flow blockage of the insert.
III. Offset strip fins for plate-fin heat exchangers	$\frac{h}{c_pG} = 0.6522\left(\frac{GD_h}{\mu}\right)^{-0.5403}\alpha^{-0.1541}\delta^{0.1499}\gamma^{-0.0678}\left[1 + 5.269 \times 10^{-5}\left(\frac{GD_h}{\mu}\right)^{1.340}\alpha^{0.504}\delta^{0.456}\gamma^{-1.055}\right]^{0.1}$
	$f_h = 9.6243\left(\frac{GD_h}{\mu}\right)^{-0.7422}\alpha^{-0.1856}\delta^{-0.3053}\gamma^{-0.2659}\left[1 + 7.669 \times 10^{-8}\left(\frac{GD_h}{\mu}\right)^{4.429}\alpha^{0.920}\delta^{3.767}\gamma^{0.236}\right]^{0.1}$
	where h/c_pG, f_h, and GD_h/μ are based on the hydraulic diameter, given by $D_h = 4shl/[2(sl + hl + th) + ts]$

References:
[a]Ravigururajan and Bergles (1985)
[b]Carnavos (1979)
[c]Lopina and Bergles (1969)
[d]Junkhan et al. (1985)
[e]Manglik and Bergles (1990)

Table 8 Electrohydrodynamic Heat Transfer Enhancement in Heat Exchangers

Source	Maximum Reported Enhancement, %	Test Fluid	Heat Transfer Wall/ Electrode Configuration	Process
Fernandez and Poulter (1987)	2,300	Transformer oil	Tube/wire	Forced convection
Ohadi et al. (1991)	320	Air	Tube/wire or rod	Forced convection
Ohadi et al. (1992)	480	R-123	Tube/wire	Boiling
Sunada et al. (1991)	600	R-123	Vertical wall/plate	Condensation
Uemura et al. (1990)	1,400	R-113	Plate/wire mesh	Film boiling
Yabe and Maki (1988)	10,000	96% (by mass) R-113, 4% ethanol	Plate/ring	Natural convection

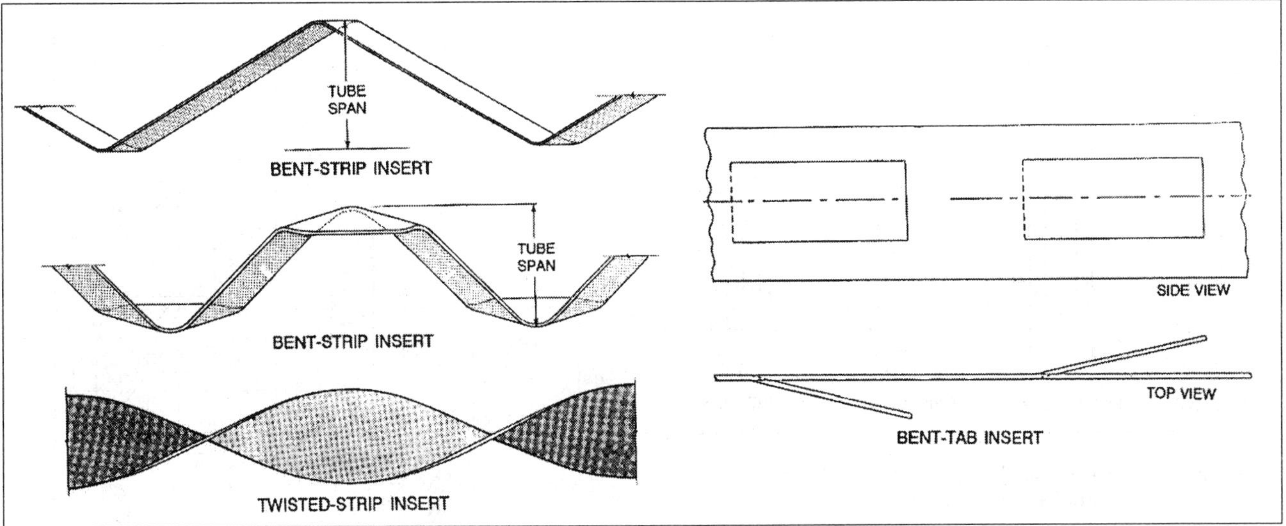

Fig. 13 Turbulators for Fire-Tube Boilers

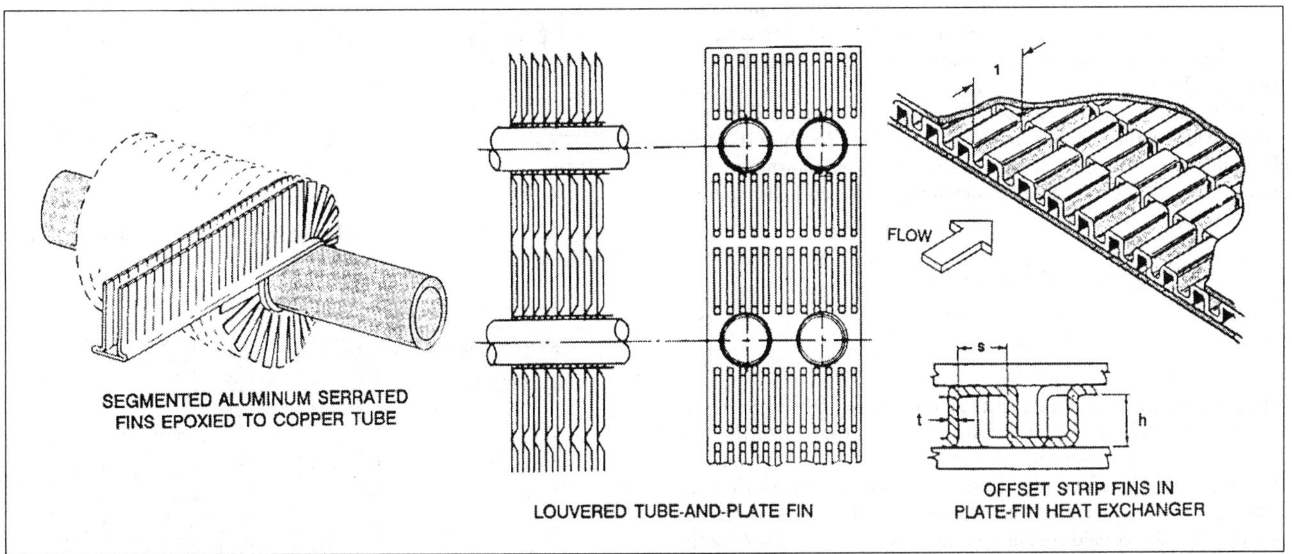

Fig. 14 Enhanced Surfaces for Gases

Several important advantages have contributed to the progress of the EHD technique. The mechanically complex rotation, injection, and vibration devices all serve to promote turbulence and better mixing of the bulk flow. However, these devices are generally cumbersome to manufacture, require excessive external power, and are not cost-effective. Furthermore, these systems often require a significant fraction of the power needed to pump the fluid. In contrast, even though high voltages are employed, the amount of electrical power consumed by the EHD process is extremely small (a few watts or less) due to the very small currents (1 mA or less). The additional manufacturing costs are minimal because the EHD process requires a small transformer and simple wire or plate electrodes. As is the case with most augmentation techniques, the increase in heat transfer coefficients is associated with a corresponding increase in pressure drop coefficients. However, unlike most other techniques where the pressure drop is substantial, the rise in pressure drops in the EHD technique is usually much less than the corresponding increase in heat transfer coefficients.

The EHD technique is limited in that it is only effective for fluids with low electrical conductivity. This includes air, certain industrial fluids, and many refrigerants. Table 8 provides a summary of selected studies involving single-phase and phase-change processes; it demonstrates that the EHD technique is particularly effective for heat transfer enhancement of refrigerants, including the ozone-safe refrigerant substitutes. Ohadi (1991) gives further details on the fundamentals, applicability, and limitations of the EHD technique.

EXTENDED SURFACE

Heat transfer from a prime surface can be increased by attaching **fins** or **extended surfaces** to increase the area available for heat transfer. Fins provide a more compact heat exchanger with lower material costs for a given performance. To achieve optimum design, fins are generally located on the side of the heat exchanger where the heat transfer coefficients are low (such as the air side of an air-to-water coil). Equipment with an extended surface includes natural- and forced-convection coils and shell-and-tube evaporators and condensers. Fins are also used inside tubes in condensers and dry expansion evaporators.

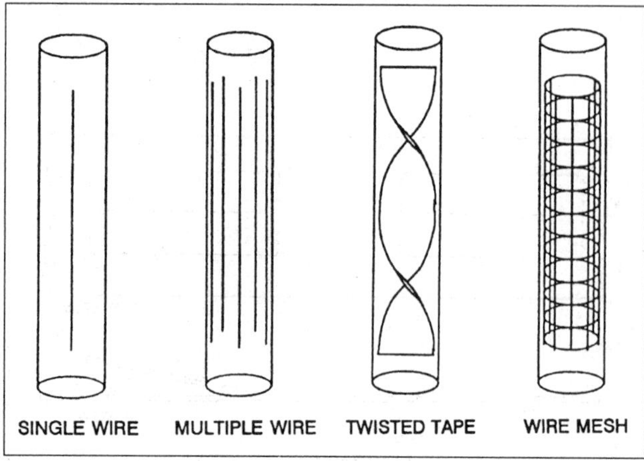

Fig. 15 Electrode Configurations for Internal Forced-Convection Flow

Fin Efficiency

As heat flows from the root of a fin to its tip, temperature drops because of the thermal resistance of the fin material. The temperature difference between the fin and the surrounding fluid is therefore greater at the root than at the tip, causing a corresponding variation in the heat flux. Therefore, increases in fin length result in proportionately less additional heat transfer. To account for this effect, **fin efficiency** ϕ is defined as the ratio of the actual heat transferred from the fin to the heat that would be transferred if the entire fin were at its root or base temperature:

$$\phi = \frac{\int h(t - t_e)dA}{\int h(t_r - t_e)dA} \tag{50}$$

where ϕ is the fin efficiency, t_e is the temperature of the surrounding environment, and t_r is the temperature at the fin root. Fin efficiency is low for long fins, thin fins, or fins made of low thermal conductivity material. Fin efficiency decreases as the heat transfer coefficient increases because of the increased heat flow. For natural convection in air-cooled condensers and evaporators, where h for the air side is low, fins can be fairly large and fabricated from low-conductivity materials such as steel instead of from copper or aluminum. For condensing and boiling, where large heat transfer coefficients are involved, fins must be very short for optimum use of material.

The heat transfer from a finned surface, such as a tube, which includes both finned or secondary area A_s and unfinned or prime area A_p is given by the following equation:

$$q = (h_p A_p + \phi h_s A_s)(t_r - t_e) \tag{51}$$

Assuming the heat transfer coefficients for the finned surface and prime surface are equal, a *surface efficiency* ϕ_s can be derived for use in Equation (52).

$$\phi_s = 1 - \left(\frac{A_s}{A}\right)(1 - \phi) \tag{52}$$

$$q = \phi_s hA(t_r - t_e) \tag{53}$$

where A is the total surface area, equal to the sum of the finned and prime areas ($A = A_s + A_p$).

Temperature distribution and fin efficiencies for various fin shapes are derived in most heat transfer texts. Figures 16 through

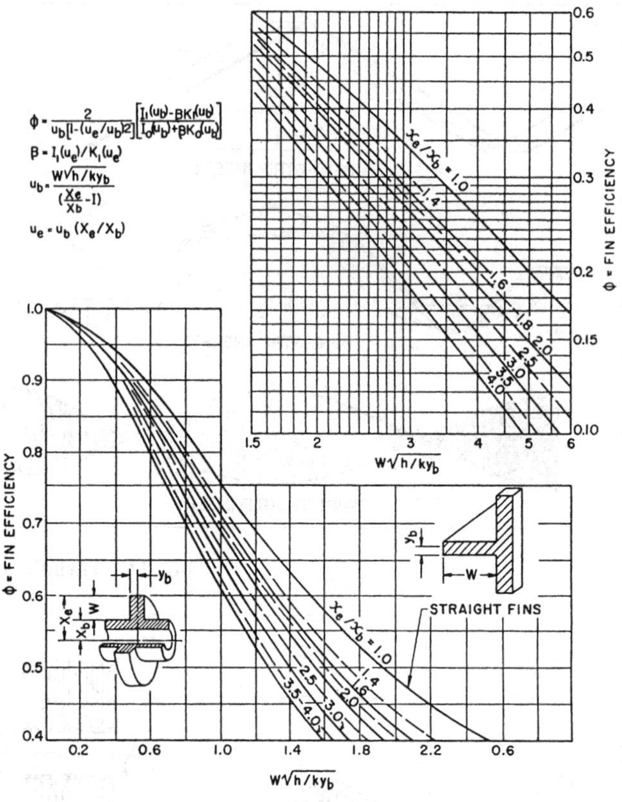

Fig. 16 Efficiency of Annular Fins of Constant Thickness

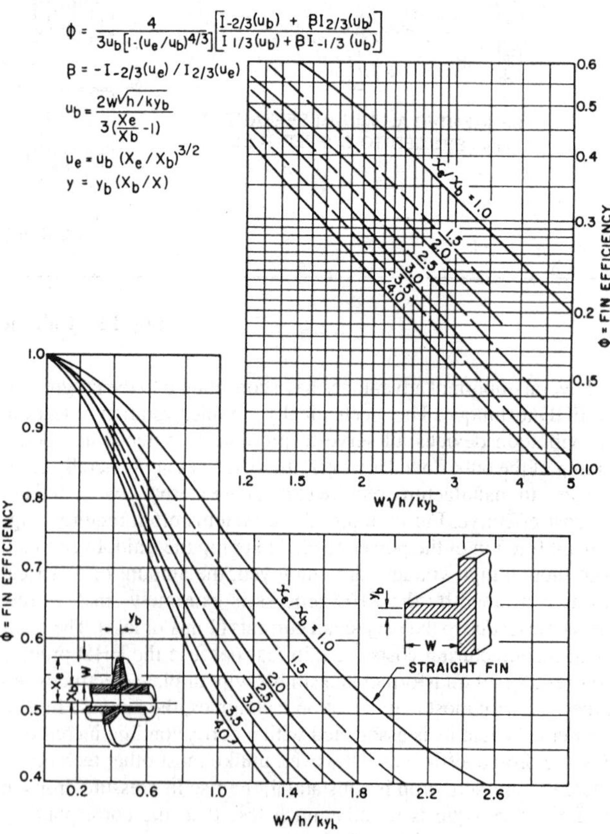

Fig. 17 Efficiency of Annular Fins with Constant Metal Area for Heat Flow

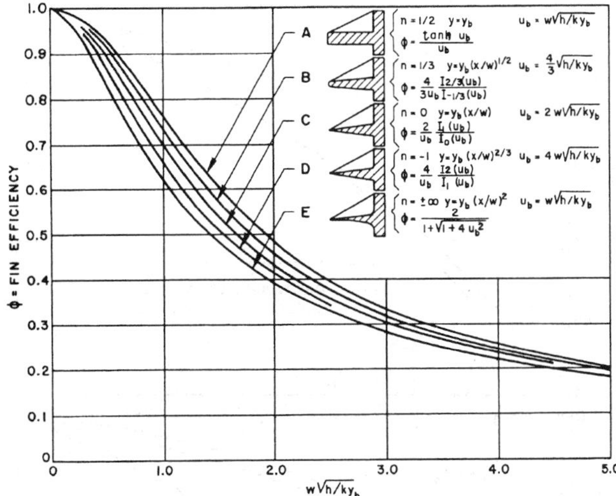

Fig. 18 Efficiency of Several Types of Straight Fin

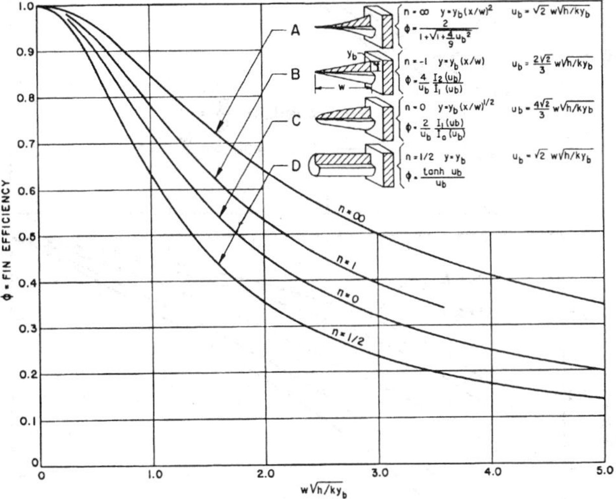

Fig. 19 Efficiency of Four Types of Spine

19 show curves and equations for annular fins, straight fins, and spines. For constant thickness square fins, the efficiency of a constant thickness annular fin of the same area can be used. More accuracy, particularly with rectangular fins of large aspect ratio, can be obtained by dividing the fin into circular sectors (Rich 1966).

Rich (1966) presents results for a wide range of geometries in a compact form for equipment designers by defining a dimensionless unit thermal resistance Φ:

$$\Phi = \frac{R_f t_o k}{l^2} \tag{54}$$

$$R_f = \left(\frac{1}{h}\right)\left(\frac{1}{\phi} - 1\right) \tag{54a}$$

where

Φ = dimensionless thermal resistance
ϕ = fin efficiency
t_o = fin thickness at fin base
l = length dimension = $r_t - r_o$ for annular fins
= W for rectangular fins

Rich (1966) also developed expressions for Φ_{max}, the maximum limiting value of Φ. Figure 20 gives Φ_{max} for annular fins of constant and tapered cross section as a function of $R = r_t/r_o$ (i.e., the ratio of the fin tip-to-root radii). Figure 21 gives Φ_{max} for rectangular fins of a given geometry as determined by the sector method. Figure 22 gives correction factors (Φ/Φ_{max}) for the determination of Φ from Φ_{max} for both annular and rectangular fins.

Example. This example illustrates the use of the fin resistance number for a rectangular fin typical of that for an air-conditioning coil.

Given: L = 0.75 in. t_o = 0.006 in.
W = 0.50 in. h = 10 Btu/h·ft²·°F
r_o = 0.25 in. k = 100 Btu·ft/h·ft²·°F

Solution: From Figure 21 at $W/r_o = 2.0$ and $L/W = 1.5$,

$$\Phi_{max} = R_{f(max)} t_o k / W^2 = 1.12$$

$$R_{f(max)} = \frac{1.12 \times 0.50^2}{0.006 \times 100 \times 12} = 0.0389 \text{ ft}^2 \cdot {}^\circ\text{F} \cdot \text{h/Btu}$$

The correction factor Φ/Φ_{max}, which is multiplied by $R_{f(max)}$ to give R_f, is given in Figure 22 as a function of the fin efficiency. As a first approximation, the fin efficiency is calculated from Equation (54a) assuming $R_f = R_{f(max)}$.

$$\phi = 1/(1 + hR_f) \approx 0.72$$

Interpolating between $L/W = 1$ and $L/W = 2$ at $W/r_o = 2$ gives

$$\Phi/\Phi_{max} = 0.88$$

Therefore,

$$R_f = 0.88 \times 0.0389 = 0.0342 \text{ ft}^2 \cdot {}^\circ\text{F} \cdot \text{h/Btu}$$

The above steps may now be repeated using the corrected value of fin resistance.

$$\phi = 0.745$$
$$\Phi/\Phi_{max} = 0.9$$
$$R_f = 0.035 \text{ ft}^2 \cdot {}^\circ\text{F} \cdot \text{h/Btu}$$

Note that the improvement in accuracy by reevaluating Φ/Φ_{max} is less than 1% of the overall thermal resistance (environment to fin base). The error produced by using $R_{f(max)}$ without correction is less than 3%. For many practical cases where greater accuracy is not warranted, a single value of R_f, obtained by estimating Φ/Φ_{max}, can be used over a range of heat transfer coefficients for a given fin. For approximate calculations, the fin resistance for other values of k and t_o can be obtained by simple proportion if the range covered is not excessive.

Schmidt (1949) presents approximate, but reasonably accurate, analytical expressions (for computer use) for circular, rectangular, and hexagonal fins. Hexagonal fins are the representative fin shape for the common staggered tube arrangement in finned-tube heat exchangers.

Schmidt's empirical solution is given by

$$\phi = \frac{\tanh(mr_i\Phi)}{mr_i\Phi}$$

where $m = \sqrt{2h/kt}$ and Φ is given by

$$\Phi = [(r_e/r_i) - 1][1 + 0.35\ln(r_e/r_i)]$$

For **circular fins**,

$$r_e/r_i = r_o/r_i$$

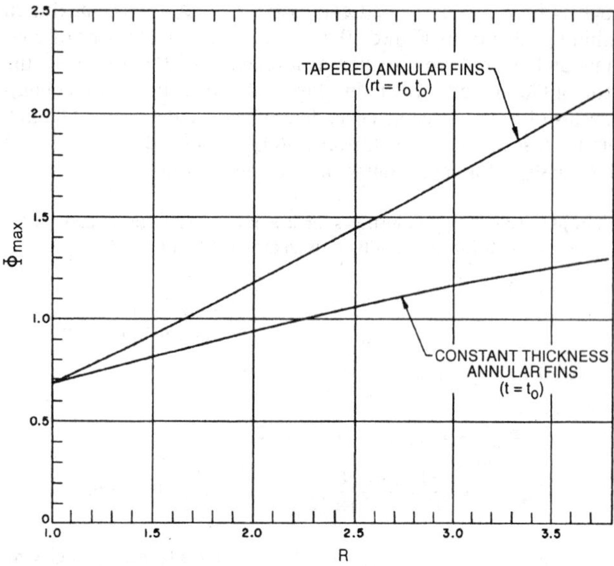

Fig. 20 Maximum Fin Resistance Number of Annular Fins
(Gardner 1945)

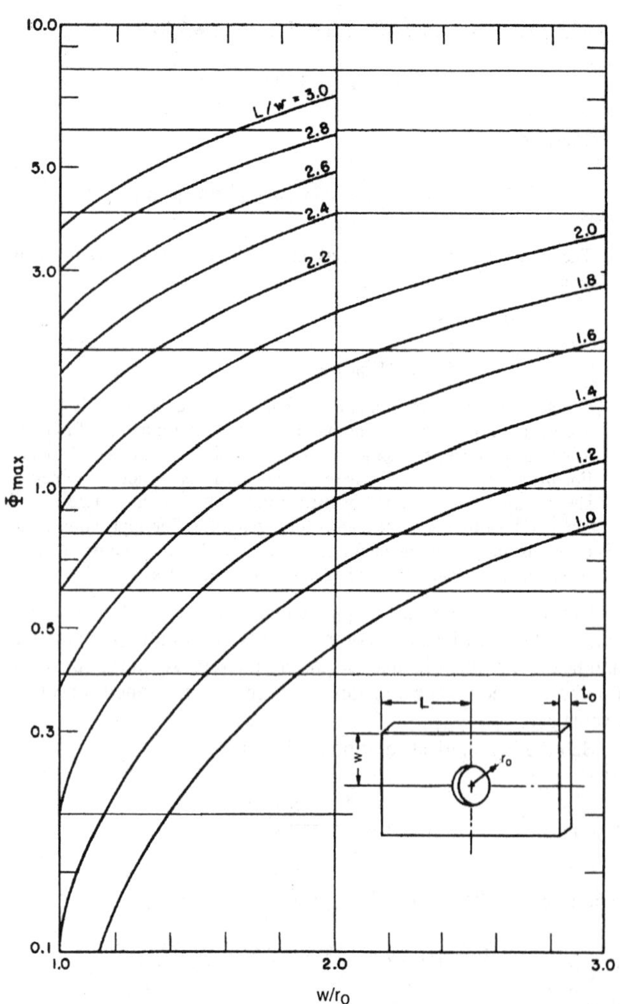

Fig. 21 Maximum Fin Resistance Number of Rectangular Fins Determined by Sector Method

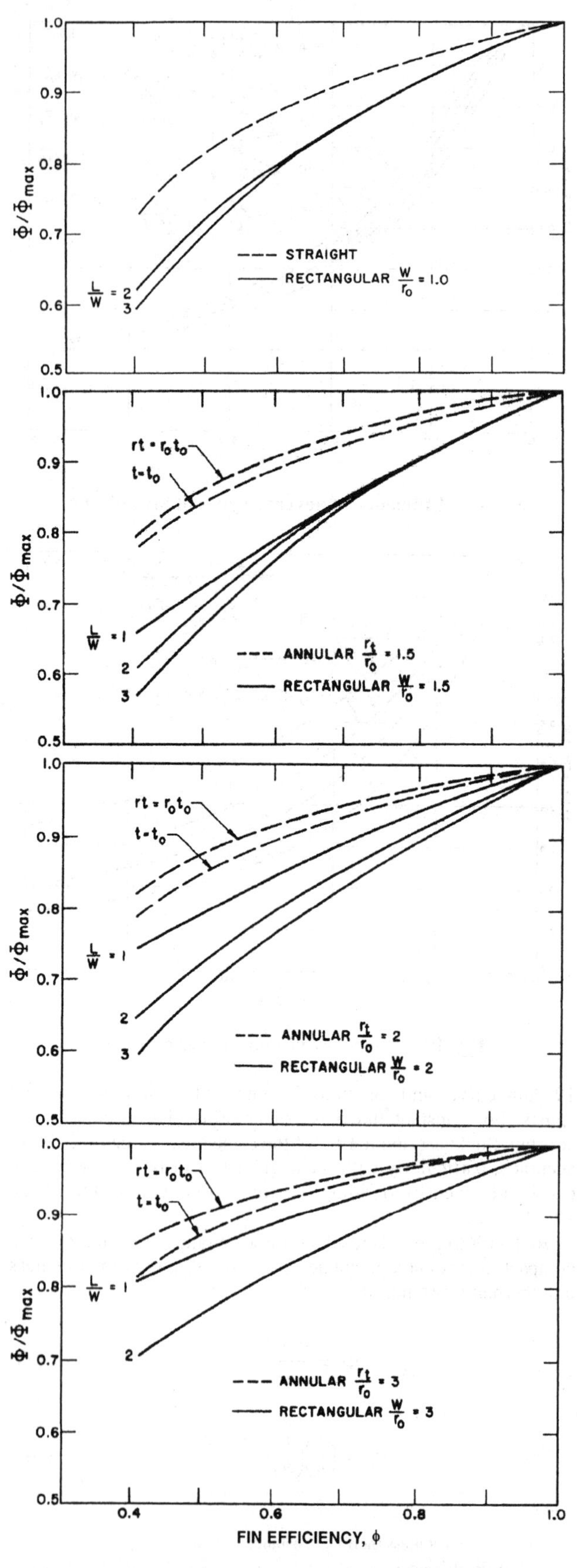

Fig. 22 Variation of Fin Resistance Number with Efficiency for Annular and Rectangular Fins
(Gardner 1945)

For **rectangular fins**,

$$r_e/r_i = 1.28\psi\sqrt{\beta - 0.2}, \quad \psi = M/r_i, \quad \beta = L/M \geq 1$$

where M and L are defined by Figure 23 as $a/2$ or $b/2$, depending on which is greater.

For **hexagonal fins**,

$$r_e/r_i = 1.27\psi\sqrt{\beta - 0.3}$$

where ψ and β are defined as above and M and L are defined by Figure 24 as $a/2$ or b (whichever is less) and $0.5\sqrt{(a^2/2)^2 + b^2}$, respectively.

The section on Bibliography lists other sources of information on finned surfaces.

Thermal Contact Resistance

Fins can be extruded from the prime surface (e.g., the short fins on the tubes in flooded evaporators or water-cooled condensers) or they can be fabricated separately, sometimes of a different material, and bonded to the prime surface. Metallurgical bonds are achieved by furnace-brazing, dip-brazing, or soldering. Nonmetallic bonding materials, such as epoxy resin, are also used. Mechanical bonds are obtained by tension-winding fins around tubes (spiral fins) or expanding the tubes into the fins (plate fins). Metallurgical bonding, properly done, leaves negligible thermal resistance at the joint but is not always economical. Thermal resistance of a mechanical bond may or may not be negligible, depending on the application, quality of manufacture, materials, and temperatures involved. Tests of plate fin coils with expanded tubes have indicated that substantial losses in performance can occur with fins that have cracked collars; but negligible thermal resistance was found in coils with continuous collars and properly expanded tubes (Dart 1959).

Thermal resistance at an interface between two solid materials is largely a function of the surface properties and characteristics of the solids, the contact pressure, and the fluid in the interface, if any. Eckels (1977) models the influence of fin density, fin thickness, and tube diameter on contact pressure and compared it to data for wet and dry coils. Shlykov (1964) shows that the range of attainable contact resistances is large. Sonokama (1964) presents data on the effects of contact pressure, surface roughness, hardness, void material, and the pressure of the gas in the voids. Lewis and Sauer (1965) show the resistance of adhesive bonds, and Kaspareck (1964) and Clausing (1964) give data on the contact resistance in a vacuum environment.

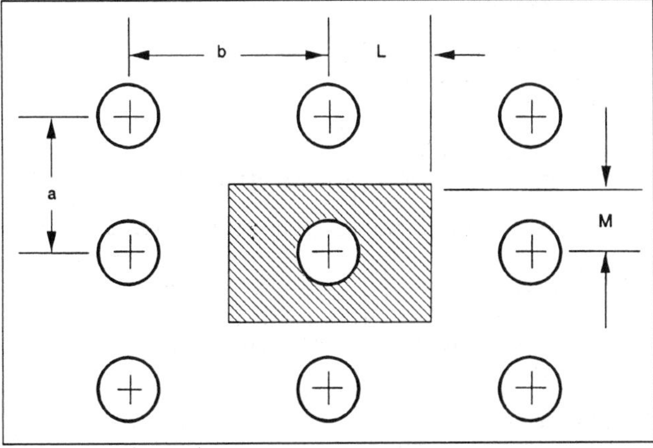

Fig. 23 Rectangular Tube Array

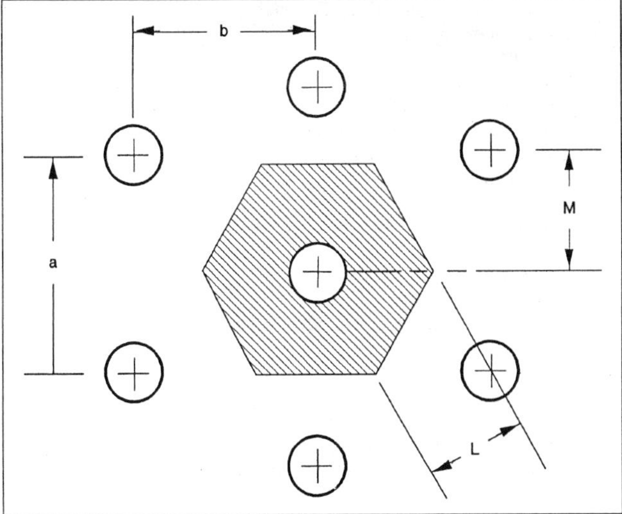

Fig. 24 Hexagonal Tube Array

Finned-Tube Heat Transfer

The heat transfer coefficients for finned coils follow the basic equations of convection, condensation, and evaporation. The arrangement of the fins affects the values of constants and the exponential powers in the equations. It is generally necessary to refer to test data for the exact coefficients.

For natural-convection finned coils (gravity coils), approximate coefficients can be obtained by considering the coil to be made of tubular and vertical fin surfaces at different temperatures and then applying the natural-convection equations to each. This calculation is difficult because the natural-convection coefficient depends on the temperature difference, which varies at different points on the fin.

Fin efficiency should be high (80 to 90%) for optimum natural-convection heat transfer. A low fin efficiency reduces the temperature near the tip. This reduces Δt near the tip and also the coefficient h, which in natural convection depends on Δt. The coefficient of heat transfer also decreases as the fin spacing decreases because of interfering convection currents from adjacent fins and reduced free-flow passage; 2 to 4 in. spacing is common. Generally, high coefficients result from large temperature differences and small flow restriction.

Edwards and Chaddock (1963) give coefficients for several circular fin-on-tube arrangements, using fin spacing δ as the characteristic length and in the form $\mathrm{Nu} = f(\mathrm{GrPr}\delta/D_o)$, where D_o is the fin diameter.

Forced-convection finned coils are used extensively in a wide variety of equipment. The fin efficiency for optimum performance is smaller than that for gravity coils because the forced-convection coefficient is almost independent of the temperature difference between the surface and the fluid. Very low fin efficiencies should be avoided because an inefficient surface gives a high (uneconomical) pressure drop. An efficiency of 70 to 90% is often used.

As fin spacing is decreased to obtain a large surface area for heat transfer, the coefficient generally increases because of higher air velocity between fins at the same face velocity and reduced equivalent diameter. The limit is reached when the boundary layer formed on one fin surface (Figure 8) begins to interfere with the boundary layer formed on the adjacent fin surface, resulting in a decrease of the heat transfer coefficient, which may offset the advantage of larger surface area.

Selection of the fin spacing for forced-convection finned coils usually depends on economic and practical considerations, such as fouling, frost formation, condensate drainage, cost, weight, and volume. Fins for conventional coils generally are spaced 14 to 6 per inch, except where factors such as frost formation necessitate wider spacing.

Several means are used to obtain higher coefficients with a given air velocity and surface, usually by creating air turbulence, generally

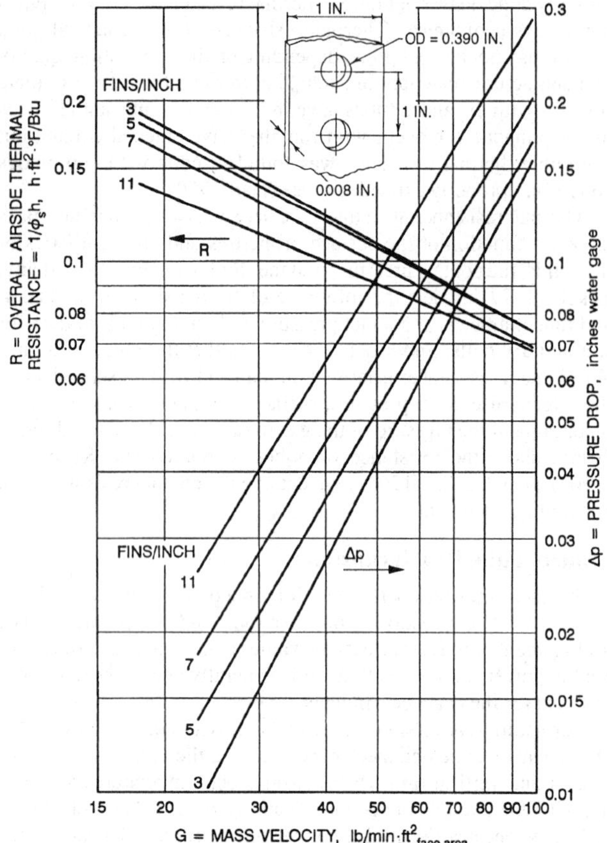

Fig. 25 Overall Air-Side Thermal Resistance and Pressure Drop for 1-Row Coils
(Shepherd 1946)

with a higher pressure drop: (1) staggered tubes instead of in-line tubes for multiple-row coils; (2) artificial additional tubes, or collars or fingers made by suitably forming the fin materials; (3) corrugated fins instead of plane fins; and (4) louvered or interrupted fins.

Figure 25 shows data for one-row coils. The thermal resistances plotted include the temperature drop through the fins, based on one square foot of total external surface area.

The section on Bibliography lists other sources of information on fins.

SYMBOLS

A = surface area for heat transfer
A_F = cross-sectional flow area
C = conductance; or fluid capacity rate
C_1, C_2 = Planck's law constants [see Equation (33)]
c = coefficient or constant
c_p = specific heat at constant pressure
c_v = specific heat at constant volume
D = diameter
d = diameter; or prefix meaning differential
e = emissivity; or protuberance height
F = angle factor [see Equations (40) and (41)]; or fin factor
Fo = Fourier number (see Table 1 and Figures 2, 3, and 4)
f = friction factor for single-phase flow
G = mass velocity; or irradiation
g = gravitational acceleration
h = heat transfer coefficient; or offset strip fin height
I = modified Bessel function
J = mechanical equivalent of heat; or radiosity
j = heat transfer factor [see Equation (4), Table 6]
k = thermal conductivity
L = length
l = length; length of one module of offset strip fins
M = mass; or molecular weight
m = general exponent
$\dot{m}$ = mass rate of flow
n = general number [see Equation (2) in Table 5]; or ratio r/r_m (see Figures 2, 3, and 4)
NTU = number of exchanger heat transfer units [see Equation (17)]
p = pressure; or fin pitch; or repeated rib pitch
q = rate of heat transfer
q'' = heat flux
R = thermal resistance
r = radius
s = lateral spacing of offset fin strips
T = absolute temperature
t = temperature; or fin thickness at base
U = overall heat transfer coefficient
V = linear velocity
W = work; or total rate of energy emission; or fin dimension
W_λ = monochromatic emissive power
x, y, z = lengths along principal coordinate axes
Y = temperature ratio (see Figures 2, 3, and 4)
Z = ratio of fluid capacity rates [see Equation (18)]
α = thermal diffusivity = $k/\rho c_p$ [Equation (28)]; or absorptance; or spiral angle for helical fins; or aspect ratio of offset strip fins
β = coefficient of thermal expansion; or contact angle of rib profile
γ = ratio, l/s
Δ = difference between values
δ = distance between fins; or ratio t/s
ε = hemispherical emittance; or exchanger heat transfer effectiveness [see Equation (16)]
λ = wavelength
μ = absolute viscosity
ν = kinematic viscosity
ρ = density; or reflectance
σ = Stefan-Boltzmann constant
τ = time; or transmittance [see Equation (39)]
Φ = fin resistance number defined by Equation (54); Φ_{max} is maximum limiting value of Φ
ϕ = fin efficiency [see Equation (50)]; or angle [see Equation (41)]

Subscripts

a = augmented
b = blackbody; or based on bulk fluid temperature
c = convection; or critical; or cold (fluid)
e = equivalent; or environment
f = film; or fin
fc = finite cylinder
frs = finite rectangular solid
g = gas
h = horizontal; or hot (fluid); or hydraulic
i = inlet; or inside; or particular surface (radiation); or based on maximum inside (envelope) diameter
ic = infinite cylinder
if = interface
is = infinite slab
iso = isothermal conditions
j = particular surface (radiation)
k = particular surface (radiation)
L = thickness
l = liquid
m = mean
max = maximum
min = minimum
n = counter variable
o = outside; or outlet; or overall; or at base of fin
p = prime heat transfer surface
r = radiation; or root (fin); or reduced
s = surface; or secondary heat transfer surface; or straight or plain; or accounting for flow blockage of twisted tape
st = static (pressure)
t = temperature; or terminal temperature; or tip (fin)
v = vapor; or vertical
w = wall; or wafer
λ = monochromatic
∞ = bulk

REFERENCES

Adams, J.A. and D.F. Rogers. 1973. *Computer aided heat transfer analysis*. McGraw-Hill, New York.

Afgan, N.H. and E.U. Schlunder. 1974. *Heat exchangers: Design and theory sourcebook*. McGraw-Hill, New York.

Altmayer, E.F., A.J. Gadgil, F.S. Bauman, and R.C. Kammerud. 1983. Correlations for convective heat transfer from room surfaces. *ASHRAE Transactions* 89(2A):61-77.

Bauman, F., A. Gadgil, R. Kammerud, E. Altmayer, and M. Nansteel. 1983. Convective heat transfer in buildings. *ASHRAE Transactions* 89(1A): 215-33.

Beckermann, C. and V. Goldschmidt. 1986. Heat transfer augmentation in the flueway of a water heater. *ASHRAE Transactions* 92(2B):485-95.

Bergles, A.E. 1985. Techniques to augment heat transfer. In *Handbook of heat transfer applications*. McGraw-Hill, New York, 3-1-3-80.

Brown, A.I. and S.M. Marco. 1958. *Introduction to heat transfer*, 3rd ed. McGraw-Hill, New York.

Burmeister, L.C. 1983. *Convective heat transfer*. John Wiley and Sons, New York.

Carnavos, T.C. 1979. Heat transfer performance of internally finned tubes in turbulent flow. *Advances in enhanced heat transfer*, pp. 61-67. American Society of Mechanical Engineers, New York.

Carslaw, H.S. and J.C. Jaeger. 1959. *Conduction of heat in solids*. Oxford University Press, England.

Clausing, A.M. 1964. Thermal contact resistance in a vacuum environment. ASME *Paper* 64-HT-16, Seventh National Heat Transfer Conference.

Croft, D.R. and D.G. Lilley. 1977. *Heat transfer calculations using finite difference equations*. Applied Science Publishers, Ltd., London.

Dart, D.M. 1959. Effect of fin bond on heat transfer. *ASHRAE Journal* 5:67.

Eckels, P.W. 1977. Contact conductance of mechanically expanded plate finned tube heat exchangers. AIChE-ASME Heat Transfer Conference, Salt Lake City, UT.

Edwards, J.A. and J.B. Chaddock. 1963. An experimental investigation of the radiation and free-convection heat transfer from a cylindrical disk extended surface. *ASHRAE Transactions* 69:313.

Fernandez, J. and R. Poulter. 1987. Radial mass flow in electrohydrodynamically-enhanced forced heat transfer in tubes. *International Journal of Heat and Mass Transfer* 80:2125-36.

Gardner, K.A. 1945. Efficiency of extended surface. *ASME Transactions* 67:621.

Grigull, U. et al. 1982. Heat transfer. Proceedings of the Seventh International Heat Transfer Conference, Munich, Vol. 3. Hemisphere Publishing, New York.

Hottel, H.C. and A.F. Sarofim. 1967. *Radiation transfer*. McGraw-Hill, New York.

Incropera, F.P. and D.P. DeWitt. 1996. *Fundamentals of heat transfer*. John Wiley and Sons, New York.

Jakob, M. 1949, 1957. *Heat transfer*, Vols. I and II. John Wiley and Sons, New York.

Junkhan, G.H. et al. 1985. Investigation of turbulence for tube flow boilers. *Journal of Heat Transfer* 107:354-60.

Junkhan, G.H., A.E. Bergles, V. Nirmalan, and W. Hanno. 1988. Performance evaluation of the effects of a group of turbulator inserts on heat transfer from gases in tubes. *ASHRAE Transactions* 94(2):1195-1212.

Kaspareck, W.E. 1964. Measurement of thermal contact conductance between dissimilar metals in a vacuum. ASME *Paper* 64-HT-38, Seventh National Heat Transfer Conference.

Kays, W.M. and A.L. London. 1984. *Compact heat exchangers*, 3rd ed. McGraw-Hill, New York.

Kays, W.M. and M. Crawford. 1980. *Convective heat and mass transfer*, 2nd ed. McGraw-Hill, New York.

Knudsen, J.G. and B.V. Roy. 1983. Studies on scaling of cooling tower water. *Fouling of heat enhancement surfaces*, pp. 517-30. Engineering Foundation, New York.

Lewis, D.M. and H.J. Sauer, Jr. 1965. The thermal resistance of adhesive bonds. ASME *Journal of Heat Transfer* 5:310.

Lopina, R.F. and A.E. Bergles. 1969. Heat transfer and pressure drop in tape generated swirl flow of single-phase water. *Journal of Heat Transfer* 91:434-42.

Manglik, R.M. and A.E. Bergles. 1990. The thermal-hydraulic design of the rectangular offset-strip-fin-compact heat exchanger. In *Compact heat exchangers*, pp. 123-49. Hemisphere Publishing, New York.

McAdams, W.H. 1954. *Heat transmission*, 3rd ed. McGraw-Hill, New York.

Metais, B. and E.R.G. Eckert. 1964. Forced, mixed and free convection regimes. ASME *Journal of Heat Transfer* 86(C2)(5):295.

Modest, M.F. 1993. *Radiation heat transfer*. McGraw-Hill, New York.

Myers, G.E. 1971. *Analytical methods in conduction heat transfer*. McGraw-Hill, New York.

Nelson, R.M. and A.E. Bergles. 1986. Performance evaluation for tubeside heat transfer enhancement of a flooded evaporative water chiller. *ASHRAE Transactions* 92(1B):739-55.

Ohadi, M.M. 1991. Heat transfer enhancement in heat exchangers. *ASHRAE Journal* (December):42-50.

Ohadi, M.M. et al. 1991. Electrohydrodynamic enhancement of heat transfer in a shell-and-tube heat exchanger. *Experimental Heat Transfer* 4(1):19-39.

Ohadi, M.M., R. Papar, T.L. Ng, M. Faani, and R. Radermacher. 1992. EHD enhancement of shell-side boiling heat transfer coefficients of R-123/oil mixture. *ASHRAE Transactions* 98(2):427-34.

Parker, J.D., J.H. Boggs, and E.F. Blick. 1969. *Introduction to fluid mechanics and heat transfer*. Addison Wesley Publishing, Reading, MA.

Patankar, S.V. 1980. *Numerical heat transfer and fluid flow*. McGraw-Hill, New York.

Ravigururajan, T.S. and A.E. Bergles. 1985. General correlations for pressure drop and heat transfer for single-phase turbulent flow in internally ribbed tubes. *Augmentation of heat transfer in energy systems*, 52:9-20. American Society of Mechanical Engineers, New York.

Rich, D.G. 1966. The efficiency and thermal resistance of annular and rectangular fins. Proceedings of the Third International Heat Transfer Conference, AIChE 111:281-89.

Schmidt, T.E. 1949. Heat transfer calculations for extended surfaces. *Refrigerating Engineering* 4:351-57.

Schneider, P.J. 1964. *Temperature response charges*. John Wiley and Sons, New York.

Shepherd, D.G. 1946. Performance of one-row tube coils with thin plate fins, low velocity forced convection. *Heating, Piping, and Air Conditioning* (April).

Shlykov, Y.P. 1964. Thermal resistance of metallic contacts. *International Journal of Heat and Mass Transfer* 7(8):921.

Siegel, R. and J.R. Howell. 1981. *Thermal radiation heat transfer*. McGraw-Hill, New York.

Somerscales, E.F.C. et al. 1991. Particulate fouling of heat transfer tubes enhanced on their inner surface, fouling and enhancement interactions. HTD 164:17-28. American Society of Mechanical Engineers, New York.

Sonokama, K. 1964. Contact thermal resistance. *Journal of the Japan Society of Mechanical Engineers* 63(505):240. English translation in RSIC-215, AD-443429.

Sunada, K., A. Yabe, T. Taketani, and Y. Yoshizawa. 1991. Experimental study of EHD pseudo-dropwise condensation. Proceedings of the ASME-JSME Thermal Engineering Joint Conference 3:47-53.

Uemura, M., S. Nishio, and I. Tanasawa. 1990. Enhancement of pool boiling heat transfer by static electric field. Ninth International Heat Transfer Conference, 75-80.

Yabe, A. and H. Maki. 1988. Augmentation of convective and boiling heat transfer by applying an electrohydrodynamical liquid jet. *International Journal of Heat Mass Transfer* 31(2):407-17.

BIBLIOGRAPHY

Fins

General

Gunter, A.Y. and A.W. Shaw. 1945. A general correlation of friction factors for various types of surfaces in cross flow. *ASME Transactions* 11:643.

Shah, R.K. and R.L. Webb. 1981. *Compact and enhanced heat exchangers, heat exchangers, theory and practice*, pp. 425-68. J. Taborek et al., eds. Hemisphere Publishing, New York.

Webb, R.L. 1980. Air-side heat transfer in finned tube heat exchangers. *Heat Transfer Engineering* 1(3):33-49.

Smooth

Clarke, L. and R.E. Winston. 1955. Calculation of finside coefficients in longitudinal finned heat exchangers. *Chemical Engineering Progress* 3:147.

Elmahdy, A.H. and R.C. Biggs. 1979. Finned tube heat exchanger: Correlation of dry surface heat transfer data. ASHRAE *Transactions* 85:2.

Ghai, M.L. 1951. Heat transfer in straight fins. General discussion on heat transfer. London Conference, September.

Gray, D.L. and R.L. Webb. 1986. Heat transfer and friction correlations for plate finned-tube heat exchangers having plain fins. Proceedings of Eighth International Heat Transfer Conference, San Francisco, CA.

Wavy

Beecher, D.T. and T.J. Fagan. 1987. Fin patternization effects in plate finned tube heat exchangers. ASHRAE *Transactions* 93:2.

Yashu, T. 1972. Transient testing technique for heat exchanger fin. *Reito* 47(531):23-29.

Spine

Abbott, R.W., R.H. Norris, and W.A. Spofford. 1980. Compact heat exchangers for general electric products—Sixty years of advances in design and manufacturing technologies. Compact heat exchangers—History, technological advancement and mechanical design problems. R.K. Shah, C.F. McDonald, and C.P. Howard, eds. Book No. G00183, pp. 37-55. American Society of Mechanical Engineers, New York.

Moore, F.K. 1975. Analysis of large dry cooling towers with spine-fin heat exchanger elements. ASME *Paper* No. 75-WA/HT-46. American Society of Mechanical Engineers, New York.

Rabas, T.J. and P.W. Eckels. 1975. Heat transfer and pressure drop performance of segmented surface tube bundles. ASME *Paper* No. 75-HT-45. American Society of Mechanical Engineers, New York.

Weierman, C. 1976. Correlations ease the selection of finned tubes. *Oil and Gas Journal* 9:94-100.

Louvered

Hosoda, T. et al. 1977. Louver fin type heat exchangers. *Heat Transfer Japanese Research* 6(2):69-77.

Mahaymam, W. and L.P. Xu. 1983. Enhanced fins for air-cooled heat exchangers—Heat transfer and friction factor correlations. Y. Mori and W. Yang, eds. Proceedings of the ASME-JSME Thermal Engineering Joint Conference, Hawaii.

Senshu, T. et al. 1979. Surface heat transfer coefficient of fins utilized in air-cooled heat exchangers. *Reito* 54(615):11-17.

Circular

Jameson, S.L. 1945. Tube spacing in finned tube banks. *ASME Transactions* 11:633.

Katz, D.L. and Associates. 1954-55. Finned tubes in heat exchangers; Cooling liquids with finned coils; Condensing vapors on finned coils; and Boiling outside finned tubes. Bulletin reprinted from *Petroleum Refiner*.

Heat Exchangers

Gartner, J.R. and H.L. Harrison. 1963. Frequency response transfer functions for a tube in crossflow. ASHRAE *Transactions* 69:323.

Gartner, J.R. and H.L. Harrison. 1965. Dynamic characteristics of water-to-air crossflow heat exchangers. ASHRAE *Transactions* 71:212.

McQuiston, F.C. 1981. Finned tube heat exchangers: State of the art for the air side. ASHRAE *Transactions* 87:1.

Myers, G.E., J.W. Mitchell, and R. Nagaoka. 1965. A method of estimating crossflow heat exchangers transients. ASHRAE *Transactions* 71:225.

Stermole, F.J. and M.H. Carson. 1964. Dynamics of flow forced distributed parameter heat exchangers. *AIChE Journal* 10(5):9.

Thomasson, R.K. 1964. Frequency response of linear counterflow heat exchangers. *Journal of Mechanical Engineering Science* 6(1):3.

Wyngaard, J.C. and F.W. Schmidt. Comparison of methods for determining transient response of shell and tube heat exchangers. ASME *Paper* 64-WA/HT-20. American Society of Mechanical Engineers, New York.

Yang, W.J. Frequency response of multipass shell and tube heat exchangers to timewise variant flow perturbance. ASME *Paper* 64-HT-18. American Society of Mechanical Engineers, New York.

Heat Transfer, General

Bennet, C.O. and J.E. Myers. 1984. *Momentum, heat and mass transfer*, 3rd ed. McGraw-Hill, New York.

Chapman, A.J. 1981. *Heat transfer*, 4th ed. Macmillan, New York.

Holman, J.D. 1981. *Heat transfer*, 5th ed. McGraw-Hill, New York.

Kern, D.Q. and A.D. Kraus. 1972. *Extended surface heat transfer*. McGraw-Hill, New York.

Kreith, F. and W.Z. Black. 1980. *Basic heat transfer*. Harper and Row, New York.

Lienhard, J.H. 1981. *A heat transfer textbook*. Prentice Hall, Englewood Cliffs, NJ.

McQuiston, F.C. and J.D. Parker. 1988. *Heating, ventilating and air-conditioning, analysis and design*, 4th ed. John Wiley and Sons, New York.

Rohsenow, W.M. and J.P. Hartnett, eds. 1973. *Handbook of heat transfer*. McGraw-Hill, New York.

Sissom, L.E. and D.R. Pitts. 1972. *Elements of transport phenomena*. McGraw-Hill, New York.

Todd, J.P. and H.B. Ellis. 1982. *Applied heat transfer*. Harper and Row, New York.

Webb, R.L. and A.E. Bergles. 1983. Heat transfer enhancement, second generation technology. *Mechanical Engineering* 6:60-67.

Welty, J.R. 1974. *Engineering heat transfer*. John Wiley and Sons, New York.

Welty, J.R., C.E. Wicks, and R.E. Wilson. 1972. *Fundamentals of momentum, heat and mass transfer*. John Wiley and Sons, New York.

Wolf, H. 1983. *Heat transfer*. Harper and Row, New York.

CHAPTER 4

TWO-PHASE FLOW

TWO-PHASE flow is encountered extensively in the air-conditioning, heating, and refrigeration industries. A combination of liquid and vapor refrigerant exists in flooded coolers, direct-expansion coolers, thermosiphon coolers, brazed and gasketed plate evaporators and condensers, and tube-in-tube evaporators and condensers, as well as in air-cooled evaporators and condensers. In the pipes of heating systems, steam and liquid water may both be present. Because the hydrodynamic and heat transfer aspects of two-phase flow are not as well understood as those of single-phase flow, no single set of correlations can be used to predict pressure drops or heat transfer rates. Instead, the correlations are for specific thermal and hydrodynamic operating conditions.

This chapter presents the basic principles of two-phase flow and provides information on the vast number of correlations that have been developed to predict heat transfer coefficients and pressure drops in these systems.

BOILING

Commonly used refrigeration evaporators are (1) flooded evaporators, where refrigerants at low fluid velocities boil outside or inside tubes; and (2) dry expansion shell-and-tube evaporators, where refrigerants at substantial fluid velocities boil outside or inside tubes.

Two-phase heat and mass transport are characterized by various flow and thermal regimes, whether vaporization takes place under natural convection or in forced flow. As in single-phase flow systems, the heat transfer coefficient for a two-phase mixture depends on the flow regime, the thermodynamic and transport properties of the vapor and the liquid, the roughness of the heating surface, the wetting characteristics of the surface-liquid pair, and other parameters. Therefore, it is necessary to consider each flow and boiling regime separately to determine the heat transfer coefficient.

Accurate data defining limits of regimes and determining the effects of various parameters are not available. The accuracy of correlations in predicting the heat transfer coefficient for two-phase flow is in most cases not known beyond the range of the test data.

Boiling and Pool Boiling in Natural Convection Systems

Regimes of Boiling. The different regimes of pool boiling described by Farber and Scorah (1948) verified those suggested by Nukiyama (1934). The regimes are illustrated in Figure 1. When the temperature of the heating surface is near the fluid saturation temperature, heat is transferred by convection currents to the free surface where evaporation occurs (Region I). Transition to nucleate boiling occurs when the surface temperature exceeds saturation by a few degrees (Region II).

In **nucleate boiling** (Region III), a thin layer of superheated liquid is formed adjacent to the heating surface. In this layer, bubbles nucleate and grow from spots on the surface. The thermal resistance of the superheated liquid film is greatly reduced by bubble-induced

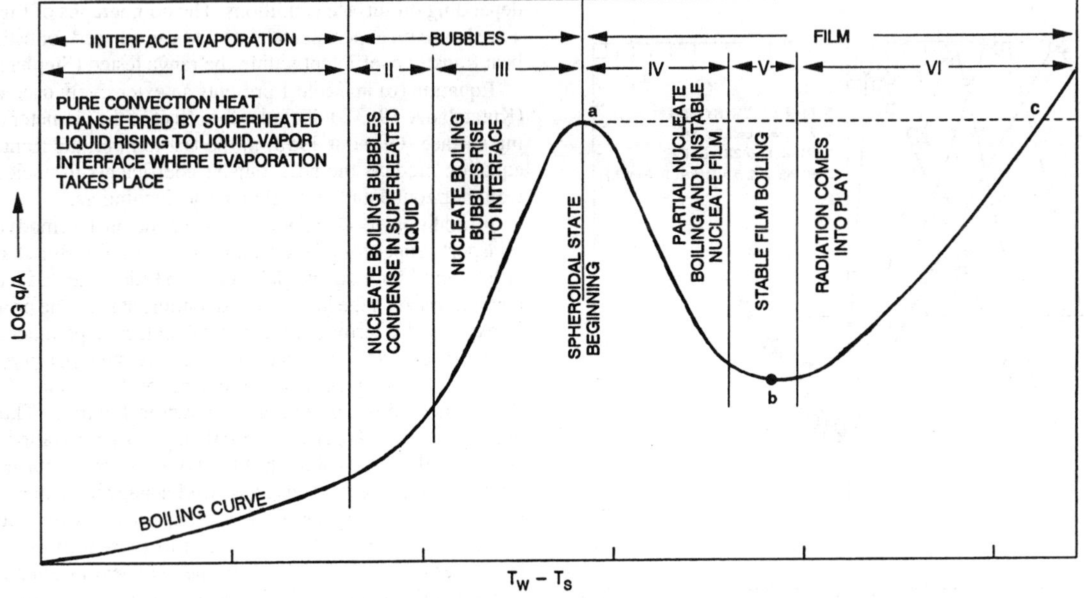

Fig. 1 Characteristic Pool Boiling Curve

The preparation of this chapter is assigned to TC 1.3, Heat Transfer and Fluid Flow.

agitation and vaporization. Increased wall temperature increases bubble population, causing a large increase in heat flux.

As heat flux or temperature difference increases further and as more vapor forms, the flow of the liquid toward the surface is interrupted, and a vapor blanket forms. This gives the **maximum** or **critical heat flux** (CHF) in nucleate boiling (point *a*, Figure 1). This flux is often termed the **burnout heat flux** or **boiling crisis** because, for constant power-generating systems, an increase of heat flux beyond this point results in a jump of the heater temperature (to point *c*, Figure 1), often beyond the melting point of a metal heating surface.

In systems with controllable surface temperature, an increase beyond the temperature for CHF causes a decrease of heat flux density. This is the **transitional boiling regime** (Region IV); liquid alternately falls onto the surface and is repulsed by an explosive burst of vapor.

At sufficiently high surface temperature, a stable vapor film forms at the heater surface; this is the **film boiling regime** (Regions V and VI). Because heat transfer is by conduction (and some radiation) across the vapor film, the heater temperature is much higher than for comparable heat flux densities in the nucleate boiling regime.

Free Surface Evaporation. In Region I, where surface temperature exceeds liquid saturation temperature by less than a few degrees, no bubbles form. Evaporation occurs at the free surface by convection of superheated liquid from the heated surface. Correlations of heat transfer coefficients for this region are similar to those for fluids under ordinary natural convection [Equations (1) through (4) in Table 1].

Nucleate Boiling. Much information is available on boiling heat transfer coefficients, but no universally reliable method is available for correlating the data. In the nucleate boiling regime, heat flux density is not a single, valued function of the temperature but depends also on the nucleating characteristics of the surface, as illustrated by Figure 2 (Berenson 1962).

The equations proposed for correlating nucleate boiling data can be put in a form that relates heat transfer coefficient *h* to temperature difference $(t_w - t_{sat})$:

$$h = \text{constant}\,(t_w - t_{sat})^a \quad (1)$$

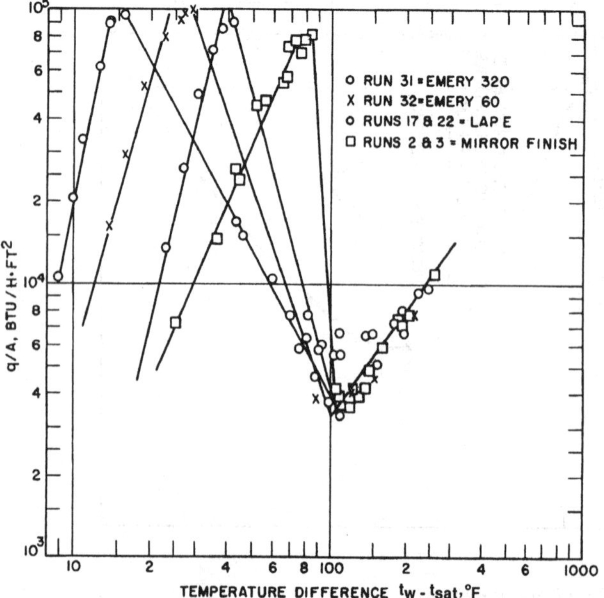

Fig. 2 Effect of Surface Roughness on Temperature in Pool Boiling of Pentane

Exponent *a* is normally between 1 and 3; the constant depends on the thermodynamic and transport properties of the vapor and the liquid. Nucleating characteristics of the surface, including the size distribution of surface cavities and the wetting characteristics of the surface-liquid pair, affect the value of the multiplying constant and the value of the exponent *a* in Equation (1). For example, variations in exponent *a* from 1 to 25 can be produced by polishing the surface with different grades of emery paper.

A generalized correlation cannot be expected without consideration of the nucleating characteristics of the heating surface. A statistical analysis of data for 25 liquids by Hughmark (1962) shows that in a correlation not considering surface condition, deviations of more than 100% are common.

In the following sections, correlations and nomographs for prediction of nucleate and flow boiling of various refrigerants are given. For most cases, these correlations have been tested for refrigerants, such as R-11, R-12, R-113, and R-114, that have now been identified as environmentally harmful and are no longer being used in new equipment. Although extensive research on the thermal and fluid characteristics of alternative refrigerants/refrigerant mixtures has taken place in recent years and some correlations have been suggested, the test databases are not yet comprehensive enough to recommend any particular equations among those recently developed.

In the absence of quantitative nucleating characteristics, Rohsenow (1951) devised a test that evaluated surface effects for a given surface-liquid combination, with the liquid at atmospheric pressure. The effect of pressure can be determined by using the dimensionless groups in Equation (5) in Table 1. Values of the coefficient C_{sf} found by Blatt and Adt (1963) for some liquid-solid combinations are presented in Section II of Table 1.

The nomographs of Figures 3 and 4 (Stephan 1963b) can be used to estimate the heat transfer coefficients for various refrigerants in nucleate boiling from a horizontal plate (Figure 3) and from the outside of a horizontal cylinder with OD = 1.18 in. (Figure 4). Pressures range from 1 to 3 atm.

Stephan's correlation (Stephan 1963c) is subject to previously mentioned limitations (particularly the heat transfer surface microstructure and nucleation characteristics) because its form is that of Equation (1), with exponent *a* equal to 4 for horizontal plates and 2.33 for horizontal cylinders. Data show variations of *a* from 2 to 25, depending on surface conditions. The nomographs of Figures 3 and 4 are based on experimental data and can be used for estimating the heat transfer coefficient within the range tested (Stephan 1963a).

Equation (6) in Table 1 presents an extensively used correlation (Kutateladze 1963). It includes the effect of the diameter of the heating surface (Gilmour 1958) in the last term on the right side. This equation predicts the heat transfer coefficients in nucleate boiling from horizontal and vertical plates and cylinders.

In addition to correlations dependent on thermodynamic and transport properties of the vapor and the liquid, Borishansky et al. (1962) and Lienhard and Schrock (1963) documented a correlating method based on the law of corresponding states. The properties can be expressed in terms of fundamental molecular parameters, leading to scaling criteria based on the reduced pressure, $p_r = p/p_c$, where p_c is the critical thermodynamic pressure for the coolant. An example of this method of correlation is shown in Figure 5. The reference pressure p^* was chosen as $p^* = 0.029p_c$. This correlation provides a simple method for scaling the effect of pressure if data are available for one pressure level. It also has an advantage if the thermodynamic and particularly the transport properties used in several equations in Table 1 are not accurately known. In its present form, this correlation gives a value of $a = 2.33$ for the exponent in Equation (1) and consequently should apply for typical aged metal surfaces.

There are explicit heat transfer coefficient correlations based on the law of corresponding states for various substances (Borishansky and Kosyrev 1966), halogenated refrigerants (Danilova 1965), and flooded evaporators (Starczewski 1965). Other investigations examined the

Table 1 Equations for Boiling Heat Transfer

Description	References	Equations	
Free convection	Jakob (1949 and 1957)		
Free convection boiling, or boiling without bubbles for low Δt and GrPr < 10^8 (all properties to be based on liquid state)		$\text{Nu} = C(\text{Gr})^m(\text{Pr})^n$	(1)
Vertical submerged surface		$\text{Nu} = 0.61(\text{Gr})^{0.25}(\text{Pr})^{0.25}$	(2)
Horizontal submerged surface		$\text{Nu} = 0.16(\text{Gr})^{1/3}(\text{Pr})^{1/3}$	(3)
Simplified equation for water		$h \sim 80(\Delta t)^{1/3}$, where h is in Btu/h·ft²·°F, Δt in °F	(4)
Nucleate boiling	Rohsenow (1951)	$(\text{Nu})_b = (\text{Re})_b^{2/3}(\text{Pr})_l^{-0.7}/C_{sf}$	(5)
	Kutateladze (1963) Gilmour (1958)	$(\text{Nu})_b = C_k\left[\frac{\rho_l}{\rho_v}\text{Re}\right]_b^{0.7}(\text{Pr})_l^{0.4}N_k^m\left[\frac{D_b}{D}\right]^n$	(6)
		where $(\text{Nu})_b = hD_b/k_l$, $(\text{Pr})_l = [\mu c_p/k]_l$; $(\text{Re})_b = qD_b/Ah_{fg}\mu_l$; $D_b = \sqrt{\sigma_t/(\rho_l-\rho_v)g}$; $N_k = p/[\sigma_t g(\rho_l-\rho_v)]^{0.5}$	
	Blatt and Adt (1963)	C_{sf} = constant that depends on solid-liquid combination	

Refrigerant	Stainless Steel	Copper
11	0.016	0.022
113	0.09	0.013

Description	References	Equations	
	Jakob (1949 and 1957)	$C_k = 7.0 \times 10^{-4}$, $m = 0.7$, $n = 0$; $C_k = 4.37 \times 10^{-3}$, $m = 0.95$, $n = 1/3$	
Critical heat flux	Kutateladze (1951) Zuber et al. (1962)	$\frac{q/A}{\rho_v h_{fg}}\left[\frac{\rho_v^2}{\sigma_t g(\rho_l-\rho_v)}\right]^{0.25} = K_D$	(7)
		For many liquids, K_D varies from 0.12 to 0.16. Recommended average value is 0.13.	
Minimum heat flux in film boiling from horizontal plate	Zuber (1959)	$\frac{q/A}{\rho_v h_{fg}}\left[\frac{(\rho_l+\rho_v)}{\sigma_t g(\rho_l-\rho_v)}\right]^{0.25} = 0.09$	(8)
Minimum heat flux in film boiling from horizontal cylinders	Lienhard and Wong (1963)	$\frac{q/A}{\rho_v h_{fg}}\left[\frac{(\rho_l+\rho_v)^2}{\sigma_t g(\rho_l-\rho_v)}\right]^{0.25} = 0.114\frac{\left[\frac{2\sigma_t}{g(\rho_l-\rho_v)D^2}\right]^{0.5}}{\left[1+\frac{2\sigma_t}{g(\rho_l-\rho_v)D^2}\right]^{0.25}}$	(9)
Minimum temperature difference for film boiling from horizontal plate	Berenson (1961)	$(t_w - t_{sat}) = 0.127\frac{\rho_v h_{fg}}{k_v}\left[\frac{g(\rho_l-\rho_v)}{\rho_l+\rho_v}\right]^{2/3} \times \left[\frac{\sigma_t}{g(\rho_l-\rho_v)}\right]^{0.5}\left[\frac{\mu_v}{\rho_l-\rho_v}\right]^{1/3}$	(10)
Film boiling from horizontal plate	Berenson (1961)	$h = 0.425\left[\frac{k_v^3\rho_v h_{fg}g(\rho_l-\rho_v)}{\mu_v(t_w-t_{sat})\sqrt{\phi_t/g(\rho_l-\rho_v)}}\right]^{0.25}$	(11)
Film boiling from horizontal cylinders	Anderson et al. (1966)	$h = 0.62\left[\frac{k_v^3\rho_v g(\rho_l-\rho_v)h_{fg}}{D\mu_v(t_w-t_{sat})}\right]^{0.25}$	(12)
Effect of radiation	Anderson et al. (1966)	Substitute $h'_{fg} = h_{fg}\left[1+0.4c_p\frac{t_w-t_b}{h_{fg}}\right]$	
Effect of surface tension and of pipe diameter	Breen and Westwater (1962)	$\Lambda/D < 0.8$: $h(\Lambda)^{0.25}/F = 0.60$	(13)
		$0.8 < \Lambda/D < 8$: $hD^{0.25}/F = 0.62$	(14)
		$8 < \Lambda/D$: $h(\Lambda)^{0.25}/F = 0.016\,(\Lambda/D)^{0.83}$	(15)
		where $\Lambda = 2\pi\left[\frac{\sigma_t}{g(\rho_l-\rho_v)}\right]^{0.25}$; $F = \left[\frac{\rho_v h_{fg}g(\rho_l-\rho_v)k_v^3}{\mu_v(t_w-t_{sat})}\right]^{0.25}$	
Turbulent film	Frederking and Clark (1962)	$\text{Nu} = 0.15\,(\text{Ra})^{1/3}$ for $\text{Ra} > 5 \times 10^7$	(16)
		$\text{Ra} = \left[\frac{D^3 g(\rho_l-\rho_v)}{v_v^2\rho_v}\left(\frac{c_p\mu}{k}\right)_v\left(\frac{h_{fg}}{c_p(t_w-t_{sat})}+0.4\right)\frac{a}{g}\right]^{1/3}$	
		a = local acceleration	

effects of oil on boiling heat transfer from diverse configurations, including boiling from a flat plate (Stephan 1963b); a 0.55 in. OD horizontal tube using an oil-R-12 mixture (Tschernobyiski and Ratiani 1955); inside horizontal tubes using an oil-R-12 mixture (Breber et al. 1980, Worsoe-Schmidt 1959, Green and Furse 1963); and commercial copper tubing using R-11 and R-113 with oil content to 10% (Dougherty and Sauer 1974). Additionally, Furse (1965) examined R-11 and R-12 boiling over a flat horizontal copper surface.

Maximum Heat Flux and Film Boiling

Maximum or critical heat flux and the film boiling region are not as strongly affected by conditions of the heating surface as the heat flux in the nucleate boiling region, making analysis of CHF and of film boiling more tractable.

Carey (1992) provides a review of the mechanisms that have been postulated to cause the CHF phenomenon in pool boiling. Each model is based on the scenario that vapor blankets, which lead to an increased thermal resistance, exist on portions of the heat transfer surface. It has been proposed that these blankets may result from Helmholtz instabilities.

When CHF (point a, Figure 1) is assumed to be a hydrodynamic instability phenomenon, a simple relation, Equation (7) in Table 1, can be derived to predict this flux for pure, wetting liquids (Kutateladze 1951, Zuber et al. 1962). The dimensionless constant K varies from approximately 0.12 to 0.16 for a large variety of liquids. The effect of wettability is still in question. Van Stralen (1959) found that for liquid mixtures, critical heat flux is a function of the concentration.

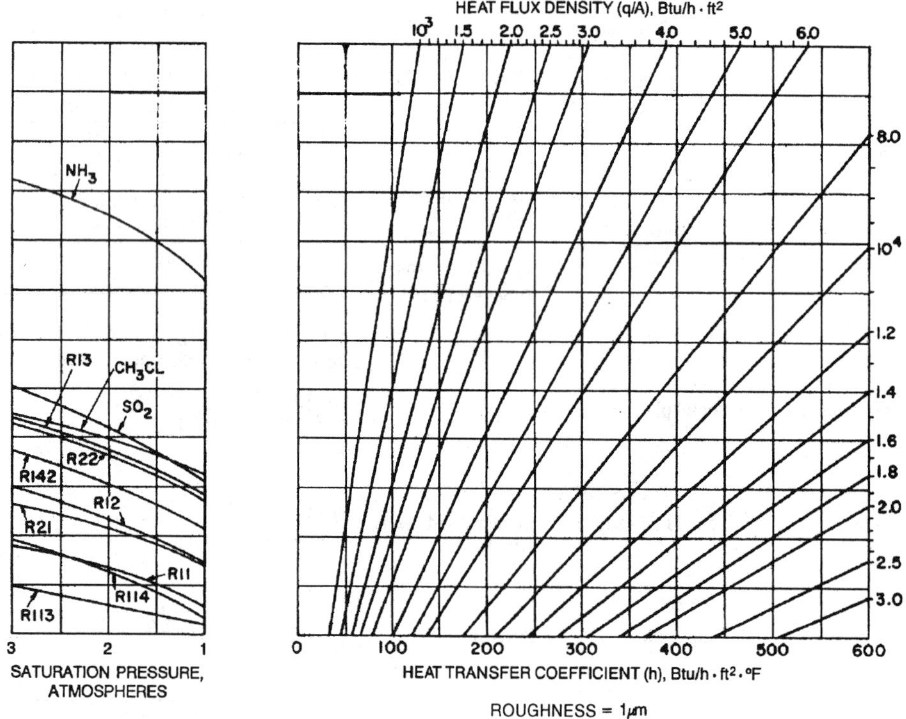

Fig. 3 Heat Transfer Coefficient for Pool Boiling from Horizontal Plate

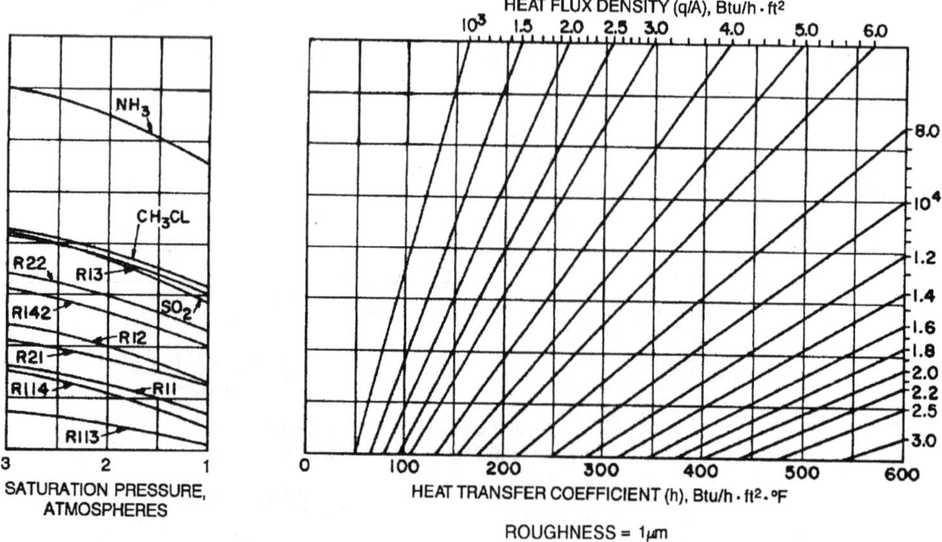

Fig. 4 Heat Transfer Coefficient for Pool Boiling from Horizontal Cylinder

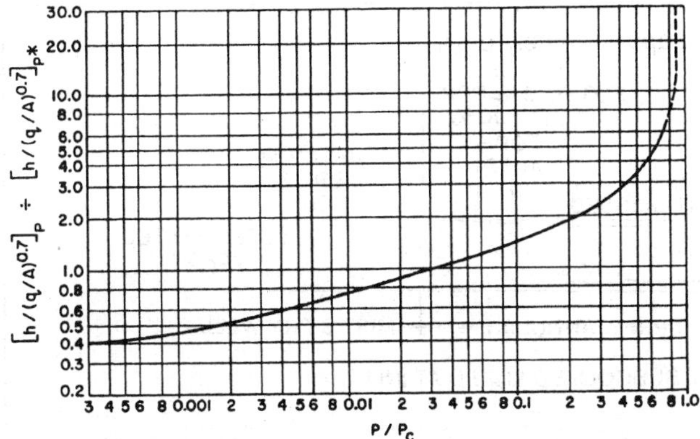

Fig. 5 Correlation of Pool Boiling Data in Terms of Reduced Pressure

The minimum heat flux density (point *b*, Figure 1) in film boiling from a horizontal surface and a horizontal cylinder can be predicted by Equations (8) and (9) in Table 1. The numerical factors 0.09 and 0.114 were adjusted to fit experimental data; values predicted by two analyses were approximately 30% higher. Equation (10) in Table 1 predicts the temperature difference at minimum heat flux of film boiling.

The heat transfer coefficient in film boiling from a horizontal surface can be predicted by Equation (11) in Table 1; and from a horizontal cylinder by Equation (12) in Table 1 (Bromley 1950), which has been generalized to include the effect of surface tension and cylinder diameter, as shown in Equations (13), (14), and (15) in Table 1 (Breen and Westwater 1962).

Frederking and Clark (1962) found that for turbulent film boiling, Equation (16) in Table 1 agrees with data from experiments at reduced gravity (Rohsenow 1963, Westwater 1963, Kutateladze 1963, Jakob 1949 and 1957).

Flooded Evaporators

Equations in Table 1 merely approximate heat transfer rates in flooded evaporators. One reason is that vapor entering the evaporator combined with vapor generated within the evaporator can produce significant forced convection effects superimposed on those caused by nucleation. Nonuniform distribution of the two-phase, vapor-liquid flow within the tube bundle of shell-and-tube evaporators or the tubes of vertical-tube flooded evaporators is also important.

Myers and Katz (1952) investigated the effect of vapor generated by the bottom rows of a tube bundle on the heat transfer coefficient for the upper rows. Improvement in coefficients for the upper tube rows is greatest at low temperature differences where nucleation effects are less pronounced. Hofmann (1957) summarizes other data for flooded tube bundles.

Typical performance of vertical tube natural circulation evaporators, based on data for water, is shown in Figure 6 (Perry 1950). Low coefficients are at low liquid levels because insufficient liquid covers the heating surface. The lower coefficient at high levels is the result of an adverse effect of hydrostatic head on temperature difference and circulation rate. Perry (1950) noted similar effects in horizontal shell-and-tube evaporators.

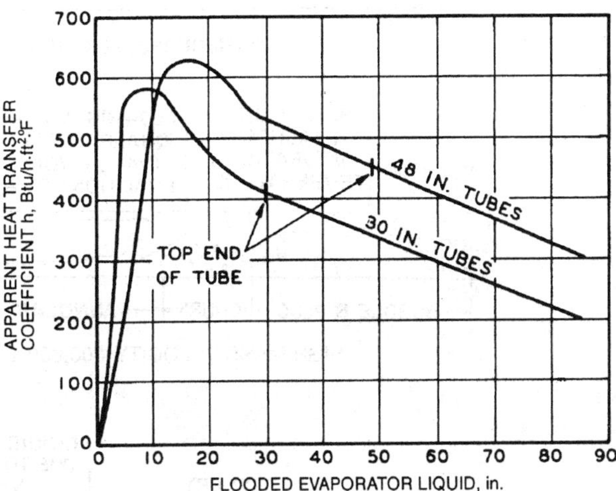

Fig. 6 Boiling Heat Transfer Coefficients for Flooded Evaporator

Forced-Convection Evaporation in Tubes

Flow Mechanics. When a mixture of liquid and vapor flows inside a tube, a number of flow patterns occur, depending on the mass fraction of liquid, the fluid properties of each phase, and the flow rate. In an evaporator tube, the mass fraction of liquid decreases along the circuit length, resulting in a series of changing vapor-liquid flow patterns. If the fluid enters as a subcooled liquid, the first indications of vapor generation are bubbles forming at the heated tube wall (nucleation). Subsequently, bubble, plug, churn (or semiannular), annular, spray annular, and mist flows can occur as the vapor content increases for two-phase flows in horizontal tubes. Idealized flow patterns are illustrated in Figure 7A for a horizontal tube evaporator.

Because nucleation occurs at the heated surface in a thin sublayer of superheated liquid, boiling in forced convection may begin while the bulk of the liquid is subcooled. Depending on the nature of the fluid and the amount of subcooling, the bubbles formed can either collapse or continue to grow and coalesce (Figure 7A), as Gouse and Coumou (1965) observed for R-113. Bergles and Rohsenow (1964) developed a method to determine the point of incipient surface boiling.

After nucleation begins, bubbles quickly agglomerate to form vapor plugs at the center of a vertical tube, or, as shown in Figure 7A, vapor plugs form along the top surface of a horizontal tube. At the point where the bulk of the fluid reaches saturation temperature, which corresponds to local static pressure, there will be up to 1% vapor quality because of the preceding surface boiling (Guerrieri and Talty 1956).

Further coalescence of vapor bubbles and plugs results in churn, or semiannular flow. If the fluid velocity is high enough, a continuous vapor core surrounded by a liquid annulus at the tube wall soon forms. This annular flow occurs when the ratio of the tube cross section filled with vapor to the total cross section is approximately 85%. With common refrigerants, this equals a vapor quality of about 3 to 5%. **Vapor quality** is the ratio of mass (or mass flow rate) of vapor to total mass (or mass flow rate) of the mixture. The usual flowing vapor quality or vapor fraction is referred to throughout this discussion. Static vapor quality is smaller because the vapor in the core flows at a higher average velocity than the liquid at the walls (see Chapter 2).

If two-phase mass velocity is high (greater than 150,000 $lb_m/h \cdot ft^2$ for a 0.5 in. tube), annular flow with small drops of entrained liquid in the vapor core (spray) can persist over a vapor quality range from a few percentage points to more than 90%. Refrigerant evaporators are fed from an expansion device at vapor qualities of approximately 20%, so that annular and spray annular flow predominate in most tube lengths. In a vertical tube, the liquid annulus is distributed uniformly over the periphery, but it is somewhat asymmetric in a horizontal tube (Figure 7A). As vapor quality reaches about 90%, the surface dries out, although there are still entrained droplets of liquid in the vapor (mist). Chaddock and Noerager (1966) found that in a horizontal tube, dryout occurs first at the top of the tube and later at the bottom (Figure 7A).

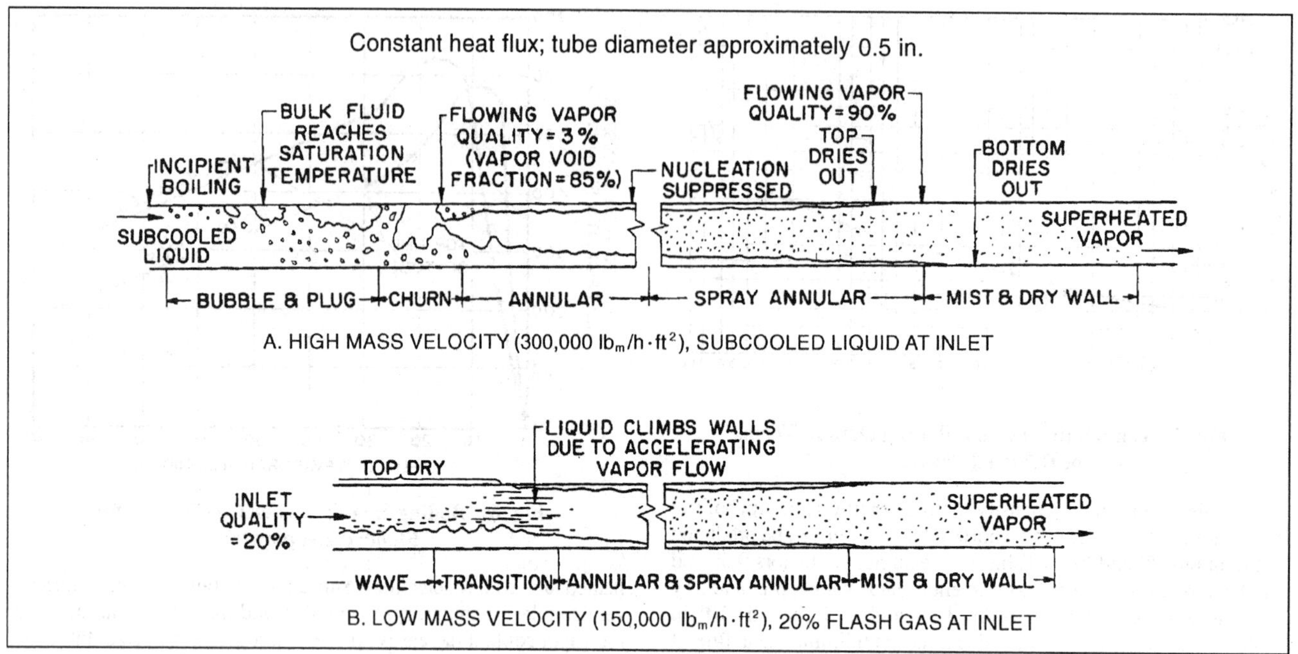

Fig. 7 Flow Regimes in Typical Smooth Horizontal Tube Evaporator

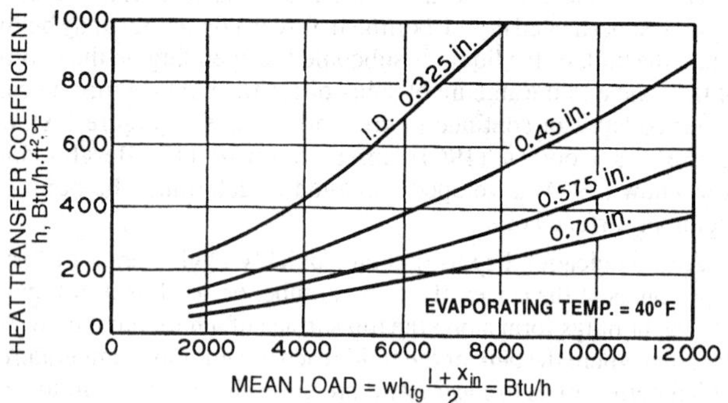

Fig. 8 Boiling Heat Transfer Coefficients for R-12 Inside Horizontal Tubes

If two-phase mass velocity is low (less than 150,000 $lb_m/h \cdot ft^2$ for a 0.5 in. horizontal tube), liquid occupies only the lower cross section of the tube. This causes a wavy type of flow at vapor qualities above about 5%. As the vapor accelerates with increasing evaporation, the interface is disturbed sufficiently to develop annular flow (Figure 7B). Liquid slugging can be superimposed on the flow configurations illustrated; the liquid forms a continuous, or nearly continuous, sheet over the tube cross section. The slugs move rapidly and at irregular intervals.

Heat Transfer. It is difficult to develop a single relation to describe the heat transfer performance for evaporation in a tube over the full quality range. For refrigerant evaporators with several percentage points of flash gas at entrance, it is less difficult because annular flow occurs in most of the tube length. The reported data are accurate only within geometry, flow, and refrigerant conditions tested; therefore, a large number of methods for calculating heat transfer coefficients for evaporation in tubes is presented in Table 2 (also see Figures 8 through 11).

Figure 8 gives heat transfer data obtained for R-12 evaporating in a 0.575 in. copper tube (Ashley 1942). The curves for the other tube diameters shown are approximations based on an assumed dependence, as in Table 2. Heat transfer coefficient dependence on the vapor fraction can be understood better from the data in

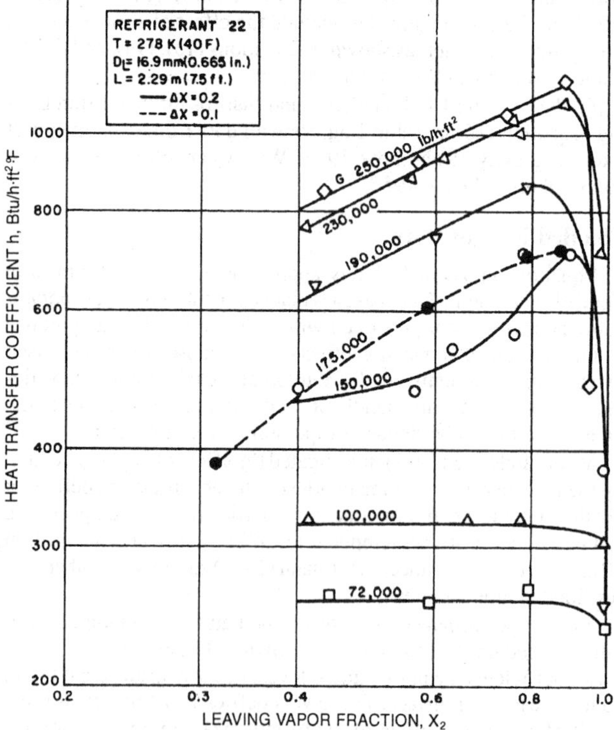

Fig. 9 Heat Transfer Coefficient Versus Vapor Fraction for Partial Evaporation

Figure 9 (Gouse and Coumou 1965). At low mass velocities (below 150,000 $lb_m/h \cdot ft^2$), the wavy flow regime shown in Figure 7B probably exists, and the heat transfer coefficient is nearly constant along the tube length, dropping at the tube exit as complete vaporization occurs. At higher mass velocities, the flow pattern is usually annular, and the coefficient increases as vapor accelerates. As the surface dries and the flow reaches a 90% vapor quality, the coefficient drops sharply.

Table 2 Equations for Forced Convection Evaporation in Tubes

Equations		Comments and References
HORIZONTAL TUBES		
Graphical presentation in Figure 8 h versus q_m where $q_m = wh_{fg}(1+x)/2$		Average coefficients for complete evaporation of R-12 at 40°F in a 0.575 in. ID copper tube, 33 ft long; the curves for other diameters in Figure 8 are based on the assumption that h varies inversely as the square of the tube diameter (Ashley 1942).
Graphical presentation in Figure 9 h versus x_2 where x_2 = leaving vapor fraction		Average coefficients for R-22 evaporating at 40°F in a 0.665 in. ID copper tube, 7.5 ft long. Vapor fraction varied from 20% to 100%. Average coefficients plotted are for vapor fraction changes of 0.20 (or 0.10). For average coefficients (at the same heat flux) over larger vapor fraction ranges, the curves can be integrated (Anderson et al. 1966).
$h = C_1\left(\frac{k_l}{d}\right)\left[\left(\frac{GD}{\mu_l}\right)^2\left(\frac{J\Delta x h_{fg}}{L}\right)\right]^n$ *where* $C_1 = 0.0009$ and $n = 0.5$ for exit qualities $\leq 90\%$; and $C_1 = 0.0082$ and $n = 0.4$ for 11°F superheat at exit	(1)	Average coefficients for R-12 and R-22 evaporating in copper tubes of 0.472 and 0.709 in. ID, from 13.4 to 31.2 ft long, and at evaporating temperatures from −4 to 32°F. Vapor fraction varied from 0.15 to 11°F superheat. Note that the term $J\Delta xh_{fg}/L$ is not dimensionless, but has the units of pound force per pound mass (Pierre 1955, 1957).
Equation (1) with $c_1 = 0.0225$ and $n = 0.375$		Average coefficients for R-22 evaporating at temperatures from 40 to 80°F in a 0.343 in. ID tube, 8 ft long. Coefficients were determined for approximately 15% vapor quality changes. The range investigated was $x = 0.20$ to superheat (Altman et al. 1960b).
$h = C_2 h_l\left(\frac{1+x}{1-x}\right)^{1.16}\left(\frac{q}{Gh_{fg}}\right)^{0.1}$ *where* $C_2 = 6.59$	(2)	Local coefficients for R-12 and R-22 evaporating in a 0.732 in. ID tube 1 ft long at saturation temperatures from 75 to 90°F. Location of transition from annular to mist flow is established, and a heat transfer equation for the mist flow regime is presented (Lavin and Young 1964).
$h_l = \frac{0.023k_l}{d}\left[\frac{DG(1-x)}{\mu_l}\right]^{0.8}(\mathrm{Pr})_l^{0.4}$	(3)	
$h = 1.85\,h_L[B_o \times 10^4 + (1/X_{tt})^{0.67}]^{0.6}$ *where*	(4)	Local coefficients for R-12 evaporating in a 0.460 in. ID stainless steel tube with a uniform wall heat flux (electric heating) over a length of 6.344 ft, and an evaporating temperature of 53°F. Vapor fraction range was 0.20 to 0.88. Equation (4) is a modified form of the Schrock and Grossman equation for vertical tube evaporation [Equation (10)] (Chaddock and Noerager 1966).
$B_o = q/Gh_{fg}$	(5)	
$h_L = \frac{0.023k_l}{d}\left(\frac{DG}{\mu_l}\right)^{0.8}(\mathrm{Pr})_l^{0.4}$	(6)	
$X_{tt} = \left(\frac{1-x}{x}\right)^{0.9}\left(\frac{\rho_v}{\rho_l}\right)^{0.5}\left(\frac{\mu_l}{\mu_v}\right)^{0.1}$	(7)	
Best agreement was with Equation (11) for vertical tubes.		Local coefficients for R-113 evaporating in a 0.430 in. ID transparent tube with a uniform wall heat flux over a length of 12.5 ft; evaporating temperature approximately 120°F. Report includes photographs of subcooled surface boiling, bubble, plug, and annular flow evaporation regimes (Gouse and Coumou 1965).
VERTICAL TUBES		
$h = 3.4h_l(1/X_{tt})^{0.45}$	(8)	Equations (8) and (9) were fitted to experimental data for vertical upflow in tubes. Both relate to forced-convection evaporation regions where nucleate boiling is suppressed (Guerrieri and Talty 1956, Dengler and Addoms 1956). A multiplying factor is recommended when nucleation is present.
$h = 3.5h_L(1/X_{tt})^{0.5}$ *where* h_l is from (3), X_{tt} from (7), h_L from (6)	(9)	
$h = 0.74h_L[B_o \times 10^4 + (1/X_{tt})^{0.67}]$ *where* B_o is from (5), h_L from (6), X_{tt} from (7)	(10)	Local coefficients for water in vertical upflow in tubes with diameters from 0.1162 to 0.4317 in. and lengths of 15 to 40 in. The boiling number B_o accounts for nucleation effects, and the Martinelli parameter X_{tt}, for forced-convection effects (Schrock and Grossman 1962).
$h = h_{mic} + h_{mac}$ *where*	(11)	Chen developed this correlation reasoning that the nucleation transfer mechanism (represented by h_{mic}) and the convective transfer mechanism (represented by h_{mac}) are additive. h_{mac} is expressed as a function of the two-phase Reynolds number after Martinelli, and h_{mic} is obtained from the nucleate boiling correlation of Forster and Zuber (1955). S_c is a suppression factor for nucleate boiling (Chen 1963).
$h_{mac} = h_l F_c$	(12)	
$h_{mic} = 0.00122\,(S_c)(E)(\Delta t)^{0.24}(\Delta p)^{0.75}$ F_c and S_c from Figures 10 and 11 $E = \frac{k_l^{0.79}(c_p)_l^{0.45}\rho_l^{0.49}g_c^{0.25}}{\sigma_l^{0.50}\mu_l^{0.29}h_{fg}^{0.24}\rho_v^{0.24}}$	(13)	
Equation (2) with $C_2 = 3.79$	(14)	See comments for Equation (2). Note the superior performance of the horizontal versus vertical configuration ($C_2 = 6.59$ versus 3.79) from this investigation, which used the same apparatus and test techniques for both orientations (Lavin and Young 1964).

Note: Except for dimensionless equations, units are lb_m, h, ft, °F, and Btu.

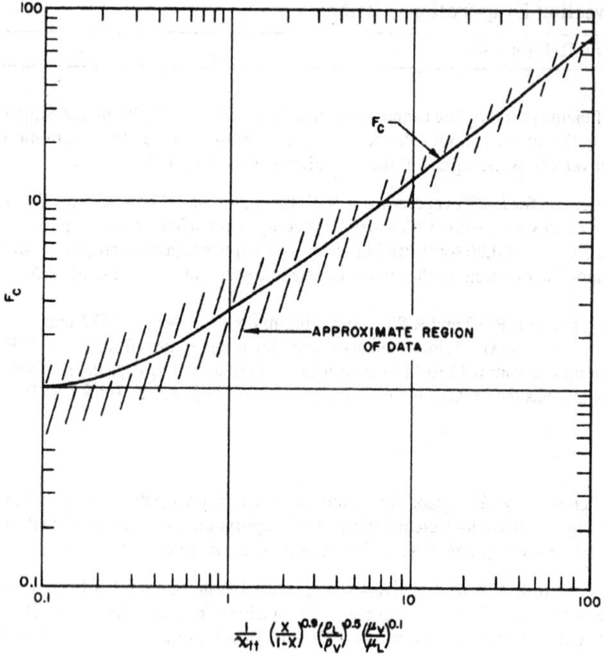

Fig. 10 Reynolds Number Factor F_c

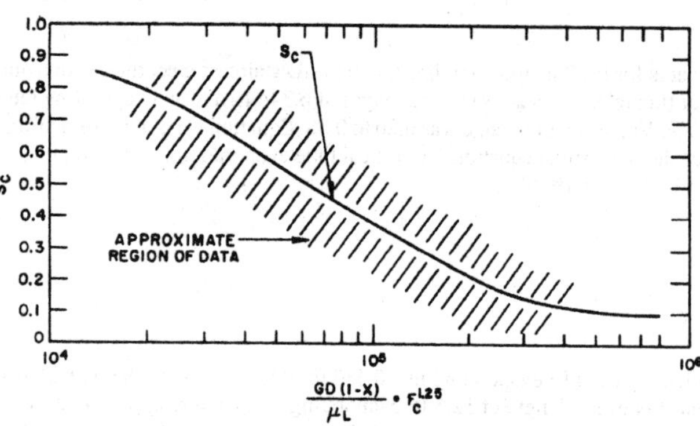

Fig. 11 Suppression Factor S_c

Equation (1) in Table 2 is recommended for refrigerant evaporation in horizontal tubes (Pierre 1955, 1957). It fits a wide range of R-12 and R-22 data.

Equations (2), (10), and (11) in Table 2 include terms for velocity effect (convection) and heat flux (nucleation) and produce local heat transfer coefficients as a function of local vapor quality x and heat transfer rate q. Local rather than average coefficients are used for accurate design.

The effect of oil on forced-convection evaporation has not been clearly determined. Increases occur in the *average* heat transfer coefficient for R-12 up to 10% oil by weight, with a maximum of about 4% (Green and Furse 1963, Worsoe-Schmidt 1960). Oil quantities greater than 10% cause reduction in heat transfer. However, oil can increase the pressure drop, offsetting possible gains in the heat transfer coefficient.

CONDENSING

In most applications that use the condensation process, condensation is initiated by removing heat at a solid-vapor interface, either through the walls of the vessel containing the saturated vapor or through the solid surface of a cooling mechanism placed within the saturated vapor. If a sufficient amount of energy is removed, the local temperature of the vapor near the interface will drop below its equilibrium saturation temperature. Because the heat removal process creates a temperature gradient with the lowest temperature near the interface, vapor droplets most likely form at this location. This defines one type of heterogeneous nucleation that can result in either dropwise condensation or film condensation, depending on the physical characteristics of the solid surface and the working fluid.

Dropwise condensation occurs on the cooling solid surface when its surface free energy is relatively low compared to that of the liquid. Examples of this type of interface include highly polished or fatty acid-impregnated surfaces in contact with steam. **Film condensation** occurs when a cooling surface having relatively high surface free energy contacts a fluid having lower surface free energy (see Isrealachvili 1991). This is the type of condensation that occurs in most systems.

The rate of heat transport depends on the condensate film thickness, which depends on the rate of vapor condensation and the rate of condensate removal. At high reduced pressures, the heat transfer coefficients for dropwise condensation are higher than those available in the presence of film condensation at the same surface loading. At low reduced pressures, the reverse is true. For example, there is a reduction of 6 to 1 in the dropwise condensation coefficient of steam when saturation pressure is decreased from 0.9 to 0.16 atm. One method for correlating the dropwise condensation heat transfer coefficient employs nondimensional parameters, including the effect of surface tension gradient, temperature difference, and fluid properties.

When condensation occurs on horizontal tubes and short vertical plates, the condensate film motion is laminar. On vertical tubes and long vertical plates, the film motion can become turbulent. Grober et al. (1961) suggest using a Reynolds number (Re) of 1600 as the critical point at which the flow pattern changes from laminar to turbulent. This Reynolds number is based on condensate flow rate divided by the breadth of the condensing surface. For a vertical tube, the breadth is the circumference of the tube; for a horizontal tube, the breadth is twice the length of the tube. $Re = 4\Gamma/\mu_f$, where Γ is the mass flow of condensate per unit of breadth, and μ_f is the absolute (dynamic) viscosity of the condensate at the film temperature t_f. In practice, condensation is usually laminar in shell-and-tube condensers with the vapor outside horizontal tubes.

Vapor velocity also affects the condensing coefficient. When this is small, condensate flows primarily by gravity and is resisted by the viscosity of the liquid. When vapor velocity is high relative to the condensate film, there is appreciable drag at the vapor-liquid interface. The thickness of the condensate film, and hence the heat transfer coefficient, is affected. When vapor flow is upward, a retarding force is added to the viscous shear, increasing the film thickness. When vapor flow is downward, the film thickness decreases and the heat transfer coefficient increases. For condensation inside horizontal tubes, the force of the vapor velocity causes the condensate to flow. When the vapor velocity is high, the transition from laminar to turbulent flow occurs at Reynolds numbers lower than previously described [i.e., 1600 according to Grober et al. (1961)].

When **superheated** vapor is condensed, the heat transfer coefficient depends on the surface temperature. When the surface temperature is *below* saturation temperature, using the value of h for condensation of saturated vapor that incorporates the difference between the *saturation* temperature and the surface temperature leads to insignificant error (McAdams 1954). If the surface temperature is *above* the saturation temperature, there is no condensation and the equations for gas convection apply.

Correlation equations for condensing heat transfer are given in Table 3. Factors F_1 and F_2, which depend only on the physical properties of the working fluid and which occur often in these equations, have been computed for some commonly used refrigerants in Table 4. Refrigerant properties used in the calculations may be found in Chapter 18.

Table 3 Heat Transfer Coefficients for Film-Type Condensation

Description	References	Equations	
1. Vertical surfaces, height L			
Laminar condensate flow, $\text{Re} = 4\Gamma/\mu_f < 1800$	McAdams (1954)	$h = 1.13F_1(h_{fg}/L\Delta t)^{0.25}$	(1)
	McAdams (1954)	$h = 1.11F_2(b/w_l)^{1/3}$	(2)
	Grigull (1952)	$h = 0.003(F_1)^2(\Delta tL/\mu_f^2 h_{fg})^{0.5}$	(3)
Turbulent flow, $\text{Re} = 4\Gamma/\mu_f > 1800$	McAdams (1954)	$h = 0.0077F_2(\text{Re})^{0.4}(1/\mu_f)^{1/3}$	(4)
2. Outside horizontal tubes, N rows in a vertical plane, length L, laminar flow	McAdams (1954)	$h = 0.79F_1(h_{fg}/Nd\Delta t)^{0.25}$	(5)
	McAdams (1954)	$h = 1.05F_2(L/w_l)^{1/3}$	(6)
Finned tubes	Beatty and Katz (1948)	$h = 0.689F_1(h_{fg}/\Delta tD_e)^{0.25}$ where D_e is determined from $\frac{1}{(D_e)^{0.25}} = 1.30\frac{A_s\phi}{A_{eff}(L_{mf})^{0.25}} + \frac{A_p}{A_{eff}(D)^{0.25}}$ with $A_{eff} = A_s\phi + A_p$ and $L_{mf} = a_f/D_o$	(7)
3. Simplified equations for steam			
Outside vertical tubes, $\text{Re} = 4\Gamma/\mu_f < 2100$	McAdams (1954)	$h = 4000/(L)^{0.25}(\Delta t)^{1/3}$	(8)
Outside horizontal tubes, $\text{Re} = 4\Gamma/\mu_f < 1800$	McAdams (1954)		
Single tube		$h = 3100/(d')^{0.25}(\Delta t)^{1/3}$	(9a)
Multiple tubes		$h = 3100/(Nd')^{0.25}(\Delta t)^{1/3}$	(9b)
4. Inside vertical tubes	Carpenter and Colburn (1949)	$h = 0.065\left(\frac{c_{pf}k_f\rho_f f'}{2\mu_f\rho_v}\right)$ where $G_m = \left(\frac{G_i^2 + G_iG_o + G_o^2}{3}\right)^{0.5}$	(10)
5. Inside horizontal tubes, $\frac{DG_l}{\mu_l} < 5000$			
$1000 < \frac{DG_v}{\mu_l}\left(\frac{\rho_l}{\rho_v}\right)^{0.5} < 20{,}000$	Ackers and Rosson (1960)	$\frac{hD}{k_l} = 13.8\left(\frac{c_p\mu_l}{k_l}\right)^{1/3}\left(\frac{h_{fg}}{c_p\Delta t}\right)^{1/6}\left[\frac{DG_v}{\mu_l}\left(\frac{\rho_l}{\rho_v}\right)^{0.5}\right]^{0.2}$	(11)
$20{,}000 < \frac{DG_v}{\mu_l}\left(\frac{\rho_l}{\rho_v}\right)^{0.5} < 100{,}000$	Ackers and Rosson (1960)	$\frac{hD}{k_l} = 0.1\left(\frac{c_p\mu_l}{k_l}\right)^{1/3}\left(\frac{h_{fg}}{c_p\Delta t}\right)^{1/6}\left[\frac{DG_v}{\mu_l}\left(\frac{\rho_l}{\rho_v}\right)^{0.5}\right]^{2/3}$	(12)
For $\frac{DG_l}{\mu_l} > 5000$ $\frac{DG_v}{\mu_l}\left(\frac{\rho_l}{\rho_v}\right)^{0.5} > 20{,}000$	Ackers et al. (1959)	$\frac{hD}{k_l} = 0.026\left(\frac{c_p\mu_l}{k_l}\right)^{1/3}\left(\frac{DG_E}{\mu_l}\right)^{0.8}$ where $G_e = G_v(\rho_l/\rho_v)^{0.5} + G_l$	(13)
	Altman et al. (1960a)	$h = 0.057\left(\frac{c_pk_f\rho_f}{\mu_f}\right)^{0.5}F^{0.5}$ where $F = \Delta p_{TPF}\left(\frac{g_oD}{4L}\right)$	(14)
	Forster and Zuber (1955)	Δp_{TPF} = frictional two-phase pressure drop	

Notes: 1. Equations (1) through (10) and Equation (14) are dimensional with units of Btu, h, ft, °F, and lb_m. 2. t_f = liquid film temperature = $t_{sat} - 0.75\Delta t$

In some cases, the equations are given in two forms: one is convenient when the amount of refrigerant to be condensed or the condensing load is known; the second is useful when the difference between the vapor temperature and the condensing surface temperature is known.

Condensation on Outside Surface of Vertical Tubes

For film-type condensation on the outside surface of vertical tubes and on vertical surfaces, Equations (1) and (2) in Table 3 are recommended when $4\Gamma/\mu_f$ is less than 1800 (McAdams 1954). For these equations, fluid properties are evaluated at the mean film temperature. When $4\Gamma/\mu_f$ is greater than 1800 (tall vertical plates or tubes), use Equation (3) or (4) in Table 3. Equations (2) and (4) in Table 3 are plotted in Figure 12. The theoretical curve for laminar film-type condensation is shown for comparison. A semitheoretical relationship for turbulent film-type condensation is also shown for Pr values of 1.0 and 5.0 (Colburn 1933-34).

Condensation on Outside Surface of Horizontal Tubes

For a bank of N tubes, Nusselt's equations, increased by 10% (Jakob 1949 and 1957), are given in Equations (5) and (6) in Table 3. Experiments by Short and Brown (1951) with R-11 suggest that drops of condensation falling from row to row cause local turbulence and increase heat transfer.

For condensation on the outside surface of horizontal finned tubes, Equation (7) in Table 3 is used for liquids that drain readily from the surface (Beatty and Katz 1948). For condensing steam outside finned tubes, where liquid is retained in the spaces between the tubes, coefficients substantially lower than those given by Equation (7) in Table 3 were reported. For additional data on condensation outside finned tubes, see Katz et al. (1947).

Table 4 Values of Condensing Coefficient Factors for Different Refrigerants (from Chapter 18)

Refrigerant	Film Temperature, °F $t_f = t_{sat} - 0.75\Delta t$	F_1	F_2
Refrigerant 11	75	154	822
	100	153	815
	125	151	803
Refrigerant 12	75	133	672
	100	122	608
	125	112	538
Refrigerant 22	75	153	822
	100	144	755
	125	132	675
Sulfur Dioxide	75	290	1920
	100	299	2000
	125	318	2170
Ammonia	75	409	3040
	100	408	3035
	125	408	3030
Propane	75	159	850
	100	157	845
	125	154	836
Butane	75	156	840
	100	156	843
	125	157	845

$$F_1 = \left(\frac{k_f^3 \rho_f^2 g}{\mu_f}\right)^{0.25} \quad \text{Units: } \left[\frac{(\text{Btu})^3(\text{lb}_m)}{(\text{h})^4(\text{ft})^7(°\text{F})^3}\right]^{0.25}$$

$$F_2 = \left(\frac{k_f^3 \rho_f^2 g}{\mu_f}\right)^{1/3} \quad \text{Units: } \left[\frac{(\text{Btu})^3(\text{lb}_m)}{(\text{h})^4(\text{ft})^7(°\text{F})^3}\right]^{1/3}$$

Simplified Equations for Steam

For film-type steam condensation at atmospheric pressure and film temperature drops of 10 to 150°F, McAdams (1954) recommends Equations (8) and (9) in Table 3.

Condensation on Inside Surface of Vertical Tubes

Condensation on the inside surface of tubes is generally affected by appreciable vapor velocity. The measured heat transfer coefficients are as much as 10 times those predicted by Equation (4) in Table 3. For vertical tubes, Jakob (1949 and 1957) gives theoretical derivations for upward and downward vapor flow. For downward vapor flow, Carpenter and Colburn (1949) suggest Equation (10) in Table 3. The friction factor f' for vapor in a pipe containing condensate should be taken from Figure 13.

Condensation on Inside Surface of Horizontal Tubes

For condensation on the inside surface of horizontal tubes (as in air-cooled condensers, evaporative condensers, and some shell-and-tube condensers), the vapor velocity and resulting shear at the vapor-liquid interface are major factors in analyzing heat transfer. Hoogendoorn (1959) identified seven types of two-phase flow patterns. For semistratified and laminar annular flow, use Equations (11) and (12) in Table 3 (Ackers and Rosson 1960). Ackers et al. (1959) recommend Equation (13) in Table 3 for turbulent annular flow (vapor Reynolds number greater than 20,000 and liquid Reynolds number greater than 5000). Equation (14) in Table 3 correlates the local heat transfer coefficients for R-22 condensing on the inside surface of pipes (Altman et al. 1960b); R-22 and several other fluids take the same form. The two-phase pressure drop in Equation (14) is determined by the method proposed by Martinelli and Nelson (1948); see also Altman et al. (1960a). A method for using a flow regime map to predict the heat transfer coefficient for condensation of pure components in a horizontal tube is presented in Breber et al. (1980).

Noncondensable Gases

Condensation heat transfer rates reduce drastically if one or more noncondensable gases are present in the condensing vapor/gas mixture. In mixtures, the condensable component is termed **vapor** and the noncondensable component is called **gas**. As the mass fraction of gas increases, the heat transfer coefficient decreases in an

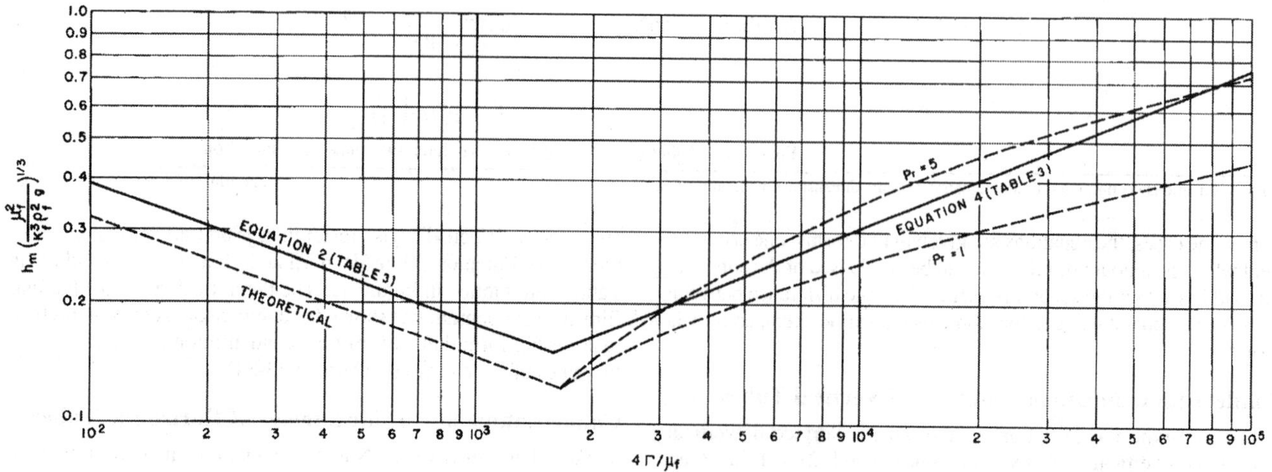

Fig. 12 Film-Type Condensation

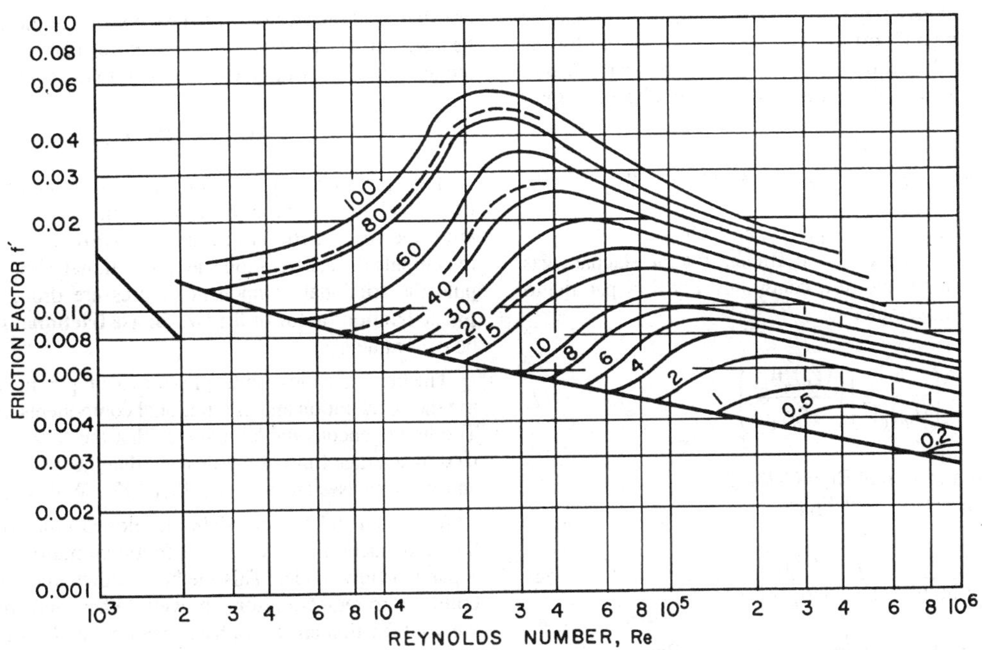

Curve parameter = Γ/ρs, where Γ = liquid flow rate, ρ = liquid density, and s = surface tension of liquid relative to water; values of gas velocity used in calculating f and Re are calculated as though no liquid were present.

Fig. 13 Friction Factors for Gas Flow Inside Pipes with Wetted Walls

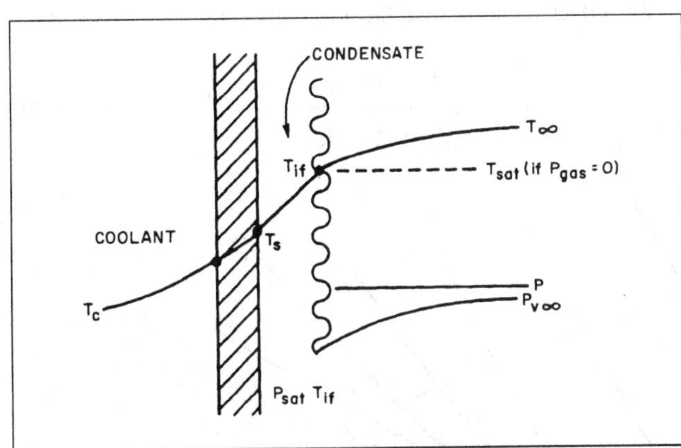

Fig. 14 Origin of Noncondensable Resistance

approximately linear manner. In a steam chest with 2.89% air by volume, Othmer (1929) found that the heat transfer coefficient dropped from about 2000 to about 600 Btu/h·ft²·°F. Consider a surface cooled to some temperature t_s below the saturation temperature of the vapor (Figure 14). In this system, accumulated condensate falls or is driven across the condenser surface. At a finite heat transfer rate, a temperature profile develops across the condensate that can be estimated from Table 3; the interface of the condensate is at a temperature $t_{if} > t_s$. In the absence of gas, the interface temperature is the vapor saturation temperature at the pressure of the condenser.

The presence of noncondensable gas lowers the vapor partial pressure and hence the saturation temperature of the vapor in equilibrium with the condensate. Further, the movement of the vapor toward the cooled surface implies similar bulk motion of the gas. At the condensing interface, the vapor is condensed at temperature t_{if} and is then swept out of the system as a liquid. The gas concentration rises to ultimately diffuse away from the cooled surface at the same rate as it is convected toward the surface (Figure 14). If gas (mole fraction) concentration is Y_g and total pressure of the system is p, the partial pressure of the bulk gas is

$$p_{g\infty} = Y_{g\infty}p \tag{2}$$

The partial pressure of the bulk vapor is

$$p_{v\infty} = (1 - Y_{g\infty})p = Y_{v\infty}p \tag{3}$$

As opposing fluxes of convection and diffusion of the gas increase, the partial pressure of gas at the condensing interface is $p_{gif} > p_{g\infty}$. By Dalton's law, assuming isobaric condition,

$$p_{gif} + p_{vif} = p \tag{4}$$

Hence, $p_{vif} < p_{v\infty}$.

Sparrow et al. (1967) noted that thermodynamic equilibrium exists at the interface, except in the case of very low pressures or liquid metal condensation, so that

$$p_{vif} = p_{sat}(t_{if}) \tag{5}$$

where $p_{sat}(t)$ is the saturation pressure of the vapor at temperature t. The available Δt for condensation across the condensate film is reduced from $(t_\infty - t_s)$ to $(t_{if} - t_s)$, where t_∞ is the bulk temperature of the condensing vapor-gas mixture, caused by the additional noncondensable resistance.

The equations in Table 3 are still valid for the condensate resistance, but the interface temperature t_{if} must be found. The noncondensable resistance, which accounts for the temperature difference $(t_\infty - t_{if})$, depends on the heat flux (through the convecting flow to the interface) and the diffusion of gas away from the interface.

In simple cases, Sparrow et al. (1967), Rose (1969), and Sparrow and Lin (1964) found solutions to the combined energy, diffusion, and momentum problem of noncondensables, but they are cumbersome.

A general method given by Colburn and Hougen (1934) can be used over a wide range if correct expressions are provided for the rate equations—add the contributions of the sensible heat transport through the noncondensable gas film and the latent heat transport via condensation:

$$h_g(t_\infty - t_{if}) + K_D M_v h_{lv}(p_{v\infty} - p_{vif}) = h(t_{if} - t_s) = U(t_{if} - t_c) \quad (6)$$

where h is from the appropriate equation in Table 3.

The value of the heat transfer coefficient for the stagnant gas depends on the geometry and flow conditions. For flow parallel to a condenser tube, for example,

$$j = \left(\frac{h_g}{(c_p)_g G}\right)\left(\frac{(c_p)_g \mu_{gv}}{K_{Dg}}\right)^{2/3} \quad (7)$$

where j is a known function of $Re = GD/\mu_{gv}$.

The mass transfer coefficient K_D is

$$\frac{K_D}{M_m}\left[\frac{p_{g\infty} - p_{gif}}{\ln(p_{g\infty}/p_{gif})}\right]\left(\frac{\mu_{gv}}{\rho_g D}\right)^{2/3} = j \quad (8)$$

The calculation method requires substitution of Equation (8) into Equation (6). For a given flow condition, G, Re, j, M_m, $p_{g\infty}$, h_g, and h (or U) are known. Assume values of t_{if}; calculate $p_{sat}(t_{if}) = p_{vif}$ and hence p_{gif}. If t_s is not known, use the overall coefficient U to the coolant and t_c in place of h and t_s in Equation (6). For either case, at each location in the condenser, iterate Equation (6) until it balances, giving the condensing interface temperature and, hence, the thermal load to that point (Colburn and Hougen 1934, Colburn 1951).

Other Impurities

Vapor entering the condenser often contains a small percentage of impurities such as oil. Oil forms a film on the condensing surfaces, creating additional resistance to heat transfer. Some allowance should be made for this, especially in the absence of an oil separator or when the discharge line from the compressor to the condenser is short.

PRESSURE DROP

Total pressure drop for two-phase flow in tubes consists of friction, acceleration, and gravitational components. It is necessary to know the **void fraction** (the ratio of gas flow area to total flow area) to compute the acceleration and gravitational components. To compute the frictional component of pressure drop, either the **two-phase friction factor** or the **two-phase frictional multiplier** must be determined.

The homogeneous model provides a simple method for computing the acceleration and gravitational components of pressure drop. The homogeneous model assumes that the flow can be characterized by average fluid properties and that the velocities of the liquid and vapor phases are equal (Collier 1972, Wallis 1969).

Martinelli and Nelson (1948) developed a method for predicting the void fraction and two-phase frictional multiplier to use with a separated flow model. This method predicts the pressure drops of boiling refrigerants reasonably well. Other methods of computing the void fraction and two-phase frictional multiplier used in a separated flow model are given in Collier (1972) and Wallis (1969).

The general nature of annular gas-liquid flow in vertical, and to some extent horizontal, pipe is indicated in Figure 15 (Wallis 1970), which plots the effective gas friction factor versus the liquid fraction $(1 - a)$. Here a is the void fraction, or fraction of the pipe cross section taken up by the gas or vapor.

The effective gas friction factor is defined as

$$f_{eff} = \left[\frac{a^{5/2} D}{2\rho_g (4Q_g/\pi D^2)^2}\right]\left(-\frac{dp}{ds}\right) \quad (9)$$

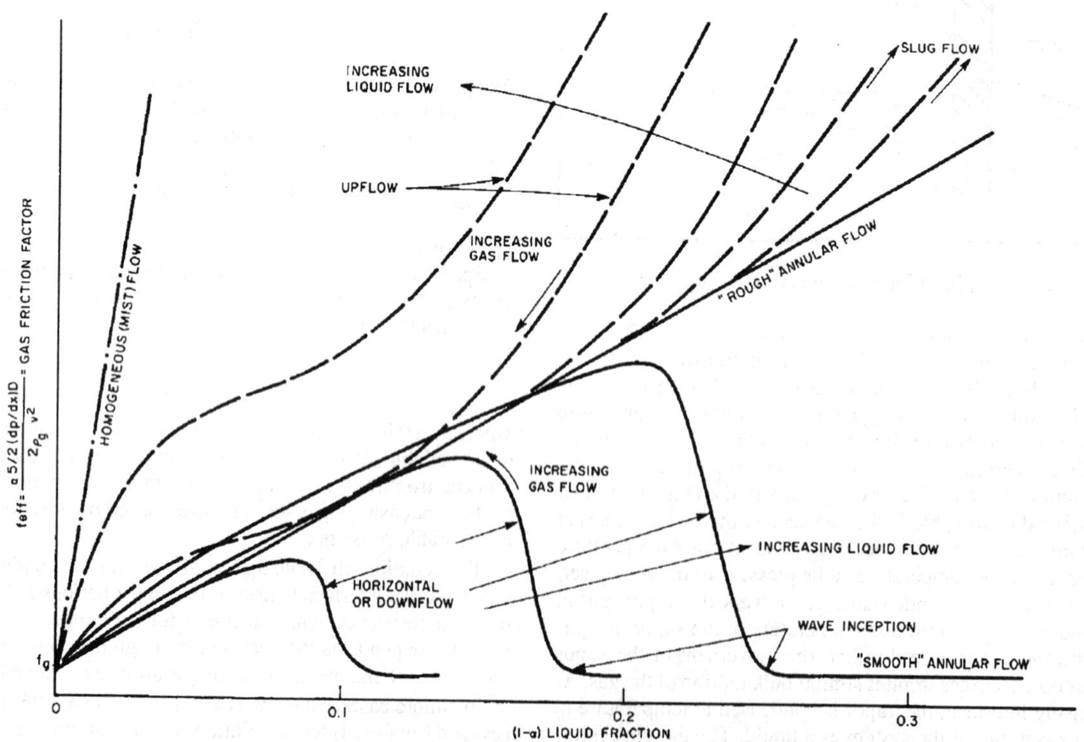

Fig. 15 Qualitative Pressure Drop Characteristics of Two-Phase Flow Regime

where D is the pipe diameter, ρ_g the gas density, and Q_g the gas volumetric flow rate. The friction factor of gas flowing by itself in the pipe (presumed smooth) is denoted by f_g. Wallis' analysis of the flow occurrences is based on interfacial friction between the gas and liquid. The wavy film corresponds to a conduit of relative roughness ε/D, about four times the liquid film thickness. Thus, the pressure drop relation of vertical flow is

$$-\frac{dp}{ds} + \rho_g g = 0.01\left(\frac{\rho_g}{D^5}\right)\left(\frac{4Q_g}{\pi}\right)^2 \frac{1+75(1-a)}{a^{5/2}} \tag{10}$$

This corresponds to the Martinelli-type analysis with

$$f_{two\text{-}phase} = \phi_g^2 f_g$$

when

$$\phi_g^2 = \frac{1+75(1-a)}{a^{5/2}} \tag{11}$$

The friction factor f_g (of the gas alone) is taken as 0.02, an appropriate turbulent flow value. This calculation can be modified for more detailed consideration of factors such as Reynolds number variation in friction, gas compressibility, and entrainment (Wallis 1970).

In two-phase flow inside horizontal tubes, the pressure gradient is written as the sum of frictional and momentum terms. Thus,

$$\frac{dp}{dz} = \left(\frac{dp}{dz}\right)_f + \left(\frac{dp}{dz}\right)_m \tag{12}$$

In adiabatic two-phase flow, the contribution of the momentum transfer to the overall pressure drop is negligibly small; theoretically, it is nonexistent if the flow is fully developed. In condensation heat transfer, the momentum transfer term contributes to the overall pressure drop due to the mass transfer that occurs at the liquid-vapor interface.

Two basic models were used in developing frictional pressure drop correlations for two-phase adiabatic flow. In the first, the flow of both phases is assumed to be homogeneous; the gas and liquid velocities are assumed equal. The frictional pressure drop is computed as if the flow were single phase, except for introducing modifiers to the single-phase friction coefficient. In the second model, the two phases are considered separate, and the velocities may differ. Two correlations used to predict the frictional pressure drop are those of Lockhart and Martinelli (1949) and Dukler et al. (1964).

In the Lockhart-Martinelli correlation, a parameter X was defined as

$$X = \left[\left(\frac{dp}{dz}\right)_l \div \left(\frac{dp}{dz}\right)_v\right]^{0.5} \tag{13}$$

where

$\left(\frac{dp}{dz}\right)_l$ = frictional pressure gradient, assuming that liquid alone flows in pipe

$\left(\frac{dp}{dz}\right)_v$ = frictional pressure gradient, assuming that gas (or vapor in case of condensation) alone flows in pipe

The frictional pressure gradient due to the single-phase flow of the liquid or vapor depends on the type of flow of each phase, laminar or turbulent. For turbulent flow during condensation, replace X by X_{tt}. Thus,

$$X_{tt} = \left(\frac{1-x}{x}\right)^{0.9}\left(\frac{\mu_l}{\mu_v}\right)^{0.1}\left(\frac{\rho_v}{\rho_l}\right)^{0.5} \tag{14}$$

Lockhart and Martinelli (1949) also defined ϕ_v as

$$\phi_v = \left[\left(\frac{dp}{dz}\right)_f \div \left(\frac{dp}{dz}\right)_v\right]^{0.5} \tag{15}$$

For condensation,

$$\left(\frac{dp}{dz}\right)_v = -\frac{2f_o(xG)^2}{\rho_v D_i} \tag{16}$$

where

$$f_o = \frac{0.045}{(GxD_i/\mu_v)^{0.2}} \tag{17}$$

Here f_o is the friction factor for adiabatic two-phase flow.

By analyzing the pressure drop data of simultaneous adiabatic flow of air and various liquids, Lockhart and Martinelli (1949) correlated the parameters ϕ_v and X and reported the results graphically. Soliman et al. (1968) approximated the graphical results of ϕ_v versus X_{tt} by

$$\phi_v = 1 + 2.85X_{tt}^{0.523} \tag{18}$$

In the correlation of Dukler et al. (1964), the frictional pressure gradient is given by

$$\left(\frac{dp}{dz}\right)_f = -\frac{2G^2 f_o \alpha(\lambda)\beta}{D_i \rho_{NS}} \tag{19}$$

where

f_o = single-phase friction coefficient evaluated at two-phase Reynolds number

$$= 0.0014 + 0.125\left(\frac{4\dot{m}_t\beta}{\pi D_i \mu_{NS}}\right)^{-0.32} \tag{20}$$

$$\alpha(\lambda) = 1 - (\ln\lambda)/[1.281 + 0.478\ln\lambda + 0.444(\ln\lambda)^2 + 0.094(\ln\lambda)^3 + 0.00843(\ln\lambda)^4] \tag{21}$$

$$\beta = \left(\frac{\rho_l}{\rho_{NS}}\right)\frac{\lambda^2}{(1-\psi)} + \left(\frac{\rho_v}{\rho_{NS}}\right)\frac{(1-\lambda)^2}{\psi} \tag{22}$$

$$\rho_{NS} = \rho_l\lambda + \rho_v(1-\lambda) \tag{23}$$

$$\mu_{NS} = \mu_l\lambda + \mu_v(1-\lambda) \tag{24}$$

$$\lambda = 1\Big/\left(1 + \frac{x}{(1-x)}\frac{\rho_v}{\rho_e}\right) \tag{25}$$

Because the correlations mentioned here were originally developed for adiabatic two-phase flow, Luu and Bergles (1980) modified the friction coefficients in Equations (16) and (19), using the modifier suggested by Silver and Wallis (1965-66). The modification

Table 5 Constants in Equation (29) for Different Void Fraction Correlations

Model	A_l	q_l	r_l	S_l
Homogeneous (Collier 1972)	1.0	1.0	1.0	0
Lockhart-Martinelli (1949)	0.28	0.64	0.36	0.07
Baroczy (1963)	1.0	0.74	0.65	0.13
Thom (1964)	1.0	1.0	0.89	0.18
Zivi (1964)	1.0	1.0	0.67	0
Turner-Wallis (1965)	1.0	0.72	0.40	0.08

replaced the friction coefficient f_o with the friction coefficient f_{co}. These terms are related by

$$\left(\frac{f_{co}}{f_o}\right) = \exp\left(\frac{\varepsilon}{2f_o}\right) - \left(\frac{\xi}{f_o}\right) \tag{26}$$

where

$$\xi = \left(\frac{D_i \psi}{2x}\right)\frac{dx}{dz} \tag{27}$$

Because the Lockhart-Martinelli and Dukler correlations for the frictional pressure gradient were based on the separated flow model, the momentum pressure gradient should be as well. Thus,

$$\left(\frac{dp}{dz}\right)_m = -G^2\left(\frac{dx}{dz}\right)\left\{\frac{2x}{\rho_v \psi} - \frac{2(1-x)}{\rho_l(1-\psi)} + q_l\left[\frac{\psi}{x}\frac{1-x}{1-\psi}\frac{1}{\rho_l} - \frac{x(1-\psi)}{\psi(1-x)\rho_v}\right]\right\} \tag{28}$$

To determine $(dp/dz)_m$, the void fraction ψ and the quality gradient must be known. A generalized expression for ψ was suggested by Butterworth (1975):

$$\psi = \frac{1}{1 + A_l[(1-x)/x]^{q_l}(\rho_v/\rho_l)^{r_l}(\mu_l/\mu_v)^{S_l}} \tag{29}$$

where A_l, q_l, r_l, and S_l are constants and are listed for the various correlations in Table 5.

The quality gradient dx/dz in Equation (28) can be estimated by assuming a constant rate of cooling. In the case of complete condensation, its value is $-1/L$, where L is the length of the condenser tube.

Evaporators and condensers often have valves, tees, bends, and other fittings that contribute to the overall pressure drop of the heat exchanger. Collier (1972) summarizes methods predicting the two-phase pressure drop in these fittings.

ENHANCED SURFACES

Enhanced heat transfer surfaces are used in heat exchangers to improve performance and decrease cost. Condensing heat transfer is often enhanced with circular fins attached to the external surfaces of tubes to increase the heat transfer area. Other enhancement methods, such as porous coatings, integral fins, and reentrant cavities, are used to augment boiling heat transfer on the external surfaces of evaporator tubes. Webb (1981) surveys external boiling surfaces and compares the performances of several enhanced surfaces with the performance of smooth tubes. For heat exchangers, the heat transfer coefficient for the refrigerant side is often smaller than the coefficient for the water side. Thus, enhancing the refrigerant-side surface can reduce the size of the heat exchanger and improve its performance.

Internal fins increase the heat transfer coefficients during evaporation or condensation in tubes. However, internal fins increase the refrigerant pressure drop and reduce the heat transfer rate by decreasing the available temperature difference between hot and cold fluids. Designers should carefully determine the number of parallel refrigerant passes that give optimum loading for best overall heat transfer.

For additional information on enhancement methods in two-phase flow, consult Bergles' comprehensive surveys (Bergles 1976, 1985).

SYMBOLS

A = area
A_{eff} = total effective area [Equation (7) in Table 3]
a = local acceleration [Equation (16) in Table 1]; void fraction [Equations (9) and (10)]
a_f = area of one side of one film
B_o = boiling number [Equation (10) in Table 2]
b = breadth of a condensing surface. For vertical tube, $b = \pi d$; for horizontal tube, $b = 2L$
C = a coefficient or constant
c_p = specific heat at constant pressure
c_v = specific heat at constant volume
C_1, C_2 = special constants (see Table 2)
C_{sf}, C_k = special constants (see Table 1)
D = diameter
D_i = inside tube diameter
D_o = outside tube diameter
d = diameter; or prefix meaning differential
(dp/dz) = pressure gradient
$(dp/dz)_f$ = frictional pressure gradient
$(dp/dz)_l$ = frictional pressure gradient, assuming that liquid alone is flowing in pipe
$(dp/dz)_m$ = momentum pressure gradient
$(dp/dz)_v$ = frictional pressure gradient, assuming that gas (or vapor) alone is flowing in pipe
F = special coefficient [Equation (14) in Table 3]
F_c = Reynolds number factor [Equation (12) in Table 2 and Figure 10]
F_1, F_2 = condensing coefficient factors [Equations (1) through (7) in Table 3 and Table 4]
f = friction factor for single-phase flow
f' = friction factor for gas flow inside pipes with wetted walls (Figure 13)
f_{co} = friction factor in presence of condensation [Equation (26)]
f_o = friction factor [Equations (17) and (19)]
G = mass velocity
Gr = Grashof number
g = gravitational acceleration
g_c = gravitational constant
h = heat transfer coefficient
h_{fg} = latent heat of vaporization or of condensation
j = Colburn j-factor
K_D = mass transfer coefficient
k = thermal conductivity
L = length
L_{mf} = mean length of fin [Equation (7) in Table 3]
ln = natural logarithm
M = mass; or molecular weight
M_m = mean molecular weight of vapor-gas mixture
M_v = molecular weight of condensing vapor
m = general exponent [Equations (1) and (6) in Table 1]
$\dot{m}$ = mass rate of flow
N = number of tubes in vertical tier
Nu = Nusselt number
n = general exponent [Equations (1) and (6) in Table 1 and Equation (1) in Table 2]
p = pressure
p_c = critical thermodynamic pressure for coolant
Pr = Prandtl number
Q = total heat transfer
q = rate of heat transfer
r = radius
Ra = Rayleigh number
Re = Reynolds number

S = distance along flow direction
S_c = suppression factor (Table 2 and Figure 11)
t = temperature
U = overall heat transfer coefficient
V = linear velocity
x = quality (i.e., vapor fraction = M_v/M); or distance in dt/dx
X_{tt} = Martinelli parameter [Figure 10, Table 2, and Equation (14)]
x,y,z = lengths along principal coordinate axes
Y_g = mole fraction of gas [Equations (2) and (3)]
Y_v = mole fraction of vapor [Equation (3)]
α = thermal diffusivity = $k/\rho c_p$
$\alpha(\lambda)$ = ratio of two-phase friction factor to single-phase friction factor at two-phase Reynolds number [Equation (21)]
β = ratio of two-phase density to no-slip density [Equation (22)]
Δ = difference between values
ε = roughness of interface
Γ = mass rate of flow of condensate per unit of breadth (see section on Condensing)
Λ = special coefficient [Equations (13) through (15) in Table 1]
λ = ratio of liquid volumetric flow rate to total volumetric flow rate [Equation (25)]
μ = absolute (dynamic) viscosity
μ_l = dynamic viscosity of saturated liquid
μ_{NS} = dynamic viscosity of two-phase homogeneous mixture [Equation (24)]
μ_v = dynamic viscosity of saturated vapor
ν = kinematic viscosity
ρ = density
ρ_l = density of saturated liquid
ρ_{NS} = density of two-phase homogeneous mixture [Equation (23)]
ρ_v = density of saturated vapor phase
σ = surface tension
ϕ_g = fin efficiency, Martinelli factor [Equation (11)]
ϕ_v = Lockhart-Martinelli parameter [Equation (15)]
ψ = void fraction

Subscripts and Superscripts

a = exponent in Equation (1)
b = bubble
c = critical or cold (fluid)
cg = condensing
e = equivalent
eff = effective
f = film or fin
g = gas
h = horizontal or hot (fluid) or hydraulic
i = inlet or inside
if = interface
L = liquid
l = liquid
m = mean
mac = convective mechanism [Equations (11) through (13) in Table 2]
max = maximum
mic = nucleation mechanism [Equations (11) through (13) in Table 2]
min = minimum
o = outside or outlet or overall
r = root (fin) or reduced pressure
s = surface or secondary heat transfer surface
sat = saturation (pressure)
t = temperature or terminal temperature of tip (fin)
v = vapor or vertical
w = wall
∞ = bulk
$*$ = reference

REFERENCES

Ackers, W.W., H.A. Deans, and O.K. Crosser. 1959. Condensing heat transfer within horizontal tubes. *Chemical Engineering Progress Symposium Series* 55(29):171-176.

Ackers, W.W. and H.F. Rosson. 1960. Condensation inside a horizontal tube. *Chemical Engineering Progress Symposium Series* 56(30):145-50.

Altman, M., R.H. Norris, and F.W. Staub. 1960a. Local and average heat transfer and pressure drop for refrigerants. *ASHRAE Transactions* (August):189.

Altman, M., F.W. Staub, and R.H. Norris. 1960b. Local heat transfer and pressure drop for Refrigerant-22 condensing to horizontal tubes. *Chemical Engineering Progress Symposium Series* 56(30):151-60.

Anderson, W., D.G. Rich, and D.F. Geary. 1966. Evaporation of Refrigerant 22 in a horizontal 3/4-in. OD tube. *ASHRAE Transactions* 72(1):28.

Ashley, C.M. 1942. The heat transfer of evaporating Freon. *Refrigerating Engineering* (February):89.

Baroczy, C.J. 1963. Correlation of liquid fraction in two-phase flow with application to liquid metals. *North American Aviation Report* SR-8171, El-Segundo, CA.

Beatty, K.O. and D.L. Katz. 1948. Condensation of vapors on outside of finned tubes. *Chemical Engineering Progress* 44(1):55.

Berenson, P.J. 1961. Film boiling heat transfer from a horizontal surface. *ASME Journal of Heat Transfer* 85:351.

Berenson, P.J. 1962. Experiments on pool boiling heat transfer. *International Journal of Heat and Mass Transfer* 5:985.

Bergles, A.E. 1976. Survey and augmentation of two-phase heat transfer. *ASHRAE Transactions* 82(1):891-905.

Bergles, A.E. 1985. Techniques to augment heat transfer. In *Handbook of heat transfer application*, 2nd ed. McGraw-Hill, New York.

Bergles, A.E. and W.M. Rohsenow. 1964. The determination of forced convection surface-boiling heat transfer. *ASME Journal of Heat Transfer*, Series C, 86(August):365.

Blatt, T.A. and R.R. Adt. 1963. Boiling heat transfer and pressure drop characteristics of Freon 11 and Freon 113 refrigerants. *Paper* No. 132, AIChE 50th National Meeting, May 5-8, Buffalo.

Borishansky, W. and A. Kosyrev. 1966. Generalization of experimental data for the heat transfer coefficient in nucleate boiling. *ASHRAE Journal* (May):74.

Borishansky, V.M., I.I. Novikov, and S.S. Kutateladze. 1962. Use of thermodynamic similarity in generalizing experimental data on heat transfer. Proceedings of the International Heat Transfer Conference.

Breber, G., J.W. Palen, and J. Taborek. 1980. Prediction of the horizontal tubeside condensation of pure components using flow regime criteria. *ASME Journal of Heat Transfer* 102(3):471-76.

Breen, B.P. and J.W. Westwater. 1962. Effects of diameter of horizontal tubes on film boiling heat transfer. AIChE Preprint No. 19, Fifth National Heat Transfer Conference, Houston, TX. *Chemical Engineering Progress* 58(7):67-72.

Bromley, L.A. 1950. Heat transfer in stable film boiling. *Chemical Engineering Progress* (46):221.

Butterworth, D. 1975. A comparison of some void-fraction relationships for co-current gas-liquid flow. *International Journal of Multiphase Flow* 1:845-50.

Carey, V.P. 1992. *Liquid-vapor phase change phenomena: An introduction to the thermophysics of vaporization and condensation processes in heat transfer equipment.* Hemisphere Publishing Corporation, Washington, D.C.

Carpenter, E.F. and A.P. Colburn. 1949. The effect of vapor velocity on condensation inside tubes. General discussion on Heat Transfer and Fluid Mechanics Institute, American Society of Mechanical Engineers, New York.

Chaddock, J.B. and J.A. Noerager. 1966. Evaporation of Refrigerant 12 in a horizontal tube with constant wall heat flux. *ASHRAE Transactions* 72(1):90.

Chen, J.C. 1963. A correlation for boiling heat transfer to saturated fluids on convective flow. ASME *Paper* 63-HT-34. American Society of Mechanical Engineers, New York.

Colburn, A.P. 1933-34. Note on the calculation of condensation when a portion of the condensate layer is in turbulent motion. *AIChE Transactions* No. 30.

Colburn, A.P. 1951. Problems in design and research on condensers of vapours and vapour mixtures. Proceedings of the Institute of Mechanical Engineers, 164:448, London.

Colburn, A.P. and O.A. Hougen. 1934. Design of cooler condensers for mixtures of vapors with noncondensing gases. *Industrial and Engineering Chemistry* 26(November):1178.

Collier, J.G. 1972. *Convective boiling and condensation.* McGraw-Hill Book Co. (UK), Ltd., London.

Danilova, G. 1965. Influence of pressure and temperature on heat exchange in the boiling of halogenated hydrocarbons. *Kholodilnaya Teknika*, No. 2. English abstract, *Modern Refrigeration* (December).

Dengler, C.E. and J.N. Addoms. 1956. Heat transfer mechanism for vaporization of water in a vertical tube. *Chemical Engineering Progress Symposium Series* 52(18):95.

Dougherty, R.L. and H.J. Sauer, Jr. 1974. Nucleate pool boiling of refrigerant-oil mixtures from tubes. *ASHRAE Transactions* 80(2):175.

Dukler, A.E., M. Wicks, III, and R.G. Cleveland. 1964. Frictional pressure drop in two-phase flow: An approach through similarity analysis. *AIChE Journal* 10(January):44-51.

Farber, E.A. and R.L. Scorah. 1948. Heat transfer to water boiling under pressure. ASME *Transactions* (May):373.

Forster, H.K. and N. Zuber. 1955. Dynamics of vapor bubbles and boiling heat transfer. *AIChE Journal* 1(4):531-35.

Frederking, T.H.K. and J.A. Clark. 1962. Natural convection film boiling on a sphere. In *Advances in cryogenic engineering*, ed. K.D. Timmerhouse, Plenum Press, New York.

Furse, F.G. 1965. Heat transfer to Refrigerants 11 and 12 boiling over a horizontal copper surface. *ASHRAE Transactions* 71(1):231.

Gilmour, C.H. 1958. Nucleate boiling—A correlation. *Chemical Engineering Progress* 54(October):77.

Gouse, S.W., Jr. and K.G. Coumou. 1965. Heat transfer and fluid flow inside a horizontal tube evaporator, Phase I. *ASHRAE Transactions* 71(2):152.

Green, G.H. and F.G. Furse. 1963. Effect of oil on heat transfer from a horizontal tube to boiling Refrigerant 12-oil mixtures. *ASHRAE Journal* (October):63.

Grigull, U. 1952. Wärmeübergang bei Filmkondensation. *Forsch, Gebiete Ingenieurw.* 18.

Grober, H., S. Erk, and U. Grigull. 1961. *Fundamentals of heat transfer*. McGraw-Hill, New York.

Guerrieri, S.A. and R.D. Talty. 1956. A study of heat transfer to organic liquids in single tube boilers. *Chemical Engineering Progress Symposium Series* 52(18):69.

Hofmann, E. 1957. Heat transfer coefficients for evaporating refrigerants. *Kältetechnik* 9(1). Argonne National Laboratory translation, Lemont, IL (September) 1958.

Hoogendoorn, C.J. 1959. Gas-liquid flow in horizontal pipes. *Chemical Engineering Sciences* IX(1).

Hughmark, G.A. 1962. A statistical analysis of nucleate pool boiling data. *International Journal of Heat and Mass Transfer* 5:667.

Isrealachvili, J.N. 1991. *Intermolecular surface forces*. Academic Press, New York.

Jakob, M. 1949 and 1957. *Heat transfer*, Vols. I and II. John Wiley and Sons, New York.

Katz, D.L., P.E. Hope, S.C. Datsko, and D.B. Robinson. 1947. Condensation of Freon-12 with finned tubes. Part I, Single horizontal tubes; Part II, Multitube condensers. *Refrigerating Engineering* (March):211, (April): 315.

Kutateladze, S.S. 1951. A hydrodynamic theory of changes in the boiling process under free convection. Izvestia Akademii Nauk, USSR, Otdelenie Tekhnicheski Nauk 4:529.

Kutateladze, S.S. 1963. *Fundamentals of heat transfer*. E. Arnold Press, London.

Lavin, J.G. and E.H. Young. 1964. Heat transfer to evaporating refrigerants in two-phase flow. AIChE *Preprint* 21e (February), Symposium on Two-Phase Flow and Heat Transfer.

Lienhard, J.H. and V.E. Schrock. 1963. The effect of pressure, geometry and the equation of state upon peak and minimum boiling heat flux. *ASME Journal of Heat Transfer* 85:261.

Lienhard, J.H. and P.T.Y. Wong. 1963. The dominant unstable wave length and minimum heat flux during film boiling on a horizontal cylinder. ASME *Paper* No. 63-HT-3. ASME-AIChE Heat Transfer Conference, Boston, August.

Lockhart, R.W. and R.C. Martinelli. 1949. Proposed correlation of data for isothermal two-phase, two-component flow in pipes. *Chemical Engineering Progress* 45(1):39-48.

Luu, M. and A.E. Bergles. 1980. Augmentation of in-tube condensation of R-113. ASHRAE *Research Project* RP-219.

Martinelli, R.C. and D.B. Nelson. 1948. Prediction of pressure drops during forced circulation boiling of water. *ASME Transactions* 70:695.

McAdams, W.H. 1954. *Heat transmission*, 3rd ed. McGraw-Hill, New York.

Myers, J.E. and D.L. Katz. 1952. Boiling coefficients outside horizontal plain, and finned tubes. *Refrigerating Engineering* (January):56.

Nukiyama, S. 1934. The maximum and minimum values of heat transmitted from metal to boiling water under atmospheric pressure. *Journal of the Japanese Society of Mechanical Engineers* 37:367.

Othmer, D.F. 1929. The condensation of steam. *Industrial and Engineering Chemistry* 21(June):576.

Perry, J.H. 1950. *Chemical engineers handbook*, 3rd ed. McGraw-Hill, New York.

Pierre, B. 1955. S.F. Review. *A.B. Svenska Flaktafabriken*, Stockholm, Sweden 2(1):55.

Pierre, B. 1957. *Kylteknisk Tidskrift* 3 (May):129.

Pierre, B. 1964. Flow resistance with boiling refrigerant. *ASHRAE Journal* (September through October).

Rohsenow, W.M. 1951. A method of correlating heat transfer for surface boiling of liquids. ASME *Transactions* 73:609.

Rohsenow, W.M. 1963. Boiling heat transfer. In *Modern developments in heat transfer*, ed. W. Ibele. Academic Press, New York.

Rose, J.W. 1969. Condensation of a vapour in the presence of a noncondensable gas. *International Journal of Heat and Mass Transfer* 12:233.

Schrock, V.E. and L.M. Grossman. 1962. Forced convection boiling in tubes. *Nuclear Science and Engineering* 12:474.

Short, B.E. and H.E. Brown. 1951. Condensation of vapors on vertical banks of horizontal tubes. American Society of Mechanical Engineers, New York.

Silver, R.S. and G.B. Wallis. 1965-66. A simple theory for longitudinal pressure drop in the presence of lateral condensation. Proceedings of Institute of Mechanical Engineering, 180 Part I(1):36-42.

Soliman, M., J.R. Schuster, and P.J. Berenson. 1968. A general heat transfer correlation for annular flow condensation. *Journal of Heat Transfer* 90:267-76.

Sparrow, E.M. and S.H. Lin. 1964. Condensation in the presence of a noncondensable gas. ASME *Transactions, Journal of Heat Transfer* 86C:430.

Sparrow, E.M., W.J. Minkowycz, and M. Saddy. 1967. Forced convection condensation in the presence of noncondensables and interfacial resistance. *International Journal of Heat and Mass Transfer* 10:1829.

Starczewski, J. 1965. Generalized design of evaporation heat transfer to nucleate boiling liquids. *British Chemical Engineering* (August).

Stephan, K. 1963a. The computation of heat transfer to boiling refrigerants. *Kältetechnik* 15:231.

Stephan, K. 1963b. Influence of oil on heat transfer of boiling Freon-12 and Freon-22. Eleventh International Congress of Refrigeration, I.I.R. *Bulletin* No. 3.

Stephan, K. 1963c. A mechanism and picture of the processes involved in heat transfer during bubble evaporation. *Chemic. Ingenieur Technik* 35:775.

Thom, J.R.S. 1964. Prediction of pressure drop during forced circulation boiling water. *International Journal of Heat and Mass Transfer* 7:709-24.

Tschernobyiski, I. and G. Ratiani. 1955. *Kholodilnaya Teknika* 32.

Turner, J.M. and G.B. Wallis. 1965. The separate-cylinders model of two-phase flow. *Report* No. NYO-3114-6. Thayer's School of Engineering, Dartmouth College, Hanover, NH.

Van Stralen, S.J. 1959. Heat transfer to boiling binary liquid mixtures. *Chemical Engineering* (British) 4(January):78.

Wallis, G.B. 1969. *One-dimensional two-phase flow*. McGraw-Hill, New York.

Wallis, G.C. 1970. Annular two-phase flow, Part I: A simple theory, Part II: Additional effect. *ASME Transactions, Journal of Basic Engineering* 92D:59 and 73.

Webb, R.L. 1981. The evolution of enhanced surface geometrics for nucleate boiling. *Heat Transfer Engineering* 2(3-4):46-69.

Westwater, J.W. 1963. Things we don't know about boiling. In *Research in Heat Transfer*, ed. J. Clark. Pergamon Press, New York.

Worsoe-Schmidt, P. 1959. Some characteristics of flow-pattern and heat transfer of Freon-12 evaporating in horizontal tubes. *Ingenieren*, International edition, 3(3).

Worsoe-Schmidt, P. 1960. *ASME Transactions* (August):197.

Zivi, S.M. 1964. Estimation of steady-state steam void-fraction by means of the principle of minimum entropy production. *Journal of Heat Transfer* 86:247-52.

Zuber, N. 1959. Hydrodynamic aspects of boiling heat transfer. U.S. Atomic Energy Commission, Technical Information Service, *Report* AECU 4439. Oak Ridge, TN.

Zuber, N., M. Tribus, and J.W. Westwater. 1962. The hydrodynamic crisis in pool boiling of saturated and subcooled liquids. Proceedings of the International Heat Transfer Conference 2:230, and discussion of the papers, Vol. 6.

CHAPTER 5

MASS TRANSFER

MASS transfer by either molecular diffusion or convection is the transport of one component of a mixture relative to the motion of the mixture and is the result of a **concentration gradient**. In an air-conditioning process, water vapor is added or removed from the air, with a simultaneous transfer of heat and mass (water vapor) between the airstream and a wetted surface. The wetted surface can be water droplets in an air washer, wetted slats of a cooling tower, condensate on the surface of a dehumidifying coil, surface presented by a spray of liquid absorbent, or wetted surfaces of an evaporative condenser. The performance of equipment with these phenomena must be calculated carefully because of the simultaneous heat and mass transfer.

This chapter addresses the principles of mass transfer and provides methods of solving a simultaneous heat and mass transfer problem involving air and water vapor. Emphasis is on air-conditioning processes involving mass transfer. The formulations presented can help in analyzing the performance of specific equipment. For a discussion on the performance of air washers, cooling coils, evaporative condensers, and cooling towers, see Chapters 19, 21, 35, and 36, respectively, of the 1996 *ASHRAE Handbook—Systems and Equipment.*

This chapter is divided into (1) the principles of molecular diffusion, (2) a discussion on the convection of mass, and (3) simultaneous heat and mass transfer and its application to specific equipment.

MOLECULAR DIFFUSION

Most mass transfer problems can be analyzed by considering the diffusion of a gas into a second gas, a liquid, or a solid. In this chapter, the diffusing or dilute component is designated as component B, and the other component as component A. For example, when water vapor diffuses into air, the water vapor is component B and dry air is component A. Properties with subscripts A or B are local properties of that component. Properties without subscripts are local properties of the mixture.

The primary mechanism of mass diffusion at ordinary temperature and pressure conditions is **molecular diffusion**, a result of density gradient. In a binary gas mixture, the presence of a concentration gradient causes transport of matter by molecular diffusion; that is, because of random molecular motion, gas B diffuses through the mixture of gases A and B in a direction that reduces the concentration gradient.

Fick's Law

The basic equation for molecular diffusion is Fick's law. Expressing the concentration of component B of a binary mixture in terms of the mass fraction ρ_B/ρ or mole fraction C_B/C, Fick's law is

$$J_B = -\rho D_v \frac{d(\rho_B/\rho)}{dy} \tag{1a}$$

$$J_B^* = -CD_v \frac{d(C_B/C)}{dy} \tag{1b}$$

The minus sign indicates that the concentration gradient is negative in the direction of diffusion. The proportionality factor D_v is the **mass diffusivity** or the **diffusion coefficient**. The diffusive mass flux J_B and the diffusive molar flux J_B^* are

$$J_B \equiv \rho_B(v_B - v) \tag{2a}$$

$$J_B^* \equiv C_B(v_B - v^*) \tag{2b}$$

where $(v_B - v)$ is the velocity of component B relative to the velocity of the mixture and v^* is the molar average velocity.

Bird et al. (1960) present an analysis of Equations (1a) and (1b). Equations (1a) and (1b) are equivalent forms of Fick's law. The equation used depends on the problem and individual preference. This chapter emphasizes mass analysis rather than molar analysis. However, all results can be converted to the molar form using the relation $C_B \equiv \rho_B / M_B$.

Fick's Law for Dilute Mixtures

In many mass diffusion problems, component B is dilute; the density of component B is small compared to the density of the mixture, and the variation in the density of the mixture throughout the problem is about ρ_B or less. In this case, Equation (1a) can be written as

$$J_B = -D_v \frac{d\rho_B}{dy} \tag{3}$$

when $\rho_B << \rho$, $\Delta\rho < \rho_B$.

Equation (3) can be used without significant error for water vapor diffusing through air at atmospheric pressure and a temperature less than 80°F. In this case, $\rho_B < 0.02\rho$, where ρ_B is the density of water vapor and ρ is the density of moist air (air and water vapor mixture). The error in J_B caused by replacing $\rho[d(\rho_B/\rho)/dy]$ with $d\rho_B/dy$ is less than 2%. At temperatures below 140°F where $\rho_B < 0.10\rho$, Equation (3) can still be used if errors in J_B as great as 10% are tolerable.

Fick's Law for Mass Diffusion Through Solids or Stagnant Fluids

Fick's law can be simplified for cases of dilute mass diffusion in solids, stagnant liquids, or stagnant gases. In these cases, $\rho_B << \rho$ and $v = 0$, which yields the following approximate result:

$$J_B = \rho_B(v_B - v) = \rho_B\left(v_B - \frac{\rho_B v_B}{\rho}\right) \approx \rho_B v_B = \dot{m}_B'' \tag{4}$$

Therefore, Fick's law reduces to

$$\dot{m}_B'' = -D_v \frac{d\rho_B}{dy} \tag{5}$$

when $\rho_B << \rho$, $\Delta\rho < \rho_B$, $v_A = 0$.

The preparation of this chapter is assigned to TC 1.3, Heat Transfer and Fluid Flow.

Fick's Law for Ideal Gases with Negligible Temperature Gradient

For cases of dilute mass diffusion, Fick's law can be written in terms of pressure gradient instead of concentration gradient when gas B can be approximated as ideal:

$$p_B = \frac{\rho_B R_U T}{M_B} \tag{6}$$

and when the gradient in T is small. Under these conditions, Equation (3) can be written as

$$J_B = -\left(\frac{M_B D_v}{R_U T}\right)\frac{dp_B}{dy} \tag{7a}$$

or

$$J_B^* = -\left(\frac{D_v}{R_U T}\right)\frac{dp_B}{dy} \tag{7b}$$

when B is dilute and ideal and T is constant. Also, under the above conditions, Equation (5) may be written as

$$\dot{m}_B'' = -\left(\frac{M_B D_v}{R_U T}\right)\frac{dp_B}{dy} \tag{8a}$$

or

$$\dot{m}_B''^* = -\left(\frac{D_v}{R_U T}\right)\frac{dp_B}{dy} \tag{8b}$$

when B is dilute and ideal, $v_A = 0$, and T is constant.

The pressure gradient formulation for mass transfer analysis has been used extensively; this is unfortunate because the pressure formulation [Equations (7) and (8)] applies only to cases where one component is dilute, the fluid closely approximates an ideal gas, and the temperature gradient has a negligible effect. The density (or concentration) gradient formulation expressed in Equations (1) through (5) is the more general formulation and can be applied to a wider range of mass transfer problems, including cases where neither component is dilute [Equation (1)]. The gases need not be ideal, nor the temperature gradient negligible. Consequently, this chapter emphasizes the density formulation.

Diffusion Coefficient

For a binary mixture, the diffusion coefficient D_v is a function of temperature, pressure, and composition. Experimental measurements of D_v for most binary mixtures are limited in range and accuracy. Table 1 gives a few experimental values for diffusion of some gases in air. For more detailed tables, see the section on Bibliography at the end of this chapter.

In the absence of data, use equations developed from (1) theory or (2) theory with constants adjusted from limited experimental data. For binary gas mixtures at low pressure, D_v is inversely proportional

Table 1 Mass Diffusivities for Gases in Air[a]

Gas	D_v, ft²/h
Ammonia	1.08
Benzene	0.34
Carbon dioxide	0.64
Ethanol	0.46
Hydrogen	1.60
Oxygen	0.80
Water vapor	0.99

[a]Gases at 77°F and 14.696 psi.

to pressure, increases with increasing temperature, and is almost independent of composition for a given gas pair. The following equation for estimating D_v at pressures less than $0.1p_{c\,min}$ was developed by Bird et al. (1960) from kinetic theory and corresponding-states arguments.

$$D_v = a\left(\frac{T}{\sqrt{T_{cA} + T_{cB}}}\right)^b \sqrt{\frac{1}{M_A} + \frac{1}{M_B}} \times \frac{(p_{cA} p_{cB})^{1/3} (T_{cA} T_{cB})^{5/12}}{p} \tag{9}$$

where

D_v = diffusion coefficient, ft²/h
a = constant, dimensionless
b = constant, dimensionless
T = absolute temperature, K
p = pressure, psi
M = molecular weight, lb_m/lb mol

The subscripts cA and cB refer to the critical states of the two gases. Analysis of experimental data gives the following values of the constants a and b:

For nonpolar gas pairs,

$$a = 7.266 \times 10^{-3}$$
$$b = 1.823$$

For water vapor with a nonpolar gas,

$$a = 9.635 \times 10^{-3}$$
$$b = 2.334$$

A *nonpolar gas* is one for which the intermolecular forces are independent of the relative orientation of molecules, depending only on the separation distance from each other. Air, composed of nonpolar gases O_2 and N_2, is nonpolar.

Equation (9) is stated to agree with experimental data at atmospheric pressure to within about 8% (Bird et al. 1960).

The mass diffusivity D_v for binary mixtures at low pressure is predictable within about 10% by kinetic theory (Reid et al. 1987).

$$D_v = 0.0072\frac{T^{1.5}}{p(\sigma_{AB})^2 \Omega_{D,AB}}\sqrt{\frac{1}{M_A} + \frac{1}{M_B}} \tag{10}$$

where

σ_{AB} = characteristic molecular diameter, nm
$\Omega_{D,AB}$ = temperature function, dimensionless

D_v is in ft²/h, p in atmospheres, and T in kelvins. If the gas molecules of A and B are considered rigid spheres having diameters σ_A and σ_B [and $\sigma_{AB} = (\sigma_A/2) + (\sigma_B/2)$], all expressed in nanometers, the dimensionless function $\Omega_{D,AB}$ equals unity. More realistic models for the molecules having intermolecular forces of attraction and repulsion lead to values of $\Omega_{D,AB}$ that are functions of temperature. Reid et al. (1987) present tabulations of this quantity. These results show that D_v increases as the 2.0 power of T at low temperatures and as the 1.65 power of T at very high temperatures.

The diffusion coefficient of moist air has been calculated for Equation (9) using a simplified intermolecular potential field function for water vapor and air (Mason and Monchick 1965).

The following is an empirical equation for mass diffusivity of water vapor in air up to 2000°F (Sherwood and Pigford 1952):

$$D_v = \frac{0.00215}{p}\left(\frac{T^{2.5}}{T+441}\right) \qquad (11)$$

where D_v is in ft²/h, p in psi, and T in °R.

Analogy Between Heat Conduction and Mass Diffusion

Molecular diffusion is, in some cases, directly analogous to conduction heat transfer. Both result from random molecular mixing in a stagnant fluid or in a fluid in laminar (streamline) flow. The equation governing each process can be expressed as

$$\text{Flux} = \text{Diffusivity} \times \text{Concentration gradient}$$

Examination of Equation (5) reveals that Fick's law applied to mass diffusion of a dilute gas through solids and stagnant fluids is analogous to Fourier's law (Chapter 3):

$$q'' = -k\frac{dt}{dx} \qquad (12)$$

The usefulness of the analogy can be proven as follows. Transient mass diffusion in a dilute mixture region where D_v is constant and component A is stagnant is governed by the following system of equations:

$$\frac{\partial \rho_B}{\partial \tau} = D_v\left(\frac{\partial^2 \rho_B}{\partial x^2} + \frac{\partial^2 \rho_B}{\partial y^2} + \frac{\partial^2 \rho_B}{\partial z^2}\right) \qquad (13a)$$

Initial condition (IC):

$$\rho_B = \rho_{Bi}(x, y, z) \text{ at time } \tau = 0 \qquad (13b)$$

Boundary condition (BC):

$$\rho_B = f(x, y, z, \tau) \text{ on boundary} \qquad (13c)$$

Transient heat diffusion in a constant k and constant ρc_p region is governed by

$$\frac{\partial t}{\partial \bar{\tau}} = \alpha\left(\frac{\partial^2 t}{\partial \bar{x}^2} + \frac{\partial^2 t}{\partial \bar{y}^2} + \frac{\partial^2 t}{\partial \bar{z}^2}\right) \qquad (14a)$$

$$\text{IC:} \quad t = t_i(\bar{x}, \bar{y}, \bar{z}) \text{ at } \bar{\tau} = 0 \qquad (14b)$$

$$\text{BC:} \quad t = \bar{f}(\bar{x}, \bar{y}, \bar{z}, \bar{\tau}) \text{ on boundary} \qquad (14c)$$

where α = thermal diffusivity ($k/\rho c_p$), ft²/h.

These two systems of equations can be put in identical form by selecting the following dimensionless parameters:

$$\theta \equiv \frac{t - t_l}{t_h - t_l}, \quad \bar{\theta} \equiv \frac{\rho_B - \rho_{Bl}}{\rho_{Bh} - \rho_{Bl}} \qquad (15a)...$$

$$X \equiv \frac{x}{L}, \quad Y \equiv \frac{y}{L}, \quad Z \equiv \frac{z}{L}, \quad \bar{X} \equiv \frac{\bar{x}}{L}, \quad \bar{Y} \equiv \frac{\bar{y}}{L}, \quad \bar{Z} \equiv \frac{\bar{z}}{L}$$

$$\text{Fo} \equiv \frac{\alpha\tau}{L^2}, \quad \text{Fo}_{\bar{m}} \equiv \frac{D_v\tau}{L^2} \qquad ...(15j)$$

Using these parameters, the heat and mass diffusion equations become identical.

Heat diffusion:

$$\frac{\partial \theta}{\partial \text{Fo}} = \frac{\partial^2 \theta}{\partial \bar{x}^2} + \frac{\partial^2 \theta}{\partial \bar{y}^2} + \frac{\partial^2 \theta}{\partial \bar{z}^2} \qquad (16a)$$

$$\text{IC:} \quad \theta = \theta_i(\bar{X}, \bar{Y}, \bar{Z}) \text{ at Fo} = 0 \qquad (16b)$$

$$\text{BC:} \quad \theta = F(\bar{X}, \bar{Y}, \bar{Z}, \text{Fo}) \text{ on boundary} \qquad (16c)$$

Mass diffusion:

$$\frac{\partial \bar{\theta}}{\partial \text{Fo}_{\bar{m}}} = \frac{\partial^2 \bar{\theta}}{\partial x^2} + \frac{\partial^2 \bar{\theta}}{\partial y^2} + \frac{\partial^2 \bar{\theta}}{\partial z^2} \qquad (17a)$$

$$\text{IC:} \quad \theta = \theta_i(X, Y, Z) \text{ at Fo}_{\bar{m}} = 0 \qquad (17b)$$

$$\text{BC:} \quad \bar{\theta} = \bar{F}(X, Y, Z, \text{Fo}_m) \text{ on boundary} \qquad (17c)$$

Therefore, if $\theta_i = \bar{\theta}_i$ and $F = \bar{F}$, then $\theta = \bar{\theta}$. That is, if the two geometries are of the same shape, the initial potential distributions θ_i and $\bar{\theta}_i$ are the same, and the boundary conditions F and $\bar{F}$ are the same, then the solutions θ and $\bar{\theta}$ are the same. Therefore, all of the heat transfer solutions for steady-state and transient conduction are available for solving analogous steady-state and transient mass transfer problems when $\rho_B << \rho$, $\Delta\rho < \rho_B$, and $v_B = 0$. For further details on the analogy between heat conduction and mass transfer, refer to Eckert and Drake (1972).

Diffusion of One Gas Through a Second Stagnant Gas

Figure 1 shows diffusion of one gas through a second stagnant gas. Water vapor diffuses from the liquid surface into surrounding stationary air. It is assumed that local equilibrium exists through the gas mixture, that the gases are ideal, and that the Gibbs-Dalton law is valid, which implies that the temperature gradient has a negligible effect. Diffusion of water vapor is due to concentration gradient and is given by Equation (7a). There is a continuous gas phase, so the mixture pressure p is constant, and the Gibbs-Dalton law yields

$$p_A + p_B = p = \text{constant} \qquad (18a)$$

or

$$\frac{\rho_A}{M_A} + \frac{\rho_B}{M_B} = \frac{p}{R_U T} = \text{constant} \qquad (18b)$$

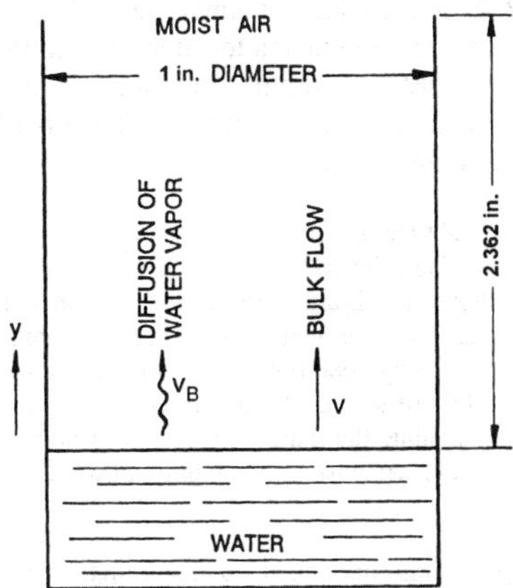

Fig. 1 Diffusion of Water Vapor Through Stagnant Air

The partial pressure gradient of the water vapor causes a partial pressure gradient of the air such that

$$\frac{dp_A}{dy} = -\frac{dp_B}{dy}$$

or

$$\left(\frac{1}{M_A}\right)\frac{d\rho_A}{dy} = -\left(\frac{1}{M_B}\right)\frac{d\rho_B}{dy} \qquad (19)$$

Air, then, diffuses toward the liquid water interface. Because it cannot be absorbed there, a bulk velocity v of the gas mixture is established in a direction away from the liquid surface, so that the net transport of air is zero (i.e., the air is stagnant).

$$\dot{m}_A'' = -D_v\frac{d\rho_A}{dy} + \rho_A v = 0 \qquad (20)$$

The bulk velocity v transports not only air but also water vapor away from the interface. Therefore, the total rate of water vapor diffusion is

$$\dot{m}_B'' = -D_v\frac{d\rho_B}{dy} + \rho_B v \qquad (21)$$

Substituting for the velocity v from Equation (20) and using Equations (18b) and (19) gives

$$\dot{m}_B'' = \left(\frac{D_v M_B p}{\rho_A R_U T}\right)\frac{d\rho_A}{dy} \qquad (22)$$

Integration yields

$$\dot{m}_B'' = \frac{D_v M_B p}{R_U T}\left[\frac{\ln(\rho_{AL}/\rho_{A0})}{y_L - y_0}\right] \qquad (23a)$$

or

$$\dot{m}_B'' = -D_v P_{Am}\left(\frac{\rho_{BL}-\rho_{B0}}{y_L - y_0}\right) \qquad (23b)$$

where

$$P_{Am} \equiv \frac{p}{p_{AL}}\rho_{AL}\left[\frac{\ln(\rho_{AL}/\rho_{A0})}{\rho_{AL}-\rho_{A0}}\right] \qquad (24)$$

P_{Am} is the logarithmic mean density factor of the stagnant air. The pressure distribution for this type of diffusion is illustrated in Figure 2. *Stagnant* refers to the net behavior of the air; it does not move because the bulk flow exactly offsets diffusion. The term P_{Am} in Equation (23b) approximately equals unity for dilute mixtures such as water vapor in air at near atmospheric conditions. This condition makes it possible to simplify Equation (23) and implies that in the case of dilute mixtures, the partial pressure distribution curves in Figure 2 are straight lines.

Example 1. A vertical tube of 1 in. diameter is partially filled with water so that the distance from the water surface to the open end of the tube is 2.362 in., as shown in Figure 1. Perfectly dried air is blown over the open tube end, and the complete system is at a constant temperature of 59°F. In 200 h of steady operation, 0.00474 lb of water evaporates from the tube. The total pressure of the system is 14.696 psia (1 atm). Using these data, (a) calculate the mass diffusivity of water vapor in air, and (b) compare this experimental result with that from Equation (11).

Solution:

(a) The mass diffusion flux of water vapor from the water surface is

$$\dot{m}_B = 0.00474/200 = 0.0000237 \text{ lb/h}$$

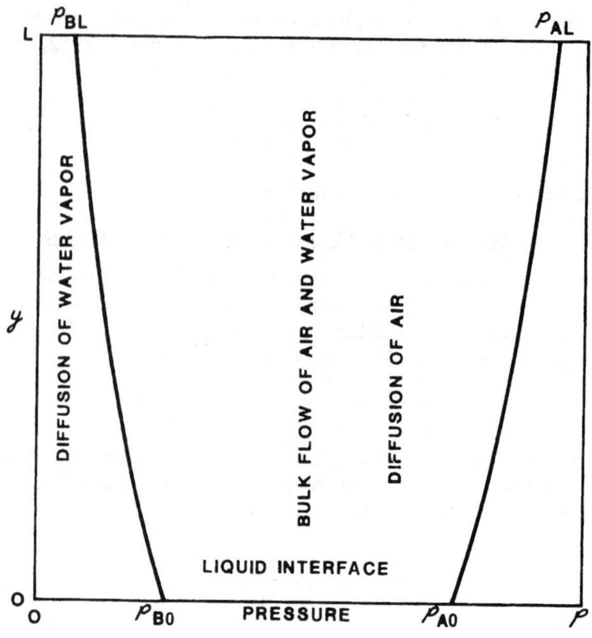

Fig. 2 Pressure Profiles for Diffusion of Water Vapor Through Stagnant Air

The cross-sectional area of a 1 in. diameter tube is $\pi(0.5)^2/144$ = 0.005454 ft². Therefore, $\dot{m}_B''$ = 0.004345 lb/ft²·h. The partial densities are determined with the aid of the psychrometric tables.

$$\rho_{BL} = 0; \quad \rho_{B0} = 0.000801 \text{ lb/ft}^3$$

$$\rho_{AL} = 0.0765 \text{ lb/ft}^3; \quad \rho_{A0} = 0.0752 \text{ lb/ft}^3$$

Since $p = p_{AL}$ = 1 atm, the logarithmic mean density factor [Equation (24)] is

$$P_{Am} = 0.0765\left[\frac{\ln(0.0765/0.0752)}{0.0765-0.0752}\right] = 1.009$$

The mass diffusivity is now computed from Equation (23b) as

$$D_v = \frac{-\dot{m}_B''(y_L - y_0)}{P_{Am}(\rho_{BL}-\rho_{B0})} = \frac{-(0.004345)(2.362)}{(1.009)(0 - 0.000801)(12)}$$
$$= 1.058 \text{ ft}^2/\text{h}$$

(b) By Equation (11), with p = 14.696 psi and T = 59 + 460 = 519°R,

$$D_v = \frac{0.00215}{14.696}\left(\frac{519^{2.5}}{519+441}\right) = 0.935 \text{ ft}^2/\text{h}$$

Neglecting the correction factor P_{Am} for this example gives a difference of less than 1% between the calculated experimental and empirically predicted values of D_v.

Molecular Diffusion in Liquids and Solids

Because of the greater density, diffusion is slower in liquids than in gases. No satisfactory molecular theories have been developed for calculating diffusion coefficients. The limited measured values of D_v show that, unlike for gas mixtures at low pressures, the diffusion coefficient for liquids varies appreciably with concentration.

Reasoning largely from analogy to the case of one-dimensional diffusion in gases and employing Fick's law as expressed by Equation (5),

$$\dot{m}_B''^* = D_v\left(\frac{C_{B1}-C_{B2}}{y_1 - y_2}\right) \qquad (25)$$

where C_B = molal concentration of solute in solvent, lb mol/ft³.

Equation (25) expresses the steady-state diffusion of the solute B through the solvent A in terms of the molal concentration difference of the solute at two locations separated by the distance $\Delta y = y_1 - y_2$. Bird et al. (1960), Hirschfelder et al. (1954), Sherwood and Pigford (1952), Reid and Sherwood (1966), Treybal (1980), and Eckert and Drake (1972) provide equations and tables for evaluating D_v. Hirschfelder et al. (1954) provide comprehensive treatment of the molecular developments.

Diffusion through a solid when the solute is dissolved to form a homogeneous solid solution is known as **structure-insensitive diffusion** (Treybal 1980). This solid diffusion closely parallels diffusion through fluids, and Equation (25) can be applied to one-dimensional steady-state problems. Values of mass diffusivity are generally lower than they are for liquids and vary with temperature.

The flow of a liquid or gas through the interstices and capillaries of a porous or granular solid is a concern. The fundamental mechanism of transport differs for gaseous diffusion, which is considered **structure-sensitive diffusion**. Experimental measurements are important because of the complex geometry of the flow passages. Generally a factor $\bar{\mu}$, called permeability, is defined by the following equation:

$$\dot{m}_B'' = -D_v \frac{d\rho_B}{d(RT)} = \bar{\mu}\left(\frac{\Delta p_B}{\Delta y}\right) \tag{26}$$

For moisture transfer through a porous building material of thickness Δy, the water vapor pressure gradient $\Delta p_B/\Delta y$ is expressed as in. Hg per inch of thickness, and the mass flux $\dot{m}_B$ as grains/h·ft^2, so that the permeability $\bar{\mu}$ has the units of grains·in/h·ft^2·in. Hg. Chapter 22 has further information.

CONVECTION OF MASS

Convection of mass involves the mass transfer mechanisms of molecular diffusion and bulk fluid motion. Fluid motion in the region adjacent to a mass transfer surface may be laminar or turbulent, depending on geometry and flow conditions.

Mass Transfer Coefficient

Convective mass transfer is analogous to convective heat transfer where geometry and boundary conditions are similar. The analogy holds for both laminar and turbulent flows and applies to both external and internal flow problems.

Mass Transfer Coefficients for External Flows. Most external convective mass transfer problems can be solved with an appropriate formulation that relates the mass transfer flux (to or from an interfacial surface) to the concentration difference across the boundary layer illustrated in Figure 3. This formulation gives rise to the convective mass transfer coefficient, defined as

$$h_M \equiv \frac{\dot{m}_B''}{\rho_{Bi} - \rho_{B\infty}} \tag{27}$$

where

h_M = local external mass transfer coefficient, ft/h
$\dot{m}_B''$ = mass flux of gas B from surface, lb$_m$/ft^2·h
ρ_{Bi} = density of gas B at interface (saturation density), lb$_m$/ft^3
$\rho_{B\infty}$ = density of component B outside boundary layer, lb$_m$/ft^3

If ρ_{Bi} and $\rho_{B\infty}$ are constant over the entire interfacial surface, the mass transfer rate from the surface can be expressed as

$$\dot{m}_B'' = \bar{h}_M(\rho_{Bi} - \rho_{B\infty}) \tag{28}$$

where $\bar{h}_M$ is the average mass transfer coefficient, defined as

$$\bar{h}_M \equiv \frac{1}{A}\int_A h_m \, dA \tag{29a}$$

Mass Transfer Coefficients for Internal Flows. Most internal convective mass transfer problems, such as those that occur in channels or in the cores of dehumidification coils, can be solved if an appropriate expression is available to relate the mass transfer flux (to or from the interfacial surface) to the difference between the concentration at the surface and the bulk concentration in the channel, as illustrated in Figure 4. This formulation leads to the definition of the mass transfer coefficient for internal flows:

$$h_M \equiv \frac{\dot{m}_B''}{\rho_{Bi} - \rho_{Bb}} \tag{29b}$$

where

h_M = internal mass transfer coefficient, ft/h
$\dot{m}_B''$ = mass flux of gas B at interfacial surface, lb$_m$/ft^2·h
ρ_{Bi} = density of gas B at interfacial surface, lb$_m$/ft^3
$\rho_{Bb} \equiv (1/\bar{u}_B A_{cs})\int_{A_{cs}} u_B \rho_B \, dA_{cs}$ = bulk density of gas B at location x
$\bar{u}_B \equiv (1/A_{cs})\int_A u_B \, dA_{cs}$ = average velocity of gas B at location x, fpm
A_{cs} = cross-sectional area of channel at station x, ft^2
u_B = velocity of component B in x direction, fpm
ρ_B = density distribution of component B at station x, lb$_m$/ft^3

Often, it is easier to obtain the bulk density of gas B from

$$\rho_{Bb} = \frac{\dot{m}_{Bo} + \int_A \dot{m}_B'' \, dA}{\bar{u}_B A_{cs}} \tag{30}$$

where

$\dot{m}_{Bo}$ = mass flow rate of component B at station $x = 0$, lb$_m$/h
A = interfacial area of channel between station $x = 0$ and station $x = x$, ft^2

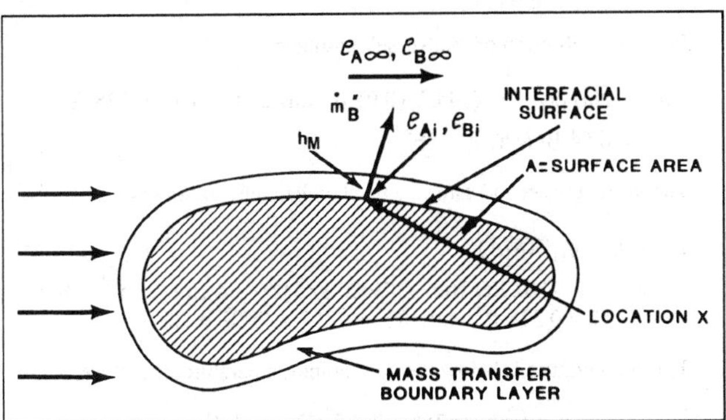

Fig. 3 Nomenclature for Convective Mass Transfer from External Surface at Location *x* Where Surface Is Impermeable to Gas A

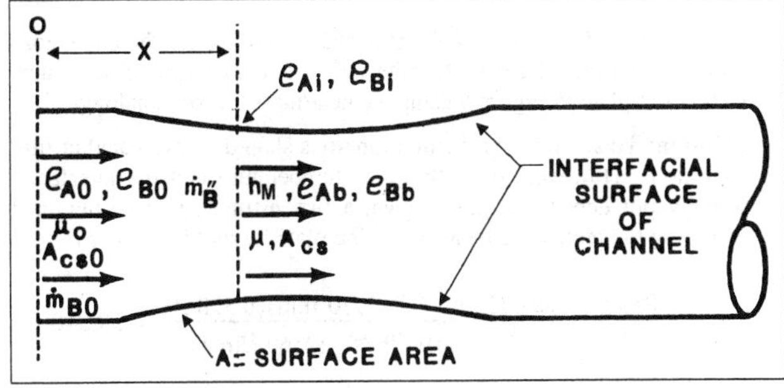

Fig. 4 Nomenclature for Convective Mass Transfer from Internal Surface Impermeable to Gas A

Equation (30) can be derived from the preceding definitions. The major problem is the determination of $\bar{u}_B$. If, however, the analysis is restricted to cases where B is dilute and concentration gradients of B in the x direction are negligibly small, $\bar{u}_B \equiv \bar{u}$. Component B is swept along in the x direction with an average velocity equal to the average velocity of the dilute mixture.

Analogy Between Convective Heat and Mass Transfer

Most expressions for the convective mass transfer coefficient h_M are determined from expressions for the convective heat transfer coefficient h.

For problems in internal and external flow where mass transfer occurs at the convective surface and where component B is dilute, it is shown by Bird et al. (1960) and Incropera and DeWitt (1996) that the Nusselt and Sherwood numbers are defined as follows:

$$\mathrm{Nu} = f(X, Y, Z, \mathrm{Pr}, \mathrm{Re}) \quad (31)$$

$$\overline{\mathrm{Nu}} = g(\mathrm{Pr}, \mathrm{Re}) \quad (32)$$

and

$$\mathrm{Sh} = f(X, Y, Z, \mathrm{Sc}, \mathrm{Re}) \quad (33)$$

$$\overline{\mathrm{Sh}} = g(\mathrm{Sc}, \mathrm{Re}) \quad (34)$$

where the function f is the same in Equations (31) and (33), and the function g is the same in Equations (32) and (34). The quantities Pr and Sc are dimensionless Prandtl and Schmidt numbers, respectively, as defined in the section on Symbols. The primary restrictions on the analogy are that the surface shapes are the same and that the dimensionless temperature boundary conditions are analogous to the dimensionless density distribution boundary conditions for component B, as indicated in Equations (16) and (17). Several primary factors prevent the analogy from being perfect. In some cases, the Nusselt number was derived for smooth surfaces. Many mass transfer problems involve wavy, droplet-like, or roughened surfaces. Many Nusselt number relations are obtained for constant temperature surfaces. Sometimes ρ_{Bi} is not constant over the entire surface because of varying saturation conditions and the possibility of surface dryout.

In all mass transfer problems, there is some blowing or suction at the surface because of the condensation, evaporation, or transpiration of component B. In most cases, this blowing/suction phenomenon has little effect on the Sherwood number, but the analogy should be examined closely if $v_i/u_\infty > 0.01$ or $v_i/\bar{u} > 0.01$, especially if the Reynolds number is large.

Example 2. Use the analogy expressed in Equations (32) and (34) to solve the following problem. An expression for heat transfer from a constant temperature flat plate in laminar flow is

$$\overline{\mathrm{Nu}}_L = 0.664\mathrm{Pr}^{1/3}\mathrm{Re}_L^{1/2} \quad (35)$$

$\mathrm{Sc} = 0.35$, $D_v = 3.87 \times 10^{-4}$ ft²/s, and $\mathrm{Pr} = 0.708$ for the given conditions; determine the mass transfer rate and temperature of the water-wetted flat plate in Figure 5 using the heat/mass transfer analogy.

Solution: To solve the problem, properties should be evaluated at film conditions. However, since the plate temperature and the interfacial water vapor density are not known, a first estimate will be obtained assuming the plate t_{i1} to be at 77°F. The plate Reynolds number is

$$\mathrm{Re}_{L1} = \frac{\rho u_\infty L}{\mu} = \frac{(0.0728\ \mathrm{lb/ft^3})(1970\ \mathrm{fpm})(0.328\ \mathrm{ft})}{(1.32 \times 10^{-5}\mathrm{lb/ft \cdot s})(60\ \mathrm{s/min})} = 59{,}340$$

The plate is entirely in laminar flow, since the transitional Reynolds number is about 5×10^5. Using the mass transfer analogy, Equation (35) yields

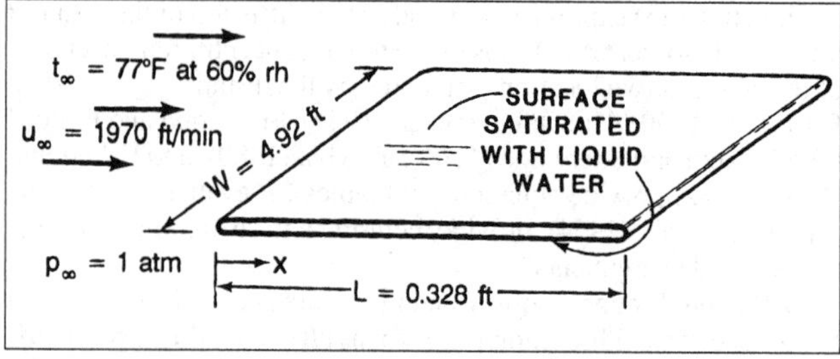

Fig. 5 Water-Saturated Flat Plate in Flowing Airstream

$$\overline{\mathrm{Sh}}_{L1} = 0.664\ \mathrm{Sc}^{1/3}\mathrm{Re}_L^{1/2}$$
$$= 0.664(0.35)^{1/3}(59{,}340)^{1/2} = 114$$

From the definition of the Sherwood number,

$$\bar{h}_{M1} = \overline{\mathrm{Sh}}_{L1}D_v/L = (114)(3.87 \times 10^{-4}\ \mathrm{ft^2/s})/0.328\ \mathrm{ft} = 0.135\ \mathrm{fps}$$

The psychrometric tables give a humidity ratio W of 0.0121 at 77°F and 60% rh. Therefore [see Equation (51)],

$$\rho_{B\infty} = 0.0121\rho_{A\infty} = (0.0121)(0.0728\ \mathrm{lb/ft^3}) = 0.000881\ \mathrm{lb/ft^3}$$

Psychrometric tables give the saturation density for water at 77°F as

$$\rho_{Bi1} = 0.02017\rho_{A\infty} = (0.02017)(0.0728\ \mathrm{lb/ft^3}) = 0.001468\ \mathrm{lb/ft^3}$$

Therefore, the mass transfer rate from the double-sided plate is

$$\dot{m}_{B1} = \bar{h}_{M1}A(\rho_{Bi} - \rho_{B\infty})$$
$$= (0.135\ \mathrm{fps})(0.328\ \mathrm{ft} \times 4.92\ \mathrm{ft} \times 2)(0.001468 - 0.000881\ \mathrm{lb/ft^3})$$
$$= 0.000256\ \mathrm{lb/s}$$

This mass rate, transformed from the liquid state to the vapor state, requires the following heat rate to the plate to maintain the evaporation:

$$q_{i1} = \dot{m}_{B1}h_{fg} = (0.000256\ \mathrm{lb/s})(1050\ \mathrm{Btu/lb}) = 0.269\ \mathrm{Btu/s}$$

To obtain a second estimate of the wetted plate temperature in this type of problem, the following criteria are used. Calculate the t_i necessary to provide a heat rate of q_{i1}. If this temperature t_{iq1} is above the dew-point temperature t_{id}, set the second estimate at $t_{i2} = (t_{iq1} + t_{i1})/2$. If t_{iq1} is below the dew-point temperature, set $t_{i2} = (t_{id} + t_{i1})/2$. For this problem, the dew point is $t_{id} = 57$°F.

Obtaining the second estimate of the plate temperature requires an approximate value of the heat transfer coefficient.

$$\overline{\mathrm{Nu}}_{L1} = 0.664\mathrm{Pr}^{1/3}\mathrm{Re}_L^{1/2} = 0.664(0.708)^{1/3}(59{,}340)^{1/2}$$
$$= 144.2$$

From the definition of the Nusselt number:

$$\bar{h}_1 = \overline{\mathrm{Nu}}_{L1}k/L = (144.2)(0.0151\ \mathrm{Btu/h \cdot ft \cdot °F})/(0.328\ \mathrm{ft})$$
$$= 6.64\ \mathrm{Btu/h \cdot ft^2 \cdot °F}$$

Therefore, the second estimate for the plate temperature is

$$t_{iq1} = t_\infty - q_{i1}/(\bar{h}_1 A)$$
$$= 77°\mathrm{F} - [(0.269\ \mathrm{Btu/s})(3600\ \mathrm{s/h})/(6.64\ \mathrm{Btu/h \cdot ft^2 \cdot °F})]$$
$$(2 \times 0.328\ \mathrm{ft} \times 4.92\ \mathrm{ft}) = 77°\mathrm{F} - 45°\mathrm{F} = 32°\mathrm{F}$$

This temperature is below the dew-point temperature; therefore,

$$t_{i2} = (57°\mathrm{F} + 77°\mathrm{F})/2 = 67°\mathrm{F}$$

The second estimate of the film temperature is

$$t_{f2} = (t_{i2} + t_\infty)/2 = (67°\mathrm{F} + 77°\mathrm{F})/2 = 72°\mathrm{F}$$

The next iteration on the solution is as follows:

$$\mathrm{Re}_{L2} = 61{,}010$$
$$\overline{\mathrm{Sh}}_{L2} = 0.664(0.393)^{1/3}(61{,}010)^{1/2} = 120$$
$$\bar{h}_{M2} = (120)(3.63\times10^{-4})/0.328 = 0.133 \text{ fps}$$

The free stream density of the water vapor has been evaluated. The density of the water vapor at the plate surface is the saturation density at 67°F.

$$\rho_{Bi2} = (0.01374)(0.0739 \text{ lb/ft}^3) = 0.00101 \text{ lb/ft}^3$$
$$A = 2\times4.92\times0.328 = 3.228 \text{ ft}^2$$
$$\dot{m}_{B2} = (0.133 \text{ fps})(3.228 \text{ ft}^2)(0.00101 \text{ lb/ft}^3 - 0.000881 \text{ lb/ft}^3) = 0.0000554 \text{ lb/s}$$
$$q_{i2} = (0.0000554)(1056) = 0.0585 \text{ Btu/s}$$
$$\overline{\mathrm{Nu}}_{L2} = 0.664(0.709)^{1/3}(61{,}010)^{1/2} = 146$$
$$\bar{h}_2 = (146)(0.01493)/0.328 = 6.65 \text{ Btu/h}\cdot\text{ft}^2\cdot{}^\circ\text{F}$$
$$t_{iq2} = 77^\circ\text{F} - (0.0585)(3600)/(6.65\times3.228) = 67.2^\circ\text{F}$$

This temperature is above the dew-point temperature; therefore

$$t_{i3} = (t_{i2} + t_{iq2})/2 = (67 + 67.2)/2 = 67.1^\circ\text{F}$$

This is approximately the same result as that obtained in the previous iteration. Therefore, the problem solution is

$$t_i = 67^\circ\text{F}$$
$$\dot{m}_B = 0.199 \text{ lb/h}$$

Eddy Diffusion

Turbulent flow is characterized by random velocity fluctuations superimposed on the time-averaged velocity. The fluctuations occur in the direction of flow and normal to it. Small mixing actions or **eddy currents** are established within the turbulent flow field. The eddies cause an exchange of momentum between different layers of the moving fluid. If the fluid has a temperature gradient, energy is exchanged by this mixing action in much the same way. If a mass concentration gradient exists, a similar mass exchange known as *eddy diffusion* occurs, which is relatively faster than molecular diffusion and depends on the intensity of the velocity fluctuations, or turbulence. Because intensity of turbulence is determined by the Reynolds number of the flow, the rate of eddy diffusion depends on the Reynolds number (Sherwood and Pigford 1952). The equation for **eddy diffusivity** ε_D has the same form as that for mass diffusivity in molecular diffusion [Equation (5)]:

$$\dot{m}_B'' = \varepsilon_D \frac{d\rho_B}{dy} \tag{36}$$

where ε_D = eddy diffusivity, ft²/h.

Because data on eddy diffusivities are difficult to obtain, the mass transfer coefficient, analogous to the heat transfer coefficient in convective heat transfer, is usually defined and determined experimentally.

Application to Turbulent Flow

Consider an airstream in steady turbulent flow over a wetted surface (Figure 6). The liquid-vapor interface is assumed to be at zero velocity, which results in a slow-moving layer of fluid in laminar flow next to the surface. Between this *laminar sublayer* and the main body of the turbulent stream, a transition region or buffer layer exists in which the fluid may be alternately in laminar flow and in turbulent flow. Within the laminar sublayer, only molecular diffusion can occur; in the buffer layer, both molecular and eddy diffusion contribute to mass transfer. In the turbulent region, eddy diffusion predominates and is so rapid that it almost equalizes the concentration gradient.

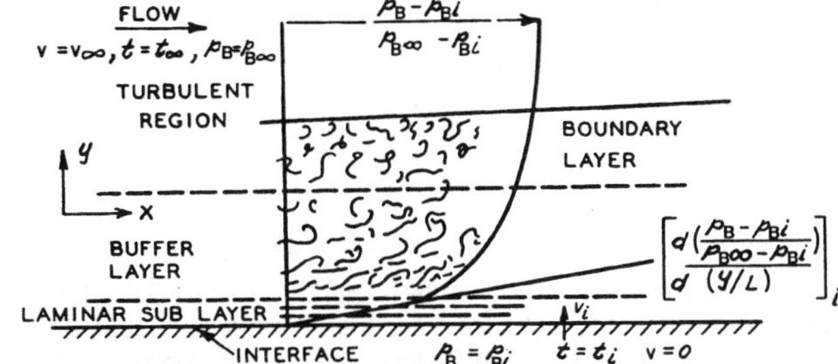

Fig. 6 Turbulent Diffusion Boundary Layer on Flat Surface

Because of the presence of the laminar sublayer, the rate of molecular mass diffusion from the wetted surface to the airstream is [from Equation (6)]

$$J_B = -D_v\left(\frac{d\rho_B}{dy}\right)_i \tag{37}$$

where $(d\rho_B/dy)_i$ is the density gradient at the interface. Assuming that the gases are ideal and obey the Gibbs-Dalton law and that the total pressure is constant, then a partial pressure gradient must also exist in the air. The wetted surface is impermeable to air, so a convective or bulk velocity must be established to counter the air diffusion rate. The total mass transfer from the wetted surface to the airstream must then be given by

$$\dot{m}_B'' = -D_v\left(\frac{d\rho_B}{dy}\right)_i + \rho_{Bi}v_i \tag{38}$$

where v_i is the convective velocity of the component B vapor at the interface (Figure 6), and the partial mass density of the water vapor at the interface is

$$\rho_{Bi} = \frac{M_B p_{Bi}}{R_U T_i}$$

For the diffusion of one gas through a second stagnant gas, the simple molecular diffusion equation corrected by the factor P_{Am} [Equation (24)] accounts for the mass transfer contribution of the convective velocity. For dilute mixtures, as in Example 1, this contribution is small. The same correction factor can be used for this forced convection mass transfer process—at least at low mass transfer rates where v_i is small. The total mass transfer rate from the interface is

$$\dot{m}_B'' = -D_v P_{Am}\left(\frac{d\rho_B}{dy}\right)_i \tag{39}$$

The concentration gradient $(d\rho_B/dy)_i$ must be evaluated experimentally. Instead of this gradient, a mass transfer coefficient h_M may be defined as in Equation (27). It follows that

$$\dot{m}_B'' = \left(\frac{h_M M_B}{R_U T_i}\right)(p_{Bi} - p_{B\infty}) = -\left(\frac{M_B D_v P_{Am}}{R_U T_i}\right)\left(\frac{dp_B}{dy}\right)_i$$

By mathematical rearrangement, this becomes

$$\mathrm{Sh}_L = h_M L/D_v$$
$$= P_{Am}\left[d\left(\frac{p_B - p_{Bi}}{p_{B\infty} - p_{Bi}}\right)/d(y/L)\right]_i \quad (40)$$

where Sh_L is the dimensionless Sherwood number and L is some characteristic dimension of the mass transfer surface, such as the length of a plate or diameter of a cylinder. Equation (40) gives a simple, physical interpretation of the dimensionless mass transfer coefficient (i.e., the Sherwood number). It is a measure of the dimensionless concentration gradient at the interface or mass transfer boundary (Figure 6).

Bird et al. (1960), Hirschfelder et al. (1954), Sherwood and Pigford (1952), Incropera and DeWitt (1996), and Reid et al. (1987) experimentally established mass transfer coefficients for a number of flow geometries. Because of the analogy between these transfer processes, heat transfer data have been used to predict mass transfer coefficients. The reliability of such similar relations has been well established at low mass transfer rates for some flow geometries (Ohadi and Sparrow 1989).

Analogy Relations for Convective Mass Transfer

Heat transfer from the wetted surface of Figure 6 to the airstream can be expressed in a form analogous to that of mass transfer in terms of the heat transfer coefficient h. Therefore, the heat flux per unit area is

$$q'' = -k\left(\frac{dt}{dy}\right)_i = h(t_i - t_\infty) \quad (41a)$$

or, by mathematical rearrangement,

$$\mathrm{Nu} = hL/k = \left[d\left(\frac{t - t_i}{t_\infty - t_i}\right)/d(y/L)\right]_i \quad (41b)$$

The Nusselt number Nu of heat transfer is the dimensionless temperature gradient at the heat transfer boundary (interface), and the Sherwood and Nusselt numbers are analogous expressions for dimensionless mass transfer and heat transfer coefficients, respectively.

For momentum transfer, a friction factor f is defined so that the momentum flux at the interface per unit area is

$$\tau_i g_c = \mu\left(\frac{du}{dy}\right)_i = \frac{f\rho\bar{u}^2}{2} \quad (42a)$$

or

$$\left(\frac{f}{2}\right)\left(\frac{\rho\bar{u}L}{\mu}\right) = \left[\frac{d(u/\bar{u})}{d(y/L)}\right]_i \quad (42b)$$

where

τ_i = sheer stress in the x-y coordinate plane, $\mathrm{lb_f/ft^2}$
g_c = gravitational constant, $\mathrm{ft \cdot lb_m/h^2 \cdot lb_f}$
$\bar{u}$ = mean free stream velocity of gas, ft/h

Equation (42b) states that one-half the friction factor times the Reynolds number ($\mathrm{Re} = \rho\bar{u}L/\mu$) equals the dimensionless velocity gradient at the stationary surface or interface.

Equations (40), (41b), and (42b) show the similarity of mass, heat, and momentum transport in a turbulent boundary layer (with both eddy and molecular diffusion), and Equations (11), (6), and (7) demonstrate the similarity for a laminar boundary layer (with molecular diffusion only). Again, assume a gaseous system in which, by chance, $D_v = \alpha = \nu$ (i.e., the molecular diffusion constants for mass, energy, and momentum are identical). Theoretically, the eddy diffusivities for these three rate processes are identical, since experiments verify that the processes rely on the same particle mixing action for the transfer operation. Therefore, the dimensionless partial pressure, temperature, and velocity profiles must be identical. This permits equating the left side of Equations (40), (41b), and (42b) to obtain

$$\mathrm{Sh}_L = \mathrm{Nu}_L = \frac{f}{2}(\mathrm{Re}_L) \quad (43a)$$

or

$$\frac{h_M L}{D_v P_{Am}} = \frac{hL}{k} = \frac{f}{2}\left(\frac{\rho\bar{u}L}{\mu}\right) \quad (43b)$$

This is equivalent to the statement that the Prandtl number ($\mathrm{Pr} = c_p\mu/k$) and Schmidt number ($\mathrm{Sc} = \mu/\rho D_v$) are unity. Using this statement, Equation (43b) can be rearranged to give

$$\frac{\mathrm{Sh}}{\mathrm{Re\,Sc}} = \frac{\mathrm{Nu}}{\mathrm{Re\,Pr}} = \frac{f}{2} \quad (44a)$$

or

$$\frac{h_M}{\bar{u}P_{Am}} = \frac{h}{\rho c_p \bar{u}} = \frac{f}{2} \quad (44b)$$

The two right-hand terms of Equation (44b) are known as the **Reynolds analogy**. The left-hand term is an extension of the Reynolds analogy to include mass transfer. The Reynolds analogy gives a plausible correlation between friction factors and heat transfer coefficients for common gases (where $\mathrm{Pr} \cong 1$) with moderate temperature potential. Since Reynolds developed this analogy, at least five modified analogies have been suggested to account for the effect of the Prandtl number. One of the simplest and most often used is the one suggested by Chilton and Colburn (1934), who showed that the correlation of heat transfer data with friction data could be improved by substituting $(\mathrm{Pr})^{1/3}$ for Pr in Equation (44a), which leads to

$$j_H = \frac{h}{\rho c_p \bar{u}}\left(\frac{c_p\mu}{k}\right)^{2/3}$$
$$= \mathrm{St\,Pr}^{2/3} = \frac{f}{2} \quad (45)$$

By analogy, Chilton and Colburn suggested that mass transfer data could be represented with good accuracy by a similar change in the Schmidt number parameter:

$$j_D = \frac{h_M P_{Am}}{\bar{u}}\left(\frac{\mu}{\rho D_v}\right)^{2/3}$$
$$= \mathrm{St}_m \mathrm{Sc}^{2/3} = \frac{f}{2} \quad (46)$$

where St and St_m refer to the dimensionless Stanton numbers for heat and mass transfer, respectively (see the section on Symbols for the definition of St).

Equations (45) and (46) are known as the **Chilton-Colburn *j*-factor analogy** and are widely used for plotting and predicting heat and mass transfer data.

The power of the Chilton-Colburn j-factor analogy is represented in Figures 7 through 10. Figure 7 plots various experimental values of j_D from a flat plate with flow parallel to the plate surface. The solid line, which represents the data to near perfection, is actually $f/2$ from Blasius' solution of laminar flow on a flat plate (left-hand portion of the solid line) and Goldstein's solution for a turbulent boundary layer (right-hand portion). The right-hand portion of the solid line also represents McAdams' (1954) correlation of turbulent flow heat transfer coefficient for a flat plate.

A *wetted-wall column* is a vertical tube in which a thin liquid film adheres to the tube surface and exchanges mass by evaporation or absorption with a gas flowing through the tube. Figure 8 illustrates typical data on vaporization in wetted-wall columns, plotted as j_D versus Re. The spread of the points with variation in $\mu/\rho D_v$ results from Gilliland's finding of an exponent of 0.56, not 2/3, representing the effect of the Schmidt number. Gilliland's equation can be written as follows:

$$j_D = 0.023\text{Re}^{-0.17}\left(\frac{\mu}{\rho D_v}\right)^{0.11} \tag{47}$$

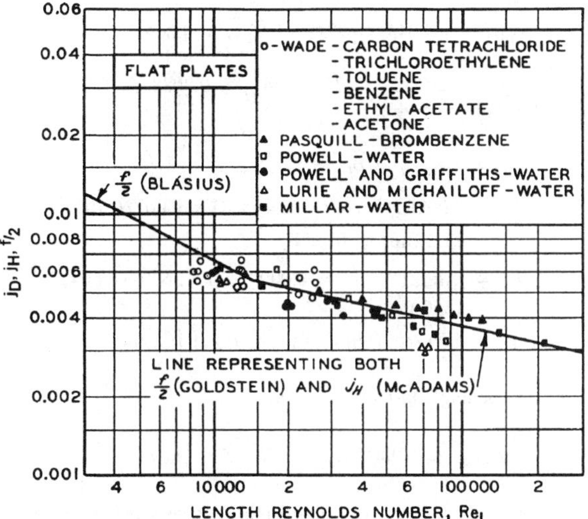

Fig. 7 Mass Transfer from Flat Plate

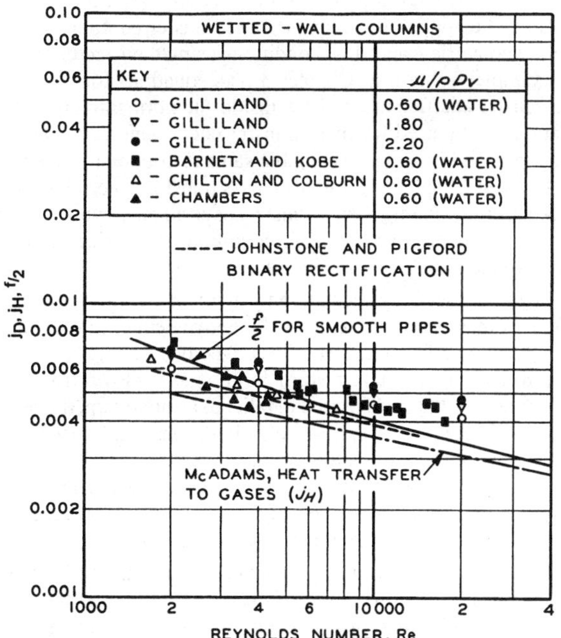

Fig. 8 Vaporization and Absorption in Wetted-Wall Column

Similarly, McAdams' (1954) equation for heat transfer in pipes can be expressed as

$$j_H = 0.023\text{Re}^{-0.20}\left(\frac{c_p\mu}{k}\right)^{0.07} \tag{48}$$

This is represented by the dash-dot curve in Figure 8, which falls below the mass transfer data. The curve $f/2$ representing friction in smooth tubes is the upper, solid curve.

Data for the evaporation of liquids from single cylinders into gas streams flowing transversely to the cylinders' axes are shown in Figure 9. Although the dash-dot line on Figure 9 represents the data, it is actually taken from McAdams (1954) as representative of a large collection of data on heat transfer to single cylinders placed transverse to airstreams. To compare these data with friction, it is necessary to distinguish between total drag and skin friction. Since the analogies are based on skin friction, the normal pressure drag must be subtracted from the measured total drag. At Re = 1000, the skin friction is 12.6% of the total drag; at Re = 31,600, it is only 1.9%. Consequently, the values of $f/2$ at a high Reynolds number, obtained by the difference, are subject to considerable error.

In Figure 10, data on the evaporation of water into air for single spheres are presented. The solid line, which best represents these data, agrees with the dashed line representing McAdams' correlation for heat transfer to spheres. These results cannot be compared

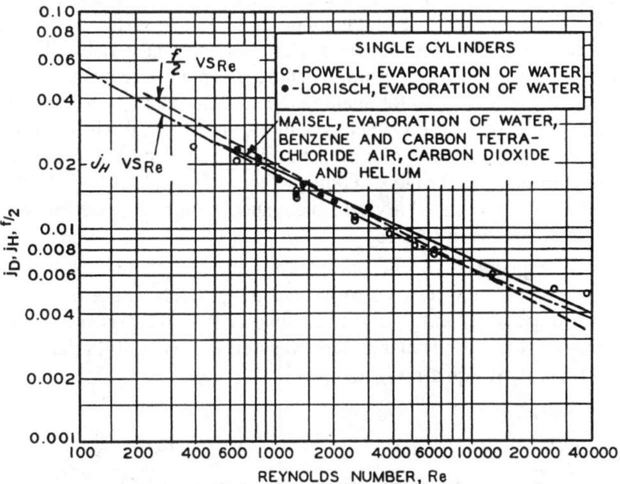

Fig. 9 Mass Transfer from Single Cylinders in Crossflow

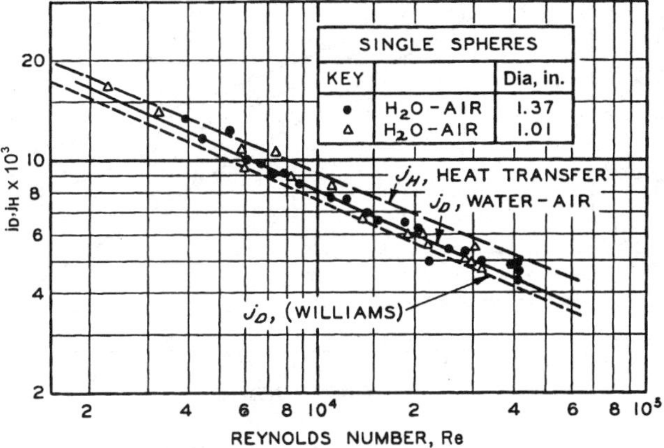

Fig. 10 Mass Transfer from Single Spheres

with friction or momentum transfer because total drag has not been allocated to skin friction and normal pressure drag. Application of these data to air-water contacting devices such as air washers and spray cooling towers is well substantiated.

When the temperature of the heat exchanger surface in contact with moist air is below the dew-point temperature of the air, vapor condensation occurs. Typically, the air dry-bulb temperature and humidity ratio both decrease as the air flows through the exchanger. Therefore, sensible and latent heat transfer occur simultaneously. This process is similar to one that occurs in a spray dehumidifier and can be analyzed using the same procedure; however, this is not generally done.

Cooling coil analysis and design are complicated by the problem of determining transport coefficients h, h_M, and f. It would be convenient if heat transfer and friction data for dry heating coils could be used with the Colburn analogy to obtain the mass transfer coefficients. However, this approach is not always reliable, and work by Guillory and McQuiston (1973) and Helmer (1974) shows that the analogy is not consistently true. Figure 11 shows j-factors for a simple parallel plate exchanger for different surface conditions with sensible heat transfer. Mass transfer j-factors and the friction factors exhibit the same behavior. Dry surface j-factors fall below those obtained under dehumidifying conditions with the surface wet. At low Reynolds numbers, the boundary layer grows quickly; the droplets are soon covered and have little effect on the flow field. As the Reynolds number is increased, the boundary layer becomes thin and more of the total flow field is exposed to the droplets. The roughness caused by the droplets induces mixing and larger j-factors. The data in Figure 11 cannot be applied to all surfaces because the length of the flow channel is also an important variable. However, the water collecting on the surface is mainly responsible for breakdown of the j-factor analogy. The j-factor analogy is approximately true when the surface conditions are identical. Under some conditions, it is possible to obtain a film of condensate on the surface instead of droplets. Guillory and McQuiston (1973) and Helmer (1974) related dry sensible j- and f-factors to those for wetted dehumidifying surfaces.

The equality of j_H, j_D, and $f/2$ for certain streamline shapes at low mass transfer rates has experimental verification. For flow past bluff objects, j_H and j_D are much smaller than $f/2$, based on total pressure drag. The heat and mass transfer, however, still relate in a useful way by equating j_H and j_D.

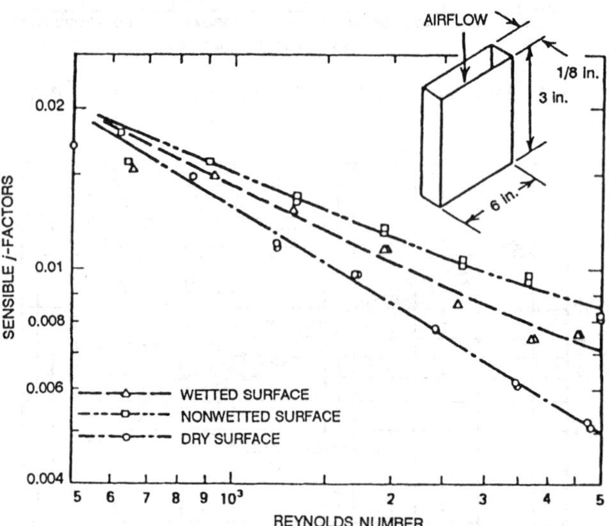

Fig. 11 Sensible Heat Transfer j-Factors for Parallel Plate Exchanger

Example 3. Using solid cylinders of volatile solids (e.g., naphthalene, camphor, dichlorobenzene) with airflow normal to these cylinders, Bedingfield and Drew (1950) found that the ratio between the heat and mass transfer coefficients could be closely correlated by the following relation:

$$\frac{h}{\rho h_M} = (0.294\ \text{Btu/lb}_m \cdot {}^\circ\text{F})\left(\frac{\mu}{\rho D_v}\right)^{0.56}$$

For completely dry air at 70°F flowing at a velocity of 31 fps over a wet-bulb thermometer of diameter d = 0.300 in., determine the heat and mass transfer coefficients from Figure 9 and compare their ratio with the Bedingfield-Drew relation.

Solution: For dry air at 70°F and standard pressure, $\rho = 0.075\ \text{lb}_m/\text{ft}^3$, $\mu = 0.044\ \text{lb}_m/\text{h}\cdot\text{ft}$, $k = 0.0149\ \text{Btu/h}\cdot\text{ft}\cdot{}^\circ\text{F}$, and $c_p = 0.240\ \text{Btu/lb}_m\cdot{}^\circ\text{F}$. From Equation (11), $D_v = 0.973\ \text{ft}^2/\text{h}$. Therefore,

$$\text{Re}_{da} = \rho u_\infty d/\mu = 0.0749 \times 31 \times 3600 \times 0.300/(12 \times 0.044) = 4750$$

$$\text{Pr} = c_p\mu/k = 0.240 \times 0.044/0.0149 = 0.709$$

$$\text{Sc} = \mu/\rho D_v = 0.044/(0.0749 \times 0.973) = 0.604$$

From Figure 9 at Re_{da} = 4750, read j_H = 0.0088, j_D = 0.0099. From Equations (45) and (46),

$$h = j_H\rho c_p u_\infty/(\text{Pr})^{2/3}$$
$$= 0.0088 \times 0.0749 \times 0.240 \times 31 \times 3600/(0.709)^{2/3}$$
$$= 22.2\ \text{Btu/h}\cdot\text{ft}^2\cdot{}^\circ\text{F}$$

$$h_M = j_D u_\infty/(\text{Sc})^{2/3} = 0.0099 \times 31 \times 3600/(0.604)^{2/3}$$
$$= 1550\ \text{ft/h}$$

$$h/\rho h_M = 22.2/(0.0749 \times 1550) = 0.191\ \text{Btu/lb}_m\cdot{}^\circ\text{F}$$

From the Bedingfield-Drew relation,

$$h/\rho h_M = 0.294(0.604)^{0.56} = 0.222\ \text{Btu/lb}_m\cdot{}^\circ\text{F}$$

The Reynolds analogy, Equation (44b), suggests that $h/\rho h_M = c_p$ = 0.240 $\text{Btu/lb}_m\cdot{}^\circ\text{F}$. This close agreement is because the ratio Sc/Pr is 0.604/0.709 or 0.85, so that the exponent of these numbers has little effect on the ratio of the transfer coefficients.

The extensive developments for calculating heat transfer coefficients can be applied to calculate mass transfer coefficients under similar geometrical and flow conditions using the j-factor analogy. For example, Table 6 of Chapter 3 lists equations for calculating heat transfer coefficients for flow inside and normal to pipes. Each equation can be used for mass transfer coefficient calculations by equating j_H and j_D and imposing the same restriction to each stated in Table 6 of Chapter 3. Similarly, mass transfer experiments often replace corresponding heat transfer experiments with complex geometries where exact boundary conditions are difficult to model (Sparrow and Ohadi 1987a, 1987b).

The j-factor analogy is useful only at low mass transfer rates. As the rate of mass transfer increases, the movement of matter normal to the transfer surface increases the convective velocity [v_i in Equation (38) and Figure 6]. For example, if a gas is blown from many small holes in a flat plate placed parallel to an airstream, the boundary layer thickens, and resistance to both mass and heat transfer increases with increasing blowing rate. Heat transfer data are usually collected at zero or, at least, insignificant mass transfer rates. Therefore, if such data are to be valid for a mass transfer process, the mass transfer rate (i.e., the blowing) must be low.

The j-factor relationship $j_H = j_D$ can still be valid at high mass transfer rates, but neither j_H nor j_D can be represented by data at zero mass transfer conditions. Eckert and Drake (1972) and Chapter 24 of Bird et al. (1960) have detailed information on high mass transfer rates.

Lewis Relation

Heat and mass transfer coefficients are satisfactorily related at the same Reynolds number by equating the Chilton-Colburn j-factors. From Equations (45) and (46), this leads to

$$\frac{h}{\rho c_p \bar{u}}\left(\frac{c_p \mu}{k}\right)^{2/3} = \frac{h_M P_{Am}}{\bar{u}}\left(\frac{\mu}{\rho D_v}\right)^{2/3}$$

or

$$\frac{h}{h_M \rho c_p} = P_{Am}\left[\frac{(\mu/\rho D_v)}{(c_p \mu/k)}\right]^{2/3}$$

$$= P_{Am}(\alpha/D_v)^{2/3} \quad (49)$$

The quantity α/D_v is the Lewis number Le. Its magnitude expresses relative rates of propagation of energy and mass within a system. It is fairly insensitive to temperature variation. For air and water vapor mixtures, the ratio is (0.60/0.71) or 0.845, and $(0.845)^{2/3}$ is 0.894. At low diffusion rates, where the heat-mass transfer analogy is valid, P_{Am} is essentially unity. Therefore, for air and water vapor mixtures,

$$\frac{h}{h_M \rho c_p} \approx 1 \quad (50)$$

The ratio of the heat transfer coefficient to the mass transfer coefficient is equal to the specific heat per unit volume of the mixture at constant pressure. This relation [Equation (50)] is usually called the Lewis relation and is nearly true for air and water vapor at low mass transfer rates. It is generally not true for other gas mixtures because the ratio Le of thermal to vapor diffusivity can differ from unity. The agreement between wet-bulb temperature and adiabatic saturation temperature is a direct result of the nearness of the Lewis number to unity for air and water vapor.

The Lewis relation is valid in turbulent flow whether or not α/D_v equals 1 because eddy diffusion in turbulent flow involves the same mixing action for heat exchange as for mass exchange, and this action overwhelms any molecular diffusion. Deviations from the Lewis relation are, therefore, due to a laminar boundary layer or a laminar sublayer and buffer zone, as in Figure 6, where molecular transport phenomena are the controlling factors.

SIMULTANEOUS HEAT AND MASS TRANSFER BETWEEN WATER-WETTED SURFACES AND AIR

A simplified method used to solve simultaneous heat and mass transfer problems was developed using the Lewis relation, and it gives satisfactory results for most air-conditioning processes. Extrapolation to very high mass transfer rates, where the simple heat-mass transfer analogy is not valid, will lead to erroneous results.

Enthalpy Potential

The water vapor concentration in the air is the humidity ratio W, defined as

$$W \equiv \frac{\rho_B}{\rho_A} \quad (51)$$

A mass transfer coefficient is defined using W as the driving potential:

$$\dot{m}_B'' = K_M(W_i - W_\infty) \quad (52)$$

where the coefficient K_M is in lb$_m$/h·ft^2. For dilute mixtures, $\rho_{Ai} \cong \rho_{A\infty}$; that is, the partial mass density of dry air changes by only a small percentage between interface and free stream conditions. Therefore,

$$\dot{m}_B'' = \frac{K_M}{\rho_{Am}}(\rho_{Bi} - \rho_\infty) \quad (53)$$

where ρ_{Am} = *mean* density of dry air, lb$_m$/ft^3. Comparing Equation (53) with Equation (27) shows that

$$h_M = \frac{K_M}{\rho_{Am}} \quad (54)$$

The *humid* specific heat c_{pm} of the airstream is, by definition (Mason and Monchick 1965),

$$c_{pm} = (1 + W_\infty)c_p \quad (55a)$$

or

$$c_{pm} = \left(\frac{\rho}{\rho_{A\infty}}\right)c_p \quad (55b)$$

where c_{pm} is in Btu/lb$_{da}$·°F.

Substituting from Equations (54) and (55b) into Equation (50) gives

$$\frac{h \rho_{Am}}{K_M \rho_{A\infty} c_{pm}} = 1 \approx \frac{h}{K_M c_{pm}} \quad (56)$$

since $\rho_{Am} \cong \rho_{A\infty}$ because of the small change in dry-air density. Using a mass transfer coefficient with the humidity ratio as the driving force, the Lewis relation becomes ratio of heat to mass transfer coefficient equals humid specific heat.

For the plate humidifier illustrated in Figure 5, the total heat transfer from liquid to interface is

$$q'' = q_A'' + \dot{m}_B'' h_{fg} \quad (57)$$

Using the definitions of the transfer coefficients [Equations (41b) and (52)],

$$q'' = h(t_i - t_\infty) + K_M(W_i - W_\infty)h_{fg} \quad (58)$$

Assuming Equation (56) is valid,

$$q'' = K_M\left[c_{pm}(t_i - t_\infty) + (W_i - W_\infty)h_{fg}\right] \quad (59)$$

The enthalpy of the air is approximately

$$h = c_{pa}t + Wh_s \quad (60)$$

The enthalpy h_s of the water vapor can be expressed by the ideal gas law as

$$h_s = c_{ps}(t - t_o) + h_{fgo} \quad (61)$$

where the base of enthalpy is taken as saturated water at temperature t_o. Choosing t_o = 0°F to correspond with the base of the dry-air enthalpy gives

$$h = (c_{pa} + Wc_{ps})t + Wh_{fgo} = c_{pm}t + Wh_{fgo} \quad (62)$$

If small changes in the latent heat of vaporization of water with temperature are neglected when comparing Equations (60) and (62), the total heat transfer can be written as

$$q'' = K_M(h_i - h_\infty) \quad (63)$$

Where the driving potential for heat transfer is temperature difference and the driving potential for mass transfer is mass concentration or partial pressure, the driving potential for simultaneous transfer of heat and mass in an air water-vapor mixture is, to a close approximation, enthalpy.

Basic Equations for Direct-Contact Equipment

Air-conditioning equipment can be classified as (1) having direct contact between air and water used as a cooling or heating fluid or (2) having the heating or cooling fluid separated from the airstream by a solid wall. Examples of the former are air washers and cooling towers; an example of the latter is a direct-expansion refrigerant (or water) cooling and dehumidifying coil. In both cases, the airstream is in contact with a water surface. Direct contact implies contact directly with the cooling (or heating) fluid. In the dehumidifying coil, the contact with the condensate removed from the airstream is direct, but it is indirect with the refrigerant flowing inside the tubes of the coil. These two cases are treated separately because the surface areas of direct-contact equipment cannot be evaluated.

For the direct-contact spray chamber air washer of cross-sectional area A_{cs} and length l (Figure 12), the steady mass flow rate of dry air per unit cross-sectional area is

$$\dot{m}_a / A_{cs} = G_a \quad (64)$$

and the corresponding mass flux of water flowing parallel with the air is

$$\dot{m}_L / A_{cs} = G_L \quad (65)$$

where

$\dot{m}_a$ = mass flow rate of air, lb/h
G_a = mass flux, or flow rate per unit cross-sectional area for air, lb/h·ft²
$\dot{m}_L$ = mass flow rate of liquid, lb/h
G_L = mass flux, or flow rate per unit cross-sectional area, for liquid, lb/h·ft²

Because water is evaporating or condensing, G_L changes by an amount dG_L in a differential length dl of the chamber. Similar changes occur in temperature, humidity ratio, enthalpy, and other properties.

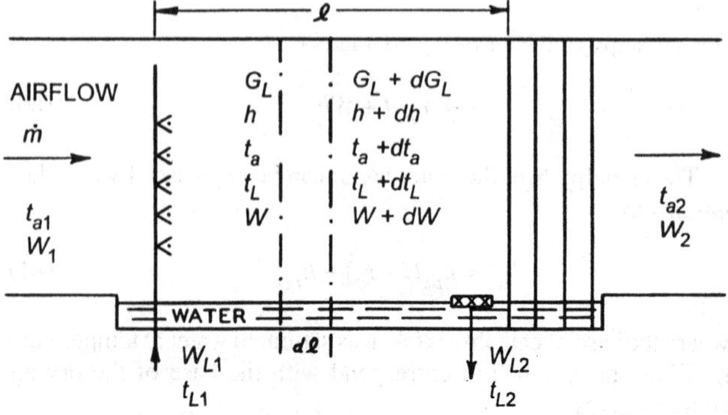

Fig. 12 Air Washer Spray Chamber

Because evaluating the true surface area in direct-contact equipment is difficult, it is common to work on a unit volume basis. If a_H and a_M are the area of heat transfer and mass transfer surface per unit of chamber volume, respectively, the total surface areas for heat and mass transfer are

$$A_H = a_H A_{cs} l \quad \text{and} \quad A_M = a_M A_{cs} l \quad (66)$$

The basic equations for the process occurring in the differential length dl can be written for

1. *Mass transfer*

$$-dG_L = G_a dW = K_M a_M (W_i - W) dl \quad (67)$$

That is, the water evaporated, the moisture increase of the air, and the mass transfer rate are all equal.

2. *Heat transfer to air*

$$G_a c_{pm} dt_a = h_a a_H (t_i - t_a) dl \quad (68)$$

3. *Total energy transfer to air*

$$G_a(c_{pm} dt_a + h_{fgo} dW) = [K_M a_M (W_i - W) h_{fg} + h_a a_H (t_i - t_a)] dl \quad (69)$$

Assuming $a_H = a_M$ and Le = 1, and neglecting small variations in h_{fg}, Equation (69) reduces to

$$G_a dh = K_M a_M (h_i - h) dl \quad (70)$$

The heat and mass transfer areas of spray chambers are assumed to be identical ($a_H = a_M$). Where packing materials, such as wood slats or Raschig rings, are used, the two areas may be considerably different because the packing may not be wet uniformly. The validity of the Lewis relation was discussed previously. It is not necessary to account for the small changes in latent heat h_{fg} after making the two previous assumptions.

4. *Energy balance*

$$G_a dh = \pm G_L c_L dt_L \quad (71)$$

A minus sign refers to parallel flow of air and water; a plus sign refers to counterflow (water flow in the opposite direction from airflow).

The water flow rate changes between inlet and outlet as a result of the mass transfer. For exact energy balance, the term $(c_L t_L dG_L)$ should be added to the right side of Equation (71). The percentage change in G_L is quite small in usual applications of air-conditioning equipment and, therefore, can be ignored.

5. *Heat transfer to water*

$$\pm G_L c_L dt_L = h_L a_H (t_L - t_i) dl \quad (72)$$

Equations (67) to (72) are the basic relations for solution of simultaneous heat and mass transfer processes in direct-contact air-conditioning equipment.

To facilitate the use of these relations in equipment design or performance, three other equations can be extracted from the above set. Combining Equations (70), (71), and (72) gives

$$\frac{h - h_i}{t_L - t_i} = -\frac{h_L a_H}{K_M a_M} = -\frac{h_L}{K_M} \quad (73)$$

Equation (73) relates the enthalpy potential for the total heat transfer through the gas film to the temperature potential for this same transfer through the liquid film. Physical reasoning leads to the conclusion that this ratio is proportional to the ratio of gas film resistance ($1/K_M$) to liquid film resistance ($1/h_L$). Combining Equations (68), (70), and (56) gives

$$\frac{dh}{dt_a} = \frac{h - h_i}{t_a - t_i} \tag{74}$$

Similarly, combining Equations (67), (68), and (56) gives

$$\frac{dW}{dt_a} = \frac{W - W_i}{t_a - t_i} \tag{75}$$

Equation (75) indicates that at any cross section in the spray chamber, the instantaneous slope of the air path dW/dt_a on a psychrometric chart is determined by a straight line connecting the air state with the interface saturation state at that cross section. In Figure 13, state 1 represents the state of the air entering the parallel flow air washer chamber of Figure 12. The washer is operating as a heating and humidifying apparatus so that the interface saturation state of the water at air inlet is the state designated 1_i. Therefore, the initial slope of the air path is along a line directed from state 1 to state 1_i. As the air is heated, the water cools and the interface temperature drops. Corresponding air states and interface saturation states are indicated by the letters *a*, *b*, *c*, and *d* in Figure 13. In each instance, the air path is directed toward the associated interface state. The interface states are derived from Equations (71) and (73). Equation (71) describes how the air enthalpy changes with water temperature; Equation (73) describes how the interface saturation state changes to accommodate this change in air and water conditions. The solution for the interface state on the normal psychrometric chart of Figure 13 can be determined either by trial and error from Equations (71) and (73) or by a complex graphical procedure (Kusuda 1957).

Air Washers

Air washers are direct-contact apparatus used to (1) simultaneously change the temperature and humidity content of air passing through the chamber and (2) remove air contaminants such as dust and odors. Adiabatic spray washers, which have no external heating or chilling source, are used to cool and humidify air. Chilled spray air washers have an external chiller to cool and dehumidify air. Heated spray air washers, whose external heating source provides additional energy for evaporation of water, are used to humidify and possibly heat air.

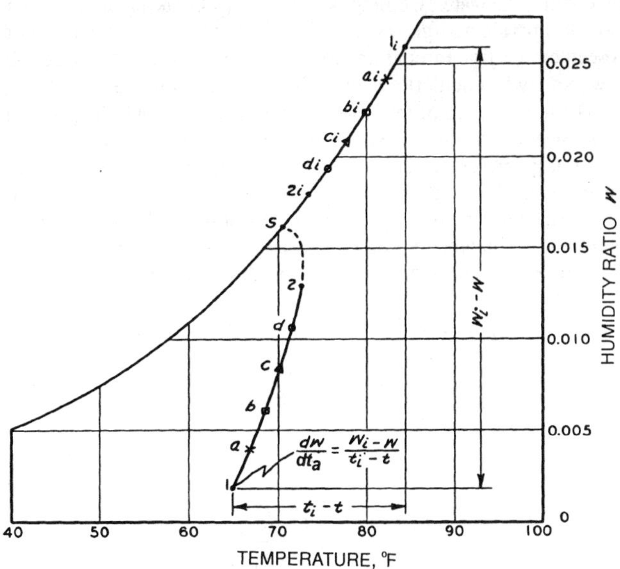

Fig. 13 Air Washer Humidification Process on Psychrometric Chart

Example 4. A parallel flow air washer with the following design conditions is to be designed (see Figure 12).

Water temperature at inlet $t_{L1} = 95°F$
Water temperature at outlet $t_{L2} = 75°F$
Air temperature at inlet $t_{a1} = 65°F$
Air wet-bulb at inlet $t'_{a1} = 45°F$
Air mass flow rate per unit area $G_a = 1200$ lb/h·ft²
Spray ratio $G_L/G_a = 0.70$
Air heat transfer coefficient per cubic foot of chamber volume $h_a a_H = 72$ Btu/h·°F·ft³
Liquid heat transfer coefficient per cubic foot of chamber volume $h_L a_H = 900$ Btu/h·°F·ft³
Air volumetric flow rate $Q = 6500$ cfm

Solution: The air mass flow rate $\dot{m}_a = 6500 \times 0.075 = 490$ lb/min; the required spray chamber cross-sectional area is, then, $A_{cs} = \dot{m}_a/G_a = 490 \times 60/1200 = 24.5$ ft². The mass transfer coefficient is given by the Lewis relation [Equation (56)] as

$$K_M a_M = (h_a a_H)/c_{pm} = 72/0.24 = 300 \text{ lb/h} \cdot \text{ft}^3$$

Figure 14 shows the enthalpy-temperature psychrometric chart with the graphical solution for the interface states and the air path through the washer spray chamber. The solution proceeds as follows:

1. Enter bottom of chart with t'_{a1} of 45°F, and follow up to saturation curve to establish air enthalpy h_1 of 17.65 Btu/lb. Extend this enthalpy line to intersect initial air temperature t_{a1} of 65°F (state 1 of air) and initial water temperature t_{L1} of 95°F at point A. (Note that the temperature scale is used for both air and water temperatures.)
2. Through point A, construct the *energy balance* line A-B with a slope of

$$\frac{dh}{dt_L} = -\frac{c_L G_L}{G_a} = -0.7$$

Point B is determined by intersection with the leaving water temperature $t_{L2} = 75°F$. The negative slope here is a consequence of the parallel flow, which results in the air-water mixture's approaching, but not reaching, the common saturation state *s*. (The line A-B has no physical significance in representing any *air state* on the psychrometric chart. It is merely a construction line in the graphical solution.)

3. Through point A, construct the *tie-line* A-1_i having a slope of

$$\frac{h - h_i}{t_L - t_i} = -\frac{h_L a_H}{K_M a_M} = -\frac{900}{300} = -3$$

The intersection of this line with the saturation curve gives the initial interface state 1_i at the chamber inlet. (Note how the energy balance line and tie-line, representing Equations (71) and (73), combine for a simple graphical solution on Figure 14 for the interface state.)

4. The initial slope of the air path can now be constructed, according to Equation (74), drawing line 1-*a* toward the initial interface state 1_i. (The length of the line 1-*a* will depend on the degree of accuracy required in the solution and the rate at which the slope of the air path is changing.)
5. Construct the horizontal line *a*-M locating the point M on the energy-balance line. Draw a new tie-line (slope of −3 as before) from M to a_i locating interface state a_i. Continue the air path from *a* to *b* by directing it toward the new interface state a_i. (Note that the change in slope of the air path from 1-*a* to *a-b* is quite small, justifying the path incremental lengths used.)

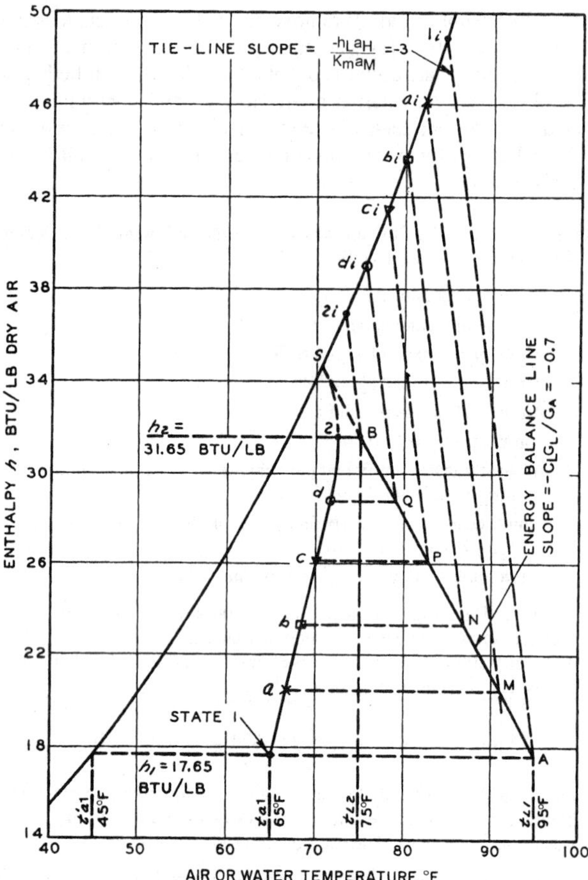

Fig. 14 Graphical Solution for Air-State Path in Parallel Flow Air Washer

6. Continue in the manner of step 5 until point 2, the final state of the air leaving the chamber, is reached. In this example, six steps are used in the graphical construction with the following results:

State	1	*a*	*b*	*c*	*d*	2
t_L	95.0	91.0	87.0	83.0	79.0	75.0
h	17.65	20.45	23.25	26.05	28.85	31.65
t_i	84.5	82.3	80.1	77.8	75.6	73.2
h_i	49.00	46.25	43.80	41.50	39.10	37.00
t_a	65.0	66.8	68.5	70.0	71.4	72.4

The final state of the air leaving the washer is t_{a2} = 72.4°F and h_2 = 31.65 Btu/lb (wet-bulb temperature t'_{a2} = 67°F).

7. The final step involves calculating the required length of the spray chamber. From Equation (70),

$$l = \frac{G_a}{K_M a_M} \int_1^2 \frac{dh}{(h_i - h)}$$

The integral is evaluated graphically by plotting $1/(h_i - h)$ versus h as shown in Figure 15. Any satisfactory graphical method can be used to evaluate the area under the curve. Simpson's rule with four equal increments of Δh equal to 3.5 gives

$$N = \int_1^2 \frac{dh}{(h_i - h)} \approx (\Delta h/3)(y_1 + 4y_2 + 2y_3 + 4y_4 + y_5)$$

$$N = (3.5/3)[0.0319 + (4 \times 0.0400) + (2 \times 0.0553) + (4 \times 0.0865) + 0.1870] = 0.975$$

The design length is, therefore, l = (1200/300)(0.975) = 3.9 ft.

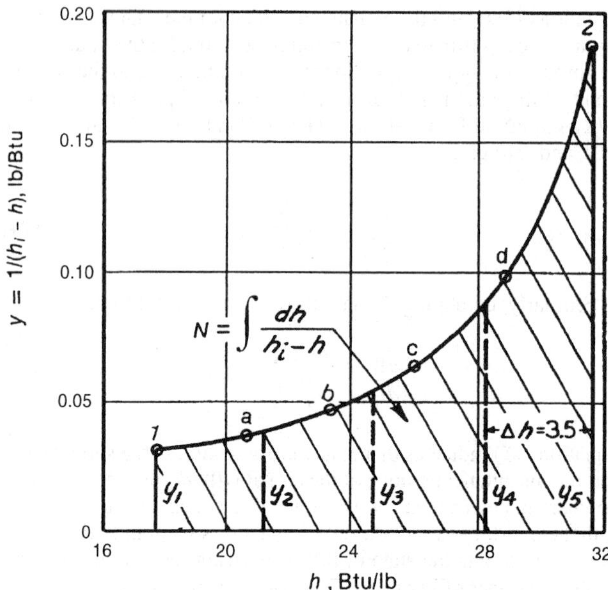

Fig. 15 Graphical Solution of $\int dh/(h_i - h)$

The method used in Example 4 can also be used to predict the performance of existing direct-contact equipment and can determine the transfer coefficients when performance data from test runs are available. By knowing the water and air temperatures entering and leaving the chamber and the spray ratio, it is possible, by trial and error, to determine the proper slope of the tie-line necessary to achieve the measured final air state. The tie-line slope gives the ratio $h_L a_H / K_M a_M$; $K_M a_M$ is found from the integral relationship in Example 4 from the known chamber length l.

Additional descriptions of air spray washers and general performance criteria are given in Chapter 19 of the 1996 *ASHRAE Handbook—Systems and Equipment.*

Cooling Towers

A cooling tower is a direct-contact heat exchanger in which waste heat picked up by the cooling water from a refrigerator, air conditioner, or industrial process is transferred to atmospheric air by cooling the water. Cooling is achieved by breaking up the water flow to provide a large water surface for air, moving by natural or forced convection through the tower, to contact the water. Cooling towers may be counterflow, crossflow, or a combination of both.

The temperature of the water leaving the tower and the packing depth needed to achieve the desired leaving water temperature are of primary interest for design. Therefore, the mass and energy balance equations are based on an overall coefficient K, which is based on (1) the enthalpy driving force due to h at the bulk water temperature and (2) neglecting the film resistance. Combining Equations (70) and (71) and using the parameters described above yields

$$G_L c_L dt = K_M a_M (h_i - h) dl = G_a dh$$
$$= \frac{K_a dV(h' - h_a)}{A_{cs}} \quad (76)$$

or

$$\frac{K_a V}{\dot{m}_L} = \int_{t_1}^{t_2} \frac{c_L\, dt}{(h' - h_a)} \quad (77)$$

Chapter 36 of the 1996 *ASHRAE Handbook—Systems and Equipment* covers cooling tower design in detail.

Cooling and Dehumidifying Coils

When water vapor is condensed out of an airstream onto an extended surface (finned) cooling coil, the simultaneous heat and mass transfer problem can be solved by the same procedure set forth for direct-contact equipment. The basic equations are the same, except that the true surface area of the coil A is known and the problem does not have to be solved on a unit volume basis. Therefore, if in Equations (67), (68), and (70) $a_M dl$ or $a_H dl$ is replaced by dA/A_{cs}, these equations become the basic heat, mass, and total energy transfer equations for indirect-contact equipment such as dehumidifying coils. The energy balance shown by Equation (71) remains unchanged. The heat transfer from the interface to the refrigerant now encounters the combined resistances of the condensate film ($R_L = 1/h_L$); the metal wall and fins, if any (R_m); and the refrigerant film ($R_r = A/h_r A_r$). If this combined resistance is designated as $R_i = R_L + R_m + R_r = 1/U_i$, Equation (72) becomes, for a coil dehumidifier,

$$\pm \dot{m}_L c_L dt_L = U_i(t_L - t_i)dA \qquad (78)$$

(plus sign for counterflow, minus sign for parallel flow).

The tie-line slope is then

$$\frac{h - h_i}{t_L - t_i} = \mp \frac{U_i}{K_M} \qquad (79)$$

Figure 16 illustrates the graphical solution on a psychrometric chart for the air path through a dehumidifying coil with a constant refrigerant temperature. Because the tie-line slope is infinite in this case, the energy balance line is vertical. The corresponding interface states and air states are denoted by the same letter symbols, and the solution follows the same procedure as in Example 4.

If the problem is to determine the required coil surface area for a given performance, the area is computed by the following relation:

$$A = \frac{\dot{m}_a}{K_M} \int_1^2 \frac{dh}{(h_i - h)} \qquad (80)$$

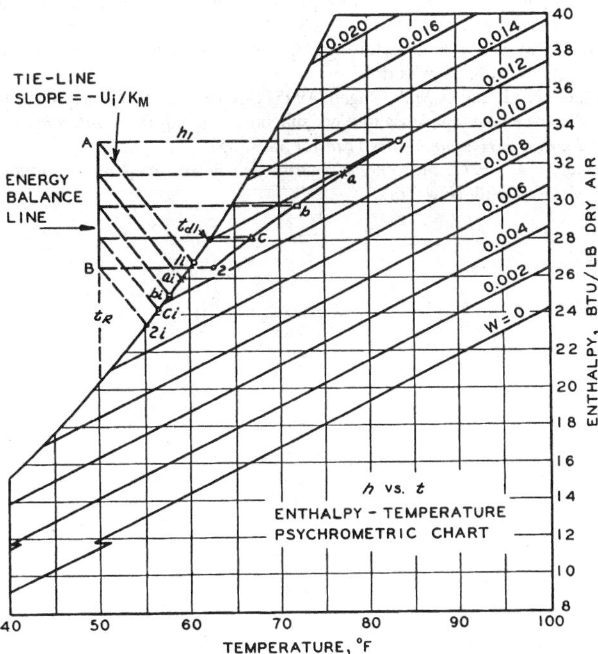

Fig. 16 Graphical Solution for Air-State Path in Dehumidifying Coil with Constant Refrigerant Temperature

This graphical solution on the psychrometric chart automatically determines whether any part of the coil is dry. Thus, in the example illustrated in Figure 16, the entering air at state 1 initially encounters an interface saturation state 1_i, clearly below its dew-point temperature t_{d1}, so the coil immediately becomes wet. Had the graphical technique resulted in an initial interface state above the dew-point temperature of the entering air, the coil would be initially dry. The air would then follow a constant humidity ratio line (the sloping W = constant lines on the chart) until the interface state reached the air dew-point temperature.

Mizushina et al. (1959) developed this method not only for water vapor and air, but also for other vapor-gas mixtures. Chapter 21 of the 1996 *ASHRAE Handbook—Systems and Equipment* shows another related method, based on ARI *Standard* 410, of determining air-cooling and dehumidifying coil performance.

SYMBOLS

a = constant, dimensionless; or surface area per unit volume, ft^2/ft^3
A = surface area, ft^2
A_{cs} = cross-sectional area, ft^2
b = exponent, dimensionless
C = molal concentration of solute in solvent, lb mol/ft^2
c_L = specific heat of liquid, Btu/lb·°F
c_p = specific heat at constant pressure, Btu/lb·°F
c_{pm} = specific heat of moist air at constant pressure, Btu/lb_{da}·°F
d = diameter, ft
D_v = diffusion coefficient (mass diffusivity), ft^2/h
f = Fanning friction factor, dimensionless
Fo = Fourier number = $\alpha\tau/L^2$, dimensionless
Fo_m = mass transfer Fourier number $D_v\tau/L^2$
g_c = gravitational constant, $ft \cdot lb_m/h^2 \cdot lb_f$
G = mass flux, flow rate per unit of cross-sectional area, $lb_m/h \cdot ft^2$
h = enthalpy, Btu/lb; or heat transfer coefficient, $Btu/h \cdot ft^2 \cdot °F$
h_{fg} = enthalpy of vaporization, Btu/lb_m
h_M = mass transfer coefficient, ft/h
J = diffusive mass flux, $lb_m/h \cdot ft^2$
J^* = diffusive molar flux, lb $mol/h \cdot ft^2$
j_D = Colburn mass transfer group = $Sh/(Re \cdot Pr^{1/3})$, dimensionless
j_H = Colburn heat transfer group = $Nu/(Re \cdot Pr^{1/3})$, dimensionless
k = thermal conductivity, Btu/h·ft·°F
K_M = mass transfer coefficient, $lb/h \cdot ft^2$
l = length, ft
L = characteristic length, ft
L/G = liquid-to-air mass flow ratio
Le = Lewis number = α/D_v, dimensionless
$\dot{m}$ = rate of mass transfer, lb/h
m'' = mass flux, $lb/h \cdot ft^2$
$m''*$ = molar flux, lb $mol/h \cdot ft^2$
M = molecular weight, lb/lb mol
Nu = Nusselt number = hL/k, dimensionless
p = pressure, atmospheres or psi
P_{Am} = logarithmic mean density factor, Equation (24)
Pr = Prandtl number = $c_p\mu/k$, dimensionless
q = rate of heat transfer, Btu/h
q'' = heat flux per unit area, $Btu/h \cdot ft^2$
Q = volumetric flow rate, cfm
Re = Reynolds number = $\rho u L/\mu$, dimensionless
R_U = universal gas constant = 1545 $lb_f \cdot ft/lb$ mol·°R
R_i = combined thermal resistance, $ft^2 \cdot °F \cdot h/Btu$
R_L = thermal resistance of condensate film, $ft^2 \cdot °F \cdot h/Btu$
R_m = thermal resistance across metal wall and fins, $ft^2 \cdot °F \cdot h/Btu$
R_r = thermal resistance of refrigerant film, $ft^2 \cdot °F \cdot h/Btu$
Sc = Schmidt number = $\mu/\rho D_v$, dimensionless
Sh = Sherwood number = $h_M L/D_v$, dimensionless
St = Stanton number = $h/\rho c_p \bar{u}$, dimensionless
St_m = mass transfer Stanton number = $h_M P_{Am}/\bar{u}$, dimensionless
t = temperature, °F
T = absolute temperature, °R
u = velocity in x direction, fpm
U_i = overall conductance from refrigerant to air-water interface for dehumidifying coil, $Btu/h \cdot ft^2 \cdot °F$
v = velocity in y direction, fpm

v_i = velocity normal to mass transfer surface for component i, ft/h
V = fluid stream velocity, fpm
W = humidity ratio, lb of water vapor per lb of dry air, lb(w)/lb(da)
x, y, z = coordinate direction, ft
X, Y, Z = coordinate direction, dimensionless
α = thermal diffusivity = $k/\rho c_p$, ft^2/h
ε_D = eddy mass diffusivity, ft^2/h
μ = absolute (dynamic) viscosity, lb$_m$/ft·h
$\bar{\mu}$ = permeability, grains·in/h·ft^2·in. Hg
ν = kinematic viscosity, ft^2/h
σ = characteristic molecular diameter, nm
θ = dimensionless time parameter
ρ = mass density or concentration, lb$_m$/ft^3
τ = time
τ_i = shear stress in the x-y coordinate plane, lb$_f$/ft^2
ω = mass fraction, lb/lb
$\Omega_{D,AB}$ = temperature function in Equation (10)

Subscripts

a = air property
da = dry air property or air-side transfer quantity
Am = logarithmic mean
A = gas component of binary mixture
B = the more dilute gas component of binary mixture
c = critical state
H = heat transfer quantity
i = air-water interface value
L = liquid
m = mean value or metal
M = mass transfer quantity
o = property evaluated at 0°F
s = water vapor property or transport quantity
w = water vapor
∞ = property of main fluid stream

Superscripts

* = on molar basis
¯ = average value
′ = wet bulb

REFERENCES

Bedingfield, G.H., Jr. and T.B. Drew. 1950. Analogy between heat transfer and mass transfer—A psychrometric study. *Industrial and Engineering Chemistry* 42:1164.

Bird, R.B., W.E. Stewart, and E.N. Lightfoot. 1960. *Transport phenomena.* John Wiley and Sons, New York.

Chilton, T.H. and A.P. Colburn. 1934. Mass transfer (absorption) coefficients—Prediction from data on heat transfer and fluid friction. *Industrial and Engineering Chemistry* 26 (November):1183.

Eckert, E.R.G. and R.M. Drake, Jr. 1972. *Analysis of heat and mass transfer.* McGraw-Hill, New York.

Guillory, J.L. and F.C. McQuiston. 1973. An experimental investigation of air dehumidification in a parallel plate heat exchanger. *ASHRAE Transactions* 79(2):146.

Helmer, W.A. 1974. *Condensing water vapor—Airflow in a parallel plate heat exchanger.* Ph.D. thesis, Purdue University, West Lafayette, IN.

Hirschfelder, J.O., C.F. Curtiss, and R.B. Bird. 1954. *Molecular theory of gases and liquids.* John Wiley and Sons, New York.

Incropera, F.P. and D.P. DeWitt. 1996. *Fundamentals of heat and mass transfer*, 4th ed. John Wiley and Sons, New York.

Kusuda, T. 1957. Graphical method simplifies determination of aircoil, wet-heat-transfer surface temperature. *Refrigerating Engineering* 65:41.

Mason, E.A. and L. Monchick. 1965. Survey of the equation of state and transport properties of moist gases. *Humidity and Moisture* 3. Reinhold Publishing Corpration, New York.

McAdams, W.H. 1954. *Heat transmission*, 3rd ed. McGraw-Hill, New York.

Mizushina, T., N. Hashimoto, and M. Nakajima. 1959. Design of cooler condensers for gas-vapour mixtures. *Chemical Engineering Science* 9:195.

Ohadi, M.M. and E.M. Sparrow. 1989. Heat transfer in a straight tube situated downstream of a bend. *International Journal of Heat and Mass Transfer* 32(2):201-12.

Reid, R.C. and T.K. Sherwood. 1966. *The properties of gases and liquids: Their estimation and correlation*, 2nd ed. McGraw-Hill, New York, pp. 520-43.

Reid, R.C., J.M. Prausnitz, and B.E. Poling. 1987. *The properties of gases and liquids*, 4th ed. McGraw-Hill, New York, pp.21-78.

Sherwood, T.K. and R.L. Pigford. 1952. *Absorption and extraction.* McGraw-Hill, New York, pp. 1-28.

Sparrow, E.M. and M.M. Ohadi. 1987a. Comparison of turbulent thermal entrance regions for pipe flows with developed velocity and velocity developing from a sharp-edged inlet. *ASME Transactions, Journal of Heat Transfer* 109:1028-30.

Sparrow, E.M. and M.M. Ohadi. 1987b. Numerical and experimental studies of turbulent flow in a tube. *Numerical Heat Transfer* 11:461-76.

Treybal, R.E. 1980. *Mass transfer operations*, 3rd ed. McGraw-Hill, New York.

BIBLIOGRAPHY

Bennett, C.O. and J.E. Myers. 1982. *Momentum, heat and mass transfer*, 3rd ed. McGraw-Hill, New York.

DeWitt, D.P. and E.L. Cussler. 1984. *Diffusion, mass transfer in fluid systems.* Cambridge University Press, UK.

Geankopolis, C.J. 1993. *Transport processes and unit operations*, 3rd ed. Prentice Hall, Englewood Cliffs, NJ

Kays, W.M. and M.E. Crawford. 1993. *Convective heat and mass transfer.* McGraw-Hill, New York.

Mikielviez, J. and A.M.A. Rageb. 1995. Simple theroretical approach to direct-contact condensation on subcooled liquid film. *International Journal of Heat and Mass Transfer* 38(3):557.

Slattery, J.C. 1972. *Momentum, energy and mass transfer in continua.* McGraw-Hill, New York.

CHAPTER 6

PSYCHROMETRICS

PSYCHROMETRICS deals with thermodynamic properties of moist air and uses these properties to analyze conditions and processes involving moist air. Hyland and Wexler (1983a, 1983b) developed formulas for thermodynamic properties of moist air and water. Perfect gas relations can be used in most air-conditioning problems instead of these formulas. Threlkeld (1970) showed that errors are less than 0.7% in calculating humidity ratio, enthalpy, and specific volume of saturated air at standard atmospheric pressure for a temperature range of −60 to 120°F. Furthermore, these errors decrease with decreasing pressure.

This chapter discusses perfect gas relations and describes their use in common air-conditioning problems. The formulas developed by Hyland and Wexler (1983a) may be used where greater precision is required.

COMPOSITION OF DRY AND MOIST AIR

Atmospheric air contains many gaseous components as well as water vapor and miscellaneous contaminants (e.g., smoke, pollen, and gaseous pollutants not normally present in free air far from pollution sources).

Dry air exists when all water vapor and contaminants have been removed from atmospheric air. The composition of dry air is relatively constant, but small variations in the amounts of individual components occur with time, geographic location, and altitude. Harrison (1965) lists the approximate percentage composition of dry air by volume as: nitrogen, 78.084; oxygen, 20.9476; argon, 0.934; carbon dioxide, 0.0314; neon, 0.001818; helium, 0.000524; methane, 0.00015; sulfur dioxide, 0 to 0.0001; hydrogen, 0.00005; and minor components such as krypton, xenon, and ozone, 0.0002. The apparent molecular mass or weighted average molecular weight of all components, for *dry air* is 28.9645, based on the carbon-12 scale (Harrison 1965). The gas constant for dry air, based on the carbon-12 scale, is:

$$R_a = 1545.32/28.9645 = 53.352 \text{ ft}\cdot\text{lb}_f/\text{lb}_m\cdot°\text{R} \quad (1)$$

Moist air is a binary (or two-component) mixture of dry air and water vapor. The amount of water vapor in moist air varies from zero (dry air) to a maximum that depends on temperature and pressure. The latter condition refers to **saturation**, a state of neutral equilibrium between moist air and the condensed water phase (liquid or solid). Unless otherwise stated, saturation refers to a flat interface surface between the moist air and the condensed phase. The molecular weight of water is 18.01528 on the carbon-12 scale. The gas constant for water vapor is:

$$R_a = 1545.32/18.01528 = 85.778 \text{ ft}\cdot\text{lb}_f/\text{lb}_m\cdot°\text{R} \quad (2)$$

UNITED STATES STANDARD ATMOSPHERE

The temperature and barometric pressure of atmospheric air vary considerably with altitude as well as with local geographic and weather conditions. The standard atmosphere gives a standard of reference for estimating properties at various altitudes. At sea level, standard temperature is 59°F; standard barometric pressure is 29.921 in. Hg. The temperature is assumed to decrease linearly with increasing altitude throughout the troposphere (lower atmosphere), and to be constant in the lower reaches of the stratosphere. The lower atmosphere is assumed to consist of dry air that behaves as a perfect gas. Gravity is also assumed constant at the standard value, 32.1740 ft/s^2. Table 1 summarizes property data for altitudes to 60,000 ft.

The values in Table 1 may be calculated from the equation

$$p = 29.92(1 - 6.8753 \times 10^{-6} Z)^{5.2559} \quad (3)$$

Table 1 Standard Atmospheric Data for Altitudes to 60,000 ft

Altitude, ft	Temperature, °F	Pressure	
		in. Hg	psia
−1000	62.6	31.02	15.236
−500	60.8	30.47	14.966
0	59.0	29.921	14.696
500	57.2	29.38	14.430
1000	55.4	28.86	14.175
2000	51.9	27.82	13.664
3000	48.3	26.82	13.173
4000	44.7	25.82	12.682
5000	41.2	24.90	12.230
6000	37.6	23.98	11.778
7000	34.0	23.09	11.341
8000	30.5	22.22	10.914
9000	26.9	21.39	10.506
10,000	23.4	20.58	10.108
15,000	5.5	16.89	8.296
20,000	−12.3	13.76	6.758
30,000	−47.8	8.90	4.371
40,000	−69.7	5.56	2.731
50,000	−69.7	3.44	1.690
60,000	−69.7	2.14	1.051

Data adapted from NASA (1976).

The preparation of this chapter is assigned to TC 1.1, Thermodynamics and Psychrometrics.

The equation for temperature as a function of altitude is given as:

$$t = 59 - 0.00356616Z \tag{4}$$

where

Z = altitude, ft
p = barometric pressure, in. Hg.
t = temperature, °F

Equations (3) and (4) are accurate from −16500 ft to 36000 ft. For higher altitudes, comprehensive tables of barometric pressure and other physical properties of the standard atmosphere, in both SI and inch-pound units, can be found in NASA (1976 U.S. Standard atmosphere).

THERMODYNAMIC PROPERTIES OF MOIST AIR

Table 2, developed from formulas by Hyland and Wexler (1983a, 1983b), shows values of thermodynamic properties based on the *thermodynamic temperature scale*. This ideal scale differs slightly from practical temperature scales used for physical measurements. For example, the standard boiling point for water (at 14.696 psia or 29.921 in. Hg) occurs at 211.95°F on this scale rather than at the traditional value of 212°F. Most measurements are currently based on the International Practical Temperature Scale of 1990 (IPTS-90). The following paragraphs briefly describe each column of Table 2.

t = Fahrenheit temperature, based on thermodynamic temperature scale and expressed relative to absolute temperature T in degrees Rankine (°R) by the relation:

$$T = t + 459.67$$

W_s = humidity ratio *at saturation*, condition at which gaseous phase (moist air) exists in equilibrium with condensed phase (liquid or solid) at given temperature and pressure (standard atmospheric pressure). At given values of temperature and pressure, humidity ratio W can have any value from zero to W_s.

v_a = specific volume of dry air, ft³/lb

v_{as} = $v_s - v_a$, difference between volume of moist air *at saturation*, per pound of dry air, and specific volume of dry air itself, ft³/lb of dry air, at same pressure and temperature.

v_s = volume of moist air *at saturation* per pound of dry air, ft³/lb of dry air.

h_a = specific enthalpy of dry air, Btu/lb of dry air. Specific enthalpy of dry air has been assigned a value of zero at 0°F and standard atmospheric pressure in Table 2.

h_{as} = $h_s - h_a$, difference between enthalpy of moist air at saturation, per pound of dry air, and specific enthalpy of dry air itself, Btu/lb of dry air, at same pressure and temperature.

h_s = enthalpy of moist air *at saturation* of dry air, Btu/lb of dry air.

h_w = specific enthalpy of condensed water (liquid or solid) in equilibrium with saturated air at specified temperature and pressure, Btu per pound of water. Specific enthalpy of liquid water is assigned a value of zero at its triple point (32.018°F) and saturation pressure.

Note that h_w is greater than the steam-table enthalpy of saturated pure condensed phase by the amount of enthalpy increase governed by the pressure increase from saturation pressure to 1 atmosphere plus influences from presence of air.

s_a = specific entropy of dry air, Btu/lb·°R. In Table 2, specific entropy of dry air has been assigned a value of zero at 0°F and standard atmospheric pressure.

s_{as} = $s_s - s_a$, difference between entropy of moist air *at saturation*, per pound of dry air, and specific entropy of dry air itself, Btu/(lb·°R), at same pressure and temperature.

s_s = entropy of moist air *at saturation* per pound of dry air, Btu/(lb·°R).

s_w = specific entropy per pound of condensed water (liquid or solid) in equilibrium with saturated air, Btu/(lb·°R); s_w differs from entropy of pure water at saturation pressure, similar to h_w.

p_s = vapor pressure of water in saturated moist air, in. Hg. Pressure p_s differs negligibly from saturation vapor pressure of pure water p_{ws} at least for conditions shown. Consequently, values of p_s can be used at same pressure and temperature in equations where p_{ws} appears. Pressure p_s is defined as $p_s = x_{ws}p$, where x_{ws} is mole fraction of water vapor in moist air saturated with water at temperature t and pressure p, and where p is total barometric pressure of moist air.

THERMODYNAMIC PROPERTIES OF WATER AT SATURATION

Table 3 shows thermodynamic properties of water at saturation for temperatures from −80 to 300°F, calculated by the formulations described by Hyland and Wexler (1983b). Symbols in the table follow standard steam table nomenclature. These properties are based on the thermodynamic temperature scale. The enthalpy and entropy of saturated liquid water are both assigned the value zero at the triple point, 32.018°F. Between the triple-point and critical-point temperatures of water, two states—liquid and vapor—may coexist in equilibrium. These states are called saturated liquid and saturated vapor.

In determining a number of moist air properties, principally the saturation humidity ratio, the *water vapor saturation pressure* is required. Values may be obtained from Table 3 or calculated from the following formulas (Hyland and Wexler 1983b).

The saturation pressure over *ice* for the temperature range of −148 to 32°F is given by:

$$\ln(p_{ws}) = C_1/T + C_2 + C_3T + C_4T^2 + C_5T^3 + C_6T^4 + C_7\ln T \tag{5}$$

where

C_1 = −1.021 416 5 E+04
C_2 = −4.893 242 8 E+00
C_3 = −5.376 579 4 E−03
C_4 = 1.920 237 7 E−07
C_5 = 3.557 583 2 E−10
C_6 = −9.034 468 8 E−14
C_7 = 4.163 501 9 E+00

The saturation pressure over *liquid water* for the temperature range of 32 to 392°F is given by:

$$\ln(p_{ws}) = C_8/T + C_9 + C_{10}T + C_{11}T^2 + C_{12}T^3 + C_{13}\ln T \tag{6}$$

where

C_8 = −1.044 039 7 E+04
C_9 = −1.129 465 0 E+01
C_{10} = −2.702 235 5 E−02
C_{11} = 1.289 036 0 E−05
C_{12} = −2.478 068 1 E−09
C_{13} = 6.545 967 3 E+00

In both Equations (5) and (6),

ln = natural logarithm
p_{ws} = saturation pressure, psia
T = absolute temperature, °R = °F + 459.67

The coefficients of Equations (5) and (6) have been derived from the Hyland-Wexler equations, which are given in SI units. Due to rounding errors in the derivations and in some computers' calculating precision, the results obtained from Equations (5) and (6) may not agree precisely with Table 3 values.

Table 2 Thermodynamic Properties of Moist Air, Standard Atmospheric Pressure, 14.696 psi (29.921 in. Hg)

Temp. t, °F	Humidity Ratio, lb_w/lb_{da} W_s	Volume, ft^3/lb dry air			Enthalpy, Btu/lb dry air			Entropy, Btu/(lb dry air) · °F			Condensed Water			Temp., °F
											Enthalpy Btu/lb	Entropy, Btu/lb · °F	Vapor Press., in. Hg	
		v_a	v_{as}	v_s	h_a	h_{as}	h_s	s_a	s_{as}	s_s	h_w	s_w	p_s	
−80	0.0000049	9.553	0.000	9.553	−19.221	0.005	−19.215	−0.04594	0.00001	−0.04592	−193.45	−0.4067	0.000236	−80
−79	0.0000053	9.579	0.000	9.579	−18.980	0.005	−18.975	−0.04531	0.00002	−0.04529	−193.06	−0.4056	0.000255	−79
−78	0.0000057	9.604	0.000	9.604	−18.740	0.006	−18.734	−0.04468	0.00002	−0.04466	−192.66	−0.4046	0.000275	−78
−77	0.0000062	9.629	0.000	9.629	−18.500	0.007	−18.493	−0.04405	0.00002	−0.04403	−192.27	−0.4036	0.000296	−77
−76	0.0000067	9.655	0.000	9.655	−18.259	0.007	−18.252	−0.04342	0.00002	−0.04340	−191.87	−0.4025	0.000319	−76
−75	0.0000072	9.680	0.000	9.680	−18.019	0.007	−18.011	−0.04279	0.00002	−0.04277	−191.47	−0.4015	0.000344	−75
−74	0.0000078	9.705	0.000	9.705	−17.778	0.008	−17.770	−0.04217	0.00002	−0.04215	−191.07	−0.4005	0.000371	−74
−73	0.0000084	9.731	0.000	9.731	−17.538	0.009	−17.529	−0.04155	0.00002	−0.04152	−190.68	−0.3994	0.000400	−73
−72	0.0000090	9.756	0.000	9.756	−17.298	0.010	−17.288	−0.04093	0.00003	−0.04090	−190.27	−0.3984	0.000430	−72
−71	0.0000097	9.781	0.000	9.782	−17.057	0.010	−17.047	−0.04031	0.00003	−0.04028	−189.87	−0.3974	0.000463	−71
−70	0.0000104	9.807	0.000	9.807	−16.806	0.011	−16.817	−0.03969	0.00003	−0.03966	−189.47	−0.3963	0.000498	−70
−69	0.0000112	9.832	0.000	9.832	−16.577	0.012	−16.565	−0.03907	0.00003	−0.03904	−189.07	−0.3953	0.000536	−69
−68	0.0000120	9.857	0.000	9.858	−16.336	0.013	−16.324	−0.03846	0.00003	−0.03843	−188.66	−0.3943	0.000576	−68
−67	0.0000129	9.883	0.000	9.883	−16.096	0.013	−16.083	−0.03785	0.00004	−0.03781	−188.26	−0.3932	0.000619	−67
−66	0.0000139	9.908	0.000	9.908	−15.856	0.015	−15.841	−0.03724	0.00004	−0.03720	−187.85	−0.3922	0.000665	−66
−65	0.0000149	9.933	0.000	9.934	−15.616	0.015	−15.600	−0.03663	0.00004	−0.03659	−187.44	−0.3912	0.000714	−65
−64	0.0000160	9.959	0.000	9.959	−15.375	0.017	−15.359	−0.03602	0.00005	−0.03597	−187.04	−0.3901	0.000766	−64
−63	0.0000172	9.984	0.000	9.984	−15.117	0.018	−15.135	−0.03541	0.00005	−0.03536	−186.63	−0.3891	0.000822	−63
−62	0.0000184	10.009	0.000	10.010	−14.895	0.019	−14.876	−0.03481	0.00005	−0.03476	−186.22	−0.3881	0.000882	−62
−61	0.0000198	10.035	0.000	10.035	−14.654	0.021	−14.634	−0.03420	0.00006	−0.03415	−185.81	−0.3870	0.000945	−61
−60	0.0000212	10.060	0.000	10.060	−14.414	0.022	−14.392	−0.03360	0.00006	−0.03354	−185.39	−0.3860	0.001013	−60
−59	0.0000227	10.085	0.000	10.086	−14.174	0.024	−14.150	−0.03300	0.00006	−0.03294	−184.98	−0.3850	0.001086	−59
−58	0.0000243	10.111	0.000	10.111	−13.933	0.025	−13.908	−0.03240	0.00007	−0.03233	−184.57	−0.3839	0.001163	−58
−57	0.0000260	10.136	0.000	10.137	−13.693	0.027	−13.666	−0.03180	0.00007	−0.03173	−184.15	−0.3829	0.001246	−57
−56	0.0000279	10.161	0.000	10.162	−13.453	0.029	−13.424	−0.03121	0.00008	−0.03113	−183.74	−0.3819	0.001333	−56
−55	0.0000298	10.187	0.000	10.187	−13.213	0.031	−13.182	−0.03061	0.00008	−0.03053	−183.32	−0.3808	0.001427	−55
−54	0.0000319	10.212	0.001	10.213	−12.972	0.033	−12.939	−0.03002	0.00009	−0.02993	−182.90	−0.3798	0.001526	−54
−53	0.0000341	10.237	0.001	10.238	−12.732	0.035	−12.697	−0.02943	0.00009	−0.02934	−182.48	−0.3788	0.001632	−53
−52	0.0000365	10.263	0.001	10.263	−12.492	0.038	−12.454	−0.02884	0.00010	−0.02874	−182.06	−0.3778	0.001745	−52
−51	0.0000390	10.288	0.001	10.289	−12.251	0.041	−12.211	−0.02825	0.00011	−0.02814	−181.64	−0.3767	0.001865	−51
−50	0.0000416	10.313	0.001	10.314	−12.011	0.043	−11.968	−0.02766	0.00011	−0.02755	−181.22	−0.3757	0.001992	−50
−49	0.0000445	10.339	0.001	10.340	−11.771	0.046	−11.725	−0.02708	0.00012	−0.02696	−180.80	−0.3747	0.002128	−49
−48	0.0000475	10.364	0.001	10.365	−11.531	0.050	−11.481	−0.02649	0.00013	−0.02636	−180.37	−0.3736	0.002272	−48
−47	0.0000507	10.389	0.001	10.390	−11.290	0.053	−11.237	−0.02591	0.00014	−0.02577	−179.95	−0.3726	0.002425	−47
−46	0.0000541	10.415	0.001	10.416	−11.050	0.056	−10.994	−0.02533	0.00015	−0.02518	−179.52	−0.3716	0.002587	−46
−45	0.0000577	10.440	0.001	10.441	−10.810	0.060	−10.750	−0.02475	0.00016	−0.02459	−179.10	−0.3705	0.002760	−45
−44	0.0000615	10.465	0.001	10.466	−10.570	0.064	−10.505	−0.02417	0.00017	−0.02400	−178.67	−0.3695	0.002943	−44
−43	0.0000656	10.491	0.001	10.492	−10.329	0.068	−10.261	−0.02359	0.00018	−0.02342	−178.24	−0.3685	0.003137	−43
−42	0.0000699	10.516	0.001	10.517	−10.089	0.073	−10.016	−0.02302	0.00019	−0.02283	−177.81	−0.3675	0.003343	−42
−41	0.0000744	10.541	0.001	10.543	−9.849	0.078	−9.771	−0.02244	0.00020	−0.02224	−177.38	−0.3664	0.003562	−41
−40	0.0000793	10.567	0.001	10.568	−9.609	0.083	−9.526	−0.02187	0.00021	−0.02166	−176.95	−0.3654	0.003793	−40
−39	0.0000844	10.592	0.001	10.593	−9.368	0.088	−9.280	−0.02130	0.00022	−0.02107	−176.52	−0.3644	0.004039	−39
−38	0.0000898	10.617	0.002	10.619	−9.128	0.094	−9.034	−0.02073	0.00024	−0.02049	−176.08	−0.3633	0.004299	−38
−37	0.0000956	10.643	0.002	10.644	−8.888	0.100	−8.788	−0.02016	0.00025	−0.01991	−175.65	−0.3623	0.004575	−37
−36	0.0001017	10.668	0.002	10.670	−8.648	0.106	−8.541	−0.01959	0.00027	−0.01932	−175.21	−0.3613	0.004866	−36
−35	0.0001081	10.693	0.002	10.695	−8.407	0.113	−8.294	−0.01902	0.00028	−0.01874	−174.78	−0.3603	0.005175	−35
−34	0.0001150	10.719	0.002	10.721	−8.167	0.120	−8.047	−0.01846	0.00030	−0.01816	−174.34	−0.3529	0.005502	−34
−33	0.0001222	10.744	0.002	10.746	−7.927	0.128	−7.799	−0.01790	0.00032	−0.01758	−173.90	−0.3582	0.005848	−33
−32	0.0001298	10.769	0.002	10.772	−7.687	0.136	−7.551	−0.01733	0.00034	−0.01699	−173.46	−0.3572	0.006214	−32
−31	0.0001379	10.795	0.002	10.797	−7.447	0.145	−7.302	−0.01677	0.00036	−0.01641	−173.02	−0.3561	0.006601	−31
−30	0.0001465	10.820	0.003	10.822	−7.206	0.154	−7.053	−0.01621	0.00038	−0.01583	−172.58	−0.3551	0.007009	−30
−29	0.0001555	10.845	0.003	10.848	−6.966	0.163	−6.803	−0.01565	0.00040	−0.01525	−172.14	−0.3541	0.007442	−29
−28	0.0001650	10.871	0.003	10.873	−6.726	0.173	−6.553	−0.01510	0.00043	−0.01467	−171.70	−0.3531	0.007898	−28
−27	0.0001751	10.896	0.003	10.899	−6.486	0.184	−6.302	−0.01454	0.00045	−0.01409	−171.25	−0.3520	0.008381	−27
−26	0.0001858	10.921	0.003	10.924	−6.245	0.195	−6.051	−0.01399	0.00048	−0.01351	−170.81	−0.3510	0.008890	−26
−25	0.0001970	10.947	0.003	10.950	−6.005	0.207	−5.798	−0.01343	0.00051	−0.01293	−170.36	−0.3500	0.009428	−25
−24	0.0002088	10.972	0.004	10.976	−5.765	0.220	−5.545	−0.01288	0.00054	−0.01235	−169.92	−0.3489	0.009995	−24
−23	0.0002214	10.997	0.004	11.001	−5.525	0.233	−5.292	−0.01233	0.00057	−0.01176	−169.47	−0.3479	0.010594	−23
−22	0.0002346	11.022	0.004	11.027	−5.284	0.247	−5.038	−0.01178	0.00060	−0.01118	−169.02	−0.3469	0.011226	−22
−21	0.0002485	11.048	0.004	11.052	−5.044	0.261	−4.783	−0.01123	0.00063	−0.01060	−168.57	−0.3459	0.011893	−21
−20	0.0002632	11.073	0.005	11.078	−4.804	0.277	−4.527	−0.01069	0.00067	−0.01002	−168.12	−0.3448	0.012595	−20
−19	0.0002786	11.098	0.005	11.103	−4.564	0.293	−4.271	−0.01014	0.00071	−0.00943	−167.67	−0.3438	0.013336	−19
−18	0.0002950	11.124	0.005	11.129	−4.324	0.311	−4.013	−0.00960	0.00075	−0.00885	−167.21	−0.3428	0.014117	−18
−17	0.0003121	11.149	0.006	11.155	−4.084	0.329	−3.754	−0.00905	0.00079	−0.00826	−166.76	−0.3418	0.014939	−17
−16	0.0003303	11.174	0.006	11.180	−3.843	0.348	−3.495	−0.00851	0.00083	−0.00768	−166.30	−0.3407	0.015806	−16
−15	0.0003493	11.200	0.006	11.206	−3.603	0.368	−3.235	−0.00797	0.00088	−0.00709	−165.85	−0.3397	0.016718	−15
−14	0.0003694	11.225	0.007	11.232	−3.363	0.390	−2.973	−0.00743	0.00093	−0.00650	−165.39	−0.3387	0.017679	−14
−13	0.0003905	11.250	0.007	11.257	−3.123	0.412	−2.710	−0.00689	0.00098	−0.00591	−164.93	−0.3377	0.018690	−13
−12	0.0004128	11.276	0.007	11.283	−2.882	0.436	−2.447	−0.00635	0.00103	−0.00532	−164.47	−0.3366	0.019754	−12
−11	0.0004362	11.301	0.008	11.309	−2.642	0.460	−2.182	−0.00582	0.00109	−0.00473	−164.01	−0.3356	0.020873	−11

Table 2 Thermodynamic Properties of Moist Air, Standard Atmospheric Pressure, 14.696 psi (29.921 in. Hg) (*Continued*)

Temp. t, °F	Humidity Ratio, lb_w/lb_{da} W_s	Volume, ft³/lb dry air v_a	v_{as}	v_s	Enthalpy, Btu/lb dry air h_a	h_{as}	h_s	Entropy, Btu/(lb dry air) · °F s_a	s_{as}	s_s	Condensed Water: Enthalpy Btu/lb h_w	Condensed Water: Entropy, Btu/lb · °F s_w	Condensed Water: Vapor Press., in. Hg p_s	Temp., °F
−10	0.0004608	11.326	0.008	11.335	−2.402	0.487	−1.915	−0.00528	0.00115	−0.00414	−163.55	−0.3346	0.022050	−10
−9	0.0004867	11.351	0.009	11.360	−2.162	0.514	−1.647	−0.00475	0.00121	−0.00354	−163.09	−0.3335	0.023289	−9
−8	0.0005139	11.377	0.009	11.386	−1.922	0.543	−1.378	−0.00422	0.00127	−0.00294	−162.63	−0.3325	0.024591	−8
−7	0.0005425	11.402	0.010	11.412	−1.681	0.574	−1.108	−0.00369	0.00134	−0.00234	−162.17	−0.3315	0.025959	−7
−6	0.0005726	11.427	0.010	11.438	−1.441	0.606	−0.835	−0.00316	0.00141	−0.00174	−161.70	−0.3305	0.027397	−6
−5	0.0006041	11.453	0.011	11.464	−1.201	0.640	−0.561	−0.00263	0.00149	−0.00114	−161.23	−0.3294	0.028907	−5
−4	0.0006373	11.478	0.012	11.490	−0.961	0.675	−0.286	−0.00210	0.00157	−0.00053	−160.77	−0.3284	0.030494	−4
−3	0.0006722	11.503	0.012	11.516	−0.721	0.712	−0.008	−0.00157	0.00165	0.00008	−160.30	−0.3274	0.032160	−3
−2	0.0007088	11.529	0.013	11.542	−0.480	0.751	0.271	−0.00105	0.00174	0.00069	−159.83	−0.3264	0.033909	−2
−1	0.0007472	11.554	0.014	11.568	−0.240	0.792	0.552	−0.00052	0.00183	0.00130	−159.36	−0.3253	0.035744	−1
0	0.0007875	11.579	0.015	11.594	0.0	0.835	0.835	0.00000	0.00192	0.00192	−158.89	−0.3243	0.037671	0
1	0.0008298	11.604	0.015	11.620	0.240	0.880	1.121	0.00052	0.00202	0.00254	−158.42	−0.3233	0.039694	1
2	0.0008742	11.630	0.016	11.646	0.480	0.928	1.408	0.00104	0.00212	0.00317	−157.95	−0.3223	0.041814	2
3	0.0009207	11.655	0.017	11.672	0.721	0.978	1.699	0.00156	0.00223	0.00380	−157.47	−0.3212	0.044037	3
4	0.0009695	11.680	0.018	11.699	0.961	1.030	1.991	0.00208	0.00235	0.00443	−157.00	−0.3202	0.046370	4
5	0.0010207	11.706	0.019	11.725	1.201	1.085	2.286	0.00260	0.00247	0.00506	−156.52	−0.3192	0.048814	5
6	0.0010743	11.731	0.020	11.751	1.441	1.143	2.584	0.00311	0.00259	0.00570	−156.05	−0.3182	0.051375	6
7	0.0011306	11.756	0.021	11.778	1.681	1.203	2.884	0.00363	0.00635	0.00272	−155.57	−0.3171	0.054060	7
8	0.0011895	11.782	0.022	11.804	1.922	1.266	3.188	0.00414	0.00286	0.00700	−155.09	−0.3161	0.056872	8
9	0.0012512	11.807	0.024	11.831	2.162	1.332	3.494	0.00466	0.00300	0.00766	−154.61	−0.3151	0.059819	9
10	0.0013158	11.832	0.025	11.857	2.402	1.402	3.804	0.00517	0.00315	0.00832	−154.13	−0.3141	0.062901	10
11	0.0013835	11.857	0.026	11.884	2.642	1.474	4.117	0.00568	0.00330	0.00898	−153.65	−0.3130	0.066131	11
12	0.0014544	11.883	0.028	11.910	2.882	1.550	4.433	0.00619	0.00347	0.00966	−153.17	−0.3120	0.069511	12
13	0.0015286	11.908	0.029	11.937	3.123	1.630	4.753	0.00670	0.00364	0.01033	−152.68	−0.3110	0.073049	13
14	0.0016062	11.933	0.031	11.964	3.363	1.714	5.077	0.00721	0.00381	0.01102	−152.20	−0.3100	0.076751	14
15	0.0016874	11.959	0.032	11.991	3.603	1.801	5.404	0.00771	0.00400	0.01171	−151.71	−0.3089	0.080623	15
16	0.0017724	11.984	0.034	12.018	3.843	1.892	5.736	0.00822	0.00419	0.01241	−151.22	−0.3079	0.084673	16
17	0.0018613	12.009	0.036	12.045	4.084	1.988	6.072	0.00872	0.00439	0.01312	−150.74	−0.3069	0.088907	17
18	0.0019543	12.035	0.038	12.072	4.324	2.088	6.412	0.00923	0.00460	0.01383	−150.25	−0.3059	0.093334	18
19	0.0020515	12.060	0.040	12.099	4.564	2.193	6.757	0.00973	0.00482	0.01455	−149.76	−0.3049	0.097962	19
20	0.0021531	12.085	0.042	12.127	4.804	2.303	7.107	0.01023	0.00505	0.01528	−149.27	−0.3038	0.102798	20
21	0.0022592	12.110	0.044	12.154	5.044	2.417	7.462	0.01073	0.00529	0.01602	−148.78	−0.3028	0.107849	21
22	0.0023703	12.136	0.046	12.182	5.285	2.537	7.822	0.01123	0.00554	0.01677	−148.28	−0.3018	0.113130	22
23	0.0024863	12.161	0.048	12.209	5.525	2.662	8.187	0.01173	0.00580	0.01753	−147.79	−0.3008	0.118645	23
24	0.0026073	12.186	0.051	12.237	5.765	2.793	8.558	0.01223	0.00607	0.01830	−147.30	−0.2997	0.124396	24
25	0.0027339	12.212	0.054	12.265	6.005	2.930	8.935	0.01272	0.00636	0.01908	−146.80	−0.2987	0.130413	25
26	0.0028660	12.237	0.056	12.293	6.246	3.073	9.318	0.01322	0.00665	0.01987	−146.30	−0.2977	0.136684	26
27	0.0030039	12.262	0.059	12.321	6.486	3.222	9.708	0.01371	0.00696	0.02067	−145.81	−0.2967	0.143233	27
28	0.0031480	12.287	0.062	12.349	6.726	3.378	10.104	0.01420	0.00728	0.02148	−145.31	−0.2956	0.150066	28
29	0.0032984	12.313	0.065	12.378	6.966	3.541	10.507	0.01470	0.00761	0.02231	−144.81	−0.2946	0.157198	29
30	0.0034552	12.338	0.068	12.406	7.206	3.711	10.917	0.01519	0.00796	0.02315	−144.31	−0.2936	0.164631	30
31	0.0036190	12.363	0.072	12.435	7.447	3.888	11.335	0.01568	0.00832	0.02400	−143.80	−0.2926	0.172390	31
32	0.0037895	12.389	0.075	12.464	7.687	4.073	11.760	0.01617	0.00870	0.02487	−143.30	−0.2915	0.180479	32
32*	0.003790	12.389	0.075	12.464	7.687	4.073	11.760	0.01617	0.00870	0.02487	0.02	0.0000	0.18050	32
33	0.003947	12.414	0.079	12.492	7.927	4.243	12.170	0.01665	0.00905	0.02570	1.03	0.0020	0.18791	33
34	0.004109	12.439	0.082	12.521	8.167	4.420	12.587	0.01714	0.00940	0.02655	2.04	0.0041	0.19559	34
35	0.004277	12.464	0.085	12.550	8.408	4.603	13.010	0.01763	0.00977	0.02740	3.05	0.0061	0.20356	35
36	0.004452	12.490	0.089	12.579	8.648	4.793	13.441	0.01811	0.01016	0.02827	4.05	0.0081	0.21181	36
37	0.004633	12.515	0.093	12.608	8.888	4.990	13.878	0.01860	0.01055	0.02915	5.06	0.0102	0.22035	37
38	0.004820	12.540	0.097	12.637	9.128	5.194	14.322	0.01908	0.01096	0.03004	6.06	0.0122	0.22920	38
39	0.005014	12.566	0.101	12.667	9.369	5.405	14.773	0.01956	0.01139	0.03095	7.07	0.0142	0.23835	39
40	0.005216	12.591	0.105	12.696	9.609	5.624	15.233	0.02004	0.01183	0.03187	8.07	0.0162	0.24784	40
41	0.005424	12.616	0.110	12.726	9.849	5.851	15.700	0.02052	0.01228	0.03281	9.08	0.0182	0.25765	41
42	0.005640	12.641	0.114	12.756	10.089	6.086	16.175	0.02100	0.01275	0.03375	10.08	0.0202	0.26781	42
43	0.005863	12.667	0.119	12.786	10.330	6.330	16.660	0.02148	0.01324	0.03472	11.09	0.0222	0.27831	43
44	0.006094	12.692	0.124	12.816	10.570	6.582	17.152	0.02196	0.01374	0.03570	12.09	0.0242	0.28918	44
45	0.006334	12.717	0.129	12.846	10.810	6.843	17.653	0.02244	0.01426	0.03669	13.09	0.0262	0.30042	45
46	0.006581	12.743	0.134	12.877	11.050	7.114	18.164	0.02291	0.01479	0.03770	14.10	0.0282	0.31206	46
47	0.006838	12.768	0.140	12.908	11.291	7.394	18.685	0.02339	0.01534	0.03873	15.10	0.0302	0.32408	47
48	0.007103	12.793	0.146	12.939	11.531	7.684	19.215	0.02386	0.01592	0.03978	16.10	0.0321	0.33651	48
49	0.007378	12.818	0.152	12.970	11.771	7.984	19.756	0.02433	0.01651	0.04084	17.10	0.0341	0.34937	49
50	0.007661	12.844	0.158	13.001	12.012	8.295	20.306	0.02480	0.01712	0.04192	18.11	0.0361	0.36264	50
51	0.007955	12.869	0.164	13.033	12.252	8.616	20.868	0.02528	0.01775	0.04302	19.11	0.0381	0.37636	51
52	0.008259	12.894	0.171	13.065	12.492	8.949	21.441	0.02575	0.01840	0.04415	20.11	0.0400	0.39054	52
53	0.008573	12.920	0.178	13.097	12.732	9.293	22.025	0.02622	0.01907	0.04529	21.11	0.0420	0.40518	53
54	0.008897	12.945	0.185	13.129	12.973	9.648	22.621	0.02668	0.01976	0.04645	22.11	0.0439	0.42030	54
55	0.009233	12.970	0.192	13.162	13.213	10.016	23.229	0.02715	0.02048	0.04763	23.11	0.0459	0.43592	55
56	0.009580	12.995	0.200	13.195	13.453	10.397	23.850	0.02762	0.02122	0.04884	24.11	0.0478	0.45205	56
57	0.009938	13.021	0.207	13.228	13.694	10.790	24.484	0.02808	0.02198	0.05006	25.11	0.0497	0.46870	57
58	0.010309	13.046	0.216	13.262	13.934	11.197	25.131	0.02855	0.02277	0.05132	26.11	0.0517	0.48589	58
59	0.010692	13.071	0.224	13.295	14.174	11.618	25.792	0.02901	0.02358	0.05259	27.11	0.0536	0.50363	59

*Extrapolated to represent metastable equilibrium with undercooled liquid.

Table 2 Thermodynamic Properties of Moist Air, Standard Atmospheric Pressure, 14.696 psi (29.921 in. Hg) (*Continued*)

Temp. t, °F	Humidity Ratio, lb$_w$/lb$_{da}$ W_s	Volume, ft³/lb dry air v_a	v_{as}	v_s	Enthalpy, Btu/lb dry air h_a	h_{as}	h_s	Entropy, Btu/(lb dry air)·°F s_a	s_{as}	s_s	Condensed Water: Enthalpy, Btu/lb h_w	Condensed Water: Entropy, Btu/lb·°F s_w	Condensed Water: Vapor Press., in. Hg p_s	Temp., °F
60	0.011087	13.096	0.233	13.329	14.415	12.052	26.467	0.02947	0.02442	0.05389	28.11	0.0555	0.52193	60
61	0.011496	13.122	0.242	13.364	14.655	12.502	27.157	0.02994	0.02528	0.05522	29.12	0.0575	0.54082	61
62	0.011919	13.147	0.251	13.398	14.895	12.966	27.862	0.03040	0.02617	0.05657	30.11	0.0594	0.56032	62
63	0.012355	13.172	0.261	13.433	15.135	13.446	28.582	0.03086	0.02709	0.05795	31.11	0.0613	0.58041	63
64	0.012805	13.198	0.271	13.468	15.376	13.942	29.318	0.03132	0.02804	0.05936	32.11	0.0632	0.60113	64
65	0.013270	13.223	0.281	13.504	15.616	14.454	30.071	0.03178	0.02902	0.06080	33.11	0.0651	0.62252	65
66	0.013750	13.248	0.292	13.540	15.856	14.983	30.840	0.03223	0.03003	0.06226	34.11	0.0670	0.64454	66
67	0.014246	13.273	0.303	13.577	16.097	15.530	31.626	0.03269	0.03107	0.06376	35.11	0.0689	0.66725	67
68	0.014758	13.299	0.315	13.613	16.337	16.094	32.431	0.03315	0.03214	0.06529	36.11	0.0708	0.69065	68
69	0.015286	13.324	0.326	13.650	16.577	16.677	33.254	0.03360	0.03325	0.06685	37.11	0.0727	0.71479	69
70	0.015832	13.349	0.339	13.688	16.818	17.279	34.097	0.03406	0.03438	0.06844	38.11	0.0746	0.73966	70
71	0.016395	13.375	0.351	13.726	17.058	17.901	34.959	0.03451	0.03556	0.07007	39.11	0.0765	0.76567	71
72	0.016976	13.400	0.365	13.764	17.299	18.543	35.841	0.03496	0.03677	0.07173	40.11	0.0783	0.79167	72
73	0.017575	13.425	0.378	13.803	17.539	19.204	36.743	0.03541	0.03801	0.07343	41.11	0.0802	0.81882	73
74	0.018194	13.450	0.392	13.843	17.779	19.889	37.668	0.03586	0.03930	0.07516	42.11	0.0821	0.84684	74
75	0.018833	13.476	0.407	13.882	18.020	20.595	38.615	0.03631	0.04062	0.07694	43.11	0.0840	0.87567	75
76	0.019491	13.501	0.422	13.923	18.260	21.323	39.583	0.03676	0.04199	0.07875	44.10	0.0858	0.90533	76
77	0.020170	13.526	0.437	13.963	18.500	22.075	40.576	0.03721	0.04339	0.08060	45.10	0.0877	0.93589	77
78	0.020871	13.551	0.453	14.005	18.741	22.851	41.592	0.03766	0.04484	0.08250	46.10	0.0896	0.96733	78
79	0.021594	13.577	0.470	14.046	18.981	23.652	42.633	0.03811	0.04633	0.08444	47.10	0.0914	0.99970	79
80	0.022340	13.602	0.487	14.089	19.222	24.479	43.701	0.03855	0.04787	0.08642	48.10	0.0933	1.03302	80
81	0.023109	13.627	0.505	14.132	19.462	25.332	44.794	0.03900	0.04945	0.08844	49.10	0.0951	1.06728	81
82	0.023902	13.653	0.523	14.175	19.702	26.211	45.913	0.03944	0.05108	0.09052	50.10	0.0970	1.10252	82
83	0.024720	13.678	0.542	14.220	19.943	27.120	47.062	0.03988	0.05276	0.09264	51.09	0.0988	1.13882	83
84	0.025563	13.703	0.561	14.264	20.183	28.055	48.238	0.04033	0.05448	0.09481	52.09	0.1006	1.17608	84
85	0.026433	13.728	0.581	14.310	20.424	29.021	49.445	0.04077	0.05626	0.09703	53.09	0.1025	1.21445	85
86	0.027329	13.754	0.602	14.356	20.664	30.017	50.681	0.04121	0.05809	0.09930	54.09	0.1043	1.25388	86
87	0.028254	13.779	0.624	14.403	20.905	31.045	51.949	0.04165	0.05998	0.10163	55.09	0.1061	1.29443	87
88	0.029208	13.804	0.646	14.450	21.145	32.105	53.250	0.04209	0.06192	0.10401	56.09	0.1080	1.33613	88
89	0.030189	13.829	0.669	14.498	21.385	33.197	54.582	0.04253	0.06392	0.10645	57.09	0.1098	1.37893	89
90	0.031203	13.855	0.692	14.547	21.626	34.325	55.951	0.04297	0.06598	0.10895	58.08	0.1116	1.42298	90
91	0.032247	13.880	0.717	14.597	21.866	35.489	57.355	0.06810	0.04340	0.11150	59.08	0.1134	1.46824	91
92	0.033323	13.905	0.742	14.647	22.107	36.687	58.794	0.04384	0.07028	0.11412	60.08	0.1152	1.51471	92
93	0.034433	13.930	0.768	14.699	22.347	37.924	60.271	0.04427	0.07253	0.11680	61.08	0.1170	1.56248	93
94	0.035577	13.956	0.795	14.751	22.588	39.199	61.787	0.04471	0.07484	0.11955	62.08	0.1188	1.61154	94
95	0.036757	13.981	0.823	14.804	22.828	40.515	63.343	0.04514	0.07722	0.12237	63.08	0.1206	1.66196	95
96	0.037972	14.006	0.852	14.858	23.069	41.871	64.940	0.04558	0.07968	0.12525	64.07	0.1224	1.71372	96
97	0.039225	14.032	0.881	14.913	23.309	43.269	66.578	0.04601	0.08220	0.12821	65.07	0.1242	1.76685	97
98	0.040516	14.057	0.912	14.969	23.550	44.711	68.260	0.04644	0.08480	0.13124	66.07	0.1260	1.82141	98
99	0.041848	14.082	0.944	15.026	23.790	46.198	69.988	0.04687	0.08747	0.13434	67.07	0.1278	1.87745	99
100	0.043219	14.107	0.976	15.084	24.031	47.730	71.761	0.04730	0.09022	0.13752	68.07	0.1296	1.93492	100
101	0.044634	14.133	1.010	15.143	24.271	49.312	73.583	0.04773	0.09306	0.14079	69.07	0.1314	1.99396	101
102	0.046090	14.158	1.045	15.203	24.512	50.940	75.452	0.04816	0.09597	0.14413	70.06	0.1332	2.05447	102
103	0.047592	14.183	1.081	15.264	24.752	52.621	77.373	0.04859	0.09897	0.14756	71.06	0.1349	2.11661	103
104	0.049140	14.208	1.118	15.326	24.993	54.354	79.346	0.04901	0.10206	0.15108	72.06	0.1367	2.18037	104
105	0.050737	14.234	1.156	15.390	25.233	56.142	81.375	0.04944	0.10525	0.15469	73.06	0.1385	2.24581	105
106	0.052383	14.259	1.196	15.455	25.474	57.986	83.460	0.04987	0.10852	0.15839	74.06	0.1402	2.31297	106
107	0.054077	14.284	1.236	15.521	25.714	59.884	85.599	0.05029	0.11189	0.16218	75.06	0.1420	2.38173	107
108	0.055826	14.309	1.279	15.588	25.955	61.844	87.799	0.05071	0.11537	0.16608	76.05	0.1438	2.45232	108
109	0.057628	14.335	1.322	15.657	26.195	63.866	90.061	0.05114	0.11894	0.17008	77.05	0.1455	2.52473	109
110	0.059486	14.360	1.367	15.727	26.436	65.950	92.386	0.05156	0.12262	0.17418	78.05	0.1473	2.59891	110
111	0.061401	14.385	1.414	15.799	26.677	68.099	94.776	0.05198	0.12641	0.17839	79.05	0.1490	2.67500	111
112	0.063378	14.411	1.462	15.872	26.917	70.319	97.237	0.05240	0.13032	0.18272	80.05	0.1508	2.75310	112
113	0.065411	14.436	1.511	15.947	27.158	72.603	99.760	0.05282	0.13434	0.18716	81.05	0.1525	2.83291	113
114	0.067512	14.461	1.562	16.023	27.398	74.964	102.362	0.05324	0.13847	0.19172	82.04	0.1543	2.91491	114
115	0.069676	14.486	1.615	16.101	27.639	77.396	105.035	0.05366	0.14274	0.19640	83.04	0.1560	2.99883	115
116	0.071908	14.512	1.670	16.181	27.879	79.906	107.786	0.05408	0.14713	0.20121	84.04	0.1577	3.08488	116
117	0.074211	14.537	1.726	16.263	28.120	82.497	110.617	0.05450	0.15165	0.20615	85.04	0.1595	3.17305	117
118	0.076586	14.562	1.784	16.346	28.361	85.169	113.530	0.05492	0.15631	0.21122	86.04	0.1612	3.26335	118
119	0.079036	14.587	1.844	16.432	28.601	87.927	116.528	0.05533	0.16111	0.21644	87.04	0.1629	3.35586	119
120	0.081560	14.613	1.906	16.519	28.842	90.770	119.612	0.05575	0.16605	0.22180	88.04	0.1647	3.45052	120
121	0.084169	14.638	1.971	16.609	29.083	93.709	122.792	0.05616	0.17115	0.22731	89.04	0.1664	3.54764	121
122	0.086860	14.663	2.037	16.700	29.323	96.742	126.065	0.05658	0.17640	0.23298	90.03	0.1681	3.64704	122
123	0.089633	14.688	2.106	16.794	29.564	99.868	129.432	0.05699	0.18181	0.23880	91.03	0.1698	3.74871	123
124	0.092500	14.714	2.176	16.890	29.805	103.102	132.907	0.05740	0.18739	0.24480	92.03	0.1715	3.85298	124
125	0.095456	14.739	2.250	16.989	30.045	106.437	136.482	0.05781	0.19314	0.25096	93.03	0.1732	3.95961	125
126	0.098504	14.764	2.325	17.090	30.286	109.877	140.163	0.05823	0.19907	0.25729	94.03	0.1749	4.06863	126
127	0.101657	14.789	2.404	17.193	30.527	113.438	143.965	0.05864	0.20519	0.26382	95.03	0.1766	4.18046	127
128	0.104910	14.815	2.485	17.299	30.767	117.111	147.878	0.05905	0.21149	0.27054	96.03	0.1783	4.29477	128
129	0.108270	14.840	2.569	17.409	31.008	120.908	151.916	0.21800	0.21810	0.27745	97.03	0.1800	4.41181	129

Table 2 Thermodynamic Properties of Moist Air, Standard Atmospheric Pressure, 14.696 psi (29.921 in. Hg) (*Continued*)

Temp. t, °F	Humidity Ratio, lb_w/lb_{da} W_s	Volume, ft³/lb dry air			Enthalpy, Btu/lb dry air			Entropy, Btu/(lb dry air)·°F			Condensed Water			Temp., °F
											Enthalpy, Btu/lb	Entropy, Btu/lb·°F	Vapor Press., in. Hg	
		v_a	v_{as}	v_s	h_a	h_{as}	h_s	s_a	s_{as}	s_s	h_w	s_w	p_s	
130	0.111738	14.865	2.655	17.520	31.249	124.828	156.076	0.05986	0.22470	0.28457	98.03	0.1817	4.53148	130
131	0.115322	14.891	2.745	17.635	31.489	128.880	160.370	0.06027	0.23162	0.29190	99.02	0.1834	4.65397	131
132	0.119023	14.916	2.837	17.753	31.730	133.066	164.796	0.06068	0.23876	0.29944	100.02	0.1851	4.77919	132
133	0.122855	14.941	2.934	17.875	31.971	137.403	169.374	0.06109	0.24615	0.30723	101.02	0.1868	4.90755	133
134	0.126804	14.966	3.033	17.999	32.212	141.873	174.084	0.06149	0.25375	0.31524	102.02	0.1885	5.03844	134
135	0.130895	14.992	3.136	18.127	32.452	146.504	178.957	0.06190	0.26161	0.32351	103.02	0.1902	5.17258	135
136	0.135124	15.017	3.242	18.259	32.693	151.294	183.987	0.06230	0.26973	0.33203	104.02	0.1919	5.30973	136
137	0.139494	15.042	3.352	18.394	32.934	156.245	189.179	0.06271	0.27811	0.34082	105.02	0.1935	5.44985	137
138	0.144019	15.067	3.467	18.534	33.175	161.374	194.548	0.06311	0.28707	0.35018	106.02	0.1952	5.59324	138
139	0.148696	15.093	3.585	18.678	33.415	166.677	200.092	0.06351	0.29602	0.35954	107.02	0.1969	5.73970	139
140	0.153538	15.118	3.708	18.825	33.656	172.168	205.824	0.06391	0.30498	0.36890	108.02	0.1985	5.88945	140
141	0.158643	15.143	3.835	18.978	33.897	177.857	211.754	0.06431	0.31456	0.37887	109.02	0.2002	6.04256	141
142	0.163748	15.168	3.967	19.135	34.138	183.754	217.892	0.06471	0.32446	0.38918	110.02	0.2019	6.19918	142
143	0.169122	15.194	4.103	19.297	34.379	189.855	224.233	0.06511	0.33470	0.39981	111.02	0.2035	6.35898	143
144	0.174694	15.219	4.245	19.464	34.620	196.183	230.802	0.06551	0.34530	0.41081	112.02	0.2052	6.52241	144
145	0.180467	15.244	4.392	19.637	34.860	202.740	237.600	0.06591	0.35626	0.42218	113.02	0.2068	6.68932	145
146	0.186460	15.269	4.545	19.815	35.101	209.550	244.651	0.06631	0.36764	0.43395	114.02	0.2085	6.86009	146
147	0.192668	15.295	4.704	19.999	35.342	216.607	251.949	0.06671	0.37941	0.44611	115.02	0.2101	7.03435	147
148	0.199110	15.320	4.869	20.189	35.583	223.932	259.514	0.06710	0.39160	0.45871	116.02	0.2118	7.21239	148
149	0.205792	15.345	5.040	20.385	35.824	231.533	267.356	0.06750	0.40424	0.47174	117.02	0.2134	7.39413	149
150	0.212730	15.370	5.218	20.589	36.064	239.426	275.490	0.06790	0.41735	0.48524	118.02	0.2151	7.57977	150
151	0.219945	15.396	5.404	20.799	36.305	247.638	283.943	0.06829	0.43096	0.49925	119.02	0.2167	7.76958	151
152	0.227429	15.421	5.596	21.017	36.546	256.158	292.705	0.06868	0.44507	0.51375	120.02	0.2184	7.96306	152
153	0.235218	15.446	5.797	21.243	36.787	265.028	301.816	0.06908	0.45973	0.52881	121.02	0.2200	8.16087	153
154	0.243309	15.471	6.005	21.477	37.028	274.245	311.273	0.06947	0.47494	0.54441	122.02	0.2216	8.36256	154
155	0.251738	15.497	6.223	21.720	37.269	283.849	321.118	0.06986	0.49077	0.56064	123.02	0.2233	8.56871	155
156	0.260512	15.522	6.450	21.972	37.510	293.849	331.359	0.07025	0.50723	0.57749	124.02	0.2249	8.77915	156
157	0.269644	15.547	6.686	22.233	37.751	304.261	342.012	0.07065	0.52434	0.59499	125.02	0.2265	8.99378	157
158	0.279166	15.572	6.933	22.505	37.992	315.120	353.112	0.07104	0.54217	0.61320	126.02	0.2281	9.21297	158
159	0.289101	15.598	7.190	22.788	38.233	326.452	364.685	0.07143	0.56074	0.63216	127.02	0.2297	9.43677	159
160	0.29945	15.623	7.459	23.082	38.474	338.263	376.737	0.07181	0.58007	0.65188	128.02	0.2314	9.6648	160
161	0.31027	15.648	7.740	23.388	38.715	350.610	389.325	0.07220	0.60025	0.67245	129.02	0.2330	9.8978	161
162	0.32156	15.673	8.034	23.707	38.956	363.501	402.457	0.07259	0.62128	0.69388	130.03	0.2346	10.1353	162
163	0.33336	15.699	8.341	24.040	39.197	376.979	416.175	0.07298	0.64325	0.71623	131.03	0.2362	10.3776	163
164	0.34572	15.724	8.664	24.388	39.438	391.095	430.533	0.07337	0.66622	0.73959	132.03	0.2378	10.6250	164
165	0.35865	15.749	9.001	24.750	39.679	405.865	445.544	0.07375	0.69022	0.76397	133.03	0.2394	10.8771	165
166	0.37220	15.774	9.355	25.129	39.920	421.352	461.271	0.07414	0.71535	0.78949	134.03	0.2410	11.1343	166
167	0.38639	15.800	9.726	25.526	40.161	437.578	477.739	0.07452	0.74165	0.81617	135.03	0.2426	11.3965	167
168	0.40131	15.825	10.117	25.942	40.402	454.630	495.032	0.07491	0.76925	0.84415	136.03	0.2442	11.6641	168
169	0.41698	15.850	10.527	26.377	40.643	472.554	513.197	0.07529	0.79821	0.87350	137.04	0.2458	11.9370	169
170	0.43343	15.875	10.959	26.834	40.884	491.372	532.256	0.07567	0.82858	0.90425	138.04	0.2474	12.2149	170
171	0.45079	15.901	11.414	27.315	41.125	511.231	552.356	0.07606	0.86058	0.93664	139.04	0.2490	12.4988	171
172	0.46905	15.926	11.894	27.820	41.366	532.138	573.504	0.07644	0.89423	0.97067	140.04	0.2506	12.7880	172
173	0.48829	15.951	12.400	28.352	41.607	554.160	595.767	0.07682	0.92962	1.00644	141.04	0.2521	13.0823	173
174	0.50867	15.976	12.937	28.913	41.848	577.489	619.337	0.07720	0.96707	1.04427	142.04	0.2537	13.3831	174
175	0.53019	16.002	13.504	29.505	42.089	602.139	644.229	0.07758	1.00657	1.08416	143.05	0.2553	13.6894	175
176	0.55294	16.027	14.103	30.130	42.331	628.197	670.528	0.07796	1.04828	1.12624	144.05	0.2569	14.0010	176
177	0.57710	16.052	14.741	30.793	42.572	655.876	698.448	0.07834	1.09253	1.17087	145.05	0.2585	14.3191	177
178	0.60274	16.078	15.418	31.496	42.813	685.260	728.073	0.07872	1.13943	1.21815	146.05	0.2600	14.6430	178
179	0.63002	16.103	16.139	32.242	43.054	716.524	759.579	0.07910	1.18927	1.26837	147.06	0.2616	14.9731	179
180	0.65911	16.128	16.909	33.037	43.295	749.871	793.166	0.07947	1.24236	1.32183	148.06	0.2632	15.3097	180
181	0.69012	16.153	17.730	33.883	43.536	785.426	828.962	0.07985	1.29888	1.37873	149.06	0.2647	15.6522	181
182	0.72331	16.178	18.609	34.787	43.778	823.487	867.265	0.08023	1.35932	1.43954	150.06	0.2663	16.0014	182
183	0.75885	16.204	19.551	35.755	44.019	864.259	908.278	0.08060	1.42396	1.50457	151.07	0.2679	16.3569	183
184	0.79703	16.229	20.564	36.793	44.260	908.061	952.321	0.08098	1.49332	1.57430	152.07	0.2694	16.7190	184
185	0.83817	16.254	21.656	37.910	44.501	955.261	999.763	0.08135	1.56797	1.64932	153.07	0.2710	17.0880	185
186	0.88251	16.280	22.834	39.113	44.742	1006.149	1050.892	0.08172	1.64834	1.73006	154.08	0.2725	17.4634	186
187	0.93057	16.305	24.111	40.416	44.984	1061.314	1106.298	0.08210	1.73534	1.81744	155.08	0.2741	17.8462	187
188	0.98272	16.330	25.498	41.828	45.225	1121.174	1166.399	0.08247	1.82963	1.91210	156.08	0.2756	18.2357	188
189	1.03951	16.355	27.010	43.365	45.466	1186.382	1231.848	0.08284	1.93221	2.01505	157.09	0.2772	18.6323	189
190	1.10154	16.381	28.661	45.042	45.707	1257.614	1303.321	0.08321	2.04412	2.12733	158.09	0.2787	19.0358	190
191	1.16965	16.406	30.476	46.882	45.949	1335.834	1381.783	0.08359	2.16684	2.25043	159.09	0.2803	19.4468	191
192	1.24471	16.431	32.477	48.908	46.190	1422.047	1468.238	0.08396	2.30193	2.38589	160.10	0.2818	19.8652	192
193	1.32788	16.456	34.695	51.151	46.431	1517.581	1564.013	0.08433	2.45144	2.53576	161.10	0.2834	20.2913	193
194	1.42029	16.481	37.161	53.642	46.673	1623.758	1670.430	0.08470	2.61738	2.70208	162.11	0.2849	20.7244	194
195	1.52396	16.507	39.928	56.435	46.914	1742.879	1789.793	0.08506	2.80332	2.88838	163.11	0.2864	21.1661	195
196	1.64070	16.532	43.046	59.578	47.155	1877.032	1924.188	0.08543	3.01244	3.09787	164.12	0.2880	21.6152	196
197	1.77299	16.557	46.580	63.137	47.397	2029.069	2076.466	0.08580	3.24914	3.33494	165.12	0.2895	22.0714	197
198	1.92472	16.583	50.636	67.218	47.638	2203.464	2251.102	0.08617	3.52030	3.60647	166.13	0.2910	22.5367	198
199	2.09975	16.608	55.316	71.923	47.879	2404.668	2452.547	0.08653	3.83275	3.91929	167.13	0.2926	23.0092	199
200	2.30454	16.633	60.793	77.426	48.121	2640.084	2688.205	0.08690	4.19787	4.28477	168.13	0.2941	23.4906	200

Table 3 Thermodynamic Properties of Water at Saturation

Temp. t, °F	Absolute Pressure p		Specific Volume, ft³/lb			Enthalpy, Btu/lb			Entropy, Btu/lb·°F			Temp., °F
	psi	in. Hg	Sat. Solid/Liq. v_i	Evap. v_{ig}	Sat. Vapor v_g	Sat. Solid/Liq. h_i	Evap. h_{ig}	Sat. Vapor h_g	Sat. Solid/Liq. s_i	Evap. s_{ig}	Sat. Vapor s_g	
−80	0.000116	0.000236	0.01732	1953234	1953234	−193.50	1219.19	1025.69	−0.4067	3.2112	2.8045	−80
−79	0.000125	0.000254	0.01732	1814052	1814052	−193.11	1219.24	1026.13	−0.4056	3.2029	2.7972	−79
−78	0.000135	0.000275	0.01732	1685445	1685445	−192.71	1219.28	1026.57	−0.4046	3.1946	2.7900	−78
−77	0.000145	0.000296	0.01732	1566663	1566663	−192.31	1219.33	1027.02	−0.4036	3.1964	2.7828	−77
−76	0.000157	0.000319	0.01732	1456752	1456752	−191.92	1219.38	1027.46	−0.4025	3.1782	2.7757	−76
−75	0.000169	0.000344	0.01733	1355059	1355059	−191.52	1219.42	1027.90	−0.4015	3.1701	2.7685	−75
−74	0.000182	0.000371	0.01733	1260977	1260977	−191.12	1219.47	1028.34	−0.4005	3.1619	2.7615	−74
−73	0.000196	0.000399	0.01733	1173848	1173848	−190.72	1219.51	1028.79	−0.3994	3.1539	2.7544	−73
−72	0.000211	0.000430	0.01733	1093149	1093149	−190.32	1219.55	1029.23	−0.3984	3.1459	2.7475	−72
−71	0.000227	0.000463	0.01733	1018381	1018381	−189.92	1219.59	1029.67	−0.3974	3.1379	2.7405	−71
−70	0.000245	0.000498	0.01733	949067	949067	−189.52	1219.63	1030.11	−0.3963	3.1299	2.7336	−70
−69	0.000263	0.000536	0.01733	884803	884803	−189.11	1219.67	1030.55	−0.3953	3.1220	2.7267	−69
−68	0.000283	0.000576	0.01733	825187	825187	−188.71	1219.71	1031.00	−0.3943	3.1141	2.7199	−68
−67	0.000304	0.000619	0.01734	769864	769864	−188.30	1219.74	1031.44	−0.3932	3.1063	2.7131	−67
−66	0.000326	0.000664	0.01734	718508	718508	−187.90	1219.78	1031.88	−0.3922	3.0985	2.7063	−66
−65	0.000350	0.000714	0.01734	670800	670800	−187.49	1219.82	1032.32	−0.3912	3.0907	2.6996	−65
−64	0.000376	0.000766	0.01734	626503	626503	−187.08	1219.85	1032.77	−0.3901	3.0830	2.6929	−64
−63	0.000404	0.000822	0.01734	585316	585316	−186.67	1219.88	1033.21	−0.3891	3.0753	2.6862	−63
−62	0.000433	0.000882	0.01734	548041	547041	−186.26	1219.91	1033.65	−0.3881	3.0677	2.6730	−62
−61	0.000464	0.000945	0.01734	511446	511446	−185.85	1219.95	1034.09	−0.3870	3.0601	2.6730	−61
−60	0.000498	0.001013	0.01734	478317	478317	−185.44	1219.98	1034.54	−0.3860	3.0525	2.6665	−60
−59	0.000533	0.001086	0.01735	447495	447495	−185.03	1220.01	1034.98	−0.3850	3.0449	2.6600	−59
−58	0.000571	0.001163	0.01735	418803	418803	−184.61	1220.03	1035.42	−0.3839	3.0374	2.6535	−58
−57	0.000612	0.001246	0.01735	392068	392068	−184.20	1220.06	1035.86	−0.3829	3.0299	2.6470	−57
−56	0.000655	0.001333	0.01735	367172	367172	−183.78	1220.09	1036.30	−0.3819	3.0225	2.6406	−56
−55	0.000701	0.001427	0.01735	343970	343970	−183.37	1220.11	1036.75	−0.3808	3.0151	2.6342	−55
−54	0.000750	0.001526	0.01735	322336	322336	−182.95	1220.14	1037.19	−0.3798	3.0077	2.6279	−54
−53	0.000802	0.001632	0.01735	302157	302157	−182.53	1220.16	1037.63	−0.3788	3.0004	2.6216	−53
−52	0.000857	0.001745	0.01735	283335	283335	−182.11	1220.18	1038.07	−0.3778	2.9931	2.6153	−52
−51	0.000916	0.001865	0.01736	265773	265773	−181.69	1220.21	1038.52	−0.3767	2.9858	2.6091	−51
−50	0.000979	0.001992	0.01736	249381	249381	−181.27	1220.23	1038.96	−0.3757	2.9786	2.6029	−50
−49	0.001045	0.002128	0.01736	234067	234067	−180.85	1220.25	1039.40	−0.3747	2.9714	2.5967	−49
−48	0.001116	0.002272	0.01736	219766	219766	−180.42	1220.26	1039.84	−0.3736	2.9642	2.5906	−48
−47	0.001191	0.002425	0.01736	206398	206398	−180.00	1220.28	1040.28	−0.3726	2.9570	2.5844	−47
−46	0.001271	0.002587	0.01736	193909	193909	−179.57	1220.30	1040.73	−0.3716	2.9499	2.5784	−46
−45	0.001355	0.002760	0.01736	182231	182231	−179.14	1220.31	1041.17	−0.3705	2.9429	2.5723	−45
−44	0.001445	0.002943	0.01736	171304	171304	−178.72	1220.33	1041.61	−0.3695	2.9358	2.5663	−44
−43	0.001541	0.003137	0.01737	161084	161084	−178.79	1220.34	1042.05	−0.3685	2.9288	2.5603	−43
−42	0.001642	0.003343	0.01737	151518	151518	−177.86	1220.36	1042.50	−0.3675	2.9218	2.5544	−42
−41	0.001749	0.003562	0.01737	142566	142566	−177.43	1220.37	1042.94	−0.3664	2.9149	2.5485	−41
−40	0.001863	0.003793	0.01737	134176	134176	−177.00	1220.38	1043.38	−0.3654	2.9080	2.5426	−40
−39	0.001984	0.004039	0.01737	126322	126322	−176.57	1220.39	1043.82	−0.3644	2.9011	2.5367	−39
−38	0.002111	0.004299	0.01737	118959	118959	−176.13	1220.40	1044.27	−0.3633	2.8942	2.5309	−38
−37	0.002247	0.004574	0.01737	112058	112058	−175.70	1220.40	1044.71	−0.3623	2.8874	2.5251	−37
−36	0.002390	0.004866	0.01738	105592	105592	−175.26	1220.41	1045.15	−0.3613	2.8806	2.5193	−36
−35	0.002542	0.005175	0.01738	99522	99522	−174.83	1220.42	1045.59	−0.3603	2.8738	2.5136	−35
−34	0.002702	0.005502	0.01738	93828	93828	−174.39	1220.42	1046.03	−0.3592	2.8671	2.5078	−34
−33	0.002872	0.005848	0.01738	88489	88489	−173.95	1220.43	1046.48	−0.3582	2.8604	2.5022	−33
−32	0.003052	0.006213	0.01738	83474	83474	−173.51	1220.43	1046.92	−0.3572	2.8537	2.4965	−32
−31	0.003242	0.006600	0.01738	78763	78763	−173.07	1220.43	1047.36	−0.3561	2.8470	2.4909	−31
−30	0.003443	0.007009	0.01738	74341	74341	−172.63	1220.43	1047.80	−0.3551	2.8404	2.4853	−30
−29	0.003655	0.007441	0.01738	70187	70187	−172.19	1220.43	1048.25	−0.3541	2.8338	2.4797	−29
−28	0.003879	0.007898	0.01739	66282	66282	−171.74	1220.43	1048.69	−0.3531	2.8272	2.4742	−28
−27	0.004116	0.008380	0.01739	62613	62613	−171.30	1220.43	1049.13	−0.3520	2.8207	2.4687	−27
−26	0.004366	0.008890	0.01739	59161	59161	−170.86	1220.43	1049.57	−0.3510	2.8142	2.4632	−26
−25	0.004630	0.009428	0.01739	55915	55915	−170.41	1220.42	1050.01	−0.3500	2.8077	2.4577	−25
−24	0.004909	0.009995	0.01739	52861	52861	−169.96	1220.42	1050.46	−0.3489	2.8013	2.4523	−24
−23	0.005203	0.010594	0.01739	49986	49986	−169.51	1220.41	1050.90	−0.3479	2.7948	2.4469	−23
−22	0.005514	0.011226	0.01739	47281	47281	−169.07	1220.41	1051.34	−0.3469	2.7884	2.4415	−22
−21	0.005841	0.011892	0.01740	44733	44733	−168.62	1220.40	1051.78	−0.3459	2.7820	2.4362	−21
−20	0.006186	0.012595	0.01740	42333	42333	−168.16	1220.39	1052.22	−0.3448	2.7757	2.4309	−20
−19	0.006550	0.013336	0.01740	40073	40073	−167.71	1220.38	1052.67	−0.3438	2.7694	2.4256	−19
−18	0.006933	0.014117	0.01740	37943	37943	−167.26	1220.37	1053.11	−0.3428	2.7631	2.4203	−18
−17	0.007337	0.014939	0.01740	35934	35934	−166.81	1220.36	1053.55	−0.3418	2.7568	2.4151	−17
−16	0.007763	0.015806	0.01740	34041	34041	−166.35	1220.34	1053.99	−0.3407	2.7506	2.4098	−16
−15	0.008211	0.016718	0.01740	32256	32256	−165.90	1220.33	1054.43	−0.3397	2.7444	2.4046	−15
−14	0.008683	0.017678	0.01741	30572	30572	−165.44	1220.31	1054.87	−0.3387	2.7382	2.3995	−14

Table 3 Thermodynamic Properties of Water at Saturation (*Continued*)

Temp. t, °F	Absolute Pressure p psi	Absolute Pressure p in. Hg	Specific Volume, ft³/lb: Sat. Solid/Liq. v_i	Evap. v_{ig}	Sat. Vapor v_g	Enthalpy, Btu/lb: Sat. Solid/Liq. h_i	Evap. h_{ig}	Sat. Vapor h_g	Entropy, Btu/lb·°F: Sat. Solid/Liq. s_i	Evap. s_{ig}	Sat. Vapor s_g	Temp., °F
−13	0.009179	0.018689	0.01741	28983	28983	−164.98	1220.30	1055.32	−0.3377	2.7320	2.3943	−13
−12	0.009702	0.019753	0.01741	27483	27483	−164.52	1220.28	1055.76	−0.3366	2.7259	2.3892	−12
−11	0.010252	0.020873	0.01741	26067	26067	−164.06	1220.26	1056.20	−0.3356	2.7197	2.3841	−11
−10	0.010830	0.022050	0.01741	24730	24730	−163.60	1220.24	1056.64	−0.3346	2.7136	2.3791	−10
−9	0.011438	0.023288	0.01741	23467	23467	−163.14	1220.22	1057.08	−0.3335	2.7076	2.3740	−9
−8	0.012077	0.024590	0.01741	22274	22274	−162.68	1220.20	1057.53	−0.3325	2.7015	2.3690	−8
−7	0.012749	0.025958	0.01742	21147	21147	−162.21	1220.18	1057.97	−0.3315	2.6955	2.3640	−7
−6	0.013456	0.027396	0.01742	20081	20081	−162.75	1220.16	1058.41	−0.3305	2.6895	2.3591	−6
−5	0.014197	0.028906	0.01742	19074	19074	−161.28	1220.13	1058.85	−0.3294	2.6836	2.3541	−5
−4	0.014977	0.030493	0.01742	18121	18121	−160.82	1220.11	1059.29	−0.3284	2.6776	2.3492	−4
−3	0.015795	0.032159	0.01742	17220	17220	−160.35	1220.08	1059.73	−0.3274	2.6717	2.3443	−3
−2	0.016654	0.033908	0.01742	16367	16367	−159.88	1220.05	1060.17	−0.3264	2.6658	2.3394	−2
−1	0.017556	0.035744	0.01742	15561	15561	−159.41	1220.02	1060.62	−0.3253	2.6599	2.3346	−1
0	0.018502	0.037671	0.01743	14797	14797	−158.94	1220.00	1061.06	−0.3243	2.6541	2.3298	0
1	0.019495	0.039693	0.01743	14073	14073	−158.47	1219.96	1061.50	−0.3233	2.6482	2.3249	1
2	0.020537	0.041813	0.01743	13388	13388	−157.99	1219.93	1061.94	−0.3223	2.6424	2.3202	2
3	0.021629	0.044037	0.01743	12740	12740	−157.52	1219.90	1062.38	−0.3212	2.6367	2.3154	3
4	0.022774	0.046369	0.01743	12125	12125	−157.05	1219.87	1062.82	−0.3202	2.6309	2.3107	4
5	0.023975	0.048813	0.01743	11543	11543	−156.57	1219.83	1063.26	−0.3192	2.6252	2.3060	5
6	0.025233	0.051375	0.01743	10991	10991	−156.09	1219.80	1063.70	−0.3182	2.6194	2.3013	6
7	0.026552	0.054059	0.01744	10468	10468	−155.62	1219.76	1064.14	−0.3171	2.6138	2.2966	7
8	0.027933	0.056872	0.01744	9971	9971	−155.14	1219.72	1064.58	−0.3161	2.6081	2.2920	8
9	0.029379	0.059817	0.01744	9500	9500	−154.66	−1219.68	−1065.03	−0.3151	−2.6024	−2.2873	9
10	0.030894	0.062901	0.01744	9054	9054	−154.18	1219.64	1065.47	−0.3141	2.5968	2.2827	10
11	0.032480	0.066131	0.01744	8630	8630	−153.70	1219.60	1065.91	−0.3130	2.5912	2.2782	11
12	0.034140	0.069511	0.01744	8228	8228	−153.21	1219.56	1066.35	−0.3120	2.5856	2.2736	12
13	0.035878	0.073047	0.01745	7846	7846	−152.73	1219.52	1066.79	−0.3110	2.5801	2.2691	13
14	0.037696	0.076748	0.01745	7483	7483	−152.24	1219.47	1067.23	−0.3100	2.5745	2.2645	14
15	0.039597	0.080621	0.01745	7139	7139	−151.76	1219.43	1067.67	−0.3089	2.5690	2.2600	15
16	0.041586	0.084671	0.01745	6811	6811	−151.27	1219.38	1068.11	−0.3079	2.5635	2.2556	16
17	0.043666	0.088905	0.01745	6501	6501	−150.78	1219.33	1068.55	−0.3069	2.5580	2.2511	17
18	0.045841	0.093332	0.01745	6205	6205	−150.30	1219.28	1068.99	−0.3059	2.5526	2.2467	18
19	0.048113	0.097960	0.01745	5924	5924	−149.81	1219.23	1069.43	−0.3049	2.5471	2.2423	19
20	0.050489	0.102796	0.01746	5657	5657	−149.32	1219.18	1069.87	−0.3038	2.5417	2.2379	20
21	0.052970	0.107849	0.01746	5404	5404	−148.82	1219.13	1070.31	−0.3028	2.5363	2.2335	21
22	0.055563	0.113128	0.01746	5162	5162	−148.33	1219.08	1070.75	−0.3018	2.5309	2.2292	22
23	0.058271	0.118641	0.01746	4932	4932	−147.84	1219.02	1071.19	−0.3008	2.5256	2.2248	23
24	0.061099	0.124398	0.01746	4714	4714	−147.34	1218.97	1071.63	−0.2997	2.5203	2.2205	24
25	0.064051	0.130408	0.01746	4506	4506	−146.85	1218.91	1072.07	−0.2987	2.5149	2.2162	25
26	0.067133	0.136684	0.01747	4308	4308	−146.35	1218.85	1072.50	−0.2977	2.5096	2.2119	26
27	0.070349	0.143233	0.01747	4119	4119	−145.85	1218.80	1072.94	−0.2967	2.5044	2.2077	27
28	0.073706	0.150066	0.01747	3940	3940	−145.35	1218.74	1073.38	−0.2956	2.4991	2.2035	28
29	0.077207	0.157195	0.01747	3769	3769	−144.85	1218.68	1073.82	−0.2946	2.4939	2.1992	29
30	0.080860	0.164632	0.01747	3606	3606	−144.35	1218.61	1074.26	−0.2936	2.4886	2.1951	30
31	0.084669	0.172387	0.01747	3450	3450	−143.85	1218.55	1074.70	−0.2926	2.4834	2.1909	31
32	0.088640	0.180474	0.01747	3302	3302	−143.35	1218.49	1075.14	−0.2915	2.4783	2.1867	32
32*	0.08865	0.18049	0.01602	3302.07	3302.09	−0.02	1075.15	1075.14	0.0000	2.1867	2.1867	32
33	0.09229	0.18791	0.01602	3178.15	3178.16	0.99	1074.59	1075.58	0.0020	2.1811	2.1832	33
34	0.09607	0.19559	0.01602	3059.47	3059.49	2.00	1074.02	1076.01	0.0041	2.1756	2.1796	34
35	0.09998	0.20355	0.01602	2945.66	2945.68	3.00	1073.45	1076.45	0.0061	2.1700	2.1761	35
36	0.10403	0.21180	0.01602	2836.60	2836.61	4.01	1072.88	1076.89	0.0081	2.1645	2.1726	36
37	0.10822	0.22035	0.01602	2732.13	2732.15	5.02	1072.32	1077.33	0.0102	2.1590	2.1692	37
38	0.11257	0.22919	0.01602	2631.88	2631.89	6.02	1071.75	1077.77	0.0122	2.1535	2.1657	38
39	0.11707	0.23835	0.01602	2535.86	2535.88	7.03	1071.18	1078.21	0.0142	2.1481	2.1623	39
40	0.12172	0.24783	0.01602	2443.67	2443.69	8.03	1070.62	1078.65	0.0162	2.1426	2.1589	40
41	0.12654	0.25765	0.01602	2355.22	2355.24	9.04	1070.05	1079.09	0.0182	2.1372	2.1554	41
42	0.13153	0.26780	0.01602	2270.42	2270.43	10.04	1069.48	1079.52	0.0202	2.1318	2.1521	42
43	0.13669	0.27831	0.01602	2189.02	2189.04	11.04	1068.92	1079.96	0.0222	2.1265	2.1487	43
44	0.14203	0.28918	0.01602	2110.92	2110.94	12.05	1068.35	1080.40	0.0242	2.1211	2.1454	44
45	0.14755	0.30042	0.01602	2035.91	2035.92	13.05	1067.79	1080.84	0.0262	2.1158	2.1420	45
46	0.15326	0.31205	0.01602	1963.85	1963.87	14.05	1067.22	1081.28	0.0282	2.1105	2.1387	46
47	0.15917	0.32407	0.01602	1894.71	1894.73	15.06	1066.66	1081.71	0.0302	2.1052	2.1354	47
48	0.16527	0.33650	0.01602	1828.28	1828.30	16.06	1066.09	1082.15	0.0321	2.1000	2.1321	48
49	0.17158	0.34935	0.01602	1764.44	1764.46	17.06	1065.53	1082.59	0.0341	2.0947	2.1288	49
50	0.17811	0.36263	0.01602	1703.18	1703.20	18.06	1064.96	1083.03	0.0361	2.0895	2.1256	50
51	0.18484	0.37635	0.01602	1644.25	1644.26	19.06	1064.40	1083.46	0.0381	2.0843	2.1224	51
52	0.19181	0.39053	0.01603	1587.64	1587.65	20.07	1063.83	1083.90	0.0400	2.0791	2.1191	52

*Extrapolated to represent metastable equilibrium with undercooled liquid.

Table 3 Thermodynamic Properties of Water at Saturation (*Continued*)

Temp.	Absolute Pressure p		Specific Volume, ft³/lb			Enthalpy, Btu/lb			Entropy, Btu/lb·°F			Temp.,
			Sat. Solid/Liq.	Evap.	Sat. Vapor	Sat. Solid/Liq.	Evap.	Sat. Vapor	Sat. Solid/Liq.	Evap.	Sat. Vapor	
t, °F	psi	in. Hg	v_i	v_{ig}	v_g	h_i	h_{ig}	h_g	s_i	s_{ig}	s_g	°F
53	0.19900	0.40516	0.01603	1533.22	1533.24	21.07	1063.27	1084.34	0.0420	2.0740	2.1159	53
54	0.20643	0.42029	0.01603	1480.89	1480.91	22.07	1062.71	1084.77	0.0439	2.0689	2.1128	54
55	0.21410	0.43591	0.01603	1430.61	1430.62	23.07	1062.14	1085.21	0.0459	2.0637	2.1096	55
56	0.22202	0.45204	0.01603	1382.19	1382.21	24.07	1061.58	1085.65	0.0478	2.0586	2,1064	56
57	0.23020	0.46869	0.01603	1335.65	1335.67	25.07	1061.01	1086.08	0.0497	2.0536	2.1033	57
58	0.23864	0.48588	0.01603	1290.85	1290.87	26.07	1060.45	1086.52	0.0517	2.0485	2.0002	58
59	0.24735	0.50362	0.01603	1247.76	1247.78	27.07	1059.89	1086.96	0.0536	2.0435	2.0971	59
60	0.25635	0.52192	0.01604	1206.30	1206.32	28.07	1059.32	1087.39	0.0555	2.0385	2.0940	60
61	0.26562	0.54081	0.01604	1166.38	1166.40	29.07	1058.76	1087.83	0.0575	2.0334	2.0909	61
62	0.27519	0.56029	0.01604	1127.93	1127.95	30.07	1058.19	1088.27	0.0594	2.0285	2.0878	62
63	0.28506	0.58039	0.01604	1090.94	1090.96	31.07	1057.63	1088.70	0.0613	2.0235	2.0848	63
64	0.29524	0.60112	0.01604	1055.32	1055.33	32.07	1057.07	1089.14	0.0632	2.0186	2.0818	64
65	0.30574	0.62249	0.01604	1020.98	1021.00	33.07	1056.50	1089.57	0.0651	2.0136	2.0787	65
66	0.31656	0.64452	0.01604	987.95	987.97	34.07	1055.94	1090.01	0.0670	2.0087	2.0758	66
67	0.32772	0.66724	0.01605	956.11	956.12	35.07	1055.37	1090.44	0.0689	2.0039	2.0728	67
68	0.33921	0.69065	0.01605	925.44	925.45	36.07	1054.81	1090.88	0.0708	1.9990	2.0698	68
69	0.35107	0.71478	0.01605	895.86	895.87	37.07	1054.24	1091.31	0.0727	1.9941	2.0668	69
70	0.36328	0.73964	0.01605	867.34	867.36	38.07	1053.68	1091.75	0.0746	1.9893	2.0639	70
71	0.37586	0.76526	0.01605	839.87	839.88	39.07	1053.11	1092.18	0.0765	1.9845	2.0610	71
72	0.38882	0.79164	0.01606	813.37	813.39	40.07	1052.55	1092.61	0.0783	1.9797	2.0580	72
73	0.40217	0.81883	0.01606	787.85	787.87	41.07	1051.98	1093.05	0.0802	1.9749	2.0552	73
74	0.41592	0.84682	0.01606	763.19	763.21	42.06	1051.42	1093.48	0.0821	1.9702	2.0523	74
75	0.43008	0.87564	0.01606	739.42	739.44	43.06	1050.85	1093.92	0.0840	1.9654	2.0494	75
76	0.44465	0.90532	0.01606	716.51	726.53	44.06	1050.29	1094.35	0.0858	1.9607	2.0465	76
77	0.45966	0.93587	0.01607	694.38	794.40	45.06	1049.72	1094.78	0.0877	1.9560	2.0437	77
78	0.47510	0.96732	0.01607	673.05	673.06	46.06	1049.16	1095.22	0.0896	1.9513	2.0409	78
79	0.49100	0.99968	0.01607	652.44	652.46	47.06	1048.59	1095.65	0.0914	1.9466	2.0380	79
80	0.50736	1.03298	0.01607	632.54	632.56	48.06	1048.03	1096.08	0.0933	1.9420	2.0352	80
81	0.52419	1.06725	0.01608	613.35	613.37	49.06	1047.46	1096.51	0.0951	1.9373	2.0324	81
82	0.54150	1.10250	0.01608	594.82	594.84	50.05	1046.89	1096.95	0.0970	1.9327	2.0297	82
83	0.55931	1.13877	0.01608	576.90	576.92	51.05	1046.33	1097.38	0.0988	1.9281	2.0269	83
84	0.57763	1.17606	0.01608	559.63	559.65	52.05	1045.76	1097.81	0.1006	1.9235	2.0242	84
85	0.59647	1.21442	0.01609	542.93	542.94	53.05	1045.19	1098.24	0.1025	1.9189	2.0214	85
86	0.61584	1.25385	0.01609	526.80	526.81	54.05	1044.63	1098.67	0.1043	1.9144	2.0187	86
87	0.63575	1.29440	0.01609	511.21	511.22	55.05	1044.06	1099.11	0.1061	1.9098	2.0160	87
88	0.65622	1.33608	0.01609	496.14	496.15	56.05	1043.49	1099.54	0.1080	1.9053	2.0133	88
89	0.67726	1.37892	0.01610	481.60	481.61	57.04	1042.92	1099.97	0.1098	1.9008	2.0106	89
90	0.69889	1.42295	0.01610	467.52	467.53	58.04	1042.36	1100.40	0.1116	1.8963	2.0079	90
91	0.72111	1.46820	0.01610	453.91	453.93	59.04	1041.79	1100.83	0.1134	1.8918	2.0053	91
92	0.74394	1.51468	0.01611	440.76	440.78	60.04	1041.22	1101.26	0.1152	1.8874	2.0026	92
93	0.76740	1.56244	0.01611	428.04	428.06	61.04	1040.65	1101.69	0.1170	1.8829	2.0000	93
94	0.79150	1.61151	0.01611	415.74	415.76	62.04	1040.08	1102.12	0.1188	1.8785	1.9973	94
95	0.81625	1.66189	0.01612	403.84	403.86	63.03	1039.51	1102.55	0.1206	1.8741	1.9947	95
96	0.84166	1.71364	0.01612	392.33	392.34	64.03	1038.95	1102.98	0.1224	1.8697	1.9921	96
97	0.86776	1.76678	0.01612	381.20	381.21	65.03	1038.38	1103.41	0.1242	1.8653	1.9895	97
98	0.89456	1.82134	0.01612	370.42	370.44	66.03	1037.81	1103.84	0.1260	1.8610	1.9870	98
99	0.92207	1.87736	0.01613	359.99	360.01	67.03	1037.24	1104.26	0.1278	1.8566	1.9844	99
100	0.95031	1.93485	0.01613	349.91	349.92	68.03	1036.67	1104.69	0.1296	1.8523	1.9819	100
101	0.97930	1.99387	0.01613	340.14	340.15	69.03	1036.10	1105.12	0.1314	1.8479	1.9793	101
102	1.00904	2.05443	0.01614	330.69	330.71	70.02	1035.53	1105.55	0.1332	1.8436	1.9768	102
103	1.03956	2.11667	0.01614	321.53	321.55	71.02	1034.95	1105.98	0.1349	1.8393	1.9743	103
104	1.07088	2.18034	0.01614	312.67	312.69	72.02	1034.38	1106.40	0.1367	1.8351	1.9718	104
105	1.10301	2.24575	0.01615	304.08	304.10	73.02	1033.81	1106.83	0.1385	1.8308	1.9693	105
106	1.13597	2.31285	0.01615	295.76	295.77	74.02	1033.24	1107.26	0.1402	1.8266	1.9668	106
107	1.16977	2.38168	0.01616	287.71	287.73	75.01	1032.67	1107.68	0.1420	1.8223	1.9643	107
108	1.20444	2.45226	0.01616	279.91	279.92	76.01	1032.10	1108.11	0.1438	1.8181	1.9619	108
109	1.23999	2.52464	0.01616	272.34	272.36	77.01	1031.52	1108.54	0.1455	1.8139	1.9594	109
110	1.27644	2.59885	0.01617	265.02	265.03	78.01	1030.95	1108.96	0.1473	1.8097	1.9570	110
111	1.31381	2.67494	0.01617	257.91	257.93	79.01	1030.38	1109.39	0.1490	1.8055	1.9546	111
112	1.35212	2.75293	0.01617	251.02	251.04	80.01	1029.80	1109.81	0.1508	1.8014	1.9521	112
113	1.39138	2.83288	0.01618	244.36	244.38	81.01	1029.23	1110.24	0.1525	1.7972	1.9497	113
114	1.43162	2.91481	0.01618	237.89	237.90	82.00	1028.66	1110.66	0.1543	1.7931	1.9474	114
115	1.47286	2.99878	0.01619	231.62	231.63	83.00	1028.08	1111.09	0.1560	1.7890	1.9450	115
116	1.51512	3.08481	0.01619	225.53	225.55	84.00	1027.51	1111.51	0.1577	1.7849	1.9426	116
117	1.55842	3.17296	0.01619	219.63	219.65	85.00	1026.93	1111.93	0.1595	1.7808	1.9402	117
118	1.60277	3.26327	0.01620	213.91	213.93	86.00	1026.36	1112.36	0.1612	1.7767	1.9379	118
119	1.64820	3.35577	0.01620	208.36	208.37	87.00	1025.78	1112.78	0.1629	1.7726	1.9356	119
120	1.69474	3.45052	0.01620	202.98	202.99	88.00	1025.20	1113.20	0.1647	1.7686	1.9332	120

Table 3 Thermodynamic Properties of Water at Saturation (*Continued*)

Temp.	Absolute Pressure p		Specific Volume, ft³/lb			Enthalpy, Btu/lb			Entropy, Btu/lb·°F			Temp.,
t, °F	psi	in. Hg	Sat. Solid/Liq. v_i	Evap. v_{ig}	Sat. Vapor v_g	Sat. Solid/Liq. h_i	Evap. h_{ig}	Sat. Vapor h_g	Sat. Solid/Liq. s_i	Evap. s_{ig}	Sat. Vapor s_g	°F
121	1.74240	3.54755	0.01621	197.76	197.76	89.00	1024.63	1113.62	0.1664	1.7645	1.9309	121
122	1.79117	3.64691	0.01621	192.69	192.69	90.00	1024.05	1114.05	0.1681	1.7605	1.9286	122
123	1.84117	3.74863	0.01622	187.78	187.78	90.99	1023.47	1114.47	0.1698	1.7565	1.9263	123
124	1.89233	3.85282	0.01622	182.98	182.99	91.99	1022.90	1114.89	0.1715	1.7525	1.9240	124
125	1.94470	3.95945	0.01623	178.34	178.36	92.99	1022.32	1115.31	0.1732	1.7485	1.9217	125
126	1.99831	4.06860	0.01623	173.85	173.86	93.99	1021.74	1115.73	0.1749	1.7445	1.9195	126
127	2.05318	4.18032	0.01623	169.47	169.49	94.99	1021.16	1116.15	0.1766	1.7406	1.9172	127
128	2.10934	4.29465	0.01624	165.23	165.25	95.99	1020.58	1116.57	0.1783	1.7366	1.9150	128
129	2.16680	4.41165	0.01624	161.11	161.12	96.99	1020.00	1116.99	0.1800	1.7327	1.9127	129
130	2.22560	4.53136	0.01625	157.11	157.12	97.99	1019.42	1117.41	0.1817	1.7288	1.9105	130
131	2.28576	4.65384	0.01625	153.22	153.23	98.99	1018.84	1117.83	0.1834	1.7249	1.9083	131
132	2.34730	4.77914	0.01626	149.44	149.46	99.99	1018.26	1118.25	0.1851	1.7210	1.9061	132
133	2.41025	4.90730	0.01626	145.77	145.78	100.99	1017.68	1118.67	0.1868	1.7171	1.9039	133
134	2.47463	5.03839	0.01627	142.21	142.23	101.99	1017.10	1119.08	0.1885	1.7132	1.9017	134
135	2.54048	5.17246	0.01627	138.74	138.76	102.99	1016.52	1119.50	0.1902	1.7093	1.8995	135
136	2.60782	5.30956	0.01627	135.37	135.39	103.98	1015.93	1119.92	0.1919	1.7055	1.8974	136
137	2.67667	5.44975	0.01628	132.10	132.12	104.98	1015.35	1120.34	0.1935	1.7017	1.8952	137
138	2.74707	5.59308	0.01628	128.92	128.94	105.98	1014.77	1120.75	0.1952	1.6978	1.8930	138
139	2.81903	5.73961	0.01629	125.83	125.85	106.98	1014.18	1121.17	0.1969	1.6940	1.8909	139
140	2.89260	5.88939	0.01629	122.82	122.84	107.98	1013.60	1121.58	0.1985	1.6902	1.8888	140
141	2.96780	6.04250	0.01630	119.90	119.92	108.98	1013.01	1122.00	0.2002	1.6864	1.8867	141
142	3.04465	6.19897	0.01630	117.05	117.07	109.98	1012.43	1122.41	0.2019	1.6827	1.8845	142
143	3.12320	6.35888	0.01631	114.29	114.31	110.98	1011.84	1122.83	0.2035	1.6789	1.8824	143
144	3.20345	6.52229	0.01631	111.60	111.62	111.98	1011.26	1123.24	0.2052	1.6752	1.8803	144
145	3.28546	6.68926	0.01632	108.99	109.00	112.98	1010.67	1123.66	0.2068	1.6714	1.8783	145
146	3.36924	6.85984	0.01632	106.44	106.45	113.98	1010.09	1124.07	0.2085	1.6677	1.8762	146
147	3.45483	7.03410	0.01633	103.96	103.98	114.98	1009.50	1124.48	0.2101	1.6640	1.8741	147
148	3.54226	7.21211	0.01633	101.55	101.57	115.98	1008.91	1124.89	0.2118	1.6603	1.8721	148
149	3.63156	7.39393	0.01634	99.21	99.22	116.98	1008.32	1125.31	0.2134	1.6566	1.8700	149
150	3.72277	7.57962	0.01634	96.93	96.94	117.98	1007.73	1125.72	0.2151	1.6529	1.8680	150
151	3.81591	7.76925	0.01635	94.70	94.72	118.99	1007.14	1126.13	0.2167	1.6492	1.8659	151
152	3.91101	7.96289	0.01635	92.54	92.56	119.99	1006.55	1126.54	0.2184	1.6455	1.8639	152
153	4.00812	8.16061	0.01636	90.44	90.46	120.99	1005.96	1126.95	0.2200	1.6419	1.8619	153
154	4.10727	8.36247	0.01636	88.39	88.41	121.99	1005.37	1127.36	0.2216	1.6383	1.8599	154
155	4.20848	8.56854	0.01637	86.40	86.41	122.99	1004.78	1127.77	0.2233	1.6346	1.8579	155
156	4.31180	8.77890	0.01637	84.45	84.47	123.99	1004.19	1128.18	0.2249	1.6310	1.8559	156
157	4.41725	8.99360	0.01638	82.56	82.58	124.99	1003.60	1128.59	0.2265	1.6274	1.8539	157
158	4.52488	9.21274	0.01638	80.72	80.73	125.99	1003.00	1128.99	0.2281	1.6238	1.8519	158
159	4.63472	9.43637	0.01639	78.92	78.94	126.99	1002.41	1129.40	0.2297	1.6202	1.8500	159
160	4.7468	9.6646	0.01639	77.175	77.192	127.99	1001.82	1129.81	0.2314	1.6167	1.8480	160
161	4.8612	9.8974	0.01640	75.471	75.488	128.99	1001.22	1130.22	0.2330	1.6131	1.8461	161
162	4.9778	10.1350	0.01640	73.812	73.829	130.00	1000.63	1130.62	0.2346	1.6095	1.8441	162
163	5.0969	10.3774	0.01641	72.196	72.213	131.00	1000.03	1131.03	0.2362	1.6060	1.8422	163
164	5.2183	10.6246	0.01642	70.619	70.636	132.00	999.43	1131.43	0.2378	1.6025	1.8403	164
165	5.3422	10.8768	0.01642	69.084	69.101	133.00	998.84	1131.84	0.2394	1.5989	1.8383	165
166	5.4685	11.1340	0.01643	67.587	67.604	134.00	998.24	1132.24	0.2410	1.5954	1.8364	166
167	5.5974	11.3963	0.01643	66.130	66.146	135.00	997.64	1132.64	0.2426	1.5919	1.8345	167
168	5.7287	11.6638	0.01644	64.707	64.723	136.01	997.04	1133.05	0.2442	1.5884	1.8326	168
169	5.8627	11.9366	0.01644	63.320	63.336	137.01	996.44	1133.45	0.2458	1.5850	1.8308	169
170	5.9993	12.2148	0.01645	61.969	61.986	138.01	995.84	1133.85	0.2474	1.5815	1.8289	170
171	6.1386	12.4983	0.01646	60.649	60.666	139.01	995.24	1134.25	0.2490	1.5780	1.8270	171
172	6.2806	12.7874	0.01646	59.363	59.380	140.01	994.64	1134.66	0.2506	1.5746	1.8251	172
173	6.4253	13.0821	0.01647	58.112	58.128	141.02	994.04	1135.06	0.2521	1.5711	1.8233	173
174	6.5729	13.3825	0.01647	56.887	56.904	142.02	993.44	1135.46	0.2537	1.5677	1.8214	174
175	6.7232	13.6886	0.01648	55.694	55.711	143.02	992.83	1135.86	0.2553	1.5643	1.8196	175
176	6.8765	14.0006	0.01648	54.532	54.549	144.02	992.23	1136.26	0.2569	1.5609	1.8178	176
177	7.0327	14.3186	0.01649	53.397	53.414	145.03	991.63	1136.65	0.2585	1.5575	1.8159	177
178	7.1918	14.6426	0.01650	52.290	52.307	146.03	991.02	1137.05	0.2600	1.5541	1.8141	178
179	7.3539	14.9727	0.01650	51.210	51.226	147.03	990.42	1137.45	0.2616	1.5507	1.8123	179
180	7.5191	15.3091	0.01651	50.155	50.171	148.04	989.81	1137.85	0.2632	1.5473	1.8105	180
181	7.6874	15.6518	0.01651	49.126	49.143	149.04	989.20	1138.24	0.2647	1.5440	1.8087	181
182	7.8589	16.0008	0.01652	48.122	48.138	150.04	988.60	1138.64	0.2663	1.5406	1.8069	182
183	8.0335	16.3564	0.01653	47.142	47.158	151.05	987.99	1139.03	0.2679	1.5373	1.8051	183
184	8.2114	16.7185	0.01653	46.185	46.202	152.05	987.38	1139.43	0.2694	1.5339	1.8034	184
185	8.3926	17.0874	0.01654	45.251	45.267	153.05	986.77	1139.82	0.2710	1.5306	1.8016	185
186	8.5770	17.4630	0.01654	44.339	44.356	154.06	986.16	1140.22	0.2725	1.5273	1.7998	186
187	8.7649	17.8455	0.01655	43.448	43.465	155.06	985.55	1140.61	0.2741	1.5240	1.7981	187
188	8.9562	18.2350	0.01656	42.579	42.595	156.07	984.94	1141.00	0.2756	1.5207	1.7963	188
189	9.1510	18.6316	0.01656	41.730	41.746	157.07	984.32	1141.39	0.2772	1.5174	1.7946	189

Table 3 Thermodynamic Properties of Water at Saturation (*Continued*)

Temp. t, °F	Absolute Pressure p psi	Absolute Pressure p in. Hg	Specific Volume, ft³/lb: Sat. Solid/Liq. v_i	Evap. v_{ig}	Sat. Vapor v_g	Enthalpy, Btu/lb: Sat. Solid/Liq. h_i	Evap. h_{ig}	Sat. Vapor h_g	Entropy, Btu/lb·°F: Sat. Solid/Liq. s_i	Evap. s_{ig}	Sat. Vapor s_g	Temp., °F
190	9.3493	19.0353	0.01657	40.901	40.918	158.07	983.71	1141.78	0.2787	1.5141	1.7929	190
191	9.5512	19.4464	0.01658	40.092	40.108	159.08	983.10	1142.18	0.2803	1.5109	1.7911	191
192	9.7567	19.8648	0.01658	39.301	39.317	160.08	982.48	1142.57	0.2818	1.5076	1.7894	192
193	9.9659	20.2907	0.01659	38.528	38.544	161.09	981.87	1142.95	0.2834	1.5043	1.7877	193
194	10.1788	20.7242	0.01659	37.774	37.790	162.09	981.25	1143.34	0.2849	1.5011	1.7860	194
195	10.3955	21.1653	0.01660	37.035	37.052	163.10	980.63	1143.73	0.2864	1.4979	1.7843	195
196	10.6160	21.6143	0.01661	36.314	36.331	164.10	980.02	1144.12	0.2880	1.4946	1.7826	196
197	10.8404	22.0712	0.01661	35.611	35.628	165.11	979.40	1144.51	0.2895	1.4914	1.7809	197
198	11.0687	22.5361	0.01662	34.923	34.940	166.11	978.78	1144.89	0.2910	1.4882	1.7792	198
199	11.3010	23.0091	0.01663	34.251	34.268	167.12	978.16	1145.28	0.2926	1.4850	1.7776	199
200	11.5374	23.4904	0.01663	33.594	33.610	168.13	977.54	1145.66	0.2941	1.4818	1.7759	200
201	11.7779	23.9800	0.01664	32.951	32.968	169.13	976.92	1146.05	0.2956	1.4786	1.7742	201
202	12.0225	24.4780	0.01665	32.324	32.340	170.14	976.29	1146.43	0.2971	1.4755	1.7726	202
203	12.2713	24.9847	0.01665	31.710	31.726	171.14	975.67	1146.81	0.2986	1.4723	1.7709	203
204	12.5244	25.5000	0.01666	31.110	31.127	172.15	975.05	1147.20	0.3002	1.4691	1.7693	204
205	12.7819	26.0241	0.01667	30.523	30.540	173.16	974.42	1147.58	0.3017	1.4660	1.7677	205
206	13.0436	26.5571	0.01667	29.949	29.965	174.16	973.80	1147.96	0.3032	1.4628	1.7660	206
207	13.3099	27.0991	0.01668	29.388	29.404	175.17	973.17	1148.34	0.3047	1.4597	1.7644	207
208	13.5806	27.6503	0.01669	28.839	28.856	176.18	972.54	1148.72	0.3062	1.4566	1.7628	208
209	13.8558	28.2108	0.01669	28.303	28.319	177.18	971.92	1149.10	0.3077	1.4535	1.7612	209
210	14.1357	28.7806	0.01670	27.778	27.795	178.19	971.29	1149.48	0.3092	1.4503	1.7596	210
212	14.7096	29.9489	0.01671	26.763	26.780	180.20	970.03	1150.23	0.3122	1.4442	1.7564	212
214	15.3025	31.1563	0.01673	25.790	25.807	182.22	968.76	1150.98	0.3152	1.4380	1.7532	214
216	15.9152	32.4036	0.01674	24.861	24.878	184.24	967.50	1151.73	0.3182	1.4319	1.7501	216
218	16.5479	33.6919	0.01676	23.970	23.987	186.25	966.23	1152.48	0.3212	1.4258	1.7469	218
220	17.2013	35.0218	0.01677	23.118	23.134	188.27	964.95	1153.22	0.3241	1.4197	1.7438	220
222	17.8759	36.3956	0.01679	22.299	22.316	190.29	963.67	1153.96	0.3271	1.4136	1.7407	222
224	18.5721	37.8131	0.01680	21.516	21.533	192.31	962.39	1154.70	0.3301	1.4076	1.7377	224
226	19.2905	39.2758	0.01682	20.765	20.782	194.33	961.11	1155.43	0.3330	1.4016	1.7347	226
228	20.0316	40.7848	0.01683	20.045	20.062	196.35	959.82	1156.16	0.3359	1.3957	1.7316	228
230	20.7961	42.3412	0.01684	19.355	19.372	198.37	958.52	1156.89	0.3389	1.3898	1.7287	230
232	21.5843	43.9461	0.01686	18.692	18.709	200.39	957.22	1157.62	0.3418	1.3839	1.7257	232
234	22.3970	45.6006	0.01688	18.056	18.073	202.41	955.92	1158.34	0.3447	1.3780	1.7227	234
236	23.2345	47.3060	0.01689	17.446	17.463	204.44	954.62	1159.06	0.3476	1.3722	1.7198	236
238	24.0977	49.0633	0.01691	16.860	16.877	206.46	953.31	1159.77	0.3505	1.3664	1.7169	238
240	24.9869	50.8738	0.01692	16.298	16.314	208.49	952.00	1160.48	0.3534	1.3606	1.7140	240
242	25.9028	52.7386	0.01694	15.757	15.774	210.51	950.68	1161.19	0.3563	1.3548	1.7111	242
244	26.8461	54.6591	0.01695	15.238	15.255	212.54	949.35	1161.90	0.3592	1.3491	1.7083	244
246	27.8172	56.6364	0.01697	14.739	14.756	214.57	948.03	1162.60	0.3621	1.3434	1.7055	246
248	28.8169	58.6717	0.01698	14.259	14.276	216.60	946.70	1163.29	0.3649	1.3377	1.7026	248
250	29.8457	60.7664	0.01700	13.798	13.815	218.63	945.36	1163.99	0.3678	1.3321	1.6998	250
252	30.9043	62.9218	0.01702	13.355	13.372	220.66	944.02	1164.68	0.3706	1.3264	1.6971	252
254	31.9934	65.1391	0.01703	12.928	12.945	222.69	942.68	1165.37	0.3735	1.3208	1.6943	254
256	33.1135	67.4197	0.01705	12.526	12.147	226.73	939.99	1166.72	0.3764	1.3153	1.6691	256
258	34.2653	69.7649	0.01707	12.123	12.140	226.76	939.97	1166.73	0.3792	1.3097	1.6889	258
260	35.4496	72.1760	0.01708	11.742	11.759	228.79	938.61	1167.40	0.3820	1.3042	1.6862	260
262	36.6669	74.6545	0.01710	11.376	11.393	230.83	937.25	1168.08	0.3848	1.2987	1.6835	262
264	37.9180	77.2017	0.01712	11.024	11.041	232.87	935.88	1168.74	0.3876	1.2932	1.6808	264
266	39.2035	79.8190	0.01714	10.684	10.701	234.90	934.50	1169.41	0.3904	1.2877	1.6781	266
268	40.5241	82.5078	0.01715	10.357	10.374	236.94	933.12	1170.07	0.3932	1.2823	1.6755	268
270	41.8806	85.2697	0.01717	10.042	10.059	238.98	931.74	1170.72	0.3960	1.2769	1.6729	270
272	43.2736	88.1059	0.01719	9.737	9.755	241.03	930.35	1171.38	0.3988	1.2715	1.6703	272
274	44.7040	91.0181	0.01721	9.445	9.462	243.07	928.95	1172.02	0.4016	1.2661	1.6677	274
276	46.1723	94.0076	0.01722	9.162	9.179	245.11	927.55	1172.67	0.4044	1.2608	1.6651	276
278	47.6794	97.0761	0.01724	8.890	8.907	247.16	926.15	1173.31	0.4071	1.2554	1.6626	278
280	49.2260	100.2250	0.01726	8.627	8.644	249.20	924.74	1173.94	0.4099	1.2501	1.6600	280
282	50.8128	103.4558	0.01728	8.373	8.390	251.25	923.32	1174.57	0.4127	1.2448	1.6575	282
284	52.4406	106.7701	0.01730	8.128	8.146	253.30	921.90	1175.20	0.4154	1.2396	1.6550	284
286	54.1103	110.1695	0.01731	7.892	7.910	255.35	920.47	1175.82	0.4182	1.2343	1.6525	286
288	55.8225	113.6556	0.01733	7.664	7.681	257.40	919.03	1176.44	0.4209	1.2291	1.6500	288
290	57.5780	117.2299	0.01735	7.444	7.461	259.45	917.59	1177.05	0.4236	1.2239	1.6476	290
292	59.3777	120.8941	0.01737	7.231	7.248	261.51	916.15	1177.66	0.4264	1.2187	1.6451	292
294	61.2224	124.6498	0.01739	7.026	7.043	263.56	914.69	1178.26	0.4291	1.2136	1.6427	294
296	63.1128	128.4987	0.01741	6.827	6.844	265.62	913.24	1178.86	0.4318	1.2084	1.6402	296
298	65.0498	132.4425	0.01743	6.635	6.652	267.68	911.77	1179.45	0.4345	1.2033	1.6378	298
300	67.0341	136.4827	0.01745	6.450	6.467	269.74	910.30	1180.04	0.4372	1.1982	1.6354	300

HUMIDITY PARAMETERS

Humidity ratio (alternatively, the moisture content or mixing ratio) ***W*** of a given moist air sample is defined as the ratio of the mass of water vapor to the mass of dry air contained in the sample:

$$W = M_w/M_a \tag{7}$$

The humidity ratio W is equal to the mole fraction ratio x_w/x_a multiplied by the ratio of molecular masses; namely, 18.01528/28.9645 = 0.62198, i.e.:

$$W = 0.62198x_w/x_a \tag{8}$$

Specific humidity ***q*** is the ratio of the mass of water vapor to the total mass of the moist air sample:

$$q = M_w/(M_w + M_a) \tag{9a}$$

In terms of the humidity ratio:

$$q = W/(1 + W) \tag{9b}$$

Absolute humidity (alternatively, water vapor density) $\boldsymbol{d_v}$ is the ratio of the mass of water vapor to the total volume of the sample:

$$d_v = M_w/V \tag{10}$$

The **density** ρ of a moist air mixture is the ratio of the total mass to the total volume:

$$\rho = (M_a + M_w)/V = (1/v)(1 + W) \tag{11}$$

where v is the moist air specific volume, ft³/lb (dry air), as defined by Equation (27).

HUMIDITY PARAMETERS INVOLVING SATURATION

The following definitions of humidity parameters involve the concept of moist air saturation:

Saturation humidity ratio $\boldsymbol{W_s(t, p)}$ is the humidity ratio of moist air saturated with respect to water (or ice) at the same temperature t and pressure p.

Degree of saturation μ is the ratio of the air humidity ratio W to the humidity ratio W_s of saturated air at the same temperature and pressure:

$$\mu = \left.\frac{W}{W_s}\right|_{t,p} \tag{12}$$

Relative humidity ϕ is the ratio of the mole fraction of water vapor x_w in a given moist air sample to the mole fraction x_{ws} in an air sample, saturated at the same temperature and pressure:

$$\phi = \left.\frac{x_w}{x_{ws}}\right|_{t,p} \tag{13}$$

Combining Equations (8), (12), and (13):

$$\mu = \frac{\phi}{1 + (1 - \phi)W_s/0.62198} \tag{14}$$

Dew-point temperature $\boldsymbol{t_d}$ is the temperature of moist air saturated at the same pressure p, with the same humidity ratio W as that of the given sample of moist air. It is defined as the solution $t_d(p, W)$ of the equation:

$$W_s(p, t_d) = W \tag{15}$$

Thermodynamic wet-bulb temperature $\boldsymbol{t^*}$ is the temperature at which water (liquid or solid), by evaporating into moist air at a given dry-bulb temperature t and humidity ratio W, can bring air to saturation adiabatically at the same temperature t^* while the pressure p is maintained constant. This parameter is considered separately in a later section.

PERFECT GAS RELATIONSHIPS FOR DRY AND MOIST AIR

When moist air is considered a mixture of independent perfect gases, dry air, and water vapor, each is assumed to obey the perfect gas equation of state as follows:

$$\text{Dry air } p_aV = n_aRT \tag{16}$$

$$\text{Water vapor } p_wV = n_wRT \tag{17}$$

where

p_a = partial pressure of dry air
p_w = partial pressure of water vapor
V = total mixture volume
n_a = number of moles of dry air
n_w = number of moles of water vapor
R = universal gas constant 1545.32 ft·lb$_f$/(lb mol·°R)
T = absolute temperature, °R

The mixture also obeys the perfect gas equation:

$$pV = nRT \tag{18}$$

or

$$(p_a + p_w)V = (n_a + n_w)RT \tag{19}$$

where $p = p_a + p_w$ is the total mixture pressure and $n = n_a + n_w$ is the total number of moles in the mixture. From Equations (16) through (19), the mole fractions of dry air and water vapor are, respectively:

$$x_a = p_a/(p_a + p_w) = p_a/p \tag{20}$$

and

$$x_w = p_w/(p_a + p_w) = p_w/p \tag{21}$$

From Equations (8), (20), and (21), the **humidity ratio** ***W*** is given by:

$$W = 0.62198\frac{p_w}{p - p_w} \tag{22}$$

The degree of saturation μ is, by definition, Equation (12):

$$\mu = \left.\frac{W}{W_s}\right|_{t,p}$$

where

$$W_s = 0.62198\frac{p_{ws}}{p - p_{ws}} \tag{23}$$

The term p_{ws} represents the saturation pressure of water vapor in the absence of air at the given temperature t. This pressure p_{ws} is a function only of temperature and differs slightly from the vapor pressure of water in saturated moist air.

The **relative humidity** ϕ is, by definition, Equation (13):

$$\phi = \left.\frac{x_w}{x_{ws}}\right|_{t,p}$$

Substituting Equation (21) for x_w and x_{ws}:

$$\phi = \left.\frac{p_w}{p_{ws}}\right|_{t,p} \tag{24}$$

Substituting Equation (21) for x_{ws} into Equation (14):

$$\phi = \frac{\mu}{1-(1-\mu)(p_{ws}/p)} \tag{25}$$

Both ϕ and μ are zero for dry air and unity for saturated moist air. At intermediate states their values differ, substantially so at higher temperatures.

The **specific volume** v of a moist air mixture is expressed in terms of a unit mass of dry air, i.e.:

$$v = V/M_a = V/(28.9645 n_a) \tag{26}$$

where V is the total volume of the mixture, M_a is the total mass of dry air, and n_a is the number of moles of dry air. By Equations (16) and (26), with the relation $p = p_a + p_w$:

$$v = \frac{RT}{28.9645(p-p_w)} = \frac{R_a T}{p-p_w} \tag{27}$$

Using Equation (22):

$$v = \frac{RT(1+1.6078W)}{28.964p} = \frac{R_a T(1+1.6078W)}{p} \tag{28}$$

In Equations (27) and (28), v is specific volume, T is absolute temperature, p is total pressure, p_w is the partial pressure of water vapor, and W is the humidity ratio.

In specific units, Equation (28) may be expressed as

$$v = 0.7543(t+459.67)(1+1.6078W)/p$$

where

v = specific volume, ft^3/lb
t = dry-bulb temperature, °F
W = humidity ratio, lb (water)/lb (dry air)
p = total pressure, in. Hg

The **enthalpy** of a mixture of perfect gases equals the sum of the individual partial enthalpies of the components. Therefore, the enthalpy of moist air can be written:

$$h = h_a + Wh_g \tag{29}$$

where h_a is the specific enthalpy for dry air and h_g is the specific enthalpy for saturated water vapor at the temperature of the mixture. Approximately:

$$h_a = 0.240t \qquad \text{(Btu/lb)} \tag{30}$$

$$h_g = 1061 + 0.444t \qquad \text{(Btu/lb)} \tag{31}$$

where t is the dry-bulb temperature, °F. The moist air enthalpy then becomes:

$$h = 0.240t + W(1061 + 0.444t) \qquad \text{(Btu/lb)} \tag{32}$$

THERMODYNAMIC WET-BULB TEMPERATURE AND DEW-POINT TEMPERATURE

For any state of moist air, a temperature t^* exists at which liquid (or solid) water evaporates into the air to bring it to saturation at exactly this same temperature and pressure (Harrison 1965). During the adiabatic saturation process, the saturated air is expelled at a temperature equal to that of the injected water (Figures 8 and 9). In the constant pressure process, the humidity ratio is increased from a given initial value W to the value W_s^*, corresponding to saturation at the temperature t^*; the enthalpy is increased from a given initial value h to the value h_s^*, corresponding to saturation at the temperature t^*; the mass of water added per unit mass of dry air is $(W_s^* - W)$, which adds energy to the moist air of amount $(W_s^* - W)h_w^*$, where h_w^* denotes the specific enthalpy of the water added at the temperature t^*. Therefore, if the process is strictly adiabatic, conservation of enthalpy at constant pressure requires that:

$$h + (W_s^* - W)h_w^* = h_s^* \tag{33}$$

The properties W_s^*, h_w^*, and h_s^* are functions only of the temperature t^* for a fixed value of pressure. The value of t^*, which satisfies Equation (33) for given values of h, W, and p, is the **thermodynamic wet-bulb temperature**.

The **psychrometer** consists of two thermometers; one thermometer's bulb is covered by a wick that has been thoroughly wetted with water. When the wet bulb is placed in an airstream, water evaporates from the wick, eventually reaching an equilibrium temperature called the **wet-bulb temperature**. This process is not one of adiabatic saturation, which defines the thermodynamic wet-bulb temperature, but is one of simultaneous heat and mass transfer from the wet bulb. The fundamental mechanism of this process is described by the Lewis relation (Chapter 5). Fortunately, only small corrections must be applied to wet-bulb thermometer readings to obtain the thermodynamic wet-bulb temperature.

As defined, thermodynamic wet-bulb temperature is a unique property of a given moist air sample independent of measurement techniques.

Equation (33) is exact since it defines the thermodynamic wet-bulb temperature t^*. Substituting the approximate perfect gas relation [Equation (32)] for h, the corresponding expression for h_s^*, and the approximate relation

$$h_w^* = t^* - 32 \qquad \text{(Btu/lb)} \tag{34}$$

into Equation (33), and solving for the humidity ratio:

$$W = \frac{(1093 - 0.556t^*)W_s^* - 0.240(t - t^*)}{1093 + 0.444t - t^*} \tag{35}$$

where t and t^* are in °F.

The **dew-point temperature t_d** of moist air with humidity ratio W and pressure p was defined earlier as the solution $t_d(p, w)$ of $W_s(p, t_d)$. For perfect gases, this reduces to:

$$p_{ws}(t_d) = p_w = (pW)/(0.62198 + W) \tag{36}$$

where p_w is the water vapor partial pressure for the moist air sample and $p_{ws}(t_d)$ is the saturation vapor pressure at temperature t_d. The saturation vapor pressure is derived from Table 3 or from Equations (5) or (6). Alternatively, the dew-point temperature can be calculated directly by one of the following equations (Peppers 1988):

For the dew-point temperature range of 32 to 200°F:

$$t_d = C_{14} + C_{15}\alpha + C_{16}\alpha^2 + C_{17}\alpha^3 + C_{18}(p_w)^{0.1984} \quad (37)$$

and for temperatures below 32°F:

$$t_d = 90.12 + 26.412\alpha + 0.8927\alpha^2 \quad (38)$$

where

t_d = dew-point temperature, °F
α = ln (p_w)
p_w = water vapor partial pressure, psia
C_{14} = 100.45
C_{15} = 33.193
C_{16} = 2.319
C_{17} = 0.17074
C_{18} = 1.2063

NUMERICAL CALCULATION OF MOIST AIR PROPERTIES

The following are outlines, citing equations and tables already presented, for calculating moist air properties using perfect gas relations. These relations are sufficiently accurate for most engineering calculations in air-conditioning practice, and are readily adapted to either hand or computer calculating methods. For more details, refer to Tables 15 through 18 in Chapter 1 of Olivieri (1996). Graphical procedures are discussed in the section on psychrometric charts.

Situation 1.

Given: Dry-bulb temperature t, Wet-bulb temperature t^*, Pressure p

To Obtain	Use	Comments
$p_{ws}(t^*)$	Table 3 or Eq. (5) or (6)	Sat. press. for temp. t^*
W_s^*	Eq. (23)	Using $p_{ws}(t^*)$
W	Eq. (35)	
$p_{ws}(t)$	Table 3 or Eq. (5) or (6)	Sat. press. for temp. t
W_s	Eq. (23)	Using $p_{ws}(t)$
μ	Eq. (12)	Using W_s
ϕ	Eq. (25)	Using $p_{ws}(t)$
v	Eq. (28)	
h	Eq. (32)	
p_w	Eq. (36)	
t_d	Table 3 with Eq. (36), (37), or (38)	

Situation 2.

Given: Dry-bulb temperature t, Dew-point temperature t_d, Pressure p

To Obtain	Use	Comments
$p_w = p_{ws}(t_d)$	Table 3 or Eq. (5) or (6)	Sat. press. for temp. t_d
W	Eq. (22)	
$p_{ws}(t)$	Table 3 or Eq. (5) or (6)	Sat. press. for temp. t_d
W_s	Eq. (23)	Using $p_{ws}(t)$
μ	Eq. (12)	Using W_s
ϕ	Eq. (25)	Using $p_{ws}(t)$
v	Eq. (28)	
h	Eq. (32)	
t^*	Eq. (23) and (35) with Table 3 or with Eq. (5) or (6)	Requires trial-and-error or numerical solution method

Situation 3.

Given: Dry-bulb temperature t, Relative humidity ϕ, Pressure p

To Obtain	Use	Comments
$p_{ws}(t)$	Table 3 or Eq. (5) or (6)	Sat. press. for temp. t
p_w	Eq. (24)	
W	Eq. (22)	
W_s	Eq. (23)	Using $p_{ws}(t)$
μ	Eq. (12)	Using W_s
v	Eq. (28)	
h	Eq. (32)	
t_d	Table 3 with Eq. (36), (37), or (38)	
t^*	Eq. (23) and (35) with Table 3 or with Eq. (5) or (6)	Requires trial-and-error or numerical solution method

EXACT RELATIONS FOR COMPUTING W_s AND ϕ

Corrections that account for (1) the effect of dissolved gases on properties of condensed phase; (2) the effect of pressure on properties of condensed phase; and (3) the effect of intermolecular force on properties of moisture itself, can be applied to Equations (23) or (25):

$$W_s = 0.62198\frac{f_s p_{ws}}{p - f_s p_{ws}} \quad (23a)$$

$$\phi = \frac{\mu}{1 - (1 - \mu)(f_s p_{ws}/p)} \quad (23b)$$

Table 4 lists f_s values for a number of pressure and temperature combinations. Hyland and Wexler (1983a) give additional values.

Table 4 Values of f_s and Estimated Maximum Uncertainties (EMU)

	14.50 psia		72.52 psia		145.04 psia	
T, °R	f	EMU E+04	f	EMU E+04	f	EMU E+04
311.67	1.0105	134	1.0540	66	1.1130	136
491.67	1.0039	2	1.0177	10	1.0353	19
671.67	1.0039	0.1	1.0180	4	1.0284	11

MOIST AIR PROPERTY TABLES FOR STANDARD PRESSURE

Table 2 shows values of thermodynamic properties for standard atmospheric pressure at temperatures from –80 to 200°F. The properties of intermediate moist air states can be calculated using the degree of saturation μ:

Volume $$v = v_a + \mu v_{as} \quad (39)$$

Enthalpy $$h = h_s + \mu h_{as} \quad (40)$$

Entropy $$s = s_a + \mu s_{as} \quad (41)$$

These equations are accurate to about 160°F. At higher temperatures, the errors can be significant. Hyland and Wexler (1983a) include charts that can be used to estimate errors for v, h, and s for standard barometric pressure.

PSYCHROMETRIC CHARTS

A psychrometric chart graphically represents the thermodynamic properties of moist air.

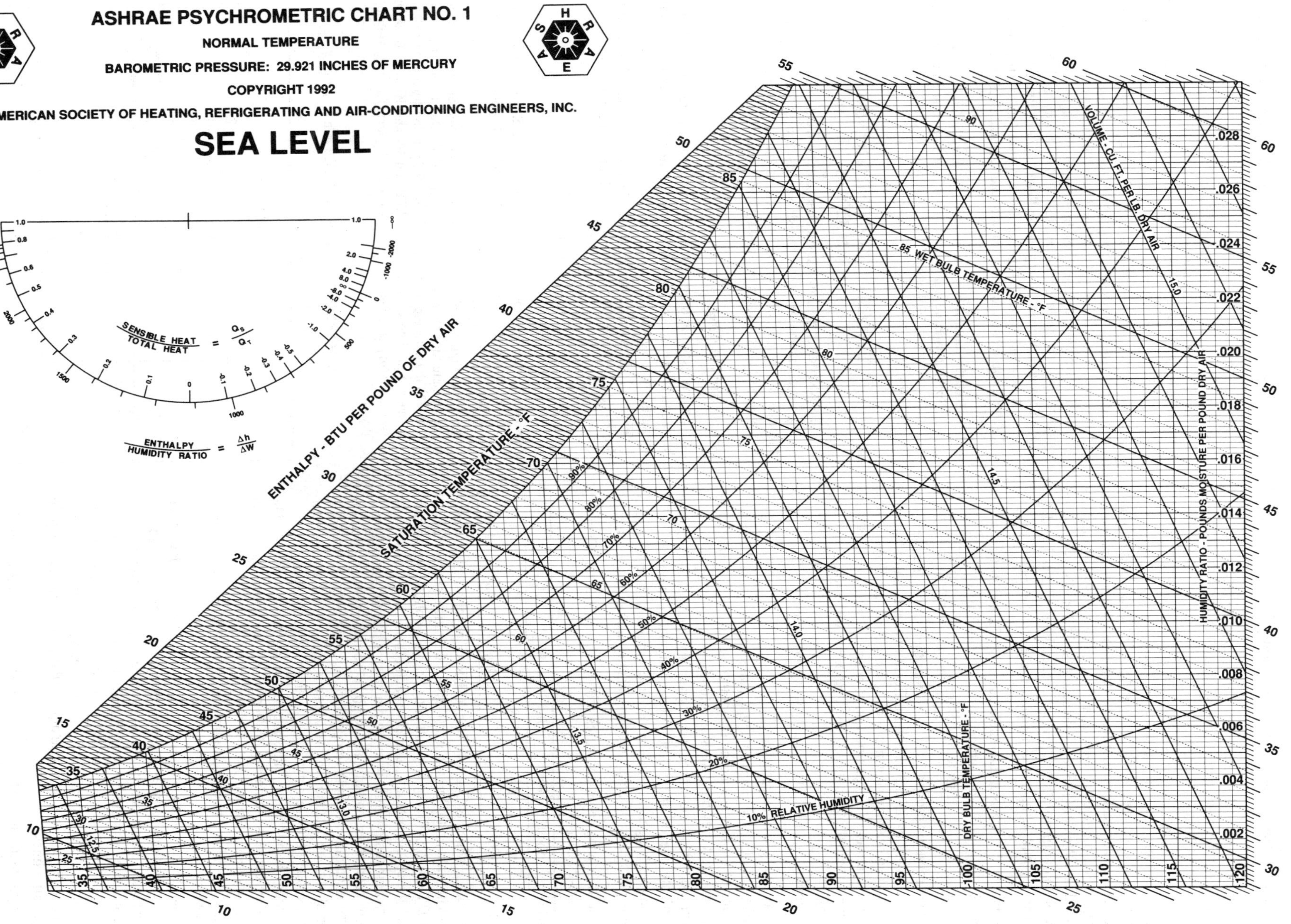

Fig. 1 ASHRAE Psychrometric Chart No. 1

The choice of coordinates for a psychrometric chart is arbitrary. A chart with coordinates of enthalpy and humidity ratio provides convenient graphical solutions of many moist air problems with a minimum of thermodynamic approximations. ASHRAE developed five such psychrometric charts.

Charts 1, 2, and 3 are for sea level pressure. Chart 4 is for 5000-ft altitude (24.89 in. Hg). Chart 5 is for 7500-ft altitude (22.65 in. Hg). All charts use oblique-angle coordinates of enthalpy and humidity ratio, and are consistent with the data of Table 2 and the properties computation methods of Goff and Gratch (1945, 1949) as well as Hyland and Wexler (1983a). Palmatier (1963) describes the geometry of chart construction applying specifically to Charts 1 and 4.

The dry-bulb temperature ranges covered by the charts are:

Charts 1, 4, 5	Normal temperature	32 to 120°F
Chart 2	Low temperature	−40 to 50°F
Chart 3	High temperature	60 to 250°F

Psychrometric properties or charts for other barometric pressures can be derived by interpolation. Sufficiently exact values for most purposes can be derived by methods described in the section on perfect gas relations. The construction of charts for altitude conditions has been treated by Haines (1961), Rohsenow (1946), and Karig (1946).

Comparison of Charts 1 and 4 by overlay reveals:

1. The dry-bulb lines coincide.
2. Wet-bulb lines for a given temperature originate at the intersections of the corresponding dry-bulb line and the two saturation curves, and they have the same slope.
3. Humidity ratio and enthalpy for a given dry- and wet-bulb increase with altitude, but there is little change in relative humidity.
4. Volume changes rapidly; for a given dry-bulb and humidity ratio, it is practically inversely proportional to barometric pressure.

The following table compares properties at sea level (Chart 1) and 5000-ft (Chart 4):

Chart No.	db	wb	h	W	rh	v
1	100	81	44.6	0.0186	45	14.5
4	100	81	49.8	0.0234	46	17.6

Figure 1, which is Chart 1 of the ASHRAE psychrometric charts, shows humidity ratio lines (horizontal) for the range from 0 (dry air) to 0.03 lb (water)/lb (dry air). Enthalpy lines are oblique lines drawn across the chart precisely parallel to each other.

Dry-bulb temperature lines are drawn straight, not precisely parallel to each other, and inclined slightly from the vertical position. Thermodynamic wet-bulb temperature lines are oblique lines that differ slightly in direction from that of enthalpy lines. They are identically straight but are not precisely parallel to each other.

Relative humidity (rh) lines are shown in intervals of 10%. The saturation curve is the line of 100% rh, while the horizontal line for $W = 0$ (dry air) is the line for 0% rh.

Specific volume lines are straight but are not precisely parallel to each other.

A narrow region above the saturation curve has been developed for fog conditions of moist air. This two-phase region represents a mechanical mixture of saturated moist air and liquid water, with the two components in thermal equilibrium. Isothermal lines in the fog region coincide with extensions of thermodynamic wet-bulb temperature lines. If required, the fog region can be further expanded by extension of humidity ratio, enthalpy, and thermodynamic wet-bulb temperature lines.

The protractor to the left of the chart shows two scales—one for sensible-total heat ratio, and one for the ratio of enthalpy difference to humidity ratio difference. The protractor is used to establish the direction of a condition line on the psychrometric chart.

Example 1 illustrates use of the ASHRAE psychrometric chart to determine moist air properties.

Example 1. Moist air exists at 100°F dry-bulb temperature, 65°F thermodynamic wet-bulb temperature, and 29.921 in. Hg pressure. Determine the humidity ratio, enthalpy, dew-point temperature, relative humidity, and volume.

Solution: Locate state point on Chart 1 (Figure 1) at the intersection of 100°F dry-bulb temperature and 65°F thermodynamic wet-bulb temperature lines. Read $W = 0.00523$ lb (water)/lb (dry air).

The **enthalpy** can be found by using two triangles to draw a line parallel to the nearest enthalpy line [30 Btu/lb (dry air)] through the state point to the nearest edge scale. Read $h = 29.80$ Btu/lb (dry air).

Dew-point temperature can be read at the intersection of W = 0.00523 lb (water)/lb (dry air) with the saturation curve. Thus, t_d = 40°F.

Relative humidity ϕ can be estimated directly. Thus, $\phi = 13\%$.

Specific volume can be found by linear interpolation between the volume lines for 14.0 and 14.5 ft³/lb (dry air). Thus, $v = 14.22$ ft³/lb (dry air).

TYPICAL AIR-CONDITIONING PROCESSES

The ASHRAE psychrometric chart can be used to solve numerous process problems with moist air. Its use is best explained through illustrative examples. In each of the following examples, the process takes place at a constant pressure of 29.921 in. Hg.

Moist Air Heating

The process of adding heat alone to moist air is represented by a horizontal line on the ASHRAE chart, since the humidity ratio remains unchanged.

Figure 2 shows a device that adds heat to a stream of moist air. For steady flow conditions, the required rate of heat addition is:

$$ {}_1q_2 = m_a(h_2 - h_1) \tag{42}$$

Example 2. Moist air, saturated at 35°F, enters a heating coil at a rate of 20,000 cfm. Air leaves the coil at 100°F. Find the required rate of heat addition.

Solution: Figure 3 schematically shows the solution. State 1 is located on the saturation curve at 35°F. Thus, $h_1 = 13.01$ Btu/lb (dry air), W_1 = 0.00428 lb (water)/lb (dry air), and $v_1 = 12.55$ ft³/lb (dry air). State 2 is located at the intersection of $t = 100$°F and W_2 and $W_1 = 0.00428$ lb (water)/lb (dry air). Thus, $h_2 = 28.77$ Btu/lb (dry air). The mass flow of dry air is:

$$ m_a = [(20{,}000)(60)/12.55] = 95{,}620 \text{ lb/h (dry air)} $$

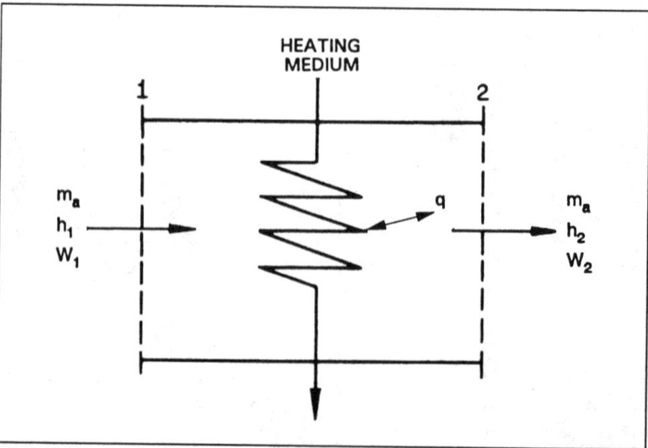

Fig. 2 Schematic of Device for Heating Moist Air

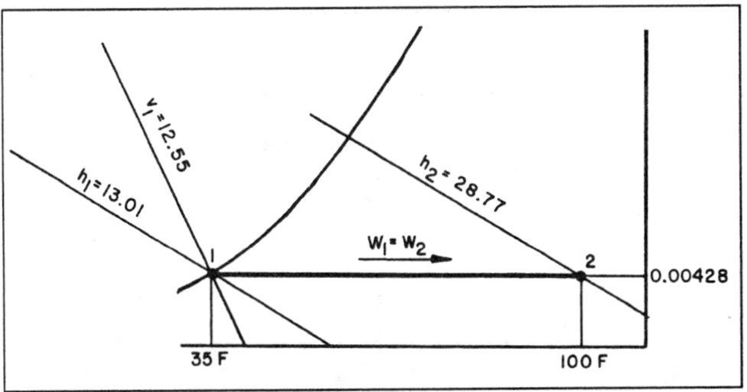

Fig. 3 Schematic Solution for Example 2

From Equation (42):

$$_1q_2 = (95{,}620)(28.77 - 13.01) = 1{,}507{,}000 \text{ Btu/h}$$

Moist Air Cooling

Moisture separation occurs when moist air is cooled to a temperature below its initial dew point. Figure 4 shows a schematic cooling coil where moist air is assumed to be uniformly processed. Although water can be separated at various temperatures ranging from the initial dew point to the final saturation temperature, it is assumed that condensed water is cooled to the final air temperature t_2 before it drains from the system.

For the system of Figure 4, the steady flow energy and material balance equations are:

$$m_a h_1 = m_a h_2 + {}_1q_2 + m_w h_{w2}$$

$$m_a W_1 = m_a W_2 + m_w$$

Thus:

$$m_w = m_a(W_1 - W_2) \tag{43}$$

$$_1q_2 = m_a[(h_1 - h_2) - (W_1 - W_2)h_{w2}] \tag{44}$$

Example 3. Moist air at 85°F dry-bulb temperature and 50% rh enters a cooling coil at 10,000 cfm and is processed to a final saturation condition at 50°F. Find the tons of refrigeration required.

Solution: Figure 5 shows the schematic solution. State 1 is located at the intersection of t = 85°F and ϕ = 50%. Thus, h_1 = 34.62 Btu/lb (dry air), W_1 = 0.01292 lb (water)/lb (dry air), and v_1 = 14.01 ft³/lb (dry air). State 2 is located on the saturation curve at 50°F. Thus, h_2 = 20.30 Btu/lb (dry air) and W_2 = 0.00766 lb (water)/lb (dry air). From Table 2, h_{w2} = 18.11 Btu/lb (water). The mass flow of dry air is:

$$m_a = 10{,}000/14.01 = 713.8 \text{ lb/min (dry air)}$$

From Equation (44):

$$_1q_2 = 713.8[(34.62 - 20.30) - (0.01292 - 0.00766)(18.11)]$$
$$= 10{,}150 \text{ Btu/min}$$

Since one ton of refrigeration equals a heat withdrawal of 200 Btu/min, the required refrigerating capacity is 50.75 tons.

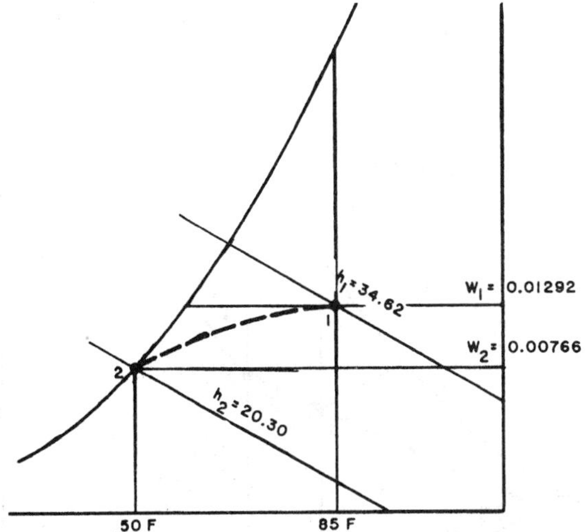

Fig. 5 Schematic Solution for Example 3

Adiabatic Mixing of Two Moist Airstreams

A common process in air-conditioning systems is the adiabatic mixing of two moist airstreams. Figure 6 schematically shows the problem. Adiabatic mixing is governed by three equations:

$$m_{a1}h_1 + m_{a2}h_2 = m_{a3}h_3$$

$$m_{a1} + m_{a2} = m_{a3}$$

$$m_{a1}W_1 + m_{a2}W_2 = m_{a3}W_3$$

Eliminating m_{a3} gives:

$$\frac{h_2 - h_3}{h_3 - h_1} = \frac{W_2 - W_3}{W_3 - W_1} = \frac{m_{a1}}{m_{a2}} \tag{45}$$

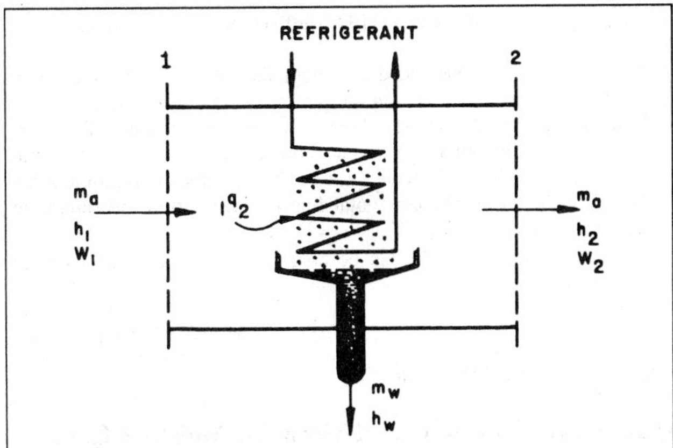

Fig. 4 Schematic of Device for Cooling Moist Air

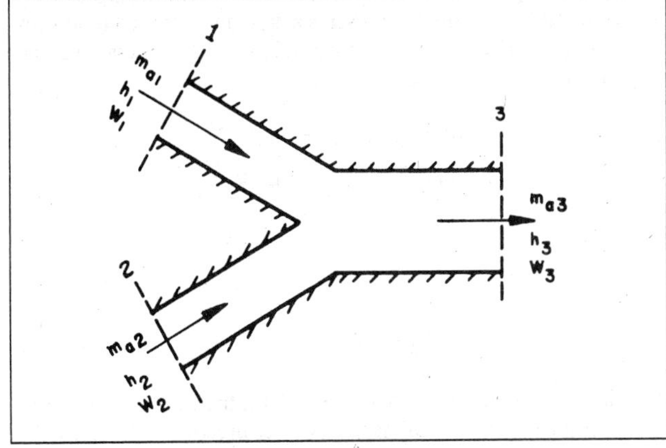

Fig. 6 Adiabatic Mixing of Two Moist Airstreams

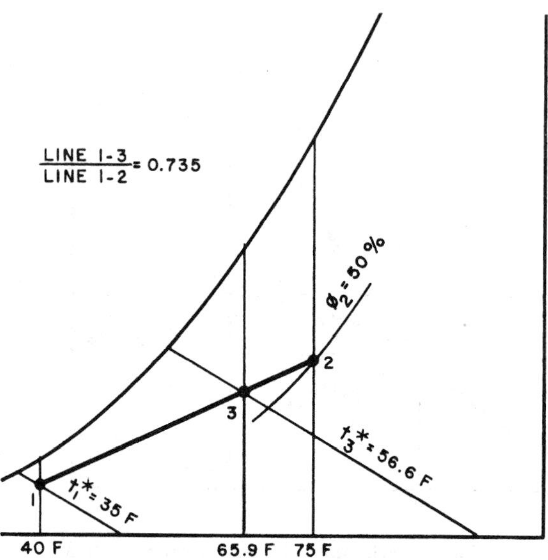

Fig. 7 Schematic Solution for Example 4

according to which, on the ASHRAE chart, the state point of the resulting mixture lies on the straight line connecting the state points of the two streams being mixed, and divides the line into two segments, in the same ratio as the masses of dry air in the two streams.

Example 4. A stream of 5000 cfm of outdoor air at 40°F dry-bulb temperature and 35°F thermodynamic wet-bulb temperature is adiabatically mixed with 15,000 cfm of recirculated air at 75°F dry-bulb temperature and 50% rh. Find the dry-bulb temperature and thermodynamic wet-bulb temperature of the resulting mixture.

Solution: Figure 7 shows the schematic solution. States 1 and 2 are located on the ASHRAE chart, revealing that v_1 = 12.65 ft³/lb (dry air), and v_2 = 13.68 ft³/lb (dry air). Therefore:

$$m_{a1} = 5000/12.65 = 395 \text{ lb/min (dry air)}$$

$$m_{a2} = 15{,}000/13.68 = 1096 \text{ lb/min (dry air)}$$

According to Equation (45):

$$\frac{\text{Line 3—2}}{\text{Line 1—3}} = \frac{m_{a1}}{m_{a2}} \text{ or } \frac{\text{Line 1—3}}{\text{Line 1—2}} = \frac{m_{a2}}{m_{a3}} = \frac{1096}{1491} = 0.735$$

Consequently, the length of line segment 1—3 is 0.735 times the length of entire line 1—2. Using a ruler, State 3 is located, and the values t_3 = 65.9°F and t_3^* = 56.6°F found.

Adiabatic Mixing of Water Injected into Moist Air

Steam or liquid water can be injected into a moist airstream to raise its humidity. Figure 8 represents a diagram of this common air-conditioning process. If the mixing is adiabatic, the following equations apply:

$$m_a h_1 + m_w h_w = m_a h_2$$

$$m_a W_1 + m_w = m_a W_2$$

Therefore,

$$\frac{h_2 - h_1}{W_2 - W_1} = h_w \qquad (46)$$

according to which, on the ASHRAE chart, the final state point of the moist air lies on a straight line whose direction is fixed by the specific enthalpy of the injected water, drawn through the initial state point of the moist air.

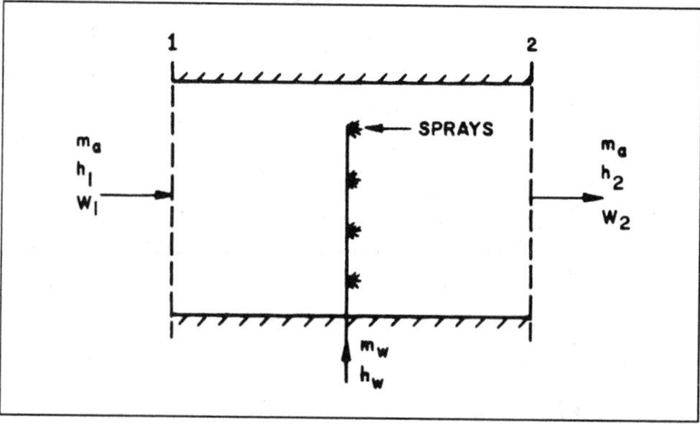

Fig. 8 Schematic Showing Injection of Water into Moist Air

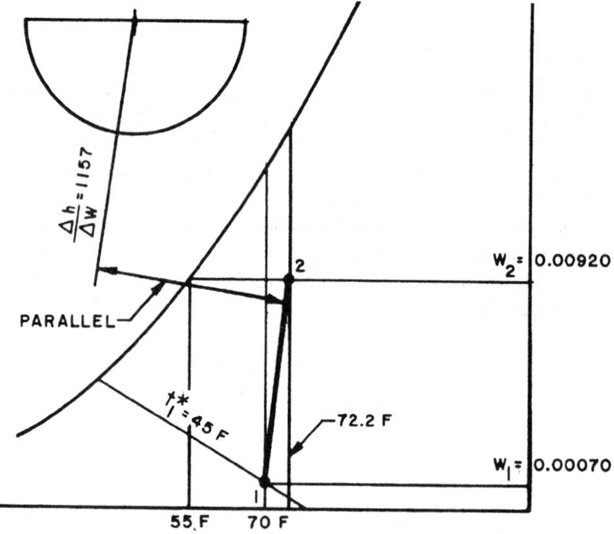

Fig. 9 Schematic Solution for Example 5

Example 5. Moist air at 70°F dry-bulb and 45°F thermodynamic wet-bulb temperature is to be processed to a final dew-point temperature of 55°F by adiabatic injection of saturated steam at 230°F. The rate of dry airflow is 200 lb/min. Find the final dry-bulb temperature of the moist air and the rate of steam flow required.

Solution: Figure 9 shows the schematic solution. By Table 3, the enthalpy of the steam h_g = 1157 Btu/lb (water). Therefore, according to Equation (46), the condition line on the ASHRAE chart connecting States 1 and 2 must have a direction:

$$\Delta h/\Delta W = 1157 \text{ Btu/lb (water)}$$

The condition line can be drawn with the $\Delta h/\Delta W$ protractor. First, establish the reference line on the protractor by connecting the origin with the value $\Delta h/\Delta W$ = 1157. Draw a second line parallel to the reference line and through the initial state point of the moist air. This second line is the condition line. State 2 is established at the intersection of the condition line with the horizontal line extended from the saturation curve at 55°F (t_{d2} = 55°F). Thus, t_2 = 72.2°F.

Values of W_2 and W_1 can be read from the chart. The required steam flow is:

$$m_w = m_a(W_2 - W_1) = (200)(60)(0.00920 - 0.00070)$$

$$= 102 \text{ lb/h (steam)}$$

Space Heat Absorption and Moist Air Moisture Gains

The problem of air conditioning a space is usually determined by (1) the quantity of moist air to be supplied, and (2) the air condition

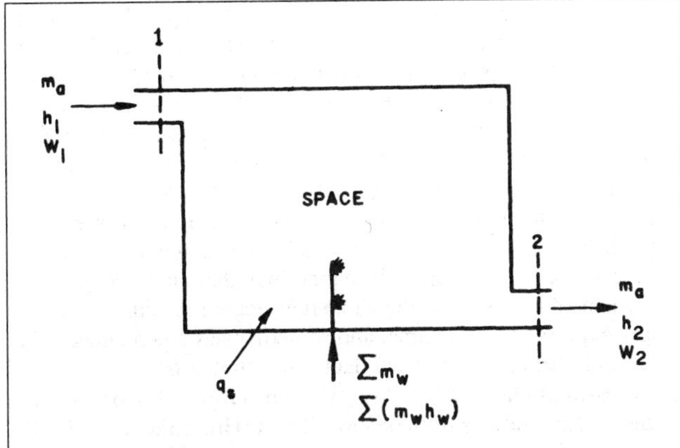

Fig. 10 Schematic of Air Conditioned Space

necessary to remove given amounts of energy and water from the space and be withdrawn at a specified condition.

Figure 10 schematically shows a space with incident rates of energy and moisture gains. The quantity q_s denotes the net sum of all rates of heat gain in the space, arising from transfers through boundaries and from sources within the space. This heat gain involves addition of energy alone and does not include energy contributions due to addition of water (or water vapor). It is usually called the *sensible heat gain*. The quantity Σm_w denotes the net sum of all rates of moisture gain on the space arising from transfers through boundaries and from sources within the space. Each pound of moisture injected into the space adds an amount of energy equal to its specific enthalpy.

Assuming steady-state conditions, governing equations are:

$$m_a h_1 + q_s + \sum(m_w h_w) = m_a h_2$$

$$m_a W_1 + \sum m_w = m_a W_2$$

or

$$q_s + \sum(m_w h_w) = m_a(h_2 - h_1) \qquad (47)$$

$$\sum m_w = m_a(W_2 - W_1) \qquad (48)$$

The left side of Equation (47) represents the total rate of energy addition to the space from all sources. By Equations (47) and (48):

$$\frac{h_2 - h_1}{W_2 - W_1} = \frac{q_s + \sum(m_w h_w)}{\sum m_w} \qquad (49)$$

according to which, on the ASHRAE chart and for a given state of the withdrawn air, all possible states (conditions) for the supply air must lie on a straight line drawn through the state point of the withdrawn air, that has a direction specified by the numerical value of $[q_s + \Sigma(m_w h_w)]/\Sigma m_w$. This line is the condition line for the given problem.

Example 6. Moist air is withdrawn from a room at 80°F dry-bulb temperature and 66°F thermodynamic wet-bulb temperature. The sensible rate of heat gain for the space is 30,000 Btu/h. A rate of moisture gain of 10 lb/h occurs from the space occupants. This moisture is assumed as saturated water vapor at 90°F. Moist air is introduced into the room at a dry-bulb temperature of 60°F. Find the required thermodynamic wet-bulb temperature and volume flow rate of the supply air.

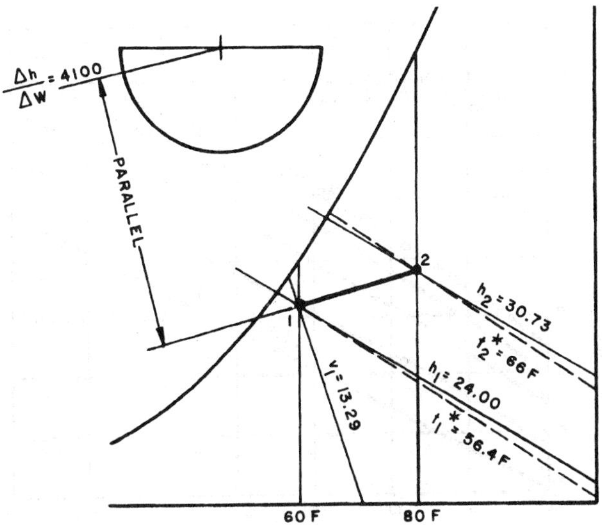

Fig. 11 Schematic Solution for Example 6

Solution: Figure 11 shows the schematic solution. State 2 is located on the ASHRAE chart. From Table 3, specific enthalpy of added water vapor is h_g = 1100.40 Btu/lb. From Equation (49):

$$\frac{\Delta h}{\Delta W} = \frac{30{,}000 + (10)(1100.40)}{10} = 4100 \text{ Btu/lb (water)}$$

With the $\Delta h/\Delta W$ protractor, establish a reference line of direction $\Delta h/\Delta W$ = 4100 Btu/lb water. Parallel to this reference line, draw a straight line on the chart through State 2. The intersection of this line with the 60°F dry-bulb temperature line is State 1. Thus, t_1^* = 56.4°F.

An alternate (and approximately correct) procedure in establishing the condition line is to use the protractor's sensible-total heat ratio scale instead of the $\Delta h/\Delta W$ scale. The quantity $\Delta H_s/\Delta H_t$ is the ratio of the rate of sensible heat gain for the space to the rate of total energy gain for the space. Therefore:

$$\frac{\Delta H_s}{\Delta H_t} = \frac{q_s}{q_s + \Sigma(m_w h_w)} = \frac{30{,}000}{30{,}000 + (10 \times 1100.44)} = 0.732$$

Note that $\Delta H_s/\Delta H_t$ = 0.732 on the protractor coincides closely with $\Delta h/\Delta W$ = 4100 Btu/lb (water).

The flow of dry air can be calculated from either Equation (47) or (48). From Equation (47):

$$m_a = \frac{q_s + \Sigma(m_w h_w)}{h_2 - h_1} = \frac{30{,}000 + (10 \times 1100.44)}{(60)(30.73 - 24.00)}$$

$$= 101.5 \text{ lb (dry air)/min}$$

At State 1, v_1 = 13.29 ft³/lb (dry air).

Therefore, supply volume = $m_a v_1$ = 101.5 × 13.29 = 1349 cfm

TRANSPORT PROPERTIES OF MOIST AIR

For certain scientific and experimental work, particularly in the heat transfer field, many other moist air properties are important. Generally classified as transport properties, these include diffusion coefficient, viscosity, thermal conductivity, and thermal diffusion factor. Mason and Monchick (1965) derive these properties by calculation. Table 5 and Figures 12 and 13 summarize the authors' results on the first three properties listed. Note that, within the boundaries of Charts 1, 2, and 3, the viscosity varies little from that of dry air at normal atmospheric pressure, and the thermal conductivity is essentially identical.

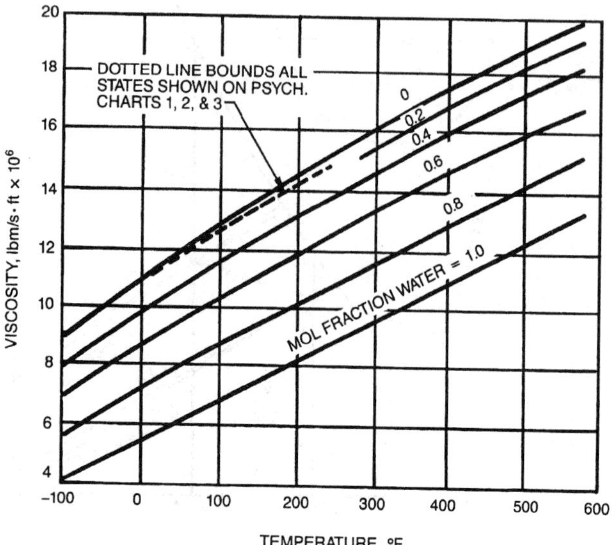

Fig. 12 Viscosity of Moist Air

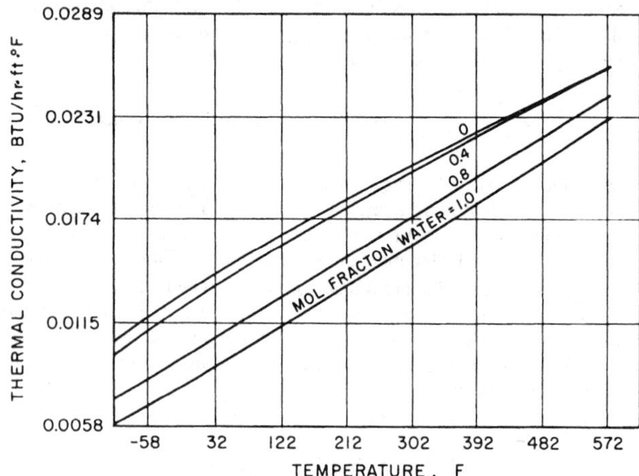

Fig. 13 Thermal Conductivity of Moist Air

Table 5 Calculated Diffusion Coefficients for Water–Air at 29.921 in. Hg Barometric Pressure

Temp., °F	ft²/h	Temp., °F	ft²/h	Temp., °F	ft²/h
−100	0.504	40	0.884	140	1.205
−50	0.600	50	0.915	150	1.240
−40	0.655	60	0.942	200	1.414
−30	0.682	70	0.973	250	1.600
−20	0.709	80	1.008	300	1.794
−10	0.736	90	1.042	350	1.996
0	0.767	100	1.073	400	2.205
10	0.794	110	1.104	450	2.422
20	0.825	120	1.139	500	2.647
30	0.853	130	1.170		

AIR, WATER, AND STEAM PROPERTIES

Coefficient f_w (or f_s) (over water) at pressures from 0.145 to 31.799 in. Hg for temperatures from −58 to 140°F (Smithsonian Institution).

Coefficient f_i (over ice) at pressures from 0.145 to 31.799 in. Hg for temperatures from 32 to 212°F (Smithsonian Institution).

Compressibility factor of dry air at pressures from 0.147 to 1470 psi and at temperatures from 90 to 5400°R (Hilsenrath et al. 1960).

Compressibility factor of moist air at pressures from 25 to 31 in. Hg at humidity ratios from 0.006 to 0.30 lb of water per lb of dry air, and for temperatures from 40 to 100°F. At pressures from 0 to 31.799 in. Hg, at values of degree of saturation from 0 to 100, and for temperatures from 32 to 140°F (Smithsonian Institution). [Note: At the time the Smithsonian Meteorological Tables were published, the value $\mu = W/W_s$ was known as relative humidity, in terms of a percentage. Since that time, there has been general agreement to designate the value μ as degree of saturation, usually expressed as a decimal and sometimes as a percentage. See Goff (1949) for more recent data and formulations.]

Compressibility factor for steam at pressures from 14.696 to 440 psi and at temperatures from 684 to 1530°R (Hilsenrath et al. 1960).

Density, enthalpy, entropy, Prandtl number, specific heat, specific heat ratio, and viscosity of dry air (Hilsenrath et al. 1960).

Density, enthalpy, entropy, specific heat, viscosity, thermal conductivity, and free energy of steam (Hilsenrath et al. 1960).

Dry air. Thermodynamic properties over a wide range of temperature (Keenan and Kaye 1945).

Enthalpy of saturated steam (Osborne et al. 1939).

Ideal-gas thermodynamic functions of dry air at temperatures from 18 to 5400°R (Hilsenrath et al. 1960).

Ideal-gas thermodynamic functions of steam at temperatures from 90 to 9000°R. Functions included are specific heat, enthalpy, free energy, and entropy (Hilsenrath et al. 1960).

Moist air properties from tabulated virial coefficients (Chaddock 1965).

Saturation humidity ratio over ice at pressures from 8.672 to 28.908 in. Hg and for temperatures from −128 to 32°F (Smithsonian Institution).

Saturation humidity ratio over water at pressures from 1.73 to 30.35 in. Hg and for temperatures from −58 to 138.2°F (Smithsonian Institution).

Saturation vapor pressure over water in in. Hg and for temperatures from −60 to 212°F (Smithsonian Institution).

Speed of sound in dry air at pressures from 0.147 to 1470 psi for temperatures from 90 to 5400°R (Hilsenrath et al. 1960). At atmospheric pressure for temperatures from −130 to 140°F (Smithsonian Institution).

Speed of sound in moist air. Relations using the formulation of Goff and Gratch and studies by Hardy et al. (1942) give methods for calculating this speed (Smithsonian Institution).

Steam tables covering the range from 32 to 1472°F.

Transport properties of moist air. Diffusion coefficient, viscosity, thermal conductivity, and thermal diffusion factor of moist air are listed (Mason and Monchick 1965). The authors' results are summarized in Table 5 and Figures 12 and 13.

Virial coefficients and other information for use with Goff and Gratch formulation (Goff 1949).

Volume of water in cubic feet for temperatures from −14 to 482°F (Smithsonian Institution 1954).

Water properties. Includes properties of ordinary water substance for the gaseous, liquid, and solid phases (Dorsey 1940).

SYMBOLS

α = $\ln(p_w)$, parameter used in Equations (37) and (38)
μ = degree of saturation W/W_s, dimensionless
ρ = moist air density, lb/ft³
ϕ = relative humidity, dimensionless
C_1 to C_{18} = constants in Equations (5), (6), and (37)
d_v = absolute humidity of moist air, mass of water per unit volume of mixture
D = enthalpy deviation, Btu/lb (dry air)
f_s = enhancement factor, used in Equations (23a) and (25a)
h = enthalpy of moist air, Btu/lb (dry air)

h_a = specific enthalpy of dry air, Btu/lb
h_{as} = $h_g - h_a$
h_f = specific enthalpy of saturated liquid water
h_{fg} = $h_g - h_f$ = enthalpy of vaporization
h_g = specific enthalpy of saturated water vapor
h_s = enthalpy of moist air at saturation per unit mass of dry air
h_s^* = enthalpy of moist air at saturation at thermodynamic wet-bulb temperature per unit mass of dry air
h_w = specific enthalpy of water (any phase) added to or removed from moist air in a process
h_w^* = specific enthalpy of condensed water (liquid or solid) at thermodynamic wet-bulb temperature and pressure of 29.921 in. Hg
H_s = rate of sensible heat gain for space
H_t = rate of total energy gain for space
m_a = mass flow of dry air, per unit time
m_w = mass flow of water (any phase), per unit time
M_a = mass of dry air in moist air sample
M_w = mass of water vapor in moist air sample
n = $n_a + n_w$, total number of moles in moist air sample
n_a = moles of dry air
n_w = moles of water vapor
p = total pressure of moist air
p_a = partial pressure of dry air
p_s = vapor pressure of water in moist air at saturation. Differs from saturation pressure of pure water because of presence of air.
p_w = partial pressure of water vapor in moist air
p_{ws} = pressure of saturated pure water
q = specific humidity of moist air, mass of water per unit mass of mixture
q_s = rate of addition (or withdrawal) of sensible heat
R = universal gas constant, 1545.32 $(lb_f/ft^2)ft^3/(lb\ mole \cdot °R)$
R_a = gas constant for dry air
R_w = gas constant for water vapor
s = entropy of moist air per unit mass of dry air
s_a = specific entropy of dry air
s_{as} = $s_s - s_a$
s_f = specific entropy of saturated liquid water
s_{fg} = $s_g - s_f$
s_g = specific entropy of saturated water vapor
s_s = specific entropy of moist air at saturation per unit mass of dry air
s_w = specific entropy of condensed water (liquid or solid) at pressure of 29.921 in. Hg
t = dry-bulb temperature of moist air, °F
t_d = dew-point temperature of moist air, °F
t^* = thermodynamic wet-bulb temperature of moist air, °F
T = absolute temperature, °R
v = volume of moist air, per unit mass of dry air
v_a = specific volume of dry air
v_{as} = $v_s - v_a$
v_f = specific volume of saturated liquid water
v_{fg} = $v_g - v_f$
v_g = specific volume of saturated water vapor
v_s = volume of moist air at saturation, per unit mass of dry air
v_T = total gas volume
V = total volume of moist air sample
W = humidity ratio of moist air, mass of water per unit mass of dry air
W_s = humidity ratio of moist air at saturation
W_s^* = humidity ratio of moist air at saturation at thermodynamic wet-bulb temperature
x_a = mole-fraction of dry air, moles of dry air per mole of mixture
x_w = mole-fraction of water, moles per mole of mixture
x_{ws} = mole-fraction of water vapor under saturated conditions, moles of vapor per mole of saturated mixture
Z = altitude, ft

REFERENCES

Chaddock, J.B. 1965. Moist air properties from tabulated virial coefficients. *Humidity and moisture measurement and control in science and industry.* 3:273. A. Wexler and W.A. Wildhack, eds. Reinhold Publishing Corp., New York.

Dorsey, N.E. 1940. *Properties of ordinary water substance.* Reinhold Publishing Corp., New York.

Goff, J.A. 1949. Standardization of thermodynamic properties of moist air. Heating, Piping, and Air Conditioning 21(11):118.

Goff, J.A. and S. Gratch. 1945. Thermodynamic properties of moist air. ASHVE *Transactions* 51:125.

Goff, J.A., J.R. Anderson, and S. Gratch. 1943. Final values of the interaction constant for moist air. ASHVE *Transactions* 49:269.

Haines, R.W. 1961. How to construct high altitude psychrometric charts. Heating, Piping, and Air Conditioning 33(10):144.

Hardy, H.C., D. Telfair, and W.H. Pielemeier. 1942. The velocity of sound in air. *Journal of the Acoustical Society of America* 13:226.

Harrison, L.P. 1965. Fundamental concepts and definitions relating to humidity. In science and industry 3:289. A. Wexler and W.H. Wildhack, eds. Reinhold Publishing Corp., New York.

Hilsenrath, J. et al. 1960. Tables of thermodynamic and transport properties of air, argon, carbon dioxide, carbon monoxide, hydrogen, nitrogen, oxygen, and steam. National Bureau of Standards. *Circular* 564, Pergamon Press, New York.

Hyland, R.W. and A. Wexler. 1983a. Formulations for the thermodynamic properties of dry air from 173.15 K to 473.15 K, and of saturated moist air from 173.15 K to 372.15 K, at pressures to 5 MPa. ASHRAE Transactions 89(2A):520-35.

Hyland, R.W. and A. Wexler. 1983b. Formulations for the thermodynamic properties of the saturated phases of H_2O from 173.15 K to 473.15 K. *ASHRAE Transactions* 89(2A):500-519.

Karig, H.E. 1946. Psychrometric charts for high altitude calculations. *Refrigerating Engineering* 52(11):433.

Keenan, J.H. and J. Kaye. 1945. *Gas tables.* John Wiley and Sons, New York.

Kusuda, T. 1970. Algorithms for psychrometric calculations. NBS Publication BSS21 (January) for sale by Superintendent of Documents, U.S. Government Printing Office, Washington, D.C.

Mason, E.A. and L. Monchick. 1965. Humidity and moisture measurement and control in science and industry. *Survey of the Equation of State and Transport Properties of Moist Gases* 3:257. Reinhold Publishing Corp., New York.

NASA. 1976. U.S. Standard atmosphere, 1976. National Oceanic and Atmospheric Administration, National Aeronautics and Space Administration, and the United States Air Force, Superintendent of Documents. U.S. Government Printing Office, Washington, D.C.

NIST. 1990. Guidelines for realizing the international temperature scale of 1990 (ITS-90). NIST Technical Note 1265. National Institute of Technology and Standards, Gaithersburg, MD.

Osborne, N.S. 1939. Stimson and Ginnings. Thermal properties of saturated steam. *Journal of Research*, National Bureau of Standards, 23(8):261.

Olivieri, J. 1996. *Psychrometrics—Theory and practice.* ASHRAE, Atlanta.

Palmatier, E.P. 1963. Construction of the normal temperature. ASHRAE psychrometric chart. *ASHRAE Journal* 5:55.

Peppers, V.W. 1988. Unpublished paper. Available from ASHRAE.

Rohsenow, W.M. 1946. Psychrometric determination of absolute humidity at elevated pressures. *Refrigerating Engineering* 51(5):423.

Smithsonian Institution. 1954. *Smithsonian physical tables*, 9th rev. ed. Available from the Smithsonian Institution, Washington, D.C.

Smithsonian Institution. *Smithsonian meteorological tables*, 6th rev. ed. Out of print, but available in many libraries. Washington, D.C.

The international temperature scale of 1990 (ITS-90). Metrologia 27:3-10.

Threlkeld, J.L. 1970. *Thermal environmental engineering*, 2nd ed. Prentice-Hall, New York, 175.

CHAPTER 7

SOUND AND VIBRATION

WITHOUT attention to sound and vibration control, the design, installation, and use of HVAC and refrigeration systems can result in complaints due to an unacceptable acoustical environment. These problems can be avoided if fundamental principles of sound and vibration control are applied. This chapter introduces these principles, including characteristics of sound; basic terminology; human response to sound; acoustical design goals; sound transmission paths; and the fundamentals of vibration isolation and measurement. Chapter 43 of the 1995 *ASHRAE Handbook—Applications* and the references listed at the end of this chapter provide further information on the subject.

ACOUSTICAL DESIGN OBJECTIVE

The primary objective of HVAC system and equipment acoustical design is to create an appropriate acoustical environment for a given space. Sound and vibration are created by a **source**, are transmitted along one or more **paths**, and reach a **receiver**. Treatments and modifications can be applied to any or all of these elements to achieve a proper acoustical environment that is free of noise and vibration.

SOUND

CHARACTERISTICS OF SOUND

Sound is a traveling oscillation in a medium exhibiting the properties of both elasticity and inertia. In fluid media (air or water), the disturbance travels as a longitudinal compression wave. Sound is generated by a vibrating surface or a turbulent fluid stream. In HVAC system design, both airborne and structure-borne sound propagation are of concern.

Speed

The speed of a longitudinal wave in a fluid medium is a function of the medium's density and modulus of elasticity. In air at room temperature, the speed of sound is about 1100 fps; in water, about 5000 fps.

Frequency

Frequency is the number of oscillations (or cycles) per unit time completed by a vibrating object. The international unit for frequency is cycles/s or hertz (Hz).

Wavelength

Wavelength is the distance between successive rarefactions or compressions of the propagation medium. Wavelength, speed, and frequency are interrelated by the following equation:

$$\lambda = c/f \tag{1}$$

where

λ = wavelength, ft
c = speed of sound, fps
f = frequency, Hz

Frequency Spectrum and Bandwidths

The **audible frequency range** for humans extends from about 20 Hz to 20 kHz. In some cases, infrasound (<20 Hz) or ultrasound (>20 kHz) is important, but methods and instrumentation for these frequency regions are specialized and are not considered here. Within the frequency range of interest, a sound source is characterized by its sound power output in octave or 1/3 octave bands, although narrower bandwidths may be appropriate for certain analyses. An **octave** is a frequency band having an upper band limit twice the frequency of its lower band limit. Table 1 lists the preferred series of octave bands and the upper and lower band limit frequencies. An octave band can, on a logarithmic frequency scale, be divided into three equally wide 1/3 octave bands with upper and lower frequency limits in the ratio of the cube root of two to one. One-third octave band center frequencies and upper and lower band limits are also listed in Table 1. The center frequency of an octave or 1/3 octave band is the geometric mean of its upper and lower band limits. Octave and 1/3 octave bands are identified by their center frequencies, not by their upper and lower band limit frequencies (ANSI *Standard* S1.11). While analysis in octave bands is usually acceptable for rating acoustical environments in rooms, 1/3 octave band analysis is often useful in product development and troubleshooting investigations.

Noise

The first and simplest definition of noise is any unwanted sound. The second definition of noise is broadband sound without distinguishable frequency characteristics, such as the sound of a waterfall. This definition is appropriate when one sound is used to mask another, as when controlled sound radiating into a room from a well-designed air-conditioning system is used to mask or hide low-level intrusive sounds from adjacent spaces to increase privacy. This controlled sound is called noise, but not in the context of unwanted sound; rather, it is a broadband, neutral sound that is frequently unobtrusive.

Three types of noise in this second context are frequently encountered in acoustics:

1. **Random noise** is an oscillation, the instantaneous magnitude of which is not specified for any given instant. The instantaneous magnitudes of a random noise are specified only by probability distributions, giving the fraction of the total time that the magnitude, or some sequence of magnitudes, lies within a specified range (ANSI *Standard* S1.1).

The preparation of this chapter is assigned to TC 2.6, Sound and Vibration Control.

Table 1 Center Approximate Cutoff Frequencies for Octave and 1/3 Octave Band Series

Octave Bands, Hz			1/3 Octave Bands, Hz		
Lower	Center	Upper	Lower	Center	Upper
			22.4	25	28
22.4	31.5	45	28	31.5	35.5
			35.5	40	45
			45	50	56
45	63	90	56	63	71
			71	80	90
			90	100	112
90	125	180	112	125	140
			140	160	180
			180	200	224
180	250	355	224	250	280
			280	315	355
			355	400	450
355	500	710	450	500	560
			560	630	710
			710	800	900
710	1,000	1,400	900	1,000	1,120
			1,120	1,250	1,400
			1,400	1,600	1,800
1,400	2,000	2,800	1,800	2,000	2,240
			2,240	2,500	2,800
			2,800	3,150	3,550
2,800	4,000	5,600	3,550	4,000	4,500
			4,500	5,000	5,600
			5,600	6,300	7,100
5,600	8,000	11,200	7,100	8,000	9,000
			9,000	10,000	11,200
			11,200	12,500	14,000
11,200	16,000	22,400	14,000	16,000	18,000
			18,000	20,000	22,400

Source: ANSI *Standard* S1.6.

2. **White noise** has a continuous frequency spectrum with equal energy/Hz over a specified frequency range. In this sense, it is like white light. White noise is not necessarily random.
3. **Pink noise** also has a continuous frequency spectrum but has equal energy per constant-percentage bandwidth, such as per octave or 1/3 octave band. Because the octave bands double in width for each successive band, the energy of white noise also doubles in each successive band. Pink noise, a constant energy per bandwidth noise, is obtained by sloping the energy profile by 3 dB per octave.

TERMINOLOGY

Decibel

The decibel (dB) is a basic unit of measurement in acoustics. It can be confusing because it is used to quantify many different descriptors relating to sound source strength, sound level, and sound attenuation. For this reason, it is important to know the context in which the term is used.

Numerically, the decibel is 10 times the base 10 logarithm of the ratio of two like quantities proportional to acoustical power or energy. The term **level**, when used in relation to sound power, sound intensity, or sound pressure, indicates that dB notation is being used. A reference quantity is always implied if it does not appear.

Sound Power and Sound Power Level

A fundamental characteristic of an acoustic source is its ability to radiate energy, whether it is weak and small in size (a cricket) or strong and large (a compressor). An energy input excites the source, which radiates some fraction of this energy in the form of sound. Because unit power radiated through a unit sphere yields unit intensity, the power reference base, established by international agreement, is 1 picowatt (pW) (10^{-12} W). The reference quantity used should be stated explicitly. A definition of sound power level is, therefore,

Table 2 Typical Sound Power Outputs and Sound Power Levels

	Approximate Power Output	
Source	Watts	Decibel re 10^{-12} W
Saturn rocket	10^8	200
Turbojet engine[a]	10^5	170
Jet aircraft at takeoff[b]	10^4	160
Turboprop at takeoff	1000	150
Prop aircraft at takeoff[c]	100	140
Large pipe organ	10	130
Small aircraft engine	1	120
Blaring radio	0.1	110
Automobile at highway speed	0.01	100
Voice, shouting	0.001	90
Garbage disposal unit	10^{-4}	80
Voice, conversation level	10^{-5}	70
Electronic equipment ventilation fan	10^{-6}	60
Office air diffuser	10^{-7}	50
Small electric clock	10^{-8}	40
Voice, soft whisper	10^{-9}	30
Rustling leaves	10^{-10}	20
Human breath	10^{-11}	10
Threshold of hearing	10^{-12}	0

[a]With afterburner. [b]Four jet engines. [c]Four propeller engines.

$$L_w = 10\ \log(w/10^{-12}\ \text{W})\ \text{dB re 1 pW}$$

or

$$L_w = 10\ \log w + 120\ \text{dB re 1 pW} \qquad (2)$$

Sound power outputs for common sources are shown in Table 2. Most mechanical equipment is rated in sound power rather than sound pressure, so that all equipment can be compared according to a common reference independent of distance and room acoustic conditions. AMCA *Application Guide* 303 provides guidelines for the application of sound power level ratings.

Sound Intensity and Sound Intensity Level

If a sphere is circumscribed around a source at an arbitrary distance from it, all the energy radiated by the source must pass through the sphere. Power flow through a unit area of the sphere is intensity, expressed in watts per square metre. (The SI units are used here rather than I-P system units because of international agreement on the definition.) Sound intensity follows the inverse square law; that is, sound intensity varies inversely as the square of distance from the source. This is true for outdoor acoustical sources and, to a limited extent, for indoor sources. Sound intensity level is expressed in dB with a reference quantity of 10^{-12} W/m^2.

Sound Pressure and Sound Pressure Level

Sound intensity is difficult to measure directly, but sound pressure is relatively easy to measure because the human ear and microphones are pressure-sensitive devices. It is for this reason that all noise codes and criteria are reported in sound pressure. It can be shown that pressure squared is proportional to intensity. A decibel

Table 3 Typical Sound Pressures and Sound Pressure Levels

Source	Sound Pressure, Pa	Sound Pressure Level, dB re 20 μPa	Subjective Reaction
Military jet takeoff at 100 ft	200	140	Extreme danger
Artillery fire at 10 ft	63.2	130	
Passenger jet takeoff at 100 ft	20	120	Threshold of pain
Loud rock band[a]	6.3	110	Threshold of discomfort
Platform of subway station (steel wheels)	2	100	
Unmuffled large diesel engine at 130 ft	0.6	90	Very loud
Computer printout room[a]	0.2	80	
Freight train at 100 ft	0.06	70	
Conversational speech at 3 ft	0.02	60	
Window air conditioner[a]	0.006	50	Moderate
Quiet residential area[a]	0.002	40	
Whispered conversation at 6 ft	0.0006	30	
Buzzing insect at 3 ft	0.0002	20	
Threshold of good hearing	0.00006	10	Faint
Threshold of excellent youthful hearing	0.00002	0	Threshold of hearing

[a]Ambient.

scale for sound pressure can be created in a manner analogous to the decibel scale for sound intensity. In this case, the reference pressure is 20 μPa, which corresponds to the approximate threshold of hearing. Because pressure squared is proportional to intensity, sound pressure level L_p is

$$L_p = 10\log(p/p_{ref})^2 \text{ re } p_{ref} \tag{3}$$

Because p_{ref} is 20 μPa (2×10^{-5} Pa), and $10 \log p^2 = 20 \log p$,

$$L_p = 20\log(p/2\times10^{-5}) \text{ re } 20\ \mu\text{Pa} \tag{4}$$

or

$$L_p = 20\log p + 94 \text{ dB re } 20\ \mu\text{Pa} \tag{5}$$

where p = root mean square (rms) value of pressure, μPa.

The human ear responds across a broad range of sound pressures; threshold of hearing to threshold of pain covers a range of approximately 10^{14}:1. Table 3 gives the sound pressure levels of various typical sources. The linear range scale for sound pressure in Table 3 is awkward in this form; therefore, the equivalent notations in the third column should be used.

Because the decibel is a logarithmic unit, two sound levels cannot be added arithmetically. Two common methods for adding decibels are used.

For the more accurate method of adding sound pressure levels, divide each by 10 and take the antilogarithm. Then add the antilogarithms, take the logarithm, and multiply by 10 to obtain the combined level. The process can be used to add a series of decibel levels, such as the octave band analysis of a noise (Table 4), to reach the overall level. Alternatively, add the largest and next largest, and repeat this process until the next addition has little or no influence.

The second method is simpler and slightly less accurate; simply refer to Table 5 to perform the desired addition. This method, although not exact, results in errors of 1 dB or less.

Table 4 Combining Decibels to Determine Overall Sound Pressure Level

Octave Band Frequency, Hz	Octave Band Level L_p, dB	Antilog of $L_p/10$		
63	85	3.2×10^8	=	0.32×10^9
125	90	1.0×10^9	=	1.0×10^9
250	92	1.6×10^9	=	1.6×10^9
500	87	5.0×10^8	=	0.5×10^9
1000	82	1.6×10^8	=	0.16×10^9
2000	78	6.3×10^7	=	0.06×10^9
4000	65	3.2×10^6	=	0.003×10^9
8000	54	2.5×10^5	=	0.0002×10^9
				3.6432×10^9
		$10 \log (3.6 \times 10^9) = 96$ dB		

Table 5 Combining Two Sound Levels

Difference Between Two Levels to Be Combined, dB	0 to 1	2 to 4	5 to 9	10 and More
Number of decibels to be added to highest level to obtain combined level	3	2	1	0

HUMAN RESPONSE TO SOUND

What is Noise?

Sound becomes noise when

- It is too loud (the sound makes speech or speech intelligibility difficult, or it is uncomfortable)
- It is unexpected (e.g., the sound of breaking glass)
- It is uncontrolled (e.g., a neighbor's lawn mower)
- It happens at the wrong time (e.g., a door slamming in the middle of the night)
- It contains pure tones (e.g., a whine, whistle, or hum)
- It contains unwanted information (e.g., an adjacent telephone conversation or undesirable music)
- It is unpleasant (e.g., a dripping faucet)
- It connotes unpleasant experiences (e.g., a mosquito buzz or a siren wail)
- It is any combination of the above examples

To be noise, sound does not have to be loud, just unwanted. In addition to being annoying, loud noise can cause hearing loss, and depending on other factors, it could affect stress level and heart rate.

Predicting Human Response to Noise

Unfortunately, predicting the response of people to any given sound is at best only a statistical concept and at worst very inaccurate. This is because response to sound is not only physiological but also psychological and depends on the state of attitude of the listener, which can vary. Hence, the effect of noise is often unpredictable. We do know, however, that people's response to noise is adverse if it is considered too loud for the situation or if it sounds "wrong." As a result, most criteria are based on descriptors that take into account level and spectrum shape.

The sound pressure level in an occupied space can be measured directly with a sound level meter or estimated from published sound power data after accounting for room volume, distance from the source, and other acoustical factors. Sound level meters measure the sound pressure at the microphone location. Estimation techniques calculate sound pressure at a specified point in an occupied space. Measured or estimated sound pressure levels in frequency bands can then be plotted, analyzed, and compared with established criteria for acceptance.

Table 6 Subjective Effect of Changes in Sound Pressure Level, Broadband Sounds

Subjective Change	Objective Change in Sound Level (Approximate)
Much quieter	Less than −10 dB
Half as loud	−10 dB
Quieter	−5 dB
Just perceptibly quieter	−3 dB
Just perceptibly louder	+3 dB
Louder	+5 dB
Twice as loud	+10 dB
Much louder	More than +10 dB

Sound Quality

It is necessary but often not sufficient to know the sound pressure levels in order to determine the acoustic acceptability of a space to occupants. The sound quality is important as well. Factors influencing sound quality include (1) loudness, (2) tone perception, (3) frequency spectrum, (4) harshness, (5) time and frequency fluctuation, and (6) vibration.

Loudness

Objectively determining the subjective reaction to sound in terms of loudness is complex; the primary method for determining this reaction is to present sounds to a statistically representative sample of human observers. Data indicate that human hearing is most sensitive in the mid-frequency range.

Most humans can say that one sound is louder than another, higher or lower pitched, and so forth. Under carefully controlled experimental conditions, humans can detect small changes in sound level. However, the human reaction describing the halving or doubling of loudness requires a change in sound pressure level of about 10 dB. For many people, the minimum perceptible change is 3 dB. This means that halving the power output of the source causes a barely noticeable change in sound pressure level, and the power output must be reduced by a factor of 10 before humans determine that loudness has been halved. Table 6 summarizes the effect of changes in sound levels for simple noises of frequency 250 Hz and higher.

Tone Perception

Although high-level sounds are generally considered noisy, people perceive sounds having tones (e.g., a whine or hum) as particularly annoying. The effect of the tone can often make a low-level sound turn to noise. Studies have been done to characterize sounds with and without pure tones.

Frequency Spectrum

The most acceptable frequency spectrum for HVAC noise is a balanced spectrum. This means that it sounds neutral and is not too "hissy" (high-frequency content) or too "rumbly" (low-frequency content). The room criterion (RC) curves approximate the balanced spectrum deemed acceptable by most people. They are straight lines sloped at −5 dB per octave band on a sound level versus frequency graph. Although still widely in use, the noise criterion (NC) curves are beginning to be replaced by the RC curves because an RC rating provides more information about sound quality.

Figure 1 shows an indoor sound spectrum and the manner in which fan noise and diffuser noise contribute to that spectrum at various frequencies. The fan noise has been attenuated such that it approaches the sound criterion (in this case, an RC-35 contour) only in the lower octave bands. If this were the only noise present in the space, it would be considered rumbly by most listeners. However, the diffusers have been selected to balance the spectrum by filling in the higher frequencies so that the quality of the sound is more pleasant. Unfortunately, achieving a balanced sound spectrum is not usually this easy—there may be a multiplicity of sound sources to consider. As a guide for the designer, Figure 2 shows the more common mechanical and electrical noise sources and frequency regions that control the indoor noise spectrum. Chapter 43 of the 1995 *ASHRAE Handbook—Applications* provides more detailed information on treating some of these noise sources.

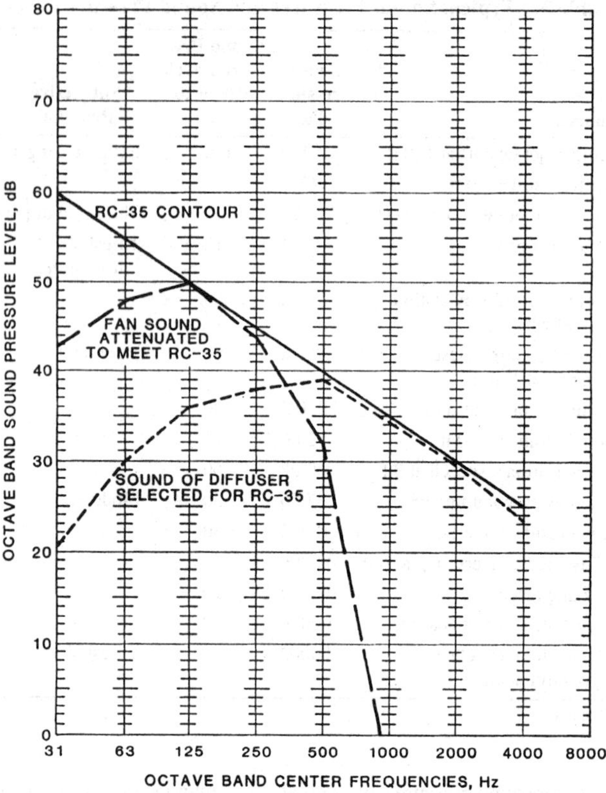

Fig. 1 Well-Balanced Sound Spectrum Resulting from Proper Selection of Air Outlets and Adequate Fan Noise Attenuation

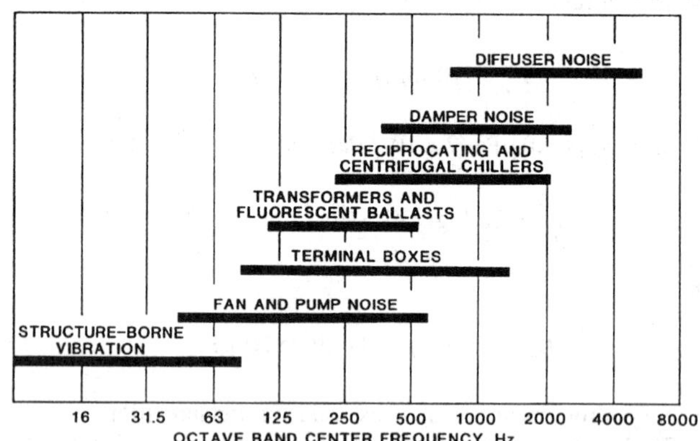

Fig. 2 Frequencies at Which Various Types of Mechanical and Electrical Equipment Generally Control Sound Spectra

ACOUSTICAL DESIGN GOALS AND RATING SYSTEMS

Establishing Design Goals

The primary acoustical design goal for air-conditioning systems is the achievement of a level of background sound that is unobtrusive in quality and low enough in level that it does not interfere with the occupancy requirements of the space being served.

For example, large conference rooms, auditoriums, and recording studios can tolerate only a low level of background sound before interference problems develop. On the other hand, higher levels of background sound are acceptable and even desirable in certain situations, such as in open-plan offices, where a certain amount of speech and activity masking is essential. Therefore, the system noise control goal depends on the required use of the space.

The degree of occupancy satisfaction achieved with a given level of background sound has many factors. To be unobtrusive, the background sound should be steady in level, neutral in character, and free of identifiable machinery noises. Background sound should have the following properties:

- A balanced distribution of sound energy over a broad frequency range
- No audible tonal or other characteristics such as a whine, whistle, hum, or rumble
- No noticeable time-varying levels from beats or other system-induced aerodynamic instability

Four types of acoustical design criteria are used by the air-conditioning industry: A-weighted sound level (dBA), noise criterion (NC) curves, room criterion (RC) curves, and loudness (sones). A more detailed discussion of acoustical design goals, as well as a table of recommended design goals (Table 2), can be found in Chapter 43 of the 1995 *ASHRAE Handbook—Applications*.

A-Weighted Sound Level

The A-weighted sound level (L_A) is widely used to state design goals as a single number, but its usefulness is limited because it gives no information on spectrum content. The measuring method is simple because the L_A can be obtained from a single reading on a hand-held instrument. The standard sound level meter includes an electronic weighting network that deemphasizes the low-frequency portions of a noise spectrum, automatically compensating for the lower sensitivity of the human ear to low-frequency sounds. Figure 3 shows the weighting characteristic of the A-weighting and other networks.

The A-weighted sound level has the advantage of identifying the desirable level as a single-valued number. However, it has a disadvantage in that it does not necessarily correlate well with the loudness or annoyance caused by the noise.

The A-weighted level comparison is best used with noises that sound alike but differ in level. It should not be used to compare sounds with distinctly different spectral characteristics; that is, two sounds at the same sound level but with different spectral content are likely to be judged differently by the listener in terms of acceptability as a background sound. One of the noises might be completely acceptable, while the other could be objectionable because its spectrum shape was rumbly, hissy, or tonal in character.

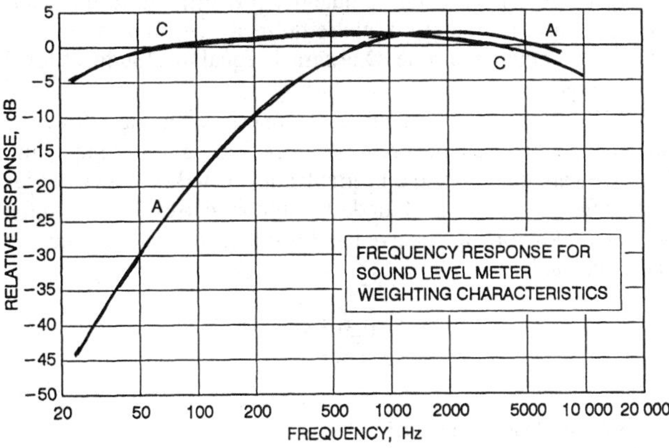

Fig. 3 Curves Showing A- and C-Weighting Responses

Noise Criterion Curves

The noise criterion curves shown in Figure 4 (Beranek 1957) are widely used. These curves define the limits that the octave-band spectrum of a noise source must not exceed in order to achieve a level of occupant acceptance. For example, an NC-35 design goal is commonly used for private offices; the background noise level meets this goal, provided that no portion of its spectrum lies above the designated NC-35 curve.

Two problems occur in using the NC design goal: (1) when the NC-level is determined by a singular tangent peak, the actual level of resulting background sound may be quieter than that desired for masking unwanted speech and activity noises because the spectrum on either side of the tangent peak drops off too rapidly; and (2) when the shape of the NC curve is closely matched, the resulting sound will be either rumbly or hissy.

Because the shape of the NC curve is not that of a well-balanced, neutral sounding noise, these curves should be used with caution in situations in which the background sound of an air-conditioning system is required to mask speech and activity noise.

Room Criterion Curves

Room criterion curves, shown in Figure 5, are designed specifically for establishing HVAC system design goals. The shape of these curves differs from that of the NC curves at both low and high frequencies.

The shape of the RC curve is a close approximation of a well-balanced, neutral-sounding spectrum. It provides guidance whenever the space requirements dictate that a certain level of background sound be maintained for masking or other purposes. Sound that

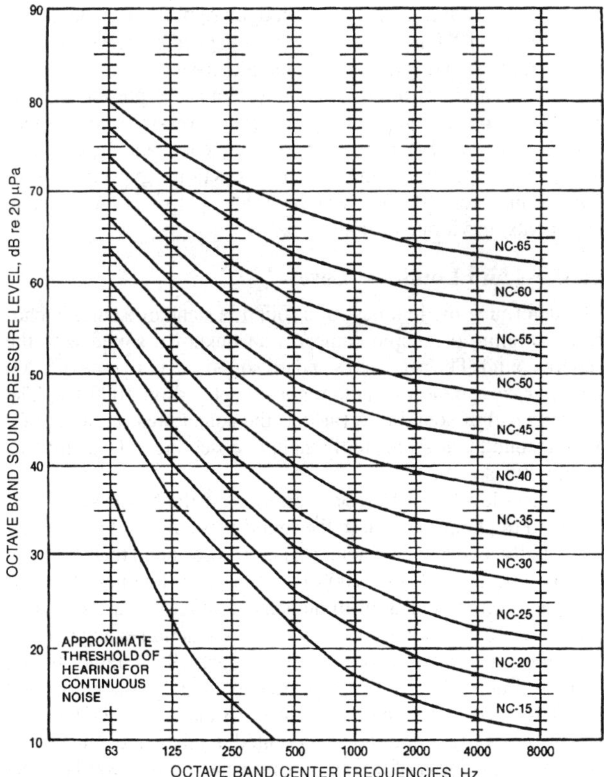

Fig. 4 NC Curves for Specifying Design Level in Terms of Maximum Permissible Sound Pressure Level for Each Frequency Band

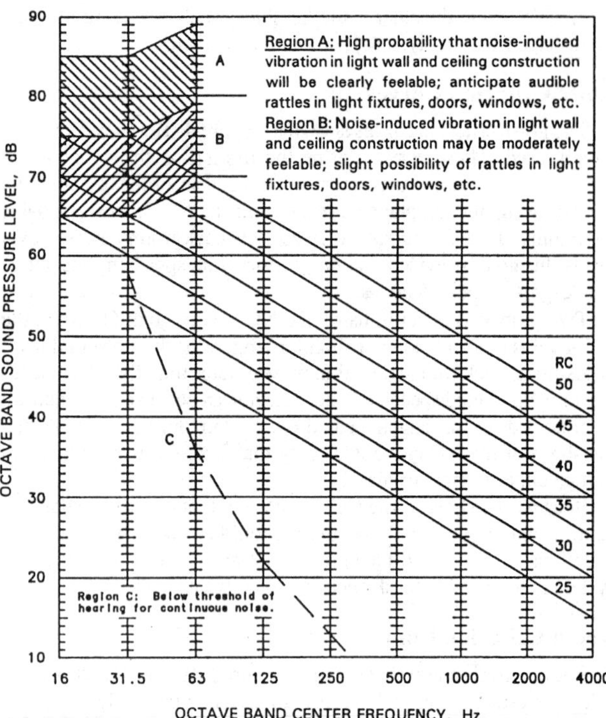

Fig. 5 RC Curves for Specifying Design Level in Terms of Balanced Spectrum Shape
(Blazier 1981)

approximates the shape of the curve within 2 dB over the entire frequency range achieves an optimum balance in quality. If the low-frequency levels (31.5 to 500 Hz) exceed the design curve by more than 5 dB, the sound is likely to be judged rumbly; if the high-frequency levels (1000 to 4000 Hz) exceed the design curve by more than 3 dB, the sound is likely to be judged hissy.

Another feature of the RC rating system is the presence of the shaded areas on the graph at high levels of low-frequency noise. These regions predict conditions of frequency and level under which noise may induce lightweight constructions (gypsum walls or ceilings) and their attached services (electrical or HVAC system components) to vibrate or rattle.

Loudness and Loudness Level

To determine the loudness of a sound, a standard sound is chosen; a sampling of people compares an unknown sound with the standard sound. The accepted standard sound is a pure tone of 1000 Hz or a narrow band of random noise centered on 1000 Hz. The loudness level of sound is defined as the sound pressure level of a standard sound that seems to the sample observers to be as loud as the unknown sound.

Loudness level is expressed in **phons**, and the loudness level of any sound in phons is equal to the sound pressure level in decibels of an equally loud standard sound. Therefore, a sound that is judged to be as loud as a 40 dB, 1000 Hz tone has a loudness level of 40 phons. Average reactions of humans to pure tones are shown in Figure 6 (Robinson and Dadson 1956). The reaction changes when the sound is a band of random noise (Pollack 1952), rather than a pure tone (Figure 7). These curves are equal loudness contours and were developed by asking the subjects to match the loudness of tones or bands of random noise presented at other frequencies and levels to the loudness of a tone or band centered on 1000 Hz. The figures indicate that the human perception of sound is most sensitive in the mid-frequency range.

Although the logarithmic phon scale covers the large dynamic range of the ear, it does not fit a subjective linear loudness scale.

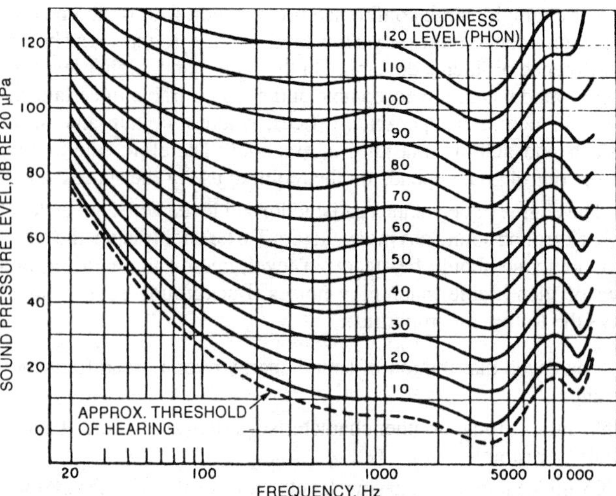

Fig. 6 Free-Field Loudness Contours for Pure Tones

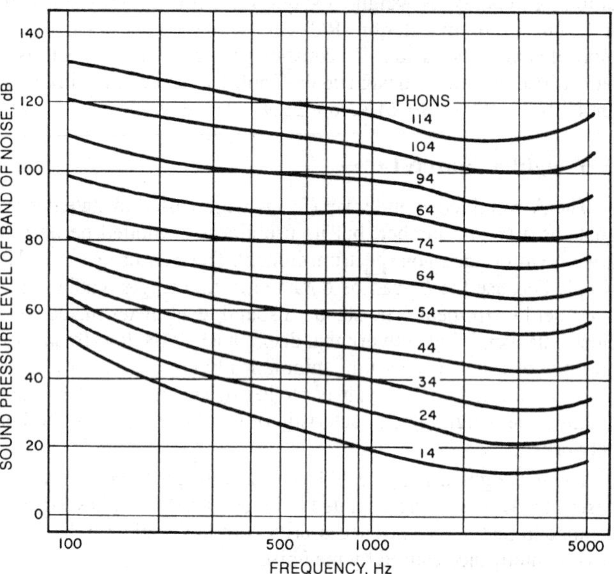

Fig. 7 Equal Loudness Contours for Relatively Narrow Bands of Random Noise

A sound that is twice as loud as another sound does not double the number of phons; as previously noted, over most of the audible range, a doubling of loudness corresponds to a change of approximately 10 phons. To obtain a quantity proportional to the loudness sensation, a loudness scale is defined in which the unit of loudness is known as a sone. One **sone** is equal to a loudness level of 40 phons.

ANSI *Standard* S3.4 calculates loudness or loudness level by using octave-band sound pressure level data as a starting point. These data are then plotted on a graph constructed as shown in Figure 8. After this graph is plotted, the number corresponding to the loudness index for each octave band is obtained. Total loudness is then calculated as

$$S = S_m + 0.3(\Sigma S - S_m) \quad (6)$$

where

S = total loudness in sones
S_m = greatest loudness index
ΣS = sum of all loudness indices

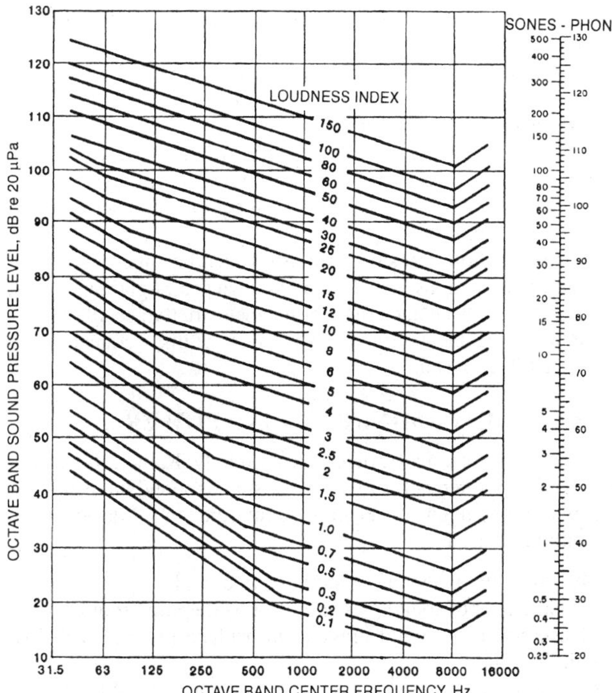

Fig. 8 Contours of Equal Loudness Index

After the loudness in sones is determined, the loudness level in phons can be found on the right side of Figure 8. In interpreting loudness data, different authors have reported values based on differing loudness contours; also, several methods are available to calculate the single-number loudness or loudness level of a complex sound. Due to their complexity, loudness measurements have not been widely used in engineering practice in the past. However, due to an increased awareness of sound quality and the availability of software for calculating loudness, this measure is now being used more frequently.

AMCA *Application Guide* 302 describes how the sone method is applied to rating the relative loudness of fans and ventilators. This calculation method is usually acceptable when there are no strong tonal components in the measured sound spectrum. A more complex calculation method using 1/3 octave band sound pressure levels by Zwicker (ISO *Standard* 532, German DIN *Standard* 45631) is more accurate in predicting loudness of sound spectra with tones.

Design Guidelines

A detailed discussion on the selection of sound design goals, as well as a table of recommended design guidelines (Table 2), can be found in Chapter 43 of the 1995 *ASHRAE Handbook—Applications*. In general, however, the following basic guidelines are important to note:

- Sound levels below NC-35 or RC-35 contribute to good speech intelligibility, while those at or above NC-35 or RC-35 may interfere with or mask speech.
- In mechanical systems with variable air volume, it may not be possible to fill in the higher frequencies to create a balanced spectrum when the quantity of supply air is moderate to low. In these cases, if acoustic privacy is important, it may be necessary to provide controlled amounts of electronic masking noise or to request the building designer to improve the sound-isolating construction.
- Even if the occupancy noise will be significantly higher than the anticipated background mechanical equipment-generated noise level, the sound design goal should not necessarily be raised to levels approaching the occupancy noise. This avoids the scenario of occupants' having to raise their voices uncomfortably to be heard over the noise.

MEASURING SOUND

Sound Level Meters

The basic instrument for measuring sound is a sound level meter. This battery-operated, electromechanical meter is a lightweight (as little as 1 lb) hand-held device that converts sound pressure at a point to sound pressure level.

All sound level meters can show the time fluctuations of the sound pressure level (no sounds are truly constant) or can hold the maximum or minimum level, but there is a type of sound level meter called an **integrating sound level meter** that takes an average of the sound pressure level over a user-definable time period. The advantages of this meter are that it is easier to read and measurements are more repeatable (especially if the time is great). The quantity measured is L_{eq}, the equivalent continuous sound pressure level.

A sound level meter can have a set of filters used to determine the frequency content of the sound. The filters are either 1/3 octave band or full octave band and are necessary for most HVAC noise control based on measurements.

Analysis in octave bands is usually acceptable for rating acoustical environments in rooms, especially for determining RC ratings. Often, however, 1/3 octave band analysis is useful in product development and troubleshooting investigations. For HVAC noise measurements, mid-frequency filters for 31.5 Hz to 8000 Hz are usually adequate.

Using filters to cover the frequency range from 31.5 Hz to 8000 Hz can require an operator to manually switch up or down through the frequencies, a time-consuming task. A sound level meter with filters that gives all filter readings simultaneously is called a **real-time analyzer** (RTA). The RTA speeds up data measurement, collection, and analysis significantly, and on most models, the information can be saved to an internal or external digital storage device.

Sound level meters are available in several accuracy grades. A Type 1 meter has a tolerance of about ±1 dB. The general-purpose meter, which is less expensive, is designated Type 2; it has a tolerance of about ±2 dB and is useful for most HVAC noise measurements.

Sound Measurement Basics

HVAC noise in general is steady. It does not vary significantly over time, except at low frequencies (250 Hz and lower), so measurements are relatively easy, but they need to be done carefully in order to assure repeatable and accurate results.

Sophisticated sound measurements should be carried out by experienced sound professionals. At present, there are only a handful of noise standards for measuring the interior noise from mechanical equipment (ASTM *Standard* E 1574, ASTM *Standard* E 1573). Most manuals for sound level meters include sections on how to measure sound. Included here are the basic methods that can help the reader make acceptable measurements.

Determining the sound spectrum in a room or troubleshooting a noise complaint usually requires measuring the octave band L_p values in the octave bands from 31.5 to 8000 Hz. In cases where tonal noise or rumble is the complaint, 1/3 octave band measurements are recommended because of their ability to discriminate narrower frequency bands of noise. A sound level meter or RTA can record the spectrum. Whatever the measurement method, L_p is measured at a point. In a room, each point often provides a different value for sound pressure level; thus, the actual location of the measurement, detailed in the report, is very important. One might survey the room, find the point that shows the highest level at, for example, 4 ft above the ground, and record that. Alternately, one might measure at 2 ft

Table 7 Guidelines for Determining Equipment Sound Levels in Presence of Contaminating Background Noise

Measurement A minus Measurement B	Correction to Measurement A to Obtain Equipment Sound Level
10 dB or more	0 dB
6 to 9 dB	−1 dB
4 to 5 dB	−2 dB
3 dB	−3 dB
2 dB	−4 dB
1 dB	−7 dB
0 dB	Equipment sound level is at least 9 dB below Measurement A

Measurement A = Tested equipment plus background noise.
Measurement B = Background noise alone.

below a register, or in the exact geometric center of the room, midway between the floor and the ceiling. Wherever the location, it must be defined and recorded. If the meter has an averaging function, one can walk the room while the meter records the continuous average of the sound pressure. This is acceptable provided there are no extreme readings in portions of the room.

When measuring, it is necessary to determine the contribution of other sources (plumbing noise, business machines, nearby traffic, etc.) to the HVAC source you are measuring. All sources other than the source you want to measure are designated background noise. Sometimes the noise due to a particular piece of HVAC equipment must be measured in the presence of background noise from other sources that cannot be turned off, such as automobile traffic or certain business machines. Determining the sound level due to the selected equipment alone requires making two sets of measurements, one set with both the HVAC equipment noise and the background noise and another set with only the background noise (with the equipment turned off). This situation might occur, for example, when determining whether the property line noise exposure from a cooling tower meets a local noise ordinance. The guidelines given in Table 7 will help to determine the sound level of a particular machine in the presence of background noise.

Measuring noise emissions from a particular piece of equipment or group of equipment requires a measurement plan specific to the situation. The Air-Conditioning and Refrigeration Institute (ARI), the Air Movement and Control Association (AMCA), the American Society of Testing and Materials (ASTM), the American National Standards Institute (ANSI), the Acoustical Society of America (ASA), and ASHRAE all publish sound level measurement procedures for various laboratory and field noise measurement situations.

Outdoor measurements are somewhat easier than indoor because there are no boundary surfaces to affect sound buildup or absorption. Important issues to address are the effect of large sound-reflecting surfaces nearby and of weather conditions such as wind, temperature, and precipitation. These effects can be estimated through guidelines in many sources such as Harris (1991).

Typical Sources of Sound

Whenever mechanical power is generated or transmitted, a fraction of the power is converted into sound power and is radiated into the air. Therefore, under certain circumstances, virtually any major component of an HVAC system could be considered a sound source (e.g., fans, pumps, ductwork, piping, motors, etc.). The component's sound source characteristics depend upon its construction, its form of mechanical power and its integration with associated system components. The most important sound source characteristics include total sound power output L_w, frequency distribution, and radiation directivity Q.

Sound sources in HVAC systems are so numerous that it is impractical to provide a complete listing herein. Typical sources of noise and vibration in HVAC systems are

- Rotating and reciprocating equipment such as fans, motors, pumps, and chillers.
- Air and fluid noises, such as those associated with flow through ductwork, piping systems, grilles, diffusers, terminal boxes, manifolds, and pressure-reducing stations.
- Excitation of surfaces—for example, friction; movement of mechanical linkages; turbulent flow impacts on ducts, plenum panels, and pipes; and impacts within equipment, such as cams and valve slap.
- Magnetostriction (transformer hum), which becomes significant in motor laminations, transformers, switch gear, lighting ballasts, and dimmers. A characteristic of magnetostrictive oscillations is that their fundamental frequency is twice the line frequency (120 Hz in a 60 Hz system).

DETERMINING SOUND POWER

Sound power cannot be determined directly. Rather, it is calculated from measurements of sound pressure or sound intensity created by a source in one of several test environments. Four methods are commonly used:

1. Measurement of sound pressure in a free field
2. Measurement of sound pressure in a reverberation room
3. Measurement of sound pressure in a plane progressive wave field in a duct
4. Measurement of sound intensity in a nonreactive field

Free-Field Method

A sound field in which the effects of the boundaries are negligible over the frequency range of interest is a **free field**. In a free field or in spaces where free-field conditions can be approximated (e.g., by sound-absorbing walls, floor, and ceiling), the sound power of a sound source can be determined from a number of measurements of sound pressure level on an imaginary spherical surface centered on and surrounding the source. This method is based on the fact that, because absorption of sound in air can be practically neglected at small distances from the sound source, the sound power generated by a source must flow through an imagined sphere with the source at its center. The intensity of the sound is determined at each of the measuring points around the source and multiplied by the area of the imagined sphere associated with the measuring points. Total sound power is the sum of these products for each point.

ANSI *Standard* S12.35 describes various methods used to calculate the sound power level under free-field conditions. Measurement accuracy is limited at the lower frequencies by the difficulty of obtaining room surface treatments that have high sound absorption coefficients at low frequencies. For example, a glass fiber wedge structure that gives useful absorption at 70 Hz must be at least 4 ft long.

Using values for the speed of sound c for air at 68°F and 14.7 psi (standard atmosphere), the relationship between sound power level and sound pressure level for a nondirectional sound source, measured in a free field at a distance r, can be written as

$$L_w = L_p + 20 \log r + 0.5 \text{ dB} \tag{7}$$

where

L_w = sound power level, dB re 10^{-12} W
L_p = sound pressure level, dB re 20 μPa

Actual sound sources radiate different amounts of sound power in different directions because the source is a certain size and the various areas of its surface do not necessarily vibrate at the same level or in phase. A directivity pattern can be established by measuring sound pressure under free-field conditions, either in an anechoic room or over a reflecting plane in a semianechoic space. Most sound sources have directional characteristics. This is why sound power and directivity measurements must be made at several points around the source. The directivity factor Q is defined as the ratio of sound pressure at a given angle from the sound source to the sound pressure that would be produced by the same source radiating uniformly in all directions. Q is a function of frequency and direction. Figure 30 in Chapter 43 of the 1995 *ASHRAE Handbook—Applications* provides more detailed information on sound source directivity.

In many cases, a fully free field is not available, and measurements have to be made in a free field over a reflecting plane. This means that the sound source is placed on a hard floor (in an otherwise sound-absorbing room) or on pavement outdoors. Because the sound is then radiated into a hemisphere rather than a full sphere, the relationship for L_w and L_p for a nondirectional sound source is

$$L_w = L_p + 20 \log r - 2.5 \text{ dB} \tag{8}$$

Reverberant Field Method

A sound field in which the sound intensity is the same in all directions and at every point due to reflections from the room surfaces is called a **reverberant field**. A widely used method for determining sound power places the sound source in a reverberation room that has sound-reflecting walls, floor, and ceiling. The sound pressure level is measured at some distance from the source and the surfaces of the room. If the rate of decay of sound (in dB/s) and the volume of the reverberation room are known, the sound power level can then be calculated from the sound pressure level (direct method). Alternately, the sound pressure level of a **reference sound source** (RSS) with known sound power output may be measured in the reverberation room, and then the unknown sound source may be measured to obtain its sound power output through comparison (substitution method).

Standardized methods for determining the sound power of HVAC equipment in reverberation rooms are given in ANSI *Standard* S12.31 for when the sound source contains mostly broadband noise; in ANSI *Standard* S12.32 for when tonal noise is prominent; and in AMCA *Standard* 300 for testing fans.

The relationship between sound power level and sound pressure level in a reverberation room is given by

$$L_w = L_p + 10 \log V + 10 \log D - 47.3 \text{ dB} \quad \text{(direct method)} \tag{9}$$

and

$$L_w = L_p + (L_w - L_p)_{ref} \quad \text{(substitution method)} \tag{10}$$

where

L_p = sound pressure level averaged over room, dB re 20 μPa
V = volume of room, ft^3
D = decay rate, dB/s
$(L_w - L_p)_{ref}$ = difference between sound power level and sound pressure level of reference sound source

Some sound sources that can be measured by these methods are room air conditioners, refrigeration compressors, components of central HVAC systems, and air terminal devices. AMCA *Standard* 300 and ARI *Standard* 880 establish special measuring procedures for some of these units. Large equipment that can operate on a large paved area, such as a parking lot, can also be measured under free-field conditions on a reflecting plane. Determining the sound power of large equipment is difficult; however, data may be available from some manufacturers.

Progressive Wave Method

One form of sound propagation is the progressive movement of a sound wave along a duct. Fan sound power can be determined in a progressive wave field by measuring the sound pressure level inside a duct. The method is described in detail in ASHRAE *Standard* 68 (AMCA *Standard* 330) for in-duct testing of fans.

Sound Intensity Method

Advances in acoustical instrumentation now permit the direct determination of sound intensity, defined as the sound power flowing through a unit area of a surface surrounding a source. The spatially averaged sound power radiated by the source can be determined by measuring the sound intensity over the sphere or hemisphere surrounding a sound source. One of the advantages of this method is that, with a few limitations, sound intensity (and, therefore, sound power) measurements can be made in the presence of background noise in ordinary rooms, thereby eliminating the need for a special testing environment. Another advantage is that, by measuring sound intensity over restricted areas around a sound source, sound directivity can be determined. This procedure can be particularly useful in reducing the noise of products during development.

International (ISO *Standard* 9614-1, ISO *Standard* 9614-2) and United States (ANSI *Standard* S12.12) standards that prescribe methods for making sound power measurements with sound intensity probes have been issued. In some situations, the sound fields may be so complex that measurements become impractical. A particular concern is that small test rooms or those having somewhat flexible boundaries (sheet metal or thin drywall) can permit a reactive sound field (one in which the room's acoustical characteristics cause it to affect the sound power output of the source) to exist.

Measurement Bandwidths

Sound power is normally determined in octave or 1/3 octave bands. Occasionally, a more exact determination of the sound source spectrum is required. In these cases, narrow-band analysis, using either constant relative bandwidth (1/12 octave) or constant absolute bandwidth (5 Hz) can be applied. The most frequently used analyzer types are digital filter analyzers for constant relative bandwidth measurements and fast Fourier transform (FFT) analyzers for constant absolute bandwidth measurements. Narrow-band analysis results are used to determine the exact frequencies of pure tones and their harmonics in a sound spectrum.

CONVERTING SOUND POWER TO SOUND PRESSURE

The designer is often required to convert the sound power level information on a source to predict the sound pressure level at a given location.

The sound pressure level at a given location in a room relative to a source of known sound power level depends on (1) room volume, (2) room furnishings and surface treatments, (3) magnitude of the sound source(s), and (4) distance from the sound source(s) to the point of observation.

The classic conversion relationship between sound power level and sound pressure level is

$$L_p = L_w + 10 \log(Q/4\pi r^2 + 4/R) + 10.5 \tag{11}$$

where

L_p = sound pressure level, dB re 20 μPa
L_w = sound power level, dB re 10^{-12} W
Q = directivity of sound source (dimensionless)
r = distance from source, ft
R = sum of all surface areas and their corresponding absorption coefficients $[S\alpha/(1-\alpha)]$, ft^2

If the source is outside, this relationship simplifies to

$$L_p = L_w + 10\log(Q/4\pi r^2) + 10.5 \tag{12}$$

Equation (12) does not account for atmospheric absorption, weather effects and barriers.

In most typical rooms, the presence of acoustically absorbent surfaces and sound-scattering elements, such as furniture, creates a relationship between sound power and sound pressure level that is almost independent of the absorptive properties of the space. For example, hospital rooms, which have only a small amount of absorption, and executive offices, which have substantial absorption, are similar when the comparison is based on the same room volume and distance between the source and point of observation.

Equation (13) can be used to estimate the sound pressure level at a chosen observation point in a normally furnished room. The estimate is accurate to ±2 dB (Schultz 1985).

$$L_p = L_w - 5\log V - 3\log f - 10\log r + 25\text{ dB} \tag{13}$$

where

L_p = room sound pressure level at chosen reference point, dB re 20 μPa
L_w = source sound power level, dB re 10^{-12} W
V = room volume, ft^3
f = octave band center frequency, Hz
r = distance from source to observation point, ft

Equation (13) applies to a single sound source in the room. With more than one source, total sound pressure level at the observation point is obtained by adding (on an energy basis) the individual contribution of each source, using the corresponding L_w and r for each source.

SOUND TRANSMISSION PATHS

Sound from a source is transmitted via a path to a receiver. Airborne and structure-borne transmission paths are of principal concern for the HVAC system designer. Airborne paths can be via the atmosphere or through ductwork; structure-borne paths are via solid materials, such as a building's floor slabs and structural elements. Sound transmission between rooms occurs via both airborne and structure-borne transmission paths. Chapter 43 of the 1995 *ASHRAE Handbook—Applications* has additional information on transmission paths.

Airborne Transmission

Atmospheric Transmission. Sound transmits readily through air, both indoors and outdoors. Indoor sound transmission paths include the direct (line-of-sight) path between the source and the receiver, as well as reflected paths introduced by the room's walls, floor, ceiling and furnishings. Sound propagation indoors does not follow the inverse square law (i.e., 6 dB attenuation per doubling of distance from a reference level).

Outdoors, the effects of the reflections are small, provided that the source is not located near large reflecting surfaces. However, sound outdoors can refract and change propagation direction due to wind and temperature gradient effects. Sound propagation outdoors follows the inverse square law. Therefore, Equations (7) and (8) can generally be used to calculate the relationship between sound power level and sound pressure level for fully free-field and hemispherical free-field conditions, respectively.

Duct-Borne Transmission. Ductwork can provide an effective sound transmission path because the sound is readily contained within the boundaries of the ductwork. Sound can transmit both upstream and downstream from the source. A special case of duct-borne transmission is crosstalk, where sound is transmitted from one room to another via the duct path.

Room-to-Room Transmission. Room-to-room sound transmission generally involves both airborne and structure-borne sound paths. The sound power incident on a room surface element undergoes three phenomena: (1) some of the sound energy is reflected from the surface element back into the room; (2) a portion of the sound energy is lost as energy transfer into the element, and (3) the remainder of the sound energy is transmitted through the element to the other room. Airborne sound is radiated as the element vibrates, and structure-borne sound can be transmitted via the studs of a partition or the floor and ceiling surfaces.

Structure-Borne Transmission

Solid structures are efficient transmission paths for sound, which frequently originates as a vibration imposed on the transmitting structure. The vibration can be either a single impulse or a steady-state energy input. Typically, only a small amount of the input energy is radiated by the structure as airborne sound. A lightweight structure with little inherent damping radiates more sound than a massive structure with greater damping.

Flanking Transmission

Sound from the source room can bypass the primary separating element and enter the receiving room along flanking paths. Common sound flanking paths include return air plenums, doors, and windows. Less obvious paths are along floor and adjoining wall structures. Such flanking paths can seriously reduce the sound isolation between the rooms. Flanking can explain poor sound isolation between spaces when the partition between spaces is known to be a very good sound insulator. It can also explain sounds generated in one room being heard in another a great distance away. Determining sound flanking paths can be difficult and should be done by a trained listener, experienced in hearing typical sound flanking. Sound intensity methods can also be used to determine flanking paths.

CONTROLLING NOISE

Terminology

The following noninterchangeable terms are used to describe the acoustical performance of many system components. ASTM *Standard* C 634 defines additional terms used to describe acoustical performance parameters.

Sound attenuation is a general term describing the reduction of the level of sound as it travels from a source to a receiver.

Insertion loss (IL) of a silencer or other sound-attenuating element is expressed in dB and is defined as the decrease in sound pressure level or sound intensity level measured at a receiver when the silencer or a sound-attenuating element is inserted into the path between the source and the receiver. For example, if a straight, unlined piece of ductwork were replaced with a duct silencer, the sound level difference would be considered the silencer's insertion loss. Measurements are typically made in either octave or 1/3 octave bands.

Transmission loss (TL) of a partition or other building element is expressed in dB and is the loss equal to 10 times the base 10 logarithm of the ratio of the airborne sound power incident on the partition to the sound power transmitted by the partition and radiated on the other side. Measurements are typically made in octave or 1/3 octave bands. Chapter 43 of the 1995 *ASHRAE Handbook—Applications* defines the special case of breakout transmission loss through duct walls.

Noise reduction (NR) is also expressed in dB and is the difference between the space-time averaged sound pressure levels of two enclosed spaces. Usually these two spaces are two adjacent room called, respectively, the source room and the receiving room. Measurements are typically made in octave or 1/3 octave bands.

Sound absorption coefficient α is the fraction of the incident sound energy that is absorbed by a construction material or assembly. It is measured using 1/3 octave bands of broadband noise and is commonly reported at the full octave band center frequencies. The α value of a material in a specific octave band depends on the material's thickness, airflow resistivity, stiffness, and method of attachment to the supporting structure. Duct and plenum liners are usually fibrous materials with high sound absorption values at mid to high frequencies.

Spherical spreading is the decrease in sound intensity due to the effects of inverse square law propagation from a point source. It occurs when the sound source is either located in free space or positioned on a reflecting plane.

Scattering is the change in direction of sound propagation due to an obstacle or inhomogeneity in the transmission medium. It causes the incident sound energy to disperse in many directions.

Enclosures and Barriers

Enclosing a noise source is a common means of controlling airborne sound transmission. A homogeneous, single panel, such as a single layer of drywall, has a certain surface weight or mass per unit area and a certain stiffness and internal resistance; the TL of such a panel has a frequency characteristic like the one shown in Figure 9. For most common materials, audible frequencies are in the mass-controlled part of the curve. The TL of a limp panel without stiffness, with sound impinging perpendicular to the panel, is designated as TL_o and is modeled by the mass law equation for sound at normal incidence, giving

$$TL_o = 20 \log w_s f - 28 \text{ dB} \quad (14)$$

where

w_s = surface mass of panel, lb/ft²
f = frequency, Hz

For homogeneous panels, heavy materials attenuate sound better than light materials. Transmission loss in a diffuse sound field is averaged over all angles of incidence to yield

$$TL_{random} = TL_o - 10 \log (0.23 TL_o) \text{ dB} \quad (15)$$

Thus, the random incidence transmission loss of common building materials is often typically about 5 dB less than the normal incidence TL. It has become common practice to use the field incidence mass law, which is an empirical expression that gives better agreement with measured data:

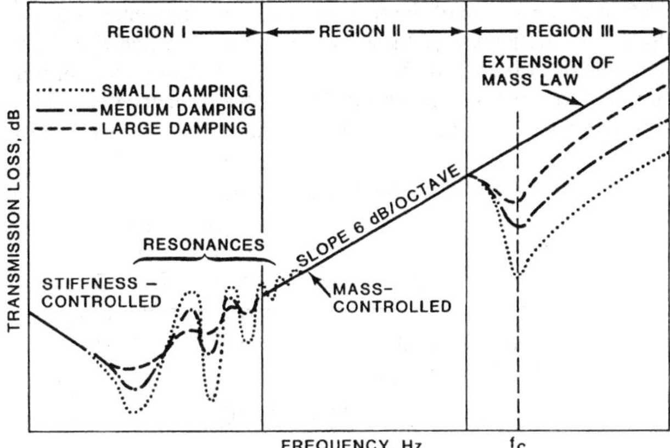

Fig. 9 Typical Transmission Loss of Panel

$$TL_o = 20 \log w_s f - 33 \text{ dB} \quad (16)$$

If the sound fields on both sides of a panel are diffuse, the panel's NR is a function of its area S_p, and the total sound absorption a_r, in the receiving space, according to the following equation:

$$\text{NR} = TL + 10 \log a_r - 10 \log S_p \text{ dB} \quad (17)$$

Because the total sound absorption in a room is expressed as the equivalent area of perfect sound absorption, both S_p and a_r are expressed in consistent units, usually square feet.

The **sound transmission class** (STC) rating of a partition or assembly is often used in architecture to classify sound isolation performance. However, the STC rating should not be used as an indicator of an assembly's ability to control noise of any source that is rich in low frequencies. This is because the STC rating system was developed to deal with speech frequency sound sources (125 to 4000 Hz). Chiller noise spectra are similar in shape to speech spectra, so a wall's STC rating can be used to select a chiller room wall or slab. However, most fan noise spectra have dominant low-frequency noise; therefore, to isolate fan noise, walls and slabs should be selected only on the basis of octave band sound transmission loss performance, particularly at low frequencies.

It is possible for a drywall partition to have a higher STC rating than a masonry wall, yet the masonry wall is likely to isolate most HVAC noise sources more effectively. Because of the limited frequency range and selective frequency discrimination of most single number rating systems, the designer should base partition and slab selections on octave band sound transmission loss values rather than on single number ratings.

The noise reduction of an enclosure may be severely compromised by openings or leaks in the enclosure. Ducts that lead into or through a noisy space can carry sound to many areas of a building. Designers should consider this factor when designing duct, piping, and electrical systems.

Attenuation of Noise in Ducts and Plenums

All ductwork, even a sheet metal duct without acoustical lining or sound attenuators, reduces sound to some degree. The natural attenuation of unlined ductwork is minimal but can, for long runs of rectangular ductwork, significantly reduce duct-borne sound. Acoustic lining of ductwork can greatly attenuate the propagation of sound through ducts, particularly at mid to high frequencies. Chapter 43 of the 1995 *ASHRAE Handbook—Applications* has a detailed discussion of lined and unlined ductwork attenuation.

If analysis shows that lined ductwork will not reduce sound propagation adequately, commercially available sound attenuators (also known as sound traps or duct silencers) can be used. There are three types: dissipative, reactive, and active. The first two are commonly known as **passive attenuators**.

Dissipative attenuators contain thick, perforated sheet metal baffles that restrict the air passage width within the attenuator housing. The baffles are filled with low-density mineral fiber insulation. This type of attenuator is most effective in reducing mid- and high-frequency sound energy.

Reactive attenuators are similar to dissipative attenuators except that the baffles are not filled with fibrous materials, but instead are subdivided into a series of cavities that absorb incident sound energy via the Helmholtz resonator concept. This attenuator type is typically used in HVAC systems serving hospitals, laboratories, or other areas with strict air quality standards. Because reactive attenuators do not contain fibrous materials, they are not as effective as dissipative attenuators. Therefore, for equal performance, a reactive attenuator must be longer than a dissipative attenuator.

The first **active attenuators** for commercial HVAC systems have been installed for only a short time. Controlled laboratory experiments have shown that active attenuators can effectively reduce both broadband and tonal noise in the 31 through 250 Hz octave bands; insertion losses of as much as 30 dB have been achieved under controlled conditions. Active attenuator systems use microphones, loudspeakers, and appropriate electronics to reduce in-duct noise by generating inverse phase sound waves that destructively interfere with the incident sound energy. Because the system's microphones and loudspeakers are mounted flush with the duct wall, there is no obstruction to airflow and, therefore, a negligible pressure drop. Because active attenuators are not effective in the presence of excessively turbulent airflow, their use is limited to relatively long, straight duct sections with an air velocity less than about 1500 fpm.

Attenuators and duct liner materials are tested according to ASTM *Standard* E 477, which defines acoustical and aerodynamic performance in terms of insertion loss, self-generated noise (or self-noise), and airflow pressure drop. Insertion loss performance is measured in the presence of both forward and reverse flows. Forward flow occurs when the air and sound move in the same direction, as in a supply air or fan discharge system; reverse flow occurs when the air and sound travel in opposite directions, as in the case of a return air or fan intake system. Self-noise can limit an attenuator's effective insertion loss for air velocities in excess of about 2000 fpm. Use extreme caution when reviewing manufacturers' performance data for attenuators and duct liner materials to be sure that the test conditions are comparable to the specific design conditions.

End reflection losses due to abrupt area changes in duct cross-section are sometimes useful in controlling low frequencies. The end reflection effect can be maximized at the end of a duct run by designing the last few feet of duct with the characteristic dimension of less than 15 in. Low-frequency noise reduction is inversely proportional to the characteristic dimension.

Where space is available, a lined plenum can provide excellent attenuation across a broad frequency range. The combination of end reflection at the plenum's entrance and exit, a large distance between the entrance and exit, and sound-absorbing lining on the plenum walls can be as effective as a sound attenuator, but with less pressure drop.

Chapter 43 of the 1995 *ASHRAE Handbook—Applications* has additional information on the control of noise.

VIBRATION

FUNDAMENTALS OF VIBRATION ISOLATION

A rigidly mounted machine transmits all vibratory forces to its supporting structure. This vibration can be isolated or, more realistically, reduced to a fraction of the original force by inserting resilient mounts between the equipment and the building structure. This section, the section on Vibration Measurement Basics, and Chapter 43 of the 1995 *ASHRAE Handbook—Applications* are useful in selecting vibration isolators and analyzing and correcting field vibration problems.

The following terms apply to vibration isolation:

M = mass of equipment, $lb_f \cdot s^2/in.$
M_f = mass of floor, $lb_f \cdot s^2/in.$
k = stiffness of isolator, $lb_f/in.$
k_f = stiffness of floor, $lb_f/in.$
F = vibratory force, lb_f
f_d = frequency of vibratory force (disturbing frequency), Hz
f_n = natural frequency of vibration isolation, Hz
f_f = natural frequency of floor, Hz
∂_{st} = static deflection of vibration isolator, in.
T = transmissibility, dimensionless ratio
$X''(\theta)$ = acceleration of system at time θ, in/s^2
$X(\theta)$ = displacement of equipment at time θ, in.
$X_f(\theta)$ = displacement of floor at time θ, in.

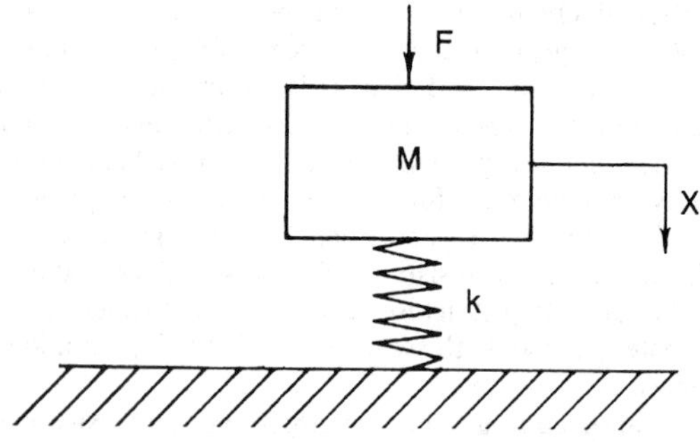

Fig. 10 Single Degree of Freedom System

Vibration evaluation requires the study of system motion. Most applications can be treated as single or two degree of freedom systems where only motion along the vertical axis is considered, and damping is disregarded. Figure 10 schematically shows a single degree of freedom system where the equation of motion is

$$MX'' + kX(\theta) = F\sin(f_d)\theta \tag{18}$$

Displacement (absolute value) is given by

$$X = \left|\frac{F/k}{1-(f_d/f_n)^2}\right| \tag{19}$$

where

$$f_n = \frac{1}{2\pi}\sqrt{\frac{k}{M}}$$

Transmissibility (absolute value) to building structure is given by

$$T = \left|\frac{1}{1-(f_d/f_n)^2}\right| \tag{20}$$

The disturbing vibration can be caused by unbalance, misalignment, eccentricities in rotating components, or defective bearings; if the frequency of the disturbing vibration is not known, it can be considered to be the lowest rotating frequency. Transmissibility T is the ratio of force transmitted by the isolator (to the floor) to force applied by the equipment. For spring isolators commonly used with HVAC equipment, the natural frequency f_n of the isolator is a function of the isolator static deflection (the distance the spring compresses under the supported equipment), as shown in Figure 11.

Equation (20) indicates that the amount of transmitted energy is a function of the square of the ratio of disturbing frequency f_d to isolator natural frequency f_n. As shown in Figure 12, when $f_d = f_n$, the denominator of Equation (20) equals zero and theoretically infinite transmission exists—a condition known as resonance. In actual resonant conditions, some limit to the transmission always exists

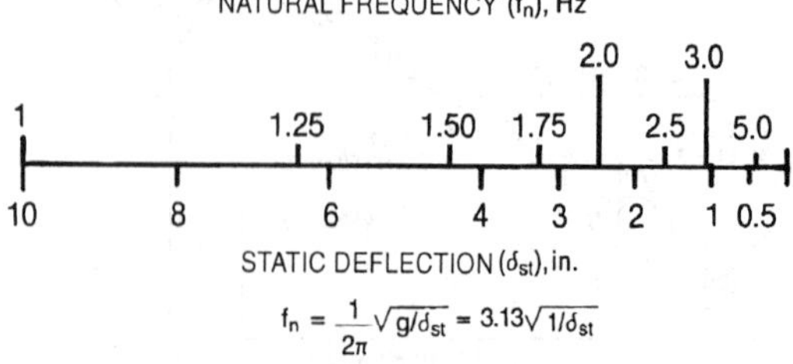

Fig. 11 Natural Frequency of Vibration Isolators

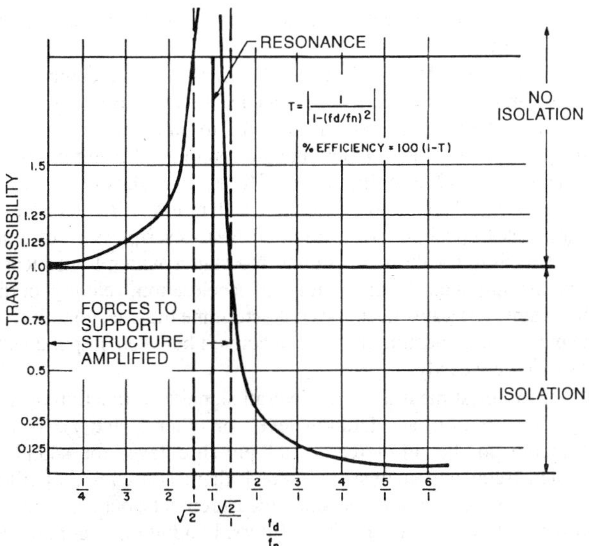

Fig. 12 Vibration Transmissibility T as Function of f_d/f_n

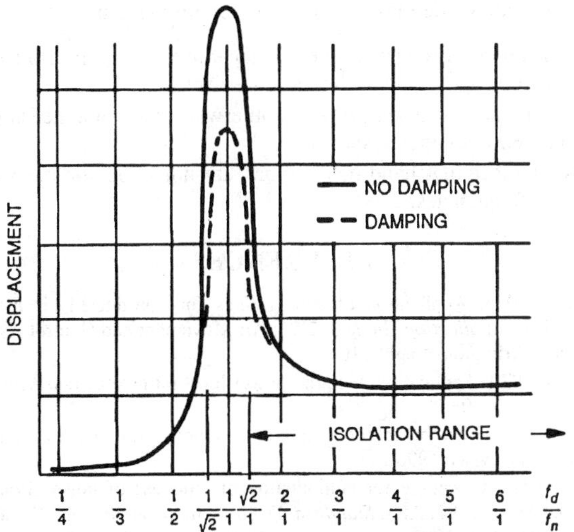

Fig. 13 Displacement of Equipment on Vibration Isolators

because of damping characteristics of isolators and structures. When $f_d/f_n \geq \sqrt{2}$, isolation occurs. The transmissibility curve becomes nearly asymptotic above a f_d/f_n ratio of about 5:1; a large increase in the f_d/f_n ratio is required to further reduce the transmissibility.

Figure 13 plots equipment displacement in accordance with Equation (19). This is similar to the transmissibility curve (Figure 12) in that equipment displacement is very great at resonance and decreases when f_d is greater than f_n and isolation occurs. Note the following:

1. Equipment displacement is greater with isolators than without isolators. Without isolators, displacement is a function of vibratory forces acting against the stiffness of equipment; with isolators, the equipment displacement varies linearly with the total system mass.
2. Once in the effective isolation range, the curve flattens out, indicating that the stiffness of the isolator has little effect on equipment displacement. This is because as isolator stiffness k decreases, the isolator deflection δ_{st} increases, resulting in a lower natural frequency f_n (see Figure 11). Therefore, as k decreases, the denominator of Equation (19) increases faster than the numerator, and equipment displacement actually decreases.

The mass of equipment has no effect on transmission through isolators to the structure (Figure 14). As shown, a 1000 lb piece of

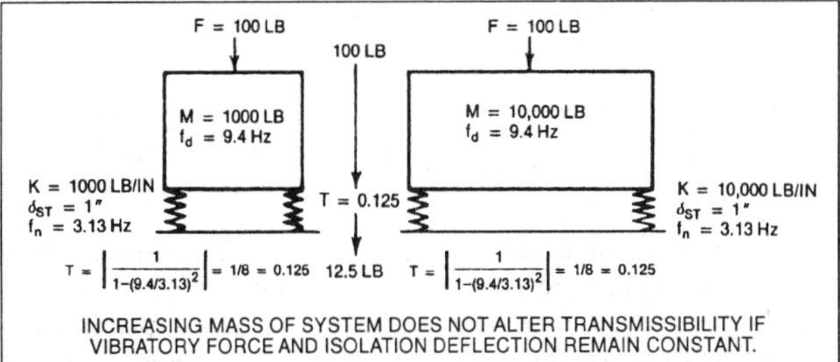

Fig. 14 Effect of Mass on Transmissibility

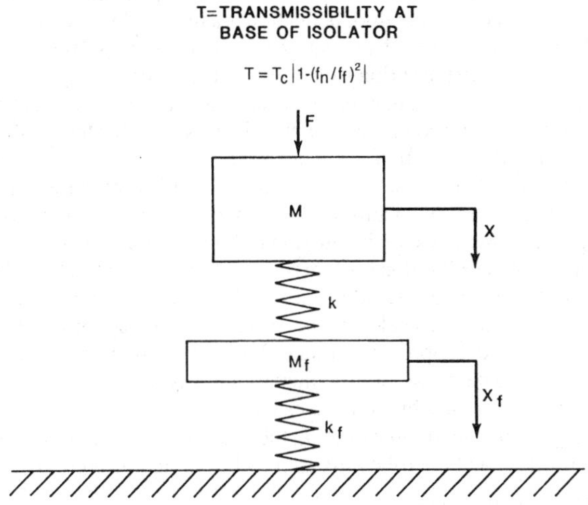

T=TRANSMISSIBILITY AT BASE OF ISOLATOR

$T = T_c \left| 1-(f_n/f_f)^2 \right|$

F M X k M_f X_f k_f

Fig. 15 Two Degree of Freedom System

equipment installed on isolators with stiffness k of 1000 lb_f/in. results in a 1 in. deflection for a system natural frequency f_n of 3.13 Hz. If equipment is operated at 564 rpm (9.4 Hz) and develops a force of 100 lb_f, a 12.5 lb_f force is transmitted to the structure. If the mass of the equipment is increased to 10,000 lb by placing it on a concrete inertia base, and the springs are changed to provide a stiffness k of 10,000 lb_f/in., the deflection is still 1 in., the natural frequency of the system is maintained at 3.13 Hz, and the force transmitted to the structure remains at 12.5 lb_f.

The increased mass of the second example reduces the equipment displacement x. Solving Equation (19) for both examples results in a 0.0125 in. displacement for the 1000 lb unit and a 0.00125 in. displacement for the 10,000 lb unit/inertia base system. This is one of the purposes of placing some high-horsepower or highly eccentric equipment on inertia pads. The greater mass permits stiffer springs and less movement of the equipment.

The foregoing discussion assumes an infinitely rigid supporting structure, which is not the case for most upper-floor locations. This condition can be represented by a two degree of freedom system, shown schematically in Figure 15, where equipment displacement is in accordance with the following equation:

$$X = \left| \frac{(F/k)\left[1 + (k/k_f) - (f_d/f_f)^2\right]}{\left[1 - (f_d/f_n)^2\right]\left[1 + (k/k_f) - (f_d/f_f)^2\right] - k/k_f} \right| \tag{21}$$

Floor displacement is in accordance with

$$X_f = \left| \frac{(F/k_f)}{\left[1 - (f_d/f_n)^2\right]\left[1 + (k/k_f) - (f_d/f_f)^2\right] - k/k_f} \right| \tag{22}$$

Transmissibility at the column supports is

$$T_c = \left| \frac{1}{\left[1 - (f_d/f_n)^2\right]\left[1 + (k/k_f) - (f_d/f_f)^2\right] - k/k_f} \right| \quad (23)$$

The denominator of Equations (21), (22), and (23) is the same; and as it approaches zero, resonance occurs, resulting in excessive transmissibility as well as excessive equipment and floor displacement.

Floor stiffness is included in the ratio k/k_f of isolator stiffness to floor stiffness. If total floor deflection and isolator deflection are the same, it does not mean that the floor and the isolators are of equal stiffness. Isolator deflection is a function of equipment mass and isolator stiffness, while floor deflection is a function of its own mass and floor spring rate. If the equipment room floor is 10 or more times stiffer than the spring isolators used to support the equipment, then k/k_f becomes so small that it is not significant. Often, this is not the case. For large equipment, refer to Table 42 in Chapter 43 of the 1995 *ASHRAE Handbook—Applications*.

The natural frequency f_f of the floor is critical because if it is equal or close to the disturbing frequency f_d, a resonant condition exists, resulting in excessive motion of both the equipment and the floor, as well as vibration transmission to the building structure. The natural frequency of the floor is a function of its mass and flexural stiffness, which depends on the material, type of construction, geometry, and support conditions. Floor deflection can indicate potential problems; as floor deflection increases, the fundamental frequency of floor vibration decreases.

If f_d is not at least three times f_f, the possibility of a resonant condition exists, and the following approaches should be considered:

1. Relocate the equipment.
2. Change the operating speed of the equipment to achieve an f_d/f_f ratio of at least 3:1.
3. Stiffen the structure to increase f_f.
4. Increase isolator deflection, thereby decreasing f_n.
5. Increase the mass M_f under the isolators.

Of these options, increasing the isolator deflection is usually the most economical, in which case the selected isolator should provide an f_d/f_n ratio of about 10:1. For example, if f_d is 15 Hz, an isolator with an f_n of about 1.5 Hz is desired, which can be provided by a 4 in. deflection isolation system. In certain situations, especially for large, low-speed equipment operating below 10 Hz, one of the other options might be more useful or cost-effective.

VIBRATION MEASUREMENT BASICS

While the control of HVAC system noise and vibration are of equal importance, the measurement of vibration is not usually necessary for determining the sources or transmission paths of disturbing noise. In addition, the techniques and instrumentation used for vibration measurement and analysis are specialized, and a thorough description is beyond the scope of this book. Therefore, designers should consult other sources (e.g., Harris 1991) for thorough descriptions of vibration measurement and analysis methods.

The typical vibrations measured are periodic motions of a surface. This surface displacement oscillates with one or more frequencies produced by mechanical means (e.g., gears), thermal means (e.g., combustion), or fluid-dynamic means (e.g., airflow through a duct or fan interactions with air). The displacement in general is inversely proportional to the frequency. In other words, if the displacements are high, the frequency is low. For most vibration measurements, we are interested in frequencies between 30 Hz and 500 Hz.

The component that detects the surface motion and converts it to electrical signals for analysis is called a **transducer**. The transducer can detect displacement, velocity, or acceleration of a surface. Displacement is the basic measure, and it is good for low frequencies. Velocity is good for overall measurements, but the transducers are usually large.

For most HVAC applications the transducer of choice is an **accelerometer**, a device that detects acceleration. It is rugged and compact. The accelerometer is attached to an amplifier, which in turn connects to a meter much like the sound level meter discussed in the section on Measuring Sound. The readouts, however, may be in acceleration level (in decibels) or acceleration (with modifiers of peak, peak-to-peak, or rms). Each of these descriptors gives a different reading for the same signal. No matter which transducer is used, the output can be in any form of acceleration, velocity, or displacement. The transducer does not determine the final quantity, but the meter and indicating device can convert between output quantities relatively easily.

The simplest measure is the overall signal as a function of time (e.g., acceleration, acceleration level, or other quantity). This is analogous to the unfiltered sound pressure level for sound. If detailed frequency analysis is needed, there is a choice of filters similar to those available for sound measurements: octave band, 1/3 octave band, or 1/12 octave band filters. In addition, there are narrow-band analyzers that use fast Fourier transforms (FFTs) to analyze and filter a signal. They are widely used but are recommended only for specialists.

Important guidelines in vibration measurement are

1. Pay attention to the frequency capability of the transducer; it must be appropriate to the measurement.
2. Mount the transducer properly; otherwise, it will not measure the claimed frequency response.
3. Remember that hand-held probes are not good for measuring high frequencies.

REFERENCES

AMCA. 1973. Application of sone ratings for non-ducted air moving devices. *Application Guide* 302-73. Air Movement and Control Association, Arlington Heights, IL.

AMCA. 1979. Application of sound power level ratings for fans. *Application Guide* 303-79.

AMCA. 1994. Reverberant room method for sound testing of fans. *Standard* 300-94 (revised 1996).

ANSI. 1980. Procedure for computation of loudness of noise. *Standard* S3.4-80 (R 1992; ASA *Standard* 37-80). American National Standards Institute, New York.

ANSI. 1986. Specifications for octave-band and fractional octave-band analog and digital filters. *Standard* S1.11-86 (R 1993; ASA *Standard* 65-86).

ANSI. 1990. Precision methods for the determination of sound power levels of broad-band noise sources in reverberation rooms. *Standard* S12.31-90 (ASA *Standard* 92-90).

ANSI. 1990. Precision methods for the determination of sound power levels of discrete-frequency and narrow-band noise sources in reverberation rooms. *Standard* S12.32-90 (ASA *Standard* 92-90).

ANSI. 1990. Precision methods for the determination of sound power levels of noise sources in anechoic and hemi-anechoic rooms. *Standard* S12.35-90 (ASA *Standard* 90-90).

ANSI. 1994. Engineering method for the determination of sound power levels of noise sources using sound intensity. *Standard* S12.12-94 (ASA *Standard* 104-94).

ANSI. 1994. Acoustical terminology. *Standard* S1.1-94.

ARI. 1994. Air terminals. *Standard* 880-94. Air-Conditioning and Refrigeration Institute, Arlington, VA.

ASHRAE. 1986. Laboratory method of testing in-duct sound power measurement procedure for fans. ANSI/ASHRAE *Standard* 68-1986 (AMCA *Standard* 330-86).

ASTM. 1989. Standard terminology relating to environmental acoustics. *Standard* C 634-89.

ASTM. 1990. Standard test method for measuring acoustical and airflow performance of duct liner materials and prefabricated silencers. *Standard* E 477-90.

ASTM. 1993. Standard test method for evaluating masking sound in open offices using A-weighted and one-third octave band sound pressure levels. *Standard* E 1573-93.

ASTM. 1995. Standard test method for measurement of sound in residential spaces. *Standard* E 1574-95.

Beranek, L.L. 1957. Revised criteria for noise in buildings. *Noise Control* 1:19.

Blazier, W.E., Jr. 1981. Revised noise criteria for design and rating of HVAC systems. *ASHRAE Transactions* 87(1).

DIN. German *Standard* 45631. Method for predicting loudness of sound spectra with tonal qualities. Deutsches Institut für Normung, Berlin.

Harris, C.M. 1991. *Handbook of acoustical measurements and noise control.* McGraw-Hill, New York.

ISO. 1975. Methods for calculating loudness level. *Standard* 532-1975. American National Standards Institute, New York.

ISO. 1993. Determination of sound power levels of noise sources using sound intensity measurements at discrete points. *Standard* 9614-1. American National Standards Institute, New York.

ISO. 1993. Draft determination of sound power levels of noise sources using sound intensity measurements by scanning. *Standard* 9614-2. American National Standards Institute, New York.

Pollack, I. 1952. The loudness of bands of noise. *Journal of the Acoustical Society of America* 24(9):533.

Robinson, D.W. and R.S. Dadson. 1956. A redetermination of the equal loudness relations for pure tones. *British Journal of Applied Physics* 7(5):166.

Schultz, T.J. 1985. Relationship between sound power level and sound pressure level in dwellings and offices. *ASHRAE Transactions* 91(1):124-53.

BIBLIOGRAPHY

AMCA. 1990. Methods for calculating fan sound ratings from laboratory test data. *Standard* 301-90. Air Movement and Control Association, Arlington Heights, IL.

ANSI. 1983. Specifications for sound level meters. *Standard* S1.4-83 (R 1994; ASA *Standard* 47-83).

ANSI. 1984. Preferred frequencies, frequency levels, and band numbers for acoustical measurements. *Standard* S1.6-84 (R 1994; ASA *Standard* 53-84).

ANSI. 1990. Precision methods for the determination of sound power levels of broad-band noise sources in reverberation rooms. *Standard* S12.31-90 (ASA *Standard* 93-90).

ASTM. 1987. Classification for rating sound insulation. *Standard* E 413-87 (R 1994). American Society for Testing and Materials, West Conshohocken, PA.

Beranek, L.L. 1960. *Noise reduction.* McGraw-Hill, New York.

Beranek, L.L. 1988. *Noise and vibration control,* revised ed. Institute of Noise Control Engineering, Washington, D.C.

Beranek, L.L. 1986. *Acoustics,* revised ed. American Institute of Physics, Acoustical Society of America, New York.

Crede, C.E. 1951. *Vibration and shock isolation.* John Wiley and Sons, New York.

Harris, C.M. and C.E. Crede. 1979. *Shock and vibration handbook,* 2nd ed. McGraw-Hill, New York.

Peterson, A.P.G. and E.E. Gross, Jr. 1974. *Handbook of noise measurement.* GenRad, Inc., Concord, MA.

Schaffer, M.E. 1991. *A practical guide to noise and vibration control for HVAC systems.* ASHRAE, Atlanta.

CHAPTER 8

THERMAL COMFORT

A PRINCIPAL purpose of heating, ventilating and air conditioning systems is to provide conditions for human thermal comfort. A widely accepted definition is, "Thermal Comfort is that condition of mind that expresses satisfaction with the thermal environment" (ASHRAE 1992). The conscious mind appears to reach conclusions about thermal comfort and discomfort from direct temperature and moisture sensations from the skin, deep body temperatures, and the efforts necessary to regulate body temperatures (Hensel 1973, 1981; Hardy 1971; Gagge 1937; Berglund 1995). In general, comfort occurs when body temperatures are held within narrow ranges, skin moisture is low, and the physiological effort of regulation is minimized.

Comfort also includes behavioral actions initiated by the conscious mind and guided by thermal and moisture sensations to reduce discomfort. For example, altering clothing, altering activity, changing posture or location, changing the thermostat setting, opening a window, complaining, or leaving the space are some of the possible behavioral actions to reduce discomfort.

Surprisingly, though regional climate conditions, living conditions, and cultures differ widely throughout the world the preferred temperature that people choose for comfort under like conditions of clothing, activity, humidity, and air movement has been found to be similar (Fanger 1972, de Dear et al. 1991, Busch 1992).

This chapter summarizes the fundamentals of human thermoregulation and comfort in terms useful to the engineer for operating systems and preparing designs and applications for the comfort and health of building occupants.

HUMAN THERMOREGULATION

The metabolic activities of the body result almost completely in heat that must be continuously dissipated and regulated to prevent abnormal body temperatures. Insufficient heat loss leads to overheating also called hyperthermia, and excessive heat loss results in body cooling also called hypothermia. Skin temperatures greater than 45°C or less than 18°C cause pain (Hardy 1952). Skin temperatures associated with comfort at sedentary activities are 91.5 to 93°F and decrease with increasing activity (Fanger 1968). In contrast internal temperatures rise with activity. The temperature regulatory center in the brain is about 98.2°F at rest in comfort and increases to about 99.3°F when walking and 100.2°F when jogging. An internal temperature less than about 82°F can lead to serious cardiac arrhythmia and death and temperatures greater than 115°F can cause irreversible brain damage. Therefore, the careful regulation of body temperature is critical to comfort and health.

The heat produced by a resting adult is about 340 Btu/h. Because most of this heat is transferred to the environment through the skin it is often convenient to characterize metabolic activity in terms of heat production per unit area of skin. For the resting person this is about 18.4 Btu/h·ft^2 (50 kcal/(h·m^2) (average person has a skin surface area of about 19.6 ft^2) and is called 1 met. Higher metabolic rates are often described in terms the resting rate. Thus a person working at metabolic rate five times the resting rate would have a metabolic rate of 5 met.

The hypothalamus section of the brain is the central control of body temperature. It has hot and cold temperature sensors, which are bathed by arterial blood. Because blood circulates rapidly and returning blood is mixed in the heart before returning to the body, arterial blood is indicative of the average internal body temperature. The hypothalamus also receives thermal information from temperature sensors in the skin and perhaps other locations as well (spinal cord, gut) (summarized by Hensel 1981).

The hypothalamus controls various physiological processes of the body to regulate body temperature. Its control behavior is primarily proportional to deviations from set-point temperatures with some integral and derivative responses. The most important and often used of the physiological processes is regulating blood flow to the skin. When internal temperatures rise above a set point an increasing proportion of the total blood is directed to the skin. This vasodilation of skin blood vessels can increase skin blood flow by 15 times to carry internal heat to the skin for transfer to the environment. When body temperatures fall below the set point, skin blood flow is reduced (vasoconstricted) to conserve body heat. The insulating effect of maximum vasoconstriction is equivalent to putting on a heavy sweater. At temperatures below the set point muscle tension increases to generate additional heat and may increase to visible shivering. Shivering can double the resting rate of heat production.

At elevated internal temperatures sweating and its control is necessary for comfort at metabolic rates above resting (Fanger 1968). Sweat glands pump perspiration onto the skin for evaporation. If conditions are good for evaporation the skin remains relatively dry with little perception of sweating. At skin conditions less favorable for evaporation the sweat must spread on the skin until its area is sufficient to evaporate the sweat coming to the surface. The fraction of the skin that is covered with water to account for the observed total evaporation rate is termed skin wettedness (Gagge 1937).

Humans are quite good at sensing skin moisture from perspiration (Berglund 1986, 1994) and skin moisture correlates well with warm discomfort and unpleasantness (Winslow 1937). It is rare for a sedentary or slightly active person to be comfortable with a skin wettedness greater than 25%. In addition to the perception of skin moisture, skin wettedness increases the friction between skin and fabrics making clothing feel less pleasant and fabrics feel more coarse (Gwosdow 1987).

With repeated intermittent heat exposure the set point for the onset of sweating decreases and the proportional gain or temperature sensitivity of the sweating system increases (Gonzalez 1978, Hensel 1981). However, under long term exposure to hot conditions the set point increases perhaps to reduce the physiological effort of sweating. Perspiration as secreted has a lower salt concentration

The preparation of this chapter is assigned to TC 2.1, Physiology and Human Environment.

than interstitial body fluid or blood plasma. After prolonged heat exposure, sweat glands acted upon by the endocrine system further reduce the salt concentration of sweat to conserve salt.

At the surface the water in sweat evaporates while the dissolved salt and other constituents remain and accumulate. Since salt lowers the vapor pressure of water and thereby impedes its evaporation, the accumulating salt results in increased skin wettedness with time. Some of the relief and pleasure of washing after a warm day is related to the restoration of a hypotonic sweat film and decreased skin wettedness. Other examples of adaptation to heat are increased blood flow and sweating in peripheral regions where heat transfer is better. Such adaptations are examples of integral control.

The role of regulatory effort in comfort is highlighted by experiments of Chatonnet and Cabanac (1965) and observations of Kuno et al. (1995). In Chatonnet's experiments the sensation of placing the hand for 30 s in relatively hot or cold water (86 to 100°F) were compared with the subject at different thermal states. When the person was over-heated or hyperthermic the cold water was pleasant and the hot water was very unpleasant; but when the persons was in a cold or hypothermic state the hand felt pleasant in hot water and unpleasant in cold water. Kuno (1995) describes similar observations during transient whole body exposures to hot and cold environment. When in a state of thermal discomfort any move away from the thermal stress of the uncomfortable environment is perceived as pleasant during the transition.

ENERGY BALANCE

The total metabolic rate of work M produced within the body is the metabolic rate required for the person's activity M_{act} plus the metabolic level required for shivering M_{shiv} (should shivering occur). A portion of the body's energy production may be expended as external work done by the muscles W; the net heat production $M - W$ is either stored (S), causing the body's temperature to rise, or is dissipated to the environment through the skin surface (Q_{sk})and respiratory tract (Q_{res}). (See Figure 1)

$$\begin{aligned} M - W &= Q_{sk} + Q_{res} + S \\ &= (C + R + E_{sk}) + (C_{res} + E_{res}) + S_{sk} + S_c \end{aligned} \quad (1)$$

where

M = rate of metabolic heat production, Btu/(h·ft²)
W = rate of mechanical work accomplished, Btu/(h·ft²)
Q_{res} = total rate of heat loss through respiration, Btu/(h·ft²)
Q_{sk} = total rate of heat loss from skin, Btu/(h·ft²)

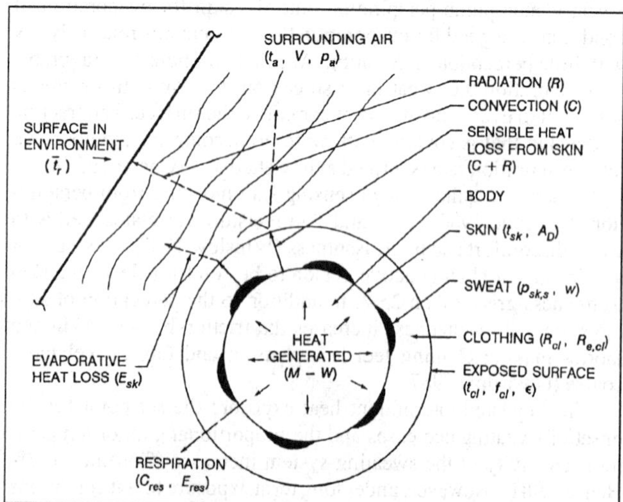

Fig. 1 Thermal Interaction of Human Body and Environment

C_{res} = rate of convective heat loss from respiration, Btu/(h·ft²)
E_{res} = rate of evaporative heat loss from respiration, Btu/(h·ft²)
$C + R$ = sensible heat loss from skin, Btu/(h·ft²)
E_{sk} = rate of total evaporative heat loss from skin, Btu/(h·ft²)
S_{cr} = rate of heat storage in core compartment, Btu/(h·ft²)
S_{sk} = rate of heat storage in skin compartment, Btu/(h·ft²)

Heat dissipation from the body to the immediate surroundings occurs by several modes of heat exchange: sensible heat flow from the skin $C + R$; latent heat flow from the evaporation of sweat E_{rsw} and from evaporation of moisture diffused through the skin E_{dif}; sensible heat flow during respiration C_{res}; and latent heat flow due to evaporation of moisture during respiration E_{res}. Sensible heat flow from the skin may be a complex mixture of conduction, convection, and radiation for a clothed person; however, it is equal to the sum of the convection C and radiation R heat transfer at the outer clothing surface (or exposed skin).

Sensible and latent heat losses from the skin are typically expressed in terms of environmental factors, skin temperature t_{sk}, and skin wettedness w. The expressions also incorporate factors that account for the thermal insulation and moisture permeability of clothing. The independent environmental variables can be summarized as: air temperature t_a, mean radiant temperature $\bar{t}_r$, relative air velocity v, and ambient water vapor pressure p_a. The independent personal variables that influence thermal comfort are activity and clothing.

The rate of heat storage in the body equals the rate of increase in internal energy. The rate of storage can be written separately for each compartment in terms of thermal capacity and time rate of change of temperature in each compartment:

$$S_{cr} = \frac{(1-\alpha)mc_{p,b}}{A_D}\frac{dt_{cr}}{d\theta} \quad (2)$$

$$S_{sk} = \frac{\alpha m c_{p,b}}{A_D}\frac{dt_{sk}}{d\theta} \quad (3)$$

where

α = fraction of body mass concentrated in skin compartment
m = body mass, lb
$c_{p,b}$ = specific heat capacity of body = 0.834 Btu/(lb·°F)
θ = time, h

The fractional skin mass α depends on the rate of blood flowing to the skin surface $\dot{m}_{bl}$.

THERMAL EXCHANGES WITH THE ENVIRONMENT

Fanger (1967, 1970), Hardy (1949), Rapp and Gagge (1967), and Gagge and Hardy (1967) give quantitative information on calculating the heat exchange between people and the environment. A summary of the mathematical statements for various terms of heat exchange used in the heat balance equations (C, R, E_{sk}, C_{res}, E_{res}) follows. Terms describing the heat exchanges associated with the thermoregulatory control mechanisms ($Q_{cr,sk}$, M_{shiv}, E_{rsw}) and values for the coefficients and appropriate equations for M_{act} and A_D are presented in later sections. The mathematical description of the energy balance of the human body represents a combined rational/empirical approach to describing the thermal exchanges with the environment. Fundamental heat transfer theory is used to describe the various mechanisms of sensible and latent heat exchange, while empirical expressions are used to determine the values of the coefficients describing these rates of heat exchange. Empirical equations are also used to describe the thermophysiological control mechanisms as a function of skin and core temperatures in the body.

Body Surface Area

The terms in the heat balance equations previously presented have units of power per unit area and refer to the surface area of the nude body. The most useful measure of nude body surface area, originally proposed by DuBois (1916), is described by:

$$A_D = 0.108 m^{0.425} l^{0.725} \quad (4)$$

where

A_D = DuBois surface area, ft^2
m = mass, lb
l = height, in.

A correction factor $f_{cl} = A_{cl}/A_D$ must be applied to the heat transfer terms from the skin (C, R, and E_{sk}) to account for the actual surface area of the clothed body A_{cl}. This factor can be found in Table 7 for various clothing ensembles. For a 68 in. tall, 154 lb man, A_D = 19.6 ft^2. All terms in the basic heat balance equations are expressed per unit DuBois surface area.

Sensible Heat Loss from Skin

Sensible heat exchange from the skin surface must pass through clothing to the surrounding environment. These paths are treated in series and can be described in terms of: (1) heat transfer from the skin surface, through the clothing insulation, to the outer clothing surface, and (2) from the outer clothing surface to the environment.

Both convective C and radiative R heat losses from the outer surface of a clothed body can be expressed in terms of a heat transfer coefficient, and the difference between the mean temperature of the outer surface of the clothed body t_{cl} and the appropriate environmental temperature:

$$C = f_{cl} h_c (t_{cl} - t_a) \quad (5)$$

$$R = f_{cl} h_r (t_{cl} - \bar{t}_r) \quad (6)$$

where

h_c = convective heat transfer coefficient, Btu/(h·ft^2·°F)
h_r = linear radiative heat transfer coefficient, Btu/(h·ft^2·°F)
f_{cl} = clothing area factor A_{cl}/A_D

The coefficients h_c and h_r are both evaluated at the clothing surface. Equations (5) and (6) are commonly combined to describe the total sensible heat exchange by these two mechanisms in terms of an operative temperature t_o and a combined heat transfer coefficient h:

$$(C + R) = f_{cl} h (t_{cl} - t_o) \quad (7)$$

where

$$t_o = \frac{h_r \bar{t}_r + h_c t_a}{h_r + h_c} \quad (8)$$

$$h = h_r + h_c \quad (9)$$

Based on Equation (8), operative temperature t_o can be defined as the average of the mean radiant and ambient air temperatures, weighted by their respective heat transfer coefficients.

The actual transport of sensible heat through clothing involves conduction, convection, and radiation. It is usually most convenient to combine these into a single thermal resistance value R_{cl}, defined by:

$$(C + R) = (t_{sk} - t_{cl})/R_{cl} \quad (10)$$

where R_{cl} = thermal resistance of clothing, h·ft^2·°F/Btu.

Since it is often inconvenient to include the clothing surface temperature in calculations, Equations (7) and (10) can be combined to eliminate t_{cl}:

$$(C + R) = \frac{t_{sk} - t_o}{R_{cl} + 1/(f_{cl} h)} \quad (11)$$

where t_o is defined in Equation (8).

Evaporative Heat Loss from Skin

Evaporative heat loss from skin E_{sk} depends on the difference between the water vapor pressure at the skin and in the ambient environment, and the amount of moisture on the skin:

$$E_{sk} = \frac{w(p_{sk,s} - p_a)}{R_{e,cl} + 1/(f_{cl} h_e)} \quad (12)$$

where

p_a = water vapor pressure in ambient air, psi
$p_{sk,s}$ = water vapor pressure at skin, normally assumed to be that of saturated water vapor at t_{sk}, psi
$R_{e,cl}$ = evaporative heat transfer resistance of clothing layer (analogous to R_{cl}), ft^2·psi·h/Btu
h_e = evaporative heat transfer coefficient (analogous to h_c), Btu/(h·ft^2·psi)
w = wet skin fraction, dimensionless

Procedures for calculating $R_{e,cl}$ and h_e are given in the Engineering Data and Measurements section. The wet skin fraction is the ratio of the actual evaporative heat loss to the maximum possible evaporative heat loss E_{max} with the same conditions and a completely wet skin (w = 1). The fraction of wet skin is important in determining evaporative heat loss. Maximum evaporative potential E_{max} occurs when the skin surface is completely wet, or w = 1.0.

Evaporative heat loss from the skin is a combination of the evaporation of sweat secreted due to thermoregulatory control mechanisms E_{rsw} and the natural diffusion of water through the skin E_{dif}:

$$E_{sk} = E_{rsw} + E_{dif} \quad (13)$$

Evaporative heat loss by regulatory sweating is directly proportional to the regulatory sweat generated:

$$E_{rsw} = \dot{m}_{rsw} h_{fg} \quad (14)$$

where

h_{fg} = heat of vaporization of water = 1045 Btu/lb at 86°F
$\dot{m}_{rsw}$ = rate at which regulatory sweat is generated, lb/h·ft^2 [see Equation (67)]

The portion of a body that must be wetted to evaporate the regulatory sweat w_{rsw} is:

$$w_{rsw} = E_{rsw}/E_{max} \quad (15)$$

With no regulatory sweating, skin wettedness due to diffusion is approximately 0.06 for normal conditions. For large values of E_{max} or long exposures to low humidities, the value may drop to as low as 0.02, since dehydration of the outer skin layers alters its diffusive characteristics. With regulatory sweating, the 0.06 value applies only to the portion of skin not covered with sweat $(1 - w_{rsw})$; the diffusion evaporative heat loss is:

$$E_{dif} = (1 - w_{rsw}) 0.06 E_{max} \quad (16)$$

These equations can be solved to give a value of w, given the maximum evaporative potential E_{max} and the regulatory sweat generation E_{rsw}:

$$w = w_{rsw} + 0.06(1 - w_{rsw}) = 0.06 + 0.94E_{rsw}/E_{max} \quad (17)$$

Once skin wettedness is determined, evaporative heat loss from the skin is calculated from Equation (12), or by:

$$E_{sk} = wE_{max} \quad (18)$$

To summarize, the following calculations determine w and E_{sk}:

- E_{max} Equation (12), with $w = 1.0$
- E_{rsw} Equation (14)
- w Equation (17)
- E_{sk} Equation (18) or (12)

Although evaporation from the skin E_{sk} as described in Equation (12) depends on w, the body does not directly regulate skin wettedness but, rather, regulates sweat rate $\dot{m}_{rsw}$, [Equation (14)]. Skin wettedness is then an indirect result of the relative activity of the sweat glands and the evaporative potential of the environment. Skin wettedness of 1.0 is the upper theoretical limit. If the aforementioned calculations yield a wettedness of more than 1.0, then Equation (14) is no longer valid because all the sweat is not evaporated. In this case, $E_{sk} = E_{max}$.

Skin wettedness is strongly correlated with warm discomfort and is also a good measure of thermal stress. Theoretically, skin wettedness can approach 1.0 while the body still maintains thermoregulatory control. In most situations, it is difficult to exceed 0.8 (Berglund 1977). Azer (1982) recommends 0.5 as a practical upper limit for sustained activity for a healthy acclimatized person.

Respiratory Losses

During respiration, the body loses both sensible and latent heat by convection and evaporation of heat and water vapor from the respiratory tract to the inhaled air. A significant amount of heat can be associated with respiration because the air is inspired at ambient conditions and expired nearly saturated at a temperature only slightly cooler than t_{cr}.

Sensible C_{res} and latent E_{res} heat losses due to respiration are:

$$C_{res} = \dot{m}_{res}c_{p,a}(t_{ex} - t_a)/A_D \quad (19)$$

$$E_{res} = \dot{m}_{res}h_{fg}(W_{ex} - W_a)/A_D \quad (20)$$

where

$\dot{m}_{res}$ = pulmonary ventilation rate, lb/h
t_{ex} = temperature of exhaled air, °F
W_{ex} = humidity ratio of exhaled air, lb (water vapor)/lb (dry air)
W_a = humidity ratio of inhaled (ambient) air, lb (water vapor)/lb (dry air)
$c_{p,a}$ = specific heat of air, Btu/(lb·°F)

These equations can be reduced by using approximations and empirical relationships to estimate the values of the parameters. Under normal circumstances, pulmonary ventilation rate is primarily a function of metabolic rate (Fanger 1970):

$$\dot{m}_{res} = K_{res}M \quad (21)$$

where

M = metabolic rate, Btu/h
K_{res} = proportionality constant (0.0645 lb·ft^2/Btu)

Respiratory air is nearly saturated and near the body temperature when exhaled. The following empirical equations developed by Fanger (1970) can be used to estimate these conditions for typical indoor environments:

$$t_{ex} = 88.6 + 0.066\,t_a + 57.6\,W_a \quad (22)$$

$$W_{ex} - W_a = 0.0265 + 0.000036\,t_a - 0.80\,W_a \quad (23)$$

where t_a and t_{ex} are in °F. For more extreme conditions, such as outdoor winter environments, different relationships may be required (see Holmer 1984).

The humidity ratio of ambient air can be expressed in terms of the total or barometric pressure p_t and ambient partial vapor pressure p_a:

$$W_a = 0.622\frac{p_a}{p_t - p_a} \quad (24)$$

Two approximations are commonly used to simplify Equations (19) and (20). First, because the dry respiratory heat loss is relatively small compared to the other terms in the heat balance, an average value for t_{ex} is determined by evaluating Equation (22) at standard conditions of 68°F, 50% rh, sea level. Second, noting in Equation (23) that there is only a weak dependence on ambient air temperature, the second term in Equation (23) and the denominator in Equation (24) are evaluated at standard conditions. Using these approximations, and substituting values of h_{fg} and $c_{p,a}$ at standard conditions, Equations (19) and (20) can be combined and written as:

$$C_{res} + E_{res} = (h_{ex} - h_a)\dot{m}_{res} \quad (25)$$

$$C_{res} + E_{res} = 0.0084M(93.2 - t_a) + 1.28M(0.851 - p_a) \quad (26)$$

where p_a is expressed in psi and t_a is in °F.

Alternative Formulations

Equations (11) and (12) describe heat loss from skin for clothed people in terms of clothing parameters R_{cl}, $R_{e,cl}$, and f_{cl}; parameters h and h_e describe outer surface resistances. Other parameters and definitions are also used. Although these alternate parameters and definitions may be confusing, note that the information presented in one form can be converted to another form. Common parameters and their qualitative descriptions are presented in Table 1. Equations showing their relation to each other are presented in Table 2. Generally, parameters related to dry or evaporative heat flows are not independent because they both rely, in part, on the same physical processes. The Lewis relation describes the relations between convective heat transfer and mass transfer coefficients for a surface (see Chapter 5). The Lewis relation can be used to relate convective and evaporative heat transfer coefficients defined in Equations (5) and (12) according to:

$$h_e/h_c = \mathrm{LR} \quad (27)$$

where LR is referred to as the "Lewis ratio" and, at typical indoor conditions, equals approximately 205°F/psi. The Lewis relation applies to surface convection coefficients. Heat transfer coefficients that include the effects of insulation layers and/or radiation will still be coupled, but the relationship may deviate significantly from that for a surface. The i terms in Tables 1 and 2 describe how the actual ratios of these parameters deviate from the ideal Lewis ratio (Woodcock 1962, Oohori et al. 1985).

Table 1 Parameters Used to Describe Clothing

Sensible Heat Flow

R_{cl} = intrinsic clothing insulation, the thermal resistance of a uniform layer of insulation covering the entire body that has the same effect on sensible heat flow as the actual clothing.

R_t = total insulation, the total equivalent uniform thermal resistance between the body and the environment: clothing and boundary resistance.

R_{cle} = effective clothing insulation, the increased body insulation due to clothing as compared to the nude state.

R_a = boundary insulation, the thermal resistance at the skin boundary for a nude body.

$R_{a,cl}$ = outer boundary insulation, the thermal resistance at the outer boundary (skin or clothing).

h' overall sensible heat transfer coefficient, the overall equivalent uniform conductance between the body and the environment.

h'_{cl} = clothing conductance, the thermal conductance of a uniform layer of insulation covering the entire body that has the same effect on sensible heat flow as the actual clothing.

F_{cle} = effective clothing thermal efficiency, the ratio of the actual sensible heat loss to that of a nude body at the same conditions.

F_{cl} = intrinsic clothing thermal efficiency, the ratio of the actual sensible heat loss to that of a nude body at the same conditions including an adjustment for the increase in surface area due to the clothing.

Evaporative Heat Flow

$R_{e,cl}$ = evaporative heat transfer resistance of the clothing, the impedance to transport of water vapor of a uniform layer of insulation covering the entire body that has the same effect on evaporative heat flow as the actual clothing.

$R_{e,t}$ = total evaporative resistance, the total equivalent uniform impedance to the transport of water vapor from the skin to the environment.

F_{pcl} = permeation efficiency, the ratio of the actual evaporative heat loss to that of a nude body at the same conditions, including an adjustment for the increase in surface area due to the clothing.

Parameters Relating Sensible and Evaporative Heat Flow

i_{cl} = clothing vapor permeation efficiency, the ratio of the actual evaporative heat flow capability through the clothing to the sensible heat flow capability as compared to the Lewis ratio.

i_m = moisture permeability index, the ratio of the actual evaporative heat flow capability between the skin and the environment to the sensible heat flow capability as compared to the Lewis ratio.

i_a = air layer vapor permeation efficiency, the ratio of the actual evaporative heat flow capability through the outer air layer to the sensible heat flow capability as compared to the Lewis ratio.

Table 2 Relationships between Clothing Parameters

Sensible Heat Flow

$$R_t = R_{cl} + 1/(hf_{cl}) = R_{cl} + R_a/f_{cl}$$
$$R_t = R_{cle} + 1/h = R_{cle} + R_a$$
$$h'_{cl} = 1/R_{cl}$$
$$h' = 1/R_t$$
$$h = 1/R_a$$
$$R_{a,cl} = R_a/f_{cl}$$
$$F_{cl} = h'/(hf_{cl}) = 1/(1 + f_{cl}hR_{cl})$$
$$F_{cle} = h'/h = f_{cl}/(1 + f_{cl}hR_{cl}) = f_{cl}F_{cl}$$

Evaporative Heat Flow

$$R_{e,t} = R_{e,cl} + 1/(h_e f_{cl}) = R_{e,cl} + R_{e,a}/f_{cl}$$
$$h_e = 1/R_{e,a}$$
$$h'_{e,cl} = 1/R_{e,cl}$$
$$h'_e = 1/R_{e,t} = f_{cl}\, F_{pcl}\, h_e$$
$$F_{pcl} = 1/(1 + f_{cl}h_e R_{e,cl})$$

Parameters Relating Sensible and Evaporative Heat Flows

$$i_{cl}\text{LR} = h'_{e,cl}/h'_{cl} = R_{cl}/R_{e,cl}$$
$$i_m\text{LR} = h'_e/h' = R_t/R_{e,t}$$
$$i_m = (R_{cl} + R_{a,cl})/[(R_{cl}/i_{cl}) + (R_{a,cl}/i_a)]$$
$$i_a\text{LR} = h_e/h$$
$$i_a = h_c/(h_c + h_r)$$

Depending on the combination of parameters used, heat transfer from the skin can be calculated using several different formulations (see Tables 2 and 3). If the parameters are used correctly, the end result will be the same regardless of the formulation used.

Total Skin Heat Loss

Total skin heat loss—sensible heat plus evaporative heat—can be calculated from any combination of the equations presented in Table 3. Total skin heat loss is used as a measure of the thermal environment; two combinations of parameters that yield the same total heat loss for a given set of body conditions (t_{sk} and w) are considered to be approximately equivalent. The fully expanded skin heat loss equation, showing each parameter that must be known or specified, is as follows:

Table 3 Skin Heat Loss Equations

Sensible Heat Loss

$$C + R = (t_{sk} - t_o)/[R_{cl} + 1/(f_{cl}h)]$$
$$C + R = (t_{sk} - t_o)/R_t$$
$$C + R = F_{cle}h(t_{sk} - t_o)$$
$$C + R = F_{cl}f_{cl}h(t_{sk} - t_o)$$
$$C + R = h'(t_{sk} - t_o)$$

Evaporative Heat Loss

$$E_{sk} = w\,(p_{sk,s} - p_a)/[R_{e,cl} + 1/(f_{cl}h_e)]$$
$$E_{sk} = w\,(p_{sk,s} - p_a)/R_{e,t}$$
$$E_{sk} = w\,F_{pcl}f_{cl}h_e\,(p_{sk,s} - p_a)$$
$$E_{sk} = h'_e\,w\,(p_{sk,s} - p_a)$$
$$E_{sk} = h'\,w_{im}\,\text{LR}\,(p_{sk,s} - p_a)$$

$$Q_{sk} = \frac{t_{sk} - t_o}{R_{cl} + R_{acl}} + \frac{w(p_{sks} - p_a)}{R_{ecl} + 1/(\text{LR}h_c f_{cl})} \tag{28}$$

where t_o is the operative temperature and represents the temperature of a uniform environment ($t_a = t_r$) that will transfer dry heat at the same rate as in the actual environment [$t_o = T_r h_r + t_a h_c)/(h_c + h_r)$]. After rearranging Equation (28) becomes:

$$Q_{sk} = F_{cl}f_{cl}h(t_{sk} - t_o) + w\text{LR}\,F_{pcl}h_c(p_{sks} - p_a) \tag{29}$$

This equation allows the trade-off between any two, or more, parameters to be evaluated under given conditions. If the trade-off between two specific variables is to be examined, then a simplified form of the equation suffices. The trade-off between operative temperature and humidity is often of interest. Equation (28) can be written in a simpler form for this purpose (Fobelets and Gagge 1988):

$$Q_{sk} = h'[(t_{sk} + wi_m\text{LR}p_{sk,s}) - (t_o + wi_m\text{LR}p_a)] \tag{30}$$

Equation (30) can be used to define a combined temperature t_{com}, which reflects the combined effect of operative temperature and humidity for an actual environment:

$$t_{com} + wi_m\mathrm{LR}p(t_{com}) = t_o + wi_m\mathrm{LR}p_a$$

or

$$t_{com} = t_o + wi_m\mathrm{LR}p_a - wi_m\mathrm{LR}p(t_{com}) \quad (31)$$

where $p(t_{com})$ is a vapor pressure related in some fixed way to t_{com} and is analogous to $p_{wb,s}$ for t_{wb}. The term $wi_m\mathrm{LR}p(t_{com})$ is constant, to the extent that i_m is constant and any combination of t_o and p_a that gives the same t_{com} will result in the same total heat loss.

Environmental indices are discussed in a later section of this chapter. Two of these, the humid operative temperature t_{oh} and the effective temperature ET*, can be represented in terms of Equation (31). The humid operative temperature is that temperature which at 100% rh yields the same total heat loss as for the actual environment:

$$t_{oh} = t_o + wi_m\mathrm{LR}(p_a - p_{oh,s}) \quad (32)$$

where $p_{oh,s}$ = saturated vapor pressure at t_{oh} in psi.

The effective temperature is the temperature at 50% rh that yields the same total heat loss from the skin as for the actual environment:

$$\mathrm{ET}^* = t_o + wi_m\mathrm{LR}(p_a - 0.5\, p_{\mathrm{ET}^*,s}) \quad (33)$$

where $p_{\mathrm{ET}^*,s}$ saturated vapor pressure at ET* in psi.

The psychrometric chart in Figure 2 shows a constant total heat loss line and the relationship between these indices. This line represents only one specific skin wettedness and moisture permeability index. The relationship between indices depends on these two parameters (see the Environmental Indices section).

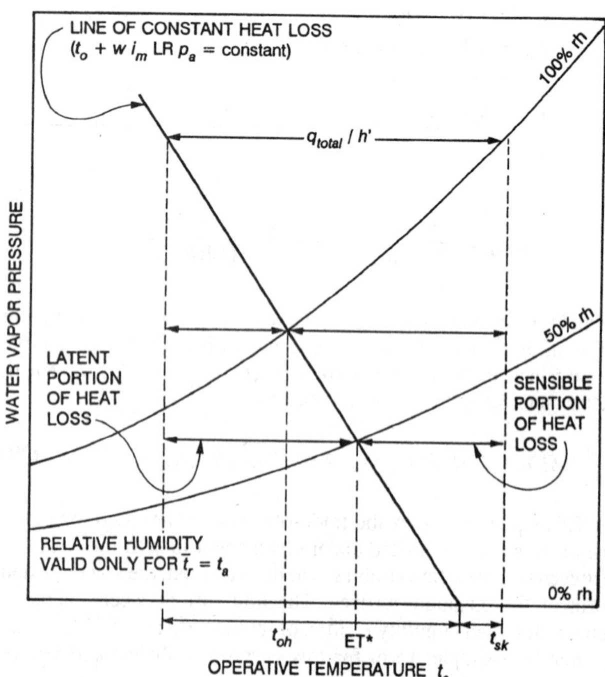

Fig. 2 Constant Skin Heat Loss Line and Its Relationship to t_{oh} and ET*

ENGINEERING DATA AND MEASUREMENTS

Applying the preceding basic equations to practical problems of the thermal environment requires quantitative estimates of the body's surface area, metabolic requirements for a given activity and the mechanical efficiency for the work accomplished, evaluation of the heat transfer coefficient h_r and h_c, and the general nature of the clothing insulation used. This section provides the necessary data and describes methods used to measure the parameters of the heat balance equation.

Metabolic Rate and Mechanical Efficiency

Maximum Capacity. In choosing optimal conditions for comfort and health, the rate of work done during routine physical activities must be known, because metabolic power increases in proportion to exercise intensity. Metabolic rate varies over a wide range, depending on the activity, the person, and the conditions under which the activity is performed. Table 4 lists typical metabolic rates for an average adult (A_D = 19.6 ft²) for activities performed continuously. The highest power a person can maintain for any continuous period is approximately 50% of the maximal capacity to use oxygen (maximum energy capacity).

Table 4 Typical Metabolic Heat Generation for Various Activities

	Btu/(h·ft²)	met[a]
Resting		
Sleeping	13	0.7
Reclining	15	0.8
Seated, quiet	18	1.0
Standing, relaxed	22	1.2
Walking (on level surface)		
2.9 ft/s (2 mph)	37	2.0
4.4 ft/s (3 mph)	48	2.6
5.9 ft/s (4 mph)	70	3.8
Office Activities		
Reading, seated	18	1.0
Writing	18	1.0
Typing	20	1.1
Filing, seated	22	1.2
Filing, standing	26	1.4
Walking about	31	1.7
Lifting/packing	39	2.1
Driving/Flying		
Car	18 to 37	1.0 to 2.0
Aircraft, routine	22	1.2
Aircraft, instrument landing	33	1.8
Aircraft, combat	44	2.4
Heavy vehicle	59	3.2
Miscellaneous Occupational Activities		
Cooking	29 to 37	1.6 to 2.0
Housecleaning	37 to 63	2.0 to 3.4
Seated, heavy limb movement	41	2.2
Machine work		
sawing (table saw)	33	1.8
light (electrical industry)	37 to 44	2.0 to 2.4
heavy	74	4.0
Handling 110 lb bags	74	4.0
Pick and shovel work	74 to 88	4.0 to 4.8
Miscellaneous Leisure Activities		
Dancing, social	44 to 81	2.4 to 4.4
Calisthenics/exercise	55 to 74	3.0 to 4.0
Tennis, singles	66 to 74	3.6 to 4.0
Basketball	90 to 140	5.0 to 7.6
Wrestling, competitive	130 to 160	7.0 to 8.7

Compiled from various sources. For additional information, see Buskirk (1960), Passmore and Durnin (1967), and Webb (1964).
[a] 1 met = 18.4 Btu/(h·ft²)

A unit used to express the metabolic rate per unit DuBois area is the met, defined as the metabolic rate of a sedentary person (seated, quiet): 1 met = 18.4 Btu/(h·ft^2) = 50 kcal/(h·m^2). A normal, healthy man has a maximum capacity of approximately M_{act} = 12 met at age 20, which drops to 7 met at age 70. Maximum rates for women are about 30% lower. Long-distance runners and trained athletes have maximum rates as high as 20 met. An average 35 year-old that does not exercise has a maximum rate of about 10 met and activities with M_{act} > 5 met are likely to prove exhausting.

Intermittent Activity. The activity of many people consists of a mixture of activities or a combination of work-rest periods. A weighted average metabolic rate is generally satisfactory, provided that activities alternate frequently (several times per hour). For example, a person typing 50% of the time, filing while seated 25% of the time, and walking about 25% of the time would have an average metabolic rate of 0.50 × 20 + 0.25 × 22 + 0.25 × 31 = 23 Btu/(h·ft^2) (see Table 4).

Accuracy. Estimating metabolic rates is difficult. The values given in Table 4 only indicate metabolic rates for the specific activities listed. Some entries give a range and some a single value, depending on the source of the data. The level of accuracy depends on the value of M_{act} and how well the activity can be defined. For well-defined activities with M_{act} < 1.5 met (e.g., reading), Table 4 is sufficiently accurate for most engineering purposes. For values of M_{act} > 3, where a task is poorly defined or where there are a variety of ways of performing a task (e.g., heavy machine work), the values may be in error by as much as ±50% for a given application. Engineering calculations should thus allow for potential variations.

Measurement. When metabolic rates must be determined more accurately than is possible with tabulated data, physiological measurements with human subjects may be necessary. The rate of metabolic heat produced by the body is most accurately measured by the rate of respiratory oxygen consumption and carbon dioxide production. An empirical equation for metabolic rate is given by Nishi (1981):

$$M = 567(0.23\ \mathrm{RQ} + 0.77)V_{O_2}/A_D,\ \text{in Btu/(h}\cdot\text{ft}^2) \tag{34}$$

where

RQ = respiratory quotient; molar ratio of V_{CO_2} exhaled to V_{O_2} inhaled, dimensionless

V_{O_2} = volumetric rate of oxygen consumption at conditions (STPD) of 32°F, 14.7 psi, ft^3/h

The exact value of the respiratory quotient used in Equation (34) RQ depends on a person's activity, diet, and physical condition. It can be determined by measuring both carbon dioxide and oxygen in the respiratory airflows, or it can be estimated with reasonable accuracy. A good estimate for the average adult is RQ = 0.83 for light or sedentary activities (M < 1.5 met), increasing proportionately to RQ = 1.0 for heavy exertion (M = 5.0 met). In extreme cases, the maximum range is 0.7 < RQ < 1.0. Estimation of RQ is generally sufficient for all except precision laboratory measurements since it does not strongly affect the value of the metabolic rate. A 10% error in estimating the respiratory quotient results in an error of less than 3% in the metabolic rate.

A second, much less accurate, method of estimating metabolic rate physiologically is to measure the heart rate. Table 5 shows the relationship between heart rate and oxygen consumption at different levels of physical exertion for a typical person. Once oxygen consumption is estimated from heart rate information, Equation (34) can be used to estimate the metabolic rate. A number of factors other than metabolic rate affect heart rate, such as physical condition, heat, emotional factors, muscles used, etc. Astrand and Rodahl (1977) show that heart rate is only a very approximate measure of metabolic rate and should not be the only source of information where accuracy is required.

Table 5 Heart Rate and Oxygen Consumption at Different Activity Levels[a]

Level of Exertion	Oxygen Consumed, ft^3/h	Heart Rate, beat/min
Light work	< 0.1	< 90
Moderate work	1.0 to 2.0	90 to 110
Heavy work	2.0 to 3.0	110 to 130
Very heavy work	3.0 to 4.0	130 to 150
Extremely heavy work	> 4.0	150 to 170

[a]Astrand and Rodahl (1977)

Mechanical Efficiency. In the heat balance equation, the rate of work accomplished W must be in the same units as metabolism M and expressed in terms of A_D in Btu/(h·ft^2). The mechanical work done by the muscles for a given task is often expressed in terms of the body's mechanical efficiency $\mu = W/M$. It is unusual for μ to be more than 5 to 10%; for most activities, it is close to zero. The maximum value under optimal conditions (e.g., bicycle ergometer) is μ = 20 to 24% (Nishi 1981). It is common to assume that mechanical work is zero for several reasons: (1) the mechanical work produced is small compared to metabolic rate, especially for office activities; (2) estimates for metabolic rates can often be inaccurate; and (3) this assumption results in a more conservative estimate when designing air-conditioning equipment for upper comfort and health limits. More accurate calculation of heat generation may require estimation of the mechanical work produced for activities where it is significant (walking on a grade, climbing a ladder, bicycling, lifting, etc.). In some cases, it is possible to either estimate or measure the mechanical work. For example, a 200-lb person walking up a 5% grade at 4.4 ft/s (3 mph) would lift a 200-lb weight a height of 0.22 ft every second, for a work rate of 44 ft·lb$_f$/s, or 204 Btu/h. This rate of mechanical work would then be subtracted from M to determine the net heat generated.

Heat Transfer Coefficients

Values for the linearized radiative heat transfer coefficient, convective heat transfer coefficient, and evaporative heat transfer coefficient are required to solve the equations describing heat transfer from the body.

Radiative heat transfer coefficient. The linearized radiative heat transfer coefficient can be calculated by:

$$h_r = 4\varepsilon\sigma\frac{A_r}{A_D}\left[459.7 + \frac{t_{cl} + \bar{t}_r}{2}\right]^3 \tag{35}$$

where

ε = average emissivity of clothing or body surface, dimensionless

σ = Stefan-Boltzmann constant, 0.1714 × 10-8 Btu/(h·ft^2·°R^4)

A_r = effective radiation area of body, ft^2

The ratio A_r/A_D is 0.70 for a sitting person and 0.73 for a standing person (Fanger 1967). The emissivity is close to unity (typically 0.95), unless special reflective materials are used or high-temperature sources are involved. It is not always possible to solve Equation (35) explicitly for h_r, since t_{cl} may also be an unknown. Some form of iteration may be required if a precise solution is required. Fortunately, h_r is nearly constant for typical indoor temperatures, and a value of 0.83 Btu/(h·ft^2·°F) suffices for most calculations. If the emissivity is significantly less than unity, the value should be adjusted by:

$$h_r = 0.83\ \varepsilon\ \text{Btu/(h}\cdot\text{ft}^2\cdot{}^\circ\text{F)} \tag{36}$$

where ε represents the area-weighted average emissivity for the clothing/body surface.

Table 6 Equations for Convection Heat Transfer Coefficients

Equation	Limits	Condition	Remarks/Sources
$h_c = 0.061\ V^{0.6}$ $h_c = 0.55$	$40 < V < 800$ $0 < V < 40$	Seated with moving air	Mitchell (1974)
$h_c = 0.475 + 0.044\ V^{0.67}$ $h_c = 0.90$	$30 < V < 300$ $0 < V < 30$	Reclining with moving air	Colin and Houdas (1967)
$h_c = 0.092\ V^{0.53}$	$100 < V < 400$	Walking in still air	V is walking speed (Nishi and Gagge 1970)
$h_c = (M - 0.85)^{0.39}$	$1.1 < M < 3.0$	Active in still air	Gagge et al. (1976)
$h_c = 0.146\ V^{0.39}$	$100 < V < 400$	Walking on treadmill in still air	V is treadmill speed (Nishi and Gagge 1970)
$h_c = 0.068\ V^{0.69}$ $h_c = 0.70$	$30 < V < 300$ $0 < V < 30$	Standing person in moving air	Developed from data presented by Seppenan et al. (1972)

Note: h_c in Btu/(h·ft²·°F), V in fpm, and M in met units, where 1 met = 18.4 Btu/h·ft².

Convective Heat Transfer Coefficient. Heat transfer by convection is usually caused by air movement within the living space or by body movements. Equations for estimating h_c under various conditions are presented in Table 6. Where two conditions apply (e.g., walking in moving air), a reasonable estimate can be obtained by taking the larger of the two values for h_c. Limits have been given to all equations. If no limits were given in the source, reasonable limits have been estimated. Care should be exercised in using these values for seated and reclining persons. The heat transfer coefficients may be accurate, but the effective heat transfer area may be substantially reduced due to body contact with a padded chair or bed.

Quantitative values of h_c are important, not only in estimating convection loss, but in evaluating (1) operative temperature t_o, (2) clothing parameters I_t and i_m, and (3) rational effective temperatures t_{oh} and ET*. All heat transfer coefficients in Table 6 were evaluated at or near 14.7 psia. These coefficients should be corrected as follows for atmospheric pressure:

$$h_{cc} = h_c(p_t/14.7)^{0.55} \tag{37}$$

where

h_{cc} = corrected convective heat transfer coefficient, Btu/(h·ft²·°F)
h_c = local atmospheric pressure, psia

The combined coefficient h is the sum of h_r and h_c described in Equation (35) and Table 6, respectively. The coefficient h governs exchange by radiation and convection from the exposed body surface to the surrounding environment.

Evaporative Heat Transfer Coefficient. The evaporative heat transfer coefficient h_e for the outer air layer of a nude or clothed person can be estimated from the convective heat transfer coefficient using the Lewis relationship given in Equation (27). If the atmospheric pressure is significantly different from standard (14.7 psia), the correction to the value obtained from Equation (27) is:

$$h_{ec} = h_e(14.7/p_t)^{0.45} \tag{38}$$

where h_{ec} is the corrected evaporative heat transfer coefficient, Btu/(h·ft²·°F).

Clothing Insulation and Moisture Permeability

Thermal Insulation. The most accurate methods for determining clothing insulation are: (1) measurements on heated manikins (McCullough and Jones 1984, Olesen and Nielsen 1983) and (2) measurements on active subjects (Nishi et al. 1975). For most routine engineering work, estimates based on tables and equations presented in this section are sufficient. Thermal manikins can measure the sensible heat loss from the "skin" ($C + R$) in a given environment. Equation (11) can then be used to evaluate R_{cl} if the environmental conditions are well defined and f_{cl} is measured. Evaluation of clothing insulation on subjects requires measurement of t_{sk}, t_{cl}, and t_o. The clothing thermal efficiency is calculated by:

$$F_{cl} = \frac{t_{cl} - t_o}{t_{sk} - t_o} \tag{39}$$

The intrinsic clothing insulation can then be calculated from manikin measurements by the following relationship, provided f_{cl} is measured and conditions are sufficiently well-defined to make an accurate determination of h:

$$R_{cl} = \frac{t_{sk} - t_o}{q} - \frac{1}{hf_{cl}} \tag{40}$$

where q = heat loss from the manikin, Btu/(h·ft²).

Clothing insulation value may be expressed in clo units. In order to avoid confusion, the symbol I is used with the clo unit instead of the symbol R. The relationship between the two is:

$$R = 0.88I \tag{41}$$

or 1.0 clo is equivalent to 0.88 ft²·h·°F/Btu.

Since clothing insulation cannot be measured for most routine engineering applications, tables of measured values for various clothing ensembles can be used to select an ensemble comparable to the one(s) in question. Table 7 gives values for typical indoor clothing

Table 7 Typical Insulation and Permeability Values for Clothing Ensembles[a]

Ensemble Description[b]	I_{cl} (clo)	I_t[c] (clo)	f_{cl}	i_{cl}	i_m[c]
Walking shorts, short-sleeve shirt	0.36	1.02	1.10	0.34	0.42
Trousers, short-sleeve shirt	0.57	1.20	1.15	0.36	0.43
Trousers, long-sleeve shirt	0.61	1.21	1.20	0.41	0.45
Same as above, plus suit jacket	0.96	1.54	1.23		
Same as above, plus vest and T-shirt	1.14	1.69	1.32	0.32	0.37
Trousers, long-sleeve shirt, long-sleeve sweater, T-shirt	1.01	1.56	1.28		
Same as above, plus suit jacket and long underwear bottoms	1.30	1.83	1.33		
Sweat pants, sweat shirt	0.74	1.35	1.19	0.41	0.45
Long-sleeve pajama top, long pajama trousers, short 3/4 sleeve robe, slippers (no socks)	0.96	1.50	1.32	0.37	0.41
Knee-length skirt, short-sleeve shirt, panty hose, sandals	0.54	1.10	1.26		
Knee-length skirt, long-sleeve shirt, full slip, panty hose	0.67	1.22	1.29		
Knee-length skirt, long-sleeve shirt, half slip, panty hose, long-sleeve sweater	1.10	1.59	1.46		
Same as above, replace sweater with suit jacket	1.04	1.60	1.30	0.35	0.40
Ankle-length skirt, long-sleeve shirt, suit jacket, panty hose	1.10	1.59	1.46		
Long-sleeve coveralls, T-shirt	0.72	1.30	1.23		
Overalls, long-sleeve shirt, T-shirt	0.89	1.46	1.27	0.35	0.40
Insulated coveralls, long-sleeve thermal underwear, long underwear bottoms	1.37	1.94	1.26	0.35	0.39

[a]From McCullough and Jones (1984) and McCullough et al. (1989)
[b]All ensembles include shoes and briefs or panties. All ensembles except those with panty hose include socks unless otherwise noted.
[c]For $\bar{t}_r = t_a$ and air velocity less than 40 fpm (I_a = 0.72 clo and i_m = 0.48 when nude)

Table 8 Garment Insulation Values

Garment Description[a]	$I_{clu,i}$, clo[b]	Garment Description[a]	$I_{clu,i}$, clo[b]	Garment Description[a]	$I_{clu,i}$, clo[b]
Underwear		Long-sleeve, flannel shirt	0.34	Long-sleeve (thin)	0.25
Men's briefs	0.04	Short-sleeve, knit sport shirt	0.17	Long-sleeve (thick)	0.36
Panties	0.03	Long-sleeve, sweat shirt	0.34	**Dresses and skirts**[c]	
Bra	0.01	**Trousers and Coveralls**	0.06	Skirt (thin)	0.14
T-shirt	0.08	Short shorts	0.08	Skirt (thick)	0.23
Full slip	0.16	Walking shorts	0.15	Long-sleeve shirtdress (thin)	0.33
Half slip	0.14	Straight trousers (thin)	0.24	Long-sleeve shirtdress (thick)	0.47
Long underwear top	0.20	Straight trousers (thick)	0.28	Short-sleeve shirtdress (thin)	0.29
Long underwear bottoms	0.15	Sweatpants	0.30	Sleeveless, scoop neck (thin)	0.23
Footwear		Overalls	0.49	Sleeveless, scoop neck (thick), i.e., jumper	0.27
Ankle-length athletic socks	0.02	Coveralls			
Calf-length socks	0.03	**Suit jackets and vests (lined)**		**Sleepwear and Robes**	
Knee socks (thick)	0.06	Single-breasted (thin)	0.36	Sleeveless, short gown (thin)	0.18
Panty hose	0.02	Single-breasted (thick)	0.44	Sleeveless, long gown (thin)	0.20
Sandals/thongs	0.02	Double-breasted (thin)	0.42	Short-sleeve hospital gown	0.31
Slippers (quilted, pile-lined)	0.03	Double-breasted (thick)	0.48	Long-sleeve, long gown (thick)	0.46
Boots	0.10	Sleeveless vest (thin)	0.10	Long-sleeve pajamas (thick)	0.57
Shirts and Blouses		Sleeveless vest (thick)	0.17	Short-sleeve pajamas (thin)	0.42
Sleeveless, scoop-neck blouse	0.12	**Sweaters**		Long-sleeve, long wrap robe (thick)	0.69
Short-sleeve, dress shirt	0.19	Sleeveless vest (thin)	0.13	Long-sleeve, short wrap robe (thick)	0.48
Long-sleeve, dress shirt	0.25	Sleeveless vest (thick)	0.22	Short-sleeve, short robe (thin)	0.34

[a]"Thin" garments are made of light, thin fabrics worn in summer; "thick" garments are made of heavy, thick fabrics worn in winter.

[b]1 clo = 0.880°F·ft²·h/Btu

[c]Knee-length

ensembles. More detailed tables are presented by McCullough and Jones (1984) and Olesen and Nielsen (1983). Accuracies for I_{cl} on the order of ±20% are typical if good matches between ensembles are found.

Often it is not possible to find an already measured clothing ensemble that matches the one in question. In this case, the ensemble insulation can be estimated from the insulation of individual garments. Table 8 gives a list of individual garments commonly worn. The insulation of an ensemble is estimated from the individual values using a summation formula (McCullough and Jones 1984):

$$I_{cl} = 0.835\sum_i I_{clu,i} + 0.161 \tag{42}$$

where $I_{clu,i}$ is the effective insulation of garment i, and I_{cl}, as before, is the insulation for the entire ensemble. A simpler and nearly as accurate summation formula is (Olesen 1985):

$$I_{cl} = \sum_i I_{clu,i} \tag{43}$$

Either Equation (42) or (43) gives acceptable accuracy for typical indoor clothing. The main source of inaccuracy is in determining the appropriate values for individual garments. Overall accuracies are on the order of ±25% if the tables are used carefully. If it is important to include a specific garment that is not included in Table 8, its insulation can be estimated by (McCullough and Jones 1984):

$$I_{clu,i} = (0.534 + 3.43x_f)(A_G/A_D) - 0.0549 \tag{44}$$

where

x_f = fabric thickness, in.
A_G = body surface area covered by garment, ft²

Values in Table 7 may be adjusted by information in Table 8 and a summation formula. Using this method, values of $I_{clu,i}$ for the selected items in Table 8 are then added to or subtracted from the ensemble value of I_{cl} in Table 7.

When sitting, the chair generally has the effect of increasing clothing insulation by up to 0.15 clo depending on the contact area (CSAC) between the chair and body (McCullough et al. 1994). A string webbed or beach chair has little or no contact area and the insulation actually decreases by about 0.1 clo due likely to compression of the clothing in the contact area. In contrast, a cushioned executive chair has a large contact area that can increase the intrinsic clothing insulation by 0.15 clo. For other chairs, the increase in intrinsic insulation (ΔI_{cl}) can be estimated from:

$$\Delta I_{cl} = 6.95 \times 10^{-6}\text{CSAC} - 0.1 \tag{45}$$

For example, a desk chair with a body contact area of 2.9 ft² has a ΔI_{cl} of 0.1 clo. This amount should be added to the intrinsic insulation of the standing clothing ensemble to obtain the insulation of the ensemble when sitting in the desk chair.

While sitting has the effect of increasing clothing insulation walking decreases it (McCullough and Hong 1994). The change in clothing insulation (ΔI_{cl}) can be estimated from the standing intrinsic insulation of the ensemble (I_{cl}) and the walking speed (Walkspeed) in steps per minute:

$$\Delta I_{cl} = 0.504 I_{cl} + 0.00281\,\text{Walkspeed} - 0.24 \tag{46}$$

For example, the clothing insulation of a person wearing a winter business suit with a standing intrinsic insulation of 1 clo would decrease by 0.52 clo when the person walks at 90 steps per minute (about 2.3 mph). Thus the ensembles intrinsic insulation when walking would be 0.48 clo.

Moisture Permeability. Moisture permeability data for some clothing ensembles are presented in terms of i_{cl} and i_m in Table 7. The values of i_m can be used to calculate $R_{e,t}$ using the relationships in Table 2. Ensembles worn indoors generally fall in the range $0.3 < i_m < 0.5$ and assuming $i_m = 0.4$ is reasonably accurate

(McCullough et al. 1989). This latter value may be used if a good match to ensembles in Table 7 cannot be made. The value of i_m or $R_{e,t}$ may be substituted directly into equations for body heat loss calculations (see Table 3). However, i_m for a given clothing ensemble is a function of the environment as well as the clothing properties. Unless i_m is evaluated at conditions very similar to the intended application, it is more rigorous to use i_{cl} to describe the moisture permeability of the clothing. The value of i_{cl} is not as sensitive to environmental conditions; thus, given data are more accurate over a wider range of air velocity and radiant and air temperature combinations for i_{cl} than for i_m. The relationships in Table 2 can be used to determine $R_{e,cl}$ from i_{cl}, and i_{cl} or $R_{e,cl}$ can then be used for body heat loss calculations (see Table 3). McCullough et al. (1989) found an average value of i_{cl} = 0.34 for common indoor clothing; this value can be used when other data are not available.

Measurements of i_m or i_{cl} may be necessary if unusual clothing (e.g., impermeable or metalized) and/or extreme environments (e.g., high radiant temperatures or high air velocities) are to be addressed. There are three different methods for measuring the moisture permeability of clothing: the first uses a wet manikin to measure the effect of sweat evaporation on heat loss (McCullough 1986); the second uses moisture permeability measurements on component fabrics as well as dry manikin measurements (Umbach 1980); the third uses measurements from sweating subjects (Nishi et al. 1975, Holmer 1984).

Clothing Surface Area. Many clothing heat transfer calculations require the clothing area factor f_{cl} to be known. The most reliable approach is to measure it using photographic methods (Olesen et al. 1982). Other than actual measurements, the best method is to use previously tabulated data for similar clothing ensembles. Table 7 is adequate for most indoor clothing ensembles. No good method of estimating f_{cl} for a clothing ensemble from other information is available, although a rough estimate can be made by (McCullough and Jones 1984):

$$f_{cl} = 1.0 + 0.3 I_{cl} \qquad (47)$$

Total Evaporative Heat Loss

The total evaporative heat loss (latent heat) from the human body due to both respirative losses and skin losses, $E_{sk} + E_{res}$, can be measured directly from the body's rate of mass loss as observed by a sensitive scale:

$$E_{sk} + E_{res} = \frac{h_{fg}}{A_D}\frac{dm}{d\theta} \qquad (48)$$

where

h_{fg} = latent heat of vaporization of water, Btu/lb
m = body mass, lb
θ = time, h

When using Equation (48), adjustments should be made for any materials consumed (e.g., food and drink) and body effluents (e.g., wastes). The fuel burned by the body also contributes slightly to weight loss and can be accounted for with the following relationship (Astrand and Rodahl 1977):

$$\frac{dm_{ge}}{d\theta} = 2.2 V_{O_2}(0.1225\ \text{RQ} - 0.0891) \qquad (49)$$

where

$dm_{ge}/d\theta$ = mass loss due to respiratory gas exchange, lb/h
V_{O_2} = uptake at STPD, ft³/h
RQ = respiratory quotient
0.1225 = density of CO_2 at STPD, lb/ft³
0.0891 = density of O_2 at STPD, lb/ft³
STPD = standard temperature and pressure of dry air at 32°F and 14.7 psia

Environmental Parameters

The parameters describing the thermal environment that must be measured or otherwise quantified if accurate estimates of human thermal response are to be made are divided into two groups—those that can be measured directly and those that are calculated from other measurements.

Directly Measured. Seven of the parameters frequently used to describe the thermal environment are psychrometric and include (1) air temperature t_a; (2) wet-bulb temperature t_{wb}; (3) dew-point temperature t_{dp}; (4) water vapor pressure p_a; (5) total atmospheric pressure p_t; (6) relative humidity (rh); and (7) humidity ratio W. These parameters are discussed in detail in Chapter 6, and methods for measuring them are discussed in Chapter 14. Two other important parameters that can be measured directly include air velocity V and mean radiant temperature $\bar{t}_r$. Air velocity measurements are also discussed in Chapter 14. The radiant temperature is the temperature of an exposed surface in the environment. The temperatures of individual surfaces are usually combined into a mean radiant temperature $\bar{t}_r$. Finally, globe temperature t_g, which can also be measured directly, is a good approximation of the operative temperature t_o and is also used with other measurements to calculate the mean radiant temperature.

Calculated Parameters. The **mean radiant temperature** $\bar{t}_r$ is a key variable in making thermal calculations for the human body. It is the uniform temperature of an imaginary enclosure in which radiant heat transfer from the human body equals the radiant heat transfer in the actual nonuniform enclosure. Measurements of the globe temperature, air temperature, and the air velocity can be combined to estimate the mean radiant temperature (see Chapter 14). The accuracy of the mean radiant temperature determined this way varies considerably depending on the type of environment and the accuracy of the individual measurements. Since the mean radiant temperature is defined with respect to the human body, the shape of the sensor is also a factor. The spherical shape of the globe thermometer gives a reasonable approximation for a seated person; an ellipsoid-shaped sensor gives a better approximation to the shape of a human, both upright and seated.

The mean radiant temperature can also be calculated from measured values of the temperature of the surrounding walls and surfaces and their positions with respect to the person. As most building materials have a high emittance ε, all the surfaces in the room can be assumed to be black. The following equation is then used:

$$\bar{T}_r^4 = T_1^4 F_{p-1} + T_2^4 F_{p-2} + \ldots + T_N^4 F_{p-N} \qquad (50)$$

where

$\bar{T}_r$ = mean radiant temperature, °R
T_N = surface temperature of surface N, °R
F_{p-N} = angle factor between a person and surface N

As the sum of the angle factors is unity, the fourth power of mean radiant temperature equals the mean value of the surrounding surface temperatures to the fourth power, weighted by the respective angle factors. In general, angle factors are difficult to determine, although Figures 3A and 3B may be used to estimate them for rectangular surfaces. The angle factor normally depends on the position and orientation of the person (Fanger 1982).

If relatively small temperature differences exist between the surfaces of the enclosure, Equation (50) can be simplified to a linear form:

$$\bar{t}_r = t_1 F_{p-1} + t_2 F_{p-2} + \ldots + t_N F_{p-N} \qquad (51)$$

Equation (51) always gives a slightly lower mean radiant temperature than Equation (50), but in many cases the difference is small. If, for example, half the surroundings (F_{p-N} = 0.5) has a temperature

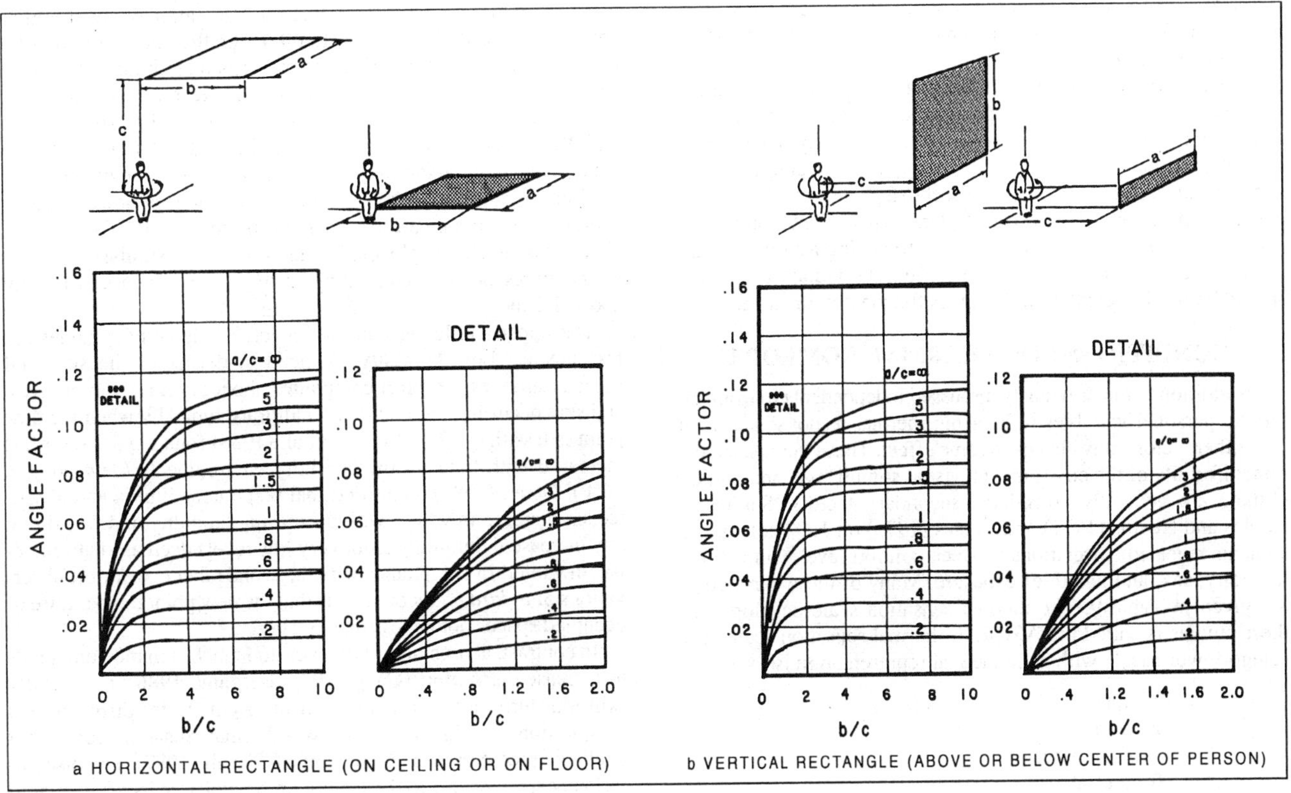

Fig. 3 Mean Value of Angle Factor between Seated Person and Horizontal or Vertical Rectangle when Person is Rotated around Vertical Axis (Fanger 1982)

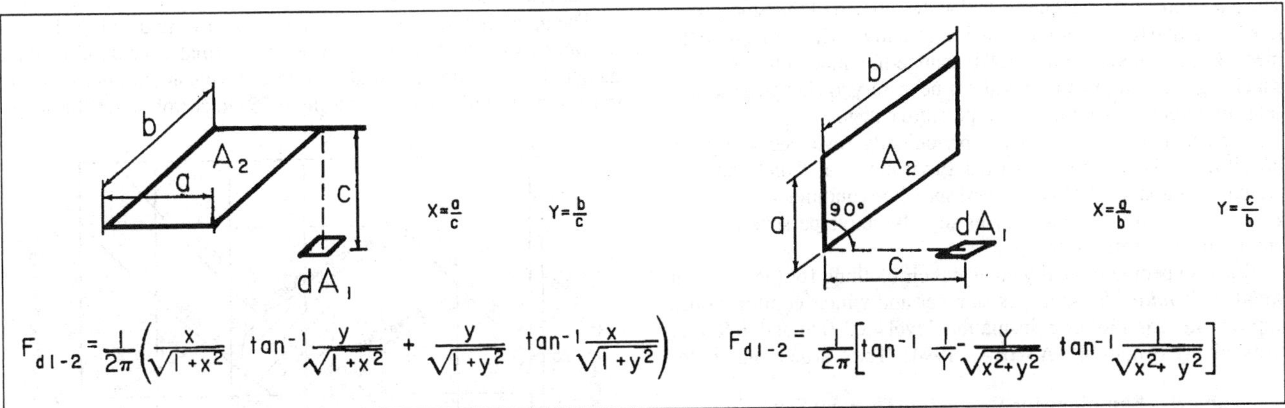

$$F_{d1-2} = \frac{1}{2\pi}\left(\frac{X}{\sqrt{1+X^2}}\tan^{-1}\frac{Y}{\sqrt{1+X^2}} + \frac{Y}{\sqrt{1+Y^2}}\tan^{-1}\frac{X}{\sqrt{1+Y^2}}\right)$$

$$F_{d1-2} = \frac{1}{2\pi}\left[\tan^{-1}\frac{1}{Y} - \frac{Y}{\sqrt{X^2+Y^2}}\tan^{-1}\frac{1}{\sqrt{X^2+Y^2}}\right]$$

Fig. 3C Analytical Formulae for Calculating Angle Factor for Small Plane Element

10°F higher than the other half, the difference between the calculated mean radiant temperatures—according to Equations (50) and (51)—is only 0.4°F. If, however, this difference is 200°F, the mean radiant temperature calculated by Equation (51) is 20°F too low.

The mean radiant temperature may also be calculated from the plane radiant temperature t_{pr} (defined below) in six directions (up, down, left, right, front, back) and for the projected area factors of a person in the same six directions. For a standing person, the mean radiant temperature may be estimated as:

$$\bar{t}_r = \{0.08[t_{pr}(\text{up}) + t_{pr}(\text{down})] + 0.23[t_{pr}(\text{right}) + t_{pr}(\text{left})] + 0.35[t_{pr}(\text{front}) + t_{pr}(\text{back})]\} / [2(0.08 + 0.23 + 0.35)] \quad (52)$$

For a seated person, the mean radiant temperature may be estimated as:

$$\bar{t}_r = \{0.18[t_{pr}(\text{up}) + t_{pr}(\text{down})] + 0.22[t_{pr}(\text{right}) + t_{pr}(\text{left})] + 0.30[t_{pr}(\text{front}) + t_{pr}(\text{back})]\} / [2(0.18 + 0.22 + 0.30)] \quad (53)$$

The **plane radiant temperature** t_{pr}, first introduced by Korsgaard (1949), is the uniform temperature of an enclosure in which the incident radiant flux on one side of a small plane element is the same as that in the actual environment. The plane radiant temperature describes the thermal radiation in one direction and its value thus depends on the direction. In comparison, the mean radiant temperature describes the thermal radiation for the human body from all directions. The plane radiant temperature can be calculated using Equations (50) and (51) with the same limitations. Area factors are determined from Figure 3C.

The **radiant temperature asymmetry** Δt_{pr} is the difference between the plane radiant temperature of the opposite sides of a small plane element. This parameter describes the asymmetry of the radiant environment, and is especially important in comfort conditions. Because it is defined with respect to a plane element, its value depends on the orientation of that plane. This orientation may be specified in some situations (e.g., floor to ceiling asymmetry) and not in others. If direction is not specified, the radiant asymmetry should be for the orientation that gives the maximum value.

CONDITIONS FOR THERMAL COMFORT

In addition to the previously discussed independent environmental and personal variables influencing thermal response and comfort, other factors may also have some effect. These factors, such as nonuniformity of the environment, visual stimuli, age, and outdoor climate are generally considered secondary factors. Studies by Rohles and Nevins (1971) and Rohles (1973) on 1600 college-age students revealed correlations between comfort level, temperature, humidity, sex, and length of exposure. Many of these correlations are given in Table 9. The thermal sensation scale developed for these studies is called the ASHRAE thermal sensation scale. This category word scale with its numerical representation is:

+3 hot
+2 warm
+1 slightly warm
0 neutral
−1 slightly cool
−2 cool
−3 cold

The equations in Table 9 indicate that the women of this study were more sensitive to temperature and less sensitive to humidity than the men. But in general about a 5.4°F change in temperature or a 0.44 psi change in water vapor pressure is necessary to change a thermal sensation vote by one unit or temperature category.

Current and past studies are periodically reviewed to update *ASHRAE Standard* 55, "Thermal Environmental Conditions for Human Occupancy." This standard specifies conditions or comfort zones where 80% of sedentary or slightly active persons find the environment thermally acceptable.

Because people typically change their clothing for the seasonal weather *Standard* 55 specifies summer and winter comfort zones appropriate for clothing insulation levels of 0.5 and 0.9 clo, respectively (Figure 4) (ASHRAE 1994). The warmer and cooler temperature borders of the comfort zones are affected by humidity and coincide with lines of constant ET*. In the middle region of a zone a typical person wearing the prescribed clothing would have a thermal sensation at or very near neutral. Nearer the boundary of the warmer zone a person would feel about +0.5 warmer on the ASHRAE thermal sensation scale and near the boundary of the cooler zone that person may have a thermal sensation of −0.5.

Comfort zones for other clothing levels can be approximated by decreasing the temperature borders of the zone by 1°F for each 0.1 clo increase in clothing insulation and visa-versa. Similarly a zone's temperatures can be decreased by 2.5°F per met increase in activity above 1.2 met.

The upper and lower humidity levels of the comfort zones are less precise. Low humidity can lead to drying of the skin and mucous surfaces. Comfort complaints about dry nose, throat, eyes, and skin occur in low humidity conditions, typically when the dew point is less than 32°F. Liviana et al. (1988) found eye discomfort increased with time in low humidity environments (dew point < 36°F). Green (1982) quantified that respiratory illness and absenteeism increase in winter with decreasing humidity and found that any increase in humidity from very low levels decreased absenteeism in winter. In compliance with these and other discomfort observations *Standard* 55 recommends that the dew point temperature of occupied spaces not be less than 36°F.

In contrast, at high humidity levels too much skin moisture tends to increase discomfort (Gagge 1937, Berglund 1986), particularly skin moisture that is physiological in origin (water diffusion and perspiration). At high humidity levels thermal sensation alone is not a reliable predictor of thermal comfort (Tanabe 1987). The discomfort appears to be due to the feeling of the moisture itself, increased friction between skin and clothing with skin moisture (Gwosdow 1986) and other factors. To prevent warm discomfort Nevins (1975) recommended that on the warm side of the comfort zone the relative humidity not exceed 60%.

The upper humidity limits of *Standard* 55 were developed theoretically from limited data. However, thermal acceptability data gathered at medium and high humidity levels at summer comfort temperatures with subjects wearing 0.55 clo corroborates the shape

Table 9 Equations for Predicting Thermal Sensation (Y)[a] of Men, Women, and Men and Women Combined[b]

Exposure Period, h	Sex	Regression Equations: t = dry-bulb temperature, °F; p = vapor pressure, psi
1.0	Male	$Y = 0.122\,t + 1.61\,p - 9.584$
	Female	$Y = 0.151\,t + 1.71\,p - 12.080$
	Combined	$Y = 0.136\,t + 1.71\,p - 10.880$
2.0	Male	$Y = 0.123\,t + 1.86\,p - 9.953$
	Female	$Y = 0.157\,t + 1.45\,p - 12.725$
	Combined	$Y = 0.140\,t + 1.65\,p - 11.339$
3.0	Male	$Y = 0.118\,t + 2.02\,p - 9.718$
	Female	$Y = 0.153\,t + 1.76\,p - 13.511$
	Combined	$Y = 0.135\,t + 1.92\,p - 11.122$

[a] Y values refer to the ASHRAE thermal sensation scale.
[b] For young adult subjects with sedentary activity and wearing clothing with a thermal resistance of approximately 0.5 clo, $\bar{t}_r \approx \bar{t}_a$ and air velocities are < 40 fpm.

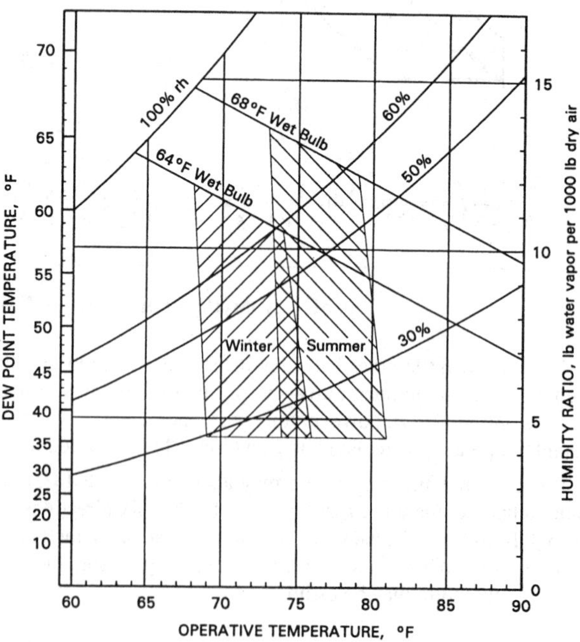

Fig. 4 ASHRAE Summer and Winter Comfort Zones
(Acceptable ranges of operative temperature and humidity for people in typical summer and winter clothing during primarily sedentary activity.)

of the upper limit and found it corresponded to an 80% thermal acceptability level (Berglund 1995).

THERMAL NONUNIFORM CONDITIONS AND LOCAL DISCOMFORT

A person may feel thermally neutral for the body as a whole, but might not be comfortable if one or more parts of the body are too warm or too cold. Nonuniformities may be due to a cold window, hot surface, draft, or a temporal variation of these. Even small variations in heat flow cause the thermal regulatory system to compensate, thus increasing the physiological effort of maintaining body temperatures. The boundaries of the comfort zones (Figure 4) of *Standard* 55 provide a thermal acceptability level of 90% if the environment is thermally uniform. Because the standard's objective is to specify conditions for 80% acceptability, the standard permits nonuniformities to decrease acceptability by 10%. Fortunately for the designer and user, the effect of common thermal nonuniformities on comfort is quantifiable and predictable as discussed in the following sections. Further, most humans are fairly insensitive to small nonuniformities.

Asymmetric Thermal Radiation

Asymmetric or nonuniform thermal radiation in a space may be caused by cold windows, uninsulated walls, cold products, cold or warm machinery, or by improperly sized panels on the wall or ceiling. In residential buildings, offices, restaurants, etc., the most common reasons for discomfort due to asymmetric thermal radiation are large windows in the winter or improperly sized or installed ceiling heating panels. At industrial workplaces, the reasons include cold or warm products, cold or warm equipment, etc.

The recommendations in ISO *Standard* 7730 and ASHRAE *Standard* 55 are based primarily on studies reported by Fanger et al. (1980) and Fanger and Christensen (1985). These standards include guidelines regarding the radiant temperature asymmetry from an overhead warm surface (heated ceiling) and a vertical cold surface (cold window). Among the studies conducted on the influence of asymmetric thermal radiation are those by McIntyre (1974, 1976), McIntyre and Griffiths (1975), Fanger and Langkilde (1975), McNall and Biddison (1970), and Olesen et al. (1972). These studies all used seated subjects. In these studies, the subjects were always in thermal neutrality and exposed only to the discomfort resulting from excessive asymmetry.

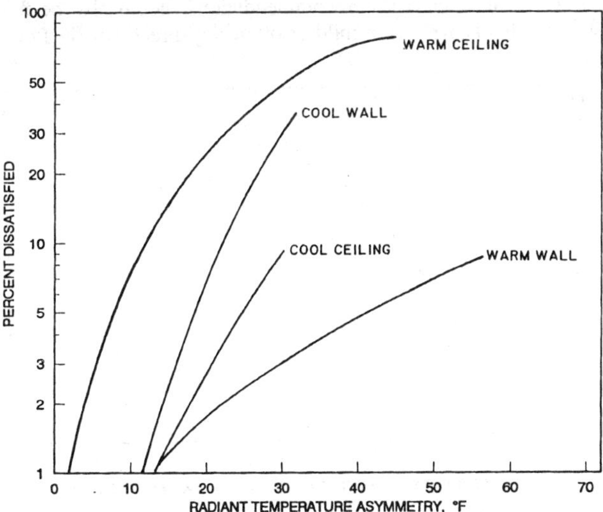

Fig. 5 Percentage of People Expressing Discomfort Due to Asymmetric Radiation

The subjects gave their reactions on their comfort sensation, and a relationship between the radiant temperature asymmetry and the number of subjects feeling dissatisfied was established (Figure 5). Radiant asymmetry, as defined in the section Environmental Parameters, is the difference in radiant temperatures of the environment seen by the person on opposite sides. More precisely, radiant asymmetry is the difference in radiant temperatures seen by a small flat element looking in opposite directions.

Figure 5 shows that people are more sensitive to asymmetry caused by an overhead warm surface than by a vertical cold surface. The influence of an overhead cold surface and a vertical warm surface is much less. These data are particularly important when applying radiant panels to provide comfort in spaces with large cold surfaces or cold windows.

Other studies of clothed persons in neutral environments found thermal acceptability unaffected by radiant temperature asymmetries of 18°F or less (Berglund 1987) and discomfort unaffected by radiant temperature asymmetries of 36°F or less (McIntyre 1975).

Draft

Draft is an undesired local cooling of the human body caused by air movement. This is a serious problem, not only in many ventilated buildings but also in automobiles, trains, and aircraft. Draft has been identified as one of the most annoying factors in offices. When people sense draft, it often results in a demand for higher air temperatures in the room or for stopping ventilation systems.

Fanger and Christensen (1985) aimed to establish the percentage of the population feeling draft when exposed to a given mean velocity. Figure 6 shows the percentage of subjects who felt draft on the head region (the dissatisfied) as a function of the mean velocity at the neck. The head region comprises head, neck, shoulders, and back. The air temperature had a significant influence on the percentage of dissatisfied. There was no significant difference between responses of men and women to draft. The data in Figure 6 applies, therefore, only to persons wearing normal indoor clothing and performing light, mainly sedentary work. Persons with higher activity levels are not so sensitive to draft (Jones et al. 1986).

A study of the effect of air velocity over the whole body found thermal acceptability unaffected in neutral environments by air speeds of 50 fpm or less (Berglund 1987). This study also found no interaction between air speed and radiant temperature asymmetry on subjective responses. This means that acceptability changes or the percent dissatisfied due to draft and radiant asymmetry are independent and additive.

Fanger et al. (1989) investigated the effect of turbulence intensity on sensation of draft. The turbulence intensity had a significant effect on the occurrence of draft sensation. The following model

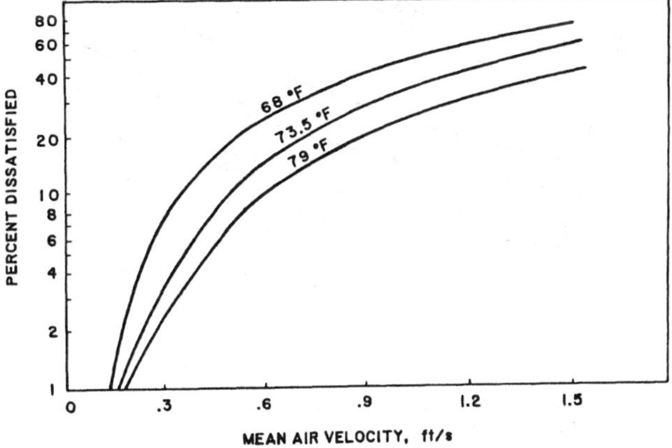

Fig. 6 Percentage of People Dissatisfied as Function of Mean Air Velocity

predicts the percentage of people dissatisfied because of draft intensity. The model can be used for quantifying draft risk in spaces and for developing air distribution systems with a low draft risk.

$$PD = 0.021(93.2 - t_a)(V - 9.8)^{0.62}(0.0019 V \text{Tu} + 3.14) \quad (54)$$

For $V < 9.8$ fpm, insert $V = 9.8$ and for PD > 100%, insert PD 100% where Tu is the turbulence intensity defined by:

$$\text{Tu} = 100 \frac{V_{sd}}{V} \quad (55)$$

and V_{sd} is the standard deviation of the velocity measured with an omnidirectional anemometer having a 0.2 s time constant.

The model extends the Fanger and Christensen draft chart model to include turbulence intensity. In this study, Tu decreases when V increases. This means that the effect of V for the experimental data to which the model is fitted are: $68 < t_a < 79$°F, $10 < V < 100$ fpm, and $0 < \text{Tu} < 70\%$. Figure 7 gives more precisely the curves that result from intersections between planes of constant Tu and the surfaces of PD = 15%.

Vertical Air Temperature Difference

In most spaces in buildings, the air temperature normally increases with height above the floor. If the gradient is sufficiently large, local warm discomfort can occur at the head, and/or cold discomfort can occur at the feet, although the body as a whole is thermally neutral. Among the few studies of vertical air temperature differences and the influence of thermal comfort reported are Olesen et al. (1979), McNair (1973), McNair and Fishman (1974), and Eriksson et al. (1975). Subjects were seated in a climatic chamber so they were individually exposed to different air temperature differences between head and ankles (Olesen et al. 1979). During the tests, the subjects were in thermal neutrality because they were allowed to change the temperature level in the test room whenever they desired; the vertical temperature difference, however, was kept unchanged. The subjects gave subjective reactions to their thermal sensation, and Figure 8 shows the percentage of dissatisfied as a function of the vertical air temperature difference between head (43 in. above the floor) and ankles (4 in. above the floor).

The case where the air temperature at head level is lower than that at ankle level will not be as critical for the occupants. Eriksson (1975) indicated that his subjects could tolerate much greater differences if the head were cooler. This observation is verified in the experiments with asymmetric thermal radiation from a cooled ceiling (Fanger et al. 1985).

Warm or Cold Floors

Due to the direct contact between the feet and the floor, local discomfort of the feet can often be caused by a too-high or too-low floor temperature. Also, the floor temperature has a significant influence on the mean radiant temperature in a room. The floor temperature is greatly influenced by the way a building is constructed (e.g., insulation of the floor, above a basement, directly on the ground, above another room, use of floor heating, floors in radiant heated areas). If a floor is too cold and the occupants feel cold discomfort in their feet, a common reaction is to increase the temperature level in the room; in the heating season, this also increases energy consumption. A radiant system, which radiates heat from the floor, is also used to avoid discomfort from cold floors.

The most extensive studies of the influence of floor temperature on feet comfort were performed by Olesen (1977a, 1977b), who, based on his own experiments and reanalysis of the data from Nevins et al. (1958, 1964, 1967), recorded the following results. For floors occupied by people with bare feet (in swimming halls, gymnasiums, dressing rooms, bathrooms, and bedrooms), flooring material is important. Ranges for some typical floor materials are as follows:

Textiles (rugs)	70 to 82°F
Pine floor	72.5 to 82°F
Oak floor	76 to 82°F
Hard linoleum	75 to 82°F
Concrete	79 to 83°F

To save energy, flooring materials with a low contact coefficient (cork, wood, carpets), radiant heated floors, or floor heating systems can be used to eliminate a desire for higher ambient temperatures caused by cold feet. These recommendations should also be followed in schools, where the children often play directly on the floor.

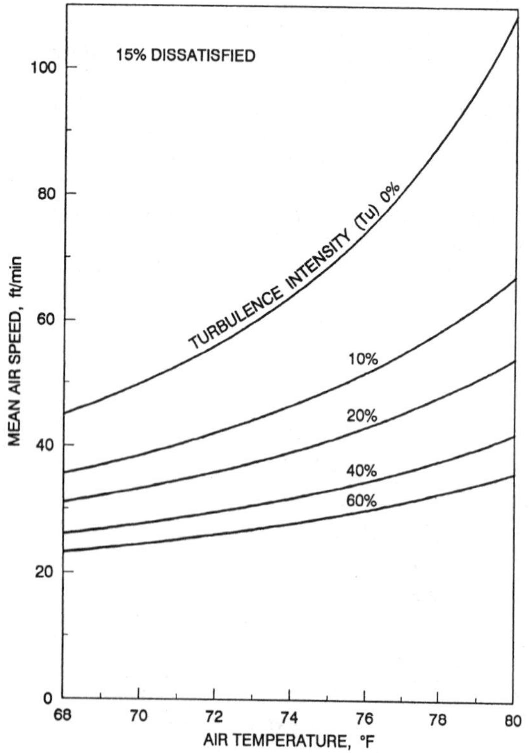

Fig. 7 Draft Conditions Dissatisfying 15% of Population

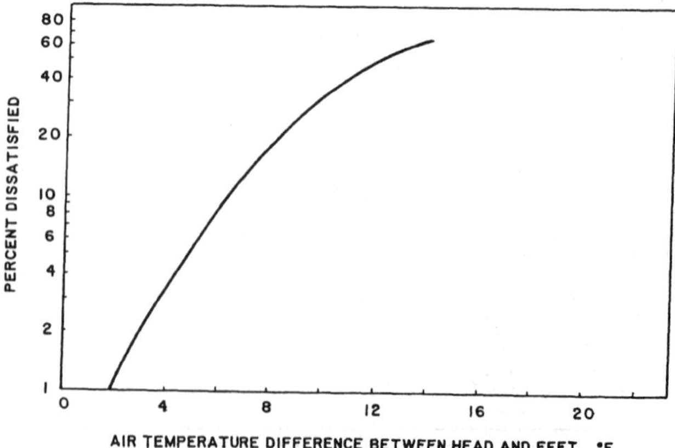

Fig. 8 Percentage of People Dissatisfied as Function of Vertical Air Temperature Difference between Head and Ankles

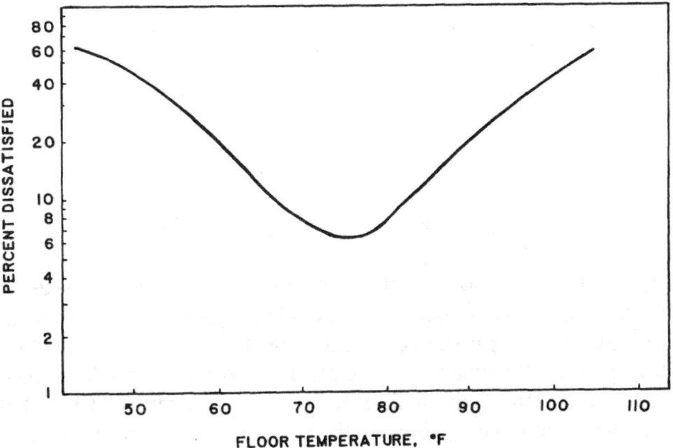

Fig. 9 Percentage of People Dissatisfied as Function of Floor Temperature

For floors occupied by people with normal indoor footwear, flooring material is insignificant. Olesen (1977b) found an optimal temperature of 77°F for sedentary and 73.5°F for standing or walking persons. At the optimal temperature, 6% of the occupants felt warm or cold discomfort in the feet. Figure 9 shows the relationship between floor temperature and percentage of dissatisfied, combining data from experiments with seated and standing subjects. In all experiments, the subjects were in thermal neutrality; thus, the percentage of dissatisfied is only related to the discomfort due to cold or warm feet. Again, no significant difference in floor temperature was preferred by females and males.

SECONDARY FACTORS AFFECTING COMFORT

The temperature, air speed, humidity of the environment, their variation, and personal parameters of metabolism and clothing insulation are primary factors that directly affect energy flow and thermal comfort. However many secondary factors, some of which are discussed in this section, may more subtly influence comfort.

Day-to-Day Variations

Fanger (1973) conducted an experiment with a group of subjects, where the preferred ambient temperature for each subject under identical conditions was determined on four different days. Since the standard deviation was only 1.0°F, Fanger concluded that the comfort conditions for the individual can be reproduced and will vary only slightly from day to day.

Age

Because metabolism decreases slightly with age, many have stated comfort conditions based on experiments with young and healthy subjects cannot be used for other age groups. Fanger (1982), Fanger and Langkilde (1975), Langkilde (1979), Nevins et al. (1966), and Rohles and Johnson (1972) conducted comfort studies in Denmark and the United States on different age groups (mean age 21 to 84). The studies revealed that the thermal environments preferred by older people do not differ from those preferred by younger people. The lower metabolism in older people is compensated for by a lower evaporative loss. Collins and Hoinville (1980) confirmed these results.

The fact that young and old people prefer the same thermal environment does not necessarily mean they are equally sensitive when exposed to cold or heat. In practice, the ambient temperature level in the homes of older people is often higher than that for younger people. This may be explained by the lower activity of elderly people, who are normally sedentary for a greater part of the day.

Adaptation

Many believe that people can acclimatize themselves by exposure to hot or cold surroundings, so that they prefer other thermal environments. Fanger (1982) conducted experiments involving subjects from the United States, Denmark, and tropical countries. The latter group was tested in Copenhagen immediately after their arrival by plane from the tropics where they had lived all their lives. Other experiments were conducted for two groups exposed to cold daily. One group comprised subjects who for 8 h daily for at least one year were doing sedentary work in cold surroundings (in the meat-packing industry). The other group consisted of winter swimmers who bathed in the sea daily.

Only slight differences regarding both the preferred ambient temperature and the physiological parameters in the comfort conditions were reported for the various groups. These results indicate that people cannot adapt to preferring warmer or colder environments. It is therefore likely that the same comfort conditions can be applied throughout the world. However, in determining the preferred ambient temperature from the comfort equations, a clo-value that corresponds to the local clothing habits should be used. A comparison of field comfort studies from different parts of the world shows significant differences in clothing habits depending on, among other things, the outdoor climate (Nichol and Humphreys 1972). According to these results, adaptation has little influence on the preferred ambient temperatures. In uncomfortable warm or cold environments, there will, however, often be an influence of adaptation. People used to working and living in warm climates can more easily accept and maintain a higher work performance in hot environments than people from colder climates.

Sex

Previously cited experiments by Fanger (1982), Fanger and Langkilde (1975), and Nevins et al. (1966) use equal numbers of male and female subjects, so comfort conditions for the two sexes can be compared. The experiments show that men and women prefer almost the same thermal environments. Women's skin temperature and evaporative loss are slightly lower than those for men, and this balances the somewhat lower metabolism of women. The reason that women often prefer higher ambient temperatures than men may be partly explained by the lighter clothing normally worn by women.

Seasonal and Circadian Rhythms

Since people cannot adapt to prefer warmer or colder environments, it follows that there is no difference between comfort conditions in winter and in summer. McNall et al. (1968) confirmed this in an investigation where results of winter and summer experiments showed no difference. On the other hand, it is reasonable to expect the comfort conditions to alter during the day as the internal body temperature has a daily rhythm—a maximum occurring late in the afternoon, and a minimum early in the morning.

In determining the preferred ambient temperature for each of 16 subjects both in the morning and in the evening, Fanger et al. (1974) and Ostberg and McNicholl (1973) observed no difference. Furthermore, Fanger et al. (1973) found only small fluctuations in the preferred ambient temperature during a simulated 8-h workday (sedentary work). There is a slight tendency to prefer somewhat warmer surroundings before lunch, but none of the fluctuations are significant.

PREDICTION OF THERMAL COMFORT

Thermal comfort and thermal sensation can be predicted several ways. One way is to use Figure 4 and Table 9 adjust for clothing and activity levels that differ from those of figure. More numerical and rigorous predictions are possible by using the PMV and Two-Node models described in the following section.

Steady-State Energy Balance

Fanger (1982) related the comfort data to physiological variables. At a given level of metabolic activity M, and when the body is not far from thermal neutrality, the mean skin temperature t_{sk} and sweat rate E_{rsw} are the only physiological parameters influencing the heat balance. However, heat balance alone is not sufficient to establish thermal comfort. In the wide range of environmental conditions where heat balance can be obtained, only a narrow range provides thermal comfort. The following linear regression equations based on data from Rohles and Nevins indicate values of t_{sk} and E_{rsw} that provide thermal comfort.

$$t_{sk,req} = 96.3 - 0.156(M - W) \tag{56}$$

$$E_{rsw,req} = 0.42(M - W - 18.43) \tag{57}$$

At higher activity levels, sweat loss increases and the mean skin temperature decreases. Both reactions increase the heat loss from the body core to the environment. These two empirical relationships link the physiological and heat flow equations and thermal comfort perceptions. By substituting these values into Equation (11) for $(C + R)$, and into Equations (17) and (18) for E_{sk}, the energy balance Equation (1) can be used to determine combinations of the six environmental and personal parameters that optimize comfort for steady-state conditions.

Fanger (1982) reduced these relationships to a single equation, which assumed all sweat generated is evaporated, eliminating clothing moisture permeability as a factor in the equation. This assumption is valid for normal indoor clothing worn in typical indoor environments with low or moderate activity levels. At higher activity levels ($M_{act} > 3$ met), where a significant amount of sweating occurs even at optimum comfort conditions, this assumption may limit accuracy. The reduced equation is slightly different from the heat transfer equations developed here. The radiant heat exchange is expressed in terms of the Stefan-Boltzmann law (instead of using h_r), and diffusion of water vapor through the skin is expressed as a diffusivity coefficient and a linear approximation for saturated vapor pressure evaluated at t_{sk}. The combination of environmental and personal variables that produce a neutral sensation may be expressed as follows:

$$\begin{aligned}(M - W) = {} & 1.196 \times 10^{-9} f_{cl}[(t_{cl} + 460)^4 - (\bar{t}_r + 460)^4] \\ & + f_{cl} h_c (t_{cl} - t_a) \\ & + 0.97[5.73 - 0.022(M - W) - 6.9 p_a] \\ & + 0.42[(M - W) - 18.43] \\ & + 0.0173 M(5.87 - 6.9 p_a) \\ & + 0.00077 M(93.2 - t_a)\end{aligned} \tag{58}$$

where

$$\begin{aligned}t_{cl} = {} & 96.3 - 0.156(M - W) \\ & - R_{cl}\{(M - W) \\ & - 0.97[5.73 - 0.022(M - W) - 6.9 p_a] \\ & - 0.42[(M - W) - 18.43] - 0.0173 M(5.87 - 6.9 p_a) \\ & - 0.00077 M(93.2 - t_a)\}\end{aligned} \tag{59}$$

The values of h_c and f_{cl} can be estimated from tables and equations given in the Engineering Data and Measurements section. Fanger used the following relationships:

$$h_c = \begin{cases} 0.361(t_{cl} - t_a)^{0.25} & 0.361(t_{cl} - t_a)^{0.25} > 0.151\sqrt{V} \\ 0.151\sqrt{V} & 0.361(t_{cl} - t_a)^{0.25} < 0.151\sqrt{V} \end{cases} \tag{60}$$

$$f_{cl} = \begin{cases} 1.0 + 0.2\, I_{cl} & I_{cl} < 0.5 \text{ clo} \\ 1.05 + 0.1\, I_{cl} & I_{cl} > 0.5 \text{ clo} \end{cases} \tag{61}$$

Figures 10 and 11 show examples of how Equation (58) can be used.

Equation (58) is expanded to include a range of thermal sensations by using a **predicted mean vote** (PMV) index. The PMV index predicts the mean response of a large group of people according to the ASHRAE thermal sensation scale. Fanger (1970) related PMV to the imbalance between the actual heat flow from the body in a given environment and the heat flow required for optimum comfort at the specified activity by the following equation.

$$\text{PMV} = 3.155[0.303\exp(-0.114M) + 0.028]L \tag{62}$$

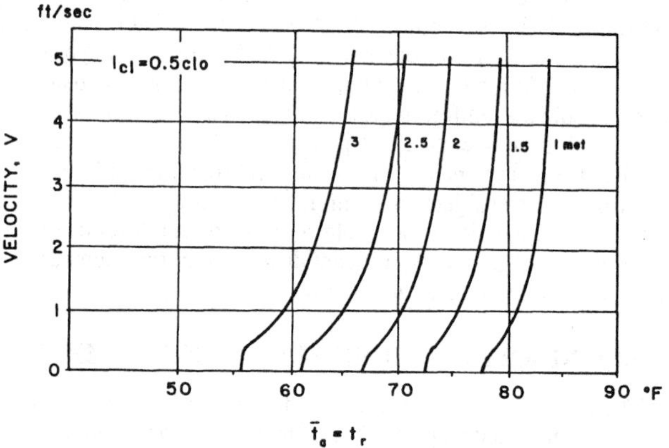

Fig. 10 Air Velocity and Operative Temperatures at 50% rh Necessary for Comfort (PMV = 0) of Persons in Summer Clothing at Various Levels of Activity

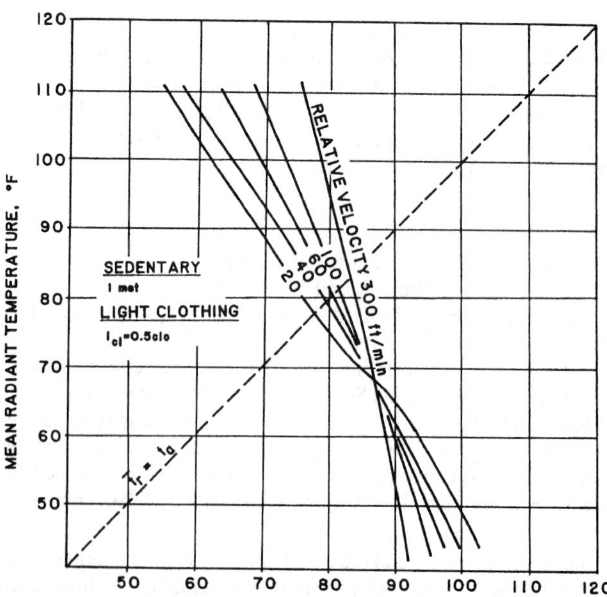

Fig. 11 Air Temperatures and Mean Radiant Temperatures Necessary for Comfort (PMV = 0) of Sedentary Persons in Summer Clothing at 50% rh

where L is the thermal load on the body, defined as the difference between the internal heat production and the heat loss to the actual environment for a person hypothetically kept at comfort values of t_{sk} and E_{rsw} at the actual activity level. Thermal load L is then the difference between the left and right sides of Equation (58) calculated for the actual values of the environmental conditions. As part of this calculation the clothing temperature t_{cl} is found by iteration as

$$t_{cl} = 96.3 - 0.156(M - W)$$
$$-R_{cl}\{1.196 \times 10^{-9} f_{cl}[(t_{cl} + 460)^4 - (\bar{t}_r + 460)^4] + f_{cl}h_c(t_{cl} - t_a)\} \quad (63)$$

After estimating the PMV with Equation (62) or some other way, the **predicted percent dissatisfied** (PPD) with a condition can also be estimated. Fanger (1982) related the PPD to the PMV as follows:

$$\text{PPD} = 100 - 95\exp[-(0.03353\,\text{PMV}^4 + 0.2179\,\text{PMV}^2)] \quad (64)$$

where dissatisfied is defined as anybody not voting either −1, +1, or 0. This relationship is shown in Figure 12. A PPD of 10% corresponds to the PMV range of ±0.5, and even with PMV = 0, about 5% of the people are dissatisfied.

The **PMV-PPD model** is widely used and accepted for design and field assessment of comfort conditions. ISO *Standard* 7730 (ISO 1984) includes a short computer listing that facilitates computing PMV and PPD for a wide range of parameters.

Two Node Model

The PMV model is only useful for predicting steady state comfort responses. The two node model (2NM) can be used to predict physiological responses or responses to transient situations, at least for low and moderate activity levels in cool to very hot environments (Gagge 1971, 1986). The 2NM is a simplification of more complex thermoregulatory models developed by Stolwijk and Hardy (1966). The simple, lumped parameter model considers a human as two concentric thermal compartments that represent the skin and core of the body.

The skin compartment simulates the epidermis and dermis and is about 1/16 in. thick. Its mass, which is about 10% of the total body, depends on the amount of blood flowing through it for thermoregulation. The temperature in a compartment is assumed to be uniform so that the only temperature gradients are between compartments. In a cold environment blood flow to the extremities may be reduced to conserve the heat of vital organs resulting in axial temperatures gradients in the arms, legs, hands, and feet. Heavy exercise with

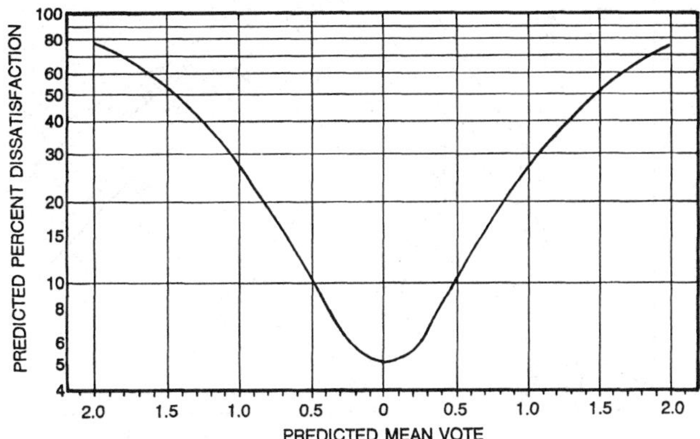

Fig. 12 Predicted Percentage of Dissatisfied (PPD) as Function of Predicted Mean Vote (PMV)

certain muscle groups or asymmetric environmental conditions may also cause nonuniform compartment temperatures and limit the accuracy of the model.

All of the heat is assumed to be generated in the core compartment. In the cold, shivering and muscle tension may generate additional metabolic heat. This increase is related to skin and core temperature depressions from their set point values, or

$$M_{shiv} = 1.9(93.2 - t_{sk})(98.6 - t_{cr}) \quad (65)$$

where the deviation terms are set to zero if they become negative.

The core loses energy when the muscles do work on the surroundings. Heat is also lost from the core through respiration. The rate of respiratory heat loss is due to sensible and latent changes in the respired air and the ventilation rate and in Equations (19) and (20).

In addition, heat is conducted passively from the core to the skin. This is modeled as a massless thermal conductor [K = 0.93 Btu/h·ft²·°F]. A controllable heat loss path from the core consists of pumping variable amounts of warm blood (m_{bl}) to the skin for cooling. This peripheral blood flow m_{bl} in L/(h·m²) depends on skin and core temperature deviations from their respective set points as:

$$m_{bl} = \text{BFN} + \frac{c_{dil}(t_{cr} - 98.6)}{1 + S_{tr}(93.2 - t_{sk})} \quad (66)$$

The bracketed temperature terms can only be > 0. If the deviation is negative the term is set to zero. For average persons the coefficients BFN, c_{dil}, and S_{tr} are 0.585, 9, and 0.28. Further, skin blood flow is limited to a maximum of 8.4 L/(h·ft²).

Dry (sensible) heat loss Q_{dry} from the skin flows through the clothing by conduction and then by parallel paths to the air and surrounding surfaces. Evaporative heat follows a similar path, flowing through the clothing and through the air boundary layer. Maximum evaporation E_{max} occurs if the skin is completely covered with sweat. The actual evaporation rate E_{sw} depends on the size of the sweat film w.

$$E_{sw} = wE_{max} \quad (67)$$

where $w = E_{rsw}/E_{max}$.

The rate of regulatory sweating E_{rsw} or rate at which water is brought to the surface of the skin in Btu/h·ft² can be predicted by skin and core temperature deviations from their set points.

$$E_{rsw} = \frac{c_{sw}(t_b - t_{bset})\exp[-(t_{sk} - 93.2)]}{19.3} \quad (68)$$

where $t_{mb} = (1 - \alpha)t_{cr} + \alpha t_{sk}$ is the mean body temperature and c_{sw} = 30 Btu/h·ft²·°F. The temperature deviation terms are set to zero when negative. α is the fraction of the total body mass that is considered to be thermally in the skin compartment.

$$\alpha = 0.0418 + 0.745/(10.8m_{bl} - 0.585) \quad (69)$$

Regulatory sweating in the model is limited to 0.1 L/h·ft² or 200 Btu/h·ft². E_{rsw} evaporates from the skin but if E_{rsw} is greater than E_{max} the excess drips off.

An energy balance on the core yields

$$M + M_{shiv} = \text{Work} + Q_{res} + (K + m_{bl}c_{p,bl})(t_{cr} - t_{sk}) + W_{cr}c_{cr}\frac{dt_{cr}}{d\theta} \quad (70)$$

and for the skin:

$$(K + m_{bl}c_{p,bl})(t_{cr} - t_{sk}) = Q_{dry} + Q_{evap} + W_{sk}c_{sk}\frac{dt_{sk}}{d\theta} \quad (71)$$

where c_{cr}, c_{sk}, and $c_{p,bl}$ are specific heats of core, skin, and blood [0.83, 0.83, and 1.0 Btu/lb·°F, respectively].

Equations (70) and (71) can be rearranged in terms of $dt_{sk}/d\theta$ and $dt_{cr}/d\theta$ and numerically integrated with small time steps (10 to 60 s) from initial conditions or previous values to find t_{cr} and t_{sk} at any time.

After calculating values of t_{sk}, t_{cr}, and w, the model uses empirical expressions to predict thermal sensation (TSENS) and thermal discomfort (DISC). These indices are based on 11-point numerical scales, where positive values represent the warm side of neutral sensation or comfort, and negative values represent the cool side. TSENS is based on the same scale as PMV, but with extra terms for ±4 (very hot/cold) and ±5 (intolerably hot/cold). Recognizing the same positive/negative convention for warm/cold discomfort, DISC is defined as

5 intolerable
4 limited tolerance
3 very uncomfortable
2 uncomfortable and unpleasant
1 slightly uncomfortable but acceptable
0 comfortable

TSENS is defined in terms of deviations of mean body temperature t_b from cold and hot set points representing the lower and upper limits for the zone of evaporative regulation: $t_{b,c}$ and $t_{b,h}$, respectively. The values of these set points depend on the net rate of internal heat production and are calculated by

$$t_{b,c} = \frac{0.34}{58.15}(M - W) + 97.34 \quad (72)$$

$$t_{b,h} = \frac{0.608}{58.15}(M - W) + 98.0 \quad (73)$$

TSENS is then determined by

$$\text{TSENS} = \begin{cases} 0.26(t_b - t_{b,c}) & t_b < t_{b,c} \\ 4.7\eta_{ev}(t_b - t_{b,c})/(t_{b,h} - t_{b,c}) & t_{b,c} \le t_b \le t_{b,h} \\ 4.7\eta_{ev} + 0.26(t_b - t_{b,h}) & t_{b,h} < t_b \end{cases} \quad (74)$$

where η_{ev}, is the evaporative efficiency (assumed to be 0.85).

Thermal discomfort is numerically equal to TSENS when t_b is below its cold set point $t_{b,c}$ and is related to skin wettedness when body temperature is regulated by sweating as

$$\text{DISC} = \begin{cases} 0.26(t_b - t_{b,c}) & t_b < t_{b,c} \\ \dfrac{4.7(E_{rsw} - E_{rsw,req})}{E_{max} - E_{rsw,req} - E_{dif}} & t_{b,c} \le t_b \end{cases} \quad (75)$$

where $E_{rsw,req}$ is calculated as in Fanger's model, using Equation (57).

Zones of Comfort and Discomfort

The preceding section shows that comfort and thermal sensation are not necessarily the same variable, especially for a person in the zone of evaporative thermal regulation. Figures 13 and 14 show this difference for the standard combination of met-clo-air movement used in the standard effective temperature. Figure 13 demonstrates that practically all basic physiological variables predicted by the two-node model are functions of ambient temperature and are relatively independent of vapor pressure. All exceptions occur at relative humidities above 80% and as the isotherms reach the ET* = 106.5°F line, where regulation by evaporation fails. Figure 14 shows that lines of constant ET* and wettedness are functions of both ambient temperature and vapor pressure. Thus, human thermal responses are divided into two classes—those in Figure 13, which respond only to heat stress from the environment, and those in Figure 14, which respond to both the heat stress from the environment and the resultant heat strain (Stolwijk et al. 1968).

For warm environments, any index with isotherms parallel to skin temperature is a reliable index of thermal sensation alone, and not of discomfort caused by increased humidity. Indices with isotherms parallel to ET* are reliable indicators of discomfort or dissatisfaction with thermal environments. For a fixed exposure time to cold, lines of constant t_{sk}, ET*, and t_o are essentially identical, and cold sensation is no different from cold discomfort. For a state

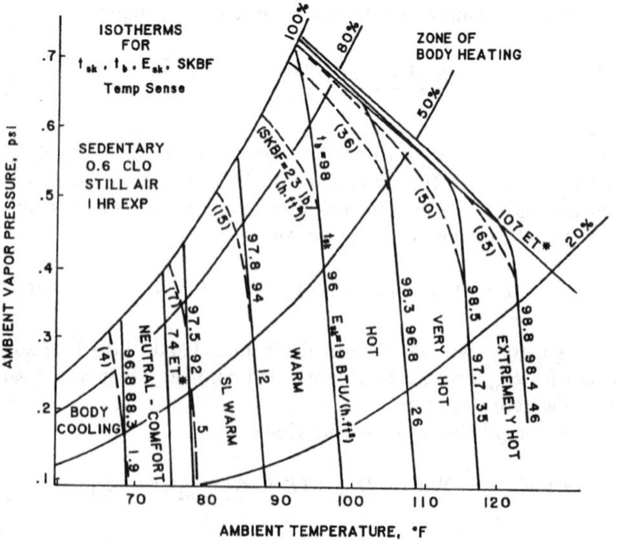

Fig. 13 Effect of Environmental Conditions on Physiological Variables

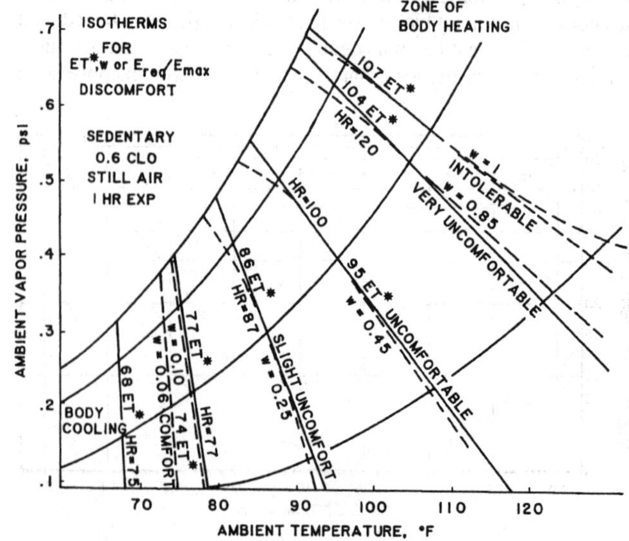

Fig. 14 Effect of Thermal Environment on Discomfort

of comfort with sedentary or light activity, lines of constant t_{sk} and ET* coincide. Thus comfort and thermal sensations coincide in this region as well. The upper and lower temperature limits for comfort at these levels can be specified either by thermal sensation (Fanger 1982) or by ET*, as is done in ASHRAE *Standard* 55, since lines of constant comfort and lines of constant thermal sensation should be identical.

ENVIRONMENTAL INDICES

An environmental index combines two or more parameters, such as air temperature, mean radiant temperature, humidity, or air velocity, into a single variable. Indices simplify the description of the thermal environment and the stress imposed by an environment. Environmental indices may be classified according to how they are developed. Rational indices are based on the theoretical concepts presented earlier. Empirical indices are based on measurements with subjects or on simplified relationships that do not necessarily follow theory. Indices may also be classified according to their application, generally either heat stress or cold stress.

Effective Temperature

The **effective temperature** ET* is probably the most common environmental index and has the widest range of application. It combines temperature and humidity into a single index, so two environments with the same ET* should evoke the same thermal response even though they have different temperatures and humidities; but they must have the same air velocities.

The original empirical effective temperature was developed by Houghten and Yaglou (1923). Gagge et al. (1971) defined a new effective temperature using a rational approach. Defined mathematically in Equation (33), this is the temperature of an environment at 50% rh that results in the same total heat loss from the skin E_{sk} as in the actual environment.

Because the index is defined in terms of operative temperature t_o, it combines the effect of three parameters ($\bar{t}_r$, t_a, and p_a) into a single index. Skin wettedness w and the permeability index i_m must be specified and are constant for a given ET* line for a particular situation. The two-node model is used to determine skin wettedness in the zone of evaporative regulation. At the upper limit of regulation, w approaches 1.0, and at the lower limit, w approaches 0.06; skin wettedness equals one of these values when the body is outside the zone of evaporative regulation. Since the slope of a constant ET* line depends on skin wettedness and clothing moisture permeability, effective temperature for a given temperature and humidity may depend on the clothing of the person and their activity. This difference is shown in Figure 15. At low skin wettedness, the air humidity has little influence, and lines of constant ET* are nearly vertical. As skin wettedness increases due to activity and/or heat stress, the lines become more horizontal and the influence of humidity is much more pronounced. The ASHRAE comfort envelope shown in Figure 4 is described in terms of ET*.

Since ET* depends on clothing and activity, it is not possible to generate a universal ET* chart. Calculation of ET* can also be tedious, requiring the solution of multiple coupled equations to determine skin wettedness. A standard set of conditions representative of typical indoor applications is used to define a **standard effective temperature** (SET*). The standard effective temperature is then defined as the equivalent air temperature of an isothermal environment at 50% rh in which a subject, while wearing clothing standardized for the activity concerned, has the same heat stress (skin temperature t_{sk}) and thermoregulatory strain (skin wettedness w) as in the actual environment.

Humid Operative Temperature

The **humid operative temperature** t_{oh} is the temperature of a uniform environment at 100% rh in which a person loses the same total amount of heat from the skin as in the actual environment. This index is defined mathematically in Equation (32). It is analogous to ET*, the only difference being that it is defined at 100% rh and 0% rh rather than at 50% rh. Figures 2 and 15 indicate that lines of constant ET* are also lines of constant t_{oh}. However, the values of these two indices differ for a given environment.

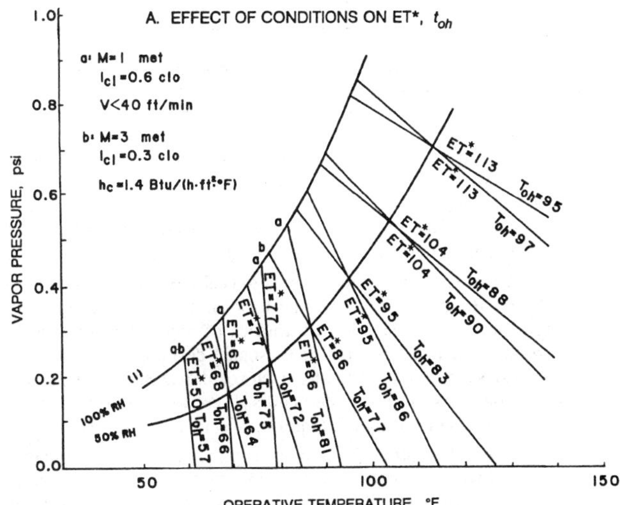

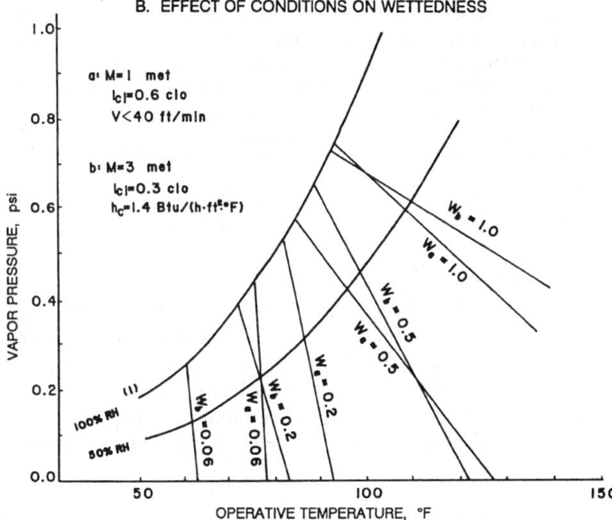

Fig. 15 Effective Temperature (ET*) and Skin Wettedness (W)
(Adapted from Nishi et al. 1975 and Gonzalez et al. 1978)

Heat Stress Index

Originally proposed by Belding and Hatch (1955), this rational index is the ratio of the total evaporative heat loss E_{sk} required for thermal equilibrium (the sum of metabolism plus dry heat load) to the maximum evaporative heat loss E_{max} possible for the environment, multiplied by 100, for steady-state conditions (S_{sk} and S_{cr} are zero), and with t_{sk} held constant at 95°F. The ratio E_{sk}/E_{max} equals skin wettedness as is shown in Equation (18). When HSI > 100, body heating occurs; when HSI < 0, body cooling occurs. Belding and Hatch (1955) limited E_{max} to 220 Btu/(h·ft²), which corresponds to a sweat rate of approximately 0.21 lb/(h·ft²). When t_{sk} is constant, loci of constant HSI coincide with lines of constant ET* on a psychrometric chart. Other indices based on wettedness of the original HSI concept have the same practical applications (Gonzalez et al. 1978, Belding 1970, ISO *Standard* 7933), but differ in their

Table 10 Evaluation of Heat Stress Index

Heat Stress Index	Physiological and Hygienic Implications of 8-h Exposures to Various Heat Stresses
0	No thermal strain.
10 20 30	Mild to moderate heat strain. If job involves higher intellectual functions, dexterity, or alertness, subtle to substantial decrements in performance may be expected. In performing heavy physical work, little decrement is expected, unless ability of individuals to perform such work under no thermal stress is marginal.
40 50 60	Severe heat strain involving a threat to health unless men are physically fit. Break-in period required for men not previously acclimatized. Some decrement in performance of physical work is to be expected. Medical selection of personnel desirable, because these conditions are unsuitable for those with cardiovascular or respiratory impairment or with chronic dermatitis. These working conditions are also unsuitable for activities requiring sustained mental effort.
70 80 90	Very severe heat strain. Only a small percentage of the population may be expected to qualify for this work. Personnel should be selected: (a) by medical examination, and (b) by trial on the job (after acclimatization). Special measures are needed to assure adequate water and salt intake. Amelioration of working conditions by any feasible means is highly desirable, and may be expected to decrease the health hazard while increasing job efficiency. Slight "indisposition," which in most jobs would be insufficient to affect performance, may render workers unfit for this exposure.
100	The maximum strain tolerated daily by fit, acclimatized young men.

treatment of E_{max} and the effect of clothing. Table 10 describes physiological factors associated with HSI values.

Index of Skin Wettedness

This index is the ratio of observed skin sweating E_{sk} to the E_{max} of the environment as defined by t_{sk}, t_a, humidity, air movement, and clothing in Equation (12). Except for the factor of 100, it is essentially the same as the heat stress index. Skin wettedness is more closely related to the sense of discomfort or unpleasantness than to temperature sensation (Gagge et al. 1969a, 1969b; Gonzalez et al. 1978).

Wet-Bulb Globe Temperature

The WBGT is an environmental heat stress index that combines dry-bulb temperature t_{db}, a *naturally ventilated* (not aspirated) wet-bulb temperature t_{nwb}, and black globe temperature t_g, according to the relation (Dukes-Dobos and Henschel 1971, 1973):

$$\text{WBGT} = 0.7\ t_{nwb} + 0.2\ t_g + 0.1\ t_a \qquad (76)$$

This form of the equation is usually used where solar radiation is present. The naturally ventilated wet-bulb thermometer is left exposed to the sunlight, but the air temperature t_a sensor is shaded. In enclosed environments, Equation (76) is simplified by dropping the t_a term and using a 0.3 weighting factor for t_g.

The black globe thermometer is responsive to air temperature, mean radiant temperature, and air movement, while the naturally ventilated wet-bulb thermometer responds to air humidity, air movement, radiant temperature, and air temperature. Thus, WBGT is a function of all four environmental factors affecting human environmental heat stress.

The WBGT is a better index of heat stress than the old ET; it shows almost as good a correlation with sweat rate as do the later Corrected Effective Temperature (CET) and the Effective Temperature with Radiation (ETR) indices (Minard 1961); the CET and ETR both require direct measurement of wind velocity which, for accuracy, requires special instruments and trained technicians.

The WBGT index is widely used for estimating the heat stress potential of industrial environments (Davis 1976). In the United

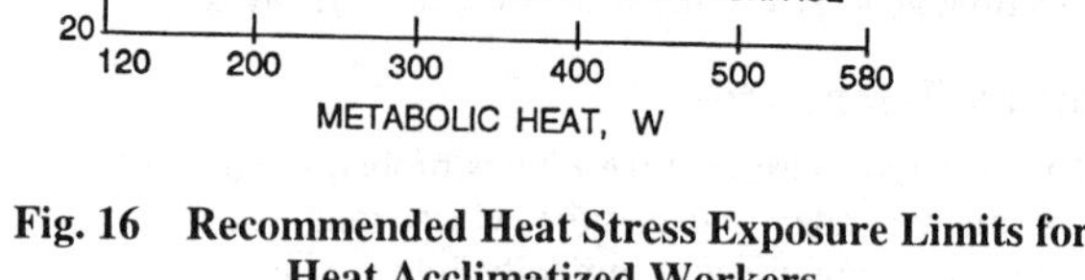

Fig. 16 Recommended Heat Stress Exposure Limits for Heat Acclimatized Workers
(From United States DHHS (NIOSH) *Publication* 86-113, 1986)

States, the National Institute of Occupational Safety and Health (NIOSH) developed a criteria document for a heat-stress limiting standard (NIOSH 1986). ISO *Standard* 7243 (ISO 1982) also uses the WBGT. Figure 16 graphically summarizes the permissible heat exposure limits, expressed as working time per hour, for a fit individual, as specified for various WBGT levels. Values apply for normal permeable clothing (0.6 clo) and must be adjusted for heavy or partly vapor-permeable clothing. USAF (1980) recommends adjusting the measured WBGT upwards by 10°F for personnel wearing chemical protective clothing or body armor. This type of clothing increases the resistance to sweat evaporation about threefold (higher if it is totally impermeable), requiring an adjustment in WBGT level to compensate for reduced evaporative cooling at the skin.

Several mathematical models are available for predicting WBGT from the environmental factors: air temperature, psychrometric wet-bulb temperature, mean radiant temperature, and air motion (Azer and Hsu 1977, Sullivan and Gorton 1976). A simpler approach, involving plotting WBGT lines on a psychrometric chart, is recommended. Isotherms of WBGT are parallel and have negative slopes varying from 0.014 psi/°F for still air to 0.016 psi/°F for air motion greater than 200 fpm. By comparison, psychrometric wet-bulb lines have negative slopes of about 0.0056 psi/°F, or are about 35% as steep.

Wet-Globe Temperature

The WGT, introduced by Botsford (1971), is a simpler approach to measuring environmental heat stress than the WBGT. The measurement is made with a wetted globe thermometer called a Botsball, which consists of a 2.5 in. black copper sphere covered with a fitted wet black mesh fabric, into which the sensor of a dial thermometer is inserted. A polished stem attached to the sphere supports the thermometer and contains a water reservoir for keeping the sphere covering wet. This instrument is suspended by the stem at the indoor (or outdoor) site to be measured.

Onkaram et al. (1980) has shown that WBGT can be predicted with reasonable accuracy from WGT for temperate to warm environments with medium to high humidities. With air temperatures between 68 and 95°F, dew points ranging from 45 to 77°F (relative humidities above 30%), and wind speeds of 15 mph or less, the

Table 11 Equivalent Wind Chill Temperatures of Cold Environments

Wind Speed, mph	Actual Thermometer Reading, °F											
	50	40	30	20	10	0	−10	−20	−30	−40	−50	−60
	Equivalent Chill Temperature, °F											
0	50	40	30	20	10	0	−10	−20	−30	−40	−50	−60
5	48	37	27	16	6	−5	−15	−26	−36	−47	−57	−68
10	40	28	16	3	−9	−21	−34	−46	−58	−71	−83	−95
15	36	22	9	−5	−18	−32	−45	−59	−72	−86	−99	−113
20	32	18	4	−11	−25	−39	−53	−68	−82	−96	−110	−125
25	30	15	0	−15	−30	−44	−59	−74	−89	−104	−119	−134
30	28	13	−3	−18	−33	−48	−64	−79	−94	−110	−125	−140
35	27	11	−4	−20	−36	−51	−67	−83	−98	−114	−129	−145
40	26	10	−6	−22	−38	−53	−69	−85	−101	−117	−133	−148
	Little danger: In less than 5 h, with dry skin. Maximum danger from false sense of security. (WCI less than 1400)				**Increasing danger:** Danger of freezing exposed flesh within one minute. (WCI between 1400 and 2000)			**Great danger:** Flesh may freeze within 30 seconds. (WCI greater than 2000)				

Notes: Cooling power of environment expressed as an equivalent temperature under calm conditions [Equation (79)].

Winds greater than 43 mph have little added chilling effect.

Source: U.S. Army Research Institute of Environmental Medicine.

experimental regression equation ($r = 0.98$) in °F for an outdoor environment is:

$$\text{WBGT} = 1.044\ \text{WGT} - 1.745 \qquad (77)$$

This equation should not be used outside the experimental range just given, since data from hot-dry desert environments show differences between WBGT and WGT that are too large (10°F and above) to be adjusted by Equation (77) (Matthew 1986). With very low humidity combined with high wind, WGT approaches the psychrometric wet-bulb temperature, which is greatly depressed below t_a. However, in the WBGT, t_{nwb} accounts for only 70% of the index value, with the remaining 30% at or above t_a.

Ciriello and Snook (1977) handle the problem by providing a series of regression equations, the choice depending on the levels of wind speed, humidity, and radiant heat. They report an accuracy of conversion from WGT to WBGT within 0.7°F (90% confidence level), if good estimates of wind speed, humidity, and radiation level are available.

Wind Chill Index

The wind chill index (WCI) is an empirical index developed from cooling measurements obtained in Antarctica on a cylindrical flask partly filled with water (Siple and Passel 1945). The index describes the rate of heat loss from the cylinder by radiation and convection for a surface temperature of 91.4°F, as a function of ambient temperature and wind velocity. As originally proposed:

$$\text{WCI} = (10.45 - 0.447\,V + 6.686\sqrt{V})(91.4 - t_a)/1.8 \qquad (78)$$

where V and t_a are in mph and °F, respectively. (WCI units are kcal/m^2·h.) The 91.4°F surface temperature was chosen to be representative of the mean skin temperature of a resting human in comfortable surroundings.

A number of valid objections have been raised about this formulation. Cooling rate data from which it was derived were measured on a 2.24 in. diameter plastic cylinder, making it unlikely that WCI would be an accurate measure of heat loss from exposed flesh, which has different characteristics than the plastic (curvature, roughness, and radiation exchange properties) and is invariably below 91.4°F in a cold environment. Moreover, values given by the equation peak at 56 mph, then decrease with increasing velocity.

Nevertheless, for velocities below 50 mph, this index reliably expresses combined effects of temperature and wind on subjective discomfort. For example, if the calculated WCI is less than 1400 and actual air temperature is above 14°F, there is little risk of frostbite during brief exposures (1 h or less), even for bare skin. However, at a WCI of 2000 or more, the probability is high that exposed flesh will begin to freeze in 1 min or less unless preventive measures are taken to shield the exposed skin (such as a fur ruff to break up the wind around the face).

Rather than using the WCI to express the severity of a cold environment, meteorologists use an index derived from the WCI called the **equivalent wind chill temperature**. This is the ambient temperature that would produce, in a calm wind (defined for this application as 4 mph), the same WCI as the actual combination of air temperature and wind velocity. Equivalent wind chill temperature $t_{eq,wc}$ in °F can be calculated by:

$$t_{eq,wc} = -0.0818(\text{WCI}) + 91.4 \qquad (79)$$

where $t_{eq,wc}$ is expressed as a temperature (and frequently referred to as a wind chill factor), thus distinguishing it from WCI, which is given either as a cooling rate or as a plain number with no units. For velocities less than 4 mph, Equation (79) does not apply, and the wind chill temperature is equal to the air temperature.

Equation (79) does not imply cooling to below ambient temperature, but recognizes that, because of wind, the cooling rate is increased as though it were occurring at the lower equivalent wind chill temperature. Wind accelerates the rate of heat loss, so that the skin surface is cooling faster toward the ambient temperature. Table 11 shows a typical wind chill chart, expressed in equivalent wind chill temperature.

SPECIAL ENVIRONMENTS

Infrared Heating

Optical and thermal properties of skin must be considered in studies concerning the effects of infrared radiation in (1) producing changes in skin temperature and skin blood flow, and (2) evoking sensations of temperature and comfort (Hardy 1961). Although the body can be considered to have the properties of water, thermal sensation and heat transfer with the environment require a study of the skin and its interaction with visible and infrared radiation.

Figure 17 shows how skin reflectance and absorptance vary for a blackbody heat source at the temperature (in °R) indicated. These curves show that darkly pigmented skin is heated more by direct radiation from a high-intensity heater at 4500°R than is lightly

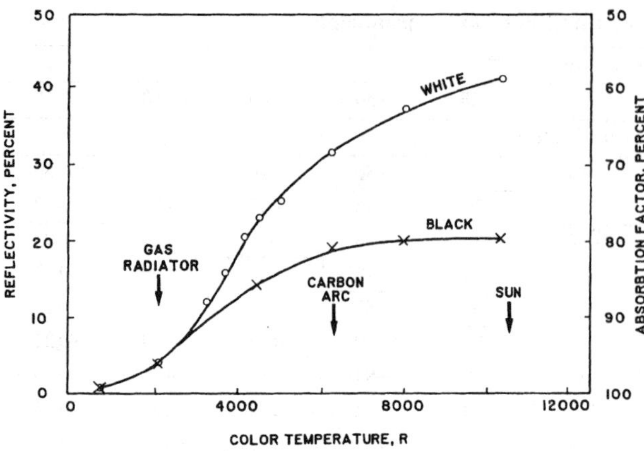

Fig. 17 Variation in Skin Reflection and Absorptivity for Blackbody Heat Sources

pigmented skin. With low-temperature and low-intensity heating equipment used for total area heating, there is minimal, if any, difference. Also, in practice, clothing minimizes differences.

Changes in skin temperature caused by high-intensity infrared radiation depend on the thermal conductivity, density, and specific heat of the living skin (Lipkin and Hardy 1954). Modeling of skin heating with the heat transfer theory yields a parabolic relation between exposure time and skin temperature rise for nonpenetrating radiation:

$$t_{sf} = t_{si} = \Delta t = 2J\varepsilon\sqrt{\theta/(\pi k\rho c_p)} \qquad (80)$$

where

t_{sf} = final skin temperature, °F
t_{si} = initial skin temperature, °F
J = irradiation intensity from source at °R, Btu/(h·ft²)
θ = time, h
k = specific thermal conductivity of tissue, Btu/(h·ft·°R)
ρ = density, lb/ft³
c_p = specific heat, Btu/(lb·°R)
ε = skin absorptance for radiation at °R, dimensionless

Product $k\rho c_p$ is the physiologically important quantity that determines temperature elevation of skin or other tissue on exposure to nonpenetrating radiation. Fatty tissue, because of its relatively low specific heat, is heated more rapidly than moist skin or bone. Experimentally, $k\rho c_p$ values can be determined by plotting Δt^2 against 1.13 $J^2\theta$ (Figure 18). Lines are linear and their slopes are inversely proportional to the $k\rho c_p$ of the specimen. Comparing leather and water with body tissues suggests that thermal inertia values depend largely on tissue water content.

Living tissues do not conform strictly to this simple mathematical formula. Figure 19 compares excised skin with living skin with normal blood flow, and skin with blood flow occluded. For short exposure times, the $k\rho c_p$ of normal skin is the same as that in which blood flow has been stopped; excised skin heats more rapidly due to unavoidable dehydration that occurs postmortem. However, with longer exposure to thermal radiation, vasodilation increases blood flow, cooling the skin. For the first 20 s of irradiation, skin with normally constricted blood vessels has a $k\rho c_p$ value of one-fourth that for skin with fully dilated vessels.

Skin temperature is the best single index of thermal comfort. The most rapid changes in skin temperature occur during the first 60 s of exposure to infrared radiation. During this initial period, thermal sensation and the heating rate of the skin vary with the quality of infrared radiation (color temperature in degrees Rankine). Since radiant heat from a gas-fired heater is absorbed at the skin surface,

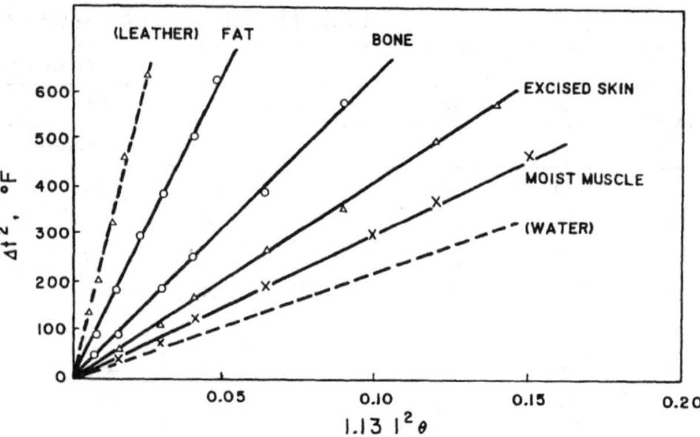

Fig. 18 Comparing Thermal Inertia of Fat, Bone, Moist Muscle, and Excised Skin to that of Leather and Water

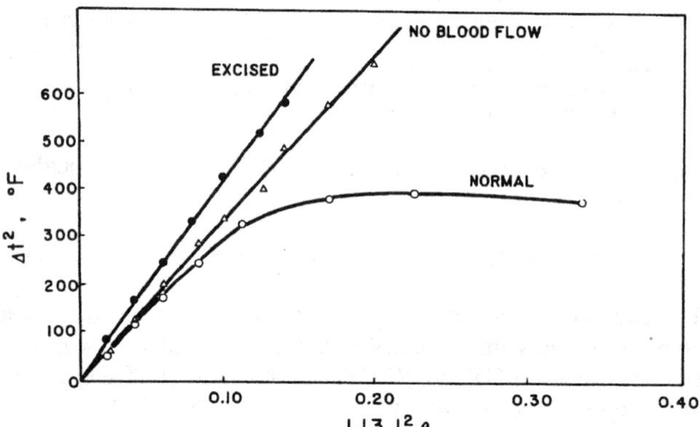

Fig. 19 Thermal Inertias of Excised, Bloodless, and Normal Living Skin

the same unit level of absorbed radiation during the first 60 s of exposure can cause an even warmer initial sensation than penetrating solar radiation. Because skin heating curves tend to level off after a 60 s exposure (Figure 19), a relative balance is quickly created between heat absorbed, heat flow to the skin surface, and heat loss to the ambient environment. Therefore, the effects of radiant heating on thermal comfort should be examined for conditions approaching thermal equilibrium.

Stolwijk and Hardy (1966) described an unclothed subject's response for a 2-h exposure to temperatures of 41 to 95°F. Nevins et al. (1966) showed a relation between ambient temperatures and thermal comfort of clothed, resting subjects. For any given uniform environmental temperature, both initial physiological response and degree of comfort can be determined for a subject at rest.

Physiological implications for radiant heating can be defined by two environmental temperatures: (1) mean radiant temperature or $\bar{t}_r$, and (2) ambient air temperature t_a. For this discussion on radiant heat, assume that (1) relative humidity is less than 50%, and (2) air movement is low and constant, with an equivalent convection coefficient of 0.51 Btu/(h·ft²·°F).

The equilibrium equation, describing heat exchange between skin surface at mean temperature t_{sk} and the radiant environment, is given in Equation (28), and can be transformed to give (see Table 2):

$$M' - E_{sk} - F_{cle}[h_r(t_{sk} - \bar{t}_r) + h_c(t_{sk} - t_o)] = 0 \qquad (81)$$

where M' is the net heat production $(M - W)$ less respiratory losses.

By algebraic transformation, Equation (81) can be rewritten:

$$M' + \mathrm{ERF} \cdot F_{cle} = E_{sk} + (h_r + h_c)(t_{sk} - t_a)F_{cle} \qquad (82)$$

where $\text{ERF} = h_r(t_r - t_a)$ is the effective radiant field and represents the additional radiant exchange with the body due to t_r differing from t_a.

The last term in Equation (82) describes heat exchange with an environment uniformly heated to temperature t_a. The term h_r, evaluated in Equation (35), is also a function of posture, for which factor (A_r/A_D) can vary from 0.67 for crouching to 0.73 for standing. For preliminary analysis, a useful value for h_r is 0.83 Btu/(h·ft²·°F), which corresponds to a normally clothed (at 75°F) sedentary subject. Ambient air movement affects h_c, which appears only in the right-hand term of Equation (82).

Although the linear radiation coefficient h_r is used in Equations (81) and (82), the same definition of ERF follows if the fourth power radiation law is used. By this law, assuming emissivity of the body surface is unity, the ERF term in Equation (82) is:

$$\text{ERF} = \sigma(A_r/A_D)[(\bar{t}_r + 460)^4 - (t_a + 460)^4]F_{cle} \qquad (83)$$

Because $\bar{t}_r$ equals the radiation of several surfaces at different temperatures ($T_1, T_2, \ldots, T_j$):

$$\begin{aligned}\text{ERF} = \sigma(A_r/A_D)[&\varepsilon_1 F_{m-1}(T_1^4 - T_a^4) \\ &+ \varepsilon_2 F_{m-2}(T_2^4 - T_a^4) \\ &+ \ldots + \varepsilon_j F_{m-j}(T_j^4 - T_a^4)]F_{cle}\end{aligned} \qquad (84)$$

where

σ = Stefan-Boltzmann constant, 0.1714×10^{-8} Btu/(h·ft²·°R⁴)
ε_j = emissivity of skin or clothing surface for source radiating at temperature T_j
F_{m-j} = angle factor to subject m from source j
T_a = ambient air temperature, °R

$$\text{ERF} = (\text{ERF})_1 + (\text{ERF})_2 + \ldots + (\text{ERF})_j \qquad (85)$$

where any ERF is given by:

$$(\text{ERF})_j = \sigma(A_r/A_D)\varepsilon_j F_{m-j}(T_j^4 - T_a^4)F_{cle}$$

ERF is the sum of the fields caused by each surface T_j [e.g., T_1 may be an infrared beam heater; T_2, a heated floor; T_3, a warm ceiling; T_4, a cold plate glass window ($T_4 < T_a$); etc.]. Only surfaces with temperature T_j differing from T_a contribute to the ERF.

Comfort Equations for Radiant Heating

The **comfort equation for radiant heat** (Gagge et al. 1967a, b) follows from definition of ERF and Equation (8):

$$t_o \text{ (for comfort)} = t_a + \text{ERF (for comfort)}/h \qquad (86)$$

Thus, operative temperature for comfort is the temperature of the ambient air plus a temperature increment ERF/h, a ratio that measures the effectiveness of the incident radiant heating on occupants. Higher air movement (which increases the values of h or h_c) reduces the effectiveness of radiant heating systems. Clothing lowers t_o for comfort and for thermal neutrality.

Values for ERF and h must be determined to apply the comfort equation for radiant heating. Table 3 may be used to estimate h. One method of determining ERF is to calculate it directly from radiometric data that give, (1) radiation emission spectrum of the source, (2) concentration of the beam, (3) radiation from the floor, ceiling, and windows, and (4) corresponding angle factors involved. This analytical approach is described in Chapter 49 of the 1995 *ASHRAE Handbook—Applications.*

For direct measurement, a skin-colored or black globe, 6 in. in diameter, can measure the radiant field ERF for comfort, in terms of the uncorrected globe temperature t_g in °F and air movement in fpm, by the following relation:

$$\text{ERF} = (A_r/A_D)[1.07 + 0.169\sqrt{V}](t_g - t_a) \qquad (87)$$

The average value of A_r/A_D is 0.7. For a skin-colored globe, no correction is needed for the quality of radiation. For a black globe, ERF must be multiplied by ε for the exposed clothing/skin surface. For a subject with 0.6 to 1.0 clo, t_o for comfort should agree numerically with t_a for comfort in Figure 4. When t_o replaces t_a in Figure 4, humidity is measured in vapor pressure rather than relative humidity, which refers only to air temperature.

Other methods may be used to measure ERF. The most accurate is by physiological means. In Equation (82), when M, $t_{sk} - t_a$, and the associated transfer coefficients are experimentally held constant:

$$\Delta\text{E} = \Delta\text{ERF} \qquad (88)$$

The variation in evaporative heat loss (rate of weight loss) caused by changing the wattage of two T-3 infrared lamps is a measure in absolute terms of the radiant heat received by the body.

A third method uses a directional radiometer to measure ERF directly. For example, radiation absorbed at the body surface [in Btu/h·ft²] is:

$$\text{ERF} = \varepsilon(A_i/A_D)J \qquad (89)$$

where irradiance J can be measured by a directional (Hardy-type) radiometer; ε is the surface absorptance effective for the source used; and A_i is the projection area of the body normal to the directional irradiance. Equation (89) can be used to calculate ERF only for the simplest geometrical arrangements. For a human subject lying supine and irradiated uniformly from above, A_i/A_D is 0.3. Figure 17 shows variance of ε for human skin with blackbody temperature (in °R) of the radiating source. When irradiance J is uneven and coming from many directions, as is usually the case, the previous physiological method can be used to obtain an effective A_i/A_D from the observed ΔE and $\Delta(\varepsilon J)$.

Hot and Humid Environments

Tolerance limits to high temperature vary with the ability to (1) sense temperature, (2) lose heat by regulatory sweating, and (3) move heat from the body core by blood flow to the skin surface, where cooling is the most effective. Many interrelating processes are involved in heat stress (Figure 20).

Skin surface temperatures of 115°F trigger pain receptors in the skin; direct contact with metal at this temperature is painful. However, since thermal insulation of the air layer around the skin is high, much higher dry air temperatures can be tolerated. For lightly clothed subjects at rest, tolerance times of nearly 50 min have been reported at 180°F dry-bulb temperature; 33 min at 200°F; 26 min at 220°F; and 24 min at 240°F. In each case, dew points were lower than 86°F. Many individuals are stimulated by brief periods of exposure to 185°F dry air in a sauna. Short exposures to these extremely hot environments are tolerable because of cooling by sweat evaporation. However, when ambient vapor pressure approaches 0.87 psi (97°F dew point, typically found on sweating skin), tolerance is drastically reduced. Temperatures of 122°F can be intolerable if the dew-point temperature is greater than 77°F and both deep body temperature and heart rate rise within minutes (Gonzalez et al. 1978).

The rate and length of time a body can sweat are limited. The maximum rate of sweating for an average man is about 4 lb/h. If

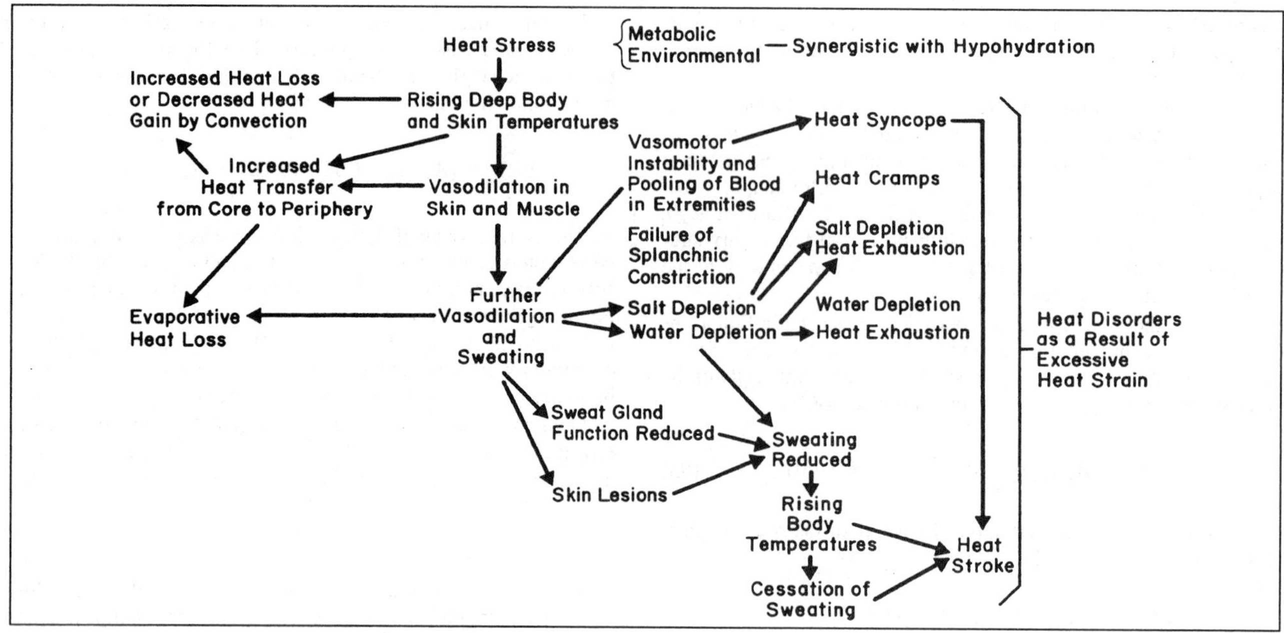

Fig. 20 Schematic Design of Heat Stress and Heat Disorders
[Modified by Buskirk from scale diagram by Belding (1967) and Leithead and Lind (1964)]

all this sweat evaporates from the skin surface under conditions of low humidity and air movement, maximum cooling is about 214 Btu/(h·°F). However, this value does not normally occur because sweat rolls off the skin surface without evaporative cooling or is absorbed by or evaporated within clothing. A more typical cooling limit is 6 mets, 110 Btu/(h·°F), representing approximately 2.4 lb/h of sweating for the average man.

Thermal equilibrium is maintained by dissipation of resting heat production (1 met) plus any radiant and convective load. If the environment does not limit heat loss from the body during heavy activity, decreasing skin temperature compensates for the core temperature rise. Therefore, mean body temperature is maintained, although the gradient from core to skin is increased. Blood flow through the skin is reduced, but muscle blood flow necessary for exercise is preserved. The upper limit of skin blood flow ($\dot{m}_{bl}$) is about 200 lb/h (Burton and Bazett 1936).

Body heat storage of 318 Btu (or a rise in t_b of 2.5°F) for an average-sized man represents an average voluntary tolerance limit. Continuing work beyond this limit increases the risk of heat exhaustion. Collapse can occur at about 635 Btu of storage (5°F); few individuals can tolerate heat storage of 872 Btu (6.8°F above normal).

The cardiovascular system affects tolerance limits. In normal, healthy subjects exposed to extreme heat, heart rate and cardiac output increase in an attempt to maintain blood pressure and supply of blood to the brain. At a heart rate of about 180 bpm, the short time between contractions prevents adequate blood supply to the heart chambers. As heart rate continues to increase, cardiac output drops, causing inadequate convective blood exchange with the skin and, perhaps more important, inadequate blood supply to the brain. Victims of this heat exhaustion faint or black out. Accelerated heart rate can also result from inadequate venous return to the heart caused by pooling of blood in the skin and lower extremities. In this case, cardiac output is limited because not enough blood is available to refill the heart between beats. This occurs most frequently when an overheated individual, having worked hard in the heat, suddenly stops working. The muscles no longer massage the blood back past the valves in the veins toward the heart. Dehydration compounds the problem, since fluid volume in the vascular system is reduced.

If core temperature increases above 106°F, critical hypothalamic proteins can be damaged, resulting in inappropriate vasoconstriction, cessation of sweating, increased heat production by shivering, or some combination of these. Heat stroke damage is frequently irreversible and carries a high risk of fatality.

A final problem, hyperventilation, occurs predominantly in hot-wet conditions, when too much CO_2 is washed from the blood. This can lead to tingling sensations, skin numbness, and vasoconstriction in the brain with occasional loss of consciousness.

Since a rise in heart rate or rectal temperature is essentially linear with ambient vapor pressure above a dew point of 77°F, these two changes can measure severe heat stress. Although individual heart rate and rectal temperature responses to mild heat stress vary, severe heat stress saturates physiological regulating systems, producing uniform increases in heart rate and rectal temperature. In contrast, sweat production measures stress under milder conditions but becomes less useful under more severe stress. The maximal sweat rate compatible with body cooling varies with (1) degree of heat acclimatization, (2) duration of sweating, and (3) whether the sweat evaporates or merely saturates the skin and drips off. Total sweat rates in excess of 4.4 lb/h can occur in short exposures, but about 2.2 lb/h is an average maximum level sustainable for an acclimatized man.

Figure 21 illustrates the decline in heart rate, rectal temperature, and skin temperature when exercising subjects are exposed to 104°F over a period of days. Acclimatization can be achieved by working in the heat for 100 min each day—30% improvement occurs after the first day, 50% after 3 days, and 95% after 6 or 7 days. Increased sweat secretion while working in the heat can be induced by rest. Although reducing salt intake during the first few days in the heat can conserve sodium, heat cramps may result. Working regularly in the heat improves cardiovascular efficiency, sweat secretion, and sodium conservation. Once induced, heat acclimatization can be maintained by as few as once-a-week workouts in the heat; otherwise, it diminishes slowly over a 2- to 3-week period and disappears.

Extreme Cold Environments

Human performance in extreme cold ultimately depends on maintaining thermal balance. Subjective discomfort is reported by a 154-lb man with 2800 in^2 of body surface area, when a heat debt of about 100 Btu is incurred. A heat debt of about 600 Btu is acutely uncomfortable; this represents a drop of approximately 4.7°F (or about 7% of total heat content) in mean body temperature.

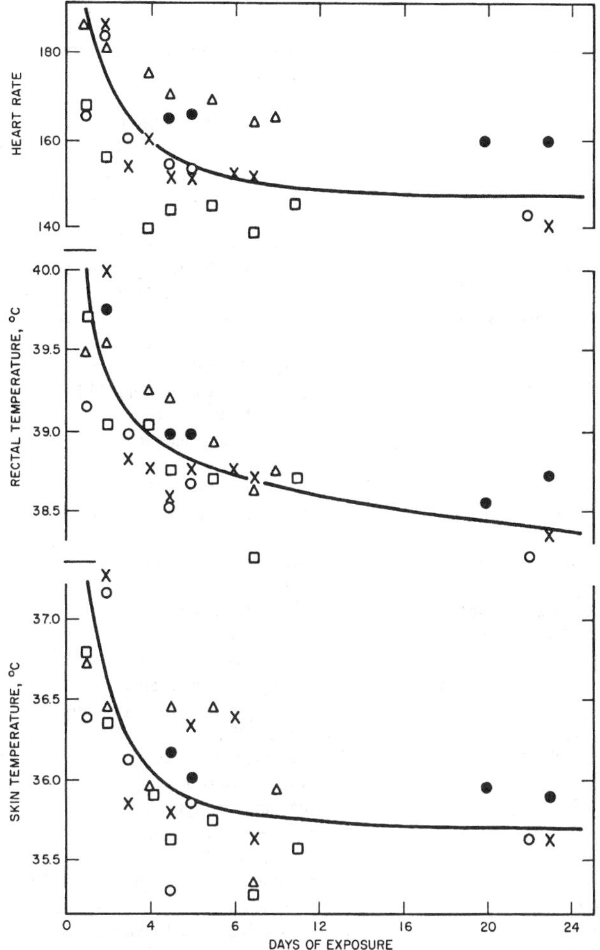

Fig. 21 Acclimatization to Heat Resulting from Daily Exposure of Five Subjects to Room Maintained at 40°C with 23% rh
(Robinson et al. 1943)

This loss can occur during 1 to 2 h of sedentary activity outdoors. A sleeping individual will awake after losing about 300 Btu, decreasing mean skin temperature by about 5.5°F and deep body temperature by about 1°F, using the Burton relationship [$\alpha = 0.67$, in Equation (41)]. A drop in deep body temperature (e.g., rectal temperature) below 95°F threatens a loss of body temperature regulation, while 82.4°F is considered critical for survival, despite recorded survival from a deep body temperature of 64.4°F. Temperature is more crucial than rate of temperature change; Witherspoon et al. (1971) observed a rate of fall in the core temperature of 5.4°F per hour in subjects immersed in 50°F water, without residual effect.

Activity level also affects human performance. Subjective sensations, reported by sendentary subjects at a mean skin temperature of 92°F, are *comfortable*; at 88°F, *uncomfortably cold*; at 86°F, *shivering cold*; and at 84°F, *extremely cold*. The critical subjective tolerance limit (without numbing) for mean skin temperature appears to be about 77°F. However, during moderate to heavy activity, subjects reported the same skin temperatures as comfortable. Although mean skin temperature is significant, the temperature of the extremities is more frequently the critical factor for comfort in the cold. Consistent with this, one of the first responses to cold exposure is vasoconstriction, which reduces circulatory heat input to the hands and feet. A hand-skin temperature of 68°F causes a report of *uncomfortably cold*; 59°F, *extreme cold*; and 41°F, *painful*. Identical verbal responses for the foot surface occur at approximately 2.7 to 3.5°F warmer temperatures.

An ambient temperature of −30°F is the lower limit for useful outdoor activity, even with adequate insulative clothing. At −60°F, almost all outdoor effort becomes exceedingly difficult; even with appropriate protective equipment, only limited exposure is possible. Reported exposures of 30 min at −103°F have occurred in the Antarctic without injury.

In response to extreme heat loss, maximal heat production becomes very important. When the less efficient vasoconstriction cannot prevent body heat loss, shivering is an automatic, more efficient defense against cold. This can be triggered by low deep body temperature, low skin temperature, rapid change of skin temperature, or some combination of all three. Shivering is usually preceded by an imperceptible increase in muscle tension and by noticeable gooseflesh produced by muscle contraction in the skin. It begins slowly in small muscle groups, initially increasing total heat production by 1.5 to 2 times resting levels. As body cooling increases, the reaction spreads to additional body segments. Ultimately violent, whole body shivering causes maximum heat production of about 6 times resting levels, rendering the individual totally ineffective.

Given sufficient cold exposure, the body undergoes changes that indicate cold acclimatization. These physiological changes include, (1) endocrine changes (e.g., sensitivity to norepinephrine), causing nonshivering heat production by metabolism of free fatty acids released from adipose tissue; (2) improved circulatory heat flow to skin, causing an overall sensation of greater comfort; and (3) improved circulatory heat flow to the extremities, reducing the risk of injury and permitting activities at what ordinarily would be severely uncomfortable temperatures in the extremities. Generally, these physiological changes are minor and are induced only by repeated extreme exposures. Nonphysiological factors, including training, experience, and selection of adequate protective clothing, are more useful and may be safer than dependence on physiological changes.

The caloric requirement for adequately clothed subjects in extreme cold is only slightly greater than that for subjects living and working in temperate climates. This greater requirement results from added work caused by (1) carrying the weight of heavy clothing (energy cost for heavy protective footwear may be six times that of an equivalent weight on the torso); and (2) the inefficiency of walking in snow, snowshoeing, or skiing, which can increase energy cost up to 300%.

To achieve proper protection in low temperatures, a person must either maintain high metabolic heat production by activity or reduce heat loss by controlling the body's microclimate with clothing. Other protective measures include spot radiant heating, showers of hot air for work at a fixed site, and warm air ventilated or electrically heated clothing. The extremities, such as fingers and toes, pose more of a problem than the torso because, as thin cylinders, they are particularly susceptible to heat loss and difficult to insulate without increasing the surface for heat loss. Vasoconstriction can reduce circulatory heat input to extremities by over 90%.

Although there is no ideal insulating material for protective clothing, radiation-reflective materials are promising. Insulation is primarily a function of clothing thickness; the thickness of trapped air, rather than fibers used, determines insulation effectiveness.

Protection for the respiratory tract seems unnecessary in healthy individuals, even at 50°F. However, asthmatics or individuals with mild cardiovascular problems may benefit from a face mask that warms inspired air. Masks are unnecessary for protecting the face since heat to facial skin is not reduced by local vasoconstriction, as it is for hands. If wind chill is great, there is always a risk of cold injury caused by freezing of exposed skin. Using properly designed torso clothing, such as a parka with fur-lined hood to minimize wind penetration to the face, and 35 Btu/h of auxiliary heat to each hand and foot, inactive people can tolerate −67°F with a 10 mph wind for more than 6 h. As long as skin temperature of fingers remains above 60°F, manual dexterity can be maintained and useful work performed without difficulty.

SYMBOLS

A = area, ft^2
A_{cl} = surface area of clothed body
A_D = DuBois surface area of nude body
A_G = body surface area covered by garment
A_r = effective radiation area of body
BFN = neutral skin blood flow, $L/h \cdot ft^2$
c_{cr} = specific heat of body core, Btu/(lb·°F)
c_{sk} = specific heat of skin
c_p = constant pressure specific heat
$c_{p,a}$ = of air
$c_{p,b}$ = of body tissue
$c_{p,bl}$ = of blood
c_{dil} = constant for skin blood flow
C_{res} = sensible heat loss due to respiration, $Btu/(h \cdot ft^2)$
c_{sw} = proportionality constant for sweat control
$C + R$ = total sensible heat loss from skin, $Btu/(h \cdot ft^2)$
E = evaporative heat loss, $Btu/(h \cdot ft^2)$
E_{dif} = due to moisture diffusion through skin
E_{max} = maximum possible
E_{rsw} = due to regulatory sweating
E_{res} = due to respiration
$E_{rsw,req}$ = required for comfort
E_{sw} = actual evaporation rate
E_{sk} = total from skin
ERF = effective radiant field, $Btu/(h \cdot ft^2)$
ET* = effective temperature, °F
f_{cl} = clothing area factor, A_{cl}/A_D, dimensionless
F_{cl} = intrinsic clothing thermal efficiency, dimensionless
F_{cle} = effective clothing thermal efficiency, dimensionless
F_{pcl} = permeation efficiency, dimensionless
F_{m-j} = angle factor to person from source j, dimensionless
F_{p-N} = angle factor from person to source N, dimensionless
h = sensible heat transfer coefficient, $Btu/(h \cdot ft^2 \cdot °F)$
h = total at surface
h' = overall including clothing
h_c = convection at surface
h_{cc} = corrected convection at surface
h_r = radiation
h_e = evaporative heat transfer coefficient, $Btu/(h \cdot ft^2 \cdot psi)$
h_e = at surface
h_{ec} = at surface, corrected for atmospheric pressure
$h_{e'}$ = overall including clothing
h_{fg} = heat of vaporization of water, Btu/lb
i = vapor permeation efficiency, dimensionless
i_a = air layer
i_{cl} = clothing
i_m = total
I = thermal resistance in clo units, clo
All subscripts given for symbol R apply to symbol I
J = irradiance, $Btu/(h \cdot ft^2)$
k = thermal conductivity of body tissue, Btu/(h·ft·°F)
K = effective conductance between core and skin, $Btu/(h \cdot ft^2 \cdot °F)$
l = height, ft
L = thermal load on body, $Btu/(h \cdot ft^2)$
LR = Lewis ratio, °F/psi
m = body mass, lb
m_{ge} = mass to gas exchange, lb
$\dot{m}$ = mass flow, $lb/(h \cdot ft^2)$
$\dot{m}_{bl}$ = blood circulation between core and skin
$\dot{m}_{rsw}$ = rate of regulatory sweat generation
$\dot{m}_{res}$ = pulmonary ventilation rate flow, lb/h
M = metabolic heat production, $Btu/(h \cdot ft^2)$
M = total
M' = net
M_{act} = due to activity
M_{shiv} = due to shivering
p = water vapor pressure, psi
p_a = in ambient air
$p_{ET^*,s}$ = saturated at ET*
$p_{oh,s}$ = saturated at t_{oh}
$p_{sk,s}$ = saturated at t_{sk}
p_t = atmospheric pressure, psi
Q = heat flow, $Btu/(h \cdot ft^2)$
Q_{dry} = sensible from skin
Q_{evap} = latent from skin
Q_{crsk} = from core to skin
Q_{res} = total due to respiration
Q_{sk} = total from the skin
R = thermal insulation, $h \cdot ft^2 \cdot °F/Btu$
R_a = air layer on nude skin
$R_{a,cl}$ = air layer at outer surface
R_{cl} = clothing
R_{cle} = change due to clothing
R_t = total
R_e = evaporative resistance, $h \cdot ft^2 \cdot psi/Btu$
$R_{e,cl}$ = clothing
R_t = total
RQ = respiratory quotient, dimensionless
S = heat storage, $Btu/(h \cdot ft^2)$
S_{cr} = in core compartment
S_{sk} = in skin compartment
S_{tr} = constriction constant for skin blood flow
SET* = standard effective temperature, °F
t = temperature, °F
t_a = ambient air
t_b = average of body
$t_{b,c}$ = lower limit for evaporative regulation zone
$t_{b,h}$ = upper limit for evaporative regulation zone
$t_{b,n}$ = at neutrality
t_{com} = combined temperature
t_{cl} = clothing surface
t_{cr} = core
$t_{cr,n}$ = at neutrality
t_{db} = dry bulb
$t_{eq,wc}$ = equivalent wind chill temperature
t_{ex} = of exhaled air
t_g = globe
t_N = of surface N
t_{nwb} = naturally ventilated wet bulb
t_o = operative
t_{pr} = plane radiant
t_s = surface
t_{sf} = final skin
t_{si} = initial skin
t_{sk} = skin
$t_{sk,n}$ = at neutrality
$t_{sk,req}$ = required for comfort
$\bar{t}_r$ = mean radiant
T = absolute temperature, °R
All subscripts that apply to symbol t may apply to symbol T.
Tu = turbulence intensity, %
V = air velocity, fpm
V_{CO_2} = volume rate of CO_2 produced, ft^3/h
V_{O_2} = volume rate of O_2 consumed, ft^3/h
w = skin wettedness, dimensionless
w_{rsw} = required to evaporate regulatory sweat
W = external work accomplished, $Btu/(h \cdot ft^2)$
W_a = humidity ratio of ambient air, lb (H_2O)/lb (dry air)
W_{ex} = humidity ratio of exhaled air, lb (H_2O)/lb (dry air)
W_{sk} = mass of skin
W_{cr} = mass of body core
α = fraction of total body mass concentrated in skin compartment, dimensionless
σ = Stefan-Boltzmann = n constant, $Btu/(h \cdot ft^2 \cdot °F^4)$
ε = emissivity, dimensionless
η = evaporative efficiency, dimensionless
ρ = density, lb/ft^3
θ = time, h

REFERENCES

ASHRAE. 1992. Thermal environmental conditions for human occupancy. ANSI/ASHRAE *Standard* 55-1992.

ASHRAE. 1994. Addendum 55a. ANSI/ASHRAE *Standard* 55-1992.

Astrand, P. and K. Rodahl. 1977. *Textbook of work physiology: Physiological bases of exercise*. McGraw-Hill, New York.

Azer, N.Z. 1982. Design guidelines for spot cooling systems: Part I—Assessing the acceptability of the environment. *ASHRAE Transactions* 88:1.

Azer, N.Z. and S. Hsu. 1977. OSHA heat stress standards and the WBGT index. *ASHRAE Transactions* 83(2):30.

Belding, H.S. and T.F. Hatch. 1955. Index for evaluating heat stress in terms of resulting physiological strains. *Heating, Piping and Air Conditioning* 207:239.

Belding, H.D. 1967. "Heat stress." In *Thermobiology*, ed. A.H. Rose. Academic Press, New York.

Belding, H.S. 1970. The search for a universal heat stress index. *Physiological and Behavioral Temperature Regulation*, eds. J.D. Hardy et al. Springfield, IL.

Berglund, L.G and D.J. Cunningham. 1986. Parameters of human discomfort in warm environments. *ASHRAE Transactions* 92(2):732-46.

Berglund, L.G. and A. Fobelets. 1987. A subjective human response to low level air currents and asymmetric radiation. *ASHRAE Transactions* 93(1):497-523.

Berglund, L.G. 1994. Common elements in the design and operation of thermal comfort and ventilation systems. *ASHRAE Transactions* 100 (1):?-?.

Berglund, L.G. 1995. Comfort Criteria: Humidity and Standards. Proceedings of Pan Pacific Symposium on Building and Urban Environmental Conditioning in Asia, 2: 369-82. Architecture Department, University of Nagoya, Japan.

Botsford, J.H. 1971. A wet globe thermometer for environmental heat measurement. *American Industrial Hygiene Association Journal* 32:1-10.

Burton, A.C. and H.C. Bazett. 1936. A study of the average temperature of the tissues, the exchange of heat and vasomotor responses in man, by a bath calorimeter. *American Journal of Physiology* 117:36.

Busch, J.F. 1992. A tale of two populations: Thermal comfort in air-conditioned and naturally ventilated offices in Thailand. *Energy and Buildings*, 18:235-49.

Buskirk, E.R. 1960. Problems related to the caloric cost of living. Bulletin of the New York Academy of Medicine 26:365.

Chatonnet, J. and M. Cabanac. 1965. The perception of thermal comfort. Int. *J. Biometeorology* 9:183-93.

Ciriello, V.M. and S.H. Snook. 1977. The prediction of WBGT from the Botsball. *American Industrial Hygiene Association Journal* 38:264.

Colin, J. and Y. Houdas. 1967. Experimental determination of coefficient of heat exchange by convection of the human body. *Journal of Applied Physiology* 22:31.

Collins, K.J. and E. Hoinville. 1980. Temperature requirements in old age. *Building Services Engineering Research and Technology* 1(4):165-72.

Davis, W.J. 1976. Typical WBGT indexes in various industrial environments. *ASHRAE Transactions* 82(2):303.

de Dear, R., K. Leow, and A. Ameen. 1991. Thermal comfort in the humid tropics— Part I, *ASHRAE Transactions* 97(1): 874-879.

DuBois, D. and E.F. DuBois. 1916. A formula to estimate approximate surface area, if height and weight are known. *Archives of Internal Medicine* 17:863-71.

Dukes-Dobos, F. and A. Henschel. 1971. The modification of the WBGT index for establishing permissible heat exposure limits in occupational work. HEW, USPHE, NIOSH, TR-69.

Dukes-Dobos, F. and A. Henschel. 1973. Development of permissible heat exposure limits for occupational work. *ASHRAE Journal* 9:57.

Eriksson, H.A. 1975. Heating and ventilating of tractor cabs. Presented at the 1975 Winter Meeting, American Society of Agricultural Engineers, Chicago.

Fanger, P.O. 1967. Calculation of thermal comfort: introduction of a basic comfort equation. *ASHRAE Transactions* 73(2):III.4.1.

Fanger, P.O. 1970. *Thermal comfort analysis and applications in environmental engineering*. McGraw-Hill, New York.

Fanger, P.O. 1972. *Thermal comfort*. McGraw-Hill, New York.

Fanger, P.O. 1973. The variability of man's preferred ambient temperature from day to day. Archives des Sciences Physiologiques 27(4):A403.

Fanger, P.O. 1982. *Thermal comfort*. Robert E. Krieger Publishing Company, Malabar, FL.

Fanger, P.O., L. Banhidi, B.W. Olesen, and G. Langkilde. 1980. Comfort limits for heated ceilings. *ASHRAE Transactions* 86.

Fanger, P.O. and N.K. Christensen. 1985. Perception of draught in ventilated spaces. *Ergonomics* 29(2):215-35.

Fanger, P.O., J. Hojbjerre, and J.O.B. Thomsen. 1974. Thermal comfort conditions in the morning and the evening. *International Journal of Biometeorology* 18(1):16.

Fanger, P.O., J. Hojbjerre, and J.O.B. Thomsen. 1973. Man's preferred ambient temperature during the day. *Arch. Science Physiology* 27(4): A395-A402.

Fanger, P.O., B.M. Ipsen, G. Langkilde, B.W. Olesen, N.K. Christensen, and S. Tanabe. 1985. Comfort limits for asymmetric thermal radiation. *Energy and Buildings*.

Fanger, P.O. and G. Langkilde. 1975. Interindividual differences in ambient temperature preferred by seated persons. *ASHRAE Transactions* 81(2):140-47.

Fanger, P.O., A. Melikov, H. Hanzawa, and J. Ring. 1987. Air turbulence and sensation of draught. *Energy and Buildings*.

Fanger, P.O., A.K. Melikov, H. Hanzawa, and J. Ring, J. 1989. Turbulence and draft. *ASHRAE Journal*, April.

Fanger, P.O., O. Ostberg, A.G.M. Nichell, N.O. Breum, and E. Jerking. 1974. Thermal comfort conditions during day and night. *European Journal of Physiology* 33:225-63.

Fobelets, A.P.R. and A.P. Gagge. 1988. Rationalization of the ET* as a measure of the enthalpy of the human environment. *ASHRAE Transactions* 94:1.

Gagge, A.P. 1937. A new physiological variable associated with sensible and insensible perspiration. *American Journal of Physiology* 20(2):277-87.

Gagge, A.P., A.C. Burton, and H.D. Bazett. 1971. A practical system of units for the description of heat exchange of man with his environment. *Science* 94:428-30.

Gagge, A.P. and J.D. Hardy. 1967. Thermal radiation exchange of the human by partitional calorimetry. *Journal of Applied Physiology*. 23:248.

Gagge, A.P., A.P. Fobelets, and L.G. Berglund. 1986. A standard predictive index of human response to the thermal environment. *ASHRAE Transactions* 92(1).

Gagge, A.P., Y. Nishi, and R.G. Nevins. 1976. The role of clothing in meeting FEA energy conservation guidelines. *ASHRAE Transactions* 82(2):234.

Gagge, A.P., G.M. Rapp, and J.D. Hardy. 1967a. The effective radiant field and operative temperature necessary for comfort with radiant heating. *ASHRAE Transactions* 73(1):I.2.1.

Gagge, A.P., G.M. Rapp, and J.D. Hardy. 1967b. The effective radiant field and operative temperature necessary for comfort with radiant heating. *ASHRAE Journal* 9(5):63.

Gagge, A.P., J.Stolwijk and Y. Nishi. 1971. An effective temperature scale based on a simple model of human physiological regulatory response. *ASHRAE Transactions* 77(1):247-62.

Gagge, A.P., J.A.J. Stolwijk, and B. Saltin. 1969a. Comfort and thermal sensation and associated physiological responses during exercise at various ambient temperatures. *Environmental Research* 2:209.

Gagge, A.P., J.A.J. Stolwijk, and Y. Nishi. 1969b. The prediction of thermal comfort when thermal equilibrium is maintained by sweating. *ASHRAE Transactions* 75(2):108.

Gonzalez, R.R., L.G. Berglund, and A.P. Gagge. 1978. Indices of thermoregulatory strain for moderate exercise in the heat. *Journal of Applied Physiology* 44:889.

Green, G.H. 1982. Positive and negative effects of building humidification. *ASHRAE Transactions* 88(1):1049-61.

Gwosdow, A.R., J.C. Stevens, L. Berglund and J.A.J. Stolwijk. 1986. Skin friction and fabric sensations in neutral and warm environments. *Textile Research Journal* 56:574-80.

Hanzawa, H., A.K. Melikov, and P.O. Fanger. 1987. Airflow characteristics in the occupied zone of ventilated spaces. *ASHRAE Transactions* 93(1).

Hardy, J.D. 1949. Heat transfer, In *Physiology of heat regulation and science of clothing*. eds. L.H. Newburgh and W.B. Saunders Ltd., London, 78.

Hardy, J.D. 1961. Physiological effects of high intensity infrared heating. *ASHRAE Journal* 4:11.

Hardy, J.D. 1961. Physiology of temperature regulation. *Physiological Reviews* 41:521-606.

Hardy, J.D., J.A.J. Stolwijk, and A.P. Gagge. 1971. Man. In *Comparative physiology of thermoregulation*, Chapter 5, Charles C. Thomas, Springfield, Ill.

Hardy, J.D., H.G. Wolf and H. Goodell. 1952. *Pain Sensations and Reactions*, Baltimore, Williams and Wilkins.

Hensel, H. 1973. Temperature reception and thermal comfort. *Archiv. des Sciences Physiologiques* 27:A359-A370.

Hensel, H. 1981. *Thermoreception and Temperature Regulation*. Academic Press, London.

Holmer, I. 1984. Required clothing insulation (IREQ) as an analytical index of cold stress. *ASHRAE Transactions* 90(1).

Horton, R.J.M. 1976. Help wanted. *ASHRAE Journal* 18:75.

Houghten, F.C. and C.P. Yaglou. 1923. ASHVE Research Report No. 673, Determination of the Comfort Zone. *ASHVE Transactions* 29:361.

ISO. 1982. Hot environments—Estimation of the heat stress on working man, based on the WBGT-index (wet-bulb globe temperature). ISO (International Organization for Standardization) *Standard* 7243, 1st ed.

ISO. 1984. Moderate thermal environments—Determination of the PMV and PPD indices and specification of the conditions for thermal comfort. ISO *Standard* 7730.

Jones, B.W., K. Hsieh, and M. Hashinaga. 1986. The effect of air velocity on thermal comfort at moderate activity levels. *ASHRAE Transactions* 92:2.

Korsgaard, V. 1949. Necessity of using a directional mean radiant temperature to describe thermal conditions in rooms. *Heat. Pipe. Air Condit.* 21:117.

Kuno, S; 1995. Comfort and pleasantness. *Proceedings of Pan Pacific Symposium on Building and Urban Environmental Conditioning in Asia*, Vol. 2: 383-92. Architecture Department, University of Nagoya, Japan.

Langkilde, G. 1979. Thermal comfort for people of high age. In *Comfort thermique: Aspects physiologiques et psychologiques*, INSERM, Paris 75:187-93.

Leithead, C.S. and A.R. Lind. 1964. *Heat stress and heat disorders*. Cassell & Co., London, England, 108. (F.A. Davis, Philadelphia, 1964).

Lipkin, M. and J.D. Hardy 1954. Measurement of some thermal properties of human tissues. *Journal of Applied Physiology* 7:212.

Liviana, J.E., F.H. Rohles and O.D. Bullock. 1988. Humidity, comfort and contact lenses. *ASHRAE Transactions* 94(1):3-11.

Matthew, W.H., et al. 1986. Botsball (WGT) performance characteristics and their impact on the implementation of existing military hot weather doctrine. U.S. Army Reserves Institute of Environmental Medicine Technical Report T 9/86, April.

McCullough, E.A. 1986. An insulation data base for military clothing. Institute for Environmental Research Report 86-01, Kansas State University, Manhattan, KS.

McCullough, E.A. and B.W. Jones. 1984. A comprehensive data base for estimating clothing insulation. IER Technical *Report* 84-01, Institute for Environmental Research, Kansas State University, Manhattan, KS. (Final report to *ASHRAE research project* 411-RP).

McCullough, E.A. and S. Hong, S. 1994. A data base for determining the decrease in clothing insulation due to body motion. *ASHRAE Transactions* 100(1).

McCullough, E.A., B.W. Jones, and T. Tamura. 1989. A data base for determining the evaporative resistance of clothing. *ASHRAE Transactions* 95(2).

McCullough, E.A., B.W. Olesen and S.W. Hong. 1994. Thermal insulation provided by chairs. *ASHRAE Transactions* 100(1).

McIntyre, D.A. 1974. The thermal radiation field. *Building Science* 9:247-62.

McIntyre, D.A. 1976. Overhead radiation and comfort. *The Building Services Engineer* 44:226-32.

McIntyre, D.A. and I.D. Griffiths. 1975. The effects of uniform and asymmetric thermal radiation on comfort. *Clima 2000, 6th International Congress of Climatritics*, Milan.

McNall, P.E., Jr. and R.E. Biddison. 1970. Thermal and comfort sensations of sedentary persons exposed to asymmetric radiant fields. *ASHRAE Transactions* 76(1):123.

McNall, P.E., P.W. Ryan, and J. Jaax. 1968. Seasonal variation in comfort conditions for college-age persons in the Middle West. *ASHRAE Transactions* 74(1):IV.2.1—IV.2.9.

McNair, H.P. 1973. A preliminary study of the subjective effects of vertical air temperature gradients. British Gas Corporation *Report* No. WH/T/R&D/73/94, London.

McNair, H.P. and D.S. Fishman. 1974. A further study of the subjective effects of vertical air temperature gradients. British Gas Corporation *Report* No. WH/T/R&D/73/94, London.

Minard, D. 1961. Prevention of heat casualties in marine corps recruits. *Military Medicine* 126:261.

Mitchell, D. 1974. Convective heat transfer in man and other animals, heat loss from animals and man. Butterworth Publishing Inc., London, 59.

Nevins, R., R.R. Gonzalez, Y. Nishi and A.P. Gagge. 1975. Effect of changes in ambient temperature and level of humidity on comfort and thermal sensations. *ASHRAE Transactions* 81(2).

Nevins, R.G. and A.M. Feyerherm. 1967. Effect of floor surface temperature on comfort: Part IV, Cold floors. *ASHRAE Transactions* 73(2):III.2.1.

Nevins, R.G. and A.O. Flinner. 1958. Effect of heated-floor temperatures on comfort. *ASHRAE Transactions* 64:175.

Nevins, R.G., K.B. Michaels, and A.M. Feyerherm. 1964. The effect of floor surface temperature on comfort: Part 1, College age males; Part II, College age females. *ASHRAE Transactions* 70:29.

Nevins, R.G., F.H. Rohles, Jr., W.E. Springer, and A.M. Feyerherm. 1966. Temperature-humidity chart for thermal comfort of seated persons. *ASHRAE Transactions* 72(1):283.

Nicol, J.F. and M.A. Humphreys. 1972. Thermal comfort as part of a self-regulating system. Proceedings of CIB symposium on thermal comfort, Building Research Station, London.

NIOSH. 1986. Criteria for a recommended standard—Occupational exposure to hot environments, revised criteria. U.S. Dept. of Health and Human Services, USDHHS (NIOSH) *Publication* 86-113.

Nishi, Y. 1981. Measurement of thermal balance of man. Bioengineering Thermal Physiology and Comfort, K. Cena and J.A. Clark, eds. Elsevier, New York.

Nishi, Y. and A.P. Gagge. 1970. Direct evaluation of convective heat transfer coefficient by naphthalene sublimation. *Journal of Applied Physiology* 29:830.

Nishi, Y., R.R. Gonzalez, and A.P. Gagge. 1975. Direct measurement of clothing heat transfer properties during sensible and insensible heat exchange with thermal environment. *ASHRAE Transactions* 81(2):183.

Olesen, B.W. 1977a. Thermal comfort requirements for floors. Proceedings of Commissions B1, B2, E1 of the IIR, Belgrade, 337-43.

Olesen, B.W. 1977b. Thermal comfort requirements for floors occupied by people with bare feet. *ASHRAE Transactions* 83(2).

Olesen, B.W. and R. Nielsen. 1983. Thermal insulation of clothing measured on a moveable manikin and on human subjects. Technical University of Denmark, Lyngby, Denmark.

Olesen, B.W., M. Scholer, and P.O. Fanger. 1979. Vertical air temperature differences and comfort. In *Indoor climate,* P.O. Fanger and O. Valbjorn, eds. Danish Building Research Institute, Copenhagen 561-79.

Olesen, S., J.J. Bassing, and P.O. Fanger. 1972. Physiological comfort conditions at sixteen combinations of activity, clothing, air velocity and ambient temperature. *ASHRAE Transactions* 78(2)199.

Onkaram, B. Stroschein, and R.F. Goldman. 1980. Three instruments for assessment of WBGT and a comparison with WGT (Botsball). *American Industrial Hygiene Association* 41:634-41.

Oohori, T., L.G. Berglund, and A.P. Gagge. 1984. Comparison of current two-parameter indices of vapor permeation of clothing—As factors governing thermal equilibrium and human comfort. *ASHRAE Transactions* 90(2).

Ostberg, O. and A.G. McNicholl. 1973. The preferred thermal conditions for "morning" and "evening" types of subjects during day and night—Preliminary results. *Build International* 6(1):147-57.

Passmore, R. and J.V.G. Durnin. 1967. *Energy, work and leisure.* Heinemann Educational Books, Ltd., London.

Rapp, G. and A.P. Gagge. 1967. Configuration factors and comfort design in radiant beam heating of man by high temperature infrared sources. *ASHRAE Transactions* 73(2):III.1.1.

Robinson, et al. 1943. *American Journal of Physiology* 140:168.

Rohles, F.H., Jr. 1973. The revised modal comfort envelope. *ASHRAE Transactions* 79(2):52.

Rohles, F.H., Jr. 1980. The preferred indoor comfort temperatures. *Report* No. 80-02, Institute for Environmental Research, Kansas State University, Manhattan, KS.

Rohles, F.H., Jr. and M.A. Johnson. 1972. Thermal comfort in the elderly. *ASHRAE Transactions* 78(1):131.

Rohles, F.H., Jr. and R.G. Nevins. 1971. The nature of thermal comfort for sedentary man. *ASHRAE Transactions* 77(1):239.

Seppanen, O., P.E. McNall, D.M. Munson, and C.H. Sprague. 1972. Thermal insulating values for typical indoor clothing ensembles. *ASHRAE Transactions* 78(1):120-30.

Siple, P.A. and C.F. Passel. 1945. Measurements of dry atmospheric cooling in subfreezing temperatures. Proceedings of the American Philosophical Society 89:177.

Stolwijk, J.A.J., A.P. Gagge, and B. Saltin. 1968. Physiological factors associated with sweating during exercise. *Journal of Aerospace Medicine* 39:1101.

Stolwijk, J.A.J. and J.D. Hardy. 1966. Partitional calorimetric studies of response of man to thermal transients. *Journal of Applied Physiology* 21:967.

Sullivan, C.D. and R.L. Gorton. 1976. A method of calculating WBGT from environmental factors. *ASHRAE Transactions* 82(2):279.

Tanabe, S., K. Kimura and T. Hara. 1987. Thermal comfort requirements during the summer season in Japan. *ASHRAE Transactions* 93(1):564-77.

Umbach, K.H. 1980. Measuring the physiological properties of textiles for clothing. *Melliand Textilberichte* (English Edition) G1:543-48.

Webb, P. 1964. *Bioastronautics Data Base*, NASA.

Winslow, C.-E.A., L.P. Herrington and A.P. Gagge. 1937. Relations between atmospheric conditions, physiological reactions and sensations of pleasantness. *American Journal of Hygiene* 26(1):103-15.

Witherspoon, J.M., R.F. Goldman, and J.R. Breckenridge. 1971. Heat transfer coefficients of humans in cold water. *Journal de Physiologie*, Paris, 63:459.

Woodcock, A.H. 1962. Moisture transfer in textile systems. *Textile Research Journal* 8:628-33.

CHAPTER 9

INDOOR ENVIRONMENTAL HEALTH

THIS chapter introduces the field of environmental health as it pertains to the indoor environment of buildings. In many cases, architectural, structural, cleaning, maintenance, materials use, and other activities that affect the environment are outside the control of the HVAC designer. Nevertheless, whenever possible, the designer should encourage features and decisions that create a healthy building environment.

TERMINOLOGY

Operational definitions of health, disease, and discomfort are controversial (Cain et al. 1995). The World Health Organization (WHO) defines health as "a state of complete physical, mental, and social well-being and not merely the absence of disease or disability." Last (1983) defines health as "a state characterized by anatomic integrity, ability to perform personally valued family, work, and social roles; ability to deal with physical, biologic, and social stress; a feeling of well-being; and freedom from the risk of disease and untimely death." Higgins (1983) defines an adverse health effect as a biological change that reduces the level of well-being or functional capacity. These definitions indicate that good health is a function of freedom from active ill health or disease (i.e., short and long-term disability or impairment, freedom from risk of disease in the future resulting from current exposures, and current subjective well-being).

Definitions of comfort are also controversial. Traditionally comfort refers to immediate satisfaction. It encompasses perception of the environment (hot, cold, noisy, etc.) and a value rating of affective implications (too hot, too cold, etc.). Rohles et al. (1989) noted that acceptability may represent a more useful concept as it allows progression towards a concrete goal. This serves as the foundation of a number of standards including thermal comfort, acoustics, etc. Nevertheless, acceptability may change over time (secular drift) as expectations change.

STANDARDS

Occupational safety and health standards are set by a variety of agencies. The United States Environmental Protection Agency (EPA) established **National Ambient Air Quality Standards**, which are summarized in Table 1.

The American Conference of Governmental Industrial Hygienists (ACGIH) reviews data on a yearly basis and publishes **Threshold Limit Values** (TLV)—levels that are frequently agreed on to be reasonably safe. The National Institute for Occupational Safety and Health (NIOSH) has developed criteria documents for specific substances and issued **Recommended Exposure Limits** (REL) that are generally below those of the ACGIH. The Occupational Safety and Health Administration (OSHA) sets **Permissible Exposure Limits** (PEL), which are the only regulatory standards in the United States. In the Federal Republic of Germany the Deutsche Forschungsgemeinschaft established **Maximale Arbeitsplatz Konzentrationen** (MAK values). However, all of these criteria were developed for 8-h exposures in industrial settings and usually exceed levels acceptable to occupants in office, residential, and other spaces. Elsewhere, a Canadian National Task Force developed similar criteria for indoor environments. Similarly, the World Health Organization has published air quality guidelines for Europe. Table 2 contrasts these various levels.

In contrast, Table 3 summarizes diseases that have been associated with specific aspects of indoor environments. For these diseases, diagnostic criteria may be used to distinguish between presence or absence of disease. These health events come about because of the presence of an exposure, a susceptible host, and a vector of transmission. Dose-response relationships are inadequately defined to establish a safe or a dangerous level of pollutants.

Normal interactions of the human body with its surrounding environment occur in predictable fashions. At extremes of the exposure range for light, heat, cold, and sound, organ dysfunction is measurable, and disease, such as frost bite, burns, and noise-induced hearing loss occur. Some transitions between normal and disease states are more difficult to delineate. Pain from bright light, erythema from heat, and nausea from vibration represent reversible effects but are interpreted by health professionals as abnormal.

DESCRIPTIONS OF SELECTED HEALTH SCIENCES

EPIDEMIOLOGY AND BIOSTATISTICS

Epidemiology is the study of distributions and determinants of disease. It represents the application of quantitative methods to evaluate diseases or conditions of interest. The subjects may be humans, animals, or even buildings. Epidemiology is traditionally subdivided into observational and analytical components. It may be primarily descriptive, or it may attempt to identify causal associations. Some classical criteria for determining causal relationships in

The preparation of this chapter is assigned to the Environmental Health Committee.

Table 1 Primary Ambient-Air Quality Standards of the United States and Their Physiological Effects

Contaminant	Long Term Concentration, μg/m³ (ppm)	Long Term Averaging Period	Short Term Concentration, μg/m³ (ppm)	Short Term Averaging Period	Physiological Effects
Sulfur dioxide	80 (0.03)	1 year	365 (0.14)	24 hours	Chronic respiratory disease
Carbon monoxide	10,000 (9)	8 hour	40,000 (35)	1 hour	Heart disease
Nitrogen dioxide	100 (0.055)	1 year			Chronic respiratory disease
Ozone			235 (0.12)[b]	1 hour	Chronic respiratory disease
Hydrocarbons			160 (0.24)	3 hours	
Total particulate (PM10)	50[a]	1 year	150	24 hours	Chronic respiratory disease
Lead	1.5	3 months[c]			Lead poisoning; central nervous system effects in children

[a]Arithmetic mean

[b]Standard is attained when expected number of days per calendar year with maximal hourly average concentrations above 0.12 ppm (235 μg/m³) is equal to or less than 1, as determined by Appendix H to subchapter C, 40 CFR 50

[c]Three-month period is a calendar quarter.

Table 2 Comparison of Standards Pertinent to Indoor Environments

	Canadian	WHO/Europe	NAAQS/EPA	NIOSH REL	OSHA	ACGIH	MAK
Aldehydes							
Acrolein	0.02 ppm[a]			0.1 ppm 0.25 ppm (15 min)	0.1 ppm 0.3 ppm (15 min)	0.1 ppm 0.3 ppm (15 min)	0.1 ppm 0.2 ppm (15 min)
Acetaldehyde	5.0 ppm			ALARA[b]	100 ppm 150 ppm (15 min)	100 ppm 150 ppm (15 min)	50 ppm
Formaldehyde	0.1 ppm[c]	0.081 ppm		0.016 ppm 0.1 ppm (15 min)	0.75 ppm 2 ppm (15 min)	0.3 ppm	0.3 ppm
Carbon Dioxide	3500 ppm			5000 ppm 30 000 ppm (15 min)	10 000 ppm 30 000 ppm (15 min)	5000 ppm 9 000 ppm (15 min)	5000 ppm 9 000 ppm (15 min)
Carbon Monoxide	11 ppm (8 h) 25 ppm (1 h)	8.6 ppm (8 h) 25 ppm (1 h) 51 ppm (30 min) 86 ppm (15 min)	9 ppm (8 h) 35 ppm (1 h)	35 ppm (8 h) 200 ppm (15 min)	35 ppm (8 h) 200 ppm (15 min)	25 ppm (8 h)	30 ppm
Nitrogen Dioxide	0.05 ppm 0.25 ppm (1 h)	0.08 ppm (24 h) 0.2 ppm (1 h)	0.053 ppm (1yr)		1 ppm (15 min)	3 ppm 5 ppm (15 min)	5 ppm
Ozone	0.12 ppm (1 h) no long-term level	0.08 ppm (8 h) 0.1 ppm (1 h)	0.12 ppm (1 h) 0.085 ppm (8 h)	0.1 ppm (15 min)	0.1 ppm (8 h) 0.3 ppm (15 min)	0.05 ppm (8 h) 0.2 ppm (15 min)	0.1 ppm
Particulate < 2.5 MMAD[d]	40 μg/m³ (8 h) 100 μg/m³ (1 h)		50 g/m³ (1 yr)		5 mg/m³ (8 h) (respirable dust)	3 mg/m³ (8 h) (no asbestos, <1% crystalline silica)	
Sulphur Dioxide	0.019 ppm 0.38 ppm (5 min)			2 ppm (8 h) 5 ppm (15 min)	2 ppm (8 h) 5 ppm (15 min)	2 ppm (8 h) 5 ppm (15 min)	2 ppm
Radon	800 Bq/m³ [e]						
Relative humidity	30-80% (summer) 30-55% (winter)						

( ) numbers in parentheses represent averaging periods

[a] parts per million (10^6)

[b] as low as reasonably achievable

[c] target level of 0.05 ppm because of its carcinogenic effects

[d] mass median aerodynamic diameter

[e] mean in normal living areas

epidemiology are consistency, temporality, plausibility, specificity, strength of the association, and dose-response relationships.

Observational studies are generally performed by defining some group of interest because of a specific exposure or risk factor. A control group is selected on the basis of similar criteria, but without the factor of interest. Observations conducted at one point in time are considered cross-sectional studies. On the other hand, a group may be defined by some criteria at a specific time, for example, all employees who worked in a certain building for at least one month in 1982. They may then be followed over time, leading to an observational cohort study.

Analytical studies may be either experimental or case-control studies. In experimental studies, individuals are selectively exposed to specific agents or conditions. Such studies are generally performed with the consent of the participants, unless the conditions are part of their usual working conditions and known to be harmless. Sometimes exposures cannot be controlled on an individual basis, and the intervention must be applied to entire groups. Control groups must be observed in parallel. Case-control studies are conducted by identifying individuals with the condition of interest and comparing risk factors between these people and individuals without that condition.

Measurement

All factors of interest must be measured in an unbiased fashion to avoid a subconscious influence of the investigator. The method

Table 3 Diseases Related to Buildings

Disease	History and Physical Examination	Laboratory Testing	Linkage	Causes
Rhinitis Sinusitis	Stuffy/ runny nose, post-nasal drip, pale or erythematous mucosa	Anterior and posterior rhinomanometry, acoustic rhinometry, nasal lavage, biopsy, rhinoscopy, RAST or skin prick testing	Immunologic skin prick or RAST testing, bradketted physiology	Direct occupational exposures; molds in the workplace; specific occupational factors (laser toners, carbonless copy paper, cleaning agents), secondary occupational exposures; pet (e.g. cat) danders brought from home
Asthma	Coughing, wheezing, episodic dyspnea, wheezing on examination, chest tightness, temporal pattern at work	Spirometry before and after work on Monday, peak expiratory flow diary, methacholine challenge	Immunologic: skin prick or RAST testing, Physiologic: related to work*	See rhinitis/sinusitis
Hypersensitivity pneumonitis	Cough, dyspnea, myalgia, weakness, rales, clubbing, feverishness	DLCO, FVC, TLC, CKR, lung biopsy	Immunologic: IgG ab to agents present, challenge testing Physiologic (in acute forms): spirometry, DLCO	Molds, moisture
Organic dust toxic syndrome	Cough, dyspnea, chest tightness, feverishness	DLCO, TLC, WBC	Temporal pattern related to work	Gram negative bacteria
Contact dermatitis (allergic)	Dry skin, itching, scaling skin	Scaling rash, eczema biopsy	Patch testing	Molds, carbonless copy paper, laser toners
Contact urticaria	Hives	Inspection biopsy	Provocative testing	Office products (carbonless copy paper)
Eye irritation	Eye itching, irritation, dryness	Tear-film break-up time, conjunctival staining (fluorescein)	Temporal pattern	Low relative humidity, volatile organic compounds (allergic conjunctivitis), particulates
Nasal irritation	Stuffy, congested nose, rhinitis	Acoustic rhinometry, posterior and anterior rhinomanometry, nasal lavage, nasal biopsy	Temporal pattern	Low relative humidity, volatile organic compounds (allergic conjunctivitis), particulates
Central nervous system symptoms	Headache, fatigue, irritability, difficulty concentrating	Neuropsychological testing	Temporal pattern (epidemiology)	Volatile organic compounds, noise, lighting, work stress, carbon monoxide, cytokines from bioaerosol exposure
Legionnaire's Disease	Pneumonic illness	History, *Legionella* culture from biopsy fluids	1) Organism isolated from patient and source, 2) immunologic watch	Aerosols from contaminated water sources, shower heads, water faucet aerators, humidifiers at home and at work, potable water sources (hot water heaters, etc.)

* (1) 10% decrement in FEV_1 across workday,
(2) peak flow changes suggestive of work relatedness, and
(3) methacholine reactivity resolving after six weeks away from exposure
RAST = radio allergen sorbent test
DLCO = single breath carbon monoxide diffusing capacity
FVC = forced vital capacity
TLC = total lung capacity
CXR = chest x-ray
IgG = class G immune globulins
FEV_1 = forced expiratory volume in the first second

of measurement should be repeatable and the technique meaningful. Statistical methods for data analysis follow standard procedures. Tests of hypotheses are performed at a specific probability level; they must have adequate power, i.e., if a sample size or a measurement difference between the factors or groups is too small, a statistically significant association may not be found even if one is present.

Results obtained in a specific situation, i.e., in a sample of exposed individuals, may be generalized to others only if they share the same characteristics. For example, it may not be legitimate to assume that all individuals have the same tolerance of thermal conditions irrespective of their heritage. Therefore, the results of studies and groups must be evaluated as they apply to a specific situation. Results obtained from a specific problem may not apply to another problem.

TOXICOLOGY

Toxicology is the study of poisons. Most researchers hold that essentially all substances may function as poisons and that only smaller doses prevent them from becoming harmful. Of fundamental importance is defining which component of the chemical structure predicts the harmful effect. The second issue is defining the dose-response relationships. The definition of dose may refer to delivered dose (exposure that is presented to the lungs) or absorbed dose (the dose that is actually absorbed through the lungs into the body and available for metabolism). Measures of exposure may be quite distinct from measures of effect, because of internal dose modifiers; for example, the delayed metabolism of some poisons because of lack of enzymes to degrade them. In addition, the mathematical characteristics of a dose may vary, depending on whether a peak dose, a geometric or arithmetic mean dose, or an integral under the dose curve is used.

Because humans often cannot be exposed in experimental conditions, most toxicological literature is based on animal studies. Recent studies suggest that it is not easy to extrapolate between dose level effects from animals to man. Isolated animal systems, such as homogenized rat livers, purified enzyme systems, or other isolated living tissues, may be used to study the impact of chemicals.

MOLECULAR BIOLOGY

Molecular biology is the branch of science that studies the chemical and physical structure of biological macromolecules. (DNA, proteins, carbohydrates, etc.). It is interested in processes on a molecular level, identifying actual mechanisms of effect or toxicity, rather on the level of cells.

CELLULAR BIOLOGY

Cellular biology is the branch of science that studies cellular organelles, activities, and processes. Little indoor air quality related research has been done at this mechanistic level, but it offers the final evidence in postulated cause and effect relationships.

GENETICS

Genetics is a branch of science that examines heredity and variation among organisms at population, individual, and chromosomal levels. Newer studies in genetics appear to indicate that some individuals are more susceptible to or are at greater risk from certain exposures than the rest of the population. This susceptibility would explain why not all individuals react the same way to the same exposure or lifestyle.

ERGONOMICS

Ergonomics may be defined as the scientific study of the relationship between man and his work environment to achieve optimum adjustment in terms of efficiency, health, and well-being. Ergonomic designs help workers interact more comfortably and efficiently with their environment. In jobs that were ergonomically designed productivity typically increased and the worker enjoyed a healthier working experience. More recently, researchers have distinguished intrinsic ergonomics from extrinsic, or traditional, ergonomics. Intrinsic ergonomics considers how the interface between an individual and the environment affects and relies on specific body parts, i.e., muscles, tendons, and bones; and work practices such as force of application, relaxation intervals, work practices and styles, and strength reserves that are not adequately considered in simple analyses of the physical environment.

The goal of ergonomics programs ranges from making work safe and humane, to increasing human efficiency, to creating human well-being. The successful application of ergonomic factors is measured by improved productivity, efficiency, safety, and acceptance of the resultant system design. The design engineer uses not only engineering skills but also the sciences and principles of anatomy, orthopedics, physiology, medicine, psychology, and sociology to apply ergonomics to a design.

Implementing ergonomic principals in the workplace helps minimize on-the-job stress and strain, and prevents cumulative trauma disorders or CTDs. These disorders are subtle injuries that can affect the muscles, tendons, and nerves at body joints, especially the hands, wrists, elbows, shoulders, neck, back, and knees. Carpal tunnel syndrome is an example of a CTD. CTDs most frequently occur as a result of strain from performing the same task on a continuous or repetitive basis. This strain can slowly build over time, until the worker experiences pain and difficulty using the injured part of the body. Higher risks of developing CTDs are encountered when the work task requires repetitive motions, excessive force, or awkward postures. The ergonomics engineer addresses these risk factors by analyzing the task thoroughly and minimizing the repetitive motion, excessive force, and awkward posture.

INDUSTRIAL HYGIENE

Industrial hygiene is the science of anticipating, recognizing, evaluating, and controlling workplace conditions that may cause worker illness or injury. Important aspects of industrial hygiene included under these principles are: (1) identification of toxic chemicals; (2) evaluation of the importance of the physical state to absorption by the lungs; evaluation of the importance of airborne particle size to absorption by the lungs and the physical state of individuals; (3) evaluation of the importance of skin absorption and ingestion to exposure and absorption; (4) identification of chemicals to be collected and analyzed; (5) determination of methods for collection of air samples; (6) identification of analytic methods to be used or collaboration with a chemist to develop methods to be used; (7) evaluation of results of measurements; (8) identification of physical stressors, including noise, heat stress, ionizing radiation, nonionizing radiation, ergonomics, and illumination; and (9) development of control measures. In addition to examining the environment, interpreting collected data, and implementing control measures, the industrial hygienist has responsibilities related to the creation of regulatory standards for the work environment, preparing programs to comply with regulations, and collaborating in epidemiologic studies to document exposures and potential exposures to help determine occupational-related illness.

Hazard Recognition

An industrial hygienist must be effective in recognizing occupational hazards. Potential hazards can include air contaminants, and chemical, microbiological, physical, and ergonomic hazards.

Air Contaminants. Air contaminants are commonly classified as either particulate or gas and vapor contaminants. Common particulate contaminants include dusts, fumes, mists, aerosols, and fibers. **Dusts** are solid particles generated by handling, crushing, or grinding. Any process that produces dust fine enough to remain in the air long enough to be inhaled or ingested should be regarded as hazardous until proven otherwise.

Fumes are formed when material from a volatilized solid condenses in cool air. In most cases, the solid particles resulting from the condensation react with air to form an oxide.

Mist is a liquid suspended in air. Mists are generated by liquids condensing from a vapor back to a liquid or by a liquid being dispersed by splashing or atomizing. **Aerosols** are also a form of a mist characterized by highly respirable, minute liquid particles.

Fibers are solid particles whose length is several times greater than their diameter, such as asbestos.

Gases are formless fluids that expand to occupy the space or enclosure in which they are confined. They are atomic, diatomic, or molecular in nature as opposed to droplets or particles which are made up of millions of atoms or molecules.

Through evaporation, liquids change into vapors and mix with the surrounding atmosphere. **Vapors** are the volatile form of substances that are normally in a solid or liquid state at room temperature and pressure. Vapors are gases in that the true vapors are atomic or molecular in nature.

Chemical Hazards. Airborne chemical hazards exist as concentrations of mists, vapors, gases, fumes, or solids. Some are toxic through inhalation, some can irritate the skin on contact, some can be toxic by absorption through the skin, or through ingestion, and some are corrosive to living tissue. The degree of risk from exposure to any given substance depends on the nature and potency of the toxic effects and the magnitude and duration of exposure.

Biological Hazards. These include bacteria, viruses, fungi, and other living organisms that can cause acute and chronic infections by entering the body either directly or through breaks in the skin.

Physical Hazards. These include excessive levels of ionizing and nonionizing electromagnetic radiation, noise, vibration, illumination, temperature, and force.

Hazard Evaluation

An industrial hygiene hazard evaluation determines the sources of potential problems. The industrial hygienist researches, inspects, and analyzes how the particular hazard affects worker health. Assessment of such exposures relies on qualitative, semi-quantitative, or quantitative approaches. In many situations air sampling will determine whether a hazardous condition exists. An appropriate sampling strategy must be used to ensure the validity of collected samples, determining worst-case (for compliance) or usual (average) exposures. Air sampling can be conducted to determine **time-weighted average** (TWA) exposures, which would cover an entire work shift; or **short-term exposures**, which would determine the magnitude of exposures to materials that are acutely hazardous. Samples may be collected for a single substance or a multicomponent mixture. Hazard evaluation also characterizes the workplace with respect to potential skin absorption or ingestion hazards. Analysis of bulk material samples and surface wipe samples could determine whether hazardous conditions exist. Physical agent characterization may require direct-reading sampling methods. After collection and analysis, the industrial hygienist must interpret the results and determine appropriate control strategies.

Hazard Control

The principles for controlling the occupational environment are substitution, isolation, and ventilation. Not all of these principles may be applied to all types of hazards, but all hazards can be controlled by using one of these principles. Engineering controls, work practices controls, administrative controls and personal protective equipment are used to apply these principles. Engineering controls are usually the most effective because they generally minimize the need for human intervention. Examples of engineering controls are automation and ventilation. Local exhaust ventilation is more effective for controlling point-source contaminants than is general ventilation. A building HVAC system is an example of a general ventilation system.

INDOOR AIR QUALITY

Indoor air contaminant materials are particulates, gases, and vapors that may be generated due to the nature of the indoor space, by occupants and their activities in a space, or brought in from the outdoors. Airborne materials may (1) occur from emissions and/or shedding of building materials and systems, (2) originate in outside air, and/or (3) be from building operating and maintenance programs and procedures. These materials include bioaerosols, particles (synthetic vitreous fibers, combustion nuclei, and others), volatile organic compounds, and inorganic and organic gases.

BIOAEROSOLS

Aerobiology is the study of airborne microorganisms and the effects of these aerosols on other living organisms (people, animals, vegetation, etc.). Bioaerosols are airborne microbiological particulate matter derived from viruses, bacteria, protozoa, algae, mites, pollen, and their cellular or cell mass components. Bioaerosols are present in both indoor and outdoor environments.

Much attention has been given to fungi, which include yeasts, molds, and mildews, as well as large mushrooms, puffballs, and bracket fungi. All fungi depend on external sources of organic material for both energy requirements and carbon skeletons. Viruses, many bacteria, algae, and protozoa are more difficult to culture, and air sampling methodology for these agents is less well known and defined (ASTM 1990).

Sources

Microorganisms break down complex molecules found in dead organic materials to simple substances such as carbon dioxide, water, and nitrates. These components are then used by photosynthetic organisms such as plants and algae. Thus the presence of bacteria and fungi in soil, water, and atmospheric habitats is a normal occurrence. Contagious diseases occur when living organisms overcome the defense of the host and establish an infection in the host that may in turn infect another human. Infected humans are the primary sources for contagious disease, and the primary disseminators as well. Virulent agents can also be released from human skin when disease produces skin lesions, or dispersed from respiratory tract infection during coughing, sneezing, or talking. Other means for release directly from infected humans include sprays of saliva and other respiratory secretions during dental and respiratory therapy procedures. Blood sprays that occur during dental and surgical procedures are of potential concern for aerosol transmission of blood-borne diseases, including HIV and hepatitis viruses. Large droplets can transmit infectious particles to those close to the disseminator, while smaller particles can remain airborne for short or very long distances (Moser et el. 1979, Fliermans 1978).

Floors in hospitals can be reservoirs for organisms that are subsequently resuspended into the air. While carpeting appears to trap microorganisms firmly, conditions within the carpet may promote their survival and dissemination. Carpet cleaning is rarely effective and may even promote resuspension (Cox 1987). Viruses may persist up to 8 weeks on nonporous surfaces (Mbithi et al. 1991).

Water is a well known source of infective agents even by aerosolization. Baylor et al. (1977) demonstrated the sequestering of small particles by foam and their subsequent dispersal through a bubble burst phenomenon. Such dispersal may take place in surf, river sprays, or man-made sources such as whirlpools. Spores of *Cladosporium*, a fungus commonly found on dead vegetation, are almost always found in outdoor air samples. Additionally, *Cladosporium* spores are found in indoor air depending on the amount of outdoor air that infiltrates into interior spaces or is brought in by the HVAC system. Bacteria and viruses that are shed from humans are also commonly present in indoor air.

Both the physical and biological properties of the bioaerosols need to be understood. For a microorganism to cause a building-related illness, it must be transported in sufficient dose to the breathing zone of a susceptible occupant. Airborne infectious particles behave physically in the same way as any other aerosol-containing particles with similar size, density, electrostatic charge, etc. The major difference with bioaerosols is that they must remain viable to cause infection; although nonviable particles may promote an immunological response. An organism that does not remain virulent in the airborne state cannot cause infection, regardless of how many units of organisms are deposited in the human respiratory tract. The greater the distance the microbial aerosols travel the less virulent they become. Virulence depends on such factors as relative humidity, temperature, oxygen, pollutants, ozone, ultraviolet light (Burge 1995). The effect of any one factor on survival and virulence can be different for different organisms.

Although microorganisms are normally present in indoor environments, the presence of abundant moisture and nutrients in interior niches amplifies the growth of some microbial agents. Thus certain types of humidifiers, water spray systems, and wet porous surfaces can be reservoirs and sites for growth of fungi, bacteria, protozoa, algae, or even nematodes (Strindehag, et al. 1988, Arnow et al. 1978, Morey et al. 1986, Morey and Jenkins 1989). Excessive air moisture (Burge 1995) and floods (Hodgson et al. 1985) may cause the proliferation of microorganisms indoors. The turbulence associated with the start-up of air-handling unit plenums may also elevate concentrations of bacteria and fungi in occupied spaces (Yoshizawa et al. 1987).

Health Effects

Exposure to airborne fungal spores, hyphal fragments, or metabolites can cause a variety of respiratory diseases. These range from allergic diseases including allergic rhinitis, asthma, and hypersensitivity pneumonitis to infectious diseases such as histoplamosis, balstomycosis and aspergillosis. In addition, acute toxicosis and cancer have been ascribed to respiratory exposure to mycotoxins (Levetin 1995). A large body of literature supports an association between moisture indicators in the home and symptoms of coughing and wheezing (Spengler et al. 1992).

The presence of microorganisms in indoor environments may cause infective and/or allergic building-related illnesses (Morey and Feeley 1988, Burge 1989). Some microorganisms under certain conditions may produce volatile chemicals (Hyppel 1984) that are malodorous or irritative (Holmber 1987), thus contributing to the development of what is called sick building syndrome.

The diseases produced by the *Legionella* genus of bacteria are collectively called Legionelloses. Presently more than 34 species of the *Legionella* family have been identified, of which over 20 have been isolated from both environmental and clinical sources. The diseases produced by *Legionella pneumophila* include the pneumonia form, Legionnaires' Disease, and the flu-like form, Pontiac fever. *L. pneumophila* serogroup 1 is the most frequently isolated from nature and most frequently associated with disease. Because *Legionella* requires special nutrients for growth and does not produce resistant spores, this bacterium is difficult to recover from air.

The survival of *Legionella* is enhanced by a variety of parameters including but not limited to warm temperatures, particular algal and protozoan associations and symbiotic relationships with certain aquatic plants (Fliermans 1985). Evidence has indicated that amoebae and other protozoa act as natural hosts and amplifiers for Legionella in the environment (Barbaree et al. 1986). It has also been suggested that the host relationship affects the virulence of *Legionella spp*.

Cooling towers, evaporative condensers, and domestic water service systems all provide water and nutrients for amplification of microorganisms such as *Legionella*. Growth of microbial populations to excessive concentrations is generally associated with inadequate preventative maintenance of these systems, at least in cooling towers. A body of literature has identified characteristics of indoor plumbing and heating systems associated with ore frequent isolation of Legionella species, including blind ends, scale, upright electric water heaters, and lower water temperatures. ASHRAE's Legionellosis Position Paper has further information on the topic (ASHRAE 1989).

Outbreaks of infective illness in the indoor air may be caused by other types of microorganisms, such as viruses. For example, most of the passengers in an airline cabin developed influenza following exposure to one acutely ill person (Moser et al. 1979). In this case, the plane had been parked on a runway for several hours with the ventilation system turned off.

Histoplasmosis, an infective illness caused by the fungus, *Histoplasma capsulatum*, has occurred (rarely) as a building-related illness among individuals involved in the removal of bat or bird droppings in abandoned buildings (Bartlett et al. 1982) and among chicken coop cleaners. Presumably asexual spores (conidia) from this fungus were inhaled by workers who removed the droppings without adequate respiratory protection.

The genus *Aspergillus* is widely distributed and is common in the soil and on decaying vegetation, dust and other organic debris (Levitin 1995). The small spores are buoyant and remain airborne for long periods (Streifel et al. 1989). Most opportunistic infections are caused by *Aspergillus fumigatus*. The literature on aspergillosis is extensive, and in many cases the environmental source of the infection has been located.

Microorganisms may cause building-related illness in indoor environments by affecting the immune system. Thus, allergic respiratory illness may develop due to inhalation of particulates containing microorganisms or their components, such as spores, enzymes, and cell wall fragments. Numerous cases of allergic respiratory illness (humidifier fever, hypersensitivity pneumonitis) report affected people manifesting acute symptoms such as malaise, fever, chills, shortness of breath, and coughing (Edwards 1980, Morey 1988). In buildings, these illnesses may occur as a response to microbiological contaminants originating from HVAC system components such as humidifiers and water spray systems, or other mechanical components that have been damaged by chronic water exposure (Hodgson et al. 1985, 1987). Affected individuals usually experience relief only after having left the building for an extended period in contrast to occupants with sick building syndrome, where relief is relatively rapid.

Wallingford and Carpenter (1986) reported that microbiological contamination was important in about 5% of the indoor air quality evaluations carried out by NIOSH. This level is likely an underestimate because neither microorganisms nor the possible importance of microbial volatiles, endotoxins, mycotoxins, nonviable microbial particulates, and viruses were routinely investigated. Crandall and Sieber (1996) showed that 47 of 104 problem buildings evaluated had water damage in occupied building areas. Other studies suggest that microorganisms in indoor air are important (Burge et al. 1987, Brundage et al. 1988, Burge 1995).

Sampling

Sampling for biological agents can include visual observation, collection of bulk or surface samples, or air sampling. The principles of sampling and analysis for microorganisms are reviewed by Chatigny (1983). ACGIH (1989) has developed assessment guidelines for the collection of microbiological particulates.

Preassessment. Sampling for microorganisms should be undertaken when medical evidence indicates the occurrence of diseases such as humidifier fever, hypersensitivity pneumonitis, allergic asthma, and allergic rhinitis. A walk-through examination of the indoor environment for visual detection of possible microbial reservoirs and amplification sites should be performed before sampling. If a reservoir or amplifier is visually identified, it is useful to obtain bulk or source samples from them. Also, removal of clearly identified reservoirs and amplifiers are preferable to complicated and costly air sampling procedures.

Air Sampling. The same principles that affect the collection of an inert particulate aerosol also govern air sampling for microorganisms. Air sampling is not likely to yield useful data and information unless the sample collected is representative of exposure. The most representative samples are those that are collected in breathing zones collected over the range of aerosol concentrations. Presently, no personal sampling method has been proposed that is sensitive enough for any bioaerosol (Burge 1995). Thus, ambient sampling designs that obtain reasonable estimates of exposures of given populations over representative periods are necessary.

The viability and/or antigenicity of the microbial particulates must be protected during sampling. In general, culture plate impactors, including multiple-and single-stage devices as well as slit-to-agar samplers, are most useful in office environments where low concentrations of bacteria and fungi are expected. Because not all microorganisms will grow on the same media, liquid impingement subculturing may be more suitable. Filter cassette samplers are useful for microorganisms or components of microorganisms (i.e., endotoxins), although binding to glass and plastic has been reported (Milton et al. 1990). Filter cassettes can also be used for spore counts. Area sampling is often used. Some investigators attempt to replicate exposure conditions through disturbance of the environment (semi- aggressive sampling) such as occurs through walking on carpets, slamming doors, and opening books or file cabinets.

Data Interpretation. Rank order assessment is a method used to interpret air sampling data for microorganisms (ACGIH 1989). Individual organisms are listed in descending order of abundance for a complainant indoor site and for one or more control locations. The predominance of one or more microbes in the complainant site, but not in the control sites or outdoors, suggests the presence of an amplifier for that organism. In the example in Table 4, *Tritirachium* and *Aspergillus* were the predominant fungi represented in complainant locations in an office building, where *Claosporium* and *Fusarium* dominated outdoor collections. In this case, *Tritirachium*

Table 4 Example Case of Airborne Fungi in Building and in Outdoor Air

Location	cfu/m³	Rank Order Taxa
Outdoors	210	Cladosporium > Fusarium > Epicoccum > Aspergillus
Complaint Office #1	2500	Tritirachium > Aspergillus > Cladosporium
Complaint Office #2	3000	Tritirachium > Aspergillus > Cladosporium

cfu/m^3 = Colony-forming units per cubic metre of air.
Culture media was malt extract agar (ACGIH 1989).

and *Aspergillus* were being amplified in the building. In addition, to comparing individual organisms, indoor-outdoor ratios of overall quantities of culturable microorganisms are useful.

Guidelines

At present, numerical guidelines for bioaerosol exposure in indoor environments are not available for the following reasons (Morey 1990):

- Incomplete data on concentrations and types of microbial particulate indoors, especially as affected by geographical, seasonal, and type-of-building parameters
- Absence of data relating bioaerosol exposure to building-related illness
- Enormous variability in kinds of microbial particulate including viable cells, dead spores, toxins, antigens, and viruses
- Large variation in human susceptibility to microbial particulate, making estimates of health risk difficult

However, even in the absence of numerical guidelines, bioaerosol sampling data can be interpreted based on such factors as

- Rank order assessment of the kinds (genera species) of microbial agents present in complainant and control locations (ACGIH 1989)
- Medical or laboratory evidence that a building-related illness is caused by a microorganism (ACGIH 1989)
- Indoor/outdoor concentration ratios for various microbial agents (Morey and Jenkins 1989, AGCIH 1989)

For a microorganism to cause a building-related illness, it must be transported in sufficient dose to the breathing zone of a susceptible occupant. Thus, the concepts of reservoir, amplifier, and disseminator need to be considered in interpreting data. Reservoirs allow microorganisms to survive, amplifiers allow microorganisms to proliferate, and disseminators effectively distribute bioaerosols. Some factors and systems may be all or only one of these. A cooling tower is all three for *Legionella*; that is, a cooling tower can harbor microorganisms in scale, allow them to proliferate, and generate an aerosol.

PARTICULATES

Dusts

Exposures and Exposure Sources. Particulates are defined as mixtures or dispersions of solid or liquid particles. Typical examples of particulates include dust, smoke, fumes, and mists. Dusts range in size from 0.1 to 25 µm, smoke particulate is typically around 0.25 µm, and fumes are usually < 0.1 µm in diameter (Zenz 1988).

Particulates found in the work environment are generated as a result of work related activities, i.e., adding batch ingredients for a manufacturing process, applying asphalt in a roofing operation, or drilling an ore deposit in preparation for blasting. Recognizing sources of particulate generation is essential for the engineer to appropriately address exposure concerns. In the indoor environment, particulate sources may include cleaning, dirt accumulation in carpets and on other fleecing surfaces, construction and renovation debris, paper dust, and deteriorated insulation.

Health Effects of Exposure. The health effects of airborne particulates depend on several factors that include particle dimension, durability, and dose. A particulate must first be respirable to be potentially hazardous. Respirable particulates vary in size from < 1 to 10 µm (Alpaugh and Hogan 1988) depending on the source of the particulate. Durability or how long the particle can live in the biological system before it dissolves can determine relative toxicity. Lastly, the dose or the amount of exposure encountered by the worker must be considered. In some instances very small exposures can cause adverse health effects (hazardous exposures) and in others seemingly large exposures do not cause any adverse health effects (nuisance exposures).

Mechanisms of Health Effects. The most common form of disease associated with the inhalation of particulates is pneumoconiosis, a fibrous hardening of the lungs caused by the irritation created from the inhalation of dust. The most commonly known pneumoconioses are asbestosis, silicosis, and coal worker's pneumoconiosis.

Asbestosis results from the inhalation of asbestos fibers (chrysotile, crocidolite, amosite, actinolite, anthophyllite, and tremolite) found in the work environment. The onset of symptomatic asbestosis is uncommon under exposures encountered in the last 45 years before at least 20 to 30 years of exposure (Selikoff et al. 1965, Smith 1955). The asbestos fibers cause fibrosis (scarring) of the lung tissue, which clinically manifests itself as dyspnea (shortness of breath) and a non-productive, irritating cough. Asbestos fiber is both dimensionally respirable as well as durable in the respiratory system.

Although asbestos is encountered in insulation in many buildings, it generally does not represent a respiratory hazard except to individuals who actively disturb it in the course of maintenance and construction. School custodians, therefore, are recognized to be at risk for asbestos-related changes, although school air is generally thought to pose no hazard to occupants. Anderson et al. (1991) and Lilienfeld (1991) do raise questions about risk to teachers.

Silicosis is probably the most common of all occupational diseases of the lung. The hazard is created by inhalation of silica dust. The worker with silicosis usually is asymtomatic, and even the early stages of massive fibrosis are not associated with signs and symptoms (Leathart 1972). It is not considered a problem in indoor environments.

Coal worker's pneumoconiosis (CWP) results from the inhalation of dust generated in coal mining operations. The dust is composed of a combination of carbon and varying percentages of silica (usually <10%) (Alpaugh and Hogan 1988). Due to the confined underground work environment, exposures have the potential to be very high at times thus creating very high doses. Data meanwhile show that workers may develop CWP at exposure below the current dust standard of 1 mg/m^3.

Exposure Control Strategies. Particulate or dust control strategies include enclosure, local exhaust, general ventilation, and the use of personal protective devices such as respirators.

The most effective means of controlling exposures to a particulate is to totally eliminate it from the work environment. The best dust-control method is a total enclosure of the dust-producing process. A negative pressure is maintained inside the entire enclosure by exhaust ventilation (Alpaugh and Hogan 1988). This control strategy is typically found in manufacturing operations.

Local exhaust ventilation as an exposure control strategy is most frequently used where particulate is generated either at large volumes or with high velocities, i.e., lathe and grinding operations. In this situation, high velocity air movement captures the particulate and removes it from the work environment. Local exhaust ventilation is one of the more costly control strategies used due to the need for engineering design and maintenance.

Table 5 OSHA Permissible Exposure Limits (PEL) for Particulates (29 CFR 1910.1000, 29 CFR 1926.1101)

Substance	CAS #	PEL
Cadmium	7440-43-9	0.05 mg/m^3
Manganese fume	7439-96-5	1.0 mg/m^3
Plaster of Paris	Nuisance	10.0 mg/m^3
Emery	Nuisance	10.0 mg/m^3
Grain dust	Nuisance	10.0 mg/m^3
Crystalline silica (as quartz)	14808-60-7	0.1 mg/m^3
Asbestos	1332-21-4	0.1 fibers/ cm^3
Total dust	Nuisance	15.0 mg/m^3
Respirable dust	Nuisance	5.0 mg/m^3

General ventilation control of the work environment is defined as a dilution approach to reducing exposures. This type of ventilation is used when particulate sources are numerous and widely distributed over a large area. In this control strategy the work environment is exhausted outside the plant and resupplied with fresh air, thus diluting the work environment. Unfortunately this strategy is the least effective means of control and very costly because conditioned (warm or cold) air is exhausted and non-conditioned air is introduced.

The last strategy that used to control exposures is the use of personal protective equipment—a respirator. Respirators are appropriate as a primary control during intermittent maintenance or cleaning activities when fixed engineering (local or general ventilation) controls may not be feasible. Respirators can also be used as a supplement to good engineering and work practice controls to increase employee protection and comfort (Alpaugh and Hogan 1988).

Few dusts pose respiratory hazards in indoor environments. In general, source control is the preferred method. If a dust problem is identified, characterization of the nature of the dust will allow the development of an appropriate intervention strategy.

Exposure Standards and Criteria. In the United States, OSHA has established Permissible Exposure Limits (PELs), which are published in the Code of Federal Regulations (CFR 1989a,b) under the authority of the Department of Labor. Table 5 lists PELs for several particulates commonly encountered in the workplace.

Synthetic Vitreous Fibers (SVF)

Exposures and Exposure Sources. A fiber can be defined as a slender, elongated structure with substantially parallel sides. These parameters distinguish this form of particulate from a dust, which is more spherical shaped. Synthetic vitreous fibers comprise a large number of important manufactured products, which are known as textile fibers; insulation and ceiling tile wool, including glass fibers, slag, and rock wool fibers; refractory ceramic fibers; and certain specialty glass fibers.

Exposures to SVF primarily occur during manufacture, fabrication and installation, and demolition. Simultaneous exposures to other dusts (asbestos during manufacture, demolition products and bioaerosols during demolition) may be important as well. Facilities generally manufacture only one form. Generally, only spun glass and refractory ceramic fibers are in the respirable range. Manufacturing operations are most easily designed to assure a clean work environment, while product application operations are more difficult to control. Good engineering practice maintains work environment exposures at levels < 0.1 fibers/cm^3 (f/cc) of air. Some data exist on exposure likely to occur in buildings show that background levels are almost uniformly below 0.0001 f/cc.

Health Effects of Exposure. The possible effects of SVFs on health include the following:

Cancer. Respirable SVF are considered to have the potential to cause carcinogenic and non-carcinogenic health effects. Although implantation studies have suggested the potential for carcinogenesis, this route of exposure is generally not pertinent for humans. Therefore, although SVF are often classified as potential human carcinogens by regulatory and professional agencies and organizations, reviews of epidemiology studies generally fail to find convincing evidence that they are associated with excess rates of human cancer. Some mortality studies have identified mild excesses of respiratory cancer. These have been attributed to concurrent asbestos exposure and to smoking. Only refractory ceramic fibers are currently considered likely to represent true human carcinogens, although other very hard fibers are likely to have similar effects. Inadequate data are currently available to recommend regulations.

Non-malignant respiratory disease. Cross-sectional surveys have suggested that few measurable adverse health effects are attributable to SVF alone. The strongest evidence suggests that SVF may exacerbate smoking-induced obstructive lung disease; some authors consider fiberglass, no different than any other dust, to cause excess rates of chest symptoms.

Dermatitis. SVF may cause an irritant contact dermatitis through embedding in the skin or conjunctivae with local inflammation. Resin binders sometimes used to tie fibers together have, on rare occasions, been associated with allergic contact dermatitis.

Exposure Control Strategies. As with other particulates, SVF control strategies include enclosure, local exhaust, general ventilation and the use of personal protective devices such as respirators. In indoor environments, SVF are primarily encountered when duct linings deteriorate. They may be identified in surface wipe samples. Appropriate intervention strategies focus on source control.

Exposure Standards and Criteria. At present, SVF is regulated by the OSHA as a "nuisance dust" with an 8-hour time-weighted average of 15 mg/m^3 for total dust and 5 mg/m^3 for respirable dust.

Combustion Nuclei

Exposures and Exposure Sources. Combustion nuclei can be defined as the gaseous and particulate products of the combustion process. Combustion products from a material include water vapor, carbon dioxide, heat, and air contaminants known as combustion nuclei. In many situations the combustion nuclei can be hazardous. These air contaminants can include carbon monoxide, nitrogen oxides, sulfur oxides, and polycyclic aromatic hydrocarbons (PAHs).

Polycyclic aromatic compounds (PAC) are the nitrogen-, sulfur-, and oxygen-heterocyclic analogues of PAH and other related PAH derivatives. Depending on their relative molecular mass and vapor pressure, PACs are distributed between vapor and particulate phases. Because PACs are generally produced by combustion, the particulate compounds are found exclusively in the respirable fraction (i.e., < 2.5 μm aerodynamic diameter).

Typical sources of combustion nuclei are tobacco smoke (cigarettes, pipes, and cigars), fossil-fuel-based heating devices such as unvented space heaters and gas ranges and flue gas from improperly vented gas- or oil-fired furnaces and wood-burning fireplaces or stoves. Infiltration of outdoor combustion contaminants can also be a significant source of such contaminants in indoor air.

The combustion nuclei of environmental tobacco smoke (ETS) consists of exhaled mainstream smoke from the smoker and sidestream smoke that is emitted from the smoldering tobacco. ETS consists of between 70 and 90% sidestream smoke and consists of the same chemical compounds of mainstream smoke. More than 4700 compounds have been identified in laboratory-based studies, including known human toxic and carcinogenic compounds such as carbon monoxide, ammonia, formaldehyde, nicotine, tobacco-specific nitrosamines, benzo(a)pyrene, benzene, cadmium, nickel, and aromatic amines. Many of these toxic constituents are more concentrated in sidestream than in mainstream smoke (Glantz and Parmley

1991). In studies conducted in residences and office buildings with tobacco smoking, ETS was a substantial source of many gas and particulate PACs (Offermann et al. 1991).

Health Effects of Exposure. The health effects of exposure to combustion nuclei depend on many factors including concentration, toxicity, and individual susceptibility or sensitivity to the particular substance. Polycyclic aromatic compounds generated by combustion processes include many PAH, nitro-PAH and aza-arenes that have been shown to be carcinogenic in animals (NAS 1983). Other PACs are biologically active as tumor promoters and/or co-carcinogens. Mumford et al. (1987) reported high exposures to PAH and aza-arenes for a population in China with very high lung cancer rates. The estimated cancer risk for exposure to benzo(a)pyrene, a classical indicator compound for PAH, is 1 ppm, for a lifetime exposure to 0.3 ng/m^3 (Offermann et al. 1991).

ETS has been shown to be causally associated with lung cancer in adults (NRC 1986, DHHS 1986), and respiratory infections, asthma exacerbations, middle ear effusion (NRC 1986, DHHS 1986), and low birth weight in children (Martin and Bracken 1986). The U.S. Environmental Protection Agency classifies ETS as a known human carcinogen (EPA 1992a), while the National Institute for Occupational Safety and Health has determined that ETS is potentially carcinogenic to occupationally exposed workers (NIOSH 1991). Health effects can also include headache, nausea, drowsiness, coma, death by asphyxiation, and in the case of mainstream tobacco smoke, chronic obstructive pulmonary disease, and cancers of the lung and other organs. Controversy exists on whether ETS is associated with the development of cardiovascular disease. ETS is also a cause of sensory irritation and annoyance (odors and eye irritation).

Exposure Control Strategies. Exposure control strategies for combustion nuclei are in many ways similar to those applied for general ventilation. For combustion nuclei derived from heating spaces, air contamination can be avoided by proper installation and ventilation of equipment to ensure that these contaminants cannot enter the work or personal environment. Proper equipment maintenance is also essential to minimize exposures to combustion nuclei. Changing makeup air availability, through the addition of enclosures, may be equally important.

Control of ETS is somewhat different in that it has been done primarily through regulatory mandates controlling the practice of tobacco smoking. Most states in the United States have passed laws to control tobacco smoking in public places such as restaurants and workplaces, and airlines have prohibited tobacco smoking on flights lasting six hours or less. Where tobacco smoking is allowed, appropriate local and general ventilation can be used for control. OSHA has proposed that tobacco smoke in indoor environments be controlled through the use of separately ventilated and exhausted smoking lounges (OSHA 1993). These lounges are kept under negative pressure relative to all adjacent and communicating indoor spaces.

Exposure Standards and Criteria. OSHA has established exposure limits for several of the carcinogens categorized as combustion nuclei (benzo(a)pyrene, cadmium, nickel, benzene, N-Nitrosodimethylamine). These limits are established for industrial work environments and are not directly applicable to indoor air situations. However, underlying atherosclerotic heart disease in individuals may be exacerbated by CO exposures below the current PEL of 35 ppm.

VOLATILE ORGANIC COMPOUNDS

Volatile organic compounds (VOCs) are air pollutants found in all nonindustrial indoor environments. After ventilation, VOCs are probably the first concern when diagnosing an IAQ problem (Morey and Singh 1991). VOCs have vapor pressures greater than about 10^{-3} to 10^{-4} mm Hg (torr). This includes 4- to 16-carbon alkanes, chlorinated hydrocarbons, alcohols, aldehydes, ketones, esters, terpenes, ethers, aromatic hydrocarbons (such as benzene and toluene), and heterocyclic hydrocarbons. The entire range of organic indoor pollutants has been categorized as indicated in Table 6 (WHO 1989). No sharp limits exist between the categories, which were defined by boiling-point ranges.

Table 6 Classification of Indoor Organic Pollutants
(WHO 1989)

Description	Abbreviation	Boiling Point Range, °F
Very volatile (gaseous) organic compounds	VVOC	<32 to 120–212
Volatile organic compounds	VOC	120–212 to 460–500
Semivolatile organics (pesticides, polynuclear aromatic compounds, plasticizers)	SVOC	460–500 to 720–750

Note: Polar compounds and higher molecular weight VOCs with appear at the higher end of each boiling-point range.

Exposures

Wallace et al. (1991) showed that individual VOC concentrations in homes and buildings are 2 to 5 times those of outdoors, and personal total VOC (TVOC) exposures were estimated to be 2 to 3 times greater than general indoor air concentrations. Normal daily activities carried out by individuals are the cause of these higher personal exposures. Personal activities frequently bring individuals close to air contaminant sources. The degree of exposure also depends on how air flows around the body due to convective forces, air turbulence, and obstructions nearby (Rodes et al. 1991).

Normally 50 to 300 volatile organic compounds are found in air samples from most nonindustrial indoor environments, using standard sorbent sampling techniques. Some researchers have used a quantity, Total Volatile Organic Compounds (TVOC), to sum all of the individuals VOCs (Brown 1994, Molhave 1992). Recent data demonstrate the limitations of that approach both theoretically and practically (Hodgson 1995). Individual compounds seldom exceed 50 μg/m^3 of air. An upper extreme average concentration of TVOCs in normally occupied houses is approximately 20 mg/m^3. A Large Buildings Study by the U.S. EPA developed the VOC sample target list in Table 7.

Table 7 Example Sample Contaminant Target List

benzene	styrene
m-,*p*-xylene	*p*-dichlorobenzene
1,2,4-trimethylbenzene	*n*-undecane
n-octane	*n*-nonane
n-decane	ethylacetate
n-dodecane	dichloromethane
butylacetate	1,1,1-trichloroethane
chloroform	tetrachloroethylene
trichloroethylene	carbon disulfide
trichlorofluoromethane	acetone
dimethyl disulfide	2-butanone
4-methyl-2-pentanone	methyl tertiary butyl ether
limonene	naphthalene
α-,β-pinene	4-phenylcyclohexene
propane	butane
butyl cellosolve	ethanol
isopropanol	phenol
formaldehyde	siloxanes
toluene	

Sources

Berglund et al. (1988) found that the sources of VOCs in nonindustrial indoor environments are confounded by the variable nature of emissions from potential sources. Daily concentrations of VOCs from outdoor air can vary drastically (Ekberg 1994). These variations are due to vehicle traffic density, wind direction, industrial emissions, and photochemical reactions. Emissions from indoor sources can be classified by their presence and rate patterns. For example, emissions are continuous and regular from building materials and furnishings (e.g., carpet and composite-wood furniture); whereas emissions from other sources can be continuous, but irregular (e.g., paints used in renovation work), intermittent and regular (e.g., VOCs in combustion products from gas stoves or cleaning products), or intermittent and irregular (e.g., VOCs from carpet shampoos) (Morey and Singh 1991).

Many "wet" emission sources that have very high emission rates initially after application (paints and adhesives) can be present in newly constructed buildings. Other new materials (carpets, wall coverings, and furnishings) also emit chemicals at higher rates until aged. Decay of these elevated VOC concentrations to what can be considered normal constant-source levels can take from weeks to months, depending on emission rates, surface areas of materials, and ventilation protocols. Renovation activities can cause similar increases of somewhat lower magnitude. The total VOC concentration in new office buildings at the time of initial occupancy can be 50 to 100 times that present in outdoor air (Sheldon et al. 1988a,b). In new office buildings with adequate outdoor air ventilation, these ratios often fall to less than 5:1, after 4 or 5 months of aging. In older buildings with continuous, regular, and irregular emission sources, indoor/outdoor ratios of total VOCs may vary from nearly 1 when maximum amounts of outdoor air are being used in HVAC systems to greater than 10 during winter and summer months when minimum amounts of outdoor air are being used (Morey and Jenkins 1989, Morey and Singh 1991).

While direct VOC emissions from primary sources are predominant, some materials act as sinks for emissions and then become secondary sources as they re-emit adsorbed chemicals (Berglund et al. 1988). Adsorption may lower the peak concentrations achieved, but the subsequent desorption prolongs the presence of indoor air pollutants. Sink materials include carpet, fabric partitions, and other fleecy materials, as well as ceiling tiles and wallboard. The type of material and compound affects the rate of adsorption and desorption (Colombo et.al. 1991). Indoor air quality models using empirically derived adsorption and desorption rates have been developed to predict the behavior of sinks. Experiments conducted in an IAQ test house confirmed the importance of sinks when trying to control the level of indoor VOCs (Tichenor 1991). Longer periods of increased ventilation will lessen sink and reemission effects.

A VOC enrichment factor (VEF) has been described to quantify VOC concentrations due to building sources beyond that expected as bioeffluents (Batterman and Peng 1995). A VEF >1 describes the abundance, while a VEF <1 the depletion of VOCs compared with VOCs expected as bioeffluents. Bioeffluents alone should produce a VEF of 1. The VEF ranged from 0.6 to 17.1 for the 20 office building studies reviewed.

While ventilation systems can transport outdoor pollutants to the indoor environment, they also may be a source of VOCs (Molhave and Thorsen 1990). The interior of the HVAC system can have large areas of porous material used as acoustical liner that can adsorb odorous compounds. This material can also hold nutrients and, with moisture, can become a reservoir for microorganisms. Microbial contaminants produce characteristic VOCs associated with their metabolism. Other HVAC components may support microbiological life, such as condensate drain pans, fouled cooling coils, and filter media. Deodorants, sealants, and encapsulants are also sources of VOCs in HVAC systems.

Floor dust, which is different from the dust in the air, has been found to be a sink (adsorption medium) and secondary emission source for VOCs. Floor dust is a mixture of organic and inorganic particles, hair and skin scales, and textile fibers. The fiber portion of floor dust has been shown to contain 169 mg/kg TVOC and the particle portion 148 mg/kg (Gyntelberg et al. 1994). These VOCs were correlated to the prevalence of irritative (sore throat) and cognitive symptoms (concentration problems) among building occupants. One hundred eighty-eight compounds were identified from thermal desorption of office dust at 250°F (Wilkens et al. 1993). Household dust was found to be similar in composition (Wolkoff and Wilkens 1994).

Human activity, including maintenance, cleaning, and cooking, is another potential source, as are human metabolism and smoking. Additional sources include liquid process copy machines, printing machines, glue, spray cans, cosmetics, and so forth (Miksch et al. 1982). Some contaminants from other sources include chloroform in water; tetrachloroethylene and 1,1,1-trichloroethane from cleaning solvents; methylene chloride from paint strippers, fresheners, cleaners, and polishers; -pinene and limonene from floor waxes; and 1-methoxy-2-propanol from spray carpet cleaners. Formaldehyde, a major VOC, has many sources, but pressed wood products appear to be the most significant.

Health Effects

Adverse health effects potentially caused by VOCs in nonindustrial indoor environments fall into three categories: (1) irritant effects, including perception of unpleasant odors, mucous membrane irritation, and exacerbation of asthma; (2) systemic effects, such as fatigue and difficulty concentrating; and (3) toxic, chronic effects, such as carcinogenicity (Girman 1989). Although allergic sensitization is a well-known additional effect, it is generally not of primary importance for the VOCs encountered in indoor spaces.

The chronic adverse health effects due to VOC exposure are of concern because some VOCs commonly found in indoor air are human (benzene) or animal (chloroform, trichloroethylene, carbon tetrachloride, p-dichlorobenzene) carcinogens. Some other VOCs are also genotoxic. Theoretical risk assessment studies suggest that chronic exposure risk due to VOCs in residential indoor air is greater than that associated with exposure to VOCs in the outdoor air or in drinking water (McCann et al. 1987, Tancrede et al. 1987).

A biological model for acute human response to low levels of VOCs indoors is based on three mechanisms: sensory perception of the environment, weak inflammatory reactions, and environmental stress reaction (Molhave 1991). A growing body of literature summarizes measurement techniques for the effects of VOCs on nasal (Koren 1990, Koren et al. 1992, Meggs 1994, Ohm et al. 1992, Molhave et al. 1993) and ocular (Kjaergard et al. 1991, 1992; Franck et al. 1993) mucosa. These observations do, however, indicate that indoor VOCs may cause discomfort due to odors, irritative symptoms in the eyes, nose, and throat, and headaches at levels well below TLVs in a dose-dependent fashion. The syndrome may include other related effects, such as decreased productivity and performance, although such effects have not yet been well characterized. It is not known how different sensory receptions are combined into perceived comfort and the sensation of air quality. This perception is apparently inter-related to stimulation of the olfactory sense in the nasal cavity, to the gustatory sense on the tongue, and the common chemical sense (Molhave 1991, Cain 1989).

Cometto-Muñiz and Cain (1994a,b) addressed the independent contribution of the trigeminal and olfactory nerves to the detection of airborne chemicals. The sense of smell is experienced through receptors in the nose of the olfactory nerve. Nasal pungency, described as common chemical sensations including prickling, irritation, tingling, freshness, stinging, and burning among others, is experienced through the non-specialized receptors of the trigeminal nerve in the face. Odor and pungency thresholds follow different patterns related to chemical concentration. Odor is detected at much lower levels. A linear correlation between pungency thresholds of homologous series—of alcohols, acetates, ketones, and alkylbenzenes, relatively non-reactive agents—suggests that nasal pungency relies on a physicochemical interaction with a susceptible biophase within the cell membrane. It is postulated that through this nonspecific mechanism, low, subthreshold levels of a wide variety of VOCs—as found in many polluted indoor environments—can be additive in their sensory impact to produce noticeable sensory irritation.

Formaldehyde is a very reactive small molecule that requires different analytical techniques than those usually employed in VOC assessment. Primary sources include phenol-fomaldehyde resin-based particle and chipboard products used in indoor spaces. It is frequently encountered in indoor spaces in concentrations between 0.04 ppm, a frequently encountered lower limit of detection, and 0.1 ppm (Liu et al. 1991, Ritchie and Lehnen 1987). Many studies have demonstrated its ability to trigger mucous membrane irritation at levels below the ACGIH TLV, and even at levels below 0.1 ppm.

Standards

No standards for exposure to VOCs relevant to nonindustrial indoor environments are in place. NIOSH, OSHA, and the ACGIH have published regulatory standards or recommended limits for industrial occupational exposures (NIOSH 1992, ACGIH annual). With few exceptions, concentrations observed in nonindustrial indoor environments fall well below (100 to 1000 times lower) these published pollutant-specific occupational standards or recommended exposure limits. However, standards for the industrial workplace are higher than would be appropriate for the general population, which includes the elderly, children, and people who are more sensitive to VOCs than the average industrial worker.

The TVOC indicator, a summation of measured concentrations of air contaminants, originated as a practical way of reporting measured VOCs. It has been used, questionably, as an indicator of the potency of VOCs to cause health effects. Because of the variety of laboratory methods that have been used to measure the TVOC, few reported TVOC measurements are comparable (Hodgson 1995).

Molhave (1991) tentatively concluded that no irritation should be expected as a result of exposure to TVOC below about 0.2 mg/m^3 for normally encountered indoor air. Outdoor air concentrations are about 0.1 mg/m^3 or below. At concentrations higher than about 3 mg/m^3 complaints occurred in all investigated buildings where occupants experienced symptoms. In controlled exposure experiments, odors are significant at 3 mg/m^3. At 5 mg/m^3, objective effects were seen in addition to the subjective irritation. Exposures for 50 minutes to 8 mg/m^3 of synthetic mixtures of 20 VOCs lead to significant irritation of mucous membranes in the eyes, nose, and throat.

Both OSHA and the ACGIH have set 8 hour standards for formaldehyde as a ceiling level. The State of California issued a residential air quality guideline of 0.1 ppm. In the setting of occupant complaints, the target guideline is 0.05 ppm.

Controlling Exposures

Much can be done to reduce building occupants' exposures to emissions to VOCs from building materials and products, and to prevent outdoor VOC contaminants from being brought into buildings. The industrial hygiene hazard control principles of substitution, isolation, and ventilation, apply. Control measures include careful planning, specifications, and selection, modification and treatment of products, as well as special installation procedures and proper ventilation system operation.

Levin (1989, 1991) has written extensively about designing new buildings for good indoor air quality. Reducing VOC emissions by careful selection and installation of building materials and furnishings is a most effective strategy for controlling IAQ. Advances in product formulation and emission testing, are leading to products claimed to be low-polluting, non-toxic, and environmentally safe. Requiring the submission of emission testing data by manufacturers for building products, whether for a new building, a building renovation or remodeling, or for substitution of a consumable product (housekeeping supplies), is becoming accepted practice. Eliminating the sources of VOCs prevents them from becoming a problem in the first place.

Gas-phase air filtration has been applied to control industrial gaseous contaminants for many years. The application of this technology to non-industrial building HVAC is of interest for improving IAQ, whether it is to provide ventilation without the need to use more outdoor air, or to help clean poor quality outdoor air. Activated charcoal and potassium permanganate-impregnated alumina are effective adsorbents that can be used, based upon the contaminant mixture present (Liu and Huza 1995, Muller and England 1995, Van Osdell and Sparks 1995). Portable air cleaners with sorbent sections are only marginally effective (Shaughnessy et. al. 1994). Photocatalytic reactors capable of destroying VOCs are being studied. These reactors use ultraviolet light and a catalytic surface, such as titanium dioxide, to convert organic pollutants to CO_2 and water.

Ventilation has traditionally been considered the primary means for controlling indoor VOC contaminants. Dilution ventilation is an effective way to control normal constant emission sources present in buildings, assuming no unusually strong sources. Compliance with *ASHRAE Standard* 62-1989 should satisfy indoor dilution ventilation requirements. Local exhaust ventilation is effective for controlling known, unavoidable point emissions sources. It is prudent to isolate office machines, such as photocopiers and laser printers, food service equipment, such as microwave ovens and coffee makers, and work areas, such as graphics and photographic labs using dedicated local exhaust systems that vent to the outside and away from outdoor air intakes.

A good ventilation protocol during renovation or remodeling includes using a single-pass (100% outdoor air) system during and at the finish of these activities, continuing until enough time has passed to lower emitted concentrations to near background. This practice minimizes sink effects and secondary emissions.

Prudent practice and administrative control should be used to minimize the generation of VOCs in indoor air during occupied hours whenever possible. Scheduling the use of volatile organic products, housekeeping activities, and pesticide application when occupant density is lowest should be considered. VOC containing supplies should be stored in well ventilated areas other than HVAC mechanical rooms or plenums.

INORGANIC GASES (OXIDES OF SULFUR, NITROGEN, AND OZONE)

Exposure and Exposure Sources

Inhalation exposures to the gaseous oxides of nitrogen, sulfur (NO_x, and SO_2) and ozone (O_3) can and do occur in residential and commercial buildings. These air pollutants are of considerable concern due to the potential for acute and chronic respiratory tract health effects in exposed individuals, particularly individuals with preexisting pulmonary disease. Nitrogen and sulfur oxides are combustion products resulting from both natural and man-made emissions. Ozone is a photochemical oxidant that forms at ground level when hydrocarbons, and oxides of nitrogen react with ultraviolet light to produce photochemical smog. Ozone can also occur from the electrical or coronal dischargers from office equipment including laser printers and photocopiers. Ozone can also be a pollutant generated when ozone-generating devices—often marketed as portable air cleaners—are used in the indoor environment (Esswein and Boeniger 1994).

Oxides of nitrogen result mainly from cooking appliances, pilot lights, and unvented heaters. Sources generating carbon monoxide (CO) often produce nitric oxide (NO) as well. Underground or attached parking garages can also contribute to concentrations of indoor NO_x. An unvented gas cook stove contributes approximately 0.025 ppm of nitrogen dioxide (NO_2) to a home. During cooking 0.2 ppm to 0.4 ppm peak levels may be reached (Samet et al. 1987). Ambient air pollution in urban locations can contribute NO_x to the indoor environment in makeup air. Oxides of nitrogen also are present in mainstream and in sidestream tobacco smoke. Nitric oxide and nitrogen dioxide are of most concern.

Sulfur dioxide (SO_2) can result from the emissions of kerosene space heaters, the combustion of fossil fuels, or burning any material containing sulfur. Sulfur dioxide is a common ambient air pollutant in many urban areas.

Health Effects and Mechanisms

Nitric oxide is an odorless, tasteless, colorless gas. Inhalation of NO causes the formation of methemoglobin, which adversely affects the body by interfering with oxygen transport at the cellular level. NO exposures of 3 ppm have been compared to carbon monoxide exposures of 10 to 15 ppm (Case et al. 1979 in EPA 1991). Acute toxicity is seldom seen from NO_2 produced by unvented indoor combustion because of the insufficient quantities of NO_2 produced. Chronic pulmonary effects from exposure to combinations of low-level combustion pollutants are possible however. (Bascom et al. 1996).

Nitrogen dioxide (NO_2) is a corrosive gas with a pungent odor, the odor threshold of which is reported to be between 0.11 and 0.22 ppm (WHO 1987). NO_2 has a low water solubility and therefore can be inhaled into the deep lung where it causes a delayed inflammatory response. Increased airway resistance has been reported at 1.5 to 2 ppm (Bascom 1996). NO_2 is reported to be a potential carcinogen by way of free radical production (Burgess and Crutchfield 1995). At high concentrations NO_2 causes lung damage directly by its oxidant properties and may cause health effects indirectly by increasing host susceptibility to respiratory infections. Health effects from exposures to ambient outdoor concentrations or in residential situations show inconsistency, especially studies relating to exposures from gas cooking stoves (Samet et al. 1987). Indoor concentrations of NO_2 often exceed ambient concentrations due to the presence of strong indoor sources and a trend toward more energy efficient (tighter) homes.

SO_2 is a colorless gas with a pungent odor detected at about 0.5 ppm (EPA 1991). Because SO_2 is quite soluble in water, it can react with moisture in the upper respiratory tract to produce irritant effects on the upper respiratory mucosa. Concomitant exposure to fine particulate, an individual's depth and rate of breathing and the presence of pre-existing disease can influence the degree of SO_2 toxicity.

Ozone is a pulmonary irritant and causes changes in human pulmonary function at concentrations of approximately 0.12 ppm (Bates 1989). Exposure to ozone at 60 to 80 ppb causes inflammation, bronchoconstriction, and increased airway responsiveness.

Exposure Control Strategies

Three methods of control for inorganic gaseous contaminates should be considered: source control, ventilation control, and removal of gaseous contaminants. Source control involves limiting (or removing) the source of the problem, for example gas cooking stoves should not be used for space heating (often a problem in low-income urban residences). Another example is limiting automobile parking around building makeup air intakes. Source control should always be the primary consideration. But source control is not always feasible when there are many diverse contaminant sources as in new buildings where the building itself, or building furnishings may be the prime contributors to the problem.

Ventilation control involves bringing clean dilution air into the occupied space or direct exhausting of air contaminants at the point of generation. *ASHRAE Standard* 62 provides guidance in applying the Ventilation Rate Method and the Indoor Air Quality Method for ventilation control.

Where neither source control nor ventilation control appear likely to control gaseous air contaminants, removal by air filtration should be investigated. Gas-phase air filtration involves dry scrubbing to remove contaminants by adsorption onto several sorbents including granular activated charcoal (GAC), potassium permanganate impregnated alumina (PIA), and alkaline-impregnated carbon filters. Muller and England (1995), Liu and Huza (1995), VanOsdell (1995), and Coutant (1994) review various filtration procedures.

No one media is effective for the broad range of gaseous contaminants found indoors. Granular activated charcoal is generally an agent of choice for non-polar compounds and is a suitable choice for O_3 and NO_2 but not for SO_x and NO. Permanganate impregnated alumina is more appropriate for SO_x and NO.

Ozone can be best controlled by local exhaust ventilation for demonstrated sources of ozone, such as photocopiers and equipment creating coronal discharges. Routine cleaning of attractor plates in an electrostatic precipitator and ensuring adequate prefilters can reduce ozone generation and limit arcing in this type of particulate removal equipment. The use of ozone generating devices as a means of air cleaning or purification has not been documented as a prudent means of air contaminant control when considering the potential health effects of the use of ozone indoors (Esswein and Boeniger 1994). The Food and Drug Administration (FDA 1990) specifically limits the use of ozone in concentrations greater than 50 ppb in areas intended for continuous occupancy such as residences, offices, schools, and hospitals.

Exposure Standards and Criteria

Currently, no specific United States government standards relatie to nonindustrial occupational exposures to air contaminants. Occupational exposure criteria are health-based, that is, they consider healthy workers in an environment, and not necessarily individuals who may be unusually responsive to the effects of chemical exposures. The EPA National Ambient Air Quality Standards (NAAQS) are also health-based standards designed to protect the general public health from the effects of hazardous airborne pollutants. However, there is a debate as to whether these standards truly represent health-based thresholds. Two of the six criteria, ozone and carbon monoxide, involve toxicologically-based research for the development of the standards. The criteria (Table 8) are not meant to be health-based guidelines for the evaluation of exposures to inorganic gasses in the indoor environment; rather Table 8 is included for comparative use and consideration by investigators of the indoor environment with the understanding that these criteria may not be completely protective to all industrial workers.

Table 8 Inorganic Gas Comparative Criteria

Contaminant	OSHA/NIOSH TWA[a]	EPA NAAQS 1 Std.
Nitric oxide	1 h 2 ppm (5 mg/m^3) 24 h 25 ppm (30 mg/m^3)	None
Nitrogen dioxide	1 h 5 ppm (9 mg/m^3) 24 h 1 ppm (1.8 mg/m^3)	0.053 ppm (100 μg/m^3)
Sulfur dioxide	1 h 5 ppm (13 mg/m^3) 24 h 2 ppm (5 mg/m^3)	0.014 ppm (365 μg/m^3)
Ozone	1 h 0.1 ppm (0.2 mg/m^3) 24 h 0.1 ppm (0.2 mg/m^3)	0.12 ppm (235 μg/m^3)

[a]The values listed are the annual arithmetic mean unless otherwise listed. The first value listed is the 24 hour average and the second value is the maximum 1 hour average. (TWA = time weighted average)

ORGANIC GASES (CARBON MONOXIDE, CARBON DIOXIDE, AND CHLOROFLUOROCARBONS)

Exposure and Exposure Sources

Carbon monoxide (CO) is an odorless, colorless, and tasteless gas produced by the incomplete combustion of hydrocarbons. It is a common ambient air pollutant. Common indoor sources of CO include gas stoves, kerosene lanterns and heaters, mainstream and sidestream tobacco smoke, woodstoves, and unvented or improperly vented combustion sources. Building makeup air intakes located at street level of near parking garages can entrain CO from automobiles and carry it to the indoor environment.

The major predictors of indoor concentrations are indoor fossil fuel sources, such as gas furnaces, hot water heaters, and other combustion appliances; attached garages; and weather inversions (Colume 1994). Levels in such homes only rarely exceed 5 ppm. In one sample of randomly selected homes, 10% failed a backdrafting test (Conibear et al. 1996). Under backdrafting conditions, indoor CO sources may contribute to much higher levels of CO.

Carbon dioxide (CO_2) or carbonic acid gas is produced by human respiration. It is not normally considered to be a toxic air contaminant, but it can be a simple asphyxiant (oxygen displacement). CO_2 is found in the ambient environment at 325 to 350 ppm. Levels in the urban environment may be higher due to emissions from gasoline and, more often, diesel fuel engines. Measurement of CO_2 in occupied spaces has been widely used to evaluate the amount of outdoor air supplied to indoor spaces. A level of 1000 ppm (or 650 ppm above outdoor air) has been suggested as being representative of delivery rates of 20 cfm per person of outside air when CO_2 is measured at equilibrium concentrations and at occupant densities of 10 people per 1000 ft^2 of floor space. Measuring CO_2 level before it has reached steady-state conditions can lead to inaccurate conclusions regarding the amount of outside air used in the building.

Chlorofluorocarbons (CFCs) are halogenated alkane gases that have been used as heat transfer gases in refrigeration applications, blowing agents, and propellants in aerosol products (including medications and consumer products), and as expanders in plastic foams. CFCs are discussed in Chapter 18 of this volume and in *ASHRAE Standards* 15 and 34.

Exposure to CFCs occur mainly through inhalation and can occur from leaks in refrigeration equipment or during servicing of HVAC systems. Of significant concern is the association between the use of CFCs and the depletion of the earth's stratospheric ozone layer. CFCs add chlorine and bromine atoms to the atmosphere which accelerate the natural ozone destruction rate. Ozone depletion is linked to increases in the amount of ultraviolet radiation (UV-B) reaching the earth's surface. Increasing UV-B can adversely affect human health, most notably by causing malignant melanoma and cataracts, cause adverse ecological impacts, and decrease crop production.

Health Effects and Mechanisms

Carbon monoxide is a chemical asphyxiant. Inhalation of CO causes a throbbing headache brought about by CO having a competitive preference for hemoglobin (about 240 times that of oxygen) and also a shift in the oxygen dissociation curve. Carbon monoxide inhibits oxygen transport in the blood through the formation of carboxyhemoglobin and inhibition of cytochrome oxidase at the cellular level. Deaths and other adverse health effects from over exposures are attributed primarily to motor vehicles. Cobb and Etzel (1991) suggested that CO poisoning at home represented a major preventable disease. Moolenaar et al. (1995) subsequently identified similar data and suggested motor vehicles and home furnaces were primary causes of mortality. Girman et al. (1996) identified both fatal outcomes and episodes. Respectively, 35.9 and 30.6% resulted from motor vehicles, 34.8% and 39.9% from appliance combustion, 4.5% and 5.2% from small appliances, 2.2% and 2.3% from camping equipment, 5.6% and 5.0% from fires, 13.4% and 13.3% from grills and hibachis, and the remainder were unknown.

Carbon dioxide can become dangerous not as a toxic agent but as a secondary asphyxiant. When concentrations exceed 35,000 ppm, central breathing receptors are triggered and cause the sensation of shortness of breath. At progressively higher concentrations, central nervous system dysfunction begins due to simple displacement of oxygen. Concentrations of CO_2 in the non-industrial environment (office buildings) are often measured in the range of 475 to 800 ppm depending on occupant density, ventilation distribution, and amount of outside air supplied to the occupied spaces.

ASHRAE Standard 34 assigns **refrigerants** to one of two toxicity classes (A or B) based on allowable exposure. Fatalities have been reported following acute exposure to fluorocarbon refrigerants. Inhalation exposures to CFCs can cause cardiotoxicity at chronic, low level exposures. Some are thought to be cardiac sensitizers to epinephrine and put occupants at risk for arrhythmias. Central nervous system (CNS) depression has been found at very high concentrations along with asphyxia. Proctor and Hughes (1991) found that volunteers exposed to 200,000 ppm of R-12 experienced significant eye irritation and CNS effects. Chronic exposure to 1000 ppm for 8 hours per day for up to 17 days caused no subjective symptoms or changes in pulmonary function.

Carbon tetrachloride (CCl_4) causes CNS depression and significant liver and kidney damage. Carbon tetrachloride has also been shown to be an animal carcinogen and is classified as a potential human carcinogen.

A significant hazard exists when chlorinated hydrocarbons (R-11 for example) are used in the vicinity of open flame or heated surfaces. Phosgene gas (carbonyl chloride), an extreme irritant to the lungs, is generated when chlorinated solvents or gasses decompose in the presence of heat.

Exposure Control Strategies

Carbon monoxide exposure control strategies primarily involve identification and control of CO emissions directly at their source. Local exhaust ventilation is an appropriate and effective control in most cases. For example, automobile repair garages commonly use a tailpipe exhaust extension to control CO exposure to the mechanics working in the repair bays. Relocating building makeup air intakes or limiting vehicle access are reasonable means to prevent entrainment of automobile exhausts into building HVAC systems.

Carbon monoxide, however, is a common pollutant of ambient air. As a result, direct control by dilution may not be feasible if ambient air is heavily contaminated with CO. Diesel or natural gas may be substituted for gasoline engines to reduce CO where specific sources from engine exhaust are identified or are a concern. Adequate venting of any combustion sources are critical to prevent the buildup of CO indoors. CO may be monitored by a properly calibrated, direct reading CO monitor, colorimetric indicator tubes, or passive diffusion sampling badges.

Exposure controls for **carbon dioxide** are generally limited to situations where exposure concentrations are expected to exceed 3 to 5%. CO_2 is not encountered at levels harmful to humans in the ambient environment. It is normally present at 350 to 375 ppm, and slightly higher in congested cities. With the exception of an intentional or an accidental CO_2 "dump" from a fire suppression system or in a dry ice manufacturing facility, CO_2 is not encountered in significant concentrations that require specific engineering controls. However, CO_2 is denser than air and can persist for some time in low areas such as trenches, depressions, and pits. This characteristic creates a simple asphyxiation hazard because it displaces oxygen. Some literature suggests CO_2 may be involved in sudden infant death syndrome, through local accumulation and respiratory reflex depression.

CFC-containing systems may only be serviced by certified technicians. Controls for preventing exposures include selection and use of appropriate fittings and valves, and insuring that compressed gas cylinders are secured when in use, in transport, and in storage. When repairs are made to leaking or defective components in HVAC equipment adequate dilution ventilation should be provided to the work area. CFCs should never be used in the vicinity of open flame or heated materials due to the potential for the formation of phosgene gas. *ASHRAE Standard* 15 establishes specific requirements for designing, installing, operating, and servicing mechanical refrigeration equipment.

Exposure Standards and Criteria

Table 9 lists time-weighted average exposure limits for organic gases from pertinent standards.

Table 9 Organic Gases Exposure Standards and Criteria

Contaminant	OSHA/NIOSH TWAs	EPA NAAQS 1 Standards
Carbon monoxide	50 ppm/35 ppm IDLH 1200 ppm	8 h 9 ppm 1 h 35 ppm
Carbon dioxide	5000 ppm IDLH 40,000 ppm	n/a
Chlorofluorocarbons	Various	n/a

IDLH = immediately dangerous to life and health

SOIL GASES

Gaseous contaminants other than radon (Rn) may enter buildings from surrounding soil. Methane from landfills has reached explosive levels in some buildings. Potentially toxic or carcinogenic volatile organic compounds in the soil as a consequence of spills, improper disposal, leaks from storage tanks, and disposal in land fills can also be transported into buildings (Wood and Porter 1987, Hodgson et al. 1992, Garbesi and Sextro 1989, and Kullman and Hill 1990). Pesticides applied to the soil beneath or adjacent to houses have also been detected in indoor air (Livingston and Jones 1981, Wright and Leidy 1982) and pressure-driven flow is a suspected entry mechanism. The broad significance of the health effects due to exposure to these soil contaminants is not well understood. Techniques that reduce Rn entry from the soil (discussed in a later section) should also be effective in reducing the entry of these contaminants into buildings. Another approach may be to increase ventilation in the building, such as by slightly opening a window. That change in home-use behavior can also help reduce the negative pressure in the house, with respect to the soil gas pressure, created by the stack effect.

PHYSICAL AGENTS

Physical factors in the indoor environment include thermal conditions (temperature, moisture, air velocity, and radiant energy), mechanical energy (noise and vibration), and electromagnetic radiation [ionizing (radon) and non-ionizing (light, radio-frequency, and extremely low frequency magnetic and electric fields)]. Physical agents can act directly upon building occupants, interact with indoor air quality factors, or affect the way humans respond to the indoor environment. Physical agents, while not categorized as indoor air quality factors, often affect human perception of the quality of indoor air.

THERMAL ENVIRONMENT

The thermal environment affects human health in that it affects body temperatures, both internally and externally (of the skin). In the normal, healthy, resting adult, internal or core body temperatures are very stable, with variations seldom exceeding 1°F. The internal temperature of a resting adult, measured orally, averages about 98.6 F; measured rectally, it is about 1°F higher (Guyton 1972). The temperature of the core is carefully modulated by an elaborate physiological control system. In contrast, the temperature of the skin is basically unregulated and can vary from about 88 to 96°F in normal environments and activities, and also varies over different parts of the skin, with the greatest variation in the hands and feet.

Range of Healthy Living Conditions

The environmental conditions for thermal comfort are those that minimize the effort of the physiological control systems to maintain the internal temperature. The control system regulates the internal body temperature by varying the amount of blood flowing to different skin areas, thus increasing or decreasing heat lost to the environment, by secreting and evaporating sweat from the skin in warm or hot environments, and by increasing the metabolic heat production by shivering in the cold. For a resting person wearing trousers and a long-sleeved shirt, thermal comfort is experienced in a still air environment at 75°F. A zone of comfort extends about 3°F above and below this optimum level. An individual can minimize the need for physiological (involuntary) responses to the thermal environment that are generally perceived as uncomfortable, by a variety of behavioral responses. In a cool or cold environment, such responses include increased clothing, increased activity, or seeking or creating an environment that is warmer. In a warm or hot environment, the amount of clothing can be reduced, the level of physical activity can be reduced, or an environment that is more conducive to increased heat loss can be created. Some of the human responses to the thermal environment are shown in Figure 1.

Cardiovascular and other diseases and the inevitable processes of aging can reduce the capacity or ability of physiological processes to maintain internal body temperature through the balancing of heat gains and heat losses. Thus, some persons are less able to deal with thermal challenges and deviations from comfortable thermally neutral conditions. Metabolic heat production tends to decrease with age, as a result of decreasing basal metabolism together with decreased physical activity. Metabolic heat produc-

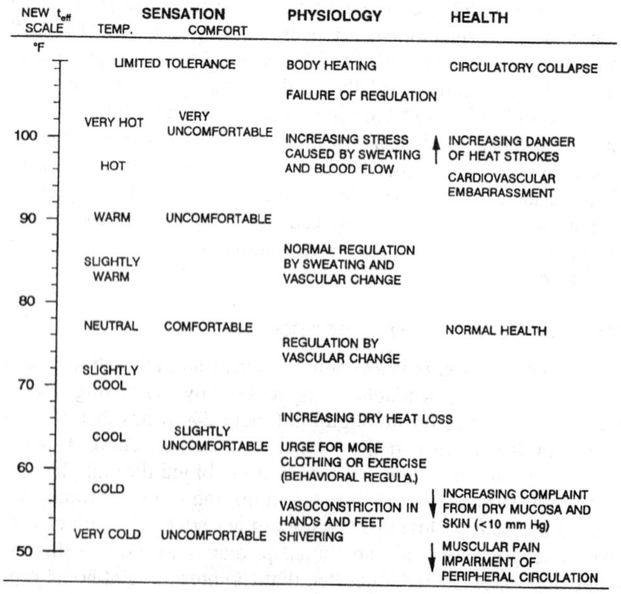

Fig. 1 Related Human Sensory, Physiological, and Health Responses for Prolonged Exposure

tion at age 80 is about 20% less than that at 20 years old, for comparable size and weight. Persons in their eighties, therefore prefer and environmental temperature about 3°F warmer than persons in their twenties. In any given environment near thermally neutral temperature an older person is likely to have a lower core and skin temperature. Older people may have reduced capacity to secrete and evaporate sweat and to increase their skin blood flow and they are, therefore, more likely to experience greater strain in warm and hot conditions as well as in cool and cold conditions.

Hypothermia

Hypothermia refers to conditions where body temperatures are below the normal range, and the normal thermoregulatory defense against cold exposure (peripheral vasoconstriction and shivering) is overwhelmed. The resulting body temperature decrease reduces the chemical activity of cells and metabolism. Reduced activity of the nervous system causes sleepiness, irritability, and affects judgement. While mild hypothermia may not cause disease, it impairs the activity of the immune system, increasing the susceptibility to infections and colds. At core temperatures below 94°F thermoregulation is impaired. Consciousness is lost at about 92°F, and thermoregulation fails completely at 86°F. Cardiac arrhythmia can develop at temperatures around 82°F and may lead to fibrillation and death. In very cold conditions, the skin and peripheral areas of the body can freeze (frostbite). If the tissue is thawed immediately there may not be permanent damage. Prolonged freezing can permanently damage local circulation and tissues may be lost.

Hyperthermia

Hyperthermia refers to the condition where body temperatures are above normal. A deep body temperature increase of 4°F above the normal range does not generally impair body function. For example, it is not unusual for runners to have rectal temperatures of 104°F after a long race. An elevated body temperature increases metabolism. Central nervous system function deteriorates at deep body temperatures above 106 to 108°F. Convulsions may occur above such temperatures and cells may be damaged. This condition is particularly dangerous for the brain, because lost neurons are not replaced. Thermoregulatory functions of sweating and peripheral vasodilation cease at about 110°F, after which body temperatures tend to rise rapidly if external cooling is not imposed.

Seasonal Patterns

Ordinary season changes in temperate climates affect the prevalence of illness. Most acute and several chronic diseases vary in frequency or severity with time of year, and some are present only in certain seasons. Minor respiratory infections, such as colds and sore throats, occur mainly in fall and winter. More serious infections, such as pneumonia, have a somewhat shorter season in winter. Intestinal infections, such as dysentery and typhoid fever, are more prevalent in summer. Diseases transmitted by insects such as encephalitis and endemic typhus are limited to summer since insects are active in warm temperatures only.

Martinez et al. (1989), Hryhorczuk et al. (1992), and others describe a correlation between weather and seasonal illnesses; but such correlations do not necessarily establish a causal relationship. Daily or weekly mortality and heat stress in heat waves have a strong physiological basis directly linked to outdoor temperature. In indoor environments which are well controlled with respect to temperature and humidity, such temperature extremes and the possible adverse effects on health are strongly attenuated.

Increased Deaths in Heat Waves

The role of ambient temperature extremes produced by weather conditions in producing discomfort, incapacity, and death has been studied extensively (Katayama 1970). Military personnel, deep mine workers, and other workers occupationally exposed to extremes of high and low temperature have been studied, but the importance of thermal stress affecting both the sick and healthy is not sufficiently appreciated. Collins and Lehmann (1953) studied weekly deaths over many years in large cities in the United States and demonstrated the impact of heat waves in producing conspicuous periods of excess mortality. Excess mortality due to heat waves was of the same amplitude as that to influenza epidemics, but tended to last one week instead of the 4 to 6 weeks duration of influenza epidemics.

Ellis (1972) reviewed heat wave related excess mortality in the United States. Mortality increases of 30% over background are commonly seen, especially in heat waves that occur early in the summer. Much of the increase occurs in the population over age 65, more of it in women than men, and many deaths are due to cardiovascular and cerebrovascular causes. Oeschli and Buechley (1970) studied heat-related deaths in Los Angeles heat waves of 1939, 1955, and 1963. Kilbourne et al. (1982) suggested that the same risk factors, i.e., age, low income, and African-American derivation, persist in the heat death epidemics that continue to occur.

Hardy (1971) showed the relationship of health data to comfort on a psychrometric diagram (Figure 2). The diagram contains ASHRAE effective temperature ET* lines and lines of constant skin moisture level or skin wettedness *w*. Skin wettedness is defined as that fraction of the skin covered with water to account for the observed evaporation rate. The ET* lines are loci of constant physiological strain, and also correspond to constant levels of physiological discomfort—slightly uncomfortable, comfortable, and very comfortable (Gonzalez et al. 1978). Skin wettedness, as an indicator of strain (Berglund and Gonzalez 1977, Berglund and Cunningham 1986) and the fraction of the skin wet with perspiration, is fairly constant along an ET* line. Numerically ET* is the equivalent temperature at 50% rh that produces the strain and discomfort of the actual condition. The summer comfort range is between an ET* of 73 and 79°F. In this region, skin wettedness is less than 0.2. Heat strokes occur generally when ET* exceeds 93°F (Bridger and Helfand 1966). Thus, the ET* line of 95°F is generally considered dangerous. At this point, skin wettedness will be 0.4 or higher.

The black dots in Figure 2 correspond to heat stroke deaths of healthy male U.S. soldiers assigned to sedentary duties in midwestern army camp offices (Shickele 1947). It is to be expected that older persons respond less well to thermal challenges than do healthy soldiers. This was apparently the case in the Illinois heat wave study mentioned earlier, where the first wave with a 33% increase in death rate and an ET* of 85°F affected mainly the over 65-year-old group. The studies suggest that the "danger line" represents a threshold of significant risk for young healthy people, and that the danger tends to move to lower values of ET* with increasing age.

Effects of Thermal Environment on Specific Diseases

Cardiovascular diseases are largely responsible for excess mortality during heat waves. For example, Burch and DePasquale (1962) found that the heart disease cases in whom decompensation is present are extremely sensitive to high temperatures and particularly to moist heat. However, both cold and hot temperature extremes are associated with increased coronary heart disease deaths and anginal symptoms (Teng and Heyer 1955).

Both acute and chronic respiratory diseases frequently increase in frequency and severity during extreme cold weather. No increase in these diseases has been noted in extreme heat. Additional studies of hospital admissions for acute respiratory illness show a negative correlation with temperature after removal of seasonal trends (Holland 1961). The symptoms of patients with chronic respiratory disease (bronchitis, emphysema) increase in cold weather. This is thought to be due to reflex constriction of the bronchi, adding to the obstruction already present. Greenburg (1964) revealed evidence of cold

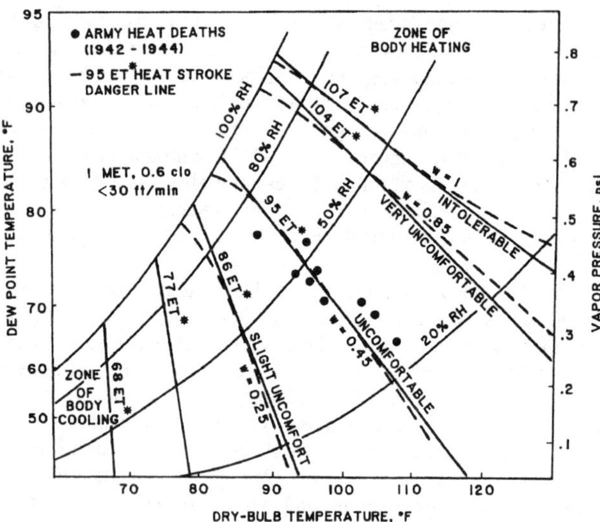

Fig. 2 Isotherms for Comfort, Discomfort, Physiological Strain, Effective Temperature (ET*), and Heat Stroke Danger Threshold

sensitivity in asthmatics; emergency room treatments for asthma increased abruptly in local hospitals with early and severe autumn cold spells. Later cold waves with even lower temperatures produced no such effects, and years without early extreme cold had no asthma epidemics of this type. Resuspension of allergens caused by resumed indoor heating may account for this phenomenon. Patients with cystic fibrosis are extremely sensitive to heat because their diminished sweat gland function greatly diminishes their ability to cope with increased temperature (Kessler and Anderson 1951).

Itching and chapping of the skin is influenced by (1) atmospheric factors, particularly cold and dry air, (2) frequent washing or wetting of skin, and (3) low indoor humidities. Although itching of the skin is usually a winter cold climate illness in the general population, it can be caused by excessive summer air conditioning (Susskind and Ishihara 1965, Gaul and Underwood 1952).

People suffering from chronic illness (heart disease) or serious acute illnesses that require hospitalization often manage to avoid serious thermal stress. Katayama et. al. (1970) found that countries with the most carefully regulated indoor climates (such as the Scandinavian countries and the United States) have had only small seasonal fluctuations in mortality in recent decades, while countries with less space heating and cooling exhibit greater seasonal swings in seasonal mortality. For example, mandatory air-conditioning in homes for the aged in the southwest United States has virtually eliminated previously observed mortality increases during heat waves.

Summer cooling reduces heat stress by removing both sensible and latent heat from the occupied space, but winter heating has a mixed effect. It reduces cold stress, but it usually does not increase the low water vapor pressure that occurs outdoors during the winter. This results in very low relative humidity in the heated space, which can contribute to dehydration and discomfort and cause injury to skin, eyes, nose, throat, and mucous membranes. These dry tissues may be less resistant to infection. Animal experiments also show that infection rates increase with low levels of either ventilation or relative humidity (Schulman and Kilbourne 1962).

In various tests conducted under identical conditions except humidity level, mechanical humidification raised the relative humidity in one space above that in the matched space; no humidified room was higher than 50% rh (Green 1979 and 1982, Gelperin 1973, Serati and Wuthrich 1969). In each investigation, the humidified rooms showed a reduction in absenteeism and upper respiratory infection—49% reduction in kindergarten children, 6% and 18% in office workers, and 8% and 18% in army recruits. Since occupants in each pair of spaces were subject to the same outdoor conditions and the same indoor air temperature, reductions were attributed to differences in humidity or a related factor (*e.g.*, reduced dust levels and coughing). Therefore, while low humidity does not have a direct pathological effect, it is a factor contributing to disease. A more direct effect has been indicated among users of contact lenses on long airline flights in cabins at low humidity. Here, dehydration of the eyes has been blamed for causing irritation and corneal edema or even ulceration of the corneal epithelium (Laviana et al. 1988).

Injury from Hot and Cold Surfaces

The skin has cold, warm, and pain sensors to feed back thermal information about surface contacts. When the skin temperature rises above 113°F or falls below about 59°F, sensations from the skin's warm and cold receptors are replaced by those from the pain receptors to warn of thermal injury to the tissue (Guyton 1968). The temperature of the skin depends on the temperature of the contact surface, its conductivity, and the contact time. Table 10 gives approximate temperature limits to avoid pain and injury when contacting three classes of conductors for various contact times (CEN).

Table 10 Approximate Surface Temperature Limits to Avoid Pain and Injury

	Contact Time				
Material	**1 s**	**10 s**	**1 min**	**10 min**	**8 h**
Metal, water	149°F	133°F	124°F	118°F	109°F
Glass, concrete	176°F	151°F	129°F	118°F	109°F
Wood	248°F	190°F	140°F	118°F	109°F

ELECTRICAL HAZARDS

Electrical current can cause burns, neural disturbances, and cardiac fibrillation (Billings 1975). The threshold of perception is about 5 mA for direct current, with a feeling of warmth at the contact site. The threshold is 1 mA for alternating current, which causes a tingling sensation.

The resistance of the current pathway through the body is a combination of core and skin resistance. The core is basically a saline volume conductor with very little resistance, therefore, the skin resistance provides the largest component of the resistance. The skin resistance decreases with moisture. If the skin is moist, voltages as low as 2 V (ac) or 5 V (dc) are sufficient to be detected, and voltages as low as 20 V (ac) or 100 V (dc) can cause a 50% loss in muscular control.

The dangerous aspect of alternating electrical current is its ability to cause cardiac arrest by ventricular fibrillation. If a weak alternating current (100 mA for 2 s) passes through the heart (as it would in going from hand to foot), the current can force the heart muscle to fibrillate and lose the rhythmic contractions of the ventricles necessary to pump blood. Unconsciousness and death will soon follow if medical aid cannot rapidly restore normal rhythm.

MECHANICAL ENERGIES

Vibration

Vibration in a building originates from both outside and inside the building. Sources outside a building include blasting operations, road traffic, overhead aircraft, underground railways, earth movements, and weather conditions. Sources inside a building include doors closing, foot traffic, moving machinery, elevators, HVAC systems, and other building services. Vibration is an omnipresent, integral part of the built environment. The effects

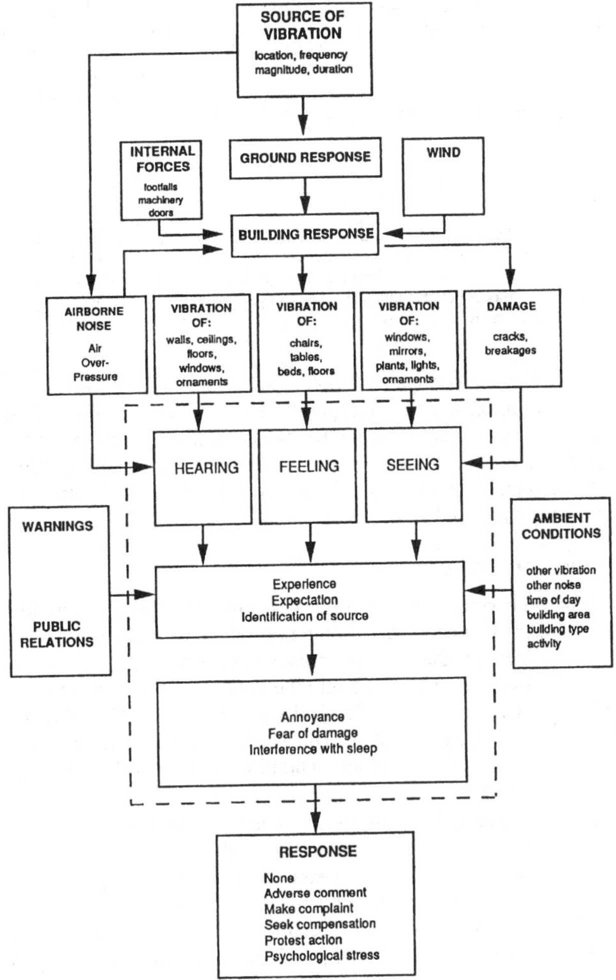

Fig. 3 Factors Affecting Acceptability of Building Vibration

of the vibration on building occupants depend on whether it is perceived by those persons and factors related to the building, the location of the building, the activities of the occupants in the building, and the perceived source and magnitude of the vibration. Factors influencing the acceptability of building vibration are presented in Figure 3.

The combination of hearing, seeing, or feeling vibration determines human response. Components concerned with hearing and seeing are part of the visual environment of a room and can be assessed as such. The perception of mechanical vibration by feeling is generally through the cutaneous and kinesthetic senses at high frequencies, and through the vestibular and visceral senses at low frequencies. Because of this and the nature of vibration sources and building responses, building vibration may be conveniently considered in two categories—low-frequency vibrations less than 1 Hz and high-frequency vibrations of 1 to 80 Hz.

Measurement and Assessment. Human response to vibration depends on the vibration of the body. The main vibrational characteristics are vibration level, frequency, axis (and area of the body), and exposure time. A root-mean-square (rms) averaging procedure (over the time of interest) is often used to represent vibration acceleration (ft/s^2 rms). Vibration frequency is measured in cycles per second (Hz), and the vibration axis is usually considered in three orthogonal, human-centered translational directions (up-and-down, side-to-side, and fore-and-aft). Although the coordinate system is centered inside the body, in practice, vibration is measured at the human surface and measurements are directly compared with relevant limit values or other data concerning human response.

Rotational motions of a building in roll, pitch, and yaw are usually about an axis of rotation some distance from the building occupants. For most purposes, these motions can be considered as the translational motions of the person. For example, a roll motion in a building about an axis of rotation some distance from a seated person will have a similar effect as side-to-side translational motions of that person, etc.

Most methods assess building vibrations with rms averaging and frequency analysis. However, human response is related to the time-varying characteristics of vibration as well. For example, many stimuli are transient, such as those caused by a train passing a building. The vibration event builds to a peak followed by a decay in level over a total period of about 10 s. The nature of the time varying-event and the number of occasions it occurs during a day are important factors that might be overlooked if data are treated as steady-state and continuous.

Standard Limits.

Low-frequency motion (1 Hz). The most commonly experienced form of slow vibration in buildings is building sway. This motion can be alarming to occupants if there is fear of building damage or injury. While occupants of two-story wood frame houses accept occasional creaks and motion from wind storms or a passing heavy vehicle, such events are not as accepted by occupants of high-rise buildings. Detected motion in tall buildings can cause discomfort and alarm. The perception thresholds of normal sensitive humans to low-frequency horizontal motion are given in Figure 12 (ISO 1984, Chen and Robertson 1972). The frequency range is from 0.06 to 1 Hz or, conversely, for oscillations with periods of 1 to 17 s. The natural frequency of sway of the Empire State Building in New York City, for example, has a period of 8.3 s (Davenport 1988). The thresholds are expressed in terms of relative acceleration which is the actual acceleration divided by the standard acceleration of gravity (g = 32.2 ft/s^2). The perception threshold to sway in terms of building accelerations decreases with increasing frequency and ranges from 0.16 to 0.06 ft/s^2.

For tall buildings, the highest horizontal accelerations generally occur near the top at the building's natural frequency of oscillation. Other parts of the building may have high accelerations at multiples of the natural frequency. Tall buildings always oscillate at their natural frequency, but the deflection is small and the motion undetectable. In general, short buildings have a higher natural frequency of vibration than taller ones. However, strong wind forces energize the oscillation and increase the horizontal deflection, speed, and accelerations of the structure.

ISO (1984) states that building motions are not to produce alarm and adverse comment from more than 2% of the building's occupants. The level of alarm depends on the interval between events. If noticeable building sway occurs for at least 10 min at intervals of 5 years or more, the accept able acceleration limit is higher than if this sway occurs annually (Figure 4). For annual intervals, the acceptable limit is only slightly above the normal person's threshold of perception. Motion at the 5-year limit level is estimated to cause 12% to complain if it occurred annually. The recommended limits are for purely horizontal motion; rotational oscillations, wind noise, and/or visual cues of the building's motion exaggerate the sensation of motion and, for such factors, the acceleration limit would be lower.

The upper line in Figure 4 is intended for offshore fixed structures such as oil drilling platforms. The line indicates the level of horizontal acceleration above which routine tasks by experienced personnel would be difficult to accomplish on the structure.

High-frequency motion (1 to 80 Hz). Higher frequency vibrations in buildings are caused by machinery, elevators, foot traffic, fans, pumps, and HVAC equipment. Further, the steel structures of modern buildings are good transmitters of high-frequency vibrations. The sensitivity to these higher frequency vibrations is indicated in

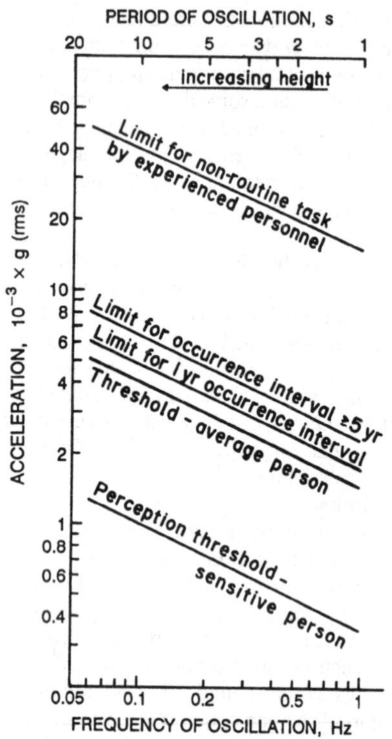

Fig. 4 Acceleration Perception Thresholds and Acceptability Limits for Horizontal Oscillations

Figure 5 (Parsons and Griffin 1988). Displayed are the median perception thresholds to vertical and horizontal vibrations in the 2 to 100-Hz frequency range. The average perception threshold for vibrations of this type is from 0.03 to 0.3 ft/s^2, depending on the frequency and on whether the person is standing, sitting, or lying down.

People detect horizontal vibrations at lower acceleration levels when lying down than when standing. However, a soft bed decouples and isolates a person fairly well from the vibrations of the structure. The threshold to vertical vibrations is nearly constant at approximately 0.04 ft/s^2 for both sitting and standing positions from 2 to 100 Hz. This agrees with earlier observations by Reiher and Meister (1931).

Many building spaces with critical work areas (surgery, precision laboratory work) are considered unacceptable if vibration is

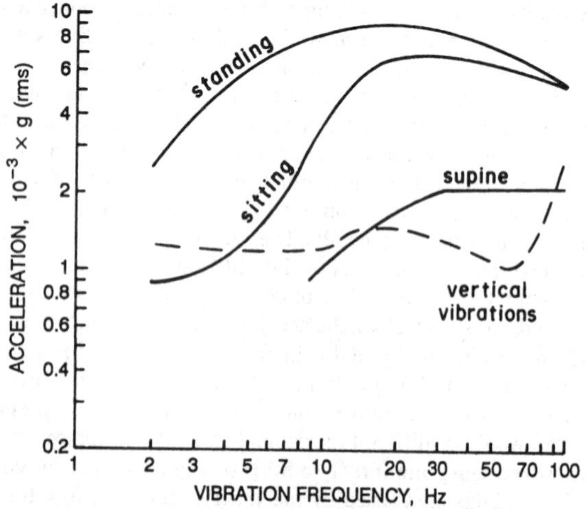

Fig. 5 Median Perception Thresholds to Horizontal (Solid Lines) and Vertical (Dashed Line) Vibrations

Table 11 Acceptable to Threshold Vibration Level Ratios

Place	Time	Continuous or Intermittent Vibration	Impulse or Transient Vibration Several Times per Day
Critical work areas	Day or night	1	1
Residential	Day/Night	2 to 4 / 1.4	30 to 90 / 1.4 to 20
Office	Day or night	4	60 to 128
Workshop	Day or night	8	90 to 128

Note: The ratios for continuous or intermittent vibration and repeated impulse shock are in the range of 0.7 to 1.0 for hospital operating theaters (room) and critical working areas. In other situations, impulse shock can generally be much higher than when the vibration is more continuous.

perceived by the occupants. In other situations and activities, perceived vibration may be acceptable. Parsons and Griffin (1988) found that accelerations twice the threshold level would be unacceptable to occupants in their homes. A method of assessing vibrational acceptability in buildings is to compare the vibration with perception threshold values (Table 11).

Sound and Noise

When the vibration of an object is transmitted to air particles, making them vibrate, a variation in normal atmospheric pressure is created. When this disturbance spreads to the eardrum it is vibrated and this vibration is translated into the sensation called "sound." In general terms, sound in the physical sense is the vibration of particles in a gas, a liquid, or a solid. The entire mechanical energy spectrum includes include infrasound and ultrasound as well as audible sound (Figure 6).

Health Effects. Hearing loss is generally considered the most undesirable effect of noise exposure, although there are other effects. Tinnitus, a ringing in the ears, is really the hearing of sounds that do not exist. It often accompanies hearing loss. Paracusis is a disorder where a sound is heard incorrectly; that is a tone is heard, but has an inappropriate pitch. Speech misperception occurs when an individual mistakenly hears one sound for another; for example when the sound for "t" is heard as a "p."

Hearing loss can be categorized as conductive, sensory, or neural. Conductive hearing loss results from a general decrease in the

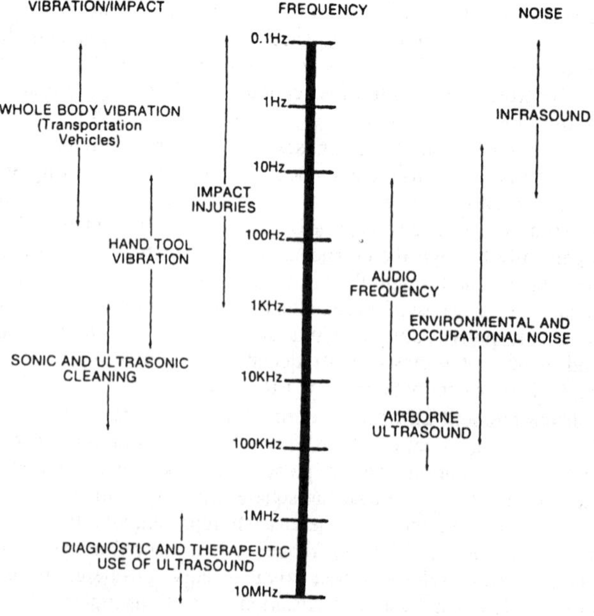

Fig. 6 Mechanical Energy Spectrum

amount of sound transmitted to the inner ear. Excessive ear wax, a ruptured ear drum, fluid in the middle ear or missing elements of the bone structures in the middle ear are all associated with conductive hearing loss. These are generally not occupationally related and are generally reversible by medical or surgical means. Sensory hearing losses are associated with irreversible damage to the inner ear. Sensory hearing loss is further classified as (1) presbyccusis, loss caused as the result of aging; (2) noise-induced hearing loss (industrial hearing loss and sociacusis, which is caused by noise in everyday life); and (3) nosoacusis, losses attributed to all other causes. Neural deficits are related to damage to higher centers of the auditory system.

Noise-induced hearing loss is believed to occur, in the most sensitive individuals, in those exposed for 8 hours per day over a working lifetime at levels of 75 dBA and for most people similarly exposed to 85 dBA.

ELECTROMAGNETIC RADIATION

Radiation energy is energy that is emitted, transmitted, or absorbed in wave or particulate form. This energy consists of electric and magnetic forces which, when disturbed in some manner, produce electromagnetic radiation. Electromagnetic radiation is grouped into a spectrum arranged by frequency and/or wavelength. The product of frequency and wavelength is the speed of light (3×10^8 m/s). The spectrum includes the ionizing, ultraviolet, visible, infrared, microwave, radio-frequency, and extremely low frequency (Figure 7). Table 12 presents these electromagnetic radiations by their range of energies, frequencies, and wavelengths. The regions are not sharply delineated from each other and, in fact, often overlap. It is convenient to divide these regions as listed in Table 12, due to the nature of the physical and biological effects.

Ionizing Radiation

Ionizing radiation is that part of the electromagnetic spectrum that has very short wave lengths and high frequencies, and it has the ability to ionize matter. Such ionizations tend to be very damaging to living matter. Background radiation that occurs naturally in the environment is from cosmic rays and naturally occurring radionuclides. It has not been established whether exposure at the

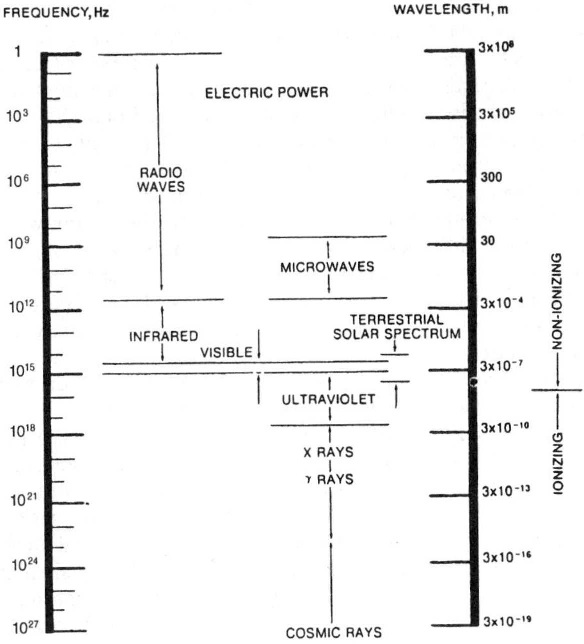

Fig. 7 Electromagnetic Spectrum

Table 12 Energy, Wavelength, and Frequency Ranges for Electromagnetic Radiation

Radiation Type	Energy Range	Wavelength Range	Frequency Range
Ionizing	> 12.4 eV	< 100 nm	> 3.00 PHz
Ultraviolet (UV)	12.40 – 3.10 eV	100 – 400 nm	3.00 PHz – 0.75 PHz
Visible	3.10 – 1.63 eV	400 – 760 nm	750 THz – 395 THz
Infrared (IR)	1.63 – 1.24 meV	760 nm – 1 mm	395 THz – 0.30 THz
Microwave (MW)	1.24 meV – 1.24 eV	1 mm – 1 m	300 GHz – 300 MHz
Radio-frequency (RF)	1.24 eV – 1.24 peV	1 m – 1 Mm	300 MHz – 300 Hz
Extremely low frequency (ELF)	< 1.24 peV	> 1 Mm	< 300 Hz

low dose rate of average background levels is harmful to humans. An important naturally occurring radionuclide is radon (^{226}Rn), a decay product of uranium in the soil (^{238}U).

Radon. Radon (Rn) is a naturally occurring, chemically inert, colorless, odorless, tasteless radioactive gas. It is produced from the radioactive decay of radium, which is formed through several intermediate steps of the decay of uranium and thorium. Radon is widely found in the natural environment. Radon-222, the gaseous decay product of radium-226 from the uranium decay series is the most common isotope of radon. As with other noble gases, radon is not chemically bound or attached to other materials. Before it decays, radon can move limited distances through very small spaces, such as those between particles of soil and rock, and enter indoor environments (Nazaroff et al. 1988, Tanner 1980). Additional but secondary sources of indoor radon include ground-water and radium-containing building materials.

Radon can be transported through soil and into buildings by pressure driven flow of Rn-containing soil gas and by molecular diffusion. Pressure driven flow of soil gas into buildings occurs through cracks, joints, and other holes in concrete foundations; directly through porous concrete blocks; through the joints and openings in crawl space ceilings; and through leakage points in HVAC ductwork that is embedded in slab floors or located in crawl spaces. Pressure driven flow is the dominant radon entry mechanism in houses with elevated radon concentrations (Nazaroff et al. 1987). The pressure differences that drive soil gas and Rn entry are caused by several factors including the thermal stack effect, wind, and operation of HVAC equipment. In addition to pressure-driven radon entry, Rn can also diffuse directly through substructural materials (e.g., concrete). The diffusive Rn entry rate is often a significant portion of the total entry rate in houses with low Rn concentrations.

Radon concentration is not the same as radon exposure. Exposure includes the time spent at a particular concentration level. The outdoor radon concentration is about 15 Bq/m^3 (0.4 pCi/L). The annual average concentration of radon in homes in the United States is about 46 Bq/m^3 (1.25 pCi/L) (EPA 1989). Only about 6% of homes in the United States (5.8 million homes) have annual average radon concentrations exceeding the action level of 147 Bq/m^3 (4 pCi/L) set by the U.S. Environmental Protection Agency (Marcinowski et al 1994). While several sources of radon may contribute to the annual indoor average, pressure-driven flow of soil gas constitutes the principal source for elevated concentrations; non-municipal water supplies can be a source of elevated indoor radon, but only in isolated instances. International action levels for radon are listed in Table 13.

Table 13 Action Levels for Radon Concentration Indoors
(DOE 1995)

Country/Agency	Action Level	
	Bq/m^3	pCi/L
Australia	200	5.4
Austria	400	10.8
Belgium	400	10.8
CEC	400	10.8
Canada	800	21.6
Czech Republic	400	10.8
P.R. China	200	5.4
Finland	400	10.8
Germany	250	6.7
ICRP	200	5.4
Ireland	200	5.4
Italy	400	10.8
Norway	400	10.8
Sweden	400	10.8
United Kingdom	200	5.4
United States	148	4.0
World Health Organization	200	5.4

Health Effects. Studies of workers in uranium and other underground mines form the principal basis for knowledge about health risks due to radon. The radioactive decay of radon produces a series of radioactive isotopes of polonium, bismuth, and lead. Unlike their chemically inert radon parent, these progeny are chemically active. They can attach to airborne particles that subsequently deposit in the lung or deposit directly in the lung without prior attachment to particles. Some of these progeny, like radon, are alpha-particle emitters, and the passage of these alpha particles through lung cells can lead to cellular changes that may initiate lung cancer (Samet 1989). Thus, adverse health effects associated with radon are due to exposures to radon decay products and the amount of risk is assumed to be directly related to the total exposure.

Measurement. Even though it is the radon progeny that present the possibility of adverse health risks, radon itself is usually measured and used as a surrogate for progeny measurements because of the expense involved in accurate measurements of radon progeny. Indoor concentrations of radon can vary hourly, daily, and seasonally, in some cases by as much as a factor of 10 to 20 on a daily basis (Turk et al. 1990b). Thus, long-term measurements (3 months to 1 year) made during normal home activities generally provide more reliable estimates of the average indoor concentration. Two techniques widely used for homeowner measurements are the short-term charcoal canister (up to 7 days), and the long-term alpha-track methods (90 days to 1 year). Generally, short-term measurements should only be used as a screening technique to see if a long-term measurement is necessary. The great uncertainties in measurement accuracy, up to 50% at the radon levels typically found in homes, as well as the natural variability of radon concentrations should be considered in interpreting the results.

Ideally, long-term measurements should be the basis for decisions on installation of radon mitigation systems and short-term measurements should only be used as a screening method to identify buildings with Rn concentrations that are very high, justifying immediate remedial action. In practice, short-term measurements at the time a building is sold are the basis for most decisions about remedial action.

Control. Exposure to indoor Rn may be reduced by: (1) inhibiting Rn entry into the building or (2) removing or diluting Rn decay products in indoor air. The most effective and energy efficient control measures are generally those that reduce Rn entry rates (Henschel 1993). One of the most common effective techniques is active subslab depressurization in which a fan and piping system draw soil gas from beneath the slab and exhausts the gas outside. This technique reduces or reverses the pressure gradient that normally draws soil gas and Rn into the building and often reduces indoor Rn concentrations by a large factor (e.g., 5 to 10). Passive control methods such as Rn-resistant construction techniques and/or passive stack subslab depressurization systems are also used; however, the performance of these control methods is not well characterized and average reductions in Rn concentrations may be 50% or less. Sealing cracks and joints in slab floor improves the performance of subslab depressurization systems. Sealing by itself is often not very effective in reducing indoor Rn.

In houses with crawl spaces, active (fan-forced) or passive crawl space ventilation is often effective in maintaining low indoor Rn concentrations although other techniques are also used (Henschel 1988, 1993).

Nonionizing Radiation

Ultraviolet radiation, visible light, and infrared radiation are components of sunlight, and of all artificial light sources. Microwave radiation and radio-frequency radiation are essential in a wide range of communication technologies and are also in widespread use for heating as in microwave ovens and heat sealers, and for heat treatments of a variety of products. Power frequency fields are an essential and unavoidable consequence of the generation, transmission, distribution, and use of electrical power.

Optical Radiation. Ultraviolet (UV), visible, and infrared (IR) radiation compose the optical radiation region of the electromagnetic spectrum. The wavelengths range from 100 nm in the UV to 1 mm in the IR, with 100 nm generally considered to be the boundary between ionizing and nonionizing. The UV region wavelengths range from 100 to 400 nm, the visible region from 400 to 760 nm, and the IR from 760 nm to 1 mm.

Optical radiation can interact with a medium by reflection, absorption, or transmission. The skin and eyes are the organs at risk in humans. Optical radiation from any of the spectral regions can cause acute and/or chronic biologic effects given appropriate energy characteristics and exposure. These effects include tanning, burning (erythema), premature "aging," and cancer of the skin; dryness, irritation, cataracts, and blindness in the eyes.

The region of the electromagnetic spectrum visible to humans is known as light. There can be biological, behavioral, psychological, and health effects from exposure to light. Assessment of these effects depends on the purpose and application of the illumination. Individual susceptibility varies, with other environmental factors (air quality, noise, chemical exposures, and diet) acting as modifiers. It is difficult, therefore, to generalize potential hazards. Light pollution results from the presence of unwanted light.

Light penetrating the retina not only allows the exterior world to be seen, but, like food and water, it is used in a variety of metabolic processes. Light stimulates the pineal gland to secrete melatonin, which regulates the human biological clock. This, in turn, influences reproductive cycles, sleeping, eating patterns, activity levels, and moods. The color of light has an affect the way the objects appear. The distortion of color rendition may result in disorientation, headache, dizziness, nausea, and fatigue.

As the daylight shortens, the human body may experience a gradual slowing down, loss of energy, and a need for more sleep. It becomes harder to get to work and depression or even withdrawal may take place. This type of seasonal depression, brought on by changes in light duration and intensity is called "seasonal affective disorder" (SAD). Sufferers of this syndrome also complain of anxiety, irritability, headache, weight gain, and lack of concentration and motivation. Treatment of this problem is through the manipulation of environmental lighting (exposure to full-spectrum lighting for extended periods, 12 hours/day).

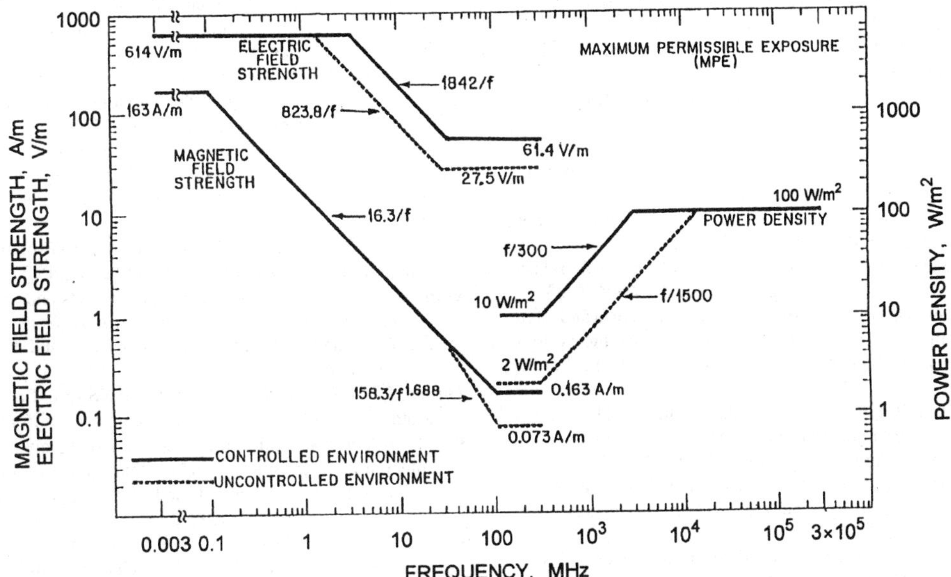

Fig. 8 Maximum Permissible Levels of Radio Frequency Radiation for Human Exposure

Radio Frequency Radiation. Just as the body absorbs infrared and light energy, which can affect thermal balance, it can also absorb other longer wavelength electromagnetic radiation. For comparison, visible light has wavelengths in the range 0.4 to 0.7 μm and infrared from 0.7 to 10 μm, while the wavelength of K and X band radar is 12 and 28.6 mm. The wavelength of radiation in a typical microwave oven is 120 mm. Infrared is absorbed within 1 mm of the surface (Murray 1995).

The heat of the absorbed radiation raises the skin temperature and, if sufficient, is detected by the skin's thermoreceptors, warning the person of the possible thermal danger. With increasing wavelength, the radiation penetrates deeper into the body. The energy can thus be deposited well beneath the skin's thermoreceptors making the person less able or slower to detect and be warned of the radiation (Justesen et al. 1982). Physiologically, these longer waves only heat the tissue and, because the heat may be deeper and less detectable, the maximum power density of such waves in occupied areas is regulated (ANSI 1991) (Figure 8). The maximum permitted power densities are less than half of sensory threshold values.

REFERENCES

ACGIH. 1989. Guidelines for the assessment of bioaerosols in the indoor environment. American Conference of Government Industrial Hygienists, Cincinnati, OH.

ACGIH. Annual. Threshold Limit Values and Biological Exposure Indices for 1995-1996.

Alpaugh, E.L. and T.J. Hogan. 1988. *Fundamentals of industrial hygiene: Particulates*, 3rd Edition. National Safety Council.

Anderson, H.A., D. Higgins, L.P. Hanrahan, P. Sarow, and J. Schirmer. 1991. Mesothelioma among employees with likely contact with in-place asbestos-containing building materials. *Ann. N.Y. Acad. Sci.* 643:550-72.

ANSI. 1991. Safety level with respect to human exposure to radio frequency electromagnetic radiation of 3KH-300GH *Standard* C95.1-1991. American National Standards Institute, New York.

Arnow, P.M., J.N. Fink, D.P. Schlueter, J.J. Barboriak, G. Mallison, S.I. Said, S. Martin, G.F. Unger, G.T. Scanlon, and V.P. Kurup. 1978. *American Journal of Medicine* 64: 237.

ASHRAE. 1989. Legionellosis position paper.

ASTM. 1990. *Biological contaminants in indoor environments*. STP 1071. American Society for Testing and Materials, W. Conshohoken, PA.

Barbaree, J.M., B.S. Fields, J.C. Feeley, G.W. Gorman and W.T. Martin. 1986. Isolation of protozoa from water associated with a Legionellosis outbreak and demonstration of intracellular multiplication of *Legionella pneumophila*. *Appl. Environ. Microbiol.* 51: 422-424.

Bartlett, P.C., L.A. Vonbehren, R.P. Tewari, R.J. Martin, L. Eagleton, M.J. Isaac, and P.S. Kulkarni. 1982. American Journal of Public Health 72: 1369.

Bascom, R.A. 1996. Environmental factors and respiratory hypersensitivity: The Americas. *Toxicol. Lett.* 86:115-30.

Bascom, R., P. Bromberg, D.L. Costa, et al. 1996. Health effects of outdoor pollution, Parts I and II. *American Journal Respir. Crit. Care Med.* 153:3-50, 477-89.

Bates, D.V. 1989. Ozone—Myth and reality. *Environmental Research* 50: 230-237.

Batterman, S. and C. Peng. 1995. TVOC and CO_2 as indicators in indoor air quality studies. *Am. Ind. Hyg. Assoc. J.* 56(1):55-65.

Baylor, E.R., V. Peters, and M.B. Baylor. 1977. Water-to-air transfer of virus. *Science*. 1977: 763.

Berglund, B., I. Johansson, and T. Lindvall. 1988. Healthy Buildings 88, 3:299-309. Stockholm, Sweden.

Berglund, L.G. and D. Cunningham. 1986. Parameters of human discomfort in warm environments. *ASHRAE Transactions* 92(2).

Berglund, L.G. and R.R. Gonzalez. 1977. Evaporation of sweat from sedentary man in humid environments. *Journal of Applied Physiology, Respiratory, Environmental and Exercise Physiology* 42(5):767-72.

Billings, C.E. 1975. Electrical shock. In *Textbook of medicine*. Saunders, Philadelphia, 72-73.

Bridger, C.A. and L.A. Helfand. 1968. Mortality from heat during July 1966 in Illinois. *International Journal of Biometeorology* 12:51.

Brundage, J.F., R.M. Scott, W.M. Lodnar, D.W. Smith, and R.N. Miller. 1988. JAMA 259:2108.

Burch, G.E. and N.P. DePasquale. 1962. Hot climates, man and his heart. Charles C. Thomas, Springfield, IL.

Burge, H.A. 1995. *Bioaerosols*. Lewis Publishers, Chelsea, MI.

Burge, H.A. 1989. Occupational medicine: State of the art reviews. J.E. Cone and M.J. Hodgson, eds. 4:713-21. Hanley and Belfus, Philadelphia.

Burge, S., A. Hedge, S. Wilson, J.H. Bass, and A. Robertson. 1987. *Annals of Occupational Hygiene* 31:493.

Burgess, J.L. and C.D. Crutchfield. 1995. Quantitative respirator fit tests of Tuscon fire fighters and measures of negative pressure excursions during exertion. *Appl. Occup. Env. Hyg.* 10(1):29-36.

Burri, G. 1984. A practical ergonomics program in place and how it works. *Cumulative Trauma Disorders and Ergonomics.* National Safety Council, Itasca, IL.

Cain, W.S. 1989. Perceptual characteristics of nasal irritation. National Danish Institute on Occupational Health, Copenhagen.

Cain, W.S., J.M. Samet, and M.J. Hodgson. 1995. The quest for negligible health risk from indoor air. *ASHRAE Journal* 37(7):38.

CEN. Surface temperatures of touchable parts, a draft proposal. TC 114 N 122 D/E. European Standards Group.

CFR. 1989a. 29 CFR 1910.1000. U.S. Government Printing Office, Washington, DC.

CFR. 1989b. 29 CFR 1926.1101.

Chatigny, M. 1983. *Air sampling instruments*, 6th ed. American Conference of Governmental Industrial Hygienists. Cincinnati, OH.

Chen, P.W. and L.E. Robertson. 1972. Human perception thresholds of horizontal motion. *ASCE Journal*, Structure Division, August.

Cobb, N. and R.A. Etzel. 1991. Unintentional carbon monoxide-related deaths in the United States, 1979 through 1988. JAMA 266:659-63.

Collins, S.D. and J. Lehmann. 1953. Excess deaths from influenza and pneumonia and from important chronic diseases during epidemic periods, 1918-51. Public Health Monograph 20.10, U.S. Public Health Service *Publication No.* 213.

Colombo, A., M. DeBortoli, H. Knöppel, H. Schauenburg, and H. Vissers. 1991. Small chamber tests and headspace analysis of volatile organic compounds emitted from household products. *Indoor Air* 1:13-21.

Cometto-Muñiz, J.E. and W.S. Cain. 1994a. Sensory reactions of nasal pungency and odor to volatile organic compounds: The alkylbenzenes. *Am. Ind. Hyg. Assoc. J.* 55(9):811-817.

Cometto-Muñiz, J.E. and W.S. Cain. 1994b. Perception of odor and nasal pungency from homologous series of volatile organic compounds. *Indoor Air* 4:140-145.

Conibear, S., S. Geneser, and B.W. Carnow. 1996. Carbon monoxide levels and sources found in a random sample of households in Chicago during the 1994-1995 heating season. *Proceedings of* IAQ 95. ASHRAE, 111-118.

Coutant, R.W., G.F. Ward, C.W. Spicer, et.al. 1994. Control of NO_2 and nitrogen acids using alkaline-impregnated carbon filters. *Proceedings of* IAQ 94. ASHRAE, 139-145.

Cox, C.S. 1987. *The aerobiological pathway of microorganisms.* John Wiley & Sons, New York.

Crandall, M.S. and W.K. Sieber. 1996. The NIOSH indoor environmental evaluation experience: Part one, building evaluations. *Applied Occupational and Environmental Hygiene.*

Davenport, A.G. 1988. The response of supertall buildings to wind. In *Second century of the skyscraper.* Van Nostrand Reinhold, New York, 705-26.

Derian, J. 1983. Dunlop company redesigns workstations, boosts productivity in high-volume inspection. *Manufacturing Ergonomics* IBM.

DHHS. 1986. The health consequences of involuntary smoking. A report of the surgeon general. *DHHS Publication No.* (PHS) 87-8398. U.S. Department of Health and Human Services, Public Health Services, Office of the Asst. Secretary for Health, Office of Smoking and Health.

DOE. 1995. *Symposium Series* CONF-77114, pp. 663-690. U.S. Dept. of Energy.

Edwards, J.H. 1980. Microbial and immunological investigations and remedial action after an outbreak of humidifier fever. *British Journal of Industrial Medicine* 37:55-62.

Ekberg, L.E. 1994. Outdoor air contaminants and indoor air quality under transient conditions. *Indoor Air* 4:189-196.

Ellis, F.P. 1972. Mortality from heat illness and heat aggravated illness in the United States. *Environmental Research* 5.

EPA. 1989. Radon and radon reduction technology. EPA-600/9-89/006a, 1:4-15. U.S. Environmental Protection Agency, Washington, DC.

EPA. 1991. Introduction to indoor air quality, a reference manual. *Report* EPA 400/3-91/003

EPA. 1992a. Respiratory health effects of passive smoking: Lung cancer and other disorders, review draft. EPA/6O0-6-90/006B. Office of Research and Development, Washington, D.C.

EPA 1992b. Technical support document for the 1992 citizen's guide to radon. EPA 400-R-92-011.

Esswein, E.J. and M.F. Boeniger. 1994. Effect of an ozone generating air purifying device on reducing concentrations of formaldehyde in air. *Appl. Occup. Environ. Hyg.* 9(2).

FDA. 1990. Maximum Acceptable Level of Ozone. 21 CFR Ch. 1. Sec. 801.415: 24-25. U.S. Food and Drug Administration.

Feeley, J.C. 1988. *Architectural design and indoor microbial pollution.* R.B. Kundsin, ed. Oxford University Press, New York, 218-27.

Fliermans, C.B. 1985. Ecological niche of *Legionella pneumophila.* In: Critical Reviews of Microbiology, 75-116.

Franck, C., P. Skov, and O. Bach. 1993. Prevalence of objective eye manifestations in people working in office buildings with different prevalences of the sick building syndrome compared with the general population. *Int. Arch. Occup. Environ. Health* 65:65-69.

Garbesi, K. and R.G. Sextro. 1989. Modeling and field evidence of pressure-driven entry of soil gas into a home through permeable below-grade walls. *Environment Science Technology* 23:1481-87.

Gaul, L.E. and G.B. Underwood. 1952. Relation of dew point and barometric pressure to chapping of normal skin. *Journal of Investigative Dermatology* 19:9.

Gelperin, A. 1973. Humidification and upper respiratory infection incidence. *Heating, Piping and Air Conditioning* 45:77.

Girman, J.R. 1989. Volatile organic compounds and building bake-out. In *Occupational medicine: State of the art reviews—Problem buildings.* J.E. Cone and M.J. Hodgson, eds. Hanley and Belfus, Philadelphia. 4(4):695-712.

Girman J, Y-L Chang, S.B. Hayward, and KS Liu.. Causes of unintentional deaths from carbon monoxide poisonings in California. Unpublished.

Glantz, S.A. and W.W. Parmley. 1991. Passive smoking and heart disease epidemiology, physiology, and biochemistry. *Circulation* 83:633-42.

Gonzalez, R.R., L.G. Berglund, and A.P. Gagge. 1978. Indices of thermoregulatory strain for moderate exercise. *Journal of Applied Physiology: Respiratory Environmental and Exercise Physiology* 44(6):889-99.

Green, G.H. 1979. The effect of indoor relative humidity on colds. *ASHRAE Transactions* 85(1).

Green, G.H. 1982. The positive and negative effects of building humidification. *ASHRAE Transactions* 88(1):1049.

Greenberg, L. 1964. Asthma and temperature change. *Archives of Environmental Health* 8:642.

Guyton, A.C. 1968. *Textbook of medical physiology.* Saunders, Philadelphia.

Gyntelberg, F., P. Suadicami, J.W. Nielsen, P. Skov, O. Valbjorn, T. Nielsen, T.O. Schneider, O. Jorgenson, P. Wolkoff, C. Wilkins, S. Gravesen, and S. Nom. 1994. Dust and the sick-building syndrome. *Indoor Air* 4:223-238.

Hardy, J.D. 1971. Thermal comfort and health. *ASHRAE Journal* 13:43.

Henschel, D.B. 1988. Radon reduction techniques for detached houses—Technical guidance, 2nd ed. EPA/625/5-87/019. U.S. Environmental Protection Agency, Washington, DC.

Henschel, D.B. 1993. Radon reduction techniques for existing detached houses—Technical guidance, 3rd ed. EPA/625/R-93/011.

Higgins, I.T.T. 1983. What is an adverse heatlth effect? *APCA Journal* 33:661-63.

Hodgson, A.T., K. Garbesi, R.G. Sextro, and J.M. Daisey. 1988. Transport of volatile organic compounds from soil into a residential basement. Paper 88-95B.1, Proceedings of the 81st Annual Meeting of the Air Pollution Control Association. Also LBL *Report No.* 25267.

Hodgson, A.T., K. Garbesi, R.G. Sextro, and J.M. Daisey 1992. Soil gas contamination and entry of volatile organic compounds into a house near a landfill. *J. Air and Waste Mgmt. Assoc.* 42:277-283.

Hodgson, A.T. 1995. A review and a limited comparison of methods for measuring total volatile organic compounds in indoor air. *Indoor Air* 5(4): 247.

Hodgson, M.J., P.R. Morey, J.S. Simon, T.D. Waters, and J.N. Fink. 1987. *American Journal of Epidemiology* 125:631.

Hodgson, M.J., P.R. Morey, M. Attfield, W. Sorenson, J.N. Fink, W.W. Rhodes, and G.S. Visvesvara. 1985. *Archives of Environmental Health* 40: 96.

Holland, W.W. 1961. Influence of the weather on respiratory and heart disease. *Lancet* 2:338.

Holmberg, K. 1987. Proceedings of the 4th International Conference on Indoor Air Quality and Climate, Berlin 1:637-42.

Houk, V. 1980. Spread of tuberculosis via recirculated air in naval vessel: The Byrd study. *An. N.Y. Acad. Sci.* 353:10.

Hryhorczuk, D.O., L.J. Frateschi, J.W. Lipscomb, and R. Zhang. 1992. Use of the scan statistic to detect temporal clustering of poisonings. *J. Tox-Clinical Toxicology* 30:459-65.

Hyppel, A. 1984. Proceedings of the 3rd International Conference on Indoor Air Quality and Climate, Stockholm, Sweden, 3:443-47.

ISO. 1984. Guideline to the evaluation of responses of occupants of fixed structures, especially buildings and off-shore structures, to low frequency horizontal motion, 0.063 to 1 Hz. ISO 6897. International Organization for Standardization, Geneva.

Justesen, D.R., E.R. Adair, J.C. Stevens and V. Bruce-Wolfe. 1982. A comparative study of human sensory thresholds: 2450 MHz Microwaves vs. far-infrared radiation. *Bioelectromagnetics* 3:117-25.

Katayama, K. and M. Momiyana-Sakamoto. 1970. A biometeorological study of mortality from stroke and heart diseases. *Meteorological Geophysics* 21:127.

Kessler, W.R. and W.R. Anderson. 1951. Heat prostration in fibrocystic disease of pancreas and other conditions. *Pediatrics* 8:648.

Kilbourne, E.M., T.S. Jones, K. Choi, and S.B. Thacker. 1982. Risk factors for heatstroke: A case-control study. *JAMA* 247(24):3332-36.

Kjaergard, S., L. Molhave, and O.F. Pedersen. 1991. Human reactions to a mixture of indoor pollutants. *Atmos. Environ.* 25:1417-26.

Kjaergard, S. 1992. Assessment methods and causes of eye irritation in humans in indoor environments. H. Knoeppel and P. Wolkoff, eds. *Chemical, Microbiological, Health, and Comfort Aspects of Indoor Air Quality.* ECSC, EEC, EAEC, Brussels 115-127.

Koren, H. 1990. The inflammatory response of the human upper airways to volatile organic compounds. In: *Proceedings of Indoor Air 90* 1:325-330.

Koren, H., D.E. Graham, and R.B. Devlin. 1992. Exposure of humans to a volatile organic mixture III: Inflammatory response. *Arch. Environ. Health* 47:39-44.

Kroemer, K.H.E. 1988. Fundamentals of Industrial Hygiene. *Ergonomics*, 3rd ed. National Safety Council.

Kullman, G.J. and R.A. Hill 1990. Indoor air quality affected by abandoned gasoline tanks. *Applied Occupational Environmental Hygiene* 5:36-37.

Last, J.M. 1983. *Public health and human ecology.* Appleton & Lange, New York.

Laviana, J.E., F.H. Rohles, and P.E. Bullock. 1988. Humidity, comfort and contact lenses. *ASHRAE Transactions* 94(1).

Leathart, G.L. 1972. Clinical Aspects of Respiratory Disease Due to Mining. *Medicine in the Mining Industry.*

Levetin, E. 1995. Fungi. In: Burge, H.A. 1995. *Bioaerosols.* CRC Press. Lewis Publishers. Boca Raton, Fl.

Levin, H. 1989. Building materials and indoor air quality. J.E. Cone and M.J. Hodgson, eds. *Occupational Medicine: State of the Art Reviews—Problem Buildings.* Hanley & Belfus, Philadelphia. 4(4):667-694.

Levin, H. 1991. Critical building design factors for indoor air quality and climate: current status and predicted trends. *Indoor Air* 1(1):79-92.

Lilienfeld, D.E. 1991. Asbestos-associated pleural mesothelioma in school teachers: A discussion of four cases. *Ann. N.Y. Acad. Sci.* 643:454-86.

Liu, K.S., J. Wesolowski, F.Y. Huang, K. Sexton, and S.B. Hayward. 1991. Irritant effects of formaldehyde exposure in mobile homes. *Environ. Health Perspect.* 94:91-94.

Liu, R. and M.A. Huza. 1995. Filtration and indoor air quality: A practical approach. *ASHRAE Journal* 37(2):18-23.

Livingston, J.M. and C.R. Jones. 1981. Living area contamination by chlordane used for termite treatment. *Bulletin of Environmental Contaminant Toxicology* 27:406-11.

Marcinowski, F., R.M. Lucas, and W.M. Yeager 1994. National and regional distributions of airborne radon concentrations in U.S. homes. *Health Physics* 66(6):699-706.

Martin, T.R. and M.B. Braken. 1986. Association of low birth weight passive smoke and exposure in pregnancy. *American Journal of Epidemiology* 124:633-42.

Martinez, B.F., M.L. Kirk, J.L. Annest, K.J. Lui, E.M. Kilbourne, and S.M. Smith. 1989. Geographic distribution of heat-related deaths among elderly persons. Use of county-level dot maps for injury surveillance and epidemiologic research. *JAMA* 262:2246-50.

Mbithi, J.N., V.S. Springthorpe, and S.A. Sattar. 1991. Effect of relative humidity and air temerature on survival of hepatitis A virus on environmental surfaces. *Appl. Environ. Microbiol.* 57: 1394.

McCann, J., L. Horn, J. Girman, and A.V. Nero. 1987. Short-term bioassays in the analysis of complex mixtures. Plenum Press, New York 5:325-54.

Meggs, W.J. 1994. RADS and RUDS—The toxic induction of asthma and rhinitis. *J. Toxicol. Clin. Toxicol.* 32:487-501.

Miksch, R.R., C.D. Hollowell, and H.E. Schmidt. 1982. Trace organic chemical contaminants in office spaces. *Atmospheric Environment* 8:129-37.

Milton, D.K., R.J. Gere, H.A. Feldman and I.A. Greaves. 1990. Endotoxin measurment: Aerosol sampling and application of a new limulus method. *Am. Ind. Hyg. Assoc. J.* 51:331.

Molhave, L. 1991. Volatile organic compounds, indoor air quality and health. *Indoor Air* 1(4):357-76.

Molhave, L., Z. Liu, A.H. Jorgensen, O.F. Pederson, and S. Kjaergard. 1993. Sensory and physiologic effects on humans of combined exposures to air temperatures and vol;atile organic compounds. *Indoor Air* 3:155-69.

Molhave, L. and M. Thorsen. 1990. A model for investigations of ventilation systems as sources for volatile organic compounds in indoor climate. *Atmospheric Environment* 25A:241-49.

Moolenaar, R.L., R.A. Etzel, and R.G. Parrish. 1995. Unintentional deaths from carbon monoxide poisoning in New Mexico, 1980 to 1988. A comparison of medical examiner and national mortality data. *Western Journal of Medicine* 163(5):431-4.

Morey, P.R. 1988. Architectural design and indoor microbial pollution. Oxford University Press, New York, 40-80.

Morey, P.R. 1990. The practitioner's approach to indoor air quality investigations. *Proceedings* of the Indoor Air Quality International Symposium. American Industrial Hygiene Association, Akron, OH.

Morey, P.R. and J.C. Feeley. 1988. *ASTM Standardization News* 16:54.

Morey, P.R., M.J. Hodgson, W.G. Sorenson, G.J. Kullman, W.W. Rhodes, and G.S. Visvesvara. 1986 Environmental studies in moldy office buildings. *ASHRAE Transactions* 93(1B):399-419.

Morey, P.R. and B.A. Jenkins. 1989. What are typical concentrations of fungi, total volatile organic compounds, and nitrogen dioxide in an office environment. Proceedings of IAQ '89, The Human Equation: Health and Comfort. ASHRAE, Atlanta, 67-71.

Morey, P.R. and J. Singh. 1991. Indoor air quality in non-industrial occupational environments. In: *Patty's industrial hygiene and toxicology*, 4th ed. 1(1).

Moser, M.R., T. R. Bender, H.S. Margolis, et al. 1979. An outbreak of influenza aboard a commercial airliner. *Am. J. Epidemiol.* 110:1.

Muller, C.O. and W.G. England. 1995. Achieving your indoor air quality goals: Which filtration system works best? *ASHRAE Journal* 27(2):24-32.

Mumford, J.L., X.Z. He, R.S. Chapman, S.R. Cao, D.B. Harris, K.M. Li, Y.L. Xian, W.Z. Jiang, C.W. Xu, J.C. Chang, W.E. Wilson, and M. Cooke. 1987. Lung cancer and indoor air pollution in Xuan Wei, China. *Science* 235:217-20.

Murray, W. 1995. Non-ionizing electromagnetic energies (chapter 14). *Patty's Industrial Hygiene and Toxicology* 3B:623-727 (644).

NAS. 1983. Polycyclic aromatic hydrocarbons: Evaluation of sources and effects. National Academy Press, Washington, D.C.

Nazaroff, W.W., S.R. Lewis, S.M. Doyle, B.A. Moed, and A.V. Nero. 1987. Experiments on pollutant transport from soil into residential basements by pressure-driven air flow. *Environment Science and Technology* 21:459.

Nazaroff, W.W., B.A. Moed, and R.G. Sextro. 1988. Soil as a source of indoor radon: Generation, migration and entry. In: *Radon and its decay products in indoor air.* Wiley, New York, 57-112.

NIOSH. 1991. Current Intelligence Bulletin 54: Environmental tobacco smoke in the workplace; Lung cancer and other health effects. DHHS (NIOSH) *Publication* 91-108. U.S. Dept. of Health and Human Services, Public Health Service, Centers for Disease Control, National Institute for Occupational Safety and Health, Cincinnati.

NIOSH. 1992. NIOSH recommendations for occupational safety and health compendium of policy documents and statements. DHHS (NIOSH) *Publication* 92-100. U.S. Dept. of Health and Human Services, Public Health Service, Centers for Disease Control, National Institute for Occupational Safety and Health, Atlanta.

NRC. 1986. Environmental tobacco smoke: Measuring exposures and assessing health effects. National Research Council. National Academy Press, Washington, D.C.

Oeschli, F.W. and R.W. Beuchley. 1970. Excess mortality associated with three Los Angeles September hot spells. *Environmental Research* 3:277.

Ohm, M., J.E. Juto, and K. Andersson. 1993. Nasal hyperreactivity and sick building syndrome. *IAQ 92: Environments for People.* ASHRAE, Atlanta.

Olishifski, J.B. 1971. Ergonomic stresses: Physical and mental. In: *Fundamentals of Industrial Hygiene*, 1st ed. National Safety Council, Itasca, IL.

Offermann, F.J., S.A. Loiselle, A.T. Hodgson, L.A. Gundel, and J.M. Daisey. 1991. A pilot study to measure indoor concentrations and emission rates of polycyclic aromatic hydrocarbons. *Indoor Air* 4:497-512.

Parsons, K.C. and M.J. Griffin. 1988. Whole-body vibration perception thresholds. *Journal of Sound and Vibration* 121(2):237-58.

Proctor and Hughes' chemical hazards in the workplace, 3rd ed. 1991. G.J. Hathaway et. al. eds. Van Nostrand Reinhold, New York.

Reiher, H. and F.J. Meister. 1931. Translation of Report Fts616RE 1946. Forshung VDI. 2,381-386, "The sensitivities of the human body to vibrations." Headquarters Air Material Command, Wright Field, Dayton, OH.

Ritchie, I.M. and R.E. Lehnen. 1987. Irritation from formaldehyde in mobile and regular homes. *Am J. Public Health* 77:323-328.

Rodes, C.E., R.M. Kamens, and R.W. Wiener. 1991. The significance and characteristics of the personal activity cloud on exposure assessment methods for indoor contaminants. *Indoor Air* 2:123-145.

Rohles, F.H., J.A. Woods, and P.R. Morey. 1989. Indoor environmental acceptability: Development of a rating scale. *ASHRAE Transactions* 95(1):23-27.

Samet, J.M., M.C. Marbury, and J.D. Spengler. 1987. Health effects and sources of indoor air pollution. *Am. Rev. Respir. Disease* 136:1486-1508.

Samet, J.M. 1989. Radon and lung cancer. *J. Natl. Cancer Institute* 81:145.

Schulman, J.H. and E.M. Kilbourne. 1962. Airborne transmission of influenza virus infection in mice. *Nature* 195:1129.

Selikoff, I.J., J. Churg, and E.C. Hammond. 1965. The occurrence of asbestosis among insulation workers in the United States. *Ann. Rpt. NY Academy of Science*.

Serati, A. and M. Wuthrich. 1969. Luftfreughtigkeit und saison Krankenheit. *Schweizerische Medizinische Wochenscrift* 99:46.

Shaughnessy, R.J., E. Levetin, J. Glocker, and K.L. Sublette. 1994. Effectiveness of portable indoor air cleaners: Sensory testing results. *Indoor Air* 4:179-188.

Sheldon, L., R.W. Handy, T. Hartwell, R.W. Whitmore, H. Zelon, and E.D. Pellizzari. 1988a. Indoor air quality in public buildings, Vol. I. EPA/600/S6-88/009a. Environmental Protection Agency, Washington, D.C.

Sheldon, L., H. Zelon, J. Sickles, C. Easton, T. Hartwell, and L. Wallace. 1988b. Indoor air quality in public buildings, Vol. II. EPA/600/S6-88/009b. Environmental Protection Agency, Research Triangle Park, NC.

Shickele, E. 1947. Environment and fatal heat stroke. *Military Surgeon* 100:235.

Smith, K.W. 1955. Pulmonary disability in asbestos workers. *Archives of Industrial Health.*

Spengler, J.D., H.A. Burge, and H.J. Su. 1992. Biological agents and the home environment. In: Bugs, Mold and Rot (I): Proceedings of the Moisture Control Workshop, E. Bales and W.B. Rose (eds). 11-18. Building Thermal Envelope Council, National Institute of Building Sciences, Washington, DC.

Stingdehag, O., I. Josefsson, and E. Hennington. 1988. Healthy Buildings '88, 3:611-20. Stockholm, Sweden.

Streifel, A.J., D. Vesley, F.S. Rhame, and B. Murray. 1989. Control of airborne fungal spores in a university hospital. *Environ. Int.* 15: 221.

Susskind, R.R. and M. Ishihara. 1965. The effects of wetting on cutaneous vulnerability. *Archives of Environmental Health* 11:529.

Tancrede, M., R. Wilson, L. Ziese, and E.A.C. Crouch. 1987. *Atmospheric Environment* 21:2187.

Tanner, A.B. 1980. Radon migration in the ground: A supplementary review. In: Natural radiation environment III. U.S. Department of Commerce, NTIS, Springfield, VA.

Teng, H.C. and H.E. Heyer, eds. 1955. The relationship between sudden changes in the weather and acute myocardial infarction. *American Heart Journal* 49:9.

Thompson, T. 1980. Innovative assembly and modular materials handling. *Assembly Engineering*.

Tichenor, B.A., G. Guo, J.E. Dunn, L.E. Sparks, and M.A. Mason. 1991. The interaction of vapour phase organic compounds with indoor sinks. *Indoor Air* 1:23-35.

Turk, B.H., R.J. Prill, D.T. Grimsrud, B.A. Moed, and R.G. Sextro. 1990. Characterizing the occurrence, sources and variability of radon in Pacific Northwest homes. *J. Air and Waste Management Assoc.* 40:498-506.

VanOsdell, D.W. and L.E. Sparks. 1995. Carbon absorption for indoor air cleaning. *ASHRAE Journal* 27(2):34-40.

Wallace, L.A., E. Pellizzari, and C. Wendel. 1991. Total volatile organic concentrations in 2700 personal, indoor, and outdoor air samples collected in the US EPA Team Studies. *Indoor Air* 4:465-477.

Wallingford, K.M. and J. Carpenter. 1986. Proceedings of IAQ '86, Managing indoor air for health and energy conservation. ASHRAE, 448-53.

WHO. 1987. Air Quality Guidelines for Europe. European Series No. 23. WHO, Copenhagen.

WHO. 1989. Indoor air quality: Organic pollutants. Report on a WHO-meeting, Euro Report and Studies III. WHO Regional Office for Europe, Copenhagen.

Wilkins, C.K., P. Wolkoff, F. Gyntelberg, P. Skov, and O. Valbjørn. 1993. Characterization of office dust by VOCs and TVOC release—Identification of potential irritant VOCs by partial least squares analysis. *Indoor Air* 3:283-290.

Wood, J.A. and M.L. Porter. 1987. Hazardous pollutants in Class II landfills. *J. Air Pollution Control Association* 37:609-15.

Wolkoff, P. and C.K. Wilkins. 1994. Indoor VOCs from household floor dust: Comparison of headspace with desorbed VOCs; method for VOC release determination. *Indoor Air* 4:248-254.

Wright, C.G. and R.B. Leidy. 1982. Chlorodane and heptachlor in the ambient air of houses treated for termites. *Bulletin of Environmental Contaminant Toxicology* 28:617-23.

Yoshizawa, S., F. Surgawa, S. Ozawo, Y. Kohsaka, and A. Matsumae. 1987. *Proceedings* 4th International Conference on Indoor Air Quality and Climate, Berlin 1:627-31.

Zenz, C. 1988. Occupational safety in industry, occupational medicine, principles and practical applications.

BIBLIOGRAPHY

Hodgson, A.T. and J.R. Girman. 1989. Design and protocol for monitoring indoor air quality. ASTM STP 1002. American Society for Testing and Materials, W. Conshohoken, PA, 244-56.

Meitz, A. 1986. Clean cooling systems minimize *Legionella* exposure. *Heating, Piping and Air Conditioning* 58(8): 99-102.

Ota, E. and E. Mulberg.1990. Exposure to formaldehyde from indoor air. California Air Resources Board ARB/RD-90-01.

Taylor, A.E., D.C. Johnson, and H. Kazemi. 1992. Environmental tobacco smoke and cardiovascular disease: A position paper from the council on cardiopulmonary and critical care. American Heart Association, Dallas.

Wallace, L.A. 1987. The total exposure assessment methodology TEAM. Study: Summary and analysis, Vol. 1. Office of Research and Development, U.S. Environmental Protection Agency.

Wallace, L.A. and C.A. Clayton. 1987. Volatile organic compounds in 600 U.S. houses: Major sources of personal exposure. *Proceedings* 4th International Conference on Indoor Air Quality and Climate, Berlin 1:183-87.

CHAPTER 10

ENVIRONMENTAL CONTROL FOR ANIMALS AND PLANTS

THERMAL conditions, air quality, lighting, noise, ion concentration, and crowding are important in designing structures for animals and plants. Thermal environment influences heat dissipation by animals and chemical process rates in plants. Lighting influences photoperiodism in animals and plants, and photosynthesis and regulation in plants. Air quality, noise, ion concentrations, and crowding can affect the health and/or productivity of animals or plants. This chapter summarizes the published results from various research projects and provides a concept of the physiological factors involved in controlling the environment.

ANIMALS

Animal performance (growth, egg or milk production, wool growth, and reproduction) and their conversion of feed to useful products are closely tied to the thermal environment. For each homeothermic species, an optimum thermal environment permits necessary and desirable body functions with minimum energetic input (Figure 1A). The optimal thermal environment—in terms of an effective temperature that integrates the effects of dry-bulb temperature, humidity, air movement, and radiation—is less important to the designer than the range of conditions that provides acceptable animal performance, efficiency, well-being, and economic return for a given species. Figure 1A depicts this range as the *zone of nominal losses*, selected to limit losses in performance to a level acceptable to the livestock manager. Researchers have found that the zone of nominal losses corresponds to the welfare plateau (i.e., welfare is enhanced by maintaining environmental conditions within the zone of nominal losses). Milk and egg production by mature animals also shows an optional thermal environment zone, or zone of nominal losses (Figure 2).

Developed from actual measurements of swine growth, Figure 1B shows the relationships of energy, growth, and efficiency with air temperature. In the case of growing pigs in Figure 1B, the range of temperatures from 60 to 72°F, which includes both optimal productivity and efficiency levels, represents acceptable design conditions to achieve maximum performance and efficiency. Even beyond that temperature range, performance and efficiency do not markedly decline in the growing pig until near the lower critical temperature (LCT) or upper critical temperature (UCT), and potential performance losses within the temperature range from 50 to 76°F may be acceptable. Response relationships, as shown in Figure 1B, allow environmental selection and design criteria to be based on penalties to performance (i.e, economic costs) and animal well-being—particularly when used with climatological information to evaluate risks for a particular situation (Hahn et al. 1983). Choosing housing requires caution, because research indicates that factors such as group versus individual penning, feed intake, and floor type can affect the LCT by 10°F.

The limits of acceptable values of the LCT and UCT depend on such effects as the species, breed, genetic characteristics of an individual animal's age, weight, sex, level of feeding and type of feed, prior conditioning, parasites, disease, social factors such as space allocation, lactation or gestation, and physical features of the environment. The LCT and UCT vary among individuals; data reported are for group means. As a result, the limits become statistical values based on animal population and altered by time-dependent factors.

Acceptable conditions are most commonly established based on temperature because an animal's sensible heat dissipation is largely influenced by the temperature difference between the animal's surface and ambient air. Humidity and air movement are sometimes included as modifiers for an effective temperature. This has been a logical development. Air movement is a secondary but influential factor in sensible heat dissipation. The importance of air velocities is species and age dependent (e.g., swine under 8 weeks of age experience slower gains and increased disease susceptibility when air velocity is increased from 25 to 50 fpm).

With warm or hot ambient temperature, elevated humidity can restrict performance severely. Relative humidity has little effect on the animal's heat dissipation during cold temperatures, and it is usually only moderately important to thermal comfort during moderate temperatures. (Information is limited on such interactions, as well as on the effects of barometric pressure, air composition, and thermal radiation.)

Animals housed in a closed environment alter air composition by reducing oxygen content and increasing carbon dioxide and vapor content. Decomposing waste products add methane, hydrogen sulfide, and ammonia. Animal activities and air movement add microscopic particles of dust from feed, bedding, and fecal material. Generally, a ventilation rate sufficient to remove water vapor adequately controls gases. However, improper air movement patterns, certain waste-handling methods, and special circumstances (e.g., disease outbreak) may indicate that more ventilation is necessary. In many cases, ventilation is not as effective for dust control as for gas control. Alternative dust control strategies may be needed.

ANIMAL CARE/WELFARE

Animal facilities that facilitate good animal care and welfare must be designed considering a wide range of environmental factors beyond thermal conditions. These include space requirements, flooring type, lighting, feed and water requirements, animal handling;

The preparation of this chapter is assigned to TC 2.2, Plant and Animal Environment, with cooperation of Committees SE-302 and SE-303 of the American Society of Agricultural Engineers.

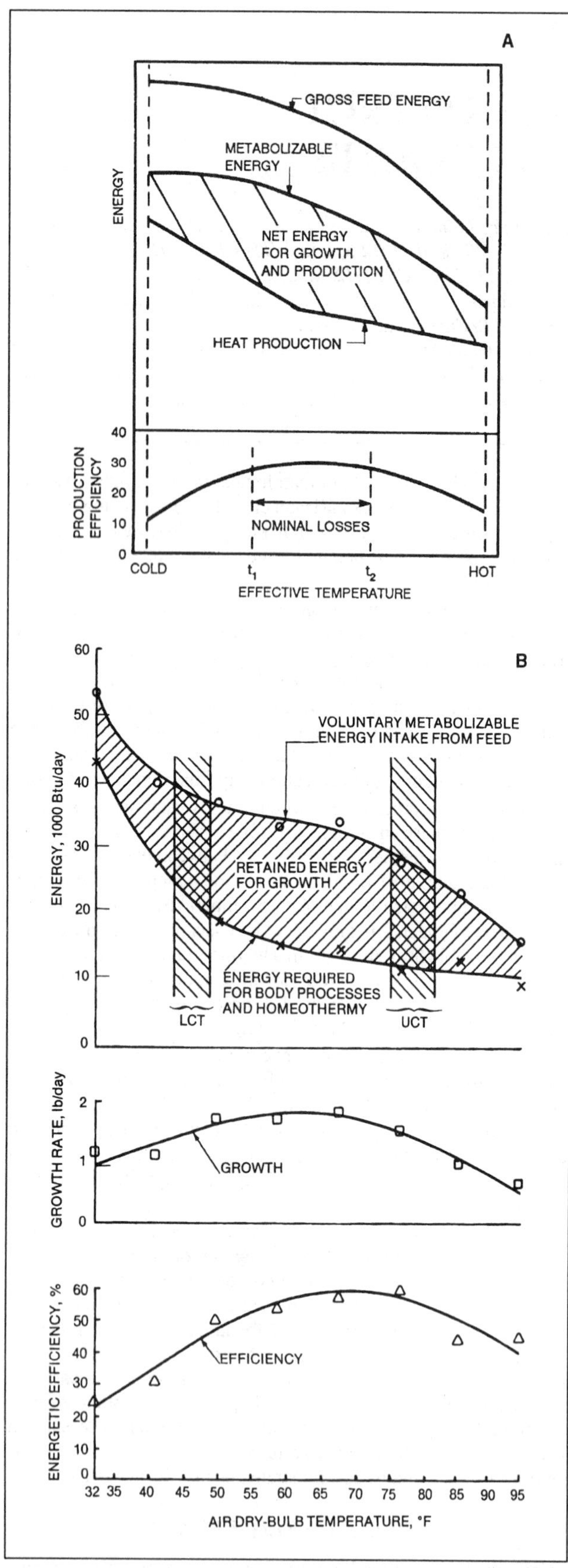

Fig. 1 Energetic and Performance Relationships Typical for Animals as Affected by Effective Environmental Temperatures
(Hahn et al. 1983)

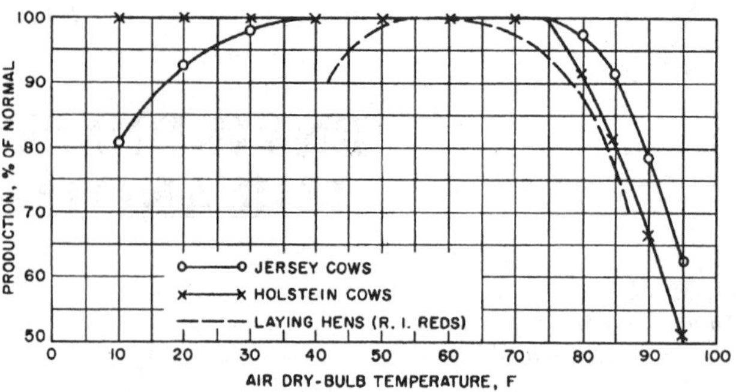

Fig. 2 Comparative Effect of Air Temperature in Mature Animal Production
(Hahn et al. 1983, Tienhoven et al. 1979)

and waste management. Facilities that meet animal care and welfare needs vary considerably by purpose, animal species, and geographic location. For additional information and guidelines for designing animal facilities to meet animal care and welfare needs, refer to the bibliography.

PHYSIOLOGICAL CONTROL SYSTEMS

An animal has a phenomenally stable control system. Despite wide variations in environmental, nutritional, or pathological conditions, animals can control blood pressure and composition, body temperature, respiration, and cardiac output without conscious effort. Physiological control systems react to unfavorable conditions to ensure survival of the animal at the expense of production or reproduction.

One physiological control system important for air-conditioning design is the homeothermic system—the means by which an animal adapts to its thermal environment. The animal strives to control body temperature by adjusting both heat production in the tissues and transfer of heat to the environment from exposed surfaces. The homeothermic system in domestic animals is a closed-loop system and can be analyzed like any closed-loop control system.

Heat Production

Heat production data have been measured for many farm animals (Yeck and Stewart 1959, Longhouse et al. 1960, 1968, Bond et al. 1959). Much of the early data were obtained for a basal condition (i.e., all life processes at a minimum level). Heat production under conditions of normal metabolic activity is more useful to the engineer. Such data are now available for most farm animals (Figures 3 and 4). Where possible, the data show total or sensible and latent heat for animals fed in a typical housing situation. Thus, the sensible heat used to evaporate urine moisture (assuming no other moisture) appears as latent heat. For design purposes, such data are better reflections of total heat partitioning than metabolic heat production obtained by calorimetry. Scott et al. (1983) provide additional information on specific situations.

The rate of heat production is primarily a function of temperature, animal species, and animal size. Heat production varies diurnally, depending on animal activity and eating times, and may change dramatically under special circumstances (e.g., when animals are disease-stressed). Building type and waste-handling system can affect the conversion of sensible to latent heat by more than 50%. Heat and moisture production of animals changes with other variables such as their genotype and growth rate. For example, a broiler chicken grows to 9 lb in 7 weeks compared to 12 weeks 30 years ago. Faster growth rate is usually associated with higher heat production, so updated data is needed. Therefore, the ranges of ventilation require careful analysis.

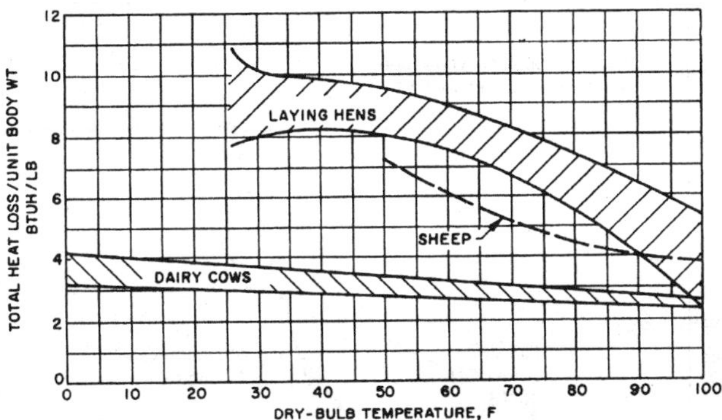

Fig. 3 Comparative Heat Loss of Mature, Producing Animals
(Tienhoven et al. 1979)

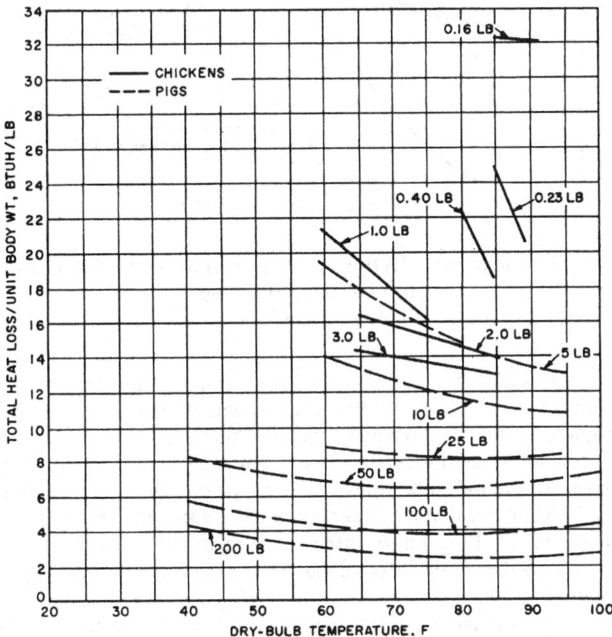

Fig. 4 Comparative Heat Loss of Growing Animals
(Bond et al. 1959)

Heat Transfer to Environment

An animal can control, to some extent, the amount of heat transferred from its body by chemical and physical regulations. In a cold environment, the animal's metabolism increases, which increases the amount of heat production, offsetting heat transfer to the environment. (The principal result of such chemical regulation is inefficient feed utilization.) Physical regulation is also used in a cold environment; blood circulation to subcutaneous capillaries decreases, hair or feathers are erected, and animals huddle together in an attempt to reduce sensible heat loss. In a hot environment, opposite physical and physiological responses generally occur to enable transfer to the environment of heat associated with necessary and productive life processes. In addition, as sensible heat dissipation becomes more difficult, evaporative heat loss increases and air moisture content becomes a factor in heat loss. Since the production of heat is a necessary by-product of growth and useful production, environmental limits to dissipating this heat cause a decrease in feed consumption and subsequent decrease in growth and production.

Most sensible heat of domestic animals is dissipated through the skin. Birds, sheep, and swine transfer most of their latent heat through the respiratory tract; cattle and horses transfer most of their latent heat through the skin. Blood transfers the heat produced by metabolism to the skin or to the lung surfaces, where it is dissipated by evaporation from the mucous layer coating the inside of the alveoli. Inspired air reaching the alveoli is heated almost to body temperature and may become nearly saturated with moisture. Expired air may not be saturated at body temperature—especially during periods of heat stress—because it is a combination of air, some of which has not reached deeply into the lungs. Minor amounts of heat are also transferred by ingestion of feed and water and through excretion.

Design factors affecting animal heat loss are (1) air temperature; (2) air vapor pressure; (3) air movement; (4) configuration, emissivity, absorptivity, and surface temperature of the surrounding shelter; and (5) temperature and conductivity of surfaces (e.g., floors with which the animal may be in contact).

Cyclic Conditions

The physiologic sensing elements respond both to environmental conditions and to changes in those conditions. Cycles of temperature, pressure, light, nutrients, parasites, magnetic fields, ionization, and other factors frequently occur with little engineering control.

Light is perhaps the earliest discovered and most important controlled environmental variable affecting reproductive processes (Farner 1961). However, the effects vary widely among animal species and age. Some animals grow and remain healthy with or without light (e.g., growing swine), while for others, lighting management is important. Studies have shown that (1) short photoperiods induce or accelerate estrus development in sheep; (2) day length affects semen production in sheep and horses (Farner 1961); (3) continuous white incandescent light during incubation of White Leghorn eggs caused eggs to hatch from 16 to 24 h earlier than eggs incubated in darkness (Shutze et al. 1962); and (4) red, yellow, or blue lights gave comparable results. Use of lights for only one of the three weeks also reduced the time required for hatching, but differences were not as marked. Percent hatchability was not affected by lighting treatments.

Light is used to delay sexual maturity in hens, which enhances subsequent production. This is done by gradually decreasing day length from hatching to 22 weeks of age or by abruptly decreasing day length to 9 h at 14 to 16 weeks of age. If pullets have previously been exposed to an increasing day length, the change should take place at 14 weeks of age; if exposed to a constant or decreasing schedule, a change at 16 weeks is adequate. Light intensities of 1 to 2 footcandles measured at bird height were adequate in all cases. Light can then be abruptly increased at 21 weeks of age to 14 or 16 h of light. The economic value of increasing day length beyond 14 h in a windowless poultry house, and 16 h in an open poultry house, has not been proved. Photoperiods such as 8 h of light (L), 10 h of darkness (D), 2 L and 4 D, and other cycles have improved feed utilization, egg production, and poultry growth (Buckland 1975, Riskowski et al. 1977).

Continuous light from hatching through 20 to 21 weeks of age markedly depressed subsequent egg production and caused a severe eye abnormality, but did not depress egg mass. (Light intensity was 1 to 3 footcandles, measured at bird height.)

Temperature cycles have been studied in cattle (Brody et al. 1955, Kibler and Brody 1956), swine (Bond et al. 1963, Nienaber et al. 1987), and poultry (Squibb 1959). The results differ somewhat among animals. Heat loss from cattle and swine can usually be calculated from average daily environmental temperatures with sufficient accuracy for design load calculation.

Productivity is only slightly different for averaged temperatures when cyclic conditions with a range less than 20°F are experienced (Squibb 1959). Above that range, productivity is depressed below that expected from the averaged temperature as determined in constant temperature tests, but under diurnally varying air conditions (68 to 77°F temperature and widely varying humidities), near-normal egg production is maintained.

Air Composition and Contaminants

The major contaminants in livestock housing are (1) respirable dusts from feed; manure; and animal skins, hair, and feathers; (2) microbes, both pathogenic and nonpathogenic, hosted in the respiratory tracts or animal wastes; and (3) gases of many types. Respirable dust particles have diameters between approximately 0.5 and 5 μm. Gases are produced from the metabolic processes of animals and from the anaerobic microbial degradation of wastes (Muehling 1970). The gases of most concern are ammonia, hydrogen sulfide, and carbon monoxide.

Dust levels in animal housing are high enough to create a nuisance in and near animal buildings, increase labor requirements for building and equipment maintenance, and interfere with the performance of heating and ventilating equipment. Dust has been implicated in poultry building fires. Dust is generated primarily by feed handling and increases in animal activity and air movement causing reentrainment of settled dust. Airborne microbes in a calf nursery are generated through defecation, urination, and coughing by the calves (Van Wicklen and Albright 1987).

Contaminants are a concern because they predispose animals to disease and poorer performance and affect operator health. Common agricultural animals can tolerate higher levels of most air contaminants than most humans without adverse health effects. However, animals experiencing stress (e.g., newborns, hot or cold, nutritionally limited, and sick animals) are more sensitive, and the presence of low levels of contaminants can have adverse effects on them.

The United States established minimum standards for exposure to air contaminants that are indicators of the maximum safe level for operators (Table 1) (OSHA 1985). These levels are time-dependent—people experiencing contaminants for less than a 40-h week can tolerate higher levels, while those exposed for longer periods can tolerate less. They are also additive; for example, someone exposed to 75% of the maximum dust level may be exposed to no more than 25% of the maximum for any other contaminant or combination of contaminants. Dust in animal buildings is biologically active, hence a lower threshold limit may be warranted. For example, Donham et al. (1989) proposed a more stringent threshold limit value for an 8 h day worker exposure of 6.5 μg/ft^3 in farm animal buildings.

The main adverse effects of dust are that it is a carrier for micro-organisms, odor, and contaminant gases; and it is an irritant that increases animal susceptibility to other contaminants. High levels of dust overload the lung clearance ability of animals. Dust may have several types of microbes that can affect animal performance. For example, Zhang (1994) found flora of micro-organisms, bacteria, and fungi in settled swine building dust. Pneumonia and other respiratory tract pathogenic bacteria can be transferred between animals from the air. Nonpathogenic bacteria may cause health problems if present in sufficient quantities to overload the body's immune system. Sick animals are more susceptible to pathogenic or nonpathogenic bacteria. Normally, when ventilation is adequate for moisture removal (Figure 5), that rate is sufficient to prevent problems due to airborne microbes; however, contaminants reach undesirable levels in confinement housing that is underventilated during cold weather (Bundy 1984). Air filtration is suggested as a means of reducing the incidence of pneumonia in an intensive calf housing (Hillman et al. 1992).

Airborne dust and microbes have been linked to respiratory disease in cattle, poultry, swine, horses, and laboratory animals.

Table 1 OSHA Standards for 8-h Day Worker Exposure

Dust	Total:	15 mg/m^3
	Respirable:	5 mg/m^3 (<5μm)
Ammonia (NH_4)	25 ppm	
Hydrogen Sulfide (H_2S)	10 ppm	
Carbon monoxide (CO)	50 ppm	

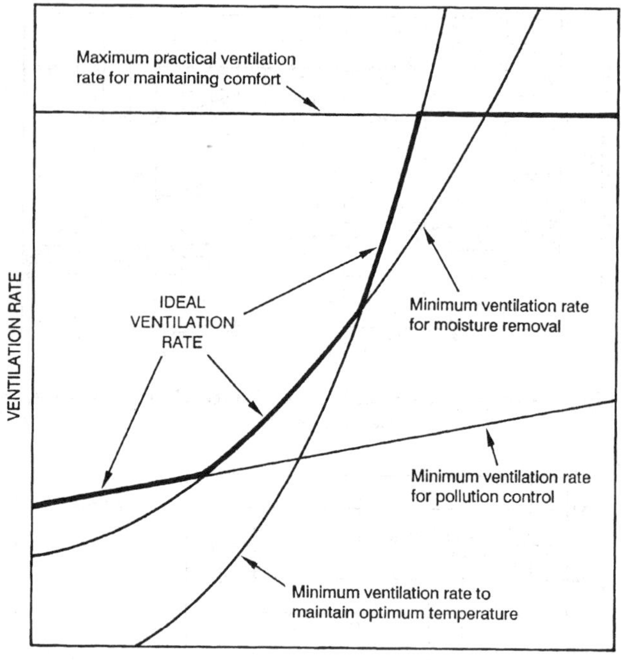

Fig. 5 Logic for Selecting the Appropriate Ventilation Rate in Livestock Buildings
(Christianson and Fehr 1983)

Infectious diseases, such as viral pneumonia and diarrhea, in calf barns have caused mortality rates between 20 and 80%, with the highest mortality occurring in enclosed barns that were underventilated during cold weather. Many respiratory diseases of poultry have been found to be transmitted via pathogenic microbes that can travel over 30 miles and remain infectious for months (Siegmund 1979). Respiratory diseases are also costly to the swine industry—the incidence of enzootic pneumonia in pigs has ranged from 30 to 75%.

Gaseous ammonia is frequently a contaminant causing serious problems in swine and poultry housing. Ammonia levels of 50 ppm have been shown to reduce body mass gain rates of swine (Curtis 1983), and many researchers suspect that adverse health effects begin at levels below 25 ppm. Some physiologists and veterinary science researchers suggest that the level is between 5 and 10 ppm (Donham 1987). Ammonia at concentrations greater than 60 ppm has been implicated in reduced body weight and feed consumption, and in increased ocular and respiratory problems with broiler chickens (Carr and Nicholson 1980). Ammonia levels are influenced by animal diet, the ventilation rate, sanitation practices, and the waste-handling system. Ventilating for moisture control is usually adequate to control ammonia below 10 ppm in buildings sanitized monthly, where wastes are not allowed to accumulate for more than two weeks and where manure solids are covered by water.

Hydrogen sulfide is produced mainly from the wastes and is present at toxic levels that usually cause problems only when liquid wastes are agitated or pumped and emitted gases can filter into the animal occupied area. Hydrogen sulfide production can rise to lethal levels when manure in a pit below the animals is agitated, so extra precautions are necessary to ensure that ventilation is sufficient to remove the contamination.

Carbon monoxide is usually produced by malfunctioning fuel-burning heaters and internal combustion engines used for spray washing equipment. Many heaters are unventilated, which, when combined with the corrosive environment in livestock housing, makes this equipment susceptible to failure. Heaters should be vented to the outdoors or sensors should be installed to ensure complete combustion.

Air contaminants are controlled primarily by proper ventilation design and management and by air cleaning. High contaminant levels are a greater problem during cold weather when ventilation rates are lowered to conserve heat (Sutton et al. 1987, Heber et al. 1988). Dust removal systems such as settling systems, cyclones, fibrous filters, and electrostatic precipitation have been considered (Bundy 1986). Proper design and management of waste-handling facilities, enhanced building hygiene, adding fat and oil to feed (Heber and Martin 1988), and reducing animal and human activity have also been used to control contaminants. A light spray of vegetable oil on the floor can reduce dust levels by 50%. Respirators have been recommended to provide protection for operators.

Air contaminants, the thermal environment, or both can reach lethal levels if the ventilation equipment is stopped by a power outage or equipment failure. Emergency power generation equipment should be available and alarm systems installed on important ventilation equipment.

Air Ionization

Charging the atmosphere with negative ions shows the beneficial results of reduced air contaminant levels and bacteria levels. Although negative ions kill bacteria, research results are inconclusive in determining whether the mode of action is air purification, a physiological effect on blood pH, ciliary stimulation, or a tranquilizing effect. Some researchers suggest that negative ionization is not beneficial because it can increase particle retention in the respiratory tract (Janni et al. 1984).

Of the various methods of ion generation investigated, those most used in the environmental field are thermionic, alpha particle emission, and high-voltage corona discharge. The latter operates more satisfactorily when discharge points are distributed over the ionized area, which helps overcome (1) ozone and nitrogen oxide production, which can occur when sufficient ion production for a large space is generated at one source and (2) much of the neutralizing effect caused by ions attached to dust particles, due to the short life of air ions. The entire enclosure acts as a low-level air cleaner because of electrostatic dust precipitation to grounded surfaces. Air bacteria counts drop about 50%, and even further reduction may be possible.

Interactive Stressors

Thermal (air temperature, air humidity, air velocity, and surrounding surface temperatures) and air quality (relative absence of dust, microbes, and contaminant gases) are the two stressors of most concern to ventilating a livestock structure. However, health status, animal age and stage of production, nutrition, and social conditions interact with the thermal and air quality conditions to determine animal health and performance.

Whether stressors are linearly additive or otherwise is an important research area. McFarlane (1987) found that ammonia, heat, acoustic noise, disease, and beak-trimming stressor effects on chickens were linearly additive in effect on the feed efficiencies and daily gain rates.

Comparing Individual Animal and Room Heat Production Data

Liquid wastes from urine, manure, waterer spillage, and cleaning affect the sensible and latent heat fraction in a room (ASAE 1991). More water on the floors requires more thermal energy from the animals or the building heating system to evaporate the moisture. This evaporated moisture must then be removed by ventilation to prevent condensation on building surfaces and equipment and minimize adverse health effects, which can occur when humidities exceed 80%.

Floor flushing or excessive water spillage by animals can increase latent heat production by one-third from animal moisture production data, with a corresponding reduction in sensible heat production on a per-animal basis (ASAE 1991). Partially slatted floors with underfloor storage of wastes reduce the room moisture production by approximately 35%, compared to solid floor systems. Rooms with slatted floors throughout the building may have moisture production rates as much as 50% lower than solid floors. Therefore, slatted floors reduce the ventilation rate required for moisture removal, while excessive water spillage increases the moisture removal ventilation rate.

Influence of Genetic Change and Breed on Heat Production

Researchers have noted significant genetic correlations with performance. For example, swine breeds with high fat content characteristics outgained low fat content breeds, but the low fat content lines used feed more efficiently (Bereskin et al. 1975). Barrows (male hogs castrated before sexual maturity) ate 6% more than gilts (young sows that have not farrowed) to sustain a 7% faster growth rate.

Genetic improvements within breeds suggest that performance data should be updated periodically. Between 1962 and 1977, for example, the average gain rate for swine increased 0.15 lb per day, and feed efficiency improved 15% for Missouri hogs (Thomeczek et al. 1977).

CATTLE

Growth

Figure 6 shows the general growth rate for both beef and dairy breeds. Efficiency of beef calf growth is of economic importance; the dairy calf is developed for adult productive and reproductive capacity. Fattening calves and yearlings could be fed increasing amounts of grain and hay, with an expected gain of about 2 lb per day. Figure 7 shows the effect of temperature on the growth rate of several breeds of beef calves.

Food and energy requirements for growth can be computed by taking the difference between the energy in the available ration consumed and the energy required for maintenance and growth. Energy requirements for maintenance at an air temperature of 68°F has been reported to be 8.65 Btu/h per pound of body weight (Blaxter and Wood 1951), 7.98 Btu/h (Bryant et al. 1967), and 8.34 Btu/h

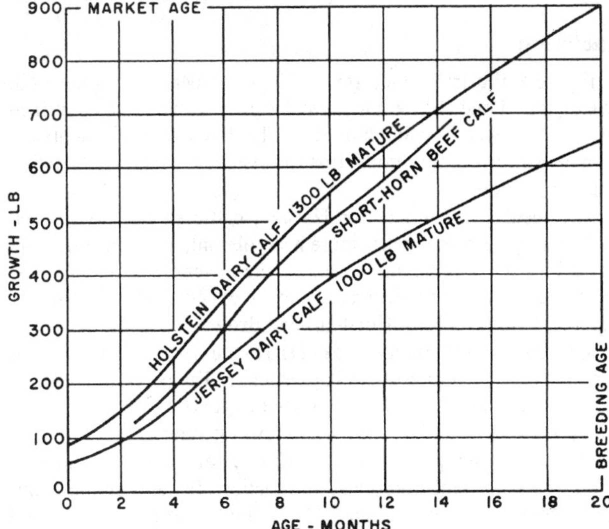

Fig. 6 General Growth Curves for Calves Fed Grain
(Johnson et al. 1958, Ragsdale 1960)

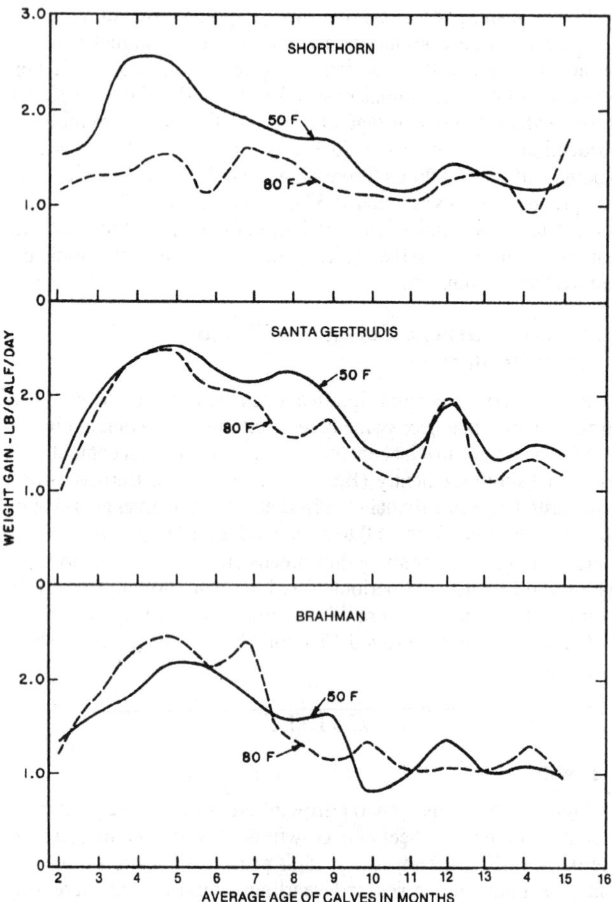

Fig. 7 Daily Weight Gain of Beef Calves
(Longhouse et al. 1960)

(Gebremedhin et al. 1983). Using 370 kcal/lb body tissue for conversion of calories to body tissue (Brody 1945), the growth rate of a calf at 70°F (5.38×10^{-4} lb gain per hour per pound of body weight) can be nearly 30 times that of a calf being kept at 37°F [0.18×10^{-4} lb/h·lb]. Alternatively, to get the same growth rate at 37°F as at 70°F, the same ration must be fed at the rate of 11.6% of body weight each day, a 16% increase (Gebremedhin et al. 1981).

Lactation

Figure 8 illustrates the effect of high temperature on the milk production of one Holstein in a test designed to study the combined effects of temperature and humidity. The dotted line represents the normal decline in milk production (persistency) from an advancing stage of lactation.

The ideal environment for Holstein cattle should not exceed 75°F. Jerseys are somewhat more heat tolerant, and their limit can be 80°F (Yeck and Stewart 1959). At the lower end of the temperature scale, production decreases may be expected at 30°F for Jerseys and below 10°F for Holsteins and Brown Swiss.

A temperature-humidity index (THI) expresses the relationship of temperature and humidity to milk production. Temperature-humidity indexes have a greater effect on cows with a genetic potential for high milk production than for those with a lower potential (Figure 9).

Under high temperatures, a cooling system can help maintain milk production and reproductive function. In field studies conducted by Bucklin and Turner (1991) in both warm and hot, humid climates, a system providing evaporative cooling by forced air movement and direct sprinkling of lactating cows increased milk production by a range of 7 to 15% over cows not exposed to cooling.

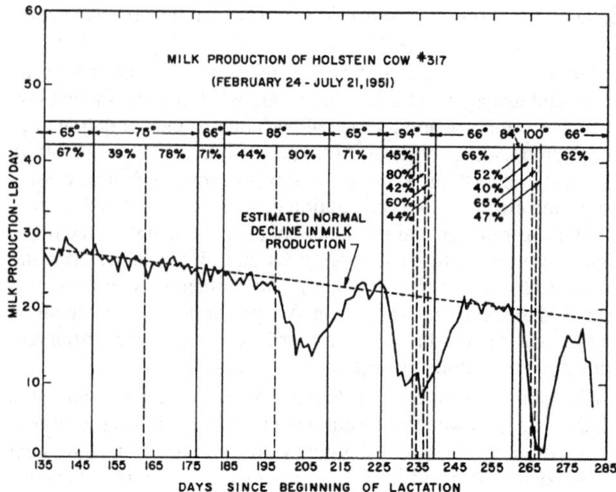

Fig. 8 Milk Production Decline of One Holstein during Temperature-Humidity Test
(Yeck and Stewart 1959)

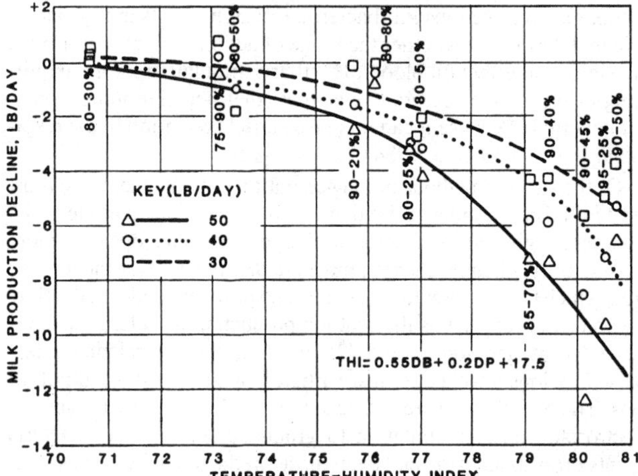

Fig. 9 Effect of Milk Production Level on Average Decline in Milk Production, versus Temperature-Humidity Index
(Johnson et al. 1962)

Reproduction

Prolonged low temperatures, even those well below freezing, do not affect the reproductive performance of farm livestock, but breeding efficiency of both sexes decreases under summer conditions. Sustained temperatures above 85°F may decrease fertility, sperm production, and semen quality of males, and increase anestrus and embryonic death in females. In bulls, temperatures above 75°F decrease spermatogenesis, and long exposures at 85°F or above cause temporary sterility. The extent of the reaction depends on temperature rise and exposure duration.

In females, Guazdauskas (1985) observed that conception rates decreased from a range of 40 to 80% conception in thermoneutral environments (50 to 72°F) to a range of 10 to 51% conception in hot environments (>81.5°F). A cooling system involving a cooled shade and artificially induced air movement increased breeding efficiency by approximately 100% in a hot climate (Wiersma and Stott 1969).

Heat and Moisture Production

Sensible and latent heat production from individual animals (Figure 10) differ from the stable heat and moisture production of

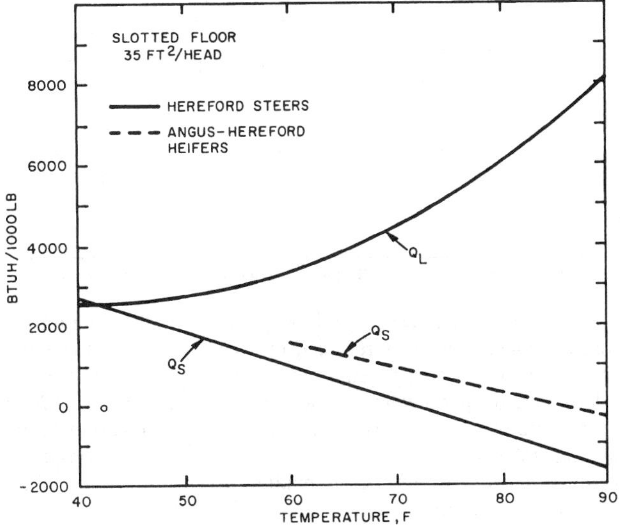

Fig. 10 Sensible and Latent Heat Production per Unit Livestock Mass (Weight)
(Hellickson et al. 1974)

Fig. 11 Stable Heat and Moisture Dissipation Rates Dairy Cattle Stanchioned in Enclosed Stables
(Yeck and Stewart 1959)

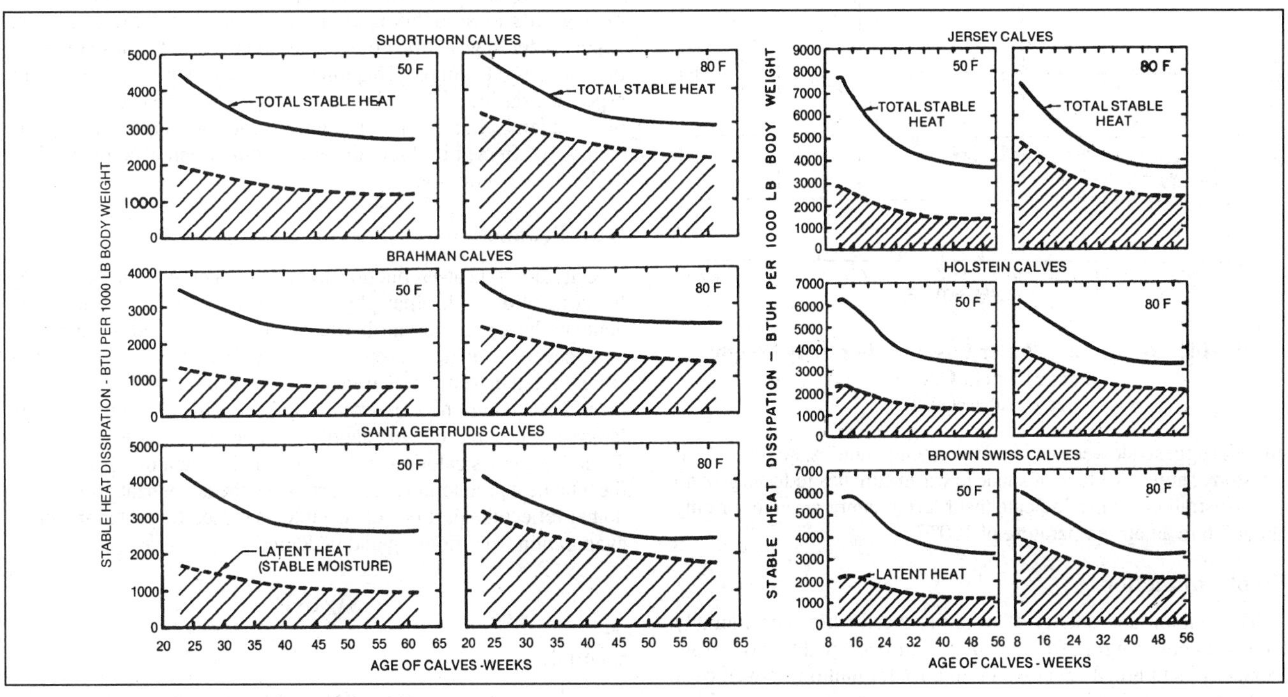

Fig. 12 Stable Heat and Moisture Loads for Beef and Dairy Calves in Pens
(Yeck 1957, Yeck and Stewart 1960)

animals (Figures 11 and 12). The data in Figures 11 and 12 were obtained while ambient conditions were at constant temperatures and relative humidities were between 55 and 70%. The effects of evaporation from feces and urine are included in these data.

The rate of cutaneous water loss is very small at colder temperatures but rises sharply above 64°F. Cutaneous evaporation as a means of heat loss in calves becomes increasingly important as the air temperature rises above 75°F. As the air temperature rises, the proportion of nonevaporative cooling decreases. Above 86°F, about 80% of the heat transferred is by evaporative cooling. Gebremedhin et al. (1981) observed that cutaneous water loss from calves varied between 6×10^{-5} and 40×10^{-5} lb/h per pound of body weight for air temperatures between 32 and 64°F and increases steadily beyond 64°F (Figure 13). Water loss by respiration, the other avenue of loss, is shown in Figure 14.

SHEEP

Growth

In normal environments, an average daily gain of 0.5 lb per day can be expected of lambs marketed at 80 to 120 lb, depending on breed. Hampshires often gain more than 1 lb per day, and lambs can be fed to weigh over 200 lb while still under a year old.

Variations exist among breeds and strains of sheep in their ability to adjust to environmental changes. Temperature effects on sheep

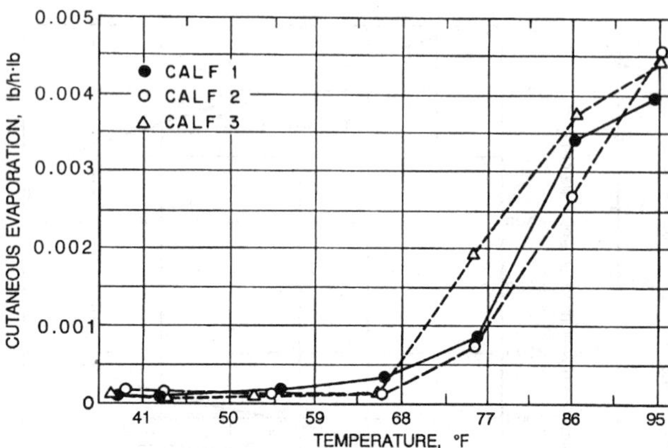

Fig. 13 Cutaneous Water Loss per Unit Body Weight of Holstein Calves
(Gebremedhin et al. 1981)

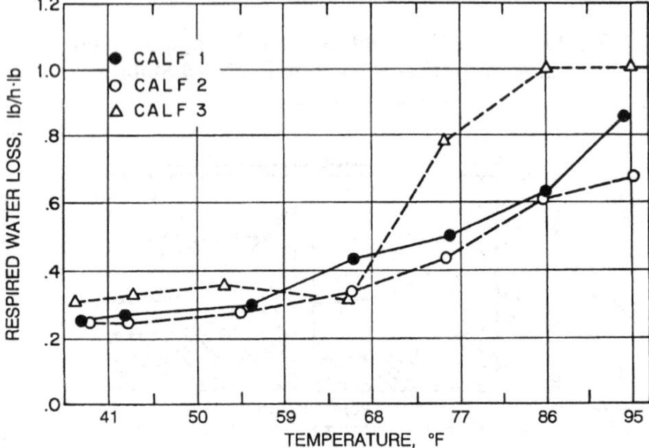

Fig. 14 Respiratory Water Loss per Unit Body Weight of Holstein Calves
(Gebremedhin et al. 1981)

growth suggest a lower rate at elevated temperatures. A South African study shows that lack of shade in warm climates reduces growth rate. Australian studies indicate that Merino lambs survive for only about 2 h in an air temperature of 100°F.

Wool Production

The amount and quality of wool produced varies considerably among breeds, ranging from about 3 lb of poor quality wool from Dorset ewes (a breed developed primarily for mutton) to well over 10 lb of high quality fleece from dual-purpose breeds. Rambouillet, Merino, and Columbia-Southdale breeds supplied 7 lb per year, and about 4.5 lb was grown by a Hampshire, where monthly shearing increased wool production by about 1 lb per year.

Environmental factors such as photoperiod, nutritional level, and temperature also affect wool growth. Skin temperature is considered a dominant factor; high wool growth is associated with high skin temperatures. In a related study, low subcutaneous blood circulation limited wool growth. A thick fleece appears to limit radiant heat loads.

Reproduction

Sheep mate only during certain periods of the year, but they can sometimes be induced to mate outside the normal season if the natural environment is modified. Ewes may breed if exposed to an air temperature of 45°F before mating is attempted.

Table 2 Heat Production of Sheep

Fleece Length, in.	At 46°F Total[a]	At 46°F Latent[b]	At 68°F Total[a]	At 68°F Latent[b]	At 90°F Total[a]	At 90°F Latent[b]
Mature, maintenance-fed						
Shorn	4.0	8%	2.7	12%	2.0	38%
1.2	2.2	29%	2.0	28%	2.0	65%
2.4	2.0	23%	1.9	43%	1.9	76%
Lambs, 1 to 14 days						
Normal	10.5	—	8.0	—	—	—

Source: Scott et al. (1983). [a]Btu/h·lb of body mass [b]Percent of total heat

In studies on the effects of high temperatures on sheep reproduction, air temperature was reported to affect the spermatogenesis rate. Rams kept at 90°F environmental temperature showed a reduced rate of spermatogenesis, although a 3-week period at 70°F effected recovery to the normal rate. Later reports suggest that high air temperatures may cause lowered fertility for other reasons. Subjecting ewes to 100°F just before mating causes fertilization failure because of some form of ovum structure degeneration.

Under practical conditions, early embryonic death appears to be the greatest loss in potential offspring conceived in a high-temperature environment. Degree of susceptibility to high temperature is greatest near mating time but generally decreases as the length of time after mating increases. A temperature of 90°F and 65% rh at mating kills most embryos at an early age. The same conditions applied later in gestation do not cause death, but do cause the birth of small, weak lambs. The longer the high-temperature period, especially during the last third of gestation, the greater the number of weak lambs born. All reproductive processes appear to be more adversely affected by high air temperature when accompanied by high relative humidity.

Heat Production

Only a few heat-production tests have been conducted, mostly for correlation with other physiological data. A lamb's ability to generate heat is important because it is normally born during the most severe climatic season. One study indicated newborn lambs were limited in heat production to 30 Btu/h·lb—five times the basal level. Test conditions were from 14 to 32°F with a 12 mph wind, but lambs are capable of withstanding temperatures as low as 40°F. Table 2 provides calorimeter heat production data for sheep; while the total heat production data are reliable, the latent heat proportions do not reflect the portions of sensible heat used for water and urine evaporation from flooring and bedding.

SWINE

Growth

Ambient air temperature affects the feed conversion and daily weight of growing swine. As shown in Figure 1B, a temperature range of 60 to 72°F produces maximum rates of gain and feed use for 155 to 220 lb hogs, while a broader range of 50 to 76°F reduces performance only slightly. For 45 to 130 lb animals, the optimal and nominal loss ranges are about 62 to 73°F and 55 to 76°F, respectively (Kibler and Brody 1956). Younger animals require temperatures of 73 to 82°F for best performance, and piglets from 3 days to 2 weeks of age should have 86 to 90°F conditions (Hahn 1983).

Daily air temperature cycles of more than 10°F on either side of 70°F result in reduced daily gain by pigs and an increased feed requirement per unit of gain (Nienaber et al. 1987). Reasonably constant air temperatures are desirable.

The level of air temperature in which swine grow affects deposition and retention of protein (carcass quality). Lean meat formation is reportedly highest in pigs raised between 60 and 70°F (Mount

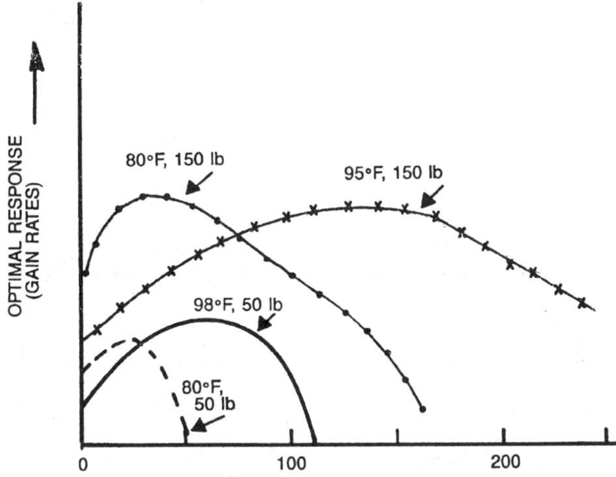

Fig. 15 Swine Response to Air Velocity
(Riskowski and Bundy 1986, Bond et al. 1965)

1963). However, the ratio of protein to fat decreases at air temperatures above 59°F.

Swine gains and feed conversion rates are highly sensitive to air velocities and, in many cases, are affected adversely at velocities as low as 50 fpm. Swine less than 8 weeks of age should not be exposed to velocities greater than 50 fpm; a velocity lower than 25 fpm is preferred when temperatures are in the recommended range. Even in hot conditions, pigs are affected adversely by air velocities greater than 200 fpm (Bond et al. 1965, Gunnarson et al. 1967, Riskowski and Bundy 1991). Feed utilization and gain are much better at low air velocities (35 fpm) than at high (300 fpm) air velocities when temperatures are optimal (Figure 15).

Reproduction

Merkle and Hazen (1967) and Heard et al. (1986) have shown that sows benefit from some type of cooling in hot weather. In these studies, cool, dry air was directed at the sow to relieve heat stress by increasing evaporative and convective heat dissipation. Field studies of breeding problems and resultant small litters have established that both sexes suffer losses from high temperature. Sprinkling the sow during hot weather at breeding time and shortly after resulted in more live births than with unsprinkled sows. Sprinkling boars before mating also increased the number of live births per litter. Conception rate varied from about 100% of normal at 70°F to only 70% at 90°F. Breeding difficulties and a decrease in live embryos were observed in tests with controlled temperatures and relative humidities (Roller and Teague 1966). In some species, spontaneous abortion under severe heat stress may save the mother's life; however, sows appear to die of heat prostration, due to extra metabolic heat generation in late pregnancy, before spontaneous abortion occurs.

Little information is available on the value of temperature control for sows during cold weather. Forcing sows to produce the necessary heat to maintain body temperature may require rigid nutritional management to avoid the animals from becoming overweight—a condition resulting in poorer conception and smaller litters.

Heat and Moisture Production

Table 3 (Butchbaker and Shanklin 1964, Ota et al. 1982) shows direct calorimetry heat production for piglets. Figures 16 and 17 (Bond et al. 1959) show heat and moisture that must be accounted for in ventilating or air conditioning older swine housing for swine between 50 and 400 lb, housed at temperatures from 50 to 90°F. These data are the sensible and latent heat levels measured in a room containing hogs. For design purposes, the room heat data reflect

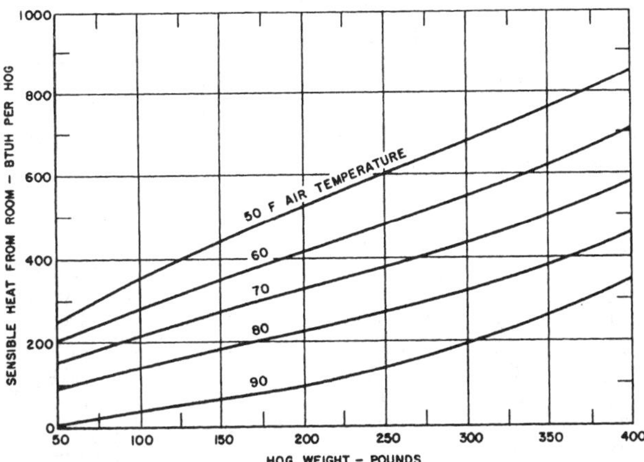

Fig. 16 Room Sensible Heat in Hog House
(Bond et al. 1959)

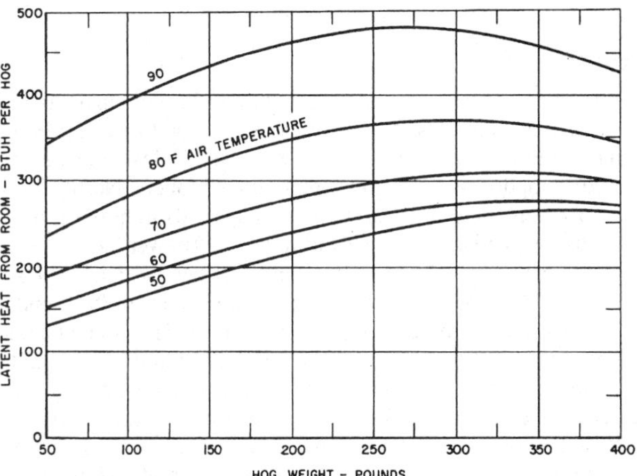

Fig. 17 Room Latent Heat in Hog House
(Bond et al. 1959)

Table 3 Heat Production of Grouped Nursery Pigs

Weight Range, lb	Temperature, °F	Total Heat Production, Btu/h · lb	Latent Heat, % of Total
9 to 13	85	5.1	33
13 to 24	75	7.0	31
24 to 37	65	7.8	30

Source: Ota et al. (1982).

swine housing conditions rather than metabolic heat production from an individual animal.

The same data can be applied even where building temperatures are cycling, if the average air temperature during the cycle is used for design. The total animal heat load and the latent load that the ventilation system must remove are greater than the values in Figures 16 and 17, if the air velocity around the animals is more than about 50 fpm.

The lower critical temperature (the ambient temperature below which heat production increases) is about 77 to 86°F for a group of newborn pigs; for a single piglet, the critical temperature is about 93 to 95°F. As the pigs grow, the critical temperature falls—for 4.5 to 9 lb pigs, it is between 86 and 95°F; for 9 to 18 lb pigs, between 77 and 86°F. The floor bedding materials for growing piglets influence these critical temperatures. For a group of nine pigs averaging 88 lb,

Table 4 Mass Grain and Feed Conversion for Meat-Type Growing Pullets

	Restricted Feeding		Full Feeding	
Age, Weeks	Gain in Weight for Week, %	Feed Conversion for Week	Gain in Weight for Week, %	Feed Conversion for Week
4	22.2	3.32	44.4	2.21
6	15.4	3.85	25.7	2.32
8	11.8	4.24	20.0	2.71
10	9.5	4.66	13.2	3.54
12	8.0	5.15	9.5	4.88
14	6.9	5.64	7.6	6.50
16	6.1	6.13	5.9	7.56
18	5.4	6.62	4.3	9.03
20	4.9	7.11	3.3	11.00
22	4.4	6.04	2.5	14.10
24	3.9	5.39	1.8	14.44

Source: North (1984).

the critical temperature is 53 to 55°F on straw, 57 to 59°F on asphalt, and 66 to 68°F on concrete slabs (Butchbaker and Shanklin 1964).

CHICKENS

Growth

The tremendous worldwide production of broiler chickens is a result of improvements in genetics, nutrition, and housing. Four-pound broilers are produced in about 8 weeks. Table 4 indicates the feed consumption and growth of one strain of broilers (North 1984).

Intermittent light and darkness (photoperiod) improve growth and conserve feed and energy (McDaniel and Brewer 1975).

Supplemental heat is generally needed for newly hatched chicks. Broiler strains respond well to brooder temperatures of 92°F under the heaters. Thereafter, air temperatures may be decreased at a rate of 7°F per week until 70°F is reached. Relative humidities of 65 to 70% promote good feathering and market-quality broilers. Air velocity effects have received limited attention. Some benefit from increased velocity (up to 500 fpm) is obtained at air temperatures from 75 to 95°F (Drury 1966). High-velocity air, at a temperature above the feather temperatures, causes more, not less, heat stress. Observations showed that 28-day-old broilers rested in areas with airflow temperature combinations of 55 fpm at 60°F, 100 fpm at 70°F, and 150 fpm at 75°F.

Reproduction

Adverse effects of high thermal environments (above 85°F) on egg production include fewer eggs, reduced egg weight, and thinner shells. Hens over 1 year of age are more adversely affected by high thermal environments than younger hens. Larger hens, such as Rhode Island Reds, are more adversely affected than smaller hens, such as White Leghorns. Hatchability also declines as temperatures increase. Although the fertility rate is good at 70°F, it declines at 86°F. Feed requirements increase markedly below 45°F; activity and productivity decline below 32°F. The suggested ideal environment is between 55 and 75°F.

Increasing relative humidity above 79%, at 85°F dry-bulb temperature, produces an increased respiration rate and drooping wings. After 7 to 10 days, acclimatization may reduce the respiration rate, but production will still be less than optimal. Generally, 70 to 75% rh is recommended to maintain conditions that are neither too dusty nor too restrictive of latent heat dissipation.

Heat and Moisture Production

Figure 18 shows the total sensible and latent heat produced by laying hens during day and night at various temperatures. Figure 19 shows the same data for chicks. Figures 20 and 21 show sensible and latent heat production for broilers grown on litter at typical

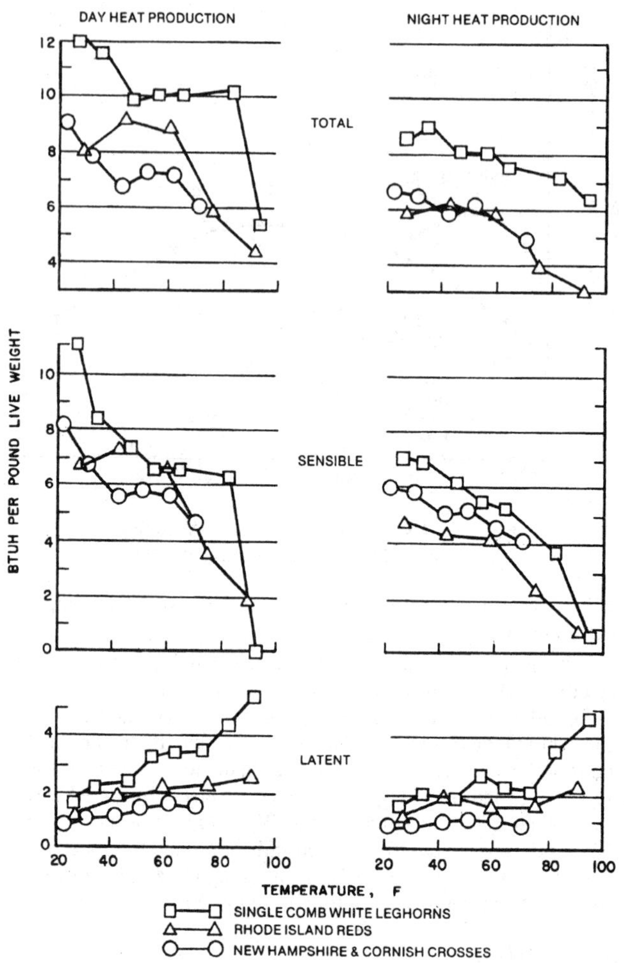

Fig. 18 Heat and Moisture Loads for Caged Laying Hens at Various Air Temperatures
(Ota and McNally 1961)

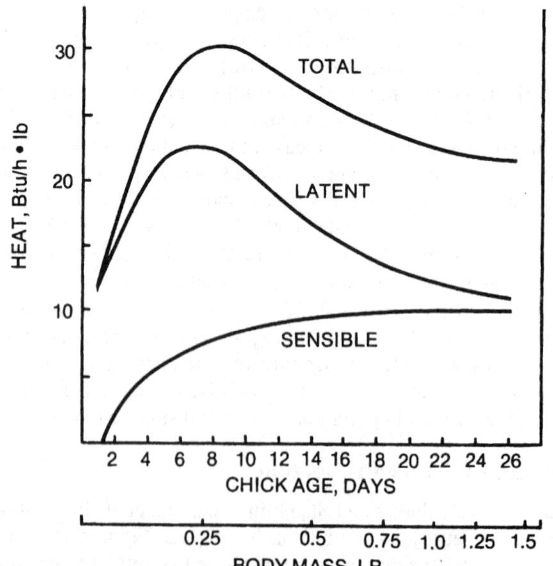

Fig. 19 Sensible, Latent, and Total Heat for Chicks Brooded on Litter
(Reece and Lott 1982)

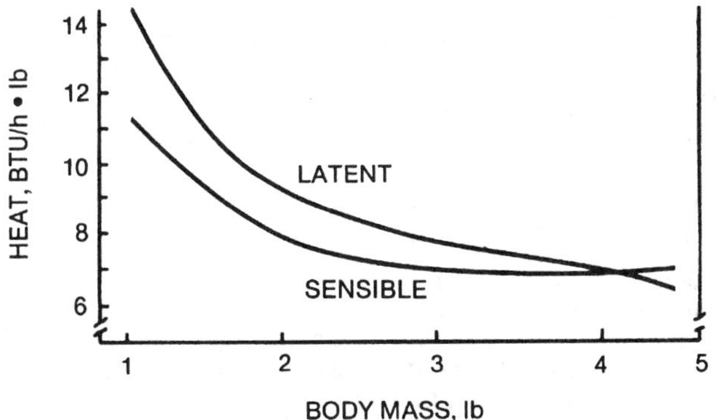

Fig. 20 Sensible and Latent Heat for Broilers Raised at 60°F on Litter
(Reece and Lott 1982)

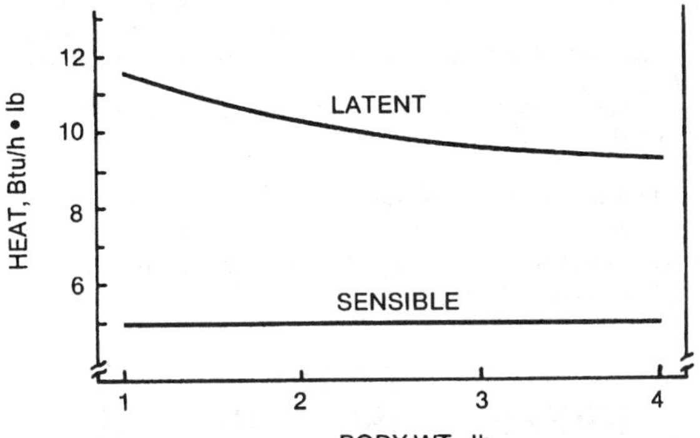

Fig. 21 Sensible and Latent Heat for Broilers Raised at 80°F on Litter
(Reece and Lott 1982)

stocking rates at two temperatures. The effects of heat and moisture absorption or release by the litter are included in these figures.

The sensible heat loss of a layer pullet is similar to that of a broiler at the same weight (Zulovich et al. 1987). However, the latent heat loss of the pullet is approximately 50% that of the broiler. Layer pullets consume 20 to 25% less feed energy than broilers. Thus, the total heat production is decreased with the reduction being in latent heat loss.

TURKEYS

Growth

The turkey poult has the best rate and efficiency of gain at a brooder room temperature of 70 to 75°F for the first 2 weeks, and 65°F thereafter, with about 70% rh. Initial brooder temperature should be 100°F, reduced 5°F per week until room temperature is reached. Proper control of photoperiod stimulates the growth rate in turkeys. Table 5 shows weekly cumulative average live weight for both sexes (Sell 1990). Figure 22 provides calorimetric heat production data for medium breed turkeys.

Reproduction

Mature turkeys tolerate temperature and humidity over a range of at least 20 to 90°F and 35 to 85% rh. Since mature birds are kept only for fertile egg production, lighting for off-season egg production becomes very important. The young stock are raised to 20 weeks on natural light. At 20 weeks, females are placed in a totally darkened pen and given 8 h per day illumination at an intensity of 2 to 5 footcandles at bird's-eye level. At 30 weeks, the sexes are mixed, and the lighting period is increased to 13 to 15 h. The breeding season then continues for 12 to 26 weeks.

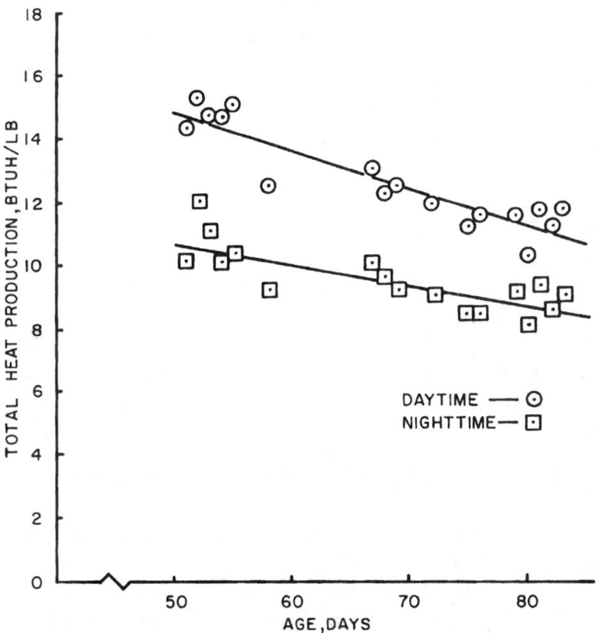

Fig. 22 Total Heat Production of Medium Breed Turkeys versus Age during Daytime and Nighttime
(Buffington et al. 1974)

Table 5 Weekly Average Live Mass of Male and Female Turkeys

Age, Weeks	Male, lb	Female, lb
0	0.07	0.07
1	0.33	0.31
2	0.66	0.64
3	1.15	1.10
4	1.85	1.76
5	2.80	2.58
6	3.92	3.55
7	5.27	4.65
8	6.75	5.89
9	8.33	7.19
10	10.10	8.56
11	11.95	9.97
12	14.00	11.40
13	16.07	12.81
14	18.26	14.18
15	20.46	15.52
16	22.67	16.80
17	24.87	17.99
18	27.06	19.14
19	29.24	20.20
20	31.40	21.17
21	33.52	
22	35.54	
23	37.51	
24	39.42	

Source: Sell (1990).

Heat and Moisture Production

Table 6 gives limited calorimetric heat loss data on large breed growing turkeys (DeShazer et al. 1974); however, an estimate from data on heavy chickens, applied to turkeys on a live-weight basis,

Table 6 Heat Loss of Growing Turkeys at Various Air Temperatures and Relative Humidities

Age, Days	Weight of Poultry, lb	Dry Bulb, °F	Relative Humidity, %	Heat Loss, Btu/h · lb Sensible	Heat Loss, Btu/h · lb Latent	Total Live Weight, lb
15	0.487	100	23	2.3	16.5	18.8
6	0.233	95	26	7.7	18.2	25.9
19	0.802	95	26	3.3	10.1	13.4
14	0.518	90	31	9.2	10.9	20.1
29	1.629	90	31	5.6	6.1	11.7
7	0.244	85	36	12.2	14.1	26.3
21	0.923	85	36	7.7	8.0	15.7
36	2.000	85	36	6.8	4.3	11.1
28	1.385	80	42	9.0	4.1	13.1
23	0.963	75	50	10.8	6.2	17.0
27	1.250	75	50	11.7	3.1	14.8
35	2.120	75	50	9.5	2.6	12.1

Source: DeShazer et al. (1974).

should be satisfactory for design purposes. The reduction in heat loss during the dark period of the day is 25% for large breeds (Buffington et al. 1974) and between 5 and 40% for small breeds.

The latent heat production of turkeys may be significantly higher than the 1970s data show if the growth rate or feed consumption is significantly higher in the 1990s than it was in the 1970s (see Chickens, Heat and Moisture Production).

LABORATORY ANIMALS

Significant environmental conditions for facilities that house laboratory animals include temperature, humidity, air motion, illumination, noise, and gaseous and viable particulate contaminants (Moreland 1975). Design conditions vary widely, depending on whether the animals are experiencing disease-induced stress (which alters environmental needs), subjected to test environments, or simply housed (Besch 1975, Murakami 1971, Nienaber and Hahn 1983). This fact reflects differing housing and ventilation guidelines for animals used in research (NIH 1985), compared to recommendations by the ASAE (1991) and Midwest Plan Service (MWPS 1983) for production agriculture. Little is known about the influence of disease on environmental requirements, animal performance, and well-being. Since significantly different conditions may exist between animal cage and animal room (macroenvironment), control of cage microenvironments is essential to ensure the animal's physiological well-being.

Heat and Moisture Production

Table 7 approximates heat released by laboratory animals at rest and during normal activity. Temperature and velocity gradient controls require low supply air-to-room air temperature differential, overhead high induction diffusion, uniform horizontal and vertical air distribution, and low return outlets. For load calculation purposes, heat gain from all laboratory animal species can be estimated (Wood et al. 1972, Gordon et al. 1976) with an acceptable level of error from:

$$ATHG = 2.5M$$

$$M = 6.6\ W^{0.75}$$

where

ATHG = average total heat gain, Btu/h per animal
M = metabolic rate of animal, Btu/h per animal
W = mass of animal, lb

Conditions in animal rooms must be maintained continuously. This requires year-round availability of refrigeration and, in some cases, dual air-conditioning facilities and emergency power for motor drives and control instruments. Chapter 13 of the 1995 *ASHRAE Handbook—HVAC Applications* has additional information on laboratory animal facilities.

Table 7 Heat Generated by Laboratory Animals

Animal	Weight, lb	Heat Generation, Btu/h per Animal: Basal[a]	Normally Active[b,c]: Sensible	Latent	Total
Mouse	0.046	0.66	1.11	0.54	1.65
Hamster	0.260	2.40	4.02	1.98	6.00
Rat	0.62	4.64	7.77	3.83	11.60
Guinea pig	0.90	6.10	10.22	5.03	15.25
Rabbit	5.41	23.41	39.22	19.31	58.53
Cat	6.61	27.21	45.57	22.45	68.02
Primate	12	42.55	71.27	35.10	106.38
Dog	22.7	68.64	104.8	56.4	161.2
Dog	50.0	124.10	230.7	124.2	354.9
Goat	79.3	175.39	293.78	144.70	438.48
Sheep	99	207.14	346.96	170.89	517.85
Pig	150	282.09	371	292	663
Chicken	4.0	18.67	12.9	21.9	34.7

[a]Based on standard metabolic rate $M = 6.6W^{0.75}$ Btu/h per animal (Kleiber 1961) or appropriate reference (W = animal mass, lb).
[b]Referenced according to availability of heat generation data. Otherwise, heat generations is calculated on basis of ATHG = 2.5M (Gordon et al. 1976). Latent heat is assumed to be 33% of total heat and sensible heat is 67% of total heat (Besch 1973, Woods et al. 1972).
[c]Data taken from Runkle (1964), Kleiber (1961), Besch (1973), Woods and Besch (1974), Woods et al. (1972), Bond et al. (1959), and Ota and McNally (1961).

PLANTS: GREENHOUSES, GROWTH CHAMBERS, AND OTHER FACILITIES

Most agronomically important plant crops are produced outdoors in favorable climates and seasons. Greenhouses and other indoor facilities are used for the out-of-season production of horticultural crops for both commercial sales and research purposes, and for producing food, floricultural, and other crops in conditions that permit the highest quality by buffering the crops from the vagaries of weather. The industry that produces crops in greenhouses may be termed controlled environment agriculture (CEA).

Historically, many cold-climate commercial greenhouses were operated only from late winter into early summer, and during autumn. Greenhouses were too warm during midsummer; during winter in some cold-climate locations, light levels were too low and the day length inadequate for many crops. Mechanical ventilation, evaporative cooling, centralized heating systems, movable insulations, carbon dioxide enrichment, and supplemental lighting have extended the use of greenhouses to year-round cropping on a relatively large scale.

Growth chambers, growth rooms, and propagation units are environmentally controlled spaces used for either research or commercial crop production. Environmentally controlled chambers may include highly sophisticated facilities used for micropropagation (e.g., tissue culture), or may be simple boxes in which air temperature and lights are controlled. Indoor facilities having controlled temperature and humidity environments may be used as warehouses to hold plants and plant products prior to commercial sale. Often these are simple refrigerated storage rooms or chambers.

Primary atmospheric requirements for plant production include: (1) favorable temperatures, (2) adequate light intensity and suitable radiation spectrum, and (3) favorable air composition and circulation. Engineering design to meet these requirements typically is based on steady-state assumptions. The thermal and ventilation time

Table 8 Recommended Night Temperatures for Greenhouse Crops

Crop Species	Night Temperatures, °F	Remarks
Aster *Callistephus chinensis*	50-55	Long days during early stages of growth
Azalea *Rhododendron* spp.	61-64	Vegetative growth and forcing specific temperatures required for flower initiation and development
Calceolaria *C. herbeohydrida*	61 50	Vegetative growth Flower initiation and development; initiation also occurs with long days and high temperatures if photon flux density is high
Calendula *C. officinallis*	39-45	
Calla *Zantedeschia* spp.	55-61	Decrease to 55°F as plants bloom
Carnation *Dianthus caryophyllus*	50-52 winter 55 spring 55-61 summer	Night temperatures adjusted seasonally in relation to photon energy flux density
Chrysanthemum *C. morifolium*	61 cut flowers 63-64 pot plants	Temperatures during flower initiation especially critical; uniform initiation very important for pot mums; cultivars classified on basis of temperature for development
Cineraria *Senecio cruentus*	61 48-52	Vegetative growth Flower initiation and development; plant quality best at low temperatures
Crossandra *C. infundibuliforms*	75-81 64	Germination Growth and flowering
Cyclamen *C. indicum*	61-64 55 50-52	Germination Seedlings Growth and flowering
Foliage plants	64-70	Species differ in their temperature and radiant energy requirements
Fuehsia *F. hydrida*	52-61	Long days for flower initiation
Geranium *Pelargonium x hortorum*	55-61	61 to 64°F for fast crops at high radiant energy flux
Gardenia *G. grandiflora* *G. jasminoides*	61-63 61-63	Lower temperatures result in iron chlorosis; higher temperatures increase bud abscission
Gloxina *Sinningia speciosa*	64-70	Lower temperatures increase bud brittleness
Hydrangea *H. macrophylla*	55-61 61-63 (forcing)	Specific temperature for flower initiation and development
Iris *I. tingitana* (*Wedgewood*)	45-61 (forcing)	Forcing temperature 55 to 57°F for 10/11 bulbs; 50 to 54°F for 9/10 bulbs
Kalanchoe (*K. Blossfeldiana*)	61	Temperatures influence rate of flower development and incidence of powdery mildew
Lily *Lilium longiflorum*	61	Temperatures manipulated to alter rate of flower development; specific temperatures for flower initiation
Orchida *Cattleya* spp.	61	Temperature requirement of hybrids related in parental species
Orchids *Phalaenopsis* spp. *Cymbidium* spp. *Cypripedium*	64 50 50-55	
Poinsettia *Euphorbia pulcherrima*	64 61-63	Vegetative growth Photoperiod requirement changes with temperature; bract development influenced by temperature
Roses *Rosa* spp.	61-63	
Saintpaulia *S. ionantha*	64-70	Below 61°F, growth is slow, hard and brittle
Snapdragon *Antirrhinum maias*	48-50 55-61	Winter Spring and Fall seedlings benefit from 61 to 64°F temperatures
Stock *Matthiola incana*	45-50	Buds fail to set if temperatures are above 64°F for 6 h or more per day. Grown mainly as field crop
Tomato	61-66	Dry temperatures from 70 to 81°F on sunny days
Lettuce	55	63 to 64°F on cloudy days 70 to 79°F on sunny days
Cucumber	64	75°F on cloudy days 81°F on sunny days

constants of most greenhouses are sufficiently short that transient conditions are seldom considered.

TEMPERATURE

Plant Requirements

Leaf and root temperatures are dominant environmental factors for plant growth and flowering. Factors in the energy balance of a plant canopy include air temperature, relative humidity, air movement, thermal radiation exchange, and convective exchange coefficients for sensible and latent heat. Therefore, leaf temperature is affected by environmental influences such as the type of heating and ventilating systems, supplemental lighting, light transmittance characteristics of the greenhouse cover, misting or evaporative cooling, location of the leaf on the plant, and the geometry of the surrounding leaf canopy.

Most information on plant responses to temperature is based on air temperature rather than plant temperature. Leaf temperature is difficult to measure, and one or several leaves represent neither the average nor the extreme temperatures of the plant. Since plants cannot actively regulate their cell and tissue temperatures in response to changing ambient conditions (passive regulation by opening and closing leaf stomata, which controls evapotranspiration, provides a small degree of control), their leaves and stems are usually within a few degrees of the surrounding air temperature (above during times of solar insolation, below at other times due to thermal reradiation and evapotranspiration).

All plants have minimum, optimum, and maximum temperatures for growth. Optimum temperature depends on the physiological process desired. Thermoperiodic species have different optimum day and night temperatures for each stage of growth, and each stage of plant growth may have its own unique optimum temperature influenced by radiant flux density, the ambient carbon dioxide level, and water and nutrient availability.

Historically, plants have been grown with night temperatures lower than day temperatures. In practice, many greenhouse crops are grown at standard (blueprint) night temperatures. Day temperatures are increased from 10 to 20°F (depending on solar intensity) above night temperatures. Table 8 presents recommended ranges of night temperatures for a selection of greenhouse crops. New practices have been developed in which night temperature is kept higher than day temperature, providing a nonchemical means of height control for some plant species (e.g., Easter lilies). However, thus far this technique has had limited application.

Heating Greenhouses

Heat loss from greenhouses is caused primarily by conduction through the structural cover and infiltration of outdoor air. The heating system is designed to meet the sum of the two. Perimeter heat loss is generally only a few percent of the total and is often neglected in design. When movable night insulation is used to conserve energy, heating systems are still designed to match conduction and infiltration losses without insulation because movable insulation may be opened early in the morning when outdoor air temperature is near its minimum, and excess heating capacity may be useful to melt snow in climatic regions where this occurs. Greenhouses are not designed to carry heavy snow loads.

Energy Balance

Radiation energy exchange. Solar gain can be estimated using procedures presented in Chapter 29. Not all insolation appears as sensible heat, however. As a general rule, from two-thirds to three-quarters of ambient insolation is available inside a typical commercial greenhouse. (Highly detailed models for calculating solar transmittance may be found in the literature.) If a greenhouse is filled with mature plants, approximately one-half of the available insolation (transmitted) may be converted to latent heat, one-quarter to one-third released as sensible heat, and the rest either reflected back outdoors or converted through photosynthesis (perhaps 3%).

Supplemental lighting can add significantly to the thermal load in a greenhouse. If movable night insulation is used, venting may be required during times of lighting, even during cold weather. The components of heat addition from supplemental lighting are divided between sensible and latent loads, with approximately one-quarter to one-third of the total heat load in latent form.

Reradiation heat loss from greenhouses comprises complex processes and may involve both reradiation from the structural cover and reradiation from inside the greenhouse if the cover is not thermally opaque. Glass is nearly thermally opaque, but many plastics are not. Newer plastic films may contain IR-inhibiting substances and can save a significant amount of heating energy, while adding only slightly to summer ventilation needs. Condensation on plastic films also reduces transmittance of thermal radiation, while diffusing but not seriously impairing transmittance of solar radiation. Generally, heat loss coefficients used in greenhouse design include the effects of thermal radiation exchange by the structural cover.

Structural heat loss. Conduction q_c plus infiltration q_i determine total heating requirements q_t. While infiltration heat loss is most accurately calculated using enthalpy differences, in practice only air temperature changes are considered, but with the apparent specific heat of air adjusted upward to account for the latent heat component.

$$q_t = q_c + q_i$$

$$q_c = \sum UA\Delta t$$

$$q_i = c_p VN\Delta t$$

where

U = heat loss coefficient, Btu/h·ft²·°F) (Table 9)
A = exposed surface area, ft²
Δt = inside minus outside air temperature, °F
c_p = volumetric specific heat of air (adjusted upward to account for latent heat component, 0.03 Btu/ft³·°F
V = greenhouse internal volume, ft³
N = number of air exchanges per hour (Table 10)
Σ = summation of all exposed surfaces of the greenhouse and perimeter heat losses

When design conditions are assumed for indoor and outdoor air temperatures and air exchange rate, the resulting heat loss may be assumed equal to the peak heating requirement for the greenhouse.

Table 9 Suggested Heat Transmission Coefficients

		U, Btu/h·ft²·°F
Glass		
	Single-glazing	1.13
	Double-glazing	0.70
Plastic film		Manufacturer's Data
	Single film[a]	1.20
	Double film, inflated	0.70
	Single film over glass	0.85
	Double film over glass	0.60
Corrugated glass fiber		
	Reinforced panels	1.20
Plastic structured sheet[b]		
	16 mm thick	0.58
	8 mm thick	0.65
	6 mm thick	0.72

[a]Infrared barrier polyethylene films reduce heat loss; however, use this coefficient when designing heating systems because the structure could occasionally be covered with non-IR materials.
[b]Plastic structured sheets are double-walled, rigid plastic panels.

Table 10 Construction U-Value Multipliers

Metal frame and glazing system, 16 to 24 in. spacing	1.08
Metal frame and glazing system, 48 in. spacing	1.05
Fiberglass on metal frame	1.03
Film plastic on metal frame	1.02
Film or fiberglass on wood	1.00

No universally accepted method exists to determine season-long heating needs for greenhouses. The heating degree-day method may be applied, but heating degree-day data must be adjusted to a base lower than 65°F because of the significant passive solar heating effect in greenhouses. The proper base must be determined locally to reflect the expected solar climate of the region and the expected greenhouse operating temperature. These difficulties often lead designers to obtain season-long heating data from comparable, existing greenhouses in the region, and apply them to new designs.

LIGHT AND RADIATION

Plant Requirements

Light (400 to 700 nm) is essential for plant vegetative growth and reproduction (Figures 23 and 24). Intensity integrated over time provides the energy for growth and development, while duration (either long or short, depending on species) may be essential for certain physiological processes such as flowering. High light intensity may exceed the ability of individual leaves to photosynthesize. However, if there is a dense canopy, excess light may be beneficial to lower leaves even when upper leaves are light saturated. The intensity at which light saturates a leaf depends on various environmental factors, such as the concentration of carbon dioxide in the ambient air, as well as biological factors (Figure 25).

Spectral distribution of light can affect plant development, but sunlight's spectral distribution need not be duplicated by artificial lighting to have suitable growth and development. Certain reproductive changes are initiated by red (660 nm) and far red (730 nm) light (Figure 26), and excessive ultraviolet light (290 to 390 nm) may be detrimental to growth.

Plants that respond to the durations of light and dark periods are termed photoperiodic (photoperiodic effects generally relate to flowering). Some plant species are long-day obligates, some are short-day obligates, some are day length-intermediate, and others are day-neutral. Such responses are usually (relatively) independent of light intensity. Photoperiodic effects can be initiated by very low light levels (less than 0.09 W/ft^2), such as that provided to chrysanthemums by incandescent lights for a short period during the middle of the night to promote vegetative growth and inhibit flowering (until a suitable size has been attained) during the winter. Some plant species can tolerate continuous light, but others require some period of darkness for proper growth and development.

Sunlight is the most common source of photosynthetically active radiation (PAR, 400 to 700 nm). Although specially designed lamp

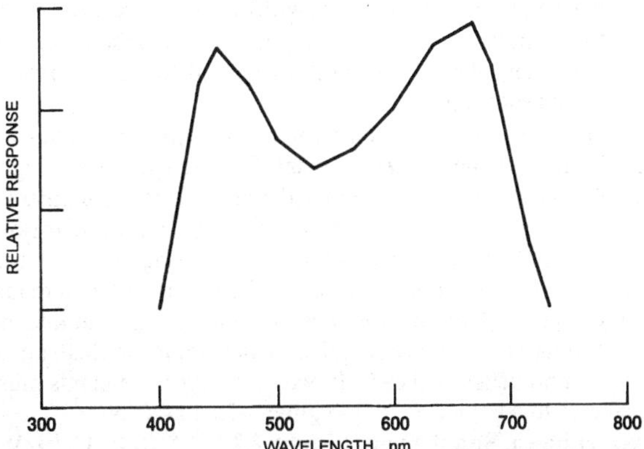

Fig. 23 Traditional Photosynthesis Action Spectra Based on Chlorophyll Absorption

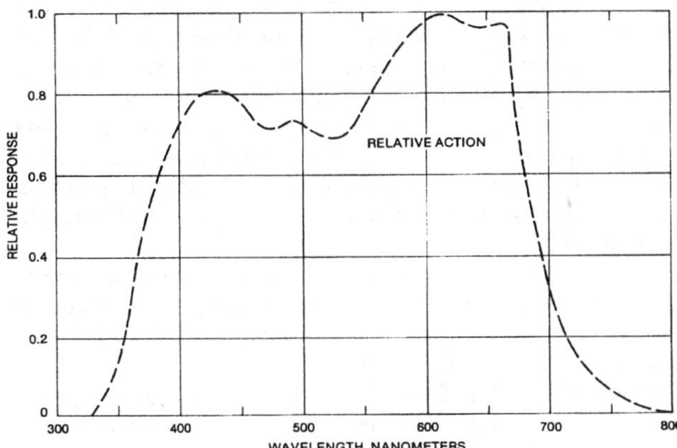

Fig. 24 Relative Photosynthetic Response

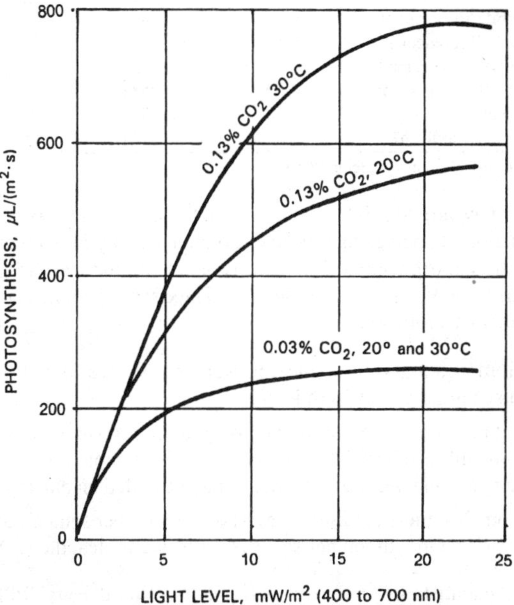

Fig. 25 Photosynthesis of Cucumber Leaf at Limiting and Saturating Carbon Dioxide Concentrations under Incandescent Light

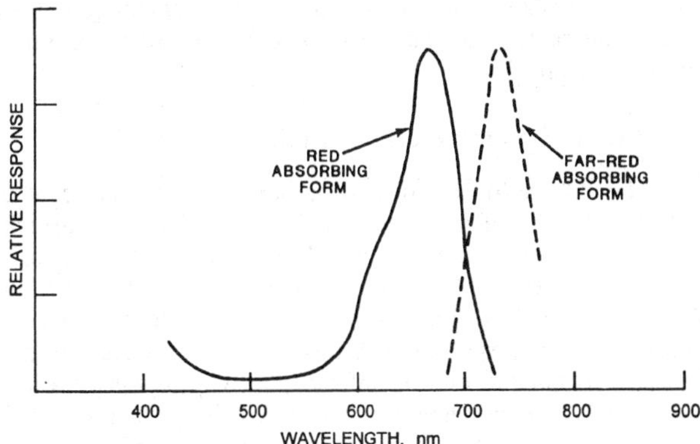

Fig. 26 Phytochrome Action Spectra

Table 11 Radiation Power Distribution of Light Sources

Light Sources	UV 300-400 nm	PAR + FAR 400-850 nm	IR 850-2700 nm	Thermal 2700 + nm	Total Radiation
FCW	2	36	1	61	100
HG/DX	3	19	18	60	100
MH	4	41	8	47	100
HPS	0.4	50	12	38	100
LPS	0.1	56	3	41	100
INC	0.2	17	74	9	100
SUN	6	59	33	2	100

Values are in watts per 100 W of radiation

Table 12 Light Conversion Factors

Light Source	Multiply W/ft² (400-850 nm) to Obtain: μmol/(s · ft²) (400-700 nm)	Divide lux by Constant to Obtain: μmol/(s · ft²) (400-700 nm)	Divide lux by Constant to Obtain: μmol/(s · ft²) (400-700 nm)
Sun and sky, daylight	4.57	581	388
Blue sky only	4.24	560	441
High-pressure sodium	4.98	883	581
Metal halide	4.59	764	657
Mercury deluxe	4.52	904	829
Warm white fluorescent	4.67	818	797
Cool white fluorescent	4.59	797	775
Plant growth fluorescent A	4.80	355	334
Plant growth fluorescent B	4.69	581	506
Incandescent	5.00	538	215
Low-pressure sodium	4.92	1141	958

Adapted from Thimijan and Heins (1983).

sources may provide light similar to sunlight, no single source or combination of sources has spectral radiation exactly like the emission of the sun from 300 to 2700 nm. Table 11 summarizes the spectral distribution of various light sources. Three systems of measuring radiation are as follows:

1. Radiometric units (irradiance) in watts per square foot (W/ft²), with specified wavelength intervals.
2. Quantum units as photon flux density in μmol/(s·m²) (at 400 to 700 nm unless otherwise specified). A mole of photons delivered in one second over a square meter may be called an einstein.
3. Photometric units (illuminance) as one lumen per square meter, or lux (lx). One lumen per square foot is one footcandle (fc).

Plant scientists use photosynthetic photon flux density (PPFD) in μmol/(s·m²) (400 to 700 nm). Engineering organizations and manufacturers of light sources use photometric and radiometric units. Because of the variation in spectral power distribution, conversion from one system of units to another must be made individually for each light source for the wavelength interval included (Table 12). To obtain comparable plant growth from different light sources, the same radiation levels (PAR) and red/far-red ratios must be maintained.

Radiation Levels for Plant Growth

Display 0.03 W/ft². For display purposes, plants can exist at an irradiance of 0.03 W/ft². The preferred lamp has changed with technological advances in efficiency and distribution. The emphasis, however, has always been on color rendering and the type of atmosphere created in the display space. Low-wattage incandescent and fluorescent lamps are preferred. At this irradiance, plants can be displayed (seen), but little or no significant positive effect on plants can be expected. Extended holding in such low light conditions will have a negative effect on many plant species. Timing (light-dark durations) and temperature interaction are not a concern.

Photoperiod Response 0.08 W/ft² (4 to 6 h). For a photoperiod response, plant growth can be regulated at an irradiance of 0.08 W/ft² for as little time as 1 h. This irradiance is called a low light intensity system. The range of plant responses (promote or delay flowering, promote growth) that can be regulated is extensive, and this lighting is widely used by commercial growers.

Survival 0.3 W/ft² (8 h). Plants can survive at an irradiance of 0.3 W/ft² for 8 or more hours daily. This level enables many green plants to maintain their color. However, stem lengthening (etteliolation) and reduction in new leaf size and thickness occur under this irradiance level. In time, the overall development of the plants falls behind that of other plants grown under higher radiation levels. Photoperiod responses do not function well at this irradiance. However, strong interactions occur between this irradiance and temperature, watering frequency, and nutrition. Cooler temperatures (less than 63°F) help conserve previously stored material, while frequent watering and fertilization aggravate stem lengthening and senescence of older foliage.

Growth Maintenance 0.8 W/ft² (12 h). Plants maintain growth over many months when exposed to an irradiance of 0.8 W/ft² of 12-h duration daily. This is the intensity at which many indoor gardeners (professional or hobbyists) grow their plants when starting them from seeds, cuttings, or meristems. Interactions with the environment (temperature, airflow, relative humidity, and pollutants) can vary among installations. Simple facilities with good air exchange and limited lamp concentration can grow a wide range of plant species. The rate of development, particularly as the plants grow in size, can be slow compared to plants grown at higher irradiances.

Propagation 1.7 W/ft² (6 to 8 h). Plants propagate rapidly when exposed to an irradiance of 1.7 W/ft² for a minimum of 6 to 8 h daily, but they prefer 12 h. Above this level, many propagators attempt to shade their greenhouses with one or several layers of neutral filters (painted films on glazing, or movable or semipermanent plastic or other fabric shade cloth materials) to restrict light (and heat) in the propagation area.

Cuttings rooted at this intensity maintain a growth rate much like that of similar tissue on a stock plant. Stem length, branching, and leaf color, however, can be regulated by manipulating temperature, moisture, stress, and nutrients. Most plants grown for their flowers and fruits can be brought to maturity by increasing the day length to 16 to 18 h for flower initiation (or rapid growth) and then reducing the day length to 8 to 12 h for development. The growth rate, however, is relatively slow. For quickest development (leaf number, number of branches, and early flower initiation), the plants must be transferred to a higher lighting regime—2.2 to 4.6 W/ft².

Greenhouse Supplemental Light 2.2 W/ft² (8 to 16 h). When natural light is inadequate, it may be supplemented up to approximately 2.2 W/ft² for 8 to 16 h daily. When coupled with the ambient sunlight (shaded by clouds, greenhouse structures, and lamp fixtures), this irradiance simulates many of the growth responses and rates associated with growth chamber studies. Plants grown in greenhouses without supplemental lighting grow slower and flower later than lighted ones in cloudy regions or in northern areas during winter. Duration (in hours) and timing (day-night) of lighting is critical.

Supplemental lighting for 8 h, particularly during the day (0800 to 1600) may not be as cost-effective as lighting at night (2000 to 0400) if off-peak electric rates are available. Neither of these lighting regimes, however, is as effective as lighting for 16 h from morning to midnight.

Lighting short-day plants, such as chrysanthemums and poinsettias, is relatively inefficient because they can be lighted only during the 8- to 12-h day, followed by an obligatory 12- to 16-h daily dark period.

Growth Chambers 4.6 W/ft² (8 to 24 h). Plants grow in growth chambers or growth rooms if the light irradiance is a minimum of 4.6 W/ft² for 8 to 24 h daily. This irradiance is approximately one-fourth that of outdoor sunlight. Cool, white fluorescent lamps, combined with incandescent lamps, are widely used. More recently, HID lamps have been substituted for fluorescent lamps. For consistent results, all require a barrier of glass or other material between the lamp and the plants, and a separate ventilating system to remove the heat from such enclosed spaces.

Since filters cannot remove infrared completely, chambers are difficult to standardize. This often leads to confusing information on plant growth and flowering of plants grown in greenhouses and outdoors. When the total irradiance is 4.6 W/ft^2 and 10 to 20% of the total radiation is provided by incandescent lamps, most kinds of plants can be grown. In typical plant forms, flowering and fruiting responses occur when the plants are subjected to the following parameters:

- Day length, 8 to 24 h
- Temperature, 50 to 95°F
- Carbon dioxide, 300 to 2000 ppm (540 to 3600 mg/m^3)
- Relative humidity, 20 to 80%

Photoperiod

Day length affects the performance of some plants. There are four basic day length plant groups:

1. *Short-day plants* flower only when the length of the daily light period is less than the critical number of hours. Daily light periods longer than the critical length inhibit flowering.
2. *Long-day plants* flower only when the daily light period is longer than the critical number of hours. They become dormant or remain vegetative when the daily light period is shorter than the critical length.
3. *Day length-intermediate plants* flower only within a narrow range of day length, usually between about 10 and 14 h. If the day length is shorter than the optimum day length for flowering, the plants stop growing.
4. *Day-neutral plants* continue in vegetative growth or flower regardless of the day length.

Continuous light inhibits flowering and promotes vegetative growth of short-day plants, but encourages continued vegetative growth and early flowering of long-day plants, blocks the flowering of day length-intermediate plants, and in many instances, increases the stem length of day-neutral plants. Plants vary in their responsiveness to light source, duration, and intensity. The technology that has evolved to control the photoperiod of plants is based primarily on the incandescent-filament lamp. Of all the light sources available, this lamp creates the regulating mechanism most similar to that of sunlight. This is because the red/far-red wavelength ratio of light from an incandescent lamp is similar to the ratio of sunlight.

The effectiveness for photoperiod response in plants peaks at wavelengths of 660 nm (red) and 730 nm (far-red). The relative order of activity in regulating photoperiod responses by lamp type is as follows: Incandescent (INC) > High-pressure sodium (HPS) >> Metal halide (MN) = Cool white fluorescent (F) = Low-pressure sodium (LPS) >> Clear mercury (Hg). Photoperiod lighting is always used in combination with daylight or another main light source. Short days (less than normal day length) are created in the greenhouse with opaque materials that surround the plants.

RELATIVE HUMIDITY

Relative humidity affects the rate at which plants take water up, the rate of latent heat transfer, and certain diseases. Normal plant growth generally occurs at relative humidities between 20 and 80% if the plants have a well-developed root system, although relative humidities above 40% are preferred to avoid water stress conditions.

Transpiration, the movement of water vapor and gases from the plant to its surroundings, is controlled by the plant's stomatal openings. It is a function of air velocity and the vapor pressure difference between water at saturation at the leaf temperature and the actual water vapor partial pressure in the air. Generally, as relative humidity decreases, transpiration increases. Very low relative humidities (less than 20%) can cause wilting, since evaporation losses may be higher than the plant can replace, especially when light intensity is high.

High humidity provides a good environment for pathogenic organisms. Many pathogenic spores do not germinate unless relative humidity is 96% or more and many require a film of water on the leaves. Somewhat lower relative humidities may support other pathogen growth stages.

Still air surrounding a plant may be much wetter than the general atmosphere because evapotranspiration from the leaves raises the relative humidity in interfoliage air. The lower leaves, which stay moist longer, are more susceptible to disease. The upper leaves are dried by radiation and air currents.

AIR COMPOSITION

Carbon dioxide, which comprises about 0.035% of ambient air, is essential for plant growth. There are basically three ways to obtain carbon dioxide for greenhouse enrichment: pure in solid, liquid, or gaseous form; from burning fuels such as propane, natural gas, or kerosene; and by the aerobic breakdown of organic matter. The three ways are listed in order of purity and reliability. Carbon dioxide enters plants through stomata and is converted to carbohydrates through photosynthesis. The carbon dioxide concentration in air surrounding a plant, as well as light level, affects the rate of photosynthesis. The concentration for maximum growth depends on many factors, including the stage of growth, leaf area, light intensity, temperature, and air velocity past the stomatal openings.

An important relationship exists between light level and carbon dioxide uptake (Figure 25). As light level increases, carbon dioxide concentration must increase concurrently to take maximum advantage of the greater photosynthetic potential. In plastic greenhouses, and in glass greenhouses sealed against infiltration, the carbon dioxide level can drop below 200 ppm when the weather is cold, light levels are moderate, and the greenhouse is not ventilated. Carbon dioxide enrichment just to maintain normal levels can then be beneficial. During times of high light levels, carbon dioxide enrichment gives maximum benefit from the available light and may even be economically desirable when greenhouse ventilation is modest. However, carbon dioxide concentrations above 1500 ppm are seldom recommended; levels between 800 and 1200 ppm are typically used.

The effects of enrichment are not always positive. Without proper crop management, the yield, quality, or both may decrease, and timing of crop maturity may change.

Pollutants

Plants are sensitive to atmospheric pollutants such as ethylene, ammonia, gaseous fuels, ozone, fluorides, photochemical smog, and oxidants (from nitrogen and sulfur). Pollution damage can range from small spots on leaves, to yellowing of leaves, to severe foliage burn, and, ultimately, to plant death in extreme but not rare situations. The effect occurs both outdoors and in greenhouses; however, this is more common in greenhouses, because of their closed nature. Pollutants indoors can be removed by activated charcoal filters in the ventilation system; however, these are seldom used in commercial greenhouses. Economically, the more feasible approach is to limit pollutant production within, or introduction into, the greenhouse air space.

Ethylene is produced naturally by plants and leads to flower and whole plant senescence. It is also produced by combustion of gaseous and liquid fuels and can rapidly cause plant damage. Concentrations above 0.2 ppm can have a detrimental effect on plant growth. Unvented heaters, air currents that bring vented combustion products back into the greenhouse, and burners for carbon dioxide production are common sources of ethylene injury. Liquefied carbon dioxide may be used to supplement natural levels rather than combustion, specifically to avoid introducing ethylene into the greenhouse air, but even liquefied carbon dioxide should be carefully selected to avoid residual amounts of ethylene that may be contained within it.

Nitrogen oxides, common components of air pollution, can cause serious plant damage. Greenhouse locations near highways, nearby industrial complexes, and even a truck left running for an extended time near a greenhouse air intake vent may lead to leaf damage from NO and NO_2.

Sulfur dioxide, produced by the burning of sulfur containing fuels, causes injury to many plants within a short time. Sources of sulfur dioxide may be nearby, such as an industrial area, or may be within the greenhouse complex, such as the vented combustion products from a central heating facility, combustion products from carbon dioxide burners (using kerosene as a fuel, for example), and sulfur burned for mildew control.

Ozone is widely recognized as a serious pollutant affecting the production of many agronomic crops. Although few research results exist to quantify the effect of ozone on greenhouse crops, damage is likely to occur when greenhouses are located near ozone sources.

Phenolics and certain other organic vapors are phytotoxic. Phenolics, as volatiles from certain wood preservatives (creosote and pentachlorophenol), can cause leaf and petal damage. Vapors from some paints can also be damaging. Misuse of herbicides and pesticides can lead to plant injury, either through spray drift or volatilization.

Covering and sealing greenhouses for energy conservation can increase concentrations of ethylene and other air pollutants if their sources are within the air space, since infiltration and ventilation are decreased. Sealing to reduce infiltration can also lead to rapid carbon dioxide depletion and inhibited plant growth during cold temperatures when, even with bright light, ventilation is not required.

AIR MOVEMENT

Air movement influences transpiration, evaporation, and the availability of carbon dioxide. Air speed affects the thickness of the boundary layer at the leaf surface, which in turn influences the transport resistance between the ambient air and the leaf stomatal cavities. Air speed of 100 to 150 fpm is commonly accepted as suitable for plant growth under CEA conditions. Air speeds across the leaf of 6 to 20 fpm are needed to facilitate carbon dioxide uptake. Air speeds above 200 fpm can induce excessive transpiration, cause the stomatal guard cells to close, reduce carbon dioxide uptake, and inhibit plant growth. Air speeds above 1000 fpm may cause physical damage to plants. Generally, if plants within a greenhouse move noticeably due to ventilation, air speed is excessive.

Air circulation within greenhouses may be created to reduce thermal stratification and maintain suitable levels of carbon dioxide within the leaf canopy. Horizontal air flow, produced by small propeller fans that move air around the greenhouse in a racetrack pattern, has been found to be effective. Such fans are approximately 14 in. in diameter, with approximately one-sixth horsepower motors, spaced at approximately 50 ft intervals. Total fan capacity in cubic feet per minute (cfm) should equal about 25% of the greenhouse volume in cubic feet.

REFERENCES

Animals

ASAE. 1991. Design of ventilation systems for poultry and livestock shelters. *ASAE Standard* D270.5, American Society of Agricultural Engineers, St. Joseph, MI.

Bereskin, B., R.J. Davey, W.H. Peters, and H.O. Hetzer. 1975. Genetic and environmental effects and interactions in swine growth and feed utilization. *Journal of Animal Science* 40(1):53.

Besch, E.L. 1973. Final report to Animal Resources Branch, Division of Research Resources, National Institutes of Health. NIH Contract 71-2511.

Besch, E.L. 1975. Animal cage room dry-bulb and dew point temperature differentials. *ASHRAE Transactions* 81(2):549.

Blaxter, K.L.and W.A. Wood. 1951. The nutrition of the young Ayrshire calf, IV some factors affecting the biological value of protein determined by nitrogen-balance methods. *Brit. J. Nutrition*, 5:55.

Bond, T.E., H. Heitman, Jr., and C.F. Kelly. 1965. Effects of increased air velocities on heat and moisture loss and growth of swine. *Transactions of ASAE* 8:167. American Society of Agricultural Engineers, St. Joseph, MI.

Bond, T.E., C.F. Kelly, and H. Heitman, Jr. 1959. Hog house air conditioning and ventilation data. *Transactions of ASAE* 2:1.

Bond, T.E., C.F. Kelly, and H. Heitman, Jr. 1963. Effect of diurnal temperature upon swine heat loss and well-being. *Transactions of ASAE* 6:132.

Brody, S. 1945. *Bioenergetics and growth*. Reinhold Publishing Co., New York.

Brody, S., A.C. Ragsdale, R.G. Yeck, and D. Worstell. 1955. Milk production, feed and water consumption, and body weight of Jersey and Holstein cows in relation to several diurnal temperature rhythms. University of Missouri Agricultural Experiment Station Research *Bulletin* No. 578.

Bryant, J.M., C.F. Foreman, N.L. Jacobson, and A.D. McGilliard. 1967. Protein and energy requirementsof the young calf. *J. Dairy Sci*, 50(10):1645-1653.

Buckland, R.B. 1975. The effect of intermittent lighting programmes on the production of market chickens and turkeys. *World Poultry Science Journal* 31(4):262.

Bucklin, R.A. and L.W. Turner. 1991. Methods to relieve heat stress in hot, humid climates. *Applied Engineering in Agriculture 7*. American Society of Agricultural Engineers, St. Joseph, MI.

Buffington, D.E., K.A. Jordan, W.A. Junnila, and L.L. Boyd. 1974. Heat production of active, growing turkeys. *Transactions of ASAE* 17:542.

Bundy, D.S. 1984. Rate of dust decay as affected by relative humidity, ionization and air movement. *Transactions of ASAE* 27(3):865-70.

Bundy, D.S. 1986. Sound preventive measures to follow when working in confinement buildings. Presented at the American Pork Congress, St. Louis, MO.

Butchbaker, A.F. and M.D. Shanklin. 1964. Partitional heat losses of newborn pigs as affected by air temperature, absolute humidity, age and body weight. *Transactions of ASAE* 7(4):380.

Carr, L.E. and J.L. Nicholson. 1980. Broiler response to three ventilation ranges. *Transactions of ASAE* 22(2):414-18.

Christianson, L.L. and R.L. Fehr. 1983. Ventilation energy and economics. In *Ventilation of agricultural structures*, M.A. Hellickson and J.N. Walker, eds. *ASAE Nomograph* No. 6, 336.

Curtis, S.E. 1983. *Environmental management in animal agriculture*. Iowa State University Press, Ames, IA.

DeShazer, J.A., L.L. Olson, and F.B. Mather. 1974. Heat losses of large white turkeys—6 to 36 days of age. *Poultry Science* 53(6):2047.

Donham, K.J. 1987. Human health and safety for workers in livestock housing. CIGR Proceedings. *ASAE Special Public.* 6-87.

Donham, K.J., P. Haglind, Y. Peterson, R. Rylander, and L. Belin. 1989. Environmental and health studies of workers in Swedish swine confinement buildings. *British J. Ind. Med.* 40:31-37.

Drury, L.N. 1966. The effect of air velocity of broiler growth in a diurnally cycling hot humid environment. *Transactions of ASAE* 9:329.

Farner, D.S. 1961. Comparative physiology: Photoperiodicity. *Annual Review of Physiology* 23:71.

Gebremedhin, K.G., C.O. Cramer, and W.P. Porter. 1981. Predictions and measurements of heat production and food and water requirements of Holstein calves in different environments. *Transactions of ASAE* 24(3):715-20.

Gebremedhin, K G., W.P. Porter, and C.O. Cramer. 1983. Quantitative analysis of the heat exchange through the fur layer of Holstein calves. *Transactions of ASAE* 26(1):188-93.

Gordon, R.L., J.E. Woods, and E.L. Besch. 1976. System load characteristics and estimation of animal heat loads for laboratory animal facilities. *ASHRAE Transactions* 82(1):107.

Guazdauskas, F.C. 1985. Effects of climate on reproduction in cattle. *Journal of Dairy Science* 68:1568-78.

Gunnarson, H.J. et al. 1967. Effect of air velocity, air temperatures and mean radiant temperature on performance of growing-finishing swine. *Transactions of ASAE* 10:715.

Hahn, G.L. 1983. Management and housing of farm animals environments. *Stress physiology in livestock*. CRC Press, Boca Raton, FL.

Hahn, G.L., A. Nygaard, and E. Simensen. 1983. Toward establishing rational criteria for selection and design of livestock environments. *ASAE Paper* 83-4517.

Heard, L., D. Froehlich, L. Christianson, R. Woerman, and W. Witmer. 1986. Snout cooling effects on sows and litters. *Transactions of ASAE* 29(4):1097.

Heber, A.J., M. Stroik, J.L. Nelssen, and D.A. Nichols. 1988. Influence of environmental factors on concentrations and inorganic content of aerial dust in swine finishing buildings. *Transactions of ASAE* 31(3):875-881.

Heber, A.J. and C.R. Martin. 1988. Effect of additives on aerodynamic segregation of dust from swine feed. *Transactions of ASAE* 31(2):558-563.

Hellickson, M.A., H.G. Young, and W.B. Witmer. 1974. Ventilation design for closed beef buildings. Proceedings of the International Livestock Environment Symposium, SP-0174, 123.

Hillman, P.E., K.G. Gebremedhin, and R.G. Warner. 1992. Ventilation system to minimize airborne bacteria, dust, humidity, and ammonia in calf nurseries. *Journal of Dairy Science* 75:1305-12.

Janni, K.A., P.T. Redig, J. Newmen, and J. Mulhausen. 1984. Respirable aerosol concentration in turkey grower building. *ASAE Paper* No. 84-4522.

Johnson, H.D., A.C. Ragsdale, and R.G. Yeck. 1958. Effects of constant environmental temperature of 50°F and 80°F on the feed and water consumption of Brahman, Santa Gertrudis and Shorthorn calves during growth. Univ. of Missouri Ag. Exp. Sta. *Research Bulletin* 683.

Johnson, H.D., A.C. Ragsdale, I.L. Berry, and M.D. Shanklin. 1962. Effect of various temperature-humidity combinations on milk production of Holstein cattle. Univ. of Missouri Ag. Exp. Sta. *Research Bulletin* 791.

Kibler, H.H. and S. Brody. 1956. Influence of diurnal temperature cycles on heat production and cardiorespirativities in Holstein and Jersey cows. Univ. of Missouri Ag. Exp. Sta. *Research Bulletin* 601.

Kleiber, M. 1961. *The fire of life: An introduction to animal energetics*. John Wiley and Sons, New York.

Longhouse, A.D. et al. 1968. Heat and moisture design data for broiler houses. *Transactions of ASAE* 41(5):694.

Longhouse, A.D., H. Ota, and W. Ashby. 1960. Heat and moisture design data for poultry housing. *Agricultural Engineering* 41(9):567.

McDaniel, G.R. and R.N. Brewer. 1975. Intermittent light speeds broiler growth and improves efficiency. *Highlights of Agricultural Research* 4:9. Auburn University, Auburn, AL.

McFarlane, J. 1987. Linear additivity of multiple concurrent environmental stressors effects on chick performance, physiology, histopathology and behavior. PhD thesis, Animal Sciences Dept., Univ. of Illinois, Urbana.

Merkle, J.A. and T.E. Hazen. 1967. Zone cooling for lactating sows.*Transactions of ASAE* 10:444.

Moreland, A.F. 1975. Characteristics of the research animal bioenvironment. *ASHRAE Transactions* 81(2):542.

Mount, L.E. 1963. Food, meat, and heat conservation. Pig Industry Development Authority Conference Circulat. Buxton, Derbyshire, England.

Muehling, A.J. 1970. Gases and odors from stored swine wastes. *Journal of Animal Science* 30:526-31.

Murakami, H. 1971. Differences between internal and external environments of the mouse cage. *Laboratory Animal Science* 21:680.

MWPS. 1983. *Structures and environment handbook*. Midwest Plan Service, Ames, IA.

Nienaber, J.A. and G.L. Hahn. 1983. Temperature distribution within controlled-environment animal rooms. *Transactions of ASAE* 26:895.

Nienaber, J.A., G.L. Hahn, H.G. Klencke, B.A. Becker, and F. Blecha. 1987. Cyclic temperature effects on growing-finishing swine. CIGR Proceedings. *ASAE Special Public.* 687:312.

NIH. 1985. Guide for the care and use of laboratory animals. National Institutes of Health Publication 85-23. Bethesda, MD.

North, M.O. 1984. *Commercial chicken production manual*, 3rd ed. AVI Publishing, Westport, CT.

OSHA. 1985. OSHA Safety and Health Standard. U.S. Department of Labor Code 1910.1000, 653-59.

Ota, H.J. and E.H. McNally. 1961. Poultry respiration calorimetric studies of laying hens. ARS-USDA 42-13, June.

Ota, H., J.A. Whitehead, and R.J. Davey. 1982. Heat production of male and female piglets. Proceedings of Second International Livestock Environment Symposium, SP-03-82. ASAE, St. Joseph, MI.

Reece, F.N. and B.D. Lott. 1982. Heat and moisture production of broiler chickens. Proceedings of the Second International Livestock Environment Symposium, SP-03-82. ASAE, St. Joseph, MI.

Riskowski, G.L., J.A. DeShazer, and F.B. Mather. 1977. Heat losses of White Leghorn hens as affected by intermittent lighting schedules. *Transactions of ASAE* 20(4):727-731.

Riskowski, G.L. and D.S. Bundy. 1991. Response of young pigs to various air temperatures and velocities. *ASHRAE Transactions* 97(2):543-549.

Roller, W.L. and H.S. Teague. 1966. Effect of controlled thermal environment on reproductive performance of swine. Proceedings of the Fourth International Biometeorological Congress. Rutgers University, New Brunswick, NJ.

Runkle, R.S. 1964. Laboratory animal housing, Part II. AIA *Journal* 4:73.

Scott, N.R., J.A. DeShazer, and W.L. Roller. 1983. Effects of the thermal and gaseous environment on livestock. *ASAE Monograph No.* 6. ASAE, St. Joseph, MI.

Sell, J.L. 1990. Faster growing, more efficient turkeys in 1989. 66(1):12.

Shutze, J.V., J.K. Lauber, M. Kato, and W.O. Wilson. 1962. Influence of incandescent and colored light on chicken embryos during incubation. *Nature* 196(4854):594.

Siegmund, H., ed. 1979. *The Merck veterinary manual, A handbook of diagnosis and therapy for the veterinarian.* Merck & Co., Inc., Rahway, NJ.

Squibb, R.L. 1959. Relation of diurnal temperature and humidity ranges to egg production and feed efficiency of New Hampshire hens. *Journal of Agricultural Science* 52(2):217.

Sutton, A.L., S.R. Nichols, D.D. Jones, D.T. Kelley, and A.B. Scheidt. 1987. Survey of seasonal atmospheric changes in confinement farrowing houses. In *Latest developments in livestock housing*. Intl. Commission of Agric. Engin. (CIGR).

Thomeczek, F.J., M.R. Ellersieck, R.K. Leavitt, and J.F. Lasley. 1977. Trends in economic traits of production tested boars in the Missouri Evaluation Station. University of Missouri Research Bulletin No. 1021.

Tienhoven, A.W., N.R. Scott, and P.E. Hillman. 1979. The hypothalmus and thermoregulation: A review. *Poultry Science* 52(6):1633.

Van Wicklen, G.L. and L.D. Albright. 1987. Removal mechanisms for calf barn aerosol particles. *Transactions of ASAE* 30(6):1758-63.

Wiersma, F. and G.H. Stott. 1969. New concepts in the physiology of heat stress in dairy cattle of interest of engineers. *Transactions of ASAE* 12(1):130-32.

Woods, J.E. and E.L. Besch. 1974. Influence of group size on heat dissipation from dogs in a controlled environment. *Laboratory Animal Science* 24:72.

Woods, J.E., E.L. Besch, and R.G. Nevins. 1972. A direct calorimetric analysis of heat and moisture dissipated from dogs. *ASHRAE Transactions* 78(2):170-83.

Yeck, R.G. 1957. Stable heat and moisture dissipation with beef calves at temperatures of 50° and 80°F. *Missouri Agricultural Exp. Sta. Research Bulletin* 645.

Yeck, R.G. and R.E. Stewart. 1959. A ten-year summary of the psychroenergetic laboratory dairy cattle research at the University of Missouri. *Transactions of ASAE* 2(1):71.

Yeck, R.G. and R.E. Stewart. 1960. Stable heat and moisture dissipation with dairy calves at temperatures of 50° and 80°F. University of Missouri *Research Bulletin No.* 759.

Zhang, Y. 1994. *Swine building ventilation*. Prairie Swine Centre. Saskatoon, SK, Canada pp.14-15.

Zulovich, J.M., M.B. Manbeck, and W.B. Roush. 1987. Whole-house heat and moisture production of young floor brood layer pullets. *Transactions of ASAE* 30(2)455-58.

BIBLIOGRAPHY

Animals

ASAE. 1991. Design of ventilation systems for poultry and livestock shelters. ASAE *Standard* D270.5. American Society of Agricultural Engineers, St. Joseph, MI.

Consortium for Developing a Guide for the Care and Use of Agricultural Animals in Agricultural Research and Teaching. 1988. *Guide for the care and use of agricultural animals in agricultural research and teaching.* Agricultural Animal Care Guide Division of Agriculture, NASULGC, Washington, D.C. 20036-1191.

Esmay, M.E. and J.E. Dixon. 1986. *Environmental control for agricultural buildings.* AVI Publications, Westport, CT.

Hellickson, M.A. and J.N. Walker, eds. 1983. Ventilation of agricultural structures. *ASAE Monograph No.* 6. American Society of Agricultural Engineers, St. Joseph, MI.

MWPS. 1995. The Midwest Plan Service handbooks and plans series. Midwest Plan Service, Ames, IA.

Plants

Cathey, H.M. and L.E. Campbell. 1975. Effectiveness of five vision lighting sources on photo-regulation of 22 species of ornamental plants. *Journal of American Society of Horticultural Science* 100(1):65.

Downs, R.J. 1975. *Controlled environments for plant research*. Columbia University Press, New York.

Mastalerz. J.W. 1977. *The greenhouse environment*. John Wiley & Sons, New York.

Tibbits, T.W. and T.T. Kozlowski. 1979. *Controlled environment guidelines for plant research*. Academic Press, New York.

Light and Radiation

Austin, R.B. and J.A. Edrich. 1974. A comparison of six sources of supplementary light for growing cereals in glasshouses during winter time. *Journal of Agricultural Research* 19:339.

Biran, I. and A.M. Kofranek. 1976. Evaluation of fluorescent lamps as an energy source for plant growth. *Journal of American Society of Horticultural Science* 101(6):625.

Campbell, L.E., R.W. Thimijan, and H.M. Cathey. 1975. Spectral radiant power of lamps used in horticulture. *Transactions of ASAE* 18(5):952.

Carpenter, W.J. and G.A. Anderson. 1972. High intensity supplementary lighting increases yields of greenhouse roses. *Journal of American Society of Horticultural Science* 101:331.

Cathey, H.M. and L.E. Campbell. 1977a. Lamps and lighting: A horticultural view. *Lighting Design & Application* 4(2):41.

Cathey, H.M. and L.E. Campbell. 1977b. Plant productivity: New approaches to efficient sources and environmental control. *Transactions of ASAE* 20(2):360.

Cathey, H.M. and L.E. Campbell. 1980. Horticultural reviews, Vol. II. AVI Publishing Co., Westport, CT, Chapter 10, 491.

Cathey, H.M., L.E. Campbell, and R.W. Thimijan. 1978. Comparative development of 11 plants grown under various fluorescent lamps and different duration of irradiation with and without additional incandescent lighting. *Journal of ASHS* 103:781.

Duke, W.B. et al. 1975. Metal halide lamps for supplemental lighting in greenhouses, crop response and spectral distribution. *Agronomics Journal* 67:49.

Gates, D.M. 1968. Transpiration and leaf temperature. *Annual Review of Plant Physiology* 19:211.

McCree, K.J. 1972a. Significance of enhancement for calculation based on the action spectrum for photosynthesis. *Plant Physiology* 49:704.

McCree, K.J. 1972b. Test of current definitions of photosynthetically active radiation against leaf photosynthesis data. *Agricultural Metrord* 10:443.

Meijer, G. 1971. Some aspects of plant irradiation. *Acta Horticulture* 22:103.

Parker, M.W. and H.A. Borthwick. 1950. Influence of light on plant growth. *Annual Review of Plant Physiology*, 43.

Stoutmeyer, V.T. and A.W. Close. 1946. Rooting cuttings and germinating seeds under fluorescent and cold cathode lighting. Proceedings of American Society of Horticultural Science 48:309.

Thimijam, R.W. and R.D. Heins. 1983. Photometric, radiometric, and quantum light units of measure: A review of procedures for interconversion. *HortScience* 18(6):818-22.

Photoperiod

Cathey, H.M. and H.A. Borthwick. 1961. Cyclic lighting for controlling flowering of chrysanthemums. *Proceedings of ASAE* 78:545.

Downs, R.J., H.A. Borthwick, and A.A. Piringer, Jr. 1958. Comparison of incandescent and fluorescent lamps for lengthening photoperiods. *Proceedings of ASHS* 71:568.

Hillman, W.S. 1962. *The physiology of flowering*. Holt, Rinehart & Winston, Inc., New York.

Jose, A.M. and D. Vince-Prue. 1978. Phytochrome action. A reappraisal. Photochemistry & Photobiology 27:209.

Temperature

Joffe, A. 1962. An evaluation of controlled temperature environments for plant growth investigations. *Nature* 195:1043.

Humidity, Carbon Dioxide, and Air Composition

Holley, W.D. 1970. CO_2 enrichment for flower production. *Transactions of ASAE* 13(3):257.

Kretchman, J. and F.S. Howlett. 1970. CO_2 enrichment for vegetable production. *Transactions of ASAE* 13(2):22.

Tibbits, T.W., J.C. McFarlane, D.T. Krizek, W.L. Berry, P.A. Hammer, R.H. Hodgsen, and R.W. Langhans. 1977. Contaminants in plant growth chambers. *Horticulture Science* 12:310.

CHAPTER 11

PHYSIOLOGICAL FACTORS IN DRYING AND STORING FARM CROPS

THIS chapter focuses on the drying and storage of grains, oilseeds, hay, cotton, and tobacco. However, the primary focus is on grains and oilseeds, collectively referred to as **grain**. Major causes of postharvest losses in these products are fungi, insects, and rodents. Substantial deterioration of grain can occur in storage. However, where the principles of good grain storage are applied, losses are minimal.

Preharvest invasion of grains by storage insects is usually not a problem in the midwestern United States. Field infestations can occur in grains when they are dried in the field at warm temperatures during harvest. Preharvest invasion by storage fungi is possible and does occur if appropriate weather conditions prevail when the grain is ripening. For example, preharvest invasion of corn by *Aspergillus flavus* occurs when hot weather is prevalent during grain ripening; it is, therefore, more common in the southeastern United States (McMillan et al. 1985). Invasion of wheat, soybeans, and corn by other fungi can occur when high ambient relative humidities prevail during grain ripening (Christensen and Meronuck 1986). However, the great majority of damage occurs during storage, due to improper conditions that permit storage fungi or insects to develop.

Deterioration from fungi during storage is prevented or minimized by (1) reduction of grain moisture content to below limits for growth of fungi, (2) maintenance of low grain temperatures throughout the storage period, (3) chemical treatment to prevent the development of fungi or to reduce the rate of fungal growth while the grain moisture content is being lowered to a safe level, and (4) airtight storage in which initial microbial and seed respiration reduces the oxygen level so that further activity by potentially harmful aerobic fungi is prevented.

Reduction of moisture by artificial drying is the most commonly used technique. The longer grain is stored, the lower its storage moisture should be. Some of the basic principles of grain drying and a summary of methods for predicting grain drying rate are included in the section on Drying Theory.

Reduction of grain temperature by aeration is practical in temperate climates and for grains that are harvested during cooler seasons. Fans are operated when ambient temperature is lower than grain temperature. Basic information on aeration is summarized in the section on Drying Theory. Use of refrigeration systems to reduce temperature is not generally cost-effective for feed grains but may have application for higher value food grains.

Chemical treatment of grain is becoming more common and is briefly described in the section on Prevention of Deterioration.

When grain is placed in airtight silos, the oxygen level is rapidly reduced, and carbon dioxide increases. Although many fungi will not grow under ideal hermetic conditions, some will grow initially in imperfectly sealed bins, and this growth can reduce the feeding value of the grain for some animals. Partially emptied bins may support harmful mold, yeast, and bacterial growth, which makes the grain unsuitable for human consumption. Airtight storage is briefly addressed in the section on Oxygen and Carbon Dioxide under Factors Determining Safe Storage.

Deterioration from insects can also be prevented by a combination of reducing moisture and lowering temperature. Lowering of temperatures is best achieved by aeration with cool ambient air during cool nights and periods of cool weather. Both the use of clean storage structures and the segregation of new crop grain from carryover grain or grain contaminated with insects are important. If insect infestation has already occurred, fumigation is often used to kill the insects. Aeration with cold air may retard the development of the insect population. Prevention and control of insect infestations are addressed in the section on Prevention of Deterioration.

For information on rodent problems, see the section on Prevention of Deterioration.

Moisture content is the most important factor determining successful storage. Although some grains are harvested at safe storage moistures, other grains (notably corn, rice, and most oilseeds) must usually be artificially dried prior to storage. During some harvest seasons, wheat and soybeans are harvested at moistures above those safe for storage and, therefore, also require drying.

Sauer (1992), Brooker et al. (1992), Hall (1980), Christensen and Meronuck (1986), and Gunasekaran (1986) summarize the basic aspects of grain storage and grain drying. Chapter 21 of the 1995 *ASHRAE Handbook—Applications* covers crop-drying equipment and aeration systems.

FACTORS DETERMINING SAFE STORAGE

Moisture Content

Grain is bought and sold on the basis of characteristics of representative samples. Probes or samplers, such as diverters, are used to obtain representative subsamples. Often representative subsamples must be taken from a large quantity (several tons) of grain. Manis (1992) summarizes sampling procedures and equipment. For safe storage, it is necessary to know the range in moisture content within a given bulk and whether any of the grain in the bulk has a moisture content high enough to permit damaging fungal growth. This range can be determined by taking probe samples from different portions of the bulk. Commonly, in large quantities of bulk-stored grain, some portions have moisture contents 2 to 3% higher than the average (Brusewitz 1987). If the moisture content anywhere in the bulk is too high, fungi will grow, regardless of the average. Therefore, the moisture content of a single representative sample is not a reliable measure of storage risk or spoilage hazard. Measurement of moisture content and the precision of various moisture-measuring methods are covered in the section on Moisture Measurement.

The preparation of this chapter is assigned to TC 2.2, Plant and Animal Environment.

Table 1 summarizes recommended safe storage moistures for several common grains. Note that for long-term storage, lower moistures are recommended. Most storage fungi will not grow in environments where the relative humidity of the air between kernels is lower than 60%. The relationship between grain moisture and the relative humidity of air between kernels is addressed in the section on Equilibrium Moisture. Table 2 summarizes the relative humidities and temperatures that permit the growth of common storage fungi. Table 3 summarizes the relative humidities and temperatures that permit growth of common storage insects.

Moisture Transfer

If temperatures vary within bulk-stored grain, moisture migrates from warmer to cooler portions. The rate of movement depends on the gradients in moisture content and temperature. Sellam and Christensen (1976) studied moisture transfer in a sample of one cubic foot of shelled corn initially at 15.5% moisture. They used heat lamps to produce a temperature differential of 18°F along the length of a sealed plastic container 1.22 ft long. After 2 days, this gradient (approximately 15°F/ft) caused the moisture content at the cool end to increase by 1.2% and the moisture content at the warm end to decrease by 1.1%.

Table 1 Safe Storage Moisture for Aerated Good-Quality Grain

Grain	Maximum Safe Moisture Content, % wet basis
Shelled corn and sorghum	
To be sold as #2 grain or equivalent by spring	15
To be stored up to 1 year	14
To be stored more than 1 year	13
Soybeans	
To be sold by spring	14
To be stored up to 1 year	12
Wheat	13
Small grain (oats, barley, etc.)	13
Sunflower	
To be stored up to 6 months	10
To be stored up to 1 year	8

Source: McKenzie (1980).

Thorpe (1982) developed an equation to describe moisture transfer caused by a temperature gradient. The equation was solved numerically, and laboratory experiments of moisture transfer in wheat were successfully modeled initially at 12% moisture content. In the experiments, an 18°F temperature gradient was developed across a column of wheat 0.66 ft thick (equivalent to a gradient of 27°F/ft). After one month, the moisture content of the warmest grain dropped to 10.6%, while the moisture content of the coolest grain increased to 14%.

Smith and Sokhansanj (1990a) provided a method of approximate analysis of the energy and velocity equations of the natural convection in grain bins. They showed that for small cereals such as wheat, the influence of convection on temperature gradients may not be significant, whereas for larger cereals such as corn, the effect of convection is more noticeable. Smith and Sokhansanj (1990b) also showed that convection flows in a grain bin are significant if the radius of the storage bin is approximately equal to the height of the bin.

Christensen and Meronuck (1986) cite an example of heating that developed in wheat initially at 13.2% in a nonaerated bin. Specially prepared samples were placed at various positions in the bin at the time the bin was filled. After 3 months, the grain began to heat from fungal activity. Moisture content in some of the samples had increased to 18%, while in others it had decreased to 10%.

These examples illustrate the importance of aeration in long-term storage. Aeration is generally required for storage structures with capacities exceeding 2000 bu or 50 tons. Moisture migration can initiate fungal and insect growth, and the heat of respiration generated by these organisms accelerates their growth and leads to spoilage. Studies suggest that temperature gradients could promote spoilage of grain loaded into a ship or barge—even if the grain is initially at a uniform moisture. Most shipments do not spoil because they remain in the ship or barge for a short time and because large temperature gradients do not develop. Christensen and Meronuck (1986) report studies of grain quality in barges and ships.

Table 2 Approximate Temperature and Relative Humidity Requirements for Spore Germination and Growth of Fungi Common on Corn Kernels

Fungus	Minimum Relative Humidity for Spore Germination,[b] %	Grain Moisture,[a] % w.b.	Growth Temperature, °F Lower Limit	Growth Temperature, °F Optimum	Growth Temperature, °F Upper Limit
Alternaria	91	19	25	68	97 to 104
Aspergillus glaucus	70 to 72	13.5 to 14	46	75	100
Aspergillus flavus	82	16 to 17	42 to 43	97 to 100	111 to 115
Aspergillus fumigatus	82	16 to 17	54	104 to 108	131
Cephalosporium acremonium[c]	97	22	46	75	104
Cladosporium	88	18	23	75 to 77	86 to 90
Epicoccum	91	19	25	75	82
Fusarium moniliforme	91	19	39	82	97
Fusarium graminearum, *Fusarium roseum* (*Gibberella zeae*)	94	20 to 21	39	75	90
Mucor	91	19	25	82	97
Nigrospora oryzae[c]	91	19	39	82	90
Penicillium funiculosum[c] (field)	91	19	46	86	97
Penicillium oxalicum[c] (field)	86	17 to 18	46	86	97
Penicillium brevicompactum (storage)	81[b]	16	28	73	86
Penicillium cyclopium (storage)	81[b]	16	28	73	86
Penicillium viridicatum (storage)	81[b]	16	28	73	97

Source: Stroshine et al. (1984).

[a] Approximate corn moisture content at 77°F, which gives an interseed relative humidity equal to the minimum at which fungus can germinate. It is probably *below* the moisture content at which the fungus would be able to compete with other fungi on grain, except for *Aspergillus glaucus*. The latter has no real competitor at 72% rh, except occasionally *Aspergillus restrictus*.

[b] Approximately 5% or more of the population can germinate at this relative humidity.

[c] Rarely found growing in stored grain, regardless of moisture and temperature.

Temperature

Most processes that cause spoilage in stored grains are accompanied by a temperature rise. Therefore, temperatures should be monitored throughout the bulk. Temperature monitoring is commonly done by attaching thermocouples to cables that extend through the bulk from the top to the bottom, with thermocouples about 3 to 6 ft apart on each cable. Single cables are used in the center of circular bins up to 25 ft in diameter. In large bins or flat storage structures, cables are spaced 20 to 25 ft apart. Relatively dry grain is a good insulator, so a **hot spot** can develop without being detected immediately (Foster and Mayes 1962). However, when these thermocouple spacings are used, extensive spoilage can usually be detected by a temperature rise at a nearby thermocouple. A temperature rise of even a few degrees is evidence that grain has spoiled or is spoiling. Forced aeration maintains a uniform and preferably low temperature throughout the bulk.

Table 2 summarizes minimum, optimum, and maximum temperatures for the growth of some common storage fungi. Storage molds grow slowly at 32 to 40°F. However, at higher moisture contents, some species of *Penicillium* will grow when the temperature is slightly below freezing. Grains with a moisture content high enough for invasion by *Aspergillus glaucus* will deteriorate rapidly at temperatures of 75 to 85°F but can be kept for months without damage at 40 to 50°F. Most grain-infesting insects become inactive below about 50°F. Mites remain active but cannot develop rapidly below about 40°F. Control of fungi and insects is described further in the section on Prevention of Deterioration.

Oxygen and Carbon Dioxide

Only a few fungi that cause stored grain deterioration can grow in an atmosphere containing only 0.1 to 0.2% oxygen or more than 60% carbon dioxide. Some yeasts can grow in grain stored in airtight storage at moisture contents above 18 to 19% and temperatures above 40°F, producing flavors that make the grain unsuitable as food. However, the grain remains suitable feed for cattle and swine (Bell and Armitage 1992), and its nutritional value may be enhanced (Beeson and Perry 1958).

Airtight storage of dry grain in underground or earth-sheltered structures is employed in many parts of the world (Dunkel 1985, Bell and Armitage 1992). Insects present when dry grain is put into storage usually die when oxygen has been depleted and will not usually reproduce if grain is sufficiently dry and in good condition. Bell and Armitage (1992) and Shejbal (1980) cover controlled atmosphere storage in more detail.

Insects can also be controlled in conventional storage structures by forcing carbon dioxide or other gases such as nitrogen through the grain (Jay 1980, Ripp 1984). However, the costs of controlled-atmosphere storage may be high unless the structure can be inexpensively sealed or the gases can be easily generated or purchased at a low price.

Grain Condition

Grain that has been stored for several months may already be invaded by storage fungi and partly deteriorated, whether or not this is evident to the naked eye. Molding occurs more rapidly in partially deteriorated grain than in sound grain when the grain is exposed to conditions favorable to mold growth. Microscopic examination and plating techniques can often reveal the fungal infection of grain in its early stages (Sauer et al. 1992, Christensen and Meronuck 1986, Stroshine et al. 1984). Accelerated storage tests, in which samples of grain are stored at a moisture content in equilibrium with air at 80% rh and 85°F and examined periodically, are useful in evaluating storability. These tests enable a manager to estimate the risk of spoilage during storage and to take appropriate action.

Equilibrium Moisture

If air remains in contact with a product for sufficient time, the partial pressure of the water vapor in the air reaches equilibrium with the partial pressure of the water vapor in the material. The relative humidity of the air at equilibrium with a material of a given moisture is the **equilibrium relative humidity**. The moisture content of a hygroscopic material in equilibrium with air of a given relative humidity is the **equilibrium moisture content** M_e.

Table 3 Estimates of Optimum and Minimum Temperatures and Relative Humidity Conditions for Population Increase of Grain-Infesting Insects

Insect Type: In Regard to Temperature	Insect Type: In Regard to Relative Humidity	Species	Temperature, °F: Minimum	Temperature, °F: Optimum	Minimum Relative Humidity, %
		Species Needing High Temperatures			
Cold hardy	Tolerant of low	*Trogoderma granarium*	75	91 to 99	1
		Cryptolestes ferrugineus	73	90 to 95	10
		Oryzaephilus surinamensis	70	88 to 93	10
	Need moderate	*Plodia interpunctella*	64	82 to 90	40
	Need high	*Cryptolestes turcicus*	70	86 to 91	50
Moderately cold hardy	Tolerant of low	*Tribolium confusum*	70	86 to 91	1
	Need moderate	*Rhyzopertha dominica*	73	90 to 95	30
		Lasioderma serricorne	72	90 to 95	30
Cold susceptible	Tolerant of low	*Tribolium castaneum*	72	90 to 95	1
		Oryzaephilus mercator	70	88 to 93	10
	Need high	*Cryptolestes pusillus*	72	82 to 91	60
		Species Thriving at Moderate Temperatures			
Cold hardy	Need moderate	*Sitotroga cerealella*	61	79 to 86	30
	Need high	*Sitophilus granarius*	59	79 to 86	50
		Stegobium paniceum	63	77 to 82	60
		Acarus siro	45	70 to 81	65
Moderately cold hardy	Need high	*Sitophilus oryzae*	63	81 to 88	60

Source: Pederson (1992). Reprinted with permission.

Several theoretical, semitheoretical, and empirical models have been proposed for calculating the M_e of grains. Morey et al. (1978) report that the modified Henderson equation is among the best equations available:

$$M_e = \frac{1}{100}\left[\frac{\ln(1.0-\phi)}{-K(t+C)}\right]^{1/N} \quad (1)$$

where

M_e = equilibrium moisture content, decimal, dry basis
t = temperature, °C = (°F − 32)/1.8
ϕ = relative humidity, decimal equivalent
K, N, C = empirical constants

Table 4 lists values for K, N, and C for various crops. ASAE *Standard* D245.4 also gives the Chung-Pfost equation, another equation used to predict M_e. Figure 1, based on the Chung-Pfost equation, shows equilibrium moisture content curves for shelled corn, wheat, soybeans, and rice. Note that equilibrium moisture depends strongly on temperature. ASAE *Standard* D245.4 gives additional curves drawn from the Chung-Pfost equation and tabulated experimental data. Locklair et al. (1957) give data for tobacco.

The modified Henderson and the Chung-Pfost equations give only approximate values of M_e and are for desorption. When grain is rewetted after it has been dried to a low moisture, the value of M_e is generally lower for a given relative humidity. Variations of as much as 0.5 to 1.0% can result from differences in variety; maturity; and relative starch, protein, and oil content. High-temperature drying can decrease the M_e of shelled corn by 0.5 to 1.0% for a given relative humidity (Tuite and Foster 1963). Therefore, to prevent mold development, shelled corn dried at high temperatures should be stored at a moisture content 0.5 to 1.0% below that of naturally dried corn. Pfost et al. (1976) summarize variations in reported values of M_e for several grains.

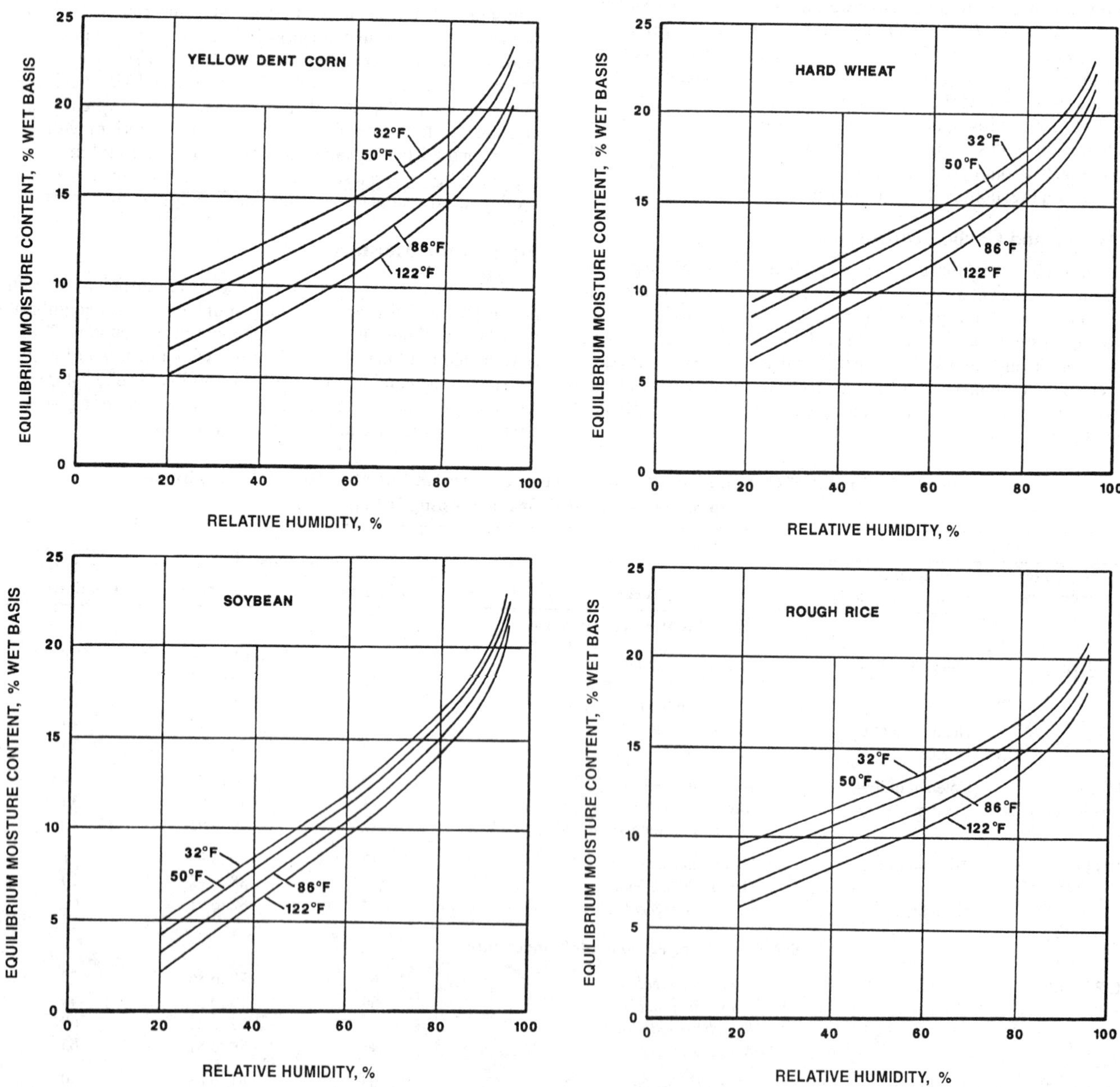

Fig. 1 Equilibrium Moisture Relationships for Certain Crops
(ASAE *Standard* D245.4)

Table 4 Desorption Equilibrium Moisture Constants for Modified Henderson Equation [Equation (1)] for Various Crops

Product	*K*	*N*	*C*
Barley	0.000022919	2.0123	195.267
Beans, edible	0.000020899	1.8812	254.230
Canola (rapeseed)[a]	0.000505600	1.5702	40.1205
Corn, yellow dent	0.000086541	1.8634	49.810
Peanut, kernel	0.000650413	1.4984	50.561
Peanut, pod	0.000066587	2.5362	23.318
Rice, rough	0.000019187	2.4451	51.161
Sorghum	0.000085320	2.4757	113.725
Soybean	0.000305327	1.2164	134.136
Wheat, durum	0.000025738	2.2110	70.318
Wheat, hard	0.000023007	2.2857	55.815
Wheat, soft	0.000012299	2.5558	64.346

Source: ASAE *Standard* D245.4.
[a] From 41 to 86°F (Sokhansanj et al. 1986).

MOISTURE MEASUREMENT

Rapid and accurate measurement of moisture of grains, seeds, and other farm crops determines whether they can be safely stored. Allowable upper limits for moisture are set by the market, and discounts and/or drying charges are usually imposed for higher moistures. Drying the grain to moistures below the accepted market limit or the limit for safe storage moisture results in additional drying expense and may actually decrease the value of the grain.

If shelled corn is dried to 12% moisture, it becomes brittle and breaks more easily during handling. The moisture removal also reduces the total weight of grain. After drying 1 bu (56.0 lb) of shelled corn at 15% moisture to 12%, only 0.966 bu (54.1 lb) remains, and the market value has decreased by 3.4%. However, at high moistures, seed respiration and fungal growth can cause greater loss in value.

Moisture content can be expressed on a wet or dry basis. The wet basis is used by farmers and the grain trade, while dry-basis moistures are often used by engineers and scientists to describe drying rates. Unless otherwise noted, moisture contents in this chapter are on a **wet basis** and are calculated by dividing the weight of water in the material by the total weight. The **dry basis** is calculated by dividing the weight of water by the weight of dry matter.

$$\text{\% Moisture (wet basis): } M_w = \frac{100 W_w}{W_w + W_d} \quad (2)$$

$$\text{\% Moisture (dry basis): } M_d = \frac{100 W_w}{W_d} \quad (3)$$

where

W_w = weight of water
W_d = weight of dry matter

Percent moisture on a wet basis M_w can be converted to percent moisture on a dry basis M_d and vice versa by the following formulas:

$$M_d = \frac{100 M_w}{100 - M_w} \quad (4)$$

$$M_w = \frac{100 M_d}{100 + M_d} \quad (5)$$

The weight change resulting from a change in moisture can be determined by assuming that the weight of dry matter is constant. The dry matter is calculated by multiplying the weight of grain by the quantity $(1 - M_w/100)$. For example, if 1000 lb of wheat at 15% moisture is dried to 13%, the new weight W_n can be calculated by equating the amount of dry matter at 15% to that at 13% as follows:

$$1000(1 - 0.15) = W_n(1 - 0.13)$$

$$W_n = 1000(0.85/0.87) = 977 \text{ lb}$$

If the final weight W_n is known, the final moisture can be calculated by making it the unknown value in the above calculation.

If two quantities of grain at differing moistures are mixed, the final moisture of the mixture can be determined by calculating the weight of water in each, adding these together, and dividing by the total weight. The weight of water is the product of the decimal equivalent of M_w and the total weight. For example, if 500 lb of shelled corn at 16% moisture is mixed with 1000 lb of shelled corn at 14% moisture, the weight of water in each sample is

$$(500 \text{ lb})(0.16) = 80 \text{ lb}$$

$$(1000 \text{ lb})(0.14) = 140 \text{ lb}$$

The moisture content after mixing will be

$$\left[\frac{80 + 140}{500 + 1000}\right] \times 100 = 14.7\%$$

Methods for determining moisture content are either *direct* or *indirect*. Direct methods involving the use of an oven determine moisture content based on the loss in product weight caused by evaporation of the water. The Karl Fischer method, a basic reference method involving a chemical reaction of water and a reagent, is classified as a direct method.

Indirect methods such as moisture meters measure the properties of the material that are functions of the moisture content. Moisture meters are used in commercial practice, while direct methods are used in research and for calibration of the indirect methods. Christensen et al. (1992) summarize approved methods used in Europe and the United States.

Direct Methods

Christensen et al. (1992) describe the fundamental or basic methods of moisture determination as (1) drying in a vacuum with a desiccant and (2) titration with a Karl Fischer reagent. It is assumed that these methods measure the true water content and can be used to verify measurements obtained with routine reference methods, including oven drying and the Brown-Duval distillation method. The Brown-Duval method, not commonly used, involves heating the grain in a special apparatus and condensing and collecting the vaporized water.

Oven techniques use either forced-convection air ovens or vacuum ovens and either ground or whole kernels. Drying times and temperatures vary considerably, and the different techniques can give significantly different results. Oven techniques are used to calibrate moisture meters (see the section on Indirect Methods). As a result, during the export of grain, the meter moisture measurements can vary between arrival and destination if the importing country uses a different standard oven technique than the exporting country.

ASAE *Standard* S352.2 is a widely used standard that recommends heating temperatures and times for various grains. The temperatures may be either 217°F (shelled corn, soybeans, sunflower) or 266°F (wheat, barley, onion). Heating times vary between 50 min (onion seeds) and 72 h (soybeans, shelled corn).

Grinding samples and using a vacuum oven reduce heating time. When initial moistures are high, a two-stage method may be used (USDA 1971). A weighed sample of whole grain is partially dried to a moisture content of 13% or below, weighed, and then ground and completely dried as in the one-stage method. The moisture lost in both stages is used to calculate moisture content.

Indirect Methods

Electronic moisture meters are simple to operate and give readings within minutes. Direct methods of moisture measurement are used to calibrate the meters for each type of grain. Meters are sensitive to grain temperature, and calibration must include a temperature correction factor. The newer automatic meters or **moisture computers** sense and correct for sample temperature and print or display the corrected moisture.

Near infrared reflectance (NIR) instruments have been developed that measure moisture, protein, starch, and oil content of ground samples (Butler 1983, Cooper 1983, Watson 1977). **Near infrared transmittance (NIT) instruments** measure the properties of whole seeds.

Conductance meters measure resistance, which varies with grain moisture. The practical range of moisture content measurable by conductance meters is approximately 7 to 23%. For up to 72 h after moisture addition or removal, the moisture at the surface of the kernels differs from the moisture in the interior. Therefore, recently dried grain reads low and recently wetted grain reads high. Mixing wet and dry grain and mixing good grain with partially deteriorated grain also result in erroneous readings. Martin et al. (1986) measured the signal from the conductance Tag-Heppenstall meter, analyzed the standard deviation of the alternating current (ac) component generated during measurement, and related this to individual kernel moisture variations in mixtures of wet and dry corn.

The dielectric properties of products depend largely on moisture content. The **capacitance meter** uses this relationship by introducing grain as the dielectric in a capacitor in a high-frequency electrical circuit. Although the capacitive reactance is the primary portion of the overall impedance measured, the resistive component is also significant in many capacitance meters. At higher frequencies and in instruments with insulated electrodes, the relative effect of the resistance is reduced, which is important in reducing errors introduced by unusual product surface conditions. Capacitance meters are affected less than conductance meters by uneven moisture distribution within kernels. Sokhansanj and Nelson (1988a) showed that the capacitance meters give low and high readings, respectively, on recently dried or rewetted grain. The range of measurable moisture content is slightly wider than that for conductance meters.

Moisture measurement by capacitance meters is sensitive to temperature, product weight, and product density (Sokhansanj and Nelson 1988b). To reduce these sources of error, a weighed sample is introduced into the measuring cell by reproducible mechanical means. Calibration, including temperature correction, is required. At least one commercially available unit measures bulk density and corrects for this factor as well as temperature. Tests of moisture meter accuracy have been reported by Hurburgh et al. (1985, 1986). Accuracy of moisture readings can be improved by taking multiple samples from a grain lot and averaging the meter measurements. Equipment for continuous measurement of moisture in flowing grain is available commercially but is not widely used in the grain trade.

Equilibrium relative humidity (described in the section on Equilibrium Moisture) can be used to indicate moisture content. It also indicates storability independent of the actual moisture content because the equilibrium relative humidity of the air surrounding the grain, to a large extent, determines whether mold growth can occur (see Table 2). Measurement of equilibrium relative humidity at specific points within a grain mass requires specialized sampling equipment and has been used primarily for research.

Determination of hay moisture content does not receive the consideration devoted to grains. Oven methods (ASAE *Standard* S358.2) are used extensively, but several conductance moisture meters are available for both hay and forages. The extreme variability of the moisture and density of the material tested lead to great variability in the readings obtained. A reasonable indication of the average moisture content of a mass of hay can be obtained if many (25 or more) measurements are taken and averaged.

PREVENTION OF DETERIORATION

Fungal Growth and Mycotoxins

Fungal growth is the most important limitation on the successful storage of grain. In cases where high-temperature drying is used, it is sometimes impossible to dry grain immediately because the harvesting capacity often exceeds the capacity of the dryer. Low-temperature or ambient drying techniques may also be used as an alternative to high-temperature drying. In this case, wet grain is placed in the bin immediately after harvest, and the drying air is blown into a plenum in the bottom of the bin. Molding may occur before the grain can be dried, so the allowable storage time at the harvest moisture and temperature must be known.

Drying in most farm bins begins in the lower layers and proceeds upward through the grain. The layer of grain that is drying is called the **drying zone**, and the upper boundary of the zone, where drying is just beginning, is called the **drying front**. The drying front may not reach the top of the bin for several days or weeks after drying begins. A risk associated with this type of drying is molding of the upper grain layers. Increasing the airflow increases the rate of drying but also increases costs. Therefore, the designer of low-temperature drying systems must know the maximum time that the grain at the top can be held at its harvest moisture and storage temperature before molding is significant.

In the United States, corn is one of the major crops that must be harvested above safe storage moistures. Shelled corn can be held at these higher moistures for a limited time before it must be dried. Mold growth produces carbon dioxide (CO_2). Allowable storage time at moistures above those for safe storage can be estimated by measuring CO_2 production of samples. By assuming that a simple sugar is being oxidized by microbial respiration, CO_2 production can be expressed in terms of dry matter loss in percent by weight.

Saul and Steele (1966) and Steele et al. (1969) studied the production of CO_2 in shelled corn, mostly on samples above 18%. Based on changes in the official grade of shelled corn, Saul and Steele (1966) established a criterion for acceptable deterioration of quality as 0.5% dry matter loss. This is equivalent to the production of 0.00735 lb of CO_2 per pound of dry matter. Thompson (1972) expressed Saul's data on dry matter loss per pound of dry matter as a function of moisture, time, and temperature using the following mathematical expression:

$$\mathrm{DML} = 1.3\left[\exp\left(\frac{0.006\theta}{K_m K_t}\right) - 1.0\right] + \frac{0.015\theta}{K_m K_t} \tag{6}$$

with

$$K_m = 0.103\left[\exp\left(\frac{455}{M_d^{1.53}}\right) - 0.00845 M_d + 1.558\right] \tag{7}$$

$$K_t = A\ \exp\left(\frac{Bt}{60}\right) + C\ \exp\left[0.61\left(\frac{t-60}{60}\right)\right] \tag{8}$$

where

DML = dry matter loss per pound of dry matter, lb/lb
θ = time in storage, h
t = grain temperature, °F
M_d = moisture content, % dry basis

Table 5 Constants for Dry Matter Loss of Shelled Corn [Equation (8)]

Temperature Range, °F	Moisture Range, % w.b.	*A*	*B*	*C*
$t < 60$	All moistures	128.76	−4.68	0
$t \geq 60$	$M_w \leq 19$	32.3	−3.48	0
$t \geq 60$	$19 < M_w \leq 28$	32.3	−3.48	$(M_w - 19)/100$
$t \geq 60$	$M_w > 28$	32.3	−3.48	0.09

Table 5 lists values for *A*, *B*, and *C*. According to Steele (1967), the damage level effect can be determined for dry matter losses of 0.1, 0.5, and 1.0% by multiplying θ from Equation (6) by K_d, where K_d is calculated as follows:

$$0.1\%\text{ DML: } K_d = 1.82\exp(-0.0143d) \quad (9a)$$

$$0.5\%\text{ DML: } K_d = 2.08\exp(-0.0239d) \quad (9b)$$

$$1.0\%\text{ DML: } K_d = 2.17\exp(-0.0254d) \quad (9c)$$

where d = mechanical damage, % by weight.

Based on a simulation, Thompson (1972) concluded that for airflow rates between 0.5 and 2.0 cfm/bu, grain deterioration in the top layer during low-temperature drying is doubled when the airflow rate is halved. Thompson also concluded that weather variations during harvest and storage seasons can cause up to a twofold difference in deterioration.

Seitz et al. (1982a, 1982b) found unacceptable levels of aflatoxin production prior to the time when 0.5% dry matter loss occurred. Nevertheless, Equations (6) through (9) give approximate predictions of mold activity, and they have been used in several computer simulation studies (Pierce and Thompson 1979 and Brooker and Duggal 1982). Pierce and Thompson (1979) give recommended airflow rates for several common low-temperature drying systems and for various locations in the midwestern United States.

Acceptable dry matter losses for wheat and barley are much lower than those for shelled corn—0.085% and 0.104%, respectively (Brook 1987). Brook found reasonable agreement with published experimental data for the following equation (Frazer and Muir 1981) for allowable storage time as a function of percent wet basis moisture and temperature based on the development of visible mold:

$$\log\theta_D = A + BM_w + Ct + G \quad (10)$$

where

θ_D = allowable storage time, days
t = temperature, °F
A,B,C,G = empirical constants, defined as follows:

Moisture Range, % w.b.	*A*	*B*	*C*	*G*
$12.0 < M_w \leq 19.0$	6.234	−0.2118	−0.0293	0.937
$19.0 < M_w < 24.0$	4.129	−0.0997	−0.0315	1.008

Brook (1987) also reported that an adaptation of Equation (6) by Morey et al. (1981) gave reasonable results for storage time of wheat. Morey's method predicts dry matter loss by adjusting M_d for differences between corn and wheat equilibrium relative humidities.

Table 2 can be used to gain insight into the deterioration of stored grain. *Aspergillus* and *Penicillium* sp. are primarily responsible for deterioration because some of their species can grow at storage moistures and temperatures frequently encountered in commercial storage. In temperate climates, shelled corn is often harvested at relatively high moistures; during the harvest and storage season, ambient temperatures can be relatively low. Aeration of the grain during cold weather and cool nights can reduce the temperature of the grain to 40 to 60°F. This is below the optimum temperature for growth of *Aspergillus* sp. (Table 2). However, *Penicillium* sp. can still grow if grain moisture is above 16 to 17%; therefore, its growth is a persistent problem in temperate climates. If hot weather prevails prior to harvest, *Aspergillus flavus*, which competes effectively at warmer temperatures and higher moistures, can begin to grow in the field and continue to grow in stored shelled corn. In growing seasons when shelled corn must be harvested at moistures above 22%, *Fusaium*, *Alternaria*, *Epiccocum* and *Mucor* can compete with *Penicillium* sp.

Chemical Treatment. Application of chemicals slows deterioration until grain can be either dried or fed to animals. Preservatives include propionic acid, acetic acid, isobutyric acid, butyric acid (Sauer and Burroughs 1974), a combination of sorbic acid and carbon dioxide (Danziger et al. 1973), ammonia (Peplinski et al. 1978), and sulfur dioxide (Eckhoff et al. 1984). Propionic acid (Hall et al. 1974) or propionic-acetic acid mixtures, although not extensively used, are perhaps the most popular in the United States with high-moisture corn. Acetic acid and formic acid are most popular in Europe. Grain treated with propionic acid can be used only as animal feed. Inert carriers have been used to lessen corrosiveness.

Hertung and Drury (1974) summarize fungicidal levels needed to preserve grain at various moistures. Both ammonia (Nofsinger et al. 1979, Nofsinger 1982) and sulfur dioxide (Eckhoff et al. 1984, Tuite et al. 1986) treatments require considerable management. Attention must be given to uniform application of the chemicals to the entire quantity of stored grain.

Insect Infestation

Insects cause major losses of stored grain. Grain containing live insects or insect fragments in sufficient numbers is unsuitable for human food. When grain is stored for long periods (a year or more), insects can infest the grain and cause significant amounts of deterioration. Traps and chemical attractants have been developed that monitor insects in storage facilities (Barak and Harein 1982, Barak and Burkholder 1985, Burkholder and Ma 1985). Detection in samples of grain taken for grading and inspection is often difficult. Many of the insects are relatively small and can be seen easily only with a magnifying lens. Many of the insect larvae develop within the kernels and cannot be detected without staining techniques or grinding of the grain sample. Infested grain mixed with good grain in marketing channels compounds the infestation problem.

Sanitation is one of the most effective methods of insect control. Cleaning of bins after removal of old-crop grain and prior to filling with new-crop grain is essential. In bins containing perforated floors, fine material that collects beneath the floors can harbor insects, which infest new-crop grain when it is added. Control by aeration is feasible in temperate climates because insect activity is reduced greatly at temperatures below 50°F. The effectiveness of temperature control has been documented by Bloome and Cuperus (1984) and Epperly et al. (1987). Chemicals have frequently been used to control live insects in grain, and methods are described by Harein and Davis (1982). Recently, thermal treatments have also been investigated (Lapp et al. 1986). Pederson (1992) summarizes the types of grain insects, the ecology of insect growth, and the methods of detecting insects in samples of grain. Control of insects in farm-stored grain is detailed by Storey et al. (1979), Quinlan (1982), and Harein and Davis (1992).

Rodents

The shift from ear corn harvesting and storage to field shelling and the introduction of metal bins have helped to reduce rodent problems. However, significant problems can arise when rodents consume grain and contaminate it with their hair and droppings. Storage structures should be made rodent-proof whenever possible. Rats can reach 13 in. up a wall, so storage structures should have concrete foundations and metal sides that resist gnawing.

In some countries, smaller on-farm storage structures are often elevated 18 in. to give protection from rodents. Double-wall construction and false ceilings should be avoided, and vents and holes should be covered with wire grates. Proper sanitation can help prevent rodent problems by eliminating areas where rodents can nest and hide. Rodents need water to survive, so elimination of available water is also effective. Techniques for killing rodents include trapping, poisoning with bait, and fumigation. Harris and Bauer (1992) address rodent problems and control in more detail.

DRYING THEORY

In ordinary applications, drying is a heat and mass transfer process that vaporizes liquid water, mixes the vapor with the drying air, and removes the vapor by carrying away the mixture mechanically. In forced-convection drying, sufficient heat for vaporization of product moisture (about 1100 Btu/lb of water) comes from the sensible heat in the drying air. A few types of dryers, mostly experimental types, have been developed to apply heat directly to the product by conduction, radiation, or dielectric heating.

The most common mode of drying uses the sensible heat content of the air. The method can be diagrammed on the psychrometric chart by locating the state points for the air as it is heated from ambient temperature to plenum temperature and then exhausted from the grain. The process is assumed to be adiabatic (i.e., all the sensible heat lost by the air is used for moisture vaporization and converted to latent heat of the water vapor in the drying air). Therefore, the state point of the air can be considered to move along adiabatic saturation lines on the psychrometric chart. In the simplified psychrometric chart in Figure 2, the ambient air at dry-bulb temperature t_a and dew-point temperature t_{dp} is heated to drying air temperature t_d, where it has a relative humidity ϕ_1. As the air passes through the grain, its sensible heat provides the latent heat of vaporization of the water. When the air exits from the grain, its temperature has dropped to t_e, and its relative humidity has increased to ϕ_2. The moisture gained by each pound of drying air is the difference $W_2 - W_1$ in humidity ratio. If the air has sufficient contact time with the grain, the value for ϕ_2 will be the equilibrium relative humidity of the grain at that moisture and temperature t_e.

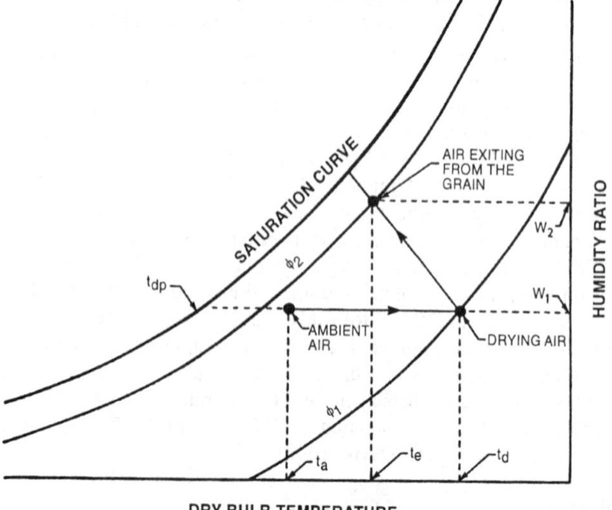

Fig. 2 Drying Process Diagrammed on Psychrometric Chart Showing Adiabatic Evaporation of Moisture from Grain

Example 1. Shelled corn at 20% moisture content is dried with air heated to 160°F. The air has an ambient temperature of 68°F with a dew point of 50°F. The air is observed to exhaust from the shelled corn at 86°F. Find the amount of energy needed to heat the air and the amount of water removed per pound of dry air.

Solution: Estimate the psychrometric air conditions, using information contained in Chapter 6 and assuming a standard atmospheric pressure of 14.696 psi.

At 68°F, the enthalpy of the dry air is h_a = 16.337 Btu/lb (Table 2 in Chapter 6) and the saturation vapor pressure p_{ws} = 0.33921 psi (Table 3 in Chapter 6). At 50°F dew point, the vapor pressure p_w = 0.17811 psi, and the enthalpy of the water vapor h_g = 1083.03 Btu/lb (Table 3 in Chapter 6). The relative humidity is then ϕ = 53% [Equation (24) in Chapter 6]; humidity ratio W = 0.0076 lb/lb [Equation (22) in Chapter 6]; and enthalpy h = 24.6 Btu/lb [Equation (29) in Chapter 6]. As the air is heated, the humidity ratio is assumed to remain constant. At 160°F, the enthalpy of the dry air is h_a = 38.474 Btu/lb (Table 2 in Chapter 6) and the saturation vapor pressure p_{ws} = 4.7468 psi (Table 3 in Chapter 6). The relative humidity has been reduced to ϕ = 4% [Equation (24) in Chapter 6]; enthalpy increased to h = 46.8 Btu/lb [Equation (29) in Chapter 6]; and the wet-bulb temperature of the drying air is t^* = 84°F [iterative solution to Equation (35) in Chapter 6]. The amount of energy needed to heat each pound of dry air is then 46.8 – 24.6 = 22.2 Btu.

As the heated air passes through the grain, it increases in moisture and decreases in temperature until it comes into equilibrium with the corn at the point of air exhaust (initially 20%). The exhaust air relative humidity can be estimated by reading the equilibrium relative humidity from the curve for shelled corn shown in Figure 1. Enter the curve for shelled corn at 20% equilibrium moisture content and a temperature of 86°F. The equilibrium relative humidity is approximately 92%. At 86°F, the saturation vapor pressure of the air p_{ws} = 0.61584 psi (Table 3 in Chapter 6); the vapor pressure p_w = 0.5666 psi [Equation (24) in Chapter 6]; and the humidity ratio W = 0.0249 lb/lb [Equation (22) in Chapter 6]. Each pound of dry air carries with it 0.0249 – 0.0076 = 0.0173 lb of water from the grain.

After the grain at the air exhaust has dried to 15%, the equilibrium moisture content curve from Figure 1 can be used to estimate the exhaust air relative humidity. If the temperature of the air were 86°F, then the equilibrium relative humidity would be approximately 76%; if the temperature of the air were 122°F, then the equilibrium relative humidity would be approximately 81%. From Equation (35) in Chapter 6, the wet-bulb temperatures associated with these two points are 80°F and 115°F, respectively. A linear interpolation between these two points results in an air temperature of 90°F and an equilibrium relative humidity of 77%. At 90°F, the saturation vapor pressure of the air p_{ws} = 0.69889 psi (Table 3 in Chapter 6); the vapor pressure p_w = 0.5381 psi [Equation (24) in Chapter 6]; and the humidity ratio W = 0.0236 lb/lb [Equation (22) in Chapter 6]. Each pound of dry air carries with it 0.0236 – 0.0076 = 0.016 lb of water from the grain.

Thin Layer Drying

A *thin layer* of grain is a layer of grain no more than several kernels deep. The ratio of grain to air is such that there is only a small change in temperature and relative humidity of the drying air when it exits the grain. The maximum rate ($dM/d\theta$) at which a thin layer of a granular hygroscopic material (such as grain) transfers moisture to or from air can be approximated by the following equation (Hukill 1947):

$$\frac{dM}{d\theta} = -C(p_g - p_a) \tag{11}$$

where

C = constant representing vapor conductivity of kernel and surrounding air film
p_g = partial pressure of water vapor in grain
p_a = partial pressure of water vapor in drying air

If $p_g > p_a$, drying takes place. If $p_g = p_a$, moisture equilibrium exists and no moisture transfer occurs. If $p_g < p_a$, wetting occurs. The assumption of a linear relationship between (1) water vapor pressure and equilibrium relative humidity and (2) equilibrium relative humidity and moisture content over the range in which drying occurs lead to the following equation:

$$\frac{dM}{d\theta} = -k(M - M_e) \tag{12}$$

where

M = moisture content (dry basis) of material at time θ
M_e = equilibrium moisture content (dry basis) of material in reference to drying air
k = constant dependent on material

The solution to this differential equation is

$$\frac{M - M_e}{M_o - M_e} = \exp(-k\theta) \tag{13}$$

where M_o = moisture content, dry basis, when $\theta = 0$.

In later work (Hukill and Schmidt 1960, Troeger and Hukill 1971), Hukill recognized that Equation (13) did not describe the drying rate of grain adequately. Misra and Brooker (1980) identified the following model as more promising for shelled corn:

$$\frac{M - M_e}{M_o - M_e} = \exp(-K\theta^N) \tag{14}$$

They give an equation for K, which is a function of drying air temperature and velocity, and another equation for N as a function of drying air relative humidity and initial grain moisture. Their equations are valid for drying air temperatures of 36 to 160°F, drying air relative humidities of 3 to 83%, drying air velocities of 5 to 459 fpm, and initial moistures of 18 to 60% (dry basis).

Li and Morey (1984) also fit their data to Equation (14) and found that within the limits of drying airflow rates and air relative humidities used, K and N can be expressed as functions of air temperature and initial grain moisture only. Their equations for K and N apply to air temperatures ranging from 80 to 240°F, initial grain moistures of 23 to 36% dry basis, airflows of 20 to 100 cfm/bu, and air relative humidities of 5 to 40%.

Other forms of the thin layer drying equation have also been proposed. Thompson et al. (1968) fitted data for shelled corn to the following equation, which is applicable in the range of 140 to 300°F:

$$\theta = A\ln MR + B(\ln MR)^2 \tag{15}$$

where

$A = -1.86178 + 0.00488t$
$B = 427.3740 \exp(-0.03301t)$
$MR = (M - M_e)/(M_o - M_e)$
θ = time, h
t = temperature, °F

Martins and Stroshine (1987) describe the effects of hybrid and damage on the thin layer drying rate and give values for constants A and B in Equation (15) for several hybrids and damage levels.

Results of thin layer drying tests for other grains have also been reported. Data are available for wheat (Watson and Bhargava 1974, Sokhansanj et al. 1984, Bruce and Sykes 1983), soybeans (Hukill and Schmidt 1960, Overhults et al. 1973, Sabbah et al. 1976), barley (O'Callaghan et al. 1971, Sokhansanj et al. 1984, Bruce 1985), sorghum (Hukill and Schmidt 1960, Paulsen and Thompson 1973), rice (Agrawal and Singh 1977, Noomhorm and Verma 1986, Banaszek and Siebenmorgen 1990), sunflower (Syarief et al. 1984, Li et al. 1987), canola (Sokhansanj et al. 1984), oats (Hukill and Schmidt 1960), and lentil seeds (Tang et al. 1989). Sokhansanj and Bruce (1987) developed more rigorous thin layer drying equations based on simultaneous heat and mass transfer through a single kernel and demonstrated that such a model accurately predicts the temperature and moisture content of the grain throughout the drying process.

Equations (13) through (15) do not describe the usual drying process, where grain is in a deep bed and where drying air changes condition but does not necessarily reach moisture equilibrium with the grain. Those models, which are formulated using thin layer drying equations such as these, are summarized in the section on Deep Bed Drying.

Airflow Resistance

Data on resistance of grain to airflow are used for a variety of design calculations such as selecting fans, determining optimum depths for drying bins, predicting airflow paths in bins with aeration ducts, and determining the practical limitations on airflow caused by fan power requirements. For a given fan and dryer or bin, airflow resistance can change with the type of grain being dried, the depth of grain, and the amount of fine material in the grain. In many grain-drying applications, such as when air is forced through a grain bin that has a uniform grain depth and a full perforated floor, airflow is one-dimensional and the pressure drop per unit depth of grain can be assumed to be constant. Shedd (1953) determined the data on pressure drop per unit depth versus airflow for a number of grains and seeds and summarized by plotting them on logarithmic axes. These curves are commonly referred to as *Shedd's curves* and are included in ASAE *Standard* D272.2. They can also be calculated from the following equation (ASAE 1994b):

$$\frac{\Delta p}{L} = \frac{aQ^2}{\ln(1 + bQ)} \tag{16}$$

where

Δp = pressure, in. of water
L = bed depth, ft
Q = airflow rate, cfm/ft^2
a, b = empirical constants

Table 6 summarizes the constants for Equation (16) for some of the more common grains. Constants for grass seeds and some vegetables are included in ASAE *Standard* D272.2. The pressure drop per foot is routinely increased by multiplying the value from Equation (16) by a packing factor. A factor of 1.5 is used for corn, 1.2 for other grains.

Equation (16) gives the airflow resistance for clean, dry grain when the bin is loaded by allowing the grain to flow into the bin through a chute from a relatively low height. In the case of shelled corn, predictions of Equation (16) can be corrected for fine material (fines) using the following equation (ASAE *Standard* D272.2, Haque et al. 1978), which is valid for airflows of 15 to 40 cfm/ft^2 and on fine material fractions of 0.0 to 0.2.

$$\left(\frac{\Delta p}{L}\right)_{ctd} = \left(\frac{\Delta p}{L}\right)_{\text{Eq. (16)}}[1 + (14.5566 - 0.1342Q)W_{fm}] \tag{17}$$

where W_{fm} = decimal fraction of fine material, by weight. Fine material is defined as broken corn and other matter that will pass through a 12/64 in. round-hole sieve.

Grama et al. (1984) reported the effect of fine material particle size distribution on resistance. They also report the effect of the increased resistance from fines on fan power requirements. Kumar and Muir (1986) report the effects of fines in wheat and barley.

Bulk density can have a significant effect on airflow resistance. For moderate heights of 14 to 24 ft, drop height does not affect bulk density in bins filled with a spout (Chang et al. 1986). Bern et al. (1982) reported that auger stirring can decrease the bulk density of

Table 6 Constants for Airflow Resistance [Equation (16)]

Material	Value of a, in. of water·min²/ft²	Value of b, ft²/cfm	Range of Q, cfm/ft²
Barley	6.76×10^{-4}	6.71×10^{-2}	1.1 to 40
Canola (rapeseed)[a]	1.65×10^{-3}	3.69×10^{-2}	4.77 to 52
Ear corn	3.29×10^{-4}	1.65	10 to 69
Lentils[b]	1.72×10^{-3}	1.87×10^{-1}	0.55 to 116
Oats	7.62×10^{-4}	7.06×10^{-2}	1.1 to 40
Peanuts	1.20×10^{-4}	5.64×10^{-1}	6 to 60
Popcorn, white	6.92×10^{-4}	5.99×10^{-2}	1.1 to 40
Popcorn, yellow	5.63×10^{-4}	8.94×10^{-2}	1.1 to 40
Rice, rough	8.12×10^{-4}	6.71×10^{-2}	1.1 to 30
Rice, long brown	6.48×10^{-4}	3.93×10^{-2}	1.1 to 32
Rice, long milled	6.89×10^{-4}	4.24×10^{-2}	1.1 to 32
Rice, medium brown	1.10×10^{-3}	5.53×10^{-2}	1.1 to 32
Rice, medium milled	9.16×10^{-4}	5.38×10^{-2}	1.1 to 32
Shelled corn	6.54×10^{-4}	1.54×10^{-1}	1.1 to 60
Shelled corn, low airflow	3.09×10^{-4}	4.34×10^{-2}	0.05 to 4
Sorghum	6.70×10^{-4}	4.09×10^{-2}	1.1 to 40
Soybeans	3.22×10^{-4}	8.13×10^{-2}	1.1 to 60
Sunflower, confectionery	3.48×10^{-4}	9.19×10^{-2}	1.1 to 35
Sunflower, oil	7.87×10^{-4}	1.20×10^{-1}	5 to 112
Wheat	8.53×10^{-4}	4.45×10^{-2}	1.1 to 40
Wheat, low airflow	2.66×10^{-4}	1.38×10^{-2}	0.05 to 4

Source: ASAE *Standard* D272.2.
[a] Jayas and Sokhansanj (1989).
[b] Sokhansanj et al. (1990).

bins filled with a grain spreader but has no effect on or increases bulk density in bins filled by gravity. Magnitudes of the increase in bulk density caused by grain spreaders have been reported by Stephens and Foster (1976b, 1978) and Chang et al. (1983). If the bulk density is known or can be accurately estimated, the pressure drop for shelled corn can be estimated using the following equation (ASAE *Standard* D272.2):

$$\frac{\Delta p}{L} = X_1 + X_2\frac{(\rho_b/\rho_k)^2 Q}{[1-(\rho_b/\rho_k)]^3} + X_3\frac{(\rho_b/\rho_k)Q^2}{[1-(\rho_b/\rho_k)]^3} \qquad (18)$$

where

Δp = pressure drop, in. of water
L = bed depth, ft
ρ_b = corn bulk density, lb/ft³
ρ_k = corn kernel density, lb/ft³
Q = airflow, cfm/ft²
X_1, X_2, X_3 = constants (Table 7)

Kumar and Muir (1986) report on the effect of filling method on the airflow resistance of wheat and barley. Jayas et al. (1987) showed that the resistance of canola to airflow in a horizontal direction was 0.5 to 0.7 times the resistance to airflow for the vertical direction.

Table 7 Constants for Equation (18) for Airflow Resistance in Shelled Corn as Function of Bulk Density, Kernel Density, and Airflow

Airflow Range, cfm/ft²	X_1	X_2	X_3
$5.3 \le Q \le 26.3$	−0.0012	5.53×10^{-4}	1.62×10^{-5}
$26.3 < Q \le 52.5$	−0.013	6.94×10^{-4}	1.39×10^{-5}
$52.5 < Q \le 117$	−0.094	1.02×10^{-3}	1.23×10^{-5}

Range of applicability: corn bulk density of 45.7 to 49.9 lb/ft³ and airflow of 5.3 to 117 cfm/ft².
Sources: ASAE *Standard* D272.2, Bern and Charity (1975).

Moisture content also affects airflow resistance. Its effect may, in part, be caused by its influence on bulk density. Shedd's curves include a footnote recommending that for loose fill of clean grain, airflow resistance should be multiplied by 0.80 if the grain is in equilibrium with air at relative humidities greater than 85% (ASAE *Standard* D272.2). At 70°F, this corresponds to a moisture of 18% or more for shelled corn (Figure 1). Haque et.al. (1982) give equations that correct for the effects of moisture content of shelled corn, sorghum, and wheat.

When the flow lines are parallel and airflow is linear (as is the case in a drying bin with a full perforated floor), calculation of the airflow is a straightforward application of Equation (16). For a given fan attached to a particular bin filled to a uniform depth with grain, the operating point of the fan can be determined as follows. A curve is plotted showing the total static pressure in the bin plenum versus airflow to the bin. Airflow rate is calculated by dividing the total air volume supplied to the plenum by the cross-sectional area of the bin. Using Equation (16), the pressure drop per unit depth can be calculated and multiplied by the total depth of grain in the bin to give total static pressure in the plenum. The fan curve showing air delivery volume versus static pressure can be plotted on the same axes. The intersection of the curves is the operating point for the fan. These calculations can also be done on a computer, and the point of intersection of the curves can be determined using appropriate numerical methods. McKenzie et al. (1980) and Hellevang (1983) summarize airflow resistances for various bin and fan combinations in tabular and graphical form. Sokhansanj and Woodward (1991) developed a design procedure for use on personal computers to select fans for near-ambient drying of grain.

In cases where airflow is nonlinear, as in conical piles or systems with air ducts, computation is complex (Miketinac and Sokhansanj 1985). Numerical methods for predicting airflow patterns have been developed and applied to bins aerated with ducts (Brooker 1969, Segerlind 1982, Khompos et al. 1984), conical-shaped piles (Jindal and Thompson 1972), and bins in which porosity varies within the bed (Lai 1980). Lai's study applies to bins in which filling methods have created differences in bulk density within the bin or where fine material is unevenly distributed.

Analysis of Deep Bed Drying

The ability to predict the rate at which grain dries in a given type of dryer operating in specific weather conditions with a specified airflow and air temperature can assist designers in developing dryers for maximum efficiency. It can also guide operators in finding the optimum way to operate their particular dryers for given weather conditions. Computer simulations have helped researchers understand the mechanisms and processes involved in drying.

Two relatively simple prediction equations can be solved on a hand calculator. Hukill (1947) developed a widely known and used method that predicts the moisture distribution in a bed of grain during drying. A graphical presentation of one of the equations, which further simplifies calculations, is available. Hukill's method is summarized by Brooker et al. (1992), who give an example calculation for shelled corn drying. Barre et al. (1971) made further adaptations of Hukill's method, and Foster (1986) gives a historical perspective on the development and utility of the method. Brooker et al. (1992) also present a technique called the *heat balance equation*, which equates the heat available in the air for drying with the amount of heat needed to evaporate the desired amount of water from the grain. Both of the above methods take into account airflow, drying air temperature and relative humidity, exit air conditions, grain moisture, and the amount of grain to be dried.

Thompson et al. (1968) considered a deep bed of grain as a series of thin layers of grain stacked one on top of another. Algebraic heat and mass balances were applied to each layer, with the exit air conditions of one layer becoming the input conditions of the next layer. Thompson et al. (1969) used the model to predict concurrent-flow, crossflow, and counterflow drying of shelled corn. Paulsen and Thompson (1973) used it to evaluate crossflow drying of sorghum. Stephens and Thompson (1976) and Pierce and Thompson (1981) used the model to make recommendations about optimum design of high-temperature grain dryers.

Bakker-Arkema et al. (1978) used simultaneous heat and mass transfer equations in a series of coupled partial differential equations to describe deep bed drying. The equations, solved using a finite difference technique, predict grain temperature, grain moisture content, and air temperature and humidity ratio. Bakker-Arkema et al. (1979, 1984) give solutions for in-bin, batch, continuous crossflow, and continuous concurrent-flow dryers. Morey et al. (1976) used the model to evaluate energy requirements for drying. Morey and Li (1984) and Bakker-Arkema et al. (1983) demonstrated the effect of thin layer drying rate on the model predictions.

Other researchers have also developed simulation models. Hamdy and Barre (1970) developed a hybrid computer simulation of high-temperature drying and compared predictions to data on moisture and temperature measurements in a laboratory crossflow dryer. Bridges et al. (1980) used the Thompson model for simulation of batch-in-bin drying. Morey et al. (1978) and Parry (1985) review many of the mathematical models used for high-temperature grain drying.

Computer simulations have also been developed for low-temperature and solar drying. Some of these models have been referenced in the section on Fungal Growth and Mycotoxins under Prevention of Deterioration. Thompson (1972) developed a model that was later used by Pierce and Thompson (1979) to make recommendations on airflow in solar grain drying and by Pierce (1986) to evaluate natural air drying. Sabbah et al. (1979) used the logarithmic model of Barre et al. (1971) for simulation of solar grain drying. Bridges et al. (1984) used a model to evaluate the economics of stirring devices in in-bin drying systems. Morey et al. (1979), Frazer and Muir (1981), Bowden et al. (1983), and Smith and Bailey (1983) have also modeled low-temperature drying. Sharp (1982) reviewed low-temperature drying simulation models.

Aeration of Grain

Aeration involves forcing small amounts of air through the stored grain to maintain a uniform temperature. Prior to the development of this concept, grain was turned by moving it from one storage bin to another. Foster (1986) credits Hukill (1953) with developing the concept of aeration. As mentioned in the sections on Fungal Growth and Mycotoxins and Insect Infestation, lowering of the grain temperature during winter in temperate climates can reduce the rate of deterioration from molds and insects. Aeration can also prevent temperature gradients from developing within the grain mass. Such gradients can cause moisture migration, which results in unacceptably high moistures in certain portions of the bin.

Aeration is used to cool stored grain in the fall. A typical practice is to aerate the grain when the difference between grain temperature and the average daily outside temperature exceeds 10°F. In the United States, grain is usually not warmed in the spring unless it is to be stored past early June. Foster and McKenzie (1979) and McKenzie (1980) give practical recommendations for aeration of grain. Airflow rates of 0.025 to 0.5 cfm per bushel are normally used. Air is usually distributed through the bottom of the bin using ducts. Duct spacing and fan selection are related to bin size and shape and to the airflow rate. Foster and Tuite (1992) give an overview of the topic and include information and charts used for design of such systems. Peterson (1982) gives recommendations for duct spacing in flat storages.

Several computer simulations have been developed to study the effects of heat buildup from microbial activity with and without aeration (Thompson 1972, Brooker and Duggal 1982, Metzger and Muir 1983, Lissik 1986). Aldis and Foster (1977) and Schultz et al. (1984) studied the effect of aeration on grain moisture changes.

DRYING SPECIFIC CROPS

Hay

Forage crops can be either harvested, dried, and stored as hay or harvested and stored under anaerobic conditions as silage. Hay quality can be judged by its color, leafiness, and appearance. Laboratory tests and feeding trials give a more detailed picture of hay quality. The traditional method of making hay is to mow the forage and allow it to field cure or dry in the swath and windrow. Harvesting at higher moistures with subsequent artificial drying may be economically feasible, depending on the local weather conditions.

Basic principles of hay drying and storage are covered by Hall (1980), FEC (1985), and Schuler et al. (1986). Forage must be harvested in the proper stage of maturity to attain maximum feeding value. Leaf loss from alfalfa is high when it is handled at moistures below 39%. Therefore, if it is baled at 40% moisture and dried artificially to the recommended storage moisture of 20% (Schuler et al. (1986), a significantly higher feeding value can be achieved. Both Schuler et al. (1986) and Hall (1980) give sketches for batch and in-storage hay dryers. They recommend airflows of 15 to 20 cfm per square foot of mow floor area or 200 to 500 cfm per ton.

Dehydrated alfalfa meal supplies provitamin A (carotene), vitamin E, xanthophylls (poultry pigmenting factors), vitamin K, vitamin C, and B vitamins. Figure 3 shows losses from field drying of hay found in tests conducted by Shepherd (1954). The rapid loss of carotene immediately after the forage is cut indicates the need for rapid transport to the dehydrator when alfalfa meal with high vitamin content is desired.

Several factors influence retention of vitamins during storage, including the starting plant material, dehydration conditions, addition of stabilizers, and storage conditions. Lowering the temperature reduces the loss rate. Inert gas atmosphere in storage also reduces losses (Hoffman et al. 1945). According to Shepherd (1954), blanching of fresh alfalfa before drying does not alter the storage stability of carotene. Table 8 shows Shepherd's results on the effect of prolonged heating at 212°F. The alfalfa was dried after 45 min; heating beyond this time represented excessive exposure to this temperature. Carotene retention in the intact meal at 149°F storage temperature was considered a measure of storage stability. Normal storage moisture is 8 to 9%. Thompson et al. (1960) summarize the effects of over- and underdrying on carotene stability.

Drying and handling of large round bales has been researched. These bales may weigh from 850 to 1500 lb and are handled individually with forklifts. Verma and Nelson (1983) studied storage of large round bales and found that dry matter loss was the primary component of the total storage losses. Bales stored so that they were protected from the weather had lower losses of dry matter than bales exposed to the weather. They were also higher in total protein. Jones et al. (1985) found significant dry matter loss in large round bales of

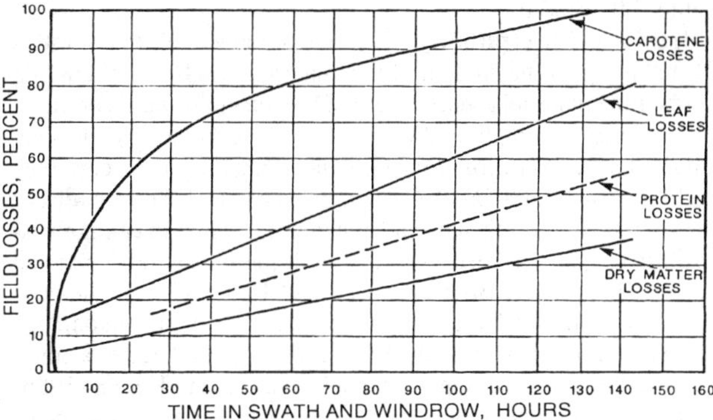

Fig. 3 Time in Swath and Windrow Versus Field Losses of Leaves, Dry Matter, Protein, and Carotene for Hay Drying

Table 8 Effect of Heating Chopped Alfalfa on Carotene Loss During Subsequent Storage of Meal

Hours in Oven at 212°F	Initial Carotene, ppm	Carotene Retained 7 Days at 149°F, %
0.75	229	37
1	228	37
2	197	37
3	176	28
4	149	21
5.5	112	18
7.5	86	15

Source: Thompson et al. (1960).
Note: Alfalfa is fresh frozen from Ryer Island, CA.

mature fescue hay. Harrison (1985) found that addition of sulfur dioxide at the rate of 1% of dry matter had little effect on dry matter loss and nutrient contents for a mixture of alfalfa and bromegrass. However, bales protected with plastic bags did have significantly lower dry matter loss. Jones et al. (1985) found that bales of mature fescue hay stored inside and bales treated with ammonia had less dry matter loss and higher in vitro dry matter digestibility. Henry et al. (1977) and Frisby et al. (1985) developed and tested solar dryers for large round bales.

Grain

The physiological factors involved in drying and storing grain are different from those of forages. Grain is the end product of plant growth, and most physiological activity within the grain or seed is approaching a low level when harvested. With forage, the biological activity within the plant is at or near its peak at the time of harvest.

Both the deterioration of grain harvested at moistures above those safe for storage and the chemical preservation of grain are addressed in the section on Fungal Growth and Mycotoxins. Preservation by ensiling or airtight storage is addressed in the section on Oxygen and Carbon Dioxide.

For more information on grain drying, see Chapter 21 of the 1995 *ASHRAE Handbook—Applications* and Brook (1992).

Corn

Shelled field corn is used primarily as livestock feed, but some is used by milling or processing industries for manufacturing starch, corn oil, and other products. Little information is available on the relationship between the drying method and the feed value of corn. Market grade, as established by the Agricultural Marketing Service of the United States Department of Agriculture (USDA), is the primary criterion for determining corn value. Tests by Cabell et al. (1958) indicated that shelled corn with a moisture content of 29 to 32% can be dried without loss of protein nutritive value by air with temperatures as high as 240°F, provided the airflow rate is approximately 110 cfm/bu.

Breakage Susceptibility. The market grade of dry corn is affected more by the amount of fine material than by other grading factors. Fine material is defined as the broken grain and other material that passes through a 12/64 in. round-hole sieve. The physical damage done to wet corn or the brittleness imparted to the corn during drying causes it to break each time it is handled. The propensity of corn to break during subsequent handling, called **breakage susceptibility**, can be measured with a multiple-impact device called the *Stein breakage tester*. Stephens and Foster (1976a) demonstrated that corn breakage in the tester was correlated with damage during handling. Watson et al. (1986) give a standardized procedure for using the Stein breakage tester, and Watson and Herum (1986) describe and compare other devices developed for measurement of breakage susceptibility. They concluded that a device developed by Singh and Finner (1983) offers great potential for testing of grain for breakage susceptibility in commercial situations.

Paulsen et al. (1983) found significant variations in breakage susceptibility among hybrids. Corn dried with air at high temperatures (140°F) was two to six times more susceptible to breakage than corn dried at near-ambient temperatures. Gustafson and Morey (1979) found that delayed cooling (maintaining the corn at or near its temperature at the end of drying for 6 to 12 h) reduced breakage susceptibility and improved the test weight.

In a study of combination drying, Gustafson et al. (1978) found that combination drying (high-temperature drying to 18% followed by low-temperature drying to 16.6% moisture or below) significantly reduced the increase in breakage susceptibility normally caused by high-temperature drying. Morey and Cloud (1980) summarized principles, energy savings, and advantages of combination drying.

Quality. Both drying temperature and corn hybrid can affect the quality of shelled corn for specific end uses. Brekke et al. (1973) found that drying at temperatures above 140°F reduced the quality of the corn for dry milling. Peplinski et al. (1982) found that optimum dry milling quality could be achieved by harvesting corn at moistures below 25%, minimizing machinery-induced damage to the kernels, and drying at air temperatures below 180°F. Paulsen and Hill (1985) found that the yield of flaking grits from dry milling of corn was significantly greater for corn that had a high test weight and relatively low breakage susceptibility. Weller et al. (1987) found that corn variety affected wet milling quality. At drying temperatures between 120 and 160°F, protein conformational changes occurred and decreased the ethanol soluble protein. Hybrids differ in resistance to storage mold (Tuite and Foster 1979), thin layer drying rate, and dry milling quality (Stroshine et al. 1986). Watson (1987) gives an extensive summary of measurement and maintenance of quality of corn, and Foster (1975) summarizes approaches to reducing damage during harvesting, handling, and drying.

Cotton

The lint moisture content for best results in ginning cotton appears to be 5 to 7%, with an optimum moisture content of 6% (Franks and Shaw 1962). Cotton, like grain, is hygroscopic and should be dried just prior to ginning. The wide variation in incoming moisture content usually requires different amounts of drying for each load. Rapid changes in the amount of drying required can best be handled by using a multipath drying tower in which the cotton is exposed for various lengths of time (2 to 10 s) at temperatures not exceeding 350°F. The air-to-cotton ratio can range from 40 to 100 cfm/lb of cotton (Franks and Shaw 1962). Laird and Baker (1983) found that substantial amounts of heat could be reclaimed and used for drying in commercial cotton gin plants. Equilibrium moisture content data for newly harvested cotton fibers are given by Griffin (1974). Anthony (1982) studied moisture gain of cotton bales during storage.

Cottonseed removed from the fibers is also dried. The germination of cottonseed is unimpaired by drying if the internal cottonseed temperatures do not exceed 140°F (Shaw and Franks 1962). This temperature is not exceeded in the tower dryer described previously. However, the moisture content of the seed can be above the recommended level of 12% following the multipath tower drying. Drying seed in a triple-pass drum at 250 to 300°F with an exposure time of 4 min, followed by cooling, reduces moisture content, inhibits the formation of free fatty acids, and improves germination compared to undried seed. Anthony (1983) dried cottonseed in a vacuum microwave dryer. The cottonseed would not germinate, but its oil properties were not harmed as long as lower temperatures were used. The drying rate was increased by reducing pressure below atmospheric. Rayburn et al. (1978) studied preservation of high-moisture cottonseed with propionic acid.

Peanuts

Peanuts in the shell normally have a moisture content of about 50% at the time of digging. Allowing peanuts to dry on the vines in the windrows for a few days removes much of this water. However, peanuts normally contain 20 to 30% moisture when removed from the vines, and some artificial drying in the shell is necessary. Drying should begin within 6 h after harvesting in order to prevent peanuts from self-heating. The maximum temperature and rate of drying must be controlled to maintain quality. High temperatures result in off-flavor or bitterness. Overly rapid drying without high temperatures results in blandness or inability to develop flavor when roasted (Bailey et al. 1954). High temperatures and rapid or excessive drying also cause the skin to slip easily and the kernels to become brittle. These conditions result in damage in the shelling operation and can be avoided if the moisture removal rate does not exceed 0.5% per hour. Because of these limitations, continuous-flow drying is not usually recommended.

Young (1984) found energy savings up to 26% when comparing recirculating dryers with conventional peanut dryers. Smith and Davidson (1982) and Smith et al. (1985) address the aeration of peanuts during warehouse storage.

Rice

Of all grains, rice is possibly the most difficult to process without quality loss. Rice containing more than 12.5% moisture cannot be stored safely for long periods, yet the recommended harvest moisture content for best milling and germination ranges from 20 to 26% (Kramer 1951). If the rice is harvested at this moisture content, drying must begin promptly to prevent heat-related damage, which can result in "stack-burn," a yellowing of the kernel. To prevent excessive internal fissuring, which results in broken kernels during milling, multiple-pass drying is usually necessary (Calderwood and Webb 1971). Kunze and Calderwood (1980) summarize rice-drying techniques.

Because the market demands polished whole kernels of rice, it is necessary to prevent damage in the form of fissures. Rapid moisture removal or addition can create moisture gradients within kernels. According to Kunze (1984), gradients can develop in the field on a humid night before harvest, in a hopper containing a mixture of rice kernels at varying moistures, and in certain types of dryers. Banaszek and Siebenmorgen (1990) quantified the rate at which moisture absorption reduces head rice yields. Velupillai and Verma (1986) report that drying at 200°F followed by tempering in a sealed container for 24 h gave good kernel strength and head rice yields. They also found that storing the rice after drying for 3 weeks gave optimum grain quality. Bakker-Arkema et al. (1984) achieved good rice quality with concurrent-flow drying of rice.

Soybean, Sunflower, and Edible Beans

Prolonged periods of extremely wet weather during the harvest season can make artificial drying of soybeans necessary. Like peanuts and other oilseeds, soybeans cannot be dried satisfactorily with the high-temperature, high-speed methods used for cereal grains. Because of the different seed structure, rapid drying splits the seed coat and reduces quality and storage life. Overhults et al. (1975) reported a significant decrease in the quality of oil extracted from soybeans dried at temperatures above 160°F. Soybeans have one of the slowest thin layer drying rates of commonly grown cereals and oilseeds (Bakker-Arkema et al. 1983). Therefore, they dry more slowly and require more energy when dried in continuous-flow dryers.

Sunflower is a major crop in some areas of the United States. Hellevang (1987) recommends maximum drying temperatures of 200°F for continuous-flow drying of oil sunflower and 180°F for nonoil sunflower to prevent scorching of the seed meat. Schuler (1974) gives data on equilibrium moisture, airflow resistance, and specific heat of sunflower seeds. Because sunflower is about half the density of shelled corn, moisture can be removed more rapidly, and there is a tendency to overdry. This factor, along with accumulation of foreign material when drying, causes an increased fire hazard (Hellevang 1982). Schmidt and Backer (1980) attribute most of the problems encountered with storage of sunflower seed to improper drying and/or aeration.

Edible beans, a major crop in several states, should be dried with air at relative humidities above 40% to prevent stress cracking. Natural air or low-temperature drying is best (Hellevang 1987). If dried at high rates, seed coats may crack, and beans may split during subsequent handling (Otten et al. 1984, Radajewski et al. 1992). Broken beans can develop a bitter or undesirable flavor and spoil more easily during storage (Uebersax and Bedford 1980).

Wheat and Barley

In northern regions of the United States, wheat and barley may be harvested above safe storage moistures to prevent excessive field losses. Moilanen et al. (1973) recommended that hard red spring wheat be dried at temperatures below 160, 140, and 120°F, respectively, for harvest moistures of 16, 20, and 24% wet basis. These data assumed airflow of 100 to 150 cfm/ft^2. For airflow of 50 cfm/ft^2, the authors recommended that the drying air temperature be reduced by 10 to 15°F. In the case of barley used for malting, the seed must be able to germinate. Therefore, the maximum recommended drying air temperature is 110 to 120°F (Hellevang 1987). Watson et al. (1962) studied the effects of harvest moisture and drying temperature on barley malting quality and recommended harvesting below 20% moisture. If wheat or barley is used for seed, the maximum recommended drying air temperature is 110°F.

In regions where soft wheat is grown, it may be economical to harvest at 20 to 24% moisture to allow double cropping with soybeans; this allows wheat harvest to begin 5 to 7 days earlier than normal and increases the yield of the soybeans (Swearingen 1979). In areas where double cropping is feasible, soft wheat can be dried using low-temperature solar drying or ambient drying with intermittent fan operation (Barrett et al. 1981). High-speed and continuous-flow systems with reduced drying air temperatures can also be used (Parsons et al. 1979). Kirleis et al. (1982) harvested soft red winter wheat at moistures of 25% or below and dried with air temperatures of 150°F or below without adverse effects on milling or cookie baking quality.

In high-temperature continuous-flow dryers, wheat and barley reduce airflow because they have a high airflow resistance. Bakker-Arkema et al. (1983) report that thin layer drying rates for barley and wheat are much faster than for corn. Barley dries more slowly than wheat, presumably because the kernels are larger. In their computer simulations of a concurrent-flow dryer, wet bushel capacity for wheat was about 80% of the capacity for shelled corn when moisture content was reduced by 4.7%. The drying capacity difference was probably caused by a decrease in airflow.

Tobacco (Curing)

Tobacco leaves normally have a moisture content of about 85% at harvest. The major methods of tobacco drying are air curing and flue curing (Johnson et al. 1960).

For **air curing**, whole plants are cut and allowed to wilt in the field until the leaves reach about 70% moisture. The plants are then hung in open barns, where temperatures range from 60 to 90°F and humidities from 65 to 70%. The curing period is 28 to 56 days (Jefries 1940). The desired end product for air curing is a tan leaf. Overdrying at low temperatures results in green color and low sugar content; overdrying at high temperatures results in yellow color (Walton and Henson 1971). Both conditions are undesirable because the normal chemical changes are arrested prematurely. Subsequent drying at optimum rates can reverse some damage. Underdrying at all temperatures results in undesirable dark color and damage from mold and bacterial growth (Walton et al. 1973).

Flue curing uses artificial heat. The leaves are harvested and hung in closed barns where temperatures are increased gradually during the curing period. Normally, 3 days of drying at temperatures of 90 to 120°F brings about yellowing. For the next 2 days, temperatures of 120 to 140°F are used for leaf drying; then, stems are dried at 170°F for 1 to 2 days. A bright yellow to orange color is desirable in flue-cured or bright-leaf tobacco.

REFERENCES

Agrawal, Y.C. and R.P. Singh. 1977. Thin-layer drying studies on short grain rice. *Paper* 77-3531. American Society of Agricultural Engineers, St. Joseph, MI.

Aldis, D.F. and G.H. Foster. 1977. Moisture changes in grain from exposure to ambient air. *Paper* 77-3524. American Society of Agricultural Engineers, St. Joseph, MI.

Anthony, W.S. 1982. Moisture gain and resilient forces of cotton bales during equilibration. *Transactions of ASAE* 25(4):1066-70.

Anthony, W.S. 1983. Vacuum microwave drying of cotton: Effect on cottonseed. *Transactions of ASAE* 26(1):275-78.

ASAE. 1992. Moisture measurement—Unground grain and seeds. *Standard* S352.2. American Society of Agricultural Engineers, St. Joseph, MI.

ASAE. 1993. Moisture measurement—Forages. *Standard* S358.2.

ASAE. 1994. Moisture relationships of grains. *Standard* D245.4.

ASAE. 1994. Resistance to airflow of grains, seeds, other agricultural products, and perforated metal sheets. *Standard* D272.2.

Bailey, W.K., T.A. Pickett, and J.G. Futral. 1954. Rapid curing adversely affects quality of peanuts. *Peanut Journal and Nut World* 33(8):37-39.

Bakker-Arkema, F.W., R.C. Brook, and L.E. Lerew. 1978. Cereal grain drying. In *Advances in cereal science and technology*, ed. Y. Pomeranz, pp. 1-90. American Association of Cereal Chemists, St. Paul, MN.

Bakker-Arkema, F.W., S. Fosdick, and J. Naylor. 1979. Testing of commercial crossflow dryers. *Paper* 79-3521. American Society of Agricultural Engineers, St. Joseph, MI.

Bakker-Arkema, F.W., C. Fontana, R.C. Brook, and C.W. Westlake. 1984. Concurrent flow rice drying. *Drying Technology* 1(2):171-91.

Bakker-Arkema, F.W., C. Fontana, G.L. Fedewa, and I.P. Schisler. 1983. A comparison of drying rates of different grains. *Paper* 83-3009. American Society of Agricultural Engineers, St. Joseph, MI.

Banaszek, M.M. and T.J. Siebenmorgen. 1990. Head rice yield reduction rate caused by moisture absorption. *Transactions of ASAE* 33(4):1263-69.

Barak, A.V. and W.E. Burkholder. 1985. A versatile and effective trap for detecting and monitoring stored-product coleoptera. *Agricultural Ecosystems and Environment* 12:207-18.

Barak, A.V. and P.K. Harein. 1982. Trap detection of stored-grain insects in farm-stored shelled corn. *Journal of Economic Entomology* 75(1):108-11.

Barre, H.J., G.R. Baughman, and M.Y. Hamdy. 1971. Application of the logarithmic model to cross-flow deep-bed grain drying. *Transactions of ASAE* 14(6):1061-64.

Barrett, Jr., J.R., M.R. Okos, and J.B. Stevens. 1981. Simulation of low temperature wheat drying. *Transactions of ASAE* 24(4):1042-46.

Beeson, W.M. and T.W. Perry. 1958. The comparative feeding value of high moisture corn and low moisture corn with different feed additives for fattening beef cattle. *Journal of Animal Science* 17(2):368-73.

Bell, C.H. and D.M. Armitage. 1992. Alternative storage practices. In *Storage of cereal grains and their products*, pp. 249-312. D.B. Sauer, ed.

Bern, C.J. and L.F. Charity. 1975. Airflow resistance characteristics of corn as influenced by bulk density. *Paper* 75-3510. American Society of Agricultural Engineers, St. Joseph, MI.

Bern, C.J., M.E. Anderson, W.F. Wilcke, and C.R. Hurburgh. 1982. Auger-stirring wet and dry corn—Airflow resistance and bulk density effects. *Transactions of ASAE* 25(1):217-20.

Bloome, P.D. and G.W. Cuperus. 1984. Aeration for management of stored grain insects in wheat. *Paper* 84-3517. American Society of Agricultural Engineers, St. Joseph, MI.

Bowden, P.J., W.J. Lamond, and E.A. Smith. 1983. Simulation of near-ambient grain drying: I, Comparison of simulations with experimental results. *Journal of Agricultural Engineering Research* 28:279-300.

Brekke, O.L., E.L. Griffin, Jr., and G.C. Shove. 1973. Dry milling of corn artificially dried at various temperatures. *Transactions of ASAE* 16(4):761-65.

Bridges, T.C., D.G. Colliver, G.M. White, and O.J. Loewer. 1984. A computer aid for evaluation of on-farm stir drying systems. *Transactions of ASAE* 27(5):1549-55.

Bridges, T.C., I.J. Ross, G.M. White, and O.J. Loewer. 1980. Determination of optimum drying depth for batch-in bin corn drying systems. *Transactions of ASAE* 23(1):228-33.

Brook, R.C. 1987. Modelling grain spoilage during near-ambient grain drying. *Divisional Note* DN 1388, AFRC Institute of Engineering Research, Wrest Park, Silsoe, Bedford, MK45 4HS, England, 20 p.

Brook, R.C. 1992. Drying cereal grains. In *Storage of cereal grains and their products*, pp. 183-218. D.B. Sauer, ed.

Brooker, D.B. 1969. Computing air pressure and velocity distribution when air flows through a porous medium and nonlinear velocity-pressure relationships exist. *Transactions of ASAE* 12(1):118-20.

Brooker, D.B. and A.K. Duggal. 1982. Allowable storage time of corn as affected by heat buildup, natural convection and aeration. *Transactions of ASAE* 25(3):806-10.

Brooker, D.B., F.W. Bakker-Arkema, and C.W. Hall. 1992. *Drying and storage of grains and oilseeds*. Van Nostrand Reinhold, New York.

Bruce, D.M. 1985. Exposed-layer barley drying: Three models fitted to new data up to 150°C. *Journal of Agricultural Engineering Research* 32:337-47.

Bruce, D.M. and R.A. Sykes. 1983. Apparatus for determining mass transfer coefficients at high temperatures for exposed particulate crops, with initial results for wheat and hops. *Journal of Agricultural Engineering Research* 28:385-400.

Brusewitz, G.H. 1987. Corn moisture variability during drying, mixing and storage. *Journal of Agricultural Engineering Research* 38:281-88.

Burkholder, W.E. and M. Ma. 1985. Pheromones for monitoring and control of stored-product insects. *Annual Review of Entomology* 30:257-72.

Butler, L.A. 1983. The history and background of NIR. *Cereal Foods World* 28(4):238-40.

Cabell, C.A., R.E. Davis, and R.A. Saul. 1958. Relation of drying air temperature, time and air flow rate to the nutritive value of field-shelled corn. *Technical Progress Report* 1957-58 ARS 44-41. USDA, Washington, D.C.

Calderwood, D.L. and B.D. Webb. 1971. Effect of the method of dryer operation on performance and on the milling and cooking characteristics of rice. *Transactions of ASAE* 14(1):142-46.

Chang, C.S., H.H. Converse, and F.S. Lai. 1986. Technical Notes: Distribution of fines and bulk density of corn as affected by choke-flow, spout-flow, and drop-height. *Transactions of ASAE* 29(2):618-20.

Chang, C.S., H.H. Converse, and C.R. Martin. 1983. Bulk properties of grain as affected by self-propelled rotational type grain spreaders. *Transactions of ASAE* 26(5):1543-50.

Christensen, C.M. and R.A. Meronuck. 1986. *Quality maintenance in stored grains and seeds*. University of Minnesota Press, Minneapolis.

Christensen, C.M., B.S. Miller, and J.A. Johnston. 1992. Moisture and its measurement. In *Storage of cereal grains and their products*, pp. 39-54. D.B. Sauer, ed.

Colliver, D.G., R.M. Peart, R.C. Brook, and J.R. Barrett, Jr. 1983. Energy usage for low temperature grain drying with optimized management. *Transactions of ASAE* 26(2):594-600.

Cooper, P.J. 1983. NIR analysis for process control. *Cereal Foods World* 28(4):241-45.

Danziger, M.T., M.P. Steinberg, and A.I. Nelson. 1973. Effect of CO_2, moisture content, and sorbate on safe storage of wet corn. *Transactions of ASAE* 16(4):679-82.

Dunkel, F.V. 1985. Underground and earth sheltered food storage: Historical, geographic, and economic considerations. *Underground Space* 9:310-15.

Eckhoff, S.R., J. Tuite, G.H. Foster, R.A. Anderson, and M.R. Okos. 1984. Inhibition of microbial growth during ambient air corn drying using sulfur dioxide. *Transactions of ASAE* 27(3):907-14.

Epperly, D.R., R.T. Noyes, G.W. Cuperus, and B.L. Clary. 1987. Control stored grain insects by grain temperature management. *Paper* 87-6035. American Society of Agricultural Engineers, St. Joseph, MI.

FEC. 1985. Hay drying: A guide to the practical design of installations. Farm Electric Center, Kenilworth, Warwickshire, England.

Foster, G.H. 1975. Causes and cures of physical damage to corn. In *Corn quality in world markets*, L.D. Hill, ed. Interstate Printers and Publishers, Danville, IL.

Foster, G.H. 1986. William V. Hukill, a pioneer in crop drying and storage. *Drying Technology* 4(3):461-71.

Foster, G.H. and H.F. Mayes. 1962. Temperature effects of an artificial hotspot embedded in stored grain. AMS-479. U.S. Department of Agriculture, Washington, D.C.

Foster, G.H. and B.A. McKenzie. 1979. Managing grain for year-round storage. AE-90. Cooperative Extension Service, Purdue University, West Lafayette, IN.

Foster, G.H. and J. Tuite. 1992. Aeration and stored grain management. In *Storage of cereal grains and their products*, pp. 219-48. D.B. Sauer, ed.

Franks, G.N. and C.S. Shaw. 1962. Multipath drying for controlling moisture in cotton. ARS 42-69. USDA, Washington, D.C.

Frazer, B.M. and W.E. Muir. 1981. Airflow requirements for drying grain with ambient and solar-heated air in Canada. *Transactions of ASAE* 24(1):208-10.

Frisby, J.C., J.T. Everett, and R.M. George. 1985. A solar dryer for large, round alfalfa bales. *Applied Engineering in Agriculture* 1(2):50-52.

Grama, S.N., C.J. Bern, and C.R. Hurburgh, Jr. 1984. Airflow resistance of moistures of shelled corn and fines. *Transactions of ASAE* 27(1):268-72.

Griffin, A.C., Jr. 1974. The equilibrium moisture content of newly harvested cotton fibers. *Transactions of ASAE* 17(2):327-28.

Gunasekaran, S. 1986. Optimal energy management in grain drying. *Critical Reviews in Food Science and Nutrition* 25(1):1-48.

Gustafson, R.J. and R.V. Morey. 1979. Study of factors affecting quality changes during high-temperature drying. *Transactions of ASAE* 22(4): 926-32.

Gustafson, R.J., R.V. Morey, C.M. Christensen, and R.A. Meronuck. 1978. Quality changes during high-low temperature drying. *Transactions of ASAE* 21(1):162-69.

Hall, C.W. 1980. *Drying and storage of agricultural crops*. AVI Publishing Company, Westport, CT.

Hall, G.E., L.D. Hill, E.E. Hatfield, and A.H. Jenson. 1974. Propionic-acetic acid for high-moisture preservation. *Transactions of ASAE* 17(2):379-82, 387.

Hamdy, M.Y. and H.J. Barre. 1970. Analysis and hybrid simulation of deep-bed drying of grain. *Transactions of ASAE* 13(6):752.

Haque, E., Y.N. Ahmed, and C.W. Deyoe. 1982. Static pressure drop in a fixed bed of grain as affected by grain moisture content. *Transactions of ASAE* 25(4):1095-98.

Haque, E., G.H. Foster, D.S. Chung, and F.S. Lai. 1978. Static pressure drop across a bed of corn mixed with fines. *Transactions of ASAE* 21(5):997-1000.

Harein, P.K. and R. Davis. 1992. Control of stored grain insects. In *Storage of cereal grains and their products*, pp. 491-534. D.B. Sauer, ed.

Harris, K.L. and F.J. Bauer. 1992. Rodents. In *Storage of cereal grains and their products*, pp. 393-434. D.B. Sauer, ed.

Harrison, H.P. 1985. Preservation of large round bales at high moisture. *Transactions of ASAE* 28(3):675-79, 686.

Hellevang, K.J. 1982. Crop dryer fires while drying sunflower. *Paper* 82-3563. American Society of Agricultural Engineers, St. Joseph, MI.

Hellevang, K.J. 1983. Natural air/low temperature crop drying. *Bulletin* 35. Cooperative Extension Service, North Dakota State University, Fargo, ND.

Hellevang, K.J. 1987. Grain drying. *Publication* AE-701. Cooperative Extension Service, North Dakota State University, Fargo, ND.

Henry, Z.A., B.L. Bledsoe, and D.D. Eller. 1977. Drying of large hay packages with solar heated air. *Paper* 77-3001. American Society of Agricultural Engineers, St. Joseph, MI.

Hertung, D.C. and E.E. Drury. 1974. Antifungal activity of volatile fatty acids on grains. *Cereal Chemistry* 51(1):74-83.

Hoffman, E.J., G.F. Lum, and A.L. Pitman. 1945. Retention of carotene in alfalfa stored in atmospheres of low oxygen content. *Journal of Agricultural Research* 71:361-73.

Hukill, W.V. 1947. Basic principles in drying corn and grain sorghum. *Agricultural Engineering* 28(8):335-38, 340.

Hukill, W.V. 1953. Grain cooling by air. *Agricultural Engineering* 34(7): 456-58.

Hukill, W.V. and J.L. Schmidt. 1960. Drying rate of fully exposed grain kernels. *Transactions of ASAE* 3(2): 71-77, 80.

Hurburgh, C.R., T.E. Hazen, and C.J. Bern. 1985. Corn moisture measurement accuracy. *Transactions of ASAE* 28(2):634-40.

Hurburgh, C.R., L.N. Paynter, S.G. Schmitt, and C.J. Bern. 1986. Performance of farm-type moisture meters. *Transactions of ASAE* 29(4): 1118-23.

Jay, E. 1980. Methods of applying carbon dioxide for insect control in stored grain. Science and Education Administration, Advances in Agricultural Technology, Southern Series, AAT-S-13, Agricultural Research (Southern Region), SEA, USDA, P.O. Box 53326, New Orleans, LA 70153.

Jayas, D.S. and S. Sokhansanj. 1989. Design data on the airflow resistance to canola (rapeseed). *Transactions of ASAE* 32(1):295-96.

Jayas, D.S., S. Sokhansanj, E.B. Moysey, and E.M. Barber. 1987. The effect of airflow direction on the resistance of canola (rapeseed) to airflow. *Canadian Agricultural Engineering* 29(2):189-92.

Jefries, R.N. 1940. The effect of temperature and relative humidity during curing upon the quality of white burley tobacco. *Bulletin* No. 407. Kentucky Agricultural Experiment Station, Lexington, KY.

Jindal, V.K. and T.L. Thompson. 1972. Air pressure patterns and flow paths in two-dimensional triangular-shaped piles of sorghum using forced convection. *Transactions of ASAE* 15(4):737-44.

Johnson, W.H., W.H. Henson, Jr., F.J. Hassler, and R.W. Watkins. 1960. Bulk curing of bright-leaf tobacco. Agricultural Engineering 41(8):511-15, 517.

Jones, A.L., R.E. Morrow, W.G. Hires, G.B. Garner, and J.E. Williams. 1985. Quality evaluation of large round bales treated with sodium diacetate or anhydrous ammonia. *Transactions of ASAE* 28(4):1043-45.

Khompos, V., L.J. Segerlind, and R.C. Brook. 1984. Pressure patterns in cylindrical grain storages. *Paper* 84-3011. American Society of Agricultural Engineers, St. Joseph, MI.

Kirleis, A.W., T.L. Housley, A.M. Emam, F.L. Patterson, and M.R. Okos. 1982. Effect of preripe harvest and artificial drying on the milling and baking quality of soft red winter wheat. *Crop Science* 22:871-76.

Kramer, H.A. 1951. Engineering aspects of rice drying. *Agricultural Engineering* 32(1):44-45, 50.

Kumar, A. and W.E. Muir. 1986. Airflow resistance of wheat and barley affected by airflow direction, filling method and dockage. *Transactions of ASAE* 29(5):1423-26.

Kunze, O.R. 1984. Physical properties of rice related to drying the grain. *Drying Technology* 2(3):369-87.

Kunze, O.R. and D.L. Calderwood. 1980. Systems for drying of rice. In *Drying and storage of agriculture crops*, C.W. Hall. AVI Publishing Company, Westport, CT.

Lai, F.S. 1980. Three dimensional flow of air through nonuniform grain beds. *Transactions of ASAE* 23(3):729-34.

Laird, W. and R.V. Baker. 1983. Heat recapture for cotton gin drying systems. *Transactions of ASAE* 26(3):912-17.

Lapp, H.M., F.J. Madrid, and L.B. Smith. 1986. A continuous thermal treatment to eradicate insects from stored wheat. *Paper* 86-3008. American Society of Agricultural Engineers, St. Joseph, MI.

Li, H. and R.V. Morey. 1984. Thin-layer drying of yellow dent corn. *Transactions of ASAE* 27(2):581-85.

Li, Y., R.V. Morey, and M. Afinrud. 1987. Thin-layer drying rates of oilseed sunflower. *Transactions of ASAE* 30(4):1172-75, 1180.

Lissik, E.A. 1986. A model for the removal of heat in respiring grains. *Paper* 86-6509. American Society of Agricultural Engineers, St. Joseph, MI.

Locklair, E.E., L.G. Veasey, and M. Samfield. 1957. Equilibrium desorption of water vapor on tobacco. *Journal of Agricultural and Food Chemistry* 5:294-98.

Manis, J.M. 1992. Sampling, inspection and grading. In *Storage of cereal grains and their products*, pp. 563-88. D.B. Sauer, ed.

Martin, C.R., Z. Czuchajowska, and Y. Pomeranz. 1986. Aquagram standard deviations of moisture in mixtures of wet and dry corn. *Cereal Chemistry* 63(5):442-45.

Martins, J. and R.L. Stroshine. 1987. Difference in drying efficiencies among corn hybrids dried in a high-temperature column-batch dryer. *Paper* 87-6559. American Society of Agricultural Engineers, St. Joseph, MI.

McKenzie, B.A. 1980. Managing dry grain in storage. AED-20. Midwest Plan Service, Iowa State University, Ames, IA.

McKenzie, B.A., G.H. Foster, and S.S. DeForest. 1980. Fan sizing and application for bin drying/cooling of grain. AE-106. Cooperative Extension Service, Purdue University, West Lafayette, IN.

McMillan, W.W., D.M. Wilson, and N.W. Widstrom. 1985. Aflatoxin contamination of preharvest corn in Georgia—A six-year study of insect damage and visible *Aspergillus flavus*. *Journal of Environmental Quality* 14:200-02.

Metzger, J.F. and W.E. Muir. 1983. Computer model of two-dimensional conduction and forced convection in stored grain. *Canadian Agricultural Engineering* 25:119-25.

Miketinac, M.J. and S. Sokhansanj. 1985. Velocity-pressure distribution in grain bins—Brooker model. *International Journal of Applied Numerical Analysis in Engineering* 21:1067-75.

Misra, M.K. and D.B. Brooker. 1980. Thin-layer drying and rewetting equations for shelled yellow corn. *Transactions of ASAE* 23(5):1254-60.

Moilanen, C.W., R.T. Schuler, and E.R. Miller. 1973. Effect on wheat quality of air flow and temperatures in mechanical dryers. *North Dakota Farm Research* 30(6):15-19.

Morey, R.V. and H.A. Cloud. 1980. Combination high-speed, natural-air corn drying. M-163. Agricultural Extension Service, University of Minnesota, St. Paul.

Morey, R.V. and H. Li. 1984. Thin-layer equation effects on deep-bed drying prediction. *Transactions of ASAE* 27(6):1924-28.

Morey, R.V., H.A. Cloud, and D.J. Hansen. 1981. Ambient air wheat drying. *Transactions of ASAE* 24(5):1312-16.

Morey, R.V., H.A. Cloud, and W.E. Lueschen. 1976. Practices for the efficient utilization of energy from drying corn. *Transactions of ASAE* 19(1):151-55.

Morey, R.V., H.A. Cloud, R.J. Gustafson, and D.W. Peterson. 1979. Management of ambient air drying systems. *Transactions of ASAE* 22(6):1418-25.

Morey, R.V., H.M. Keener, T.L. Thompson, G.M. White, and F.W. Bakker-Arkema. 1978. The present status of grain drying simulation. *Paper* 78-3009. American Society of Agricultural Engineers, St. Joseph, MI.

Nofsinger, G.W. 1982. The trickle ammonia process—An update. Grain Conditioning Conference Proceedings, Agricultural Engineering Department, University of Illinois, Champaign-Urbana.

Nofsinger, G.W., R.J. Bothast, and R.A. Anderson. 1979. Field trials using extenders for ambient-conditioning high-moisture corn. *Transactions of ASAE* 22(5):1208-13.

Noomhorm, A. and L.R. Verma. 1986. Generalized single-layer rice drying models. *Transactions of ASAE* 29(2):587-91.

O'Callaghan, J.R., D.J. Menzies, and P.H. Bailey. 1971. Digital simulation of agricultural dryer performance. *Journal of Agricultural Engineering Research* 16:223-44.

Otten, L., R. Brown, and W.S. Reid. 1984. Drying of white beans—Effects of temperature and relative humidity on seed coat damage. *Canadian Agricultural Engineering* 26(2):101-04.

Overhults, D.G., G.M. White, M.E. Hamilton, and I.J. Ross. 1973. Drying soybeans with heated air. *Transactions of ASAE* 16(1):112-13.

Overhults, D.G., G.M. White, M.E. Hamilton, I.J. Ross, and J.D. Fox. 1975. Effect of heated air drying on soybean oil quality. *Transactions of ASAE* 16(1):112-13.

Parry, J.L. 1985. Mathematical modelling and computer simulation of heat and mass transfer in agricultural grain drying: A review. *Journal of Agricultural Engineering Research* 32:1-29.

Parsons, S.D., B.A. McKenzie, and J.R. Barrett, Jr. 1979. Harvesting and drying high-moisture wheat. In *Double cropping winter wheat and soybeans in Indiana*. ID 96, Cooperative Extension Service, Purdue University, West Lafayette, IN.

Paulsen, M.R. and L.D. Hill. 1985. Corn quality factors affecting dry milling performance. *Journal of Agricultural Engineering Research* 31:255-63.

Paulsen, M.R. and T.L. Thompson. 1973. Drying analysis of grain sorghum. *Transactions of ASAE* 16(3):537-40.

Paulsen, M.R., L.D. Hill, D.G. White, and G.F. Sprague. 1983. Breakage susceptibility of corn-belt genotypes. *Transactions of ASAE* 26(6):1830-36.

Pederson, J.R. 1992. Insects: Identification, damage, and detection. In *Storage of cereal grains and their products*, ed. D.B. Sauer, pp. 435-90.

Peplinski, A.J., R.A. Anderson, and O.L. Brekke. 1982. Corn dry milling as influenced by harvest and drying conditions. *Transactions of ASAE* 25(4):1114-17.

Peplinski, A.J., O.L. Brekke, R.J. Bothast, and L.T. Black. 1978. High moisture corn—An extended preservation trial with ammonia. *Transactions of ASAE* 21(4): 773-76, 781.

Peterson, W.H. 1982. Design principles for grain aeration in flat storages. Illinois Farm Electrification Council *Fact Sheet* No. 9. University of Illinois, Agricultural Engineering Department, Urbana.

Pfost, H.B., S.G. Mauer, D.S. Chung, and G.A. Milliken. 1976. Summarizing and reporting equilibrium moisture data for grains. *Paper* 76-3520. American Society of Agricultural Engineers, St. Joseph, MI.

Pierce, R.O. 1986. Economic consideration for natural air corn drying in Nebraska. *Transactions of ASAE* 29(4):1131-35.

Pierce, R.O. and T.L. Thompson. 1979. Solar grain drying in the North Central Region—Simulation results. *Transactions of ASAE* 15(1):178-87.

Pierce, R.O. and T.L. Thompson. 1981. Energy use and performance related to crossflow dryer design. *Transactions of ASAE* 24 (1):216-20.

Quinlan, J.K. 1982. Grain protectants for insect control. *Marketing Bulletin* 72. Agricultural Research Service, U.S. Department of Agriculture, Washington, D.C.

Radajewski, W., T. Jensen, G.Y. Abawi, and E.J. McGahan. 1992. Drying rate and damage to navy beans. *Transactions of ASAE* 35(2):583-90.

Rayburn, S.T., A.C. Griffin, Jr., and M.E. Whitten. 1978. Storing cottonseed with propionic acid. *Transactions of ASAE* 21(5):990-92.

Ripp, B.E., ed. 1984. Controlled atmosphere and fumigation in grain storages. Proceedings of an International Symposium, Practical Aspects of Controlled Atmosphere and Fumigation in Grain Storages, in Perth, Western Australia. Elsevier Science Publishing Company, New York.

Sabbah, M.A., H.M. Keener, and G.E. Meyer. 1979. Simulation of solar drying of shelled corn using the logarithmic model. *Transactions of ASAE* 22(3):637-43.

Sabbah, M.A., G.E. Meyer, H.M. Keener, and W.L. Roller. 1976. Reversed-air drying for fixed bed of soybean seed. *Paper* 76-3023. American Society of Agricultural Engineers, St. Joseph, MI.

Sauer, D.B., ed. 1992. *Storage of cereal grains and their products*. American Association of Cereal Chemists, St. Paul, MN.

Sauer, D.B. and R. Burroughs. 1974. Efficacy of various chemicals as grain mold inhibitors. *Transactions of ASAE* 17(3):557-59.

Sauer, D.B., R.A. Meronuck, and C.M. Christensen. 1992. Microflora. In *Storage of cereal grains and their products*, ed. D.B. Sauer, pp. 313-40.

Saul, R.A. and J.L. Steele. 1966. Why damaged shelled corn costs more to dry. *Agricultural Engineering* 47:326-29, 337.

Schmidt, B.J. and L.F. Backer. 1980. Results of a sunflower storage monitoring program in North Dakota. *Paper* 80-6033, American Society of Agricultural Engineers, St. Joseph, MI.

Schuler, R.T. 1974. Drying related properties of sunflower seeds. *Paper* 74-3534. American Society of Agricultural Engineers, St. Joseph, MI.

Schuler, R.T., B.J. Holmes, R.J. Straub, and D.A. Rohweder. 1986. Hay drying. *Publication* A3380. Cooperative Extension Service, University of Wisconsin, Madison.

Schultz, L.J., M.L. Stone, and P.D. Bloome. 1984. A comparison of simulation techniques for wheat aeration. *Paper* 84-3012. American Society of Agricultural Engineers, St. Joseph, MI.

Segerlind, L.J. 1982. Solving the nonlinear airflow equation. *Paper* 82-3017. American Society of Agricultural Engineers, St. Joseph, MI.

Seitz, L.M., D.B. Sauer, and H.E. Mohr. 1982a. Storage of high-moisture corn: Fungal growth and dry matter loss. *Cereal Chemistry* 59(2):100-105.

Seitz, L.M., D.B. Sauer, H.E. Mohr, and D.F. Aldis. 1982b. Fungal growth and dry matter loss during bin storage of high-moisture corn. *Cereal Chemistry* 59(1):9-14.

Sellam, M.A. and C.M. Christensen. 1976. Temperature differences, moisture transfer and spoilage in stored corn. *Feedstuffs* 48(36):28, 33.

Sharp, J.R. 1982. A review of low-temperature drying simulation models. *Journal of Agricultural Engineering Research* 27(3):169-90.

Shaw, C.S. and G.N. Franks. 1962. Cottonseed drying and storage at cotton gins. *Technical Bulletin* 1262. USDA, ARS, Washington, D.C.

Shedd, C.K. 1953. Resistance of grains and seeds to air flow. *Agricultural Engineering* 34(9):616-18.

Shejbal, J., ed. 1980. *Controlled atmosphere storage of grains*. Elsevier Science Publishing Company, New York.

Shepherd, J.B. 1954. Experiments in harvesting and preserving alfalfa for dairy cattle feed. *Technical Bulletin* 1079. USDA, Washington, D.C.

Singh, S.S. and M.F. Finner. 1983. A centrifugal impacter for damage susceptibility evaluation of shelled corn. *Transactions of ASAE* 26(6):1858-63.

Smith, E.A. and P.H. Bailey. 1983. Simulation of near-ambient grain drying. II, Control strategies for drying barley in Northern Britain. *Journal of Agricultural Engineering Research* 28:301-17.

Smith, E.A. and S. Sokhansanj. 1990a. Moisture transport due to natural convection in grain stores. *Journal of Agricultural Engineering Research* 47:23-34.

Smith, E.A. and S. Sokhansanj. 1990b. Natural convection and temperature of stored products—A theoretical analysis. *Canadian Agricultural Engineering* 32(1):91-97.

Smith, J.S., Jr. and J.I. Davidson, Jr. 1982. Psychrometrics and kernel moisture content as related to peanut storage. *Transactions of ASAE* 25(1):231-36.

Smith, J.S., Jr., J.I. Davidson, Jr., T.H. Sanders, and R.J. Cole. 1985. Storage environment in a mechanically ventilated peanut warehouse. *Transactions of ASAE* 28(4):1248-52.

Sokhansanj, S. and D.M. Bruce. 1987. A conduction model to predict grain drying simulation. *Transactions of ASAE* 30(4):1181-84.

Sokhansanj, S. and S.O. Nelson. 1988a. Dependence of dielectric properties of whole-grain wheat on bulk density. *Journal of Agricultural Engineering Research* 39:173-79.

Sokhansanj, S. and S.O. Nelson. 1988b. Transient dielectric properties of wheat associated with non-equilibrium kernel moisture conditions. *Transactions of ASAE* 31(4):1251-54.

Sokhansanj, S. and G.E. Woodward. 1991. Computer assisted fan selection for natural grain drying—A teaching and extension tool. *Applied Engineering in Agriculture* 6(6):782-84.

Sokhansanj, S., D. Singh, and J.D. Wasserman. 1984. Drying characteristics of wheat, barley and canola subjected to repetitive wetting and drying cycles. *Transactions of ASAE* 27(3):903-906, 914.

Sokhansanj, S., W. Zhijie, D.S. Jayas, and T. Kameoka. 1986. Equilibrium relative humidity moisture content of rapeseed from 5 to 25°C. *Transactions of ASAE* 29(3):837-39.

Sokhansanj, S., A.A. Falacinski, F.W. Sosulski, D.S. Jayas, and J. Tang. 1990. Resistance of bulk lentils to airflow. *Transactions of ASAE* 33(4):1281-85.

Steele, J.L. 1967. Deterioration of damaged shelled corn as measured by carbon dioxide production. Unpublished Ph.D. diss., Department of Agricultural Engineering, Iowa State University, Ames, IA.

Steele, J.L., R.A. Saul, and W.V. Hukill. 1969. Deterioration of shelled corn as measured by carbon dioxide production. *Transactions of ASAE* 12(5):685-89.

Stephens, G.R. and T.L. Thompson. 1976. Improving crossflow grain dryer design using simulation. *Transactions of ASAE* 19(4):778-81.

Stephens, L.E. and G.H. Foster. 1976a. Breakage tester predicts handling damage in corn. ARS-NC-49. ARS, USDA, Washington, D.C.

Stephens, L.E. and G.H. Foster. 1976b. Grain bulk properties as affected by mechanical grain spreaders. *Transactions of ASAE* 19(2):354-58.

Stephens, L.E. and G.H. Foster. 1978. Bulk properties of wheat and grain sorghum as affected by a mechanical grain spreader. *Transactions of ASAE* 21(6):1217-18.

Storey, C.L., R.D. Speirs, and L.S. Henderson. 1979. Insect control in farm-stored grain. *Farmers Bulletin* 2269. USDA-SEA (Available for sale from Superintendent of Documents, U.S. Government Printing Office, Washington, D.C. 20402).

Stroshine, R.L., A.W. Kirleis, J.F. Tuite, L.F. Bauman, and A. Emam. 1986. Differences in grain quality among selected corn hybrids. *Cereal Foods World* 31(4):311-16.

Stroshine, R.L., J. Tuite, G.H. Foster, and K. Baker. 1984. Self-study guide for grain drying and storage. Department of Agricultural Engineering, Purdue University, West Lafayette, IN.

Swearingen, M.L. 1979. A practical guide to no-till double cropping. In *Double cropping winter wheat and soybeans in Indiana*. ID 96. Cooperative Extension Service, Purdue University, West Lafayette, IN.

Syarief, A.M., R.V. Morey, and R.J. Gustafson. 1984. Thin-layer drying rates of sunflower seed. *Transactions of ASAE* 27(1):195-200.

Tang, J., S. Sokhansanj, and F.W. Sosulski. 1989. Thin-layer drying of lentil. *Paper* 89-6607. American Society of Agricultural Engineers, St. Joseph, MI.

Thompson, C.R., E.M. Bickoff, G.R. VanAtta, G.O. Kohler, J. Guggolz, and A.L. Livingston. 1960. Carotene stability in alfalfa as affected by laboratory- and industrial-scale processing. *Technical Bulletin* 1232. ARS, USDA, Washington, D.C.

Thompson, T.L. 1972. Temporary storage of high-moisture shelled corn using continuous aeration. *Transactions of ASAE* 15(2):333-37.

Thompson, T.L., G.H. Foster, and R.M. Peart. 1969. Comparison of concurrent-flow, crossflow and counterflow grain drying methods. *Marketing Research Report* 841. USDA-ARS, Washington, D.C.

Thompson, T.L., R.M. Peart, and G.H. Foster. 1968. Mathematical simulation of corn drying—A new model. *Transactions of ASAE* 11(4):582-86.

Thorpe, G.R. 1982. Moisture diffusion through bulk grain subjected to a temperature gradient. *Journal of Stored Products Research* 18:9-12.

Troeger, J.M. and W.V. Hukill. 1971. Mathematical description of the drying rate of fully exposed corn. *Transactions of ASAE* 14(6):1153-56, 1162.

Tuite, J. and G.H. Foster. 1963. Effect of artificial drying on the hygroscopic properties of corn. *Cereal Chemistry* 40:630-37.

Tuite, J. and G.H. Foster. 1979. Control of storage diseases of grain. *Annual Review of Phytopathology* 17:343.

Tuite, J., G.H. Foster, S.R. Eckhoff, and O.L. Shotwell. 1986. Sulfur dioxide treatment to extend corn drying time (note). *Cereal Chemistry* 63(5): 462-64.

Uebersax, M.A. and C.L. Bedford. 1980. Navy bean processing: Effect of storage and soaking methods on quality of canned beans. *Research Report* 410. Agricultural Experiment Station, Michigan State University, East Lansing.

USDA. 1971. Oven methods for determining moisture content of grain and related agricultural commodities, Chapter 12. *Equipment manual*, GR Instruction 916-6. U.S. Department of Agriculture, Consumer and Marketing Service, Grain Division, Hyattville, MD.

Velupillai, L. and L.R. Verma. 1986. Drying and tempering effects on parboiled rice quality. *Transactions of ASAE* 29(1):312-19.

Verma, L.R. and B.D. Nelson. 1983. Changes in round bales during storage. *Transactions of ASAE* 26(2):328-32.

Walton, L.R. and W.H. Henson, Jr. 1971. Effect of environment during curing on the quality of burley tobacco: Effect of low humidity curing on support price. *Tobacco Science* 15:54-57.

Walton, L.R., W.H. Henson, Jr., and J.M. Bunn. 1973. Effect of environment during curing on the quality of burley tobacco: Effect of high humidity curing on support price. *Tobacco Science* 17:25-27.

Watson, C.A. 1977. Near infrared reflectance spectro-photometric analysis of agricultural products. *Analytical Chemistry* 49(9):835A-40A.

Watson, C.A., O.J. Banasick, and G.L. Pratt. 1962. Effect of drying temperature on barley malting quality. *Brewers Digest* 37(7):44-48.

Watson, E.L. and V.K. Bhargava. 1974. Thin-layer drying studies on wheat. *Canadian Agricultural Engineering* 16(1):18-22.

Watson, S.A. 1987. Measurement and maintenance of quality. In *Corn: Chemistry and technology*, S.A. Watson and P.E. Ramstad, eds., pp. 125-83. American Association of Cereal Chemists, St. Paul, MN.

Watson, S.A. and F.L. Herum. 1986. Comparison of eight devices for measuring breakage susceptibility of shelled corn. *Cereal Chemistry* 63(2):139-42.

Watson, S.A. and P.E. Ramstad. 1987. *Corn: Chemistry and technology*. American Association of Cereal Chemists, St. Paul, MN.

Watson, S.A., L.L. Darrah, and F.L. Herum. 1986. Measurement of corn breakage susceptibility with the Stein breakage tester: A collaborative study. *Cereal Foods World* 31(5):366-72.

Weller, C.L., M.R. Paulsen, and M.P. Steinberg. 1987. Varietal, harvest moisture and drying air temperature effects on quality factors affecting corn wet milling. *Paper* 87-6046. American Society of Agricultural Engineers, St. Joseph, MI.

Young, J.H. 1984. Energy conservation by partial recirculation of peanut drying air. *Transactions of ASAE* 27(3):928-34.

CHAPTER 12

AIR CONTAMINANTS

AIR is composed of many gases. The gaseous composition of clean, dry air near sea level is approximately 21% oxygen, 78% nitrogen, 1% argon, and 0.03% carbon dioxide. Also included are trace amounts of hydrogen, neon, krypton, helium, ozone, and xenon, in addition to varying amounts of water vapor and small quantities of microscopic and submicroscopic solid matter called permanent atmospheric impurities.

Air composition may be changed accidentally or deliberately. In sewers, sewage treatment plants, tunnels, and mines, the oxygen content of air can become so low that people cannot remain conscious or survive. In confined spaces with concentrations of people (theaters, survival shelters, submarines), carbon dioxide given off by normal respirations must be removed and replaced with oxygen. Pilots of high-altitude aircraft, breathing at greatly reduced pressure, require systems that increase oxygen concentration. Conversely, for divers working at extreme depths, it is common to increase the percentage of helium in the air and reduce nitrogen and sometimes oxygen concentrations.

At atmospheric pressure, oxygen concentrations of less than 12% or carbon dioxide concentrations greater than 5% are dangerous, even for short periods.

Lesser deviations from normal composition can be hazardous under prolonged exposure. Chapter 11 of the 1991 *ASHRAE Handbook—Applications* details acceptable variations in air composition. Chapter 9 of this volume further discusses environmental health issues.

CLASSIFICATION OF AIR CONTAMINANTS

Normal air contains varying amounts of foreign materials (**permanent atmospheric impurities**). These materials can be generated by natural processes such as wind erosion, sea spray evaporation, and volcanic eruption. Natural concentrations in the air vary but are usually less than the concentrations caused by human activity.

Man-made contaminants are many and varied, originating from numerous areas of human activity. Electric power-generating plants, various modes of transportation, industrial processes, mining and smelting, construction, and agriculture generate large amounts of contaminants. Contaminants that present particular problems in the indoor environment include, among others, tobacco smoke, radon, and formaldehyde.

Air contaminants can be classified as follows:

- Particulate or gaseous
- Organic or inorganic
- Visible or invisible
- Submicroscopic, microscopic, or macroscopic
- Toxic or harmless
- Stable or unstable

Loose classifications based on the phase of the suspended contaminant (solid, liquid, or gas) and the method of formation of the contaminant are as follows:

- Dusts, fumes, and smokes that are solid particulate matter, although smoke often contains liquid particulates
- Mists, fogs, and smokes that are suspended liquid particulates
- Vapors and gases

Dusts, Fumes, and Smokes

Dusts. Dusts are solid particles projected into the air by natural forces such as wind, volcanic eruption, or earthquakes; or by mechanical processes including crushing, grinding, demolition, blasting, drilling, shoveling, screening, and sweeping. Some of these forces produce dusts by reducing larger masses, while others disperse materials that have already been reduced.

Particles are not dust unless they are smaller than about 100 μm in diameter. Dusts can be mineral, such as rock, metal, or clay; vegetable, such as grain, flour, wood, cotton, or pollen; or animal, including wool, hair, silk, feathers, and leather.

Fumes. Fumes are solid particles formed by condensation of vapors of solid materials. Metallic fumes are generated from molten metals and usually occur as oxides because of the highly reactive nature of finely divided matter. Fumes can also be formed by sublimation, distillation, or chemical reaction. Such processes create airborne particles smaller than 1 μm. Fumes permitted to age may agglomerate into larger clusters.

Smokes. Smokes are small solid and/or liquid particles produced by incomplete combustion of organic substances such as tobacco, wood, coal, oil, and other carbonaceous materials. The term smoke is applied to a mixture of solid, liquid, and gaseous products, although technical literature distinguishes between such components as soot or carbon particles, fly ash, cinders, tarry matter, unburned gases, and gaseous combustion products. Smoke particles vary in size, the smallest being much less than 1 μm. The average is often in the range of 0.1 to 0.3 μm.

Environmental tobacco smoke consists of a suspension of small 0.01 to 1.0 μm (mass median diameter of 0.5 μm) liquid particles that form as the superheated vapors leaving the burning tobacco condense. Also produced are numerous gaseous contaminants including carbon monoxide.

Bioaerosols. Airborne viruses, bacteria, pollen, and fungus spores are sometimes referred to as bioaerosols. **Viruses** range in size from 0.003 to 0.06 μm, although they usually occur in colonies or attached to other particles. Most **bacteria** range between 0.4 and 5 μm and are usually associated with large particles. **Fungus** spores are usually from 10 to 30 μm, while **pollen** grains are from 10 to 100 μm, with many common varieties in the 20 to 40 μm range.

The preparation of this chapter is assigned to TC 2.3, Gaseous Air Contaminants and Gas Contaminant Removal Equipment, in conjunction with TC 2.4, Particulate Air Contaminants and Particulate Contaminant Removal Equipment.

Mists and Fogs

Mists are small airborne droplets of materials that are ordinarily liquid at normal temperatures and pressure. They can be formed by atomizing, spraying, mixing, violent chemical reactions, evolution of gas from liquid, or escape as a dissolved gas when pressure is released. Small droplets expelled or atomized by sneezing constitute mists containing microorganisms that become air contaminants.

Fogs are fine airborne droplets usually formed by condensation of vapor. Fog nozzles are named for their ability to produce extra fine droplets, as compared with mists from ordinary spray devices. Many droplets in fogs or clouds are microscopic and submicroscopic and serve as a transition stage between mists and vapors.

The volatile nature of most liquids reduces the size of their airborne droplets from the mist range to the fog range and eventually to the vapor phase, until the air becomes saturated with that liquid. If solid material is suspended or dissolved in the liquid droplet, it remains in the air as particulate contamination. For example, sea spray evaporates fairly rapidly, generating a large number of fine salt particles that remain suspended in the atmosphere.

Smog commonly refers to air pollution; it implies an air mixture of smoke particles, mists, and fog droplets of such concentration and composition as to impair visibility, in addition to being irritating or harmful. The composition varies among different locations and over time. The term is often applied to haze caused by a sunlight-induced photochemical reaction involving the materials in automobile exhausts. Smog is often associated with temperature inversions in the atmosphere that prevent normal dispersion of contaminants.

Vapors and Gases

The terms gas and vapor are often used to describe a common state of a substance. Gas is normally used to describe any mixture, except atmospheric air, that exists in the gaseous state under normal atmospheric conditions. Examples are oxygen, helium, and nitrogen. Vapor is used to describe a substance in the gaseous state that can also exist as a liquid or solid under normal atmospheric conditions. Examples include gasoline, benzine, carbon tetrachloride, and water.

NATURE OF AIRBORNE CONTAMINANTS

Sizes of Airborne Particles

Figure 1 shows the sizes and characteristics of airborne solids and liquids. Particles less than 0.1 µm in diameter begin to behave similarly to gas molecules, traveling with Brownian movement and with no predictable or measurable settling velocity. Particles from 0.1 to 1 µm have settling velocities that can be calculated but that are so low that settling is usually negligible because normal air currents counteract any settling. On a particle count basis, over 99% of the particles in a typical atmosphere are below 1 µm.

Particles in the 1 to 10 µm range settle in still air at constant and appreciable velocity. However, normal air currents keep them in suspension for appreciable periods.

Industrial hygienists are primarily concerned with particles less than 2 µm in diameter because it is particles in this size range that are most likely to be retained in the lungs (Morrow 1964). Particles larger than 8 to 10 µm in diameter are separated and retained by the upper respiratory tract. Intermediate sizes are deposited mainly in the conducting airways of the lungs, from which they are rapidly cleared and swallowed or coughed out. About 50% or fewer of the particles in inhaled air settle in the respiratory tract.

Particles larger than 10 µm settle fairly rapidly and can be found suspended in air only near their source or under strong wind conditions. Exceptions are lint and other light fibrous materials such as portions of certain weed seeds, which remain suspended longer. Most individual particles 10 µm or larger are visible to the naked eye under favorable conditions of lighting and contrast. Smaller particles are visible in high concentrations. Cigarette smoke (with an average particle size of less than 0.5 µm) and clouds are common examples.

Direct fallout in the vicinity of the dispersing stack or flue and other nuisance problems of air pollution involve larger particles. Smaller particles, as well as generated mists, fogs, and fumes, remain in suspension longer. In this size range, meteorology and topography are more important than physical characteristics of the particles. Because settling velocities are low, the ability of the atmosphere to disperse these small particles depends largely on local weather conditions.

Comparison is often made to screen sizes used for grading industrial dusts and granular materials. Table 1 illustrates the relation of U.S. Standard sieve mesh to particle size in micrometres. Particles above 40 µm are the screen sizes, and those below are the subscreen or microscopic sizes.

Table 1 Relation of Screen Mesh to Particle Size

U.S. Standard sieve mesh	400	325	200	140	100	60	35	18
Nominal sieve opening in µm	37	44	74	105	149	250	500	1000

Particle Size Distribution

The particle size distribution in any sample can be expressed as the percentage of particles smaller than a specified size. The upper curve of Figure 2 shows these data plotted for typical atmospheric contamination (Whitby et al. 1955, 1957). The middle curve shows the percentage of the total projected area of the particles contributed by particles smaller than a specified size. The lower curve shows the percentage of the total particle mass contributed by those particles smaller than a given size.

The differences among values presented by the three curves should be noted. For example, particles 0.1 µm or less in diameter (but still above electron microscope minimum detection size of about 0.005 µm) make up 80% of the number of particles in the atmosphere but contribute only 1% of the mass. Also, the 0.1% of particles larger than 1 µm carry 70% of the total mass, which is the direct result of the fact that mass of a spherical particle increases as the cube of its diameter. Although most of the mass is contributed by intermediate and larger particles, about 80% of the contamination is supplied by particles less than 5 µm in diameter. Most of the surface staining effect is caused by deposited particles less than 1 µm in diameter. Suspended particles in urban air are predominantly smaller than 1 µm (expressed as equivalent spheres of unit density) and have a distribution that is approximately log-normal.

SUSPENDED PARTICULATES

The total amount of suspended particulate matter in the atmosphere can influence the loading rate of air filters and their selection. The amount of soot that falls in cities in the United States ranges from 20 to 200 ton/mile^2 per month. Soot fall data indicate effectiveness of smoke abatement and proper combustion methods and serve as comparative indices of such control programs. However, these values have little significance to the ventilating and air-conditioning engineer because they do not express any measure of the suspended material that must be cleaned from the ventilation air before it can be used.

The number of suspended particulates is enormous. A room with heavy cigarette smoke has a total particle concentration of 30×10^6 particles per cubic foot; even clean air contains over 10^6 particles/ft^3. If smaller particles detectable by other means, such an electron microscope or condensation nuclei counter, were also included, the total particle count would be greater than the above concentrations by a factor of 10 to 100.

Extensive measurements have been made of outdoor pollution, but limited data have been gathered on indoor pollution not associated with specific industrial processes. Indoor levels are influenced

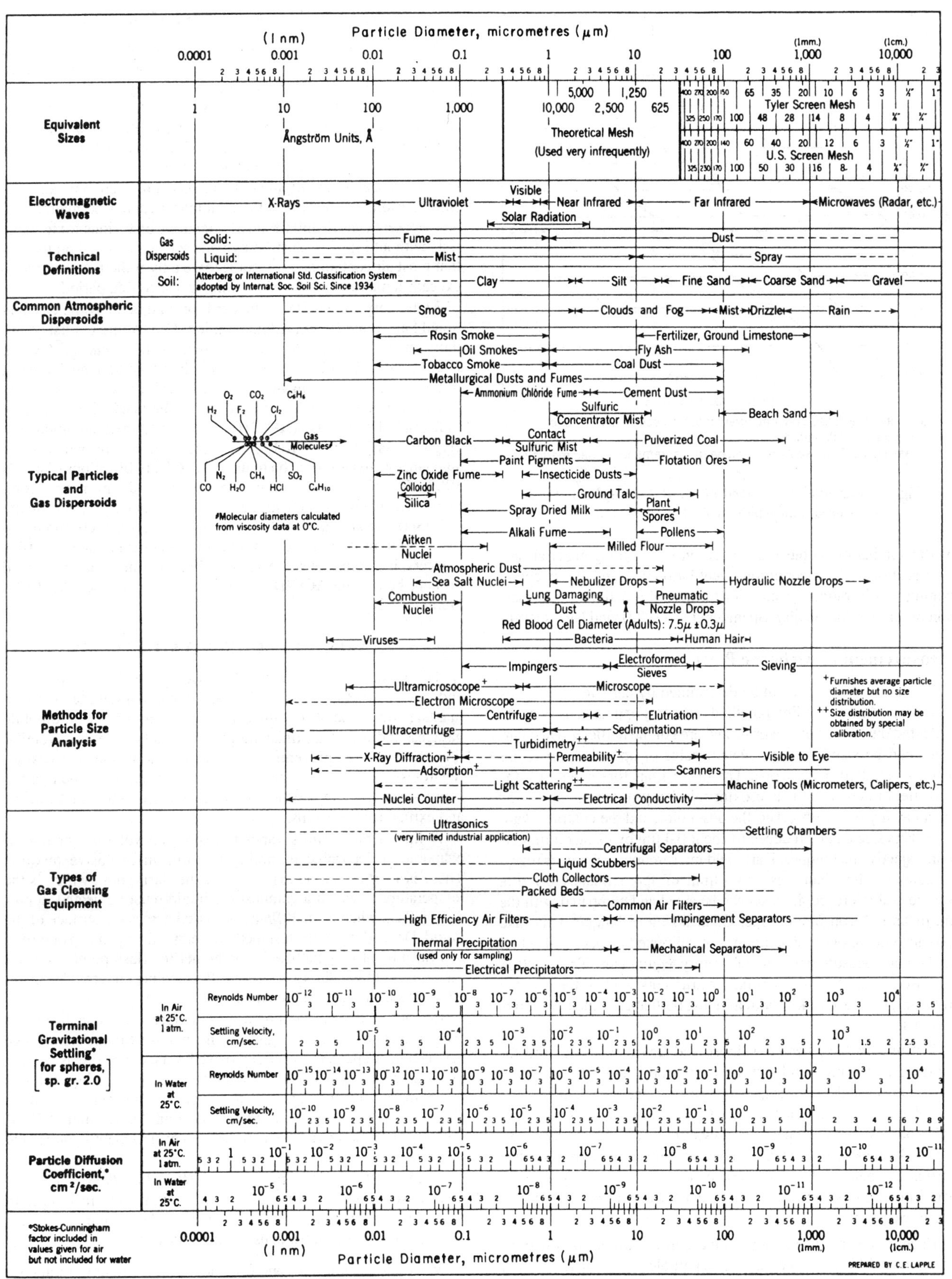

Fig. 1 Characteristics of Particles and Particle Dispersoids
(Courtesy of Stanford Research Institute)

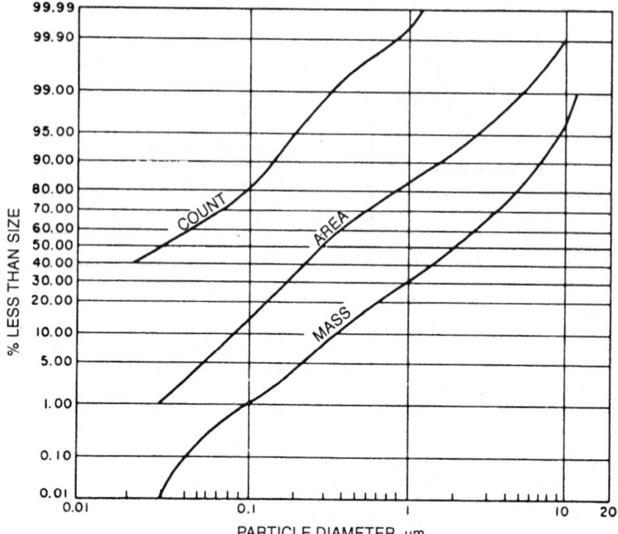

Count curve: Based on measurements by electron microscope.
Area curve: Calculated.
Mass curve: Solid section based on measurements by sedimentation.

Fig. 2 Particle Size Distribution of Atmospheric Dust
[Reproduced from Whitby et al. (1955) by permission.]

by the number of people and their activities, building materials and construction, outside conditions, ventilation rate, and the air-conditioning and filtration system. For further information, see the section on Indoor Air Quality, Spengler et al. (1982), and NRC (1981).

Measurement of Airborne Particles

The quantity of dust can be determined by particle count or by total mass. Another indirect method measures optical density of the collected dust based on the projected area of the particles. Dust particles can be sized with graduated scales or optical comparisons using a standard microscope. The lower limit for sizing with the light field microscope is approximately 0.9 µm, depending on the visual acuity of the observer, the dust color, and the contrast available. This size can be reduced to about 0.4 µm by using oil immersion objective techniques. Dark field microscopic techniques reveal particles smaller than these, to a limit of approximately 0.1 µm. Smaller submicroscopic dusts can be sized and compared with the aid of an electron microscope. Other sizing techniques may take into account velocity of samplings in calibrated devices and actual settlement measurements in laboratory equipment. The electron microscope and various sampling instruments, such as the cascade impacter, have been successful in sizing particulates, including fogs and mists.

Instruments have been developed and test methods devised for continuous counting and sizing of airborne contaminants. These instruments use the light-scattering principle and, depending on the model, can detect particles down to 0.3 µm. Instruments using laser light sources can detect particles down to 0.1 µm (ASTM *Standard* F 50). A condensation nucleus counter can count submicrometric particles to below 0.01 µm. These particles, present in great numbers in the atmosphere, serve as nuclei for condensation of water vapor (Scala 1963).

Each of the various methods of measuring particle size distribution gives a different value for a given particle size because different properties are actually measured. For example, a microscopic technique may measure the longest dimension, while cascade impacter results are based on aerodynamic behavior (ACGIH 1995).

INDUSTRIAL AIR CONTAMINANTS

Many industrial processes produce air contaminants in the form of dusts, fumes, smokes, mists, vapors, and gases. Contaminants should be controlled at the source so that they are not dispersed through the factory or allowed to increase to toxic concentration levels. Control methods are covered in Chapter 25 of the 1996 *ASHRAE Handbook—Systems and Equipment* and Chapters 24 and 26 of the 1995 *ASHRAE Handbook—Applications*.

Zero concentration of all contaminants is not economically feasible. Absolute control of all contaminants cannot be maintained, and workers can assimilate small quantities of various toxic materials without injury. Industrial hygiene science is based on the fact that most air contaminants become toxic only if their concentration exceeds a maximum allowable limit for a specified period.

Mandatory United States standards promulgated by the Occupational Safety and Health Administration (OSHA) are published initially in the *Federal Register* and then in the *Code of Federal Regulations* (29 CFR 1910.1000), which is revised annually. Most of the standards promulgated initially by OSHA were those previously recommended by the American Conference of Governmental Industrial Hygienists (ACGIH) and the American Industrial Hygiene Association (AIHA). The health effects on which these standards were based can be found in ACGIH publications (ACGIH 1987). Revisions of the initial OSHA standards are now recommended by the National Institute of Occupational Safety and Health (NIOSH) and are published in a series of individual criteria documents. For protection of nonindustrial occupants such as office workers, including sensitive persons, indoor concentrations should be kept below the ACGIH threshold limit values (see ASHRAE *Standard* 62).

FLAMMABLE GASES AND VAPORS

The use of flammable materials is widespread. Flammable gases and vapors (NFPA *Standard* 325) can be found in sewage treatment plants, sewage and utility tunnels, dry-cleaning plants, automobile garages, and industrial finishing process plants. Adequate ventilation minimizes or prevents fires and explosions and is necessary regardless of other precautions, such as elimination of the ignition sources, safe building construction, and the use of automatic alarm and extinguisher systems.

A flammable liquid's vapor pressure and volatility or rate of evaporation determine its ability to form an explosive mixture. These properties can be expressed by the flash point, which is the temperature to which a combustible liquid must be heated to produce a flash when a small flame is passed across the surface of the liquid. Depending on the test methods, either the open cup or closed cup flash point may be listed. The higher the flash point, the safer the liquid is to handle. Liquids with flash points under 70°F should be regarded as highly flammable.

In practice, the air-vapor or air-gas mixture must be in the **explosive range** before it can be ignited. The explosive range is the range between the upper and lower explosive limits, expressed as percent by volume in air. Concentrations of material above the higher range or below the lower range will not explode. The range for many chemicals is found National Fire Protection Association (NFPA) *Standard* 325. Some representative limits of flammability are listed in Table 2.

Chapters 26 and 41 of the 1995 *ASHRAE Handbook—Applications* covers equipment for control of combustible materials. In designing ventilation systems to control flammable gases and vapors, the engineer must consider the following:

1. Most safety authorities and fire underwriters prefer to limit concentrations to 20 to 25% of the lower explosive limit of a material. The resulting safety factor of 4 or 5 allows latitude for imperfections in air distribution and variations of temperature or

Table 2 Flammable Limits of Some Gases and Vapors

Gas or Vapor	Flash Point, °F	Flammable Limits, % by Volume Lower	Upper
Acetone	0	2.6	12.8
Ammonia	Gas	16	25
Benzene (benzol)	12	1.3	7.1
n-Butane	Gas	1.9	8.5
Carbon disulfide	−22	1.3	44
Carbon monoxide	Gas	12.5	74
1,2-Dichloroethylene	43	9.7	12
Diethyl ether	−49	1.9	48
Ethyl alcohol	55	4.3	19
Ethylene	Gas	3.1	32
Gasoline	−45	1.4	7.6
Hydrogen	Gas	4.0	75
Hydrogen sulfide	Gas	4.3	45
Isopropyl alcohol	53	2.0	12
Methyl alcohol	52	7.3	36
Methyl ethyl ketone	21	1.8	10
Natural gas (variable)	Gas	3.8	17
Naphtha (benzine)	50	0.9	6.7
Propane	Gas	2.2	9.5
Toluene (toluol)	40	1.2	7.1
o-Xylene	90	1.0	6.0

mixture, and guards against unpredictable or unrecognized sources of ignition. Operation at concentrations above the upper explosive limit should be resorted to only in rare instances. To reach the upper explosive limit, the flammable gas or vapor must pass through the active explosive range, in which any source of ignition causes an explosion. In addition, a drop in gas concentration due to unforeseen dilution or reduced evaporation rate results in operation in the dangerous explosive range.

2. In occupied places where ventilation is applied for proper health control, the danger of an explosion is minimized. In most instances, flammable gases and vapors are also toxic, and the maximum allowable concentrations are far below the lower explosive limit of the material. For example, proper ventilation for acetone vapors keeps the concentration below 1000 ppm. This is equivalent to 0.1% by mass. The lower explosive limit for acetone is 2.5% by volume.
3. Proper location of exhaust and supply ventilation equipment depends primarily on how the contaminant is given off and on other problems of the process, and secondarily on the relative density of flammable vapor.

The specific density of the explosive mixture is the same as that of air. Cross drafts, moving equipment, and temperature differentials may cause sufficient mixing to produce explosive concentrations and disperse these throughout the atmosphere, without regard to heavier-than-air vapors. Therefore, the engineer must either provide proper exhaust and supply air patterns to control the hazardous material, preferably at its source, or offset the effects of drafts, equipment movement, and convective forces by providing good distribution of exhaust and supply air for general dilution and exhaust. The intake duct should be positioned so that it does not bring in exhaust gases or emissions from ambient sources.

COMBUSTIBLE DUSTS

Many organic and some mineral dusts can produce dust explosions (Hartmann 1958). Often, a primary explosion results from a small amount of dust in suspension that has been exposed to a source of ignition; the pressure and vibration created can dislodge large accumulations of dust on horizontal surfaces, creating a larger secondary explosion.

For ignition, dust clouds require high temperatures and sufficient dust concentration. These temperatures and concentrations and the minimum spark energy can be found in Avallone and Baumeister (1987).

Explosive dusts are potential hazards whenever uncontrolled dust escapes, dispersing in the atmosphere or settling on horizontal surfaces such as beams and ledges. Proper exhaust ventilation design involves the principles covered in Chapter 26 of the 1995 *ASHRAE Handbook—Applications*. The ventilation systems and equipment chosen must prevent the pocketing of dust inside the equipment. When local exhaust ventilation is used, separation equipment should be installed as close to the dust source as possible to prevent transport of dust in the exhaust system.

AIR POLLUTION

Concentration levels of outdoor pollutants vary. Table 3 presents a sample of the annual median of several pollutants in selected urban centers. See Table 1 in Chapter 9 for ambient air quality standards set by the U.S. Environmental Protection Agency (EPA) for several pollutants known as **criteria pollutants**.

Odorants are also objectionable in occupied spaces. In many cases (see Chapter 41 of the 1995 *ASHRAE Handbook—Applications*), equipment that removes odorants also removes some nonodorous gases. The engineer must (1) cope with existing pollution by selecting equipment that provides clean air and (2) reduce pollution by providing adequate controls to accompany the design. This includes selecting the appropriate equipment for satisfactory combustion, as well as controls that enable the operator to run the equipment according to good practice and applicable air pollution ordinances. When dealing

Table 3 Annual Median Concentrations for Total Suspended Particles, NO_2, O_3, and CO for 1979[a]

	Concentration			
	μg/m³			mg/m³
Location	TSP (Annual Average)[b]	NO_2 (1 h Average)	O_3 (1 h Average)	CO (1 h Average)
Baltimore	43-102	45	20	1.5
Boston	67	75	—	3.5
Burbank, CA	—	124	39	3.5
Charleston, WV	43-70	37	14	1.2
Chicago	56-125	63	29	2.9
Cincinnati	47-87	60	24	1.0
Cleveland	58-155	89[c]	26	2.0
Dallas	43-73	59[c]	39	1.4
Denver	80-194	89	37	4.6
Detroit	52-135	68	14	1.8
Houston	51-147	90[c]	39[d]	1.0
Indianapolis	48-81	91[c]	33	2.7
Los Angeles	90	85	117	2.6
Louisville	60-102	70[c]	31	1.5
Milwaukee	47-105	86[c]	41	1.4
Minneapolis	45-87	65[c]	—	1.8
Nashville	41-82	62[c]	49[d]	2.6
New York	40-77	57	35	5.5
Philadelphia	51-109	85	39	3.2
Pittsburgh	88-162	—	29[d]	3.9
St. Louis	63-107	90[d]	22[d]	2.3[d]
San Diego	57-75	69	39	1.1
San Francisco	51	46	20[e]	2.1
Washington, D.C.	47-70	52	29	1.6

Source: Wadden and Schiff (1983).
[a] EPA (1980)
[b] Annual geometric mean of 24 h averages
[c] 24 h averages
[d] Not a full year
[e] Total oxidants

with industrial contaminants, the engineer must design local exhaust systems that minimize the removal of useful materials and provide collection equipment that reduces the amount of material discharged to the atmosphere as required by good practice and existing codes and standards.

RADIOACTIVE AIR CONTAMINANTS

Radioactive contaminants (Jacobson and Morris 1976-77) can be particulate or gaseous and are similar to ordinary industrial contaminants. A major source of human exposure to airborne radioactivity is radon from soil gas. Many radioactive materials would be chemically toxic if present in high concentrations; however, in most cases, their radioactivity is the factor that necessitates limiting their concentration in air.

Most radioactive air contaminants affect the body when they are absorbed and retained. This is known as the **internal radiation hazard**. Inert gases may be hazardous because the whole body comes in contact with radiation from the gas in the surrounding air. These gases are, therefore, **external radiation hazards**. Particulate contaminants may settle to the ground, where they contaminate plants and eventually enter the food chain and the human body. Material deposited on the ground increases external radiation exposure. However, except for fallout from nuclear weapons or a serious reactor accident, such exposure is insignificant.

Radioactive air contaminants can emit alpha, beta, or gamma rays. The alpha rays have very low penetrating power and present no hazard except when the material is deposited inside or on the body. Beta rays are somewhat more penetrating and can be both an internal and an external hazard. The penetrating ability of gamma rays depends on their energy, which varies from one type of radioactive element or isotope to another. A distinction should be made between the properties of the radioactive material and the radiation given off by this material. Radioactive particles and gases can be removed from air by devices such as filters and absorption traps, but the gamma radiation from such material is capable of penetrating solid walls. This distinction is frequently overlooked. The amount of radioactive material in air is measured in becquerels per cubic metre (1 becquerel equals 2.702702×10^{-11} curies), while the dose of radiation from deposited material is measured in rads.

Radioactive materials present problems that make them distinctive. The contaminants can generate enough heat to damage filtration equipment or ignite spontaneously. The concentrations at which most radioactive materials are hazardous are much lower than those of ordinary materials; as a result, special electronic instruments that respond to radioactivity must be used to detect these hazardous levels.

Radon is a colorless, odorless, radioactive gas found in all soils in various concentrations. Radon gas enters a house or building primarily through leakage paths in the foundation. Radon can then decay through a succession of decay products, producing metallic ions. These decay products then attach to particles suspended in the air or deposit onto surfaces within an occupied space; they can then be inhaled and deposited in the respiratory tract. Their subsequent decay produces alpha particles, which may increase the health risk.

The ventilation engineer faces difficulty in dealing with radioactive air contamination because of the extremely low permissible concentrations for radioactive materials. Contaminants must be kept from entering certain sensitive industrial plants, such as those in the photographic industry. If radioactive materials are handled inside the plant, the problem is to collect the contaminated air as close to the source as possible, and then remove the contaminant from the air with a high degree of efficiency, before releasing the air to the outdoors. Filters are generally used for particulate materials, but venturi scrubbers, wet washers, and other devices can be used as prefilters to meet special needs.

The design of equipment and systems for control of radioactive particulates and gases in nuclear laboratories, power plants, and fuel-processing facilities is highly specialized. Careful attention must be given to the reliability and contaminant-removal ability of the equipment under the special environmental stresses involved. Various publications of the U.S. Department of Energy (e.g., Boronsted et al. 1976) can provide guidance in this field.

The basic standards for permissible air concentrations are those of the National Committee on Radiation Protection, published by the National Institute of Standards and Technology (NIST) as NBS *Handbook* No. 69 (NIST 1959). Industries operating under licenses from the U.S. Nuclear Regulatory Commission or state licensing agencies must meet requirements of the *Code of Federal Regulations*, Title 10, Part 20. Some states have additional requirements.

ATMOSPHERIC POLLEN

Pollen grains discharged by weeds, grasses, and trees (Hewson et al. 1967, Jacobson and Morris 1976-77, Solomon and Mathews 1978) cause hay fever and have properties of special interest to air-cleaning equipment designers (see Chapter 24 of the 1996 *ASHRAE Handbook—Systems and Equipment*). Whole grains and fragments transported by air range between 10 and 50 μm; however, some measure as small as 5 μm, and others measure over 100 μm in diameter. Ragweed pollen grains are fairly uniform in size, ranging from 15 to 25 μm. Pollen grains can be removed from the air more readily than the dust particles prevalent in outdoor air or those produced by dusty processes because the latter predominate in the size range from 0.1 to 10 μm.

Most grains are hygroscopic and, therefore, vary in mass with the humidity. Illustrations and data on pollen grains are available in botanical literature. Geographical distribution of plants that produce hay fever is also recorded.

The quantity of pollen grains in the air is generally estimated by exposing an adhesive-coated glass plate outdoors for 24 h, then counting calibrated areas under the microscope. Pollen counts measured by this technique usually represent the number of grains found on 1.8 cm^2 of a 24 h gravity slide. Methods are available for determining the number of grains in a measured volume of air, but, despite their greater accuracy, they have not replaced the simpler gravity slide method used for most pollen counts. Counting techniques vary.

Hay fever sufferers may experience the first symptoms when the pollen count is 10 to 25; in some localities, maximum figures for the seasonal peak may exceed 4000 grains per cubic metre for a 24-h period, depending on the sampling and reporting methods used by the laboratory. These levels are usually given for specific pollens. When such information is important, it should be obtained directly by a volumetric instrument. The number of pollen grains per cubic metre of air varies from 2 to 20 times the number found on 1 cm^2 of a 24 h gravity slide, depending on grain diameter, shape, specific gravity, wind velocity, humidity, and physical placement of the collecting plate.

Whole-grain pollens are easily removed from the outside air entering a ventilation system with medium-efficiency filters selected to remove 99% of particles 10 μm and greater. Once they have entered a building, the rapid settling rate of whole-grain pollens makes it difficult for further air cleaning to reduce the concentrations of these particles. On resuspension from occupant activities, the whole-grain pollens may be broken into particles, which may then be controlled effectively with a high-efficiency filter capable of removing a high percentage of particles a few micrometres in diameter.

BIOAEROSOLS

Interest in airborne microorganisms and their ability to subsist in different environments increased with the need to assemble and launch space probes under sterile conditions, as well as to prevent cross-infection in hospitals. Methods for contamination control are similar to those used in clean rooms to protect critical manufacturing areas from dust contamination. For more information on clean rooms, see Chapter 15 of the 1995 *ASHRAE Handbook—Applications*.

Public interest focuses on airborne microorganisms, primarily bacteria, responsible for diseases and infections. However, a variety of airborne microorganisms besides bacteria are of economic significance and can cause product contamination or loss. In the food-processing industry, yeast and mold can reduce the shelf life of some products. Refined syrups can be damaged by mold scums. Wild yeast can destroy a batch of beer. Antibiotic yields can be reduced by foreign organisms in the culture mix.

Most microorganisms become airborne by attachment to dust particles. Bacteria from the soil are likely to be spore-formers and are capable of surviving in hostile environments. Other airborne bacteria, especially within closed occupied spaces, originate from droplet nuclei caused by actions such as sneezing. Concentrations of microorganisms in the atmosphere vary from a few to several hundred per cubic metre, depending on many factors. The sampling method for microorganisms has an effect on the measured count. Collection on dry filter paper can cause count degradation because of the dehydration loss of some organisms. Glass impingers may give high counts because agitation can cause clusters to break up into smaller individual organisms. Slit samples may give a more accurate colony count.

When maximum removal of airborne microorganisms is necessary or desirable, superinterception, high-efficiency particulate air (HEPA) filters are used. These filters create essentially sterile atmospheres and are preferred over chemical scrubbers and ultraviolet radiation, which are also used to control airborne microorganisms.

In many situations, total control of airborne microorganisms is not required. For these situations, there are different types of high-efficiency dry media extended surface filters that provide the efficiency necessary for certain applications. These filters have lower pressure drops than HEPA filters and can manage or filter out the contaminant of concern.

INDOOR AIR QUALITY

Indoor air quality is a concern in residences, offices, and other indoor, nonindustrial environments (Spengler et al. 1982, NRC 1981). Exposure to indoor pollutants can be as important as exposure to outdoor pollutants because a large portion of the population spends up to 90% of their time indoors and because indoor pollutant concentrations are frequently higher than corresponding outdoor pollutant levels.

Characterization of the indoor air quality has been the subject of numerous recent studies. ASHRAE Indoor Air Quality (IAQ) Conference proceedings discuss indoor air quality problems and some practical controls. ASHRAE *Standard* 62 addresses many indoor air quality concerns. Table 4 illustrates the sources, levels, and indoor-to-outdoor concentration ratios of several pollutants found in indoor environments. Chapter 9 has further information.

Table 4 Sources, Possible Concentrations, and Indoor-to-Outdoor Concentration Ratios of Some Indoor Pollutants

Pollutant	Sources of Indoor Pollution	Possible Indoor Concentration[a]	I/O Concentration Ratio	Location
Carbon monoxide	Combustion equipment, engines, faulty heating systems	100 ppm	>>1	Skating rinks, offices, homes, cars, shops
Respirable particles	Stoves, fireplaces, cigarettes, condensation of volatiles, aerosol sprays, resuspension, cooking	100 to 500 μg/m^3	>>1	Homes, offices, cars, public facilities, bars, restaurants
Organic vapors	Combustion, solvents, resin products, pesticides, aerosol sprays	NA	>1	Homes, restaurants, public facilities, offices, hospitals
Nitrogen dioxide	Combustion, gas stoves, water heaters, dryers, cigarettes, engines	200 to 1000 μg/m^3	>>1	Homes, skating rinks
Sulfur dioxide	Heating system	20 μg/m^3	<1	
Total suspended particles without smoking	Combustion, resuspension, heating system	100 μg/m^3	1	Homes, offices, transportation, restaurants
Sulfate	Matches, gas stoves	5 μg/m^3	<1	Homes, offices
Formaldehyde	Insulation, product binders, particleboard	0.05 to 1.0 ppm	>1	Homes, offices
Radon and progeny	Building materials, groundwater, soil	0.1 to 200 nCi/m^3	>>1	Homes, buildings
Asbestos	Fireproofing, insulation	<10^6 fiber/m^3	1	Homes, schools, offices
Mineral and synthetic fibers	Cloth, rugs, wallboard	NA	—	Homes, schools, offices
Carbon dioxide	Combustion, humans, pets	3000 ppm	>>1	Homes, schools, offices
Viable organisms	Humans, pets, rodents, insects, plants, fungi, humidifiers, air conditioners	NA	>1	Homes, hospitals, schools, offices, public facilities
Ozone	Electric arcing	20 ppb	<1	Airplanes
	Ultraviolet light sources	200 ppb	>1	Offices

Source: NRC (1981).
[a]Concentrations listed are only those reported indoors. Both higher and lower concentrations have been measured. No averaging times are given. NA indicates it is not appropriate to list a concentration.

REFERENCES

ACGIH. 1995. *Air sampling instruments*, 8th ed. American Conference of Governmental Industrial Hygienists, Cincinnati, OH.

ASHRAE. 1989. Ventilation for acceptable indoor air quality. ANSI/ASHRAE *Standard* 62-1989.

ASTM. 1992. Practice for continuous sizing and counting of airborne particles in dust-controlled areas and clean rooms using instruments capable of detecting single sub-micrometre and larger particles. ASTM *Standard* F 50.

Avallone, E.A. and T. Baumeister. 1987. *Marks' standard handbook for mechanical engineers*. McGraw-Hill, New York.

Boronsted, C.A., J.E. Kahn, and A.B. Fuller. 1976. *Nuclear air cleaning handbook*. ORNL *Report* ERDA-76-21. Oak Ridge National Laboratory, Oak Ridge, TN.

Hartmann, I. 1958. Explosion and fire hazards of combustible dusts. *Industrial Hygiene and Toxicology* 1:2. Interscience Publishers, New York.

Hewson, E.W. et al. 1967. Air pollution by ragweed pollen. *Journal of the Air Pollution Control Association* 17(10):651.

Jacobson, A.R. and S.C. Morris. 1976-77. The primary pollutants, viable particulates, their occurrence, sources and effects. In *Air pollution*, 3rd ed., I(4):169. Academic Press, New York.

Morrow, P.E. 1964. Evaluation of inhalation hazards based upon the respirable dust concept and the philosophy and application of selective sampling. *AIHA Journal* 25:213.

National Primary and Secondary Ambient Air Quality Standards. *Code of Federal Regulations*, Title 40, Part 50, 40CFR50.

NFPA. 1994. Guide to fire hazard properties of flammable liquids, gases and volatile solids. ANSI/NFPA *Standard* 325-94. National Fire Protection Association, Quincy, MA.

NIST. 1959. Maximum permissible body burdens and maximum permissible concentrations of radionuclides in air and water for occupational exposure. NBS *Handbook* No. 69. National Institute of Standards and Technology, Gaithersburg, MD.

NRC. 1981. *Indoor pollutants*. 1981. National Research Council, National Academy Press, Washington, D.C.

Scala, G.F. 1963. A new instrument for the continuous measurement of condensation nuclei. *Analytical Chemistry* 35(5):702.

Solomon, W.R. and K.P. Mathews. 1978. Aerobiology and inhalant allergens. *Allergy, principles and practices*, E. Middleton, Jr., C.E. Reed, and E.F. Ellis, eds. G.V. Mosely, St. Louis, MO.

Spengler, J., C. Hallowell, D. Moschandreas, and O. Fanger. 1982. Indoor air pollution. *Environmental International*. Pergamon Press, Oxford, England.

Wadden, A. and P.A. Schiff. 1983. *Indoor air pollution: Characterization, prediction and control*. John Wiley and Sons, New York.

Whitby, K.T., A.B. Algren, and R.C. Jordan. 1955. Size distribution and concentration of airborne dust. *Heating, Piping and Air Conditioning* 27:121.

Whitby, K.T., A.B. Algren, and R.C. Jordan. 1957. The ASHRAE airborne dust survey. *Heating, Piping and Air Conditioning* 29(11):185.

CHAPTER 13

ODORS

VARIOUS factors make odor control a primary consideration in ventilation engineering: (1) modern buildings permit less air infiltration through walls and have more indoor sources of odors associated with building materials, furnishings, and office equipment; (2) outdoor air is often polluted; and (3) energy costs have encouraged ventilation rate reductions at a time when requirements for a relatively odor-free environment are greater than ever.

This chapter reviews how odoriferous substances are perceived. Chapter 41 of the 1995 *ASHRAE Handbook—Applications* has information on control methods.

SENSE OF SMELL

Olfactory Stimuli

Among organic substances, those with molecular weights greater than 300 are generally odorless. Some substances with molecular weights less than 300 are such potent olfactory stimuli that they can be perceived at concentrations too low to be detected with direct-reading instruments. Trimethylamine, for example, can be recognized by a human observer at a concentration of about 10^{-4} ppm.

Table 1 shows **threshold concentrations** for selected compounds. These threshold values are not precise numbers and may vary by more than one order of magnitude. The **threshold limit value** (TLV) is the concentration of a compound that should have no adverse health consequences if a worker is regularly exposed for 8-hour periods (OSHA, revised annually). Table 1 also includes the ratio of the TLV to the odor threshold for each compound. For ratios greater than one, most occupants can detect the odor and leave the area long before the compound becomes a health risk. As the ratio increases, the safety factor provided by the odor also increases. Table 1 is not a comprehensive list of the chemicals found in indoor air. AIHA (1989), Cain et al. (1992), and Van Germent and Mettenbreijer (1977) provide lists of odor thresholds. ACGIH (annually) and Moore and Houtala (1983) provide TLVs or relations between odor thresholds and TLVs.

Olfactory sensitivity often makes it possible to detect potentially harmful substances at concentrations below dangerous levels so that they can be eliminated. Foul-smelling air is often assumed to be unhealthy. When symptoms such as nausea, headache, and loss of appetite are caused by an unpleasant odor, it may not matter whether the air is toxic. The magnitude of the symptoms is related to the magnitude of the odor. Even a room with a low but recognizable odor can arouse uneasiness among occupants. Cometto-Muniz and Cain (1992) review sensory irritation and its relation to indoor air pollution.

Anatomy and Physiology

The olfactory receptors lie in the *olfactory cleft*, which is high in the nasal passages. The surrounding tissue contains other diffusely distributed receptors—free nerve endings that also respond to airborne vapors. These receptors of common chemical sense mediate the tickling, burning, cooling, and, occasionally, painful sensations that accompany olfactory sensations. Ammonia stimulates these free nerve endings, as do other substances that have any degree of pungency. Olfaction and the common chemical sense operate as a single perceptual system (Cain 1976).

About five million olfactory neurons, a small cluster of nerve cells inside the nasal cavity above the bridge of the nose, determine whether the air is odoriferous. These olfactory receptors are connected by a nerve fiber into the olfactory bulb of the brain. The bulb passes information it receives from the receptors to various central structures of the brain, including an area called the *seat of emotion*. One sniff of an odorant can often evoke a complex, emotion-laden memory, such as a scene from childhood.

Hormonal factors, which often mediate emotional states, can modulate olfactory sensitivity. Although the evidence is not uniformly compelling, research has found that (1) the sensitivity of women varies during the menstrual cycle, reaching a peak just before and during ovulation (Schneider 1974); (2) women are generally more sensitive than men, but this difference only emerges around the time of sexual maturity (Koelega and Koster 1974); (3) sensitivity is altered by certain diseases (Schneider 1974); and (4) various hormones and drugs (e.g., estrogen and alcohol) alter sensitivity (Schneider 1974, Engen et al. 1975). Other factors that may affect olfactory perception include the olfactory acuity of an individual, the magnitude of the flow rate toward the olfactory receptors, the temperature, and the relative humidity. Humans are able to perceive over 4000 different odors, yet individuals are able to name only a small number of odors (Ruth 1986).

SENSORY MEASUREMENT

Sensory measurements may be made using one or more of the following categories of data:

- Nominal data that indicate different categories, but not in any order of magnitude; for example, the numbers used in a team or panel of odor sniffers
- Ordinal data that fall into two or more classes belonging to an ordered series; for example, the odor is either slight, moderate, or strong
- Interval data that are placed in ordered classes separated by a meaningful measure or interval between them; for example, first, second, third
- Ratio data that indicate how many times a quantity is larger than a reference quantity

Odor Sensation Attributes

Odor sensation has four components or attributes: detectability, intensity, character, and hedonic tone. **Detectability** or **threshold** is the minimum concentration of an odorant that provokes detection by some predetermined segment of the population. Two types of thresholds exist: the **detection threshold** and the **recognition**

The preparation of this chapter is assigned to TC 2.3, Gaseous Air Contaminants and Gas Contaminant Removal Equipment, and TC 2.4, Particulate Air Contaminants and Particulate Contaminant Removal Equipment.

Table 1 Odor Thresholds and Safety Factors of Selected Gaseous Air Pollutants

Pollutant	Odor Threshold, mg/m^3	TLV/Odor Threshold	Pollutant	Odor Threshold, mg/m^3	TLV/Odor Threshold
Acetaldehyde	1.2	150	Hydrogen cyanide	1	10
Acetone	47	38	Hydrogen fluoride	2.7	0.9
Acetonitrile	>0		Hydrogen sulfide	0.007	2000
Acrolein	0.35	0.7	Mercury	Infinite	0.0
Acrylonitrile	50		Methanol	130	2
Allyl chloride	1.4	2	Methyl chloride	595	0.2
Ammonia	33	0.5	Methylene chloride	750	0.2
Benzene	15	2	Nitric oxide	>0	
Benzyl chloride	0.2	25	Nitrogen dioxide	51	0.1
2-Butanone (MEK)	30	20	Ozone	0.2	1
Carbon dioxide	Infinite	0.0	Phenol	0.18	106
Carbon monoxide	Infinite	0.0	Phosgene	4	0.1
Carbon disulfide	0.6	50	Propane	1800	
Carbon tetrachloride	130	0.2	Sulfur dioxide	1.2	4
Chlorine	0.007	430	Sulfuric acid	1	
Chloroform	1.5	33	Tetrachloroethane	24	0.3
p-Cresol	0.056	390	Tetrachloroethylene	140	2
Dichlorodifluoromethane	5400	0.9	*o*-Toluidene	24	
Dioxane	304	0.3	Toluene	8	47
Ethylene dichloride	25	1.6	Toluene diisocyanate	15	0.003
Ethylene oxide	196	0.01	1,1,1-Trichloroethane	1.1	1730
Formaldehyde	1.2	1.3	Trichloroethylene	120	2
n-Heptane	2.4	670	Vinyl chloride monomer	1400	0.007
Hydrogen chloride	12	0.6	Xylene	2	220

threshold. The detection threshold is the lowest level that elicits response by a segment of the population. If that segment is 50%, the detection threshold is denoted by ED_{50}. Thresholds can be attributed to 100%, which includes all olfactory sensitivities, or to 10%, which includes only the most sensitive segment of the population. Threshold values are not physical constants; rather they are statistical measurements of best estimates.

Intensity is a quantitative aspect of a descriptive analysis, stating the degree to which a characteristic odor is present. Intensity of the perceived odor is, therefore, the strength of the odoriferous sensation. Detection threshold values and, most often, odor intensity determine the need for indoor odor controls.

Character defines the odor as similar to some familiar smell, for example, fishy, sour, flowery, and the like. **Hedonics**, or the hedonic tone of an odor, is the degree to which an odor is perceived as pleasant or unpleasant. Hedonic judgments include both *category* judgments (pleasant, neutral, unpleasant) and a *magnitude* judgment (very unpleasant, slightly pleasant).

Important questions are (1) What is the minimum concentration of odorant that can be detected? (2) How does perceived odor magnitude grow with concentration above the threshold? No universal method has been accepted to measure either the threshold or the perceived magnitude of the odor above threshold. However, certain guidelines and conventions simplify the choice of methods.

Olfactory Acuity

The olfactory acuity of the population is normally distributed—that is, most people have an average ability to smell substances or to respond to odoriferous stimuli, a few people are very sensitive or hypersensitive, and a few others are insensitive, including some who are totally unable to smell (anosmic). The olfactory acuity of an individual varies as a function of the odorant. For instance, repeated exposure to an odorant may lead to anosmia only to that odorant.

Threshold

The perception of weak odoriferous signals is probabilistic; at one moment the signal may make a discernible impact, at the next moment it may not. Factors affecting this phenomenon include the moment-to-moment variability in the number of molecules striking the receptors, the variability in which receptors are contacted, and a person's state of readiness or fatigue. The combined effect of these factors prevents an individual from perceiving an odor during the entire time of the stimulus. Dilution to detection (recognition) threshold values allows determination of the largest number of dilutions that still permits half of the panelists to detect (recognize) the odor.

Dravnieks Olfactometer

Measurement of odor attributes is complex. Several methods, such as those described by Dravnieks (1975) and NAS (1979), are used to measure odor threshold values. ASTM STP 434 describes the forced-choice dynamic olfactometer method. The most frequently used dynamic olfactometer has six ports or sets. Within each port, three glass sniffers are arranged in a circular pattern to achieve blind presentation to the subject. Two of the sniffers supply nonodoriferous air, and the third sniffer provides the diluted odorous air. The subject is asked to identify or guess which sniffer has the odoriferous air.

The odoriferous air is supplied to each port at various levels of dilution. Samples are diluted either by compressed air or room air using a peristaltic pump and are presented in decreasing levels of dilution. The dilution factors from port one to port six are 3600x, 1200x, 400x, 45x, and 15x (IITRI 1980). The odorant is presented in increasing rather than random order because exposure to one of the higher concentrations can temporarily impair the ability to detect weaker concentrations.

The subject can be tested with additional ascending series. The concentration of interest in each series is the lowest in the string of three correct detections. The geometric mean taken across the various series provides a good index of the subject's threshold. In a similar manner, the geometric mean taken across several subjects provides a good index of the threshold for a group.

Three issues of technical and practical importance are (1) the rate of airflow to the sniffers, (2) the size of panel, and (3) the design of the sniffers. The flow rate usually varies from 8 to 50 mL/s but, in some cases, is much greater. Test practices use as few as five panelists or as many as ten. Although a small panel is representative only of itself, too large a panel is difficult to manage. A panel of ten provides statistically sound data.

Sniffers used with dynamic olfactometers are small glass ports that allow the panelist to sniff as close to or as far from as desired. An alternative design is a cup arrangement that requires the panelist to insert the nose into the cup and eliminates variation in the dilution between the delivered air and the air actually sniffed by the panelist.

In all sensory testing, it is necessary to guard against subtle biases and clues that may influence results. Also, adequate statistical procedures should be used to detect biases such as whether there is a reliable psychophysical difference between odorants, between means of air sampling, etc.

For more details regarding psychophysical procedures, ways to sample odoriferous air, handling samples, means of stimulus presentation, and statistical procedures, consult ASTM STP 434.

Suprathreshold Intensity

The relation between **perceived odor magnitude** S and concentration C conforms to a power function

$$S = kC^n \tag{1}$$

This exemplifies the **psychophysical power law**, also called **Stevens' law** (Stevens 1957). The law applies throughout the sensory realm and has almost completely replaced the Weber-Fechner logarithmic law in modern psychophysics. In the olfactory realm, the exponent n of the power function is less than 1.0. Accordingly, a given percentage change in concentration causes a smaller percentage change in perceived magnitude. The size of the exponent varies from one odorant to another, ranging from less than 0.2 to about 0.7 (Cain and Moskowitz 1974), which has important consequences for malodor control. An exponent of 0.7 implies that in order to reduce perceived intensity by a factor of 5, for instance, concentration must be reduced by a factor of 10. An exponent of 0.2 implies that a fivefold reduction in perceived magnitude can be achieved by slightly more than a three-thousandfold reduction in concentration.

Of the various ways to scale perceived magnitude, a **category scale**, which can be either number or word categorized, is commonly used. Numerical values on this scale do not reflect ratio relations among odor magnitudes (e.g., a value of 2 does not represent a perceived magnitude twice as great as a value of 1). Table 2 gives four examples of category scales (Meilgaard et al. 1987).

Although category scaling procedures can be advantageous in field situations, **ratio scaling** techniques are used frequently in the laboratory (Cain and Moskowitz 1974). Ratio scaling procedures require observers to assign numbers proportional to perceived magnitude. For example, if the observer is instructed to assign the number 10 to one concentration and a subsequently presented concentration seems three times as strong, the observer calls it 30; if another seems one-half as strong, the observer assigns it 5. This ratio scaling procedure, called **magnitude estimation**, was used to derive the power function for butanol (Figure 1). Equation (1) in this case becomes $\log S = n \log C + \log k$, which has the form $y = ax + b$. By means of this transformation, the exponent becomes the slope of the line. The function for butanol conforms to the equation $S = 0.26\,C^{0.66}$, which represents a consensus of results obtained in various laboratories (Moskowitz et al. 1974). According to this equation, a concentration of 250 ppm has a perceived magnitude of 10. This corresponds to *moderate intensity*. A concentration near threshold (2 to 5 ppm) has a perceived magnitude of about 0.5.

Table 2 Examples of Category Scales

Number Category		Word Category	
Scale I	Scale II	Scale I	Scale II
0	0	None	None at all
1	1	Threshold	Just detectable
2	2.5	Very slight	Very mild
3	5	Slight	Mild
4	7.5	Slight-moderate	Mild-distinct
5	10	Moderate	Distinct
6	12.5	Moderate-strong	Distinct-strong
7	15	Strong	Strong

A third way to measure suprathreshold odor intensity is to **match the intensity of odorants**. An observer can be given a concentration series of a matching odorant (e.g., 1-butanol) to choose the member that matches most closely the intensity of an *unknown* odorant. The matching odorant can be generated by a relatively inexpensive olfactometer such as that shown in Figure 2 (Dravnieks 1975). Figure 3 shows, in logarithmic coordinates, functions for various odorants obtained by the matching method (Dravnieks and Laffort 1972). The left-hand ordinate expresses intensity in terms of concentration of butanol, and the right-hand ordinate expresses intensity in terms of perceived magnitude. The two ordinates are related by the function in Figure 1, the standardized function for butanol. The matching method illustrated here has been incorporated into ASTM *Standard* E 544-75, *Standard Practices for Referencing Suprathreshold Odor Intensity*.

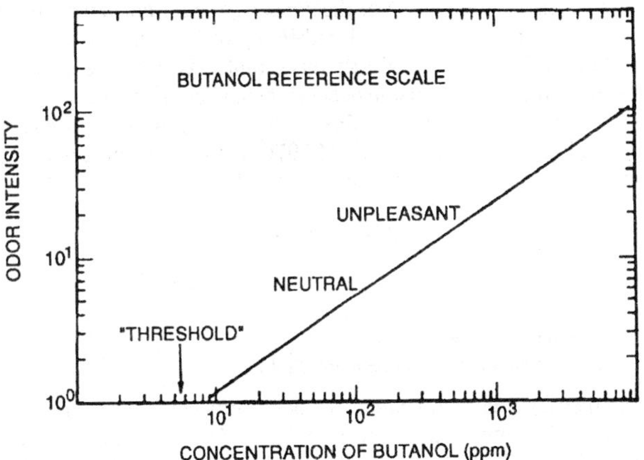

Fig. 1 Standardized Function Relating Perceived Magnitude to Concentration of 1-Butanol
(Moskowitz et al. 1974)

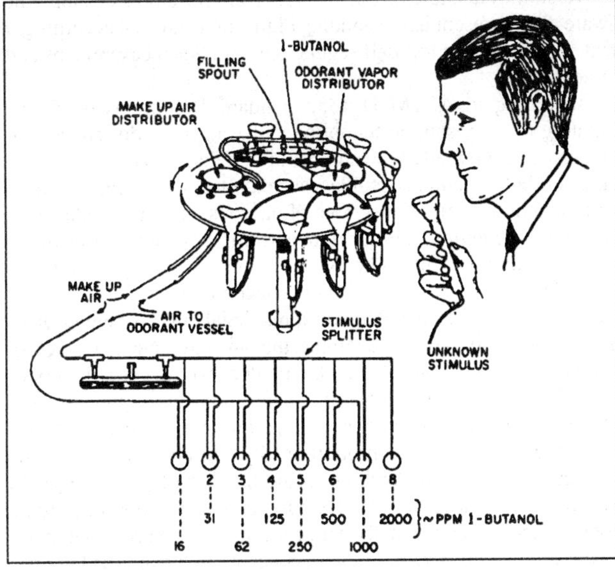

Fig. 2 Panelist Using Dravnieks Binary Dilution Olfactometer

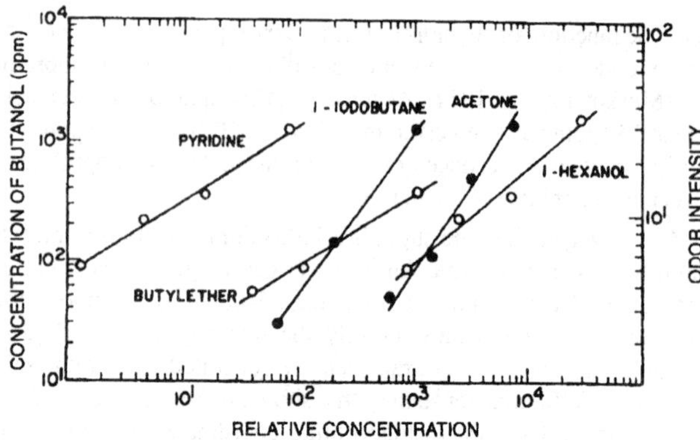

Fig. 3 Matching Functions Obtained with Dravnieks Olfactometer
(Dravnieks and Laffort 1972, Cain 1978)

Odor Units

A National Academy of Sciences document (NAS 1979) elaborates on attitudes regarding odor units. The objection to the odor unit is associated with the fact that odor is a sensation, not a substance. This is an example of the concept of odor concentration: if 1 ft^3 of odorous air is diluted with N ft^3 of fresh air to reach the threshold value, then the odoriferous sample has an odor concentration of N units/ft^3. Confusion occurs when other dimensions for volume are used. In response to this confusion, the American Society for Testing and Materials (ASTM 1976) defined the dimensionless factor Z for the dilution to detection threshold:

$$Z = C/C_{thr} \tag{2}$$

where

C = odorant concentration
C_{thr} = odorant concentration at threshold

Note that Z numerically equals the odor units per cubic foot.

The following example from NAS (1979) illustrates the concepts. $Z = 1000$ when one volume of odorous sample requires 1000 volumes to reach the dilution threshold. Furthermore, if the sampled air is emitted at 50 m^3/min, the odorous emission rate is $1000 \times 50 = 50{,}000$ m^3/min, which is the clean air demand of the odorous emission. This value is equivalent to odor units per minute after conversion from cubic meters to cubic feet. The engineer should be aware of the present understanding of this unit and deal accordingly with existing and proposed regulations on odor measurement and removal.

According to ASTM D 135, Standard Definitions of Terms Relating to Atmospheric Sampling and Analysis, odor concentration can be expressed as the number of unit volumes that a unit volume of odorous sample occupies when diluted to the odor threshold with nonodorous air. If a sample of odorous air can be reduced to threshold by a tenfold dilution with pure air, the concentration of the original sample is said to be 10 odor units. Hence, odor units are equivalent to multiples of threshold concentrations.

Odor units are used widely to express legal limits for emission of odoriferous materials. For example, the law may state that a factory operation may not cause the ambient odor level to exceed 15 odor units. It should be recognized that odor units are *not* units of perceived magnitude. For every odorant, odor units and parts-per-million are proportional. The proportionality constant varies from one odorant to another, depending on the number of ppm needed to evoke a threshold response. Because perceived odor magnitude does not grow proportionally with concentration expressed in ppm, it cannot grow proportionally with concentration expressed in odor units. According to the results of psychophysical scaling, a sample of 20 odor units is always perceived as less than twice as strong as a sample of 10 odor units. Moreover, because the rate of growth of the psychophysical function varies from one odorant to another, samples of two odorants, each at 20 odor units, may have unequal perceived intensities.

Fig. 4 Hedonic Scale Used for Evaluation of Odors in Population Surveys

Although odor units are not equivalent to units of perceived magnitude, they can be useful. Most indoor and outdoor contaminants are complex mixtures; the actual concentration of the odoriferous portion of a sample cannot be expressed with certainty. Thus, the odor unit is a useful measure of *concentration* when evaluating, for example, the efficiency of a filter or ventilation system.

Odor Acceptability

The acceptability or pleasantness of an odor can be measured psychophysically in the same way as odor intensity. Both ratio and category scaling procedures can be adapted to odor acceptability.

Odors do not always cause adverse reactions. Products are manufactured to elicit favorable responses. Acceptance tests may involve product comparison (frequently used in the perfume industry) or a hedonic scale (Figure 4). The premise of acceptance tests is that the larger the segment of subjects accepting the odor, the better the odor. A hedonic scale that allows for negative as well as positive responses is likely to be more responsive to the question of how acceptable the odor is.

All persons exposed to a given odor are not likely to agree whether the odor acceptable or unacceptable. Acceptability of a given odor by a person is due to a complex combination of associations and is not simply a characteristic of the odor itself (Beck and Day 1991).

Odor Quality

The quality or character of an odor is difficult to assess quantitatively. A primary difficulty is that odors can vary along many dimensions. One way to assess quality is to ask panelists to judge the similarity between a test sample and various reference samples, using a 5-point category scale. For certain applications, reference odorants can be chosen to represent only that portion of the qualitative range relevant to the odor problem under investigation (e.g., animal odors). Another procedure is to ask panelists to assess the degree of association between a test sample's quality and certain verbal descriptors (e.g., sweaty, woody, chalky, sour).

The number of odorant descriptors and the descriptors to be used have been subjects of disagreement (Harper et al. 1968). The number of descriptors varies from a minimum of seven (Amoore 1962) to as many as 830 used by an ASTM subcommittee. An atlas of odor characters, containing 146 descriptors, was compiled for 180 chemicals by ASTM (1985).

An odor can be characterized either by an open-ended word description or by multidimensional scaling. Multidimensional scaling is based on similarity and dissimilarity judgments in comparison to a set of standard odors or to various descriptors.

In some instances, the interest merely may be whether an odor's quality has changed as a result of some treatment (e.g., use of a bacteriostat). Under these circumstances, samples of air taken before and after treatment can be compared directly (using a simple scale of similarity) or indirectly (with appropriate verbal descriptors).

Olf Unit

Sometimes indoor air quality scientists cannot successfully resolve complaints about the air in offices, schools, and other nonindustrial environments. Customarily, complaints are attributed to elevated pollutant concentrations; frequently, however, such high concentrations are not found indoors, yet complaints persist. The inability to determine cause and effect relationships between complaints and indoor pollutant concentrations has led investigators to define the cause of these complaints as **sick building syndrome**.

Based on the assumption that the inability to find a difference between air pollutant levels in buildings with registered complaints and those without any complaints is due to inadequacies of prevailing measurement techniques, Fanger and others changed the focus from chemical analysis to sensory analysis (Fanger 1987, Fanger et al. 1988). Fanger quantified air pollution sources by comparing them with a well-known source—a sedentary person in thermal comfort. A new unit, the olf, was introduced. An **olf** is defined as the emission rate of air pollutants (bioeffluents) from a standard person.

To use this unit, Fanger generated curves that relate the percentage of persons dissatisfied with the emissions of one person in a chamber as a function of fresh air ventilation rate, and obtained the following expression:

$$\begin{aligned} D &= 395\exp(-1.83q^{0.26}) \quad \text{for} \quad q \geq 0.322 \\ D &= 100 \quad \text{for} \quad q < 0.322 \end{aligned} \tag{3}$$

where

D = percentage dissatisfied persons
q = ventilation-emission ratio, L/s × olf

The curves can be used in the study of indoor nonindustrial environments. A large number of judges (more than 50) is used in a study. They visit the test environment under three different conditions: when the test indoor environment is (1) unoccupied and unventilated; (2) unoccupied but ventilated; and (3) occupied and ventilated. These measurements help in quantifying the sensory impact from such sources as building materials, the ventilation system, and occupant activity. Results are obtained by analyzing the responses to a questionnaire (Fanger et al. 1988).

ANALYTICAL MEASUREMENT

Performance data on the control of specific odorants can be obtained using suitable analytical methods. Detectors permit detection of substances in amounts as little as one nanogram. Air contains many minor components, so gas chromatographic separation of the components must precede detection. Because odor thresholds for some compounds are low, preconcentration of the minor components is necessary. Preconcentration consists of adsorption or absorption by a stable, sufficiently nonvolatile material, followed by thermal desorption or extraction.

Mass spectrometry can be used with gas chromatography to identify constituents of complex mixtures. The chromatograph resolves a mixture into its constituents, and the spectrometer provides identification and concentration of selected constituents.

Several other detectors are sufficiently sensitive and specific to detect resolved components. Hydrogen flame ionization detectors respond adequately and nearly mass proportionally to almost all organic compounds. Flame photometric detectors can pinpoint, with equal sensitivity, compounds that contain sulfur; many sulfur compounds are strongly odorous and are of interest in odor work. A Coulson conductometric detector is specifically and adequately sensitive to ammonia and organic nitrogen compounds. Thermal conductivity detectors are generally not sensitive enough for analytical work on odors.

Airborne volatile organic compounds (VOCs) cause odors, but the correlation between indoor VOC concentrations and odor complaints in indoor environments is poor. The state of the art for sampling and analysis of VOCs in indoor air is reviewed in NIOSH (1993). The nose is, however, usually the most sensitive detector available and can perceive odors from complex mixtures of VOCs at low indoor concentrations.

Frequently, a sniffing port (Dravnieks and O'Donnell 1971, Dravnieks and Krotoszynski 1969) is installed in parallel with the detector(s). Part of the resolved effluent exhausts through the port and allows the components that are particularly odorous or carry some relevant odor quality to be annotated. Usually, only a fraction of all components studied exhibit odors—particularly those of interest in odor evaluation. A change in odorant concentration indicated by detector responses may result from odorant removal treatment by ventilation or another process (Dravnieks and O'Donnell 1971). Odor control may not influence all odorant concentrations to the same extent, and new odorants may appear.

Because the physicochemical correlates of olfaction are poorly understood, no simple analytical means to predict the perceived quality and intensity of an odorant exists. Moreover, since acceptability of an odorant depends strongly on context, it is unlikely that analytical instruments will supplant human evaluation.

ODOR SOURCES

Outdoor sources of odors include automotive and diesel exhausts, hazardous waste sites, sewage treatment plants, compost piles, refuse facilities, abattoirs, printing plants, refineries, chemical plants, and many other stationary and mobile sources. These sources produce both inorganic compounds (e.g., ammonia and hydrogen sulfide) and VOCs, including some which evaporate from solid or liquid particulate matter. Odors emitted by sources in the ambient environment eventually enter the indoor environment.

Indoor sources also emit odors. Such sources include tobacco products; bathrooms and toilets; building materials (adhesives, paints, caulks, processed wood, carpets, plastic sheeting, and insulation board); consumer products such as food, toiletries, cleaning materials, and polishes; hobby materials; fabrics; and foam cushions. In offices, offset printing processes, copiers, and computer printers may produce odors. If electrostatic processes are involved, ozone may be emitted.

Mildew and other processes of decay often produce odors in occupied spaces (home and office), damp basements, and ventilation systems (e.g., from wetted air-conditioning coils and spray dehumidifiers).

Chapter 41 of the 1995 *ASHRAE Handbook—Applications* gives further information on contaminant sources and generation rates.

FACTORS AFFECTING ODORS

Humidity and Temperature

Odor perception of cigarette smoke (which is a suspension of tobacco tar droplets plus vapor) and pure vapors is affected by temperature and humidity. An increase in humidity, at constant dry-bulb temperature, lowers the intensity level of cigarette smoke odor, as well as that of pure vapors. This effect is more pronounced for some odorants than for others. An increase in temperature at constant specific humidity lowers the odor level of cigarette smoke slightly. Adaptation to odors takes place more rapidly during the initial stages of exposure. Although the perceptible odor level of cigarette smoke decreases with exposure time, irritation to the eyes and nose generally increases. The irritation is greatest at low relative humidities.

To keep odor perception and irritation at a minimum, the air-conditioned space should be operated at about 45 to 60% rh. Because temperature has only a slight effect on odor level at constant specific humidity, it generally can be ignored; temperature should be maintained at conditions desired.

Sorption and Release of Odors

Frequently, spaces develop normal occupancy odor levels long after occupancy has ceased. This results from sorption of odors by the furnishings during occupancy, with later desorption. This is observed when furnaces or radiators, after a long shutdown, are heated at winter start-up. It also can be observed when evaporator coils warm up. The rate of desorption can be decreased by decreasing temperature and relative humidity, and increased (as for cleaning) by the reverse.

Perception of smoke, cooking, and body odors decreases as humidity increases. Where the odor source is intrinsic with the materials (as in linoleum, paint, rubber, and upholstery) reducing the relative humidity is beneficial and decreases the rate of odor release. Quantitative values should not be used without considering the sorption-desorption phenomenon.

Dilution of Odors by Ventilation

Cain et al. (1983, 1987) compared occupancy odor with smoke odor by relating perceived odor intensity and odor acceptability during smoking and nonsmoking under controlled chamber conditions. When smoking took place, they found that the odor intensity was nearly five times as high as the odor intensity perceived under severe occupancy conditions (hot, humid, and low ventilation) but with no smoking. This study also determined the ventilation rates required to control occupancy odor and tobacco smoke odor to acceptable levels.

REFERENCES

ACGIH. Annually. *Threshold limit values for chemical substances and physical agents and biological exposure indices.* American Conference of Governmental Industrial Hygienists, Cincinnati, OH.

AIHA. 1989. Odor thresholds for chemicals with established occupational health standards. American Industrial Hygiene Association, Akron, OH.

Amoore, J.E. 1962. The stereochemical theory of olfaction 1, Identification of seven primary odors. *Proc. Sci. Sect. Toilet Goods Assoc.* 37:1-12.

ASTM. 1975. Standard practices for referencing suprathreshold odor intensity. ASTM E 544-75 (R 1993). American Society for Testing and Materials, Conshohocken, PA.

ASTM. 1976. Standard method for measurement of odor in atmosphere (dilution method). American Society for Testing and Materials, Conshohocken, PA.

ASTM. 1982. Manual on sensory testing methods. *Special Technical Publication* STP 434.

ASTM. 1985. Atlas of odor character profiles. *Data Series* 61.

Beck, L. and V. Day. 1991. New Jersey's approach to odor problems. From *Transactions: Recent developments and current practices and odor regulations control and technology*, D.R. Derenzo and A. Gnyp, eds. Air and Waste Management Association, Pittsburgh, PA.

Cain, W.S. 1976. Olfaction and the common chemical sense; Some psychophysical contrasts. *Sensory Processes* 1:57.

Cain, W.S. 1978. The odoriferous environment and the application of olfactory research. Carterette and Friedman, eds. *Handbook of perception*, Vol. 6, Tasting, smelling, feeling, and hurting. Academic Press, New York.

Cain, W.S., et al. 1992. Reference guide to odor thresholds. EPA/600/R-92/047. National Institute of Science and Technology (NIST), Springfield, VA.

Cain, W.S. and H.R. Moskowitz. 1974. Psychophysical scaling of odor. Turk, Johnson, and Moulton, eds. *Human responses to environmental odors.* Academic Press, New York.

Cain, W.S., T. Tosun, L.C. See, and B.P. Leaderer. 1987. Environmental tobacco smoke: Sensory reactions of occupants. *Atmospheric Environment* 21:347-53.

Cain, W.S., B.P. Leaderer, R. Isseroff, L.G. Berglund, R.I. Huey, E.D. Lipsitt, and D. Perlman. 1983. Ventilation requirements in buildings I. Control of occupancy odor and tobacco smoke odor. *Atmospheric Environment* 17:1183-97.

Cometto-Muniz, J.E. and W.S. Cain. 1992. Sensory irritation, relation to indoor air pollution in sources of indoor air contaminants—characterizing emissions and health effects. *Annals of the New York Academy of Sciences*, Vol. 641.

Dravnieks, A. 1975. Evaluation of human body odors, methods and interpretations. *Journal of the Society of Cosmetic Chemists* 26:551.

Dravnieks, A. and B. Krotoszynski. 1969. Analysis and systematization of data for odorous compounds in air. *ASHRAE Symposium Bulletin*, Odor and odorants: The engineering view.

Dravnieks, A. and P. Laffort. 1972. Physicochemical basis of quantitative and qualitative odor discrimination in humans. Schneider, ed. Olfaction and Taste IV, p. 142, Wissenschaftliche Verlagsgesellschaft mBH, Stuttgart.

Dravnieks, A. and A. O'Donnell. 1971. Principles and some techniques of high resolution headspace analysis. *Journal of Agricultural and Food Chemistry* 19:1049.

Engen, T., R.A. Kilduff, and N.J. Rummo. 1975. The influence of alcohol on odor detection. *Chemical Senses and Flavor* 1:323.

Fanger, P.O. 1987. A solution to the sick building mystery. Indoor Air '87, Proceedings of the International Conference on Indoor Air and Climate. Institute of Water, Soil and Air Hygiene, Berlin.

Fanger, P.O., J. Lautidsen, P. Bluyssen, and G. Clausen. 1988. Air pollution sources in offices and assembly halls quantified by the olf unit. *Energy and Buildings* 12:7-19.

Harper, R., E.C. Bate Smith, and D.G. Land. 1968. *Odour description and odour classification.* American Elsevier Publishing, New York. Distributors for Churchill Livingston Publishing, Edinburgh, Scotland.

IITRI. 1980. *Instruction manual: Dynamic dilution forced-choice triangle olfactometer, model 101 for emission odors.* Illinois Institute of Technology Research Institute, Chicago.

Koelega, H.S. and E.P. Koster. 1974. Some experiments on sex differences in odor perception. *Annals of the New York Academy of Sciences* 237:234.

Meilgaard, M., G.V. Civille, and B.T. Carr. 1987. *Sensory evaluation techniques.* CRC Press, Boca Raton, FL.

Moore, J.E. and E. Houtala. 1983. Odor as an aid to chemical safety. *Journal of Applied Toxicology* 3:272-90.

Moskowitz, H.R., A. Dravnieks, W.S. Cain, and A. Turk. 1974. Standardized procedure for expressing odor intensity. *Chemical Senses and Flavor* 1:235.

NAS. 1979. *Odors from stationary and mobile sources.* National Academy of Sciences, Washington, D.C.

NIOSH. 1993. Case Studies—Indoor Environmental Quality "From the ground up." *Applied Occupational and Environmental Hygiene* 8:677-80.

OSHA. Revised annually. Code of Federal Regulations, 29 CFR 1900. U.S. Government Printing Office, Washington, D.C.

Ruth, J.H. 1986. Odor thresholds and irritation levels of several chemical substances: A review. *American Industrial Hygiene Association Journal* 47:142-51.

Schneider, R.A. 1974. Newer insights into the role and modifications of olfaction in man through clinical studies. *Annals of the New York Academy of Sciences* 237:217.

Stevens, S.S. 1957. On the psychophysical law. *Psychological Review* 64:153.

Van Germent, L.J. and A.H. Mettenbreijer. 1977. Odor threshold values in air and water. Central Institute for Nutrition and Food Research, TNO, Netherlands.

BIBLIOGRAPHY

ACGIH. 1988. *Advances in air sampling.* American Conference of Governmental Industrial Hygienists, Cincinnati, OH.

Clemens, J.B. and R.G. Lewis. 1988. Sampling for organic compounds. In *Principles of environmental sampling* 20:147-57. American Chemical Society, Washington, D.C.

Moschandreas, D.J. and S.M. Gordon. 1991. Volatile organic compounds in the indoor environment: Review of characterization methods and indoor air quality studies. In *Organic chemistry of the atmosphere.* CRC Press, Boca Raton, FL.

CHAPTER 14

MEASUREMENT AND INSTRUMENTS

HEATING, refrigerating, and air-conditioning engineers and technicians require instruments for both laboratory work and fieldwork. Precision is more essential in the laboratory, where research and development are undertaken, than in the field, where acceptance and adjustment tests are conducted. This chapter describes the characteristics and uses of some of these instruments.

TERMINOLOGY

The following definitions are generally accepted.

Accuracy. The capability of an instrument to indicate the true value of measured quantity. This is often confused with inaccuracy, which is the departure from the true value to which all causes of error (e.g., hysteresis, nonlinearity, drift, temperature effect, and other sources) contribute.

Amplitude. The magnitude of variation from its zero value in an alternating quantity.

Average. The sum of a number of values divided by the number of values.

Bandwidth. The range of frequencies over which a given device is designed to operate within specified limits.

Bias. The tendency of an estimate to deviate in one direction from a true value (a systematic error).

Calibration. (1) The process of comparing a set of discrete magnitudes or the characteristic curve of a continuously varying magnitude with another set or curve previously established as a standard. Deviation between indicated values and their corresponding standard values constitutes the correction (or calibration curve) for inferring true magnitude from indicated magnitude thereafter; (2) the process of adjusting an instrument to fix, reduce, or eliminate the deviation defined in (1).

Calibration curve. (1) The path or locus of a point that moves so that its coordinates on a graph correspond to values of input signals and output deflections; (2) the plot of error versus input (or output).

Confidence. The degree to which a statement (measurement) is believed to be true.

Dead band. The range of values of the measured variable to which an instrument will not effectively respond.

Deviate. Any item of a statistical distribution that differs from the selected measure of control tendency (average, median, mode).

Deviation, standard. The square root of the average of the squares of the deviations from the mean (root mean square deviation). A measure of dispersion of a population.

The preparation of this chapter is assigned to TC 1.2, Instruments and Measurements.

Distortion. An unwanted change in wave form. Principal forms of distortion are inherent nonlinearity of the device, nonuniform response at different frequencies, and lack of constant proportionality between phase-shift and frequency. (A wanted or intentional change might be identical, but it is called **modulation.**)

Drift. A gradual, undesired change in output over a period of time that is unrelated to input, environment, or load. Drift is gradual; if variation is rapid, the fluctuation is referred to as **cycling.**

Dynamic error band. The spread or band of output-amplitude deviation incurred by a constant amplitude sine wave as its frequency is varied over a specified portion of the frequency spectrum (see *Static error band*).

Error. The difference between the true or actual value to be measured (input signal) and the indicated value (output) from the measuring system. Errors can be systematic or random.

Error, accuracy. See *Error, systematic.*

Error, fixed. See *Error, systematic.*

Error, instrument. The error of an instrument's measured value that includes random or systematic errors.

Error, precision. See *Error, random.*

Error, probable. An error with a 50% or higher chance of occurrence. A statement of probable error is of little value.

Error, random. A statistical error caused by chance and not recurring. This term is a general category for errors that can take values on either side of an average value. To describe a random error, its distribution must be known.

Error, root mean square, or **RMS.** An accuracy statement of a system comprising several items. For example, a laboratory potentiometer, volt box, null detector, and reference voltage source have individual accuracy statements assigned to them. These errors are generally independent of one another, so a system of these units displays an accuracy given by the square root of the sum of the squares of the individual limits of error. For example, four individual errors of 0.1% yield a calibrated accuracy of 0.4% but an RMS error of only 0.2%.

Error, systematic. A persistent error not due to chance; systematic errors are causal. A systematic error is likely to have the same magnitude and sign for every instrument constructed with the same components and procedures. Errors in calibrating equipment cause systematic errors because all instruments calibrated are biased in the direction of the calibrating equipment error. Voltage and resistance drifts over time are generally in one direction and are classed as systematic errors.

Frequency response (flat). The portion of the frequency spectrum over which the measuring system has a constant value of amplitude response and a constant value of time lag. Input signals that have frequency components within this range are indicated by the measuring system (without distortion).

Hysteresis. The summation of all effects, under constant environmental conditions, that cause the output of an instrument to assume different values at a given stimulus point when that point is approached with increasing stimulus and with decreasing stimulus. Hysteresis includes backlash. It is usually measured as a percent of full scale when input varies over the full increasing and decreasing range. In instrumentation, hysteresis and dead band are similar quantities.

Linearity. The straight-lineness of the transfer curve between an input and an output; that condition prevailing when output is directly proportional to input (see *Nonlinearity*).

Loading error. A loss of output signal from a device caused by a current drawn from its output. It increases the voltage drop across the internal impedance, where no voltage drop is desired.

Mean. See *Average*.

Median. The middle value in a distribution, above and below which lie an equal number of values.

Noise. Any unwanted disturbance or spurious signal that modifies the transmission, measurement, or recording of desired data.

Nonlinearity. The prevailing condition (and the extent of its measurement) under which the input-output relationship (known as the input-output curve, transfer characteristic, calibration curve, or response curve) fails to be a straight line. Nonlinearity is measured and reported in several ways, and the way, along with the magnitude, must be stated in any specification.

Minimum-deviation-based nonlinearity. The maximum departure between the calibration curve and a straight line drawn to give the greatest accuracy; expressed as a percent of full-scale deflection.

Slope-based nonlinearity. The ratio of maximum slope error anywhere on the calibration curve to the slope of the nominal sensitivity line; usually expressed as a percent of nominal slope.

Most variations beyond these two definitions result from the many ways in which the straight line can be arbitrarily drawn. All are valid as long as construction of the straight line is explicit.

Population. A group of individual persons, objects, or items from which samples may be taken for statistical measurement.

Precision. The repeatability of measurements of the same quantity under the same conditions; not a measure of absolute accuracy. The precision of a measurement is used here to describe the relative tightness of the distribution of measurements of a quantity about their mean value. Therefore, precision of a measurement is associated more with its repeatability than its accuracy. It combines uncertainty caused by random differences in a number of identical measurements and the smallest readable increment of the scale or chart. Precision is given in terms of deviation from a mean value.

Primary calibration. A calibration procedure in which the instrument output is observed and recorded while the input stimulus is applied under precise conditions—usually from a primary external standard traceable directly to the National Institute of Standards and Technology (NIST).

Range. A statement of the upper and lower limits between which an instrument's input can be received and for which the instrument is calibrated.

Reliability. The probability that an instrument's precision and accuracy will continue to fall within specified limits.

Repeatability. See *Precision*.

Reproducibility. In instrumentation, the closeness of agreement among repeated measurements of the output for the same value of input made under the same operating conditions over a period of time, approaching from both directions; it is usually measured as a nonreproducibility and expressed as reproducibility in percent of span for a specified time period. Normally, this implies a long period of time, but under certain conditions, the period may be a short time so that drift is not included. Reproducibility includes hysteresis, dead band, drift, and repeatability. Between repeated measurements, the input may vary over the range, and operating conditions may vary within normal limits.

Resolution. The smallest change in input that produces a detectable change in instrument output. Resolution differs from precision in that it is a psychophysical term referring to the smallest increment of humanly perceptible output (rated in terms of the corresponding increment of input). The precision, the resolution, or both may be better than the accuracy. An ordinary six-digit (or dial) instrument has a resolution of one part per million (ppm) of full scale; however, it is possible that the accuracy is no better than 25 ppm (0.0025%). Note that the practical resolution of an instrument cannot be any better than the resolution of the indicator or detector, whether internal or external.

Scale factor. (1) The amount by which a measured quantity must change to produce unity output; (2) the ratio of real to analog values.

Sensitivity. The property of an instrument that determines scale factor. The word is often short for maximum sensitivity or the minimum scale factor with which an instrument can respond. The minimum input signal strength required to produce a desired value of output signal (e.g., full scale or unit output or the ratio of output to input values).

Sensitivity inaccuracy. The maximum error in sensitivity displayed as a result of the summation of the following: frequency response; attenuator inaccuracy; hysteresis or dead band; amplitude distortion (sensitivity nonlinearity); phase distortion (change in phase relationship between input signal and output deflection); and gain instability. Only by taking into account all these factors can nominal sensitivity, as indicated by the numeral on the attenuator readout, be discounted for accurate interpretation.

Stability. (1) Independence or freedom from changes in one quantity as the result of a change in another; (2) the absence of drift.

Static error band. (1) The spread of error present if the indicator (pen, needle) stopped at some value (e.g., at one-half of full scale). It is normally reported as a percent of full scale; (2) a specification or rating of maximum departure from the point where the indicator must be when an on-scale signal is stopped and held at a given signal level. This definition stipulates that the stopped position can be approached from either direction in following any random waveform. Therefore, it is a quantity that includes hysteresis and nonlinearity but excludes items such as chart paper accuracy or electrical drift (see *Dynamic error band*).

Step-function response. The characteristic curve or output plotted against time resulting from the input application of a step function (a function that is zero for all values of time before a certain instant, and a constant for all values of time thereafter).

Threshold. The smallest stimulus or signal that results in a detectable output.

Time constant. The time required for an exponential quantity to change by an amount equal to 0.632 times the total change required to reach steady state for first-order systems.

Transducer. A device for translating the changing magnitude of one kind of quantity into corresponding changes of another kind of quantity. The second quantity often has dimensions different from the first and serves as the source of a useful signal. The first quantity may be considered an input and the second an output. Significant energy may or may not transfer from the transducer's input to output.

Uncertainty. An estimated value for the error (i.e., what an error might be if it were measured by calibration). Although uncertainty may be the result of both systematic and precision errors, only precision error can be treated by statistical methods.

Zero shift. Drift in the zero indication of an instrument without any change in the measured variable.

UNCERTAINTY ANALYSIS

Uncertainty Sources

Measurement generally consists of a sequence of operations or steps. Virtually every step introduces a conceivable source of uncertainty, the effect of which must be assessed. The following list is representative of the most common, but not all, sources of uncertainty.

- Inaccuracy in the mathematical model that describes the physical quantity
- Inherent stochastic variability of the measurement process
- Uncertainties in measurement standards and calibrated instrumentation
- Time-dependent instabilities due to gradual changes in standards and instrumentation
- Effects of environmental factors such as temperature, humidity, and pressure
- Values of constants and other parameters obtained from outside sources
- Uncertainties arising from interferences, impurities, inhomogeneity, inadequate resolution, and incomplete discrimination
- Computational uncertainties and data analysis
- Incorrect specifications and procedural errors
- Laboratory practice, including handling techniques, cleanliness, and operator techniques, etc.
- Uncertainty in corrections made for known effects, such as installation effect corrections

Uncertainty of a Measured Variable

For a measured variable X_i, the total error is caused by both **precision (random)** and **systematic (bias) errors**. This relationship is shown in Figure 1. The possible measurement values of the variable are scattered in a distribution around the parent population mean μ_i (Figure 1A). The curve is the normal or Gaussian distribution and is the theoretical distribution function for the infinite population of measurements that generated X_i. The parent population mean differs from $(X_i)_{true}$ by an amount called the systematic (or bias) error β_i (Figure 1B). The quantity β_i is the total fixed error that remains after all calibration corrections have been made. In general, there are several sources of bias error, such as calibration standard errors, data acquisition errors, data reduction errors, test technique errors, and so forth. There is usually no direct way to measure these errors. These errors are unknown and are assumed to be zero; otherwise, an additional correction would be applied to reduce them to zero to the best of one's ability.

The **precision uncertainty** for a variable, which is an estimate of the possible error associated with the repeatability of a particular measurement, is determined from the sample standard deviation, or the estimate of the error associated with the repeatability of a particular measurement. Unlike the systematic error, the precision error varies from reading to reading. As the number of readings of a particular variable tends to infinity, the distribution of these possible errors becomes Gaussian.

For each bias error source, the experimenter must estimate a **systematic uncertainty**. Systematic uncertainties are usually estimated from previous experience, calibration data, analytical models, and engineering judgment. For a discussion on estimating systematic uncertainties (bias limits), see Coleman and Steele (1989).

For further information on measurement uncertainty, see ASME *Standards* MFC-2M and PTC 19.1 and Coleman and Steele (1995).

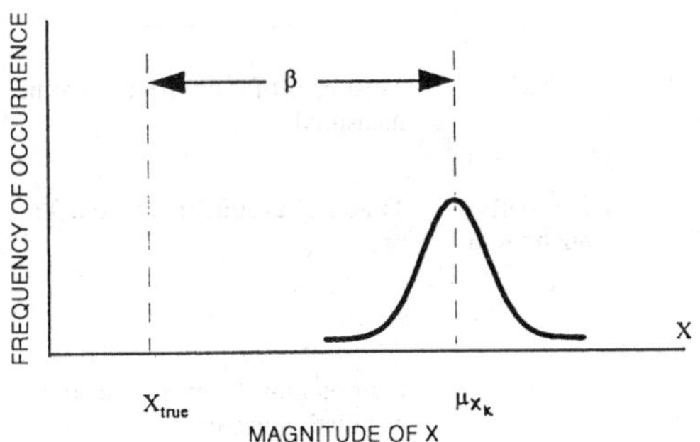

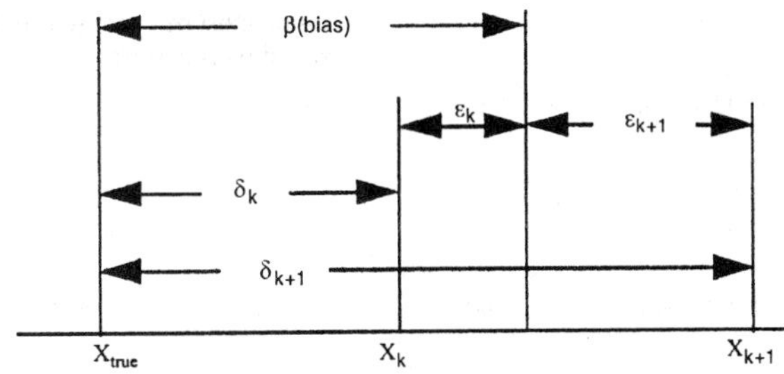

Fig. 1 Errors in the Measurement of a Variable *X*

TEMPERATURE MEASUREMENT

Instruments for measuring temperature are listed in Table 1. Temperature sensor output must be related to an accepted temperature scale. This is achieved by manufacturing the instrument according to certain specifications or by calibrating it against a temperature standard. To help users conform to standard temperatures and temperature measurements, the International Committee of Weights and Measures (CIPM) has adopted the International Temperature Scale of 1990 (ITS-90).

The unit of temperature of the ITS-90 is the kelvin (K) and has a size equal to the fraction 1/273.16 of the thermodynamic temperature of the triple point of water.

The ITS-90 is maintained in the United States by the National Institute of Standards and Technology (NIST), and any laboratory may obtain calibrations from NIST based on this scale.

Benedict (1984), Considine (1985), Quinn (1990), Schooley (1986, 1992), and DeWitt and Nutter (1988) cover temperature measurement in more detail.

Sampling and Averaging

Although temperature is usually measured within, and is associated with, a relatively small volume (depending on the size of the thermometer), it can also be associated with an area (e.g., on a surface or in a flowing stream). To determine average stream temperature, the cross section must be divided into smaller areas and the temperature of each area measured. The temperatures measured are then combined into a weighted mass flow average by either (1) using equal areas and multiplying each temperature by the fraction of total mass flow in its area or (2) using areas of size inversely proportional to mass flow and taking a simple arithmetic average of the temperatures in each. A means of mixing or selective sampling may be preferable to these cumbersome procedures. While mixing can occur from turbulence alone, **transposition** is much more effective. In transposition, the stream is divided into parts determined by the type of stratification, and alternate parts pass through one another.

Static Temperature Versus Total Temperature

When a fluid stream impinges on a temperature-sensing element such as a thermometer or thermocouple, the element is at a temperature greater than the true stream temperature. The difference is a fraction of the temperature equivalent of the stream velocity t_e.

$$t_e = \frac{3600V^2}{2g_c J c_p} \qquad (1)$$

where

t_e = temperature equivalent of stream velocity, °F
V = velocity of stream, fpm
g_c = gravitational constant = 32.174 $lb_m \cdot ft/lb_f \cdot s^2$
J = mechanical equivalent of heat, 778.3 $ft \cdot lb_f$/Btu
c_p = specific heat of stream at constant pressure, $Btu/lb_m \cdot °F$

Table 1 Temperature Measurement

Measurement Means	Application	Approximate Range, °F	Uncertainty, °F	Limitations
Liquid-in-glass thermometers				
Mercury-in-glass	Temperature of gases and liquids by contact	−36/1000	0.05 to 3.6	In gases, accuracy affected by radiation
Organic	Temperature of gases and liquids by contact	−330/400	0.05 to 3.6	In gases, accuracy affected by radiation
Gas thermometer	Primary standard	−456/1200	Less than 0.02	Requires considerable skill to use
Resistance thermometers				
Platinum	Precision; remote readings; temperature of fluids or solids by contact	−430/1800	Less than 0.0002 to 0.2	High cost; accuracy affected by radiation in gases
Rhodium-iron	Transfer standard for cryogenic applications	−460/−400	0.0002 to 0.2	High cost
Nickel	Remote readings; temperature by contact	−420/400	0.02 to 2	Accuracy affected by radiation in gases
Germanium	Remote readings; temperature by contact	−460/−400	0.0002 to 0.2	
Thermistors	Remote readings; temperature by contact	Up to 400	0.0002 to 0.2	
Thermocouples				
Pt-Rh/Pt (type S)	Standard for thermocouples on IPTS-68, not on ITS-90	32/2650	0.2 to 5	High cost
Au/Pt	Highly accurate reference thermometer for laboratory applications	−60/1800	0.1 to 2	High cost
Types K and N	General testing of high temperature; remote rapid readings by direct contact	Up to 2300	0.2 to 18	Less accurate than listed above thermocouples
Iron/Constantan (type J)	Same as above	Up to 1400	0.2 to 10	Subject to oxidation
Copper/Constantan (type T)	Same as above, especially suited for low temperature	Up to 660	0.2 to 5	
Ni-Cr/Constantan (type E)	Same as above, especially suited for low temperature	Up to 1650	0.2 to 13	
Beckman thermometers (metastatic)	For differential temperature in same applications as in glass-stem thermometer	10°F scale, used 32 to 212°F	0.01	Must be set for temperature to be measured
Bimetallic thermometers	For approximate temperature	−4/1200	2, usually much more	Time lag; unsuitable for remote use
Pressure-bulb thermometers				
Gas-filled bulb	Remote testing	−100/1200	4	Caution must be exercised so that installation is correct
Vapor-filled bulb	Remote testing	−25/500	4	Caution must be exercised so that installation is correct
Liquid-filled bulb	Remote testing	−60/2100	4	Caution must be exercised so that installation is correct
Optical pyrometers	For intensity of narrow spectral band of high-temperature radiation (remote)	1500 and up	30	
Radiation pyrometers	For intensity of total high-temperature radiation (remote)	Any range		
Seger cones (fusion pyrometers)	Approximate temperature (within temperature source)	1200/3600	90	
Triple points, freezing/melting points, and boiling points of materials	Standards	All except extremely high temperatures	Extremely precise	For laboratory use only

This fraction of the temperature equivalent of the velocity is the **recovery factor**, and it varies from 0.6 to 0.8°F for bare thermometers to 1.0°F for aerodynamically shielded thermocouples. For precise temperature measurement, each temperature sensor must be calibrated to determine its recovery factor. However, for most applications where air velocities are below 2000 fpm, the recovery factor can be omitted.

Various temperature sensors are available for temperature measurement in fluid streams. The principal sensors are the **static temperature thermometer**, which indicates true stream temperature but is cumbersome, and the **thermistor**, used for accurate temperature measurement within a limited range.

LIQUID-IN-GLASS THERMOMETERS

Any device that changes monotonically with temperature is a thermometer; however, the term usually signifies an ordinary liquid-in-glass temperature-indicating device. Mercury-filled thermometers have a useful range from −37.8°F, the freezing point of mercury, to about 1000°F, near which the glass usually softens. Lower temperatures can be measured with organic-liquid-filled thermometers (e.g., alcohol-filled), with ranges of −330 to 400°F. During manufacture, thermometers are roughly calibrated for at least two temperatures, often the freezing and boiling points of water; space between the calibration points is divided into desired scale divisions. Thermometers that are intended for precise measurement applications have scales etched into the glass that forms their stems. The probable error for as-manufactured, etched-stem thermometers is ±1 scale division. The highest quality mercury thermometers may have uncertainties of ±0.06 to ±4°F if they have been calibrated by comparison against primary reference standards.

Liquid-in-glass thermometers are used for many applications within the HVAC industry. Some of these uses include local temperature indication of process fluids related to HVAC systems, such as cooling and heating fluids and air.

The use of mercury-in-glass thermometers as temperature measurement standards is fairly common because of their relatively high accuracy and low cost. Such thermometers used as references must be calibrated on the ITS-90 by comparison in a uniform bath with a standard platinum resistance thermometer that has been calibrated either by the appropriate standards agency or by a laboratory that has direct traceability to the standards agency and the ITS-90. Such a calibration is necessary in order to determine the proper corrections to be applied to the scale readings. For application and calibration of liquid-in-glass thermometers, refer to NIST (1976, 1986).

Liquid-in-glass thermometers are calibrated by the manufacturer for total or partial stem immersion. If a thermometer calibrated for total immersion is used at partial immersion (i.e., with a portion of the liquid column at a temperature different from that of the bath), an emergent stem correction must be made. This correction can be calculated as follows:

$$\text{Stem correction} = Kn(t_b - t_s) \tag{2}$$

where

K = differential expansion coefficient of mercury or other liquid in glass. K is 0.00009 for Fahrenheit mercurial thermometers. For K values for other liquids and specific glasses, refer to Schooley (1992).
n = number of degrees that liquid column emerges from bath
t_b = temperature of bath, °F
t_s = average temperature of emergent liquid column of n degrees, °F

Sources of Thermometer Errors

A thermometer measuring gas temperatures can be affected by radiation from surrounding surfaces. If the gas temperature is approximately the same as that of the surrounding surfaces, radiation effects can be ignored. If the temperature differs considerably from that of the surroundings, radiation effects should be minimized by shielding or aspiration (ASME *Standard* PTC 193). **Shielding** may be provided by highly reflective surfaces placed between the thermometer bulb and the surrounding surfaces such that air movement around the bulb is not appreciably restricted (Parmelee and Huebscher 1946). Improper shielding can increase errors. **Aspiration** results from passing a high-velocity stream of air or gas over the thermometer bulb.

When a **thermometer well** within a container or pipe under pressure is required, the thermometer should fit snugly and be surrounded with a high thermal conductivity material (oil, water, or mercury, if suitable). Liquid in a long, thin-walled well is advantageous for rapid response to temperature changes. The surface of the pipe or container around the well should be insulated to eliminate heat transfer to or from the well.

Industrial thermometers are available for permanent installation in pipes or ducts. These instruments are fitted with metal guards to prevent breakage and are useful for many other purposes. The considerable heat capacity and conductance of the guards or shields can cause errors, however.

Allowing ample time for the thermometer to attain temperature equilibrium with the surrounding fluid prevents excessive errors in temperature measurements. When reading a liquid-in-glass thermometer, the eye should be kept at the same level as the top of the liquid column to avoid parallax.

RESISTANCE THERMOMETERS

Resistance thermometers depend on a change of the electrical resistance of a sensing element (usually metal) with a change in temperature; resistance increases with increasing temperature. The use of resistance thermometers largely parallels that of thermocouples, although readings are usually unstable above about 1000°F. Two-lead temperature elements are not recommended because they do not permit correction for lead resistance. Three leads to each resistor are necessary to obtain consistent readings, and four leads are preferred. Wheatstone bridge circuits or 6-1/2-digit multimeters can be used for measurements.

A typical circuit used by several manufacturers is shown in Figure 2. In this design, a differential galvanometer is used in which coils L and H exert opposing forces on the indicating needle. Coil L is in series with the thermometer resistance AB, and coil H is in series with the constant resistance *R*. As the temperature falls, the resistance of AB decreases, allowing more current to flow through coil L than through coil H. This causes an increase in the force exerted by coil L, pulling the needle down to a lower reading. Likewise, as the temperature rises, the resistance of AB increases, causing less current to flow through coil L than through coil H. This forces the indicating needle to a higher reading. Rheostat S must be adjusted occasionally to maintain a constant current.

The resistance thermometer is more costly to make and likely to have considerably longer response times than thermocouples. A resistance thermometer gives best results when used to measure steady or slowly changing temperature.

Resistance Temperature Devices

Resistance temperature devices (RTDs) are typically constructed from platinum, rhodium-iron, nickel, nickel-iron, tungsten, or copper. These devices are further characterized by their simple circuit designs, high degree of linearity, good sensitivity, and excellent stability. The choice of materials for an RTD usually depends on the intended application; temperature range, corrosion protection, mechanical stability, and cost are some of the selection criteria.

Platinum RTDs. Presently, for HVAC applications, RTDs constructed of platinum are the most widely used. Platinum is extremely stable and corrosion-resistant. Platinum RTDs are highly malleable and can thus be drawn into fine wires; they can also be manufactured at low cost as thin films. They have a high melting point and can be refined to a high degree of purity, thus attaining highly reproducible results. Due to these properties, platinum RTDs are used to define the ITS-90 for the range of 13.8033 K (triple point of equilibrium hydrogen) to 1234.93 K (freezing point of silver).

Platinum resistance temperature devices can measure the widest range of temperatures and are the most accurate and stable temperature sensors. Their resistance-temperature relationship is one of the most linear. The higher the purity of the platinum, the more stable and accurate the sensor. With high-purity platinum, primary grade platinum RTDs are capable of achieving reproducibility of ±0.00002°F, whereas the minimum uncertainty of a recently calibrated thermocouple is ±0.4°F.

Platinum RTD Design. The most widely used RTD is designed with a resistance of 100 Ω at 32°F (R_0 = 100 Ω). Other RTDs are available that use lower resistances at temperatures above 1100°F. The lower the resistance value, the faster the response time for sensors of the same size.

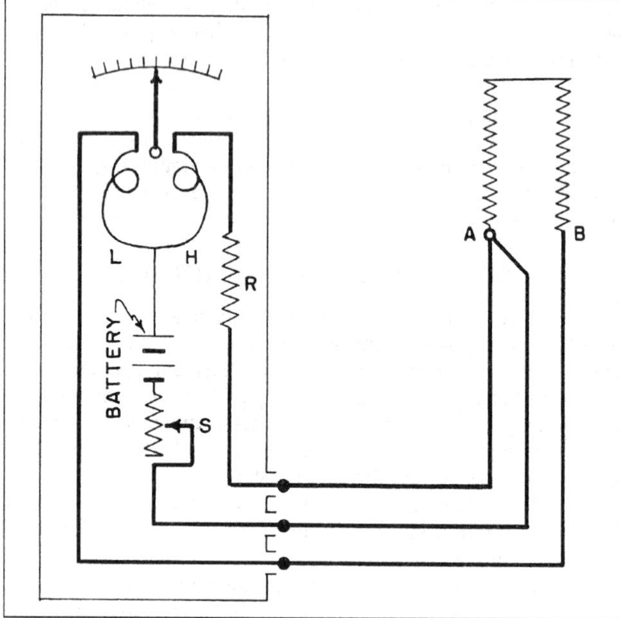

Fig. 2 Typical Resistance Thermometer Circuit

Thin-Film Platinum RTDs. Thin-film 1000 Ω platinum RTDs are readily available. They have the excellent linear properties of lower resistance platinum RTDs and are more cost-effective because they are mass produced and have lower platinum purity. However, the problem with many platinum RTDs with R_0 values of greater than 100 Ω is the difficulty in obtaining transmitters or electronic interface boards from sources other than the RTD manufacturer. In addition to a nonstandard interface, higher R_0 value platinum RTDs may have higher self-heating losses if the excitation current is not controlled properly.

Thin-film RTDs have the advantages of lower cost and smaller sensor size. They are specifically adapted to surface mounting. Thin-film sensors tend to have an accuracy limitation of ±0.1% or ±0.2°F. This may prove to be adequate for most HVAC applications; only in tightly controlled facilities may users wish to install the standard wire-wound platinum RTDs with accuracies of 0.01% or ±0.02°F (these are available upon special request for certain temperature ranges).

Assembly and Construction. Regardless of the R_0 resistance value of RTDs, their assembly and construction are relatively simple. The electrical connections come in three basic types, depending on the number of wires to be connected to the resistance measurement circuitry. Two, three, or four wires are used for electrical connection using a Wheatstone bridge or a variation of it (Figure 3).

In the basic two-wire configuration, the resistance of the RTD is measured through the two connecting wires. Because the connecting wires extend from the site of the temperature measurement, any additional changes in resistivity due to a change in temperature may affect the measured resistance. Three- and four-wire assemblies are built to compensate for the connecting lead resistance values. The original three-wire circuit improved the resistance measurement by adding a compensating wire to the voltage side of the circuit. This helps reduce part of the connecting wire resistance. When more accurate measurements (better than ±0.2°F) are required, the four-wire bridge is recommended. The four-wire bridge eliminates all connecting wire resistance errors.

All the bridges discussed here are direct current (dc) circuits and were used extensively until the advent of precision alternating current (ac) circuits using microprocessor-controlled ratio transformers, dedicated analog-to-digital converters, and other solid-state

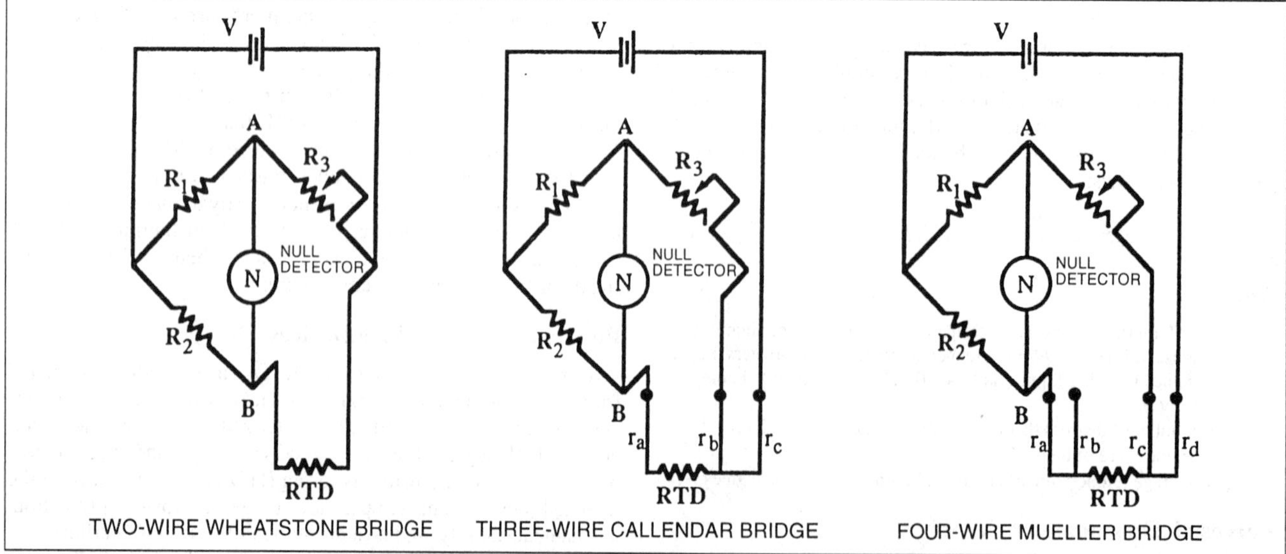

Fig. 3 Typical Resistance Temperature Device Bridge Circuits

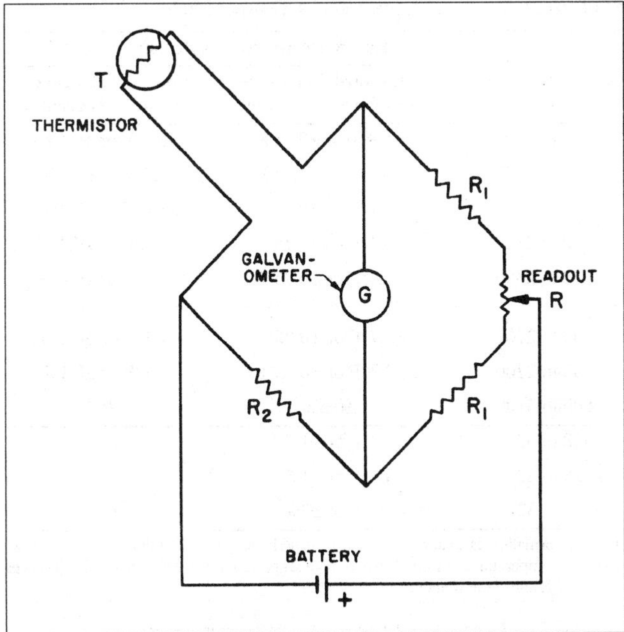

Fig. 4 Basic Thermistor Circuit

devices that measure resistance with uncertainties of less than 1 ppm. Resistance measurement technology now allows more portable thermometers, lower cost, ease of use, and high-precision temperature measurement in industrial uses.

Thermistors

Certain semiconductor compounds (usually sintered metallic oxides) exhibit large changes in resistance with temperature, usually decreasing as the temperature increases. For use, the thermistor element may be connected by lead wires into a galvanometer bridge circuit and calibrated. Alternatively, a 6-1/2-digit multimeter and a constant-current source with a means for reversing the current to eliminate thermal electromotive force (emf) effects may also be used. This method of measurement is easier and faster, and it may be more precise and accurate. Thermistors are usually applied to electronic temperature compensation circuits, such as thermocouple reference junction compensation, or to other applications where high resolution and limited operating temperature ranges exist. Figure 4 illustrates a typical thermistor circuit.

Semiconductor Devices

In addition to the positive resistance coefficient RTDs and the negative resistance coefficient thermistor, there are two other types of devices that vary resistance or impedance with temperature. Although the principle of their operation has long been known, their reliability was questioned due to imprecise manufacturing techniques. Improvements in silicon microelectronics manufacturing techniques have brought semiconductors to the point where low-cost, precise temperature sensors are commercially available.

Elemental Semiconductors. Due to controlled doping of impurities into elemental germanium, a germanium semiconductor is a reliable temperature sensor for cryogenic temperature measurement in the range of 1.8 to 150°R.

Junction Semiconductors. The first simple junction semiconductor device consisted of a single diode or transistor, in which the forward-connected base emitter voltage is very sensitive to temperature. Today the more common form is a pair of diode-connected transistors, which make the device suitable for ambient temperature measurement. Applications include thermocouple reference junction compensation.

The primary advantages of silicon transistor temperature sensors are their extreme linearity and exact R_0 value. Another advantage is the incorporation of signal conditioning circuitry into the same device as the sensor element. As with thermocouples, these semiconductors require highly precise manufacturing techniques, extremely precise voltage measurements, multiple-point calibration, and temperature compensation to achieve an accuracy as high as ±0.02°F, but with a much higher cost. Lower cost devices achieve accuracies of ±0.2°F using mass manufacturing techniques and single-point calibration. A mass-produced silicon temperature sensor can be interchanged easily. If one device fails, only the sensor element need be changed. Electronic circuitry can be used to recalibrate the new device.

Winding Temperature. The winding temperature of electrical operating equipment is usually determined from the resistance change of these windings in operation. With copper windings, the relation between these parameters is

$$\frac{R_1}{R_2} = \frac{148 + t_1}{148 + t_2} \tag{3}$$

where

R_1 = winding resistance at temperature t_1, Ω
R_2 = winding resistance at temperature t_2, Ω
t_1, t_2 = winding temperatures, °F

The classical method of determining winding temperature is to measure the equipment when it is inoperative and temperature-stabilized at room temperature. After the equipment has operated sufficiently to cause temperature stabilization under load conditions, the winding resistance should be measured again. The latter value is obtained by taking resistance measurements at known short time intervals after shutdown. These values may be extrapolated to zero time to indicate the winding resistance at the time of shutdown. The obvious disadvantage of this method is that the device must be shut down to determine winding temperature. A circuit described by Seely (1955), however, makes it possible to measure resistances while the device is operating.

THERMOCOUPLES

When two wires of dissimilar metals are joined by soldering, welding, or twisting, they form a thermocouple junction or **thermojunction**. An emf that depends on the wire materials and the junction temperature exists between the wires. This is known as the **Seebeck voltage**.

Thermocouples for temperature measurement yield less precise results than platinum resistance thermometers, but except for glass thermometers, thermocouples are the most common instruments of temperature measurement for the range of 32 to 1800°F. Due to their low cost, moderate reliability, and ease of use, thermocouples continue to maintain widespread acceptance.

The most commonly used thermocouples in industrial applications are assigned letter designations. The tolerances of such commercially available thermocouples are given in Table 2.

Because the measured emf is a function of the difference in temperature and the type of dissimilar metals used, a known temperature at one junction is required, whereas the remaining junction temperature may be calculated. It is common practice to call the one with a known temperature the (cold) **reference** junction and the one with the unknown temperature the (hot) **measured** junction. The reference junction is typically kept at a reproducible temperature, such as the ice point of water.

Various systems are used to maintain the reference junction temperature—a mixture of ice and water contained in an insulated flask or commercially available thermoelectric coolers to maintain the ice point temperature automatically within a reference chamber. When

Table 2 Thermocouple Tolerances on Initial Values of Electromotive Force Versus Temperature

Thermocouple Type	Material Identification	Temperature Range, °F	Reference Junction Tolerance at 32°F[a]: Standard Tolerance (whichever is greater)	Reference Junction Tolerance at 32°F[a]: Special Tolerance (whichever is greater)
T	Copper versus Constantan	32 to 700	±1.8°F or ±0.75%	±0.9°F or ±0.4%
J	Iron versus Constantan	32 to 1400	±4°F or ±0.75%	±2°F or ±0.4%
E	Nickel-10% Chromium versus Constantan	32 to 1600	±3.1°F or ±0.5%	±1.8°F or ±0.4%
K	Nickel-10% Chromium versus 5% Aluminum, Silicon	32 to 2300	±4°F or ±0.75%	±2°F or ±0.4%
N	Nickel-14% Chromium, 1.5% Silicon versus Nickel-4.5% Silicon, 0.1% Magnesium	32 to 2300	±4°F or ±0.75%	±2°F or ±0.4%
R	Platinum-13% Rhodium versus Platinum	32 to 2700	±2.7°F or ±0.25%	±1.1°F or ±0.1%
S	Platinum-10% Rhodium versus Platinum	32 to 2700	±2.7°F or ±0.25%	±1.1°F or ±0.1%
B	Platinum-30% Rhodium versus Platinum-6% Rhodium	1600 to 3100	±0.5%	±0.25%
T[b]	Copper versus Constantan	−328 to 32	±1.8°F or ±1.5%	c
E[b]	Nickel-10% Chromium versus Constantan	−328 to 32	±3.1°F or ±1%	c
K[b]	Nickel-10% Chromium versus 5% Aluminum, Silicon	−328 to 32	±4°F or ±2%	c

Source: ASTM *Standard* E 230, Temperature-Electromotive Force (EMF) Tables for Standardized Thermocouples.

[a]Tolerances in this table apply to new thermocouple wire, normally in the size range of 0.01 to 0.1 in. diameter and used at temperatures not exceeding the recommended limits. Thermocouple wire is available in two grades: standard and special.

[b]Thermocouples and thermocouple materials are normally supplied to meet the tolerance specified in the table for temperatures above 32°F. The same materials, however, may not fall within the tolerances given in the second section of the table when operated below freezing (32°F). If materials are required to meet tolerances at subfreezing temperatures, the purchase order must state so.

[c]Little information is available to justify establishing special tolerances for below-freezing temperatures. Limited experience suggests the following special tolerances for types E and T thermocouples:

Type E −328 to 32°F; ±2°F or ±0.5% (whichever is greater)

Type T −328 to 32°F; ±1°F or ±0.8% (whichever is greater)

These tolerances are given only as a guide for discussion between purchaser and supplier.

these systems cannot be used in an application, measuring instruments with automatic reference junction temperature compensation may be used.

As previously described, the principle for measuring temperature with a thermocouple is based on the accurate measurement of the Seebeck voltage. The acceptable dc voltage measurement methods are (1) millivoltmeter, (2) millivolt potentiometer, and (3) a high-input impedance digital voltmeter. Many digital voltmeters include built-in software routines for the direct calculation and display of temperature. Regardless of the method selected, many options to simplify the measurement process are available.

Solid-state digital readout devices in combination with a milli-volt- or microvoltmeter, as well as packaged thermocouple readouts with built-in cold junction and linearization circuits, are available. The latter requires a proper thermocouple to provide direct meter reading of temperature. Accuracy approaching or surpassing that of potentiometers can be attained, depending on the instrument quality. This method is popular because it eliminates the null balancing requirement and reads temperature directly in a digital readout.

Wire Diameter and Composition

Thermocouple wire is selected by considering the temperature to be measured, the corrosion protection afforded to the thermocouple, and the precision and service life required. Type T thermocouples are suitable for temperatures up to 700°F; type J, up to 1400°F; and types K and N, up to 2300°F. Higher temperatures require noble metal thermocouples (type S, R, or B), which have a higher initial cost and do not develop as high an emf as the base metal thermocouples. Thermocouple wires of the same type have small compositional variation from lot to lot from the same manufacturer and especially among different manufacturers. Consequently, calibrating samples from each wire spool is essential for precision. Calibration data on wire may be obtained from the manufacturer.

Reference functions are available for relating temperature and emf of letter-designated thermocouple types. Such functions are easy to use with computers. The functions depend on thermocouple type and on the range of temperature; they are used to generate reference tables of emf as a function of temperature but are not well suited for calculating temperatures directly from values of emf. Approximate inverse functions are available, however, for calculating temperature and are of the form

$$t = \sum_{i=0}^{n} a_i E^i \tag{4}$$

where t = temperature, a = thermocouple constant, and E = voltage. Burns et al. (1992) give the reference functions and approximate inverses for all letter-designated thermocouples.

The emf of a thermocouple, as measured with a high-input impedance device, is independent of the diameters of its constituent wires. Thermocouples with small-diameter wires respond faster to temperature changes and are less affected by radiation than larger ones. Large-diameter wire thermocouples, however, are necessary for high-temperature work when wire corrosion is a problem. For use in heated air or gases, thermocouples are often shielded and sometimes aspirated. An arrangement for avoiding error due to radiation involves using several thermocouples of different wire sizes and estimating the true temperature by extrapolating readings to zero diameter.

With thermocouples, temperatures can be indicated or recorded remotely on conveniently located instruments. Because thermocouples can be made of small-diameter wire, they can be used to measure temperatures within thin materials, within narrow spaces, or in otherwise inaccessible locations.

Multiple Thermocouples

Thermocouples in series, with alternate junctions maintained at a common temperature, produce an emf that, when divided by the number of thermocouples, gives the average emf corresponding to the temperature difference between two sets of junctions. This series arrangement of thermocouples, often called a **thermopile**, is used to increase sensitivity and is often used for measuring small temperature changes and differences.

Connecting a number of thermocouples of the same type in parallel with a common reference junction is useful for obtaining an average temperature of an object or volume. In such measurements, however, it is important that the electrical resistances of the individual thermocouples be the same. The use of thermocouples in series and parallel arrangements is discussed in ASTM *Manual* 12.

Surface Temperature Measurement

The thermocouple is useful in determining surface temperature. It can be attached to a metal surface in several ways. For permanent installations, soldering, brazing, or peening is suggested. For peening, a small hole is drilled and the thermocouple measuring junction is driven into it. For temporary arrangements, thermocouples can be attached by tape, adhesive, or putty-like material. For boiler or furnace surfaces, furnace cement should be used. To minimize the possibility of error caused by heat conduction along wires, a surface thermocouple should be made of fine wires placed in close contact with the surface being measured for about an inch from the junction to ensure good thermal contact. The wires must be insulated electrically from each other and from the metal surface (except at the junction).

Thermocouple Construction

The thermocouple wires are typically insulated with fibrous glass, fluorocarbon resin, or ceramic insulators. In another form of thermocouple, the thermocouple wires are insulated with compacted ceramic insulation inside a metal sheath. This form of thermocouple provides both mechanical protection and protection from stray electromagnetic fields. The measuring junction may be exposed or enclosed within the metal sheath. An enclosed junction may be either grounded or ungrounded to the metal sheath.

For the exposed junction type, the measuring junction is in direct contact with the process stream; it is therefore subject to corrosion or contamination but provides a fast temperature response. The grounded enclosed junction type, in which the thermocouple wires are welded to the metal sheath, provides electrical grounding, as well as mechanical and corrosion protection. This type, however, has a slower response time than the exposed junction type. With the ungrounded enclosed junction construction, the response time is even slower, but the thermocouple wires are isolated electrically and are less susceptible to some forms of mechanical strain than those with grounded construction.

INFRARED RADIOMETERS

Infrared radiation thermometers, also known as remote temperature sensors (Hudson 1969), permit noncontact measurement of surface temperature over a wide range. In these instruments, radiant flux from the observed object is focused by an optical system onto an infrared detector that generates an output signal proportional to the incident radiation that can be read from a meter or display unit. Point and scanning radiometers are available; the latter are able to display the temperature variation existing within the field of view.

Radiometers are usually classified according to the detector used—either thermal or photon. In thermal detectors, a change in electrical property is caused by the heating effect of the incident radiation. Examples of thermal detectors are the thermocouple, the thermopile, and metallic and semiconductor bolometers. In photon detectors, a change in electrical property is caused by the surface absorption of incident photons. Because these detectors do not require an increase in temperature for activation, their response time is much shorter than that of thermal detectors. Scanning radiometers usually use photon detectors.

A radiometer only measures the power level of the radiation incident on the detector; this incident radiation is a combination of the thermal radiation emitted by the object and the surrounding background radiation reflected from the surface of the object. An accurate measurement of the temperature, therefore, requires knowledge of the long-wavelength emissivity of the object as well as the effective temperature of the thermal radiation field surrounding the object. Calibration against an internal or external source of known temperature and emissivity is required in order to obtain true surface temperature from the radiation measurements.

The temperature resolution of a radiometer decreases as the object temperature decreases. For example, a radiometer that can resolve a temperature difference of 0.5°F on an object near 70°F may only resolve a difference of 2°F on an object at 32°F.

INFRARED THERMOGRAPHY

Infrared thermography is the discipline concerned with the acquisition and analysis of thermal information in the form of images from an infrared imaging system. An infrared imaging system consists of (1) an infrared television camera and (2) a display unit. The infrared television camera scans a surface and senses the self-emitted and reflected radiation viewed from the surface. The display unit contains either a cathode-ray tube (CRT) that displays a gray-tone or color-coded thermal image of the surface or a color liquid crystal display (LCD) screen. A photograph of the image on the CRT is called a **thermogram**. An introductory treatise on infrared thermography is given by Paljak and Pettersson (1972).

Thermography has been used successfully to detect missing insulation and air infiltration paths in building envelopes (Burch and Hunt 1978). Standard practices for conducting thermographic inspections of buildings are given in ASTM *Standard* C 1060. A technique for quantitatively mapping the heat loss in building envelopes is given by Mack (1986).

Aerial infrared thermography of buildings is effective in identifying regions of an individual built-up roof that have wet insulation (Tobiasson and Korhonen 1985), but it is ineffective in ranking a group of roofs according to their thermal resistance (Goldstein 1978, Burch 1980). In this latter application, the emittances of the separate roofs and outdoor climate (i.e., temperature and wind speed) throughout the microclimate often produce changes in the thermal image that may be incorrectly attributed to differences in thermal resistance.

Industrial applications include locating defective or missing pipe insulation in buried heat distribution systems, surveys of manufacturing plants to quantify energy loss from equipment, and locating defects in coatings (Bentz and Martin 1987).

HUMIDITY MEASUREMENT

Any instrument capable of measuring the humidity or psychrometric state of air is a hygrometer, and many are available. The indication sensors used on the instruments respond to different moisture property contents. These responses are related to factors such as wet-bulb temperature, relative humidity, humidity (mixing) ratio, dew point, and frost point.

Table 3 lists instruments for measuring humidity. Each is capable of accurate measurement under certain conditions and within specific limitations. The following sections describe various instruments used to measure humidity.

PSYCHROMETERS

A typical industrial psychrometer consists of a pair of matched electrical or mechanical temperature sensors, one of which is kept wet with a moistened wick. A blower aspirates the sensor, which lowers the temperature at the moistened temperature sensor. The lowest temperature depression occurs when the evaporation rate required to saturate the moist air adjacent to the wick is constant. This is a steady-state, open-loop, nonequilibrium process, which depends on the purity of the water, the cleanliness of the wick, the ventilation rate, radiation effects, the size and accuracy of the temperature sensors, and the transport properties of the gas.

Table 3 Humidity Sensor Properties

Type of Sensor	Sensor Category	Method of Operation	Approximate Range	Some Uses	Approximate Accuracy
Psychrometer	Evaporative cooling	Temperature measurement of wet bulb	32 to 180°F	Measurement, standard	±3 to ±7% rh
Adiabatic saturation psychrometer	Evaporative cooling	Temperature measurement of thermodynamic wet bulb	40 to 85°F	Measurement, standard	±0.2 to ±2% rh
Chilled mirror	Dew point	Optical determination of moisture formation	−110 to 200°F dp	Measurement, control, meteorology	±0.4 to ±4°F
Heated saturated salt solution	Water vapor pressure	Vapor pressure depression in salt solution	−20 to 160°F dp	Measurement, control, meteorology	±3°F
Hair	Mechanical	Dimensional change	5 to 100% rh	Measurement, control	±5% rh
Nylon	Mechanical	Dimensional change	5 to 100% rh	Measurement, control	±5% rh
Dacron thread	Mechanical	Dimensional change	5 to 100% rh	Measurement	±7% rh
Goldbeater's skin	Mechanical	Dimensional change	5 to 100% rh	Measurement	±7% rh
Cellulosic materials	Mechanical	Dimensional change	5 to 100% rh	Measurement, control	±5% rh
Carbon	Mechanical	Dimensional change	5 to 100% rh	Measurement	±5% rh
Dunmore type	Electrical	Impedance	7 to 98% rh at 40 to 140°F	Measurement, control	±1.5% rh
Ion exchange resin	Electrical	Impedance or capacitance	10 to 100% rh at −40 to 190°F	Measurement, control	±5% rh
Porous ceramic	Electrical	Impedance or capacitance	Up to 400°F	Measurement, control	±1 to ±1.5% rh
Aluminum oxide	Electrical	Capacitance	5 to 100% rh	Measurement, control	±3% rh
Aluminum oxide	Electrical	Capacitance	−110 to 140°F dp	Trace moisture measurement, control	±2°F dp
Electrolytic hygrometer	Electrical	Capacitance			
Coulometric	Electrolytic cell	Electrolyzes due to adsorbed moisture	1 to 1000 ppm	Measurement	
Infrared laser diode	Electrical	Optical diodes	0.1 to 100 ppm	Trace moisture measurement	±0.1 ppm
Surface acoustic wave	Electrical	SAW attenuation	85 to 98% rh	Measurement, control	±1% rh
Piezoelectric	Mass sensitive	Mass changes due to adsorbed moisture	−100 to 0°F	Trace moisture measurement, control	±2 to ±10°F dp
Radiation absorption	Moisture absorption	Moisture absorption of UV or IR radiation	0 to 180°F dp	Measurement, control, meteorology	±4°F dp, ±5% rh
Gravimetric	Direct measurement of mixing ratio	Comparison of sample gas with dry airstream	120 to 20,000 ppm mixing ratio	Primary standard, research and laboratory	±0.13% of reading
Color change	Physical	Color changes	10 to 80% rh	Warning device	±10% rh

Notes:
1. This table does not encompass all of the available technology for the measurement of humidity.
2. The approximate range for the device types listed is based on surveys of device manufacturers.
3. The approximate accuracy is based on manufacturers' data.
4. Presently, the National Institute of Standards and Technology (NIST) will only certify instruments whose operating range is within −103 to 212°F dew point.

ASHRAE *Standard* 41.6 recommends an airflow over both the wet and dry bulbs of 600 to 1000 fpm for transverse ventilation and 300 to 500 fpm for axial ventilation.

The **sling psychrometer** consists of two thermometers mounted side by side in a frame fitted with a handle for whirling the device through the air. The thermometers are spun until their readings become steady. In the **ventilated** or **aspirated psychrometer**, the thermometers remain stationary, and a small fan, blower, or syringe moves the air across the thermometer bulbs. Various designs are used in the laboratory, and commercial models are available.

Other temperature sensors, such as thermocouples and thermistors, are also used and can be adapted for recording temperatures or for use where a small instrument is required. Small-diameter wet-bulb sensors operate with low ventilation rates.

Charts and tables showing the relationship between the temperatures and humidity are available. Data are usually based on a barometric pressure equal to one standard atmosphere. To meet special needs, charts can be produced that apply to nonstandard pressure (e.g., the ASHRAE 7500 ft psychrometric chart). Alternatively, mathematical calculations can be made (Kusuda 1965). Uncertainties of 3 to 7% rh are typical for psychrometer-based derivation. The degree of uncertainty is a function of the accuracy of the temperature measurements, wet and dry bulb, knowledge of the barometric pressure, and conformance to accepted operational procedures such as those outlined in ASHRAE *Standard* 41.6.

In air temperatures below 32°F, the water on the wick may either freeze or supercool. Because the wet-bulb temperature is different for ice and water, the state must be known and the proper chart or table used. Some operators remove the wick from the wet-bulb for freezing conditions and dip the bulb in water a few times; this allows water to freeze on the bulb between dips, forming a film of ice. Because the wet-bulb depression is slight at low temperatures, precise temperature readings are essential. A psychrometer can be used at high temperatures, but if the wet-bulb depression is large, the wick must remain wet and water supplied to the wick must be cooled so as not to influence the wet-bulb temperature by carrying sensible heat to it (Richardson 1965, Worrall 1965).

Greenspan and Wexler (1968) and Wentzel (1961) developed devices to measure adiabatic saturation temperature.

DEW-POINT HYGROMETERS

Condensation Dew-Point Hygrometers

The condensation (chilled mirror) dew-point hygrometer is an accurate and reliable instrument with a wide humidity range. However, these features are obtained through an increase in complexity and cost compared to the psychrometer. In the condensation hygrometer, a surface is cooled (thermoelectrically, mechanically, or chemically) until dew or frost begins to condense out. The condensate surface is maintained electronically in vapor pressure equilibrium with the surrounding gas, while surface condensation is detected by optical, electrical, or nuclear techniques. The measured surface temperature is then the dew-point temperature.

The largest source of error in a condensation hygrometer stems from the difficulty in measuring condensate surface temperature accurately. Typical industrial versions of the instrument are accurate to ±1.0°F over wide temperature spans. With proper attention to the condensate surface temperature measuring system, errors can be reduced to about ±0.4°F. Condensation hygrometers can be made surprisingly compact using solid-state optics and thermoelectric cooling.

Wide span and minimal errors are two of the main features of this instrument. A properly designed condensation hygrometer can measure dew points from 200°F down to frost points of −100°F. Typical condensation hygrometers can cool to 150°F below the ambient temperature, establishing lower limits of the instrument to dew points corresponding to approximately 0.5% rh. Accuracies for measurements above −40°F can be ±2°F or better, deteriorating to ±4°F at lower temperatures.

The response time of a condensation dew-point hygrometer is usually specified in terms of its cooling/heating rate, typically 4°F/s for thermoelectric cooled mirrors. This makes it somewhat faster than a heated salt hygrometer. Perhaps the most significant feature of the condensation hygrometer is its fundamental measuring technique, which essentially renders the instrument self-calibrating. For calibration, it is necessary only to manually override the surface cooling control loop, causing the surface to heat, and witness that the instrument recools to the same dew point when the loop is closed. Assuming that the surface temperature measuring system is correct, this is a reasonable check on the instrument's performance.

Although condensation hygrometers can become contaminated, they can easily be cleaned and returned to service with no impairment to performance.

Salt-Phase Heated Hygrometers

Another instrument in which the temperature varies with ambient dew-point temperature is variously designated as a self-heating salt-phase transition hygrometer or a heated electrical hygrometer. This device usually consists of a tubular substrate covered by glass fiber fabric, with a spiral bifilar winding for electrodes. The surface is covered with a salt solution, usually lithium chloride. The sensor is connected in series with a ballast and a 24 V (ac) supply. When the instrument is in operation, electrical current flowing through the salt film heats the sensor. The electrical resistance characteristics of the salt are such that a balance is reached with the salt at a critical moisture content corresponding to a saturated solution. The sensor temperature adjusts automatically so that the water vapor pressures of the salt film and ambient atmosphere are equal.

With lithium chloride, this sensor cannot be used to measure relative humidity below approximately 12% (the equilibrium relative humidity of this salt), and it has an upper dew-point limit of about 160°F. The regions of highest precision are between −10 and 93°F, and above 105°F dew point. Another problem is that the lithium chloride solution can be washed off when exposed to water. In addition, this type of sensor is subject to contamination problems, which limits its accuracy. Its response time is also very slow; it takes approximately 2 min for a 67% step change.

MECHANICAL HYGROMETERS

Many organic materials change in dimension with changes in humidity; this action is used in a number of simple and effective humidity indicators, recorders, and controllers (see Chapter 42 of the 1995 *ASHRAE Handbook—Applications*). They are coupled to pneumatic leak ports, mechanical linkages, or electrical transduction elements to form hygrometers.

Commonly used organic materials are human hair, nylon, Dacron, animal membrane, animal horn, wood, and paper. Their inherent nonlinearity and hysteresis must be compensated for within the hygrometer. These devices are generally unreliable below 32°F. The response is generally inadequate for monitoring a changing process. Responses can be affected significantly by exposure to extremes of humidity. Mechanical hygrometers require initial calibration and frequent recalibration; however, they are useful because they can be arranged to read relative humidity directly, and they are simpler and less expensive than most other types.

ELECTRICAL IMPEDANCE AND CAPACITANCE HYGROMETERS

Many substances adsorb or lose moisture with changing relative humidity and exhibit corresponding changes in electrical impedance or capacitance.

Dunmore Hygrometers

This sensor consists of dual electrodes on a tubular or flat substrate; it is coated with a film containing salt, such as lithium chloride, in a binder to form an electrical connection between windings. The relation of sensor resistance to humidity is usually represented by graphs. Because the sensor is highly sensitive, the graphs are a series of curves, each for a given temperature, with intermediate values found by interpolation. Several resistance elements, called Dunmore elements, cover a standard range. Systematic calibration is essential because the resistance grid varies with time and contamination as well as with exposure to temperature and humidity extremes.

Polymer Film Electronic Hygrometers

These devices consist of a hygroscopic organic polymer deposited by means of thin or thick film processing technology on a water-permeable substrate. Both capacitance and impedance sensors are available. The impedance devices may be either ionic or electronic conduction types. These hygrometers typically have integrated circuits that provide temperature correction and signal conditioning. The primary advantages of this sensor technology are small size; low cost; fast response times (on the order of 1 to 120 s for 64% change in relative humidity); and good accuracy over the full range, including the low end (1 to 15% h), where most other devices are less accurate.

Ion Exchange Resin Electric Hygrometers

A conventional ion exchange resin consists of a polymer having a high relative molecular mass and polar groups of positive or negative charge in cross-link structure. Associated with these polar groups are ions of opposite charge that are held by electrostatic forces to the fixed polar groups. In the presence of water or water vapor, the electrostatically held ions become mobile; thus, when a voltage is impressed across the resin, the ions are capable of electrolytic conduction. The **Pope cell** is one example of an ion exchange element. It is a wide-range sensor, typically covering 15 to 95% rh; therefore, one sensor can be used where several Dunmore elements would be required. The Pope cell, however, has a nonlinear characteristic from approximately 1000 Ω at 100% rh to several megohms at 10% rh.

Impedance-Based Porous Ceramic Electronic Hygrometers

Using the adsorption characteristics of oxides, humidity-sensitive ceramic oxide devices employ either ionic or electronic measurement techniques to relate adsorbed water to relative humidity. Ionic conduction is produced by dissociation of water molecules forming surface hydroxyls. The dissociation causes migration of protons such that the impedance of the device decreases with increasing water content. The ceramic oxide is sandwiched between porous metal electrodes that connect the device to an impedance-measuring circuit for linearizing and signal conditioning. These sensors have excellent sensitivity, are resistant to contamination and high temperature (up to 400°F), and may get fully wet without sensor degradation. These sensors are accurate to about ±1.5% rh, and ±1% rh when temperature compensated. These sensors have a moderate cost.

Aluminum Oxide Capacitive Sensor

This sensor consists of an aluminum strip that is anodized by a process that forms a porous oxide layer. A very thin coating of cracked chromium or gold is then evaporated over this structure. The aluminum base and the cracked chromium or gold layer form the two electrodes of what is essentially an aluminum oxide capacitor.

Water vapor is rapidly transported through the cracked chromium or gold layer and equilibrates on the walls of the oxide pores in a manner functionally related to the vapor pressure of water in the atmosphere surrounding the sensor. The number of water molecules adsorbed on the oxide structure determines the capacitance between the two electrodes.

ELECTROLYTIC HYGROMETERS

In electrolytic hygrometers, air is passed through a tube, where moisture is adsorbed by a highly effective desiccant (usually phosphorous pentoxide) and electrolyzed. The airflow is regulated to 0.0035 cfm at a standard temperature and pressure. As the incoming water vapor is absorbed by the desiccant and electrolyzed into hydrogen and oxygen, the current of electrolysis determines the mass of water vapor entering the sensor. The flow rate of the entering gas is controlled precisely to maintain a standard sample mass flow rate into the sensor. The instrument is usually designed for use with moisture-air ratios in the range of less than 1 ppm to 1000 ppm but can be used with higher humidities.

PIEZOELECTRIC SORPTION

This hygrometer compares the changes in frequency of two hygroscopically coated quartz crystal oscillators. As the mass of the crystal changes due to the absorption of water vapor, the frequency changes. The amount of water sorbed on the sensor is a function of relative humidity (i.e., partial pressure of water as well as ambient temperature).

A commercial version uses a hygroscopic polymer coating on the crystal. The humidity is measured by monitoring the change in the vibration frequency of the quartz crystal when the crystal is alternately exposed to wet and dry gas.

SPECTROSCOPIC (RADIATION ABSORPTION) HYGROMETERS

Radiation absorption devices operate on the principle that selective absorption of radiation is a function of frequency for different media. Water vapor absorbs **infrared** radiation at 2 to 3 μm wavelengths and **ultraviolet** radiation centered about the Lyman-alpha line at 0.122 μm. The amount of absorbed radiation is directly related to the absolute humidity or water vapor content in the gas mixture according to Beer's law. The basic unit consists of an energy source and optical system for isolating wavelengths in the spectral region of interest and a measurement system for determining the attenuation of radiant energy caused by the water vapor in the optical path. The absorbed radiation is measured extremely quickly and independent of the degree of saturation of the gas mixture. Response times of 0.1 to 1 s for 90% change in moisture content are common. Spectroscopic hygrometers are primarily used where a noncontact application is required; this may include atmospheric studies, industrial drying ovens, and harsh environments. The primary disadvantages of this device are its high cost and relatively large size.

GRAVIMETRIC HYGROMETERS

Humidity levels can be measured by extracting and finding the mass of water vapor in a known quantity or atmosphere. For precise laboratory work, powerful desiccants, such as phosphorous pentoxide and magnesium perchlorate, are used for the extraction process; for other purposes, calcium chloride or silica gel is satisfactory.

When the highest level of accuracy is required, the gravimetric hygrometer, developed and maintained by NIST, is the ultimate in the measurement hierarchy. The gravimetric hygrometer gives the absolute water vapor content, where the mass of the absorbed water and the precise measurement of the gas volume associated with the water vapor determine the mixing ratio or absolute humidity of the sample. This system has been chosen as the primary standard because the required measurements of mass, temperature, pressure, and volume can be made with extreme precision. However, its complexity and required attention to detail limit the usefulness of the gravimetric hygrometer.

CALIBRATION

For many hygrometers, the need for recalibration depends on the accuracy required, the stability of the sensor, and the conditions to which the sensor is being subjected. Many hygrometers should be calibrated regularly by exposure to an atmosphere maintained at a known humidity and temperature, or by comparison with a transfer standard hygrometer. Complete calibration usually requires observation of a series of temperatures and humidities. Methods for producing known humidities include saturated salt solutions (Greenspan 1977, Huang and Whetstone 1985); sulfuric acid solutions, and mechanical systems, such as the divided flow, two-pressure (Amdur 1965); two-temperature (Till and Handegord 1960); and NIST two-pressure humidity generator (Hasegawa 1976). All these systems rely on precise methods of temperature and pressure control within a controlled environment to produce a known humidity, usually with accuracies of 0.5 to 1.0%. The operating range for the precision generator is typically 5 to 95% rh.

PRESSURE MEASUREMENT

Pressure is the force exerted per unit area by a medium, generally a liquid or gas. Pressure so defined is sometimes called **absolute pressure**. Thermodynamic and material properties are expressed in terms of absolute pressures; thus, the properties of a refrigerant will be given in terms of absolute pressures. **Vacuum** refers to pressures below atmospheric.

Differential pressure is the difference between two absolute pressures. In many cases, the differential pressure can be very small compared to either of the absolute pressures (these are often referred to as low-range, high-line differential pressures). A common example of differential pressure is the pressure drop, or difference between inlet and outlet pressures, across a filter or flow element.

Gage pressure is a special case of differential pressure where one of the pressures (the reference pressure) is atmospheric pressure. Many pressure gages, including most refrigeration test sets, are designed to make gage pressure measurements, and there are probably more gage pressure measurements made than any other. Gage pressure measurements are often used as surrogates for absolute pressures. However, because of variations in atmospheric pressure due to elevation (atmospheric pressure in Denver, Colorado, is about 81% of sea-level pressure) and weather changes, the measurement of gage pressures to determine absolute pressures can significantly restrict the accuracy of the measured pressure, unless corrections are made for the local atmospheric pressure at the time of the measurement.

Pressures can be further classified as static or dynamic. **Static pressures** have a small or undetectable change with time; **dynamic pressures** include a significant pulsed, oscillatory, or other time-dependent component. Static pressure measurements are the most common, but equipment such as blowers and compressors can generate significant oscillatory pressures at discrete frequencies. Flow in pipes and ducts can generate resonant pressure changes, as well as turbulent "noise" that can span a wide range of frequencies.

Units

A plethora of pressure units, many of them poorly defined, are in common use. The international (SI) unit is the newton per square metre, called the pascal (Pa). The bar is an acceptable alternate unit, as is the standard atmosphere, but they should not be introduced where they are not used at present. Although not internationally recognized, the pound per square inch (psi) is a widely used and properly defined unit. Also widely used, but not as rigorously defined and, therefore, a potential source of error, are units based on the length of liquid columns, including inches of mercury (in. Hg), mm of mercury (mm Hg), and inches of water (in. of water). The latter is often used for low-range differential pressure measurements. In the case of pounds per square inch, the type of pressure measurement is often indicated by a modification of the unit (i.e., both psi and psia are used to indicate absolute pressure measurements, psid indicates a differential measurement, and psig indicates a gage measurement). No such standard convention exists for other units, and unless explicitly stated, reported values are assumed to be absolute pressures. Conversion factors for different pressure units can be found in Chapter 35.

The difference between the conversion factors for inches of mercury and inches of water at the different temperatures is indicative of the errors that can arise from uncertainties about the definitions of these units.

Types of Pressure-Measuring Instruments

Broadly speaking, pressure instruments can be divided into three different categories—standards, mechanical gages, and electromechanical transducers. Standards instruments are used for the most accurate calibrations. The liquid-column manometer, which is the most common and potentially the most accurate standard, is used for a variety of applications, including field applications. Mechanical pressure gages are generally the least expensive and the most common pressure instruments. However, electromechanical transducers have become much less expensive and are easier to use, so they are being used more often.

PRESSURE STANDARDS

Liquid-column manometers measure pressure by determining the vertical displacement of a liquid of known density in a known gravitational field. Typically they are constructed as a U-tube of transparent material (glass or plastic). The pressure to be measured is applied to one side of the U-tube. If the other (reference) side is evacuated (zero pressure), the manometer measures absolute pressure; if the reference side is open to the atmosphere, it measures gage pressure; if the reference side is connected to some other pressure, the manometer measures the differential between the two pressures. Manometers filled with water and different oils are often used to measure low-range differential pressures. In some low-range instruments, one tube of the manometer is inclined in order to enhance the readability. Mercury-filled manometers are used for higher range differential and absolute pressure measurements. In the latter case, the reference side is evacuated, generally with a mechanical vacuum pump. Typical full-scale ranges for manometers vary from 10 in. of water to 3 atm.

For pressures above the range of manometers, standards are generally of the piston-gage, pressure-balance, or deadweight-tester type. These instruments apply pressure to the bottom of a vertical piston, which is surrounded by a close-fitting cylinder (typical clearances are millionths of an inch). The pressure generates a force approximately equal to the pressure times the area of the piston. This force is balanced by weights stacked on the top of the piston. If the mass of the weights, the local acceleration of gravity, and the area of the piston (or more properly, the "effective area" of the piston and cylinder assembly) are known, the applied pressure can be calculated. Piston gages generally generate gage pressures with respect to the atmospheric pressure above the piston. They can be used to measure absolute pressures either indirectly by separately measuring the atmospheric pressure and adding it to the gage pressure determined by the piston gage, or directly by surrounding the top of the piston and weights with an evacuated bell jar. Piston gage full-scale ranges vary from 5 to 200,000 psi.

At the other extreme, very low absolute pressures (below about 0.4 in. of water), a number of different types of standards are used. These tend to be specialized and expensive instruments found only in major standards laboratories. However, one low-pressure standard, the **McLeod gage**, has been used for field applications. Unfortunately, although the theory of the McLeod gage is simple and straightforward, it is difficult to make accurate measurements with this instrument, and major errors can occur when it is used to measure gases that condense or are adsorbed (e.g., water). In general, gages other than the McLeod gage should be used for most low-pressure or vacuum applications.

MECHANICAL PRESSURE GAGES

Mechanical pressure gages couple a pressure sensor to a mechanical readout, typically a pointer and dial. The most common type employs a **Bourdon tube** sensor, which is essentially a coiled metal tube of circular or elliptical cross section. Increasing pressure applied to the inside of the tube causes it to uncoil. A mechanical linkage translates the motion of the end of the tube to the rotation of a pointer. In most cases, the Bourdon tube is surrounded by atmospheric pressure, so that the gages measure gage pressure. A few instruments surround the Bourdon tube with a sealed enclosure that can be evacuated for absolute measurements or connected to another pressure for differential measurements. Available instruments vary widely in cost, size, pressure range, and accuracy. Full-scale ranges can vary from 5 to 100,000 psi. Accuracy of properly calibrated and used instruments can vary from 0.1 to 10% of full scale. Generally there is a strong correlation between size, accuracy, and price; larger instruments are more accurate and expensive.

To achieve better sensitivity, some low-range mechanical gages, sometimes called **aneroid gages**, employ corrugated diaphragms or capsules as sensors. The capsule is basically a short bellows sealed with end caps. These sensors are more compliant than a Bourdon tube, and a given applied pressure will cause a larger deflection of the sensor. The inside of a capsule can be evacuated and sealed in order to measure absolute pressures or connected to an external fitting to allow differential pressures to be measured. Typically, these gages are used for low-range measurements of 1 atm or less. In instruments of better quality, accuracies of 0.1% of reading or better can be achieved.

ELECTROMECHANICAL TRANSDUCERS

Mechanical pressure gages are generally limited by inelastic behavior of the sensing element, friction in the readout mechanism, and limited resolution of the pointer and dial. These effects can be eliminated or reduced by using electronic techniques to sense the distortion or stress of a mechanical sensing element and electronically convert that stress or distortion to a pressure reading. A wide variety of sensors is used, including Bourdon tubes, capsules, diaphragms, and different resonant structures whose vibration frequency varies with the applied pressure. Capacitive, inductive, and optical lever sensors are used to measure the displacement of the sensor element. In some cases, feedback techniques may be used to constrain the sensor in a null position, minimizing distortion and hysteresis of the sensing element. Temperature control or compensation is often included. Readout may be in the form of a digital display, analog voltage or current, or a digital code. Size varies, but in the case of transducers employing a diaphragm fabricated as part of a silicon chip, the sensor and signal-conditioning electronics can be contained in a small transistor package, and the largest part of the device is the pressure fitting. The best of these instruments achieve long-term instabilities of 0.01% or less of full scale, and corresponding accuracies when properly calibrated. Performance of the less expensive instruments can be more on the order of several percent.

While the dynamic response of most mechanical gages is limited by the sensor and readout, the response of some electromechanical transducers can be much faster, allowing measurements of dynamic pressures at frequencies up to 1 kHz and beyond in the case of transducers specifically designed for dynamic measurements. Manufacturers' literature should be consulted as a guide to the dynamic response of specific instruments.

As the measured pressure is reduced below about 1.5 psia, it becomes increasingly difficult to sense mechanically. A variety of gages have been developed that measure some other property of the gas that is related to the pressure. In particular, thermal conductivity gages, known as thermocouple, thermistor, Pirani, and convection gages, are used for pressures down to about 0.0004 in. of water. These gages have a sensor tube with a small heated element and a temperature sensor; the temperature of the heated element is determined by the thermal conductivity of the gas, and the output of the temperature sensor is displayed on an analog or digital electrical meter contained in an attached electronics unit. The accuracy of thermal conductivity gages is limited by their nonlinearity, dependence on gas species, and tendency to read high when contaminated. Oil contamination is a particular problem. However, these gages are small, reasonably rugged, and relatively inexpensive; in the hands of a typical user, they will give far more reliable results than a McLeod gage. They can be used to check the base pressure in a system that is being evacuated prior to being filled with refrigerant. They should be checked periodically for contamination by comparing the reading with that from a new, clean sensor tube.

GENERAL CONSIDERATIONS

Accurate values of atmospheric or barometric pressure are required for weather prediction and aircraft altimetry. In the United States, a network of calibrated instruments, generally accurate to within 0.1% of reading and located at airports, is maintained by the National Weather Service, the Federal Aviation Administration, and local airport operating authorities. These agencies are generally cooperative in providing current values of atmospheric pressure that can be used to check the calibration of absolute pressure gages or to correct gage pressure readings to absolute pressures. However, the pressure readings generally reported for weather and altimetry purposes are not the true atmospheric pressure, but rather a value adjusted to an equivalent sea level pressure. Therefore, unless the location is near sea level, it is important to ask for the station or true atmospheric pressure rather than using the adjusted values broadcast by radio stations. Further, the atmospheric pressure decreases with increasing elevation at a rate (near sea level) of about 0.001 in. Hg/ft, and corresponding corrections should be made to account for the difference in elevation between the instruments being compared.

As noted before, gage-pressure instruments are sometimes used to measure absolute pressures, and the accuracy of these measurements can be compromised by uncertainties in the atmospheric pressure. This error can be particularly serious when gage-pressure instruments are used to measure a vacuum (negative gage pressures). For all but the most crude measurements, absolute-pressure gages should be used for vacuum measurements; for pressures below about 0.4 in. of water, a thermal conductivity gage should be used.

All pressure gages are susceptible to temperature errors. Several techniques are used to minimize these errors—sensor materials are generally chosen to minimize temperature effects, mechanical readouts can include temperature compensation elements, electromechanical transducers may include a temperature sensor and compensation circuit, and some transducers are operated at a controlled temperature. Clearly, temperature effects are of greater concern for field applications, and it is prudent to check the manufacturers' literature for the temperature range over which the specified accuracy can be maintained. Abrupt temperature changes can also cause large transient errors that may take some time to decay.

The readings of some electromechanical transducers with a resonant or vibrating sensor can depend on the gas species. Although some of these units can achieve calibrated accuracies of the order of 0.01% of reading, they are typically calibrated with dry air or nitrogen, and the readings for other gases can be in error by several percent, quite possibly much more for refrigerants and other high-density gases. High-accuracy readings can be maintained by calibrating these devices with the gas to be measured. Manufacturer's literature should be consulted.

The measurement of dynamic pressures is limited not just by the frequency response of the pressure gage, but also by the hydraulic or pneumatic time constant of the connection between the gage and the system to be monitored. As a general rule, the longer the connecting lines and the smaller their diameter, the lower the frequency response of the system. Further, even if only the static component of the pressure is of interest, and a gage with a low-frequency response is used, a significant pulsating or oscillating pressure component can cause significant errors in pressure gage readings and, in some cases, can damage the gage, particularly gages with a mechanical readout mechanism. In these cases, a filter or snubber should be used to reduce the higher frequency components.

VELOCITY MEASUREMENT

Heating and air-conditioning engineers measure the flow of air more often than any other gases, and the air is usually measured at or near atmospheric pressure. Under this condition, the air can be treated as an incompressible fluid, and simple formulas give sufficient precision to solve many problems. Instruments that measure

Table 4 Velocity Measurement

Measurement Means	Application	Range, fpm	Precision	Limitations
Smoke puff or airborne solid tracer	Low air velocities in rooms; highly directional	5 to 50	10 to 20%	Awkward to use but valuable in tracing air movement
Deflecting vane anemometer	Air velocities in rooms, at outlets, etc.; directional	30 to 24,000	5%	Needs periodic check calibration
Revolving vane anemometer	Moderate air velocities in ducts and rooms; somewhat directional	100 to 3000	2 to 5%	Extremely subject to error with variations in velocities with space or time; easily damaged; needs periodic calibration
Hot-wire anemometer	a. Low air velocities; directional and nondirectional available	1 to 2000	2 to 5%	Requires accurate calibration at frequent intervals. Some are relatively costly.
	b. High air velocities	Up to 60,000	0.2 to 5%	
	c. Transient velocity and turbulence			
Pitot tube	Standard instrument for measuring duct velocities	180 to 10,000 with micromanometer; 600 to 10,000 with draft gages; 10,000 up with manometer	1 to 5%	Accuracy falls off at low end of range
Impact tube and sidewall or other static tap	High velocities, small tubes and where air direction may be variable	120 to 10,000 with micromanometer; 600 to 10,000 with draft gages; 10,000 up with manometer	1 to 5%	Accuracy depends on constancy of static pressure across stream section
Cup anemometer	Meteorological	Up to 12,000	2 to 5%	Poor accuracy at low air velocity (<500 fpm)
Laser Doppler velocimeter	Calibration of air velocity instruments	1 to 6000	1 to 3%	High cost and complexity limit LDVs to laboratory applications

fluid velocity and their application range and precision are listed in Table 4.

AIRBORNE TRACER TECHNIQUES

Tracer techniques are suitable for measuring velocity in an open space. Typical tracers include smoke, feathers, pieces of lint, and radioactive or nonradioactive gases. Measurements are made by timing the rate of movement of solid tracers or by monitoring the change in concentration level of gas tracers.

Smoke is a useful qualitative tool in studying air movements. Smoke can be obtained from titanium tetrachloride (irritating to nasal membranes) or by mixing potassium chlorate and powdered sugar (a nonirritating smoke) and firing the mixture with a match. The latter process produces considerable heat and should be confined to a pan away from flammable materials. Titanium tetrachloride smoke works well for spot tests, particularly for leakage through casings and ducts, because it can be handled easily in a small, pistol-like ejector.

The fumes of ammonia water and sulfuric acid, if permitted to mix, form a white precipitate. Two bottles, one containing ammonia water and the other containing acid, are connected to a common nozzle by rubber tubing. A syringe forces air over the liquid surfaces in the bottles; the two streams mix at the nozzle and form a white cloud.

A satisfactory test smoke also can be made by bubbling an airstream through ammonium hydroxide and then hydrochloric acid (Nottage et al. 1952). Smoke tubes, smoke candles, and smoke bombs are available for studying airflow patterns.

ANEMOMETERS

Deflecting Vane Anemometers

The deflecting vane anemometer consists of a pivoted vane enclosed in a case. Air exerts pressure on the vane as it passes through the instrument from an upstream to a downstream opening. A hair spring and a damping magnet resist vane movement. The instrument gives instantaneous readings of directional velocities on an indicating scale. With fluctuating velocities, it is necessary to average the needle swings visually to obtain average velocities. This instrument is useful for studying air motion in a room; locating objectionable drafts; measuring air velocities at supply and return diffusers and grilles; and measuring laboratory hood face velocities.

Propeller or Revolving Vane Anemometers

The propeller anemometer consists of a light, revolving wind-driven wheel connected through a gear train to a set of recording dials that read linear feet of air passing in a measured length of time. It is made in various sizes—3, 4, and 6 in. are the most common. Each instrument requires individual calibration. At low velocities, the friction drag of the mechanism is considerable. To compensate for this, a gear train that overspeeds is commonly used. For this reason, the correction is often additive at the lower range and subtractive at the upper range, with the least correction in the middle range of velocities. The best of these instruments have starting speeds of 50 fpm or higher; therefore, they cannot be used below that air speed. Electronic revolving vane anemometers, with optical or magnetic pickups to sense the rotation of the vane, are available. Sizes for the vanes range as small as 1/2 in. in diameter for the electronic versions.

Cup Anemometers

The cup anemometer is primarily used to measure outdoor, meteorological wind speeds. It consists of three or four hemispherical cups mounted radially from a vertical shaft. Wind from any direction with a vector component in the plane of cup rotation causes the cups and shaft to rotate. Because the primary use of this anemometer is to make meteorological wind speed measurements, the instrument is usually constructed so that wind speeds can be recorded or indicated electrically at a remote point.

Thermal Anemometers

The thermal or hot-wire anemometer consists of a heated RTD, thermocouple junction, or thermistor sensor constructed at the end of a probe; it is designed to provide a direct, simple method of determining air velocity at a point in the flow field. The probe is placed into an airstream, and the movement of air past the electrically heated velocity sensor tends to cool the sensor in proportion to the speed of the airflow. The electronics and sensor are commonly combined into a portable, hand-held device that interprets the sensor signal and provides a direct reading of air velocity in either analog or digital display format. Often the sensor probe also incorporates an ambient temperature-sensing RTD or thermistor, in which case the indicated air velocity is "temperature compensated" to "standard" air density conditions (typically 0.0748 lb/ft^3).

Hot-wire anemometers have long been used in the fluid flow research field. Research anemometer sensors have been constructed using very fine wires in configurations that allow the researcher to characterize fluid flows in one, two, and three dimensions with sensor/electronics response rates up to several hundred kilohertz. This technology has been incorporated into more ruggedized sensors suitable for measurements in the HVAC field, primarily for unidirectional airflow measurement. Omnidirectional sensing instruments suitable for thermal comfort studies are also available.

The principal advantages of thermal anemometers are their wide dynamic range and their ability to sense extremely low velocities. Typical accuracy (including repeatability) of 2 to 5% of reading over the entire velocity range is often achieved in commercially available portable instruments.

Among the limitations of thermal anemometers are the following: (1) the unidirectional sensor must be carefully aligned in the airstream (typically to within ±20° rotation) to achieve accurate results; (2) the velocity sensor must be kept clean because contaminant buildup will cause the calibration to change; and (3) due to the inherent high speed of response of thermal anemometers, measurements in turbulent flows can yield fluctuating velocity measurements. Electronically controlled time-integrated functions are now available in many digital air velocity meters to help smooth these turbulent flow measurements.

In the HVAC field, thermal anemometers are suitable for use in a variety of applications. They are particularly well-suited to the low velocities associated with laboratory fume hood face velocity measurements (typically in the 60 to 200 fpm range). Thermal anemometers can also be used for taking multipoint traverse measurements in ventilation ductwork.

Laser Doppler Velocimeters (or Anemometers)

The laser Doppler velocimeter (LDV) or laser Doppler anemometer (LDA) is an extremely complex system that collects scattered light produced by a particle passing through the intersection volume of two intersecting laser beams of the same light frequency (Mease et al. 1992). The scattered light consists of bursts containing a regularly spaced fringe pattern whose frequency is linearly proportional to the speed of the particle. Due to the cost and complexity of these systems, they are usually not suitable for in situ field measurements. Rather, the primary application of LDV systems in the HVAC industry is the calibration of systems used to calibrate other air velocity instruments.

The greatest advantage of an LDV is its performance at low air speeds. It is capable of reading air speeds as low as 15 fpm with uncertainty levels of 1% or less (Mease et al. 1992). In addition, it is nonintrusive in the flow—only optical access is required. It can be used to measure fluctuating components as well as mean speeds and is available in one-, two-, and even three-dimensional configurations. Its biggest disadvantages are its high cost and extreme technological complexity, which requires highly skilled operators. Modern fiber optic systems require less-skilled operators but at a considerable increase in cost.

PITOT-STATIC TUBES

The pitot-static tube, in conjunction with a suitable manometer or differential pressure transducer, provides a simple method of determining air velocity at a point in a flow field. Figure 5 shows the construction of a standard pitot tube (ASHRAE *Standard* 51) and the method of connecting it with inclined manometers to display both static pressure and velocity pressure. The equation for determining air velocity from measured velocity pressure is

$$V = C\sqrt{\frac{2p_w g_c}{\rho}} \tag{5}$$

where

V = velocity, fpm
p_w = velocity pressure (pitot-tube manometer reading), in. of water
ρ = density of air, lb$_m$/ft^3
g_c = gravitational constant = 32.174 lb$_m$·ft/lb$_f$·s^2
C = unit conversion factor = 136.8

The type of manometer or differential pressure transducer used with a pitot-static tube depends on the magnitude of velocity pressure being measured and on the desired accuracy. At velocities greater than 1500 fpm, a draft gage of appropriate range is usually satisfactory. If the pitot-static tube is used to measure air velocities lower than 1500 fpm, a precision manometer or comparable pressure differential transducer is essential.

Other pitot-static tubes have been used and calibrated. To meet special conditions, various sizes of pitot-static tubes geometrically similar to the standard tube can be used. For relatively high velocities in ducts of small cross-sectional area, total pressure readings can be obtained with an impact (pitot) tube. Where static pressure across the stream is relatively constant, as in turbulent flow in a straight duct, a sidewall tap to obtain static pressure can be used with the impact tube to obtain the velocity pressure head. One form of impact tube is a small streamlined tube with a fine hole in its upstream end and with its axis placed parallel to the stream.

If the Mach number of the flow is greater than about 0.3, the effects of compressibility should be included in the computation of the air speed from pitot-static and impact (stagnation or pitot) tube measurements (Mease et al. 1992).

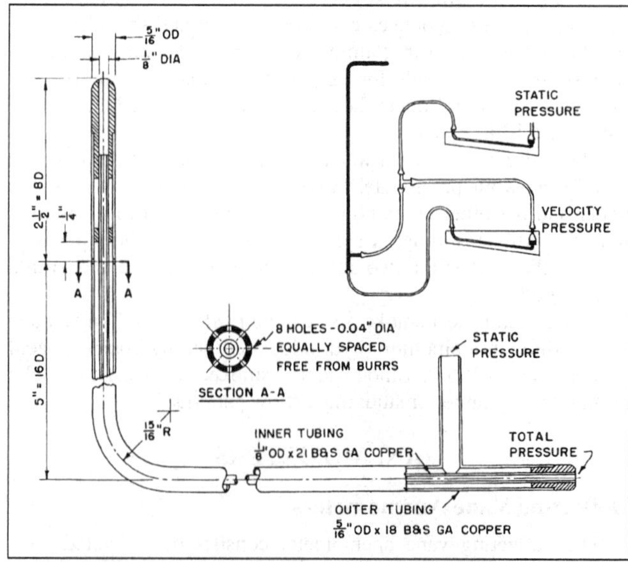

Fig. 5 Standard Pitot Tube

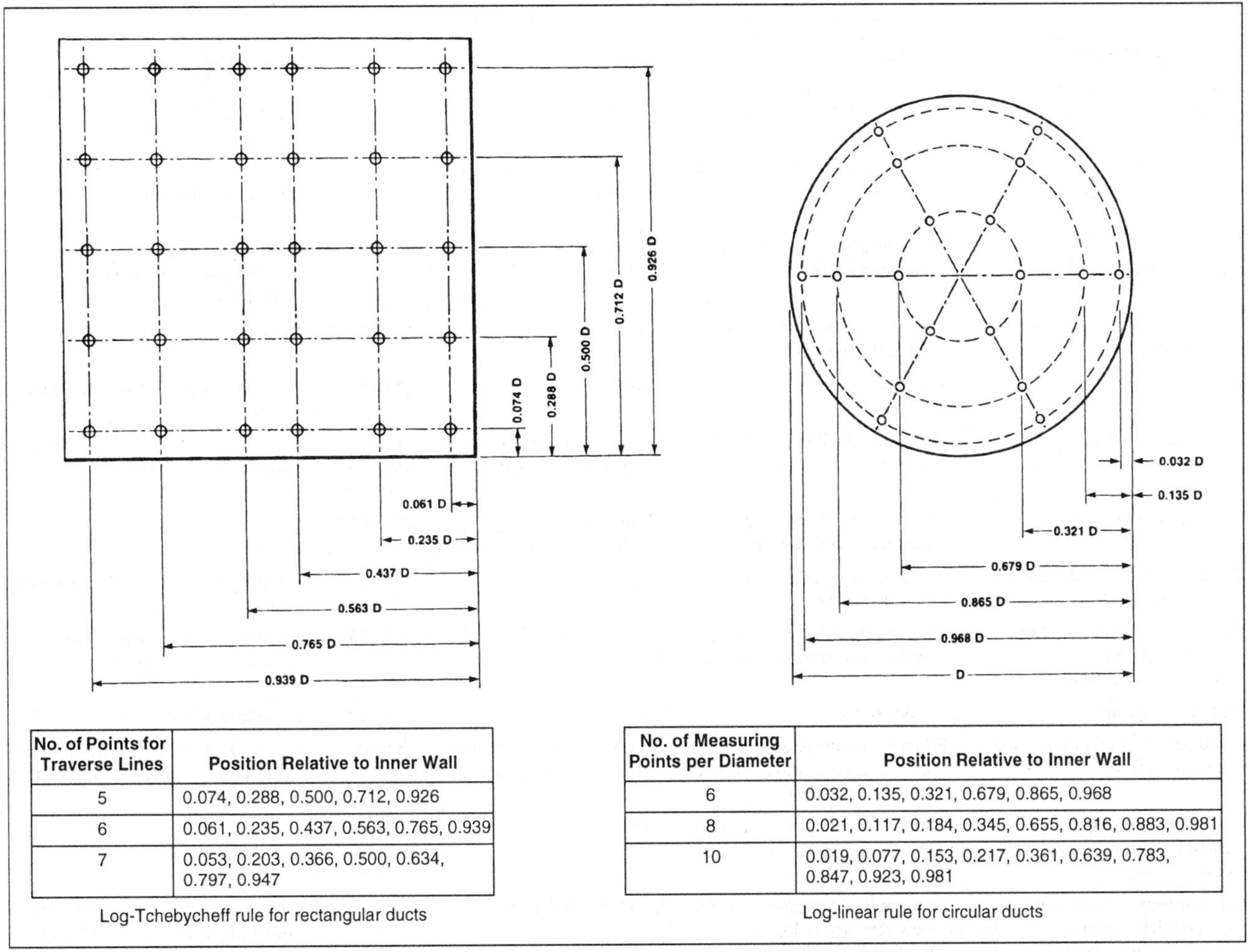

No. of Points for Traverse Lines	Position Relative to Inner Wall
5	0.074, 0.288, 0.500, 0.712, 0.926
6	0.061, 0.235, 0.437, 0.563, 0.765, 0.939
7	0.053, 0.203, 0.366, 0.500, 0.634, 0.797, 0.947

Log-Tchebycheff rule for rectangular ducts

No. of Measuring Points per Diameter	Position Relative to Inner Wall
6	0.032, 0.135, 0.321, 0.679, 0.865, 0.968
8	0.021, 0.117, 0.184, 0.345, 0.655, 0.816, 0.883, 0.981
10	0.019, 0.077, 0.153, 0.217, 0.361, 0.639, 0.783, 0.847, 0.923, 0.981

Log-linear rule for circular ducts

Fig. 6 Measuring Points for Rectangular and Round Duct Traverse

MEASURING FLOW IN DUCTS

Because velocity in a duct is seldom uniform across any section, and a pitot tube reading or thermal anemometer indicates velocity at only one location, a traverse is usually made to determine average velocity. Generally, velocity is lowest near the edges or corners and greatest at or near the center.

To determine the velocity in the traverse plane, a straight average of individual point velocities will give satisfactory results when point velocities are determined by the log-Tchebycheff rule (ISO *Standard* 3966). Figure 6 shows suggested sensor locations for traversing round and rectangular ducts (ASHRAE *Standard* 111). The log-Tchebycheff rule provides the greatest accuracy because its location of traverse points accounts for the effect of wall friction and the fall-off of velocity near the duct walls. For circular ducts, the log-Tchebycheff and log-linear traverse methods are similar. Log-Tchebycheff is now recommended for rectangular ducts as well. It minimizes the positive error (measured greater than actual) caused by the failure to account for losses at the duct wall. This error can occur when using the older method of equal subareas to traverse rectangular ducts.

For a rectangular duct traverse, a minimum of 25 points should be measured. For a duct side less than 18 in., the points should be located at the center of equal areas not more than 6 in. apart, and a minimum of 2 points per side should be used. For a duct side greater than 56 in., the maximum distance between points is 8 in. For a circular duct traverse, the log-linear rule and three symmetrically disposed diameters may be used (Figure 6). Points on two perpendicular diameters may be used where access is limited.

If possible, measuring points should be located at least 7.5 diameters downstream and 3 diameters upstream from a disturbance (e.g., caused by a turn). Compromised traverses as close as 2 diameters downstream and 1 diameter upstream can be performed with an increase in measurement error. Because field-measured airflows are rarely steady and uniform, particularly near disturbances, accuracy can be improved by increasing the number of measuring points. Straightening vanes (ASHRAE *Standard* 51) located 1.5 duct diameters ahead of the traverse plane improve measurement precision.

When velocities at a traverse plane fluctuate, the readings should be averaged on a time-weighted basis. Two traverse readings in short succession also help to average out velocity variations that occur with time. If negative velocity pressure readings are encountered, they are considered a measurement value of zero and calculated in the average velocity pressure. ASHRAE *Standard* 111 has further information on measuring flow in ducts.

FLOW RATE MEASUREMENT

Various means of measuring fluid flow rate are listed in Table 5. The values for volumetric or mass flow rate measurement (ASME *Standard* PTC 19.5, Benedict 1984) are often

Table 5 Volumetric or Mass Flow Rate Measurement

Measurement Means	Application	Range	Precision	Limitations
Orifice and differential pressure measurement system	Flow through pipes, ducts, and plenums for all fluids	Above Reynolds number of 5000	1 to 5%	Discharge coefficient and accuracy influenced by installation conditions
Nozzle and differential pressure measurement system	Flow through pipes, ducts, and plenums for all fluids	Above Reynolds number of 5000	0.5 to 2.0%	Discharge coefficient and accuracy influenced by installation conditions
Venturi tube and differential pressure measurement system	Flow through pipes, ducts, and plenums for all fluids	Above Reynolds number of 5000	0.5 to 2.0%	Discharge coefficient and accuracy influenced by installation conditions
Timing given mass or volumetric flow	Liquids or gases; used to calibrate other flowmeters	Any	0.1 to 0.5%	System is bulky and slow
Rotameters	Liquids or gases	Any	0.5 to 5.0%	Should be calibrated for fluid being metered
Displacement meter	Relatively small volumetric flow with high pressure loss	As high as 1000 cfm depending on type	0.1 to 2.0% depending on type	Most types require calibration with fluid being metered
Gasometer or volume displacement	Short-duration tests; used to calibrate other flowmeters	Total flow limited by available volume of containers	0.5 to 1.0%	—
Thomas meter (temperature rise of stream due to electrical heating)	Elaborate setup justified by need for good accuracy	Any	1%	Uniform velocity; usually used with gases
Element of resistance to flow and differential pressure measurement system	Used for check where system has calibrated resistance element	Lower limit set by readable pressure drop	1 to 5%	Secondary reading depends on accuracy of calibration
Turbine flowmeters	Liquids or gases	Any	0.25 to 2.0%	Uses electronic readout
Instrument for measuring velocity at point in flow	Primarily for installed systems with no special provision for flow measurement	Lower limit set by accuracy of velocity measurement	2 to 4%	Accuracy depends on uniformity of flow and completeness of traverse
Heat input and temperature changes with steam and water coil	Check value in heater or cooler tests	Any	1 to 3%	—
Laminar flow element and differential pressure measurement system	Measure liquid or gas volumetric flow rate; nearly linear relationship with pressure drop; simple and easy to use	0.0001 to 2000 cfm	1%	Fluid must be free of dirt, oil, and other impurities that could plug meter or affect its calibration
Magnetohydrodynamic flowmeter (electromagnetic)	Measures electrically conductive fluids, slurries; meter does not obstruct flow; no moving parts	0.1 to 10,000 gpm	1%	At present state of the art, conductivity of fluid must be greater than 5 μmho/cm
Swirl flowmeter and vortex shedding meter	Measure liquid or gas flow in pipe; no moving parts	Above Reynolds number of 10^4	1%	—

determined by measuring pressure difference across an orifice, nozzle, or venturi tube. The various meters have different advantages and disadvantages. For example, the orifice plate is more easily changed than the complete nozzle or venturi tube assembly. However, the nozzle is often preferred to the orifice because its discharge coefficient is more precise. The venturi tube is a nozzle followed by an expanding recovery section to reduce net pressure loss. Differential pressure-type flow measurement has benefited through workshops addressing fundamental issues, textbooks, research, and improved standards (Miller 1983, DeCarlo 1984, Mattingly 1984, ASME *Standard* B40.1, ASME *Standard* MFC-9M, ASME *Standard* MFC-10M, ASME *Standard* MFC-1M).

Fluid meters use a wide variety of physical techniques to make flow measurements (ASME *Standard* PTC 19.5, Miller 1983, DeCarlo 1984); those more prevalently used are described in this section. To assure and validate the accuracy of flow rate measurement instruments, appropriate calibration procedures should include documentation of traceability to the calibration facility. The calibration facility should, in turn, provide documentation of traceability to national standards.

Flow Measurement Methods

Direct. Both gas and liquid flow can be measured quite accurately by timing a collected amount of fluid that is measured gravimetrically or volumetrically. While this method is commonly used for calibrating other metering devices, it is particularly useful where the flow rate is low or intermittent and where a high degree of accuracy is required. These systems are generally large and slow, but in their simplicity, they can be considered primary devices.

The variable area meter or rotameter is a convenient direct-reading flowmeter for liquids and gases. This is a vertical, tapered tube in which the flow rate is indicated by the position of a float suspended in the upward flow. The position of the float is determined by its buoyancy and the upwardly directed fluid drag.

Displacement meters measure total liquid or gas flow over time. The two major types of displacement meters used for gases are the conventional gas meter, which uses a set of bellows, and the wet test meter, which uses a water displacement principle.

Indirect. The **Thomas meter** is used in laboratories to measure high gas flow rates with low pressure losses. The gas is heated by electric heaters, and the temperature rise is measured by two resistance thermometer grids. When the heat input and temperature rise are known, the mass flow of gas is calculated as the quantity of gas that will remove the equivalent heat at the same temperature rise.

A velocity traverse (made using a pitot tube or other velocity-measuring instrument) measures airflow rates in the field or calibrates large nozzles. This method can be imprecise at low velocities and impracticable where many test runs are in progress.

Another field-estimating method measures the pressure drop across elements with known pressure drop characteristics, such as heating and cooling coils or fans. If the pressure drop/flow rate relationship has been calibrated, the results can be precise. If the method depends on rating data, it should be used for check purposes only.

VENTURI, NOZZLE, AND ORIFICE FLOWMETERS

Flow in a pipeline can be measured by a venturi meter (Figure 7), flow nozzle (Figure 8), or orifice plate (Figure 9). American Society of Mechanical Engineers (ASME) *Standard* MFC-3M describes measurement of fluid flow in pipes using the orifice, nozzle, and venturi; ASME *Standard* PTC 19.5 specifies their construction.

Assuming an incompressible fluid (liquid or slow-moving gas), uniform velocity profile, frictionless flow, and no gravitational effects, the principle of conservation of mass and energy can be applied to the venturi and nozzle geometries to give

$$w = \rho V_1 A_1 = \rho V_2 A_2 = A_2 \sqrt{\frac{2g_c \rho (p_1 - p_2)}{1 - \beta^4}} \tag{6}$$

where

w = mass flow rate, lb_m/s
V = velocity of stream, fps
A = flow area, ft^2
g_c = gravitational constant = 32.174 $lb_m \cdot ft/lb_f \cdot s^2$
ρ = density of fluid, lb_m/ft^3
p = absolute pressure, lb_f/ft^2
β = (D_2/D_1) for venturi and sharp edge orifice and d/D for flow nozzle

Note: Subscript 1 refers to the entering conditions; subscript 2 refers to the throat conditions.

Because the flow through the meter is not frictionless, a correction factor C is defined to account for friction losses. If the fluid is at a high temperature, an additional correction factor F_a should be included to account for thermal expansion of the primary element. Because this amounts to less than 1% at 500°F, it can usually be omitted. Equation (6) then becomes

$$w = CA_2 \sqrt{\frac{2g_c \rho (p_1 - p_2)}{1 - \beta^4}} \tag{7}$$

where C is the friction loss correction factor.

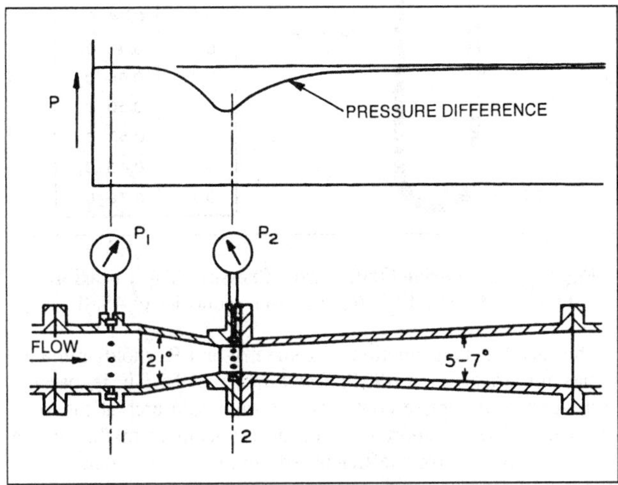

Fig. 7 Typical Herschel Type Venturi Meter

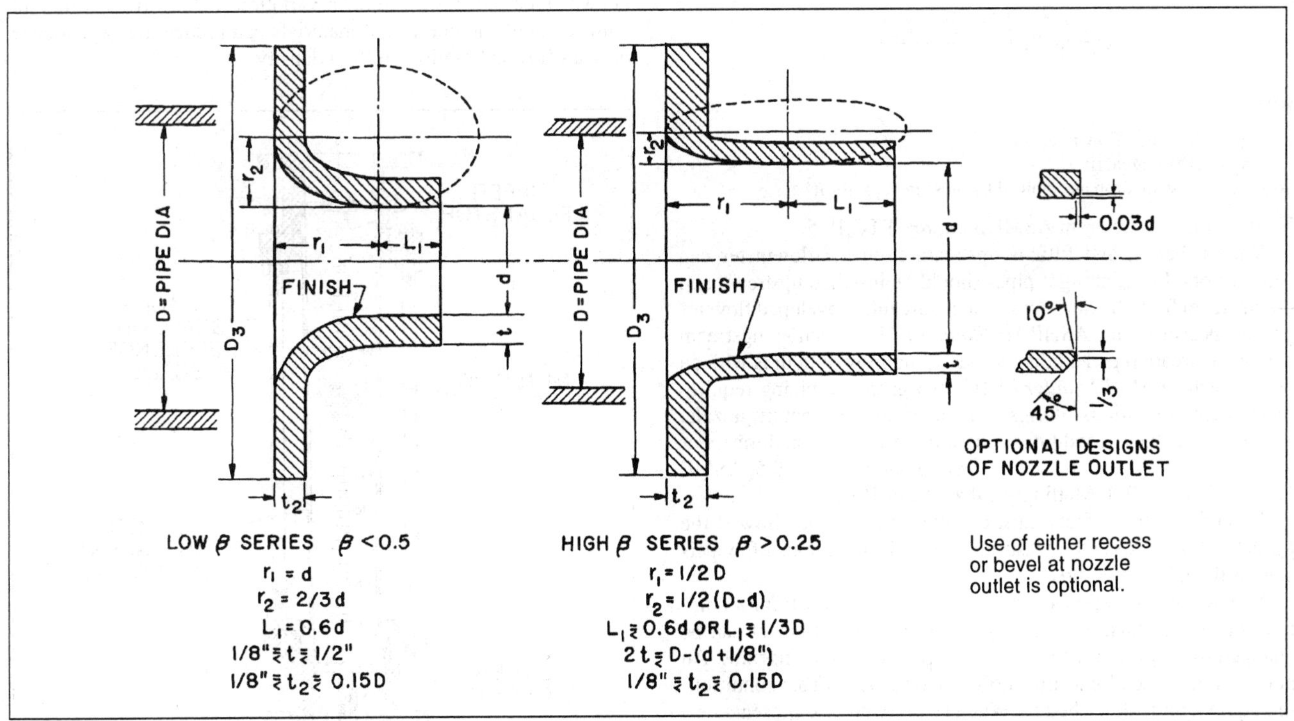

Fig. 8 Dimensions of ASME Long-Radius Flow Nozzles
From ASME PTC 19.5. Reprinted with permission of ASME.

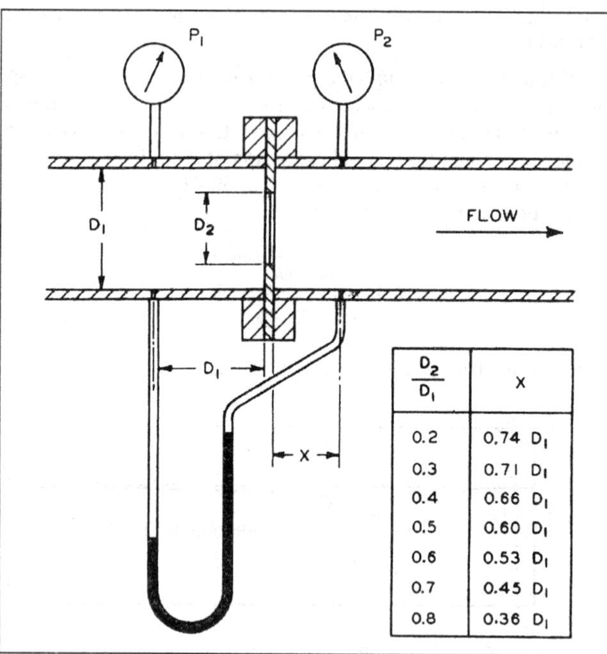

D_2/D_1	X
0.2	0.74 D_1
0.3	0.71 D_1
0.4	0.66 D_1
0.5	0.60 D_1
0.6	0.53 D_1
0.7	0.45 D_1
0.8	0.36 D_1

Fig. 9 Sharp Edge Orifice with Pressure Tap Locations
From ASME PTC 19.5. Reprinted with permission of ASME.

The factor C is a function of geometry and Reynolds number. Values of C are given in ASME *Standard* PTC 19.5. The jet passing through an orifice plate contracts to a minimum area at the vena contracta located a short distance downstream from the orifice plate. The contraction coefficient, friction loss coefficient C, and approach factor $1/(1-\beta^4)^{0.5}$ can be combined into a single constant K, which is a function of geometry and Reynolds number. The orifice flow rate equations then become

$$Q = KA_2\sqrt{\frac{2g_c(p_1-p_2)}{\rho}} \tag{8}$$

where

Q = discharge flow rate, cfs
A_2 = orifice area, ft^2
$p_1 - p_2$ = pressure drop as obtained by pressure taps, lb$_f$/ft^2

Values of K are shown in ASME *Standard* PTC 19.5.

Valves, bends, and fittings upstream from the flowmeter can cause errors. Long, straight pipes should be installed upstream and downstream from the flow devices to assure fully developed flow for proper measurement. ASHRAE *Standard* 41.8 specifies upstream and downstream pipe lengths for measuring flow of liquids with an orifice plate. ASME *Standard* PTC 19.5 gives the piping requirements between various fittings and valves and the venturi, nozzle, and orifice. If these conditions cannot be met, flow conditioners or straightening vanes can be used (ASME *Standard* PTC 19.5, ASME *Standard* MFC-10M, Mattingly 1984, Miller 1983).

Compressibility effects must be considered for gas flow if the pressure drop across the measuring device is more than a few percent of the initial pressure.

Nozzles are sometimes arranged in parallel pipes from a common manifold; thus, the capacity of the testing equipment can be changed by shutting off the flow through one or more nozzles. An apparatus designed for testing airflow and capacity of air-conditioning equipment is described by Wile (1947), who also presents pertinent information on nozzle discharge coefficients, Reynolds numbers, and resistance of perforated plates. Some laboratories refer to this apparatus as a code tester.

VARIABLE AREA FLOWMETERS (ROTAMETERS)

In permanent installations where high precision, ruggedness, and operational ease are important, the variable area flowmeter is satisfactory. It is frequently used to measure liquids or gases in small-diameter pipes. For ducts or pipes over 6 in. in diameter, the expense of this meter may not be warranted. In larger systems, however, the meter can be placed in a bypass line and used with an orifice.

The variable area meter (Figure 10) commonly consists of a float that is free to move vertically in a transparent tapered tube. The fluid to be metered enters at the narrow bottom end of the tube and moves upward, passing at some point through the annulus formed between the float and the inside wall of the tube. At any particular flow rate, the float assumes a definite position in the tube; a calibrated scale on the tube shows the float's location and the fluid flow rate.

The position of the float is established by a balance between the fluid pressure forces across the annulus and gravity on the float. The buoyant force supporting the float, $v_f(\rho_f-\rho)g/g_c$, is balanced by the pressure difference acting on the cross-sectional area of the float $A_f\Delta p$, where ρ_f, A_f, and v_f are, respectively, the float density, float cross-sectional area, and float volume. The pressure difference across the annulus is

$$\Delta p = \frac{v_f(\rho_f-\rho)g}{A_f g_c} \tag{9}$$

The mass flow follows from Equation (8) as

$$w = KA_2\sqrt{\frac{2v_f(\rho_f-\rho)g\rho}{A_f}} \tag{10}$$

The flow for any selected fluid is nearly proportional to the area, so that calibration of the tube is convenient. To use the meter for different fluids, the flow coefficient variation for any float must be known. Float design can reduce variation of the flow coefficient with Reynolds number; float materials can reduce the dependence of mass flow calibration on fluid density.

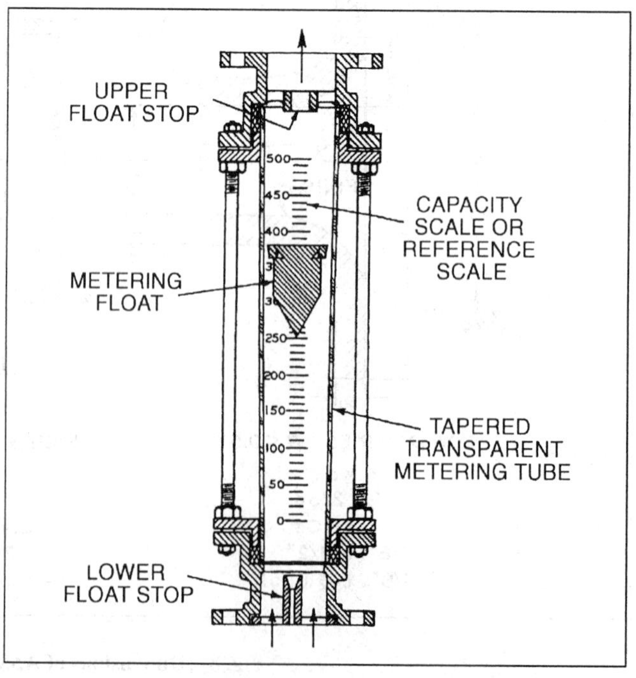

Fig. 10 Variable Area Flowmeter

POSITIVE DISPLACEMENT METERS

Many positive displacement meters are available for measuring total liquid or gas volumetric flow rates. The fluid measured in these meters flows progressively into compartments of definite size. As the compartments are filled, they are rotated so that the fluid discharges from the meter. The flow rate through the meter is equal to the product of the compartment volume, the number of compartments, and the rotation rate of the rotor. Most of these meters have a mechanical register calibrated to show total flow.

TURBINE FLOWMETERS

Turbine flowmeters are volumetric flow rate sensing meters with a magnetic stainless steel turbine rotor suspended in the flow stream of a nonmagnetic meter body. The fluid stream exerts a force on the blades of the turbine rotor, setting it in motion and converting the fluid's linear velocity to an angular velocity. Design motivation for turbine meters is to have the rotational speed of the turbine proportional to the average fluid velocity and thus to the volume rate of fluid flow (Miller 1983, DeCarlo 1984, Mattingly 1992).

The rotational speed of the rotor is monitored by an externally mounted pickoff assembly. Magnetic and radio frequency are the most commonly used pickoffs. The magnetic pickoff contains a permanent magnet and coil. As the turbine rotor blades pass through the field produced by the permanent magnet, a shunting action induces ac voltage in the winding of the coil wrapped around the magnet. A sine wave with a frequency proportional to the flow rate develops. With the radio frequency pickoff, an oscillator applies a high-frequency carrier signal to a coil in the pickoff assembly. The rotor blades pass through the field generated by the coil and modulate the carrier signal by shunting action on the field shape. The carrier signal is modulated at a rate corresponding to the rotor speed, which is proportional to the flow rate. With both pickoffs, frequency of the pulses generated becomes a measure of flow rate, and the total number of pulses measures total volume (Woodring 1969, Shafer 1961, Mattingly 1992).

Because output frequency of the turbine flowmeter is proportional to flow rate, every pulse from the turbine meter is equivalent to a known volume of fluid that has passed through the meter; the sum of these pulses yields total volumetric flow. Summation is accomplished by electronic counters designed for use with turbine flowmeters; they combine a mechanical or electronic register with the basic electronic counter.

Turbine flowmeters should be installed with straight lengths of pipe upstream and downstream from the meter. The length of the inlet and outlet pipes should be according to manufacturers' recommendations or pertinent standards. Where recommendations of standards cannot be accommodated, the meter installation should be calibrated. Some turbine flowmeters can be used in bidirectional flow applications. A fluid strainer, used with liquids of poor or marginal lubricity, minimizes bearing wear.

The lubricity of the process fluid and the type and quality of rotor bearings determine whether the meter is satisfactory for the particular application. When choosing turbine flowmeters for use with fluorocarbon refrigerants, attention must be paid to the type of bearings used in the meter and to the oil content of the refrigerant. For these applications, sleeve-type rather than standard ball bearings are recommended. The amount of oil in the refrigerant can severely affect calibration and bearing life.

In metering liquid fluorocarbon refrigerants, the liquid must not flash to a vapor (cavitate). This would cause a tremendous increase in flow volume. Flashing results in erroneous measurements and rotor speeds that can damage the bearings or cause a failure. Flashing can be avoided by maintaining an adequate back pressure on the downstream side of the meter (Liptak 1972).

AIRFLOW-MEASURING HOODS

Flow-measuring hoods are portable instruments designed to measure supply or exhaust airflow through diffusers and grilles in HVAC systems. A flow-measuring hood assembly typically consists of a fabric hood section, a plastic or metal base, an airflow-measuring manifold, a meter, and handles for carrying and holding the hood in place.

For volumetric airflow measurements, the flow-measuring hood is placed over a diffuser or grille. The fabric hood captures and directs airflow from the outlet or inlet across the flow-sensing manifold in the base of the instrument. The manifold consists of a number of tubes containing upstream and downstream holes in a grid pattern designed to simultaneously sense and average multiple velocity points across the base of the flow-measuring hood. Air from the upstream holes flows through the tubes past a sensor and then exits through the downstream holes. Sensors employed by different manufacturers include swinging vane anemometers, electronic micromanometers, and thermal anemometers. In the case of the electronic micromanometer sensor, air does not actually flow through the manifold, but the airtight sensor senses the pressure differential from the upstream to downstream series of holes. The meter on the base of the flow-measuring hood interprets the signal from the sensor and provides a direct reading of volumetric flow in either an analog or digital display format.

As a performance check in the field, the indicated flow of a measuring hood can be compared to a duct traverse flow measurement (using a pitot-tube or a thermal anemometer). All flow-measuring hoods induce some back pressure on the air-handling system because the hood restricts the flow coming out of the diffuser. This added resistance alters the true amount of air coming out of the diffuser. In most cases, this error is negligible and is less than the accuracy of the instrument. For proportional balancing, this error need not be taken into account because all similar diffusers will have about the same amount of back pressure. To determine whether back pressure is significant, a velocity traverse can be made in the duct ahead of the diffuser with and without the flow-measuring hood in place. The difference in the average velocity of the traverse indicates the degree of back-pressure compensation required on similar diffusers in the system. For example, if the average velocity is 800 fpm with the hood in place and 820 fpm without the hood, the indicated flow reading can be multiplied by 1.025 on similar diffusers in the system (820/800 = 1.025). As an alternative, the designer of the air-handling system can predict the reduction in airflow due to the additional pressure of the hood by using a curve supplied by the flow-measuring hood manufacturer. This curve indicates the pressure drop through the hood for different flow rates.

ELECTRIC MEASUREMENT

Ammeters

Ammeters are low-resistance instruments for measuring current. They should be connected in series with the circuit being measured (Figure 11). Ideally, they have the appearance of a short circuit, but in practice, all ammeters have a nonzero input impedance that influences the measurement to some extent.

Ammeters often have several ranges, and it is good practice when measuring unknown currents to start with the highest range and then reduce the range to the appropriate value to obtain the most sensitive reading. Ammeters with range switches maintain circuit continuity during switching. On some older instruments, it may be necessary to short-circuit the ammeter terminals when changing the range.

Current transformers are often used to increase the operating range of ammeters. They may also provide isolation and thus protection from a high-voltage line. Current transformers have at least two separate windings on a magnetic core (Figure 12).

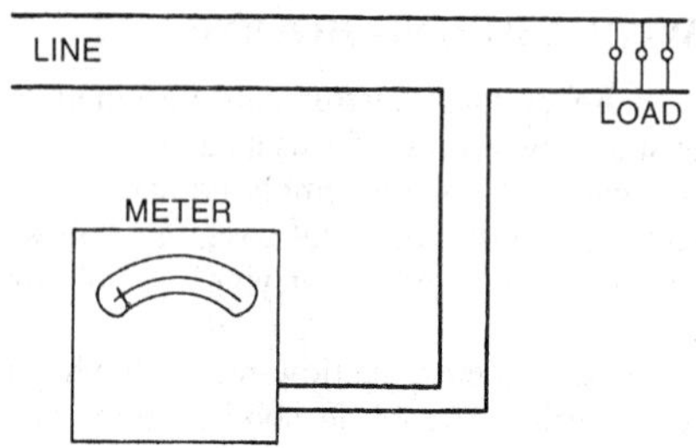

Fig. 11 Ammeter Connected in Power Circuit

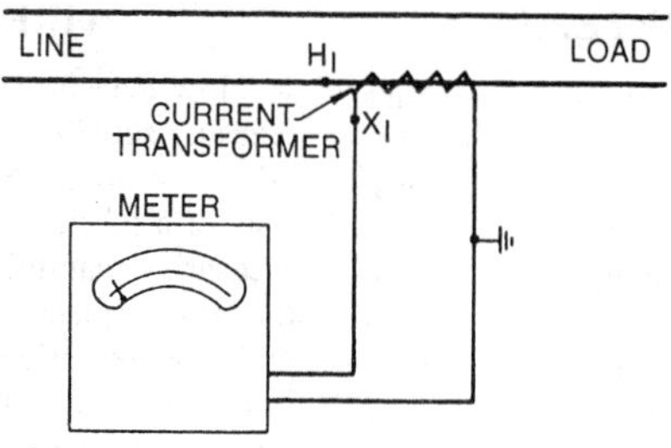

Fig. 12 Ammeter with Current Transformer

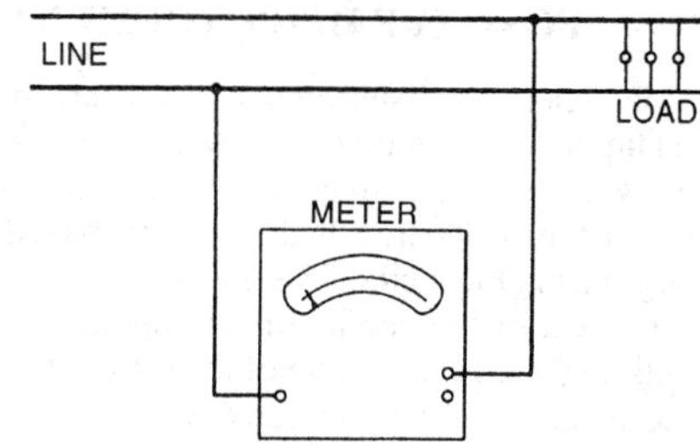

Fig. 13 Voltmeter Connected Across Load

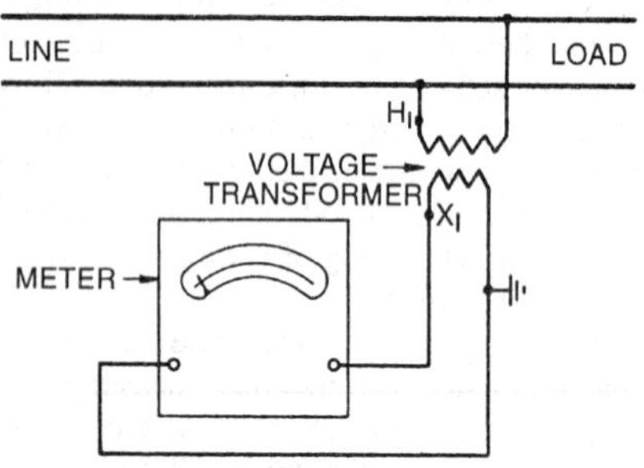

Fig. 14 Voltmeter with Potential Transformer

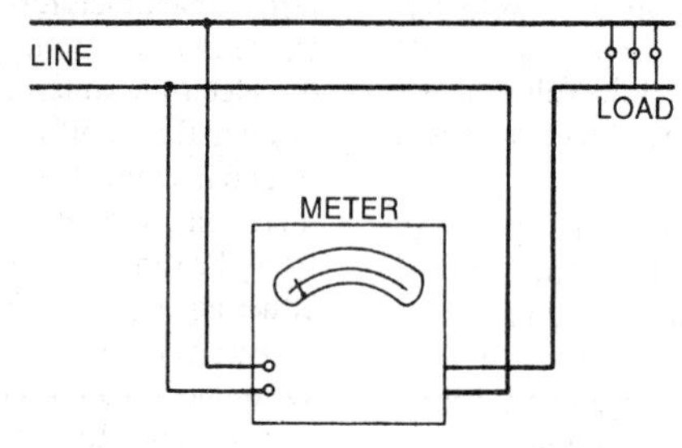

Fig. 15 Wattmeter in Single-Phase Circuit Measuring Power Load plus Loss in Current-Coil Circuit

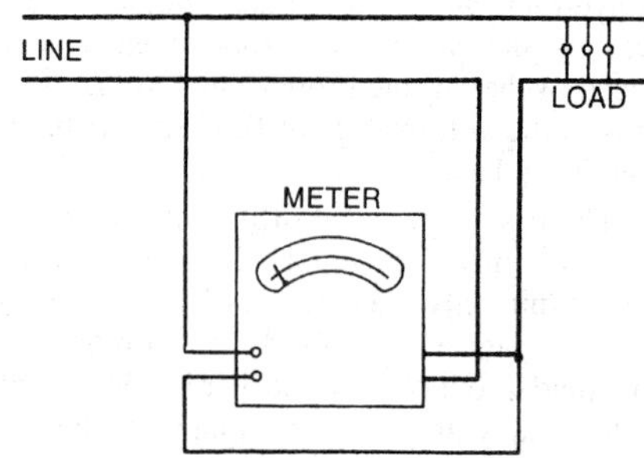

Fig. 16 Wattmeter in Single-Phase Circuit Measuring Power Load plus Loss in Potential-Coil Circuit

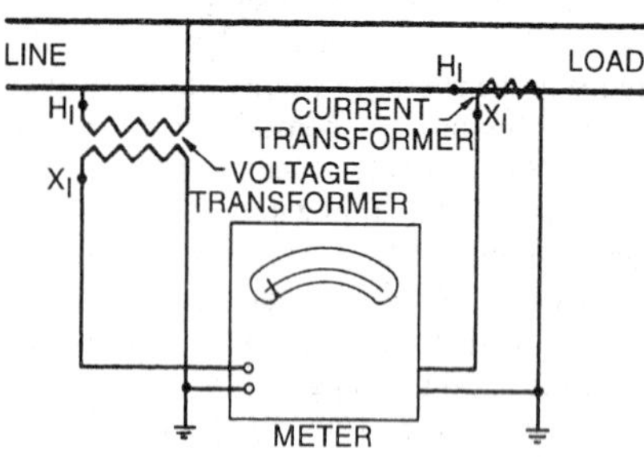

Fig. 17 Wattmeter with Current and Potential Transformer

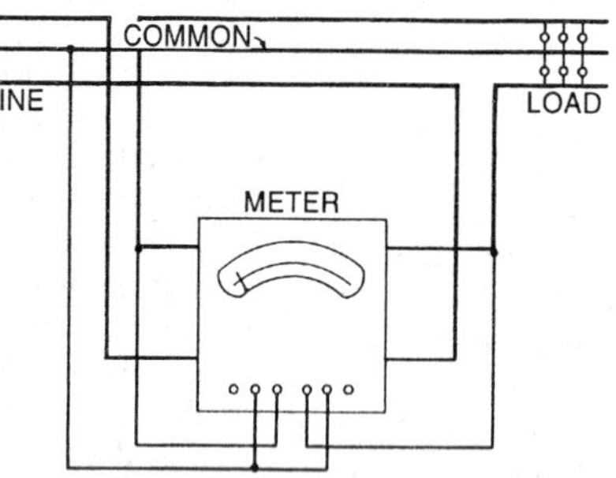

Fig. 18 Polyphase Wattmeter in Two-Phase, Three-Wire Circuit with Balanced or Unbalanced Voltage or Load

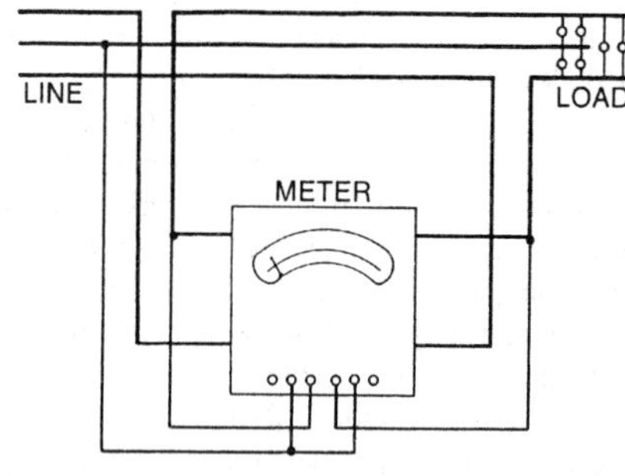

Fig. 19 Polyphase Wattmeter in Three-Phase, Three-Wire Circuit

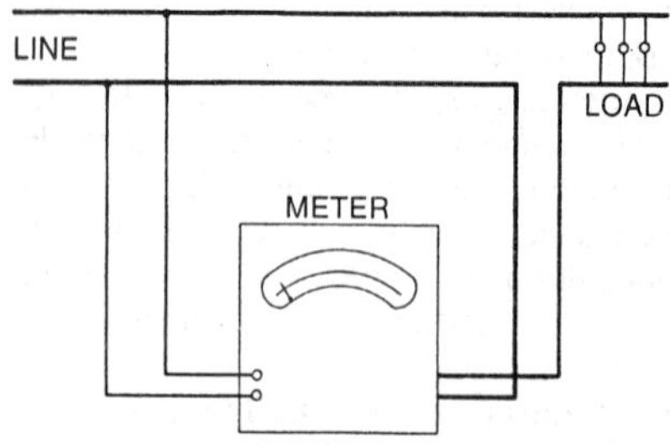

Fig. 20 Single-Phase Power-Factor Meter

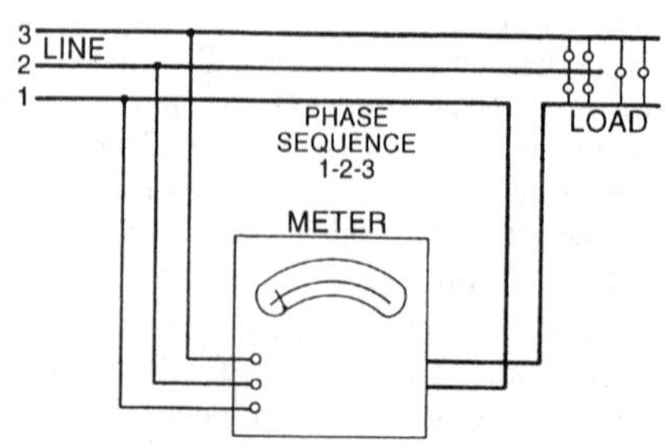

Fig. 21 Three-Wire, Three-Phase Power-Factor Meter

The primary winding is connected in series with the circuit in which the current is measured. In the case of a clamp-on probe, the transformer core is actually opened and then connected around a single conductor carrying the current to be measured. That conductor serves as the primary winding. The secondary winding carries a scaled-down version of the primary current, which is connected to an ammeter. Depending on the type of instrument, the ammeter reading may have to be multiplied by the ratio of the transformer.

When using an auxiliary current transformer, the secondary circuit must not be open when current is flowing in the primary winding; dangerous high voltage may exist across the secondary terminals. A short-circuiting blade between the secondary terminals should be closed before the secondary circuit is opened at any point.

Transformer accuracy can be impaired by the residual magnetism in the core when the primary circuit is opened at an instant when the flux is large. The transformer core may be left magnetized, resulting in ratio and phase angle errors. The primary and secondary windings should be short-circuited before making changes.

Voltmeters

Voltmeters are high-resistance instruments that should be connected across the load (in parallel), as shown in Figure 13. Ideally they have the appearance of an open circuit, but in practice all voltmeters have some finite impedance that influences the measurement to some extent.

Voltage transformers are often used to increase the operating range of a voltmeter (Figure 14). They also provide isolation from high voltages and prevent injury to the operator. Like current transformers, voltage transformers consist of two or more windings on a magnetic core. The primary winding is generally connected across the high voltage to be measured, and the secondary winding is connected to the voltmeter. It is important not to short-circuit the secondary winding of a voltage transformer.

Wattmeters

Wattmeters are instruments that measure the active power of an ac circuit, which equals the voltage multiplied by that part of the current in phase with the voltage. There are generally two sets of terminals—one to connect the load voltage and the other to connect in series with the load current. Current and voltage transformers can be used to extend the range of a wattmeter or to isolate it from high voltage. Figure 15 and Figure 16 show connections for single-phase wattmeters, and Figure 17 shows use of current and voltage transformers with a single-phase wattmeter.

Wattmeters with multiple current and voltage elements are available to measure polyphase power. Polyphase wattmeter connections are shown in Figure 18 and Figure 19.

Power-Factor Meters

Power-factor meters measure the ratio of the active power to the apparent power (product of the voltage and current). The connections for power-factor meters and wattmeters are similar, and current and voltage transformers can be used to extend their range. Connections for single-phase and polyphase power-factor meters are shown in Figure 20 and Figure 21, respectively.

ROTATIVE SPEED MEASUREMENT

Tachometers

Tachometers, or direct-measuring rpm counters, vary from handheld mechanical or electric meters to shaft-driven and electronic pulse counters. They are used in general laboratory and shop work to check the rotative speeds of motors, engines, and turbines.

Stroboscopes

Optical rpm counters work by producing a controlled high-speed electronic flashing light. The operator directs the light on a rotating member and increases the rate of flashes until the optical effect of stopping rotation of the member is achieved. At this point, the rpm measured is equal to the flashes per minute emitted by the strobe unit. Care must be taken to start at the bottom of the instrument scale and work up because multiples of the rpm produce almost the same optical effect as true synchronism. Multiples can be indicated by positioning suitable marks on the shaft, such as a bar on one side and a circle on the opposite side. If, for example, the two are seen superimposed, then the strobe light is flashing at an even multiple of the true rpm.

AC Tachometer-Generators

A tachometer-generator consists of a rotor and a stator. The rotor is a permanent magnet driven by the equipment. The stator is a winding with a hole through the center for the rotor. Concentricity is not critical; bearings are not required between rotor and stator. The output can be a single-cycle-per-revolution signal whose voltage is a linear function of rotor speed. The polypole configuration that generates 10 cycles per revolution permits measurement of speeds as low as 20 rpm without causing the indicating needle to flutter. The output of the ac tachometer-generator is rectified and connected to a dc voltmeter.

SOUND AND VIBRATION MEASUREMENT

Measurement systems for determining sound pressure, sound intensity, and mechanical vibration generally involve the use of transducers to convert mechanical signals into electrical signals, which are then processed electronically in order to characterize the measured mechanical signals. These measurement systems contain one or more of the following elements, which may or may not be contained in a single instrument:

1. A transducer, or an assembly of transducers, to convert sound pressure, sound intensity, or mechanical vibration (time-varying strain, displacement, velocity, acceleration, or force) into an electrical signal that is quantitatively related to the mechanical quantity being measured.
2. Amplifiers and networks to provide such functions as electrical impedance matching, signal conditioning, integration, differentiation, frequency weighting, and gain.
3. Signal-processing equipment to detect and quantify those aspects of the signal that are being measured (peak value, rms value, time-weighted average level, power spectral density, or magnitude or phase of a complex linear spectrum or transfer function).
4. A device such as a meter, oscilloscope, digital display, or level recorder to display the signal or the aspects of it that are being quantified.

The relevant range of sound and vibration signals can vary over more than 12 orders of magnitude in amplitude and more than 8 orders of magnitude in frequency, depending on the application. References on instrumentation, measurement procedures, and signal analysis are given in the section on Bibliography. Product and application notes, technical reviews, and books published by instrumentation manufacturers are an excellent source of additional reference material. See Chapter 43 of the 1995 *ASHRAE Handbook—Applications* and Chapter 7 of this volume for further information on sound and vibration.

SOUND MEASUREMENT

Microphones

A microphone is an electroacoustical transducer that transforms an acoustical signal into an electrical signal. The two predominant transduction principles used in the measurement of sound (as opposed to broadcasting or recording) are the electrostatic and the piezoelectric. **Electrostatic (capacitor) microphones** are available either as electric microphones, which do not require an external polarizing voltage, or as condenser microphones, which do require an external polarizing voltage, typically in the range of 28 to 200 V (dc). **Piezoelectric microphones** may be manufactured using either natural piezoelectric crystals or poled ferroelectric crystals. The types of response characteristics of measuring microphones are pressure, free field, and random incidence (diffuse field).

A microphone with a uniform pressure response characteristic maintains uniform sensitivity over its operating frequency range when exposed to a sound pressure that is uniform over the surface of the sensing element. A microphone with a uniform free-field response characteristic maintains uniform sensitivity over its operating frequency range when exposed to a plane progressive sound wave at a specified angle of incidence to the surface of the sensing element. A microphone with a uniform random-incidence response characteristic maintains uniform sensitivity over its operating frequency range when exposed to a diffuse sound field.

The sensitivity and the frequency range over which the microphone has uniform sensitivity (flat frequency response) vary with both the diameter (surface area) of the sensing element and the microphone type. Other factors that may critically affect the performance or response of a measuring microphone and preamplifier in a given measurement application are atmospheric pressure, temperature, relative humidity, external magnetic and electrostatic fields, mechanical vibration, and radiation. A microphone should be selected based on its long- and short-term stability; the match between its performance characteristics (e.g., sensitivity, frequency response, amplitude linearity, self-noise) and the expected amplitude of sound pressure, frequency, range of analysis, and expected environmental conditions of measurement; and any other pertinent considerations, such as size and directional characteristics.

Sound Measurement Systems

Microphone preamplifiers, amplifiers, weighting networks (see Chapter 7), filters, and displays are available either separately or integrated into a measuring instrument such as a sound level meter, personal noise exposure meter, measuring amplifier, or real-time fractional octave or Fourier [e.g., fast Fourier transform (FFT)] signal analyzer. The instrument(s) included in a sound measurement system depends on the purpose of the measurement and the frequency range and resolution of signal analysis. In the case of community and industrial noise measurements for regulatory purposes, the instrument, signal processing, and quantity to be measured are usually dictated by the pertinent regulation. The optimal set of instruments generally varies for measurement of different characteristics such as sound power in HVAC ducts, sound power emitted by machinery, noise criteria (NC) numbers, sound absorption coefficients, sound transmission loss of building partitions, and reverberation times (T_{60}).

Frequency Analysis

Measurement criteria often dictate the use of filters to analyze the signal in order to indicate the spectrum of the sound being measured. Filters of different bandwidths for different purposes include fractional octave band (one, one-third, one-twelfth, etc.), constant percentage bandwidth, and constant (typically narrow) bandwidth. The filters may be analog or digital and, if digital, may or may not be capable of real-time data acquisition during the measurement period, depending on the bandwidth of frequency analysis. FFT signal analyzers are generally used in situations that require very narrow band signal analysis when the amplitudes of the sound spectra vary significantly with respect to frequency. This may occur in regions of resonance or when it is necessary to identify narrow-band or discrete sine-wave signal components of a spectrum in the presence of other such components or of broadband noise.

Sound Chambers

Special rooms and procedures are required in order to characterize and calibrate sound sources and receivers. The rooms are generally classified into three types—anechoic, semianechoic, and reverberant. The ideal **anechoic** room or chamber would have boundary surfaces that completely absorb sound energy at all frequencies. The ideal **semianechoic** room or chamber would be identical to the ideal anechoic room, except that one surface would totally reflect sound energy at all frequencies. The ideal **reverberant** room or chamber would have boundary surfaces that totally reflect sound energy at all frequencies.

Anechoic chambers are used to perform measurements under conditions approximating those of a free sound field. They can be used in calibrating and characterizing individual microphones, microphone arrays, acoustic intensity probes, reference sound power sources, loudspeakers, sirens, and other individual or complex sources of sound.

Semianechoic chambers are built with a hard reflecting floor in order to accommodate heavy machinery or to simulate large factory floor or outdoor conditions. They can be used in calibrating and characterizing reference sound power sources, obtaining sound power levels of noise sources, and characterizing the sound output of emergency vehicle sirens when mounted on an emergency motor vehicle.

Reverberation chambers are used to perform measurements under conditions approximating those of a diffuse sound field. They can be used in calibrating and characterizing random-incidence microphones and reference sound power sources, obtaining sound power ratings of equipment and sound power levels of noise sources, measuring sound absorption coefficients of building materials and panels, and measuring the transmission loss through building partitions and components such as doors and windows.

Calibration

A measurement system should be calibrated as a system from microphone or probe to indicating device before it is used to perform absolute measurements of sound. Acoustic calibrators and pistonphones of fixed or variable frequency and amplitude are available for this purpose. These calibrators should be used at a frequency low enough that the pressure, free-field, and random-incidence response characteristics of the measuring microphone(s) are, for practical purposes, equivalent, or at least related in a known quantitative manner for that specific measurement system. In general, the sound pressure produced by these calibrators may vary, depending on the microphone type, whether the microphone has a protective grid, atmospheric pressure, temperature, and relative humidity. Correction factors and coefficients are required for conditions of use that differ from those existing during the calibration of the acoustic calibrator or pistonphone. For demanding applications, precision sound sources and measuring microphones should periodically be sent to the manufacturer, a private testing laboratory, or a national standards laboratory for calibration.

VIBRATION MEASUREMENT

With the exception of seismic instruments that record or indicate vibration directly via a mechanical or optomechanical device connected to the test surface, vibration measurements involve the use of an electromechanical or interferometric vibration transducer. Here, the term vibration transducer refers to a generic mechanical vibration transducer. Electromechanical and interferometric vibration transducers belong to a large and varied group of transducers that detect mechanical motion and furnish an electrical signal that is quantitatively related to a particular physical characteristic of the motion. Depending on the design of the transducer, the electrical signal may be related to mechanical strain, displacement, velocity, acceleration, or force. The operating principles of vibration transducers may involve optical interference; electrodynamic coupling; piezoelectric (including poled ferroelectric) or piezoresistive crystals; or variable capacitance, inductance, reluctance, or resistance. A considerable variety of vibration transducers with a wide range of sensitivities and bandwidths is commercially available. Vibration transducers may be contacting (e.g., seismic transducers) or non-contacting (e.g., interferometric or capacitive).

Transducers

Seismic transducers use a spring mass resonator within the transducer. At frequencies much greater than the fundamental natural frequency of the mechanical resonator, the relative displacement between the base and the seismic mass of the transducer is nearly proportional to the displacement of the transducer base. At frequencies much lower than the fundamental resonant frequency, the relative displacement between the base and the seismic mass of the transducer is nearly proportional to the acceleration of the transducer base. Therefore, seismic displacement transducers and seismic electrodynamic velocity transducers tend to have a relatively compliant suspension with a low resonant frequency; piezoelectric accelerometers and force transducers have a relatively stiff suspension with a high resonant frequency.

Strain transducers include the metallic resistance gage and the piezoresistive strain gage. For dynamic strain measurements, these are usually of the bonded type, where the gages are bonded directly to the test surface. The accuracy with which a bonded strain gage replicates strain occurring in the test structure is largely a function of how well the strain gage was oriented and bonded to the test surface.

Displacement transducers include the capacitance gage, fringe-counting interferometer, seismic displacement transducer, and linear variable differential transformer (LVDT). Velocity transducers include the reluctance (magnetic) gage, laser Doppler interferometer, and seismic electrodynamic velocity transducer. Accelerometers and force transducers include the piezoelectric, piezoresistive, and force-balance servo.

Vibration Measurement Systems

The sensitivity, frequency limitations, bandwidth, and amplitude linearity of vibration transducers vary greatly with the transduction mechanism and the manner in which the transducer is applied in a given measurement apparatus. The performance of contacting transducers can be significantly affected by the mechanical mounting methods and points of attachment of the transducer and connecting cable and by the mechanical impedance of the structure loading the transducer. Amplitude linearity varies significantly over the operating range of the transducer, with some transducer types or configurations being inherently more linear than others. Other factors that may critically affect the performance or response of a vibration transducer in a given measurement application are temperature; relative humidity; external acoustic, magnetic, and electrostatic fields; transverse vibration; base strain; chemicals; and radiation. A vibration transducer should be selected based on its long- and short-term stability; the match between its performance characteristics (e.g., sensitivity, frequency response, amplitude linearity, self-noise) and the expected amplitude of vibration, frequency range of analysis, and expected environmental conditions of measurement; and any other pertinent considerations (e.g., size, mass, and resonant frequency).

Vibration exciters, or **shakers**, are used in structural analysis, vibration analysis of machinery, fatigue testing, mechanical impedance measurements, and vibration calibration systems. Vibration exciters have a table or moving element with a drive mechanism that may be mechanical, electrodynamic, piezoelectric, or hydraulic. They range from relatively small, low-power units for calibrating transducers such as accelerometers to relatively large, high-power units for structural and fatigue testing.

Conditioning amplifiers, power supplies, preamplifiers, charge amplifiers, voltage amplifiers, power amplifiers, filters, controllers, and displays are available either separately or integrated into a measuring instrument or system, such as a structural analysis system, vibration analyzer, vibration monitoring system, vibration meter, measuring amplifier, multichannel data-acquisition and modal analysis system, or real-time fractional-octave or FFT signal analyzer. The choice of instrument(s) to include in a vibration measurement system depends on the mechanical quantity to be determined, the purpose of the measurement, and the frequency range and resolution of signal analysis. In the case of vibration measurements, the signal analysis is relatively narrow in bandwidth and may be relatively low in frequency in order to accurately characterize structural resonances. Accelerometers with internal integrated circuitry are available to provide impedance matching or servo control for measuring very low frequency acceleration (servo accelerometers). Analog integration and differentiation of vibration signals is available through integrating and differentiating networks and amplifiers, and digital is available through FFT analyzers. Vibration measurements made for different purposes (e.g., machinery diagnostics and health monitoring, balancing rotating machinery, analysis of torsional vibration, analysis of machine-tool vibration, modal analysis, analysis of vibration isolation, stress monitoring, industrial control) will generally each dictate different mechanical measurement requirements and a different optimal set of instrumentation.

Calibration

Because of their inherent long- and short-term stability, amplitude linearity, wide bandwidth, wide dynamic range, low noise, and wide range of sensitivities, seismic accelerometers have traditionally been used as a reference standard for dynamic mechanical measurements. A measurement system should be calibrated as a system from transducer to indicating device before it is used to perform absolute dynamic measurements of mechanical quantities. Calibrated reference vibration exciters, standard reference accelerometers, precision conditioning amplifiers, and precision calibration exciters are available for this purpose. These exciters and standard reference accelerometers can be used to transfer a calibration to another transducer. For demanding applications, either a calibrated exciter or a standard reference accelerometer with connecting cable and conditioning amplifier should periodically be sent to the manufacturer, a private testing laboratory, or a national standards laboratory for calibration.

LIGHTING MEASUREMENT

Light level, or illuminance, is usually measured with a photocell made from a semiconductor such as silicon or selenium. Such photocells produce an output current proportional to incident luminous flux; when linked with a microammeter, color- and cosine-corrected filters, and multirange switches, they are used in inexpensive handheld light meters and more precise instruments. Different cell heads allow multirange use in precision meters.

Cadmium sulfide photocells, in which the resistance varies with illumination, are also used in light meters. Both gas-filled and vacuum photoelectric cells are in use.

Small survey-type meters are not as accurate as laboratory meters; their readings should be considered approximate, although consistent, for a given condition. Their range is usually from 5 to 5000 footcandles. Precision low-level meters have cell heads with ranges down to 0 to 2 footcandles.

A photometer installed in a revolving head is called a goniophotometer and is used to measure the distribution of light sources or luminaires. To measure total luminous flux, the luminaire is placed in the center of a sphere painted inside with a high-reflectance white with a near perfect diffusing matte surface. Total light output is measured through a small baffled window in the sphere wall.

To measure irradiation from germicidal lamps, a filter of fused quartz with fluorescent phosphor is placed over the light meter cell.

If meters are used to measure the number of lumens per unit area diffusely leaving a surface, luminance (cd/in^2) instead of illumination (footcandles) is read. Light meters can be used to measure luminance; or electronic lux meters containing a phototube, an amplifier, and a microammeter can read luminance directly. In the case of a perfectly diffuse reflecting surface, which has a constant luminance regardless of viewing angle, the unit of footlamberts in lumens/ft^2 is sometimes applied.

THERMAL COMFORT MEASUREMENT

Thermal comfort depends on the combined influence of clothing, activity, air temperature, air velocity, mean radiant temperature, and air humidity. Thermal comfort is influenced by heating or cooling of particular body parts. This is due to radiant temperature asymmetry (plane radiant temperature), draft (air temperature, air velocity, turbulence), vertical air temperature differences, and floor temperature (surface temperature).

A general description of thermal comfort is given in Chapter 8, and guidelines for an acceptable thermal environment are given in ASHRAE *Standard* 55 and ISO *Standard* 7730. ASHRAE *Standard* 55 also includes required measuring accuracy. In addition to specified accuracy, ISO *Standard* 7726 includes recommended measuring locations and a detailed description of instruments and methods.

Clothing and Activity Level

These values are estimated from tables (Chapter 8, ISO *Standard* 9920, ISO *Standard* 8996). The thermal insulation of clothing (clo-value) can be measured on a thermal mannequin (McCullough et al. 1985, Olesen 1985). The activity (met-value) can be estimated from measuring CO_2 and O_2 in a person's expired air.

Air Temperature

Various types of thermometers may be used to measure air temperature. Placed in a room, the sensor registers a temperature between air temperature and mean radiant temperature. One way of reducing the radiant error is to make the sensor as small as possible because the convective heat transfer coefficient increases as the size decreases while the radiant heat transfer coefficient is constant. A smaller sensor also provides a favorably low time constant. The radiant error can also be reduced by using a shield (an open, polished aluminum cylinder) around the sensor, by using a sensor with a low-emittance surface, or by artificially increasing the air velocity around the sensor (aspirating air through a tube in which the sensor is placed).

Air Velocity

In occupied zones, air velocities are usually small (0 to 100 fpm) but have an effect on human thermal sensation. Because the velocity fluctuates, the mean value should be measured over a suitable period, typically 3 min. Velocity fluctuations with frequencies up to 1 Hz significantly increase human discomfort due to draft, which is a function of air temperature, mean air velocity, and turbulence (see Chapter 8). The fluctuations can be given as the standard deviation of the air velocity over the measuring period (3 min) or as the turbulence intensity (standard deviation divided by mean air velocity). Velocity direction may change and is difficult to identify at low air velocities. An omnidirectional sensor with a short response time should be used. A thermal anemometer is suitable. If a hot-wire anemometer is used, the direction of the flow being measured must be perpendicular to the hot wire. Smoke puffs can be used to identify the direction.

Plane Radiant Temperature

This refers to the uniform temperature of an enclosure in which the radiant flux on one side of a small plane element is the same as in the actual nonuniform environment. It describes the radiation in one direction. The plane radiant temperature can be calculated from the surface temperatures of the environment (half-room) and the angle factors between the surfaces and a plane element (ASHRAE *Standard* 55). The plane radiant temperature may also be measured by a net-radiometer or a radiometer with a sensor consisting of a reflective disk (polished) and an absorbent disk (painted black) (Olesen et al. 1989).

Mean Radiant Temperature

This is the uniform temperature of an imaginary black enclosure in which an occupant would exchange the same amount of radiant heat as in the actual nonuniform enclosure. The mean radiant temperature can be calculated from measured surface temperatures and the corresponding angle factors between the person and the surfaces (Chapter 8).

The mean radiant temperature can also be determined from the plane radiant temperature in six opposite directions weighted according to the projected area factors for a person (Chapter 8).

Because of its simplicity, the instrument most commonly used to determine the mean radiant temperature is a **black globe thermometer** (Vernon 1932, Bedford and Warmer 1935). This thermometer consists of a hollow sphere usually 6 in. in diameter coated in flat black paint with a thermocouple or thermometer bulb at its center. The temperature assumed by the globe at equilibrium results from a balance between heat gained and lost by radiation and convection.

The mean radiant temperatures are calculated from

$$\bar{t}_r = \left[(t_g + 460)^4 + \frac{4.74 \times 10^7 V_a^{0.6}}{\varepsilon D^{0.4}} (t_g - t_a) \right]^{1/4} - 460 \qquad (11)$$

where

$\bar{t}_r$ = mean radiant temperature, °F
t_g = globe temperature, °F
V_a = air velocity, fpm
t_a = air temperature, °F
D = globe diameter, ft
ε = emissivity (0.95 for black globe)

According to Equation (11), air temperature and air velocity around the globe must also be determined. The globe thermometer is spherical, while mean radiant temperature is defined in relation to the human body. For sedentary people, the globe represents a good approximation. For people who are standing, the globe, in a radiant nonuniform environment, overestimates the radiation from floor or ceiling. An ellipsoid-shaped sensor gives a closer approximation to the human shape. A black globe will also overestimate the influence of short-wave radiation (e.g., sunshine). A flat gray color better represents the radiant characteristic of normal clothing (Olesen et al. 1989). The hollow sphere is usually made of copper, which results in an undesirable high time constant. This can be overcome by using lighter materials (e.g., a thin plastic bubble).

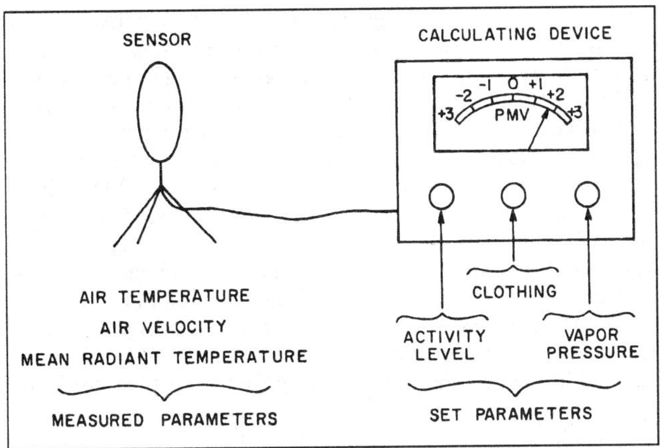

Fig. 22 Madsen's Comfort Meter
(Madsen 1976)

Air Humidity

The water vapor pressure (absolute humidity) is usually uniform in the occupied zone of a space; therefore, it is sufficient to measure absolute humidity at one location. Many of the instruments listed in Table 3 are applicable. At ambient temperatures that provide comfort or slight discomfort, the thermal effect of humidity is only moderate, and highly accurate humidity measurements are unnecessary.

CALCULATING THERMAL COMFORT

When the thermal parameters have been measured, their combined effect can be calculated by the thermal indices in Chapter 8. For example, the effective temperature (Gagge et al. 1971) can be determined from air temperature and humidity. Based on the four environmental parameters and an estimation of clothing and activity, the **predicted mean vote** (PMV) can be determined with the aid of tables (Fanger 1982, ISO *Standard* 7730, Chapter 8). The PMV is an index predicting the average thermal sensation that a group of occupants may experience in a given space.

For certain types of normal activity and clothing, the environmental parameters measured can be compared directly with those described in ASHRAE *Standard* 55 or ISO *Standard* 7730.

INTEGRATING INSTRUMENTS

Several instruments have been developed to evaluate the combined effect of two or more thermal parameters on human comfort. Madsen (1976) developed an instrument that gives information on the occupants' expected thermal sensation by direct measurement of the PMV value. The comfort meter has a heated ellipsoid-shaped sensor that simulates the body (Figure 22). The estimated clothing (insulation value), activity in the actual space, and humidity are set on the instrument. The sensor then integrates the thermal effect of the air temperature, mean radiant temperature, and air velocity in approximately the same way the body does. The electronic instrument gives the measured operative and equivalent temperature, calculated PMV, and predicted percentage of dissatisfied (PPD).

MOISTURE CONTENT AND TRANSFER MEASUREMENT

Little off-the-shelf instrumentation exists to measure the moisture content of porous materials or the moisture transfer through those materials. However, many measurements can be set up with a small investment of time and money. Three moisture properties are most commonly sought—(1) the sorption isotherm, a measure of the amount of water vapor a hygroscopic material will adsorb from humid air; (2) vapor permeability, a measure of the rate at which water vapor will pass through a given material; and (3) liquid diffusivity, a measure of the rate at which liquid water will pass through a porous material.

Sorption Isotherm

A sorption isotherm relates the **equilibrium moisture content** (EMC) of a hygroscopic material to the ambient relative humidity under conditions of constant temperature. Moisture content is the ratio of the total mass of water in a sample to the dry mass of the sample. Determining a sorption isotherm involves exposing a sample of material to a known relative humidity at a known temperature and then measuring the sample's moisture content after a sufficient period of time has elapsed for the sample to reach equilibrium with its surroundings. Hysteresis in the sorption behavior of most hygroscopic materials requires that measurements be made both for increasing relative humidity (the adsorption isotherm) and for decreasing relative humidity (the desorption isotherm).

The ambient relative humidity can be controlled using saturated salt solutions or mechanical refrigeration equipment (Tveit 1966, Cunningham and Sprott 1984, Carotenuto et al. 1991). Precise measurements of the relative humidity produced by various salt solutions have been reported by Greenspan (1977). ASTM *Standard* E 104 describes the use of saturated salt solutions. The EMC of a sample is usually determined gravimetrically using a precision balance. The sample dry mass, necessary to calculate moisture content, can be found by oven drying or desiccant drying. Oven dry mass may be lower than desiccant dry mass because of the loss of volatiles other than water in the oven (Richards et al. 1992).

A major difficulty in the measurement of the sorption isotherms of engineering materials is the long time required for many materials to reach equilibrium—often as long as weeks or months. The rate-limiting mechanism for these measurements is usually the slow process of vapor diffusion into the pores of the material. The use of smaller samples can help reduce the diffusion time.

Vapor Permeability

The diffusive transfer of water vapor through porous materials is often described by a modified form of Fick's law:

$$w''_v = -\mu \frac{dp}{dx} \quad (12)$$

where

w''_v = mass of vapor diffusing through unit area in unit time, gr/h·ft^2
dp/dx = vapor pressure gradient, in. Hg/in.
μ = vapor permeability, gr·in/h·ft^2·in. Hg

In engineering practice, permeance may be used instead of permeability. **Permeance** is simply permeability divided by the material thickness in the direction of the flow of vapor; thus, while permeability is a material property, permeance depends on thickness.

Measurement of permeability is made with wet-cup, dry-cup, or modified cup tests. Wet- and dry-cup tests are described in Chapter 22.

For many engineering materials, vapor permeability is a strong function of mean relative humidity. Wet and dry cups cannot adequately characterize this dependence on relative humidity. Instead, a modified cup method can be used (McLean et al. 1990, Burch et al. 1992). In the modified cup method, the pure water or desiccant within a cup is replaced with a saturated salt solution. A second saturated salt solution is used to condition the environment external to the cup. With such an arrangement, the relative humidities on both sides of the sample material can be varied from 0 to 100%. Several cups with a range of mean relative humidities are used to map out the dependence of vapor permeability on relative humidity.

In measuring materials of high permeability, the finite rate of vapor diffusion through air in the cup may become a factor. The air-film resistance could then be a significant fraction of the resistance to vapor flow presented by the sample material. An accurate measurement of high-permeability materials may require an accounting of diffusive rates across all air gaps (Fanney et al. 1991).

Liquid Diffusivity

The transfer of liquid water through porous materials may be characterized as a diffusion-like process:

$$w_l'' = -\rho D_l \frac{d\gamma}{dx} \tag{13}$$

where

w_l'' = mass of liquid transferred through unit area per unit time, lb/h·ft^2
ρ = liquid density, lb/ft^3
D_l = liquid diffusivity, ft^2/h
$d\gamma/dx$ = moisture content gradient, ft^{-1}

D_l typically shows a strong dependence on moisture content.

Transient measurement methods deduce the functional form of $D_l\gamma$ by observing the evolution of a one-dimensional moisture content profile over time. An initially dry specimen is brought into contact with liquid water. The free water will migrate into the specimen, drawn in by surface tension. The resulting moisture content profile, which changes with time, must be differentiated to find the liquid diffusivity of the material (Bruce and Klute 1956).

Determining the transient moisture content profile typically involves the use of a noninvasive and nondestructive method of measuring local moisture content. Gamma ray absorption (Freitas et al. 1991, Kumaran and Bomberg 1985, Quenard and Sallee 1989), X-ray radiography (Ambrose et al. 1990), neutron radiography (Prazak et al. 1990), and nuclear magnetic resonance (NMR) (Gummerson et al. 1979) have all been employed.

Uncertainty in the resulting measurement of the liquid diffusivity is often large because of the necessity to differentiate noisy experimental data.

HEAT TRANSFER THROUGH BUILDING MATERIALS

Thermal Conductivity

The thermal conductivity of a heat insulator, as defined in Chapter 22, is a unit heat transfer factor. Two methods of determining the thermal conductivity of flat insulation are the **guarded hot plate** and the **heat flow meter apparatus**, according to ASTM *Standards* C 177 and C 518, respectively. Both methods use parallel isothermal plates to induce a steady temperature gradient across the thickness of the specimen(s). The guarded hot plate is considered an absolute method for determining thermal conductivity. The heat flow meter apparatus requires calibration with a specimen having a known thermal conductivity, usually determined in the guarded hot plate. The heat flow meter apparatus is calibrated by determining the voltage output of its heat flux transducer(s) as a function of the heat flux through the transducer(s).

The basic design of the guarded hot plate consists of an electrically heated plate and two liquid-cooled plates. Two similar specimens of a material are required for a test; one is mounted on each side of the hot plate. A cold plate is then pressed against the outside of each specimen by a clamp screw. The heated plate consists of two sections separated by a small gap. During tests, the central (metering) section and the outer (guard) section are maintained at the same temperature to minimize errors caused by edge effects. The electric energy required to heat the metering section is measured carefully and converted to heat flow. The thermal conductivity of the material can be calculated under steady-state conditions using this heat flow quantity, the area of the metering section, the temperature gradient, and the specimen thickness. The thermal conductivity of cylindrical or pipe insulation (Chapter 22) is determined in a similar manner, but an equivalent thickness must be calculated to account for the cylindrical shape (ASTM *Standard* C 335). Transient methods have been developed by D'Eustachio and Schreiner (1952), Hooper and Lepper (1950), and Hooper and Chang (1953) using a line heat source within a slender probe. These instruments are available commercially and have the advantages of rapidity and a small test specimen requirement. The probe is a useful research and development tool, but it has not been as accepted as the guarded hot plate, heat flow meter apparatus, or pipe insulation apparatus.

Thermal Conductance and Resistance

Thermal conductances (C-factors) and resistances (R-values) of many building assemblies can be calculated from the conductivities and dimensions of their components, as described in Chapter 24. Test values can also be determined experimentally using large specimens representative of the building assemblies tested in the hot box apparatus described in ASTM *Standards* C 236 and C 976. This laboratory apparatus allows measurement of heat transfer through a specimen under controlled air temperature, air velocity, and radiation conditions. It is specially suited for large nonhomogeneous specimens.

For in situ measurements, heat flux and temperature transducers are useful in measuring the dynamic or steady-state behavior of opaque building components (ASTM *Standard* C 1046). A heat flux transducer is simply a differential thermopile within a core or substrate material. There are two types of construction: (1) multiple thermocouple junctions wrapped around a core material, or (2) printed circuits with a uniform array of thermocouple junctions. The transducer is calibrated by determining its voltage output as a function of the heat flux through the transducer. For in situ measurements, the transducer is installed in either the wall or roof construction, or mounted on an exterior surface with tape or glue. The data obtained from the heat flux and temperature transducers can be used to compute the thermal conductance or resistance of the building component (ASTM *Standard* C 1255).

AIR INFILTRATION AND AIRTIGHTNESS MEASUREMENT

Two major characteristics describe air infiltration in buildings—air exchange rates and envelope air leakage. The measurement approaches used to determine these factors are described in Chapter 25. The air exchange rate of a building refers to the rate at which outdoor air enters the building under normal conditions of weather and ventilation system operation. In general, the air change rate includes both outdoor air taken in through the air-handling system and air leakage through the building envelope (infiltration). The outdoor air intake rate is determined by the design, installation, and operation of the mechanical ventilation system. Infiltration is determined by the extent and distribution of leaks over the building envelope and the pressure differences across these leaks. These pressure differences are induced by wind, inside-outside temperature differences, and the operation of building mechanical equipment. Building air exchange rates can only be measured by injecting a tracer gas into a building and monitoring and analyzing the tracer gas concentration response. The instrumentation required for tracer gas testing includes (1) a means of injecting the tracer gas and sampling the air in the building and (2) a tracer gas monitor. There is a variety of tracer gas techniques, distinguished by the injection strategy and analysis approach employed. Carbon dioxide is often used as the tracer gas, with occupant-generated CO_2 precluding the need to inject additional tracer gas. To fully characterize the air exchange performance of a building, the air exchange rate must be measured over a range of weather and equipment operation.

The airtightness of a building envelope can be measured relatively quickly using building pressurization techniques, which are described in Chapter 25. In the pressurization technique, a large fan or blower mounted in a door or window induces a large and roughly uniform pressure difference across the building shell. The airflow required to maintain this pressure difference is then measured. The more leakage in the building, the more airflow is required to induce a specific indoor-outdoor pressure difference. The building airtightness is characterized by the airflow rate at a reference pressure, normalized by the building volume or surface area. Under proper test conditions, the results of a pressurization test are independent of weather conditions. The instrumentation requirements for pressurization testing include air-moving equipment, a device to measure airflow, and a differential pressure gage.

AIR CONTAMINANT MEASUREMENT

Three measures of particulate air contamination include the number, projected area, and mass of particles per unit volume of air (ASTM Volume 11.03, ASTM *Special Technical Publication* STP 957). Each requires an appropriate sampling technique.

Particles are counted by capturing them in impingers, impactors, membrane filters, or thermal or electrostatic precipitators. Counting may be done by microscope, using stage counts if the sample covers a broad range of sizes.

Electronic particle counters can give rapid data on particle size distribution and concentration. However, their accuracy depends on careful calibration, appropriate maintenance, and proper application. Particle counters have been used in indoor office environments as well as in clean rooms.

Projected area determinations are usually made by sampling onto a filter paper and comparing the light transmitted or scattered by this filter to a standard filter. The staining ability of dusts depends on the projected area and refractive index per unit volume. For sampling, filters must collect the minimum sized particle of interest. In this respect, membrane or glass fiber filters are recommended.

To determine particle mass, a measured quantity of air is drawn through filters, preferably of membrane or glass fiber, and the filter mass is compared to the mass before sampling. Electrostatic or thermal precipitators and various impactors have also been used. For further information, see ACGIH (1983), Lundgren et al. (1979), and Lodge (1989).

Chapter 41 of the 1995 *ASHRAE Handbook—Applications* presents information on measuring and monitoring gaseous contaminants. Relatively costly analytical equipment, which must be calibrated and operated carefully by experienced personnel, is needed. Numerous methods of sampling the contaminants, as well as the laboratory analysis techniques used after sampling, are specified. Some of the analytical methods are specific to a single pollutant; others are capable of presenting a concentration spectrum for many compounds simultaneously.

CARBON DIOXIDE MEASUREMENT

Carbon dioxide has become an important measurement parameter for air-conditioning, heating, and refrigerating engineers, particularly for use in indoor air quality (IAQ) applications. Although CO_2 is generally not of concern as a specific toxin in indoor air, it is used as a surrogate indicator of odor related to human occupancy. ASHRAE *Standard* 62 states that maintaining CO_2 concentrations below 1000 ppm (based on a differential of 700 ppm between indoor and outdoor CO_2 concentrations) usually results in conditions conducive to comfort and reduced odor from human-generated pollutants. *Standard* 62 also recommends specific minimum outdoor air ventilation rates to ensure adequate indoor air quality. Carbon dioxide is often used as a tracer gas when quantifying outdoor air ventilation rates. Carbon dioxide sensors are also used in building control strategies to optimize ventilation as a function of occupancy.

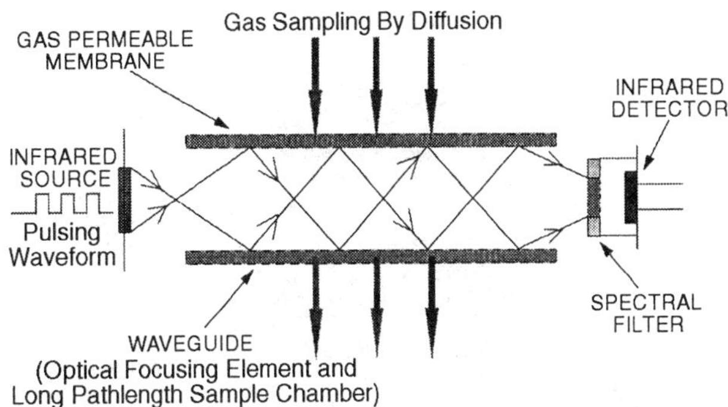

Fig. 23 Nondispersive Infrared Carbon Dioxide Sensor

NONDISPERSIVE INFRARED CO_2 DETECTORS

The technology in most widespread use for IAQ applications is the nondispersive infrared (NDIR) sensor (Figure 23). This device makes use of the strong absorption band that CO_2 produces at 4.2 μm when excited by an infrared light source. Indoor air quality-specific NDIR instruments, when calibrated between 0 and 5000 ppm, are typically accurate within 150 ppm, but the accuracy of some sensors can be improved to within 50 ppm if the instrument is calibrated for a narrower range. Portable NDIR meters are available with direct-reading digital displays; however, response time varies significantly among different instruments. While most NDIR cell designs facilitate very rapid CO_2 sample diffusion, some of the instruments now in widespread use for indoor air quality measurement exhibit slower sensor response, resulting in stabilization times greater than 5 min (up to 15 min), which may complicate walk-through inspections.

Calibration

In a clean, stable environment, NDIR sensors can hold calibration for months, but condensation, dust, dirt, and mechanical shock may offset calibration. As with all other CO_2 sensor technologies, NDIR sensor readings are proportional to pressure due to the change in density of the gas molecules that results from a change in the sample pressure. This leads to errors in CO_2 readings when the barometric pressure changes from the calibration pressure. Weather-induced errors will be small, but all CO_2 instruments should be recalibrated if used at an altitude that is significantly different from the calibration altitude. Most NDIR sensors are sensitive to cooling effects when placed in an airstream. This is an important consideration when locating a fixed sensor or when using a portable system to evaluate air-handling system performance because airflow in supply and return ducts may significantly shift readings.

Applications

Nondispersive infrared sensors are well suited for equilibrium tracer and tracer decay ventilation studies, and faster response models are ideal for a quick, basic evaluation of human-generated pollution and ventilation adequacy. When properly located, these sensors are also appropriate for continuous monitoring and for control strategies using equilibrium tracer and air fraction tracer calculations.

AMPEROMETRIC ELECTROCHEMICAL CO_2 DETECTORS

Amperometric electrochemical CO_2 sensors (Figure 24), a newly available technology, offer repeatability and response time optimized for portable diagnostic work. A measured current is

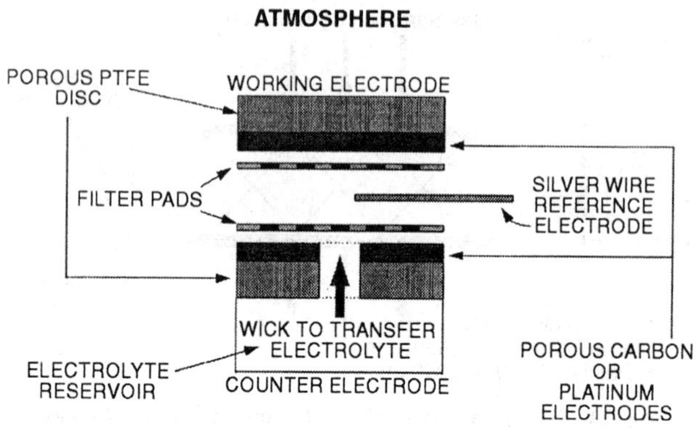

Fig. 24 Amperometric Carbon Dioxide Sensor

driven between two electrodes by the reduction of CO_2 that has diffused across a porous membrane. Electrochemical CO_2 sensors may exhibit a change in electrolyte chemistry over time (typically 18 to 36 months), so sensors should be replaced, unlike NDIR sensors, which normally last the lifetime of the instrument (although NDIR lamps may need periodic replacement). These sensors typically hold their calibration for several weeks, but over the longer term they may exhibit more drift if exposed to low humidities; this drift often makes them less suitable for continuous monitoring applications. At low humidity (below 20% rh), the sensors must be kept moist (with a simple accessory) to maintain specified accuracy.

Advantages to this technology are a very quick response (90% of response in less than 2 minutes) and no required warm-up period. Amperometric electrochemical sensors have a far lower power requirement than NDIR detectors, usually operating continuously for weeks where NDIR instruments typically operate for only 3 to 12 h on a new set of internal batteries or on a full charge. The longer battery life is advantageous for spot checks and walk-throughs and for measuring CO_2 distribution throughout a building and within a zone. Unlike most NDIR sensors, amperometric electrochemical sensors are not affected by very high humidity or shock, although readings may be affected while high levels of condensate are on the sensor. Another important advantage is that these detectors are insensitive to the cooling effects of airflow, making them appropriate for spot measurements of CO_2 in supply ducts and in return airflows when determining total outdoor air supplied by the HVAC system and when surveying HVAC system performance.

PHOTOACOUSTIC CO_2 DETECTORS

Both open- and closed-cell photoacoustic sensors are available.

Open-Cell Sensors

Open-cell photoacoustic CO_2 sensors (Figure 25) operate as air diffuses through a permeable membrane into a chamber that is pulsed with filtered light at the characteristic CO_2 absorption frequency of 4.2 µm. The light energy absorbed by the CO_2 heats the sample chamber, causing a pressure pulse, which is sensed by a piezoresistor. Open-cell photoacoustic CO_2 sensors are presently unavailable in portable instruments, in part because any vibration that might occur during transportation would affect calibration and might affect the signal obtained for a given concentration of CO_2. Ambient acoustical noise may also influence readings. For continuous monitoring, vibration is a concern, as are temperature and airflow cooling effects. However, if a sensor is located properly and if the optical filter is kept relatively clean, photoacoustic CO_2 sensors may be very stable. Commercially available open-cell photoacoustic transmitters do not allow recalibration to adjust for pressure differences, so an offset should be incorporated in any control system using these sensors at an altitude or duct pressure other than calibration conditions.

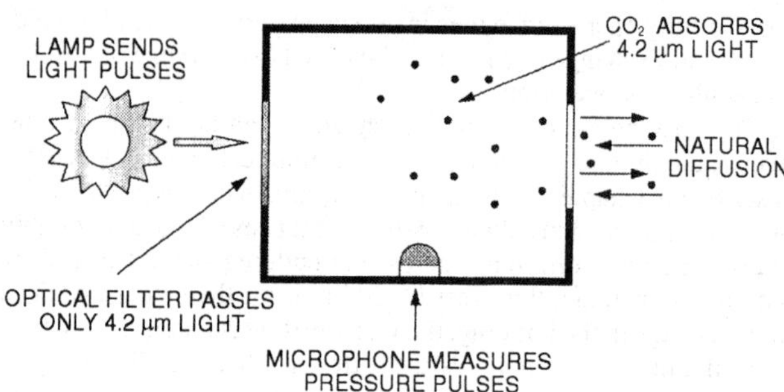

Fig. 25 Open-Cell Photoacoustic Carbon Dioxide Sensor

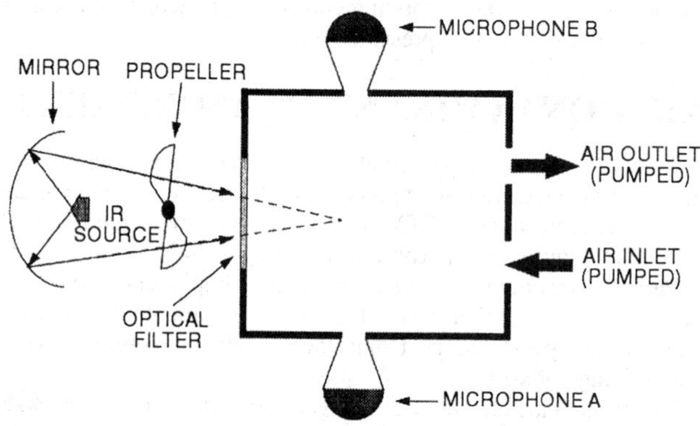

Fig. 26 Closed-Cell Photoacoustic Carbon Dioxide Sensor

Closed-Cell Sensors

Closed-cell photoacoustic sensors (Figure 26) operate under the same principle as the open-cell version, except that samples are pumped into a sample chamber that is sealed and environmentally stabilized. Two acoustic sensors are sometimes used in the chamber to minimize vibration effects. Closed-cell units, available as portable or fixed monitors, come with particle filters that are easily replaced (typically at 3- to 6-month intervals) if dirt or dust accumulates on them. Closed-cell photoacoustic monitors permit recalibration to correct for drift, pressure effects, or other environmental factors that might influence accuracy.

POTENTIOMETRIC ELECTROCHEMICAL CO_2 DETECTORS

Potentiometric electrochemical CO_2 sensors use a porous fluorocarbon membrane that is permeable to CO_2, which diffuses into a carbonic acid electrolyte, changing the electrolyte pH. This pH change is monitored by a pH electrode inside the cell. The pH electrode isopotential drift prohibits long-term monitoring to the accuracy and resolution required for continuous measurement or control or for detailed IAQ evaluations, although accuracy within 100 ppm, achievable short-term over the 2000 ppm range, may be adequate for basic ventilation and odor evaluations. In addition, this type of sensor exhibits slow response, which increases the operator time necessary for field applications or for performing a walk-through of a building.

COLORIMETRIC DETECTOR TUBES

Colorimetric detector tubes contain a chemical compound that discolors in the presence of CO_2 gas, with the amount of discoloration related to the CO_2 concentration. These detector tubes are often used to spot check CO_2 levels; when used properly, they are accurate to within 25%. If numerous samples are taken (i.e., six or more), uncertainty may be reduced; this may render the tubes, if used in the late afternoon, adequate as a very basic determination of odor and discomfort related to human occupation. However, CO_2 detector tubes are generally not appropriate for specific ventilation assessment because of their inaccuracy and inability to record concentration changes over time.

LABORATORY MEASUREMENTS

Laboratory techniques for measuring CO_2 concentration include mass spectroscopy, thermal conductivity, infrared spectroscopy, and gas chromatography. These techniques typically require taking on-site **grab samples** for laboratory analysis. Capital costs for each piece of equipment are high, and significant training is required. A considerable drawback to grab sampling is that CO_2 levels change significantly during the day and over the course of a week, making it sensible to place sensors on site with an instrument capable of recording or data logging measurements continuously over the course of a workweek. An automated grab sampling system capturing many samples of data would be quite cumbersome and expensive if designed to provide CO_2 trend information over time. However, an advantage to laboratory techniques is that they can be highly accurate. A mass spectrometer, for example, can measure CO_2 concentration to within 5 ppm from 0 to 2000 ppm. All laboratory measurement techniques are subject to errors resulting from interfering agents. A gas chromatograph is typically used in conjunction with the mass spectrometer to eliminate interference from nitrous oxide (N_2O), which has an equivalent mass, if samples are collected in a hospital or in another location where N_2O might be present.

COMBUSTION ANALYSIS

Two approaches are used to measure the thermal output or capacity of a boiler, furnace, or other fuel-burning device. The direct or **calorimetric test** measures change in enthalpy or heat content of the fluid, air, or water heated by the device, and multiplies this by the flow rate to arrive at the unit's capacity. The indirect test or **flue gas analysis** method determines the heat losses in the flue gases and the jacket and deducts them from the heat content (higher heating value) of the measured fuel input to the appliance. A **heat balance** simultaneously applies both tests to the same device. The indirect test usually indicates the greater capacity, and the difference is credited to radiation from the casing or jacket and unaccounted-for losses.

With small equipment, the expense of the direct test is usually not justified, and the indirect test is used with an arbitrary radiation and unaccounted-for loss factor.

FLUE GAS ANALYSIS

The flue gases from burning fossil fuels generally contain carbon dioxide (CO_2), water, and hydrogen (H_2) with some small amounts of carbon monoxide (CO), nitrogen oxides (NO_x), sulfur oxides (SO_x), and unburned hydrocarbons. However, generally only CO_2 (or O_2) and CO are measured to determine completeness of combustion and efficiency.

In the laboratory, the instruments most commonly used to measure CO and CO_2 are nondispersive infrared (NDIR) analyzers. The NDIR instruments have several advantages: (1) they are not very sensitive to flow rate, (2) no wet chemicals are required, (3) they have a relatively fast response, (4) measurements can be made over a wide range of concentrations, and (5) they are not sensitive to the presence of contaminants in the ambient air.

In the laboratory, oxygen is generally measured with an instrument that makes use of the paramagnetic properties of oxygen. The paramagnetic instruments are generally used because of their excellent accuracy and because they can be made specific to the measurement of oxygen.

For field testing and burner adjustment, portable combustion testing equipment is available. These instruments generally measure O_2 and CO with electrochemical cells. The CO_2 is then calculated by an on-board microprocessor and, together with temperature, is used to calculate thermal efficiency. If a less expensive approach is required, a portable Orsat apparatus can be used to measure CO_2, and a length-of-stain tube to measure CO.

DATA ACQUISITION AND RECORDING

Almost every type of transducer and sensor is available with the necessary interface system to make it computer-compatible. The transducer itself begins to lose its identity when integrated into a system that incorporates such features as linearization, offset correction, self-calibration, and so forth. This has eliminated the concern regarding the details of signal conditioning and amplification of basic transducer outputs. The personal computer is integrated into every aspect of data recording, including sophisticated graphics, acquisition and control, and analysis. Modems connected to the Internet or an internal network allow easy access to remote personal computer-based data-recording systems from virtually any locale.

Other means of recording, such as chart recorders, which can be either multipurpose or specifically designed for a given sensor, are available. Chart recorders provide a visual indication and a hard copy record of the data. Rarely is the output of a chart recorder used to process data. Simple indicators and readouts are used mostly to monitor the output of a sensor visually. In most situations, analog indicators such as d'Arsonval movement meters have been rendered obsolete by modern digital indicators. Industrial environments commonly employ signal transmitters for control or computer data-handling systems to convert the signal output of the primary sensor into a compatible common signal span (e.g., the standard 4-20 mA current loop). All signal conditioning (ranging, zero suppression, reference-junction compensation) is provided at the transmitter. Thus, all recorders and controllers in the system can have an identical electrical span, with variations only in charts and scales offering the advantages of interchangeability and economy in equipment cost. Long signal transmission lines can be used, and receiving devices can be added to the loop without degrading performance.

The vast selection of available hardware, an often confusing set of terminology, and the challenge of optimizing the performance/cost ratio for a specific application make the task of configuring a data acquisition system difficult. A system specifically configured to meet a particular measurement need can quickly become obsolete if it has inadequate flexibility. Memory size, recording speed, and signal processing capability are major considerations in determining the correct recording system. Thermal, mechanical, electromagnetic interference, portability, and meteorological factors also influence the selection.

Digital Recording

A digital data acquisition system must contain an interface, which is a system involving one or several analog-to-digital converters, and in the case of multichannel inputs, circuitry for multiplexing. The interface may also provide excitation for transducers, calibration, and conversion of units. The digital data are arranged into one or several standard digital bus formats. Many data acquisition systems are designed to acquire data rapidly and store large records of data for later recording and analysis. Once the input signals have been digitized, the digital data are essentially immune to noise and can be transmitted over great distances.

Information is transferred to a computer/recorder from the interface as a pulse train, which can be transmitted as 4-, 8-, 12-, 16-, or 32-bit words. An 8-bit word is a byte; many communications methods are rated according to their bytes per second transfer rate. Digital data are transferred in either serial or parallel mode. Serial transmission means that the data are sent as a series of pulses, one bit at a time. Although slower than parallel systems, serial interfaces require only two wires, which lowers their cabling cost. The speed of serial transmissions is rated according to the symbols per second rate, or baud rate. In parallel transmission, the entire data word is transmitted at one time. To do this, each bit of a data word has to have its own transmission line; other lines are needed for clocking and control. Parallel mode is used for short distances or when high data transmission rates are required. Serial mode must be used for long-distance communications where wiring costs are prohibitive.

The two most popular interface bus standards currently used for data transmission are the IEEE 488, or general-purpose interface bus (GPIB), and the RS232 serial interface. The IEEE 488 bus system feeds data down eight parallel wires, one data byte at a time. This parallel operation allows it to transfer data rapidly at up to 1 million characters per second. However, the IEEE 488 bus is limited to a cable length of 65 ft and requires an interface connection on every meter for proper termination. The RS232 system feeds data serially down two wires, one bit at a time. An RS232 line may be over 1000 ft long. For longer distances, it may feed a modem to send data over standard telephone lines. A local area network (LAN) may be available in a facility for transmitting information. With appropriate interfacing, transducer data are available to any computer connected to the network.

Bus measurements can greatly simplify three basic applications—data gathering, automated limit testing, and computer-controlled processes. In data gathering applications, readings are collected over time. The most common applications include aging tests in quality control, temperature tests in quality assurance, and testing for intermittents in service. A controller can monitor any output indefinitely and then display the data directly on its screen or record it on magnetic tape or disks for future use.

In automated limit testing, the computer simply compares each measurement with programmed limits. The controller converts the readings to a good/bad readout. Automatic limit testing becomes highly cost-effective when working with large number of parameters of a particular unit under test.

In computer-controlled processes, the IEEE 488 bus system becomes a permanent part of a larger, completely automated system. For example, a large industrial process may require many electrical sensors that feed a central computer controlling many parts of the manufacturing process. An IEEE 488 bus controller collects readings from several sensors and saves the data until asked to dump an entire batch of readings to a larger central computer at one time. Used in this manner, the IEEE 488 bus controller serves as a slave of the central computer.

Dynamic range and accuracy are two important parameters that must be considered in a digital recording system. **Dynamic range** refers to the ratio of the maximum input signal for which the system is useful to the noise floor of the system. The **accuracy** figure for a system is impacted by the signal noise level, nonlinearity, temperature, time, crosstalk, and so forth. In selecting an 8-, 12-, or 16-bit analog-to-digital converter, the designer cannot assume that the system accuracy will necessarily be determined by the resolution of the encoders (i.e., 0.4%, 0.025%, and 0.0016%, respectively. If the sensor preceding the converter is limited to 1% full-scale accuracy, for example, no significant benefits are gained by using a 12-bit system over an 8-bit system and suppressing the least significant bit. However, a greater number of bits may be required to cover a larger dynamic range.

Data Logging Devices

Data loggers digitally store electrical signals (analog or digital) to an internal memory storage component. The signal from connected sensors is typically stored to memory at timed intervals ranging from MHz to hourly sampling. Some data loggers store data based on an event (e.g., button push, contact closure). Many data loggers can perform linearization, scaling, or other signal conditioning and permit logged readings to be either instantaneous or averaged values. Most data loggers have built-in clocks that record the time and date together with transducer signal information. Data loggers range from single-channel input to 256 or more channels. Some are general-purpose devices that will accept a multitude of analog and/or digital inputs, while others are more specialized to a specific measurement (e.g., a portable anemometer with built-in data logging capability) or for a specific application (e.g., a temperature, relative humidity, CO_2, and CO monitor with data logging for IAQ applications). Stored data are generally downloaded using a serial interface with a temporary direct connection to a personal computer. Remote data loggers may also download via modem through telephone lines. Some data loggers are designed to allow downloading directly to a printer.

Chart Recorders

Chart recorders convert electrical signals (analog or digital) to records of the data versus time on a hard copy, usually paper. Mechanical styluses use ink, hot wire, pressure, or electrically sensitive paper to provide a continuous trace. They are useful up to a few hundred hertz. Thermal and ink recorders are confined to chart speeds of a few inches per second for recording relatively slow processes. Newer advances in portable recorders provide multichannel inputs and up to 25 kHz real-time frequency response without using a pen or pen motor. Both *x-y* recorders and plotters allow two variables to be recorded with respect to one another. Their response times are generally limited to that of thermal and ink recorders. Oscillographic recorders have largely been made obsolete by digital oscilloscopes.

REFERENCES

ACGIH. 1983. *Air sampling instruments for evaluation of atmospheric contaminants*, 6th ed. American Conference of Governmental Industrial Hygienists, Cincinnati, OH.

Ambrose, J.H., L.C. Chow, and J.E. Beam. 1990. Capillary flow properties of mesh wicks. AIAA *Journal of Thermophysics* 4:318-24.

Amdur, E.J. 1965. Two-pressure relative humidity standards. In *Humidity and Moisture* 3:445. Reinhold Publishing Corporation, New York.

ASHRAE. 1985. Laboratory methods of testing fans for rating. ANSI/ASHRAE *Standard* 51-1985, also ANSI/AMCA *Standard* 210-85.

ASHRAE. 1988. Practices for measurement, testing, adjusting, and balancing of building heating, ventilation, air-conditioning and refrigeration systems. ANSI/ASHRAE *Standard* 111-1988.

ASHRAE. 1989. Standard methods of measurement of flow of liquids in pipes using orifice flowmeters. ANSI/ASHRAE *Standard* 41.8-1989.

ASHRAE. 1992. Thermal environmental conditions for human occupancy. ANSI/ASHRAE *Standard* 55-1992.

ASHRAE. 1994. Standard method for measurement of moist air properties. ANSI/ASHRAE *Standard* 41.6-1994.

ASME. 1972. Application of fluid meters, Part II, 6th ed. (Interim Supplement 19.5 on Instruments and Apparatus) *Standard* PTC 19.5-72. American Society of Mechanical Engineers, New York.

ASME. 1974. Temperature measurement instruments and apparatus. ANSI/ASME *Standard* PTC 19.3-74 (R 1986).

ASME. 1983. Measurement uncertainty for fluid flow in closed conduits. ANSI/ASME *Standard* MFC-2M-83.

ASME. 1985. Part I: Measurement uncertainty instruments and apparatus. ANSI/ASME *Standard* PTC 19.1-85.

ASME. 1988. Measurement of liquid flow in closed conduits by weighing methods. ANSI/ASME *Standard* MFC-9M-88.

ASME. 1989. Measurement of fluid flow in pipes using orifice, nozzle, and venturi. *Standard* MFC-3M-85.

ASME. 1991. Gauges—Pressure indicating dial type—Elastic elements. ANSI/ASME *Standard* B40.1-91.

ASME. 1991. Glossary of terms used in the measurement of fluid flow in pipes. ANSI/ASME *Standard* MFC-1M-91.

ASME. 1994. Method for establishing installation effects on flowmeters. ANSI/ASME *Standard* MFC-10M-94.

ASTM. 1985. Standard practice for maintaining constant relative humidity by means of aqueous solutions. *Standard* E 104-85 (R 1991). American Society for Testing and Materials, West Conshohocken, PA.

ASTM. 1985. Test method for steady-state heat flux measurements and thermal transmission properties by means of the guarded-hot-plate apparatus. *Standard* C 177-85.

ASTM. 1987. Sampling and calibration for atmospheric measurements. *Special Technical Publication* STP 957.

ASTM. 1989. Test method for steady-state thermal performance of building assemblies by means of a guarded hot box. *Standard* C 236-89.

ASTM. 1990. Standard practice for thermographic inspection of insulation installations in envelope cavities of frame buildings. *Standard* C 1060-90.

ASTM. 1990. Test method for thermal performance of building assemblies by means of a calibrated hot box. *Standard* C 976-90.

ASTM. 1991. Test method for steady-state heat flux measurements and thermal transmission properties by means of the heat flow meter apparatus. *Standard* C 518-91.

ASTM. 1993. *Manual on the use of thermocouples in temperature measurement. Manual* 12.

ASTM. 1993. Standard practice for use of the International System of Units (SI) (the modernized metric system). *Standard* E 380-93.

ASTM. 1993. Temperature-electromotive force (EMF) tables for standardized thermocouples. *Standard* E 230-93.

ASTM. 1995. Standard practice for determining thermal resistance of building envelope components from in-situ data. *Standard* C 1155-95.

ASTM. 1995. Standard practice for in-situ measurement of heat flux and temperature on building envelope components. *Standard* C 1046-95.

ASTM. 1995. Test method for steady-state heat transfer properties of horizontal pipe insulation. *Standard* C 335-95.

ASTM. 1995. Test methods for water vapor transmission of materials. *Standard* E 96-95.

ASTM. 1996. *Atmospheric analysis: Occupational health and safety*. Vol. 11.03.

Bedford, T. and C.G. Warmer. 1935. The globe thermometer in studies of heating and ventilating. *Journal of the Institution of Heating and Ventilating Engineers* 2:544.

Benedict, R.P. 1984. *Fundamentals of temperature, pressure and flow measurements*, 3rd ed. John Wiley and Sons, New York.

Bentz, D.P. and J.W. Martin. 1987. Using the computer to analyze coating defects. *Journal of Protective Coatings and Linings* 4(5).

BIPM. 1981. Report of the BIPM Working Group on the Statement of Uncertainties to the Comité International des Poids et Mésures, Kaarls R. Rapporteur.

Bruce, R.R. and A. Klute. 1956. The measurement of soil moisture diffusivity. Proceedings of the Soil Science Society of America 20:458-62.

Burch, D.M. 1980. Infrared audits of roof heat loss. *ASHRAE Transactions* 86(2).

Burch, D.M. and C.M. Hunt. 1978. Retrofitting an existing residence for energy conservation—An experimental study. *Building Science Series* 105. National Institute of Standards and Technology, Gaithersburg, MD.

Burch, D.M., W.C. Thomas, and A.H. Fanney. 1992. Water vapor permeability measurements of common building materials. *ASHRAE Transactions* 98(1).

Burns, G.W., M.G. Scroger, G.F. Strouse, M.C. Croarkin, and W.F. Guthrie. 1992. Temperature-electromotive force reference functions and tables for the letter-designated thermocouple types based on the ITS-90. NIST *Monograph* 175. U.S. Government Printing Office, Washington, D.C.

Carotenuto, A., F. Fucci, and G. LaFianzi. 1991. Adsorption phenomena in porous media in the presence of moist air. *International Journal of Heat and Mass Transfer* 18:71-81.

Cohen, E.R. 1990. The expression of uncertainty in physical measurements. 1990 Measurement Science Conference Proceedings, Anaheim, CA.

Coleman, H.W. and W.G. Steele. 1989. *Experimentation and uncertainty analysis for engineers*. John Wiley and Sons, New York.

Coleman, H.W. and W.G. Steele. 1995. Engineering application of experimental uncertainty analysis. *AIAA Journal* 33(10).

Considine, D.M. 1985. *Process instruments and controls handbook*, 3rd ed. McGraw-Hill, New York.

Cunningham, M.J. and T.J. Sprott. 1984. Sorption properties of New Zealand building materials. Building Research Association of New Zealand *Research Report* No. 45, Judgeford.

D'Eustachio, D. and R.E. Schreiner. 1952. A study of transient heat method for measuring thermal conductivity. *ASHVE Transactions* 58:331.

DeCarlo, J.P. 1984. *Fundamentals of flow measurement*. Instrumentation Society of America, Research Triangle Park, NC.

DeWitt, D.P. and G.D. Nutter. 1988. *Theory and practice of radiation thermometry*. John Wiley and Sons, New York.

Fanger, P.O. 1982. *Thermal comfort*. Robert E. Krieger, Malabar, FL.

Fanney, A.H., W.C. Thomas, D.M. Burch, and L.R. Mathena. 1991. Measurements of moisture diffusion in building materials. *ASHRAE Transactions* 97:99-113.

Freitas, V., P. Crausse, and V. Abrantes. 1991. Moisture diffusion in thermal insulating materials. In *Insulation materials: Testing and applications*, Vol. 2. ASTM *Special Technical Publication* STP 1116. American Society for Testing and Materials, West Conshohocken, PA.

Gagge, A.P., J.A.J. Stolwijk, and Y. Nishi. 1971. An effective temperature scale based on a simple model of human physiological regulatory response. *ASHRAE Transactions* 77(1).

Goldstein, R.J. 1978. Application of aerial infrared thermography. *ASHRAE Transactions* 84(1).

Greenspan, L. 1977. Humidity fixed points of binary saturated aqueous solutions. *Journal of Research of the National Bureau of Standards* 81A:89-95.

Greenspan, L. and A. Wexler. 1968. An adiabatic saturation psychrometer. *Journal of Research of the National Bureau of Standards* 72C(1):33.

Gummerson, R.J., C. Hall, W.D. Hoff, R. Hawkes, G.N. Holland, and W.S. Moore. 1979. Unsaturated water flow within porous materials observed by NMR imaging. *Nature* 281:56-57.

Hasegawa, S. 1976. The NBS two-pressure humidity generator, Mark 2. *Journal of Research of the National Bureau of Standards* 81A:81.

Hooper, F.C. and F.C. Lepper. 1950. Transient heat flow apparatus for the determination of thermal conductivity. *ASHVE Transactions* 56:309.

Hooper, F.C. and S.C. Chang. 1953. Development of thermal conductivity probe. *ASHVE Transactions* 59:463.

Hudson, R.D., Jr. 1969. *Infrared system engineering*. John Wiley & Sons, NY.

ISO. 1977. Measurement of fluid flow in closed conduits—Velocity area method using Pitot static tubes. *Standard* 3966. International Organization for Standardization, Geneva.

ISO. 1985. Thermal environments—Instruments and methods for measuring physical quantities. *Standard* 7726.

ISO. 1990. Ergonomics—Determination of metabolic heat production. *Standard* 8996.

ISO. 1994. Moderate thermal environments—Determination of the PMV and PPD indices and specification of the conditions for thermal comfort. *Standard* 7730.

ISO. 1995. Ergonomics of the thermal environment—Estimation of the thermal insulation and evaporative resistance of a clothing ensemble. *Standard* 9920.

Kumaran, M.K. and M. Bomberg. 1985. A gamma-spectrometer for determination of density distribution and moisture distribution in building materials. Proceedings of the International Symposium on Moisture and Humidity, Washington, D.C., pp. 485-90.

Kusuda, T. 1965. Calculation of the temperature of a flat-plate wet surface under adiabatic conditions with respect to the Lewis relation. *Humidity and Moisture* I:16. Reinhold Publishing Corporation, New York.

Liptak, B.G., ed. 1972. *Instrument engineers handbook*, Vol. 1. Chilton Book Company, Philadelphia, PA.

Lodge, J.P., ed. 1989. *Methods of air sampling and analysis*, 3rd ed. Lewis Publishers, MI.

Lundgren, D.A. et al., eds. 1979. *Aerosol measurement*. University Presses of Florida, Gainesville.

Mack, R.T. 1986. Energy loss profiles: Foundation for future profit in thermal imager, sales, and service. Proceedings of the 5th Infrared Information Exchange, Book 1, AGEMA Infrared Systems, Secaucus, NJ.

Madsen, T.L. 1976. Thermal comfort measurements. *ASHRAE Transactions* 82(1).

Mattingly, G.E. 1984. Workshop on fundamental research issues in orifice metering. GRI *Report* 84/0190. Gas Research Institute, Chicago, IL.

Mattingly, G.E. 1992. The characterization of a piston displacement-type flowmeter calibration facility and the calibration and use of pulsed output type flowmeters. *Journal of Research of the National Institute of Standards and Technology* 97(5):509.

McCullough, E.A., B.W. Jones, and J. Huck. 1985. A comprehensive data base for estimating clothing insulation. *ASHRAE Transactions* 92:29-47.

McLean, R.C., G.H. Galbraith, and C.H. Sanders. 1990. Moisture transmission testing of building materials and the presentation of vapour permeability values. *Building Research and Practice*. 82-103.

Mease, N.E., W.G. Cleveland, Jr., G.E. Mattingly, J.M. Hall. 1992. Air speed calibrations at the National Institute of Standards and Technology. Proceedings of the 1992 Measurement Science Conference, Anaheim, CA.

Miller, R.W. 1983. Measurement engineering handbook. McGraw-Hill, NY.

NIST. 1976. Liquid-in-glass thermometry. NIST *Monograph* 150. National Institute of Standards and Technology, Gaithersburg, MD.

NIST. 1986. Thermometer calibrations. NIST *Monograph* 174.

Nottage, H.B., J.G. Slaby, and W.P. Gojsza. 1952. A smoke-filament technique for experimental research in room air distribution. *ASHVE Transactions* 58:399.

Olesen, B.W. 1985. A new and simpler method for estimating the thermal insulation of a clothing ensemble. *ASHRAE Transactions* 92:478-92.

Olesen, B.W., J. Rosendahl, L.N. Kalisperis, L.H. Summers, and M. Steinman. 1989. Methods for measuring and evaluating the thermal radiation in a room. *ASHRAE Transactions* 95(1).

Paljak, I. and B. Pettersson. 1972. *Thermography of buildings*. National Swedish Institute for Materials Testing, Stockholm.

Parmelee, G.V. and R.G. Huebscher. 1946. The shielding of thermocouples from the effects of radiation. *ASHVE Transactions* 52:183.

Prazak, J., J. Tywoniak, F. Peterka, and T. Slonc. 1990. Description of transport of liquid in porous media—A study based on neutron radiography data. *International Journal of Heat and Mass Transfer* 33:1105-20.

Quenard, D. and H. Sallee. 1989. A gamma-ray spectrometer for measurement of the water diffusivity of cementitious materials. Proceedings of the Materials Research Society Symposium, Vol. 137.

Quinn, T.J. 1990. *Temperature*, 2nd ed. Academic Press, New York.

Richards, R.F., D.M. Burch, W.C. Thomas. 1992. Water vapor sorption measurements of common building materials. *ASHRAE Transactions* 98(1).

Richardson, L. 1965. A thermocouple recording psychrometer for measurement of relative humidity in hot, arid atmosphere. In *Humidity and Moisture* I:101. Reinhold Publishing Corporation, New York.

Schooley, J.F. 1986. *Thermometry*. CRC Press, Boca Raton, FL.

Schooley, J.F., ed. 1992. *Temperature: Its measurement and control in science and in industry*, Vol. 6. American Institute of Physics, New York.

Seely, R.E. 1955. A circuit for measuring the resistance of energized A-C windings. *AIEE Transactions*, 214.

Shafer, M.R. 1961. Performance characteristics of turbine flowmeters. Proceedings of the Winter Annual Meeting, *Paper* No. 61-WA-25. American Society of Mechanical Engineers, New York.

Till, C.E. and G.E. Handegord. 1960. Proposed humidity standard. *ASHRAE Transactions* 66:288.

Tobiasson, W. and C. Korhonen. 1985. Roofing moisture surveys: Yesterday, today, and tomorrow. Proceedings of the Second International Symposium on Roofing Technology, Gaithersburg, MD.

Tveit, A. 1966. Measurement of moisture sorption and moisture permeability of porous materials. *Report* No. 45. Norwegian Building Research Institute, Oslo.

Vernon, H.M. 1932. The globe thermometer. *Proceedings of the Institution of Heating and Ventilating Engineers* 39:100.

Wentzel, J.D. 1961. An instrument for measurement of the humidity of air. *ASHRAE Journal* 11:67.

Wile, D.D. 1947. Air flow measurement in the laboratory. *Refrigerating Engineering* 6:515.

Woodring, E.D. 1969. Magnetic turbine flowmeters. *Instruments and Control Systems* 6:133.

Worrall, R.W. 1965. Psychrometric determination of relative humidities in air with dry-bulb temperatures exceeding 212°F. In *Humidity and Moisture* I:105. Reinhold Publishing Corporation, New York.

BIBLIOGRAPHY

ASA. 1980. Techniques of machinery vibration measurement. ANSI *Standard* S2.17-80 (R 1986). Acoustical Society of America, New York.

ASA. 1984. Mechanical vibration of rotating and reciprocating machinery—Requirements for instruments for measuring vibration severity. ANSI *Standard* S2.40-84 (R 1990).

ASA. 1984. Specification for acoustical calibrators. ANSI *Standard* S1.40-84 (R 1994).

ASA. 1985. Statistical methods for determining and verifying stated noise emission values of machinery and equipment. ANSI *Standard* S12.3-85 (R 1990).

ASA. 1987. Methods for determination of insertion loss of outdoor noise barriers. ANSI *Standard* S12.8-87.

ASA. 1989. Guide to the mechanical mounting of accelerometers. ANSI *Standard* S2.61-89 (R 1991).

ASA. 1989. Mechanical vibration—Balance quality requirements of rigid rotors, Part 1: Determination of permissible residual unbalance. ANSI *Standard* S2.19-89.

ASA. 1989. Method for the designation of sound power emitted by machinery and equipment. ANSI *Standard* S12.23-89.

ASA. 1989. Reference quantities for acoustical levels. ANSI *Standard* S1.8-89.

ASA. 1990. Survey methods for the determination of sound power levels of noise sources. ANSI *Standard* S12.36-90.

ASA. 1990. Vibrations of buildings—Guidelines for the measurements of vibrations and evaluation of their effects on buildings. ANSI *Standard* S2.47-90.

ASA. 1995. Measurement of sound pressure levels in air. ANSI *Standard* S1.13-95.

ASHRAE. 1984. Standard method for measurement of flow of gas. ANSI/ASHRAE *Standard* 41.7-1984 (RA 91).

ASHRAE. 1986. Engineering analysis of experimental data. *Guideline* 2-1986 (RA 96).

ASHRAE. 1986. Laboratory method of testing in-duct sound power measurement procedure for fans. ANSI/ASHRAE *Standard* 68-1986, also ANSI/AMCA *Standard* 330-86.

ASHRAE. 1986. Standard method for temperature measurement. ANSI/ASHRAE *Standard* 41.1-1986 (RA 91).

ASHRAE. 1987. Standard methods for laboratory air flow measurement. ANSI/ASHRAE *Standard* 41.2-1987 (RA 92).

ASHRAE. 1988. A standard calorimeter test method for flow measurement of a volatile refrigerant. ANSI/ASHRAE *Standard* 41.9-1988.

ASHRAE. 1989. Standard method for pressure measurement. ANSI/ASHRAE *Standard* 41.3-1989.

ASHRAE. 1996. Method for measurement of proportion of lubricant in liquid refrigerant. *Standard* 41.4-1996.

ASTM. 1995. Standard specifications for ASTM thermometers. *Standard* E 1-95. American Society for Testing and Materials, West Conshohocken, PA.

Beranek, L.L. 1988. *Acoustical measurements*. Published for the Acoustical Society of America by the American Institute of Physics, New York.

Beranek, L.L. 1989. *Noise and vibration control*. Institute of Noise Control Engineering, Poughkeepsie, NY.

Dally, J.W., W.F. Riley, and K.G. McConnell. 1984. *Instrumentation for engineering measurements*. John Wiley and Sons, New York.

Den Hartog, J.P. 1985. *Mechanical vibrations*. Dover Publications, Inc., New York.

EIA. 1991. Interface between data terminal equipment and data circuit-terminating equipment employing serial binary data interchange. ANSI/EIA *Standard* TIA-232-E-91.

EPA. 1991. Introduction to indoor air quality: A self-paced learning module. EPA/400/3-91/002, U.S. Environmental Protection Agency, Washington.

Harris, C.M. 1987. *Shock and vibration handbook*, 3rd ed. McGraw-Hill, New York.

IEC. 1984. Advance edition of the International electrotechnical vocabulary, Chapter 801: Acoustics and electro-acoustics. IEC *Standard* 50 (801)-84.

IEC. 1990. Rotating electrical machines, Part 9: Noise limits. IEC *Standard* 34/9-90 including Corrigendum 1991, 2nd ed.

IEEE. 1987. Standard digital interface for programmable instrumentation. ANSI/IEEE *Standard* 488.1-87 (R 1994).

IEEE. 1992. Codes, formats, protocols, and common commands for use with IEEE 488.1-87. ANSI/IEEE *Standard* 488.2-92.

IESNA. 1993. *Lighting handbook reference and application*, 8th ed. Illuminating Engineering Society of North America, New York.

Lord, H.W. et al. 1987. *Noise control for engineers*. Krieger Publishing, Melbourne, FL.

Morrison, R. 1986. *Grounding and shielding techniques in instrumentation*, 3rd ed. John Wiley and Sons, New York.

Papoulis, A. 1987. *The Fourier integral and its applications*. McGraw-Hill, New York.

Spitzer, D.W., ed. 1991. *Flow measurement*. Instrumentation Society of America, Research Triangle Park, NC.

Steele, W.G., R.A. Ferguson, R.P. Taylor, and H.W. Coleman. 1994. Comparison of ANSI/ASME and ISO models for calculation of uncertainty. *ISA Transactions* 33:339-52.

Tilford, C.R. 1992. Pressure and vacuum measurements. In *Physical methods of chemistry*, 2nd ed., Vol. 6, pp. 106-73. B.W. Rossiter and J.F. Hamilton, eds. John Wiley and Sons, New York.

CHAPTER 15

AIRFLOW AROUND BUILDINGS

AIRFLOW around buildings affects worker safety, process and building equipment operation, weather and pollution protection at inlets, and the ability to control environmental factors of temperature, humidity, air motion, and contaminants. Wind causes surface pressures that vary around buildings, changing intake and exhaust system flow rates, natural ventilation, infiltration and exfiltration, and interior pressure. The mean flow patterns and turbulence of wind passing over a building can cause recirculation of exhaust gases to air intakes. This chapter contains information for evaluating flow patterns, estimating wind pressures and air intake contamination, and solving problems caused by the effects of wind on intakes, exhausts, and equipment. Related information can be found in Chapters 12, 14, 25, and 26 of this volume; in Chapters 24, 26, and 48 of the 1995 *ASHRAE Handbook—Applications*; and in Chapters 25, 30, 35, and 36 of the 1996 *ASHRAE Handbook—Systems and Equipment.*

FLOW PATTERNS

Buildings having even moderately complex shapes, such as L- or U-shaped structures formed by two or three rectangular blocks, can generate flow patterns too complex to generalize for design. To determine flow conditions influenced by surrounding buildings or topography, wind tunnel or water channel tests of scale models or tests of existing buildings are required. However, if a building is oriented perpendicular to the wind, it can be considered as consisting of several independent rectangular blocks. Only isolated rectangular block buildings are discussed here. Hosker (1984, 1985) reviews the effects of nearby buildings.

The mean speed U_H of wind approaching a building increases with height H above the ground (Figure 1). Both the upwind velocity profile shape and its turbulence level strongly influence flow patterns and surface pressures. A stagnation zone exists on the upwind wall. The flow separates at the sharp edges to generate recirculating flow zones that cover the downwind surfaces of the building (roof, sides, and leeward walls) and extend for some distance into the wake. If the building has sufficient length L in the windward direction, the flow will reattach to the building (Figure 2) and may generate two distinct regions of separated recirculating flow—on the building and in its wake.

Surface flow patterns on the upwind wall are largely influenced by approach wind characteristics. Higher wind speed at roof level causes a larger stagnation pressure on the upper part of the wall than near the ground, which leads to downwash on the lower one-half to two-thirds of the building (Figure 1). On the upper one-quarter to one-third of the building, the surface flow is directed upward over the roof. For a building whose height H is three or more times the width W of the upwind face, an intermediate zone can exist between the upwash and downwash regions, where the surface streamlines pass horizontally around the building. The downwash on the lower surface of the upwind face separates from the building before it reaches ground level and moves upwind to form a vortex that can generate high velocities close to the ground. This ground level upwind vortex is carried around the sides of the building in a U shape (Figure 1) and is responsible for the suspension of dust and debris that can contaminate air intakes close to ground level.

For wind perpendicular to a building wall, the height H and width W of the upwind building face determine the flow patterns shown in Figure 3. According to Wilson (1979), the **scaling length** R (ft) is:

$$R = B_s^{0.67} B_L^{0.33} \tag{1}$$

where

B_s = smaller of upwind building face dimensions H and W, ft
B_L = larger of upwind building face dimensions H and W, ft

When B_L is larger than $8B_s$, use $B_L = 8B_s$ in Equation (1). For buildings with varying roof levels or with wings separated by at least a distance B_s, only the height and width of the building face below the portion of the roof in question should be used to calculate R.

Streamline patterns are independent of wind speed and depend mainly on building shape and upwind conditions. Because of the three-dimensional flow around a building, the shape and size of the recirculation airflow is not constant over the surface; instead, it reattaches closer to the upwind face of the building along the edges at the roof and at ground level (Figure 2). The height H_c of the recirculation region (Figure 3) also decreases near roof edges.

The wind above the roof recirculation region is affected by the presence of the building; the flow accelerates as the streamlines curve upward over the roof and decelerates as they curve downward into the wake on the lee side of the building. The distance above roof level where a building influences the flow is approximately $1.5R$ (Figure 1). The roof pitch begins to affect flow when it exceeds about 15° (1:4). When roof pitch reaches 20° (1:3), the flow remains attached to the upwind pitched roof and produces a recirculation region downwind from the roof ridge that is larger than that for a flat roof.

The downwind wall of a building faces a region of low average velocity and high turbulence. Velocities near the downwind wall are typically one-quarter of those at the corresponding upwind wall location. Figures 1 and 2 show that an upward flow exists over most of the downwind walls. A flow recirculation region extends for an approximate distance $L_r = 1.0R$ downwind.

If the angle of the approach wind is not perpendicular to the upwind face, complex flow patterns result. Strong vortices develop from the upwind edges of a roof, causing a strong downwash into the building wake above the roof. High speeds in these vortices cause large negative pressures near roof corners that can be a hazard to roof-mounted equipment during high winds. When the angle between the wind direction and the upwind face of the building is less than about 70°, the downwash-upwash patterns on the upwind face of the building are less pronounced, as is the ground level vortex

The preparation of this chapter is assigned to TC 2.5, Air Flow Around Buildings.

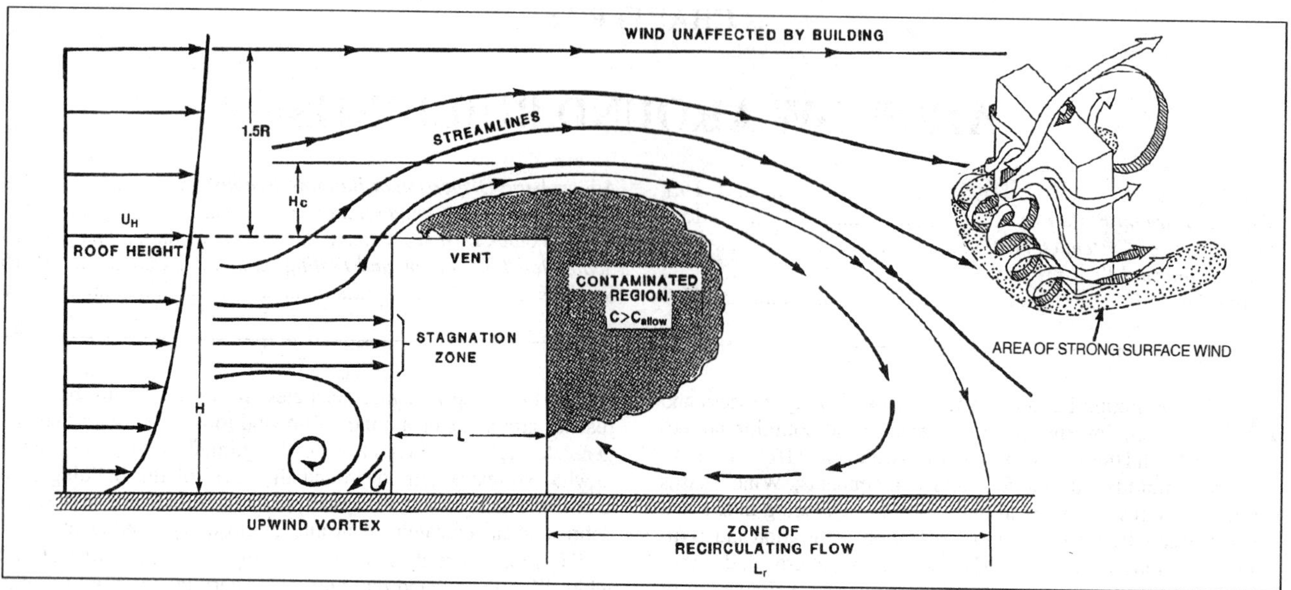

Fig. 1 Flow Patterns Around Rectangular Building

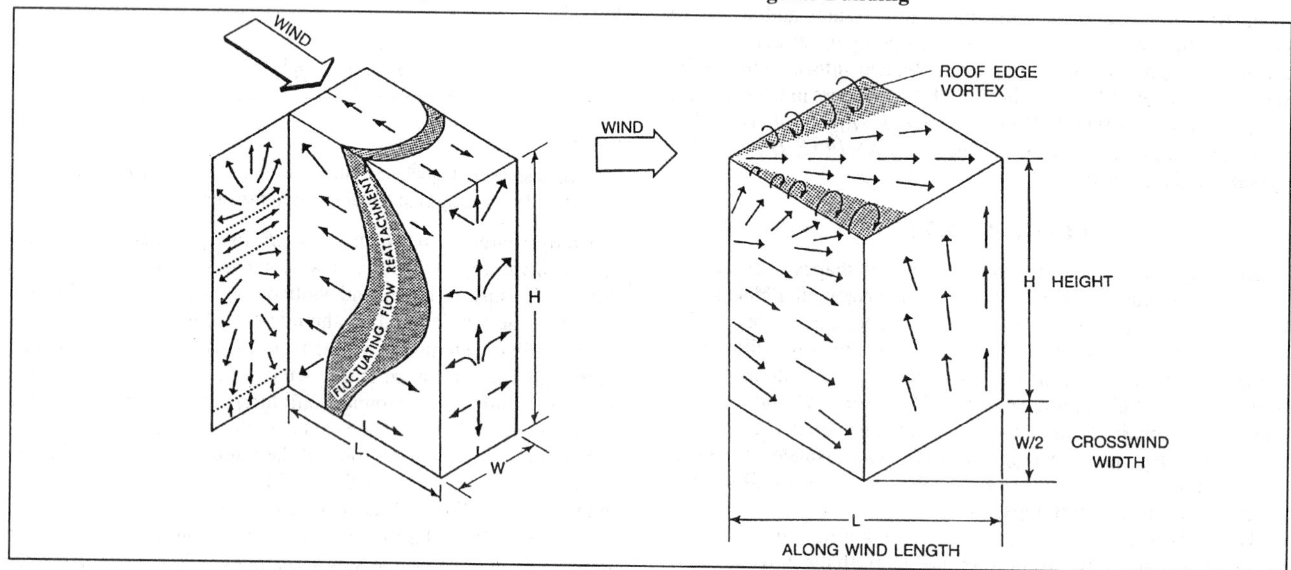

Fig. 2 Surface Flow Patterns and Building Dimensions

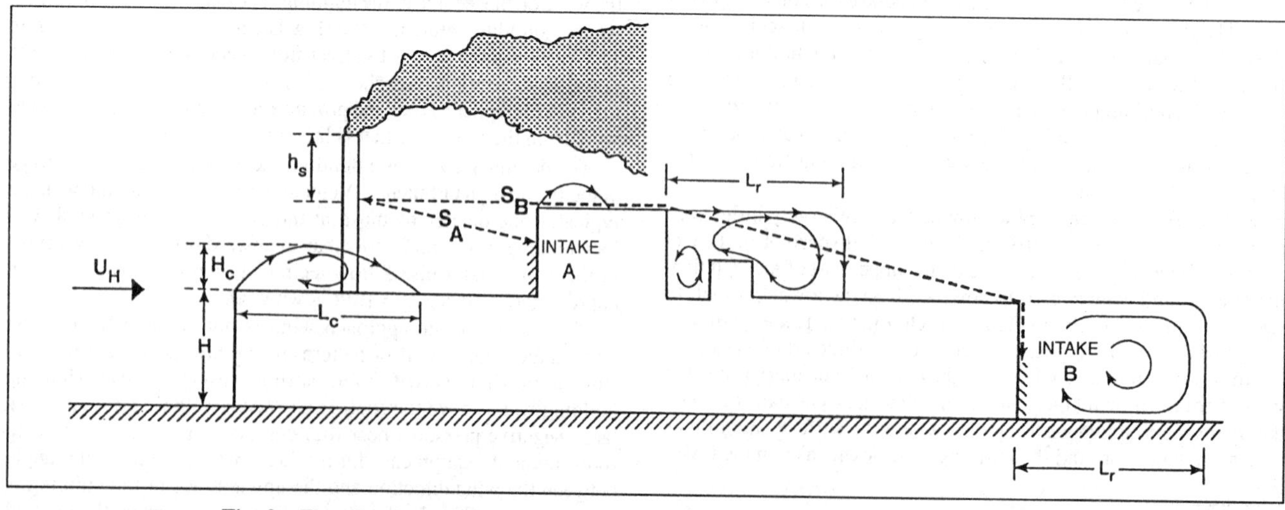

Fig. 3 Flow Recirculation Regions and Exhaust-to-Intake Stretched-String Distances
(Wilson 1982)

shown in Figure 1. For an approach flow angle of 45°, streamlines remain close to the horizontal in their passage around the sides of the building (Figures 1 and 2), except near roof level where the flow is sucked upward into the roof edge vortices.

WIND PRESSURE ON BUILDINGS

In addition to the flow patterns described previously, the turbulence or gustiness of the approaching wind and the unsteady character of the separated flows cause surface pressures to fluctuate. The pressures discussed here are time-averaged values, with an averaging period of about 600 s. Instantaneous pressures may vary significantly above and below these averages, and peak pressures two or three times the mean values are possible. Although peak pressures are important with regard to structural loads, mean values are more appropriate for computing infiltration and ventilation rates. The time-averaged surface pressures are proportional to the wind velocity pressure p_v (lb/ft^2) given by **Bernoulli's equation**

$$p_v = \rho_a U_H^2 / 2g \tag{2}$$

where

U_H = approach wind speed at upwind wall height H, ft/s
ρ_a = outdoor air density, lb/ft^3
g = gravitational acceleration, ft/s^2

The difference p_s (lb/ft^2) between the pressure on the building surface and the local outdoor atmospheric pressure at the same level in an undisturbed wind approaching the building is

$$p_s = C_p p_v \tag{3}$$

where C_p is the local wind pressure coefficient for the building surface.

The local wind speed U_H at the top of the wall that is required for Equation (2) is estimated by applying terrain and height corrections to the **hourly wind speed U_{met} from a nearby meteorological station.** U_{met} is generally measured in flat, open terrain. The anemometer that records U_{met} is located at a height H_{met}, usually 33 ft above ground level. The hourly average wind speed U_H at wall height H in the undisturbed wind approaching a building in its local terrain (Figures 1 and 3) can be calculated from U_{met} as follows:

$$U_H = U_{met}\left(\frac{\delta_{met}}{H_{met}}\right)^{a_{met}}\left(\frac{H}{\delta}\right)^{a} \tag{4}$$

The wind boundary layer thickness δ and exponent a for the local building terrain and a_{met} and δ_{met} for the meteorological station are determined from Table 1. Typical values for meteorological stations located in flat, open terrain (Category 3 in Table 1) are $a_{met} = 0.14$ and $\delta_{met} = 900$ ft. The values and terrain categories in Table 1 are consistent with those adopted in other engineering applications, for example ASCE *Standard* 7.

Equation (4) gives the wind speed at height H above the plan area-weighted average height of local obstacles, such as buildings and vegetation. At heights at or below this average obstacle height (e.g., at roof height in densely built-up suburbs), the speed depends on the geometrical arrangement of the buildings, and Equation (4) is less reliable.

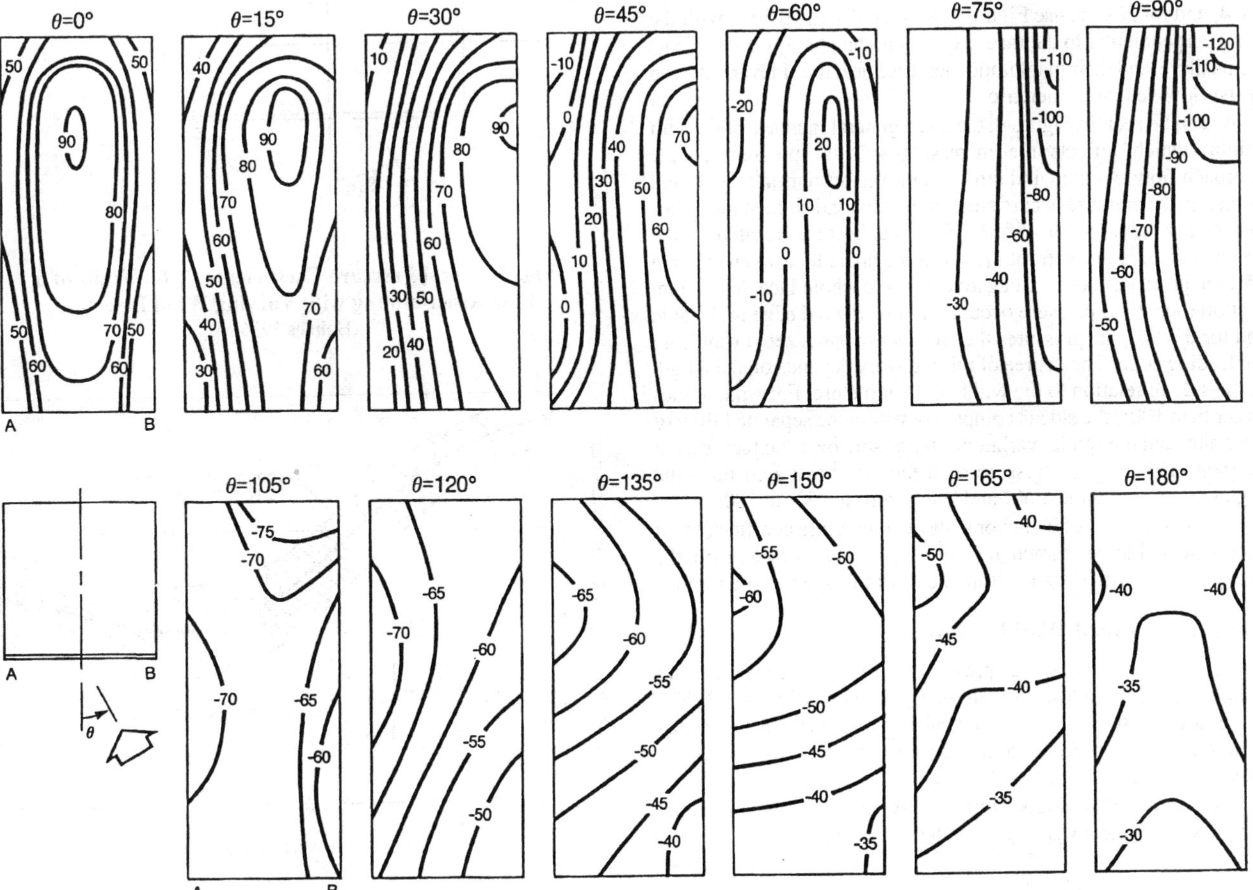

Fig. 4 Local Pressure Coefficients ($C_p \times 100$) for a Tall Building with Varying Wind Direction
(Davenport and Hui 1982)

Table 1 Atmospheric Boundary Layer Parameters

Terrain Category	Description	Exponent a	Layer Thickness δ, ft
1	Large city centers, in which at least 50% of buildings are higher than 70 ft, over a distance of at least 6500 ft upwind	0.33	1500
2	Urban, suburban, wooded areas, and other areas with closely spaced obstructions compared to or larger than single-family dwellings (over a distance of at least 6500 ft upwind)	0.22	1200
3	Open terrain with scattered obstacles generally less than 30 ft high	0.14	900
4	Flat, unobstructed areas exposed to wind flowing over a large water body (no more than 1600 ft inland)	0.10	700

Local Wind Pressure Coefficients

Values of the local wind pressure coefficient C_p used in Equation (3) depend on building shape, wind direction, and the influence of nearby buildings, vegetation, and terrain features. Accurate determination of C_p can be obtained only from wind tunnel model tests of the specific site and building. Ventilation rate calculations for single, unshielded rectangular buildings can be reasonably estimated using existing wind tunnel data.

Figure 4 shows pressure coefficients for walls of a tall building with a square cross section for simulated urban terrain (Davenport and Hui 1982), and Figure 5 for walls of a low-rise building (Holmes 1986). Generally, high-rise buildings are those where the height H is more than three times the crosswind width W. For $H > 3W$, use Figure 4, and for $H < W$, use Figure 5. At a wind angle $\theta = 0°$, with the wind perpendicular to the face in question, the pressure coefficients are positive, and their magnitudes decrease near the sides and the top as the flow velocities increase.

As can be seen in Figure 4, there is a general increase of C_p with height, which reflects the increasing velocity pressure in the approach flow. As the wind direction moves off normal, the region of maximum pressure occurs closer to the upwind edge of the building. At a wind angle of 45°, the pressures become negative at the downwind edge of the front face; at some angle θ between 60° and 75°, the pressures become negative over the whole face. Maximum suction (negative) pressure occurs near the upwind edge of the side and for $\theta = 90°$, the pressures then recover towards zero away from the leading edge. The degree of this recovery depends on the length of the side in relation to the width of the structure. For wind angles larger than 100°, the side is completely within the separated flow of the wake, and the spatial variations in pressure over the face are not as great. The average pressure on a face is above zero for wind angles from 0° to almost 60° and below zero for 60° to 180°.

A similar pattern of behavior in the wall pressure coefficients for a low-rise building is shown in Figure 5. Here the recovery from the strong suction with distance from the upwind edge is more rapid.

Surface Averaged Wall Pressures

Figure 6 shows the surface pressure coefficient C_s averaged over a complete wall of a low-rise building (Swami and Chandra 1987). The figure also includes the values calculated from the pressure distributions shown in Figure 5. Similar results for a tall building are shown in Figure 7 (Akins et al. 1979).

The wind-induced indoor-outdoor pressure difference is found using the coefficient $C_{p(in\text{-}out)}$, which is defined as

$$C_{p(in\text{-}out)} = C_p - C_{in} \quad (5)$$

where C_{in} is the internal wind-induced pressure coefficient.

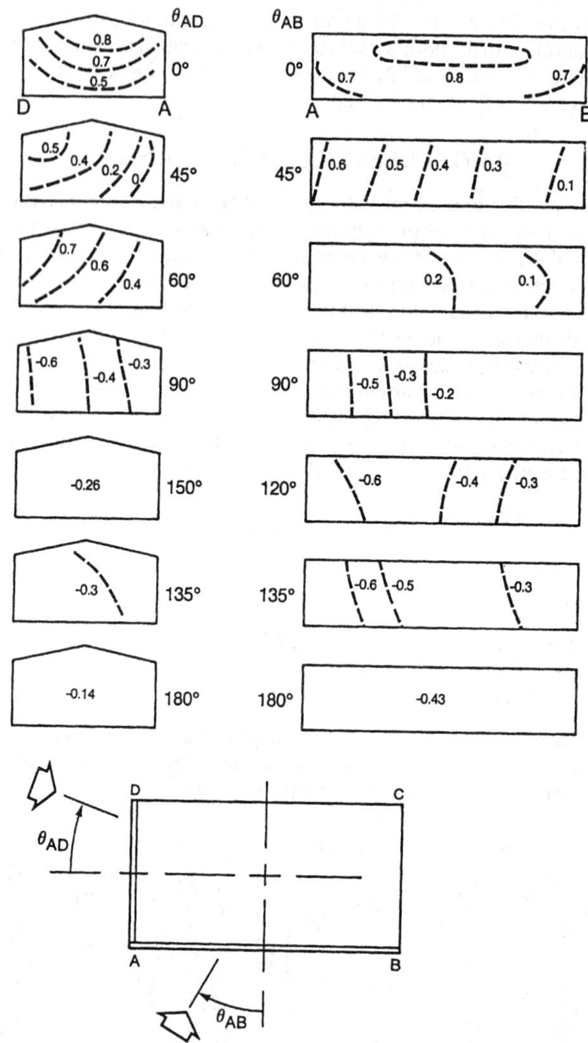

Fig. 5 Local Pressure Coefficients C_p for Walls of a Low-Rise Building with Varying Wind Direction
(Holmes 1986)

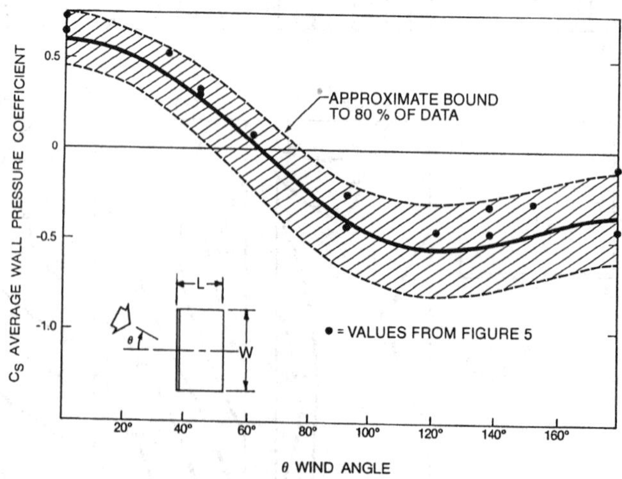

Fig. 6 Variation of Surface Averaged Wall Pressure Coefficients for Low-Rise Buildings
(Swami and Chandra 1987)

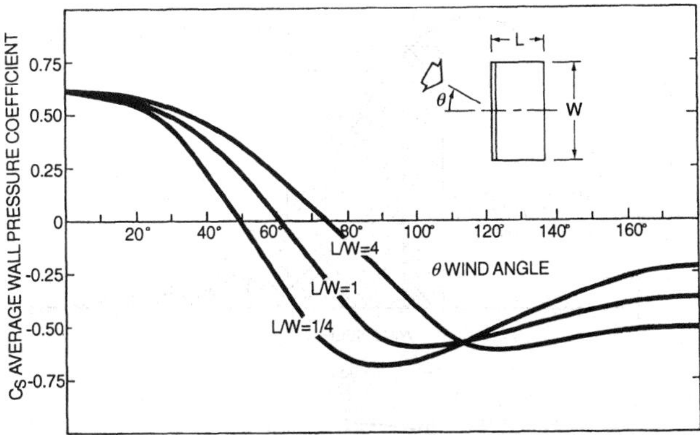

Fig. 7 Surface Averaged Wall Pressure Coefficients for Tall Buildings

The mean of the spatially averaged wall pressure coefficients C_s in Figures 6 and 7 for the four building walls is about −0.2 for most wind directions. This implies that for uniformly distributed air leakage sites in all the walls, C_{in} is about −0.2. Chapter 25 explains how stack effect pressures induced by the indoor-outdoor temperature difference combine with wind-induced pressures.

Roof Pressures

Surface pressures on the roof of a low-rise building depend strongly on roof slope. Figure 8 shows typical distributions for a wind direction normal to a side of the building. For very low slopes, the pressures are negative over the whole roof surface. The magnitude is greatest within the separated flow zone near the leading edge and recovers toward the free stream pressure away from the edge. For steeper slopes, the pressures are weakly positive on the windward slope and negative within the separated flow over the leeward slope. With a wind angle of about 45°, the vortices originating at the leading corner of a roof with a low slope can induce very large localized negative pressures. Figure 9 shows the average pressure coefficient over the roof of a tall building (Akins et al. 1979).

Interference and Shielding Effects on Pressures

Nearby structures strongly influence surface pressures on both high- and low-rise buildings. These effects are very strong for spacing-to-height ratios less than five, where the distributions of pressure shown in Figures 5 through 9 do not apply. Although the effect of shielding is still significant at larger spacings, for low-rise buildings it is largely accounted for by the reduction in p_v with increased terrain roughness. Chapter 25 gives shielding classes for air infiltration and ventilation applications.

Sources of Wind Data

In order to design for the effects of airflow around buildings, wind speed and direction frequency data should be obtained. The simplest forms of wind data are tables or charts of climatic normals, which give hourly average wind speeds, prevailing wind directions, and peak gust wind speeds for each month of the year. This information can be found in sources such as *The Weather Almanac* (Bair 1992) and the *Climatic Atlas of the United States* (DOC 1968). A more up-to-date source, which contains information on wind speed and direction frequencies, is the *International Station Meteorological Climatic Summary* available in CD-ROM format from the National Climatic Data Center (NCDC) in Asheville, North Carolina. Where more detailed information is required, digital records of hourly winds and other meteorological parameters are available (on magnetic tape or CD-ROM) from the NCDC for stations throughout the world. Most countries also have weather services that provide data. For example, in Canada, the Atmospheric Environment Service in Downsview, Ontario, provides hourly meteorological data and summaries.

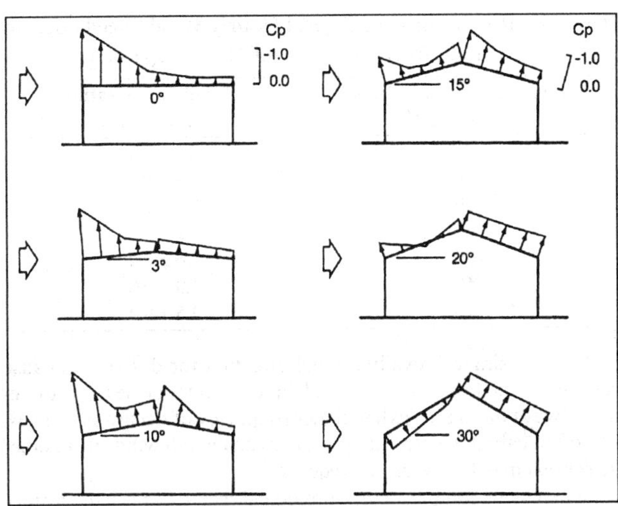

Fig. 8 Local Roof Pressure Coefficients C_p for Roof of Low-Rise Buildings
(Holmes 1986)

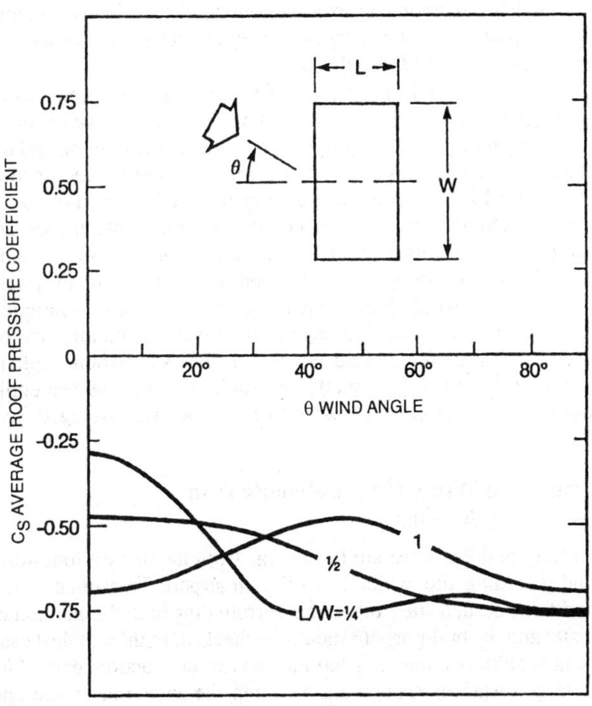

Fig. 9 Surface Averaged Roof Pressure Coefficients for Tall Buildings

If only the annual average U_{annual} of the hourly wind speeds U_{met} is available for a meteorological station, the percentage of time that U_{met} exceeds U_{annual} can be estimated from Table 2. The ratios U_{met}/U_{annual} are based on long-term data from 24 weather stations widely distributed over North America. At these stations, U_{met} ranges from 600 to 1200 fpm. The uncertainty ranges listed in Table 2 are one standard deviation of the wind speed ratios. The following example demonstrates the use of Table 2.

Example 1. The wind speed U_{met} for a building pressure or exhaust dilution calculation should be exceeded less than 88 h each year (1% of the total hours). If U_{annual} = 800 fpm, find U_{met}.

Solution: From Table 2, the 1% exceedance speed is (2.5 ± 0.4) times U_{annual}. For U_{annual} = 800 fpm, U_{met} is 2000 fpm for 1% exceedance, with an uncertainty range of 1700 to 2300 fpm for one standard deviation.

Table 2 Relationship of Typical Hourly Wind Speeds U_{met} to Annual Average U_{annual} of Hourly Speeds

Percentage of Hourly Values Exceeded by U_{met}	Wind Speed Ratio U_{met}/U_{annual}
90%	0.2 ± 0.1
75%	0.5 ± 0.1
50%	0.8 ± 0.1
25%	1.2 ± 0.15
10%	1.6 ± 0.2
5%	1.9 ± 0.3
1%	2.5 ± 0.4

Using a single prevailing wind direction for design can cause serious errors. For any set of wind direction frequencies, one direction always has a somewhat higher frequency of occurrence. Thus, it is often called the prevailing wind, even though winds from other directions may be almost as frequent.

When using long-term meteorological records, check the history of mounting conditions for the anemometer, because the instrument may have been relocated and its height varied. This can affect its directional exposure and the recorded wind speeds. Equation (4) can be used to correct wind data collected at different mounting heights. Poor anemometer exposure due to obstructions or mounting on top of a building cannot be easily corrected, and the records for that period should be deleted.

If an estimate of the probability of an extreme wind speed outside the range of the recorded values at a site is required, the observations may be fitted to an appropriate probability distribution and the particular probabilities calculated from the fitted curve function (see Figure 10). This process is usually repeated for each of the 16 wind directions. Where estimates of extreme probabilities are required, curve fitting at the tail of the distribution is very important.

Accurate extrapolation of frequency distributions to rarely occurring high wind speeds requires special statistical techniques. Building codes for wind loading on structures contain information on estimating extreme wind conditions. For ventilation applications, extreme winds are usually not required, and the 99 percentile limit can be accurately estimated from airport data averaged over less than 10 years.

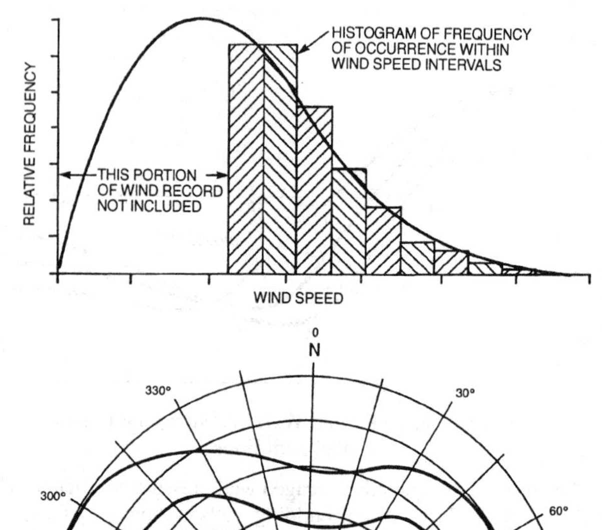

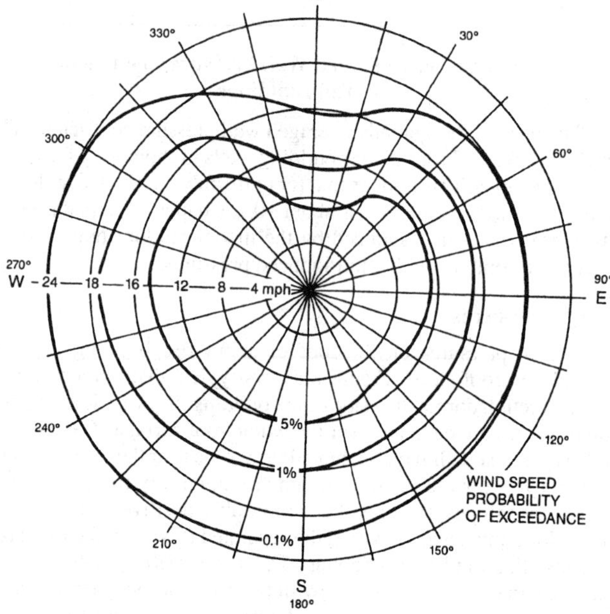

Fig. 10 Frequency Distribution of Wind Speed and Direction

Estimating Wind at Sites Remote from Recording Stations

Many building sites are located far from the nearest long-term wind recording site, which is usually an airport. To estimate wind conditions at such sites, the terrain surrounding both the anemometer site and the building site should be checked. In the simplest case of flat or slightly undulating terrain with few obstructions extending for large distances around and between the anemometer site and building site, recorded wind data can be assumed to be representative of that at the building site. Wind direction occurrence frequency at a building site should be inferred from airport data only if the two locations are on the same terrain, with no terrain features that could alter wind direction between them.

In cases where the only significant difference between the airport recording site terrain and the building site terrain is surface roughness, the mean wind speed can be adjusted, using Equation (4) and Table 1, to yield approximate wind velocities at the building site.

In using Equation (4), cases may be encountered where, for a given wind direction, the terrain upwind of either the building site or the recording site does not fall into just one of the categories in Table 1. The terrain immediately upwind of the site may fall into one category, while that somewhat further upwind falls into a different category. For example, at a downtown airport the terrain may be flat and open (Category 3) immediately around the recording instrument, but urban or suburban (Category 2) a relatively short distance away. This difference in terrains also occurs when a building site or recording site is in an urban area near open water or at the edge of town. In these cases, the suggested approach is to use the terrain category that is most representative of the average condition within approximately 1 mile upwind of the site (Deaves 1981). If the average condition is somewhere between two categories described in Table 1, the values of a and δ can be interpolated from those given in the table.

A rough guideline is that only wind speeds U_H of 800 fpm (9 mph) or greater at the building site can be estimated reliably using Equation (4) and Table 2 if the building and meteorological station are in different terrain categories of Table 1.

In addition to changes in surface roughness, several other factors are important in causing the wind speed and direction at a building site to differ from values recorded at a nearby airport. Wind speeds for buildings on hill crests or in valleys where the wind is channeled can be 50% higher than airport data. Shelter in the lee of hills and escarpments can reduce speeds to half the values at nearby flat airport terrain.

Solar heating of valley slopes can cause light winds of 200 to 800 fpm to occur as warm air flows upslope. At night, cooling of the ground by heat radiation can produce these speeds as cold air drains downslope. In general, rolling terrain experiences a smaller fraction of low speeds than nearly flat terrain.

When the wind is calm or light in the rural area surrounding a city, urban air tends to rise in a buoyant plume over the city center. This rising air, heated by man-made sources and higher solar absorption in the city, is replaced by air sucked toward the city center from the edges. In this way, the urban heat island can produce light wind speeds and direction frequencies significantly different than those at a rural airport.

In more complex terrain, both wind speed and direction may be significantly different from those at the distant recording site. In these cases, building site wind conditions should not be estimated from airport data. Options are either to establish an on-site wind recording station or to commission a detailed wind tunnel correlation study between the building site and long-term airport wind observations.

WIND EFFECTS ON SYSTEM OPERATION

With few exceptions, building intakes and exhausts cannot be located or oriented such that a prevailing wind ensures ventilation and air-conditioning system operation. Wind can assist or hinder inlet and exhaust fans, depending on their positions on the building, but even in locations with a predominant wind direction, the ventilating system must perform adequately for all other directions. To avoid variability in system flow rates, use Figures 4, 5, and 8 as a guide to placing inlets and exhausts in locations where the surface pressure coefficients do not vary greatly with the wind direction.

Cooling towers and similar equipment should be oriented to take advantage of prevailing wind directions, based on careful study of the meteorological data and flow patterns on the building for the area and time of year involved.

A building with only upwind openings is under a positive pressure (Figure 11). Building pressures are negative when there are only downwind openings. A building with internal partitions and openings is under various pressures depending on the relative sizes of the openings and the wind direction. With larger openings on the windward face, the building interior tends to remain under positive pressure; the reverse is also true (see Figures 4 through 9, and Chapter 25).

Airflow through a wall opening results from differential pressures, which may exceed 0.5 in. of water during high winds. Supply and exhaust systems, and openings, dampers, louvers, doors, and windows make the building flow conditions too complex for direct calculation. Iterative calculations are required because of the nonlinear dependence of volume flow rate on the differential pressure across an opening. Several multizone airflow models are available for these iterative calculations (Feustel and Dieris 1992). The opening and closing of doors and windows by building occupants add further complications. In determining $C_{p(in\text{-}out)}$ from Equation (5), the wind direction is more important than the position of an opening on a wall, as shown in Figures 4 and 5.

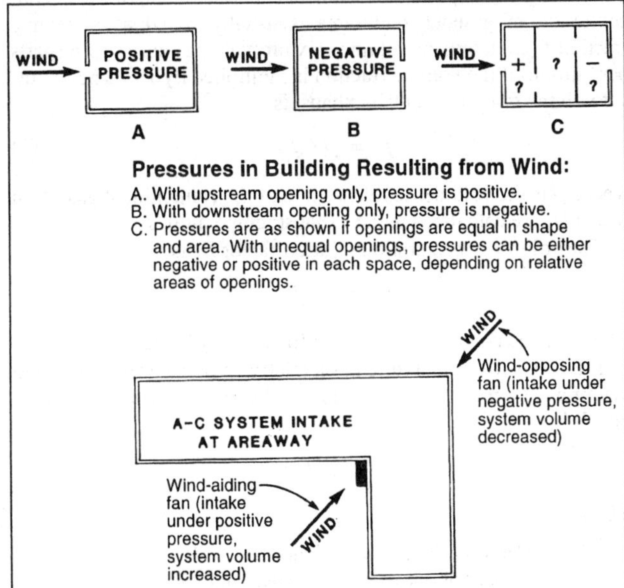

Fig. 11 Sensitivity of System Volume to Locations of Building Openings, Intakes, and Exhausts

Natural and Mechanical Ventilation

With natural ventilation, wind may augment, impede, or sometimes reverse the airflow through a building. For large roof areas (Figure 2), the wind can reattach to the roof downstream from the cavity and thus reverse natural ventilation discharging out of monitor or similar windows. These reversals can be avoided by using stacks, continuous roof ventilators, or other exhaust devices in which the flow is augmented by the wind.

Mechanical ventilation is also affected by wind conditions. A low-pressure wall exhaust fan, 0.05 to 0.1 in. of water, can suffer drastic reduction in capacity. Flow can be reversed by wind pressures on windward walls, or its rate can be increased substantially when subjected to negative pressures on the lee and other sides. Clarke (1967), when measuring medium-pressure air-conditioning systems, 1 to 1.5 in. of water, found flow rate changes of 25% for wind blowing into intakes on an L-shaped building compared to wind blowing away from intakes. Such changes in flow rate can cause noise at the supply outlets and drafts in the space served.

For mechanical systems, the wind can be thought of as an additional pressure source in series with a system fan, either assisting or opposing it (Houlihan 1965). Where system stability is essential, the supply and exhaust systems must be designed for high pressures (about 3 to 4 in. of water) or must use devices to actively minimize unacceptable variations in flow rate. To conserve energy, the system pressure selected should be consistent with system needs.

Building Pressure Balance

Proper building pressure balance avoids flow conditions that make doors hard to open, cause drafts, and prevent the confinement of contaminants to specific areas. Although the supply and exhaust systems in an area may be in nominal balance, wind can upset this balance, not only because of the changes in fan capacity but also by superimposing infiltrated or exfiltrated air or both on the area. The effects can make it impossible to control environmental conditions. Where building balance and minimum infiltration are important, consider the following:

- Fan system design with pressure adequate to minimize wind effects
- Controls to regulate flow rate or pressure or both
- Separate supply and exhaust systems to serve each building area requiring control or balance
- Doors (possibly self-closing) or double-door air locks to noncontrolled adjacent areas, particularly outside doors
- Sealing windows and other leakage sources and closing natural vent openings

System volume and pressure control is described in Chapter 42 of the 1995 *ASHRAE Handbook—Applications*. This control is not possible without adequate system pressure for both the supply and exhaust systems to overcome wind effects. Such a control system may require fan inlet or discharge dampers, fan speed or pitch control, or both.

Fume Hood Operation

Wind effects can interfere with safe hood operation. Supply (makeup) volume variations can cause disturbances at hood faces or a lack of adequate makeup air. Volume surges, due to fluctuating wind pressures acting on the exhaust system, can cause momentary inadequate hood exhaust. If highly toxic contaminants are involved, surging is unacceptable. The system should be designed to eliminate this condition. On low-pressure exhaust systems, it is impossible to test the hoods under wind-induced, surging conditions. These systems should be tested during calm conditions for safe flow into the hood faces; they should be rechecked by smoke tests during high wind conditions. For more information, see Chapter 13, Laboratory Systems, in the 1995 *ASHRAE Handbook—Applications*.

Minimizing Wind Effect on System Volume

Wind effect can be reduced by careful selection of inlet and outlet locations. Because wall surfaces are subject to a wide variety of positive and negative pressures, wall openings should be avoided when possible. When they are required, wall openings should be away from corners formed by building wings (Figure 11). Mechanical ventilation systems should operate at a pressure high enough to minimize wind effect. Low-pressure systems and propeller exhaust fans should not be used with wall openings unless their ventilation rates are small or they are used in noncritical services such as storage areas.

Although roof air intakes within flow recirculation zones best minimize wind effect on system flow rates, current and future air quality in these zones must be considered. These locations should be avoided if a source of contamination exists or may be added in the future. The best area is near the middle of the roof because the negative pressure there is small and least affected by changes in wind direction (Figure 8). Avoid the edges of the roof and walls, where large pressure fluctuations occur. Either vertical or horizontal (mushroom) openings can be used. On roofs having large areas, where the intake may be outside the roof recirculation zone, mushroom or 180° gooseneck designs minimize impact pressure from wind flow. The 135° gooseneck that is frequently used or vertical louvered openings are undesirable for this purpose or for rain protection.

Heated air or industrial contaminants should be exhausted vertically through stacks, above the roof recirculation zone. Horizontal, louvered (45° down), and 135° gooseneck discharges are undesirable, even for heat removal systems because of their sensitivity to wind effect. A 180° gooseneck for systems handling hot air may be undesirable because of air impingement on tar and felt roofs. Vertically discharging stacks located in a recirculation region (except near a wall) have the advantage of being subjected only to negative pressure created by wind flow over the tip of the stack.

BUILDING INTERNAL PRESSURE AND FLOW CONTROL

In air-conditioning and ventilation systems for a building containing airborne contaminants, the correct airflow is toward the contaminated areas. Airflow direction is maintained by controlling pressure differentials between spaces. In a laboratory building, for example, peripheral rooms such as offices and conference rooms are maintained at a positive pressure, and laboratories at a negative pressure, both with reference to corridor pressure. Pressure differentials between spaces are normally obtained by balancing the air-conditioning and ventilation supply system airflows in the spaces in conjunction with the exhaust systems in the laboratories, with differential pressure instrumentation to control the airflow. Chapter 42 of the 1995 *ASHRAE Handbook—Applications* has further information on controls.

Airflow in corridors is sometimes controlled by an outdoor reference probe that senses static pressure at doorways and air intakes. The differential pressure measured between the corridor and the outside may then signal a controller to increase or decrease airflow to the corridor. Unfortunately, it is difficult to locate an external probe where it will sense the proper external static pressure. High wind velocity and resulting pressure changes around entrances can cause great variations in pressure. Care must be taken to ensure that the probe is unaffected by wind pressure.

The pressure differential for a room adjacent to a corridor can be controlled using the corridor pressure as the reference. Outdoor pressure cannot control pressure differentials within rooms, even during periods of relatively constant wind velocity and pressure. A single pressure sensor can measure the outside pressure at one point only and may not be representative of pressures elsewhere.

ATMOSPHERIC DISPERSION OF BUILDING EXHAUST

A building exhaust system is frequently used to release a mixture of building air and pollutant gas at concentration C_e (mass of pollutant per volume of air) into the atmosphere through a stack or vent on the building. The exhaust mixes with atmospheric air to produce a pollutant concentration C, which may contaminate an air intake or receptor if the concentration is larger than some specified allowable value C_{allow} (Figure 1). The **dilution factor** D between source and receptor mass concentrations is defined as

$$D = C_e/C \tag{6}$$

where

C_e = contaminant mass concentration in exhaust, lb/ft³
C = contaminant mass concentration at receptor, lb/ft³

The dilution increases with distance from the source, starting from its initial value of unity. If C is replaced by C_{allow} in Equation (6), the atmospheric dilution D_{req} required to meet the allowable concentration at the intake (receptor) is

$$D_{req} = C_e/C_{allow} \tag{7}$$

The exhaust (source) concentration is given by

$$C_e = \dot{m}/Q_e = \dot{m}/(A_e V_e) \tag{8}$$

where

$\dot{m}$ = contaminant mass release rate, lb/s
$Q_e = A_e V_e$ = total exhaust volumetric flow rate, ft³/s
A_e = exhaust face area, ft²
V_e = exhaust face velocity, ft/s

The concentration units of mass per mixture volume are appropriate for gaseous pollutants, aerosols, dusts, and vapors. The concentration of gaseous pollutants is usually stated as a volume fraction f (contaminant volume/mixture volume), or as ppm (parts per million) if the volume fraction is multiplied by 10^6. The pollutant volume fraction f_e in the exhaust is

$$f_e = Q/Q_e \tag{9}$$

where Q is the volumetric release rate of the contaminant gas. Both Q and Q_e are calculated at the exhaust temperature T_e.

The volume concentration dilution factor D_v is

$$D_v = f_e/f \tag{10}$$

where f is the contaminant volume fraction at the receptor.

If the exhaust gas mixture has a relative molecular mass close to that of air, D_v may be calculated from the mass concentration dilution D by

$$D_v = (T_e/T_a)D \tag{11}$$

where

T_e = exhaust air absolute temperature, °R
T_a = outdoor ambient air absolute temperature, °R

Many building exhausts are close enough to ambient temperature to assume that volume fraction and mass concentration dilutions D_v and D are equal.

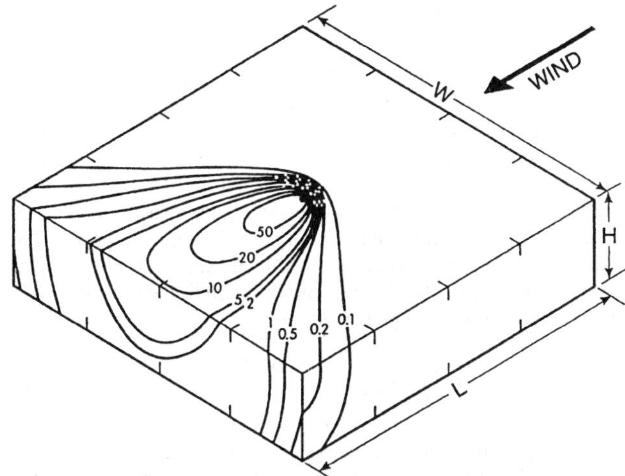

Fig. 12 Normalized Concentration Contours for Central Vent on Roof of a Low-Rise Building
(Wilson 1976)

Exhaust Gas Concentrations at Air Intakes

The dispersion of pollutants from building exhaust depends on the combined effect of atmospheric turbulence in the wind approaching the building and turbulence generated by the building itself. This building-generated turbulence is most intense in and near the flow recirculation zones that occur on the upwind edges of the building (Figures 2 and 3). Because of turbulence and distortion of wind streamlines by the building, the concentration caused by a source near a building cannot be estimated accurately by the design procedures developed for tall isolated stacks. Meroney (1982), Wilson and Britter (1982), Hosker (1984), and Halitsky (1982) review gas diffusion near buildings.

When the exhaust source is in or near a flow recirculation zone, contours of the **normalized concentration coefficient** K_c may be used (see Figure 12). For C in contaminant mass per volume,

$$K_c = CU_H HW/\dot{m} \tag{12}$$

The contours of K_c are developed from wind tunnel model studies on buildings of similar shape. Halitsky (1985) gives details of this technique. Surface concentrations for block buildings with uncapped exhaust vents and short stacks in a uniform nonturbulent airstream are reported by Halitsky (1963), for situations where jet diameters and emission velocities were large enough to project the exhaust plume above the roof and sometimes through the recirculation cavity.

ESTIMATING INTAKE CONTAMINATION

If a representative configuration with published surface concentration coefficients K_c cannot be found to match the existing building, available data may be used to estimate the minimum dilution observed at the same receptor distance from the source.

The most important variables determining minimum dilution are the exhaust-to-intake stretched-string distance S and the effective stack height h_s (Figure 3). Only the stack height extending above large rooftop obstacles such as penthouses and architectural barriers should be used to define the **effective stack height**.

The **stretched-string distance** S is defined as the shortest length of string connecting the point on a stack where $h_s = 0$ to the nearest point on an air intake. The point where $h_s = 0$ is defined as the height of obstacles close to the stack, as shown in Figure 3, or the boundary of any roof recirculation region through which the stack passes.

All the following equations for **minimum dilution** D_{min} were developed for exposures equivalent to 10-min averaging times in the atmosphere, and with nonbuoyant exhaust jets from roof vents. If the exhaust gases are hot, buoyancy increases the rise of the exhaust gas mixture and produces lower concentrations at roof level. By neglecting buoyant plume rise, the D_{min} equations have an inherent safety factor, particularly at low wind speed, where buoyancy is most important.

The **averaging time** t_a over which exhaust gas concentration exposures are measured is also important in determining minimum dilution. As the averaging time increases, the exhaust gas plume meanders more from side to side, thus reducing the time-averaged concentration at an intake location. The effect of changing the averaging time over a range of about 3 min to 3 h can be estimated by adjusting the 10-min values given in Equations (14) to (21) using the following equation:

$$D_{min,1}/D_{min,2} = (t_{a1}/t_{a2})^{0.2} \tag{13}$$

See Wollenweber and Panofsky (1989).

The ratio of minimum dilution at any two averaging times t_{a1} and t_{a2} applies for $S/A_e^{0.5} > 10$ to allow the exhaust jet to begin meandering. If the exhaust and intake are both located in the same flow recirculation region, dilution is less sensitive to averaging time than predicted by Equation (13). In this case, assume the D_{min} values for 3-min averages also apply for averaging times from 3 to 60 min. Use Equation (13) to adjust the 10-min values in Equations (14) to (21) to 3-min values for this flow recirculation situation.

Strong Jets in Flow Recirculation Cavity

For block buildings with surface vents or short stacks that produce significant jet rise because of their large diameter, high emission velocity, and free discharge, the minimum dilution at a surface intake on the same building or on the jet centerline above the building roof is given by Halitsky (1963) as

$$D_{min} = \left[\alpha + 0.11(1 + 0.2\alpha)S/A_e^{0.5}\right]^2 \tag{14}$$

where α is a numerical constant related to building shape, emission velocity ratio V_e/U_H, building orientation to the wind, and stack height. Values of α can be found in Halitsky (1963). The smallest D_{min} occurs along the elevated centerline of the jet plume, where α = 1.0 is the appropriate value. Larger values of D_{min} occur on the building surfaces, where α ranges from 2.0 to 20. The larger values are associated with greater stack height and buoyant exhaust.

Strong Jets on Multiwinged Buildings

For a strong vertical exhaust with $V_e/U_H \approx 2$ from surface vents or short stacks on the roof of a multiwinged building with different wing roof heights, Halitsky (1962, 1982) recommends a minimum dilution of

$$D_{min} = M\left[3.16 + 0.1S/A_e^{0.5}\right]^2 \tag{15}$$

where M is an intake location factor with values of $M = 1.5$ when the intake is on the same roof as the source, $M = 2.0$ when the source and intake are on different wings separated by an air space, and $M = 4.0$ when the intake is substantially lower than the source.

Surface Vents on Flat-Roofed Buildings

For buildings with exhausts of zero stack height on a flat roof, the minimum dilution $D_{min,o}$ at a roof or wall intake is given by Wilson and Lamb (1994), Wilson and Chui (1985, 1987), and Chui and Wilson (1988) as

$$D_{min,o} = \left[D_o^{0.5} + D_s^{0.5}\right]^2 \quad (16)$$

where

$$D_o = 1 + 13.0\beta(V_e/U_H) \quad (17)$$

and

$$D_s = B_1(U_H/V_e)(S^2/A_e) \quad (18)$$

D_o is the apparent initial dilution at roof level caused by internal turbulence in the exhaust jet (Halitsky 1962, 1966). D_s is the distance dilution caused by the combined action of building and atmospheric turbulence. The wind speed U_H at wall height H is calculated from local airport weather data using Equation (4) and Table 1. For vertically directed uncapped exhausts, the capping factor $\beta = 1.0$; for capped, louvered, or downward-facing exhausts, $\beta = 0$. The distance dilution parameter B_1 depends on the exhaust jet trajectory and on the intensity of turbulence in the approach wind and generated by the building. This upwind turbulence is represented by σ_θ, which is the standard deviation (in degrees) of wind direction fluctuations averaged over 10-min periods. The full-scale data of Wilson and Lamb (1994) suggest that

$$B_1 = 0.027 + 0.0021\sigma_\theta \quad (19)$$

Typically, σ_θ varies between 0° and 30° for 10-min averages. For buildings located in urban terrain (Category 2 in Table 1), the recommended design value is $\sigma_\theta = 15°$, for which $B_1 = 0.059$.

Equations (16), (17), and (18) imply that minimum dilution does not depend on the location of either the exhaust or intake, only on the distance S between them. This is true when exhaust and intake locations are on the same building wall or on the roof. The dilution may increase if the intake and exhaust are located on different faces, as indicated by the M factor in Equation (15). For roof exhausts with wall intakes, the results of Li and Meroney (1983) suggest that the first term in Equation (19) may be increased from 0.027 to 0.10.

For buildings less than about 300 ft high and also less than twice as high as the surrounding buildings, atmospheric turbulence makes a significant contribution to exhaust gas dilution. Wilson (1976, 1977) gives surface concentration contours (see Figure 12) for flat-roofed buildings in a simulated approach wind that is typical of an urban area. Flush vents with small exhaust velocity make these results suitable for estimates for capped exhaust stacks or louvered exhaust vents.

The effect of atmospheric turbulence is relatively insignificant for isolated high-rises that are taller than 300 ft and also twice the average height of buildings for 3000 ft upwind. On buildings where the effects of atmospheric turbulence are small, Wilson and Chui (1987) found that maximum surface concentrations for 10-min exposures were two to ten times higher than on an equivalent low-rise building. For these high-rise buildings, use $\sigma_\theta = 0°$ in Equation (19), for which $B_1 = 0.027$.

When exhaust from several collecting stations is combined in a single vent or in a tight cluster of stacks, the effective exhaust flow area A_e in Equation (18) increases, which causes the minimum dilution in Equation (16) to decrease. To qualify as a cluster, the stacks must all lie within a two-stack diameter radius of the middle of the group. Stacks lined up in a row do not act as a single stack, as shown by Gregoric et al. (1982). However, the exhaust concentration C_e of each contaminant decreases by mixing with other exhaust streams, and the plume rise increases due to the higher momentum in the combined jets. For combined vertical exhaust jets, the rooflevel intake concentration C is almost always lower than the intake concentration caused by separate exhausts. Where possible, exhausts should be combined before release to take advantage of this increase in overall dilution.

Critical Wind Speed and Dilution

At very low wind speed, the exhaust jet from an uncapped stack rises high above roof level, producing a large exhaust dilution D_{min} at a given intake location. Likewise, at high wind speed, the dilution is also large because of longitudinal stretching of the plume by the wind. Between these extremes, a critical wind speed exists at which the least dilution occurs for a given exhaust and intake location. This critical, absolute minimum dilution D_{crit} may be used to determine if an exhaust vent will be safe under all wind conditions. The critical wind speed for an uncapped vertical exhaust ($\beta = 1.0$) can be evaluated by finding the absolute minimum in Equations (16), (17), and (18). It is closely approximated by

$$U_{crit,o}/V_e = \frac{3.6}{S}\left(\frac{A_e}{B_1}\right)^{0.5} \quad (20)$$

where $U_{crit,o}$ is the critical wind speed producing the smallest minimum dilution for an uncapped vertical exhaust with negligible stack height. This critical dilution $D_{crit,o}$ is

$$D_{crit,o} = \frac{(1 + 26V_e/U_{crit,o})^2}{1 + 13V_e/U_{crit,o}} \quad (21)$$

To assess the severity of the hazard caused by intake contamination, it is useful to know how often the worst-case D_{crit} is likely to occur. The number of hours per year during which the dilution is no more than twice the critical minimum value may be estimated from weather records by finding the fraction of time that the wind speed is between 0.5 $U_{crit,o}$ and 3.0 $U_{crit,o}$ (Wilson 1982, 1983). This fraction is then multiplied by the fraction of time the local wind direction lies in a sector 22.5° on each side of the line joining the exhaust and intake location.

EXHAUST STACK DESIGN

This section describes strategies for designing exhaust stack systems and for locating air intakes on a building emitting toxic or odorous exhausts. A geometric method is presented for calculating stack heights to avoid reentry of toxic or odorous exhausts into the emitting building. In addition, a method is given for computing minimum dilution at nearby air intakes and other receptors.

Stack Design Strategies

The dilution a stack exhaust system can provide is limited. Therefore, before discharge, exhaust contamination should be reduced by filters, collectors, and scrubbers.

Central exhausts that combine flows from many collecting stations should always be used where safe and practical. By combining several exhaust streams, central systems dilute intermittent bursts of contamination from a single station. Also, the combined flow forms an exhaust plume that rises a greater distance above the emitting building. Additional air volume can be added to the exhaust near the exit with a makeup air unit to increase the initial dilution and exhaust plume rise. This added air volume does not need heating or cooling, saving on energy costs.

In some cases, separate exhaust systems are mandatory. The nature of the contaminants to be combined, recommended industrial hygiene practice, and applicable safety codes need to be considered. Separate exhaust stacks should be grouped in a tight cluster to take advantage of the larger plume rise of the resulting combined jet. In addition, a single stack location for a central exhaust system or a tight cluster of stacks allows building air intakes to be positioned as far as possible from the exhaust location. For a tight cluster to be considered as a single stack in dilution calculations, the stacks must be uncapped and all lie within a two-stack diameter radius of the middle of the group.

As shown in Figure 3, the effective stack height h_s is the portion of the exhaust stack that extends above local recirculation zones and upwind and downwind obstacles. Wilson and Winkel (1982) demonstrated that stacks terminating below the level of adjacent walls and architectural enclosures do not effectively reduce roof-level exhaust contamination. To take full advantage of their height, stacks should be located on the highest roof of a building.

Architectural screens used to mask rooftop equipment can adversely affect exhaust dilution, depending on such variables as porosity, relative height, and distance from the stack. Stacks should extend above the architectural screen by the height H_c of the flow recirculation zone due to the screen itself, in order to prevent exhaust contamination of equipment within the enclosure.

Large buildings, structures, and terrain close to the emitting building can have adverse effects on dilution of stack exhaust, because the emitting building can be within the recirculation flow zones downwind of these nearby flow obstacles. In addition, an air intake located on a nearby taller building can be contaminated by exhausts from the shorter building. Wherever possible, facilities emitting toxic or highly odorous contaminants should not be located near taller buildings or at the base of steep terrain.

As shown in Figure 13, stacks should be vertically directed and uncapped. Stack caps that deflect the exhaust jet have a detrimental effect on both exhaust-to-intake dilution and exhaust plume rise. Conical stack caps often do not exclude rain, because rain does not usually fall straight down. Changnon (1966) shows that periods of heavy rainfall are often accompanied by high winds that deflect raindrops under the cap and into the stack. A stack exhaust velocity V_e of about 2500 fpm prevents condensed moisture from draining down the stack and keeps rain from entering the stack. For intermittently operated systems, protection from rain and snow should be provided by stack drains, as shown in Figures 13F through 13J, rather than stack caps.

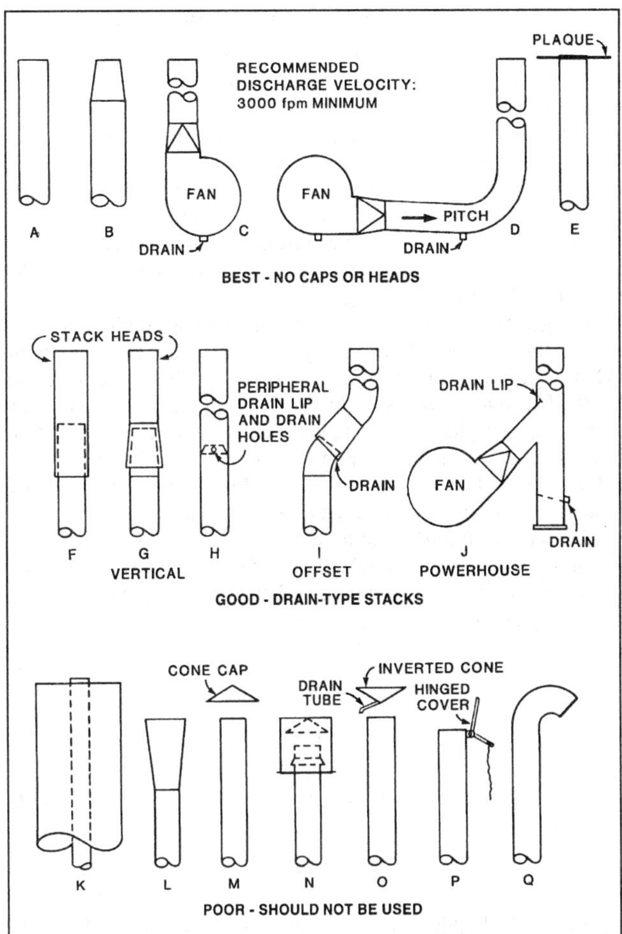

Fig. 13 Stack Designs Providing Vertical Discharge and Rain Protection

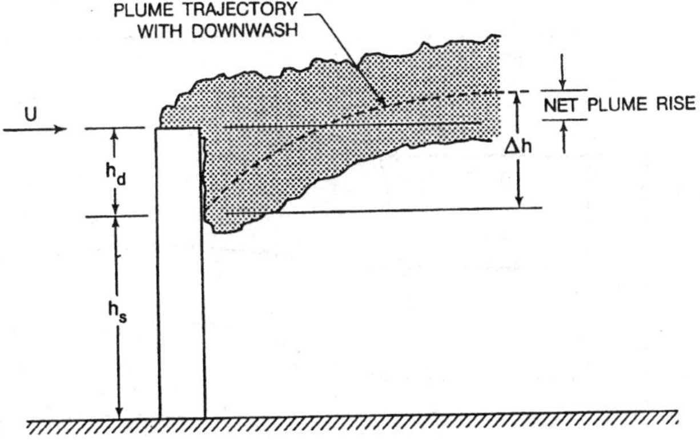

Fig. 14 Reduction of Effective Stack Height by Stack Wake Downwash

Recommended Stack Exhaust Velocity

High stack exhaust velocity and temperature increase plume rise and reduce intake contamination by increasing plume dilution and elevating the plume trajectory. The exhaust velocity V_e should be maintained above 2000 fpm (even when there are drains in the stack) to provide adequate plume rise and jet dilution. Velocities above 2000 fpm provide still more plume rise and dilution, but above 3000 to 4000 fpm, noise and vibration from exhaust fans become important considerations. The stack tip may have an exit nozzle (Figure 13B) to increase the exhaust velocity and plume rise.

An exception to these exhaust velocity recommendations may be needed when corrosive condensate droplets are discharged. In this case, a 1000 fpm in-stack velocity and a condensate drain are recommended to reduce droplet emission. An exit nozzle (Figure 13B) should be used to attain the exhaust velocity needed to produce plume rise and avoid downwash.

Stack wake downwash occurs where low-momentum exhausts are pulled downward by negative pressures immediately downwind of the stack, as shown in Figure 14. V_e should be 1.5 times the design wind speed U_H at roof level to avoid stack wake downwash. For design purposes, a meteorological station design wind speed U_{met} that is exceeded less than 1% to 5% of the time can be used, computed from climatological records or estimated from Table 2. Then, Equation 4 is used to calculate U_H from U_{met}, and V_e can be calculated ($V_e = 1.5U_H$).

Other Stack Design Standards

Minimum heights for chimneys and other flues are discussed in the *Uniform Building Code* (ICBO 1994). AIHA *Standard* 29.5, the American National Standard for Laboratory Ventilation, recommends a minimum stack height of 10 ft above the adjacent roof line; an exhaust velocity V_e of 3000 fpm, or 2000 fpm when internal condensation may occur; and a stack height extending one stack diameter above any architectural screen. The stack must also be situated to avoid reentry (reentrainment) into the laboratory or nearby buildings at concentrations above 20% of allowable concentrations within the laboratory under any atmospheric condition. NFPA *Standard* 45, Fire Protection for Laboratories Using Chemicals, specifies a minimum stack height of 7 ft to protect rooftop workers. Toxic chemical emissions may also be regulated by federal, state, and local air quality agencies.

Stack Height to Avoid Exhaust Reentrainment

This section presents a geometric method of specifying stack height h_s so that the lower edge of the exhaust plume is above intakes and recirculation zones on the emitting building roof. This stack height should nearly eliminate reentry of exhausts into the emitting building provided there are no large nearby buildings, structures, or terrain to disturb the approaching wind. This geometric method considers only intakes on the emitting building. Additional stack height or an exhaust-to-intake minimum dilution calculation should be used if the exhaust plume can impinge on the air intake of a nearby building.

The geometric design procedure starts by considering the flow recirculation regions shown in Figure 15. To avoid entrainment of exhaust gases, the stack plume must rise above the recirculation height H_c. Where stacks or exhaust vents discharge within this region, gases rapidly diffuse to the roof and may enter ventilation intakes or other openings. Figure 1 shows that this effluent is entrained into the zone of recirculating flow behind the downwind face and is, in some cases, brought back up to the roof.

Wilson (1976) found that for a flat-roofed building, the recirculation region maximum height H_c at location X_c, and recirculation lengths L_c and L_r (shown in Figures 3 and 15) are given by the following equations:

$$H_c = 0.22R \tag{22}$$

$$X_c = 0.5R \tag{23}$$

$$L_c = 0.9R \tag{24}$$

$$L_r = 1.0R \tag{25}$$

where R is the building scaling length from Equation (1). The high turbulence boundary downwind of the rooftop recirculation region can be approximated by a straight line sloping downward from H_c to the roof at L_c. The dimensions of the recirculating zones are somewhat sensitive to the intensity and scale of turbulence in the approaching wind. High levels of turbulence from upwind obstacles may decrease the coefficients in Equations (22) through (25) by up to half. Turbulence in the recirculation region and in the approaching wind causes the reattachment locations shown in Figure 2 to fluctuate.

To account for changes in roof-level, penthouses, and equipment-housings, the scale length R for each of these rooftop obstacles should be calculated from Equation (1) using the upwind face dimensions of the obstacle. The recirculation region for each obstacle is calculated from Equations (22), (23), and (24). The length L_r of the recirculation region downwind from the obstacle, or from the entire building, is given by Equation (25), with R based on the dimensions of the downwind face of the obstacle. The high turbulence region boundary Z_2 in Figure 15 follows a 10:1 (5.7°) downward slope from the top of the recirculation regions at X_c or L_r. When an obstacle is close to the upwind edge of a roof or near another obstacle, the flow recirculation zones interact. Wilson (1976) gives methods for dealing with these situations.

Building-generated turbulence is confined to the roof wake region, whose upper boundary Z_3 in Figure 15 is given by the following equation:

$$Z_3/R = 0.28(X/R)^{0.33} \tag{26}$$

where X is the distance from the upwind roof edge where the recirculation region forms. Building-generated turbulence decreases with height above roof level. At the edge of the rooftop wake boundary Z_3, the turbulence intensity is close to the background level in the approach wind. The high levels of turbulence below the boundary Z_2 in Figure 15 rapidly diffuse exhaust gases downward to contaminate roof-level intakes.

The next step in the geometric design procedure for stack height is to calculate the height h_{sc} of a stack with a rain cap and, therefore, no plume rise. The h_{sc} required to avoid excessive exhaust gas reentry is estimated by assuming that the plume spreads upward and downward from h_{sc} with a 5:1 slope (11.3°), as shown in Figure 15. (This slope represents a downward spread of approximately two standard deviations of a Gaussian plume concentration distribution.) Then, h_{sc} is raised until the lower edge of the plume avoids contact with all recirculation (zone 1) and high turbulence (zone 2)

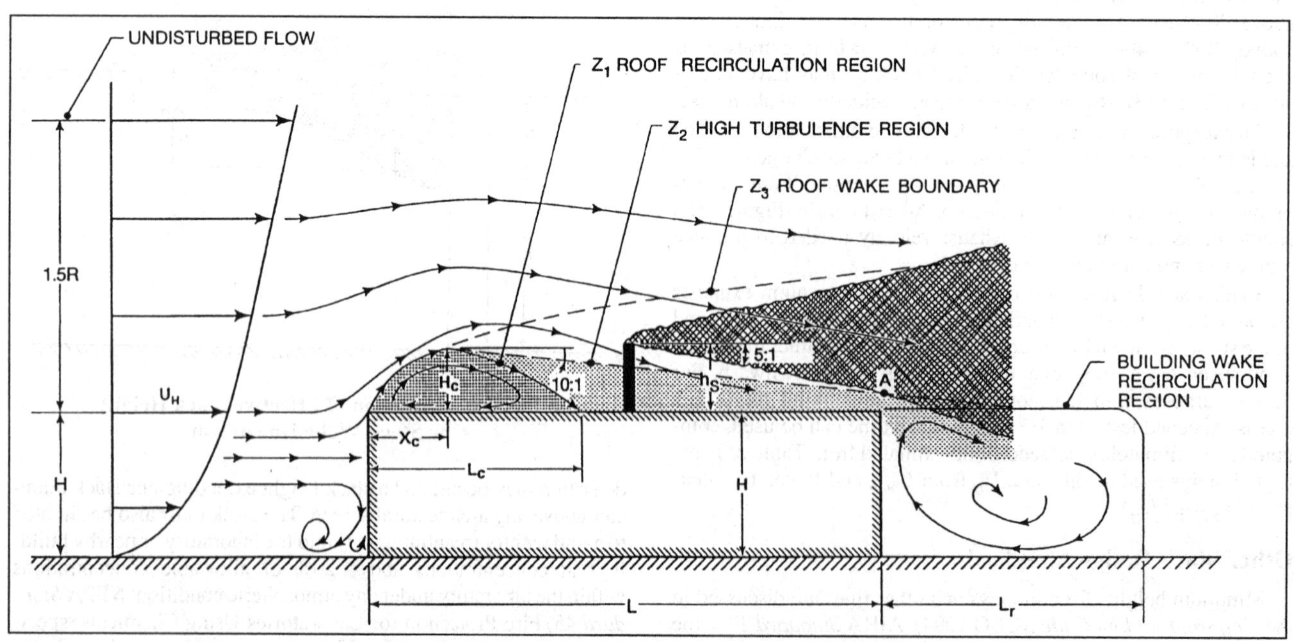

Fig. 15 Design Procedure for Required Stack Height to Avoid Contamination
(Wilson 1979)

boundaries from rooftop obstacles such as air intake housings, architectural screens, or penthouses. The sizes of the recirculation zones are given by Equations (22), (23), and (24).

If air intakes are located on the downwind wall, the lower edge of the plume, sloping down at 5:1 must lie above recirculation and high turbulence zone boundaries (point A in Figure 15) at the downwind edge of the roof. For a highly toxic contaminant that requires a large dilution factor at a wall intake, the lower edge of the plume should lie above the recirculation zone in the wake downwind of the building. The boundary of the building wake recirculation, shown in Figures 1, 3, and 15, is defined by the horizontal line extending a distance L_r from the downwind edge of the roof. The recirculation length L_r is calculated from Equation (25).

The final step in the geometric stack height design procedure is to reduce the stack height to give credit for plume rise from uncapped stacks, and to increase stack height to account for stack wake downwash caused by low exhaust velocity. Only jet momentum rise is used; buoyancy rise is considered as a safety factor. For an uncapped stack of diameter d, the **plume rise** h_r due to the vertical momentum of the exhaust is estimated from Briggs (1984) as

$$h_r = 3.0\beta d(V_e/U_H) \qquad (27)$$

where

$$d = (4A_e/\pi)^{0.5} \qquad (28)$$

For an uncapped stack, the capping factor is $\beta = 1.0$. For a capped stack, $\beta = 0$, so $h_r = 0$, and no credit is given for plume rise. U_H is the maximum design wind speed for which air intake contamination must be avoided.

Stack wake downwash is described in the section on Recommended Stack Exhaust Velocity. For a vertically directed jet from an uncapped stack ($\beta = 1.0$), Briggs (1973) recommends a stack wake **downwash adjustment** h_d of

$$h_d = 2.0d(1.5 - \beta V_e/U_H) \qquad (29)$$

for $V_e/U_H < 1.5$. For $V_e/U_H > 1.5$, there is no downwash and $h_d = 0$. Rain caps are frequently used on stacks of gas- and oil-fired furnaces and packaged ventilation units, for which $\beta = 0$ and $h_d = 3.0d$.

The stack height h_s recommended by this geometric method is

$$h_s = h_{sc} - h_r + h_d \qquad (30)$$

The advantage of using an uncapped stack instead of a capped stack is considerable. If the minimum recommended exhaust velocity V_e of $1.5U_H$ is maintained for an uncapped stack ($\beta = 1.0$), plume downwash $h_d = 0$ and $h_r = 4.5d$. For a capped stack ($\beta = 0$), $h_d = 3.0d$ and $h_r = 0$. Using these values in Equation (30), an uncapped stack can be made $7.5d$ shorter than a capped stack.

The largest flow recirculation, high turbulence, and wake regions occur when the wind is normal to the upwind wall of the building. The required stack height is the largest of the heights calculated for the four (or more) wind directions that are normal to an upwind wall.

Estimating Critical Dilution for Exhaust Stacks

The geometric stack design procedure described in the previous section does not give any estimate of the worst-case critical dilution factor D_{crit} between the stack and an air intake. If D_{crit} can be specified from knowledge of stack emissions and required health limits or odor thresholds, the computation of critical dilutions can be an alternative method for specifying stack heights. Smeaton et al. (1991) and Petersen and Ratcliff (1991) discuss the use of emission information and the formulation of dilution requirements in more detail. Air quality regulations may also play a role. Exhaust from a single-source dedicated stack may require more atmospheric dilution than a stack with manifolded exhausts because emissions are diluted within the manifold.

In this section, a method is presented for estimating D_{crit} for a predetermined stack height. The method has been shown to give accurate estimates of minimum dilution compared to field data (Wilson and Lamb 1994).

An increase in stack height or in exhaust velocity ratio V_e/U_H reduces roof-level contamination by keeping the high concentrations on the plume centerline far enough above the roof so that the intakes see only intermittent concentrations in the fringes of the plume. An increase in stack height or exhaust velocity increases the critical wind speed at which the absolute minimum dilution occurs. This higher critical wind speed often significantly reduces the number of hours per year that high intake contamination (i.e., low dilution) is observed.

Using a Gaussian plume dispersion equation, with a 10-min averaged vertical plume spread standard deviation $\sigma_z = 0.093S$, a crosswind plume spread standard deviation $\sigma_y = (0.093 + 0.0072\sigma_\theta)S$ where σ_θ is defined for Equation (19), and an uncapped vertical exhaust jet with no buoyancy and with plume rise inversely proportional to wind speed, the critical wind speed U_{crit} at which the smallest minimum dilution D_{crit} is observed is

$$\frac{U_{crit,o}}{U_{crit}} = (Y+1)^{0.5} - Y^{0.5} \qquad (31)$$

$U_{crit,o}$ is the critical wind speed for a flush (zero stack height) vertical exhaust, computed from Equation (20). The influence of stack height on the worst-case critical dilution for the standard 10-min exposure time may be calculated as follows:

$$\frac{D_{crit}}{D_{crit,o}} = \frac{U_{crit}}{U_{crit,o}} \exp\left[Y + Y^{0.5}(Y+1)^{0.5}\right] \qquad (32)$$

where Y is the height-to-spread parameter:

$$Y = 0.025h_s^2/\sigma_z^2 \qquad (33)$$

For 10-min averages, $\sigma_z = 0.093S$, and

$$Y = 28.9h_s^2/S^2 \qquad (34)$$

$D_{crit,o}$ in Equation (32) is the dilution at critical wind speed for a flush vertical roof exhaust with no stack height, from Equation (21). Equations (31) and (32) are reliable only for $Y < 2.0$. Close to the stack, where $Y > 2.0$, use $Y = 2.0$ in Equations (31) and (32). Because both wind speed and turbulence intensity vary strongly with height above the building roof, the plume rise h_r of the exhaust jet may not be inversely proportional to wind speed; normally h_r is proportional to $U^{-0.4}$ to $U^{-1.0}$. Thus, Equations (31) and (32) are only approximations. Because buoyancy is not included, the added rise due to buoyancy provides a safety factor, particularly at low wind speed.

Because Equations (31) and (32) give the effect of a stack relative to a flush exhaust with $h_s = 0$, they are useful for assessing the advantages of increasing stack height as a remedial measure. By comparing two different heights, this calculation allows the relative benefits of a stack to be estimated without knowing any details of the contaminant concentrations or exhaust velocity in the existing stack. For example, the stack height required using the simple geometrical design procedure (see Figure 15) has an h_s/S of at least 0.2. Equations (31) and (32) show that the critical wind speed U_{crit} for this stack height is about twice as large, and the critical dilution D_{crit} about eight times as large, as the U_{crit} and D_{crit} for the vertical jet from an uncapped exhaust with zero effective stack height.

Example 2. The stack height h_s of the uncapped vertical exhaust on the building in Figure 3 must be specified to avoid excessive contamination of air intakes A and B by stack gases. The stack has a diameter d of 1.64 ft and an exhaust velocity V_e of 1770 fpm. It is located 52.5 ft from the upwind edge of the roof. The penthouse's upwind wall (with intake A) is located 98.4 ft from the upwind edge of the roof, a height of 13.1 ft, and a length of 23.0 ft in the wind direction. The top of intake A is 6.56 ft below the penthouse roof. The building has a height H of 49.2 ft and a length of 203 ft. The top of intake B is 19.7 ft below roof level. The width (measured into the page) of the building is 164 ft, and the penthouse is 29.5 ft wide. What is the required stack height h_s for a design wind speed specified as twice the annual average hourly wind speed of 7.95 mph at a nearby airport with an anemometer height H_{met} of 32.8 ft? The building is located in suburban terrain (Category 2 in Table 1).

Solution: The first step is to set the height h_{sc} of a capped stack by projecting lines with 5:1 slopes upwind from points of potential plume impact. For intake A, the highest point of impact is the top of the recirculation zone on the roof of the penthouse. To find the height of this recirculation zone, start with Equation (1):

$$R = (13.1)^{0.67}(29.5)^{0.33} = 17.1 \text{ ft}$$

Then use Equations (22) and (23):

$$H_c = 0.22(17.1) = 3.76 \text{ ft}$$

$$X_c = 0.5(17.1) = 8.55 \text{ ft}$$

With the 5:1 slope of the lower plume boundary shown in Figure 15, the capped stack height in Figure 3 must be

$$h_{sc} = 0.2(98.4 - 52.5 + 8.55) + 3.76 = 14.7 \text{ ft}$$

above the penthouse roof to avoid intake A. For intake B on the downwind wall, the plume boundary from the stack in Figure 3 must lie above the end of the roof. To avoid intake B, the capped stack height must be

$$h_{sc} = 0.2(203 - 52.5) - 13.1 = 17.1 \text{ ft}$$

The design stack height is set by the condition of avoiding contamination of intake B, because intake A requires only a 14.7 ft capped stack. Credit for plume rise h_r from the uncapped stack requires calculation of the building wind speed U_H at H = 49.2 ft. The design wind speed for H_{met} = 32.8 ft at the airport meteorological station is U_{met} = 2(7.95) = 15.9 mph = 1400 fpm. With the airport in open terrain (Category 3 of Table 1), and the building in urban terrain (Category 2), the wind speed adjustment parameters are a_{met} = 0.14 and δ_{met} = 900 ft at the airport, and a = 0.22 and δ = 1200 ft at the building. Using Equation (4), with the building height H = 49.2 ft,

$$U_H = 1400\left(\frac{900}{32.8}\right)^{0.14}\left(\frac{49.2}{1200}\right)^{0.22} = 1102 \text{ fpm}$$

Because V_e/U_H = 1770/1102 = 1.61 is greater than 1.5, there is no plume downwash, and h_d = 0 from Equation (29). Using Equation (27), the plume rise at the design wind speed is

$$h_r = 3.0(1.64)(1.61) = 7.92 \text{ ft}$$

Deducting this rise from the uncapped height h_{sc} for intake B,

$$h_s = 17.1 - 7.92 = 9.18 \text{ ft}$$

As shown in Figure 3, this stack height is measured above the roof of the nearby penthouse. Adding the penthouse height of 13.1 ft sets the required stack height at 22.2 ft above roof level.

Example 3. The uncapped stack from Example 2 (with h_s = 9.18 ft above the penthouse roof) is used to exhaust a toxic contaminant. The stack gas concentration must be diluted by 1000:1 to be considered safe for 60-min exposures. (This required dilution is determined from occupational health standards.) Calculate the critical minimum dilution factor D_{crit} and the critical wind speed U_{crit} for contamination for intakes A and B. Is the stack height of 9.18 ft sized in Example 1 sufficient to handle this toxic substance?

Solution: At intake A, the stretched-string distance is

$$S_A = \left[(98.4 - 52.5)^2 + (6.56)^2\right]^{0.5} = 46.3 \text{ ft}$$

and the exhaust area of the 1.64 ft diameter stack is A_e = 2.11 ft². The first step is to determine the critical wind speed and dilution for an uncapped stack with h_s = 0. The normalized exhaust-to-intake distance is

$$S_A/A_e^{0.5} = 46.3/(2.11)^{0.5} = 31.8$$

For dilution calculations, B_1 = 0.059 from the text following Equation (19). The critical wind speed for a vent without a stack is calculated as follows from Equation (20), with V_e = 1770 fpm:

$$U_{crit,o} = \frac{1770(3.6)}{31.8(0.059)^{0.5}} = 825 \text{ fpm}$$

From Equation (21), the worst-case minimum dilution at intake A for this critical wind speed is

$$D_{crit,o} = \frac{[1 + 26(1770)/825]^2}{1 + 13(1770)/825} = 112$$

The effect of stack height h_s = 9.18 ft is calculated from Equations (31) and (32). The height-to-spread parameter Y in Equation (34) is

$$Y = 28.9(9.18/46.3)^2 = 1.12$$

From Equation (31),

$$U_{crit} = \frac{825}{(1.12 + 1)^{0.5} - (1.12)^{0.5}} = 2070 \text{ fpm}$$

From Equation (32),

$$D_{crit} = 112(2070/825)\exp\left[1.12 + (1.12)^{0.5}(1.12 + 1)^{0.5}\right] = 4020$$

The stack height has increased U_{crit} to a high, infrequently occurring 2070 fpm (23.5 mph), and increased the worst-case critical dilution by more than a factor of 35 from the zero stack height baseline case $D_{crit,o}$ = 112 to D_{crit} = 4020.

The next step is conversion of the dilution from the 10-min exposure averaging time on which all the design equations are based, to the 60-min exposure required for the specified occupational health standard. Using Equation (13),

$$D_{crit} = 4020(60/10)^{0.2} = 5750$$

Thus, the 9.18 ft stack height produces much more than the required 1000:1 dilution at intake A.

Repeating the calculations for intake B, the stretched-string distance in Figure 3 is

$$S_B = (98.4 - 52.5) + 23 + (81.6^2 + 13.1^2)^{0.5} + 19.7 = 172 \text{ ft}$$

and the normalized exhaust-to-intake distance is

$$S_B/A_e^{0.5} = 172/(2.11)^{0.5} = 118$$

For the zero stack height baseline case, B_1 = 0.059 and Equation (20) at intake B gives

$$U_{crit,o} = \frac{1770(3.6)}{118(0.059)^{0.5}} = 222 \text{ fpm}$$

From Equation (21), the zero stack height critical dilution for intake B is

$$D_{crit,o} = \frac{[1 + 26(1770)/222]^2}{1 + 13(1770)/222} = 414$$

For the specified stack height h_s = 9.18 ft, the height-to-spread parameter Y in Equation (34) is

$$Y = 28.9(9.18/172)^2 = 0.082$$

Then, from Equations (31) and (32),

$$U_{crit} = \frac{222}{(0.082 + 1)^{0.5} - (0.082)^{0.5}} = 295 \text{ fpm}$$

$$D_{crit} = 414(295/222)\exp\left[0.082 + (0.082)^{0.5}(0.082 + 1)^{0.5}\right] = 804$$

For intake B, which is further from the stack, the 9.18 ft stack increases the critical wind speed by a factor of 1.3 (from 222 fpm to 295 fpm), and the critical worst-case dilution increases only by a factor of 1.9 (from 414 to 804). For intake A, which is closer to the stack, the 9.18 ft stack increases the critical wind speed by a factor of 2.5 (from 825 fpm to 2070 fpm) and increases critical dilution by a factor of 36 (from 112 to 4020). This shows that a stack has a diminishing benefit as the exhaust-to-intake distance increases. Using Equation (11), for a 60-min exposure at intake B,

$$D_{crit} = 804(60/10)^{0.2} = 1150$$

This meets the required 1000:1 minimum hourly averaged dilution specified in the design requirement.

Air Intake Location to Minimize Contamination

Stack height requirements can sometimes be reduced by careful location of air intakes. Even in the absence of toxic building exhausts, intakes should be located to avoid other sources of contamination such as automobile traffic, kitchens, dust from streets and plants, evaporative cooling towers, emergency generators, and plumbing vents. Diesel odors from loading docks and emergency generators can be nuisances (Smeaton et al. 1991, Ratcliff et al. 1994). Kitchen exhaust can be a source of odors and can cause plugging and corrosion of heat exchangers.

Evaporative cooling towers located too close to air intakes can have several effects: water vapor can increase air-conditioning loads, freezing water vapor can damage equipment, and ice can block intake grilles and filters. The effect on indoor air quality of escaping droplets of cooling tower water (drift) containing chemicals added to retard scaling and biological contamination is discussed by Vanderhayden and Schuyler (1994).

Intakes should not be located within the same architectural screen enclosure as contaminated exhaust outlets. If contaminated exhaust is discharged from several locations on a roof, intakes should be sited to minimize contamination. Where all exhausts of concern are emitted from a single relatively tall stack or tight cluster of stacks, a good intake location may be at the base of the stack, provided this location is not adversely affected by exhaust from nearby buildings.

Intakes near loading zones should be avoided. Overhead canopies on vehicle docks do not prevent hot vehicle exhaust from rising to intakes above the canopy. When the loading zone is within the flow recirculation region downwind of the building, vehicle exhausts may spread upwind over the entire building surface (Ratcliff et al. 1994).

When the wind is normal to the upwind wall, the streamline patterns in Figures 1 and 2 show that air flows up and down the wall, dividing at about two-thirds up the wall. The downward flow creates ground-level swirl (shown in Figure 1) that stirs up dust and debris. To take advantage of the natural separation of wind flow over the upper and lower half of a building, toxic or nuisance exhausts should be located on the roof and intakes located on the lower one-third of the building, but high enough to avoid wind-blown dust, debris, and vehicle exhaust. If ground-level sources such as wind-blown dust and vehicle exhaust are the major sources of contamination, a rooftop intake is desirable. If possible, inlet designs should take into account future sources of contamination, future changes to building geometries, and planned construction of buildings nearby.

Cooling towers and similar heat-rejection devices are very sensitive to airflow around buildings. This equipment is frequently roof-mounted, with intakes close to the roof where air is considerably hotter and at a higher wet-bulb temperature than air that is not affected by the roof. This can reduce the capacity of cooling towers and air-cooled condensers.

Heat exchangers often take in air on one side and discharge heated, moist air horizontally from the other side. For these horizontal-flow cooling towers, changes in wind direction and velocity caused by obstructions that are immediately adjacent can drastically reduce equipment performance by reducing the airflow rate. Recirculation is even more serious than reduction in airflow rate for such devices. Recirculation of warm moist exhaust raises the inlet wet-bulb temperature, which reduces performance. Recirculation can be caused by local disturbance of the airflow by an upwind obstruction or by a close downwind obstruction. The vertical discharge ducts may need to be extended to reduce recirculation and improve the effectiveness of this equipment.

SCALE MODEL SIMULATION AND TESTING

For many routine design applications, the flow patterns, wind pressures, and exhaust dilution can be estimated using the data and equations presented in the previous sections. However, in critical applications, such as where health and safety are of concern, physical modeling or full-scale field evaluations may be required to obtain more accurate estimates. Measurements on small-scale models in wind tunnels or water channels can provide information for design prior to construction and an economical method of performance evaluation for existing facilities. Full-scale testing is not generally useful in the initial design process because of the time and expense required to obtain meaningful information. On the other hand, full-scale testing is useful for verifying data derived from physical modeling and for planning remedial changes to improve existing facilities.

Detailed accounts of physical modeling, field measurements and applications, and engineering problems resulting from atmospheric flow around buildings are available in the proceedings of conferences on wind engineering (see Bibliography).

The wind tunnel is the main tool used to assess and understand the airflow around buildings. A water channel can also be used. Models of buildings, complexes, and the local surrounding topography are constructed and tested in a simulated turbulent atmospheric boundary layer. The airflow, wind pressures, snow loads, structural response, or pollutant concentrations can then be measured directly by properly scaling the wind and exhaust flow characteristics. Weil et al. (1981), Petersen (1987a), and Dagliesh (1975) found generally good agreement between the results of wind tunnel simulations and corresponding full-scale data.

Similarity Requirements

Physical modeling is most appropriate for applications involving small-scale atmospheric motions, such as recirculation of exhaust downwind of a laboratory, wind loads on structures, wind speeds around building clusters, snow loads on roofs, and airflow over hills. At present, winds associated with tornadoes, thunderstorms, and large-scale atmospheric motion cannot be simulated.

Snyder (1981) gives guidelines for fluid modeling of atmospheric diffusion. This report contains explicit directions and should be used whenever designing wind tunnel studies to assess concentration levels due to air pollutants. ASCE *Standard* 7 also provides guidance when wind tunnels are used for evaluating wind effects on structures.

A complete and exact simulation of the airflow over buildings and the resulting concentration or pressure distributions cannot be achieved in a physical model. This is not a serious limitation, however. Cermak (1971, 1976a,b), Snyder (1981), and Petersen (1987a,b) found that an accurate simulation of the transport and dispersion of laboratory exhaust can be achieved if the following criteria are met in the model and full scale:

1. Match exhaust velocity to wind speed ratios, V_e/U_H.
2. Match exhaust to ambient air density ratios ρ_e/ρ_a.
3. Match exhaust Froude numbers $Fr^2 = \rho_a V_e^2/[(\rho_e - \rho_a)gd]$, where d is the effective exhaust stack diameter.
4. Ensure fully turbulent stack gas flow by ensuring stack Reynolds number $Re_s = V_e d/\nu$ greater than 2000 (where ν is the kinematic viscosity of outdoor air, ft^2/s), or by placing an obstruction inside the stack to enhance turbulence.
5. Ensure fully turbulent wind flow.
6. Scale all dimensions and roughness by a common factor.
7. Match atmospheric stability by the bulk Richardson number (see Cermak 1975). For most applications related to airflow around buildings, neutral stratification is assumed, and no Richardson number matching is required.
8. Match mean velocity and turbulence distributions in the wind.
9. Ensure building wind Reynolds number $U_H R/\nu$ is greater than 11,000 for sharp-edged structures, or greater than 90,000 for round-edged structures.
10. Ensure less than 5% blockage of wind tunnel cross section.

For wind speeds, flow patterns, or pressure distributions around buildings, only conditions 5 through 10 are necessary. Usually, each wind tunnel study requires a detailed assessment to determine which appropriate parameters to match in the model and full scale.

In wind tunnel simulation of exhaust gas recirculation, the buoyancy of the exhaust gas (condition 3) is often not modeled. This allows using a high wind tunnel speed or a smaller model to achieve a high enough Reynolds number (conditions 4, 5, and 9). Neglecting buoyancy is justified if the density of building exhaust air is within 10% of the outdoor ambient air. Also, critical minimum dilution D_{crit} occurs at wind speeds high enough to produce a well-mixed, neutrally stable atmosphere, allowing stability matching (condition 7) to be neglected. Omission of conditions 3 and 7 simplifies the test procedure considerably, reducing both testing time and cost.

Buoyancy should be properly simulated for high-temperature exhausts from boilers and diesel generators. Equality of model and prototype Froude numbers (condition 3) requires tunnel speeds of less than 100 fpm for testing. However, greater tunnel speeds may be needed to meet the minimum building Reynolds number requirement (condition 4).

Wind Simulation Facilities

Boundary layer wind tunnels are required for conducting most wind studies. The wind tunnel test section should be long enough so that a deep boundary layer that slowly changes with downwind distance can be established upwind of the model building.

Other important wind tunnel characteristics include the width and height of the test section, range of wind speeds, roof adjustability, and temperature control. Larger models can be used in tunnels that are wider and taller, which, in turn, gives better measurement resolution. Model blockage effects can be minimized by an adjustable roof height. Temperature control of the tunnel surface and airflow is required when atmospheric conditions other than neutral stability are to be simulated. Boundary layer characteristics appropriate for the site are established by using roughness elements on the tunnel floor that produce mean velocity and turbulence intensity profiles characteristic of the full scale.

Water as well as air can be used for the modeling fluid if an appropriate flow facility is available. Flow facilities may be in the form of a tunnel, tank, or open channel. Water tanks with a free surface ranging in size up to that of a wind tunnel test section have been used by towing a model (upside down) through the nonflowing fluid. Stable stratification can be obtained by adding a salt solution at the tank bottom. This technique does not permit development of a boundary layer and yields only approximate qualitative information on flow around buildings. Water channels can be designed to develop thick turbulent boundary layers similar to those developed in the wind tunnel. One advantage of such a flow system is ease of flow visualization, but this is offset by a greater difficulty in measurement of flow variables and concentrations.

Designing Model Test Programs

The first step in planning a test program is selection of the model length scale. Choice of this scale depends on cross-sectional dimensions of the test section, dimensions of the building to be studied and nearby neighboring buildings, and/or topographic features and thickness of the simulated atmospheric boundary layer. Typical geometric scales range from about 150:1 to 1000:1.

Because a large model size is desirable to meet minimum Reynolds number and Froude number requirements, a wide test section is advantageous. In general, the model at any section should be small compared to the test section area so that blockage is less than 5%.

The test program must include specifications of the meteorological variables to be considered. These include wind direction, wind speed, and thermal stability. Data taken at the nearest meteorological station should be reviewed to obtain a realistic assessment of wind climate for a particular site. Ordinarily, local winds around a building, pressures, and/or concentrations are measured for 16 wind directions, in 22.5° intervals. This is easily accomplished by mounting the building model and its nearby surroundings on a turntable. More than 16 wind directions are required for highly toxic exhausts or for finding peak fluctuating pressures on a building. If only local wind information and pressures are of interest, testing at one wind speed with neutral stability is sufficient.

SYMBOLS

a = exponent in power law wind speed profile for local building terrain, Equation (4) and Table (1)
a_{met} = exponent a for the meteorological station, Equation (4) and Table 1
A_e = stack or exhaust exit face area, ft^2
B_1 = air entrainment parameter in distance dilution D_s, Equation (19)
B_L = larger of the two upwind building face dimensions H and W, ft
B_s = smaller of the two upwind building face dimensions H and W, ft
C = contaminant mass concentration at receptor at ambient air temperature T_e, Equation (6), lb/ft^3
C_{allow} = allowable concentration of contaminant at receptor, Equation (7)
C_e = contaminant mass concentration in exhaust at exhaust temperature T_e, Equation (6), lb/ft^3
C_{in} = internal wind-induced pressure coefficient, Equation (5)
C_p = local wind pressure coefficient for building surface, Equation (3)
$C_{p(in-out)}$ = difference between outdoor and indoor pressure coefficients, Equation (5)
C_s = wall-averaged pressure coefficient, Figure 6
d = effective exhaust stack diameter, Equation (28), ft
D = dilution factor between source and receptor mass concentrations, Equation (6)
D_o = apparent initial dilution factor for exhaust jet, Equation (16)
D_{crit} = critical dilution factor at roof level for uncapped vertical exhaust at critical wind speed U_{crit} that produces smallest value of D_{min} for given exhaust-to-intake distance S and stack height h_s, Equation (31)
$D_{crit,o}$ = critical dilution factor D_{crit} at roof level for uncapped vertical exhaust with zero stack height ($h_s = 0$), Equation (21)

D_{min} = minimum dilution factor D at given wind speed for all exhaust locations at same fixed distance S from intake, Equation (13)

$D_{min,o}$ = minimum dilution factor D_{min} at roof level for flush vent with zero stack height ($h_s = 0$), Equation (16)

D_{req} = atmospheric dilution required to meet allowable concentration of contaminant C_{allow}, Equation (7)

D_s = distance dilution factor at fixed wind speed, Equation (18)

D_v = dilution factor between source and receptor volume fraction concentrations, Equation (10)

f = contaminant volume concentration fraction at receptor; ratio of contaminant gas volume to total mixture volume, Equation (10), ppm $\times 10^{-6}$

f_e = contaminant volume concentration fraction in exhaust gas; ratio of contaminant gas volume to total mixture volume, Equation (9), ppm $\times 10^{-6}$

g = gravitational acceleration, Equation (2), ft/s^2

h_d = downwash correction to be subtracted from stack height, Equation (29), ft

h_r = plume rise of uncapped vertical exhaust jet, Equation (27), ft

h_s = effective exhaust stack height above rooftop obstacles and enclosures, Equation (30), ft

h_{sc} = required height of capped exhaust stack to avoid excessive intake contamination, Equation (30), ft

H = wall height above ground on upwind building face, Equation (4), ft

H_c = maximum height above roof level of upwind roof edge flow recirculation zone, Equation (22), ft

H_{met} = height of wind anemometer at meteorological station, Equation (4), ft

K_c = normalized concentration coefficient, Equation (12)

L = length of building in wind direction, Figure 2, ft

L_c = length of upwind roof edge recirculation zone, Equation (24), ft

L_r = length of flow recirculation zone behind rooftop obstacle or building, Equation (25), ft

$\dot{m}$ = contaminant mass release rate, Equation (8), lb/s

M = configuration factor, Equation (15)

p_s = wind pressure difference between exterior building surface location and local outdoor atmospheric pressure at same level in undisturbed approach wind, Equation (3), lb/ft^2

p_v = velocity pressure of wind at roof level, Equation (2), lb/ft^2

Q = contaminant volumetric release rate, Equation (9), ft^3/s

Q_e = total exhaust volumetric flow rate, Equation (8), ft^3/s

R = scaling length for roof flow patterns, Equation (1), ft

S = stretched-string distance; the shortest distance from exhaust to intake over and along building surface, Equation (14), ft

t_a = time interval over which receptor (intake) concentrations are averaged in computing dilution, Equation (13), s

T_a = outdoor ambient air absolute temperature, Equation (11), °R

T_e = exhaust air mixture absolute temperature, Equation (11), °R

U_{annual} = annual average of hourly wind speeds U_{met}, Table 2, ft/s

U_{crit} = critical wind speed that produces smallest minimum dilution factor D_{crit} for uncapped vertical exhaust at given S and h_s, Equation (31), ft/s

$U_{crit,o}$ = critical wind speed for smallest minimum dilution factor $D_{crit,o}$ for flush uncapped exhaust with zero stack height ($h_s = 0$), Equation (20), ft/s

U_H = mean wind speed at height H of upwind wall in undisturbed flow approaching building, Equation (2), ft/s

U_{met} = meteorological station hourly wind speed, measured at height H_{met} above ground in smooth terrain, Equation (4) and Table 2, ft/s

V_e = exhaust face velocity, Equation (8), ft/s

W = width of upwind building face, Equation (12), ft

X = distance from upwind roof edge, Equation (26), ft

X_c = distance from upwind roof edge to H_c, Equation (23), ft

Y = height-to-spread parameter, Equation (33)

Z_1 = height of flow recirculation zone boundary above roof, Figure 15, ft

Z_2 = height of high turbulence zone boundary above roof, Figure 15, ft

Z_3 = height of roof edge wake boundary above the roof, Equation (26) and Figure 15, ft

α = configuration parameter, Equation (14)

β = capping factor; $\beta = 1.0$ for vertical uncapped roof exhaust; $\beta = 0$ for capped, louvered, or downward-facing exhaust, Equation (17)

δ = fully developed strong wind atmospheric boundary layer thickness, Equation (4) and Table 1, ft

δ_{met} = atmospheric boundary layer thickness at the meteorological station, Equation (4) and Table 1, ft

ν = kinematic viscosity of outdoor air, ft^2/s

ρ_a = density of outdoor air, Equation (2), lb/ft^3

ρ_e = density of exhaust gas mixture, lb/ft^3

σ_y = standard deviation of crosswind plume spread, above Equation (31), ft

σ_z = standard deviation of vertical plume spread, Equation (33), ft

σ_θ = standard deviation of wind direction and fluctuations in time t_a, Equation (19) and above Equation (31), degrees

θ = wind angle between perpendicular line from the upwind building face to wind direction, Figure 5, degrees

REFERENCES

AIHA. 1992. American national standard for laboratory ventilation. ANSI/AIHA *Standard* Z9.5-1992. American Industrial Hygiene Association, Fairfax, VA.

Akins, R.E., J.A. Peterka, and J.E. Cermak. 1979. "Averaged pressure coefficients for rectangular buildings." In *Wind Engineering*. Proceedings of the Fifth International Conference 7:369-80, Fort Collins, CO. Pergamon Press, NY.

ASCE. 1995. Minimum design loads for buildings and other structures. ANSI/ASCE *Standard* 7-1995. American Society of Civil Engineers, New York.

ASHRAE. 1986. Bin and degree hour weather data for simplified energy calculations.

Bair, F.E. 1992. *The Weather Almanac*, 6th ed. Gale Research Inc., Detroit, MI.

Briggs. 1973. Diffusion estimates for small emissions. Oak Ridge Atmospheric Turbulence and Diffusion Laboratory, Draft *Report* No. 79.

Briggs. 1984. "Plume rise and buoyancy effects." In *Atmospheric science and power production*. D. Randerson, ed. U.S. Department of Energy DOE/TIC-27601 (DE 84005177).

Cermak, J.E. 1971. Laboratory simulation of the atmospheric boundary layer. *AIAA Journal* 9(9):1746.

Cermak, J.E. 1975. Applications of fluid mechanics to wind engineering. *Journal of Fluid Engineering*, Transactions of ASME 97:9.

Cermak, J.E. 1976a. Nature of airflow around buildings. *ASHRAE Transactions* 82(1):1044-60.

Cermak, J.E. 1976b. Aerodynamics of buildings. *Annual Review of Fluid Mechanics* 8:75.

Changnon, S.A. 1966. Selected rain-wind relations applicable to stack design. *Heating Piping and Air Conditioning* 38(3):93.

Chui, E.H. and D.J. Wilson. 1988. Effects of varying wind direction on exhaust gas dilution. *Journal of Wind Engineering and Industrial Aerodynamics* 31:87-104.

Clarke, J.H. 1967. Airflow around buildings. *Heating Piping and Air Conditioning* 39(5):145.

Dagliesh, W.A. 1975. Comparison of model/full-scale wind pressures on a high-rise building. *Journal of Industrial Aerodynamics* 1:55-66.

Davenport, A.G. and H.Y.L. Hui. 1982. External and internal wind pressures on cladding of buildings. Boundary Layer Wind Tunnel Laboratory, University of Western Ontario, London, Ontario, Canada. BLWT-820133.

Deaves, D.M. 1981. Computations of wind flow over changes in surface roughness. *Journal of Wind Engineering and Industrial Aerodynamics* 7:65-94.

DOC. 1968. Climatic Atlas of the United States. U.S. Department of Commerce, Washington, D.C.

Feustel, H.E. and J. Dieris. 1992. A survey of airflow models for multizone buildings. *Energy and Buildings* 18: 79-100.

Gregoric, M., L.R. Davis, and D.J. Bushnell. 1982. An experimental investigation of merging buoyant jets in a crossflow. *Journal of Heat Transfer*, Transactions of ASME 104:236-40.

Halitsky, J. 1962. Diffusion of vented gas around buildings. *Journal of the Air Pollution Control Association* 12:74-80.

Halitsky, J. 1963. Gas diffusion near buildings. *ASHRAE Transactions* 69:464-84.

Halitsky, J. 1966. A method of estimating concentrations in transverse jet plumes. *International Journal of Air and Water Pollution* 10:821-43.

Halitsky, J. 1982. Atmospheric dilution of fume hood exhaust gases. *American Industrial Hygiene Association Journal* 43(3):185-89.

Halitsky, J. 1985. Concentration coefficients in atmospheric dispersion calculations. *ASHRAE Transactions* 91(2B):1722-36.

Holmes, J.D. 1986. *Wind loads on low-rise buildings: The structural and environmental effects of wind on buildings and structures*, Chapter 12. Faculty of Engineering, Monash University, Melbourne, Australia.

Hosker, R.P. 1984. "Flow and diffusion near obstacles." In *Atmospheric science and power production*. D. Randerson, ed. U.S. Department of Energy DOE/TIC-27601 (DE 84005177).

Hosker, R.P. 1985. Flow around isolated structures and building clusters: A review. *ASHRAE Transactions* 91(2b):1671-92.

Houlihan, T.F. 1965. Effects of relative wind on supply air systems. *ASHRAE Journal* 7(7):28.

ICBO. 1994. *Uniform building code*. International Conference of Building Officials, Whittier, CA.

Li, W.W. and R.N. Meroney. 1983. Gas dispersion near a cubical building. *Journal of Wind Engineering and Industrial Aerodynamics* 12:15-33.

Meroney, R.N. 1982. "Turbulent diffusion near buildings." In *Engineering meteorology* 48:525. E.J. Plate, ed. Elsevier, Amsterdam.

NCDC. Updated periodically. *International station meteorological climatic summary* (CD-ROM). National Climatic Data Center, Asheville, NC. Published jointly with U.S. Air Force and U.S. Navy.

NFPA. 1991. Fire protection for laboratories using chemicals. ANSI/NPPA *Standard* 45-91. National Fire Protection Association, Quincy, MA.

Petersen, R.L. 1987a. Wind tunnel investigation of the effect of platform-type structures on dispersion of effluents from short stacks. *Journal of Air Pollution Control Association* 36:1347-52.

Petersen, R.L. 1987b. Designing building exhausts to achieve acceptable concentrations of toxic effluent. *ASHRAE Transactions* 93(2):2165-85.

Petersen, R.L. and M.A. Ratcliff. 1991. An objective approach to laboratory stack design. *ASHRAE Transactions* 97(2):553-62.

Ratcliff, M.A., R.L. Petersen, and B.C. Cochran. 1994. Wind tunnel modeling of diesel motors for fresh air intake design. *ASHRAE Transactions* 100(2):603-11.

Smeaton, W.H., M.F. Lepage, and G.D. Schuyler. 1991. Using wind tunnel data and other criteria to judge acceptability of exhaust stacks. *ASHRAE Transactions* 97(2):583-88.

Snyder, W.H. 1981. Guideline for fluid modelling of atmospheric diffusion. Environmental Protection Agency *Report*, EPA-600/881-009.

Swami, H.V. and S. Chandra. 1987. Procedures for calculating natural ventilation airflow rates in buildings. Final *Report* FSEC-CR-163-86. Florida Solar Energy Center, Cape Canaveral, FL.

Vanderhayden, M.D. and G.D. Schuyler. 1994. Evaluation and quantification of the impact of cooling tower emissions on indoor air quality. *ASHRAE Transactions* 100(2):612-20.

Weil, J.C., J.E. Cermak, and R.L. Petersen. 1981. Plume dispersion about the windward side of a hill at short range: Wind tunnel versus field measurements. Paper presented at Fifth American Meteorological Society Symposium, Atlanta, GA.

Wilson, D.J. 1976. Contamination of air intakes from roof exhaust vents. *ASHRAE Transactions* 82:1024-38.

Wilson, D.J. 1977. Dilution of exhaust gases from building surface vents. *ASHRAE Transactions* 83(1):168-76.

Wilson, D.J. 1979. Flow patterns over flat roofed buildings and application to exhaust stack design. *ASHRAE Transactions* 85:284-95.

Wilson, D.J. 1982. Critical wind speeds for maximum exhaust gas reentry from flush vents at roof level intakes. *ASHRAE Transactions* 88(1):503-13.

Wilson, D.J. 1983. A design procedure for estimating air intake contamination from nearby exhaust vents. *ASHRAE Transactions* 89(2):136-52.

Wilson, D.J. 1985. Ventilation intake air contamination by nearby exhausts. Proceedings of the Air Pollution Control Association Conference, Ottawa, Canada.

Wilson, D.J. and R.E. Britter. 1982. Estimates of building surface concentrations from nearby point sources. *Atmospheric Environment* 16:2631-46.

Wilson, D.J. and E.H. Chui. 1985. Influence of exhaust velocity and wind incidence angle on dilution from roof vents. *ASHRAE Transactions* 91(2b):1693-1706.

Wilson, D.J. and E.H. Chui. 1987. Effect of turbulence from upwind buildings on dilution of exhaust gases. *ASHRAE Transactions* 93(2):2186-97.

Wilson, D.J. and G. Winkel. 1982. The effect of varying exhaust stack height on contaminant concentration at roof level. *ASHRAE Transactions* 88(1):513-33.

Wilson, D.J. and B.K. Lamb. 1994. Dispersion of exhaust gases from roof-level stacks and vents on a laboratory building. *Atmospheric Environment* 28:3099-111.

Wollenweber, G.C. and H.A. Panofsky. 1989. Dependence of velocity variance on sampling time. *Boundary Layer Meteorology* 47:205-15.

BIBLIOGRAPHY

Anonymous. 1964. How to design drain-type stacks. *Heating Piping and Air Conditioning* 36(6):143.

ASCE. 1987. Wind tunnel model studies of building and structures. ASCE Manuals and Reports on Engineering Practice No. 67. American Society of Civil Engineers, NY.

Cermak, J.E. 1977. Wind-tunnel testing of structures. *Journal of the Engineering Mechanics Division*, ASCE 103, EM6:1125.

Cermak, J.E., ed. 1979. *Wind engineering*. Proceedings of Fifth International Conference, Fort Collins, CO. Pergamon Press, New York.

Clarke, J.H. 1965. The design and location of building inlets and outlets to minimize wind effect and building reentry of exhaust fumes. *Journal of American Industrial Hygiene Association* 26:242.

Cochran, L.S. and Cermak, J.E. 1992. Full and model-scale cladding pressures on the Texas Tech University experimental building. *Journal of Wind Engineering and Industrial Aerodynamics* 41-44, 1589-1600.

Defant, F. 1951. "Local winds." In *Compendium of meteorology*, pp. 655-72. American Meteorology Society, Boston.

Elliot, W.P. 1958. The growth of the atmospheric internal boundary layer. *Transactions of the American Geophysical Union* 39:1048-54.

ESDU. 1990. Strong winds in the atmospheric boundary layer. Part 1: Mean hourly windspeeds, pp. 15-17. Engineering Science Data Unit, Item 82-26, London, UK.

Geiger, R. 1966. *The climate near the ground*. Harvard University Press, Cambridge, MA.

Houghton, E.L. and N.B. Carruthers. 1976. *Wind forces on buildings and structures: An introduction*. Edward Arnold, London.

Landsberg, H. 1981. *The urban climate*. Academic Press, New York.

Panofsky, H.A. and J.A. Dutton. 1984. *Atmospheric turbulence: Models and methods for engineering applications*. John Wiley and Sons, New York.

Pasquill, F. and F. Smith. 1983. *Atmospheric diffusion*, 3rd ed. Halstead Press, New York.

Proceedings of the Fifth National Conference on Wind Engineering. 1985. Texas Tech University, Lubbock, TX.

Simiu, V. and R. Scanlan. 1986. *Wind effects on structures: An introduction to wind engineering*, 2nd ed. Wiley Interscience, NY.

CHAPTER 16

ENERGY RESOURCES

BUILDINGS and facilities of various types may be heated, ventilated, air conditioned, and refrigerated—using systems and equipment designed for that purpose and using the site energy forms commonly available—without concern for the original **energy resources** from whence those energy forms came. Since the energy used in buildings and facilities comprises a significant amount of the total energy used for all purposes, and since the use of this energy has an impact on energy resources, ASHRAE recognizes the "effect of its technology on the environment and natural resources to protect the welfare of posterity" (ASHRAE 1990).

Many governmental agencies regulate energy conservation legislation for obtaining building permits (Conover 1984). The application of specific values to building energy use situations has a considerable effect on the selection of HVAC&R systems and equipment and how they are applied.

CHARACTERISTICS OF ENERGY AND ENERGY RESOURCE FORMS

The HVAC&R industry deals with energy forms as they occur on or arrive at a building site. Generally, these energy forms are fossil fuels (natural gas, oil, and coal) and electricity. Solar energy and wind energy are also available at most sites, as is low-level geothermal energy (**energy source** for heat pumps). Direct-use (high-temperature) geothermal energy is available at some. These are the prime forms of energy used to power or heat the improvements on a site.

Forms of On-Site Energy

Fossil fuels and electricity are commodities that are usually metered or measured for payment by the facility owner or operator. On the other hand, solar or wind, each of which might be considered a dispersed energy form in its natural state (i.e., requiring neither central processing nor a distribution network) costs nothing for the commodity itself, but does incur cost for the means to make use of it. High-temperature geothermal energy, which is not universally available, may or may not be a sold commodity, depending on the particular locale and local regulations (Chapter 29).

Some prime on-site energy forms require further processing or conversion into other forms more directly suited for the particular systems and equipment needed in a building or facility. For instance, natural gas or oil is burned in a boiler to produce steam or hot water, a form of thermal energy which is then distributed to various use points (such as heating coils in air-handling systems, unit heaters, convectors, fin-tube elements, steam-powered cooling units, humidifiers, and kitchen equipment) throughout the building. Although electricity is not converted in form on-site, it is nevertheless used in a variety of ways, including lighting, running motors for fans and pumps, powering electronic equipment and office machinery, and space heating. While the methods and efficiencies with which these processes take place fall within the scope of the HVAC&R designer, the process by which a prime energy source arrives at a given facility site is not under direct control of the professional. On-site energy choices, if available, may be controlled by the designer based in part on the present and future availability of the associated resource commodities.

The basic energy source for heating may be natural gas, oil, coal, or electricity. Cooling may be produced by electricity, thermal energy, or natural gas. If electricity is generated on-site, the generator may be turned by an engine using natural gas or oil, or by a turbine using steam or gas directly.

The term *energy source* refers to on-site energy in the form in which it arrives at or occurs on a site (e.g., electricity, gas, oil, or coal). *Energy resource* refers to the raw energy, which (1) is extracted from the earth (wellhead or mine-mouth), (2) is used in the generation of the energy source delivered to a building site (coal used to generate electricity), or (3) occurs naturally and is available at a site (solar, wind, or geothermal energy).

Nonrenewable and Renewable Energy Resources

From the standpoint of energy conservation, energy resources may be classified in two broad categories: (1) nonrenewable (or discontinuous) resources, which have definite, although sometimes unknown, limitations; and (2) renewable (or continuous) resources, which can generally be freely used without depletion or have the potential to renew in a reasonable period. Resources used most in industrialized countries, both now and in the past, are nonrenewable (Gleeson 1951).

Nonrenewable resources of energy include

- Coal
- Crude Oil
- Natural gas
- Uranium 235 (atomic energy)

Renewable resources of energy include

- Hydropower
- Solar
- Wind
- Earth heat (geothermal)
- Biomass (wood, wood wastes and municipal solid waste)
- Tidal power
- Ocean thermal
- Atmosphere or large body of water (as used by the heat pump)
- Crops (for alcohol production)

Characteristics of Fossil Fuels and Electricity

Most on-site energy for buildings in developed countries involves electricity and fossil fuels as the prime on-site energy sources. Both fossil fuels and electricity can be described in terms of their energy content (Btu). This implies that the two energy forms

The preparation of this chapter is assigned to TC 1.10, Energy Resources.

are comparable and that an equivalence can be established. In reality, however, fossil fuels and electricity are only comparable in energy terms when they are used to generate heat. Fossil fuels, for example, cannot directly drive motors or energize light bulbs. Conversely, electricity gives off heat as a by-product regardless of whether it is used for running a motor or lighting a light bulb, and regardless of whether that heat is needed. Thus, electricity and fossil fuels have different characteristics, uses, and capabilities aside from any differences relating to their derivation.

Beyond the building site, further differences between these energy forms may be observed, such as methods of extraction, transformation, transportation, and delivery, and the characteristics of the resource itself. Natural gas arrives at the site in virtually the same form in which it was extracted from the earth. Oil is processed (distilled) before arriving at the site; having been extracted as crude oil, it arrives at a given site as, for example, No. 2 oil or diesel fuel. Electricity is created (converted) from a different energy form, often a fossil fuel, which itself may first be converted to a thermal form. The total electricity conversion, or generation, process includes energy losses governed largely by the laws of thermodynamics.

Fuel cells, which are used only on a small scale, convert a fossil fuel to electricity by chemical means.

Fossil fuels undergo a conversion process by combustion (oxidation) and heat transfer to thermal energy in the form of steam or hot water. The conversion equipment used is a boiler or a furnace in lieu of a generator, and it usually occurs on a project site rather than off-site. (District heating is an exception.) Inefficiencies of the fossil fuel conversion occur on-site, while the inefficiencies of most electricity generation occur off-site, before the electricity arrives at the building site. (Cogeneration is an exception.)

WORLD ENERGY RESOURCES

Production

Energy production trends for the world, leading producers, and world areas from 1984 to 1993 are shown on Figure 1. World primary energy production in 1993 was up 17.5% since 1982, but it has been essentially flat since 1990. The largest total energy producers in 1993 were the United States (20%), three constituent republics of the former U.S.S.R. (14%), and China (9%), with China exhibiting the greatest growth in absolute terms of 10^{16} Btu (10 quads). Saudi Arabia (6%) was the world's fourth largest producer. Together these producers accounted just under half of the world's energy. (Note that the U.S.S.R. officially dissolved on January 1, 1992.) Total world energy production by resource type is shown in Figure 2.

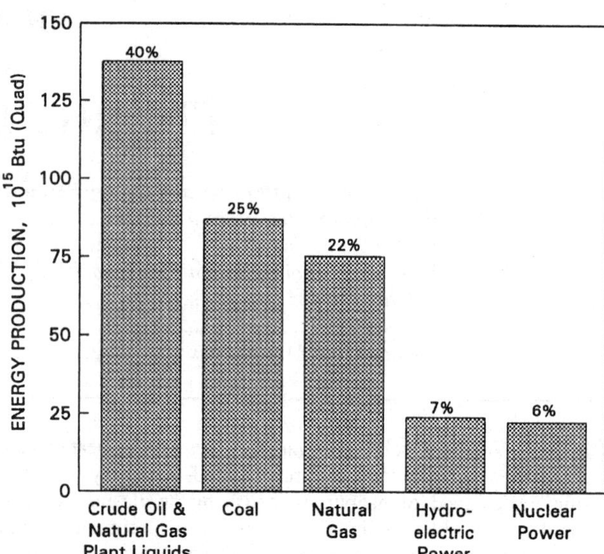

Fig. 2 World Primary Energy Production by Resource Type: 1993

Crude Oil. World crude oil production was 61 million barrels per day in 1994—up 10% since 1973. The biggest crude oil producers in 1994 were the eight nations comprising the Organization of Petroleum Exporting Countries (OPEC) at 43% (including Saudi Arabia at 13%), Russia (10%), and the United States (11%). Oil production declined most noticeably since 1993 in Russia (10%) and in the United States (3%). The other primary non-OPEC producers were China, Mexico, the United Kingdom, and Canada.

Natural Gas. World production reached 76×10^{12} ft^3 in 1991—up 28% from the 1984 level. The biggest producers in 1993 were Russia (29%) and the United States (24%).

Coal. At 4.9×10^9 short tons in 1993, coal production had risen just 6% since 1984 and comprised 25% of the world's energy production. The leading producers of coal were China (26%), the United States (19%), Russia (7%), and Germany (6%).

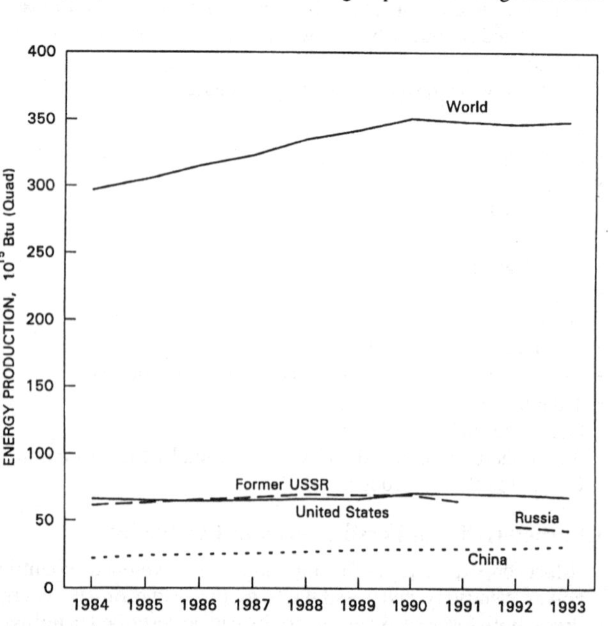

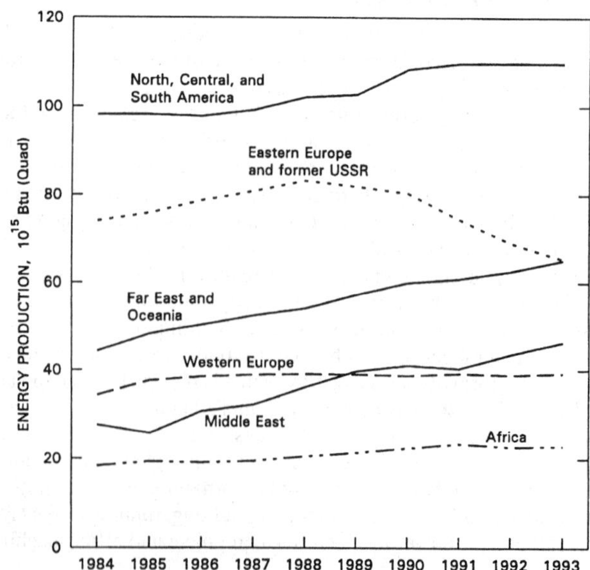

Fig. 1 World Primary Energy Production Trends

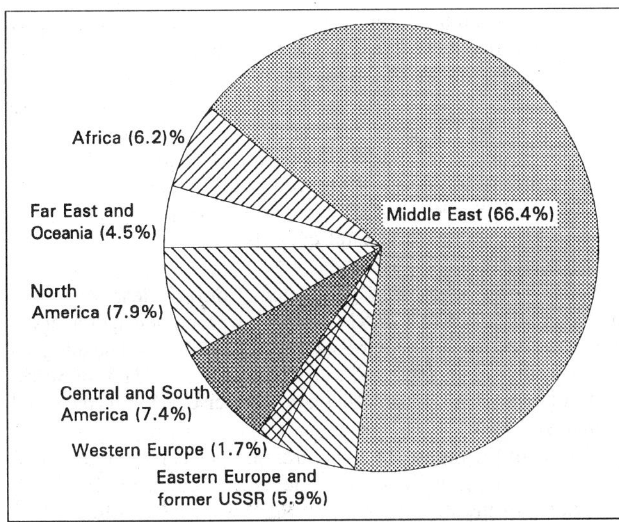

Fig. 3 World Crude Oil Reserves: January 1, 1994
(Basis: *Oil and Gas Journal*)

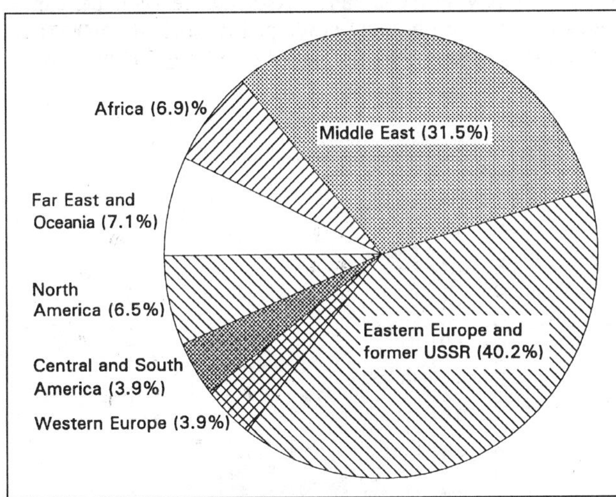

Fig. 4 World Natural Gas Reserves: January 1, 1994
(Basis: *Oil and Gas Journal*)

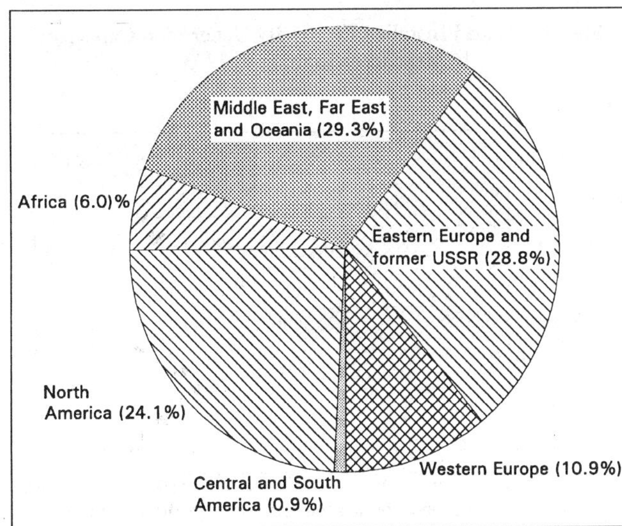

Fig. 5 World Recoverable Coal Reserves: January 1, 1994
(Basis: World Energy Council)

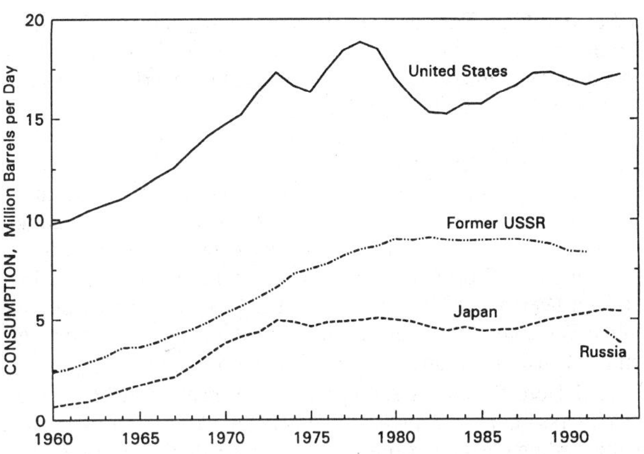

Fig. 6 Petroleum Consumption Trends of Leading Consumers: 1960 - 1993

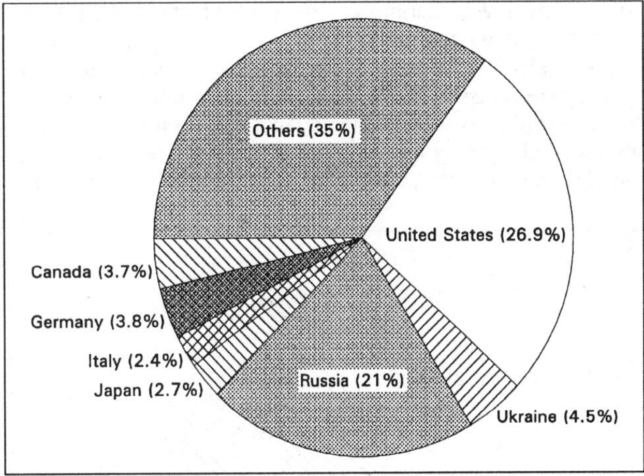

Fig. 7 World Natural Gas Consumption: 1993

Reserves

On January 1, 1994 the estimated world reserves of crude oil and gas were distributed by world region according to Figures 3 and 4. Saudi Arabia was estimated to have 39% of the Middle Eastern crude oil reserves. Iraq, the United Arab Emirates, Kuwait, and Iran were each estimated to have more crude oil reserves than any world region outside the Middle East. Outside of the Middle East, the biggest reserves were estimated to be in the former U.S.S.R., Venezuela, and Mexico. The single area with the largest gas reserves by far was the former U.S.S.R.

World coal reserves as of January 1, 1994 are shown by world area in Figure 5. The countries with the most plentiful reserves, as a percent of total, were the former U.S.S.R. (23%), the United States (23%), China (11%), and Australia (9%).

Consumption

Data on world energy consumption are available only by type of resource rather than by total energy consumed.

Petroleum. The consumption trends of the leading consumers from 1960 to 1993 are depicted in Figure 6. In 1993, the United States consumed far more than any other country—26% of world consumption and 44% of the consumption of Organization for Economic Cooperation and Development (OECD) countries. By contrast, Japan, another OECD country, consumed just 8% of the world total and 14% of that of the OECD countries. Of the non-OECD countries, Russia and China were the biggest consumers (6% and 5%, respectively, of world consumption).

Natural Gas. Transport pipelines notwithstanding, this energy resource tends to be consumed close to the site of production; and indeed, in 1993 the two biggest natural gas producers were also the two biggest consumers. Figure 7 depicts natural gas consumption by the leading consumer countries as a percentage of world consumption. Of the major consumers, the United States consumed more than it produced (111%), and the former U.S.S.R. consumed less (73%). The Ukraine and Germany, the third and fourth largest consumers, produced very little. Canadian consumption was 56% of its production. World consumption of natural gas increased 43% between 1980 and 1993, with the former U.S.S.R. up 54% and the United States up 2%. After the United States and Russia, no single country consumed more than 5% of the world total.

Coal. Here, the three largest producers were also the three largest consumers. Figure 8 depicts the percentage of world consumption by the leading consumers during 1993. Since 1980, world coal consumption had increased 18% (a slight drop-off of 8% since its peak in 1989). In the same period, China's consumption increased 87%, the United States' 32%, and India's 29%. In 1991 the leading coal exporters by a wide margin were the United States and Australia, while the leading importer was Japan.

Electricity. Figure 9 shows the world's electricity generation by energy resource in 1992. Figure 10 shows installed capacity for the same resources at the beginning of 1993. Both net generation and installed capacity were dominated by the United States (26% and 25%, respectively), the former U.S.S.R. (13% for both), and Japan (7% for both). China was close to Japan's net electricity generation capacity, with 6% of the world total.

Electricity generated by hydroelectric means increased in the world by 16% between 1984 and 1993, with the largest increase occurring in the Far East/Oceania. (It decreased in the United States over that same period, however.) The top countries for hydroelectric generation in 1992 were Canada, the United States, the former U.S.S.R., and Brazil—collectively accounting for close to half of the world total quantity of electricity generated by that means.

Total world electricity generation from nuclear resources increased 303% between 1981 and 1994, with higher-than-average increases occurring in the Far East and Africa (355%). The top-generating countries in 1992 were the United States (31% of world total), France (16%), Japan (10%), the former U.S.S.R. (10%), and Germany (7.5%).

Per Capita. Figure 11 compares the per capita commercial energy consumption by selected developed countries for 1989. According to the data source, World Resources Institute (WRI 1993), commercial energy consumption includes "primary electricity and solid, liquid, and gaseous fuels produced domestically or imported—but excludes traditional fuels such as fuelwood, charcoal, crop wastes, and dung." As is apparent, the per capita energy consumption in more developed is vastly different from that in less developed countries, and differs even among the more developed countries.

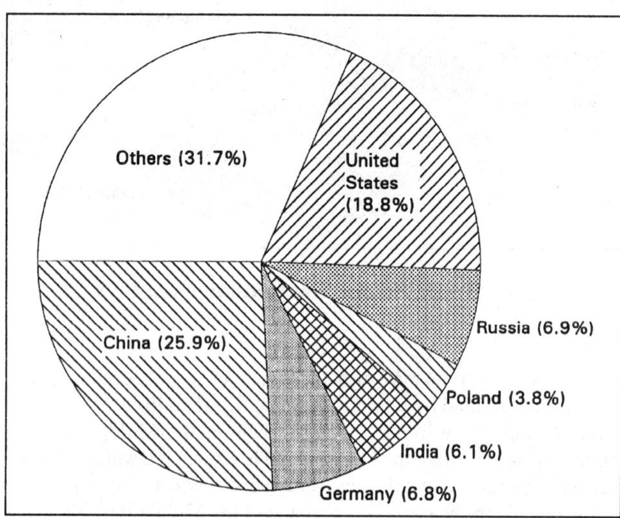

Fig. 8 World Coal Consumption: 1993

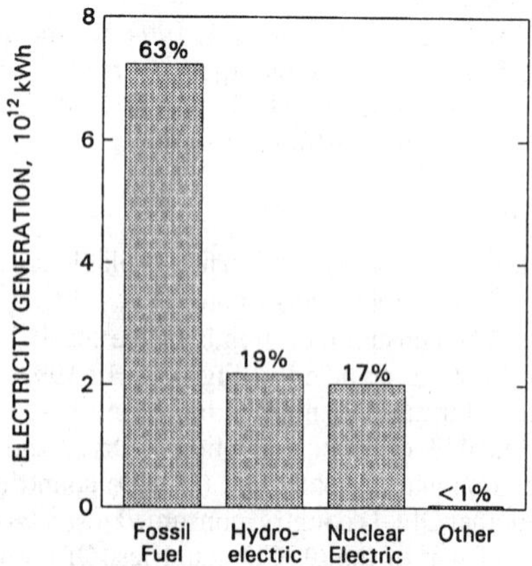

Fig. 9 World Electricity Generation by Resource: 1992

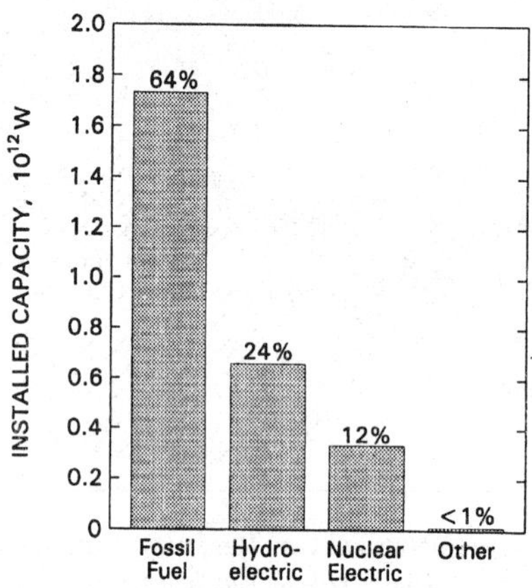

Fig. 10 World Installed Electricity Generation Capacity by Resource: January 1, 1993

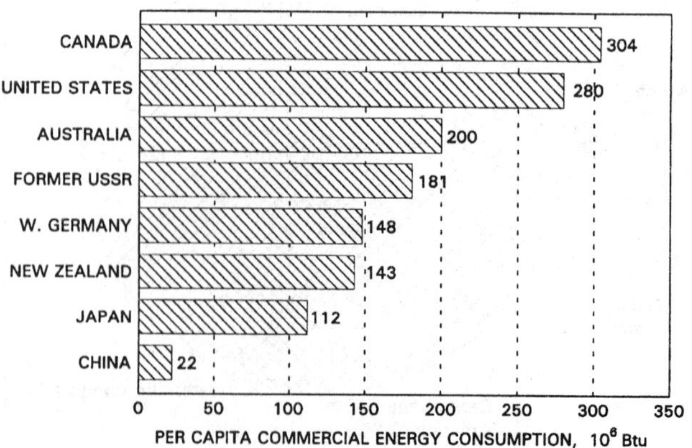

Fig. 11 Per Capita Commercial Consumption by Selected Developed Countries in 1989

UNITED STATES ENERGY USE

Per Capita Energy Consumption

Figure 12, which is based on EIA (1994), presents a capsule overview of past U.S. energy use intensity by relating per capita energy use since 1950 to population growth. Per capita energy use has varied significantly. The 1960s experienced a sharp increase in per capita energy use, which leveled off during the 1970s due to higher energy prices and the emphasis on energy conservation. In the early 1980s, however, a significant drop in per capita energy use occurred as industrial output decreased, efficiency of use improved, and global economic pressures mounted. The last half of the 1980s indicates an increase in per capita energy use at a rate paralleling that of the population—but at a considerably lower level.

The *Annual Energy Outlook* is the basic source of data for projecting the use of energy in the United States (EIA 1996). Figures 13 and 14 are summaries of data from this source.

EIA (1996) presents forecasts of energy trends that are based on three macroeconomic growth scenarios prepared by Data Resources, Inc. (DRI). EIA has also added a low and a high world oil price scenario to DRI's mid-growth economic scenario. Thus, the National Energy Modeling System (NEMS), which EIA uses for these forecasts, has yielded a total of five scenarios. Figures 13 and 14 present the baseline scenario or reference case. It assumes average annual growth of the Gross Domestic Product (GDP) at 2.0%, of the labor force at 1.0%, and of labor productivity at just under 1%. The forecast, in order to be policy-neutral, also assumes that all federal, state, and local laws and regulations in effect as of October 1, 1995 remain unchanged through 2015, except for the lifting of the ban on exports of Alaskan crude oil. Further, any potential impact of the evolving restructuring of the electricity industry is not included in these forecasts.

Projected Overall Energy Consumption

Figure 13 shows energy use by end-use sector, with the major end-use sectors being residential, commercial, industrial, and transportation. HVAC&R engineers are primarily concerned with the first three sectors. Figure 14 shows energy consumption by type of resource. Figure 13 shows less total energy consumption than Figure 14 primarily because it excludes the thermodynamic losses of elec-

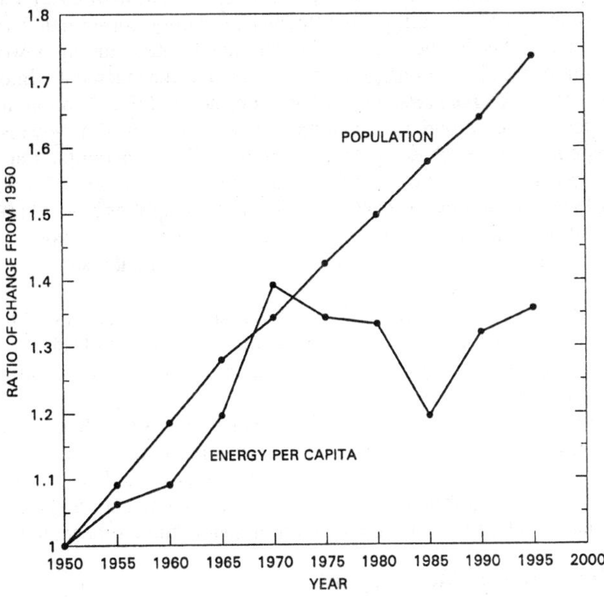

Fig. 12 Per Capita End-Use Energy Consumption Trends in the United States

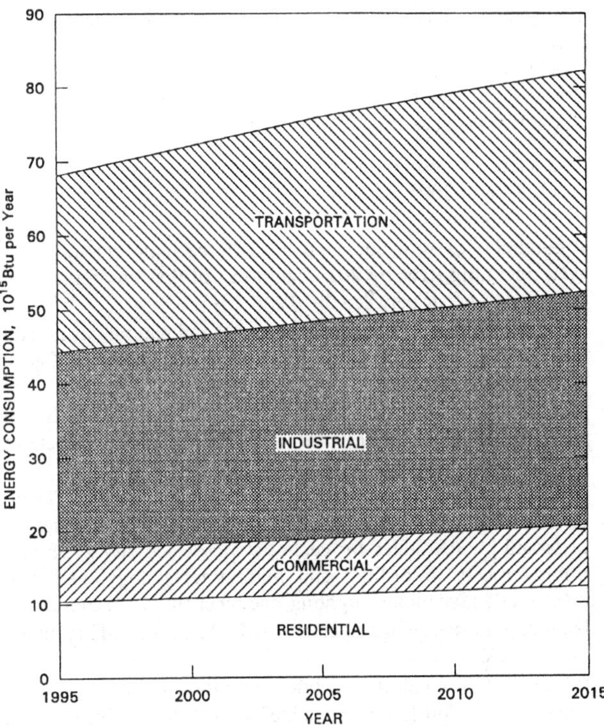

Fig. 13 Projected Total U.S. Energy Consumption by End-Use Sector

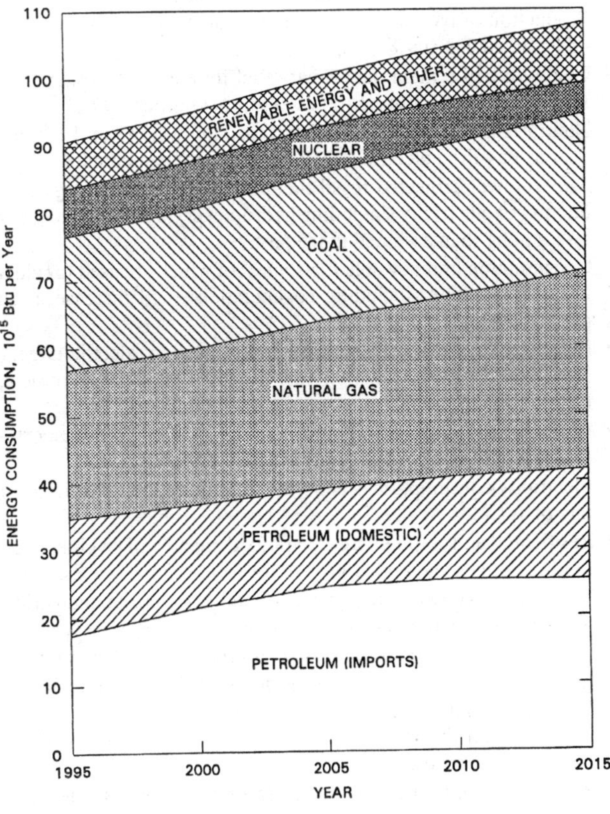

Fig. 14 Projected Total U.S. Energy Consumption by Resource

tricity generation and the processing and delivery burdens of various energy forms.

The following background comments and observations apply to the overall picture of projected energy use in the United States over the next two decades (Figures 13 and 14):

- The share of petroleum consumption met by net imports will reach 57% in 2005 and remain at about that level through 2015. (This is up from 45% in 1994.)
- A number of existing nuclear plants will reach the end of their 40-year operating lives by 2015, and most of these retirements are assumed to occur between 2010 and 2015.
- Generation of electricity from both coal and natural gas is likely to expand significantly to fill the resulting gap and to meet growing customer demand.
- After 2010, there will be a notable shift of "baby boomers" into retirement. The resulting reduced workforce size is expected to slow economic growth and shift patterns of energy use.
- Technology developments are a key factor in the forecast. The mix of advanced technologies influencing energy trends includes fuel cells and photovoltaics for central station electricity generation, improvements in drilling and mining technology, and more energy-effective motors, lighting, and other end-use technologies.
- Natural gas prices will rise more slowly than assumed in previous projections.
- Inflation-adjusted electricity prices will remain flat.
- Coal prices are projected to decline based on continuing improvements in mining productivity and flat wage rates for miners.
- Higher oil production is expected from OPEC countries, and oil prices are projected slightly lower than assumed in previous projections.
- Because of lower fuel prices, renewable energy sources will penetrate into use more slowly than in previous projections.
- Production of natural gas will increase at an annual average rate of 1.3%, coal at 0.8%.
- Coal will still remain the primary fuel for electricity generation, with 90% of its consumption used for that purpose.
- Natural gas consumption will increase an average of 1.6% per year as gas-fired electricity generation more than doubles.
- Due to lower projected fossil fuel prices, renewable energy production of electricity (including hydropower) is expected to grow less rapidly than assumed in the past.
- Consumption of renewable fuels, most of which are used for electricity generation (including cogeneration), will increase at an average annual rate of 1.5%.
- Hydropower, the main renewable source used for electricity generation, will increase only slightly because of a shortage of new, large sites and because of regulatory actions.
- With two-thirds of petroleum consumption used for transportation, growth will average approximately 0.9% a year.
- Electricity consumption is expected to grow at an average annual rate of 1.4%.
- Residential and commercial electricity demand will be higher due to lower prices and higher growth for new electricity uses.
- With the slowdown in the growth of energy-intensive industries, industrial demand will drop.
- Demand for transportation energy will remain about the same.
- Energy intensity (energy used per dollar of GDP) will decline at an average annual rate of 1.1% due to moderate energy price increases and the projected slow growth rate of more energy-intensive industries.
- Per capita energy use is expected to remain nearly stable—and well below the record highs of the early 1970s—due to increasingly efficient technologies offsetting growing demand for energy services. Because lower prices result in more consumption of electricity per capita, demand for electricity will increase at an annual average rate of 0.5%, while the per capita demand for other energy sources will remain flat.
- Carbon emissions from fossil fuel combustion will continue to grow and will contribute to global climate changes, increasing the "greenhouse gas effect."

Outlook Summary

In general, the following key issues will dominate energy matters in the next two decades:

- Rising dependency of the United States on imported oil
- Expected retirement of a large portion of existing nuclear power plants
- Role of technology developments, including energy conservation and energy efficiency as alternatives to energy production
- Deregulation of the utility industry and the growth of independent power producers
- Growth of population, coupled with the shift of "baby boomers" into retirement

INTEGRATED RESOURCE PLANNING/ DEMAND-SIDE MANAGEMENT

The desire to improve economic effectiveness and achieve environmental and societal goals has led to two techniques for improving the selection of an energy resource and influencing energy consumption by the end user. Although these techniques are commonly practiced by electric utilities, they could be effectively implemented by any entity acting *like* a utility in that it chooses energy resources to provide one or more converted energy forms to multiple users.

Integrated Resource Planning

Integrated resource planning (IRP) is a technique that has gained acceptance among utility management and regulators. It goes by many names, including "least cost utility planning," and it has many definitions. Although a single definition that includes all the elements considered in discussions and regulations related to utility resource planning does not exist, several common elements distinguish IRP from traditional approaches to utility investment planning.

- IRP includes a broader range of resource options than those traditionally considered in electricity generation planning, particularly energy conservation measures. Originally, least cost planning considered utility investments in energy conservation as alternatives to the construction of new central-station power plants (similar to one goal of demand-side management). Integrated resource planning considers such nontraditional options as renewable resource generating technologies, customer-owned generation, purchase contracts from nonutility generators, transactions with other utilities, and many others.
- IRP considers the costs and benefits of these options beyond traditional measures of revenue requirements. Regulators typically require determining costs and benefits from a "total resource" or societal perspective.
- IRP attempts to optimize resource decisions by determining the costs and benefits of all options on the same basis and selecting the options that provide the most favorable balance of benefits against costs.
- IRP addresses a wider population of stakeholders than traditional utility planning. Many state regulatory commissions involve the public in the review of utility resource plans—and sometimes even in their formation. Customers, environmentalists, and other public interest groups are often prominent in these proceedings.

Demand-Side Management

Broadly defined, **demand-side management** (DSM) includes any actions taken by an electric utility or industrial plant that provide the customer (or end user) with the same level of energy services, but at lower overall costs, and the energy provider with economic or other desirable benefits. Some public utility commissions have requested that gas utilities apply these techniques as well. The fundamental objectives of DSM are to reduce demand, conserve energy, and reduce pollutants. DSM may be integrated into an energy provider's integrated resource plan. Many (but not all) DSM plans offer some form of economic incentive in exchange for a modified energy use pattern.

Supporters of DSM may be primarily interested in one of its benefits. For example, industry may use DSM to reduce operating expenses. A public service commission may require a utility to incorporate DSM as part of its IRP to minimize the need for new electric generating capacity and subsequent rate increases. Environmental advocates may encourage DSM to reduce energy consumption. However, supporters may also share reasons they support DSM. For example, industry and environmental advocates may both be interested in delaying new electric generation and in reducing contaminants. Commissions and environmental advocates are concerned with the economic health of industry (Pritchett et al. 1993).

RELATIONSHIPS

In designing the systems required for a facility, an HVAC&R designer sooner or later must consider the use of one or more forms of prime energy. Most likely, these would be nonrenewable energy sources (fossil fuels and electricity), although installations are sometimes designed using a single energy source (e.g., only a fossil fuel or only electricity).

Solar energy normally impinges on the site (and on the facilities to be put there), so it will have an impact on the energy consumption of the facility. The designer must account for this impact as well as decide whether to make active use of solar energy. When solar energy is used beneficially, it can reduce the requirements for nonrenewable energy forms. Other naturally occurring and distributed renewable forms such as wind power and earth heat (if available) might also be considered.

If they are to understand and be concerned with the earth's energy resources, designers must be aware of the relationship between on-site energy sources and raw energy resources—including how these resources are used and what they are used for. The relationship between energy sources and energy resources involves two parts: (1) quantifying the energy resource units expended and (2) considering the societal impact of the depletion of one energy resource (caused by on-site energy use) with respect to others. The following two sections describe those parts in more specific terms.

Quantifiable Relationships

As on-site energy sources are consumed, a corresponding amount of resources are consumed to produce that on-site energy. For instance, for every 1000 gal of No. 2 oil consumed by a boiler at a building site, some greater number of gallons of crude oil is extracted from the earth. On leaving the well, the crude oil is transported and processed into its final form, perhaps stored, and then transported to the site where it will be used.

Even though gas requires no significant processing, it is transported, often over long distances, to reach its final destination, which causes some energy loss. Electricity may have as its raw energy resource a fossil fuel, uranium, or an elevated body of water (hydroelectric generating plant).

Data to assist in determining the amount of resource use per delivered on-site energy source unit are available. In the United States, data are available from entities within the U.S. Department of Energy and from the agencies and associations listed at the end of this chapter.

A **resource utilization factor** (RUF) is the ratio of resources consumed to energy delivered (for each form of energy) to a building site. Specific RUFs may be determined for various energy sources normally consumed on-site, including nonrenewable sources such as coal, gas, oil, and electricity, and renewable sources such as solar, geothermal, waste, and wood energy. With electricity, which may derive from several resources depending on the particular fuel mix of the generating stations in the region served, the overall RUF is the weighted combination of individual factors applicable to electricity and a particular energy resource. Grumman (1984) gives specific formulas for calculating RUFs.

While a designer does not need to determine the amount of energy resources attributable to a given building or building site for its design or operation, this information may be helpful when planning the long-range availability of energy for a building or the building's impact on energy resources. Currently, factors or fuel-quantity-to-energy resource ratios are used, which suggests that energy resources are of concern to the HVAC&R industry.

Intangible Relationships

Energy resources should not simply be converted into common energy units [e.g., quadrillion (10^{15}) Btu or quad] because the commonality thus established gives a misleading picture of the equivalence of these resources. Other differences and limitations of each of the resources defy easy quantification but are nonetheless real.

For instance, consider electricity that arrives and is used on a site and the resources from which it is derived. Electricity can be generated from coal, oil, natural gas, uranium, or elevated bodies of water (hydropower). The end result is the same: electricity at *X* kilovolts, *Y* hertz. However, is a kilowatt-hour of electricity generated by hydropower equal in societal impact to that same kilowatt-hour generated by coal? by uranium? by domestic oil? or by imported oil? In other words, electricity generated by hydropower, though identical in quantity with the electricity generated by imported oil, might be considered more desirable from a societal impact standpoint.

Intangible factors such as safety, environmental acceptability, availability, and national interest also are affected in different ways by the consumption of each resource. Heiman (1984) proposes a procedure for weighting the following intangible factors:

National/Global Considerations

- Balance of trade
- Environmental impacts
- International policy
- Employment
- Minority employment
- Availability of supply
- Alternative uses
- National defense
- Domestic policy
- Effect on capital markets

Local Considerations

- Exterior environmental impact—air
- Exterior environmental impact—solid waste
- Exterior environmental impact—water resources
- Local employment
- Local balance of trade
- Use of distribution infrastructure
- Local energy independence
- Land use
- Exterior safety

Site Considerations

- Reliability of supply
- Indoor air quality
- Aesthetics
- Interior safety
- Anticipated changes in energy resource prices

SUMMARY

In designing HVAC&R systems, the need to address immediate issues such as economics, performance, and space constraints often prevents designers from fully considering the energy resources affected. Today's energy resources are less certain because of issues such as availability, safety, national interest, environmental concerns, and the world political situation. As a result, the reliability, economics, and continuity of many common energy resources over the potential life of a building being designed are unclear. For this reason, the designer of building energy systems must consider the energy resources on which the long-term operation of the building will depend. If the continued viability of those resources is reason for concern, the design should provide for, account for, or address such an eventuality.

AGENCIES AND ASSOCIATIONS

American Gas Association (AGA), Arlington, VA
American Petroleum Institute (API), Washington, D.C.
Bureau of Mines, Department of Interior, Washington, D.C.
Council on Environmental Quality (CEQ), Washington, D.C.
Edison Electric Institute (EEI), Washington, D.C.
Electric Power Research Institute (EPRI), Palo Alto, CA
Energy Information Administration (EIA), Washington, D.C.
Gas Research Institute (GRI), Chicago, IL
National Coal Association (NCA), Washington, D.C.
North American Electric Reliability Council (NAERC), Princeton, NJ
Organization of Petroleum Exporting Countries (OPEC), Vienna, Austria

REFERENCES

ASHRAE. 1990. Energy position statement (June).

Bartlett, A.A. 1983. *Forgotten fundamentals of the energy crisis.* Boulder, CO.

Conover, D.R. 1984. Accounting for energy resource use in building regulations. *ASHRAE Transactions* 90(1B):547-63.

EIA. 1994. *Annual energy review.* DOE/EIA—0384(94). Energy Information Administration, Office of Energy Markets and End Use, U.S. Department of Energy, Washington, D.C.

EIA. 1996. *Annual energy outlook 1996.* DOE/EIA—0383(96). Energy Information Administration, U.S. Department of Energy, Washington, D.C.

Gleeson, G.W. 1951. Energy—Choose it wisely today for safety tomorrow. *ASHVE Transactions* 57:523-40.

Grumman, D.L. 1984. Energy resource accounting: ASHRAE *Standard* 90C-1977R. *ASHRAE Transactions* 90(1B):531-46.

Heiman, J.L. 1984. Proposal for a simple method for determining resource impact factors. *ASHRAE Transactions* 90(1B):564-70.

Pritchett, T., L. Moody, and M. Brubaker. 1993. Why industry demand-side management programs should be self-directed. *Electricity Journal.* Seattle, WA.

WRI. 1993. *World resources 1992-93.* World Resources Institute, Washington, DC.

BIBLIOGRAPHY

Anderson, R.J. 1984. The energy resource picture in 1984 in the U.S. and abroad. *ASHRAE Transactions* 90(1B):521-30.

DOE. 1979. *Impact assessment of a mandatory source-energy approach to energy conservation in new construction.* U.S. Department of Energy, Washington, D.C.

Pacific Northwest Laboratory. 1987. *Development of whole-building energy design targets for commercial buildings phase 1 planning.* PNL-5854, Vol. 2. U.S. Department of Energy, Washington, D.C.

CHAPTER 17

COMBUSTION AND FUELS

PRINCIPLES OF COMBUSTION

COMBUSTION is the chemical reaction in which an oxidant reacts rapidly with a fuel to liberate stored energy as thermal energy, generally in the form of high-temperature gases. Small amounts of electromagnetic energy (light), electric energy (free ions and electrons), and mechanical energy (noise) are also released during combustion. Except in special applications, the oxidant for combustion is oxygen in the air.

Conventional hydrocarbon fuels contain primarily hydrogen and carbon, in elemental form or in various compounds. Their complete combustion produces mainly carbon dioxide (CO_2) and water (H_2O); however, small quantities of carbon monoxide (CO) and partially reacted flue gas constituents (gases and liquid or solid aerosols) may form. Most conventional fuels also contain small amounts of sulfur, which is oxidized to sulfur dioxide (SO_2) or sulfur trioxide (SO_3) during combustion, and noncombustible substances such as mineral matter (ash), water, and inert gases. Flue gas is the product of complete or incomplete combustion and includes excess air (if present), but not dilution air.

Fuel combustion rate depends on (1) the rate of the chemical reaction of the combustible fuel constituents with oxygen, (2) the rate at which oxygen is supplied to the fuel (the mixing of air and fuel), and (3) the temperature in the combustion region. The reaction rate is fixed by fuel selection. Increasing the mixing rate or temperature increases the combustion rate.

With **complete combustion** of hydrocarbon fuels, all hydrogen and carbon in the fuel are oxidized to H_2O and CO_2. Generally, for complete combustion, excess oxygen or excess air must be supplied beyond the amount theoretically required to oxidize the fuel. Excess air is usually expressed as a percentage of the air required to completely oxidize the fuel.

In **stoichiometric combustion** of a hydrocarbon fuel, fuel is reacted with the exact amount of oxygen required to oxidize all carbon, hydrogen, and sulfur in the fuel to CO_2, H_2O, and SO_2. Therefore, exhaust gas from stoichiometric combustion theoretically contains no incompletely oxidized fuel constituents and no unreacted oxygen (i.e., no carbon monoxide and no excess air or oxygen). The percentage of CO_2 contained in products of stoichiometric combustion is the maximum attainable and is referred to as the **stoichiometric CO_2, ultimate CO_2,** or **maximum theoretical percentage of CO_2.**

Stoichiometric combustion is seldom realized in practice because of imperfect mixing and finite reaction rates. For economy and safety, most combustion equipment should operate with some excess air. This ensures that fuel is not wasted and that combustion is complete despite variations in fuel properties and in the supply rates of fuel and air. The amount of excess air supplied to any particular piece of combustion equipment depends on such factors as (1) expected variations in fuel properties and in fuel and air supply rates, (2) equipment application, (3) degree of operator supervision required or available, and (4) control requirements. For maximum efficiency, combustion at low excess air is desirable.

Incomplete combustion occurs when a fuel element is not completely oxidized in the combustion process. For example, a hydrocarbon may not completely oxidize to carbon dioxide and water but may form partially oxidized compounds, such as carbon monoxide, aldehydes, and ketones. Conditions that promote incomplete combustion include (1) insufficient air and fuel mixing (causing local fuel-rich and fuel-lean zones), (2) insufficient air supply to the flame (providing less than the required quantity of oxygen), (3) insufficient reactant residence time in the flame (preventing completion of combustion reactions), (4) flame impingement on a cold surface (quenching combustion reactions), or (5) flame temperature that is too low (slowing combustion reactions).

Incomplete combustion uses fuel inefficiently, can be hazardous because of carbon monoxide production, and contributes to air pollution.

Combustion Reactions

The reaction of oxygen with the combustible elements and compounds in fuels occurs according to fixed chemical principles, including

- Chemical reaction equations
- Law of matter conservation: the mass of each element in the reaction products must equal the mass of that element in the reactants
- Law of combining masses: chemical compounds are formed by elements combining in fixed mass relationships
- Chemical reaction rates

Oxygen for combustion is normally obtained from air, which is a physical mixture of nitrogen, oxygen, small amounts of water vapor, carbon dioxide, and inert gases. For practical combustion calculations, dry air consists of 20.95% oxygen and 79.05% inert gases (nitrogen, argon, and so forth) by volume, or 23.15% oxygen and 76.85% inert gases by mass. For calculation purposes, nitrogen is assumed to pass through the combustion process unchanged (although small quantities of nitrogen oxides are known to form). Table 1 lists oxygen and air requirements for stoichiometric combustion of some pure combustible materials (or constituents) found in common fuels. Table 2 lists the products of stoichiometric combustion of the same pure combustible materials in Table 1.

Flammability Limits

Fuel burns in a self-sustained reaction only when the volume percentages of fuel and air in a mixture at standard temperature and pressure are within specific limits: the upper and lower flammability limits or explosive limits (UEL and LEL). See Table 3. Both temperature and pressure affect these limits. As the temperature of the mixture increases, the upper limit increases and the lower limit decreases. As the pressure of the mixture decreases below atmospheric pressure, the upper limit decreases and the lower limit increases. However, as pressure increases above atmospheric pressure, the upper limit increases and the lower limit is relatively constant.

The preparation of this chapter is assigned to TC 6.10, Fuels and Combustion.

Table 1 Combustion Reactions of Common Fuel Constituents

Constituent	Molecular Formula	Combustion Reactions	Stoichiometric Oxygen and Air Requirements: lb/lb Fuel[a] O_2	lb/lb Fuel[a] Air	ft³/ft³ Fuel O_2	ft³/ft³ Fuel Air
Carbon (to CO)	C	$C + 0.5O_2 \rightarrow CO$	1.33	5.75	b	b
Carbon (to CO_2)	C	$C + O_2 \rightarrow CO_2$	2.66	11.51	b	b
Carbon monoxide	CO	$CO + 0.5O_2 \rightarrow CO$	0.57	2.47	0.50	2.39
Hydrogen	H_2	$H_2 + 0.5O_2 \rightarrow H_2O$	7.94	34.28	0.50	2.39
Methane	CH_4	$CH_4 + 2O_2 \rightarrow CO_2 + 2H_2O$	3.99	17.24	2.00	9.57
Ethane	C_2H_6	$C_2H_6 + 3.5O_2 \rightarrow 2CO_2 + 3H_2O$	3.72	16.09	3.50	16.75
Propane	C_3H_8	$C_3H_8 + 5O_2 \rightarrow 3CO_2 + 4H_2O$	3.63	15.68	5.00	23.95
Butane	C_4H_{10}	$C_4H_{10} + 6.5O_2 \rightarrow 4CO_2 + 5H_2O$	3.58	15.47	6.50	31.14
Alkanes	C_nH_{2n+2}	$C_nH_{2n+2} + (1.5n + 0.5)O_2 \rightarrow nCO_2 + (n+1)H_2O$	—	—	$1.5n + 0.5$	$7.18n + 2.39$
Ethylene	C_2H_4	$C_2H_4 + 3O_2 \rightarrow 2CO_2 + 2H_2O$	3.42	14.78	3.00	14.38
Propylene	C_3H_6	$C_3H_6 + 4.5O_2 \rightarrow 3CO_2 + 3H_2O$	3.42	14.78	4.50	21.53
Alkenes	C_nH_{2n}	$C_nH_{2n} + 1.5nO_2 \rightarrow nCO_2 + nH_2O$	3.42	14.78	$1.50n$	$7.18n$
Acetylene	C_2H_2	$C_2H_2 + 2.5O_2 \rightarrow 2CO_2 + H_2O$	3.07	13.27	2.50	11.96
Alkynes	C_nH_{2m}	$C_nH_{2m} + (n + 0.5m)O_2 \rightarrow nCO_2 + mH_2O$	—	—	$n + 0.5m$	$4.78n + 2.39m$
Sulfur (to SO_2)	S	$S + O_2 \rightarrow SO_2$	1.00	4.31	b	b
Sulfur (to SO_3)	S	$S + 1.5O_2 \rightarrow SO_3$	1.50	6.47	b	b
Hydrogen sulfide	H_2S	$H_2S + 1.5O_2 \rightarrow SO_2 + H_2O$	1.41	6.08	1.50	7.18

[a]Atomic masses: H = 1.008, C = 12.01, O = 16.00, S = 32.06.
[b]Volume ratios are not given for fuels that do not exist in vapor form at reasonable temperatures or pressure.

Table 2 Combustion Reactions of Common Fuel Constituents

Constituent	Molecular Formula	Ultimate CO_2, %	Dew Point, °F	Flue Gas from Stoichiometric Combustion: Unit Vol./Unit Vol. Fuel CO_2	Unit Vol./Unit Vol. Fuel H_2O	Unit Mass/Unit Mass Fuel CO_2	Unit Mass/Unit Mass Fuel H_2O
Carbon (to CO)	C	—	—	—	—	—	—
Carbon (to CO_2)	C	29.30	—	—	—	3.664	—
Carbon monoxide	CO	34.70	—	1.0	—	1.571	—
Hydrogen	H_2	—	162	—	1.0	—	8.937
Methane	CH_4	11.73	139	1.0	2.0	2.744	2.246
Ethane	C_2H_6	13.18	134	2.0	3.0	2.927	1.798
Propane	C_3H_8	13.75	131	3.0	4.0	2.994	1.634
Butane	C_4H_{10}	14.05	129	4.0	5.0	3.029	1.550
Alkanes	C_nH_{2n+2}	—	128 to 127	n	$n + 1$	$\frac{44.01n}{14.026n + 2.016}$	$\frac{18.01(n+1)}{14.026n + 2.016}$
Ethylene	C_2H_4	15.05	125	2.0	2.0	3.138	1.285
Propylene	C_3H_6	15.05	125	3.0	3.0	3.138	1.285
Alkenes	C_nH_{2n}	15.05	125	n	n	3.138	1.285
Acetylene	C_2H_2	17.53	103	2.0	1.0	3.834	0.692
Alkynes	C_nH_{2m}	—	—	n	m	$\frac{22.005n}{6.005n + 1.008m}$	$\frac{9.008m}{6.005n + 1.008m}$
				SO_x	H_2O	SO_x	H_2O
Sulfur (to SO_2)	S	—	—	1.0 SO_2	—	1.998 (SO_2)	—
Sulfur (to SO_3)	S	—	—	1.0 SO_3	—	2.497 (SO_3)	—
Hydrogen sulfide	H_2S	—	125	1.0 SO_2	1.0	1.880 (SO_2)	0.528

Adapted, in part, from *Gas Engineers Handbook* (1965).
Note: Dew point is determined from Figure 2.

Ignition Temperature

Ignition temperature is the lowest temperature at which heat is generated by combustion faster than heat is lost to the surroundings and combustion becomes self-propagating. See Table 3. The fuel-air mixture will not burn freely and continuously below the ignition temperature unless heat is supplied, but chemical reaction between the fuel and air may occur. Ignition temperature is affected by a large number of factors.

The ignition temperature and flammability limits of a fuel-air mixture, together, are a measure of the potential for ignition (*Gas Engineers Handbook* 1965).

Combustion Modes

Combustion reactions occur in either continuous or pulse flame modes. **Continuous combustion** burns fuel in a sustained manner as long as fuel and air are continuously fed to the combustion zone and the fuel-air mixture is within the flammability limits. Continuous combustion is more common than pulse combustion and is used in most fuel-burning equipment.

Pulse combustion is an acoustically resonant process that burns various fuels in small, discrete fuel-air mixture volumes in a very rapid series of combustions.

The introduction of fuel and air into the pulse combustor is controlled by mechanical or aerodynamic valves. Typical combustors

Table 3 Flammability Limits and Ignition Temperatures of Common Fuels in Fuel-Air Mixtures

Substance	Molecular Formula	Lower Flammability Limit, %	Upper Flammability Limit, %	Ignition Temperature, °F	References
Carbon (activated coke)	C	—	—	1220	Hartman (1958)
Carbon monoxide	CO	12.5	74	1128	Scott et al. (1948)
Hydrogen	H_2	4.0	75.0	968	Zabetakis (1956)
Methane	CH_4	5.0	15.0	1301	*Gas Engineers Handbook* (1965)
Ethane	C_2H_6	3.0	12.5	968 to 1166	Trinks (1947)
Propane	C_3H_8	2.1	10.1	871	NFPA (1962)
n-Butane	C_4H_{10}	1.86	8.41	761	NFPA (1962)
Ethylene	C_2H_4	2.75	28.6	914	Scott et al. (1948)
Propylene	C_3H_6	2.00	11.1	856	Scott et al. (1948)
Acetylene	C_2H_2	2.50	81	763 to 824	Trinks (1947)
Sulfur	S	—	—	374	Hartman (1958)
Hydrogen sulfide	H_2S	4.3	45.50	558	Scott et al. (1948)

Flammability limits adapted from Coward and Jones (1952). All values corrected to 60°F, 30 in. Hg, dry.

consist of one or more valves, a combustion chamber, an exit pipe, and a control system (ignition means, fuel-metering devices, etc.). Typically, combustors for warm air furnace, hot water boiler, and commercial cooking equipment use mechanical valves. Aerodynamic valves are usually used in higher pressure applications, such as thrust engines. Separate valves for air and fuel, a single valve for premixed air and fuel, or multiple valves of either type can be used. Premix valve systems may require a flame trap at the combustion chamber entrance to prevent flashback.

In a mechanically valved pulse combustor, air and fuel are forced into the combustion chamber through the valves under pressures less than 0.5 psi. An ignition source, such as a spark, ignites the fuel-air mixture, causing a positive pressure buildup in the combustion chamber. The positive pressure causes the valves to close, leaving only the exit pipe of the combustion chamber as a pressure relief opening. The combustion chamber and exit pipe geometry determine the resonant frequency of the combustor. The pressure wave from the initial combustion travels down the exit pipe at sonic velocity. As this wave exits the combustion chamber, most of the flue gases present in the chamber are carried with it into the exit pipe. Flue gases remaining in the combustion chamber begin to cool immediately. The contraction of the cooling gases and the momentum of gases in the exit pipe create a vacuum inside the chamber that opens the valves and allows more fuel and air into the chamber. While the fresh charge of fuel-air enters the chamber, the pressure wave reaches the end of the exit pipe and is partially reflected from the open end of the pipe. The fresh fuel-air charge is ignited by residual combustion and/or heat. The resulting combustion starts another cycle.

Typical pulse combustors operate at 30 to 100 cycles per second and emit resonant sound, which must be considered in their application. The pulses produce high convective heat transfer rates.

Heating Value

Combustion releases thermal energy or heat. The quantity of heat generated by complete combustion of a unit of specific fuel is constant and is termed the **heating value, heat of combustion,** or **caloric value** of that fuel. The heating value of a fuel can be determined by measuring the heat evolved during combustion of a known quantity of the fuel in a calorimeter, or it can be estimated from chemical analysis of the fuel and the heating values of the various chemical elements in the fuel. For information on calculating heating values, see the sections on Characteristics of Fuel Oils and Characteristics of Coals.

Higher heating value, **gross heating value**, or **total heating value** includes the latent heat of vaporization and is determined when water vapor in the fuel combustion products is condensed. Conversely, **lower heating value** or **net heating value** is obtained when the latent heat of vaporization is *not* included. When the heating value of a fuel is specified without designating higher or lower, it generally means the higher heating value in the United States. (Lower heating value is mainly used for internal combustion engine fuels.)

Table 4 Heating Values of Substances Occurring in Common Fuels

Substance	Molecular Formula	Higher Heating Values,[a] Btu/lb	Lower Heating Values,[a] Btu/lb	Specific Volume,[b] ft^3/lb
Carbon (to CO)	C	3,950	3,950	—
Carbon (to CO_2)	C	14,093	14,093	—
Carbon monoxide	CO	4,347	4,347	13.5
Hydrogen	H_2	61,095	51,623	188.0
Methane	CH_4	23,875	21,495	23.6
Ethane	C_2H_6	22,323	20,418	12.5
Propane	C_3H_8	21,669	19,937	8.36
Butane	C_4H_{10}	21,321	19,678	6.32
Ethylene	C_2H_4	21,636	20,275	—
Propylene	C_3H_6	21,048	19,687	9.01
Acetylene	C_2H_2	21,502	20,769	14.3
Sulfur (to SO_2)	S	3,980	3,980	—
Sulfur (to SO_3)	S	5,940	5,940	—
Hydrogen Sulfide	H_2S	7,097	6,537	11.0

Adapted from *Gas Engineers Handbook* (1965).
[a]All values corrected to 60°F, 30 in. Hg, dry. For gases saturated with water vapor at 60°F, deduct 1.74% of the value to adjust for gas volume displaced by water vapor.
[b]At 32°F and 29.92 in. Hg.

Heating values are usually expressed in Btu/ft^3 for gaseous fuels, Btu/gal for liquid fuels, and Btu/lb for solid fuels. Heating values are always given in relation to a certain reference temperature and pressure, usually 60, 68, or 77°F and 14.696 psia, depending on the particular industry practice. Heating values of several substances in common fuels are listed in Table 4.

With incomplete combustion, not all fuel is completely oxidized, and the heat released is less than the heating value of the fuel. Therefore, the quantity of heat produced per unit of fuel consumed decreases, implying lower combustion efficiency.

Not all heat released during combustion can be used effectively. The greatest heat loss is in the form of the increased temperature (thermal energy) of hot exhaust gases above the temperature of incoming air and fuel. Other heat losses include radiation and convection heat transfer from the outer walls of combustion equipment to the environment.

Altitude Compensation

Air at altitudes above sea level is less dense and has less oxygen per unit volume. Therefore, combustion at altitudes above sea level has less available oxygen to burn with the fuel unless compensation is made for the altitude. Combustion occurs, but the amount of excess air is reduced. If excess air is reduced enough by an increase in altitude, combustion is incomplete or ceases.

Altitude compensation is achieved by matching the fuel and air supply rates to attain complete combustion without too much excess air or too much fuel. Fuel and air supply rates can be matched by increasing the air supply rate to the combustion zone or by decreasing the fuel supply rate to the combustion zone. The air supply rate can be increased with a combustion air blower, and the fuel supply rate can be reduced by decreasing the fuel input (derating).

Power burners use combustion air blowers and can increase the air supply rate to compensate for altitude. The combustion zone can be pressurized to attain the same air density in the combustion chamber as that attained at sea level.

Derating can be used as an alternative to power combustion. In the United States, the fuel gas codes generally do not require derating of nonpower burners at altitudes up to 2000 ft. At altitudes above 2000 ft, burners should be derated 4% for each 1000 ft above sea level (NFPA/IAS National Fuel Gas Code). Chimney or vent operation also must be considered at high altitudes (see Chapter 30 of the 1996 *ASHRAE Handbook—Systems and Equipment*).

FUEL CLASSIFICATION

Generally, hydrocarbon fuels are classified according to physical state (gas, liquid, or solid). Different types of combustion equipment are usually needed to burn fuels in the different physical states. Gaseous fuels can be burned in premix or diffusion burners. Liquid fuel burners must include a means for atomizing or vaporizing fuel into small droplets or a vapor and must provide adequate mixing of fuel and air. Solid fuel combustion equipment must (1) heat fuel to vaporize sufficient volatiles to initiate and sustain combustion, (2) provide residence time to complete combustion, and (3) provide space for ash containment.

Principal fuel applications include space heating and cooling of residential, commercial, industrial, and institutional buildings; service water heating; steam generation; and refrigeration. Major fuels for these applications are natural and liquefied petroleum gases, fuel oils, diesel and gas turbine fuels (for total energy applications), and coal.

Fuels of limited use, such as manufactured gases, kerosene, briquettes, wood, and coke, are not discussed here. Fuel choice is based on one or more of the following:

1. Fuel factors
   - Availability, including dependability of supply
   - Convenience of use and storage
   - Economy
   - Cleanliness
2. Combustion equipment factors
   - Operating requirements
   - Cost
   - Service requirements
   - Ease of control

GASEOUS FUELS

Although various gaseous fuels have been used as energy sources in the past, heating and cooling applications are presently limited to natural gas and liquefied petroleum gases.

Types and Properties

Natural gas is a nearly odorless and colorless gas that accumulates in the upper parts of oil and gas wells. Raw natural gas is a mixture of methane (55 to 98%), higher hydrocarbons (primarily ethane), and noncombustible gases. Some constituents, principally water vapor, hydrogen sulfide, helium, and gases for liquefied petroleum gases and gasoline are removed prior to distribution.

Natural gas used as fuel typically contains methane, CH_4 (70 to 96%); ethane, C_2H_6 (1 to 14%); propane, C_3H_8 (0 to 4%); butane, C_4H_{10} (0 to 2%); pentane, C_5H_{12} (0 to 0.5%); hexane, C_6H_{14} (0 to 2%); carbon dioxide, CO_2 (0 to 2%); oxygen, O_2 (0 to 1.2%); and nitrogen, N_2 (0.4 to 17%).

The composition of natural gas depends on its geographical source. Because the gas is drawn from various sources, the composition of gas distributed in a given location can vary slightly, but a fairly constant heating value is usually maintained for control and safety. Local gas utilities are the best sources of current gas composition data for a particular area.

Heating values of natural gases vary from 900 to 1200 Btu/ft^3; the usual range is 1000 to 1050 Btu/ft^3 at sea level. The heating value for a particular gas can be calculated from the composition data and values in Table 4.

For safety purposes, odorants (such as mercaptans) are added to natural gas and LPG to give them noticeable odors.

Liquefied petroleum gases (LPG) consist primarily of propane and butane, and are usually obtained as a byproduct of oil refinery operations or by stripping natural gas. Propane and butane are gaseous under usual atmospheric conditions, but can be liquefied under moderate pressures at normal temperatures.

Three liquefied petroleum gases—butane, propane, and a mixture of the two—are commercially available as fuels.

Commercial propane consists primarily of propane but generally contains about 5 to 10% propylene. It has a heating value of about 21,560 Btu/lb or about 2500 Btu/ft^3 of gas. At atmospheric pressure, commercial propane has a boiling point of about −40°F. The low boiling point of propane allows it to be used during winter in the northern United States and in Canada. Tank heaters and vaporizers permit its use in colder climates and where high fuel flow rates are required. Propane is available in cylinders, bottles, tank trucks, or tank cars.

Propane-air mixtures are used in place of natural gas in small communities and by natural gas companies at peak loads. Table 5 lists heating values and specific gravities for various fuel-air ratios.

Commercial butane consists primarily of butane but may contain up to 5% butylene. It has a heating value of about 21,180 Btu/lb or about 3200 Btu/ft^3. At atmospheric pressure, commercial butane has a relatively high boiling point of about 32°F. Therefore, butane cannot be used in cold weather unless the gas temperature is maintained above 32°F or the partial pressure is decreased by dilution with a gas having a lower boiling point. Butane is usually available in bottles, tank trucks, or tank cars, but not in cylinders.

Table 5 Propane-Air and Butane-Air Gas Mixtures

Heating Value, Btu/ft^3	Propane-Air[a]			Butane-Air[b]		
	% Gas	% Air	Sp Gr	% Gas	% Air	Sp Gr
500	19.8	80.2	1.103	15.3	84.7	1.155
600	23.8	76.2	1.124	18.4	81.6	1.186
700	27.8	72.2	1.144	21.5	78.5	1.216
800	31.7	68.3	1.165	24.5	75.5	1.248
900	35.7	64.3	1.185	27.6	72.4	1.278
1000	39.7	60.3	1.206	30.7	69.3	1.310
1100	43.6	56.4	1.227	33.7	66.3	1.341
1200	47.5	52.5	1.248	36.8	63.2	1.372
1300	51.5	48.5	1.268	39.8	60.2	1.402
1400	55.5	44.5	1.288	42.9	57.1	1.433
1500	59.4	40.6	1.309	46.0	54.0	1.464
1600	63.4	36.6	1.330	49.0	51.0	1.495
1700	67.4	32.6	1.350	52.1	47.9	1.526
1800	71.3	28.7	1.371	55.2	44.8	1.557

Adapted from *Gas Engineers Handbook* (1965).
[a]Values used for calculation: 2522 Btu/ft^3; 1.52 specific gravity.
[b]Values used for calculation: 3261 Btu/ft^3; 2.01 specific gravity.

Butane-air mixtures are used in place of natural gas in small communities and by natural gas companies at peak loads. Table 5 lists heating values and specific gravities for various fuel-air ratios.

Commercial propane-butane mixtures with various ratios of propane and butane are available. Their properties generally fall between those of the unmixed fuels.

Manufactured gases are combustible gases produced from coal, coke, oil, liquefied petroleum gases, or natural gas. For more detailed information, see *Gas Engineers Handbook* (1965). These fuels are used primarily for industrial in-plant operations or as specialty fuels (e.g., acetylene for welding).

LIQUID FUELS

Significant liquid fuels include various fuel oils for firing combustion equipment and engine fuels for total energy systems. Liquid fuels, with few exceptions, are mixtures of hydrocarbons derived by refining crude petroleum. In addition to hydrocarbons, crude petroleum usually contains small quantities of sulfur, oxygen, nitrogen, vanadium, other trace metals, and impurities such as water and sediment. Refining produces a variety of fuels and other products. Nearly all lighter hydrocarbons are refined into fuels (e.g., liquefied petroleum gases, gasoline, kerosene, jet fuels, diesel fuels, and light heating oils). Heavy hydrocarbons are refined into residual fuel oils and other products (e.g., lubricating oils, waxes, petroleum coke, and asphalt).

Crude petroleums from different oil fields vary in hydrocarbon molecular structure. Crude is paraffin-base (principally chain-structured paraffin hydrocarbons), naphthene- or asphaltic-base (containing relatively large quantities of saturated ring-structural naphthenes), aromatic-base (containing relatively large quantities of unsaturated, ring-structural aromatics), or mixed- or intermediate-base (between paraffin- and naphthene-base crudes). Except for heavy fuel oils, the crude type has little significant effect on resultant products and combustion applications.

Types of Fuel Oils

Fuel oils for heating are broadly classified as **distillate fuel oils** (lighter oils) or **residual fuel oils** (heavier oils). ASTM *Standard* D 396 has specifications for fuel oil properties that subdivide the oils into various grades. Grades No. 1 and 2 are distillate fuel oils. Grades 4, 5 (Light), 5 (Heavy), and 6 are residual fuel oils. Specifications for the grades are based on required characteristics of fuel oils for use in different types of burners.

Grade No. 1 is a light distillate intended for vaporizing-type burners. High volatility is essential to continued evaporation of the fuel oil with minimum residue.

Grade No. 2 is a heavier distillate than No. 1. It is used primarily with pressure-atomizing (gun) burners that spray the oil into a combustion chamber. The atomized oil vapor mixes with air and burns. This grade is used in most domestic burners and many medium-capacity commercial-industrial burners. A dewaxed No. 2 oil with a pour point of −58°F is supplied only to areas where regular No. 2 oil would jell.

Grade No. 4 is an intermediate fuel that is considered either a heavy distillate or a light residual. Intended for burners that atomize oils of higher viscosity than domestic burners can handle, its permissible viscosity range allows it to be pumped and atomized at relatively low storage temperatures.

Grade No. 5 (Light) is a residual fuel of intermediate viscosity for burners that handle fuel more viscous than No. 4 without preheating. Preheating may be necessary in some equipment for burning and, in colder climates, for handling.

Grade No. 5 (Heavy) is a residual fuel more viscous than No. 5 (Light), but intended for similar purposes. Preheating is usually necessary for burning and, in colder climates, for handling.

Grade No. 6, sometimes referred to as Bunker C, is a high-viscosity oil used mostly in commercial and industrial heating. It requires preheating in the storage tank to permit pumping, and additional preheating at the burner to permit atomizing.

Low-sulfur residual oils are marketed in many areas to permit users to meet sulfur dioxide emission regulations. These fuel oils are produced (1) by refinery processes that remove sulfur from the oil (hydrodesulfurization), (2) by blending high-sulfur residual oils with low-sulfur distillate oils, or (3) by a combination of these methods. These oils have significantly different characteristics than regular residual oils. For example, the viscosity-temperature relationship can be such that low-sulfur fuel oils have viscosities of No. 6 fuel oils when cold, and of No. 4 when heated. Therefore, normal guidelines for fuel handling and burning can be altered when using these fuels.

Fuel oil grade selection for a particular application is usually based on availability and economic factors, including fuel cost, clean air requirements, preheating and handling costs, and equipment cost. Installations with low firing rates and low annual fuel consumption cannot justify the cost of preheating and other methods that use residual fuel oils. Large installations with high annual fuel consumption cannot justify the premium cost of distillate fuel oils. Disagreements on economy occur somewhere in between.

Characteristics of Fuel Oils

Characteristics that determine grade classification and suitability for given applications are (1) viscosity, (2) flash point, (3) pour point, (4) water and sediment content, (5) carbon residue, (6) ash, (7) distillation qualities or distillation temperature ranges, (8) specific gravity, (9) sulfur content, (10) heating value, and (11) carbon-hydrogen content. Not all of these are included in ASTM *Standard* D 396.

Viscosity is an oil's resistance to flow. It is significant because it indicates the ease at which oil flows or can be pumped and the ease of atomization. Differences in fuel oil viscosities are caused by variations in the concentrations of fuel oil constituents and different refining methods. Approximate viscosities of fuel oils are shown in Figure 1.

Flash point is the lowest temperature to which an oil must be heated for its vapors to ignite in a flame. Minimum permissible flash point is usually prescribed by state and municipal laws.

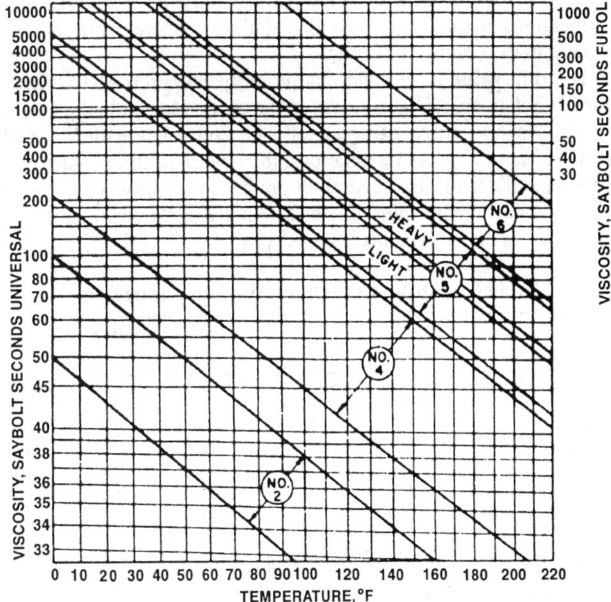

Fig. 1 Approximate Viscosity of Fuel Oils

Pour point is the lowest temperature at which a fuel can be stored and handled. Fuels with higher pour points can be used when heated storage and piping facilities are provided.

Water and **sediment content** should be low to prevent fouling the facilities. Sediment accumulates on filter screens and burner parts. Water in distillate fuels can cause tanks to corrode and emulsions to form in residual oil.

Carbon residue is obtained by a test in which the oil sample is destructively distilled in the absence of air. When commercial fuels are used in proper burners, this residue has almost no relationship to soot deposits, except indirectly when deposits are formed by vaporizing burners.

Ash is the noncombustible material in an oil. An excessive amount indicates the presence of materials that cause high wear on burner pumps.

The **distillation** test shows the volatility and ease of vaporization of a fuel.

Specific gravity is the ratio of the of a fuel oil to the density of water at a specific temperature. **Specific gravities** cover a range in each grade, with some overlap between distillate and residual grades. **API gravity** (developed by the American Petroleum Institute) is a parameter widely used in place of specific gravity. It is obtained by the following formula:

$$\text{Degrees API} = \frac{141.5}{\text{Sp Gr at }60/60°\text{F}} - 131.5 \qquad (1)$$

where Sp Gr at 60/60°F is the ratio of the mass of a given volume of oil at 60°F to the mass of the same volume of water at 60°F.

The API gravity of water at 60°F is 10.0.

Air pollution considerations are important in determining the allowable **sulfur content** of fuel oils. Sulfur content is frequently limited by legislation aimed at reducing sulfur oxide emissions from combustion equipment. These laws require sulfur content to be below a certain level, usually 1.0, 0.5, or 0.3%. Table 6 lists sulfur levels of some marketed fuel oils.

Sulfur in fuel oils is also undesirable because of the corrosiveness of sulfur compounds in the flue gas. Although low-temperature corrosion can be minimized by maintaining the stack at temperatures above the dew point of the flue gas, this limits the overall thermal efficiency of combustion equipment.

For certain industrial applications, the sulfur content of a fuel must be limited because of adverse effects on product quality. The applications include direct-fired metallurgy where work is performed in the combustion zone.

Heating value is an important property, although ASTM *Standard* D 396 does not list it as one of the criteria for fuel oil classification. Heating value can generally be correlated with the API gravity. Table 7 shows the relationship between heating value, API gravity, and density for several oil grades. In the absence of more specific data, heating values can be calculated as shown in the *North American Combustion Handbook* (1965):

Table 6 Sulfur Content of Marketed Fuel Oils

Grade of Oil	No. 1	No. 2	No. 4	No. 5 (Light)	No. 5 (Heavy)	No. 6
Total fuel samples	123	158	13	15	16	96
Sulfur content,						
% mass min.	0.002	0.03	0.46	0.90	0.57	0.32
max.	0.380	0.64	1.44	3.50	2.92	4.00
No. samples with S						
over 0.3%	1	32	13	15	16	96
over 0.5%	0	1	11	15	16	93
over 1.0%	0	0	3	9	11	60
over 3.0%	0	0	0	2	0	8

Derived from Shelton (1974).

Table 7 Typical API Gravity, Density, and Heating Value of Standard Grades of Fuel Oil

Grade No.	API Gravity	Density, lb/gal	Heating Value, Btu/gal
1	38 to 45	6.950 to 6.675	137,000 to 132,900
2	30 to 38	7.296 to 6.960	141,800 to 137,000
4	20 to 28	7.787 to 7.396	148,100 to 143,100
5L	17 to 22	7.940 to 7.686	150,000 to 146,800
5H	14 to 18	8.080 to 7.890	152,000 to 149,400
6	8 to 15	8.448 to 8.053	155,900 to 151,300

$$\text{Higher heating value, Btu/lb} = 22{,}320 - 3{,}780(\text{Specific gravity})^2 \qquad (2)$$

Distillate fuel oils (Grades 1 and 2) have a carbon-hydrogen content of 84 to 86% carbon, with the remainder predominantly hydrogen. The heavier residual fuel oils (Grades 4, 5, and 6) may contain up to 88% carbon and as little as 11% hydrogen. An approximate relationship for determining the hydrogen content of fuel oils is

$$\text{Hydrogen, \%} = 26 - (15 \times \text{Specific gravity}) \qquad (3)$$

ASTM *Standard* D 396 is more a classification than a specification, distinguishing between six generally nonoverlapping grades, one of which characterizes any commercial fuel oil. Quality is not defined, as a refiner might control it; for example, the standard lists the distillation temperature 90% point for Grade No. 2 as having a maximum of 640°F, whereas commercial practice rarely exceeds 600°F.

Types and Properties of Liquid Fuels for Engines

The primary stationary engine fuels are diesel and gas turbine oils, natural gases, and liquefied petroleum gases. Other fuels include sewage gas, manufactured gas, and gas mixtures. Gasoline and the JP series of gas turbine fuels are rarely used for stationary engines.

Only properties of diesel and gas turbine fuel oils are covered here; properties of natural and liquefied petroleum gases are found in the section on Gaseous Fuels. For properties of gasolines and JP turbine fuel, consult texts on internal combustion engines and gas turbines (Bibliography). Properties of currently marketed gasolines can be found in the latest volumes of *Mineral Industry Surveys, Motor Gasolines*, issued semiannually by the U.S. Bureau of Mines.

Properties of the three grades of diesel fuel oils (1-D, 2-D, and 4-D) are listed in ASTM *Standard* D 975.

Grade No. 1-D includes the class of volatile fuel oils from kerosene to intermediate distillates. These fuels are used in high-speed engines with frequent and relatively wide variations in loads and speeds and where abnormally low fuel temperatures are encountered.

Grade No. 2-D includes the class of lower volatility distillate gas oils. These fuels are used in high-speed engines with relatively high loads and uniform speeds, or in engines not requiring fuels with the higher volatility or other properties specified for Grade No. 1-D.

Grade No. 4-D covers the class of more viscous distillates and blends of these distillates with residual fuel oils. These fuels are used in low- and medium-speed engines involving sustained loads at essentially constant speed.

Property specifications and test methods for Grade No. 1-D, 2-D, and 4-D diesel fuel oils are essentially identical to specifications of Grade No. 1, 2, and 4 fuel oils, respectively. However, diesel fuel oils have an additional specification for **cetane number**, which measures ignition quality and influences combustion roughness. Cetane number requirements depend on engine design, size, speed and load variations, and starting and atmospheric conditions. An increase in cetane number over values actually required does not improve engine performance. Thus, the cetane number should be as low as possible to assure maximum fuel availability. ASTM *Standard* D 975 provides

Table 8 Classification of Coals by Rank[a]

Class	Group	Limits of Fixed Carbon or Energy Content, Mineral-Matter-Free Basis	Requisite Physical Properties
I Anthracite	1. Metaanthracite	Dry FC, 98% or more (Dry VM, 2% or less)	Nonagglomerating
	2. Anthracite	Dry FC, 92% or more, and less than 98% (Dry VM, 8% or less, and more than 2%)	
	3. Semianthracite	Dry FC, 86% or more, and less than 92% (Dry VM, 14% or less, and more than 8%)	
II Bituminous[d]	1. Low-volatile bituminous coal	Dry FC, 78% or more, and less than 86% (Dry VM, 22% or less, and more than 14%)	Either agglomerating[b] or nonweathering[f]
	2. Medium-volatile bituminous coal	Dry FC, 69% or more, and less than 78% (Dry VM, 31% or less, and more than 22%)	
	3. High-volatile *A* bituminous coal	Dry FC, less than 69% (Dry VM, more than 31%), and moist[c], about 14,000 Btu/lb[e] or more	
	4. High-volatile *B* bituminous coal	Moist[c], about 13,000 Btu/lb or more, and less than 14,000 Btu/lb[e]	
	5. High-volatile *C* bituminous coal	Moist[c], about 11,000 Btu/lb or more, and less than 13,000 Btu/lb[e]	
III Subbituminous	1. Subbituminous *A* coal	Moist[c], about 11,000 Btu/lb or more, and less than 13,000 Btu/lb[e]	Both weathering and nonagglomerating[b]
	2. Subbituminous *B* coal	Moist[c], about 9,500 Btu/lb or more, and less than 11,000 Btu/lb[e]	
	3. Subbituminous *C* coal	Moist[c], about 8,300 Btu/lb or more, and less than 9,500 Btu/lb[e]	
IV Lignitic	1. Lignite	Moist[c], less than 8,300 Btu/lb	Consolidated
	2. Brown coal	Moist[c], less than 8,300 Btu/lb	Unconsolidated

Source: Adapted from ASTM *Standard* D 388, Standard Classification of Coals by Rank.

FC = Fixed Carbon; VM = Volatile Matter.

[a]This classification does not include a few coals of unusual physical and chemical properties which come within the limits of fixed carbon or Btu of high-volatile bituminous and subbituminous ranks. All these coals either contain less than 48% dry, mineral-matter-free fixed carbon, or have more than about 15,500 moist, mineral-matter-free Btu/lb.

[b]If agglomerating, classify in low-volatile group of the bituminous class.

[c]Moist (Btu/lb) refers to coal containing its natural bed moisture but not including visible water on the coal surface.

[d]There may be noncaking varieties in each group of the bituminous class.

[e]Coals having 69% or more fixed carbon on the dry, mineral-matter-free basis shall be classified according to fixed carbon, regardless of energy content.

[f]There are three varieties of coal in the high-volatile *C* bituminous coal group: Variety 1, agglomerating and nonweathering; Variety 2, agglomerating and weathering; and Variety 3, nonagglomerating and nonweathering.

several methods for estimating cetane number from other fuel oil properties.

ASTM *Standard* D 2880 for gas turbine fuel oils relates gas turbine fuel oil grades to fuel and diesel fuel oil grades. Test methods for determining properties of gas turbine fuel oils are essentially identical to those for fuel oils. However, gas turbine specifications contain quantity limits on some trace elements that may be present. These limits are intended to prevent excessive corrosion in gas turbine engines. For a detailed discussion of fuels for gas turbines and combustion in gas turbines, see Chapters 5 and 9, respectively, in Hazard (1971).

SOLID FUELS

Solid fuels include coal, coke, wood, and waste products of industrial and agricultural operations. Of these, only coal is widely used for heating and cooling applications.

The complex composition of coal makes classification difficult. Chemically, coal consists of carbon, hydrogen, oxygen, nitrogen, sulfur, and a mineral residue, ash. Chemical analysis provides some indication of coal quality, but does not define its burning characteristics sufficiently. The coal user is principally interested in the available heat per unit mass of coal and the amount of ash and dust produced, but is also interested in burning characteristics and handling and storing properties. A description of coal qualities and their characteristics can be obtained from the U.S. Bureau of Mines.

Types of Coals

Commonly accepted definitions for classifying coals are listed in Table 8. This classification is arbitrary because there are no distinct demarcation lines between coal types.

Anthracite is a clean, dense, hard coal that creates little dust in handling. It is comparatively hard to ignite, but burns freely once started. It is noncaking and burns uniformly and smokelessly with a short flame.

Semianthracite has a higher volatile content than anthracite. It is not as hard and ignites more easily. Otherwise, its properties are similar to those of anthracite.

Bituminous coal includes many types of coal with distinctly different compositions, properties, and burning characteristics. Coals range from high-grade bituminous, such as those found in the eastern United States, to low-rank coals, such as those found in the western United States. Caking properties range from coals that melt or become fully plastic, to those from which volatiles and tars are distilled without changing form (classed as noncaking or free-burning). Most bituminous coals are strong and nonfriable enough to permit screened sizes to be delivered free of fines. Generally, they ignite easily and burn freely. Flame length is long and varies with different coals. If improperly fired, much smoke and soot are possible, especially at low burning rates.

Semibituminous coal is soft and friable, and handling creates fines and dust. It ignites slowly and burns with a medium-length flame. Its caking properties increase as volatile matter increases, but the coke formed is weak. With only half the volatile matter content of bituminous coals, burning produces less smoke; hence, it is sometimes called smokeless coal.

Subbituminous coal, such as that found in the western United States, is high in moisture when mined and tends to break up as it dries or is exposed to the weather; it is likely to ignite spontaneously when piled or stored. It ignites easily and quickly, has a medium-length flame, and is noncaking and free-burning. The lumps tend to break into small pieces if poked. Very little smoke and soot are formed.

Lignite is woody in structure, very high in moisture when mined, of low heating value, and clean to handle. It has a greater tendency than subbituminous coals to disintegrate as it dries and is also more likely to ignite spontaneously. Because of its high moisture, freshly mined lignite ignites slowly and is noncaking. The char left after moisture and volatile matter are driven off burns very easily, like charcoal. The lumps tend to break up in the fuel bed and pieces of char that fall into the ash pit continue to burn. Very little smoke or soot forms.

Characteristics of Coal

The characteristics of coals that determine classification and suitability for given applications are the proportions of (1) volatile matter, (2) fixed carbon, (3) moisture, (4) sulfur, and (5) ash. Each of these is reported in the proximate analysis. Coal analyses can be reported on several bases: as-received, moisture-free (dry) or dry, and mineral-matter-free (or ash-free). As-received is applicable for combustion calculations; dry and mineral-matter-free, for classification purposes.

Volatile matter is driven off as gas or vapor when the coal is heated according to a standardized temperature test. It consists of a variety of organic gases, generally resulting from distillation and decomposition. Volatile products given off by coals when heated differ materially in the ratios by mass of the gases to oils and tars. No heavy oils or tars are given off by anthracite, and very small quantities are given off by semianthracite. As volatile matter in the coal increases to as much as 40% of the coal (dry and ash-free basis), increasing amounts of oils and tars are released. However, for coals of higher volatile content, the quantity of oils and tars decreases and is relatively low in the subbituminous coals and in lignite.

Fixed carbon is the combustible residue left after the volatile matter is driven off. It is not all carbon. Its form and hardness are an indication of fuel coking properties and, therefore, guide the choice of combustion equipment. Generally, fixed carbon represents that portion of fuel that must be burned in the solid state.

Moisture is difficult to determine accurately because a sample can lose moisture on exposure to the atmosphere, particularly when reducing the sample size for analysis. To correct for this loss, total moisture content of a sample is customarily determined by adding the moisture loss obtained when air-drying the sample to the measured moisture content of the dried sample. Moisture does not represent all of the water present in coal; water of decomposition (combined water) and of hydration are not given off under standardized test conditions.

Ash is the noncombustible residue remaining after complete coal combustion. Generally, the mass of ash is slightly less than that of mineral matter before burning.

Sulfur is an undesirable constituent in coal, because the sulfur oxides formed when it burns contribute to air pollution and cause combustion system corrosion. Table 9 lists the sulfur content of typical coals. Legislation has limited the sulfur content of coals burned in certain locations.

Table 9 Typical Ultimate Analyses for Coals

Rank	As Received, Btu/lb	Constituents, Percent by Mass: Oxygen	Hydrogen	Carbon	Nitrogen	Sulfur	Ash
Anthracite	12,700	5.0	2.9	80.0	0.9	0.7	10.5
Semianthracite	13,600	5.0	3.9	80.4	1.1	1.1	8.5
Low-volatile bituminous	14,350	5.0	4.7	81.7	1.4	1.2	6.0
Medium-volatile bituminous	14,000	5.0	5.0	81.4	1.4	1.5	6.0
High-volatile bituminous *A*	13,800	9.3	5.3	75.9	1.5	1.5	6.5
High-volatile bituminous *B*	12,500	13.8	5.5	67.8	1.4	3.0	8.5
High-volatile bituminous *C*	11,000	20.6	5.8	59.6	1.1	3.5	9.4
Subbituminous *B*	9,000	29.5	6.2	52.5	1.0	1.0	9.8
Subbituminous *C*	8,500	35.7	6.5	46.4	0.8	1.0	9.6
Lignite	6,900	44.0	6.9	40.1	0.7	1.0	7.3

Heating value may be reported on an as-received, dry, dry and mineral-matter-free, or moist and mineral-matter-free basis. Higher heating values of coals are frequently reported with their proximate analysis. When more specific data are lacking, the higher heating value of higher quality coals can be calculated by the Dulong formula:

$$\text{Higher heating value, Btu/lb} = 14{,}544C + 62{,}028[H - (O/8)] + 4{,}050S \qquad (4)$$

where C, H, O, and S are the mass fractions of carbon, hydrogen, oxygen, and sulfur in the coal.

Other important parameters in judging coal suitability include (1) ultimate analysis, (2) ash-fusion temperature, (3) grindability, and (4) free-swelling index.

Ultimate analysis is another method of reporting coal composition. Percentages of C, H, O, N, S, and ash in the coal sample are reported. Ultimate analysis is used for detailed fuel studies and for computing a heat balance when required in heating device testing. Typical ultimate analyses of various coals are shown in Table 9.

Ash-fusion temperature indicates the fluidity of the ash at elevated temperatures. It is helpful in selecting coal to be burned in a particular furnace and in estimating the possibility of ash handling and slagging problems.

The **grindability index** indicates the ease with which a coal can be pulverized and is helpful in estimating ball mill capacity with various coals. There are two common methods for determining the index—Hardgrove and ball mill.

The **free-swelling index** denotes the extent of coal swelling on combustion on a fuel bed and indicates the coking characteristics of coal.

COMBUSTION CALCULATIONS

Calculations of the quantity of air required for combustion and the quantity of flue gas products generated during combustion are frequently needed for sizing system components and as input to efficiency calculations. Other calculations, such as values for excess air and theoretical CO_2, are useful in estimating combustion system performance.

Frequently, combustion calculations can be simplified by using molecular mass. The molecular mass of a compound equals the sum of the atomic masses of the elements in the compound. Molecular mass can be expressed in any mass units. The pound molecular weight or pound mole is the molecular weight of the compound expressed in pounds. The molecular mass of any substance contains the same number of molecules as the molecular mass of any other substance.

Corresponding to measurement standards common to the industries, calculations involving gaseous fuels are generally based on volume, and calculations involving liquid and solid fuels are generally based on mass.

Some calculations described here require data on concentrations of carbon dioxide, carbon monoxide, and oxygen in the flue gas. Gas analyses for CO_2, CO, and O_2 can be obtained by volumetric chemical analysis and other analytical techniques, including electromechanical cells used in portable electronic flue gas analyzers.

Air Required for Combustion

Stoichiometric or theoretical air is the exact quantity of air required to provide oxygen for complete combustion.

The three most prevalent components in hydrocarbon fuels (C, H_2, and S) are completely combusted as in the following reactions:

$$C + O_2 \rightarrow CO_2$$
$$H_2 + 0.5O_2 \rightarrow H_2O$$
$$S + O_2 \rightarrow SO_2$$

In the reactions, C, H_2, and S can be taken to represent 1 lb mole of carbon, hydrogen, and sulfur, respectively. Using approximate atomic masses (C = 12, H = 1, S = 32, and O = 16), 12 lb of C are oxidized by 32 lb of O_2 to form 44 lb of CO_2, 2 lb of H_2 are oxidized by 16 lb of O_2 to form 18 lb of H_2O, and 32 lb of S are oxidized by 32 lb of O_2 to form 64 lb of SO_2. These relationships can be extended to include hydrocarbons.

The mass of dry air required to supply a given quantity of oxygen is 4.32 times the mass of the oxygen. The mass of air required to oxidize the fuel constituents listed in Table 1 was calculated on this basis. Oxygen contained in the fuel, except that contained in ash, should be deducted from the amount of oxygen required, because this oxygen is already combined with fuel components. In addition, when calculating the mass of air to be supplied for combustion, allowance should be made for water vapor, which is always present in atmospheric air.

As stated previously, combustion calculations for gaseous fuels are based on volume. Avogadro's law states that, for any gas, one mole occupies the same volume at a given temperature and pressure. Therefore, in reactions involving gaseous compounds, the gases react in volume ratios identical to the pound mole ratios. That is, for the oxidation of hydrogen in the above reaction, one volume (or one lb mole) of hydrogen reacts with one-half volume (or one-half lb mole) of oxygen to form one volume (or one lb mole) of water vapor.

The volume of air required to supply a given volume of oxygen is 4.78 times the volume of oxygen. The volumes of dry air required to oxidize the fuel constituents listed in Table 1 were calculated on this basis. Volume ratios are not given for fuels that do not exist in vapor form at reasonable temperatures or pressures. Again, oxygen contained in the fuel should be deducted from the quantity of oxygen required, because this oxygen is already combined with fuel components. Allowance should be made for water vapor, which increases the volume of dry air by 1 to 3%.

From the relationships just described, the theoretical mass m_a of dry air required for stoichiometric combustion of a unit mass of any hydrocarbon fuel is

$$m_a = 0.0144(8C + 24H + 3S - 3O) \quad (5)$$

where C, H, S, and O are the mass percentages of carbon, hydrogen, sulfur, and oxygen in the fuel.

Analyses of gaseous fuels are generally based on hydrocarbon components rather than elemental content.

If the fuel analysis is based on mass, the theoretical mass m_a of dry air required for stoichiometric combustion of a unit mass of gaseous fuel is

$$m_a = 2.47CO + 34.28H_2 + 17.24CH_4 + 16.09C_2H_6 + 15.68C_3H_8 + 15.47C_4H_{10} + 13.27C_2H_2 + 14.78C_2H_4 + 6.08H_2S - 4.32O_2 \quad (6)$$

If the fuel analysis is reported on a volumetric or molecular basis, it is simplest to calculate air requirements based on volume and, if necessary, convert to mass. The theoretical volume V_a of air required for stoichiometric combustion of a unit volume of gaseous fuels is

$$V_a = 2.39CO + 2.39H_2 + 9.57CH_4 + 16.75C_2H_6 + 23.95C_3H_8 + 31.14C_4H_{10} + 11.96C_2H_2 + 14.38C_2H_4 + 7.18H_2S - 4.78O_2 + 30.47 \text{ illuminants} \quad (7)$$

where CO, H_2, and so forth are the volumetric fractions of each constituent in the fuel gas.

Illuminants include a variety of compounds not separated by usual gas analysis. In addition to ethylene (C_2H_4) and acetylene (C_2H_2), the principal illuminants included in Equation (7), and the dry air required for combustion, per unit volume of each gas, are: propylene (C_3H_6), 21.44; butylene (C_4H_8), 28.58; pentene (C_5H_{10}), 35.73; benzene (C_6H_6); 35.73, toluene (C_7H_8), 42.88; and xylene (C_8H_{10}), 50.02. Because toluene and xylene are normally scrubbed from the gas before distribution, they can be disregarded in computing air required for combustion of gaseous fuels. The percentage of illuminants present in gaseous fuels is small, so the values can be lumped together, and an approximate value of 30 unit volumes of dry air per unit volume of gas can be used. If ethylene and acetylene are included as illuminants, a value of 20 unit volumes of dry air per unit volume of gaseous illuminants can be used.

For many combustion calculations, only approximate values of air requirements are necessary. If approximate values for theoretical air are sufficient, or if complete information on the fuel is not available, the values in Tables 10 and 11 can be used. Another frequently used value for estimating air requirements is 0.9 ft³ of air for 100 Btu of fuel.

Table 10 Approximate Air Requirements for Stoichiometric Combustion of Fuels

Type of Fuel	Air Required		Approx. Precision,	Exceptions
	lb/lb Fuel	ft³/Unit Fuel[a]	%	
Solid	Btu/lb × 0.00073	Btu/lb × 0.0097	3	Fuels containing more than 30% water
Liquid	Btu/lb × 0.00071	Btu/lb × 0.0094	3	Results low for gasoline and kerosene
Gas	Btu/lb × 0.00067	Btu/ft³ × 0.0089	5	300 Btu/ft³ or less

Source: Data based on Shnidman (1954).
[a]Unit fuel for solid and liquid fuels in lb, for gas in ft³.

Table 11 Approximate Air Requirements for Stoichiometric Combustion of Various Fuels

Type of Fuel	Theoretical Air Required for Combustion
Solid fuels	lb/lb fuel
Anthracite	9.6
Semibituminous	11.2
Bituminous	10.3
Lignite	6.2
Coke	11.2
Liquid fuels	lb/gal fuel
No. 1 fuel oil	103
No. 2 fuel oil	106
No. 5 fuel oil	112
No. 6 fuel oil	114
Gaseous fuels	ft³/ft³ fuel
Natural gas	9.6
Butane	31.1
Propane	24.0

In addition to the amount theoretically required for combustion, **excess air** must be supplied to most practical combustion systems to ensure complete combustion.

$$\text{Excess air, \%} = \frac{\text{Air supplied} - \text{Theoretical air}}{\text{Theoretical air}} \quad (8)$$

The excess air level at which a combustion process operates significantly affects its overall efficiency. Too much excess air dilutes flue gas excessively, lowering its heat transfer temperature and increasing sensible flue gas loss. Conversely, if the level of excess air is too low, incomplete combustion and loss of unburned combustible gases from the equipment can result. The highest combustion efficiency is usually obtained when just enough excess air is supplied and properly mixed with combustible gases to ensure complete combustion. The general practice is to supply from 5 to 50% excess air, the exact amount depending on the type of fuel burned, combustion equipment, and other factors.

The amount of dry air supplied per unit mass of fuel burned can be obtained from the following equation, which is reasonably precise for most solid and liquid fuels.

$$\text{Dry air supplied} = \frac{C\,(3.04N_2)}{CO_2 + CO} \quad (9)$$

where

Dry air supplied = unit mass per unit mass of fuel
C = unit mass of carbon burned per unit mass of fuel, corrected for carbon in the ash
CO_2, CO, N_2 = percentages by volume from the flue gas analysis

These values of dry air supplied and theoretical air can be used in Equation (8) to determine excess air.

Excess air can also be calculated from unit volumes of stoichiometric combustion products and air, and from volumetric analysis of the flue gas:

$$\text{Excess air, \%} = 100\left(\frac{P}{A}\right)\left(\frac{U - CO_2}{CO_2}\right) \quad (10)$$

where

U = ultimate carbon dioxide of flue gases resulting from stoichiometric combustion, %
CO_2 = carbon dioxide content of flue gases, %
P = dry products from stoichiometric combustion, unit volume per unit volume of gas burned
A = air required for stoichiometric combustion, unit volume per unit volume of gas burned

As the ratio P/A is approximately 0.9 for most natural gases, a value of 90 can be substituted for 100 (P/A) in Equation (10) for rough calculation.

Because excess air calculations are almost invariably made from flue gas analysis results and theoretical air requirements are not always known, another convenient method of expressing the relation of Equation (8) is

$$\text{Excess air, \%} = \frac{100[O_2 - (CO/2)]}{0.264N_2 - [O_2 - (CO/2)]} \quad (11)$$

where O_2, CO, and N_2 are percentages by volume from the flue gas analysis.

Theoretical CO_2

The theoretical CO_2, ultimate CO_2, stoichiometric CO_2, or maximum CO_2 concentration attainable in the products from the combustion of a hydrocarbon fuel with air is obtained when the fuel is completely burned with the theoretical quantity of air and zero excess air. Theoretical CO_2 varies with the carbon-hydrogen ratio of the fuel. For combustion with excess air present, theoretical CO_2 values can be calculated from the flue gas analysis:

$$\text{Theoretical } CO_2,\ \% = U = \frac{CO_2}{1 - (O_2/20.95)} \quad (12)$$

where CO_2 and O_2 are percentages by volume from the flue gas analysis.

Table 12 gives approximate theoretical CO_2 values for stoichiometric combustion of several common types of fuel, as well as CO_2 values attained with different amounts of excess air. In practice, desirable CO_2 values depend on the excess air, fuel, firing method, and other considerations.

Table 12 Approximate Maximum Theoretical (Stoichiometric) CO_2 Values, and CO_2 Values of Various Fuels with Different Percentages of Excess Air

Type of Fuel	Theoretical or Maximum CO_2, %	Percent CO_2 at Given Excess Air Values		
		20%	40%	60%
Gaseous fuels				
Natural gas	12.1	9.9	8.4	7.3
Propane gas (Commercial)	13.9	11.4	9.6	8.4
Butane gas (Commercial)	14.1	11.6	9.8	8.5
Mixed gas (Natural and Carbureted water gas)	11.2	12.5	10.5	9.1
Carbureted water gas	17.2	14.2	12.1	10.6
Coke oven gas	11.2	9.2	7.8	6.8
Liquid fuels				
No. 1 and 2 fuel oil	15.0	12.3	10.5	9.1
No. 6 fuel oil	16.5	13.6	11.6	10.1
Solid fuels				
Bituminous coal	18.2	15.1	12.9	11.3
Anthracite	20.2	16.8	14.4	12.6
Coke	21.0	17.5	15.0	13.0

Quantity of Flue Gas Produced

The mass of dry flue gas produced per mass of fuel burned is required in heat loss and efficiency calculations. This mass is equal to the sum of the mass of (1) fuel (minus ash retained in the furnace), (2) air theoretically required for combustion, and (3) excess air. For solid fuels, this mass, determined from the flue gas analysis, is

$$\text{Dry flue gas} = \frac{11CO_2 + 8O_2 + 7(CO + N_2)}{3(CO_2 + CO)}\,C \quad (13)$$

where

Dry flue gas = lb/lb of fuel
C = lb of carbon burned per lb of fuel, corrected for carbon in the ash
CO_2, O_2, CO, N_2 = percentages by volume from the flue gas analysis

The total dry gas volume of flue gases resulting from combustion of one unit volume of gaseous fuels for various percentages of CO_2 is

$$\text{Dry flue gas} = \left(\frac{\text{vol. of } CO_2 \text{ produced}}{\text{unit vol. of gas burned}}\right)\left(\frac{100}{CO_2}\right) \quad (14)$$

where

Dry flue gas = unit volume per unit volume of gaseous fuel
CO_2 = percentage by volume from the flue gas analysis

Excess air quantity can be estimated by subtracting the quantity of dry flue gases resulting from stoichiometric combustion from the total volume of flue gas.

Water Vapor and Dew Point of Flue Gas

Water vapor in flue gas is the total of (1) the water contained in the fuel; (2) the water contained in the stoichiometric, excess, and dilution air; and (3) the water produced from the combustion

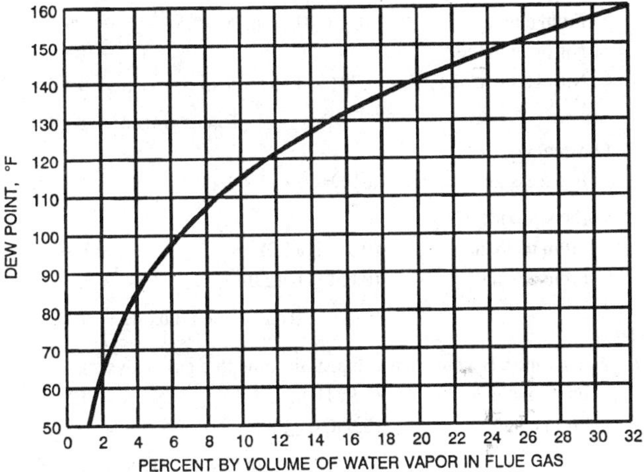

Fig. 2 Water Vapor and Dew Point of Flue Gas
Adapted from *Gas Engineers Handbook* (1965). Printed with permission of Industrial Press and American Gas Association.

of hydrogen or hydrocarbons in the fuel. The amount of water vapor in the stoichiometric combustion products may be calculated from the fuel burned by using the water data in Table 2.

The dew point is the temperature at which condensation begins and can be determined using Figure 2. The volume fraction of water vapor in the flue gas can be determined as follows:

$$P_{wv} = \frac{V_w}{(100\,V_c/P_c) + V_w} \tag{15}$$

where

V_w = total water vapor volume (from fuel; from stoichiometric, excess, and dilution air; and from combustion)
V_c = unit volume of CO_2 produced per unit volume of gaseous fuel
P_c = percent CO_2 in flue gas

Using Figure 3, the dew points of solid, liquid, or gaseous fuels may be estimated. For example, to find the dew point of flue gas resulting from the combustion of a solid fuel with a weight ratio (hydrogen to carbon-plus-sulfur) of 0.088 and sufficient excess air to produce 11.4% oxygen in the flue gas, start with the weight ratio of 0.088. Proceed vertically to the intersection of the solid fuels curve and then to the theoretical dew point of 115°F on the dew-point scale (see dotted lines in Figure 3). Follow the curve fixed by this point (down and to the right) to 11.4% oxygen in the flue gas (on the abscissa). The actual dew point is 93°F and is found on the dew-point scale.

An estimation can be made of the dew point of the flue gas from natural gas having a higher heating value (HHV) of 1020 Btu/ft^3 with 6.3% oxygen or 31.5% air. Start with 1020 Btu/ft^3 and proceed vertically to the intersection of the gaseous fuels curve and then to the theoretical dew point of 139°F on the dew-point scale. Follow

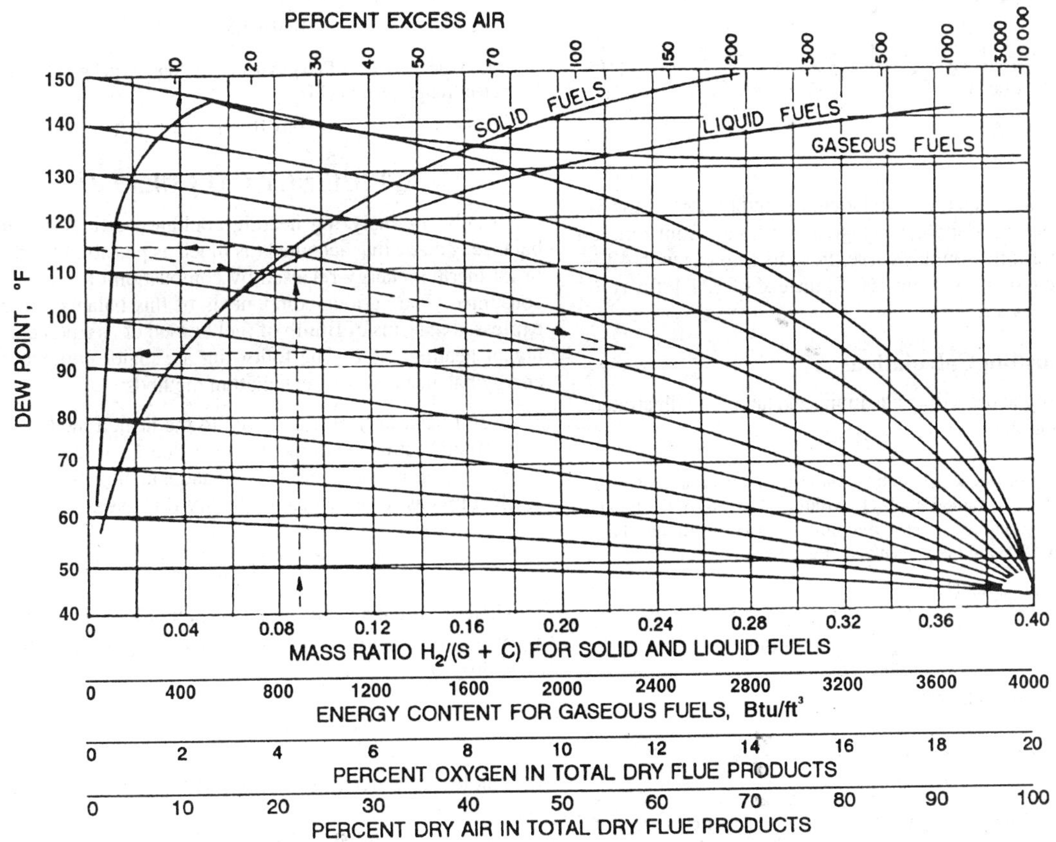

Fig. 3 Theoretical Dew Points of Combustion Products of Industrial Fuels
Adapted from *Gas Engineers Handbook* (1965). Printed with permission of Industrial Press and American Gas Association.

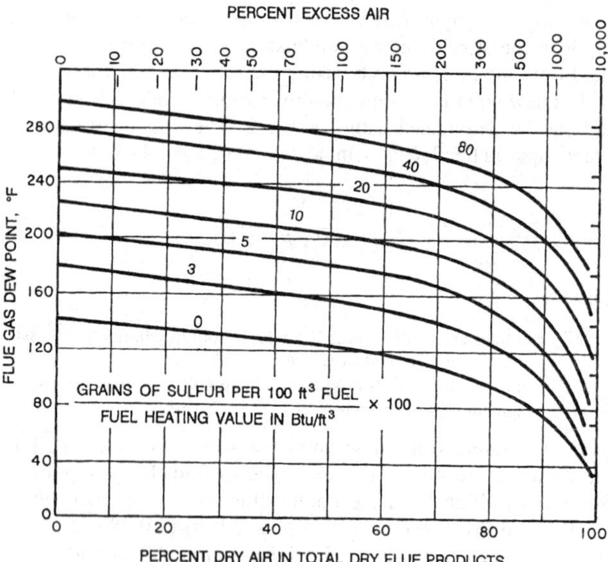

Fig. 4 Influence of Sulfur Oxides on Flue Gas Dew Point
(Stone 1969)

the curve fixed by this point to 6.3% oxygen or 31.5% air in the flue gas. The actual dew point is 127°F.

The presence of sulfur dioxide, and particularly sulfur trioxide, influences the vapor pressure of condensate in flue gas, and the dew point can be raised by as much as 25 to 75°F, as shown in Figure 4. To illustrate the use of Figure 4, for a manufactured gas with a HHV of 550 Btu/ft^3 containing 15 grains of sulfur per 100 ft^3 being burned with 40% excess air, the proper curve in Figure 4 is determined as follows:

$$\frac{\text{Grains of S per 100 ft}^3\text{ of fuel}}{\text{Btu per ft}^3\text{ of fuel}} \times 100 \qquad (16)$$

$$= \frac{15}{550} \times 100 = 2.73$$

This curve lies between the 0- and 3-curves and is close to the 3-curve. The dew point for any percentage of excess air from zero to 100% can be determined on this curve. For this flue gas with 40% excess air, the dew point is about 160°F, instead of 127°F for zero sulfur at 40% excess air.

Sample Combustion Calculations

Applications of the preceding equations and tables are illustrated by Examples 1 and 2.

Example 1. Analysis of flue gases from the burning of a natural gas shows 10.0% CO_2, 3.1% O_2, and 86.9% N_2 by volume. Analysis of the fuel is 90% CH_4, 5% N_2, and 5% C_2H_6 by volume. Find U (maximum theoretical percent CO_2), and the percentage of excess air.

Solution: From Equation (12),

$$U = \frac{10.0}{1-(3.1/20.95)} = 11.74\%\ CO_2$$

From Equation (10), using 100 (P/A) = 90,

$$\text{Excess air} = \frac{(11.74 - 10.0)90}{10} = 15.7\%$$

Example 2. For the same analysis as in Example 1, find, per cubic foot of fuel gas, the volume of dry air required for combustion, the volume of each constituent in the flue gases, and the total volume of dry and wet flue gases.

Solution: From Equation (7), the volume of dry air required for combustion is:

$$9.57CH_4 + 16.75C_2H_6 = (9.57 \times 0.90) + (16.75 \times 0.05)$$
$$= 9.45\text{ ft}^3\text{ per ft}^3\text{ of fuel gas}$$

(The volume of dry air may also be calculated using Table 11.)

From Table 1, the cubic feet of flue gas constituents per cubic foot of fuel gas are

Nitrogen, N_2

From methane	$(0.9CH_4)(9.57 - 2.0)$	= 6.81
From ethane	$(0.05C_2H_6)(16.75 - 3.5)$	= 0.66
Nitrogen in fuel		= 0.05
Nitrogen in excess air	$0.791 \times 0.157 \times 9.45$	= 1.17
	Total nitrogen	= 8.69 ft^3

Oxygen, O_2

In excess air	$0.209 \times 0.157 \times 9.45$	= 0.31 ft^3

Carbon dioxide, CO_2

From methane	$(0.9CH_4)(1.0)$	= 0.90
From ethane	$(0.05C_2H_6)(2.0)$	= 0.10
	Total carbon dioxide	= 1.00 ft^3

Water vapor, H_2O (does not appear in some flue gas analyses)

From methane	$(0.9CH_4)(2.0)$	= 1.8
From ethane	$(0.05C_2H_6)(3.0)$	= 0.15
	Total water vapor	= 1.95 ft^3

Total volume of dry gas per cubic foot of fuel gas

$$8.69 + 0.31 + 1.00 = 10.0\text{ ft}^3$$

Total volume of wet gases per cubic foot of fuel gas (neglecting water vapor in combustion air)

$$10.0 + 1.95 = 11.95\text{ ft}^3$$

The cubic feet of dry flue gas per cubic foot of fuel gas can also be computed from Equation (14):

$$(1.00)(100)/10.0 = 10.0\text{ ft}^3$$

EFFICIENCY CALCULATIONS

Usually, in analyzing heating appliance efficiency, an energy balance is made that accounts (as much as possible) for disposition of all thermal energy released by combustion of the fuel quantity consumed. The various components of this balance are generally expressed in terms of Btu/lb of fuel burned or as a percentage of its higher heating value. The following are major components of an energy balance and their calculation methods:

1. Useful heat, or heat transferred to the heated medium; for convection heating equipment, this value q_1 is computed as the product of the mass rate of flow and enthalpy change.
2. Heat loss as sensible heat in the dry flue gases

$$q_2 = m_g c_{pg}(t_g - t_a) \qquad (17)$$

where m_g is calculated as in Equation (13).

3. Heat loss in water vapor in products formed by combustion of hydrogen

$$q_3 = (9H_2/100)[(h)_{tg} - (h_f)_{ta}] \qquad (18)$$

4. Heat loss in water vapor in the combustion air

$$q_4 = Mm_a[(h)_{tg} - (h_g)_{ta}] \qquad (19)$$

where m_a is calculated as in Equations (5) and (6).

5. Heat loss from incomplete combustion of carbon

$$q_5 = 10{,}143\ \mathrm{C}\left(\frac{\mathrm{CO}}{\mathrm{CO_2 + CO}}\right) \tag{20}$$

6. Heat loss from unburned carbon in the ash or refuse

$$q_6 = 14{,}600[(C_u/100) - \mathrm{C}] \tag{21}$$

7. Unaccounted-for heat losses, q_7

The following symbols are used in Equations (17) through (21):

q_1 = useful heat, Btu/lb of fuel
q_2 = heat loss in dry flue gases, Btu/lb of fuel
q_3 = heat loss in water vapor from combustion of hydrogen, Btu/lb of fuel
q_4 = heat loss in water vapor in combustion air, Btu/lb of fuel
q_5 = heat loss from incomplete combustion of carbon, Btu/lb of fuel
q_6 = heat loss from unburned carbon in ash, Btu/lb of fuel
q_7 = unaccounted-for heat losses, Btu/lb of fuel
c_{pg} = mean specific heat of flue gases at constant pressure [c_{pg} ranges from 0.242 to 0.254 Btu/lb·°F for flue gas temperatures from 300 to 1000°F], Btu/lb·°F
$(h)_{tg}$ = enthalpy of superheated steam at flue gas temperature and 1 psia, Btu/lb
$(h_f)_{ta}$ = enthalpy of saturated water vapor at air temperature, Btu/lb
$(h_g)_{ta}$ = enthalpy of saturated steam at combustion air temperature, Btu/lb
m_a = mass of combustion air per mass of fuel used, lb/lb of fuel
m_g = mass of dry flue gas per mass of fuel, lb/lb of fuel
t_a = temperature of combustion air, °F
t_g = temperature of flue gases at exit of heating device, °F
H_2 = hydrogen in fuel, % by mass (from ultimate analysis of fuel)
M = humidity ratio of combustion air, mass of water vapor per mass of dry air
CO, CO_2 = carbon monoxide and carbon dioxide in flue gases, % by volume
C = mass of carbon burned per unit of mass of fuel, corrected for carbon in ash, lb/lb of fuel

$$\mathrm{C} = \frac{WC_u - W_aC_a}{100W} \tag{22}$$

where

C_u = percentage of carbon in fuel by mass from ultimate analysis
W_a = mass of ash and refuse
C_a = percent of combustible in ash by mass (combustible in ash is usually considered to be carbon)
W = mass of fuel used

Useful heat (item 1) is generally measured for a particular piece of combustion equipment.

Flue gas loss is the sum of items 2 through 6. However, for clean-burning gas- and oil-fired equipment, items 5 and 6 are usually negligible and flue gas loss is the sum of items 2, 3, and 4.

Flue gas losses (the sum of items 2, 3, and 4) can be determined with sufficient precision for most purposes from the curves in Figure 5, if O_2 content and flue gas temperature are known. Values of the losses were computed from typical ultimate analyses, assuming 1% water vapor (by mass) in the combustion air. Curves for medium-volatile bituminous coal can be used for high-volatile bituminous coal with no appreciable error.

Generally, item 5 is negligible for modern combustion equipment in good operating condition.

Item 6 is generally negligible for gas and oil firing, but should be determined for coal-firing applications.

Item 7 consists primarily of radiation and convection losses from combustion equipment surfaces and losses caused by incomplete combustion not included in items 5 and 6. Heat loss from incomplete combustion is determined by subtracting the sum of items 1 through 6 from the fuel heating value.

Radiation and convection losses are not usually determined by direct measurement. But if the heating appliance is located within the heated space, radiation and convection losses can be considered useful heat rather than lost heat and can be omitted from heat loss calculations or added to item 1.

If CO is present in flue gases, small amounts of unburned hydrogen and hydrocarbons may also be present. The small losses caused by incomplete combustion of these gases would be included in item 7, if item 7 was determined by subtracting items 1 through 6 from the fuel heating value.

The overall thermal efficiency of combustion equipment is defined as

$$\text{Thermal efficiency, \%} = 100\,\frac{\text{Useful heat}}{\text{Heating value of fuel}} \tag{23}$$

The following equation can be used to estimate efficiency for equipment where item 7 is small or radiation and convection are useful heat:

$$\text{Thermal efficiency, \%} = 100\,\frac{\text{Heating value of fuel} - (q_2 + q_3 + q_4 + q_5 + q_6)}{\text{Heating value of fuel}} \tag{24}$$

Using heating values based on gas volume, the thermal efficiency of a gas appliance can be computed with sufficient precision by the following equation:

$$\eta = \frac{100(Q_g - Q_{fl})}{Q_g} \tag{25}$$

where

η = thermal efficiency, %
Q_g = gross heating value of fuel gas per unit volume
Q_{fl} = flue gas losses per unit volume of fuel gas

To produce heat efficiently by burning any common fuel, flue gas losses must be minimized by (1) providing adequate heat-absorbing surface in the appliance, (2) maintaining clean heat transfer surfaces on both fire and water or air sides, and (3) reducing excess air to the minimum level consistent with complete combustion and discharge of combustion products.

Seasonal Efficiency

The method just presented is useful for calculating the steady-state efficiency of a heating system or device. Unfortunately, the seasonal efficiency of a combustion heating system can be significantly different from the steady-state efficiency. The primary factor affecting the seasonal efficiency is flue loss during the burner-off period. The warm stack that exists at the end of the firing period can cause airflow in the stack while the burner is off. This airflow can remove heat from furnace and heat exchanger components, from the structure itself, and from pilot flames. Also, if combustion air is drawn from the heated space within the structure, the heated air lost must be at least partly replaced with cold infiltrated air. For further discussion of seasonal efficiency, see Chapter 28 of the 1996 *ASHRAE Handbook—Systems and Equipment.*

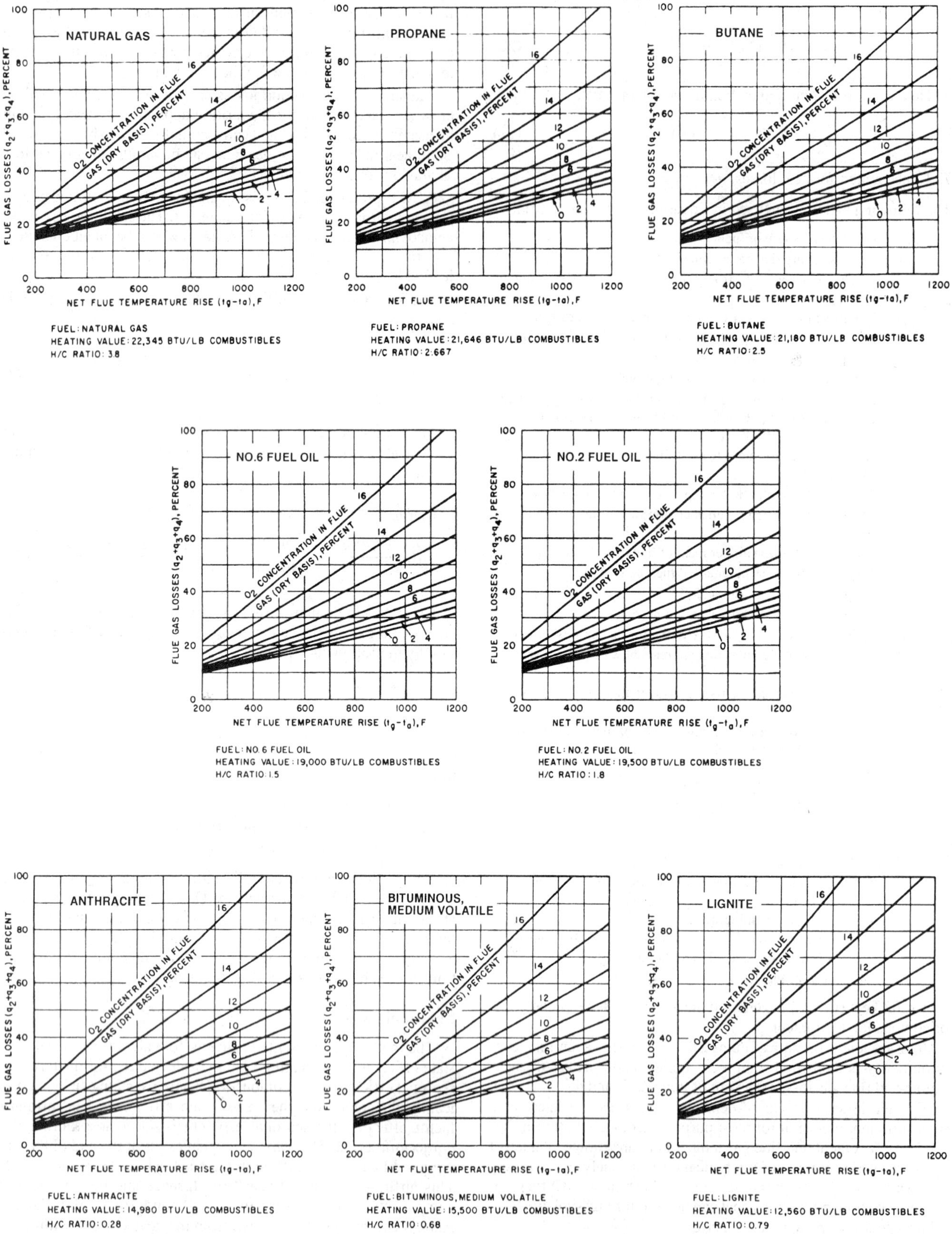
NATURAL GAS
PROPANE
BUTANE
NO.6 FUEL OIL
NO.2 FUEL OIL
ANTHRACITE
BITUMINOUS, MEDIUM VOLATILE
LIGNITE
O2 CONCENTRATION IN FLUE GAS (DRY BASIS), PERCENT
FLUE GAS LOSSES (q2+q3+q4), PERCENT
NET FLUE TEMPERATURE RISE (tg-ta), F
FUEL: NATURAL GAS
HEATING VALUE: 22,345 BTU/LB COMBUSTIBLES
H/C RATIO: 3.8
FUEL: PROPANE
HEATING VALUE: 21,646 BTU/LB COMBUSTIBLES
H/C RATIO: 2.667
FUEL: BUTANE
HEATING VALUE: 21,180 BTU/LB COMBUSTIBLES
H/C RATIO: 2.5
FUEL: NO. 6 FUEL OIL
HEATING VALUE: 19,000 BTU/LB COMBUSTIBLES
H/C RATIO: 1.5
FUEL: NO. 2 FUEL OIL
HEATING VALUE: 19,500 BTU/LB COMBUSTIBLES
H/C RATIO: 1.8
FUEL: ANTHRACITE
HEATING VALUE: 14,980 BTU/LB COMBUSTIBLES
H/C RATIO: 0.28
FUEL: BITUMINOUS, MEDIUM VOLATILE
HEATING VALUE: 15,500 BTU/LB COMBUSTIBLES
H/C RATIO: 0.68
FUEL: LIGNITE
HEATING VALUE: 12,560 BTU/LB COMBUSTIBLES
H/C RATIO: 0.79

Fig. 5 Flue Gas Losses with Various Fuels

(Flue gas temperature rise shown. Loss is based on 65°F room temperature.)

COMBUSTION CONSIDERATIONS

Air Pollution

Combustion processes constitute the largest single source of air pollution. Pollutants can be grouped into four categories:

1. Products of incomplete fuel combustion
   - Combustible aerosols (solid and liquid), including smoke, soot, and organics, but excluding ash
   - Carbon monoxide, CO
   - Gaseous hydrocarbons
2. Oxides of nitrogen (generally grouped and referred to as NO_x)
   - Nitric oxide, NO
   - Nitrogen dioxide, NO_2
3. Emissions resulting from fuel contaminants
   - Sulfur oxides, primarily sulfur dioxide, SO_2, and small quantities of sulfur trioxide, SO_3
   - Ash
   - Trace metals
4. Emissions resulting from additives
   - Combustion-controlling additives
   - Other additives

Table 13 lists NO_x emission factors for uncontrolled fuel-burning equipment (i.e., equipment that does not have exhaust gas recirculation, low-NO_x burners, or other emission controls). Differences in the NO_x emissions of fuels are caused by the flame temperature and different levels of fuel nitrogen. The data in Table 13 are adapted from EPA (1993), Compilation of Air Pollutant Emission Factors, which lists emission factors of a wide variety of equipment, as well as emission reduction options. Carbon monoxide emissions are less dependent on fuel type and typically range from 0.03 to 0.04 lb/10^6 Btu of heat input. For gas-fired commercial and industrial boilers, particulate emissions range from 0.005 to 0.006 lb/10^6 Btu. For distillate-oil-fired commercial and industrial boilers, particulates are typically 0.014 lb/10^6 Btu. For residential oil-fired equipment, particulate emission factors are 0.003 lb/10^6 Btu. For residual-oil-fired equipment, particulate emissions depend on the sulfur content. For a sulfur content of 1%, the particulate emission rate is typically 0.083 lb/10^6 Btu.

Emission levels of nitrogen oxides and products of incomplete combustion are directly related to the combustion process and can be controlled, to some extent, by process modification. Emissions due to fuel contaminants are related to fuel selection and are slightly affected by the combustion process. Emissions due to additives must be considered in the overall evaluation of the merits of using additives.

Emission levels of products of incomplete fuel combustion can be reduced by reducing burner cycling, ensuring adequate excess air, improving the mixing of air and fuel (by increasing turbulence, improving distribution, and improving liquid fuel atomization), increasing residence time in the hot combustion zone (possibly by decreasing the firing rate), increasing combustion zone temperatures (to speed reactions), and avoiding quenching the flame before reactions are completed.

Nitrogen oxides are produced during the combustion process, either (1) by thermal fixation (reaction of nitrogen and oxygen at high combustion temperatures), or (2) from fuel nitrogen (oxidation of organic nitrogen in fuel molecules). Unfortunately, high excess air and high flame temperature techniques, which ensure complete fuel combustion, tend to promote NO_x formation.

The relative contribution of each of these mechanisms to the total NO_x emissions depends on the amount of organic nitrogen in the fuel. Natural gas contains very little nitrogen. Virtually all NO_x emissions with gas firing are due to the thermal mechanism. The nitrogen content of distillate oil varies, but an average of 20 ppm of fuel NO_x is produced (about 20-30% of the total NO_x). The fuel nitrogen in residual oil can be significantly higher, with fuel NO_x contributing heavily to the total emissions.

Table 13 NO_x Emission Factors for Combustion Sources Without Emission Controls

Source	NO_x Emission Factor, lb/10^6 Btu of Heat Input
Gas-Fired Equipment	
Small industrial boilers	0.14
Commercial boilers	0.10
Residential furnaces	0.09
Distillate-oil-fired small industrial boilers, commercial boilers, and residential furnaces	0.14
Residual oil-fired small industrial boilers and commercial boilers	0.37

Thermal fixation is strongly dependent on flame maximum temperature. For example, increasing the flame temperature from 2600 to 2800°F increases thermal NO_x tenfold. Therefore, methods to control thermal NO_x are based on methods to reduce the maximum flame temperature. Flue gas recirculation is perhaps the most effective method of reducing thermal NO_x in commercial and industrial boilers. In gas-fired boilers, NO_x reductions of 70% can be realized with 15-20% recirculation of flue gas into the flame. The NO_x reduction decreases with increasing fuel nitrogen content. With distillate-oil firing, reductions of 60-70% can be achieved. In residual-oil-fired boilers, flue gas recirculation can reduce NO_x emissions by 15 to 30%. The maximum rate of flue gas recirculation is limited by combustion instability and CO production.

Two-stage or biased firing is the only technique that reduces NO_x produced both by thermal fixation and fuel nitrogen in industrial and utility applications. The fuel-rich or air-deficient primary combustion zone retards NO_x formation early in the combustion process (when NO_x forms most readily from fuel nitrogen), and avoids peak temperatures, reducing thermal NO_x. Retrofit low-NO_x burners that control air distribution and fuel air mixing in the flame zone can be used to achieve staged combustion. With oil firing, NO_x reductions of 20 to 50% can be obtained with low-NO_x burners. The application of flue gas recirculation and other control methods to residential, oil-fired warm air furnaces was reviewed by Butcher et al. (1994).

Two-stage firing is not practical in smaller gas appliances because of increased complexity and dangerous interstage CO levels. The following are some methods of reducing NO_x emissions from gas-fired appliances (Murphy and Putnam 1985):

- Burner adjustment
- Flame inserts (radiation screens or rods)
- Staged combustion and delayed mixing
- Secondary air baffling
- Catalytic and radiant burners

Radiation screens or rods (flame inserts) surrounding or inserted into the flame absorb radiation to reduce flame temperature and retard NO_x formation. Proprietary appliance burners with no flame inserts have been developed and produced to comply with the very strict NO_x emission limitations of California's Air Quality Management Districts.

The U.S. EPA sets limits on air pollutant emissions (Source Performance Standards) from boilers larger than 10 million Btu/h of heat input. In addition, states set emission regulations that are at least as strict at the federal limits and may apply to smaller equipment.

The EPA's automobile emission standard is 1.0 g of NO_2 per mile, which is equivalent to 750 ng/J of NO_x emission. California's maximum is 0.4 g/mile, equivalent to 300 ng/J. California's Air Quality Management Districts for the South Coast (Los Angeles) and the San Francisco Bay Area limit NO_x emission to 40 ng/J of useful heat for some natural gas-fired residential heating appliances.

For further discussion of air pollution aspects of fuel combustion, see EPA (1971a and 1971b).

Condensation and Corrosion

Fuel-burning systems that cycle on and off to meet demand cool down during the off-cycle. When the appliance starts again, condensate forms briefly on surfaces until they are heated above the dew-point temperature. Low-temperature corrosion occurs in system components (heat exchangers, flues, vents, chimneys) when their surfaces remain below the dew-point temperature of flue gas constituents (water vapor, sulfides, chlorides, fluorides, etc.) long enough to cause condensation. Corrosion increases as condensate dwell time increases.

Acids in the flue gas condensate are the principle substances responsible for low-temperature corrosion in fuel-fired systems. Sulfuric, hydrochloric, and other acids are formed when acidic compounds in fuel and air combustion products combine with condensed moisture in appliance heat exchangers, flues, or vents. Corrosion can be avoided by maintaining these surfaces above the flue gas dew point.

In high-efficiency, condensing-type appliances and economizers, flue gas temperatures are intentionally reduced below the flue gas dew-point temperatures to achieve efficiencies approaching 100%. In these systems, the surfaces subjected to condensate must be made of corrosion-resistant materials. The most corrosive conditions exist at the leading edge of the condensing region, especially those areas that experience evaporation during each cycle (Strickland et al. 1987). Drainage of condensate retards the concentration of acids on system surfaces. Regions from which condensate partially or completely drains away before evaporation are less severely attacked than regions from which condensate does not drain before evaporation.

The metals most resistant to condensate corrosion are stainless-steel alloys with high chromium and molybdenum content, and nickel-chromium alloys with high molybdenum content (Stickford et al. 1988). Aluminum experiences general corrosion rather than pitting when exposed to flue gas condensate. If applied in sufficiently thick cross section to allow for metal loss, aluminum can be used in condensing regions. Most ceramic and high-temperature polymer materials resist the corrosive effects of flue gas condensate. These materials may have application in the condensing regions, if they can meet the structural and temperature requirements of a particular application.

In coal-fired power plants, the rate of corrosion for carbon steel condensing surfaces by the mixed acids (primarily sulfuric and hydrochloric) is reported to be maximum at about 122 ± 18°F (Davis 1987). Mitigation techniques include (1) acid neutralization with a base such as NH_3 or $Ca(OH)_2$; (2) use of protective linings of glass-filled polyester or coal-tar epoxy; and (3) replacement of steel with molybdenum-bearing stainless steels, nickel alloys, polymers, or other corrosion-resistant materials. Other elements in residual fuel oils and coals that contribute to high-temperature corrosion include sodium, potassium, and vanadium. Each fuel-burning system component should be evaluated during installation, or when modified, to determine the potential for corrosion and the means to retard corrosion (Paul et al. 1988).

Soot

Soot deposits on flue surfaces of a boiler or heater act as an insulating layer over the surface, reducing heat transfer to the water or air. Soot can also clog flues, reduce draft and available air, and prevent proper combustion. Proper burner adjustment can minimize soot accumulation. The use of off-specification fuel can contribute to the generation of soot.

REFERENCES

ASTM. 1995. Standard specification for fuel oils. ANSI/ASTM *Standard* D 396-95. American Society for Testing and Materials, West Conshohocken, PA.

ASTM. 1996. Standard specification for diesel fuel oils. ANSI/ASTM *Standard* D 975-96.

ASTM. 1996. Standard specification for gas turbine fuel oils. ANSI/ASTM *Standard* D 2880-96.

ASTM. 1995. Standard classification of coals by rank. ASTM *Standard* D 388-95.

Butcher, T.A., L. Fisher, B. Kamath, T. Kirchstetter, and J. Batey. 1994. Nitrogen oxides (NO_x) and oil burners. Proceedings of the 1994 Oil Heat Technology Conference and Workshops. BNL *Report* No. 52430. Brookhaven National Laboratory, Upton, NY.

Coward, H.F. and G.W. Jones. 1952. Limits of flammability of gases and vapors. *Bulletin* 503. U.S. Bureau of Mines, Washington, D.C.

Davis, J.R., ed. 1987. *Metals handbook*, 9th ed., Vol. 13. ASM International, Metals Park, OH.

EPA. 1971a. Standards of performance for new stationary sources, Group I, Federal Register 36, August 17. U.S. Environmental Protection Agency, Washington, D.C.

EPA. 1971b. Standards of performance for new stationary sources, Group I, Part II, Federal Register 36, December 23. U.S. Environmental Protection Agency, Washington, D.C.

EPA. 1993. Compilation of air pollutant emission factors. *Report* AP-42. U.S. Environmental Protection Agency, Washington, D.C.

Gas engineers handbook. 1965. The Industrial Press, New York.

Hartman, I. 1958. "Dust explosions." In *Mechanical engineers' handbook*, 6th ed., Section 7, pp. 41-48. McGraw-Hill, New York.

Hazard, H.R. 1971. "Gas turbine fuels." In *Gas turbine handbook*, Gas Turbine Publications, Stamford, CT.

Murphy, M.J. and A.A. Putnam. 1985. Burner technology bulletin: Control of NO_x emissions from residential gas appliances. *Report* GRI-85/0132. Battelle Columbus Division for Gas Research Institute.

NFPA. 1962. Fire-hazard properties of flammable liquids, gases and volatile solids, Tables 6-126, pp. 6-131 ff. In *Fire protection handbook*, 12th ed., National Fire Protection Association, Quincy, MA.

NFPA/IAS. 1992. ANSI/NFPA *Standard* 54-1992. National Fuel Gas Code, Section 8.1.2. National Fire Protection Association, Quincy, MA. ANSI/IAS *Standard* Z223.1-1992. American Gas Association, Arlington, VA.

North American combustion handbook. 1965. The North American Manufacturing Co., Cleveland, OH.

Paul, D.D., A.L. Rutz, S.G. Talbert, J.J. Crisafolli, G.R. Whitacre, and R.D. Fischer. 1988. User's manual for Vent-II Ver. 3.0—A dynamic microcomputer program for analyzing gas venting systems. *Report* GRI-88/0304. Battelle Columbus Division for Gas Research Institute.

Scott, G.S., et al. 1948. Determination of ignition temperatures of combustible liquids and gases. *Analytical Chemistry* 20:238-41.

Shelton, E.M. 1974. *Burner oil fuels*. Petroleum Products Survey 86. U.S. Bureau of Mines, Washington, D.C.

Shnidman, L. 1954. *Gaseous fuels*. American Gas Association, Arlington, VA.

Stickford, G.H., S.G. Talbert, B. Hindin, and D.W. Locklin. 1988. Research on corrosion-resistant materials for condensing heat exchangers. Proceedings of the 39th Annual International Appliance Technical Conference.

Stone, R.L. 1969. Fireplace operation depends upon good chimney design. *ASHRAE Journal* 11(February):63-69.

Trinks, W. 1947. Simplified calculation of radiation from non-luminous furnace gases. *Industrial Heating* 14:40-46.

U.S. Bureau of Mines. Semiannually. *Mineral Industry Surveys, Motor Gasolines*. Washington, D.C.

Zabetakis, M.G. 1956. Research on the combustion and explosion hazards of hydrogen-water vapor-air mixtures. Division of Explosives Technology, *Progress Report* 1. U.S. Bureau of Mines, Washington, D.C.

BIBLIOGRAPHY

Bonne, U. and A. Patani. 1982. Combustion system performance analysis and simulation study. *Report* GRI-81/0093 (PB 83-161 406). Honeywell SSPL, Bloomington, MN.

Gas Appliance Technology Center, Gas Research Institute, Manufacturer update on status of GATC research on heat-exchanger corrosion, May 1984. Battelle Columbus Laboratories and American Gas Association Laboratories.

Lewis, B. and G. von Elbe. 1987. *Combustion, flames, and explosion of gases*, 3rd ed. Academic Press, New York.

Stickford, G.H., S.G. Talbert, and D.W. Locklin. 1987. Condensate corrosivity in residential condensing appliances. Proceedings of the International Symposium on Condensing Heat Exchangers, Paper 3, BNL *Report* No. 52068, 1 and 2. Brookhaven National Laboratory, Upton, NY.

CHAPTER 18

REFRIGERANTS

REFRIGERANTS are the working fluids in refrigeration, air-conditioning, and heat pumping systems. They absorb heat from one area, such as an air-conditioned space, and reject it into another, such as outdoors, usually through evaporation and condensation, respectively. These phase changes occur both in absorption and mechanical vapor compression systems, but they do not occur in systems operating on a gas cycle using a fluid such as air. (See Chapter 1 for more information on refrigeration cycles.) The design of the refrigeration equipment depends strongly on the properties of the selected refrigerant. Table 1 lists ASHRAE standard refrigerant designations from ASHRAE *Standard* 34.

Refrigerant selection involves compromises between conflicting desirable thermodynamic properties. A refrigerant must satisfy many requirements, some of which do not directly relate to its ability to transfer heat. Chemical stability under conditions of use is the most important characteristic. Safety codes may require a nonflammable refrigerant of low toxicity for some applications. Cost, availability, efficiency, and compatibility with compressor lubricants and materials with which the equipment is constructed are other concerns.

The environmental consequences of a refrigerant that leaks from a system must also be considered. Because of their great stability, fully halogenated compounds, such as **chlorofluorocarbons** (CFCs), persist in the atmosphere for many years and eventually diffuse into the stratosphere. The molecules of CFCs, such as R-11 and R-12, contain only carbon and the halogens chlorine and fluorine. Once in the upper atmosphere, CFC molecules break down and release chlorine, which destroys ozone (**ozone depletion**). In the lower atmosphere, these molecules absorb infrared radiation, which may contribute to the warming of the earth. Substitution of a hydrogen atom for one or more of the halogens in a CFC molecule greatly reduces its atmospheric lifetime and lessens its environmental impact. These compounds are called **hydrochlorofluorocarbons** (HCFCs). A similar class of compounds used as fire extinguishing agents and called halons also cause ozone depletion. **Halons** are compounds containing bromine, fluorine, and carbon. Like CFCs, halons break down, but release bromine, which is even more destructive to stratospheric ozone than chlorine.

Latent heat of vaporization is another important property. On a molar basis, fluids with similar boiling points have almost the same latent heat. Since the compressor operates on volumes of gas, refrigerants with similar boiling points produce similar capacities in a given compressor. On a mass basis, latent heat varies widely among fluids. The maximum efficiency of a theoretical vapor compression cycle is achieved by fluids with low vapor heat capacity. This property is associated with fluids having a simple molecular structure and low molecular weight.

Transport properties of thermal conductivity and viscosity affect the performance of heat exchangers and piping. High thermal conductivity and low viscosity are desirable.

The preparation of this chapter is assigned to TC 3.1, Refrigerants and Brines.

No single fluid satisfies all the attributes desired of a refrigerant; as a result, a variety of refrigerants is used. This chapter describes the basic characteristics of various refrigerants, and Chapter 19 lists thermophysical properties.

PHASEOUT OF REFRIGERANTS

The Montreal Protocol is an international treaty that controls the production of ozone-depleting substances, including refrigerants containing chlorine and/or bromine (U.N. 1994, 1996). The original Protocol was signed September 16, 1987, by the European Economic Community (currently the European Union) and 24 nations, including the United States. It entered into force on January 1, 1989, and limits the 1998 production of specified CFCs to 50% of their 1986 levels. Starting in 1992, the production of specified halons (including R-13B1) was frozen at 1986 levels. Developing countries were granted additional time to meet these deadlines.

The original Protocol contained provisions for periodic revision. Two such revisions, referred to as the London and Copenhagen Amendments, were agreed to in 1990 and 1992, respectively. As of September 1996, the Montreal Protocol had been ratified by 157 parties, the London Amendment by 110 parties, and the Copenhagen Amendment by 58 parties.

The Copenhagen Amendment entered into force on June 14, 1994. It called for a complete cessation of the production of CFCs by January 1, 1996, and of halons by January 1, 1994. Continued use from existing (reclaimed or recycled) stock is permitted. Allowance is also provided for continued production for very limited "essential uses." In addition, HCFCs (including R-22) are to be phased out relative to a 1989 reference level for developed countries. Production was frozen at the reference level on January 1, 1996. Production will be limited to 65% of the reference level by January 1, 2004; to 35% by January 1, 2010; to 10% by January 1, 2015; and to 0.5% of the reference level by January 1, 2020. Complete cessation of the production of HCFCs is called for by January 1, 2030. In addition to the international agreement, individual countries may have domestic regulations for ozone-depleting compounds.

The production and use of **hydrofluorocarbon** (HFC) refrigerants (such as R-32, R-125, R-134a, and R-143a and their mixtures, including R-404, R-407, and R-410) are not regulated by the Montreal Protocol, but may be regulated by individual countries.

REFRIGERANT PROPERTIES

Physical Properties

Table 2 lists some physical properties of commonly used refrigerants, a few very low-boiling cryogenic fluids, some newer refrigerants, and some older refrigerants of historical interest. These refrigerants are arranged in increasing order of atmospheric boiling point, from helium at −452.1°F to water at 212°F.

Table 2 also includes the freezing point, critical properties, and refractive index. Of these properties, the boiling point is most important because it is a direct indicator of the temperature level at

Table 1 Standard Designation of Refrigerants (ASHRAE *Standard* 34)

Refrigerant Number	Chemical Name or Composition (% by mass)	Chemical Formula
Methane Series		
10	tetrachloromethane (carbon tetrachloride)	CCl_4
11	trichlorofluoromethane	CCl_3F
12	dichlorodifluoromethane	CCl_2F_2
12B1	bromochlorodifluoromethane	$CBrClF_2$
12B2	dibromodifluoromethane	CBr_2F_2
13	chlorotrifluoromethane	$CClF_3$
13B1	bromotrifluoromethane	$CBrF_3$
14	tetrafluoromethane (carbon tetrafluoride)	CF_4
20	trichloromethane (chloroform)	$CHCl_3$
21	dichlorofluoromethane	$CHCl_2F$
22	chlorodifluoromethane	$CHClF_2$
22B1	bromodifluoromethane	$CHBrF_2$
23	trifluoromethane	CHF_3
30	dichloromethane (methylene chloride)	CH_2Cl_2
31	chlorofluoromethane	CH_2ClF
32	difluoromethane (methylene fluoride)	CH_2F_2
40	chloromethane (methyl chloride)	CH_3Cl
41	fluoromethane (methyl fluoride)	CH_3F
50	methane	CH_4
Ethane Series		
110	hexachloroethane	CCl_3CCl_3
111	pentachlorofluoroethane	CCl_3CCl_2F
112	1,1,2,2-tetrachloro-1,2-difluoroethane	CCl_2FCCl_2F
112a	1,1,1,2-tetrachloro-2,2-difluoroethane	CCl_3CClF_2
113	1,1,2-trichloro-1,2,2-trifluoroethane	CCl_2FCClF_2
113a	1,1,1-trichloro-2,2,2-trifluoroethane	CCl_3CF_3
114	1,2-dichloro-1,1,2,2-tetrafluoroethane	$CClF_2CClF_2$
114a	1,1-dichloro-1,2,2,2-tetrafluoroethane	CCl_2FCF_3
114B2	1,2-dibromo-1,1,2,2-tetrafluoroethane	$CBrF_2CBrF_2$
115	chloropentafluoroethane	$CClF_2CF_3$
116	hexafluoroethane	CF_3CF_3
120	pentachloroethane	$CHCl_2CCl_3$
123	2,2-dichloro-1,1,1-trifluoroethane	$CHCl_2CF_3$
123a	1,2-dichloro-1,1,2-trifluoroethane	$CHClFCClF_2$
124	2-chloro-1,1,1,2-tetrafluoroethane	$CHClFCF_3$
124a	1-chloro-1,1,2,2-tetrafluoroethane	CHF_2CClF_2
125	pentafluoroethane	CHF_2CF_3
133a	2-chloro-1,1,1-trifluoroethane	CH_2ClCF_3
134a	1,1,1,2-tetrafluoroethane	CH_2FCF_3
140a	1,1,1-trichloroethane (methyl chloroform)	CH_3CCl_3
141b	1,1-dichloro-1-fluoroethane	CCl_2FCH_3
142b	1-chloro-1,1-difluoroethane	$CClF_2CH_3$
143a	1,1,1-trifluoroethane	CF_3CH_3
150a	1,1-dichloroethane	$CHCl_2CH_3$
152a	1,1-difluoroethane	CHF_2CH_3
160	chloroethane (ethyl chloride)	CH_3CH_2Cl
170	ethane	CH_3CH_3
Propane Series		
216ca	1,3-dichloro-1,1,2,2,3,3-hexafluoropropane	$CClF_2CF_2CClF_2$
218	octafluoropropane	$CF_3CF_2CF_3$
245cb	1,1,1,2,2-pentafluoropropane	$CF_3CF_2CH_3$
290	propane	$CH_3CH_2CH_3$
Cyclic Organic Compounds		
C316	1,2-dichloro-1,2,3,3,4,4-hexafluorocyclobutane	$C_4Cl_2F_6$
C317	chloroheptafluorocyclobutane	C_4ClF_7
C318	octafluorocyclobutane	C_4F_8
Zeotropic Blends (% by mass)		
400	R-12/114 (must be specified)	
401A	R-22/152a/124 (53/13/34)	
401B	R-22/152a/124 (61/11/28)	
401C	R-22/152a/124 (33/15/52)	
402A	R-125/290/22 (60/2/38)	
402B	R-125/290/22 (38/2/60)	
Zeotropes (*Continued*)		
403A	R-290/22/218 (5/75/20)	
403B	R-290/22/218 (5/56/39)	
404A	R-125/143a/134a (44/52/4)	
405A	R-22/152a/142b/C318 (45/7/5.5/42.5)	
406A	R-22/600a/142b (55/4/41)	
407A	R-32/125/134a (20/40/40)	
407B	R-32/125/134a (10/70/20)	
407C	R-32/125/134a (23/25/52)	
407D	R-32/125/134a (15/15/70)	
408A	R-125/143a/22 (7/46/47)	
409A	R-22/124/142b (60/25/15)	
409B	R-22/124/142b (65/25/10)	
410A	R-32/125 (50/50)	
410B	R-32/125 (45/55)	
411A	R-1270/22/152a (1.5/87.5/11.0)	
411B	R-1270/22/152a (3/94/3)	
412A	R-22/218/142b (70/5/25)	
Azeotropic Blends (% by mass)		
500	R-12/152a (73.8/26.2)	
501	R-22/12 (75.0/25.0)*	
502	R-22/115 (48.8/51.2)	
503	R-23/13 (40.1/59.9)	
504	R-32/115 (48.2/51.8)	
505	R-12/31 (78.0/22.0)*	
506	R-31/114 (55.1/44.9)	
507A	R-125/143a (50/50)	
508A	R-23/116 (39/61)	
508B	R-23/116 (46/54)	
509A	R-22/218 (44/56)	
Miscellaneous Organic Compounds		
Hydrocarbons		
600	butane	$CH_3CH_2CH_2CH_3$
600a	2-methyl propane (isobutane)	$CH(CH_3)_3$
Oxygen Compounds		
610	ethyl ether	$C_2H_5OC_2H_5$
611	methyl formate	$HCOOCH_3$
Sulfur Compounds		
620	(Reserved for future assignment)	
Nitrogen Compounds		
630	methyl amine	CH_3NH_2
631	ethyl amine	$C_2H_5NH_2$
Inorganic Compounds		
702	hydrogen	H_2
704	helium	He
717	ammonia	NH_3
718	water	H_2O
720	neon	Ne
728	nitrogen	N_2
732	oxygen	O_2
740	argon	Ar
744	carbon dioxide	CO_2
744A	nitrous oxide	N_2O
764	sulfur dioxide	SO_2
Unsaturated Organic Compounds		
1112a	1,1-dichloro-2,2-difluoroethene	$CCl_2{=}CF_2$
1113	1-chloro-1,2,2-trifluoroethene	$CClF{=}CF_2$
1114	tetrafluoroethene	$CF_2{=}CF_2$
1120	trichloroethene	$CHCl{=}CCl_2$
1130	1,2-dichloroethene (trans)	$CHCl{=}CHCl$
1132a	1,1 difluoroethene (vinylidene fluoride)	$CF_2{=}CH_2$
1140	1-chloroethene (vinyl chloride)	$CHCl{=}CH_2$
1141	1-fluoroethene (vinyl fluoride)	$CHF{=}CH_2$
1150	ethene (ethylene)	$CH_2{=}CH_2$
1270	propene (propylene)	$CH_3CH{=}CH_2$

*The exact composition of this azeotrope is in question.

Table 2 Physical Properties of Selected Refrigerants[a]

Refrigerant No.	Chemical Name or Composition (% by mass)	Chemical Formula	Molecular Mass	Boiling Pt. (NBP) at 14.696 psia, °F	Freezing Point, °F	Critical Temperature, °F	Critical Pressure, psia	Critical Volume, ft³/lb	Refractive Index of Liquid[b,c]
704	Helium	He	4.0026	−452.1	None	−450.3	33.21	0.2311	1.021 (NBP) 5461 Å
702p	Hydrogen, para	H_2	2.0159	−423.2	−434.8	−400.3	187.5	0.5097	1.09 (NBP)[f]
702n	Hydrogen, normal	H_2	2.0159	−423.0	−434.5	−399.9	190.8	0.5320	1.097 (NBP) 5791 Å
720	Neon	Ne	20.183	−410.9	−415.5	−379.7	493.1	0.03316	—
728	Nitrogen	N_2	28.013	−320.4	−346.0	−232.4	492.9	0.05092	1.205 (83 K) 5893 Å
729	Air	—	28.97	−317.8	—	−220.95	548.9	0.0530	
						−221.1	546.3	0.05007	
740	Argon	Ar	39.948	−302.55	−308.7	−188.48	704.9	0.0301	1.233 (84 K) 5893 Å
732	Oxygen	O_2	31.9988	−297.332	−361.8	−181.424	731.4	0.03673	1.221 (92 K) 5893 Å
50	Methane	CH_4	16.04	−258.7	−296	−116.5	673.1	0.099	
14	Tetrafluoromethane	CF_4	88.01	−198.3	−299	−50.2	543	0.0256	
1150	Ethylene	C_2H_4	28.05	−154.7	−272	48.8	742.2	0.070	1.363(−148)[1]
744A[2]	Nitrous oxide	N_2O	44.02	−129.1	−152	97.7	1048	0.0355	
170	Ethane	C_2H_6	30.07	−127.85	−297	90.0	709.8	0.0830	
503	R-23/13 (40.1/59.9)	—	87.5	−127.6	—	67.1	607	0.0326	
23	Trifluoromethane	CHF_3	70.02	−115.7	−247	78.1	701.4	0.0311	
13	Chlorotrifluoromethane	$CClF_3$	104.47	−114.6	−294	83.9	561	0.0277	1.146 (77)[4]
744	Carbon dioxide	CO_2	44.01	−109.2[d]	−69.9[e]	87.9	1070.0	0.0342	1.195 (59)
13B1	Bromotrifluoromethane	$CBrF_3$	148.93	−71.95	−270	152.6	575	0.0215	1.239 (77)[4]
504	R-32/115 (48.2/51.8)	—	79.2	−71.0	—	151.5	690.5	0.0324	
32	Difluoromethane	CH_2F_2	52.02	−61.1	−213	173.14	845.6	0.03726	
125	Pentafluoroethane	C_2HF_5	120.03	−55.43	−153.67	151.34	526.57	—	—
1270	Propylene	C_3H_6	42.09	−53.86	−301	197.2	670.3	0.0720	1.3640 (−58)[1]
502[5]	R-22/115 (48.8/51.2)	—	111.63	−49.8	—	179.9	591.0	0.0286	
290	Propane	C_3H_8	44.10	−43.76	−305.8	206.1	616.1	0.0726	1.3397 (−43)
22	Chlorodifluoromethane	$CHClF_2$	86.48	−41.36	−256	204.8	721.9	0.0305	1.234 (77)[4]
115	Chloropentafluoroethane	$CClF_2CF_3$	154.48	−38.4	−159	175.9	457.6	0.0261	1.221 (77)[4]
500	R-12/152a (73.8/26.2)	—	99.31	−28.3	−254	221.9	641.9	0.0323	
717	Ammonia	NH_3	17.03	−28.0	−107.9	271.4	1657	0.068[d]	1.325 (61.7)
12	Dichlorodifluoromethane	CCl_2F_2	120.93	−21.62	−252	233.6	596.9	0.0287	1.288 (77)[4]
134a	Tetrafluoroethane	CF_3CH_2F	102.03	−15.08	−141.9	214.0	589.8	0.029	—
152a	Difluoroethane	CHF_2CH_3	66.05	−13.0	−178.6	236.3	652	0.0439	
40[2]	Methyl chloride	CH_3Cl	50.49	−11.6	−144	289.6	968.7	0.0454	
124	Chlorotetrafluoroethane	$CHClFCF_3$	136.47	8.26	−326.47	252.5	530.84	—	—
600a	Isobutane	C_4H_{10}	58.13	10.89	−255.5	275.0	529.1	0.0725	1.3514 (−13)[1]
764[6]	Sulfur dioxide	SO_2	64.07	14.0	−103.9	315.5	1143	0.0306	
142b	Chlorodifluoroethane	$CClF_2CH_3$	100.5	14.4	−204	278.8	598	0.0368	
630[6]	Methyl amine	CH_3NH_2	31.06	19.9	−134.5	314.4	1082		1.432 (63.5)
C318	Octafluorocyclobutane	C_4F_8	200.04	21.5	−42.5	239.6	403.6	0.0258	
600	Butane	C_4H_{10}	58.13	31.1	−217.3	305.6	550.7	0.0702	1.3562 (5)[1]
114	Dichlorotetrafluoroethane	$CClF_2CClF_2$	170.94	38.8	−137	294.3	473	0.0275	1.294 (77)
21[7]	Dichlorofluoromethane	$CHCl_2F$	102.92	47.8	−211	353.3	750	0.0307	1.332 (77)[4]
160[2]	Ethyl chloride	C_2H_5Cl	64.52	54.32	−216.9	369.0	764.4	0.0485	
631[6]	Ethyl amine	$C_2H_5NH_2$	45.08	61.88	−113	361.4	815.6		
11	Trichlorofluoromethane	CCl_3F	137.38	74.87	−168	388.4	639.5	0.0289	1.362 (77)[4]
123	Dichlorotrifluoroethane	$CHCl_2CF_3$	152.93	82.17	−160.87	362.82	532.87	—	—
611[6]	Methyl formate	$C_2H_4O_2$	60.05	89.2	−146	417.2	870	0.0459	
141b	Dichlorofluoroethane	CCl_2FCH_3	116.95	89.6	—	399.6	616.4	—	
610[6]	Ethyl ether	$C_4H_{10}O$	74.12	94.3	−177.3	381.2	523	0.0607	1.3526 (68)
216ca	Dichlorohexafluoropropane	$C_3Cl_2F_6$	220.93	96.24	−193.7	356.0	399.5	0.0279	
30[6]	Methylene chloride	CH_2Cl_2	84.93	104.4	−142	458.6	882		1.4244 (68)[3]
113	Trichlorotrifluoroethane	CCl_2FCClF_2	187.39	117.63	−31	417.4	498.9	0.0278	1.357 (77)[4]
1130[8]	Dichloroethylene	$CHCl{=}CHCl$	96.95	118	−58	470	795		
1120[6]	Trichloroethylene	$CHCl{=}CCl_2$	131.39	189.0	−99	520	728		1.4782(68)[3]
718[6]	Water	H_2O	18.02	212	32	705.18	3200	0.0498	

Notes:

[a]Data from ASHRAE *Thermodynamic Properties of Refrigerants* (Stewart et al. 1986) or from McLinden (1990), unless otherwise noted.

[b]Temperature of measurement (°F, unless kelvin is noted) shown in parentheses. Data from CRC *Handbook of Chemistry and Physics* (CRC 1987), unless otherwise noted.

[c]For the sodium D line.

[d]Sublimes.

[e]At 76.4 psia.

[f]Dielectric constant data.

References:

[1]Kirk and Othmer (1956).

[2]*Matheson Gas Data Book* (1966).

[3]Electrochemicals Department, E.I. duPont de Nemours & Co.

[4]*Bulletin* B-32A (duPont).

[5]*Bulletin* T-502 (duPont 1980).

[6]*Handbook of Chemistry* (1967).

[7]*Bulletin* G-1 (duPont).

[8]CRC *Handbook of Chemistry and Physics* (CRC 1987).

which a refrigerant can be used. The freezing point must be lower than any contemplated usage. The critical properties describe a material at the point where the distinction between liquid and gas is lost. At higher temperatures, no separate liquid phase is possible. In refrigeration cycles involving condensation, a refrigerant must be chosen that allows this change of state to occur at a temperature somewhat below the critical. Cycles that reject heat at supercritical temperatures (such as cycles using carbon dioxide) are also possible.

Lithium Bromide-Water and Ammonia-Water Solutions. These are the most commonly used working fluids in absorption refrigeration systems. Figure 1 shows specific gravity, Figure 2 shows specific heat, and Figure 3 shows viscosity of lithium bromide-water solutions. Chapter 19 has an enthalpy-concentration diagram and a vapor pressure diagram for lithium bromide-water solutions. Chapter 19 also has equilibrium properties of water-ammonia solutions.

Electrical Properties

Tables 3 and 4 list the electrical characteristics of refrigerants that are especially important in hermetic systems.

Sound Velocity

Table 5 gives examples of the velocity of sound in the vapor phase of various fluorinated refrigerants. Chapter 19 has sound velocity data for many refrigerants. The velocity increases when the temperature is increased and decreases when the pressure is increased. The velocity of sound can be calculated from the equation

$$V_a = \left(g_c \frac{dp}{d\rho}\right)_S^{0.5} = \left[\gamma g_c \left(\frac{dp}{d\rho}\right)_T\right]^{0.5} \qquad (1)$$

where

V_a = sound velocity, ft/s
g_c = gravitational constant = 32.1740 $lb_m \cdot ft/lb_f \cdot s^2$
p = absolute pressure, lb_f/ft^2
ρ = density, lb_m/ft^2
γ = c_p/c_v = ratio of specific heats
S = entropy, Btu/lb · °R
T = temperature, °R

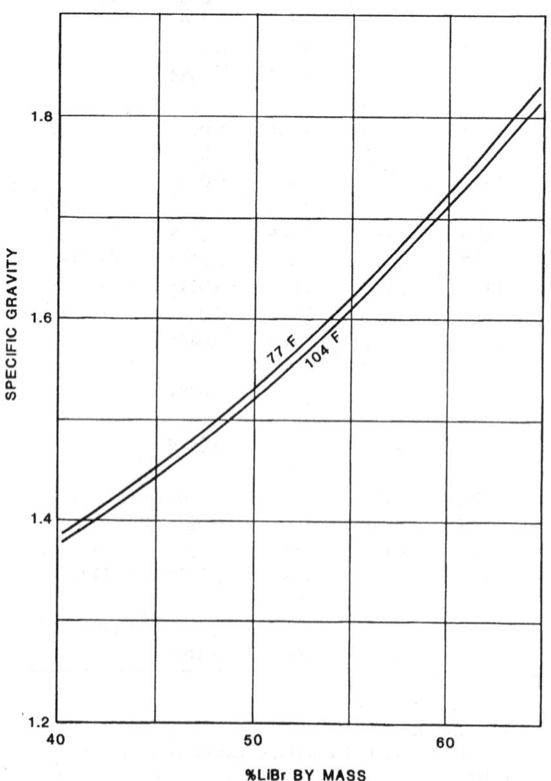

Fig. 1 Specific Gravity of Aqueous Solutions of Lithium Bromide

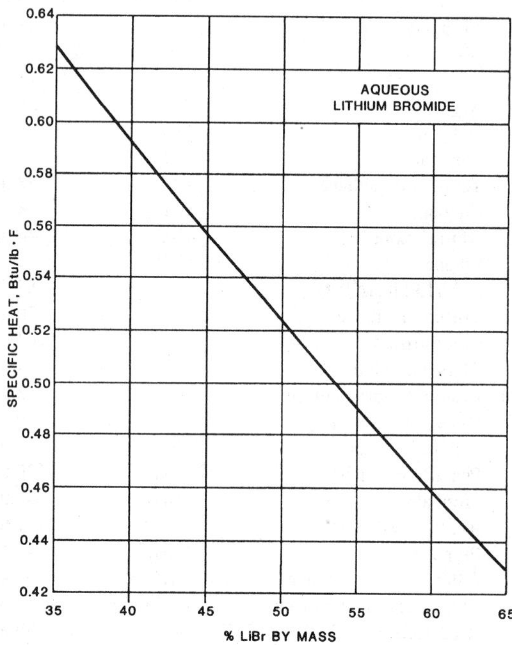

Fig. 2 Specific Heat of Aqueous Lithium Bromide Solutions

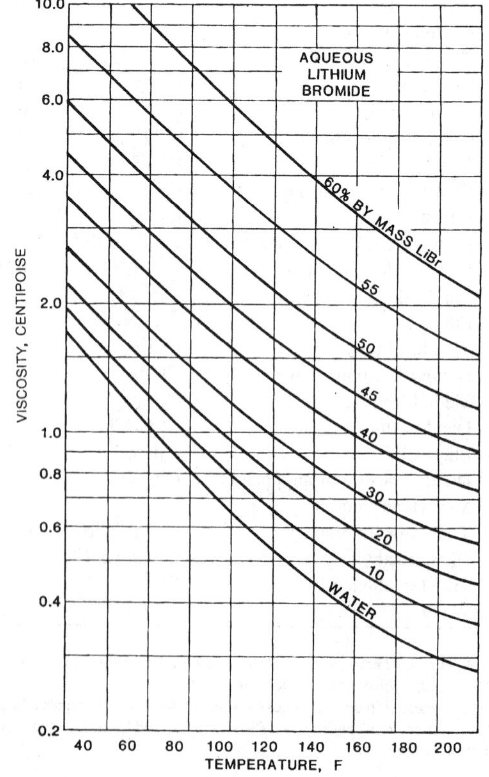

Fig. 3 Viscosity of Aqueous Solutions of Lithium Bromide

Table 3 Electrical Properties of Liquid Refrigerants

Refrigerant No.	Chemical Name or Composition (% by mass)	Temp., °F	Dielectric Constant	Volume Resistivity, MΩ·m	Ref.
11	Trichlorofluoromethane	84	2.28		1
		a	1.92	63680	2
		77	2.5	90	3
12	Dichlorodifluoromethane	84	2.13		1
		a	1.74	53900	2
		77	2.1	> 120	3
		77	2.100		4
13	Chlorotrifluoromethane	−22	2.3	120	4
		68	1.64		
22	Chlorodifluoromethane	75	6.11		1
		a	6.12	0.83	2
		77	6.6	75	3
23	Trifluoromethane	−22	6.3		3
		68	5.51		4
113	Trichlorotrifluoroethane	86	2.44		1
		a	1.68	45490	2
		77	2.6	> 120	3
114	Dichlorotetrafluoroethane	88	2.17		1
		a	1.83	66470	2
		77	2.2	> 70	3
124a	Chlorotetrafluoroethane	77	4.0	50	3
290	Propane	a	1.27	73840	2
500	R-12/152a (73.8/26.2)	a	1.80	55750	2
717	Ammonia	69	15.5		5
744	Carbon dioxide	32	1.59		5

[a]Ambient temperature.

References:
1 Data from E.I. duPont de Nemours & Co., Inc. Used by permission.
2 Beacham and Divers (1955).
3 Eiseman (1955).
4 Makita et al. (1976).
5 CRC *Handbook of Chemistry and Physics* (CRC 1987).

The sound velocity can be estimated from the tables of thermodynamic properties. The change in pressure with a change in density ($dp/d\rho$) can be estimated either at constant entropy or at constant temperature. It is simpler to estimate at constant temperature but then the ratio of specific heats must also be known. The practical velocity of a gas in piping or through openings is limited by the velocity of sound in the gas.

Latent Heat of Vaporization

An empirical rule of chemistry (Trouton's rule) states that the latent heat of vaporization at the boiling point on a molar basis, divided by the temperature in absolute units, is a constant for most materials. This rule is applied to refrigerants in Table 6. It applies fairly well to these refrigerants, although the result is not entirely constant. The rule helps in comparing different refrigerants and in understanding the operation of refrigeration systems.

REFRIGERANT PERFORMANCE

Chapter 1 describes several methods of calculating refrigerant performance, and Chapter 19 includes tables of thermodynamic properties of the various refrigerants.

Table 7 shows the theoretical calculated performance of a number of refrigerants for the U.S. standard cycle of 5°F evaporation and 86°F condensation. Calculated data for other conditions are given in Table 8. The tables can be used to compare the properties

Table 4 Electrical Properties of Refrigerant Vapors

Refrigerant No.	Chemical Name or Composition (% by mass)	Pressure, atm.	Temperature, °F	Dielectric Constant	Relative Dielectric Strength, Nitrogen = 1	Volume Resistivity, GΩ·m	Ref.
11	Trichlorofluoromethane	0.5	79	1.0019			3
		a	b	1.009		74.35	2
		1.0	73		3.1		4
12	Dichlorodifluoromethane	0.5	84	1.0016			3
		a	b	1.012	452[c]	72.77	2
		1.0	73		2.4		4
		4.9	68	1.019			5
13	Chlorotrifluoromethane	0.5	84	1.0013			3
		1.0	73		1.4		4
		4.9	68	1.013			5
		19.5	90	1.055			6
14	Tetrafluoromethane	0.5	76	1.0006			3
		1.0	73		1.0		4
22	Chlorodifluoromethane	0.5	78	1.0035			3
		a	b	1.004	460[c]	2113	2
		1.0	73		1.3		4
		4.9	68	1.033			5
23	Trifluoromethane	4.9	68	1.042			5
113	Trichlorotrifluoroethane	a	b	1.010	440[c]	94.18	2
		0.4	73		2.6		4
114	Dichlorotetrafluoroethane	0.5	80	1.0021			3
		a	b	1.002	295[c]	148.3	2
		1.0	73		2.8		4
116	Hexafluoroethane	0.94	73	1.002			3
133a	Chlorotrifluoroethane	0.94	80	1.010			3
142b	Chlorodifluoroethane	0.93	81	1.013			3
143a	Trifluoroethane	0.85	77	1.013			3
170	Ethane	1.0	32	1.0015			1
290	Propane	a	b	1.009	440[c]	105.3	2
500	R-12/152a (73.8/26.2)	a	b	1.024	470[c]	76.45	2
717	Ammonia	1.0	32	1.0072			1
		a	b		0.82		4
729	Air	1.0	32	1.00059			1
744	Carbon dioxide	1.0	32	1.00099			1
		1.0	b		0.88		4
1150	Ethylene	1.0	32	1.00144			1
		1.0	73		1.21		4

Notes:
[a]Saturation vapor pressure.
[b]Ambient temperature.
[c]Measured breakdown voltage, volts/mil.

References:
1 CRC *Handbook of Chemistry and Physics* (CRC 1987).
2 Beacham and Divers (1955).
3 Fuoss (1938).
4 Charlton and Cooper (1937).
5 Makita et al. (1976).
6 Hess et al. (1962).

of different refrigerants, but actual operating conditions are somewhat different from the calculated data. In most cases, the suction vapor is assumed to be saturated, and the compression is assumed adiabatic or at constant entropy. For R-113 and R-114, these assumptions would cause some liquid in the discharge vapor. In these cases, it is assumed that the discharge vapor is saturated and that the suction vapor is slightly superheated. In Section F of Table 8, the temperature of the suction gas is assumed to be 65°F (−10°F saturated evaporating plus 75°F superheat). Comparison with Section E illustrates the effect of suction gas superheating on refrigerant performance.

SAFETY

Table 9 summarizes the toxicity and flammability characteristics of many refrigerants. In ASHRAE *Standard* 34, refrigerants are classified according to the hazard involved in their use. The toxicity and flammability classifications yield six safety groups (A1, A2, A3, B1, B2, and B3) for refrigerants. Group A1 refrigerants are the least hazardous, Group B3 the most hazardous.

The safety classification in ASHRAE *Standard* 34 consists of a capital letter and a numeral. The capital letter designates the toxicity of the refrigerant at concentrations below 400 ppm by volume:

- Class A Toxicity not identified
- Class B Evidence of toxicity identified

The numeral denotes the flammability of the refrigerant:

- Class 1 No flame propagation in air at 65°F and 14.7 psia
- Class 2 Lower flammability limit (LFL) greater than 0.00625 lb/ft^3 at 70°F and 14.7 psia **and** heat of combustion less than 8174 Btu/lb
- Class 3 Highly flammable as defined by LFL less than or equal to 0.00625 lb/ft^3 at 70°F and 14.7 psia **or** heat of combustion greater than or equal to 8174 Btu/lb

Table 5 Velocity of Sound in Refrigerant Vapors[a]

Refrigerant	Pressure, psia	Temperature, °F: 50	100	150
		Velocity of Sound, ft/s		
11	10	b	469	490
12	10	480	503	525
22	10	583	610	635
113	10	b	435	456
114	10	391	411	430
502	10	501	525	547
123	10	b	435	456
124	10	443	465	486
125	10	477	500	521
134a	10	517	543	566
12	100	b	457	490
22	100	b	574	607
502	100	450	488	519
124	100	b	b	440
125	100	b	466	497
134a	100	b	490	528
12	200	b	b	442
22	200	b	523	572
502	200	b	435	483
125	200	b	420	467
134a	200	b	b	476

[a]Data in I-P units from E.I. duPont de Nemours & Co., Inc. Used by permission. Some data from NIST Standard Reference Database 23.
[b]Below saturation temperature.

LEAK DETECTION

Leak detection in refrigeration equipment is a major problem for manufacturers and service engineers. The following sections describe several leak detection methods.

Electronic Detection

The electronic detector is widely used in the manufacture and assembly of refrigeration equipment. Instrument operation depends on the variation in current flow caused by ionization of decomposed refrigerant between two oppositely charged platinum electrodes. This instrument can detect any of the halogenated refrigerants except R-14; however, *it is not recommended for use in atmospheres*

Table 6 Latent Heat of Vaporization Versus Boiling Point

Refrigerant No.	Chemical Name or Composition (% by mass)	Normal Boiling Pt., °F	Latent Heat λ at NBP, Btu/lb·mol	Trouton Constant, λ/°R[b]	Ref.
717	Ammonia	−28.0	10,036	23.256	1
630	Methyl amine[a]	23.0	11,141	23.086	4
764	Sulfur dioxide	13.6	10,705	22.626	2
631	Ethyl amine	68.0	11,645	22.076	4
611	Methyl formate[a]	100.0	12,094	21.616	4
134a	Tetrafluoroethane	−15.07	9,531	21.44	5
504	R-32/115 (48.2/51.8)	−71.0	8,282	21.316	1
23	Trifluoromethane	−115.7	7,325	21.29	1
124	Chlorotetrafluoroethane	8.26	9,742	20.82	5
C318	Octafluorocyclobutane	21.5	10,017	20.81	1
21	Dichlorofluoromethane	47.8	10,557	20.80	3
22	Chlorodifluoromethane	−41.4	8,687	20.76	1
40	Methyl chloride	−10.8	9,305	20.73	3
123	Dichlorotrifluoroethane	82.17	11,215	20.70	5
506	R-31/114 (55.1/44.9)	9.9	9,644	20.54	3
125	Pentafluoroethane	−55.43	8,295	20.52	5
113	Trichlorotrifluoroethane	117.6	11,828	20.49	1
152a	Difluoroethane	−13.0	9,045	20.25	1
502	R-22/115 (48.8/51.2)	−49.9	8,280	20.21	3
114	Dichlorotetrafluoroethane	38.8	10,005	20.07	1
216ca	Dichlorohexafluoropropane	96.2	11,154	20.07	1
505	R-12/31 (78.0/22.0)[c]	−21.8	8,735	19.95	3
11	Trichlorofluoromethane	74.9	10,648	19.92	1
500	R-12/152a (73.8/26.2)	−28.3	8,588	19.91	1
14	Tetrafluoromethane	−198.3	5,146	19.69	1
30	Methylene chloride[a]	120.0	11,398	19.66	4
600	Butane	31.1	9,641	19.64	1
13B1	Bromotrifluoromethane	−72.0	7,607	19.62	1
12	Dichlorodifluoromethane	−21.6	8,591	19.61	1
142b	Chlorodifluoroethane	14.4	9,297	19.61	1
115	Chloropentafluoroethane	−38.4	8,245	19.57	1
1270	Propylene	−53.9	7,931	19.55	1
503	R-23/13 (40.1/59.9)	−126.1	6,483	19.43	1
600a	Isobutane	10.9	9,103	19.34	1
13	Chlorotrifluoromethane	−114.6	6,670	19.33	1
290	Propane	−43.7	8,026	19.29	1
1150	Ethylene	−154.7	5,793	19.00	1
170	Ethane	−127.9	6,296	18.98	1
50	Methane	−258.7	3,521	17.52	1

Notes:
[a]Not at normal atmospheric pressure.
[b]Normal boiling temperatures.
[c]The exact composition of this azeotrope is in question.

References:
1 ASHRAE *Thermodynamic Properties of Refrigerants* (Stewart et al. 1986).
2 CRC *Handbook of Chemistry and Physics* (CRC 1987).
3 ASHRAE (1977).
4 *Chemical Engineer's Handbook* (1973).
5 NIST Standard Reference Database 23.

Table 7 Comparative Refrigerant Performance per Ton of Refrigeration[a]

Refrigerant No.	Chemical Name or Composition (% by mass)	Evaporator Pressure, psia	Condenser Pressure, psia	Compression Ratio	Net Refrigerating Effect, Btu/lb_m	Refrigerant Circulated, lb_m/min	Liquid Circulated, in^3/min	Specific Volume of Suction Gas, ft^3/lb_m	Compressor Displacement, cfm	Power Consumption, hp	Coefficient of Performance	Comp. Discharge Temp., °F
170	Ethane	236.410	674.710	2.85	69.27	2.88704	289.1266	0.5344	1.543	1.733	2.72	123
744	Carbon dioxide	332.375	1045.360	3.15	57.75	3.46320	158.5272	0.2639	0.914	1.678	2.81	156
13B1	Bromotrifluoromethane	77.820	264.128	3.39	28.45	7.02901	129.7814	0.3798	2.669	1.134	4.16	104
125	Pentafluoroethane	58.870	228.110	3.87	37.69	5.30645	126.8148	0.6281	3.333	1.283	3.67	108
1270	Propylene	52.704	189.440	3.59	123.15	1.62401	90.7048	2.0487	3.327	1.035	4.56	108
290	Propane	42.37	156.820	3.70	120.30	1.66251	95.0386	2.4589	4.088	1.031	4.57	98
502	R-22/115 (48.8/51.2)	50.561	191.290	3.78	44.91	4.45305	103.3499	0.8015	3.569	1.067	4.42	98
22	Chlorodifluoromethane	42.963	172.899	4.02	69.90	2.86144	67.6465	1.2394	3.546	1.011	4.67	128
717	Ammonia	34.170	168.795	4.94	474.20	0.42177	19.6087	8.1790	3.450	0.989	4.77	210
500	R-12/152a (73.8/26.2)	31.064	127.504	4.10	60.64	3.29834	80.1925	1.5022	4.955	1.005	4.69	105
12	Dichlorodifluoromethane	26.505	107.991	4.07	50.25	3.97981	85.2280	1.4649	5.830	0.992	4.75	100
134a	Tetrafluoroethane	23.790	111.630	4.69	64.77	3.08785	71.8199	1.9500	6.021	1.070	4.41	108
124	Chlorotetrafluoroethane	12.960	64.590	4.98	50.93	3.92696	81.1580	2.7140	10.658	1.054	4.47	90
600a	Isobutane	12.924	59.286	4.59	113.00	1.76991	90.0059	6.4189	11.361	1.070	4.41	80
600	Butane	8.176	41.191	5.04	125.55	1.59299	77.7772	10.2058	16.258	0.952	4.95	88
114	Dichlorotetrafluoroethane[b]	6.747	36.493	5.41	43.02	4.64889	89.5631	4.3400	20.176	1.015	4.65	86
11	Trichlorofluoromethane	2.937	18.318	6.24	67.21	2.97592	56.2578	12.2400	36.425	0.939	5.02	110
123	Dichlorotrifluoroethane	2.290	15.900	6.94	61.19	3.26829	62.3495	14.0800	46.018	0.974	4.84	94
113	Trichlorotrifluoroethane[b]	1.006	7.884	7.83	52.08	3.84047	68.5997	26.2845	100.945	1.105	4.27	86

Notes: [a]Based on 5°F evaporation and 86°F condensation. [b]Saturated suction except R-113 and R-114. Enough superheat was added to give saturated discharge.

Table 8 Comparative Refrigerant Performance per Ton at Various Evaporating and Condensing Temperatures

Refrigerant No.	Chemical Name or Composition (% by mass)	Suction Temp., °F	Evaporator Pressure, psia	Condenser Pressure, psia	Compression Ratio	Net Refrigerating Effect, Btu/lb_m	Refrigerant Circulated, lb_m/min	Specific Volume of Suction Gas, ft^3/lb_m	Compressor Displacement, cfm	Power Consumption, hp
A. −130°F Saturated Evaporating, 0°F Suction Superheat, −40°F Saturated Condensing										
1150	Ethylene	−130	30.887	210.670	6.82	142.01	1.40835	3.8529	5.426	1.756
170	Ethane	−130	13.620	112.790	8.28	156.58	1.27730	8.3575	10.675	1.633
13	Chlorotrifluoromethane	−130	9.059	88.037	9.72	45.82	4.36529	3.6245	15.822	1.685
23	Trifluoromethane	−130	9.06	103.03	11.37	79.38	2.51953	5.4580	13.752	1.753
B. −100°F Saturated Evaporating, 0°F Suction Superheat, −30°F Saturated Condensing										
170	Ethane	−100	31.267	134.730	4.31	157.76	1.26775	3.8671	4.903	1.118
13	Chlorotrifluoromethane	−100	22.276	106.290	4.77	46.23	4.32581	1.5631	6.762	1.153
125	Pentafluoroethane	−100	3.780	27.760	7.34	56.43	3.54403	8.3900	29.734	1.101
22	Chlorodifluoromethane	−100	2.380	19.629	8.25	90.75	2.20397	18.5580	40.901	1.074
23	Trifluoromethane	−100	23.74	125.99	5.31	79.37	2.51984	2.219	5.592	1.178
C. −76°F Saturated Evaporating, 0°F Suction Superheat, 5°F Saturated Condensing										
1150	Ethylene	−76	109.370	416.235	3.81	116.95	1.71021	1.1617	1.987	1.478
170	Ethane	−76	54.634	235.440	4.31	322.65	0.61987	2.2906	1.420	0.566
23	Trifluoromethane	−76	45.410	237.180	5.22	69.60	2.87356	1.2030	3.457	1.394
13	Chlorotrifluoromethane	−76	40.872	192.135	4.70	39.42	5.07389	0.8801	4.465	1.382
13B1	Bromotrifluoromethane	−76	13.173	77.820	5.91	37.80	5.29128	2.0329	10.757	1.253
125	Pentafluoroethane	−76	8.210	58.870	7.17	50.62	3.95101	4.0720	16.089	1.277
290	Propane	−76	6.150	42.367	6.89	147.39	1.35699	14.8560	20.159	1.196
22	Chlorodifluoromethane	−76	5.438	42.963	7.90	84.24	2.37425	8.5925	20.401	1.195
717	Ammonia	−76	3.18	34.26	10.79	540.63	0.37	75.7838	28.04	1.247
12	Dichlorodifluoromethane	−76	3.277	26.501	8.09	58.61	3.41219	10.2448	34.957	1.191
134a	Tetrafluoroethane	−76	2.3	23.77	10.32	78.1	2.561	17.3038	44.315	1.182
D. −40°F Saturated Evaporating, 0°F Suction Superheat, 68°F Saturated Condensing										
744	Carbon dioxide	−40	145.770	830.530	5.70	77.22	2.59000	0.6128	1.587	2.208
23	Trifluoromethane	−40	103.030	597.900	5.80	45.67	4.37924	0.5448	2.386	2.442
13B1	Bromotrifluoromethane	−40	31.855	207.854	6.53	28.81	6.94155	0.8915	6.189	1.855
125	Pentafluoroethane	−40	21.840	175.100	8.02	37.44	5.34188	1.6250	8.681	1.962
290	Propane	−40	16.099	121.560	7.55	119.33	1.67602	6.0829	10.195	1.670
22	Chlorodifluoromethane	−40	15.268	131.997	8.65	70.65	2.83106	3.2805	9.287	1.606
717	Ammonia	−40	10.4	124.31	11.95	486.55	0.411	25.1436	10.334	1.576
500	R-12/152a (73.8/26.2)	−40	10.959	96.948	8.85	60.24	3.31989	3.9895	13.245	1.583
12	Dichlorodifluoromethane	−40	9.304	82.295	8.84	49.44	4.04572	3.8868	15.725	1.596
134a	Tetrafluoroethane	−40	7.42	83.0	11.19	63.17	3.166	5.7899	18.331	1.597

Table 8 Comparative Refrigerant Performance per Ton at Various Evaporating and Condensing Temperatures (*Concluded*)

Refrigerant No.	Refrigerant Chemical Name or Composition (% by mass)	Suction Temp., °F	Evaporator Pressure, psia	Condenser Pressure, psia	Compression Ratio	Net Refrigerating Effect, Btu/lb_m	Refrigerant Circulated, lb_m/min	Specific Volume of Suction Gas, ft^3/lb_m	Compressor Displacement, cfm	Power Consumption, hp
	E. −10°F Saturated Evaporating, 0°F Suction Superheat, 100°F Saturated Condensing									
123	Dichlorotrifluoroethane	−10	1.48	20.8	14.07	55.64	3.594	21.1405	75.979	1.436
11	Trichlorofluoromethane	−10	1.92	23.37	12.2	61.82	3.235	18.1691	58.777	1.398
124	Chlorotetrafluoroethane	−10	8.950	80.920	9.04	44.99	4.44543	3.8410	17.075	1.649
134a	Tetrafluoroethane	−10	16.62	138.98	8.36	56.57	3.535	2.7114	9.585	1.589
12	Dichlorodifluoromethane	−10	19.197	131.720	6.86	44.89	4.45563	1.9803	8.823	1.606
717	Ammonia	−10	23.73	211.96	8.93	461.25	0.434	11.6774	5.068	1.494
22	Chlorodifluoromethane	−10	31.231	210.670	6.75	64.07	3.12173	1.6757	5.231	1.602
502	R-22/115 (48.8/51.2)	−10	37.256	230.890	6.20	39.05	5.12177	1.0727	5.494	1.904
125	Pentafluoroethane	−10	43.320	276.950	6.39	31.09	6.43294	0.8459	5.442	2.172
	F. −10°F Saturated Evaporating, 75°F Suction Superheat (Not Included in Refrigeration Effect), 100°F Saturated Condensing									
123	Dichlorotrifluoroethane	65	1.48	20.8	14.07	55.64	3.594	24.8022	89.139	1.678
11	Trichlorofluoromethane	65	1.92	23.37	12.2	61.82	3.235	21.2804	68.842	1.632
124	Chlorotetrafluoroethane	65	8.950	80.920	9.04	44.99	4.44543	4.5310	20.142	1.919
134a	Tetrafluoroethane	65	16.62	138.98	8.36	56.57	3.535	3.2359	11.439	1.906
12	Dichlorodifluoromethane	65	19.197	131.720	6.86	44.89	4.45563	2.3597	10.514	1.914
717	Ammonia	65	23.73	211.96	8.93	461.25	0.434	13.7281	5.958	1.742
22	Chlorodifluoromethane	65	31.231	210.670	6.75	64.07	3.12173	2.0121	6.281	1.924
502	R-22/115 (48.8/51.2)	65	37.256	230.890	6.20	39.05	5.12177	1.3015	6.666	2.310
125	Pentafluoroethane	65	43.320	276.950	6.39	31.09	6.43294	1.0280	6.613	2.573
	G. −10°F Saturated Evaporating, 75°F Suction Superheat (Included in Refrigeration Effect), 100°F Saturated Condensing									
123	Dichlorotrifluoroethane	65	1.48	20.8	14.07	67.3	2.972	24.7971	73.697	1.387
11	Trichlorofluoromethane	65	1.92	23.37	12.2	71.88	2.783	21.2763	59.212	1.403
124	Chlorotetrafluoroethane	65	8.950	80.920	9.04	57.33	3.48857	4.5310	15.807	1.506
134a	Tetrafluoroethane	65	16.62	138.98	8.36	71.25	2.807	3.2358	9.083	1.513
12	Dichlorodifluoromethane	65	19.197	131.720	6.86	55.83	3.58251	2.3597	8.454	1.539
717	Ammonia	65	23.73	211.96	8.93	498.44	0.401	13.7506	5.514	1.612
22	Chlorodifluoromethane	65	31.231	210.670	6.75	75.95	2.63326	2.0121	5.298	1.623
502	R-22/115 (48.8/51.2)	65	37.256	230.890	6.20	51.23	3.90362	1.3015	5.081	1.761
125	Pentafluoroethane	65	43.320	276.950	6.39	45.13	4.43164	1.0280	4.556	1.773
	H. 20°F Saturated Evaporating, 0°F Suction Superheat, 80°F Saturated Condensing									
125	Pentafluoroethane	20	78.400	209.270	2.67	41.47	4.82276	0.4735	2.284	0.831
290	Propane	20	55.931	144.330	2.58	128.39	1.55775	1.8873	2.940	0.721
22	Chlorodifluoromethane	20	57.786	158.360	2.74	73.12	2.73512	0.9334	2.553	0.707
717	Ammonia	20	48.19	153.06	3.18	497.1	0.402	6.0498	2.432	0.677
500	R-12/152a (73.8/26.2)	20	41.936	116.620	2.78	64.15	3.11784	1.1294	3.521	0.702
12	Dichlorodifluoromethane	20	35.765	98.850	2.76	53.22	3.75827	1.1045	4.151	0.701
134a	Tetrafluoroethane	20	33.13	101.49	3.06	67.91	2.945	1.4088	4.149	0.693
124	Chlorotetrafluoroethane	20	18.290	58.410	3.19	54.67	3.65831	1.9640	7.185	0.710
600a	Isobutane	20	17.916	53.907	3.01	121.45	1.64677	4.7361	7.799	0.706
600	Butane	20	11.557	37.225	3.22	134.18	1.49054	7.3947	11.022	0.686
123	Dichlorotrifluoroethane	20	3.48	14.07	4.04	64.75	3.089	9.5073	29.368	0.656
11	Trichlorofluoromethane	20	4.33	16.17	3.74	69.78	2.866	8.5213	24.422	0.649
	I. 40°F Saturated Evaporating, 0°F Suction Superheat, 100°F Saturated Condensing									
125	Pentafluoroethane	40	111.710	276.950	2.48	37.10	5.39084	0.3312	1.785	0.860
290	Propane	40	78.782	189.040	2.40	114.96	1.73974	1.3563	2.360	0.750
22	Chlorodifluoromethane	40	83.246	210.670	2.53	68.71	2.91091	0.6557	1.909	0.696
717	Ammonia	40	73.3	211.96	2.89	480.33	0.416	4.0841	1.699	0.653
500	R-12/152a (73.8/26.2)	40	60.722	155.790	2.57	60.54	3.30344	0.7920	2.616	0.692
12	Dichlorodifluoromethane	40	51.705	131.720	2.55	50.50	3.96024	0.7784	3.083	0.689
134a	Tetrafluoroethane	40	49.77	138.98	2.79	63.72	3.139	0.9522	2.989	0.679
124	Chlorotetrafluoroethane	40	27.890	80.920	2.90	52.06	3.84172	1.3180	5.063	0.698
600a	Isobutane	40	26.750	73.364	2.74	115.83	1.72667	3.2564	5.623	0.693
600	Butane	40	17.679	51.683	2.92	129.22	1.54775	4.9754	7.701	0.669
11	Trichlorofluoromethane	40	6.99	23.37	3.34	68.04	2.939	5.4546	16.031	0.624
123	Dichlorotrifluoroethane	40	5.79	20.8	3.59	62.82	3.184	5.9212	18.853	0.635
113	Trichlorotrifluoroethane	47	2.695	10.494	3.89	54.14	3.69433	10.7059	39.551	0.710

Table 9 Comparison of Safety Group Classifications in ASHRAE *Standard* 34-1989 and ASHRAE *Standard* 34-1992

Refrigerant Number	Chemical Formula	Safety Group Old	Safety Group New
10	CCl_4	2	B1
11	CCl_3F	1	A1
12	CCl_2F_2	1	A1
13	$CClF_3$	1	A1
13B1	$CBrF_3$	1	A1
14	CF_4	1	A1
21	$CHCl_2F$	2	B1
22	$CHClF_2$	1	A1
23	CHF_3		A1
30	CH_2Cl_2	2	B2
32	CH_2F_2		A2
40	CH_3Cl	2	B2
50	CH_4	3a	A3
113	CCl_2FCClF_2	1	A1
114	$CClF_2CClF_2$	1	A1
115	$CClF_2CF_3$	1	A1
116	CF_3CF_3		A1
123	$CHCl_2CF_3$		B1
124	$CHClFCF_3$		A1
125	CHF_2CF_3		A1
134a	CF_3CH_2F		A1
142b	$CClF_2CH_3$	3b	A2
143a	CF_3CH_3		A2
152a	CHF_2CH_3	3b	A2
170	CH_3CH_3	3a	A3
218	$CF_3CF_2CF_3$		A1
290	$CH_3CH_2CH_3$	3a	A3
C318	C_4F_8	1	A1
400	R-12/114 (must be specified)	1	A1/A1
500	R-12/152a (73.8/26.2)	1	A1
501	R-22/12 (75.0/25.0)*	1	A1
502	R-22/115 (48.8/51.2)	1	A1
507A	R-125/143a (50/50)		A1
508A	R-23/116 (39/61)		A1
508B	R-23/116 (46/54)		A1/A1
509A	R-22/218 (44/56)		A1
600	$CH_3CH_2CH_2CH_3$	3a	A3
600a	$CH(CH_3)_3$	3a	A3
611	$HCOOCH_3$	2	B2
702	H_2		A3
704	He		A1
717	NH_3	2	B2
718	H_2O		A1
720	Ne		A1
728	N_2		A1
740	Ar		A1
744	CO_2	1	A1
764	SO_2	2	B1
1140	$CHCl{=}CH_2$		B3
1150	$CH_2{=}CH_2$	3a	A3
1270	$CH_3CH{=}CH_2$	3a	A3

*The exact composition of this azeotrope is in question.

that contain explosive or flammable vapors. Other vapors, such as alcohol and carbon monoxide, may interfere with the test.

The electronic detector is the most sensitive of the various leak detection methods, reportedly capable of sensing a leak of 1/100 oz of R-12 per year.

A portable model is available for field testing. Other models are available with automatic balancing systems that correct for refrigerant vapors that might be present in the atmosphere around the test area.

Halide Torch

The halide torch is a fast and reliable method of detecting leaks of chlorinated refrigerants. Air is drawn over a copper element heated by a methyl alcohol or hydrocarbon flame. If halogenated vapors are present, they decompose, and the color of the flame changes to bluish-green. Although not as sensitive as the electronic detector, this method is suitable for most purposes.

Bubble Method

The object to be tested is pressurized with air or nitrogen. A pressure corresponding to operating conditions is generally used. The object is immersed in water, and any leaks are detected by observing the formation of bubbles in the liquid. Adding a detergent to the water decreases the surface tension, prevents escaping gas from clinging to the side of the object, and promotes the formation of a regular stream of small bubbles. Kerosene or other organic liquids are sometimes used for the same reason. A solution of soap or detergent can be brushed or poured onto joints or other spots where leakage is suspected. Leaking gas forms soap bubbles that can be readily detected.

Leaks can also be determined by pressurizing or evacuating and observing the change in pressure or vacuum over a period of time. This is effective in checking the tightness of the system but does not locate the point of leakage.

Ammonia and Sulfur Dioxide Leaks

Ammonia can be detected by burning a sulfur candle in the vicinity of the suspected leak or by bringing a solution of hydrochloric acid near the object. If ammonia vapor is present, a white cloud or smoke of ammonium sulfite or ammonium chloride forms. Ammonia can also be detected with indicator paper that changes color in the presence of a base.

Sulfur dioxide can be detected by the appearance of white smoke when aqueous ammonia is brought near the leak.

EFFECT ON CONSTRUCTION MATERIALS

Metals

Halogenated refrigerants can be used satisfactorily under normal conditions with most common metals, such as steel, cast iron, brass, copper, tin, lead, and aluminum. Under more severe conditions, various metals affect such properties as hydrolysis and thermal decomposition in varying degrees. The tendency of metals to promote thermal decomposition of halogenated compounds is in the following order:

(least decomposition) Inconel < 18-8 stainless steel < nickel < copper < 1340 steel < aluminum < bronze < brass < zinc < silver (most decomposition)

This order is only approximate, and exceptions may be found for individual compounds or for special use conditions. The effect of metals on hydrolysis is probably similar.

Magnesium, zinc, and aluminum alloys containing more than 2% magnesium are not recommended for use with halogenated compounds where even trace amounts of water may be present.

Warning: Never use methyl chloride with aluminum in any form. A highly flammable gas is formed, and the explosion hazard is great.

Ammonia should never be used with copper, brass, or other alloys containing copper. When water is present in sulfur dioxide systems, sulfurous acid is formed and can attack iron or steel rapidly and other metals at a slower rate.

Further discussion of the compatibility of refrigerants and lubricants with construction materials may be found in Chapter 5 of the 1994 *ASHRAE Handbook—Refrigeration*.

Elastomers

The linear swelling of some elastomers in the liquid phase of various refrigerants is shown in Table 10. Swelling data can be used to a limited extent in comparing the effect of refrigerants on elastomers. However, other factors, such as the amount of extraction, tensile strength, and degree of hardness of the exposed elastomer must be

Table 10 Swelling of Elastomers in Liquid Refrigerants at Room Temperature

Refrigerant		Linear Swell, %							
No.	Chemical Name or Composition (% by mass)	Buna N	Buna S (GR-S)	Butyl (GR-1)	Natural Rubber	Neoprene GN	Thiokol FA	Viton B	Silicone
11	Trichlorofluoromethane	6	21	41	23	17	2	6	38
12	Dichlorodifluoromethane	2	3	6	6	0	1	9	—
13	Chlorotrifluoromethane	1	1	0	1	0	0	4	—
13B1	Bromotrifluoromethane	1	1	2	1	2	—	7	—
21	Dichlorofluoromethane	48	49	24	34	28	28	22	—
22	Chlorodifluoromethane	26	4	1	6	2	4	20	20
30	Methylene chloride	52	26	23	34	37	59	—	—
40	Methyl chloride	35	20	16	26	22	11	—	—
113	Trichlorotrifluoroethane	1	9	21	17	3	1	7	34
114	Dichlorotetrafluoroethane	0	2	2	2	0	0	9	—
502	R-22/115 (48.8/51.2)	7	3	—	4	1	—	—	—
600	Butane	1	8	20	16	3	0	—	—

Adapted from Eiseman (1949).

Table 11 Diffusion of Water and R-22 Through Elastomers

Elastomer	Diffusion Rate	
	Water[a]	R-22[b]
Neoprene	0.717	1.31
Buna N	0.109	19.7
Hypalon 40	0.457	0.52
Butyl	0.043	0.30
Viton	—	3.61
Polyethylene	0.123	—
Natural	1.428	—

Adapted from Eiseman (1966).
[a]0.003 in. film, 100% rh at 100°F. Diffusion rate per hour in pounds of water per 1000 ft^2 of elastomer.
[b]Film thickness = 0.001 in.; temperature = 77°F. Gas at 1 atm and 32°F. Diffusion rate per day in ft^3 of gas per ft^2 of elastomer.

Table 12 Swelling of Plastics in Liquid Refrigerants at Room Temperature

Plastic	Linear Swell, % Refrigerant						
	11	12	21	30	113	114a	22
Phenol formaldehyde resin	0	0	0	0	−0.2	−0.2	a
Cellulose acetate	0.4	0	b	b	0	−0.1	a
Cellulose nitrate	0.6	0	b	b	0	−0.1	a
Nylon	0	0	0	0	0	−0.2	1
Methyl methacrylate resin	0	−0.1	b	b	−0.2	−0.2	b
Polyethylene	6.7	0.4	4.5	4.6	2.3	0.6	2
Polystyrene	b	−0.1	b	b	−0.2	−0.2	a
Polyvinyl alcohol	0.3	−0.7	12.9	9.1	−0.1	0.2	a
Polyvinyl chloride	0	0	15.1	b	0	0.1	a
Polyvinylidene chloride	−0.2	0	1.0	2.4	−0.1	0	4
Polytetrafluoroethylene	0	−0.7	0.1	0	0	−0.3	1

Adapted from Brown (1960).
[a]Data not available.
[b]Sample completely disintegrated.

considered. When other fluids are present in addition to the refrigerant, the combined effect on elastomers should be determined. In some instances, somewhat higher swelling of elastomers is found in mixtures of R-22 and lubricating oil than in either fluid alone. Table 11 shows the diffusion rate of water and R-22 through elastomers.

Plastics

The effect of a refrigerant on a plastic material should be thoroughly examined under the conditions of intended use. Plastics are often mixtures of two or more basic types, and it is difficult to predict the effect of the refrigerant. The linear swelling of some plastic materials in refrigerants is shown in Table 12. Swelling data can be used as a guide but, as with elastomers, the effect on the properties of the plastic should also be examined. Comparable data for R-22 is limited, but the effect on plastics is generally more severe than that of R-12, but not as severe as that of R-21. The effect of R-114 is very similar to that of R-114a.

REFERENCES

ASHRAE. 1977. *ASHRAE Handbook—Fundamentals,* Chapter 16.

ASHRAE. 1992. Number designation and safety classification of refrigerants. ANSI/ASHRAE *Standard* 34-1992.

Beacham, E.A. and R.T. Divers. 1955. Some aspects of the dielectric properties of refrigerants. *Refrigerating Engineering* 7:33.

Brown, J.A. 1960. Effect of propellants on plastic valve components. *Soap and Chemical Specialties* 3:87.

Charlton, E.E. and F.S. Cooper. 1937. Dielectric strengths of insulating fluids. *General Electric Review* 865(9):438.

Chemical engineer's handbook, 5th ed. 1973. McGraw-Hill, New York.

CRC Handbook of chemistry and physics, 68th ed. 1987. CRC Press, Boca Raton, FL.

duPont. *Bulletin* B-32A. Freon Products Division. E.I. duPont de Nemours & Co., Inc., Wilmington, DE.

duPont. *Bulletin* G-1. Freon Products Division. E.I. duPont de Nemours & Co., Inc., Wilmington, DE.

duPont. 1980. *Bulletin* T-502. Freon Products Division. E.I. duPont de Nemours & Co., Inc., Wilmington, DE.

Eiseman, B.J., Jr. 1949. Effect on elastomers of Freon compounds and other halohydrocarbons. *Refrigerating Engineering* 12:1171.

Eiseman, B.J., Jr. 1955. How electrical properties of Freon compounds affect hermetic system's insulation. *Refrigerating Engineering* 4:61.

Fellows, B.R., R.G. Richard, and I.R. Shankland. 1991. Electrical characterization of alternate refrigerants. Proceedings of the 18th International Congress of Refrigeration. International Institute of Refrigeration, Paris, France.

Fuoss, R.M. 1938. Dielectric constants of some fluorine compounds. *Journal of the American Chemical Society*, 1633.

Handbook of chemistry, 10th ed. 1967. McGraw-Hill, New York.

Handbook of chemistry and physics, 41st ed. 1959-60. The Chemical Rubber Publishing Co., Cleveland, OH.

Kirk and Othmer. 1956. *The encyclopedia of chemical technology*. The Interscience Encyclopedia, Inc., New York.

Matheson gas data book. 1966. The Matheson Company, Inc., East Rutherford, NJ.

McLinden, M.O. 1990. *International Journal of Refrigeration* 13:149-62.

NIST Standard Reference Database 23. National Institute of Standards and Technology, Gaithersburg, MD.

Stewart, R.B., R.T. Jacobsen, and S.G. Penoncello. 1986. *ASHRAE Thermodynamic properties of refrigerants*. ASHRAE.

U.N. 1994. *1994 Report of the refrigeration, air conditioning, and heat pumps technical options committee.* United Nations Environment Programme, Nairobi, Kenya. ISBN 92-807-1455-4.

U.N. 1996. OzonAction (The Newsletter of the United Nations Environment Programme Industry and Environment OzonAction Programme). October (No. 20):10.

CHAPTER 19

THERMOPHYSICAL PROPERTIES OF REFRIGERANTS

THIS chapter presents tabular data for the thermodynamic and transport properties of refrigerants. Most of the refrigerants have a thermodynamic property chart on pressure-enthalpy coordinates with an abbreviated set of tabular data for the saturated liquid and vapor on the facing page. In addition, tabular data in the superheated vapor region are given for R-134a to assist students working on compression cycle examples. For each of the cryogenic fluids, a second table of properties is provided for the vapor at a pressure of one standard atmosphere; these tables provide data needed when these gases are used in heat-transfer or purge-gas applications.

New for the 1997 Handbook are data for several of the "new" zeotropic refrigerant blends, namely, R-404A, R-407C, and R-410A, and the azeotropic blend R-507A. Most of the CFC refrigerants have been retained to assist in making comparisons. For the zeotropic blends and R-729 (air), tables are incremented in pressure with properties given for the liquid on the bubble line and vapor on the dew line. This arrangement is chosen because pressure is more commonly measured in the field while servicing equipment; it also highlights the difference between the bubble and dew point temperatures—the so-called "temperature glide" experienced with blends. Revised formulations have been used for most of the hydrocarbon refrigerants and the cryogenic fluids; the formulations used are detailed in the section on References.

The reference states used for most of the refrigerants correspond to the international convention of 200 kJ/kg for enthalpy and 1 kJ/(kg·K) for entropy, both for the saturated liquid at 0 °C. For the I-P edition, the traditional American convention of 0 Btu/lb and 0 Btu/lb·°F for enthalpy and entropy, respectively, for the saturated liquid at −40 °F is used. The exceptions are water and fluids that have very low critical temperatures, such as ethylene and the cryogens.

These data are intended to help engineers make preliminary comparisons among unfamiliar fluids. For greater detail and a wider range of data, consult the sources listed in the section on References at the end of the chapter and/or *ASHRAE Thermodynamic Properties of Refrigerants* (1986) and *ASHRAE Thermophysical Properties of Refrigerants* (1993).

The preparation of this chapter is assigned to TC 3.1, Refrigerants and Brines.

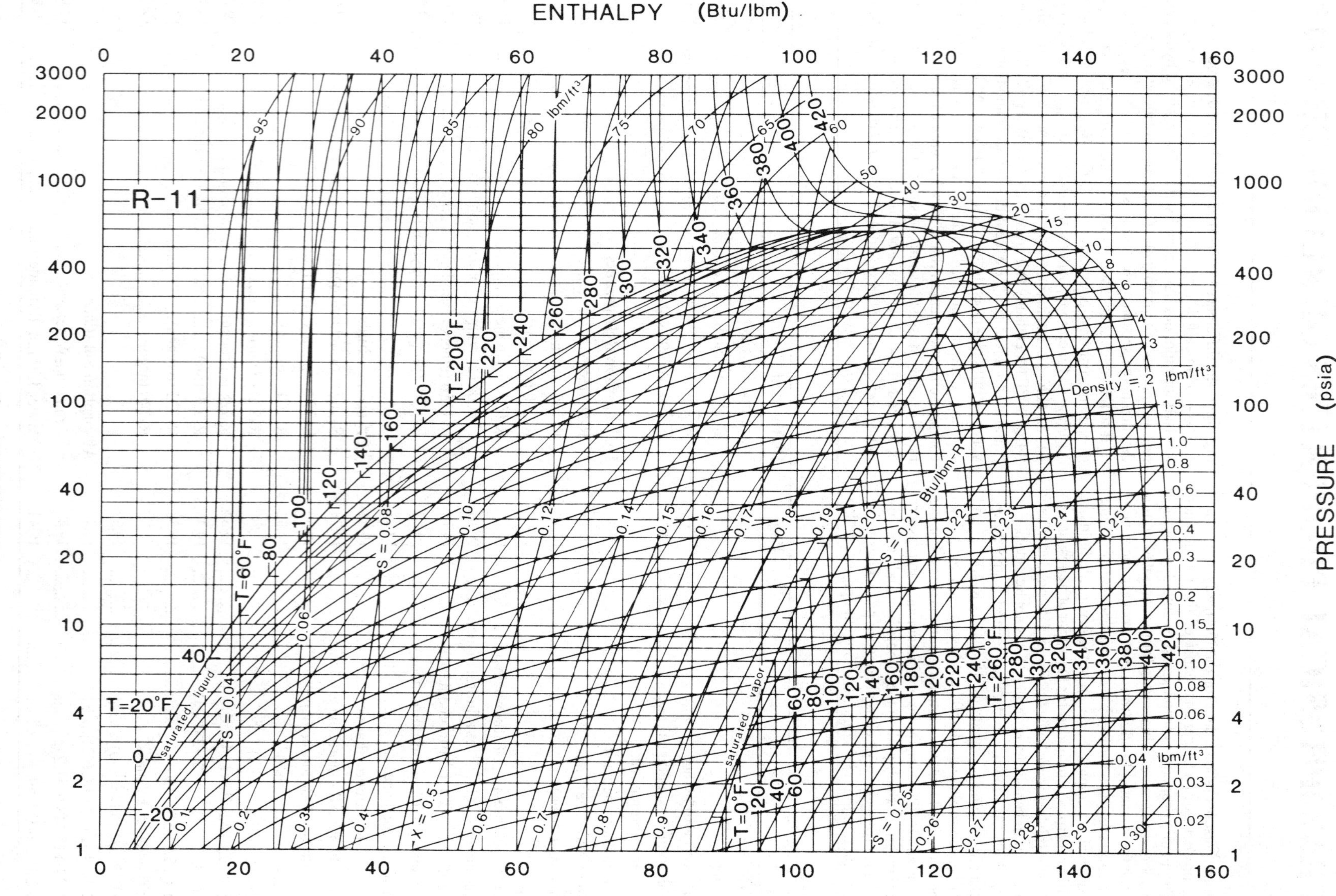

Prepared by: CENTER FOR APPLIED THERMODYNAMIC STUDIES, University of Idaho

Fig. 1 Pressure-Enthalpy Diagram for Refrigerant 11

Refrigerant 11 (Trichlorofluoromethane) Properties of Saturated Liquid and Saturated Vapor

Temp,* °F	Pressure, psia	Density, lb/ft³ Liquid	Volume, ft³/lb Vapor	Enthalpy, Btu/lb Liquid	Enthalpy, Btu/lb Vapor	Entropy, Btu/lb·°F Liquid	Entropy, Btu/lb·°F Vapor	Specific Heat c_p, Btu/lb·°F Liquid	Specific Heat c_p, Btu/lb·°F Vapor	c_p/c_v Vapor	Velocity of Sound, ft/s Liquid	Velocity of Sound, ft/s Vapor	Viscosity, lb_m/ft·h Liquid	Viscosity, lb_m/ft·h Vapor	Thermal Cond, Btu/h·ft·°F Liquid	Thermal Cond, Btu/h·ft·°F Vapor	Surface Tension, dyne/cm	Temp,* °F
−166.85a	0.001	110.41	24230.	−24.922	73.328	−0.07085	0.26467	–	0.1005	1.1680	–	352.	–	–	–	–	–	−166.85
−160.00	0.002	109.93	14650.	−23.116	74.019	−0.06474	0.25940	0.1840	0.1019	1.1655	–	356.	–	–	–	–	–	−160.00
−150.00	0.003	109.23	7355.6	−21.267	75.045	−0.05867	0.25235	0.1860	0.1038	1.1620	3953.	361.	–	–	–	–	35.26	−150.00
−140.00	0.006	108.52	3882.0	−19.398	76.088	−0.05273	0.24597	0.1878	0.1058	1.1588	3895.	366.	–	–	–	–	34.44	−140.00
−130.00	0.012	107.81	2143.2	−17.511	77.149	−0.04692	0.24022	0.1894	0.1077	1.1558	3829.	371.	–	–	–	–	33.63	−130.00
−120.00	0.022	107.09	1232.5	−15.610	78.226	−0.04123	0.23502	0.1909	0.1095	1.1531	3758.	376.	–	–	–	–	32.82	−120.00
−110.00	0.037	106.37	735.62	−13.694	79.318	−0.03568	0.23032	0.1922	0.1114	1.1506	3684.	381.	–	–	–	–	32.01	−110.00
−100.00	0.062	105.65	454.14	−11.767	80.425	−0.03024	0.22608	0.1933	0.1133	1.1482	3609.	386.	–	–	–	–	31.21	−100.00
−90.00	0.100	104.92	289.15	−9.829	81.546	−0.02493	0.22225	0.1943	0.1151	1.1461	3533.	391.	–	–	–	–	30.41	−90.00
−80.00	0.156	104.19	189.37	−7.880	82.681	−0.01973	0.21880	0.1953	0.1170	1.1442	3457.	396.	–	–	–	–	29.62	−80.00
−70.00	0.239	103.46	127.27	−5.923	83.827	−0.01464	0.21568	0.1962	0.1188	1.1425	3382.	400.	–	–	–	–	28.83	−70.00
−60.00	0.355	102.73	87.583	−3.957	84.986	−0.00966	0.21288	0.1970	0.1206	1.1410	3308.	405.	–	–	–	–	28.04	−60.00
−50.00	0.517	101.99	61.601	−1.983	86.155	−0.00478	0.21036	0.1978	0.1225	1.1396	3235.	409.	–	–	–	–	27.26	−50.00
−40.00	0.738	101.25	44.203	0.000	87.335	0.00000	0.20810	0.1986	0.1243	1.1384	3163.	414.	–	–	–	–	26.49	−40.00
−30.00	1.032	100.50	32.310	1.991	88.524	0.00469	0.20608	0.1994	0.1262	1.1374	3093.	418.	–	–	–	–	25.72	−30.00
−20.00	1.419	99.75	24.022	3.989	89.721	0.00928	0.20427	0.2002	0.1280	1.1366	3024.	422.	1.889	0.0222	0.0595	–	24.95	−20.00
−10.00	1.918	99.00	18.143	5.997	90.927	0.01380	0.20267	0.2011	0.1298	1.1360	2955.	426.	1.782	0.0227	0.0587	–	24.19	−10.00
0.00	2.554	98.24	13.903	8.013	92.139	0.01823	0.20124	0.2020	0.1316	1.1356	2888.	430.	1.678	0.0231	0.0579	–	23.43	0.00
5.00	2.931	97.86	12.233	9.024	92.747	0.02041	0.20059	0.2024	0.1325	1.1354	2855.	432.	1.627	0.0233	0.0575	–	23.05	5.00
10.00	3.352	97.48	10.798	10.038	93.357	0.02258	0.19998	0.2029	0.1334	1.1354	2822.	433.	1.578	0.0236	0.0571	–	22.68	10.00
15.00	3.822	97.09	9.5606	11.054	93.967	0.02473	0.19941	0.2033	0.1343	1.1353	2789.	435.	1.529	0.0238	0.0567	–	22.30	15.00
20.00	4.343	96.71	8.4906	12.072	94.579	0.02686	0.19887	0.2038	0.1352	1.1354	2757.	437.	1.482	0.0240	0.0563	–	21.93	20.00
25.00	4.920	96.32	7.5621	13.093	95.192	0.02898	0.19837	0.2043	0.1361	1.1355	2725.	438.	1.436	0.0243	0.0559	–	21.56	25.00
30.00	5.557	95.93	6.7536	14.117	95.806	0.03108	0.19790	0.2048	0.1370	1.1356	2693.	440.	1.390	0.0245	0.0555	–	21.19	30.00
35.00	6.259	95.54	6.0477	15.143	96.420	0.03316	0.19747	0.2053	0.1379	1.1358	2661.	442.	1.347	0.0247	0.0551	0.00470	20.82	35.00
40.00	7.031	95.14	5.4294	16.172	97.035	0.03523	0.19706	0.2059	0.1388	1.1361	2629.	443.	1.304	0.0250	0.0548	0.00475	20.45	40.00
45.00	7.876	94.75	4.8863	17.203	97.650	0.03728	0.19668	0.2064	0.1397	1.1364	2598.	444.	1.262	0.0252	0.0544	0.00480	20.08	45.00
50.00	8.800	94.35	4.4080	18.238	98.265	0.03931	0.19633	0.2070	0.1406	1.1368	2567.	446.	1.222	0.0254	0.0540	0.00485	19.72	50.00
55.00	9.809	93.95	3.9856	19.275	98.880	0.04134	0.19601	0.2075	0.1415	1.1373	2536.	447.	1.183	0.0256	0.0536	0.00490	19.35	55.00
60.00	10.907	93.55	3.6116	20.315	99.495	0.04334	0.19571	0.2081	0.1424	1.1378	2505.	448.	1.145	0.0259	0.0532	0.00495	18.99	60.00
65.00	12.099	93.14	3.2795	21.358	100.109	0.04534	0.19543	0.2087	0.1433	1.1385	2474.	450.	1.109	0.0261	0.0528	0.00501	18.63	65.00
70.00	13.392	92.73	2.9841	22.405	100.723	0.04732	0.19518	0.2093	0.1442	1.1392	2444.	451.	1.073	0.0264	0.0524	0.00506	18.27	70.00
74.67b	14.696	92.35	2.7369	23.386	101.297	0.04916	0.19496	0.2099	0.1451	1.1399	2415.	452.	1.041	0.0266	0.0521	0.00511	17.94	74.67
75.00	14.790	92.33	2.7206	23.455	101.337	0.04928	0.19495	0.2100	0.1452	1.1400	2413.	452.	1.039	0.0266	0.0521	0.00512	17.91	75.00
80.00	16.301	91.91	2.4851	24.507	101.949	0.05124	0.19473	0.2106	0.1461	1.1409	2383.	453.	1.006	0.0268	0.0517	0.00517	17.56	80.00
85.00	17.929	91.50	2.2741	25.564	102.560	0.05318	0.19454	0.2113	0.1470	1.1419	2353.	454.	0.974	0.0271	0.0513	0.00523	17.20	85.00
90.00	19.681	91.08	2.0846	26.624	103.170	0.05511	0.19437	0.2119	0.1479	1.1429	2323.	455.	0.943	0.0273	0.0509	0.00528	16.85	90.00
95.00	21.563	90.66	1.9142	27.687	103.778	0.05703	0.19421	0.2126	0.1489	1.1441	2293.	456.	0.913	0.0276	0.0505	0.00534	16.49	95.00
100.00	23.581	90.23	1.7605	28.754	104.384	0.05894	0.19407	0.2134	0.1498	1.1454	2263.	456.	0.884	0.0278	0.0502	0.00540	16.14	100.00
105.00	25.743	89.81	1.6217	29.825	104.989	0.06083	0.19394	0.2141	0.1508	1.1468	2234.	457.	0.856	0.0281	0.0498	0.00546	15.79	105.00
110.00	28.053	89.38	1.4960	30.900	105.591	0.06272	0.19383	0.2149	0.1518	1.1483	2204.	458.	0.829	0.0283	0.0494	0.00552	15.45	110.00
115.00	30.520	88.94	1.3821	31.978	106.191	0.06459	0.19374	0.2156	0.1528	1.1500	2174.	458.	0.804	0.0286	0.0490	0.00559	–	115.00
120.00	33.150	88.51	1.2787	33.060	106.788	0.06646	0.19365	0.2164	0.1538	1.1517	2145.	459.	0.778	0.0288	0.0486	0.00565	–	120.00
125.00	35.950	88.07	1.1846	34.147	107.382	0.06832	0.19358	0.2172	0.1548	1.1536	2115.	459.	0.754	0.0291	0.0483	0.00572	–	125.00
130.00	38.926	87.62	1.0988	35.238	107.974	0.07017	0.19352	0.2181	0.1558	1.1557	2086.	459.	0.731	0.0293	0.0479	0.00578	–	130.00
135.00	42.087	87.17	1.0205	36.333	108.562	0.07200	0.19346	0.2189	0.1569	1.1579	2057.	459.	0.708	0.0296	0.0475	0.00585	–	135.00
140.00	45.439	86.72	0.9489	37.433	109.146	0.07383	0.19342	0.2198	0.1580	1.1602	2027.	459.	0.687	0.0299	0.0471	0.00592	–	140.00
145.00	48.989	86.26	0.8833	38.537	109.727	0.07565	0.19339	0.2207	0.1591	1.1628	1998.	460.	0.666	0.0302	0.0468	0.00599	–	145.00
150.00	52.745	85.80	0.8232	39.646	110.304	0.07747	0.19336	0.2217	0.1602	1.1655	1969.	459.	0.646	0.0304	0.0464	0.00606	–	150.00
160.00	60.905	84.86	0.7172	41.878	111.445	0.08107	0.19333	0.2236	0.1626	1.1715	1910.	459.	0.607	0.0310	0.0456	0.00621	–	160.00
170.00	69.979	83.91	0.6273	44.130	112.566	0.08464	0.19333	0.2257	0.1650	1.1783	1851.	458.	0.572	0.0316	0.0449	0.00637	–	170.00
180.00	80.030	82.93	0.5506	46.403	113.666	0.08819	0.19334	0.2279	0.1677	1.1861	1793.	457.	0.539	0.0322	0.0442	0.00654	–	180.00
190.00	91.120	81.93	0.4849	48.699	114.743	0.09171	0.19337	0.2303	0.1705	1.1950	1734.	456.	0.508	0.0328	0.0434	0.00671	–	190.00
200.00	103.31	80.90	0.4283	51.019	115.793	0.09521	0.19341	0.2328	0.1735	1.2052	1675.	454.	0.479	0.0334	0.0427	0.00689	–	200.00
210.00	116.68	79.85	0.3793	53.364	116.815	0.09870	0.19344	0.2356	0.1768	1.2169	1615.	451.	0.453	0.0341	0.0419	0.00707	–	210.00
220.00	131.28	78.77	0.3367	55.736	117.805	0.10216	0.19348	0.2385	0.1804	1.2302	1555.	448.	0.428	0.0347	0.0412	0.00727	–	220.00
230.00	147.18	77.65	0.2996	58.136	118.760	0.10561	0.19351	0.2418	0.1844	1.2456	1495.	445.	0.404	0.0354	0.0404	0.00748	–	230.00
240.00	164.46	76.50	0.2670	60.567	119.677	0.10905	0.19353	0.2453	0.1888	1.2634	1434.	441.	0.383	0.0362	0.0397	0.00769	–	240.00
250.00	183.19	75.30	0.2384	63.032	120.551	0.11248	0.19353	0.2493	0.1937	1.2842	1372.	437.	0.363	0.0369	0.0390	0.00792	–	250.00
260.00	203.43	74.06	0.2131	65.533	121.378	0.11591	0.19351	0.2537	0.1993	1.3086	1309.	432.	0.344	0.0377	0.0382	0.00815	–	260.00
270.00	225.26	72.78	0.1906	68.074	122.152	0.11934	0.19346	0.2587	0.2057	1.3375	1245.	426.	0.326	0.0385	0.0375	0.00840	–	270.00
280.00	248.77	71.43	0.1706	70.659	122.867	0.12278	0.19336	0.2645	0.2131	1.3721	1180.	420.	0.310	0.0393	0.0367	0.00866	–	280.00
290.00	274.03	70.01	0.1528	73.294	123.513	0.12623	0.19322	0.2713	0.2219	1.4142	1114.	413.	0.294	0.0401	0.0360	0.00893	–	290.00
300.00	301.12	68.51	0.1367	75.985	124.082	0.12970	0.19301	0.2794	0.2326	1.4663	1046.	406.	0.280	0.0410	0.0352	0.00921	–	300.00
310.00	330.14	66.92	0.1222	78.742	124.559	0.13320	0.19273	0.2894	0.2457	1.5322	976.	397.	–	–	–	–	–	310.00
320.00	361.18	65.21	0.1090	81.578	124.926	0.13675	0.19235	0.3020	0.2625	1.6179	904.	388.	–	–	–	–	–	320.00
330.00	394.36	63.35	0.0970	84.510	125.159	0.14037	0.19184	0.3188	0.2848	1.7336	830.	378.	–	–	–	–	–	330.00
340.00	429.78	61.29	0.0859	87.565	125.220	0.14408	0.19117	0.3423	0.3159	1.8976	752.	366.	–	–	–	–	–	340.00
350.00	467.60	58.97	0.0755	90.787	125.054	0.14794	0.19026	0.3778	0.3628	2.1476	671.	354.	–	–	–	–	–	350.00
360.00	507.98	56.24	0.0657	94.248	124.561	0.15203	0.18901	0.4376	0.4418	2.5730	586.	340.	–	–	–	–	–	360.00
370.00	551.15	52.86	0.0561	98.092	123.541	0.15651	0.18718	0.5570	0.6042	3.4525	499.	325.	–	–	–	–	–	370.00
380.00	597.49	48.25	0.0458	102.678	121.417	0.16180	0.18412	–	–	–	–	–	–	–	–	–	–	380.00
388.33c	639.27	34.59	0.0289	112.749	112.749	0.17350	0.17350	∞	∞	∞	0.	0.	–	–	∞	∞	0.00	388.33

*temperatures are on the ITS-90 scale a = triple point b = normal boiling point c = critical point

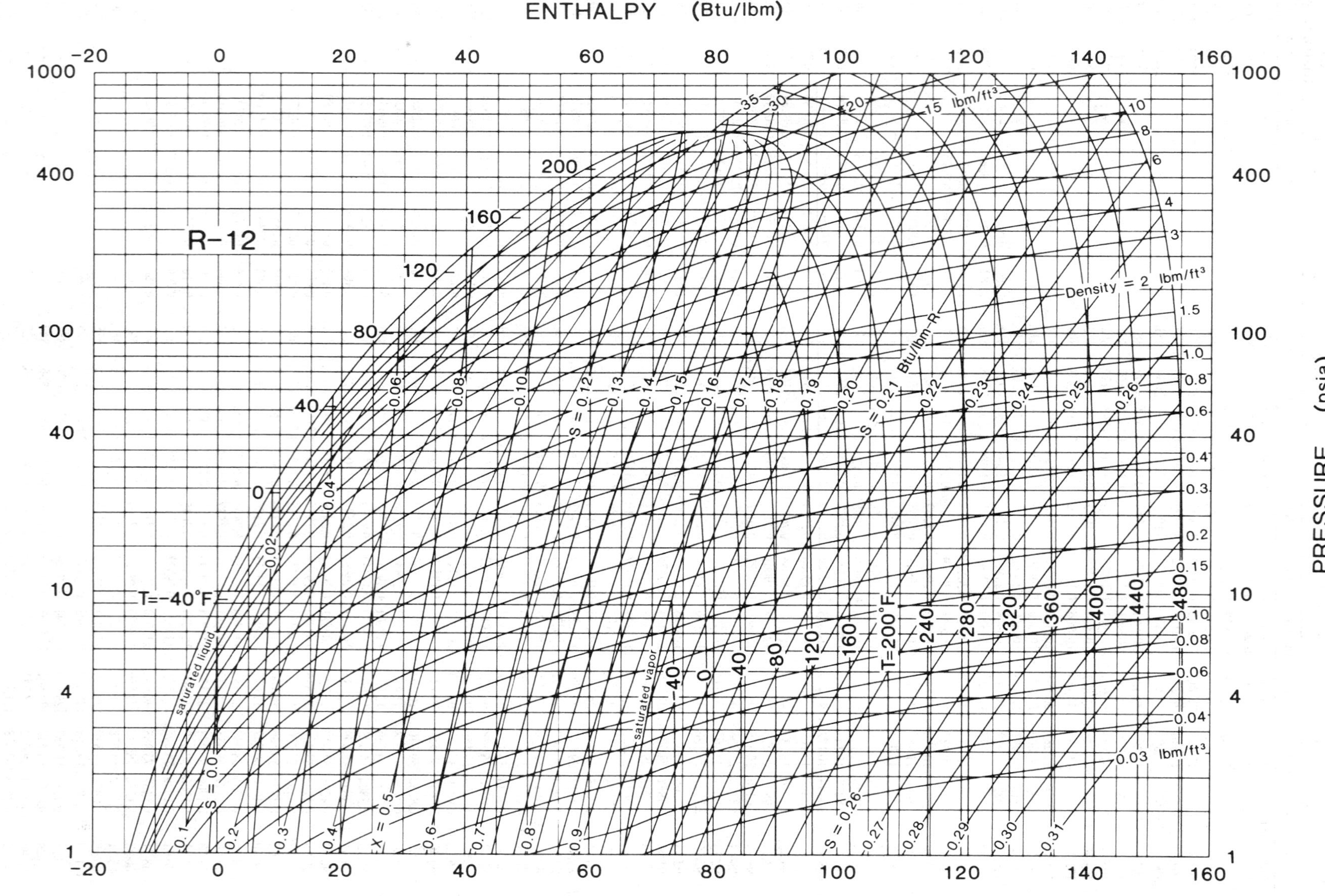

Prepared by CENTER FOR APPLIED THERMODYNAMIC STUDIES, University of Idaho

Fig. 2 Pressure-Enthalpy Diagram for Refrigerant 12

Refrigerant 12 (Dichlorodifluoromethane) Properties of Saturated Liquid and Saturated Vapor

Temp,* °F	Pressure, psia	Density, lb/ft³ Liquid	Volume, ft³/lb Vapor	Enthalpy, Btu/lb Liquid	Enthalpy, Btu/lb Vapor	Entropy, Btu/lb·°F Liquid	Entropy, Btu/lb·°F Vapor	Specific Heat c_p, Btu/lb·°F Liquid	Specific Heat c_p, Btu/lb·°F Vapor	c_p/c_v Vapor	Velocity of Sound, ft/s Liquid	Velocity of Sound, ft/s Vapor	Viscosity, lb_m/ft·h Liquid	Viscosity, lb_m/ft·h Vapor	Thermal Cond, Btu/h·ft·°F Liquid	Thermal Cond, Btu/h·ft·°F Vapor	Surface Tension, dyne/cm	Temp,* °F
−140.00	0.260	104.06	108.98	−20.324	61.801	−0.05521	0.20169	0.1978	0.1099	1.1791	2676.	393.	–	–	–	–	25.79	−140.00
−130.00	0.417	103.16	69.954	−18.338	62.880	−0.04910	0.19727	0.1994	0.1122	1.1759	2665.	398.	–	–	–	–	24.91	−130.00
−120.00	0.648	102.24	46.326	−16.337	63.974	−0.04312	0.19332	0.2005	0.1145	1.1732	2648.	403.	–	–	–	–	24.04	−120.00
−110.00	0.979	101.32	31.556	−14.327	65.081	−0.03729	0.18981	0.2014	0.1168	1.1708	2625.	408.	–	–	–	–	23.18	−110.00
−100.00	1.438	100.40	22.049	−12.308	66.199	−0.03160	0.18668	0.2023	0.1190	1.1688	2597.	413.	–	–	–	–	22.33	−100.00
−95.00	1.727	99.93	18.594	−11.295	66.762	−0.02880	0.18525	0.2027	0.1202	1.1680	2581.	415.	–	–	–	–	21.90	−95.00
−90.00	2.063	99.46	15.766	−10.280	67.327	−0.02604	0.18390	0.2031	0.1213	1.1673	2563.	417.	1.343	0.0200	0.0594	0.00294	21.48	−90.00
−85.00	2.450	98.99	13.438	−9.263	67.894	−0.02331	0.18262	0.2035	0.1224	1.1666	2545.	420.	1.302	0.0204	0.0588	0.00303	21.06	−85.00
−80.00	2.894	98.52	11.511	−8.244	68.462	−0.02061	0.18142	0.2039	0.1235	1.1661	2526.	422.	1.262	0.0209	0.0582	0.00311	20.64	−80.00
−75.00	3.402	98.04	9.9073	−7.223	69.032	−0.01794	0.18029	0.2044	0.1245	1.1657	2505.	424.	1.223	0.0213	0.0576	0.00320	20.22	−75.00
−70.00	3.980	97.56	8.5655	−6.200	69.603	−0.01530	0.17923	0.2048	0.1256	1.1654	2484.	426.	1.185	0.0217	0.0569	0.00328	19.80	−70.00
−65.00	4.634	97.08	7.4374	−5.173	70.175	−0.01269	0.17823	0.2053	0.1267	1.1653	2462.	428.	1.148	0.0222	0.0563	0.00337	19.39	−65.00
−60.00	5.372	96.60	6.4844	−4.145	70.747	−0.01010	0.17728	0.2059	0.1278	1.1652	2440.	429.	1.112	0.0226	0.0557	0.00345	18.98	−60.00
−55.00	6.202	96.12	5.6756	−3.113	71.320	−0.00754	0.17640	0.2064	0.1289	1.1653	2416.	431.	1.077	0.0230	0.0551	0.00353	18.57	−55.00
−50.00	7.131	95.63	4.9863	−2.078	71.893	−0.00500	0.17556	0.2070	0.1300	1.1655	2392.	433.	1.043	0.0233	0.0545	0.00361	18.16	−50.00
−45.00	8.167	95.14	4.3963	−1.041	72.465	−0.00249	0.17478	0.2076	0.1311	1.1658	2367.	435.	1.010	0.0237	0.0538	0.00369	17.75	−45.00
−40.00	9.319	94.65	3.8894	0.000	73.038	0.00000	0.17404	0.2082	0.1322	1.1663	2342.	436.	0.978	0.0241	0.0532	0.00377	17.35	−40.00
−35.00	10.597	94.15	3.4520	1.044	73.609	0.00247	0.17334	0.2089	0.1334	1.1669	2316.	438.	0.947	0.0244	0.0526	0.00385	16.95	−35.00
−30.00	12.008	93.65	3.0734	2.092	74.179	0.00492	0.17269	0.2096	0.1345	1.1676	2289.	439.	0.917	0.0248	0.0520	0.00393	16.55	−30.00
−25.00	13.562	93.15	2.7443	3.144	74.749	0.00734	0.17207	0.2104	0.1357	1.1685	2262.	440.	0.888	0.0251	0.0514	0.00401	16.15	−25.00
−21.63b	14.696	92.81	2.5467	3.855	75.131	0.00897	0.17168	0.2109	0.1365	1.1692	2244.	441.	0.869	0.0254	0.0510	0.00407	15.88	−21.63
−20.00	15.270	92.64	2.4574	4.200	75.316	0.00975	0.17150	0.2112	0.1369	1.1696	2235.	442.	0.860	0.0255	0.0508	0.00409	15.75	−20.00
−15.00	17.141	92.13	2.2065	5.260	75.881	0.01214	0.17096	0.2120	0.1381	1.1708	2207.	443.	0.834	0.0258	0.0502	0.00417	15.36	−15.00
−10.00	19.186	91.62	1.9863	6.324	76.445	0.01451	0.17045	0.2129	0.1393	1.1722	2178.	444.	0.808	0.0262	0.0496	0.00426	14.97	−10.00
−5.00	21.414	91.10	1.7924	7.392	77.006	0.01686	0.16997	0.2138	0.1406	1.1737	2149.	445.	0.783	0.0265	0.0490	0.00434	14.58	−5.00
0.00	23.837	90.58	1.6213	8.466	77.564	0.01920	0.16952	0.2147	0.1419	1.1755	2120.	446.	0.758	0.0268	0.0484	0.00442	14.19	0.00
5.00	26.466	90.05	1.4698	9.544	78.119	0.02152	0.16910	0.2157	0.1432	1.1774	2091.	446.	0.735	0.0272	0.0478	0.00450	–	5.00
10.00	29.313	89.52	1.3353	10.627	78.671	0.02383	0.16870	0.2167	0.1446	1.1796	2061.	447.	0.713	0.0275	0.0472	0.00458	–	10.00
15.00	32.387	88.99	1.2156	11.716	79.219	0.02612	0.16833	0.2177	0.1460	1.1819	2030.	447.	0.691	0.0278	0.0466	0.00467	–	15.00
20.00	35.702	88.44	1.1088	12.810	79.763	0.02840	0.16798	0.2188	0.1474	1.1845	2000.	448.	0.670	0.0282	0.0460	0.00475	–	20.00
25.00	39.269	87.90	1.0133	13.910	80.303	0.03066	0.16765	0.2199	0.1489	1.1873	1969.	448.	0.650	0.0285	0.0454	0.00484	–	25.00
30.00	43.101	87.35	0.9276	15.016	80.838	0.03291	0.16734	0.2211	0.1504	1.1904	1938.	448.	0.631	0.0289	0.0448	0.00492	–	30.00
35.00	47.209	86.79	0.8506	16.127	81.369	0.03516	0.16704	0.2223	0.1520	1.1938	1906.	448.	0.612	0.0292	0.0442	0.00501	–	35.00
40.00	51.605	86.23	0.7813	17.245	81.894	0.03739	0.16677	0.2236	0.1537	1.1974	1875.	448.	0.595	0.0296	0.0436	0.00510	–	40.00
45.00	56.303	85.66	0.7187	18.370	82.413	0.03961	0.16651	0.2249	0.1554	1.2014	1843.	448.	0.577	0.0299	0.0431	0.00519	–	45.00
50.00	61.316	85.08	0.6621	19.501	82.927	0.04181	0.16626	0.2263	0.1571	1.2057	1810.	448.	0.561	0.0303	0.0425	0.00528	–	50.00
55.00	66.656	84.49	0.6109	20.640	83.434	0.04401	0.16602	0.2278	0.1589	1.2103	1778.	448.	0.545	0.0306	0.0419	0.00537	–	55.00
60.00	72.336	83.90	0.5643	21.786	83.935	0.04621	0.16580	0.2293	0.1608	1.2154	1745.	447.	0.529	0.0310	0.0413	0.00546	–	60.00
65.00	78.369	83.30	0.5219	22.939	84.428	0.04839	0.16559	0.2308	0.1628	1.2209	1712.	446.	0.514	0.0314	0.0407	0.00556	–	65.00
70.00	84.770	82.70	0.4833	24.100	84.914	0.05056	0.16538	0.2324	0.1648	1.2268	1679.	445.	0.500	0.0318	0.0401	0.00566	–	70.00
75.00	91.551	82.08	0.4481	25.269	85.392	0.05273	0.16518	0.2342	0.1670	1.2332	1645.	444.	0.486	0.0322	0.0396	0.00576	–	75.00
80.00	98.727	81.45	0.4158	26.447	85.861	0.05489	0.16499	0.2359	0.1692	1.2402	1611.	443.	0.473	0.0326	0.0390	0.00586	–	80.00
85.00	106.31	80.82	0.3863	27.633	86.321	0.05705	0.16480	0.2378	0.1716	1.2478	1577.	442.	0.460	0.0330	0.0384	0.00596	–	85.00
90.00	114.32	80.17	0.3592	28.829	86.772	0.05920	0.16462	0.2398	0.1740	1.2560	1543.	441.	0.448	0.0335	0.0378	0.00606	–	90.00
95.00	122.76	79.51	0.3342	30.034	87.212	0.06135	0.16444	0.2419	0.1766	1.2650	1508.	439.	0.436	0.0339	0.0373	0.00617	–	95.00
100.00	131.65	78.84	0.3113	31.249	87.641	0.06349	0.16425	0.2441	0.1794	1.2749	1473.	437.	0.424	0.0344	0.0367	0.00628	–	100.00
105.00	141.01	78.16	0.2901	32.474	88.059	0.06563	0.16407	0.2464	0.1823	1.2856	1438.	435.	0.413	0.0349	0.0361	0.00639	–	105.00
110.00	150.85	77.46	0.2706	33.711	88.464	0.06777	0.16389	0.2489	0.1855	1.2974	1402.	433.	0.402	0.0354	0.0356	0.00650	–	110.00
115.00	161.19	76.75	0.2525	34.959	88.856	0.06991	0.16370	0.2515	0.1888	1.3103	1367.	431.	0.392	0.0359	0.0350	0.00662	–	115.00
120.00	172.04	76.02	0.2358	36.219	89.233	0.07205	0.16351	0.2543	0.1924	1.3246	1331.	429.	0.382	0.0364	0.0345	0.00674	–	120.00
125.00	183.41	75.28	0.2203	37.492	89.596	0.07419	0.16330	0.2573	0.1963	1.3404	1294.	426.	0.372	0.0370	0.0339	0.00686	–	125.00
130.00	195.33	74.51	0.2058	38.778	89.942	0.07633	0.16309	0.2606	0.2005	1.3579	1258.	423.	0.363	0.0375	0.0333	0.00698	–	130.00
135.00	207.80	73.73	0.1924	40.079	90.270	0.07847	0.16287	0.2641	0.2050	1.3774	1220.	420.	0.354	0.0381	0.0328	0.00711	–	135.00
140.00	220.86	72.93	0.1799	41.394	90.578	0.08062	0.16264	0.2679	0.2100	1.3992	1183.	417.	0.345	0.0387	0.0322	0.00724	–	140.00
145.00	234.50	72.10	0.1682	42.727	90.866	0.08277	0.16239	0.2721	0.2155	1.4238	1145.	413.	0.337	0.0394	0.0317	0.00737	–	145.00
150.00	248.75	71.25	0.1573	44.076	91.131	0.08494	0.16212	0.2767	0.2217	1.4516	1107.	410.	0.329	0.0400	0.0311	0.00750	–	150.00
155.00	263.64	70.37	0.1471	45.444	91.371	0.08711	0.16183	0.2817	0.2285	1.4833	1069.	406.	0.321	0.0407	0.0306	0.00764	–	155.00
160.00	279.17	69.45	0.1375	46.833	91.583	0.08929	0.16151	0.2874	0.2363	1.5197	1030.	401.	–	–	–	–	–	160.00
165.00	295.37	68.51	0.1285	48.244	91.765	0.09149	0.16116	0.2937	0.2451	1.5619	990.	397.	–	–	–	–	–	165.00
170.00	312.25	67.52	0.1200	49.680	91.912	0.09371	0.16078	0.3009	0.2553	1.6113	950.	392.	–	–	–	–	–	170.00
175.00	329.85	66.50	0.1120	51.142	92.021	0.09594	0.16035	0.3092	0.2673	1.6697	910.	387.	–	–	–	–	–	175.00
180.00	348.18	65.42	0.1044	52.635	92.085	0.09820	0.15988	0.3188	0.2814	1.7397	868.	382.	–	–	–	–	–	180.00
185.00	367.27	64.28	0.0973	54.162	92.099	0.10050	0.15935	0.3302	0.2986	1.8253	826.	376.	–	–	–	–	–	185.00
190.00	387.14	63.08	0.0904	55.727	92.054	0.10283	0.15874	0.3439	0.3197	1.9317	783.	370.	–	–	–	–	–	190.00
195.00	407.83	61.80	0.0839	57.339	91.938	0.10520	0.15805	0.3607	0.3465	2.0675	739.	364.	–	–	–	–	–	195.00
200.00	429.37	60.43	0.0776	59.004	91.736	0.10764	0.15726	0.3822	0.3817	2.2464	693.	357.	–	–	–	–	–	200.00
210.00	475.17	57.29	0.0656	62.550	90.978	0.11276	0.15521	0.4508	0.4997	2.8478	595.	343.	–	–	–	–	–	210.00
220.00	524.96	53.26	0.0537	66.578	89.410	0.11848	0.15207	0.6212	0.8091	4.4129	483.	326.	–	–	–	–	–	220.00
230.00	579.55	46.30	0.0397	72.122	85.410	0.12628	0.14554	–	–	–	–	–	–	–	–	–	–	230.00
233.20c	598.27	35.25	0.0284	78.775	78.775	0.13578	0.13578	∞	∞	∞	0.	0.	–	–	∞	∞	0.00	233.20

*temperatures are on the ITS–90 scale   b = normal boiling point   c = critical point

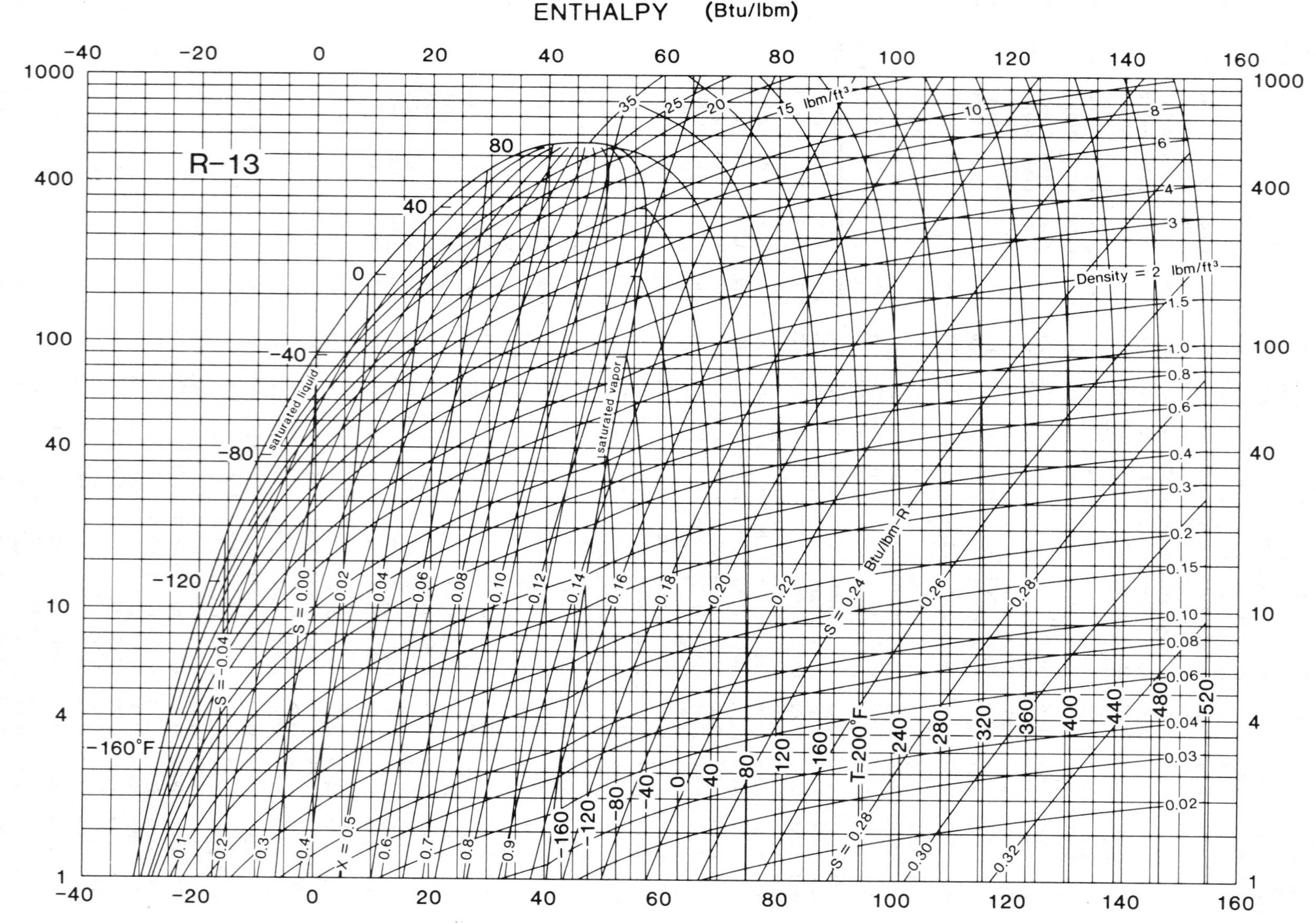

Prepared by: CENTER FOR APPLIED THERMODYNAMIC STUDIES, University of Idaho

Fig. 3 Pressure-Enthalpy Diagram for Refrigerant 13

Refrigerant 13 (Chlorotrifluoromethane) Properties of Saturated Liquid and Saturated Vapor

Temp,* °F	Pressure, psia	Density, lb/ft³ Liquid	Volume, ft³/lb Vapor	Enthalpy, Btu/lb Liquid	Enthalpy, Btu/lb Vapor	Entropy, Btu/lb·°F Liquid	Entropy, Btu/lb·°F Vapor	Specific Heat c_p, Btu/lb·°F Liquid	Specific Heat c_p, Btu/lb·°F Vapor	c_p/c_v Vapor	Velocity of Sound, ft/s Liquid	Velocity of Sound, ft/s Vapor	Viscosity, lb_m/ft·h Liquid	Viscosity, lb_m/ft·h Vapor	Thermal Cond, Btu/h·ft·°F Liquid	Thermal Cond, Btu/h·ft·°F Vapor	Surface Tension, dyne/cm	Temp,* °F
−150.00	4.452	99.55	7.0160	−24.352	43.797	−0.06649	0.15358	–	0.1138	1.2216	–	417.	–	–	–	–	17.33	−150.00
−145.00	5.368	98.92	5.8959	−23.355	44.297	−0.06330	0.15169	–	0.1152	1.2216	–	419.	–	–	–	–	16.86	−145.00
−140.00	6.431	98.29	4.9849	−22.347	44.796	−0.06013	0.14991	–	0.1167	1.2218	–	421.	–	–	–	–	16.39	−140.00
−135.00	7.655	97.65	4.2390	−21.329	45.292	−0.05698	0.14822	–	0.1182	1.2223	–	423.	–	–	–	–	15.93	−135.00
−130.00	9.059	97.00	3.6245	−20.301	45.786	−0.05384	0.14662	–	0.1198	1.2232	–	425.	–	–	–	–	15.46	−130.00
−125.00	10.659	96.35	3.1150	−19.262	46.276	−0.05072	0.14511	–	0.1214	1.2244	–	426.	–	–	–	–	15.00	−125.00
−120.00	12.474	95.70	2.6902	−18.212	46.762	−0.04762	0.14367	–	0.1231	1.2259	–	428.	–	–	–	–	14.55	−120.00
−115.00	14.523	95.03	2.3340	−17.151	47.245	−0.04453	0.14230	–	0.1248	1.2277	–	429.	–	–	–	–	14.09	−115.00
−114.60b	14.696	94.98	2.3083	−17.067	47.283	−0.04429	0.14220	–	0.1250	1.2279	–	429.	–	–	–	–	14.06	−114.60
−110.00	16.826	94.36	2.0338	−16.080	47.722	−0.04146	0.14100	–	0.1266	1.2299	–	431.	–	–	–	–	13.64	−110.00
−105.00	19.403	93.69	1.7795	−14.998	48.194	−0.03840	0.13977	–	0.1285	1.2325	–	432.	–	–	–	–	13.20	−105.00
−100.00	22.276	93.01	1.5632	−13.905	48.660	−0.03536	0.13860	–	0.1304	1.2355	–	433.	–	–	–	–	12.75	−100.00
−95.00	25.466	92.32	1.3781	−12.802	49.121	−0.03233	0.13748	–	0.1324	1.2390	–	434.	–	–	–	–	12.31	−95.00
−90.00	28.995	91.62	1.2192	−11.689	49.574	−0.02932	0.13641	0.1721	0.1344	1.2428	–	434.	0.634	0.0243	0.0468	0.00383	11.87	−90.00
−85.00	32.885	90.91	1.0822	−10.565	50.021	−0.02632	0.13539	0.1761	0.1365	1.2472	–	435.	0.600	0.0247	0.0459	0.00395	11.44	−85.00
−80.00	37.160	90.20	0.9636	−9.431	50.459	−0.02333	0.13441	0.1800	0.1388	1.2520	–	435.	0.569	0.0251	0.0450	0.00407	11.01	−80.00
−75.00	41.842	89.47	0.8606	−8.287	50.890	−0.02037	0.13347	0.1839	0.1411	1.2574	–	435.	0.540	0.0255	0.0442	0.00418	10.58	−75.00
−70.00	46.955	88.74	0.7706	−7.133	51.312	−0.01741	0.13257	0.1878	0.1435	1.2634	–	436.	0.514	0.0259	0.0434	0.00430	10.16	−70.00
−68.00	49.126	88.44	0.7379	−6.668	51.478	−0.01623	0.13222	0.1893	0.1445	1.2659	–	436.	0.504	0.0261	0.0431	0.00435	9.99	−68.00
−66.00	51.372	88.15	0.7069	−6.202	51.643	−0.01506	0.13188	0.1909	0.1455	1.2686	–	435.	0.494	0.0262	0.0428	0.00440	9.82	−66.00
−64.00	53.694	87.85	0.6774	−5.734	51.806	−0.01389	0.13154	0.1924	0.1466	1.2713	–	435.	0.484	0.0264	0.0425	0.00444	9.66	−64.00
−62.00	56.092	87.55	0.6494	−5.265	51.968	−0.01272	0.13120	0.1940	0.1476	1.2742	–	435.	0.475	0.0265	0.0422	0.00449	9.49	−62.00
−60.00	58.570	87.24	0.6228	−4.794	52.128	−0.01155	0.13087	0.1956	0.1487	1.2772	–	435.	0.466	0.0267	0.0418	0.00454	9.33	−60.00
−58.00	61.129	86.94	0.5975	−4.322	52.287	−0.01038	0.13055	0.1971	0.1498	1.2803	–	435.	0.458	0.0269	0.0416	0.00459	9.16	−58.00
−56.00	63.769	86.63	0.5735	−3.848	52.443	−0.00922	0.13023	0.1987	0.1509	1.2835	–	435.	0.449	0.0270	0.0413	0.00464	9.00	−56.00
−54.00	66.493	86.32	0.5506	−3.373	52.599	−0.00806	0.12991	0.2003	0.1520	1.2869	–	435.	0.441	0.0272	0.0410	0.00468	8.83	−54.00
−52.00	69.302	86.01	0.5288	−2.895	52.752	−0.00690	0.12960	0.2019	0.1532	1.2904	–	434.	0.433	0.0273	0.0407	0.00473	8.67	−52.00
−50.00	72.199	85.70	0.5081	−2.417	52.903	−0.00575	0.12929	0.2035	0.1543	1.2940	–	434.	0.426	0.0275	0.0404	0.00478	8.51	−50.00
−48.00	75.183	85.38	0.4883	−1.937	53.053	−0.00459	0.12898	0.2051	0.1555	1.2977	–	434.	0.418	0.0276	0.0401	0.00484	8.35	−48.00
−46.00	78.258	85.06	0.4694	−1.455	53.201	−0.00344	0.12868	0.2068	0.1568	1.3016	–	433.	0.411	0.0278	0.0398	0.00489	8.18	−46.00
−44.00	81.424	84.74	0.4515	−0.972	53.347	−0.00229	0.12838	0.2085	0.1580	1.3057	–	433.	0.404	0.0279	0.0395	0.00494	8.02	−44.00
−42.00	84.683	84.42	0.4343	−0.487	53.491	−0.00114	0.12809	0.2102	0.1593	1.3099	–	433.	0.397	0.0281	0.0393	0.00499	7.86	−42.00
−40.00	88.037	84.10	0.4179	0.000	53.633	0.00000	0.12780	0.2119	0.1606	1.3143	–	432.	0.391	0.0282	0.0390	0.00505	7.70	−40.00
−38.00	91.487	83.77	0.4022	0.488	53.773	0.00114	0.12751	0.2136	0.1620	1.3188	–	432.	0.384	0.0284	0.0387	0.00510	7.55	−38.00
−36.00	95.036	83.44	0.3873	0.978	53.911	0.00228	0.12722	0.2154	0.1633	1.3236	–	431.	0.378	0.0286	0.0385	0.00516	7.39	−36.00
−34.00	98.684	83.11	0.3730	1.469	54.046	0.00342	0.12694	0.2172	0.1647	1.3285	–	431.	0.372	0.0287	0.0382	0.00521	7.23	−34.00
−32.00	102.43	82.77	0.3593	1.962	54.180	0.00456	0.12665	0.2190	0.1662	1.3336	–	430.	0.366	0.0289	0.0379	0.00527	7.08	−32.00
−30.00	106.29	82.43	0.3462	2.457	54.311	0.00569	0.12637	0.2208	0.1676	1.3389	–	429.	0.360	0.0291	0.0377	0.00533	6.92	−30.00
−28.00	110.24	82.09	0.3337	2.953	54.440	0.00682	0.12609	0.2227	0.1691	1.3445	–	429.	0.354	0.0292	0.0374	0.00539	6.77	−28.00
−26.00	114.31	81.75	0.3217	3.451	54.566	0.00795	0.12582	0.2246	0.1707	1.3502	–	428.	0.349	0.0294	0.0372	0.00545	6.61	−26.00
−24.00	118.48	81.40	0.3102	3.951	54.691	0.00908	0.12554	0.2266	0.1723	1.3562	–	428.	0.343	0.0296	0.0369	0.00551	6.46	−24.00
−22.00	122.76	81.05	0.2992	4.452	54.812	0.01020	0.12527	0.2286	0.1739	1.3625	–	427.	0.338	0.0298	0.0367	0.00558	6.31	−22.00
−20.00	127.15	80.70	0.2886	4.955	54.931	0.01133	0.12500	0.2306	0.1756	1.3690	–	426.	0.333	0.0299	0.0364	0.00564	6.16	−20.00
−18.00	131.65	80.34	0.2785	5.460	55.048	0.01245	0.12472	0.2327	0.1773	1.3758	–	425.	0.328	0.0301	0.0362	0.00571	6.00	−18.00
−16.00	136.27	79.98	0.2688	5.967	55.162	0.01357	0.12445	0.2348	0.1791	1.3829	–	424.	0.323	0.0303	0.0359	0.00578	5.85	−16.00
−14.00	141.01	79.62	0.2594	6.476	55.273	0.01469	0.12418	0.2370	0.1809	1.3903	–	423.	0.318	0.0305	0.0357	0.00585	5.71	−14.00
−12.00	145.86	79.25	0.2505	6.986	55.381	0.01581	0.12391	0.2392	0.1827	1.3980	–	423.	0.314	0.0307	0.0354	0.00592	5.56	−12.00
−10.00	150.83	78.88	0.2418	7.498	55.486	0.01692	0.12364	0.2414	0.1847	1.4061	–	422.	0.309	0.0309	0.0352	0.00599	5.41	−10.00
−8.00	155.93	78.51	0.2336	8.013	55.588	0.01804	0.12337	0.2437	0.1867	1.4145	–	421.	0.305	0.0311	0.0350	0.00606	5.26	−8.00
−6.00	161.14	78.13	0.2256	8.529	55.687	0.01915	0.12310	0.2461	0.1887	1.4234	–	420.	0.301	0.0313	0.0347	0.00614	5.12	−6.00
−4.00	166.48	77.74	0.2179	9.048	55.783	0.02027	0.12283	0.2485	0.1909	1.4326	–	419.	0.296	0.0315	0.0345	0.00622	4.97	−4.00
−2.00	171.95	77.36	0.2106	9.568	55.875	0.02138	0.12256	0.2509	0.1931	1.4423	–	417.	0.292	0.0317	0.0343	0.00629	4.83	−2.00
0.00	177.55	76.96	0.2035	10.091	55.965	0.02249	0.12228	0.2534	0.1954	1.4525	–	416.	0.288	0.0319	0.0340	0.00638	4.69	0.00
5.00	192.12	75.96	0.1869	11.408	56.172	0.02526	0.12159	0.2599	0.2015	1.4802	–	413.	0.279	0.0325	0.0335	0.00659	4.33	5.00
10.00	207.53	74.92	0.1717	12.740	56.354	0.02803	0.12089	0.2668	0.2082	1.5117	–	410.	0.270	0.0331	0.0329	0.00681	3.99	10.00
15.00	223.81	73.85	0.1579	14.089	56.510	0.03080	0.12017	0.2742	0.2157	1.5478	–	406.	0.261	0.0337	0.0324	0.00705	3.65	15.00
20.00	241.00	72.73	0.1452	15.457	56.637	0.03358	0.11943	0.2819	0.2241	1.5894	–	402.	0.253	0.0343	0.0318	0.00730	3.31	20.00
25.00	259.13	71.56	0.1335	16.846	56.730	0.03636	0.11865	0.2901	0.2337	1.6380	–	398.	0.245	0.0350	0.0313	0.00756	2.98	25.00
30.00	278.23	70.34	0.1227	18.260	56.787	0.03916	0.11784	0.2988	0.2447	1.6953	–	394.	0.238	0.0358	0.0308	0.00784	2.67	30.00
35.00	298.34	69.06	0.1127	19.702	56.800	0.04198	0.11698	–	0.2577	1.7640	–	389.	–	–	–	–	2.35	35.00
40.00	319.50	67.69	0.1035	21.177	56.765	0.04483	0.11606	–	0.2731	1.8477	–	383.	–	–	–	–	2.05	40.00
45.00	341.75	66.24	0.0948	22.691	56.671	0.04773	0.11506	–	0.2920	1.9522	–	378.	–	–	–	–	1.76	45.00
50.00	365.15	64.68	0.0867	24.256	56.506	0.05068	0.11396	–	0.3158	2.0862	–	371.	–	–	–	–	1.48	50.00
55.00	389.75	62.98	0.0791	25.883	56.254	0.05372	0.11273	–	0.3470	2.2644	–	365.	–	–	–	–	1.20	55.00
60.00	415.62	61.10	0.0718	27.593	55.889	0.05688	0.11133	–	0.3900	2.5141	–	357.	–	–	–	–	0.95	60.00
65.00	442.85	58.97	0.0648	29.418	55.370	0.06021	0.10967	–	0.4541	2.8901	–	349.	–	–	–	–	0.70	65.00
70.00	471.55	56.47	0.0579	31.418	54.624	0.06383	0.10764	–	0.5612	3.5254	–	340.	–	–	–	–	0.47	70.00
75.00	501.88	53.36	0.0507	33.723	53.494	0.06797	0.10495	–	0.7824	4.8478	–	331.	–	–	–	–	0.27	75.00
80.00	534.14	48.85	0.0426	36.745	51.495	0.07338	0.10071	–	–	–	–	–	–	–	–	–	0.09	80.00
83.93c	561.23	36.07	0.0277	44.270	44.270	0.08704	0.08704	∞	∞	∞	0.	0.	–	–	∞	∞	0.00	83.93

*temperatures are on the IPTS–68 scale b = normal boiling point c = critical point

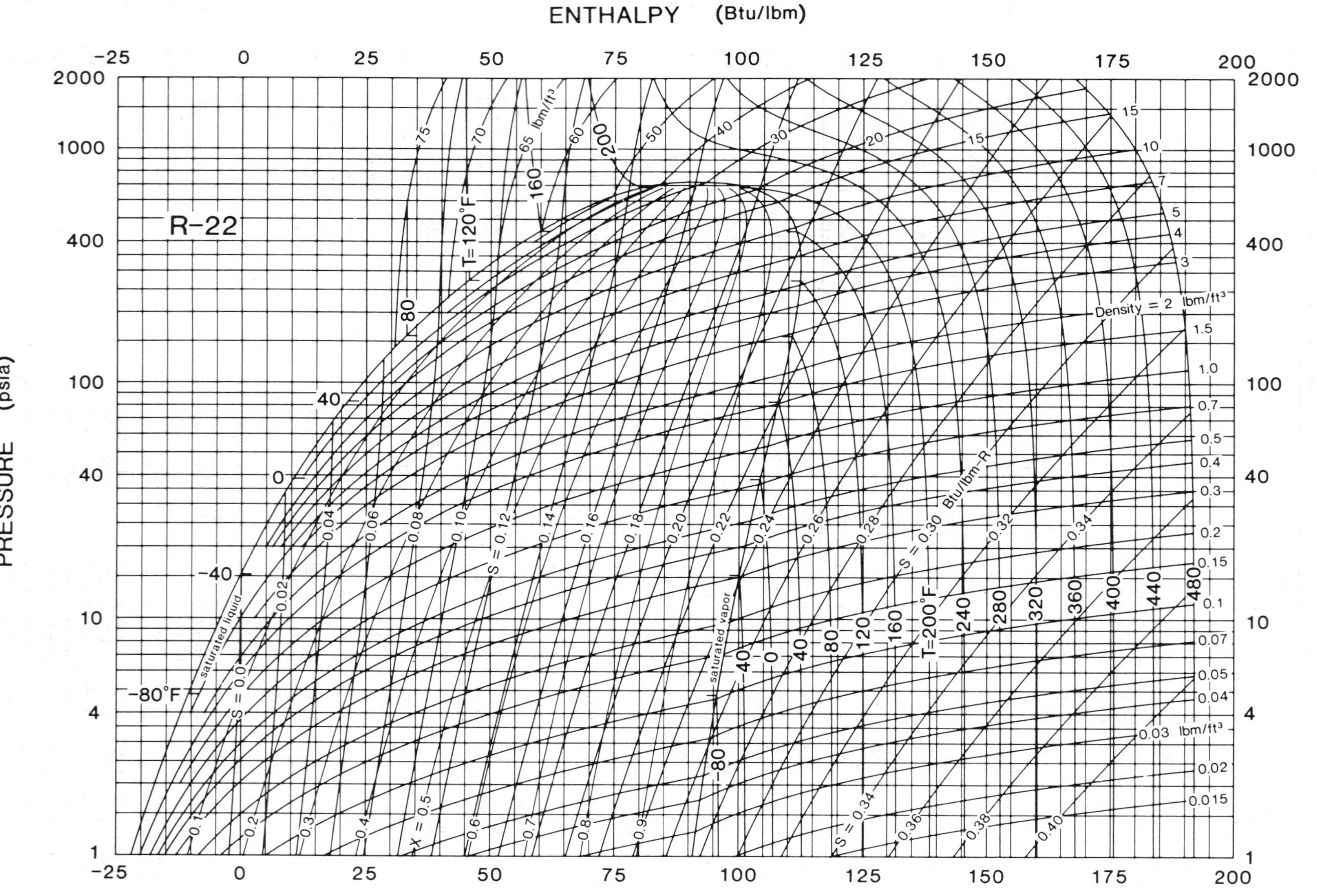

Prepared by: CENTER FOR APPLIED THERMODYNAMIC STUDIES, University of Idaho

Fig. 4 Pressure-Enthalpy Diagram for Refrigerant 22

Refrigerant 22 (Chlorodifluoromethane) Properties of Saturated Liquid and Saturated Vapor

Temp,* °F	Pressure, psia	Density, lb/ft³ Liquid	Volume, ft³/lb Vapor	Enthalpy, Btu/lb Liquid	Enthalpy, Btu/lb Vapor	Entropy, Btu/lb·°F Liquid	Entropy, Btu/lb·°F Vapor	Specific Heat c_p, Btu/lb·°F Liquid	Specific Heat c_p, Btu/lb·°F Vapor	c_p/c_v Vapor	Velocity of Sound, ft/s Liquid	Velocity of Sound, ft/s Vapor	Viscosity, lb_m/ft·h Liquid	Viscosity, lb_m/ft·h Vapor	Thermal Cond, Btu/h·ft·°F Liquid	Thermal Cond, Btu/h·ft·°F Vapor	Surface Tension, dyne/cm	Temp,* °F
−250.00	—	107.37	—	−63.169	76.604	−0.21914	0.44952	—	0.1018	1.2914	—	395.	—	—	—	—	—	−250.00
−240.00	—	106.41	—	−56.462	77.629	−0.18786	0.42332	—	0.1033	1.2860	—	403.	—	—	—	—	—	−240.00
−230.00	—	105.48	—	−51.569	78.669	−0.16605	0.40101	—	0.1048	1.2807	—	411.	—	—	—	—	36.75	−230.00
−220.00	0.002	104.58	16805.	−47.705	79.724	−0.14958	0.38211	—	0.1064	1.2754	—	419.	—	—	—	—	35.70	−220.00
−210.00	0.004	103.70	6982.6	−44.426	80.796	−0.13616	0.36538	—	0.1080	1.2703	—	427.	—	—	—	—	34.67	−210.00
−200.00	0.010	102.81	3151.5	−41.474	81.882	−0.12457	0.35048	—	0.1096	1.2653	—	435.	—	—	—	—	33.63	−200.00
−190.00	0.022	101.92	1527.4	−38.706	82.984	−0.11411	0.33715	—	0.1113	1.2604	—	442.	—	—	—	—	32.61	−190.00
−180.00	0.044	101.03	787.79	−36.038	84.100	−0.10439	0.32518	—	0.1130	1.2558	—	449.	—	—	—	—	31.59	−180.00
−170.00	0.084	100.12	429.22	−33.424	85.230	−0.09521	0.31441	—	0.1147	1.2515	—	456.	—	—	—	—	30.58	−170.00
−160.00	0.151	99.22	245.51	−30.839	86.373	−0.08644	0.30470	—	0.1165	1.2474	—	463.	—	—	—	—	29.57	−160.00
−150.00	0.262	98.30	146.65	−28.269	87.528	−0.07800	0.29594	—	0.1183	1.2437	—	470.	—	—	—	—	28.57	−150.00
−140.00	0.435	97.38	91.059	−25.708	88.692	−0.06986	0.28801	—	0.1201	1.2403	—	476.	—	—	—	—	27.57	−140.00
−130.00	0.696	96.46	58.544	−23.150	89.864	−0.06198	0.28082	—	0.1221	1.2374	—	482.	—	—	—	—	26.59	−130.00
−120.00	1.080	95.53	38.833	−20.594	91.040	−0.05435	0.27430	0.2555	0.1241	1.2349	3483.	488.	—	—	—	—	25.61	−120.00
−110.00	1.626	94.60	26.494	−18.038	92.218	−0.04694	0.26838	0.2555	0.1262	1.2329	3384.	494.	—	—	0.0765	—	24.64	−110.00
−100.00	2.384	93.66	18.540	−15.481	93.397	−0.03973	0.26298	0.2557	0.1285	1.2315	3290.	500.	—	—	0.0749	—	23.67	−100.00
−90.00	3.413	92.71	13.275	−12.921	94.572	−0.03271	0.25807	0.2561	0.1308	1.2307	3198.	505.	—	—	0.0734	0.00292	22.71	−90.00
−80.00	4.778	91.75	9.7044	−10.355	95.741	−0.02587	0.25357	0.2567	0.1334	1.2305	3110.	510.	—	—	0.0718	0.00315	21.76	−80.00
−70.00	6.555	90.79	7.2285	−7.783	96.901	−0.01919	0.24945	0.2574	0.1361	1.2310	3023.	514.	—	—	0.0703	0.00338	20.82	−70.00
−60.00	8.830	89.81	5.4766	−5.201	98.049	−0.01266	0.24567	0.2584	0.1389	1.2323	2937.	519.	—	—	0.0688	0.00360	19.89	−60.00
−50.00	11.696	88.83	4.2138	−2.608	99.182	−0.00627	0.24220	0.2596	0.1420	1.2344	2852.	522.	—	—	0.0673	0.00382	18.96	−50.00
−45.00	13.383	88.33	3.7160	−1.306	99.742	−0.00312	0.24056	0.2604	0.1436	1.2358	2810.	524.	—	—	0.0665	0.00393	18.50	−45.00
−41.44b	14.696	87.97	3.4048	−0.377	100.138	−0.00090	0.23944	0.2609	0.1448	1.2369	2780.	525.	—	—	0.0660	0.00401	18.18	−41.44
−40.00	15.255	87.82	3.2880	0.000	100.296	0.00000	0.23899	0.2611	0.1453	1.2374	2768.	526.	—	—	0.0658	0.00404	18.05	−40.00
−35.00	17.329	87.32	2.9185	1.310	100.847	0.00309	0.23748	0.2620	0.1471	1.2393	2725.	527.	—	—	0.0651	0.00414	17.59	−35.00
−30.00	19.617	86.81	2.5984	2.624	101.391	0.00616	0.23602	0.2629	0.1489	1.2414	2683.	529.	—	—	0.0643	0.00425	17.14	−30.00
−25.00	22.136	86.29	2.3202	3.944	101.928	0.00920	0.23462	0.2638	0.1507	1.2437	2641.	530.	—	—	0.0636	0.00435	16.69	−25.00
−20.00	24.899	85.77	2.0774	5.268	102.461	0.01222	0.23327	0.2648	0.1527	1.2463	2599.	531.	—	—	0.0629	0.00445	16.24	−20.00
−15.00	27.924	85.25	1.8650	6.598	102.986	0.01521	0.23197	0.2659	0.1547	1.2493	2557.	532.	—	—	0.0622	0.00456	15.79	−15.00
−10.00	31.226	84.72	1.6784	7.934	103.503	0.01818	0.23071	0.2671	0.1567	1.2525	2515.	533.	—	—	0.0614	0.00466	—	−10.00
−5.00	34.821	84.18	1.5142	9.276	104.013	0.02113	0.22949	0.2684	0.1589	1.2560	2473.	534.	—	—	0.0607	0.00476	—	−5.00
0.00	38.726	83.64	1.3691	10.624	104.515	0.02406	0.22832	0.2697	0.1611	1.2599	2431.	535.	0.615	0.0268	0.0600	0.00486	—	0.00
5.00	42.960	83.09	1.2406	11.979	105.009	0.02697	0.22718	0.2710	0.1634	1.2641	2389.	535.	0.597	0.0271	0.0593	0.00496	—	5.00
10.00	47.538	82.54	1.1265	13.342	105.493	0.02987	0.22607	0.2725	0.1658	1.2687	2346.	535.	0.580	0.0274	0.0586	0.00506	—	10.00
15.00	52.480	81.98	1.0250	14.712	105.968	0.03275	0.22500	0.2740	0.1683	1.2737	2304.	536.	0.563	0.0276	0.0579	0.00516	—	15.00
20.00	57.803	81.41	0.9343	16.090	106.434	0.03561	0.22395	0.2756	0.1709	1.2792	2262.	536.	0.546	0.0279	0.0572	0.00526	—	20.00
25.00	63.526	80.84	0.8532	17.476	106.891	0.03846	0.22294	0.2773	0.1737	1.2851	2219.	536.	0.530	0.0282	0.0566	0.00536	—	25.00
30.00	69.667	80.26	0.7804	18.871	107.336	0.04129	0.22195	0.2791	0.1765	1.2915	2177.	536.	0.515	0.0284	0.0559	0.00546	—	30.00
35.00	76.245	79.67	0.7150	20.275	107.769	0.04411	0.22098	0.2809	0.1794	1.2984	2134.	535.	0.499	0.0287	0.0552	0.00555	—	35.00
40.00	83.280	79.07	0.6561	21.688	108.191	0.04692	0.22004	0.2829	0.1825	1.3059	2091.	535.	0.484	0.0290	0.0545	0.00565	—	40.00
45.00	90.791	78.46	0.6029	23.111	108.600	0.04972	0.21912	0.2849	0.1857	1.3141	2048.	534.	0.470	0.0292	0.0538	0.00575	—	45.00
50.00	98.799	77.84	0.5548	24.544	108.997	0.05251	0.21821	0.2870	0.1891	1.3229	2005.	533.	0.456	0.0295	0.0532	0.00584	—	50.00
55.00	107.32	77.22	0.5111	25.988	109.379	0.05529	0.21732	0.2893	0.1927	1.3324	1962.	532.	0.442	0.0298	0.0525	0.00594	—	55.00
60.00	116.38	76.58	0.4715	27.443	109.748	0.05806	0.21644	0.2916	0.1964	1.3428	1919.	531.	0.429	0.0301	0.0518	0.00604	—	60.00
65.00	126.00	75.93	0.4355	28.909	110.103	0.06082	0.21557	0.2941	0.2003	1.3540	1876.	530.	0.416	0.0303	0.0512	0.00613	—	65.00
70.00	136.19	75.27	0.4026	30.387	110.441	0.06358	0.21472	0.2967	0.2045	1.3663	1832.	528.	0.404	—	0.0505	0.00623	—	70.00
75.00	146.98	74.60	0.3726	31.877	110.761	0.06633	0.21387	0.2994	0.2089	1.3796	1788.	527.	0.392	—	0.0499	0.00632	—	75.00
80.00	158.40	73.92	0.3451	33.381	111.066	0.06907	0.21302	0.3024	0.2135	1.3941	1744.	525.	0.380	—	0.0492	0.00642	—	80.00
85.00	170.45	73.22	0.3199	34.898	111.350	0.07182	0.21218	0.3055	0.2185	1.4100	1700.	523.	0.369	—	0.0486	0.00652	—	85.00
90.00	183.17	72.51	0.2968	36.430	111.616	0.07456	0.21134	0.3088	0.2238	1.4275	1655.	520.	0.358	—	0.0479	0.00661	—	90.00
95.00	196.57	71.79	0.2756	37.977	111.859	0.07730	0.21050	0.3123	0.2295	1.4467	1611.	518.	0.348	—	0.0473	0.00671	—	95.00
100.00	210.69	71.05	0.2560	39.538	112.081	0.08003	0.20965	0.3162	0.2356	1.4678	1566.	515.	0.338	—	0.0466	0.00680	—	100.00
105.00	225.53	70.29	0.2379	41.119	112.278	0.08277	0.20879	0.3203	0.2422	1.4912	1520.	512.	—	—	0.0460	0.00690	—	105.00
110.00	241.14	69.51	0.2212	42.717	112.448	0.08552	0.20793	0.3248	0.2495	1.5173	1474.	509.	—	—	0.0454	0.00699	—	110.00
115.00	257.52	68.71	0.2058	44.334	112.591	0.08827	0.20705	0.3298	0.2573	1.5464	1428.	506.	—	—	0.0447	0.00709	—	115.00
120.00	274.71	67.89	0.1914	45.972	112.704	0.09103	0.20615	0.3353	0.2660	1.5791	1382.	502.	—	—	0.0441	0.00719	—	120.00
125.00	292.73	67.05	0.1781	47.633	112.783	0.09379	0.20522	0.3413	0.2756	1.6160	1334.	498.	—	—	—	—	—	125.00
130.00	311.61	66.17	0.1657	49.319	112.825	0.09657	0.20427	0.3482	0.2864	1.6581	1287.	494.	—	—	—	—	—	130.00
135.00	331.38	65.27	0.1542	51.032	112.826	0.09937	0.20329	0.3559	0.2985	1.7063	1238.	489.	—	—	—	—	—	135.00
140.00	352.07	64.33	0.1434	52.775	112.784	0.10220	0.20227	0.3648	0.3123	1.7621	1189.	485.	—	—	—	—	—	140.00
145.00	373.71	63.35	0.1332	54.553	112.692	0.10504	0.20119	0.3752	0.3282	1.8275	1139.	479.	—	—	—	—	—	145.00
150.00	396.32	62.33	0.1237	56.370	112.541	0.10793	0.20006	0.3873	0.3468	1.9050	1088.	474.	—	—	—	—	—	150.00
160.00	444.65	60.12	0.1063	60.145	112.035	0.11383	0.19757	0.4198	0.3957	2.1126	983.	462.	—	—	—	—	—	160.00
170.00	497.35	57.59	0.0907	64.175	111.165	0.12001	0.19464	0.4711	0.4716	2.4409	873.	448.	—	—	—	—	—	170.00
180.00	554.82	54.57	0.0763	68.597	109.753	0.12668	0.19102	0.5657	0.6073	3.0349	752.	433.	—	—	—	—	—	180.00
190.00	617.53	50.62	0.0625	73.742	107.398	0.13432	0.18613	0.7952	0.9222	4.4150	616.	415.	—	—	—	—	—	190.00
200.00	686.11	44.44	0.0478	80.558	102.809	0.14432	0.17805	—	—	—	—	—	—	—	—	—	—	200.00
205.06c	723.74	32.70	0.0306	91.052	91.052	0.15989	0.15989	∞	∞	∞	0.	0.	—	—	∞	∞	0.00	205.06

*temperatures are on the ITS–90 scale

b = normal boiling point

c = critical point

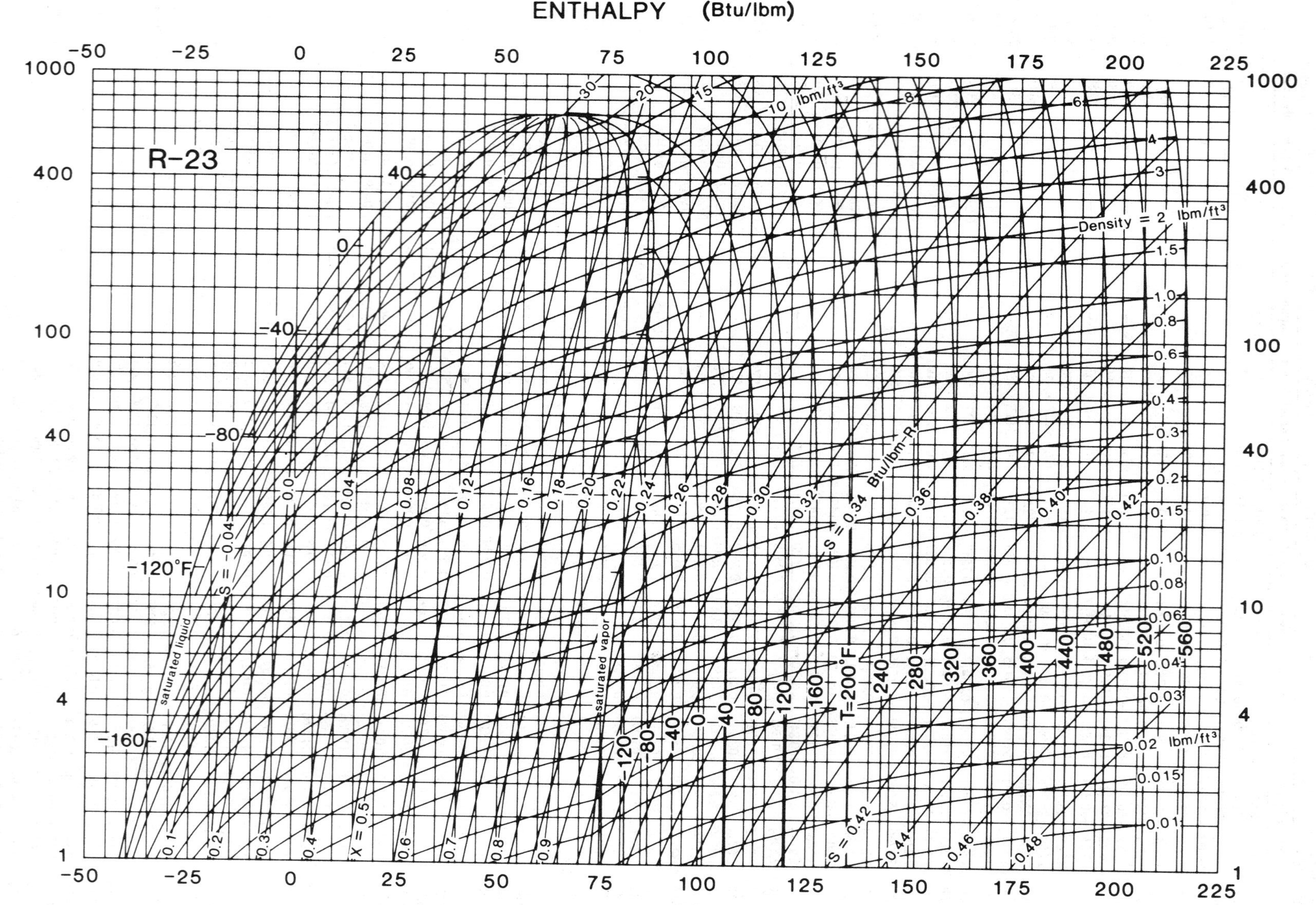

Fig. 5 Pressure-Enthalpy Diagram for Refrigerant 23

Refrigerant 23 (Trifluoromethane) Properties of Saturated Liquid and Saturated Vapor

Temp,* °F	Pressure, psia	Density, lb/ft³ Liquid	Volume, ft³/lb Vapor	Enthalpy, Btu/lb Liquid	Enthalpy, Btu/lb Vapor	Entropy, Btu/lb·°F Liquid	Entropy, Btu/lb·°F Vapor	Specific Heat c_p, Btu/lb·°F Liquid	Specific Heat c_p, Btu/lb·°F Vapor	c_p/c_v Vapor	Velocity of Sound, ft/s Liquid	Velocity of Sound, ft/s Vapor	Viscosity, lb$_m$/ft·h Liquid	Viscosity, lb$_m$/ft·h Vapor	Thermal Cond, Btu/h·ft·°F Liquid	Thermal Cond, Btu/h·ft·°F Vapor	Surface Tension, dyne/cm	Temp,* °F
−160.00	2.774	94.42	16.315	−35.171	75.195	−0.09766	0.27063	–	0.1363	1.2960	–	517.	–	–	–	–	–	−160.00
−155.00	3.442	93.95	13.334	−33.755	75.756	−0.09298	0.26646	–	0.1380	1.2969	–	521.	–	–	–	–	–	−155.00
−150.00	4.237	93.46	10.981	−32.354	76.309	−0.08842	0.26247	–	0.1399	1.2980	–	524.	–	–	–	–	–	−150.00
−145.00	5.177	92.97	9.1062	−30.964	76.853	−0.08398	0.25866	–	0.1419	1.2995	–	527.	–	–	–	–	–	−145.00
−140.00	6.279	92.46	7.6019	−29.581	77.388	−0.07962	0.25500	–	0.1441	1.3014	–	529.	–	–	–	–	–	−140.00
−135.00	7.566	91.95	6.3857	−28.201	77.913	−0.07535	0.25149	–	0.1464	1.3036	–	532.	–	–	–	–	–	−135.00
−130.00	9.058	91.42	5.3956	−26.822	78.427	−0.07114	0.24811	–	0.1488	1.3062	–	535.	–	–	–	–	–	−130.00
−125.00	10.778	90.88	4.5844	−25.440	78.930	−0.06699	0.24487	–	0.1514	1.3091	–	537.	–	–	–	–	–	−125.00
−120.00	12.751	90.32	3.9155	−24.052	79.421	−0.06289	0.24174	–	0.1542	1.3125	–	539.	–	–	–	–	–	−120.00
−115.64b	14.696	89.83	3.4267	−22.837	79.839	−0.05935	0.23911	–	0.1567	1.3157	–	541.	–	–	–	–	–	−115.64
−115.00	15.003	89.76	3.3607	−22.657	79.900	−0.05882	0.23873	–	0.1571	1.3162	–	541.	–	–	–	–	–	−115.00
−110.00	17.561	89.18	2.8981	−21.252	80.366	−0.05479	0.23582	–	0.1602	1.3204	–	543.	–	–	–	–	–	−110.00
−105.00	20.452	88.59	2.5102	−19.835	80.818	−0.05078	0.23301	–	0.1634	1.3251	–	545.	–	–	–	–	–	−105.00
−100.00	23.708	87.98	2.1834	−18.405	81.256	−0.04680	0.23029	–	0.1669	1.3302	–	546.	–	–	–	–	–	−100.00
−98.00	25.118	87.74	2.0672	−17.829	81.428	−0.04521	0.22923	–	0.1683	1.3324	–	547.	–	–	–	–	–	−98.00
−96.00	26.594	87.49	1.9584	−17.251	81.597	−0.04363	0.22818	–	0.1698	1.3346	–	547.	–	–	–	–	–	−96.00
−94.00	28.137	87.24	1.8565	−16.670	81.764	−0.04204	0.22714	–	0.1713	1.3370	–	548.	–	–	–	–	–	−94.00
−92.00	29.748	86.98	1.7609	−16.087	81.928	−0.04046	0.22612	–	0.1728	1.3394	–	548.	–	–	–	–	–	−92.00
−90.00	31.431	86.73	1.6712	−15.501	82.090	−0.03888	0.22511	–	0.1744	1.3419	–	549.	–	–	–	–	–	−90.00
−88.00	33.187	86.47	1.5871	−14.912	82.249	−0.03730	0.22411	–	0.1760	1.3445	–	549.	–	–	–	–	–	−88.00
−86.00	35.019	86.21	1.5079	−14.321	82.407	−0.03573	0.22313	–	0.1776	1.3472	–	550.	–	–	–	–	–	−86.00
−84.00	36.927	85.95	1.4336	−13.728	82.561	−0.03416	0.22216	–	0.1792	1.3500	–	550.	–	–	–	–	–	−84.00
−82.00	38.916	85.68	1.3636	−13.132	82.713	−0.03258	0.22119	–	0.1809	1.3529	–	550.	–	–	–	–	–	−82.00
−80.00	40.986	85.42	1.2977	−12.533	82.863	−0.03101	0.22025	–	0.1826	1.3559	–	551.	–	–	–	–	–	−80.00
−78.00	43.140	85.15	1.2356	−11.931	83.010	−0.02945	0.21930	–	0.1844	1.3589	–	551.	–	–	–	–	–	−78.00
−76.00	45.380	84.87	1.1771	−11.327	83.155	−0.02788	0.21838	–	0.1862	1.3621	–	551.	–	–	–	–	13.68	−76.00
−74.00	47.709	84.60	1.1219	−10.720	83.297	−0.02632	0.21746	–	0.1880	1.3654	–	551.	–	–	–	–	13.43	−74.00
−72.00	50.128	84.32	1.0698	−10.110	83.436	−0.02475	0.21655	–	0.1899	1.3688	–	551.	–	–	–	–	13.18	−72.00
−70.00	52.639	84.04	1.0207	−9.497	83.573	−0.02319	0.21565	–	0.1918	1.3722	–	551.	–	–	–	–	12.93	−70.00
−68.00	55.246	83.76	0.9742	−8.882	83.707	−0.02163	0.21477	–	0.1937	1.3758	–	552.	–	–	–	–	12.69	−68.00
−66.00	57.951	83.47	0.9302	−8.264	83.839	−0.02007	0.21389	–	0.1957	1.3796	–	552.	–	–	–	–	12.44	−66.00
−64.00	60.755	83.18	0.8887	−7.644	83.968	−0.01852	0.21302	–	0.1977	1.3834	–	552.	–	–	–	–	12.20	−64.00
−62.00	63.660	82.89	0.8493	−7.021	84.094	−0.01696	0.21216	–	0.1997	1.3874	–	552.	–	–	–	–	11.96	−62.00
−60.00	66.671	82.59	0.8121	−6.395	84.217	−0.01541	0.21131	–	0.2018	1.3915	–	552.	–	–	–	–	11.72	−60.00
−58.00	69.787	82.30	0.7768	−5.767	84.337	−0.01386	0.21047	–	0.2039	1.3957	–	551.	–	–	–	–	11.48	−58.00
−56.00	73.013	82.00	0.7433	−5.136	84.455	−0.01231	0.20963	–	0.2061	1.4000	–	551.	–	–	–	–	11.24	−56.00
−54.00	76.350	81.69	0.7116	−4.502	84.570	−0.01076	0.20881	–	0.2083	1.4045	–	551.	–	–	–	–	11.01	−54.00
−52.00	79.800	81.38	0.6814	−3.866	84.682	−0.00922	0.20799	–	0.2106	1.4092	–	551.	–	–	–	–	10.77	−52.00
−50.00	83.367	81.07	0.6528	−3.228	84.791	−0.00767	0.20718	–	0.2129	1.4140	–	551.	–	–	–	–	10.54	−50.00
−48.00	87.052	80.76	0.6257	−2.587	84.897	−0.00614	0.20637	–	0.2152	1.4190	–	550.	–	–	–	–	10.31	−48.00
−46.00	90.858	80.44	0.5998	−1.944	85.000	−0.00460	0.20558	–	0.2176	1.4241	–	550.	–	–	–	–	10.08	−46.00
−44.00	94.787	80.12	0.5753	−1.298	85.100	−0.00306	0.20479	–	0.2200	1.4294	–	550.	–	–	–	–	9.85	−44.00
−42.00	98.842	79.80	0.5519	−0.650	85.197	−0.00153	0.20401	–	0.2225	1.4349	–	549.	–	–	–	–	9.62	−42.00
−40.00	103.03	79.47	0.5296	0.000	85.291	0.00000	0.20323	–	0.2251	1.4405	738.	549.	0.396	–	0.0595	0.00654	9.40	−40.00
−38.00	107.34	79.14	0.5085	0.652	85.382	0.00153	0.20246	–	0.2277	1.4464	719.	548.	0.391	–	0.0590	0.00661	9.17	−38.00
−36.00	111.79	78.80	0.4883	1.307	85.469	0.00305	0.20170	–	0.2303	1.4525	701.	548.	0.386	–	0.0585	0.00668	8.95	−36.00
−34.00	116.37	78.47	0.4690	1.964	85.553	0.00457	0.20094	–	0.2330	1.4588	684.	547.	0.380	–	0.0581	0.00675	8.73	−34.00
−32.00	121.09	78.12	0.4507	2.623	85.634	0.00609	0.20019	–	0.2358	1.4653	667.	547.	0.375	–	0.0576	0.00683	8.51	−32.00
−30.00	125.95	77.78	0.4332	3.284	85.711	0.00761	0.19944	–	0.2386	1.4720	651.	546.	0.370	–	0.0572	0.00691	8.29	−30.00
−25.00	138.74	76.89	0.3928	4.947	85.888	0.01138	0.19760	–	0.2460	1.4901	613.	544.	0.357	–	0.0561	0.00711	7.76	−25.00
−20.00	152.46	75.98	0.3567	6.622	86.042	0.01514	0.19577	–	0.2538	1.5099	578.	542.	0.344	0.0299	0.0550	0.00732	7.23	−20.00
−15.00	167.17	75.04	0.3244	8.311	86.171	0.01888	0.19397	–	0.2622	1.5318	546.	540.	0.331	0.0304	0.0540	0.00755	6.72	−15.00
−10.00	182.90	74.07	0.2954	10.013	86.274	0.02260	0.19219	–	0.2711	1.5561	517.	537.	0.318	0.0309	0.0530	0.00778	6.22	−10.00
−5.00	199.70	73.07	0.2692	11.730	86.346	0.02630	0.19041	–	0.2808	1.5833	491.	534.	0.306	0.0314	0.0519	0.00803	–	−5.00
0.00	217.62	72.03	0.2457	13.462	86.387	0.02999	0.18863	–	0.2912	1.6138	466.	530.	0.293	0.0319	0.0508	0.00829	–	0.00
5.00	236.71	70.96	0.2243	15.213	86.392	0.03367	0.18685	–	0.3027	1.6484	444.	527.	0.281	0.0324	0.0497	0.00856	–	5.00
10.00	257.02	69.84	0.2049	16.985	86.358	0.03735	0.18505	–	0.3153	1.6879	423.	523.	0.269	0.0330	0.0484	0.00884	–	10.00
15.00	278.61	68.67	0.1872	18.783	86.278	0.04103	0.18323	–	0.3293	1.7334	403.	518.	–	0.0337	–	–	–	15.00
20.00	301.55	67.46	0.1710	20.611	86.148	0.04474	0.18136	–	0.3452	1.7866	385.	513.	–	0.0343	–	–	–	20.00
25.00	325.90	66.19	0.1562	22.479	85.959	0.04847	0.17944	–	0.3634	1.8495	367.	508.	–	0.0351	–	–	–	25.00
30.00	351.74	64.85	0.1426	24.396	85.701	0.05225	0.17745	–	0.3847	1.9250	349.	502.	–	0.0359	–	–	–	30.00
35.00	379.14	63.43	0.1300	26.375	85.361	0.05611	0.17536	–	0.4100	2.0176	332.	495.	–	0.0368	–	–	–	35.00
40.00	408.19	61.93	0.1183	28.434	84.923	0.06008	0.17314	–	0.4411	2.1338	314.	488.	–	0.0379	–	–	–	40.00
45.00	438.99	60.32	0.1074	30.598	84.365	0.06421	0.17074	–	0.4803	2.2840	296.	481.	–	0.0390	–	–	–	45.00
50.00	471.64	58.58	0.0971	32.901	83.654	0.06855	0.16813	–	0.5319	2.4859	277.	472.	–	0.0402	–	–	–	50.00
55.00	506.27	56.68	0.0874	35.389	82.745	0.07319	0.16520	–	0.6039	2.7721	257.	463.	–	0.0416	–	–	–	55.00
60.00	543.01	54.55	0.0780	38.138	81.561	0.07827	0.16182	–	0.7125	3.2100	236.	452.	–	–	–	–	–	60.00
65.00	581.99	52.08	0.0688	41.270	79.972	0.08400	0.15777	–	0.8981	3.9663	213.	441.	–	–	–	–	–	65.00
70.00	623.39	49.05	0.0594	45.032	77.695	0.09085	0.15252	–	1.2953	5.5959	–	–	–	–	–	–	–	70.00
75.00	667.39	44.71	0.0488	50.138	73.842	0.10012	0.14445	–	–	–	–	–	–	–	–	–	–	75.00
78.66c	701.40	32.78	0.0305	61.455	61.455	0.12090	0.12090	∞	∞	∞	0.	0.	–	–	∞	∞	0.00	78.66

*temperatures are on the IPTS-68 scale

b = normal boiling point

c = critical point

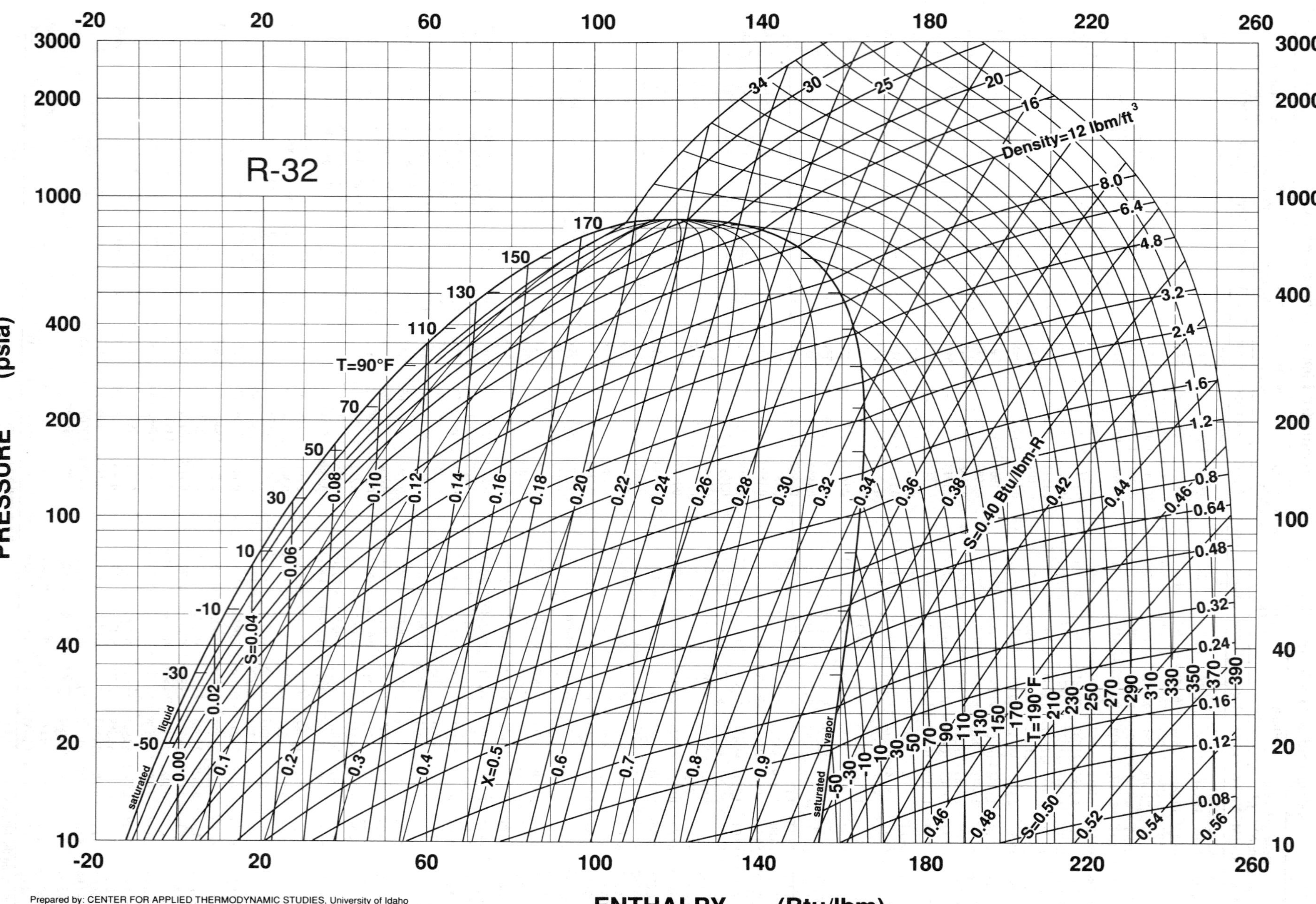

Fig. 6 Pressure–Enthalpy Diagram for Refrigerant 32

Refrigerant 32 (Difluoromethane) Properties of Saturated Liquid and Saturated Vapor

Temp,* °F	Pressure, psia	Density, lb/ft³ Liquid	Volume, ft³/lb Vapor	Enthalpy, Btu/lb Liquid	Enthalpy, Btu/lb Vapor	Entropy, Btu/lb·°F Liquid	Entropy, Btu/lb·°F Vapor	Specific Heat c_p, Btu/lb·°F Liquid	Specific Heat c_p, Btu/lb·°F Vapor	c_p/c_v Vapor	Velocity of Sound, ft/s Liquid	Velocity of Sound, ft/s Vapor	Viscosity, lb$_m$/ft·h Liquid	Viscosity, lb$_m$/ft·h Vapor	Thermal Cond, Btu/h·ft·°F Liquid	Thermal Cond, Btu/h·ft·°F Vapor	Surface Tension, dyne/cm	Temp,* °F
−70.00	11.318	75.87	6.8747	−11.246	155.538	−0.02771	0.40031	–	0.1862	1.3250	–	680.	0.993	–	0.1115	–	–	−70.00
−65.00	13.108	75.41	5.9913	−9.405	156.182	−0.02302	0.39653	–	0.1885	1.3278	–	682.	0.952	–	0.1102	–	–	−65.00
−61.00b	14.696	75.04	5.3818	−7.923	156.687	−0.01930	0.39360	–	0.1904	1.3302	–	684.	0.921	–	0.1092	–	–	−61.00
−60.00	15.117	74.95	5.2410	−7.551	156.812	−0.01837	0.39288	–	0.1909	1.3309	–	685.	0.914	–	0.1089	–	–	−60.00
−55.00	17.365	74.49	4.6010	−5.683	157.427	−0.01374	0.38933	–	0.1935	1.3343	–	687.	0.878	–	0.1076	–	–	−55.00
−50.00	19.872	74.02	4.0529	−3.802	158.026	−0.00913	0.38589	–	0.1962	1.3380	–	689.	0.844	–	0.1063	–	–	−50.00
−45.00	22.657	73.56	3.5817	−1.907	158.608	−0.00455	0.38254	–	0.1991	1.3421	–	691.	0.811	–	0.1051	–	–	−45.00
−40.00	25.741	73.08	3.1751	0.000	159.174	0.00000	0.37928	–	0.2021	1.3466	–	693.	0.781	–	0.1038	–	–	−40.00
−35.00	29.147	72.61	2.8229	1.920	159.721	0.00453	0.37611	–	0.2053	1.3515	–	694.	0.752	–	0.1025	–	–	−35.00
−30.00	32.897	72.12	2.5168	3.852	160.251	0.00903	0.37302	–	0.2086	1.3568	–	696.	0.725	–	0.1012	–	–	−30.00
−25.00	37.014	71.64	2.2499	5.797	160.761	0.01350	0.37001	–	0.2122	1.3626	–	697.	0.699	–	0.0999	–	–	−25.00
−20.00	41.521	71.14	2.0165	7.753	161.251	0.01795	0.36707	–	0.2159	1.3688	–	698.	0.674	–	0.0986	–	–	−20.00
−15.00	46.444	70.65	1.8116	9.721	161.721	0.02237	0.36420	–	0.2197	1.3756	–	699.	0.650	–	0.0973	–	–	−15.00
−10.00	51.807	70.14	1.6313	11.701	162.170	0.02677	0.36139	–	0.2238	1.3829	–	700.	0.627	–	0.0960	–	–	−10.00
−5.00	57.636	69.64	1.4721	13.693	162.596	0.03114	0.35864	–	0.2281	1.3908	–	700.	0.605	–	0.0948	–	–	−5.00
0.00	63.956	69.12	1.3312	15.696	163.000	0.03549	0.35594	–	0.2326	1.3993	–	700.	0.584	–	0.0935	–	–	0.00
2.00	66.629	68.91	1.2794	16.500	163.155	0.03722	0.35488	–	0.2345	1.4029	–	701.	0.576	–	0.0929	–	–	2.00
4.00	69.385	68.70	1.2300	17.306	163.306	0.03894	0.35382	–	0.2364	1.4066	–	701.	0.568	–	0.0924	–	–	4.00
6.00	72.229	68.49	1.1829	18.115	163.453	0.04066	0.35277	–	0.2384	1.4105	–	701.	0.560	–	0.0919	–	–	6.00
8.00	75.161	68.28	1.1379	18.925	163.596	0.04238	0.35173	–	0.2404	1.4145	–	700.	0.552	–	0.0914	–	–	8.00
10.00	78.183	68.07	1.0949	19.736	163.735	0.04410	0.35069	–	0.2424	1.4186	–	700.	0.545	–	0.0909	–	–	10.00
12.00	81.297	67.85	1.0538	20.550	163.870	0.04581	0.34967	–	0.2445	1.4228	–	700.	0.537	–	0.0904	–	–	12.00
14.00	84.504	67.63	1.0146	21.365	164.001	0.04752	0.34864	–	0.2466	1.4271	–	700.	0.530	–	0.0899	–	–	14.00
16.00	87.808	67.42	0.9771	22.183	164.128	0.04922	0.34763	–	0.2487	1.4316	–	700.	0.522	–	0.0893	–	–	16.00
18.00	91.208	67.20	0.9412	23.002	164.250	0.05092	0.34662	–	0.2509	1.4363	–	700.	0.515	–	0.0888	–	–	18.00
20.00	94.709	66.98	0.9069	23.823	164.368	0.05261	0.34562	–	0.2532	1.4411	–	699.	0.508	–	0.0883	–	–	20.00
22.00	98.310	66.76	0.8740	24.646	164.481	0.05430	0.34462	–	0.2555	1.4460	–	699.	0.501	–	0.0878	–	–	22.00
24.00	102.02	66.53	0.8425	25.471	164.590	0.05599	0.34362	–	0.2578	1.4512	–	699.	0.494	–	0.0873	–	–	24.00
26.00	105.82	66.31	0.8124	26.299	164.695	0.05768	0.34264	–	0.2602	1.4565	–	698.	0.487	–	0.0868	–	–	26.00
28.00	109.74	66.08	0.7835	27.128	164.794	0.05936	0.34165	–	0.2627	1.4619	–	698.	0.481	–	0.0863	–	–	28.00
30.00	113.77	65.86	0.7558	27.959	164.889	0.06104	0.34067	–	0.2652	1.4676	–	698.	0.474	–	0.0857	–	–	30.00
32.00	117.90	65.63	0.7292	28.792	164.979	0.06271	0.33970	–	0.2678	1.4734	–	697.	0.467	–	0.0852	–	–	32.00
34.00	122.15	65.40	0.7038	29.627	165.064	0.06438	0.33873	–	0.2704	1.4795	–	697.	0.461	–	0.0847	–	–	34.00
36.00	126.52	65.16	0.6793	30.465	165.144	0.06605	0.33776	–	0.2731	1.4858	–	696.	0.455	–	0.0842	–	–	36.00
38.00	131.00	64.93	0.6559	31.305	165.218	0.06771	0.33680	–	0.2758	1.4922	–	695.	0.448	–	0.0837	–	–	38.00
40.00	135.60	64.69	0.6333	32.146	165.288	0.06938	0.33583	–	0.2787	1.4989	–	695.	0.442	–	0.0832	–	–	40.00
42.00	140.32	64.45	0.6117	32.991	165.352	0.07104	0.33488	–	0.2816	1.5059	–	694.	0.436	–	0.0826	–	–	42.00
44.00	145.17	64.21	0.5909	33.837	165.410	0.07269	0.33392	–	0.2845	1.5131	–	693.	0.430	–	0.0821	–	–	44.00
46.00	150.14	63.97	0.5709	34.686	165.463	0.07435	0.33297	–	0.2876	1.5206	–	693.	0.424	–	0.0816	–	–	46.00
48.00	155.24	63.73	0.5517	35.538	165.510	0.07600	0.33201	–	0.2907	1.5283	–	692.	0.418	–	0.0811	–	–	48.00
50.00	160.46	63.48	0.5332	36.392	165.552	0.07765	0.33106	–	0.2939	1.5363	–	691.	0.412	–	0.0806	–	–	50.00
52.00	165.82	63.23	0.5154	37.248	165.587	0.07929	0.33012	–	0.2972	1.5447	–	690.	0.407	–	0.0801	–	–	52.00
54.00	171.32	62.98	0.4983	38.107	165.616	0.08094	0.32917	–	0.3006	1.5533	–	689.	0.401	–	0.0796	–	–	54.00
56.00	176.95	62.73	0.4818	38.970	165.639	0.08258	0.32822	–	0.3040	1.5623	–	688.	0.395	–	0.0790	–	–	56.00
58.00	182.72	62.48	0.4659	39.834	165.655	0.08422	0.32727	–	0.3076	1.5716	–	687.	0.390	–	0.0785	–	–	58.00
60.00	188.62	62.22	0.4506	40.702	165.665	0.08586	0.32633	–	0.3113	1.5813	–	686.	–	–	0.0780	–	–	60.00
62.00	194.68	61.96	0.4359	41.573	165.667	0.08750	0.32538	–	0.3151	1.5914	–	685.	–	–	0.0775	–	–	62.00
64.00	200.88	61.70	0.4217	42.447	165.663	0.08914	0.32443	–	0.3190	1.6020	–	684.	–	–	0.0770	–	–	64.00
66.00	207.22	61.43	0.4080	43.324	165.652	0.09077	0.32348	–	0.3231	1.6129	–	683.	–	–	0.0765	–	–	66.00
68.00	213.72	61.16	0.3948	44.205	165.633	0.09241	0.32253	–	0.3272	1.6243	–	682.	–	–	0.0760	–	–	68.00
70.00	220.37	60.89	0.3820	45.089	165.607	0.09404	0.32157	–	0.3315	1.6362	–	681.	–	–	0.0754	–	–	70.00
75.00	237.68	60.20	0.3520	47.316	165.507	0.09812	0.31918	–	0.3429	1.6683	–	677.	–	–	0.0741	–	7.11	75.00
80.00	255.99	59.49	0.3246	49.567	165.355	0.10221	0.31676	–	0.3554	1.7042	–	674.	–	–	–	–	6.68	80.00
85.00	275.36	58.76	0.2993	51.847	165.145	0.10630	0.31432	–	0.3691	1.7446	–	670.	–	–	–	–	6.26	85.00
90.00	295.83	58.01	0.2761	54.157	164.874	0.11041	0.31183	–	0.3843	1.7902	–	666.	–	–	–	–	5.84	90.00
95.00	317.43	57.23	0.2547	56.502	164.536	0.11453	0.30930	–	0.4012	1.8420	–	661.	–	–	–	–	5.42	95.00
100.00	340.22	56.42	0.2349	58.885	164.124	0.11867	0.30671	–	0.4203	1.9015	–	656.	–	–	–	–	5.00	100.00
105.00	364.24	55.59	0.2166	61.313	163.631	0.12285	0.30405	–	0.4419	1.9702	–	651.	–	–	–	–	4.59	105.00
110.00	389.54	54.71	0.1996	63.791	163.046	0.12707	0.30130	–	0.4667	2.0504	–	646.	–	–	–	–	4.19	110.00
115.00	416.19	53.80	0.1838	66.326	162.360	0.13134	0.29845	–	0.4956	2.1450	–	640.	–	–	–	–	3.79	115.00
120.00	444.23	52.84	0.1690	68.928	161.558	0.13568	0.29548	–	0.5297	2.2583	–	634.	–	–	–	–	3.39	120.00
125.00	473.74	51.83	0.1553	71.607	160.624	0.14010	0.29235	–	0.5706	2.3959	–	628.	–	–	–	–	3.01	125.00
130.00	504.76	50.75	0.1424	74.378	159.537	0.14463	0.28905	–	0.6208	2.5665	–	621.	–	–	–	–	2.63	130.00
135.00	537.39	49.61	0.1302	77.258	158.269	0.14929	0.28552	–	0.6839	2.7830	–	614.	–	–	–	–	2.26	135.00
140.00	571.69	48.37	0.1187	80.273	156.784	0.15412	0.28171	–	0.7659	3.0661	–	606.	–	–	–	–	1.90	140.00
145.00	607.76	47.02	0.1078	83.455	155.029	0.15918	0.27754	–	0.8769	3.4516	–	598.	–	–	–	–	1.55	145.00
150.00	645.68	45.52	0.0973	86.856	152.930	0.16453	0.27290	–	1.0357	4.0057	–	590.	–	–	–	–	1.21	150.00
155.00	685.57	43.82	0.0871	90.554	150.366	0.17030	0.26760	–	1.2823	4.8681	–	580.	–	–	–	–	0.89	155.00
160.00	727.57	41.81	0.0769	94.694	147.123	0.17671	0.26132	–	1.7181	6.3925	–	570.	–	–	–	–	0.59	160.00
165.00	771.85	39.28	0.0665	99.588	142.741	0.18425	0.25333	–	2.6971	9.8124	–	559.	–	–	–	–	0.32	165.00
170.00	818.67	35.45	0.0544	106.291	135.718	0.19457	0.24130	–	–	–	–	–	–	–	–	–	0.09	170.00
173.14c	849.61	26.20	0.0382	120.726	120.726	0.21713	0.21713	∞	∞	∞	0.	0.	–	–	∞	∞	0.00	173.14

*temperatures are on the IPTS-68 scale

b = normal boiling point

c = critical point

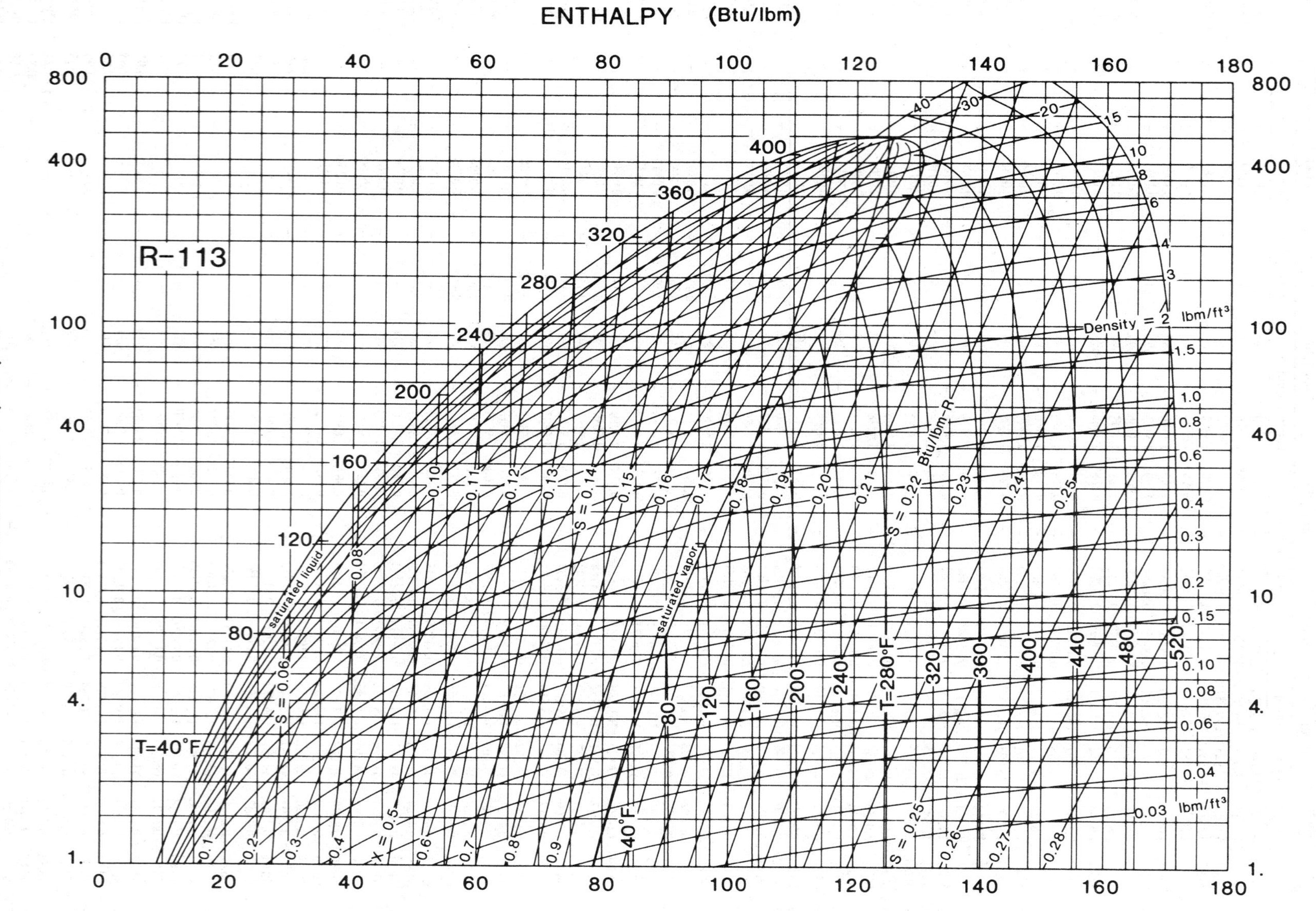

Prepared by: CENTER FOR APPLIED THERMODYNAMIC STUDIES, University of Idaho

Fig. 7 Pressure-Enthalpy Diagram for Refrigerant 113

Refrigerant 113 (1,1,2-Trichloro-1,2,2-trifluoroethane) Properties of Saturated Liquid and Saturated Vapor

Temp,* °F	Pressure, psia	Density, lb/ft³ Liquid	Volume, ft³/lb Vapor	Enthalpy, Btu/lb Liquid	Enthalpy, Btu/lb Vapor	Entropy, Btu/lb·°F Liquid	Entropy, Btu/lb·°F Vapor	Specific Heat c_p, Btu/lb·°F Liquid	Specific Heat c_p, Btu/lb·°F Vapor	c_p/c_v Vapor	Velocity of Sound, ft/s Liquid	Velocity of Sound, ft/s Vapor	Viscosity, lb_m/ft·h Liquid	Viscosity, lb_m/ft·h Vapor	Thermal Cond, Btu/h·ft·°F Liquid	Thermal Cond, Btu/h·ft·°F Vapor	Surface Tension, dyne/cm	Temp,* °F
−20.00	0.445	105.08	56.424	4.000	75.095	0.00931	0.17101	–	–	–	–	–	–	–	–	–	22.33	−20.00
−10.00	0.624	104.32	41.061	6.035	76.554	0.01388	0.17071	–	–	–	–	–	–	–	–	–	21.69	−10.00
0.00	0.861	103.56	30.372	8.092	78.023	0.01841	0.17054	–	–	–	–	–	–	–	–	–	21.06	0.00
10.00	1.170	102.78	22.808	10.172	79.503	0.02288	0.17050	–	–	–	–	–	–	–	–	–	20.43	10.00
20.00	1.566	102.01	17.370	12.274	80.991	0.02731	0.17057	–	–	–	–	–	–	–	–	–	19.81	20.00
30.00	2.068	101.23	13.403	14.398	82.488	0.03169	0.17074	–	–	–	–	–	–	–	–	–	19.19	30.00
40.00	2.695	100.44	10.469	16.544	83.991	0.03603	0.17101	0.2220	–	–	–	–	2.173	0.0206	0.0469	0.00457	18.57	40.00
50.00	3.470	99.65	8.2704	18.712	85.500	0.04032	0.17136	0.2240	–	–	–	–	2.016	0.0215	0.0464	0.00472	17.96	50.00
60.00	4.417	98.85	6.6030	20.900	87.014	0.04457	0.17179	0.2258	–	–	–	–	1.873	0.0224	0.0458	0.00487	17.35	60.00
70.00	5.563	98.05	5.3241	23.109	88.531	0.04877	0.17229	0.2275	–	–	–	–	1.744	0.0232	0.0452	0.00503	16.74	70.00
75.00	6.220	97.64	4.7975	24.221	89.291	0.05086	0.17256	0.2283	–	–	–	–	1.683	0.0236	0.0449	0.00510	16.44	75.00
80.00	6.937	97.23	4.3326	25.338	90.051	0.05294	0.17285	0.2291	–	–	–	–	1.625	0.0239	0.0447	0.00518	16.14	80.00
85.00	7.719	96.82	3.9212	26.460	90.812	0.05500	0.17315	0.2298	–	–	–	–	1.570	0.0243	0.0444	0.00526	15.84	85.00
90.00	8.570	96.41	3.5562	27.587	91.573	0.05706	0.17347	0.2305	–	–	–	–	1.518	0.0246	0.0441	0.00534	15.55	90.00
95.00	9.494	96.00	3.2316	28.718	92.334	0.05911	0.17380	0.2312	–	–	–	–	1.467	0.0249	0.0438	0.00541	15.25	95.00
100.00	10.494	95.58	2.9424	29.854	93.095	0.06114	0.17414	0.2319	–	–	–	–	1.419	0.0252	0.0435	0.00549	14.95	100.00
102.00	10.917	95.42	2.8356	30.310	93.400	0.06195	0.17428	0.2322	–	–	–	–	1.400	0.0253	0.0434	0.00552	14.84	102.00
104.00	11.353	95.25	2.7335	30.766	93.705	0.06276	0.17442	0.2325	0.1690	–	–	–	1.382	0.0254	0.0433	0.00555	14.72	104.00
106.00	11.803	95.08	2.6358	31.223	94.009	0.06357	0.17456	0.2328	0.1697	–	–	–	1.364	0.0256	0.0432	0.00559	14.60	106.00
108.00	12.267	94.91	2.5423	31.681	94.313	0.06438	0.17471	0.2330	0.1704	–	–	–	1.346	0.0257	0.0431	0.00562	14.48	108.00
110.00	12.745	94.74	2.4529	32.140	94.618	0.06518	0.17486	0.2333	0.1710	–	–	–	1.329	0.0258	0.0430	0.00565	14.37	110.00
112.00	13.237	94.58	2.3672	32.599	94.922	0.06599	0.17500	0.2336	0.1716	–	–	–	1.312	0.0259	0.0428	0.00568	14.25	112.00
114.00	13.744	94.41	2.2852	33.059	95.227	0.06679	0.17515	0.2338	0.1722	–	–	–	1.295	0.0260	0.0427	0.00571	14.13	114.00
116.00	14.266	94.24	2.2066	33.520	95.531	0.06759	0.17531	0.2341	0.1727	–	–	–	1.279	0.0261	0.0426	0.00574	14.02	116.00
117.60b	14.696	94.10	2.1460	33.890	95.775	0.06823	0.17543	0.2343	0.1732	–	–	–	1.266	0.0262	0.0425	0.00577	13.93	117.60
118.00	14.804	94.07	2.1313	33.981	95.835	0.06838	0.17546	0.2344	0.1733	–	–	–	1.262	0.0262	0.0425	0.00577	13.90	118.00
120.00	15.357	93.90	2.0591	34.443	96.139	0.06918	0.17561	0.2346	0.1738	–	–	–	1.246	0.0263	0.0424	0.00581	13.79	120.00
122.00	15.926	93.73	1.9899	34.906	96.443	0.06998	0.17577	0.2349	0.1743	–	–	–	1.231	0.0264	0.0423	0.00584	13.67	122.00
124.00	16.512	93.56	1.9235	35.369	96.747	0.07077	0.17593	0.2352	0.1748	–	–	–	1.215	0.0265	0.0422	0.00587	13.55	124.00
126.00	17.114	93.38	1.8598	35.833	97.051	0.07156	0.17609	0.2355	0.1752	–	–	–	1.200	0.0266	0.0421	0.00590	13.44	126.00
128.00	17.733	93.21	1.7987	36.298	97.355	0.07235	0.17625	0.2357	0.1757	–	–	–	1.185	0.0266	0.0419	0.00593	13.32	128.00
130.00	18.369	93.04	1.7400	36.763	97.659	0.07314	0.17641	0.2360	0.1761	–	–	–	1.171	0.0267	0.0418	0.00596	13.21	130.00
132.00	19.023	92.87	1.6836	37.230	97.962	0.07393	0.17657	0.2363	0.1765	–	–	–	1.156	0.0268	0.0417	0.00600	13.09	132.00
134.00	19.695	92.69	1.6294	37.696	98.266	0.07471	0.17674	0.2365	0.1769	–	–	–	1.142	0.0269	0.0416	0.00603	12.98	134.00
136.00	20.385	92.52	1.5774	38.164	98.569	0.07550	0.17690	0.2368	0.1773	–	–	–	1.128	0.0270	0.0415	0.00606	12.87	136.00
138.00	21.093	92.34	1.5273	38.632	98.872	0.07628	0.17707	0.2371	0.1777	–	–	–	1.114	0.0271	0.0414	0.00609	12.75	138.00
140.00	21.820	92.17	1.4792	39.100	99.176	0.07706	0.17724	0.2374	0.1780	–	–	–	1.101	0.0272	0.0413	0.00612	12.64	140.00
142.00	22.567	91.99	1.4330	39.570	99.478	0.07784	0.17741	0.2376	0.1784	–	–	–	1.088	0.0272	0.0412	0.00616	12.52	142.00
144.00	23.333	91.82	1.3885	40.040	99.781	0.07861	0.17758	0.2379	0.1787	–	–	–	1.075	0.0273	0.0410	0.00619	12.41	144.00
146.00	24.119	91.64	1.3457	40.510	100.080	0.07939	0.17775	0.2382	0.1791	–	–	–	1.062	0.0274	0.0409	0.00622	12.30	146.00
148.00	24.925	91.47	1.3044	40.981	100.390	0.08016	0.17792	0.2385	0.1794	–	–	–	1.049	0.0275	0.0408	0.00625	12.18	148.00
150.00	25.752	91.29	1.2648	41.453	100.690	0.08094	0.17810	0.2388	0.1797	–	–	–	1.037	0.0276	0.0407	0.00629	12.07	150.00
155.00	27.910	90.84	1.1718	42.636	101.440	0.08286	0.17853	0.2395	0.1805	–	–	–	1.007	0.0278	0.0404	0.00637	11.79	155.00
160.00	30.204	90.39	1.0871	43.822	102.200	0.08477	0.17898	0.2403	0.1812	–	–	–	0.978	0.0279	0.0401	0.00645	11.51	160.00
165.00	32.640	89.94	1.0098	45.012	102.950	0.08668	0.17943	0.2411	0.1819	–	–	–	0.950	0.0281	0.0399	0.00653	11.23	165.00
170.00	35.222	89.48	0.9391	46.205	103.700	0.08857	0.17988	0.2419	0.1826	–	–	–	0.923	0.0283	0.0396	0.00661	10.95	170.00
175.00	37.956	89.03	0.8743	47.403	104.450	0.09046	0.18034	0.2427	0.1833	–	–	–	0.897	0.0285	0.0393	0.00670	10.68	175.00
180.00	40.848	88.56	0.8149	48.604	105.200	0.09233	0.18081	0.2436	0.1841	–	–	–	0.873	0.0286	0.0390	0.00678	10.40	180.00
185.00	43.903	88.09	0.7604	49.808	105.940	0.09420	0.18127	0.2445	0.1848	–	–	–	0.849	0.0288	0.0387	0.00687	–	185.00
190.00	47.127	87.62	0.7102	51.017	106.690	0.09606	0.18175	0.2455	0.1856	–	–	–	0.826	0.0290	0.0385	0.00695	–	190.00
195.00	50.527	87.14	0.6640	52.229	107.430	0.09790	0.18222	0.2465	0.1865	–	–	–	0.804	0.0291	0.0382	0.00704	–	195.00
200.00	54.107	86.66	0.6214	53.445	108.170	0.09974	0.18270	0.2475	0.1874	–	–	–	0.783	0.0293	0.0379	0.00712	–	200.00
210.00	61.835	85.68	0.5457	55.888	109.640	0.10339	0.18366	0.2498	0.1895	–	–	–	0.743	0.0297	0.0373	0.00730	–	210.00
220.00	70.358	84.68	0.4808	58.346	111.100	0.10701	0.18462	0.2522	0.1921	–	–	–	0.705	0.0300	0.0368	0.00747	–	220.00
230.00	79.726	83.66	0.4249	60.820	112.550	0.11059	0.18559	0.2550	0.1952	–	–	–	0.671	0.0304	0.0362	0.00765	–	230.00
240.00	89.990	82.61	0.3766	63.312	113.980	0.11415	0.18656	0.2579	0.1989	–	–	–	0.638	0.0308	0.0357	0.00783	–	240.00
250.00	101.20	81.53	0.3346	65.820	115.400	0.11767	0.18753	0.2612	0.2033	–	–	–	0.608	0.0312	0.0351	0.00802	–	250.00
260.00	113.42	80.42	0.2979	68.349	116.790	0.12117	0.18849	0.2648	0.2086	–	–	–	0.580	0.0317	0.0345	0.00821	–	260.00
270.00	126.69	79.27	0.2658	70.898	118.170	0.12464	0.18943	0.2687	0.2148	–	–	–	0.554	0.0322	0.0340	0.00840	–	270.00
280.00	141.08	78.08	0.2375	73.472	119.530	0.12810	0.19037	0.2730	0.2220	–	–	–	0.529	0.0327	0.0334	0.00859	–	280.00
290.00	156.64	76.84	0.2125	76.072	120.850	0.13154	0.19128	0.2777	0.2303	–	–	–	0.506	0.0333	0.0329	0.00879	–	290.00
300.00	173.45	75.55	0.1903	78.703	122.150	0.13497	0.19216	0.2828	0.2399	–	–	–	0.485	0.0340	0.0323	0.00899	–	300.00
310.00	191.55	74.20	0.1705	81.369	123.410	0.13840	0.19302	0.2883	–	–	–	–	0.465	0.0347	0.0318	0.00920	–	310.00
320.00	211.04	72.77	0.1528	84.076	124.620	0.14183	0.19383	0.2943	–	–	–	–	0.446	0.0355	0.0312	0.00941	–	320.00
330.00	231.97	71.26	0.1369	86.832	125.780	0.14527	0.19460	0.3008	–	–	–	–	0.428	0.0364	0.0307	0.00962	–	330.00
340.00	254.43	69.65	0.1225	89.647	126.880	0.14874	0.19530	0.3078	–	–	–	–	0.411	0.0373	–	–	–	340.00
350.00	278.52	67.91	0.1094	92.533	127.900	0.15224	0.19592	0.3153	–	–	–	–	0.395	0.0384	–	–	–	350.00
360.00	304.31	66.01	0.0974	95.510	128.830	0.15580	0.19645	–	–	–	–	–	–	–	–	–	–	360.00
370.00	331.93	63.90	0.0863	98.602	129.630	0.15944	0.19684	–	–	–	–	–	–	–	–	–	–	370.00
380.00	361.50	61.50	0.0760	101.850	130.260	0.16322	0.19704	–	–	–	–	–	–	–	–	–	–	380.00
390.00	393.16	58.67	0.0661	105.330	130.630	0.16720	0.19697	–	–	–	–	–	–	–	–	–	–	390.00
417.80c	494.70	35.58	0.0281	123.900	123.900	0.18710	0.18710	∞	∞	∞	0	0	–	–	∞	∞	0	417.80

*temperatures are on the IPTS-68 scale

b = normal boiling point

c = critical point

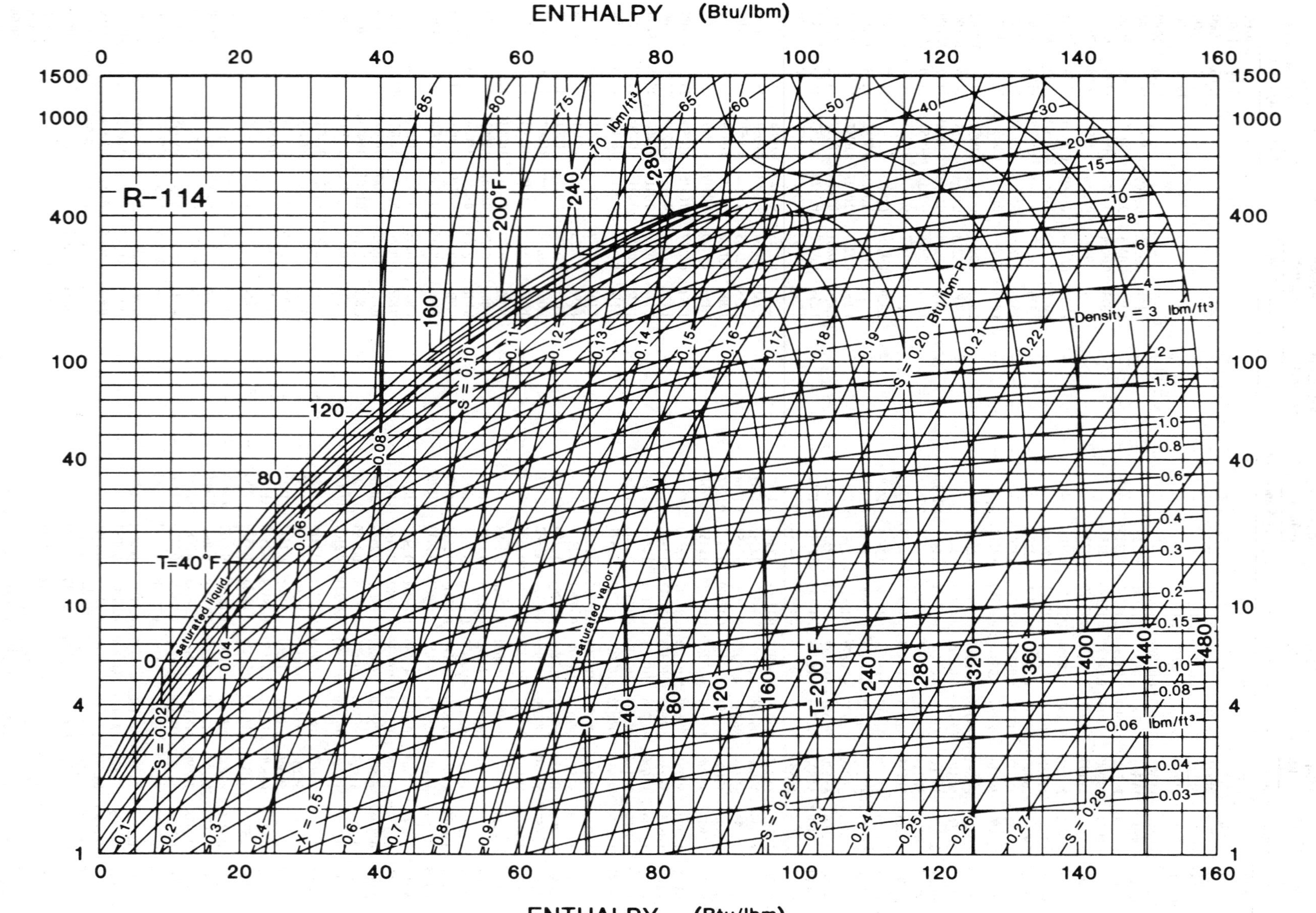

Prepared by: CENTER FOR APPLIED THERMODYNAMIC STUDIES, University of Idaho

Fig. 8 Pressure-Enthalpy Diagram for Refrigerant 114

Refrigerant 114 (1,2-Dichloro-1,1,2,2-tetrafluoroethane) Properties of Saturated Liquid and Saturated Vapor

Temp,* °F	Pressure, psia	Density, lb/ft³ Liquid	Volume, ft³/lb Vapor	Enthalpy, Btu/lb Liquid	Enthalpy, Btu/lb Vapor	Entropy, Btu/lb·°F Liquid	Entropy, Btu/lb·°F Vapor	Specific Heat c_p, Btu/lb·°F Liquid	Specific Heat c_p, Btu/lb·°F Vapor	c_p/c_v Vapor	Velocity of Sound, ft/s Liquid	Velocity of Sound, ft/s Vapor	Viscosity, lb_m/ft·h Liquid	Viscosity, lb_m/ft·h Vapor	Thermal Cond, Btu/h·ft·°F Liquid	Thermal Cond, Btu/h·ft·°F Vapor	Surface Tension, dyne/cm	Temp,* °F
−100.00	0.196	98.40	115.22	−14.466	55.647	−0.03723	0.15880	–	–	–	–	–	–	–	–	–	–	−100.00
−80.00	0.455	98.04	52.141	−9.506	58.230	−0.02381	0.15599	–	–	–	–	–	–	–	–	–	–	−80.00
−60.00	0.966	97.60	25.809	−4.693	60.887	−0.01145	0.15412	–	–	–	–	–	–	–	–	–	–	−60.00
−40.00	1.890	97.07	13.787	0.000	63.604	0.00000	0.15301	0.2072	0.1388	–	2617.	362.	2.012	–	0.0466	–	–	−40.00
−20.00	3.452	96.42	7.8606	4.603	66.367	0.01071	0.15253	0.2128	0.1485	–	2437.	366.	1.744	–	0.0449	–	–	−20.00
−15.00	3.975	96.24	6.8926	5.744	67.063	0.01329	0.15250	0.2140	0.1505	–	2395.	368.	1.682	–	0.0445	–	–	−15.00
−10.00	4.560	96.05	6.0640	6.881	67.761	0.01583	0.15249	0.2151	0.1523	–	2353.	369.	1.622	–	0.0441	–	–	−10.00
−5.00	5.213	95.85	5.3523	8.017	68.460	0.01834	0.15252	0.2161	0.1539	–	2313.	370.	1.563	–	0.0436	–	–	−5.00
0.00	5.939	95.64	4.7389	9.151	69.161	0.02081	0.15257	0.2171	0.1554	–	2273.	371.	1.506	–	0.0432	–	–	0.00
5.00	6.745	95.41	4.2083	10.283	69.863	0.02326	0.15265	0.2181	0.1568	–	2234.	372.	1.452	–	0.0428	–	–	5.00
10.00	7.636	95.18	3.7479	11.416	70.567	0.02568	0.15275	0.2190	0.1581	–	2196.	373.	1.399	–	0.0424	–	–	10.00
12.00	8.018	95.08	3.5810	11.869	70.848	0.02664	0.15280	0.2193	0.1585	–	2181.	374.	1.378	–	0.0423	–	–	12.00
14.00	8.415	94.98	3.4230	12.322	71.130	0.02760	0.15285	0.2196	0.1590	–	2166.	374.	1.358	–	0.0421	–	–	14.00
16.00	8.827	94.88	3.2734	12.775	71.412	0.02855	0.15291	0.2200	0.1594	–	2151.	374.	1.338	–	0.0420	–	–	16.00
18.00	9.255	94.78	3.1317	13.228	71.694	0.02950	0.15297	0.2203	0.1599	–	2136.	375.	1.318	–	0.0418	–	–	18.00
20.00	9.699	94.67	2.9973	13.682	71.976	0.03045	0.15303	0.2206	0.1603	–	2122.	375.	1.298	–	0.0417	–	–	20.00
22.00	10.160	94.56	2.8699	14.136	72.258	0.03139	0.15310	0.2209	0.1607	–	2107.	376.	1.279	–	0.0415	–	–	22.00
24.00	10.638	94.45	2.7490	14.590	72.540	0.03233	0.15317	0.2212	0.1611	–	2093.	376.	1.260	–	0.0414	–	–	24.00
26.00	11.134	94.33	2.6343	15.044	72.822	0.03326	0.15324	0.2216	0.1615	–	2078.	376.	1.242	–	0.0412	–	–	26.00
28.00	11.648	94.22	2.5253	15.499	73.104	0.03420	0.15331	0.2219	0.1618	–	2064.	377.	1.224	–	0.0410	–	–	28.00
30.00	12.180	94.10	2.4218	15.954	73.386	0.03513	0.15339	0.2222	0.1622	–	2050.	377.	1.206	–	0.0409	–	–	30.00
32.00	12.732	93.98	2.3235	16.410	73.669	0.03605	0.15347	0.2225	0.1626	–	2036.	377.	1.188	0.0254	0.0407	–	13.80	32.00
34.00	13.302	93.86	2.2299	16.866	73.951	0.03698	0.15356	0.2228	0.1629	–	2022.	378.	1.170	0.0255	0.0406	–	13.67	34.00
36.00	13.892	93.73	2.1410	17.322	74.233	0.03789	0.15365	0.2231	0.1632	–	2009.	378.	1.153	0.0256	0.0404	–	13.54	36.00
38.00	14.503	93.61	2.0563	17.780	74.516	0.03881	0.15374	0.2234	0.1636	–	1995.	378.	1.136	0.0257	0.0403	–	13.41	38.00
38.62b	14.696	93.56	2.0309	17.921	74.603	0.03910	0.15377	0.2235	0.1637	–	1991.	378.	1.131	0.0258	0.0402	–	13.37	38.62
40.00	15.134	93.47	1.9757	18.237	74.798	0.03973	0.15383	0.2237	0.1639	–	1982.	379.	1.119	0.0258	0.0401	–	13.28	40.00
42.00	15.787	93.34	1.8989	18.696	75.080	0.04064	0.15393	0.2240	0.1642	–	1968.	379.	1.103	0.0260	0.0400	–	13.15	42.00
44.00	16.461	93.20	1.8257	19.155	75.362	0.04155	0.15403	0.2242	0.1645	–	1955.	379.	1.087	0.0261	0.0398	–	13.02	44.00
46.00	17.158	93.07	1.7560	19.614	75.644	0.04246	0.15413	0.2245	0.1648	–	1941.	379.	1.071	0.0262	0.0397	–	12.89	46.00
48.00	17.877	92.92	1.6895	20.075	75.927	0.04337	0.15423	0.2248	0.1652	–	1928.	380.	1.055	0.0263	0.0395	–	12.76	48.00
50.00	18.619	92.78	1.6260	20.536	76.209	0.04427	0.15434	0.2251	0.1655	–	1915.	380.	1.040	0.0264	0.0394	–	12.63	50.00
52.00	19.384	92.63	1.5655	20.998	76.491	0.04517	0.15445	0.2254	0.1658	–	1902.	380.	1.025	0.0265	0.0392	–	12.51	52.00
54.00	20.174	92.48	1.5076	21.460	76.772	0.04607	0.15456	0.2257	0.1661	–	1889.	380.	1.010	0.0266	0.0391	–	12.38	54.00
56.00	20.988	92.33	1.4524	21.924	77.054	0.04697	0.15467	0.2260	0.1664	–	1876.	381.	0.995	0.0267	0.0389	–	12.25	56.00
58.00	21.828	92.17	1.3997	22.388	77.336	0.04786	0.15479	0.2263	0.1667	–	1863.	381.	0.981	0.0269	0.0388	–	12.12	58.00
60.00	22.693	92.02	1.3492	22.853	77.617	0.04876	0.15491	0.2266	0.1670	–	1850.	381.	0.966	0.0270	0.0386	–	12.00	60.00
62.00	23.584	91.85	1.3010	23.319	77.899	0.04965	0.15503	0.2270	0.1673	–	1838.	381.	0.952	0.0271	0.0385	–	11.87	62.00
64.00	24.501	91.69	1.2549	23.786	78.180	0.05054	0.15515	0.2273	0.1676	–	1825.	381.	0.939	0.0272	0.0383	–	11.74	64.00
66.00	25.446	91.52	1.2108	24.254	78.461	0.05143	0.15528	0.2276	0.1679	–	1812.	381.	0.925	0.0273	0.0382	–	11.62	66.00
68.00	26.418	91.35	1.1685	24.722	78.742	0.05231	0.15540	0.2279	0.1682	–	1800.	381.	0.912	0.0274	0.0380	–	11.49	68.00
70.00	27.418	91.18	1.1281	25.192	79.023	0.05320	0.15553	0.2282	0.1686	–	1787.	382.	0.899	0.0275	0.0379	–	11.36	70.00
72.00	28.447	91.00	1.0894	25.662	79.303	0.05408	0.15566	0.2286	0.1689	–	1775.	382.	0.886	0.0276	0.0377	–	11.24	72.00
74.00	29.505	90.83	1.0522	26.133	79.584	0.05496	0.15580	0.2289	0.1692	–	1763.	382.	0.873	0.0277	0.0376	–	11.11	74.00
76.00	30.593	90.64	1.0167	26.605	79.864	0.05584	0.15593	0.2292	0.1696	–	1750.	382.	0.861	0.0278	0.0374	–	10.99	76.00
78.00	31.710	90.46	0.9826	27.078	80.144	0.05672	0.15607	0.2296	0.1699	–	1738.	382.	0.848	0.0279	0.0373	–	10.86	78.00
85.00	35.868	89.79	0.8737	28.740	81.122	0.05977	0.15656	0.2309	0.1713	–	1696.	382.	0.807	0.0283	0.0367	–	10.43	85.00
90.00	39.080	89.29	0.8049	29.934	81.818	0.06194	0.15692	0.2319	0.1723	–	1666.	382.	0.779	0.0285	0.0364	–	10.12	90.00
95.00	42.504	88.78	0.7426	31.132	82.513	0.06410	0.15729	0.2329	0.1734	–	1636.	382.	0.752	0.0288	0.0360	–	9.81	95.00
100.00	46.148	88.25	0.6861	32.335	83.205	0.06624	0.15767	0.2340	0.1746	–	1606.	381.	0.726	0.0290	0.0356	–	9.51	100.00
105.00	50.021	87.70	0.6347	33.541	83.895	0.06837	0.15805	0.2352	0.1759	–	1577.	381.	0.701	0.0293	0.0353	–	9.21	105.00
110.00	54.132	87.14	0.5880	34.752	84.583	0.07049	0.15845	0.2364	0.1773	–	1547.	380.	0.677	0.0295	0.0349	–	8.90	110.00
120.00	63.103	85.98	0.5063	37.185	85.949	0.07469	0.15925	0.2391	0.1806	–	1489.	378.	0.632	0.0301	0.0341	–	8.31	120.00
130.00	73.134	84.76	0.4380	39.630	87.302	0.07884	0.16007	0.2423	0.1845	–	1431.	376.	0.591	0.0306	0.0334	–	7.72	130.00
140.00	84.299	83.50	0.3803	42.088	88.638	0.08293	0.16089	0.2458	0.1892	–	1372.	373.	0.552	0.0312	0.0326	–	7.14	140.00
150.00	96.678	82.19	0.3314	44.561	89.955	0.08697	0.16173	0.2498	0.1946	–	1313.	370.	0.517	0.0318	0.0319	–	6.57	150.00
160.00	110.35	80.84	0.2896	47.050	91.250	0.09097	0.16256	0.2543	0.2009	–	1253.	366.	0.485	0.0324	0.0311	–	6.01	160.00
170.00	125.41	79.44	0.2537	49.560	92.518	0.09494	0.16338	0.2594	0.2082	–	1192.	361.	0.455	0.0331	0.0304	–	–	170.00
180.00	141.93	77.99	0.2227	52.096	93.753	0.09887	0.16418	0.2652	0.2167	–	1130.	355.	0.427	0.0339	0.0296	–	–	180.00
190.00	160.03	76.48	0.1958	54.666	94.950	0.10279	0.16495	0.2715	0.2263	–	1067.	348.	0.401	0.0347	0.0288	–	–	190.00
200.00	179.78	74.90	0.1722	57.275	96.101	0.10670	0.16568	0.2787	0.2372	–	1001.	341.	0.378	0.0356	0.0280	–	–	200.00
210.00	201.32	73.23	0.1515	59.935	97.195	0.11062	0.16636	0.2865	0.2494	–	934.	332.	0.356	0.0366	0.0272	–	–	210.00
220.00	224.75	71.46	0.1332	62.655	98.218	0.11456	0.16696	0.2952	0.2632	–	864.	323.	0.335	–	0.0263	–	–	220.00
230.00	250.19	69.55	0.1168	65.449	99.154	0.11855	0.16747	0.3048	0.2785	–	792.	313.	0.316	–	0.0255	–	–	230.00
240.00	277.77	67.47	0.1021	68.337	99.977	0.12260	0.16785	0.3152	0.2954	–	716.	302.	0.299	–	0.0246	–	–	240.00
250.00	307.65	65.15	0.0888	71.346	100.650	0.12675	0.16806	–	–	–	638.	289.	–	–	–	–	–	250.00
260.00	339.97	62.49	0.0766	74.519	101.110	0.13106	0.16801	–	–	–	556.	276.	–	–	–	–	–	260.00
270.00	374.90	59.28	0.0650	77.941	101.240	0.13563	0.16757	–	–	–	470.	261.	–	–	–	–	–	270.00
280.00	412.62	55.02	0.0537	81.819	100.810	0.14074	0.16641	–	–	–	381.	245.	–	–	–	–	–	280.00
290.00	453.31	47.62	0.0409	87.049	98.880	0.14757	0.16335	–	–	–	–	–	–	–	–	–	–	290.00
294.60c	473.00	34.75	0.0288	93.94	93.94	0.1566	0.1566	∞	∞	∞	0.	0.	–	–	∞	∞	0	294.60

*temperatures are on the IPTS-68 scale b = normal boiling point c = critical point

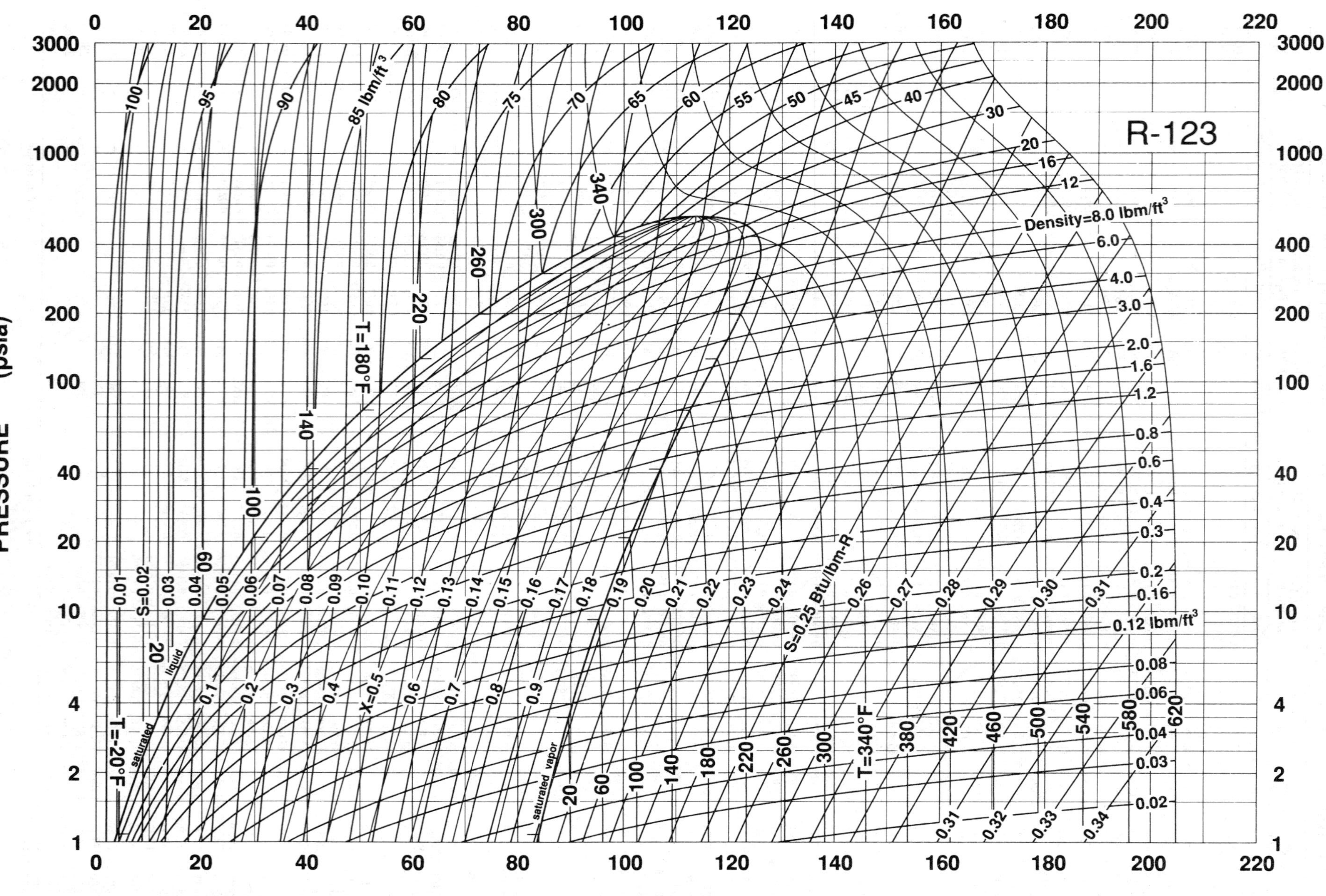

Prepared by: CENTER FOR APPLIED THERMODYNAMIC STUDIES, University of Idaho

Fig. 9 Pressure-Enthalpy Diagram for Refrigerant 123

Refrigerant 123 (2,2-Dichloro-1,1,1-trifluoroethane) Properties of Saturated Liquid and Saturated Vapor

Temp,* °F	Pressure, psia	Density, lb/ft³ Liquid	Volume, ft³/lb Vapor	Enthalpy, Btu/lb Liquid	Enthalpy, Btu/lb Vapor	Entropy, Btu/lb·°F Liquid	Entropy, Btu/lb·°F Vapor	Specific Heat c_p, Btu/lb·°F Liquid	Specific Heat c_p, Btu/lb·°F Vapor	c_p/c_v Vapor	Velocity of Sound, ft/s Liquid	Velocity of Sound, ft/s Vapor	Viscosity, lb_m/ft·h Liquid	Viscosity, lb_m/ft·h Vapor	Thermal Cond, Btu/h·ft·°F Liquid	Thermal Cond, Btu/h·ft·°F Vapor	Surface Tension, dyne/cm	Temp,* °F
–40.00	0.548	100.71	53.445	0.000	81.340	0.00000	0.19382	—	0.1419	1.1064	—	386.	2.537	—	—	—	23.49	–40.00
–30.00	0.776	99.96	38.593	2.005	82.707	0.00472	0.19254	—	0.1442	1.1057	—	390.	2.303	—	—	—	22.78	–30.00
–20.00	1.080	99.20	28.345	3.923	84.085	0.00913	0.19146	—	0.1466	1.1052	—	394.	2.100	—	—	—	22.06	–20.00
–10.00	1.478	98.44	21.141	5.812	85.476	0.01338	0.19054	—	0.1490	1.1047	—	398.	1.922	—	—	—	21.35	–10.00
0.00	1.993	97.67	15.993	7.710	86.877	0.01755	0.18978	—	0.1514	1.1045	—	401.	1.766	—	0.0515	—	20.65	0.00
5.00	2.303	97.28	13.979	8.672	87.581	0.01963	0.18945	—	0.1526	1.1044	—	403.	1.695	—	0.0510	—	20.30	5.00
10.00	2.651	96.89	12.258	9.645	88.289	0.02171	0.18916	—	0.1539	1.1043	—	405.	1.629	—	0.0506	—	19.95	10.00
15.00	3.042	96.49	10.782	10.631	88.998	0.02380	0.18890	—	0.1551	1.1043	—	406.	1.566	—	0.0501	—	19.61	15.00
20.00	3.480	96.09	9.5110	11.632	89.710	0.02590	0.18867	—	0.1563	1.1043	—	408.	1.507	—	0.0496	—	19.26	20.00
25.00	3.969	95.69	8.4141	12.647	90.424	0.02800	0.18847	0.2046	0.1576	1.1043	2636.	409.	1.451	—	0.0492	—	18.92	25.00
30.00	4.513	95.29	7.4644	13.679	91.140	0.03012	0.18831	0.2079	0.1588	1.1044	2596.	411.	1.398	—	0.0487	—	18.57	30.00
35.00	5.116	94.88	6.6397	14.727	91.859	0.03224	0.18817	0.2111	0.1601	1.1046	2558.	412.	1.348	—	0.0482	—	18.23	35.00
40.00	5.785	94.47	5.9215	15.792	92.579	0.03438	0.18806	0.2144	0.1613	1.1047	2521.	414.	1.301	—	0.0478	—	17.89	40.00
45.00	6.522	94.06	5.2943	16.873	93.301	0.03653	0.18797	0.2177	0.1626	1.1050	2486.	415.	1.257	—	0.0473	—	17.55	45.00
50.00	7.334	93.64	4.7450	17.971	94.025	0.03869	0.18791	0.2209	0.1639	1.1052	2452.	416.	1.214	—	0.0468	—	17.21	50.00
55.00	8.226	93.22	4.2628	19.084	94.750	0.04086	0.18788	0.2240	0.1651	1.1056	2419.	417.	1.174	—	0.0464	—	16.87	55.00
60.00	9.203	92.80	3.8384	20.213	95.477	0.04304	0.18787	0.2271	0.1664	1.1059	2387.	418.	1.136	—	0.0459	—	16.53	60.00
65.00	10.271	92.38	3.4639	21.356	96.205	0.04523	0.18789	0.2300	0.1676	1.1063	2355.	419.	1.099	—	0.0454	—	16.20	65.00
70.00	11.436	91.95	3.1326	22.515	96.933	0.04742	0.18792	0.2329	0.1689	1.1068	2323.	420.	1.065	—	0.0450	0.00572	15.86	70.00
75.00	12.704	91.52	2.8389	23.687	97.663	0.04962	0.18798	0.2356	0.1701	1.1074	2292.	421.	1.032	—	0.0445	0.00583	15.53	75.00
80.00	14.080	91.08	2.5779	24.873	98.394	0.05182	0.18805	0.2382	0.1714	1.1080	2261.	422.	1.000	—	0.0440	0.00594	15.20	80.00
82.11b	14.696	90.89	2.4764	25.378	98.702	0.05275	0.18809	0.2392	0.1719	1.1083	2248.	422.	0.987	—	0.0438	0.00598	15.06	82.11
85.00	15.572	90.64	2.3454	26.072	99.124	0.05403	0.18815	0.2406	0.1726	1.1087	2231.	423.	0.970	—	0.0436	0.00605	14.87	85.00
90.00	17.185	90.20	2.1379	27.282	99.855	0.05623	0.18826	0.2430	0.1738	1.1095	2200.	423.	0.941	—	0.0431	0.00616	14.54	90.00
95.00	18.927	89.75	1.9522	28.504	100.586	0.05844	0.18839	0.2452	0.1751	1.1103	2169.	424.	0.913	—	0.0426	0.00627	14.21	95.00
100.00	20.803	89.30	1.7858	29.737	101.317	0.06065	0.18854	0.2473	0.1763	1.1113	2138.	425.	0.887	—	0.0422	0.00638	13.89	100.00
105.00	22.821	88.85	1.6363	30.980	102.047	0.06285	0.18871	0.2493	0.1775	1.1124	2107.	425.	0.861	—	0.0417	0.00649	13.56	105.00
110.00	24.988	88.39	1.5018	32.233	102.777	0.06505	0.18888	0.2511	0.1788	1.1135	2075.	425.	0.837	—	0.0412	0.00659	13.24	110.00
115.00	27.310	87.92	1.3804	33.495	103.505	0.06725	0.18907	0.2529	0.1800	1.1148	2044.	425.	0.813	—	0.0408	0.00670	12.92	115.00
120.00	29.796	87.46	1.2708	34.766	104.231	0.06944	0.18928	0.2545	0.1812	1.1162	2012.	426.	0.790	—	0.0403	0.00681	12.59	120.00
125.00	32.451	86.98	1.1716	36.044	104.956	0.07163	0.18949	0.2560	0.1825	1.1177	1980.	426.	0.768	0.0285	0.0398	0.00692	12.27	125.00
130.00	35.285	86.51	1.0816	37.330	105.679	0.07381	0.18972	0.2575	0.1837	1.1194	1948.	426.	0.747	0.0287	0.0394	0.00703	11.96	130.00
135.00	38.304	86.03	0.9999	38.623	106.399	0.07598	0.18995	0.2588	0.1850	1.1212	1915.	425.	0.727	0.0289	0.0389	0.00714	11.64	135.00
140.00	41.515	85.54	0.9255	39.922	107.116	0.07814	0.19019	0.2601	0.1862	1.1232	1882.	425.	0.707	0.0292	0.0384	0.00725	11.32	140.00
145.00	44.928	85.05	0.8577	41.228	107.830	0.08030	0.19044	0.2613	0.1875	1.1254	1849.	425.	0.687	0.0294	0.0380	0.00736	11.01	145.00
150.00	48.549	84.55	0.7958	42.540	108.540	0.08245	0.19070	0.2624	0.1888	1.1277	1816.	425.	0.669	0.0296	0.0375	0.00747	10.70	150.00
155.00	52.388	84.05	0.7393	43.857	109.246	0.08458	0.19097	0.2635	0.1901	1.1303	1782.	424.	0.651	0.0298	0.0370	0.00758	10.39	155.00
160.00	56.451	83.54	0.6874	45.179	109.948	0.08671	0.19123	0.2645	0.1914	1.1331	1749.	423.	0.633	0.0301	0.0366	0.00769	10.08	160.00
165.00	60.747	83.02	0.6399	46.506	110.645	0.08883	0.19151	0.2655	0.1927	1.1361	1715.	423.	0.616	0.0303	0.0361	0.00780	9.77	165.00
170.00	65.285	82.50	0.5963	47.838	111.337	0.09094	0.19178	0.2664	0.1941	1.1393	1681.	422.	0.599	0.0305	0.0356	0.00791	9.47	170.00
175.00	70.073	81.98	0.5561	49.174	112.022	0.09303	0.19206	0.2673	0.1955	1.1429	1646.	421.	0.583	0.0307	0.0352	0.00802	9.16	175.00
180.00	75.119	81.44	0.5191	50.515	112.702	0.09512	0.19234	0.2682	0.1969	1.1467	1612.	420.	—	0.0310	0.0347	0.00813	8.86	180.00
185.00	80.434	80.90	0.4850	51.860	113.374	0.09719	0.19261	0.2691	0.1984	1.1509	1578.	419.	—	0.0312	0.0342	0.00824	8.56	185.00
190.00	86.024	80.35	0.4535	53.209	114.039	0.09926	0.19289	0.2700	0.1999	1.1554	1543.	417.	—	0.0314	0.0338	0.00835	8.26	190.00
195.00	91.900	79.79	0.4244	54.562	114.697	0.10131	0.19317	0.2709	0.2015	1.1604	1509.	416.	—	0.0316	0.0333	0.00846	7.96	195.00
200.00	98.071	79.23	0.3974	55.920	115.346	0.10336	0.19344	0.2719	0.2032	1.1657	1474.	415.	—	0.0318	0.0328	—	7.67	200.00
205.00	104.55	78.65	0.3724	57.281	115.986	0.10539	0.19371	0.2728	0.2049	1.1715	1439.	413.	—	0.0321	0.0324	—	7.38	205.00
210.00	111.33	78.07	0.3492	58.647	116.617	0.10741	0.19398	0.2738	0.2067	1.1777	1405.	411.	—	0.0323	0.0319	—	7.09	210.00
215.00	118.44	77.48	0.3277	60.018	117.237	0.10943	0.19424	0.2749	0.2086	1.1846	1370.	409.	—	0.0325	—	—	6.80	215.00
220.00	125.89	76.88	0.3076	61.393	117.847	0.11143	0.19449	0.2760	0.2106	1.1920	1336.	407.	—	0.0327	—	—	6.51	220.00
225.00	133.67	76.26	0.2889	62.774	118.445	0.11343	0.19474	0.2772	0.2127	1.2001	1301.	405.	—	0.0329	—	—	6.23	225.00
230.00	141.81	75.64	0.2715	64.159	119.032	0.11542	0.19498	0.2786	0.2150	1.2089	1267.	403.	—	0.0332	—	—	5.94	230.00
235.00	150.31	75.00	0.2552	65.551	119.605	0.11740	0.19521	0.2800	0.2174	1.2185	1233.	401.	—	0.0334	—	—	5.67	235.00
240.00	159.19	74.35	0.2400	66.948	120.165	0.11937	0.19543	0.2816	0.2200	1.2291	1198.	398.	—	0.0336	—	—	5.39	240.00
245.00	168.44	73.69	0.2258	68.353	120.711	0.12134	0.19564	0.2834	0.2228	1.2407	1164.	396.	—	0.0338	—	—	5.11	245.00
250.00	178.10	73.01	0.2125	69.764	121.241	0.12330	0.19583	0.2853	0.2258	1.2534	1130.	393.	—	0.0340	—	—	4.84	250.00
260.00	198.63	71.61	0.1882	72.613	122.250	0.12721	0.19618	0.2899	0.2327	1.2831	1061.	387.	—	—	—	—	4.31	260.00
270.00	220.87	70.13	0.1668	75.501	123.184	0.13112	0.19646	0.2957	0.2411	1.3198	992.	380.	—	—	—	—	3.79	270.00
280.00	244.92	68.56	0.1478	78.437	124.031	0.13503	0.19667	0.3031	0.2515	1.3663	922.	372.	—	—	—	—	3.28	280.00
290.00	270.89	66.88	0.1308	81.434	124.773	0.13895	0.19676	0.3127	0.2649	1.4268	851.	363.	—	—	—	—	2.79	290.00
300.00	298.88	65.06	0.1154	84.507	125.389	0.14292	0.19674	0.3254	0.2829	1.5085	778.	353.	—	—	—	—	2.31	300.00
310.00	329.01	63.06	0.1015	87.677	125.845	0.14695	0.19654	0.3431	0.3083	1.6250	701.	342.	—	—	—	—	1.85	310.00
320.00	361.45	60.82	0.0887	90.977	126.088	0.15109	0.19612	0.3691	0.3471	1.8034	621.	329.	—	—	—	—	1.42	320.00
330.00	396.37	58.22	0.0768	94.457	126.025	0.15538	0.19536	0.4109	0.4137	2.1089	534.	315.	—	—	—	—	1.01	330.00
340.00	434.02	55.08	0.0652	98.209	125.478	0.15995	0.19405	0.4911	0.5508	2.7349	441.	300.	—	—	—	—	0.63	340.00
350.00	474.80	50.88	0.0534	102.454	124.000	0.16505	0.19166	—	—	—	—	—	—	—	—	—	0.29	350.00
360.00	519.44	43.25	0.0389	108.347	119.584	0.17206	0.18577	—	—	—	—	—	—	—	—	—	0.04	360.00
362.63c	532.00	34.34	0.0291	113.806	113.806	0.17788	0.17788	∞	∞	∞	0.	0.	—	—	∞	∞	0.00	362.63

*temperatures are on the ITS–90 scale b = normal boiling point c = critical point

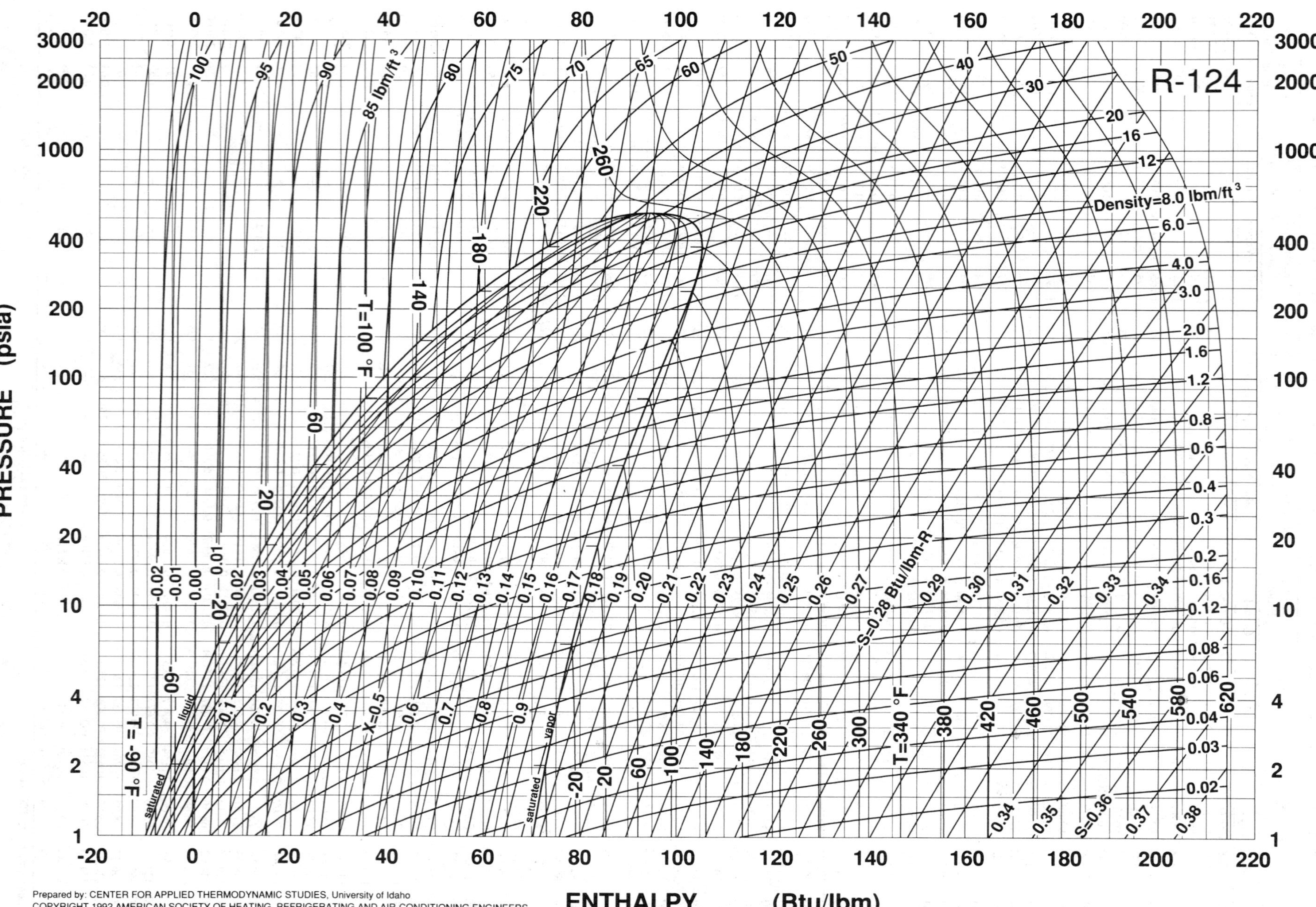

Fig. 10 Pressure–Enthalpy Diagram for Refrigerant 124

Refrigerant 124 (2-Chloro-1,1,1,2-tetrafluoroethane) Properties of Saturated Liquid and Saturated Vapor

Temp,* °F	Pressure, psia	Density, lb/ft³ Liquid	Volume, ft³/lb Vapor	Enthalpy, Btu/lb Liquid	Enthalpy, Btu/lb Vapor	Entropy, Btu/lb·°F Liquid	Entropy, Btu/lb·°F Vapor	Specific Heat c_p, Btu/lb·°F Liquid	Specific Heat c_p, Btu/lb·°F Vapor	c_p/c_v Vapor	Velocity of Sound, ft/s Liquid	Velocity of Sound, ft/s Vapor	Viscosity, lb_m/ft·h Liquid	Viscosity, lb_m/ft·h Vapor	Thermal Cond, Btu/h·ft·°F Liquid	Thermal Cond, Btu/h·ft·°F Vapor	Surface Tension, dyne/cm	Temp,* °F
−100.00	0.438	100.79	64.170	−16.341	67.816	−0.03222	0.19140	0.2290	0.1382	1.1232	3019.	382.	2.721	—	0.0672	—	22.10	−100.00
−90.00	0.666	100.79	43.371	−12.650	69.146	−0.02797	0.18901	0.2294	0.1408	1.1224	2778.	386.	2.390	—	0.0661	—	21.53	−90.00
−80.00	0.988	100.64	29.950	−9.306	70.485	−0.02328	0.18687	0.2296	0.1436	1.1219	2654.	390.	2.118	—	0.0649	—	20.92	−80.00
−70.00	1.433	99.74	21.136	−7.002	71.832	−0.01730	0.18501	0.2311	0.1464	1.1217	2585.	394.	1.902	—	0.0638	—	20.16	−70.00
−60.00	2.034	98.83	15.228	−4.684	73.184	−0.01143	0.18340	0.2325	0.1493	1.1218	2523.	398.	1.719	—	0.0626	—	19.41	−60.00
−50.00	2.830	97.91	11.182	−2.350	74.542	−0.00566	0.18203	0.2340	0.1524	1.1223	2464.	402.	1.563	—	0.0615	—	18.66	−50.00
−45.00	3.315	97.44	9.6449	−1.177	75.223	−0.00282	0.18142	0.2349	0.1539	1.1226	2436.	404.	1.493	—	0.0609	—	18.29	−45.00
−40.00	3.865	96.97	8.3539	0.000	75.904	0.00000	0.18087	0.2357	0.1555	1.1231	2409.	405.	1.428	—	0.0603	—	17.92	−40.00
−35.00	4.487	96.50	7.2644	1.181	76.586	0.00280	0.18036	0.2366	0.1572	1.1236	2381.	407.	1.368	—	0.0598	—	17.55	−35.00
−30.00	5.189	96.02	6.3410	2.368	77.268	0.00557	0.17989	0.2375	0.1588	1.1243	2354.	408.	1.311	—	0.0592	—	17.18	−30.00
−25.00	5.977	95.54	5.5550	3.559	77.950	0.00832	0.17947	0.2385	0.1605	1.1250	2327.	410.	1.259	—	0.0586	—	16.81	−25.00
−20.00	6.858	95.06	4.8834	4.755	78.633	0.01105	0.17909	0.2395	0.1622	1.1258	2300.	411.	1.209	—	0.0580	—	16.45	−20.00
−15.00	7.842	94.57	4.3073	5.956	79.316	0.01377	0.17874	0.2406	0.1640	1.1267	2273.	412.	1.163	—	0.0575	—	16.09	−15.00
−10.00	8.935	94.08	3.8113	7.163	79.999	0.01646	0.17844	0.2417	0.1658	1.1278	2245.	414.	1.120	—	0.0569	—	15.72	−10.00
−5.00	10.146	93.58	3.3827	8.376	80.681	0.01914	0.17817	0.2429	0.1676	1.1289	2218.	415.	1.079	—	0.0563	0.00447	15.36	−5.00
0.00	11.485	93.08	3.0110	9.594	81.363	0.02180	0.17793	0.2441	0.1695	1.1301	2190.	416.	1.040	—	0.0557	0.00456	15.00	0.00
5.00	12.960	92.58	2.6877	10.819	82.045	0.02444	0.17772	0.2453	0.1713	1.1314	2162.	417.	1.004	—	0.0552	0.00466	14.65	5.00
10.00	14.580	92.07	2.4056	12.051	82.726	0.02707	0.17755	0.2466	0.1732	1.1329	2134.	418.	0.969	—	0.0546	0.00476	14.29	10.00
10.34b	14.696	92.03	2.3877	12.135	82.772	0.02725	0.17754	0.2467	0.1734	1.1330	2132.	418.	0.967	—	0.0546	0.00476	14.27	10.34
15.00	16.356	91.55	2.1586	13.289	83.406	0.02968	0.17740	0.2479	0.1752	1.1344	2105.	418.	0.937	—	0.0540	0.00486	13.94	15.00
20.00	18.298	91.03	1.9417	14.533	84.086	0.03228	0.17728	0.2493	0.1772	1.1361	2076.	419.	0.905	—	0.0535	0.00495	13.58	20.00
25.00	20.416	90.51	1.7507	15.785	84.764	0.03487	0.17719	0.2507	0.1792	1.1379	2047.	420.	0.876	—	0.0529	0.00505	13.23	25.00
30.00	22.720	89.98	1.5821	17.044	85.441	0.03745	0.17713	0.2521	0.1812	1.1398	2017.	420.	0.848	—	0.0523	0.00515	12.88	30.00
35.00	25.221	89.44	1.4328	18.310	86.116	0.04001	0.17708	0.2536	0.1833	1.1419	1987.	420.	0.821	—	0.0517	0.00524	12.53	35.00
40.00	27.930	88.90	1.3003	19.584	86.790	0.04256	0.17706	0.2551	0.1854	1.1441	1957.	421.	0.795	—	0.0512	0.00534	12.18	40.00
45.00	30.859	88.35	1.1823	20.866	87.462	0.04510	0.17706	0.2567	0.1875	1.1465	1926.	421.	0.770	—	0.0506	0.00544	11.84	45.00
50.00	34.020	87.80	1.0771	22.155	88.132	0.04763	0.17708	0.2582	0.1897	1.1491	1895.	421.	0.746	—	0.0500	0.00553	11.49	50.00
55.00	37.423	87.24	0.9831	23.453	88.800	0.05015	0.17712	0.2599	0.1919	1.1518	1863.	421.	0.723	—	0.0494	0.00563	11.15	55.00
60.00	41.081	86.67	0.8988	24.759	89.465	0.05266	0.17717	0.2616	0.1942	1.1547	1831.	420.	0.701	—	0.0489	0.00573	10.81	60.00
65.00	45.006	86.10	0.8231	26.074	90.127	0.05516	0.17724	0.2633	0.1965	1.1579	1799.	420.	0.680	—	0.0483	0.00583	10.47	65.00
70.00	49.210	85.52	0.7549	27.397	90.787	0.05765	0.17733	0.2650	0.1989	1.1612	1766.	420.	0.659	—	0.0477	0.00592	10.13	70.00
75.00	53.706	84.93	0.6935	28.729	91.443	0.06014	0.17743	0.2669	0.2013	1.1648	1733.	419.	0.639	—	0.0472	0.00602	9.79	75.00
80.00	58.506	84.33	0.6379	30.070	92.095	0.06262	0.17755	0.2687	0.2037	1.1686	1700.	418.	0.619	0.0283	0.0466	0.00611	9.46	80.00
85.00	63.624	83.72	0.5876	31.421	92.744	0.06509	0.17767	0.2706	0.2063	1.1728	1666.	417.	0.600	0.0285	0.0460	0.00621	9.13	85.00
90.00	69.073	83.11	0.5420	32.781	93.388	0.06755	0.17781	0.2726	0.2089	1.1772	1632.	416.	0.582	0.0288	0.0454	0.00631	8.80	90.00
95.00	74.865	82.49	0.5005	34.151	94.027	0.07001	0.17796	0.2746	0.2115	1.1820	1597.	415.	0.564	0.0290	0.0449	0.00640	8.47	95.00
100.00	81.015	81.85	0.4627	35.531	94.661	0.07246	0.17811	0.2768	0.2143	1.1872	1562.	414.	0.546	0.0292	0.0443	0.00649	8.14	100.00
105.00	87.535	81.21	0.4282	36.922	95.290	0.07491	0.17827	0.2789	0.2171	1.1928	1527.	412.	0.529	0.0295	0.0437	0.00659	7.82	105.00
110.00	94.441	80.55	0.3967	38.323	95.912	0.07735	0.17844	0.2812	0.2200	1.1988	1491.	411.	0.512	0.0297	0.0432	0.00668	7.49	110.00
115.00	101.75	79.88	0.3679	39.735	96.527	0.07979	0.17861	0.2836	0.2231	1.2054	1456.	409.	0.496	0.0300	0.0426	0.00677	7.17	115.00
120.00	109.46	79.20	0.3415	41.158	97.135	0.08222	0.17879	0.2860	0.2262	1.2125	1419.	407.	0.479	0.0302	0.0420	0.00687	6.85	120.00
125.00	117.61	78.51	0.3172	42.594	97.734	0.08466	0.17897	0.2886	0.2296	1.2203	1383.	405.	0.463	0.0305	0.0415	0.00696	6.54	125.00
130.00	126.20	77.80	0.2949	44.041	98.325	0.08709	0.17914	0.2913	0.2330	1.2288	1346.	403.	0.448	0.0307	0.0409	0.00705	6.22	130.00
135.00	135.24	77.08	0.2743	45.501	98.906	0.08951	0.17932	0.2942	0.2367	1.2382	1309.	400.	0.432	0.0310	0.0403	0.00713	5.91	135.00
140.00	144.76	76.34	0.2553	46.974	99.476	0.09194	0.17949	0.2972	0.2406	1.2484	1271.	397.	0.417	0.0313	0.0398	0.00722	5.61	140.00
145.00	154.77	75.59	0.2378	48.461	100.035	0.09437	0.17966	0.3004	0.2447	1.2598	1233.	394.	0.403	0.0316	0.0392	0.00731	5.30	145.00
150.00	165.28	74.81	0.2215	49.963	100.580	0.09680	0.17983	0.3038	0.2491	1.2724	1195.	391.	0.388	0.0319	0.0386	0.00739	5.00	150.00
155.00	176.31	74.01	0.2065	51.480	101.112	0.09924	0.17998	0.3075	0.2538	1.2864	1156.	388.	0.374	0.0322	0.0381	0.00747	4.70	155.00
160.00	187.88	73.20	0.1925	53.013	101.627	0.10167	0.18012	0.3114	0.2590	1.3022	1117.	384.	0.360	0.0325	0.0375	0.00755	4.40	160.00
165.00	200.00	72.35	0.1795	54.563	102.125	0.10411	0.18025	0.3157	0.2646	1.3199	1077.	381.	0.346	0.0329	0.0369	0.00763	4.11	165.00
170.00	212.70	71.48	0.1674	56.132	102.604	0.10656	0.18037	0.3204	0.2707	1.3400	1038.	376.	0.333	0.0332	0.0364	0.00770	3.82	170.00
175.00	225.99	70.59	0.1561	57.720	103.061	0.10902	0.18046	0.3256	0.2775	1.3629	997.	372.	0.319	0.0337	0.0358	0.00777	3.54	175.00
180.00	239.89	69.66	0.1455	59.329	103.494	0.11149	0.18053	0.3314	0.2852	1.3893	957.	368.	0.306	0.0341	0.0352	0.00783	3.26	180.00
185.00	254.41	68.69	0.1357	60.962	103.900	0.11397	0.18058	0.3379	0.2938	1.4200	916.	363.	0.294	0.0346	0.0347	0.00789	2.98	185.00
190.00	269.60	67.68	0.1264	62.620	104.274	0.11647	0.18059	0.3452	0.3038	1.4560	874.	357.	0.282	0.0352	0.0341	0.00795	2.71	190.00
195.00	285.45	66.63	0.1176	64.306	104.613	0.11899	0.18056	0.3537	0.3153	1.4987	832.	352.	0.270	0.0358	0.0336	0.00800	2.44	195.00
200.00	302.01	65.52	0.1094	66.024	104.910	0.12153	0.18048	0.3636	0.3289	1.5502	789.	346.	0.258	0.0364	0.0330	0.00803	2.18	200.00
205.00	319.28	64.35	0.1016	67.778	105.160	0.12410	0.18035	0.3754	0.3453	1.6133	746.	339.	0.247	0.0372	0.0325	0.00806	1.92	205.00
210.00	337.32	63.11	0.0942	69.573	105.353	0.12672	0.18015	0.3897	0.3654	1.6921	701.	333.	0.236	0.0381	0.0319	0.00808	1.67	210.00
215.00	356.13	61.78	0.0871	71.416	105.477	0.12938	0.17986	0.4076	0.3909	1.7932	656.	325.	0.226	—	0.0314	0.00808	1.43	215.00
220.00	375.76	60.34	0.0804	73.319	105.518	0.13210	0.17947	0.4307	0.4243	1.9270	610.	318.	0.216	—	0.0309	0.00806	1.19	220.00
225.00	396.25	58.77	0.0738	75.295	105.451	0.13490	0.17895	0.4619	0.4698	2.1119	563.	310.	0.207	—	0.0303	0.00801	0.97	225.00
230.00	417.64	57.01	0.0675	77.367	105.245	0.13782	0.17824	0.5069	0.5360	2.3825	514.	301.	0.199	—	0.0298	0.00793	0.75	230.00
235.00	440.00	55.01	0.0612	79.572	104.843	0.14090	0.17727	0.5779	0.6410	2.8142	463.	292.	—	—	—	—	0.55	235.00
240.00	463.38	52.61	0.0547	81.981	104.143	0.14424	0.17591	0.7078	0.8330	3.6065	410.	281.	—	—	—	—	0.36	240.00
245.00	487.91	49.49	0.0479	84.761	102.909	0.14807	0.17382	1.0242	1.2942	5.5136	353.	271.	—	—	—	—	0.19	245.00
250.00	513.78	44.38	0.0392	88.581	100.240	0.15333	0.16975	—	—	—	—	—	—	—	—	—	0.04	250.00
252.45c	527.07	34.57	0.0289	94.596	94.596	0.16087	0.16087	∞	∞	∞	0.	0.	—	—	∞	∞	0.00	252.45

*temperatures are on the ITS–90 scale b = normal boiling point c = critical point

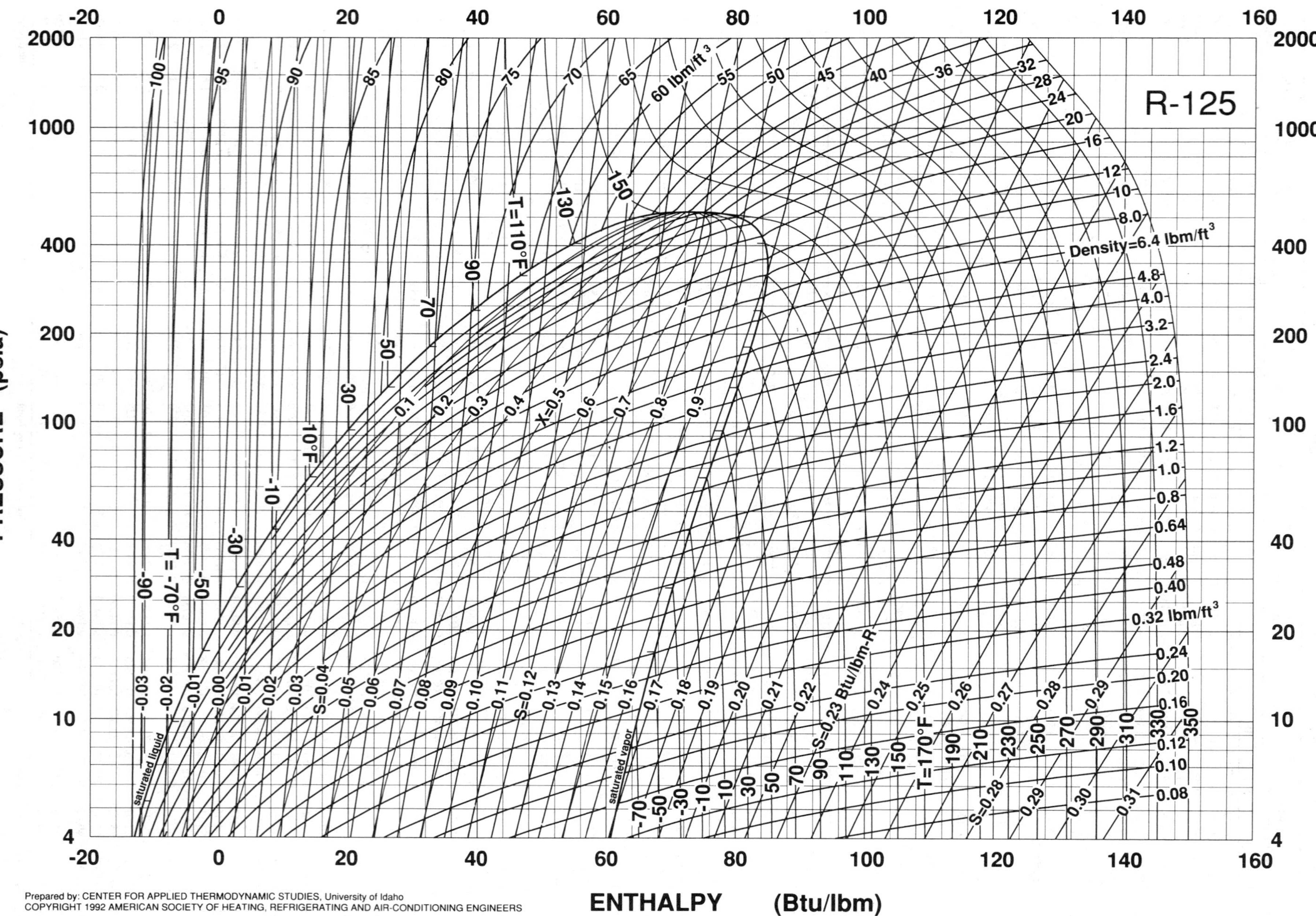

Fig. 11 Pressure–Enthalpy Diagram for Refrigerant 125

Refrigerant 125 (Pentafluoroethane) Properties of Saturated Liquid and Saturated Vapor

Temp,* °F	Pressure, psia	Density, lb/ft³ Liquid	Volume, ft³/lb Vapor	Enthalpy, Btu/lb Liquid	Enthalpy, Btu/lb Vapor	Entropy, Btu/lb·°F Liquid	Entropy, Btu/lb·°F Vapor	Specific Heat c_p, Btu/lb·°F Liquid	Specific Heat c_p, Btu/lb·°F Vapor	c_p/c_v Vapor	Velocity of Sound, ft/s Liquid	Velocity of Sound, ft/s Vapor	Viscosity, lb_m/ft·h Liquid	Viscosity, lb_m/ft·h Vapor	Thermal Cond, Btu/h·ft·°F Liquid	Thermal Cond, Btu/h·ft·°F Vapor	Surface Tension, dyne/cm	Temp,* °F
−100.00	3.691	99.92	8.5830	−13.439	60.483	−0.03294	0.17155	—	0.1511	1.1333	—	405.	1.557	—	—	—	—	−100.00
−95.00	4.376	99.34	7.3248	−12.593	61.186	−0.03088	0.17071	—	0.1526	1.1333	—	407.	1.474	—	—	—	—	−95.00
−90.00	5.164	98.76	6.2782	−11.675	61.890	−0.02861	0.16994	—	0.1542	1.1333	—	409.	1.398	—	—	—	—	−90.00
−85.00	6.067	98.17	5.4034	−10.694	62.595	−0.02617	0.16922	—	0.1558	1.1335	—	410.	1.327	—	—	—	—	−85.00
−80.00	7.096	97.58	4.6691	−9.658	63.301	−0.02359	0.16856	—	0.1574	1.1339	—	412.	1.262	—	—	—	—	−80.00
−75.00	8.264	96.99	4.0501	−8.573	64.006	−0.02089	0.16795	—	0.1590	1.1344	—	414.	1.201	—	—	—	—	−75.00
−70.00	9.586	96.39	3.5261	−7.444	64.710	−0.01809	0.16738	—	0.1607	1.1350	—	415.	1.144	—	—	—	—	−70.00
−65.00	11.075	95.78	3.0809	−6.278	65.414	−0.01521	0.16687	—	0.1624	1.1359	—	416.	1.092	—	—	—	—	−65.00
−60.00	12.746	95.17	2.7010	−5.077	66.116	−0.01226	0.16639	—	0.1642	1.1368	—	418.	1.042	—	—	—	—	−60.00
−55.00	14.615	94.55	2.3758	−3.846	66.817	−0.00925	0.16595	0.2473	0.1660	1.1380	2413.	419.	0.997	—	—	—	—	−55.00
−54.79b	14.696	94.52	2.3634	−3.794	66.846	−0.00912	0.16594	0.2476	0.1661	1.1381	2409.	419.	0.995	—	—	—	—	−54.79
−50.00	16.698	93.92	2.0963	−2.587	67.515	−0.00620	0.16556	0.2530	0.1678	1.1394	2338.	420.	0.953	—	—	—	—	−50.00
−45.00	19.012	93.29	1.8553	−1.305	68.211	−0.00311	0.16520	0.2581	0.1697	1.1409	2269.	421.	0.913	—	—	—	—	−45.00
−40.00	21.574	92.65	1.6468	0.000	68.904	0.00000	0.16487	0.2626	0.1716	1.1427	2204.	421.	0.875	—	—	—	—	−40.00
−35.00	24.402	92.01	1.4658	1.325	69.593	0.00313	0.16458	0.2667	0.1736	1.1447	2141.	422.	0.839	—	—	—	—	−35.00
−30.00	27.514	91.35	1.3083	2.668	70.278	0.00628	0.16432	0.2705	0.1757	1.1469	2082.	423.	0.805	—	—	—	—	−30.00
−25.00	30.928	90.69	1.1706	4.027	70.959	0.00943	0.16409	0.2739	0.1778	1.1494	2024.	423.	0.773	—	—	—	—	−25.00
−20.00	34.664	90.02	1.0500	5.403	71.636	0.01259	0.16388	0.2770	0.1799	1.1522	1969.	423.	0.742	—	—	—	—	−20.00
−15.00	38.742	89.35	0.9441	6.793	72.307	0.01575	0.16370	0.2799	0.1822	1.1552	1914.	423.	0.713	—	—	—	—	−15.00
−10.00	43.179	88.66	0.8508	8.197	72.972	0.01890	0.16354	0.2826	0.1845	1.1586	1861.	423.	0.686	—	—	—	—	−10.00
−5.00	47.997	87.96	0.7684	9.614	73.632	0.02205	0.16341	0.2852	0.1868	1.1623	1810.	423.	0.659	—	—	—	—	−5.00
0.00	53.215	87.25	0.6954	11.044	74.284	0.02520	0.16329	0.2877	0.1893	1.1663	1759.	422.	0.634	—	—	—	—	0.00
2.00	55.419	86.97	0.6685	11.620	74.543	0.02645	0.16325	0.2886	0.1903	1.1680	1739.	422.	0.624	—	—	—	—	2.00
4.00	57.691	86.68	0.6429	12.197	74.801	0.02771	0.16321	0.2896	0.1913	1.1698	1719.	422.	0.615	—	—	—	—	4.00
6.00	60.034	86.39	0.6184	12.777	75.058	0.02896	0.16318	0.2906	0.1924	1.1717	1699.	421.	0.605	—	—	—	—	6.00
8.00	62.448	86.10	0.5951	13.358	75.314	0.03021	0.16315	0.2915	0.1934	1.1736	1679.	421.	0.596	—	—	—	—	8.00
10.00	64.934	85.80	0.5728	13.942	75.568	0.03147	0.16311	0.2925	0.1945	1.1756	1659.	421.	0.587	—	—	—	—	10.00
12.00	67.495	85.51	0.5515	14.527	75.821	0.03271	0.16309	0.2934	0.1956	1.1776	1640.	420.	0.578	—	—	—	—	12.00
14.00	70.131	85.21	0.5311	15.114	76.073	0.03396	0.16306	0.2944	0.1967	1.1798	1621.	420.	0.569	—	—	—	—	14.00
16.00	72.843	84.91	0.5116	15.704	76.323	0.03521	0.16304	0.2953	0.1978	1.1820	1601.	420.	0.560	—	—	—	—	16.00
18.00	75.633	84.61	0.4930	16.295	76.572	0.03645	0.16301	0.2963	0.1989	1.1843	1582.	419.	0.552	—	—	—	—	18.00
20.00	78.503	84.30	0.4752	16.888	76.819	0.03769	0.16299	0.2972	0.2001	1.1867	1563.	419.	0.543	—	—	—	—	20.00
22.00	81.454	84.00	0.4581	17.483	77.065	0.03893	0.16298	0.2982	0.2013	1.1892	1544.	418.	0.535	—	—	—	—	22.00
24.00	84.487	83.69	0.4418	18.081	77.309	0.04017	0.16296	0.2992	0.2025	1.1918	1525.	418.	0.526	—	—	—	—	24.00
26.00	87.603	83.38	0.4261	18.680	77.552	0.04141	0.16294	0.3002	0.2037	1.1944	1506.	417.	0.518	—	—	—	—	26.00
28.00	90.805	83.06	0.4111	19.281	77.793	0.04265	0.16293	0.3012	0.2049	1.1972	1487.	416.	0.510	—	—	—	—	28.00
30.00	94.092	82.75	0.3967	19.885	78.033	0.04388	0.16292	0.3023	0.2062	1.2001	1468.	416.	0.502	—	—	—	—	30.00
32.00	97.468	82.43	0.3829	20.491	78.270	0.04512	0.16291	0.3033	0.2075	1.2030	1450.	415.	0.494	—	—	—	—	32.00
34.00	100.93	82.11	0.3696	21.098	78.506	0.04635	0.16289	0.3044	0.2088	1.2061	1431.	414.	0.487	—	—	—	—	34.00
36.00	104.49	81.78	0.3569	21.708	78.741	0.04758	0.16288	0.3055	0.2102	1.2094	1412.	414.	0.479	—	—	—	—	36.00
38.00	108.14	81.45	0.3447	22.320	78.973	0.04881	0.16288	0.3066	0.2116	1.2127	1394.	413.	0.472	—	—	—	—	38.00
40.00	111.88	81.12	0.3330	22.935	79.203	0.05003	0.16287	0.3078	0.2130	1.2162	1375.	412.	0.464	—	—	—	—	40.00
42.00	115.72	80.79	0.3217	23.551	79.431	0.05126	0.16286	0.3090	0.2144	1.2198	1357.	411.	0.457	—	—	—	—	42.00
44.00	119.65	80.45	0.3109	24.170	79.658	0.05248	0.16285	0.3102	0.2159	1.2236	1338.	410.	0.449	—	—	—	—	44.00
46.00	123.68	80.11	0.3005	24.792	79.882	0.05371	0.16284	0.3115	0.2174	1.2275	1320.	409.	0.442	—	—	—	—	46.00
48.00	127.81	79.77	0.2905	25.415	80.104	0.05493	0.16284	0.3128	0.2189	1.2316	1302.	408.	0.435	—	—	—	—	48.00
50.00	132.05	79.42	0.2808	26.042	80.323	0.05615	0.16283	0.3141	0.2205	1.2358	1283.	407.	0.428	—	—	—	—	50.00
55.00	143.09	78.54	0.2583	27.618	80.862	0.05920	0.16281	0.3177	0.2247	1.2473	1238.	405.	0.411	—	—	—	—	55.00
60.00	154.80	77.63	0.2377	29.212	81.385	0.06225	0.16278	0.3217	0.2291	1.2600	1192.	402.	0.394	—	—	—	—	60.00
65.00	167.20	76.69	0.2190	30.824	81.889	0.06530	0.16274	0.3260	0.2339	1.2743	1147.	399.	0.378	—	—	—	—	65.00
70.00	180.34	75.73	0.2018	32.456	82.375	0.06835	0.16269	0.3309	0.2391	1.2904	1102.	395.	0.362	—	—	—	—	70.00
75.00	194.23	74.73	0.1861	34.108	82.839	0.07140	0.16263	0.3364	0.2447	1.3087	1057.	392.	0.346	—	—	—	3.99	75.00
80.00	208.91	73.69	0.1716	35.784	83.279	0.07445	0.16255	0.3426	0.2510	1.3295	1012.	387.	0.331	—	—	—	3.65	80.00
85.00	224.41	72.61	0.1582	37.485	83.692	0.07752	0.16244	0.3497	0.2579	1.3535	967.	383.	0.316	—	—	—	3.31	85.00
90.00	240.78	71.49	0.1459	39.215	84.075	0.08061	0.16229	0.3579	0.2657	1.3814	922.	378.	0.301	—	—	—	2.99	90.00
95.00	258.04	70.31	0.1345	40.976	84.423	0.08371	0.16211	0.3674	0.2746	1.4142	878.	373.	0.287	—	—	—	2.67	95.00
100.00	276.24	69.07	0.1239	42.772	84.730	0.08684	0.16189	0.3787	0.2850	1.4534	833.	368.	0.273	—	—	—	2.36	100.00
105.00	295.41	67.75	0.1140	44.610	84.991	0.09001	0.16160	0.3922	0.2971	1.5009	788.	362.	0.260	—	—	—	2.05	105.00
110.00	315.62	66.35	0.1048	46.495	85.198	0.09323	0.16124	0.4085	0.3118	1.5598	743.	355.	0.247	—	—	—	1.76	110.00
115.00	336.89	64.84	0.0961	48.438	85.338	0.09652	0.16080	0.4288	0.3300	1.6345	698.	348.	0.234	—	—	—	1.48	115.00
120.00	359.28	63.21	0.0879	50.450	85.397	0.09988	0.16024	0.4547	0.3534	1.7326	653.	340.	0.222	—	—	—	1.21	120.00
125.00	382.82	61.41	0.0801	52.549	85.356	0.10337	0.15954	0.4887	0.3846	1.8667	607.	332.	0.210	—	—	—	0.95	125.00
130.00	407.55	59.39	0.0726	54.765	85.184	0.10701	0.15866	0.5353	0.4292	2.0609	561.	323.	0.198	—	—	—	0.71	130.00
135.00	433.46	57.06	0.0654	57.146	84.837	0.11090	0.15751	0.6037	0.4984	2.3666	515.	312.	—	—	—	—	0.48	135.00
140.00	460.51	54.23	0.0583	59.789	84.238	0.11519	0.15599	0.7158	0.6211	2.9150	466.	302.	—	—	—	—	0.27	140.00
145.00	488.50	50.45	0.0511	62.944	83.233	0.12030	0.15385	—	—	—	—	—	—	—	—	—	0.10	145.00
150.00	516.72	43.11	0.0344	68.072	76.834	0.12864	0.14293	—	—	—	—	—	—	—	—	—	—	150.00
150.87c	521.41	35.66	0.0280	72.653	72.653	0.13611	0.13611	∞	∞	∞	0.	0.	—	—	∞	∞	0.00	150.87

*temperatures are on the IPTS–68 scale

b = normal boiling point

c = critical point

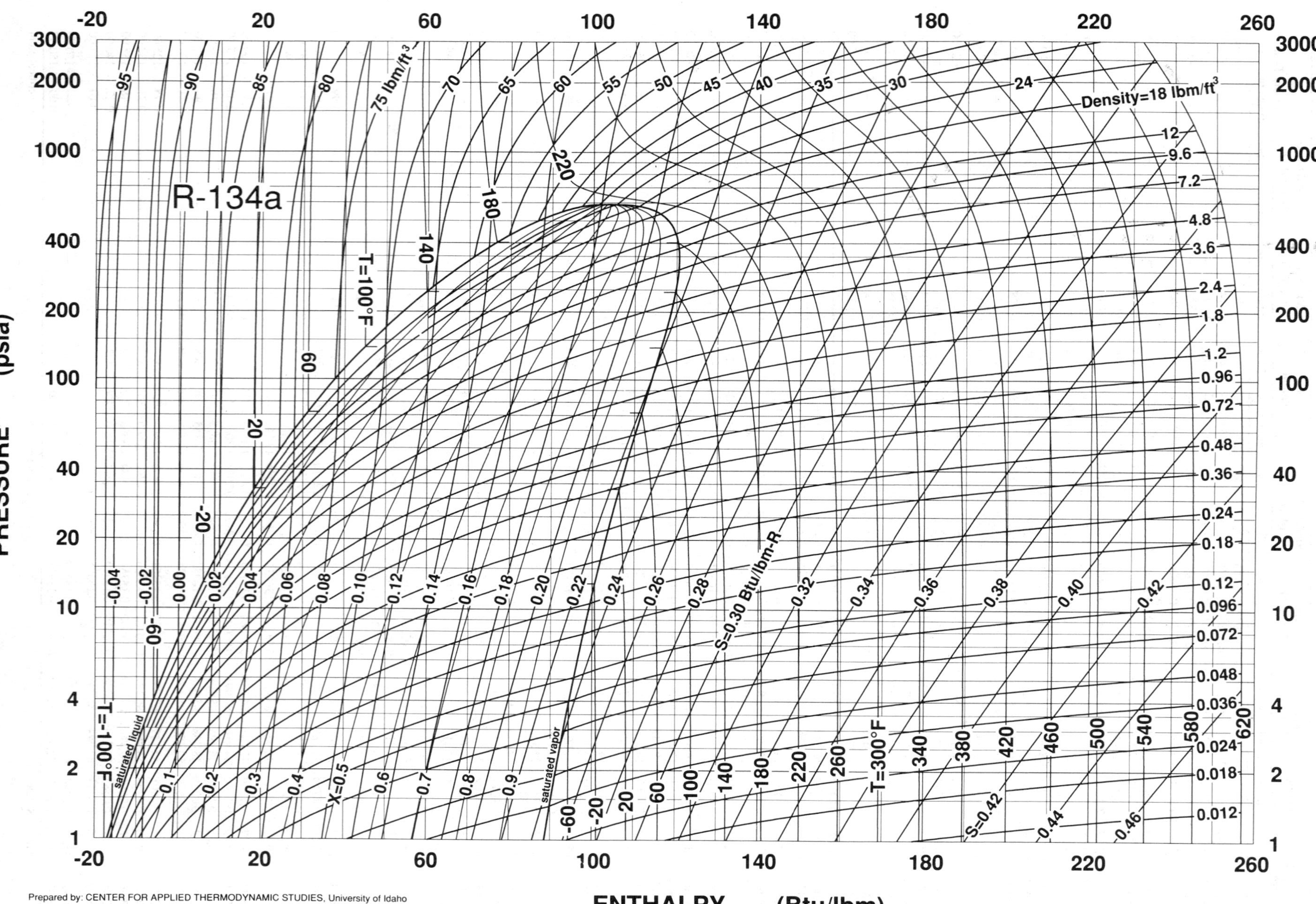

Fig. 12 Pressure–Enthalpy Diagram for Refrigerant 134a

Refrigerant 134a (1,1,1,2-Tetrafluoroethane) Properties of Saturated Liquid and Saturated Vapor

Temp,* °F	Pressure, psia	Density, lb/ft³ Liquid	Volume, ft³/lb Vapor	Enthalpy, Btu/lb Liquid	Enthalpy, Btu/lb Vapor	Entropy, Btu/lb·°F Liquid	Entropy, Btu/lb·°F Vapor	Specific Heat c_p, Btu/lb·°F Liquid	Specific Heat c_p, Btu/lb·°F Vapor	c_p/c_v Vapor	Velocity of Sound, ft/s Liquid	Velocity of Sound, ft/s Vapor	Viscosity, lb_m/ft·h Liquid	Viscosity, lb_m/ft·h Vapor	Thermal Cond, Btu/h·ft·°F Liquid	Thermal Cond, Btu/h·ft·°F Vapor	Surface Tension, dyne/cm	Temp,* °F
−153.94a	0.057	99.34	564.85	32.989	80.235	0.09154	0.27880	0.2740	0.1397	1.1628	3723.	416.	5.289	0.0160	—	—	28.15	−153.94
−150.00	0.072	98.95	449.29	31.902	80.783	0.08801	0.27588	0.2776	0.1409	1.1615	3670.	418.	4.913	0.0163	—	—	27.76	−150.00
−140.00	0.130	97.98	259.15	29.093	82.190	0.07908	0.26903	0.2837	0.1438	1.1583	3552.	424.	4.117	0.0168	—	—	26.78	−140.00
−130.00	0.222	97.01	155.69	26.238	83.618	0.07029	0.26294	0.2870	0.1467	1.1554	3448.	430.	3.497	0.0173	—	—	25.81	−130.00
−120.00	0.367	96.05	97.027	23.359	85.066	0.06169	0.25752	0.2886	0.1497	1.1528	3352.	436.	3.006	0.0179	—	—	24.86	−120.00
−110.00	0.586	95.09	62.509	20.467	86.531	0.05330	0.25270	0.2895	0.1527	1.1507	3261.	441.	2.613	0.0184	—	—	23.92	−110.00
−100.00	0.906	94.13	41.496	17.569	88.011	0.04513	0.24842	0.2900	0.1559	1.1489	3173.	446.	2.292	0.0190	—	—	22.99	−100.00
−90.00	1.363	93.17	28.303	14.665	89.504	0.03717	0.24462	0.2906	0.1591	1.1475	3087.	451.	2.028	0.0195	0.0721	—	22.08	−90.00
−80.00	1.997	92.21	19.783	11.755	91.005	0.02940	0.24125	0.2913	0.1624	1.1466	3001.	456.	1.808	0.0200	0.0706	—	21.17	−80.00
−75.00	2.396	91.73	16.680	10.297	91.759	0.02559	0.23972	0.2917	0.1641	1.1463	2959.	458.	1.712	0.0203	0.0699	—	20.73	−75.00
−70.00	2.859	91.25	14.138	8.837	92.514	0.02182	0.23827	0.2922	0.1658	1.1462	2916.	460.	1.623	0.0206	0.0691	—	20.28	−70.00
−65.00	3.393	90.77	12.045	7.374	93.270	0.01809	0.23691	0.2928	0.1676	1.1461	2874.	462.	1.542	0.0209	0.0684	—	19.84	−65.00
−60.00	4.006	90.28	10.310	5.907	94.026	0.01440	0.23563	0.2935	0.1694	1.1462	2832.	464.	1.466	0.0212	0.0677	0.00405	19.40	−60.00
−55.00	4.707	89.80	8.8656	4.437	94.783	0.01075	0.23443	0.2943	0.1712	1.1465	2790.	466.	1.396	0.0214	0.0669	0.00423	18.96	−55.00
−50.00	5.505	89.31	7.6569	2.963	95.539	0.00713	0.23331	0.2951	0.1731	1.1468	2747.	468.	1.331	0.0217	0.0662	0.00440	18.52	−50.00
−45.00	6.409	88.82	6.6405	1.484	96.295	0.00355	0.23225	0.2960	0.1750	1.1473	2705.	470.	1.271	0.0220	0.0654	0.00457	18.09	−45.00
−40.00	7.429	88.32	5.7819	0.000	97.050	0.00000	0.23125	0.2970	0.1769	1.1479	2663.	471.	1.215	0.0223	0.0647	0.00473	17.66	−40.00
−35.00	8.577	87.83	5.0533	1.489	97.804	0.00352	0.23032	0.2981	0.1789	1.1487	2621.	473.	1.163	0.0225	0.0639	0.00489	17.23	−35.00
−30.00	9.862	87.33	4.4325	2.984	98.556	0.00701	0.22945	0.2992	0.1810	1.1497	2579.	474.	1.113	0.0228	0.0632	0.00505	16.81	−30.00
−25.00	11.297	86.82	3.9014	4.484	99.306	0.01048	0.22863	0.3004	0.1831	1.1508	2536.	476.	1.068	0.0231	0.0625	0.00521	16.38	−25.00
−20.00	12.895	86.32	3.4452	5.991	100.054	0.01392	0.22786	0.3017	0.1852	1.1521	2494.	477.	1.024	0.0234	0.0617	0.00536	15.96	−20.00
−15.00	14.667	85.81	3.0519	7.505	100.799	0.01733	0.22714	0.3031	0.1874	1.1535	2452.	478.	0.984	0.0237	0.0610	0.00550	15.54	−15.00
−14.92b	14.696	85.80	3.0462	7.529	100.811	0.01739	0.22713	0.3031	0.1874	1.1535	2451.	478.	0.983	0.0237	0.0610	0.00551	15.54	−14.92
−10.00	16.626	85.29	2.7116	9.026	101.542	0.02073	0.22647	0.3045	0.1897	1.1552	2410.	479.	0.946	0.0240	0.0602	0.00565	15.13	−10.00
−5.00	18.787	84.77	2.4161	10.554	102.280	0.02409	0.22584	0.3060	0.1920	1.1570	2367.	480.	0.910	0.0243	0.0595	0.00579	14.71	−5.00
0.00	21.162	84.25	2.1587	12.090	103.015	0.02744	0.22525	0.3075	0.1943	1.1590	2325.	481.	0.876	0.0245	0.0588	0.00593	14.30	0.00
5.00	23.767	83.72	1.9337	13.634	103.745	0.03077	0.22470	0.3091	0.1968	1.1613	2283.	481.	0.843	0.0248	0.0580	0.00607	13.89	5.00
10.00	26.617	83.18	1.7365	15.187	104.471	0.03408	0.22418	0.3108	0.1993	1.1637	2240.	482.	0.813	0.0251	0.0573	0.00621	13.48	10.00
15.00	29.726	82.64	1.5630	16.748	105.192	0.03737	0.22370	0.3126	0.2018	1.1664	2198.	482.	0.784	0.0254	0.0565	0.00635	13.08	15.00
20.00	33.110	82.10	1.4101	18.318	105.907	0.04065	0.22325	0.3144	0.2045	1.1694	2155.	482.	0.756	0.0257	0.0558	0.00649	12.67	20.00
25.00	36.785	81.55	1.2749	19.897	106.617	0.04391	0.22283	0.3162	0.2072	1.1726	2113.	482.	0.730	0.0260	0.0550	0.00662	12.27	25.00
30.00	40.768	80.99	1.1550	21.486	107.320	0.04715	0.22244	0.3182	0.2100	1.1761	2070.	482.	0.705	0.0263	0.0543	0.00676	11.87	30.00
35.00	45.075	80.42	1.0484	23.085	108.016	0.05038	0.22207	0.3202	0.2129	1.1799	2027.	482.	0.681	0.0267	0.0536	0.00690	11.48	35.00
40.00	49.724	79.85	0.9534	24.694	108.705	0.05359	0.22172	0.3223	0.2159	1.1841	1985.	482.	0.658	0.0270	0.0528	0.00704	11.08	40.00
45.00	54.732	79.26	0.8685	26.314	109.386	0.05679	0.22140	0.3244	0.2190	1.1886	1942.	481.	0.636	0.0273	0.0521	0.00718	10.69	45.00
50.00	60.116	78.67	0.7925	27.944	110.058	0.05998	0.22110	0.3267	0.2222	1.1935	1899.	481.	0.615	0.0276	0.0513	0.00732	10.30	50.00
55.00	65.895	78.07	0.7243	29.586	110.722	0.06316	0.22081	0.3290	0.2255	1.1988	1856.	480.	0.595	0.0280	0.0506	0.00746	9.91	55.00
60.00	72.087	77.46	0.6630	31.239	111.376	0.06633	0.22054	0.3314	0.2289	1.2046	1813.	479.	0.576	0.0283	0.0499	0.00761	9.53	60.00
65.00	78.712	76.84	0.6077	32.905	112.019	0.06949	0.22028	0.3339	0.2325	1.2109	1770.	477.	0.557	0.0286	0.0491	0.00776	9.15	65.00
70.00	85.787	76.21	0.5577	34.583	112.652	0.07264	0.22003	0.3366	0.2363	1.2178	1726.	476.	0.539	0.0290	0.0484	0.00791	8.77	70.00
75.00	93.333	75.57	0.5125	36.274	113.272	0.07578	0.21979	0.3393	0.2402	1.2252	1683.	474.	0.522	0.0294	0.0476	0.00806	8.39	75.00
80.00	101.37	74.91	0.4715	37.978	113.880	0.07892	0.21957	0.3422	0.2444	1.2334	1640.	472.	0.505	0.0297	0.0469	0.00822	8.02	80.00
85.00	109.92	74.25	0.4343	39.697	114.475	0.08205	0.21934	0.3453	0.2487	1.2424	1596.	470.	0.489	0.0301	0.0462	0.00838	7.65	85.00
90.00	119.00	73.57	0.4004	41.430	115.055	0.08518	0.21912	0.3485	0.2533	1.2522	1552.	468.	0.473	0.0305	0.0454	0.00855	7.28	90.00
95.00	128.63	72.87	0.3694	43.179	115.619	0.08830	0.21890	0.3519	0.2582	1.2630	1509.	466.	0.458	0.0309	0.0447	0.00872	6.91	95.00
100.00	138.83	72.16	0.3411	44.943	116.166	0.09142	0.21868	0.3555	0.2633	1.2748	1465.	463.	0.443	0.0313	0.0439	0.00890	6.55	100.00
105.00	149.63	71.43	0.3153	46.725	116.694	0.09454	0.21845	0.3594	0.2689	1.2880	1421.	460.	0.428	0.0318	0.0432	0.00908	6.20	105.00
110.00	161.05	70.68	0.2915	48.524	117.203	0.09766	0.21822	0.3635	0.2748	1.3026	1376.	457.	0.414	0.0322	0.0425	0.00926	5.84	110.00
115.00	173.11	69.91	0.2697	50.343	117.690	0.10078	0.21797	0.3680	0.2812	1.3189	1332.	454.	0.400	0.0327	0.0417	0.00946	5.49	115.00
120.00	185.84	69.12	0.2497	52.181	118.153	0.10391	0.21772	0.3728	0.2881	1.3372	1288.	450.	0.387	0.0332	0.0410	0.00965	5.15	120.00
125.00	199.25	68.31	0.2312	54.040	118.591	0.10704	0.21744	0.3781	0.2957	1.3577	1243.	446.	0.374	0.0338	0.0403	0.00986	4.80	125.00
130.00	213.38	67.47	0.2141	55.923	119.000	0.11018	0.21715	0.3839	0.3040	1.3810	1198.	442.	0.361	0.0343	0.0395	0.01007	4.47	130.00
135.00	228.25	66.60	0.1983	57.830	119.377	0.11333	0.21683	0.3903	0.3133	1.4075	1153.	437.	0.348	0.0349	0.0388	0.01029	4.13	135.00
140.00	243.88	65.70	0.1836	59.764	119.720	0.11650	0.21648	0.3974	0.3236	1.4379	1108.	432.	0.335	0.0356	0.0380	0.01052	3.81	140.00
145.00	260.31	64.77	0.1700	61.727	120.024	0.11968	0.21609	0.4053	0.3353	1.4731	1062.	427.	0.323	0.0363	0.0373	0.01075	3.48	145.00
150.00	277.57	63.80	0.1574	63.722	120.284	0.12288	0.21566	0.4144	0.3486	1.5143	1017.	421.	0.311	0.0370	0.0366	0.01100	3.17	150.00
155.00	295.69	62.78	0.1455	65.752	120.495	0.12611	0.21517	0.4247	0.3640	1.5630	971.	416.	0.298	0.0378	0.0358	0.01125	2.86	155.00
160.00	314.69	61.72	0.1345	67.823	120.650	0.12938	0.21463	0.4368	0.3821	1.6213	924.	409.	0.286	0.0387	0.0351	0.01151	2.55	160.00
165.00	334.62	60.60	0.1241	69.939	120.739	0.13268	0.21400	0.4511	0.4036	1.6921	877.	403.	0.274	0.0397	0.0343	0.01178	2.26	165.00
170.00	355.51	59.42	0.1144	72.106	120.753	0.13603	0.21329	0.4683	0.4299	1.7798	829.	396.	0.262	0.0407	0.0336	0.01206	1.97	170.00
175.00	377.40	58.16	0.1052	74.335	120.677	0.13945	0.21247	0.4896	0.4627	1.8908	781.	388.	0.249	0.0420	0.0329	0.01235	1.69	175.00
180.00	400.34	56.80	0.0965	76.636	120.493	0.14295	0.21151	0.5168	0.5048	2.0354	731.	380.	0.237	0.0434	0.0321	0.01265	1.41	180.00
185.00	424.37	55.33	0.0881	79.027	120.175	0.14655	0.21037	0.5527	0.5612	2.2309	680.	372.	0.224	0.0450	0.0314	0.01296	1.15	185.00
190.00	449.55	53.70	0.0801	81.534	119.684	0.15029	0.20901	0.6031	0.6406	2.5087	627.	363.	0.211	0.0469	—	—	0.90	190.00
195.00	475.95	51.86	0.0723	84.196	118.963	0.15423	0.20733	0.6794	0.7612	2.9338	572.	353.	0.197	0.0493	—	—	0.67	195.00
200.00	503.64	49.70	0.0646	87.088	117.906	0.15847	0.20519	0.8100	0.9673	3.6640	514.	343.	0.182	0.0523	—	—	0.45	200.00
205.00	532.72	47.00	0.0566	90.368	116.289	0.16326	0.20226	1.0906	1.4042	5.2174	450.	331.	0.164	0.0566	—	—	0.25	205.00
210.00	563.34	43.03	0.0474	94.548	113.411	0.16933	0.19750	—	—	—	378.	316.	0.142	0.0639	—	—	0.09	210.00
213.85c	588.27	32.04	0.0312	103.775	103.775	0.18128	0.18128	∞	∞	∞	0.	0.	—	—	∞	∞	0.00	213.85

*temperatures are on the ITS–90 scale a = triple point b = normal boiling point c = critical point

Refrigerant 134a Properties of Superheated Vapor

Pressure = 14.696 psia
Saturation temperature = −14.92°F

Temp,* °F	Density, lb/ft³	Enthalpy, Btu/lb	Entropy, Btu/lb·°F	Vel. Sound, ft/s
Saturated				
Liquid	85.7972	7.53	0.01739	2451.2
Vapor	0.3283	100.81	0.22713	478.0
0.00	0.3158	103.62	0.23335	487.2
20.00	0.3008	107.45	0.24149	499.0
40.00	0.2874	111.34	0.24944	510.2
60.00	0.2753	115.31	0.25723	521.0
80.00	0.2642	119.35	0.26486	531.5
100.00	0.2541	123.47	0.27236	541.6
120.00	0.2448	127.68	0.27974	551.4
140.00	0.2362	131.96	0.28700	561.0
160.00	0.2282	136.32	0.29416	570.4
180.00	0.2208	140.77	0.30122	579.5
200.00	0.2139	145.30	0.30819	588.5
220.00	0.2074	149.90	0.31507	597.3
240.00	0.2013	154.59	0.32187	606.0
260.00	0.1955	159.36	0.32858	614.5
280.00	0.1901	164.20	0.33522	622.8
300.00	0.1850	169.12	0.34178	631.1

Pressure = 25.00 psia
Saturation temperature = 7.22°F

Temp,* °F	Density, lb/ft³	Enthalpy, Btu/lb	Entropy, Btu/lb·°F	Vel. Sound, ft/s
Saturated				
Liquid	83.4823	14.32	0.03224	2263.9
Vapor	0.5426	104.07	0.22446	481.5
20.00	0.5245	106.60	0.22982	489.9
40.00	0.4991	110.61	0.23800	502.4
60.00	0.4765	114.66	0.24596	514.1
80.00	0.4563	118.78	0.25373	525.4
100.00	0.4379	122.96	0.26135	536.2
120.00	0.4212	127.22	0.26881	546.6
140.00	0.4058	131.55	0.27615	556.7
160.00	0.3916	135.95	0.28337	566.5
180.00	0.3786	140.43	0.29048	576.0
200.00	0.3663	144.98	0.29750	585.3
220.00	0.3549	149.61	0.30441	594.4
240.00	0.3443	154.32	0.31124	603.3
260.00	0.3343	159.10	0.31798	612.0
280.00	0.3248	163.96	0.32464	620.6
300.00	0.3160	168.90	0.33122	629.0

Pressure = 50.00 psia
Saturation temperature = 40.29°F

Temp,* °F	Density, lb/ft³	Enthalpy, Btu/lb	Entropy, Btu/lb·°F	Vel. Sound, ft/s
Saturated				
Liquid	79.8125	24.79	0.05377	1982.3
Vapor	1.0545	108.74	0.22170	481.7
60.00	0.9982	113.00	0.23005	496.2
80.00	0.9489	117.32	0.23822	509.8
100.00	0.9055	121.68	0.24614	522.5
120.00	0.8670	126.07	0.25385	534.5
140.00	0.8322	130.51	0.26139	545.8
160.00	0.8008	135.01	0.26877	556.7
180.00	0.7718	139.57	0.27601	567.2
200.00	0.7454	144.20	0.28313	577.4
220.00	0.7208	148.89	0.29014	587.2
240.00	0.6980	153.65	0.29704	596.8
260.00	0.6768	158.48	0.30385	606.1
280.00	0.6569	163.38	0.31056	615.2
300.00	0.6383	168.35	0.31719	624.1

Pressure = 75.00 psia
Saturation temperature = 62.24°F

Temp,* °F	Density, lb/ft³	Enthalpy, Btu/lb	Entropy, Btu/lb·°F	Vel. Sound, ft/s
Saturated				
Liquid	77.1862	31.98	0.06775	1793.6
Vapor	1.5686	111.67	0.22042	478.1
80.00	1.4873	115.74	0.22809	492.7
100.00	1.4092	120.30	0.23639	507.7
120.00	1.3416	124.85	0.24439	521.6
140.00	1.2822	129.43	0.25215	534.5
160.00	1.2294	134.04	0.25971	546.6
180.00	1.1817	138.69	0.26710	558.2
200.00	1.1383	143.39	0.27434	569.2
220.00	1.0984	148.15	0.28145	579.8
240.00	1.0620	152.97	0.28843	590.1
260.00	1.0280	157.85	0.29531	600.0
280.00	0.9966	162.79	0.30208	609.7
300.00	0.9671	167.80	0.30876	619.0
320.00	0.9398	172.87	0.31535	628.2
340.00	0.9138	178.01	0.32186	637.1
360.00	0.8895	183.21	0.32828	645.9
380.00	0.8665	188.48	0.33463	654.5
400.00	0.8448	193.82	0.34091	662.9

Pressure = 100.00 psia
Saturation temperature = 79.17°F

Temp,* °F	Density, lb/ft³	Enthalpy, Btu/lb	Entropy, Btu/lb·°F	Vel. Sound, ft/s
Saturated				
Liquid	75.0245	37.69	0.07840	1646.8
Vapor	2.0917	113.78	0.21960	472.8
80.00	2.0858	113.98	0.21998	473.6
100.00	1.9576	118.80	0.22874	491.6
120.00	1.8509	123.55	0.23709	507.8
140.00	1.7597	128.29	0.24512	522.4
160.00	1.6800	133.02	0.25288	536.0
180.00	1.6094	137.78	0.26044	548.8
200.00	1.5463	142.57	0.26781	560.8
220.00	1.4891	147.40	0.27502	572.3
240.00	1.4368	152.27	0.28210	583.3
260.00	1.3886	157.21	0.28905	593.9
280.00	1.3444	162.19	0.29588	604.1
300.00	1.3031	167.24	0.30261	614.0
320.00	1.2647	172.35	0.30925	623.6
340.00	1.2287	177.52	0.31579	632.9
360.00	1.1950	182.75	0.32225	642.1
380.00	1.1633	188.04	0.32863	651.0
400.00	1.1334	193.39	0.33494	659.8

Pressure = 125.00 psia
Saturation temperature = 93.15°F

Temp,* °F	Density, lb/ft³	Enthalpy, Btu/lb	Entropy, Btu/lb·°F	Vel. Sound, ft/s
Saturated				
Liquid	73.1279	42.53	0.08715	1524.7
Vapor	2.6279	115.41	0.21898	466.7
100.00	2.5638	117.16	0.22212	473.9
120.00	2.4025	122.16	0.23090	492.9
140.00	2.2694	127.08	0.23924	509.7
160.00	2.1561	131.96	0.24725	525.0
180.00	2.0577	136.83	0.25498	539.0
200.00	1.9710	141.71	0.26250	552.2
220.00	1.8935	146.62	0.26983	564.6
240.00	1.8233	151.56	0.27700	576.4
260.00	1.7592	156.55	0.28402	587.6
280.00	1.7006	161.59	0.29093	598.4
300.00	1.6463	166.67	0.29771	608.8
320.00	1.5959	171.82	0.30440	618.9
340.00	1.5492	177.02	0.31098	628.7
360.00	1.5055	182.27	0.31747	638.2
380.00	1.4644	187.59	0.32388	647.5
400.00	1.4258	192.97	0.33021	656.6

Pressure = 150.00 psia
Saturation temperature = 105.17°F

Temp,* °F	Density, lb/ft³	Enthalpy, Btu/lb	Entropy, Btu/lb·°F	Vel. Sound, ft/s
Saturated				
Liquid	71.4013	46.78	0.09464	1419.1
Vapor	3.1801	116.71	0.21844	460.0
120.00	3.0077	120.64	0.22530	476.6
140.00	2.8181	125.78	0.23403	496.0
160.00	2.6620	130.83	0.24231	513.3
180.00	2.5295	135.83	0.25026	528.9
200.00	2.4146	140.82	0.25794	543.3
220.00	2.3132	145.82	0.26539	556.7
240.00	2.2223	150.83	0.27267	569.3
260.00	2.1401	155.88	0.27978	581.3
280.00	2.0658	160.97	0.28675	592.7
300.00	1.9971	166.10	0.29360	603.7
320.00	1.9338	171.28	0.30033	614.2
340.00	1.8751	176.51	0.30696	624.5
360.00	1.8208	181.80	0.31349	634.4
380.00	1.7695	187.14	0.31993	644.0
400.00	1.7216	192.54	0.32628	653.4
420.00	1.6766	198.00	0.33256	662.6
440.00	1.6341	203.51	0.33876	671.6
460.00	1.5940	209.08	0.34488	680.4
480.00	1.5558	214.71	0.35094	689.0
500.00	1.5197	220.40	0.35692	697.4

Pressure = 175.00 psia
Saturation temperature = 115.76°F

Temp,* °F	Density, lb/ft³	Enthalpy, Btu/lb	Entropy, Btu/lb·°F	Vel. Sound, ft/s
Saturated				
Liquid	69.7902	50.62	0.10126	1325.3
Vapor	3.7511	117.76	0.21794	453.0
120.00	3.6836	118.95	0.21999	458.4
140.00	3.4148	124.38	0.22921	481.3
160.00	3.2025	129.64	0.23783	500.9
180.00	3.0271	134.79	0.24602	518.3
200.00	2.8785	139.90	0.25388	534.0
220.00	2.7494	144.99	0.26148	548.5
240.00	2.6349	150.08	0.26887	562.1
260.00	2.5328	155.19	0.27607	574.8
280.00	2.4403	160.34	0.28312	586.9
300.00	2.3558	165.51	0.29003	598.5
320.00	2.2785	170.73	0.29681	609.5
340.00	2.2071	176.00	0.30348	620.2
360.00	2.1411	181.32	0.31004	630.5
380.00	2.0795	186.69	0.31651	640.5
400.00	2.0216	192.11	0.32290	650.3
420.00	1.9675	197.59	0.32920	659.7
440.00	1.9164	203.12	0.33542	669.0
460.00	1.8683	208.71	0.34156	678.0
480.00	1.8228	214.36	0.34763	686.9
500.00	1.7797	220.05	0.35363	695.6

Pressure = 200.00 psia
Saturation temperature = 125.27°F

Temp,* °F	Density, lb/ft³	Enthalpy, Btu/lb	Entropy, Btu/lb·°F	Vel. Sound, ft/s
Saturated				
Liquid	68.2602	54.14	0.10721	1240.5
Vapor	4.3437	118.61	0.21743	445.6
140.00	4.0726	122.86	0.22460	465.2
160.00	3.7850	128.36	0.23363	487.8
180.00	3.5561	133.70	0.24210	507.2
200.00	3.3656	138.94	0.25018	524.5
220.00	3.2036	144.14	0.25793	540.2
240.00	3.0623	149.31	0.26544	554.7
260.00	2.9371	154.50	0.27274	568.3
280.00	2.8247	159.69	0.27987	581.1
300.00	2.7234	164.92	0.28684	593.2
320.00	2.6305	170.18	0.29368	604.8
340.00	2.5455	175.49	0.30039	615.9
360.00	2.4668	180.83	0.30700	626.7
380.00	2.3934	186.23	0.31350	637.0
400.00	2.3254	191.68	0.31991	647.1
420.00	2.2614	197.18	0.32624	656.9
440.00	2.2017	202.73	0.33248	666.4
460.00	2.1453	208.34	0.33864	675.7
480.00	2.0920	214.00	0.34473	684.8
500.00	2.0417	219.71	0.35075	693.7

*temperatures are on the ITS-90 scale

Refrigerant 134a Properties of Superheated Vapor (*Concluded*)

Pressure = 225.00 psia
Saturation temperature = 133.93°F

Temp,* °F	Density, lb/ft³	Enthalpy, Btu/lb	Entropy, Btu/lb·°F	Vel. Sound, ft/s
Saturated				
Liquid	66.7870	57.42	0.11266	1162.8
Vapor	4.9609	119.30	0.21690	438.1
140.00	4.8123	121.16	0.22002	447.3
160.00	4.4191	126.99	0.22959	473.6
180.00	4.1206	132.54	0.23840	495.5
200.00	3.8796	137.94	0.24671	514.6
220.00	3.6784	143.25	0.25465	531.6
240.00	3.5058	148.52	0.26229	547.2
260.00	3.3542	153.78	0.26970	561.6
280.00	3.2202	159.04	0.27691	575.1
300.00	3.0995	164.32	0.28395	587.9
320.00	2.9899	169.62	0.29084	600.0
340.00	2.8897	174.97	0.29761	611.7
360.00	2.7978	180.35	0.30425	622.8
380.00	2.7122	185.77	0.31079	633.6
400.00	2.6330	191.24	0.31723	644.0
420.00	2.5592	196.77	0.32358	654.0
440.00	2.4900	202.34	0.32984	663.9
460.00	2.4249	207.96	0.33603	673.4
480.00	2.3636	213.64	0.34213	682.8
500.00	2.3057	219.36	0.34816	691.9

Pressure = 250.00 psia
Saturation temperature = 141.89°F

Temp,* °F	Density, lb/ft³	Enthalpy, Btu/lb	Entropy, Btu/lb·°F	Vel. Sound, ft/s
Saturated				
Liquid	65.3526	60.50	0.11770	1090.7
Vapor	5.6060	119.84	0.21634	430.3
160.00	5.1189	125.49	0.22560	458.2
180.00	4.7275	131.31	0.23484	483.1
200.00	4.4239	136.89	0.24343	504.2
220.00	4.1756	142.34	0.25156	522.8
240.00	3.9664	147.71	0.25935	539.5
260.00	3.7854	153.05	0.26688	554.9
280.00	3.6265	158.37	0.27418	569.2
300.00	3.4847	163.71	0.28129	582.6
320.00	3.3571	169.06	0.28824	595.3
340.00	3.2408	174.44	0.29506	607.4
360.00	3.1342	179.85	0.30175	618.9
380.00	3.0359	185.31	0.30832	630.1
400.00	2.9451	190.81	0.31479	640.8
420.00	2.8604	196.35	0.32117	651.2
440.00	2.7813	201.94	0.32745	661.3
460.00	2.7072	207.58	0.33365	671.1
480.00	2.6374	213.27	0.33977	680.7
500.00	2.5717	219.02	0.34582	690.1

Pressure = 275.00 psia
Saturation temperature = 149.27°F

Temp,* °F	Density, lb/ft³	Enthalpy, Btu/lb	Entropy, Btu/lb·°F	Vel. Sound, ft/s
Saturated				
Liquid	63.9423	63.43	0.12241	1023.4
Vapor	6.2831	120.25	0.21572	422.3
160.00	5.9060	123.82	0.22155	441.2
180.00	5.3869	129.98	0.23133	469.9
200.00	5.0031	135.78	0.24026	493.4
220.00	4.6978	141.38	0.24862	513.7
240.00	4.4465	146.87	0.25658	531.7
260.00	4.2314	152.30	0.26423	548.0
280.00	4.0446	157.69	0.27162	563.1
300.00	3.8803	163.08	0.27881	577.2
320.00	3.7317	168.49	0.28583	590.5
340.00	3.5987	173.91	0.29270	603.1
360.00	3.4764	179.36	0.29943	615.1
380.00	3.3646	184.84	0.30604	626.6
400.00	3.2612	190.37	0.31254	637.7
420.00	3.1653	195.93	0.31894	648.4
440.00	3.0758	201.55	0.32525	658.8
460.00	2.9922	207.20	0.33147	668.9
480.00	2.9136	212.91	0.33761	678.7
500.00	2.8397	218.67	0.34368	688.3

Pressure = 300.00 psia
Saturation temperature = 156.16°F

Temp,* °F	Density, lb/ft³	Enthalpy, Btu/lb	Entropy, Btu/lb·°F	Vel. Sound, ft/s
Saturated				
Liquid	62.5436	66.23	0.12686	959.8
Vapor	6.9967	120.54	0.21505	414.2
160.00	6.8168	121.92	0.21730	422.0
180.00	6.1118	128.55	0.22782	455.8
200.00	5.6239	134.61	0.23715	482.1
220.00	5.2494	140.39	0.24578	504.3
240.00	4.9472	146.00	0.25393	523.7
260.00	4.6939	151.52	0.26171	541.1
280.00	4.4758	157.00	0.26921	557.0
300.00	4.2852	162.45	0.27649	571.8
320.00	4.1160	167.90	0.28357	585.7
340.00	3.9631	173.37	0.29049	598.8
360.00	3.8247	178.85	0.29727	611.2
380.00	3.6981	184.37	0.30392	623.1
400.00	3.5816	189.92	0.31045	634.6
420.00	3.4737	195.51	0.31688	645.6
440.00	3.3735	201.15	0.32321	656.3
460.00	3.2799	206.83	0.32945	666.6
480.00	3.1922	212.55	0.33561	676.7
500.00	3.1098	218.32	0.34169	686.5

Pressure = 325.00 psia
Saturation temperature = 162.62°F

Temp,* °F	Density, lb/ft³	Enthalpy, Btu/lb	Entropy, Btu/lb·°F	Vel. Sound, ft/s
Saturated				
Liquid	61.1446	68.92	0.13110	899.5
Vapor	7.7526	120.71	0.21431	405.9
180.00	6.9220	126.96	0.22423	440.3
200.00	6.2928	133.36	0.23408	470.2
220.00	5.8341	139.34	0.24301	494.5
240.00	5.4723	145.10	0.25136	515.5
260.00	5.1741	150.73	0.25930	534.0
280.00	4.9208	156.29	0.26692	550.9
300.00	4.7017	161.81	0.27428	566.4
320.00	4.5082	167.31	0.28144	580.9
340.00	4.3352	172.82	0.28841	594.5
360.00	4.1790	178.35	0.29524	607.4
380.00	4.0368	183.90	0.30193	619.7
400.00	3.9063	189.48	0.30849	631.5
420.00	3.7859	195.09	0.31495	642.8
440.00	3.6743	200.75	0.32131	653.8
460.00	3.5703	206.44	0.32757	664.4
480.00	3.4731	212.19	0.33375	674.7
500.00	3.3819	217.97	0.33984	684.7

Pressure = 350.00 psia
Saturation temperature = 168.71°F

Temp,* °F	Density, lb/ft³	Enthalpy, Btu/lb	Entropy, Btu/lb·°F	Vel. Sound, ft/s
Saturated				
Liquid	59.7334	71.54	0.13516	841.7
Vapor	8.5577	120.76	0.21349	397.5
180.00	7.8491	125.18	0.22046	423.2
200.00	7.0242	132.01	0.23098	457.6
220.00	6.4561	138.24	0.24029	484.4
240.00	6.0219	144.17	0.24888	507.1
260.00	5.6728	149.91	0.25698	526.9
280.00	5.3805	155.56	0.26472	544.7
300.00	5.1295	161.15	0.27218	561.0
320.00	4.9098	166.72	0.27941	576.1
340.00	4.7148	172.27	0.28644	590.2
360.00	4.5396	177.84	0.29332	603.6
380.00	4.3807	183.42	0.30005	616.3
400.00	4.2355	189.03	0.30665	628.4
420.00	4.1019	194.67	0.31313	640.1
440.00	3.9784	200.35	0.31952	651.4
460.00	3.8636	206.06	0.32580	662.2
480.00	3.7564	211.82	0.33199	672.8
500.00	3.6561	217.62	0.33811	683.0

Pressure = 375.00 psia
Saturation temperature= 174.46°F

Temp,* °F	Density, lb/ft³	Enthalpy, Btu/lb	Entropy, Btu/lb·°F	Vel. Sound, ft/s
Saturated				
Liquid	58.2974	74.09	0.13908	785.9
Vapor	9.4209	120.69	0.21256	389.0
180.00	8.9498	123.10	0.21634	403.8
200.00	7.8311	130.54	0.22781	444.1
220.00	7.1211	137.08	0.23758	474.0
240.00	6.6028	143.19	0.24644	498.5
260.00	6.1926	149.07	0.25473	519.6
280.00	5.8555	154.82	0.26260	538.4
300.00	5.5694	160.49	0.27016	555.5
320.00	5.3212	166.11	0.27747	571.3
340.00	5.1022	171.72	0.28457	586.0
360.00	4.9066	177.32	0.29149	599.8
380.00	4.7300	182.94	0.29826	612.9
400.00	4.5692	188.58	0.30490	625.4
420.00	4.4217	194.24	0.31141	637.4
440.00	4.2857	199.94	0.31782	648.9
460.00	4.1596	205.68	0.32413	660.1
480.00	4.0421	211.46	0.33034	670.8
500.00	3.9323	217.28	0.33647	681.3

Pressure = 400.00 psia
Saturation temperature = 197.93°F

Temp,* °F	Density, lb/ft³	Enthalpy, Btu/lb	Entropy, Btu/lb·°F	Vel. Sound, ft/s
Saturated				
Liquid	56.8213	76.60	0.14289	731.8
Vapor	10.3541	120.50	0.21152	380.4
180.00	10.3454	120.53	0.21158	380.6
200.00	8.7370	128.93	0.22451	429.5
220.00	7.8399	135.85	0.23484	463.0
240.00	7.2145	142.18	0.24403	489.7
260.00	6.7351	148.21	0.25252	512.3
280.00	6.3472	154.06	0.26055	532.1
300.00	6.0221	159.81	0.26821	550.0
320.00	5.7425	165.49	0.27560	566.5
340.00	5.4977	171.15	0.28277	581.7
360.00	5.2802	176.80	0.28975	596.0
380.00	5.0848	182.45	0.29656	609.5
400.00	4.9075	188.12	0.30323	622.4
420.00	4.7454	193.82	0.30978	634.7
440.00	4.5964	199.54	0.31621	646.5
460.00	4.4584	205.30	0.32254	657.9
480.00	4.3303	211.09	0.32878	669.0
500.00	4.2107	216.93	0.33492	679.6

Pressure = 600.00 psia
Saturation temperature = n/a (supercritical)

Temp,* °F	Density, lb/ft³	Enthalpy, Btu/lb	Entropy, Btu/lb·°F	Vel. Sound, ft/s
220.00	19.6784	118.27	0.20421	340.3
240.00	14.2159	131.50	0.22343	409.1
260.00	12.2674	139.92	0.23530	449.7
280.00	11.0672	147.15	0.24522	480.8
300.00	10.2049	153.83	0.25413	506.7
320.00	9.5351	160.21	0.26241	529.2
340.00	8.9895	166.39	0.27024	549.2
360.00	8.5305	172.45	0.27774	567.5
380.00	8.1351	178.45	0.28496	584.4
400.00	7.7885	184.40	0.29197	600.1
420.00	7.4804	190.34	0.29879	615.0
440.00	7.2035	196.26	0.30546	629.0
460.00	6.9523	202.20	0.31198	642.4
480.00	6.7229	208.15	0.31838	655.3
500.00	6.5118	214.12	0.32467	667.6

*temperatures are on the ITS-90 scale

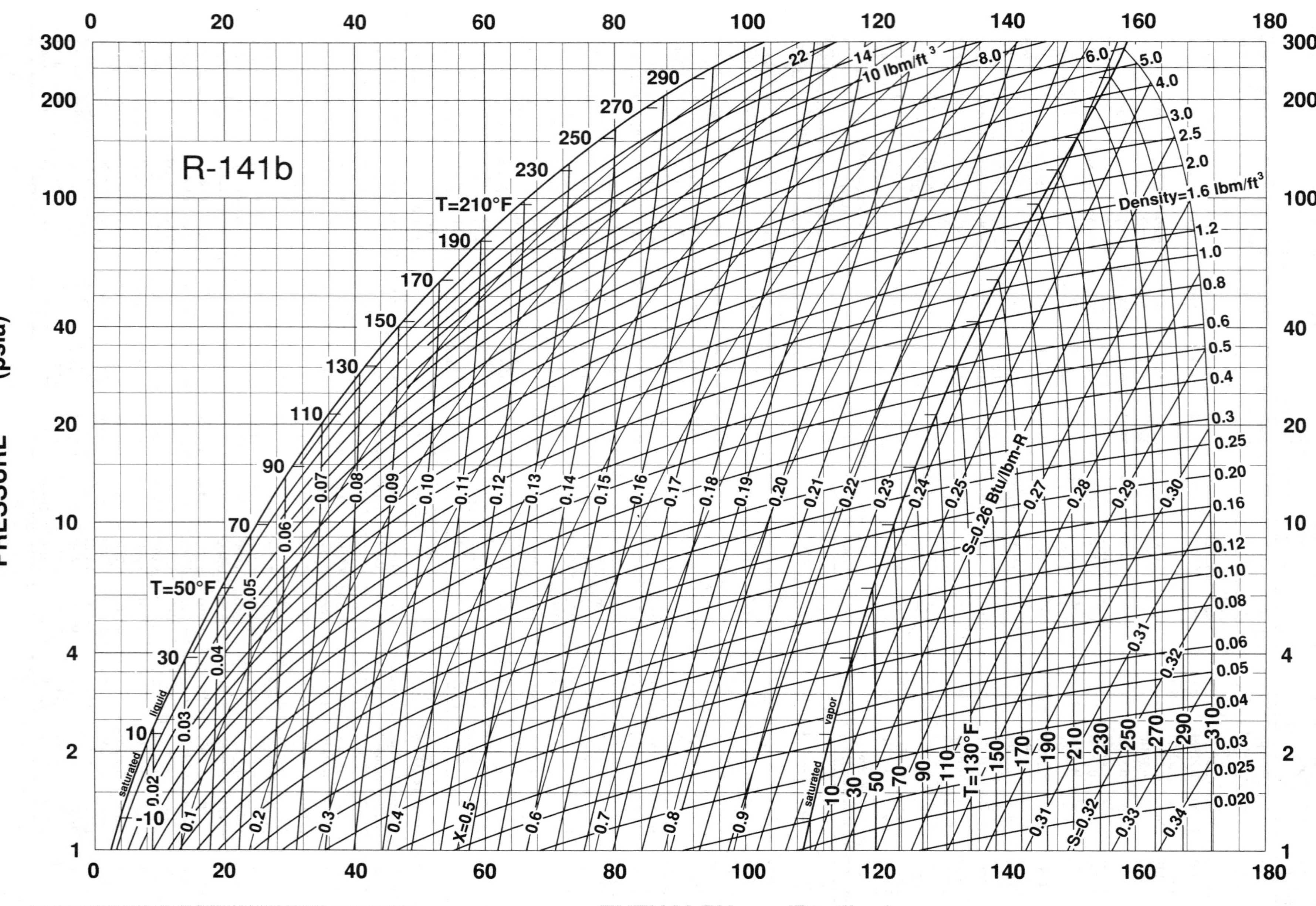

Fig. 13 Pressure–Enthalpy Diagram for Refrigerant 141b

Refrigerant 141b (1,1-Dichloro-1-fluoroethane) Properties of Saturated Liquid and Saturated Vapor

Temp,* °F	Pressure, psia	Density, lb/ft³ Liquid	Volume, ft³/lb Vapor	Enthalpy, Btu/lb Liquid	Enthalpy, Btu/lb Vapor	Entropy, Btu/lb·°F Liquid	Entropy, Btu/lb·°F Vapor	Specific Heat c_p, Btu/lb·°F Liquid	Specific Heat c_p, Btu/lb·°F Vapor	c_p/c_v Vapor	Velocity of Sound, ft/s Liquid	Velocity of Sound, ft/s Vapor	Viscosity, lb_m/ft·h Liquid	Viscosity, lb_m/ft·h Vapor	Thermal Cond, Btu/h·ft·°F Liquid	Thermal Cond, Btu/h·ft·°F Vapor	Surface Tension, dyne/cm	Temp,* °F
0.00	1.695	81.68	24.691	6.101	111.614	0.01372	0.24326	—	0.1655	1.1197	—	464.	1.712	—	0.0623	—	—	0.00
5.00	1.962	81.36	21.538	7.331	112.410	0.01638	0.24252	—	0.1667	1.1193	—	466.	1.645	—	0.0617	—	—	5.00
10.00	2.263	81.05	18.852	8.606	113.208	0.01911	0.24182	—	0.1679	1.1191	—	468.	1.582	—	0.0610	—	—	10.00
15.00	2.602	80.73	16.555	9.915	114.008	0.02188	0.24118	—	0.1691	1.1188	—	470.	1.524	—	0.0604	—	—	15.00
20.00	2.981	80.41	14.586	11.248	114.811	0.02467	0.24058	—	0.1703	1.1187	—	472.	1.469	—	0.0598	—	—	20.00
25.00	3.404	80.10	12.890	12.599	115.615	0.02747	0.24002	—	0.1716	1.1186	—	474.	1.417	—	0.0591	—	—	25.00
30.00	3.874	79.78	11.427	13.963	116.422	0.03027	0.23951	—	0.1728	1.1185	—	475.	1.369	—	0.0585	—	—	30.00
35.00	4.396	79.46	10.159	15.333	117.230	0.03305	0.23904	—	0.1741	1.1186	—	477.	1.323	—	0.0579	—	21.12	35.00
40.00	4.972	79.13	9.0571	16.708	118.039	0.03581	0.23861	—	0.1753	1.1186	—	479.	1.280	—	0.0572	—	20.77	40.00
45.00	5.609	78.81	8.0968	18.084	118.850	0.03855	0.23822	—	0.1766	1.1188	—	480.	1.239	—	0.0566	—	20.41	45.00
50.00	6.309	78.49	7.2572	19.460	119.662	0.04126	0.23786	—	0.1779	1.1190	—	482.	1.200	—	0.0560	—	20.06	50.00
55.00	7.077	78.16	6.5209	20.836	120.475	0.04394	0.23754	—	0.1792	1.1193	—	483.	1.163	—	0.0554	—	19.70	55.00
60.00	7.918	77.84	5.8735	22.210	121.288	0.04660	0.23725	—	0.1805	1.1196	—	485.	1.128	—	0.0548	—	19.35	60.00
65.00	8.836	77.51	5.3026	23.583	122.102	0.04922	0.23700	—	0.1819	1.1200	—	486.	1.095	—	0.0542	—	19.00	65.00
70.00	9.838	77.18	4.7979	24.955	122.916	0.05182	0.23677	—	0.1832	1.1205	—	488.	1.063	—	0.0536	—	18.65	70.00
75.00	10.927	76.85	4.3506	26.327	123.731	0.05439	0.23657	—	0.1846	1.1210	—	489.	1.033	—	0.0530	—	18.30	75.00
80.00	12.109	76.52	3.9531	27.700	124.546	0.05694	0.23640	—	0.1860	1.1216	—	490.	1.003	—	0.0524	—	17.95	80.00
85.00	13.390	76.18	3.5990	29.073	125.360	0.05947	0.23625	—	0.1874	1.1223	—	491.	0.975	—	0.0518	—	17.61	85.00
89.72b	14.696	75.86	3.2994	30.374	126.130	0.06184	0.23614	—	0.1888	1.1230	—	492.	0.949	—	0.0512	—	17.28	89.72
90.00	14.775	75.84	3.2829	30.450	126.174	0.06198	0.23613	—	0.1888	1.1231	—	492.	0.948	—	0.0512	—	17.26	90.00
95.00	16.270	75.51	3.0001	31.830	126.988	0.06447	0.23603	—	0.1903	1.1239	—	493.	0.922	—	0.0506	—	16.92	95.00
100.00	17.881	75.17	2.7465	33.214	127.801	0.06695	0.23596	—	0.1918	1.1248	—	494.	0.896	—	0.0500	—	16.58	100.00
105.00	19.614	74.82	2.5187	34.603	128.614	0.06941	0.23590	—	0.1933	1.1258	—	495.	0.872	—	0.0495	—	16.24	105.00
110.00	21.474	74.48	2.3136	35.998	129.425	0.07187	0.23587	—	0.1948	1.1269	—	496.	0.848	—	0.0489	—	15.90	110.00
115.00	23.469	74.13	2.1285	37.400	130.236	0.07431	0.23585	—	0.1963	1.1281	—	497.	0.825	—	0.0483	—	15.56	115.00
120.00	25.604	73.78	1.9613	38.808	131.045	0.07674	0.23586	—	0.1979	1.1293	—	497.	0.802	—	0.0477	—	15.22	120.00
125.00	27.886	73.43	1.8099	40.225	131.853	0.07916	0.23588	—	0.1995	1.1306	—	498.	0.780	—	0.0472	—	14.89	125.00
130.00	30.321	73.08	1.6726	41.648	132.660	0.08158	0.23592	—	0.2011	1.1321	—	498.	0.759	—	0.0466	—	14.55	130.00
135.00	32.916	72.72	1.5479	43.080	133.464	0.08398	0.23597	—	0.2028	1.1336	—	499.	0.737	—	0.0461	—	14.22	135.00
140.00	35.677	72.36	1.4343	44.520	134.267	0.08638	0.23604	—	0.2045	1.1352	—	499.	0.717	—	0.0455	—	13.89	140.00
145.00	38.611	72.00	1.3308	45.968	135.069	0.08877	0.23613	—	0.2062	1.1369	—	499.	0.696	—	0.0450	—	13.56	145.00
150.00	41.726	71.63	1.2364	47.423	135.868	0.09116	0.23623	—	0.2079	1.1387	—	499.	0.676	—	0.0444	—	13.23	150.00
155.00	45.027	71.26	1.1500	48.886	136.665	0.09353	0.23634	—	0.2097	1.1407	—	499.	0.656	—	0.0439	—	12.90	155.00
160.00	48.522	70.89	1.0708	50.357	137.459	0.09590	0.23646	—	0.2116	1.1427	—	499.	0.637	—	0.0433	—	12.58	160.00
165.00	52.218	70.51	0.9983	51.834	138.252	0.09826	0.23660	—	0.2134	1.1449	—	499.	0.618	—	0.0428	—	12.26	165.00
170.00	56.121	70.13	0.9316	53.319	139.041	0.10061	0.23675	—	0.2153	1.1472	—	499.	0.599	—	0.0422	—	—	170.00
175.00	60.240	69.75	0.8703	54.810	139.828	0.10295	0.23691	—	0.2173	1.1496	—	498.	0.580	—	0.0417	—	—	175.00
180.00	64.581	69.36	0.8138	56.307	140.612	0.10528	0.23708	—	0.2193	1.1521	—	498.	—	—	0.0412	—	—	180.00
185.00	69.152	68.97	0.7617	57.811	141.392	0.10761	0.23726	—	0.2214	1.1548	—	497.	—	—	0.0407	—	—	185.00
190.00	73.960	68.57	0.7136	59.321	142.169	0.10992	0.23744	—	0.2235	1.1577	—	496.	—	—	0.0401	—	—	190.00
195.00	79.013	68.17	0.6691	60.837	142.943	0.11222	0.23764	—	0.2256	1.1607	—	496.	—	—	0.0396	—	—	195.00
200.00	84.320	67.76	0.6278	62.359	143.713	0.11452	0.23784	—	0.2279	1.1639	—	494.	—	—	0.0391	—	—	200.00
205.00	89.887	67.35	0.5896	63.888	144.479	0.11680	0.23805	—	0.2302	1.1673	—	493.	—	—	0.0386	—	—	205.00
210.00	95.723	66.93	0.5541	65.424	145.240	0.11908	0.23827	—	0.2325	1.1709	—	492.	—	—	0.0381	—	—	210.00
215.00	101.84	66.51	0.5211	66.968	145.997	0.12135	0.23849	—	0.2350	1.1747	—	491.	—	—	0.0376	—	—	215.00
220.00	108.24	66.08	0.4903	68.519	146.748	0.12362	0.23872	—	0.2375	1.1788	—	489.	—	—	0.0371	—	—	220.00
225.00	114.93	65.65	0.4616	70.080	147.494	0.12588	0.23895	—	0.2401	1.1832	—	487.	—	—	0.0366	—	—	225.00
230.00	121.93	65.21	0.4349	71.651	148.234	0.12813	0.23918	—	0.2428	1.1878	—	485.	—	—	0.0361	—	—	230.00
235.00	129.24	64.76	0.4099	73.233	148.968	0.13039	0.23941	—	0.2457	1.1928	—	483.	—	—	0.0356	—	—	235.00
240.00	136.87	64.31	0.3865	74.829	149.694	0.13265	0.23965	—	0.2486	1.1982	—	480.	—	—	0.0351	—	—	240.00
245.00	144.83	63.84	0.3645	76.439	150.413	0.13491	0.23989	—	0.2517	1.2040	—	478.	—	—	0.0346	—	—	245.00
250.00	153.13	63.37	0.3439	78.067	151.123	0.13718	0.24012	—	0.2550	1.2104	—	475.	—	—	0.0341	—	—	250.00
255.00	161.78	62.89	0.3246	79.713	151.823	0.13945	0.24035	—	0.2584	1.2172	—	472.	—	—	—	—	—	255.00
260.00	170.80	62.41	0.3064	81.382	152.513	0.14174	0.24058	—	0.2620	1.2248	—	468.	—	—	—	—	—	260.00
265.00	180.18	61.91	0.2892	83.075	153.191	0.14405	0.24080	—	0.2659	1.2331	—	464.	—	—	—	—	—	265.00
270.00	189.95	61.40	0.2730	84.797	153.855	0.14638	0.24102	—	0.2701	1.2424	—	460.	—	—	—	—	—	270.00
275.00	200.11	60.88	0.2577	86.551	154.505	0.14873	0.24122	—	0.2746	1.2528	—	455.	—	—	—	—	—	275.00
280.00	210.67	60.35	0.2432	88.341	155.136	0.15111	0.24142	—	0.2795	1.2645	—	450.	—	—	—	—	—	280.00
285.00	221.65	59.81	0.2294	90.174	155.748	0.15354	0.24160	—	0.2849	1.2779	—	444.	—	—	—	—	—	285.00
290.00	233.05	59.26	0.2163	92.055	156.337	0.15601	0.24175	—	0.2909	1.2935	—	438.	—	—	—	—	—	290.00
295.00	244.90	58.69	0.2037	93.993	156.897	0.15853	0.24189	—	0.2977	1.3120	—	431.	—	—	—	—	—	295.00
300.00	257.19	58.10	0.1917	95.996	157.424	0.16113	0.24199	—	0.3056	1.3342	—	423.	—	—	—	—	—	300.00

*temperatures are on the IPTS–68 scale

b = normal boiling point

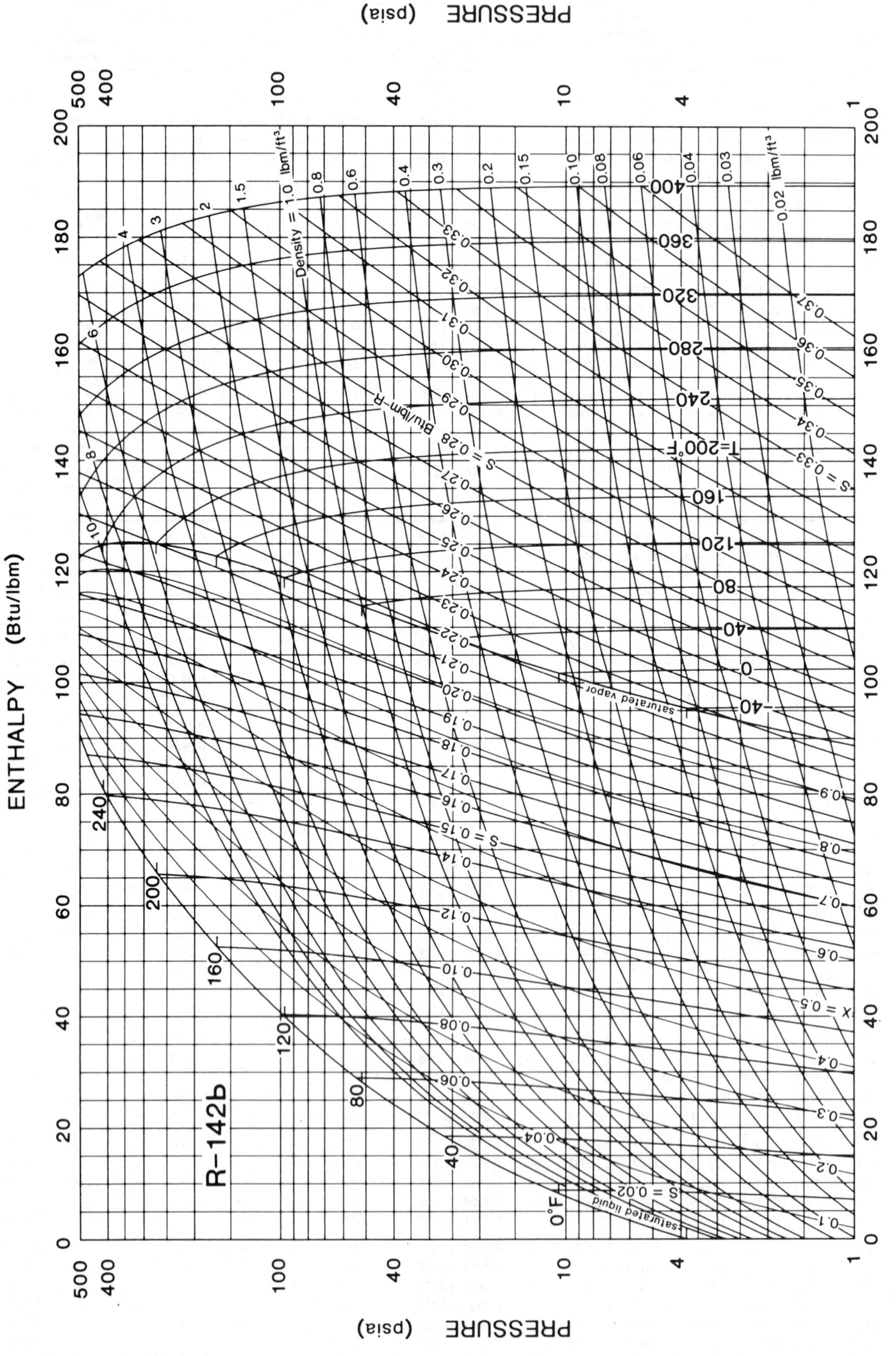

Prepared by: CENTER FOR APPLIED THERMODYNAMIC STUDIES, University of Idaho

Fig. 14 Pressure–Enthalpy Diagram for Refrigerant 142b

Refrigerant 142b (1-Chloro-1,1-difluoroethane) Properties of Saturated Liquid and Saturated Vapor

Temp,* °F	Pressure, psia	Density, lb/ft³ Liquid	Volume, ft³/lb Vapor	Enthalpy, Btu/lb Liquid	Enthalpy, Btu/lb Vapor	Entropy, Btu/lb·°F Liquid	Entropy, Btu/lb·°F Vapor	Specific Heat c_p, Btu/lb·°F Liquid	Specific Heat c_p, Btu/lb·°F Vapor	c_p/c_v Vapor	Velocity of Sound, ft/s Liquid	Velocity of Sound, ft/s Vapor	Viscosity, lb_m/ft·h Liquid	Viscosity, lb_m/ft·h Vapor	Thermal Cond, Btu/h·ft·°F Liquid	Thermal Cond, Btu/h·ft·°F Vapor	Surface Tension, dyne/cm	Temp,* °F
−60.00	2.135	80.26	19.796	−3.732	92.263	−0.00910	0.23109	—	0.1623	1.1441	—	471.	—	—	—	—	—	−60.00
−55.00	2.485	79.89	17.199	−2.822	93.043	−0.00684	0.23006	—	0.1636	1.1435	—	473.	—	—	—	—	—	−55.00
−50.00	2.881	79.52	14.995	−1.897	93.826	−0.00457	0.22909	—	0.1649	1.1430	—	475.	—	—	—	—	—	−50.00
−45.00	3.329	79.15	13.116	−0.956	94.611	−0.00229	0.22818	—	0.1663	1.1425	—	477.	—	—	—	—	—	−45.00
−40.00	3.833	78.77	11.509	0.000	95.399	0.00000	0.22732	—	0.1676	1.1422	—	479.	—	—	—	—	—	−40.00
−35.00	4.399	78.40	10.130	0.972	96.187	0.00230	0.22651	—	0.1690	1.1420	—	481.	—	—	—	—	—	−35.00
−30.00	5.033	78.02	8.9418	1.961	96.977	0.00461	0.22575	—	0.1704	1.1419	—	483.	—	—	—	—	—	−30.00
−25.00	5.740	77.64	7.9154	2.967	97.768	0.00693	0.22503	—	0.1718	1.1419	—	485.	—	—	—	—	—	−25.00
−20.00	6.526	77.26	7.0257	3.990	98.560	0.00927	0.22436	—	0.1732	1.1420	—	487.	—	—	—	—	—	−20.00
−15.00	7.400	76.88	6.2523	5.030	99.351	0.01162	0.22373	—	0.1746	1.1422	—	488.	—	—	—	—	—	−15.00
−10.00	8.366	76.49	5.5778	6.088	100.143	0.01398	0.22314	—	0.1760	1.1425	—	490.	—	—	—	—	—	−10.00
−5.00	9.433	76.11	4.9880	7.163	100.933	0.01635	0.22259	—	0.1775	1.1430	—	491.	—	—	0.0561	—	—	−5.00
0.00	10.609	75.72	4.4709	8.257	101.722	0.01874	0.22207	—	0.1789	1.1436	—	493.	—	—	0.0555	—	—	0.00
5.00	11.901	75.33	4.0163	9.370	102.510	0.02114	0.22158	—	0.1804	1.1444	—	494.	—	—	0.0548	—	—	5.00
10.00	13.318	74.93	3.6156	10.501	103.295	0.02355	0.22113	—	0.1819	1.1453	—	495.	—	—	0.0542	—	16.17	10.00
14.47b	14.696	74.58	3.2973	11.527	103.995	0.02572	0.22074	—	0.1833	1.1463	—	496.	—	—	0.0537	—	15.85	14.47
15.00	14.868	74.53	3.2616	11.651	104.078	0.02598	0.22070	—	0.1834	1.1464	—	496.	—	—	0.0536	—	15.81	15.00
20.00	16.560	74.13	2.9480	12.819	104.857	0.02842	0.22030	—	0.1850	1.1477	—	497.	—	—	0.0530	—	15.44	20.00
25.00	18.404	73.73	2.6696	14.007	105.633	0.03087	0.21992	—	0.1866	1.1492	—	498.	—	—	0.0524	—	15.08	25.00
30.00	20.410	73.32	2.4219	15.213	106.404	0.03334	0.21957	—	0.1882	1.1509	—	499.	—	—	0.0519	—	14.72	30.00
35.00	22.587	72.91	2.2010	16.439	107.171	0.03582	0.21924	0.2993	0.1898	1.1528	—	499.	—	0.0236	0.0513	—	14.37	35.00
40.00	24.945	72.50	2.0037	17.684	107.932	0.03831	0.21892	0.3009	0.1915	1.1549	—	500.	—	0.0238	0.0507	—	14.01	40.00
45.00	27.495	72.08	1.8270	18.948	108.687	0.04081	0.21863	0.3026	0.1932	1.1572	—	500.	—	0.0241	0.0501	—	13.65	45.00
50.00	30.248	71.66	1.6684	20.230	109.436	0.04333	0.21835	0.3043	0.1950	1.1599	—	500.	—	0.0243	0.0495	—	13.30	50.00
55.00	33.216	71.24	1.5260	21.531	110.177	0.04585	0.21809	0.3059	0.1968	1.1628	—	500.	—	0.0245	0.0489	—	12.95	55.00
60.00	36.408	70.80	1.3976	22.852	110.910	0.04839	0.21784	0.3076	0.1986	1.1660	—	501.	—	0.0247	0.0483	—	12.60	60.00
65.00	39.838	70.37	1.2819	24.190	111.635	0.05094	0.21760	0.3093	0.2005	1.1696	—	500.	—	0.0250	0.0478	—	12.25	65.00
70.00	43.517	69.93	1.1773	25.547	112.350	0.05349	0.21737	0.3111	0.2025	1.1735	—	500.	—	0.0252	0.0472	—	11.91	70.00
75.00	47.458	69.48	1.0826	26.922	113.055	0.05606	0.21715	0.3128	0.2046	1.1778	—	500.	—	0.0254	0.0466	—	11.56	75.00
80.00	51.672	69.03	0.9967	28.314	113.750	0.05863	0.21694	0.3146	0.2067	1.1825	—	500.	—	0.0256	0.0460	—	11.22	80.00
85.00	56.172	68.57	0.9187	29.724	114.433	0.06121	0.21673	0.3164	0.2089	1.1876	—	499.	—	0.0258	0.0455	—	10.88	85.00
90.00	60.971	68.11	0.8478	31.151	115.103	0.06379	0.21652	0.3183	0.2112	1.1933	—	498.	—	0.0260	0.0449	—	10.54	90.00
95.00	66.083	67.64	0.7832	32.595	115.761	0.06638	0.21632	0.3202	0.2136	1.1994	—	498.	—	0.0262	0.0443	—	10.20	95.00
100.00	71.521	67.16	0.7242	34.054	116.405	0.06897	0.21611	0.3222	0.2161	1.2062	—	497.	—	0.0265	0.0438	—	9.86	100.00
105.00	77.297	66.67	0.6704	35.530	117.034	0.07157	0.21591	0.3242	0.2187	1.2135	—	496.	—	0.0267	0.0432	—	9.53	105.00
110.00	83.427	66.18	0.6211	37.021	117.647	0.07416	0.21570	0.3263	0.2214	1.2216	—	495.	—	0.0269	0.0427	—	9.20	110.00
115.00	89.923	65.68	0.5760	38.526	118.245	0.07676	0.21548	0.3284	0.2243	1.2304	—	493.	—	0.0271	0.0421	—	8.87	115.00
120.00	96.799	65.17	0.5346	40.046	118.824	0.07936	0.21527	0.3307	0.2274	1.2400	—	492.	—	0.0274	0.0415	—	8.54	120.00
125.00	104.07	64.64	0.4966	41.579	119.386	0.08196	0.21504	0.3330	0.2306	1.2505	—	491.	—	0.0276	0.0410	—	8.22	125.00
130.00	111.75	64.11	0.4617	43.126	119.929	0.08456	0.21481	0.3354	0.2340	1.2620	—	489.	—	0.0279	0.0404	—	7.90	130.00
135.00	119.85	63.57	0.4295	44.685	120.451	0.08715	0.21456	0.3378	0.2377	1.2746	—	487.	—	0.0281	0.0399	—	7.58	135.00
140.00	128.39	63.01	0.3999	46.256	120.952	0.08974	0.21430	0.3404	0.2415	1.2883	—	486.	—	0.0284	0.0394	—	7.26	140.00
145.00	137.39	62.44	0.3726	47.838	121.432	0.09232	0.21403	0.3431	0.2457	1.3033	—	484.	—	0.0287	0.0388	—	6.94	145.00
150.00	146.85	61.86	0.3473	49.432	121.888	0.09490	0.21375	0.3458	0.2501	1.3198	—	482.	—	0.0290	0.0383	—	6.63	150.00
155.00	156.79	61.26	0.3239	51.036	122.321	0.09747	0.21345	0.3487	0.2549	1.3379	—	480.	—	0.0293	0.0377	—	6.32	155.00
160.00	167.22	60.65	0.3023	52.650	122.728	0.10004	0.21313	0.3517	0.2600	1.3576	—	478.	—	0.0296	0.0372	—	6.01	160.00
165.00	178.17	60.02	0.2823	54.275	123.109	0.10259	0.21279	0.3548	0.2655	1.3794	—	476.	—	0.0299	0.0367	—	5.70	165.00
170.00	189.64	59.37	0.2637	55.909	123.463	0.10514	0.21243	0.3580	0.2714	1.4032	—	473.	—	0.0303	0.0361	—	5.40	170.00
175.00	201.66	58.71	0.2465	57.552	123.789	0.10768	0.21205	—	0.2779	1.4294	—	471.	—	0.0307	—	—	5.10	175.00
180.00	214.23	58.02	0.2305	59.206	124.085	0.11022	0.21164	—	0.2849	1.4582	—	469.	—	0.0311	—	—	4.81	180.00
185.00	227.37	57.30	0.2156	60.871	124.351	0.11274	0.21121	—	0.2925	1.4900	—	466.	—	0.0315	—	—	4.51	185.00
190.00	241.09	56.56	0.2017	62.547	124.585	0.11527	0.21076	—	0.3009	1.5251	—	463.	—	0.0319	—	—	4.22	190.00
195.00	255.41	55.80	0.1888	64.236	124.786	0.11778	0.21027	—	0.3100	1.5638	—	461.	—	0.0324	—	—	3.94	195.00
200.00	270.35	55.00	0.1767	65.940	124.952	0.12030	0.20976	—	0.3200	1.6067	—	458.	—	0.0329	—	—	3.65	200.00
205.00	285.92	54.16	0.1654	67.661	125.082	0.12282	0.20921	—	0.3310	1.6544	—	455.	—	0.0334	—	—	3.38	205.00
210.00	302.13	53.29	0.1548	69.404	125.174	0.12535	0.20863	—	0.3432	1.7076	—	451.	—	0.0340	—	—	3.10	210.00
215.00	319.00	52.38	0.1449	71.173	125.225	0.12789	0.20801	—	0.3568	1.7673	—	448.	—	0.0345	—	—	2.83	215.00
220.00	336.54	51.41	0.1356	72.975	125.232	0.13046	0.20735	—	0.3721	1.8347	—	444.	—	0.0352	—	—	2.56	220.00
225.00	354.77	50.40	0.1269	74.819	125.192	0.13307	0.20664	—	0.3894	1.9116	—	439.	—	0.0358	—	—	2.30	225.00
230.00	373.70	49.31	0.1186	76.717	125.099	0.13573	0.20588	—	0.4093	2.0006	—	434.	—	0.0365	—	—	2.05	230.00
235.00	393.34	48.16	0.1107	78.686	124.943	0.13846	0.20505	—	0.4327	2.1059	—	428.	—	—	—	—	1.80	235.00
240.00	413.72	46.92	0.1032	80.750	124.714	0.14131	0.20415	—	0.4611	2.2343	—	421.	—	—	—	—	1.55	240.00
245.00	434.84	45.57	0.0959	82.943	124.391	0.14431	0.20313	—	0.4970	2.3984	—	411.	—	—	—	—	1.31	245.00
250.00	456.71	44.10	0.0887	85.320	123.937	0.14755	0.20196	—	0.5463	2.6252	—	399.	—	—	—	—	1.08	250.00

*temperatures are on the IPTS–68 scale

b = normal boiling point

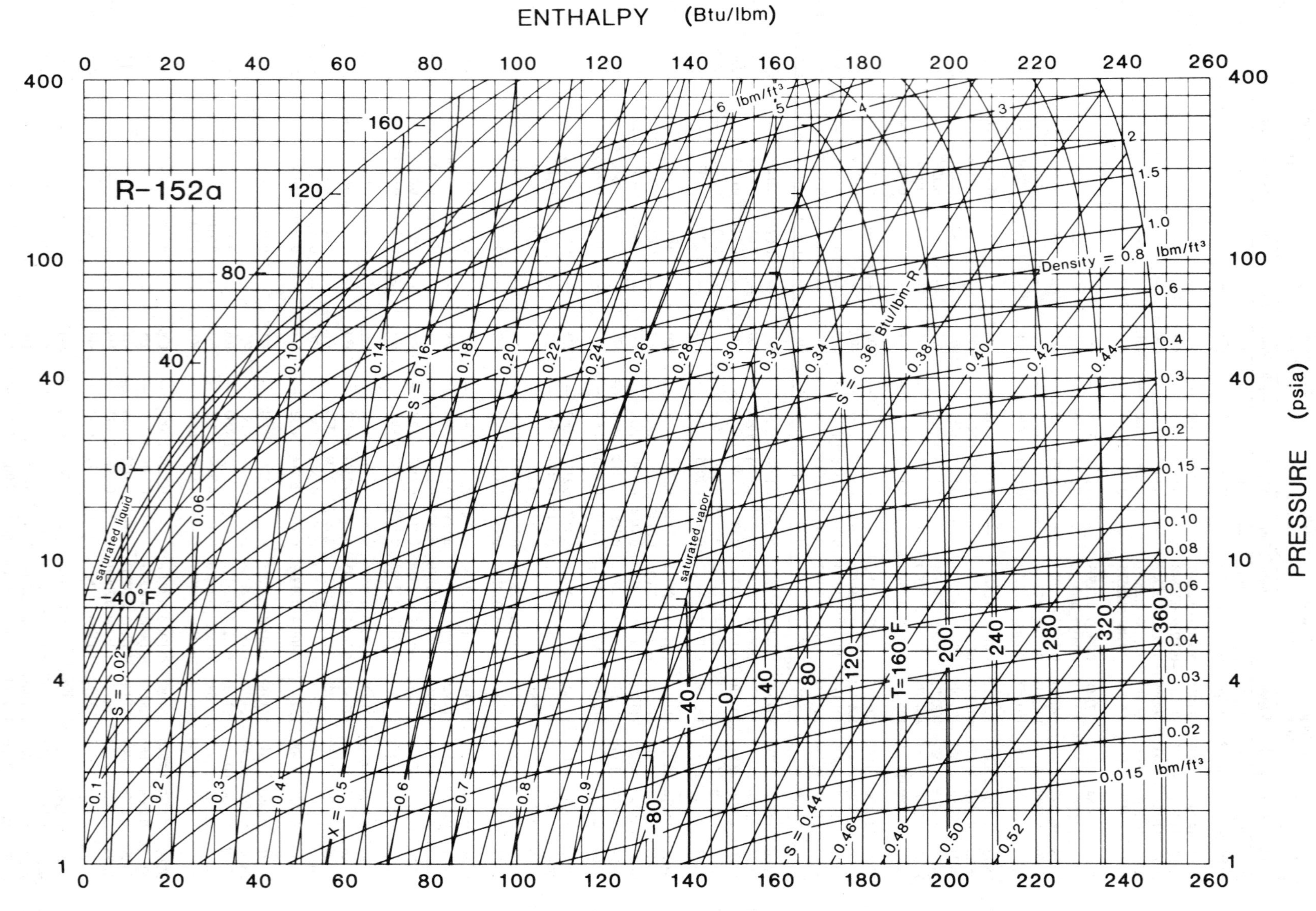

Prepared by: CENTER FOR APPLIED THERMODYNAMIC STUDIES, University of Idaho

Fig. 15 Pressure–Enthalpy Diagram for Refrigerant 152a

Refrigerant 152a (1,1-Difluoroethane) Properties of Saturated Liquid and Saturated Vapor

Temp,* °F	Pressure, psia	Density, lb/ft³ Liquid	Volume, ft³/lb Vapor	Enthalpy, Btu/lb Liquid	Enthalpy, Btu/lb Vapor	Entropy, Btu/lb·°F Liquid	Entropy, Btu/lb·°F Vapor	Specific Heat c_p, Btu/lb·°F Liquid	Specific Heat c_p, Btu/lb·°F Vapor	c_p/c_v Vapor	Velocity of Sound, ft/s Liquid	Velocity of Sound, ft/s Vapor	Viscosity, lb_m/ft·h Liquid	Viscosity, lb_m/ft·h Vapor	Thermal Cond, Btu/h·ft·°F Liquid	Thermal Cond, Btu/h·ft·°F Vapor	Surface Tension, dyne/cm	Temp,* °F
−181.46a	0.012	74.28	3784.5	—	122.659	—	0.46931	—	0.1626	1.2274	—	507.	—	—	—	—	—	−181.46
−180.00	0.013	74.19	3483.2	—	122.896	—	0.46751	—	0.1632	1.2266	—	508.	—	—	—	—	—	−180.00
−170.00	0.024	73.58	1962.1	—	124.536	—	0.45497	—	0.1668	1.2208	—	516.	—	—	—	—	—	−170.00
−160.00	0.044	72.96	1113.9	—	126.207	—	0.44262	—	0.1705	1.2156	—	523.	—	—	—	—	—	−160.00
−150.00	0.078	72.34	646.96	—	127.907	—	0.43090	—	0.1742	1.2107	—	531.	—	—	—	—	—	−150.00
−140.00	0.134	71.72	386.82	—	129.635	—	0.42001	—	0.1780	1.2063	—	538.	—	—	—	—	—	−140.00
−130.00	0.224	71.09	238.52	—	131.387	—	0.41000	—	0.1819	1.2023	—	545.	—	—	—	—	—	−130.00
−120.00	0.363	70.45	151.60	—	133.160	—	0.40084	—	0.1858	1.1988	—	552.	—	—	—	—	—	−120.00
−110.00	0.570	69.81	99.179	—	134.952	—	0.39250	—	0.1899	1.1958	—	558.	—	—	—	—	—	−110.00
−100.00	0.871	69.16	66.660	−21.865	136.759	−0.05611	0.38492	—	0.1942	1.1932	—	565.	—	—	—	—	—	−100.00
−90.00	1.297	68.51	45.934	−18.413	138.576	−0.04664	0.37803	—	0.1987	1.1912	—	571.	1.301	—	—	—	—	−90.00
−80.00	1.884	67.85	32.386	−14.852	140.402	−0.03714	0.37178	—	0.2035	1.1898	—	577.	1.187	—	—	—	—	−80.00
−70.00	2.678	67.19	23.317	−11.216	142.230	−0.02769	0.36609	—	0.2086	1.1890	—	582.	1.088	—	—	—	—	−70.00
−60.00	3.729	66.52	17.113	−7.522	144.058	−0.01834	0.36092	—	0.2140	1.1889	—	587.	1.000	—	—	—	—	−60.00
−50.00	5.096	65.84	12.781	−3.781	145.882	−0.00911	0.35622	0.3759	0.2199	1.1894	3248.	592.	0.921	—	—	—	—	−50.00
−45.00	5.918	65.49	11.113	−1.895	146.790	−0.00454	0.35403	0.3778	0.2229	1.1900	3202.	594.	0.886	—	—	—	—	−45.00
−40.00	6.845	65.15	9.7001	0.000	147.696	0.00000	0.35193	0.3797	0.2261	1.1907	3157.	597.	0.852	—	—	—	—	−40.00
−35.00	7.885	64.80	8.4974	1.905	148.599	0.00450	0.34994	0.3815	0.2294	1.1916	3112.	599.	0.820	—	—	—	—	−35.00
−30.00	9.049	64.45	7.4699	3.819	149.499	0.00898	0.34803	0.3833	0.2328	1.1928	3068.	601.	0.790	—	—	—	—	−30.00
−25.00	10.347	64.10	6.5885	5.742	150.394	0.01342	0.34621	0.3851	0.2363	1.1941	3023.	602.	0.761	—	—	—	—	−25.00
−20.00	11.789	63.74	5.8297	7.674	151.285	0.01783	0.34446	0.3869	0.2399	1.1956	2979.	604.	0.734	—	—	—	—	−20.00
−15.00	13.387	63.38	5.1740	9.616	152.171	0.02221	0.34280	0.3887	0.2436	1.1973	2935.	606.	0.708	—	—	—	—	−15.00
−11.24b	14.696	63.11	4.7396	11.080	152.833	0.02548	0.34159	0.3901	0.2464	1.1988	2901.	607.	0.689	—	—	—	—	−11.24
−10.00	15.151	63.02	4.6055	11.566	153.052	0.02656	0.34120	0.3906	0.2474	1.1993	2890.	607.	0.683	—	—	—	—	−10.00
−5.00	17.095	62.66	4.1108	13.526	153.927	0.03088	0.33968	0.3924	0.2513	1.2015	2846.	608.	0.660	—	0.0737	—	—	−5.00
0.00	19.231	62.29	3.6791	15.496	154.796	0.03518	0.33822	0.3943	0.2553	1.2039	2802.	609.	0.638	—	0.0728	—	—	0.00
5.00	21.570	61.92	3.3010	17.476	155.658	0.03944	0.33682	0.3963	0.2594	1.2066	2758.	611.	0.617	—	0.0718	—	—	5.00
10.00	24.127	61.55	2.9691	19.465	156.513	0.04369	0.33548	0.3982	0.2636	1.2094	2714.	611.	0.597	—	0.0709	—	14.98	10.00
15.00	26.915	61.17	2.6767	21.465	157.361	0.04790	0.33420	0.4003	0.2679	1.2126	2670.	612.	0.578	—	0.0700	—	14.57	15.00
20.00	29.948	60.79	2.4184	23.475	158.201	0.05210	0.33297	0.4024	0.2723	1.2160	2626.	613.	0.559	—	0.0690	—	14.17	20.00
25.00	33.239	60.41	2.1898	25.496	159.033	0.05627	0.33179	0.4045	0.2768	1.2197	2582.	613.	0.542	—	0.0681	—	13.77	25.00
30.00	36.804	60.02	1.9867	27.529	159.856	0.06042	0.33066	0.4068	0.2814	1.2237	2537.	613.	0.525	—	0.0672	—	13.37	30.00
35.00	40.657	59.62	1.8060	29.572	160.669	0.06455	0.32957	0.4091	0.2860	1.2280	2493.	614.	0.509	—	0.0662	—	12.98	35.00
40.00	44.813	59.23	1.6447	31.628	161.473	0.06865	0.32852	0.4114	0.2908	1.2326	2448.	613.	0.494	—	0.0653	—	12.59	40.00
45.00	49.289	58.83	1.5005	33.695	162.266	0.07274	0.32751	0.4139	0.2957	1.2375	2404.	613.	0.479	—	0.0643	—	12.20	45.00
50.00	54.099	58.42	1.3712	35.775	163.048	0.07681	0.32653	0.4164	0.3007	1.2429	2359.	613.	0.464	0.0232	0.0634	—	11.81	50.00
55.00	59.260	58.01	1.2550	37.868	163.818	0.08087	0.32559	0.4190	0.3058	1.2486	2314.	612.	0.451	0.0235	0.0625	—	11.42	55.00
60.00	64.788	57.59	1.1505	39.974	164.576	0.08491	0.32468	0.4218	0.3110	1.2547	2269.	612.	0.438	0.0237	0.0615	—	11.04	60.00
65.00	70.700	57.17	1.0561	42.094	165.320	0.08893	0.32380	0.4246	0.3163	1.2613	2224.	611.	0.425	0.0239	0.0606	—	10.65	65.00
70.00	77.014	56.74	0.9708	44.228	166.051	0.09294	0.32294	0.4276	0.3218	1.2684	2179.	609.	0.413	0.0241	0.0597	—	10.28	70.00
75.00	83.746	56.31	0.8936	46.377	166.767	0.09694	0.32210	0.4306	0.3275	1.2760	2134.	608.	0.401	0.0244	0.0588	—	9.90	75.00
80.00	90.913	55.87	0.8235	48.542	167.468	0.10092	0.32129	0.4339	0.3333	1.2842	2088.	607.	0.390	0.0246	0.0579	—	9.52	80.00
85.00	98.535	55.42	0.7597	50.722	168.151	0.10490	0.32049	0.4372	0.3394	1.2931	2042.	605.	0.379	0.0248	0.0570	—	9.15	85.00
90.00	106.63	54.97	0.7016	52.919	168.818	0.10886	0.31971	0.4408	0.3456	1.3026	1996.	603.	0.368	0.0251	0.0561	—	8.78	90.00
95.00	115.21	54.51	0.6487	55.133	169.465	0.11282	0.31895	0.4445	0.3522	1.3129	1950.	601.	0.358	0.0253	0.0552	—	8.42	95.00
100.00	124.31	54.04	0.6002	57.366	170.092	0.11677	0.31819	0.4485	0.3590	1.3241	1904.	598.	0.348	0.0256	0.0543	—	8.06	100.00
105.00	133.93	53.56	0.5559	59.617	170.698	0.12072	0.31744	0.4526	0.3661	1.3363	1857.	596.	0.338	0.0259	0.0534	—	7.70	105.00
110.00	144.10	53.07	0.5153	61.887	171.281	0.12466	0.31669	0.4570	0.3736	1.3496	1811.	593.	0.328	0.0261	0.0525	—	7.34	110.00
115.00	154.84	52.57	0.4779	64.179	171.839	0.12860	0.31594	0.4617	0.3815	1.3640	1763.	590.	0.319	0.0264	0.0516	—	6.98	115.00
120.00	166.17	52.06	0.4436	66.493	172.371	0.13254	0.31519	0.4667	0.3900	1.3799	1716.	587.	0.310	0.0267	0.0508	—	6.63	120.00
125.00	178.11	51.54	0.4119	68.829	172.874	0.13648	0.31443	0.4721	0.3989	1.3973	1668.	583.	0.302	0.0270	0.0499	—	6.29	125.00
130.00	190.68	51.00	0.3827	71.190	173.347	0.14042	0.31367	0.4779	0.4085	1.4165	1620.	579.	0.293	0.0273	0.0491	—	5.94	130.00
135.00	203.90	50.46	0.3557	73.577	173.787	0.14437	0.31288	0.4841	0.4189	1.4377	1572.	575.	0.285	0.0276	0.0483	—	5.60	135.00
140.00	217.79	49.90	0.3307	75.991	174.192	0.14833	0.31208	0.4909	0.4301	1.4613	1523.	571.	0.277	0.0280	0.0474	—	5.26	140.00
145.00	232.39	49.32	0.3075	78.435	174.557	0.15230	0.31126	0.4983	0.4424	1.4877	1474.	566.	0.269	0.0283	0.0466	—	4.93	145.00
150.00	247.71	48.72	0.2860	80.911	174.880	0.15628	0.31041	0.5064	0.4559	1.5174	1425.	561.	0.261	0.0287	0.0458	—	4.60	150.00
155.00	263.77	48.11	0.2660	83.421	175.156	0.16028	0.30952	0.5154	0.4708	1.5509	1375.	555.	0.254	0.0291	0.0450	—	4.28	155.00
160.00	280.61	47.48	0.2473	85.968	175.382	0.16430	0.30859	0.5254	0.4874	1.5890	1324.	550.	—	0.0296	0.0443	—	3.96	160.00
165.00	298.24	46.82	0.2299	88.555	175.551	0.16835	0.30761	0.5366	0.5062	1.6328	1273.	544.	—	0.0300	0.0435	—	3.64	165.00
170.00	316.70	46.13	0.2136	91.187	175.657	0.17243	0.30657	0.5493	0.5276	1.6834	1221.	537.	—	0.0305	0.0428	—	3.33	170.00
175.00	336.02	45.42	0.1984	93.869	175.692	0.17654	0.30547	0.5639	0.5523	1.7427	1169.	530.	—	0.0311	—	—	3.02	175.00
180.00	356.22	44.67	0.1840	96.605	175.647	0.18071	0.30427	0.5808	0.5812	1.8128	1115.	523.	—	0.0317	—	—	2.72	180.00
185.00	377.34	43.89	0.1705	99.403	175.511	0.18493	0.30299	0.6007	0.6154	1.8970	1061.	516.	—	0.0323	—	—	2.43	185.00
190.00	399.41	43.06	0.1578	102.272	175.270	0.18921	0.30158	0.6245	0.6568	1.9999	1006.	507.	—	0.0330	—	—	2.14	190.00
195.00	422.47	42.17	0.1457	105.223	174.905	0.19359	0.30002	0.6537	0.7081	2.1282	949.	499.	—	—	—	—	1.86	195.00
200.00	446.56	41.23	0.1343	108.272	174.392	0.19806	0.29829	0.6904	0.7733	2.2925	891.	490.	—	—	—	—	1.59	200.00
210.00	498.01	39.08	0.1127	114.753	172.776	0.20746	0.29410	0.8040	0.9778	2.8110	769.	469.	—	—	—	—	1.06	210.00
220.00	554.22	36.38	0.0922	122.043	169.915	0.21785	0.28829	—	—	—	636.	446.	—	—	—	—	0.59	220.00
230.00	606.14	29.31	0.0341	135.228	135.236	0.23666	0.23667	—	—	—	—	—	—	—	—	—	0.17	230.00
235.87c	655.07	22.97	0.0435	147.302	147.302	0.25359	0.25359	∞	∞	∞	0.	0.	—	—	∞	∞	0.00	235.87

*temperatures are on the ITS–90 scale a = triple point b = normal boiling point c = critical point

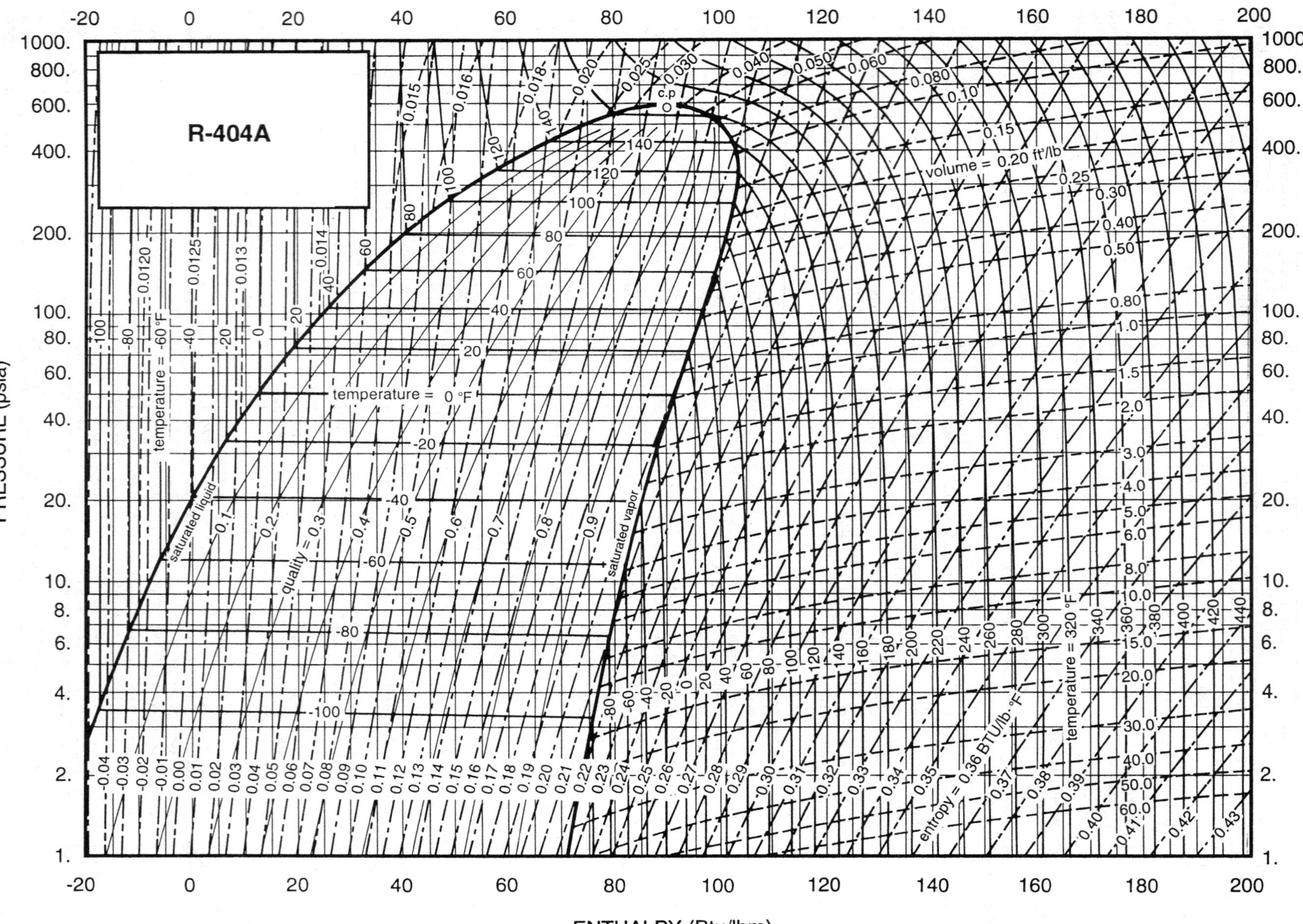

Fig. 16 Pressure–Enthalpy Diagram for Refrigerant 404A
Reprinted with permission from E.I. DuPont de Nemours.

Refrigerant 404A [R-125/143a/134a (44/52/4)] Properties of Liquid on the Bubble Line and Vapor on the Dew Line

Pressure,	Temperature*, °F		Density, lb/ft³	Volume, ft³/lb	Enthalpy, Btu/lb		Entropy, Btu/lb·°F		Specific Heat c_p, Btu/lb·°F		c_p/c_v	Velocity of Sound, ft/s		Pressure,
psia	Bubble	Dew	Liquid	Vapor	Liquid	Vapor	Liquid	Vapor	Liquid	Vapor	Vapor	Liquid	Vapor	psia
1.00	−130.49	−128.46	89.51	36.17	−25.81	71.15	−0.06905	0.22497	0.2733	0.1535	1.160	3217.	439.5	1.00
1.50	−120.93	−119.01	88.60	24.74	−23.19	72.53	−0.06119	0.22089	0.2756	0.1568	1.159	3113.	444.4	1.50
2.00	−113.74	−111.89	87.91	18.90	−21.20	73.57	−0.05539	0.21811	0.2774	0.1593	1.158	3037.	447.9	2.00
2.50	−107.91	−106.11	87.35	15.34	−19.58	74.42	−0.05074	0.21604	0.2788	0.1614	1.158	2978.	450.6	2.50
3.00	−102.97	−101.22	86.86	12.94	−18.20	75.14	−0.04685	0.21440	0.2801	0.1633	1.158	2928.	452.8	3.00
4.00	−94.84	−93.16	86.06	9.888	−15.91	76.32	−0.04051	0.21192	0.2821	0.1664	1.158	2848.	456.3	4.00
5.00	−88.22	−86.59	85.41	8.027	−14.04	77.29	−0.03543	0.21008	0.2838	0.1690	1.158	2785.	458.9	5.00
6.00	−82.60	−81.02	84.84	6.768	−12.44	78.11	−0.03116	0.20863	0.2853	0.1712	1.159	2731.	461.0	6.00
7.00	−77.69	−76.15	84.35	5.859	−11.03	78.83	−0.02746	0.20746	0.2867	0.1733	1.159	2685.	462.7	7.00
8.00	−73.32	−71.80	83.90	5.171	−9.77	79.46	−0.02419	0.20647	0.2879	0.1751	1.160	2643.	464.1	8.00
10.00	−65.74	−64.28	83.12	4.195	−7.58	80.56	−0.01858	0.20488	0.2901	0.1784	1.161	2573.	466.4	10.00
12.00	−59.29	−57.86	82.45	3.535	−5.70	81.50	−0.01386	0.20365	0.2921	0.1813	1.163	2513.	468.1	12.00
14.00	−53.64	−52.25	81.86	3.059	−4.04	82.31	−0.00976	0.20266	0.2938	0.1839	1.165	2461.	469.5	14.00
14.696b	−51.82	−50.44	81.67	2.922	−3.51	82.57	−0.00845	0.20235	0.2944	0.1848	1.165	2445.	469.9	14.70
16.00	−48.59	−47.23	81.32	2.697	−2.55	83.03	−0.00613	0.20183	0.2954	0.1864	1.166	2415.	470.5	16.00
18.00	−44.02	−42.69	80.84	2.414	−1.20	83.69	−0.00286	0.20113	0.2969	0.1886	1.168	2373.	471.4	18.00
20.00	−39.83	−38.52	80.38	2.185	0.05	84.28	0.00012	0.20052	0.2983	0.1907	1.170	2335.	472.0	20.00
22.00	−35.95	−34.66	79.96	1.997	1.21	84.83	0.00286	0.19999	0.2997	0.1927	1.172	2300.	472.6	22.00
24.00	−32.34	−31.07	79.57	1.839	2.30	85.34	0.00541	0.19951	0.3009	0.1945	1.174	2267.	473.0	24.00
26.00	−28.95	−27.71	79.19	1.704	3.32	85.81	0.00778	0.19909	0.3022	0.1963	1.175	2237.	473.3	26.00
28.00	−25.77	−24.54	78.84	1.588	4.29	86.26	0.01001	0.19871	0.3033	0.1981	1.177	2208.	473.5	28.00
30.00	−22.75	−21.54	78.50	1.487	5.21	86.68	0.01210	0.19837	0.3045	0.1997	1.179	2181.	473.7	30.00
32.00	−19.88	−18.68	78.18	1.398	6.08	87.07	0.01409	0.19805	0.3056	0.2013	1.181	2155.	473.8	32.00
34.00	−17.15	−15.97	77.87	1.319	6.92	87.45	0.01598	0.19776	0.3066	0.2029	1.183	2130.	473.8	34.00
36.00	−14.54	−13.37	77.57	1.249	7.72	87.80	0.01778	0.19750	0.3077	0.2044	1.185	2106.	473.8	36.00
38.00	−12.04	−10.88	77.28	1.186	8.50	88.14	0.01950	0.19725	0.3087	0.2059	1.187	2084.	473.8	38.00
40.00	−9.64	−8.49	77.00	1.129	9.24	88.47	0.02115	0.19702	0.3097	0.2073	1.188	2062.	473.7	40.00
42.00	−7.33	−6.19	76.73	1.077	9.96	88.78	0.02273	0.19681	0.3106	0.2087	1.190	2041.	473.5	42.00
44.00	−5.10	−3.97	76.47	1.030	10.65	89.08	0.02425	0.19661	0.3116	0.2101	1.192	2021.	473.4	44.00
46.00	−2.94	−1.83	76.22	0.9864	11.33	89.37	0.02572	0.19642	0.3125	0.2115	1.194	2002.	473.2	46.00
48.00	−0.86	0.24	75.97	0.9465	11.98	89.64	0.02713	0.19624	0.3134	0.2128	1.196	1983.	473.0	48.00
50.00	1.16	2.25	75.73	0.9098	12.62	89.91	0.02850	0.19607	0.3143	0.2141	1.198	1965.	472.8	50.00
55.00	5.94	7.01	75.16	0.8292	14.13	90.53	0.03174	0.19569	0.3166	0.2172	1.203	1921.	472.1	55.00
60.00	10.41	11.46	74.61	0.7616	15.55	91.11	0.03476	0.19535	0.3187	0.2203	1.208	1881.	471.3	60.00
65.00	14.61	15.64	74.09	0.7040	16.89	91.65	0.03758	0.19505	0.3208	0.2233	1.213	1843.	470.5	65.00
70.00	18.57	19.58	73.60	0.6543	18.17	92.14	0.04023	0.19478	0.3229	0.2262	1.218	1807.	469.5	70.00
75.00	22.32	23.31	73.12	0.6111	19.39	92.61	0.04274	0.19453	0.3249	0.2290	1.223	1773.	468.5	75.00
80.00	25.88	26.86	72.66	0.5731	20.55	93.05	0.04511	0.19430	0.3269	0.2318	1.228	1740.	467.4	80.00
85.00	29.28	30.25	72.22	0.5393	21.67	93.46	0.04738	0.19409	0.3288	0.2345	1.233	1709.	466.3	85.00
90.00	32.54	33.49	71.79	0.5092	22.74	93.85	0.04954	0.19389	0.3308	0.2372	1.239	1679.	465.1	90.00
95.00	35.66	36.59	71.37	0.4822	23.78	94.21	0.05162	0.19370	0.3327	0.2399	1.244	1651.	463.9	95.00
100.00	38.66	39.58	70.97	0.4577	24.78	94.56	0.05361	0.19353	0.3347	0.2426	1.250	1623.	462.7	100.00
110.00	44.34	45.24	70.19	0.4152	26.69	95.20	0.05737	0.19320	0.3385	0.2479	1.261	1571.	460.1	110.00
120.00	49.65	50.52	69.44	0.3796	28.50	95.79	0.06088	0.19290	0.3424	0.2531	1.273	1521.	457.4	120.00
130.00	54.63	55.48	68.72	0.3492	30.21	96.31	0.06417	0.19262	0.3463	0.2584	1.285	1475.	454.6	130.00
140.00	59.34	60.17	68.02	0.3230	31.84	96.80	0.06728	0.19235	0.3502	0.2638	1.298	1431.	451.7	140.00
150.00	63.80	64.61	67.35	0.3001	33.41	97.24	0.07023	0.19209	0.3542	0.2692	1.312	1389.	448.8	150.00
160.00	68.05	68.83	66.69	0.2800	34.91	97.64	0.07304	0.19184	0.3583	0.2747	1.326	1349.	445.7	160.00
170.00	72.10	72.87	66.05	0.2622	36.36	98.01	0.07573	0.19159	0.3624	0.2803	1.341	1311.	442.7	170.00
180.00	75.97	76.73	65.41	0.2462	37.76	98.35	0.07830	0.19134	0.3667	0.2861	1.357	1274.	439.6	180.00
190.00	79.69	80.43	64.79	0.2319	39.12	98.66	0.08077	0.19109	0.3711	0.2921	1.374	1238.	436.4	190.00
200.00	83.27	83.99	64.18	0.2189	40.44	98.94	0.08316	0.19084	0.3757	0.2982	1.391	1203.	433.2	200.00
220.00	90.04	90.72	62.99	0.1963	42.98	99.42	0.08769	0.19032	0.3853	0.3113	1.430	1138.	426.7	220.00
240.00	96.36	97.01	61.81	0.1772	45.39	99.82	0.09196	0.18978	0.3957	0.3256	1.474	1076.	420.1	240.00
260.00	102.29	102.92	60.64	0.1610	47.71	100.12	0.09600	0.18921	0.4072	0.3414	1.524	1016.	413.3	260.00
280.00	107.90	108.49	59.48	0.1468	49.95	100.35	0.09985	0.18860	0.4200	0.3592	1.582	960.	406.4	280.00
300.00	113.20	113.77	58.32	0.1344	52.12	100.49	0.10355	0.18794	0.4345	0.3796	1.650	906.	399.4	300.00
320.00	118.24	118.78	57.15	0.1234	54.24	100.55	0.10712	0.18722	0.4511	0.4032	1.730	853.	392.2	320.00
340.00	123.04	123.56	55.96	0.1136	56.31	100.53	0.11058	0.18643	0.4705	0.4310	1.827	803.	385.0	340.00
360.00	127.63	128.12	54.75	0.1047	58.35	100.43	0.11395	0.18557	0.4937	0.4646	1.945	753.	377.7	360.00
380.00	132.02	132.49	53.49	0.0965	60.37	100.23	0.11726	0.18461	0.5220	0.5061	2.092	705.	370.2	380.00
400.00	136.24	136.67	52.19	0.0890	62.38	99.94	0.12053	0.18353	0.5579	0.5588	2.282	657.	362.6	400.00
450.00	146.06	146.42	48.56	0.0722	67.49	98.62	0.12873	0.18011	0.7133	0.7888	3.120	541.	343.1	450.00
500.00	154.99	155.25	43.76	0.0565	73.20	95.88	0.13774	0.17463	1.2479	1.5641	5.974	422.	322.1	500.00
548.18c	162.50	162.50	35.84	0.0279	80.83	80.83	0.14972	0.14972	∞	∞	∞	0.	0.0	548.18

*temperatures are on the ITS-90 scale b = one standard atmosphere c = critical point

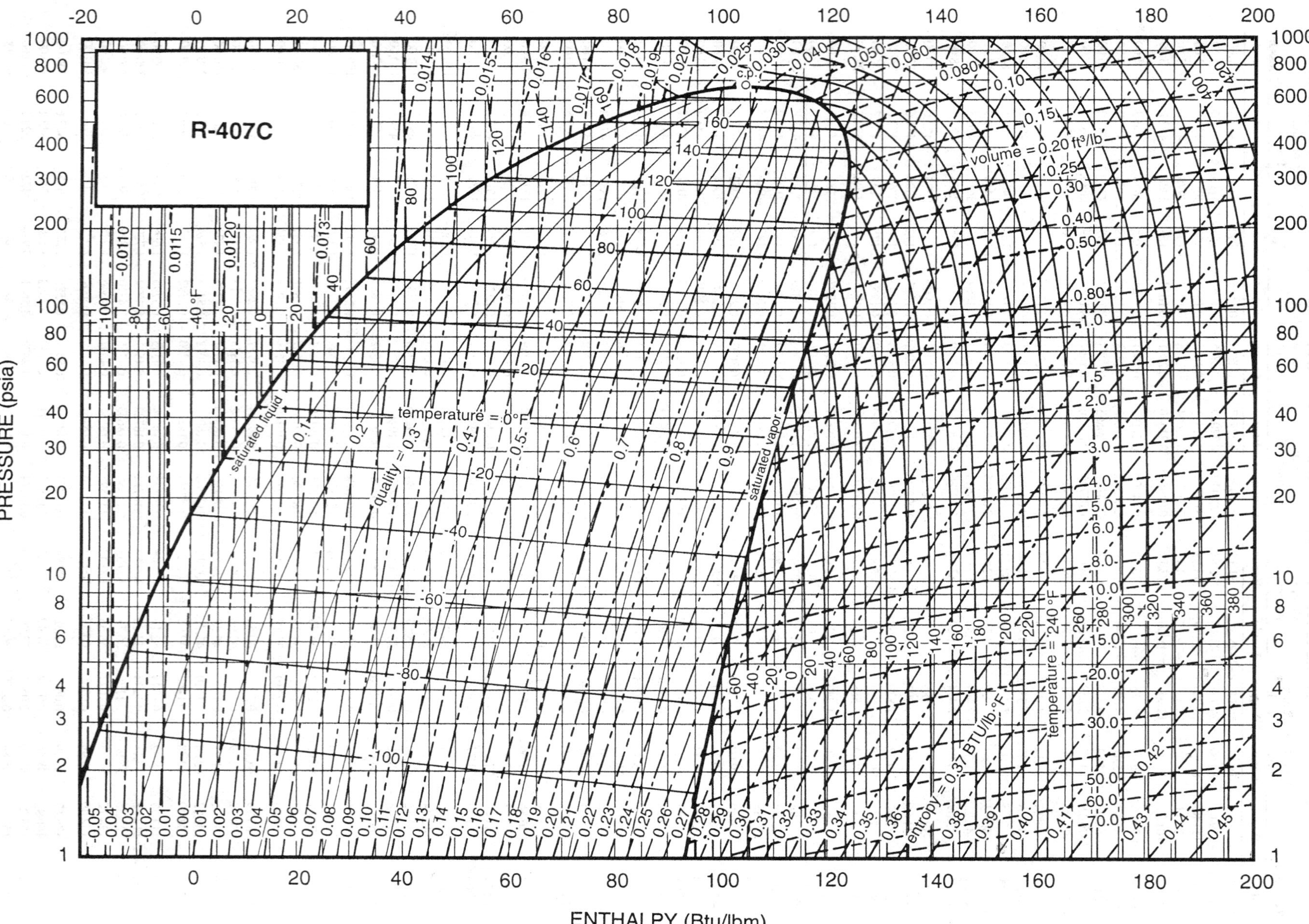

Fig. 17 Pressure–Enthalpy Diagram for Refrigerant 407C

Reprinted with permission from E.I. DuPont de Nemours.

Refrigerant 407C [R-32/125/134a (23/25/52)] Properties of Liquid on the Bubble Line and Vapor on the Dew Line

Pressure, psia	Temperature*, °F Bubble	Temperature*, °F Dew	Density, lb/ft³ Liquid	Volume, ft³/lb Vapor	Enthalpy, Btu/lb Liquid	Enthalpy, Btu/lb Vapor	Entropy, Btu/lb·°F Liquid	Entropy, Btu/lb·°F Vapor	Specific Heat c_p, Btu/lb·°F Liquid	Specific Heat c_p, Btu/lb·°F Vapor	c_p/c_v Vapor	Velocity of Sound, ft/s Liquid	Velocity of Sound, ft/s Vapor	Pressure, psia
1.00	−125.86	−111.56	94.31	43.09	−26.07	93.44	−0.06935	0.28147	0.2961	0.1557	1.182	3459.	484.3	1.00
1.50	−116.20	−102.09	93.37	29.45	−23.20	94.84	−0.06088	0.27613	0.2977	0.1585	1.180	3354.	489.6	1.50
2.00	−108.95	−94.98	92.66	22.49	−21.04	95.89	−0.05464	0.27246	0.2988	0.1606	1.180	3278.	493.4	2.00
2.50	−103.07	−89.22	92.07	18.24	−19.28	96.74	−0.04967	0.26968	0.2998	0.1625	1.179	3218.	496.3	2.50
3.00	−98.09	−84.34	91.58	15.38	−17.78	97.46	−0.04551	0.26746	0.3005	0.1640	1.179	3168.	498.7	3.00
4.00	−89.91	−76.32	90.76	11.75	−15.32	98.64	−0.03877	0.26404	0.3019	0.1667	1.179	3086.	502.5	4.00
5.00	−83.26	−69.80	90.09	9.530	−13.31	99.59	−0.03338	0.26145	0.3030	0.1690	1.179	3022.	505.4	5.00
6.00	−77.61	−64.27	89.51	8.033	−11.59	100.40	−0.02887	0.25939	0.3040	0.1710	1.180	2967.	507.8	6.00
7.00	−72.69	−59.44	89.01	6.953	−10.09	101.11	−0.02497	0.25769	0.3049	0.1728	1.180	2920.	509.7	7.00
8.00	−68.30	−55.15	88.55	6.134	−8.75	101.73	−0.02153	0.25624	0.3058	0.1744	1.181	2879.	511.4	8.00
10.00	−60.71	−47.71	87.77	4.975	−6.42	102.81	−0.01565	0.25386	0.3073	0.1774	1.183	2808.	514.0	10.00
12.00	−54.25	−41.38	87.09	4.192	−4.43	103.71	−0.01071	0.25198	0.3087	0.1800	1.184	2747.	516.0	12.00
14.00	−48.60	−35.84	86.49	3.626	−2.68	104.50	−0.00643	0.25042	0.3100	0.1823	1.186	2695.	517.7	14.00
14.696b	−46.79	−34.06	86.30	3.464	−2.11	104.75	−0.00507	0.24994	0.3104	0.1831	1.187	2678.	518.2	14.70
16.00	−43.56	−30.90	85.95	3.198	−1.11	105.19	−0.00265	0.24910	0.3112	0.1845	1.188	2648.	519.0	16.00
18.00	−38.99	−26.43	85.46	2.861	0.32	105.82	0.00075	0.24795	0.3123	0.1866	1.190	2606.	520.1	18.00
20.00	−34.81	−22.33	85.01	2.590	1.63	106.39	0.00384	0.24694	0.3134	0.1885	1.192	2568.	521.0	20.00
22.00	−30.94	−18.55	84.59	2.367	2.84	106.91	0.00667	0.24603	0.3144	0.1903	1.193	2533.	521.7	22.00
24.00	−27.35	−15.02	84.19	2.180	3.98	107.39	0.00930	0.24522	0.3154	0.1921	1.195	2500.	522.3	24.00
26.00	−23.97	−11.72	83.82	2.020	5.04	107.83	0.01175	0.24448	0.3163	0.1937	1.197	2469.	522.8	26.00
28.00	−20.80	−8.61	83.46	1.883	6.05	108.25	0.01404	0.24381	0.3172	0.1953	1.199	2440.	523.3	28.00
30.00	−17.80	−5.67	83.13	1.763	7.01	108.64	0.01620	0.24318	0.3181	0.1969	1.201	2413.	523.6	30.00
32.00	−14.95	−2.88	82.81	1.658	7.92	109.01	0.01824	0.24260	0.3190	0.1984	1.203	2387.	523.9	32.00
34.00	−12.23	−0.23	82.50	1.565	8.78	109.36	0.02018	0.24206	0.3198	0.1999	1.205	2362.	524.1	34.00
36.00	−9.64	2.31	82.21	1.482	9.62	109.69	0.02203	0.24156	0.3207	0.2013	1.207	2339.	524.3	36.00
38.00	−7.15	4.74	81.92	1.407	10.42	110.01	0.02379	0.24108	0.3215	0.2027	1.209	2316.	524.4	38.00
40.00	−4.76	7.08	81.65	1.339	11.19	110.31	0.02548	0.24064	0.3223	0.2040	1.211	2295.	524.5	40.00
42.00	−2.47	9.32	81.38	1.278	11.93	110.60	0.02709	0.24021	0.3231	0.2054	1.212	2274.	524.6	42.00
44.00	−0.26	11.49	81.12	1.222	12.65	110.87	0.02865	0.23981	0.3238	0.2067	1.214	2254.	524.6	44.00
46.00	1.88	13.58	80.87	1.171	13.34	111.13	0.03015	0.23943	0.3246	0.2079	1.216	2234.	524.6	46.00
48.00	3.95	15.60	80.63	1.124	14.01	111.39	0.03159	0.23907	0.3253	0.2092	1.218	2216.	524.5	48.00
50.00	5.95	17.55	80.40	1.080	14.67	111.63	0.03299	0.23872	0.3261	0.2104	1.220	2198.	524.4	50.00
55.00	10.70	22.20	79.83	0.9850	16.22	112.20	0.03629	0.23791	0.3279	0.2134	1.225	2155.	524.2	55.00
60.00	15.13	26.52	79.30	0.9050	17.68	112.72	0.03935	0.23718	0.3296	0.2163	1.230	2114.	523.8	60.00
65.00	19.28	30.58	78.79	0.8370	19.06	113.20	0.04221	0.23652	0.3313	0.2191	1.235	2077.	523.3	65.00
70.00	23.20	34.41	78.31	0.7784	20.36	113.64	0.04490	0.23590	0.3330	0.2219	1.240	2041.	522.7	70.00
75.00	26.91	38.03	77.85	0.7273	21.60	114.05	0.04744	0.23533	0.3346	0.2246	1.245	2008.	522.1	75.00
80.00	30.44	41.47	77.41	0.6824	22.79	114.44	0.04984	0.23479	0.3363	0.2272	1.250	1976.	521.4	80.00
85.00	33.81	44.75	76.98	0.6426	23.93	114.80	0.05213	0.23429	0.3379	0.2298	1.255	1945.	520.6	85.00
90.00	37.03	47.89	76.57	0.6071	25.02	115.14	0.05431	0.23382	0.3395	0.2324	1.260	1916.	519.8	90.00
95.00	40.11	50.90	76.17	0.5751	26.07	115.45	0.05640	0.23337	0.3411	0.2349	1.266	1888.	519.0	95.00
100.00	43.08	53.79	75.78	0.5463	27.09	115.75	0.05840	0.23294	0.3426	0.2375	1.271	1861.	518.1	100.00
110.00	48.69	59.25	75.03	0.4962	29.02	116.30	0.06217	0.23214	0.3458	0.2424	1.282	1809.	516.2	110.00
120.00	53.93	64.35	74.33	0.4542	30.84	116.80	0.06569	0.23140	0.3489	0.2473	1.293	1761.	514.2	120.00
130.00	58.85	69.14	73.65	0.4185	32.57	117.24	0.06898	0.23071	0.3520	0.2522	1.304	1716.	512.1	130.00
140.00	63.50	73.65	72.99	0.3876	34.21	117.64	0.07209	0.23006	0.3552	0.2571	1.316	1674.	510.0	140.00
150.00	67.90	77.92	72.36	0.3608	35.78	118.00	0.07503	0.22944	0.3583	0.2621	1.328	1633.	507.8	150.00
160.00	72.10	81.99	71.75	0.3371	37.28	118.33	0.07782	0.22885	0.3615	0.2670	1.341	1594.	505.5	160.00
170.00	76.10	85.86	71.16	0.3162	38.73	118.63	0.08048	0.22828	0.3648	0.2720	1.354	1557.	503.2	170.00
180.00	79.92	89.57	70.58	0.2975	40.13	118.90	0.08303	0.22774	0.3681	0.2771	1.367	1522.	500.8	180.00
190.00	83.60	93.12	70.01	0.2806	41.48	119.14	0.08548	0.22720	0.3714	0.2823	1.381	1488.	498.4	190.00
200.00	87.13	96.53	69.46	0.2654	42.79	119.36	0.08783	0.22669	0.3748	0.2875	1.396	1455.	496.0	200.00
220.00	93.81	102.98	68.38	0.2390	45.30	119.73	0.09230	0.22568	0.3819	0.2984	1.427	1392.	491.0	220.00
240.00	100.06	108.99	67.33	0.2168	47.68	120.01	0.09648	0.22470	0.3893	0.3098	1.460	1334.	486.0	240.00
260.00	105.93	114.64	66.31	0.1979	49.95	120.23	0.10042	0.22374	0.3972	0.3220	1.497	1278.	480.8	260.00
280.00	111.48	119.95	65.31	0.1815	52.14	120.38	0.10417	0.22278	0.4056	0.3350	1.537	1225.	475.6	280.00
300.00	116.74	124.98	64.33	0.1672	54.25	120.47	0.10774	0.22183	0.4146	0.3491	1.581	1174.	470.3	300.00
320.00	121.74	129.75	63.35	0.1546	56.29	120.50	0.11116	0.22087	0.4244	0.3644	1.630	1125.	464.9	320.00
340.00	126.52	134.30	62.39	0.1434	58.27	120.48	0.11446	0.21989	0.4350	0.3812	1.685	1078.	459.5	340.00
360.00	131.09	138.63	61.42	0.1333	60.20	120.40	0.11765	0.21890	0.4468	0.3998	1.746	1033.	454.0	360.00
380.00	135.48	142.78	60.45	0.1242	62.10	120.26	0.12074	0.21787	0.4599	0.4205	1.815	988.	448.5	380.00
400.00	139.70	146.75	59.47	0.1160	63.96	120.06	0.12375	0.21681	0.4745	0.4439	1.894	944.	442.9	400.00
450.00	149.59	156.00	56.97	0.0981	68.50	119.31	0.13100	0.21397	0.5212	0.5187	2.151	838.	428.6	450.00
500.00	158.69	164.41	54.31	0.0832	72.98	118.13	0.13803	0.21072	0.5918	0.6320	2.546	734.	413.9	500.00
550.00	167.13	172.06	51.33	0.0703	77.55	116.38	0.14509	0.20679	0.7152	0.8279	3.234	630.	398.5	550.00
600.00	175.04	179.00	47.71	0.0583	82.49	113.66	0.15262	0.20158	0.9958	1.2600	4.754	524.	381.7	600.00
650.00	182.58	185.04	42.09	0.0448	88.90	108.29	0.16233	0.19247	2.3366	3.0952	11.150	410.	360.9	650.00
674.84c	186.94	186.94	31.59	0.0317	98.82	98.82	0.17754	0.17754	∞	∞	∞	0.	0.0	674.84

*temperatures are on the ITS-90 scale; b = one standard atmosphere; c = critical point

Refrigerant 410A [R-32/125 (50/50)] Properties of Liquid on the Bubble Line and Vapor on the Dew Line

Pressure, psia	Temperature*, °F Bubble	Temperature*, °F Dew	Density, lb/ft³ Liquid	Volume, ft³/lb Vapor	Enthalpy, Btu/lb Liquid	Enthalpy, Btu/lb Vapor	Entropy, Btu/lb·°F Liquid	Entropy, Btu/lb·°F Vapor	Specific Heat c_p, Btu/lb·°F Liquid	Specific Heat c_p, Btu/lb·°F Vapor	c_p/c_v Vapor	Velocity of Sound, ft/s Liquid	Velocity of Sound, ft/s Vapor	Pressure, psia
1.00	−135.76	−135.67	92.14	47.62	−30.61	99.89	−0.08257	0.32027	0.3116	0.1538	1.226	3726.	518.5	1.00
1.50	−126.59	−126.50	91.23	32.58	−27.75	101.21	−0.07385	0.31327	0.3134	0.1561	1.225	3604.	524.5	1.50
2.00	−119.71	−119.62	90.54	24.89	−25.58	102.20	−0.06743	0.30840	0.3144	0.1579	1.225	3518.	528.8	2.00
2.50	−114.13	−114.04	89.97	20.20	−23.83	103.00	−0.06231	0.30468	0.3152	0.1594	1.225	3451.	532.2	2.50
3.00	−109.41	−109.32	89.49	17.04	−22.34	103.67	−0.05803	0.30169	0.3158	0.1608	1.225	3396.	534.9	3.00
4.00	−101.65	−101.56	88.69	13.02	−19.88	104.77	−0.05110	0.29703	0.3169	0.1631	1.225	3308.	539.3	4.00
5.00	−95.35	−95.25	88.04	10.57	−17.88	105.65	−0.04556	0.29347	0.3177	0.1652	1.226	3239.	542.6	5.00
6.00	−90.00	−89.90	87.48	8.915	−16.18	106.40	−0.04093	0.29061	0.3185	0.1670	1.227	3181.	545.3	6.00
7.00	−85.33	−85.24	86.99	7.718	−14.69	107.04	−0.03693	0.28822	0.3192	0.1686	1.228	3132.	547.5	7.00
8.00	−81.18	−81.08	86.55	6.812	−13.36	107.61	−0.03340	0.28617	0.3199	0.1701	1.229	3088.	549.4	8.00
10.00	−73.99	−73.89	85.79	5.527	−11.05	108.58	−0.02738	0.28278	0.3212	0.1729	1.231	3013.	552.5	10.00
12.00	−67.88	−67.77	85.13	4.659	−9.09	109.40	−0.02233	0.28005	0.3223	0.1753	1.233	2950.	554.8	12.00
14.00	−62.53	−62.42	84.55	4.031	−7.36	110.10	−0.01796	0.27777	0.3234	0.1776	1.236	2895.	556.8	14.00
14.696b	−60.81	−60.71	84.36	3.852	−6.80	110.33	−0.01656	0.27706	0.3238	0.1783	1.236	2877.	557.4	14.70
16.00	−57.76	−57.65	84.02	3.556	−5.81	110.72	−0.01410	0.27582	0.3245	0.1797	1.238	2846.	558.3	16.00
18.00	−53.45	−53.33	83.55	3.182	−4.41	111.28	−0.01063	0.27411	0.3255	0.1817	1.240	2803.	559.7	18.00
20.00	−49.49	−49.38	83.11	2.881	−3.12	111.78	−0.00748	0.27259	0.3264	0.1835	1.243	2763.	560.8	20.00
22.00	−45.84	−45.72	82.70	2.633	−1.92	112.23	−0.00459	0.27122	0.3274	0.1853	1.245	2726.	561.7	22.00
24.00	−42.44	−42.32	82.32	2.425	−0.80	112.65	−0.00191	0.26998	0.3283	0.1870	1.247	2691.	562.5	24.00
26.00	−39.26	−39.13	81.96	2.248	0.24	113.04	0.00058	0.26884	0.3291	0.1887	1.250	2659.	563.2	26.00
28.00	−36.26	−36.13	81.61	2.096	1.23	113.40	0.00291	0.26779	0.3300	0.1903	1.252	2629.	563.7	28.00
30.00	−33.43	−33.30	81.29	1.963	2.17	113.74	0.00511	0.26682	0.3309	0.1918	1.255	2601.	564.2	30.00
32.00	−30.74	−30.61	80.98	1.846	3.07	114.06	0.00719	0.26591	0.3317	0.1933	1.257	2574.	564.6	32.00
34.00	−28.17	−28.04	80.68	1.742	3.92	114.35	0.00916	0.26506	0.3325	0.1948	1.259	2548.	565.0	34.00
36.00	−25.73	−25.59	80.40	1.650	4.73	114.64	0.01104	0.26426	0.3333	0.1962	1.262	2523.	565.3	36.00
38.00	−23.38	−23.25	80.12	1.567	5.52	114.90	0.01283	0.26351	0.3341	0.1976	1.264	2500.	565.5	38.00
40.00	−21.13	−20.99	79.86	1.492	6.27	115.16	0.01454	0.26279	0.3349	0.1990	1.266	2477.	565.7	40.00
42.00	−18.97	−18.83	79.60	1.423	7.00	115.40	0.01618	0.26211	0.3356	0.2003	1.269	2455.	565.8	42.00
44.00	−16.88	−16.74	79.35	1.361	7.70	115.63	0.01776	0.26147	0.3364	0.2016	1.271	2434.	566.0	44.00
46.00	−14.87	−14.73	79.11	1.304	8.38	115.85	0.01928	0.26085	0.3371	0.2029	1.274	2414.	566.0	46.00
48.00	−12.92	−12.78	78.88	1.252	9.04	116.06	0.02075	0.26026	0.3379	0.2042	1.276	2395.	566.1	48.00
50.00	−11.04	−10.89	78.65	1.204	9.68	116.26	0.02217	0.25969	0.3386	0.2054	1.278	2376.	566.1	50.00
55.00	−6.57	−6.41	78.10	1.098	11.20	116.72	0.02552	0.25837	0.3404	0.2085	1.284	2331.	566.0	55.00
60.00	−2.40	−2.24	77.59	1.009	12.63	117.15	0.02862	0.25716	0.3422	0.2115	1.290	2289.	565.8	60.00
65.00	1.51	1.67	77.10	0.9328	13.97	117.53	0.03152	0.25605	0.3439	0.2144	1.296	2249.	565.5	65.00
70.00	5.20	5.36	76.64	0.8676	15.24	117.89	0.03425	0.25502	0.3456	0.2173	1.302	2212.	565.1	70.00
75.00	8.68	8.85	76.20	0.8107	16.45	118.21	0.03682	0.25405	0.3473	0.2201	1.309	2176.	564.6	75.00
80.00	12.00	12.17	75.77	0.7607	17.61	118.52	0.03925	0.25315	0.3490	0.2229	1.315	2143.	564.1	80.00
85.00	15.16	15.33	75.36	0.7164	18.72	118.79	0.04156	0.25229	0.3506	0.2256	1.321	2111.	563.4	85.00
90.00	18.18	18.36	74.96	0.6769	19.78	119.05	0.04377	0.25148	0.3523	0.2283	1.327	2080.	562.8	90.00
95.00	21.08	21.26	74.58	0.6413	20.81	119.29	0.04588	0.25071	0.3539	0.2310	1.334	2051.	562.1	95.00
100.00	23.86	24.04	74.21	0.6092	21.79	119.52	0.04791	0.24998	0.3555	0.2337	1.340	2022.	561.3	100.00
110.00	29.12	29.31	73.49	0.5535	23.68	119.92	0.05173	0.24860	0.3588	0.2390	1.353	1968.	559.7	110.00
120.00	34.03	34.22	72.81	0.5067	25.45	120.28	0.05528	0.24732	0.3620	0.2442	1.367	1918.	557.9	120.00
130.00	38.64	38.84	72.16	0.4669	27.12	120.59	0.05861	0.24613	0.3653	0.2495	1.381	1871.	556.1	130.00
140.00	42.99	43.19	71.54	0.4325	28.72	120.86	0.06174	0.24501	0.3686	0.2548	1.395	1826.	554.2	140.00
150.00	47.11	47.31	70.94	0.4026	30.24	121.09	0.06470	0.24395	0.3719	0.2601	1.410	1783.	552.2	150.00
160.00	51.02	51.23	70.35	0.3763	31.70	121.30	0.06752	0.24294	0.3753	0.2655	1.425	1743.	550.1	160.00
170.00	54.76	54.96	69.79	0.3529	33.10	121.48	0.07021	0.24197	0.3787	0.2709	1.440	1704.	548.0	170.00
180.00	58.33	58.54	69.24	0.3321	34.45	121.63	0.07278	0.24104	0.3821	0.2764	1.456	1667.	545.8	180.00
190.00	61.75	61.97	68.70	0.3134	35.76	121.76	0.07524	0.24014	0.3856	0.2821	1.473	1631.	543.7	190.00
200.00	65.04	65.26	68.18	0.2964	37.03	121.87	0.07761	0.23927	0.3892	0.2878	1.491	1596.	541.4	200.00
220.00	71.27	71.49	67.16	0.2670	39.45	122.03	0.08210	0.23759	0.3966	0.2997	1.527	1531.	536.9	220.00
240.00	77.08	77.30	66.17	0.2423	41.76	122.11	0.08631	0.23599	0.4044	0.3123	1.567	1469.	532.2	240.00
260.00	82.53	82.75	65.21	0.2212	43.95	122.13	0.09027	0.23444	0.4127	0.3256	1.610	1411.	527.5	260.00
280.00	87.67	87.89	64.27	0.2030	46.05	122.10	0.09403	0.23293	0.4214	0.3399	1.657	1356.	522.7	280.00
300.00	92.54	92.76	63.36	0.1872	48.08	122.00	0.09761	0.23145	0.4308	0.3553	1.709	1303.	517.8	300.00
320.00	97.17	97.39	62.45	0.1731	50.04	121.86	0.10104	0.22999	0.4410	0.3720	1.766	1253.	512.9	320.00
340.00	101.58	101.79	61.55	0.1607	51.95	121.66	0.10434	0.22853	0.4519	0.3903	1.829	1205.	507.9	340.00
360.00	105.80	106.01	60.66	0.1495	53.80	121.41	0.10752	0.22708	0.4640	0.4106	1.900	1158.	502.9	360.00
380.00	109.83	110.04	59.77	0.1394	55.61	121.12	0.11061	0.22561	0.4772	0.4330	1.979	1113.	497.8	380.00
400.00	113.71	113.91	58.88	0.1302	57.39	120.77	0.11360	0.22413	0.4919	0.4583	2.069	1069.	492.7	400.00
450.00	122.77	122.95	56.63	0.1105	61.70	119.66	0.12080	0.22028	0.5375	0.5379	2.357	963.	479.7	450.00
500.00	131.05	131.22	54.29	0.0941	65.92	118.16	0.12770	0.21612	0.6032	0.6559	2.790	862.	466.2	500.00
550.00	138.68	138.82	51.77	0.0800	70.15	116.15	0.13453	0.21140	0.7091	0.8525	3.518	764.	452.0	550.00
600.00	145.72	145.83	48.90	0.0672	74.56	113.36	0.14154	0.20563	0.9150	1.2550	5.014	663.	436.5	600.00
650.00	152.22	152.29	45.22	0.0542	79.55	108.88	0.14942	0.19735	1.5215	2.6002	10.010	557.	418.1	650.00
703.78c	158.40	158.40	34.18	0.0293	91.19	91.19	0.16792	0.16792	∞	∞	∞	0.	0.0	703.78

*temperatures are on the ITS-90 scale b = one standard atmosphere c = critical point

Refrigerant 507A [R-125/143a (50/50)] Properties of Saturated Liquid and Saturated Vapor

Temp,* °F	Pressure,** psia	Density, lb/ft³ Liquid	Volume, ft³/lb Vapor	Enthalpy, Btu/lb Liquid	Enthalpy, Btu/lb Vapor	Entropy, Btu/lb·°F Liquid	Entropy, Btu/lb·°F Vapor	Specific Heat c_p, Btu/lb·°F Liquid	Specific Heat c_p, Btu/lb·°F Vapor	c_p/c_v Vapor	Velocity of Sound, ft/s Liquid	Velocity of Sound, ft/s Vapor	Temp,* °F
−150.00	0.41	92.04	82.03	−30.87	66.51	−0.08495	0.22952	0.2655	0.1459	1.164	3445.	424.2	−150.00
−145.00	0.52	91.56	64.93	−29.53	67.22	−0.08068	0.22681	0.2671	0.1475	1.163	3379.	427.1	−145.00
−140.00	0.67	91.09	51.85	−28.19	67.93	−0.07646	0.22425	0.2686	0.1491	1.162	3317.	429.9	−140.00
−135.00	0.84	90.61	41.76	−26.85	68.64	−0.07228	0.22186	0.2700	0.1508	1.161	3257.	432.7	−135.00
−130.00	1.05	90.14	33.89	−25.49	69.36	−0.06814	0.21960	0.2713	0.1525	1.160	3199.	435.3	−130.00
−125.00	1.30	89.65	27.72	−24.13	70.09	−0.06405	0.21749	0.2725	0.1542	1.159	3143.	437.9	−125.00
−120.00	1.60	89.17	22.83	−22.77	70.81	−0.06000	0.21550	0.2738	0.1559	1.158	3088.	440.5	−120.00
−115.00	1.95	88.68	18.93	−21.40	71.53	−0.05599	0.21363	0.2750	0.1577	1.158	3035.	442.9	−115.00
−110.00	2.37	88.19	15.79	−20.02	72.26	−0.05202	0.21188	0.2762	0.1595	1.157	2983.	445.3	−110.00
−105.00	2.86	87.70	13.26	−18.63	72.99	−0.04809	0.21024	0.2775	0.1613	1.157	2932.	447.6	−105.00
−100.00	3.43	87.20	11.20	−17.24	73.72	−0.04420	0.20870	0.2787	0.1632	1.157	2882.	449.8	−100.00
−95.00	4.09	86.70	9.507	−15.84	74.45	−0.04035	0.20726	0.2800	0.1651	1.157	2832.	451.9	−95.00
−90.00	4.84	86.20	8.114	−14.44	75.18	−0.03653	0.20591	0.2813	0.1671	1.157	2783.	453.9	−90.00
−85.00	5.70	85.69	6.959	−13.03	75.91	−0.03274	0.20464	0.2826	0.1691	1.158	2735.	455.8	−85.00
−80.00	6.69	85.18	5.997	−11.61	76.64	−0.02899	0.20345	0.2840	0.1712	1.158	2687.	457.6	−80.00
−75.00	7.81	84.67	5.190	−10.19	77.37	−0.02527	0.20234	0.2854	0.1732	1.159	2640.	459.3	−75.00
−70.00	9.07	84.15	4.512	−8.75	78.10	−0.02157	0.20130	0.2868	0.1754	1.160	2592.	460.8	−70.00
−65.00	10.49	83.63	3.937	−7.31	78.82	−0.01791	0.20033	0.2883	0.1776	1.161	2546.	462.3	−65.00
−60.00	12.07	83.10	3.449	−5.87	79.54	−0.01428	0.19942	0.2898	0.1798	1.162	2499.	463.6	−60.00
−55.00	13.85	82.57	3.033	−4.41	80.26	−0.01067	0.19858	0.2913	0.1822	1.164	2453.	464.8	−55.00
−52.79b	14.696b	82.33	2.868	−3.77	80.58	−0.00908	0.19822	0.2920	0.1832	1.165	2432.	465.3	−52.79
−50.00	15.82	82.03	2.676	−2.95	80.98	−0.00709	0.19778	0.2929	0.1845	1.166	2407.	465.8	−50.00
−45.00	18.01	81.49	2.368	−1.48	81.69	−0.00353	0.19704	0.2946	0.1869	1.168	2361.	466.8	−45.00
−40.00	20.42	80.94	2.103	0.00	82.40	0.00000	0.19635	0.2962	0.1894	1.170	2315.	467.6	−40.00
−35.00	23.09	80.39	1.873	1.49	83.11	0.00351	0.19571	0.2980	0.1920	1.172	2269.	468.2	−35.00
−30.00	26.01	79.83	1.672	2.99	83.81	0.00700	0.19511	0.2998	0.1946	1.175	2224.	468.7	−30.00
−25.00	29.21	79.27	1.497	4.49	84.50	0.01047	0.19455	0.3016	0.1973	1.178	2178.	469.0	−25.00
−20.00	32.71	78.70	1.344	6.01	85.19	0.01392	0.19403	0.3035	0.2001	1.181	2133.	469.2	−20.00
−15.00	36.53	78.12	1.210	7.53	85.88	0.01735	0.19354	0.3055	0.2030	1.184	2087.	469.2	−15.00
−10.00	40.67	77.53	1.091	9.07	86.56	0.02076	0.19308	0.3076	0.2059	1.188	2042.	469.0	−10.00
−5.00	45.17	76.94	0.9861	10.61	87.23	0.02415	0.19266	0.3097	0.2090	1.193	1996.	468.7	−5.00
0.00	50.03	76.34	0.8931	12.17	87.89	0.02753	0.19226	0.3119	0.2122	1.197	1951.	468.2	0.00
5.00	55.29	75.73	0.8104	13.74	88.54	0.03090	0.19188	0.3142	0.2155	1.202	1905.	467.5	5.00
10.00	60.95	75.11	0.7367	15.32	89.19	0.03425	0.19153	0.3166	0.2189	1.208	1860.	466.6	10.00
15.00	67.04	74.48	0.6709	16.91	89.82	0.03759	0.19120	0.3192	0.2224	1.214	1814.	465.5	15.00
20.00	73.58	73.84	0.6119	18.52	90.45	0.04092	0.19089	0.3218	0.2261	1.220	1769.	464.2	20.00
25.00	80.59	73.18	0.5589	20.14	91.06	0.04424	0.19059	0.3246	0.2300	1.228	1723.	462.7	25.00
30.00	88.08	72.52	0.5112	21.77	91.66	0.04755	0.19030	0.3276	0.2340	1.236	1677.	461.0	30.00
35.00	96.10	71.84	0.4682	23.41	92.25	0.05086	0.19003	0.3307	0.2383	1.244	1631.	459.1	35.00
40.00	104.6	71.15	0.4294	25.08	92.83	0.05415	0.18976	0.3340	0.2428	1.254	1585.	456.9	40.00
45.00	113.7	70.45	0.3942	26.75	93.39	0.05745	0.18949	0.3375	0.2475	1.264	1538.	454.5	45.00
50.00	123.4	69.73	0.3622	28.45	93.93	0.06074	0.18923	0.3412	0.2526	1.276	1492.	451.9	50.00
55.00	133.7	68.99	0.3331	30.16	94.46	0.06403	0.18897	0.3452	0.2580	1.289	1445.	449.0	55.00
60.00	144.6	68.23	0.3066	31.89	94.97	0.06731	0.18870	0.3495	0.2638	1.304	1398.	445.8	60.00
65.00	156.2	67.45	0.2824	33.64	95.46	0.07061	0.18843	0.3542	0.2701	1.320	1351.	442.4	65.00
70.00	168.4	66.65	0.2602	35.41	95.92	0.07390	0.18815	0.3593	0.2769	1.338	1303.	438.6	70.00
75.00	181.4	65.82	0.2399	37.21	96.36	0.07721	0.18785	0.3648	0.2843	1.359	1255.	434.6	75.00
80.00	195.0	64.97	0.2212	39.03	96.77	0.08052	0.18752	0.3709	0.2925	1.382	1207.	430.3	80.00
85.00	209.5	64.08	0.2040	40.87	97.15	0.08385	0.18718	0.3776	0.3016	1.409	1159.	425.6	85.00
90.00	224.7	63.17	0.1881	42.74	97.49	0.08719	0.18680	0.3852	0.3118	1.440	1110.	420.6	90.00
95.00	240.7	62.21	0.1735	44.65	97.80	0.09056	0.18639	0.3937	0.3234	1.476	1060.	415.3	95.00
100.00	257.6	61.21	0.1599	46.59	98.06	0.09395	0.18593	0.4034	0.3367	1.519	1011.	409.6	100.00
105.00	275.4	60.17	0.1472	48.57	98.28	0.09737	0.18541	0.4146	0.3523	1.570	960.	403.5	105.00
110.00	294.0	59.07	0.1355	50.59	98.43	0.10083	0.18482	0.4278	0.3707	1.631	909.	396.9	110.00
115.00	313.7	57.91	0.1245	52.66	98.52	0.10434	0.18416	0.4436	0.3929	1.708	857.	390.0	115.00
120.00	334.3	56.67	0.1141	54.79	98.53	0.10791	0.18338	0.4630	0.4206	1.804	805.	382.5	120.00
125.00	355.9	55.34	0.1044	56.98	98.45	0.11156	0.18248	0.4876	0.4559	1.929	751.	374.6	125.00
130.00	378.7	53.89	0.0951	59.26	98.24	0.11531	0.18141	0.5198	0.5028	2.098	696.	366.1	130.00
135.00	402.6	52.29	0.0863	61.65	97.88	0.11920	0.18013	0.5647	0.5686	2.337	640.	357.1	135.00
140.00	427.7	50.50	0.0778	64.17	97.31	0.12327	0.17853	0.6322	0.6679	2.702	581.	347.3	140.00
145.00	454.2	48.40	0.0693	66.90	96.43	0.12765	0.17648	0.7465	0.8361	3.325	519.	336.8	145.00
150.00	482.0	45.80	0.0607	69.98	95.05	0.13253	0.17366	0.9864	1.1857	4.626	454.	325.2	150.00
159.34c	538.8	30.74	0.0325	83.13	83.13	0.15345	0.15345	∞	∞	∞	0.	0.0	159.34

*temperatures are on the ITS-90 scale

**Small deviations from azeotropic behavior occur at some conditions–tabulated pressures are the average of the bubble and dew point pressures.

b = normal boiling point

c = critical point

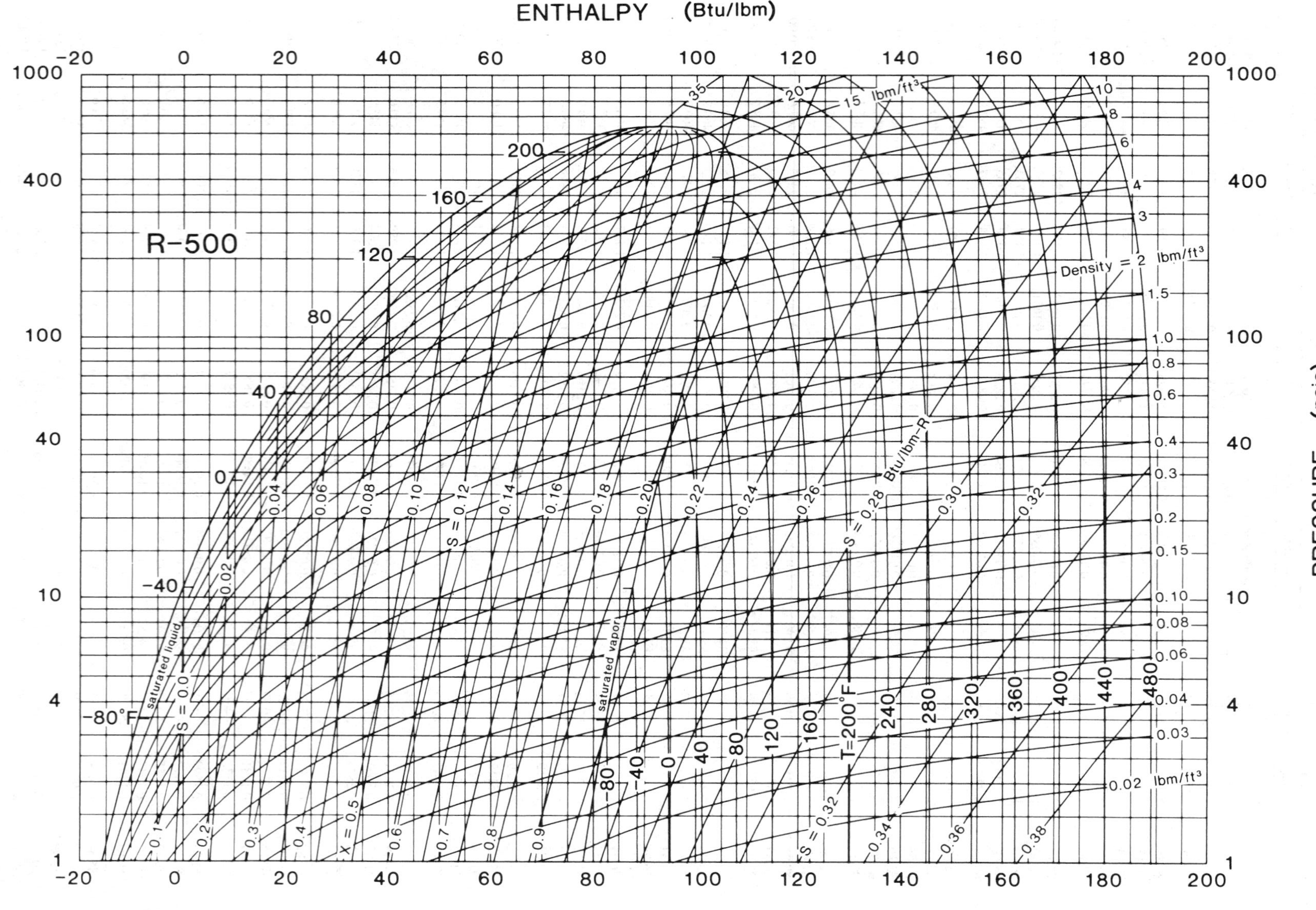

Prepared by: CENTER FOR APPLIED THERMODYNAMIC STUDIES, University of Idaho

Fig. 18 Pressure–Enthalpy Diagram for Refrigerant 500

Refrigerant 500 [R-12/152a (73.8/26.2)] Properties of Saturated Liquid and Saturated Vapor

Temp,* °F	Pressure, psia	Density, lb/ft³ Liquid	Volume, ft³/lb Vapor	Enthalpy, Btu/lb Liquid	Enthalpy, Btu/lb Vapor	Entropy, Btu/lb·°F Liquid	Entropy, Btu/lb·°F Vapor	Specific Heat c_p, Btu/lb·°F Liquid	Specific Heat c_p, Btu/lb·°F Vapor	c_p/c_v Vapor	Velocity of Sound, ft/s Liquid	Velocity of Sound, ft/s Vapor	Viscosity, lb_m/ft·h Liquid	Viscosity, lb_m/ft·h Vapor	Thermal Cond, Btu/h·ft·°F Liquid	Thermal Cond, Btu/h·ft·°F Vapor	Surface Tension, dyne/cm	Temp,* °F
−90.00	2.482	88.78	15.887	−10.763	81.207	−0.02723	0.22156	0.2451	0.1425	—	—	—	1.336	—	0.0643	—	—	−90.00
−80.00	3.456	87.90	11.676	−8.700	82.493	−0.02173	0.21846	0.2422	0.1456	—	—	—	1.231	—	0.0630	—	—	−80.00
−70.00	4.725	87.01	8.7298	−6.593	83.779	−0.01626	0.21566	0.2398	0.1486	—	—	—	1.137	—	0.0617	—	—	−70.00
−60.00	6.351	86.11	6.6307	−4.442	85.060	−0.01082	0.21312	0.2381	0.1517	—	—	—	1.053	—	0.0605	—	—	−60.00
−50.00	8.404	85.21	5.1094	−2.244	86.337	−0.00540	0.21083	0.2371	0.1549	—	—	—	0.978	—	0.0592	—	—	−50.00
−40.00	10.959	84.29	3.9895	0.000	87.604	0.00000	0.20875	0.2366	0.1582	—	—	—	0.911	—	0.0579	—	—	−40.00
−38.00	11.538	84.10	3.8025	0.455	87.857	0.00108	0.20835	0.2366	0.1589	—	—	—	0.898	—	0.0577	—	—	−38.00
−36.00	12.141	83.92	3.6260	0.911	88.109	0.00215	0.20797	0.2366	0.1595	—	—	—	0.886	—	0.0574	—	—	−36.00
−34.00	12.768	83.73	3.4594	1.369	88.360	0.00323	0.20759	0.2367	0.1602	—	—	—	0.874	—	0.0572	—	—	−34.00
−32.00	13.421	83.54	3.3019	1.829	88.612	0.00430	0.20722	0.2368	0.1609	—	—	—	0.862	—	0.0569	—	—	−32.00
−30.00	14.101	83.35	3.1530	2.291	88.862	0.00538	0.20686	0.2369	0.1616	—	—	—	0.850	—	0.0567	—	—	−30.00
−28.31b	14.696	83.19	3.0335	2.683	89.074	0.00629	0.20656	0.2370	0.1622	—	—	—	0.841	—	0.0565	—	—	−28.31
−28.00	14.807	83.17	3.0122	2.755	89.112	0.00645	0.20651	0.2370	0.1623	—	—	—	0.839	—	0.0564	—	—	−28.00
−26.00	15.541	82.98	2.8789	3.221	89.362	0.00753	0.20616	0.2371	0.1630	—	—	—	0.828	—	0.0562	—	—	−26.00
−24.00	16.304	82.79	2.7527	3.689	89.611	0.00860	0.20582	0.2373	0.1637	—	—	—	0.817	—	0.0559	—	—	−24.00
−22.00	17.096	82.60	2.6331	4.159	89.859	0.00967	0.20548	0.2375	0.1644	—	—	—	0.806	—	0.0557	—	—	−22.00
−20.00	17.918	82.41	2.5198	4.631	90.107	0.01074	0.20515	0.2378	0.1652	—	—	—	0.795	—	0.0554	—	—	−20.00
−18.00	18.771	82.22	2.4124	5.104	90.355	0.01181	0.20483	0.2380	0.1659	—	—	—	0.785	—	0.0552	—	—	−18.00
−16.00	19.656	82.02	2.3104	5.580	90.602	0.01288	0.20451	0.2383	0.1667	—	—	—	0.775	—	0.0550	—	—	−16.00
−14.00	20.573	81.83	2.2137	6.057	90.848	0.01395	0.20420	0.2386	0.1674	—	—	—	0.765	—	0.0547	—	—	−14.00
−12.00	21.523	81.64	2.1218	6.537	91.093	0.01502	0.20390	0.2390	0.1682	—	—	—	0.755	—	0.0545	—	—	−12.00
−10.00	22.507	81.44	2.0345	7.018	91.338	0.01609	0.20360	0.2393	0.1689	—	—	—	0.746	—	0.0542	—	—	−10.00
−8.00	23.526	81.25	1.9515	7.501	91.583	0.01716	0.20331	0.2397	0.1697	—	—	—	0.736	—	0.0540	—	—	−8.00
−6.00	24.581	81.06	1.8726	7.987	91.826	0.01822	0.20302	0.2402	0.1705	—	—	—	0.727	—	0.0537	—	—	−6.00
−4.00	25.672	80.86	1.7975	8.474	92.069	0.01929	0.20274	0.2406	0.1713	—	—	—	0.718	—	0.0535	—	—	−4.00
−2.00	26.801	80.66	1.7261	8.963	92.311	0.02035	0.20247	0.2411	0.1721	—	—	—	0.709	—	0.0532	—	—	−2.00
0.00	27.968	80.47	1.6581	9.454	92.553	0.02142	0.20220	0.2416	0.1730	—	—	—	0.701	—	0.0530	—	—	0.00
2.00	29.174	80.27	1.5933	9.947	92.794	0.02248	0.20193	0.2422	0.1738	—	—	—	0.692	—	0.0527	—	—	2.00
4.00	30.420	80.07	1.5316	10.443	93.034	0.02355	0.20167	0.2427	0.1746	—	—	—	0.684	—	0.0525	—	—	4.00
6.00	31.707	79.87	1.4727	10.940	93.273	0.02461	0.20142	0.2433	0.1755	—	—	—	0.676	—	0.0522	—	—	6.00
8.00	33.036	79.67	1.4166	11.439	93.511	0.02567	0.20116	0.2439	0.1764	—	—	—	0.667	—	0.0520	—	—	8.00
10.00	34.408	79.47	1.3631	11.940	93.749	0.02673	0.20092	0.2446	0.1773	—	—	—	0.660	—	0.0517	—	—	10.00
12.00	35.823	79.27	1.3120	12.443	93.986	0.02780	0.20068	0.2453	0.1782	—	—	—	0.652	—	0.0515	—	—	12.00
14.00	37.282	79.07	1.2632	12.948	94.221	0.02886	0.20044	0.2460	0.1791	—	—	—	0.644	—	0.0512	—	—	14.00
16.00	38.787	78.86	1.2165	13.455	94.457	0.02992	0.20021	0.2467	0.1800	—	—	—	0.637	—	0.0510	—	—	16.00
18.00	40.338	78.66	1.1720	13.964	94.691	0.03098	0.19998	0.2475	0.1809	—	—	—	0.629	—	0.0507	—	—	18.00
20.00	41.936	78.45	1.1294	14.475	94.924	0.03204	0.19976	0.2483	0.1819	—	—	—	0.622	—	0.0505	—	—	20.00
22.00	43.583	78.25	1.0887	14.988	95.157	0.03310	0.19954	0.2491	0.1828	—	—	—	0.615	—	0.0503	—	—	22.00
24.00	45.278	78.04	1.0498	15.503	95.388	0.03416	0.19932	0.2500	0.1838	—	—	—	0.608	—	0.0500	—	—	24.00
26.00	47.024	77.83	1.0125	16.020	95.619	0.03522	0.19911	0.2509	0.1848	—	—	—	0.601	—	0.0498	—	—	26.00
28.00	48.820	77.62	0.9768	16.539	95.848	0.03627	0.19890	0.2518	0.1858	—	—	—	0.595	—	0.0495	—	—	28.00
30.00	50.669	77.42	0.9426	17.060	96.077	0.03733	0.19870	0.2528	0.1868	—	—	—	0.588	—	0.0493	—	—	30.00
32.00	52.570	77.20	0.9099	17.583	96.304	0.03839	0.19850	0.2537	0.1878	—	—	—	0.581	—	0.0490	—	—	32.00
34.00	54.525	76.99	0.8785	18.109	96.531	0.03944	0.19830	0.2547	0.1889	—	—	—	0.575	—	0.0488	—	—	34.00
36.00	56.535	76.78	0.8485	18.636	96.757	0.04050	0.19811	0.2558	0.1900	—	—	—	0.569	—	0.0485	—	—	36.00
38.00	58.600	76.57	0.8196	19.165	96.981	0.04156	0.19792	0.2568	0.1910	—	—	—	0.563	—	0.0483	—	—	38.00
40.00	60.722	76.35	0.7920	19.697	97.204	0.04261	0.19773	0.2579	0.1921	—	—	—	0.557	—	0.0480	—	—	40.00
45.00	66.283	75.81	0.7276	21.034	97.758	0.04525	0.19727	0.2608	0.1950	—	—	—	0.542	—	0.0474	—	—	45.00
50.00	72.219	75.26	0.6694	22.385	98.304	0.04788	0.19684	0.2639	0.1979	—	—	—	0.528	—	0.0468	—	—	50.00
55.00	78.548	74.71	0.6168	23.749	98.843	0.05051	0.19642	0.2672	0.2009	—	—	—	0.514	—	0.0462	—	—	55.00
60.00	85.283	74.15	0.5690	25.127	99.374	0.05315	0.19602	0.2706	0.2041	—	—	—	0.501	—	0.0455	—	—	60.00
65.00	92.443	73.58	0.5255	26.519	99.895	0.05578	0.19563	0.2743	0.2074	—	—	—	0.489	—	0.0449	—	—	65.00
70.00	100.04	73.00	0.4860	27.924	100.41	0.05841	0.19526	0.2782	0.2108	—	—	—	0.477	—	0.0443	—	—	70.00
75.00	108.10	72.41	0.4499	29.344	100.91	0.06103	0.19489	0.2822	0.2143	—	—	—	0.465	—	0.0436	—	—	75.00
80.00	116.62	71.81	0.4169	30.777	101.41	0.06366	0.19454	0.2865	0.2180	—	—	—	0.454	—	0.0430	—	—	80.00
85.00	125.64	71.20	0.3867	32.225	101.89	0.06629	0.19419	0.2909	0.2218	—	—	—	0.444	—	0.0424	—	—	85.00
90.00	135.16	70.58	0.3591	33.688	102.36	0.06892	0.19385	0.2956	0.2257	—	—	—	0.433	—	0.0417	—	—	90.00
100.00	155.79	69.30	0.3103	36.661	103.27	0.07418	0.19319	0.3056	0.2341	—	—	—	0.414	—	0.0404	—	—	100.00
110.00	178.65	67.96	0.2689	39.699	104.12	0.07945	0.19253	0.3164	0.2431	—	—	—	0.396	—	0.0391	—	—	110.00
120.00	203.89	66.56	0.2335	42.807	104.90	0.08474	0.19186	0.3281	0.2527	—	—	—	0.379	—	0.0378	—	—	120.00
130.00	231.65	65.09	0.2031	45.992	105.61	0.09005	0.19116	0.3406	0.2630	—	—	—	0.364	—	0.0365	—	—	130.00
140.00	262.10	63.53	0.1768	49.263	106.23	0.09540	0.19040	0.3540	0.2741	—	—	—	—	—	0.0352	—	—	140.00
150.00	295.41	61.85	0.1539	52.633	106.75	0.10080	0.18956	—	—	—	—	—	—	—	—	—	—	150.00
160.00	331.75	60.04	0.1338	56.119	107.12	0.10629	0.18859	—	—	—	—	—	—	—	—	—	—	160.00
170.00	371.35	58.06	0.1160	59.752	107.32	0.11189	0.18744	—	—	—	—	—	—	—	—	—	—	170.00
180.00	414.45	55.84	0.1000	63.577	107.29	0.11768	0.18602	—	—	—	—	—	—	—	—	—	—	180.00
190.00	461.34	53.28	0.0855	67.676	106.94	0.12376	0.18419	—	—	—	—	—	—	—	—	—	—	190.00
200.00	512.40	50.19	0.0719	72.219	106.07	0.13036	0.18168	—	—	—	—	—	—	—	—	—	—	200.00
210.00	568.16	46.10	0.0584	77.651	104.25	0.13809	0.17781	—	—	—	—	—	—	—	—	—	—	210.00
220.00	629.50	38.52	0.0416	86.497	99.24	0.15030	0.16905	—	—	—	—	—	—	—	—	—	—	220.00
222.00c	641.9	31.10	0.0321	94.40	94.40	0.1608	0.1608	∞	∞	∞	0	0	—	—	∞	∞	0	222.00

*temperatures are on the IPTS–68 scale

b = normal boiling point

c = critical point

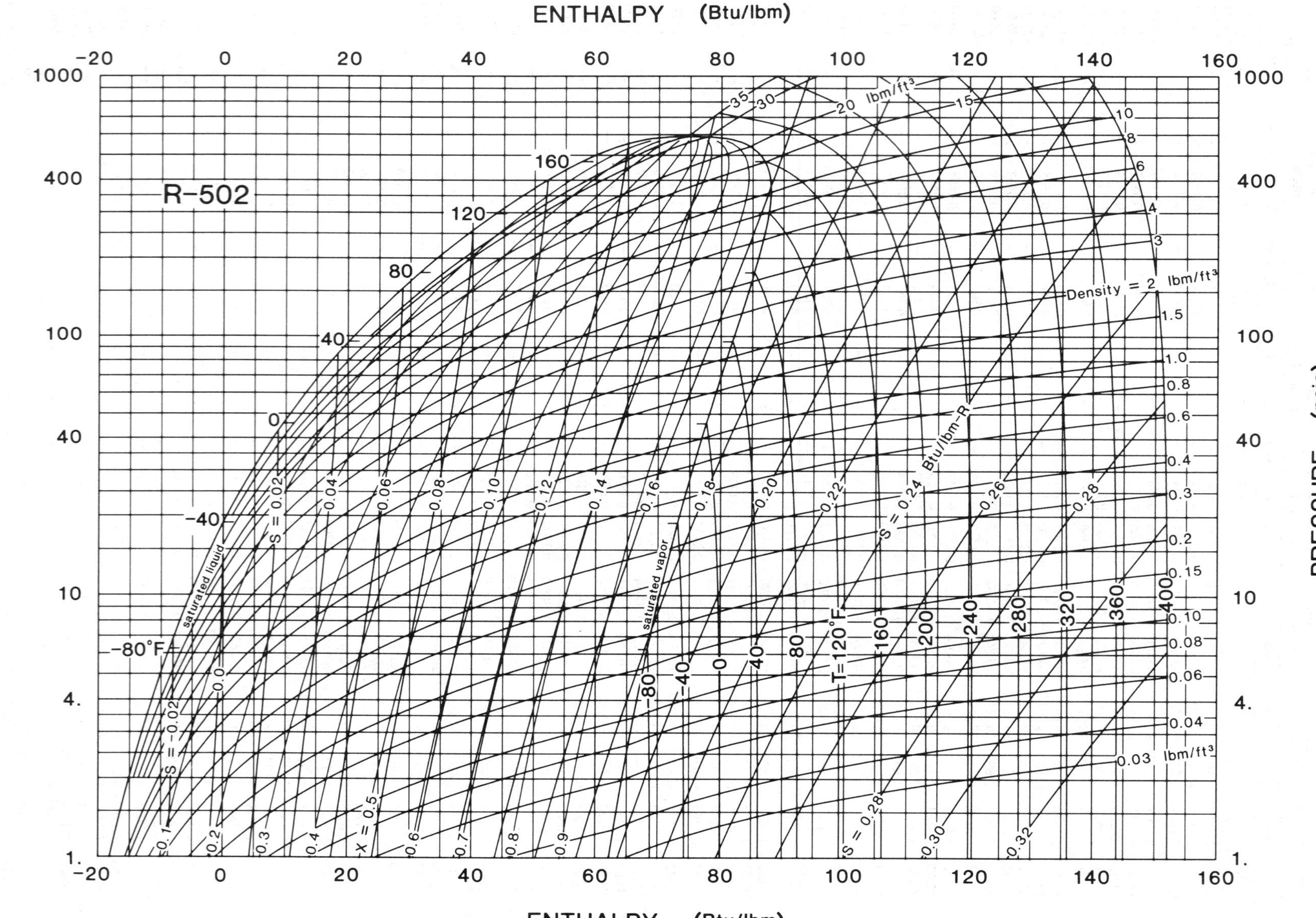

Fig. 19 Pressure-Enthalpy Diagram for Refrigerant 502

Refrigerant 502 [R-22/115 (48.8/51.2)] Properties of Saturated Liquid and Saturated Vapor

Temp,* °F	Pressure, psia	Density, lb/ft³ Liquid	Volume, ft³/lb Vapor	Enthalpy, Btu/lb Liquid	Enthalpy, Btu/lb Vapor	Entropy, Btu/lb·°F Liquid	Entropy, Btu/lb·°F Vapor	Specific Heat c_p, Btu/lb·°F Liquid	Specific Heat c_p, Btu/lb·°F Vapor	c_p/c_v Vapor	Velocity of Sound, ft/s Liquid	Velocity of Sound, ft/s Vapor	Viscosity, lb_m/ft·h Liquid	Viscosity, lb_m/ft·h Vapor	Thermal Cond, Btu/h·ft·°F Liquid	Thermal Cond, Btu/h·ft·°F Vapor	Surface Tension, dyne/cm	Temp,* °F
−70.00	8.434	94.70	4.3241	−6.564	69.570	−0.01617	0.17921	0.2506	0.1387	—	—	—	1.081	—	0.0558	—	16.89	−70.00
−65.00	9.731	94.16	3.7849	−5.511	70.174	−0.01349	0.17828	0.2518	0.1399	—	—	—	1.040	—	0.0552	—	16.46	−65.00
−60.00	11.182	93.61	3.3248	−4.441	70.777	−0.01080	0.17740	0.2532	0.1410	—	—	—	1.002	—	0.0545	—	16.03	−60.00
−55.00	12.802	93.07	2.9306	−3.355	71.377	−0.00811	0.17656	0.2545	0.1422	—	—	—	0.965	—	0.0539	—	15.61	−55.00
−50.00	14.602	92.51	2.5915	−2.252	71.975	−0.00541	0.17578	0.2559	0.1433	—	—	—	0.930	—	0.0533	—	15.19	−50.00
−49.75b	14.696	92.49	2.5761	−2.198	72.004	−0.00528	0.17574	0.2559	0.1433	—	—	—	0.929	—	0.0533	—	15.16	−49.75
−48.00	15.376	92.29	2.4693	−1.807	72.213	−0.00433	0.17547	0.2564	0.1437	—	—	—	0.917	—	0.0530	—	15.02	−48.00
−46.00	16.183	92.07	2.3541	−1.359	72.451	−0.00325	0.17518	0.2570	0.1442	—	—	—	0.904	—	0.0528	—	14.85	−46.00
−44.00	17.022	91.84	2.2452	−0.909	72.689	−0.00217	0.17489	0.2575	0.1446	—	—	—	0.891	—	0.0525	—	14.68	−44.00
−42.00	17.895	91.62	2.1425	−0.456	72.925	−0.00108	0.17461	0.2581	0.1451	—	—	—	0.879	—	0.0523	—	14.51	−42.00
−40.00	18.802	91.39	2.0453	0.000	73.162	0.00000	0.17433	0.2587	0.1456	—	—	—	0.866	—	0.0520	0.00411	14.35	−40.00
−38.00	19.746	91.17	1.9535	0.458	73.398	0.00109	0.17406	0.2592	0.1460	—	—	—	0.854	—	0.0518	0.00415	14.18	−38.00
−36.00	20.726	90.94	1.8666	0.919	73.633	0.00217	0.17380	0.2598	0.1465	—	—	—	0.843	—	0.0516	0.00420	14.01	−36.00
−34.00	21.744	90.71	1.7844	1.382	73.867	0.00326	0.17354	0.2604	0.1470	—	—	—	0.831	—	0.0513	0.00424	13.85	−34.00
−32.00	22.801	90.48	1.7066	1.848	74.101	0.00434	0.17329	0.2610	0.1474	—	—	—	0.820	—	0.0511	0.00429	13.68	−32.00
−30.00	23.897	90.25	1.6328	2.317	74.335	0.00543	0.17304	0.2616	0.1479	—	—	—	0.809	—	0.0508	0.00433	13.52	−30.00
−28.00	25.034	90.02	1.5629	2.787	74.567	0.00652	0.17280	0.2622	0.1484	—	—	—	0.798	—	0.0506	0.00438	13.35	−28.00
−26.00	26.212	89.79	1.4966	3.261	74.799	0.00761	0.17257	0.2627	0.1489	—	—	—	0.787	—	0.0503	0.00442	13.19	−26.00
−24.00	27.433	89.56	1.4336	3.737	75.031	0.00870	0.17234	0.2633	0.1494	—	—	—	0.777	—	0.0501	0.00447	13.03	−24.00
−22.00	28.697	89.32	1.3739	4.215	75.261	0.00979	0.17211	0.2639	0.1499	—	—	—	0.767	—	0.0498	0.00451	12.86	−22.00
−20.00	30.006	89.09	1.3172	4.696	75.491	0.01088	0.17189	0.2646	0.1504	—	—	—	0.757	—	0.0496	0.00455	12.70	−20.00
−18.00	31.361	88.85	1.2633	5.179	75.720	0.01197	0.17168	0.2652	0.1510	—	—	—	0.747	—	0.0494	0.00460	12.54	−18.00
−16.00	32.762	88.62	1.2120	5.665	75.948	0.01306	0.17147	0.2658	0.1515	—	—	—	0.737	—	0.0491	0.00464	12.38	−16.00
−14.00	34.211	88.38	1.1633	6.154	76.175	0.01415	0.17126	0.2664	0.1521	—	—	—	0.728	—	0.0489	0.00468	12.22	−14.00
−12.00	35.709	88.14	1.1169	6.644	76.402	0.01524	0.17106	0.2670	0.1526	—	—	—	0.719	—	0.0486	0.00473	12.05	−12.00
−10.00	37.256	87.90	1.0727	7.138	76.627	0.01633	0.17087	0.2676	0.1532	—	—	—	0.710	—	0.0484	0.00477	11.89	−10.00
−8.00	38.854	87.66	1.0307	7.633	76.852	0.01742	0.17067	0.2683	0.1538	—	—	—	0.701	—	0.0481	0.00481	11.73	−8.00
−6.00	40.504	87.42	0.9907	8.131	77.075	0.01852	0.17049	0.2689	0.1544	—	—	—	0.692	—	0.0479	0.00485	11.58	−6.00
−4.00	42.207	87.17	0.9525	8.632	77.298	0.01961	0.17030	0.2695	0.1550	—	—	—	0.684	—	0.0476	0.00490	11.42	−4.00
−2.00	43.964	86.93	0.9161	9.135	77.520	0.02070	0.17012	0.2702	0.1557	—	—	—	0.675	—	0.0474	0.00494	11.26	−2.00
0.00	45.776	86.68	0.8813	9.640	77.741	0.02180	0.16995	0.2708	0.1563	—	—	—	0.667	—	0.0472	0.00498	11.10	0.00
2.00	47.644	86.44	0.8482	10.147	77.960	0.02289	0.16978	0.2715	0.1570	—	—	—	0.659	—	0.0469	0.00502	10.94	2.00
4.00	49.569	86.19	0.8166	10.657	78.179	0.02398	0.16961	0.2721	0.1577	—	—	—	0.651	—	0.0467	0.00507	10.79	4.00
6.00	51.552	85.94	0.7864	11.169	78.397	0.02508	0.16944	0.2728	0.1584	—	—	—	0.643	—	0.0464	0.00511	10.63	6.00
8.00	53.594	85.69	0.7575	11.684	78.613	0.02617	0.16928	0.2734	0.1591	—	—	—	0.636	—	0.0462	0.00515	10.47	8.00
10.00	55.697	85.43	0.7299	12.200	78.828	0.02726	0.16912	0.2741	0.1598	—	—	—	0.628	—	0.0459	0.00519	10.32	10.00
15.00	61.226	84.80	0.6661	13.502	79.362	0.02999	0.16874	0.2758	0.1618	—	—	—	0.610	—	0.0453	0.00530	9.93	15.00
20.00	67.155	84.15	0.6088	14.818	79.887	0.03272	0.16838	0.2775	0.1638	—	—	—	0.593	—	0.0447	0.00541	9.55	20.00
25.00	73.503	83.50	0.5575	16.147	80.405	0.03545	0.16803	0.2792	0.1660	—	—	—	0.576	—	0.0441	0.00552	9.17	25.00
30.00	80.287	82.83	0.5112	17.490	80.913	0.03818	0.16770	0.2809	0.1684	—	—	—	0.561	—	0.0435	0.00562	8.79	30.00
35.00	87.523	82.16	0.4695	18.846	81.413	0.04090	0.16738	0.2827	0.1709	—	—	—	0.545	0.0285	0.0429	0.00573	8.42	35.00
40.00	95.229	81.47	0.4318	20.216	81.903	0.04362	0.16707	0.2845	0.1735	—	—	—	0.531	0.0288	0.0423	0.00584	8.05	40.00
45.00	103.42	80.77	0.3976	21.597	82.383	0.04633	0.16678	0.2863	0.1764	—	—	—	0.517	0.0292	0.0417	0.00595	7.68	45.00
50.00	112.12	80.06	0.3666	22.991	82.852	0.04904	0.16649	0.2882	0.1794	—	—	—	—	0.0296	0.0411	0.00607	—	50.00
55.00	121.34	79.33	0.3383	24.397	83.310	0.05174	0.16621	0.2900	0.1827	—	—	—	—	0.0300	0.0404	0.00618	—	55.00
60.00	131.10	78.59	0.3126	25.814	83.755	0.05444	0.16594	0.2919	0.1861	—	—	—	—	0.0304	0.0398	0.00629	—	60.00
65.00	141.42	77.83	0.2892	27.244	84.187	0.05713	0.16566	0.2938	0.1898	—	—	—	—	0.0308	0.0392	0.00641	—	65.00
70.00	152.32	77.06	0.2677	28.685	84.606	0.05982	0.16539	0.2958	0.1937	—	—	—	—	0.0312	—	0.00653	—	70.00
75.00	163.81	76.27	0.2480	30.138	85.009	0.06249	0.16512	0.2977	0.1978	—	—	—	—	0.0316	—	0.00665	—	75.00
80.00	175.92	75.46	0.2299	31.602	85.397	0.06517	0.16484	0.2997	0.2022	—	—	—	—	0.0320	—	0.00678	—	80.00
85.00	188.66	74.62	0.2133	33.078	85.767	0.06783	0.16456	0.3017	0.2068	—	—	—	—	0.0324	—	0.00690	—	85.00
90.00	202.06	73.77	0.1980	34.566	86.118	0.07049	0.16427	0.3037	0.2117	—	—	—	—	0.0328	—	0.00703	—	90.00
95.00	216.13	72.88	0.1839	36.066	86.449	0.07314	0.16397	0.3058	0.2168	—	—	—	—	0.0332	—	0.00716	—	95.00
100.00	230.89	71.97	0.1708	37.578	86.758	0.07579	0.16366	0.3079	0.2223	—	—	—	—	0.0336	—	0.00730	—	100.00
105.00	246.38	71.02	0.1587	39.104	87.042	0.07843	0.16332	—	—	—	—	—	—	—	—	—	—	105.00
110.00	262.61	70.04	0.1474	40.644	87.298	0.08107	0.16297	—	—	—	—	—	—	—	—	—	—	110.00
115.00	279.61	69.02	0.1369	42.201	87.524	0.08371	0.16258	—	—	—	—	—	—	—	—	—	—	115.00
120.00	297.41	67.96	0.1271	43.774	87.716	0.08635	0.16216	—	—	—	—	—	—	—	—	—	—	120.00
125.00	316.05	66.84	0.1179	45.369	87.869	0.08901	0.16170	—	—	—	—	—	—	—	—	—	—	125.00
130.00	335.54	65.66	0.1094	46.987	87.977	0.09167	0.16118	—	—	—	—	—	—	—	—	—	—	130.00
135.00	355.94	64.41	0.1013	48.634	88.032	0.09435	0.16060	—	—	—	—	—	—	—	—	—	—	135.00
140.00	377.30	63.08	0.0936	50.316	88.024	0.09706	0.15994	—	—	—	—	—	—	—	—	—	—	140.00
145.00	399.65	61.65	0.0863	52.045	87.941	0.09982	0.15919	—	—	—	—	—	—	—	—	—	—	145.00
150.00	423.06	60.09	0.0793	53.834	87.763	0.10265	0.15830	—	—	—	—	—	—	—	—	—	—	150.00
155.00	447.61	58.37	0.0726	55.705	87.463	0.10558	0.15724	—	—	—	—	—	—	—	—	—	—	155.00
160.00	473.39	56.43	0.0660	57.698	86.997	0.10867	0.15595	—	—	—	—	—	—	—	—	—	—	160.00
165.00	500.50	54.17	0.0595	59.878	86.293	0.11202	0.15430	—	—	—	—	—	—	—	—	—	—	165.00
170.00	529.11	51.39	0.0527	62.386	85.198	0.11584	0.15207	—	—	—	—	—	—	—	—	—	—	170.00
175.00	559.42	47.55	0.0451	65.610	83.303	0.12073	0.14861	—	—	—	—	—	—	—	—	—	—	175.00
179.90c	591.0	35.00	0.0286	74.81	74.81	0.1346	0.1346	∞	∞	∞	0	0	—	—	∞	∞	0	179.90

*temperatures are on the IPTS–68 scale

b = normal boiling point

c = critical point

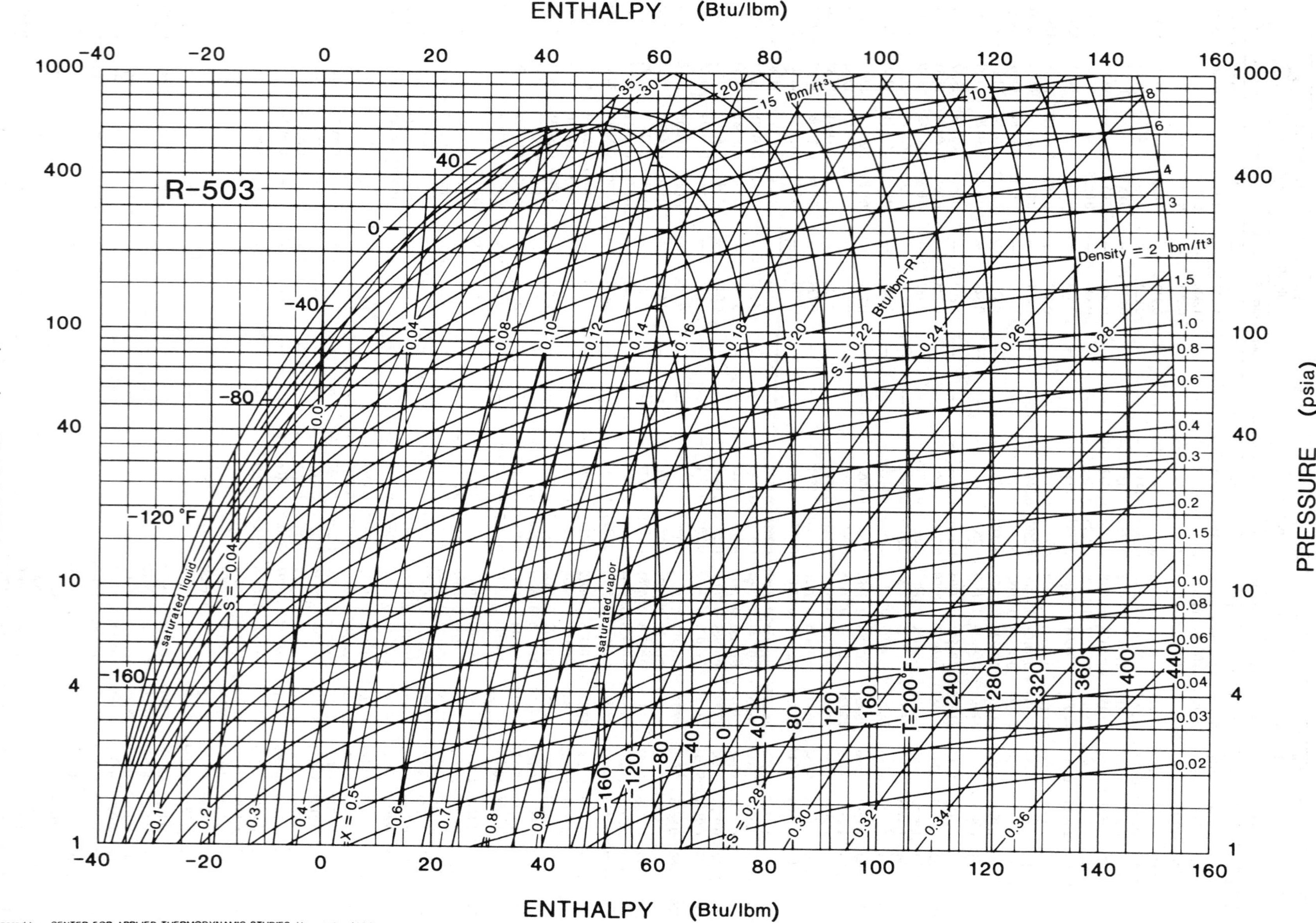

Prepared by CENTER FOR APPLIED THERMODYNAMIC STUDIES, University of Idaho

Fig. 20 Pressure–Enthalpy Diagram for Refrigerant 503

Refrigerant 503 [R-23/13 (40.1/59.9)] Properties of Saturated Liquid and Saturated Vapor

Temp,* °F	Pressure, psia	Density, lb/ft³ Liquid	Volume, ft³/lb Vapor	Enthalpy, Btu/lb Liquid	Enthalpy, Btu/lb Vapor	Entropy, Btu/lb·°F Liquid	Entropy, Btu/lb·°F Vapor	Specific Heat c_p, Btu/lb·°F Liquid	Specific Heat c_p, Btu/lb·°F Vapor	c_p/c_v Vapor	Velocity of Sound, ft/s Liquid	Velocity of Sound, ft/s Vapor	Viscosity, lb_m/ft·h Liquid	Viscosity, lb_m/ft·h Vapor	Thermal Cond, Btu/h·ft·°F Liquid	Thermal Cond, Btu/h·ft·°F Vapor	Surface Tension, dyne/cm	Temp,* °F
−190.00	1.069	97.81	30.827	−38.625	47.727	−0.11247	0.20774	—	—	—	—	—	—	—	—	—	—	−190.00
−180.00	1.768	97.04	19.265	−36.283	48.778	−0.10395	0.20020	0.1241	—	—	—	—	—	—	0.0724	—	—	−180.00
−170.00	2.809	96.23	12.506	−33.900	49.821	−0.09559	0.19344	0.1341	—	—	—	—	—	—	0.0705	—	—	−170.00
−160.00	4.307	95.37	8.3963	−31.475	50.851	−0.08737	0.18736	0.1435	0.1203	—	—	—	—	—	0.0686	—	—	−160.00
−150.00	6.396	94.46	5.8079	−29.011	51.864	−0.07929	0.18187	0.1524	0.1225	—	—	—	—	—	0.0668	—	—	−150.00
−140.00	9.227	93.49	4.1253	−26.510	52.854	−0.07136	0.17691	0.1611	0.1247	—	—	—	0.871	—	0.0649	—	—	−140.00
−138.00	9.897	93.29	3.8638	−26.006	53.050	−0.06979	0.17597	0.1628	0.1251	—	—	—	0.857	—	0.0645	—	—	−138.00
−136.00	10.605	93.09	3.6222	−25.500	53.244	−0.06823	0.17505	0.1645	0.1256	—	—	—	0.842	—	0.0642	—	—	−136.00
−134.00	11.353	92.88	3.3987	−24.993	53.437	−0.06667	0.17415	0.1662	0.1260	—	—	—	0.828	—	0.0638	—	—	−134.00
−132.00	12.141	92.68	3.1918	−24.484	53.629	−0.06512	0.17327	0.1679	0.1265	—	—	—	0.814	—	0.0634	—	—	−132.00
−130.00	12.972	92.47	3.0001	−23.975	53.819	−0.06358	0.17240	0.1697	0.1269	—	—	—	0.801	—	0.0630	—	—	−130.00
−128.00	13.847	92.26	2.8223	−23.464	54.009	−0.06204	0.17155	0.1714	0.1274	—	—	—	0.787	—	0.0627	—	—	−128.00
−126.15b	14.696	92.06	2.6693	−22.991	54.183	−0.06062	0.17077	0.1730	0.1278	—	—	—	0.775	—	0.0623	—	—	−126.15
−126.00	14.767	92.04	2.6572	−22.952	54.197	−0.06050	0.17071	0.1731	0.1279	—	—	—	0.774	—	0.0623	—	—	−126.00
−124.00	15.735	91.83	2.5037	−22.439	54.385	−0.05897	0.16989	0.1748	0.1284	—	—	—	0.761	—	0.0619	—	—	−124.00
−122.00	16.751	91.61	2.3610	−21.924	54.570	−0.05745	0.16908	0.1765	0.1289	—	—	—	0.749	—	0.0615	—	—	−122.00
−120.00	17.818	91.39	2.2281	−21.409	54.755	−0.05594	0.16829	0.1783	0.1293	—	—	—	0.737	—	0.0612	—	—	−120.00
−118.00	18.936	91.17	2.1042	−20.892	54.938	−0.05443	0.16751	0.1800	0.1298	—	—	—	0.724	—	0.0608	—	—	−118.00
−116.00	20.109	90.94	1.9887	−20.375	55.120	−0.05292	0.16675	0.1818	0.1303	—	—	—	0.713	—	0.0604	—	—	−116.00
−114.00	21.336	90.71	1.8809	−19.856	55.301	−0.05143	0.16600	0.1836	0.1309	—	—	—	0.701	—	0.0601	—	—	−114.00
−112.00	22.621	90.48	1.7801	−19.336	55.480	−0.04993	0.16526	0.1854	0.1314	—	—	—	0.689	—	0.0597	—	—	−112.00
−110.00	23.965	90.25	1.6859	−18.815	55.658	−0.04845	0.16453	0.1872	0.1319	—	—	—	0.678	—	0.0593	—	—	−110.00
−108.00	25.369	90.01	1.5977	−18.293	55.834	−0.04697	0.16382	0.1890	0.1325	—	—	—	0.667	—	0.0590	—	—	−108.00
−106.00	26.835	89.78	1.5152	−17.771	56.009	−0.04550	0.16312	0.1909	0.1330	—	—	—	0.657	—	0.0586	—	—	−106.00
−104.00	28.365	89.53	1.4378	−17.247	56.183	−0.04403	0.16243	0.1927	0.1336	—	—	—	0.646	—	0.0582	—	—	−104.00
−102.00	29.962	89.29	1.3652	−16.722	56.354	−0.04256	0.16175	0.1946	0.1341	—	—	—	0.636	—	0.0578	—	—	−102.00
−100.00	31.626	89.05	1.2971	−16.197	56.525	−0.04111	0.16108	0.1966	0.1347	—	—	—	0.626	—	0.0575	—	—	−100.00
−98.00	33.359	88.80	1.2331	−15.670	56.694	−0.03966	0.16042	0.1985	0.1353	—	—	—	0.616	—	0.0571	—	—	−98.00
−96.00	35.164	88.55	1.1729	−15.143	56.861	−0.03822	0.15978	0.2005	0.1359	—	—	—	0.606	—	0.0567	—	—	−96.00
−94.00	37.041	88.29	1.1163	−14.614	57.026	−0.03678	0.15914	0.2024	0.1365	—	—	—	0.596	—	0.0564	—	—	−94.00
−92.00	38.994	88.03	1.0630	−14.085	57.190	−0.03535	0.15851	0.2045	0.1371	—	—	—	0.587	—	0.0560	—	—	−92.00
−90.00	41.024	87.78	1.0128	−13.555	57.352	−0.03392	0.15789	0.2065	0.1378	—	—	—	0.578	—	0.0556	—	—	−90.00
−88.00	43.133	87.51	0.9654	−13.024	57.513	−0.03250	0.15728	0.2086	0.1384	—	—	—	0.569	—	0.0553	—	—	−88.00
−86.00	45.322	87.25	0.9208	−12.493	57.671	−0.03109	0.15668	0.2107	0.1391	—	—	—	0.560	—	0.0549	—	—	−86.00
−84.00	47.594	86.98	0.8786	−11.960	57.828	−0.02968	0.15609	0.2129	0.1397	—	—	—	0.551	—	0.0545	—	—	−84.00
−82.00	49.950	86.71	0.8388	−11.427	57.984	−0.02827	0.15551	0.2151	0.1404	—	—	—	0.543	—	0.0542	—	—	−82.00
−80.00	52.393	86.43	0.8012	−10.892	58.137	−0.02688	0.15494	0.2173	0.1411	—	—	—	0.535	—	0.0538	—	—	−80.00
−78.00	54.925	86.16	0.7656	−10.357	58.288	−0.02549	0.15437	0.2196	0.1419	—	—	—	0.526	—	0.0534	—	—	−78.00
−76.00	57.546	85.87	0.7320	−9.821	58.438	−0.02410	0.15381	0.2219	0.1426	—	—	—	0.518	—	0.0531	—	11.24	−76.00
−74.00	60.261	85.59	0.7001	−9.284	58.585	−0.02272	0.15326	0.2242	0.1433	—	—	—	0.511	—	0.0527	—	11.04	−74.00
−72.00	63.069	85.30	0.6699	−8.747	58.731	−0.02134	0.15271	0.2266	0.1441	—	—	—	0.503	—	0.0523	—	10.84	−72.00
−70.00	65.974	85.01	0.6412	−8.208	58.874	−0.01997	0.15218	0.2291	0.1449	—	—	—	0.495	—	0.0519	—	10.63	−70.00
−65.00	73.672	84.28	0.5759	−6.857	59.224	−0.01657	0.15086	0.2354	0.1469	—	—	—	0.477	—	0.0510	—	10.13	−65.00
−60.00	82.017	83.51	0.5184	−5.500	59.560	−0.01320	0.14958	0.2420	0.1490	—	—	—	0.460	—	0.0501	—	9.64	−60.00
−55.00	91.041	82.73	0.4677	−4.138	59.882	−0.00987	0.14834	0.2489	0.1513	—	—	—	0.443	—	0.0492	—	9.15	−55.00
−50.00	100.78	81.93	0.4228	−2.766	60.188	−0.00655	0.14712	0.2562	0.1536	—	—	—	0.428	—	0.0483	—	8.67	−50.00
−45.00	111.26	81.10	0.3830	−1.387	60.478	−0.00326	0.14593	0.2639	0.1561	—	—	—	0.413	—	0.0473	—	8.19	−45.00
−40.00	122.53	80.24	0.3475	0.000	60.751	0.00000	0.14476	0.2720	0.1588	—	—	—	0.398	—	0.0464	—	7.72	−40.00
−35.00	134.61	79.36	0.3158	1.398	61.005	0.00325	0.14361	0.2805	0.1615	—	—	—	0.385	—	0.0455	—	7.25	−35.00
−30.00	147.54	78.45	0.2873	2.808	61.238	0.00647	0.14246	0.2895	0.1644	—	—	—	0.372	—	0.0445	—	6.80	−30.00
−25.00	161.36	77.52	0.2618	4.231	61.450	0.00969	0.14133	0.2989	0.1675	—	—	—	0.360	—	0.0436	—	6.35	−25.00
−20.00	176.11	76.55	0.2388	5.670	61.638	0.01290	0.14020	0.3088	0.1707	—	—	—	0.348	—	0.0427	—	5.90	−20.00
−15.00	191.83	75.55	0.2180	7.126	61.799	0.01611	0.13906	0.3192	0.1741	—	—	—	0.336	—	0.0417	—	5.47	−15.00
−10.00	208.55	74.51	0.1992	8.601	61.932	0.01931	0.13791	0.3302	0.1776	—	—	—	0.326	—	0.0408	—	5.04	−10.00
−5.00	226.32	73.43	0.1820	10.101	62.032	0.02253	0.13674	0.3417	0.1814	—	—	—	0.315	—	0.0399	—	—	−5.00
0.00	245.19	72.31	0.1664	11.626	62.097	0.02576	0.13555	—	—	—	—	—	—	—	—	—	—	0.00
5.00	265.19	71.15	0.1522	13.183	62.121	0.02901	0.13433	—	—	—	—	—	—	—	—	—	—	5.00
10.00	286.37	69.94	0.1391	14.776	62.100	0.03230	0.13306	—	—	—	—	—	—	—	—	—	—	10.00
15.00	308.79	68.67	0.1271	16.411	62.026	0.03563	0.13173	—	—	—	—	—	—	—	—	—	—	15.00
20.00	332.49	67.33	0.1161	18.095	61.892	0.03902	0.13032	—	—	—	—	—	—	—	—	—	—	20.00
25.00	357.53	65.92	0.1058	19.838	61.687	0.04248	0.12882	—	—	—	—	—	—	—	—	—	—	25.00
30.00	383.97	64.43	0.0963	21.649	61.398	0.04603	0.12720	—	—	—	—	—	—	—	—	—	—	30.00
35.00	411.85	62.83	0.0873	23.544	61.006	0.04970	0.12543	—	—	—	—	—	—	—	—	—	—	35.00
40.00	441.25	61.10	0.0789	25.543	60.487	0.05352	0.12345	—	—	—	—	—	—	—	—	—	—	40.00
45.00	472.23	59.20	0.0710	27.673	59.804	0.05754	0.12120	—	—	—	—	—	—	—	—	—	—	45.00
50.00	504.86	57.05	0.0633	29.977	58.900	0.06182	0.11857	—	—	—	—	—	—	—	—	—	—	50.00
55.00	539.21	54.55	0.0558	32.533	57.678	0.06651	0.11537	—	—	—	—	—	—	—	—	—	—	55.00
60.00	575.35	51.35	0.0481	35.513	55.945	0.07188	0.11120	—	—	—	—	—	—	—	—	—	—	60.00
65.00	613.38	46.08	0.0397	39.596	53.255	0.07909	0.10513	—	—	—	—	—	—	—	—	—	—	65.00
67.00c	631.90	35.21	0.0284	47.41	47.41	0.0924	0.0924	∞	∞	∞	0	0	—	—	∞	∞	0	67.00

*temperatures are on the IPTS–68 scale b = normal boiling point c = critical point

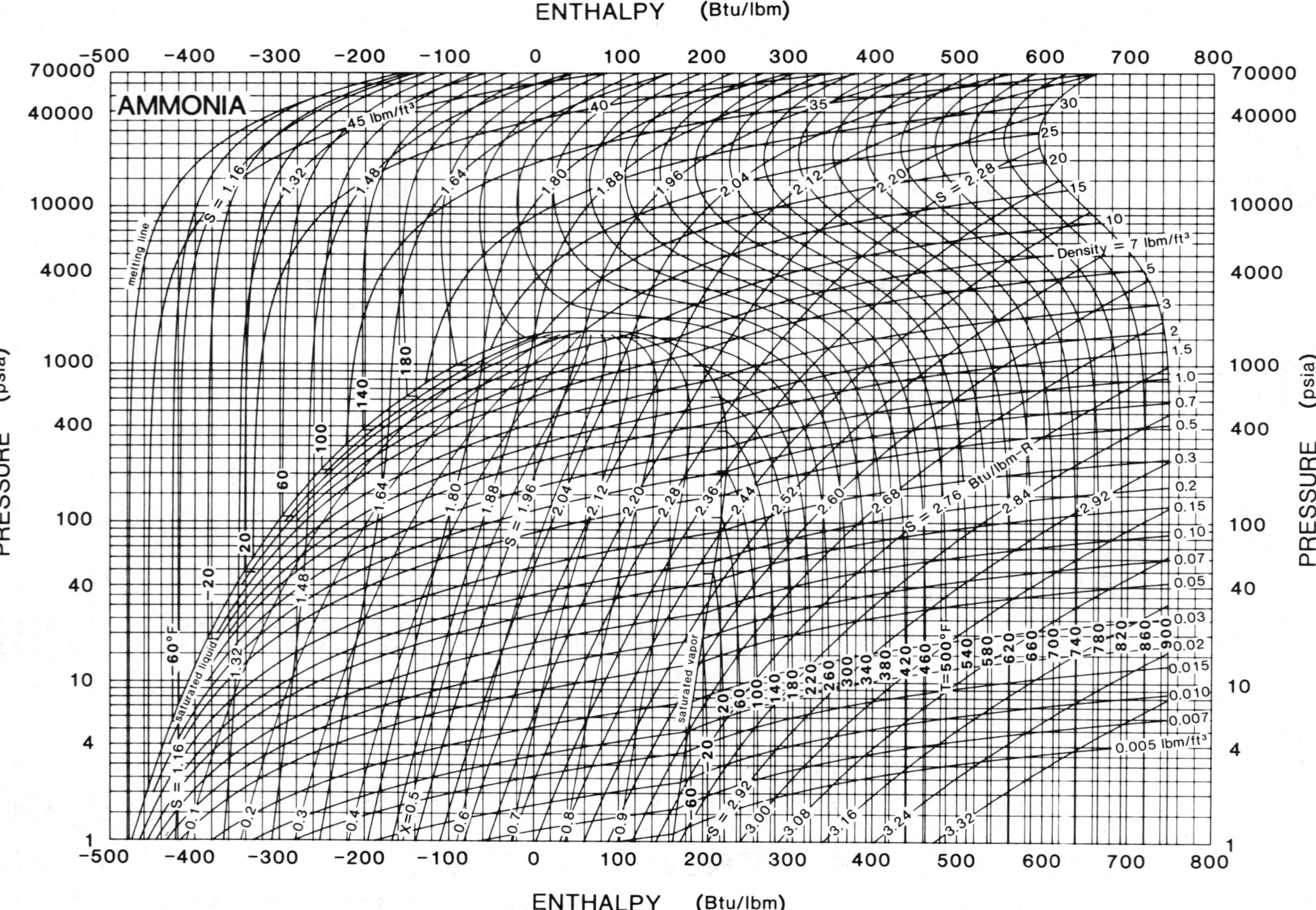

Fig. 21 Pressure–Enthalpy Diagram for Refrigerant 717 (Ammonia)

Note: The reference states for enthalpy and entropy differ from those in the table.

Refrigerant 717 (Ammonia) Properties of Saturated Liquid and Saturated Vapor

Temp,* °F	Pressure, psia	Density, lb/ft³ Liquid	Volume, ft³/lb Vapor	Enthalpy, Btu/lb Liquid	Enthalpy, Btu/lb Vapor	Entropy, Btu/lb·°F Liquid	Entropy, Btu/lb·°F Vapor	Specific Heat c_p, Btu/lb·°F Liquid	Specific Heat c_p, Btu/lb·°F Vapor	c_p/c_v Vapor	Velocity of Sound, ft/s Liquid	Velocity of Sound, ft/s Vapor	Viscosity, lb_m/ft·h Liquid	Viscosity, lb_m/ft·h Vapor	Thermal Cond, Btu/h·ft·°F Liquid	Thermal Cond, Btu/h·ft·°F Vapor	Surface Tension, dyne/cm	Temp,* °F
−107.78a	0.876	45.81	252.01	−71.827	569.276	−0.18678	1.63586	—	0.4752	1.3352	—	1167.	1.224	0.0166	—	0.00742	—	−107.78
−100.00	1.230	45.51	183.31	−63.032	572.742	−0.16206	1.60621	—	0.4777	1.3362	—	1179.	1.167	0.0169	—	0.00769	—	−100.00
−90.00	1.856	45.13	124.64	−52.230	577.125	−0.13244	1.57049	—	0.4817	1.3379	—	1194.	1.076	0.0172	—	0.00802	41.94	−90.00
−80.00	2.733	44.73	86.784	−41.713	581.419	−0.10437	1.53719	—	0.4869	1.3399	—	1208.	0.983	0.0176	—	0.00836	40.67	−80.00
−70.00	3.931	44.32	61.753	−31.305	585.612	−0.07732	1.50607	—	0.4932	1.3423	—	1222.	0.897	0.0180	—	0.00869	39.41	−70.00
−60.00	5.539	43.91	44.818	−20.912	589.693	−0.05100	1.47691	—	0.5007	1.3450	—	1235.	0.820	0.0183	—	0.00902	38.16	−60.00
−50.00	7.656	43.49	33.121	−10.484	593.653	−0.02526	1.44953	—	0.5097	1.3481	—	1247.	0.753	0.0187	0.3543	0.00936	36.90	−50.00
−45.00	8.941	43.28	28.652	−5.250	595.584	−0.01257	1.43646	—	0.5146	1.3499	—	1253.	0.723	0.0189	0.3510	0.00953	36.28	−45.00
−40.00	10.396	43.07	24.885	0.000	597.482	0.00000	1.42376	1.0508	0.5200	1.3518	5046.	1259.	0.694	0.0191	0.3477	0.00971	35.65	−40.00
−35.00	12.040	42.85	21.694	5.267	599.344	0.01246	1.41144	1.0541	0.5256	1.3538	5025.	1264.	0.668	0.0193	0.3444	0.00987	35.03	−35.00
−30.00	13.890	42.64	18.981	10.551	601.171	0.02481	1.39946	1.0575	0.5317	1.3559	5006.	1269.	0.643	0.0195	0.3411	0.01004	34.41	−30.00
−27.99b	14.696	42.55	18.006	12.682	601.896	0.02975	1.39473	1.0589	0.5342	1.3568	4999.	1272.	0.634	0.0196	0.3398	0.01011	34.16	−27.99
−25.00	15.964	42.42	16.665	15.854	602.962	0.03706	1.38781	1.0609	0.5380	1.3582	4988.	1275.	0.620	0.0197	0.3379	0.01021	33.78	−25.00
−20.00	18.281	42.21	14.680	21.174	604.714	0.04921	1.37648	1.0644	0.5448	1.3607	4971.	1279.	0.598	0.0199	0.3346	0.01038	33.16	−20.00
−15.00	20.861	41.99	12.973	26.511	606.428	0.06125	1.36545	1.0679	0.5518	1.3633	4954.	1284.	0.577	0.0201	0.3313	0.01055	32.54	−15.00
−10.00	23.727	41.77	11.500	31.867	608.102	0.07320	1.35471	1.0714	0.5593	1.3661	4936.	1288.	0.558	0.0203	0.3281	0.01073	31.92	−10.00
−5.00	26.900	41.55	10.223	37.242	609.736	0.08506	1.34424	1.0750	0.5671	1.3692	4916.	1293.	0.539	0.0205	0.3248	0.01092	31.31	−5.00
0.00	30.402	41.33	9.1135	42.635	611.327	0.09682	1.33403	1.0786	0.5752	1.3724	4894.	1297.	0.521	0.0207	0.3216	0.01112	30.69	0.00
5.00	34.258	41.11	8.1464	48.046	612.876	0.10849	1.32408	1.0822	0.5837	1.3759	4870.	1300.	0.504	0.0209	0.3183	0.01133	30.07	5.00
10.00	38.492	40.88	7.3005	53.477	614.382	0.12008	1.31437	1.0860	0.5925	1.3797	4844.	1304.	0.488	0.0211	0.3151	0.01155	29.46	10.00
15.00	43.131	40.65	6.5585	58.928	615.842	0.13158	1.30488	1.0898	0.6017	1.3838	4816.	1307.	0.473	0.0213	0.3119	0.01178	28.84	15.00
20.00	48.199	40.43	5.9059	64.398	617.256	0.14299	1.29561	1.0937	0.6113	1.3881	4785.	1310.	0.458	0.0215	0.3086	0.01202	28.23	20.00
25.00	53.724	40.20	5.3303	69.888	618.623	0.15432	1.28654	1.0977	0.6212	1.3928	4751.	1313.	0.444	0.0217	0.3053	0.01227	27.62	25.00
30.00	59.734	39.96	4.8211	75.400	619.941	0.16558	1.27767	1.1018	0.6315	1.3979	4716.	1316.	0.431	0.0219	0.3021	0.01252	27.01	30.00
35.00	66.258	39.73	4.3695	80.933	621.208	0.17676	1.26898	1.1060	0.6422	1.4033	4677.	1318.	0.417	0.0221	0.2988	0.01278	26.40	35.00
40.00	73.324	39.49	3.9682	86.488	622.424	0.18787	1.26047	1.1103	0.6533	1.4092	4637.	1320.	0.405	0.0223	0.2956	0.01305	25.79	40.00
45.00	80.962	39.25	3.6105	92.066	623.585	0.19891	1.25213	1.1148	0.6648	1.4155	4594.	1322.	0.393	0.0225	0.2923	0.01333	25.18	45.00
50.00	89.204	39.01	3.2910	97.666	624.692	0.20987	1.24394	1.1194	0.6767	1.4222	4549.	1323.	0.381	0.0227	0.2890	0.01361	24.57	50.00
55.00	98.081	38.76	3.0050	103.291	625.741	0.22077	1.23590	1.1241	0.6890	1.4295	4502.	1325.	0.370	0.0229	0.2857	0.01390	23.97	55.00
60.00	107.62	38.51	2.7484	108.941	626.731	0.23161	1.22800	1.1290	0.7018	1.4374	4452.	1326.	0.359	0.0231	0.2824	0.01419	23.36	60.00
65.00	117.87	38.26	2.5177	114.615	627.659	0.24238	1.22022	1.1341	0.7152	1.4458	4401.	1327.	0.349	0.0234	0.2791	0.01449	22.76	65.00
70.00	128.84	38.00	2.3098	120.315	628.523	0.25309	1.21257	1.1393	0.7290	1.4549	4348.	1327.	0.339	0.0236	0.2757	0.01479	22.16	70.00
75.00	140.58	37.75	2.1222	126.041	629.321	0.26374	1.20503	1.1448	0.7434	1.4647	4293.	1327.	0.329	0.0238	0.2724	0.01510	21.56	75.00
80.00	153.12	37.49	1.9525	131.795	630.051	0.27434	1.19759	1.1504	0.7584	1.4752	4236.	1327.	0.320	0.0240	0.2690	0.01541	20.96	80.00
85.00	166.50	37.22	1.7987	137.577	630.709	0.28488	1.19024	1.1563	0.7740	1.4865	4178.	1327.	0.310	0.0242	0.2657	0.01572	20.36	85.00
90.00	180.74	36.95	1.6591	143.389	631.292	0.29537	1.18298	1.1625	0.7903	1.4987	4118.	1327.	0.302	0.0244	0.2623	0.01604	19.76	90.00
95.00	195.90	36.68	1.5321	149.232	631.799	0.30581	1.17580	1.1689	0.8074	1.5117	4057.	1326.	0.293	0.0246	0.2589	0.01636	19.17	95.00
100.00	211.99	36.40	1.4165	155.106	632.224	0.31621	1.16868	1.1757	0.8253	1.5258	3994.	1325.	0.285	0.0249	0.2555	0.01669	18.57	100.00
105.00	229.07	36.12	1.3109	161.013	632.567	0.32656	1.16163	1.1829	0.8441	1.5410	3931.	1324.	0.277	0.0251	0.2520	0.01703	17.98	105.00
110.00	247.17	35.84	1.2144	166.955	632.821	0.33687	1.15462	1.1905	0.8639	1.5574	3865.	1322.	0.270	0.0253	0.2486	0.01738	17.39	110.00
115.00	266.32	35.55	1.1261	172.933	632.985	0.34715	1.14766	1.1986	0.8847	1.5750	3799.	1320.	0.262	0.0256	0.2451	0.01773	16.80	115.00
120.00	286.57	35.26	1.0451	178.950	633.053	0.35739	1.14073	1.2072	0.9067	1.5941	3731.	1318.	0.255	0.0258	0.2417	0.01809	16.22	120.00
125.00	307.96	34.96	0.9708	185.007	633.023	0.36760	1.13383	1.2165	0.9299	1.6146	3663.	1315.	0.248	0.0260	0.2382	0.01846	15.63	125.00
130.00	330.52	34.66	0.9024	191.107	632.888	0.37778	1.12694	1.2264	0.9546	1.6368	3593.	1312.	0.241	0.0263	0.2347	0.01885	15.05	130.00
135.00	354.30	34.35	0.8394	197.253	632.646	0.38795	1.12006	1.2372	0.9808	1.6609	3523.	1309.	0.235	0.0265	0.2311	0.01925	14.46	135.00
140.00	379.34	34.03	0.7813	203.447	632.289	0.39809	1.11318	1.2489	1.0088	1.6869	3451.	1306.	0.229	0.0268	0.2276	0.01966	13.88	140.00
145.00	405.67	33.71	0.7277	209.694	631.813	0.40823	1.10628	1.2616	1.0387	1.7152	3378.	1302.	0.222	0.0271	0.2240	0.02009	13.31	145.00
150.00	433.35	33.39	0.6781	215.997	631.213	0.41836	1.09937	1.2754	1.0707	1.7460	3305.	1298.	0.216	0.0274	0.2204	0.02054	12.73	150.00
155.00	462.42	33.05	0.6322	222.361	630.480	0.42849	1.09241	1.2907	1.1051	1.7796	3230.	1293.	0.211	0.0276	0.2167	0.02102	12.16	155.00
160.00	492.92	32.71	0.5896	228.792	629.609	0.43863	1.08541	1.3074	1.1423	1.8162	3154.	1288.	0.205	0.0279	0.2130	0.02151	11.58	160.00
165.00	524.91	32.36	0.5500	235.294	628.590	0.44878	1.07836	1.3259	1.1826	1.8564	3078.	1283.	0.199	0.0282	0.2093	0.02204	11.01	165.00
170.00	558.42	32.01	0.5133	241.874	627.416	0.45897	1.07123	1.3464	1.2264	1.9007	3000.	1277.	0.194	0.0286	0.2056	0.02259	10.45	170.00
175.00	593.51	31.64	0.4790	248.540	626.075	0.46919	1.06401	1.3692	1.2743	1.9495	2922.	1271.	0.189	0.0289	0.2018	0.02318	9.88	175.00
180.00	630.22	31.26	0.4471	255.298	624.558	0.47946	1.05669	1.3946	1.3268	2.0037	2842.	1264.	0.184	0.0293	0.1979	0.02380	9.32	180.00
185.00	668.61	30.87	0.4173	262.159	622.851	0.48978	1.04925	1.4232	1.3848	2.0640	2761.	1257.	0.178	0.0296	0.1940	0.02446	8.76	185.00
190.00	708.73	30.47	0.3894	269.133	620.938	0.50018	1.04167	1.4555	1.4493	2.1317	2679.	1249.	0.173	0.0300	0.1900	0.02517	8.20	190.00
195.00	750.64	30.06	0.3632	276.232	618.804	0.51067	1.03393	1.4921	1.5214	2.2081	2595.	1240.	0.168	0.0305	0.1860	0.02593	7.65	195.00
200.00	794.38	29.63	0.3387	283.469	616.428	0.52127	1.02599	1.5339	1.6028	2.2950	2511.	1231.	0.163	0.0309	0.1819	0.02674	7.10	200.00
205.00	840.03	29.18	0.3157	290.862	613.785	0.53200	1.01782	1.5821	1.6955	2.3947	2424.	1222.	0.159	0.0314	0.1777	0.02762	6.55	205.00
210.00	887.64	28.71	0.2940	298.429	610.846	0.54289	1.00939	1.6382	1.8022	2.5104	2336.	1211.	0.154	0.0320	0.1734	0.02857	6.00	210.00
215.00	937.28	28.23	0.2735	306.194	607.577	0.55396	1.00065	1.7041	1.9265	2.6462	2247.	1200.	0.149	0.0326	0.1689	0.02960	5.46	215.00
220.00	989.03	27.71	0.2541	314.185	603.934	0.56525	0.99154	1.7826	2.0737	2.8080	2155.	1188.	0.144	0.0332	0.1644	0.03073	4.93	220.00
225.00	1043.0	27.17	0.2357	322.439	599.861	0.57682	0.98199	1.8778	2.2509	3.0042	2061.	1175.	0.139	0.0339	0.1597	0.03196	4.40	225.00
230.00	1099.1	26.59	0.2182	330.998	595.287	0.58871	0.97190	1.9956	2.4691	3.2471	1966.	1162.	0.134	0.0348	—	—	3.87	230.00
235.00	1157.7	25.98	0.2015	339.924	590.119	0.60101	0.96115	2.1451	2.7450	3.5558	1867.	1147.	0.129	0.0357	—	—	3.35	235.00
240.00	1218.6	25.31	0.1854	349.296	584.228	0.61382	0.94957	2.3415	3.1056	3.9616	1766.	1130.	0.124	0.0368	—	—	2.84	240.00
245.00	1282.1	24.58	0.1698	359.229	577.434	0.62729	0.93692	2.6113	3.5985	4.5184	1662.	1113.	0.119	0.0381	—	—	2.33	245.00
250.00	1348.3	23.76	0.1545	369.899	569.465	0.64166	0.92285	3.0058	4.3134	5.3293	1555.	1094.	0.114	0.0396	—	—	1.83	250.00
255.00	1417.3	22.83	0.1393	381.593	559.887	0.65731	0.90677	3.6381	5.4427	6.6150	1443.	1073.	0.108	0.0416	—	—	1.34	255.00
260.00	1489.4	21.73	0.1239	394.854	547.928	0.67496	0.88765	4.8147	7.4812	8.9443	1327.	1049.	0.101	0.0443	—	—	0.87	260.00
269.99c	1643.7	14.67	0.0682	467.152	467.152	0.77240	0.77240	∞	∞	∞	0.	0.	—	—	∞	∞	0.00	269.99

*temperatures have been converted from the IPTS-68 scale of the original formulation to the ITS-90 scale a = triple point b = normal boiling point c = critical point

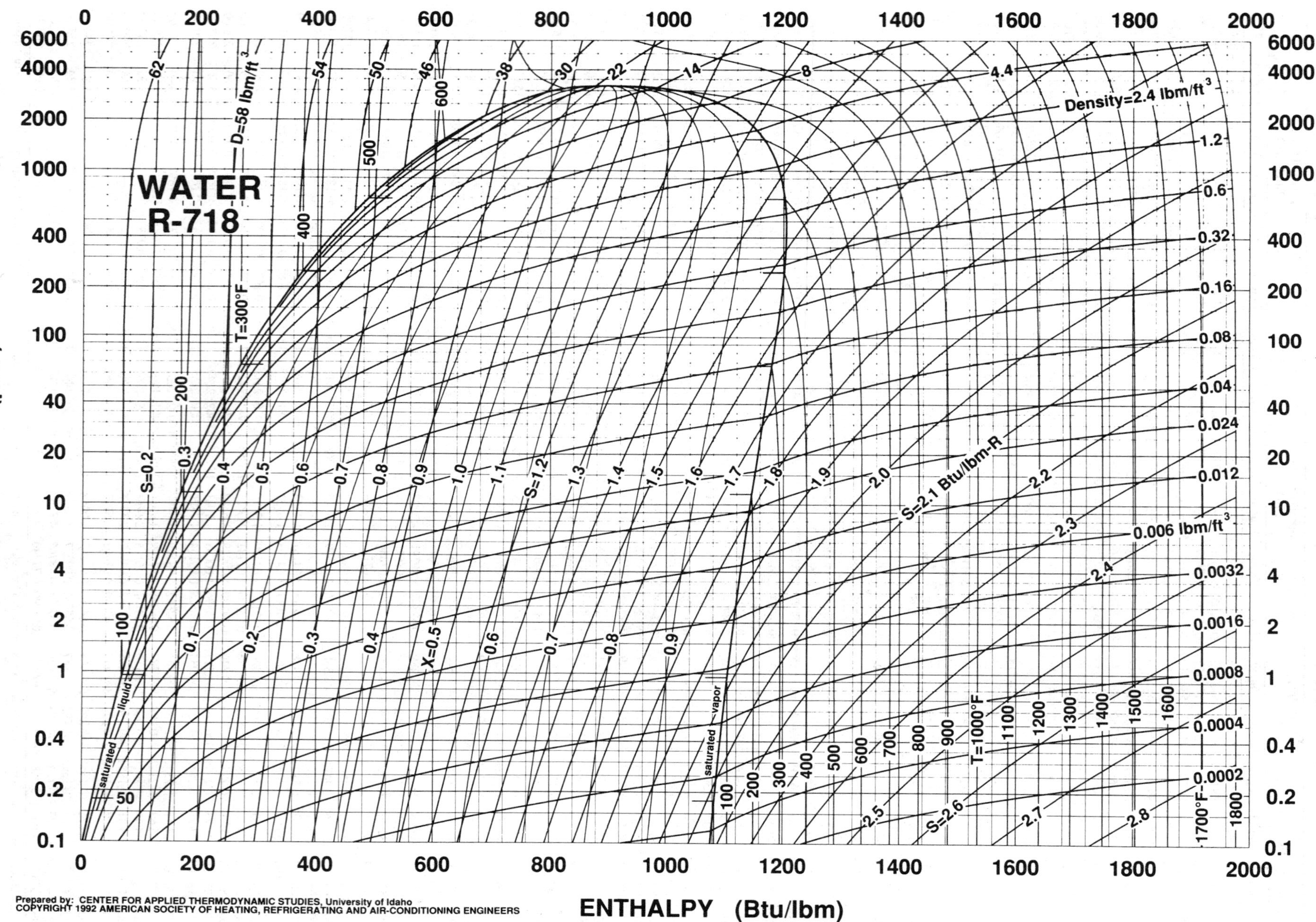

Fig. 22 Pressure–Enthalpy Diagram for Refrigerant 718 (Water/Steam)

Refrigerant 718 (Water/Steam) Properties of Saturated Liquid and Saturated Vapor

Temp,* °F	Pressure, psia	Density, lb/ft³ Liquid	Volume, ft³/lb Vapor	Enthalpy, Btu/lb Liquid	Enthalpy, Btu/lb Vapor	Entropy, Btu/lb·°F Liquid	Entropy, Btu/lb·°F Vapor	Specific Heat c_p, Btu/lb·°F Liquid	Specific Heat c_p, Btu/lb·°F Vapor	c_p/c_v Vapor	Velocity of Sound, ft/s Liquid	Velocity of Sound, ft/s Vapor	Viscosity, lb_m/ft·h Liquid	Viscosity, lb_m/ft·h Vapor	Thermal Cond, Btu/h·ft·°F Liquid	Thermal Cond, Btu/h·ft·°F Vapor	Surface Tension, dyne/cm	Temp,* °F
32.02a	0.089	62.41	3299.6	0.0	1075.0	0.0000	2.1864	1.0100	0.4461	1.3305	4596.	1343.	4.336	0.0223	0.3241	0.00986	75.65	32.02
40.00	0.122	62.42	2443.5	8.0	1078.5	0.0162	2.1586	1.0037	0.4467	1.3303	4670.	1354.	3.740	0.0226	0.3290	0.01000	75.02	40.00
50.00	0.178	62.41	1703.1	18.1	1082.9	0.0361	2.1254	1.0003	0.4476	1.3302	4750.	1367.	3.161	0.0229	0.3351	0.01018	74.22	50.00
60.00	0.256	62.37	1206.3	28.0	1087.3	0.0555	2.0938	0.9993	0.4486	1.3301	4818.	1380.	2.713	0.0232	0.3411	0.01037	73.41	60.00
70.00	0.363	62.30	867.45	38.0	1091.7	0.0745	2.0637	0.9992	0.4497	1.3299	4876.	1393.	2.360	0.0236	0.3469	0.01057	72.57	70.00
80.00	0.507	62.21	632.67	48.0	1096.0	0.0932	2.0351	0.9992	0.4511	1.3299	4926.	1405.	2.075	0.0240	0.3524	0.01078	71.72	80.00
90.00	0.699	62.11	467.66	58.0	1100.3	0.1116	2.0078	0.9991	0.4525	1.3298	4967.	1418.	1.842	0.0244	0.3576	0.01100	70.85	90.00
100.00	0.950	61.99	350.05	68.0	1104.6	0.1296	1.9818	0.9990	0.4542	1.3299	5002.	1430.	1.648	0.0248	0.3625	0.01123	69.96	100.00
110.00	1.276	61.86	265.16	78.0	1108.9	0.1473	1.9569	0.9989	0.4560	1.3300	5030.	1442.	1.486	0.0252	0.3670	0.01147	69.06	110.00
120.00	1.694	61.71	203.11	88.0	1113.2	0.1646	1.9332	0.9988	0.4580	1.3303	5053.	1454.	1.348	0.0256	0.3710	0.01171	68.13	120.00
130.00	2.224	61.55	157.23	98.0	1117.4	0.1817	1.9105	0.9988	0.4602	1.3305	5070.	1465.	1.230	0.0260	0.3747	0.01197	67.20	130.00
140.00	2.891	61.38	122.93	108.0	1121.6	0.1985	1.8888	0.9990	0.4627	1.3310	5083.	1476.	1.129	0.0264	0.3781	0.01224	66.24	140.00
150.00	3.720	61.19	97.021	118.0	1125.7	0.2150	1.8680	0.9995	0.4654	1.3315	5091.	1488.	1.040	0.0269	0.3810	0.01252	65.27	150.00
160.00	4.743	61.00	77.262	128.0	1129.8	0.2313	1.8481	1.0002	0.4683	1.3322	5096.	1498.	0.963	0.0273	0.3836	0.01281	64.29	160.00
170.00	5.994	60.80	62.046	138.0	1133.9	0.2473	1.8290	1.0011	0.4715	1.3332	5096.	1509.	0.894	0.0278	0.3859	0.01311	63.29	170.00
180.00	7.513	60.58	50.224	148.0	1137.9	0.2631	1.8107	1.0023	0.4750	1.3343	5092.	1520.	0.834	0.0282	0.3879	0.01342	62.27	180.00
190.00	9.341	60.36	40.962	158.0	1141.9	0.2787	1.7930	1.0036	0.4788	1.3356	5085.	1530.	0.781	0.0287	0.3896	0.01374	61.24	190.00
200.00	11.526	60.12	33.646	168.1	1145.8	0.2940	1.7761	1.0052	0.4829	1.3372	5075.	1540.	0.733	0.0291	0.3910	0.01408	60.19	200.00
210.00	14.122	59.88	27.826	178.2	1149.6	0.3091	1.7598	1.0069	0.4873	1.3390	5061.	1549.	0.690	0.0296	0.3922	0.01443	59.13	210.00
212.00b	14.696	59.83	26.809	180.2	1150.4	0.3121	1.7566	1.0072	0.4882	1.3394	5058.	1551.	0.682	0.0297	0.3924	0.01450	58.92	212.00
220.00	17.184	59.63	23.159	188.2	1153.4	0.3241	1.7441	1.0087	0.4921	1.3411	5044.	1558.	0.651	0.0300	0.3931	0.01479	58.06	220.00
230.00	20.774	59.37	19.393	198.3	1157.0	0.3388	1.7289	1.0108	0.4973	1.3435	5025.	1568.	0.616	0.0305	0.3939	0.01516	56.97	230.00
240.00	24.960	59.10	16.332	208.5	1160.6	0.3534	1.7143	1.0129	0.5029	1.3463	5002.	1576.	0.585	0.0310	0.3944	0.01555	55.87	240.00
250.00	29.814	58.82	13.830	218.6	1164.1	0.3678	1.7001	1.0153	0.5089	1.3495	4976.	1585.	0.556	0.0314	0.3948	0.01595	54.75	250.00
260.00	35.411	58.54	11.772	228.8	1167.6	0.3820	1.6864	1.0177	0.5155	1.3530	4948.	1593.	0.530	0.0319	0.3950	0.01636	53.62	260.00
270.00	41.835	58.24	10.069	239.0	1170.9	0.3960	1.6732	1.0203	0.5225	1.3570	4918.	1600.	0.506	0.0324	0.3950	0.01679	52.48	270.00
280.00	49.173	57.94	8.6520	249.2	1174.1	0.4099	1.6603	1.0231	0.5301	1.3614	4884.	1608.	0.484	0.0328	0.3949	0.01723	51.33	280.00
290.00	57.516	57.63	7.4677	259.5	1177.2	0.4237	1.6478	1.0261	0.5383	1.3663	4849.	1615.	0.464	0.0333	0.3946	0.01769	50.16	290.00
300.00	66.963	57.31	6.4725	269.8	1180.2	0.4373	1.6356	1.0292	0.5472	1.3718	4811.	1621.	0.445	0.0338	0.3942	0.01816	48.99	300.00
310.00	77.615	56.99	5.6322	280.1	1183.0	0.4507	1.6238	1.0326	0.5567	1.3778	4771.	1628.	0.428	0.0342	0.3936	0.01864	47.80	310.00
320.00	89.580	56.65	4.9191	290.5	1185.7	0.4641	1.6123	1.0362	0.5670	1.3845	4728.	1633.	0.412	0.0347	0.3929	0.01914	46.60	320.00
330.00	102.97	56.31	4.3118	300.9	1188.3	0.4773	1.6010	1.0401	0.5780	1.3917	4684.	1639.	0.397	0.0351	0.3920	0.01965	45.39	330.00
340.00	117.90	55.96	3.7920	311.3	1190.7	0.4904	1.5900	1.0443	0.5899	1.3997	4637.	1644.	0.383	0.0356	0.3910	0.02018	44.17	340.00
350.00	134.50	55.60	3.3457	321.8	1193.0	0.5033	1.5792	1.0488	0.6028	1.4084	4588.	1648.	0.371	0.0361	0.3898	0.02072	42.94	350.00
360.00	152.89	55.23	2.9608	332.4	1195.1	0.5162	1.5687	1.0537	0.6166	1.4180	4537.	1652.	0.359	0.0365	0.3885	0.02128	41.70	360.00
370.00	173.20	54.85	2.6276	342.9	1197.0	0.5289	1.5583	1.0590	0.6314	1.4284	4484.	1656.	0.347	0.0370	0.3871	0.02185	40.46	370.00
380.00	195.57	54.46	2.3382	353.6	1198.7	0.5416	1.5481	1.0647	0.6473	1.4398	4430.	1658.	0.337	0.0375	0.3854	0.02244	39.20	380.00
390.00	220.14	54.06	2.0859	364.3	1200.3	0.5542	1.5381	1.0709	0.6645	1.4522	4373.	1661.	0.327	0.0379	0.3837	0.02305	37.94	390.00
400.00	247.05	53.66	1.8654	375.1	1201.6	0.5667	1.5282	1.0776	0.6830	1.4657	4314.	1663.	0.318	0.0384	0.3817	0.02367	36.67	400.00
410.00	276.45	53.24	1.6720	385.9	1202.8	0.5791	1.5184	1.0849	0.7028	1.4804	4253.	1664.	0.309	0.0389	0.3796	0.02431	35.39	410.00
420.00	308.50	52.81	1.5017	396.8	1203.7	0.5915	1.5087	1.0927	0.7242	1.4965	4190.	1665.	0.300	0.0393	0.3774	0.02497	34.11	420.00
430.00	343.36	52.37	1.3515	407.8	1204.5	0.6038	1.4992	1.1013	0.7473	1.5140	4126.	1665.	0.293	0.0398	0.3749	0.02565	32.82	430.00
440.00	381.18	51.92	1.2186	418.9	1204.9	0.6160	1.4897	1.1106	0.7722	1.5332	4059.	1664.	0.285	0.0403	0.3723	0.02635	31.52	440.00
450.00	422.14	51.46	1.1006	430.1	1205.2	0.6282	1.4802	1.1208	0.7991	1.5541	3990.	1663.	0.278	0.0407	0.3695	0.02708	30.23	450.00
460.00	466.41	50.98	0.9957	441.4	1205.1	0.6403	1.4708	1.1319	0.8281	1.5770	3920.	1662.	0.271	0.0412	0.3664	0.02783	28.93	460.00
470.00	514.15	50.49	0.9020	452.8	1204.8	0.6525	1.4614	1.1440	0.8596	1.6022	3847.	1659.	0.264	0.0417	0.3632	0.02861	27.62	470.00
480.00	565.56	49.99	0.8183	464.3	1204.3	0.6646	1.4521	1.1572	0.8938	1.6299	3772.	1656.	0.258	0.0422	0.3597	0.02943	26.31	480.00
490.00	620.82	49.47	0.7433	475.9	1203.4	0.6767	1.4427	1.1717	0.9311	1.6604	3695.	1652.	0.252	0.0427	0.3560	0.03029	25.01	490.00
500.00	680.11	48.93	0.6759	487.7	1202.2	0.6888	1.4333	1.1877	0.9717	1.6942	3616.	1647.	0.246	0.0432	0.3521	0.03119	23.70	500.00
510.00	743.64	48.38	0.6152	499.6	1200.6	0.7009	1.4238	1.2053	1.0163	1.7317	3534.	1641.	0.240	0.0438	0.3480	0.03215	22.39	510.00
520.00	811.60	47.81	0.5603	511.7	1198.7	0.7130	1.4143	1.2249	1.0653	1.7735	3450.	1634.	0.235	0.0443	0.3435	0.03317	21.09	520.00
530.00	884.21	47.22	0.5107	524.0	1196.4	0.7252	1.4047	1.2466	1.1196	1.8205	3364.	1627.	0.229	0.0449	0.3388	0.03427	19.78	530.00
540.00	961.69	46.60	0.4657	536.4	1193.7	0.7374	1.3949	1.2709	1.1798	1.8732	3275.	1618.	0.224	0.0455	0.3339	0.03546	18.48	540.00
550.00	1044.2	45.97	0.4248	549.1	1190.5	0.7497	1.3850	1.2982	1.2473	1.9332	3184.	1609.	0.219	0.0461	0.3287	0.03676	17.19	550.00
560.00	1132.1	45.31	0.3875	562.0	1186.9	0.7621	1.3749	1.3292	1.3233	2.0016	3089.	1598.	0.214	0.0467	0.3233	0.03820	15.90	560.00
570.00	1225.5	44.62	0.3535	575.2	1182.7	0.7745	1.3645	1.3646	1.4097	2.0803	2992.	1586.	0.209	0.0474	0.3176	0.03980	14.63	570.00
580.00	1324.8	43.89	0.3224	588.6	1177.9	0.7872	1.3539	1.4054	1.5089	2.1718	2891.	1573.	0.204	0.0481	0.3117	0.04162	13.36	580.00
590.00	1430.1	43.13	0.2938	602.4	1172.4	0.7999	1.3429	1.4531	1.6240	2.2791	2787.	1559.	0.199	0.0489	0.3056	0.04369	12.10	590.00
600.00	1541.7	42.34	0.2675	616.6	1166.2	0.8129	1.3316	1.5094	1.7594	2.4067	2679.	1543.	0.194	0.0497	0.2994	0.04609	10.86	600.00
610.00	1660.0	41.49	0.2432	631.2	1159.2	0.8261	1.3197	1.5772	1.9212	2.5606	2567.	1526.	0.189	0.0506	0.2930	0.04890	9.63	610.00
620.00	1785.3	40.59	0.2207	646.3	1151.2	0.8396	1.3073	1.6603	2.1184	2.7500	2451.	1507.	0.184	0.0516	0.2866	0.05225	8.43	620.00
630.00	1917.9	39.62	0.1998	662.0	1142.1	0.8536	1.2941	1.7650	2.3644	2.9882	2331.	1486.	0.178	0.0527	0.2800	0.05631	7.24	630.00
640.00	2058.2	38.57	0.1803	678.4	1131.6	0.8679	1.2801	1.9010	2.6805	3.2965	2204.	1462.	0.173	0.0540	0.2734	0.06133	6.09	640.00
650.00	2206.6	37.43	0.1619	695.8	1119.5	0.8830	1.2648	2.0855	3.1030	3.7114	2072.	1436.	0.167	0.0555	0.2667	0.06765	4.97	650.00
660.00	2363.6	36.15	0.1445	714.3	1105.2	0.8989	1.2481	2.3508	3.6981	4.2993	1934.	1406.	0.161	0.0572	0.2600	0.07586	3.89	660.00
670.00	2529.7	34.70	0.1279	734.5	1088.2	0.9161	1.2292	2.7658	4.6038	5.1991	1786.	1372.	0.154	0.0594	0.2532	0.08693	2.85	670.00
680.00	2705.6	32.97	0.1115	757.1	1067.1	0.9352	1.2072	3.5086	6.1610	6.7546	1629.	1332.	0.146	0.0622	0.2468	0.10285	1.89	680.00
690.00	2892.1	30.78	0.0948	784.0	1039.1	0.9577	1.1796	5.2151	9.5419	10.1841	1459.	1284.	0.137	0.0662	0.2424	0.12873	1.00	690.00
700.00	3090.8	27.42	0.0756	821.9	994.3	0.9894	1.1381	13.7960	25.9020	26.2094	1245.	1209.	0.123	0.0736	0.2543	0.19639	0.26	700.00
705.18c	3199.2	20.10	0.0497	896.8	896.8	1.0531	1.0531	∞	∞	∞	0.	0.	0.104	0.1043	∞	∞	0.00	705.18

*temperatures are on the IPTS–68 scale a = triple point b = normal boiling point c = critical point

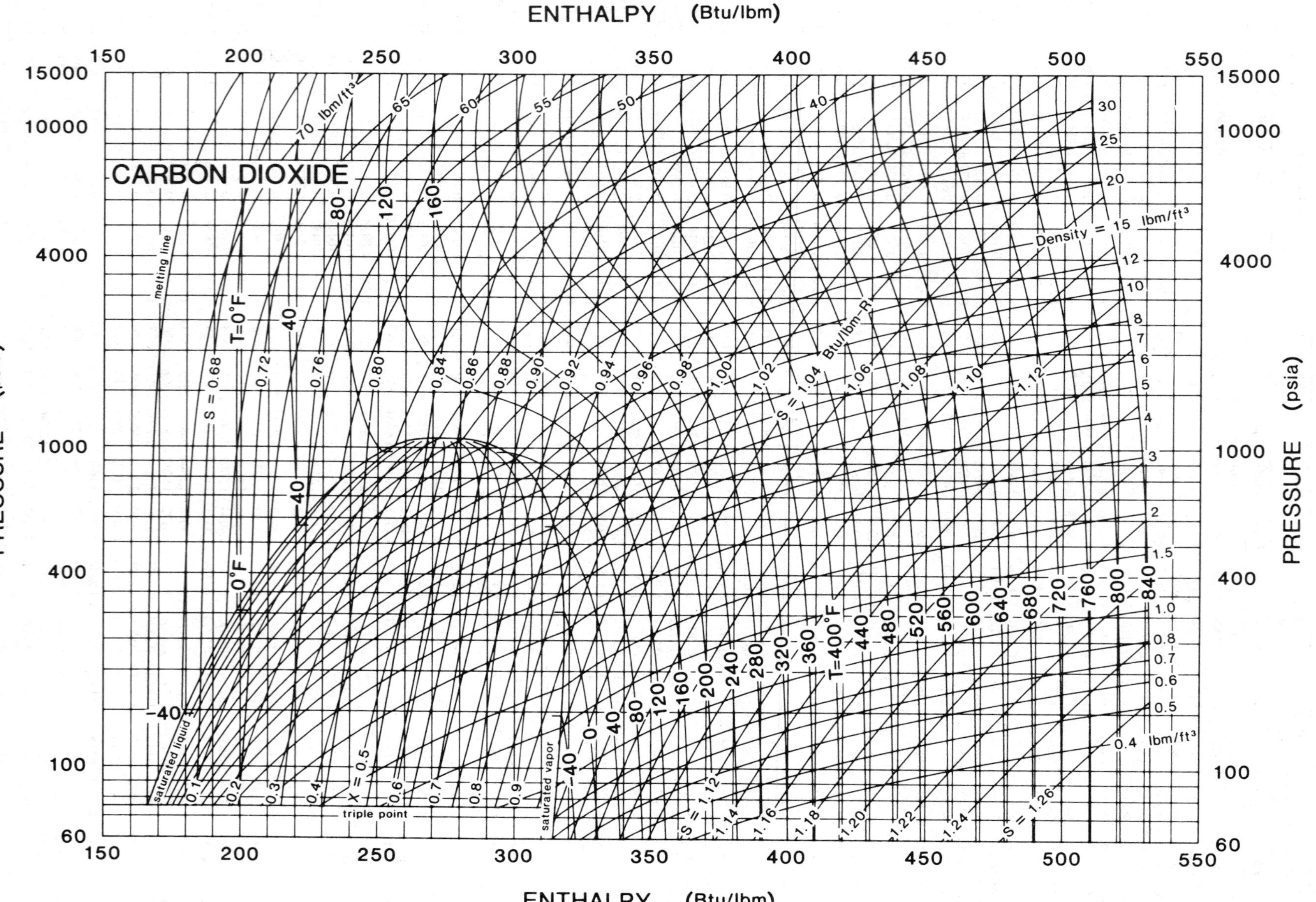

Prepared by: CENTER FOR APPLIED THERMODYNAMIC STUDIES, University of Idaho

Fig. 23 Pressure–Enthalpy Diagram for Refrigerant 744 (Carbon Dioxide)

Note: The reference states for enthalpy and entropy differ from those in the table.

Refrigerant 744 (Carbon Dioxide) Properties of Saturated Liquid and Saturated Vapor

Temp.,* °F	Pressure, psia	Density, lb/ft³ Liquid	Volume, ft³/lb Vapor	Enthalpy, Btu/lb Liquid	Enthalpy, Btu/lb Vapor	Entropy, Btu/lb·°F Liquid	Entropy, Btu/lb·°F Vapor	Specific Heat c_p, Btu/lb·°F Liquid	Specific Heat c_p, Btu/lb·°F Vapor	c_p/c_v Vapor	Velocity of Sound, ft/s Liquid	Velocity of Sound, ft/s Vapor	Viscosity, lb_m/ft·h Liquid	Viscosity, lb_m/ft·h Vapor	Thermal Cond, Btu/h·ft·°F Liquid	Thermal Cond, Btu/h·ft·°F Vapor	Surface Tension, dyne/cm	Temp,* °F
−69.83a	75.138	73.56	1.1640	−14.225	136.515	−0.03471	0.35196	0.4764	0.2052	1.4669	3224.	737.	0.625	0.0267	0.1044	0.00654	17.09	−69.83
−65.00	84.296	72.95	1.0427	−11.922	136.902	−0.02890	0.34819	0.4751	0.2106	1.4750	3164.	737.	0.597	0.0271	0.1024	0.00667	16.41	−65.00
−60.00	94.643	72.30	0.9328	−9.541	137.274	−0.02297	0.34437	0.4745	0.2166	1.4845	3101.	738.	0.571	0.0274	0.1004	0.00681	15.72	−60.00
−55.00	105.91	71.65	0.8368	−7.161	137.616	−0.01712	0.34065	0.4745	0.2228	1.4951	3037.	738.	0.545	0.0277	0.0984	0.00694	15.03	−55.00
−50.00	118.16	70.99	0.7525	−4.780	137.927	−0.01135	0.33699	0.4752	0.2293	1.5069	2972.	738.	0.521	0.0281	0.0963	0.00708	14.35	−50.00
−48.00	123.34	70.73	0.7216	−3.826	138.041	−0.00906	0.33555	0.4757	0.2321	1.5121	2946.	737.	0.512	0.0282	0.0955	0.00714	14.08	−48.00
−46.00	128.69	70.46	0.6923	−2.871	138.151	−0.00678	0.33412	0.4762	0.2348	1.5174	2920.	737.	0.503	0.0283	0.0947	0.00719	13.81	−46.00
−44.00	134.21	70.20	0.6644	−1.915	138.255	−0.00451	0.33270	0.4769	0.2377	1.5230	2893.	737.	0.494	0.0285	0.0939	0.00725	13.54	−44.00
−42.00	139.90	69.93	0.6379	−0.958	138.353	−0.00225	0.33129	0.4776	0.2406	1.5288	2867.	737.	0.485	0.0286	0.0931	0.00731	13.28	−42.00
−40.00	145.77	69.65	0.6126	0.000	138.445	0.00000	0.32989	0.4784	0.2436	1.5349	2840.	737.	0.477	0.0287	0.0923	0.00737	13.01	−40.00
−38.00	151.82	69.38	0.5885	0.961	138.532	0.00224	0.32850	0.4794	0.2466	1.5413	2814.	736.	0.469	0.0289	0.0915	0.00743	12.75	−38.00
−36.00	158.06	69.11	0.5655	1.923	138.612	0.00448	0.32711	0.4804	0.2498	1.5479	2787.	736.	0.460	0.0290	0.0907	0.00749	12.48	−36.00
−34.00	164.48	68.83	0.5436	2.887	138.686	0.00671	0.32574	0.4816	0.2530	1.5549	2760.	735.	0.453	0.0291	0.0899	0.00756	12.22	−34.00
−32.00	171.09	68.55	0.5227	3.853	138.754	0.00893	0.32437	0.4828	0.2563	1.5621	2732.	735.	0.445	0.0293	0.0891	0.00762	11.96	−32.00
−30.00	177.90	68.27	0.5027	4.822	138.816	0.01115	0.32300	0.4841	0.2596	1.5697	2705.	734.	0.437	0.0294	0.0883	0.00769	11.70	−30.00
−28.00	184.90	67.99	0.4836	5.793	138.871	0.01336	0.32165	0.4856	0.2631	1.5776	2678.	734.	0.430	0.0295	0.0875	0.00776	11.45	−28.00
−26.00	192.11	67.71	0.4654	6.767	138.919	0.01557	0.32030	0.4871	0.2667	1.5859	2650.	733.	0.422	0.0297	0.0867	0.00783	11.19	−26.00
−24.00	199.52	67.42	0.4480	7.743	138.960	0.01777	0.31895	0.4888	0.2703	1.5945	2622.	733.	0.415	0.0298	0.0859	0.00790	10.93	−24.00
−22.00	207.14	67.13	0.4313	8.723	138.995	0.01996	0.31761	0.4905	0.2741	1.6036	2594.	732.	0.408	0.0300	0.0851	0.00797	10.68	−22.00
−20.00	214.97	66.84	0.4154	9.706	139.021	0.02215	0.31627	0.4924	0.2779	1.6130	2566.	731.	0.401	0.0301	0.0843	0.00805	10.43	−20.00
−18.00	223.02	66.54	0.4001	10.693	139.041	0.02434	0.31494	0.4944	0.2819	1.6229	2538.	730.	0.394	0.0302	0.0835	0.00813	10.18	−18.00
−16.00	231.29	66.25	0.3855	11.683	139.053	0.02653	0.31361	0.4965	0.2860	1.6333	2509.	730.	0.388	0.0304	0.0827	0.00822	9.93	−16.00
−14.00	239.78	65.95	0.3714	12.677	139.056	0.02871	0.31228	0.4987	0.2903	1.6441	2481.	729.	0.381	0.0305	0.0819	0.00830	9.68	−14.00
−12.00	248.49	65.65	0.3580	13.675	139.052	0.03089	0.31095	0.5011	0.2947	1.6554	2452.	728.	0.375	0.0307	0.0811	0.00839	9.43	−12.00
−10.00	257.44	65.34	0.3451	14.677	139.040	0.03307	0.30963	0.5036	0.2992	1.6673	2423.	727.	0.368	0.0309	0.0804	0.00849	9.19	−10.00
−8.00	266.62	65.03	0.3328	15.683	139.019	0.03524	0.30831	0.5062	0.3039	1.6798	2394.	726.	0.362	0.0310	0.0796	0.00858	8.95	−8.00
−6.00	276.04	64.72	0.3209	16.694	138.989	0.03741	0.30698	0.5090	0.3088	1.6929	2364.	725.	0.356	0.0312	0.0788	0.00869	8.70	−6.00
−4.00	285.70	64.41	0.3095	17.710	138.950	0.03959	0.30566	0.5119	0.3138	1.7067	2335.	724.	0.350	0.0314	0.0780	0.00879	8.46	−4.00
−2.00	295.60	64.09	0.2986	18.732	138.901	0.04176	0.30433	0.5150	0.3190	1.7211	2305.	722.	0.344	0.0316	0.0772	0.00890	8.22	−2.00
0.00	305.76	63.76	0.2880	19.758	138.843	0.04394	0.30300	0.5183	0.3245	1.7363	2275.	721.	0.338	0.0317	0.0764	0.00902	7.99	0.00
2.00	316.17	63.44	0.2779	20.790	138.775	0.04611	0.30167	0.5218	0.3301	1.7523	2245.	720.	0.333	0.0319	0.0756	0.00914	7.75	2.00
4.00	326.83	63.11	0.2682	21.828	138.697	0.04829	0.30034	0.5254	0.3360	1.7691	2215.	719.	0.327	0.0321	0.0749	0.00927	7.52	4.00
6.00	337.76	62.77	0.2589	22.872	138.608	0.05046	0.29900	0.5292	0.3422	1.7868	2185.	717.	0.322	0.0323	0.0741	0.00940	7.29	6.00
8.00	348.95	62.43	0.2498	23.923	138.508	0.05264	0.29766	0.5333	0.3486	1.8055	2154.	716.	0.316	0.0325	0.0733	0.00954	7.05	8.00
10.00	360.41	62.09	0.2412	24.980	138.397	0.05483	0.29631	0.5376	0.3553	1.8253	2123.	714.	0.311	0.0327	0.0725	0.00968	6.83	10.00
12.00	372.14	61.74	0.2328	26.044	138.273	0.05702	0.29495	0.5421	0.3624	1.8461	2092.	713.	0.305	0.0330	0.0718	0.00984	6.60	12.00
14.00	384.15	61.39	0.2248	27.116	138.138	0.05920	0.29359	0.5469	0.3698	1.8682	2061.	711.	0.300	0.0332	0.0710	0.01000	6.37	14.00
16.00	396.44	61.03	0.2170	28.195	137.990	0.06140	0.29222	0.5519	0.3776	1.8917	2029.	709.	0.295	0.0334	0.0702	0.01016	6.15	16.00
18.00	409.02	60.67	0.2095	29.283	137.828	0.06360	0.29084	0.5573	0.3857	1.9165	1997.	708.	0.290	0.0337	0.0694	0.01034	5.93	18.00
20.00	421.89	60.30	0.2023	30.379	137.653	0.06581	0.28945	0.5630	0.3944	1.9429	1966.	706.	0.284	0.0339	0.0687	0.01052	5.71	20.00
22.00	435.05	59.92	0.1953	31.484	137.463	0.06802	0.28805	0.5690	0.4035	1.9710	1933.	704.	0.279	0.0342	0.0679	0.01072	5.49	22.00
24.00	448.52	59.54	0.1886	32.598	137.258	0.07025	0.28663	0.5755	0.4131	2.0009	1901.	702.	0.274	0.0345	0.0671	0.01092	5.28	24.00
26.00	462.28	59.15	0.1821	33.722	137.037	0.07248	0.28520	0.5823	0.4234	2.0329	1869.	700.	0.269	0.0347	0.0663	0.01114	5.06	26.00
28.00	476.36	58.75	0.1758	34.857	136.800	0.07472	0.28376	0.5896	0.4342	2.0670	1836.	698.	0.264	0.0350	0.0656	0.01136	4.85	28.00
30.00	490.74	58.35	0.1697	36.002	136.545	0.07697	0.28230	0.5974	0.4458	2.1037	1803.	696.	0.259	0.0353	0.0648	0.01160	4.64	30.00
32.00	505.45	57.94	0.1638	37.159	136.273	0.07923	0.28082	0.6057	0.4582	2.1430	1770.	694.	0.254	0.0356	0.0640	0.01185	4.43	32.00
34.00	520.47	57.52	0.1581	38.328	135.981	0.08151	0.27932	0.6146	0.4715	2.1853	1736.	692.	0.249	0.0360	0.0633	0.01211	4.23	34.00
36.00	535.83	57.09	0.1526	39.511	135.669	0.08380	0.27779	0.6242	0.4858	2.2310	1702.	689.	0.244	0.0363	0.0625	0.01239	4.03	36.00
38.00	551.51	56.65	0.1473	40.706	135.336	0.08610	0.27625	0.6345	0.5011	2.2804	1668.	687.	0.239	0.0366	0.0617	0.01268	3.82	38.00
40.00	567.54	56.20	0.1421	41.917	134.980	0.08842	0.27467	0.6457	0.5177	2.3339	1634.	684.	0.235	0.0370	0.0610	0.01299	3.63	40.00
42.00	583.91	55.74	0.1371	43.143	134.600	0.09076	0.27307	0.6577	0.5358	2.3923	1599.	682.	0.230	0.0374	0.0602	0.01332	3.43	42.00
44.00	600.62	55.27	0.1322	44.385	134.195	0.09312	0.27144	0.6708	0.5554	2.4559	1565.	679.	0.225	0.0378	0.0595	0.01366	3.24	44.00
46.00	617.69	54.79	0.1274	45.645	133.763	0.09551	0.26977	0.6851	0.5769	2.5257	1529.	677.	0.220	0.0382	0.0587	0.01402	3.05	46.00
48.00	635.13	54.29	0.1228	46.924	133.302	0.09791	0.26806	0.7007	0.6005	2.6025	1494.	674.	0.215	0.0386	0.0579	0.01441	2.86	48.00
50.00	652.92	53.78	0.1183	48.223	132.809	0.10035	0.26631	0.7179	0.6265	2.6875	1458.	671.	0.210	0.0390	0.0572	0.01481	2.67	50.00
52.00	671.09	53.25	0.1140	49.544	132.283	0.10281	0.26451	0.7369	0.6554	2.7818	1421.	668.	0.205	0.0395	0.0564	0.01525	2.49	52.00
54.00	689.64	52.71	0.1097	50.888	131.720	0.10531	0.26267	0.7580	0.6877	2.8873	1384.	665.	0.200	0.0400	0.0557	0.01570	2.31	54.00
56.00	708.58	52.15	0.1056	52.259	131.117	0.10784	0.26077	0.7815	0.7240	3.0059	1347.	662.	0.194	0.0404	0.0549	0.01618	2.14	56.00
58.00	727.91	51.57	0.1015	53.657	130.471	0.11041	0.25880	0.8081	0.7652	3.1402	1309.	658.	0.189	0.0410	0.0542	0.01669	1.96	58.00
60.00	747.64	50.97	0.0975	55.087	129.777	0.11303	0.25676	0.8383	0.8122	3.2935	1270.	655.	0.184	0.0415	0.0534	0.01724	1.80	60.00
62.00	767.78	50.34	0.0937	56.551	129.029	0.11570	0.25464	0.8729	0.8665	3.4703	1230.	652.	0.179	0.0420	0.0526	0.01781	1.63	62.00
64.00	788.33	49.68	0.0899	58.054	128.222	0.11843	0.25243	0.9132	0.9299	3.6763	1189.	648.	0.173	0.0426	0.0519	0.01842	1.47	64.00
66.00	809.32	49.00	0.0861	59.600	127.348	0.12123	0.25011	0.9605	1.0050	3.9194	1148.	644.	0.168	0.0432	0.0511	0.01907	1.31	66.00
68.00	830.74	48.28	0.0824	61.197	126.396	0.12411	0.24767	1.0173	1.0952	4.2107	1105.	640.	0.163	0.0438	0.0504	0.01977	1.16	68.00
70.00	852.61	47.51	0.0788	62.851	125.355	0.12708	0.24508	1.0866	1.2057	4.5663	1060.	636.	0.157	0.0445	0.0496	0.02050	1.01	70.00
75.00	909.34	45.36	0.0697	67.321	122.243	0.13505	0.23777	1.3554	1.6345	5.9316	941.	625.	0.143	—	—	—	0.66	75.00
80.00	969.22	42.64	0.0604	72.565	117.967	0.14434	0.22847	1.9829	2.6237	9.0199	809.	610.	0.128	—	—	—	0.35	80.00
85.00	1032.7	38.47	0.0494	79.825	110.771	0.15720	0.21401	—	—	—	—	—	—	—	—	—	0.10	85.00
87.76c	1069.6	29.20	0.0342	94.234	94.234	0.18320	0.18320	∞	∞	∞	0.	0.	—	—	∞	∞	0.00	87.76

*temperatures are on the IPTS–68 scale a = triple point c = critical point

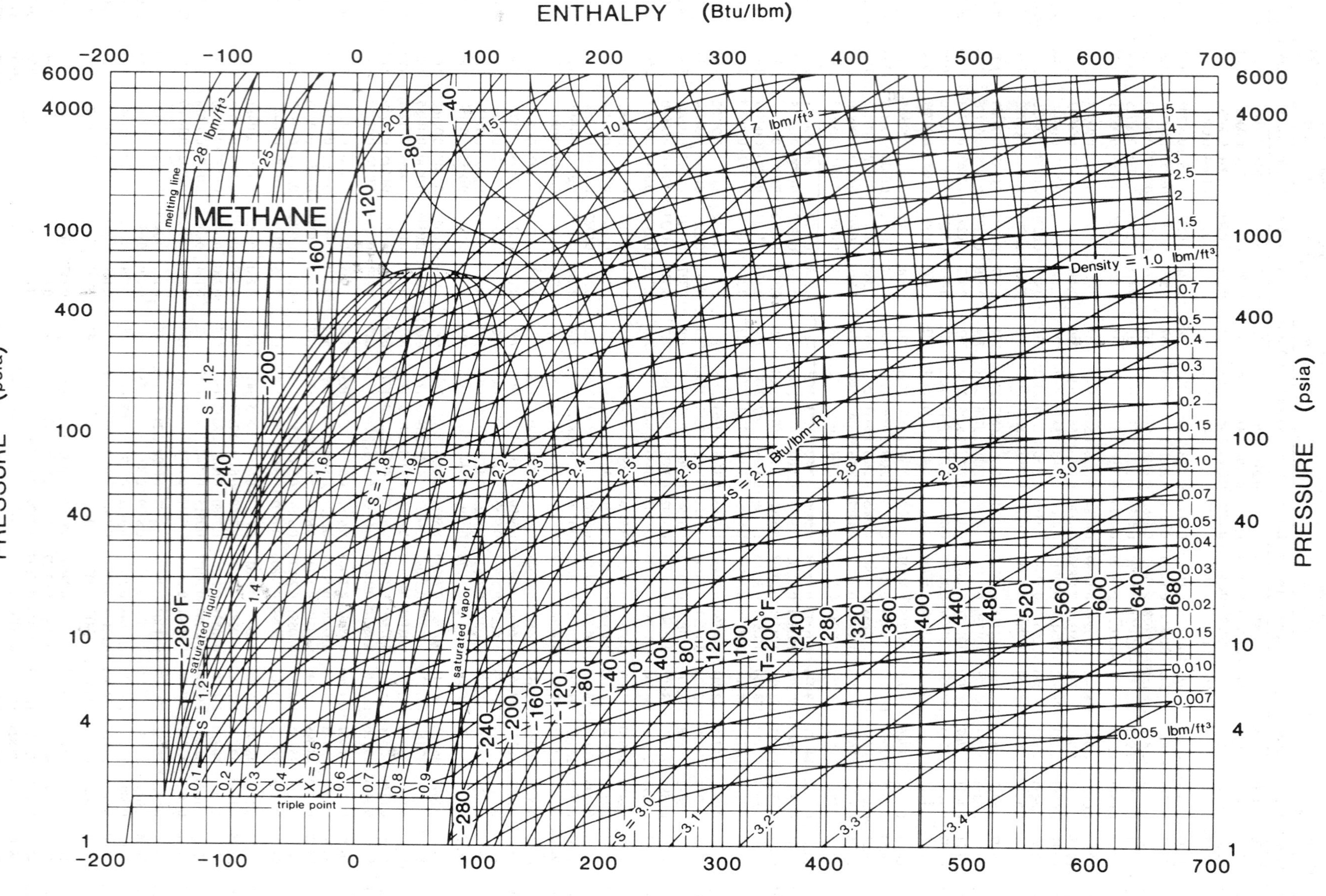

Prepared by CENTER FOR APPLIED THERMODYNAMIC STUDIES, University of Idaho

Fig. 24 Pressure–Enthalpy Diagram for Refrigerant 50 (Methane)

Refrigerant 50 (Methane) Properties of Saturated Liquid and Saturated Vapor

Temp,** °F	Pressure psia	Pressure psig	Volume, ft³/lb$_m$ Vapor	Density, lb$_m$/ft³ Liquid	Enthalpy, Btu/lb$_m$ Liquid	Enthalpy, Btu/lb$_m$ Vapor	Entropy, Btu/lb$_m$·°F Liquid	Entropy, Btu/lb$_m$·°F Vapor	Temp,** °F	Pressure psia	Pressure psig	Volume, ft³/lb$_m$ Vapor	Density, lb$_m$/ft³ Liquid	Enthalpy, Btu/lb$_m$ Liquid	Enthalpy, Btu/lb$_m$ Vapor	Entropy, Btu/lb$_m$·°F Liquid	Entropy, Btu/lb$_m$·°F Vapor
−296.45a	1.6997	26.461*	63.723	28.169	−153.88	79.912	1.0252	2.4575	−210	87.030	72.334	1.6822	23.688	−81.027	111.84	1.3788	2.1513
−295	1.8831	26.087*	57.992	28.104	−152.75	80.584	1.0321	2.4490	−208	92.229	77.533	1.5909	23.565	−79.203	112.28	1.3859	2.1467
−290	2.6448	24.536*	42.441	27.878	−148.81	82.886	1.0556	2.4211	−206	97.649	82.953	1.5057	23.441	−77.368	112.70	1.3930	2.1423
−285	3.6375	22.515*	31.678	27.648	−144.82	85.152	1.0788	2.3953	−204	103.30	88.601	1.4259	23.315	−75.523	113.09	1.4001	2.1378
−280	4.9082	19.928*	24.070	27.414	−140.79	87.377	1.1015	2.3713	−202	109.18	94.481	1.3511	23.188	−73.668	113.47	1.4071	2.1334
−270	8.4934	12.629*	14.564	26.936	−132.63	91.690	1.1455	2.3282	−200	115.50	100.60	1.2810	23.059	−71.801	113.83	1.4141	2.1290
−268	9.4086	10.765*	13.261	26.838	−130.98	92.528	1.1541	2.3202	−198	121.66	106.96	1.2152	22.928	−69.922	114.16	1.4212	2.1246
−266	10.399	8.7491*	12.099	26.740	−129.33	93.358	1.1626	2.3125	−196	128.27	113.58	1.1534	22.796	−68.032	114.48	1.4281	2.1203
−264	11.468	6.5716*	11.061	26.641	−127.68	94.179	1.1711	2.3049	−194	135.14	120.44	1.0953	22.662	−66.128	114.76	1.4351	2.1160
−262	12.621	4.2244*	10.132	26.542	−126.02	94.990	1.1795	2.2976	−192	142.27	127.57	1.0406	22.526	−64.211	115.03	1.4421	2.1117
−260	13.862	1.6988*	9.2970	26.442	−124.35	95.792	1.1878	2.2904	−190	149.66	134.97	0.98903	22.389	−62.280	115.26	1.4491	2.1074
−258.73b	14.696	0.0000	8.8117	26.378	−123.30	96.296	1.1931	2.2859	−188	157.33	142.63	0.94044	22.249	−60.334	115.48	1.4560	2.1246
−258	15.194	0.4980	8.5463	26.341	−122.68	96.584	1.1961	2.2834	−186	165.28	150.58	0.89458	22.107	−58.373	115.66	1.4630	2.0989
−256	16.623	1.9267	7.8696	26.239	−121.01	97.366	1.2043	2.2765	−184	173.50	158.81	0.85127	21.962	−56.395	115.82	1.4699	2.0946
−254	18.152	3.4562	7.2585	26.137	−119.33	98.138	1.2125	2.2699	−182	182.02	167.33	0.81032	21.816	−54.400	115.94	1.4769	2.0903
−252	19.787	5.0910	6.7054	26.034	−117.65	98.898	1.2205	2.2633	−180	190.84	176.14	0.77158	21.666	−52.388	116.04	1.4838	2.0860
−250	21.532	6.8357	6.2040	25.931	−115.97	99.648	1.2286	2.2569	−178	199.95	185.26	0.73489	21.514	−50.357	116.10	1.4908	2.0817
−248	23.391	8.6951	5.7485	25.827	−114.27	100.39	1.2365	2.2507	−176	209.38	194.68	0.70012	21.359	−48.305	116.13	1.4977	2.0774
−246	25.370	10.674	5.3340	25.722	−112.58	101.11	1.2444	2.2446	−174	219.12	204.42	0.66715	21.202	−46.234	116.13	1.5047	2.0731
−244	27.472	12.776	4.9562	25.616	−110.88	101.83	1.2523	2.2386	−172	229.18	214.48	0.63584	21.041	−44.140	116.09	1.5117	2.0687
−242	29.704	15.008	4.6113	25.510	−109.17	102.53	1.2601	2.2327	−170	239.56	224.87	0.60610	20.877	−42.023	116.01	1.5187	2.0643
−240	32.069	17.373	4.2959	25.403	−107.46	103.22	1.2678	2.2269	−168	250.28	235.58	0.57782	20.709	−39.882	115.89	1.5258	2.0598
−238	34.573	19.877	4.0069	25.295	−105.74	103.90	1.2755	2.2213	−166	261.34	246.64	0.55090	20.537	−37.715	115.73	1.5328	2.0553
−236	37.221	22.525	3.7418	25.186	−104.02	104.57	1.2832	2.2158	−164	272.75	258.05	0.52527	20.362	−35.521	115.53	1.5399	2.0508
−234	40.017	25.321	3.4983	25.076	−102.29	105.22	1.2908	2.2103	−162	284.51	269.81	0.50083	20.182	−33.298	115.28	1.5471	2.0462
−232	42.967	28.271	3.2742	24.966	−100.56	105.85	1.2983	2.2050	−160	296.63	281.93	0.47752	19.998	−31.045	114.98	1.5542	2.0415
−230	46.075	31.379	3.0677	24.855	−98.817	106.48	1.3059	2.1997	−155	328.55	313.86	0.42370	19.516	−25.265	114.00	1.5724	2.0295
−228	49.348	34.652	2.8772	24.743	−97.071	107.09	1.3133	2.1946	−150	362.90	348.21	0.37547	18.998	−19.246	112.63	1.5909	2.0168
−226	52.789	38.093	2.7012	24.630	−95.318	107.68	1.3207	2.1895	−145	399.81	385.11	0.33196	18.437	−12.937	110.81	1.6099	2.0032
−224	56.405	41.709	2.5384	24.516	−93.558	108.26	1.3281	2.1845	−140	439.41	424.71	0.29239	17.821	−6.2642	108.43	1.6297	1.9885
−222	60.200	45.504	2.3876	24.401	−91.792	108.82	1.3355	2.1795	−135	481.88	467.18	0.25597	17.132	0.8920	105.30	1.6505	1.9721
−220	64.179	49.483	2.2477	24.285	−90.018	109.37	1.3428	2.1747	−130	527.42	512.73	0.22182	16.335	8.7538	101.12	1.6730	1.9532
−218	68.349	53.653	2.1178	24.168	−88.236	109.89	1.3500	2.1699	−125	576.50	561.60	0.18855	15.350	17.826	95.192	1.6986	1.9298
−216	72.713	58.017	1.9970	24.050	−86.447	110.41	1.3573	2.1651	−120	628.91	614.22	0.15232	13.902	29.880	85.179	1.7323	1.8951
−214	77.278	62.582	1.8846	23.930	−84.649	110.90	1.3645	2.1605	−116.7c	666.4	651.8	0.09877	10.12	56.91	56.91	1.810	1.810
−212	82.049	67.353	1.7799	23.810	−82.843	111.38	1.3716	2.1558									

*in. Hg vacuum **temperatures are on the IPTS-68 scale a = triple point b = normal boiling point c = critical point

Temp, °F	Viscosity, lb$_m$/ft·h Sat. Liquid	Viscosity Sat. Vapor	Viscosity Gas at $p = 1$ atm × 10^{-2}*	Thermal Conductivity, Btu/h·ft·°F Sat. Liquid	Thermal Conductivity Sat. Vapor	Thermal Conductivity Gas at $p = 1$ atm × 10^{-3}*	Specific Heat c_p, Btu/lb$_m$·°F Sat. Liquid	Specific Heat Sat. Vapor	Specific Heat Gas at $p = 0$ atm	Specific Heat Gas at $p = 1$ atm	Temp., °F
−290	0.436	0.0094		0.127	0.0055		0.797				−290
−280	0.372	0.0099		0.122	0.0059		0.804		0.497		−280
−270	0.324	0.0103		0.117	0.0064		0.811		0.496		−270
−260	0.285	0.0108		0.112	0.0069		0.820	0.509	0.496		−260
−240	0.229	0.0117	1.18	0.102	0.0078	7.6	0.841	0.586	0.496		−240
−220	0.191	0.0127	1.27	0.093	0.0088	8.3	0.871	0.644	0.495		−220
−200	0.165	0.0139	1.37	0.083	0.0099	9.0	0.912	0.698	0.495		−200
−180	0.145	0.0156	1.47	0.074	0.0110	9.6	0.966	0.764	0.496		−180
−160	0.120	0.0177	1.57	0.064	0.0126	10.4	1.074	0.858	0.497		−160
−140	0.086	0.021	1.67	0.054	0.015	11.1	1.30	1.03	0.497	0.505	−140
−120	0.057	0.026	1.77	0.038	0.023	11.8	1.71	1.85	0.498	0.505	−120
−116.7c	0.040	0.040	1.79	0.029	0.029	12.0			0.499	0.505	−116.7
−110			1.82			12.2			0.499	0.505	−110
−100			1.87			12.6			0.499	0.505	−100
− 60			2.07			14.1			0.503	0.507	− 60
−40			2.17			14.9			0.506	0.509	−40
−20			2.26			15.7			0.508	0.512	−20
0			2.35			16.5			0.511	0.514	0
20			2.44			17.3			0.515	0.517	20
40			2.53			18.1			0.519	0.522	40
60			2.62			19.0			0.526	0.528	60
80			2.71			19.9			0.533	0.535	80
100			2.79			20.7			0.540	0.542	100
120			2.88			21.5			0.548	0.549	120
140			2.96			22.4			0.556	0.557	140
160			3.05			23.3			0.564	0.565	160
180			3.13			24.2			0.572	0.573	180
200			3.21			25.1			0.580	0.581	200
300			3.60			30.4			0.624	0.625	300
400			3.96			36.4			0.671	0.672	400
500			4.31			42.5			0.720	0.720	500

c = Critical temperature. Tabulated properties ignore critical region effects. *Actual value = (Table value) × (Indicated multiplier).

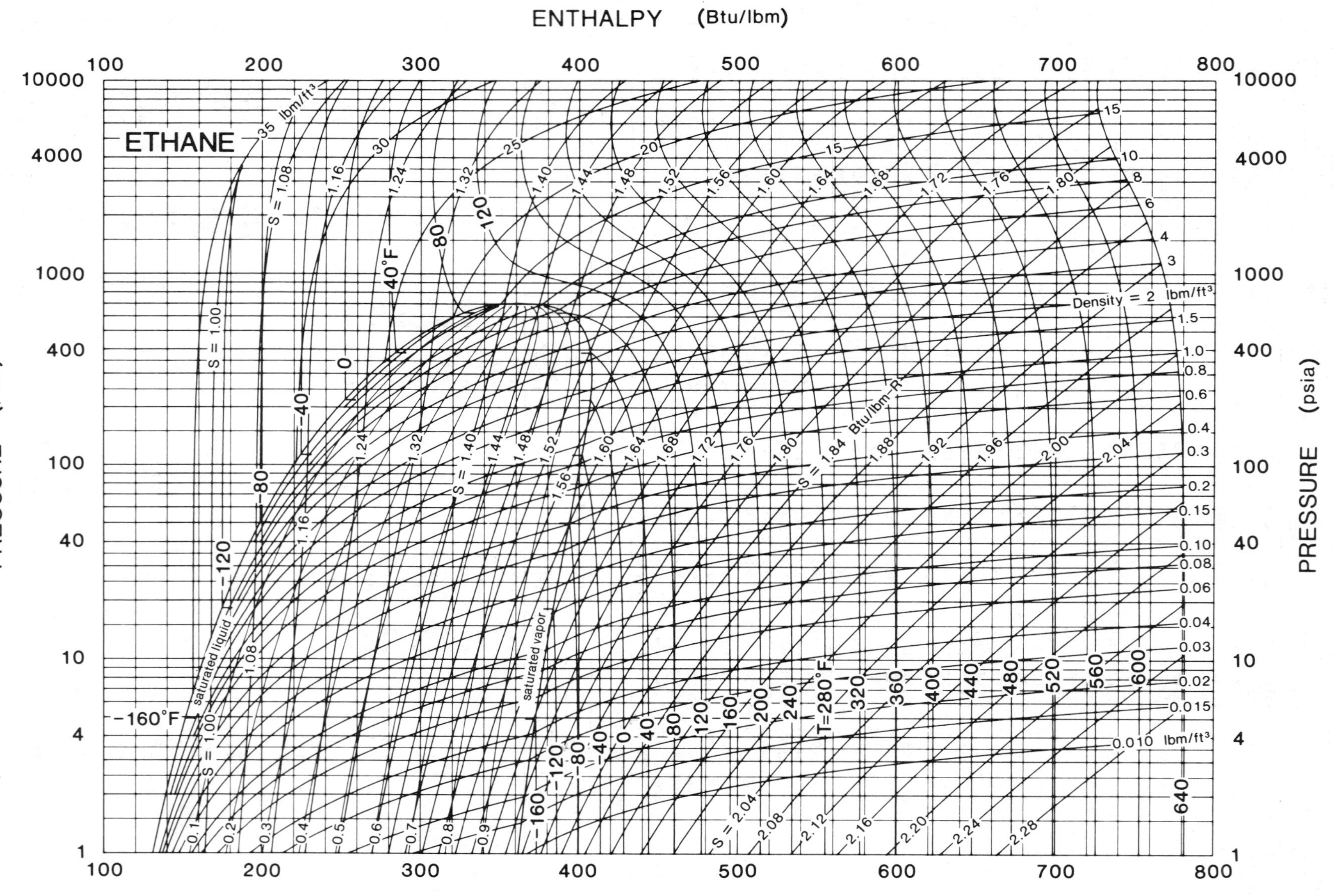

Fig. 25 Pressure–Enthalpy Diagram for Refrigerant 170 (Ethane)

Refrigerant 170 (Ethane) Properties of Saturated Liquid and Saturated Vapor

Temp,** °F	Pressure psia	Pressure psig	Volume, ft^3/lb_m Vapor	Density, lb_m/ft^3 Liquid	Enthalpy, Btu/lb_m Liquid	Enthalpy, Btu/lb_m Vapor	Entropy, $Btu/lb_m \cdot °F$ Liquid	Entropy, $Btu/lb_m \cdot °F$ Vapor
−297.04a	0.00016	29.921*	351536.	40.698	76.078	332.08	0.61190	2.1861
−280	0.00155	29.918*	41296.	40.050	85.317	337.11	0.66592	2.0661
−270	0.00471	29.912*	14350.	39.670	90.758	340.06	0.69540	2.0081
−260	0.01268	29.895*	5617.1	39.288	96.212	343.01	0.72343	1.9575
−250	0.03075	29.859*	2433.0	38.904	101.68	345.97	0.75013	1.9131
−240	0.06822	29.782*	1149.3	38.520	107.15	348.93	0.77565	1.8741
−230	0.14010	29.636*	584.95	38.133	112.64	351.90	0.80008	1.8397
−220	0.26909	29.373*	317.59	37.744	118.15	354.87	0.82354	1.8091
−210	0.48753	28.929*	182.41	37.352	123.67	357.84	0.84610	1.7820
−200	0.83922	28.213*	110.04	36.958	129.21	360.80	0.86785	1.7578
−190	1.3809	27.110*	69.301	36.560	134.77	363.76	0.88887	1.7362
−180	2.1836	25.475*	45.331	36.158	140.37	366.70	0.90921	1.7169
−170	3.3327	23.136*	30.660	35.752	145.99	369.62	0.92895	1.6995
−160	4.9289	19.886*	21.360	35.341	151.65	372.51	0.94812	1.6838
−150	7.0872	15.492*	15.276	34.925	157.34	375.37	0.96679	1.6696
−140	9.9369	9.6895*	11.183	34.503	163.09	378.18	0.98499	1.6567
−130	13.620	2.1903*	8.3575	34.074	168.88	380.94	1.0028	1.6450
−128	14.470	0.4595*	7.9027	33.988	170.04	381.49	1.0063	1.6428
−127.48b	14.696	0.0000	7.7904	33.966	170.34	381.63	1.0072	1.6422
−126	15.361	0.6651	7.4780	33.901	171.21	382.03	1.0098	1.6406
−124	16.294	1.5980	7.0811	33.814	172.38	382.57	1.0133	1.6385
−122	17.270	2.5743	6.7099	33.726	173.55	383.11	1.0167	1.6364
−120	18.291	3.5954	6.3625	33.638	174.73	383.64	1.0202	1.6343
−118	19.358	4.6625	6.0370	33.550	175.90	384.17	1.0236	1.6323
−116	20.473	5.7771	5.7318	33.462	177.08	384.70	1.0270	1.6303
−114	21.636	6.9405	5.4455	33.373	178.27	385.23	1.0304	1.6283
−112	22.850	8.1541	5.1767	33.284	179.45	385.75	1.0338	1.6264
−110	24.115	9.4194	4.9240	33.194	180.64	386.27	1.0372	1.6245
−108	25.434	10.738	4.6865	33.104	181.83	386.79	1.0406	1.6226
−106	26.807	12.111	4.4630	33.014	183.02	387.30	1.0440	1.6208
−104	28.236	13.540	4.2525	32.923	184.22	387.81	1.0473	1.6190
−102	29.722	15.026	4.0541	32.832	185.41	388.32	1.0506	1.6172
−100	31.267	16.571	3.8671	32.741	186.61	388.82	1.0540	1.6154
−98	32.873	18.177	3.6906	32.649	187.82	389.32	1.0573	1.6137
−96	34.540	19.844	3.5240	32.556	189.03	389.82	1.0606	1.6120
−94	36.271	21.575	3.3666	32.464	190.24	390.31	1.0639	1.6103
−92	38.067	23.371	3.2178	32.370	191.45	390.80	1.0672	1.6087
−90	39.929	25.233	3.0770	32.277	192.67	391.29	1.0704	1.6071
−88	41.860	27.164	2.9437	32.183	193.89	391.77	1.0737	1.6055
−86	43.860	29.164	2.8175	32.088	195.11	392.24	1.0769	1.6039
−84	45.931	31.235	2.6980	31.993	196.34	392.72	1.0802	1.6023
−82	48.075	33.379	2.5846	31.898	197.57	393.18	1.0834	1.6008
−80	50.293	35.597	2.4770	31.802	198.80	393.65	1.0866	1.5993
−78	52.587	37.891	2.3748	31.705	200.04	394.11	1.0899	1.5978
−76	54.959	40.263	2.2778	31.608	201.28	394.56	1.0931	1.5963
−74	57.410	42.714	2.1857	31.510	202.53	395.01	1.0963	1.5948
−72	59.942	45.246	2.0980	31.412	203.78	395.46	1.0995	1.5934
−70	62.556	47.860	2.0147	31.314	205.03	395.90	1.1026	1.5919
−65	69.462	54.766	1.8233	31.064	208.18	396.97	1.1106	1.5884
−60	76.919	62.223	1.6537	30.811	211.36	398.02	1.1185	1.5850
−55	84.954	70.258	1.5029	30.554	214.56	399.03	1.1263	1.5817
−50	93.593	78.897	1.3684	30.293	217.79	400.00	1.1341	1.5785
−45	102.86	88.167	1.2482	30.027	221.06	400.93	1.1419	1.5753
−40	112.79	98.094	1.1404	29.756	224.36	401.82	1.1497	1.5722
−35	123.40	108.71	1.0435	29.479	227.69	402.67	1.1574	1.5691
−30	134.73	120.03	0.95614	29.198	231.06	403.47	1.1651	1.5661
−25	146.80	132.10	0.87726	28.910	234.47	404.22	1.1728	1.5631
−20	159.63	144.94	0.80583	28.615	237.91	404.92	1.1805	1.5601
−15	173.27	158.57	0.74102	28.314	241.41	405.56	1.1882	1.5571
−10	187.73	173.04	0.68207	28.005	246.95	406.14	1.1959	1.5542
−5	203.05	188.36	0.62833	27.688	248.54	406.65	1.2036	1.5512
0	219.26	204.57	0.57924	27.362	252.18	407.10	1.2114	1.5482
10	254.47	239.77	0.49307	26.680	259.68	407.76	1.2270	1.5422
20	293.61	278.91	0.42022	25.950	267.47	408.06	1.2428	1.5359
30	336.96	322.26	0.35806	25.161	275.58	407.93	1.2589	1.5292
40	384.81	370.11	0.30448	24.296	284.10	407.25	1.2754	1.5219
50	437.50	422.80	0.25771	23.329	293.15	405.86	1.2925	1.5137
60	495.41	480.71	0.21615	22.215	302.94	403.46	1.3106	1.5041
70	559.03	544.33	0.17813	20.862	313.88	399.45	1.3304	1.4921
80	629.07	614.37	0.14098	19.023	327.13	392.28	1.3540	1.4748
89.9c	706.5	691.8	0.1120	8.928	384.3	384.3	1.457	1.457

*in. Hg vacuum **temperatures are on the IPTS-68 scale a = triple point b = normal boiling point c = critical point

Temp, °F	Viscosity, $lb_m/ft \cdot h$ Sat. Liquid	Viscosity Sat. Vapor	Viscosity Gas at $p = 1$ atm $\times 10^{-2}$*	Thermal Conductivity, $Btu/h \cdot ft \cdot °F$ Sat. Liquid	Thermal Conductivity Sat. Vapor	Thermal Conductivity Gas at $p = 1$ atm $\times 10^{-3}$*	Specific Heat c_p, $Btu/lb_m \cdot °F$ Sat. Liquid	Specific Heat Sat. Vapor	Specific Heat Gas at $p = 0$ atm	Specific Heat Gas at $p = 1$ atm	Temp, °F
−160	0.48			0.0940	0.0039		0.5665		0.3176		−160
−140	0.41			0.0896	0.0049		0.5736		0.3245		−140
−120	0.36	0.0147	1.47	0.0852	0.0057	5.41	0.5825		0.3317	0.346	−120
−100	0.32	0.0160	1.55	0.0808	0.0064	5.99	0.5934	0.3808	0.3391	0.352	−100
−80	0.29	0.0172	1.64	0.0765	0.0070	6.61	0.6067	04147	0.3468	0.357	−80
−60	0.26	0.0180	1.72	0.0721	0.0076	7.25	0.6227	0.4340	0.3549	0.364	−60
−40	0.24	0.0187	1.80	0.0677	0.0083	7.93	0.6511	0.4508	0.3630	0.371	−40
−20	0.21	0.0196	1.88	0.0633	0.0093	8.63	0.6702	0.4771	0.3715	0.379	−20
0	0.181	0.0208	1.96	0.0589	0.0106	9.35	0.6931	0.5248	0.3803	0.387	0
20	0.158	0.023	2.05	0.0545	0.0122	10.11	0.7345	0.6060	0.3893	0.396	20
40	0.135	0.025	2.13	0.0501	0.0144	10.9	0.8091	0.7413	0.3984	0.405	40
60	0.114	0.028	2.21	0.0458	0.0169	11.7	0.9314	0.9040	0.4097	0.415	60
70	0.102	0.030	2.24	0.0428	0.0197	11.9		1.041	0.4153	0.420	70
80	0.088	0.034	2.28	0.0391	0.0228	12.6			0.4210	0.424	80
89.9c	0.052	0.052	2.32	0.0312	0.0312	13.0			0.4266	0.429	89.9
100			2.36			13.4			0.4323	0.435	100
120			2.44			14.3			0.4436	0.445	120
140			2.52			15.2			0.4548	0.456	140
180			2.67			17.0			0.4773		180
220			2.82			18.9			0.4996		220
260			2.96			20.8			0.5218		260
300			3.11			22.7			0.5439		300
400			3.45			27.8			0.5979		400
500			3.78			33.0			0.6501		500
600			4.10			38.4			0.6998		600
700			4.40			43.9			0.7467		700
800			4.69			49.7			0.7901		800
1000			5.23			63.6			0.8662		1000

c = Critical temperature. Tabulated properties ignore critical region effects. *Actual value = (Table value) × (Indicated multiplier).

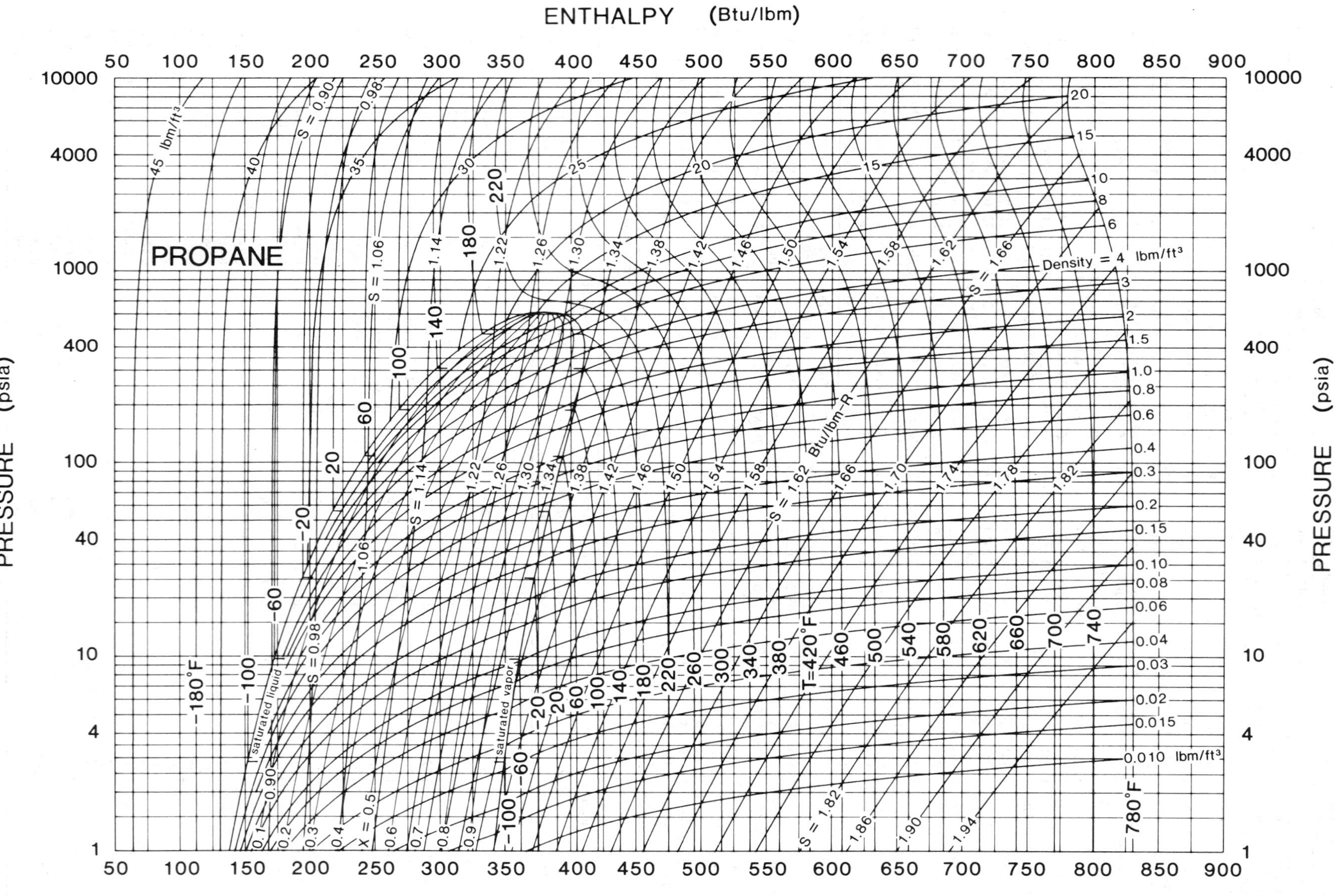

Prepared by: CENTER FOR APPLIED THERMODYNAMIC STUDIES, University of Idaho

Fig. 26 Pressure–Enthalpy Diagram for Refrigerant 290 (Propane)
Note: The reference states for enthalpy and entropy differ from those in the table.

Refrigerant 290 (Propane) Properties of Saturated Liquid and Saturated Vapor

Temp,* °F	Pressure, psia	Density, lb/ft³ Liquid	Volume, ft³/lb Vapor	Enthalpy, Btu/lb Liquid	Enthalpy, Btu/lb Vapor	Entropy, Btu/lb·°F Liquid	Entropy, Btu/lb·°F Vapor	Specific Heat c_p, Btu/lb·°F Liquid	Specific Heat c_p, Btu/lb·°F Vapor	c_p/c_v Vapor	Velocity of Sound, ft/s Liquid	Velocity of Sound, ft/s Vapor	Viscosity, lb_m/ft·h Liquid	Viscosity, lb_m/ft·h Vapor	Thermal Cond, Btu/h·ft·°F Liquid	Thermal Cond, Btu/h·ft·°F Vapor	Surface Tension, dyne/cm	Temp,* °F
−200.00	0.020	42.07	3133.0	−80.222	137.497	−0.23927	0.59918	0.4746	0.2596	1.2104	5653.	595.	1.800	0.0100	0.1072	0.00313	28.59	−200.00
−190.00	0.040	41.72	1636.4	−75.465	140.102	−0.22129	0.57808	0.4768	0.2638	1.2066	5531.	605.	1.589	0.0104	0.1053	0.00330	27.74	−190.00
−180.00	0.075	41.36	901.99	−70.685	142.743	−0.20389	0.55925	0.4792	0.2681	1.2031	5410.	615.	1.416	0.0107	0.1033	0.00348	26.90	−180.00
−170.00	0.135	41.01	521.52	−65.881	145.416	−0.18701	0.54243	0.4817	0.2725	1.1998	5289.	625.	1.272	0.0111	0.1013	0.00366	26.06	−170.00
−160.00	0.231	40.65	314.68	−61.050	148.122	−0.17062	0.52739	0.4845	0.2770	1.1967	5169.	635.	1.150	0.0114	0.0992	0.00384	25.22	−160.00
−150.00	0.381	40.29	197.25	−56.189	150.856	−0.15466	0.51393	0.4875	0.2818	1.1938	5049.	644.	1.046	0.0117	0.0971	0.00404	24.39	−150.00
−145.00	0.482	40.11	158.22	−53.747	152.233	−0.14684	0.50775	0.4891	0.2842	1.1925	4990.	648.	1.000	0.0119	0.0960	0.00414	23.98	−145.00
−140.00	0.605	39.93	127.95	−51.297	153.616	−0.13912	0.50190	0.4908	0.2867	1.1912	4931.	653.	0.956	0.0121	0.0949	0.00424	23.56	−140.00
−135.00	0.754	39.75	104.26	−48.838	155.005	−0.13149	0.49636	0.4925	0.2892	1.1900	4871.	657.	0.916	0.0123	0.0939	0.00434	23.15	−135.00
−130.00	0.932	39.57	85.583	−46.370	156.400	−0.12395	0.49112	0.4943	0.2918	1.1889	4812.	661.	0.878	0.0124	0.0928	0.00444	22.74	−130.00
−125.00	1.144	39.39	70.737	−43.893	157.800	−0.11649	0.48617	0.4962	0.2945	1.1879	4753.	665.	0.843	0.0126	0.0917	0.00455	22.33	−125.00
−120.00	1.394	39.20	58.851	−41.406	159.205	−0.10912	0.48148	0.4981	0.2972	1.1869	4694.	669.	0.810	0.0128	0.0906	0.00466	21.92	−120.00
−115.00	1.688	39.02	49.269	−38.910	160.614	−0.10183	0.47705	0.5001	0.3000	1.1860	4635.	673.	0.779	0.0130	0.0896	0.00477	21.51	−115.00
−110.00	2.031	38.84	41.493	−36.403	162.028	−0.09461	0.47287	0.5022	0.3029	1.1853	4576.	677.	0.749	0.0132	0.0885	0.00488	21.10	−110.00
−105.00	2.429	38.65	35.141	−33.885	163.444	−0.08747	0.46891	0.5043	0.3059	1.1846	4517.	681.	0.722	0.0133	0.0874	0.00499	20.70	−105.00
−100.00	2.889	38.46	29.922	−31.357	164.864	−0.08040	0.46516	0.5066	0.3089	1.1840	4458.	684.	0.695	0.0135	0.0864	0.00511	20.29	−100.00
−95.00	3.417	38.28	25.608	−28.817	166.287	−0.07339	0.46163	0.5088	0.3120	1.1834	4400.	688.	0.671	0.0137	0.0853	0.00523	19.89	−95.00
−90.00	4.020	38.09	22.023	−26.265	167.713	−0.06645	0.45828	0.5112	0.3152	1.1830	4341.	691.	0.647	0.0139	0.0842	0.00535	19.48	−90.00
−85.00	4.708	37.90	19.027	−23.700	169.140	−0.05957	0.45513	0.5137	0.3184	1.1827	4282.	694.	0.625	0.0141	0.0832	0.00547	19.08	−85.00
−80.00	5.486	37.71	16.512	−21.124	170.569	−0.05274	0.45215	0.5162	0.3218	1.1825	4224.	697.	0.604	0.0143	0.0821	0.00559	18.68	−80.00
−75.00	6.364	37.52	14.389	−18.533	171.999	−0.04598	0.44934	0.5188	0.3252	1.1825	4165.	700.	0.584	0.0144	0.0811	0.00572	18.28	−75.00
−70.00	7.351	37.33	12.589	−15.930	173.430	−0.03927	0.44668	0.5215	0.3287	1.1825	4107.	703.	0.564	0.0146	0.0801	0.00585	17.89	−70.00
−65.00	8.456	37.13	11.057	−13.313	174.861	−0.03261	0.44418	0.5242	0.3323	1.1826	4048.	706.	0.546	0.0148	0.0790	0.00598	17.49	−65.00
−60.00	9.689	36.94	9.7463	−10.681	176.292	−0.02599	0.44182	0.5271	0.3360	1.1829	3989.	709.	0.529	0.0150	0.0780	0.00612	17.10	−60.00
−55.00	11.059	36.74	8.6209	−8.034	177.722	−0.01943	0.43960	0.5300	0.3398	1.1833	3931.	711.	0.512	0.0152	0.0770	0.00625	16.70	−55.00
−50.00	12.577	36.55	7.6507	−5.372	179.151	−0.01291	0.43751	0.5331	0.3436	1.1838	3872.	713.	0.496	0.0154	0.0760	0.00639	16.31	−50.00
−45.00	14.253	36.35	6.8111	−2.694	180.579	−0.00644	0.43554	0.5362	0.3476	1.1845	3814.	715.	0.481	0.0156	0.0750	0.00653	15.92	−45.00
−43.76b	14.696	36.30	6.6203	−2.026	180.934	−0.00483	0.43507	0.5370	0.3486	1.1847	3799.	716.	0.477	0.0156	0.0748	0.00657	15.82	−43.76
−40.00	16.099	36.15	6.0819	0.000	182.005	0.00000	0.43369	0.5394	0.3517	1.1853	3755.	717.	0.466	0.0158	0.0741	0.00668	15.53	−40.00
−35.00	18.126	35.94	5.4464	2.711	183.429	0.00640	0.43195	0.5427	0.3558	1.1863	3696.	719.	0.452	0.0160	0.0731	0.00683	15.14	−35.00
−30.00	20.344	35.74	4.8906	5.439	184.849	0.01276	0.43031	0.5462	0.3601	1.1874	3638.	721.	0.438	0.0162	0.0721	0.00698	14.76	−30.00
−25.00	22.767	35.54	4.4031	8.185	186.267	0.01908	0.42878	0.5497	0.3644	1.1887	3579.	722.	0.425	0.0164	0.0712	0.00713	14.37	−25.00
−20.00	25.405	35.33	3.9740	10.949	187.680	0.02537	0.42733	0.5533	0.3689	1.1902	3520.	724.	0.413	0.0166	0.0702	0.00729	13.99	−20.00
−15.00	28.272	35.12	3.5952	13.732	189.089	0.03163	0.42598	0.5571	0.3735	1.1918	3462.	725.	0.401	0.0168	0.0693	0.00745	13.61	−15.00
−10.00	31.380	34.91	3.2600	16.535	190.493	0.03786	0.42472	0.5610	0.3782	1.1936	3403.	726.	0.389	0.0170	0.0684	0.00761	13.23	−10.00
−5.00	34.742	34.69	2.9624	19.357	191.891	0.04407	0.42354	0.5650	0.3830	1.1957	3344.	726.	0.378	0.0172	0.0675	0.00778	12.86	−5.00
0.00	38.370	34.48	2.6975	22.201	193.283	0.05024	0.42243	0.5691	0.3880	1.1979	3285.	727.	0.367	0.0174	0.0666	0.00795	12.48	0.00
5.00	42.278	34.26	2.4611	25.065	194.668	0.05639	0.42139	0.5733	0.3931	1.2004	3226.	727.	0.357	0.0176	0.0657	0.00812	12.11	5.00
10.00	46.480	34.04	2.2496	27.951	196.045	0.06252	0.42042	0.5777	0.3983	1.2032	3167.	727.	0.346	0.0179	0.0648	0.00829	11.74	10.00
15.00	50.988	33.81	2.0600	30.860	197.414	0.06863	0.41952	0.5823	0.4037	1.2061	3107.	727.	0.337	0.0181	0.0639	0.00847	11.37	15.00
20.00	55.817	33.59	1.8895	33.793	198.774	0.07472	0.41867	0.5870	0.4092	1.2094	3048.	727.	0.327	0.0183	0.0630	0.00866	11.00	20.00
25.00	60.980	33.36	1.7359	36.749	200.124	0.08079	0.41788	0.5918	0.4149	1.2130	2989.	726.	0.318	0.0185	0.0622	0.00884	10.64	25.00
30.00	66.492	33.13	1.5973	39.730	201.463	0.08685	0.41714	0.5969	0.4208	1.2169	2929.	725.	0.309	0.0188	0.0613	0.00903	10.27	30.00
35.00	72.367	32.89	1.4718	42.736	202.790	0.09289	0.41645	0.6021	0.4269	1.2212	2869.	724.	0.300	0.0190	0.0605	0.00923	9.91	35.00
40.00	78.619	32.65	1.3580	45.769	204.105	0.09892	0.41580	0.6075	0.4332	1.2258	2810.	723.	0.292	0.0192	0.0597	0.00943	9.55	40.00
45.00	85.263	32.41	1.2547	48.829	205.406	0.10494	0.41520	0.6131	0.4397	1.2309	2750.	722.	0.283	0.0195	0.0588	0.00963	9.20	45.00
50.00	92.314	32.16	1.1606	51.918	206.691	0.11095	0.41462	0.6190	0.4465	1.2364	2690.	720.	0.275	0.0197	0.0580	0.00984	8.85	50.00
55.00	99.786	31.91	1.0749	55.035	207.961	0.11695	0.41409	0.6251	0.4535	1.2425	2630.	718.	0.267	0.0200	0.0572	0.01005	8.49	55.00
60.00	107.70	31.66	0.9965	58.183	209.212	0.12295	0.41358	0.6314	0.4609	1.2491	2570.	715.	0.260	0.0202	0.0564	0.01027	8.15	60.00
65.00	116.06	31.40	0.9247	61.362	210.445	0.12894	0.41309	0.6381	0.4685	1.2563	2509.	713.	0.252	0.0205	0.0556	0.01049	7.80	65.00
70.00	124.89	31.13	0.8590	64.573	211.657	0.13494	0.41263	0.6450	0.4765	1.2642	2449.	710.	0.245	0.0208	0.0548	0.01072	7.46	70.00
75.00	134.20	30.86	0.7985	67.818	212.846	0.14093	0.41218	0.6523	0.4850	1.2728	2388.	707.	0.238	0.0210	0.0540	0.01095	7.12	75.00
80.00	144.01	30.59	0.7430	71.098	214.011	0.14692	0.41174	0.6599	0.4938	1.2823	2327.	703.	0.231	0.0213	0.0533	0.01119	6.78	80.00
85.00	154.34	30.31	0.6917	74.414	215.149	0.15293	0.41131	0.6680	0.5032	1.2928	2266.	699.	0.224	0.0216	0.0525	0.01144	6.45	85.00
90.00	165.20	30.02	0.6445	77.768	216.259	0.15893	0.41089	0.6765	0.5131	1.3043	2205.	695.	0.218	0.0219	0.0517	0.01169	6.12	90.00
95.00	176.61	29.73	0.6008	81.162	217.337	0.16495	0.41046	0.6855	0.5237	1.3171	2144.	691.	0.211	0.0222	0.0510	0.01195	5.79	95.00
100.00	188.58	29.43	0.5603	84.598	218.380	0.17098	0.41002	0.6951	0.5350	1.3313	2082.	686.	0.205	0.0225	0.0502	0.01222	5.47	100.00
110.00	214.30	28.80	0.4879	91.602	220.352	0.18310	0.40911	0.7163	0.5604	1.3648	1958.	675.	0.192	0.0232	0.0487	0.01279	4.83	110.00
120.00	242.50	28.14	0.4252	98.799	222.142	0.19530	0.40808	0.7408	0.5903	1.4071	1832.	663.	0.180	0.0239	0.0472	0.01340	4.21	120.00
130.00	273.33	27.44	0.3706	106.214	223.711	0.20763	0.40689	0.7699	0.6266	1.4618	1705.	649.	0.168	0.0247	0.0457	0.01408	3.60	130.00
140.00	306.96	26.68	0.3227	113.879	225.009	0.22014	0.40545	0.8055	0.6723	1.5342	1576.	634.	0.157	0.0256	0.0442	0.01485	3.02	140.00
150.00	343.57	25.87	0.2803	121.839	225.962	0.23287	0.40366	0.8509	0.7322	1.6340	1444.	616.	0.145	0.0266	0.0428	0.01576	2.46	150.00
160.00	383.36	24.97	0.2425	130.161	226.466	0.24594	0.40135	0.9123	0.8154	1.7785	1308.	597.	0.134	0.0278	0.0414	0.01688	1.92	160.00
170.00	426.55	23.96	0.2084	138.952	226.358	0.25949	0.39830	1.0029	0.9410	2.0040	1165.	575.	0.123	0.0292	0.0403	0.01836	1.41	170.00
180.00	473.42	22.78	0.1770	148.412	225.346	0.27381	0.39408	1.1563	1.1564	2.4021	1013.	551.	0.111	0.0311	0.0394	0.02050	0.94	180.00
190.00	524.30	21.30	0.1469	158.996	222.796	0.28956	0.38776	1.4915	1.6272	3.2904	847.	523.	0.098	0.0336	0.0392	0.02399	0.51	190.00
200.00	579.74	19.07	0.1147	172.340	216.510	0.30916	0.37612	—	—	—	—	—	0.082	0.0382	—	—	0.15	200.00
206.06c	616.07	13.77	0.0726	195.358	195.358	0.34325	0.34325	∞	∞	∞	0.	0.	—	—	∞	∞	0.00	206.06

*temperatures are on the IPTS-68 scale b = normal boiling point c = critical point

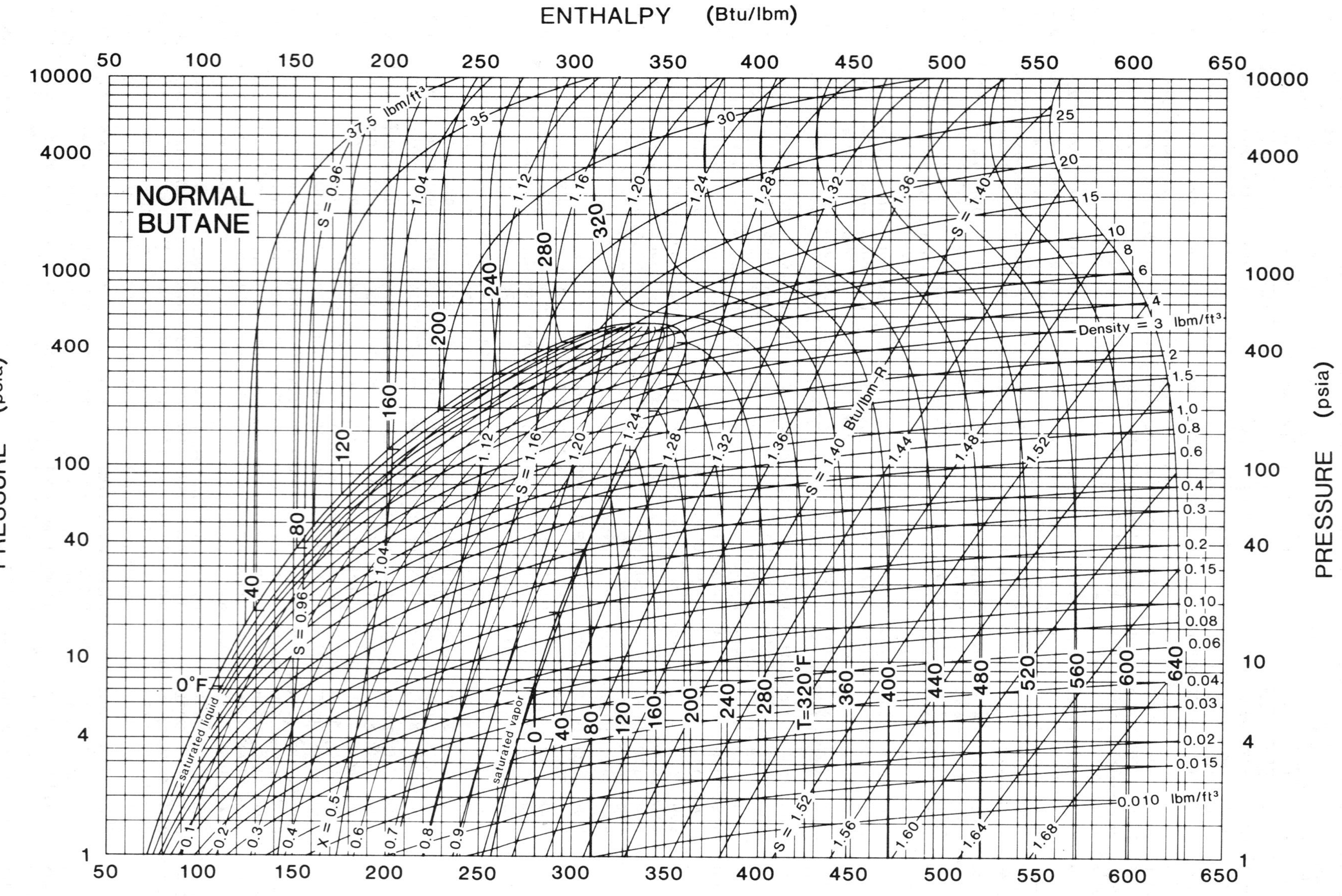

Fig. 27 Pressure–Enthalpy Diagram for Refrigerant 600 (*n*-Butane)

Note: The reference states for enthalpy and entropy differ from those in the table.

Refrigerant 600 (*n*-Butane) Properties of Saturated Liquid and Saturated Vapor

Temp,* °F	Pressure, psia	Density, lb/ft³ Liquid	Volume, ft³/lb Vapor	Enthalpy, Btu/lb Liquid	Enthalpy, Btu/lb Vapor	Entropy, Btu/lb·°F Liquid	Entropy, Btu/lb·°F Vapor	Specific Heat c_p, Btu/lb·°F Liquid	Specific Heat c_p, Btu/lb·°F Vapor	c_p/c_v Vapor	Velocity of Sound, ft/s Liquid	Velocity of Sound, ft/s Vapor	Viscosity, lb_m/ft·h Liquid	Viscosity, lb_m/ft·h Vapor	Thermal Cond, Btu/h·ft·°F Liquid	Thermal Cond, Btu/h·ft·°F Vapor	Surface Tension, dyne/cm	Temp,* °F
−150.00	0.021	43.72	2680.0	−54.034	145.702	−0.14904	0.49595	0.4804	0.2930	1.1323	4976.	548.	1.945	0.0109	0.0991	0.00362	28.19	−150.00
−140.00	0.038	43.39	1536.9	−49.229	148.643	−0.13377	0.48522	0.4807	0.2971	1.1304	4906.	556.	1.737	0.0112	0.0967	0.00381	27.41	−140.00
−130.00	0.066	43.06	917.01	−44.419	151.622	−0.11895	0.47570	0.4815	0.3012	1.1286	4833.	564.	1.564	0.0115	0.0943	0.00399	26.64	−130.00
−120.00	0.110	42.74	567.05	−39.597	154.636	−0.10455	0.46728	0.4829	0.3055	1.1268	4757.	572.	1.416	0.0118	0.0921	0.00419	25.87	−120.00
−110.00	0.178	42.41	362.20	−34.759	157.687	−0.09051	0.45985	0.4848	0.3099	1.1251	4678.	579.	1.290	0.0121	0.0901	0.00439	25.11	−110.00
−100.00	0.278	42.08	238.26	−29.899	160.772	−0.07681	0.45332	0.4873	0.3145	1.1236	4597.	587.	1.182	0.0124	0.0881	0.00459	24.35	−100.00
−90.00	0.423	41.75	160.98	−25.011	163.890	−0.06340	0.44759	0.4903	0.3192	1.1221	4514.	594.	1.087	0.0127	0.0862	0.00480	23.59	−90.00
−80.00	0.626	41.41	111.45	−20.090	167.042	−0.05027	0.44261	0.4938	0.3241	1.1208	4428.	601.	1.003	0.0130	0.0843	0.00502	22.84	−80.00
−70.00	0.907	41.08	78.895	−15.133	170.225	−0.03739	0.43829	0.4976	0.3292	1.1196	4341.	608.	0.930	0.0133	0.0826	0.00524	22.10	−70.00
−60.00	1.285	40.74	56.999	−10.134	173.438	−0.02473	0.43458	0.5019	0.3344	1.1185	4252.	614.	0.865	0.0136	0.0809	0.00547	21.36	−60.00
−50.00	1.786	40.40	41.955	−5.091	176.679	−0.01227	0.43143	0.5065	0.3399	1.1177	4161.	620.	0.806	0.0139	0.0793	0.00571	20.63	−50.00
−40.00	2.438	40.06	31.414	0.000	179.946	0.00000	0.42878	0.5114	0.3456	1.1170	4070.	626.	0.754	0.0143	0.0777	0.00595	19.90	−40.00
−30.00	3.273	39.71	23.893	5.142	183.238	0.01210	0.42659	0.5165	0.3515	1.1166	3977.	631.	0.706	0.0146	0.0762	0.00621	19.18	−30.00
−20.00	4.327	39.36	18.436	10.337	186.552	0.02404	0.42483	0.5219	0.3576	1.1163	3883.	637.	0.663	0.0149	0.0747	0.00647	18.46	−20.00
−10.00	5.638	39.01	14.416	15.587	189.887	0.03583	0.42345	0.5275	0.3640	1.1163	3788.	641.	0.623	0.0153	0.0733	0.00674	17.75	−10.00
0.00	7.251	38.66	11.410	20.896	193.240	0.04749	0.42242	0.5334	0.3706	1.1166	3693.	646.	0.587	0.0156	0.0719	0.00701	17.04	0.00
5.00	8.184	38.48	10.194	23.573	194.922	0.05328	0.42203	0.5364	0.3740	1.1168	3645.	648.	0.570	0.0158	0.0712	0.00716	16.69	5.00
10.00	9.211	38.30	9.1329	26.266	196.608	0.05903	0.42171	0.5394	0.3774	1.1171	3597.	650.	0.554	0.0159	0.0705	0.00730	16.34	10.00
15.00	10.337	38.12	8.2031	28.974	198.298	0.06475	0.42147	0.5426	0.3810	1.1175	3549.	651.	0.539	0.0161	0.0698	0.00745	16.00	15.00
20.00	11.569	37.93	7.3863	31.698	199.991	0.07045	0.42130	0.5457	0.3845	1.1180	3500.	653.	0.524	0.0163	0.0691	0.00760	15.65	20.00
25.00	12.914	37.75	6.6666	34.438	201.686	0.07612	0.42119	0.5489	0.3882	1.1185	3452.	655.	0.509	0.0165	0.0685	0.00775	15.31	25.00
30.00	14.378	37.56	6.0309	37.194	203.384	0.08176	0.42115	0.5522	0.3919	1.1191	3403.	656.	0.495	0.0166	0.0678	0.00790	14.96	30.00
31.03b	14.696	37.52	5.9090	37.765	203.735	0.08292	0.42115	0.5528	0.3927	1.1193	3393.	656.	0.492	0.0167	0.0677	0.00793	14.89	31.03
35.00	15.969	37.38	5.4677	39.967	205.084	0.08738	0.42117	0.5555	0.3957	1.1198	3354.	657.	0.482	0.0168	0.0672	0.00806	14.62	35.00
40.00	17.693	37.19	4.9676	42.757	206.786	0.09297	0.42125	0.5588	0.3995	1.1206	3305.	659.	0.469	0.0170	0.0665	0.00822	14.28	40.00
45.00	19.559	37.00	4.5224	45.564	208.490	0.09854	0.42138	0.5622	0.4035	1.1215	3256.	660.	0.456	0.0172	0.0659	0.00838	13.94	45.00
50.00	21.574	36.81	4.1251	48.389	210.194	0.10409	0.42156	0.5657	0.4074	1.1225	3207.	661.	0.444	0.0174	0.0652	0.00854	13.61	50.00
55.00	23.746	36.62	3.7697	51.231	211.899	0.10962	0.42180	0.5692	0.4115	1.1236	3158.	661.	0.432	0.0176	0.0646	0.00871	13.27	55.00
60.00	26.081	36.42	3.4512	54.091	213.605	0.11513	0.42208	0.5728	0.4157	1.1248	3108.	662.	0.421	0.0178	0.0639	0.00888	12.94	60.00
65.00	28.590	36.23	3.1649	56.970	215.311	0.12062	0.42241	0.5764	0.4199	1.1261	3059.	662.	0.410	0.0179	0.0633	0.00905	12.61	65.00
70.00	31.279	36.03	2.9072	59.867	217.017	0.12609	0.42278	0.5801	0.4242	1.1276	3009.	663.	0.400	0.0181	0.0627	0.00923	12.28	70.00
75.00	34.157	35.83	2.6747	62.783	218.721	0.13154	0.42319	0.5839	0.4286	1.1291	2960.	663.	0.389	0.0183	0.0621	0.00941	11.95	75.00
80.00	37.232	35.63	2.4646	65.719	220.425	0.13697	0.42364	0.5877	0.4331	1.1308	2910.	663.	0.379	0.0185	0.0614	0.00959	11.63	80.00
85.00	40.513	35.42	2.2742	68.673	222.127	0.14239	0.42413	0.5916	0.4377	1.1326	2860.	663.	0.370	0.0187	0.0608	0.00977	11.30	85.00
90.00	44.009	35.22	2.1015	71.648	223.827	0.14780	0.42465	0.5956	0.4424	1.1346	2810.	663.	0.360	0.0189	0.0602	0.00996	10.98	90.00
95.00	47.728	35.01	1.9445	74.643	225.524	0.15318	0.42520	0.5997	0.4472	1.1367	2760.	662.	0.351	0.0191	0.0596	0.01015	10.66	95.00
100.00	51.679	34.80	1.8015	77.658	227.219	0.15856	0.42579	0.6038	0.4521	1.1390	2710.	662.	0.342	0.0194	0.0590	0.01035	10.34	100.00
105.00	55.871	34.59	1.6711	80.694	228.909	0.16392	0.42640	0.6080	0.4571	1.1415	2660.	661.	0.333	0.0196	0.0584	0.01055	10.03	105.00
110.00	60.314	34.37	1.5519	83.752	230.596	0.16927	0.42704	0.6123	0.4622	1.1442	2610.	660.	0.325	0.0198	0.0578	0.01075	9.71	110.00
115.00	65.015	34.15	1.4428	86.831	232.278	0.17461	0.42770	0.6168	0.4675	1.1470	2559.	659.	0.316	0.0200	0.0572	0.01096	9.40	115.00
120.00	69.986	33.93	1.3429	89.933	233.954	0.17993	0.42839	0.6213	0.4729	1.1501	2509.	657.	0.308	0.0202	0.0566	0.01117	9.09	120.00
125.00	75.234	33.71	1.2511	93.057	235.625	0.18525	0.42909	0.6259	0.4784	1.1534	2459.	656.	0.301	0.0204	0.0560	0.01138	8.78	125.00
130.00	80.770	33.48	1.1667	96.204	237.288	0.19056	0.42982	0.6307	0.4841	1.1570	2408.	654.	0.293	0.0207	0.0554	0.01160	8.48	130.00
135.00	86.603	33.25	1.0889	99.375	238.945	0.19586	0.43056	0.6356	0.4900	1.1609	2357.	652.	0.285	0.0209	0.0548	0.01182	8.18	135.00
140.00	92.743	33.02	1.0173	102.570	240.592	0.20115	0.43131	0.6406	0.4961	1.1650	2307.	650.	0.278	0.0211	0.0542	0.01205	7.88	140.00
145.00	99.199	32.78	0.9511	105.790	242.231	0.20644	0.43208	0.6458	0.5023	1.1695	2256.	647.	0.271	0.0214	0.0536	0.01228	7.58	145.00
150.00	105.98	32.54	0.8899	109.035	243.859	0.21172	0.43286	0.6512	0.5088	1.1744	2205.	645.	0.264	0.0216	0.0530	0.01251	7.28	150.00
155.00	113.10	32.30	0.8332	112.306	245.476	0.21700	0.43365	0.6567	0.5156	1.1796	2154.	642.	0.257	0.0219	0.0524	0.01275	6.99	155.00
160.00	120.57	32.05	0.7807	115.604	247.081	0.22227	0.43444	0.6625	0.5226	1.1853	2103.	639.	0.250	0.0221	0.0519	0.01300	6.70	160.00
165.00	128.39	31.80	0.7319	118.930	248.672	0.22754	0.43524	0.6685	0.5298	1.1915	2052.	635.	0.243	0.0224	0.0513	0.01325	6.41	165.00
170.00	136.59	31.54	0.6865	122.284	250.249	0.23281	0.43604	0.6747	0.5374	1.1982	2001.	632.	0.237	0.0227	0.0507	0.01350	6.12	170.00
175.00	145.16	31.28	0.6443	125.668	251.809	0.23809	0.43684	0.6812	0.5454	1.2055	1949.	628.	0.230	0.0229	0.0501	0.01376	5.84	175.00
180.00	154.12	31.01	0.6049	129.082	253.351	0.24336	0.43763	0.6880	0.5537	1.2135	1898.	624.	0.224	0.0232	0.0496	0.01403	5.56	180.00
185.00	163.48	30.73	0.5682	132.527	254.873	0.24864	0.43842	0.6951	0.5626	1.2222	1846.	619.	0.218	0.0235	0.0490	0.01430	5.28	185.00
190.00	173.25	30.46	0.5339	136.006	256.374	0.25392	0.43920	0.7026	0.5719	1.2318	1794.	615.	0.212	0.0238	0.0484	0.01458	5.01	190.00
195.00	183.45	30.17	0.5018	139.518	257.851	0.25921	0.43996	0.7106	0.5817	1.2424	1742.	610.	0.206	0.0241	0.0479	0.01486	4.74	195.00
200.00	194.09	29.88	0.4717	143.066	259.302	0.26451	0.44071	0.7190	0.5923	1.2540	1690.	604.	0.200	0.0244	0.0473	0.01516	4.47	200.00
210.00	216.71	29.27	0.4170	150.277	262.113	0.27515	0.44215	0.7377	0.6157	1.2814	1585.	593.	0.188	0.0251	0.0462	0.01577	3.94	210.00
220.00	241.23	28.62	0.3687	157.654	264.783	0.28585	0.44347	0.7594	0.6433	1.3158	1480.	579.	0.177	0.0258	0.0451	0.01641	3.43	220.00
230.00	267.76	27.94	0.3256	165.218	267.276	0.29664	0.44462	0.7853	0.6766	1.3599	1372.	565.	0.165	0.0266	0.0440	0.01710	2.93	230.00
240.00	296.42	27.20	0.2871	173.001	269.548	0.30757	0.44556	0.8173	0.7182	1.4183	1263.	548.	0.154	0.0274	0.0430	0.01784	2.45	240.00
250.00	327.34	26.40	0.2523	181.042	271.533	0.31868	0.44619	0.8587	0.7726	1.4988	1152.	529.	0.143	0.0284	0.0419	0.01866	1.99	250.00
260.00	360.69	25.51	0.2207	189.406	273.135	0.33005	0.44639	0.9155	0.8484	1.6157	1038.	509.	0.132	0.0295	0.0408	0.01962	1.55	260.00
270.00	396.64	24.50	0.1915	198.197	274.203	0.34181	0.44597	1.0008	0.9633	1.8000	918.	485.	0.121	0.0308	0.0396	0.02082	1.14	270.00
280.00	435.43	23.32	0.1640	207.608	274.458	0.35421	0.44459	1.1490	1.1632	2.1306	792.	459.	0.109	0.0325	0.0387	0.02256	0.75	280.00
290.00	477.36	21.80	0.1370	218.088	273.282	0.36782	0.44144	1.4852	1.6136	2.8921	655.	428.	0.096	0.0348	0.0384	0.02571	0.40	290.00
300.00	522.95	19.41	0.1068	231.321	268.472	0.38481	0.43371	—	—	—	—	—	0.079	0.0392	—	—	0.11	300.00
305.62c	550.56	14.22	0.0703	251.554	251.554	0.41093	0.41093	∞	∞	∞	0.	0.	—	—	∞	∞	0.00	305.62

*temperatures are on the IPTS-68 scale b = normal boiling point c = critical point

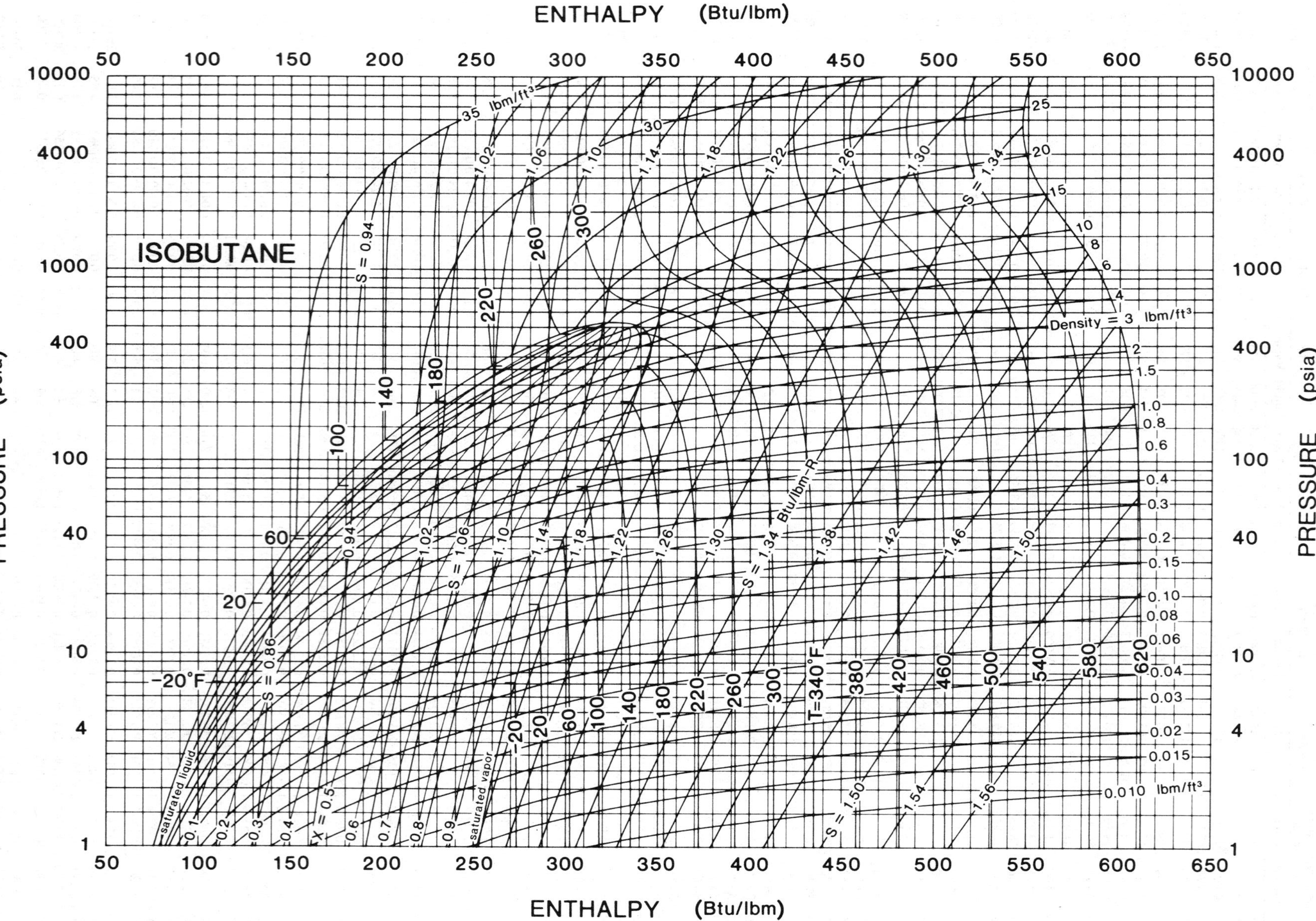

Prepared by: CENTER FOR APPLIED THERMODYNAMIC STUDIES, University of Idaho

Fig. 28 Pressure–Enthalpy Diagram for Refrigerant 600a (Isobutane)

Note: The reference states for enthalpy and entropy differ from those in the table.

Refrigerant 600a (Isobutane) Properties of Saturated Liquid and Saturated Vapor

Temp,* °F	Pressure, psia	Density, lb/ft³ Liquid	Volume, ft³/lb Vapor	Enthalpy, Btu/lb Liquid	Enthalpy, Btu/lb Vapor	Entropy, Btu/lb·°F Liquid	Entropy, Btu/lb·°F Vapor	Specific Heat c_p, Btu/lb·°F Liquid	Specific Heat c_p, Btu/lb·°F Vapor	c_p/c_v Vapor	Velocity of Sound, ft/s Liquid	Velocity of Sound, ft/s Vapor	Viscosity, lb_m/ft·h Liquid	Viscosity, lb_m/ft·h Vapor	Thermal Cond, Btu/h·ft·°F Liquid	Thermal Cond, Btu/h·ft·°F Vapor	Surface Tension, dyne/cm	Temp,* °F
−150.00	0.047	42.76	1219.7	−52.320	136.451	−0.14409	0.46549	0.4486	0.2694	1.1460	4760.	551.	2.313	0.0108	0.0799	0.00345	25.67	−150.00
−140.00	0.082	42.42	722.16	−47.811	139.153	−0.12976	0.45510	0.4532	0.2747	1.1431	4675.	559.	2.040	0.0111	0.0800	0.00363	24.95	−140.00
−130.00	0.137	42.08	443.87	−43.255	141.900	−0.11573	0.44590	0.4579	0.2801	1.1404	4589.	566.	1.813	0.0114	0.0798	0.00382	24.24	−130.00
−120.00	0.222	41.74	282.21	−38.653	144.691	−0.10198	0.43779	0.4626	0.2856	1.1379	4503.	574.	1.623	0.0117	0.0794	0.00402	23.53	−120.00
−110.00	0.348	41.40	185.01	−34.002	147.523	−0.08849	0.43064	0.4675	0.2911	1.1356	4416.	581.	1.461	0.0120	0.0789	0.00422	22.82	−110.00
−100.00	0.530	41.06	124.71	−29.302	150.396	−0.07524	0.42438	0.4724	0.2968	1.1335	4329.	588.	1.323	0.0123	0.0783	0.00443	22.11	−100.00
−90.00	0.787	40.72	86.220	−24.551	153.306	−0.06222	0.41891	0.4775	0.3026	1.1317	4241.	595.	1.203	0.0126	0.0775	0.00464	21.40	−90.00
−80.00	1.141	40.37	60.998	−19.749	156.251	−0.04940	0.41416	0.4827	0.3086	1.1301	4153.	602.	1.099	0.0130	0.0765	0.00486	20.70	−80.00
−70.00	1.618	40.02	44.072	−14.895	159.230	−0.03679	0.41006	0.4880	0.3148	1.1288	4064.	608.	1.008	0.0133	0.0755	0.00509	20.00	−70.00
−60.00	2.248	39.67	32.460	−9.986	162.240	−0.02436	0.40656	0.4934	0.3212	1.1277	3973.	614.	0.928	0.0136	0.0744	0.00533	19.30	−60.00
−50.00	3.065	39.32	24.331	−5.021	165.280	−0.01210	0.40361	0.4990	0.3278	1.1269	3883.	620.	0.856	0.0139	0.0732	0.00557	18.60	−50.00
−40.00	4.109	38.96	18.534	0.000	168.346	0.00000	0.40114	0.5047	0.3346	1.1263	3791.	625.	0.792	0.0143	0.0719	0.00583	17.91	−40.00
−30.00	5.421	38.60	14.328	5.080	171.438	0.01195	0.39912	0.5106	0.3416	1.1261	3698.	630.	0.735	0.0146	0.0706	0.00609	17.22	−30.00
−20.00	7.049	38.23	11.226	10.220	174.552	0.02376	0.39752	0.5166	0.3489	1.1261	3604.	634.	0.684	0.0150	0.0693	0.00636	16.53	−20.00
−10.00	9.042	37.86	8.9062	15.423	177.688	0.03543	0.39629	0.5228	0.3564	1.1264	3510.	638.	0.637	0.0153	0.0678	0.00664	15.85	−10.00
0.00	11.456	37.48	7.1465	20.690	180.841	0.04699	0.39540	0.5292	0.3642	1.1271	3415.	642.	0.595	0.0157	0.0664	0.00693	15.17	0.00
5.00	12.838	37.29	6.4271	23.348	182.425	0.05273	0.39507	0.5325	0.3682	1.1276	3367.	643.	0.575	0.0158	0.0657	0.00708	14.84	5.00
10.00	14.347	37.10	5.7947	26.022	184.012	0.05844	0.39482	0.5358	0.3722	1.1281	3319.	645.	0.556	0.0160	0.0650	0.00724	14.50	10.00
11.10b	14.696	37.06	5.6663	26.611	184.361	0.05969	0.39478	0.5366	0.3731	1.1283	3308.	645.	0.552	0.0161	0.0648	0.00727	14.43	11.10
15.00	15.990	36.91	5.2370	28.714	185.603	0.06412	0.39464	0.5392	0.3764	1.1288	3271.	646.	0.538	0.0162	0.0642	0.00739	14.16	15.00
20.00	17.776	36.72	4.7439	31.423	187.196	0.06978	0.39453	0.5427	0.3806	1.1295	3222.	647.	0.521	0.0164	0.0635	0.00755	13.83	20.00
25.00	19.711	36.52	4.3068	34.150	188.793	0.07541	0.39448	0.5462	0.3849	1.1304	3174.	648.	0.504	0.0166	0.0628	0.00771	13.50	25.00
30.00	21.805	36.32	3.9181	36.895	190.393	0.08103	0.39450	0.5497	0.3892	1.1313	3125.	649.	0.488	0.0168	0.0620	0.00787	13.16	30.00
35.00	24.065	36.13	3.5718	39.658	191.994	0.08662	0.39457	0.5534	0.3936	1.1324	3076.	650.	0.473	0.0170	0.0613	0.00803	12.83	35.00
40.00	26.500	35.92	3.2624	42.439	193.598	0.09219	0.39470	0.5571	0.3982	1.1336	3027.	651.	0.458	0.0172	0.0606	0.00820	12.50	40.00
45.00	29.119	35.72	2.9853	45.240	195.203	0.09774	0.39489	0.5608	0.4028	1.1349	2978.	651.	0.444	0.0174	0.0598	0.00837	12.17	45.00
50.00	31.929	35.52	2.7366	48.060	196.809	0.10327	0.39512	0.5647	0.4074	1.1363	2928.	651.	0.431	0.0176	0.0591	0.00855	11.85	50.00
55.00	34.941	35.31	2.5129	50.899	198.416	0.10878	0.39541	0.5686	0.4122	1.1379	2879.	652.	0.418	0.0178	0.0584	0.00872	11.52	55.00
60.00	38.162	35.10	2.3112	53.758	200.024	0.11428	0.39574	0.5726	0.4171	1.1396	2829.	651.	0.405	0.0180	0.0577	0.00890	11.20	60.00
65.00	41.602	34.89	2.1290	56.638	201.631	0.11976	0.39611	0.5766	0.4221	1.1415	2779.	651.	0.393	0.0182	0.0570	0.00908	10.87	65.00
70.00	45.270	34.68	1.9641	59.538	203.237	0.12522	0.39652	0.5808	0.4271	1.1435	2729.	651.	0.381	0.0184	0.0562	0.00927	10.55	70.00
75.00	49.176	34.46	1.8145	62.459	204.843	0.13067	0.39697	0.5851	0.4323	1.1457	2679.	650.	0.370	0.0186	0.0555	0.00946	10.23	75.00
80.00	53.328	34.25	1.6786	65.402	206.446	0.13611	0.39746	0.5894	0.4376	1.1481	2629.	650.	0.359	0.0189	0.0548	0.00965	9.92	80.00
85.00	57.736	34.03	1.5549	68.367	208.047	0.14153	0.39798	0.5939	0.4431	1.1507	2578.	649.	0.348	0.0191	0.0541	0.00984	9.60	85.00
90.00	62.409	33.80	1.4420	71.355	209.646	0.14695	0.39853	0.5985	0.4486	1.1536	2528.	647.	0.338	0.0193	0.0535	0.01004	9.29	90.00
95.00	67.358	33.58	1.3389	74.365	211.240	0.15235	0.39912	0.6032	0.4544	1.1566	2477.	646.	0.328	0.0195	0.0528	0.01024	8.97	95.00
100.00	72.592	33.35	1.2445	77.399	212.830	0.15774	0.39973	0.6080	0.4602	1.1600	2426.	644.	0.319	0.0198	0.0521	0.01044	8.66	100.00
105.00	78.121	33.11	1.1580	80.458	214.415	0.16313	0.40036	0.6129	0.4663	1.1636	2376.	643.	0.309	0.0200	0.0514	0.01065	8.35	105.00
110.00	83.955	32.88	1.0785	83.540	215.993	0.16851	0.40101	0.6181	0.4725	1.1675	2325.	641.	0.300	0.0203	0.0508	0.01086	8.05	110.00
115.00	90.104	32.64	1.0054	86.649	217.565	0.17388	0.40169	0.6233	0.4789	1.1718	2273.	638.	0.291	0.0205	0.0501	0.01108	7.74	115.00
120.00	96.578	32.40	0.9381	89.783	219.128	0.17924	0.40238	0.6288	0.4855	1.1764	2222.	636.	0.283	0.0208	0.0495	0.01129	7.44	120.00
125.00	103.39	32.15	0.8760	92.943	220.682	0.18461	0.40308	0.6344	0.4924	1.1815	2171.	633.	0.275	0.0211	0.0488	0.01152	7.14	125.00
130.00	110.55	31.90	0.8186	96.132	222.225	0.18996	0.40380	0.6403	0.4995	1.1870	2119.	630.	0.267	0.0213	0.0482	0.01174	6.84	130.00
135.00	118.06	31.64	0.7655	99.348	223.757	0.19532	0.40453	0.6463	0.5070	1.1930	2068.	627.	0.259	0.0216	0.0476	0.01197	6.54	135.00
140.00	125.94	31.38	0.7163	102.594	225.275	0.20068	0.40526	0.6527	0.5147	1.1996	2016.	623.	0.251	0.0219	0.0470	0.01221	6.25	140.00
145.00	134.21	31.11	0.6706	105.870	226.778	0.20604	0.40600	0.6593	0.5229	1.2068	1964.	619.	0.243	0.0222	0.0464	0.01244	5.96	145.00
150.00	142.86	30.84	0.6282	109.177	228.265	0.21140	0.40673	0.6662	0.5314	1.2148	1912.	615.	0.236	0.0225	0.0458	0.01269	5.67	150.00
155.00	151.92	30.57	0.5887	112.516	229.732	0.21677	0.40746	0.6734	0.5404	1.2235	1859.	610.	0.229	0.0228	0.0452	0.01293	5.38	155.00
160.00	161.39	30.28	0.5518	115.889	231.179	0.22214	0.40819	0.6811	0.5500	1.2332	1807.	606.	0.222	0.0232	0.0446	0.01319	5.10	160.00
165.00	171.29	29.99	0.5175	119.297	232.603	0.22752	0.40890	0.6891	0.5601	1.2440	1754.	601.	0.215	0.0235	0.0440	0.01345	4.82	165.00
170.00	181.62	29.70	0.4854	122.741	234.000	0.23291	0.40960	0.6977	0.5710	1.2560	1701.	595.	0.209	0.0239	0.0434	0.01371	4.54	170.00
175.00	192.42	29.39	0.4553	126.224	235.369	0.23831	0.41028	0.7068	0.5827	1.2694	1648.	589.	0.202	0.0242	0.0428	0.01398	4.27	175.00
180.00	203.67	29.08	0.4271	129.747	236.705	0.24373	0.41093	0.7166	0.5953	1.2845	1595.	583.	0.196	0.0246	0.0423	0.01426	4.00	180.00
185.00	215.41	28.76	0.4007	133.312	238.005	0.24916	0.41156	0.7271	0.6090	1.3015	1541.	577.	0.189	0.0250	0.0417	0.01455	3.73	185.00
190.00	227.64	28.43	0.3758	136.922	239.264	0.25462	0.41215	0.7385	0.6240	1.3208	1486.	570.	0.183	0.0254	0.0411	0.01484	3.47	190.00
195.00	240.39	28.08	0.3524	140.580	240.477	0.26010	0.41269	0.7510	0.6406	1.3429	1432.	562.	0.177	0.0259	0.0406	0.01515	3.21	195.00
200.00	253.66	27.73	0.3304	144.290	241.639	0.26561	0.41318	0.7647	0.6592	1.3683	1377.	554.	0.171	0.0263	0.0400	0.01546	2.96	200.00
210.00	281.84	26.98	0.2898	151.878	243.781	0.27674	0.41397	0.7970	0.7037	1.4322	1265.	538.	0.159	0.0273	0.0389	0.01614	2.46	210.00
220.00	312.35	26.16	0.2534	159.731	245.618	0.28806	0.41443	0.8388	0.7629	1.5218	1150.	519.	0.148	0.0285	0.0378	0.01690	1.98	220.00
230.00	345.33	25.25	0.2203	167.910	247.050	0.29966	0.41441	0.8964	0.8467	1.6544	1031.	498.	0.136	0.0299	0.0367	0.01782	1.53	230.00
240.00	380.98	24.22	0.1900	176.521	247.915	0.31167	0.41371	0.9837	0.9765	1.8676	906.	475.	0.124	0.0315	0.0356	0.01903	1.10	240.00
250.00	419.52	22.99	0.1615	185.763	247.926	0.32436	0.41195	1.1383	1.2081	2.2589	773.	450.	0.112	0.0337	0.0347	0.02084	0.71	250.00
260.00	461.23	21.40	0.1337	196.109	246.440	0.33835	0.40828	1.5073	1.7513	3.1957	628.	420.	0.098	0.0367	0.0345	0.02405	0.36	260.00
270.00	506.48	18.76	0.1025	209.553	240.979	0.35632	0.39938	—	—	—	—	—	0.080	0.0427	—	—	0.08	270.00
274.46c	527.94	14.01	0.0714	226.821	226.821	0.37957	0.37957	∞	∞	∞	0.	0.	—	—	∞	∞	0.00	274.46

*temperatures are on the IPTS-68 scale b = normal boiling point c = critical point

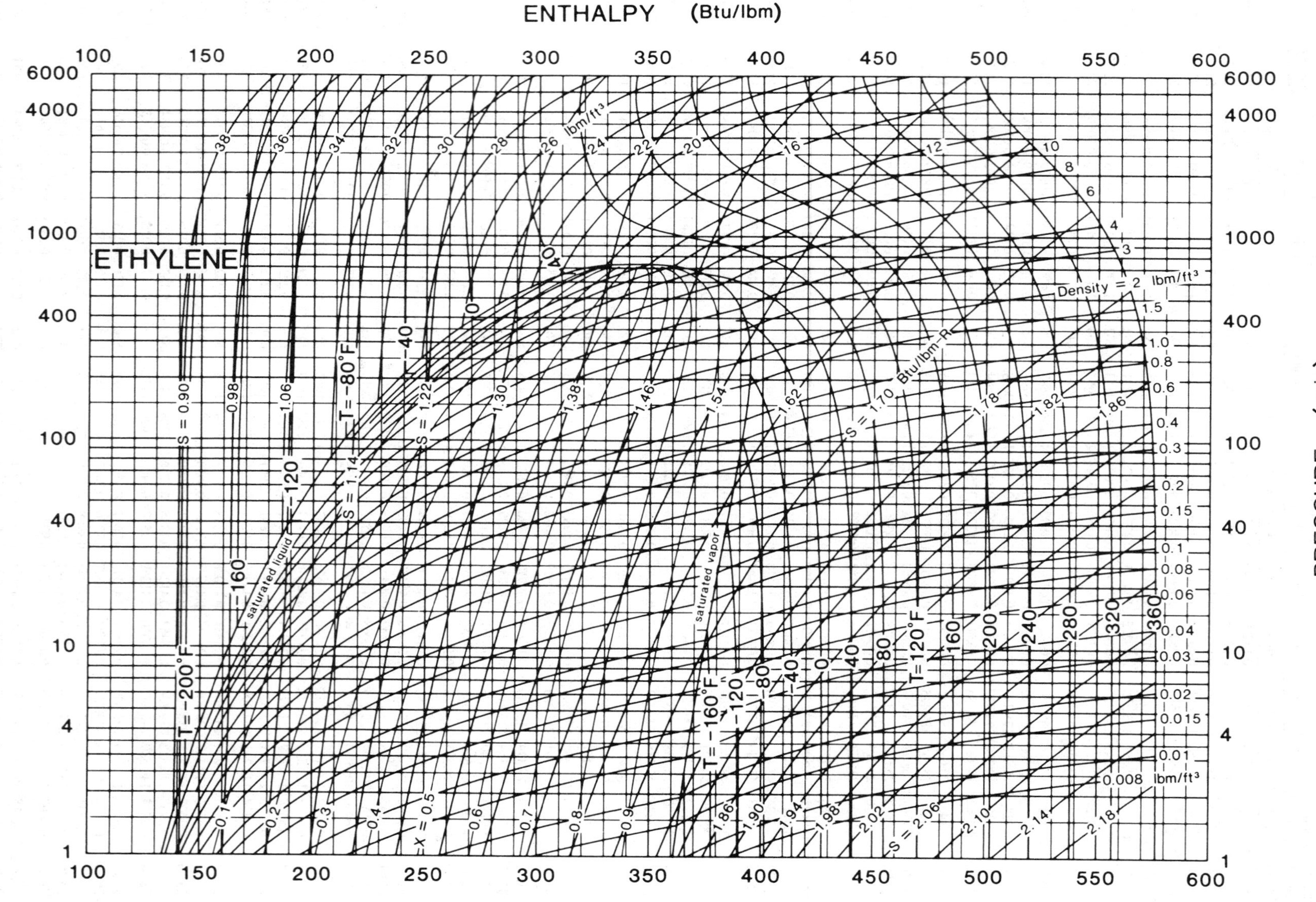

Fig. 29 Pressure–Enthalpy Diagram for Refrigerant 1150 (Ethylene)

Refrigerant 1150 (Ethylene) Properties of Saturated Liquid and Saturated Vapor

Temp,**	Pressure		Volume, ft³/lbm	Density, lbm/ft³	Enthalpy, Btu/lbm		Entropy, Btu/lbm·°F	
°F	psia	psig	Vapor	Liquid	Liquid	Vapor	Liquid	Vapor
−230	0.49203	28.919*	178.02	38.930	126.39	357.70	0.83928	1.8464
−220	0.88923	28.111*	102.63	38.493	132.09	360.44	0.86357	1.8163
−210	1.5232	26.820*	62.275	38.052	137.78	363.14	0.88682	1.7895
−200	2.4897	24.852*	39.511	37.604	143.47	365.80	0.90913	1.7654
−190	3.9056	21.969*	26.060	37.148	149.17	368.40	0.93063	1.7437
−185	4.8243	20.099*	21.442	36.916	152.02	369.68	0.94110	1.7337
−180	5.9086	17.891*	17.783	36.682	154.88	370.95	0.95139	1.7241
−175	7.1789	15.305*	14.858	36.445	157.74	372.19	0.96152	1.7150
−170	8.6568	12.296*	12.501	36.205	160.61	373.42	0.97149	1.7063
−165	10.365	8.8182*	10.587	35.962	163.49	374.63	0.98132	1.6980
−160	12.327	4.8232*	9.0215	35.717	166.38	375.82	0.99100	1.6900
−158	13.189	3.0692*	8.4759	35.618	167.54	376.28	0.99483	1.6870
−156	14.096	1.2210*	7.9705	35.519	168.70	376.75	0.99864	1.6839
−154.73b	14.696	0.0000	7.6695	35.456	169.43	377.04	1.0010	1.6820
−154	15.052	0.3559	7.5018	35.419	169.86	377.21	1.0024	1.6810
−152	16.057	1.3611	7.0666	35.319	171.02	377.67	1.0062	1.6780
−150	17.114	2.4176	6.6623	35.219	172.18	378.13	1.0100	1.6752
−148	18.223	3.5271	6.2862	35.118	173.34	378.58	1.0137	1.6723
−146	19.387	4.6913	5.9360	35.016	174.51	379.02	1.0174	1.6696
−144	20.608	5.9119	5.6097	34.914	175.68	379.47	1.0211	1.6668
−142	21.887	7.1909	5.3053	34.812	176.85	379.90	1.0248	1.6641
−140	23.226	8.5299	5.0212	34.709	178.02	380.34	1.0284	1.6615
−138	24.627	9.9308	4.7556	34.606	179.19	380.77	1.0320	1.6589
−136	26.091	11.395	4.5073	34.502	180.37	381.20	1.0357	1.6563
−134	27.622	12.926	4.2748	34.398	181.54	381.62	1.0393	1.6538
−132	29.219	14.523	4.0571	34.293	182.72	382.03	1.0428	1.6513
−130	30.887	16.191	3.8529	34.188	183.90	382.45	1.0464	1.6488
−128	32.625	17.929	3.6614	34.082	185.08	382.85	1.0499	1.6464
−126	34.436	19.740	3.4815	33.976	186.27	383.26	1.0535	1.6440
−124	36.323	21.627	3.3124	33.870	187.45	383.66	1.0570	1.6416
−122	38.287	23.591	3.1534	33.763	188.64	384.05	1.0605	1.6393
−120	40.329	25.633	3.0038	33.655	189.83	384.44	1.0640	1.6370
−118	42.453	27.757	2.8629	33.547	191.03	384.82	1.0674	1.6348
−116	44.660	29.964	2.7300	33.438	192.22	385.20	1.0709	1.6325
−114	46.951	32.255	2.6048	33.329	193.42	385.57	1.0743	1.6303
−112	49.330	34.634	2.4865	33.220	194.62	385.94	1.0778	1.6282

Temp,**	Pressure		Volume, ft³/lbm	Density, lbm/ft³	Enthalpy, Btu/lbm		Entropy, Btu/lbm·°F	
°F	psia	psig	Vapor	Liquid	Liquid	Vapor	Liquid	Vapor
−110	51.797	37.101	2.3749	33.110	195.82	386.30	1.0812	1.6260
−108	54.356	39.660	2.2693	32.999	197.03	386.66	1.0846	1.6239
−106	57.008	42.312	2.1695	32.888	198.24	387.01	1.0879	1.6218
−104	59.755	45.059	2.0751	32.776	199.45	387.36	1.0913	1.6197
−102	62.599	47.903	1.9857	32.663	200.66	387.70	1.0947	1.6177
−100	65.542	50.846	1.9010	32.550	201.88	388.03	1.0980	1.6157
−98	68.587	53.891	1.8206	32.437	203.10	388.36	1.1014	1.6137
−96	71.735	57.039	1.7445	32.323	204.32	388.68	1.1047	1.6117
−94	74.988	60.292	1.6722	32.208	205.55	388.99	1.1080	1.6097
−92	78.350	63.654	1.6035	32.092	206.78	389.30	1.1113	1.6078
−90	81.821	67.125	1.5382	31.976	208.01	389.60	1.1146	1.6059
−85	90.993	76.297	1.3888	31.682	211.11	390.32	1.1228	1.6011
−80	100.90	86.201	1.2566	31.384	214.24	391.00	1.1309	1.5965
−75	111.57	96.873	1.1393	31.080	217.39	391.63	1.1390	1.5920
−70	123.04	108.35	1.0349	30.771	220.57	392.20	1.1470	1.5875
−60	148.53	133.83	0.85834	30.135	227.03	393.20	1.1630	1.5788
−55	162.61	147.92	0.78345	29.806	230.31	393.60	1.1710	1.5745
−50	177.64	162.94	0.71603	29.470	233.64	393.94	1.1789	1.5702
−45	193.64	178.95	0.65516	29.124	237.02	394.21	1.1868	1.5659
−35	228.74	214.04	0.55006	28.404	243.92	394.52	1.2027	1.5574
−30	247.91	233.21	0.50456	28.027	247.45	394.55	1.2107	1.5530
−25	268.21	253.51	0.46305	27.636	251.06	394.48	1.2187	1.5487
−20	289.68	274.98	0.42509	27.231	254.73	394.30	1.2268	1.5442
−10	336.32	321.62	0.35826	26.370	262.34	393.57	1.2432	1.5350
−5	361.57	346.88	0.32872	25.909	266.29	393.00	1.2515	1.5302
0	388.19	373.49	0.30141	25.423	270.35	392.25	1.2600	1.5252
5	416.20	401.51	0.27605	24.909	274.55	391.32	1.2686	1.5199
20	509.30	494.60	0.20947	23.138	288.18	386.98	1.2960	1.5019
30	579.59	564.89	0.17067	21.650	298.61	382.08	1.3163	1.4868
40	657.38	642.68	0.13295	19.595	311.46	373.66	1.3409	1.4653
48.5c	731.0	716.3	0.07480	13.37	342.2	342.2	1.400	1.400

*in. Hg vacuum **temperatures are on the IPTS-68 scale b = normal boiling point c = critical point

Temp, °F	Viscosity, lbm/ft·h Sat. Liquid	Sat. Vapor	Gas at p = 1 atm × 10^{-2}*	Thermal Conductivity, Btu/h·ft·°F Sat. Liquid	Sat. Vapor	Gas at p = 1 atm × 10^{-2}*	Specific Heat c_p, Btu/lbm·°F Sat. Liquid	Sat. Vapor	Gas at p = 0 atm	Gas at p = 1 atm	Temp, °F
−180	0.487			0.119			0.572				−180
−160	0.412			0.113			0.573	0.288			−160
−140	0.356			0.107	0.0049	0.49	0.576	0.299			−140
−120	0.314	0.016	1.62	0.100	0.0055	0.55	0.582	0.313			−120
−100	0.279	0.017	1.71	0.094	0.0061	0.61	0.592	0.335			−100
−80	0.252	0.018	1.80	0.087	0.0068	0.67	0.607	0.364			−80
−60	0.230	0.019	1.89	0.081	0.0075	0.73	0.642	0.403			−60
−40	0.211	0.020	1.98	0.074	0.0085	0.79	0.705	0.454			−40
−20	0.187	0.023	2.07	0.068	0.0097	0.85	0.804	0.529			−20
0	0.157	0.026	2.16	0.061	0.0110	0.91	0.946	0.657	0.331	0.336	0
20	0.123	0.029	2.25	0.053	0.013	0.98	1.139	0.882	0.342	0.346	20
30	0.103	0.031	2.30	0.048	0.015	1.01	1.25	1.014	0.347	0.351	30
40	0.082	0.036	2.34	0.041	0.018	1.04	1.39	1.35	0.352	0.356	40
48.5c	0.053	0.053	2.38	0.027	0.027	1.07			0.357	0.361	48.5
60			2.43			1.11			0.362	0.366	60
80			2.51			1.18			0.373	0.376	80
100			2.60			1.25			0.383	0.386	100
120			2.68			1.33			0.393	0.395	120
140			2.77			1.41			0.403	0.405	140
180			2.93			1.58			0.423	0.424	180
220			3.10			1.77			0.442	0.443	220
260			3.25			1.96			0.461	0.462	260
300			3.41			2.14			0.479	0.480	300
400			3.79			2.64			0.523	0.524	400
500			4.15			3.15			0.564	0.565	500
600			4.49			3.67			0.602	0.602	600
700			4.81			4.19			0.637		700
800			5.13			4.73			0.670		800
1000			5.72			5.87			0.726		1000

c = Critical temperature. Tabulated properties ignore critical region effects. *Actual value = (Table value) × (Indicated multiplier).

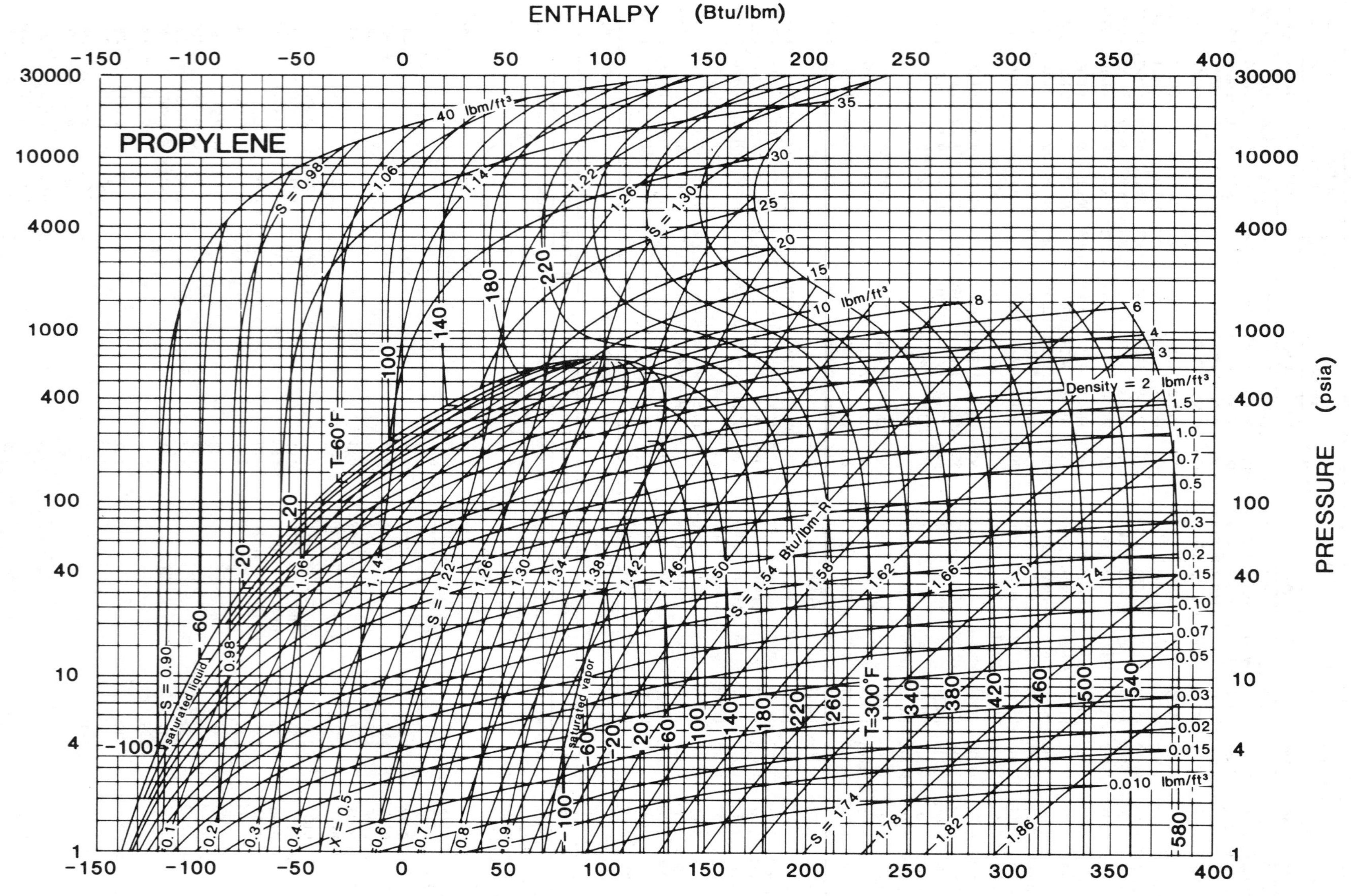

Prepared by: CENTER FOR APPLIED THERMODYNAMIC STUDIES, University of Idaho

Fig. 30 Pressure–Enthalpy Diagram for Refrigerant 1270 (Propylene)

Refrigerant 1270 (Propylene) Properties of Saturated Liquid and Saturated Vapor

Temp,** °F	Pressure psia	Pressure psig	Volume, ft³/lb$_m$ Vapor	Density, lb$_m$/ft³ Liquid	Enthalpy, Btu/lb$_m$ Liquid	Enthalpy, Btu/lb$_m$ Vapor	Entropy, Btu/lb$_m$·°F Liquid	Entropy, Btu/lb$_m$·°F Vapor
−301.47a	2.3×10^{-7}	29.921*	2.9×10^{8}	47.998	−213.80	31.094	0.51019	2.0581
−300	3.0×10^{-7}	29.921*	2.2×10^{8}	47.940	−212.95	31.407	0.51552	2.0459
−280	8.3×10^{-6}	29.921*	6.9×10^{6}	47.136	−203.33	35.763	0.57244	1.9032
−260	1.2×10^{-4}	29.921*	4.7×10^{5}	46.325	−194.87	40.275	0.61709	1.7948
−240	0.00108	29.919*	5.4×10^{4}	45.519	−186.29	44.930	0.65801	1.7106
−220	0.00663	29.908*	9340.1	44.720	−177.40	49.714	0.69676	1.6444
−200	0.03031	29.859*	2187.1	43.928	−168.22	54.619	0.73353	1.5917
−180	0.10956	29.698*	649.45	43.140	−158.83	59.637	0.76834	1.5495
−160	0.32751	29.254*	232.62	42.353	−149.29	64.755	0.80129	1.5156
−140	0.83912	28.213*	96.766	41.562	−139.62	69.956	0.83253	1.4881
−120	1.8952	26.063*	45.404	40.763	−129.81	75.215	0.86226	1.4659
−100	3.8587	22.065*	23.490	39.954	−119.87	80.504	0.89067	1.4478
−95	4.5452	20.667*	20.184	39.750	−117.36	81.827	0.89759	1.4438
−90	5.3266	19.076*	17.425	39.544	−114.85	83.149	0.90444	1.4400
−85	6.2120	17.274*	15.111	39.338	−112.32	84.470	0.91123	1.4365
−80	7.2109	15.240*	13.160	39.130	−109.78	85.789	0.91795	1.4330
−75	8.3334	12.954*	11.508	38.921	−107.22	87.105	0.92461	1.4298
−70	9.5899	10.396*	10.102	38.711	−104.66	88.418	0.93121	1.4267
−65	10.991	7.5434*	8.9010	38.499	−102.09	89.728	0.93776	1.4238
−60	12.548	4.3732*	7.8702	38.286	−99.499	91.034	0.94426	1.4210
−55	14.273	0.8620*	6.9821	38.072	−96.899	92.334	0.95070	1.4183
−53.85b	14.696	0.0000	6.7948	38.022	−96.297	92.634	0.95219	1.4177
−50	16.177	1.4805	6.2140	37.856	−94.285	93.630	0.95710	1.4158
−48	16.991	2.2951	5.9360	37.769	−93.236	94.146	0.95965	1.4148
−46	17.837	3.1411	5.6732	37.682	−92.184	94.662	0.96219	1.4139
−44	18.715	4.0194	5.4246	37.594	−91.130	95.176	0.96472	1.4129
−42	19.627	4.9308	5.1893	37.507	−90.074	95.689	0.96724	1.4120
−40	20.572	5.8760	4.9663	37.419	−89.015	96.201	0.96976	1.4111
−38	21.552	6.8561	4.7551	37.330	−87.954	96.712	0.97227	1.4102
−36	22.568	7.8717	4.5548	37.242	−86.891	97.222	0.97477	1.4093
−34	23.620	8.9238	4.3648	37.153	−85.825	97.731	0.97727	1.4085
−32	24.709	10.013	4.1845	37.064	−84.757	98.238	0.97976	1.4076
−30	25.837	11.141	4.0133	36.974	−83.686	98.744	0.98225	1.4068
−28	27.003	12.307	3.8506	36.884	−82.612	99.249	0.98472	1.4060
−26	28.210	13.514	3.6960	36.794	−81.536	99.752	0.98720	1.4052
−24	29.457	14.761	3.5489	36.704	−80.458	100.25	0.98967	1.4045
−22	30.746	16.050	3.4090	36.613	−79.376	100.75	0.99213	1.4037
−20	32.078	17.382	3.2759	36.521	−78.292	101.25	0.99458	1.4029
−18	33.453	18.757	3.1490	36.430	−77.206	101.75	0.99703	1.4022
−16	34.873	20.177	3.0282	36.338	−76.116	102.24	0.99948	1.4015
−14	36.338	21.642	2.9130	36.246	−75.024	102.74	1.0019	1.4008
−12	37.849	23.153	2.8032	36.153	−73.929	103.23	1.0044	1.4001
−10	39.407	24.711	2.6984	36.060	−72.831	103.72	1.0068	1.3994
−8	41.014	26.318	2.5984	35.966	−71.730	104.21	1.0092	1.3987
−6	42.670	27.974	2.5029	35.872	−70.626	104.69	1.0116	1.3981
−4	44.375	29.679	2.4117	35.778	−69.519	105.18	1.0140	1.3974
−2	46.132	31.436	2.3245	35.683	−68.409	105.66	1.0165	1.3968
0	47.941	33.245	2.2412	35.588	−67.296	106.14	1.0189	1.3962
2	49.803	35.107	2.1615	35.493	−66.180	106.62	1.0213	1.3956
4	51.718	37.022	2.0852	35.397	−65.060	107.10	1.0237	1.3949
6	53.689	38.993	2.0122	35.300	−63.938	107.57	1.0260	1.3944
8	55.716	41.020	1.9424	35.203	−62.812	108.04	1.0284	1.3938
10	57.799	43.103	1.8754	35.106	−61.683	108.51	1.0308	1.3932
12	59.940	45.244	1.8113	35.008	−60.551	108.98	1.0332	1.3926
14	62.140	47.444	1.7498	34.910	−59.415	109.44	1.0356	1.3921
16	64.401	49.705	1.6908	34.811	−58.276	109.91	1.0380	1.3915
18	66.722	52.026	1.6343	34.712	−57.133	110.37	1.0403	1.3910
25	75.340	60.644	1.4536	34.360	−53.106	111.96	1.0486	1.3892
30	81.984	67.288	1.3393	34.104	−50.202	113.07	1.0545	1.3879
35	89.055	74.359	1.2356	33.845	−47.274	114.16	1.0604	1.3867
40	96.569	81.873	1.1414	33.582	−44.321	115.24	1.0662	1.3855
45	104.54	89.845	1.0557	33.315	−41.343	116.29	1.0721	1.3844
50	112.99	98.294	0.97747	33.043	−38.338	117.33	1.0779	1.3833
60	131.38	116.69	0.84066	32.487	−32.245	119.32	1.0895	1.3812
70	151.89	137.19	0.72565	31.910	−26.033	121.20	1.1011	1.3791
80	174.64	159.95	0.62850	31.310	−19.690	122.96	1.1128	1.3771
90	199.80	185.10	0.54535	50.685	−13.206	124.57	1.1244	1.3750
100	227.50	212.80	0.47422	30.029	−6.5635	126.03	1.1361	1.3730
120	291.16	276.46	0.35950	28.606	7.2746	128.32	1.1596	1.3685
140	366.99	352.29	0.27172	26.978	22.063	129.48	1.1839	1.3630
160	456.56	441.87	0.20226	25.009	38.298	128.83	1.2094	1.3555
180	562.06	547.36	0.14370	22.290	57.444	124.53	1.2385	1.3434
198.4c	676.5	661.8	0.07171	13.95	97.42	97.42	1.298	1.298

*in. Hg vacuum **temperatures are on the IPTS-68 scale a = triple point b = normal boiling point c = critical point

Temp, °F	Viscosity, lb$_m$/ft·h Sat. Liquid	Viscosity Sat. Vapor	Viscosity Gas at $p = 1$ atm $\times 10^{-2}$*	Thermal Conductivity, Btu/h·ft·°F Sat. Liquid	Thermal Conductivity Sat. Vapor	Thermal Conductivity Gas at $p = 1$ atm $\times 10^{-3}$*	Specific Heat c_p, Btu/lb$_m$·°F Sat. Liquid	Specific Heat Sat. Vapor	Specific Heat Gas at $p = 0$ atm	Specific Heat Gas at $p = 1$ atm	Temp, °F
−100	0.595			0.094	0.0035		0.485		0.287		−100
−80	0.511	0.0150		0.091	0.0044		0.483		0.295		−80
−60	0.443	0.0157		0.087	0.0052		0.484		0.303		−60
−40	0.388	0.0165	1.65	0.084	0.0058	0.56	0.486	0.321	0.312		−40
−20	0.344	0.0173	1.73	0.081	0.0064	0.63	0.492	0.334	0.321		−20
0	0.307	0.0182	1.82	0.078	0.0069	0.69	0.501	0.344	0.329	0.333	0
20	0.276	0.0191	1.90	0.074	0.0076	0.76	0.513	0.359	0.338	0.343	20
40	0.250	0.0197	1.98	0.071	0.0083	0.83	0.530	0.375	0.347	0.353	40
60	0.220	0.0206	2.06	0.067	0.0091	0.90	0.551	0.408	0.356	0.363	60
80	0.191	0.0218	2.14	0.064	0.0102	0.97	0.577	0.458	0.366	0.373	80
100	0.163	0.0233	2.21	0.061	0.0116	1.05	0.609	0.540	0.375	0.383	100
120	0.135	0.026	2.29	0.057	0.0133	1.12	0.647	0.644	0.384	0.393	120
140	0.111	0.028	2.37	0.054	0.0153	1.19	0.692	0.798	0.394	0.403	140
160	0.089	0.032	2.44	0.050	0.018	1.27	0.743		0.403	0.413	160
180	0.071	0.037	2.52	0.045	0.022	1.35			0.413	0.422	180
190	0.065	0.045	2.55	0.040	0.025	1.39			0.418	0.427	190
198.4c	0.058	0.058	2.58	0.029	0.029	1.42			0.421	0.430	198.4
200			2.59			1.42			0.423	0.432	200
220						1.50			0.432	0.441	220
240						1.58			0.442	0.450	240
260						1.67			0.452	0.459	260
280						1.75			0.461	0.468	280
300						1.84			0.471	0.477	300
320						1.92			0.480	0.486	320
340						2.01			0.490	0.495	340
400						2.29			0.518	0.521	400
500						2.75			0.563	0.563	500
600						3.25			0.605		600
700						3.77			0.642		700
800						4.27			0.674		800

c = Critical temperature. Tabulated properties ignore critical region effects.

*Actual value = (Table value) × (Indicated multiplier).

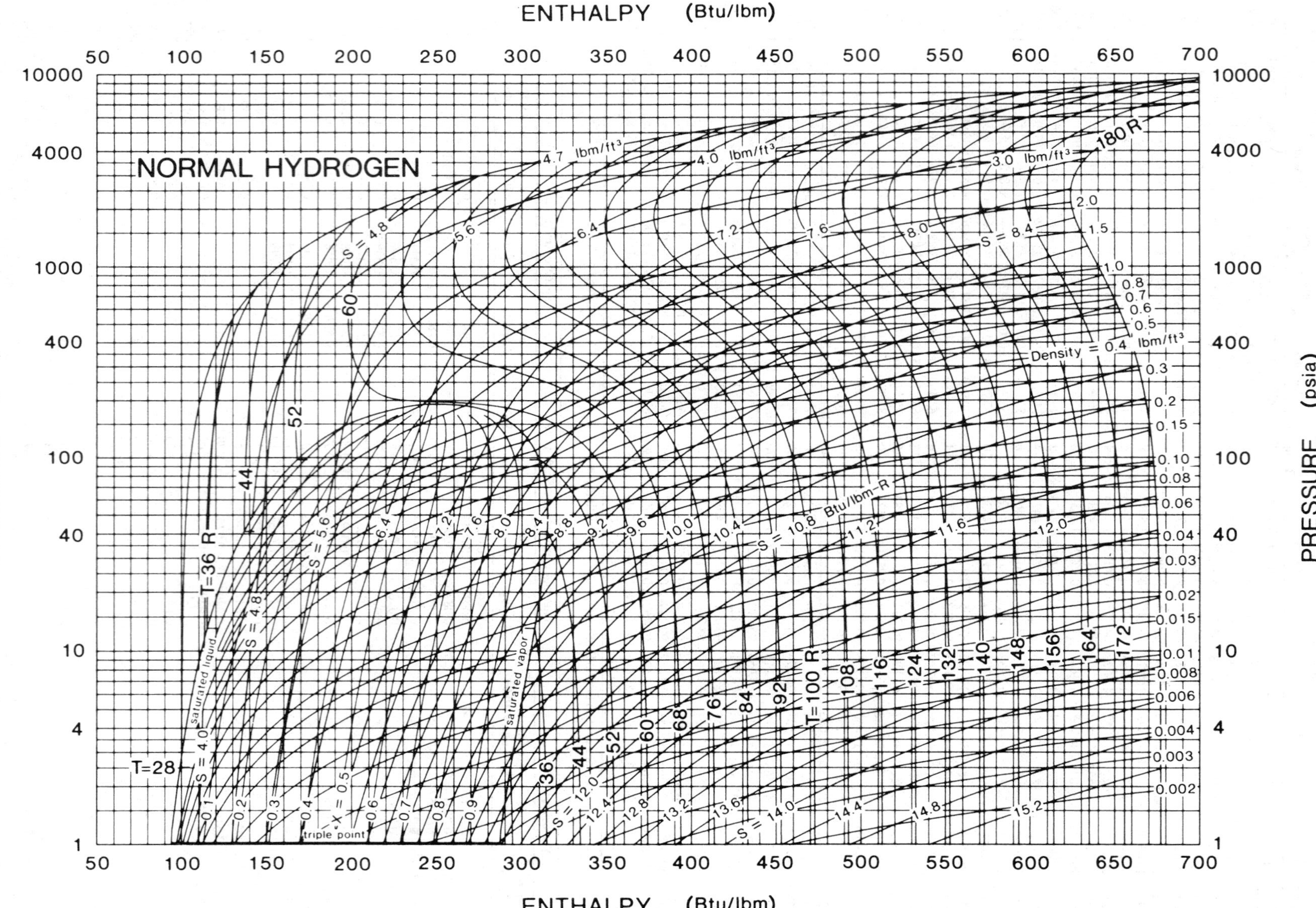

Fig. 31 Pressure–Enthalpy Diagram for Refrigerant 702 (Normal Hydrogen)

Refrigerant 702 (Normal Hydrogen) Properties of Saturated Liquid and Saturated Vapor

Temp,* °R	Pressure, psia	Density, lb/ft³ Liquid	Volume, ft³/lb Vapor	Enthalpy, Btu/lb Liquid	Enthalpy, Btu/lb Vapor	Entropy, Btu/lb·°R Liquid	Entropy, Btu/lb·°R Vapor	Specific Heat c_p, Btu/lb·°R Liquid	Specific Heat c_p, Btu/lb·°R Vapor	c_p/c_v Vapor	Velocity of Sound, ft/s Liquid	Velocity of Sound, ft/s Vapor	Viscosity, lb_m/ft·h Liquid	Viscosity, lb_m/ft·h Vapor	Thermal Cond, Btu/h·ft·°R Liquid	Thermal Cond, Btu/h·ft·°R Vapor	Surface Tension, dyne/cm	Temp,* °R
25.11a	1.125	4.801	116.73	93.84	287.11	3.3656	11.0477	1.858	2.605	1.669	4467.	998.9	0.0617	0.00160	0.0440	0.00599	3.181	25.11
26.00	1.464	4.776	92.561	95.60	289.08	3.4270	10.8701	1.751	2.590	1.688	4405.	1018.5	0.0577	0.00170	0.0463	0.00635	3.081	26.00
28.00	2.513	4.715	57.548	99.07	293.31	3.5540	10.5077	1.716	2.611	1.711	4240.	1054.2	0.0503	0.00189	0.0506	0.00706	2.859	28.00
30.00	4.060	4.651	37.736	102.54	297.29	3.6755	10.1889	1.807	2.658	1.732	4072.	1084.8	0.0446	0.00206	0.0539	0.00771	2.640	30.00
31.00	5.062	4.618	31.072	104.38	299.18	3.7357	10.0431	1.872	2.686	1.744	3992.	1098.9	0.0422	0.00215	0.0553	0.00803	2.532	31.00
32.00	6.239	4.584	25.834	106.31	301.00	3.7960	9.9051	1.944	2.717	1.758	3917.	1112.3	0.0400	0.00223	0.0564	0.00835	2.424	32.00
33.00	7.611	4.549	21.668	108.34	302.74	3.8565	9.7740	2.020	2.752	1.773	3845.	1125.0	0.0380	0.00231	0.0574	0.00867	2.318	33.00
34.00	9.195	4.513	18.319	110.47	304.39	3.9173	9.6491	2.100	2.790	1.791	3776.	1137.0	0.0361	0.00239	0.0583	0.00899	2.212	34.00
35.00	11.013	4.477	15.600	112.68	305.96	3.9785	9.5297	2.182	2.832	1.811	3709.	1148.3	0.0345	0.00247	0.0590	0.00932	2.107	35.00
36.00	13.082	4.439	13.372	114.98	307.42	4.0401	9.4152	2.267	2.879	1.833	3644.	1159.0	0.0329	0.00256	0.0595	0.00966	2.003	36.00
36.70b	14.696	4.412	12.045	116.66	308.39	4.0836	9.3376	2.328	2.915	1.851	3598.	1166.1	0.0319	0.00261	0.0598	0.00990	1.931	36.70
37.00	15.425	4.400	11.531	117.38	308.79	4.1023	9.3051	2.355	2.931	1.858	3579.	1169.0	0.0315	0.00264	0.0600	0.01000	1.900	37.00
38.00	18.061	4.359	9.9976	119.86	310.04	4.1650	9.1989	2.445	2.989	1.887	3514.	1178.2	0.0301	0.00272	0.0603	0.01036	1.798	38.00
39.00	21.010	4.317	8.7112	122.43	311.18	4.2284	9.0961	2.540	3.054	1.919	3448.	1186.8	0.0289	0.00281	0.0605	0.01073	1.698	39.00
40.00	24.295	4.273	7.6245	125.10	312.19	4.2924	8.9963	2.640	3.127	1.955	3380.	1194.8	0.0277	0.00289	0.0606	0.01111	1.598	40.00
41.00	27.935	4.227	6.7006	127.87	313.06	4.3573	8.8991	2.746	3.209	1.995	3311.	1202.0	0.0266	0.00298	0.0607	0.01150	1.499	41.00
42.00	31.952	4.179	5.9104	130.75	313.80	4.4229	8.8041	2.860	3.302	2.042	3239.	1208.7	0.0255	0.00307	0.0606	0.01191	1.402	42.00
43.00	36.367	4.129	5.2308	133.75	314.38	4.4894	8.7109	2.982	3.408	2.094	3166.	1214.7	0.0245	0.00316	0.0605	0.01233	1.306	43.00
44.00	41.202	4.077	4.6430	136.88	314.80	4.5567	8.6192	3.116	3.528	2.154	3089.	1220.0	0.0236	0.00326	0.0603	0.01278	1.211	44.00
45.00	46.477	4.022	4.1320	140.14	315.04	4.6250	8.5285	3.262	3.666	2.224	3011.	1224.7	0.0227	0.00335	0.0600	0.01324	1.117	45.00
46.00	52.214	3.965	3.6857	143.56	315.10	4.6944	8.4386	3.425	3.827	2.304	2929.	1228.8	0.0218	0.00345	0.0596	0.01373	1.025	46.00
47.00	58.434	3.905	3.2939	147.13	314.94	4.7649	8.3490	3.609	4.014	2.398	2845.	1232.3	0.0209	0.00356	0.0591	0.01425	0.935	47.00
48.00	65.158	3.842	2.9484	150.87	314.56	4.8368	8.2592	3.817	4.235	2.509	2757.	1235.2	0.0201	0.00367	0.0586	0.01480	0.846	48.00
49.00	72.407	3.776	2.6423	154.81	313.92	4.9103	8.1690	4.059	4.499	2.642	2666.	1237.4	0.0193	0.00378	0.0580	0.01540	0.758	49.00
50.00	80.202	3.705	2.3698	158.95	313.01	4.9858	8.0776	4.344	4.820	2.803	2572.	1239.1	0.0185	0.00390	0.0573	0.01603	0.673	50.00
52.00	97.512	3.547	1.9068	167.99	310.19	5.1449	7.8888	5.123	5.718	3.252	2367.	1240.8	0.0170	0.00417	0.0556	0.01750	0.508	52.00
54.00	117.25	3.356	1.5282	178.43	305.67	5.3220	7.6857	6.484	7.246	4.013	2133.	1240.5	0.0154	0.00451	0.0535	0.01933	0.352	54.00
56.00	139.59	3.102	1.2092	191.38	298.64	5.5351	7.4556	9.791	10.325	5.537	1851.	1238.9	0.0138	0.00500	0.0507	0.02187	0.208	56.00
58.00	164.68	2.702	0.9281	210.22	287.39	5.8358	7.1718	—	—	—	—	—	—	—	—	—	0.082	58.00
59.74c	190.75	1.880	0.5320	248.30	248.30	6.4441	6.4441	∞	∞	∞	0.	0.0	—	—	∞	∞	0.000	59.74

*temperatures are on the IPTS-68 scale   a = triple point   b = normal boiling point   c = critical point

Refrigerant 702 (Normal Hydrogen) Properties of Gas at 14.696 psia (one standard atmosphere)

Temp, °F	Density, lb/ft³	Enthalpy, Btu/lb	Entropy, Btu/lb·°F	c_p, Btu/lb·°F	c_p/c_v	Vel. Sound, ft/s	Viscosity, lb/ft·h	Thermal Cond, Btu/h·ft·°F	Temp, °F	Density, lb/ft³	Enthalpy, Btu/lb	Entropy, Btu/lb·°F	c_p, Btu/lb·°F	c_p/c_v	Vel. Sound, ft/s	Viscosity, lb/ft·h	Thermal Cond, Btu/h·ft·°F
−423.0a	0.08302	308.39	9.3376	2.915	1.851	1166.1	0.0026	0.0099	0.0	0.00600	1545.27	16.3039	3.367	1.414	4006.9	0.0194	0.0944
−420.0	0.07529	316.85	9.5592	2.792	1.812	1226.5	0.0028	0.0106	10.0	0.00587	1578.98	16.3765	3.376	1.413	4048.1	0.0197	0.0961
−410.0	0.05801	343.71	10.1639	2.617	1.748	1401.0	0.0035	0.0132	20.0	0.00575	1612.78	16.4477	3.384	1.411	4088.9	0.0200	0.0978
−400.0	0.04750	369.53	10.6376	2.555	1.720	1549.3	0.0042	0.0156	30.0	0.00563	1646.66	16.5176	3.391	1.410	4129.4	0.0203	0.0994
−390.0	0.04032	394.93	11.0312	2.527	1.704	1681.7	0.0048	0.0179	40.0	0.00552	1680.61	16.5862	3.398	1.409	4169.5	0.0205	0.1010
−380.0	0.03507	420.11	11.3690	2.512	1.694	1802.7	0.0054	0.0201	50.0	0.00541	1714.62	16.6536	3.405	1.408	4209.4	0.0208	0.1026
−370.0	0.03105	445.19	11.6655	2.505	1.686	1914.8	0.0060	0.0221	60.0	0.00531	1748.69	16.7198	3.410	1.407	4248.9	0.0211	0.1042
−360.0	0.02787	470.22	11.9302	2.503	1.679	2019.3	0.0065	0.0241	70.0	0.00521	1782.83	16.7849	3.416	1.406	4288.2	0.0214	0.1057
−340.0	0.02314	520.39	12.3889	2.518	1.663	2208.2	0.0075	0.0281	80.0	0.00511	1817.01	16.8488	3.421	1.405	4327.1	0.0216	0.1073
−320.0	0.01980	571.09	12.7806	2.556	1.641	2373.7	0.0084	0.0318	90.0	0.00502	1851.24	16.9117	3.425	1.404	4365.8	0.0219	0.1088
−300.0	0.01730	622.76	13.1262	2.613	1.615	2520.1	0.0093	0.0355	100.0	0.00493	1885.51	16.9735	3.429	1.404	4404.2	0.0222	0.1104
−280.0	0.01537	675.70	13.4385	2.683	1.588	2651.9	0.0101	0.0394	120.0	0.00476	1954.17	17.0940	3.437	1.402	4480.2	0.0227	0.1134
−260.0	0.01382	730.11	13.7256	2.759	1.561	2773.0	0.0109	0.0437	140.0	0.00460	2022.97	17.2107	3.443	1.401	4555.2	0.0233	0.1166
−240.0	0.01256	786.05	13.9925	2.835	1.537	2886.5	0.0117	0.0479	160.0	0.00445	2091.87	17.3237	3.447	1.401	4629.1	0.0238	0.1199
−220.0	0.01151	843.50	14.2428	2.910	1.515	2994.3	0.0124	0.0521	180.0	0.00431	2160.85	17.4333	3.451	1.400	4702.1	0.0243	0.1231
−200.0	0.01063	902.40	14.4788	2.979	1.497	3098.0	0.0131	0.0562	200.0	0.00418	2229.91	17.5396	3.454	1.399	4774.1	0.0248	0.1261
−180.0	0.00986	962.63	14.7022	3.043	1.481	3198.3	0.0138	0.0604	220.0	0.00406	2299.02	17.6428	3.457	1.399	4845.2	0.0253	0.1292
−160.0	0.00921	1024.07	14.9143	3.100	1.468	3295.9	0.0145	0.0645	240.0	0.00394	2368.17	17.7431	3.459	1.399	4915.3	0.0258	0.1322
−140.0	0.00863	1086.59	15.1163	3.151	1.457	3391.1	0.0151	0.0686	260.0	0.00383	2437.36	17.8406	3.460	1.398	4984.6	0.0263	0.1352
−120.0	0.00812	1150.06	15.3089	3.196	1.447	3484.2	0.0158	0.0726	280.0	0.00373	2506.57	17.9354	3.461	1.398	5053.0	—	—
−100.0	0.00767	1214.38	15.4929	3.235	1.439	3575.5	0.0164	0.0765	300.0	0.00363	2575.80	18.0278	3.462	1.398	5120.4	—	—
−80.0	0.00727	1279.43	15.6689	3.269	1.432	3664.9	0.0170	0.0802	320.0	0.00354	2645.06	18.1178	3.463	1.398	5187.1	—	—
−60.0	0.00690	1345.13	15.8375	3.299	1.427	3752.7	0.0176	0.0839	340.0	0.00345	2714.33	18.2055	3.464	1.398	5252.8	—	—
−40.0	0.00657	1411.38	15.9992	3.325	1.422	3838.9	0.0182	0.0875	360.0	0.00337	2783.62	18.2911	3.465	1.398	5317.8	—	—
−20.0	0.00627	1478.11	16.1546	3.348	1.418	3923.7	0.0188	0.0910	380.0	0.00329	2852.93	18.3746	3.466	1.397	5381.9	—	—
									400.0	0.00321	2922.25	18.4562	3.467	1.397	5445.2	—	—

a = saturated vapor at normal boiling point

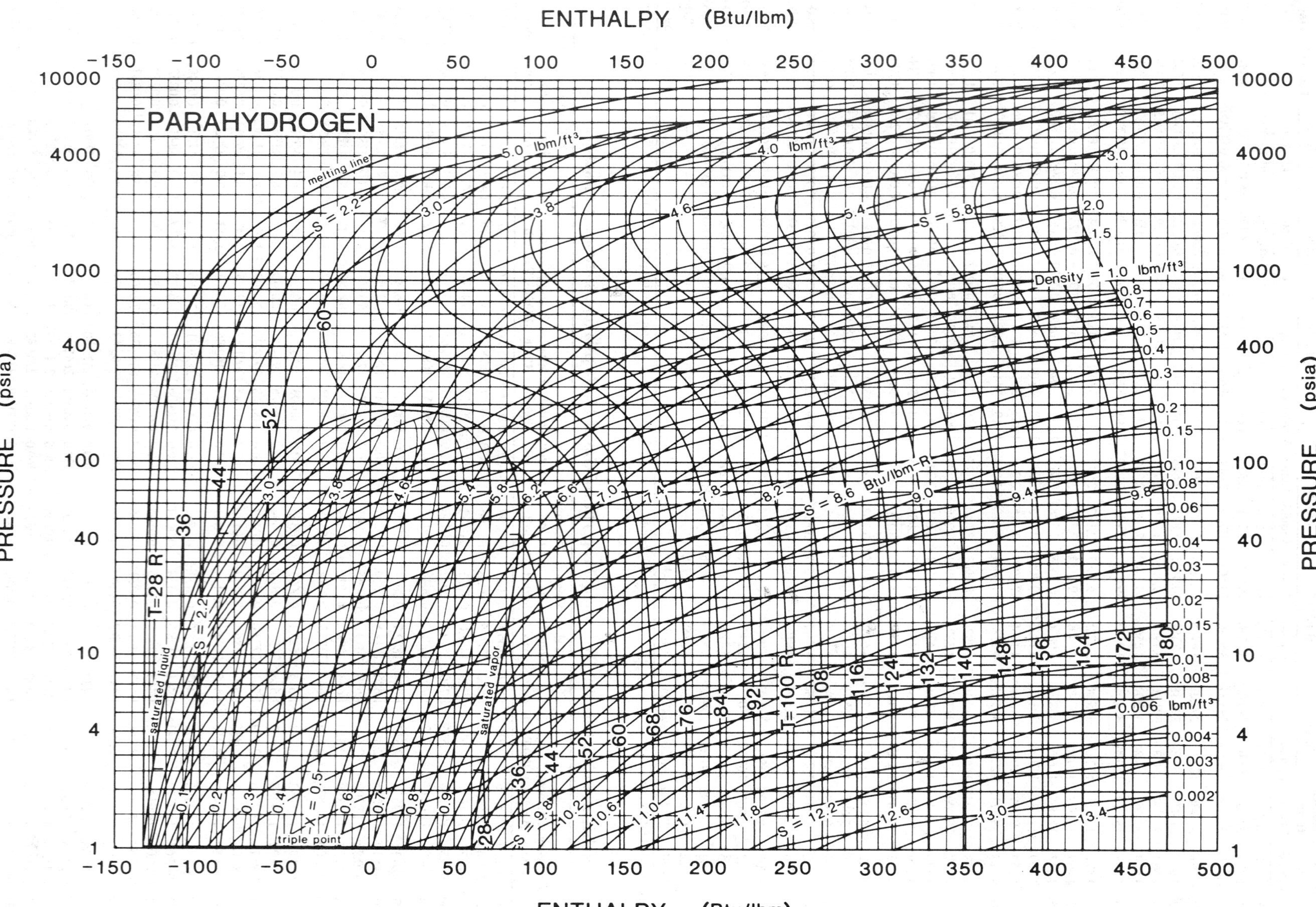

Prepared by: CENTER FOR APPLIED THERMODYNAMIC STUDIES, University of Idaho

Fig. 32 Pressure–Enthalpy Diagram for Refrigerant 702p (Parahydrogen)

Refrigerant 702p (Parahydrogen) Properties of Saturated Liquid and Saturated Vapor

Temp,* °R	Pressure, psia	Density, lb/ft³ Liquid	Volume, ft³/lb Vapor	Enthalpy, Btu/lb Liquid	Enthalpy, Btu/lb Vapor	Entropy, Btu/lb·°R Liquid	Entropy, Btu/lb·°R Vapor	Specific Heat c_p, Btu/lb·°R Liquid	Specific Heat c_p, Btu/lb·°R Vapor	c_p/c_v Vapor	Velocity of Sound, ft/s Liquid	Velocity of Sound, ft/s Vapor	Viscosity, lb_m/ft·h Liquid	Viscosity, lb_m/ft·h Vapor	Thermal Cond, Btu/h·ft·°R Liquid	Thermal Cond, Btu/h·ft·°R Vapor	Surface Tension, dyne/cm	Temp,* °R
24.84a	1.022	4.809	125.6440	−132.18	60.28	1.1902	8.9456	1.845	2.552	1.690	4503.	1000.6	0.0630	0.00157	0.0435	0.00604	3.124	24.84
25.00	1.075	4.805	120.3520	−131.88	60.63	1.2018	8.9123	1.823	2.554	1.690	4490.	1003.5	0.0622	0.00159	0.0439	0.00609	3.107	25.00
26.00	1.462	4.775	92.4333	−130.06	62.81	1.2710	8.7083	1.733	2.571	1.696	4406.	1021.2	0.0577	0.00170	0.0464	0.00640	3.001	26.00
27.00	1.948	4.745	71.9555	−128.33	64.94	1.3356	8.5153	1.705	2.590	1.704	4319.	1038.1	0.0538	0.00180	0.0486	0.00672	2.894	27.00
28.00	2.548	4.715	56.8038	−126.61	67.00	1.3976	8.3337	1.716	2.612	1.712	4231.	1054.2	0.0503	0.00189	0.0506	0.00703	2.789	28.00
29.00	3.279	4.683	45.4567	−124.86	69.01	1.4583	8.1635	1.754	2.636	1.722	4146.	1069.6	0.0473	0.00198	0.0523	0.00735	2.684	29.00
30.00	4.158	4.651	36.8429	−123.07	70.94	1.5185	8.0038	1.807	2.663	1.733	4063.	1084.2	0.0446	0.00206	0.0539	0.00766	2.579	30.00
31.00	5.201	4.618	30.2129	−121.21	72.81	1.5786	7.8539	1.872	2.693	1.746	3985.	1098.1	0.0422	0.00215	0.0552	0.00798	2.475	31.00
32.00	6.427	4.585	25.0403	−119.26	74.59	1.6388	7.7125	1.943	2.726	1.761	3910.	1111.3	0.0400	0.00223	0.0564	0.00831	2.371	32.00
33.00	7.854	4.550	20.9531	−117.24	76.30	1.6993	7.5789	2.019	2.762	1.777	3839.	1123.7	0.0380	0.00231	0.0574	0.00863	2.268	33.00
34.00	9.500	4.514	17.6851	−115.13	77.92	1.7602	7.4522	2.099	2.803	1.796	3770.	1135.5	0.0361	0.00239	0.0582	0.00897	2.166	34.00
35.00	11.384	4.477	15.0434	−112.93	79.44	1.8215	7.3315	2.181	2.848	1.817	3703.	1146.6	0.0345	0.00247	0.0589	0.00931	2.064	35.00
36.00	13.526	4.439	12.8866	−110.64	80.87	1.8833	7.2162	2.266	2.897	1.840	3637.	1157.0	0.0329	0.00256	0.0595	0.00965	1.963	36.00
36.50b	14.696	4.420	11.9588	−109.47	81.54	1.9143	7.1606	2.310	2.924	1.853	3604.	1161.9	0.0322	0.00260	0.0597	0.00983	1.913	36.50
37.00	15.945	4.400	11.1093	−108.26	82.19	1.9456	7.1057	2.354	2.952	1.867	3572.	1166.7	0.0315	0.00264	0.0600	0.01001	1.863	37.00
38.00	18.660	4.359	9.6324	−105.79	83.40	2.0084	6.9993	2.445	3.014	1.897	3506.	1175.7	0.0301	0.00272	0.0603	0.01037	1.763	38.00
39.00	21.691	4.317	8.3954	−103.21	84.49	2.0718	6.8966	2.540	3.083	1.930	3440.	1184.0	0.0289	0.00281	0.0605	0.01074	1.664	39.00
40.00	25.058	4.273	7.3516	−100.54	85.46	2.1357	6.7970	2.639	3.160	1.968	3373.	1191.8	0.0277	0.00289	0.0607	0.01112	1.565	40.00
41.00	28.781	4.227	6.4648	−97.76	86.29	2.2003	6.7002	2.744	3.246	2.011	3305.	1198.8	0.0266	0.00298	0.0607	0.01152	1.468	41.00
42.00	32.880	4.180	5.7066	−94.87	86.98	2.2656	6.6057	2.856	3.344	2.059	3235.	1205.2	0.0255	0.00307	0.0606	0.01193	1.371	42.00
43.00	37.375	4.130	5.0544	−91.87	87.52	2.3317	6.5131	2.977	3.455	2.115	3163.	1211.0	0.0245	0.00316	0.0605	0.01235	1.275	43.00
44.00	42.287	4.079	4.4901	−88.74	87.89	2.3986	6.4219	3.108	3.582	2.178	3088.	1216.2	0.0236	0.00326	0.0603	0.01280	1.180	44.00
45.00	47.637	4.025	3.9992	−85.48	88.08	2.4665	6.3319	3.252	3.728	2.251	3011.	1220.7	0.0227	0.00335	0.0600	0.01326	1.086	45.00
46.00	53.446	3.968	3.5699	−82.07	88.09	2.5354	6.2424	3.412	3.897	2.336	2931.	1224.6	0.0218	0.00345	0.0596	0.01375	0.993	46.00
47.00	59.734	3.909	3.1926	−78.51	87.87	2.6056	6.1532	3.593	4.096	2.435	2847.	1227.9	0.0209	0.00356	0.0592	0.01427	0.901	47.00
48.00	66.525	3.846	2.8593	−74.78	87.43	2.6773	6.0638	3.798	4.332	2.553	2760.	1230.6	0.0201	0.00367	0.0586	0.01482	0.811	48.00
49.00	73.840	3.779	2.5634	−70.87	86.73	2.7507	5.9735	4.037	4.615	2.695	2670.	1232.6	0.0193	0.00378	0.0580	0.01541	0.721	49.00
50.00	81.703	3.708	2.2994	−66.74	85.73	2.8262	5.8818	4.320	4.962	2.868	2575.	1234.0	0.0185	0.00390	0.0573	0.01604	0.633	50.00
52.00	99.171	3.550	1.8483	−57.74	82.68	2.9853	5.6909	5.089	5.952	3.361	2370.	1235.0	0.0170	0.00417	0.0556	0.01750	0.462	52.00
54.00	119.139	3.362	1.4749	−47.40	77.70	3.1604	5.4812	6.388	7.721	4.238	2142.	1233.4	0.0154	0.00451	0.0535	0.01933	0.299	54.00
56.00	141.856	3.124	1.1525	−34.92	69.57	3.3637	5.2331	9.186	11.753	6.220	1882.	1229.1	0.0138	0.00500	0.0507	0.02186	0.146	56.00
58.00	167.654	2.772	0.8435	−17.82	54.28	3.6351	4.8811	20.040	29.226	14.714	1576.	1223.1	0.0118	0.00593	0.0466	0.02617	0.015	58.00
59.29c	186.195	1.958	0.5108	17.32	17.32	4.2101	4.2101	∞	∞	∞	0.	0.0	—	—	∞	∞	0.000	59.29

*temperatures are on the IPTS-68 scale a = triple point b = normal boiling point c = critical point

Refrigerant 702p (Parahydrogen) Properties of Gas at 14.696 psia (one standard atmosphere)

Temp, °F	Density, lb/ft³	Enthalpy, Btu/lb	Entropy, Btu/lb·°F	c_p, Btu/lb·°F	c_p/c_v	Vel. Sound, ft/s	Viscosity, lb/ft·h	Thermal Cond, Btu/h·ft·°F
−423.2a	0.08362	81.54	7.1606	2.924	1.853	1161.9	0.0026	0.0098
−420.0	0.07527	90.59	7.3986	2.791	1.811	1226.7	0.0028	0.0107
−410.0	0.05800	117.45	8.0031	2.616	1.748	1401.1	0.0035	0.0132
−400.0	0.04750	143.26	8.4768	2.555	1.720	1549.3	0.0042	0.0156
−390.0	0.04032	168.65	8.8703	2.526	1.704	1681.8	0.0048	0.0179
−380.0	0.03507	193.86	9.2084	2.517	1.691	1801.4	0.0054	0.0201
−370.0	0.03105	219.03	9.5060	2.519	1.680	1911.1	0.0060	0.0222
−360.0	0.02787	244.28	9.7729	2.534	1.665	2011.1	0.0065	0.0244
−340.0	0.02314	295.67	10.2426	2.618	1.622	2180.8	0.0075	0.0291
−320.0	0.01980	349.47	10.6580	2.773	1.563	2316.2	0.0084	0.0343
−300.0	0.01730	406.94	11.0422	2.979	1.502	2430.0	0.0093	0.0400
−280.0	0.01537	468.76	11.4068	3.205	1.449	2533.3	0.0101	0.0464
−260.0	0.01382	535.09	11.7566	3.424	1.408	2633.2	0.0109	0.0532
−240.0	0.01256	605.50	12.0925	3.610	1.378	2733.3	0.0117	0.0597
−220.0	0.01151	679.18	12.4135	3.749	1.359	2835.2	0.0124	0.0657
−200.0	0.01063	755.20	12.7180	3.844	1.346	2938.1	0.0131	0.0712
−180.0	0.00986	832.67	13.0054	3.896	1.340	3042.0	0.0138	0.0760
−160.0	0.00921	910.82	13.2753	3.913	1.338	3146.5	0.0145	0.0802
−140.0	0.00863	989.03	13.5280	3.905	1.339	3250.9	0.0151	0.0839
−120.0	0.00812	1066.89	13.7642	3.879	1.341	3354.7	0.0158	0.0870
−100.0	0.00767	1144.11	13.9851	3.842	1.346	3457.5	0.0164	0.0899
−80.0	0.00727	1220.56	14.1920	3.801	1.351	3558.8	0.0170	0.0924
−60.0	0.00690	1296.16	14.3861	3.759	1.356	3658.5	0.0176	0.0948
−40.0	0.00657	1370.93	14.5686	3.718	1.361	3756.1	0.0182	0.0972
−20.0	0.00627	1444.91	14.7408	3.680	1.366	3851.6	0.0188	0.0995
0.0	0.00600	1518.17	14.9038	3.646	1.371	3944.9	0.0194	0.1017
10.0	0.00587	1554.55	14.9821	3.630	1.373	3990.7	0.0197	0.1029
20.0	0.00575	1590.78	15.0584	3.616	1.375	4036.0	0.0200	0.1040
30.0	0.00563	1626.87	15.1329	3.602	1.377	4080.6	0.0203	0.1052
40.0	0.00552	1662.83	15.2056	3.590	1.379	4124.8	0.0205	0.1063
50.0	0.00541	1698.67	15.2766	3.578	1.381	4168.3	0.0208	0.1075
60.0	0.00531	1734.40	15.3460	3.567	1.382	4211.4	0.0211	0.1087
70.0	0.00521	1770.02	15.4139	3.558	1.383	4253.9	0.0214	0.1099
80.0	0.00511	1805.55	15.4804	3.549	1.385	4295.9	0.0216	0.1110
90.0	0.00502	1841.00	15.5455	3.541	1.386	4337.4	0.0219	0.1123
100.0	0.00493	1876.37	15.6092	3.533	1.387	4378.3	0.0222	0.1135
110.0	0.00484	1911.66	15.6717	3.526	1.388	4418.9	0.0225	0.1148
120.0	0.00476	1946.90	15.7330	3.520	1.389	4458.9	0.0227	0.1160
130.0	0.00468	1982.07	15.7932	3.515	1.390	4498.5	0.0230	0.1173
140.0	0.00460	2017.20	15.8523	3.510	1.391	4537.7	0.0233	0.1186
150.0	0.00453	2052.27	15.9103	3.506	1.391	4576.5	0.0235	0.1199
160.0	0.00445	2087.31	15.9673	3.501	1.392	4614.9	0.0238	0.1212
170.0	0.00438	2122.30	16.0233	3.498	1.393	4652.8	0.0241	0.1226
180.0	0.00431	2157.27	16.0784	3.495	1.393	4690.4	0.0243	0.1239
190.0	0.00425	2192.20	16.1326	3.492	1.393	4727.7	0.0246	0.1252
200.0	0.00418	2227.10	16.1859	3.489	1.394	4764.6	0.0248	0.1265
210.0	0.00412	2261.98	16.2384	3.487	1.394	4801.2	0.0251	0.1278
220.0	0.00406	2296.84	16.2900	3.485	1.395	4837.5	0.0253	0.1292
230.0	0.00400	2331.67	16.3409	3.483	1.395	4873.4	0.0256	0.1306
240.0	0.00394	2366.49	16.3910	3.481	1.395	4909.1	0.0258	0.1319
250.0	0.00389	2401.29	16.4404	3.479	1.395	4944.5	0.0261	0.1333

a = saturated vapor at normal boiling point

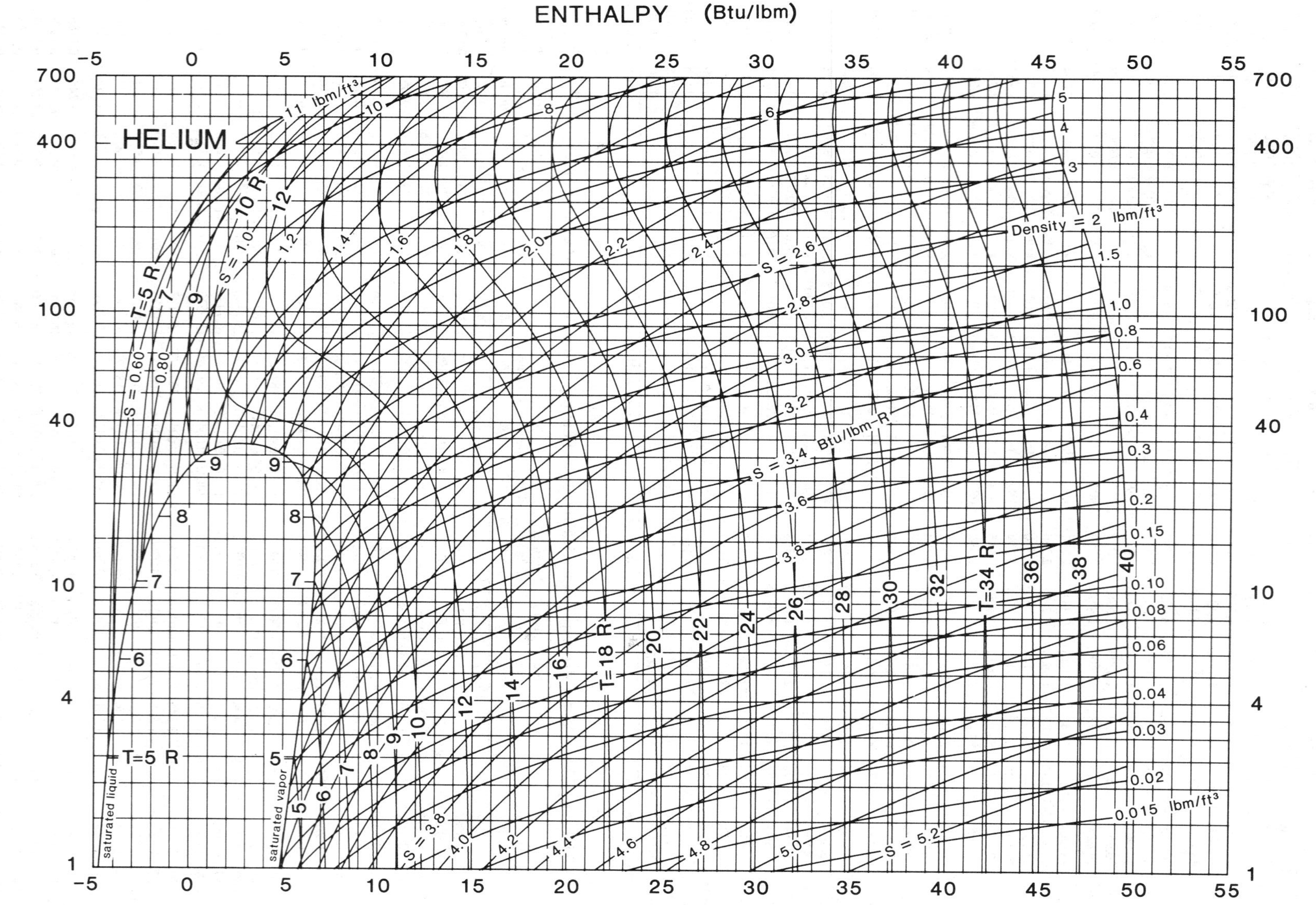

Prepared by: CENTER FOR APPLIED THERMODYNAMIC STUDIES, University of Idaho

Fig. 33 Pressure–Enthalpy Diagram for Refrigerant 704 (Helium)

Note: The reference states for enthalpy and entropy differ from those in the table.

Refrigerant 704 (Helium) Properties of Saturated Liquid and Saturated Vapor

Temp,* °R	Pressure, psia	Density, lb/ft³ Liquid	Volume, ft³/lb Vapor	Enthalpy, Btu/lb Liquid	Enthalpy, Btu/lb Vapor	Entropy, Btu/lb·°R Liquid	Entropy, Btu/lb·°R Vapor	Specific Heat c_p, Btu/lb·°R Liquid	Specific Heat c_p, Btu/lb·°R Vapor	c_p/c_v Vapor	Velocity of Sound, ft/s Liquid	Velocity of Sound, ft/s Vapor	Viscosity, lb_m/ft·h Liquid	Viscosity, lb_m/ft·h Vapor	Thermal Cond, Btu/h·ft·°R Liquid	Thermal Cond, Btu/h·ft·°R Vapor	Surface Tension, dyne/cm	Temp,* °R
3.92a	0.704	9.130	13.9837	1.005	10.998	0.33557	2.88589	1.5100	1.4485	1.747	711.	273.0	—	—	—	—	0.388	3.92
4.00	0.789	9.122	12.7008	1.121	11.076	0.36428	2.85306	1.2809	1.4556	1.752	710.	275.2	—	—	—	—	0.382	4.00
4.20	1.021	9.099	10.1926	1.340	11.263	0.41671	2.77927	0.9051	1.4720	1.768	708.	280.3	—	—	—	—	0.366	4.20
4.40	1.296	9.068	8.3274	1.504	11.445	0.45355	2.71288	0.7039	1.4879	1.786	708.	285.1	—	—	—	—	0.350	4.40
4.60	1.617	9.030	6.9024	1.640	11.621	0.48225	2.65223	0.6064	1.5037	1.805	710.	289.7	—	—	—	—	0.334	4.60
4.80	1.988	8.987	5.7905	1.763	11.792	0.50693	2.59623	0.5700	1.5199	1.827	711.	294.0	—	—	—	—	0.319	4.80
5.00	2.413	8.938	4.9078	1.884	11.955	0.52983	2.54407	0.5694	1.5371	1.851	711.	298.0	—	—	—	—	0.303	5.00
5.20	2.896	8.883	4.1970	2.008	12.111	0.55216	2.49514	0.5893	1.5554	1.879	708.	301.7	—	—	—	—	0.287	5.20
5.40	3.442	8.824	3.6172	2.138	12.260	0.57454	2.44894	0.6209	1.5755	1.910	703.	305.2	—	—	—	—	0.272	5.40
5.60	4.053	8.760	3.1392	2.276	12.400	0.59727	2.40509	0.6590	1.5978	1.944	696.	308.5	—	—	—	—	0.256	5.60
5.80	4.734	8.690	2.7410	2.423	12.531	0.62047	2.36323	0.7009	1.6229	1.984	688.	311.5	—	—	—	—	0.241	5.80
6.00	5.488	8.616	2.4063	2.579	12.652	0.64419	2.32304	0.7453	1.6513	2.029	679.	314.3	—	—	—	—	0.226	6.00
6.20	6.319	8.536	2.1226	2.744	12.763	0.66842	2.28427	0.7918	1.6839	2.080	669.	316.8	—	—	—	—	0.211	6.20
6.40	7.230	8.451	1.8803	2.920	12.862	0.69316	2.24664	0.8407	1.7217	2.139	659.	319.2	0.00844	0.00238	0.0103	0.00416	0.196	6.40
6.60	8.224	8.360	1.6717	3.106	12.950	0.71841	2.20990	0.8927	1.7657	2.207	647.	321.3	0.00832	0.00248	0.0104	0.00432	0.181	6.60
6.80	9.306	8.262	1.4910	3.303	13.025	0.74419	2.17383	0.9489	1.8177	2.286	635.	323.2	0.00819	0.00258	0.0105	0.00448	0.166	6.80
7.00	10.480	8.158	1.3333	3.511	13.085	0.77053	2.13816	1.0109	1.8798	2.379	623.	324.9	0.00807	0.00268	0.0106	0.00466	0.152	7.00
7.20	11.748	8.047	1.1948	3.732	13.129	0.79748	2.10264	1.0810	1.9547	2.491	609.	326.3	0.00793	0.00279	0.0107	0.00483	0.138	7.20
7.40	13.114	7.927	1.0724	3.965	13.155	0.82515	2.06700	1.1621	2.0466	2.626	595.	327.6	0.00780	0.00290	0.0107	0.00502	0.123	7.40
7.60	14.584	7.797	0.9634	4.214	13.161	0.85365	2.03091	1.2586	2.1613	2.792	580.	328.7	0.00766	0.00301	0.0108	0.00522	0.109	7.60
7.61b	14.696	7.787	0.9559	4.233	13.160	0.85578	2.02824	1.2664	2.1708	2.806	579.	328.8	0.00765	0.00302	0.0108	0.00524	0.108	7.61
7.80	16.161	7.656	0.8657	4.479	13.143	0.88319	1.99398	1.3768	2.3077	3.003	564.	329.6	0.00752	0.00313	0.0108	0.00544	0.095	7.80
8.00	17.850	7.501	0.7774	4.764	13.098	0.91401	1.95573	1.5272	2.5001	3.278	547.	330.3	0.00737	0.00326	0.0108	0.00568	0.082	8.00
8.20	19.657	7.329	0.6969	5.072	13.018	0.94651	1.91551	1.7271	2.7626	3.650	529.	330.8	0.00721	0.00340	0.0109	0.00596	0.068	8.20
8.40	21.587	7.135	0.6226	5.410	12.895	0.98128	1.87237	2.0092	3.1400	4.181	509.	331.2	0.00704	0.00354	0.0109	0.00629	0.055	8.40
8.60	23.650	6.911	0.5529	5.788	12.716	1.01927	1.82486	2.4415	3.7248	5.003	487.	331.6	0.00686	0.00371	0.0110	0.00670	0.042	8.60
8.80	25.854	6.641	0.4861	6.222	12.453	1.06226	1.77037	3.1940	4.7457	6.434	463.	332.2	0.00664	0.00389	0.0110	0.00724	0.030	8.80
9.00	28.215	6.294	0.4192	6.751	12.055	1.11412	1.70345	4.8375	6.9536	9.531	436.	333.4	0.00638	0.00411	0.0112	0.00801	0.018	9.00
9.20	30.755	5.771	0.3451	7.493	11.364	1.18708	1.60775	11.0204	14.9798	20.845	404.	336.9	0.00603	0.00440	0.0115	0.00932	0.007	9.20
9.35c	32.990	4.348	0.2300	9.339	9.339	1.37760	1.37760	∞	∞	∞	0.	0.0	—	—	∞	∞	0.000	9.35

*temperatures are on the EPT-76 scale a = lower lambda point b = normal boiling point c = critical point

Refrigerant 704 (Helium) Properties of Gas at 14.696 psia (one standard atmosphere)

Temp, °F	Density, lb/ft³	Enthalpy, Btu/lb	Entropy, Btu/lb·°F	c_p, Btu/lb·°F	c_p/c_v	Vel. Sound, ft/s	Viscosity, lb/ft·h	Thermal Cond, Btu/h·ft·°F
−452.1a	1.04613	13.16	2.0282	2.1708	2.806	328.8	0.0030	0.00523
−450.0	0.66840	16.73	2.4459	1.5277	2.037	415.4	0.0035	0.00624
−440.0	0.28462	30.24	3.4138	1.2880	1.729	636.4	0.0058	0.01035
−430.0	0.18546	42.96	3.9370	1.2616	1.691	786.5	0.0076	0.01343
−420.0	0.13809	55.52	4.3020	1.2525	1.679	910.2	0.0092	0.01608
−400.0	0.09163	80.49	4.8118	1.2459	1.671	1115.8	0.0119	0.02074
−380.0	0.06862	105.39	5.1716	1.2437	1.669	1288.4	0.0143	0.02493
−360.0	0.05486	130.25	5.4500	1.2427	1.668	1440.3	0.0164	0.02882
−340.0	0.04571	155.10	5.6772	1.2421	1.667	1577.5	0.0183	0.03248
−320.0	0.03917	179.94	5.8691	1.2418	1.667	1703.7	0.0202	0.03596
−300.0	0.03427	204.77	6.0353	1.2416	1.667	1821.2	0.0219	0.03931
−280.0	0.03046	229.60	6.1818	1.2415	1.667	1931.5	0.0236	0.04254
−260.0	0.02741	254.43	6.3129	1.2414	1.667	2035.8	0.0248	0.04566
−240.0	0.02492	279.26	6.4314	1.2413	1.667	2135.1	0.0264	0.04870
−220.0	0.02285	304.08	6.5395	1.2413	1.667	2229.9	0.0280	0.05166
−200.0	0.02109	328.91	6.6390	1.2413	1.666	2320.9	0.0295	0.05454
−180.0	0.01958	353.73	6.7311	1.2412	1.666	2408.4	0.0310	0.05737
−160.0	0.01827	378.56	6.8168	1.2412	1.666	2492.9	0.0324	0.06013
−140.0	0.01713	403.38	6.8970	1.2412	1.666	2574.6	0.0338	0.06284
−120.0	0.01613	428.21	6.9724	1.2412	1.666	2653.7	0.0352	0.06551
−100.0	0.01523	453.03	7.0434	1.2412	1.666	2730.6	0.0366	0.06813
−80.0	0.01443	477.85	7.1105	1.2412	1.667	2805.4	0.0380	0.07070
−60.0	0.01371	502.68	7.1743	1.2412	1.667	2878.2	0.0393	0.07324
−40.0	0.01305	527.50	7.2349	1.2412	1.667	2949.3	0.0406	0.07574
−20.0	0.01246	552.32	7.2927	1.2412	1.667	3018.7	0.0419	0.07821
0.0	0.01192	577.15	7.3479	1.2412	1.667	3086.5	0.0432	0.08064
20.0	0.01142	601.97	7.4007	1.2412	1.667	3152.8	0.0445	0.08304
40.0	0.01096	626.79	7.4514	1.2412	1.667	3217.8	0.0457	0.08542
60.0	0.01054	651.62	7.5001	1.2412	1.667	3281.5	0.0470	0.08776
80.0	0.01015	676.44	7.5470	1.2411	1.667	3344.0	0.0482	0.09008
100.0	0.00979	701.26	7.5922	1.2411	1.667	3405.3	0.0494	0.09238
120.0	0.00945	726.09	7.6357	1.2411	1.667	3465.6	0.0506	0.09465
140.0	0.00914	750.91	7.6778	1.2411	1.667	3524.8	0.0518	0.09690
160.0	0.00884	775.73	7.7186	1.2411	1.667	3583.1	0.0530	0.09912
180.0	0.00857	800.55	7.7580	1.2411	1.667	3640.4	0.0542	0.10133
200.0	0.00831	825.38	7.7962	1.2411	1.667	3696.8	0.0553	0.10351
240.0	0.00783	875.02	7.8693	1.2411	1.667	3807.1	0.0576	0.10783
280.0	0.00741	924.67	7.9383	1.2411	1.667	3914.4	0.0599	0.11207
320.0	0.00703	974.32	8.0036	1.2411	1.667	4018.8	0.0621	0.11624
360.0	0.00669	1023.96	8.0657	1.2411	1.667	4120.5	0.0643	0.12036
400.0	0.00637	1073.61	8.1249	1.2411	1.667	4219.8	0.0665	0.12441
440.0	0.00609	1123.25	8.1813	1.2411	1.667	4316.8	0.0686	0.12841
480.0	0.00583	1172.90	8.2353	1.2411	1.667	4411.6	0.0707	0.13236
520.0	0.00559	1222.55	8.2871	1.2411	1.667	4504.5	0.0728	0.13626
560.0	0.00537	1272.19	8.3367	1.2411	1.667	4595.5	0.0749	0.14011
600.0	0.00517	1321.84	8.3845	1.2411	1.667	4684.7	0.0769	0.14392
640.0	0.00498	1371.48	8.4305	1.2412	1.667	4772.3	0.0789	0.14768
680.0	0.00481	1421.13	8.4748	1.2412	1.667	4858.3	0.0809	0.15141
720.0	0.00465	1470.78	8.5176	1.2412	1.667	4942.8	0.0829	0.15509
760.0	0.00449	1520.42	8.5590	1.2412	1.667	5025.8	0.0849	0.15874
800.0	0.00435	1570.07	8.5991	1.2412	1.667	5107.6	0.0868	0.16236

a = saturated vapor at normal boiling point

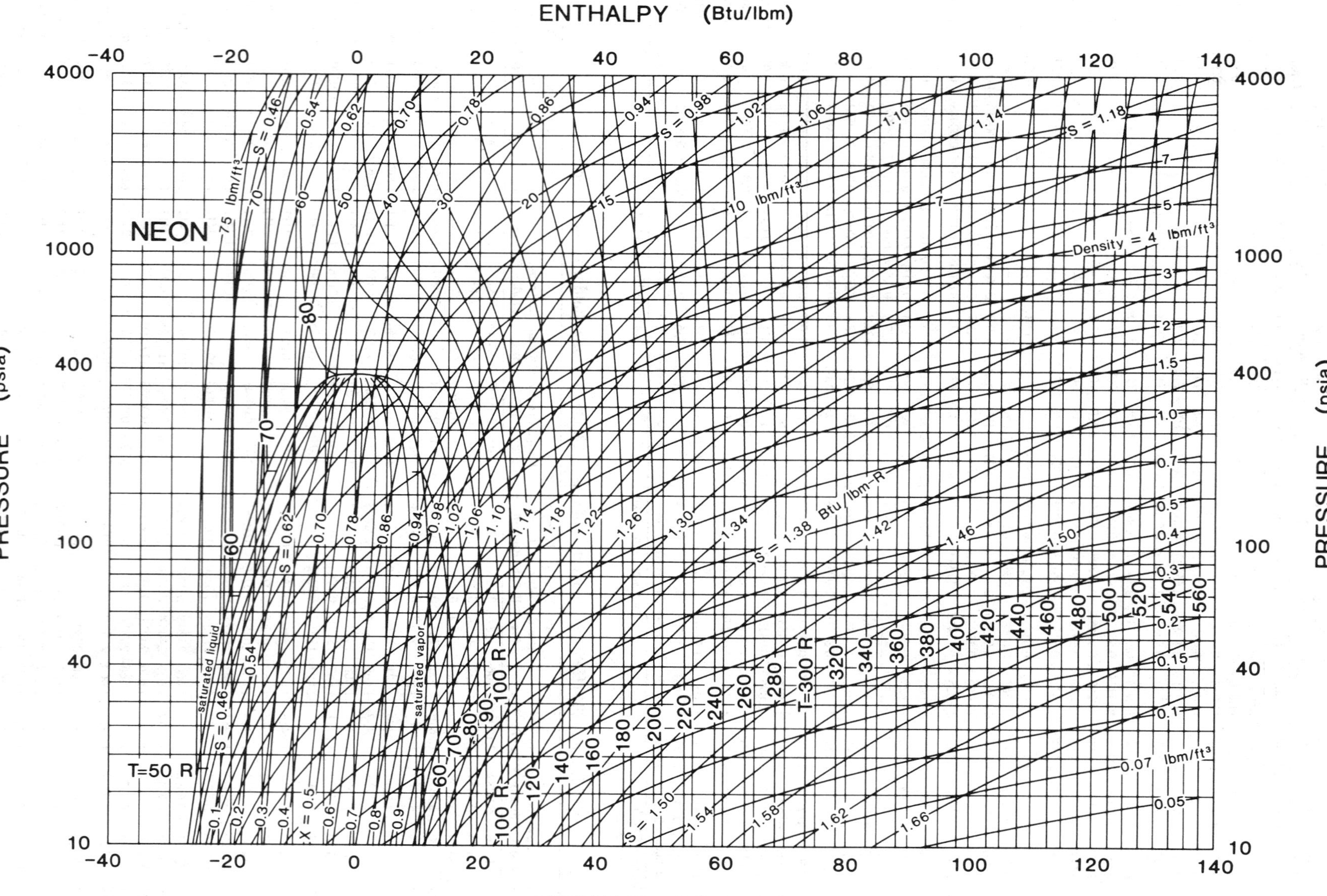

Fig. 34 Pressure–Enthalpy Diagram for Refrigerant 720 (Neon)

Refrigerant 720 (Neon) Properties of Saturated Liquid and Saturated Vapor

Temp,* °R	Pressure, psia	Density, lb/ft³ Liquid	Volume, ft³/lb Vapor	Enthalpy, Btu/lb Liquid	Enthalpy, Btu/lb Vapor	Entropy, Btu/lb·°R Liquid	Entropy, Btu/lb·°R Vapor	Specific Heat c_p, Btu/lb·°R Liquid	Specific Heat c_p, Btu/lb·°R Vapor	c_p/c_v Vapor	Velocity of Sound, ft/s Liquid	Velocity of Sound, ft/s Vapor	Surface Tension, dyne/cm	Temp,* °R
44.21a	6.30	78.18	3.6041	−27.77	10.14	0.3294	1.1868	0.510	0.568	1.362	2213.	370.8	5.705	44.21
45.00	7.39	77.66	3.1245	−27.38	10.38	0.3381	1.1771	0.479	0.495	1.401	2139.	378.9	5.549	45.00
46.00	8.96	77.03	2.6245	−26.91	10.64	0.3483	1.1647	0.457	0.429	1.456	2069.	389.3	5.353	46.00
47.00	10.77	76.42	2.2201	−26.46	10.88	0.3580	1.1523	0.447	0.384	1.513	2017.	399.3	5.158	47.00
48.00	12.85	75.82	1.8903	−26.01	11.09	0.3673	1.1401	0.444	0.354	1.568	1976.	408.8	4.964	48.00
48.79b	14.696	75.35	1.6725	−25.66	11.23	0.3745	1.1306	0.445	0.338	1.609	1949.	415.5	4.813	48.79
49.00	15.22	75.22	1.6194	−25.56	11.27	0.3764	1.1281	0.446	0.335	1.620	1943.	417.2	4.772	49.00
50.00	17.91	74.61	1.3952	−25.11	11.44	0.3854	1.1164	0.450	0.322	1.666	1913.	424.6	4.582	50.00
51.00	20.95	73.99	1.2085	−24.65	11.59	0.3943	1.1050	0.456	0.315	1.709	1885.	430.9	4.393	51.00
52.00	24.35	73.36	1.0519	−24.19	11.73	0.4032	1.0938	0.464	0.312	1.747	1858.	436.3	4.205	52.00
53.00	28.14	72.71	0.9199	−23.72	11.85	0.4120	1.0830	0.472	0.312	1.782	1830.	440.9	4.020	53.00
54.00	32.35	72.04	0.8079	−23.23	11.96	0.4208	1.0724	0.481	0.314	1.816	1801.	444.7	3.836	54.00
55.00	37.01	71.35	0.7123	−22.74	12.05	0.4296	1.0621	0.490	0.318	1.849	1771.	448.0	3.653	55.00
56.00	42.15	70.65	0.6304	−22.24	12.13	0.4384	1.0521	0.499	0.323	1.882	1740.	450.8	3.473	56.00
57.00	47.78	69.93	0.5598	−21.73	12.19	0.4471	1.0422	0.508	0.330	1.916	1707.	453.2	3.294	57.00
58.00	53.94	69.19	0.4987	−21.21	12.24	0.4558	1.0326	0.517	0.338	1.953	1673.	455.3	3.117	58.00
59.00	60.65	68.43	0.4455	−20.69	12.27	0.4646	1.0231	0.527	0.347	1.993	1638.	457.1	2.942	59.00
60.00	67.95	67.65	0.3991	−20.15	12.28	0.4732	1.0138	0.536	0.357	2.036	1601.	458.6	2.769	60.00
62.00	84.40	66.03	0.3224	−19.05	12.26	0.4906	0.9955	0.558	0.383	2.138	1525.	461.0	2.430	62.00
64.00	103.52	64.32	0.2625	−17.90	12.15	0.5079	0.9775	0.583	0.417	2.267	1444.	462.6	2.100	64.00
66.00	125.56	62.50	0.2149	−16.70	11.95	0.5253	0.9595	0.616	0.461	2.436	1359.	463.3	1.780	66.00
68.00	150.76	60.55	0.1765	−15.45	11.63	0.5430	0.9412	0.660	0.521	2.665	1269.	463.3	1.471	68.00
70.00	179.41	58.41	0.1451	−14.11	11.17	0.5611	0.9222	0.723	0.607	2.993	1173.	462.5	1.174	70.00
72.00	211.81	56.01	0.1190	−12.65	10.53	0.5801	0.9020	0.823	0.739	3.494	1068.	460.9	0.891	72.00
74.00	248.30	53.19	0.0967	−11.02	9.62	0.6008	0.8797	1.001	0.966	4.345	950.	458.5	0.625	74.00
76.00	289.29	49.60	0.0772	−9.08	8.32	0.6247	0.8536	1.398	1.430	6.073	812.	455.1	0.381	76.00
78.00	335.11	44.27	0.0592	−6.46	6.30	0.6563	0.8199	2.650	2.814	11.168	652.	451.0	0.165	78.00
80.09c	388.50	30.09	0.0332	−0.08	−0.08	0.7335	0.7335	∞	∞	∞	0.	0.0	0.000	80.09

*temperatures are on the IPTS-68 scale a = triple point b = normal boiling point c = critical point

Refrigerant 720 (Neon) Properties of Gas at 14.696 psia (one standard atmosphere)

Temp, °F	Density, lb/ft³	Enthalpy, Btu/lb	Entropy, Btu/lb·°F	c_p, Btu/lb·°F	c_p/c_v	Vel. Sound, ft/s
−410.9a	0.59789	11.23	1.1306	0.3382	1.609	415.5
−400.0	0.47681	14.23	1.1862	0.2601	1.721	488.3
−380.0	0.35134	19.33	1.2600	0.2519	1.694	569.2
−360.0	0.27910	24.33	1.3160	0.2493	1.683	638.6
−340.0	0.23177	29.31	1.3615	0.2482	1.677	700.7
−320.0	0.19826	34.26	1.3998	0.2476	1.674	757.5
−300.0	0.17327	39.21	1.4329	0.2472	1.672	810.2
−280.0	0.15389	44.15	1.4621	0.2470	1.671	859.6
−260.0	0.13842	49.09	1.4882	0.2468	1.670	906.2
−240.0	0.12579	54.03	1.5117	0.2467	1.669	950.6
−200.0	0.10638	63.89	1.5530	0.2465	1.668	1033.5
−180.0	0.09877	68.82	1.5713	0.2465	1.668	1072.6
−160.0	0.09217	73.75	1.5883	0.2465	1.668	1110.2
−140.0	0.08640	78.68	1.6042	0.2464	1.668	1146.7
−120.0	0.08131	83.61	1.6192	0.2464	1.668	1182.0
−100.0	0.07679	88.54	1.6333	0.2464	1.667	1216.2
−80.0	0.07275	93.46	1.6466	0.2463	1.667	1249.5
−60.0	0.06910	98.39	1.6592	0.2463	1.667	1282.0
−40.0	0.06581	103.32	1.6713	0.2463	1.667	1313.7
−20.0	0.06282	108.24	1.6827	0.2463	1.667	1344.5
0.0	0.06009	113.17	1.6937	0.2463	1.667	1374.8
20.0	0.05758	118.09	1.7042	0.2463	1.667	1404.3
40.0	0.05528	123.02	1.7142	0.2463	1.667	1433.3
60.0	0.05315	127.94	1.7239	0.2463	1.667	1461.6
80.0	0.05118	132.87	1.7332	0.2463	1.667	1489.5

Temp, °F	Density, lb/ft³	Enthalpy, Btu/lb	Entropy, Btu/lb·°F	c_p, Btu/lb·°F	c_p/c_v	Vel. Sound, ft/s
100.0	0.04935	137.80	1.7422	0.2463	1.667	1516.8
120.0	0.04765	142.72	1.7508	0.2462	1.667	1543.6
140.0	0.04606	147.65	1.7592	0.2462	1.667	1570.0
160.0	0.04457	152.57	1.7672	0.2462	1.667	1596.0
180.0	0.04318	157.50	1.7751	0.2462	1.667	1621.5
200.0	0.04187	162.42	1.7827	0.2462	1.667	1646.6
220.0	0.04064	167.34	1.7900	0.2462	1.667	1671.4
240.0	0.03948	172.27	1.7972	0.2462	1.667	1695.8
260.0	0.03838	177.19	1.8041	0.2462	1.667	1719.8
280.0	0.03735	182.12	1.8108	0.2462	1.667	1743.5
300.0	0.03636	187.04	1.8174	0.2462	1.667	1766.9
320.0	0.03543	191.97	1.8238	0.2462	1.667	1790.0
340.0	0.03454	196.89	1.8300	0.2462	1.667	1812.8
360.0	0.03370	201.82	1.8361	0.2462	1.667	1835.3
380.0	0.03290	206.74	1.8421	0.2462	1.667	1857.5
400.0	0.03213	211.66	1.8478	0.2462	1.667	1879.5
440.0	0.03070	221.51	1.8591	0.2462	1.667	1922.7
480.0	0.02940	231.36	1.8698	0.2462	1.667	1965.0
520.0	0.02820	241.21	1.8800	0.2462	1.667	2006.3
560.0	0.02709	251.06	1.8899	0.2462	1.667	2046.9
600.0	0.02607	260.91	1.8994	0.2462	1.667	2086.6
640.0	0.02512	270.75	1.9085	0.2462	1.667	2125.6
680.0	0.02424	280.60	1.9173	0.2462	1.667	2163.9
720.0	0.02342	290.45	1.9258	0.2462	1.667	2201.5
760.0	0.02265	300.30	1.9340	0.2462	1.667	2238.5
800.0	0.02193	310.15	1.9419	0.2462	1.667	2274.9

a = saturated vapor at normal boiling point

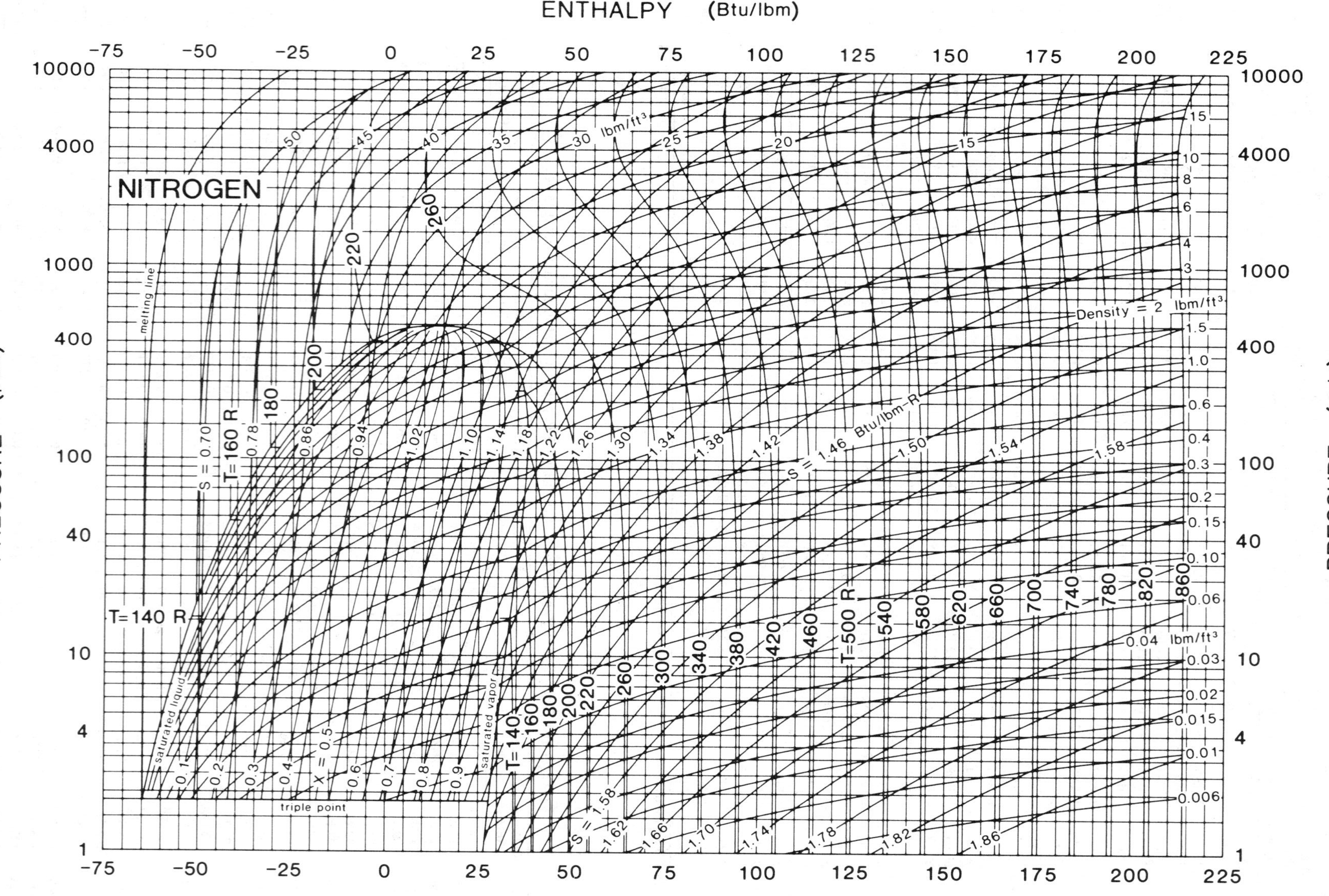

Prepared by CENTER FOR APPLIED THERMODYNAMIC STUDIES, University of Idaho

Fig. 35 Pressure–Enthalpy Diagram for Refrigerant 728 (Nitrogen)

Refrigerant 728 (Nitrogen) Properties of Saturated Liquid and Saturated Vapor

Temp,* °R	Pressure, psia	Density, lb/ft³ Liquid	Volume, ft³/lb Vapor	Enthalpy, Btu/lb Liquid	Enthalpy, Btu/lb Vapor	Entropy, Btu/lb·°R Liquid	Entropy, Btu/lb·°R Vapor	Specific Heat c_p, Btu/lb·°R Liquid	Specific Heat c_p, Btu/lb·°R Vapor	c_p/c_v Vapor	Velocity of Sound, ft/s Liquid	Velocity of Sound, ft/s Vapor	Viscosity, lb_m/ft·h Liquid	Viscosity, lb_m/ft·h Vapor	Thermal Cond, Btu/h·ft·°R Liquid	Thermal Cond, Btu/h·ft·°R Vapor	Surface Tension, dyne/cm	Temp,* °R
113.67a	1.82	54.29	23.738	−64.91	27.76	0.57825	1.39348	0.4826	0.2838	1.390	3352.	523.2	0.6916	0.0101	0.1009	0.00311	12.24	113.67
115.00	2.08	54.09	20.970	−64.26	28.06	0.58387	1.38665	0.4821	0.2857	1.390	3318.	525.7	0.6639	0.0102	0.1000	0.00315	12.06	115.00
120.00	3.34	53.34	13.542	−61.85	29.16	0.60435	1.36278	0.4811	0.2929	1.394	3197.	535.0	0.5749	0.0107	0.0963	0.00333	11.39	120.00
125.00	5.15	52.58	9.0919	−59.44	30.22	0.62397	1.34131	0.4815	0.3002	1.400	3084.	543.7	0.5044	0.0111	0.0928	0.00350	10.72	125.00
130.00	7.66	51.81	6.3130	−57.03	31.25	0.64286	1.32189	0.4830	0.3075	1.408	2976.	551.7	0.4472	0.0116	0.0893	0.00369	10.06	130.00
135.00	11.04	51.03	4.5130	−54.60	32.22	0.66109	1.30423	0.4854	0.3146	1.419	2873.	559.2	0.3999	0.0121	0.0859	0.00388	9.42	135.00
139.23b	14.696	50.36	3.4662	−52.53	33.01	0.67606	1.29047	0.4881	0.3205	1.430	2787.	565.0	0.3659	0.0125	0.0831	0.00404	8.88	139.23
140.00	15.45	50.23	3.3089	−52.16	33.15	0.67875	1.28806	0.4887	0.3215	1.432	2772.	566.0	0.3602	0.0126	0.0826	0.00407	8.78	140.00
145.00	21.11	49.42	2.4801	−49.69	34.01	0.69590	1.27315	0.4928	0.3285	1.449	2673.	572.1	0.3262	0.0131	0.0793	0.00428	8.15	145.00
150.00	28.20	48.58	1.8949	−47.20	34.81	0.71259	1.25931	0.4977	0.3355	1.470	2575.	577.5	0.2968	0.0137	0.0761	0.00449	7.53	150.00
155.00	36.93	47.73	1.4722	−44.69	35.53	0.72888	1.24637	0.5035	0.3430	1.495	2478.	582.1	0.2710	0.0142	0.0729	0.00472	6.92	155.00
160.00	47.53	46.85	1.1605	−42.13	36.16	0.74482	1.23417	0.5104	0.3513	1.526	2380.	585.9	0.2482	0.0148	0.0698	0.00496	6.32	160.00
165.00	60.20	45.94	0.9264	−39.54	36.71	0.76044	1.22256	0.5186	0.3610	1.563	2281.	588.9	0.2277	0.0153	0.0667	0.00521	5.74	165.00
170.00	75.19	44.99	0.7476	−36.91	37.14	0.77580	1.21141	0.5284	0.3726	1.608	2180.	591.1	0.2093	0.0159	0.0636	0.00548	5.16	170.00
175.00	92.71	44.02	0.6089	−34.22	37.46	0.79095	1.20059	0.5404	0.3870	1.663	2078.	592.3	0.1925	0.0166	0.0605	0.00577	4.60	175.00
180.00	113.01	42.99	0.4997	−31.48	37.65	0.80595	1.18999	0.5553	0.4051	1.731	1972.	592.6	0.1770	0.0172	0.0575	0.00609	4.05	180.00
185.00	136.32	41.92	0.4127	−28.65	37.69	0.82086	1.17946	0.5741	0.4284	1.817	1862.	592.0	0.1627	0.0179	0.0545	0.00644	3.52	185.00
190.00	162.89	40.78	0.3424	−25.74	37.55	0.83577	1.16887	0.5984	0.4591	1.927	1748.	590.3	0.1493	0.0187	0.0515	0.00683	3.00	190.00
195.00	192.99	39.56	0.2850	−22.71	37.21	0.85077	1.15805	0.6309	0.5007	2.074	1627.	587.5	0.1366	0.0195	0.0484	0.00727	2.50	195.00
200.00	226.88	38.23	0.2374	−19.54	36.62	0.86603	1.14680	0.6762	0.5590	2.275	1499.	583.5	0.1245	0.0204	0.0453	0.00778	2.02	200.00
205.00	264.86	36.76	0.1975	−16.16	35.71	0.88175	1.13481	0.7430	0.6455	2.570	1361.	578.2	0.1127	0.0215	0.0421	0.00841	1.56	205.00
210.00	307.25	35.07	0.1634	−12.52	34.39	0.89828	1.12165	0.8496	0.7847	3.035	1212.	571.5	0.1009	0.0228	0.0388	0.00925	1.13	210.00
215.00	354.43	33.06	0.1336	−8.45	32.47	0.91622	1.10651	1.0416	1.0417	3.877	1049.	563.0	0.0889	0.0244	0.0354	0.01052	0.73	215.00
220.00	406.84	30.49	0.1065	−3.66	29.52	0.93683	1.08763	1.4809	1.6619	5.856	867.	552.1	0.0761	0.0266	0.0326	0.01310	0.38	220.00
225.00	465.28	26.51	0.0784	2.91	23.94	0.96462	1.05807	3.8790	5.1786	16.586	654.	535.3	0.0606	0.0309	0.0358	0.02362	0.08	225.00
227.15c	492.81	19.55	0.0512	12.83	12.83	1.00750	1.00750	∞	∞	∞	0.	0.0	—	—	∞	∞	0.00	227.15

*temperatures are on the IPTS-68 scale a = triple point b = normal boiling point c = critical point

Refrigerant 728 (Nitrogen) Properties of Gas at 14.696 psia (one standard atmosphere)

Temp, °F	Density, lb/ft³	Enthalpy, Btu/lb	Entropy, Btu/lb·°F	c_p, Btu/lb·°F	c_p/c_v	Vel. Sound, ft/s	Viscosity, lb/ft·h	Thermal Cond, Btu/h·ft·°F
−320.4a	0.2885	33.01	1.2905	0.3205	1.430	565.0	0.0125	0.00404
−300.0	0.2471	38.64	1.3283	0.2595	1.436	619.8	0.0144	0.00470
−275.0	0.2116	45.05	1.3656	0.2546	1.425	670.7	0.0167	0.00550
−250.0	0.1853	51.39	1.3978	0.2527	1.418	717.1	0.0189	0.00631
−225.0	0.1650	57.69	1.4262	0.2516	1.413	760.2	0.0211	0.00711
−200.0	0.1487	63.97	1.4516	0.2508	1.410	800.8	0.0232	0.00783
−180.0	0.1379	68.98	1.4702	0.2504	1.408	831.7	0.0248	0.00841
−160.0	0.1286	73.99	1.4875	0.2500	1.407	861.4	0.0264	0.00898
−140.0	0.1204	78.99	1.5036	0.2498	1.406	890.1	0.0279	0.00954
−120.0	0.1132	83.98	1.5188	0.2496	1.405	917.9	0.0294	0.01008
−100.0	0.1069	88.97	1.5331	0.2494	1.404	944.8	0.0309	0.01061
−80.0	0.1012	93.96	1.5466	0.2493	1.404	970.9	0.0323	0.01113
−60.0	0.0961	98.94	1.5594	0.2492	1.403	996.3	0.0337	0.01163
−40.0	0.0915	103.92	1.5715	0.2491	1.403	1021.1	0.0351	0.01213
−20.0	0.0873	108.90	1.5831	0.2490	1.403	1045.2	0.0364	0.01261
0.0	0.0835	113.88	1.5942	0.2490	1.402	1068.8	0.0378	0.01309
10.0	0.0817	116.37	1.5996	0.2489	1.402	1080.4	0.0384	0.01332
20.0	0.0800	118.86	1.6048	0.2489	1.402	1091.9	0.0390	0.01355
30.0	0.0784	121.35	1.6099	0.2489	1.402	1103.3	0.0397	0.01378
40.0	0.0768	123.84	1.6150	0.2489	1.402	1114.5	0.0403	0.01401
50.0	0.0753	126.33	1.6199	0.2489	1.402	1125.6	0.0409	0.01424
60.0	0.0738	128.82	1.6247	0.2489	1.402	1136.6	0.0416	0.01446
70.0	0.0724	131.31	1.6295	0.2489	1.401	1147.5	0.0422	0.01469
80.0	0.0711	133.80	1.6341	0.2489	1.401	1158.3	0.0428	0.01491
90.0	0.0698	136.28	1.6387	0.2489	1.401	1169.0	0.0434	0.01513
100.0	0.0686	138.77	1.6432	0.2489	1.401	1179.6	0.0440	0.01534
120.0	0.0662	143.75	1.6519	0.2489	1.401	1200.4	0.0452	0.01578
140.0	0.0640	148.73	1.6604	0.2490	1.400	1220.9	0.0463	0.01620
160.0	0.0619	153.71	1.6685	0.2491	1.400	1241.0	0.0475	0.01662
180.0	0.0600	158.69	1.6765	0.2492	1.400	1260.8	0.0486	0.01704

Temp, °F	Density, lb/ft³	Enthalpy, Btu/lb	Entropy, Btu/lb·°F	c_p, Btu/lb·°F	c_p/c_v	Vel. Sound, ft/s	Viscosity, lb/ft·h	Thermal Cond, Btu/h·ft·°F
200.0	0.0581	163.68	1.6841	0.2493	1.399	1280.2	0.0497	0.01745
220.0	0.0564	168.66	1.6916	0.2494	1.399	1299.4	0.0508	0.01786
240.0	0.0548	173.65	1.6988	0.2496	1.399	1318.2	0.0519	0.01826
260.0	0.0533	178.65	1.7059	0.2497	1.398	1336.7	0.0530	0.01866
280.0	0.0518	183.64	1.7127	0.2499	1.398	1354.9	0.0540	0.01906
300.0	0.0505	188.65	1.7194	0.2502	1.397	1372.8	0.0551	0.01945
320.0	0.0492	193.65	1.7259	0.2504	1.396	1390.5	0.0561	0.01985
340.0	0.0480	198.66	1.7322	0.2507	1.396	1407.9	0.0571	0.02024
360.0	0.0468	203.68	1.7384	0.2510	1.395	1425.0	0.0581	0.02062
380.0	0.0457	208.70	1.7445	0.2514	1.394	1441.9	0.0591	0.02101
400.0	0.0446	213.73	1.7504	0.2517	1.393	1458.6	0.0601	0.02140
420.0	0.0436	218.77	1.7562	0.2521	1.392	1475.0	0.0611	0.02178
440.0	0.0426	223.82	1.7619	0.2525	1.391	1491.2	0.0620	0.02216
460.0	0.0417	228.87	1.7674	0.2529	1.391	1507.2	0.0630	0.02254
480.0	0.0408	233.93	1.7729	0.2534	1.390	1522.9	0.0639	0.02292
500.0	0.0400	239.01	1.7782	0.2538	1.389	1538.5	0.0648	0.02330
520.0	0.0391	244.09	1.7834	0.2543	1.387	1553.9	0.0658	0.02368
540.0	0.0384	249.18	1.7886	0.2548	1.386	1569.0	0.0667	0.02406
560.0	0.0376	254.28	1.7936	0.2553	1.385	1584.0	0.0676	0.02444
580.0	0.0369	259.39	1.7986	0.2559	1.384	1598.8	0.0685	0.02482
600.0	0.0362	264.51	1.8035	0.2564	1.383	1613.5	0.0693	0.02519
620.0	0.0355	269.65	1.8083	0.2570	1.382	1627.9	0.0702	0.02557
640.0	0.0349	274.79	1.8130	0.2575	1.381	1642.2	0.0711	0.02594
660.0	0.0342	279.95	1.8176	0.2581	1.379	1656.3	0.0720	0.02632
680.0	0.0336	285.12	1.8222	0.2587	1.378	1670.3	0.0728	0.02670
700.0	0.0331	290.30	1.8267	0.2593	1.377	1684.2	0.0737	0.02707
720.0	0.0325	295.49	1.8312	0.2599	1.376	1697.9	0.0745	0.02745
740.0	0.0320	300.69	1.8355	0.2606	1.374	1711.4	0.0754	0.02782
760.0	0.0314	305.91	1.8399	0.2612	1.373	1724.8	0.0762	0.02820
780.0	0.0309	311.14	1.8441	0.2618	1.372	1738.1	0.0770	0.02857
800.0	0.0304	316.38	1.8483	0.2624	1.371	1751.3	0.0778	0.02895

a = saturated vapor at normal boiling point

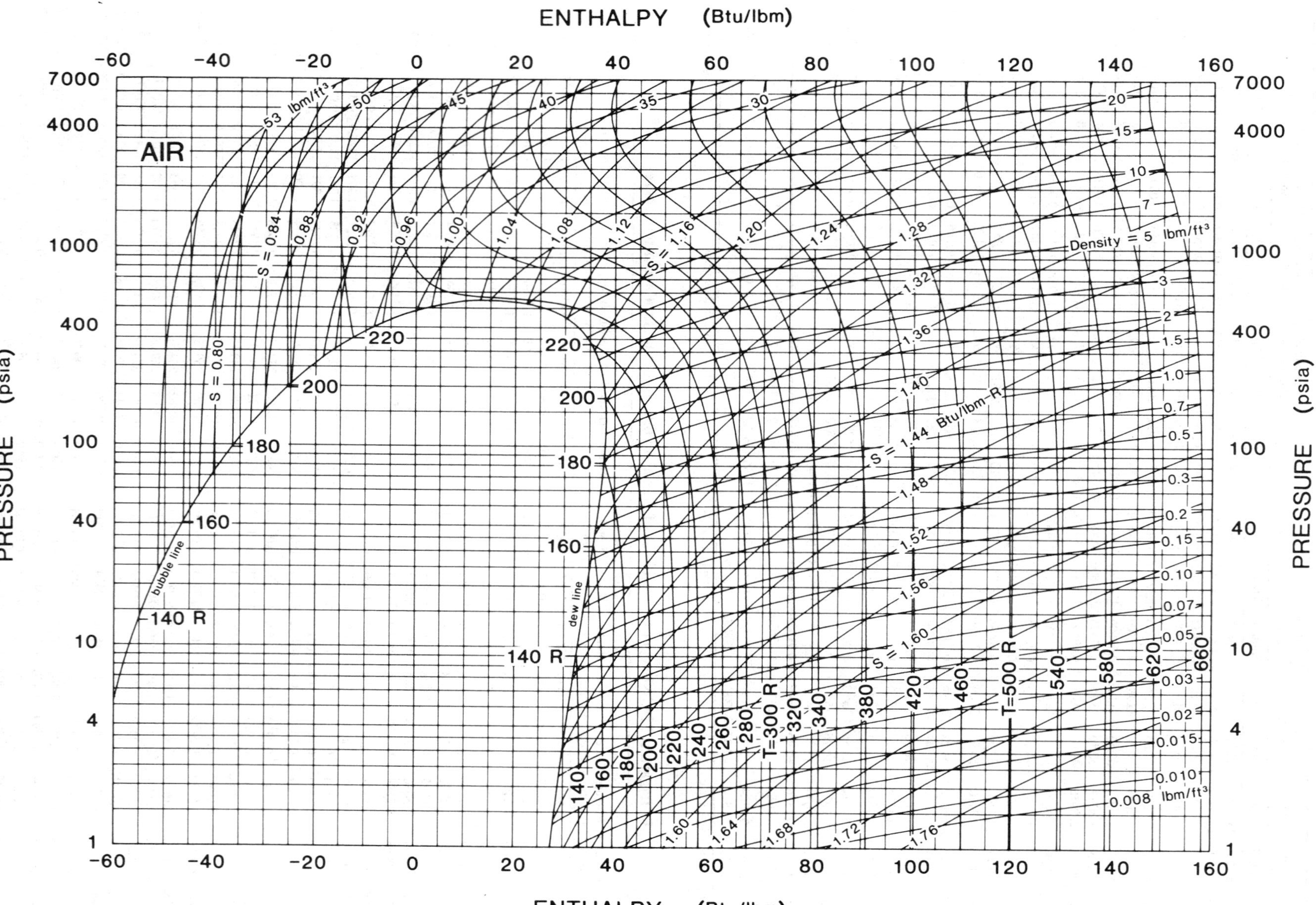

Prepared by: CENTER FOR APPLIED THERMODYNAMIC STUDIES, University of Idaho

Fig. 36 Pressure-Enthalpy Diagram for Refrigerant 729 (Air)

Refrigerant 729 (Air) Properties of Liquid on the Bubble Line and Vapor on the Dew Line

Pressure, psia	Temp*, °R Bubble	Temp*, °R Dew	Density, lb/ft³ Liquid	Volume, ft³/lb Vapor	Enthalpy, Btu/lb Liquid	Enthalpy, Btu/lb Vapor	Entropy, Btu/lb·°R Liquid	Entropy, Btu/lb·°R Vapor	Specific Heat c_p, Btu/lb·°R Liquid	Specific Heat c_p, Btu/lb·°R Vapor	c_p/c_v Vapor	Vel. of Sound, ft/s Liquid	Vel. of Sound, ft/s Vapor	Viscosity, lb_m/ft·h Liquid	Viscosity, lb_m/ft·h Vapor	Thermal Cond, Btu/h·ft·°R Liquid	Thermal Cond, Btu/h·ft·°R Vapor	Surface Tension, dyne/cm
0.91a	107.55	115.19	59.79	46.8083	−70.04	27.29	0.58484	1.46086	0.4591	0.2447	1.406	3322.	524.	1.0650	0.0106	0.1009	0.00325	14.12
1.00	108.49	116.03	59.65	42.7531	−69.61	27.48	0.58883	1.45585	0.4588	0.2450	1.406	3312.	526.	1.0248	0.0107	0.1003	0.00328	14.00
2.00	115.58	122.47	58.62	22.4796	−66.36	28.92	0.61781	1.42069	0.4573	0.2475	1.410	3225.	539.	0.7885	0.0114	0.0961	0.00350	13.04
4.00	123.53	129.89	57.44	11.8472	−62.72	30.51	0.64820	1.38636	0.4573	0.2511	1.417	3115.	553.	0.6173	0.0121	0.0914	0.00375	11.98
6.00	128.66	134.75	56.66	8.1498	−60.37	31.51	0.66682	1.36662	0.4581	0.2539	1.423	3039.	562.	0.5382	0.0126	0.0885	0.00393	11.31
8.00	132.56	138.46	56.07	6.2499	−58.57	32.24	0.68050	1.35274	0.4591	0.2563	1.429	2979.	568.	0.4892	0.0129	0.0863	0.00406	10.80
10.00	135.75	141.49	55.58	5.0865	−57.10	32.83	0.69142	1.34203	0.4602	0.2585	1.435	2928.	572.	0.4546	0.0133	0.0845	0.00417	10.39
12.00	138.47	144.09	55.15	4.2980	−55.84	33.31	0.70054	1.33332	0.4613	0.2605	1.440	2885.	576.	0.4283	0.0135	0.0830	0.00427	10.04
14.696b	141.63	147.10	54.65	3.5634	−54.38	33.86	0.71094	1.32367	0.4627	0.2630	1.448	2833.	580.	0.4009	0.0138	0.0812	0.00439	9.64
20.00	146.73	151.96	53.84	2.6779	−52.00	34.69	0.72728	1.30905	0.4657	0.2675	1.461	2749.	586.	0.3624	0.0144	0.0784	0.00458	9.00
40.00	159.68	164.35	51.68	1.4015	−45.87	36.53	0.76687	1.27616	0.4765	0.2819	1.511	2524.	599.	0.2875	0.0157	0.0715	0.00510	7.42
60.00	168.40	172.73	50.13	0.9542	−41.64	37.50	0.79222	1.25669	0.4875	0.2952	1.561	2364.	605.	0.2492	0.0167	0.0669	0.00549	6.39
80.00	175.19	179.27	48.87	0.7234	−38.27	38.08	0.81141	1.24256	0.4988	0.3085	1.612	2235.	608.	0.2239	0.0175	0.0633	0.00582	5.61
100.00	180.84	184.71	47.77	0.5815	−35.40	38.43	0.82710	1.23128	0.5107	0.3222	1.666	2124.	609.	0.2051	0.0182	0.0604	0.00612	4.98
150.00	192.04	195.53	45.40	0.3865	−29.49	38.66	0.85774	1.20957	0.5440	0.3609	1.819	1893.	609.	0.1724	0.0197	0.0544	0.00680	3.77
200.00	200.80	203.97	43.34	0.2851	−24.59	38.32	0.88164	1.19252	0.5846	0.4089	2.009	1701.	606.	0.1497	0.0210	0.0497	0.00745	2.88
250.00	208.12	210.98	41.41	0.2222	−20.23	37.56	0.90188	1.17767	0.6368	0.4719	2.258	1530.	601.	0.1322	0.0223	0.0457	0.00815	2.18
300.00	214.47	217.03	39.51	0.1787	−16.17	36.45	0.91999	1.16383	0.7075	0.5593	2.602	1372.	596.	0.1175	0.0236	0.0420	0.00895	1.60
350.00	220.11	222.38	37.57	0.1464	−12.26	34.96	0.93692	1.15024	0.8101	0.6893	3.109	1223.	590.	0.1046	0.0250	0.0387	0.00998	1.12
400.00	225.21	227.19	35.51	0.1210	−8.33	33.03	0.95342	1.13618	0.9738	0.9040	3.934	1079.	584.	0.0927	0.0266	0.0357	0.01149	0.72
450.00	229.89	231.56	33.18	0.0998	−4.19	30.48	0.97040	1.12061	1.2881	1.3302	5.534	936.	578.	0.0813	0.0287	0.0335	0.01417	0.39
500.00	234.23	235.51	30.19	0.0802	0.65	26.68	0.99001	1.10078	2.2460	2.6522	10.317	773.	571.	0.0694	0.0315	0.0336	0.02069	0.12
548.97c	238.71	238.71	18.89	0.0529	16.38	16.38	1.05489	1.05489	—	—	—	—	—	—	—	—	—	0.00

*temperatures are on the IPTS-68 scale   a = triple point   b = bubble and dew points at 14.696 psia   c = critical point

Refrigerant 729 (Air) Properties of Gas at 14.696 psia (one standard atmosphere)

Temp, °F	Density, lb/ft³	Enthalpy, Btu/lb	Entropy, Btu/lb·°F	c_p, Btu/lb·°F	c_p/c_v	Vel. Sound, ft/s	Viscosity, lb/ft·h	Thermal Cond, Btu/ft·h·°F
−312.6a	0.2806	33.86	1.3237	0.2630	1.448	580.3	0.0138	0.00439
−300.0	0.2561	37.11	1.3449	0.2547	1.438	608.6	0.0150	0.00480
−280.0	0.2255	42.13	1.3745	0.2483	1.428	649.6	0.0169	0.00545
−260.0	0.2017	47.06	1.4005	0.2454	1.422	687.3	0.0188	0.00611
−240.0	0.1827	51.95	1.4239	0.2437	1.417	722.5	0.0206	0.00677
−220.0	0.1670	56.81	1.4451	0.2427	1.414	755.9	0.0224	0.00764
−200.0	0.1538	61.66	1.4645	0.2420	1.412	787.7	0.0241	0.00799
−180.0	0.1426	66.50	1.4824	0.2415	1.410	818.2	0.0258	0.00856
−160.0	0.1329	71.32	1.4991	0.2412	1.409	847.5	0.0275	0.00913
−140.0	0.1245	76.14	1.5147	0.2409	1.408	875.7	0.0291	0.00968
−120.0	0.1171	80.96	1.5293	0.2407	1.407	903.0	0.0307	0.01023
−100.0	0.1105	85.77	1.5431	0.2405	1.406	929.5	0.0322	0.01076
−80.0	0.1047	90.58	1.5561	0.2404	1.405	955.2	0.0337	0.01128
−60.0	0.0994	95.39	1.5684	0.2403	1.405	980.2	0.0351	0.01179
−40.0	0.0946	100.19	1.5801	0.2403	1.404	1004.6	0.0366	0.01229
−20.0	0.0903	105.00	1.5913	0.2402	1.404	1028.3	0.0380	0.01278
0.0	0.0863	109.81	1.6020	0.2402	1.403	1051.5	0.0393	0.01326
20.0	0.0827	114.61	1.6122	0.2402	1.403	1074.2	0.0407	0.01372
40.0	0.0794	119.42	1.6221	0.2403	1.403	1096.3	0.0420	0.01419
60.0	0.0763	124.22	1.6315	0.2403	1.402	1118.0	0.0433	0.01465
80.0	0.0735	129.03	1.6406	0.2404	1.402	1139.3	0.0446	0.01510
100.0	0.0709	133.84	1.6493	0.2405	1.401	1160.1	0.0458	0.01554
120.0	0.0684	138.65	1.6578	0.2407	1.401	1180.6	0.0470	0.01599
140.0	0.0661	143.47	1.6659	0.2408	1.400	1200.6	0.0483	0.01642
160.0	0.0640	148.28	1.6738	0.2410	1.400	1220.3	0.0495	0.01685
180.0	0.0620	153.11	1.6815	0.2412	1.399	1239.6	0.0506	0.01728
200.0	0.0601	157.93	1.6889	0.2414	1.398	1258.6	0.0518	0.01771
220.0	0.0583	162.76	1.6961	0.2417	1.398	1277.3	0.0529	0.01813
240.0	0.0567	167.60	1.7032	0.2420	1.397	1295.6	0.0541	0.01854
260.0	0.0551	172.44	1.7100	0.2423	1.396	1313.7	0.0552	0.01896
280.0	0.0536	177.29	1.7166	0.2426	1.395	1331.5	0.0563	0.01937
300.0	0.0522	182.15	1.7231	0.2430	1.394	1348.9	0.0574	0.01978
320.0	0.0508	187.01	1.7294	0.2433	1.394	1366.2	0.0584	0.02019
340.0	0.0496	191.88	1.7356	0.2437	1.393	1383.1	0.0595	0.02059
360.0	0.0484	196.76	1.7416	0.2442	1.392	1399.8	0.0605	0.02099
380.0	0.0472	201.65	1.7475	0.2446	1.391	1416.3	0.0616	0.02140
400.0	0.0461	206.54	1.7533	0.2451	1.389	1432.5	0.0626	0.02180
420.0	0.0451	211.45	1.7589	0.2455	1.388	1448.5	0.0636	0.02220
440.0	0.0441	216.36	1.7644	0.2460	1.387	1464.3	0.0646	0.02260
460.0	0.0431	221.29	1.7698	0.2465	1.386	1479.9	0.0656	0.02299
480.0	0.0422	226.22	1.7752	0.2471	1.385	1495.3	0.0666	0.02339
500.0	0.0413	231.17	1.7804	0.2476	1.384	1510.5	0.0675	0.02378
520.0	0.0405	236.13	1.7855	0.2482	1.382	1525.5	0.0685	0.02418
540.0	0.0397	241.10	1.7905	0.2487	1.381	1540.2	0.0694	0.02457
560.0	0.0389	246.08	1.7954	0.2493	1.380	1554.9	0.0704	0.02496
580.0	0.0381	251.07	1.8003	0.2499	1.379	1569.3	0.0713	0.02536
600.0	0.0374	256.08	1.8051	0.2505	1.377	1583.6	0.0722	0.02575
620.0	0.0367	261.09	1.8097	0.2511	1.376	1597.8	0.0732	0.02614
640.0	0.0360	266.12	1.8144	0.2517	1.375	1611.8	0.0741	0.02653
660.0	0.0354	271.16	1.8189	0.2524	1.374	1625.6	0.0750	0.02692
680.0	0.0348	276.22	1.8234	0.2530	1.372	1639.3	0.0759	0.02731
700.0	0.0342	281.28	1.8278	0.2536	1.371	1652.8	0.0768	0.02770
720.0	0.0336	286.36	1.8321	0.2543	1.370	1666.2	0.0776	0.02808
740.0	0.0330	291.45	1.8364	0.2549	1.368	1679.5	0.0785	0.02847
760.0	0.0325	296.56	1.8406	0.2555	1.367	1692.7	0.0794	0.02885
780.0	0.0320	301.68	1.8448	0.2562	1.366	1705.7	0.0802	0.02924
800.0	0.0315	306.81	1.8489	0.2568	1.365	1718.6	0.0811	0.02962
820.0	0.0310	311.95	1.8529	0.2574	1.363	1731.4	0.0820	0.03001
840.0	0.0305	317.10	1.8569	0.2581	1.362	1744.1	0.0828	0.03039
860.0	0.0300	322.27	1.8609	0.2587	1.361	1756.7	0.0836	0.03078
880.0	0.0296	327.45	1.8648	0.2594	1.360	1769.1	0.0845	0.03116
900.0	0.0292	332.64	1.8686	0.2600	1.359	1781.5	0.0853	0.03154

a = dew point temperature

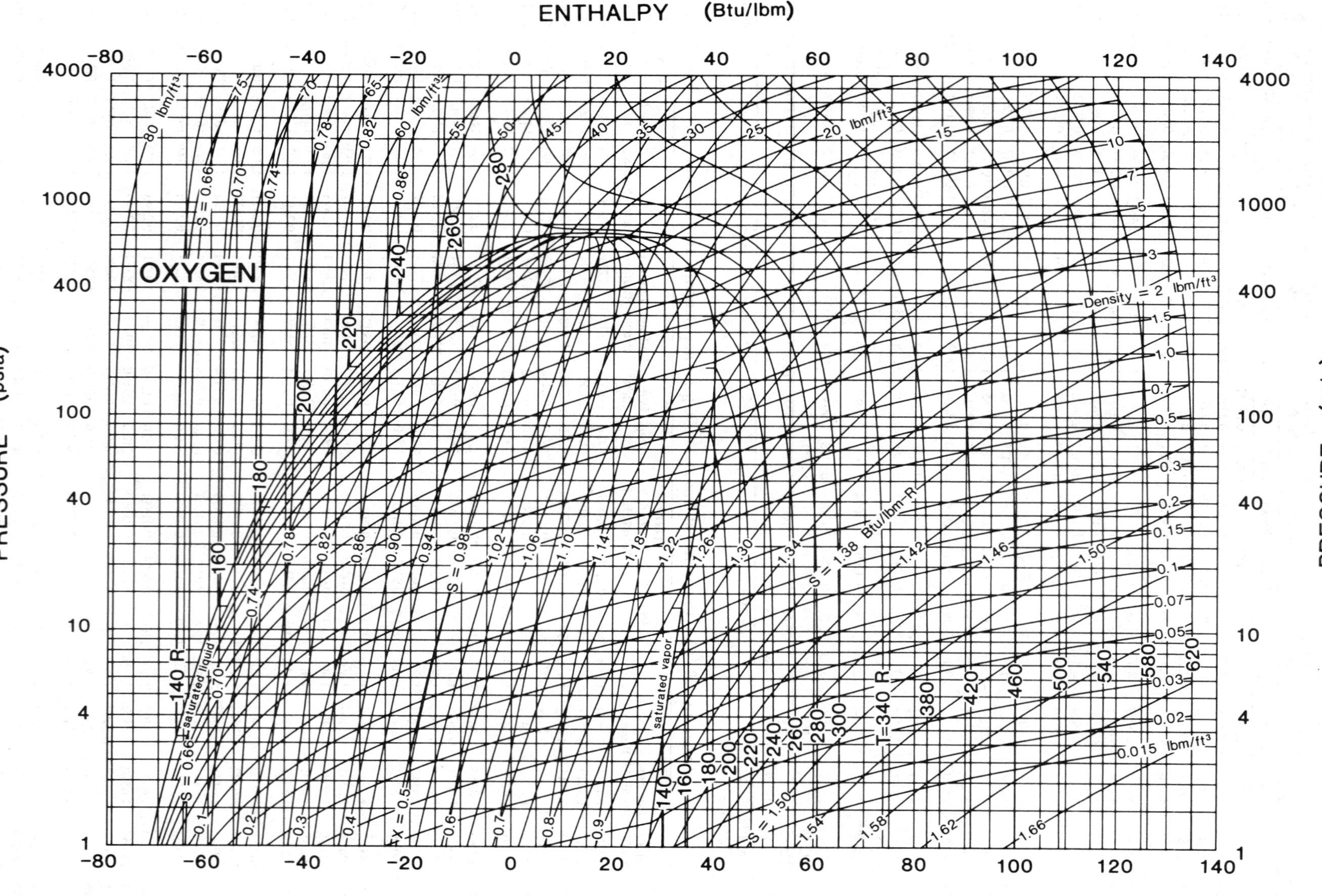

Prepared by: CENTER FOR APPLIED THERMODYNAMIC STUDIES, University of Idaho

Fig. 37 Pressure–Enthalpy Diagram for Refrigerant 732 (Oxygen)

Refrigerant 732 (Oxygen) Properties of Saturated Liquid and Saturated Vapor

Temp,* °R	Pressure, psia	Density, lb/ft³ Liquid	Volume, ft³/lb Vapor	Enthalpy, Btu/lb Liquid	Enthalpy, Btu/lb Vapor	Entropy, Btu/lb·°R Liquid	Entropy, Btu/lb·°R Vapor	Specific Heat c_p, Btu/lb·°R Liquid	Specific Heat c_p, Btu/lb·°R Vapor	c_p/c_v Vapor	Velocity of Sound, ft/s Liquid	Velocity of Sound, ft/s Vapor	Viscosity, lb_m/ft·h Liquid	Viscosity, lb_m/ft·h Vapor	Thermal Cond, Btu/h·ft·°R Liquid	Thermal Cond, Btu/h·ft·°R Vapor	Surface Tension, dyne/cm	Temp,* °R
97.85a	0.02	81.54	1546.5	−83.30	21.13	0.49922	1.56638	0.3999	0.2213	1.395	3686.	460.4	2.3755	0.0097	0.1173	0.00291	22.68	97.85
100.00	0.03	81.23	1092.6	−82.44	21.59	0.50790	1.54816	0.3993	0.2223	1.394	3704.	465.2	2.1422	0.0099	0.1164	0.00297	22.35	100.00
110.00	0.14	79.73	265.18	−78.44	23.74	0.54599	1.47490	0.4002	0.2276	1.389	3685.	486.6	1.4145	0.0109	0.1119	0.00326	20.82	110.00
120.00	0.48	78.17	83.783	−74.43	25.87	0.58085	1.41673	0.4010	0.2322	1.386	3578.	506.8	1.0216	0.0119	0.1074	0.00356	19.31	120.00
130.00	1.34	76.58	32.306	−70.42	27.98	0.61295	1.36988	0.4011	0.2341	1.388	3443.	526.3	0.7868	0.0129	0.1028	0.00386	17.82	130.00
140.00	3.20	74.97	14.5114	−66.40	30.04	0.64267	1.33157	0.4015	0.2335	1.397	3298.	545.1	0.6339	0.0139	0.0982	0.00417	16.36	140.00
150.00	6.73	73.33	7.3371	−62.38	32.02	0.67039	1.29969	0.4028	0.2320	1.411	3151.	562.8	0.5270	0.0149	0.0937	0.00449	14.92	150.00
155.00	9.39	72.49	5.4088	−60.36	32.96	0.68360	1.28566	0.4039	0.2316	1.421	3077.	571.0	0.4848	0.0154	0.0914	0.00465	14.20	155.00
160.00	12.80	71.64	4.0699	−58.33	33.87	0.69643	1.27267	0.4054	0.2317	1.431	3002.	578.8	0.4480	0.0160	0.0891	0.00482	13.50	160.00
162.34b	14.696	71.24	3.5859	−57.38	34.28	0.70230	1.26692	0.4062	0.2320	1.437	2967.	582.2	0.4324	0.0162	0.0881	0.00490	13.17	162.34
165.00	17.11	70.78	3.1187	−56.29	34.74	0.70890	1.26060	0.4072	0.2325	1.444	2927.	586.0	0.4156	0.0165	0.0869	0.00499	12.80	165.00
170.00	22.45	69.90	2.4290	−54.24	35.56	0.72105	1.24932	0.4094	0.2342	1.457	2851.	592.6	0.3868	0.0170	0.0846	0.00517	12.11	170.00
175.00	28.97	69.01	1.9195	−52.18	36.34	0.73291	1.23873	0.4121	0.2369	1.473	2775.	598.5	0.3609	0.0176	0.0823	0.00535	11.43	175.00
180.00	36.84	68.10	1.5366	−50.10	37.06	0.74451	1.22875	0.4153	0.2405	1.491	2697.	603.9	0.3376	0.0181	0.0801	0.00554	10.75	180.00
185.00	46.21	67.17	1.2444	−48.00	37.73	0.75587	1.21928	0.4190	0.2453	1.510	2619.	608.5	0.3163	0.0187	0.0778	0.00573	10.09	185.00
190.00	57.25	66.22	1.0182	−45.88	38.34	0.76701	1.21026	0.4234	0.2512	1.533	2540.	612.4	0.2968	0.0193	0.0755	0.00593	9.43	190.00
195.00	70.13	65.24	0.8409	−43.74	38.88	0.77796	1.20163	0.4284	0.2583	1.559	2460.	615.7	0.2787	0.0198	0.0732	0.00615	8.77	195.00
200.00	85.02	64.24	0.7002	−41.56	39.35	0.78874	1.19333	0.4343	0.2668	1.589	2378.	618.2	0.2620	0.0204	0.0709	0.00637	8.13	200.00
205.00	102.10	63.20	0.5873	−39.36	39.75	0.79938	1.18529	0.4411	0.2768	1.623	2294.	620.0	0.2464	0.0210	0.0686	0.00661	7.50	205.00
210.00	121.55	62.13	0.4958	−37.12	40.07	0.80990	1.17746	0.4491	0.2884	1.663	2209.	621.1	0.2318	0.0216	0.0663	0.00686	6.87	210.00
215.00	143.55	61.02	0.4208	−34.84	40.30	0.82033	1.16980	0.4585	0.3020	1.711	2122.	621.4	0.2180	0.0223	0.0640	0.00714	6.26	215.00
220.00	168.28	59.87	0.3590	−32.51	40.43	0.83069	1.16224	0.4696	0.3181	1.767	2033.	621.1	0.2049	0.0229	0.0616	0.00743	5.66	220.00
225.00	195.93	58.66	0.3074	−30.12	40.46	0.84101	1.15474	0.4830	0.3372	1.835	1942.	619.9	0.1924	0.0236	0.0593	0.00776	5.07	225.00
230.00	226.69	57.40	0.2641	−27.68	40.38	0.85134	1.14722	0.4991	0.3603	1.919	1848.	618.1	0.1805	0.0243	0.0569	0.00813	4.49	230.00
235.00	260.77	56.07	0.2274	−25.16	40.15	0.86170	1.13962	0.5190	0.3886	2.023	1750.	615.4	0.1690	0.0251	0.0545	0.00855	3.93	235.00
240.00	298.37	54.65	0.1961	−22.55	39.78	0.87216	1.13185	0.5442	0.4244	2.155	1649.	611.9	0.1578	0.0259	0.0521	0.00903	3.38	240.00
245.00	339.72	53.13	0.1692	−19.83	39.22	0.88279	1.12381	0.5768	0.4710	2.328	1544.	607.5	0.1470	0.0268	0.0496	0.00962	2.84	245.00
250.00	385.04	51.47	0.1457	−16.97	38.45	0.89367	1.11535	0.6208	0.5342	2.563	1434.	602.2	0.1363	0.0278	0.0471	0.01034	2.33	250.00
260.00	488.67	47.59	0.1068	−10.68	35.97	0.91683	1.09626	0.7794	0.7669	3.421	1192.	588.2	0.1148	0.0303	0.0423	0.01257	1.36	260.00
270.00	611.86	42.17	0.0745	−2.87	31.22	0.94436	1.07061	1.3059	1.5835	6.314	898.	567.0	0.0914	0.0342	0.0383	0.01751	0.51	270.00
278.25c	731.42	27.23	0.0367	13.95	13.95	1.00320	1.00320	∞	∞	∞	0.	0.0	—	—	∞	∞	0.00	278.25

*temperatures are on the IPTS-68 scale a = triple point b = normal boiling point c = critical point

Refrigerant 732 (Oxygen) Properties of Gas at 14.696 psia (one standard atmosphere)

Temp, °F	Density, lb/ft³	Enthalpy, Btu/lb	Entropy, Btu/lb·°F	c_p, Btu/lb·°F	c_p/c_v	Vel. Sound, ft/s	Viscosity, lb/ft·h	Thermal Cond, Btu/h·ft·°F
−297.3a	0.2789	34.28	1.2669	0.2320	1.437	582.2	0.0162	0.00490
−280.0	0.2499	38.20	1.2898	0.2236	1.433	617.3	0.0180	0.00542
−260.0	0.2234	42.66	1.3134	0.2226	1.424	652.8	0.0200	0.00603
−240.0	0.2022	47.10	1.3346	0.2217	1.418	686.2	0.0221	0.00664
−220.0	0.1848	51.53	1.3539	0.2208	1.414	718.0	0.0241	0.00724
−200.0	0.1702	55.94	1.3716	0.2201	1.411	748.3	0.0261	0.00783
−180.0	0.1577	60.33	1.3879	0.2196	1.409	777.3	0.0280	0.00843
−160.0	0.1470	64.72	1.4030	0.2192	1.407	805.2	0.0300	0.00899
−140.0	0.1377	69.10	1.4172	0.2189	1.406	832.1	0.0318	0.00954
−120.0	0.1295	73.48	1.4304	0.2187	1.405	858.1	0.0337	0.01009
−100.0	0.1222	77.85	1.4430	0.2186	1.404	883.2	0.0355	0.01062
−80.0	0.1157	82.23	1.4548	0.2185	1.404	907.7	0.0372	0.01115
−60.0	0.1099	86.60	1.4660	0.2185	1.403	931.4	0.0389	0.01167
−40.0	0.1046	90.97	1.4767	0.2185	1.402	954.5	0.0406	0.01217
−20.0	0.0998	95.34	1.4869	0.2186	1.401	977.0	0.0423	0.01267
0.0	0.0955	99.71	1.4966	0.2188	1.401	998.9	0.0439	0.01316
20.0	0.0915	104.09	1.5059	0.2190	1.400	1020.3	0.0455	0.01365
40.0	0.0878	108.47	1.5149	0.2192	1.399	1041.1	0.0470	0.01413
60.0	0.0844	112.86	1.5235	0.2195	1.398	1061.5	0.0485	0.01460
80.0	0.0812	117.25	1.5318	0.2199	1.396	1081.4	0.0500	0.01507
100.0	0.0783	121.65	1.5398	0.2203	1.395	1100.9	0.0515	0.01553
120.0	0.0756	126.06	1.5475	0.2207	1.394	1120.0	0.0530	0.01600
140.0	0.0731	130.48	1.5550	0.2212	1.392	1138.7	0.0544	0.01646
160.0	0.0707	134.91	1.5623	0.2218	1.391	1157.0	0.0558	0.01692
180.0	0.0685	139.35	1.5693	0.2223	1.389	1174.9	0.0572	0.01738
200.0	0.0664	143.80	1.5762	0.2230	1.388	1192.5	0.0586	0.01784
220.0	0.0645	148.27	1.5828	0.2236	1.386	1209.7	0.0599	0.01829
240.0	0.0626	152.75	1.5893	0.2243	1.384	1226.7	0.0613	0.01874
260.0	0.0609	157.24	1.5957	0.2251	1.382	1243.3	0.0626	0.01920
280.0	0.0592	161.75	1.6018	0.2258	1.381	1259.7	0.0639	0.01964
300.0	0.0577	166.28	1.6079	0.2266	1.379	1275.8	0.0652	0.02009
320.0	0.0562	170.81	1.6138	0.2274	1.377	1291.6	0.0665	0.02054
340.0	0.0548	175.37	1.6196	0.2282	1.375	1307.2	0.0677	0.02099
360.0	0.0535	179.94	1.6252	0.2290	1.373	1322.6	0.0690	0.02143
380.0	0.0522	184.53	1.6307	0.2298	1.371	1337.7	0.0702	0.02188
400.0	0.0510	189.13	1.6361	0.2307	1.369	1352.6	0.0714	0.02232
420.0	0.0498	193.75	1.6415	0.2315	1.367	1367.3	0.0726	0.02276
440.0	0.0487	198.39	1.6467	0.2323	1.366	1381.9	0.0738	0.02320
460.0	0.0476	203.05	1.6518	0.2332	1.364	1396.2	0.0750	0.02364
480.0	0.0466	207.72	1.6568	0.2340	1.362	1410.4	0.0761	0.02407
500.0	0.0457	212.41	1.6618	0.2349	1.360	1424.4	0.0773	0.02451
520.0	0.0447	217.12	1.6666	0.2357	1.358	1438.3	0.0784	0.02494
540.0	0.0438	221.84	1.6714	0.2365	1.357	1452.0	0.0795	0.02537
560.0	0.0430	226.58	1.6761	0.2374	1.355	1465.5	0.0807	0.02580
580.0	0.0421	231.33	1.6807	0.2382	1.353	1478.9	0.0818	0.02622
600.0	0.0413	236.10	1.6852	0.2390	1.352	1492.2	0.0829	0.02665
650.0	0.0395	248.10	1.6963	0.2409	1.348	1524.8	0.0856	0.02770
700.0	0.0378	260.20	1.7070	0.2428	1.344	1556.6	0.0882	0.02873
750.0	0.0362	272.38	1.7172	0.2447	1.340	1587.7	0.0908	0.02975
800.0	0.0348	284.66	1.7272	0.2464	1.337	1618.3	0.0934	0.03076

a = saturated vapor at normal boiling point

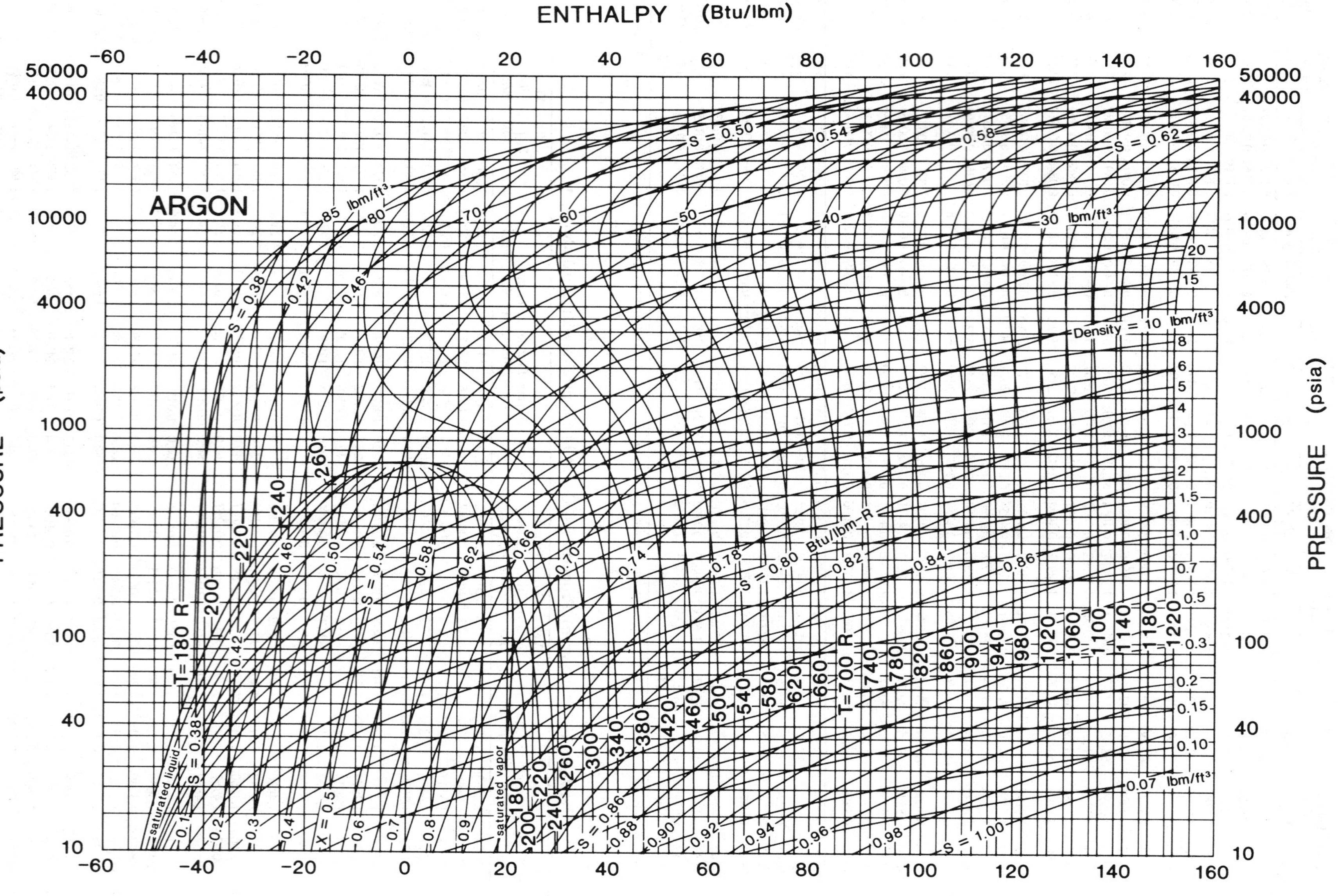

Prepared by: CENTER FOR APPLIED THERMODYNAMIC STUDIES, University of Idaho

Fig. 38 Pressure-Enthalpy Diagram for Refrigerant 740 (Argon)

Refrigerant 740 (Argon) Properties of Saturated Liquid and Saturated Vapor

Temp,* °R	Pressure, psia	Density, lb/ft³ Liquid	Volume, ft³/lb Vapor	Enthalpy, Btu/lb Liquid	Enthalpy, Btu/lb Vapor	Entropy, Btu/lb·°R Liquid	Entropy, Btu/lb·°R Vapor	Specific Heat c_p, Btu/lb·°R Liquid	Specific Heat c_p, Btu/lb·°R Vapor	c_p/c_v Vapor	Velocity of Sound, ft/s Liquid	Velocity of Sound, ft/s Vapor	Viscosity, lb_m/ft·h Liquid	Viscosity, lb_m/ft·h Vapor	Thermal Cond, Btu/h·ft·°R Liquid	Thermal Cond, Btu/h·ft·°R Vapor	Surface Tension, dyne/cm	Temp,* °R
150.85a	10.00	88.47	3.9491	−52.08	18.32	0.31822	0.78492	0.2549	—	—	2799.	685.0	0.6840	0.0162	0.0784	0.00304	13.42	150.85
155.00	12.95	87.61	3.1154	−51.01	18.64	0.32515	0.77453	0.2568	—	—	2749.	644.0	0.6382	0.0166	0.0764	0.00313	12.83	155.00
157.13b	14.696	87.17	2.7731	−50.46	18.80	0.32865	0.76944	0.2577	—	—	2723.	630.8	0.6166	0.0169	0.0754	0.00318	12.53	157.13
160.00	17.34	86.56	2.3822	−49.72	19.00	0.33332	0.76282	0.2592	—	—	2688.	618.0	0.5889	0.0172	0.0740	0.00325	12.13	160.00
165.00	22.78	85.49	1.8525	−48.41	19.34	0.34130	0.75192	0.2619	—	—	2626.	604.2	0.5449	0.0178	0.0716	0.00337	11.44	165.00
170.00	29.43	84.38	1.4624	−47.08	19.66	0.34913	0.74173	0.2649	—	—	2563.	596.8	0.5053	0.0184	0.0693	0.00350	10.75	170.00
175.00	37.45	83.25	1.1701	−45.74	19.95	0.35681	0.73217	0.2683	0.1379	1.866	2500.	593.1	0.4696	0.0190	0.0670	0.00364	10.08	175.00
180.00	46.99	82.09	0.9474	−44.38	20.21	0.36436	0.72316	0.2721	0.1474	1.854	2436.	591.6	0.4371	0.0196	0.0647	0.00378	9.41	180.00
185.00	58.23	80.90	0.7753	−43.00	20.43	0.37179	0.71463	0.2763	0.1563	1.859	2370.	591.4	0.4074	0.0203	0.0624	0.00392	8.76	185.00
190.00	71.32	79.68	0.6405	−41.59	20.61	0.37913	0.70650	0.2810	0.1647	1.875	2304.	592.1	0.3802	0.0210	0.0601	0.00408	8.11	190.00
195.00	86.44	78.42	0.5335	−40.16	20.74	0.38638	0.69870	0.2864	0.1731	1.904	2235.	593.2	0.3551	0.0217	0.0579	0.00425	7.48	195.00
200.00	103.76	77.12	0.4477	−38.70	20.82	0.39356	0.69119	0.2924	0.1818	1.944	2165.	594.5	0.3319	0.0224	0.0557	0.00442	6.86	200.00
205.00	123.47	75.77	0.3782	−37.21	20.85	0.40069	0.68389	0.2992	0.1912	1.995	2093.	595.8	0.3103	0.0231	0.0535	0.00462	6.25	205.00
210.00	145.72	74.38	0.3212	−35.68	20.80	0.40778	0.67676	0.3072	0.2016	2.060	2018.	597.0	0.2901	0.0239	0.0514	0.00482	5.65	210.00
215.00	170.72	72.93	0.2742	−34.12	20.69	0.41485	0.66975	0.3166	0.2136	2.140	1941.	598.0	0.2711	0.0247	0.0492	0.00504	5.07	215.00
220.00	198.64	71.41	0.2349	−32.51	20.48	0.42193	0.66279	0.3278	0.2279	2.239	1860.	598.6	0.2532	0.0256	0.0471	0.00529	4.50	220.00
225.00	229.66	69.82	0.2019	−30.84	20.19	0.42905	0.65583	0.3415	0.2453	2.363	1775.	598.9	0.2362	0.0266	0.0450	0.00556	3.95	225.00
230.00	264.00	68.13	0.1739	−29.11	19.78	0.43624	0.64881	0.3586	0.2671	2.519	1686.	598.6	0.2200	0.0276	0.0428	0.00587	3.41	230.00
235.00	301.85	66.34	0.1499	−27.31	19.25	0.44355	0.64165	0.3805	0.2952	2.721	1591.	597.8	0.2043	0.0287	0.0407	0.00623	2.89	235.00
240.00	343.44	64.40	0.1292	−25.41	18.56	0.45104	0.63424	0.4097	0.3328	2.989	1490.	596.4	0.1892	0.0300	0.0385	0.00665	2.39	240.00
245.00	388.99	62.27	0.1111	−23.39	17.68	0.45881	0.62646	0.4505	0.3853	3.360	1380.	594.2	0.1742	0.0314	0.0363	0.00717	1.91	245.00
250.00	438.76	59.88	0.0952	−21.21	16.56	0.46701	0.61811	0.5110	0.4636	3.904	1260.	591.2	0.1593	0.0331	0.0341	0.00785	1.46	250.00
255.00	493.05	57.12	0.0809	−18.81	15.12	0.47586	0.60889	0.6095	0.5914	4.778	1129.	587.3	0.1440	0.0352	0.0319	0.00882	1.03	255.00
260.00	552.18	53.78	0.0678	−16.05	13.19	0.48581	0.59826	0.7963	0.8335	6.393	984.	582.1	0.1278	0.0379	0.0299	0.01043	0.64	260.00
265.00	616.56	49.38	0.0553	−12.65	10.43	0.49785	0.58497	1.2601	1.4494	10.381	824.	574.9	0.1099	0.0418	0.0293	0.01386	0.30	265.00
271.19c	704.88	33.14	0.0302	−1.53	−1.53	0.53776	0.53776	∞	∞	∞	0.	0.0	—	—	∞	∞	0.00	271.19

*temperatures are on the IPTS-68 scale a = triple point b = normal boiling point c = critical point

Refrigerant 740 (Argon) Properties of Gas at 14.696 psia (one standard atmosphere)

Temp, °F	Density, lb/ft³	Enthalpy, Btu/lb	Entropy, Btu/lb·°F	c_p Btu/lb·°F	c_p/c_v	Vel. Sound, ft/s	Viscosity, lb/ft·h	Thermal Cond, Btu/h·ft·°F
−302.5a	0.3606	18.80	0.7694	—	—	630.8	0.0169	0.00318
−300.0	0.3545	19.06	0.7711	—	—	597.4	0.0171	0.00323
−275.0	0.3028	22.35	0.7902	0.1335	1.689	609.4	0.0199	0.00373
−250.0	0.2648	25.62	0.8068	0.1291	1.689	654.0	0.0226	0.00425
−225.0	0.2356	28.82	0.8213	0.1273	1.686	694.2	0.0254	0.00478
−200.0	0.2124	31.99	0.8341	0.1264	1.682	731.6	0.0281	0.00534
−180.0	0.1969	34.52	0.8435	0.1260	1.680	759.9	0.0303	0.00578
−160.0	0.1835	37.03	0.8521	0.1257	1.678	787.0	0.0324	0.00609
−140.0	0.1719	39.54	0.8603	0.1255	1.676	813.2	0.0345	0.00647
−120.0	0.1616	42.05	0.8679	0.1253	1.675	838.6	0.0366	0.00684
−100.0	0.1525	44.56	0.8750	0.1252	1.674	863.1	0.0386	0.00721
−80.0	0.1444	47.06	0.8818	0.1251	1.673	887.0	0.0406	0.00759
−60.0	0.1372	49.56	0.8882	0.1250	1.672	910.2	0.0426	0.00795
−40.0	0.1306	52.06	0.8943	0.1249	1.672	932.8	0.0445	0.00831
−20.0	0.1246	54.56	0.9001	0.1248	1.671	954.9	0.0464	0.00866
0.0	0.1192	57.05	0.9057	0.1248	1.671	976.5	0.0482	0.00900
10.0	0.1166	58.30	0.9084	0.1248	1.671	987.1	0.0491	0.00917
20.0	0.1142	59.55	0.9110	0.1248	1.670	997.5	0.0500	0.00934
30.0	0.1118	60.79	0.9136	0.1247	1.670	1007.9	0.0509	0.00950
40.0	0.1096	62.04	0.9161	0.1247	1.670	1018.2	0.0518	0.00967
50.0	0.1074	63.29	0.9186	0.1247	1.670	1028.4	0.0527	0.00983
60.0	0.1053	64.54	0.9210	0.1247	1.670	1038.4	0.0535	0.00999
70.0	0.1034	65.78	0.9234	0.1247	1.670	1048.4	0.0544	0.01016
80.0	0.1014	67.03	0.9257	0.1247	1.669	1058.3	0.0553	0.01031
90.0	0.0996	68.28	0.9280	0.1246	1.669	1068.0	0.0561	0.01047
100.0	0.0978	69.52	0.9302	0.1246	1.669	1077.7	0.0569	0.01063
120.0	0.0944	72.01	0.9346	0.1246	1.669	1096.8	0.0586	0.01094
140.0	0.0913	74.51	0.9388	0.1246	1.669	1115.6	0.0602	0.01125
160.0	0.0883	77.00	0.9429	0.1246	1.669	1134.1	0.0619	0.01155
180.0	0.0855	79.49	0.9469	0.1246	1.668	1152.3	0.0635	0.01184
200.0	0.0829	81.98	0.9507	0.1245	1.668	1170.2	0.0650	0.01214
220.0	0.0805	84.47	0.9544	0.1245	1.668	1187.8	0.0666	0.01243
240.0	0.0782	86.96	0.9580	0.1245	1.668	1205.2	0.0681	0.01271
260.0	0.0760	89.45	0.9616	0.1245	1.668	1222.3	0.0696	0.01299
280.0	0.0740	91.94	0.9650	0.1245	1.668	1239.2	0.0711	0.01327
300.0	0.0720	94.43	0.9683	0.1245	1.668	1255.8	0.0726	0.01354
320.0	0.0702	96.92	0.9715	0.1245	1.668	1272.2	0.0740	0.01382
340.0	0.0684	99.41	0.9747	0.1245	1.668	1288.4	0.0755	0.01408
360.0	0.0667	101.90	0.9778	0.1245	1.668	1304.5	0.0769	0.01435
380.0	0.0651	104.39	0.9808	0.1245	1.668	1320.3	0.0783	0.01461
400.0	0.0636	106.88	0.9837	0.1245	1.668	1335.9	0.0797	0.01487
420.0	0.0622	109.37	0.9865	0.1244	1.668	1351.3	0.0811	0.01513
440.0	0.0608	111.86	0.9893	0.1244	1.667	1366.6	0.0824	0.01538
460.0	0.0595	114.34	0.9921	0.1244	1.667	1381.7	0.0838	0.01563
480.0	0.0582	116.83	0.9948	0.1244	1.667	1396.7	0.0851	0.01588
500.0	0.0570	119.32	0.9974	0.1244	1.667	1411.5	0.0864	0.01613
520.0	0.0558	121.81	0.9999	0.1244	1.667	1426.1	0.0878	0.01637
540.0	0.0547	124.30	1.0025	0.1244	1.667	1440.6	0.0891	0.01662
560.0	0.0536	126.79	1.0049	0.1244	1.667	1454.9	0.0903	0.01685
580.0	0.0526	129.28	1.0073	0.1244	1.667	1469.1	0.0916	0.01709
600.0	0.0516	131.76	1.0097	0.1244	1.667	1483.2	0.0929	0.01733
620.0	0.0507	134.25	1.0120	0.1244	1.667	1497.1	0.0941	0.01756
640.0	0.0497	136.74	1.0143	0.1244	1.667	1510.9	0.0954	0.01779
660.0	0.0488	139.23	1.0166	0.1244	1.667	1524.5	0.0966	0.01802
680.0	0.0480	141.72	1.0188	0.1244	1.667	1538.1	0.0978	0.01825
700.0	0.0472	144.21	1.0209	0.1244	1.667	1551.5	0.0990	0.01847
720.0	0.0464	146.69	1.0231	0.1244	1.667	1564.8	0.1002	0.01870
740.0	0.0456	149.18	1.0252	0.1244	1.667	1578.0	0.1014	0.01892
760.0	0.0448	151.67	1.0272	0.1244	1.667	1591.2	0.1026	0.01914
780.0	0.0441	154.16	1.0292	0.1244	1.667	1604.1	0.1037	0.01936
800.0	0.0434	156.65	1.0312	0.1244	1.667	1617.0	0.1049	0.01957

a = saturated vapor at normal boiling point

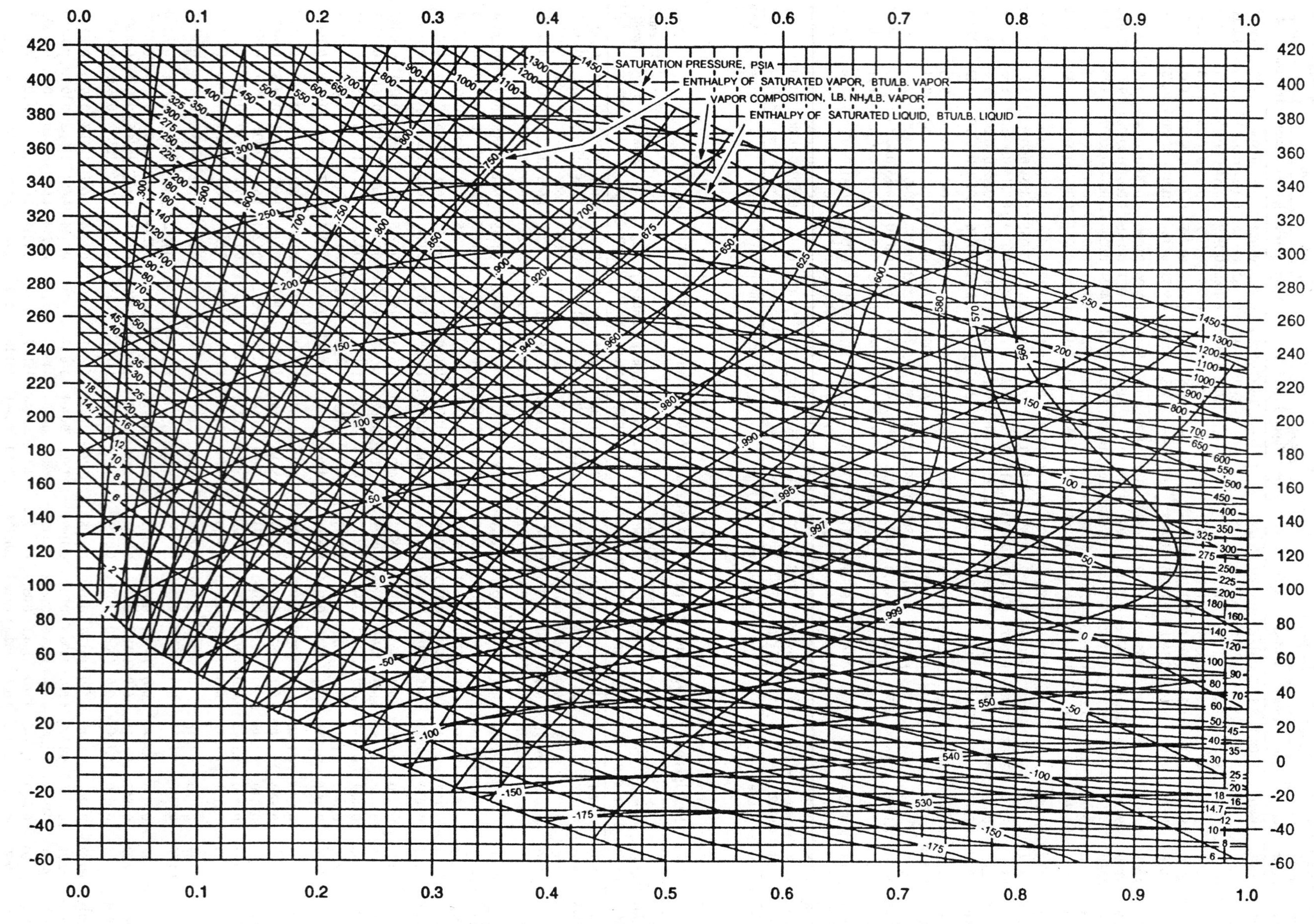

Prepared by: Kwang Kim and Keith Herold
CENTER FOR ENVIRONMENTAL ENERGY ENGINEERING
University of Maryland at College Park

Fig. 39 Enthalpy-Concentration Diagram for Ammonia-Water Solutions

Specific Volume of Saturated Ammonia Solutions, ft³/lb

Temp, °F	Concentration, Ammonia (Mass basis) 0	10	20	30	40	50	60	70	80	90	100	Temp, °F
20	0.0160	0.0165	0.0170	0.0176	0.0182	0.0190	0.0197	0.0207	0.0217	0.0230	0.0245	20
40	0.0160	0.0165	0.0171	0.0177	0.0184	0.0191	0.0200	0.0209	0.0221	0.0236	0.0253	40
60	0.0160	0.0166	0.0172	0.0178	0.0186	0.0193	0.0202	0.0212	0.0225	0.0241	0.0260	60
80	0.0161	0.0167	0.0173	0.0180	0.0188	0.0196	0.0205	0.0216	0.0230	0.0247	0.0267	80
100	0.0161	0.0168	0.0174	0.0182	0.0190	0.0198	0.0208	0.0220	0.0235	0.0254	0.0275	100
120	0.0162	0.0169	0.0176	0.0184	0.0192	0.0201	0.0211	0.0224	0.0241	0.0261	0.0284	120
140	0.0163	0.0170	0.0177	0.0185	0.0194	0.0203	0.0215	0.0229	0.0247	0.0268	0.0294	140
160	0.0164	0.0172	0.0179	0.0187	0.0196	0.0206	0.0219	0.0235	0.0254	0.0277	0.0306	160
180	0.0165	0.0173	0.0181	0.0190	0.0199	0.0210	0.0223	0.0241	0.0262	0.0286	0.0320	180
200	0.0166	0.0175	0.0183	0.0192	0.0202	0.0213	0.0228	0.0247	0.0270	0.0298	0.0338	200
220	0.0168	0.0176	0.0185	0.0194	0.0205	0.0217	0.0234	0.0255	0.0279	0.0312	0.0361	220

Prepared under Research Project No. 271-RP, sponsored by TC 8.3.
Data reference: B.H. Jennings, Ammonia water properties (paper presented at ASHRAE meeting, January 1965).

Refrigerant Temperature (t' = °F) and Enthalpy (h = Btu/lb) of Lithium Bromide Solutions

Temp, (t = °F)		Percent LiBr 0	10	20	30	40	45	50	55	60	65	70
80	t'	80.0	78.2	75.6	70.5	60.9	53.5	42.1	28.6	13.8	−0.2#	−11.6#
	h	48.0	39.2	31.8	25.6	21.6	21.2	23.0	28.7	38.9	52.7#	67.1#
100	t'	100.0	98.1	95.3	89.9	79.6	71.8	60.0	46.1	30.9	16.2#	3.8#
	h	68.0	56.6	47.0	38.7	33.2	32.1	33.2	38.2	47.8	61.1#	75.1#
120	t'	120.0	117.9	114.9	109.2	98.3	90.1	77.9	63.6	48.1	32.7	19.1#
	h	87.9	73.6	61.7	51.7	44.7	43.0	43.6	48.0	56.9	69.4	83.0#
140	t'	140.0	137.8	134.6	128.5	117.1	108.5	95.8	81.2	65.2	49.1	34.4#
	h	107.9	91.0	77.0	65.1	56.5	54.1	54.1	57.9	66.1	78.0	91.1#
160	t'	160.0	157.7	154.3	147.9	135.8	126.8	113.8	98.7	82.3	65.6	49.7#
	h	127.9	108.2	92.0	78.2	68.1	65.1	64.7	67.9	75.4	86.6	99.2#
180	t'	180.0	177.5	173.9	167.2	154.5	145.1	131.7	116.2	99.5	82.0	65.1#
	h	147.9	125.4	107.9	91.9	80.4	76.6	75.3	77.7	84.6	95.1	107.2#
200	t'	200.0	197.4	193.6	186.5	173.3	163.5	149.6	133.7	116.6	98.5	80.4#
	h	168.0	143.4	123.3	105.3	92.1	87.4	85.9	87.8	94.1	104.0	115.6#
220	t'	220.0	217.2	213.3	205.8	192.0	181.8	167.5	151.3	133.7	114.9	95.7
	h	188.1	160.7	138.2	119.0	104.1	99.0	96.5	97.8	103.3	112.5	123.6
240	t'	240.0*	237.1*	232.9	225.2	210.7	200.2	185.4	168.8	150.9	131.4	111.0
	h	208.3*	178.4*	154.0	132.6	116.0	110.3	107.1	107.7	112.5	121.1	131.6
260	t'	260.0*	256.9*	252.6*	244.5*	229.4	218.5	203.3	186.3	168.0	147.9	126.4
	h	228.6*	195.7*	169.1*	146.2*	128.1	121.6	117.6	117.6	121.6	129.5	139.5
280	t'	280.0*	276.8*	272.3*	263.8*	248.2*	236.8*	221.2	203.9	185.1	164.3	141.7
	h	249.1*	213.8*	185.1*	159.7*	140.0*	132.8*	128.1	127.5	130.6	137.9	147.6
300	t'	300.0*	296.7*	291.9*	283.1*	266.9*	255.2*	239.2*	221.4	202.3	180.8	157.0
	h	269.6*	231.6*	200.7*	173.5*	152.1*	144.1*	138.9*	137.3	139.8	146.5	155.5
320	t'	320.0*	316.5*	311.6*	302.5*	285.6*	273.5*	257.1*	238.9*	219.4	197.2	172.4
	h	290.3*	249.7*	216.3*	187.2*	164.2*	155.3*	149.5*	147.1*	148.8	154.9	163.4
340	t'	340.0*	336.4*	331.3*	321.8*	304.4*	291.9*	275.0*	256.4*	236.5*	213.7	187.7
	h	311.1*	267.9*	232.1*	201.0*	176.1*	166.6*	160.1*	157.0*	158.0*	163.5	171.0
360	t'	360.0*	356.2*	350.9*	341.1*	323.1*	310.2*	292.9*	274.0*	253.7*	230.1	203.0
	h	332.2*	286.1*	248.0*	214.9*	188.2*	178.0*	170.6*	166.8*	167.0*	171.9	178.3

*Extensions of data above 235°F are well above the original data and should be used with care.
#Supersaturated solution.

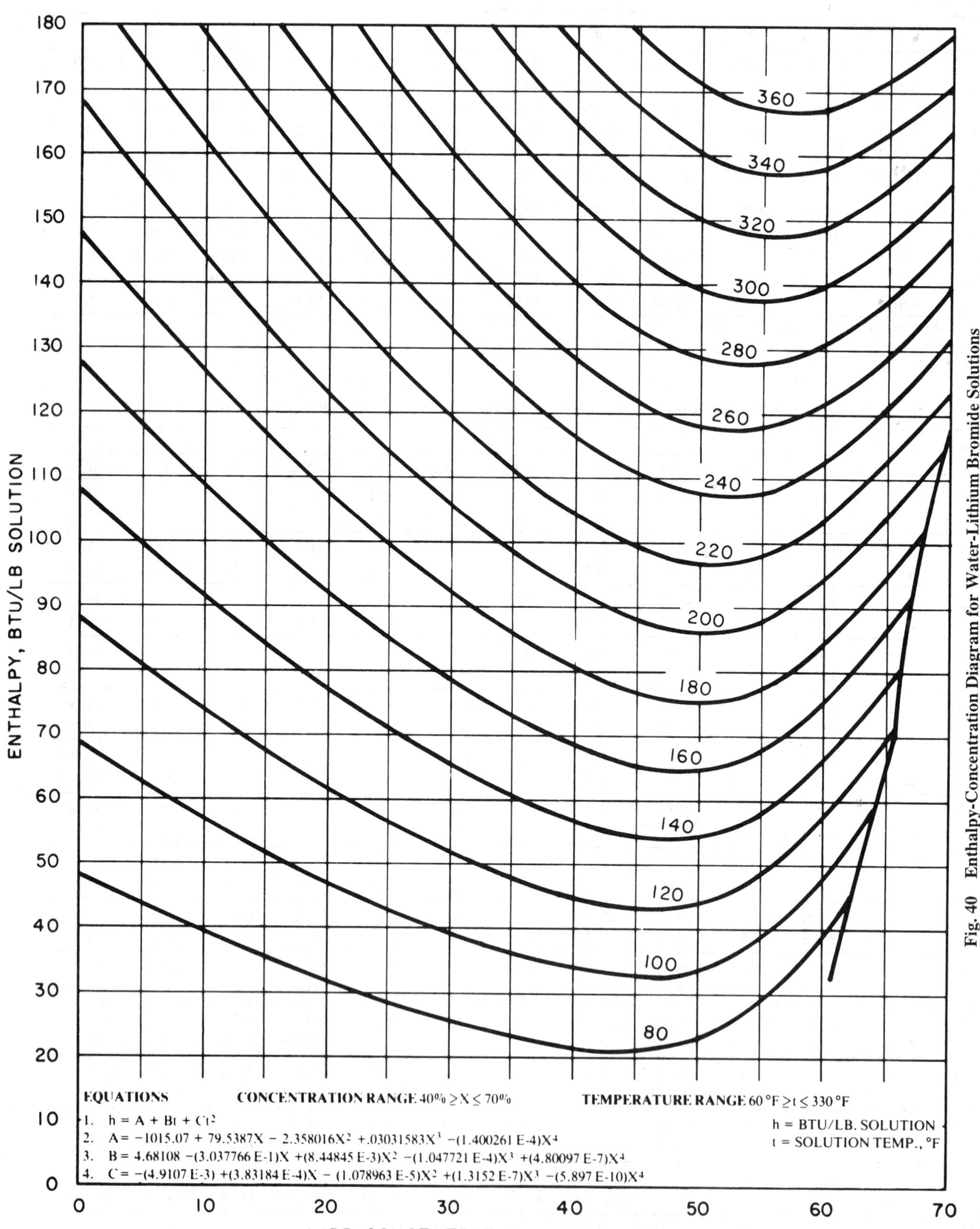

Fig. 40 Enthalpy-Concentration Diagram for Water-Lithium Bromide Solutions

EQUATIONS

1. $t = At' + B$
2. $t' = (t - B)/A$
3. $A = -2.00755 + .16976X - (3.133362\ E\text{-}3)X^2 + (1.97668\ E\text{-}5)X^3$
4. $B = 321.128 - 19.322X + .374382X^2 - (2.0637\ E\text{-}3)X^3$
5. $\text{Log}_{10} P = C + D/(t' + 459.72)^2 + E/(t' + 459.72)^2$
6. $t' = \dfrac{-2E}{D + [D^2 - 4E(C - \log_{10}P)]^{0.5}} - 459.72$

TEMP. RANGE (REFRIGERANT)	$0 < t' \leq 230°F$
TEMP. RANGE (SOLUTION)	$40 < t \leq 350°F$
CONCENTRATION RANGE	$45\% < X \leq 70\%$

C = 6.21147
D = −2886.373
E = −337269.46
t′ = REFRIGERANT TEMP. °F
t = SOLUTION TEMP. °F
X = PERCENT LIBR
P = PSIA

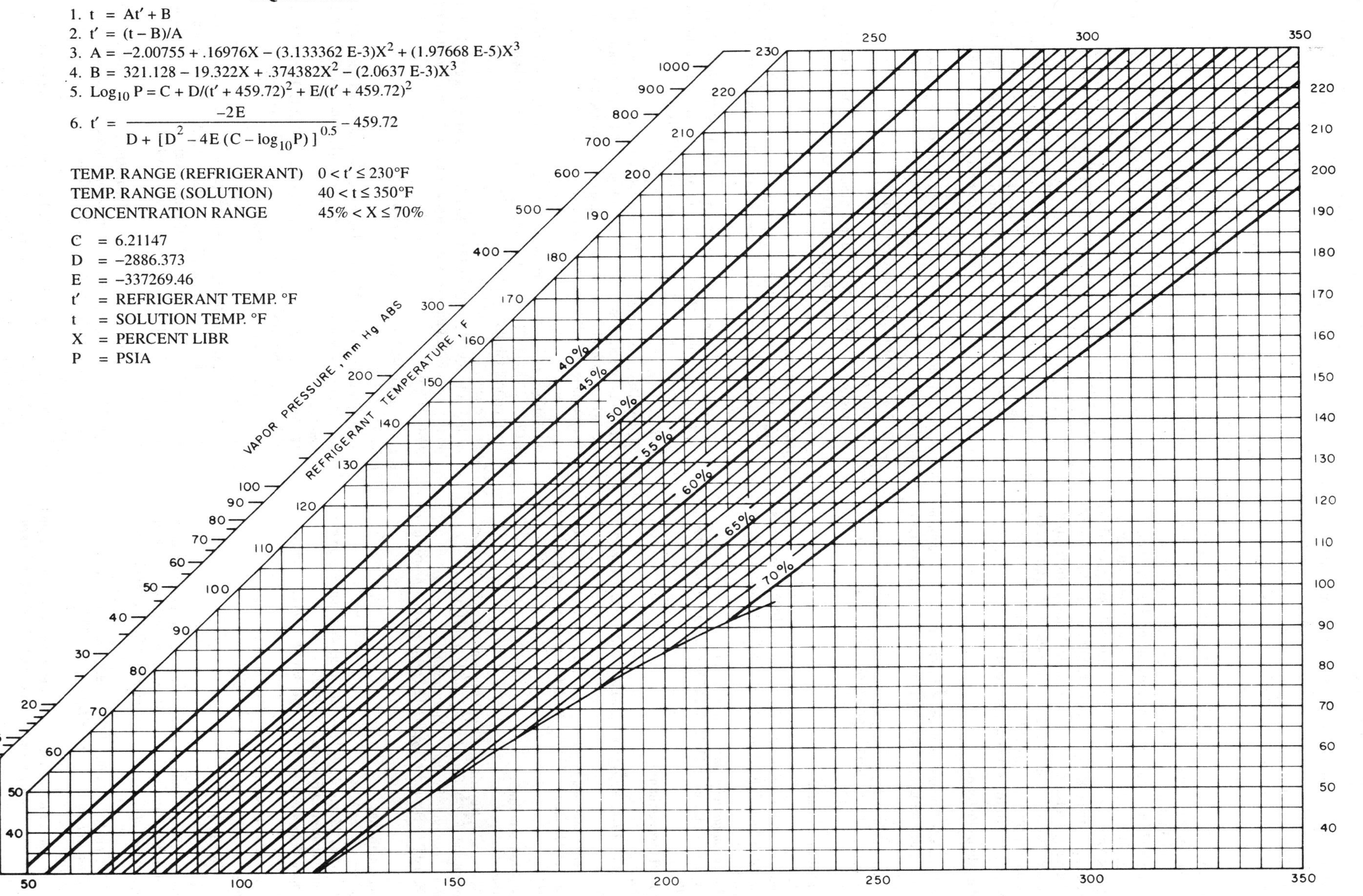

[a]*Reprinted by permission of Carrier Corp.*

Fig. 41 Equilibrium Chart for Aqueous Lithium Bromide Solutions

REFERENCES

The sources used for the tabular data presented in this chapter are listed by fluid and property. The reference listed under "*Equation of state*" was used for vapor pressure, liquid density, vapor volume, enthalpy, entropy, specific heat, and the velocity of sound, unless indicated otherwise by additional entries for one or more of these properties. Transport properties for many of the refrigerants have been updated by P.E. Liley and P.D. Desai of Purdue University as part of *ASHRAE Thermophysical properties of refrigerants* (1993). The values in the tables in this Handbook were calculated from the equations given by Liley and Desai and may differ slightly from the tabular values given by Liley. Tables that are new or revised for the 1997 Handbook were calculated using several computerized fluid properties packages; the package is indicated under the fluid name, with the underlying source for the data noted. Data for some fluids have been taken directly from previous editions of the Handbook; these are indicated by a (*) preceding the citation to indicate the 1989 edition or a (†) to indicate the 1993 edition.

R-11 (Trichlorofluoromethane)

Equation of state:

†Jacobsen, R.T., S.G. Penoncello, and E.W. Lemmon. 1992. A fundamental equation for trichlorofluoromethane (R-11). *Fluid Phase Equilibria* 80(11):45-56.

Viscosity, thermal conductivity, and surface tension:

†Liley, P.E. and P.D. Desai. 1993. *ASHRAE Thermophysical properties of refrigerants.*

R-12 (Dichlorodifluoromethane)

Equation of state:

†Penoncello, S.G., R.T. Jacobsen, and E.W. Lemmon. 1992. A fundamental equation for dichlorodifluoromethane (R-12). *Fluid Phase Equilibria* 80(11):57-70.

Viscosity, thermal conductivity, and surface tension:

†Liley, P.E. and P.D. Desai. 1993. *ASHRAE Thermophysical properties of refrigerants.*

R-13 (Chlorotrifluoromethane)

Equation of state:

†Downing, R.C. 1974. Refrigerant equations. *ASHRAE Transactions* 80(2):159.

Liquid heat capacity, viscosity, thermal conductivity, and surface tension:

†Liley, P.E. and P.D. Desai. 1993. *ASHRAE Thermophysical properties of refrigerants.*

R-22 (Chlorodifluoromethane)

Equation of state:

†Kamei, A. and S.W. Beyerlein. 1992. A fundamental equation for chlorodifluoromethane (R-22). *Fluid Phase Equilibria* 80(11):71-86.

Viscosity, thermal conductivity, and surface tension:

†Liley, P.E. and P.D. Desai. 1993. *ASHRAE Thermophysical properties of refrigerants.*

R-23 (Trifluoromethane)

Equation of state:

†Hou, Y.C. and J.J. Martin. 1959. Physical and thermodynamic properties of trifluoromethane. AIChE *Journal* 5:125-29.

Liquid heat capacity, liquid sound speed, viscosity, thermal conductivity, and surface tension:

†Liley, P.E. and P.D. Desai. 1993. *ASHRAE Thermophysical properties of refrigerants.*

R-32 (Difluoromethane)

Equation of state:

†Malbrunot, P.F., P.A. Meunier, G.M. Scatena, W.H. Mears, K.P. Murphy, and J.V. Sinka. 1968. Pressure-volume-temperature behavior of difluoromethane. *Journal of Chemical and Engineering Data* 13:16-21.

Viscosity:

†Phillips, T.W. and K.P. Murphy. 1970. Liquid viscosity of halocarbons. *Journal of Chemical and Engineering Data* 15:304-307.

Thermal conductivity:

†Tauscher, W. 1969. Measurement of the thermal conductivity of liquid refrigerants by an unsteady-state hot wire method. *ASHRAE Journal* 11(1):97-104.

Surface tension:

†Schmidt, J.W. and M.R. Moldover. 1994. Alternative refrigerants CH_2F_2 and C_2HF_5: Critical temperature, refractive index, surface tension, and estimates of liquid, vapor and critical densities. *Journal of Chemical and Engineering Data* 39:39-44.

R-113 (1,1,2-Trichloro-1,2,2-trifluoroethane)

Equation of state:

*Mastroianni, M.J., R.F. Stahl, and P.N. Sheldon. 1978. Physical and thermodynamic properties of 1,1,2-trifluorotrichloroethane (R-113). *Journal of Chemical and Engineering Data* 23:113-18.

Heat capacity, viscosity, thermal conductivity, and surface tension:

†Liley, P.E. and P.D. Desai. 1993. *ASHRAE Thermophysical properties of refrigerants.*

R-114 (1,2-Dichloro-1,1,2,2-tetrafluoroethane)

Equation of state:

*Hules, K.R. and D.P. Wilson. 1982. An interim engineering model of the thermodynamic properties of 1,2-dichlorotetrafluoroethane, Refrigerant 114. Proceedings of the 8th ASME Thermophysical Properties Symposium 2:370-79. American Society of Mechanical Engineers, New York.

Heat capacity, velocity of sound, viscosity, thermal conductivity, and surface tension:

†Liley, P.E. and P.D. Desai. 1993. *ASHRAE Thermophysical properties of refrigerants.*

R-123 (2,2-Dichloro-1,1,1-trifluoroethane)

Equation of state:

†Younglove, B.A. 1992. National Institute of Standards and Technology, Boulder, CO. Personal communication to M.O. McLinden, January 16, 1992.

Liquid thermal conductivity:

†Yata, J. 1990. Thermal conductivity of alternatives to CFCs. Proceedings of the 11th Japan Symposium on Thermophysical Properties, pp. 143-46.

Vapor thermal conductivity:

†Gross, U., Y.W. Song, J. Kallweit, and E. Hahne. 1990. Thermal conductivity of saturated R-123 and R-134a—Transient hot wire measurements. Proceedings of the meeting of IIR Commission B1, Tel Aviv.

Liquid viscosity:

†Diller, D.E., A.S. Aragon, and A. Laesecke. 1993. Measurements of the viscosities of saturated and compressed liquid 1,1,1,2-tetrafluoroethane (R-134a), 2,2-dichloro-1,1,1-trifluoroethane (R-123) and 1,1-dichloro-1-fluoroethane (R-141b). *Fluid Phase Equilibria* 88:251-62.

Vapor viscosity:

†Takahashi, M., C. Yokoyama, and S. Takahashi. 1990. Gas viscosities of HCFC-123 and HCFC-123a. Proceedings of the 11th Japan Symposium on Thermophysical Properties, pp. 115-18.

Surface tension:

†Chae, H.B., J.W. Schmidt, and M.R. Moldover. 1990. Surface Tension of Refrigerants R-123 and R-134a. *Journal of Chemical and Engineering Data* 35:6-8.

R-124 (2-Chloro-1,1,1,2-tetrafluoroethane)

Equation of state:

†Younglove, B.A. 1992. National Institute of Standards and Technology, Boulder, Co. Personal communication to M.O. McLinden, April 1, 1992.

Thermal conductivity:

†Perkins, R.A., L.J. van Poolen, J.B. Howley, and M.L. Huber. 1993. Thermal conductivity of 2-chloro-1,1,1,2-tetrafluoroethane (R-124). *International Journal of Thermophysics.*

Liquid viscosity:

†Diller, D.E. and S.M. Peterson. 1993. Measurements of the viscosities of saturated and compressed fluid 1-chloro-1,2,2,2-tetrafluoroethane (R-124) and pentafluoroethane (R-125) at temperatures between 120 and 420 K. *International Journal of Thermophysics* 14:55-66.

Vapor viscosity:

†Takahashi, M., C. Yokoyama, and S. Takahashi. 1992. Gas viscosity of HFC-124 [sic] at high pressures. Proceedings of the 13th Japan Symposium on Thermophysical Properties, pp. 347-50.

Surface tension:

†Okada, M. and Y. Higashi. 1992. Measurements of the surface tension for HCFC-124 and HCFC-141b. Proceedings of the 13th Japan Symposium on Thermophysical Properties, pp. 73-76.

R-125 (Pentafluoroethane)

Equation of state:

†Wilson, L.C., W.V. Wilding, G.M. Wilson, R.L. Rowley, V.M. Felix, and T. Chisolm-Carter. 1992. Thermophysical properties of HFC-125. *Fluid Phase Equilibria* 80(11):167-78.

Liquid viscosity:

†Diller, D.E. and S.M. Peterson. 1993. Measurements of the viscosities of saturated and compressed fluid 1-chloro-1,2,2,2-tetrafluoroethane (R-124) and pentafluoroethane (R-125) at temperatures between 120 and 420 K. *International Journal of Thermophysics* 14:55-66.

Surface tension:

†Schmidt, J.W. and M.R. Moldover. 1994. Alternative refrigerants CH_2F_2 and C_2HF_5: Critical temperature, refractive index, surface tension, and estimates of liquid, vapor and critical densities. *Journal of Chemical and Engineering Data* 39:39-44.

R-134a (1,1,1,2-Tetrafluoroethane)

Equation of state:

†Huber, M.L. and M.O. McLinden. 1992. Thermodynamic properties of R-134a (1,1,1,2-tetrafluoroethane). Proceedings of the International Refrigeration Conference, pp. 453-62. Purdue University, West Lafayette, Indiana. July 14-17.

Thermal conductivity:

†Laesecke, A., R.A. Perkins, and C.A. Nieto de Castro. 1992. Thermal conductivity of R-134a. *Fluid Phase Equilibria* 80(11):263-74.

Viscosity:

†Huber, M.L. and J.F. Ely. 1992. Prediction of viscosity of refrigerants and refrigerant mixtures. *Fluid Phase Equilibria* 80(11):249-62.

Surface tension:

†Chae, H.B., J.W. Schmidt, and M.R. Moldover. 1990. Surface tension of refrigerants R-123 and R-134a. *Journal of Chemical and Engineering Data* 35:6-8.

R-141b (1,1-Dichloro-1-fluoroethane)

Equation of state:

†Weber, L.A. 1991. PVT and thermodynamic properties of R-141b in the gas phase. Proceedings of the 18th International Congress of Refrigeration. Paper No. 69.

Liquid thermal conductivity:

†Yata, J., M. Hori, T. Kurahashi, and T. Minamiyama. 1992. Thermal conductivity of alternative fluorocarbons in liquid phase. *Fluid Phase Equilibria* 80(11):287-96.

Liquid viscosity:

†Diller, D.E., A.S. Aragon, and A. Laesecke. 1993. Measurements of the viscosities of saturated and compressed liquid 1,1,1,2-tetrafluoroethane (R-134a), 2,2-dichloro-1,1,1-trifluoroethane (R-123) and 1,1-dichloro-1-fluoroethane (R-141b). *Fluid Phase Equilibria* 88:251-62.

Surface tension:

†Okada, M. and Y. Higashi. 1992. Measurements of the surface tension for HCFC-124 and HCFC-141b. Proceedings of the 13th Japan Symposium on Thermophysical Properties, pp. 73-76.

R-142b (1-Chloro-1,1-difluoroethane)

Equation of state:

†Mears, W.H., R.F. Stahl, S.R. Orfeo, R.C. Shair, L.F. Kells, W. Thompson, and H. McCann. 1955. Thermodynamic properties of halogenated ethanes and ethylenes. *Industrial and Engineering Chemistry* 47:1449-53.

Liquid heat capacity:

†Nakagawa, S., H. Sato, and K. Watanabe. 1990. Measurements of isobaric heat capacity of liquid HCFC-142b. Proceedings of the 11th Japan Symposium on Thermophysical Properties, pp. 103-106.

Liquid thermal conductivity:

†Yata, J., M. Hori, T. Kurahashi, and T. Minamiyama. 1992. Thermal conductivity of alternative fluorocarbons in liquid phase. *Fluid Phase Equilibria* 80(11):287-96.

Vapor viscosity:

†Takahashi, M., C. Yokoyama, and S. Takahashi. 1987. Viscosities of gaseous R-13B1, R-142b, and R-152a. *Journal of Chemical and Engineering Data* 32:98-103.

Surface tension:

†Okada, M., Y. Higashi, T. Ikeda, and T. Kuwana. 1991. Measurements of the surface tension for HCFC-142b and HFC-152a. Proceedings of the 12th Japan Symposium on Thermophysical Properties, pp. 105-108.

R-152a (1,1-Difluoroethane)

Equation of state:

†Tillner-Roth, R. 1993. Die thermodynamischen Eigenschaften von R-152a, R-134a und ihren Gemischen—Messungen und Fundamentalgleichungen. Doktor-Ingenieur thesis, Universität Hannover, Germany,

Liquid thermal conductivity:

†Yata, J., M. Hori, T. Kurahashi, and T. Minamiyama. 1992. Thermal conductivity of alternative fluorocarbons in liquid phase. *Fluid Phase Equilibria* 80(11):287-96.

Liquid viscosity:

†Correlation based on data of T.W. Phillips and K.P. Murphy (1970), *ASHRAE Transactions* 76(II):146-56, and A. Kumagai and S. Takahashi (1991), *International Journal of Thermophysics* 12:105-17.

Vapor viscosity:

†Takahashi, M., C. Yokoyama, and S. Takahashi. 1987. Viscosities of gaseous R-13B1, R-142b, and R-152a. *Journal of Chemical and Engineering Data* 32:98-103.

Surface tension:

†Okada, M., Y. Higashi, T. Ikeda, and T. Kuwana. 1991. Measurements of the surface tension for HCFC-142b and HFC-152a. Proceedings of the 12th Japan Symposium on Thermophysical Properties, pp. 105-108.

R-404A [R-125/143a/134a (44/52/4)]

Data computed using the Refprop database, version 6(b18). M.O. McLinden and S.A. Klein. 1996. A next generation refrigerant properties database. pp. 409-14. 6th International Refrigeration Conference at Purdue, West Lafayette, IN, July 23-26.

Equation of state:

Lemmon, E.W. 1996. A generalized model for the prediction of the thermodynamic properties of mixtures including vapor-liquid equilibrium. Ph.D. thesis, University of Idaho, Moscow, ID.

R-407C [R-32/125/134a (23/25/52)]

Data computed using the Refprop database, version 6(b18). M.O. McLinden and S.A. Klein. 1996. A next generation refrigerant properties database. pp. 409-14. 6th International Refrigeration Conference at Purdue, West Lafayette, IN, July 23-26.

Equation of state:

Lemmon, E.W. 1996. A generalized model for the prediction of the thermodynamic properties of mixtures including vapor-liquid equilibrium. Ph.D. thesis, University of Idaho, Moscow, ID.

R-410A [R-32/125 (50/50)]

Data computed using the Refprop database, version 6(b18). M.O. McLinden and S.A. Klein. 1996. A next generation refrigerant properties database. pp. 409-14. 6th International Refrigeration Conference at Purdue, West Lafayette, IN, July 23-26.

Equation of state:

Lemmon, E.W. 1996. A generalized model for the prediction of the thermodynamic properties of mixtures including vapor-liquid equilibrium. Ph.D. thesis, University of Idaho, Moscow, ID.

R-507A [R-125/143a (50/50)]

Data computed using the Refprop database, version 6(b18). M.O. McLinden and S.A. Klein. 1996. A next generation refrigerant properties database. pp. 409-14. 6th International Refrigeration Conference at Purdue, West Lafayette, IN, July 23-26.

Equation of state:

Lemmon, E.W. 1996. A generalized model for the prediction of the thermodynamic properties of mixtures including vapor-liquid equilibrium. Ph.D. thesis, University of Idaho, Moscow, ID.

R-500 [R-12/152a (73.8/26.2)]

Equation of state:

*Sinka, J.V. and K.P. Murphy. 1967. Pressure-volume-temperature relationship for a mixture of difluorodichloromethane and 1,1-difluoroethane. *Journal of Chemical and Engineering Data* 12:315-16.

Heat capacity, viscosity, and thermal conductivity:

†Liley, P.E. and P.D. Desai. 1993. *ASHRAE Thermophysical properties of refrigerants.*

R-502 [R-22/115 (48.8/51.2)]

Equation of state:

*Martin, J.J. and R.C. Downing. 1970. Thermodynamic properties of refrigerant 502. *ASHRAE Transactions* 76(2):129-39.

Heat capacity, viscosity, thermal conductivity, and surface tension:

†Liley, P.E. and P.D. Desai. 1993. *ASHRAE Thermophysical properties of refrigerants.*

R-503 [R-23/13 (40.1/59.9)]

Equation of state:

*Sinka, J.V., E. Rosenthal, and R.P. Dixon. 1970. Pressure-volume-temperature relationship for a mixture of monochlorotrifluoromethane and trifluoromethane. *Journal of Chemical and Engineering Data* 15:73-74.

Heat capacity, viscosity, thermal conductivity, and surface tension:

†Liley, P.E. and P.D. Desai. 1993. *ASHRAE Thermophysical properties of refrigerants.*

R-717 (Ammonia)

Equation of state:

†Haar, L. and J.S. Gallagher. 1978. Thermodynamic properties of ammonia. *Journal of Physical and Chemical Reference Data* 7:635-792.

Viscosity:

†Watson, J.T.R. 1983. The dynamic viscosity of ammonia. Paper presented to Transport Formulations Subcommittee of International Union of Pure and Applied Chemistry.

Thermal conductivity:

†Krauss, R. 1991. Section in *VDI-Warmeatlas* (Thermal Atlas of the Association of German Engineers), 6th ed., Vol. D, Section b, pp. 59-71. VDI-Verlag, Dusseldorf, Germany.

Surface tension:

†Liley, P.E. and P.D. Desai. 1993. *ASHRAE Thermophysical properties of refrigerants.*

R-718 (Water/Steam)

Equation of state:

†Haar, L., J.S. Gallagher, and G.S. Kell. 1984. NBS/NRC Steam Tables. Hemisphere Publishing Corporation, Washington, D.C.

Viscosity and thermal conductivity:

†Sengers, J.V. and J.T.R. Watson. 1986. Improved international formulations for the viscosity and thermal conductivity of water substance. *Journal of Physical and Chemical Reference Data* 15:1291-1322.

Surface tension:

†International Association for the Properties of Steam. 1975. Release on surface tension of water substance.

R-744 (Carbon dioxide)

Equation of state:

†Ely, J.F., J.W. Magee, and W.M. Haynes. 1987. Thermophysical properties for special high content mixtures. Research Report RR-110. Gas Processors Association, Tulsa, OK.

Viscosity and thermal conductivity:

†Vesovic, V., W.A. Wakeham, G.A. Olchowy, J.V. Sengers, J.T.R. Watson, and J. Millat. 1990. The transport properties of carbon dioxide. *Journal of Physical and Chemical Reference Data* 19:763-808.

Surface tension:

†Liley, P.E. and P.D. Desai. 1993. *ASHRAE Thermophysical properties of refrigerants.*

R-50 (Methane)

Equation of state:

*Angus, S., B. Armstrong, and K.M. de Reuck. 1978. International Thermodynamic Tables of the Fluid State—5: Methane. International Union of Pure and Applied Chemistry, Chemical Data Series, No. 16. Pergammon Press, Oxford, England.

Heat capacity, viscosity, and thermal conductivity:

**ASHRAE Thermophysical properties of refrigerants*. 1976.

R-170 (Ethane)

Equation of state:

*Goodwin, R.D., H.M. Roder, and G.C. Straty. 1976. Thermodynamic properties of ethane, from 90 to 600 K at pressures to 700 bar. National Bureau of Standards *Technical Note* 684. Government Printing Office, Washington, D.C.

Heat capacity, viscosity, and thermal conductivity:

**ASHRAE Thermophysical properties of refrigerants*. 1976.

R-290 (Propane)

Data computed using the Refprop database, version 6(b18). M.O. McLinden and S.A. Klein. 1996. A next generation refrigerant properties database. pp. 409-14. 6th International Refrigeration Conference at Purdue, West Lafayette, IN, July 23-26.

Equation of state, viscosity, and thermal conductivity:

Younglove, B.A. and J.F. Ely. 1987. Thermophysical properties of fluids. II. Methane, ethane, propane, isobutane and normal butane. *Journal of Physical and Chemical Reference Data* 16:577-798.

Surface tension:

Baidakov, V.G. and I.I. Sulla. 1985. Surface tension of propane and isobutane at near-critical temperatures. *Russian Journal of Physical Chemistry* 59:551-54.

R-600 (*n*-Butane)

Data computed using the Refprop database, version 6(b18). M.O. McLinden and S.A. Klein. 1996. A next generation refrigerant properties database. pp. 409-14. 6th International Refrigeration Conference at Purdue, West Lafayette, IN, July 23-26.

Equation of state, viscosity, and thermal conductivity:

Younglove, B.A. and J.F. Ely. 1987. Thermophysical properties of fluids. II. Methane, ethane, propane, isobutane and normal butane. *Journal of Physical and Chemical Reference Data* 16:577-798.

Surface tension:

Calado, J.C.G., I.A. McLure, and V.A.M. Soares. 1978. Surface tension for octafluorocyclobutane, *n*-butane and their mixtures from 233 K to 254 K, and vapour pressure, excess Gibbs function and excess volume for the mixture at 233 K. *Fluid Phase Equilibria* 2:99-213.

R-600a (Isobutane)

Data computed using the Refprop database, version 6(b18). M.O. McLinden and S.A. Klein. 1996. A next generation refrigerant properties database. pp. 409-14. 6th International Refrigeration Conference at Purdue, West Lafayette, IN, July 23-26.

Equation of state, viscosity, and thermal conductivity:

Younglove, B.A. and J.F. Ely. 1987. Thermophysical properties of fluids. II. Methane, ethane, propane, isobutane and normal butane. *Journal of Physical and Chemical Reference Data* 16:577-798.

Surface tension:

Baidakov, V.G. and I.I. Sulla. 1985. Surface tension of propane and isobutane at near-critical temperatures. *Russian Journal of Physical Chemistry* 59:551-54.

R-1150 (Ethylene)

Equation of state:

*McCarty, R.D. and R.T. Jacobsen. 1981. An equation of state for fluid ethylene. National Bureau of Standards *Technical Note*. Government Printing Office, Washington, D.C.

Heat capacity, viscosity, and thermal conductivity:

**ASHRAE Thermophysical properties of refrigerants*. 1976.

R-1270 (Propylene)

Equation of state:

*Angus, S., B. Armstrong, and K.M. de Reuck. 1980. International Thermodynamic Tables of the Fluid State—7: Propylene. International Union of Pure and Applied Chemistry. Pergammon Press, Oxford, England.

Heat capacity, viscosity, and thermal conductivity:

**ASHRAE Thermophysical properties of refrigerants*. 1976.

R-702 (Hydrogen)

Thermodynamic data computed using the ALLPROPS database, version 4.0. E.W. Lemmon, R.T. Jacobsen, S.G. Penoncello, and S.W. Beyerlein. 1994. Computer programs for the calculation of thermodynamic properties of cryogens and other fluids. *Advances in Cryogenic Engineering* 39:1891-97.

Transport data computed using the NIST12 database, version 3.0. D.G. Friend, R.D. McCarty, and V. Arp. 1992. NIST Thermophysical properties of pure fluids database, version 3.0. Standard Reference Data Program, National Institute of Standards and Technology, Gaithersburg, MD.

Equation of state, viscosity, and thermal conductivity:

McCarty, R.D. 1975. Hydrogen: Technology survey—Thermophysical properties. NASA SP-3089.

Surface tension:

Liley, P.E. and P.D. Desai. 1993. *ASHRAE Thermophysical properties of refrigerants.*

R-702p (Parahydrogen)

Thermodynamic data computed using the ALLPROPS database, version 4.0. E.W. Lemmon, R.T. Jacobsen, S.G. Penoncello, and S.W. Beyerlein. 1994. Computer programs for the calculation of thermodynamic properties of cryogens and other fluids. *Advances in Cryogenic Engineering* 39:1891-97.

Transport data computed using the NIST12 database, version 3.0. D.G. Friend, R.D. McCarty, and V. Arp. 1992. NIST Thermophysical properties of pure fluids database, version 3.0. Standard Reference Data Program, National Institute of Standards and Technology, Gaithersburg, MD.

Equation of state, viscosity, and thermal conductivity:

Younglove, B.A. 1982. Thermophysical properties of fluids. I. Argon, ethylene, parahydrogen, nitrogen, nitrogen trifluoride, and oxygen. *Journal of Physical and Chemical Reference Data* 11(Supplement No. 1).

Surface tension:

Liley, P.E. and P.D. Desai. 1993. *ASHRAE Thermophysical properties of refrigerants.*

R-704 (Helium)

Thermodynamic data computed using the ALLPROPS database, version 4.0. E.W. Lemmon, R.T. Jacobsen, S.G. Penoncello, and S.W. Beyerlein. 1994. Computer programs for the calculation of thermodynamic properties of cryogens and other fluids. *Advances in Cryogenic Engineering* 39:1891-97.

Transport data computed using the NIST12 database, version 3.0. D.G. Friend, R.D. McCarty, and V. Arp. 1992. NIST Thermophysical properties of pure fluids database, version 3.0. Standard Reference Data Program, National Institute of Standards and Technology, Gaithersburg, MD.

Equation of state, viscosity, and thermal conductivity:

Arp, V.D., R.D. McCarty, and D.G. Friend. 1995. Thermophysical properties of helium-4 from 0.8 to 1500 K with pressures to 2000 MPa. NIST *Technical Note* 1334 (revised). National Institute of Standards and Technology, Gaithersburg, MD.

Surface tension:

Liley, P.E. and P.D. Desai. 1993. *ASHRAE Thermophysical properties of refrigerants.*

R-720 (Neon)

Data computed using the ALLPROPS database, version 4.0. E.W. Lemmon, R.T. Jacobsen, S.G. Penoncello, and S.W. Beyerlein. 1994. Computer programs for the calculation of thermodynamic properties of cryogens and other fluids. *Advances in Cryogenic Engineering* 39:1891-97.

Equation of state:

Katti, R.S., R.T. Jacobsen, R.B. Stewart, and M. Jahangiri. 1986. Thermodynamic properties for neon for temperatures from the triple point to 700 K at pressures to 700 MPa. *Advances in Cryogenic Engineering* 31:1189-1197.

Surface tension:

Liley, P.E. and P.D. Desai. 1993. *ASHRAE Thermophysical properties of refrigerants.*

R-728 (Nitrogen)

Data computed using the ALLPROPS database, version 4.0. E.W. Lemmon, R.T. Jacobsen, S.G. Penoncello, and S.W. Beyerlein. 1994. Computer programs for the calculation of thermodynamic properties of cryogens and other fluids. *Advances in Cryogenic Engineering* 39:1891-97.

Equation of state:

Jacobsen, R.T., R.B. Stewart, and M. Jahangiri. 1986. Thermodynamic properties of nitrogen from the freezing line to 2000 K at pressures to 1000 MPa. *Journal of Physical and Chemical Reference Data* 15(2):735-909.

Viscosity, and thermal conductivity:

Ely, J.F. 1997. Correlation contained in the AIRPROPS database, version 1.0. National Institute of Standards and Technology, Gaithersburg, MD.

Surface tension:

Lemmon, E.W. and S.G. Penoncello. 1994. The surface tension of air and air component mixtures. *Advances in Cryogenic Engineering* 39:1927-34.

R-729 (Air)

Data computed using the AIRPROPS database, version 1.0. E.W. Lemmon. 1997. NIST Thermophysical properties of air and air component mixtures, version 1.0. Standard Reference Data Program, National Institute of Standards and Technology, Gaithersburg, MD.

Equation of state, viscosity, and thermal conductivity:

Jacobsen, R.T., S.G. Penoncello, S.W. Beyerlein, D.G. Friend, J.F. Ely, J.C. Rainwater, and W.M. Haynes. 1995. Thermophysical properties of air. National Institute of Standards and Technology, Supplement to NASP *Technical Memorandum* 1005, NASA Langley Research Center.

Surface tension:

Lemmon, E.W. and S.G. Penoncello. 1994. The surface tension of air and air component mixtures. *Advances in Cryogenic Engineering* 39:1927-34.

R-732 (Oxygen)

Data computed using the ALLPROPS database, version 4.0. E.W. Lemmon, R.T. Jacobsen, S.G. Penoncello, and S.W. Beyerlein. 1994. Computer programs for the calculation of thermodynamic properties of cryogens and other fluids. *Advances in Cryogenic Engineering* 39:1891-97.

Equation of state:

Schmidt, R. and W. Wagner. 1985. A new form of the equation of state for pure substances and its application to oxygen. Fluid Phase Equilibria 19:175-200.

Viscosity, and thermal conductivity:

Ely, J.F. 1997. Correlation contained in the AIRPROPS database, version 1.0. National Institute of Standards and Technology, Gaithersburg, MD.

Surface tension:

Lemmon, E.W. and S.G. Penoncello. 1994. The surface tension of air and air component mixtures. *Advances in Cryogenic Engineering* 39:1927-34.

R-740 (Argon)

Data computed using the ALLPROPS database, version 4.0. E.W. Lemmon, R.T. Jacobsen, S.G. Penoncello, and S.W. Beyerlein. 1994. Computer programs for the calculation of thermodynamic properties of cryogens and other fluids. *Advances in Cryogenic Engineering* 39:1891-97.

Equation of state:

Stewart, R.B. and R.T.Jacobsen. 1989. Thermodynamic properties of argon from the triple point to 1200 K with pressures to 1000 MPa. *Journal of Physical and Chemical Reference Data* 18(2):639-798.

Viscosity, and thermal conductivity:

Ely, J.F. 1997. Correlation contained in the AIRPROPS database, version 1.0. National Institute of Standards and Technology, Gaithersburg, MD.

Surface tension:

Lemmon, E.W. and S.G. Penoncello. 1994. The surface tension of air and air component mixtures. *Advances in Cryogenic Engineering* 39:1927-34.

---

*Indicates data reprinted from the 1989 Handbook.

†Indicates data reprinted from the 1993 Handbook.

CHAPTER 20

PHYSICAL PROPERTIES OF SECONDARY COOLANTS (BRINES)

IN many refrigeration applications, heat is transferred to a **secondary coolant**, which can be any liquid cooled by the refrigerant and used to transfer heat without changing state. These liquids are also known as **heat transfer fluids**, **brines**, or **secondary refrigerants**.

Other ASHRAE Handbooks describe various applications for secondary coolants. In the 1994 *ASHRAE Handbook—Refrigeration*, refrigeration systems are discussed in Chapter 5; their uses in food processing are found in Chapters 8, 11, and 13; ice rinks are discussed in Chapter 33; and environmental test facilities are covered in Chapter 36. In the 1995 *ASHRAE Handbook—Applications*, solar energy utilization is discussed in Chapter 30, thermal storage in Chapter 40, and snow melting in Chapter 46.

This chapter describes the physical properties of several secondary coolants and provides information on their use. The chapter also includes information on corrosion protection. Additional information on corrosion inhibition can be found in Chapter 44 of the 1995 *ASHRAE Handbook—Applications* and Chapter 4 of the 1994 *ASHRAE Handbook—Refrigeration.*

BRINES

Physical Properties

Water solutions of calcium chloride and sodium chloride are the most common refrigeration brines. Tables 1 and 2 list the properties of pure calcium chloride brine and sodium chloride brine. For commercial grades, use the formulas in the footnotes to these tables. Figures 1 and 5 give the specific heats for calcium chloride and sodium chloride brines and are used for computation of heat loads with ordinary brine (Carrier 1959). Figures 2 and 6 show the ratio of the mass of the solution to that of water, which is commonly used as the measure of salt concentration. Viscosities are given in Figures 3 and 7. Figures 4 and 8 show thermal conductivity of calcium and sodium brines at varying temperatures and concentrations.

Brine applications in refrigeration are mainly in the industrial machinery field and in skating rinks. Corrosion is the principal

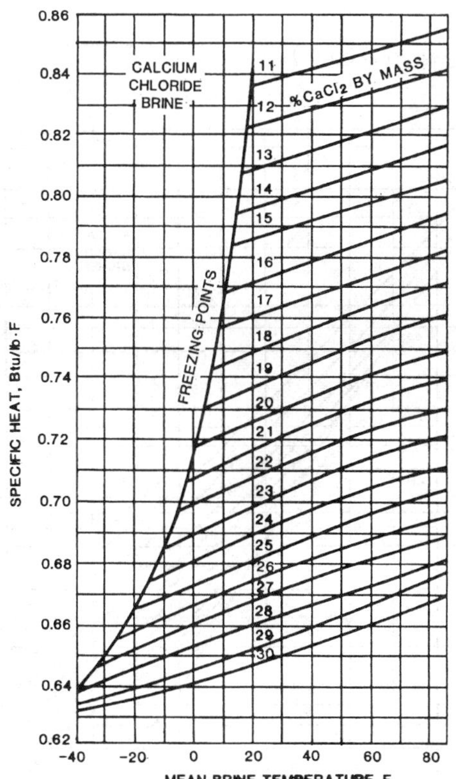

Fig. 1 Specific Heat of Calcium Chloride Brines

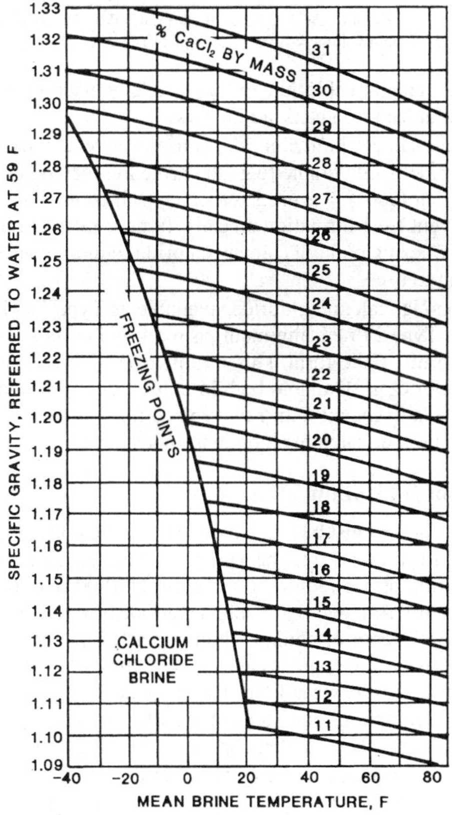

Fig. 2 Specific Gravity of Calcium Chloride Brines

The preparation of this chapter is assigned to TC 3.1, Refrigerants and Brines.

Table 1 Properties of Pure Calcium Chloride[a] Brines

Pure $CaCl_2$, % by Mass	Ratio of Mass to Water at 60°F	Baume Density at 60°F	Specific Heat at 60°F, Btu/lb·°F	Crystallization Starts, °F	Mass per Unit Volume[b] at 60°F: $CaCl_2$, lb/gal	Brine, lb/gal	$CaCl_2$, lb/ft³	Brine, lb/ft³	Ratio of Mass at Various Temperatures to Water at 60°F: −4°F	14°F	32°F	50°F
0	1.000	0.0	1.000	32.0	0.000	8.34	0.00	62.40				
5	1.044	6.1	0.924	27.7	0.436	8.717	3.26	65.15			1.043	1.042
6	1.050	7.0	0.914	26.8	0.526	8.760	3.93	65.52			1.052	1.051
7	1.060	8.2	0.898	25.9	0.620	8.851	4.63	66.14			1.061	1.060
8	1.069	9.3	0.884	24.6	0.714	8.926	5.34	66.70			1.071	1.069
9	1.078	10.4	0.869	23.5	0.810	9.001	6.05	67.27			1.080	1.078
10	1.087	11.6	0.855	22.3	0.908	9.076	6.78	67.83			1.089	1.087
11	1.096	12.6	0.842	20.8	1.006	9.143	7.52	68.33			1.098	1.096
12	1.105	13.8	0.828	19.3	1.107	9.227	8.27	68.95			1.108	1.105
13	1.114	14.8	0.816	17.6	1.209	9.302	9.04	69.51			1.117	1.115
14	1.124	15.9	0.804	15.5	1.313	9.377	9.81	70.08			1.127	1.124
15	1.133	16.9	0.793	13.5	1.418	9.452	10.60	70.64		1.139	1.137	1.134
16	1.143	18.0	0.779	11.2	1.526	9.536	11.40	71.26		1.149	1.146	1.143
17	1.152	19.1	0.767	8.6	1.635	9.619	12.22	71.89		1.159	1.156	1.153
18	1.162	20.2	0.756	5.9	1.747	9.703	13.05	72.51		1.169	1.166	1.163
19	1.172	21.3	0.746	2.8	1.859	9.786	13.90	73.13		1.180	1.176	1.173
20	1.182	22.1	0.737	−0.4	1.970	9.853	14.73	73.63		1.190	1.186	1.183
21	1.192	23.0	0.729	−3.9	2.085	9.928	15.58	74.19				
22	1.202	24.4	0.716	−7.8	2.208	10.037	16.50	75.00	1.215	1.211	1.207	1.203
23	1.212	25.5	0.707	−11.9	2.328	10.120	17.40	75.63				
24	1.223	26.4	0.697	−16.2	2.451	10.212	18.32	76.32	1.236	1.232	1.228	1.224
25	1.233	27.4	0.689	−21.0	2.574	10.295	19.24	76.94				
26	1.244	28.3	0.682	−25.8	2.699	10.379	20.17	77.56				
27	1.254	29.3	0.673	−31.2	2.827	10.471	21.13	78.25				
28	1.265	30.4	0.665	−37.8	2.958	10.563	22.10	78.94				
29	1.276	31.4	0.658	−49.4	3.090	10.655	23.09	79.62				
29.87	1.290	32.6	0.655	−67.0	3.16	10.75	23.65	80.45				
30	1.295	33.0	0.653	−50.8	3.22	10.80	24.06	80.76				
32	1.317	34.9	0.640	−19.5	3.49	10.98	26.10	82.14				
34	1.340	36.8	0.630	4.3	3.77	11.17	28.22	83.57				

[a]Mass of Type 1 (77% min.) $CaCl_2$ = (mass of pure $CaCl_2$)/(0.77). Mass of Type 2 (94% min.) $CaCl_2$ = (mass of pure $CaCl_2$)/(0.94).
[b]Mass of water per unit volume = Brine mass minus $CaCl_2$ mass.

problem for calcium chloride brines, especially in ice-making tanks where galvanized iron cans are immersed.

Ordinary salt (sodium chloride) is used where contact with calcium chloride is intolerable (e.g., the brine fog method of freezing fish and other foods). It is used as a spray in air cooling of unit coolers to prevent frost formation on coils. In most refrigerating work, the lower freezing point of calcium chloride solution makes it more convenient to use.

Commercial calcium chloride, available as Type 1 (77% minimum) and Type 2 (94% minimum), is marketed in flake, solid, and solution forms; flake form is used most extensively. Commercial sodium chloride is available both in crude (rock salt) and refined grades. Because magnesium salts tend to form sludge, their presence in sodium or calcium chloride is undesirable.

Corrosion Inhibition

Brine systems must be treated to control corrosion and deposits. The standard chromate treatment program is the most effective. Calcium chloride brines require a minimum of 1800 ppm of sodium chromate with pH 6.5 to 8.5. Sodium chloride brines require a minimum of 3600 ppm of sodium chromate and a pH of 6.5 to 8.5. Sodium nitrite at 3000 ppm in calcium brines or 4000 ppm in sodium brines controls pH between 7.0 and 8.5, and should provide adequate protection. Organic inhibitors are available that may provide adequate protection where neither chromates nor nitrites can be used.

Before using any chromate-based inhibitor package, review federal, state, and local regulations concerning the use and disposal of chromate-containing fluids. If the regulations prove too restrictive, an alternative inhibition system should be considered.

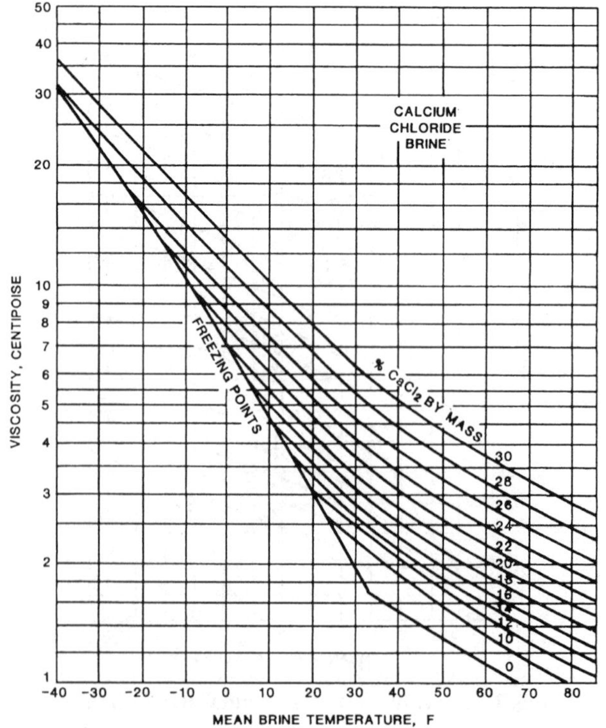

Fig. 3 Viscosity of Calcium Chloride Brines

Table 2 Properties of Pure Sodium Chloride[a] Brines

Pure NaCl, % by Mass	Ratio of Mass to Water at 59°F	Baume Density at 60°F	Specific Heat at 59°F, Btu/lb·°F	Crystallization Starts, °F	Mass per Unit Volume[b] at 60°F				Ratio of Mass at Various Temperatures to Water at 60°F			
					NaCl, lb/gal	Brine, lb/gal	NaCl, lb/ft^3	Brine, lb/ft^3	14°F	32°F	50°F	68°F
0	1.000	0.0	1.000	32.0	0.000	8.34	0.000	62.4				
5	1.035	5.1	0.938	26.7	0.432	8.65	3.230	64.6		1.0382	1.0366	1.0341
6	1.043	6.1	0.927	25.5	0.523	8.71	3.906	65.1		1.0459	1.0440	1.0413
7	1.050	7.0	0.917	24.3	0.613	8.76	4.585	65.5		1.0536	1.0515	1.0486
8	1.057	8.0	0.907	23.0	0.706	8.82	5.280	66.0		1.0613	1.0590	1.0559
9	1.065	9.0	0.897	21.6	0.800	8.89	5.985	66.5		1.0691	1.0665	1.0633
10	1.072	10.1	0.888	20.2	0.895	8.95	6.690	66.9		1.0769	1.0741	1.0707
11	1.080	10.8	0.879	18.8	0.992	9.02	7.414	67.4		1.0849	1.0817	1.0782
12	1.087	11.8	0.870	17.3	1.090	9.08	8.136	67.8		1.0925	1.0897	1.0857
13	1.095	12.7	0.862	15.7	1.188	9.14	8.879	68.3		1.1004	1.0933	1.0971
14	1.103	13.6	0.854	14.0	1.291	9.22	9.632	68.8		1.1083	1.1048	1.1009
15	1.111	14.5	0.847	12.3	1.392	9.28	10.395	69.3	1.1195	1.1163	1.1126	1.1086
16	1.118	15.4	0.840	10.5	1.493	9.33	11.168	69.8	1.1277	1.1243	1.1205	1.1163
17	1.126	16.3	0.833	8.6	1.598	9.40	11.951	70.3	1.1359	1.1323	1.1284	1.1241
18	1.134	17.2	0.826	6.6	1.705	9.47	12.744	70.8	1.1442	1.1404	1.1363	1.1319
19	1.142	18.1	0.819	4.5	1.813	9.54	13.547	71.3	1.1535	1.1486	1.1444	1.1398
20	1.150	19.0	0.813	2.3	1.920	9.60	14.360	71.8	1.1608	1.1568	1.1542	1.1478
21	1.158	19.9	0.807	0.0	2.031	9.67	15.183	72.3	1.1692	1.1651	1.1606	1.1559
22	1.166	20.8	0.802	−2.3	2.143	9.74	16.016	72.8	1.1777	1.1734	1.1688	1.1640
23	1.175	21.7	0.796	−5.1	2.256	9.81	16.854	73.3	1.1862	1.1818	1.1771	1.1721
24	1.183	22.5	0.791	3.8	2.371	9.88	17.712	73.8	1.1948	1.1902	1.1854	1.1804
25	1.191	23.4	0.786	16.1	2.488	9.95	18.575	74.3				
25.2	1.200			32.0								

[a]Mass of commercial NaCl required = (mass of pure NaCl required)/(% purity).
[b]Mass of water per unit volume = Brine mass minus NaCl mass.

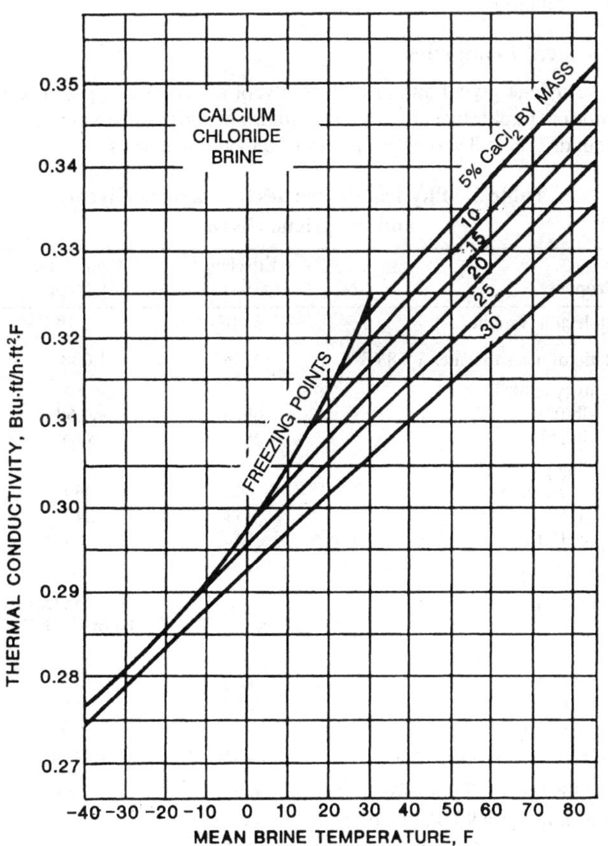

Fig. 4 Thermal Conductivity of Calcium Chloride Brines

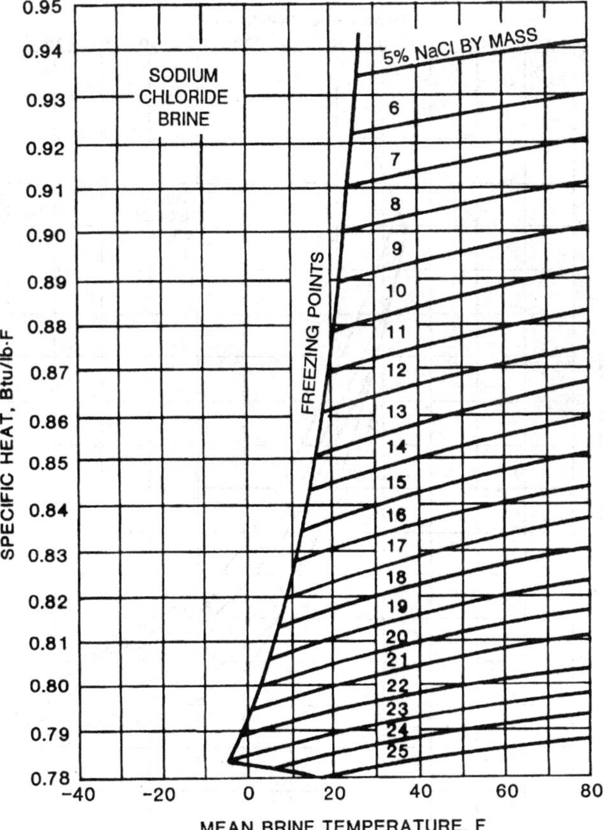

Fig. 5 Specific Heat of Sodium Chloride Brines

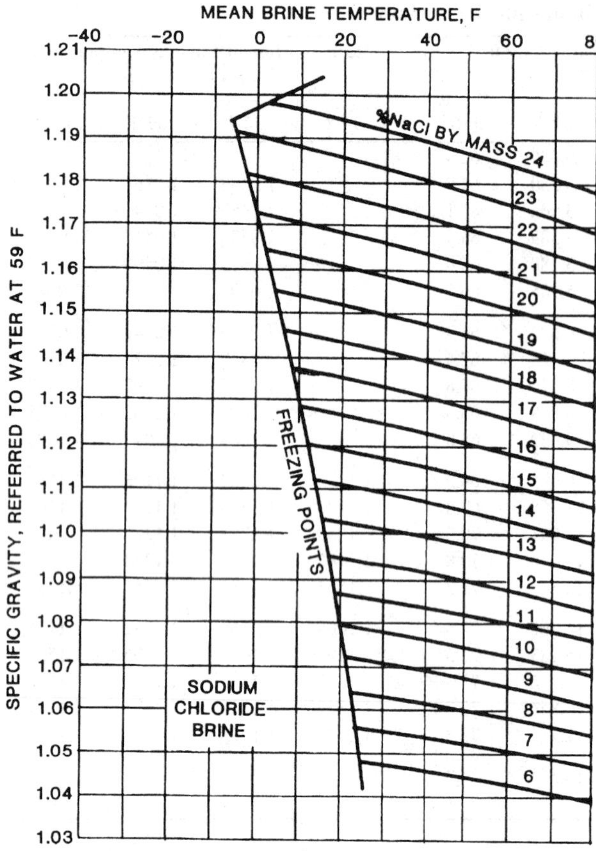

Fig. 6 Specific Gravity of Sodium Chloride Brines

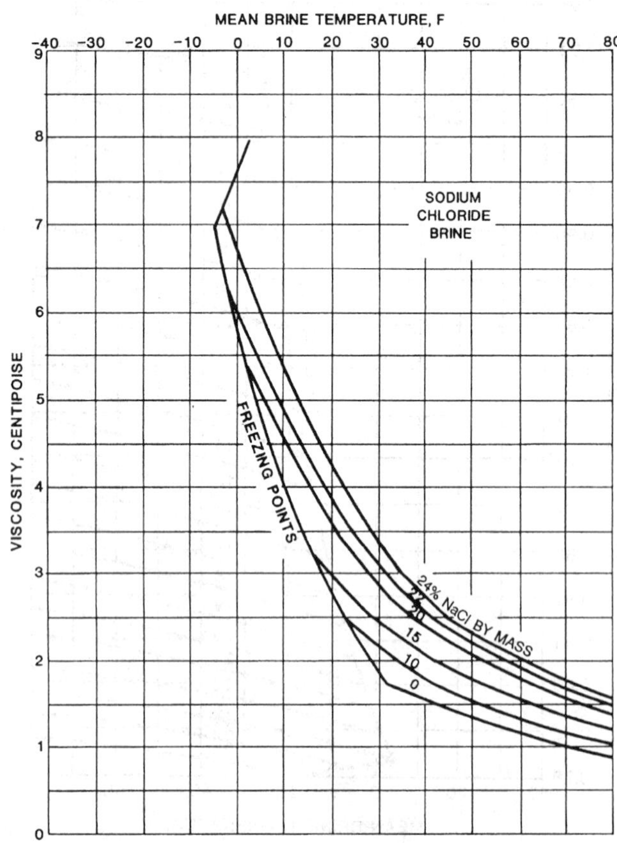

Fig. 7 Viscosity of Sodium Chloride Brines

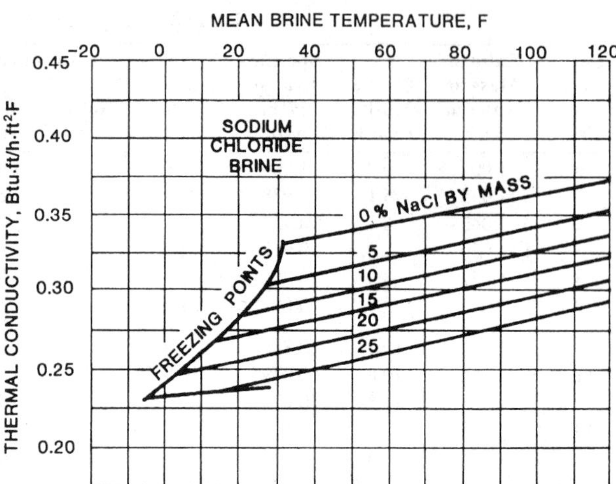

Fig. 8 Thermal Conductivity of Sodium Chloride Brines
(Carrier 1959)

INHIBITED GLYCOLS

Ethylene glycol and propylene glycol, inhibited for corrosion control, are used as aqueous freezing point depressants (antifreeze) and heat transfer media in heating and cooling systems. Their chief attributes are their ability to lower the freezing point of water, and their low volatility, and relatively low corrosivity when properly inhibited.

Inhibited ethylene glycol solutions have better physical properties than propylene glycol solutions, especially at lower temperatures. However, the less toxic propylene glycol is preferred for applications involving possible human contact or where mandated by regulations.

Physical Properties

Ethylene glycol and propylene glycol are colorless, practically odorless liquids that are miscible with water and many organic compounds. Table 3 shows properties of the pure materials.

Table 3 Physical Properties of Ethylene Glycol and Propylene Glycol

Property	Ethylene Glycol	Propylene Glycol
Molecular weight	62.07	76.10
Ratio of mass to water at 68/68°F	1.1155	1.0381
Density at 68°F		
lb/ft³	69.50	64.68
lb/gal	9.29	8.65
Boiling point, °F		
at 760 mm Hg	388	369
at 50 mm Hg	253	241
at 10 mm Hg	192	185
Vapor pressure at 68°F, mm Hg	0.05	0.07
Freezing point, °F	9.1	Sets to glass below −60°F
Viscosity, centipoise		
at 32°F	57.4	243
at 68°F	20.9	60.5
at 104°F	9.5	18.0
Refractive index n_D at 68°F	1.4319	1.4329
Specific heat at 68°F, Btu/lb·°F	0.561	0.593
Heat of fusion at 9.1°F, Btu/lb	80.5	—
Heat of vaporization at 1 atm, Btu/lb	364	296
Heat of combustion at 68°F, Btu/lb	8,280	10,312

The freezing and boiling points of aqueous solutions of ethylene glycol and propylene glycol are given in Tables 4 and 5. Note that increasing the concentration of ethylene glycol above 60% by mass causes the freezing point of the solution to increase. Propylene glycol solutions above 60% by mass do not have freezing points. Instead of freezing, propylene glycol solutions become a glass (glass being an amorphous, undercooled liquid of extremely high viscosities that has all the appearances of a solid). On the dilute side of the eutectic, ice forms on freezing; on the concentrated side, solid glycol separates from solution on freezing. The freezing velocity of such solutions is often quite slow; but, in time, they set to a hard, solid mass.

Physical properties (i.e., density, specific heat, thermal conductivity, and viscosity) for aqueous solutions of ethylene glycol can be found in Tables 6 through 9 and Figures 9 through 12; similar data for aqueous solutions of propylene glycol can be found in Tables 10 through 13 and Figures 13 through 16. Densities are for aqueous solutions of industrially inhibited glycols. These densities are somewhat higher than those for pure glycol and water alone. Typical corrosion inhibitor packages do not significantly affect the other physical properties. The physical properties for the two fluids are similar, with the exception of viscosity. At the same concentration, aqueous solutions of propylene glycol are more viscous than solutions of ethylene glycol. This higher viscosity accounts for the majority of the performance difference between the two fluids.

The choice of glycol concentration depends on the type of protection required by the application. If the fluid is being used to prevent equipment damage during idle periods in cold weather, such as winterizing coils in an HVAC system, 30% ethylene glycol or 35%

Table 4 Freezing and Boiling Points of Aqueous Solutions of Ethylene Glycol

Percent Ethylene Glycol		Freezing Point, °F	Boiling Point, °F at 14.6 psia
By Mass	By Volume		
0.0	0.0	32.0	212
5.0	4.4	29.4	213
10.0	8.9	26.2	214
15.0	13.6	22.2	215
20.0	18.1	17.9	216
21.0	19.2	16.8	216
22.0	20.1	15.9	216
23.0	21.0	14.9	217
24.0	22.0	13.7	217
25.0	22.9	12.7	218
26.0	23.9	11.4	218
27.0	24.8	10.4	218
28.0	25.8	9.2	219
29.0	26.7	8.0	219
30.0	27.7	6.7	220
31.0	28.7	5.4	220
32.0	29.6	4.2	220
33.0	30.6	2.9	220
34.0	31.6	1.4	220
35.0	32.6	–0.2	221
36.0	33.5	–1.5	221
37.0	34.5	–3.0	221
38.0	35.5	–4.5	221
39.0	36.5	–6.4	221
40.0	37.5	–8.1	222
41.0	38.5	–9.8	222
42.0	39.5	–11.7	222
43.0	40.5	–13.5	223
44.0	41.5	–15.5	223
45.0	42.5	–17.5	224
46.0	43.5	–19.8	224
47.0	44.5	–21.6	224
48.0	45.5	–23.9	224
49.0	46.6	–26.7	224
50.0	47.6	–28.9	225
51.0	48.6	–31.2	225
52.0	49.6	–33.6	225
53.0	50.6	–36.2	226
54.0	51.6	–38.8	226
55.0	52.7	–42.0	227
56.0	53.7	–44.7	227
57.0	54.7	–47.5	228
58.0	55.7	–50.0	228
59.0	56.8	–52.7	229
60.0	57.8	–54.9	230
65.0	62.8	a	235
70.0	68.3	a	242
75.0	73.6	a	248
80.0	78.9	–52.2	255
85.0	84.3	–34.5	273
90.0	89.7	–21.6	285
95.0	95.0	–3.0	317

[a]Freezing points are below –60°F.

Table 5 Freezing and Boiling Points of Aqueous Solutions of Propylene Glycol

Percent Propylene Glycol		Freezing Point, °F	Boiling Point, °F at 14.6 psia
By Mass	By Volume		
0.0	0.0	32.0	212
5.0	4.8	29.1	212
10.0	9.6	26.1	212
15.0	14.5	22.9	212
20.0	19.4	19.2	213
21.0	20.4	18.3	213
22.0	21.4	17.6	213
23.0	22.4	16.6	213
24.0	23.4	15.6	213
25.0	24.4	14.7	214
26.0	25.3	13.7	214
27.0	26.4	12.6	214
28.0	27.4	11.5	215
29.0	28.4	10.4	215
30.0	29.4	9.2	216
31.0	30.4	7.9	216
32.0	31.4	6.6	216
33.0	32.4	5.3	216
34.0	33.5	3.9	216
35.0	34.4	2.4	217
36.0	35.5	0.8	217
37.0	36.5	–0.8	217
38.0	37.5	–2.4	218
39.0	38.5	–4.2	218
40.0	39.6	–6.0	219
41.0	40.6	–7.8	219
42.0	41.6	–9.8	219
43.0	42.6	–11.8	219
44.0	43.7	–13.9	219
45.0	44.7	–16.1	220
46.0	45.7	–18.3	220
47.0	46.8	–20.7	220
48.0	47.8	–23.1	221
49.0	48.9	–25.7	221
50.0	49.9	–28.3	222
51.0	50.9	–31.0	222
52.0	51.9	–33.8	222
53.0	53.0	–36.7	223
54.0	54.0	–39.7	223
55.0	55.0	–42.8	223
56.0	56.0	–46.0	223
57.0	57.0	–49.3	224
58.0	58.0	–52.7	224
59.0	59.0	–56.2	224
60.0	60.0	–59.9	225
65.0	65.0	a	227
70.0	70.0	a	230
75.0	75.0	a	237
80.0	80.0	a	245
85.0	85.0	a	257
90.0	90.0	a	270
95.0	95.0	a	310

[a]Above 60% by mass, solutions do not freeze but become a glass.

Table 6 Density of Aqueous Solutions of Ethylene Glycol

Temperature, °F	Concentrations in Volume Percent Ethylene Glycol								
	10%	20%	30%	40%	50%	60%	70%	80%	90%
−30					68.12	69.03	69.90	70.75	
−20					68.05	68.96	69.82	70.65	71.45
−10				67.04	67.98	68.87	69.72	70.54	71.33
0				66.97	67.90	68.78	69.62	70.43	71.20
10			65.93	66.89	67.80	68.67	69.50	70.30	71.06
20		64.83	65.85	66.80	67.70	68.56	69.38	70.16	70.92
30	63.69	64.75	65.76	66.70	67.59	68.44	69.25	70.02	70.76
40	63.61	64.66	65.66	66.59	67.47	68.31	69.10	69.86	70.59
50	63.52	64.56	65.55	66.47	67.34	68.17	68.95	69.70	70.42
60	63.42	64.45	65.43	66.34	67.20	68.02	68.79	69.53	70.23
70	63.31	64.33	65.30	66.20	67.05	67.86	68.62	69.35	70.04
80	63.19	64.21	65.17	66.05	66.90	67.69	68.44	69.15	69.83
90	63.07	64.07	65.02	65.90	66.73	67.51	68.25	68.95	69.62
100	62.93	63.93	64.86	65.73	66.55	67.32	68.05	68.74	69.40
110	62.79	63.77	64.70	65.56	66.37	67.13	67.84	68.52	69.17
120	62.63	63.61	64.52	65.37	66.17	66.92	67.63	68.29	68.92
130	62.47	63.43	64.34	65.18	65.97	66.71	67.40	68.05	68.67
140	62.30	63.25	64.15	64.98	65.75	66.48	67.16	67.81	68.41
150	62.11	63.06	63.95	64.76	65.53	66.25	66.92	67.55	68.14
160	61.92	62.86	63.73	64.54	65.30	66.00	66.66	67.28	67.86
170	61.72	62.64	63.51	64.31	65.05	65.75	66.40	67.01	67.58
180	61.51	62.42	63.28	64.07	64.80	65.49	66.12	66.72	67.28
190	61.29	62.19	63.04	63.82	64.54	65.21	65.84	66.42	66.97
200	61.06	61.95	62.79	63.56	64.27	64.93	65.55	66.12	66.65
210	60.82	61.71	62.53	63.29	63.99	64.64	65.24	65.81	66.33
220	60.57	61.45	62.27	63.01	63.70	64.34	64.93	65.48	65.99
230	60.31	61.18	61.99	62.72	63.40	64.03	64.61	65.15	65.65
240	60.05	60.90	61.70	62.43	63.10	63.71	64.28	64.81	65.29
250	59.77	60.62	61.40	62.12	62.78	63.39	63.94	64.46	64.93

Note: Density in lb/ft^3.

Table 7 Specific Heat of Aqueous Solutions of Ethylene Glycol

Temperature, °F	Concentrations in Volume Percent Ethylene Glycol								
	10%	20%	30%	40%	50%	60%	70%	80%	90%
−30					0.734	0.680	0.625	0.567	
−20					0.739	0.686	0.631	0.574	0.515
−10				0.794	0.744	0.692	0.638	0.581	0.523
0				0.799	0.749	0.698	0.644	0.588	0.530
10			0.849	0.803	0.754	0.703	0.651	0.595	0.538
20		0.897	0.853	0.808	0.759	0.709	0.657	0.603	0.546
30	0.940	0.900	0.857	0.812	0.765	0.715	0.664	0.610	0.553
40	0.943	0.903	0.861	0.816	0.770	0.721	0.670	0.617	0.561
50	0.945	0.906	0.864	0.821	0.775	0.727	0.676	0.624	0.569
60	0.947	0.909	0.868	0.825	0.780	0.732	0.683	0.631	0.576
70	0.950	0.912	0.872	0.830	0.785	0.738	0.689	0.638	0.584
80	0.952	0.915	0.876	0.834	0.790	0.744	0.696	0.645	0.592
90	0.954	0.918	0.880	0.839	0.795	0.750	0.702	0.652	0.600
100	0.957	0.922	0.883	0.843	0.800	0.756	0.709	0.659	0.607
110	0.959	0.925	0.887	0.848	0.806	0.761	0.715	0.666	0.615
120	0.961	0.928	0.891	0.852	0.811	0.767	0.721	0.673	0.623
130	0.964	0.931	0.895	0.857	0.816	0.773	0.728	0.680	0.630
140	0.966	0.934	0.898	0.861	0.821	0.779	0.734	0.687	0.638
150	0.968	0.937	0.902	0.865	0.826	0.785	0.741	0.694	0.646
160	0.971	0.940	0.906	0.870	0.831	0.790	0.747	0.702	0.654
170	0.973	0.943	0.910	0.874	0.836	0.796	0.754	0.709	0.661
180	0.975	0.946	0.913	0.879	0.842	0.802	0.760	0.716	0.669
190	0.978	0.949	0.917	0.883	0.847	0.808	0.766	0.723	0.677
200	0.980	0.952	0.921	0.888	0.852	0.813	0.773	0.730	0.684
210	0.982	0.955	0.925	0.892	0.857	0.819	0.779	0.737	0.692
220	0.985	0.958	0.929	0.897	0.862	0.825	0.786	0.744	0.700
230	0.987	0.961	0.932	0.901	0.867	0.831	0.792	0.751	0.708
240	0.989	0.964	0.936	0.905	0.872	0.837	0.799	0.758	0.715
250	0.992	0.967	0.940	0.910	0.877	0.842	0.805	0.765	0.723

Note: Specific heat in Btu/lb·°F.

Table 8 Thermal Conductivity of Aqueous Solutions of Ethylene Glycol

	Concentrations in Volume Percent Ethylene Glycol								
Temperature, °F	10%	20%	30%	40%	50%	60%	70%	80%	90%
−30					0.190	0.178	0.167	0.158	
−20					0.193	0.181	0.170	0.160	0.151
−10				0.212	0.197	0.184	0.172	0.161	0.152
0				0.216	0.200	0.186	0.174	0.163	0.153
10			0.238	0.220	0.204	0.189	0.176	0.164	0.154
20		0.264	0.243	0.224	0.207	0.191	0.178	0.166	0.155
30	0.294	0.269	0.247	0.227	0.210	0.194	0.180	0.167	0.156
40	0.300	0.274	0.251	0.231	0.212	0.196	0.182	0.169	0.157
50	0.305	0.279	0.255	0.234	0.215	0.198	0.183	0.170	0.158
60	0.311	0.284	0.259	0.237	0.218	0.200	0.185	0.171	0.159
70	0.316	0.288	0.263	0.240	0.220	0.202	0.186	0.172	0.160
80	0.320	0.292	0.266	0.243	0.223	0.204	0.188	0.173	0.161
90	0.325	0.296	0.269	0.246	0.225	0.206	0.189	0.174	0.161
100	0.329	0.299	0.272	0.248	0.227	0.208	0.190	0.175	0.162
110	0.333	0.302	0.275	0.251	0.229	0.209	0.192	0.176	0.163
120	0.336	0.305	0.277	0.253	0.230	0.210	0.193	0.177	0.163
130	0.339	0.308	0.280	0.255	0.232	0.212	0.194	0.178	0.164
140	0.342	0.311	0.282	0.256	0.233	0.213	0.195	0.179	0.165
150	0.345	0.313	0.284	0.258	0.235	0.214	0.196	0.180	0.165
160	0.347	0.315	0.285	0.259	0.236	0.215	0.197	0.180	0.166
170	0.349	0.316	0.287	0.261	0.237	0.216	0.197	0.181	0.166
180	0.351	0.318	0.288	0.262	0.238	0.217	0.198	0.181	0.167
190	0.352	0.319	0.289	0.263	0.239	0.218	0.199	0.182	0.167
200	0.353	0.320	0.290	0.263	0.240	0.218	0.199	0.182	0.168
210	0.354	0.321	0.291	0.264	0.240	0.219	0.200	0.183	0.168
220	0.355	0.321	0.291	0.265	0.240	0.219	0.200	0.183	0.168
230	0.355	0.322	0.291	0.265	0.241	0.219	0.200	0.183	0.169
240	0.355	0.322	0.291	0.265	0.241	0.219	0.200	0.184	0.169
250	0.354	0.321	0.291	0.265	0.241	0.220	0.201	0.184	0.169

Note: Thermal conductivity in Btu·ft/h·ft^2·°F.

Table 9 Viscosity of Aqueous Solutions of Ethylene Glycol

	Concentrations in Volume Percent Ethylene Glycol								
Temperature, °F	10%	20%	30%	40%	50%	60%	70%	80%	90%
−30					63.69	89.67	128.79	185.22	
−20					40.38	60.46	89.93	131.32	284.48
−10				19.58	27.27	42.05	63.50	91.88	169.83
0				13.76	19.34	30.08	45.58	65.04	107.77
10			6.83	10.13	14.26	22.06	33.31	46.89	71.87
20		3.90	5.38	7.74	10.85	16.56	24.79	34.48	49.94
30	2.16	3.14	4.33	6.09	8.48	12.68	18.77	25.84	35.91
40	1.82	2.59	3.54	4.91	6.77	9.90	14.45	19.71	26.59
50	1.56	2.18	2.95	4.04	5.50	7.85	11.31	15.29	20.18
60	1.35	1.86	2.49	3.38	4.55	6.33	8.97	12.05	15.65
70	1.18	1.61	2.13	2.87	3.81	5.17	7.22	9.62	12.37
80	1.04	1.41	1.84	2.46	3.23	4.28	5.88	7.79	9.93
90	0.93	1.24	1.60	2.13	2.76	3.58	4.85	6.38	8.10
100	0.83	1.11	1.41	1.87	2.39	3.03	4.04	5.28	6.68
110	0.75	0.99	1.25	1.64	2.08	2.58	3.40	4.41	5.58
120	0.68	0.90	1.11	1.46	1.82	2.23	2.88	3.73	4.71
130	0.62	0.81	1.00	1.30	1.61	1.93	2.47	3.17	4.01
140	0.57	0.74	0.90	1.17	1.43	1.69	2.13	2.72	3.45
150	0.53	0.68	0.82	1.05	1.28	1.49	1.86	2.35	2.98
160	0.49	0.63	0.75	0.95	1.15	1.32	1.63	2.05	2.60
170	0.46	0.58	0.68	0.87	1.04	1.18	1.43	1.80	2.28
180	0.43	0.54	0.63	0.79	0.94	1.06	1.27	1.58	2.01
190	0.40	0.50	0.58	0.73	0.85	0.95	1.14	1.40	1.79
200	0.37	0.47	0.54	0.67	0.78	0.86	1.02	1.25	1.60
210	0.35	0.43	0.50	0.61	0.71	0.78	0.92	1.12	1.43
220	0.33	0.41	0.46	0.57	0.66	0.72	0.83	1.01	1.29
230	0.32	0.38	0.43	0.53	0.60	0.66	0.76	0.91	1.16
240	0.30	0.36	0.40	0.49	0.56	0.61	0.69	0.83	1.06
250	0.29	0.34	0.38	0.45	0.52	0.56	0.63	0.75	0.96

Note: Viscosity in centipoises.

Table 10 Density of Aqueous Solutions of an Industrially Inhibited Propylene Glycol

	Concentrations in Volume Percent Propylene Glycol								
Temperature, °F	10%	20%	30%	40%	50%	60%	70%	80%	90%
−30						67.05	67.47	68.38	68.25
−20					66.46	66.93	67.34	68.13	68.00
−10					66.35	66.81	67.20	67.87	67.75
0				65.71	66.23	66.68	67.05	67.62	67.49
10			65.00	65.60	66.11	66.54	66.89	67.36	67.23
20		64.23	64.90	65.48	65.97	66.38	66.72	67.10	66.97
30	63.38	64.14	64.79	65.35	65.82	66.22	66.54	66.83	66.71
40	63.30	64.03	64.67	65.21	65.67	66.05	66.35	66.57	66.44
50	63.20	63.92	64.53	65.06	65.50	65.87	66.16	66.30	66.18
60	63.10	63.79	64.39	64.90	65.33	65.68	65.95	66.04	65.91
70	62.98	63.66	64.24	64.73	65.14	65.47	65.73	65.77	65.64
80	62.86	63.52	64.08	64.55	64.95	65.26	65.51	65.49	65.37
90	62.73	63.37	63.91	64.36	64.74	65.04	65.27	65.22	65.09
100	62.59	63.20	63.73	64.16	64.53	64.81	65.03	64.95	64.82
110	62.44	63.03	63.54	63.95	64.30	64.57	64.77	64.67	64.54
120	62.28	62.85	63.33	63.74	64.06	64.32	64.51	64.39	64.26
130	62.11	62.66	63.12	63.51	63.82	64.06	64.23	64.11	63.98
140	61.93	62.46	62.90	63.27	63.57	63.79	63.95	63.83	63.70
150	61.74	62.25	62.67	63.02	63.30	63.51	63.66	63.55	63.42
160	61.54	62.03	62.43	62.76	63.03	63.22	63.35	63.26	63.13
170	61.33	61.80	62.18	62.49	62.74	62.92	63.04	62.97	62.85
180	61.11	61.56	61.92	62.22	62.45	62.61	62.72	62.68	62.56
190	60.89	61.31	61.65	61.93	62.14	62.29	62.39	62.39	62.27
200	60.65	61.05	61.37	61.63	61.83	61.97	62.05	62.10	61.97
210	60.41	60.78	61.08	61.32	61.50	61.63	61.69	61.81	61.68
220	60.15	60.50	60.78	61.00	61.17	61.28	61.33	61.51	61.38
230	59.89	60.21	60.47	60.68	60.83	60.92	60.96	61.21	61.08
240	59.61	59.91	60.15	60.34	60.47	60.55	60.58	60.91	60.78
250	59.33	59.60	59.82	59.99	60.11	60.18	60.19	60.61	60.48

Note: Density in lb/ft^3.

Table 11 Specific Heat of Aqueous Solutions of Propylene Glycol

	Concentrations in Volume Percent Propylene Glycol								
Temperature, °F	10%	20%	30%	40%	50%	60%	70%	80%	90%
−30						0.741	0.680	0.615	0.542
−20					0.799	0.746	0.687	0.623	0.550
−10					0.804	0.752	0.693	0.630	0.558
0				0.855	0.809	0.758	0.700	0.637	0.566
10			0.898	0.859	0.814	0.764	0.707	0.645	0.574
20		0.936	0.902	0.864	0.820	0.770	0.713	0.652	0.583
30	0.966	0.938	0.906	0.868	0.825	0.776	0.720	0.660	0.591
40	0.968	0.941	0.909	0.872	0.830	0.782	0.726	0.667	0.599
50	0.970	0.944	0.913	0.877	0.835	0.787	0.733	0.674	0.607
60	0.972	0.947	0.917	0.881	0.840	0.793	0.740	0.682	0.615
70	0.974	0.950	0.920	0.886	0.845	0.799	0.746	0.689	0.623
80	0.976	0.953	0.924	0.890	0.850	0.805	0.753	0.696	0.631
90	0.979	0.956	0.928	0.894	0.855	0.811	0.760	0.704	0.639
100	0.981	0.959	0.931	0.899	0.861	0.817	0.766	0.711	0.647
110	0.983	0.962	0.935	0.903	0.866	0.823	0.773	0.718	0.656
120	0.985	0.965	0.939	0.908	0.871	0.828	0.779	0.726	0.664
130	0.987	0.967	0.942	0.912	0.876	0.834	0.786	0.733	0.672
140	0.989	0.970	0.946	0.916	0.881	0.840	0.793	0.740	0.680
150	0.991	0.973	0.950	0.921	0.886	0.846	0.799	0.748	0.688
160	0.993	0.976	0.953	0.925	0.891	0.852	0.806	0.755	0.696
170	0.996	0.979	0.957	0.929	0.896	0.858	0.812	0.762	0.704
180	0.998	0.982	0.961	0.934	0.902	0.864	0.819	0.770	0.712
190	1.000	0.985	0.964	0.938	0.907	0.869	0.826	0.777	0.720
200	1.002	0.988	0.968	0.943	0.912	0.875	0.832	0.784	0.729
210	1.004	0.991	0.971	0.947	0.917	0.881	0.839	0.792	0.737
220	1.006	0.994	0.975	0.951	0.922	0.887	0.845	0.799	0.745
230	1.008	0.996	0.979	0.956	0.927	0.893	0.852	0.806	0.753
240	1.011	0.999	0.982	0.960	0.932	0.899	0.859	0.814	0.761
250	1.013	1.002	0.986	0.965	0.937	0.905	0.865	0.821	0.769

Note: Specific heat in Btu/lb·°F.

Table 12 Thermal Conductivity of Aqueous Solutions of Propylene Glycol

	Concentrations in Volume Percent Propylene Glycol								
Temperature, °F	10%	20%	30%	40%	50%	60%	70%	80%	90%
−30						0.171	0.159	0.147	0.137
−20					0.188	0.174	0.160	0.148	0.137
−10					0.191	0.176	0.161	0.148	0.136
0				0.211	0.194	0.178	0.162	0.149	0.136
10			0.235	0.215	0.196	0.179	0.163	0.149	0.136
20		0.262	0.239	0.218	0.199	0.181	0.164	0.150	0.136
30	0.293	0.267	0.243	0.222	0.201	0.183	0.165	0.150	0.135
40	0.299	0.272	0.247	0.225	0.204	0.184	0.166	0.150	0.135
50	0.304	0.277	0.251	0.227	0.206	0.186	0.167	0.150	0.135
60	0.310	0.281	0.254	0.230	0.208	0.187	0.168	0.150	0.134
70	0.315	0.285	0.258	0.233	0.210	0.188	0.168	0.151	0.134
80	0.319	0.289	0.261	0.235	0.211	0.189	0.169	0.151	0.134
90	0.323	0.292	0.263	0.237	0.213	0.190	0.169	0.151	0.133
100	0.327	0.295	0.266	0.239	0.214	0.191	0.170	0.151	0.133
110	0.331	0.298	0.268	0.241	0.215	0.192	0.170	0.151	0.132
120	0.334	0.301	0.270	0.243	0.217	0.193	0.170	0.150	0.132
130	0.338	0.304	0.272	0.244	0.218	0.193	0.170	0.150	0.131
140	0.340	0.306	0.274	0.245	0.218	0.194	0.171	0.150	0.131
150	0.343	0.308	0.276	0.246	0.219	0.194	0.171	0.150	0.130
160	0.345	0.309	0.277	0.247	0.220	0.194	0.171	0.150	0.130
170	0.347	0.311	0.278	0.248	0.220	0.195	0.171	0.149	0.129
180	0.348	0.312	0.279	0.249	0.221	0.195	0.170	0.149	0.129
190	0.350	0.313	0.280	0.249	0.221	0.195	0.170	0.148	0.128
200	0.351	0.314	0.280	0.249	0.221	0.194	0.170	0.148	0.127
210	0.351	0.314	0.280	0.249	0.221	0.194	0.169	0.147	0.127
220	0.352	0.314	0.280	0.249	0.220	0.194	0.169	0.147	0.126
230	0.352	0.314	0.280	0.249	0.220	0.193	0.168	0.146	0.125
240	0.351	0.314	0.280	0.249	0.220	0.193	0.168	0.146	0.125
250	0.351	0.314	0.279	0.248	0.219	0.192	0.167	0.145	0.124

Note: Thermal conductivity in Btu·ft/h·ft^2·°F.

Table 13 Viscosity of Aqueous Solutions of Propylene Glycol

	Concentrations in Volume Percent Propylene Glycol								
Temperature, °F	10%	20%	30%	40%	50%	60%	70%	80%	90%
−30						497.57	864.87	1363.75	3555.22
−20					156.08	298.75	493.93	820.58	1819.72
−10					95.97	182.96	291.28	495.68	983.05
0				40.99	61.32	114.90	177.73	303.94	558.32
10			13.44	27.17	40.62	74.19	112.20	190.41	332.02
20		5.36	9.91	18.64	27.83	49.29	73.22	122.30	205.91
30	2.80	4.23	7.47	13.20	19.66	33.68	49.32	80.66	132.67
40	2.28	3.41	5.75	9.63	14.28	23.65	34.22	54.64	88.51
50	1.89	2.79	4.52	7.22	10.65	17.05	24.41	37.99	60.93
60	1.60	2.32	3.61	5.55	8.13	12.59	17.86	27.10	43.16
70	1.38	1.95	2.94	4.36	6.34	9.51	13.38	19.79	31.37
80	1.20	1.66	2.43	3.50	5.04	7.34	10.25	14.79	23.35
90	1.05	1.43	2.04	2.86	4.08	5.77	8.00	11.29	17.75
100	0.93	1.25	1.73	2.37	3.35	4.62	6.37	8.79	13.76
110	0.83	1.10	1.49	2.00	2.79	3.76	5.15	6.97	10.86
120	0.75	0.97	1.30	1.71	2.36	3.11	4.23	5.62	8.71
130	0.68	0.87	1.14	1.49	2.02	2.61	3.53	4.60	7.09
140	0.62	0.78	1.01	1.30	1.75	2.22	2.98	3.82	5.85
150	0.57	0.71	0.90	1.16	1.53	1.91	2.54	3.22	4.89
160	0.52	0.64	0.82	1.03	1.35	1.66	2.19	2.75	4.13
170	0.48	0.59	0.74	0.93	1.20	1.45	1.91	2.37	3.52
180	0.44	0.54	0.68	0.85	1.08	1.29	1.69	2.07	3.04
190	0.41	0.50	0.62	0.78	0.97	1.15	1.50	1.82	2.64
200	0.38	0.46	0.58	0.72	0.88	1.04	1.34	1.61	2.31
210	0.36	0.43	0.54	0.67	0.81	0.94	1.21	1.45	2.04
220	0.34	0.40	0.50	0.62	0.74	0.86	1.10	1.31	1.82
230	0.32	0.38	0.47	0.59	0.69	0.79	1.00	1.19	1.63
240	0.30	0.36	0.45	0.55	0.64	0.73	0.92	1.09	1.47
250	0.28	0.34	0.42	0.52	0.59	0.68	0.85	1.00	1.33

Note: Viscosity in centipoises.

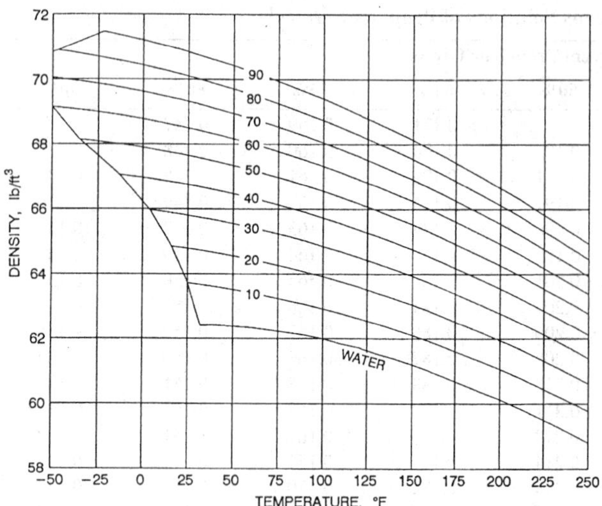

Fig. 9 Density of Aqueous Solutions of Industrially Inhibited Ethylene Glycol (vol. %)

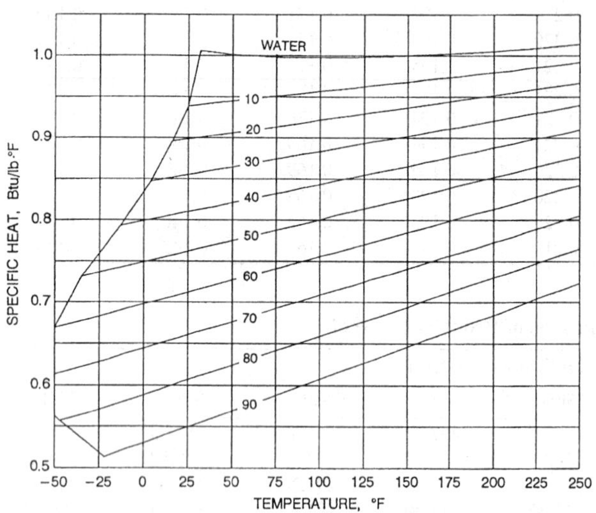

Fig. 10 Specific Heat of Aqueous Solutions of Industrially Inhibited Ethylene Glycol (vol. %)

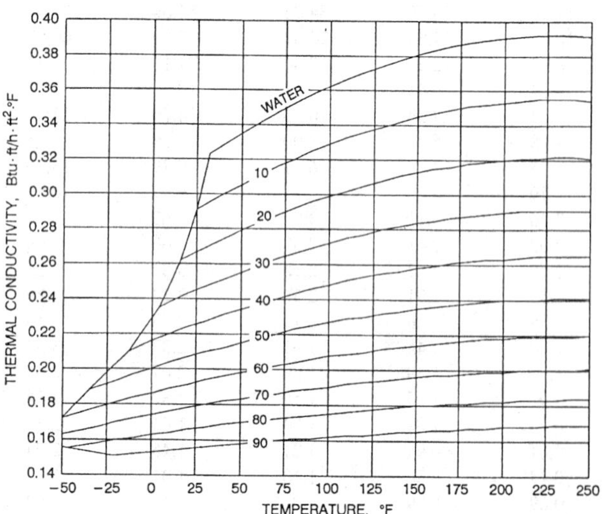

Fig. 11 Thermal Conductivity of Aqueous Solutions of Industrially Inhibited Ethylene Glycol (vol. %)

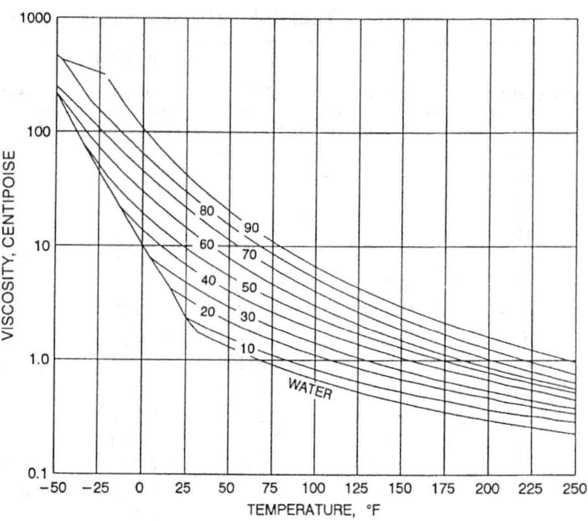

Fig. 12 Viscosity of Aqueous Solutions of Industrially Inhibited Ethylene Glycol (vol. %)

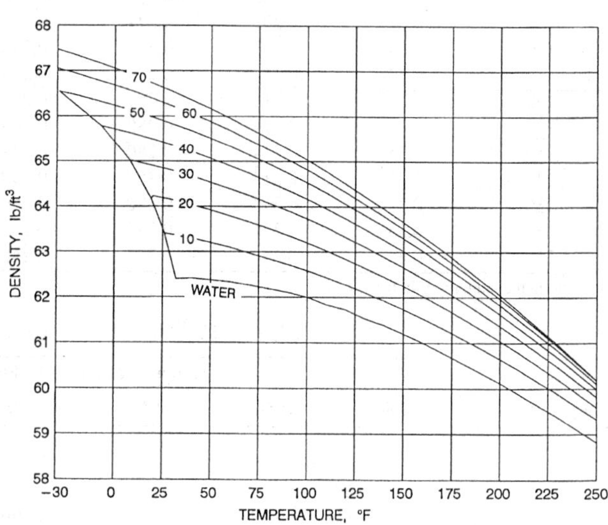

Fig. 13 Density of Aqueous Solutions of Industrially Inhibited Propylene Glycol (vol. %)

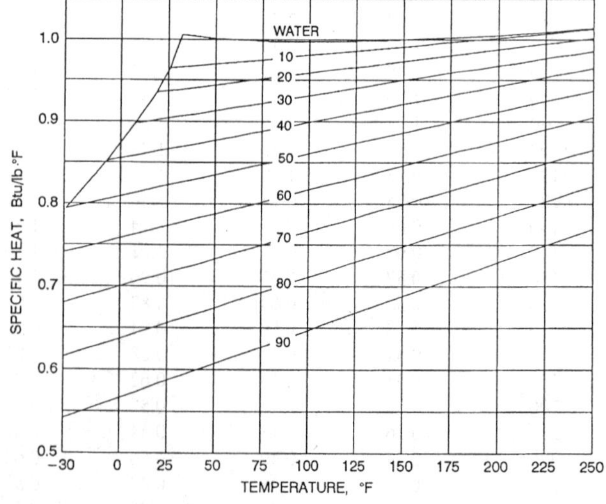

Fig. 14 Specific Heat of Aqueous Solutions of Industrially Inhibited Propylene Glycol (vol. %)

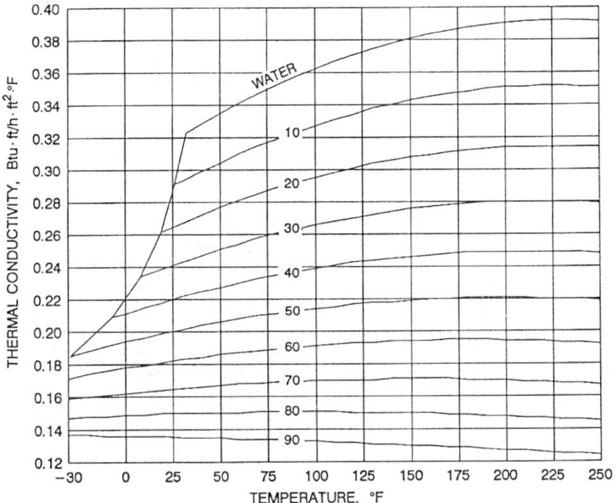

Fig. 15 Thermal Conductivity of Aqueous Solutions of Industrially Inhibited Propylene Glycol (vol. %)

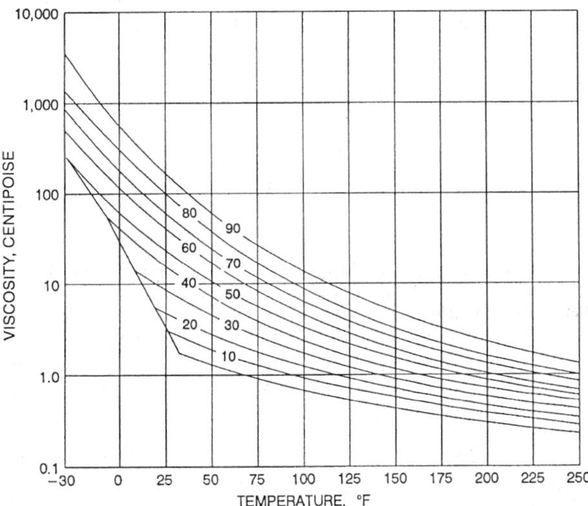

Fig. 16 Viscosity of Aqueous Solutions of Industrially Inhibited Propylene Glycol (vol. %)

propylene glycol is sufficient. These concentrations will allow the fluid to freeze. As the fluid freezes, it forms a slush that expands and flows into any available space. Therefore, expansion volume must be included with this type of protection. If the application requires that the fluid remain entirely liquid, a concentration with a freezing point 5°F below the lowest expected temperature should be chosen. Avoid excessive glycol concentration because it increases initial cost and adversely affects the physical properties of the fluid.

Additional physical property data is available from suppliers of industrially inhibited ethylene and propylene glycol.

Corrosion Inhibition

Commercial ethylene glycol or propylene glycol, when pure, is generally less corrosive than water to common metals used in construction. However, aqueous solutions of these glycols assume the corrosivity of the water from which they are prepared and can become increasingly corrosive with use if they are not properly inhibited. Without inhibitors, glycols oxidize into acidic end products. The amount of oxidation is influenced by temperature, degree of aeration, and, to some extent, the particular combination of metal components to which the glycol solution is exposed.

Corrosion inhibition can be described by classifying additives as either (1) corrosion inhibitors, or (2) environmental stabilizers and adjusters. **Corrosion inhibitors** form a surface barrier that protects the metal from attack. These barriers are usually formed by adsorption of the inhibitor by the metal, by reaction of the inhibitor with the metal, or by the incipient reaction product. In most cases, metal surfaces are covered by films of their oxides that inhibitors reinforce.

Environmental stabilizers or adjusters, while not corrosion inhibitors in the strict sense, decrease corrosion by stabilizing or favorably altering the overall environment. An alkaline buffer such as borax is an example of an environmental stabilizer, since its prime purpose is to maintain an alkaline condition (pH above 7). Some chelating agents function as stabilizers by removing from the solution certain deleterious ions that accelerate the corrosion process or mechanism; however, exercise caution in their use because improper combinations of pH and concentration may lead to excessive corrosion.

Certain oxidants, such as sodium chromate, should not be used with glycol solutions, because the glycol can oxidize prematurely. Generally, combinations of the two types of additives, inhibitors, and stabilizers offer the best corrosion resistance in a given system. Commercial inhibited glycols are available from several suppliers.

Service Considerations

Design Considerations. Inhibited glycols can be used at temperatures as high as 350°F. However, maximum-use temperatures vary from fluid to fluid. Therefore, the manufacturer's suggested temperature-use ranges should be followed. In systems with a high degree of aeration, the bulk fluid temperature should not exceed 180°F; however, temperatures up to 350°F are permissible in a pressurized system if air intake is eliminated. Maximum film temperatures should not exceed 50°F above the bulk temperature. Nitrogen blanketing minimizes oxidation when the system operates at elevated temperatures for extended periods.

Minimum operating temperatures are typically −10°F for ethylene glycol solutions and 0°F for propylene glycol solutions. Operation below these temperatures is generally impractical, because the viscosity of the fluids builds dramatically, thus increasing pumping horsepower requirements and reducing heat transfer film coefficients.

Standard materials can be used with most inhibited glycol solutions except galvanized steel, because the galvanizing material, zinc, reacts with a portion of the inhibitor package found in most formulated glycols.

Because the removal of sludge and other contaminants is critical, install suitable filters. If inhibitors are rapidly and completely adsorbed by such contamination, the fluid is ineffective for corrosion inhibition. Consider such adsorption when selecting filters.

Storage and Handling. Inhibited glycol concentrates are stable, relatively noncorrosive materials with high flash points. These fluids can be stored in mild steel, stainless steel, or aluminum vessels. However, aluminum should be used only when the fluid temperature is below 150°F. Corrosion in the vapor space of vessels may be a problem, because the fluid's inhibitor package cannot reach these surfaces to protect them. To prevent this problem, a coating may be used. Suitable coatings include novolac-based vinyl ester resins, high-bake phenolic resins, polypropylene, and polyvinylidene fluoride. To ensure the coating is suitable for a particular application and temperature, the manufacturer should be consulted. Since the chemical properties of an inhibited glycol concentrate differ from those of its dilutions, the effect of the concentrate on different containers should be known when selecting storage.

Choose transfer pumps only after considering temperature-viscosity data. Centrifugal pumps with electric motor drives are often used. Materials compatible with ethylene or propylene glycol should be used for pump packing material. Mechanical seals are also satisfactory. Welded mild steel transfer piping with a minimum diameter is normally used in conjunction with the piping, although flanged and gasketed joints are also satisfactory.

Preparation Before Application. Before an inhibited glycol is charged into a system, remove residual contaminants such as sludge, rust, brine deposits, and oil so the contained inhibitor functions properly. Avoid strong acid cleaners; if they are required, consider inhibited acids. Completely remove the cleaning agent before charging with inhibited glycol.

Use distilled, deionized, or condensate water, because water from some sources contains elements that reduce the effectiveness of the inhibited formulation. If water of this quality is unavailable, water containing less than 25 ppm chloride, less than 25 ppm sulfate, and less than 100 ppm of total hardness may be used.

Fluid Maintenance. Glycol concentrations can be determined by refractive index, gas chromatography, or Karl Fischer analysis for water (assuming that the concentration of other fluid components, such as inhibitor, is known). Using density to determine glycol concentration is unsatisfactory because (1) density measurements are temperature sensitive, (2) inhibitor concentrations can change density, (3) values for propylene glycol are close to those of water, and (4) propylene glycol values are maximum at 70 to 75% concentration.

A rigorous inhibitor monitoring and maintenance schedule is essential to maintain a glycol solution in relatively noncorrosive condition for a long period. However, a specific schedule is not always easy to establish, because inhibitor depletion rate depends on the particular conditions of use. Analysis of samples immediately after installation, after two to three months, and after six months should establish the pattern for the schedule. Visually inspecting the solution and filter residue can detect active corrosion.

Properly inhibited and maintained glycol solutions provide better corrosion protection than brine solutions in most systems. A long, though not indefinite, service life can be expected. Avoid indiscriminate mixing of inhibited formulations. Exercise caution in replacing brine systems with inhibited glycols because brine components are incompatible with glycol formulations.

HALOCARBONS

Many common refrigerants are used as secondary coolants as well as primary refrigerating media. Their favorable properties as heat transfer fluids include low freezing points, low viscosities, nonflammability, and good stability. Chapters 18 and 19 present physical and thermodynamic properties for common refrigerants. Table 14 lists two halocarbon compounds that are commonly used as secondary coolants. Table 15 gives vapor pressure, specific heat, thermal conductivity, density, and viscosity values for methylene chloride (R-30). Table 16 gives the same properties for trichloroethylene (R-1120).

Table 9 in Chapter 18 summarizes comparative safety characteristics for halocarbons. *Threshold Limit Values and Biological Exposure Indices* (ACGIH 1996) has more information on halocarbon toxicity.

Construction materials and stability factors in halocarbon use are discussed in Chapter 18 of this volume and Chapter 5 of the 1994 *ASHRAE Handbook—Refrigeration*. Note particularly that methylene chloride and trichloroethylene should not be used in contact with aluminum components.

NONHALOCARBON, NONAQUEOUS FLUIDS

In addition to the aforementioned fluids, numerous other secondary refrigerants are available. These fluids have been used primarily by the chemical processing and pharmaceutical industries. They have been used rarely in the HVAC and allied industries due to their cost and relative novelty. Before choosing these types of fluids, consider electrical classifications, disposal, potential worker exposure, process containment, and other relevant issues.

Table 14 Freezing and Boiling Points of Halocarbon Coolants

Refrigerant	Name	Freezing Point, °F	Boiling Point, °F
30	Methylene chloride	−142	103.6
1120	Trichloroethylene	−123	189

Table 15 Properties of Liquid Methylene Chloride (R-30)

Temperature, °F	Vapor Pressure, psia	Specific Heat, Btu/lb·°F	Thermal Conductivity, Btu/h·ft·°F	Density, lb/ft³	Viscosity, Centipoise
140	25.4	0.296	0.074	78.3	0.32
122	19.9	0.293	0.076	79.4	0.34
104	14.5	0.289	0.079	80.5	0.37
86	10.2	0.286	0.081	81.6	0.40
68	6.82	0.284	0.083	82.7	0.44
50	4.39	0.282	0.085	83.8	0.48
32	2.73	0.280	0.087	84.9	0.53
14	1.64	0.278	0.089	86.0	0.59
−4	0.97	0.277	0.091	87.1	0.66
−22	0.55	0.275	0.093	88.2	0.76
−40	0.32	0.274	0.094	89.3	0.88
−58	0.18	0.273	0.096	90.4	1.05
−76	0.10	0.273	0.098	91.5	1.29
−94	0.06	0.273	0.099	92.6	1.68
−112	0.03	0.272	0.101	93.7	2.50

Table 16 Properties of Liquid Trichloroethylene (R-1120)

Temperature, °F	Vapor Pressure, psia	Specific Heat, Btu/lb·°F	Thermal Conductivity, Btu/h·ft·°F	Density, lb/ft³	Viscosity, Centipoise
140	5.73	0.231	0.062	86.8	0.40
122	4.21	0.228	0.063	88.0	0.44
104	2.87	0.225	0.065	89.0	0.48
86	1.86	0.223	0.066	90.1	0.52
68	1.13	0.220	0.068	91.3	0.57
50	0.667	0.218	0.069	92.4	0.63
32	0.370	0.216	0.071	93.5	0.70
14	0.199	0.213	0.073	94.6	0.78
−4	0.102	0.211	0.074	95.6	0.87
−22	0.052	0.209	0.076	96.6	0.99
−40	0.024	0.207	0.077	97.7	1.14
−58	0.011	0.206	0.079	98.7	1.33
−76	0.005	0.204	0.080	99.7	1.60
−94	0.002	0.202	0.082	100.6	1.93
−112	0.001	0.201	0.084	101.6	2.45

Table 17 Summary of Physical Properties of Polydimethylsiloxane Mixture and d-Limonene

	Polydimethylsiloxane Mixture	d-Limonene
Flash point, °F, closed cup	116	115
Boiling point, °F	347	310
Freezing point, °F	−168	−142
Operational temperature range, °F	−100 to 500	None published

Table 18 Properties of a Polydimethylsiloxane Heat Transfer Fluid

Temperature, °F	Vapor Pressure, psia	Viscosity, Centipoise	Density, lb/ft³	Heat Capacity, Btu/lb·°F	Thermal Conductivity, Btu/h·ft·°F
−100	0.00	12.5	57.8	0.337	0.0748
−90	0.00	10.5	57.5	0.340	0.0742
−80	0.00	8.82	57.2	0.344	0.0736
−70	0.00	7.50	56.9	0.347	0.0730
−60	0.00	6.43	56.6	0.350	0.0724
−50	0.00	5.55	56.3	0.354	0.0717
−40	0.00	4.83	56.0	0.357	0.0711
−30	0.00	4.22	55.7	0.361	0.0705
−20	0.00	3.72	55.4	0.364	0.0699
−10	0.00	3.29	55.1	0.367	0.0692
0	0.00	2.93	54.8	0.371	0.0686
10	0.00	2.62	54.5	0.374	0.0679
20	0.00	2.36	54.2	0.378	0.0673
30	0.00	2.13	53.9	0.381	0.0666
40	0.01	1.93	53.6	0.384	0.0659
50	0.01	1.76	53.3	0.388	0.0652
60	0.02	1.60	53.0	0.391	0.0646
70	0.03	1.47	52.7	0.395	0.0639
80	0.04	1.35	52.4	0.398	0.0632
90	0.05	1.25	52.1	0.402	0.0625
100	0.08	1.15	51.8	0.405	0.0618
110	0.11	1.07	51.5	0.408	0.0610
120	0.15	0.993	51.1	0.412	0.0603
130	0.20	0.926	50.8	0.415	0.0596
140	0.27	0.865	50.5	0.419	0.0589
150	0.35	0.810	50.2	0.422	0.0581
160	0.46	0.760	49.8	0.425	0.0574
170	0.60	0.715	49.5	0.429	0.0567
180	0.76	0.673	49.2	0.432	0.0559
190	0.96	0.635	48.8	0.436	0.0551
200	1.20	0.601	48.5	0.439	0.0544
210	1.49	0.569	48.1	0.442	0.0536
220	1.84	0.540	47.8	0.446	0.0528
230	2.24	0.513	47.4	0.449	0.0521
240	2.72	0.488	47.0	0.453	0.0513
250	3.27	0.465	46.7	0.456	0.0505
260	3.91	0.443	46.3	0.459	0.0497
270	4.65	0.424	45.9	0.463	0.0489
280	5.50	0.405	45.5	0.466	0.0481
290	6.46	0.388	45.1	0.470	0.0473
300	7.55	0.372	44.7	0.473	0.0465
310	8.78	0.357	44.3	0.476	0.0457
320	10.16	0.343	43.9	0.480	0.0449
330	11.71	0.330	43.5	0.483	0.0441
340	13.43	0.317	43.1	0.487	0.0432
350	15.33	0.306	42.6	0.490	0.0424
360	17.45	0.295	42.2	0.494	0.0416
370	19.77	0.285	41.7	0.497	0.0407
380	22.32	0.275	41.3	0.500	0.0399
390	25.12	0.266	40.8	0.504	0.0390
400	28.17	0.257	40.4	0.507	0.0382
410	31.49	0.249	39.9	0.511	0.0373
420	35.10	0.242	39.4	0.514	0.0365
430	39.00	0.234	38.9	0.517	0.0356
440	43.21	0.227	38.4	0.521	0.0348
450	47.75	0.221	37.9	0.524	0.0339
460	52.63	0.214	37.4	0.528	0.0330
470	57.86	0.209	36.8	0.531	0.0321
480	63.46	0.203	36.3	0.534	0.0313
490	69.44	0.197	35.8	0.538	0.0304
500	75.81	0.192	35.2	0.541	0.0295

Table 19 Physical Properties of d-Limonene

Temperature, °F	Specific Heat, Btu/lb·°F	Viscosity, Centipoise	Density, lb/ft³	Thermal Conductivity, Btu/h·ft·°F
−100	0.3	3.8	57.1	0.0794
−50	0.34	2.8	55.8	0.0764
0	0.37	2.1	54.5	0.0734
50	0.41	1.6	53.2	0.0704
100	0.44	1.2	51.8	0.0674
150	0.48	0.9	50.4	0.0644
200	0.51	0.7	49	0.0614
250	0.54	0.6	47.6	0.0584
300	0.58	0.4	46	0.0554

Note: Properties are estimated or based on incomplete data.

Tables 17 through 19 contain physical property information on a mixture of dimethylsiloxane polymers of various molecular weights (Dow Corning 1989) and d-limonene. Note that the information on d-limonene is limited; it is based on measurements made over small data temperature ranges or simply on standard physical property estimation techniques. The compound is an optically active terpene (molecular formula $C_{10}H_{16}$) derived as an extract from orange and lemon oils. The "d" indicates that the material is dextrorotatory, which is a physical property of the material that does not affect the transport properties of the material significantly.

The mixture of dimethylsiloxane polymers can be used with most standard construction materials; d-limonene, however, can be quite corrosive, easily autooxidizing at ambient temperatures. This fact should be understood and considered before using d-limonene in a system.

REFERENCES

ACGIH. 1996. Threshold limit values and biological exposure indices. Published annually by the American Conference of Governmental Industrial Hygienists, Cincinnati, OH.

Carrier Air Conditioning Company. 1959. Basic data, Section 17M. Syracuse, NY.

Dow Corning USA. 1989. Syltherm heat transfer liquids. Midland, MI.

BIBLIOGRAPHY

Born, D.W. 1989. Inhibited glycols for corrosion and freeze protection in water-based heating and cooling systems. Midland, MI.

CCI. Calcium chloride for refrigeration brine. Manual RM-1. Calcium Chloride Institute.

Dow Chemical USA. 1994. Engineering manual for DOWFROST and DOWFROST HD heat transfer fluids. Midland, MI.

Dow Chemical USA. 1996. Engineering manual for Dowtherm SR-1 and Dowtherm 4000 heat transfer fluids. Midland, MI.

Fontana, M.G. 1986. *Corrosion engineering*. McGraw-Hill, New York.

NACE. 1984. *Corrosion basics: An introduction*. National Association of Corrosion Engineers, Houston, TX.

Nathan C.C. 1973. *Corrosion inhibitors*. National Association of Corrosion Engineers, Houston, TX.

Refrigeration Engineering Application Data—Section 40. *Refrigerating Engineering* 54 (November).

Sawens, R.H. 1947. Calcium chloride and sodium chloride refrigeration brines.

Sax, N.I. and R.J. Lewis. 1987. *Hawley's condensed chemical dictionary*. Van Nostrand Reinhold, New York.

Union Carbide Corporation. 1994. Ucartherm heat transfer fluids. South Charleston, WV.

CHAPTER 21

SORBENTS AND DESICCANTS

SORPTION refers to the binding of one substance to another. **Sorbents** are materials that have an ability to attract and hold other gases or liquids. They can be used to attract gases or liquids other than water vapor, a characteristic that makes them very useful in chemical separation processes. **Desiccants** are a subset of sorbents; they have a particular affinity for water.

Virtually all materials are desiccants; that is, they attract and hold water vapor. Wood, natural fibers, clays, and many synthetic materials attract and release moisture as commercial desiccants do, but they lack the holding capacity. For example, woolen carpet fibers attract up to 23% of their dry weight in water vapor, and nylon can take up almost 6% of its weight in water. In contrast, a commercial desiccant takes up between 10 and 1100% of its dry weight in water vapor, depending on its type and on the moisture available in the environment. Furthermore, commercial desiccants continue to attract moisture even when the surrounding air is quite dry, a characteristic that other materials do not share.

All desiccants behave in a similar way—they attract moisture until they reach equilibrium with the surrounding air. Moisture is usually removed from the desiccant by heating it to temperatures between 120 and 500°F and exposing it to a scavenger airstream. After the desiccant dries, it must be cooled so that it can attract moisture once again. Sorption always generates sensible heat equal to the latent heat of the water vapor taken up by the desiccant plus an additional heat of sorption that varies between 5 and 25% of the latent heat of the water vapor. This heat is transferred to the desiccant and to the surrounding air.

The process of attracting and holding moisture is described as either adsorption or absorption, depending on whether the desiccant undergoes a chemical change as it takes on moisture. **Adsorption** does not change the desiccant, except by the addition of the weight of water vapor; it is similar in some ways to a sponge soaking up water. **Absorption**, on the other hand, changes the desiccant. An example of an absorbent is table salt, which changes from a solid to a liquid as it absorbs moisture.

DESICCANT APPLICATIONS

Desiccants can dry either liquids or gases, including ambient air, and are used in many air-conditioning applications, particularly when

- The latent load is large in comparison to the sensible load.
- The cost of energy to regenerate the desiccant is low compared to the cost of energy to dehumidify the air by chilling it below its dew point.
- The moisture control level for the space would require chilling the air to subfreezing dew points if compression refrigeration alone were used to dehumidify the air.
- The temperature control level for the space or process requires continuous delivery of air at subfreezing temperatures.

In any of these situations, the cost of running a vapor compression cooling system can be very high. A desiccant process may offer considerable advantages in energy, initial cost of equipment, and maintenance.

Because desiccants are able to attract and hold more than simply water vapor, they can remove contaminants from airstreams to improve indoor air quality. Desiccants have been used to remove organic vapors, and in special circumstances, to control microbiological contaminants (Batelle 1971, Buffalo Testing Laboratory 1974). Hines et al. (1991) have also confirmed the usefulness of desiccants in removing vapors that can degrade indoor air quality. Desiccant materials are capable of adsorbing hydrocarbon vapors at the same time they are collecting moisture from air. These desiccant cosorption phenomena show promise of improving indoor air quality in typical building HVAC systems.

Desiccants are also used in drying compressed air to low dew points. In this application, moisture can be removed from the desiccant without heat. Desorption is accomplished using differences in vapor pressures compared to the total pressures of the compressed and ambient pressure airstreams.

Finally, desiccants are used to dry the refrigerant circulating in air-conditioning and refrigeration systems. This reduces corrosion in refrigerant piping and prevents valves and capillaries from becoming clogged with ice crystals. In this application, the desiccant is not regenerated; it is discarded when it has adsorbed its limit of water vapor.

This chapter discusses the water sorption characteristics of desiccant materials and explains some of the implications of those characteristics in ambient pressure air-conditioning applications. Information on other applications for desiccants can be found in Chapters 6, 26, 31, and 33 of the 1994 *ASHRAE Handbook—Refrigeration*, Chapters 11, 13 through 24, 27, and 41 of the 1995 *ASHRAE Handbook—Applications,* and Chapter 22 of the 1996 *ASHRAE Handbook—Systems and Equipment.*

DESICCANT CYCLE

Practically speaking, all desiccants function by the same mechanism—transferring moisture because of a difference between the water vapor pressure at their surface and that of the surrounding air. When the vapor pressure at the desiccant surface is lower than that of the air, the desiccant attracts moisture. When the surface vapor pressure is higher than that of the surrounding air, the desiccant releases moisture.

Figure 1 shows the relationship between the moisture content of the desiccant and its surface vapor pressure. As the moisture content of the desiccant rises, so does the water vapor pressure at its surface. At some point, the vapor pressure at the desiccant surface is the same as that of the air—the two are in equilibrium. Then moisture

---

The preparation of this chapter is assigned to TC 3.5, Desiccant and Sorption Technology.

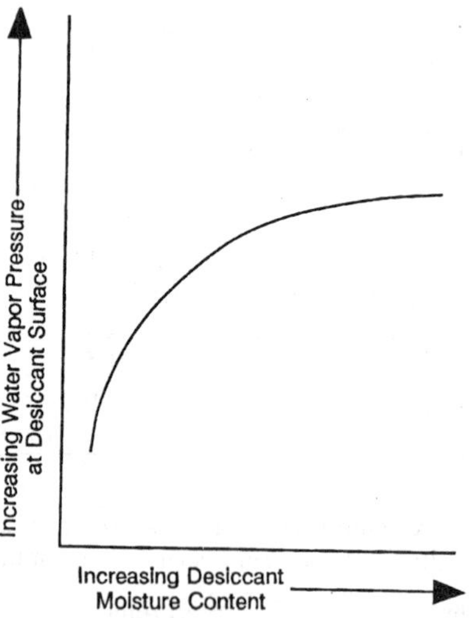

Fig. 1 Desiccant Water Vapor Pressure as Function of Moisture Content
(Harriman 1990)

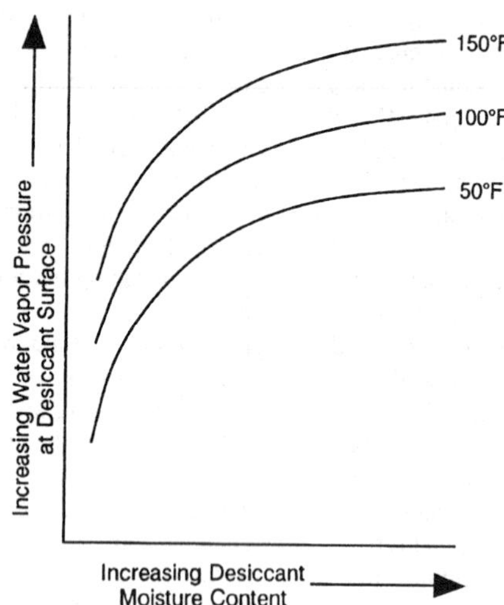

Fig. 2 Desiccant Water Vapor Pressure as Function of Desiccant Moisture Content and Temperature
(Harriman 1990)

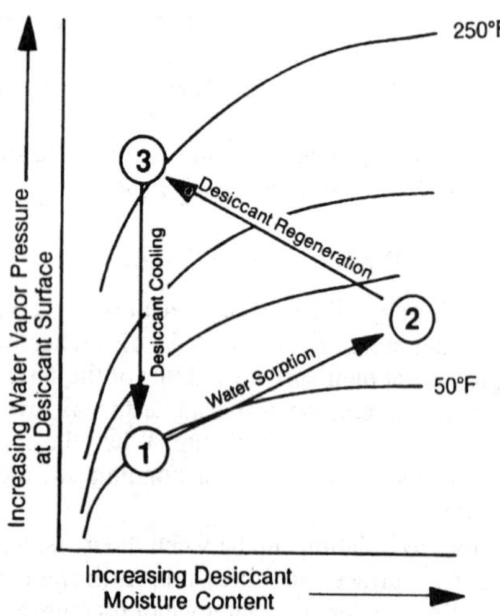

Fig. 3 Desiccant Cycle
(Harriman 1990)

cannot move in either direction until some external force changes the vapor pressure at the desiccant or in the air.

Figure 2 shows the effect of temperature on the vapor pressure at the desiccant. Both higher temperature and increased moisture content increase the vapor pressure at the surface. When the surface vapor pressure exceeds that of the surrounding air, moisture leaves the desiccant—a process called reactivation or regeneration. After the desiccant is dried (reactivated) by the heat, its vapor pressure remains high, so that it has very little ability to absorb moisture. Cooling the desiccant reduces its surface vapor pressure so that it can absorb moisture once again. The complete cycle is illustrated in Figure 3.

The economics of desiccant operation depend on the energy cost of moving a given material through this cycle. The dehumidification of air (loading the desiccant with water vapor) generally proceeds without energy input other than fan and pump costs. The major portion of energy is invested in regenerating the desiccant (moving from point 2 to point 3) and cooling the desiccant (point 3 to point 1).

Regeneration energy is equal to the sum of three variables:

1. The heat necessary to raise the desiccant to a temperature high enough to make its surface vapor pressure higher than that of the surrounding air
2. The heat necessary to vaporize the moisture it contains (about 1060 Btu/lb)
3. The small amount of heat from desorption of the water from the desiccant

The cooling energy is proportional to (1) the mass of the desiccant and (2) the difference between its temperature after regeneration and the lower temperature that allows the desiccant to remove water from the airstream once again.

The cycle is similar when desiccants are regenerated using pressure differences in a compressed air application. The desiccant is saturated in a high-pressure chamber (i.e., that of the compressed air). Then valves open, isolating the compressed air from the material, and the desiccant is exposed to air at ambient pressure. The vapor pressure of the saturated desiccant is much higher than ambient air at normal pressures; thus the moisture leaves the desiccant for the surrounding air. An alternate desorption strategy uses a small portion of the dried air, returning it to the moist desiccant bed to reabsorb the moisture, then venting the air to the atmosphere at ambient pressures.

Table 1 shows the range of vapor pressures over which the desiccant must operate in space-conditioning applications. It converts the relative humidity at 70°F to dew point and the corresponding vapor pressure. The greater the difference between the air and desiccant surface vapor pressures, the greater the ability of the material to absorb moisture from the air at that moisture content.

Table 1 Vapor Pressures of Different Relative Humidities at 70°F

Relative Humidity at 70°F, %	Dew Point, °F	Vapor Pressure, in. Hg
10	12	0.07
20	28	0.15
30	37	0.22
40	45	0.30
50	51	0.37
60	55	0.44
70	60	0.52
80	64	0.59
90	67	0.67
100	70	0.74

The ideal desiccant for a particular application depends on the range of water vapor pressures likely to occur in the air, the temperature level of the regeneration heat source, and the moisture sorption and desorption characteristics of the desiccant within those constraints. In commercial practice, however, most desiccants can be made to perform well in a wide variety of operating situations through careful engineering of the mechanical aspects of the dehumidification system. Some of these hardware issues are discussed in Chapter 22 of the 1996 *ASHRAE Handbook—Systems and Equipment*.

TYPES OF DESICCANTS

Desiccants can be liquids or solids and can hold moisture through absorption or adsorption, as described earlier. Most absorbents are liquids, and most adsorbents are solids.

Liquid Absorbents

Liquid absorption dehumidification can best be illustrated by comparison to the operation of an air washer. When air passes through an air washer, its dew point approaches the temperature of the water supplied to the machine. Air that is more humid is dehumidified and air that is less humid is humidified. In a similar manner, a liquid absorption dehumidifier brings air into contact with a liquid desiccant solution. The liquid has a vapor pressure lower than water at the same temperature, and the air passing over the solution approaches this reduced vapor pressure; it is dehumidified.

The vapor pressure of a liquid absorption solution is directly proportional to its temperature and inversely proportional to its concentration. Figure 4 illustrates the effect of increasing desiccant concentration on the water vapor pressure at its surface. The figure shows the vapor pressure of various solutions of water and triethylene glycol, a common commercial desiccant. As the glycol content of the mixture increases, the vapor pressure of the mixture decreases. This pressure difference allows the glycol solution to absorb moisture from the air whenever the vapor pressure of the air is greater than that of the solution.

From a slightly different perspective, the vapor pressure of a given concentration of absorbent solution approximates the vapor pressure values of a fixed relative humidity line on a psychrometric chart. Higher solution concentrations give lower equilibrium relative humidities, which allow the absorbent to dry air to lower levels.

Figure 5 illustrates the effect of temperature on the vapor pressure of lithium chloride (LiCl), another liquid desiccant in common use. A solution that is 25% lithium chloride has a vapor pressure of 0.37 in. Hg at a temperature of 70°F. If the same 25% solution is heated to 100°F, its vapor pressure more than doubles to 0.99 in. Hg. This can be expressed another way, as the 70°F, 25% solution is in equilibrium with air at a 51°F dew point. The same 25% solution at 100°F is at equilibrium with an airstream at a 79°F dew point. The warmer the desiccant, the less moisture it can attract from the air.

In standard practice, the behavior of a liquid desiccant is controlled by adjusting its temperature, its concentration, or both. Desiccant temperature is controlled by simple heaters and coolers. Concentration is controlled by heating the desiccant to drive moisture out into a waste airstream or directly to the ambient.

Commercially available liquid desiccants have an especially high water-holding capacity. Each molecule of lithium chloride, for example, can hold two water molecules, even in the dry state. Above two water molecules per molecule of LiCl, the desiccant becomes a liquid and continues to absorb water. If the solution is in equilibrium with air at 90% rh, approximately 26 water molecules are attached to each molecule of LiCl. This represents a water absorption of more than 1000% on a dry weight basis.

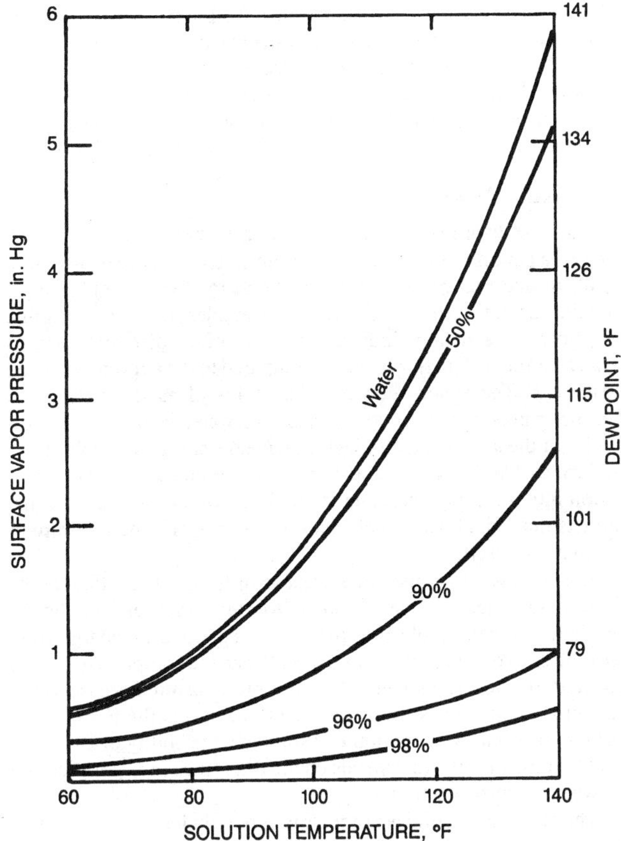

Fig. 4 Surface Vapor Pressure of Water-Triethylene Glycol Solutions
(Dow 1981)

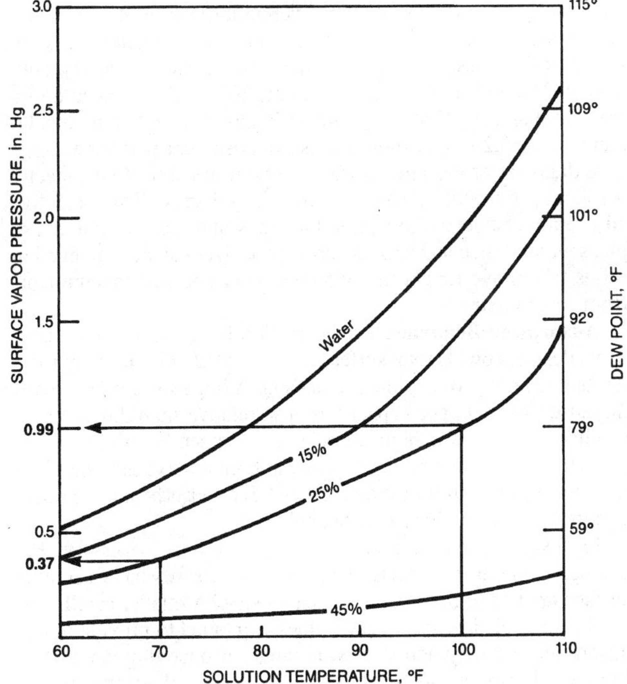

Fig. 5 Surface Vapor Pressure of Water-Lithium Chloride Solutions
(Foote Mineral 1988)

As a practical matter, however, the absorption process is limited by the exposed surface area of the desiccant and by the contact time allowed for the reaction. More surface area and more contact time allow the desiccant to approach its theoretical capacity. Commercial desiccant systems stretch these limits by spraying the desiccant onto an extended surface much like in a cooling tower.

Solid Adsorbents

Adsorbents are solid materials with a tremendous internal surface area per unit of mass; a single gram can have more than 50,000 ft^2 of surface area. Structurally, adsorbents resemble a rigid sponge, and the surface of the sponge in turn resembles the ocean coastline of a fjord. This analogy indicates the scale of the different surfaces in an adsorbent. The fjords can be compared to the **capillaries** in the adsorbent. The spaces between the grains of sand on the fjord beaches can be compared to the spaces between the individual molecules of the adsorbent, all of which have the capacity to hold water molecules. The bulk of the adsorbed water is contained by condensation into the capillaries, and the majority of the surface area that attracts individual water molecules is in the crystalline structure of the material itself.

Adsorbents attract moisture because of the electrical field at the desiccant surface. The field is not uniform in either force or charge, so it attracts water molecules that have a net opposite charge from specific sites on the desiccant surface. When the complete surface is covered, the adsorbent can hold still more moisture because vapor condenses into the first water layer and fills the capillaries throughout the material. As with liquid absorbents, the ability of an adsorbent to attract moisture depends on the difference in vapor pressure between its surface and the air.

The capacity of solid adsorbents is generally less than the capacity of liquid absorbents. For example, a typical molecular sieve adsorbent will hold 17% of its dry weight in water when the air is at 70°F and 20% rh. In contrast, lithium chloride can hold 130% of its mass at the same temperature and relative humidity. But solid adsorbents have several other favorable characteristics.

For example, molecular sieves continue to adsorb moisture even when they are quite hot, allowing dehumidification of very warm airstreams. Also, several solid adsorbents can be manufactured to precise tolerances, with pore diameters that can be closely controlled. This means they can be tailored to adsorb a molecule of a specific diameter. Water, for example, has an effective molecular diameter of 3.2 nm. A molecular sieve adsorbent with an average pore diameter of 4.0 nm adsorbs water but has almost no capacity for larger molecules, such as organic solvents. This selective adsorption characteristic is useful in many applications. For example, several desiccants with different pore sizes can be combined in series to remove first water and then other specific contaminants from an airstream.

Adsorption Behavior. The adsorption behavior of solid adsorbents depends on (1) total surface area, (2) total volume of capillaries, and (3) range of capillary diameters. A large surface area gives the adsorbent a larger capacity at low relative humidities. Large capillaries provide a high capacity for condensed water, which gives the adsorbent a higher capacity at high relative humidities. A narrow range of capillary diameters makes an adsorbent more selective in the vapor molecules it can hold.

In designing a desiccant, some trade-offs are necessary. For example, materials with large capillaries necessarily have a smaller surface area per unit of volume than those with smaller capillaries. As a result, adsorbents are sometimes combined to provide a high adsorption capacity across a wide range of operating conditions. Figure 6 illustrates this point using three silica gel adsorbents prepared for use in laboratory research. Each has a different internal structure, but since they are all silicas, they have similar surface adsorption characteristics. Gel 1 has large capillaries, making its total volume large, but its total surface area is small. It has a large adsorption capacity at high relative humidities but adsorbs a small amount at low relative humidities.

In contrast, Gel 8 has a capillary volume one-seventh the size of Gel 1, but a total surface area almost twice as large. This gives it a higher capacity at low relative humidities but a lower capacity to hold the moisture that condenses at high relative humidities.

The table in Figure 6 illustrates the wide range of performance characteristics possible within a single class of adsorbent. The table shows three noncommercial silica gels. Silica gels and most other adsorbents can be manufactured to provide optimum performance in a specific application, balancing capacity against strength, weight, and other favorable characteristics (Bry-Air 1986).

Types of Solid Adsorbents. General classes of solid adsorbents include

- Silica gels
- Zeolites
- Synthetic zeolites (molecular sieves)
- Activated aluminas
- Carbons
- Synthetic polymers

Silica gels are amorphous solid structures formed by condensing soluble silicates from solutions of water or other solvents. They have the advantages of a relatively low cost and relative simplicity of structural customizing. They are available as large as spherical beads about 3/16 in. in diameter or as small as grains of a fine powder.

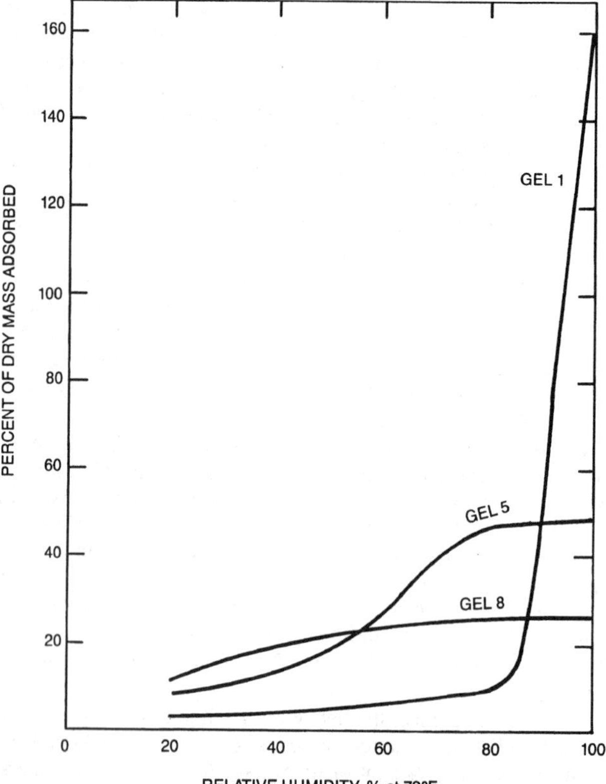

Gel Number	Total Surface Area, m^3/g	Average Capillary Diameter, nm	Total Volume of Capillaries, m^3/g	Total Volume of Capillaries, m^3/g
1	315	21	21	1700
5	575	3.8	3.8	490
8	540	2.2	2.2	250

Fig. 6 Adsorption and Structural Characteristics of Some Experimental Silica Gels
(Oscic and Cooper 1982)

Zeolites are aluminosilicate minerals. They occur in nature and are mined rather than synthesized. Zeolites have a very open crystalline lattice that allows molecules like water vapor to be held inside the crystal itself like an object in a cage. Particular atoms of an aluminosilicate determine the size of the openings between the "bars" of the cage, which in turn governs the maximum size of the molecule that can be adsorbed into the structure.

Synthetic zeolites, also called **molecular sieves**, are crystalline aluminosilicates manufactured in a thermal process. Controlling the temperature of the process and the composition of the ingredient materials allows close control of the structure and surface characteristics of the adsorbent. At a somewhat higher cost, this provides a much more uniform product than naturally occurring zeolites.

Activated aluminas are oxides and hydrides of aluminum that are also manufactured in thermal processes. Their structural characteristics can be controlled by the gases used to produce them and by the temperature and duration of the thermal process.

Carbons are most frequently used for adsorption of gases other than water vapor because they often have a greater affinity for the nonpolar molecules typical of organic solvents. Like other adsorbents, carbons have a large internal surface and especially large capillaries. This capillary volume gives them a high capacity to adsorb water vapor at relative humidities of 45 to 100%.

Synthetic polymers have potential for use as desiccants as well. Long molecules, like those found in polystyrenesulfonic acid sodium salt (PSSASS), are twisted together like the strands of string. Each of the many sodium ions in the long PSSASS molecules has the potential to bind several water molecules, and the spaces between the packed strings can also contain condensed water, giving the polymer a capacity exceeding that of many other solid adsorbents.

DESICCANT ISOTHERMS

Figure 7 shows a rough comparison of the sorption characteristics of different desiccants. Large variations from these isotherms occur because manufacturers use different methods to optimize the materials for different applications. The suitability of a given desiccant to a particular application is generally governed as much by the engineering of the mechanical system that presents the material to the airstreams as by the characteristics of the material itself.

Several sources give details of desiccant equipment design and information about desiccant isotherm characteristics. Brunauer (1945) considers five basic isotherm shape types. Each isotherm shape is determined by the dominant sorption mechanisms of the desiccant, which give rise to its specific capacity characteristics at different vapor pressures. Isotherm shape can be important in designing the optimum desiccant for applications where a narrow range of operating conditions can be expected. Collier (1986, 1988) illustrates how an optimum isotherm shape can be used to ensure a maximum coefficient of performance in one particular air-conditioning desiccant application.

DESICCANT LIFE

The useful life of desiccant materials depends largely on the quantity and type of contamination in the airstreams they dry. In commercial equipment, desiccants last from 10,000 to 100,000 h and longer before they need replacement. Normally, two mechanisms cause the loss of desiccant capacity: (1) change in desiccant sorption characteristics through reactions with contaminants and (2) loss of effective surface area through clogging or **hydrothermal degradation**.

Liquid absorbents are more susceptible to chemical reaction with airstream contaminants other than water vapor than are solid adsorbents. For example, certain sulfur compounds can react with lithium chloride to form lithium sulfate, which is not a desiccant. If the concentration of sulfur compounds in the airstream were below 10 ppm and the desiccant were in use 24 h a day, the capacity

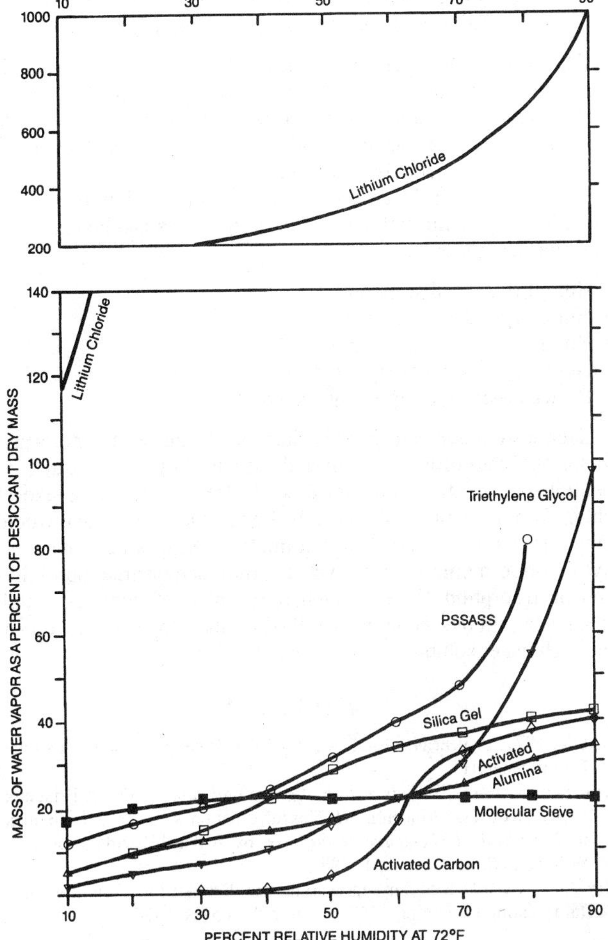

The sources for isotherms presented in the figure include
PSSASS: National Renewable Energy Laboratory *Report* No. PR-255-3308
Lithium chloride: Munters Corporation—Cargocaire Division and Kathabar, Inc.
Triethylene glycol: Dow Chemical Corporation
Silica gel: Davison Chemical Division of W.R. Grace Co.
Activated carbon: Calgon Corporation
Activated alumina: LaRoche Industries Inc.
Molecular sieve: Davison Chemical Division of W.R. Grace Co.

Fig. 7 Sorption Isotherms of Various Desiccants at 72°F

reduction would be approximately 10 to 20% after 3 years of operation. If the concentration were 30 ppm, this reduction would occur after 1 year.

Adsorbents tend to be less chemically reactive and more sensitive to clogging, a function of the type and quantity of particulate material in the airstream. In some situations, certain adsorbents are sensitive to hydrothermal stress due to the thermal expansion and contraction of the desiccant material on rapid changes in desiccant moisture content. For example, silica gel that must move between an airstream above 95% rh at low temperatures and a reactivating airstream at high temperatures six times per hour, 24 h a day can partly fracture; this may mean a 10% reduction in capacity over the course of a year. For applications where such capacity reduction is undesirable, thermally stabilized desiccants are used in place of more sensitive materials.

In air-conditioning applications, desiccant equipment is designed to minimize the need for desiccant replacement in much the same way that vapor compression cooling systems are designed to avoid the need for compressor replacement. Unlike filters, desiccants are seldom intended to be frequently replaced during normal service in an air-drying application.

COSORPTION OF WATER VAPOR AND INDOOR AIR CONTAMINANTS

Hines et al. (1991) have confirmed that many desiccant materials can collect common indoor pollutants at the same time they collect water vapor from ambient air. This characteristic promises to become useful in future air-conditioning systems where the quality of indoor air is especially important.

The behavior of different desiccant and vapor mixtures is complex, but in general, pollutant sorption reactions can be classified into five categories:

- Humidity-neutral sorption
- Humidity-reduced sorption
- Humidity-enhanced sorption
- Humidity-pollutant displacement
- Desiccant-catalyzed pollutant conversion

Humidity-reduced sorption is illustrated by the behavior of water vapor and chloroform on activated carbon. Sorption is humidity-neutral until relative humidity exceeds 45%, when the uptake of chloroform is reduced. The adsorbed water blocks sites that would otherwise attract and hold chloroform. In contrast, water and carbonyl chloride mixtures on activated carbon demonstrate humidity-enhanced sorption. Here, sorption of the pollutant increases at high relative humidities. Hines et al. (1991) attribute this phenomenon to the high water solubility of carbonyl chloride.

REFERENCES

Batelle. 1971. Project No. N-0914-5200-1971. Batelle Memorial Institute, Columbus, OH.

Brunauer, S. 1945. *The adsorption of gases and vapors*, Vol. I. Princeton University Press, Princeton, NJ. This information is quoted and expanded in *The physical chemistry of surfaces*, by Arthur W. Adamson. John Wiley and Sons, New York, 1982.

Bry-Air. 1986. MVB Series engineering data. Bry-Air Inc., Sunbury, OH.

Buffalo Testing Laboratory. 1974. *Report* No. 65711-1974.

Collier, R.K. Advanced desiccant materials assessment. *Research Report* 5084-243-1089. Phase I-1986, Phase II-1988. Gas Research Institute, Chicago.

Czanderna, A.W. 1988. Polymers as advanced materials for desiccant applications. *Research Report* NREL/PR-255-3308. National Renewable Energy Laboratory, Golden, CO.

Dow. 1981. Guide to glycols. Dow Chemical Corporation, Organic Chemicals Division, Midland, MI.

Foote Mineral. 1988. Lithium chloride technical data. *Bulletin* 151. Foote Mineral Corporation, Exton, PA.

Harriman, L.G., III. 1990. *The dehumidification handbook*, 2nd ed. Munters Cargocaire, Amesbury, MA.

Hines, A.J., T.K. Ghosh, S.K. Loyalka, and R.C. Warder, Jr. 1991. Investigation of co-sorption of gases and vapors as a means to enhance indoor air quality. ASHRAE *Research Project* 475-RP and Gas Research Institute *Project* GRI-90/0194. Gas Research Institute, Chicago.

Oscic, J. and I.L. Cooper. 1982. *Adsorption*. John Wiley and Sons, New York.

SUNY Buffalo School of Medicine. Effects of glycol solution on microbiological growth. Niagrara Blower *Report* No. 03188.

BIBLIOGRAPHY

Adamson, A.W. 1982. *The physical chemistry of surfaces*. John Wiley and Sons, New York.

Falcone, J.S., Jr., ed. 1982. Soluble silicates. *Symposium Series* 194. American Chemical Society, Washington, D.C.

Gas conditioning factbook. 1962. Dow Chemical Company, Midland, MI.

Iler, R.K. *The chemistry of silica*. 1979. John Wiley and Sons, New York.

King, C.J. 1980. *Separation processes*, 2nd ed. McGraw-Hill, New York.

McCabe, W.L. and J.C. Smith. 1978. *Unit operations of chemical engineering*. McGraw-Hill, New York.

Perry, R.H. and C.H. Chilton. 1973. *Chemical engineers' handbook*, 5th ed. McGraw-Hill, New York.

Ruthven, D.M. 1984. *Principles of adsorption and adsorption processes*. John Wiley and Sons, New York.

Strauss, W. 1975. *Industrial gas cleaning*. Pergamon Press, New York.

Valenzuela, D. and A. Myers. 1989. *Adsorption equilibrium data handbook*. Simon and Schuster/Prentice-Hall, Englewood Cliffs, NJ.

CHAPTER 22

THERMAL AND MOISTURE CONTROL IN INSULATED ASSEMBLIES—FUNDAMENTALS

PROPER design of space heating, air-conditioning, refrigeration, and other industrial systems requires knowledge of thermal insulations and thermal behavior of building structures. This chapter deals with heat and moisture transfer definitions, fundamentals and properties of thermal insulation materials, heat flow calculations, economic thickness of insulation, and the fundamentals of moisture as it relates to building components and systems.

TERMINOLOGY AND SYMBOLS

The following heat and moisture transfer definitions and symbols are commonly used in the building industry. The following I-P units are used in the definitions: British thermal units (Btu), hour (h), temperature (°F), absolute temperature (°R), temperature difference (°F), feet (ft), inches (in.), pound (lb), grains (gr), and inches of mercury (in. Hg).

Thermal transmission, heat transfer, or rate of heat flow. The flow of heat energy induced by a temperature difference. Heat may be transferred by conduction, convection, radiation, and mass transfer. These can occur separately or in combinations, depending on specific circumstances.

Thermal conductivity, *k*. The time rate of heat flow through a unit area of homogeneous material in a direction perpendicular to isothermal planes, induced by a unit temperature gradient. (ASTM *Standard* C 168 defines homogeneity.) Units are Btu·in/h·ft^2·°F or Btu/h·ft·°F. Thermal conductivity must be evaluated for a specific mean temperature, because in most materials it varies with temperature.

For porous materials, heat flows by a combination of modes and may depend on orientation, direction, or both. The measured property of such materials may be called *effective* or *apparent* thermal conductivity. The specific test conditions (i.e., sample thickness, orientation, environment, environmental pressure, surface temperature, mean temperature, and temperature difference) should be reported with the values. With thermal conductivity, the symbol k_{app} is used to note the lack of pure conduction or to indicate that all values reported are apparent.

Thermal resistivity, R_u. The reciprocal of thermal conductivity. Units are °F·ft^2·h/Btu·in.

Thermal conductance, C-factor, *C*. Time rate of heat flow through a unit area of a body induced by a unit temperature difference between the body surfaces. Units are Btu/h·ft^2·°F.

When the two defined surfaces of mass-type (i.e., nonreflective) thermal insulation have unequal areas, as in the case of radial heat flow through a curved block or through a pipe covering (see Table 2 in Chapter 3) or through materials of nonuniform thickness, an appropriate mean area and mean thickness must be given. Heat flow formulas involving materials that are not uniform slabs must contain *shape factors* to account for the area variation involved.

When heat flow is by conduction alone, thermal conductance of a material may be obtained by dividing the thermal conductivity of the material by its thickness. When several modes of heat transfer are involved, the *apparent* or *effective thermal conductance* may be obtained by dividing the apparent thermal conductivity by the thickness.

Where air circulates within or passes through insulation, as it may with low-density fibrous materials, the effective thermal conductance is affected.

Thermal conductances and thermal resistances of the more common building materials and industrial insulations are tabulated in Table 4 in Chapter 24.

Heat transfer film coefficient (or surface coefficient of heat transfer or surface film conductance), *h* or *f*. Heat transferred between a surface and a fluid in unit time through unit area induced by unit temperature difference between the surface and the fluid in contact with it, Btu/h·ft^2·°F.

Surface film resistance. The reciprocal of the heat transfer film coefficient, °F·ft^2·h/Btu.

The surrounding space must be air or other fluids for convection to take place. If the space is evacuated, the only heat flow is by radiation.

Note that subscripts *i* and *o* often denote inside and outside surface resistances and conductances, respectively.

Thermal resistance R-value, *R*. Under steady conditions, the mean temperature difference between two defined surfaces of material or construction that induces unit heat flow through a unit area, °F·ft^2·h/Btu.

Thermal transmittance, U-factor, *U*. The time rate of heat flow per unit area under steady conditions from the fluid on the warm side of a barrier to the fluid on the cold side, per unit temperature difference between the two fluids. It is determined by first evaluating the R-value and then computing its reciprocal, *U*, in Btu/h·ft^2·°F.

The U-factor is sometimes called the *overall coefficient of heat transfer*.

In building practice, the heat transfer fluid is air. The temperature of the fluid is obtained by averaging its temperature over a finite region near the surface involved.

The preparation of this chapter is assigned to TC 4.4, Thermal Insulation and Moisture Retarders.

Thermal emittance, ε. The ratio of the radiant flux emitted by a body to that which would be emitted by a blackbody at the same temperature and under the same conditions.

Effective emittance E is the combined effect of emittances from the boundary surfaces of an air space, where the boundaries are parallel and of a dimension much larger than the distance between them. Table 2 in Chapter 24 lists values of *E* for various air spaces.

Surface reflectance, ρ. The fraction of the radiant flux falling on a surface that is reflected by it.

Water vapor permeance, *M*. The rate of water vapor transmission by diffusion per unit area of a body between two specified parallel surfaces, induced by unit vapor pressure difference between the two surfaces, gr/h·ft^2·in. Hg or perm.

Water vapor permeability, μ. The rate of water vapor transmission by diffusion per unit area of flat material of unit thickness induced by unit vapor pressure difference between two surfaces, under specified temperature and humidity conditions. When permeability varies with psychrometric conditions, the spot or specific permeability defines the property at a specific condition in gr·in/h·ft^2·in. Hg or perm·in., where the vapor pressure difference is in. Hg and 1 grain = 1/7000 lb.

Water vapor resistance, Z. The reciprocal of permeance—signifies a resistance to moisture flow, in. Hg·ft^2·h/gr. A *rep* is a unit of resistance to water vapor flow, where rep = 1/perm.

THERMAL INSULATION

Thermal insulations are materials or combinations of materials that, when properly applied, retard the flow of heat energy by conductive, convective, and/or radiative transfer modes. Thermal insulations can be fibrous, particulate, film or sheet, block or monolithic, open-cell or closed-cell, or composites of these materials that can be chemically or mechanically bound or supported.

By retarding heat flow, thermal insulations can serve one or more of the following thermal functions:

1. Conserve energy by reducing heat loss or gain of piping, ducts, vessels, equipment, and structures
2. Control surface temperatures of equipment and structures for personnel protection and comfort
3. Help control the temperature of a chemical process, a piece of equipment, or a structure
4. Prevent vapor condensation on surfaces with a temperature below the dew point of the surrounding atmosphere
5. Reduce temperature fluctuations within an enclosure when heating or cooling is not needed or available
6. Reduce temperature variations within a conditioned space for increased personal comfort
7. Provide fire protection

Thermal insulation can serve additional functions, although such secondary functions should be consistent with its capabilities and primary purpose. Under certain conditions, insulations may

1. Add structural strength to a wall, ceiling, or floor section
2. Provide support for a surface finish
3. Impede water vapor transmission and air infiltration
4. Prevent or reduce damage to equipment and structures from exposure to fire and freezing conditions
5. Reduce noise and vibration
6. Reduce growth of mold and mildew

Thermal insulation is used to control heat flow at all temperatures, the limiting value being its survival temperature.

BASIC MATERIALS

Thermal insulations normally consist of the following basic materials and composites:

- Inorganic, fibrous, or cellular materials such as glass, rock, or slag wool; and calcium silicate, bonded perlite, vermiculite, and ceramic products. Asbestos insulations used to be applied, but asbestos has been shown to be a carcinogen. Extreme caution must be used if it is encountered.
- Organic fibrous materials such as cellulose, cotton, animal hair, wood, pulp, cane, or synthetic fibers, and organic cellular materials such as cork, foamed rubber, polystyrene, polyurethane, and other polymers.
- Metallic or metallized organic reflective membranes. These surfaces must face an air, gas-filled, or evacuated space to be effective.

PHYSICAL STRUCTURE AND FORM

Mass-type insulation can be cellular, granular, or fibrous solid material to retard heat flow. Reflective insulation consists of smooth-surfaced sheets of metal foil or foil-surfaced material separated by air spaces.

The physical forms of industrial and building insulations include the following:

Loose-fill insulations consist of fibers, powders, granules, or nodules that are usually poured or blown into walls or other spaces.

Insulating cement is a loose material that is mixed with water or a suitable binder to obtain plasticity and adhesion. It is troweled or blown wet on a surface and dried in place. Both loose-fill and insulating cement are suited for covering irregular spaces.

Flexible and semirigid insulations consist of organic and inorganic materials with and without binders and with varying degrees of compressibility and flexibility. These insulations are generally available as blanket, batt, or felt insulation, and in either sheets or rolls. Coverings and facings may be fastened to one or both sides and serve as reinforcing, air or vapor retarders or both, reflective surfaces, or surface finishes. These coverings include combinations of laminated foil, glass, cloth or plastics and paper, wire mesh, or metal lath. Although standard sizes are generally used, thickness and shape of insulation can be any dimension handled conveniently.

Rigid materials are available in rectangular blocks, boards, or sheets, which are preformed during manufacture to standard lengths, widths, and thicknesses. Insulation for pipes and curved surfaces is supplied in sections or segments, with radii of curvature available to suit all standard sizes of pipe and tubing.

Reflective materials are available in sheets and rolls of single layer or multilayer construction and in preformed shapes with integral air spaces.

Formed-in-place insulations are available as liquid components or expandable pellets that can be poured, frothed, or sprayed in place to form rigid or semirigid foam insulation. Fibrous materials mixed with liquid binders can also be sprayed in place, and in some products, the binder is also a foam.

Accessory materials for thermal insulation include mechanical and adhesive fasteners, exterior and interior finishes, vapor and air retarder coatings, jackets and weather coatings, sealants, lagging adhesives, membranes, and flashing compounds. ASTM *Standard* C 168 defines terms related to thermal insulating materials.

PROPERTIES

Thermal insulation selection may involve secondary criteria in addition to the primary property of low thermal conductivity. Characteristics such as resiliency or rigidity, acoustical energy absorption, water vapor permeability, airflow resistance, fire hazard and fire resistance, ease of application, applied cost, health and safety aspects, or other parameters may influence the choice among materials that have almost equal thermal performance values.

Thermal Properties

Thermal resistance is a measure of the effectiveness of thermal insulation to retard heat flow. A material with a high thermal resistance (low thermal conductance) is an effective insulator.

Heat transmission in most thermal insulations is accomplished by a combination of gas and solid conduction, radiation, and convection. Heat transfer through materials or systems is controlled by factors such as length of heat flow paths, temperature, temperature difference characteristics of the system, and environmental conditions.

Although heat transmission characteristics are usually determined by measuring thermal conductivity, this property does not strictly apply to thermal insulation. A particular sample of a material has a unique value of thermal conductivity for a particular set of conditions. This value may not be representative of the material at other conditions and should be called **apparent thermal conductivity**. For details, refer to ASTM *Standards* C 168, C 177, C 236, C 335, C 518, C 976, and C 1045.

Reflective insulations impede radiant heat transfer because the surfaces have high reflectance and low emittance values. (Table 1 and Table 2 in Chapter 24 give typical design values.) To be effective, the reflective face of both single and multiple layer reflective insulations must face an air or evacuated space. Multiple layers of reflective materials and smooth and parallel sealed air spaces increase overall thermal resistance. Air exchange and movement must be inhibited.

Mass-type insulation can be combined with reflective surfaces and air spaces to obtain a higher thermal resistance. However, each design must be evaluated because maximum thermal performance of these systems depends on such factors as condition of the insulation, shape and form of the construction, the means to avoid air leakage and movement, and the condition of the installed reflective surfaces and their aging characteristics.

Design values of effective or apparent thermal conductivity, thermal conductance, and thermal resistance for the most common insulations are listed in Table 4 in Chapter 24. These values have been selected as typical and useful for engineering calculations. The manufacturer or test results of the insulation under appropriate conditions can give values for a particular insulation.

Other thermal properties that can be important are specific heat, heat capacity, thermal diffusivity, the coefficient of thermal expansion, and the maximum temperature limit. **Heat capacity** is the product of specific heat and mass. **Thermal diffusivity** becomes important for applications where temperature varies with time, since the rate of temperature change within an insulation is proportional to its thermal diffusivity for a given thickness. Chapter 3 covers symbols, definitions, and methods of calculation in steady-state heat transfer.

Mechanical Properties

Some insulations have sufficient structural strength for load bearing. They are used occasionally to support load-bearing roofs and floors, form self-supporting partitions, or stiffen structural panels. For such applications, one or more of the following properties of an insulation may be important: strength in compression, tension, shear, impact, flexure, and resistance to vibration. These mechanical properties vary with basic composition, density, cell size, fiber diameter and orientation, type and amount of binder (if any), and temperature and environmental conditioning.

Health and Safety

Most thermal insulations have good resistance to fire, vermin, rot, objectionable odors, and vapors; some are a potential risk to health and safety. These risks can result from direct exposure to these materials and accessories while they are being stored or transported, during or after installation, or as a result of intervening or indirect actions or events, such as aging, fire, or physical disturbance. The potential health and safety effects of thermal insulation can be considered in two categories: (1) those related to storage, handling, and installation operations and (2) those that occur after installation (such as aging). Potential hazards during manufacture are not discussed. Correct handling, installation, and precautionary measures can reduce or eliminate these problems.

Potential health effects range from temporary irritation to serious changes in body functions. The principal concerns are with insulation containing asbestos. Questions have also been raised about man-made fibers. To date, research is inconclusive as to their potential hazard; however, they can be very annoying in installation, and the use of proper safeguards is desirable. Potential traumatic injury can occur from direct contact with materials that are sharp, rough, have protrusions or abrasive surfaces, permit overheating, or transmit electrical energy.

Combustion of insulation materials and accessories may release heat, hazardous gases, fibers, and particulates. Manufacturers' recommendations and applicable government codes and standards (ASTM *Standard* C 930) have more details.

Acoustics

Some thermal insulations are used as acoustical control materials, whether or not thermal performance is a design requirement. Acoustical efficiency depends on the physical structure of the material. Materials with open, porous surfaces have sound absorption capability. Those with high density and resilient characteristics can act as vibration insulators; either alone or in combination with other materials, some are effective barriers to sound transmission. Insulations for sound conditioning include flexible and semirigid, formed-in-place fibrous materials, and rigid fibrous insulation.

Sound absorption insulations are normally installed on interior surfaces or used as interior surfacing materials. Rigid insulations are fabricated into tile or blocks, edge-treated to facilitate mechanical or adhesive application, and prefinished during manufacture. Some insulation units have a natural porous surface, others include mechanical perforations to facilitate the entry of sound waves, and still others use a diaphragm or decorative film surfacing attached only to the edges of the units, which allows the sound waves to reach the fibrous backing by diaphragm action.

Flexible, semirigid, and formed-in-place fibrous materials used for sound absorption are available in a variety of thicknesses and densities that determine their sound absorption characteristics.

When density is increased by reducing the thickness of the material, sound absorption is generally reduced; however, as thickness increases, the influence of density decreases.

Thermal insulations improve sound transmission loss when installed in *discontinuous construction.* A wall of staggered stud construction that uses resilient clips or channels on one side of the stud or resilient insulation boards of special manufacture to prevent acoustical coupling mechanically between the surfaces reduces sound transmission. A sound absorption thermal insulation blanket in a wall cavity reduces sound transmission, depending on the type of construction.

In floor construction, resilient channels or separate floor and ceiling joists form a discontinuous construction; sound-absorbing thermal insulation placed within this construction further reduces sound transmission. Sound-deadening boards underlying finish flooring absorb impact sounds and improve the airtightness of the construction, thus reducing airborne sound transmission.

Thermal insulation boards can be placed under mechanical equipment to isolate vibration. The imposed loading and natural resonant frequency of materials are critical for proper design. Because material must deflect properly under load to provide isolation, the system should be neither overloaded nor underloaded.

For further information on sound and vibration control, refer to Chapter 43 of the 1995 *ASHRAE Handbook—Applications.*

Other Properties

Other properties of insulating materials that can be important, depending on the application, include density, resilience, resistance to settling, permanence, reuse or salvage value, ease of handling, dimensional uniformity and stability, resistance to chemical action and chemical change, resistance to moisture penetration, ease in fabrication, application of finishes, and sizes and thicknesses obtainable.

HEAT FLOW

FACTORS AFFECTING THERMAL PERFORMANCE

A wide variety of physical, environmental, application, and, in some cases, aging factors affect the thermal performance of insulations.

Thermal conductivity k is a property of a homogeneous material. Building materials, such as lumber, brick, and stone, are usually considered homogeneous. Most thermal insulation and many other building materials are porous and consist of combinations of solid matter with small voids.

For most insulating materials, conduction is not the only mode of heat transfer. Consequently, the term *apparent thermal conductivity* describes the heat flow properties of most materials. Some materials with low thermal conductivities are almost purely conductive (silica opacified aerogel, corkboard, etc.).

The apparent thermal conductivity of insulation varies with form and physical structure, environment, and application conditions. Form and physical structure vary with the basic material and manufacturing process. Typical variations include density, cell size, diameter and arrangement of fibers or particles, degree and extent of bonding materials, transparency to thermal radiation, and the type and pressure of gas within the insulation.

Environment and application conditions include mean temperature, temperature gradient, moisture content, air infiltration, orientation, and direction of heat flow. Thermal performance values for insulation materials and systems are usually obtained by standard methods listed in Volume 04.06 (*Thermal Insulation and Environmental Acoustics*) of the *Annual Book of ASTM Standards*. The methods apply mainly to laboratory measurements on dried or conditioned samples at specific mean temperatures and temperature gradient conditions. Although the fundamental heat transmission characteristics of a material or system can be determined accurately, actual performance in a structure may vary from that indicated in the laboratory due to application variations. Field measurement techniques continue to be developed. The design of the envelope, its construction, and the materials used may all affect the test procedure. These factors are detailed in ASTM STP 544, STP 660, STP 718, STP 789, STP 879, STP 885, STP 922, STP 1030, and STP 1116.

The effective thermal conductivity of some thermal insulation materials varies with density. Figure 1 illustrates this variation at one mean temperature for a number of materials currently used to insulate building envelopes. For most mass-type insulations, there is a minimum in the respective apparent thermal conductivity versus density. This minimum depends on the type and form of material, temperature, and direction of heat flow. For fibrous materials, the values of density at which the minimum value occurs increase as both the fiber diameter or cell size and the mean temperature increase. These effects are shown in Figure 2 (Lotz 1969) and Figure 3, respectively.

Other factors that affect thermal performance include compaction and settling of insulation, air permeability, type and amount of binder used, additives that may influence the bond or contact between fibers or particles, and, if used, the type and form of radiation transfer inhibitor. In cellular materials, the factors that influence thermal performance and strength properties are the same as

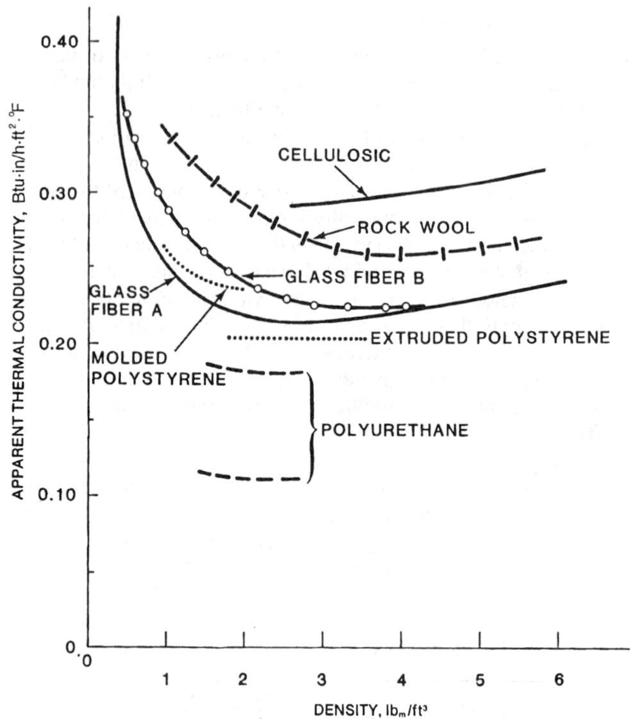

Fig. 1 Apparent Thermal Conductivity Versus Density of Several Thermal Insulations Used as Building Insulations

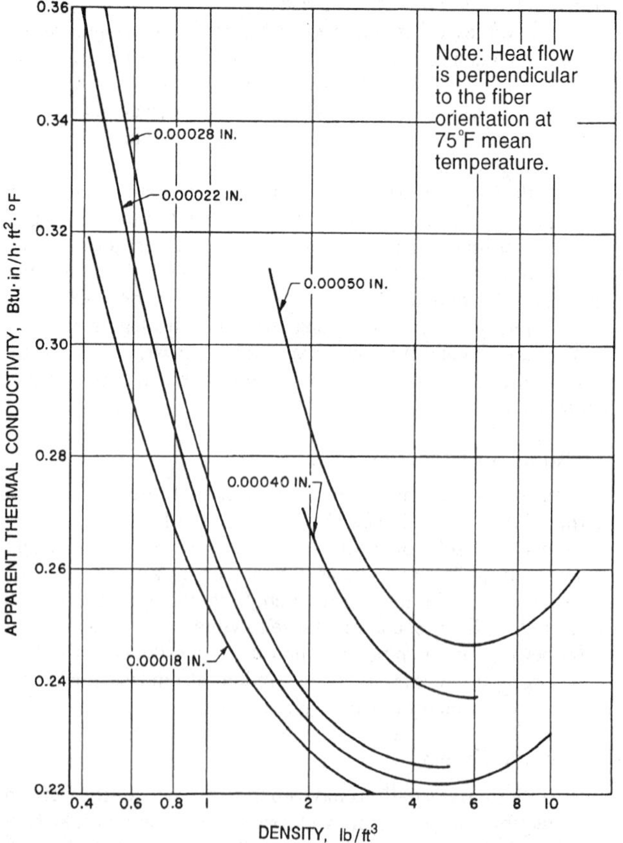

Fig. 2 Typical Variation of Apparent Thermal Conductivity with Fiber Diameter and Density

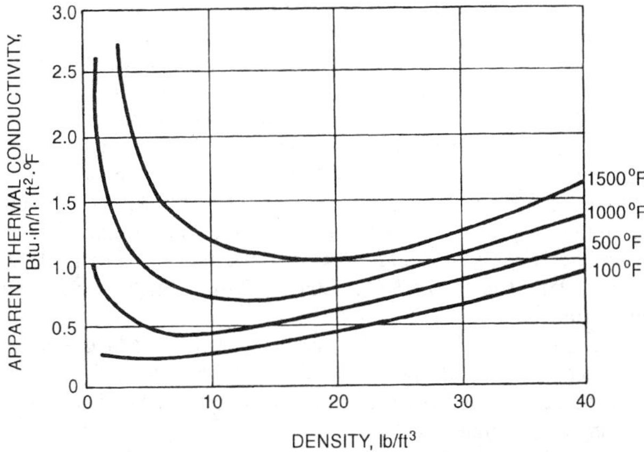

Fig. 3 Typical Variation of Apparent Thermal Conductivity with Mean Temperature and Density for Fibrous Insulations

those that control the thermal conductivity of the basic structured material: size and shape of the cells, thickness of the cell walls, gas contained in the cells, orientation of the cells, and radiation characteristics of the cell surfaces.

A change in density caused by the degree of compaction of insulation powders affects the effective thermal conductivity. Insulating concretes made from lightweight aggregates can be produced in a wide range of densities, with corresponding thermal conductances.

Fibrous insulations reach a minimum conductivity when fibers are uniformly spaced and perpendicular to the direction of heat flow. Generally, a decrease in the diameter of the fiber lowers the conductivity for the same density (Figure 2). For cellular insulation, a specific combination of cell size, density, and gas composition produces optimum thermal conductivity.

At temperatures below 400 to 570°F, a large portion of heat transfer across most insulations occurs by conduction through the air or other gas in the insulations (Rowley et al. 1952, Lander 1955, Simons 1955, Verschoor and Greebler 1952). The overall heat transfer can be closely approximated by supposition of gas conduction with other mechanisms of heat transfer, each determined separately. If the gas within the insulation is replaced by another gas with a different thermal conductivity, the apparent thermal conductivity of the insulation is changed by an amount approximately equal to the difference in conductivity of the two gases. For example, replacing air with a fluorinated hydrocarbon gas can lower the apparent thermal conductivity of the insulation by as much as 50%. Fluorocarbon-expanded cellular plastic foams with a high proportion (greater than 90%) of closed cells retain the fluorocarbon within the cells for extended periods. As these products are initially produced, they have apparent thermal conductivities of approximately 0.111 Btu·in/h·ft^2·°F at 75°F when the gas contained has a mean-free-path greater than the dimensions within the cells. However, this value increases with time as atmospheric gases diffuse into the cells and, over a long period of time, the fluorocarbon gas dissolves in the polymer or diffuses out. The rates of diffusion and increase in apparent thermal conductivity depend on several factors, including permeance of the cell walls to the gases involved, age of the foam, temperature, geometry of the insulation (thickness), and integrity of the surface protection provided. The significance of the surface protection is apparent when foams are encased in gas-impermeable membranes or some water vapor retarders (Brandreth 1986, Tye 1987). For estimating the long-term change in thermal resistance of unfaced rigid closed-cell plastic foams, a test method (ASTM *Standard* C 1303) has been developed that involves slicing and scaling under controlled laboratory conditions. Christian et al. (1995) provides an example of an application of this accelerated aging test method.

Brandreth (1986) and Tye (1988) show that the aging process of polyurethane and polyisocyanurate materials is reasonably well understood analytically and confirmed experimentally. The dominant parameters for minimum aging are as follows:

- Closed-cell content > 90%, preferably > 95%
- Small uniform cell diameter << 0.04 in., with a larger proportion of polymer in windows
- Small anisotropy in cell structure
- High density
- Increasing thickness
- High initial pressure of fluorocarbon blowing agent in cell
- Polymer highly resistant to gas diffusion and solubility
- Polymer distributed evenly in struts and windows of cells
- Low aging temperature

Aging may be further reduced, particularly for laminated and spray-applied products, with higher density polymer skins, or by well-adhered facings and coverings with low gas and moisture permeance characteristics. An oxygen diffusion rate of less than 0.02 in^3/1000 ft^2·day for 0.001 in. thickness of barrier is one criterion used by some industry organizations manufacturing laminated products. The adhesion of any facing must be continuous, and every effort must be made in the manufacturing process to eliminate or minimize the shear plane layer at the foam/substrate interface (Ostrogorsky and Glicksman 1986).

Closed-cell phenolic-type materials and products, which are blown with similar gases, age differently and much more slowly. The reasons for this are believed to be higher material density, smaller more uniform cell size with a larger proportion of polymer in windows, and a basic polymer more resistant to gas diffusion.

The average distance, or **mean free path**, that an enclosed gas molecule travels before striking another gas molecule increases as pressure within an insulation decreases. When the mean free path equals the average distance a gas molecule travels before striking a solid part of the insulation, apparent thermal conductivity of the insulation decreases with decreasing pressure. Correspondingly, for materials such as silica gel and fine carbon black, which have an average pore size smaller than the mean free path of air at atmospheric pressure, it is possible to attain thermal conductivity values lower than those for still air (Verschoor and Greebler 1952).

For homogeneous, dense materials, the primary mode of heat transfer is conduction. However, as the temperatures increase, the heat transmission by thermal radiation and possible convection becomes a greater part of the total heat transferred. The magnitude of radiation and convection transfer depends on temperature difference, direction of heat flow, the nature of materials involved, and geometric considerations. The rate of radiant heat transfer varies in proportion to the fourth power of the absolute temperature.

Because of radiation heat transfer in low-density insulation, measured apparent thermal conductivity depends on test thickness. The thickness effect increases the apparent thermal conductivity measured at installed thickness over that commonly determined at 1 in. (Pelanne 1979). From a thermal resistance standpoint, the effect is small, typically less than 10% for most 1 to 6 in. thick insulations.

Environmental and Application Conditions

The apparent conductivity of insulating materials generally increases with increasing temperature. Figure 4 shows typical variations with mean temperature. However, some materials, such as fluorocarbon-expanded, closed-cell urethanes, have an inflection in the curve over the temperature range where there is a change of phase of the fluorocarbon from gas to liquid (see Table 10 in Chapter 24).

The apparent thermal conductivity of a sample at one mean temperature (average of the two surface temperatures) only applies to the material at the particular thickness tested. Further testing is required to obtain values suitable for all thicknesses. The rate of

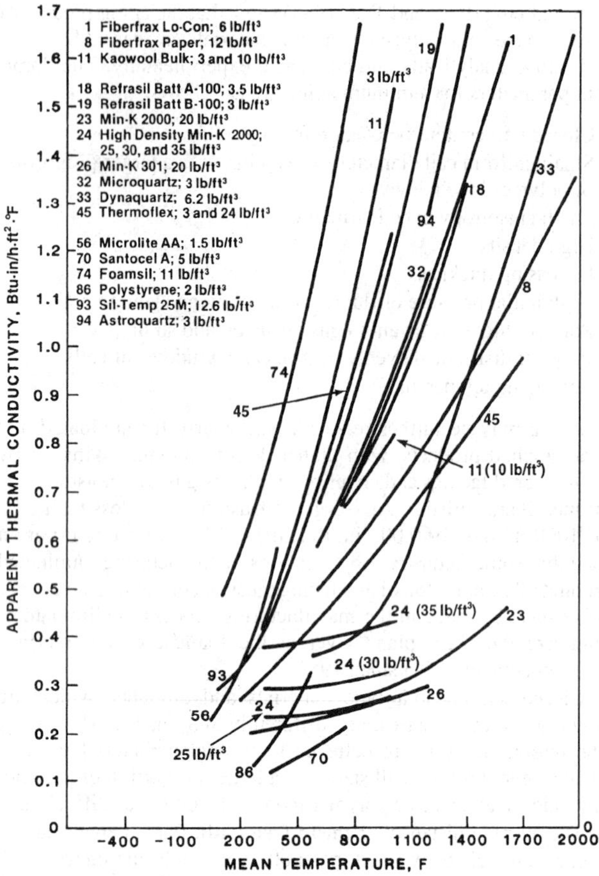

Fig. 4 Apparent Thermal Conductivity Versus Mean Temperature for Various Materials (in Air at Atmospheric Pressure)
(Glaser et al. 1967, Pelanne 1977)

change of apparent thermal conductivity with temperature and environmental conditions varies with the type and density of material.

Insulating materials that permit a large percentage of heat transfer by radiation, such as low-density fibrous and cellular products, show the greatest change in apparent thermal conductivity with changes of temperature and surrounding surface emittance. The ASTM Standard Test Methods recognize the importance of radiation in heat transmission and require that the surfaces of all plates be painted or otherwise treated to have a total emittance greater than 0.8 at operating temperature.

The effect of temperature alone on structural integrity is not ordinarily important for most materials in low-temperature insulation applications. Decomposition, excessive linear shrinkage, softening, or some other effects of temperature alone limit the maximum temperature for which a material is suited. At extreme temperatures, both high and low, selecting materials for a specific service becomes more critical and must be based on experience and actual performance data (see Table 4 and Table 10 in Chapter 24).

Convection and air infiltration in or through some insulation systems, may increase heat transfer across them. Low-density, loose-fill, large open-cell and fibrous insulations, and poorly designed or installed reflective systems are most susceptible to increased heat transfer by air filtration and convection. The temperature difference across an insulation, as well as the height, thickness, or width of the insulated space, influences the amount of convection. In some cases, natural convection may be inherent in the systems (Wilkes and Rucker 1983, Wilkes and Childs 1992); however, in many cases, the effect of convection in or through insulation can be minimized by careful design of an insulated structure (Donnelly et al. 1976).

Gaps between both board- and batt-type insulation can lower insulation effectiveness. Board-type insulation may not be perfectly square, may be installed improperly, and may be applied to uneven surfaces. Joints formed in board-type insulation allow it to fit together without air gaps. Boards and batts can be installed in two layers with joints between layers offset and staggered. For example, a 4% void area around batt insulation can produce a 50% loss in effective thermal resistance for an R-19 ceiling application (Verschoor 1977). Similar results have been obtained with different test conditions and for wall configurations (Lewis 1979, Hedlin 1985, Tye et al. 1981). Chapter 23 has further details on these effects.

If moisture condenses in the insulation, it may reduce thermal resistance, and, perhaps physically damage the system. The reduction in thermal resistance depends on the material, the moisture content, and its distribution.

More information on the effects of moisture are in the section on Effect of Moisture on Heat Flow. Section A3 of the CIBSE *Guide* (1986) and Chapter 23 cover the thermal properties of building structures affected by moisture.

THERMAL TRANSMITTANCE

The method of calculating an overall coefficient of heat transmission requires knowledge of (1) the apparent thermal conductivity and thickness of homogeneous components, (2) thermal conductance of nonhomogeneous components, (3) surface conductances of both sides of the construction, and (4) conductance of air spaces in the construction. Procedures for calculating thermal conductance and resistance and definitions of heat transfer terms and symbols are included in this chapter and in Chapter 3. In some construction, multidimensional heat flow effects are significant. Parallel heat flow paths of different resistances occur in wood frame house construction, for example. In such cases, the guarded hot box method (ASTM *Standard* C 236) or the calibrated hot box method (ASTM *Standard* C 976) may be used to determine overall thermal transmittance.

Surface Conductance

Surface conductance is the heat transfer to or from the surface by the combined effects of radiation, convection, and conduction. Each of these transport modes can vary independently. Heat transfer by radiation between two surfaces is controlled by the character of the surfaces (emittance and reflectivity), the temperature difference between them, and the solid angle through which they see each other. Heat transfer by convection and conduction is controlled by surface roughness, air movement, and temperature difference between the air and surface. Table 1 illustrates the importance of the effect of temperature of surrounding surfaces on surface heat flux caused by radiation.

In many cases, because the thermal resistance (reciprocal of conductance) of the internal parts of the wall is high compared with the surface resistance, the surface factors are of minor importance. However, surface resistances on a window with a single pane of glass constitute almost the entire resistance and are very important. Raber and Hutchinson (1945) analyzed the factors affecting surface conductance and the difference between surface and air temperatures.

Table 1 Variation in Surface Heat Flux for Vertical Surfaces at 80°F with Different Temperatures of Surrounding Surface (70°F Ambient Still Air; 0.83 Emittance)

Surrounding Surface	Surface Heat Flux, Btu/h·ft²				
Temperature, °F	75	70	65	60	50
Convection	6.6	6.6	6.6	6.6	6.6
Radiation	4.4	8.6	12.8	17.0	24.9
Total	11.0	15.2	19.4	23.6	31.5

The convective part of the surface conductance is markedly affected by the nature of the air movement on the surface, illustrated by Figure 1 in Chapter 24. On smooth surfaces, surface length (Parmelee and Huebscher 1947) also affects the convection part of conductance; the average value decreases as the surface length increases. Moreover, observations (Parmelee and Aubele 1950) of the magnitude of low-temperature radiant energy from outdoor surroundings show that only under certain conditions can outdoors be treated as a blackbody radiating at an effective air temperature. Therefore, selection of surface conductance coefficients for a building becomes a matter of judgment. Surface conductances in Table 1 in Chapter 24 are applicable to ordinary building materials. In special cases, where surface conductances become important factors in the overall rates of heat transfer, more accurate coefficients may be required. Principles and data given in Chapter 3 can be applied in such cases.

FACTORS AFFECTING HEAT TRANSFER ACROSS AIR SPACES

Heat transfer across an air space is affected by the nature of the boundary surfaces, as well as the intervening air space, the orientation of the air space, the distance between boundary surfaces, and the direction of heat flow. Air space conductance coefficients represent the total conductance from one surface bounding the air space to the other. The total conductance is the sum of a radiation component and a convection and conduction component. In all cases, the spaces are considered airtight with no through air leakage.

The radiation portion of the coefficient is affected by the temperature of the two boundary surfaces and by their respective surface properties. For surfaces that can be considered ideal graybodies, the surface properties can be characterized by emissivity. The combined effect of the emittances of the boundary surfaces of an air space is expressed by the effective emittance E of the air space. The radiation component is affected only slightly by the thickness of the space, its orientation, the direction of heat flow, or the order of emittance (hot or cold surface). The heat transfer by convection and conduction combined is affected markedly by orientation of the air space and the direction of heat flow, by the temperature difference across the space, and, in some cases, by the thickness of the space. It is also slightly affected by the mean temperature of its surfaces. For air spaces in building construction, the radiation and convection-conduction components can vary independently of each other.

Table 3 in Chapter 24 lists the thermal resistance values of sealed air spaces of uniform thickness and moderately smooth, plane, parallel surfaces. These data are based on experimental measurements (Robinson et al. 1954). Resistance values for systems with air spaces can be estimated from these results if emissivity values are corrected for field conditions. However, the resistance value of some common composite building insulation systems involving mass-type insulation with a reflective surface in conjunction with an air space may be appreciably lower than the estimated value (Palfey 1980), particularly if the air space is not sealed or of uniform thickness. The thermal resistance values for plane air spaces in Table 3 in Chapter 24 represent typical values; for critical applications, the effectiveness of a particular design should be confirmed by actual test data undertaken by using ASTM hot box methods (ASTM *Standards* C 236 and C 976). This test is especially necessary for constructions combining reflective and nonreflective thermal insulation.

For narrow air spaces, defined as those for which the product of the temperature difference (in degrees Fahrenheit) and the cube of the space thickness (in inches) is less than 3 for heat flow horizontally or downward, or less than 1 for heat flow upward, convection is practically suppressed. The conductance for these spaces is the sum of the radiative heat transfer coefficient and that for heat conduction alone through air. The radiation and conduction component can be computed by the method shown in the footnote to Table 3 in Chapter 24. Effects of different mean temperatures, temperature differences, and effective emittances may be found in this table.

To obtain high thermal resistance with reflective insulation, a series of multiple air layers bounded by reflective surfaces is needed. The total resistance equals the sum of the resistance values across each air space. All air layers or spaces must be sealed because air moving between the layers can increase the heat flow. Depending on the type of reflective insulation, one or both sides may have highly reflective surfaces. Except for thick horizontal air spaces with heat flow down, little is gained thermally by the addition of a second highly reflective surface to the same air space. If an air space has only one reflective surface, the side on which the reflective surface is placed makes no appreciable difference in the rate of heat transfer; however, placing the surface on the warm side minimizes or prevents condensation. Condensation should be prevented from forming on a reflective surface because, apart from other effects, it degrades the reflective properties of the surface. A reflective surface placed on the warm side of an air space usually is not a condensing surface and, therefore, maintains the thermal resistance of the air space and acts as a water vapor retarder if the material and its joints have sufficiently low permeance (see the section on Vapor Retarder Functions and Properties).

The emittance of a surface is the measure of its ability to emit radiant energy and, for the same temperature and wavelength, is equal to its absorptance (ratio of the radiant energy absorbed by a surface to the total radiant energy falling on it). The ratio of the energy reflected by the surface to that falling on it is the reflectance; for an opaque surface, reflectance is equal to one minus the emittance. This emittance varies with surface type and condition and radiation wavelength.

For reflective insulation used with heating, air-conditioning, and refrigeration applications, the emittance value for long-wavelength (infrared) radiation is important, not the value for the shorter wavelengths of the visible spectrum. Visible brightness is not a true measure of the reflectance for thermal radiation because the reflectance for light and for long-wavelength radiation is unrelated. Table 2 in Chapter 24 lists typical emittance values for reflective surfaces and building materials, and the corresponding emittance factors for air spaces.

Chemical action, dust or oil accumulation, or the presence of condensation or frost can change a reflective surface enough to reduce its reflectance and increase its emittance. Chemical changes include oxidation, corrosion, or tarnishing caused by air, moisture, wet plaster, or the chemical treatment of wood spacing strips or other adjoining structural members. Surface emittance values should be obtained by tests. Low-emittance windows have significantly greater thermal resistance, for example.

CALCULATING OVERALL THERMAL RESISTANCE

Using the principles of heat flow presented in Chapter 3, calculating heat flow by the overall thermal resistance method is preferred.

The total resistance to heat flow through building construction such as a flat ceiling, floor, or wall (or curved surface if the curvature is small) is the numerical sum of the resistances (R-values) of all parts of the construction in series:

$$R = R_1 + R_2 + R_3 + R_4 + \ldots + R_n \tag{1}$$

where

$R_1, R_2, \ldots, R_n$ = individual resistances of the parts
R = resistance of the construction from inside surface to outside surface

However, in buildings, to obtain the overall resistance R_T, the air film resistances R_i and R_o from Table 1 in Chapter 24 must be added to R.

$$R_T = R_i + R + R_o \tag{2}$$

The U-factor (thermal transmittance) is the reciprocal of R_T:

$$U = \frac{1}{R_T} \tag{3}$$

With the use of higher values of R_T, the corresponding values of U become very small. This is one reason why it is sometimes preferable to specify resistance rather than transmittance. Also, a whole number is more understandable to an insulation buyer than is a decimal or fraction.

Thus, for a wall with air space construction, consisting of two homogeneous materials of conductivities k_1 and k_2 and thickness x_1 and x_2, respectively, and separated by an air space of conductance C, the overall resistance is

$$R_T = \frac{1}{h_i} + \frac{x_1}{k_1} + \frac{1}{C} + \frac{x_2}{k_2} + \frac{1}{h_o} \tag{4}$$

where h_i and h_o are the heat transfer film coefficients.

Series and Parallel Heat Flow Paths

In many installations, components are arranged so that heat flows in parallel paths of different conductances. If no heat flows between lateral paths, heat flow in each path may be calculated using Equations (1) and (2). The average transmittance is then

$$U_{av} = aU_a + bU_b + \ldots + nU_n \tag{5}$$

where $a, b, \ldots, n$ are respective fractions of a typical basic area composed of several different paths with transmittances $U_a, U_b, \ldots, U_n$.

If heat can flow laterally with little resistance in any continuous layer so that transverse isothermal planes result, total average resistance $R_{T(av)}$ is the sum of the resistance of the layers between such planes, each layer being calculated by the appropriate Equation (1) or modification of Equation (4), using the resistance values. This is a series combination of layers, of which one (or more) provides parallel paths.

The calculated heat flow, assuming parallel heat flow only, is usually considerably lower than that calculated with the assumption of combined series-parallel heat flow. The actual heat flow is some value between the two calculated values. In the absence of test values for the combination, an intermediate value should be used; examination of the construction usually reveals whether a value closer to the higher or lower calculated value should be used. Generally, if the construction contains any highly conductive layer in which lateral conduction is very high compared to heat flow through the wall, a value closer to the series-parallel calculation should be used. If, however, there is no layer of high lateral conductance, a value closer to the parallel heat flow calculation should be used.

CALCULATING SURFACE TEMPERATURES

The temperature at any interface can be calculated, since the temperature drop through any component of the wall is proportional to its resistance. Thus, the temperature drop Δt through R_1 in Equation (1) is

$$\Delta t_1 = \frac{R_1(t_i - t_o)}{R_T} \tag{6}$$

where t_i and t_o are the indoor and outdoor temperatures, respectively. Hence, the temperature at the interface between R_1, and R_2 is

$$t_{1-2} = t_i - \Delta t_1 \tag{7}$$

For types of building materials having nonuniform or irregular sections such as hollow clay tile or concrete blocks, it is necessary to use the R-value of the unit as manufactured.

If the resistances of materials in a wall are highly dependent on temperature, the mean temperature must be known to assign the correct value. In such cases, it is perhaps most convenient to use a trial-and-error procedure for the calculation of the total resistance R_T. First, the mean operating temperature for each layer is estimated, and R-values for the particular materials are selected. The total resistance R_T is then calculated as in Equation (4), and the temperature at each interface is calculated using Equations (6) and (7).

The mean temperature of each component (arithmetic mean of its surface temperatures) can then be used to obtain second generation R-values. This procedure can then be repeated until the R-values have been correctly selected for the resulting mean temperatures. Generally, this can be done in two or three trial calculations.

In many heating and cooling load calculations, it is necessary to determine the inside surface temperature or the temperature of the surfaces within the structure. The resistances through any two paths of heat flow are proportional to the temperature drops through these paths and can be expressed as

$$\frac{R_1}{R_2} = \frac{t_i - t_x}{t_i - t_o} \tag{8}$$

where

R_1 = resistance from indoor air to any point in structure at which temperature is to be determined
R_2 = overall resistance of wall from indoor air to outdoor air
t_i = indoor air temperature
t_x = temperature to be determined
t_o = outdoor air temperature

HEAT FLOW CALCULATIONS

Equation (9) is used to calculate heat flow through flat surfaces; Equation (10) is used for cylindrical surfaces (Figure 5).

$$q_s = \frac{t_{is} - t_{os}}{R} \tag{9}$$

$$q_s = \frac{t_{is} - t_{os}}{\dfrac{r_s \ln(r_1/r_i)}{k_1} + \dfrac{r_s \ln(r_s/r_1)}{k_2}} \tag{10}$$

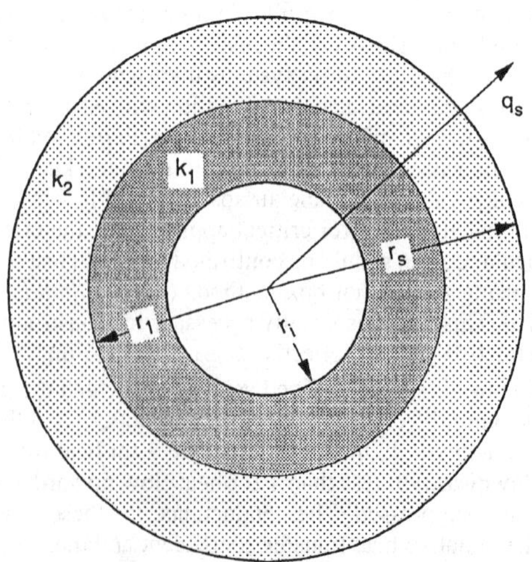

Fig. 5 Heat Flow Through Cylindrical Surfaces

where

q_s = rate of heat transfer per unit area of outer surface of insulation
R = surface-to-surface thermal resistance from Equation (1)
k = thermal conductivity of insulation at calculated mean temperature
t_{is} = temperature of inner surface
t_{os} = temperature of outer surface
r_i = inner radius of insulation
r_1 = outer radius of intermediate layer of insulation
r_s = outer radius of insulation
ln = natural or Naperian logarithm

To calculate the heat flow per unit area of pipe surface,

$$q_o = q_s(r_s/r_i) \tag{11}$$

where the units of q_o are Btu/h·ft^2.

Heat flow per unit length of pipe is the preferred unit. For steady-state conditions, heat flow through each successive material is the same. However, the temperature drop through each material is proportional to its thermal resistance. The terms that appear in the denominator of Equation (10) represent the resistances to heat flow based on log mean thickness. The thermal resistances discussed for flat surfaces and a thermal resistance of a cylindrical insulation should not be confused where the surface areas involved are never equal.

Heat flow is inversely proportional to the sum of the resistances of the system. The various temperature drops in the system are proportional to the resistances.

The following are usually assumed for calculations of heat flow:

t_{is} = temperature at inner surface of insulation = temperature of fluid in pipe or container
t_o = still air ambient temperature
r_i = inner radius of insulation = outside radius of iron pipe
r_s = outer radius of insulation = $r_i + x$
x = equivalent thickness of insulation, based on log mean thickness versus flat insulation thickness equivalent in performance

INSULATION THICKNESS

ECONOMIC THICKNESS: MECHANICAL SYSTEMS

Economics can be used (1) to select the optimum insulation thickness for a specific insulation or (2) to evaluate two or more insulation materials for least cost for a given thermal performance. In either case, economics determine the most cost-effective solution for insulating over a specific period (FEA 1976). This solution can be reached by different techniques, but only one solution exists for a given set of economic variables. This section presents the basic concepts used to determine economic insulation thickness for a single-insulation material system.

Greater than optimum insulation thickness may also require more capital investment for structure and piping. However, in some instances, limited energy availability may require more insulation than is normally justified by minimizing insulation and lost energy costs alone.

Determining the most profitable thickness of insulation is difficult. The economics of each plant (including cost of producing energy, cost of insulating, discount rate or cost of capital, and potential for energy loss) indicate the preferred amount of insulation. Various types of equipment and piping also require different economic thicknesses. This analysis is further complicated because future energy cost and the life of the facility and insulation must also be considered. For every plant, these factors dictate different solutions to the economic analysis.

Economic Analysis

The cost of installed insulation increases with thickness. This incremental cost is for both labor and material. Insulation is often applied in multiple layers (1) because materials are not manufactured in single layers of sufficient thickness and, (2) in many cases, to compensate for expansion and contraction. Figure 6 shows installed costs for a multilayer application. The average slope of the curves increases with the number of layers because labor and material costs increase at a more rapid rate as thickness increases.

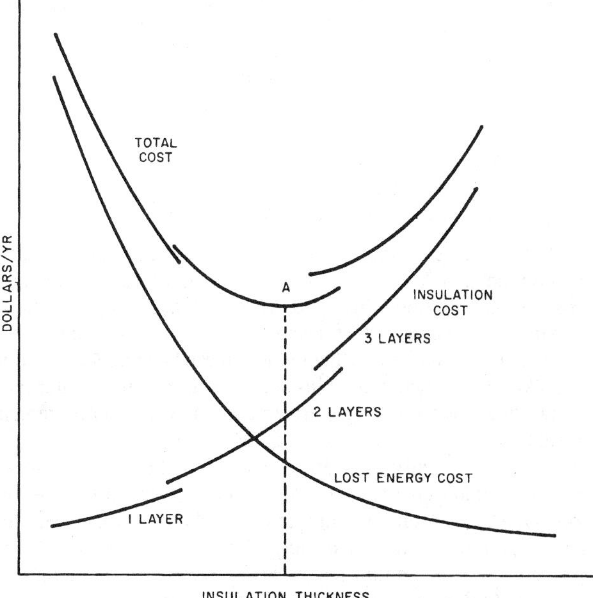

Fig. 6 Determination of Economic Thickness of Installed Insulation

Because the optimal economic thickness is the lowest total cost of lost energy and installed insulation over the life of the insulation, these two costs must be compared in similar terms. Either the annual cost of the insulation must be compared to the average annual cost of lost energy, or the cost of the energy lost each year must be expressed in present dollars and compared with the total cost of the insulation investment. The former method, annualizing the insulation cost and comparing this with the average annual cost of lost energy, is easier to compute.

Life-cycle costing spreads the initial cost of the insulation over the number of years the insulation is expected to be in service. The life cycle selected affects the economic thickness of the insulation. Thinner insulation pays back on a short life cycle, and thicker insulation pays back over a longer cycle. An insulation system designed to pay for itself in energy savings in a short payback period that stays in service longer, does not produce the lowest total cost over the service period.

Annualized cost of the installed insulation must also be adjusted for the cost of money, which can be a discount rate including the desired rate of return on the insulation investment. Insulation system maintenance costs should also be included in the annual cost.

Because fuel cost is likely to change during the depreciation period (the life of the facility or payback period), the average cost should be estimated and this value used rather than current cost.

Insulation reduces the size and capital costs of the heating and cooling equipment required for an installation because it lowers energy demand. This capital cost may be annualized by considering the plant depreciation period, cost of money, annual energy output for the plant, and operational expenses.

Total Cost

Because the total annual cost represents the sum of the annual cost of insulation and the annual cost of lost energy, for each incremental increase in insulation thickness, the corresponding change in the total cost is

$$m_t = m_c + m_s$$

where

m_t = incremental change in total annual costs
m_c = incremental change in cost of insulation
m_s = incremental change in cost of lost energy

Initially, as insulation is applied, the total cost decreases because the incremental energy savings is greater than the incremental cost of insulation. Additional insulation reduces this cost up to a thickness where the change in the total cost is equal to zero. At this point, $m_t = m_c + m_s = 0$, and no further reduction can be obtained. Beyond this thickness, the incremental insulation costs become greater than the additional energy savings derived by adding another increment of insulation.

Figure 6 shows curves of total annual costs of operation, insulation costs, and lost energy costs. Point A on the total cost curve corresponds to the economic insulation thickness, which, in this example, is in the double-layer range. Viewing the calculated economic thickness as a minimum thickness provides a hedge against unforeseen fuel price increases and conserves energy.

ECONOMIC THICKNESS: BUILDING ENVELOPES

In buildings such as residences and warehouses, the internal energy gains are insignificant compared with the heat losses and gains through the envelope. For these buildings, the heating and cooling requirements are roughly proportional to the difference between the indoor and outdoor temperature. For commercial, industrial, and institutional buildings, internal heat loads can be significant, and the heating and cooling requirements are not as directly related to the indoor/outdoor temperature difference. In both types of buildings, solar heat can be an important factor and should be evaluated.

Dominant Heat Loss and Gain Through Envelope

Thermal insulation is generally installed in building envelope components (e.g., ceilings, walls, and floors) to reduce space heating (and, in some cases, cooling) costs on a long-term basis. Additional benefits may include increased occupant comfort, reduced requirements for heating and cooling system capacity, and elimination of condensation on wall surfaces in cold climates. However, because these benefits are difficult to quantify and are likely to be negligible beyond a nominal level of thermal resistance, the following evaluation methods focus only on reductions in energy costs. The economically optimal insulation thickness (best measured in terms of thermal resistance) in an envelope component minimizes total life-cycle space heating and cooling costs attributable to that component. Total life-cycle costs are the sum of present-value heating and cooling costs over the useful lifetime of the insulation plus the installed cost of the insulation.

Generally, as the thermal resistance of the insulation is increased, space heating costs decrease, and insulation costs increase. As long as the additional reduction in present-value heating plus cooling costs (incremental savings) caused by an increase in insulation thickness exceeds the additional cost of that increased insulation thickness (incremental costs), total life-cycle costs are reduced. But, as insulation thickness is increased, the incremental savings per unit change in resistance decreases, and incremental insulation costs beyond the first resistance unit typically remain constant or increase. Because start-up costs are attributed to the first resistance unit, this may be the most expensive unit. Beyond a certain resistance level, incremental costs exceed incremental savings, so that such additions increase life-cycle costs. At the point where incremental savings equal incremental costs, life-cycle costs are minimized. However, if total insulation costs then exceed total insulation benefits, the optimal level of insulation for a given component is no insulation at all. This may occur if the start-up costs are extremely high, as for installation of insulation in a sealed-cavity wall.

If the R-value of the insulation used is continuously variable (e.g., loose-fill insulation in attics), uniformly small increments of insulation (R-1 ...) can be used to determine appropriate optimal thicknesses; or calculus can be used to determine an exact optimum. If the insulation materials used are available only in discrete levels of thermal resistance (R-12, R-18, R-27), the increment used in determining optimal thickness should be based on differences between those levels (R-12 over R-0, R-18 over R-12, R-27 over R-18). Where discrete increments of resistance are used, determining the resistance level for which incremental savings equal incremental costs may not be feasible. In such cases, the selection should be left to the judgment of the analyst based on the level of conservation desired.

If a building envelope component requires structural modifications to accommodate increased insulation thickness, this cost must be included in the installed cost of the additional insulation. Generally, such modifications should only be considered when they are less costly than the use of more efficient (i.e., lower thermal conductivity), but more expensive, insulation materials than those ordinarily used.

Typically, the incremental energy savings and insulation costs differ for each building envelope component; therefore, the optimal insulation level differs for each component in the same building. Less efficient heating plants and higher costs of heating energy necessitate higher optimal insulation levels in each building envelope component. Conversely, more efficient heating equipment lowers the optimal insulation level. The effects of climate, cooling energy costs, and cooling equipment efficiency on optimal insulation levels are less clear and differ widely, depending on overall building design and operational profile.

Computation

Optimal insulation thickness is found by examining the incremental savings and costs of each additional level of insulation included in a building envelope component. The present-value, **life-cycle savings** from each increment are calculated as follows:

$$\Delta \mathrm{LCS}_{ij} = \frac{\Delta \mathrm{AHR}_{ij}(P_h)(\mathrm{UPW}_h{}^*)}{n_h} + \frac{\Delta \mathrm{ACR}_{ij}(P_c)(\mathrm{UPW}_c{}^*)}{n_c} \quad (12)$$

where

$\Delta \mathrm{LCS}_{ij}$ = present-value, life-cycle savings in dollars from the *i*th increment of insulation in the *j*th component
$\Delta \mathrm{AHR}_{ij}$, $\Delta \mathrm{ACR}_{ij}$ = changes in annual heating and cooling requirements, respectively, of the building due to the *i*th increment of insulation in the *j*th component (treating a reduction as positive)
n_h, n_c = seasonal conversion efficiencies (or coefficients of performance) of the building's heating and cooling equipment, respectively
P_h, P_c = current costs per thermal unit for heating and cooling energy, respectively
$\mathrm{UPW}_h{}^*$, $\mathrm{UPW}_c{}^*$ = modified uniform present-worth factors for heating and cooling energy types used, respectively

The terms $\Delta \mathrm{AHR}_{ij}$, $\Delta \mathrm{ACR}_{ij}$, n_h, and n_c must be determined through thermal engineering analysis and require consideration of the overall building design, the climate, and the operational profile of the building. In some cases (floors), $\Delta \mathrm{ACR}_{ij}$ from additional insulation may be negative, that is, additional insulation may increase annual cooling requirements. Careful construction and insulation practices ensure that theoretical savings are actually realized.

The terms $\mathrm{UPW}_h{}^*$ and $\mathrm{UPW}_c{}^*$ relate present-value, life-cycle savings to annual savings calculated at current energy costs. They are modified uniform present-worth factors in the sense that they incorporate both a discount rate and a rate of energy price increase over the time horizon considered in the life-cycle cost analysis. The UPW* is calculated as follows:

$$\text{UPW*} = \frac{1+P}{D-P}\left[1+\left(\frac{1+P}{1+D}\right)^L\right] \quad \text{if } D \neq P \tag{13a}$$

or

$$\text{UPW*} = L \quad \text{if } D = P \tag{13b}$$

where

P = annual rate of energy price increase
D = discount rate
L = useful life of insulation

Because different energy types may have different projected rates of price increase, UPW_h* and UPW_c* may differ when different energy types are used for heating and cooling. However, the same discount rate and lifetime should be used to determine the appropriate UPW* factors. Generally, longer lifetimes (30 years) rather than shorter lifetimes (10 years) approach true life-cycle cost minimization more closely and avoid the high cost of retrofitting if energy prices increase.

Dominant Internal Heat Loads

In buildings with dominant internal loads, the energy requirements vary so widely that no generalizations can be made regarding insulation. This contrasts with envelope-dominated structures, in which more insulation reduces energy consumption (Hart 1981).

In internal-load-dominated buildings with cooling, for cool climates having more hours below room temperature than above, higher thermal resistance increases cooling energy consumption while reducing heating energy consumption. Therefore, the calculation of economically optimum resistance becomes quite complex and involves multiple measure or hourly methods described in Chapter 30. Spielvogel (1974), Burch and Hunt (1978), and Rudoy (1975) give more details.

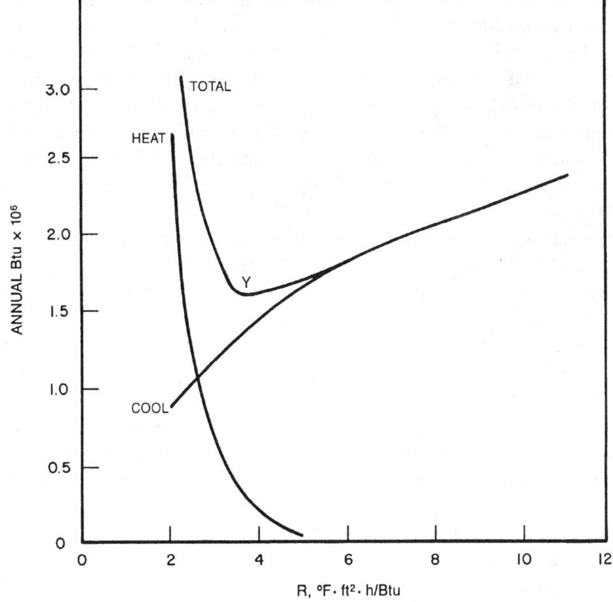

Fig. 7 Example of Optimal Thermal Resistance for Building with Internal Heat Gains
(Adapted from Spielvogel 1974)

Figure 7 shows the results of these calculations for a building in Columbus, Ohio, with 8.2 Btu/h·ft² of internal heat gains that operates 24 h per day (Spielvogel 1974). This solution is not the only one possible but illustrates problems faced by the designer.

As thermal resistance increases, the U-factor decreases, annual heating energy decreases, and annual cooling energy increases. The energy optimum exists at Point Y in Figure 7, where the total heating plus cooling energy is at its minimum. Because the cost of cooling energy differs from the cost of heating energy, the economic optimum will not be the same as the energy optimum.

These results occur in some localities that have far more hours per year with outdoor temperatures between 50 and 75°F than between 75 and 100°F. At temperatures between 75 and 100°F, low U-factors result in less energy consumption for cooling; however, for temperatures between 50 and 75°F, low U-factors inhibit the flow of internal heat from the building, thereby creating higher cooling loads and higher energy requirements than those in buildings with higher U-factors. What might be saved at outdoor temperatures over 75°F can be more than spent in additional cooling energy at temperatures below 75°F. Economizer cycles could offset these excess internal gains with ventilation air, however.

Where the hours of use or the quantity of internal heat gains vary from room to room, the optimum thermal resistance also varies. For example, in a cold climate, a hotel kitchen requires little or no insulation, because the internal heat is sufficient to heat the space almost all year. In a meeting room adjacent to the kitchen, substantially more insulation is justified. Thus, the economic thermal resistance of any envelope element, such as a roof, will not be the same throughout the entire building.

This type of analysis must include the level and duration of internal gains and the nature of the energy consumption of the heating and cooling systems. Most buildings need evaluation of walls, roofs, and floors on a room-by-room basis. Computer programs make these more complex analyses possible. Due to the wide diversity of building types, internal gains, system types, and operating conditions, no simple rules can establish U-factors for minimum energy consumption.

Effectiveness of Added Insulation

The effectiveness of added insulation varies with many factors, including climate, original insulation level, preparation costs, and predicted life, based on payback calculations. Building codes generally balance life-cycle costs between construction, financing, and energy expenditures. Figure 8 shows typical relationships between life-cycle costs and energy consumption. Individual points on the

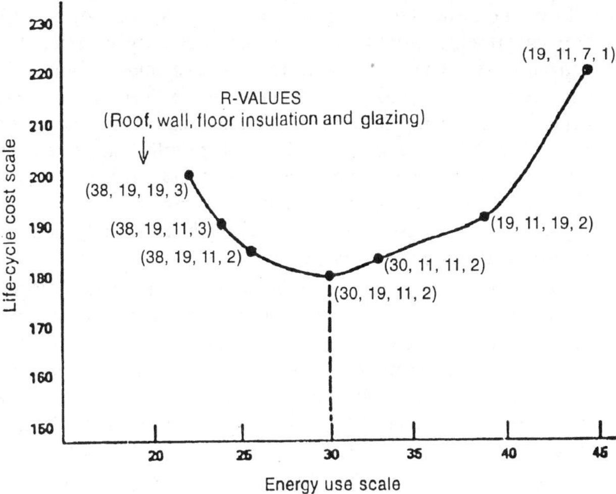

Fig. 8 Typical Relationship of Life-Cycle Cost to Energy Use

curve represent different combinations of ceiling, wall, and floor insulation in R-values and glazing types (single, double, or triple). Because life-cycle costs vary not only with construction and energy costs but also with climatic factors, the profiles of this curve vary according to locality. In Figure 8, the optimal condition for this example is attained with R-30 attic insulation, R-19 wall insulation, R-11 floor insulation, and double glazing. However, the effectiveness of added insulation can be determined only by analyzing actual conditions.

FUNDAMENTALS OF MOISTURE IN BUILDINGS

MOISTURE PROBLEMS IN BUILDINGS

Moisture control is necessary to avoid moisture-related problems with building energy performance, building maintenance and durability, and human comfort and health. Moisture degradation is the largest factor limiting the useful life of a building and can be visible or invisible. Invisible degradation includes degradation of the thermal resistance of building materials and decrease in the strength and stiffness of some materials. Visible moisture degradation may be in the form of (1) mold and mildew, (2) the decay of wood-based materials, (3) spalling of masonry and concrete caused by freeze-thaw cycles, (4) hydration of plastic materials, (5) corrosion of metals, (6) damage due to expansion of materials (e.g., buckling of wood floors), and (7) a decline in visual appearance (e.g., buckling of wood siding or efflorescence of masonry materials, which is the formation of a salt crust from the leaching of free alkalies). In addition, high moisture levels can lead to odors and mold spores in indoor air, which can seriously affect occupant health and comfort. Short summaries of such moisture conditions and related performance and health issues follow.

Mold, Mildew, Dust Mites, and Human Health

Mold and mildew in buildings are offensive, and the spores can cause respiratory problems and other allergic reactions in humans. Mold and mildew will grow on most surfaces if the relative humidity at the surface is above a critical value and the surface temperature is conducive to growth. The longer the relative humidity remains above the critical value, the more likely is visible mold growth; and the higher the humidity or temperature, the shorter is the time needed for germination. The surface relative humidity is a complex function of material moisture content, material properties, and local temperature and humidity conditions. In addition, mold growth depends on the type of surface. Fully recognizing the complexity of the issue, the International Energy Agency Annex 14 (1990) nevertheless established a surface humidity criterion for design purposes: The monthly average surface relative humidity should remain below 80%. Others have proposed more stringent criteria, the most stringent requiring that surface relative humidity remain below 70% *at all times*. Although there still is no agreement on which criterion is most appropriate, mold and mildew can usually be avoided by limiting surface moisture conditions over 80% to short time periods. These criteria should only be relaxed for nonporous surfaces that are regularly cleaned.

Most molds grow at temperatures above approximately 40°F. Moisture accumulation below 40°F may not cause mold and mildew if the material is allowed to dry out below the critical moisture content before the temperature rises above 40°F. Dust mites can trigger allergies and asthma (Burge et al. 1994). Dust mites thrive at high relative humidities (over 70%) at room temperature, but will not survive sustained relative humidities below 50% (Burge et al. 1994). Note that these relative humidities relate to local conditions in the typical places that mites tend to inhabit: mattresses, carpets, soft furniture, etc.

Paint Failure and Other Appearance Problems

Moisture trapped behind paint films may cause failure of the paint. Water or condensation may also cause streaking or staining. Excessive swings in the moisture content of wood-based panels or boards may cause buckling or warp. Excessive moisture in masonry and concrete may cause efflorescence, a white powdery area or lines, or, when combined with low temperatures, may cause freeze-thaw damage and spalling (chipping).

Structural Failures

Structural failures due to decay of wood are rare but have occurred (e.g. Merrill and TenWolde 1989). Decay generally requires a wood moisture content at fiber saturation (usually about 30%) or higher and temperatures between 50 and 100°F. Wood moisture contents above fiber saturation are only possible in green lumber or by absorption of liquid water from condensation, leaks, ground water, or other saturated materials in contact with the wood. To maintain a safety margin, 20% moisture content is sometimes used during field inspections as the maximum allowable moisture level. Because wood moisture content can vary widely with sample location, a local moisture content of 20% or higher may indicate fiber saturation elsewhere. Once established, decay fungi produce water that enables them to maintain moisture conditions conducive to their own growth.

Rusting or corrosion of nails, nail plates, or other metal building components is also a potential cause of structural failure. Corrosion may occur at high relative humidities near the metal surface or as a result of liquid water from elsewhere. Wood moisture content over 20% encourages corrosion of steel fasteners in wood, especially if the wood is treated with preservatives. In buildings, metal fasteners are often the coldest surfaces, encouraging condensation on and corrosion of the fasteners.

Effect of Moisture on Heat Flow

Moisture in the building envelope can significantly degrade the thermal performance of most insulation materials. Bomberg and Shirtliffe (1978), Epstein and Putnam (1977), Hedlin (1977, 1983, 1987, 1988a, 1988b), Jespersen (1960), Joy (1957), Knab et al. (1980), Kumaran (1989), Kyle and Desjarlais (1994), Langlais et al. (1983), Paljack (1973), Pedersen (1990), Pedersen et al. (1991), Shapiro and Motakef (1990), Thomas et al. (1983), Tobiasson and Richard (1979), Tobiasson (1987, 1991), and Tye (1987) have investigated the effect of moisture content on heat flow and show that the effect is dependent on the type of insulation material, the moisture content, the temperature of the insulation material, the insulation material's thermal history, and the building envelope's interior and exterior environments. The reported relationships between the thermal performance of the insulation material and heat flow can vary significantly; the variations are more pronounced in open-cell and fibrous insulations. Rapid vapor transfer through vapor-permeable insulations during testing accounts for these variations. Variations can also be due to the location of water in the insulation layer. Kyle and Desjarlais (1994) have estimated that water distribution can account for a difference of up to 25% in heat flow in certain specific cases.

Moisture can contribute to heat flow in both sensible and latent forms, as well as through mass transfer. Mass transfer occurs when there is a vapor pressure difference across the insulation layer; water vapor from the side with the higher vapor pressure migrates through the insulation to the side with the lower vapor pressure. Evaporation on the warm side of the insulation and condensation or adsorption on the cold side add a latent heat component to the heat flow.

Under steady-state conditions, the effect of moisture on thermal resistance may be small. Verschoor (1985) showed that an insulated residential-type stud wall panel with a poor vapor retarder on the warm side accumulated 0.3 lb of moisture per square foot when exposed to conditions continuously below freezing (steady-state) for 31 days. All of the accumulated moisture was located in the 0.5 in. layer of mineral fiber insulation immediately adjacent to the cold-side sheathing and at the interface between the insulation and the sheathing. In a continuation of the test program, to a total exposure period of 60 days, the rate of moisture gain remained constant during the entire period. Subsequently, the same test wall was subjected to diurnal outside temperature cycles through the freezing point. With this exposure, most of the accumulated moisture was found in the sheathing and the bottom of the test wall rather than in the coldest insulation layer. Except for the sheathing, there was negligible change in the overall thermal performance of the wall.

The behavior may be different for closed-cell plastic foam insulations. Field observations of low-temperature insulated tanks have shown that when no air space exists between the cold surface and the insulation, the expected accumulation of condensed moisture did not occur.

Degradation of thermal resistance is more pronounced when daily reversals in temperature across the insulation drive moisture back and forth through the insulation layer and is exacerbated when the insulation material has a high water vapor permeance. A wetted compact low-slope roof is a good example of these phenomena. During the nighttime, moisture migrates upward through the insulation layer and can condense in the upper part of the insulation layer on the underside of the membrane. The following day, the increase in ambient temperature coupled with solar radiation heats the membrane and reverses the vapor pressure difference, evaporating some or all of the condensed water and driving it downward into the roofing system, transporting heat in the process. If sufficient water vapor can be driven downward, it may condense somewhere in the lower portion of the insulation layer, releasing its heat of condensation.

Hedlin (1988b) and Shuman (1980) experimentally showed that, for building envelopes containing permeable fibrous insulations that were undergoing temperature reversals, the rate of energy transfer increased sharply as the moisture content (MC) increased to approximately 1% by volume. The rate of increase in energy transfer diminished rapidly with further increases in moisture content. Energy transfer for permeable insulation with 1% MC by volume was roughly double that for dry insulation. Pedersen et al. (1991) analytically reproduced Hedlin's results; he demonstrated the high mobility of moisture in a permeable insulation and showed that latent effects are appreciable for a wide variety of North American climates. The latent effects typically add to the energy load and can increase peak energy demand. The extra load is added to the building load in the warm afternoon, and nearly the same amount of heat is removed in the cool evening.

Certain organic insulations such as wood fiberboard and perlite are hygroscopic and can contain 1% MC by volume if installed at equilibrium conditions; water leakage into the building envelope component is *not* a prerequisite for significant increases in heat flow.

In studies of hourly temperature and moisture content variations in wood frame wall cavities, Duff (1971) showed measurable levels of daily moisture migration across the cavity. Pedersen et al. (1991) demonstrated the same phenomenon in compact low-slope roofs.

The building envelope is exposed to ever-changing exterior conditions; fluctuations in the outdoor air temperature and the amount of solar radiation affect the temperature profile through the building envelope. These temperature changes affect the magnitude and the direction of the vapor pressure gradients. It is unlikely that conditions will ever exist that prevent changes in the temperature profile of building envelope components. Water vapor transfer rates and direction are constantly changing, adjusting not only to the thermal changes but to changes in moisture concentration. However, conditions may exist where the *average* vapor pressure drive is in one direction, for example, upward in a low-slope roof during winter. In this situation, moisture accumulates in the insulation just below the impermeable roof membrane, and, assuming these conditions can be maintained for a reasonable length of time, water accumulates or frost forms unless the top layer of the insulation has adequate moister absorption capacity (e.g., perlite or wood fiberboard). Under similar conditions in the summer, moisture is driven down to the vapor retarder or deck (if the deck is less permeable than the insulation layer). If there are no layers in the system that are less permeable than the insulation, the water vapor simply diffuses into the building interior.

Under conditions where the vapor pressure is changing slowly or where the insulation layer has an extremely low water vapor permeance, little water vapor is transported. Moisture still affects the sensible heat transfer in the building envelope component. Epstein and Putnam (1977) and Larsson et al. (1977) showed a nearly linear increase in sensible energy transfer of approximately 3 to 5% for each volume percent increase in moisture content in cellular plastic insulations. For example, an insulation material with a 5% MC by volume has 15 to 25% greater energy transfer than the dry insulation material. Other field studies by Dechow and Epstein (1978) and Ovstaas et al. (1983) have shown similar results for insulations installed in below-grade applications such as foundation walls.

Moisture Effect on Heat Storage

Moisture also affects the thermal storage capacity of certain hygroscopic building materials considered composites of a given substance and water. At 10% MC, nearly 30% of the heat storage capacity of wood is in the water held in the cell walls. Since the specific heat of wood is a function of its temperature and moisture content (but almost independent of density and species), heat storage calculations must include an estimate of the equilibrium in-service moisture content of wood building components.

PROPERTIES OF WATER VAPOR IN AIR

Chapter 6 on Psychrometrics describes in detail the properties of moist air and defines the various terms associated with water vapor in air, such as dew-point temperature, dry-bulb and wet-bulb temperatures, and relative humidity. Chapter 6 also explains the use of the psychrometric chart, as well as the physics of heating and cooling moist air.

MOISTURE IN BUILDING MATERIALS

Many common building materials are porous. The pores provide a large internal surface area generally with an affinity for water. In some materials, such as wood, moisture may also be adsorbed in the cell wall itself. The amount of water in these **hygroscopic** (water-attracting) materials is related to the relative humidity (rh) of the surrounding atmosphere. When relative humidity of the surrounding air rises, hygroscopic materials gain moisture (**adsorption**), and when the relative humidity drops these materials lose moisture (**desorption**). The relationship between relative humidity and the moisture content (MC) at a particular temperature can be represented in a graph called a sorption isotherm. Often isotherms obtained by adsorption are not identical to isotherms obtained by desorption, because the material tends to retain moisture when it is drying. This difference between desorption and adsorption isotherms is called **hysteresis**. At high relative humidity, small pores become entirely filled with water. The maximum moisture content is reached when all pores are filled with water, but experimentally this can only be achieved in a vacuum, by boiling the material or by keeping the material in contact with water for an extremely long time (i.e., many years). In practice, the maximum moisture content of porous materials is lower. This lower maximum moisture content is sometimes referred to as the **capillary saturation moisture content**. Figure 9 shows typical sorption curves. It shows the **equilibrium moisture content** (EMC) as a function of relative humidity. The EMC increases with relative humidity, especially above 80% rh. The EMC decreases slightly with increasing temperature.

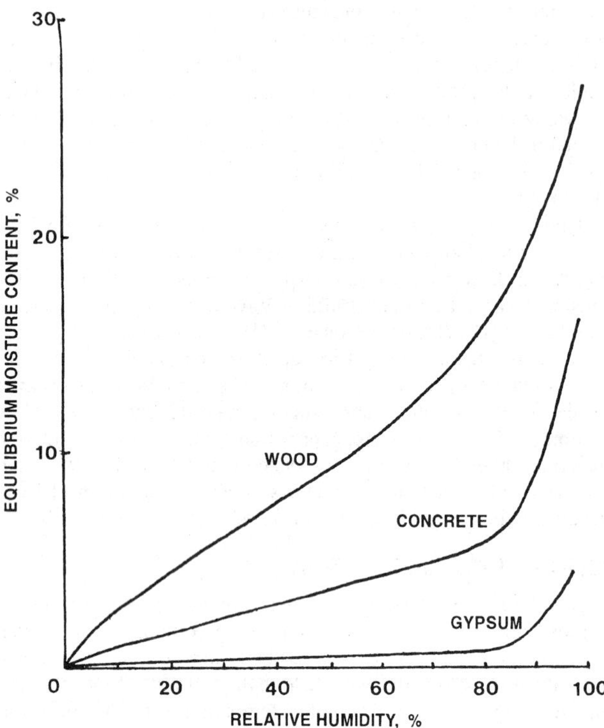

Fig. 9 Typical Sorption Isotherms for Wood, Concrete, and Gypsum (Hysteresis Is Ignored)

Table 2 Linear and Volumetric Shrinkage Values of Wood, from Green to Oven Dry Moisture Content

Wood Species	Radial %	Tangential %	Volumetric %
Hardwoods			
Birch			
Yellow	7.3	9.5	16.8
Oak			
Northern red	4.0	8.6	13.7
Southern red	4.7	11.3	16.1
White	5.6	10.5	16.3
Softwoods			
Cedar			
Northern white	2.2	4.9	7.2
Western red cedar	2.4	5.0	6.8
Douglas fir			
Coast	4.8	7.6	12.4
Fir			
White	3.3	7.0	9.8
Hemlock			
Western	4.2	7.8	12.4
Pine			
Eastern white	2.1	6.1	8.2
Longleaf	5.1	7.5	12.2
Ponderosa	3.9	6.2	9.7
Redwood			
Young growth	2.2	4.9	7.0
Imported wood			
Lauan	3.8	8.0	—

Note: Values expressed as a percentage of the green dimension (FPL 1987).

Chapter 21 describes hygroscopic substances and their use as dehumidifying agents. Table 2 in Chapter 11 of the 1995 *ASHRAE Handbook—Applications* has data on the moisture content of various common materials in equilibrium with the atmosphere at various relative humidities.

Wood and many other hygroscopic materials change dimensions with variations in moisture content. The moisture content at which cell walls of wood are saturated but no free water exists in cell cavities is the **fiber saturation point**. It represents the upper limit for moisture gain from the air as water vapor. It is also the upper limit of swelling; although more water can be absorbed in cell cavities, additional swelling does not occur. Wood moisture content is expressed as a percentage of its oven dry weight. The average fiber saturation point for most species is about 30%. The average EMC at 68°F and 45% rh (heating season indoor conditions) is about 8.5%. At 75°F and 70% rh (summer conditions), the EMC is about 13%.

The resulting dimensional changes in wood are proportional to the change in moisture content, but vary with species and direction of grain. For white oak, which is representative of interior trim material, 4.5% MC variation causes 2.5% volumetric shrinkage or swelling. Longitudinal dimensional change in straight-grained wood is insignificant, but it increases with crossgrain and other irregularities. Accordingly, residential wood trusses with top and bottom chords exposed to different temperature and moisture regimes can show measurable seasonal vertical movements (Plewes 1976).

Thermal expansion of wood is usually outweighed by shrinkage or swelling due to moisture content changes. The linear thermal expansion coefficient for wood across the grain ranges from about 7×10^{-6} for light wood species to about 25×10^{-6} for dense wood species, which is small compared to that of many other materials. The thermal expansion coefficient parallel to the grain is between 1.7×10^{-6} and 2.5×10^{-6} per °F.

Most plant and animal fibers undergo dimensional changes similar to those noted in wood. Typical values for wood can be found in Table 2; the *Wood Handbook* (FPL 1987) gives more detailed information. Related changes in other materials are not as well documented. However, different expansion rates caused by temperature and moisture changes in different materials used in composite constructions should be considered (Baker 1964, BRS 1974).

Porous materials also absorb water in liquid form when in contact with it. Liquid water may be present because of leaks, rain penetration, flooding, or surface condensation. Wetting may be so complete that the material reaches the capillary saturation moisture content.

MOISTURE MIGRATION

Liquid water and water vapor migrate by a variety of moisture transport mechanisms. The following are some of the most important mechanisms:

- Liquid flow by gravity or air pressure differences
- Capillary suction of liquid water in porous building materials
- Liquid diffusion by moisture content differences
- Movement of water vapor by air movement
- Water vapor diffusion by vapor pressure differences

Although in the past many moisture control strategies focused on control of vapor diffusion through the installation of vapor (diffusion) retarders, the other mechanisms, when present, can move far greater amounts of moisture. Therefore, liquid flow, capillary suction, and air movement should be controlled first.

Liquid flow by gravity and air pressure difference is not discussed here, but a short description of the other mechanisms follows. A more comprehensive treatment of moisture transport and storage may be found in Kumaran et al. (1994).

Capillary Suction

Within small pores of diameter less than approximately 0.1 μm, molecular attraction between the surface of the capillary and the water molecules causes a capillary suction defined by

$$s = \frac{2\sigma\cos\theta}{r} \qquad (14)$$

where

s = capillary suction
σ = surface tension of water
r = radius of the capillary
θ = contact wetting angle

The **contact wetting angle** is the angle between the water meniscus and the capillary surface. The smaller this angle is, the larger the capillary suction. In hydrophilic (water-attracting) materials, the contact wetting angle is less than 90°, and in hydrophobic (water-repelling) materials, the angle is between 90° and 180°. Moisture movement by capillary suction is governed by differences in capillary suction. Capillary suction is greater in smaller capillaries, so capillary suction moves moisture from larger to smaller capillaries. Surface tension σ is a function of temperature: the higher the temperature, the lower the surface tension. Thus, capillary suction moves moisture towards lower temperatures.

Isothermal and nonisothermal movement may occur in the liquid phase, or in the vapor phase if the capillaries are not completely filled. The transfer in the vapor phase is by vapor diffusion and is caused by a difference in vapor saturation pressure in the capillaries. **Thompson's law** states that the saturation vapor pressure in equilibrium with the water in a capillary is given by

$$p'' = p'\exp(s/\rho RT) = p'\exp(2\sigma\cos\theta/r\rho RT) \qquad (15)$$

where

p'' = saturation vapor pressure in capillary
p' = saturation vapor pressure in ambient air at same temperature as p''
ρ = density of water
R = gas constant of water vapor
T = absolute temperature

Equation (15) shows that the saturation vapor pressure is lower in smaller capillaries than in larger capillaries, which causes vapor diffusion from larger water-filled capillaries to smaller capillaries. The equation also shows that saturation vapor pressure in the capillaries is lower at lower temperatures, which causes diffusion from higher to lower temperatures.

If the vapor pressure in the ambient air is in equilibrium with the saturation vapor pressure in the capillaries, Equations (14) and (15) can be combined:

$$s = \rho RT \ln\phi \qquad (16)$$

where ϕ is the relative humidity of the ambient air.

Capillary flow of water can be expressed as a function of suction pressure gradients:

$$w_m = -k_m \frac{ds}{dx} \qquad (17)$$

where

w_m = water flux
k_m = water permeability coefficient

Air Movement

Water vapor movement by air can be represented by

$$w = W\rho v \qquad (18)$$

where

w = water vapor flux (flow per unit area)
W = humidity ratio
ρ = density of air
v = airflow velocity

Even small airflows can carry large amounts of water vapor when compared to vapor diffusion. Airflow retarders are designed to inhibit the flow of air and thereby the transport of water vapor into the construction.

Water Vapor Diffusion

Water vapor also migrates by diffusion through air and building materials. Moisture diffusion through air is rapid, resulting in minimal vapor pressure differentials between connected spaces and rapid flow to condensing surfaces, such as cold glass or dehumidifier coils. This equalization is further assisted by convective mixing of the air. Although movement by air convection, when present, usually dominates, vapor diffusion can be the an important mode of transportation in industrial applications such as cold storage facilities or built-in refrigerators. The control of diffusion also becomes more important with increasingly airtight construction.

The equation used to calculate water vapor diffusion flux through materials is based on a form of Fick's law:

$$w = -\mu \frac{dp}{dx} \qquad (19)$$

where

w = water vapor flux (flow rate per unit area)
p = water vapor pressure
x = distance along flow path
μ = water vapor permeability

According to Equation (19), water vapor flow by diffusion is proportional to the water vapor pressure gradient and closely parallels Fourier's equation for heat flow. The actual diffusion of vapor through a material is complex; the apparent water vapor permeability is a function of relative humidity and temperature and may vary by flow direction and along the flow path within a material (Chang and Hutcheon 1956, Bomberg 1989).

Standard test methods for the measurement of water vapor permeability are described in ASTM *Standard* E 96, Standard Test Methods for Water Vapor Transmission of Materials. The rate of moisture transfer through a material is determined gravimetrically while maintaining a steady temperature and vapor pressure differential across the specimen. ASTM *Standard* E 96 defines a dry-cup or desiccant method, and a wet-cup or water method. In the dry-cup method, the air on one side of the specimen is maintained close to 0% rh, while the other side is usually maintained near 50% rh. In that humidity range, moisture transport is primarily by vapor diffusion for most materials tested, so the test can produce consistent results. The wet-cup method calls for humidities between 50% and 100%, and generally produces higher permeability values. In this humidity range, part of the transport in many materials is by mechanisms other than vapor diffusion and may no longer adhere to Equation (19). Especially in strongly hygroscopic materials, such as wood, this can lead to inconsistent values for wet-cup permeability. Additional detailed information on the reliability of permeability testing can be found in Trechsel and Bomberg (1989).

Permeance is usually expressed in perms (grains/h·ft^2·in. Hg), and permeability in perm·in. Whereas permeability refers to water vapor flux per unit thickness, the term permeance is used in reference to a material of a specific thickness. For example, a material that is 2 in. thick generally is assumed to have a permeance of half its permeability, even though permeance of many materials often is not strictly proportional to thickness. In many cases, this simplification ignores the effect of moisture content variations within the material, and the effect of cracks or holes in the surface. It is inappropriate to use the term permeability in reference to inhomogeneous or composite materials, such as plywood or gypsum board with paper facings. Whenever a building product is made of two dissimilar materials, only water vapor permeance is meaningful for characterizing vapor transfer through that product.

Methods have been developed that allow measurement of water vapor transport with temperature gradients across the specimen (Douglas et al. 1992). These test methods promise more accurate data on isothermal vapor transfer through materials and will eventually allow better distinction between various transport modes.

Combined Liquid and Vapor Flow

It is nearly impossible to experimentally distinguish between liquid flows and vapor flows in hygroscopic materials. Therefore, moisture flow is often expressed as a combined moisture flow, and is most commonly expressed as a function of moisture content and temperature gradients:

$$w_m = -D_w \frac{dm}{dx} - D_T \frac{dT}{dx} \quad (20)$$

where

w_m = moisture flux (combined vapor and liquid)
D_w = moisture diffusion coefficient (T = constant)
m = moisture content, %
x = distance along flow path
D_T = thermal moisture diffusion coefficient (m = constant)
T = absolute temperature

This approach has the advantage that both temperature and moisture content are measurable quantities. However, the parameters D_w and D_T are often both functions of moisture content and, to a lesser degree, temperature. Another disadvantage is that moisture content is discontinuous at interfaces between different materials and therefore cannot be used as a moisture flow potential at those interfaces. An alternative formulation with suction as moisture flow potential has the advantage of continuity at material interfaces:

$$w_m = -k_m' \frac{ds}{dx} - D_T' \frac{dT}{dx} \quad (21)$$

where

k_m' = moisture permeability coefficient
D_T' = thermal moisture diffusion coefficient without a suction gradient

WATER VAPOR RETARDERS AND AIRFLOW RETARDERS

Water vapor retarders and airflow retarders combine to control the movement of moisture and air. Although their functions are different, a single component may serve both functions. The designer should assess the needs for moisture and air movement control in the building envelope and provide a system that combines the required vapor retarder and airflow retarder properties.

Airflow Retarder Functions and Properties

In addition to a vapor retarder, control of moisture requires an effective airflow retarder (also referred to as an air barrier or air infiltration barrier). Without effective control of airflow, vapor retarders are completely ineffective.

A vapor retarder may also be an airflow retarder, that is, an air/vapor retarder. In the past, many designs were based on this, and measures were taken to ensure that the vapor retarder was continuous in order to control airflow through it. This remains a valid approach. Some recent designs treat airflow retarders and vapor retarders as separate entities, but an airflow retarder should not be placed in a location where it can cause moisture to condense if it also has vapor retarder properties. For example, an airflow retarder placed on the cold side of a building envelope may cause condensation, particularly if the vapor retarder is ineffective and the airflow retarder is impermeable to moisture.

However, a cold-side air/vapor retarder that also has sufficient insulation may result in a lower potential for condensation by raising the temperature of the surface of the air/vapor retarder, but this requires careful installation and sealing of joints.

Air leakage characteristics of airflow retarders can be determined with ASTM *Standard* E 283 (test method for airflow retarders placed on the interior side of the building envelope) or ASTM *Standard* E 1424 (test method for exterior airflow retarders). Specific air leakage criteria for airflow retarders in heating climates (Canada) can be found in Di Lenardo et al. (1995). These specifications call for maximum permissible air leakage rates between 0.01 and 0.04 cfm/ft^2 (as measured with an air pressure difference of 0.3 in. of water), depending on the water vapor permeance of the outermost layer of the building envelope. The highest recommended maximum permissible air leakage rate of the airflow retarder applies if the permeance of the outermost layer is greater than 14 perms, and the lowest rate applies if the permeance is between 0.26 and 1 perm. Intermediate values are provided as well. The recommendations apply only to heating climates.

The effectiveness of an airflow retarder can be greatly reduced if openings, even small ones, exist in it. Such openings can be caused by poor design, poor workmanship during application, poorly sealed joints and edges, insufficient coating thickness, improper caulking and flashing, uncompensated thermal expansion, mechanical forces, aging, and other forms of degradation. Common faults or leaks occur at electrical boxes, plumbing penetrations, telephone and television wiring, and other unsealed openings in the structure.

A ceiling airflow retarder needs to be continuous at chases for plumbing, ducts, flues, and electrical wiring. In flat roofing, mechanical fasteners are sometimes used to adhere the system to the deck. These often penetrate the airflow retarder, and the resulting hole may allow air and accompanying moisture leakage into the roof.

Because it resists airflow, an airflow retarder must withstand pressures exerted by chimney (stack) effects, wind effects, or both, during construction and over the life of the building. The magnitude of the pressure varies, depending on the type of building and the sequence of construction. At one extreme, single-family dwellings may be built with the exterior cladding partly or entirely installed and insulation in place before the airflow retarder is added. Chimney effects in such buildings are small even in cold weather, so stresses on the airflow retarder during construction are small.

At the other extreme, in tall buildings, wind and chimney-effect forces are much greater than they are in single-family or other low-rise buildings. A fragile, unprotected sheet material should not be used as an airflow retarder (or vapor retarder) in a tall building because it will probably be torn by the wind before construction is completed.

In summary, an airflow retarder must

- Meet air permeability requirements
- Be continuous
  1. Tight joints in the airflow retarder must be constructible
  2. Effective bonds in the airflow retarder must be made at intersections (e.g., wall/roof)
  3. It must accommodate dimensional changes due to temperature or shrinkage without damage to joints or the airflow retarder material
- Be strong enough to support the stresses applied to it
  1. It must not be ruptured or excessively deformed by air pressures due to wind and stack effects
  2. Where an adhesive is used to complete a joint, it must be designed to withstand forces that might gradually peel it away

Note that a small penetration across an airflow retarder system may seriously affect its performance. It concentrates the air/vapor flow in such a way that large local deposits of water and ice are possible. Calculations of moisture flow and accumulation using permeance values are useless when airflow is involved.

In addition, the following properties may be important, depending on the application:

- Elasticity
- Thermal stability
- Fire and flammability resistance
- Inertness to deteriorating elements
- Ease of installation

More information on air leakage in buildings may be found in Chapter 25 (Ventilation and Infiltration).

Vapor Retarder Functions and Properties

A vapor retarder is a component that retards water vapor diffusion, but does not totally prevent its transmission. The requirements for buildings are entirely different than those for pipes and equipment. Conditions inside and outside of buildings vary continually, and air movement and ventilation can provide wetting as well as drying at various times. Moisture entering one side of a wall cavity can be stored and released at a later time, or transmitted immediately out of the cavity through the other side. Requirements for vapor retarders in building components are therefore not extremely stringent. In contrast, moisture that enters the insulation of cold storage facilities, cold pipes, or equipment is unlikely to escape, except during periods when the facilities, pipes, or equipment are not in use and are allowed to warm up. Vapor retarders for cold pipe or equipment applications must therefore have an extremely low permeance (e.g., less that than 0.05 perm, or lower for severe conditions). However, Korsgaard (1993) demonstrated that water condensing on an insulated cold pipe can be removed continuously by wicking action with a specially designed and installed wick system, as long as pipe temperatures are above freezing.

In HVAC applications, vapor retarders are applied to thermal insulation on tanks, cold pipes, ducts, refrigerated enclosures, and buildings. If conditions are conducive to condensation, water vapor retarders help (1) keep the insulation dry, thereby reducing the cooling load; (2) prevent structural damage by rot, corrosion, or the expansion of freezing water; and (3) reduce paint problems on exterior wall construction (ASTM *Standard* C 755).

In addition to vapor permeance, the following properties of vapor retarders are important, depending on the application:

- Mechanical strength in tension, shear, impact, and flexure
- Adhesion
- Elasticity
- Thermal stability
- Fire and flammability resistance
- Inertness to other deteriorating elements
- Ease of fabrication, application, and joint sealing

Any vapor retarder's effectiveness depends on its vapor permeance, installation, and location within the insulated section. The vapor retarder is usually located at or near the surface exposed to the higher water vapor pressure. For residences with heating systems, this is usually the winter-warm side of the insulation.

Under conditions of reversible water vapor flow that can occur during temperature cycling of industrial insulations or of special purpose buildings, the selection and location of water vapor retarders require special study and treatment (Stachelek 1955).

Vapor retarder material is usually a thin sheet or coating. However, a construction of several materials, some perhaps of substantial thickness, could also constitute a vapor retarder system.

Korsgaard and Pedersen (1989, 1992) describe a vapor retarder composed of a synthetic fabric sandwiched between staggered strips of plastic film. The fabric wicks free water from the building envelope, while the plastic film retards vapor flow into it. Figure 10 shows a cathedral ceiling application and the physical principles involved.

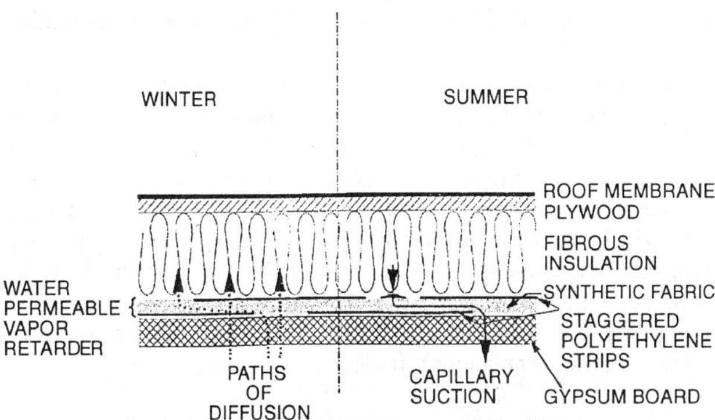

Fig. 10 Cathedral Ceiling with Water-Permeable Vapor Retarder

Water vapor permeances and permeabilities of some vapor retarders and other building materials are given in Table 9 in Chapter 24.

Vapor retarder material for pipes is usually sheet metal with soldered seams; heavy foil with wide, sealed overlaps; plastic pipe; or other very low permeance systems.

Classification of Vapor Retarders

Historically, a material or system with a permeance of 1 perm or less qualifies as a vapor retarder. More recently, further classification of vapor retarders has been proposed. For example, the Canadian General Standards Board (CGSB) has specified Type I vapor retarders as retarders with a permeance of 0.25 perm or less, and Type II as retarders with a permeance of 0.75 perm or less before aging and 1 perm or less after aging.

Water vapor retarders are classified as rigid, flexible, or coating materials. Rigid retarders include reinforced plastics, aluminum, and stainless steel. These retarders usually are mechanically fastened in place and are vapor-sealed at the joints.

Flexible retarders include metal foils, laminated foil and treated papers, coated felts and papers, and plastic films or sheets. Such retarders are supplied in roll form or as an integral part of a building material (e.g., insulation). Accessory materials are required for sealing joints.

Coating retarders may be semifluid or mastic; paint (arbitrarily called surface coatings); or hot melt, including thermofusible sheet materials. Their basic composition may be asphaltic, resinous, or polymeric, with or without pigments and solvents, as required to meet design conditions. They can be applied by spray, brush, trowel, roller, dip or mop, or in sheet form, depending on the type of coating and surface to which it is applied. Potentially, each of these materials is an airflow retarder; however, to meet airflow retarder specifications, it must satisfy the requirements for strength, continuity, and air permeance.

Designers have many options. For example, the conditions for control of airflow and moisture movement might be achieved using an interior finish, such as drywall, to provide strength and stiffness, along with a low-permeability coating, such as a vapor retarder paint, to provide the required low level of permeance. In this case, edge sealing is needed to establish continuity with adjacent airflow retarder/vapor retarder components.

Other designs may use more than one component. However, (1) any component that qualifies as a vapor retarder usually also impedes airflow, and is thus subject to pressure differences that it must resist, and (2) any component that impedes airflow often also retards vapor movement and may promote condensation or frost formation.

Additional information regarding the control of moisture and airflow through the use of vapor retarders and airflow retarders may be found in Chapters 23 and 25 and in Construction Specifications Canada (1990).

Several studies have reported a significant increase in the apparent permeance of vapor retarders as a result of small holes in the vapor retarder. For example, Seiffert (1970) reports a 100-fold increase in the permeance of aluminum foil when it is 0.014% perforated, and a 4000-fold increase when 0.22% of the surface is perforated. In general, penetrations particularly degrade a vapor retarder's effectiveness if the vapor retarder has a very low permeance (e.g., polyethylene or aluminum foil). In addition, perforations may lead to additional air leakage, which further erodes the effectiveness of the vapor retarder.

STEADY-STATE DESIGN TOOLS

Traditional methods for moisture design of the exterior building envelope all have severe limitations, and the results are sometimes difficult to interpret. However, these methods are used by design professionals and form the basis for current codes dealing with moisture control and vapor retarders.

The three best-known manual steady-state design tools for evaluating the probability of condensation within exterior envelopes (exterior walls, roofs, or ceilings) are (1) the dew-point method, (2) the Glaser diagram, and (3) the Kieper diagram. All three methods compare the vapor pressures within the envelope, as calculated by simple vapor diffusion equations, with the saturation pressures, which are based on the calculated temperatures within the envelope. If the calculated vapor pressure is above the saturation pressure at any point within the envelope, condensation is indicated.

The dew-point method, used in North America, and the Glaser diagram, commonly used in Europe and elsewhere, are almost identical. They differ slightly in the formulation of the vapor diffusion equation for flow through a building material and in definition of terms; the main difference lies in the graphical procedures. These methods are often misused, especially when condensation is present. Like the dew-point method and Glaser diagram, the Kieper diagram is based on vapor diffusion theory, although provisions can be added to include the approximate effect of cavity ventilation and air movement.

Many people advocate abandoning steady-state design tools because of their severe limitations. Perhaps the greatest limitation is that their focus is restricted to prevention of sustained interstitial condensation. Many building failures (such as mold and mildew, buckling of siding, and paint failure) are not necessarily related to surface condensation. Conversely, limited condensation can often be tolerated, depending on the materials involved, the temperature conditions, and the speed at which the material dries out. Wetting and drying cycles cannot be accurately analyzed with steady-state tools because these tools neglect moisture storage in the building materials. Another weakness is that these methods exclude all moisture transfer mechanisms other than vapor diffusion. Results obtained with any of these methods should therefore be considered as approximations and be used with extreme care. The validity and usefulness of any of these methods depend on the judicious selection of boundary conditions and material properties. Specifically, they should only be used to estimate seasonal mean conditions, rather than daily or even weekly mean conditions. Furthermore, the user should remember that permeances may vary with moisture content and that the effect of rain splash, flashing imperfections, leaky or poorly formed joints, weather exposure, and sunshine can have overriding effects.

For those who want to use these simplified tools, despite their shortcomings, short descriptions of the dew-point method and the Kieper diagram are presented, with examples of their use. The Glaser diagram parallels the dew-point method to such an extent that it is not described separately in this chapter. A comprehensive description of all three methods can be found in TenWolde (1994).

Table 3 Approximate Thermal and Vapor Diffusion Properties of Wall in Example 1

Air Film or Material	Thermal Resistance R, °F·ft²·h/Btu	Permeance M, perm	Diffusion Resistance Z, rep
Air film (still)	0.68	160[a]	0.0063[a]
Gypsum board, painted	0.45	5	0.2
Insulation	11	30	0.033
Plywood sheathing	0.62	0.5	2
Wood siding	1[b]	35[b]	0.029[b]
Air film (wind)	0.17	1000[a]	0.001[a]
Total	13.92	Not applicable	2.27

[a]Approximate values; permeances of surface air films are very large compared to those of other materials and do not affect results of calculations.
[b]Approximate values; permeance reflects limited ventilation of back of siding.

Dew-Point Method

The dew-point method is based on a slight modification of diffusion Equation (19):

$$w = -\mu \frac{\Delta p}{d} \tag{22}$$

where

w = water vapor flux through a layer of material, grains/h·ft²
μ = water vapor permeability of material, perm·in.
Δp = vapor pressure difference across the layer, in. Hg
d = thickness of the layer, in.

The term μ/d represents the permeance of the material. Water vapor resistance Z is defined as the inverse of the permeance and is expressed in reps (where rep = 1/perm). Then Equation (22) can be written as

$$w = -\frac{\Delta p}{Z} \tag{23}$$

Example 1. The dew-point method is explained and demonstrated for the frame wall construction and materials described in Table 3. Assume 70°F, 40% rh indoors, and 20°F, 50% rh outdoors.

Solution:

Step 1. Calculate the temperature drop across each material. The temperature drop is proportional to the R-value as follows:

$$\frac{\Delta T_{material}}{\Delta T_{wall}} = \frac{R_{material}}{R_{wall}}$$

Table 4 lists the resulting temperature drops and resulting temperatures at each surface.

Step 2. Find the saturation vapor pressures corresponding to the surface temperatures (Table 4). These values can be found in Table 2 in Chapter 6, or in other psychrometric tables or charts.

Step 3. Calculate the vapor pressure drops across each material. These are calculated in much the same way as the temperature drops in step 1.

$$\frac{\Delta p_{material}}{\Delta p_{wall}} = \frac{Z_{material}}{Z_{wall}}$$

where

p = vapor pressure, in. of Hg
Z = vapor diffusion resistance, rep

From Table 3, the total resistance of the wall with the vapor retarder is

Z_{wall} = 1/160 + 1/5 + 1/30 + 1/0.5 + 1/35 + 1/1000 = 2.27 reps

The vapor pressure drop across the wall is calculated from indoor and outdoor relative humidities and the indoor and outdoor saturation vapor pressures (see Table 4).

$$\Delta p_{wall} = p_{indoor} - p_{outdoor}$$
$$= (40/100)\ 0.7392 - (50/100)\ 0.1096 = 0.2409 \text{ in. Hg}$$

As with temperatures, the vapor pressures at the surfaces of each material can easily be determined from the vapor pressure drops. Table 5 lists the results for the example wall in the initial calculation column.

Step 4. Figure 11 shows the saturation and calculated vapor pressures. Comparison with saturation pressures reveals that the calculated vapor pressure on the interior surface of the sheathing (0.2702 in. Hg) is well above the saturation pressure at that location (0.1434 in. Hg). This indicates condensation, probably on the surface of the sheathing. *It does not indicate condensation within the insulation.*

If the location of the condensation or the condensation rate is of interest, additional calculations (steps 5 and 6) are necessary.

Step 5. Figure 11 shows that the calculated vapor pressure exceeds the saturation vapor pressure by the greatest amount at the interior surface of the plywood sheathing. Therefore, this is the most likely location for condensation to occur. With condensation at that surface, the vapor pressure should equal the saturation vapor pressure at that location (see the final calculation column in Table 5).

Step 6. The change of vapor pressure on the plywood sheathing alters all other vapor pressures as well as the vapor flow through the wall. The calculation of vapor pressures is similar to the calculation in step 3, but the wall is now divided into two parts: one part on the interior of the condensation plane (i.e., gypsum board and insulation) and the other part on the exterior (plywood sheathing and wood siding). The vapor pressure drop over the first part of the wall is

$$\Delta p_1 = 0.2957 - 0.1434 = 0.1523 \text{ in. Hg}$$

and that over the second part is

$$\Delta p_2 = 0.1434 - 0.0548 = 0.0886 \text{ in. Hg}$$

The vapor diffusion resistances of both parts of the wall are

$$Z_1 = 1/160 + 1/5 + 1/30 = 0.24 \text{ rep}$$

$$Z_2 = 1/0.5 + 1/35 + 1/1000 = 2.03 \text{ reps}$$

The vapor pressure drops across each material can now be calculated from

$$\frac{\Delta p_{material}}{\Delta p_i} = \frac{Z_{material}}{Z_i} \quad \text{for } i = 1, 2$$

Final calculations of vapor pressure are shown in Table 5. The vapor pressure no longer exceeds the saturation vapor pressure, which means that the condensation plane was chosen correctly. Figure 11 shows the vapor pressure profile (labeled as Vapor pressure, final calculation).

Vapor flow is no longer the same throughout the wall: vapor flow into the wall from the indoor air increases as a result of the lower vapor pressure at the plywood surface, while flow from the wall to the outside decreases. The difference between the two flows is the rate of moisture accumulation:

$$w_c = \Delta p_1/Z_1 - \Delta p_2/Z_2 = 0.1523/0.24 - 0.0886/2.03 = 0.59 \text{ gr/h} \cdot \text{ft}^2$$

This amounts to about 0.002 lb/day·ft². In our example, the plywood surface is below freezing, and this moisture would probably accumulate as frost. The moisture content of the plywood would be increased by 1% after about a week of condensation at this rate.

Table 4 Calculated Temperature Drops, Surface Temperatures, and Saturation Vapor Pressures in Example 1

Air Film or Material or Surface	Temperature Drop, °F	Surface Temperature, °F	Saturation Vapor Pressure at Surface, in. Hg
Indoor air	—	70	0.7392
Surface air film	2.4	—	—
Interior wall surface	—	67.6	0.6807
Gypsum board	1.7	—	—
Gypsum board/Insulation	—	65.9	0.6419
Insulation	39.5	—	—
Insulation/Sheathing	—	26.4	0.1434
Plywood sheathing	2.2	—	—
Sheathing/Siding	—	24.2	0.1309
Wood siding	3.6	—	—
Exterior wall surface	—	20.6	0.1124
Surface air film	0.6	—	—
Outdoor air	—	20	0.1096

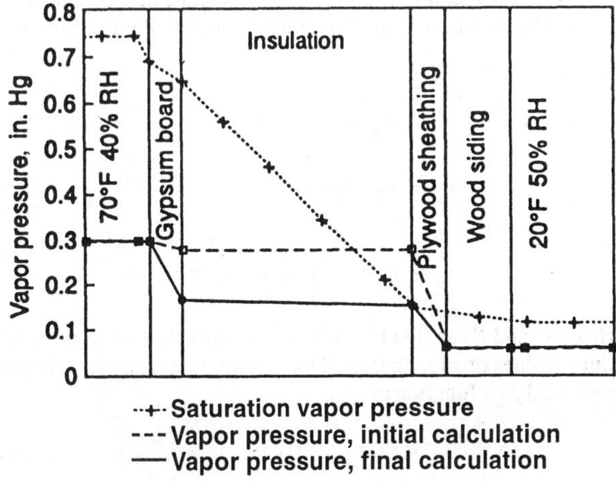

Fig. 11 Dew-Point Calculation in Example 1

Table 5 Initial and Final Calculations of Vapor Pressure Drops and Surface Vapor Pressures in Example 1

Air Film or Material or Surface	Saturation Vapor Pressure, in. Hg	Initial Calculation Vapor Pressure, in. Hg		Final Calculation Vapor Pressure, in. Hg	
		Drop	At Surface	Drop	At Surface
Indoor air (40% rh)	0.7392	—	0.2957	—	0.2957
Surface air film	—	0.0007	—	0.0040	—
Interior wall surface	0.6807	—	0.2950	—	0.2917
Gypsum board	—	0.0212	—	0.1271	—
Gypsum board/Insulation	0.6419	—	0.2738	—	0.1646
Insulation	—	0.0036	—	0.0212	—
Insulation/Sheathing	0.1434	—	0.2702	—	0.1434
Plywood sheathing	—	0.2123	—	0.0874	—
Sheathing/Siding	0.1309	—	0.0579	—	0.0560
Wood siding	—	0.0030	—	0.0012	—
Exterior wall surface	0.1124	—	0.0549	—	0.0548
Surface air film	—	0.0001	—	0.00004	—
Outdoor air (50% rh)	0.1096	—	0.0548	—	0.0548

The dew-point method can be summarized as follows:

1. Calculate temperature drops and surface temperatures.
2. Find corresponding saturation vapor pressures.
3. Calculate vapor pressure drops and vapor pressures.
4. Check whether the saturation pressure is above the vapor pressure at all surfaces; if so, no condensation is indicated. Vapor flow through the wall may be determined if desired.

If condensation is indicated, the following steps may be followed:

5. Select the condensation surface; the vapor pressure at this surface equals the saturation vapor pressure.
6. Recalculate the vapor pressures; if any vapor pressures are above saturation, steps 5 and 6 should be repeated with a different condensation surface.
7. If needed, calculate the rate of condensation.

Kieper Diagram

The Kieper diagram was first introduced by Kieper et al. (1976) and described in greater detail by Trethowen (1979) and TenWolde (1983, 1994). As with the dew-point method and the Glaser diagram, the Kieper diagram is entirely based on vapor diffusion theory. The advantages of this method are that (a) the same diagram can be used for different wall configurations, as long as indoor and outdoor conditions are not changed, and (b) the calculation does not need to be repeated if condensation is indicated. Rather than graphing vapor pressures and saturation pressures, the Kieper diagram uses two parameters, x and y, which represent normalized thermal properties and vapor diffusion properties of the materials in the wall, respectively. The normalized thermal property x is defined as follows:

$$x_1 = \frac{R_1}{R_{wall}}$$

$$x_2 = x_1 + \frac{R_2}{R_{wall}}$$

$$x_n = x_1 + \ldots + x_{n-1} + \frac{R_n}{R_{wall}} \qquad (24)$$

where R_1 and R_2 are the R-values of the individual materials and air films. Values of x range from 0 to 1. Temperature in the wall can be expressed as a function of x:

$$t(x) = t_i - x(t_i - t_o) \qquad (25)$$

where

t_i = indoor temperature
t_o = outdoor temperature

The vapor diffusion y parameter is defined similarly as

$$y_n = y_1 + \ldots + y_{n-1} + \frac{Z_n}{Z_{wall}} \qquad (26)$$

Values of y also range from 0 to 1.

Lines of equal wetting potential ($w_c Z_{wall}$ in in. Hg) can be drawn in the Kieper diagram (TenWolde 1983, 1994 has derivation):

$$w_c Z_{wall} = \frac{p_i - p_s[t(x)] - y(p_i - p_o)}{y(y-1)} \qquad (27)$$

where

w_c = moisture accumulation rate, gr/h·ft^2
p_i = indoor vapor pressure
p_o = outdoor vapor pressure
$p_s[t(x)]$ = saturation vapor pressure at temperature $t(x)$ as defined in Equation (25)

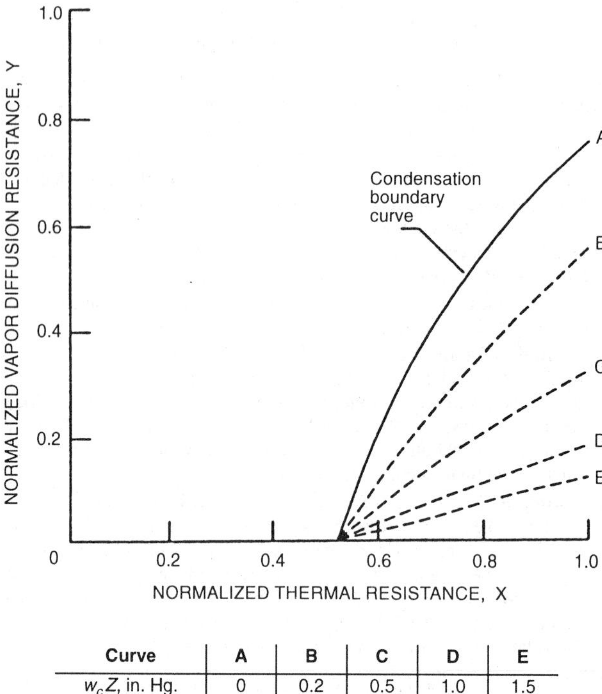

Curve	A	B	C	D	E
$w_c Z$, in. Hg.	0	0.2	0.5	1.0	1.5

Fig. 12 Kieper Diagram for 70°F, 40% rh Indoors and 20°F, 50% rh Outdoors

If w_c is positive, condensation (wetting) is indicated; if negative, evaporation (drying) takes place. The right-hand side of Equation (27) includes only x, y, and the indoor and outdoor vapor pressures; it contains no material property parameters. Thus, the curves of equal wetting potential in the Kieper diagram depend only on indoor and outdoor conditions. The curve where the wetting potential is zero is sometimes called the **condensation boundary curve**. Figure 12 shows the Kieper diagram with the curves for 70°F, 40% rh indoors and 20°F, 50% rh outdoors. For milder climates, the set of curves moves to the right.

Example 2: Table 6 shows the x and y values associated with the frame wall without a vapor retarder from Example 1. When the wall profile is entered in the Kieper diagram (Figure 13), it penetrates the condensation region (the area below the condensation boundary curve). The diagram shows that the plywood surface (point P in Figure 13) has the greatest wetting potential.

From Figure 13, the wetting potential at the plywood surface can be estimated to be

$$w_c Z = 1.4 \text{ in. Hg}$$

With Z = 2.27 reps, the estimated rate of condensation is

$$w_c = 1.4/2.27 = 0.6 \text{ gr/h} \cdot \text{ft}^2$$

When a vapor retarder is added (see Table 6), Figure 13 shows that the wall curve is entirely outside the condensation region and that no condensation is indicated.

One advantage of the Kieper approach is that Equation (27) can be used directly without use of a diagram to calculate drying or wetting rates at various locations inside a wall or ceiling. Equation (27) can also easily be incorporated into a computer program or standard computer spreadsheet.

Both the dew-point method and the Kieper diagram allow simple estimation of the effect of wall or roof cavity ventilation by representing the effect of ventilation on thermal and vapor transport through the addition of parallel thermal and vapor diffusion resistances (Trethowen 1979, TenWolde and Carll 1992). The parallel

Table 6 Kieper Diagram: x and y Values for Frame Wall With and Without Vapor Retarder in Example 2

Air Film or Material	Thermal Resistance, °F·ft²·h/Btu	Permeance, perm	Diffusion Resistance, rep	x	y Without Vapor Retarder	y With Vapor Retarder
Air film (still)	0.68	160[a]	0.006[a]	0.049	0.003	0.0003
Gypsum board, painted	0.45	5	0.2	0.081	0.091	0.011
Vapor retarder	—	0.06	16.67	0.081	—	0.891
Insulation	11	30	0.033	0.871	0.105	0.893
Plywood sheathing	0.62	0.5	2	0.916	0.986	0.998
Wood siding	1[b]	35[b]	0.029[b]	0.988	0.999	1.000
Air film (wind)	0.17	1000[a]	0.001[a]	1.000	1.000	1.000
Total						
Without vapor retarder	13.92	Not applicable	2.27	—	—	—
With vapor retarder	13.92		18.94			

[a]Approximate values; permeances of surface air films are very large compared to those of other materials and do not affect results of calculations.

[b]Approximate values; permeance reflects limited ventilation of back of siding.

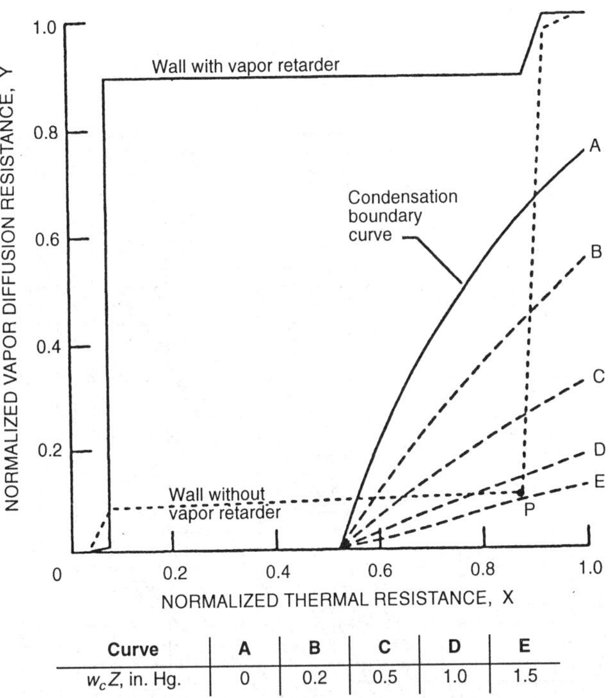

Curve	A	B	C	D	E
$w_c Z$, in. Hg.	0	0.2	0.5	1.0	1.5

Fig. 13 Kieper Diagram with Curves for Frame Walls With and Without Vapor Retarder in Example 2

resistances account for the heat and vapor bypassing the exterior material layers with outside ventilation air. The magnitude of the parallel resistances may be determined from the following equations:

$$R_{t,par} = \frac{S}{Q\rho c_p} \tag{28}$$

$$R_{v,par} = \frac{S}{Q\rho c} \tag{29}$$

where

$R_{t,par}$ = parallel thermal resistance, °F·ft²·h/Btu
$R_{v,par}$ = parallel vapor flow resistance, rep
S = surface area of the wall or ceiling, ft²
Q = cavity ventilation airflow rate, ft³/h
ρ = density of air, lb/ft³
c = ratio of humidity ratio and vapor pressure, approximately 145 gr/lb·in. Hg
c_p = specific heat, Btu/lb·°F

Although this method of including ventilation only approximates the actual effects of ventilation, it can be a useful tool.

MATHEMATICAL MODELS

In a review of the state of the art of heat and moisture transport modeling for buildings, 25 different models of various complexity were identified (Hens 1996). Most of the models are research tools that may be too complex for users other than researchers. The following models are readily available:

- Glasta (Physibel, Belgium)
- EMPTEDD (Trow, Toronto, Canada)
- Match (Technical University, Lyngby, Denmark)
- COND (Technical University, Dresden, Germany)
- MOIST (NIST, Gaithersburg, MD)

An extensive discussion of most of these models can be found in a final report of the International Energy Agency Annex 24 (Hens 1996), and a more general discussion in Ojanen et al. (1994).

Some models are steady-state, but most are transient. The advantages of most of the transient models include more complete transport equations and the ability to analyze wetting and drying phenomena. The disadvantages are that they tend to be more complex and less easily understood, slower and more difficult to use, and difficult to modify.

PREVENTING SURFACE CONDENSATION

Condensation occurs when water vapor comes in contact with a surface that has a temperature lower than the dew point of that vapor. To prevent condensation from occurring on the warm side of insulated rooms, pipes, ducts, and equipment, the insulation should be sufficiently thick to ensure that the insulation surface temperature always exceeds the dew-point temperature.

Figure 14 illustrates the steady-state heat transfer for which the following relationship can be used to calculate the data in Figure 15.

$$q = \frac{t_b - t_p}{R_s} = \frac{t_p - t_o}{R_i} \tag{30}$$

since

$$\frac{R_s}{R_i} = \frac{t_b - t_p}{t_p - t_o}$$

$$R_i = \frac{L}{k}$$

$$L = kR_i = \frac{kR_s(t_p - t_o)}{t_b - t_p} \tag{31}$$

where

q = heat flux through unit area perpendicular to the flow, Btu/h·ft²
t_b = ambient still-air dry-bulb temperature, °F
t_p = dew point, °F
t_o = operating temperature, °F
R_s = surface thermal resistance, °F·ft²·h/Btu
R_i = insulation thermal resistance, °F·ft²·h/Btu
L = thickness, in. (After calculating L, Figure 16 can be used to obtain the actual thickness as needed for pipe insulation.)
k = mean thermal conductivity, Btu·in/h·ft²·°F

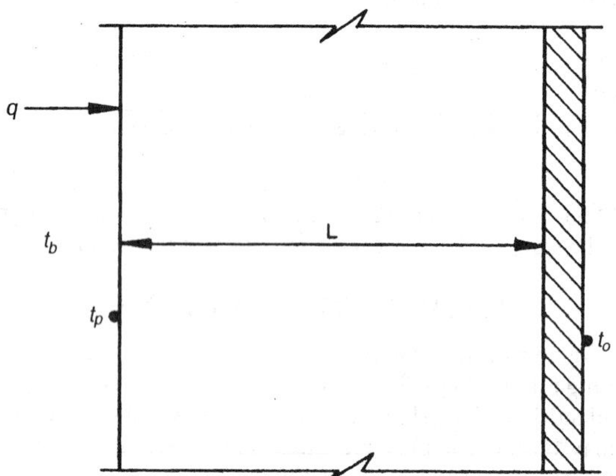

Fig. 14 Steady-State Heat Transfer

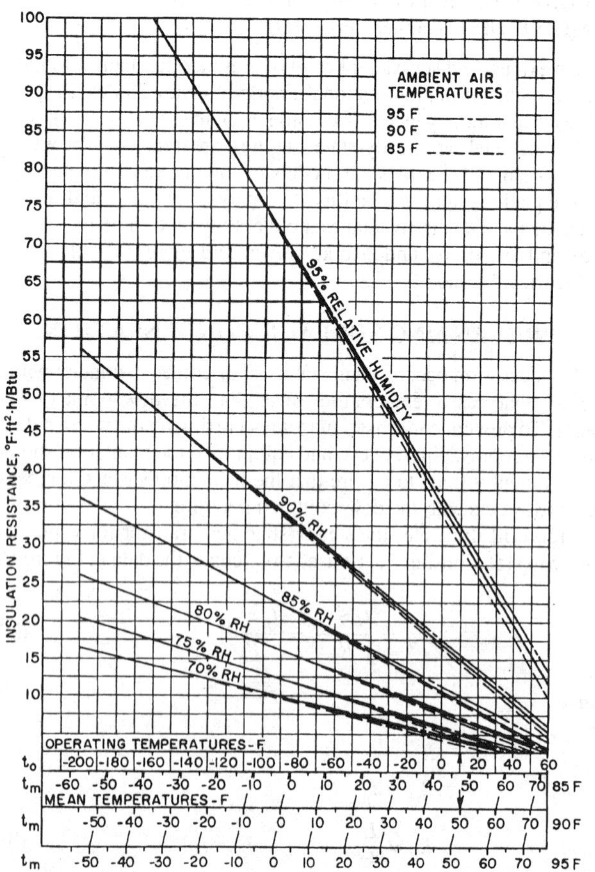

Fig. 15 Thermal Resistance of Insulation to Prevent Surface Condensation

Equivalent thickness is the thickness of insulation on a flat surface required to give the same rate of heat transmission per unit area of outer surface of insulation as on a cylinder or pipe:

$$L = r_s \ln\left(\frac{r_s}{r_i}\right) \tag{32}$$

where

r_s = outer radius of insulation
r_i = inner radius of insulation

The approximate thickness of insulation L required to prevent condensation on flat metallic surfaces can be obtained from Figure 15, which shows the approximate thermal resistance R needed to prevent condensation. The formula $L = Rk$ can then be used to find the thickness. In Figure 15, t_m is the mean temperature for a specific insulating material and is the arithmetic average of t_b and t_o:

$$t_m = \frac{t_b + t_o}{2}$$

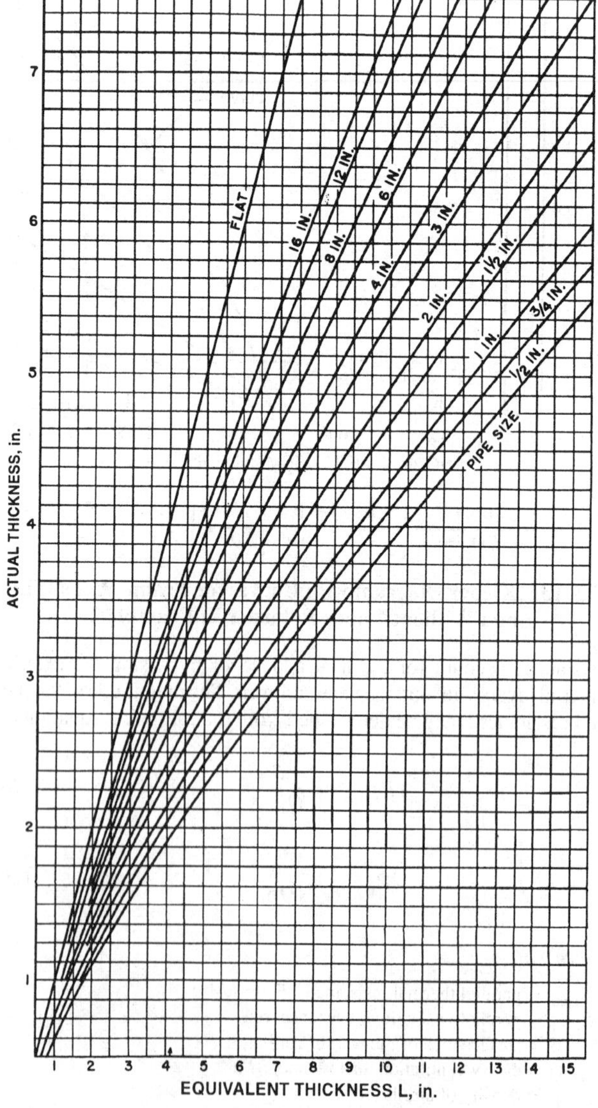

Fig. 16 Conversion of Equivalent Thickness to Actual Thickness for Pipe Insulation

A more accurate determination of the insulation thickness L needed to control condensation can be made with Equations (30) and (31). When pipes are involved, Figure 16 can be applied to relate the thickness of insulation, as determined by either of the methods discussed previously, to the actual thickness needed. By entering Figure 16 with L as the thickness needed on a flat surface (noted as equivalent thickness) and moving vertically until the curve for the pipe size involved is intersected, the actual thickness needed is found by moving horizontally to the y axis. When choosing pipe insulation, always use the next greater nominal thickness than the thickness determined.

Example 3. A nominal 4 in. pipe operates at 10°F in ambient still-air conditions of 90°F and 90% rh. The mean temperature is 50°F. The insulation specified has a thermal conductivity k of 0.28 Btu·in/h·ft^2·°F at the mean temperature, according to manufacturers' tables.

Solution: Using Figure 15, locate the insulation resistance needed to prevent condensation and obtain R = 14.8 °F·ft^2·h/Btu. Since $R = L/k$, the thickness required for a flat surface or the equivalent thickness L for any pipe size is

$$L = Rk = 14.8 \times 0.28 = 4.14 \text{ in.}$$

To find the actual thickness needed on a 4 in. pipe, read the Equivalent Thickness Scale in Figure 16 from L of 4.1 in. to pipe size of 4 in. and find the actual thickness of 2.9 in. on the ordinate. Select the next higher standard thickness of 3 in.

REFERENCES

ASHRAE. 1989. Thermal performance of the exterior envelopes of buildings IV. Proceedings of the Fourth ASHRAE/DOE/BTECC/CIBSE Conference.

ASHRAE. 1992. Thermal performance of the exterior envelopes of buildings V. Proceedings of the Fifth ASHRAE/DOE/BTECC Conference.

ASTM. 1974. Heat transmission measurements in thermal insulations. *Special Technical Publication* STP 544. American Society for Testing and Materials, West Conshohocken, PA.

ASTM. 1979. Thermal transmission measurements of insulation. *Special Technical Publication* STP 660.

ASTM. 1980. Thermal insulation performance. *Special Technical Publication* STP 718.

ASTM. 1983. Thermal insulation materials and systems for energy conservation in the 80s. *Special Technical Publication* STP 789.

ASTM. 1985. Building applications of heat flux transducers. *Special Technical Publication* STP 885.

ASTM. 1985. Guarded hot plate and heat flow meter methodology. *Special Technical Publication* STP 879.

ASTM. 1985. Standard practice for selection of vapor retarders for thermal insulation. *Standard* C 755-85 (R 1990).

ASTM. 1985. Standard test method for steady-state heat flux measurements and thermal transmission properties by means of the guarded-hot-plate apparatus. *Standard* C 177-85 (R 1993).

ASTM. 1988. Thermal insulation—Materials and systems. *Special Technical Publication* STP 922.

ASTM. 1989. Standard test method for steady-state thermal performance of building assemblies by means of guarded hot box. *Standard* C 236-89 (R 1993).

ASTM. 1990. Insulation materials, testing, and applications. *Special Technical Publication* STP 1030.

ASTM 1990. Standard practice for calculating thermal transmission properties from steady-state heat flux measurements. *Standard* C 1045-90.

ASTM. 1990. Standard terminology relating to thermal insulating materials. *Standard* C 168-90.

ASTM 1990. Standard test method for thermal performance of building assemblies by means of a calibrated hot box. *Standard* C 976-90.

ASTM. 1991. Insulation materials, testing, and applications, Second volume. *Special Technical Publication* STP 1116.

ASTM. 1991. Standard test method for steady-state heat flux measurements and thermal transmission properties by means of the heat flow meter apparatus. *Standard* C 518-91.

ASTM. 1992. Standard classification of potential health and safety concerns associated with thermal insulation materials and accessories. *Standard* C 930-92.

ASTM. 1995. Standard test method for determining the rate of air leakage through exterior windows, curtain walls, and doors under specified pressure differences across the specimen. *Standard* E 283-91.

ASTM. 1995. Standard test method for determining the rate of air leakage through exterior windows, curtain walls, and doors under specified pressure and temperature differences across the specimen. *Standard* E 1424-91.

ASTM. 1995. Standard test method for steady-state heat transfer properties of horizontal pipe insulation. *Standard* C 335-95.

ASTM. 1995. Standard test methods for water vapor transmission of materials. *Standard* E 96-95.

ASTM. 1996. *Annual book of ASTM standards*. Volume 04.06, *Thermal insulation and environmental acoustics*.

ASTM. 1996. Standard test method for estimating the long-term change in the thermal resistance of unfaced rigid closed cell plastic foams by slicing and scaling under controlled laboratory conditions. *Standard* C 1303-95.

Baker, M.C. 1964. Thermal and moisture deformation in building materials. *Canadian Building Digest* 56. National Research Council of Canada, Ottawa.

Bomberg, M. 1989. Testing water vapor transmission: Unresolved issues. In: *Water vapor transmission through materials and systems*. ASTM *Special Technical Publication* STP 1039:157-67.

Bomberg, M. and C. Shirtliffe. 1978. Influence of moisture and moisture gradients on heat transfer through porous building materials. ASTM *Special Technical Publication* STP 660:211-33.

Brandreth, D.A., ed. 1986. Advances in foam aging—A topic in energy conservation series. Caissa Editions, Yorklyn, DE.

BRS. 1974. Estimation of thermal and moisture movements and stresses, Parts 1, 2, and 3. *Building Research Station Digest* 227, 228, and 229, Watford, England.

Burch, D.M. and C.M. Hunt. 1978. Retrofitting an existing wood frame residence to reduce its heating and cooling energy requirements. *ASHRAE Transactions* 84:176.

Burge, H.A., H.J. Su, and J.D. Spengler. 1994. Moisture, organisms, and health effects. In: *Moisture control in buildings*. ASTM Manual *Series*: MNL 18, Chapter 6.

Chang, S.C. and N.B. Hutcheon. 1956. Dependence of water vapor permeability on temperature and humidity. *ASHRAE Transactions* 62:437.

Christian, J.E., A. Desjarlais, R. Graves, and T.L. Smith. 1995. Five-year field study confirms the PIMA standard for estimating polyisocyanurate insulation long-term thermal performance. 11th Conference of Roofing Technology, National Roofing Contractors Association, Rosemont, IL.

CIBSE. 1986. Thermal properties of building structures in *CIBSE Guide, Volume A, Design Data*. The Chartered Institution of Building Services Engineers, London, England.

Construction Specifications Canada. 1990. Tek-AID on Air Barrier Systems. 1 St. Clair Street W., Toronto, M4W 1K6, Canada.

Dechow, F.J. and K.A. Epstein. 1978. Laboratory and field investigations of moisture absorption and its effect on thermal performance of various insulations. ASTM *Special Technical Publication* STP 660:234-60.

Di Lenardo, B., W.C. Brown, W.A. Dalgleish, K. Kumaran, and G.F. Poirier. 1995. Technical guide for air barrier systems for exterior walls of low-rise buildings. Canadian Construction Materials Centre, National Research Council Canada, Ottawa, Ontario.

Donnelly, R.G., V.J. Tennery, D.L. McElroy, T.G. Godfrey, and J.O. Kolb. 1976. Industrial thermal insulation, An assessment. Oak Ridge National Laboratory *Report* TM-5283 and TM-5515, TID-27120.

Douglas, J.S., T.H. Kuehn, and J.W. Ramsey. 1992. A new moisture permeability measurement method and representative test data. *ASHRAE Transactions* 98(2):513-19.

Duff, J.E. 1971. The effect of air conditioning on the moisture conditions in wood walls. USDA Forest Service Research *Paper* SE-78. Madison, WI.

Epstein, K.A. and L.E. Putnam. 1977. Performance criteria for the protected membrane roof system. Proceedings of the Symposium on Roofing Technology. National Institute of Standards and Technology, Gaithersburg, MD, and National Roofing Contractors Association, Rosemont, IL.

FEA. 1976. Economic thickness for industrial insulation. GPO No. 41-018-00115-8. U.S. Government Printing Office, Washington, D.C.

FPL. 1987. *Wood handbook*. Agriculture Handbook No. 72, Forest Products Laboratory, Forest Service, U.S. Department of Agriculture.

Glaser, P.E., I.A. Black, R.S. Lindstrom, F.E. Ruccia, and A.E. Wechsler. 1967. Thermal insulation systems—A survey. NASA SP5027.

Hart, G.H. 1981. Heating the perimeter zone of an office building: An analytical study using the proposed ASHRAE comfort standard (55-74R). *ASHRAE Transactions* 87(2):529.

Hedlin, C.P. 1977. Moisture gains by foam plastic roof insulations under controlled temperature gradients. *Journal of Cellular Plastics* 13(5): 313-19.

Hedlin, C.P. 1983. Effect of moisture on thermal resistances of some insulations in a flat roof system under field-type conditions. ASTM *Special Technical Publication* STP 789:602-25.

Hedlin, C.P. 1985. Effect of insulation joints on heat loss through flat roofs. *ASHRAE Transactions* 91(2B):608-22.

Hedlin, C.P. 1987. Seasonal variations in the modes of heat transfer in a moist porous thermal insulation in a flat roof. *Journal of Thermal Insulation* 11:54-66.

Hedlin, C.P. 1988a. Heat flow through a roof insulation having moisture contents between 0 and 1% by volume, in summer. *ASHRAE Transactions* 94(2):1579-94.

Hedlin, C.P. 1988b. Heat transfer in a wet porous thermal insulation in a flat roof. *Journal of Thermal Insulation* 11:165-88.

Hens, H. 1996. Heat, air and moisture transfer in highly insulated envelope parts, task 1: Modelling. Final *Report*, Volume 1, International Energy Agency, Annex 24. Catholic University-Leuven, Laboratorium for Building Physics, Leuven, Belgium.

International Energy Agency. 1990. Guidelines & Practice, Volume 2. International Energy Agency, Annex XIV, Leuven, Belgium.

Jespersen, H.B. 1960. The effect of moisture on insulating materials made of plastic foam, impregnated mineral wool, vermiculite, and concrete. Denmark Technological Institute.

Joy, F.A. 1957. Thermal effect of moisture on insulation containing moisture. In: *Thermal conductivity measurements and applications of thermal insulations*. ASTM *Special Technical Publication* STP 217.

Kieper, G., W. Caemmerer, and A. Wagner. 1976. A new diagram to evaluate the performance of building constructions with a view to water vapor diffusion. Presented at CIB Working Group W40 Meeting, Washington, DC.

Knab, L.I., D.R. Jenkins, and R.G. Mathey. 1980. *The effect of moisture on the thermal conductance of roofing systems*. Building Science Series No. 123. National Institute of Standards and Technology, Gaithersburg, MD.

Korsgaard, V. 1993. Innovative concept to prevent moisture formation and icing of cold pipe insulation. *ASHRAE Transactions* 99(1):270-73.

Korsgaard, V. and C.R. Pedersen. 1989. Transient moisture distribution in flat roofs with hygro diode vapor retarder. Thermal Performance of Exterior Envelopes of Buildings IV. ASHRAE/DOE/BTECC/CIBSE Conference Proceedings. Available from ASHRAE.

Korsgaard, V. and C.R. Pedersen. 1992. Laboratory and practical experience with a novel water-permeable vapor retarder. In: *Thermal performance of the exterior envelopes of buildings V*, pp. 480-90. ASHRAE.

Kumaran, M.K. 1989. Experimental investigation on simultaneous heat and moisture transport through thermal insulation. Proceedings of the (Conseil International du Batiment-International Building Council) CIB 11th International Conference 2:275-84.

Kumaran, M.K., G.P. Mitalas, and M.T. Bomberg. 1994. Fundamentals of transport and storage of moisture in building materials and components. In: *Moisture control in buildings*. ASTM Manual *Series*: MNL 18, Chapter 1.

Kyle, D.M. and A.O. Desjarlais. 1994. Assessment of technologies for constructing self-drying low-slope roofs. Oak Ridge National Laboratory *Report* ORNL/CON-380. Oak Ridge, TN.

Lander, R.M. 1955. Gas is an important factor in the thermal conductivity of most insulating materials. *ASHRAE Transactions* 61:151.

Langlais, C., M. Hyrien, and S. Klarsfeld. 1983. Influence of moisture on heat transfer through fibrous insulating materials. ASTM *Special Technical Publication* STP 789:563-81.

Larsson, L.E., J. Ondrus, and B.A. Petersson. 1977. The protected membrane roof (PMR)—A study combining field and laboratory tests. Proceedings of the Symposium on Roofing Technology. National Institute of Standards and Technology, Gaithersburg, MD, and National Roofing Contractors Association, Rosemont, IL.

Lewis, J.E. 1979. Thermal evaluation of the effects of gaps between adjacent roof insulation panels. *Journal of Thermal Insulation* (October): 80-103.

Lotz, W.A. 1969. Facts about thermal insulation. *ASHRAE Journal* (June): 83-84.

Merrill, J.L. and A. TenWolde. 1989. Overview of moisture-related damage in one group of Wisconsin manufactured homes. *ASHRAE Transactions* 95(1):405-14.

Ojanen, T., R. Kohonen, and M.K Kumaran. 1994. Modeling heat, air, and moisture transport through building materials and components. In: *Moisture control in buildings*. ASTM Manual *Series*: MNL 18, Chapter 2.

Ostrogorsky, A.G. and L.R. Glicksman. 1986. Laboratory tests of effectiveness of diffusion barriers. *Journal of Cellular Plastics* 22:303.

Ovstaas, G., S. Smith, W. Strzepek, and G. Titley. 1983. Thermal performance of various insulations in below-earth-grade perimeter application. ASTM *Special Technical Publication* STP 789:435-54.

Palfey, A.J. 1980. Thermal performance of low emittance building sheathing. *Journal of Thermal Insulation* 3.

Paljak, I. 1973. Condensation in slabs of cellular plastics. *Materiaux et Constructions* 6(31):53.

Parmelee, G.V. and WW. Aubele. 1950. ASHVE Research *Report* No.1399. Heat flow through unshaded glass: Design data for load calculations. *ASHVE Transactions* 56:371.

Parmelee, G.V. and R.G. Huebscher. 1947. Forced convection heat transfer from flat surfaces. *ASHVE Transactions* 53:245.

Pedersen, C.R. 1990. Combined heat and moisture transfer in building construction. *Report* No.214. Thermal Insulation Laboratory, Technical University of Denmark, Lyngby, Denmark.

Pedersen, C.R., T.W. Petrie, G.E. Courville, P.W. Childs, and K.E. Wilkes. 1991. Moisture migration and drying rates for low slope roofs—Preliminary results. Proceedings of the 3rd International Symposium on Roofing Technology. National Roofing Contractors Association, Rosemont, IL.

Pelanne, C.M. 1977. Heat flow principles in thermal insulation. *Journal of Thermal Insulation* 1:48.

Pelanne, C.M. 1979. Thermal insulation heat flow measurements: Requirements for implementation. *ASHRAE Journal* 21(3):51.

Plewes, W.G. 1976. Upward deflection of wood trusses in winter. Building Research *Note* 107, National Research Council of Canada, Ottawa.

Raber, B.F. and F.W. Hutchinson. 1945. Radiation corrections for basic constants used in the design of all types of heating systems. *ASHVE Transactions* 51:213.

Robinson, H.E., F.J. Powlitch, and R.S. Dill. 1954. The thermal insulating value of airspaces. Housing and Home Finance Agency, Housing Research *Paper* No. 32, U.S. Government Printing Office, Washington, D.C.

Rowley, F.B., R.C. Jordan, C.E. Lund, and R.M. Lander. 1952. Gas is an important factor in the thermal conductivity of most insulating materials. *ASHVE Transactions* 58:15.

Rudoy, W. 1975. Effect of building envelope parameters on annual heating/cooling load. *ASHRAE Journal* 17(7):19.

Seiffert, K. 1970. *Damp diffusion and buildings*. Elsevier, Amsterdam, Netherlands.

Shapiro, A.P. and S. Motakef. 1990. Unsteady heat and mass transfer with phase change in a porous slab: Analytical solutions and experimental results. *International Journal of Heat and Mass Transfer* 33(1):163-73.

Shuman, E.C. 1980. Field measurement of heat flux through a roof with saturated thermal insulation and covered with black and white granules. ASTM *Special Technical Publication* STP 718:519-39.

Simons, E. 1955. In-place studies of insulated structures. *Refrigerating Engineering* 63:40 and 63:128.

Spielvogel, L.G. 1974. More insulation can increase energy consumption. *ASHRAE Journal* 16(1):61.

Stachelek, S.J. 1955. Vapor barrier requirements for cold storage structures. *Industrial Refrigeration* 34.

TenWolde, A. 1983. The Kieper and MOISTWALL moisture analysis methods for walls. In: *Thermal performance of the exterior envelopes of buildings II*, pp. 1033-51. ASHRAE.

TenWolde, A. 1994. Design Tools. In: *Moisture control in buildings*. ASTM Manual *Series*: MNL 18, Chapter 11.

TenWolde A. and C. Carll. 1992. Effect of cavity ventilation on moisture in walls and roofs. In: *Thermal performance of the exterior envelopes of buildings V*, pp. 555-62. ASHRAE.

Thomas, W.C., G.P. Bal, and R.J. Onega. 1983. Heat and moisture transfer in glass fiber roof-insulating material. ASTM *Special Technical Publication* STP 789:582-601.

Tobiasson, W. and J. Richard. 1979. Moisture gain and its thermal consequences for common roof insulations. Proceedings of the 5th Conference on Roofing Technology, National Institute of Standards and Technology, Gaithersburg, MD, and National Roofing Contractors Association, Rosemont, IL.

Tobiasson, W., A. Greatorex, and D. Van Pelt. 1987. Wetting of polystyrene and urethane roof insulations in the laboratory and on a protected membrane roof. In: *Thermal Insulation, Materials and Systems*. ASTM *Special Technical Publication* STP 922:421-30.

Tobiasson, W., A. Greatorex, and D. Van Pelt. 1991. New wetting curves for common insulations. In: *International Symposium on Roofing Technology*. National Institute of Standards and Technology, Gaithersburg, MD, and National Roofing Contractors Association, Rosemont, IL.

Trechsel, H.R. and M. Bomberg, eds. 1989. *Water vapor transmission through materials and systems*. ASTM *Special Technical Publication* STP 1039.

Trethowen, H.A. 1979. The Kieper method for building moisture design. BRANZ Reprint 12, Building Research Association of New Zealand, New Zealand.

Tye, R.P. 1987. Assessment of foam in-place urethane foam insulation used in buildings. ORNL/Sub-86/56525/1. Oak Ridge National Laboratory, Oak Ridge, TN.

Tye, R.P. 1988. Aging of cellular plastics: A comprehensive bibliography. *Journal of Thermal Insulation* 11:196-222.

Tye, R.P. and C.F. Baker. 1987. Development of experimental data on cellular plastic insulation under simulated winter exposure conditions. In: *Thermal Insulation, Materials and Systems*, ASTM *Special Technical Publication* STP 922:518-37.

Tye, R.P. and A.O. Desjarlais. 1981. Performance characteristics of foam-in-place urea formaldehyde insulation. ORNL/Sub-78/86993/1. Oak Ridge National Laboratory, Oak Ridge, TN.

Verschoor, J.D. 1977. Effectiveness of building insulation applications. USN/CEL *Report* No. CR78.006—NTIS No. AD-AO53 452/9ST.

Verschoor J.D. 1985. Measurement of water vapor migration and storage in composite building construction. *ASHRAE Transactions* 91(2):390-403.

Verschoor, J.D. and P. Greebler. 1952. Heat transfer by gas conductivity and radiation in fibrous insulations. *ASME Transactions* 74(6):961-68.

Wilkes, K.E. and P.W. Childs. 1992. Thermal performance of fiberglass and cellulose attic insulations. In: *Thermal performance of the exterior envelopes of buildings V*, pp. 357-67. ASHRAE.

Wilkes, K.E. and J.L. Rucker. 1983. Thermal performance of residential attic insulation E. *Energy and Buildings* 5:263-77.

BIBLIOGRAPHY

ASHRAE. 1979. Thermal performance of exterior envelopes of buildings. Proceedings of the First ASHRAE/DOE/BTECC Conference.

ASHRAE. 1982. Thermal performance of exterior envelopes of buildings II. Proceedings of the Second ASHRAE/DOE/BTECC Conference.

ASHRAE. 1985. Thermal performance of exterior envelopes of buildings III. Proceedings of the Third ASHRAE/DOE/BTECC Conference.

ASTM. 1989. Standard practice for determination of heat gain or loss and the surface temperature of insulated pipe and equipment systems by the use of a computer program. *Standard* C 680-89 (R 1995). American Society for Testing and Materials, West Conshohocken, PA.

ASTM. 1994. Standard test method for water vapor transmission rate of flexible barrier materials using an infrared detection technique. *Standard* F 372-94.

Bomberg, M., ed. 1991. Moisture research in North America: 25 Years of building science. Lund University, Lund, Sweden.

ECON-I. 1973. How to determine economic thickness of thermal insulation. Thermal Insulation Manufacturers Association, Mt. Kisco, NY.

FEA. 1974. Retrofitting existing housing for energy conservation: An economic analysis. Federal Energy Administration, Washington, D.C.

Hite, S.C. and J.L. Dray. 1948. Research in home humidity control. Research Series No. 106. Purdue University, Engineering Experiment Station, West Lafayette, IN.

Rose, W.B. and A. TenWolde, eds. 1993. *Bugs, Mold & Rot II: Workshop Proceedings*. National Institute of Building Sciences, Washington, DC.

Trechsel, H.R., ed. 1994. *Moisture control in buildings*. ASTM Manual *Series*: MNL 18.

CHAPTER 23

THERMAL AND MOISTURE CONTROL IN INSULATED ASSEMBLIES—APPLICATIONS

THERMAL and moisture design and long-term performance of buildings must be considered in the original planning phase. Installation of adequate insulation and moisture control assemblies during construction can be much more economical than installation later. Proper selection of thermal insulation and moisture control assemblies must be based on

- Thermal and moisture properties of the materials
- Other properties required by the location of the materials
- Space availability
- Compatibility of the materials with adjacent materials
- Interior and exterior climate

Types of thermal insulation, their properties, economic thickness, and principles of moisture control and moisture transport are discussed in Chapters 22 and 24. Insulation in various assemblies that can be used interchangeably for a given construction, as well as specific moisture control options for various climatic regions, are discussed in this chapter.

GENERAL BUILDING INSULATION PRACTICE

MAJOR CONSTRUCTION TYPES

Wood Frame Construction

Wood framing members and structural panels such as plywood, particleboard, and fiberboard only provide limited resistance to heat flow; therefore, wood frame construction is well suited to application of both cavity insulation and surface-applied insulation. The most common materials for cavity insulation are glass fiber, mineral fiber, cellulose, and spray-applied foams. For surface applications, a wide variety of sheathing insulations exist, such as rigid foam panels. Insulating sheathing placed on exterior walls may also have sufficient structural properties to provide required lateral bracing. Prefinished insulating paneling can be used as an inside finish on exterior walls, or on one or both sides of interior partitions; it can also underlay other finishes.

Roof decks of wood, metal, or preformed units may be insulated on top of or below the deck. Structural monolithic insulating decks must be at least 3 in. thick to provide an acceptable overall heat transmission coefficient. Additional thermal insulation can be placed on top of or below the deck.

Attic construction with conventional rafters and ceiling joists or roof trusses can be insulated between framing members with batt, blanket, or loose-fill insulation. In some warm climates, radiant barriers and reflective insulations can provide additional reduction of cooling loads. The United States Department of Energy's Radiant Barrier Attic Fact Sheet (U.S. Dept. of Energy 1991) provides information on climatic areas best suited for radiant barrier applications. This document also provides comparative information on the relative performance of these products versus conventional fibrous insulations.

Existing frame construction can be insulated by blowing in suitable loose-fill material. When using loose-fill materials in wall retrofit applications, extra care must be taken during the installation to eliminate voids within the wall cavity. All cavities should be checked prior to installation for obstructions such as fire stop headers and wiring that could prevent complete filling of the cavity. In addition, the material must be installed at the manufacturer's recommended density to ensure the desired thermal performance. Accessible space can be insulated by manual placement of batt, blanket, or loose-fill material.

In addition to being properly insulated, the exterior envelope of a building should be constructed to minimize airflow into or through the building envelope. Airflow may degrade the thermal performance of insulation and may cause excessive moisture accumulation in the building envelope. The use and function of airflow retarders are discussed in Chapter 22 and in this chapter.

Heavy Steel Frame Construction

The installation of preformed metal framing members is similar to that of wood frame construction, except that the physical dimensions of and spacing between members may differ. Preformed insulation material of the proper dimension may be installed if loose-fill or sprayed-in-place insulation is not used.

Structures having steel framing supports and exterior metal cladding are usually insulated between the frame and cladding with faced blanket or spray insulation as shown in Figure 1. In heating climates, the facing may serve as a combination vapor retarder and interior finish and must be protected against physical damage.

Other types of insulation can be used by adding framing or furring to the inside of frame or exterior siding. Securing insulation as shown in Figure 1 causes local reduction of thermal resistance. This reduction plus the thermal bridging caused by the screw or bolt

---

The preparation of this chapter is assigned to TC 4.4, Thermal Insulation and Moisture Retarders.

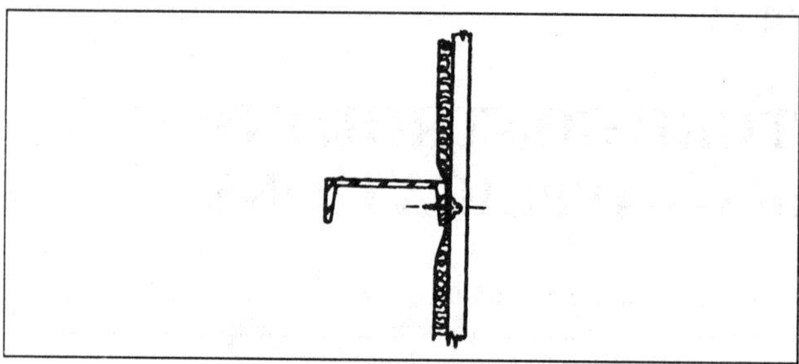

Fig. 1 Steel Frame Building Wall Insulated with Flexible Blanket

penetrating the insulation to the interior skin may result in condensation during cold weather.

Curtain walls may be (1) custom walls for specific projects, where most construction components are developed on a one-time basis; (2) commercial walls, where standard components are adapted to a particular building design; and (3) industrial-type walls, where standard metal sheets are fluted or ribbed to form field-assembled sandwiches to meet job conditions.

Curtain walls may consist of prefabricated panels or sections that may be classified as one of the following general types:

1. To withstand the elements, single-thickness facings are usually inserted in subframing with the windows to form a veneer over separate backup walls. Thermal insulation is integral with, or added to, the exterior of the backup wall and covers the edge of the floor slab or spandrel beam. In heating climates, this method reduces heat loss and keeps floors warm at their exterior edge. Insulating the exterior of the framing reduces thermal movement of the building structure (Figure 2).
2. Sandwich construction or adhesive-bonded panels are generally three-ply: exterior skin, core materials, and interior skin. This type of panel, when manufactured with concrete faces and an insulation core, comes in large sizes and can be attached directly to the exterior of the building frame. When using metal facings, the width of the panels is usually restricted to match that of the standard formed sheet. These panels are installed on the exterior of the building framing. The thermal performance required for a particular installation (including the interfaces between walls, floors, ceiling, doors, and windows) should be checked carefully because adding insulation later is difficult.
3. Mechanically fastened panels are generally the hollow box type, with the exterior facing nesting over the interior facing. The cavity can be filled with flexible or semirigid insulation (Figure 3). The edges should vent to the exterior of the building to allow the panel to serve as a rain screen. The panels are normally installed in a subframing system.
4. Industrial metal panels have an exterior facing of standard ribbed, corrugated V-beam materials and an interior facing of proprietary metal pans or standard corrugated or ribbed sheeting (Figure 4). Insulation varies from semirigid to rigid, depending on the design of the inner surfacing materials. It provides good thermal and moisture control when installed with tight inner surface construction.

All curtain wall panels with insulation in the cavity or as a core should be sufficiently tight at their edges to prevent the entrance of free water or moisture. However, because some moisture may enter the wall from the inside or outside, the wall should be capable of drying out. Panel edges should not be hermetically sealed. For cold climates, the subframing members or wall mullions should be of noncontinuous construction. The exterior to interior path should have a thermal break or insulated mullion cover on the interior or exterior (see Figure 5) to reduce the hazard of condensation.

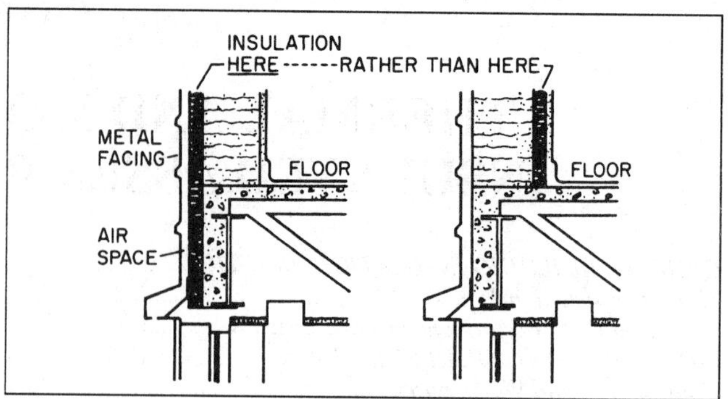

Fig. 2 Proper Placement to Insulate Structural Frame

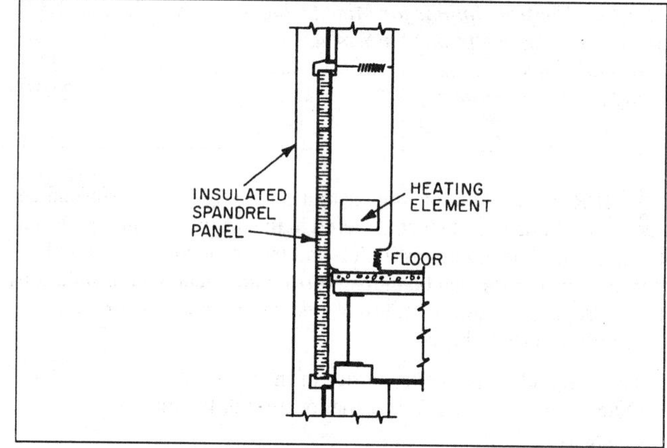

Fig. 3 Insulated Pan-Type Panel

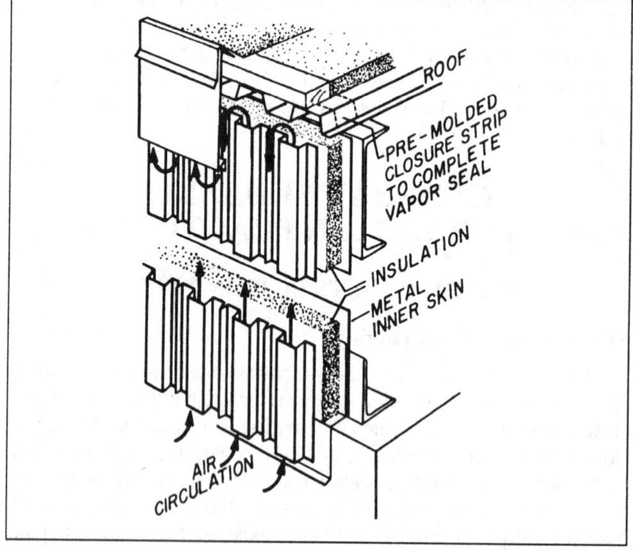

Fig. 4 Vented and Insulated Field-Fabricated Curtain Wall

Masonry Construction

Because concrete masonry unit walls and masonry cavity walls have hollow vertical cavities, insulating materials can usually be placed within the wall itself. Loose-fill insulation, such as water-repellent perlite and vermiculite, as well as foam inserts and foamed-in-place insulations, are commonly used. This insulation method is more effective with low-density masonry units. Rigid insulation can be placed between the wythes in a cavity wall or

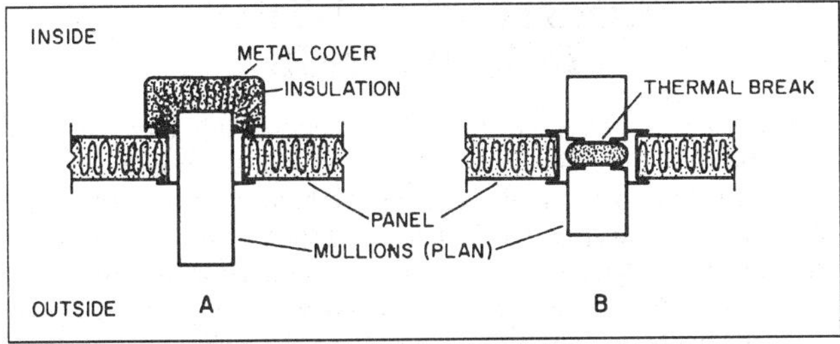

Fig. 5 Methods of Reducing Thermal Conduction Through Mullions

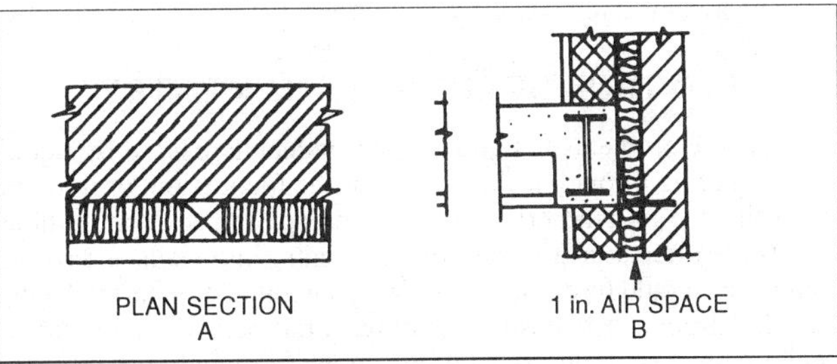

Fig. 6 Section of Insulated Walls

veneered construction (Figure 6B), and furring on the inside of the wall is still used in many areas (Figure 6A).

The thickness and density of concrete masonry wall construction coupled with its interior air cavities moderate the heat transmission. The thermal mass or inertia of such heavy construction causes a time lag in heat migration, which may lower the peak gains or losses. Insulation placed on the outside of a concrete masonry wall enhances this time lag effect and helps protect the structure from expansion and contraction caused by temperature extremes.

Resistance to water penetration in masonry walls depends on wall construction. For cavity wall construction, metal flashing at each floor and roof intersection and at windows and doors collects and diverts water out through weep holes in the masonry wall. For through-wall construction, waterproof coatings of portland cement, latex, or oil-base paints are used, as well as epoxy, silicone, and bituminous-based coatings. Depending on the severity of the moisture problem, both filler and finish coats can be used. Manufacturers' suggestions should be followed in such applications. Fill coats (fillers or primer-sealers) fill voids in coarse-textured masonry units and help smooth out surface irregularities before application of finish coats.

Whatever the construction, tooled concave or V-shaped mortar joints are preferred. These joints impede water infiltration because the joint tool compresses the mortar and forces it into holes or voids that may develop when laying the concrete masonry units. They also create a shape that sheds water.

In addition to augmenting effective thermal performance, proper insulation can enhance the fire resistance and noise-reduction qualities of masonry construction.

Precast or poured-in-place solid concrete walls are insulated similarly to solid masonry construction. The design and method of fabricating the panels dictates the type of insulation selected. Figure 7 details the joint of a typical precast concrete panel with sandwiched insulation. Note that the break in insulation (Figure 7) could cause a serious thermal short.

Foundation and Floor Systems

Perimeter insulation should be applied vertically on the inside (Figure 8A) or outside (Figure 8B) of foundation walls. Rigid mineral fiber or cellular plastic insulation are used as perimeter outside insulation (Figure 8B). Drainage (e.g., gravel-fill) may be required by

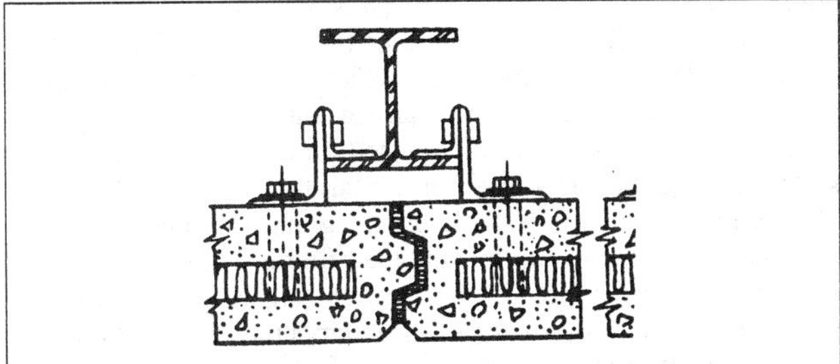

Fig. 7 Plan Section of Insulated Precast Concrete Panel

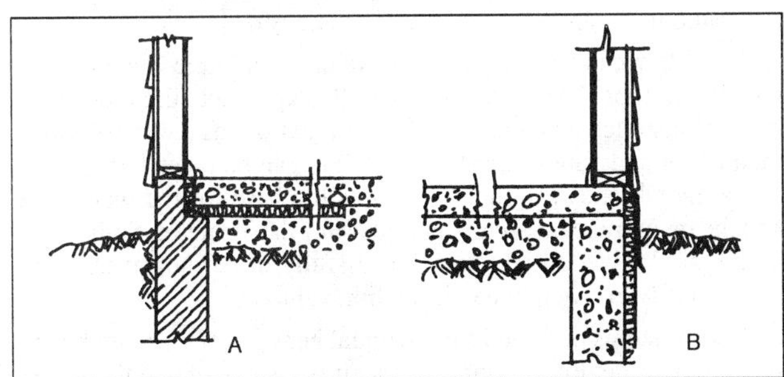

Fig. 8 Perimeter Insulation

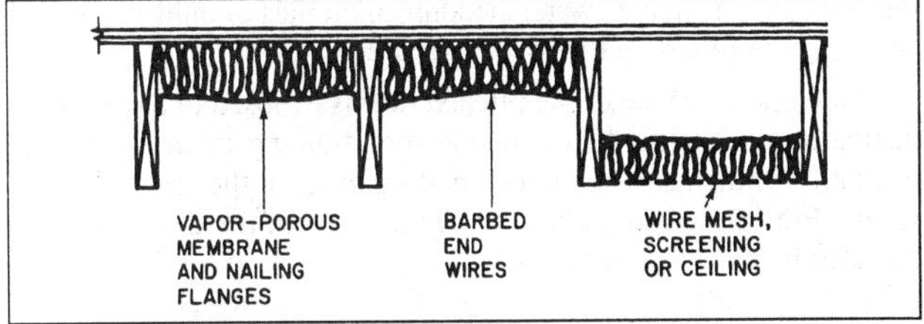

Fig. 9 Insulation Methods for Framed Floors

manufacturers for some perimeter insulations. Perimeter insulation usually needs protection from physical damage and ultraviolet radiation if it extends above grade level. Insulation can also be installed horizontally under the slab, from the foundation wall inward for a distance of 2 ft or more, often with vertically placed insulation.

Below-grade, back-filled walls enclosing usable and heated space can be insulated and finished in the same way as solid masonry or concrete walls. Unoccupied and unheated spaces below heated areas, such as crawl spaces, may have insulated walls to maintain crawl space temperatures and minimize the need for underfloor insulation. Details on below-grade insulation may be found in the *Building Foundation Design Handbook* (Labs et al. 1988).

Floor construction with joists can be insulated between the joists (Figure 9), but the insulation may need supplemental support. Material with an integral vapor retarder, installed with the vapor retarder up, can be secured by the flange provided for this purpose. Laced wires, barbed-end wires spanning the joist space, wire mesh, screening, or ceiling finish support loose-fill insulation.

Another floor insulation method uses double-sided, perforated aluminum foil draped and stapled over floor joists (BRANZ 1983). For a sag depth of about 4 in., laboratory R-values had a mean of 6.8 and ranged from 5.7 to 14.2 $ft^2 \cdot °F \cdot h/Btu$, depending on the humidity above the floor space. Field measurements have shown R-values up to 14 $ft^2 \cdot °F \cdot h/Btu$ for carpeted floors over good installations with a sag of 4 in.

Roof Deck Construction

Almost all structural roof deck construction systems require thermal insulation to economically maintain the design indoor environment. For low-slope roofs, insulation should be placed on top of the deck, on the outside of the structure. This location moderates the deck temperature, which reduces thermal movement of the deck and the potential for underdeck condensation. Traditionally, the insulation is placed under the roof membrane so that it functions as a base for the built-up roof (BUR) or single-ply membrane. For a stable base, a good bond must be established between the insulation and the roof deck, and between the insulation and the BUR. Ineffective bonds that allow the BUR to move in reaction to stresses from temperature changes often cause the roof to fail.

Single-ply membrane roofs are frequently installed with ballasted, loose-laid, or mechanically or adhesively fastened systems. Adhesion to the insulation is usually not desirable.

Insulation can also be placed above the membrane in a protected membrane roof (PMR) system. With this approach, the roof membrane is installed on the deck, where it functions as a waterproofing membrane and vapor retarder. Insulation can be placed above and below membranes to function both as the base for the membrane and the membrane protector, as well as for the insulator. Some insulation can be wetted and then successfully dried; however, while wet, it has a greatly reduced insulation value.

Roof systems designed for continual heavy traffic should have a secondary structure, supported on a curb system, above the waterproof membrane of the roof system.

Insulation for pitched roofs covered with shingles, tile, or similar mechanically fastened watershedding materials should be placed above the ceiling directly on the attic floor.

If the undersurface of the pitched roof is exposed to the interior, insulation can be applied over the roof framing or decking. The insulation is then covered with a nailing base for the roofing application. Prefinished roof planks with the required thermal resistance can also be placed over the rafters.

INSULATION FIELD PERFORMANCE CHARACTERISTICS

Convection and air infiltration in some insulation systems may increase the heat transfer across them. Low-density loose-fill, large open-cell, and fibrous insulations and poorly designed or installed reflective systems are most susceptible to increased heat transfer caused by natural and forced convection (air infiltration). A change of temperature differential across an insulation, as well as the height, thickness, or width of the insulated space, influences the amount of convection. When a membrane with low air permeance is applied to one surface or when the cavity is filled with insulation, natural convection is reduced significantly and apparent thermal conductivities measured by standard test methods apply.

If a membrane is applied to an insulation surface, care must be taken to ensure that the insulation is not compressed or disturbed significantly. The heat loss due to air convection may not be significant for many types of insulation products, such as batts and higher density loose-fill insulations. Convective heat loss potentials should be obtained from insulation manufacturers.

The effectiveness of thermal insulation is seriously impaired when it is installed incorrectly. For example, a 4% void area in R-11 $ft^2 \cdot °F \cdot h/Btu$ wall insulation increases heat loss by 15%. A 4% void area in R-19 ceiling insulation causes a 50% increase in heat loss. Verschoor (1977) found that air interchange around thin wall insulation installed vertically with air spaces on both sides increases heat loss by 60%. Other factors, including vibration, temperature cycling, and other mechanical forces, can affect thermal performance by causing settling or other dimensional changes. Section A3 of the *CIBSE Guide* (1986) gives more information on the effect of moisture on thermal properties of building structures.

MOISTURE CONTROL IN BUILDINGS

Strategies to control moisture accumulation fall into two general categories: (1) minimizing moisture entry into the building envelope and (2) removing moisture from the building envelope. Once basic moisture transport mechanisms and specific moisture control practices are understood, roof, wall, and foundation constructions for various climates can be reviewed systematically to determine whether each potentially significant moisture transport mechanism is explicitly controlled. Because it is not possible to prevent moisture migration completely, construction should include drainage, ventilation, removal by capillary suction, or other provisions to carry away unwanted water.

CONTROL OF LIQUID WATER ENTRY

Moisture problems in buildings are frequently caused by liquid water entering through leaking roofs or the foundation, or through the walls due to wind-driven rain or rain splashing. Poor flashing details are often a major cause of water entry into walls and roofs. Rainwater should be carried away from the foundation through gutters, downspouts, and positive grading. A rain screen can minimize penetration of walls due to raindrop momentum, capillarity, gravity, and air pressure difference. The rain screen wall is designed so that the air pressure difference across the exterior rain screen is nearly zero at all times. A rain screen wall contains three components: an airflow retarder system, a pressure-equalization chamber, and a rain screen. The airflow retarder must be able to resist pressures from wind, the stack effect, and mechanical ventilation. The pressure-equalization chamber separates the rain screen and the airflow retarder system. It may be an air cavity or may be filled with a self-draining material to prevent water that penetrates the rain screen from reaching the airflow retarder system. The chamber should consist of separate compartments to avoid lateral airflow, especially around corners of the building. Each chamber compartment is vented to the outside through the rain screen to provide pressure equalization and must be flashed to the outside to drain water that has penetrated the rain screen. The rain screen must contain sufficient vents to provide pressure equalization, that is, the airflow resistance of the rain screen must be much lower than that of the airflow retarder.

CONTROL OF WATER VAPOR MIGRATION

Water vapor entry into the building envelope can be limited by airflow retarders and water vapor retarders. As described in Chapter 22, airflow retarders are intended to restrict airflow, and thereby water vapor flow, whereas water vapor retarders are designed to restrict vapor flow by diffusion. The vapor retarder may also function as an airflow retarder, but this is not necessary to qualify as a vapor retarder.

Air Leakage Control

Wilson and Garden (1965), Dickens and Hutcheon (1965), Latta (1976), Handegord (1979), Tamura (1975), and numerous others have demonstrated that air movement is more effective than water vapor diffusion for transporting water vapor within the building envelope. In order to minimize moisture penetration by air leakage, the building envelope should be as airtight as possible. In the past, air leakage in residential buildings provided sufficient ventilation. However, in airtight buildings it is often necessary to provide mechanical ventilation to ensure acceptable air quality and prevent moisture and health problems caused by excessive indoor humidity. Ventilation or drainage must go to the outside of the airtight layer of construction or it will increase air leakage of the building. To avoid condensation on the airtight layer, either the temperature of the layer must be kept above the dew point by locating it on the warm side of the insulation, or the permeance of the layer must be adequate to permit vapor transmission.

The airflow retarder must also be sufficiently strong and well supported to resist wind loads.

As described in the section on Leakage Distribution in Residential Buildings in Chapter 25, air leakage through the building envelope is not confined to doors and windows. Although 6 to 22% of the air leakage occurs at windows and doors, 18 to 50% typically takes place through walls, and 3 to 30% through the ceiling. Leakage often occurs between the sill plate and the foundation, through interior walls, electrical outlets, plumbing penetrations, and cracks at the top and bottom of the exterior walls. More detailed information can be found in Chapter 25.

Not all cracks and openings can be sealed in existing buildings, nor can absolutely tight construction be achieved in new buildings. However, an effort should be made to provide as tight an enclosure as possible to reduce leakage and minimize potential condensation within the envelope. Such measures also reduce energy loss.

Moisture accumulation in the building envelope can also be minimized by controlling the dominant direction of airflow. This can be accomplished by operating the building at a small negative or positive air pressure, depending on climate. In cooling climates, the pressure should be positive to prevent the entry of humid outside air into the envelope. From a moisture control perspective, in heating climates this pressure should be neutral or slightly negative, but because of the danger of backdrafting of combustion appliances and the danger of an increase in the entry of radon gas from the soil, significant negative indoor pressures are generally not recommended. Significant positive indoor pressures should be avoided in heating climates.

MOISTURE CONTROL OPTIONS

Options for moisture control under heating conditions often differ from those under cooling conditions, even though the physical principles of moisture movement are the same. Therefore, the selection of moisture control options depends on whether the local climate is predominantly a heating or cooling climate. *The Moisture Control Handbook* (Lstiburek and Carmody 1991) recommends a three-step procedure for designing energy-efficient roofs, walls, and foundations with inherent moisture control capabilities:

1. Identify the climate: heating, cooling, or mixed
2. Determine the potential moisture transport mechanisms in each part of the exterior envelope: liquid flow, capillary suction, air movement, and vapor diffusion
3. Select the moisture control strategies: control moisture entry, control liquid moisture accumulation (condensation), or remove moisture by draining, venting, or diffusion

The definitions of climate zones are somewhat arbitrary. Lstiburek and Carmody (1991) recommend that heating climates be defined as climates with 4000 heating degree days (base 65°F) or more. Cooling climates are defined as warm, humid climates where one or both of the following conditions occur: (1) a 67°F or higher wet-bulb temperature for 3000 or more hours during the warmest six consecutive months of the year; (2) a 73°F or higher wet-bulb temperature for 1500 or more hours during the warmest six consecutive months of the year.

Mixed climates are all other climates that do not fall under the definitions of heating or cooling. Regions with heating climates in North America generally include the northern half of the United States, Alaska, and all of Canada. The climate in southeastern coastal regions of the United States generally can be characterized as cooling. However, the local climate should be evaluated to determine whether to design for heating, cooling, or mixed-climate conditions.

Not all moisture problems can be avoided at all times. Proper design can help reduce the risk and make a building more tolerant to moisture, and the recommendations in this chapter are intended to provide guidance.

MOISTURE CONTROL OPTIONS FOR HEATING CLIMATES

Surface Condensation

Heating climates are defined as climates with 4000 heating degree days (base 65°F) or more. In such climates, occasional window condensation is common in buildings during winter and fall.

Lowering indoor humidity to minimize surface condensation on windows is one approach, but increasing the interior surface temperature of the window using multiple glazing, low-emittance glazing, or gas-filled glazing may be more effective. Higher thermal resistance in windows has the added advantages of saving energy, improving occupant comfort, and reducing the possibility of condensate damage to the interior adjacent to the window (e.g., staining of the wall, rotting of the window sill, and mold growth).

Appropriate choice of glazing based on the outside design temperature should ensure that windows remain clear most of the time. Some condensation may appear around the edge of the window perimeter but should disappear with a warming trend. This criterion should be used to decide which glazing should be installed to maintain the desired humidity, or whether to reduce the humidity to avoid condensation during the coldest periods.

Local condensation and mildew growth on walls and ceilings is often the result of low inside surface temperatures due to insufficient or faulty insulation. Increasing the thermal insulation or eliminating the voids in the insulation is the obvious remedy. If the problem is due to infiltration of cold air, an attempt should be made to eliminate the air leakage. However, in existing buildings these measures are often difficult or too expensive. In these cases, the only alternatives are lowering the indoor humidity, raising the indoor temperature, or increasing the air circulation near the surface.

Indoor Humidity Control

A common cause of moisture problems during the heating season is excessive indoor humidity. This is caused by an improper balance between moisture generation and moisture removal. This balance can be changed by reducing the sources of moisture or by increasing the removal rate, usually by ventilation or dehumidification. However, it is important to avoid lowering the relative humidity too far below the lower comfort limit, which is generally about 25 to 30% rh.

The moisture content of the air in an occupied building without dehumidification is always higher than that of the outdoor air, because water vapor is introduced into the building from various sources. Christian (1994) provides a detailed discussion of various individual moisture sources, primarily in residences. In the section on Internal Moisture Gains in Chapter 20 of the 1996 *ASHRAE Handbook—Systems and Equipment*, it is stated that a family of four produces an average of 0.7 lb/h of moisture. TenWolde (1988, 1994) reports production rates between 0.3 and 0.73 lb/h for one to two adults, with an average of 0.53 lb/h. European sources report rates between 0.6 and 1.2 lb/h for families without children, and rates between 0.46 and 2.1 lb/h for families with one to three children (Christian 1994). These numbers demonstrate that moisture production rates in residences can vary widely. Moisture production rates for various kinds of livestock and plants can be found in Chapter 10.

A residential crawl space or basement can contribute significant amounts of additional water vapor. Trethowen (1994) reported an average moisture release of 0.082 $lb/ft^2 \cdot day$ from moist or wet crawl spaces. Finally, moisture released from building materials that are drying (construction moisture) in a new building can add large amounts of water vapor.

Moisture generation can be reduced by the removal of some sources. Humidifiers should have a humidistat control. Spot ventilation of kitchens, bathrooms, and attached greenhouses can reduce moisture contributions from those areas.

Exposed soil surfaces in crawl spaces or cellars should be covered with vapor retarder membranes held down by sand, gravel, or other suitable means (see the section on Crawl Spaces). Houseplants are a source of moisture, as is green or wet wood stored inside for fuel. (See Table 2 in Chapter 22 of the 1996 *ASHRAE Handbook—Systems and Equipment.*)

If indoor humidity is excessive, and source reduction is impossible, increasing the ventilation rate should be considered. An air-to-air heat recovery device can be included to reduce the heating energy penalty from an exhaust fan. Other approaches include the use of mechanical dehumidifiers and insulated vent stacks that extend from the living space through the roof.

Short-term ventilation procedures, such as occasionally opening a window or door, may lower humidity momentarily, but it will rise to its original level soon after the opening has been closed. This is due to the evaporation of stored moisture into the indoor air. When water vapor is released from showers or cooking, much of it is adsorbed by hygroscopic materials (paper, wood, fabrics, etc.) in the building. Some temporary storage in the form of surface condensation may also occur. This moisture is released more slowly at a later time. Moisture storage effectively dampens the effect of short-term (hourly or daily) changes in moisture release or weather conditions on indoor humidity. Stored moisture also slows the effect of ventilation and dehumidification because this moisture needs to be released and removed before the indoor humidity can be lowered permanently. The evaporation of moisture stored during the summer's periods of high relative humidities is the cause of high relative humidity and window condensation in early fall.

In cold or cool winter climates, house ventilation can be an effective method for moisture removal. In these climates, a ventilation level of 0.35 ACH (as recommended in ASHRAE *Standard* 62, Ventilation for Acceptable Air Quality) is generally sufficient to prevent excessive indoor humidity and most window condensation (TenWolde 1994). Ventilation is primarily required to ensure acceptable indoor air quality. If the recommended minimum ventilation levels are achieved, additional ventilation is probably unnecessary and ineffective for humidity control. In mild, humid climates, ventilation rates greater than 0.35 ACH may be needed for humidity control, but in such climates other means of moisture removal, such as dehumidifiers, should be considered.

It is often suggested that high indoor humidity is caused by vapor retarders that lock in the moisture. However, only a small fraction of the total moisture generated can be removed by vapor diffusion through the building envelope. Most high indoor humidity is due to inadequate ventilation, inadequate dehumidification and air conditioning, or an unusually large moisture source in the building. In general, only a vapor retarder that also functions as an airflow retarder can significantly affect house ventilation rates, and thereby indoor humidity conditions.

Vapor Retarders and Airflow Retarders

Vapor retarders are recommended and often mandated in heating climates, and should be placed on the interior side of the insulation. Airflow retarders are also necessary, but their placement is probably less critical and still subject to debate. Vapor retarder and airflow retarder functions may be combined in one material. Airflow retarder placement on the exterior prevents cold air from penetrating the insulation (wind washing) and therefore improves thermal performance of the building envelope. However, in heating climates airflow retarders on the exterior should have a high water vapor permeance. Special airflow retarder materials for exterior use with sufficiently high water vapor permeance are commonly available. Exterior airflow retarders do not prevent penetration and circulation of warm indoor air inside the wall or ceiling/roof cavity. Conversely, interior airflow retarders do not prevent wind washing. Interior airflow retarders do not need to have a high water vapor permeance. For additional general guidance on the placement and properties of airflow retarders, see Chapter 22.

The use of vapor retarders in compact low-slope roofing systems has been a long-standing issue for the roofing industry. Unlike other portions of the building envelope, water intrusion into a low-slope roof due to membrane failure is inevitable. Wet insulation performs below thermal performance levels specified during design. A survey by Kyle and Desjarlais (1994) has indicated that the average energy efficiency of the entire roofing inventory in the United States is reduced by approximately 40% due to moisture contamination.

Powell and Robinson (1971) studied these problems and stated that the "most practical and economical solution to the problem of moisture in insulated flat-roof constructions (is) to provide a design that would have in-service self-drying characteristics." A **self-drying roof** uses the local meteorological conditions to create a vapor drive into the building interior. Desjarlais (1995) has demonstrated that climates with up to 9000 heating degree days create annually averaged downward vapor drives. In a self-drying roof, any leakage into the roofing system is passively driven into the building interior; if the leak is repaired, the roof system will dry. A vapor retarder prevents the incorporation of self-drying characteristics into the roofing system design by placing an impermeable layer between the roof insulation and the building interior and should only be placed in a roofing system when the amount of wintertime water uptake that the roofing system will experience exceeds the moisture limit of the insulation material employed in the roof. Desjarlais (1995) offers guidelines on how to determine these limits.

Penetrations of the airflow retarder (such as electrical outlets, light fixtures or plumbing stacks) should be minimized. Care should be taken to seal around the penetration. Special airtight electrical boxes are available. Limited or minor penetrations of the vapor retarder are not of great concern if there is an effective airflow retarder elsewhere in the wall or ceiling.

Attics

The aim of current code requirements, which typically call for attic ventilation, is the prevention of condensation on the underside of the roof sheathing. Summer cooling of the attic air and extension of the service life of shingles are often cited as additional benefits of attic ventilation. Although the ventilation of attics and cathedral ceilings is often required and enforced in residential construction, ventilation of low-slope roof systems (typically in commercial and industrial construction) is uncommon. The first requirements for attic ventilation were promulgated in 1942 by the Federal Housing Administration (FHA 1942) with no supporting statement indicating a research basis. Studies on roof cavity and attic ventilation were conducted in the late 1930s and in the 1940s (Rowley et al. 1939, Britton 1948, Jordan et al. 1948), and the results gave a mixed appraisal of the effectiveness of attic ventilation. Britton (1948) was the first to note that moisture damage to attic sheathing may be due to air leakage from a wet foundation area, via air movement through chases and furred wall spaces.

Although there clearly are potential benefits from attic vents in heating climates, there are also disadvantages: vents can be prone to snow and rain entry that can wet the insulation, and cold air blowing through eave vents can degrade the thermal performance of attic insulation. Older buildings and experience in cold Canadian climates have also demonstrated that unvented attics can perform satisfactorily. Thus, the question that must be answered is whether the potential benefits of attic vents outweigh their cost and disadvantages.

For attic construction with a large unconditioned attic space and steep roof construction, vents have been shown to provide effective attic ventilation. It is relatively easy and inexpensive to install vents in such an attic without compromising the effectiveness of the ceiling insulation. In heating climates, attic ventilation usually provides a measure of protection from excessive moisture accumulation in the roof sheathing, but if indoor humidity is high and humid indoor air leaks into the attic, the use of attic vents does not guarantee that attic moisture problems will not develop. Therefore, moisture control in attics in heating climates depends primarily on maintaining low indoor humidity levels during cold weather and on assuring sufficient airtightness and vapor resistance (i.e., a vapor retarder) in the ceiling.

A ventilated attic is noticeably cooler in the summer than an unventilated attic. With standard levels of insulation, this temperature difference does not usually translate into significant cooling energy savings (Burch and Treado 1978). Ventilation may slightly reduce the temperature of shingles on a sloped roof, but it is not clear whether this slight temperature difference is a significant factor in shortening the service life of shingles.

Ventilation of roofs has been shown to reduce ice dam damage during winter by providing a roof surface with more uniform cool temperatures. However, under some conditions, large amounts of ventilation are needed to prevent ice dams, necessitating mechanical attic ventilation (Tobiasson et al. 1994). Moreover, many ice dams are due to local heating of the roof. Well installed insulation, prevention of leakage of warm indoor air into the attic, and avoidance of heat sources such as uninsulated or leaky heating ducts in the attic are probably more effective in preventing ice dams than attic ventilation. Damage due to ice damming in roof valleys and eaves can best be prevented by use of a waterproof underlayment of sufficient width beneath the shingles.

Attic vents probably provide a net benefit in most heating climates. However, with proper indoor humidity control and airtight ceiling construction, unvented attics can perform satisfactorily over many years in cold heating climates. This alternative is especially attractive in regions with frequent blowing snow, where attic vents are likely to provide snow entry into the attic.

Roofs with absorbent claddings, such as wood shingles or cement or clay tiles, are subject to solar-driven moisture penetration (Cunningham et al. 1990). This occurs when the roofing is wetted by rain or dew and subsequently exposed to sunshine, which drives the moisture into the roof. It is aggravated by the use of dark roof colors. An impermeable membrane under the shingles or tiles can greatly reduce this moisture transfer into the roof, but measures should be taken to prevent condensation on the underside of this membrane.

Cathedral Ceilings

Cathedral ceiling construction is inherently more prone to moisture damage than attic construction, because cathedral ceiling construction creates isolated conditions in each rafter cavity. Although providing effective ventilation to attics with simple geometries is relatively easy and inexpensive, providing soffit and ridge ventilation to each individual cavity in a cathedral ceiling may be impractical. Rose (1992) has shown that during winter, ridge ventilation without sufficient soffit ventilation may admit harmful amounts of humid indoor air into the cathedral ceiling cavity. Wind washing of the insulation, especially near the soffit vents, is another common problem with ventilated cathedral ceilings. An airtight ceiling plane, a vapor retarder, and foam air chutes between the sheathing and the top of the insulation have been shown to be effective for moisture control in cathedral ceilings, while the advantages of roof vents over normal soffit air leakage are slight (Rose 1995).

The recommended air space (minimum 1 in.) between the top of the insulation and the roof sheathing also restricts the amount of insulation that can be installed in the roof cavity. TenWolde and Carll (1992) also demonstrated that ventilation of wall and roof cavities may lead to increased air leakage, and that the net effect on moisture in the cavity greatly depends on the ratio of indoor air leakage into the cavity to the ventilation with outdoor air. Because vents in cathedral ceilings are less likely to provide effective ventilation, potential beneficial effects on moisture conditions, shingle life, and energy conservation during the cooling season are very limited. Therefore, the benefits of vents in cathedral ceilings do not clearly outweigh their potential drawbacks and should not be required in cases where adding vents is particularly difficult or undesirable. With proper indoor humidity control, airtight ceiling construction, and a ceiling vapor retarder, unvented cathedral ceilings can perform satisfactorily in cold heating climates.

Operating Practices

In heating climates it is very important to avoid excessive indoor humidity. Details of indoor humidity control are discussed in the section on Indoor Humidity Control. Buildings in heating climates should not be operated at substantial positive indoor air pressures, which drive moist air into the building envelope and increase the potential for moisture accumulation. Large negative pressures should also be avoided if any unsealed combustion equipment is operated in the building. Negative pressure in the basement or in slab-on-grade buildings should also be avoided when there is potential radon leakage from the soil into the building, unless a subslab depressurization system has been installed.

Other Considerations

All previously mentioned considerations require special attention in buildings where the use dictates sustained high relative humidity levels. To gain flexible and uninterrupted floor spaces, mechanical service shafts are often located outside the perimeter of the occupied area, allowing unrestricted airflow into exterior wall cavities. Potential problems may be further aggravated by duct leakage. In controlled humidity areas, such as operating rooms, the outflow of moisture to surrounding cavities should be reduced by low-permeance surface finishes such as vinyl fabrics or vapor retarder paints. Complex piping and electrical systems may not permit effective control of air and water vapor flow in the plane of the interior surface. Such interferences require careful placement of air and vapor retarders because they can be rendered ineffective by penetrations of mechanical work. Such penetrations should be kept to a minimum and should be sealed. The sealing effectiveness can be verified with the use of smoke sticks, preferably while the building is being pressurized.

Example of Residential Wall Construction for Heating Climates

Figure 10 shows an example of a residential wall detail for heating climates. Moisture control is handled in the following ways:

Rain. The brick veneer, an air space minimum of 2 in., and an airflow retarder form an effective rain screen. Sufficient airtightness can be obtained by airtight installation of the sheathing (i.e., installed vertically with joints over the studs, with sealant or caulk used at the joints). Building paper provides additional airflow resistance.

Liquid movement. The 2 in. air space behind the brick veneer provides a capillary break for any rain water absorbed by the brick and mortar. Mortar should not breach the air space and touch the building paper, as this would allow rainwater to bypass the capillary break. The building paper protects the fiberboard or gypsum from any water penetrating the rain screen.

Air movement. The sheathing and building paper serve as an airflow retarder. The interior gypsum board may be sealed or caulked to provide additional protection. The polyethylene vapor retarder can provide additional airflow resistance if there are no penetrations.

Vapor diffusion. Vapor diffusion from the inside is inhibited by the polyethylene vapor retarder.

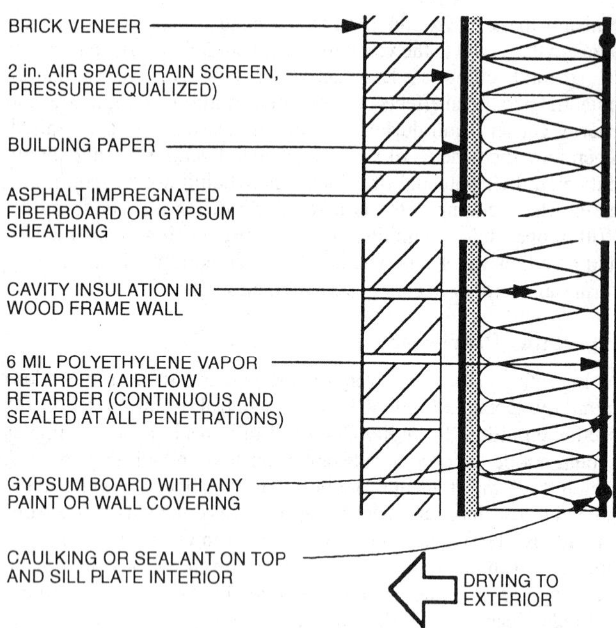

Fig. 10 Example of Residential Wall Construction for Heating Climates
Source: Lstiburek and Carmody (1991). Reprinted with permission.

MOISTURE CONTROL OPTIONS FOR MIXED CLIMATES

Mixed or temperate climates fall neither under the definition of a heating climate, nor under the definition of a hot, humid climate. Mixed climates may be heating or cooling dominated. This zone includes areas with hot and dry climates (e.g., Arizona). Buildings in mixed climates may encounter high interior levels of humidity during winter, and high exterior levels of humidity during summer.

Summer cooling or winter heating for comfort in mixed climates does not usually create serious vapor problems in exterior walls and ceilings. During summer, outdoor dew point, especially peak values, may exceed the design dew-point temperatures in common use, but they seldom exceed 75°F for any prolonged period. Condensation within exterior walls exposed to an indoor temperature of 75°F is seldom as serious as winter condensation.

In a study of a wood-sided house in Athens, Georgia, Duff (1956) showed under summer cooling conditions that temperatures were lower outside than inside long enough to prevent moisture buildup from damaging the structure. This was true regardless of whether or not a vapor retarder was placed near the inner surface. However, masonry or brick-veneered structures with a low-permeance vapor retarder (e.g., vinyl wallpaper or polyethylene) near inner surfaces do have a moisture buildup under summer cooling conditions.

Vapor Retarders and Airflow Retarders

In mixed climates the need for vapor retarders in most types of buildings is probably less pronounced than in heating climates or in warm, humid climates. However, if deemed necessary, vapor retarder placement in mixed climates presents somewhat of a dilemma.

A vapor retarder to prevent condensation within exterior walls under summer cooling conditions would normally be located on the outside of the insulation, whereas under winter conditions, it would be located on the inner side. Use of vapor retarders at both locations is undesirable because it can restrict not only the entry of moisture into the insulation but the escape of any moisture as well. In dwelling construction, the vapor retarder should be placed to protect against the more serious condensation (winter or summer). However, if indoor humidity is kept below 35% (at 70°F) during winter, a vapor retarder on the inside of the insulation is probably not necessary in mixed climates.

The choice and placement of vapor retarder, airflow retarder, and other materials should allow for drying of any accumulated moisture. For example, if a vapor retarder is installed on the interior, an exterior airflow retarder and/or sheathing should have sufficient permeance to allow drying.

The corresponding situation in cold storage buildings, in which a more serious reversal of vapor flow conditions from winter to summer may occur, requires individual analysis.

Serious wetting within walls can occur in summer under special conditions. The National Research Council of Canada tested the walls of huts of brick masonry, finished inside with furring, insulation, vapor retarder, and plasterboard. The walls were opened during a sunny period following rain. Extensive wetting was observed in the insulation, particularly on the back of the vapor retarder. The absorptive brick wall had taken on substantial quantities of water during the rainfall. Subsequent heating by the sun had driven the moisture as vapor into the wall, where it condensed and caused serious wetting. The construction had no protection in the form of parging or paper on the inside of the brick for the furring and insulation. Walls with absorptive exterior coverings capable of absorbing and storing considerable quantities of water during a rain, and providing little resistance to vapor flow into the insulation from outdoors, may experience serious interior wetting by condensation. No wetting occurred in a similar construction when a saturated sheathing paper was used between the insulation and the brick. Thus, a moderate vapor flow resistance, such as that provided by parging or a good sheathing paper on the outside of the insulation, can be effective in such cases.

Problems with wetting of the interior vapor retarder may be alleviated by the use of a water-permeable vapor retarder (Korsgaard and Pedersen 1992). This innovative type of vapor retarder is described in the section on Vapor Retarder Functions and Properties in Chapter 22. The water-permeable vapor retarder allows liquid water to escape, while maintaining its vapor retarder attributes under dry conditions. This allows drying of wall and roof cavities during summer if moisture accumulated during winter. Use of this vapor retarder may also accommodate designs with double vapor retarders, which have vapor retarders at the interior and exterior sides of the insulation.

Airtight construction is recommended in all climates. Airflow retarders provide protection from excessive moisture accumulation in the building envelope during cooling and heating and may reduce energy consumption.

Attics and Cathedral Ceilings

Venting of attics and cathedral ceilings during winter in a mixed climate has similar benefits and drawbacks as in a heating climate. Venting may provide some net benefits for moisture control in attics where effective vents can be installed relatively easily and cheaply. However, vents in cathedral ceilings are far less likely to be effective and beneficial, and therefore should not be required in cases where adding vents is particularly difficult or undesirable. With proper indoor humidity control and airtight ceiling construction, unvented cathedral ceilings can perform satisfactorily in mixed climates. More detailed discussion of ventilation of attics and cathedral ceilings can be found in the sections on Attics and Cathedral Ceilings under Moisture Control Options for Heating Climates.

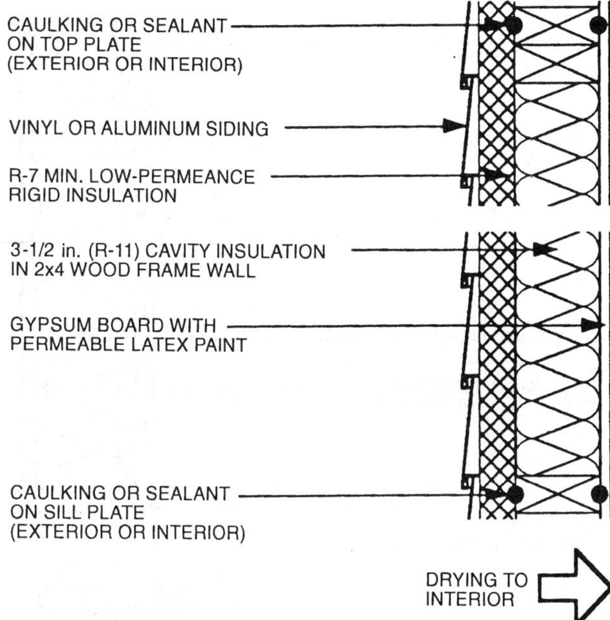

Fig. 11 Example of Residential Wall Construction for Mixed Climates

Source: Lstiburek and Carmody (1991). Reprinted with permission.

Example of Residential Wall Construction for Mixed Climates

Figure 11 shows an example of a residential wall detail for mixed climates, with rigid insulation serving as a vapor retarder and air retarder. Moisture control is handled in the following ways:

Rain. The combination of siding and airtight foam sheathing serves as a screen system and controls rain penetration. The air cavities behind the siding should be sufficient to create air equalization. If the air space is insufficient, the siding may be installed on furring strips to provide the air space. With vinyl or aluminum siding, liquid absorption and capillary moisture transfer are not a concern.

Air movement. The sheathing can be caulked at the top and bottom plates to provide an exterior airflow retarder. Alternatively, caulking of the gypsum board can provide an interior airflow retarder.

Vapor diffusion. The impermeable rigid insulation acts as a vapor retarder. During cooling periods, vapor diffusion from the outside is impeded at the exterior sheathing surface. During heating periods, vapor diffusion from the inside is inhibited at the interior surface of the foam sheathing. To keep moisture condensation to a minimum, this first condensing surface temperature should be elevated through the use of a foam sheathing with a high R-value. For mixed climates, the thermal resistance of the insulating sheathing in this example should be R-7 or greater, with R-11 thermal resistance in the cavity.

MOISTURE CONTROL OPTIONS FOR WARM, HUMID CLIMATES

Warm, humid cooling climates are defined as climates where one or both of the following conditions occur: (1) a 67°F or higher wet-bulb temperature for 3000 or more hours during the warmest six consecutive months of the year; (2) a 73°F or higher wet-bulb temperature for 1500 or more hours during the warmest six consecutive months of the year.

Depending on local experience with moisture problems, humid-climate design criteria may also be desirable in locations that do not quite meet the foregoing conditions.

In warm, humid climates ventilation is counter-productive, and dehumidification by air conditioning or other means is the most practical means of moisture removal from the conditioned space.

Air-conditioned buildings in humid climates tend to experience growth of mold and mildew unless a reliable air-conditioning system with good humidity control is used to decrease indoor humidity and provide adequate air circulation.

In high-humidity climates, the outdoor-to-indoor vapor pressure differential can be twice as much as in humidified buildings in cold climates. Thus, the latent cooling load caused by diffusion through the building envelope may need to be considered, even though the calculation is imprecise. This load is in addition to ventilation, infiltration, and internally generated latent cooling loads.

Because the latent cooling load on an air-conditioning system in high-humidity climates frequently exceeds the sensible load, a system should be capable of handling the latent load without overcooling. In residential buildings, oversized air-conditioning units do not alleviate the problem of high humidity, because they are off for longer periods or cycle too frequently.

The most successful approach to solving this problem includes proper sizing of the air-conditioning system and use of reheat. An air-conditioning system capable of operation at a low sensible heat ratio and with multiple capacities may suffice in some cases. The air-conditioning systems commonly used in other climates, that is, those with limited dehumidification capability, especially under light sensible loads (such as room fan-coil units handling the full load) should be avoided. Any air-conditioning system capable of providing comfort conditions within the range shown in the ASHRAE Comfort Zones (see Figure 4 in Chapter 8), under all sensible loads, prevents the growth of mold and mildew indoors. All-air systems with reheat generally provide satisfactory conditions.

Ventilating air should be continuously dehumidified since its vapor pressure is usually higher than that of indoor air. Similarly, infiltration should be minimized. Temperature controls for air-conditioning units, especially those handling outside air, should not allow variable temperature supply air unless humidity is also controlled. When the control system permits supply air temperature to increase with a consequent reduction in dehumidification, previously cooled surfaces (such as ceilings, diffusers, or registers) may be at temperatures below the dew point of the supply air, causing condensation.

Airflow Retarders and Water Vapor Retarders

Construction should be airtight as in all other climates. Many moisture and condensation problems in cooling climates have been found to be caused by excessive leakage of outside air into the building envelope. Airflow retarders in cooling climates are best placed on the exterior. Negative pressures of the indoor space should be avoided. Under high-humidity conditions, ambient water vapor diffuses through building materials from the outside into air-conditioned spaces. Exterior surfaces should have lower permeance than interior surfaces in high-humidity climates. Paints and finishes can provide the necessary permeance, with the lower permeance at the outside surface and the higher permeance toward the inside.

Low-permeance paints, vinyl wall paper, or any other similar low-permeance vapor retarder should not be used on the inside of walls and ceilings in warm, humid cooling climates. If practical, faced insulation should be installed with the facing toward the exterior of the cavity. However, kraft paper facing does not have a very low permeance, and insulation has been installed with the kraft paper facing towards the interior in warm, humid climates for many years without causing apparent moisture problems.

Where vapor retarders are used, they should be on the outside. Then, any water vapor that enters the construction can flow through to the inside of the building, where it can be removed by the air-conditioning system instead of accumulating in the wall or roof construction. Note that this is the reverse of the recommended practice for cold climates.

Attics and Cathedral Ceilings

The commonly stated rules for attic and cathedral ceiling construction—ventilation and vapor retarder toward the inside—pertain to cold climates and not necessarily to warm, humid climates. In warm, humid climates, venting tends to increase moisture and humidity levels in the attic rather than reduce them. When cooling ducts are located in the attic space, attic ventilation with humid outdoor air may therefore increase the chance of condensation on the ducts. When the ceiling is not airtight, attic venting may also increase the latent cooling load in the building.

As with other climates, a ventilated attic in a warm, humid climate is noticeably cooler in the summer than an unventilated attic. However, with standard levels of insulation, this temperature difference does not translate into significant cooling energy savings (Burch and Treado 1978). Ventilation may slightly reduce the temperature of shingles on a sloped roof, but it is not clear whether this slight temperature difference is a significant factor in shortening the service life of shingles. In summary, because the net benefits of attic venting do not always clearly outweigh the disadvantages, venting should not be required in a warm, humid climate. Venting should be considered a design option.

For similar reasons, ventilation of cathedral ceilings should not be required in warm, humid climates. In a warm and humid climate, design and construction of cathedral ceilings should be considered that (1) restrict the amount of outdoor air that can enter the cavity by eliminating vents to the outdoors and (2) allow cavities to dry to the interior, by not using low-permeance vapor retarders or low-permeance interior coverings such as vinyl on the ceiling.

Other Considerations

In high-humidity climates, moisture may condense on the exterior surface of a building because its temperature is frequently lower than the ambient dew point. In roof systems with high relative humidity under the roof, condensation may occur at night on the bottom surface of the roofing, wetting the insulation. This may be mistaken as a leaking roof.

To encourage drying, shaded exterior surfaces that do not benefit from the evaporative effects of sun and wind, such as inside corners, should be avoided or minimized. Special attention should be focused on building components that are prone to thermal bridging, such as exterior cantilevers, columns, foundations, or window and door frames. Although these solutions may not totally eliminate mold and mildew growth, they substantially reduce the potential.

Operation and Maintenance

Because the potential for damage to a building and its contents is substantial in an air-conditioned building in humid climates, it is more important to properly operate and maintain the building and its air-conditioning system in humid climates than it is in others. The chilled water supply temperature and flow should be reliable, and multiple chillers and pumps should be considered to ensure continuous uninterrupted dehumidification.

Raising the chilled water supply temperature to conserve energy should not be attempted under these conditions.

Lowering the cooling thermostat setting generally increases the chance for mold and condensation in exterior walls, especially in locations where the cooled air is blown directly towards the wall.

Example of Residential Wall Construction for Warm, Humid Climates

Figure 12 shows an example of a residential wall detail for warm, humid climates, with rigid insulation serving as a vapor retarder and airflow retarder. Moisture control is handled in the following ways:

Rain. The combination of airtight foam sheathing and siding serves as a screen system and controls rain penetration. The air cavities behind the siding should be sufficient to create air equalization. If the air space is insufficient, the siding may be installed on furring strips to provide the air space. With vinyl or aluminum siding, liquid absorption and capillary moisture transfer are not a concern. Wood siding may be backprimed to prevent moisture absorption through the back, and installation of wedges and clips on wood lapped siding should be considered to minimize capillary transport between the boards.

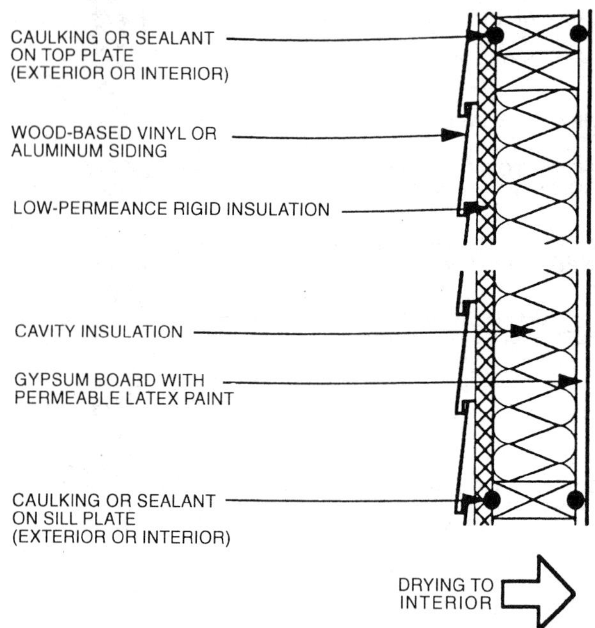

Fig. 12 Example of Residential Wall Construction for Warm, Humid Climates

Source: Lstiburek and Carmody (1991). Reprinted with permission.

Air movement. The exterior sheathing is the best location for an air seal, using either an adhesive or caulk to fasten the sheathing to the framing members.

Vapor diffusion. In warm, humid climates, the dominant source of moisture is the outside air, and moisture is typically driven toward the interior. This makes the exterior, or near the exterior, the best location for the vapor retarder. Vapor-permeable paint should be used on the interior gypsum wallboard.

MEMBRANE ROOF SYSTEMS

Since most membrane roof systems in commercial and institutional construction are highly resistant to vapor leakage, condensation must be prevented when insulation is placed between the heated interior and the roof membrane.

Wet insulation in low-slope roof construction is difficult to dry. Drainage is likely to be so slow as to be ineffective. Ventilation to the outside is not effective for drying roof insulation, because forces acting to remove the moisture are small. The vents themselves may present a hazard to the insulation by admitting moisture and drifting snow. Also, water leaks can occur where the vents penetrate the roof, unless they are properly installed. Finally, vents may allow chimney action to carry air upward through openings in the deck and ceiling. Then as air flows to the outside, further moisture is drawn into the roof with the replacement air and may condense.

A vapor retarder in a conventional flat roof can trap moisture in the roof cavity. The decision whether to use a vapor retarder depends on interior humidity and climate. The absence of a vapor retarder allows vapor to enter a roof during the heating season, but it also facilitates the removal of moisture in warm weather. This may not be acceptable in buildings with high indoor humidity or in extremely cold climates, when a large accumulation of frost or liquid condensation results in dripping. Where humidities are lower, or the climate less severe, the roof system may successfully store moisture through the heating season without problems (Baker 1980). The success of this strategy, however, also depends on the airtightness of the roof assembly. More information on this can be found in the section on Self-Drying Low-Slope Roofs.

Korsgaard and Pedersen (1992) demonstrated that the use of a water-permeable vapor retarder allows wet roof insulation and sheathing to dry out, but, at the same time, inhibits water vapor transfer into the cavity from below. One type of water-permeable vapor retarder consists of a layer of synthetic fabric with strips of polyethylene on each side with exposed fabric between them. The top and bottom strips are staggered so that liquid moisture can be wicked through the membrane, while the membrane still functions effectively as a vapor retarder (see Figure 10 in Chapter 22).

Regular inspection of the membrane and flashings helps prevent water leakage into the roof. Link (1977), Tobiasson et al. (1977), and Hedlin (1974) describe how infrared scanners or capacitance meters can help detect wet insulation, which can be removed or possibly dried out.

Inverted Roof Systems

The top layers in protected membrane or inverted roof systems are not waterproof; therefore, insulation is exposed to rainwater. To remain effective, it must be able to resist moisture penetration. Extruded polystyrene board has been used extensively. Insulation moisture content commonly ranges up to 4 or 5% by volume.

Some insulations are damaged by freezing and thawing, which fracture cell walls and allow water into an otherwise low-permeance material. When free moisture is available, the rate of entry increases rapidly as the temperature gradient increases (Hedlin 1977). Even when insulation is immersed in ponded water, moisture absorption through the edges is less than through the upper and lower surfaces, because the temperature gradient is normal to the roof surface.

Protective measures can reduce moisture gains. Roof slope performs much the same function for protected membrane roofs as it does for conventional ones. Covering the bottom surface of the insulation with a low-permeance layer inhibits moisture entry there. The upper surface should be open to the atmosphere so that water can evaporate freely. If it is trapped against the upper surface (e.g., by paving stones), solar heating may drive the water into the insulation.

Where a high thermal resistance is required, roofs may combine conventional and protected membrane systems when they are applied in two separate lifts. The protected membrane system may be applied to existing conventional roofs to increase the thermal resistance, if the roof structure can support the added weight. This addition keeps the insulation below the roof membrane warmer, so that the chance of moisture condensation entering from below is significantly reduced.

Self-Drying Low-Slope Roof Systems

A major cause of roof replacement is excessive accumulation of water in the roofing system. Historically, this accumulation has been minimized by delaying its ingress into the roofing system through the use of improved roofing membranes and periodic planned maintenance. Of course, most roofing systems eventually leak. Without periodic inspection, small leaks in a roofing system containing a vapor retarder or some other low-permeance layer (such as an asphalt mopping) can go undetected for long periods of time and lead to a major roof system failure. The self-drying roof facilitates the controlled outflow of water vapor into the building interior, preventing any long-term accumulation of water in the roofing system. Although they have not been optimized, the roofing industry has been constructing self-drying roofs for many years. A roofing system installed without a vapor retarder or a low-permeance layer is effectively a self-drying roof.

A self-drying roof should be considered whenever the average yearly vapor drive is into the building interior. Tobiasson and Harrington (1985) have produced vapor drive maps for the continental United States. Desjarlais (1995) has reported that this condition (vapor drive to the interior) is satisfied for climatic regions having less than 8800 heating degree days (65°F base).

The self-drying roof system must be carefully designed and include special features. The deck system must be reasonably permeable to water vapor so that downward drying can be maximized. The water vapor permeance of the insulation materials must be selected so that the anticipated wintertime wetting is maintained at a level that the insulation materials can tolerate. Water vapor absorptive layer(s) should be included in the roof system so that a major leak into the roof can be controlled without leakage into the building interior. The self-drying roof system must not contain a vapor retarder or any layers that are relatively impermeable. A suggested roof design procedure for self-drying roofs has been proposed by Kyle and Desjarlais (1994).

MOISTURE CONTROL IN FOUNDATIONS

Floor Slabs

Summer surface condensation may form on concrete floors on grade, especially during the first few years after construction. Carpeting tends to lower the interior slab surface temperature, increasing the condensation potential. Dehumidification and ventilation may be sufficient to avoid odors or floor cover bonding problems caused by moisture, which are generally more objectionable than actual damage to the floor covering.

As with any type of foundation, proper drainage of rainwater and runoff is critical and is aided by grading away from the foundation (minimum 5% slope). Gutters and downspouts should carry the water away from the foundation and should be properly maintained.

Entry of ground moisture can be further reduced by isolating the slab with the placement of a low-permeability membrane over the soil beneath the slab and by using coarse gravel to break the capillary moisture rise. Note that application of a membrane is difficult because of the potential damage during construction.

Sealing of floor slabs and basements against the entry of radon should also be considered. Although soil cover sheets are commonly referred to as "vapor retarders," they can also act as a waterproofing membrane when exposed to liquid water.

Control of slab surface temperatures is an important consideration in minimizing the need for mechanical dehumidification, particularly with solar-oriented designs that emphasize the effects of thermal mass. In localities with severe summer surface condensation problems, lightweight concrete should be considered for floor slabs to increase their insulating value, or insulation should be added under the slab.

Crawl Spaces

Moisture problems generally occur when improper drainage or grading around the house lead to wet soil or even standing water in the crawl space. Evaporation of this moisture then causes high humidity in the crawl space and often in the rest of the building, and sometimes leads to high moisture content in wood framing members in the floor and in the band joist (header joist). Providing proper drainage of water away from the foundation is critical for moisture control. Standard dewatering techniques (including sump pumps, drain tiles, etc.) should be used to keep the soil in the crawl space as dry as possible. In addition, ground covers that restrict evaporation of water from the soil into the crawl space provide an effective way to prevent moisture and humidity problems.

It is important to seal any ducts in the crawl space, to avoid venting clothes dryers into the crawl space, and to repair any leaking water pipes. A minimum clearance of 18 in. between the crawl space soil and the underside of any wood framing members is also recommended.

Whether or not to ventilate a crawl space has been a controversial issue. Most building codes require installation of vents to provide ventilation with outside air, but a symposium on crawl space design concluded that there is no compelling technical basis for crawl space ventilation requirements (ASHRAE 1994). There must be a distinction between conditioned and unconditioned crawl spaces. Conditioned crawl spaces have insulated perimeter walls and may contain plumbing and heating runs. These must be ventilated with indoor air or be within the house air circulation system. One way to accomplish this is by exhausting indoor air through the insulated crawl space, in conjunction with an air-to-air heat exchanger for energy efficiency (Samuelson 1994). Unconditioned crawl spaces are those with no services, and an insulated floor above the crawl space. Ventilation with outside air is permitted but not always necessary. Unvented crawl spaces must have a ground cover, which should cover 100% of the crawl space soil. Ground cover treatments for conditioned and unconditioned crawl spaces are similar, although ground cover in conditioned spaces may be upgraded to handle heavier foot traffic that is expected with quality storage space. Recommendations for grading and drainage requirements for crawl spaces are the same as for other types of foundation.

The ground cover material should have a perm rating of no more than 1.0 and must be rugged enough to withstand foot and knee traffic. Recommended materials include 6 mil (0.006 in.) polyethylene, smooth roll roofing of 45 or 55 lb per 108 ft^2 (federal specification Class A, Type 1), and 45 mil (0.045 in.) ethylene-propylene terpolymer (EPDM) membranes. Roll roofing should not be used for very wet soils, because fungi will decompose it over time. Any source of subfloor warmth (heating ducts, furnaces) is likely to seriously increase the evaporation from wet subfloor soil (Trethowen and Middlemass 1988, Trethowen 1988).

Before laying the membrane, all debris must be removed and the soil leveled. Edges need only be lapped 4 to 6 in., and no sealing is required. The membrane need not be carried up the face of the wall unless the interior grade is below the outside grade.

If control of entry of radon or other soil gases is desired, a 6 mil (0.006 in.) polyethylene ground cover is recommended. Radon control is especially important in unvented crawl spaces. Some have recommended that the polyethylene should be weighted down and sealed at the perimeter and overlaps to retard radon entry, but others argue that weighting and sealing may lead to water ponding on top of the polyethylene. If radon control is not of primary importance, the ground cover may be cut in several low spots to help drainage should ponding occur. The primary intended function of the polyethylene (i.e., moisture control or radon control) should govern the decision on this issue.

Example of Residential Foundation Construction Details

Figure 13 shows an example of a residential basement detail for mixed climates. Moisture control has been handled in the following ways:

Rain and groundwater. Rain is carried away by gutters, downspouts, grading away from the building, and a cap of low-permeance backfill material. Subgrade drainage prevents water from reaching the foundation wall by use of a drain screen (gravel and footer drain connected to daylight, sump, or storm sewer).

Liquid moisture transport. Dampproof coating is installed on the exterior of the foundation wall and over the top of the footing to control water entry. Capillary moisture movement into the slab is inhibited by a gravel pad 4 in. thick.

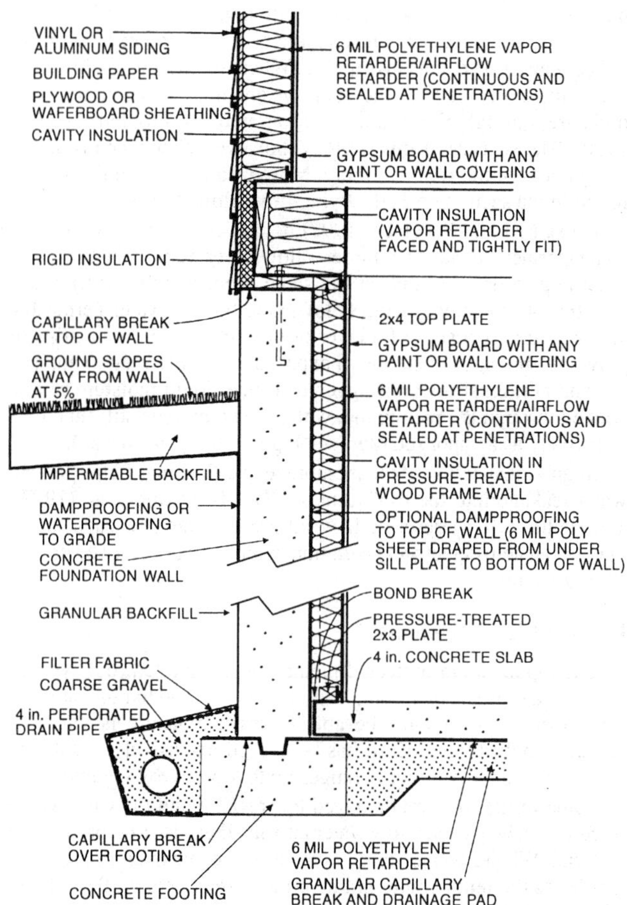

Fig. 13 Example of Residential Basement Construction for Mixed Climates

Source: Lstiburek and Carmody (1991). Reprinted with permission.

Air movement. All air leakage openings (i.e., floor slab/wall intersection, rim joist area) are caulked and sealed.

Vapor diffusion. Dampproofing on the wall and polyethylene under the slab inhibit vapor diffusion into the slab and foundation walls. During heating periods, vapor may diffuse from the interior into the rim joist framing, where it may accumulate. To reduce moisture accumulation, the temperature of the rim joist is raised through the installation of exterior insulation.

Figure 14 shows an example of a moisture-controlling residential slab-on-grade foundation detail for warm, humid climates, with insulation laid horizontally beneath the perimeter of the floor. Rigid insulation is also placed in the vertical joint between the wall and the slab. Because the rigid insulation can act as a conduit for insects into the building, additional protection such as metal flashings or other treatments may be necessary.

Rain and groundwater. The bottom of the gravel layer is the grade level adjacent to the perimeter. The ground should be graded to direct water away from the building.

Liquid moisture transport. The granular layer under the slab provides a capillary break between the soil and the slab. This pad can also be integrated into a subslab ventilation system to provide radon mitigation, if needed. Extension of the vapor diffusion retarder over the top of the foundation wall and appropriate flashing for the brick facing serve as a capillary break protecting the above-grade wall from ground moisture.

Air movement and vapor diffusion. The vapor retarder placed under the slab restricts both moist soil gas entry and vapor diffusion through the slab. Ductwork located in slabs

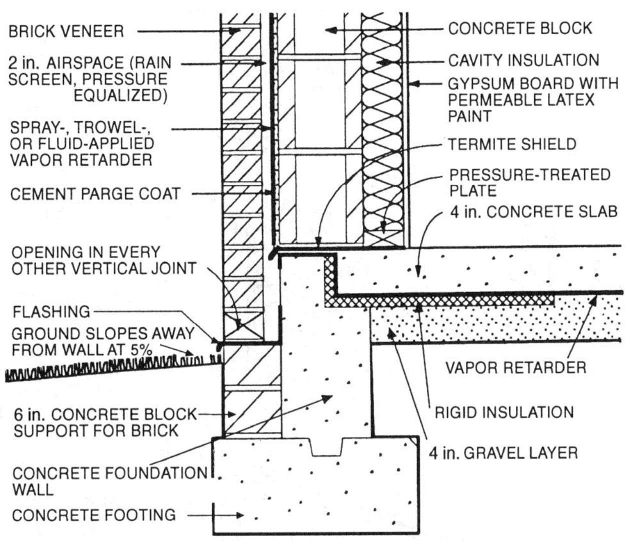

Fig. 14 Example of Residential Slab-On-Grade Construction in Warm, Humid Climates

Source: Lstiburek and Carmody (1991). Reprinted with permission.

increases the risk of ground source moisture entering the conditioned space if groundwater and soil gas are permitted to seep into the ducts.

ENVELOPE COMPONENT INTERSECTIONS

A moisture control strategy must consider not only envelope components, but also how these components come together. Component intersections are especially prone to air leakage and thermal bridging and therefore require special care.

Exterior Wall Corners. Exterior corners are common locations for mold and mildew during heating periods due to cold surfaces caused by (1) thermal bridges, where structural members penetrate the insulation and provide a low-resistance heat flow path; (2) wind washing; (3) increased heat loss due to the fin effect; and (4) poor circulation of indoor air. Figure 15 shows heat loss effects at building corners. Insulating sheathings and two-stud corners help prevent cold interior surfaces and corner moisture problems.

Wall/Window Intersections. Restricting air-transported moisture at all potential openings makes a major contribution to the overall building tightness. The airflow retarder must be continuous. Figure 16 shows several details that help form a continuous airflow retarder at a window jamb.

Wall/Roof Intersections. Exterior wall/ceiling intersections are other common cool spots during the heating season caused by reduced attic insulation at the eaves and wind washing. High-heel trusses that allow installation of more insulation, wind baffles, and rigid insulation exterior sheathing all help control moisture at these locations. Figure 17 shows the heat loss mechanisms at attic/wall intersections and how to minimize the risk of moisture problems.

Wall/Floor Intersections. Air leakage at rim joist assemblies is avoided by making sure the airflow retarder is continuous. Caulking and sealing are necessary at all polyethylene seams, as shown in Figure 13. Floor structural members penetrating the insulation can cause thermal bridging, but the use of insulating sheathing helps minimize this (see Figure 13).

Wall/Foundation Intersections. Concrete footings are frequently poured directly in contact with the ground, which occasionally becomes damp or wet. Concrete used for most residential foundations has the right degree of porosity to provide capillary suction, which draws water into the footings and then into the

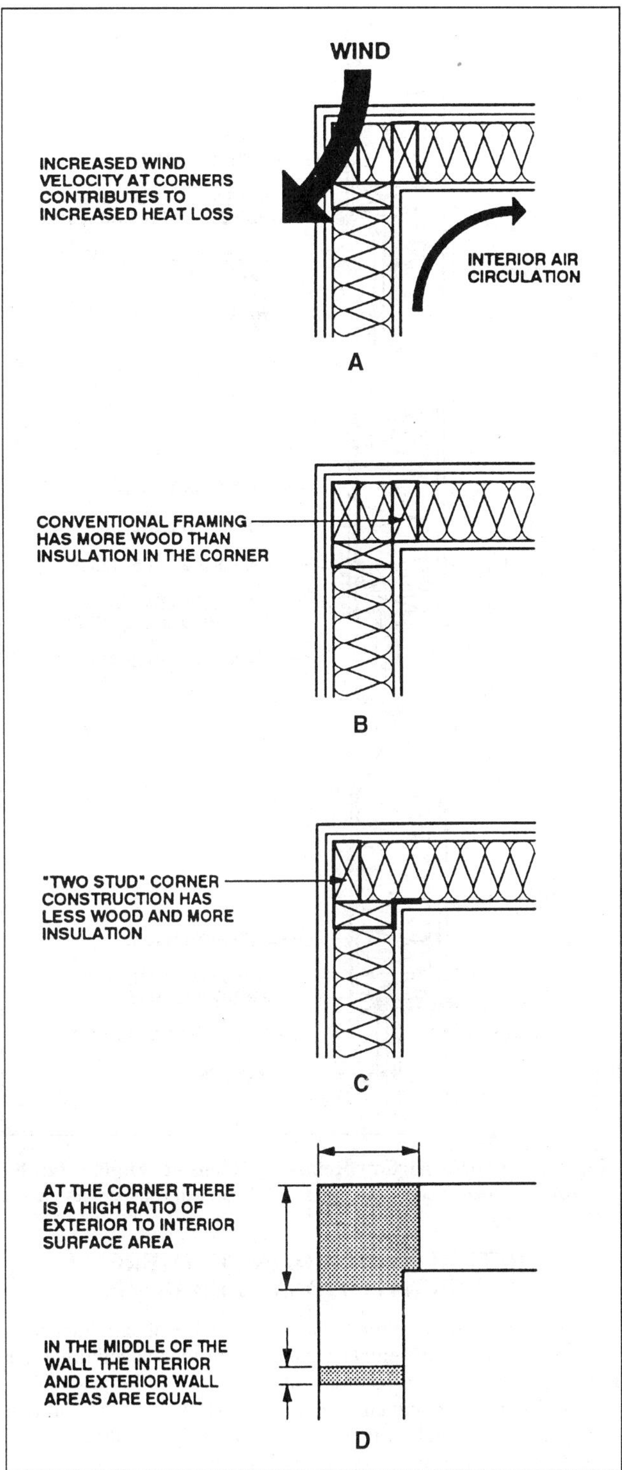

Fig. 15 Heat Loss at Building Corners

Source: Lstiburek and Carmody (1991). Reprinted with permission.

foundation wall. This water usually evaporates into the inside space undetected. However, the moisture occasionally manifests itself as a ring of dampness visible at the bottom interior surface of a basement or crawl space wall. Gravel and capillary breaks installed between the footing and the foundation wall are effective moisture control strategies in these below-grade envelope intersections. Several techniques to control capillary moisture below grade are shown in Figures 13 and 14.

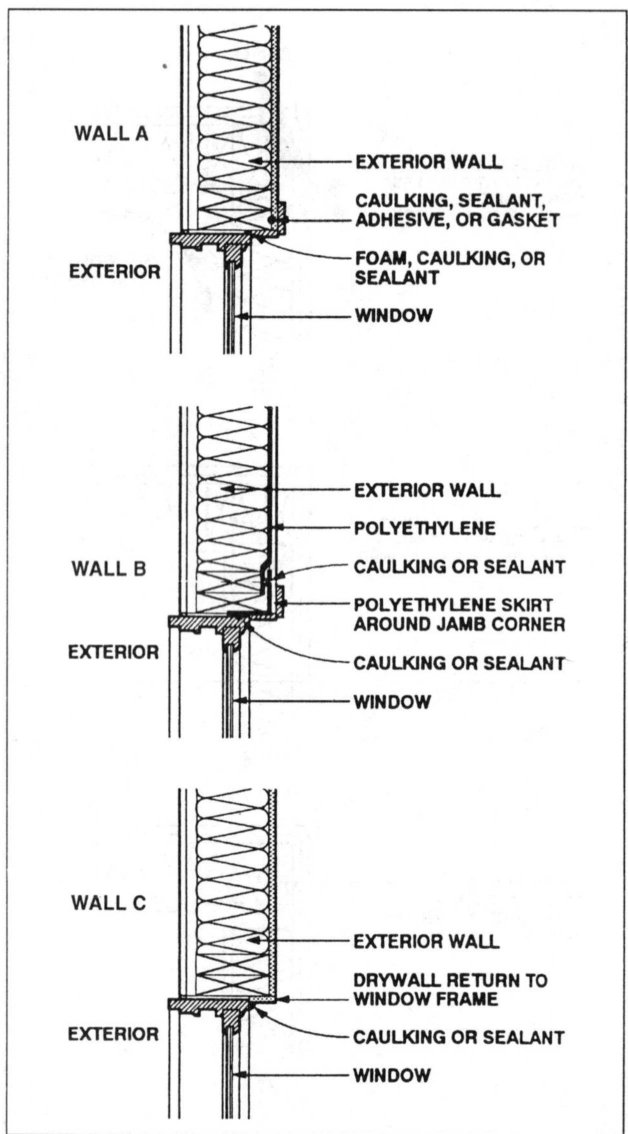

Fig. 16 Interior Airflow Retarder Details at Window Jamb
Source: Lstiburek and Carmody (1991). Reprinted with permission.

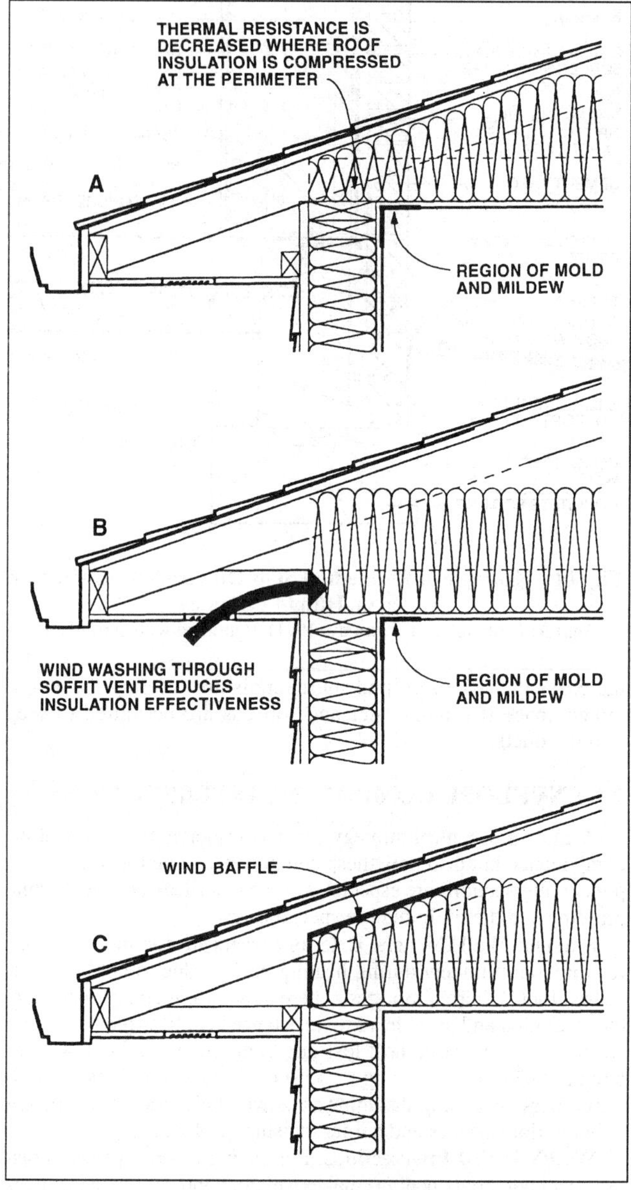

Fig. 17 Heat Loss Effect at Ceiling Edge
Source: Lstiburek and Carmody (1991). Reprinted with permission.

MOISTURE CONTROL IN COMMERCIAL AND INSTITUTIONAL BUILDINGS

Moisture control in commercial and institutional buildings often requires approaches different from those in residential buildings. As a rule, commercial and institutional buildings require mechanical heating, ventilation, and air-conditioning (HVAC) equipment that is specially sized and designed for the particular building. Indoor humidity conditions can vary greatly from one building to the next, depending on the use and requirements. The building envelope should be designed to perform well with these indoor conditions. Special thought should be given to the moisture control features of the HVAC equipment and the building envelope for certain buildings with special indoor humidity conditions or requirements, such as swimming pools, hospitals, and museums.

Materials commonly used in commercial and institutional construction tend to be somewhat more moisture-tolerant and decay-resistant than those used in residential construction. Vapor retarders in the walls of commercial buildings are less common. As a result, air leakage through hollow concrete masonry is often greater than in other types of construction. Upward airflow in the cavities is not always adequately blocked, and parallel random leakage paths are found between gypsum wallboard or other finishes and the block face. Because of such conditions, plaster or heavy-textured paints and mastics that fill the pores and holes control airflow better than sheet materials.

Air can leak through exterior walls where the structural system or services penetrate the air barrier or at joints between dissimilar materials or components. For example, masonry cannot form a tight seal with structural steel columns and beams. To reduce this problem, the structural frame should be inside and separate from the exterior wall. The resulting curtain wall can then incorporate a more continuous air barrier and be protected from the fluctuating weather by insulation applied to the outside. Exterior cladding to control rain penetration is best applied following the weather-tightening system. The interior wythe (masonry course) and air barrier should be accessible for maintenance of the air seal and joints.

The deterioration of exterior structural elements of a building and damage to the interior through condensation from air leakage have an important bearing on the operation and maintenance costs of the building. Improving the airtightness of internal floors and partitions, particularly in high-rise buildings, redistributes the pressure differences caused by the stack effect and reduces the pressure difference across the exterior wall on each floor. This approach also improves ventilation and air distribution and reduces the air circulation between occupancies on different floors. It also helps control smoke movement in the case of fire and may enable a more equitable apportioning of energy charges for space heating between individual units in apartment buildings.

Leakage in actual buildings often occurs through holes cut accidentally or deliberately in a reasonably tight membrane or component, such as the penetration of services through specified air or vapor retarders or solid components. Other leakage openings in the exterior envelope result from dimensional changes in improperly placed materials or from inadequate sealants or membranes applied to bridge joints or cracks that will eventually open.

REFRIGERATORS, COLD PIPES, AND FREEZERS

Refrigerators, cold pipes, and cold vessels require insulation and should have a vapor retarder placed on the warm side of the insulation or as close as possible to the warm surface of any enclosure. Insulation on a cold pipeline has no opportunity to dry out periodically. Also, vapor does not escape from the cold side of any metal (vaportight) pipe. Such applications require insulating materials that are highly resistant to water vapor, or a highly vapor-resistant jacket or coating whose permeance is not over 0.01 perm. Metal coverings are desirable but difficult to apply.

Similar considerations apply in cold rooms, whether constructed inside a heated building or as a separate building. Cold rooms operating above freezing may experience some periods of vapor reversal in winter, but such drying is of little help. Refrigerators lined with cement or other vapor-permeable material allow slight amounts of vapor to pass, which reduces the accumulation of moisture that penetrates the retarder. However, an adequate warm-side retarder with a permeance not greater than 0.01 perm should be installed.

Air can infiltrate through gaps in joints, tears in the retarder, or cracks where pipes or similar items penetrate the wall. Opening and closing cold room doors also causes pumping of air, which can increase the infiltration of moist air.

Insulated cold piping (and sometimes ductwork) should receive special attention when exposed to ambient or nonconditioned air. Because cold piping frequently operates year-round, a constant vapor drive exists under humid conditions. Even with vapor retarder insulation, jackets, and vapor sealing of joints and fittings, moisture inevitably accumulates in permeable insulations. This moisture not only reduces the thermal resistance of the insulation, it also accelerates condensation on the jacket surface with consequent dripping of water and growth of mold and mildew. Depending on local conditions, these problems can arise in less than 3 years, or they can take as long as 20 or 30 years. Since periodic insulation replacement is the only known solution, the piping installation should be accessible for such replacement and should have a means for draining water. As an alternative, very low-permeance insulating materials (e.g., materials not exceeding 0.1 perms) have been used to extend the life of the system and reduce replacement frequency. The lower the permeance of the insulation material, the longer its life, provided good workmanship is practiced during installation. Another alternative has been offered by Korsgaard (1993), who has demonstrated that water condensing on an insulated pipe can be removed continuously by wicking action with a specially designed and installed wick system, provided the pipe temperatures are above freezing.

INDUSTRIAL INSULATION PRACTICE

The applications of industrial insulation discussed in this section include (1) pipes; (2) tanks, vessels, and equipment; (3) ducts; and (4) refrigerated rooms and buildings.

PIPES

Small pipes are insulated with cylindrical half-sections of insulation, with factory-applied jackets that form a hinge-and-lap, or with flexible, closed-cell material. Large pipes can be insulated with flexible material or with curved, flat segmented, or cylindrical half-, third-, or quarter-sections of rigid insulation, particularly where removal for frequent servicing of the pipe is necessary. Fittings (valves, tees, crosses, and elbows) are insulated with preformed fitting insulation, fabricated fitting insulation, individual pieces cut from sectional straight pipe insulation, or insulating cements. Fittings insulation should always be consistent with pipe insulation.

Securing Methods

The method of securement varies with the size of pipe, form and weight of the insulation, and the type of jacketing (i.e., separate or factory-applied). Insulation with certain factory-applied jacketing can be secured on small piping by cementing the overlapping jacket. Large piping may require supplemental wiring or banding. Insulation on large piping requiring separate jacketing is wired or banded in place, and the jacket is cemented, wired, or banded, depending on the type. Insulation with factory-applied metal jacketing is secured by specific design of the jacket and its joint closure. The flexible closed-cell materials require no jacket for most applications and are applied using specially formulated contact adhesives.

Insulation Finish for Above-Ambient Temperatures

Pipe insulation finishes for indoor use are usually governed by location. If concealed, the simplest finish is a laminate of kraft paper, glass fiber reinforcing, and a vapor retarder. If exposed, the finishes are factory-applied jackets designed to meet fire safety requirements. For maximum fire safety, unusual exposure conditions, or appearance, factory or separately applied metal jackets may be used. Flexible closed-cell material is usually painted where exposed and left unfinished where hidden. An outdoor finish protects the insulation from the weather; chemical exposure, mechanical abuse, and appearance are additional considerations.

Insulation Finish for Below-Ambient Temperatures

Piping at temperatures below ambient is insulated to control heat gain and prevent condensation of moisture from the ambient air. Since piping is an absolute barrier to the passage of water vapor, the outer surface of the insulation must be covered by an impervious membrane or cover, which also helps protect the pipe against corrosion.

Retarder treatment should be recommended by the insulation manufacturer as established by performance testing. The insulation should be as dry as possible, and therefore should be protected from undue weather exposure.

Vapor-seal finishes for straight pipe insulation are generally designed to meet operating temperature, fire safety, and appearance requirements. Jacketing commonly consists of various combinations of laminates of paper, aluminum foil, plastic film, and glass fiber reinforcing. An important feature of such jacketing is very low permeance in a relatively thin layer, which provides flexibility for ease of cementing and sealing laps and end-joint strips. This type of jacketing is commonly used indoors without additional treatment. In some cases of subzero operating temperature, multilayer insulation and jacketing may be used. Flexible closed-cell materials must be carefully cemented to avoid openings in the insulations.

Insulation fittings are usually vapor-sealed by field application of suitable materials and may vary with the type of insulation and operating temperature. For temperatures above 10°F, the vapor seal can be a lapped spiral wrap of plastic film adhesive tape or a relatively thin coat of vapor-seal mastic. For temperatures below 10°F, common practice is two coats of vapor-seal mastic reinforced with open-weave glass or other fabric. The thickness of the mastic increases with decreasing temperature. With long lines of piping, the insulation should be sealed off every 15 or 20 ft to limit water penetration if vapor-seal damage occurs.

For dual-temperature service, where piping is alternately cold and hot, the vapor-seal finish, including mastics, must withstand pipe movement and exposure temperatures without deterioration. When flexible closed-cell insulation is used, it should be applied slightly compressed to prevent it from being strained when the piping expands.

Outdoor pipe insulation may be vapor-sealed in the same manner as indoor piping, by applying added weather protection jacketing without damage to the retarder and sealing it to keep out water. In some instances, heavy-duty weather and vapor-seal finish may be used.

Surface Temperature

In elevated temperature applications, the surface temperature of the insulation system should be below that at which personnel coming into contact with the surface could be harmed. In below-ambient temperature applications, the surface temperature of the system should be above dew point to prevent condensation.

Compared to a jacket with a nonreflective surface, a jacket with a reflective surface has a higher surface temperature for hot applications and a lower surface temperature for cold-temperature applications, because the lower emissivity reduces the rate of heat exchange. Therefore, adding a reflective jacket could produce a surface temperature capable of burning personnel on hot applications and causing condensation on cold applications. The jacketing material used also contributes to the relative safety at a given surface temperature. For example, at 175°F, a stainless steel jacket blisters skin more severely than a canvas jacket does.

Insulating Pipes to Prevent Freezing

If the surrounding air temperature remains sufficiently low for an extended period, insulation cannot prevent freezing of still water or of water flowing at a rate insufficient for the available heat content to offset the heat loss. Insulation can only prolong the time required for water to freeze and prevent freezing if water flow is maintained at a sufficient rate. The first section of Table 1 can be used to estimate the thickness of insulation necessary to prevent freezing of still water in pipes; the second section of Table 1 gives the minimum flow of water at an initial temperature of 42°F to prevent the temperature of the pipe from reaching 32°F at the end of the exposed length.

Table 1 Estimated Requirements to Prevent Freezing of Water in Pipes

Steel Pipe Nominal Diameter, in.	Insulation Thickness, in. 2	3	4	2	3	4
	Time to Cool Water to Freezing, h			Water Mass Flow Rate per Unit Length of Exposed Pipe to Prevent Freezing, lb/h·ft		
0.5	0.27	0.32	0.36	0.54	0.45	0.39
1	0.61	0.75	0.85	0.70	0.55	0.48
1.5	1.16	1.46	1.69	0.89	0.68	0.58
2	1.67	2.13	2.49	1.05	0.79	0.66
3	2.83	3.71	4.42	1.44	1.03	0.84
4	4.07	5.43	6.54	1.83	1.26	1.00
5	5.45	7.36	8.96	2.30	1.52	1.19
6	6.86	9.37	11.5	2.83	1.80	1.37
8	9.59	13.3	16.5	4.08	2.39	1.76
10	12.6	17.6	22.1	5.94	3.13	2.21
12	15.4	21.7	27.4	8.63	3.99	2.70

Design Conditions: Surrounding air temperature $t_a = -18$°F, initial water temperature $t_i = 42$°F, and insulation thermal conductivity $k_I = 0.025$ Btu/h·ft·°F. Thermal resistances of pipe and air film at surface of insulation are ignored.

Calculations are for 40ST steel pipe. See Table 2 in Chapter 40 of the 1996 *ASHRAE Handbook—Systems and Equipment* for actual pipe dimensions.

To calculate time θ (in hours) required for water to cool to 32°F, the following equation can be used:

$$\theta = \rho c_p \pi \left(\frac{D_i}{2}\right)^2 R_T \ln\left[\frac{t_i - t_a}{t_f - t_a}\right] \tag{1}$$

where

θ = time for water to cool to freezing, h
ρ = density of water = 62.4 lb/ft^3
c_p = specific heat of water = 1.0 Btu/lb·°F
D_i = inside diameter of pipe, ft
D_p = outer diameter of pipe or inner diameter of insulation, ft
D_I = outer diameter of insulation, ft
$R_T = R_p + R_I + R_a$ = combined thermal resistance of pipe wall, insulation, and exterior air film per foot of pipe, ft·°F·h/Btu
$R_a = 1/(h_a \pi D_I)$ = resistance between ambient air and outer surface of insulation per foot of pipe, ft·°F·h/Btu
h_a = air heat transfer coefficient (see Chapter 3 for values)
$R_I = \ln(D_I/D_p)/(2\pi k_I)$ = resistance of thermal insulation per foot of pipe, ft·°F·h/Btu
$R_p = \ln(D_p/D_i)/(2\pi k_p)$ = resistance of pipe per foot pipe, ft·°F·h/Btu ($R_p \approx 0$ for metal pipe)
k_I = thermal conductivity of insulation, Btu/h·ft·°F
k_p = thermal conductivity of pipe material, Btu/h·ft·°F (see Table 10 in Chapter 40 of the 1996 *ASHRAE Handbook—Systems and Equipment* for thermal conductivity of various plastic pipe materials)
t_a = ambient air temperature, °F
t_i = initial water temperature, °F
t_f = freezing temperature, °F

To calculate water flow, the following equation can be used:

$$\frac{W}{L} = \frac{1}{c_p(R_T + R_w)\ln\left[\frac{R_T}{(R_T + R_w)}\frac{(t_i - t_a)}{(t_f - t_a)}\right]} \tag{2}$$

where

W = flow rate required to keep pipe free of ice, lb/h
L = length of exposed pipe, ft
$R_w = 1/(\pi k_w \text{Nu})$ = resistance between water and inner surface of pipe per foot of pipe = 0.23 ft·°F·h/Btu
k_w = thermal conductivity of water near freezing = 0.32 Btu/h·ft·°F
Nu = Nusselt number for water = 4.36 for fully developed laminar flow and constant heat flux

When unusual conditions make it impractical to maintain protection with insulation alone, a hot trace pipe or, preferably, electric resistance heating cable is required along the bottom or top of the water pipe. The heating system then supplies the heat lost through the insulation. The insulation thickness is determined by the cost of the heating system, the insulation, and the heat loss.

Underground Pipe Insulation

Both heated and cooled underground piping systems are insulated. Protecting underground insulated piping is more difficult than protecting aboveground piping. Groundwater conditions, including chemical or electrolytic contributions by the soil and the existence of a water pressure, require a special design to protect insulated pipes from corrosion. Walk-through tunnels, conduits, or integral protective coverings are generally provided to protect the pipe and insulation from water. Examples and general design features of conduits and a description of tunnels can be found in Chapter 11 of the 1996 *ASHRAE Handbook—Systems and Equipment*.

Temperatures above Ambient. Piping for heated systems in walk-through tunnels is usually covered with sectional insulation and finished with effective mechanical protection such as metal or waterproofing jackets. The use of walk-through tunnels is declining because of cost.

Conduit systems are generally used for underground insulated piping systems. The most successful application is the use of sectional insulation with the conduit sized for drainage and adequate drying of insulation on heated piping in the event of accidental flooding. BRAB Technical *Report* 66 (1975) gives detailed design criteria for conduit systems. The criteria require that (1) all systems provide for draining and insulation drying, (2) the insulation withstand boiling and drying without physical damage and loss of insulating value, and (3) the conduit casing is watertight in the field. The insulation should be a nonconductor of electricity, verminproof, and chemically and dimensionally stable at the operating temperature of the pipe.

BRAB Technical *Report* 39 (1963) describes suitable laboratory tests for compliance with the criteria just discussed. BRAB Technical *Report* 47 (1964) describes evaluative tests and field investigations, which have shown that calcium silicate is resistant to severe boiling action. Fibrous glass (density of 4 to 7 lb/ft^3) will not withstand boiling when a conduit becomes flooded, and wet poured-in-place insulations are likely to remain partially wet for their installed life.

The thickness of insulation for underground piping is not determined on the same basis as aboveground piping.

Temperatures below Ambient. Integrally protected, insulated piping buried directly in the ground is commonly used for chilled water. However, since no heat is available to drive out moisture, an absolute protective covering against water and insulation with low permeance and water absorption is extremely important. Cellular glass with proper protection has been widely used for this type of application. The acceptance of plastic foams is increasing, but their long-term performance has not yet been established.

Conduit for chilled water piping requires a different approach than for hot piping. Insulation must have low conductivity, and conduit design must use this low conductivity and maintain continuing performance. More recent designs use low-conductivity plastic foam insulation with plastic pipe as the internal water-carrying piping and as the external conduit.

Where the temperature difference between the pipe at 40°F and the soil at 55 to 60°F is small, pipe size, length, and flow rate may economically justify direct burial without insulation. However, good piping protection may be required.

TANKS, VESSELS, AND EQUIPMENT

Flat, curved, and irregular surfaces, such as tanks, vessels, boilers, and chimney connectors, are normally insulated with flat blocks or bevelled lags, curved segments, blanket forms of insulation, or mineral fiber-inorganic binder insulations. On surfaces operating below 185°F, closed cellular insulations are used extensively. Since no general procedure can apply to all materials and conditions, manufacturers' specifications and instructions must be followed for specific applications.

Securing Methods

Insulations are secured in a variety of ways, depending on the form of insulation and contour of the surface to be insulated. On small-diameter, cylindrical vessels, insulation can be secured by banding around the circumference. On larger cylindrical vessels, banding can be supplemented by angle iron support ledges to support the insulation against slippage. Where diameters exceed 10 to 15 ft, slotted angle iron may be run lengthwise on the cylinder, at intervals around the circumference, as securement, and to avoid an excessive length of banding.

On large flat and cylindrical surfaces, banding or wiring can be supplemented by fastening to various welded studs at frequent intervals. On large flat, cylindrical, and spherical surfaces, it is often advantageous to secure the insulation by impaling it on welded studs or pins and fastening it with speed washers. Flexible closed-cell insulations are adhered directly to the surface to be insulated, using a suitable contact adhesive.

Insulation Finish for Above-Ambient Temperatures

For temperatures above ambient, insulation is finished to protect it against mechanical damage and weather, consistent with acceptable appearance. On smaller equipment indoors, insulation is commonly covered with hexagonal wire mesh tightly stretched and secured. Then a base and hard finish coat of cement is applied. This can be additionally finished by painting.

For the same equipment outdoors, the insulation can be finished with a coat of hard cement, hexagonal mesh properly secured, and a coat of weather-resistant mastic (preferably a breathing type). A variation is to apply only two coats of weather-resistant mastic reinforced with open-mesh glass or other fabric; however, this finish is limited to an operating temperature of about 300°F, because metal expansion can rupture the finish at insulation joints. Larger equipment may be finished indoors and out with suitable sheet metal. Outdoor finish of any type must be properly flashed around penetrations (e.g., manhole access openings, pipe connections, and structural supports) to maintain a weatherproof condition.

Insulation Finish for Below-Ambient Temperatures

For temperatures below ambient, insulation is finished, as required, to prevent condensation and protect against mechanical damage and weather, consistent with acceptable appearance. In accordance with the operating temperature, the finish must retard vapor to avoid moisture entry from surrounding air, which can cause an increase in thermal conductivity of the insulation, deterioration of the insulation, or corrosion of the metal equipment surface. For moderately low temperatures, insulation can be finished with hexagonal mesh properly secured, a coat of hard finish cement, and several coats of suitable paint. Where a greater degree of vapor retarding is required, two coats of mastic reinforced with open-mesh glass or other fabric can be used. For an even greater degree of vapor retarding, two coats of asphalt vapor-retarder mastic reinforced with asphalt-saturated and perforated felt with edges lapped and cemented can be used (see Table 9 in Chapter 24). If this type of vapor retarding is used indoors, standards for acceptable appearance and protection from mechanical damage may require the addition of hexagonal mesh, hard finish cement, and finish painting.

For the same equipment outdoors, the insulation can be finished with heavier or additional coats of vapor-retarder mastic reinforced with open-mesh glass or other fabric. For appearance, mastics can be painted, but since they usually contain solvents, they must be properly dried before painting. Paints must be the appropriate type recommended by paint manufacturers. In some cases, the additional finish over the mastic may be suitable sheet metal, applied carefully so as not to damage the vapor retarder (see Table 9 in Chapter 24 for building material properties).

Whenever vapor retarder finish is applied, all penetrations such as manholes, pipe connections, and structural supports must be properly vapor retarded with mastic or other sealant reinforced with open-mesh glass or other fabric. Equipment must be insulated from structural steel by isolating supports of high compressive strength and reasonably low thermal conductivity, such as hardwood block. The vapor retarding must carry over this insulation from the equipment to the supporting steel to assure proper sealing.

If the equipment rests directly on steel supports, the supports must be insulated for some distance from the points of contact. Commonly, insulation and vapor retarder are extended 8 to 10 times the thickness of the insulation applied to the equipment.

For dual-temperature service, where vessels are alternately cold and hot, the materials and a design of vapor retarder finish selected must withstand movement caused by temperature change.

Reflective insulation is commonly applied to large refrigerated tanks. In this system, the external finishing sheets of heavy-gage metal form the vapor retarder, and the joints are sealed to permit contraction and expansion. Valved outlets at the bottom of the tank allow drainage.

DUCTS

The need for duct insulation is influenced by (1) duct location, whether indoors or outdoors; (2) the effect of heat loss or gain on equipment size and operating cost; (3) the need to prevent condensation on low-temperature ducts; (4) the need to control temperature change in long duct lengths; and (5) the need to control noise with interior duct lining.

All ducts exposed to outdoor conditions, as well as those passing through unconditioned spaces, should be insulated. While analyses of temperature change, heat loss or gain, and other factors affecting the economics of thermal insulation are seldom made for residential installations, they are essential for large commercial and industrial projects.

The U-factor for uninsulated sheet metal ducts is affected by air velocity, the emittance of the metal, and the shape of the duct. An approximate value of 1.0 Btu/h·ft^2·°F may be used. For insulated ducts, U-factors of 0.24 and 0.13 Btu/h·ft^2·°F represent 1 and 2 in. thick rigid insulation with a thermal conductivity of 0.27 Btu·in/h·ft^2·°F at 75°F mean temperature. A method for determining heat loss or gain for ducts is given in Chapter 32.

Materials for Ducts, Insulations, and Liners

Ducts within buildings can be of insulated sheet metal or fibrous glass, both of which provide combined air barrier, thermal insulation, and sound absorption. Ducts embedded in or below floor slabs may be of compressed fiber, ceramic tile, or other rigid materials.

Duct insulations include semirigid boards and flexible blanket types, composed of organic and inorganic materials in fibrous, cellular, or bonded particle forms. Insulations for exterior surfaces may have attached vapor barriers or facings, or vapor barriers may be applied separately. When applied to the duct interior as a liner, insulation both insulates thermally and absorbs sound. Duct liner insulations have sound-permeable coatings or other treatment on the side facing the airstream to withstand air velocities without deterioration.

Fibrous glass ducts are available in preformed round ducts or in board form for fabrication of rectangular ductwork. Round and rectangular ducts have a minimum density of about 3 lb/ft^3, with a thermal conductivity of 0.23 Btu·in/h·ft^2·°F at 75°F mean temperature. Round ducts with a diameter of up to about 30 in. and boards in various thicknesses and sizes are used to form required sizes of rectangular ducts. The maximum recommended velocity is 2000 fpm, although tests at velocities exceeding 10,000 fpm show no fiber erosion. Primary use is for low-pressure systems tested at 1.5 times the recommended static pressure. Maximum recommended air temperature is 250°F. A complete system provides greater decibel attenuation than is usually provided by standard duct liners, with greater reduction in airborne equipment noise and crosstalk. Higher design velocities are also possible.

To satisfy most building codes, the various duct insulations and fibrous glass duct materials must meet the fire hazard requirements of the National Fire Protection Association's (1) NFPA *Standard* 90A, to restrict spread of smoke, heat, and fire through duct systems, and to minimize ignition sources; and (2) NFPA *Standard* 90B, on supply ducts, controls, clearances, heating panels, return ducts, air filters, and heat pumps. Local code authorities should also be consulted.

Where thermally insulated air-conditioning ducts pass through unconditioned spaces, such as attics, the maximum allowable heat flux should be no greater than that required by NFPA *Standard* 90A.

Securing Methods

Exterior duct insulation can be attached with adhesive, with supplemental preattached pins and clips, or with wiring or banding. Liners can be attached with adhesive and supplemental pins and clips.

Manufacturers provide information on the construction of fibrous glass duct systems. Preformed round duct for straight runs is combined with fittings fabricated from straight duct. Rectangular ducts and fittings are fabricated by grooving, folding, and taping, with metal accessories such as turning vanes, splitters, and dampers incorporated into the system. When rectangular ducts exceed predetermined dimensions for particular static pressures, ductwork must be reinforced. The Sheet Metal and Air Conditioning Contractors National Association's (SMACNA) *Fibrous Glass Duct Construction Standards* has further information.

Heating Ducts

The effect of duct insulation on residential heating system equipment size can be marginal. However, insulation can reduce operating costs significantly, depending on unit costs for heating and the extent of duct exposed to outside conditions. In addition, duct insulation maintains the supply air temperature, which may keep the air entering the conditioned space within a more comfortable range.

Vapor retarders are not required on exterior insulation of ducts used for heating alone, but they must be provided for ducts used for alternate heating and cooling.

Cooling Ducts

Insulation can reduce operating costs and cooling equipment size significantly. The advantage of adequate insulation is especially significant in areas with high dry-bulb and dew-point temperatures. Ducts for summer air conditioning are insulated with the same materials as heating ducts. Ducts in any unconditioned space should be insulated and have vapor retarders to prevent condensation. Joints and laps in the vapor retarder must be sealed. Flexible closed-cell insulation does not always need a supplemental vapor retarder, but care must be taken to form vapor-tight seams at joints.

REFRIGERATED ROOMS AND BUILDINGS

Cold storage facilities require the most attention to thermal insulation and air/vapor retarders, since they greatly influence costs. Many refrigerated spaces operate at subzero temperatures (as much as 100°F below outside temperature), which increases the severity of the service condition imposed on the insulation system. The primary source of moisture is airborne water vapor; however, soil moisture must also be considered because whenever moisture freezes (either in the insulation system or in soil under slabs-on-ground), both operating and maintenance costs increase. Control of moisture entry into the insulation is of primary importance in limiting water and ice collection in the structure. The location and integrity of the vapor retarder system is crucial and places a premium on good workmanship.

Interior Finish

The permeance of the interior finish (e.g., paint) should be greater than that of the vapor retarder so that the drywall, etc., can dry out in the direction of the finish and away from the vapor retarder.

The panels in structural panel systems fill in between structural members and, therefore, become the building surfaces. Where panels are used as a building lining, no special surface preparation is needed, assuming the surfaces are sound and reasonably smooth. Sheet metal and other specialty boards (when installed to permit passage of water vapor, if necessary) have been successfully used for walls and ceilings of rooms operating over a wide temperature range.

Insulated structural panels, with metal exteriors and metal or reinforced plastic interior faces, are commonly used for cooler and freezer rooms. These finishes keep moisture out of the insulation, so that only the joints between panels must be vapor-sealed to avoid moisture penetration.

In selecting an interior finish, the following factors should be considered: (1) fire resistance, (2) washdown requirements, (3) mechanical damage, (4) moisture and gas permeance, and (5) applicable codes in food-handling installations. All interior walls of insulated spaces should be protected by bumper guards and curbs whenever there is possibility of damage to the finish.

Thermal Resistance

Generally, the R-values in Table 2 may be used for the different types of facilities noted. The range in R-values results from variation in cost of energy, insulation materials, and climatic conditions. For more exact values, consult a designer and/or insulation supplier. R-values less than those shown should not be used.

Table 2 Recommended R-Values for Refrigerated Facilities

		Recommended R-Values, ft²·°F·h/Btu[a]		
Facility	Temperature Range, °F	Floors	Walls/ Suspended Ceilings	Roofs
Cooler	40 to 50	Perimeter insulation only	17 to 30	25 to 35
Chill cooler[b]	28 to 35	15 to 20	25 to 35	30 to 40
Holding freezer	−10 to −20	30 to 35	33 to 50	40 to 50
Blast freezer[c]	−40	35 to 40	50 to 60	50 to 60

[a]The R-values were derived using the economic thickness methods described in Chapter 22. Due to the wide range in the cost of energy and insulation materials on a thermal performance basis, a recommended R-value is a guide. For more exact values, consult a designer, insulation supplier, or both.

[b]If a cooler has the possibility of being converted to a freezer, the owner should consider insulating the facility with the higher R-values from the freezer section.

[c]R-values shown are for a blast freezer built within an unconditioned space. If built within a cooler or freezer, consult a designer and/or insulation supplier.

LAND TRANSPORT VEHICLES

Insulation installed in transport vehicles should have the same qualities desirable for most other applications (i.e., moderate cost, low density, low thermal conductivity, low moisture permeability and retention, ease of application, uniformity, resistance to breakdown at temperature extremes, and fire resistance). The insulation should resist cracking, crumbling, shifting, and pack-down from shock, vibration, and flexing of the body structure that occurs in transport.

State laws and other factors limit the outside width, height, and length of vehicles. Therefore, the designer must establish an insulation thickness to obtain the optimum cargo space and operating performance. For example, increasing insulation thickness from 3 to 4 in. in a 40 ft long highway trailer decreases cargo space by about 100 ft^3, or 5%. A decrease in insulation thickness, assuming other factors remain equal, requires an increase in refrigerating or heating unit capacity. In determining the optimum thickness of insulation, the vehicle body and its refrigeration or heating unit must be considered as one system.

Urethane foam insulations are used increasingly for transport vehicles, although glass fiber, polystyrene, and other materials continue to be used. Urethane can be formed in place between the inner and outer skin of the vehicle, or it can be preformed outside the vehicle.

The total amount of heat transferred from outside to inside a vehicle, excluding heat that enters when the vehicle door is open for loading or unloading of cargo, is composed of (1) heat transferred through insulation, (2) heat transferred through structural members, and (3) heat due to air and moisture leakage into the vehicle and insulated space.

Since air leakage can significantly affect the overall heat transfer rate of the vehicle body, some buyers specify a maximum rate of air leakage that the body can have with a given air pressure imposed in the cargo space. Furthermore, the buyer often specifies the maximum overall heat transfer rate for the vehicle, usually at conditions of 100°F and 50% rh outside and 0°F inside. Chapters 27 and 28 of the 1994 *ASHRAE Handbook—Refrigeration* have further information on refrigerated land transport vehicles.

ENVIRONMENTAL SPACES

Atmospheric environmental spaces provide environments of selected temperature, humidity, pressure, and composition. These may be desired singly or in combination, in stabilized degree or in phased fluctuating degree. The designed environment may be required for short- or long-term tests, for continued or periodic production or manufacture, or for short- or long-term storage. Design requirements and specifications for vapor retarder construction, thermal insulation construction, and protective finish of insulation construction should include performance requirements in terms of (1) permissible water vapor permeance; (2) thermal transfer; and (3) physical and structural requirements for the protective finish based on the environmental conditions, range of temperature exposure, and permeance to water vapor or other gases to which it may be exposed or for which it may be a controlling boundary.

In a tightly designed system, pressure changes caused by temperature change alone can be sufficient to rupture the insulation system. The effect of vacuum during a pulldown can also tear a barrier fastening.

Refrigeration equipment is required to maintain the proper environment in packing facilities for items such as electrical cable and in the manufacture of electronic equipment; pharmaceutical and biological chemicals; and textiles, paper, and tobaccos. The permissible water vapor permeance rates through walls, ceilings, and floors of the manufacturing space must be considered in the design load calculations for humidification equipment.

Controlled atmosphere storages for fruit and other perishable commodities are held at refrigerated temperatures and control the percentage of oxygen, carbon dioxide, and other gases in the air mixture within the space. Permeance of these gases through the internal gas barrier, which is normally also the protective finish of the thermal insulation, as well as the water vapor permeance through the external water vapor retarder, which should not exceed that of the internal gas barrier, should be specified.

General Principles

The insulation principles for refrigerated rooms or buildings given in this chapter apply only to steady-state operation. Environmental facilities used alternately for refrigerated and high temperatures are subjected to reversal of vapor flow and are usually designed with vapor retarders of equal permeance on both sides of the thermal insulation.

In altitude chambers, heavy welded steel shells form an external structure that resists the pressure differences between external atmospheric and internal vacuum conditions.

Environmental conditions at normal ambient temperatures and above, with injected high humidity, present problems of operation similar to those described for tanks and vessels. Interior protective finish on thermal insulation is usually designed with the lowest possible water vapor permeance through the protective material and its sealed joints, or with some provision for draining condensate that can be expected in the thermal insulation construction.

Altitude chambers with prestressed shells are generally designed with the insulation on the inside of the shell. This protects the shell from extremes of temperature and eliminates the effect of the shell's thermal mass on the refrigeration load during a rapid pulldown of temperature.

The insulation design must also consider the heat transfer through sleeves, pipes, conduits, door jambs, observation windows, and other openings, but the following is a general guide to insulation thickness for low temperatures:

$$x = 0.15k(t_o - t_i) \tag{3}$$

where

x = insulation thickness, in.
k = conductivity of insulation, Btu·in/h·ft^2·°F
t_o = design ambient temperature, °F
t_i = chamber temperature, °F

The use of Equation (3) results in an outside surface temperature of less than 5°F below ambient.

When the temperature-controlled space involves both heating and refrigeration, the thermal resistance of the structure should be designed to satisfy the most severe need. Thermal insulation should be selected to perform safely at all the temperatures encountered. The effect of thermal mass of insulation and protective finish and the temperature change rate should also be considered.

The thermal insulation construction must prevent the insulation material from (1) settling with ensuing noninsulated voids, (2) deteriorating through thermal shock, and (3) absorbing or retaining flammable liquids or vapors (if present in the environment of the space).

General Installation Techniques

The four types of insulation (rigid, loose-fill, flexible, and reflective) can be used in environmental space insulation construction. Requirements of vapor control, temperature range, thermal mass, and effect on temperature pulldown rate are set by the specification of environmental conditions.

Reflective insulations in the form of aluminum alloy sheets with emittances of 0.05 to 0.03, spaced to form the requisite number of dimensioned air spaces, are used extensively in large altitude chambers, low-temperature test rooms, and large all-weather rooms. The inherent low thermal mass of this insulation construction is advantageous for rapid temperature pulldown. Evacuation of air in spaces between reflective insulation sheets must be simultaneous with evacuation of air from the environmental space, either by vent openings through the protective liner of the thermal insulation, or by bleed lines exhausting the thermal areas to the main evacuation line. Bleed lines permit the use of a completely sealed protective liner between the insulation space and the environmental space, with no pressure differential created against the liner. Drainage of the spaces between reflective insulation sheets is also provided. Resistance to fire is afforded by the limited supply of oxygen in the spaces between the reflective insulation sheets.

Wood separator strips between reflective insulation sheets can be employed from the low range of temperature up to 200°F. For temperature ranges of 250 to 700°F, aluminum rods may be used as separators. For temperatures ranging from 700°F to the limits of stainless steel, a separator system of stainless steel rods and reflective insulation sheets of stainless steel may be used. The protective liner can be heavy-gage aluminum or stainless steel, depending on environmental conditions and temperatures.

Rigid cellular glass insulation can be used in combination with reflective insulation. Use cellular glass in the outer layer of thermal insulation where it does not affect rapid temperature pulldown; use reflective insulation in the inner layer where low thermal mass permits rapid temperature pulldown.

Cellular-glass rigid insulation can be used through a range of high temperatures. It is applied with adhesive or mechanical fasteners and joint sealers designed for the temperatures encountered. The protective finish must be suitable for the same temperatures. The high load-bearing value of cellular-glass permits support of floor constructions. Because the material has low conductivity, it is used to insulate piping in reflective insulation construction.

Mineral fiber insulation in flexible form, which is often used in reach-in and walk-in altitude chambers and low-temperature test facilities, is manufactured in plants and shipped assembled or in sections.

Organic rigid, foamed plastic rigid, semirigid, flexible, and loose-fill insulation, employed in the temperature range permitted by their physical characteristics, must be designed to preclude contraction cracks, settlement, and condensed water vapor collection.

REFERENCES

ASHRAE. 1989. Ventilation for acceptable indoor air quality. ANSI/ASHRAE *Standard* 62-1989.

ASHRAE. 1994. Recommended practices for controlling moisture in crawl spaces. Technical Data *Bulletin* 10(3).

Baker, M.C. 1980. *Roofs*, Chapters 6 and 8. Multiscience Publications, Montreal, Canada.

BRAB. 1963. Evaluation of components for underground heat distribution systems. Building Research Advisory Board, Federal Construction Council Technical *Report* 39. National Academy of Sciences, National Research Council, Publication No. 828. Washington, D.C.

BRAB. 1964. Field evaluation of underground heat distribution systems. Building Research Advisory Board, Federal Construction Council Technical *Report* 47. National Academy of Sciences, National Research Council, Washington, D.C.

BRAB. 1975. Criteria for underground distribution. Technical *Report* 66. Standing Committee on Mechanical Engineering of the FCC, BRAB, and NRC.

BRANZ. 1983. Insulating suspended timber frame floors. BRANZ *Bulletins* 177 and 233. Building Research Association, New Zealand.

Britton, R.R. 1948. Condensation in walls and roofs. HHFA Technical Papers *No.* 1, 2, 3 and 8. Housing and Home Finance Agency, Washington, D.C.

Burch, D.M. and S.J. Treado. 1978. Ventilating residences and their attics for energy conservation—An experimental study. In: Summer attic and whole-house ventilation. NBS Special Publication 548. National Institute of Standards and Technology, Gaithersburg, MD.

Christian, J.E. 1994. Moisture sources. In: *Moisture control in buildings*, pp. 176-82. ASTM Manual *Series*: MNL 18. American Society for Testing and Materials, West Conshohocken, PA.

CIBSE. 1986. *CIBSE Guide—Volume A* (*Design Data*). Section A3. Chartered Institution of Building Services Engineers, London.

Cunningham, M.J., G.A. Tsongas, and D. McQuade. 1990. Solar-driven moisture transfer through absorbent roofing materials. *ASHRAE Transactions* 96(2):465-71.

Desjarlais, A.O. 1995. Self-drying roofs: What! No dripping! In: *Proceedings of the ASHRAE/DOE/BETEC thermal performance of the exterior envelopes of buildings VI*, pp. 763-73. ASHRAE.

Dickens, H.B. and N.B. Hutcheon. 1965. Moisture accumulation in roof spaces under extreme winter conditions. Paper presented at RILEM/CIB symposium, Moisture Problems in Buildings, Helsinki, Finland. NRCC 9132.

Duff, J.E. 1956. The effect of air conditioning of water vapor permeability on temperature and humidity. *ASHRAE Transactions* 62:437.

FHA. 1942. Property standards and minimum construction requirements for dwelling. Federal Housing Administration, Washington, D.C.

Handegord, G.O. 1979. The need for improved air tightness in buildings. Building Research *Note* 151. National Research Council of Canada, Ottawa.

Hedlin, C.P. 1974. The use of nuclear moisture density meters to measure moisture in flat roofing insulations. National Research Council of Canada. NRCC 14593.

Hedlin, C.P. 1977. Moisture gains by foam plastic roof insulations under controlled temperature gradients. *Journal of Cellular Plastics* 13(5). NRCC 16317.

Jordan, C.A., E.C. Peck, F.A. Strange, and L.V. Teesdale. 1948. Attic condensation in tightly built houses. Housing and Home Finance Agency Technical *Bulletin* No 6.

Korsgaard, V. and C.R. Pedersen. 1992. Laboratory and practical experience with a novel water-permeable vapor retarder. In: *Thermal performance of the exterior envelopes of buildings V*, pp. 480-90. ASHRAE.

Kyle, D.M. and A.O. Desjarlais. 1994. Assessment of technologies for constructing self-drying low-slope roofs. Oak Ridge National Laboratory *Report* ORNL/CON-380. Oak Ridge, TN.

Labs, K., J. Carmody, R. Sterling, L. Shen, Y.J. Huang, and D. Parker. 1988. *Building foundation design handbook.* Oak Ridge National Laboratory *Report* ORNL/sub/86-72143/1.

Latta, J.K. 1976. Vapor barriers: What are they? Are they effective? *Canadian Building Digest* 175.

Link, R.E., Jr. 1977. Airborne thermal infrared and nuclear meter systems for detecting roof moisture. Proceedings of the NBS-NRCA Symposium, Roofing Technology.

Lstiburek, J. and J. Carmody. 1991. *The moisture control handbook—New low-rise, residential construction*, ORNL/Sub/89-SD350/1. Martin Marietta Energy Systems, Oak Ridge National Laboratory, Oak Ridge, TN.

NFPA. 1993. Standard for the installation of air conditioning and ventilating systems. *Standard* 90A-93, National Fire Protection Association, Quincy, MA.

NFPA. 1993. Standard for the installation of warm air heating and air conditioning systems. *Standard* 90B-93. National Fire Protection Association, Quincy, MA.

Powell, F.J. and H.E. Robinson. 1971. The effect of moisture on the heat transfer performance of insulated flat-roof constructions. Building Science *Series* 37. National Institute of Standards and Technology, Gaithersburg, MD.

Rose, W. 1992. Measured values of temperature and sheathing content in residential attic assemblies. In: *Thermal performance of the exterior envelopes of buildings V*, pp. 379-90. ASHRAE.

Rose, W.B. 1995. Attic construction with sheathing-applied insulation. *ASHRAE Transactions* 101(2):789-98.

Rowley, F.B., A.B. Algren, and C.E. Lund. 1939. Condensation of moisture and its relation to building construction and operation. *ASHVE Transactions*, 45:231-52.

Samuelson, I. 1994. Moisture control in crawl spaces. In: Recommended practices for controlling moisture in crawl spaces. ASHRAE Technical Data *Bulletin* 10(3):pp. 58-64.

SMACNA. 1992. Fibrous glass duct construction standards, 6th ed. Sheet Metal and Air Conditioning Contractor's National Association, Chantilly, VA.

Tamura, G.T. 1975. Measurement of air leakage characteristics of house enclosures. *ASHRAE Transactions* 81:202.

TenWolde, A. 1988. A mathematical model for indoor humidity in homes during winter. In: *Proceedings of symposium on air infiltration, ventilation, and moisture transfer*, pp. 3-32. BTECC-National Institute for Building Science, Washington D.C.

TenWolde, A. 1994. Ventilation, humidity, and condensation in manufactured houses during winter. *ASHRAE Transactions* 100(1):103-15.

TenWolde A. and C. Carll. 1992. Effect of cavity ventilation on moisture in walls and roofs. In: *Thermal performance of the exterior envelopes of buildings V*, pp. 555-62. ASHRAE.

Tobiasson, W. and M. Harrington. 1985. Vapor drive maps of the U.S. *Proceedings of the ASHRAE/DOE/BTECC thermal performance of the exterior envelopes of buildings III*, pp. 663-72.

Tobiasson, W., C. Korhonen, and A. van den Berg. 1977. Hand held infrared systems for detecting roof moisture. Proceedings of the NBS-NRCA Symposium, Roofing Technology.

Tobiasson, W., J. Buska, and A. Greatorex. 1994. Ventilating attics to minimize icing at eaves. *Energy and Buildings* 20:229-34.

Trethowen, H.A. 1988. A survey of subfloor ground evaporation rates. Building Research Association of New Zealand, BRANZ Study *Report* SR 13.

Trethowen, H.A. 1994. Three surveys of subfloor moisture in New Zealand. *ASHRAE Transactions* 100(1):1427-38.

Trethowen, H.A. and G. Middlemass. 1988. A survey of moisture damage in southern New Zealand buildings. Building Research Association of New Zealand, BRANZ Study *Report* SR 7.

U.S. Dept. of Energy. 1991. Radiant barrier attic fact sheet. DOE/CE-0335P.

Verschoor, J.D. 1977. Effectiveness of building insulation applications. USN/CEL *Report* No. CR78.006-NTIS No. AD-A053 452/9ST.

Wilson, A.G. and G.K. Garden. 1965. Moisture accumulation in walls due to air leakage. NRCC 9131. Paper presented at RILEM/CIB Symposium, Moisture Problems in Buildings, Helsinki, Finland.

BIBLIOGRAPHY

IEA. 1991. Vol. 2, Guidelines and practice. *Annex XIV, Condensation and Energy*. International Energy Agency, Leuven, Belgium.

Trechsel, H.R., ed. 1994. *Moisture control in buildings*. ASTM Manual *Series* MNL18. American Society for Testing and Materials, West Conshohocken, PA.

CHAPTER 24

THERMAL AND WATER VAPOR TRANSMISSION DATA

THIS chapter presents thermal and water vapor transmission data based on steady-state or equilibrium conditions. Chapter 3 covers heat transfer under transient or changing temperature conditions. Chapter 22 discusses selection of insulation materials and procedures for determining overall thermal resistances by simplified methods.

BUILDING ENVELOPES

Thermal Transmission Data for Building Components

The steady-state thermal resistances (R-values) of building components (walls, floors, windows, roof systems, etc.) can be calculated from the thermal properties of the materials in the component; or the heat flow through the assembled component can be measured directly with laboratory equipment such as the guarded hot box (ASTM *Standard* C 236) or the calibrated hot box (ASTM *Standard* C 976).

Tables 1 through 6 list thermal values, which may be used to calculate thermal resistances of building walls, floors, and ceilings. The values shown in these tables were developed under ideal conditions. In practice, overall thermal performance can be reduced significantly by such factors as improper installation and shrinkage, settling, or compression of the insulation (Tye and Desjarlais 1983; Tye 1985, 1986).

Most values in these tables were obtained by accepted ASTM test methods described in ASTM *Standards* C 177 and C 518 for materials and ASTM *Standards* C 236 and C 976 for building envelope components. Because commercially available materials vary, not all values apply to specific products.

The most accurate method of determining the overall thermal resistance for a combination of building materials assembled as a building envelope component is to test a representative sample by a hot box method. However, all combinations may not be conveniently or economically tested in this manner. For many simple constructions, calculated R-values agree reasonably well with values determined by hot box measurement.

The performance of materials fabricated in the field is especially subject to the quality of workmanship during construction and installation. Good workmanship becomes increasingly important as the insulation requirement becomes greater. Therefore, some engineers include additional insulation or other safety factors based on experience in their design.

Figure 1 shows how convection affects surface conductance of several materials. Other tests on smooth surfaces show that the average value of the convection part of the surface conductance decreases as the length of the surface increases.

Vapor retarders, which are discussed in Chapters 22 and 23, require special attention. Moisture from condensation or other sources may reduce the thermal resistance of insulation, but the effect of moisture must be determined for each material. For example, some materials with large air spaces are not affected significantly if the moisture content is less than 10% by weight, while the effect of moisture on other materials is approximately linear.

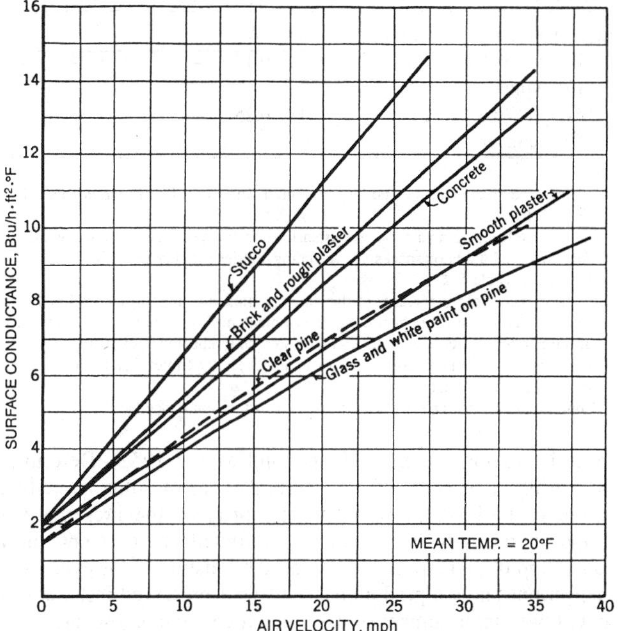

Fig. 1 Surface Conductance for Different Surfaces as Affected by Air Movement

Ideal conditions of components and installations are assumed in calculating overall R-values (i.e., insulating materials are of uniform nominal thickness and thermal resistance, air spaces are of uniform thickness and surface temperature, moisture effects are not involved, and installation details are in accordance with design). The National Institute of Standards and Technology Building Materials and Structures Report BMS 151 shows that measured values differ from calculated values for certain insulated constructions. For this reason, some engineers decrease the calculated R-values a moderate amount to account for departures of constructions from requirements and practices.

Tables 3 and 2 give values for well-sealed systems constructed with care. Field applications can differ substantially from laboratory test conditions. Air gaps in these insulation systems can seriously degrade thermal performance as a result of air movement due to both natural and forced convection. Sabine et al. (1975) found that the tabular values are not necessarily additive for multiple-layer, low-emittance air spaces, and tests on actual constructions should be conducted to accurately determine thermal resistance values.

Values for foil insulation products supplied by manufacturers must also be used with caution because they apply only to systems that are identical to the configuration in which the product was tested. In addition, surface oxidation, dust accumulation, condensation, and other factors that change the condition of the low-emittance surface can reduce the thermal effectiveness of

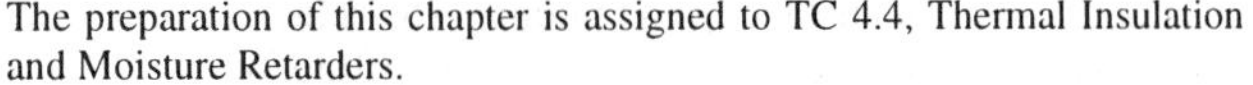

The preparation of this chapter is assigned to TC 4.4, Thermal Insulation and Moisture Retarders.

Table 1 Surface Conductances and Resistances for Air

Position of Surface	Direction of Heat Flow	Surface Emittance, ε: Non-reflective ε = 0.90 h_i	R	Reflective ε = 0.20 h_i	R	Reflective ε = 0.05 h_i	R
STILL AIR							
Horizontal	Upward	1.63	0.61	0.91	1.10	0.76	1.32
Sloping—45°	Upward	1.60	0.62	0.88	1.14	0.73	1.37
Vertical	Horizontal	1.46	0.68	0.74	1.35	0.59	1.70
Sloping—45°	Downward	1.32	0.76	0.60	1.67	0.45	2.22
Horizontal	Downward	1.08	0.92	0.37	2.70	0.22	4.55
MOVING AIR (Any position)		h_o	R				
15-mph Wind (for winter)	Any	6.00	0.17	—	—	—	—
7.5-mph Wind (for summer)	Any	4.00	0.25	—	—	—	—

Notes:
1. Surface conductance h_i and h_o measured in Btu/h·ft²·°F; resistance R in °F·ft²·h/Btu.
2. No surface has both an air space resistance value and a surface resistance value.
3. For ventilated attics or spaces above ceilings under summer conditions (heat flow down), see Table 5.
4. Conductances are for surfaces of the stated emittance facing virtual blackbody surroundings at the same temperature as the ambient air. Values are based on a surface-air temperature difference of 10°F and for surface temperatures of 70°F.
5. See Chapter 3 for more detailed information, especially Tables 5 and 6, and see Figure 1 for additional data.
6. Condensate can have a significant impact on surface emittance (see Table 2).

these insulation systems (Hooper and Moroz 1952). Deterioration results from contact with several types of solutions, either acidic or basic (e.g., wet cement mortar or the preservatives found in decay-resistant lumber). Polluted environments may cause rapid and severe material degradation. However, site inspections show a predominance of well-preserved installations and only a small number of cases in which rapid and severe deterioration has occurred. An extensive review of the reflective building insulation system performance literature is provided by Goss and Miller (1989).

CALCULATING OVERALL THERMAL RESISTANCES

Relatively small, highly conductive elements in an insulating layer called thermal bridges can substantially reduce the average thermal resistance of a component. Examples include wood and metal studs in frame walls, concrete webs in concrete masonry walls, and metal ties or other elements in insulated wall panels. The following examples illustrate the calculation of R-values and U-factors for components containing thermal bridges.

These conditions are assumed in calculating the design R-values:

- Equilibrium or steady-state heat transfer, disregarding effects of thermal storage
- Surrounding surfaces at ambient air temperature
- Exterior wind velocity of 15 mph for winter (surface with R = 0.17°F·ft²·h/Btu) and 7.5 mph for summer (surface with R = 0.25°F·ft²·h/Btu)
- Surface emittance of ordinary building materials is 0.90

Wood Frame Walls

The average overall R-values and U-factors of wood frame walls can be calculated by assuming either parallel heat flow paths through areas with different thermal resistances or by assuming isothermal planes. Equations (1) through (5) from Chapter 22 are used.

Table 2 Emittance Values of Various Surfaces and Effective Emittances of Air Spaces[a]

Surface	Average Emittance ε	Effective Emittance ε_{eff} of Air Space: One Surface Emittance ε; Other, 0.9	Both Surfaces Emittance ε
Aluminum foil, bright	0.05	0.05	0.03
Aluminum foil, with condensate just visible (> 0.7 gr/ft²)	0.30[b]	0.29	—
Aluminum foil, with condensate clearly visible (> 2.9 gr/ft²)	0.70[b]	0.65	—
Aluminum sheet	0.12	0.12	0.06
Aluminum coated paper, polished	0.20	0.20	0.11
Steel, galvanized, bright	0.25	0.24	0.15
Aluminum paint	0.50	0.47	0.35
Building materials: wood, paper, masonry, nonmetallic paints	0.90	0.82	0.82
Regular glass	0.84	0.77	0.72

[a]These values apply in the 4 to 40 μm range of the electromagnetic spectrum.
[b]Values are based on data presented by Bassett and Trethowen (1984).

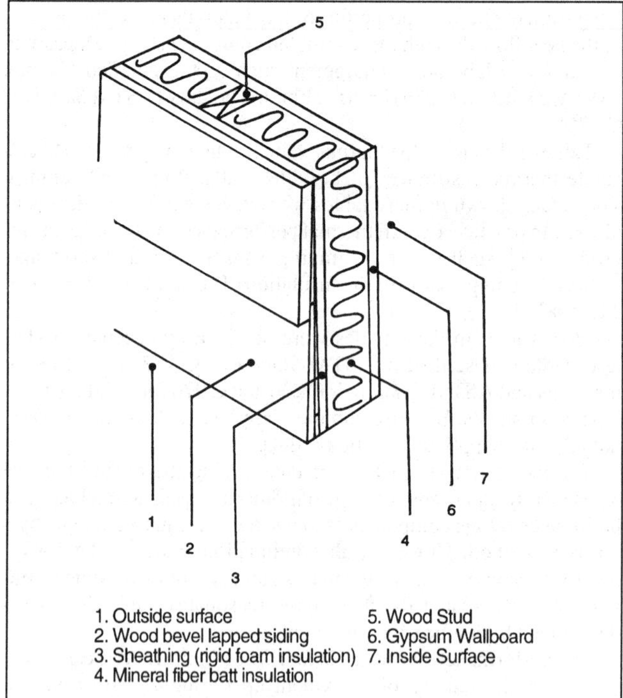

Fig. 2 Insulated Wood Frame Wall (Example 1)

The framing factor or fraction of the building component that is framing depends on the specific type of construction, and it may vary based on local construction practices—even for the same type of construction. For stud walls 16 in. on center (OC), the fraction of insulated cavity may be as low as 0.75, where the fraction of studs, plates, and sills is 0.21 and the fraction of headers is 0.04. For studs 24 in. OC, the respective values are 0.78, 0.18, and 0.04. These fractions contain an allowance for multiple studs, plates, sills, extra framing around windows, headers, and band joists. These assumed framing fractions are used in the following example, to illustrate the importance of including the effect of framing in determining the overall thermal conductance of a building. The actual framing fraction should be calculated for each specific construction.

Table 3 Thermal Resistances of Plane Air Spaces[a,b,c], °F·ft²·h/Btu

Position of Air Space	Direction of Heat Flow	Air Space: Mean Temp.[d], °F	Air Space: Temp. Diff.[d], °F	0.5-in. Air Space[c] Effective Emittance ε_{eff}[d,e] 0.03	0.05	0.2	0.5	0.82	0.75-in. Air Space[c] Effective Emittance ε_{eff}[d,e] 0.03	0.05	0.2	0.5	0.82
		90	10	2.13	2.03	1.51	0.99	0.73	2.34	2.22	1.61	1.04	0.75
		50	30	1.62	1.57	1.29	0.96	0.75	1.71	1.66	1.35	0.99	0.77
		50	10	2.13	2.05	1.60	1.11	0.84	2.30	2.21	1.70	1.16	0.87
Horiz.	Up	0	20	1.73	1.70	1.45	1.12	0.91	1.83	1.79	1.52	1.16	0.93
		0	10	2.10	2.04	1.70	1.27	1.00	2.23	2.16	1.78	1.31	1.02
		−50	20	1.69	1.66	1.49	1.23	1.04	1.77	1.74	1.55	1.27	1.07
		−50	10	2.04	2.00	1.75	1.40	1.16	2.16	2.11	1.84	1.46	1.20
		90	10	2.44	2.31	1.65	1.06	0.76	2.96	2.78	1.88	1.15	0.81
		50	30	2.06	1.98	1.56	1.10	0.83	1.99	1.92	1.52	1.08	0.82
		50	10	2.55	2.44	1.83	1.22	0.90	2.90	2.75	2.00	1.29	0.94
45° Slope	Up	0	20	2.20	2.14	1.76	1.30	1.02	2.13	2.07	1.72	1.28	1.00
		0	10	2.63	2.54	2.03	1.44	1.10	2.72	2.62	2.08	1.47	1.12
		−50	20	2.08	2.04	1.78	1.42	1.17	2.05	2.01	1.76	1.41	1.16
		−50	10	2.62	2.56	2.17	1.66	1.33	2.53	2.47	2.10	1.62	1.30
		90	10	2.47	2.34	1.67	1.06	0.77	3.50	3.24	2.08	1.22	0.84
		50	30	2.57	2.46	1.84	1.23	0.90	2.91	2.77	2.01	1.30	0.94
		50	10	2.66	2.54	1.88	1.24	0.91	3.70	3.46	2.35	1.43	1.01
Vertical	Horiz.	0	20	2.82	2.72	2.14	1.50	1.13	3.14	3.02	2.32	1.58	1.18
		0	10	2.93	2.82	2.20	1.53	1.15	3.77	3.59	2.64	1.73	1.26
		−50	20	2.90	2.82	2.35	1.76	1.39	2.90	2.83	2.36	1.77	1.39
		−50	10	3.20	3.10	2.54	1.87	1.46	3.72	3.60	2.87	2.04	1.56
		90	10	2.48	2.34	1.67	1.06	0.77	3.53	3.27	2.10	1.22	0.84
		50	30	2.64	2.52	1.87	1.24	0.91	3.43	3.23	2.24	1.39	0.99
		50	10	2.67	2.55	1.89	1.25	0.92	3.81	3.57	2.40	1.45	1.02
45° Slope	Down	0	20	2.91	2.80	2.19	1.52	1.15	3.75	3.57	2.63	1.72	1.26
		0	10	2.94	2.83	2.21	1.53	1.15	4.12	3.91	2.81	1.80	1.30
		−50	20	3.16	3.07	2.52	1.86	1.45	3.78	3.65	2.90	2.05	1.57
		−50	10	3.26	3.16	2.58	1.89	1.47	4.35	4.18	3.22	2.21	1.66
		90	10	2.48	2.34	1.67	1.06	0.77	3.55	3.29	2.10	1.22	0.85
		50	30	2.66	2.54	1.88	1.24	0.91	3.77	3.52	2.38	1.44	1.02
		50	10	2.67	2.55	1.89	1.25	0.92	3.84	3.59	2.41	1.45	1.02
Horiz.	Down	0	20	2.94	2.83	2.20	1.53	1.15	4.18	3.96	2.83	1.81	1.30
		0	10	2.96	2.85	2.22	1.53	1.16	4.25	4.02	2.87	1.82	1.31
		−50	20	3.25	3.15	2.58	1.89	1.47	4.60	4.41	3.36	2.28	1.69
		−50	10	3.28	3.18	2.60	1.90	1.47	4.71	4.51	3.42	2.30	1.71
		Air Space		1.5-in. Air Space[c]					3.5-in. Air Space[c]				
		90	10	2.55	2.41	1.71	1.08	0.77	2.84	2.66	1.83	1.13	0.80
		50	30	1.87	1.81	1.45	1.04	0.80	2.09	2.01	1.58	1.10	0.84
		50	10	2.50	2.40	1.81	1.21	0.89	2.80	2.66	1.95	1.28	0.93
Horiz.	Up	0	20	2.01	1.95	1.63	1.23	0.97	2.25	2.18	1.79	1.32	1.03
		0	10	2.43	2.35	1.90	1.38	1.06	2.71	2.62	2.07	1.47	1.12
		−50	20	1.94	1.91	1.68	1.36	1.13	2.19	2.14	1.86	1.47	1.20
		−50	10	2.37	2.31	1.99	1.55	1.26	2.65	2.58	2.18	1.67	1.33
		90	10	2.92	2.73	1.86	1.14	0.80	3.18	2.96	1.97	1.18	0.82
		50	30	2.14	2.06	1.61	1.12	0.84	2.26	2.17	1.67	1.15	0.86
		50	10	2.88	2.74	1.99	1.29	0.94	3.12	2.95	2.10	1.34	0.96
45° Slope	Up	0	20	2.30	2.23	1.82	1.34	1.04	2.42	2.35	1.90	1.38	1.06
		0	10	2.79	2.69	2.12	1.49	1.13	2.98	2.87	2.23	1.54	1.16
		−50	20	2.22	2.17	1.88	1.49	1.21	2.34	2.29	1.97	1.54	1.25
		−50	10	2.71	2.64	2.23	1.69	1.35	2.87	2.79	2.33	1.75	1.39
		90	10	3.99	3.66	2.25	1.27	0.87	3.69	3.40	2.15	1.24	0.85
		50	30	2.58	2.46	1.84	1.23	0.90	2.67	2.55	1.89	1.25	0.91
		50	10	3.79	3.55	2.39	1.45	1.02	3.63	3.40	2.32	1.42	1.01
Vertical	Horiz.	0	20	2.76	2.66	2.10	1.48	1.12	2.88	2.78	2.17	1.51	1.14
		0	10	3.51	3.35	2.51	1.67	1.23	3.49	3.33	2.50	1.67	1.23
		−50	20	2.64	2.58	2.18	1.66	1.33	2.82	2.75	2.30	1.73	1.37
		−50	10	3.31	3.21	2.62	1.91	1.48	3.40	3.30	2.67	1.94	1.50
		90	10	5.07	4.55	2.56	1.36	0.91	4.81	4.33	2.49	1.34	0.90
		50	30	3.58	3.36	2.31	1.42	1.00	3.51	3.30	2.28	1.40	1.00
		50	10	5.10	4.66	2.85	1.60	1.09	4.74	4.36	2.73	1.57	1.08
45° Slope	Down	0	20	3.85	3.66	2.68	1.74	1.27	3.81	3.63	2.66	1.74	1.27
		0	10	4.92	4.62	3.16	1.94	1.37	4.59	4.32	3.02	1.88	1.34
		−50	20	3.62	3.50	2.80	2.01	1.54	3.77	3.64	2.90	2.05	1.57
		−50	10	4.67	4.47	3.40	2.29	1.70	4.50	4.32	3.31	2.25	1.68
		90	10	6.09	5.35	2.79	1.43	0.94	10.07	8.19	3.41	1.57	1.00
		50	30	6.27	5.63	3.18	1.70	1.14	9.60	8.17	3.86	1.88	1.22
		50	10	6.61	5.90	3.27	1.73	1.15	11.15	9.27	4.09	1.93	1.24
Horiz.	Down	0	20	7.03	6.43	3.91	2.19	1.49	10.90	9.52	4.87	2.47	1.62
		0	10	7.31	6.66	4.00	2.22	1.51	11.97	10.32	5.08	2.52	1.64
		−50	20	7.73	7.20	4.77	2.85	1.99	11.64	10.49	6.02	3.25	2.18
		−50	10	8.09	7.52	4.91	2.89	2.01	12.98	11.56	6.36	3.34	2.22

[a]See Chapter 22, section Factors Affecting Heat Transfer across Air Spaces. Thermal resistance values were determined from the relation, $R = 1/C$, where $C = h_c + \varepsilon_{eff} h_r$, h_c is the conduction-convection coefficient, $\varepsilon_{eff} h_r$ is the radiation coefficient $\approx 0.0068\varepsilon_{eff}[(t_m + 460)/100]^3$, and t_m is the mean temperature of the air space. Values for h_c were determined from data developed by Robinson et al. (1954). Equations (5) through (7) in Yarbrough (1983) show the data in this table in analytic form. For extrapolation from this table to air spaces less than 0.5 in. (as in insulating window glass), assume $h_c = 0.159(1 + 0.0016\, t_m)/l$ where l is the air space thickness in inches, and h_c is heat transfer through the air space only.

[b]Values are based on data presented by Robinson et al. (1954). (Also see Chapter 3, Tables 3 and 4, and Chapter 36). Values apply for ideal conditions, i.e., air spaces of uniform thickness bounded by plane, smooth, parallel surfaces with no air leakage to or from the space. When accurate values are required, use overall U-factors determined through calibrated hot box (ASTM C 976) or guarded hot box (ASTM C 236) testing. Thermal resistance values for multiple air spaces must be based on careful estimates of mean temperature differences for each air space.

[c]A single resistance value cannot account for multiple air spaces; each air space requires a separate resistance calculation that applies only for the established boundary conditions. Resistances of horizontal spaces with heat flow downward are substantially independent of temperature difference.

[d]Interpolation is permissible for other values of mean temperature, temperature difference, and effective emittance ε_{eff}. Interpolation and moderate extrapolation for air spaces greater than 3.5 in. are also permissible.

[e]Effective emittance ε_{eff} of the air space is given by $1/\varepsilon_{eff} = 1/\varepsilon_1 + 1/\varepsilon_2 - 1$, where ε_1 and ε_2 are the emittances of the surfaces of the air space (see Table 2).

Table 4 Typical Thermal Properties of Common Building and Insulating Materials—Design Values[a]

Description	Density, lb/ft^3	Conductivity[b] (*k*), $\frac{Btu \cdot in}{h \cdot ft^2 \cdot °F}$	Conductance (*C*), $\frac{Btu}{h \cdot ft^2 \cdot °F}$	Resistance[c] (*R*) Per Inch Thickness (1/*k*), $\frac{°F \cdot ft^2 \cdot h}{Btu \cdot in}$	Resistance[c] (*R*) For Thickness Listed (1/*C*), $\frac{°F \cdot ft^2 \cdot h}{Btu}$	Specific Heat, $\frac{Btu}{lb \cdot °F}$
BUILDING BOARD						
Asbestos-cement board	120	4.0	—	0.25	—	0.24
Asbestos-cement board ... 0.125 in.	120	—	33.00	—	0.03	
Asbestos-cement board ... 0.25 in.	120	—	16.50	—	0.06	
Gypsum or plaster board ... 0.375 in.	50	—	3.10	—	0.32	0.26
Gypsum or plaster board ... 0.5 in.	50	—	2.22	—	0.45	
Gypsum or plaster board ... 0.625 in.	50	—	1.78	—	0.56	
Plywood (Douglas Fir)[d]	34	0.80	—	1.25	—	0.29
Plywood (Douglas Fir) ... 0.25 in.	34	—	3.20	—	0.31	
Plywood (Douglas Fir) ... 0.375 in.	34	—	2.13	—	0.47	
Plywood (Douglas Fir) ... 0.5 in.	34	—	1.60	—	0.62	
Plywood (Douglas Fir) ... 0.625 in.	34	—	1.29	—	0.77	
Plywood or wood panels ... 0.75 in.	34	—	1.07	—	0.93	0.29
Vegetable fiber board						
Sheathing, regular density[e] ... 0.5 in.	18	—	0.76	—	1.32	0.31
... 0.78125 in.	18	—	0.49	—	2.06	
Sheathing intermediate density[e] ... 0.5 in.	22	—	0.92	—	1.09	0.31
Nail-base sheathing[e] ... 0.5 in.	25	—	0.94	—	1.06	0.31
Shingle backer ... 0.375 in.	18	—	1.06	—	0.94	0.31
Shingle backer ... 0.3125 in.	18	—	1.28	—	0.78	
Sound deadening board ... 0.5 in.	15	—	0.74	—	1.35	0.30
Tile and lay-in panels, plain or acoustic	18	0.40	—	2.50	—	0.14
... 0.5 in.	18	—	0.80	—	1.25	
... 0.75 in.	18	—	0.53	—	1.89	
Laminated paperboard	30	0.50	—	2.00	—	0.33
Homogeneous board from repulped paper	30	0.50	—	2.00	—	0.28
Hardboard[e]						
Medium density	50	0.73	—	1.37	—	0.31
High density, service-tempered grade and service grade	55	0.82	—	1.22	—	0.32
High density, standard-tempered grade	63	1.00	—	1.00	—	0.32
Particleboard[e]						
Low density	37	0.71	—	1.41	—	0.31
Medium density	50	0.94	—	1.06	—	0.31
High density	62	.5	1.18	—	0.85	—
Underlayment ... 0.625 in.	40	—	1.22	—	0.82	0.29
Waferboard	37	0.63	—	1.59	—	—
Wood subfloor ... 0.75 in.	—	—	1.06	—	0.94	0.33
BUILDING MEMBRANE						
Vapor—permeable felt	—	—	16.70	—	0.06	
Vapor—seal, 2 layers of mopped 15-lb felt	—	—	8.35	—	0.12	
Vapor—seal, plastic film	—	—	—	—	Negl.	
FINISH FLOORING MATERIALS						
Carpet and fibrous pad	—	—	0.48	—	2.08	0.34
Carpet and rubber pad	—	—	0.81	—	1.23	0.33
Cork tile ... 0.125 in.	—	—	3.60	—	0.28	0.48
Terrazzo ... 1 in.	—	—	12.50	—	0.08	0.19
Tile—asphalt, linoleum, vinyl, rubber	—	—	20.00	—	0.05	0.30
vinyl asbestos						0.24
ceramic						0.19
Wood, hardwood finish ... 0.75 in.	—	—	1.47	—	0.68	
INSULATING MATERIALS						
Blanket and Batt[f,g]						
Mineral fiber, fibrous form processed from rock, slag, or glass						
approx. 3-4 in.	0.4-2.0	—	0.091	—	11	
approx. 3.5 in.	0.4-2.0	—	0.077	—	13	
approx. 3.5 in.	1.2-1.6	—	0.067	—	15	
approx. 5.5-6.5 in.	0.4-2.0	—	0.053	—	19	
approx. 5.5 in.	0.6-1.0	—	0.048	—	21	
approx. 6-7.5 in.	0.4-2.0	—	0.045	—	22	
approx. 8.25-10 in.	0.4-2.0	—	0.033	—	30	
approx. 10-13 in.	0.4-2.0	—	0.026	—	38	
Board and Slabs						
Cellular glass	8.0	0.33	—	3.03	—	0.18
Glass fiber, organic bonded	4.0-9.0	0.25	—	4.00	—	0.23
Expanded perlite, organic bonded	1.0	0.36	—	2.78	—	0.30
Expanded rubber (rigid)	4.5	0.22	—	4.55	—	0.40
Expanded polystyrene, extruded (smooth skin surface) (CFC-12 exp.)	1.8-3.5	0.20	—	5.00	—	0.29

Table 4 Typical Thermal Properties of Common Building and Insulating Materials—Design Values[a] (Continued)

Description	Density, lb/ft³	Conductivity[b] (*k*), Btu·in / h·ft²·°F	Conductance (*C*), Btu / h·ft²·°F	Resistance[c] (*R*) Per Inch Thickness (1/*k*), °F·ft²·h / Btu·in	Resistance[c] (*R*) For Thickness Listed (1/*C*), °F·ft²·h / Btu	Specific Heat, Btu / lb·°F
Expanded polystyrene, extruded (smooth skin surface) (HCFC-142b exp.)[h]	1.8-3.5	0.20	—	5.00	—	0.29
Expanded polystyrene, molded beads	1.0	0.26	—	3.85	—	—
	1.25	0.25	—	4.00	—	—
	1.5	0.24	—	4.17	—	—
	1.75	0.24	—	4.17	—	—
	2.0	0.23	—	4.35	—	—
Cellular polyurethane/polyisocyanurate[i,l] (CFC-11 exp.) (unfaced)	1.5	0.16-0.18	—	6.25-5.56	—	0.38
Cellular polyisocyanurate[i] (CFC-11 exp.) (gas-permeable facers)	1.5-2.5	0.16-0.18	—	6.25-5.56	—	0.22
Cellular polyisocyanurate[j] (CFC-11 exp.) (gas-impermeable facers)	2.0	0.14	—	7.04	—	0.22
Cellular phenolic (closed cell) (CFC-11, CFC-113 exp.)[k]	3.0	0.12	—	8.20	—	—
Cellular phenolic (open cell)	1.8-2.2	0.23	—	4.40	—	—
Mineral fiber with resin binder	15.0	0.29	—	3.45	—	0.17
Mineral fiberboard, wet felted						
Core or roof insulation	16-17	0.34	—	2.94	—	—
Acoustical tile	18.0	0.35	—	2.86	—	0.19
Acoustical tile	21.0	0.37	—	2.70	—	—
Mineral fiberboard, wet molded						
Acoustical tile[l]	23.0	0.42	—	2.38	—	0.14
Wood or cane fiberboard						
Acoustical tile[l] 0.5 in.	—	—	0.80	—	1.25	0.31
Acoustical tile[l] 0.75 in.	—	—	0.53	—	1.89	—
Interior finish (plank, tile)	15.0	0.35	—	2.86	—	0.32
Cement fiber slabs (shredded wood with Portland cement binder)	25-27.0	0.50-0.53		2.0-1.89	—	
Cement fiber slabs (shredded wood with magnesia oxysulfide binder)	22.0	0.57	—	1.75	—	0.31
Loose Fill						
Cellulosic insulation (milled paper or wood pulp)	2.3-3.2	0.27-0.32	—	3.70-3.13	—	0.33
Perlite, expanded	2.0-4.1	0.27-0.31	—	3.7-3.3	—	0.26
	4.1-7.4	0.31-0.36	—	3.3-2.8	—	—
	7.4-11.0	0.36-0.42	—	2.8-2.4	—	—
Mineral fiber (rock, slag, or glass)[g]						
approx. 3.75-5 in.	0.6-2.0	—	—	—	11.0	0.17
approx. 6.5-8.75 in.	0.6-2.0	—	—	—	19.0	—
approx. 7.5-10 in.	0.6-2.0	—	—	—	22.0	—
approx. 10.25-13.75 in.	0.6-2.0	—	—	—	30.0	—
Mineral fiber (rock, slag, or glass)[g]						
approx. 3.5 in. (closed sidewall application)	2.0-3.5	—	—	—	12.0-14.0	—
Vermiculite, exfoliated	7.0-8.2	0.47	—	2.13	—	0.32
	4.0-6.0	0.44	—	2.27	—	—
Spray Applied						
Polyurethane foam	1.5-2.5	0.16-0.18	—	6.25-5.56	—	—
Ureaformaldehyde foam	0.7-1.6	0.22-0.28	—	4.55-3.57	—	—
Cellulosic fiber	3.5-6.0	0.29-0.34	—	3.45-2.94	—	—
Glass fiber	3.5-4.5	0.26-0.27	—	3.85-3.70	—	—
Reflective Insulation						
Reflective material ($\varepsilon < 0.5$) in center of 3/4 in. cavity forms two 3/8 in. vertical air spaces[m]	—	—	0.31	—	3.2	—
METALS (See Chapter 36, Table 3)						
ROOFING						
Asbestos-cement shingles	120	—	4.76	—	0.21	0.24
Asphalt roll roofing	70	—	6.50	—	0.15	0.36
Asphalt shingles	70	—	2.27	—	0.44	0.30
Built-up roofing 0.375 in.	70	—	3.00	—	0.33	0.35
Slate 0.5 in.	—	—	20.00	—	0.05	0.30
Wood shingles, plain and plastic film faced	—	—	1.06	—	0.94	0.31
PLASTERING MATERIALS						
Cement plaster, sand aggregate	116	5.0	—	0.20	—	0.20
Sand aggregate 0.375 in.	—	—	13.3	—	0.08	0.20
Sand aggregate 0.75 in.	—	—	6.66	—	0.15	0.20

Table 4 Typical Thermal Properties of Common Building and Insulating Materials—Design Values[a] (Continued)

Description	Density, lb/ft³	Conductivity[b] (k), Btu·in/(h·ft²·°F)	Conductance (C), Btu/(h·ft²·°F)	Resistance[c] (R) Per Inch Thickness (1/k), °F·ft²·h/(Btu·in)	Resistance[c] (R) For Thickness Listed (1/C), °F·ft²·h/Btu	Specific Heat, Btu/(lb·°F)
Gypsum plaster:						
Lightweight aggregate ....0.5 in.	45	—	3.12	—	0.32	—
Lightweight aggregate ....0.625 in.	45	—	2.67	—	0.39	—
Lightweight aggregate on metal lath ....0.75 in.	—	—	2.13	—	0.47	—
Perlite aggregate	45	1.5	—	0.67	—	0.32
Sand aggregate	105	5.6	—	0.18	—	0.20
Sand aggregate ....0.5 in.	105	—	11.10	—	0.09	—
Sand aggregate ....0.625 in.	105	—	9.10	—	0.11	—
Sand aggregate on metal lath ....0.75 in.	—	—	7.70	—	0.13	—
Vermiculite aggregate	45	1.7	—	0.59	—	—
MASONRY MATERIALS						
Masonry Units						
Brick, fired clay	150	8.4-10.2	—	0.12-0.10	—	—
	140	7.4-9.0	—	0.14-0.11	—	—
	130	6.4-7.8	—	0.16-0.12	—	—
	120	5.6-6.8	—	0.18-0.15	—	0.19
	110	4.9-5.9	—	0.20-0.17	—	—
	100	4.2-5.1	—	0.24-0.20	—	—
	90	3.6-4.3	—	0.28-0.24	—	—
	80	3.0-3.7	—	0.33-0.27	—	—
	70	2.5-3.1	—	0.40-0.33	—	—
Clay tile, hollow						
1 cell deep ....3 in.	—	—	1.25	—	0.80	0.21
1 cell deep ....4 in.	—	—	0.90	—	1.11	—
2 cells deep ....6 in.	—	—	0.66	—	1.52	—
2 cells deep ....8 in.	—	—	0.54	—	1.85	—
2 cells deep ....10 in.	—	—	0.45	—	2.22	—
3 cells deep ....12 in.	—	—	0.40	—	2.50	—
Concrete blocks[n, o]						
Limestone aggregate						
8 in., 36 lb, 138 lb/ft³ concrete, 2 cores	—	—	—	—	—	—
Same with perlite filled cores	—	—	0.48	—	2.1	—
12 in., 55 lb, 138 lb/ft³ concrete, 2 cores	—	—	—	—	—	—
Same with perlite filled cores	—	—	0.27	—	3.7	—
Normal weight aggregate (sand and gravel)						
8 in., 33-36 lb, 126-136 lb/ft³ concrete, 2 or 3 cores	—	—	0.90-1.03	—	1.11-0.97	0.22
Same with perlite filled cores	—	—	0.50	—	2.0	—
Same with vermiculite filled cores	—	—	0.52-0.73	—	1.92-1.37	—
12 in., 50 lb, 125 lb/ft³ concrete, 2 cores	—	—	0.81	—	1.23	0.22
Medium weight aggregate (combinations of normal weight and lightweight aggregate)						
8 in., 26-29 lb, 97-112 lb/ft³ concrete, 2 or 3 cores	—	—	0.58-0.78	—	1.71-1.28	—
Same with perlite filled cores	—	—	0.27-0.44	—	3.7-2.3	—
Same with vermiculite filled cores	—	—	0.30	—	3.3	—
Same with molded EPS (beads) filled cores	—	—	0.32	—	3.2	—
Same with molded EPS inserts in cores	—	—	0.37	—	2.7	—
Lightweight aggregate (expanded shale, clay, slate or slag, pumice)						
6 in., 16-17 lb 85-87 lb/ft³ concrete, 2 or 3 cores	—	—	0.52-0.61	—	1.93-1.65	—
Same with perlite filled cores	—	—	0.24	—	4.2	—
Same with vermiculite filled cores	—	—	0.33	—	3.0	—
8 in., 19-22 lb, 72-86 lb/ft³ concrete	—	—	0.32-0.54	—	3.2-1.90	0.21
Same with perlite filled cores	—	—	0.15-0.23	—	6.8-4.4	—
Same with vermiculite filled cores	—	—	0.19-0.26	—	5.3-3.9	—
Same with molded EPS (beads) filled cores	—	—	0.21	—	4.8	—
Same with UF foam filled cores	—	—	0.22	—	4.5	—
Same with molded EPS inserts in cores	—	—	0.29	—	3.5	—
12 in., 32-36 lb, 80-90 lb/ft³ concrete, 2 or 3 cores	—	—	0.38-0.44	—	2.6-2.3	—
Same with perlite filled cores	—	—	0.11-0.16	—	9.2-6.3	—
Same with vermiculite filled cores	—	—	0.17	—	5.8	—
Stone, lime, or sand	180	72	—	0.01	—	—
Quartzitic and sandstone	160	43	—	0.02	—	—
	140	24	—	0.04	—	—
	120	13	—	0.08	—	0.19
Calcitic, dolomitic, limestone, marble, and granite	180	30	—	0.03	—	—
	160	22	—	0.05	—	—
	140	16	—	0.06	—	—
	120	11	—	0.09	—	0.19
	100	8	—	0.13	—	—

Table 4 Typical Thermal Properties of Common Building and Insulating Materials—Design Values[a] (Continued)

Description	Density, lb/ft³	Conductivity[b] (k), Btu·in/h·ft²·°F	Conductance (C), Btu/h·ft²·°F	Resistance[c] (R) Per Inch Thickness (1/k), °F·ft²·h/Btu·in	Resistance[c] (R) For Thickness Listed (1/C), °F·ft²·h/Btu	Specific Heat, Btu/lb·°F
Gypsum partition tile						
3 by 12 by 30 in., solid	—	—	0.79	—	1.26	0.19
3 by 12 by 30 in., 4 cells	—	—	0.74	—	1.35	—
4 by 12 by 30 in., 3 cells	—	—	0.60	—	1.67	—
Concretes[o]						
Sand and gravel or stone aggregate concretes (concretes with more than 50% quartz or quartzite sand have conductivities in the higher end of the range)	150	10.0-20.0	—	0.10-0.05	—	—
	140	9.0-18.0	—	0.11-0.06	—	0.19-0.24
	130	7.0-13.0	—	0.14-0.08	—	—
Limestone concretes	140	11.1	—	0.09	—	—
	120	7.9	—	0.13	—	—
	100	5.5	—	0.18	—	—
Gypsum-fiber concrete (87.5% gypsum, 12.5% wood chips)	51	1.66	—	0.60	—	0.21
Cement/lime, mortar, and stucco	120	9.7	—	0.10	—	—
	100	6.7	—	0.15	—	—
	80	4.5	—	0.22	—	—
Lightweight aggregate concretes						
Expanded shale, clay, or slate; expanded slags; cinders; pumice (with density up to 100 lb/ft³); and scoria (sanded concretes have conductivities in the higher end of the range)	120	6.4-9.1	—	0.16-0.11	—	—
	100	4.7-6.2	—	0.21-0.16	—	0.20
	80	3.3-4.1	—	0.30-0.24	—	0.20
	60	2.1-2.5	—	0.48-0.40	—	—
	40	1.3	—	0.78	—	—
Perlite, vermiculite, and polystyrene beads	50	1.8-1.9	—	0.55-0.53	—	—
	40	1.4-1.5	—	0.71-0.67	—	0.15-0.23
	30	1.1	—	0.91	—	—
	20	0.8	—	1.25	—	—
Foam concretes	120	5.4	—	0.19	—	—
	100	4.1	—	0.24	—	—
	80	3.0	—	0.33	—	—
	70	2.5	—	0.40	—	—
Foam concretes and cellular concretes	60	2.1	—	0.48	—	—
	40	1.4	—	0.71	—	—
	20	0.8	—	1.25	—	—
SIDING MATERIALS (on flat surface)						
Shingles						
Asbestos-cement	120	—	4.75	—	0.21	—
Wood, 16 in., 7.5 exposure	—	—	1.15	—	0.87	0.31
Wood, double, 16-in., 12-in. exposure	—	—	0.84	—	1.19	0.28
Wood, plus ins. backer board, 0.312 in.	—	—	0.71	—	1.40	0.31
Siding						
Asbestos-cement, 0.25 in., lapped	—	—	4.76	—	0.21	0.24
Asphalt roll siding	—	—	6.50	—	0.15	0.35
Asphalt insulating siding (0.5 in. bed.)	—	—	0.69	—	1.46	0.35
Hardboard siding, 0.4375 in.	—	—	1.49	—	0.67	0.28
Wood, drop, 1 by 8 in.	—	—	1.27	—	0.79	0.28
Wood, bevel, 0.5 by 8 in., lapped	—	—	1.23	—	0.81	0.28
Wood, bevel, 0.75 by 10 in., lapped	—	—	0.95	—	1.05	0.28
Wood, plywood, 0.375 in., lapped	—	—	1.69	—	0.59	0.29
Aluminum, steel, or vinyl[p, q], over sheathing						
Hollow-backed	—	—	1.64	—	0.61	0.29[q]
Insulating-board backed nominal 0.375 in.	—	—	0.55	—	1.82	0.32
Insulating-board backed nominal 0.375 in., foil backed	—	—	0.34	—	2.96	—
Architectural (soda-lime float) glass	158	6.9	—	—	—	0.21
WOODS (12% moisture content)[e,r]						
Hardwoods						0.39[s]
Oak	41.2-46.8	1.12-1.25	—	0.89-0.80	—	
Birch	42.6-45.4	1.16-1.22	—	0.87-0.82	—	
Maple	39.8-44.0	1.09-1.19	—	0.92-0.84	—	
Ash	38.4-41.9	1.06-1.14	—	0.94-0.88	—	
Softwoods						0.39[s]
Southern Pine	35.6-41.2	1.00-1.12	—	1.00-0.89	—	
Douglas Fir-Larch	33.5-36.3	0.95-1.01	—	1.06-0.99	—	
Southern Cypress	31.4-32.1	0.90-0.92	—	1.11-1.09	—	
Hem-Fir, Spruce-Pine-Fir	24.5-31.4	0.74-0.90	—	1.35-1.11	—	
West Coast Woods, Cedars	21.7-31.4	0.68-0.90	—	1.48-1.11	—	
California Redwood	24.5-28.0	0.74-0.82	—	1.35-1.22	—	

Notes for Table 4

[a]Values are for a mean temperature of 75°F. Representative values for dry materials are intended as design (not specification) values for materials in normal use. Thermal values of insulating materials may differ from design values depending on their in-situ properties (e.g., density and moisture content, orientation, etc.) and variability experienced during manufacture. For properties of a particular product, use the value supplied by the manufacturer or by unbiased tests.

[b]To obtain thermal conductivities in Btu/h·ft·°F, divide the *k*-factor by 12 in/ft.

[c]Resistance values are the reciprocals of *C* before rounding off *C* to two decimal places.

[d]Lewis (1967).

[e]U.S. Department of Agriculture (1974).

[f]Does not include paper backing and facing, if any. Where insulation forms a boundary (reflective or otherwise) of an airspace, see Tables 2 and 3 for the insulating value of an airspace with the appropriate effective emittance and temperature conditions of the space.

[g]Conductivity varies with fiber diameter. (See Chapter 22, Factors Affecting Thermal Performance.) Batt, blanket, and loose-fill mineral fiber insulations are manufactured to achieve specified R-values, the most common of which are listed in the table. Due to differences in manufacturing processes and materials, the product thicknesses, densities, and thermal conductivities vary over considerable ranges for a specified R-value.

[h]This material is relatively new and data are based on limited testing.

[i]For additional information, see Society of Plastics Engineers (SPI) *Bulletin* U108. Values are for aged, unfaced board stock. For change in conductivity with age of expanded polyurethane/polyisocyanurate, see Chapter 22, Factors Affecting Thermal Performance.

[j]Values are for aged products with gas-impermeable facers on the two major surfaces. An aluminum foil facer of 0.001 in. thickness or greater is generally considered impermeable to gases. For change in conductivity with age of expanded polyisocyanurate, see Chapter 22, Factors Affecting Thermal Performance, and SPI *Bulletin* U108.

[k]Cellular phenolic insulation may no longer be manufactured. The thermal conductivity and resistance values do not represent aged insulation, which may have a higher thermal conductivity and lower thermal resistance.

[l]Insulating values of acoustical tile vary, depending on density of the board and on type, size, and depth of perforations.

[m]Cavity is framed with 0.75 in. wood furring strips. Caution should be used in applying this value for other framing materials. The reported value was derived from tests and applies to the reflective path only. The effect of studs or furring strips must be included in determining the overall performance of the wall.

[n]Values for fully grouted block may be approximated using values for concrete with a similar unit weight.

[o]Values for concrete block and concrete are at moisture contents representative of normal use.

[p]Values for metal or vinyl siding applied over flat surfaces vary widely, depending on amount of ventilation of airspace beneath the siding; whether airspace is reflective or nonreflective; and on thickness, type, and application of insulating backing used. Values are averages for use as design guides, and were obtained from several guarded hot box tests (ASTM C 236) or calibrated hot box (ASTM C 976) on hollow-backed types and types made using backing-boards of wood fiber, foamed plastic, and glass fiber. Departures of ±50% or more from these values may occur.

[q]Vinyl specific heat = 0.25 Btu/lb·°F

[r]See Adams (1971), MacLean (1941), and Wilkes (1979). The conductivity values listed are for heat transfer across the grain. The thermal conductivity of wood varies linearly with the density, and the density ranges listed are those normally found for the wood species given. If the density of the wood species is not known, use the mean conductivity value. For extrapolation to other moisture contents, the following empirical equation developed by Wilkes (1979) may be used:

$$k = 0.1791 + \frac{(1.874 \times 10^{-2} + 5.753 \times 10^{-4} M)\rho}{1 + 0.01M}$$

where ρ is density of the moist wood in lb/ft³, and M is the moisture content in percent.

[s]From Wilkes (1979), an empirical equation for the specific heat of moist wood at 75°F is as follows:

$$c_p = \frac{(0.299 + 0.01M)}{(1 + 0.01M)} + \Delta c_p$$

where Δc_p accounts for the heat of sorption and is denoted by

$$\Delta c_p = M(1.921 \times 10^{-3} - 3.168 \times 10^{-5} M)$$

where M is the moisture content in percent by mass.

Example 1. Calculate the U-factor of the 2 by 4 stud wall shown in Figure 2. The studs are at 16 in. OC. There is 3.5 in. mineral fiber batt insulation (R-13) in the stud space. The inside finish is 0.5 in. gypsum wallboard; the outside is finished with rigid foam insulating sheathing (R-4) and 0.5 in. by 8 in. wood bevel lapped siding. The insulated cavity occupies approximately 75% of the transmission area; the studs, plates, and sills occupy 21%; and the headers occupy 4%.

Solution. Obtain the R-values of the various building elements from Tables 1 and 4. Assume the R = 1.25 per inch for the wood framing. Also, assume the headers are solid wood, in this case, and group them with the studs, plates, and sills.

Element	*R* (Insulated Cavity)	*R* (Studs, Plates, and Headers)
1. Outside surface, 15 mphwind	0.17	0.17
2. Wood bevel lapped siding	0.81	0.81
3. Rigid foam insulating sheathing	4.0	4.0
4. Mineral fiber batt insulation, 3.5 in.	13.0	—
5. Wood stud, nominal 2 × 4	—	4.38
6. Gypsum wallboard, 0.5 in.	0.45	0.45
7. Inside surface, still air	0.68	0.68
	$R_1 = 19.11$	$R_2 = 10.49$

Since the U-factor is the reciprocal of R-value, $U_1 = 0.052$ and $U_2 = 0.095$ Btu/h·ft²·°F.

If the wood framing (thermal bridging) is not included, Equation (3) from Chapter 22 may be used to calculate the U-factor of the wall as follows:

$$U_{av} = U_1 = \frac{1}{R_1} = 0.052 \text{ Btu/h} \cdot \text{ft}^2 \cdot {}^\circ\text{F}$$

If the wood framing is accounted for using the parallel-path flow method, the U-factor of the wall is determined using Equation (5) from Chapter 22 as follows:

$$U_{av} = (0.75 \times 0.052) + (0.25 \times 0.095) = 0.063 \text{ Btu/h} \cdot \text{ft}^2 \cdot {}^\circ\text{F}$$

If the wood framing is included using the isothermal planes method, the U-factor of the wall is determined using Equations (2) and (3) from Chapter 22 as follows:

$$R_{T(av)} = 4.98 + 1/[(0.75/13.0) + (0.25/4.38)] + 1.13$$

$$= 14.82 {}^\circ\text{F} \cdot \text{ft}^2 \cdot \text{h/Btu}$$

$$U_{av} = 0.067 \text{ Btu/h} \cdot \text{ft}^2 \cdot {}^\circ\text{F}$$

For a frame wall with a 24-in. OC stud space, the average overall R-value is 15.18°F·ft²·h/Btu. Similar calculation procedures may be used to evaluate other wall designs, except those with thermal bridges.

Masonry Walls

The average overall R-values of masonry walls can be estimated by assuming a combination of layers in series, one or more of which provides parallel paths. This method is used because heat flows laterally through block face shells so that transverse isothermal planes result. Average total resistance $R_{T(av)}$ is the sum of the resistances of the layers between such planes, each layer calculated as shown in Example 2.

Example 2. Calculate the overall thermal resistance and average U-factor of the 7-5/8-in. thick insulated concrete block wall shown in Figure 3. The two-core block has an average web thickness of 1-in. and a face shell thickness of 1-1/4-in. Overall block dimensions are 7-5/8 by 7-5/8 by 15-5/8 in. Measured thermal resistances of 112 lb/ft^3 concrete and 7 lb/ft^3 expanded perlite insulation are 0.10 and 2.90°F·ft^2·h/Btu per inch, respectively.

Solution. The equation used to determine the overall thermal resistance of the insulated concrete block wall is derived from Equations (2) and (5) from Chapter 22 and is given below:

$$R_{T(av)} = R_i + R_f + \left(\frac{a_w}{R_w} + \frac{a_c}{R_c}\right)^{-1} + R_o$$

where

$R_{T(av)}$ = overall thermal resistance based on assumption of isothermal planes
R_i = thermal resistance of inside air surface film (still air)
R_o = thermal resistance of outside air surface film (15 mph wind)
R_f = total thermal resistance of face shells
R_c = thermal resistance of cores between face shells
R_w = thermal resistance of webs between face shells
a_w = fraction of total area transverse to heat flow represented by webs of blocks
a_c = fraction of total area transverse to heat flow represented by cores of blocks

From the information given and the data in Table 1, determine the values needed to compute the overall thermal resistance.

R_i = 0.68
R_o = 0.17
R_f = (2)(1.25)(0.10) = 0.25
R_c = (5.125)(2.90) = 14.86
R_w = (5.125)(0.10) = 0.51
a_w = 3/15.625 = 0.192
a_c = 12.625/15.625 = 0.808

Using the equation given, the overall thermal resistance and average U-factor are calculated as follows:

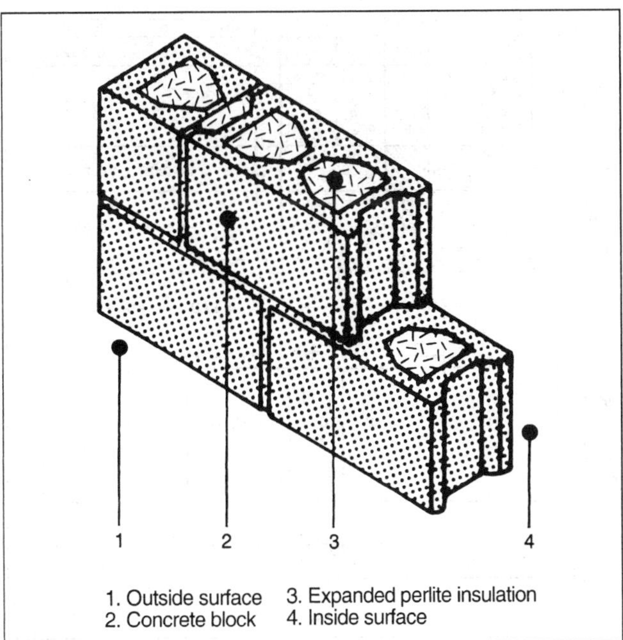

Fig. 3 Insulated Concrete Block Wall (Example 2)

$$R_{T(av)} = 0.68 + 0.25 + \frac{0.51 \times 14.86}{(0.808 \times 0.51) + (0.192 \times 14.86)} + 0.17$$

$$= 3.43\,°\text{F} \cdot \text{ft}^2 \cdot \text{h/Btu}$$

$$U_{av} = 1/3.43 = 0.29\ \text{Btu/h} \cdot \text{ft}^2 \cdot °\text{F}$$

Based on guarded hot box tests of this wall without mortar joints, Tye and Spinney (1980) measured the average R-value for this insulated concrete block wall as 3.13°F·ft^2·h/Btu.

Assuming parallel heat flow only, the calculated resistance is higher than that calculated on the assumption of isothermal planes. The actual resistance generally is some value between the two calculated values. In the absence of test values, examination of the construction usually reveals whether a value closer to the higher or lower calculated R-value should be used. Generally, if the construction contains a layer in which lateral conduction is high compared with transmittance through the construction, the calculation with isothermal planes should be used. If the construction has no layer of high lateral conductance, the parallel heat flow calculation should be used.

Hot box tests of insulated and uninsulated masonry walls constructed with block of conventional configuration show that thermal resistances calculated using the isothermal planes heat flow method agree well with measured values (Van Geem 1985, Valore 1980, Shu et al. 1979). Neglecting horizontal mortar joints in conventional block can result in thermal transmittance values up to 16% lower than actual, depending on the density and thermal properties of the masonry, and 1 to 6% lower, depending on the core insulation material (Van Geem 1985, McIntyre 1984). For aerated concrete block walls, other solid masonry, and multicore block walls with full mortar joints, neglecting mortar joints can cause errors in R-values up to 40% (Valore 1988). Horizontal mortar joints usually found in concrete block wall construction are neglected in Example 2.

Constructions Containing Metal

Curtain and metal stud-wall constructions often include metallic and other thermal bridges, which can significantly reduce the thermal resistance. However, the capacity of the adjacent facing materials to transmit heat transversely to the metal is limited, and some contact resistance between all materials in contact limits the reduction. Contact resistances in building structures are only 0.06 to 0.6°F·ft^2·h/Btu—too small to be of concern in many cases. However, the contact resistances of steel framing members may be important. Also, in many cases (as illustrated in Example 3), the area of metal in contact with the facing greatly exceeds the thickness of the metal, which mitigates the contact reistance effects.

Thermal characteristics for panels of sandwich construction can be computed by combining the thermal resistances of the various layers. However, few panels are true sandwich constructions; many have ribs and stiffeners that create complicated heat flow paths. R-values for the assembled sections should be determined on a representative sample by using a hot box method. If the sample is a wall section with air cavities on both sides of fibrous insulation, the sample must be of representative height since convective airflow can contribute significantly to heat flow through the test section. Computer modeling can also be useful, but all heat transfer mechanisms must be considered.

In Example 3, the metal member is only 0.020 in. thick, but it is in contact with adjacent facings over a 1.25 in.-wide area. The steel member is 3.50 in. deep, has a thermal resistance of approximately 0.011°F·ft^2·h/Btu, and is virtually isothermal. The calculation involves careful selection of the appropriate thickness for the steel member. If the member is assumed to be 0.020 in. thick, the fact that the flange transmits heat to the adjacent facing is ignored, and the heat flow through the steel is underestimated. If the member is assumed to be 1.25 in. thick, the heat flow through the steel is overestimated. In Example 3, the steel member behaves in much the same way as a rectangular member 1.25 in. thick and 3.50 in. deep with a thermal resistance of $(1.25/0.020) \times 0.011 = 0.69$°F·ft²·h/Btu does. The Building Research Association of New Zealand (BRANZ) commonly uses this approximation.

Example 3. Calculate the C-factor of the insulated steel frame wall shown in Figure 4. Assume that the steel member has an R-value of 0.69°F·ft²·h/Btu and that the framing behaves as though it occupies approximately 8% of the transmission area.

Solution. Obtain the R-values of the various building elements from Table 4.

Element	*R* (Insul.)	*R* (Framing)
1. 0.5-in. gypsum wallboard	0.45	0.45
2. 3.5-in. mineral fiber batt insulation	11	—
3. Steel framing member	—	0.69
4. 0.5-in. gypsum wallboard	0.45	0.45
	$R_1 = 11.90$	$R_2 = 1.59$

Therefore, $C_1 = 0.084$; $C_2 = 0.629$ Btu/h·ft²·°F.

If the steel framing (thermal bridging) is not considered, the C-factor of the wall is calculated using Equation (3) from Chapter 22 as follows:

$$C_{av} = C_1 = 1/R_1 = 0.084 \text{ Btu/h}\cdot\text{ft}^2\cdot{}^\circ\text{F}$$

If the steel framing is accounted for using the parallel flow method, the C-factor of the wall is determined using Equation (5) from Chapter 22 as follows:

$$C_{av} = (0.92 \times 0.084) + (0.08 \times 0.629)$$

$$= 0.128 \text{ Btu/h}\cdot\text{ft}^2\cdot{}^\circ\text{F}$$

$$R_{T(av)} = 7.81{}^\circ\text{F}\cdot\text{ft}^2\cdot\text{h/Btu}$$

If the steel framing is included using the isothermal planes method, the C-factor of the wall is determined using Equations (2) and (3) from Chapter 22 as follows:

$$R_{T(av)} = 0.45 + 1/[(0.92/11.00) + (0.08/0.69)] + 0.45$$

$$= 5.91{}^\circ\text{F}\cdot\text{ft}^2\cdot\text{h/Btu}$$

$$C_{av} = 0.169 \text{ Btu/h}\cdot\text{ft}^2\cdot{}^\circ\text{F}$$

For this insulated steel frame wall, Farouk and Larson (1983) measured an average R-value of 6.61°F·ft²·h/Btu.

In ASHRAE/IESNA *Standard* 90.1-1989, one method given for determining the thermal resistance of wall assemblies containing metal framing involves using a parallel path correction factor F_c, which is listed in Table 8C-2 of the standard. For 2 by 4 steel framing, 16 in. OC, $F_c = 0.50$. Using the correction factor method, an R-value of 6.40°F·ft²·h/Btu [0.45 + 11(0.50) + 0.45] is obtained for the wall described in Example 3.

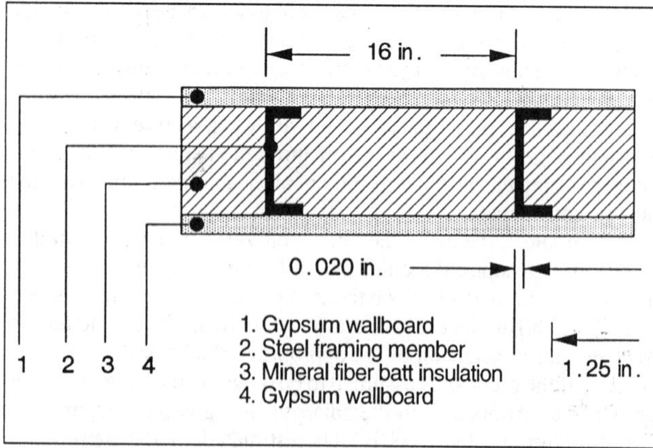

Fig. 4 Insulated Steel Frame Wall (Example 3)

Zone Method of Calculation

For structures with widely spaced metal members of substantial cross-sectional area, calculation by the isothermal planes method can result in thermal resistance values that are too low. For these constructions, the **zone method** can be used. This method involves two separate computations—one for a chosen limited portion, Zone A, containing the highly conductive element; the other for the remaining portion of simpler construction, Zone B. The two computations are then combined using the parallel flow method, and the average transmittance per unit overall area is calculated. The basic laws of heat transfer are applied by adding the area conductances *CA* of elements in parallel, and adding area resistances *R/A* of elements in series.

The surface shape of Zone A is determined by the metal element. For a metal beam (see Figure 5), the Zone A surface is a strip of width *W* that is centered on the beam. For a rod perpendicular to panel surfaces, it is a circle of diameter *W*. The value of *W* is calculated from Equation (1), which is empirical. The value of *d* should not be less than 0.5 in. for still air.

$$W = m + 2d \tag{1}$$

where

m = width or diameter of metal heat path terminal, in.
d = distance from panel surface to metal, in.

Generally, the value of *W* should be calculated using Equation (1) for each end of the metal heat path; the larger value, within the limits of the basic area, should be used as illustrated in Example 4.

Example 4. Calculate transmittance of the roof deck shown in Figure 5. Tee-bars at 24 in. OC support glass fiber form boards, gypsum concrete, and built-up roofing. Conductivities of components are: steel, 314.4 Btu·in/h·ft²·°F; gypsum concrete, 1.66 Btu·in/h·ft²·°F; and glass fiber form board, 0.25 Btu·in/h·ft²·°F. Conductance of built-up roofing is 3.00 Btu/h·ft²·°F.

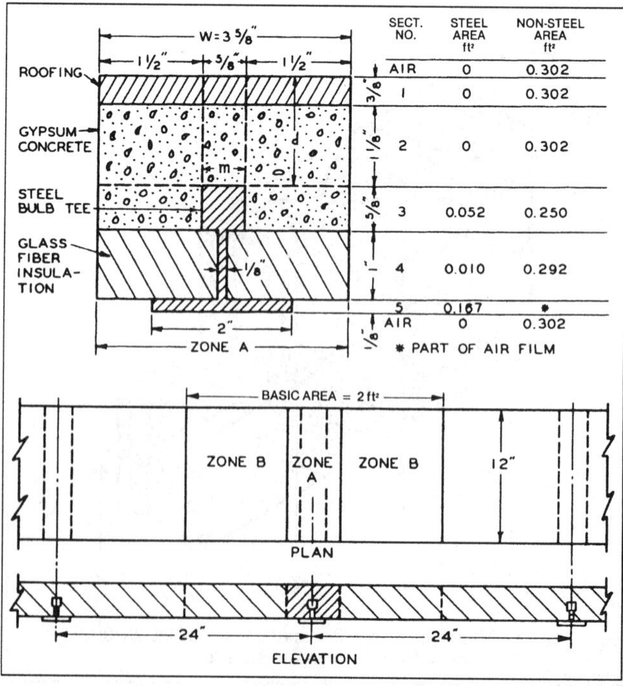

Fig. 5 Gypsum Roof Deck on Bulb Tees (Example 4)

Solution. The basic area is 2 ft^2 (24 in. by 12 in.) with a tee-bar (12 in. long) across the middle. This area is divided into Zones A and B.

Zone A is determined from Equation (1) as follows:

$$\text{Top side } W = m + 2d = 0.625 + (2 \times 1.5) = 3.625 \text{ in.}$$
$$\text{Bottom side } W = m + 2d = 2.0 + (2 \times 0.5) = 3.0 \text{ in.}$$

Using the larger value of W, the area of Zone A is (12 × 3.625)/144 = 0.302 ft^2. The area of Zone B is 2.0 – 0.302 = 1.698 ft^2.

To determine area transmittance for Zone A, divide the structure within the zone into five sections parallel to the top and bottom surfaces (Figure 5). The area conductance CA of each section is calculated by adding the area conductances of its metal and nonmetal paths. Area conductances of the sections are converted to area resistances R/A and added to obtain the total resistance of Zone A.

Section	Area × Conductance =	CA	$\frac{1}{CA} = \frac{R}{A}$
Air (outside, 15 mph)	0.302 × 6.00	1.81	0.55
No. 1, Roofing	0.302 × 3.00	0.906	1.10
No. 2, Gypsum concrete	0.302 × 1.66/1.125	0.446	2.24
No. 3, Steel	0.052 × 314.4/0.625	26.2	0.04
No. 3, Gypsum concrete	0.250 × 1.66/0.625	0.664	
No. 4, Steel	0.010 × 314.4/1.00	3.14	0.31
No. 4, Glass fiberboard	0.292 × 0.25/1.00	0.073	
No. 5, Steel	0.167 × 314.4/0.125	420.0	0.002
Air (inside)	0.302 × 1.63	0.492	2.03
		Total R/A =	6.27

Area transmittance of Zone A = 1/(R/A) = 1/6.27 = 0.159.

For Zone B, the unit resistances are added and then converted to area transmittance, as shown in the following table.

Section	Resistance, R
Air (outside, 15 mph)	1/6.00 = 0.17
Roofing	1/3.00 = 0.33
Gypsum concrete	1.75/1.66 = 1.05
Glass fiberboard	1.00/0.25 = 4.00
Air (inside)	1/1.63 = 0.61
Total resistance	= 6.16

Since unit transmittance = $1/R$ = 0.162, the total area transmittance UA is calculated as follows:

$$\begin{aligned} \text{Zone B} = 1.698 \times 0.162 &= 0.275 \\ \text{Zone A} &= 0.159 \\ \text{Total area transmittance of basic area} &= 0.434 \\ \text{Transmittance per ft}^2 = 0.434/2.0 &= 0.217 \\ \text{Resistance per ft}^2 &= 4.61 \end{aligned}$$

Overall R-values of 4.57 and 4.85°F·ft^2·h/Btu have been measured in two guarded hot box tests of a similar construction.

When the steel member represents a relatively large proportion of the total heat flow path, as in Example 4, detailed calculations of resistance in sections 3, 4, and 5 of Zone A are unnecessary; if only the steel member is considered, the final result of Example 4 is the same. However, if the heat flow path represented by the steel member is small, as for a tie rod, detailed calculations for sections 3, 4, and 5 are necessary. A panel with an internal metallic structure and bonded on one or both sides to a metal skin or covering presents special problems of lateral heat flow not covered in the zone method.

Modified Zone Method for Metal Stud Walls with Insulated Cavities

The modified zone method is similar to the parallel path method and the zone method. All three methods are based on parallel-path calculations. Figure 6 shows the width w of the zone of thermal anomalies around a metal stud. This zone can be assumed to equal the length of the stud flange L (parallel path method), or can be calculated as a sum of the length of stud flange and a distance double that from wall surface to metal Σd_i (zone method). In the modified zone method the width of the zone depends on the following three parameters:

- Ratio between thermal resistivity of sheathing material and cavity insulation
- Size (depth) of stud
- Thickness of sheathing material

The Modified Zone Method is explained in Figure 6 (which can be copied and used as a calculation form). The wall cross section shown in Figure 6, is divided into two zones: the zone of thermal anomalies around metal stud w and the cavity zone cav. Wall material layers are grouped into an exterior and interior surface sections—A (sheathing, siding) and B (wallboard)—and interstitial sections I and II (cavity insulation, metal stud flange).

Assuming that the layers or layer of wall materials in wall section A are thicker than those in wall section B, as show by the cross section in Figure 6, they can be described as follows:

$$\sum_{i=1}^{n} d_i \geq \sum_{j=1}^{m} d_j \qquad (2)$$

where

n = number of material layer (of thickness d_i) between metal stud flange and wall surface for section A
m = number of material layer (of thickness d_j) for section B

Then, the width of the zone of thermal anomalies around the metal stud w can be estimated by

$$w = L + z_f \sum_{i=1}^{n} d_i \qquad (3)$$

where

L = stud flange size,
d_i = thickness of material layer in section A
z_f = zone factor, which is shown in Figure 7 (z_f = 2 for zone method)

Kosny and Christian (1995) verified the accuracy of the Modified Zone Method for over 200 simulated cases of metal frame walls with insulated cavities. For all configurations considered the discrepancy between results were within ±2%. Hot box measured R-values for 15 metal stud walls tested by Barbour et al. (1994) were compared with results obtained by Kosny and Christian (1995) and McGowan and Desjarlais (1997). The Modified Zone Method was found to be the most accurate simple method for estimating the clear wall R-value of light-gage steel stud walls with insulated cavities. However, this analysis does not apply to construction with metal sheathing. Also, *ASHRAE Standard* 90.1 may require a different method of analysis.

Ceilings and Roofs

The overall R-value for ceilings of wood frame flat roofs can be calculated using Equations (1) through (5) from Chapter 22. Properties of the materials are found in Tables 1, 3, 2, and 4. The fraction of framing is assumed to be 0.10 for joists at 16 in. OC and 0.07 for joists at 24 in. OC. The calculation procedure is similar to that shown in Example 1. Note that if the ceiling contains plane air spaces (see Table 3), the resistance depends on the direction of heat flow, i.e., whether the calculation is for a winter (heat flow up) or summer (heat flow down) condition.

For ceilings of pitched roofs under winter conditions, calculate the R-value of the ceiling using the procedure for flat roofs. Table 5 can be used to determine the effective resistance of the

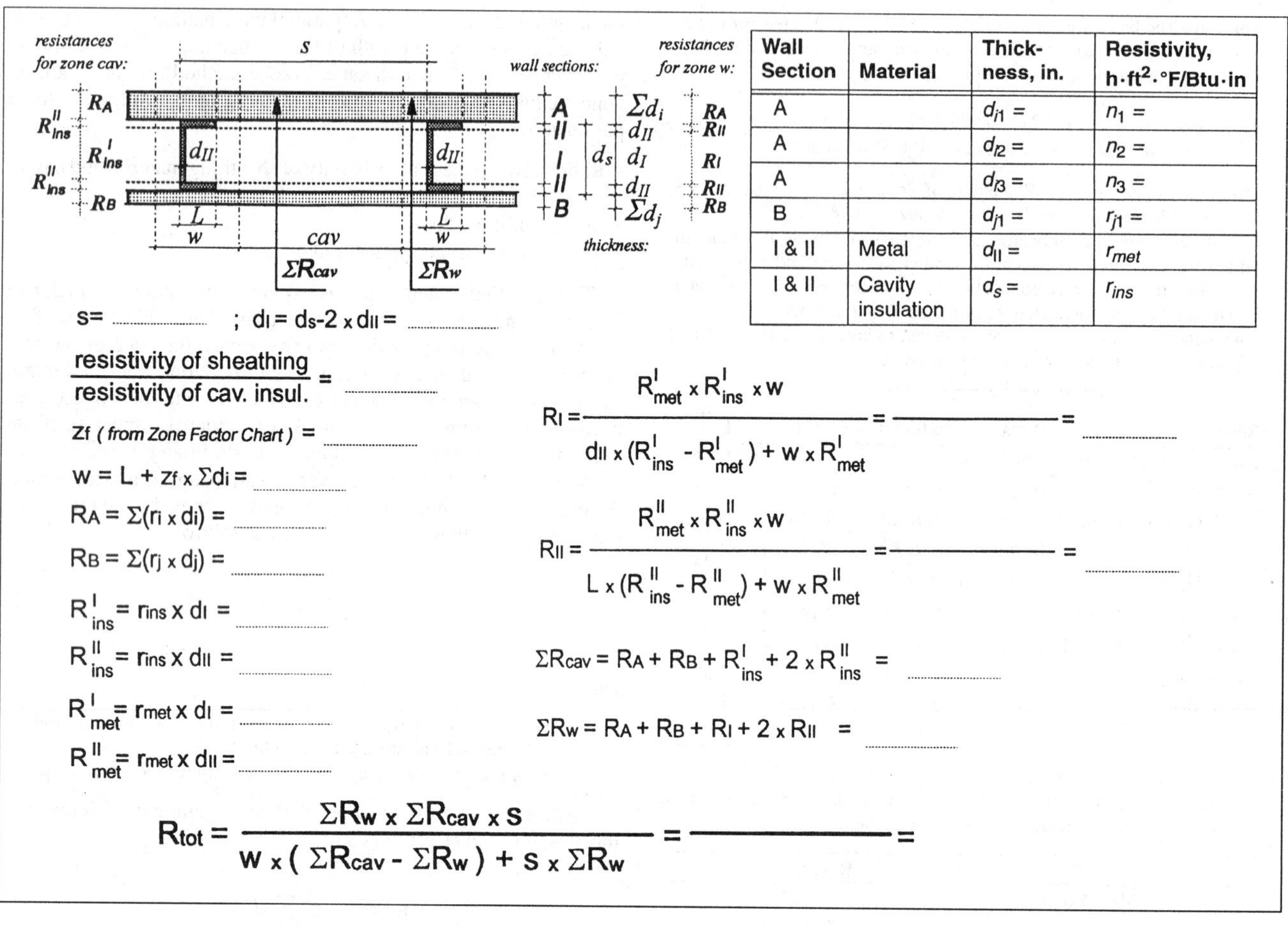

Wall Section	Material	Thick-ness, in.	Resistivity, h·ft²·°F/Btu·in
A		d_{i1} =	r_{i1} =
A		d_{i2} =	r_{i2} =
A		d_{i3} =	r_{i3} =
B		d_{j1} =	r_{j1} =
I & II	Metal	d_{II} =	r_{met}
I & II	Cavity insulation	d_s =	r_{ins}

s = ... ; $d_I = d_s - 2 \times d_{II}$ = ...

$\frac{\text{resistivity of sheathing}}{\text{resistivity of cav. insul.}}$ = ...

z_f (from Zone Factor Chart) = ...

$w = L + z_f \times \Sigma d_i$ = ...

$R_A = \Sigma(r_i \times d_i)$ = ...

$R_B = \Sigma(r_j \times d_j)$ = ...

$R^{I}_{ins} = r_{ins} \times d_I$ = ...

$R^{II}_{ins} = r_{ins} \times d_{II}$ = ...

$R^{I}_{met} = r_{met} \times d_I$ = ...

$R^{II}_{met} = r_{met} \times d_{II}$ = ...

$$R_I = \frac{R^{I}_{met} \times R^{I}_{ins} \times w}{d_{II} \times (R^{I}_{ins} - R^{I}_{met}) + w \times R^{I}_{met}} = \text{——} = \text{........}$$

$$R_{II} = \frac{R^{II}_{met} \times R^{II}_{ins} \times w}{L \times (R^{II}_{ins} - R^{II}_{met}) + w \times R^{II}_{met}} = \text{——} = \text{........}$$

$\Sigma R_{cav} = R_A + R_B + R^{I}_{ins} + 2 \times R^{II}_{ins}$ = ...

$\Sigma R_w = R_A + R_B + R_I + 2 \times R_{II}$ = ...

$$R_{tot} = \frac{\Sigma R_w \times \Sigma R_{cav} \times s}{w \times (\Sigma R_{cav} - \Sigma R_w) + s \times \Sigma R_w} = \text{——} =$$

Fig. 6 Modified Zone Method R-Value Calculation Form for Metal Stud Walls

attic space under summer conditions for varying conditions of ventilation air temperature, airflow direction and rates, ceiling resistance, roof or sol-air temperatures, and surface emittances (Joy 1958).

The R-value is the total resistance obtained by adding the ceiling and effective attic resistances. The applicable temperature difference is that difference between room air and sol-air temperatures or between room air and roof temperatures (see Table 5, footnote f). Table 5 can be used for pitched and flat residential roofs over attic spaces. When an attic has a floor, the ceiling resistance should account for the complete ceiling-floor construction.

Windows and Doors

Table 5 of Chapter 29 lists U-factors for various fenestration products. Table 6 in Chapter 29 lists U-factors for exterior wood and steel doors. All U-factors are approximate, because a significant portion of the resistance of a window or door is contained in the air film resistances, and some parameters that may have important effects are not considered. For example, the listed U-factors assume the surface temperatures of surrounding bodies are equal to the ambient air temperature. However, the indoor surface of a window or door in an actual installation may be exposed to nearby radiating surfaces, such as radiant heating panels, or opposite walls with much higher or lower temperatures than the indoor air. Air movement across the indoor surface of a window or door, such as that caused by nearby heating and cooling outlet grilles, increases the U-factor; and air movement (wind) across the outdoor surface of a window or door also increases the U-factor.

U_o Concept

U_o is the combined thermal transmittance of the respective areas of gross exterior wall, roof or ceiling or both, and floor assemblies. The U_o equation for a wall is as follows:

$$U_o = (U_{wall}A_{wall} + U_{window}A_{window} + U_{door}A_{door})/A_o \quad (4)$$

where

U_o = average thermal transmittance of gross wall area
A_o = gross area of exterior walls
U_{wall} = thermal transmittance of all elements of opaque wall area
A_{wall} = opaque wall area
U_{window} = thermal transmittance of window area (including frame)
A_{window} = window area (including frame)
U_{door} = thermal transmittance of door area
A_{door} = door area (including frame)

Where more than one type of wall, window, or door is used, the *UA* term for that exposure should be expanded into its subelements, as shown in Equation (3).

$$\begin{aligned} U_oA_o = {} & U_{wall\,1}A_{wall\,1} + U_{wall\,2}A_{wall\,2} + \dots + U_{wall\,m}A_{wall\,m} \\ & + U_{window\,1}A_{window\,1} + U_{window\,2}A_{window\,2} + \dots \\ & + U_{window\,n}A_{window\,n} + U_{door\,1}A_{door\,1} \\ & + U_{door\,2}A_{door\,2} + \dots + U_{door\,o}A_{door\,o} \end{aligned} \quad (5)$$

Table 5 Effective Thermal Resistance of Ventilated Attics[a] (Summer Condition)

		NONREFLECTIVE SURFACES									
		No Ventilation[b]		Natural Ventilation		Power Ventilation[c]					
		Ventilation Rate, cfm/ft²									
		0		0.1[d]		0.5		1.0		1.5	
		Ceiling Resistance R^e, °F·ft²·/Btu									
Ventilation Air Temperature, °F	Sol-Air[f] Temperature, °F	10	20	10	20	10	20	10	20	10	20
80	120	1.9	1.9	2.8	3.4	6.3	9.3	9.6	16	11	20
	140	1.9	1.9	2.8	3.5	6.5	10	9.8	17	12	21
	160	1.9	1.9	2.8	3.6	6.7	11	10	18	13	22
90	120	1.9	1.9	2.5	2.8	4.6	6.7	6.1	10	6.9	13
	140	1.9	1.9	2.6	3.1	5.2	7.9	7.6	12	8.6	15
	160	1.9	1.9	2.7	3.4	5.8	9.0	8.5	14	10	17
100	120	1.9	1.9	2.2	2.3	3.3	4.4	4.0	6.0	4.1	6.9
	140	1.9	1.9	2.4	2.7	4.2	6.1	5.8	8.7	6.5	10
	160	1.9	1.9	2.6	3.2	5.0	7.6	7.2	11	8.3	13
		REFLECTIVE SURFACES[g]									
80	120	6.5	6.5	8.1	8.8	13	17	17	25	19	30
	140	6.5	6.5	8.2	9.0	14	18	18	26	20	31
	160	6.5	6.5	8.3	9.2	15	18	19	27	21	32
90	120	6.5	6.5	7.5	8.0	10	13	12	17	13	19
	140	6.5	6.5	7.7	8.3	12	15	14	20	16	22
	160	6.5	6.5	7.9	8.6	13	16	16	22	18	25
100	120	6.5	6.5	7.0	7.4	8.0	10	8.5	12	8.8	12
	140	6.5	6.5	7.3	7.8	10	12	11	15	12	16
	160	6.5	6.5	7.6	8.2	11	14	13	18	15	20

[a]Although the term effective resistance is commonly used when there is attic ventilation, this table includes values for situations with no ventilation. The effective resistance of the attic added to the resistance ($1/U$) of the ceiling yields the effective resistance of this combination based on sol-air (see Chapter 28) and room temperatures. These values apply to wood frame construction with a roof deck and roofing that has a conductance of 1.0 Btu/h·ft²·°F.

[b]This condition cannot be achieved in the field unless extreme measures are taken to tightly seal the attic.

[c]Based on air discharging outward from attic.

[d]When attic ventilation meets the requirements stated in Chapter 25, 0.1 cfm/ft² is assumed as the natural summer ventilation rate.

[e]When determining ceiling resistance, do not add the effect of a reflective surface facing the attic, as it is accounted for in the Reflective Surfaces part of the table.

[f]Roof surface temperature rather than sol-air temperature (see Chapter 28) can be used if 0.25 is subtracted from the attic resistance shown.

[g]Surfaces with effective emittance ε_{eff} = 0.05 between ceiling joists facing attic space.

Table 6 Transmission Coefficients U for Wood and Steel Doors, Btu/h·ft²·°F

Nominal Door Thickness, in.	Description	No Storm Door	Wood Storm Door[c]	Metal Storm Door[d]
Wood Doors[a,b]				
1-3/8	Panel door with 7/16-in. panels[e]	0.57	0.33	0.37
1-3/8	Hollow core flush door	0.47	0.30	0.32
1-3/8	Solid core flush door	0.39	0.26	0.28
1-3/4	Panel door with 7/16-in. panels[e]	0.54	0.32	0.36
1-3/4	Hollow core flush door	0.46	0.29	0.32
1-3/4	Panel door with 1-1/8-in. panels[e]	0.39	0.26	0.28
1-3/4	Solid core flush door	0.40	—	0.26
2-1/4	Solid core flush door	0.27	0.20	0.21
Steel Doors[b]				
1-3/4	Fiberglass or mineral wool core with steel stiffeners, no thermal break[f]	0.60	—	—
1-3/4	Paper honeycomb core without thermal break[f]	0.56	—	—
1-3/4	Solid urethane foam core without thermal break[a]	0.40	—	—
1-3/4	Solid fire rated mineral fiberboard core without thermal break[f]	0.38	—	—
1-3/4	Polystyrene core without thermal break (18 gage commercial steel)[f]	0.35	—	—
1-3/4	Polyurethane core without thermal break (18 gage commercial steel)[f]	0.29	—	—
1-3/4	Polyurethane core without thermal break (24 gage residential steel)[f]	0.29	—	—
1-3/4	Polyurethane core with thermal break and wood perimeter (24 gage residential steel)[f]	0.20	—	—
1-3/4	Solid urethane foam core with thermal break[a]	0.20	—	0.16

Note: All U-factors for exterior doors in this table are for doors with no glazing, except for the storm doors which are in addition to the main exterior door. Any glazing area in exterior doors should be included with the appropriate glass type and analyzed as a window (see Chapter 29). Interpolation and moderate extrapolation are permitted for door thicknesses other than those specified.

[a]Values are based on a nominal 32 in. by 80 in. door size with no glazing.

[b]Outside air conditions: 15 mph wind speed, 0°F air temperature; inside air conditions: natural convection, 70°F air temperature.

[c]Values for wood storm door are for approximately 50% glass area.

[d]Values for metal storm door are for any percent glass area.

[e]55% panel area.

[f]ASTM C 236 hotbox data on a nominal 3 ft by 7 ft door size with no glazing.

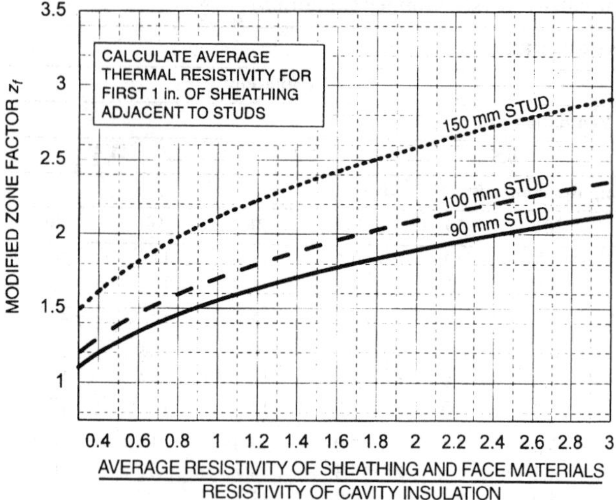

Use z_f = **–0.5** for walls when total thickness of layer of materials attached to one side of metal frame ≤ **5/8 in. and** thermal resistivity of sheathing ≤ **1.5 h·ft²·°F/Btu·in.**

Use z_f = +**0.5** for walls when total thickness of layer of materials attached to one side of metal frame ≤ **5/8 in. and** thermal resistivity of sheathing > **1.5 h·ft²·°F/Btu·in**

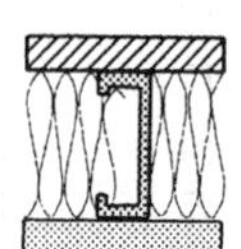

Find z_f in chart above for walls when total thickness of layer of materials attached to one side of metal frame > **5/8 in.**

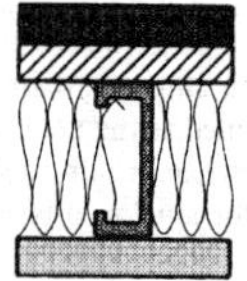

Fig. 7 Modified Zone Factor for Calculating R-Value of Metal Stud Walls with Cavity Insulation

Example 5. Calculate U_o for a wall 30 ft by 8 ft, constructed as in Example 1. The wall contains two double-glazed (0.5 in. airspace) fixed windows with wood/vinyl frames. (From Table 5 in Chapter 29, U = 0.52 Btu/h·ft²·°F.) One window is 60 in. by 34 in. and the second 36 in. by 30 in. The wall also contains a 1.75-in. solid core flush door with a metal storm door 34 in. by 80 in. (U = 0.26 Btu/h·ft²·°F from Table 6).

Solution. The U-factor for the wall was obtained in Example 1. The areas of the different components are:

$$A_{window} = [(60 \times 34) + (36 \times 30)]/144 = 21.7 \text{ ft}^2$$

$$A_{door} = (34 \times 80)/144 = 18.9 \text{ ft}^2$$

$$A_{wall} = (30 \times 8) - (21.7 + 18.9) = 199.4 \text{ ft}^2$$

Therefore, the combined thermal transmittance for the wall is:

$$U_o = \frac{(0.063 \times 199.4) + (0.52 \times 21.7) + (0.26 \times 18.9)}{(30 \times 8)}$$

$$= 0.119 \text{ Btu/h} \cdot \text{ft}^2 \cdot {}^\circ\text{F}$$

Slab-on-Grade and Below-Grade Construction

Heat transfer through basement walls and floors to the ground depends on the following factors: (1) the difference between the air temperature within the room and that of the ground and outside air, (2) the material of the walls or floor, and (3) the thermal conductivity of the surrounding earth. The latter varies with local conditions and is usually unknown. Because of the great thermal inertia of the surrounding soil, ground temperature varies with depth, and there is a substantial time lag between changes in outdoor air temperatures and corresponding changes in ground temperatures. As a result, ground-coupled heat transfer is less amenable to steady-state representation than above-grade building elements. However, several simplified procedures for estimating ground-coupled heat transfer have been developed. These fall into two principal categories: (1) those that reduce the ground heat transfer problem to a closed form solution, and (2) those that use simple regression equations developed from statistically reduced multidimensional transient analyses.

Closed form solutions, including the ASHRAE arc-length procedure discussed in Chapter 27 by Latta and Boileau (1969), generally reduce the problem to one-dimensional, steady-state heat transfer. These procedures use simple, "effective" U-factors or ground temperatures or both. Methods differ in the various parameters averaged or manipulated to obtain these effective values. Closed form solutions provide acceptable results in climates that have a single dominant season, because the dominant season persists long enough to permit a reasonable approximation of steady-state conditions at shallow depths. The large errors (percentage) that are likely during transition seasons should not seriously affect building design decisions, since these heat flows are relatively insignificant when compared with those of the principal season.

The ASHRAE arc-length procedure is a reliable method for wall heat losses in cold winter climates. Chapter 27 discusses a slab-on-grade floor model developed by one study. Although both procedures give results comparable to transient computer solutions for cold climates, their results for warmer U.S. climates differ substantially.

Research conducted by Hougten et al. (1942) and Dill et al. (1945) indicates a heat flow of approximately 2.0 Btu/h·ft² through an uninsulated concrete basement floor with a temperature difference of 20°F between the basement floor and the air 6 in. above it. A U-factor of 0.10 Btu/h·ft²·°F is sometimes used for concrete basement floors on the ground. For basement walls below grade, the temperature difference for winter design conditions is greater than for the floor. Test results indicate that at the midheight of the below-grade portion of the basement wall, the unit area heat loss is approximately twice that of the floor.

For concrete slab floors in contact with the ground at grade level, tests indicate that for small floor areas (equal to that of a 25 ft by 25 ft house) the heat loss can be calculated as proportional to the length of exposed edge rather than total area. This amounts to 0.81 Btu/h per linear foot of exposed edge per degree Fahrenheit difference between the indoor air temperature and the average outdoor air temperature. This value can be reduced appreciably by installing insulation under the ground slab and along the edge between the floor and abutting walls. In most calculations, if the perimeter loss is calculated accurately, no other floor losses need to be considered. Chapter 27 contains data for load calculations and heat loss values for below-grade walls and floors at different depths.

The second category of simplified procedures uses transient two-dimensional computer models to generate the ground heat transfer data that are then reduced to compact form by regression analysis (see Mitalas 1982 and 1983, Shipp 1983). These are the most accurate procedures available, but the database is very expensive to generate. In addition, these methods are limited to the range of climates and constructions specifically examined. Extrapolating beyond the outer bounds of the regression surfaces can produce significant errors.

Apparent Thermal Conductivity of Soil

Effective or apparent soil thermal conductivity is difficult to estimate precisely and may change substantially in the same soil at different times due to changed moisture conditions and the presence of

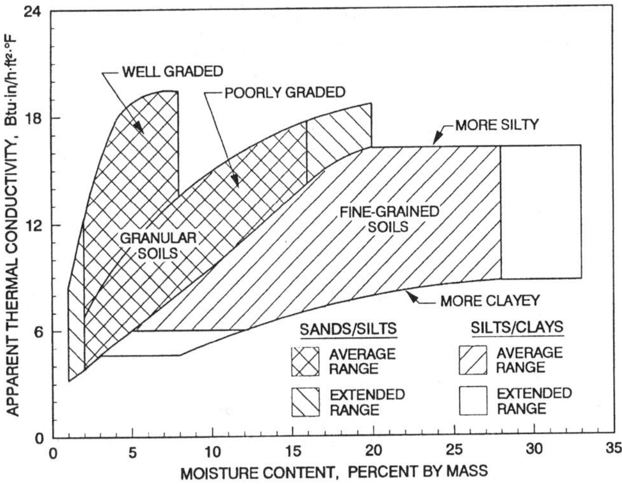

Fig. 8 Trends of Apparent Thermal Conductivity of Moist Soils

freezing temperatures in the soil. Figure 8 shows the typical apparent soil thermal conductivity as a function of moisture content for different general types of soil. The figure is based on data presented in Salomone and Marlowe (1989) using envelopes of thermal behavior coupled with field moisture content ranges for different soil types. In Figure 8, the term well-graded applies to granular soils with good representation of all particle sizes from largest to smallest. The term poorly graded refers to granular soils with either a uniform gradation, in which most particles are about the same size, or a skip (or gap) gradation, in which particles of one or more intermediate sizes are not present.

Although thermal conductivity varies greatly over the complete range of possible moisture contents for a soil, this range can be narrowed if it is assumed that the moisture contents of most field soils lie between the "wilting point" of the soil (i.e., the moisture content of a soil below which a plant cannot alleviate its wilting symptoms) and the "field capacity" of the soil (i.e., the moisture content of a soil that has been thoroughly wetted and then drained until the drainage rate has become negligibly small). After a prolonged dry spell, the moisture will be near the wilting point, and after a rainy period, the soil will have a moisture content near its field capacity. The moisture contents at these limits have been studied by many agricultural researchers, and data for different types of soil are given by Salomone and Marlowe (1989) and Kersten (1949). The shaded areas on Figure 8 approximate (1) the full range of moisture contents for different soil types and (2) a range between average values of each limit.

Table 7 gives a summary of design values for thermal conductivities of the basic soil classes. Table 8 gives ranges of thermal conductivity for some basic classes of rock. The value chosen depends on whether heat transfer is being calculated for minimum heat loss through the soil, as in a ground heat exchange system, or a maximum value, as in peak winter heat loss calculations for a basement. Hence, a high and a low value are given for each soil class.

As heat flows through the soil, the moisture tends to move away from the source of heat. This moisture migration provides initial mass transport of heat, but it also dries the soil adjacent to the heat source, hence lowering the apparent thermal conductivity in that zone of soil.

Trends typical in a soil when other factors are held constant are:

- *k* increases with moisture content
- *k* increases with increasing dry density of a soil
- *k* decreases with increasing organic content of a soil
- *k* tends to decrease for soils with uniform gradations and rounded soil grains (because the grain-to-grain contacts are reduced)
- *k* of a frozen soil may be higher or lower than that of the same unfrozen soil (because the conductivity of ice is higher than that of water but lower than that of the typical soil grains). Differences in *k* below moisture contents of 7 to 8% are quite small. At approximately 15% moisture content, differences in k-factors may vary up to 30% from unfrozen values.

Table 7 Typical Apparent Thermal Conductivity Values for Soils, Btu·in/h·ft²·°F

	Normal Range	Recommended Values for Design[a]	
		Low[b]	High[c]
Sands	4.2 to 17.4	5.4	15.6
Silts	6 to 17.4	11.4	15.6
Clays	6 to 11.4	7.8	10.8
Loams	6 to 17.4	6.6	15.6

[a]Reasonable values for use when no site- or soil-specific data are available.
[b]Moderately conservative values for minimum heat loss through soil (e.g., use in soil heat exchanger or earth-contact cooling calculations). Values are from Salomone and Marlowe (1989).
[c]Moderately conservative values for maximum heat loss through soil (e.g., use in peak winter heat loss calculations). Values are from Salomone and Marlowe (1989).

Table 8 Typical Apparent Thermal Conductivity Values for Rocks, Btu·in/h·ft²·°F

	Normal Range
Pumice, tuff, obsidian	3.6 to 15.6
Basalt	3.6 to 18.0
Shale	6 to 27.6
Granite	12 to 30
Limestone, dolomite, marble	8.4 to 30
Quartzose sandstone	9.6 to 54

When calculating annual energy use, values that represent typical site conditions as they vary during the year should be chosen. In climates where ground freezing is significant, accurate heat transfer simulations should include the effect of the latent heat of fusion of water. The energy released during this phase change significantly retards the progress of the frost front in moist soils.

Water Vapor Transmission Data for Building Components

Table 9 gives typical water vapor permeance and permeability values for common building materials. These values can be used to calculate water vapor flow through building components and assemblies using equations in Chapter 22.

MECHANICAL AND INDUSTRIAL SYSTEMS

Thermal Transmission Data

Table 10 lists the thermal conductivities of various materials used as industrial insulations. These values are functions of the arithmetic mean of the temperatures of the inner and outer surfaces for each insulation.

Heat Loss from Pipes and Flat Surfaces

Tables 11A, 11B, and 12 give heat losses from bare steel pipes and flat surfaces and bare copper tubes. These tables were calculated using ASTM *Standard* C 680. User inputs for the programs described in the standard include operating temperature, ambient temperature, pipe size, insulation type, number of insulation layers, and thickness for each layer. A program option allows the user to input a surface coefficient or surface emittance, surface orientation, and wind speed. The computer uses this information to calculate the

Table 9 Typical Water Vapor Permeance and Permeability Values for Common Building Materials[a]

Material	Thickness, in.	Permeance, Perm	Resistance[h], Rep	Permeability, Perm-in.	Resistance/in.[h], Rep/in.
Construction Materials					
Concrete (1:2:4 mix)				3.2	0.31
Brick masonry	4	0.8[f]	1.3		
Concrete block (cored, limestone aggregate)	8	2.4[f]	0.4		
Tile masonry, glazed	4	0.12[f]	8.3		
Asbestos cement board	0.12	4-8[d]	0.1-0.2		
With oil-base finishes		0.3-0.5[d]	2-3		
Plaster on metal lath	0.75	15[f]	0.067		
Plaster on wood lath		11[e]	0.091		
Plaster on plain gypsum lath (with studs)		20[f]	0.050		
Gypsum wall board (plain)	0.375	50[f]	0.020		
Gypsum sheathing (asphalt impregnated)	0.5			20[d]	0.050
Structural insulating board (sheathing quality)				20-50[f]	0.050-0.020
Structural insulating board (interior, uncoated)	0.5	50-90[f]	0.020-0.011		
Hardboard (standard)	0.125	11[f]	0.091		
Hardboard (tempered)	0.125	5[f]	0.2		
Built-up roofing (hot mopped)		0			
Wood, sugar pine				0.4-5.4[b]	2.5-0.19
Plywood (douglas fir, exterior glue)	0.25	0.7[f]	1.4		
Plywood (douglas fir, interior glue)	0.25	1.9[f]	0.53		
Acrylic, glass fiber reinforced sheet	0.056	0.12[d]	8.3		
Polyester, glass fiber reinforced sheet	0.048	0.05[d]	20		
Thermal Insulations					
Air (still)				120[f]	0.0083
Cellular glass				0[d]	∞
Corkboard				2.1-2.6[d]	0.48-0.38
				9.5[e]	0.11
Mineral wool (unprotected)				116[e]	0.0086
Expanded polyurethane (R-11 blown) board stock				0.4-1.6[d]	2.5-0.62
Expanded polystyrene—extruded				1.2[d]	0.83
Expanded polystyrene—bead				2.0-5.8[d]	0.50-0.17
Phenolic foam (covering removed)				26	0.038
Unicellular synthetic flexible rubber foam				0.02-0.15[d]	50-6.7
Plastic and Metal Foils and Films[c]					
Aluminum foil	0.001	0.0[d]	∞		
Aluminum foil	0.00035	0.05[d]	20		
Polyethylene	0.002	0.16[d]	6.3		3100
Polyethylene	0.004	0.08[d]	12.5		3100
Polyethylene	0.006	0.06[d]	17		3100
Polyethylene	0.008	0.04[d]	25		3100
Polyethylene	0.010	0.03[d]	33		3100
Polyvinylchloride, unplasticized	0.002	0.68[d]	1.5		
Polyvinylchloride, plasticized	0.004	0.8-1.4[d]	1.3-0.72		
Polyester	0.001	0.73[d]	1.4		
Polyester	0.0032	0.23[d]	4.3		
Polyester	0.0076	0.08[d]	12.5		
Cellulose acetate	0.01	4.6[d]	0.2		
Cellulose acetate	0.125	0.32[d]	3.1		

heat flow and the surface temperature. The programs calculate the surface coefficients if the user has not already supplied them.

The equations used in ASTM C 680 are:

$$h_{cv} = C\left(\frac{1}{d}\right)^{0.2}\left(\frac{1}{T_{avg}}\right)^{0.181}(\Delta T^{0.266})\sqrt{1 + 1.277(\text{Wind})} \quad (6)$$

where

h_{cv} = convection surface coefficient, Btu/h·ft²·°F
d = diameter for cylinder, in. For flat surfaces and large cylinders (d > 24 in.), use d = 24 in.
T_{avg} = average temperature of air film = $(T_a + T_s)/2$, °R
T_a = temperature of ambient air, °R
T_s = temperature of surface, °R
ΔT = surface to air temperature difference, °R
Wind = air speed, mph
C = constant depending on shape and heat flow condition
= 1.016 for horizontal cylinders
= 1.235 for longer vertical cylinders
= 1.394 for vertical plates
= 1.79 for horizontal plates, warmer than air, facing upward
= 0.89 for horizontal plates, warmer than air, facing downward
= 0.89 for horizontal plates, cooler than air, facing upward
= 1.79 for horizontal plates, cooler than air, facing downward

Table 9 Typical Water Vapor Permeance and Permeability Values for Common Building Materials (*Concluded*)[a]

Material	Weight, lb/100 ft²	Permeance, Perms			Resistance[h] Rep		
		Dry-Cup	Wet-Cup	Other	Dry-Cup	Wet-Cup	Other
Building Paper, Felts, Roofing Papers[g]							
Duplex sheet, asphalt laminated, aluminum foil one side	8.6	0.002	0.176		500	5.8	
Saturated and coated roll roofing	65	0.05	0.24		20	4.2	
Kraft paper and asphalt laminated, reinforced 30-120-30	6.8	0.3	1.8		3.3	0.55	
Blanket thermal insulation backup paper, asphalt coated	6.2	0.4	0.6-4.2		2.5	1.7-0.24	
Asphalt-saturated and coated vapor retarder paper	8.6	0.2-0.3	0.6		5.0-3.3	1.7	
Asphalt-saturated, but not coated, sheathing paper	4.4	3.3	20.2		0.3	0.05	
15-lb asphalt felt	14	1.0	5.6		1.0	0.18	
15-lb tar felt	14	4.0	18.2		0.25	0.055	
Single-kraft, double	3.2	31	42		0.032	0.024	
Liquid-Applied Coating Materials	**Thickness, in.**						
Commercial latex paints (dry film thickness)[i]							
Vapor retarder paint	0.0031			0.45			2.22
Primer-sealer	0.0012			6.28			0.16
Vinyl acetate/acrylic primer	0.002			7.42			0.13
Vinyl-acrylic primer	0.0016			8.62			0.12
Semi-gloss vinyl-acrylic enamel	0.0024			6.61			0.15
Exterior acrylic house and trim	0.0017			5.47			0.18
Paint-2 coats							
Asphalt paint on plywood			0.4			2.5	
Aluminum varnish on wood		0.3-0.5			3.3-2.0		
Enamels on smooth plaster				0.5-1.5			2.0-0.66
Primers and sealers on interior insulation board				0.9-2.1			1.1-0.48
Various primers plus 1 coat flat oil paint on plaster				1.6-3.0			0.63-0.33
Flat paint on interior insulation board				4			0.25
Water emulsion on interior insulation board				30-85			0.03-0.012
	Weight, oz/ft²						
Paint-3 coats							
Exterior paint, white lead and oil on wood siding		0.3-1.0			3.3-1.0		
Exterior paint, white lead-zinc oxide and oil on wood		0.9			1.1		
Styrene-butadiene latex coating	2	11			0.09		
Polyvinyl acetate latex coating	4	5.5			0.18		
Chlorosulfonated polyethylene mastic	3.5	1.7			0.59		
	7.0	0.06			16		
Asphalt cutback mastic, 1/16 in., dry		0.14			7.2		
3/16 in., dry		0.0			—		
Hot melt asphalt	2	0.5			2		
	3.5	0.1			10		

[a]This table permits comparisons of materials; but in the selection of vapor retarder materials, exact values for permeance or permeability should be obtained from the manufacturer or from laboratory tests. The values shown indicate variations among mean values for materials that are similar but of different density, orientation, lot, or source. The values should not be used as design or specification data. Values from dry-cup and wet-cup methods were usually obtained from investigations using ASTM E 96 and C 355; values shown under others were obtained by two-temperature, special cell, and air velocity methods. Permeance, resistance, permeability, and resistance per unit thickness values are given in the following units:

Permeance	Perm	= gr/h·ft²·in. Hg
Resistance	Rep	= in. Hg·ft²·h/gr
Permeability	Perm-in.	= gr/h·ft²·(in. Hg/in.)
Resistance/unit thickness	Rep/in.	= (in. Hg·ft²·h/gr)/in.

[b]Depending on construction and direction of vapor flow.

[c]Usually installed as vapor retarders, although sometimes used as an exterior finish and elsewhere near the cold side, where special considerations are then required for warm side barrier effectiveness.

[d]Dry-cup method.

[e]Wet-cup method.

[f]Other than dry- or wet-cup method.

[g]Low permeance sheets used as vapor retarders. High permeance used elsewhere in construction.

[h]Resistance and resistance/in. values have been calculated as the reciprocal of the permeance and permeability values.

[i]Cast at 10 mils (0.01 in.) wet film thickness.

$$h_{rad} = \frac{\varepsilon\sigma(T_a^4 - T_s^4)}{T_a - T_s} \quad (7)$$

where

h_{rad} = radiation surface coefficient, Btu/h·ft²·°F

ε = surface emittance

σ = Stefan-Boltzmann constant = 0.1713×10^{-8} Btu/h·ft²·°R⁴

Example 6. Compute the total annual heat loss from 165 ft of nominal 2-in. bare steel pipe in service 4000 h per year. The pipe is carrying steam at 10 psig and is exposed to an average air temperature of 80°F.

Solution. The pipe temperature is taken as the steam temperature, which is 239.4°F, obtained by interpolation from Steam Tables. By interpolation in Table 11A between 180°F and 280°F, heat loss from a nominal 2-in. pipe is 285 Btu/h·ft. Total annual heat loss from the entire line is 285 Btu/h·ft × 165 ft × 4000 h = 188 × 10⁶ Btu.

In calculating heat flow, Equations (9) and (10) from Chapter 22 generally are used. For dimensions of standard pipe and fitting sizes, refer to the *Piping Handbook*. For insulation product dimensions, refer to ASTM *Standard* C 585, or to the insulation manufacturers' literature.

Examples 7 and 8 illustrate how Equations (9) and (10) from Chapter 22 can be used to determine heat loss from both flat and

Table 10 Typical Thermal Conductivity for Industrial Insulations at Various Mean Temperatures—Design Values[a]

Material	Max. Temp.,[b] °F	Typical Density, lb/ft³	Typical Conductivity in Btu·in/h·ft²·°F at Mean Temp., °F −100	−75	−50	−25	0	25	50	75	100	200	300	500	700	900
BLANKETS AND FELTS																
ALUMINOSILICATE FIBER																
7 to 10 μm diameter fiber	1800	4								0.24		0.32		0.54	0.99	1.03
	2000	6-8								0.25		0.30		0.48	0.78	0.95
3 μm diameter fiber	2200	4								0.22		0.29		0.45	0.59	0.74
MINERAL FIBER (Rock, slag, or glass)																
Blanket, metal reinforced	1200	6-12									0.26	0.32	0.39	0.54		
	1000	2.5-6									0.24	0.31	0.40	0.61		
Blanket, flexible, fine-fiber	350	0.75				0.25	0.26	0.28	0.30	0.33	0.36	0.53				
organic bonded		0.75				0.24	0.25	0.27	0.29	0.32	0.34	0.48				
		1.0				0.23	0.24	0.25	0.27	0.29	0.32	0.43				
		1.5				0.21	0.22	0.23	0.25	0.27	0.28	0.37				
		2.0				0.20	0.21	0.22	0.23	0.25	0.26	0.33				
		3.0				0.19	0.20	0.21	0.22	0.23	0.24	0.31				
Blanket, flexible, textile fiber,	350	0.65				0.27	0.28	0.29	0.30	0.31	0.32	0.50	0.68			
organic bonded		0.75				0.26	0.27	0.28	0.29	0.31	0.32	0.48	0.66			
		1.0				0.24	0.25	0.26	0.27	0.29	0.31	0.45	0.60			
		1.5				0.22	0.23	0.24	0.25	0.27	0.29	0.39	0.51			
		3.0				0.20	0.21	0.22	0.23	0.24	0.25	0.32	0.41			
Felt, semirigid organic bonded	400	3-8						0.24	0.25	0.26	0.27	0.35	0.44			
Laminated and felted without binder	850	3	0.16	0.17	0.18	0.19	0.20	0.21	0.22	0.23	0.24	0.35	0.55			
	1200	7.5											0.35	0.45	0.60	
BLOCKS, BOARDS, AND PIPE INSULATION																
MAGNESIA	600	11-12									0.35	0.38	0.42			
85% CALCIUM SILICATE	1200	11-15									0.38	0.41	0.44	0.52	0.62	0.72
	1800	12-15												0.63	0.74	0.95
CELLULAR GLASS	900	7.8-8.2	0.24	0.25	0.26	0.28	0.29	0.30	0.32	0.33	0.34	0.41	0.49	0.70	1.01	
DIATOMACEOUS SILICA	1600	21-22												0.64	0.68	0.72
	1900	23-25												0.70	0.75	0.80
MINERAL FIBER (Glass)																
Organic bonded, block and boards	400	3-10	0.16	0.17	0.18	0.19	0.20	0.22	0.24	0.25	0.26	0.33	0.40			
Nonpunking binder	1000	3-10									0.26	0.31	0.38	0.52		
Pipe insulation, slag, or glass	350	3-4					0.20	0.21	0.22	0.23	0.24	0.29				
	500	3-10					0.20	0.22	0.24	0.25	0.26	0.33	0.40			
Inorganic bonded block	1000	10-15									0.33	0.38	0.45	0.55		
	1800	15-24									0.32	0.37	0.42	0.52	0.62	0.74
Pipe insulation, slag, or glass	1000	10-15									0.33	0.38	0.45	0.55		
Resin binder		15	0.23	0.24	0.25	0.26	0.28	0.29								
RIGID POLYSTYRENE																
Extruded (CFC-12 exp.)(smooth skin surface)	165	1.8-3.5	0.16	0.16	0.17	0.16	0.17	0.18	0.19	0.20						
Molded beads	165	1	0.17	0.19	0.20	0.21	0.22	0.24	0.25	0.26	0.28					
		1.25	0.17	0.18	0.19	0.20	0.22	0.23	0.24	0.25	0.27					
		1.5	0.16	0.17	0.19	0.20	0.21	0.22	0.23	0.24	0.26					
		1.75	0.16	0.17	0.18	0.19	0.20	0.22	0.23	0.24	0.25					
		2.0	0.15	0.16	0.18	0.19	0.20	0.21	0.22	0.23	0.24					
RIGID POLYURETHANE/POLYISOCYANURATE[c,d]																
Unfaced (CFC-11 exp.)	210	1.5-2.5	0.16	0.17	0.18	0.18	0.18	0.17	0.16	0.16	0.17					
RIGID POLYISOCYANURATE																
Gas-impermeable facers (CFC-11 exp.)	250	2.0						0.12	0.13	0.14	0.15					
RIGID PHENOLIC																
Closed cell (CFC-11, CFC-113 exp.)		3.0						0.11	0.115	0.12	0.125					
RUBBER, Rigid foamed	150	4.5						0.20	0.21	0.22	0.23					
VEGETABLE AND ANIMAL FIBER																
Wool felt (pipe insulation)	180	20						0.28	0.30	0.31	0.33					
INSULATING CEMENTS																
MINERAL FIBER (Rock, slag, or glass)																
With colloidal clay binder	1800	24-30									0.49	0.55	0.61	0.73	0.85	
With hydraulic setting binder	1200	30-40									0.75	0.80	0.85	0.95		
LOOSE FILL																
Cellulose insulation (milled pulverized paper or wood pulp)		2.5-3							0.26	0.27	0.29					
Mineral fiber, slag, rock, or glass		2-5			0.19	0.21	0.23	0.25	0.26	0.28	0.31					
Perlite (expanded)		3-5	0.22	0.24	0.25	0.27	0.28	0.30	0.31	0.33	0.35					
Silica aerogel		7.6			0.13	0.14	0.15	0.15	0.16	0.17	0.18					
Vermiculite (expanded)		7-8.2			0.39	0.40	0.42	0.44	0.45	0.47	0.49					
		4-6			0.34	0.35	0.38	0.40	0.42	0.44	0.46					

[a]Representative values for dry materials, which are intended as design (not specification) values for materials in normal use. Insulation materials in actual service may have thermal values that vary from design values depending on their in-situ properties (e.g., density and moisture content). For properties of a particular product, use the value supplied by the manufacturer or by unbiased tests.

[b]These temperatures are generally accepted as maximum. When operating temperature approaches these limits, follow the manufacturers' recommendations.

[c]Some polyurethane foams are formed by means that produce a stable product (with respect to *k*), but most are blown with refrigerant and will change with time.

[d]See Table 4, footnote i.

[e]See Table 4, footnote j.

Table 11A Heat Loss from Bare Steel Pipe to Still Air at 80°F[a], Btu/h·ft

Nominal Pipe Size[b], in.	Pipe Inside Temperature, °F									
	180	280	380	480	580	680	780	880	980	1080
0.50	59.3	147.2	263.2	412.3	600.9	836.8	1128.6	1485.6	1918.0	2436.8
0.75	72.5	180.1	322.6	506.2	739.2	1031.2	1392.9	1836.0	2373.5	3018.8
1.00	88.8	220.8	396.1	622.7	910.9	1272.6	1721.2	2271.5	2939.4	3741.6
1.25	109.7	272.8	490.4	772.3	1131.7	1583.8	2145.6	2835.4	3673.4	4680.9
1.50	123.9	308.5	555.1	875.1	1283.8	1798.3	2438.2	3224.6	4180.5	5330.0
2.00	151.8	378.1	681.4	1076.3	1581.5	2218.9	3012.6	3989.2	5177.2	6606.8
2.50	180.5	450.0	811.9	1284.0	1888.8	2652.6	3604.3	4775.3	6199.5	7912.5
3.00	215.9	538.8	973.5	1541.8	2271.4	3194.0	4344.9	5762.2	7486.9	9562.3
3.50	243.9	609.0	1101.4	1746.1	2574.7	3623.6	4933.0	6546.4	8510.4	10874.3
4.00	271.6	678.6	1228.2	1948.7	2875.9	4050.5	5517.5	7326.0	9528.1	12178.9
4.50	299.2	747.7	1354.4	2150.9	3176.8	4477.7	6103.8	8109.5	10553.2	13496.2
5.00	329.8	824.7	1494.8	2375.4	3510.6	4950.7	6751.3	8972.5	11678.4	14936.3
6.00	387.1	968.7	1757.8	2796.8	4138.0	5841.4	7972.7	10603.1	13808.2	17667.6
7.00	440.5	1102.8	2003.0	3189.9	4723.9	6673.5	9114.2	12127.4	15799.4	20220.8
8.00	493.3	1235.7	2246.1	3580.0	5305.5	7500.0	10248.4	13642.2	17778.2	22758.0
9.00	545.9	1368.1	2488.8	3970.2	5888.7	8331.0	11392.1	15174.5	19787.1	25343.6
10.00	604.3	1514.8	2757.2	4400.7	6530.1	9241.1	12638.6	16835.1	21949.2	28104.9
11.00	656.0	1644.8	2995.5	4783.8	7102.1	10054.9	13756.2	18328.4	23900.3	30606.1
12.00	704.0	1762.3	3203.8	5104.9	7557.3	10661.8	14524.9	19256.7	24967.6	31766.8
14.00	771.0	1934.2	3525.9	5636.0	8373.9	11862.4	16235.5	21635.6	28212.3	36120.3
16.00	872.2	2189.0	3993.2	6387.4	9495.9	13458.0	18424.8	24556.6	32021.1	40990.7
18.00	972.5	2441.7	4456.7	7132.9	10609.4	15041.3	20596.7	27453.2	35795.6	45813.1
20.00	1072.1	2692.4	4916.8	7873.2	11715.1	16613.4	22752.5	30326.8	39537.6	50590.0
24.00	1269.3	3188.9	5828.3	9339.9	13905.5	19726.9	27019.7	36010.1	46930.3	60014.7

Table 11B Heat Loss from Flat Surfaces to Still Air at 80°F, Btu/h·ft²

	Surface Inside Temperature, °F									
	180	280	380	480	580	680	780	880	980	1080
Vertical surface	212.2	533.1	973.3	1558.6	2321.2	3298.0	4530.1	6062.8	7945.5	10231.5
Horizontal surface										
Facing up	234.7	586.4	1061.1	1683.5	2484.9	3501.9	4775.4	6350.4	8276.3	10606.1
Facing down	183.6	465.3	861.4	1399.6	2112.8	3038.4	4217.8	5696.7	7524.5	9754.7

[a]Calculations from ASTM C 680; steel: $k = 314.4$ Btu·in/h·ft²·°F; $\varepsilon = 0.94$.

[b]Losses per square foot of pipe for pipes larger than 24 in. can be considered the same as losses per square foot for 24-in. pipe.

cylindrical surfaces. Figure 9 shows surface resistance as a function of heat transmission for both flat and cylindrical surfaces. The surface emittance is assumed to be 0.85 to 0.90 in still air at 80°F.

Example 7. Compute the heat loss from a boiler wall if the interior insulation surface temperature is 1100°F and ambient still air temperature is 80°F. The wall is insulated with 4.5 in. of mineral fiber block and 0.5 in. of mineral fiber insulating and finishing cement.

Solution. Assume that the mean temperature of the mineral fiber block is 700°F, the mean temperature of the insulating cement is 200°F, and the surface resistance R_s is 0.60 ft²·°F·h/Btu.

From Table 10, $k_1 = 0.62$ and $k_2 = 0.80$. Using Equation (9) from Chapter 22:

$$q_s = \frac{1100 - 80}{(4.5/0.62) + (0.5/0.80) + 0.60} = 120.2 \text{ Btu/h} \cdot \text{ft}^2$$

As a check, from Figure 9, at 120.2 Btu/h·ft², $R_s = 0.56$. The mean temperature of the mineral fiber block is:

$$4.5/0.62 = 7.26;\ 7.26/2 = 3.63$$

$$1100 - \frac{3.63}{8.48}(1020) = 663°\text{F}$$

and the mean temperature of the insulating cement is:

$$0.5/0.80 = 0.63;\ 0.63/2 = 0.31;\ 7.26 + 0.31 = 7.57$$

$$1100 - \frac{7.57}{8.48}(1020) = 189°\text{F}$$

From Table 10, at 663°F, $k_1 = 0.60$; at 189°F, $k_2 = 0.79$.

Using these adjusted values to recalculate q_s:

$$q_s = \frac{1020}{(4.5/0.60) + (0.5/0.79) + 0.56} = \frac{1020}{8.69}$$

$$= 117.4 \text{ Btu/h} \cdot \text{ft}^2$$

From Figure 9, at 117.4 Btu/h·ft², $R_s = 0.56$. The mean temperature of the mineral fiber block is:

$$4.5/0.6 = 7.50;\ 7.50/2 = 3.75$$

$$1100 - \frac{3.75}{8.69}(1020) = 660°\text{F}$$

and the mean temperature of the insulating cement is:

$$0.5/0.79 = 0.63;\ 0.63/2 = 0.31;\ 7.50 + 0.31 = 7.81$$

$$1100 - \frac{7.81}{8.69}(1020) = 183°\text{F}$$

From Table 10, at 660°F, $k_1 = 0.60$; at 183°F, $k_2 = 0.79$.

Since R_s, k_1, and k_2 do not change at these values, $q_s = 117.4$ Btu/h·ft.

Example 8. Compute heat loss per square foot of outer surface of insulation if pipe temperature is 1200°F and ambient still air temperature is 80°F. The pipe is nominal 6-in. steel pipe, insulated with a nominal 3-in. thick diatomaceous silica as the inner layer and a nominal 2-in. thick calcium silicate as the outer layer.

Table 12 Heat Loss from Bare Copper Tube to Still Air at 80°F[a], Btu/h·ft

Nominal Tube Size, in.	Tube Inside Temperature, °F							
	120	150	180	210	240	270	300	330
Dull ε = 0.44								
0.250	7.1	14.1	21.9	30.6	39.9	49.9	60.6	71.9
0.375	9.1	18.0	28.1	39.1	51.1	63.9	77.6	92.2
0.500	11.0	21.8	34.0	47.4	61.9	77.5	94.1	111.8
0.750	14.7	29.1	45.4	63.3	82.7	103.6	126.0	149.8
1.000	18.3	36.2	56.4	78.7	102.8	128.9	156.7	186.5
1.250	21.8	43.1	67.2	93.6	122.4	153.4	186.7	222.2
1.500	25.2	49.8	77.6	108.3	141.5	177.4	216.0	257.1
2.000	31.8	62.9	98.0	136.7	178.8	224.3	273.1	325.4
2.500	38.3	75.6	117.9	164.4	215.1	269.8	328.7	391.8
3.000	44.6	88.1	137.2	191.5	250.5	314.4	383.2	456.9
3.500	50.8	100.3	156.3	218.0	285.4	358.2	436.7	520.8
4.000	57.0	112.3	175.0	244.2	319.7	401.4	489.4	583.9
5.000	69.0	135.9	211.7	295.5	386.9	486.0	592.8	707.6
6.000	80.7	159.0	247.7	345.7	452.8	568.9	694.2	829.0
8.000	103.7	204.1	317.8	443.7	581.3	730.7	892.1	1066.0
10.000	126.1	247.9	386.1	539.1	706.5	888.4	1085.2	1297.4
12.000	148.0	290.9	453.0	632.5	829.2	1043.1	1274.6	1524.4
Bright ε = 0.08								
0.250	5.4	10.8	16.9	23.5	30.5	37.9	45.5	53.5
0.375	6.8	13.7	21.4	29.7	38.6	47.9	57.6	67.6
0.500	8.2	16.4	25.7	35.7	46.3	57.4	69.1	81.2
0.750	10.7	21.6	33.8	46.9	60.9	75.6	90.9	106.8
1.000	13.2	26.5	41.4	57.6	74.7	92.8	111.6	131.2
1.250	15.5	31.3	48.8	67.8	88.0	109.3	131.6	154.7
1.500	17.8	35.8	56.0	77.8	100.9	125.3	150.8	177.4
2.000	22.2	44.6	69.7	96.8	125.7	156.1	187.9	221.1
2.500	26.4	53.0	82.8	115.1	149.5	185.6	223.5	263.0
3.000	30.5	61.2	95.6	132.8	172.4	214.2	257.9	303.5
3.500	34.4	69.1	107.9	150.0	194.8	242.0	291.4	342.9
4.000	38.3	76.8	120.0	166.8	216.6	269.1	324.1	381.4
5.000	45.7	91.8	143.4	199.3	258.8	321.6	387.4	456.1
6.000	53.0	106.3	166.0	230.7	299.7	372.5	448.7	528.3
8.000	66.8	134.1	209.4	291.1	378.2	470.1	566.5	667.2
10.000	80.2	160.8	251.0	349.0	453.4	563.7	679.5	800.4
12.000	93.0	186.5	291.3	404.9	526.1	654.2	788.7	929.3

[a]Calculations from ASTM C 680; for copper: $k = 2784$ Btu·in/h·ft^2·°F.

Solution. From Chapter 40 of the 1996 *ASHRAE Handbook—Equipment*, $r_o = 3.31$ in. A nominal 3-in. thick diatomaceous silica insulation to fit a nominal 6-in. steel pipe is 3.02 in. thick. A nominal 2-in. thick calcium silicate insulation to fit over the 3.02-in. diatomaceous silica is 2.08 in. thick. Therefore, $r_i = 6.33$ in. and $r_s = 8.41$ in..

Assume that the mean temperature of the diatomaceous silica is 600°F, the mean temperature of the calcium silicate is 250°F and the surface resistance R_s is 0.50. From Table 10, $k_1 = 0.66$; $k_2 = 0.42$. By Equation (10) from Chapter 22:

$$q_s = \frac{1200 - 80}{[8.41\ln(6.33/3.31)/0.66] + [8.41\ln(8.41/3.31)/0.40] + 0.50}$$

$$= \frac{1120}{(5.45/0.66) + (2.39/0.40) + 0.50} = 76.0 \text{ Btu/h} \cdot \text{ft}^2$$

From Figure 9, at 76.0 Btu/h·ft^2, $R_s = 0.60$. The mean temperature of the diatomaceous silica is:

$$5.45/0.66 = 8.26;\ 8.26/2 = 4.13$$

$$1200 - \frac{4.13}{14.83}(1120) = 888°\text{F}$$

and the mean temperature of the calcium silicate is:

$$2.39/0.40 = 5.98;\ 5.98/2 = 2.99;\ 8.26 + 2.99 = 11.25$$

$$1200 - \frac{11.25}{14.83}(1120) = 350°\text{F}$$

From Table 10, $k_1 = 0.72$; $k_2 = 0.46$. Recalculating:

$$q_s = \frac{1120}{(5.45/0.72) + (2.39/0.46) + 0.60} = 83.8 \text{ Btu/h} \cdot \text{ft}^2$$

From Figure 9 at 83.8 Btu/h·ft^2, $R_s = 0.59$. The mean temperature of the diatomaceous silica is:

$$5.45/0.72 = 7.57;\ 7.57/2 = 3.78$$

$$1200 - \frac{3.78}{13.36}(1120) = 883°\text{F}$$

and the mean temperature of the calcium silicate is:

$$2.39/0.40 = 5.98;\ 5.98/2 = 2.99;\ 8.26 + 2.99 = 11.25$$

$$1200 - \frac{11.25}{14.83}(1120) = 350°\text{F}$$

From Table 10, $k_1 = 0.72$; $k_2 = 0.46$. Recalculating:

$$2.39/0.46 = 5.20;\ 5.20/2 = 2.60;\ 7.57 + 2.60 = 10.17$$

$$1200 - \frac{10.17}{13.36}(1120) = 347°\text{F}$$

Since R_s, k_1, and k_2 do not change at 83.8 Btu/h·ft^2, this is q_s. The heat flow per ft^2 of the inner surface of the insulation is:

$$q_o = q_s(r_s/r_o) = 83.8(8.41/3.31) = 213 \text{ Btu/h} \cdot \text{ft}^2$$

Table 13 Recommended Thicknesses for Pipe and Equipment Insulation

Nom Dia., in.		MINERAL FIBER (Fiberglass and Rock Wool) Process Temperature, °F										CALCIUM Process Temp., °F		
		150	250	350	450	550	650	750	850	950	1050	150	250	350
	Thickness	1	1½	2	2½	3	3½	4	4	4½	5½	1	1½	2
½	Heat loss	8	16	24	33	43	54	66	84	100	114	13	24	34
	Surface temp.	72	75	76	78	79	81	82	86	87	87	75	78	80
	Thickness	1	1½	2	2½	3½	4	4	4½	5	5½	1	2	2½
1	Heat loss	11	21	30	41	49	61	79	96	114	135	16	26	38
	Surface temp.	73	76	78	80	79	81	84	86	88	89	76	76	79
	Thickness	1	2	2½	3	4	4	4	5½	5½	6	1½	2½	3
1½	Heat loss	14	22	33	45	54	73	94	103	128	152	17	29	42
	Surface temp.	73	74	77	79	79	82	86	84	88	90	73	75	78
	Thickness	1½	2	3	3½	4	4	4	5½	6	6	1½	2½	3
2	Heat loss	13	25	34	47	61	81	105	114	137	168	19	32	47
	Surface temp.	71	75	75	77	79	83	87	85	87	91	74	76	79
	Thickness	1½	2½	3½	4	4	4½	4½	6	6½	7	2	3	3½
3	Heat loss	16	28	39	54	75	94	122	133	154	184	21	37	54
	Surface temp.	72	74	75	77	81	83	87	86	87	90	73	75	78
	Thickness	1½	3	4	4	4	5	5½	6	7	7½	2	3	4
4	Heat loss	19	29	42	63	88	102	126	152	174	206	25	43	58
	Surface temp.	72	73	74	78	82	86	85	87	88	90	70	76	77
	Thickness	2	3	4	4	4½	5	5½	6½	7½	8	2	3½	4
6	Heat loss	21	38	54	81	104	130	159	181	208	246	33	51	75
	Surface temp.	71	74	75	79	82	84	87	88	89	91	74	75	79
	Thickness	2	3½	4	4	5	5	5½	7	8	8½	2½	3½	4
8	Heat loss	26	42	65	97	116	155	189	204	234	277	35	62	90
	Surface temp.	71	73	76	80	81	86	89	88	89	92	73	76	79
	Thickness	2	3½	4	4	5	5½	5½	7½	8½	9	2½	4	4
10	Heat loss	32	50	77	115	136	170	220	226	259	307	41	66	106
	Surface temp.	72	74	77	81	82	85	90	87	89	91	73	75	80
	Thickness	2	3½	4	4	5	5½	5½	7½	8½	9½	2½	4	4
12	Heat loss	36	57	87	131	154	192	249	253	290	331	47	75	121
	Surface temp.	72	74	77	82	82	86	91	88	89	91	73	76	81
	Thickness	2	3½	4	4	5	5½	6½	7½	9	9½	2½	4	4
14	Heat loss	40	61	94	141	165	206	236	271	297	352	51	81	130
	Surface temp.	72	74	77	82	83	86	87	89	89	91	73	76	81
	Thickness	2½	3½	4	4	5½	5½	7	8	9	10	3	4	4
16	Heat loss	37	68	105	157	171	228	247	284	326	372	50	90	144
	Surface temp.	71	74	78	83	82	87	86	88	89	91	72	76	82
	Thickness	2½	3½	4	4	5½	5½	7	8	9	10	3	4	4
18	Heat loss	41	75	115	173	187	250	270	310	354	404	55	99	159
	Surface temp.	71	74	78	83	83	87	87	88	90	91	73	76	82
	Thickness	2½	3½	4	4	5½	5½	7	8	9	10	3	4	4
20	Heat loss	45	82	126	189	204	272	292	335	383	436	60	108	174
	Surface temp.	71	75	78	83	83	87	87	89	90	92	73	77	82
	Thickness	2½	4	4	4	5½	6	7½	8	9	10	3	4	4
24	Heat loss	53	86	147	221	237	295	320	386	439	498	71	127	203
	Surface temp.	71	74	78	83	83	86	86	89	91	93	73	77	82
	Thickness	2½	4	4	4	5½	6½	7½	8½	10	10	3	4	4
30	Heat loss	65	105	179	268	286	332	383	439	481	591	86	154	247
	Surface temp.	71	74	79	84	84	85	87	89	89	94	73	77	83
	Thickness	2½	4	4	4	5½	7	8	9	10	10	2½	4	4
36	Heat loss	77	123	211	316	335	364	422	486	556	683	119	181	291
	Surface temp.	71	74	79	84	84	84	86	88	90	94	74	77	83
	Thickness	2	3½	4	4½	5½	8½	9½	10	10	10	2½	3½	4
Flat	Heat loss	10	14	20	27	31	27	31	38	47	58	12	20	28
	Surface temp.	72	74	77	80	82	80	82	85	89	93	73	77	81

Consult manufacturer's literature for product temperature limitations. Table is based on typical operating conditions, e.g., 65°F ambient temperature and 7.5 mph wind speed, and may not represent actual conditions of use. Units for thickness, heat loss, and surface temperature are in inches, Btu/h·ft (Btu/h·ft^2 for flat surfaces), and °F, respectively.

Table 13 Recommended Thicknesses for Pipe and Equipment Insulation (*Concluded*)

Nom. Dia., in.		SILICATE							CELLULAR GLASS						
		Process Temperature, °F							Process Temperature, °F						
		450	550	650	750	850	950	1050	150	250	350	450	550	650	750
	Thickness	2½	3	3½	4	4	4	4	1½	1½	2	2½	3	3½	4
½	Heat loss	42	53	63	75	90	108	128	9	23	34	48	62	78	92
	Surface temp.	81	82	83	84	87	91	94	70	76	78	82	83	85	84
	Thickness	3	3½	4	4	4	4	4	1½	2	2½	3	3½	4	4
1	Heat loss	49	60	72	89	109	130	154	12	25	38	52	68	86	112
	Surface temp.	80	82	83	86	90	94	98	71	75	77	79	81	83	88
	Thickness	3½	4	4	4	4	5	5	1½	2½	3	4	4	4	4
1½	Heat loss	54	68	86	106	128	139	164	15	28	44	56	79	105	137
	Surface temp.	80	81	85	88	92	91	94	72	75	77	78	82	87	92
	Thickness	3½	4	4½	5	5½	6	6	1½	2½	3	4	4	4	4½
2	Heat loss	61	75	90	106	123	142	167	17	31	47	61	84	113	140
	Surface temp.	81	82	84	85	87	88	91	72	74	77	78	82	86	89
	Thickness	4	4½	5	5½	6	6	6	1½	3	3½	4	4	4½	5
3	Heat loss	71	87	105	123	143	71	202	22	35	54	75	105	132	161
	Surface temp.	80	82	84	85	87	90	94	73	74	77	79	84	86	89
	Thickness	4	4½	5	5½	6	6½	7	2	3	4	4	4	4½	5
4	Heat loss	82	101	121	142	164	187	213	22	41	59	87	122	150	185
	Surface temp.	81	83	85	87	89	90	92	71	74	76	80	85	87	90
	Thickness	4	4½	5	5½	6	7	8	2	3½	4	4	4½	5½	6
6	Heat loss	105	129	153	178	205	224	245	30	48	74	111	144	171	212
	Surface temp.	83	85	87	89	91	91	91	72	74	77	82	85	86	89
	Thickness	4½	5	5	6	7	8	8½	2½	3½	4	4	5	5½	6½
8	Heat loss	117	144	183	200	220	243	277	30	58	90	134	161	203	238
	Surface temp.	82	85	89	89	89	90	92	71	74	78	83	84	87	89
	Thickness	4	5	5½	6	7½	8½	9	2½	4	4	4	5½	5½	7
10	Heat loss	149	168	200	233	243	269	306	37	63	106	159	178	238	264
	Surface temp.	85	86	88	90	89	89	91	71	74	79	84	84	87	88
	Thickness	4	5	5½	7	8	8½	9½	2½	4	4	4	5½	5½	7½
12	Heat loss	170	191	266	236	262	300	330	42	71	121	181	201	269	284
	Surface temp.	86	86	89	88	88	90	91	71	74	79	85	84	90	88
	Thickness	4	5	5½	7	8	9	9½	2½	4	4	4	5½	5½	8
14	Heat loss	183	205	242	252	262	308	352	47	79	134	199	219	293	293
	Surface temp.	86	87	89	88	88	89	91	72	74	80	85	85	91	87
	Thickness	4	5½	6½	7½	8	9	10	2½	4	4	4	5½	5½	8
16	Heat loss	204	211	237	265	307	338	372	53	88	149	222	242	325	322
	Surface temp.	87	85	86	87	89	90	91	72	75	80	86	86	91	88
	Thickness	4	5½	6½	7½	8½	9	10	2½	4	4	4	5½	5½	8
18	Heat loss	225	232	259	289	320	367	403	59	96	164	245	266	356	351
	Surface temp.	87	86	87	87	88	90	91	72	75	80	86	86	92	88
	Thickness	4	5½	6½	7½	8½	9½	10	2½	4	4	4½	5½	5½	8
20	Heat loss	245	252	281	312	346	381	435	64	105	179	243	289	387	379
	Surface temp.	87	86	87	88	89	90	92	72	75	81	84	86	92	88
	Thickness	4	5½	6½	7½	8½	9½	10	2½	4	4	5	5½	5½	8
24	Heat loss	287	293	325	360	397	437	497	76	123	209	260	336	449	436
	Surface temp.	88	87	88	88	89	90	93	72	75	81	83	87	93	89
	Thickness	4	5½	7	8	9	10	10	2½	4	4	5½	5½	5½	8
30	Heat loss	349	353	368	409	452	498	589	93	150	254	290	405	542	521
	Surface temp.	88	87	87	88	89	90	94	72	75	81	82	87	93	90
	Thickness	4	6½	7½	8	9	10	10	2½	4	4	5½	5½	5½	8
36	Heat loss	410	359	406	475	524	576	681	110	176	229	340	474	635	606
	Surface temp.	89	84	86	88	89	91	94	73	76	81	82	88	94	90
	Thickness	5½	6½	7½	8½	9½	10	10	2½	4	4	5½	5½	7½	8½
Flat	Heat loss	29	33	36	39	43	49	58	11	17	29	31	44	43	50
	Surface temp.	81	83	84	85	87	89	93	73	76	83	84	90	90	93

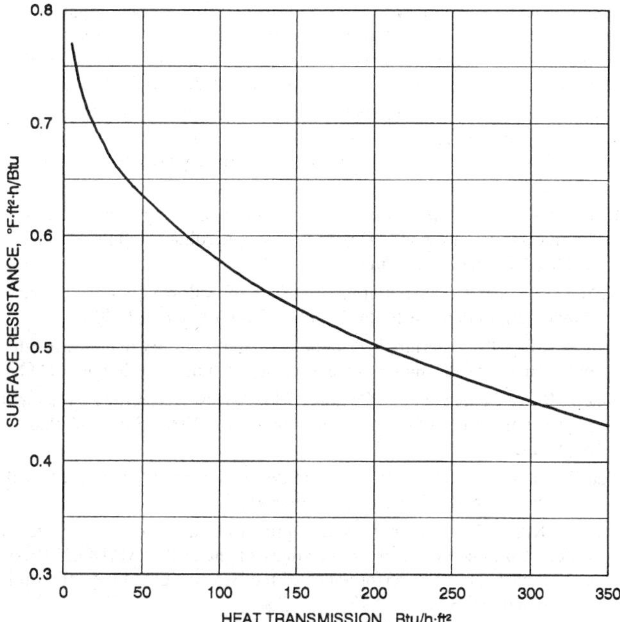

Fig. 9 Surface Resistance as Function of Heat Transmission for Flat Surfaces and Cylindrical Surfaces Greater than 24 in. in Diameter

Because trial and error techniques are tedious, the computer programs previously described should be used to estimate heat flows per unit area of flat surfaces or per unit length of piping, and interface temperatures including surface temperatures.

Several methods can be used to determine the most effective thickness of insulation for piping and equipment. Table 13 shows the recommended insulation thicknesses for three different pipe and equipment insulations. Installed cost data can be developed using procedures described by the Federal Energy Administration (1976). Computer programs capable of calculating thickness information are available from several sources. Also, manufacturers of insulations offer computerized analysis programs for designers and owners to evaluate insulation requirements. For more information on determining economic insulation thickness, see Chapter 22.

Chapters 3 and 22 give guidance concerning process control, personnel protection, condensation control, and economics. For specific information on sizes of commercially available pipe insulation, see ASTM *Standard* C 585 and consult with the North American Insulation Manufacturers Association (NAIMA) and its member companies.

CALCULATING HEAT FLOW FOR BURIED PIPELINES

In calculating heat flow to or from buried pipelines, the thermal properties of the soil must be assumed. Table 7 gives the apparent thermal conductivity values of various soil types, and Figure 8 shows the typical trends of apparent soil thermal conductivity with moisture content for various soil types. Table 8 provides ranges of apparent thermal conductivity for various types of rock. Kernsten (1949) also discusses thermal properties of soils. Carslaw and Jaeger (1959) give methods for calculating the heat flow taking place between one or more buried cylinders and the surroundings.

REFERENCES

Adams, L. 1971. Supporting cryogenic equipment with wood. *Chemical Engineering* (May):156-58.

ASTM. 1990. Standard practice for inner and outer diameters of rigid thermal insulation for nominal sizes of pipe and tubing. *Standard* C585-90. American Society for Testing and Materials, W. Conshohoken, PA.

ASTM. 1991. Standard test method for steady-state heat flux measurements and thermal transmission properties by means of the heat flow meter apparatus. *Standard* C 518-91.

ASTM. 1993. Standard test method for steady-state heat flux measurements and thermal transmission properties by means of the guarded-hot-plate apparatus. *Standard* C 177-85 (Revised 1993).

ASTM. 1993. Standard test method for steady-state thermal performance of building assemblies by means of a guarded hot box. *Standard* C 236-89 (Revised 1993).

ASTM. 1995. Standard practice for determination of heat gain or loss and the surface temperatures of insulated pipe and equipment systems by the use of a computer program. *Standard* C 680-89 (Revised 1995).

ASTM. 1996. Standard test method for thermal performance of building assemblies by means of a calibrated hot box. *Standard* C 976-90 (Revised 1996).

Barbour, E., J. Goodrow, J. Kosny, and J.E. Christian. 1994. Thermal performance of steel-framed walls. Prepared for American Iron and Steel Institute by NAHB Research Center.

Bassett, M.R. and H.A. Trethowen. 1984. Effect of condensation on emittance of reflective insulation. *Journal of Thermal Insulation* 8 (October):127.

Carslaw, H.S. and J.C. Jaeger. 1959. Conduction of heat in solids. Oxford University Press, Amen House, London, England, 449.

Dill, R.S., W.C. Robinson, and H.E. Robinson. 1945. Measurements of heat losses from slab floors. National Bureau of Standards. Building Materials and Structures Report, BMS 103.

Economic thickness for industrial insulation. 1976. GPO No. 41-018-001 15-8, Federal Energy Administration, Washington, D.C.

Farouk, B. and D.C. Larson. 1983. Thermal performance of insulated wall systems with metal studs. Proceedings of the 18th Intersociety Energy Conversion Engineering Conference, Orlando, FL.

Farouki, O.T. 1981. Thermal properties of soil. CRREL *Monograph* 81-1, United States Army Corps of Engineers Cold Regions Research and Engineering Laboratory, December.

Fishenden, M. 1962. Tables of emissivity of surfaces. *International Journal of Heat and Mass Transfer* 5:67-76.

Goss, W.P. and R.G. Miller. 1989. Literature review of measurement and prediction of reflective building insulation system performance: 1900-1989. *ASHRAE Transactions* 95(2).

Hooper, F.C. and W.J. Moroz. 1952. The impact of aging factors on the emissivity of reflective insulations. ASTM *Bulletin* (May):92-95.

Hougten, F.C., S.I. Taimuty, C. Gutberlet, and C.J. Brown. 1942. Heat loss through basement walls and floors. *ASHVE Transactions* 48:369.

Joy, F.A. 1958. Improving attic space insulating values. *ASHAE Transactions* 64:251.

Kersten, M.S. 1949. Thermal properties of soils. University of Minnesota, Engineering Experiment Station Bulletin 28, June.

Kosny, J. and J.E. Christian. 1995. Reducing the uncertainties associated with using the ASHRAE zone method for R-value calculations of metal frame walls. *ASHRAE Transactions* 101(2).

Latta, J.K. and G.G. Boileau. 1969. Heat losses from house basements. *Canadian Building* 19(10).

Lewis, W.C. 1967. Thermal conductivity of wood-base fiber and particle panel materials. Forest Products Laboratory, Research Paper FPL 77, June.

Lotz, W.A. 1964. Vapor barrier design, neglected key to freezer insulation effectiveness. *Quick Frozen Foods* (November):122.

MacLean, J.D. 1941. Thermal conductivity of wood. *ASHVE Transactions* 47:323.

McElroy, D.L., D.W. Yarbrough, and R.S. Graves. 1987. Thickness and density of loose-fill insulations after installation in residential attics. *Thermal insulation: Materials and systems.* F.J. Powell and S.L. Matthews, eds. ASTM STP 922:423-505.

McGowan, A. and A.O. Desjarlais. 1997. An investigation of common thermal bridges in walls. *ASHRAE Transactions* 103(2).

McIntyre, D.A. 1984. The increase in U-value of a wall caused by mortar joints, ECRC/M1843. The Electricity Council Research Centre, Copenhurst, England, June.

Mitalas, G.P. 1982. Basement heat loss studies at DBR/NRC, NRCC 20416. Division of Building Research, National Research Council of Canada, September.

Mitalas, G.P. 1983. Calculation of basement heat loss. *ASHRAE Transactions* 89(1B):420.

Prangnell, R.D. 1971. The water vapor resistivity of building materials—A literature survey. *Materiaux et Constructions* 4:24 (November).

Robinson, H.E., F.J. Powell, and L.A. Cosgrove. 1957. Thermal resistance of airspaces and fibrous insulations bounded by reflective surfaces. National Bureau of Standards, Building Materials and Structures Report BMS 151.

Robinson, H.E., F.J. Powlitch, and R.S. Dill. 1954. The thermal insulation value of airspaces. Housing and Home Finance Agency, Housing Research Paper No. 32.

Sabine, H.J., M.B. Lacher, D.R. Flynn, and T.L. Quindry. 1975. Acoustical and thermal performance of exterior residential walls, doors and windows. *NBS Building Science Series* 77. National Institute of Standards and Technology, Gaithersburg, MD.

Salomone, L.A. and J.I. Marlowe. 1989. Soil and rock classification according to thermal conductivity: Design of ground-coupled heat pump systems. EPRI CU-6482, Electric Power Research Institute, August.

Shipp, P.H. 1983. Basement, crawlspace and slab-on-grade thermal performance. Proceedings of the ASHRAE/DOE Conference, Thermal Performance of the Exterior Envelopes of Buildings II, ASHRAE SP 38:160-79.

Shu, L.S., A.E. Fiorato, and J.W. Howanski. 1979. Heat transmission coefficients of concrete block walls with core insulation. Proceedings of the ASHRAE/DOE-ORNL Conference, Thermal Performance of the Exterior Envelopes of Buildings, ASHRAE SP 28:421-35.

Tye, R.P. 1985. Upgrading thermal insulation performance of industrial processes. *Chemical Engineering Progress* (February):30-34.

Tye, R.P. 1986. Effects of product variability on thermal performance of thermal insulation. Proceedings of the First Asian Thermal Properties Conference, Beijing, People's Republic of China.

Tye, R.P. and A.O. Desjarlais. 1983. Factors influencing the thermal performance of thermal insulations for industrial applications. *Thermal insulation, materials, and systems for energy conservation in the '80s.* F.A. Govan, D.M. Greason, and J.D. McAllister, eds. ASTM STP 789:733-48.

Tye, R.P. and S.C. Spinney. 1980. A study of various factors affecting the thermal performance of perlite insulated masonry construction. Dynatech Report No. PII-2. Holometrix, Inc. (formerly Dynatech R/D Company), Cambridge, MA.

USDA. 1974. *Wood handbook.* Wood as an engineering material. Forest Products Laboratory, U.S. Department of Agriculture Handbook No. 72, Tables 3-7 and 4-2, and Figures 3-4 and 3-5.

Valore, R.C. 1980. Calculation of U-values of hollow concrete masonry. American Concrete Institute, *Concrete International* 2(2):40-62.

Valore, R.C. 1988. Thermophysical properties of masonry and its constituents, Parts I and II. International Masonry Institute, Washington, D.C.

Valore, R., A. Tuluca, and A. Caputo. 1988. Assessment of the thermal and physical properties of masonry block products (ORNL/Sub/86-22020/1), September.

Van Geem, M.G. 1985. Thermal transmittance of concrete block walls with core insulation. *ASHRAE Transactions* 91(2).

Wilkes, K.E. 1979. Thermophysical properties data base activities at Owens-Corning Fiberglas. Proceedings of the ASHRAE/DOE-ORNL Conference, Thermal Performance of the Exterior Envelopes of Buildings, ASHRAE SP 28:662-77.

Yarbrough, E.W. 1983. Assessment of reflective insulations for residential and commercial applications (ORNL/TM-8891), October.

Yellott, J.I. 1965. Thermal and mechanical effects of solar radiation on steel doors. *ASHRAE Transactions* 71(2):42.

CHAPTER 25

VENTILATION AND INFILTRATION

PROVIDING a comfortable and healthy indoor environment for building occupants is the primary concern of HVAC engineers. Comfort and indoor air quality (IAQ) are dependent on many factors, including thermal regulation, control of internal and external sources of pollutants, supply of acceptable air, removal of unacceptable air, occupants' activities and preferences, and proper operation and maintenance of building systems. Ventilation and infiltration are only part of the acceptable indoor air quality and thermal comfort problem. HVAC designers, occupants, and building owners must be aware of and address the many other factors as well. Reducing ventilation and infiltration to solve thermal comfort problems and to reduce energy consumption can affect indoor air quality and may be against code, so such procedures should be approached with care and be under the direction of a registered professional engineer with expertise in HVAC analysis and design.

HVAC design engineers and others concerned with building ventilation and indoor air quality should obtain a copy of ASHRAE *Standard* 62, Ventilation for Acceptable Indoor Air Quality. This standard, like all ASHRAE standards, is reviewed regularly and contains ventilation design and evaluation requirements for commercial and residential buildings. In the design of a new building or the analysis of an existing building, the version of *Standard* 62 that has been adopted by the local code authority must be determined. An existing building may be required to meet current code, or it may be grandfathered under an older code. If a project involves infiltration in residences, then ASHRAE *Standards* 119 and 136 should be consulted. The last chapter of each year's *ASHRAE Handbook* (Chapter 38 of this volume) has a list of current standards.

This chapter focuses on commercial and institutional buildings, where ventilation concerns usually dominate, and on single- and multifamily residences, where infiltration is important. The basic concepts and terminology for both are presented before more advanced analytical and design techniques are given. Ventilation of industrial buildings is covered in Chapter 24 of the 1995 *ASHRAE Handbook—Applications*. However, many of the fundamental ideas and terminology covered in this chapter can also be applied to industrial buildings.

BASIC CONCEPTS AND TERMINOLOGY

Outdoor air that flows through a building is often used to dilute and remove indoor air contaminants. However, the energy required to condition this outdoor air can be a significant portion of the total space-conditioning load. The magnitude of the outdoor airflow into the building must be known for proper sizing of the HVAC equipment and evaluation of energy consumption. For buildings without mechanical cooling and dehumidification, proper ventilation and infiltration airflows are important for providing comfort for occupants. Additionally, the flow of air into buildings and between zones will affect fires and the movement of smoke. Smoke management is addressed in Chapter 48 of the 1995 *ASHRAE Handbook—Applications*.

Ventilation and Infiltration

Air exchange of outdoor air with the air already in a building can be divided into two broad classifications: ventilation and infiltration. **Ventilation** is the intentional introduction of air from the outside into a building; it is further subdivided into natural ventilation and forced ventilation. **Natural ventilation** is the intentional flow of air through open windows, doors, grilles, and other planned building envelope penetrations, and it is driven by natural and/or artificially produced pressure differentials. **Forced ventilation**, shown in Figure 1, is the intentional movement of air into and out of a building using fans and intake and exhaust vents; it is also called **mechanical ventilation**.

Infiltration is the uncontrolled flow of outdoor air into a building through cracks and other unintentional openings and through the normal use of exterior doors for entrance and egress. Infiltration is also known as **air leakage** into a building. **Exfiltration**, depicted in Figure 1, is the leakage of indoor air out of a building. Like natural ventilation, infiltration and exfiltration are driven by natural and/or artificial pressure differences. These forces are discussed in detail in the section on Driving Mechanisms. *Transfer air* (*ta*) is air that moves from one interior space to another, either intentionally or not.

These modes of air exchange differ significantly in how they affect energy, air quality, and thermal comfort, and they can each vary with weather conditions, building operation, and use. Although one mode may be expected to dominate in a particular building, all must be considered for the proper design and operation of an HVAC system.

Modern commercial and institutional buildings are normally required to have forced ventilation and are usually pressurized somewhat to reduce or eliminate infiltration. Forced ventilation has

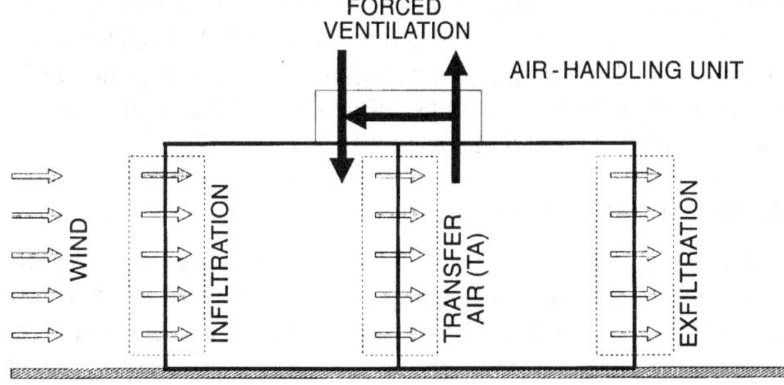

Fig. 1 Two-Space Building with Forced Ventilation, Infiltration, and Exfiltration

The preparation of this chapter is assigned to TC 4.3, Ventilation Requirements and Infiltration.

the greatest potential for control of air exchange when the system is properly designed, installed, and operated; it should provide acceptable indoor air quality when ASHRAE *Standard* 62 requirements are followed. Forced ventilation equipment and systems are described in Chapters 1, 2, and 9 of the 1996 *ASHRAE Handbook—Systems and Equipment*.

In commercial and institutional buildings, uncontrolled natural ventilation, such as through operable windows, may not be desirable from the point of view of energy conservation and comfort. In commercial and institutional buildings with mechanical cooling and forced ventilation, an air- or water-side economizer cycle may be preferable to operable windows for taking advantage of cool outdoor conditions when interior cooling is required. Infiltration may be significant in commercial and institutional buildings, especially in tall, leaky, or underpressurized buildings.

In most of the United States, residential buildings typically rely on infiltration and natural ventilation to meet their ventilation needs (ASHRAE 1989). Infiltration is not reliable for ventilation purposes because it depends on weather conditions, building construction, and maintenance. Natural ventilation, usually through operable windows, is dependent on weather and building design but allows occupants to control airborne contaminants and interior air temperature. However, natural ventilation can have a substantial energy cost if used while the residence's heating or cooling equipment is operating.

In place of operable windows, small exhaust fans may be provided for localized venting in residential spaces such as kitchens and bathrooms. Not all local building codes require that the exhaust be vented to the outside. Instead, the code may allow the air to be treated and returned to the space or to be discharged to an attic space. Poor maintenance of these treatment devices can make nonducted vents ineffective for ventilation purposes. Condensation in attics should be avoided. In northern Europe and in Canada, some building codes require general forced ventilation in residences, and heat recovery heat exchangers are popular for reducing the energy impact. Residential buildings with low rates of infiltration and natural ventilation require forced ventilation at rates given in ASHRAE *Standard* 62.

Airflow

Figure 2 shows a simple **air-handling unit** (AHU) that conditions air for a building. Air brought back to the air handler from the conditioned space is *return air* (*ra*). The portion of the return air that is discharged to the environment is *exhaust air* (*ea*), and the part of the return air that is reused is *recirculated air* (*ca*). Air brought in intentionally from the environment is *outdoor* or *outside air* (*oa*). Because outside air may need treatment to be acceptable for use in a building, it should not be called "fresh air." The outside air and the recirculated air are combined to form *mixed air* (*ma*), which is then conditioned and delivered to the thermal zone as *supply air* (*sa*). Any portion of the mixed air that intentionally or unintentionally circumvents conditioning is *bypass air* (*ba*). *Ventilation air* is air used to provide acceptable indoor air quality. It may be composed of forced or natural ventilation air, infiltration air, suitably treated recirculated air, transfer air, or an appropriate combination (ASHRAE 1989). Due to the wide variety of air-handling systems, these airflows may not all be present in a particular system as defined here. Also, more complex systems may have additional airflows.

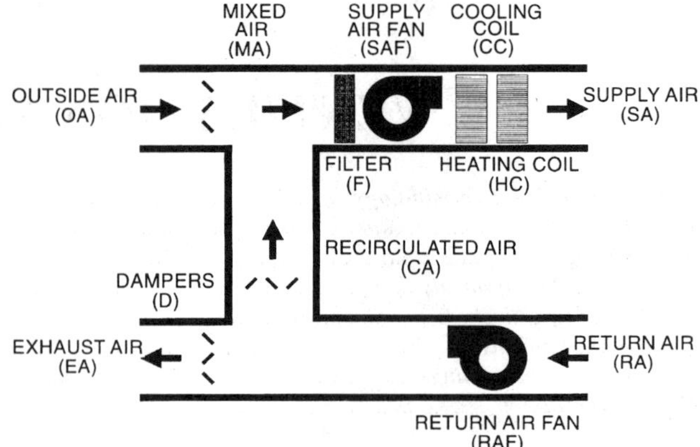

Fig. 2 Simple All-Air Air-Handling Unit with Associated Airflows

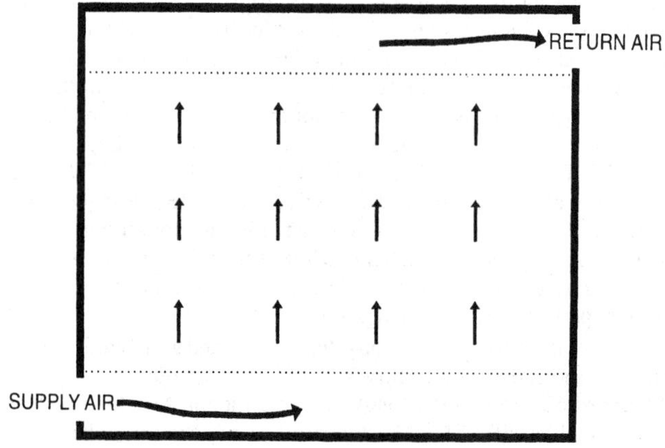

Fig. 3 Displacement Flow Within a Space

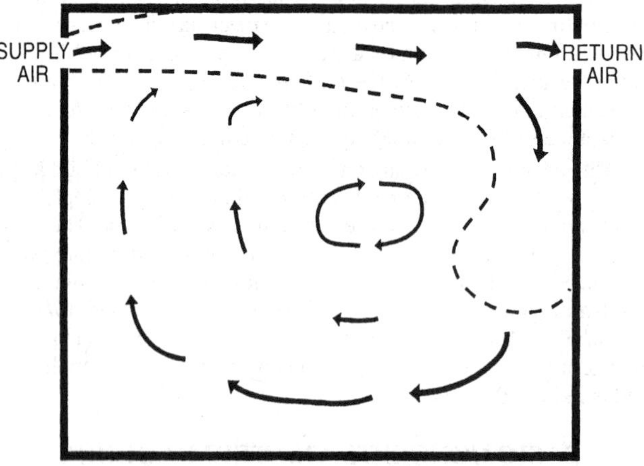

Fig. 4 Entrainment Flow Within a Space

Room Air Movement

Air movement within spaces affects the diffusion of ventilation air and therefore indoor air quality and comfort. Two distinct flow patterns are commonly used to characterize air movement in rooms: displacement flow and entrainment flow. **Displacement flow**, shown in Figure 3, is the movement of air within a space in a piston- or plug-type motion. No mixing of the room air occurs in ideal displacement flow, which is desirable for removing pollutants generated within a space. A laminar flow air distribution system that sweeps air across a space may produce displacement flow.

Entrainment flow, shown in Figure 4, is also known as *conventional mixing*. Systems with ceiling-based supply air diffusers and return air grilles are common examples of air distribution systems that produce entrainment flow. Entrainment flow with very poor mixing within the room has been called short-circuiting flow because much of the supply air leaves the room without mixing with the room air. There is little evidence that properly designed, installed, and operated air distribution systems exhibit short-circuiting behavior. There is some evidence that poorly design, installed, or operated systems can exhibit short-circuiting behavior, especially ceiling-based systems in the heating mode (Offermann and Int-Hout 1989).

Perfect mixing occurs when supply air is instantly and evenly distributed throughout a space. Perfect mixing is also known as *complete or uniform mixing*; the air may be called *well stirred* or *well mixed.* This theoretical performance is approached by entrainment flow systems that have good mixing and by displacement flow systems that allow too much mixing (Rock et al. 1995). The outdoor air requirements given in Table 2 of *Standard* 62 assume delivery of ventilation air with perfect mixing within spaces (ASHRAE 1989). For more detailed information on space air diffusion, see Chapter 31.

Air Exchange Rate

The **air exchange rate** I compares airflow to volume and is

$$I = \frac{Q}{V} \tag{1}$$

where

Q = volumetric flow rate of air into space, cfm
V = interior volume of space, ft^3

The air exchange rate has units of 1/time. When the time unit is hours, the air exchange rate is also called **air changes per hour (ACH)**. The air exchange rate may be defined for several different situations. For example, the air exchange rate for an entire building or zone served by an air-handling unit compares the amount of outside air brought into the building or zone to the total interior volume. This *nominal air exchange rate* I_N is

$$I_N = \frac{Q_{oa}}{V} \tag{2}$$

where Q_{oa} is the outdoor airflow rate including ventilation and infiltration. The nominal air exchange rate describes the outside air ventilation rate entering a building or zone. It does not describe recirculation or the distribution of the ventilation air to each space within a building or zone.

For a particular space, the *space air exchange rate* I_S compares the supply airflow rate Q_{sa} to the volume of that space:

$$I_S = \frac{Q_{sa}}{V} \tag{3}$$

The space air exchange rate for a particular space or zone includes recirculated as well as outside air in the supply air, and it is used frequently in the evaluation of supply air diffuser performance and space air mixing.

Time Constants

Time constants τ, which have units of time (usually in hours or seconds), are also used to describe ventilation and infiltration. One time constant is the time required for one air change in a building, zone, or space if ideal displacement flow existed. It is the inverse of the air exchange rate:

$$\tau = \frac{1}{I} = \frac{V}{Q} \tag{4}$$

The *nominal time constant* compares the interior volume of a building or zone to the volumetric outdoor airflow rate:

$$\tau_N = \frac{V}{Q_{oa}} \tag{5}$$

Like the nominal air exchange rate, the nominal time constant does not describe recirculation of air within a building or zone. It also does not characterize the distribution of the outside air to individual spaces within a building or zone.

The *space time constant* compares the interior volume of a particular space to the total supply airflow rate to that space. The space time constant is the inverse of the space air exchange rate:

$$\tau_S = \frac{V}{Q_{sa}} \tag{6}$$

The space time constant includes the effect of recirculated air, if present, as well as that of outside air introduced to the space through the supply air. If infiltration is significant in a space, then the infiltration flow rate should be included when determining both the space air exchange rate and the space time constant.

Outside Air Fraction

The outside airflow being introduced to a building or zone by an air-handling unit can also be described by the *outside air fraction* X_{oa}, which is the ratio of the volumetric flow rate of outside air brought in by the air handler to the total supply airflow rate:

$$X_{oa} = \frac{Q_{oa}}{Q_{sa}} = \frac{Q_{oa}}{Q_{ma}} = \frac{Q_{oa}}{Q_{oa} + Q_{ca}} \tag{7}$$

When expressed as a percentage, the outside air fraction is called the *percent outside air.* The design outside airflow rate for a building's ventilation system is found through evaluating the requirements of ASHRAE *Standard* 62. The supply airflow rate is that required to meet the thermal load. The outside air fraction and percent outside air then describe the degree of recirculation, where a low value indicates a high rate of recirculation, and a high value shows little recirculation. Conventional all-air air-handling systems for commercial and institutional buildings have approximately 10 to 40% outside air.

100% outside air means no recirculation of return air through the air-handling system. Instead, all the supply air is treated outside air, also known as *makeup air* (*ka*), and all return air is discharged directly to the outside as *relief air* (*la*). An air-handling unit that provides 100% outside air is typically called a **makeup air unit** (MAU).

THERMAL LOADS

Outdoor air introduced into a building constitutes a large part of the total space-conditioning (heating, cooling, humidification, and dehumidification) load, which is one reason to limit air exchange rates in buildings to the minimum required. Air exchange typically represents 20 to 40% of a shell-dominated building's thermal load. Chapter 27 and Chapter 28 cover thermal loads in more detail.

Air exchange increases a building's thermal load in three ways. First, the incoming air must be heated or cooled from the outdoor air temperature to the indoor air temperature. The rate of energy consumption due to this sensible heating or cooling is given by

$$q_s = 60 Q \rho c_p \Delta t \tag{8}$$

where

q_s = sensible heat load, Btu/h
Q = airflow rate, cfm
ρ = air density, lb$_m$/ft^3 (about 0.075)
c_p = specific heat of air, Btu/lb·°F (about 0.24)
Δt = indoor-outdoor temperature difference, °F

Second, air exchange modifies the moisture content of the air in a building. This is particularly important in some locations in the summer when the outdoor air must be dehumidified. In the winter, when the relative humidity of the indoor air is below 30%, humidification may be needed. The rate of energy consumption associated with these latent loads is given by

$$q_l = 60Q\rho h_{fg}\Delta W \quad (9)$$

where

q_l = latent heat load, Btu/h
h_{fg} = latent heat of vapor at appropriate air temperature, Btu/lb_m (about 1000)
ΔW = humidity ratio of indoor air minus humidity ratio of outdoor air, lb_m water/lb_m dry air

Finally, air exchange can increase a building's thermal load by decreasing the performance of the envelope insulation system. Air flowing around and through the insulation can raise heat transfer rates above design rates. The effect of such airflow on insulation system performance is difficult to quantify but should be considered. Airflow within the insulation system can also decrease the system's performance due to moisture condensation in and on the insulation.

Infiltration Degree-Days

Heating and cooling degree-days are a simple way to characterize the severity of a particular climate. Heating and cooling degree-day values are based on sensible temperature data, but infiltration loads are both sensible and latent. *Infiltration degree-days (IDDs)* more fully describe a climate and can be used to estimate heat loss or gain due to infiltration in residences (Sherman 1986). Total infiltration degree-days is the sum of the heating and cooling infiltration degree-days and is calculated from hour-by-hour weather data and base conditions using the following equation:

$$\text{IDD} = \frac{1}{24}\sum_{hours}^{t_{oa}<t_b}\frac{Q}{Q_o}(t_b - t_{oa}) + \frac{1}{24c_p}\sum_{hours}^{\substack{h_{oa}>h_b \\ t_{oa}>t_b}}\frac{Q}{Q_o}(h_{oa} - h_b) \quad (10)$$

where

IDD = total infiltration degree-days, °F·day
Q = infiltration flow rate as calculated from Equation (46), cfm
Q_o = reference infiltration flow rate, cfm
t_b = base air temperature (variable), °F
t_{oa} = hourly temperature of outside air, °F
c_p = specific heat of air, Btu/lb·°F (about 0.24)
h_{oa} = hourly specific enthalpy of outside air, Btu/lb
h_b = base air enthalpy (variable), Btu/lb

The selection of h_b and t_b is an important part of the calculation of the IDDs. ASHRAE *Standard* 119 lists IDDs for many locations with a particular set of base conditions.

INDOOR AIR QUALITY

Outdoor air requirements for acceptable indoor air quality (IAQ) have long been debated, and different rationales have produced radically different ventilation standards (Grimsrud and Teichman 1989, Janssen 1989, Klauss et al. 1970, Yaglou et al. 1936, Yaglou and Witheridge 1937). Historically, the major considerations have included the amount of outdoor air required to control moisture, carbon dioxide (CO_2), odors, and tobacco smoke generated by occupants. These considerations have led to prescriptions of a minimum rate of outdoor air supply per occupant. More recently, the maintenance of acceptable indoor concentrations of a variety of additional pollutants that are not generated primarily by occupants has been a major concern. The most common pollutants of concern and their sources are presented in Table 1. Additional information on contaminants can be found in Chapter 12; odors are covered in Chapter 13.

Table 1 Indoor Air Pollutants and Sources

Sources	Pollutant Types
OUTDOOR	
Ambient air	SO_2, NO, NO_2, O_3, hydrocarbons, CO, particulates, bioaerosols
Motor vehicles	CO, Pb, hydrocarbons, particulates
Soil	Radon, organics
INDOOR	
Building construction materials	
Concrete, stone	Radon
Particleboard, plywood	Formaldehyde
Insulation	Formaldehyde, fiberglass
Fire retardant	Asbestos
Adhesives	Organics
Paint	Mercury, organics
Building contents	
Heating and cooking combustion appliances	CO, NO, NO_2, formaldehyde, particulates, organics
Furnishings	Organics
Water service; natural gas	Radon
Human occupants	
Metabolic activity	H_2O, CO_2, NH_3, odors
Human activities	
Tobacco smoke	CO, NO_2, organics, particulates, odors
Aerosol spray devices	Vinyl chloride, organics
Cleaning and cooking products	Organics, NH_3, odors
Hobbies and crafts	Organics
Damp organic materials, stagnant water	Bioaerosols
Coil drain pans	
Humidifiers	

Indoor pollutant concentrations depend on the strength of pollutant sources and the total rate of pollutant removal. Pollutant sources include the outdoor air; indoor sources such as occupants, furnishings, and appliances; and the soil adjacent to the building. Pollutant removal processes include dilution with outside air, local exhaust ventilation, deposition on surfaces, chemical reactions, and air-cleaning processes. If (1) general building ventilation is the only significant pollutant removal process, (2) the indoor air is thoroughly mixed, and (3) the pollutant source strength and ventilation rate have been stable for a sufficient period; then the steady-state indoor pollutant concentration is given by

$$C_i = C_o + S/Q_{oa} \quad (11)$$

where

C_i = steady-state indoor concentration, μg/ft^3
C_o = outdoor concentration, μg/ft^3
S = total pollutant source strength, μg/min
Q_{oa} = ventilation rate, cfm

Variation in pollutant source strengths (rather than variation in ventilation rate) is considered the largest cause of building-to-building variation in the concentrations of pollutants that are not generated by occupants. Turk et al. (1989) found that a lack of correlation between average indoor respirable particle concentrations and whole-building outdoor ventilation rate indicated that source strength, high outdoor concentrations, building volume, and removal processes are important. Because pollutant source strengths are highly variable, maintenance of minimum ventilation rates does not ensure acceptable indoor air quality in all situations. The lack of health-based concentration standards for many indoor air pollutants, primarily due to the lack of health data, makes the specification of minimum ventilation rates even more difficult.

Regardless of these complexities and uncertainties, designers and operators need guidance on ventilation and indoor air quality.

ASHRAE *Standard* 62 provides guidance on ventilation and indoor air quality in the form of two alternative procedures. In the **Ventilation Rate Procedure**, indoor air quality is assumed to be acceptable if (1) the concentrations of six pollutants in the incoming outdoor air meet the United States national ambient air quality standards, and (2) the outdoor air supply rates meet or exceed values (which vary depending on the type of space) provided in a table. The minimum outside air supply per person for any type of space is 15 cfm. This minimum rate will maintain an indoor CO_2 concentration below 0.1% (1000 parts per million) assuming a typical CO_2 generation rate per occupant (Janssen 1989). This minimum outside air supply rate was based, in part, on research by Berg-Munch et al. (1986), which indicated that 15 cfm was required to satisfy the odor perceptions of 80% or more of visitors.

The second alternative in *Standard* 62 is the **Indoor Air Quality Procedure**. In this procedure, any outside air supply rate is acceptable if (1) the indoor concentrations of nine pollutants are maintained below specified values, and (2) the air is deemed acceptable via subjective evaluations of odor. If users of the IAQ Procedure control pollutant source strengths or use air cleaning or local exhaust ventilation, they may be able to reduce the outside air supply rates to below those specified in the ventilation rate procedure. Modest energy savings may result, as described by Eto and Meyer (1988) and Eto (1990). However, the maximum acceptable CO_2 concentration of 0.1% in the IAQ Procedure effectively limits the minimum ventilation rate to 15 cfm per occupant, unless CO_2 is removed by air cleaning, which is generally considered impractical.

In cases of high contaminant source strengths, impractically high rates of ventilation are required to control contaminant levels, and other methods of control are more effective. Removal or reduction of contaminant sources is a very effective means of control. Controlling a localized source by means of local exhaust, such as range hoods or bathroom exhaust fans, can also be effective.

Particles can be removed with various types of air filters. Gaseous contaminants with higher molecular weight can be controlled with activated carbon or alumina pellets impregnated with a substance such as potassium permanganate. Chapter 24 of the 1996 *ASHRAE Handbook—Systems and Equipment* has information on air cleaning.

A combination of source control and local exhaust, as opposed to dilution with ventilation air, is the method of choice in industrial environments. Indoor air quality problems and methods of control are covered in the proceedings of annual IAQ conferences sponsored by ASHRAE. Industrial ventilation is discussed in Chapters 24 and 26 of the 1995 *ASHRAE Handbook—Applications* and in *Industrial Ventilation: A Manual of Recommended Practice* (ACGIH 1995).

TRACER GAS MEASUREMENTS

The only reliable way to determine the air exchange rate of an existing building is to measure it. Several tracer gas measurement procedures exist, all involving an inert or nonreactive gas used to label the indoor air (Hunt 1980; Sherman et al. 1980; Harrje et al. 1981; Lagus and Persily 1985; Dietz et al. 1986; Charlesworth 1988; Persily 1988; Fisk et al. 1989; Lagus 1989; Sherman 1989a, 1989b; Fortmann et al. 1990; Harrje et al. 1990; Persily and Axley 1990; Sherman 1990). The tracer is released into the building in a specified manner, and the concentration of the tracer within the building is monitored and related to the building's air exchange rate. A variety of tracer gases and associated concentration detection devices have been used. Desirable qualities of a tracer gas are detectability, nonreactivity, nontoxicity, neutral buoyancy, relatively low concentration in ambient air, and low cost (Hunt 1980).

All tracer gas measurement techniques are based on a mass balance of the tracer gas within the building. Assuming the outdoor concentration is zero and the indoor air is well mixed, this mass balance takes the following form:

$$V\left(\frac{dC}{d\theta}\right) = F(\theta) - Q(\theta)C(\theta) \tag{12}$$

where

V = volume of space being tested, ft^3
$C(\theta)$ = tracer gas concentration at time θ
$dC/d\theta$ = time rate of change of concentration, min^{-1}
$F(\theta)$ = tracer gas injection rate at time θ, cfm
$Q(\theta)$ = airflow rate out of building at time θ, cfm
θ = time, min

In Equation (12), density differences between indoor and outdoor air are generally ignored for moderate climates; therefore, Q also refers to the airflow rate into the building. While Q is often referred to as the infiltration rate, any measurement includes both mechanical and natural ventilation in addition to envelope infiltration. The ratio of Q to the volume V being tested has units of 1/time (often converted to ACH) and is the air exchange rate I.

Equation (12) is based on the assumptions that (1) no unknown tracer gas sources exist; (2) the airflow out of the building is the dominant means of removing the tracer gas from the space (i.e., the tracer gas does not react chemically within the space and/or is not adsorbed onto interior surfaces), and (3) the tracer gas concentration within the building can be represented by a single value (i.e., the tracer gas is uniformly mixed within the space).

Three different tracer gas procedures are used to measure air exchange rates: (1) decay or growth, (2) constant concentration, and (3) constant injection.

Decay or Growth

Decay. The simplest tracer gas measurement technique is the decay method (also known as the step-down method), which is a standardized procedure (ASTM *Standard* E 741). A small amount of tracer gas is injected into the space and is allowed to mix with the interior air. After the injection, $F = 0$ and the solution to Equation (12) is

$$C(\theta) = C_o e^{-I\theta} \tag{13}$$

where C_o is the concentration at $\theta = 0$.

Equation (13) is generally used to solve for I by measuring the tracer gas concentration periodically during the decay and fitting the data to the logarithmic form of Equation (13):

$$\ln C(\theta) = \ln C_o - I\theta \tag{14}$$

Like all tracer gas techniques, the tracer gas decay method has advantages and disadvantages. One advantage is the fact that Equation (13) is an exact solution to the tracer gas mass balance equation. Also, because logarithms of concentration are taken, only relative concentrations are needed, which can simplify the calibration of the concentration-measuring equipment. Finally, the tracer gas injection rate need not be measured, although it must be controlled so that the tracer gas concentrations are within the range of the concentration-measuring device. The concentration-measuring equipment can be located on site, or building samples can be collected in suitable containers and analyzed elsewhere.

The most serious problem with the decay technique is imperfect mixing of the tracer gas with the interior air, both at initial injection and during the decay. Equations (12) and (13) employ the assumption that the tracer gas concentration within the building is uniform. If the tracer is not well mixed, this assumption is not appropriate and the determination of I will be subject to errors. It is difficult to estimate the magnitude of the errors due to poor mixing, and little analysis of this problem has been performed.

Growth. The growth or step-up method is similar to the decay method except that the initial tracer gas concentration is low and is increased during the test.

Constant Concentration

In the constant concentration technique, the tracer gas injection rate is adjusted to maintain a constant concentration within the building. If the concentration is truly constant, then Equation (12) reduces to

$$Q(\theta) = F(\theta)/C \tag{15}$$

There is less experience with this technique than with the decay procedure, but several applications exist (Kumar et al. 1979, Collet 1981, Bohac et al. 1985, Fortmann et al. 1990).

Because the tracer gas injection is continuous, no initial mixing period is required. Another advantage is that the tracer gas injection into each zone of the building can be separately controlled; thus, the amount of outdoor air flowing into each zone can be determined. This procedure has the disadvantage of requiring the measurement of absolute tracer concentrations and injection rates. Also, imperfect mixing of the tracer and the interior air causes a delay in the response of the concentration to changes in the injection rate. This delay in concentration response makes it impossible to keep the concentration constant, so Equation (15) is only an approximation. The magnitude of the error has not been well examined.

Constant Injection

In the constant injection procedure, the tracer is injected at a constant rate, and the solution to Equation (12) becomes

$$C(\theta) = (F/Q)(1 - e^{-I\theta}) \tag{16}$$

After sufficient time, the transient term reduces to zero, the concentration attains equilibrium, and Equation (16) reduces to

$$Q = F/C \tag{17}$$

Equation (17) is valid only when air exchange rate I and airflow rate Q are constant; thus, this technique is appropriate for systems at or near equilibrium. It is particularly useful in spaces with mechanical ventilation or with high air exchange rates. Constant injection requires the measurement of absolute concentrations and injection rates.

Dietz et al. (1986) introduced a special case of the constant injection technique. This technique uses permeation tubes as a tracer gas source. The tubes release the tracer at an ideally constant rate into the building being tested. A sampling tube packed with an adsorbent collects the tracer from the interior air at a constant rate by diffusion. After a sampling period of one week or more, the sampler is removed and analyzed to determine the average tracer gas concentration within the building during the sampling period.

Solving Equation (12) for C and taking the time average gives

$$\langle C\rangle = \langle F/Q\rangle = F\langle 1/Q\rangle \tag{18}$$

where < ... > denotes time average. (Note that the time average of $dC/d\theta$ is assumed to equal zero.)

Equation (18) shows that the average tracer concentration $\langle C\rangle$ and the injection rate F can be used to calculate the average of the inverse airflow rate. The average of the inverse is less than the inverse of the actual average, with the magnitude of this difference depending on the distribution of airflow rates during the measurement period. Sherman and Wilson (1986) calculated these differences to be about 20% for one-month averaging periods. Differences greater than 30% have been measured when there were large changes in air exchange rate due to occupant airing of houses; errors from 5 to 30% were measured when the variation was due to weather effects (Bohac et al. 1987). Longer averaging periods and large changes in air exchange rates during the measurement periods generally lead to larger differences between the average inverse exchange rate and the inverse of the actual average rate.

Multizone Air Exchange Measurement

Equation (12) is based on the assumption of a single, well-mixed enclosure, and the techniques described are for single-zone measurements. Airflow between internal zones and between the exterior and individual internal zones has led to the development of multizone measurement techniques (Harrje et al. 1985, Sherman and Dickerhoff 1989, Fortmann et al. 1990, Harrje et al. 1990). These techniques are important when considering the transport of pollutants from one room of a building to another. For a theoretical development, see Sinden (1978b). Multizone measurements typically use either multiple tracer gases for the different zones or the constant concentration technique. A proper error analysis is essential in all multizone flow determination (Charlesworth 1988, D'Ottavio et al. 1988).

AIR CHANGE EFFECTIVENESS

Ventilation effectiveness is a description of an air distribution system's ability to remove internally generated pollutants from a building, zone, or space. **Air change effectiveness** is a description of an air distribution system's ability to deliver ventilation air to a building, zone, or space. The HVAC design engineer does not have knowledge or control of actual pollutant sources within buildings, so Table 2 of ASHRAE *Standard* 62 defines outdoor air requirements for typical, expected building uses. For most projects, therefore, the air change effectiveness is of more relevance to HVAC system design than the ventilation effectiveness. Various definitions for air change effectiveness have been proposed. The specific measure that meets the local code requirements must be determined, if any is needed at all. The following definitions and symbols may vary depending on the application.

Age of Air

The age of air θ_{age} (Sandberg 1981) is the length of time that some quantity of outside air has been in a building, zone, or space. The "youngest" air is at the point where outside air enters the building by forced or natural ventilation or through infiltration (Grieve 1989). The "oldest" air may be at some location within the building or in the exhaust air. When the characteristics of the air distribution system are varied, a longer age of air suggests poorer outside air delivery compared to a short age of air for the same location. The age of air has units of time, usually in seconds or minutes, so it is not a true "efficiency" or "effectiveness" measure. The age of air concept, however, has gained wide acceptance in Europe and is used increasingly in North America.

The age of air can be evaluated for existing buildings using tracer gas methods. Using either the decay (step-down) or the growth (step-up) tracer gas method, the zone average or **nominal age of air** $\theta_{age,N}$ can be determined by taking concentration measurements in the exhaust air. The **local age of air** $\theta_{age,L}$ is evaluated through tracer gas measurements at any desired point in a space, such as at a worker's desk. Once time-dependent data of tracer gas concentration are available, the age of air can be calculated from

$$\theta_{age} = \int_{\theta=0}^{\infty} \frac{C_{in} - C}{C_{in} - C_o} d\theta \tag{19}$$

where C_{in} = concentration of tracer gas being injected.

Because evaluation of the age of air requires integration to infinite time, an exponential tail is usually added to the known concentration data (Farrington et al. 1990).

Air Change Effectiveness

Air change effectiveness measures ε_I are nondimensional gages of ventilation air delivery. One common definition of air change effectiveness is the ratio of a time constant to an age of air:

$$\varepsilon_I = \frac{\tau}{\theta_{age}} \tag{20}$$

The **nominal air change effectiveness** $\varepsilon_{I,N}$ shows the effectiveness of outside air delivery to the entire building, zone, or space:

$$\varepsilon_{I,N} = \frac{\tau_N}{\theta_{age,N}} \tag{21}$$

where the nominal time constant τ_N is usually calculated from measured airflow rates.

The **local air change effectiveness** $\varepsilon_{I,L}$ shows the effectiveness of outside air delivery to one specific point in a space:

$$\varepsilon_{I,L} = \frac{\tau_N}{\theta_{age,L}} \tag{22}$$

where τ_N is found either through airflow measurements or from tracer gas concentration data. An $\varepsilon_{I,L}$ value of 1.0 indicates that the air distribution system delivers air equivalent to that of a system with perfectly mixed air in the spaces. A value less than 1.0 shows less than perfect mixing with some degree of stagnation. A value of $\varepsilon_{I,L}$ greater than 1.0 suggests that a degree of plug or displacement flow is present at that point (Rock 1992).

Currently, the HVAC design engineer must assume that a properly designed, installed, operated, and maintained air distribution system provides an air change effectiveness of about 1. Therefore, the Table 2 values of ASHRAE *Standard* 62 are appropriate for the design of commercial, institutional, and residential buildings when the Ventilation Rate Procedure is used. If the Indoor Air Quality Procedure of *Standard* 62 is used, then actual pollutant sources and the air change effectiveness must be known for the successful design of HVAC systems that have fixed ventilation airflow rates.

DRIVING MECHANISMS

Natural ventilation and infiltration are driven by (1) pressure differences across the building envelope caused by wind; (2) air density differences due to temperature differences between indoor and outdoor air (buoyancy, or the stack effect); and (3) the operation of appliances, such as combustion devices, leaky forced-air thermal distribution systems, and mechanical ventilation systems. The indoor-outdoor pressure difference at a location depends on the magnitude of these driving mechanisms as well as on the characteristics of the openings in the building envelope (i.e., their locations and the relationship between pressure difference and airflow for each opening).

Pressure differences across the building envelope are based on the requirement that the mass flow of air into the building equal the mass flow out. In general, density differences between indoor and outdoor air can be neglected, so the volumetric airflow rate into the building equals the volumetric airflow rate out. Based on this assumption, the envelope pressure differences can be determined; however, such a determination requires a great deal of detailed information that is essentially impossible to obtain.

When wind impinges on a building, it creates a distribution of static pressures on the building's exterior surface that depends on the wind direction, wind speed, air density, surface orientation, and surrounding conditions. When no significant shell openings are present, this pressure distribution is independent of the pressure p_i inside the building. If (1) no other forces act on the building, (2) no indoor-outdoor temperature difference exists, and (3) no appliance forces air through the building, the pressure differences are determined by the interior static pressure according to the following equation:

$$\Delta p = p_o + p_w - p_i \tag{23}$$

where

Δp = pressure difference between outdoors and indoors at location, in. of water
p_o = static pressure at reference height in undisturbed flow, in. of water
p_w = wind pressure at location, in. of water
p_i = interior pressure at height of location, in. of water

If no indoor-outdoor temperature difference exists, the interior static pressure p_i decreases linearly with height at a rate dependent on the interior temperature. This rate of pressure decrease equals $-\rho_i g$, where ρ_i is the average interior air density and g is the acceleration due to gravity. The interior static pressure equilibrates to a value such that the total airflow into the building equals the total airflow out of the building. The interior static pressure may be determined by calculating the airflow through each opening as a function of the interior pressure, adding all these airflow rates together, setting this sum equal to zero, and solving for the interior pressure. However, to solve for the interior pressure in this way, the location of each opening in the building envelope, the value of p_w at each opening, and the relationship between airflow rate and pressure difference for each opening must be known.

When an indoor-outdoor temperature difference exists, it imposes a gradient Δp_s on the pressure difference. This stack gradient is a function of height and temperature difference and may be added to the pressure difference due to wind in Equation (23). The pressure difference is now expressed as

$$\Delta p = p_o + p_w - p_{i,r} + \Delta p_s \tag{24}$$

The parameter $p_{i,r}$ is the interior static pressure at some reference height, and this pressure again assumes a value such that the total inflow equals the total outflow. A summation of all the airflows through these openings can be set equal to zero and solved for the interior pressure at the reference height.

When an appliance such as a combustion device, leaky forced-air distribution system, or ventilation fan operates, an additional airflow is imposed on the building. The pressure difference is still calculated using Equation (24), but the interior reference pressure $p_{i,r}$ changes so that the balance between inflow and outflow is maintained. This balance necessarily includes the airflow rate(s) associated with the appliance(s).

To determine the pressure differences across the building envelope and the corresponding air exchange rates, building-specific information about the exterior pressure distribution due to wind and the location of and airflow rate/pressure difference relationship for every opening in the building envelope are needed. These inputs are difficult to obtain for any given building, which makes such a determination unrealistic.

Wind Pressure

Wind pressures are generally positive with respect to the static pressure in the undisturbed airstream on the windward side of a building and negative on the leeward side. Pressures on the other sides are negative or positive, depending on wind angle and building shape. Static pressures over building surfaces are almost proportional to the velocity head of the undisturbed airstream. The wind pressure or velocity head is given by the Bernoulli equation, assuming no height change or pressure losses:

$$p_w = C_1 C_p \rho \frac{V^2}{2} \quad (25)$$

where

p_w = wind surface pressure relative to static pressure in undisturbed flow, in. of water
ρ = air density, lb_m/ft^3 (about 0.075)
V = wind speed, mph
C_p = wind surface pressure coefficient, dimensionless
C_1 = unit conversion factor = 0.0129

Therefore, Equation (23) can be rewritten as

$$\Delta p = p_o + C_p \rho \frac{V^2}{2} - p_i \quad (26)$$

C_p is a function of location on the building envelope and wind direction. Chapter 15 provides additional information on the values of C_p. Although standard conditions are frequently used, the air density and consequently the wind pressure can vary for a given wind speed with changes in temperature or elevation. For example, for an elevation rise from sea level to 5000 ft or an air temperature change from −20 to 70°F, the air density will drop about 20%. If these elevation and temperature changes both apply, the air density will drop about 45%. Therefore, the effects of local air density cannot be ignored.

The wind speed incident on a building is generally lower than the average meteorological wind speed for a region; thus, meteorological data usually overestimate wind pressures on a building. Building wind speeds are lower because of the effects of height, terrain, and shielding (Lee et al. 1980) but could be higher in some situations. The wind speed is zero at the ground surface and increases with height up to an altitude of about 2000 ft above ground level. Meteorological measurements are typically made at a height of 33 ft in open areas. Residential buildings are generally shorter than 33 ft and are therefore subject to lower wind pressures. Tall buildings are subject to a range of wind speeds over the height of the building, exposing the exterior to varying wind pressures that are both lower and higher than estimates based on Equation (25).

A database of surface pressure coefficients from eight investigations of low-rise buildings and one of high-rise buildings was developed by Swami and Chandra (1988). The 544 average surface pressure coefficients from the database were used to develop a relationship (with a correlation coefficient of 0.80) between wind incident angle, building side ratio, and average surface pressure coefficient:

$$\begin{aligned} C_{p,n} &= \ln[1.248 - 0.703\sin(a/2) - 1.175\sin^2(a) \\ &\quad + 0.131\sin^3(2aG) + 0.769\cos(a/2) \\ &\quad + 0.07G^2\sin^2(a/2) + 0.717\cos^2(a/2)] \end{aligned} \quad (27)$$

where

$C_{p,n}$ = normalized C_p
a = angle between wind direction and outward normal of wall under consideration, degrees
G = natural log of ratio of width of wall under consideration to width of adjacent wall

They further noted that uncertainties in the estimation of site wind speed and of the effect of surrounding buildings are likely to be equal to or greater than the uncertainty in estimating C_p from Equation (27).

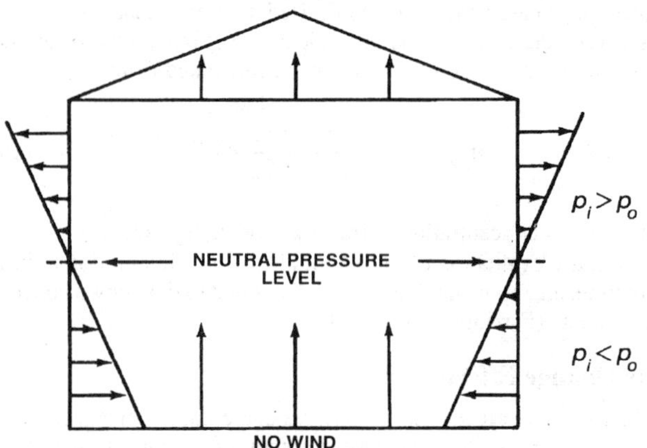

Fig. 5 Pressure Differences Caused by Stack Effect for Typical Structure (Heating)

The shielding effects of trees, shrubbery, and other buildings within several building heights of a particular building produce large-scale turbulence eddies that not only reduce effective wind speed but also alter wind direction. Thus, meteorological wind speed data must be reduced carefully when applied to low buildings. Chapter 15 provides additional guidance on estimating wind pressures.

The magnitude of the pressure differences found on the surfaces of buildings varies rapidly with time because of turbulent fluctuations in the wind (Grimsrud et al. 1979, Etheridge and Nolan 1979). However, the use of average wind pressures to calculate pressure differences is usually sufficient to calculate average infiltration values. For residential buildings, the magnitude of wind pressure differences averaged over 20 min seldom exceeds ±0.02 in. of water under typical conditions. In many cases, the averages are less than ±0.01 in. of water. For tall buildings or buildings completely exposed to open terrain, the pressure on the windward side is much closer to those calculated from average wind speeds for the site (Tamura and Wilson 1968).

Stack Pressure

Temperature differences between indoors and outdoors cause density differences (and therefore pressure differences) that drive infiltration. During the heating season, the warmer inside air rises and flows out of the building near its top. It is replaced by colder outdoor air that enters the building near its base. During the cooling season, the flow directions are reversed and the infiltration rate is generally lower because the indoor-outdoor temperature differences are smaller.

In the heating season, the pressure distribution over the building due to the stack effect takes the qualitative form shown in Figure 5. The height at which the interior and exterior pressures are equal is called the **neutral pressure level** (NPL) (Tamura and Wilson 1966, 1967a). Above this point, the interior pressure is greater than the exterior; below this point, the greater exterior pressure causes airflow into the building.

The pressure difference due to the stack effect at height H is

$$\begin{aligned} \Delta p_s &= C_2(\rho_o - \rho_i)g(H - H_{NPL}) \\ &= C_2\rho_i g(H - H_{NPL})(T_i - T_o)/T_o \end{aligned} \quad (28)$$

where

Δp_s = pressure difference due to stack effect, in. of water
ρ = air density, lb_m/ft^3 (about 0.075 for indoor conditions)

g = gravitational constant, 32.2 ft/s²
H = height of observation, ft
H_{NPL} = height of neutral pressure level, ft
T = average absolute temperature, °R
C_2 = unit conversion factor = 0.00598

Subscripts

i = indoor
o = outdoor

Chastain and Colliver (1989) showed that the average of the vertical distribution of temperature differences $T_i - T_o$ due to stratification rather than the localized temperature difference near the opening is the appropriate temperature difference to use in Equation (28).

A useful estimate is that the pressure difference induced in a building by the stack effect is 2.7×10^{-5} in. of water/ft·°R, neglecting any resistance to airflow within the structure. Therefore, in a one-story house with an 8 ft ceiling, an NPL of one-half the building height, and a temperature difference of 45°R, the stack pressure will be 0.005 in. of water between the ceiling and floor. In a tall building (e.g., 20 stories of 13 ft each) with no internal resistance to airflow, the stack pressure under these same conditions will be 0.16 in. of water. For cold conditions, these values are significantly greater due to the increased outdoor air density.

The location of the NPL at zero wind speed is a structure-dependent parameter that depends only on the vertical distribution of openings in the shell, the resistance of the openings to airflow, and the resistance to vertical airflow within the building. If the openings are uniformly distributed vertically, they have the same resistance to airflow, and there is no internal airflow resistance, the NPL is at the midheight of the building (see Figure 5). If there is only one opening or one opening that is extremely large relative to any others, the NPL is at or near the center of this opening. Foster and Down (1987) studied the location of the NPL as it relates to natural ventilation in a building with only two openings. Lee et al. (1985) studied, both experimentally and analytically, the characteristics of thermal performance of high-rise buildings, using an idealized model building with a number of openings at various locations and temperature distributions. Later, Lee et al. (1988) showed that the ratio of floor opening areas to exterior opening areas had a dominant role in pressure distribution. Chastain and Colliver (1989) investigated the location of the NPL in walls with multiple openings and found that it was independent of the temperature difference and that the mean air density difference across the building envelope should be used in Equation (28).

Internal partitions, stairwells, elevator shafts, utility ducts, chimneys, vents, and mechanical supply and exhaust systems complicate the analysis of NPL location. Chimneys and openings at or above roof height raise the NPL in small buildings. Exhaust systems increase the height of the NPL; outdoor air supply systems lower it.

Available data on the NPL in various kinds of buildings are limited. The NPL in tall buildings varies from 0.3 to 0.7 of total building height (Tamura and Wilson 1966, 1967a). For houses, especially houses with chimneys, the NPL is usually above midheight. Operating a combustion heat source with a flue raises the NPL further, sometimes above the ceiling (Shaw and Brown 1982).

Equation (28) provides a maximum stack pressure difference, given no internal airflow resistance. The sum of the pressure differences across the exterior wall at the bottom and at the top of the building, as calculated by Equation (28), equals the total theoretical draft for the building. The sum of the actual top and bottom pressure differences, divided by the total theoretical draft, equals the **thermal draft coefficient**. The value of the thermal draft coefficient depends on the airflow resistance of the exterior walls relative to the airflow resistance between floors. For a building without internal partitions, the total theoretical draft is achieved across the exterior walls (Figure 6A), and the thermal draft coefficient equals 1. In a building with airtight separations at each floor, each story acts independently, its own stack effect being unaffected by that of any other floor (Figure 6B). The ratio of the actual to the theoretical draft is minimized in this case.

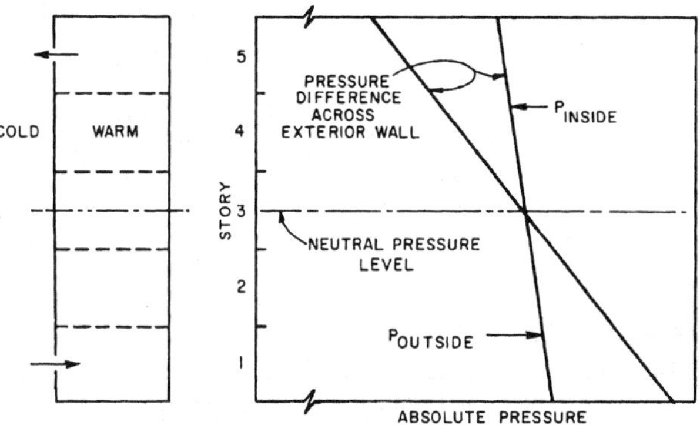

A. BUILDING WITH NO INTERNAL PARTITION

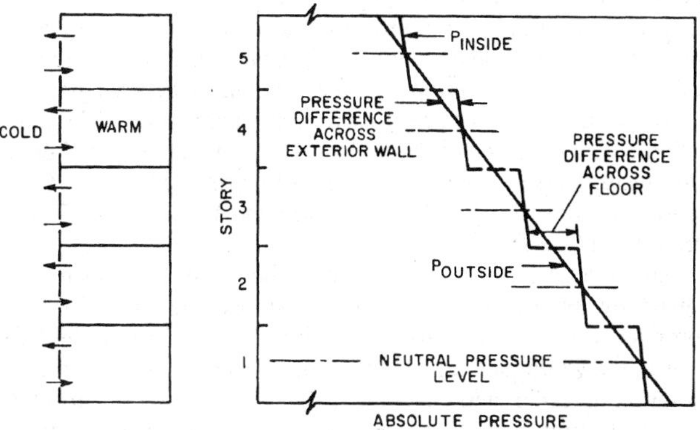

B. BUILDING WITH AIRTIGHT SEPARATION OF EACH STORY

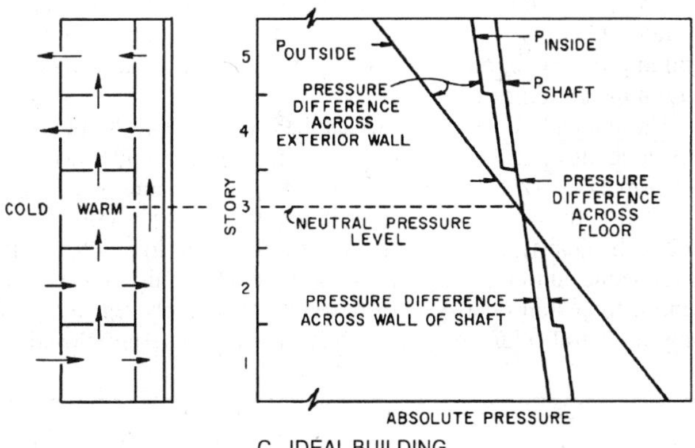

C. IDEAL BUILDING

Fig. 6 Stack Effect in Various Buildings

Real multistory buildings are neither open inside (Figure 6A), nor airtight between stories (Figure 6B). Vertical air passages, stairwells, elevators, and other service shafts allow airflow between floors. Figure 6C represents a heated building with uniform openings in the exterior wall, through each floor, and into the vertical shaft at each story. Between floors, the slope of the line representing the inside pressure is the same as that shown in Figure 6A, and the discontinuity at each floor (Figure 6B) represents the pressure difference across it. Total stack effect for the building remains the same, but some of the total pressure difference maintains flow through openings in the floors and vertical shafts. As a result, the pressure difference across the exterior wall at any level is less than it would be with no internal flow resistance.

Maintaining airtightness between floors and from floors to vertical shafts is a means of controlling indoor-outdoor pressure differences due to the stack effect and therefore infiltration. Good separation is also conducive to the proper operation of mechanical ventilation and smoke management systems. Tamura and Wilson (1967b) showed that when vertical shaft leakage is at least two times the envelope leakage, the thermal draft coefficient is almost 1. Measurements of pressure differences in three tall office buildings by Tamura and Wilson (1967a) indicated that the thermal draft coefficient ranged from 0.8 to 0.9 with the ventilation systems off.

Mechanical Systems

The operation of mechanical equipment, such as ventilation/exhaust systems and vented combustion devices, affects pressure differences across the building envelope. The interior static pressure adjusts such that the airflows through all the openings in the building envelope plus equipment-induced airflows balance to zero. These changes in pressure differences and airflow rates caused by mechanical equipment are unpredictable, unless the location of each opening in the envelope and the relationship between pressure difference and airflow rate for each opening are known. The interaction between mechanical ventilation system operation and envelope airtightness has been discussed for low-rise buildings (Nylund 1980) and for office buildings (Tamura and Wilson 1966, 1967b; Persily and Grot 1985a).

Air exhausted from a building must be balanced by increasing the airflow into the building through other openings. The airflow at some locations changes from outflow to inflow because the NPL rises in the winter and falls in the summer. For supply fans, the situation is reversed, and envelope inflows become outflows. Thus, the effects a mechanical system has on a building must be considered. Depressurization caused by an improperly designed system can increase the rate of radon entry into a building and interfere with the proper operation of combustion device venting or other exhaust systems. Overpressurization can force moist indoor air through the building envelope, and in cold climates, moisture may condense within the building envelope.

The interaction between mechanical systems and the building envelope also pertains to systems serving zones of buildings. The performance of zone-specific exhaust or pressurization systems is affected by the leakage in zone partitions as well as in exterior walls.

Mechanical systems can also create infiltration-driving forces in single-zone buildings. Specifically, some single-family houses with central forced-air duct systems have multiple supply registers, yet only a central return register. When internal doors are closed in these houses, large positive indoor-outdoor pressure differentials are created for rooms with only supply registers, whereas the room with the return duct tends to depressurize relative to outside. This is caused by the resistance of internal door undercuts to flow from the supply register to the return (Modera et al. 1991). The magnitudes of the indoor-outdoor pressure differentials created have been measured to average 0.012 to 0.024 in. of water (Modera et al. 1991).

Building envelope airtightness and interzonal airflow resistance can also affect the performance of mechanical systems. The actual airflow rate delivered by these systems, particularly ventilation systems, depends on the pressure they work against. This effect is the same as the interaction of a fan with its associated ductwork, which is discussed in Chapter 32 of this volume and Chapter 18 of the 1996 *ASHRAE Handbook—Systems and Equipment*. The building envelope and its leakage can be considered part of the ductwork in determining the pressure drop of the system.

Combining Driving Forces

The pressure differences due to wind pressure, stack pressure, and mechanical systems are considered in combination by adding them together and determining the airflow rate through each opening due to this total pressure difference. The driving forces must be combined in this manner, as opposed to adding the airflow rates due to the separate driving forces, because the airflow rate through the openings is not linearly related to pressure difference.

Figure 7 qualitatively shows the addition of driving forces for a building with uniform openings above and below midheight and without significant internal resistance to airflow. The slopes of the pressure lines are a function of the densities of the indoor and outdoor air. In Figure 7A, with inside air warmer than outside and pressure differences caused solely by thermal forces, the NPL is at midheight, with inflow through lower openings and outflow through higher openings. (Direction of flow is always from the higher to the lower pressure region.) A chimney or mechanical exhaust would decrease the inside pressure and thus shift the inside pressure line to the left, raising the NPL; an excess of outdoor supply air over exhaust would lower it. Figure 7B presents qualitative uniform pressure differences caused by wind alone, with the effect on the windward and leeward sides equal but opposite. When the temperature difference and wind effects both exist, the pressures due to each are added together to determine the total pressure difference across the building envelope. Figure 7C shows the combination, where the wind force of Figure 7B has just balanced the

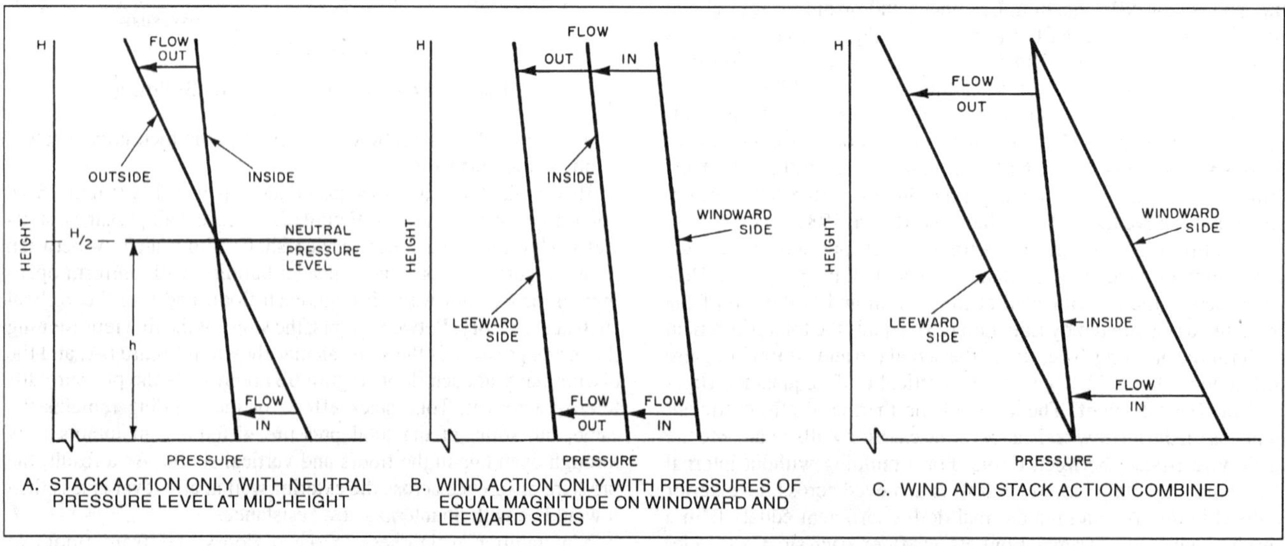

Fig. 7 Distribution of Inside and Outside Pressures over Height of Building

thermal force of Figure 7A, causing no pressure difference at the top windward or bottom leeward side. Total airflow is similar to that with the wind acting alone, but significantly larger than the airflow due to the stack effect only.

The relative importance of the wind and stack pressures in a building depends on building height, internal resistance to vertical airflow, location and flow resistance characteristics of envelope openings, local terrain, and the immediate shielding of the building. The taller the building is and the smaller its internal resistance to airflow, the stronger the stack effect will be. The more exposed a building is, the more susceptible it will be to wind. For any building, there will be ranges of wind speed and temperature difference for which the building's infiltration is dominated by the stack effect, the wind, or the driving pressures of both (Sinden 1978a). These building and terrain factors determine, for specific values of temperature difference and wind speed, in which regime the building's infiltration lies.

The effect of mechanical ventilation on envelope pressure differences is more complex and depends on both the direction of the ventilation flow (exhaust or supply) and the differences in these ventilation flows among the zones of the building. If mechanically supplied outdoor air is provided uniformly to each story, the change in the exterior wall pressure difference pattern from thermal pressures is uniform. With a nonuniform supply of outdoor air (for example, to one story only), the extent of pressurization varies from story to story and depends on the internal airflow resistance. Pressurizing all levels uniformly has little effect on the pressure differences across floors and vertical shaft enclosures, but pressurizing individual stories increases the pressure drop across these internal separations. Pressurization of the ground level is often used in tall buildings to reduce the stack pressures across entries.

The pressure fields illustrated in Figure 7 indicate that the wind and stack effects can be combined by simply adding the pressures. Because this pressure addition can cause directional changes in the resulting flow, the flows cannot be combined as simply. An accurate superposition of the two effects requires detailed knowledge of the two pressure fields. Several theoretical models and measurements have been made for combining forced and natural ventilation (Kiel and Wilson 1987, Persily and Norford 1987, Shaw 1987). A simpler model (Sherman 1992) uses the neutral pressure level and wind direction to combine the two effects. For typical buildings, the following equation can be used to combine the stack and wind effects:

$$Q_{ws} = \sqrt{Q_w^2 + Q_s^2} \tag{29}$$

where

Q_{ws} = infiltration airflow from both wind and stack effects, cfm
Q_w = infiltration airflow from wind, cfm
Q_s = infiltration airflow from stack effect, cfm

Mechanical ventilation can change the flow through the envelope by affecting the internal pressure. When both supply and exhaust fans are present, the mechanical ventilation can be broken up into the amount of ventilation flow that creates no change in pressure distribution, or the "balanced" part:

$$Q_{bal} = \text{Minimum of } (Q_{supply}, Q_{exhaust}) \tag{30}$$

and an "unbalanced" part:

$$Q_{unbal} = \text{Maximum of } (Q_{supply}, Q_{exhaust}) - Q_{bal} \tag{31}$$

If the building has only supply or only exhaust, all the flow is unbalanced.

Because the balanced part of the flow does not affect the pressure distribution across the envelope, it adds simply to the wind- and stack-driven pressures. The unbalanced part, however, does affect

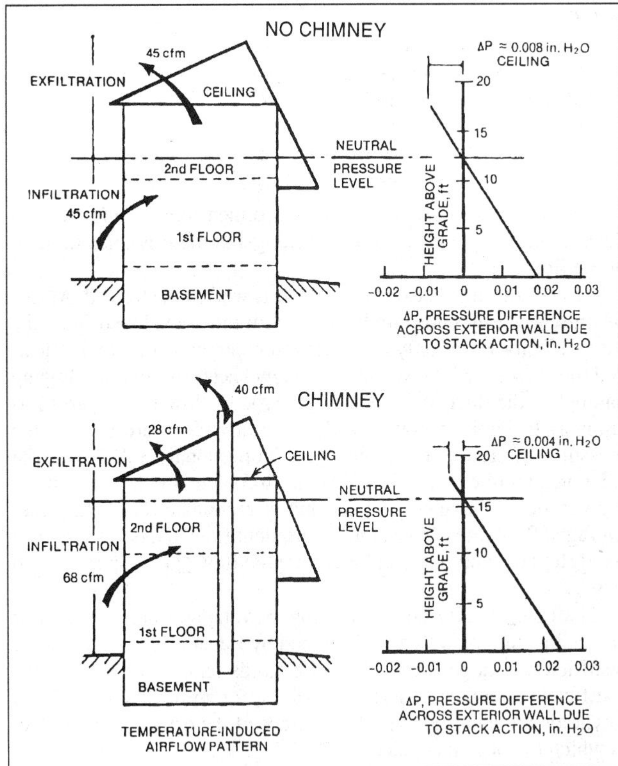

Fig. 8 Temperature-Induced Pressure and Airflow Patterns under Operation of Electric or Gas Furnace for Δt = 50°F

the pressure distribution, and therefore only part of it contributes to additional ventilation (Sherman 1992) until the fan completely dominates. The efficiency of unbalanced ventilation depends again on the NPL and wind direction, but it can be assumed to be 50% for typical cases.

The total ventilation then becomes

$$Q = Q_{bal} + \text{Maximum of } (Q_{unbal}, Q_{ws} + 0.5\, Q_{unbal}) \tag{32}$$

Levins (1982) and Kiel and Wilson (1987) further discuss the combination of mechanical ventilation airflow rates with naturally induced infiltration rates.

Shaw and Brown (1982) compared air infiltration in identical homes with and without a gas furnace with a chimney. Figure 8 presents the effects of exfiltration through the chimney and ceiling with and without the gas furnace; it also shows the impact of the chimney on the NPL.

AIRFLOW THROUGH OPENINGS

The relationship between the airflow Q through an opening in the building envelope and the pressure difference Δp across it is called the **leakage function** of the opening. The form of the leakage function depends on the geometry of the opening. Background theoretical material relevant to leakage functions may be found in Hopkins and Hansford (1974), Etheridge (1977), Kronvall (1980a), and Chastain et al. (1987).

The relationship describing the airflow through an opening is based on the Bernoulli equation with steady, incompressible flow. The most commonly used expression for the airflow rate through an opening is

$$Q = C_3 C_D A \sqrt{2\Delta p/\rho} \tag{33}$$

where

Q = airflow rate, cfm
C_D = discharge coefficient for opening, dimensionless
A = cross-sectional area of opening, ft^2
ρ = air density, lb$_m$/ft^3
Δp = pressure difference across opening, in. of water
C_3 = unit conversion factor = 776

The discharge coefficient C_D is a dimensionless number that depends on the geometry of the opening and the Reynolds number of the flow.

Airflow through constant area ducts is well characterized. At sufficiently low Reynolds numbers and with fully developed flow, the fluid velocity varies only in the direction perpendicular to the flow, and the flow may be visualized as many sheets or laminae flowing parallel to the duct walls. Thus, this type of flow is referred to as laminar. In laminar flow, C_D depends on the square root of the pressure difference; therefore, Q is proportional to Δp. At large Reynolds numbers, the flow becomes turbulent and the velocity at a given point fluctuates rapidly and at random, even if the time-averaged flow rate is constant. In turbulent flow, C_D is constant at a fixed Reynolds number, and therefore the flow Q is proportional to $\sqrt{\Delta p}$.

The case of fully developed flow traversing a hole, nozzle, or orifice in a thin plate is also described by Equation (33). Again, for a sufficiently large value of the Reynolds number, the discharge coefficient is constant. The value of C_D for an orifice depends on Reynolds number and the relative areas of the orifice and the duct in which the orifice is placed.

This discussion of laminar and turbulent flow applies to constant area ducts with fully developed flow and orifices in such ducts. The openings in a building envelope are much less uniform in geometry. Generally, the flow never becomes fully developed, thereby preventing the applicability of the simple relations between Q and Δp. Each opening in the building envelope can still be described by Equation (33), where A is an equivalent cross-sectional area, and C_D depends on opening geometry and the pressure difference across it. The following equation, commonly called the power law equation, is sometimes used instead:

$$Q = c(\Delta p)^n \qquad (34)$$

where

c = flow coefficient, cfm/(in. of water)^n
n = flow exponent, dimensionless

Equation (34) only approximates the relationship between Q and Δp. In fact, the values of c and n depend on the range of Δp over which Equation (34) is applied and do not have a physical interpretation. Honma (1975) measured Q as a function of Δp for several simple openings, and the measured data were fit to Equation (34). The cracks with larger flow resistances (i.e., greater flow lengths or narrower widths), tended to have an exponent n closer to 1 than did gaps with less resistance. For openings in the shell of a building, the value of n depends on the opening geometry, as well as on entrance and exit effects. Additional investigation of pressure/flow data for simple cracks by Chastain et al. (1987) further indicated the importance of adequately characterizing the three-dimensional geometry of openings and the entrance and exit effects. A modeling procedure was developed to determine the discharge coefficient C_D in Equation (33) for the complex openings found in the building shell.

NATURAL VENTILATION

Natural ventilation is the flow of outdoor air due to wind and thermal pressures through intentional openings in the building's shell. It can effectively control both temperature and contaminants, particularly in mild climates. Temperature control by natural ventilation is often the only means of providing cooling when mechanical air conditioning is not available. The arrangement, location, and control of ventilation openings should combine the driving forces of wind and temperature to achieve a desired ventilation rate and good distribution of ventilation air through the building.

Natural Ventilation Openings

Natural ventilation openings include (1) windows, doors, dormer (monitor) openings, and skylights; (2) roof ventilators; (3) stacks connecting to registers; and (4) specially designed inlet or outlet openings.

Windows transmit light and provide ventilation when open. They may open by sliding vertically or horizontally; by tilting on horizontal pivots at or near the center; or by swinging on pivots at the top, bottom, or side. The type of pivoting used is important for weather protection and affects airflow rate.

Roof ventilators provide a weatherproof air outlet. Capacity is determined by the ventilator's location on the roof; the resistance to airflow of the ventilator and its ductwork; the ventilator's ability to use kinetic wind energy to induce flow by centrifugal or ejector action; and the height of the draft.

Natural-draft or gravity roof ventilators can be stationary, pivoting, oscillating, or rotating. Selection criteria include ruggedness, corrosion resistance, stormproofing features, dampers and operating mechanisms, noise, cost, and maintenance. Natural ventilators can be supplemented with power-driven supply fans; the motors need only be energized when the natural exhaust capacity is too low. Gravity ventilators can have manual dampers or dampers controlled by thermostat or wind velocity.

A natural-draft roof ventilator should be positioned so that it receives the full, unrestricted wind. Turbulence created by surrounding obstructions, including higher adjacent buildings, impairs a ventilator's ejector action. The ventilator inlet should be conical or bell mounted to give a high flow coefficient. The opening area at the inlet should be increased if screens, grilles, or other structural members cause flow resistance. Building air inlets at lower levels should be larger than the combined throat areas of all roof ventilators.

Stacks or vertical flues should be located where wind can act on them from any direction. Without wind, stack effect alone removes air from the room with the inlets.

Required Flow

The ventilation airflow rate required to remove a given amount of heat from a building can be calculated from the following equation if the quantity of heat to be removed and the indoor-outdoor temperature difference are known.

$$Q = 60q/c_p\rho(t_i - t_o) \qquad (35)$$

where

Q = airflow rate required to remove heat, cfm
q = rate of heat removal, Btu/h
c_p = specific heat of air, Btu/lb·°F (about 0.024)
ρ = air density, lb$_m$/ft^3 (about 0.075)
$t_i - t_o$ = indoor-outdoor temperature difference, °F

Flow Caused by Wind

Factors due to wind forces that affect the ventilation rate include average speed, prevailing direction, seasonal and daily variation in speed and direction, and local obstructions such as nearby buildings, hills, trees, and shrubbery. Liddament (1988) reviewed the relevance of wind pressure as a driving mechanism. A multiflow path simulation model was developed and used to illustrate the effects of wind on air exchange rate.

Wind speeds may be lower in summer than in winter; directional frequency is also a function of season. Natural ventilation systems are often designed for wind speeds of one-half the seasonal average. The following equation shows the rate of air forced through ventilation inlet openings by wind or determines the proper size of openings to produce given airflow rates:

$$Q = C_4 C_v A V \tag{36}$$

where

Q = airflow rate, cfm
C_v = effectiveness of openings (C_v is assumed to be 0.5 to 0.6 for perpendicular winds and 0.25 to 0.35 for diagonal winds)
A = free area of inlet openings, ft^2
V = wind speed, mph
C_4 = unit conversion factor = 88.0

Inlets should face directly into the prevailing wind. If they are not advantageously placed, flow will be less than that predicted by Equation (36); if the inlets are unusually well placed, flow will be slightly more. Desirable outlet locations are (1) on the leeward side of the building directly opposite the inlet, (2) on the roof, in the low-pressure area caused by a flow discontinuity of the wind, (3) on the side adjacent to the windward face where low-pressure areas occur, (4) in a dormer on the leeward side, (5) in roof ventilators, or (6) by stacks. Chapter 15 gives a general description of the wind pressure distribution on a building. The inlets should be placed in the exterior high-pressure regions; the outlets should be placed in the exterior low-pressure regions.

Flow Caused by Thermal Forces

If building internal resistance is not significant, the flow caused by stack effect can be expressed by

$$Q = 60 C_D A \sqrt{2 g \Delta H_{NPL} (T_i - T_o)/T_i} \tag{37}$$

where

Q = airflow rate, cfm
C_D = discharge coefficient for opening
ΔH_{NPL} = height from midpoint of lower opening to NPL, ft
T_i = indoor temperature, °R
T_o = outdoor temperature, °R

Equation (37) applies when $T_i > T_o$. If $T_i < T_o$, replace T_i in the denominator with T_o, and replace $(T_i - T_o)$ in the numerator with $(T_o - T_i)$. An average temperature should be used for T_i if there is thermal stratification. If the building has more than one opening, the outlet and inlet areas are considered equal. The discharge coefficient C_D accounts for all viscous effects such as surface drag and interfacial mixing.

Estimation of ΔH_{NPL} is difficult. If one window or door represents a large fraction (approximately 90%) of the total opening area in the envelope, then the NPL is at the midheight of that aperture, and ΔH_{NPL} equals one-half the height of the aperture. For this condition, flow through the opening is bidirectional (i.e., air from the warmer side flows through the top of the opening, and air from the colder side flows through the bottom). Interfacial mixing occurs across the counterflow interface, and the orifice coefficient can be calculated according to the following equation (Kiel and Wilson 1986):

$$C_D = 0.40 + 0.0025 |T_i - T_o| \tag{38}$$

If enough other openings are available, the airflow through the opening will be unidirectional, and mixing cannot occur. A discharge coefficient of C_D = 0.65 should then be used. Additional information on stack-driven airflows for natural ventilation can be found in Foster and Down (1987).

Greatest flow per unit area of openings is obtained when inlet and outlet areas are equal; Equations (37) and (38) are based on this

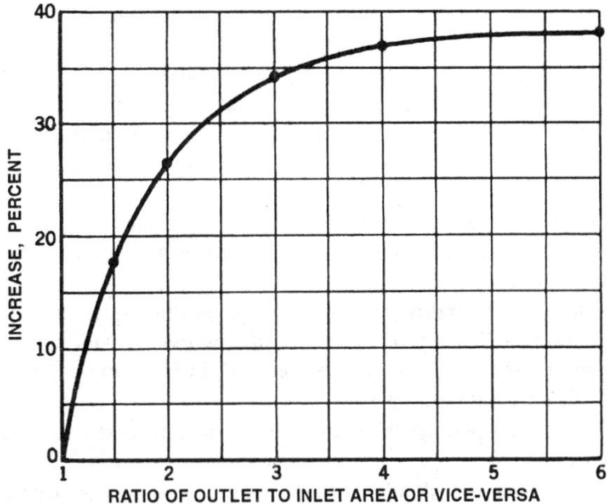

Fig. 9 Increase in Flow Caused by Excess Area of One Opening over the Other

equality. Increasing the outlet area over inlet area (or vice versa) increases airflow but not in proportion to the added area. When openings are unequal, use the smaller area in Equation (37) and add the increase as determined from Figure 9.

Natural Ventilation Guidelines

Several general guidelines should be observed in designing for natural ventilation. Some of these may conflict with other climate-responsive strategies (such as using orientation and shading devices to minimize solar gain) or other design considerations.

1. In hot, humid climates, air velocities should be maximized in the occupied zones for bodily cooling. In hot, arid climates, airflow throughout the building should be maximized for structural cooling, particularly at night when the temperature is low.
2. Topography, landscaping, and surrounding buildings should be used to redirect airflow and give maximum exposure to breezes. Vegetation can funnel breezes and avoid wind dams, which reduce the driving pressure differential around the building. Site objects should not obstruct inlet openings.
3. The building should be shaped to expose maximum shell openings to breezes.
4. Architectural elements such as wing walls, parapets, and overhangs should be used to promote airflow into the building interior.
5. The long facade of the building and the majority of the door and window openings should be oriented with respect to the prevailing summer breezes. If there is no prevailing direction, openings should be sufficient to provide ventilation regardless of wind direction.
6. Windows should be located in opposing pressure zones. Two openings on opposite sides of a space increase the ventilation flow. Openings on adjacent sides force air to change direction, providing ventilation to a greater area. The benefits of the window arrangement depend on the outlet location relative to the direction of the inlet airstream.
7. If a room has only one external wall, better airflow is achieved with two widely spaced windows.
8. If the openings are at the same level and near the ceiling, much of the flow may bypass the occupied level and be ineffective in diluting contaminants there.
9. Vertical distance between openings is required to take advantage of the stack effect; the greater the vertical distance, the greater the ventilation.
10. Openings in the vicinity of the NPL are least effective for thermally induced ventilation. If the building has only one large

opening, the NPL tends to move to that level, which reduces the pressure across the opening.

11. Greatest flow per unit area of total opening is obtained by inlet and outlet openings of nearly equal areas. An inlet window smaller than the outlet creates higher inlet velocities. An outlet smaller than the inlet creates lower but more uniform airspeed through the room.
12. Openings with areas much larger than calculated are sometimes desirable when anticipating increased occupancy or very hot weather.
13. Horizontal windows are generally better than square or vertical windows. They produce more airflow over a wider range of wind directions and are most beneficial in locations where prevailing wind patterns shift.
14. Window openings should be accessible to and operable by occupants.
15. Inlet openings should not be obstructed by indoor partitions. Partitions can be placed to split and redirect airflow but should not restrict flow between the building's inlets and outlets.
16. Vertical airshafts or open staircases can be used to increase and take advantage of stack effects. However, enclosed staircases intended for evacuation during a fire should not be used for ventilation.

RESIDENTIAL INFILTRATION

Although the terms infiltration and air leakage are sometimes used synonymously, they are different, though related, quantities. Infiltration is the rate of uncontrolled air exchange through unintentional openings that occurs under given conditions, while **air leakage area** is a measure of the airtightness of the building shell. The greater the air leakage area of a building, the greater its infiltration rate, all else (weather, exposure, and building geometry) being equal. Infiltration is the rate of air entering the structure and is equal to the exfiltration (the rate of air leaving the structure).

Infiltration may be reduced either by reducing the surface pressures driving the flow or by reducing the air leakage area of the shell. Surface pressures caused by the wind can be reduced by changing the landscaping in the vicinity of the building (Mattingly and Peters 1977). Stack pressures can be reduced by increasing the airflow resistance between floors and from floors to any vertical shafts within the building, although this is an issue almost exclusively reserved to tall buildings.

The infiltration rate of an individual building depends on weather conditions, equipment operation, and occupant activities. The rate can vary by a factor of five from weather effects alone (Malik 1978). When associating an infiltration rate with a building, it is important to either provide the corresponding weather conditions and equipment status or describe the rate as a seasonal or annual average.

Typical infiltration values in housing in North America vary by a factor of about ten, from tightly constructed housing with seasonal average air exchange rates of about 0.2 air exchanges per hour (ACH) to loosely constructed housing with air exchange rates as great as 2.0 ACH. Figure 10 and Figure 11 show histograms of infiltration rates measured in two different samples of North American housing (Grimsrud et al. 1982, Grot and Clark 1979). Figure 10 shows the average seasonal infiltration of 312 houses located in different areas in North America. The median infiltration value of this sample is 0.5 ACH. Figure 11 represents measurements in 266 houses located in 16 cities in the United States. The median value of this sample is 0.9 ACH. The group of houses contained in the Figure 10 sample is biased toward new, energy-efficient houses, while the group in Figure 11 represents older, low-income housing in the United States.

Additional studies have found average values for houses in regional areas. Palmiter and Brown (1989) and Parker et al.

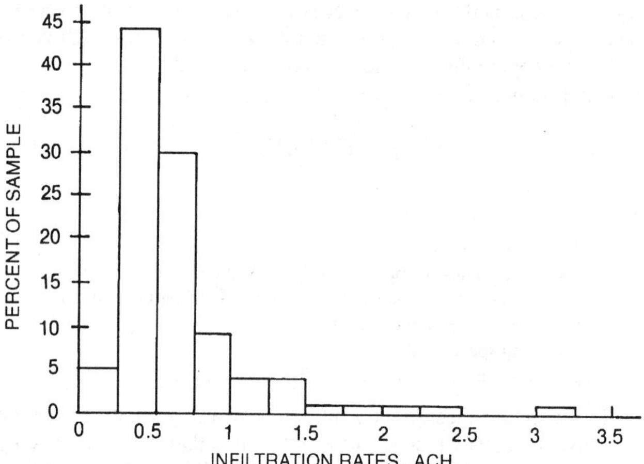

Fig. 10 Histogram of Infiltration Values—New Construction

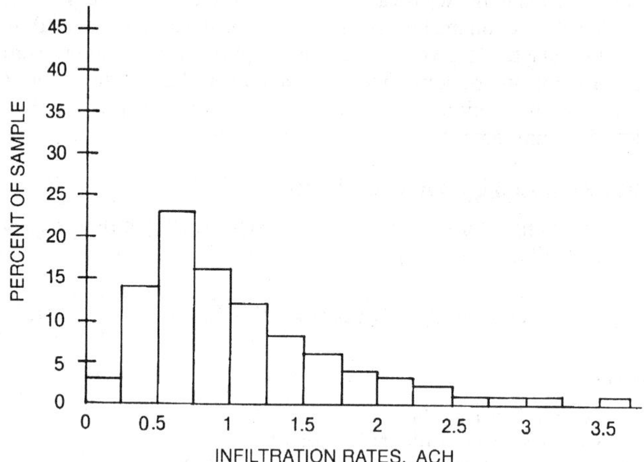

Fig. 11 Histogram of Infiltration Values—Low-Income Housing

(1990) found a heating season average of 0.40 ACH (range: 0.13 to 1.11 ACH) for 134 houses in the Pacific Northwest. In a comparison of 292 houses incorporating energy-efficient features (including measures to reduce air infiltration and provide ventilation heat recovery) with 331 control houses, Parker et al. (1990) found an average of about 0.25 ACH (range: 0.02 to 1.63 ACH) for the energy-efficient houses versus 0.49 (range: 0.05 to 1.63 ACH) for the control. Ek et al. (1990) found an average of 0.5 ACH (range: 0.26 to 1.09) for 93 double-wide manufactured homes also in the Pacific Northwest. Canadian housing stock has been characterized by Yuill and Comeau (1989) and Riley (1990). While these studies do not represent random samples of North American housing, they indicate the distribution of infiltration rates expected in a group of buildings.

Occupancy influences have not been measured directly and vary widely. Desrochers and Scott (1985) estimated that they add an average of 0.10 to 0.15 ACH to unoccupied values. Kvisgaard and Collet (1990) found that in 16 Danish dwellings, the users on average provided 63% of the total air exchange rate.

RESIDENTIAL AIR LEAKAGE

The air leakage of a building characterizes the relationship between the pressure difference across the building envelope and the airflow rate through the envelope (see the section on Airflow Through Openings). Building air leakage area is a physical property of a building determined by its design, its construction, and its deterioration over time. Although **airtightness** is just one factor in determining the air exchange rate of a building, it is useful for comparing buildings to one another or to airtightness standards, for evaluating design and construction quality, and for studying the effectiveness of airtightening retrofits. No simple relationship exists between a building's airtightness and its air exchange rate, but calculation methods do exist (see the section on Simplified Models).

Measurement

While tracer gas measurement procedures provide building air exchange rates, they are somewhat expensive and time-consuming. In many cases, it is sufficient, or preferable, to measure the air leakage of a building with pressurization testing (Stricker 1975, Tamura 1975, Kronvall 1978, Blomsterberg and Harrje 1979, Gadsby and Harrje 1985). Fan pressurization is relatively quick and inexpensive, and it characterizes building envelope airtightness independent of weather conditions. In this procedure, a large fan or blower is mounted in a door or window and induces a large and roughly uniform pressure difference across the building shell (CGSB 1986, ASTM *Standard* E 779). The airflow required to maintain this pressure difference is then measured. The leakier the building is, the more airflow is necessary to induce a specific indoor-outdoor pressure difference. The airflow rate is generally measured at a series of pressure differences ranging from about 0.04 in. to 0.30 in. of water.

The results of a pressurization test, therefore, consist of several combinations of pressure difference and airflow rate data. An example of typical data is shown in Figure 12. These data points characterize the air leakage of a building and are generally converted to a single value that serves as a measure of the building's airtightness. There are several different measures of airtightness, most of which involve fitting the data to a curve in the form of Equation (34) (i.e., $Q = c\Delta p^n$). The airtightness ratings are based on airflow rates predicted at particular reference pressures by Equation (34). The basic difference between the various airtightness ratings is the value of the reference pressure.

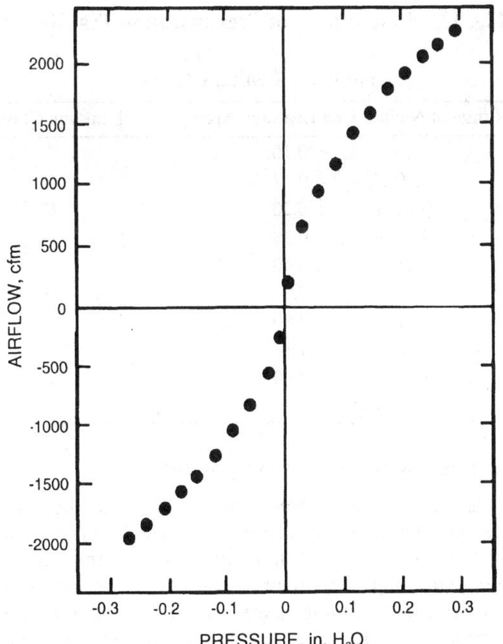

Fig. 12 Airflow Rate Versus Pressure Difference Data from Whole-House Pressurization Test

Airtightness Ratings

In some cases, the predicted airflow rate is converted to an equivalent or effective air leakage area by rearranging Equation (33):

$$A_L = C_5 Q_r \frac{\sqrt{\rho/2\Delta p_r}}{C_D} \tag{39}$$

where

A_L = equivalent or effective air leakage area, in^2
Q_r = predicted airflow rate at Δp_r (from curve fit to pressurization test data), cfm
ρ = air density, lb$_m$/ft^3
Δp_r = reference pressure difference, in. of water
C_D = discharge coefficient
C_5 = unit conversion factor = 0.186

All the openings in the building shell are combined into an overall opening area and discharge coefficient for the building when the equivalent or effective air leakage area is calculated. Some users of the leakage area approach set the discharge coefficient equal to 1. Others set $C_D \cong 0.6$ (i.e., the discharge coefficient for a sharp-edged orifice). The air leakage area of a building is, therefore, the area of an orifice (with an assumed value of C_D) that would produce the same amount of leakage as the building envelope at the reference pressure.

An airtightness rating, whether based on an air leakage area or a predicted airflow rate, is generally normalized by some factor to account for building size. Normalization factors include floor area, exterior envelope area, and building volume.

With the wide variety of possible approaches to normalization and reference pressure difference, and the use of the air leakage area concept, many different airtightness ratings are being used. Reference pressure differences include 0.016, 0.04, 0.10, 0.20, and 0.30 in. of water. Reference pressure differences of 0.016 and 0.04 in. of water are advocated because they are closer to the pressure differences that actually induce air exchange and, therefore, better model the flow characteristics of the openings. While this may be true, they are outside the range of measured values in the test; therefore, the predicted airflow rates at 0.016 and 0.04 in. of water are subject to significant uncertainty. The uncertainty in these predicted airflow rates and the implications for quantifying airtightness are discussed in Persily and Grot (1985b), Chastain (1987), and Modera and Wilson (1990). Round robin tests by Murphy et al. (1991) to determine the repeatability and reproducibility of fan pressurization devices found that subtle errors in fan calibration or operator technique are greatly exaggerated when extrapolating the pressure versus flow curve out to 0.016 in. of water, with errors as great as ±40%, mainly due to the fan calibration errors at low flow.

Some common airtightness ratings include the effective air leakage area at 0.016 in. of water assuming C_D = 1.0 (Sherman and Grimsrud 1980); the equivalent air leakage area at 0.04 in. of water assuming C_D = 0.611 (CGSB 1986); and the airflow rate at 0.20 in. of water, divided by the building volume to give units of air changes per hour (Blomsterberg and Harrje 1979).

Conversion Between Ratings

Air leakage areas at one reference pressure difference can be converted to air leakage areas at some other reference pressure difference according to

$$A_{r,2} = A_{r,1}\left(\frac{C_{D,1}}{C_{D,2}}\right)\left(\frac{\Delta p_{r,2}}{\Delta p_{r,1}}\right)^{n-0.5} \tag{40}$$

where

$A_{r,1}$ = air leakage area at reference pressure difference $\Delta p_{r,1}$, in^2
$A_{r,2}$ = air leakage area at reference pressure difference $\Delta p_{r,2}$, in^2
$C_{D,1}$ = discharge coefficient used to calculate $A_{r,1}$
$C_{D,2}$ = discharge coefficient used to calculate $A_{r,2}$
n = flow exponent from Equation (34)

An air leakage area at one reference pressure difference can be converted to an airflow rate at some other reference pressure difference according to

$$Q_{r,2} = C_6 C_{D,1} A_{r,1} (2/\rho)^{0.5} (\Delta p_{r,1})^{0.5-n} (\Delta p_{r,2})^n \qquad (41)$$

where

$Q_{r,2}$ = airflow rate at reference pressure difference $\Delta p_{r,2}$, cfm
C_6 = unit conversion factor = 5.39

Finally, an air leakage area may be converted to the flow coefficient c in Equation (34) according to

$$c = C_6 C_D A_L (2/\rho)^{0.5} (\Delta p_r)^{0.5-n} \qquad (42)$$

Equations (40) through (42) require the assumption of a value of n, unless it is reported with the measurement results. When whole-building pressurization test data are fitted to Equation (34), the value of n generally lies between 0.6 and 0.7. Therefore, using a value of n in this range is reasonable.

Building Air Leakage Data

Fan pressurization measures a building property that ideally varies little with time and weather conditions. In reality, unless the wind and temperature differences during the measurement period are sufficiently mild, the pressure differences they induce during the test will interfere with the test pressures and cause measurement errors. Persily (1982) and Modera and Wilson (1990) presented experimental studies of the effects of wind speed on pressurization test results. Several experimental studies have also shown variations on the order of 20 to 40% over a year in the measured airtightness in homes (Persily 1982, Kim and Shaw 1986, Warren and Webb 1986).

Several pressurization test results for residential buildings are presented in Figure 13 (Persily 1986). These results are in units of air changes per hour at 0.20 in. of water and show the wide range in airtightness among houses, even houses of identical design. The data for passive solar and energy-efficient houses also show that houses expected to be relatively airtight are not necessarily very tight. The houses in Sweden—which has a residential building airtightness standard of 3 air changes per hour at 0.20 in. of water for single-family detached houses (Swedish Building Code 1980)—are exceptionally tight, as are the houses in Canada.

ASHRAE *Standard* 119 establishes air leakage performance levels for residential buildings. These levels are in terms of the normalized leakage area A_n:

$$A_n = 6.944 \left(\frac{A_L}{A_f}\right) \left(\frac{H}{H_o}\right)^{0.3} \qquad (43)$$

where

A_n = normalized leakage area, dimensionless
A_L = effective leakage area at 0.016 in. of water (C_D = 1.0), in²
A_f = gross floor area (within exterior walls), ft²
H = building height, ft
H_o = reference height of one-story building = 8 ft

Table 2 presents the leakage classes of *Standard* 119. The values of A_n in this table correspond approximately to annual average building air exchange rates in units of air changes per hour. *Standard* 119 specifies appropriate leakage classes for a building based on climate.

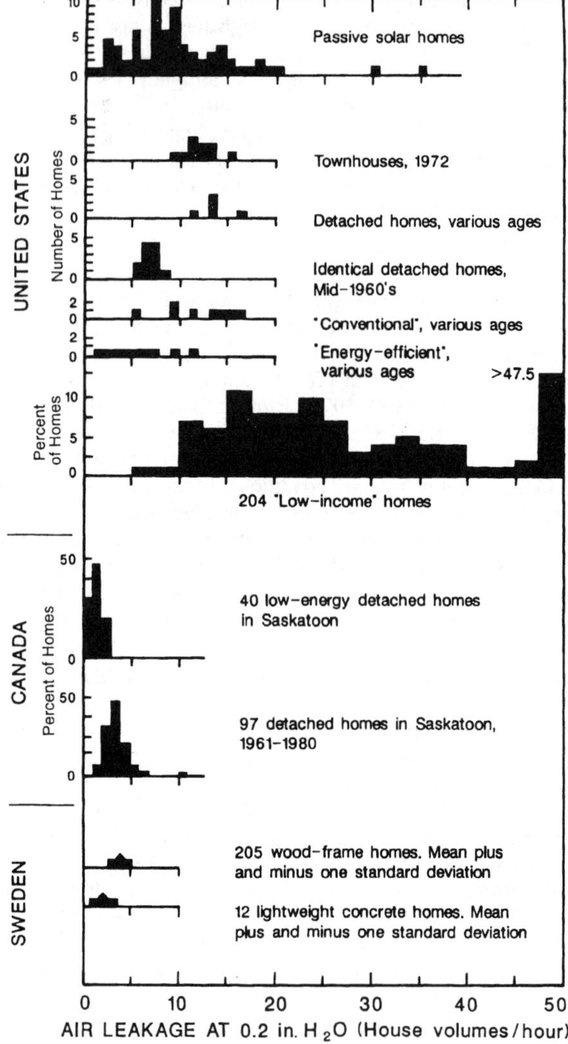

Fig. 13 Comparison of Pressurization Test Results

Table 2 Leakage Classes

Range of Normalized Leakage Area	Leakage Class
$A_n < 0.10$	A
$0.10 \le A_n < 0.14$	B
$0.14 \le A_n < 0.20$	C
$0.20 \le A_n < 0.28$	D
$0.28 \le A_n < 0.40$	E
$0.40 \le A_n < 0.57$	F
$0.57 \le A_n < 0.80$	G
$0.80 \le A_n < 1.13$	H
$1.13 \le A_n < 1.60$	I
$1.60 \le A_n$	J

Air Leakage of Building Components

The fan pressurization procedure discussed in the section on Measurement enables the measurement of whole-building air leakage. The location and size of individual openings in building envelopes are extremely important because they influence the air infiltration rate of a building as well as the heat and moisture transfer characteristics of the envelope. Additional test procedures exist for pressure-testing individual building components such as windows, walls, and doors; they are discussed in ASTM *Standards* E 283 and E 783 for laboratory and field tests, respectively.

Leakage Distribution

Dickerhoff et al. (1982) and Harrje and Born (1982) studied the air leakage of individual building components and systems. The following points summarize the percentages of whole-building air leakage area associated with various components and systems. The values in parentheses include the range determined for each component and the mean of the range.

Walls (18 to 50%; 35%). Both interior and exterior walls contribute to the leakage of the structure. Leakage between the sill plate and the foundation, cracks below the bottom of the gypsum wallboard, electrical outlets, plumbing penetrations, and leaks into the attic at the top plates of walls all occur. Because interior walls are not filled with insulation, open paths connecting these walls and the attic permit the walls to behave like heat exchanger fins within the conditioned living space of the house.

Ceiling details (3 to 30%; 18%). Leakage across the top ceiling of the heated space is particularly insidious because it reduces the effectiveness of insulation on the attic floor and contributes to infiltration heat loss. Ceiling leakage also reduces the effectiveness of ceiling insulation in buildings without attics. Recessed lighting, plumbing, and electrical penetrations leading to the attic are some particular areas of concern.

Heating system (3 to 28%; 18%). The location of the furnace or ductwork in conditioned or unconditioned spaces, the venting arrangement of a fuel-burning device, and the existence and location of a combustion air supply all affect leakage. Modera et al. (1991) and Robison and Lambert (1989), among others, have shown that the variability of leakage in ducts passing through unconditioned spaces is high, the coefficient of variation being on the order of 50%. Field studies have also shown that in-situ repairs can eliminate one-quarter to two-thirds of the observed leakage (Cummings and Tooley 1989, Cummings et al. 1990, Robison and Lambert 1989). The 18% contribution of ducts to total leakage significantly underestimates their impact because during system operation, the pressure differentials across the duct leaks are approximately ten times higher than typical pressure differences across the envelope leaks (Modera 1989, Modera et al. 1991).

Windows and doors (6 to 22%; 15%). More variation in window leakage is seen among window types (e.g., casement versus double-hung) than among new windows of the same type from different manufacturers (Weidt et al. 1979). Windows that seal by compressing the weather strip (casements, awnings) show significantly lower leakage than windows with sliding seals.

Fireplaces (0 to 30%; 12%). When a fireplace is not in use, poorly fitting dampers allow air to escape. Glass doors reduce excess air while a fire is burning but rarely seal the fireplace structure more tightly than a closed damper does. Chimney caps or fireplace plugs (with signs that warn they are in place) effectively reduce leakage through a cold fireplace.

Vents in conditioned spaces (2 to 12%; 5%). Exhaust vents in conditioned spaces frequently have either no dampers or dampers that do not close properly.

Diffusion through walls (<1%). Diffusion, in comparison to infiltration through holes and other openings in the structure, is not an important flow mechanism. At 0.02 in. of water, the permeability of building materials produces an air exchange rate of less than 0.01 ACH by wall diffusion in a typical house.

Component leakage areas. Table 3 shows effective air leakage areas for a variety of residential building components at 0.016 in. of water with C_D assumed equal to 1 (Colliver et al. 1992). The values in the table present results in terms of air leakage area per unit component. Per unit component means per component, per unit surface area, or per unit length of crack or sash, whichever is appropriate. These air leakage areas may be converted to air leakage areas at other reference pressures, airflow rates, or flow coefficients using Equations (40) through (42).

Multifamily Building Leakage

Leakage distribution is particularly important in multifamily apartment buildings. These buildings often cannot be treated as single zones due to the internal resistance between apartments. Moreover, the leakage between apartments varies widely, tending to be small in modern construction, and ranging as high as 60% of the total apartment leakage in turn-of-the-century brick walk-up apartment buildings (Modera et al. 1991, Diamond et al. 1986). Little information on interzonal leakage has been reported because of the difficulty and expense of these measurements.

Controlling Air Leakage

New Buildings. It is much easier to build a tight building than to tighten an existing building. Elmroth and Levin (1983), Eyre and Jennings (1983), Marbek Resource Consultants (1984), and Nelson et al. (1985) provide information and construction details on airtight building design for houses.

A continuous air infiltration retarder is one of the most effective means of reducing air leakage through walls, around window and door frames, and at joints between major building elements. The air infiltration retarder can be installed either on the inside of the wall framing, in which case it usually functions as a vapor retarder as well, or on the outside of the wall framing, in which case it should have a permeance rating high enough to permit diffusion of water vapor from the wall. For a discussion of moisture transfer in building envelopes, see Chapter 22 and Chapter 23.

When the air infiltration retarder is also a continuous plastic film vapor retarder, particular care must be taken to ensure its continuity at all wall, floor, and ceiling joints; at window and door frames; and at all penetrations of the retarder, such as electrical outlets and switches, plumbing connections, and utility service penetrations. Joints in the **air-vapor retarder** must be lapped and sealed. Plastic vapor retarders installed in the ceiling should be tightly sealed with the vapor retarder in the outside walls and should be continuous over the partition walls. A seal at the top of the partition walls prevents leakage into the attic; a plate on top of the studs generally gives a poor seal.

A continuous air infiltration retarder installed on the outside of wall framing can cover many difficult construction details associated with the installation of continuous air-vapor retarders. Interior air-vapor retarders must be lapped and sealed at electrical outlets and switches, at joints between walls and floors and between walls and ceilings, and at plumbing connections penetrating the wall's interior finish. The exterior air infiltration retarder can cover these problem areas continuously. Joints in the air infiltration retarder should be lapped and sealed or taped. Exterior air infiltration retarders are generally made of a material stronger than plastic film and are more likely to withstand damage during construction. Sealing the wall against air leakage at the exterior of the insulation also cuts down on convection currents within the wall cavity, allowing insulation to retain more of its effectiveness.

Existing Buildings. The air leakage sites must first be located in order to tighten the envelope of an existing building. As discussed earlier, air leakage in buildings is due not only to windows and doors, but to a wide range of unexpected and unobvious construction defects. Many important leakage sites can be very difficult to find. A variety of techniques developed to locate leakage sites are described in ASTM *Standard* E 1186 and Charlesworth (1988).

Once leakage sites are located, they can be repaired with materials and techniques appropriate to the size and location of the leak. Harrje et al. (1979), Diamond et al. (1982), and Energy Resource Center (1982) include information on airtightening in existing residential buildings. With these procedures, the air leakage of residential buildings can be reduced dramatically. Depending on the extent of the tightening effort and the experience of those doing the work, residential buildings can be tightened anywhere from 5% to more

Table 3 Effective Air Leakage Areas (Low-Rise Residential Applications)

	Units (see note)	Best Estimate	Minimum	Maximum
Ceiling				
General	in^2/ft^2	0.026	0.011	0.04
Drop	in^2/ft^2	0.0027	0.00066	0.003
Ceiling penetrations				
Whole-house fans	in^2 ea	3.1	0.25	3.3
Recessed lights	in^2 ea	1.6	0.23	3.3
Ceiling/Flue vent	in^2 ea	4.8	4.3	4.8
Surface-mounted lights	in^2 ea	0.13		
Chimney	in^2 ea	4.5	3.3	5.6
Crawl space				
General (area for exposed wall)	in^2/ft^2	0.144	0.1	0.24
8 in. by 16 in. vents	in^2 ea	20		
Door frame				
General	in^2 ea	1.9	0.37	3.9
Masonry, not caulked	in^2/ft^2	0.07	0.024	0.07
Masonry, caulked	in^2/ft^2	0.014	0.004	0.014
Wood, not caulked	in^2/ft^2	0.024	0.009	0.024
Wood, caulked	in^2/ft^2	0.004	0.001	0.004
Trim	in^2/lftc	0.05		
Jamb	in^2/lftc	0.4	0.3	0.5
Threshold	in^2/lftc	0.1	0.06	1.1
Doors				
Attic/crawl space, not weatherstripped	in^2 ea	4.6	1.6	5.7
Attic/crawl space, weatherstripped	in^2 ea	2.8	1.2	2.9
Attic fold down, not weatherstripped	in^2 ea	6.8	3.6	13
Attic fold down, weatherstripped	in^2 ea	3.4	2.2	6.7
Attic fold down, with insulated box	in^2 ea	0.6		
Attic from unconditioned garage	in^2 ea	0	0	0
Double, not weatherstripped	in^2/ft^2	0.16	0.1	0.32
Double, weatherstripped	in^2/ft^2	0.12	0.04	0.33
Elevator (passenger)	in^2 ea	0.04	0.022	0.054
General, average	in^2/lftc	0.015	0.011	0.021
Interior (pocket, on top floor)	in^2 ea	2.2		
Interior (stairs)	in^2/lftc	0.04	0.012	0.070
Mail slot	in^2/lftc	0.2		
Sliding exterior glass patio	in^2 ea	3.4	0.46	9.3
Sliding exterior glass patio	in^2/ft^2	0.079	0.009	0.22
Storm (difference between with and without)	in^2 ea	0.9	0.46	0.96
Single, not weatherstripped	in^2 ea	3.3	1.9	8.2
Single, weatherstripped	in^2 ea	1.9	0.6	4.2
Vestibule (subtract per each location)	in^2 ea	1.6		
Electrical outlets/Switches				
No gaskets	in^2 ea	0.38	0.08	0.96
With gaskets	in^2 ea	0.023	0.012	0.54
Furnace				
Sealed (or no) combustion	in^2 ea	0	0	0
Retention head or stack damper	in^2 ea	4.6	3.1	4.6
Retention head and stack damper	in^2 ea	3.7	2.8	4.6
Floors over crawl spaces				
General	in^2/ft^2	0.032	0.006	0.071
Without ductwork in crawl space	in^2/ft^2	0.0285		
With ductwork in crawl space	in^2/ft^2	0.0324		
Fireplace				
With damper closed	in^2/ft^2	0.62	0.14	1.3
With damper open	in^2/ft^2	5.04	2.09	5.47
With glass doors	in^2/ft^2	0.58	0.06	0.58
With insert and damper closed	in^2/ft^2	0.52	0.37	0.66
With insert and damper open	in^2/ft^2	0.94	0.58	1.3
Gas water heater	in^2 ea	3.1	2.3	3.9
Joints				
Ceiling-wall	in^2/lftc	0.070	0.0075	0.12
Sole plate, floor/wall, uncaulked	in^2/lftc	0.2	0.018	0.26
Sole plate, floor/wall, caulked	in^2/lftc	0.04	0.0035	0.056
Top plate, band joist	in^2/lftc	0.005	0.0035	0.018

	Units (see note)	Best Estimate	Minimum	Maximum
Piping/Plumbing/Wiring penetrations				
Uncaulked	in^2 ea	0.9	0.31	3.7
Caulked	in^2 ea	0.3	0.16	0.3
Vents				
Bathroom with damper closed	in^2 ea	1.6	0.39	3.1
Bathroom with damper open	in^2 ea	3.1	0.95	3.4
Dryer with damper	in^2 ea	0.46	0.45	1.1
Dryer without damper	in^2 ea	2.3	1.9	5.3
Kitchen with damper open	in^2 ea	6.2	2.2	11
Kitchen with damper closed	in^2 ea	0.8	0.16	1.1
Kitchen with tight gasket	in^2 ea	0.16		
Walls (exterior)				
Cast-in-place concrete	in^2/ft^2	0.007	0.0007	0.026
Clay brick cavity wall, finished	in^2/ft^2	0.0098	0.0007	0.033
Precast concrete panel	in^2/ft^2	0.017	0.0004	0.024
Lightweight concrete block, unfinished	in^2/ft^2	0.05	0.019	0.058
Lightweight concrete block, painted or stucco	in^2/ft^2	0.016	0.0075	0.016
Heavyweight concrete block, unfinished	in^2/ft^2	0.0036		
Continuous air infiltration barrier	in^2/ft^2	0.0022	0.0008	0.003
Rigid sheathing	in^2/ft^2	0.005	0.0042	0.006
Window framing				
Masonry, uncaulked	in^2/ft^2	0.094	0.082	0.148
Masonry, caulked	in^2/ft^2	0.019	0.016	0.03
Wood, uncaulked	in^2/ft^2	0.025	0.022	0.039
Wood, caulked	in^2/ft^2	0.004	0.004	0.007
Windows				
Awning, not weatherstripped	in^2/ft^2	0.023	0.011	0.035
Awning, weatherstripped	in^2/ft^2	0.012	0.006	0.017
Casement, weatherstripped	in^2/lftc	0.011	0.005	0.14
Casement, not weatherstripped	in^2/lftc	0.013		
Double horizontal slider, not weatherstripped	in^2/lftc	0.052	0.0009	0.16
Double horizontal slider, wood, weatherstripped	in^2/lftc	0.026	0.0070	0.081
Double horizontal slider, aluminum, weatherstripped	in^2/lftc	0.034	0.027	0.038
Double-hung, not weatherstripped	in^2/lftc	0.12	0.040	0.29
Double-hung, weatherstripped	in^2/lftc	0.031	0.009	0.089
Double-hung with storm, not weatherstripped	in^2/lftc	0.046	0.023	0.080
Double-hung with storm, weatherstripped	in^2/lftc	0.037	0.021	0.05
Double-hung with pressurized track, weatherstripped	in^2/lftc	0.023	0.018	0.026
Jalousie	in^2/louver	0.524		
Lumped	in^2/lfts	0.022	0.00042	0.097
Single horizontal slider, weatherstripped	in^2/lfts	0.031	0.009	0.097
Single horizontal slider, aluminum	in^2/lfts	0.04	0.013	0.097
Single horizontal slider, wood	in^2/lfts	0.021	0.013	0.047
Single horizontal slider, wood clad	in^2/lfts	0.030	0.025	0.038
Single-hung, weatherstripped	in^2/lfts	0.041	0.029	0.058
Sill	in^2/lftc	0.0099	0.0065	0.010
Storm inside, heat shrink	in^2/lfts	0.00085	0.00042	0.00085
Storm inside, rigid sheet with magnetic seal	in^2/lfts	0.0056	0.00085	0.011
Storm inside, flexible sheet with mechanical seal	in^2/lfts	0.0072	0.00085	0.039
Storm inside, rigid sheet with mechanical seal	in^2/lfts	0.019	0.0021	0.039
Storm outside, pressurized track	in^2/lftc	0.025		
Storm outside, 2-track	in^2/lftc	0.058		
Storm outside, 3-track	in^2/lftc	0.116		

Note: Air leakage areas are based on values found in the literature. The effective air leakage area (in square inches) is based on a pressure difference of 0.016 in. of water and $C_D = 1$.

Abbreviations: ft^2 = gross area in square feet; ea = each; lftc = linear foot of crack; lfts = linear foot of sash

than 50% (Blomsterberg and Harrje 1979, Harrje and Mills 1980, Jacobson et al. 1986, Verschoor and Collins 1986, Giesbrecht and Proskiw 1986). Much less information is available for airtightening large, commercial buildings, but the same general principles apply (Parekh et al. 1991, Persily 1991).

NONRESIDENTIAL AIR LEAKAGE

Commercial Building Envelope Leakage

The building envelopes of large commercial buildings are often thought to be quite airtight. The National Association of Architectural Metal Manufacturers specifies a maximum leakage per unit of exterior wall area of 0.060 cfm/ft² at a pressure difference of 0.30 in. of water exclusive of leakage through operable windows. Tamura and Shaw (1976a) found that, assuming a flow exponent of 0.65 in Equation (34), air leakage measurements in eight Canadian office buildings with sealed windows ranged from 0.120 to 0.480 cfm/ft². Persily and Grot (1986) ran whole-building pressurization tests in large office buildings that showed that pressurization airflow rate divided by building volume is relatively low compared to that of houses. However, if these airflow rates are normalized by building envelope area instead of by volume, the results indicate envelope airtightness levels similar to those in typical American houses. In a study of eight U.S. office buildings, Persily and Grot (1986) found air leakage ranging from 0.213 to 1.028 cfm/ft² at 0.30 in. of water. Therefore, office building envelopes are leakier than expected. Typical air leakage values per unit wall area at 0.30 in. of water are 0.10, 0.30, and 0.60 cfm/ft² for tight, average, and leaky walls, respectively (Tamura and Shaw 1976a).

Air Leakage Through Internal Partitions

In large buildings, the air leakage associated with internal partitions becomes very important. Elevator, stair, and service shaft walls; floors; and other interior partitions are the major separations of concern in these buildings. Their leakage characteristics are needed to determine infiltration through exterior walls and airflow within a building. These internal resistances are also important in the event of a fire to predict smoke movement patterns and evaluate smoke management systems.

Table 4 gives air leakage areas (calculated at 0.30 in. of water with C_D = 0.65) for different internal partitions of commercial buildings (Klote and Fothergill 1983). Figure 14 presents examples of measured air leakage rates of elevator shaft walls (Tamura and Shaw 1976b), the type of data used to derive the values in Table 4. Chapter 48 of the 1995 *ASHRAE Handbook—Applications* should be consulted for performance models and applications of smoke management systems.

Leakage openings at the top of elevator shafts are equivalent to orifice areas of 620 to 1550 in². Air leakage rates through stair shaft and elevator doors are shown in Figure 15 as a function of average crack width around the door. The air leakage areas associated with other openings within commercial buildings are also important for air movement calculations. These include interior doors and partitions, suspended ceilings in buildings where the space above the ceiling is used in the air distribution system, and other components of the air distribution system.

Table 4 Air Leakage Areas for Internal Partitions in Commercial Buildings (at 0.30 in. of water and C_D = 0.65)

Construction Element	Wall Tightness	Area Ratio
		A_L/A_w
Stairwell walls	Tight	0.14×10^{-4}
	Average	0.11×10^{-3}
	Loose	0.35×10^{-3}
Elevator shaft walls	Tight	0.18×10^{-3}
	Average	0.84×10^{-3}
	Loose	0.18×10^{-2}
		A_L/A_f
Floors	Average	0.52×10^{-4}

A_L = air leakage area A_w = wall area A_f = floor area

Air Leakage Through Exterior Doors

Door infiltration depends on the type of door, room, and building. In residences and small buildings where doors are used infrequently, the air exchange associated with a door can be estimated based on air leakage through cracks between the door and the frame.

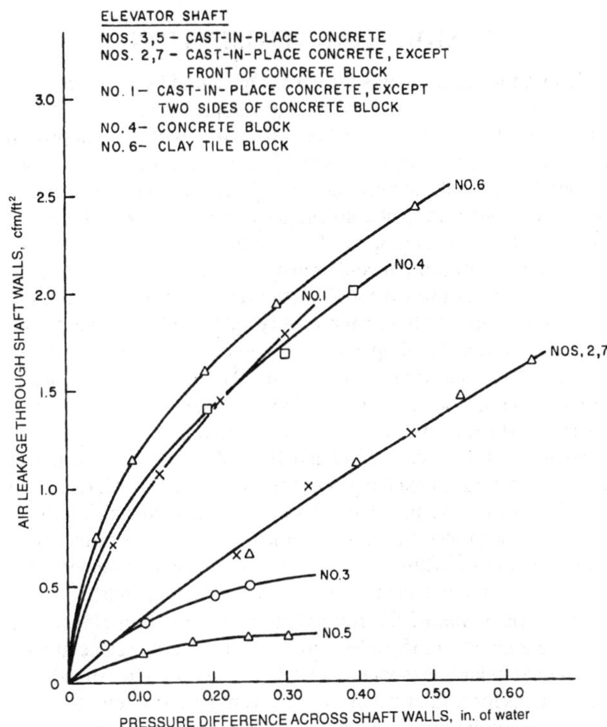

Fig. 14 Air Leakage Rates of Elevator Shaft Walls

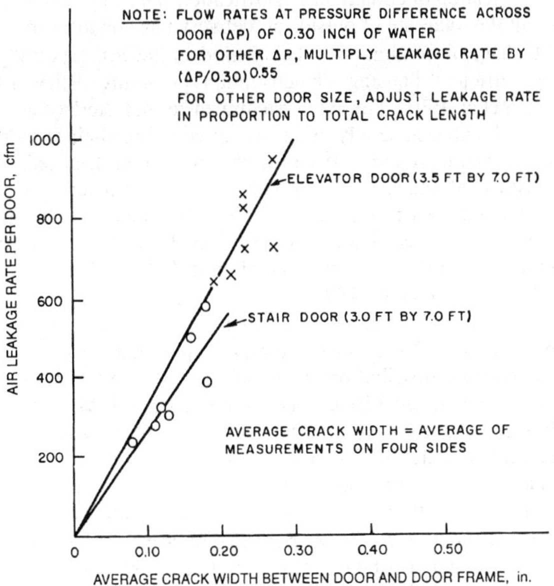

Fig. 15 Air Leakage Rate of Door Versus Average Crack Width

A frequently opened single door, as in a small retail store, has a much larger amount of airflow than a closed door. An ASHRAE research program provided data on air leakage characteristics of swinging door entrances (Min 1958, Tamura and Wilson 1966, 1967a) and revolving doors (Schutrum et al. 1961). A design chart (Min 1961) based on the data from Schutrum et al. (1961) evaluates infiltration through manual and power-operated revolving doors.

Grot and Persily (1986) also found that eight recently constructed office buildings had infiltration rates ranging from 0.1 to 0.6 ACH with no outdoor air intake. The infiltration rates of these buildings exhibited varying degrees of weather dependence, generally much lower than that measured in houses.

RESIDENTIAL VENTILATION

Ventilation requirements for houses have traditionally been met on the assumption that the building envelope is leaky enough that infiltration will suffice. Possible difficulties with this approach include low ventilation when natural forces (temperature difference and wind) are weak, unnecessary energy consumption when such forces are strong, drafts in cold climates, lack of control of ventilation rates to meet changing needs, potential for interstitial condensation from exfiltration in cold climates, and lack of opportunity to recover the energy used to condition the ventilation air. The solution to these concerns is to have a reasonably tight building envelope and a properly designed and operated mechanical ventilation system.

ASHRAE *Standard* 119 and the National Building Code of Canada (NRCC 1995) encourage the transition to tighter envelope construction. Hamlin (1991) shows a 30% increase in airtightness of tract-built Canadian houses between 1982 and 1989. Also, 82% of the newer houses had natural air exchange rates below 0.3 ACH in March. Yuill et al. (1991) derived a procedure to show the extent to which infiltration contributes toward meeting ventilation rate requirements. As a result, the National Building Code of Canada has requirements for mechanical ventilation capability in all new dwelling units.

ASHRAE *Standard* 62 gives ventilation rate requirements for houses, essentially 0.35 ACH with at least 15 cfm per occupant. Canadian Standards Association (CSA) *Standard* F326 expands the requirements for residential mechanical ventilation systems to cover air distribution within the house, thermal comfort, minimum temperatures for equipment and ductwork, system controls, pressurization and depressurization of the dwelling, installation requirements, and verification of compliance. Verification can be by design or by test, but the total rate of outside air delivery must be measured.

Mechanical ventilation is being used in houses, especially in energy-efficient housing demonstration programs (Riley 1990, Palmiter et al. 1991). Possible systems can be characterized as local or central; exhaust, supply, or balanced; with forced-air or radiant/hydronic heating/cooling systems; with or without heat recovery; and with continuous, occupant controlled, or demand controlled (i.e., by pollutant sensing) operation. Note that not all combinations are viable. Various options are described by Fisk et al. (1984), Hekmat et al. (1986), Sibbitt and Hamlin (1991), Palmiter et al. (1991), and Yuill et al. (1991).

The simplest systems use bathroom and kitchen fans to augment infiltration. Noise, installed capacity, lifetime under continuous operation, distribution to all rooms (especially bedrooms), and energy efficiency issues need to be addressed. Many present bath and kitchen fans are ineffective ventilators because of poor installation and design. However, properly specified and installed exhaust fans can form part of good whole-house ventilation systems and are so specified in some Canadian building codes.

Central supply systems use a furnace blower to induce air from the outdoors and distribute it. However, if the blower is thermostatically controlled, it will operate intermittently and provide little ventilation in mild weather. If the blower operates continuously, cold drafts can be a problem when the furnace is off.

Central exhaust systems use leakage sites and, in some cases, intentional and controllable openings in the building envelope as the supply. Such systems are suitable for retrofit in existing houses. Energy can be recovered from the exhaust airstream with a heat pump to supplement domestic hot water and/or space heating.

For new houses with tightly constructed envelopes, balanced ventilation systems with passive heat recovery (air-to-air heat exchangers or heat recovery ventilators) are appropriate. Fan-induced supply and exhaust air flows at nearly equal rates over a heat exchanger, where heat and sometimes moisture is transferred between the airstreams. This reduces the energy required to condition the ventilation air by typically 60 to 80% (Cutter 1987). It also reduces the thermal comfort problem that occurs when untempered air is introduced directly into the house. Airflow balance, leakage between streams, biological contamination of wet surfaces, and frosting are concerns associated with these systems.

Sibbitt and Hamlin (1991) found several low-cost mechanical systems that, with proper design and commissioning, met CSA *Standard* F326 requirements. Palmiter et al. (1991) found that mechanical systems provide significant ventilation only when operated continuously, met CSA *Standard* F326 requirements. A separate ventilation system (air-to-air heat exchanger or exhaust air heat pump) or an existing system integrated with multispeed fans can best accomplish continuous ventilation. Continuous central supply systems are not recommended for cold climates because they cause cold air drafts and condensation in the walls.

The type of ventilation system can be selected based on house leakage class as defined in Table 2. Balanced air-to-air systems with heat recovery are optimal for tight houses (leakage classes A-C). The leakier the house is, the larger will be the contribution from infiltration and the less effective will be heat recovery ventilation. In mild climates, these systems can also effectively be used in leakage classes D-F. Central exhaust systems should not be used for leakage classes A-C unless special provisions are made for air inlets; otherwise, internal pressure will become too low during operation and backdrafting through fossil-fueled appliances could occur. Central exhaust systems are optimal for leakage classes D-F. Ventilation systems are normally not needed for leakage classes G-J, but for those cases in which they are, central exhaust is usually the best choice.

Residential Ventilation Zones

For guidance in the selection of residential ventilation systems, Sherman (1995) developed four climatic zones for the United States. These zones are shown in Figure 16 for the continental United States. Alaska is in Zone 1, and Hawaii is in Zone 4.

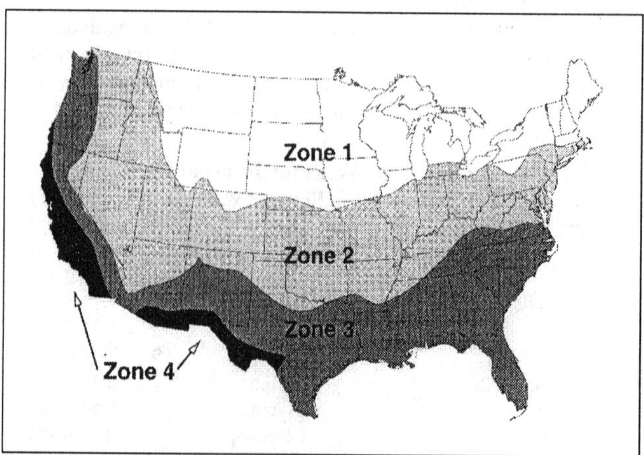

Fig. 16 Airtightness Zones for Residences in the United States
(Sherman 1995)

Zone 1 includes the severe climates of the northern tier of states. A Zone 1 residence that meets airtightness and energy conservation standards probably cannot meet its ventilation needs through infiltration and will require forced (mechanical) ventilation. Zone 2 includes the moderate climates where careful design and construction may allow buildings to simultaneously meet energy standards and ventilation needs through infiltration and mechanical exhaust. The mild climates in Zone 3 allow residences to meet both ASHRAE *Standards* 119 and 62 over a substantial range of airtightness. Zone 4 residences have relatively small energy penalties associated with infiltration or ventilation. In this zone, natural ventilation is usually preferred to forced ventilation as a technique to supplement infiltration.

Combining Residential Infiltration and Ventilation

Both significant infiltration and ventilation often occur simultaneously in residences. Usually, these flow rates are determined individually. When the leakage distribution and driving forces are known, superposition can be used to combine the flow rates (Sherman 1992). The following method is used in ASHRAE *Standard* 136.

1. If the ventilation supply and exhaust flow rates are balanced, then they are simply added to the infiltration flow rate.
2. If the ventilation supply and exhaust flow rates are unbalanced (Q_{unbal}) and much larger than the infiltration rate, then infiltration may be ignored.
3. If the unbalanced flow rate is much smaller than the infiltration flow rate, then only half is counted:

$$Q_{comb} = Q_{bal} + \frac{Q_{unbal}}{2} + Q_{infiltration} \tag{44}$$

4. If the unbalanced flow rate is not much larger or much smaller than the infiltration flow rate, then the combined flow rate is

$$Q_{comb} = Q_{bal} + \sqrt{Q^2_{unbal} + Q^2_{infiltration}} \tag{45}$$

SIMPLIFIED MODELS

Techniques for calculating building air exchange rates have improved in recent years (Liddament 1983, Liddament and Allen 1983). This section describes several calculation procedures, ranging from simple estimation techniques to more physical models. The air exchange rate of a building cannot be reliably deduced from the building's construction or age or from a simple visual inspection. Some measurement is necessary, such as a pressurization test of envelope airtightness or a detailed quantification of the leakage sites and their magnitude. As discussed in the section on Driving Mechanisms, it is straightforward to calculate the air exchange rate of a building given (1) the location and leakage function for every opening in the building envelope and between major building zones, (2) the wind pressure coefficients over the building envelope, and (3) any mechanical ventilation airflow rates. These inputs are generally unavailable for all except very simple structures or extremely well studied buildings. Therefore, assumptions as to their values must be made. The appropriateness of these assumptions determines the accuracy of predictions of air exchange rates.

Empirical Models

These models of residential infiltration are based on statistical fits of infiltration rate data for specific houses. They use pressurization test results to account for house airtightness and take the form of simple relations between infiltration rate, an airtightness rating, and, in most cases, weather conditions. Empirical models account for envelope infiltration only and do not deal with intentional ventilation. In one approach, the ACH at 0.20 in. of water from a pressurization test is simply divided by a constant approximately equal to 20 (Sherman 1987). This technique does not account for the effect of infiltration driving mechanisms on air exchange. Empirical models that do account for weather effects have been developed by Reeves et al. (1979), Kronvall (1980a), and Shaw (1981).

The latter two models account for building air leakage using the values of c and n from Equation (34). The only other inputs required are wind speed and temperature difference. Such empirical models predict long-term (one-week) infiltration rates very well in the houses from which they were developed; they do not, however, work as well in other houses due to the building-specific nature of leakage distribution, wind pressure, and internal partitioning. Persily and Linteris (1983) and Persily (1986) show comparisons between measured and predicted house infiltration rates for these and other models. The average long-term differences between measurements and predictions are generally on the order of 40%, although individual predictions can be off by 100% or more.

Single-Zone Models

Several procedures have been developed to calculate building air exchange rates that are based on physical models of the building interior as a single zone. These single-zone models are only appropriate to buildings with no internal resistance to airflow and are therefore inappropriate to large, multizone buildings. Models of this type have been developed by the Institute of Gas Technology (IGT) (Cole et al. 1980), the Building Research Establishment (Warren and Webb 1980), and the Lawrence Berkeley National Laboratory (LBNL) (Sherman and Grimsrud 1980). The LBNL model has been widely used and serves as the basis of the calculation procedure described in the section on Residential Calculation Examples. It uses pressurization test results to characterize house air leakage through the effective air leakage area A_L at 0.016 in. of water (discharge coefficient $C_D = 1$). In addition to wind speed and temperature difference, the user must input information on distribution of leakage over the building envelope, a shielding parameter, and a local terrain coefficient. The predictive accuracy of this model can be very good (±7% for weekly, ±20% for short term) when the parameters are well known for the building in question (Sherman and Modera 1986), but the predictions are not as accurate when the parameters are not known. All these single-zone models are sensitive to the values of the inputs, which are quite difficult to determine. These models have exhibited average errors on the order of 40% for many measurements on groups of houses and can be off by 100% in individual cases (Persily 1986).

Multizone Models

Multicell models of air exchange treat buildings as a series of interconnected zones and assume that the air within each zone is well mixed. Several such models have been developed by Allard and Herrlin (1989), Etheridge and Alexander (1980), Liddament and Allen (1983), Walton (1984, 1989), Herrlin (1985), and Feustel and Raynor-Hoosen (1990). They are all based on a mass balance for each zone of the building. These mass balances are used to solve for an interior static pressure within the building by requiring that the inflows and outflows for each zone balance to zero. The models require the user to input a location and leakage function for every opening in the building envelope and in relevant interior partitions, a value for the wind pressure coefficient C_p at the location of each building envelope leakage site, and any mechanical ventilation airflow rates. Such detailed information is difficult to obtain for a building. Wind pressure coefficient data in the literature, air leakage measurement results from the building or its components, and air leakage data from the literature can be used. These models not only solve for whole-building and individual zone air exchange rates, but

Table 5 Example of Calculation of Building Effective Air Leakage Area Based on Component Leakage Areas

Component	Description	Size or Number	×	A_L per unit	=	A_L, in²
Sills	Uncaulked	142 ft		0.19 in²/ft		27.0
Electrical outlets		20		0.08 in² ea		1.6
Windows	Sliding	141 ft²		0.057 in²/ft²		11.4
Framing		141 ft²		0.024 in²/ft²		
Exterior doors	Single	62 ft²		0.11 in²/ft²		8.3
Framing		62 ft²		0.024 in²/ft²		
Fireplace	Without damper	1		54.0 in² ea		54.0
Penetrations	Pipes	7		0.93 in² ea		6.5
Heating ducts	Ducts untaped, in basement	1		22.0 in² ea		22.0
Calculated total building air leakage area A_c =						130.8 in²

Table 6 Stack Coefficient C_s

	House Height (Stories)		
	One	Two	Three
Stack coefficient	0.0150	0.0299	0.0449

also determine airflow rates between zones. These interzone airflow rates are useful for predicting pollutant transport within buildings and smoke movement patterns in the event of a fire. Multizone models have the advantage of providing a physically correct determination of airflow rates, and very complex representations of buildings can be easily modeled on personal computers; however, determining the correct inputs to these models is difficult.

Residential Calculation Examples

This section presents a simple, single-zone approach to calculating air infiltration rates in houses based on the LBNL model (Sherman and Grimsrud 1980). The approach requires the effective air leakage area at 0.016 in. of water, which can be obtained from a whole-building pressurization test. If a test value is not available, the data in Table 3 can be used to estimate the air leakage area of the building. To obtain the building's total air leakage area, multiply the overall dimensions or number of occurrences of each building component by the appropriate table entry. The sum of the resulting products is the total building air leakage area.

Table 5 gives the result of an example calculation of the effective air leakage area of a residence. Each leakage component is identified in the first column and described in the second. The length, area, or number of the component is in the third column. The fourth column contains the air leakage area per unit component, from Table 3, and the fifth contains the total air leakage area associated with that component. The sum of the terms in the last column is the total air leakage area of the building, in this case 131 in².

Using the effective air leakage area, the airflow rate due to infiltration is calculated according to

$$Q = A_L \sqrt{C_s \Delta t + C_w V^2} \qquad (46)$$

where

Q = airflow rate, cfm
A_L = effective air leakage area, in²
C_s = stack coefficient, cfm²/(in⁴·°F)
Δt = average indoor-outdoor temperature difference for time interval of calculation, °F
C_w = wind coefficient, cfm²/(in⁴·mph²)
V = average wind speed measured at local weather station for time interval of calculation, mph

The air exchange rate I of the building is obtained by dividing Q by the building volume V. Table 6 presents values of C_s for one-, two-, and three-story houses. The value of the wind coefficient C_w depends on the local shielding class of the building and the building height. Table 7 lists five different shielding classes. Table 8 presents values of C_w for one-, two-, and three-story houses in shielding classes 1 through 5. In calculating the values in Table 6 and Table 8, the following assumptions were made regarding input to the LBNL model:

- Terrain class = 3 (rural area with scattered obstacles)
- R = 0.5 (half of the building leakage in the walls)
- X = 0 (equal amounts of leakage in the floor and ceiling)
- Heights of one-, two-, and three-story buildings = 8, 16, and 24 ft, respectively

Table 7 Local Shielding Classes

Class	Description
1	No obstructions or local shielding
2	Light local shielding; few obstructions, few trees, or small shed
3	Moderate local shielding; some obstructions within two house heights, thick hedge, solid fence, or one neighboring house
4	Heavy shielding; obstructions around most of perimeter, buildings or trees within 30 ft in most directions; typical suburban shielding
5	Very heavy shielding; large obstructions surrounding perimeter within two house heights; typical downtown shielding

Table 8 Wind Coefficient C_w

Shielding Class	House Height (Stories)		
	One	Two	Three
1	0.0119	0.0157	0.0184
2	0.0092	0.0121	0.0143
3	0.0065	0.0086	0.0101
4	0.0039	0.0051	0.0060
5	0.0012	0.0016	0.0018

Example 1. Estimate the infiltration at design conditions for a two-story house in Lincoln, Nebraska. The house has an effective air leakage area of 77 in² and a volume of 12,000 ft³, and it is surrounded by a thick hedge (shielding class 3). The indoor air temperature is 68°F.

Solution: The 97.5% design temperature for Lincoln is −2°F. Assume a design wind speed of 15 mph. From Equation (46), with C_s = 0.0299 from Table 6 and C_w = 0.0086 from Table 8, the airflow rate due to infiltration is

$$Q = 77\sqrt{(0.0299)(70) + (0.0086)(15)^2}$$
$$= 155 \text{ cfm} = 9300 \text{ ft}^3/\text{h}$$

From Equation (1), the air exchange rate I is equal to Q divided by the building volume:

$$I = (9300 \text{ ft}^3/\text{h})/12{,}000 \text{ ft}^3$$
$$= 0.78 \text{ h}^{-1} = 0.78 \text{ ACH}$$

Example 2. Calculate the average infiltration during a one-week period in January for a one-story house in Portland, Oregon. During this period, the average indoor-outdoor temperature difference is 30°F, and the average wind speed is 6 mph.

The house has a volume of 9000 ft³ and an effective air leakage area of 107 in², and it is located in an area with buildings and trees within 30 ft in most directions (shielding class 4).

Solution: From Equation (46), the airflow rate due to infiltration is

$$Q = 107\sqrt{(0.0150)(30) + (0.0039)(6^2)}$$
$$= 82.2 \text{ cfm} = 4930 \text{ ft}^3/\text{h}$$

The air exchange rate is therefore

$$I = 4930/9000 = 0.55\ \text{h}^{-1} = 0.55\ \text{ACH}$$

This value of air exchange rate is an estimate of the average value over the one-week interval for which the weather information was obtained and averaged.

Example 3. Estimate the average infiltration over the heating season in a two-story house with a volume of 11,000 ft^3 and the air leakage area calculated in Table 5 (131 in^2). The house is located on a lot with several large trees but no other close buildings (shielding class 3). The average wind speed during the heating season is 7 mph while the average indoor-outdoor temperature difference is 36°F.

Solution: From Equation (46), the airflow rate due to infiltration is

$$Q = 131\sqrt{(0.0299)(36) + (0.0086)(7^2)}$$

$$= 160\ \text{cfm} = 9620\ \text{ft}^3/\text{h}$$

The average air exchange rate is therefore

$$I = 9620/11{,}000 = 0.87\ \text{h}^{-1} = 0.87\ \text{ACH}$$

Again, this estimate is valid for the time interval used in computing the average values of the weather variables. Because the temperature difference and wind speed are values averaged over the entire heating season, the infiltration estimate is valid over the same interval.

NONRESIDENTIAL VENTILATION

Low-rise commercial and institutional building ventilation systems are typically designed to provide a slight pressurization to minimize infiltration. This pressurization is achieved by having the outside or makeup airflow rate higher than the exhaust or relief airflow rate. In low-rise commercial and institutional buildings, infiltration is usually neglected except in areas such as lobbies, where infiltration can be significant due to doors. As discussed in the section on Driving Mechanisms, wind and the stack effect can also cause significant infiltration and exfiltration. Ventilation airflow rates for commercial and institutional buildings are determined from ASHRAE *Standard* 62, Ventilation for Acceptable Indoor Air Quality. In the design of forced ventilation systems for commercial and institutional buildings, ventilation air credit is usually not taken for infiltration. However, the pressure differentials that drive infiltration and exfiltration, if significant, must be considered when selecting the air distribution system to ensure delivery of ventilation and conditioning air.

ASHRAE *Standard* 62 includes two procedures for obtaining acceptable indoor air quality: the Ventilation Rate Procedure and the Indoor Air Quality Procedure. The Ventilation Rate Procedure is by far the more commonly used.

Ventilation Rate Procedure

A basis for the Ventilation Rate Procedure is the concept that occupants are a significant source of indoor air pollution. The ventilation rates required to control this pollution can be expressed as a flow rate of clean outdoor air per occupant. The required outdoor airflow rates for many types of facilities are listed in Table 2 of *Standard* 62. In other types of facilities (e.g., swimming pools and pet shops), the main pollutant sources are unrelated to human occupancy; therefore, outdoor airflow rates per unit floor area are used to express the required ventilation rates. These flow rates are also listed in Table 2 of *Standard* 62.

The HVAC designer faces several challenges in developing an air distribution system to deliver outdoor air to the occupants of a building. The first is to determine whether the outdoor air is acceptable for use and to design a system for cleaning the air if it is not acceptable. A second problem for the designer is to develop an air intake and mixing system that will deliver to the building the needed amount of outdoor air. This outdoor air must be delivered not only at the design conditions but throughout the year. The task is complicated by weather-related variations in indoor-outdoor pressure difference, by pressure variations due to building components such as exhaust fans or dirty filters, and probably most severely by flow rate variations caused by the operation of variable air volume (VAV) systems (Mumma and Wong 1990, Janu et al. 1995).

Even when the outdoor air is delivered to the building, another challenge remains: delivering it to the occupants in their various locations. Unfortunately, the ratio of the required outdoor airflow rate to the supply airflow rate needed to meet the thermal load is not the same in all spaces. ASHRAE *Standard* 62 includes the multiple spaces equation [Equation (47) of this chapter] to address this difficulty.

The delivery of outdoor air, not just to the building but to the rooms where it is required, is further complicated when a VAV system is used. This is because the flow rate of supply air delivered to each room varies over time as the cooling loads in the rooms vary. Thus, the outdoor air fraction required in each of the rooms, including the critical room, will also vary. This means that the outdoor air fraction required at the central air-handling unit will vary as well, reaching a maximum when supply airflow rates are low.

Another problem the designer faces in the delivery of outdoor air to a building is the air change effectiveness within each room. It is not enough to supply the required outdoor air to the space; the ventilation air must be delivered to the occupants. This problem is discussed in the section on Air Change Effectiveness presented earlier in this chapter.

Multiple Spaces. Table 2 in ASHRAE *Standard* 62 prescribes the outdoor air requirements of the occupants of various types of spaces. In Section 6.1.3.2, *Standard* 62 describes how to deal with the problem of delivering this outdoor air to each of the spaces served by a single air-handling system. The problem is that each of these spaces will require a different fraction of outdoor air in the supply air delivered to it, but a central air-handling unit serving multiple thermal zones can only supply a single outdoor air fraction.

At first glance, it might appear that it would be necessary to find the critical space (the one with the highest required outdoor air fraction) served by the air-handling system and set the minimum outdoor air fraction of the AHU to the value the critical space requires. Such a system would overventilate all the other spaces served, use excessive energy for conditioning the outdoor air, and increase the required capacity of the heating and cooling equipment.

Fortunately, none of this is necessary. Because the return air from all the noncritical zones served by the system has an airborne contaminant level below the allowable limit, the combined recirculated air still has the capacity to remove contaminants from the critical and other spaces. This fact is recognized in the multiple spaces equation (Equation 6-1) of *Standard* 62:

$$Y = \frac{X}{[1 + X - Z]} \tag{47}$$

where

$Y = Q_{oa,req}/Q_{sa}$ = corrected fraction of outdoor air in system supply
$X = Q_{oa,nom}/Q_{sa}$ = uncorrected fraction of outdoor air in system supply
$Z = Q_{oa,crit}/Q_{sa,crit}$ = required fraction of outdoor air in critical space
$Q_{oa,req}$ = required outdoor airflow rate
Q_{sa} = total supply airflow rate of air-handling system
$Q_{oa,nom}$ = sum of outdoor airflow rates for all spaces served by system (i.e., design occupancy of each space multiplied by outdoor air required per person in that type of space, summed for all spaces)
$Q_{oa,crit}$ = total outdoor airflow rate required in critical space
$Q_{sa,crit}$ = supply airflow rate to critical space

Equation (47) expresses the corrected fraction Y of outdoor air in the system supply, which is the fraction that the system must actually deliver. The ratio X/Y can be thought of as an efficiency of the ventilation system. It is the ratio of the pollutant concentration in the relief air being exhausted from the building by the air-handling system to the pollutant concentration in a space where each occupant receives just the prescribed amount of outdoor air.

An examination of Equation (47) shows that the outdoor air fraction Y actually required by the air-handling system will always be less than the outdoor air fraction Z required in the critical space, except when the critical zone requires 100% outside air. In that case, $Z = 1.0$ and $Y = X$.

Constant Volume Systems. For most types of constant volume systems, if the designer applies Equation (47), the result should be reduced energy consumption for conditioning ventilation air. When multiple interior spaces have the same occupant density and cooling load, Y will also be equal to X. Other spaces with different ratios of outdoor air requirement to cooling load will have different required outdoor air fractions. There are several approaches the designer can use to reduce the required amount of outdoor air in large buildings. One is to group zones that require similar outdoor air fractions on the same air-handling system. Another is to provide direct ventilation for problem areas such as conference rooms. A third is to transfer air from an overventilated zone to one that requires a high outdoor air fraction.

Variable Air Volume Systems. Suppling the required amounts of outdoor air to all the spaces served by a VAV system presents the designer with a more complex problem. The flow rate to each space in the building will vary throughout the year. Therefore, the outdoor air fraction required in each space will change continually. Different spaces will be the critical one (with the highest required outdoor air fraction) at different hours of the year.

The first element of this problem is the flow rate analysis. In simple VAV systems, Equation (47) can be used, but it must be applied for each hour of the year. This means that the designer must use an hour-by-hour building energy analysis computer program to calculate the hourly cooling load for each space. Then Equation (47) can be applied in each hour, and the minimum outdoor airflow rate required for each hour can be found.

Some systems require more complex analysis. In systems with fan-powered terminal units and in dual-fan/dual-duct systems, the outdoor air fractions delivered to the individual spaces can be different from the outdoor air fraction leaving the primary air-handling unit. Warden (1995) presents a set of equations similar to Equation (47) that apply to these systems.

The second element of this problem is the development of an economical design. In cooler climates, most buildings contain spaces that will require as much cooling air as outdoor air in some hours of the year. This means that when Equation (47) is applied, the entire air-handling system will be found to require 100% outdoor air in those particular hours. If an air-mixing system with a constant "minimum outdoor air" setting were used, this setting would have to be 100%, greatly increasing the heating and cooling loads in the building. This suggests that the ventilation system should be designed to vary the outdoor airflow rate as required throughout the year.

Other means of reducing the required flow rate of outdoor air in VAV systems include those suggested for constant volume systems. In addition, limiting the minimum airflow rates of the VAV boxes in the critical zones may reduce the outdoor air requirements for the whole building. Of course, that will increase the use of reheat, but it may reduce the total energy consumption of the building. It is also possible to use a real-time optimization control technique that will either apply reheat to the critical space or increase the outdoor air supply, whichever is more economical at that time (Mumma and Bolin 1994).

Economizer Cycles. The multiple spaces ventilation problems discussed should not be encountered at moderate outdoor temperatures for air-handling systems having air-side economizer cycles. In the "economizer" range of operation, the air-handling unit supplies 100% outdoor air, but the chiller also operates. In the "free cooling" range, the chiller does not run and the outdoor airflow rate is modulated to meet the thermal loads. In the economizer range and part of the free cooling range, the outdoor airflow rate will likely be above that required by the critical zone. It is at high and low outdoor temperatures, when the air-handling system is delivering the minimum amount of outdoor air, that it becomes necessary to ensure that the amount meets the requirements of Equation (47). However, designers should be cautious about calculating the lower limit of the free cooling range. At low outdoor temperatures, perimeter and top floor spaces require reduced supply airflow rates, so that the outdoor temperature at which the free cooling outdoor air fraction falls to the minimum outdoor air fraction required by Equation (47) will be quite high.

SYMBOLS

a = angle, degrees
A = area, ft^2 or in^2
c = flow coefficient, cfm/(in. of water)^n
c_p = specific heat, Btu/lb·°F
C = concentration
C_D = discharge coefficient
C_p = pressure coefficient
C_v = effectiveness of openings
F = tracer gas injection rate, cfm
h = specific enthalpy, Btu/lb
h_{fg} = heat of vaporization, Btu/lb$_m$
H = height, ft
I = air exchange rate, 1/time
IDD = infiltration degree-days, °F·day
n = flow exponent
p = pressure, in. of water
q = heat rate, Btu/h
Q = volumetric flow rate, cfm
S = source strength, μg/mins
t = relative temperature, °F
T = absolute temperature, °R
V = volume, ft^3; or velocity, mph
W = humidity ratio, lb$_m$ water/lb$_m$ dry air
X = nominal outdoor air fraction (uncorrected)
X_{oa} = outdoor air fraction
Y = corrected outdoor air fraction
Z = required outdoor air fraction in critical space
ε_I = air change effectiveness
ρ = air density, lb$_m$/ft^3
θ = time
θ_{age} = age of air
τ = time constant

Subscripts

b = base
ba = bypass air
c = calculated
ca = recirculated air
ea = exhaust air
f = floor
i = indoor
ka = makeup air
l = latent
la = relief air
L = leakage or local
ma = mixed air
n = normalized
N = nominal
NPL = neutral pressure level
o = outdoor, initial condition, or reference
oa = outside air
p = pressure
r = reference
ra = return air
s = sensible or stack
S = space or source
sa = supply air
ta = transfer air
w = wind

REFERENCES

ACGIH. 1995. *Industrial ventilation: A manual of recommended practice*, 22nd ed. American Conference of Governmental Industrial Hygienists, Lansing, MI.

Allard, F. and M. Herrlin. 1989. Wind-induced ventilation. *ASHRAE Transactions* 95(2):722-28.

ASHRAE. 1988. Air leakage performance for detached single-family residential buildings. ANSI/ASHRAE *Standard* 119-1988 (RA 94).

ASHRAE. 1989. Ventilation for acceptable indoor air quality. ANSI/ASHRAE *Standard* 62-1989.

ASHRAE. 1993. A method of determining air change rates in detached dwellings. ANSI/ASHRAE *Standard* 136-1993.

ASTM. 1987. Practices for air leakage site detection in building envelopes. *Standard* E 1186-87(R 1992). American Society for Testing and Materials, West Conshohocken, PA.

ASTM. 1988. Standard test method for determining air leakage rate by fan pressurization. *Standard* E 779-88.

ASTM. 1991. Test method for determining the rate of air leakage through exterior windows, curtain walls, and doors under specified pressure differences across the specimen. *Standard* E 283-91.

ASTM. 1993. Test method for field measurement of air leakage through installed exterior windows and doors. *Standard* E 783-93.

ASTM. 1993. Test methods for determining air change in a single zone by means of a tracer gas dilution. *Standard* E 741-93.

Berg-Munch, B., G. Clausen, and P.O. Fanger. 1986. Ventilation requirements for the control of body odor in spaces occupied by women. *Environmental International* 12(1-4):195.

Blomsterberg, A.K. and D.T. Harrje. 1979. Approaches to evaluation of air infiltration energy losses in buildings. *ASHRAE Transactions* 85(1):797.

Bohac, D., D.T. Harrje, and G.S. Horner. 1987. Field study of constant concentration and PFT infiltration measurements. Proceedings of the 8th IEA Conference of the Air Infiltration and Ventilation Centre, Uberlingen, Germany.

Bohac, D., D.T. Harrje, and L.K. Norford. 1985. Constant concentration infiltration measurement technique: An analysis of its accuracy and field measurements, 176. Proceedings of the ASHRAE-DOE-BTECC Conference on the Thermal Performance of the Exterior Envelopes of Buildings III, Clearwater Beach, FL.

CGSB. 1986. Determination of the airtightness of building envelopes by the fan depressurization method. CGSB *Standard* 149.10-M86. Canadian General Standards Board, Ottawa.

Charlesworth, P.S. 1988. Measurement of air exchange rates. Chapter 2 in *Air exchange rate and airtightness measurement techniques—An applications guide*. Air Infiltration and Ventilation Centre, Coventry, Great Britain.

Chastain, J.P. 1987. Pressure gradients and the location of the neutral pressure axis for low-rise structures under pure stack conditions. Unpublished M.S. thesis. University of Kentucky, Lexington.

Chastain, J.P. and D.G. Colliver. 1989. Influence of temperature stratification on pressure differences resulting from the infiltration stack effect. *ASHRAE Transactions* 95(1):256-68.

Chastain, J.P., D.G. Colliver, and P.W. Winner, Jr. 1987. Computation of discharge coefficients for laminar flow in rectangular and circular openings. *ASHRAE Transactions* 93(2):2259-83.

Cole, J.T., T.S. Zawacki, R.H. Elkins, J.W. Zimmer, and R.A. Macriss. 1980. Application of a generalized model of air infiltration to existing homes. *ASHRAE Transactions* 86(2):765.

Collet, P.F. 1981. Continuous measurements of air infiltration in occupied dwellings, 147. Proceedings of the 2nd IEA Conference of the Air Infiltration Centre, Stockholm, Sweden.

Colliver, D.G., W.E. Murphy, and W. Sun. 1992. Evaluation of the techniques for the measurement of air leakage of building components. *Final Report* of ASHRAE Research Project RP-438. University of Kentucky, Lexington.

CSA. 1991. Residential mechanical ventilation systems. CAN/CSA-F326-M91. Canadian Standards Association, Toronto.

Cummings, J.B. and J.J. Tooley, Jr. 1989. Infiltration and pressure differences induced by forced air systems in Florida residences. *ASHRAE Transactions* 96(20):551-60.

Cummings, J.B., J.J. Tooley, Jr., and R. Dunsmore. 1990. Impacts of duct leakage on infiltration rates, space conditioning energy use and peak electrical demand in Florida homes. Proceedings of ACEEE Summer Study, Pacific Grove, CA. American Council for an Energy-Efficient Economy, Washington, D.C.

Cutter. 1987. Air-to-air heat exchangers. In *Energy design update*. Cutter Information Corporation, Arlington, MA.

Desrochers, D. and A.G. Scott. 1985. Residential ventilation rates and indoor radon daughter levels, 362. Transactions of the APCA Specialty Conference, Indoor Air Quality in Cold Climates: Hazards and Abatement Measures, Ottawa, Canada.

Diamond, R.C., M.P. Modera, and H.E. Feustel. 1986. Ventilation and occupant behavior in two apartment buildings. Lawrence Berkeley National Laboratory *Report* LBL-21862. Presented at the 7th AIC Conference, Stratford-upon-Avon, UK.

Diamond, R.C., J.B. Dickinson, R.D. Lipschutz, B. O'Regan, and B. Shohl. 1982. The house doctor's manual. *Report* PUB 3017. Lawrence Berkeley National Laboratory, Berkeley, CA.

Dickerhoff, D.J., D.T. Grimsrud, and R.D. Lipschutz. 1982. Component leakage testing in residential buildings. Proceedings of the American Council for an Energy-Efficient Economy, 1982 Summer Study, Santa Cruz, CA. *Report* LBL 14735. Lawrence Berkeley National Laboratory, Berkeley, CA.

Dietz, R.N., R.W. Goodrich, E.A. Cote, and R.F. Wieser. 1986. Detailed description and performance of a passive perfluorocarbon tracer system for building ventilation and air exchange measurement. In *Measured air leakage of buildings*, p. 203. ASTM STP 904. H.R. Trechsel and P.L. Lagus, eds. American Society for Testing and Materials, West Conshohocken, PA.

D'Ottavio, T.W., G.I. Senum, and R.N. Dietz. 1988. Error analysis techniques for perfluorocarbon tracer derived multizone ventilation rates. *Building and Environment* 23(40).

Ek, C.W., S.A. Anisko, and G.O. Gregg. 1990. Air leakage tests of manufactured housing in the Northwest United States. In *Air change rate and airtightness in buildings*, pp. 152-64. ASTM STP 1067. M.H. Sherman, ed. American Society for Testing and Materials, West Conshohocken, PA.

Elmroth, A. and P. Levin. 1983. Air infiltration control in housing, a guide to international practice. *Report* D2:1983. Air Infiltration Centre, Swedish Council for Building Research, Stockholm.

Energy Resource Center. 1982. How to house doctor. University of Illinois, Chicago.

Etheridge, D.W. 1977. Crack flow equations and scale effect. *Building and Environment* 12:181.

Etheridge, D.W. and D.K. Alexander. 1980. The British gas multi-cell model for calculating ventilation. *ASHRAE Transactions* 86(2):808.

Etheridge, D.W. and J.A. Nolan. 1979. Ventilation measurements at model scale in a turbulent flow. *Building and Environment* 14(1):53.

Eto, J. 1990. The HVAC costs of increased fresh air ventilation rates in office buildings. Proceedings of Indoor Air '90 4:53-58. International Conference on Indoor Air Quality and Climate, Ottawa, Canada.

Eto, J. and C. Meyer. 1988. The HVAC costs of increased fresh air ventilation rates in office buildings. *ASHRAE Transactions* 94(2):331-45.

Eyre, D. and D. Jennings. 1983. Air-vapour barriers—A general perspective and guidelines for installation. Energy, Mines, and Resources Canada, Ottawa.

Farrington, R., D. Martin, and R. Anderson. 1990. A comparison of displacement efficiency, decay time constant, and age of air for isothermal flow in an imperfectly mixed enclosure. Proceedings of the ACEEE 1990 Summer Study on Energy Efficiency in Buildings, pp. 4.35-4.43. American Council for an Energy-Efficient Economy, Washington, D.C.

Feustel, H.E. and A. Raynor-Hoosen, eds. 1990. Fundamentals of the multizone air flow model—COMIS. *Technical Note* 29. Air Infiltration and Ventilation Centre, Coventry, Great Britain.

Fisk, W.J., R.J. Prill, and O. Steppanen. 1989. A multi-tracer technique for studying rates of ventilation, air distribution patterns and air exchange efficiencies. Proceedings of Conference on Building Systems—Room Air and Air Contaminant Distribution, pp. 237-40. ASHRAE, Atlanta.

Fisk, W.J., R.K. Spencer, D.T. Grimsrud, F.J. Offermann, B. Pedersen, and R. Sextro. 1984. Indoor air quality control techniques: A critical review. *Report* LBL 16493. Lawrence Berkeley National Laboratory, Berkeley, CA.

Fortmann, R.C., N.L. Nagda, and H.E. Rector. 1990. Comparison of methods for the measurement of air change rates and interzonal airflows to two test residences. In *Air change rate and airtightness in buildings*, pp. 104-18. ASTM STP 1067. M.H. Sherman, ed., American Society for Testing and Materials, West Conshohocken, PA.

Foster, M.P. and M.J. Down. 1987. Ventilation of livestock buildings by natural convection. *Journal of Agricultural Engineering Research* 37:1.

Gadsby, K.J. and D.T. Harrje. 1985. Fan pressurization of buildings: Standards, calibration and field experience. *ASHRAE Transactions* 91(2): 95-104.

Giesbrecht, P. and G. Proskiw. 1986. An evaluation of the effectiveness of air leakage sealing. In *Measured air leakage of buildings*, p. 312. ASTM STP 904. H.R. Trechsel and P.L. Lagus, eds. American Society for Testing and Materials, West Conshohocken, PA.

Grieve, P.W. 1989. *Measuring ventilation using tracer-gases*. Brüel and Kjær, Denmark.

Grimsrud, D.T. and K.Y. Teichman. 1989. The scientific basis of *Standard* 62-1989. *ASHRAE Journal* 31(10):51-54.

Grimsrud, D.T., M.H. Sherman, and R.C. Sonderegger. 1982. Calculating infiltration: Implications for a construction quality standard. Proceedings of the ASHRAE-DOE Conference on the Thermal Performance of the Exterior Envelope of Buildings II, p. 422. Las Vegas, NV.

Grimsrud, D.T., M.H. Sherman, R.C. Diamond, P.E. Condon, and A.H. Rosenfeld. 1979. Infiltration-pressurization correlations: Detailed measurements in a California house. *ASHRAE Transactions* 85(1):851.

Grot, R.A. and R.E. Clark. 1979. Air leakage characteristics and weatherization techniques for low-income housing. Proceedings of the ASHRAE-DOE Conference on the Thermal Performance of the Exterior Envelopes of Buildings, p. 178. Orlando, FL.

Grot, R.A. and A.K. Persily. 1986. Measured air infiltration and ventilation rates in eight large office buildings. In *Measured air leakage of buildings*, p. 151. ASTM STP 904. H.R. Trechsel and P.L. Lagus, eds. American Society for Testing and Materials, West Conshohocken, PA.

Hamlin, T.L. 1991. Ventilation and airtightness in new, detached Canadian housing. *ASHRAE Transactions* 97(2):904-10.

Harrje, D.T. and G.J. Born. 1982. Cataloguing air leakage components in houses. Proceedings of the ACEEE 1982 Summer Study, Santa Cruz, CA. American Council for an Energy-Efficient Economy, Washington, D.C.

Harrje, D.T. and T.A. Mills, Jr. 1980. Air infiltration reduction through retrofitting, 89. *Building air change rate and infiltration measurements*. ASTM STP 719. C.M. Hunt, J.C. King, and H.R. Trechsel, eds. American Society for Testing and Materials, West Conshohocken, PA.

Harrje, D.T., G.S. Dutt, and J. Beyea. 1979. Locating and eliminating obscure but major energy losses in residential housing. *ASHRAE Transactions* 85(2):521.

Harrje, D.T., R.A. Grot, and D.T. Grimsrud. 1981. Air infiltration site measurement techniques. Proceedings of the 2nd IEA Conference of the Air Infiltration Centre, p. 113. Stockholm, Sweden.

Harrje, D.T., G.S. Dutt, D.L. Bohac, and K.J. Gadsby. 1985. Documenting air movements and infiltration in multicell buildings using various tracer techniques. *ASHRAE Transactions* 91(2):2012-27.

Harrje, D.T., R.N. Dietz, M. Sherman, D.L. Bohac, T.W. D'Ottavio, and D.J. Dickerhoff. 1990. Tracer gas measurement systems compared in a multifamily building. In *Air change rate and airtightness in buildings*, p. 5-12. ASTM STP 1067. M.H. Sherman, ed. American Society for Testing and Materials, West Conshohocken, PA.

Hekmat, D., H.E. Feustel, and M.P. Modera. 1986. Impacts of ventilation strategies on energy consumption and indoor air quality in single-family residences. *Energy and Buildings* 9(3):239.

Herrlin, M.K. 1985. MOVECOMP: A static-multicell-airflow-model. *ASHRAE Transactions* 91(2B):1989.

Honma, H. 1975. *Ventilation of dwellings and its disturbances*. Faibo Grafiska, Stockholm.

Hopkins, L.P. and B. Hansford. 1974. Air flow through cracks. *Building Service Engineer* 42(September):123.

Hunt, C.M. 1980. Air infiltration: A review of some existing measurement techniques and data. *Building air change rate and infiltration measurements*, p. 3. ASTM STP 719. C.M. Hunt, J.C. King, and H.R. Trechsel, eds. American Society for Testing and Materials, West Conshohocken, PA.

Jacobson, D.I., G.S. Dutt, and R.H. Socolow. 1986. Pressurization testing, infiltration reduction, and energy savings. In *Measured air leakage of buildings*, p. 265. ASTM STP 904. H.R. Trechsel and P.L. Lagus, eds. American Society for Testing and Materials, West Conshohocken, PA.

Janssen, J.E. 1989. Ventilation for acceptable indoor air quality. *ASHRAE Journal* 31(10):40-48.

Janu, G.J., J.D. Wegner, C.G. Nesler. 1995. Outdoor air flow control for VAV systems. *ASHRAE Journal* 37(4):62-68.

Kiel, D.E. and D.J. Wilson. 1986. Gravity driven airflows through open doors, 15.1. Proceedings of the 7th IEA Conference of the Air Infiltration Centre, Stratford-upon-Avon, United Kingdom.

Kiel, D.E. and D.J. Wilson. 1987. Influence of natural ventilation on total building ventilation dominated by strong fan exhaust. *ASHRAE Transactions* 93(2):1286.

Kim, A.K. and C.Y. Shaw. 1986. Seasonal variation in airtightness of two detached houses. In *Measured air leakage of buildings*, p. 17. ASTM STP 904. H.R. Trechsel and P.L. Lagus, eds. American Society for Testing and Materials, West Conshohocken, PA.

Klauss, A.K., R.H. Tull, L.M. Rootsd, and J.R. Pfafflino. 1970. History of the changing concepts in ventilation requirements. *ASHRAE Journal* 12(6):51-55.

Klote, J.H. and J.W. Fothergill, Jr. 1983. *Design of smoke control systems for buildings*. ASHRAE.

Kronvall, J. 1978. Testing of homes for air leakage using a pressure method. *ASHRAE Transactions* 84(1):72.

Kronvall, J. 1980. Correlating pressurization and infiltration rate data—Tests of an heuristic model. Lund Institute of Technology, Division of Building Technology, Lund, Sweden.

Kumar, R., A.D. Ireson, and H.W. Orr. 1979. An automated air infiltration measuring system using SF6 tracer gas in constant concentration and decay methods. *ASHRAE Transactions* 85(2):385.

Kvisgaard, B. and P.F. Collet. 1990. The user's influence on air change. In *Air change rate and airtightness in buildings*, pp. 67-76. ASTM STP 1067. M.H. Sherman, ed. American Society for Testing and Materials, West Conshohocken, PA.

Lagus, P.L. 1989. Tracer measurement instrumentation suitable for infiltration, air leakage, and air flow pattern characterization. Proceedings of Conference on Building Systems—Room Air and Air Contaminant Distribution, pp. 97-102. ASHRAE, Atlanta.

Lagus, P. and A.K. Persily. 1985. A review of tracer-gas techniques for measuring airflows in buildings. *ASHRAE Transactions* 91(2B):1075.

Lee, B.E., M. Hussain, and B. Soliman. 1980. Predicting natural ventilation forces upon low-rise buildings. *ASHRAE Journal* 22(2):35-39.

Lee, K.H., Y. Lee, and H. Tanaka. 1985. Thermal effect on pressure distribution in simulated high-rise buildings—Experiment and analysis. *ASHRAE Transactions* 91(2):530-44.

Lee, K.H., H. Tanaka, and Y. Lee. 1988. Thermally induced pressure distribution in simulated tall buildings with floor partitions. *ASHRAE Transactions* 94(1):228-42.

Levins, W.P. 1982. Measured effect of forced ventilation on house infiltration rate. Proceedings of the ASHRAE-DOE Conference on the Thermal Performance of the Exterior Envelopes of Buildings II, Las Vegas, NV.

Liddament, M.W. 1983. The Air Infiltration Centre's Program of Model Validation. *ASHRAE Transactions* 89(2):129-45.

Liddament, M.W. 1988. The calculation of wind effect on ventilation. *ASHRAE Transactions* 94(2):1645-60.

Liddament, M. and C. Allen. 1983. The validation and comparison of mathematical models of air infiltration. *Technical Note* 11. Air Infiltration Centre, Bracknell, Great Britain.

Malik, N. 1978. Field studies of dependence of air infiltration on outside temperature and wind. *Energy and Buildings* 1(3):281.

Marbek Resource Consultants. 1984. Air sealing homes for energy conservation. Energy, Mines and Resources Canada, Buildings Energy Technology Transfer Program, Ottawa.

Mattingly, G.E. and E.F. Peters. 1977. Wind and trees: Air infiltration effects on energy in housing. *Journal of Industrial Aerodynamics* 2(1):1.

Min, T.C. 1958. Winter infiltration through swinging-door entrances in multistory buildings. *ASHRAE Transactions* 64:421.

Min, T.C. 1961. Engineering concept and design of controlling infiltration and traffic through entrances in tall commercial buildings. International Conference on Heating, Ventilating and Air Conditioning, London.

Modera, M.P. 1989. Residential duct system leakage: Magnitude, impacts, and potential for reduction. *ASHRAE Transactions*. 96(2):561-69.

Modera, M.P. and D.J. Wilson. 1990. The effects of wind on residential building leakage measurements. *Air change rate and airtightness in buildings*, pp. 132-45. ASTM STP 1067. M.H. Sherman, ed. American Society for Testing and Materials, West Conshohocken, PA; Lawrence Berkeley National Laboratory *Report* LBL-24195.

Modera, M.P., D. Dickerhoff, R. Jansky, and B. Smith. 1991. Improving the energy efficiency of residential air distribution systems in California. *Report* LBL-30866. Lawrence Berkeley National Laboratory, Berkeley, CA.

Mumma, S.A., and R.J. Bolin. 1994. Real-time, on-line optimization of VAV system control to minimize the energy consumption rate and to satisfy ASHRAE *Standard* 62-1989 for all occupied zones. *ASHRAE Transactions* 94(1):168-79.

Mumma, S.A., and Y.M. Wong. 1990. Analytical evaluation of outdoor airflow rate variation vs. supply airflow rate variation in VAV systems when the outside air damper position is fixed. *ASHRAE Transactions* 90(1):1197-1208.

Murphy, W.E., D.G. Colliver, and L.R. Piercy. 1991. Repeatability and reproducibility of fan pressurization devices in measuring building air leakage. *ASHRAE Transactions* 97(2):885-95.

Nelson, B.D., D.A. Robinson, and G.D. Nelson. 1985. Designing the envelope—Guidelines for buildings. Thermal Performance of the Exterior Envelopes of Buildings III. Proceedings of ASHRAE/DOE/ BTECC Conference, Florida. ASHRAE SP 49, pp. 1117-22.

NRCC. 1995. National Building Code of Canada. National Research Council of Canada, Ottawa.

Nylund, P.O. 1980. Infiltration and ventilation. *Report* D22:1980. Swedish Council for Building Research, Stockholm.

Offermann, F., and D. Int-Hout. 1989. Ventilation effectiveness measurements of three supply/return air configurations. *Environment International* 15(1-6):585-92.

Palmiter, L. and I. Brown. 1989. The Northwest Residential Infiltration Survey, description and summary of results. Proceedings of the ASHRAE/DOE/BTECC/CIBSE Conference—Thermal Performance of the Exterior Envelopes of Buildings IV, Florida, pp. 445-57.

Palmiter, L., I.A. Brown, and T.C. Bond. 1991. Measured infiltration and ventilation in 472 all-electric homes. *ASHRAE Transactions* 97(2):979-87.

Parekh, A., K. Ruest, and M. Jacobs. 1991. Comparison of airtightness, indoor air quality and power consumption before and after air-sealing of high-rise residential buildings. Proceedings of the 12th AIVC Conference: Air Movement and Ventilation Control within Buildings, Air Infiltration and Ventilation Centre, Coventry, Great Britain.

Parker, G.B., M. McSorley, and J. Harris. 1990. The Northwest Residential Infiltration Survey: A field study of ventilation in new houses in the Pacific Northwest. In *Air change rate and airtightness in buildings*, pp. 93-103. ASTM STP 1067. M.H. Sherman, ed. American Society for Testing and Materials, West Conshohocken, PA.

Persily, A. 1982. Repeatability and accuracy of pressurization testing. Proceedings of the ASHRAE-DOE Conference, Thermal Performance of the Exterior Envelopes of Buildings II, Las Vegas, NV.

Persily, A.K. 1986. Measurements of air infiltration and airtightness in passive solar homes. In *Measured air leakage of buildings*, p. 46. ASTM STP 904. H.R. Trechsel and P.L. Lagus, eds. American Society for Testing and Materials, West Conshohocken, PA.

Persily, A.K. 1988. Tracer gas techniques for studying building air exchange. *Report* NBSIR 88-3708. National Institute of Standards and Technology, Gaithersburg, MD.

Persily, A.K. 1991. Design guidelines for thermal envelope integrity in office buildings. Proceedings of the 12th AIVC Conference: Air Movement and Ventilation Control within Buildings, Air Infiltration and Ventilation Centre, Coventry, Great Britain.

Persily, A.K. and J. Axley. 1990. Measuring airflow rates with pulse tracer techniques. In *Air change rate and airtightness in buildings*, pp. 31-51. ASTM STP 1067. M.H. Sherman, ed. American Society for Testing and Materials, West Conshohocken, PA.

Persily, A.K. and R.A. Grot. 1985a. The airtightness of office building envelopes. Thermal Performance of the Exterior Envelopes of Buildings III. Proceedings of the ASHRAE-DOE-BTECC Conference, Clearwater Beach, FL, p. 125.

Persily, A.K. and R.A. Grot. 1985b. Accuracy in pressurization data analysis. *ASHRAE Transactions* 91(2B):105.

Persily, A.K. and R.A. Grot. 1986. Pressurization testing of federal buildings. In *Measured air leakage of buildings*, p. 184. ASTM STP 904. H.R. Trechsel and P.L. Lagus, eds. American Society for Testing and Materials, West Conshohocken, PA.

Persily, A.K. and G.T. Linteris. 1983. A comparison of measured and predicted infiltration rates. *ASHRAE Transactions* 89(2):183.

Persily, A.K. and L.K. Norford. 1987. Simultaneous measurements of infiltration and intake in an office building. *ASHRAE Transactions* 93(2):42-56.

Reeves, G., M.F. McBride, and C.F. Sepsy. 1979. Air infiltration model for residences. *ASHRAE Transactions* 85(1):667.

Riley, M. 1990. Indoor air quality and energy conservation: The R-2000 home program experience. Proceedings of Indoor Air '90 5:143. International Conference on Indoor Air Quality and Climate, Ottawa, Canada.

Robison, P.E. and L.A. Lambert. 1989. Field investigation of residential infiltration and heating duct leakage. *ASHRAE Transactions* 95(2):542-50.

Rock, B.A. 1992. Characterization of transient pollutant transport, dilution, and removal for the study of indoor air quality. Ph.D. diss., University of Colorado at Boulder. University Microfilms International.

Rock, B.A., M.J. Brandemuehl, and R. Anderson. 1995. Toward a simplified design method for determining the air change effectiveness. *ASHRAE Transactions* 101(1):217-27.

Sandberg, M.H. 1981. What is ventilation efficiency? *Building and Environment* 16:123-35.

Schutrum, L.F., N. Ozisik, C.M. Humphrey, and J.T. Baker. 1961. Air infiltration through revolving doors. *ASHRAE Transactions* 67:488.

Shaw, C.Y. 1981. A correlation between air infiltration and air tightness for a house in a developed residential area. *ASHRAE Transactions* 87(2):333.

Shaw, C.Y. 1987. Methods for estimating air change rates and sizing mechanical ventilation systems for houses. *ASHRAE Transactions* 93(2):2284-2302.

Shaw, C.Y. and W.C. Brown. 1982. Effect of a gas furnace chimney on the air leakage characteristic of a two-story detached house, 12.1. Proceedings of the 3rd IEA Conference of the Air Infiltration Centre, London.

Sherman, M.H. 1986. Infiltration degree-days: A statistic for quantifying infiltration-related climate. *ASHRAE Transactions* 92(2):161-81.

Sherman, M.H. 1987. Estimation of infiltration from leakage and climate indications. *Energy and Buildings* 10(1):81.

Sherman, M.H. 1989a. Uncertainty in airflow calculations using tracer gas measurements. *Building and Environment* 24(4):347-54.

Sherman, M.H. 1989b. On the estimation of multizone ventilation rates from tracer gas measurements. *Building and Environment* 24(4):355-62.

Sherman, M.H. 1990. Tracer gas techniques for measuring ventilation in a single zone. *Building and Environment* 25(4):365-74.

Sherman, M.H. 1992. Superposition in infiltration modeling. *Indoor Air* 2:101-14.

Sherman, M.H. 1995. The use of blower door data. *Indoor Air* 5:215-24.

Sherman, M.H. and D. Dickerhoff. 1989. Description of the LBL multitracer measurement system. Proceedings of the ASHRAE/DOE/BTECC/CIBSE Conference—Thermal Performance of the Exterior Envelopes of Buildings IV, pp. 417-32.

Sherman, M.H. and D.T. Grimsrud. 1980. Infiltration-pressurization correlation: Simplified physical modeling. *ASHRAE Transactions* 86(2):778.

Sherman, M.H., D.T. Grimsrud, P.E. Condon, and B.V. Smith. 1980. Air infiltration measurement techniques, 9. Proceedings of the 1st IEA Symposium of the Air Infiltration Centre, London. *Report* LBL 10705. Lawrence Berkeley National Laboratory, Berkeley, CA.

Sherman, M.H. and M.P. Modera. 1986. Comparison of measured and predicted infiltration using the LBL infiltration model. In *Measured air leakage of buildings*, p. 325. ASTM STP 904. H.R. Trechsel and P.L. Lagus, eds. American Society for Testing and Materials, West Conshohocken, PA.

Sherman, M.H. and D.J. Wilson. 1986. Relating actual and effective ventilation in determining indoor air quality. *Building and Environment* 21(3/4):135.

Sibbitt, B.E., and T. Hamlin. 1991. *Meeting Canadian residential ventilation standard requirements with low-cost systems.* Canada Mortgage and Housing Corporation, Ottawa.

Sinden, F.W. 1978a. Wind, temperature and natural ventilation—Theoretical considerations. *Energy and Buildings* 1(3):275.

Sinden, F.W. 1978b. Multi-chamber theory of air infiltration. *Building and Environment* 13:21-28.

Stricker, S. 1975. Measurement of air-tightness of houses. *ASHRAE Transactions* 81(1):148.

Swami, M.V. and S. Chandra. 1988. Correlations for pressure distribution on buildings and calculation of natural-ventilation airflow. *ASHRAE Transactions* 94(1):243-66.

Swedish Building Code. 1980. Thermal insulation and air tightness. SBN 1980.

Tamura, G.T. 1975. Measurement of air leakage characteristics of house enclosures. *ASHRAE Transactions* 81(1):202.

Tamura, G.T. and C.Y. Shaw. 1976a. Studies on exterior wall airtightness and air infiltration of tall buildings. *ASHRAE Transactions* 82(1):122.

Tamura, G.T. and C.Y. Shaw. 1976b. Air leakage data for the design of elevator and stair shaft pressurization system. *ASHRAE Transactions* 82(2):179.

Tamura, G.T. and A.G. Wilson. 1966. Pressure differences for a nine-story building as a result of chimney effect and ventilation system operation. *ASHRAE Transactions* 72(1):180.

Tamura, G.T. and A.G. Wilson. 1967a. Pressure differences caused by chimney effect in three high buildings. *ASHRAE Transactions* 73(2): II.1.1.

Tamura, G.T. and A.G. Wilson. 1967b. Building pressures caused by chimney action and mechanical ventilation. *ASHRAE Transactions* 73(2): II.2.1.

Tamura, G.T. and A.G. Wilson. 1968. Pressure differences caused by wind on two tall buildings. *ASHRAE Transactions* 74(2):170.

Turk, B.T., D.T. Grimsrud, J.T. Brown, K.L. Geisling-Sobotka, J. Harrison, and R.J. Prill. 1989. Commercial building ventilation rates and particle concentrations. *ASHRAE Transactions* 95(1):422-33.

Verschoor, J.D. and J.O. Collins. 1986. Demonstration of air leakage reduction program in navy family housing. In Measured air leakage of buildings, p. 294. ASTM STP 904. H.R. Trechsel and P.L. Lagus, eds. American Society for Testing and Materials, West Conshohocken, PA.

Walton, G.N. 1984. A computer algorithm for predicting infiltration and interroom airflows. *ASHRAE Transactions* 90(1B):601.

Walton, G.N. 1989. Airflow network models for element-based building airflow modeling. *ASHRAE Transactions* 95(2):611-20.

Warden, D. 1995. Outdoor air: Calculation and delivery. *ASHRAE Journal* 37(6):54-63.

Warren, P.R. and B.C. Webb. 1980. The relationship between tracer gas and pressurization techniques in dwellings. Proceedings of the 1st IEA Symposium of the Air Infiltration Centre, London.

Warren, P.R. and B.C. Webb. 1986. Ventilation measurements in housing. CIBSE Symposium, Natural Ventilation by Design. Chartered Institution of Building Services Engineers, London.

Weidt, J.L., J. Weidt, and S. Selkowitz. 1979. Field air leakage of newly installed residential windows. Proceedings of the ASHRAE-DOE Conference, Thermal Performance of the Exterior Envelopes of Buildings, p. 149. Orlando, FL.

Yaglou, C.P. and W.N. Witheridge. 1937. Ventilation requirements. *ASHVE Transactions* 43: 423.

Yaglou, C.P., E.C. Riley, and D.I. Coggins. 1936. Ventilation requirements. *ASHVE Transactions* 42:133.

Yuill, G.K. and G.M. Comeau. 1989. Investigation of the indoor air quality, air tightness and air infiltration rates of a random sample of 78 houses in Winnipeg, pp. 122-27. Proceedings of IAQ '89, The Human Equation—Health and Comfort. ASHRAE, Atlanta.

Yuill, G.K., M.R. Jeanson, and C.P. Wray. 1991. Simulated performance of demand-controlled ventilation systems using carbon dioxide as an occupancy indicator. *ASHRAE Transactions* 97(2):963-68.

BIBLIOGRAPHY

AIVC. 1992. AIRBASE Bibliographic Database. International Energy Agency Air Infiltration and Ventilation Centre, Coventry, Great Britain.

Fleming, W.S. 1986. Indoor air quality, infiltration and ventilation in residential buildings. Proceedings of IAQ '86, Managing Indoor Air for Health and Energy Conservation, pp. 192-207. ASHRAE, Atlanta.

Harrje, D.T. and K.J. Gadsby. 1986. Indoor air quality—Some residential answers. *ASHRAE Journal* 28(7):32-36.

Kronvall, J. 1980. Air flow in building components. *Report* TVBH-1002. Lund Institute of Technology, Division of Building Technology, Lund, Sweden.

Parker, D.S. 1990. Air tightness characteristics of electrically heated houses in the residential standards demonstration program. In *Air change rate and airtightness in buildings*, pp. 283-93. ASTM STP 1067. M.H. Sherman, ed. American Society for Testing and Materials, West Conshohocken, PA.

Reinhold, G. and R. Sonderegger. 1983. Component leakage areas in residential buildings. Proceedings of the 4th IEA Conference of the Air Infiltration Centre, Elm, Switzerland. *Report* LBL 16221. Lawrence Berkeley National Laboratory, Berkeley, CA.

Shaw, C.Y. and G.T. Tamura. 1977. The calculation of air infiltration rates caused by wind and stack action for tall buildings. *ASHRAE Transactions* 83(2):145.

Yuill, G.K. 1991. The development of a method of determining air change rates in detached dwellings for assessing indoor air quality. *ASHRAE Transactions* 97(2):896-903.

Yuill, G.K. and J.E. Lovatt. 1986. Prediction of pollutant concentration in houses. Proceedings of IAQ '86, Managing Indoor Air for Health and Energy Conservation, pp. 383-91. ASHRAE, Atlanta.

BIBLIOGRAPHIC DATABASE

AIRBASE, a database of bibliographic references that contains abstracts in English of technical papers covering air infiltration in buildings, has been developed by the Air Infiltration and Ventilation Centre (AIVC 1992). Most of the articles are concerned with the prediction, measurement, and reduction of air infiltration and leakage rates. Abstracts are also included for selected papers relating to indoor air quality, occupant behavior, thermal comfort, ventilation efficiency, natural and mechanical ventilation, wind pressure and its influence on infiltration, energy-saving measures, and moisture and condensation.

CHAPTER 26

CLIMATIC DESIGN INFORMATION

THIS chapter provides tables of climatic conditions for 1459 locations in the United States, Canada, and around the world. These summaries include values of dry-bulb, wet-bulb, and dew-point temperature and wind speed with direction at various frequencies of occurrence. This information is commonly used for design, sizing, distribution, installation, and marketing of heating, ventilating, air-conditioning, and dehumidification equipment; as well as for other energy-related processes in residential, agricultural, commercial and industrial applications. Sources of other information such as degree days and typical weather years for energy calculations are also described.

UPDATED INFORMATION

A recent research project (ASHRAE 1997a) developed new design information for this chapter. Limited information on psychrometric conditions were provided in the 1993 *ASHRAE Handbook* through the design values of dry-bulb temperature with mean coincident wet-bulb temperature and the design wet-bulb temperature. Design values of wet-bulb and dew-point temperature with mean coincident dry-bulb temperature and humidity ratio are now included. These new data allow the designer to consider additional operational peak conditions. Motivated by a need for the design of smoke management systems, design values of wind speed have also been added (Lamming and Salmon 1994).

The design conditons in this chapter are provided for those locations for which long-term hourly observations were available (at least 12 years). Consequently, many United States locations listed in previous versions of this chapter are no longer listed because they lacked long-term data. The number of Canadian and international locations has increased significantly.

Design values of temperature and humidity have been updated from the 1993 *ASHRAE Handbook*. The temperature and humidity conditions previously provided for at the 1, 2.5, and 5 percentile frequency of occurrence during summer months have been replaced by conditions corresponding to annual percentile values of 0.4, 1 and 2. Winter month conditions for 99 and 97.5 percentiles have been replaced by conditions based on annual percentiles of 99.6 and 99. This change was made in order to provide design conditions representing the same probability of occurrence anywhere, regardless of the seasonal distribution of extreme temperature and humidity. Previously, the summer and winter months used for the calculation of design conditions varied depending on location. For instance, summer design conditions for the United States were calculated over the four month period from June through September, whereas Canadian summer design conditions were based on only the month of July. The following sections describe how the annual percentiles were chosen to yield design conditions that are similar to those previously calculated on a seasonal basis for most of the United States.

CLIMATIC DESIGN CONDITIONS

Design conditions for the United States appear in Tables 1a and 1b, for Canada in Tables 2a and 2b, and for international locations in Tables 3a and 3b. Information on station location, period analyzed, the heating design conditions, wind, mean annual extreme and standard deviation of minimum and maximum dry-bulb temperature, and mean daily temperature range are listed in Tables 1a, 2a, and 3a. Information on the design conditions for cooling and humidity control are provided in Tables 1b, 2b and 3b.

The information provided in Tables 1a, 2a, and 3a in the indicated column numbers are:

1. Name of the observing station as it appears in the data set from which it was abstracted, the World Meteorological Organization station number, latitude, longitude, elevation, standard pressure at elevation (see Chapter 6 for the equations used to calculate standard pressure), and the period analyzed
2. Dry-bulb temperature corresponding to 99.6% and 99.0% annual cumulative frequency of occurrence (cold)
3. Wind speed corresponding to 1.0%, 2.5%, and 5.0% annual cumulative frequency of occurrence
4. Wind speed corresponding to the 0.4% and 1.0% cumulative frequency of occurrence for the coldest month (lowest average dry-bulb temperature) at the location, and the mean coincident dry-bulb temperature
5. Mean wind speed coincident with the 99.6% dry-bulb temperature in Column 2 and 0.4% dry-bulb temperature from Column 2 of Tables 1b through 3b, and the wind direction most frequently occurring with the 99.6% and 0.4% dry-bulb temperatures (direction is reported in degrees true: 360 represents a north wind, 90 east, etc.)
6. Average of annual extreme maximum and minimum dry-bulb temperatures and standard deviations

Information provided in Tables 1b, 2b, and 3b includes:

1. Station name
2. Dry-bulb temperature corresponding to 0.4%, 1.0%, and 2.0% annual cumulative frequency of occurrence and the mean coincident wet-bulb temperature (warm)
3. Wet-bulb temperature corresponding to 0.4%, 1.0%, and 2.0% annual cumulative frequency of occurrence and the mean coincident dry-bulb temperature
4. Dew-point temperature corresponding to 0.4%, 1.0%, and 2.0% annual cumulative frequency of occurrence and the mean coincident dry-bulb temperature and the humidity ratio (calculated for the dew-point temperature at the standard atmospheric pressure at the elevation of the station)
5. Mean daily range, which is the mean of the difference between daily maximum and minimum temperatures for the warmest month (highest average dry-bulb temperature)

Values of Cumulative Frequency of Occurrence Representing Design Conditions

Values of ambient dry-bulb, dew-point, and wet-bulb temperature and wind speed corresponding to the various annual percentiles represent the value that is exceeded on average by the indicated percentage of the total number of hours in a year (8760). The 0.4%, 1.0%, 2.0%, and 5.0% values are exceeded on average 35, 88, 175, and 438 hours.

The preparation of this chapter is assigned to TC 4.2, Weather Information.

The 99.0% and 99.6% (cold) values are defined in the same way but are usually viewed as the values for which the corresponding weather element are less than the design condition 88 and 35 hours, respectively. Mean coincident values are the average of the indicated weather element occurring concurrently with the corresponding design value.

These design conditions were calculated from the frequency distribution analyzed from data sets observed over several years. The design values occur more frequently than the corresponding nominal percentile in some years and less frequently in others.

Data Sources

The three following primary sources of observational data sets were used for the calculation of design values:

1. Hourly weather observations from Surface Airways Meteorological and Solar Observing Network (SAMSON) data from NCDC (National Climatic Data Center) for 239 United States observing locations from 1961 through 1990 (NCDC 1991)
2. Hourly observational records in the DATSAV2 format (provided by NCDC) for 860 international locations and for 538 United States locations for the period 1982 through 1993, including the 239 SAMSON locations
3. Hourly weather records for the period 1953 through 1993 for 145 Canadian locations from the Canadian Weather Energy and Engineering Data Sets (CWEEDS) produced by Environment Canada (1993b).

Two primary periods of record were used in the calculations. The values for the United States SAMSON and Canadian CWEEDS locations are generally based on the period 1961-1993. DATSAV2 data were used for the 1991-1993 period for the SAMSON sites. The values for international locations and the rest of the United States, whose data were analysed from the DATSAV2 files, are based on the period 1982 through 1993. DATSAV2 is a comprehensive database containing hourly observations for locations around the world collected from global telecommunications circuits. It is quality-controlled and archived by the Air Force Combat Climatology Center Operating Location A at Asheville, NC. Tables 1a, 2a and 3a indicate the period of record used for each location.

In summary, data source for United States locations with the period identified as "6193" is SAMSON data supplemented with DATSAV2 data for the last 3 years. The source for United States locations with the period "8293" is DATSAV2.

Calculation of Design Conditions

Details of the analysis procedures are available in ASHRAE (1997a), including the measures used to ensure that the number and distribution of missing data, both by month and by hour of the day, did not introduce significant biases into the analysis. Generally, the annual cumulative frequency distribution was constructed from the relative frequency distributions compiled for each month. Each individual month's data was included if they met screening criteria for completeness and unbiased distribution of missing data. Although the minimum period of record chosen for this analysis was 12 years (1982 through 1993), some variation and gaps in observing programs meant that some months' data were unusable due to incompleteness.

A station's design conditions were included in this chapter only if there were data from at least 8 months that met the screening criteria from the period of record for each month of the year. For instance, there had to be 8 months each for January, February, March, etc. whose data met the completeness screening criteria. Gaps of up to 5 hours were filled. A month's data were included if the month was at least 85% complete after filling and the difference between the number of day and nighttime observations was less than 60.

Relationship between Design Conditions and Design Temperatures Published Previously

The design conditions in this chapter are calculated on a different basis compared to the design conditions published in previous editions of this handbook. Previous design conditions were based on a 4-month summer period and a 3-month winter period in the United States, on the months of July and January in Canada, and the warmest 4-month period and coldest 3-month period in international locations. Although generally suitable as design values, the different periods resulted in design temperatures representing different annual probabilities of occurrence, depending on the country; and within countries, on the distribution of temperature and humidity conditions throughout the year typical of regional climatic zones. The design conditions in this chapter explicitly represent the same annual probability of occurrence in any location, regardless of country or general climatic conditions.

The annual cumulative frequency of occurrences representing the design dry-bulb temperatures generally correspond to the seasonal design temperatures in the following fashion for locations in the mid-latitude, continental locations (characterized by a hot summer and cold winter). The 0.4% annual value is about the same as the 1% summer design temperature. The 1% annual values is about 1°F lower than the 2.5% summer design temperature in the 1993 *ASHRAE Handbook*, and the 2% annual condition corresponds approximately to the 5% summer design temperature in the 1993 Handbook.

In Canadian continental locations, the 0.4% annual condition is about the same as the 2.5% July design temperature in the 1993 *ASHRAE Handbook*. In the Pacific region and southern coastal locations of the United States, where the extremes are generally more widely distributed throughout the year, the values in this chapter represent more extreme conditions than design temperatures in the 1993 *ASHRAE Handbook*.

Annual 99.6% and 99.0% design conditions represent a slightly colder condition than the previous cold season design temperatures, although there is considerable variability in this relationship from location to location.

Further details concerning differences between the design conditions in this Chapter and previous versions are described in ASHRAE (1997a).

Applicability and Characteristics of Design Conditions

The sets of design values in this chapter represent different psychrometric conditions. Design data based on dry-bulb temperature represent peak occurrences of the sensible component of ambient outdoor conditions. Design values based on wet-bulb temperature are related to the enthalpy of the outdoor air. Conditions based on dew point relate to the peaks of the humidity ratio. The designer, engineer, or other user must decide which set(s) of conditions and probability of occurrence apply to the design situation. The addition of the psychrometric design conditions in this chapter allows for several viewpoints of operational peak loads. Additional sources of information on the frequency and duration of extremes of temperature and humidity are provided later in this chapter.

Heating Design Conditions (Winter). The 99.6% and 99.0% design conditions in Column 2 in Tables 1a, 2a, and 3a are often used in sizing of heating equipment. In cold spells, dry-bulb temperatures below the design conditions can last for a week or more.

Columns 4 and 5 of Tables 1a, 2a, and 3a provide information useful for estimating peak loads accounting for infiltration. Column 4 provides extreme wind speeds only for the coldest month, with the mean coincident dry-bulb temperature. Column 5 provides the mean wind speed and direction coincident to the corresponding percentile design dry-bulb temperature.

Cooling and Dehumidification Design Conditions (Summer). The 0.4%, 1.0%, and 2.0% dry-bulb temperatures and mean coincident wet-bulb temperatures in Column 2 of Tables 1b, 2b, and 3b often represent conditions on hot, mostly sunny days. These are useful for cooling applications, especially air-conditioning.

Design conditions based on wet-bulb temperature in Column 3 represent extremes of the total sensible plus latent heat of outdoor air. This information is useful for cooling towers, evaporative coolers, and fresh air ventilation system design.

The design conditions based on dew-point temperatures in Column 4 of Tables 1b, 2b and 3b are directly related to extremes of humidity ratio, which represent peak moisture loads from the weather. Extreme dew-point conditions may occur on days with moderate dry-bulb temperatures resulting in high relative humidity. These values are especially useful for applications involving humidity control, such as desiccant cooling and dehumidification, cooling-based dehumidification, and fresh air ventilation systems. The values are also used as a check point when analyzing the behavior of cooling systems at part load conditions, particularly when such systems are used for humidity control as a secondary function.

The humidity ratio values in Column 2 correspond to the combination of dew-point temperature and the mean coincident dry-bulb temperature calculated at the standard pressure of the elevation of the location.

Wind. Design wind speeds in Column 3 of Tables 1a, 2a, and 3a are used for design of smoke management systems. Annual percentiles of 1.0, 2.5 and 5.0 are appropriate for this application. Values for United States and Canadian locations are from Lamming and Salmon (1995), in which adjustments to the standard 33 ft anemometer height are made. Wind speed values for other locations are taken from ASHRAE (1997a), in which no adjustment for non-standard anemometer height is made.

Annual Extreme Temperatures. Column 6 of Tables 1a, 2a, and 3a provides the mean and standard deviation of the annual extreme maximum and minimum dry-bulb temperatures. The probability of occurrence of very extreme conditions can be required for the operational design of equipment to ensure continuous operation and serviceability (regardless of whether the heating or cooling loads are being met). These values were calculated from the extremes of the hourly temperature observations. The true maximum and minimum temperatures for any day generally occurs between hourly readings Thus, the mean maximum and minimum temperatures calculated in this way are about 1°F less extreme than the mean daily extreme temperatures observed with maximum and minimum thermometers.

Return period (or recurrence interval) is defined as the reciprocal of the annual probability of occurrence. For instance, the 50-year return period maximum dry-bulb temperature has a probability of occurring or being exceeded of 2.0% (i.e. 1/50) each year. This statistic does not indicate how often the condition will occur in terms of the number of hours each year (as in the design conditions based on percentiles), but describes the probability of the condition occurring at all in any year. The following method can be used to estimate the return period (recurrence interval) of extreme temperatures.

$$T_n = M + IFs$$

where

T_n = n-year return period value of extreme dry-bulb temperature to be estimated;
M = mean of annual extreme maximum or minimum dry-bulb temp.
s = standard deviation of annual extreme maximum or minimum dry-bulb temperatures
I = 1 if maximum dry-bulb temperatures are being considered
I = −1 if minimum extremes are being considered

$$F = -\frac{\sqrt{6}}{\pi}\left[0.5772 + \ln\left(\frac{n}{n-1}\right)\right]$$

For example, the 50-year return period extreme maximum dry-bulb temperature estimated for Terre Haute, Indiana is 104°F (M = 96, s = 3.2, n = 50, I = 1). Similarly, the 100-year return period extreme minimum dry-bulb temperature for Winnipeg, Manitoba is −48°F (M = −33°F, s = 4.7, n = 100, I = −1).

This calculation is based on the assumptions that the annual maxima and minima are distributed according to the Gumbel (Type 1 Extreme Value) distribution and are fitted with the method of moments (Lowery and Nash 1970). The uncertainty or standard error using this method increases with increasing standard deviation, increasing value of return period, and decreasing length of the period of record; and it can be significant. For instance, the standard error in the 50-year return period maximum dry-bulb temperature estimated at a location with a 12-year period of record can be 5°F or more. Thus, the uncertainty of return-period values estimated in this way is greater for locations from the DATSAV2 data sets compared to those from the longer SAMSON and CWEEDS data sets.

Mean Daily Range. The mean daily range is the mean difference between the daily maximum and minimum temperatures during the hottest month. These values are calculated from the extremes of the hourly temperature observations. The true daily temperature range is generally about 2°F greater, for the same reason discussed in the previous section.

Representativeness of Data and Sources of Uncertainty

The information in the tables were obtained by direct analysis of the observations from the indicated locations. The design values are provided and used as an estimate of the annual cumulative frequency of occurrence of the weather conditions at the recording station for several years into the future. Several sources of uncertainty affect the accuracy of using the design conditions to represent other locations or periods.

The most important of these factors is spatial representativeness. Data representing the psychrometric conditions are generally properties of air masses, rather than local features, and tend to vary on regional scales. As a result, a particular value often may reasonably represent an area extending several miles. However, significant variations can occur with changes in local elevation, across large metropolitan areas, or in the vicinity of large bodies of water. Judgment must always be exercised in assessing the representativeness of the design conditions. It is especially important to note the elevation of locations in the tables because design conditions vary significantly for locations whose elevations differ by a few hundred feet or more. An applied climatologist should be consulted in estimating design conditions for locations not explicitly listed in this chapter.

Weather conditions vary from year to year, and to some extent, from decade to decade, due to the inherent variability of climate. Similarly, values representing design conditions vary depending on the period of record used in the analysis. Thus, there is always some uncertainty in using the design conditions from one period to represent another due to short term climatic variability. Typically, the values of design dry-bulb temperature vary less than 2°F from decade to decade but larger variations can occur. Differing periods used in the analysis can lead to differences in design conditions between nearby locations at similar elevations in the tables in this chapter. Design conditions may show trends in areas of increasing urbanization or other regions experiencing extensive changes to land use. Longer term climatic change due to human or natural causes may also introduce trends into design conditions, but no conclusive evidence or consensus of opinion is available on the rapidity or nature of such changes.

Wind speed is very sensitive to local exposure features such as terrain and surface cover. Wind speed values in Columns 3, 4, and 5 of Tables 1a, 2a, and 3a are often representative of a flat, open exposure, such as at airports. Wind engineering methods, as described in Chapter 14, can be used to account for exposure differences between airport and building sites. But, estimating exposure is a complex procedure, best undertaken by an experienced applied climatologist or wind engineer with knowledge of the exposure of the observing and building sites and surrounding regions.

OTHER SOURCES OF CLIMATIC INFORMATION

Joint Frequency Tables of Psychrometric Conditions

Most of the design values in this chapter were developed by a research project (ASHRAE 1997a). The joint frequency tables provide the annual and monthly joint frequency of occurrence of various psychrometric combinations for each location in the tables.

The International Station Meteorological Climate Summary (ISMCS) (NCDC 1996) is a CD-ROM containing several tables of climatic summary information from over 7000 locations around the world. Only several hundred of these locations have observed hourly weather data. A table providing the joint frequency of dry-bulb temperature and wet-bulb temperature depression is provided for these locations. It can be used to assist in estimating design conditions for locations for which no other information is available.

The monthly frequency distribution of dry-bulb temperatures and mean coincident wet-bulb temperatures are available for 134 Canadian locations from Environment Canada (1983-1987).

Some ASHRAE chapters have compiled design conditions for locations in their regions. Local chapter offices can be consulted on availability of such material. State climatologists in the United States are often a valuable source of information.

Degree-Days and Climatic Normals

Degree-day summary information and climatic normals for the United States are available on CD-ROM in the Climatography of the U.S. (NCDC 1992a, 1992b, 1994); and, for Canada, in the Canadian Climate Normals, 1961 to 1990 (Environment Canada 1993a).

Typical Year Data Sets

Software exists to simulate the annual energy performance of buildings requiring a one-year data set (8760 h) of weather conditions. Many data sets in different record formats have been developed to meet this requirement. The data represent a typical year from the viewpoint of weather-induced energy loads on a building. No explicit effort was made to represent extreme conditions, so these files do not represent design conditions.

Weather Year for Energy Calculations, Version 2 (WYEC2) for 52 United States and 5 Canadian locations is available from ASHRAE. A user manual and software toolkit explains the development, format, and characteristics of this data, and provides access and toolkit software.

Typical Meteorological Year (TMY) files were produced using an objective statistical algorithm to choose the most typical month from the long term record. These files were originally intended for the design of solar energy systems, and accordingly solar radiation values were weighted heavily. A new version of these files, TMY2, is available for 239 United States locations from the National Renewable Energy Laboratory (Marion and Urban 1995).

Canadian Weather year for Energy Calculation (CWEC) files for 47 Canadian locations were developed for use with the Canadian National Energy Code, using the TMY algorithm and software (Environment Canada 1993b).

Seasonal Percentiles of Dew-Point Temperature

Seasonal percentiles of dew-point temperature are available for 381 United States and Canadian locations in Colliver et al. (1995).

Sequences of Extreme Temperature and Humidity

ASHRAE (1997b) is a compilation of extreme sequences of 1, 3, 5, and 7-day duration for 239 United States (SAMSON data) and 144 Canadian (CWEEDS data) locations based independently on the following five criteria: high dry-bulb temperature, high dew-point temperature, high enthalpy, low dry-bulb temperature, and low wet-bulb depression. The sequences are selected according to annual percentiles of 0.4, 1.0, and 2.0 for each criteria (99.6, 99.0 and 98.0, in the case of low dry-bulb temperature). The data included for each hour of a sequence are solar radiation, dry-bulb and dew-point temperature, atmospheric pressure, and wind speed and direction. Accompanying information allows the user to go back to the source SAMSON and CWEEDS data and obtain sequences with different characteristics (i.e. different probability of occurrence, windy conditions, low or high solar radiation, etc.).

These sequences were developed primarily to assist in the design of heating or cooling systems having a finite capacity before regeneration is required, or that rely on thermal mass to limit loads. The information is also useful where information on the hourly weather sequence during extreme episodes is required for design.

Observational Data Sets

For some detailed designs, a custom analysis of the most appropriate long term weather record is best. National weather services are generally the best source of long term observational data. The WMO World Data Center A at the National Climate Data Center in Asheville, NC collects and makes available a significant volume of archived weather observations from around the world. The SAMSON and CWEEDS data sets provide long term hourly data, including solar radiation values, for the United States and Canada.

Increasingly, information about weather and climate services and data sets, and the data sets themselves are becoming available through the Internet and World Wide Web.

REFERENCES:

ASHRAE. 1997a. Updating the tables of design weather conditions in the ASHRAE Handbook of Fundamentals. *Research Report* RP-890.

ASHRAE. 1997b. Sequences of extreme temperature and humidity for design calculations. *Research Report* RP-828.

Colliver, D.G., H. Zhang, R.S. Gates, and K.T. Priddy. 1995. Design data for the 1%, 2%, and 5% occurrences of extreme dew-point temperature, with mean coincident dry-bulb temperature. *Final Report of* RP-754. ASHRAE, Atlanta.

Environment Canada. 1983-1987. Principal station data. PSD 1 to 134. Atmospheric Environment Service, Downsview, Ontario.

Environment Canada. 1993a. Canadian climate normals (1961-1990). Atmospheric Environment Service, Downsview, Ontario.

Environment Canada. 1993b. Canadian weather energy and engineering data sets (CWEEDS files) and Canadian weather for energy calculations (CWEC files) User's Manual. Atmospheric Environment Service, Downsview, Ontario.

Lamming, S.D. and J.R. Salmon. 1994. Wind data for design of smoke control systems. *Final Report of* ASHRAE RP-816. ASHRAE, Atlanta.

Lowery, M.D. and J. E. Nash. 1970. A comparison of methods of fitting the double exponential distribution. *Journal of Hydrology* 10(3): 259-275.

Marion, W. and K. Urban. 1995. User's Manual for TMY2s, typical meteorological years, derived from the 1961-1990 national solar radiation data base. NREL/SP-463-7668, E95004064. National Renewable Energy Laboratory, Golden, CO.

NCDC. 1991. Surface airways meteorological and solar observing network (SAMSON) data set. National Climatic Data Center, Asheville, NC.

NCDC. 1992a. Monthly normals of temperature, precipitation, and heating and cooling degree-days. In *Climatography of the U.S.* #81. National Climatic Data Center, Asheville, NC.

NCDC. 1992b, Annual degree-days to selected bases (1961-1990). In *Climatography of the U.S.* #81. National Climatic Data Center.

NCDC. 1994. U.S. division and station climatic data and normals. National Climatic Data Center, Asheville, NC.

NCDC. 1996. International station meteorological climate summary (ISMCS). National Climatic Data Center.

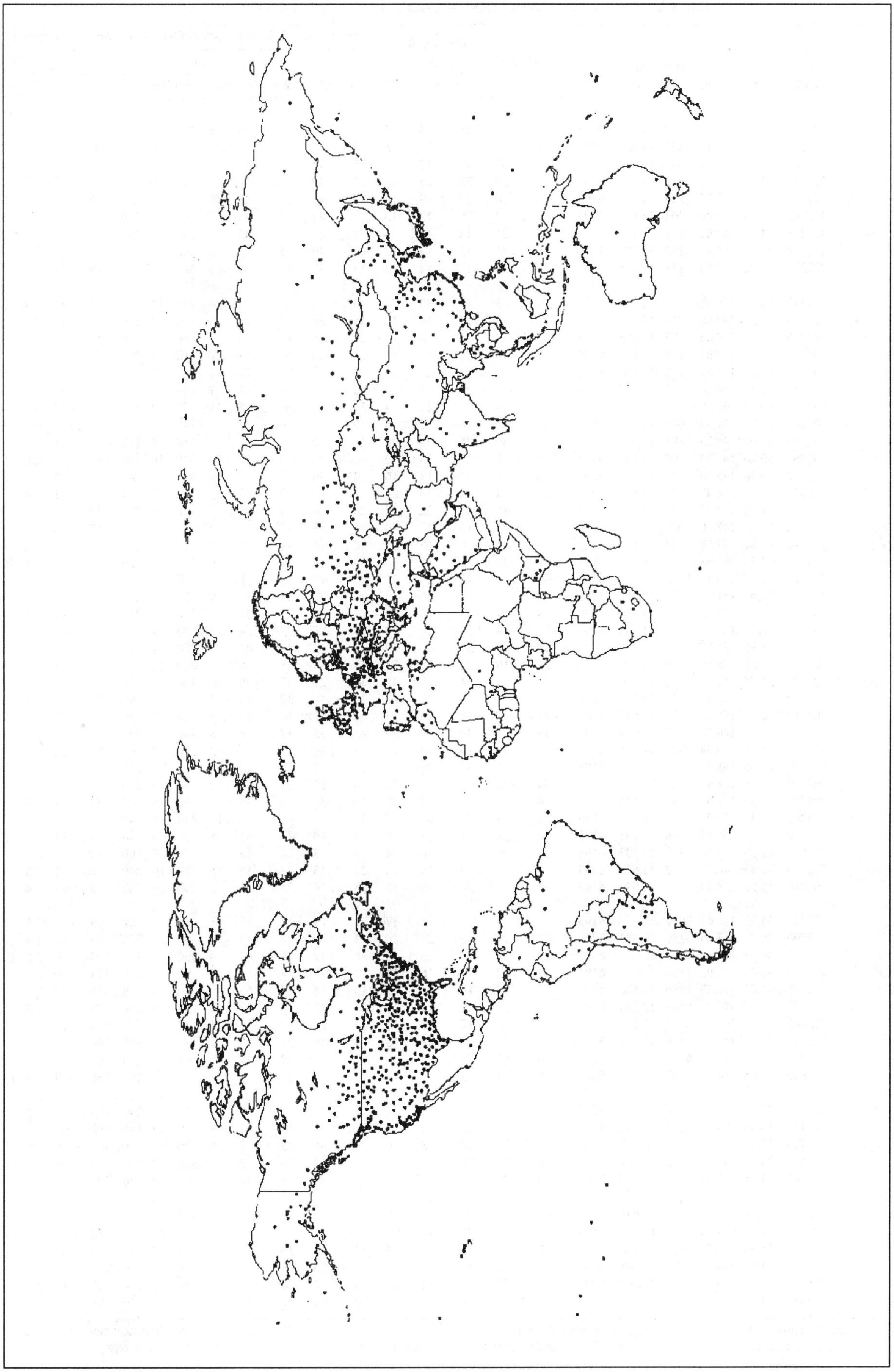

Fig. 1 Location of Weather Stations

Table 1A Heating and Wind Design Conditions—United States

Station	WMO#	Lat.	Long.	Elev. ft	StdP psia	Dates	Heating DB 99.6%	Heating DB 99%	Extreme Wind Speed 1%	Extreme Wind Speed 2.5%	Extreme Wind Speed 5%	Coldest Month 0.4% WS	Coldest Month 0.4% MDB	Coldest Month 1% WS	Coldest Month 1% MDB	MWS/MWD to DB 99.6% MWS	MWS/MWD to DB 99.6% PWD	MWS/MWD to DB 0.4% MWS	MWS/MWD to DB 0.4% PWD	Annual Extreme Daily Mean DB Max	Annual Extreme Daily Mean DB Min	Annual Extreme Daily StdD DB Max	Annual Extreme Daily StdD DB Min
ALABAMA																							
Anniston	722287	33.58	85.85	610	14.374	8293	19	24	16	14	13	18	47	15	46	6	300	7	240	98	10	3.2	7.4
Birmingham	722280	33.57	86.75	630	14.364	6193	18	23	19	17	15	20	41	18	42	7	340	9	320	98	9	3.3	6.4
Dothan	722268	31.32	85.45	400	14.484	8293	28	32	18	17	15	19	45	17	47	9	320	8	320	99	16	1.6	7.2
Huntsville	723230	34.65	86.77	643	14.357	6193	15	20	23	20	18	23	40	21	40	10	340	10	270	97	7	3.0	7.5
Mobile	722230	30.68	88.25	220	14.579	6193	26	30	22	19	17	23	48	21	48	10	360	9	320	97	18	1.9	6.3
Montgomery	722260	32.30	86.40	203	14.588	6193	24	27	20	17	15	20	45	18	45	7	360	8	270	98	15	2.9	6.3
Muscle Shoals/Florence	723235	34.75	87.62	551	14.405	8293	16	21	18	16	14	19	42	17	42	9	360	7	290	98	7	3.1	9.2
Ozark, Fort Rucker	722269	31.28	85.72	299	14.538	8293	28	31	16	13	12	17	49	15	47	5	340	5	300	99	18	2.3	5.9
Tuscaloosa	722286	33.22	87.62	171	14.605	8293	20	24	17	14	13	18	47	16	51	5	360	7	240	99	11	1.8	6.8
ALASKA																							
Adak, NAS	704540	51.88	176.65	13	14.688	8293	19	23	34	30	27	40	34	34	35	4	210	10	170	67	11	3.4	2.9
Anchorage, Elemendorf AFB	702720	61.25	149.80	213	14.583	8293	−13	−8	17	14	12	18	26	15	26	3	50	7	260	77	−18	3.2	6.5
Anchorage, Fort Richardson	702700	61.27	149.65	377	14.496	8293	−19	−13	19	14	11	20	35	15	36	3	50	5	270	80	−23	2.2	6.3
Anchorage, Int'l Airport	702730	61.17	150.02	131	14.626	6193	−14	−9	22	19	17	23	18	19	18	4	10	8	290	77	−18	2.9	7.2
Annette	703980	55.03	131.57	112	14.636	6193	13	17	31	27	23	31	41	28	40	10	40	8	320	81	10	3.8	5.4
Barrow	700260	71.30	156.78	13	14.688	6193	−41	−36	28	25	22	30	3	26	−1	7	140	12	90	65	−45	4.7	4.4
Bethel	702190	60.78	161.80	151	14.615	6193	−28	−24	31	27	24	34	8	30	5	13	20	12	360	78	−32	3.3	6.6
Bettles	701740	66.92	151.52	643	14.357	6193	−49	−44	18	16	14	19	11	16	7	2	340	8	190	85	−55	4.0	5.8
Big Delta, Ft. Greely	702670	64.00	145.73	1283	14.027	6193	−45	−39	34	29	25	38	0	33	3	3	180	9	180	84	−48	3.3	7.5
Cold Bay	703160	55.20	162.73	102	14.642	6193	6	10	38	34	30	46	34	40	34	15	340	16	140	67	2	4.0	5.3
Cordova	702960	60.50	145.50	43	14.673	8293	−4	1	22	19	16	24	40	22	38	1	340	8	240	79	−9	5	5.4
Deadhorse	700637	70.20	148.47	56	14.666	8293	−36	−34	32	28	25	34	−1	30	−7	12	240	12	60	78	−51	14	5.2
Dillingham	703210	59.05	158.52	95	14.645	8293	−20	−13	25	22	20	28	20	24	21	5	40	10	180	74	−27	3.1	9.4
Fairbanks, Eielson AFB	702650	64.67	147.10	548	14.407	8293	−33	−31	14	12	10	14	21	11	16	0	150	5	290	87	−46	3.8	7.7
Fairbanks, Int'l Airport	702610	64.82	147.87	453	14.457	6193	−47	−41	18	15	13	16	11	12	11	2	10	8	220	87	−48	3.8	7.8
Galena	702220	64.73	156.93	151	14.615	8293	−33	−31	18	15	13	19	14	16	15	0	270	5	320	84	−50	2.5	10.4
Gulkana	702710	62.15	145.45	1578	13.877	6193	−44	−39	26	24	21	22	17	19	18	3	360	7	180	82	−46	3.2	7.4
Homer	703410	59.63	151.50	72	14.657	8293	0	4	22	20	18	23	24	21	27	9	30	10	270	70	−5	4	6.8
Juneau	703810	58.37	134.58	23	14.683	8293	4	7	27	23	20	29	39	25	38	5	360	9	230	81	−1	2.5	4.9
Kenai	702590	60.57	151.25	95	14.645	8293	−22	−14	23	20	18	25	25	22	24	2	30	9	270	75	−27	3.4	7.4
Ketchikan	703950	55.35	131.70	95	14.645	8293	13	20	25	22	19	29	42	24	42	5	280	11	320	78	7	1.8	5.2
King Salmon	703260	58.68	156.65	49	14.669	6193	−24	−19	32	28	24	33	36	28	36	7	360	12	270	78	−31	3.5	7.2
Kodiak, State USCG Base	703500	57.75	152.50	112	14.636	6193	7	12	34	30	26	34	28	30	30	18	300	11	320	76	1	3.6	6.1
Kotzebue	701330	66.87	162.63	16	14.687	6193	−36	−31	35	31	28	38	14	32	14	7	70	12	300	75	−39	4.8	6.5
McGrath	702310	62.97	155.62	338	14.517	6193	−47	−42	18	16	14	18	23	14	12	1	310	7	340	83	−52	3.3	7.0
Middleton Island	703430	59.43	146.33	46	14.671	8293	18	21	40	34	30	42	35	37	36	18	330	8	260	66	15	4.9	6.8
Nenana	702600	64.55	149.08	361	14.505	8293	−51	−44	16	14	12	18	10	15	8	2	250	7	60	87	−52	4.1	7.2
Nome	702000	64.50	165.43	23	14.683	6193	−31	−26	30	26	23	31	17	28	18	4	20	12	260	76	−35	4.2	6.3
Northway	702910	62.97	141.93	1722	13.803	8293	−34	−32	15	13	12	14	−13	12	−6	0	300	7	290	83	−54	2.7	5.9
Port Heiden	703330	56.95	158.62	95	14.645	8293	−6	−2	38	32	28	38	36	32	35	17	60	15	160	74	−11	4.3	7.4
Saint Paul Island	703080	57.15	170.22	30	14.680	6193	−2	3	41	37	33	47	24	41	21	19	350	14	240	58	−3	5.2	6.9
Sitka	703710	57.07	135.35	66	14.661	8293	16	21	23	21	19	24	40	22	41	8	70	9	230	76	11	6.1	5.0
Talkeetna	702510	62.30	150.10	358	14.507	6193	−28	−21	17	16	14	19	13	17	15	4	50	8	200	82	−35	2.8	8.0
Valdez	702750	61.13	146.35	33	14.678	8293	4	7	24	19	16	28	13	22	15	15	70	10	240	76	1	3.6	6.1
Yakutat	703610	59.52	139.67	30	14.680	6193	−3	2	24	19	16	25	36	21	33	2	100	9	320	75	−8	4.0	7.0
ARIZONA																							
Flagstaff	723755	35.13	111.67	7011	11.335	6193	1	8	21	18	17	21	29	18	30	3	20	9	220	89	−10	2.5	7.3
Kingman	723700	35.27	113.95	3389	12.983	8293	22	27	26	23	20	24	46	21	43	5	90	13	240	103	15	1.8	6.8
Page	723710	36.93	111.45	4278	12.561	8293	20	24	19	16	13	16	42	12	40	4	300	7	360	104	8	3.6	12.2
Phoenix, Int'l Airport	722780	33.43	112.02	1106	14.118	6193	34	37	19	16	14	17	59	14	58	5	90	9	270	114	30	2.2	4.6
Phoenix, Luke AFB	722785	33.53	112.38	1089	14.126	8293	35	38	19	15	13	16	58	13	55	4	340	9	210	115	30	2.2	3.8
Prescott	723723	34.65	112.42	5043	12.208	6193	15	20	22	19	17	21	42	18	42	6	190	11	230	98	7	2.2	6.2
Safford, Agri Center	722747	32.82	109.68	3117	13.114	8293	21	26	17	14	12	15	50	13	48	4	110	7	310	106	11	3.8	11.5
Tucson	722740	32.12	110.93	2556	13.388	6193	31	34	24	21	18	24	56	21	56	7	140	12	300	108	25	2.8	4.0
Winslow	723740	35.02	110.73	4882	12.281	8293	10	14	26	22	19	24	46	19	45	5	140	9	250	100	3	4.9	6.3
Yuma	722800	32.65	114.60	207	14.586	8293	40	44	19	17	15	20	59	17	58	4	30	7	280	116	29	1.8	11.9
ARKANSAS																							
Blytheville, Eaker AFB	723408	35.97	89.95	256	14.560	8293	12	18	22	19	17	23	36	21	38	10	10	6	240	99	6	5	9.0
Fayetteville	723445	36.00	94.17	1250	14.044	8293	6	13	21	19	18	21	44	19	44	9	350	10	190	100	−1	3.2	9.2
Fort Smith	723440	35.33	94.37	463	14.451	6193	13	19	20	18	16	21	46	18	41	9	320	9	270	102	6	3.9	6.6
Little Rock, AFB	723405	34.92	92.15	312	14.531	8293	16	21	20	18	16	20	42	18	42	9	360	9	200	101	10	3.8	6.2
Texarkana	723418	33.45	93.98	390	14.489	8293	20	25	19	17	15	20	47	18	48	9	50	9	190	101	13	3.1	7.6
CALIFORNIA																							
Alameda, NAS	745060	37.78	122.32	13	14.688	8293	40	42	21	18	16	20	51	17	52	6	120	8	300	93	25	4.9	14.2
Arcata/Eureka	725945	40.98	124.10	217	14.581	6193	30	32	21	19	17	21	53	18	51	5	90	10	320	82	26	4.5	3.2
Bakersfield	723840	35.43	119.05	492	14.436	6193	32	35	19	16	14	19	56	14	54	5	90	12	310	108	28	2.3	3.4
Barstow/Daggett	723815	34.85	116.78	1926	13.701	6193	28	32	30	27	23	30	58	25	54	6	270	12	290	111	22	2.5	4.8
Blue Canyon	725845	39.28	120.72	5285	12.097	8293	21	24	15	12	11	16	35	14	35	5	70	6	290	89	11	3.8	18.4
Burbank/Glendale	722880	34.20	118.35	774	14.289	8293	39	41	18	14	12	20	56	17	57	2	330	8	180	106	33	3.1	4.5

WMO# = World Meteorological Organization number
Lat = latitude Long. = longitude
Elev = elevation, ft
StdP = standard pressure at station elevation, psia
DB = dry-bulb temperature, °F
WS = wind speed, mph

Table 1B Cooling and Dehumidification Design Conditions—United States

	Cooling DB/MWB						WB/MDB						DP/MDB and HR									
	0.4%		1%		2%		0.4%		1%		2%		0.4%			1%			2%			Range
Station	DB	MWB	DB	MWB	DB	MWB	WB	MDB	WB	MDB	WB	MDB	DP	HR	MDB	DP	HR	MDB	DP	HR	MCDB	of DB
ALABAMA																						
Anniston	95	76	93	76	90	75	79	90	78	88	77	86	77	143	84	76	137	82	75	133	81	19.6
Birmingham	94	75	92	75	90	74	78	89	77	88	76	87	75	135	83	74	131	82	73	127	81	18.7
Dothan	95	76	93	76	92	76	80	90	79	88	78	87	77	144	83	76	139	82	76	136	82	17.5
Huntsville	94	75	92	74	90	74	78	89	77	88	76	86	75	135	83	74	130	82	73	126	81	18.5
Mobile	94	77	92	76	91	76	79	89	79	88	78	87	77	142	83	76	139	83	76	135	82	16.5
Montgomery	95	76	93	76	91	76	79	91	78	89	78	88	76	139	85	75	134	84	75	130	83	18.7
Muscle Shoals/Florence	96	76	94	75	92	74	78	90	78	89	77	87	76	137	82	75	133	82	74	130	81	20.0
Ozark, Fort Rucker	95	77	94	77	92	76	81	90	79	89	78	88	78	146	85	77	142	84	76	138	83	18.0
Tuscaloosa	95	77	94	77	92	76	80	90	79	89	78	88	77	142	84	76	137	83	75	134	82	19.6
ALASKA																						
Adak, NAS	59	55	57	53	55	51	55	59	53	57	51	54	53	59	58	51	55	56	49	51	53	9.7
Anchorage, Elemendorf AFB	71	58	69	57	66	56	60	69	58	66	57	64	57	69	62	55	65	61	53	61	60	12.6
Anchorage, Fort Richardson	74	60	71	58	68	57	61	72	59	69	58	66	56	69	64	54	63	62	53	61	61	15.5
Anchorage, Int'l Airport	71	59	68	57	65	56	60	69	58	66	57	63	56	68	62	55	64	61	53	61	60	12.6
Annette	74	61	70	59	66	57	62	72	60	68	58	65	58	71	65	56	68	63	55	65	61	10.5
Barrow	57	51	52	49	48	46	52	56	49	52	46	48	49	53	54	46	46	51	44	42	48	10.6
Bethel	72	59	68	57	64	55	60	69	58	66	56	63	56	68	62	55	64	60	53	60	58	13.4
Bettles	79	61	75	59	72	58	63	76	61	73	59	70	58	72	66	56	67	64	54	63	63	19.4
Big Delta, Ft. Greely	78	59	75	58	71	56	61	74	59	72	58	69	56	70	65	54	65	63	52	61	61	17.3
Cold Bay	60	54	57	53	55	52	55	59	54	56	53	55	54	62	56	53	59	55	51	56	54	7.4
Cordova	70	59	67	57	63	56	60	69	58	65	56	62	56	67	63	54	62	60	53	60	59	13.5
Deadhorse	66	57	61	54	58	53	58	64	55	62	53	58	54	61	62	51	56	59	49	51	56	13.7
Dillingham	69	57	66	56	62	54	59	67	57	64	55	61	56	67	62	53	61	59	52	57	57	13.1
Fairbanks, Eielson AFB	81	61	78	60	75	59	64	78	62	75	60	72	58	74	66	56	69	66	54	63	66	19.4
Fairbanks, Int'l Airport	81	61	77	59	74	58	63	77	61	74	59	71	58	72	65	56	68	64	54	64	63	18.6
Galena	78	61	74	59	71	58	63	74	61	71	59	69	58	73	66	56	69	65	54	63	64	15.3
Gulkana	77	58	73	56	69	55	59	73	57	70	56	67	53	63	62	51	60	60	50	55	59	20.3
Homer	65	56	62	55	60	54	57	63	56	61	55	59	54	64	59	53	60	58	52	57	57	11.9
Juneau	74	60	69	58	66	57	61	71	59	68	58	64	57	70	63	56	67	61	55	64	60	13.9
Kenai	68	56	65	55	62	54	58	65	56	63	55	61	55	64	59	53	60	58	52	58	57	13.3
Ketchikan	71	60	68	59	66	58	62	69	60	67	59	64	59	74	64	57	71	62	56	68	61	10.3
King Salmon	71	58	67	56	64	55	59	68	57	65	56	62	56	66	61	54	62	60	52	58	58	15.5
Kodiak, State USCG Base	68	58	65	56	62	55	59	66	57	63	56	61	56	67	61	55	64	59	53	61	57	11.2
Kotzebue	68	59	64	58	61	56	60	67	58	64	56	61	57	70	64	55	65	61	54	62	59	8.8
McGrath	77	60	73	58	70	56	61	74	59	70	58	67	57	69	63	55	65	62	54	62	61	17.4
Middleton Island	62	54	60	51	59	51	55	61	54	59	53	57	52	57	56	51	56	56	50	54	55	5.8
Nenana	80	60	76	59	73	57	62	75	60	73	59	70	57	69	65	55	65	65	53	60	63	21.2
Nome	69	57	65	55	61	54	58	66	56	63	55	60	55	64	61	53	60	59	51	56	57	10.9
Northway	78	58	74	57	71	56	60	76	58	71	57	69	54	66	62	53	62	61	51	60	59	20.0
Port Heiden	64	54	61	52	59	51	56	62	54	60	52	58	51	57	59	50	54	57	49	51	55	9.7
Saint Paul Island	54	51	52	50	51	49	51	53	50	52	49	50	50	55	52	49	52	51	48	50	50	5.4
Sitka	66	59	64	58	61	57	60	65	59	62	58	60	58	74	62	57	71	60	56	68	59	9.2
Talkeetna	77	60	73	58	70	57	62	74	60	70	58	67	57	71	64	56	67	62	54	64	61	16.4
Valdez	69	56	66	55	62	54	58	67	56	64	55	61	53	60	59	53	59	57	52	57	56	12.2
Yakutat	66	56	63	55	60	54	58	62	57	60	56	59	56	67	58	55	64	57	54	62	57	12.0
ARIZONA																						
Flagstaff	85	56	83	55	80	55	61	74	60	73	59	72	58	93	65	56	88	64	55	83	63	27.6
Kingman	99	64	97	63	95	62	71	82	67	85	66	86	67	112	77	62	92	75	59	85	76	24.8
Page	99	62	97	62	95	61	66	85	65	86	64	86	60	92	74	58	85	74	56	80	74	23.8
Phoenix, Int'l Airport	110	70	108	70	106	70	76	97	75	96	74	95	71	118	82	69	111	84	67	104	85	23.0
Phoenix, Luke AFB	110	71	107	71	105	71	78	97	76	97	75	96	74	130	85	71	118	85	69	111	86	25.2
Prescott	94	60	91	60	89	60	67	81	66	80	64	79	63	104	71	61	98	71	60	93	70	25.4
Safford, Agri Center	102	66	99	66	97	65	71	89	71	89	69	88	67	111	77	66	106	76	64	102	77	34.7
Tucson	104	65	102	65	100	65	72	88	71	87	70	86	69	116	76	67	111	76	66	106	77	29.4
Winslow	95	60	93	60	91	59	65	80	64	81	63	80	61	95	71	59	91	69	58	85	69	27.4
Yuma	111	72	109	72	106	72	80	96	78	95	77	95	76	136	87	74	127	88	71	117	89	23.8
ARKANSAS																						
Blytheville, Eaker AFB	97	78	95	77	93	77	82	92	80	91	78	89	78	149	88	77	142	86	76	135	85	18.7
Fayetteville	95	75	93	75	90	74	78	90	77	89	76	87	75	136	85	74	132	84	72	124	81	21.4
Fort Smith	99	76	96	76	93	75	79	92	78	91	77	90	75	134	85	74	130	84	73	126	83	21.5
Little Rock, AFB	97	77	95	77	92	76	80	92	79	91	78	89	77	141	86	76	137	85	75	133	84	19.5
Texarkana	97	77	95	77	93	76	80	91	79	90	78	89	77	143	85	76	139	85	75	135	84	20.5
CALIFORNIA																						
Alameda, NAS	83	65	79	64	76	63	67	79	65	76	64	73	62	85	70	61	80	69	60	78	67	14.8
Arcata/Eureka	70	60	67	59	65	58	62	67	61	65	60	64	60	78	64	59	75	63	58	71	62	15.5
Bakersfield	104	70	101	69	99	69	73	98	71	96	70	95	64	92	84	62	85	83	60	79	83	26.5
Barstow/Daggett	107	68	105	67	102	67	72	95	71	95	69	95	66	103	81	63	91	85	59	81	85	27.8
Blue Canyon	84	59	81	57	79	56	62	80	60	78	58	75	54	74	70	52	69	70	50	64	68	16.6
Burbank/Glendale	98	69	95	69	92	68	74	90	72	89	71	86	69	108	80	67	103	78	66	98	77	23.4

MDB = mean coincident dry-bulb temperature, °F
MWB = mean coincident wet-bulb temperature, °F
MWS = mean coincident wind speed, mph
MWD = mean coincident wind direction
StdD = standard deviation
A = airport
HR = humidity ratio
DP = dew-point temperature, °F

Table 1A Heating and Wind Design Conditions—United States

Station	WMO#	Lat.	Long.	Elev. ft	StdP psia	Dates	Heating DB		Extreme Wind Speed			Coldest Month WS/MDB				MWS/MWD to DB				Annual Extreme Daily			
												0.4%		1%		99.6%		0.4%		Mean DB		StdD DB	
							99.6%	99%	1%	2.5%	5%	WS	MDB	WS	MDB	MWS	PWD	MWS	PWD	Max	Min	Max	Min
Fairfield, Travis AFB	745160	38.27	121.93	62	14.662	8293	31	34	28	24	22	26	53	22	52	4	20	9	240	105	26	3.4	4.0
Fresno	723890	36.77	119.72	328	14.522	6193	30	32	17	15	13	17	53	14	52	4	90	9	290	107	26	2.2	3.8
Lancaster/Palmdale	723816	34.73	118.22	2346	13.492	8293	22	24	30	28	25	29	48	26	49	2	260	14	240	107	15	2	5.9
Lemoore, Reeves NAS	747020	36.33	119.95	236	14.570	8293	30	32	19	16	14	20	49	16	51	4	150	7	360	110	17	4.1	10.4
Long Beach	722970	33.82	118.15	39	14.675	6193	40	43	19	16	14	19	58	16	58	4	300	10	270	102	35	4.5	2.8
Los Angeles	722950	33.93	118.40	105	14.640	6193	43	45	21	18	16	20	56	17	56	6	70	10	250	97	38	5.1	3.0
Marysville, Beale AFB	724837	39.13	121.43	112	14.636	8293	31	34	20	17	14	23	53	19	53	3	20	5	200	106	26	3.2	4.1
Merced, Castle AFB	724810	37.38	120.57	187	14.596	8293	30	32	18	15	12	21	51	17	49	2	110	9	320	104	26	2.7	3.6
Mount Shasta	725957	41.32	122.32	3543	12.909	8293	16	21	14	12	10	14	36	12	37	4	60	4	180	95	10	2.7	6.7
Mountain View, Moffet NAS	745090	37.42	122.05	39	14.675	8293	36	39	19	17	15	19	54	16	52	1	140	9	330	98	23	2.5	12.2
Ontario	722865	34.05	117.60	942	14.202	8293	35	38	22	19	17	28	62	21	57	4	10	13	240	108	29	3.4	2.5
Oxnard, Pt. Mugu NAWS	723910	34.12	119.12	7	14.692	8293	39	41	22	19	16	25	57	21	58	5	20	12	50	93	24	5	10.6
Paso Robles	723965	35.67	120.63	837	14.257	8293	26	29	22	20	18	21	52	18	51	3	110	11	300	108	21	2.2	4.9
Red Bluff	725910	40.15	122.25	354	14.508	8293	29	32	23	21	19	26	53	23	50	6	340	9	160	111	25	3.2	3.8
Riverside, March AFB	722860	33.88	117.27	1539	13.896	8293	34	36	18	15	13	22	51	18	55	1	210	9	300	107	29	2.3	3.2
Sacramento, Mather Field	724835	38.55	121.30	95	14.645	8293	30	32	20	17	14	24	53	20	51	2	120	6	310	105	26	5	4.3
Sacramento, McClellan AFB	724836	38.67	121.40	75	14.655	8293	31	34	20	16	14	23	53	19	52	2	340	5	220	107	27	2.5	4.9
Sacramento, Metro	724839	38.70	121.58	23	14.683	6193	31	33	22	19	17	23	51	20	50	3	340	8	220	107	27	6.6	3.0
Salinas	724917	36.67	121.60	85	14.650	8293	33	35	21	19	18	23	51	21	51	6	130	11	310	95	29	4.7	2.2
San Bernardino, Norton AFB	722866	34.10	117.23	1158	14.091	8293	34	36	17	13	11	21	56	16	55	2	50	8	250	109	29	2.5	2.7
San Diego, Int'l Airport	722900	32.73	117.17	30	14.680	6193	44	46	18	16	15	20	59	16	60	3	70	10	310	95	39	6.4	4.4
San Diego, Miramar NAS	722930	32.85	117.12	420	14.474	8293	39	42	13	11	9	15	59	12	59	3	90	6	310	102	27	4	13.7
San Francisco	724940	37.62	122.38	16	14.687	6193	37	39	29	26	23	27	53	22	52	5	160	13	300	94	33	4.3	3.0
San Jose Int'l Airport	724945	37.37	121.93	56	14.666	8293	35	38	20	18	17	20	56	17	56	1	160	10	320	101	27	3.1	9.0
Santa Barbara	723925	34.43	119.83	10	14.690	8293	34	37	20	17	14	19	58	16	58	1	40	10	260	97	28	6.7	6.5
Santa Maria	723940	34.90	120.45	240	14.569	6193	32	35	23	21	19	21	59	18	59	4	110	11	300	95	27	5.0	2.8
Stockton	724920	37.90	121.25	26	14.681	8293	30	32	22	19	17	24	52	21	49	4	110	11	280	106	26	3.1	3.4
Victorville, George AFB	723825	34.58	117.38	2874	13.232	8293	27	30	22	19	16	22	49	18	47	3	160	9	180	106	21	3.1	5.6
COLORADO																							
Alamosa	724620	37.45	105.87	7543	11.108	6193	−17	−11	26	23	21	23	33	20	30	3	190	12	240	88	−27	2.0	7.9
Colorado Springs	724660	38.82	104.72	6171	11.701	6193	−2	4	29	25	21	28	35	23	33	7	20	12	160	95	−9	2.0	6.9
Craig	725700	40.50	107.53	6283	11.652	8293	−20	−12	26	20	17	22	33	17	27	2	270	9	250	93	−31	2	10.6
Denver	724699	39.75	104.87	5331	12.076	8293	−3	3	24	21	18	25	39	21	40	6	180	9	160	97	−11	2	7.0
Eagle	724675	39.65	106.92	6539	11.539	6193	−13	−7	22	19	17	20	33	18	32	3	90	11	230	93	−23	3.2	7.8
Grand Junction	724760	39.12	108.53	4839	12.301	6193	2	7	22	19	17	17	33	14	30	5	70	11	290	100	−3	2.0	8.5
Limon	724665	39.27	103.67	5364	12.062	8293	−6	1	27	23	21	27	29	22	25	9	160	12	200	96	−13	2.2	6.5
Pueblo	724640	38.28	104.52	4721	12.355	6193	−1	5	32	27	24	30	44	26	43	5	270	12	140	102	−12	1.9	7.7
Trinidad	724645	37.27	104.33	5761	11.883	8293	−2	6	25	22	19	24	41	21	42	5	290	10	210	98	−10	2	6.8
CONNECTICUT																							
Bridgeport	725040	41.17	73.13	16	14.687	6193	8	12	27	23	21	34	29	30	29	14	320	14	230	93	2	2.8	4.9
Hartford, Brainard Field	725087	41.73	72.65	20	14.685	6193	2	6	23	20	18	23	25	20	26	7	320	11	250	97	−6	2.4	5.7
Windsor Locks, Bradley Fld	725080	41.93	72.68	180	14.600	8293	3	8	21	19	17	22	30	20	29	7	360	11	240	97	−5	2	5.8
DELAWARE																							
Dover, AFB	724088	39.13	75.47	30	14.680	8293	14	18	22	19	17	23	36	21	35	8	340	9	240	97	6	3.2	6.1
Wilmington	724089	39.68	75.60	79	14.654	6193	10	14	25	22	19	27	29	23	30	11	290	11	240	96	3	2.7	6.8
FLORIDA																							
Apalachicola	722200	29.73	85.03	20	14.685	8293	31	35	19	17	15	19	51	17	51	6	360	9	220	93	23	6.7	7.4
Cape Canaveral, NASA	747946	28.62	80.72	10	14.690	8293	38	42	19	17	15	21	60	19	60	8	320	8	220	96	29	1.4	6.1
Daytona Beach	722056	29.18	81.05	36	14.676	6193	34	37	21	19	17	22	61	19	61	7	310	11	240	96	27	1.9	4.4
Fort Lauderdale/Hollywood	722025	26.07	80.15	23	14.683	8293	46	50	22	20	18	22	69	20	71	9	330	11	120	97	39	1.1	6.1
Fort Myers	722106	26.58	81.87	16	14.687	8293	42	47	19	18	16	20	64	18	66	6	30	9	70	97	34	1.3	4.7
Gainesville	722146	29.68	82.27	151	14.615	8293	30	33	19	17	14	19	65	17	62	4	300	9	270	97	21	1.8	7.2
Homestead, AFB	722026	25.48	80.38	7	14.692	8293	48	52	17	15	13	17	70	15	70	6	360	7	120	95	41	2.2	5.6
Jacksonville, Cecil Field NAS	722067	30.22	81.88	82	14.652	8293	31	34	18	16	14	19	62	17	62	3	290	7	270	100	20	2	8.8
Jacksonville, Int'l Airport	722060	30.50	81.70	30	14.680	6193	29	32	21	18	17	21	54	19	55	6	310	9	230	98	22	2.1	5.1
Jacksonville, Mayport Naval	722066	30.40	81.42	16	14.687	8293	34	39	19	17	14	21	54	18	55	6	310	7	270	99	20	2.2	13.1
Key West	722010	24.55	81.75	20	14.685	6193	55	58	22	20	18	24	65	21	66	12	50	9	140	91	51	1.2	4.0
Melbourne	722040	28.10	80.65	36	14.676	8293	38	43	21	19	18	22	62	20	62	9	320	11	120	97	30	1.8	6.7
Miami, Int'l Airport	722020	25.82	80.28	13	14.688	6193	46	50	23	20	18	22	68	20	69	10	340	11	150	94	39	2.1	5.1
Miami, New Tamiami A	722029	25.65	80.43	10	14.690	8293	45	49	21	19	18	21	72	19	72	8	360	11	130	95	39	2	6.3
Milton, Whiting Field NAS	722226	30.72	87.02	200	14.589	8293	28	31	18	16	14	19	50	17	52	6	340	6	330	99	17	1.8	8.6
Orlando	722050	28.43	81.32	105	14.640	8293	37	42	20	18	16	21	66	19	65	8	330	9	290	96	29	1.6	6.5
Panama City, Tyndall AFB	747750	30.07	85.58	16	14.687	8293	33	37	18	16	14	19	52	17	52	9	360	7	240	94	24	2.3	6.3
Pensacola, Sherman AFB	722225	30.35	87.32	30	14.680	8293	28	32	23	20	18	25	43	22	48	9	360	10	200	100	15	6.3	7.6
Saint Petersburg	722116	27.92	82.68	10	14.690	8293	43	47	21	19	17	22	65	20	63	11	10	10	230	97	35	1.8	4.7
Sarasota/Bradenton	722115	27.40	82.55	30	14.680	8293	39	43	22	19	17	23	67	20	67	5	40	9	270	97	29	1.6	12.1
Tallahassee	722140	30.38	84.37	69	14.659	6193	25	28	18	16	14	19	52	17	54	3	350	8	360	98	17	2.0	4.6
Tampa, Int'l Airport	722110	27.97	82.53	10	14.690	6193	36	40	19	17	15	21	59	19	59	8	20	10	270	95	29	1.2	4.8
Valparaiso, Eglin AFB	722210	30.48	86.53	85	14.650	8293	30	33	19	16	14	18	49	16	51	6	360	7	210	97	19	2	6.1
Vero Beach	722045	27.65	80.42	26	14.681	8293	39	43	20	19	17	21	67	19	67	8	310	11	240	96	31	2	6.5

WMO# = World Meteorological Organization number
Lat = latitude
Long. = longitude
Elev = elevation, ft
StdP = standard pressure at station elevation, psia
DB = dry-bulb temperature, °F
WS = wind speed, mph

Table 1B Cooling and Dehumidification Design Conditions—United States

	Cooling DB/MWB						WB/MDB						DP/MDB and HR									
	0.4%		1%		2%		0.4%		1%		2%		0.4%			1%			2%			Range
Station	DB	MWB	DB	MWB	DB	MWB	WB	MDB	WB	MDB	WB	MDB	DP	HR	MDB	DP	HR	MDB	DP	HR	MCDB	of DB
Fairfield, Travis AFB	98	67	94	67	91	66	70	92	69	90	67	88	62	85	76	61	79	74	59	75	73	29.0
Fresno	103	71	101	70	98	69	73	98	71	96	70	94	64	92	85	62	85	84	61	80	82	30.9
Lancaster/Palmdale	101	66	98	65	96	64	70	94	69	91	67	90	62	92	80	60	84	81	58	78	81	27.9
Lemoore, Reeves NAS	103	72	101	71	98	70	75	97	73	96	71	94	67	101	89	65	94	87	62	85	86	32.9
Long Beach	92	67	88	67	84	66	71	85	70	82	69	80	67	101	76	66	96	75	65	92	75	16.7
Los Angeles	85	64	81	64	78	64	70	78	69	76	68	75	67	99	75	66	95	73	65	92	72	10.9
Marysville, Beale AFB	101	70	98	69	95	68	72	97	71	95	69	92	63	86	85	61	80	82	60	76	81	29.9
Merced, Castle AFB	99	69	97	69	94	68	72	96	71	93	69	92	64	90	81	62	82	84	60	78	81	30.2
Mount Shasta	91	62	88	61	85	60	64	87	63	84	61	82	56	76	74	53	69	73	51	64	71	32.0
Mountain View, Moffet NAS	88	65	84	65	80	64	68	82	67	80	65	78	62	83	74	61	80	73	60	76	72	18.0
Ontario	102	71	98	70	95	69	75	94	73	92	72	90	70	113	80	68	106	80	66	101	78	27.7
Oxnard, Pt. Mugu NAWS	83	62	79	64	77	64	70	77	69	75	67	74	68	103	74	66	97	73	65	93	72	14.6
Paso Robles	102	68	98	67	95	65	70	97	68	94	67	91	61	81	76	58	75	73	57	71	71	37.8
Red Bluff	105	70	102	69	98	67	72	98	71	95	69	93	65	94	82	62	85	80	61	80	78	29.5
Riverside, March AFB	101	68	98	68	95	67	72	92	71	91	70	90	67	104	79	65	97	80	62	90	79	29.0
Sacramento, Mather Field	101	69	97	68	95	67	71	97	69	94	68	92	61	80	79	60	77	77	58	74	76	33.7
Sacramento, McClellan AFB	102	70	98	69	95	68	72	97	71	95	69	92	63	85	84	61	80	81	60	77	79	29.7
Sacramento, Metro	100	69	97	69	94	68	72	96	70	94	69	91	62	84	82	61	79	80	59	76	78	33.3
Salinas	83	63	78	62	75	61	66	78	65	75	63	72	62	82	69	60	78	68	59	76	67	18.7
San Bernardino, Norton AFB	103	70	101	70	97	69	74	94	73	94	71	92	68	107	83	66	101	83	65	95	82	31.5
San Diego, Int'l Airport	85	67	81	67	79	67	73	79	71	78	70	77	70	111	77	68	104	76	67	99	74	8.9
San Diego, Miramar NAS	92	69	88	67	85	67	72	85	71	83	69	81	68	104	78	67	99	77	65	96	75	17.5
San Francisco	83	63	78	62	74	61	64	79	63	75	62	72	59	76	67	58	73	66	57	71	65	16.7
San Jose Int'l Airport	93	67	89	66	86	65	70	88	68	85	67	83	63	85	77	61	81	76	60	78	74	22.3
Santa Barbara	83	64	80	64	77	64	69	77	67	76	66	74	66	96	74	65	91	71	63	85	70	18.0
Santa Maria	86	63	82	62	78	61	66	81	65	78	64	75	61	80	70	60	77	69	59	74	68	19.4
Stockton	100	69	97	68	94	67	71	96	70	94	68	92	62	83	78	60	78	78	59	75	77	30.4
Victorville, George AFB	101	65	98	65	96	64	69	88	68	88	67	88	65	102	78	61	90	79	59	83	78	28.3
COLORADO																						
Alamosa	84	55	82	55	80	54	60	75	59	74	58	73	55	87	62	54	81	62	52	77	62	31.2
Colorado Springs	90	58	87	58	84	58	63	78	62	77	61	76	59	92	66	57	88	66	56	83	65	24.9
Craig	88	57	85	56	83	55	60	79	59	78	57	77	53	77	66	52	72	65	50	68	64	36.4
Denver	93	60	90	59	87	59	65	81	63	80	62	79	60	96	69	58	90	68	57	85	68	26.9
Eagle	88	58	86	57	83	57	62	80	60	78	59	76	57	88	64	55	82	65	53	76	65	36.1
Grand Junction	96	61	94	60	92	60	65	84	64	83	63	82	60	93	70	58	87	71	56	79	71	26.6
Limon	90	60	88	60	85	59	64	79	63	78	62	77	60	96	67	59	92	66	58	88	66	26.8
Pueblo	97	62	94	62	92	62	67	84	66	83	65	83	63	104	71	62	98	71	60	92	71	29.4
Trinidad	93	61	90	60	87	60	65	84	64	83	63	81	60	96	71	58	91	69	57	86	69	28.3
CONNECTICUT																						
Bridgeport	86	73	84	72	82	71	76	83	74	81	73	79	74	126	79	72	120	78	71	115	77	14.1
Hartford, Brainard Field	91	73	88	72	85	70	76	87	74	84	73	82	81	228	71	79	216	70	78	207	0	20.9
Windsor Locks, Bradley Fld	92	73	88	71	85	70	76	87	74	84	72	82	72	119	81	71	114	79	69	109	77	20.9
DELAWARE																						
Dover, AFB	93	76	89	75	87	74	79	88	78	86	76	84	77	141	84	76	135	82	74	129	81	16.2
Wilmington	91	75	89	74	86	73	78	87	76	85	75	83	75	132	82	74	125	81	72	120	80	17.0
FLORIDA																						
Apalachicola	92	79	90	78	89	78	81	88	80	87	79	87	79	148	85	78	145	84	77	141	84	13.3
Cape Canaveral, NASA	92	78	90	78	89	77	80	88	79	87	79	87	78	145	84	77	141	84	76	138	83	16.0
Daytona Beach	92	77	90	77	88	77	79	88	79	87	78	86	77	141	84	76	137	84	76	134	83	15.4
Fort Lauderdale/Hollywood	92	78	90	78	89	78	81	88	80	87	79	87	78	147	85	78	145	84	77	141	84	11.3
Fort Myers	94	77	93	77	92	77	80	89	80	88	79	87	78	147	84	78	144	83	77	140	83	16.9
Gainesville	94	77	92	77	90	76	80	89	79	88	78	87	77	143	84	76	139	83	76	136	82	18.7
Homestead, AFB	92	79	90	79	89	78	81	89	80	88	80	87	79	150	87	78	145	86	77	141	85	11.7
Jacksonville, Cecil Field NAS	96	76	95	76	93	76	79	91	78	90	77	89	76	138	84	75	134	83	75	130	82	20.0
Jacksonville, Int'l Airport	94	77	93	77	91	77	80	90	79	89	78	88	77	142	85	76	138	84	76	134	83	17.8
Jacksonville, Mayport Naval	95	78	92	78	90	77	81	89	80	89	79	88	78	147	86	77	142	85	77	139	85	15.3
Key West	90	79	89	79	89	79	81	87	80	87	80	87	79	149	85	78	146	85	77	143	85	8.1
Melbourne	93	79	91	79	89	79	82	89	81	88	80	87	80	155	86	79	150	85	78	146	85	15.3
Miami, Int'l Airport	91	77	90	77	89	77	80	87	79	87	78	86	78	144	83	77	141	83	76	138	82	11.4
Miami, New Tamiami A	92	78	91	78	90	77	80	89	79	88	79	87	78	145	83	77	141	83	76	138	83	15.5
Milton, Whiting Field NAS	95	78	93	77	92	76	81	90	80	89	79	88	78	148	86	77	143	85	76	138	84	18.5
Orlando	94	76	93	76	92	76	79	88	79	88	78	87	77	142	83	76	139	82	76	136	81	16.6
Panama City, Tyndall AFB	91	79	89	79	88	79	83	88	82	87	81	86	81	160	86	80	154	85	79	150	84	12.2
Pensacola, Sherman AFB	93	78	92	78	90	78	81	89	80	88	79	88	79	150	85	78	144	85	76	138	85	15.3
Saint Petersburg	94	80	93	79	92	79	82	90	82	89	81	88	80	156	86	80	153	85	79	150	85	13.5
Sarasota/Bradenton	93	80	92	79	90	79	82	90	81	89	80	88	79	153	87	79	148	86	78	146	86	15.8
Tallahassee	95	77	93	76	91	76	80	89	79	88	78	87	77	142	83	76	138	82	76	135	82	18.5
Tampa, Int'l Airport	92	77	91	77	90	77	80	88	79	88	78	87	78	144	85	77	140	84	76	137	83	15.0
Valparaiso, Eglin AFB	92	78	90	78	89	77	81	88	80	87	79	86	79	149	85	78	144	84	77	141	83	13.9
Vero Beach	92	77	90	78	89	77	80	88	79	88	79	87	77	141	85	77	139	84	76	137	84	15.7

MDB = mean coincident dry-bulb temperature, °F
MWB = mean coincident wet-bulb temperature, °F
MWS = mean coincident wind speed, mph
MWD = mean coincident wind direction
StdD = standard deviation
A = airport
HR = humidity ratio
DP = dew-point temperature, °F

Table 1A Heating and Wind Design Conditions—United States

Station	WMO#	Lat.	Long.	Elev. ft	StdP psia	Dates	Heating DB		Extreme Wind Speed			Coldest Month WS/MDB				MWS/MWD to DB				Annual Extreme Daily			
												0.4%		1%		99.6%		0.4%		Mean DB		StdD DB	
							99.6%	99%	1%	2.5%	5%	WS	MDB	WS	MDB	MWS	PWD	MWS	PWD	Max	Min	Max	Min
West Palm Beach	722030	26.68	80.12	20	14.685	6193	43	47	24	21	19	24	69	21	70	9	320	12	110	94	35	2.0	5.0
GEORGIA																							
Albany	722160	31.53	84.18	194	14.593	8293	27	30	19	17	15	19	50	18	50	4	360	9	250	100	17	2.2	7.2
Athens	723110	33.95	83.32	810	14.270	6193	20	25	19	17	15	20	40	18	40	10	290	9	270	98	11	3.5	6.6
Atlanta	722190	33.65	84.42	1033	14.155	6193	18	23	22	19	17	23	37	21	36	12	320	9	300	96	9	3.5	7.3
Augusta	722180	33.37	81.97	148	14.617	6193	21	25	20	18	15	21	45	19	46	5	290	9	250	100	13	3.7	5.6
Brunswick	722137	31.15	81.38	20	14.685	8293	30	34	18	17	16	19	49	18	49	8	350	10	250	98	22	2.5	7.7
Columbus, Fort Benning	722250	32.33	85.00	233	14.572	8293	23	27	16	13	11	17	46	15	46	3	320	5	240	100	14	2.9	6.7
Columbus, Metro Airport	722255	32.52	84.93	397	14.486	6193	23	27	17	15	14	18	44	16	46	7	310	9	310	99	14	2.3	6.1
Macon	722170	32.70	83.65	361	14.505	6193	23	27	19	17	15	20	46	18	45	7	320	9	270	100	14	2.7	6.4
Marietta, Dobbins AFB	722270	33.92	84.52	1070	14.136	8293	21	26	18	16	13	20	35	18	38	9	340	6	300	97	12	3.6	6.7
Rome	723200	34.35	85.17	643	14.357	8293	15	21	14	12	10	14	42	13	42	5	340	6	270	98	4	3.8	7.0
Savannah	722070	32.13	81.20	49	14.669	6193	26	29	20	17	15	21	49	19	49	7	270	9	270	98	18	3.0	5.4
Valdosta, Moody AFB	747810	30.97	83.20	233	14.572	8293	30	34	15	13	12	16	53	14	52	4	360	5	300	99	21	2.5	7.6
Valdosta, Regional Airport	722166	30.78	83.28	203	14.588	8293	28	31	17	15	14	18	55	16	56	4	340	8	300	99	17	3.2	7.7
Waycross	722130	31.25	82.40	151	14.615	8293	29	32	16	14	12	16	52	14	52	4	250	7	240	98	21	7	7.6
HAWAII																							
Ewa, Barbers Point NAS	911780	21.32	158.07	49	14.669	8293	59	61	20	18	16	22	73	19	75	5	40	11	60	93	35	1.6	21.4
Hilo	912850	19.72	155.07	36	14.676	6193	61	63	19	16	14	21	76	18	76	7	230	12	110	88	58	1.6	1.8
Honolulu	911820	21.35	157.93	16	14.687	6193	61	63	23	21	20	23	74	21	75	5	320	15	60	91	58	1.9	2.2
Kahului	911900	20.90	156.43	66	14.661	6193	59	61	27	25	24	32	76	28	76	6	160	19	50	92	54	1.5	4.4
Kaneohe, MCAS	911760	21.45	157.77	10	14.690	8293	67	68	20	18	17	21	74	19	74	7	190	10	70	88	40	1.4	29.0
Lihue	911650	21.98	159.35	148	14.617	6193	60	62	26	24	21	25	73	23	73	8	270	14	60	87	57	1.4	3.0
Molokai	911860	21.15	157.10	449	14.458	8293	60	61	24	22	21	22	74	21	74	4	70	13	60	92	43	4	22.0
IDAHO																							
Boise	726810	43.57	116.22	2867	13.235	6193	2	9	24	21	18	22	37	19	37	6	130	11	320	103	–4	2.7	9.1
Burley	725867	42.55	113.77	4150	12.621	8293	–5	2	23	21	19	23	30	22	28	7	60	8	280	98	–11	4	8.5
Idaho Falls	725785	43.52	112.07	4741	12.346	8293	–12	–6	27	23	21	28	32	23	29	7	360	12	180	96	–20	3.6	9.0
Lewiston	727830	46.38	117.02	1437	13.948	8293	6	15	20	17	14	24	38	20	40	5	280	7	310	103	3	2.7	9.9
Mountain Home, AFB	726815	43.05	115.87	2995	13.173	8293	0	5	23	21	18	23	33	21	31	2	90	8	350	105	–6	3.2	8.5
Mullan	727836	47.47	115.80	3317	13.017	8293	–1	7	10	10	9	11	18	9	21	2	10	4	10	92	–7	2	7.9
Pocatello	725780	42.92	112.60	4478	12.468	6193	–7	0	29	25	23	30	36	27	36	6	50	11	250	98	–15	2.3	9.1
ILLINOIS																							
Belleville, Scott AFB	724338	38.55	89.85	453	14.457	8293	3	10	21	18	15	23	32	20	31	7	360	7	190	100	–3	3.1	7.2
Chicago, Meigs Field	725340	41.78	87.75	623	14.367	8293	–4	3	23	22	19	26	17	23	30	12	240	13	220	97	–10	3.2	8.1
Chicago, O'Hare Int'l A	725300	41.98	87.90	673	14.342	6193	–6	–1	26	23	21	27	24	23	23	10	270	12	230	96	–12	2.8	6.5
Decatur	725316	39.83	88.87	682	14.337	8293	–2	3	24	22	20	27	24	24	27	13	310	12	210	99	–10	5.8	7.2
Glenview, NAS	725306	42.08	87.82	653	14.352	8293	–3	4	22	19	17	23	17	20	25	11	250	10	240	98	–10	3.1	7.7
Marseilles	744600	41.37	88.68	738	14.308	8293	–5	1	26	22	20	28	18	25	21	12	290	10	250	96	–11	4	5.9
Moline/Davenport IA	725440	41.45	90.52	594	14.383	6193	–8	–3	26	23	20	28	16	24	18	9	290	12	200	97	–14	2.7	6.0
Peoria	725320	40.67	89.68	663	14.347	6193	–6	–1	25	22	20	26	16	23	19	9	290	11	180	96	–12	3.3	6.1
Quincy	724396	39.95	91.20	768	14.292	8293	–4	2	26	23	20	28	23	24	22	12	330	12	210	97	–10	3.6	8.1
Rockford	725430	42.20	89.10	741	14.306	6193	–10	–4	26	23	21	26	18	23	20	9	290	13	200	95	–16	3.1	5.5
Springfield	724390	39.85	89.67	614	14.373	6193	–4	2	25	23	21	27	25	24	27	10	270	12	230	97	–11	2.8	5.5
West Chicago	725305	41.92	88.25	758	14.297	8293	–7	0	23	21	19	25	13	23	20	11	290	11	240	96	–14	3.2	7.7
INDIANA																							
Evansville	724320	38.05	87.53	387	14.491	6193	3	9	22	19	17	22	33	20	34	7	320	9	240	97	–4	2.7	8.5
Fort Wayne	725330	41.00	85.20	827	14.262	6193	–4	2	25	23	20	27	19	24	22	10	250	12	230	95	–11	3.6	5.2
Indianapolis	724380	39.73	86.27	807	14.272	6193	–3	3	24	21	19	25	26	22	27	8	230	11	230	94	–10	2.8	6.8
Lafayette, Purdue Univ	724386	40.42	86.93	607	14.376	8293	–5	3	22	20	18	24	26	22	27	9	270	12	220	97	–11	3.8	7.7
Peru, Grissom AFB	725335	40.65	86.15	810	14.270	8293	–3	4	24	21	18	29	20	24	22	11	270	9	210	96	–8	3.8	7.4
South Bend	725350	41.70	86.32	774	14.289	6193	–2	3	25	23	20	26	22	23	23	13	230	12	230	95	–10	3.3	5.8
Terre Haute	724373	39.45	87.32	584	14.388	8293	–3	5	23	20	18	23	31	21	32	8	150	11	230	96	–10	3.2	7.9
IOWA																							
Burlington	725455	40.78	91.13	699	14.328	8293	–4	1	21	19	17	24	12	21	18	9	310	11	200	98	–10	4	6.8
Cedar Rapids	725450	41.88	91.70	869	14.240	8293	–11	–5	25	22	20	29	12	26	14	10	300	11	180	96	–15	3.6	5.4
Des Moines	725460	41.53	93.65	965	14.190	6193	–9	–4	27	24	21	28	14	24	19	11	320	12	180	98	–15	3.4	5.1
Fort Dodge	725490	42.55	94.18	1165	14.087	8293	–13	–7	27	23	21	29	10	26	10	11	340	11	190	96	–17	4.9	4.9
Lamoni	725466	40.62	93.95	1122	14.109	8293	–6	0	19	17	15	21	23	19	20	7	320	9	210	99	–12	4.3	6.8
Mason City	725485	43.15	93.33	1214	14.062	6193	–15	–10	27	23	22	30	9	27	12	12	300	14	200	97	–23	3.6	11.4
Ottumwa	725465	41.10	92.45	846	14.251	8293	–5	0	29	26	23	31	20	28	24	13	320	15	200	98	–12	4	6.8
Sioux City	725570	42.40	96.38	1102	14.119	6193	–11	–6	29	25	22	31	14	28	16	11	320	14	180	99	–18	3.6	4.7
Spencer	726500	43.17	95.15	1339	13.998	8293	–16	–11	24	22	20	25	13	23	13	10	300	12	180	99	–20	6.3	4.0
Waterloo	725480	42.55	92.40	879	14.234	6193	–14	–9	27	24	22	29	10	25	13	9	300	13	180	96	–20	3.5	5.9
KANSAS																							
Concordia	724580	39.55	97.65	1483	13.925	8293	–4	3	28	25	22	28	32	25	32	13	360	16	10	104	–8	4	9.4
Dodge City	724510	37.77	99.97	2592	13.370	6193	0	6	30	27	24	31	31	27	32	13	10	17	200	104	–6	2.8	5.6
Ft Riley, Marshall AAF	724550	39.05	96.77	1066	14.138	8293	–2	5	21	18	16	20	39	18	37	5	350	9	180	104	–5	3.1	9.0
Garden City	724515	37.93	100.72	2890	13.224	8293	–3	4	30	26	23	29	32	25	34	12	360	16	190	104	–9	2.7	6.5
Goodland	724650	39.37	101.70	3688	12.840	6193	–3	2	32	28	24	31	27	27	30	12	270	13	180	102	–11	2.9	6.6

WMO# = World Meteorological Organization number
Lat = latitude
Long. = longitude
Elev = elevation, ft
StdP = standard pressure at station elevation, psia
DB = dry-bulb temperature, °F
WS = wind speed, mph

Table 1B Cooling and Dehumidification Design Conditions—United States

Station	Cooling DB/MWB 0.4% DB	0.4% MWB	1% DB	1% MWB	2% DB	2% MWB	WB/MDB 0.4% WB	0.4% MDB	1% WB	1% MDB	2% WB	2% MDB	DP/MDB and HR 0.4% DP	0.4% HR	0.4% MDB	1% DP	1% HR	1% MDB	2% DP	2% HR	2% MCDB	Range of DB
West Palm Beach	91	78	90	78	89	77	80	88	79	88	78	87	77	143	84	77	139	84	76	137	83	13.1
GEORGIA																						
Albany	96	76	95	76	93	75	79	90	78	89	78	88	77	141	83	76	136	82	75	133	81	19.8
Athens	94	75	92	75	90	74	78	89	77	87	76	86	75	133	82	74	129	81	73	125	80	18.4
Atlanta	93	75	91	74	88	73	77	88	76	87	75	85	74	133	82	73	128	81	72	124	80	17.3
Augusta	96	76	94	76	92	75	79	91	78	89	77	88	76	135	84	75	130	83	74	127	82	20.2
Brunswick	93	78	91	79	88	78	81	89	80	88	79	87	78	147	86	78	144	85	77	141	84	14.4
Columbus, Fort Benning	97	76	94	76	92	76	80	91	79	89	78	88	77	142	85	76	136	83	75	133	82	20.5
Columbus, Metro Airport	95	76	93	75	91	75	79	89	78	88	77	87	76	139	82	75	134	82	74	130	81	18.0
Macon	96	76	94	75	92	75	79	91	78	89	77	88	76	136	83	75	132	82	74	129	82	19.3
Marietta, Dobbins AFB	94	74	91	74	89	74	77	88	76	87	75	86	74	134	82	73	130	81	72	123	79	17.1
Rome	96	74	94	74	91	74	78	90	77	89	76	88	75	134	83	74	130	83	73	127	83	20.7
Savannah	95	77	93	76	91	76	79	90	78	89	78	87	77	139	84	76	135	83	75	132	82	17.5
Valdosta, Moody AFB	95	77	94	77	92	76	80	91	79	89	78	88	77	142	85	76	139	84	76	135	83	17.8
Valdosta, Regional Airport	95	77	94	76	92	76	80	90	79	89	78	88	77	144	83	76	139	82	76	136	82	19.4
Waycross	96	76	94	76	93	75	78	91	78	90	77	89	75	134	84	75	130	83	74	127	83	20.3
HAWAII																						
Ewa, Barbers Point NAS	92	73	90	72	89	72	76	86	75	86	75	85	74	126	83	72	118	82	71	113	82	15.8
Hilo	85	74	84	74	83	73	76	82	76	81	75	81	75	130	79	74	127	79	73	123	78	13.3
Honolulu	89	73	88	73	87	73	76	84	75	84	74	84	74	125	80	72	120	80	71	116	79	12.2
Kahului	89	74	88	74	87	73	76	85	76	85	75	84	74	127	80	73	122	80	72	118	80	15.6
Kaneohe, MCAS	86	75	85	74	84	74	78	82	77	82	76	82	76	138	81	75	133	81	74	128	80	7.4
Lihue	85	75	85	74	84	74	77	83	76	82	75	82	75	132	80	74	128	80	73	125	79	9.6
Molokai	88	73	87	73	86	72	76	85	75	83	74	83	74	128	80	73	124	79	71	118	79	13.3
IDAHO																						
Boise	96	63	94	63	91	62	66	90	64	89	63	87	58	79	72	55	72	71	53	67	71	30.3
Burley	94	63	90	62	87	61	67	86	65	84	63	83	60	90	75	58	84	72	56	78	72	29.0
Idaho Falls	92	61	89	60	86	60	64	84	63	82	61	81	58	88	71	56	81	69	54	73	68	34.0
Lewiston	97	65	93	64	90	63	67	91	65	89	64	87	58	76	72	56	71	71	54	65	71	26.5
Mountain Home, AFB	99	63	96	62	93	61	66	91	64	91	63	89	58	79	71	54	70	69	52	64	71	32.8
Mullan	87	62	84	61	80	60	65	81	63	79	62	77	60	86	69	58	80	68	56	75	66	28.1
Pocatello	93	61	90	60	87	59	64	84	62	83	61	82	57	83	70	55	76	70	53	70	69	32.1
ILLINOIS																						
Belleville, Scott AFB	95	78	93	77	90	76	80	92	78	90	77	88	77	141	87	76	136	85	74	131	84	19.8
Chicago, Meigs Field	92	74	89	73	86	71	77	88	76	85	74	83	74	132	84	72	121	80	71	115	80	16.0
Chicago, O'Hare Int'l A	91	74	88	73	86	71	77	88	75	85	74	83	74	130	84	72	123	82	71	115	80	19.6
Decatur	94	76	91	75	88	74	79	90	78	89	76	86	76	140	86	75	133	84	73	127	83	20.0
Glenview, NAS	93	75	89	73	87	71	78	90	76	87	74	84	74	130	85	72	120	82	70	113	81	17.6
Marseilles	93	74	89	73	86	71	78	89	76	86	74	84	75	135	85	73	126	82	71	117	81	19.4
Moline/Davenport IA	93	76	90	74	87	73	78	90	77	87	75	85	75	134	85	73	127	83	72	120	82	20.0
Peoria	92	76	89	74	86	73	78	89	77	86	75	84	75	137	85	74	130	83	72	123	81	19.5
Quincy	94	76	91	75	88	74	78	89	77	88	76	85	76	138	84	74	132	82	73	126	82	18.9
Rockford	91	74	88	73	85	71	77	87	75	85	74	82	74	132	84	73	124	81	71	116	79	19.8
Springfield	93	76	91	75	88	74	79	89	77	88	76	85	76	139	86	75	132	84	73	125	82	19.4
West Chicago	91	75	88	74	86	72	78	88	76	85	74	83	76	138	85	74	130	83	71	119	80	19.8
INDIANA																						
Evansville	94	77	92	76	90	75	79	90	78	89	77	87	76	137	86	75	132	84	73	126	83	19.8
Fort Wayne	90	74	88	73	85	71	77	86	75	84	74	82	74	131	83	72	124	81	71	117	79	19.9
Indianapolis	91	75	88	74	86	73	78	88	77	86	75	83	75	137	84	74	131	82	73	125	81	18.9
Lafayette, Purdue Univ	93	75	90	75	88	73	79	89	77	86	75	84	76	139	85	74	132	83	73	125	82	20.9
Peru, Grissom AFB	93	75	89	75	87	73	79	89	77	86	75	83	76	142	85	75	134	83	73	127	81	18.5
South Bend	90	73	87	72	85	71	77	86	75	84	73	81	74	130	83	72	123	80	71	116	78	18.6
Terre Haute	93	76	90	76	88	75	80	89	78	87	76	85	77	144	86	76	136	84	74	131	82	19.6
IOWA																						
Burlington	94	76	91	76	88	73	78	89	77	88	76	85	75	136	85	74	131	83	72	124	82	18.7
Cedar Rapids	93	75	89	74	86	72	78	89	76	86	74	84	75	136	84	74	129	83	71	120	80	20.0
Des Moines	93	76	90	74	87	73	78	89	76	87	75	85	74	133	85	73	126	83	71	120	81	18.5
Fort Dodge	92	75	88	73	86	71	77	88	75	86	74	83	74	133	84	72	123	82	70	116	79	18.5
Lamoni	96	74	92	74	89	72	77	89	76	87	75	85	74	134	83	73	127	82	71	120	80	18.9
Mason City	91	74	88	73	85	71	77	87	75	85	74	82	75	135	84	73	126	82	71	118	80	20.8
Ottumwa	95	75	92	75	88	73	78	90	76	88	75	86	75	136	84	74	130	83	72	121	81	18.7
Sioux City	94	75	90	74	88	72	78	89	76	87	75	85	75	135	86	73	127	84	71	120	82	20.4
Spencer	91	75	88	73	85	71	77	88	75	86	73	82	74	134	84	72	123	82	70	117	79	20.2
Waterloo	91	75	88	73	85	71	77	87	75	85	74	83	74	132	84	72	124	82	71	117	80	20.0
KANSAS																						
Concordia	100	73	96	72	93	72	77	90	76	89	74	88	74	133	84	72	123	82	70	118	81	22.5
Dodge City	100	70	97	70	94	69	74	90	73	89	71	88	70	120	79	68	114	78	67	109	77	24.3
Ft Riley, Marshall AAF	99	75	96	74	93	74	78	90	77	90	75	88	75	136	86	73	130	83	71	120	82	22.7
Garden City	100	69	97	69	94	69	73	89	72	89	71	88	69	118	79	67	113	78	66	108	77	27.5
Goodland	97	66	94	66	91	65	70	86	69	84	68	84	66	111	74	65	106	73	63	100	73	26.5

MDB = mean coincident dry-bulb temperature, °F
MWB = mean coincident wet-bulb temperature, °F
MWS = mean coincident wind speed, mph
MWD = mean coincident wind direction
StdD = standard deviation
A = airport
HR = humidity ratio
DP = dew-point temperature, °F

Table 1A Heating and Wind Design Conditions—United States

							Heating DB		Extreme Wind Speed			Coldest Month WS/MDB				MWS/MWD to DB				Annual Extreme Daily			
												0.4%		1%		99.6%		0.4%		Mean DB		StdD DB	
Station	WMO#	Lat.	Long.	Elev. ft	StdP psia	Dates	99.6%	99%	1%	2.5%	5%	WS	MDB	WS	MDB	MWS	PWD	MWS	PWD	Max	Min	Max	Min
Russell	724585	38.87	98.82	1864	13.732	8293	−4	3	29	26	23	29	33	25	35	11	10	16	190	105	−8	3.6	8.5
Salina	724586	38.80	97.65	1273	14.032	8293	−3	4	27	23	22	28	33	24	34	11	360	15	180	106	−7	2.3	9.9
Topeka	724560	39.07	95.62	886	14.231	6193	−2	4	25	22	20	25	28	22	29	9	320	12	180	100	−8	3.8	7.4
Wichita, Airport	724500	37.65	97.43	1339	13.998	6193	2	8	29	25	23	28	30	26	31	13	360	16	200	105	−4	2.9	6.3
Wichita, McConnell AFB	724505	37.62	97.27	1371	13.982	8293	2	10	25	23	20	25	38	23	36	11	360	12	190	105	−1	2.7	7.7
KENTUCKY																							
Bowling Green	746716	36.97	86.42	548	14.407	8293	7	14	20	19	17	21	40	19	40	6	220	9	230	97	−2	3.2	10.3
Covington/Cincinnati Airport	724210	39.05	84.67	876	14.236	6193	1	7	22	20	18	25	30	22	33	9	250	10	230	95	−7	3.1	8.5
Fort Campbell, AAF	746710	36.67	87.50	571	14.395	8293	9	15	19	16	14	20	40	17	43	4	330	6	240	98	0	3.1	9.5
Fort Knox, Godman AAF	724240	37.90	85.97	755	14.299	8293	9	15	17	15	13	18	42	16	39	4	290	6	270	97	0	4.1	7.9
Jackson	724236	37.60	83.32	1381	13.977	8293	7	14	17	14	13	18	44	16	40	7	230	6	230	94	−3	2.7	9.4
Lexington	724220	38.03	84.60	988	14.179	6193	4	10	21	19	17	23	38	20	38	8	270	9	240	94	−4	3.5	8.3
Louisville	724230	38.18	85.73	489	14.438	6193	6	12	22	19	17	22	40	20	34	10	290	10	250	96	−1	3.1	7.9
Paducah	724350	37.07	88.77	413	14.477	8293	7	13	22	19	17	22	45	19	42	8	40	9	180	98	−1	2.9	9.4
LOUISIANA																							
Alexandria, England AFB	747540	31.33	92.55	89	14.648	8293	27	30	16	13	12	17	53	15	49	7	360	3	180	98	20	2.2	6.3
Baton Rouge	722317	30.53	91.15	69	14.659	6193	27	30	20	18	16	21	48	19	49	8	360	8	270	97	20	2.2	5.4
Bossier City, Barksdale AFB	722485	32.50	93.67	167	14.607	8293	22	27	18	16	14	19	49	16	51	7	360	5	180	99	15	2.3	6.7
Lafayette	722405	30.20	91.98	43	14.673	8293	28	32	21	18	16	21	54	19	53	9	10	8	200	97	19	1.6	8.1
Lake Charles	722400	30.12	93.22	33	14.678	6193	29	32	22	19	17	24	50	21	49	10	20	8	230	96	23	2.3	4.7
Leesville, Fort Polk	722390	31.05	93.20	328	14.522	8293	27	30	16	13	12	16	51	14	52	4	20	4	180	98	20	2	5.9
Monroe	722486	32.52	92.03	79	14.654	8293	22	27	19	17	15	20	50	18	47	9	10	7	230	99	17	1.8	8.5
New Orleans, Int'l Airport	722310	29.98	90.25	30	14.680	6193	30	34	21	19	17	21	48	19	49	7	340	8	360	96	23	2.0	5.3
New Orleans, Lakefront A	722315	30.05	90.03	10	14.690	8293	35	39	22	19	18	21	49	20	50	14	360	9	300	94	21	8.1	12.4
Shreveport	722480	32.47	93.82	259	14.558	6193	22	26	20	18	16	22	46	19	48	9	360	8	180	99	16	3.1	5.6
MAINE																							
Augusta	726185	44.32	69.80	351	14.510	8293	−3	1	23	21	19	25	20	22	22	10	320	11	210	93	−10	3.1	3.4
Bangor	726088	44.80	68.83	194	14.593	8293	−7	−2	22	19	17	24	18	21	20	6	300	10	240	94	−16	2.9	5.9
Brunswick, NAS	743920	43.88	69.93	75	14.655	8293	−2	2	20	17	15	21	27	19	25	4	340	9	190	96	−12	7.9	6.1
Caribou	727120	46.87	68.02	623	14.367	6193	−14	−10	28	24	22	30	13	27	11	10	270	13	250	90	−23	2.8	4.5
Limestone, Loring AFB	727125	46.95	67.88	745	14.304	8293	−13	−9	23	20	18	25	12	22	11	7	300	9	260	91	−20	2.3	2.9
Portland	726060	43.65	70.32	62	14.662	6193	−3	2	24	21	18	24	26	21	25	7	320	12	270	93	−13	3.6	5.5
MARYLAND																							
Camp Springs, Andrews AFB	745940	38.82	76.87	282	14.546	8293	13	18	21	18	16	23	30	21	32	7	350	9	230	98	4	2.9	6.7
Baltimore, BWI Airport	724060	39.18	76.67	154	14.614	6193	11	15	24	21	19	25	31	22	31	10	290	11	280	97	4	2.9	5.8
Lex Park, Patuxent River NAS	724040	38.28	76.40	39	14.675	8293	16	21	20	17	15	22	30	19	35	9	340	9	270	98	8	2.3	6.1
Salisbury	724045	38.33	75.52	52	14.668	8293	13	18	20	18	16	20	35	19	37	6	10	9	240	97	4	2.7	5.8
MASSACHUSETTS																							
Boston	725090	42.37	71.03	30	14.680	6193	7	12	29	25	23	30	30	27	28	17	320	14	270	96	0	2.7	4.7
East Falmouth, Otis Angb	725060	41.65	70.52	131	14.626	8293	11	14	26	22	20	26	34	23	33	9	300	10	240	90	5	2.5	3.8
Weymouth, S Weymouth NAS	725097	42.15	70.93	161	14.610	8293	6	11	19	16	14	18	29	16	29	7	320	9	260	97	−2	3.8	3.8
Worcester	725095	42.27	71.88	1010	14.167	6193	0	5	27	23	20	29	22	26	21	14	270	10	270	90	−6	1.9	4.1
MICHIGAN																							
Alpena	726390	45.07	83.57	692	14.332	6193	−7	−1	21	19	17	22	20	19	20	5	270	11	240	93	−17	3.4	5.9
Detroit, Metro	725370	42.23	83.33	663	14.347	6193	0	5	27	23	21	28	28	24	27	11	240	13	230	95	−7	3.0	5.4
Flint	726370	42.97	83.75	764	14.294	6193	−2	3	25	22	20	27	24	23	23	8	230	13	230	93	−10	3.1	5.0
Grand Rapids	726350	42.88	85.52	804	14.274	6193	0	5	25	22	20	26	25	23	24	8	180	13	240	93	−9	2.1	5.3
Hancock	727440	47.17	88.50	1079	14.131	6193	−9	−4	21	19	18	23	18	20	16	8	270	10	250	90	−16	2.9	5.6
Harbor Beach	725386	44.02	82.80	600	14.379	8293	9	12	26	22	19	26	27	23	27	10	220	9	230	94	2	2.9	4.1
Jackson	725395	42.27	84.47	1001	14.172	8293	−3	4	20	19	17	23	22	20	23	9	240	11	210	93	−11	2.5	5.6
Lansing	725390	42.77	84.60	873	14.238	6193	−3	2	26	23	20	28	23	25	24	8	290	13	250	94	−13	2.8	5.9
Marquette, Sawyer AFB	727435	46.35	87.40	1220	14.059	8293	−11	−6	24	21	18	26	18	23	17	6	280	10	210	91	−18	4.7	4.7
Marquette/Ishpeming, A	727430	46.53	87.55	1424	13.955	8293	−13	−8	22	19	18	22	20	20	16	8	270	11	230	90	−22	4.5	4.5
Mount Clemens, Angb	725377	42.62	82.83	581	14.390	8293	3	7	21	18	16	25	21	21	24	7	280	9	230	95	−3	4	2.7
Muskegon	726360	43.17	86.25	633	14.362	6193	3	7	27	24	22	28	25	25	26	10	290	12	200	90	−5	2.7	5.0
Oscoda, Wurtsmith AFB	726395	44.45	83.40	633	14.362	8293	0	3	21	19	17	23	26	21	24	6	220	11	200	95	−7	4.1	4.7
Pellston	727347	45.57	84.80	719	14.318	8293	−9	−3	26	23	20	28	22	24	22	4	300	14	250	92	−21	3.1	4.9
Saginaw	726379	43.53	84.08	669	14.343	8293	0	4	23	21	19	25	22	22	23	10	260	13	240	96	−6	5.8	4.5
Sault Ste. Marie	727340	46.47	84.37	725	14.314	6193	−12	−7	23	20	18	24	19	21	18	7	90	10	230	89	−22	3.5	5.4
Seul Choix Point	726399	45.92	85.92	591	14.385	8293	0	4	28	24	22	30	27	26	27	9	300	8	200	82	−5	2.3	6.3
Traverse City	726387	44.73	85.58	623	14.367	6193	−3	2	21	19	18	23	23	21	23	7	180	13	230	94	−13	2.8	7.3
MINNESOTA																							
Alexandria	726557	45.87	95.40	1424	13.955	8293	−20	−15	25	22	20	28	12	24	8	10	300	14	180	96	−26	3.6	4.5
Brainerd, Pequot Lakes	727500	46.60	94.32	1280	14.029	8293	−24	−17	11	10	9	11	8	10	11	3	320	5	190	95	−30	7.9	6.8
Duluth	727450	46.83	92.18	1417	13.958	6193	−21	−16	25	22	20	25	12	22	11	10	310	12	230	90	−28	2.8	4.7
Hibbing	727455	47.38	92.83	1352	13.992	8293	−25	−20	20	19	17	20	13	19	13	6	330	11	200	92	−34	2.5	4.7
International Falls	727470	48.57	93.38	1184	14.077	6193	−29	−23	22	20	18	22	10	20	8	6	270	11	180	92	−37	3.4	3.8
Minneapolis-St. Paul	726580	44.88	93.22	837	14.257	6193	−16	−11	25	22	20	25	12	22	14	9	300	14	180	97	−22	3.5	5.4
Redwood Falls	726556	44.55	95.08	1024	14.160	8293	−17	−12	26	22	20	28	14	24	15	11	280	14	180	99	−22	4.1	5.2
Rochester	726440	43.92	92.50	1319	14.008	6193	−17	−12	29	26	24	32	12	28	12	13	300	15	200	94	−23	3.7	5.2

WMO# = World Meteorological Organization number Elev = elevation, ft DB = dry-bulb temperature, °F
Lat = latitude Long. = longitude StdP = standard pressure at station elevation, psia WS = wind speed, mph

Table 1B Cooling and Dehumidification Design Conditions—United States

Station	Cooling DB/MWB						WB/MDB						DP/MDB and HR									Range of DB
	0.4%		1%		2%		0.4%		1%		2%		0.4%			1%			2%			
	DB	MWB	DB	MWB	DB	MWB	WB	MDB	WB	MDB	WB	MDB	DP	HR	MDB	DP	HR	MDB	DP	HR	MCDB	
Russell	100	72	96	72	94	72	76	91	75	90	73	88	72	126	83	71	120	82	69	116	80	24.1
Salina	101	74	97	73	94	73	77	92	76	90	75	89	74	132	85	72	123	83	71	118	82	23.0
Topeka	96	75	93	75	90	75	79	90	78	89	76	88	76	139	87	74	132	85	73	126	83	20.3
Wichita, Airport	100	73	97	73	94	73	77	91	76	90	74	89	73	129	83	72	123	82	71	118	81	22.2
Wichita, McConnell AFB	100	73	97	73	94	73	77	92	76	90	75	89	74	133	84	72	124	83	71	119	82	21.8
KENTUCKY																						
Bowling Green	94	76	91	75	88	74	78	89	77	87	76	86	76	136	84	75	132	82	74	127	81	20.0
Covington/Cincinnati Airport	91	74	89	73	86	72	77	87	76	86	74	83	74	132	84	73	126	81	72	120	80	18.9
Fort Campbell, AAF	95	77	93	76	90	76	80	90	78	89	77	87	77	143	85	76	136	84	74	132	83	19.4
Fort Knox, Godman AAF	94	76	92	74	89	74	78	90	77	88	76	86	76	138	85	74	132	83	73	126	82	19.4
Jackson	90	74	87	73	85	72	77	87	76	85	74	83	74	135	83	73	130	81	71	122	79	18.2
Lexington	91	74	89	73	86	72	77	87	75	86	74	83	74	130	83	72	124	81	71	120	80	18.4
Louisville	93	76	90	75	88	74	78	90	77	88	76	86	75	134	85	74	129	84	73	125	82	18.2
Paducah	96	77	93	76	92	75	80	91	79	90	78	88	77	143	86	76	138	85	75	132	83	20.2
LOUISIANA																						
Alexandria, England AFB	95	78	94	78	92	77	81	90	80	90	79	89	78	147	86	77	142	85	76	138	85	18.4
Baton Rouge	94	78	92	77	91	77	80	89	79	88	78	87	78	145	84	77	141	84	76	137	83	16.7
Bossier City, Barksdale AFB	96	77	94	77	93	77	80	90	79	90	78	89	77	144	84	76	139	83	76	134	83	20.0
Lafayette	94	78	93	78	91	77	80	89	80	89	79	88	78	146	84	77	143	83	77	140	83	17.1
Lake Charles	93	78	91	78	90	77	80	88	80	88	79	87	78	148	84	78	145	84	77	141	83	16.2
Leesville, Fort Polk	95	77	94	76	92	76	79	89	79	88	78	87	77	144	83	76	140	82	76	136	82	18.2
Monroe	96	78	94	78	93	77	81	91	80	90	79	89	78	147	86	77	143	85	77	139	84	19.3
New Orleans, Int'l Airport	93	79	92	78	90	78	81	90	80	88	80	87	79	151	86	78	146	85	77	142	84	15.5
New Orleans, Lakefront A	93	78	92	78	90	77	81	88	80	87	79	87	79	150	85	78	145	84	77	141	83	11.9
Shreveport	97	77	95	77	93	76	79	91	79	90	78	89	76	139	84	76	135	83	75	132	83	19.1
MAINE																						
Augusta	87	71	84	69	80	67	73	83	71	80	69	77	70	113	77	68	106	75	67	100	74	18.4
Bangor	87	71	84	69	81	67	73	83	71	81	69	77	70	111	78	68	104	75	67	99	73	20.5
Brunswick, NAS	87	71	84	69	80	67	73	83	71	80	70	77	70	111	78	69	105	76	67	100	74	19.1
Caribou	85	69	82	67	79	66	72	81	70	77	68	76	70	112	76	68	104	75	66	97	72	19.5
Limestone, Loring AFB	84	68	80	66	78	64	71	79	69	76	67	74	68	107	75	67	101	72	65	94	71	18.7
Portland	86	71	83	70	80	68	74	83	72	80	70	77	71	114	79	69	107	76	67	101	74	18.7
MARYLAND																						
Camp Springs, Andrews AFB	94	75	91	74	88	73	78	88	77	87	75	85	75	134	83	74	129	82	73	124	80	18.7
Baltimore, BWI Airport	93	75	91	74	88	73	78	88	76	86	75	85	75	132	83	74	125	81	72	120	80	18.8
Lex Park, Patuxent River NAS	93	76	90	75	87	74	79	88	77	87	76	85	76	136	84	75	131	83	74	125	82	15.8
Salisbury	93	77	90	76	88	75	80	88	78	86	77	85	78	144	84	76	137	82	75	132	81	18.7
MASSACHUSETTS																						
Boston	91	73	87	71	84	70	75	87	74	83	72	81	72	119	80	71	113	79	69	108	78	15.3
East Falmouth, Otis Angb	85	72	82	72	79	69	75	81	74	78	72	76	74	125	78	72	118	76	71	113	75	14.6
Weymouth, S Weymouth NAS	92	73	87	72	85	71	77	87	75	84	73	81	74	129	82	72	118	79	70	111	78	19.6
Worcester	85	71	83	69	80	68	74	82	72	80	70	77	71	119	78	69	112	76	68	105	75	16.6
MICHIGAN																						
Alpena	87	71	84	69	81	67	74	83	72	81	70	78	71	116	79	69	107	76	67	100	74	22.9
Detroit, Metro	90	73	87	72	84	70	76	86	74	84	73	81	73	125	83	71	118	80	70	111	78	20.4
Flint	88	73	86	71	83	70	75	84	74	82	72	80	73	125	81	71	116	78	69	110	77	20.6
Grand Rapids	89	73	86	71	84	70	76	85	74	83	72	81	73	126	81	71	118	79	70	112	77	20.7
Hancock	86	71	83	69	80	67	73	82	71	80	70	77	70	116	79	69	109	76	67	103	74	20.6
Harbor Beach	90	71	86	69	83	68	74	86	72	83	70	80	70	113	82	68	106	80	67	100	78	14.4
Jackson	88	74	86	73	84	71	77	86	75	83	73	81	74	134	83	72	123	81	71	117	78	20.3
Lansing	89	73	86	72	84	70	76	85	74	83	73	81	73	127	81	72	120	79	70	114	78	21.7
Marquette, Sawyer AFB	86	69	83	68	79	65	72	83	70	79	68	75	69	113	77	67	106	74	66	99	73	22.1
Marquette/Ishpeming, A	85	69	82	67	78	65	72	82	70	78	68	75	69	111	77	67	104	75	65	98	72	22.1
Mount Clemens, Angb	90	74	87	72	84	71	77	87	75	83	73	80	74	131	83	72	120	81	70	113	78	19.6
Muskegon	85	71	83	70	81	69	75	82	73	80	71	78	72	122	80	70	115	77	69	109	76	18.1
Oscoda, Wurtsmith AFB	89	72	86	71	83	69	75	86	73	83	71	79	72	120	80	70	112	79	68	106	77	21.4
Pellston	87	71	85	69	81	68	74	83	72	81	70	78	70	115	78	69	108	76	67	103	75	23.9
Saginaw	90	74	87	72	84	70	77	86	75	84	73	81	74	132	83	72	120	80	70	112	78	21.2
Sault Ste. Marie	83	69	80	68	77	66	72	80	70	77	68	74	69	111	76	67	103	74	65	95	72	21.9
Seul Choix Point	78	66	76	65	74	64	70	76	68	72	66	71	68	106	74	67	101	72	65	94	70	13.9
Traverse City	89	71	86	70	83	68	74	84	72	82	70	80	71	117	80	69	109	78	67	103	76	22.0
MINNESOTA																						
Alexandria	89	72	86	70	83	69	75	86	73	82	71	80	72	123	82	70	116	79	68	109	77	19.3
Brainerd, Pequot Lakes	88	70	85	68	81	66	72	85	70	82	68	78	68	108	81	66	102	77	65	96	75	21.6
Duluth	84	69	81	67	78	65	72	81	69	78	67	75	68	110	77	66	102	75	64	94	72	20.2
Hibbing	85	70	81	68	78	66	73	82	71	78	68	75	70	116	78	68	108	76	66	101	73	23.2
International Falls	86	69	83	67	80	66	72	82	70	79	68	77	69	112	78	67	103	75	65	96	73	21.8
Minneapolis-St. Paul	91	73	88	71	85	70	76	88	74	84	72	82	73	124	83	71	116	81	69	109	79	19.1
Redwood Falls	92	74	88	72	86	70	77	89	75	85	73	82	75	135	83	72	123	81	70	116	80	20.7
Rochester	88	72	85	71	82	70	76	85	74	82	72	80	73	128	81	71	120	79	69	111	77	19.7

MDB = mean coincident dry-bulb temperature, °F
MWB = mean coincident wet-bulb temperature, °F
MWS = mean coincident wind speed, mph
MWD = mean coincident wind direction
StdD = standard deviation
A = airport
HR = humidity ratio
DP = dew-point temperature, °F

Table 1A Heating and Wind Design Conditions—United States

Station	WMO#	Lat.	Long.	Elev. ft	StdP psia	Dates	Heating DB 99.6%	Heating DB 99%	Extreme Wind Speed 1%	2.5%	5%	Coldest Month WS/MDB 0.4% WS	0.4% MDB	1% WS	1% MDB	MWS/MWD to DB 99.6% MWS	99.6% PWD	0.4% MWS	0.4% PWD	Annual Extreme Daily Mean DB Max	Mean DB Min	StdD DB Max	StdD DB Min
Saint Cloud	726550	45.55	94.07	1024	14.160	6193	−20	−14	22	20	18	23	11	20	10	8	300	12	200	95	−27	3.0	5.6
Tofte	727554	47.58	90.83	791	14.280	8293	−10	−6	24	20	17	25	16	22	18	8	260	8	330	86	−19	4.5	4.9
MISSISSIPPI																							
Biloxi, Keesler AFB	747686	30.42	88.92	33	14.678	8293	31	35	17	14	13	18	49	16	50	8	360	7	210	97	23	2	7.4
Columbus, AFB	723306	33.65	88.45	220	14.579	8293	20	25	18	15	13	19	43	16	46	6	360	6	240	100	12	2.7	6.8
Greenwood	722359	33.50	90.08	154	14.614	8293	20	24	19	17	14	19	46	18	47	6	360	6	180	99	13	2.3	7.6
Jackson	722350	32.32	90.08	331	14.520	6193	21	25	20	18	16	21	45	19	46	7	340	8	270	98	14	2.7	5.8
McComb	722358	31.18	90.47	413	14.477	8293	23	28	17	14	13	17	49	15	49	6	350	7	230	98	15	2	7.2
Meridian	722340	32.33	88.75	308	14.532	6193	21	25	19	17	15	19	43	17	46	6	360	8	360	99	13	2.9	5.9
Tupelo	723320	34.27	88.77	361	14.505	8293	18	22	19	17	15	20	44	17	44	7	10	7	260	99	10	2.9	8.5
MISSOURI																							
Cape Girardeau	723489	37.23	89.57	341	14.515	8293	6	13	21	19	18	22	35	20	36	9	360	10	200	100	−1	2.9	9.2
Columbia	724450	38.82	92.22	899	14.224	6193	−1	5	25	22	20	25	27	22	28	11	310	11	200	99	−8	4.4	6.2
Joplin	723495	37.15	94.50	981	14.182	8293	3	11	23	21	19	24	50	21	47	10	10	11	220	100	−2	4	9.4
Kansas City	724460	39.32	94.72	1024	14.160	6193	−1	4	26	23	20	26	34	23	33	10	320	13	190	100	−7	4.1	6.6
Poplar Bluff	723300	36.77	90.47	479	14.443	8293	8	13	18	15	13	17	40	15	38	7	360	7	200	101	2	6.8	9.4
Spickard/Trenton	725400	40.25	93.72	886	14.231	8293	1	6	23	20	18	25	29	22	31	8	360	11	200	100	−5	5.2	7.4
Springfield	724400	37.23	93.38	1270	14.034	6193	3	9	24	21	19	23	35	21	35	10	340	10	230	98	−4	3.5	6.4
St. Louis, Int'l Airport	724340	38.75	90.37	564	14.398	6193	2	8	26	23	20	26	26	23	27	12	290	11	240	99	−5	3.5	6.2
Warrensburg, Whiteman AFB	724467	38.73	93.55	869	14.240	8293	1	7	22	19	17	23	34	21	34	9	360	9	190	101	−5	4	7.7
MONTANA																							
Billings	726770	45.80	108.53	3570	12.896	6193	−13	−7	28	24	22	30	25	27	30	10	230	10	240	99	−19	2.8	6.2
Bozeman	726797	45.78	111.15	4475	12.469	8293	−20	−12	21	18	15	20	36	17	34	4	140	9	360	96	−29	2.9	7.7
Butte	726785	45.95	112.50	5545	11.980	6193	−22	−14	23	21	18	21	29	19	30	4	150	13	120	92	−34	2.5	7.9
Cut Bank	727796	48.60	112.37	3855	12.760	6193	−21	−16	34	30	27	40	36	34	36	7	320	13	270	93	−28	4.0	5.7
Glasgow	727680	48.22	106.62	2297	13.516	6193	−22	−17	29	26	23	28	18	25	15	8	330	13	160	99	−29	3.2	6.6
Great Falls, Int'l Airport	727750	47.48	111.37	3658	12.854	6193	−19	−13	33	29	26	34	38	31	38	7	240	12	230	98	−25	3.2	7.4
Great Falls, Malmstrom AFB	727755	47.50	111.18	3527	12.917	8293	−17	−11	28	24	21	33	38	29	38	4	240	8	260	99	−22	3.2	7.9
Havre	727770	48.55	109.77	2598	13.367	8293	−25	−19	24	21	19	26	35	23	33	6	240	9	270	102	−33	5	8.1
Helena	727720	46.60	112.00	3898	12.740	6193	−18	−10	25	22	19	25	40	22	35	5	290	12	280	96	−24	3.3	7.2
Kalispell	727790	48.30	114.27	2972	13.184	6193	−12	−3	24	20	17	25	12	21	18	7	20	9	170	95	−19	2.9	8.6
Lewistown	726776	47.05	109.47	4167	12.613	6193	−18	−12	26	23	20	29	35	25	35	7	250	11	90	95	−25	3.5	7.3
Miles City	742300	46.43	105.87	2628	13.352	6193	−19	−13	27	23	20	28	25	23	27	8	290	11	140	102	−25	2.7	6.5
Missoula	727730	46.92	114.08	3189	13.079	6193	−9	−1	22	19	17	22	17	19	21	7	120	10	290	97	−15	2.9	8.2
NEBRASKA																							
Bellevue, Offutt AFB	725540	41.12	95.92	1047	14.148	8293	−5	1	22	19	17	26	23	22	23	8	330	10	190	100	−9	4.5	6.3
Grand Island	725520	40.97	98.32	1857	13.736	6193	−8	−2	30	26	23	29	21	26	19	11	270	15	180	102	−14	3.2	5.2
Lincoln	725510	40.85	96.75	1188	14.076	8293	−7	−2	27	23	21	28	25	24	27	9	350	15	180	103	−11	6.5	8.1
Norfolk	725560	41.98	97.43	1552	13.890	6193	−11	−5	29	25	22	33	20	28	21	11	340	15	190	101	−18	3.0	5.4
North Platte	725620	41.13	100.68	2785	13.275	6193	−10	−4	29	25	22	28	24	24	26	7	320	12	180	101	−16	2.9	6.6
Omaha, Eppley Airfield	725500	41.30	95.90	981	14.182	6193	−7	−2	26	23	20	27	21	23	17	10	340	12	180	100	−14	3.3	4.8
Omaha, Wso	725530	41.37	96.02	1332	14.002	8293	−8	−2	22	20	18	25	23	22	25	10	310	11	170	98	−14	4	6.5
Scottsbluff	725660	41.87	103.60	3957	12.712	6193	−11	−3	30	26	22	32	35	27	35	8	300	11	300	101	−19	2.9	8.0
Sidney	725610	41.10	102.98	4304	12.549	8293	−8	−1	29	24	22	31	32	26	35	9	290	12	160	101	−18	4.5	8.6
Valentine	725670	42.87	100.55	2598	13.367	8293	−16	−8	27	23	21	26	25	23	28	9	250	15	180	104	−22	4.3	8.5
NEVADA																							
Elko	725825	40.83	115.78	5135	12.166	6193	−5	1	21	18	16	20	36	16	37	4	70	10	230	98	−13	3.2	8.0
Ely	724860	39.28	114.85	6263	11.660	6193	−6	0	28	24	21	26	33	22	30	11	190	13	230	93	−15	2.3	7.3
Las Vegas, Int'l Airport	723860	36.08	115.17	2178	13.575	6193	27	30	30	26	23	25	48	22	49	7	250	12	230	111	21	2.2	4.7
Mercury	723870	36.62	116.02	3310	13.021	8293	24	28	25	22	19	25	44	21	42	8	50	12	230	102	19	18	6.3
North Las Vegas, Nellis AFB	723865	36.23	115.03	1870	13.729	8293	28	31	24	21	18	23	52	19	49	2	20	9	210	112	21	2	4.5
Reno	724880	39.50	119.78	4400	12.505	6193	8	13	26	22	19	26	46	21	44	3	160	10	290	99	1	2.2	8.4
Tonopah	724855	38.05	117.08	5427	12.033	6193	7	13	25	22	20	24	37	22	36	9	340	12	180	98	1	2.0	6.5
Winnemucca	725830	40.90	117.80	4314	12.544	6193	1	7	23	19	17	21	39	18	38	5	160	11	250	101	−9	2.3	10.2
NEW HAMPSHIRE																							
Concord	726050	43.20	71.50	344	14.513	6193	−8	−2	23	20	17	23	20	20	21	4	320	10	230	95	−18	2.9	5.5
Lebanon	726116	43.63	72.30	597	14.381	8293	−7	−3	18	15	14	18	25	16	26	2	360	9	220	94	−17	2	5.2
Mount Washington	726130	44.27	71.30	6266	11.659	8293	−23	−19	88	81	73	99	−14	92	−15	73	280	21	270	65	−33	2.3	4.1
Portsmouth, Pease AFB	726055	43.08	70.82	102	14.642	8293	4	9	21	18	16	22	26	20	27	8	280	8	270	94	−2	2.2	3.2
NEW JERSEY																							
Atlantic City	724070	39.45	74.57	66	14.661	6193	8	13	27	23	20	29	36	25	34	9	310	11	250	96	0	2.9	5.7
Millville	724075	39.37	75.07	82	14.652	8293	10	15	19	18	17	20	35	19	35	7	300	11	240	96	0	2.3	7.4
Newark	725020	40.70	74.17	30	14.680	6193	10	14	26	23	20	27	28	23	29	13	260	13	230	98	4	2.6	4.8
Teterboro	725025	40.85	74.07	10	14.690	8293	10	14	21	19	17	21	29	19	30	11	280	12	240	97	2	2.5	5.6
Trenton, McGuire AFB	724096	40.02	74.60	135	14.624	8293	11	15	22	19	17	23	31	21	31	8	330	8	240	97	2	2.2	5.2
NEW MEXICO																							
Alamogordo, Holloman AFB	747320	32.85	106.10	4094	12.647	8293	20	23	20	17	14	18	50	15	48	3	10	8	250	102	13	2.9	3.6
Albuquerque	723650	35.05	106.62	5315	12.084	6193	13	18	29	25	22	26	34	22	37	8	360	10	240	100	6	2.6	7.3
Carlsbad	722687	32.33	104.27	3294	13.028	8293	19	23	25	22	19	25	57	21	54	8	340	12	150	104	9	3.6	7.0
Clayton	723600	36.45	103.15	4970	12.241	8293	1	9	30	27	24	30	40	26	39	10	40	13	200	98	−5	2.5	7.4

WMO# = World Meteorological Organization number
Lat = latitude
Long. = longitude
Elev = elevation, ft
StdP = standard pressure at station elevation, psia
DB = dry-bulb temperature, °F
WS = wind speed, mph

Table 1B Cooling and Dehumidification Design Conditions—United States

	Cooling DB/MWB						WB/MDB						DP/MDB and HR									
	0.4%		1%		2%		0.4%		1%		2%		0.4%			1%			2%			Range
Station	DB	MWB	DB	MWB	DB	MWB	WB	MDB	WB	MDB	WB	MDB	DP	HR	MDB	DP	HR	MDB	DP	HR	MCDB	of DB
Saint Cloud	91	72	88	71	85	70	76	87	74	84	72	82	72	125	83	70	116	80	68	109	78	21.5
Tofte	79	64	75	62	71	61	66	74	64	72	63	70	64	92	70	61	83	69	59	77	68	13.0
MISSISSIPPI																						
Biloxi, Keesler AFB	92	79	91	78	89	78	81	89	80	88	80	87	79	151	86	78	147	85	78	144	84	13.0
Columbus, AFB	96	77	94	76	92	76	80	91	78	89	78	88	77	141	85	76	136	83	75	132	82	19.3
Greenwood	96	78	94	78	93	77	81	91	80	90	79	89	78	148	86	77	143	85	76	139	84	19.1
Jackson	95	77	93	76	92	76	80	90	79	89	78	88	77	142	84	76	138	83	75	134	82	19.2
McComb	94	76	92	76	91	76	79	89	78	88	78	87	77	141	83	76	138	82	75	135	81	19.8
Meridian	96	77	94	76	92	76	79	91	78	90	77	88	76	139	84	75	134	83	74	130	83	20.3
Tupelo	96	76	94	76	92	75	79	89	78	89	77	88	76	137	83	75	134	83	74	131	82	18.9
MISSOURI																						
Cape Girardeau	96	77	94	77	91	76	80	92	78	90	78	88	77	141	86	76	136	85	75	132	83	19.8
Columbia	95	75	92	75	89	74	78	89	77	88	75	86	75	137	85	74	130	83	72	124	82	20.3
Joplin	96	75	94	75	91	74	78	90	77	89	76	88	75	137	85	74	132	85	72	125	83	20.0
Kansas City	96	75	93	75	90	74	78	90	77	89	76	87	75	137	86	74	130	84	73	125	83	18.8
Poplar Bluff	95	77	92	76	90	76	80	90	78	88	77	87	77	144	85	76	138	83	75	133	82	20.0
Spickard/Trenton	96	74	93	73	89	72	78	88	76	88	75	86	76	139	83	73	128	83	71	118	81	19.6
Springfield	95	74	92	74	89	74	78	89	76	88	75	86	74	134	84	73	128	83	72	124	81	20.8
St. Louis, Int'l Airport	95	76	93	75	90	74	79	90	78	88	76	87	76	138	85	75	132	83	73	127	82	18.3
Warrensburg, Whiteman AFB	96	76	93	76	90	75	79	91	78	90	76	88	76	139	86	75	134	85	73	128	83	19.3
MONTANA																						
Billings	93	63	90	62	87	61	65	86	64	84	62	83	59	83	71	57	78	71	55	74	70	25.8
Bozeman	91	61	87	60	85	59	64	83	62	82	61	81	58	83	69	56	78	67	53	71	66	31.7
Butte	86	57	84	56	80	55	60	76	58	76	57	76	54	76	61	52	70	63	50	66	62	31.5
Cut Bank	87	60	84	59	80	58	62	81	60	79	59	77	56	77	67	54	70	65	51	65	63	26.1
Glasgow	94	64	90	63	86	62	68	85	66	83	64	82	62	91	74	60	83	71	58	77	69	25.3
Great Falls, Int'l Airport	92	61	88	60	85	59	64	84	62	82	61	81	57	81	69	55	74	67	53	69	66	27.2
Great Falls, Malmstrom AFB	93	62	89	61	86	60	65	85	63	83	62	81	59	84	71	57	78	69	54	71	68	26.3
Havre	94	63	90	62	86	61	66	87	64	84	62	82	60	84	72	58	78	69	56	74	68	27.9
Helena	90	60	87	59	84	59	63	82	61	81	60	80	57	80	68	55	73	66	52	68	66	28.0
Kalispell	89	62	86	61	82	60	65	83	63	81	61	79	59	82	69	57	76	67	55	71	67	29.9
Lewistown	89	61	86	60	82	59	64	81	63	80	61	78	58	85	71	56	79	69	54	74	67	28.3
Miles City	97	66	93	65	90	64	69	89	67	86	66	84	63	95	76	61	88	75	59	82	73	25.9
Missoula	91	62	88	61	85	60	65	84	63	82	62	81	58	82	68	56	76	68	55	71	66	31.3
NEBRASKA																						
Bellevue, Offutt AFB	95	76	91	75	88	74	79	89	77	88	76	86	76	141	85	74	134	83	73	127	82	18.4
Grand Island	97	72	93	72	90	70	76	89	74	88	73	86	72	127	82	70	120	81	69	113	79	22.4
Lincoln	97	74	94	74	91	73	78	90	76	89	75	87	75	136	84	73	130	83	71	121	82	22.3
Norfolk	95	74	92	72	89	72	76	90	75	88	73	86	73	129	83	71	121	82	70	115	81	20.8
North Platte	95	69	92	69	89	68	73	87	72	86	70	85	69	118	80	67	111	78	66	105	77	25.5
Omaha, Eppley Airfield	95	75	92	75	89	73	78	90	77	88	75	86	75	136	85	73	128	84	72	121	82	19.9
Omaha, Wso	94	75	90	75	87	73	77	89	76	87	74	85	74	134	84	72	126	83	71	120	82	17.6
Scottsbluff	95	65	92	64	89	64	69	87	68	85	66	84	64	102	76	62	97	74	60	91	73	28.9
Sidney	95	63	92	63	89	63	67	84	66	84	64	84	62	97	73	60	91	72	58	86	71	27.9
Valentine	97	68	94	67	90	67	72	90	71	89	69	87	67	110	79	65	103	78	63	94	77	26.5
NEVADA																						
Elko	95	60	92	59	90	58	63	85	61	84	60	84	57	84	68	54	75	66	51	67	67	38.4
Ely	89	56	87	56	85	55	60	78	59	78	58	78	55	82	64	53	75	64	50	68	65	34.6
Las Vegas, Int'l Airport	108	66	106	66	103	65	71	95	70	93	69	93	65	102	79	63	92	81	60	84	85	24.8
Mercury	102	65	100	64	98	63	69	88	67	89	66	89	64	102	72	60	89	77	58	80	80	25.9
North Las Vegas, Nellis AFB	108	68	106	67	104	66	72	94	71	94	70	94	67	106	79	64	97	82	61	86	84	26.3
Reno	95	61	92	60	90	59	63	87	62	86	60	85	56	77	69	53	69	69	50	63	68	37.3
Tonopah	94	58	92	57	89	57	62	83	61	82	60	81	56	83	67	53	74	68	50	67	69	31.1
Winnemucca	97	61	94	60	92	59	63	88	62	87	60	86	56	79	68	53	69	67	50	62	68	37.4
NEW HAMPSHIRE																						
Concord	90	71	87	70	84	68	74	85	73	82	71	79	71	118	79	70	111	77	68	105	76	24.1
Lebanon	88	71	86	69	83	68	74	84	72	82	70	79	70	113	79	69	108	77	67	103	75	23.0
Mount Washington	60	56	58	54	56	54	58	59	56	57	54	56	58	90	58	56	84	57	54	78	55	8.5
Portsmouth, Pease AFB	89	72	85	70	83	70	75	84	73	82	72	79	73	123	85	71	113	77	69	106	76	18.2
NEW JERSEY																						
Atlantic City	91	74	88	73	86	72	77	87	76	84	75	82	75	131	81	74	125	80	72	120	79	18.1
Millville	92	75	89	74	87	73	78	87	76	86	75	83	75	134	81	74	129	80	73	125	80	18.7
Newark	93	74	90	73	87	71	77	88	76	85	74	83	74	127	81	73	121	80	71	116	80	15.9
Teterboro	92	76	89	74	87	73	78	88	77	87	75	83	76	134	84	74	128	82	72	119	81	18.4
Trenton, McGuire AFB	93	75	90	74	87	73	78	89	76	87	75	84	75	132	83	74	127	82	72	118	80	18.9
NEW MEXICO																						
Alamogordo, Holloman AFB	98	63	96	63	93	63	68	87	67	85	67	85	65	106	72	62	98	72	61	92	73	30.2
Albuquerque	96	60	93	60	91	60	65	83	64	82	64	81	61	98	68	60	93	69	58	89	69	25.4
Carlsbad	101	65	98	66	96	66	72	88	71	87	70	85	69	121	76	68	116	76	67	111	75	25.4
Clayton	94	62	91	62	88	62	67	84	65	84	65	82	61	98	72	60	94	71	59	90	70	26.1

MDB = mean coincident dry-bulb temperature, °F
MWB = mean coincident wet-bulb temperature, °F
MWS = mean coincident wind speed, mph
MWD = mean coincident wind direction
StdD = standard deviation
A = airport
HR = humidity ratio
DP = dew-point temperature, °F

Table 1A Heating and Wind Design Conditions—United States

							Heating DB		Extreme Wind Speed			Coldest Month WS/MDB				MWS/MWD to DB				Annual Extreme Daily			
												0.4%		1%		99.6%		0.4%		Mean DB		StdD DB	
Station	WMO#	Lat.	Long.	Elev. ft	StdP psia	Dates	99.6%	99%	1%	2.5%	5%	WS	MDB	WS	MDB	MWS	PWD	MWS	PWD	Max	Min	Max	Min
Clovis, Cannon AFB	722686	34.38	103.32	4295	12.554	8293	10	15	26	23	20	26	40	23	39	8	50	11	220	101	5	2.3	4.0
Farmington	723658	36.75	108.23	5502	11.999	8293	8	13	23	21	18	22	35	19	34	6	60	10	240	99	−1	3.8	7.2
Gallup	723627	35.52	108.78	6470	11.570	8293	−1	5	23	20	18	19	39	18	37	1	140	11	270	94	−12	2.3	7.9
Roswell	722680	33.30	104.53	3668	12.849	8293	14	20	22	19	17	20	51	18	48	8	360	11	140	105	6	4.7	6.5
Truth Or Consequences	722710	33.23	107.27	4859	12.292	8293	22	26	25	21	18	24	43	21	41	8	350	10	170	102	6	2.9	23.8
Tucumcari	723676	35.18	103.60	4065	12.661	6193	9	15	25	22	20	28	50	23	45	8	50	12	230	102	1	3.1	7.2
NEW YORK																							
Albany	725180	42.75	73.80	292	14.541	6193	−7	−2	24	22	19	23	20	20	22	5	300	10	230	95	−18	3.0	6.3
Binghamton	725150	42.22	75.98	1631	13.850	6193	−2	2	24	21	19	24	20	22	19	13	270	11	220	89	−9	3.1	4.4
Buffalo	725280	42.93	78.73	705	14.325	6193	2	5	29	26	23	34	25	30	24	12	270	13	240	91	−6	2.3	5.3
Central Islip	725035	40.80	73.10	98	14.643	8293	11	15	22	20	18	23	32	21	31	10	340	11	210	94	2	3.2	5.4
Elmira/Corning	725156	42.17	76.90	955	14.195	8293	−2	3	21	19	17	23	20	20	27	5	240	11	210	95	−10	3.6	6.1
Glens Falls	725185	43.33	73.62	328	14.522	8293	−10	−4	18	16	14	19	22	17	22	2	350	10	190	93	−20	2.7	4.7
Massena	726223	44.93	74.85	213	14.583	6193	−15	−10	21	18	17	23	22	21	22	4	270	10	230	92	−27	3.0	6.0
New York, JFK Airport	744860	40.65	73.78	23	14.683	6193	11	15	27	24	21	30	29	27	28	17	320	13	230	96	6	2.6	4.6
New York, La Guardia A	725030	40.77	73.90	30	14.680	8293	13	17	28	25	22	30	29	27	28	18	310	12	280	97	6	2.2	4.3
Newburgh	725038	41.50	74.10	492	14.436	8293	6	10	23	20	18	26	17	23	26	8	260	10	230	92	−4	3.1	6.5
Niagara Falls	725287	43.10	78.95	591	14.385	8293	4	7	26	22	20	30	24	27	23	11	240	13	230	91	−4	3.2	6.1
Plattsburgh, AFB	726225	44.65	73.47	236	14.570	8293	−9	−4	21	18	16	22	27	19	24	2	350	8	260	93	−17	2.7	4.7
Poughkeepsie	725036	41.63	73.88	167	14.607	8293	2	6	18	16	14	19	25	17	25	3	250	9	250	96	−8	3.1	5.9
Rochester	725290	43.12	77.67	554	14.403	6193	1	5	27	23	21	29	22	26	21	10	230	12	250	93	−7	2.8	4.9
Rome, Griffiss AFB	725196	43.23	75.40	505	14.429	8293	−5	1	22	19	16	23	22	20	22	3	330	8	260	93	−15	2.5	3.8
Syracuse	725190	43.12	76.12	407	14.481	6193	−3	2	26	22	20	28	20	25	21	7	90	11	250	93	−13	3.0	6.0
Watertown	726227	44.00	76.02	325	14.524	8293	−12	−6	21	19	18	24	24	21	25	5	80	11	240	90	−25	2.9	7.0
White Plains	725037	41.07	73.70	440	14.463	8293	7	12	19	17	15	19	29	18	29	13	310	9	260	95	0	2.9	4.5
NORTH CAROLINA																							
Asheville	723150	35.43	82.55	2169	13.580	6193	11	16	25	22	19	26	26	23	28	11	340	9	340	91	3	2.6	6.8
Cape Hatteras	723040	35.27	75.55	10	14.690	6193	26	29	26	22	20	27	47	23	47	11	340	11	230	91	20	2.0	4.9
Charlotte	723140	35.22	80.93	768	14.292	6193	18	23	20	17	15	20	44	18	45	6	50	9	240	97	10	2.9	6.0
Cherry Point, Mcas	723090	34.90	76.88	30	14.680	8293	24	28	19	16	15	19	43	17	48	5	10	7	240	100	12	2.5	8.5
Fayetteville, Fort Bragg	746930	35.13	78.93	243	14.567	8293	22	27	17	14	12	19	42	16	44	4	10	6	240	100	15	3.8	6.8
Goldsboro, Johnson AFB	723066	35.33	77.97	108	14.638	8293	22	27	17	14	12	18	46	15	44	4	270	8	260	100	14	3.1	7.4
Greensboro	723170	36.08	79.95	886	14.231	6193	15	19	19	17	15	20	40	18	40	7	290	8	230	96	7	2.7	5.0
Hickory	723145	35.73	81.38	1188	14.076	8293	18	23	17	15	13	18	41	16	41	4	320	9	240	97	8	3.2	6.8
Jacksonville, New River Mcaf	723096	34.72	77.45	26	14.681	8293	23	27	18	16	14	19	49	17	47	5	350	7	240	99	13	2	8.8
New Bern	723095	35.07	77.05	20	14.685	8293	22	27	18	16	14	19	49	17	47	6	10	8	240	99	13	1.4	8.3
Raleigh/Durham	723060	35.87	78.78	440	14.463	6193	16	20	21	18	16	21	42	19	43	8	360	9	240	96	9	2.9	5.3
Wilmington	723013	34.27	77.90	33	14.678	6193	23	27	21	19	17	22	51	20	48	7	320	10	220	97	17	2.2	5.7
Winston-Salem	723193	36.13	80.22	971	14.187	8293	18	23	19	17	15	21	38	19	38	7	290	8	240	96	8	2.7	5.8
NORTH DAKOTA																							
Bismarck	727640	46.77	100.75	1660	13.835	6193	−21	−16	29	25	22	29	13	25	16	7	290	13	180	100	−30	3.6	6.4
Devils Lake	727580	48.10	98.87	1453	13.940	8293	−23	−19	26	22	20	27	12	24	10	9	300	11	10	98	−27	7	5.0
Fargo	727530	46.90	96.80	899	14.224	6193	−22	−17	31	27	24	32	7	28	7	8	180	14	160	98	−27	3.4	4.4
Grand Forks, AFB	727575	47.97	97.40	912	14.217	8293	−20	−16	27	24	21	30	9	26	13	7	290	13	180	98	−25	4.7	4.9
Minot, AFB	727675	48.42	101.35	1667	13.832	8293	−21	−16	28	24	21	30	18	27	16	10	310	12	150	101	−25	3.8	7.4
Minot, Int'l Airport	727676	48.27	101.28	1716	13.807	6193	−20	−16	28	24	22	30	14	27	14	12	290	13	200	98	−25	3.0	4.8
Williston	727670	48.18	103.63	1906	13.711	8293	−24	−18	27	23	21	28	25	24	20	8	220	14	150	101	−30	4.5	8.3
OHIO																							
Akron/Canton	725210	40.92	81.43	1237	14.050	6193	0	5	24	21	19	25	26	22	26	11	270	10	230	92	−7	2.9	7.0
Cincinnati, Lunken Field	724297	39.10	84.42	482	14.441	8293	5	12	21	19	17	22	35	19	33	9	260	10	210	96	−3	3.2	9.4
Cleveland	725240	41.42	81.87	804	14.274	6193	1	6	26	23	20	27	28	24	28	12	230	12	230	93	−6	2.8	6.3
Columbus, Int'l Airport	724280	40.00	82.88	817	14.267	6193	1	6	23	20	18	24	30	21	25	9	270	11	270	94	−6	2.6	7.1
Columbus, Rickenbckr AFB	724285	39.82	82.93	745	14.304	8293	3	10	21	18	16	23	26	20	27	7	210	8	270	96	−4	4.5	8.6
Dayton, Int'l Airport	724290	39.90	84.20	1004	14.170	6193	−1	5	24	21	19	25	26	22	28	11	270	11	240	95	−8	2.9	7.0
Dayton, Wright-Paterson AFB	745700	39.83	84.05	823	14.263	8293	1	8	21	18	16	23	28	21	30	7	270	9	240	96	−7	3.2	7.7
Findlay	725366	41.02	83.67	810	14.270	8293	−2	4	23	20	19	25	34	22	29	11	250	12	210	94	−9	3.8	7.9
Mansfield	725246	40.82	82.52	1296	14.020	6193	−1	4	25	22	20	28	28	25	26	13	240	12	240	91	−8	2.8	6.0
Toledo	725360	41.60	83.80	692	14.332	6193	−2	3	23	20	18	25	25	22	21	10	250	11	230	95	−10	3.0	5.4
Youngstown	725250	41.27	80.67	1184	14.077	6193	−1	4	23	21	19	24	22	22	21	10	230	10	230	91	−8	2.5	5.8
Zanesville	724286	39.95	81.90	899	14.224	8293	2	9	19	18	16	21	32	19	31	7	240	9	220	94	−7	3.6	8.5
OKLAHOMA																							
Altus, AFB	723520	34.67	99.27	1378	13.978	8293	13	19	23	21	19	24	40	21	42	9	20	10	190	107	7	3.4	7.7
Enid, Vance AFB	723535	36.33	97.92	1306	14.015	8293	5	12	26	23	20	27	38	23	38	12	10	11	190	105	1	3.6	6.7
Lawton, Fort Sill/Post Field	723550	34.65	98.40	1188	14.076	8293	12	19	24	21	19	26	35	22	36	11	10	11	170	103	8	2.5	7.4
McAlester	723566	34.88	95.78	771	14.291	8293	10	17	20	18	16	21	47	19	45	9	360	9	190	102	4	4	8.3
Oklahoma City, Tinker AFB	723540	35.42	97.38	1293	14.022	8293	10	17	24	22	19	25	42	22	42	10	10	11	190	103	6	3.2	6.1
Oklahoma City, W. Rogers A	723530	35.40	97.60	1302	14.017	6193	9	15	29	25	23	29	33	26	37	15	360	13	180	103	4	3.4	4.9
Tulsa	723560	36.20	95.90	676	14.340	6193	9	14	25	23	21	24	46	22	40	11	360	12	180	103	3	3.6	5.6
OREGON																							
Astoria	727910	46.15	123.88	23	14.683	6193	25	29	25	22	19	29	51	24	49	8	90	12	320	87	20	4.5	6.1

WMO# = World Meteorological Organization number
Lat = latitude
Long. = longitude
Elev = elevation, ft
StdP = standard pressure at station elevation, psia
DB = dry-bulb temperature, °F
WS = wind speed, mph

Table 1B Cooling and Dehumidification Design Conditions—United States

Station	Cooling DB/MWB						WB/MDB						DP/MDB and HR									Range of DB
	0.4%		1%		2%		0.4%		1%		2%		0.4%			1%			2%			
	DB	MWB	DB	MWB	DB	MWB	WB	MDB	WB	MDB	WB	MDB	DP	HR	MDB	DP	HR	MDB	DP	HR	MCDB	
Clovis, Cannon AFB	96	64	93	64	91	65	70	84	69	83	68	83	66	114	75	65	109	74	64	105	73	24.5
Farmington	94	60	92	60	89	60	65	83	64	83	63	82	60	94	69	58	90	68	57	85	69	28.8
Gallup	89	57	87	56	85	56	62	76	61	76	60	75	59	94	65	57	90	64	56	85	64	30.6
Roswell	98	65	96	65	94	65	70	87	69	86	68	85	66	111	73	65	108	73	64	104	73	24.8
Truth Or Consequences	97	61	95	61	93	61	66	85	65	85	64	84	60	94	71	59	90	71	58	87	72	25.0
Tucumcari	98	64	95	65	93	64	69	87	68	85	67	84	65	109	73	64	104	73	63	100	72	24.9
NEW YORK																						
Albany	90	71	86	70	84	69	74	85	73	82	71	79	72	118	79	70	111	77	68	106	76	23.7
Binghamton	85	70	82	69	80	67	73	81	71	79	70	77	70	118	77	69	111	75	67	106	74	17.5
Buffalo	86	70	84	69	81	68	74	82	72	80	71	78	71	118	78	70	113	77	68	106	75	17.7
Central Islip	88	73	85	72	83	70	76	83	75	81	74	79	74	129	79	73	124	78	71	116	77	15.1
Elmira/Corning	90	72	87	71	84	69	75	86	73	82	72	80	72	122	81	70	116	78	69	110	76	24.1
Glens Falls	88	73	85	71	83	70	76	85	74	82	72	80	74	127	81	71	116	79	69	108	76	22.1
Massena	87	72	84	71	82	69	75	84	73	81	71	79	72	118	80	70	111	78	68	105	76	21.8
New York, JFK Airport	91	74	88	72	85	71	76	86	75	84	74	82	74	125	80	72	120	80	71	114	79	13.9
New York, La Guardia A	92	74	89	73	86	72	77	87	76	85	74	83	74	129	81	73	125	80	71	116	80	14.6
Newburgh	88	74	85	72	83	70	76	85	74	83	73	80	74	130	82	72	119	80	70	111	78	17.1
Niagara Falls	87	72	85	71	83	69	75	84	74	81	72	79	73	125	81	71	116	78	69	111	76	18.9
Plattsburgh, AFB	86	71	83	69	80	68	74	82	72	80	70	78	71	115	79	69	108	76	67	102	75	19.6
Poughkeepsie	92	75	88	72	85	71	76	87	75	85	73	82	74	126	82	71	116	80	70	111	78	23.0
Rochester	89	73	86	71	83	70	75	85	74	82	72	80	73	123	81	71	116	79	69	109	77	20.1
Rome, Griffiss AFB	88	71	86	70	83	69	74	84	73	82	71	79	71	117	80	70	111	78	68	105	76	22.9
Syracuse	88	72	85	71	83	70	75	85	73	82	72	80	72	120	80	70	113	78	69	107	77	20.3
Watertown	85	71	83	70	80	69	74	82	72	80	71	77	71	118	78	70	111	77	69	106	75	20.5
White Plains	89	74	87	72	84	71	76	86	75	83	73	80	74	128	80	72	120	80	71	114	78	18.0
NORTH CAROLINA																						
Asheville	88	72	85	71	83	70	75	84	73	82	72	80	72	128	79	71	123	78	70	118	76	19.4
Cape Hatteras	88	78	86	77	85	77	80	86	79	84	78	83	78	147	83	77	142	83	76	138	82	11.4
Charlotte	94	74	91	74	89	73	77	88	76	87	75	86	74	130	82	73	125	80	72	122	80	17.8
Cherry Point, Mcas	95	79	92	78	90	77	81	91	80	90	79	88	78	146	87	77	141	86	76	136	85	16.6
Fayetteville, Fort Bragg	96	77	94	76	92	75	79	91	78	89	77	88	76	139	84	76	135	83	75	131	83	18.2
Goldsboro, Johnson AFB	96	77	94	76	91	76	80	91	78	89	77	87	76	139	84	76	135	83	75	132	82	18.4
Greensboro	92	75	90	74	88	73	77	88	76	86	75	85	74	132	82	73	127	81	72	123	80	18.5
Hickory	94	73	91	72	88	72	76	87	75	85	74	84	74	133	80	73	128	80	71	120	78	19.6
Jacksonville, New River Mcaf	94	79	92	78	89	77	81	90	79	89	78	87	78	145	86	77	140	85	76	136	84	17.1
New Bern	94	78	92	78	90	76	81	91	79	89	78	87	78	144	86	77	139	84	76	134	83	17.1
Raleigh/Durham	93	76	90	75	88	74	78	88	77	87	76	85	75	134	82	74	130	81	73	125	80	18.8
Wilmington	93	79	91	78	89	77	80	89	79	88	78	86	78	146	85	77	141	83	76	137	83	15.7
Winston-Salem	92	74	89	74	87	73	77	86	76	86	75	85	74	134	81	73	129	80	72	121	79	17.6
NORTH DAKOTA																						
Bismarck	93	68	90	67	86	66	72	86	70	84	68	82	68	109	79	65	100	77	63	92	75	26.5
Devils Lake	91	69	87	67	84	66	72	86	70	83	68	80	68	108	78	66	100	77	63	90	75	21.1
Fargo	91	71	88	70	85	69	75	86	73	84	71	81	72	122	82	69	112	80	67	104	77	22.3
Grand Forks, AFB	91	71	88	69	85	68	75	86	72	83	70	80	71	118	81	69	109	78	67	101	76	22.9
Minot, AFB	94	68	90	67	86	66	72	87	70	85	68	82	68	109	80	65	100	78	63	90	75	24.7
Minot, Int'l Airport	92	67	88	66	84	65	71	85	69	83	67	80	67	106	78	64	96	75	62	89	73	22.9
Williston	96	67	92	66	87	65	71	87	69	86	67	83	66	103	78	63	92	76	61	85	73	25.7
OHIO																						
Akron/Canton	88	72	85	71	83	70	75	84	73	82	72	80	72	125	80	71	118	78	69	113	77	18.8
Cincinnati, Lunken Field	93	74	90	75	88	73	77	89	76	87	75	84	75	132	82	74	128	81	72	120	80	20.0
Cleveland	89	73	86	72	84	71	76	85	74	83	72	81	73	125	82	71	118	80	70	112	78	18.6
Columbus, Int'l Airport	90	74	88	73	86	71	77	87	75	85	74	82	73	128	82	72	123	81	71	117	79	19.3
Columbus, Rickenbckr AFB	92	74	89	73	87	72	77	88	75	86	74	84	74	130	83	72	120	82	70	115	79	19.8
Dayton, Int'l Airport	90	74	88	73	86	71	76	87	75	85	74	82	73	129	82	72	123	80	71	117	79	19.2
Dayton, Wright-Paterson AFB	92	74	89	74	87	73	78	88	76	86	74	84	75	136	84	73	127	83	71	119	81	19.8
Findlay	90	74	87	72	85	71	76	86	75	83	73	81	74	132	81	72	121	80	71	116	78	18.9
Mansfield	88	73	85	72	83	71	76	85	74	83	73	80	73	128	81	71	122	79	70	116	78	17.8
Toledo	90	73	87	72	85	71	77	86	75	84	73	81	74	129	82	72	122	80	70	115	78	20.9
Youngstown	88	72	85	70	83	69	74	84	73	82	71	79	72	122	80	70	116	78	69	110	76	20.6
Zanesville	90	74	88	73	86	71	76	87	75	85	74	82	74	130	82	72	120	80	71	116	78	20.7
OKLAHOMA																						
Altus, AFB	102	73	100	73	97	73	77	93	76	92	75	91	74	132	84	72	124	83	71	119	82	23.6
Enid, Vance AFB	101	74	98	74	95	73	77	92	76	91	75	90	73	130	85	71	121	83	70	116	82	21.8
Lawton, Fort Sill/Post Field	99	73	97	73	95	73	77	90	76	90	75	89	74	135	83	73	129	82	71	121	81	20.7
McAlester	98	76	96	76	93	76	79	92	78	91	77	89	76	141	85	75	137	83	74	133	83	21.8
Oklahoma City, Tinker AFB	98	74	96	75	94	74	78	92	77	91	76	89	75	138	87	74	132	85	72	123	83	19.4
Oklahoma City, W. Rogers A	99	74	96	74	94	73	77	91	76	90	75	89	73	129	83	72	125	82	71	120	81	21.0
Tulsa	100	76	97	76	94	75	79	92	78	92	77	90	76	137	87	74	132	85	73	127	84	19.5
OREGON																						
Astoria	76	64	72	62	69	61	65	75	63	71	62	68	61	81	69	60	76	66	59	74	65	14.2

MDB = mean coincident dry-bulb temperature, °F
MWB = mean coincident wet-bulb temperature, °F
MWS = mean coincident wind speed, mph
MWD = mean coincident wind direction
StdD = standard deviation
A = airport
HR = humidity ratio
DP = dew-point temperature, °F

Table 1A Heating and Wind Design Conditions—United States

Station	WMO#	Lat.	Long.	Elev. ft	StdP psia	Dates	Heating DB 99.6%	Heating DB 99%	Extreme Wind Speed 1%	Extreme Wind Speed 2.5%	Extreme Wind Speed 5%	Coldest Month WS/MDB 0.4% WS	Coldest Month WS/MDB 0.4% MDB	Coldest Month WS/MDB 1% WS	Coldest Month WS/MDB 1% MDB	MWS/MWD to DB 99.6% MWS	MWS/MWD to DB 99.6% PWD	MWS/MWD to DB 0.4% MWS	MWS/MWD to DB 0.4% PWD	Annual Extreme Daily Mean DB Max	Annual Extreme Daily Mean DB Min	Annual Extreme Daily StdD DB Max	Annual Extreme Daily StdD DB Min
Eugene	726930	44.12	123.22	374	14.498	6193	21	26	20	18	16	22	46	19	45	8	360	12	360	99	16	3.7	7.9
Hillsboro	726986	45.53	122.95	203	14.588	8293	19	24	19	17	15	23	26	19	34	8	60	9	360	100	15	3.8	6.3
Klamath Falls	725895	42.15	121.73	4091	12.649	8293	4	10	25	22	19	28	39	23	33	6	320	9	320	97	−4	4.1	8.6
Meacham	726885	45.52	118.40	4055	12.666	8293	−9	0	12	10	9	13	33	11	33	1	130	5	360	93	−21	4.7	12.2
Medford	725970	42.37	122.87	1329	14.003	6193	21	24	19	16	13	20	51	15	50	3	130	9	290	104	15	3.4	6.4
North Bend	726917	43.42	124.25	13	14.688	6193	30	32	25	23	20	23	51	20	50	7	140	14	340	82	24	4.0	5.5
Pendleton	726880	45.68	118.85	1496	13.918	6193	3	11	28	24	20	27	44	23	42	6	140	9	310	102	−1	3.6	11.2
Portland	726980	45.60	122.60	39	14.675	6193	22	27	25	21	18	28	37	25	39	13	120	11	340	99	18	4.4	6.0
Redmond	726835	44.25	121.15	3077	13.133	6193	1	9	20	17	16	20	42	18	41	6	320	11	340	98	−7	3.2	10.2
Salem	726940	44.92	123.00	200	14.589	6193	20	25	23	19	17	25	46	22	46	6	350	10	360	100	14	3.3	6.6
Sexton Summit	725975	42.62	123.37	3842	12.767	8293	21	24	24	22	19	27	37	24	38	9	120	6	340	89	16	4.9	8.5
PENNSYLVANIA																							
Allentown	725170	40.65	75.43	384	14.493	6193	5	10	27	23	21	28	26	25	24	9	270	11	240	95	−2	2.8	5.2
Altoona	725126	40.30	78.32	1503	13.915	8293	5	10	20	18	17	23	20	20	22	9	270	8	250	92	−5	3.6	7.9
Bradford	725266	41.80	78.63	2142	13.593	6193	−6	−1	19	18	16	22	22	19	21	7	270	9	240	87	−15	2.8	5.0
Du Bois	725125	41.18	78.90	1818	13.756	8293	0	5	21	19	17	23	20	21	20	11	280	10	270	90	−9	3.1	7.0
Erie	725260	42.08	80.18	738	14.308	6193	2	7	27	24	22	29	28	26	28	14	200	12	250	90	−4	3.1	6.4
Harrisburg	725115	40.20	76.77	308	14.532	6193	9	13	22	20	18	24	29	22	29	8	270	10	250	97	2	3.3	5.7
Philadelphia, Int'l Airport	724080	39.88	75.25	30	14.680	6193	11	15	24	21	19	26	31	23	30	12	290	11	230	96	5	2.8	5.6
Philadelphia, Northeast A	724085	40.08	75.02	121	14.631	8293	11	15	21	19	17	22	30	19	29	10	300	10	260	97	3	2.5	6.1
Philadelphia, Willow Gr NAS	724086	40.20	75.15	361	14.505	8293	10	14	18	15	13	19	30	16	30	5	300	6	250	99	2	5.4	5.8
Pittsburgh, Allegheny Co. A	725205	40.35	79.93	1253	14.042	8293	4	11	21	19	17	23	24	21	24	11	250	11	240	94	−4	3.1	9.4
Pittsburgh, Int'l Airport	725200	40.50	80.22	1224	14.057	6193	2	7	25	21	19	26	24	23	25	10	260	11	230	93	−6	3.1	6.7
Wilkes-Barre/Scranton	725130	41.33	75.73	948	14.199	6193	2	7	20	18	16	21	26	19	25	8	230	11	220	92	−5	2.8	4.9
Williamsport	725140	41.25	76.92	525	14.419	6193	2	7	23	20	18	24	23	21	25	8	270	10	250	94	−6	3.1	5.9
RHODE ISLAND																							
Providence	725070	41.73	71.43	62	14.662	6193	5	10	27	23	21	27	31	23	32	12	340	13	230	95	−2	3.7	5.0
SOUTH CAROLINA																							
Beaufort, Mcas	722085	32.48	80.72	39	14.675	8293	28	31	18	15	13	19	46	17	45	4	300	7	270	101	13	2.9	8.3
Charleston	722080	32.90	80.03	49	14.669	6193	25	28	22	19	17	22	52	19	51	7	20	10	230	98	18	2.3	5.6
Columbia	723100	33.95	81.12	226	14.576	6193	21	24	20	17	15	20	48	18	49	5	220	9	240	100	13	3.1	5.6
Florence	723106	34.18	79.72	148	14.617	8293	23	27	19	17	15	20	51	18	50	7	360	10	240	100	14	2.7	7.6
Greer/Greenville	723120	34.90	82.22	971	14.187	6193	19	23	20	18	16	21	45	18	44	6	50	9	230	97	11	2.6	5.5
Myrtle Beach, AFB	747910	33.68	78.93	26	14.681	8293	25	29	18	15	13	18	49	15	47	4	360	7	290	98	17	2.9	7.4
Sumter, Shaw AFB	747900	33.97	80.47	243	14.567	8293	24	29	18	16	14	19	48	17	48	5	10	8	240	100	17	3.1	6.1
SOUTH DAKOTA																							
Chamberlain	726530	43.80	99.32	1739	13.795	8293	−13	−7	27	24	21	28	18	25	20	11	270	13	190	106	−12	8.1	18.4
Huron	726540	44.38	98.22	1289	14.024	6193	−17	−12	29	25	22	29	14	25	15	9	290	14	180	102	−25	4.6	5.9
Pierre	726686	44.38	100.28	1742	13.794	6193	−14	−9	29	25	22	32	15	27	20	11	320	14	180	106	−20	3.8	5.7
Rapid City	726620	44.05	103.07	3169	13.089	6193	−11	−5	36	31	27	37	26	32	26	9	350	13	160	102	−17	3.4	5.4
Sioux Falls	726510	43.58	96.73	1427	13.953	6193	−16	−11	28	25	22	30	15	26	17	8	310	15	180	100	−23	4.1	4.9
TENNESSEE																							
Bristol	723183	36.48	82.40	1519	13.906	6193	9	14	20	17	15	21	35	19	36	6	270	8	250	92	−1	3.0	7.5
Chattanooga	723240	35.03	85.20	689	14.333	6193	15	20	19	17	15	20	37	18	38	7	360	8	280	97	7	3.6	7.0
Crossville	723265	35.95	85.08	1880	13.724	8293	7	15	16	14	13	18	33	16	36	4	310	8	270	93	−3	4	8.6
Jackson	723346	35.60	88.92	433	14.467	8293	12	18	20	18	16	21	46	19	44	9	360	8	240	98	4	2.3	8.8
Knoxville	723260	35.82	83.98	981	14.182	6193	13	19	21	18	15	21	48	19	45	7	50	8	250	95	4	3.0	7.8
Memphis	723340	35.05	90.00	285	14.544	6193	16	21	22	19	17	22	42	20	42	10	20	9	240	99	9	2.8	7.2
Nashville	723270	36.13	86.68	591	14.385	6193	10	16	22	19	17	22	46	20	42	8	340	9	230	97	1	3.3	7.8
TEXAS																							
Abilene	722660	32.42	99.68	1791	13.769	6193	16	22	27	24	22	26	48	23	46	12	0	11	140	102	10	2.8	5.9
Amarillo	723630	35.23	101.70	3606	12.879	6193	6	12	30	27	24	30	40	27	38	14	20	15	200	100	−1	2.8	5.5
Austin	722540	30.30	97.70	620	14.369	6193	25	30	23	20	18	25	41	22	43	12	10	10	180	101	20	2.4	5.9
Beaumont/Port Arthur	722410	29.95	94.02	23	14.683	6193	29	32	22	20	18	23	51	21	51	10	340	9	200	97	22	2.5	4.5
Beeville, Chase Field NAS	722556	28.37	97.67	190	14.595	8293	28	33	22	20	18	23	58	20	53	13	350	9	150	104	22	2.5	8.3
Brownsville	722500	25.90	97.43	20	14.685	6193	36	40	27	24	22	26	64	23	62	13	330	16	160	98	31	2.4	5.2
College Station/Bryan	722445	30.58	96.37	322	14.525	8293	22	29	21	19	17	21	47	19	49	12	350	9	170	101	17	2.3	8.3
Corpus Christi	722510	27.77	97.50	43	14.673	6193	32	36	28	25	23	27	59	24	58	13	360	15	140	98	25	1.9	5.2
Dallas/Fort Worth, Int'l A	722590	32.90	97.03	597	14.381	8293	17	24	26	23	21	26	46	24	47	13	350	10	170	103	14	3.1	8.3
Del Rio, Laughlin AFB	722615	29.37	100.78	1083	14.130	8293	28	32	22	19	17	22	47	19	50	7	10	9	140	105	22	3.1	5.8
El Paso	722700	31.80	106.40	3917	12.731	6193	21	25	25	21	18	24	51	21	49	5	20	8	180	104	14	3.3	6.1
Fort Worth, Carswell AFB	722595	32.77	97.45	650	14.354	8293	18	24	22	20	18	23	43	20	45	11	10	8	10	103	15	2.2	8.1
Fort Worth, Meacham Field	722596	32.82	97.37	709	14.323	6193	19	24	27	24	21	27	40	24	44	13	350	10	180	103	14	3.0	5.8
Guadalupe Pass	722620	31.83	104.80	5453	12.022	8293	13	19	51	45	41	50	39	46	37	19	70	13	250	98	10	2.9	6.8
Houston, Hobby Airport	722435	29.65	95.28	46	14.671	8293	29	34	22	20	18	23	52	21	52	13	350	7	190	98	24	2	8.1
Houston, Inter Airport	722430	29.97	95.35	108	14.638	6193	27	31	20	18	16	22	47	20	52	8	340	10	180	98	22	3.1	5.4
Junction	747400	30.50	99.77	1713	13.808	8293	19	23	19	16	15	19	53	16	53	6	360	9	150	104	12	2.3	6.7
Killeen, Fort Hood	722576	31.07	97.83	1014	14.165	8293	20	27	22	19	17	22	48	19	53	11	360	9	160	102	15	2	8.6
Kingsville, NAS	722516	27.50	97.82	49	14.669	8293	31	36	23	21	19	22	61	20	60	11	360	11	150	102	18	2	10.1
Laredo	722520	27.55	99.47	509	14.427	8293	32	36	24	22	20	22	59	20	62	9	320	13	140	106	28	2.2	6.7

WMO# = World Meteorological Organization number
Lat = latitude Long. = longitude
Elev = elevation, ft
StdP = standard pressure at station elevation, psia
DB = dry-bulb temperature, °F
WS = wind speed, mph

Table 1B Cooling and Dehumidification Design Conditions—United States

Station	Cooling DB/MWB						WB/MDB						DP/MDB and HR									Range of DB
	0.4%		1%		2%		0.4%		1%		2%		0.4%			1%			2%			
	DB	MWB	DB	MWB	DB	MWB	WB	MDB	WB	MDB	WB	MDB	DP	HR	MDB	DP	HR	MDB	DP	HR	MCDB	
Eugene	91	67	87	65	83	64	69	87	67	84	65	81	62	83	74	60	78	73	59	74	71	27.6
Hillsboro	92	69	88	67	84	65	71	89	68	86	66	82	64	90	79	61	82	75	60	78	72	26.6
Klamath Falls	91	64	87	62	85	61	66	87	64	84	63	81	58	85	74	57	80	73	55	75	71	34.2
Meacham	87	59	84	58	80	57	61	82	59	80	58	78	52	67	66	50	63	64	49	59	64	37.1
Medford	98	67	95	66	91	65	69	94	67	91	66	88	60	81	75	58	76	74	56	71	73	33.7
North Bend	71	60	69	60	67	59	62	69	61	67	60	66	60	76	65	58	73	64	57	70	63	12.8
Pendleton	97	64	93	63	90	62	66	92	64	90	63	87	57	74	71	55	68	69	53	62	70	27.2
Portland	90	67	86	66	83	64	69	87	67	84	65	80	62	83	75	60	78	72	59	75	71	21.6
Redmond	93	62	89	61	86	59	63	88	62	86	60	83	55	71	68	52	65	67	50	60	66	35.0
Salem	92	67	87	66	83	64	68	89	67	85	65	81	61	81	75	59	76	73	58	72	71	27.9
Sexton Summit	83	60	80	59	77	58	62	80	61	77	59	74	55	76	70	53	69	68	52	66	66	18.9
PENNSYLVANIA																						
Allentown	90	73	88	72	85	71	76	86	74	84	73	82	73	123	81	71	117	79	70	111	78	19.4
Altoona	89	72	86	70	83	69	74	85	72	83	71	80	71	119	79	69	113	77	68	109	76	19.4
Bradford	83	69	80	68	78	66	72	79	70	77	68	75	69	116	75	68	111	73	66	105	72	21.2
Du Bois	86	70	84	69	81	67	72	81	71	79	70	78	70	116	76	69	112	74	67	108	73	19.4
Erie	85	72	83	70	80	70	74	82	73	80	71	78	72	122	79	70	115	77	69	109	76	15.6
Harrisburg	92	74	89	73	86	72	77	87	76	85	74	83	74	130	82	73	123	80	72	118	79	18.8
Philadelphia, Int'l Airport	92	75	89	74	87	73	78	88	77	86	75	84	75	132	83	74	126	81	73	121	80	17.7
Philadelphia, Northeast A	93	76	90	74	88	73	78	88	77	87	75	84	76	135	83	74	129	82	72	121	82	19.1
Philadelphia, Willow Gr NAS	93	75	90	74	88	72	78	89	76	87	75	85	74	131	83	73	125	82	71	116	81	19.4
Pittsburgh, Allegheny Co. A	90	72	87	71	85	70	75	85	74	84	72	81	71	122	79	70	117	78	69	113	77	18.0
Pittsburgh, Int'l Airport	89	72	86	70	84	69	74	85	73	82	71	80	71	121	80	70	115	78	68	109	77	19.5
Wilkes-Barre/Scranton	88	71	85	70	83	69	74	83	73	81	71	79	71	120	79	70	115	77	69	109	76	18.8
Williamsport	90	73	87	71	84	70	76	85	74	83	73	80	73	125	80	72	118	78	70	113	77	20.3
RHODE ISLAND																						
Providence	89	73	86	71	83	70	76	85	74	82	73	80	73	124	80	72	118	78	70	112	77	17.4
SOUTH CAROLINA																						
Beaufort, Mcas	95	78	93	78	92	77	80	90	80	89	79	88	78	145	85	77	141	85	76	137	84	16.7
Charleston	94	78	92	77	90	77	80	90	79	88	78	87	78	145	84	77	139	83	76	134	83	16.2
Columbia	96	76	94	75	92	74	78	90	77	89	77	87	75	134	82	75	130	81	74	127	81	19.9
Florence	96	76	94	76	92	76	80	90	78	89	78	88	77	142	85	76	136	83	75	132	82	19.8
Greer/Greenville	93	74	91	74	88	73	77	88	76	87	75	85	74	130	81	73	126	80	72	122	80	18.2
Myrtle Beach, AFB	93	79	90	78	88	78	81	89	80	88	79	87	79	150	87	78	144	86	77	140	84	14.4
Sumter, Shaw AFB	95	76	93	75	90	75	78	89	77	88	76	86	76	136	83	75	132	82	74	129	81	18.5
SOUTH DAKOTA																						
Chamberlain	98	72	94	71	90	70	76	91	74	89	72	87	71	124	84	70	116	82	68	109	80	23.8
Huron	95	72	91	71	88	70	76	89	74	87	72	84	72	126	84	70	117	81	69	110	79	24.1
Pierre	99	70	95	69	91	68	74	90	72	89	71	86	70	116	81	68	109	80	66	102	78	25.6
Rapid City	95	65	91	65	88	64	70	85	68	84	67	82	65	104	76	63	98	75	61	92	73	25.3
Sioux Falls	94	73	90	72	87	71	76	89	75	87	73	84	73	127	84	71	119	82	69	112	80	22.1
TENNESSEE																						
Bristol	89	72	87	72	85	71	75	85	74	84	73	82	72	125	81	71	120	79	70	116	77	19.2
Chattanooga	94	75	92	75	89	74	78	89	77	88	76	86	75	134	82	74	130	82	73	125	81	19.5
Crossville	89	73	87	72	85	72	76	85	74	83	73	82	74	134	80	72	125	79	71	121	78	19.8
Jackson	95	77	93	76	91	76	80	91	78	90	78	88	76	140	85	75	135	85	75	132	84	19.8
Knoxville	92	74	90	74	87	73	77	88	76	86	75	85	74	131	82	73	127	81	72	123	80	18.1
Memphis	96	78	94	77	92	77	80	92	79	91	78	89	77	143	87	76	137	86	75	133	84	16.8
Nashville	94	76	92	75	90	74	78	89	77	88	76	86	75	134	83	74	130	82	73	126	81	19.1
TEXAS																						
Abilene	99	71	97	71	95	71	75	89	74	89	73	88	71	123	81	70	119	80	69	115	79	20.5
Amarillo	96	67	94	66	92	66	71	86	70	86	69	85	67	112	76	65	107	75	64	104	74	23.3
Austin	98	74	96	74	94	74	78	89	77	88	76	87	76	137	81	75	134	80	74	130	80	20.1
Beaumont/Port Arthur	94	79	92	79	91	78	81	90	81	89	80	88	79	152	86	79	148	85	78	145	84	15.9
Beeville, Chase Field NAS	101	77	98	77	96	77	82	91	81	91	80	90	80	155	86	78	148	85	78	144	84	21.6
Brownsville	95	78	94	77	93	77	80	89	79	88	79	88	78	146	83	77	142	83	77	140	82	16.5
College Station/Bryan	98	75	96	75	94	75	79	89	78	89	78	88	77	141	82	76	138	81	75	134	81	21.4
Corpus Christi	95	78	94	78	92	78	81	90	80	89	79	88	79	148	84	78	146	83	77	143	83	16.5
Dallas/Fort Worth, Int'l A	100	74	98	74	96	74	78	92	77	91	76	90	75	132	82	74	130	82	73	126	81	20.3
Del Rio, Laughlin AFB	101	72	98	73	96	72	78	91	77	90	76	89	75	136	82	74	131	82	72	124	81	20.9
El Paso	101	64	98	64	96	64	70	85	69	84	68	84	67	114	73	65	109	73	64	103	74	28.0
Fort Worth, Carswell AFB	100	75	97	75	96	75	79	92	78	91	77	90	76	141	85	75	135	84	74	130	84	19.3
Fort Worth, Meacham Field	100	75	98	74	96	74	78	91	77	90	76	89	75	135	83	74	131	82	73	127	82	20.0
Guadalupe Pass	92	61	89	60	87	60	66	82	65	80	64	79	62	102	71	60	96	71	59	91	71	20.9
Houston, Hobby Airport	94	77	93	77	92	77	80	89	80	88	79	87	78	147	84	78	144	83	77	141	82	16.6
Houston, Inter Airport	96	77	94	77	92	77	80	90	79	89	79	88	78	144	83	77	141	83	76	137	83	18.2
Junction	100	72	98	71	96	71	76	89	75	88	74	87	73	130	80	71	121	79	70	118	79	24.8
Killeen, Fort Hood	98	74	96	73	95	74	78	90	77	89	76	88	75	137	81	74	132	81	73	128	80	21.4
Kingsville, NAS	97	77	96	78	95	78	81	91	80	91	80	90	79	149	85	78	144	84	77	141	84	19.8
Laredo	102	73	101	74	98	74	79	92	78	91	77	89	76	138	82	75	136	81	75	132	81	21.2

MDB = mean coincident dry-bulb temperature, °F
MWB = mean coincident wet-bulb temperature, °F
MWS = mean coincident wind speed, mph
MWD = mean coincident wind direction
StdD = standard deviation
A = airport
HR = humidity ratio
DP = dew-point temperature, °F

Table 1A Heating and Wind Design Conditions—United States

							Heating DB		Extreme Wind Speed			Coldest Month WS/MDB				MWS/MWD to DB				Annual Extreme Daily			
												0.4%		1%		99.6%		0.4%		Mean DB		StdD DB	
Station	WMO#	Lat.	Long.	Elev. ft	StdP psia	Dates	99.6%	99%	1%	2.5%	5%	WS	MDB	WS	MDB	MWS	PWD	MWS	PWD	Max	Min	Max	Min
Lubbock, Int'l Airport	722670	33.65	101.82	3241	13.054	6193	11	17	30	26	23	30	43	27	44	12	0	14	160	102	4	2.6	5.6
Lubbock, Reese AFB	722675	33.60	102.05	3337	13.008	8293	11	18	25	22	19	25	48	22	44	10	20	11	170	102	6	3.1	4.9
Lufkin	722446	31.23	94.75	289	14.543	6193	23	27	18	16	14	18	44	17	46	6	330	8	230	99	17	3.2	5.3
Marfa	722640	30.37	104.02	4859	12.292	8293	15	19	24	21	18	25	44	22	45	5	360	9	220	97	5	2.3	5.0
McAllen	722506	26.18	98.23	108	14.638	8293	34	40	24	22	20	23	68	21	68	11	350	14	130	106	27	4.3	8.1
Midland/Odessa	722650	31.95	102.18	2861	13.238	6193	17	22	28	25	22	27	50	23	48	9	20	13	180	103	9	2.6	6.9
San Angelo	722630	31.37	100.50	1909	13.709	6193	20	24	26	23	21	25	52	22	51	10	20	11	160	103	13	2.8	6.1
San Antonio, Int'l Airport	722530	29.53	98.47	794	14.279	6193	26	30	22	19	17	23	43	20	45	10	350	10	160	100	19	2.9	5.2
San Antonio, Kelly AFB	722535	29.38	98.58	689	14.333	8293	27	32	19	17	15	21	51	18	52	8	360	8	160	103	22	2.9	6.5
San Antonio, Randolph AFB	722536	29.53	98.28	761	14.296	8293	27	31	19	17	15	20	45	17	48	7	340	7	150	101	20	2.2	6.7
Sanderson	747300	30.17	102.42	2838	13.250	8293	23	28	19	16	13	20	44	17	48	6	360	7	120	102	9	2.9	8.3
Victoria	722550	28.85	96.92	118	14.633	6193	29	33	26	23	21	26	50	23	51	12	360	12	180	99	23	2.5	5.2
Waco	722560	31.62	97.22	509	14.427	6193	22	26	26	23	21	29	38	25	42	13	360	12	180	104	16	2.8	6.4
Wichita Falls, Sheppard AFB	723510	33.98	98.50	1030	14.157	6193	14	19	29	25	23	28	42	25	43	12	360	13	180	107	7	3.4	6.6
UTAH																							
Cedar City	724755	37.70	113.10	5623	11.945	6193	2	8	26	22	20	24	38	21	39	4	140	12	200	97	−6	2.3	8.3
Ogden, Hill AFB	725755	41.12	111.97	4787	12.325	8293	6	11	22	19	17	22	27	19	28	9	110	6	190	96	1	2.9	6.3
Salt Lake City	725720	40.78	111.97	4226	12.586	6193	6	11	27	23	20	27	42	22	40	7	160	11	340	100	−3	1.9	6.7
VERMONT																							
Burlington	726170	44.47	73.15	341	14.515	6193	−11	−6	23	21	18	24	30	21	27	6	70	11	180	93	−19	2.7	5.6
Montpelier/Barre	726145	44.20	72.57	1165	14.087	8293	−10	−6	21	19	17	22	20	20	20	4	320	9	220	91	−18	3.6	5.9
VIRGINIA																							
Fort Belvoir	724037	38.72	77.18	69	14.659	8293	12	18	18	14	12	19	35	17	34	2	320	6	160	100	2	2.3	7.6
Hampton, Langley AFB	745980	37.08	76.37	10	14.690	8293	21	24	22	19	17	22	41	20	40	10	330	9	240	97	13	3.2	6.1
Lynchburg	724100	37.33	79.20	938	14.204	6193	12	17	19	17	15	21	35	18	35	8	360	9	230	95	5	2.9	5.8
Newport News	723086	37.13	76.50	43	14.673	8293	18	22	19	18	16	20	40	18	41	8	350	10	220	99	11	2.3	4.7
Norfolk	723080	36.90	76.20	30	14.680	6193	20	24	25	22	20	26	40	23	40	12	340	12	230	97	14	2.8	5.4
Oceana, NAS	723075	36.82	76.03	23	14.683	8293	22	25	21	19	17	21	42	19	42	8	310	9	220	98	14	1.8	6.8
Quantico, Mcas	724035	38.50	77.30	13	14.688	8293	16	21	17	14	12	19	36	15	38	6	340	5	230	100	8	3.6	5.9
Richmond	724010	37.50	77.33	177	14.602	6193	14	18	20	18	16	21	40	18	39	7	340	10	230	98	6	2.6	5.8
Roanoke	724110	37.32	79.97	1175	14.082	6193	12	17	23	20	17	27	31	23	32	10	320	10	290	96	4	3.3	5.6
Sterling	724030	38.95	77.45	322	14.525	6193	9	14	22	19	16	25	32	21	31	7	340	10	250	97	−1	3.3	7.0
Washington, National A	724050	38.85	77.03	66	14.661	8293	15	20	23	20	18	24	34	21	35	11	340	11	170	99	8	2.5	6.8
WASHINGTON																							
Bellingham	727976	48.80	122.53	157	14.612	8293	15	21	23	20	18	28	33	23	34	17	40	9	290	87	11	3.1	7.4
Hanford	727840	46.57	119.60	732	14.311	8293	5	12	25	21	18	24	44	19	44	6	20	8	20	105	2	3.1	9.0
Olympia	727920	46.97	122.90	200	14.589	6193	18	23	21	18	16	21	45	19	45	5	180	8	50	94	10	4.0	8.1
Quillayute	727970	47.95	124.55	203	14.588	6193	23	27	33	27	21	41	45	35	45	7	60	9	240	87	19	8.4	6.4
Seattle, Int'l Airport	727930	47.45	122.30	449	14.458	6193	23	28	22	19	17	24	44	21	44	10	10	10	350	92	19	3.6	6.8
Spokane, Fairchild AFB	727855	47.62	117.65	2461	13.435	6193	1	7	27	23	20	28	39	25	38	7	50	9	240	98	−7	3.2	8.7
Stampede Pass	727815	47.28	121.33	3967	12.708	8293	3	10	21	19	16	27	19	22	25	13	90	7	100	84	2	3.2	7.2
Tacoma, McChord AFB	742060	47.13	122.48	322	14.525	8293	18	24	18	15	13	22	45	18	46	2	180	7	20	94	12	2.7	6.8
Walla Walla	727846	46.10	118.28	1204	14.067	8293	4	12	22	19	17	24	49	22	47	6	180	9	300	105	1	3.2	11.7
Wenatchee	727825	47.40	120.20	1243	14.047	8293	3	9	22	19	17	17	36	12	31	3	100	9	280	101	−2	2.5	7.2
Yakima	727810	46.57	120.53	1066	14.138	6193	4	11	24	20	17	23	47	19	43	7	250	7	90	101	−2	3.2	8.5
WEST VIRGINIA																							
Bluefield	724125	37.30	81.20	2858	13.240	8293	5	12	15	13	12	18	34	15	33	6	270	6	290	88	−6	4	8.5
Charleston	724140	38.37	81.60	981	14.182	6193	6	11	18	16	14	20	38	18	34	7	250	8	240	94	−2	2.8	6.7
Elkins	724170	38.88	79.85	1998	13.665	6193	−2	5	20	18	16	22	30	19	30	4	280	8	290	88	−12	2.8	5.4
Huntington	724250	38.37	82.55	837	14.257	6193	6	11	19	16	14	20	32	17	32	8	270	8	270	94	−2	5.0	7.6
Martinsburg	724177	39.40	77.98	558	14.402	8293	8	14	21	18	16	23	33	20	34	7	270	9	290	99	−3	4	8.3
Morgantown	724176	39.65	79.92	1247	14.045	8293	4	11	18	15	13	19	32	17	33	6	210	8	240	93	−4	3.6	8.6
Parkersburg	724273	39.35	81.43	860	14.245	8293	4	11	18	16	14	20	32	18	29	7	240	8	270	95	−4	3.1	9.2
WISCONSIN																							
Eau Claire	726435	44.87	91.48	906	14.221	6193	−18	−13	22	19	17	21	14	20	13	7	250	13	220	95	−25	3.2	5.7
Green Bay	726450	44.48	88.13	702	14.326	6193	−13	−8	25	22	20	25	19	22	18	10	270	12	200	93	−19	2.8	5.6
La Crosse	726430	43.87	91.25	663	14.347	6193	−14	−8	23	20	18	23	13	21	13	7	310	12	180	97	−21	3.2	6.2
Madison	726410	43.13	89.33	866	14.241	6193	−11	−6	24	21	19	25	16	22	17	8	300	12	230	94	−18	3.2	6.0
Milwaukee	726400	42.95	87.90	692	14.332	6193	−7	−2	28	24	22	28	19	24	20	13	290	15	220	95	−12	3.2	6.7
Wausau	726463	44.93	89.63	1201	14.069	8293	−15	−9	19	17	15	19	16	17	17	7	300	10	200	93	−22	3.1	4.7
WYOMING																							
Big Piney	726710	42.57	110.10	6969	11.353	8293	−22	−15	24	20	17	22	25	19	21	3	60	11	260	87	−33	2.7	8.5
Casper	725690	42.92	106.47	5289	12.096	6193	−13	−5	34	30	27	35	35	32	32	9	260	13	240	97	−22	2.2	8.4
Cheyenne, Warren AFB	725640	41.15	104.82	6142	11.714	6193	−7	0	34	29	26	38	36	33	34	10	290	13	290	92	−15	2.2	7.5
Cody	726700	44.52	109.02	5095	12.184	8293	−14	−7	34	28	23	35	35	30	35	6	40	11	70	95	−20	4.1	9.4
Gillette	726650	44.35	105.53	4035	12.675	8293	−16	−7	28	25	22	30	34	27	33	8	260	11	140	101	−20	5.9	10.1
Lander	725760	42.82	108.73	5558	11.974	6193	−14	−7	23	19	16	25	38	19	37	3	120	10	270	95	−20	2.5	7.8
Rock Springs	725744	41.60	109.07	6759	11.444	6193	−9	−2	28	25	23	32	25	29	24	7	70	13	280	90	−17	2.0	8.0
Sheridan	726660	44.77	106.97	3967	12.708	6193	−14	−8	28	24	20	29	32	23	27	5	280	9	120	99	−22	3.0	6.4
Worland	726665	43.97	107.95	4245	12.577	8293	−22	−13	22	19	16	20	28	17	28	3	210	9	220	103	−30	2.2	10.4

WMO# = World Meteorological Organization number
Lat = latitude Long. = longitude
Elev = elevation, ft
StdP = standard pressure at station elevation, psia
DB = dry-bulb temperature, °F
WS = wind speed, mph

Table 1B Cooling and Dehumidification Design Conditions—United States

	Cooling DB/MWB						WB/MDB						DP/MDB and HR									
	0.4%		1%		2%		0.4%		1%		2%		0.4%			1%			2%			Range
Station	DB	MWB	DB	MWB	DB	MWB	WB	MDB	WB	MDB	WB	MDB	DP	HR	MDB	DP	HR	MDB	DP	HR	MCDB	of DB
Lubbock, Int'l Airport	97	67	95	67	93	67	73	87	72	86	71	85	69	120	77	68	115	76	67	111	76	22.1
Lubbock, Reese AFB	98	67	95	67	93	67	73	87	72	86	71	85	69	122	78	68	115	77	66	110	77	23.8
Lufkin	97	76	95	77	93	76	79	90	79	89	78	89	77	143	83	76	139	83	75	134	82	20.9
Marfa	94	62	92	61	89	62	68	82	67	81	66	80	65	110	72	63	103	71	62	98	71	31.3
McAllen	100	76	98	76	97	76	80	91	80	90	79	89	78	146	83	77	143	82	77	140	82	20.7
Midland/Odessa	99	67	97	67	95	67	73	87	72	86	71	86	69	120	76	68	115	75	67	111	75	23.7
San Angelo	100	70	97	70	95	70	75	90	74	89	73	88	71	123	80	70	118	79	69	116	78	22.3
San Antonio, Int'l Airport	98	73	96	73	94	74	78	87	77	87	76	86	76	139	81	75	135	81	74	132	80	19.1
San Antonio, Kelly AFB	99	74	97	74	96	74	79	89	78	88	77	88	77	145	83	76	140	82	75	136	81	20.5
San Antonio, Randolph AFB	98	74	96	74	94	74	78	90	77	89	76	88	76	138	82	75	134	81	74	132	81	22.3
Sanderson	97	67	95	68	94	68	74	86	73	86	72	86	70	123	79	69	119	78	68	114	77	20.7
Victoria	95	76	94	76	92	77	80	88	79	88	78	87	78	145	83	77	141	82	76	139	82	17.4
Waco	101	75	99	75	97	75	79	93	78	92	77	91	75	135	83	74	131	82	74	127	82	21.6
Wichita Falls, Sheppard AFB	103	74	100	73	98	73	77	93	76	92	75	91	73	129	82	72	124	82	71	120	81	23.9
UTAH																						
Cedar City	93	59	91	59	88	58	64	80	62	80	61	79	59	93	68	57	85	68	55	78	68	28.5
Ogden, Hill AFB	93	61	90	60	87	60	65	83	64	81	62	81	60	91	72	57	83	73	55	77	73	22.0
Salt Lake City	96	62	94	62	92	61	66	85	65	85	64	85	60	92	73	58	84	73	56	77	73	27.7
VERMONT																						
Burlington	87	71	84	69	82	68	74	83	72	81	70	78	71	115	79	69	109	77	67	102	75	20.4
Montpelier/Barre	85	70	83	68	80	67	72	82	70	80	69	77	69	111	78	67	106	75	66	99	73	21.1
VIRGINIA																						
Fort Belvoir	95	78	93	76	89	75	80	92	78	89	77	87	77	139	86	75	133	85	74	127	83	20.9
Hampton, Langley AFB	94	78	91	77	88	76	80	90	79	89	78	86	77	141	85	76	136	84	75	132	83	14.9
Lynchburg	93	74	90	74	88	73	77	88	76	87	75	85	74	129	81	73	125	80	72	120	79	18.2
Newport News	95	78	92	77	89	76	80	91	78	89	77	87	77	139	84	76	135	83	75	132	82	18.2
Norfolk	93	77	91	76	88	75	79	89	77	87	77	85	76	135	83	75	130	82	74	126	81	15.3
Oceana, NAS	94	77	91	76	88	75	79	89	78	87	77	86	77	139	85	76	134	83	74	129	82	15.7
Quantico, Mcas	94	77	92	76	89	74	79	91	78	89	76	87	76	136	87	75	130	85	73	125	83	18.5
Richmond	94	76	92	75	89	74	79	90	78	88	76	86	76	137	84	75	131	82	74	126	81	19.1
Roanoke	92	73	89	72	87	71	75	88	74	86	73	84	72	123	80	71	118	79	70	115	78	19.6
Sterling	93	75	90	74	88	73	77	88	76	87	75	85	74	130	83	73	125	81	72	120	80	21.0
Washington, National A	95	76	92	76	89	74	79	89	78	88	76	86	76	137	83	75	132	83	74	127	81	16.6
WASHINGTON																						
Bellingham	79	65	76	64	74	62	67	78	65	75	63	72	61	81	73	60	78	70	59	74	67	16.7
Hanford	100	67	96	65	93	64	68	96	66	94	65	90	58	73	72	56	68	75	53	62	74	26.5
Olympia	87	67	83	65	79	64	68	85	66	81	64	78	61	81	73	60	76	71	58	73	69	25.2
Quillayute	80	62	74	61	70	59	64	76	62	72	60	67	60	76	65	58	74	63	57	71	62	15.4
Seattle, Int'l Airport	85	65	81	64	78	62	66	83	65	79	63	76	60	78	71	59	74	69	57	71	68	18.3
Spokane, Fairchild AFB	92	62	89	61	85	60	65	86	63	84	61	82	57	77	68	55	71	68	53	67	67	26.1
Stampede Pass	78	57	74	56	71	54	59	74	57	71	56	69	53	70	63	51	65	61	50	62	58	16.0
Tacoma, McChord AFB	86	65	82	63	78	62	67	83	65	80	63	76	60	79	71	59	76	70	58	72	68	22.5
Walla Walla	98	66	95	65	92	64	68	92	67	91	65	88	60	82	74	58	76	72	57	71	72	27.0
Wenatchee	95	67	92	65	88	63	67	91	66	89	64	85	59	78	75	57	73	75	55	68	74	25.2
Yakima	95	65	92	64	88	63	67	90	66	89	64	86	59	78	75	57	71	74	55	67	72	31.1
WEST VIRGINIA																						
Bluefield	85	69	83	69	80	67	72	81	71	79	70	77	69	120	75	68	116	75	67	111	73	16.4
Charleston	91	73	88	73	86	71	76	86	75	85	74	82	73	129	81	72	123	80	71	118	78	19.1
Elkins	85	71	83	70	81	69	73	82	72	80	71	78	71	121	78	69	116	77	68	111	75	21.1
Huntington	91	74	89	73	86	72	77	87	76	85	74	83	74	132	82	73	127	81	72	121	79	19.1
Martinsburg	94	74	91	73	88	72	77	87	75	86	74	85	74	130	81	72	120	80	71	116	79	21.8
Morgantown	89	72	87	71	85	70	75	85	74	83	73	82	72	124	79	71	119	78	70	115	76	20.3
Parkersburg	91	74	88	72	86	72	76	87	75	85	74	82	74	132	82	72	122	80	71	118	78	19.6
WISCONSIN																						
Eau Claire	90	73	87	71	84	70	76	86	74	83	72	81	73	125	82	71	116	80	69	109	78	20.6
Green Bay	88	73	85	72	82	70	76	85	74	82	72	80	73	124	82	71	116	79	69	109	77	20.7
La Crosse	91	74	88	73	85	71	77	87	75	84	74	82	75	132	83	73	125	81	71	117	78	20.1
Madison	90	73	87	72	84	70	76	86	74	84	72	82	73	126	83	71	118	80	69	111	78	21.9
Milwaukee	89	74	86	72	83	70	76	86	74	83	72	81	73	127	83	71	119	80	70	111	78	16.6
Wausau	88	71	85	70	82	69	74	83	72	82	71	78	71	120	79	69	113	77	68	108	75	19.6
WYOMING																						
Big Piney	83	54	80	53	78	52	56	75	55	74	53	74	50	69	60	48	64	60	45	57	59	32.8
Casper	92	59	89	58	86	58	62	81	61	80	60	79	57	85	66	55	78	66	53	73	65	30.4
Cheyenne, Warren AFB	87	58	85	57	82	57	62	77	61	76	60	75	58	90	66	56	85	65	55	80	64	25.7
Cody	91	59	87	58	84	57	61	83	60	81	58	80	54	76	70	52	69	66	50	64	65	25.4
Gillette	94	61	91	61	87	60	65	84	63	83	62	82	59	88	73	57	80	69	54	73	68	28.6
Lander	90	59	87	58	85	57	62	81	61	80	59	80	56	81	69	53	74	68	51	69	67	26.7
Rock Springs	86	54	84	54	82	53	58	75	57	74	56	74	54	78	62	51	71	61	49	66	61	27.7
Sheridan	93	62	90	61	86	61	66	85	64	83	63	81	60	88	71	58	82	71	56	76	69	29.1
Worland	96	63	93	63	90	62	67	88	66	86	64	84	61	94	75	59	86	75	57	80	73	31.0

MDB = mean coincident dry-bulb temperature, °F
MWB = mean coincident wet-bulb temperature, °F
MWS = mean coincident wind speed, mph
MWD = mean coincident wind direction
StdD = standard deviation
A = airport
HR = humidity ratio
DP = dew-point temperature, °F

Table 2A Heating and Wind Design Conditions—Canada

Station	WMO#	Lat.	Long.	Elev. ft	StdP psia	Dates	Heating DB 99.6%	Heating DB 99%	Extreme Wind Speed 1%	Extreme Wind Speed 2.5%	Extreme Wind Speed 5%	Coldest Month WS/MDB 0.4% WS	Coldest Month WS/MDB 0.4% MDB	Coldest Month WS/MDB 1% WS	Coldest Month WS/MDB 1% MDB	MWS/MWD to DB 99.6% MWS	MWS/MWD to DB 99.6% PWD	MWS/MWD to DB 0.4% MWS	MWS/MWD to DB 0.4% PWD	Annual Extreme Daily Mean DB Max	Annual Extreme Daily Mean DB Min	Annual Extreme Daily StdD DB Max	Annual Extreme Daily StdD DB Min
ALBERTA																							
Calgary Intl A	718770	51.12	114.02	3556	12.90	6193	−22	−17	28	24	21	32	29	28	27	7	0	11	160	89	−28	2.7	5.6
Cold Lake A	711200	54.42	110.28	1785	13.77	6193	−31	−26	21	18	16	21	18	18	11	3	270	9	180	88	−40	3.2	6.3
Coronation	718730	52.07	111.45	2595	13.37	6193	−27	−23	25	21	19	28	12	23	13	9	320	11	160	92	−35	3.1	6.3
Edmonton Intl A	711230	53.30	113.58	2372	13.479	6193	−28.1	−22.9	24	21	18	24	12	21	11	6	180	9	180	87	−36	3.1	8.1
Fort McMurray A	719320	56.65	111.22	1211	14.06	6193	−32	−29	17	15	13	18	16	15	11	3	90	9	250	90	−42	3.6	4.9
Grande Prairie A	719400	55.18	118.88	2195	13.57	6193	−32	−27	27	23	19	29	32	24	28	3	320	8	270	87	−41	2.7	6.7
Lethbridge A	718740	49.63	112.80	3048	13.15	6193	−22	−16	36	32	28	45	39	39	38	5	250	13	270	94	−30	3.2	6.5
Medicine Hat A	718720	50.02	110.72	2349	13.49	6193	−24	−19	26	22	20	29	36	25	33	5	230	11	220	97	−32	3.6	7.2
Peace River A	710680	56.23	117.43	1873	13.727	6193	−31.5	−27.2	21	18	17	22	30	19	24	4	0	9	270	87	−42	2.9	6.7
Red Deer A	718780	52.18	113.90	2969	13.19	6193	−27	−21	22	19	17	27	13	22	13	6	200	10	180	88	−35	3.1	6.5
Rocky Mtn. House	719280	52.43	114.92	3245	13.05	6193	−25	−20	19	16	13	19	26	16	20	3	340	8	160	87	−36	2.7	5.0
Vermilion A		53.35	110.83	2028	13.65	6193	−30	−25	22	19	17	21	13	19	11	3	270	11	180	90	−43	3.6	6.7
Whitecourt	719300	54.15	115.78	2566	13.38	6193	−30	−24	17	15	14	19	22	17	18	4	270	7	90	87	−41	2.0	5.4
BRITISH COLUMBIA																							
Abbotsford A	711080	49.03	122.37	190	14.59	6193	15	20	20	17	15	29	33	25	34	12	90	7	220	92	10	4.0	6.7
Cape St. James	710310	51.93	131.02	302	14.54	6193	25	29	50	46	40	60	40	54	42	22	50	11	300	69	22	3.4	5.6
Castlegar A	718840	49.30	117.63	1624	13.85	6693	5	9	18	15	14	21	18	19	21	8	0	7	180	98	−3	2.9	7.0
Comox A	718930	49.72	124.90	79	14.65	6193	21	25	29	25	21	31	43	28	42	7	290	7	340	87	17	3.8	5.2
Cranbrook A	718800	49.60	115.78	3081	13.13	7093	−15	−8	20	18	16	20	33	18	33	2	200	10	210	94	−22	2.7	6.7
Fort Nelson A	719450	58.83	122.58	1253	14.04	6193	−33	−30	16	14	12	15	8	12	2	1	220	5	120	88	−42	3.2	6.3
Fort St. John A	719430	56.23	120.73	2280	13.52	6193	−30	−25	25	22	19	29	23	25	22	7	0	9	230	86	−36	2.9	6.3
Kamloops A	718870	50.70	120.45	1135	14.10	6693	−8	−1	23	20	18	25	26	22	27	4	90	8	270	98	−14	2.7	8.8
Penticton A	718890	49.47	119.60	1129	14.11	6193	5	10	23	20	17	28	34	25	35	8	340	9	180	96	1	2.7	7.2
Port Hardy A	711090	50.68	127.37	72	14.66	6193	22	26	28	24	21	33	38	29	39	8	110	9	340	76	18	3.4	4.9
Prince George A	718960	53.88	122.68	2267	13.53	6193	−25	−18	21	18	15	27	32	23	24	2	0	6	180	85	−37	9.2	5.9
Prince Rupert A	718980	54.30	130.43	112	14.64	6393	7	13	28	23	20	30	44	26	43	6	70	8	270	75	2	4.5	6.8
Quesnel A	711030	53.03	122.52	1788	13.77	6193	−22	−14	17	15	14	19	18	17	20	1	340	5	340	92	−31	4.0	8.3
Sandspit A	711010	53.25	131.82	20	14.68	6193	21	25	38	32	27	42	44	37	42	18	320	9	270	72	18	3.4	5.0
Smithers A	719500	54.82	127.18	1716	13.81	6193	−19	−12	17	15	13	18	23	16	19	3	140	6	320	88	−26	4.0	7.0
Spring Island	714790	50.12	127.93	322	14.53	6193	29	31	41	35	29	44	46	40	45	6	50	6	320	78	25	6.1	4.5
Terrace A	719510	54.47	128.58	712	14.32	6193	−2	2	26	23	20	32	10	29	14	19	0	8	270	89	−5	4.0	5.8
Tofino A	711060	49.08	125.77	79	14.65	6193	25	29	24	20	18	29	46	24	45	5	70	7	290	81	21	4.0	5.6
Vancouver Intl A	718920	49.18	123.17	7	14.69	6193	18	24	22	19	16	25	41	21	42	6	90	7	290	82	14	2.9	6.3
Victoria Intl A	717990	48.65	123.43	62	14.66	6193	23	26	20	16	14	24	37	20	38	10	50	6	90	87	18	3.2	5.8
Williams Lake A	711040	52.18	122.05	3084	13.13	6193	−20	−14	22	19	17	24	29	21	30	3	320	6	140	88	−29	4.0	7.9
MANITOBA																							
Brandon A	711400	49.92	99.95	1342	14.00	6193	−29	−24	27	23	20	28	2	24	2	9	270	12	160	94	−36	3.1	4.5
Churchill A	719130	58.75	94.07	95	14.64	6193	−36	−33	34	30	26	36	−11	30	−14	15	270	13	230	86	−41	4.5	4.0
Dauphin A	718550	51.10	100.05	1001	14.17	6193	−28	−23	28	25	22	31	2	28	5	9	250	13	200	93	−36	3.4	4.3
Portage La Prairie A	718510	49.90	98.27	883	14.23	6193	−25	−21	26	23	20	29	1	25	1	9	250	12	180	95	−31	3.4	4.1
The Pas A	718670	53.97	101.10	889	14.23	6193	−32	−28	24	21	19	25	−6	22	−2	6	290	11	160	89	−40	3.4	4.1
Thompson A	710790	55.80	97.87	715	14.32	6893	−38	−34	20	18	16	19	−5	17	−7	3	270	10	180	89	−48	3.8	4.3
Winnipeg Int'l A	718520	49.90	97.23	784	14.28	6193	−27	−23	29	25	23	30	5	26	5	7	320	13	180	94	−33	3.4	4.7
NEW BRUNSWICK																							
Charlo A	717110	47.98	66.33	125	14.63	6793	−14	−10	24	21	19	27	3	24	8	11	250	11	250	89	−21	2.7	4.5
Chatham A	717170	47.00	65.45	102	14.64	6193	−12	−7	24	21	18	27	16	24	16	7	270	11	230	93	−20	2.2	4.3
Fredericton A	717000	45.87	66.53	66	14.66	6193	−12	−7	23	20	17	25	17	22	18	5	270	11	230	92	−21	2.5	5.4
Moncton A	717050	46.12	64.68	233	14.57	6193	−10	−5	26	23	20	30	19	26	19	13	270	13	250	89	−17	2.0	4.5
Saint John A	716090	45.32	65.88	358	14.51	6193	−9	−4	26	23	20	32	24	28	23	9	340	11	230	84	−18	4.0	4.7
NEWFOUNDLAND																							
Battle Harbour	718170	52.30	55.83	26	14.68	6193	−14	−10	40	35	32	48	17	42	15	18	270	17	230	78	−18	4.7	5.9
Bonavista	711960	48.67	53.12	89	14.65	6193	3	7	43	38	34	48	21	42	23	24	280	17	230	81	0	2.7	5.8
Cartwright	718180	53.70	57.03	46	14.67	6493	−18	−15	36	30	27	40	18	35	19	12	220	12	210	84	−23	4.1	4.5
Daniels Harbour	711850	50.23	57.58	62	14.66	6693	−7	−3	40	35	31	45	16	39	21	12	270	15	230	75	−12	3.2	6.5
Deer Lake A	718090	49.22	57.40	72	14.66	6693	−13	−7	24	22	18	26	21	22	19	3	240	14	220	86	−23	2.3	6.7
Gander Intl A	718030	48.95	54.57	495	14.43	6193	−4	0	32	28	25	37	20	32	22	16	270	13	230	85	−8	2.9	6.3
Goose A	718160	53.32	60.37	160	14.36	6193	−23	−20	26	22	20	30	3	26	5	11	250	12	250	89	−29	4.0	3.6
Hopedale	719000	55.45	60.23	26	14.68	6493	−21	−18	36	30	27	40	13	35	11	12	250	13	250	79	−25	4.5	5.8
St. John's A	718010	47.62	52.73	459	14.45	6193	3	7	37	33	29	41	24	37	25	17	290	17	250	82	−2	2.5	4.9
Stephenville A	718150	48.53	58.55	85	14.65	6193	−2	3	33	28	23	36	21	30	21	11	50	9	250	79	−7	2.5	6.3
Wabush Lake A	718250	52.93	66.87	1808	13.76	6193	−33	−30	23	20	17	25	−5	21	−4	5	270	12	240	82	−43	3.8	4.3
NORTHWEST TERRITORIES																							
Baker Lake	719260	64.30	96.08	59	14.66	6393	−41	−39	36	32	28	42	−26	37	−26	12	0	10	270	77	−50	4.9	4.3
Cambridge Bay A	719250	69.10	105.12	89	14.65	6193	−38	−35	35	31	27	36	−19	31	−19	9	320	11	140	68	−50	4.9	3.8
Cape Parry A	719480	70.17	124.68	56	14.67	6193	−34	−33	31	28	25	34	−12	29	−12	7	270	9	110	66	−42	4.9	4.3
Chesterfield	719164	63.33	90.72	36	14.68	6393	−35	−34	33	29	26	35	−26	31	−26	14	320	13	320	73	−49	13.9	4.9
Coral Harbour A	719150	64.20	83.37	210	14.58	6193	−40	−38	37	32	27	39	−5	32	−7	9	340	13	270	71	−49	3.8	5.2
Fort Smith A	719340	60.02	111.95	666	14.35	6193	−34	−32	18	17	15	20	−3	18	−5	3	150	10	180	88	−48	3.4	5.6
Hall Beach A	710810	68.78	81.25	26	14.68	6193	−42	−38	33	29	25	34	−20	30	−21	10	320	11	180	64	−53	5.2	5.4
Inuvik UA	719570	68.30	133.48	223	14.58	7393	−43	−40	17	15	14	19	−7	17	−7	1	70	8	180	84	−52	2.5	5.0
Iqualuit A (Frobisher)	719090	63.75	68.55	108	14.64	6193	−39	−36	32	28	25	40	−12	34	−12	4	320	12	320	68	−43	4.3	4.7
Norman Wells A	710430	65.28	126.80	243	14.57	6193	−40	−36	24	21	18	29	−4	24	−7	2	170	8	140	87	−49	3.1	5.4

WMO# = World Meteorological Organization number
Lat = latitude Long. = longitude
Elev = elevation, ft
StdP = standard pressure at station elevation, psia
DB = dry-bulb temperature, °F
WS = wind speed, mph

Table 2B Cooling and Dehumidification Design Conditions—Canada

	Cooling DB/MWB						WB/MDB						DP/MDB and HR									
	0.4%		1%		2%		0.4%		1%		2%		0.4%			1%			2%			Range
Station	**DB**	**MWB**	**DB**	**MWB**	**DB**	**MWB**	**WB**	**MDB**	**WB**	**MDB**	**WB**	**MDB**	**DP**	**HR**	**MDB**	**DP**	**HR**	**MDB**	**DP**	**HR**	**MCDB**	**of DB**
ALBERTA																						
Calgary Intl A	83	60	80	59	77	57	62	78	61	75	59	73	57	79	67	55	74	65	53	69	64	22.0
Cold Lake A	82	64	78	62	75	60	66	78	64	76	62	72	62	88	71	60	82	69	58	76	67	20.0
Coronation	85	62	82	60	78	59	65	80	63	78	61	75	60	84	69	58	78	67	56	73	66	22.1
Edmonton Intl A	82	63	78	62	75	60	66	77	64	75	62	73	62	89	73	59	83	69	57	77	66	21.8
Fort McMurray A	84	64	80	62	76	60	66	79	64	76	62	73	61	84	71	59	78	68	57	74	66	22.0
Grande Prairie A	81	62	78	60	75	58	64	77	62	74	60	71	59	81	68	57	76	65	55	71	63	20.9
Lethbridge A	88	61	84	61	81	60	65	81	63	79	61	77	59	84	70	57	78	69	55	73	67	24.8
Medicine Hat A	90	63	87	62	84	61	66	84	64	83	62	80	60	83	70	58	78	69	56	72	68	25.0
Peace River A	81	62	78	60	75	59	65	77	62	75	60	71	60	82	69	58	76	67	56	71	65	21.4
Red Deer A	82	62	79	61	76	59	65	78	63	75	61	73	60	85	71	58	80	69	56	74	66	22.9
Rocky Mtn. House	80	62	78	61	75	59	64	78	63	75	61	72	60	87	70	58	81	68	56	76	66	22.5
Vermilion A	83	64	80	62	77	61	66	78	64	77	62	74	62	88	72	60	83	69	58	77	67	22.0
Whitecourt	80	61	77	60	74	59	65	77	62	74	60	71	60	86	69	58	80	67	57	75	64	23.4
BRITISH COLUMBIA																						
Abbotsford A	85	67	80	66	77	64	69	83	66	79	64	76	62	85	77	61	80	73	59	76	70	21.4
Cape St. James	64	59	62	58	60	57	60	63	59	61	58	60	59	74	61	58	71	59	57	69	59	7.6
Castlegar A	92	64	88	63	84	62	67	85	65	83	63	81	60	83	71	59	78	70	57	74	68	27.9
Comox A	80	63	76	62	73	61	65	76	63	74	62	71	60	78	68	59	75	66	58	71	65	16.4
Cranbrook A	88	61	85	60	81	59	63	83	62	80	60	78	57	77	66	55	72	66	53	67	65	24.8
Fort Nelson A	82	62	78	60	75	59	64	77	62	74	60	72	60	80	68	58	75	66	56	70	64	21.1
Fort St. John A	79	61	76	59	73	58	63	76	61	72	59	70	59	80	67	57	74	65	55	69	63	18.7
Kamloops A	93	65	88	63	85	62	66	88	65	85	63	81	59	78	70	57	74	69	56	69	68	24.7
Penticton A	90	65	87	64	83	63	67	85	65	83	64	81	60	81	73	59	76	72	57	72	71	26.5
Port Hardy A	68	59	65	58	63	57	60	66	59	64	58	62	58	72	62	57	69	61	56	67	60	12.4
Prince George A	81	60	78	59	74	58	63	78	61	74	59	71	58	77	66	56	73	64	54	67	62	23.2
Prince Rupert A	66	58	63	57	61	56	60	64	58	62	57	60	58	71	61	57	69	60	56	66	59	10.4
Quesnel A	85	62	81	60	77	59	64	80	62	77	61	74	59	81	67	57	75	65	56	71	63	25.4
Sandspit A	68	60	65	59	63	58	61	66	60	64	59	62	59	75	62	58	71	61	57	69	60	8.6
Smithers A	81	61	77	59	73	58	62	78	61	74	59	71	57	74	65	56	70	64	54	65	62	22.0
Spring Island	68	60	66	59	63	58	61	66	60	64	59	62	59	76	62	58	74	61	57	71	60	8.8
Terrace A	83	62	78	60	74	59	64	79	62	76	60	72	58	73	66	56	69	64	55	65	64	17.1
Tofino A	72	62	68	60	66	58	62	70	60	67	59	64	59	76	63	58	73	62	57	71	60	12.2
Vancouver Intl A	76	65	74	64	71	62	66	75	64	72	63	70	62	83	71	61	79	70	60	76	68	14.0
Victoria Intl A	79	63	75	62	72	61	64	77	63	74	61	71	59	75	69	58	71	67	57	69	65	18.4
Williams Lake A	83	59	79	57	75	56	60	77	59	75	57	72	55	73	62	53	67	61	51	62	60	22.0
MANITOBA																						
Brandon A	87	67	84	66	80	65	71	82	69	80	66	77	67	105	77	65	97	74	62	88	72	23.6
Churchill A	77	62	72	60	67	58	64	74	61	70	58	66	59	76	68	56	68	65	53	60	62	16.7
Dauphin A	87	67	84	66	80	64	70	82	68	80	66	77	67	102	77	64	94	74	62	86	72	22.1
Portage La Prairie A	88	68	85	67	81	65	72	83	69	80	68	78	68	108	78	66	99	75	64	92	73	20.5
The Pas A	83	66	79	64	76	62	68	79	66	76	64	73	64	93	73	62	85	70	60	80	68	18.4
Thompson A	83	64	79	62	76	60	66	78	64	75	62	73	62	85	71	60	78	69	58	73	66	23.0
Winnipeg Int'l A	87	68	84	67	81	66	72	82	70	80	68	78	68	107	78	66	99	75	64	92	73	20.5
NEW BRUNSWICK																						
Charlo A	83	68	79	66	75	65	70	78	68	76	66	73	68	102	74	66	97	72	64	90	70	18.4
Chatham A	86	69	83	67	79	65	71	81	69	78	68	75	68	104	75	67	98	73	65	92	71	20.3
Fredericton A	86	69	83	68	79	66	72	82	70	79	68	76	68	104	77	67	99	75	65	93	72	20.7
Moncton A	83	68	80	67	77	65	71	79	69	77	68	74	68	104	75	67	99	73	65	93	71	19.4
Saint John A	78	65	75	64	72	62	68	75	66	72	64	69	66	96	71	64	90	69	62	84	66	16.9
NEWFOUNDLAND																						
Battle Harbour	65	58	60	55	58	53	59	63	56	60	54	57	57	69	61	54	62	58	52	57	55	10.4
Bonavista	74	65	71	63	68	62	67	72	65	69	63	67	65	92	70	63	85	68	61	80	66	11.7
Cartwright	75	62	70	59	67	58	63	72	61	68	59	66	59	76	67	57	70	65	55	64	63	17.5
Daniels Harbour	69	63	66	62	65	61	65	67	63	66	62	64	64	90	67	62	83	65	60	78	63	9.7
Deer Lake A	81	66	77	64	74	62	68	77	66	74	65	71	66	95	73	64	88	71	61	81	69	21.4
Gander Intl A	79	65	76	63	72	62	68	75	66	72	64	70	65	95	71	63	88	70	61	83	68	17.8
Goose A	81	63	77	61	73	60	66	77	63	73	61	71	61	82	70	59	76	68	57	71	65	18.2
Hopedale	70	59	65	57	61	55	60	68	58	64	55	61	57	69	64	54	62	61	52	57	59	12.6
St. John's A	76	65	73	64	70	63	68	73	66	71	64	68	66	98	71	64	92	69	62	85	67	15.7
Stephenville A	74	64	71	64	69	63	67	71	65	69	64	68	66	95	70	64	88	68	62	83	66	12.4
Wabush Lake A	76	60	72	58	68	57	63	71	61	69	59	66	60	83	66	58	76	63	56	70	62	16.9
NORTHWEST TERRITORIES																						
Baker Lake	69	57	65	55	61	53	59	67	56	63	54	60	55	63	64	52	57	59	50	53	57	16.4
Cambridge Bay A	60	53	57	51	53	48	54	59	51	56	49	53	50	55	56	48	50	53	46	46	51	12.4
Cape Parry A	58	53	54	50	50	47	53	58	50	53	47	50	50	53	56	47	48	52	45	43	50	9.7
Chesterfield	66	54	60	52	56	50	55	65	52	60	50	56	50	53	60	48	49	55	46	46	52	13.7
Coral Harbour A	64	53	60	51	56	50	55	63	52	59	50	56	50	54	58	48	50	55	46	46	52	14.8
Fort Smith A	82	63	78	61	75	60	65	78	63	75	61	72	61	81	69	59	76	67	57	71	66	21.4
Hall Beach A	56	50	52	47	49	45	50	55	48	52	45	49	47	48	53	44	42	50	42	39	47	9.9
Inuvik Ua	78	60	75	59	71	57	62	75	60	72	58	69	56	67	67	53	61	65	52	57	64	18.4
Iqualuit A (Frobisher)	60	50	57	48	53	46	51	59	49	55	47	53	46	46	53	44	43	52	42	40	50	12.4
Norman Wells A	80	62	77	60	74	59	64	77	62	74	60	71	59	75	68	57	71	66	56	67	65	18.5

MDB = mean coincident dry-bulb temperature, °F
MWB = mean coincident wet-bulb temperature, °F
MWS = mean coincident wind speed, mph
MWD = mean coincident wind direction
StdD = standard deviation
A = airport
HR = humidity ratio
DP = dew-point temperature, °F

Table 2A Heating and Wind Design Conditions—Canada

Station	WMO#	Lat.	Long.	Elev. ft	StdP psia	Dates	Heating DB		Extreme Wind Speed			Coldest Month WS/MDB				MWS/MWD to DB				Annual Extreme Daily			
												0.4%		1%		99.6%		0.4%		Mean DB		StdD DB	
							99.6%	99%	1%	2.5%	5%	WS	MDB	WS	MDB	MWS	PWD	MWS	PWD	Max	Min	Max	Min
Resolute A	719240	74.72	94.98	220	14.58	6393	−42	−40	39	34	30	42	−15	37	−18	9	320	12	110	55	−49	3.8	4.9
Yellowknife A	719360	62.47	114.45	676	14.34	6193	−39	−36	22	19	17	23	−8	20	−9	5	50	10	160	82	−47	3.6	4.5
NOVA SCOTIA																							
Greenwood A	713970	44.98	64.92	92	14.65	6193	−2	3	29	25	22	35	25	30	23	7	300	15	250	89	−11	2.3	5.4
Halifax Intl A	713950	44.88	63.50	476	14.44	6993	−2	2	27	23	20	30	26	27	26	11	320	12	200	87	−8	2.7	4.3
Sable Island	716000	43.93	60.02	13	14.69	6193	14	17	38	34	23	43	30	39	29	24	290	13	200	73	10	2.5	4.0
Shearwater A	716010	44.63	63.50	167	14.61	6193	1	5	29	25	22	33	24	28	24	12	340	10	230	85	−5	3.2	4.1
Sydney A	717070	46.17	60.05	203	14.59	6193	−1	3	30	26	23	35	23	30	24	13	270	14	230	87	−7	2.5	4.5
Truro	713980	45.37	63.27	131	14.63	6193	−9	−4	24	21	18	30	25	26	25	5	0	11	270	86	−16	3.2	5.8
Yarmouth A	716030	43.83	66.08	141	14.62	6193	7	10	28	25	22	30	26	28	27	12	320	10	190	79	1	2.7	3.6
ONTARIO																							
Armstrong A	718410	50.30	89.03	1152	14.09	6193	−33	−30	22	19	17	21	−2	18	−3	3	270	12	0	88	−49	3.2	4.9
Atikokan	717480	48.75	91.62	1289	14.02	6793	−31	−27	16	14	12	16	6	14	6	1	270	8	230	89	−42	3.6	3.8
Big Trout Lake	718480	53.83	89.87	735	14.31	6793	−32	−30	23	21	18	24	−1	21	−2	6	290	9	200	86	−42	3.2	4.9
Earlton A	717350	47.70	79.85	797	14.28	6193	−27	−21	21	19	17	24	14	21	12	4	320	12	200	91	−38	3.8	5.0
Geraldton	718340	49.78	86.93	1152	14.09	6893	−32	−28	21	18	16	22	3	19	4	2	270	11	0	87	−45	4.1	6.7
Gore Bay A	717330	45.88	82.57	633	14.36	6193	−12	−7	27	23	21	29	23	26	22	7	0	11	180	86	−21	4.5	5.9
Kapuskasing A	718310	49.42	82.47	745	14.30	6193	−30	−25	20	17	16	21	5	19	5	4	270	10	230	90	−39	2.7	5.0
Kenora A	718500	49.78	94.37	1348	13.99	6193	−27	−22	21	18	17	21	7	19	6	8	320	11	180	89	−32	3.4	5.0
London A	716230	43.03	81.15	912	14.22	6193	−3	2	25	22	20	30	21	26	21	9	250	11	250	90	−11	3.4	5.6
Mount Forest	716310	43.98	80.75	1362	13.99	6293	−7	−3	25	21	19	28	19	25	18	6	90	10	250	87	−16	2.2	4.5
Muskoka A	716300	44.97	79.30	925	14.21	6193	−17	−11	21	19	18	23	21	20	21	7	320	9	270	87	−30	2.2	5.6
North Bay A	717310	46.35	79.43	1217	14.06	6193	−18	−13	20	17	16	23	16	20	16	6	0	10	230	86	−26	3.2	5.4
Ottawa Int'l A	716280	45.32	75.67	374	14.50	6193	−13	−8	23	20	17	27	17	23	15	9	290	10	250	91	−19	2.7	5.0
Sault Ste. Marie A	712600	46.48	84.52	630	14.36	6293	−13	−8	27	23	20	28	18	24	18	4	90	9	220	88	−25	3.2	5.9
Simcoe	715270	42.85	80.27	791	14.28	6293	−2	3	24	20	18	28	23	24	24	10	270	11	230	91	−10	2.9	4.7
Sioux Lookout A	718420	50.12	91.90	1280	14.03	6193	−30	−25	17	15	14	19	2	17	2	4	270	9	200	89	−39	3.2	4.9
Sudbury A	717300	46.62	80.80	1142	14.10	6193	−19	−14	30	26	23	30	13	27	13	10	0	14	230	89	−27	4.0	5.2
Thunder Bay A	717490	48.37	89.32	653	14.35	6193	−22	−18	24	21	18	25	9	22	6	9	250	11	200	90	−31	3.8	4.7
Timmins A	717390	48.57	81.37	968	14.19	6193	−28	−23	21	19	17	22	6	20	7	5	180	10	250	91	−39	3.6	4.7
Toronto Int'l A	716240	43.67	79.63	568	14.40	6593	−4	1	26	22	20	29	22	26	23	9	340	12	270	92	−11	2.9	5.6
Trenton A	716210	44.12	77.53	282	14.55	6193	−8	−3	26	22	19	30	23	26	24	6	50	12	230	89	−16	3.1	5.6
Wiarton A	716330	44.75	81.10	728	14.31	6193	−5	0	25	22	19	27	27	24	25	7	340	12	230	88	−15	2.3	7.0
Windsor A	715380	42.27	82.97	623	14.37	6193	2	6	28	24	21	30	22	26	22	12	230	12	250	94	−5	2.7	5.4
PRINCE EDWARD ISLAND																							
Charlottetown A	717060	46.28	63.13	177	14.60	6193	−6	−2	24	22	20	35	17	29	18	13	270	12	230	85	−12	2.3	4.7
Summerside A	717020	46.43	63.83	79	14.65	6193	−5	−1	32	28	25	40	18	35	18	15	270	13	200	85	−11	2.7	4.1
QUEBEC																							
Bagotville A	717270	48.33	71.00	522	14.42	6193	−23	−19	26	23	20	30	3	26	5	6	270	10	270	91	−30	2.5	5.0
Baie Comeau A	711870	49.13	68.20	72	14.66	6593	−19	−14	27	24	21	30	14	25	14	11	270	13	230	82	−28	3.2	6.8
Grindstone Island		47.38	61.87	194	14.59	6193	−1	3	49	43	39	55	20	49	21	24	290	18	250	79	−5	2.0	5.6
Kuujjuarapik A	719050	55.28	77.77	39	14.67	6193	−33	−30	29	25	22	26	1	24	−2	8	120	13	180	85	−43	4.0	4.7
Kuujjuaq A	719060	58.10	68.42	121	14.63	6193	−34	−31	29	25	21	31	2	27	−2	5	230	11	180	82	−41	3.4	3.6
La Grande Riviere A	718270	53.63	77.70	640	14.36	7793	−33	−30	22	20	17	23	0	20	−2	6	270	13	240	85	−39	3.6	3.4
Lake Eon A	714210	51.87	63.28	1841	13.74	6193	−31	−27	23	21	18	23	4	21	0	5	270	10	230	80	−42	2.5	4.0
Mont Joli A	717180	48.60	68.22	171	14.61	6193	−12	−8	28	25	22	35	8	30	9	15	290	14	230	87	−18	3.1	4.5
Montreal Intl A	716270	45.47	73.75	118	14.63	6193	−12	−7	23	20	18	30	19	26	18	7	250	11	230	90	−19	2.3	4.5
Montreal Mirabel A	716278	45.68	74.03	269	14.55	7693	−16	−11	21	18	16	25	13	21	11	6	240	9	240	88	−25	1.8	4.3
Nitchequon		53.20	70.90	1759	13.79	6193	−33	−31	25	22	20	29	−6	25	−5	6	270	11	230	78	−46	3.6	4.7
Quebec A	717140	46.80	71.38	240	14.57	6193	−16	−11	24	21	19	30	13	26	11	10	250	12	250	89	−23	2.3	4.9
Riviere Du Loup	717150	47.80	69.55	486	14.44	6693	−13	−10	21	19	17	24	14	21	13	10	180	11	230	85	−18	2.0	4.7
Roberval A	717280	48.52	72.27	587	14.39	6193	−23	−19	23	20	18	27	8	23	10	7	270	12	220	90	−30	3.1	4.3
Schefferville A	718280	54.80	66.82	1709	13.81	6193	−33	−31	26	23	21	30	−10	26	−9	7	320	12	270	81	−43	4.0	5.0
Sept-Iles A	718110	50.22	66.27	180	14.60	6893	−20	−15	27	23	20	30	16	25	13	8	300	11	220	82	−26	4.3	4.3
Sherbrooke A	716100	45.43	71.68	791	14.28	6393	−20	−14	19	17	15	22	15	19	14	5	110	9	250	88	−30	2.2	5.0
St. Hubert A	713710	45.52	73.42	89	14.65	6193	−12	−7	27	23	20	30	18	27	19	7	20	13	250	91	−20	2.5	4.9
Ste. Agathe Des Monts	717200	46.05	74.28	1296	14.02	6693	−19	−15	19	17	15	22	12	19	11	4	290	9	270	86	−28	2.9	4.0
Val d'Or A	717250	48.07	77.78	1106	14.12	6193	−27	−22	21	18	16	22	12	20	10	5	310	11	230	88	−37	2.9	4.9
SASKATCHEWAN																							
Broadview	718610	50.38	102.68	1975	13.68	6693	−30	−25	26	23	20	28	9	25	9	8	290	12	160	94	−37	3.6	5.8
Estevan A	718620	49.22	102.97	1906	13.71	6193	−25	−21	30	26	23	32	12	28	12	10	290	14	180	97	−32	3.2	5.4
Moose Jaw A	718640	50.33	105.55	1893	13.72	6193	−26	−21	30	26	23	35	21	30	19	9	290	13	180	97	−33	3.2	5.4
North Battleford A	718760	52.77	108.25	1798	13.77	6193	−31	−26	25	22	20	26	8	22	7	4	320	11	140	93	−38	3.1	5.4
Prince Albert A	718690	53.22	105.68	1404	13.96	6193	−34	−29	23	20	18	24	2	21	3	3	270	11	180	91	−43	3.8	5.9
Regina A	718630	50.43	104.67	1893	13.72	6193	−29	−24	30	27	23	34	8	29	10	9	270	14	180	96	−36	3.6	5.0
Saskatoon	718660	52.17	106.68	1654	13.84	6193	−31	−26	27	23	20	27	12	23	6	7	290	13	180	94	−37	3.8	5.9
Swift Current A	718700	50.28	107.68	2684	13.33	6193	−25	−21	33	28	25	36	17	30	14	14	270	13	180	95	−33	3.1	5.8
Uranium City A	710760	59.57	108.48	1043	14.15	6393	−38	−35	20	18	16	18	−7	16	−9	4	70	7	230	86	−49	4.0	3.4
Wynyard	718650	51.77	104.20	1841	13.74	6593	−29	−25	26	23	21	27	13	24	11	8	270	13	180	92	−35	3.4	6.1
Yorkton A	711380	51.27	102.47	1634	13.85	6193	−30	−26	25	22	20	27	6	23	2	7	290	12	180	93	−37	3.2	5.6
YUKON TERRITORY																							
Burwash A	719670	61.37	139.05	2644	13.34	6793	−34	−32	28	24	21	29	31	25	31	0	290	9	110	79	−55	3.4	6.5
Whitehorse A	719640	60.72	135.07	2306	13.51	6193	−34	−31	23	20	18	28	17	24	15	2	340	8	140	83	−47	4.0	5.9

WMO# = World Meteorological Organization number
Lat = latitude
Long. = longitude
Elev = elevation, ft
StdP = standard pressure at station elevation, psia
DB = dry-bulb temperature, °F
WS = wind speed, mph

Table 2B Cooling and Dehumidification Design Conditions—Canada

Station	Cooling DB/MWB 0.4% DB	MWB	1% DB	MWB	2% DB	MWB	WB/MDB 0.4% WB	MDB	1% WB	MDB	2% WB	MDB	DP/MDB and HR 0.4% DP	HR	MDB	1% DP	HR	MDB	2% DP	HR	MCDB	Range of DB
Resolute A	50	45	48	43	44	41	46	50	43	47	41	44	42	39	47	40	36	45	38	34	43	8.5
Yellowknife A	77	60	74	59	70	57	62	73	60	71	59	68	58	73	66	56	68	65	53	62	63	14.4
NOVA SCOTIA																						
Greenwood A	84	69	80	67	78	66	72	79	70	77	69	75	69	108	76	68	102	74	66	96	72	20.0
Halifax Intl A	80	68	78	66	75	64	70	77	69	74	67	72	68	106	73	67	99	71	65	95	69	16.7
Sable Island	70	67	69	66	67	65	68	69	67	68	66	67	68	101	69	66	97	68	65	92	66	8.3
Shearwater A	78	66	75	64	72	63	69	74	67	72	66	69	68	102	71	66	96	69	65	92	68	13.1
Sydney A	81	68	78	67	75	65	71	78	69	75	67	72	68	105	74	67	99	72	65	93	70	17.3
Truro	79	69	77	67	75	66	71	77	69	75	68	72	69	106	75	68	102	73	66	97	70	19.3
Yarmouth A	73	66	71	64	69	63	68	71	66	69	64	67	67	98	69	65	93	68	63	86	66	13.1
ONTARIO																						
Armstrong A	81	66	78	65	75	63	69	78	67	75	65	73	67	102	73	64	94	71	61	85	69	24.7
Atikokan	84	67	80	66	77	64	71	80	69	77	67	74	68	109	75	66	102	73	64	95	71	23.0
Big Trout Lake	79	64	75	62	72	61	67	75	65	72	63	70	64	91	70	62	84	68	60	78	67	16.4
Earlton A	85	69	81	67	78	65	71	80	69	77	67	75	69	108	76	67	101	74	65	95	72	21.6
Geraldton	81	66	78	65	75	63	69	77	67	75	65	73	67	103	74	65	97	72	62	87	69	22.1
Gore Bay A	80	68	78	67	75	65	71	77	69	75	68	73	69	108	74	67	102	72	66	97	71	16.4
Kapuskasing A	84	67	80	65	77	64	70	79	68	77	66	74	67	102	75	65	95	72	62	87	70	22.5
Kenora A	84	67	81	65	78	64	70	79	68	77	66	75	68	106	75	66	99	73	63	92	72	16.4
London A	85	71	83	70	80	69	74	82	72	80	71	77	71	119	79	70	113	77	68	108	75	19.8
Mount Forest	83	70	80	68	77	67	72	80	70	77	69	75	70	115	77	68	109	74	67	103	73	20.3
Muskoka A	84	69	80	68	78	66	72	80	70	77	68	75	70	113	76	68	106	74	66	100	72	20.7
North Bay A	81	67	78	66	76	64	70	77	68	75	67	73	68	108	74	66	102	72	65	96	70	17.1
Ottawa Int'l A	86	70	83	69	80	67	73	82	71	80	69	78	70	112	78	68	106	76	67	99	75	18.5
Sault Ste. Marie A	83	69	79	67	76	65	71	79	69	76	67	74	69	108	76	67	102	73	65	95	71	20.9
Simcoe	85	72	83	70	80	69	74	83	72	80	71	77	71	118	79	70	112	77	68	107	75	19.3
Sioux Lookout A	84	67	80	65	78	63	70	80	68	76	66	74	67	104	74	65	97	71	63	90	70	18.9
Sudbury A	84	67	81	66	78	64	70	80	68	77	66	74	68	105	74	66	99	72	64	92	71	19.1
Thunder Bay A	84	68	80	66	77	64	70	80	68	77	66	73	67	101	76	65	93	72	62	86	70	21.8
Timmins A	84	67	81	65	78	64	70	80	68	77	66	74	67	102	74	65	95	72	63	88	70	23.0
Toronto Int'l A	87	71	84	70	81	68	74	83	72	80	70	78	71	116	79	69	110	77	68	104	75	20.2
Trenton A	84	71	81	70	78	68	74	81	72	78	70	76	71	116	78	70	110	76	68	104	74	18.0
Wiarton A	82	70	80	69	77	67	72	80	70	77	69	75	70	113	77	68	106	75	67	101	73	18.0
Windsor A	89	73	86	71	83	70	76	85	74	83	73	80	73	125	82	71	118	79	70	112	77	17.5
PRINCE EDWARD ISLAND																						
Charlottetown A	79	69	77	67	74	65	71	77	69	74	67	72	68	105	75	67	99	72	65	94	71	15.1
Summerside A	79	68	77	66	74	65	71	76	69	74	67	72	68	104	74	67	99	72	65	94	70	14.4
QUEBEC																						
Bagotville A	84	67	80	65	77	63	69	79	68	76	66	74	66	99	74	65	93	72	62	85	70	19.8
Baie Comeau A	75	63	71	61	69	60	65	71	63	69	61	67	62	85	68	61	79	66	59	75	64	17.1
Grindstone Island	73	66	70	65	68	64	68	71	67	69	65	68	67	100	70	66	95	69	64	90	68	8.6
Kuujjuarapik A	75	61	70	59	66	57	63	72	61	68	58	65	60	76	67	57	69	64	54	62	62	15.8
Kuujuaq A	74	60	69	57	65	55	62	69	59	66	56	64	58	72	65	55	65	63	52	58	60	18.7
La Grande Riviere A	78	62	75	60	71	58	65	73	62	70	60	68	62	83	68	59	77	65	57	70	63	21.2
Lake Eon A	74	60	70	58	67	57	63	70	61	67	59	64	60	83	65	58	78	63	57	74	62	16.4
Mont Joli A	80	67	76	65	74	64	69	77	67	75	65	72	66	97	75	64	90	72	62	83	70	16.4
Montreal Intl A	85	71	83	70	80	68	74	82	72	80	70	77	70	112	79	69	106	77	67	101	75	17.6
Montreal Mirabel A	84	71	81	69	78	67	73	81	71	78	69	76	70	110	78	68	103	75	66	98	73	20.2
Nitchequon	72	60	69	58	66	57	63	68	61	65	59	64	61	85	65	59	79	63	57	74	62	14.0
Quebec A	84	70	80	68	78	66	73	80	70	77	68	75	70	111	77	68	103	75	66	95	73	19.1
Riviere Du Loup	79	68	76	67	74	65	70	76	69	74	67	72	68	106	74	67	99	72	65	93	70	15.5
Roberval A	83	68	79	66	76	65	71	79	69	77	67	74	68	105	76	66	98	74	64	92	72	17.8
Schefferville A	74	58	69	57	66	55	61	69	59	66	57	64	58	76	64	56	71	62	53	64	60	16.0
Sept-Iles A	72	60	69	59	67	58	64	69	62	67	60	65	61	82	66	60	77	64	58	73	62	13.9
Sherbrooke A	84	70	80	68	78	66	72	81	70	78	68	75	69	109	77	67	103	75	66	97	73	22.3
St. Hubert A	86	71	83	69	80	68	74	83	72	80	70	78	71	113	78	69	106	77	67	101	75	19.1
Ste. Agathe Des Monts	81	68	78	66	76	65	71	77	69	75	67	73	69	110	75	67	103	72	65	97	70	19.4
Val d'Or A	83	67	80	65	77	63	69	79	68	76	66	73	67	102	73	65	95	72	62	88	70	20.7
SASKATCHEWAN																						
Broadview	87	65	83	64	79	63	69	81	66	79	64	76	64	97	74	62	89	71	60	83	70	23.8
Estevan A	90	66	86	65	83	64	70	83	68	81	66	79	66	103	76	63	92	73	61	86	71	23.8
Moose Jaw A	90	64	87	64	83	62	68	83	66	81	64	79	63	92	73	61	85	71	59	79	69	23.8
North Battleford A	86	64	82	63	78	61	67	80	65	78	63	76	62	88	71	60	83	69	58	77	68	21.1
Prince Albert A	84	65	81	64	78	62	68	80	65	77	64	75	63	90	74	61	84	71	59	78	69	21.8
Regina A	89	64	85	64	82	62	68	82	66	80	64	78	63	92	75	61	85	71	59	79	69	23.6
Saskatoon	87	64	84	63	80	62	67	81	65	79	63	77	62	88	73	60	82	70	58	76	68	22.7
Swift Current A	88	63	84	62	80	61	66	81	64	79	62	77	61	88	71	59	82	68	57	76	67	23.0
Uranium City A	79	62	76	60	72	58	64	76	62	72	60	69	59	78	68	58	74	66	56	69	64	17.1
Wynyard	85	64	81	63	78	62	68	79	65	77	63	75	64	94	74	61	85	70	59	81	68	20.7
Yorkton A	86	65	82	64	79	62	68	80	66	78	64	75	64	96	74	62	87	71	60	81	69	22.0
YUKON TERRITORY																						
Burwash A	73	57	69	55	66	53	57	71	56	68	54	64	52	62	60	50	58	58	48	55	57	21.4
Whitehorse A	77	57	73	55	69	53	58	73	56	70	55	67	52	62	61	50	58	59	48	55	58	20.9

MDB = mean coincident dry-bulb temperature, °F
MWB = mean coincident wet-bulb temperature, °F
MWS = mean coincident wind speed, mph
MWD = mean coincident wind direction
StdD = standard deviation
A = airport
HR = humidity ratio
DP = dew-point temperature, °F

Table 3A Heating and Wind Design Conditions—World Locations

							Heating DB		Extreme Wind Speed			Coldest Month WS/MDB				MWS/MWD to DB				Annual Extreme Daily			
												0.4%		1%		99.6%		0.4%		Mean DB		StdD DB	
Station	WMO#	Lat.	Long.	Elev. ft	StdP psia	Dates	99.6%	99%	1%	2.5%	5%	WS	MDB	WS	MDB	MWS	PWD	MWS	PWD	Max	Min	Max	Min
ALGERIA																							
Algiers	603900	36.72 N	3.25 E	82	14.656	8293	36	38	25	21	19	26	55	22	55	2	200	12	60	105	32	4.5	2.5
Annaba	603600	36.83 N	7.82 E	13	14.693	8293	39	41	25	22	20	27	54	22	53	5	220	11	240	106	34	5.8	2.2
Biskra	605250	34.80 N	5.73 E	285	14.549	8293	41	43	34	30	26	33	56	29	58	7	10	11	180	113	36	2.7	2.3
Constantine	604190	36.28 N	6.62 E	2276	13.530	8293	31	33	24	20	17	26	46	22	46	2	320	12	240	103	27	2.5	1.4
El Golea	605900	30.57 N	2.87 E	1302	14.021	8293	33	35	26	22	19	26	58	22	57	4	320	8	180	113	28	2.3	2.7
Oran	604900	35.63 N	0.60 W	295	14.544	8293	35	38	29	24	21	33	57	26	58	2	200	14	240	104	28	8.8	12.4
Tebessa	604750	35.48 N	8.13 E	2667	13.337	8293	29	31	26	22	19	30	44	23	49	1	280	11	220	103	25	3.1	2.0
ARGENTINA																							
Buenos Aires	875760	34.82 S	58.53 W	65	14.665	8293	31	34	23	20	18	22	53	19	53	4	270	11	270	98	27	2.5	2.0
Comodoro Rivadavia	878600	45.78 S	67.50 W	150	14.620	8293	29	31	43	37	33	43	51	36	50	11	270	22	290	97	26	7.2	2.0
Cordoba	873440	31.32 S	64.22 W	1555	13.892	8293	31	34	27	24	21	28	66	24	60	3	270	16	20	100	25	3.1	3.2
Junin Airport	875480	34.55 S	60.95 W	266	14.559	8293	30	32	26	23	20	26	55	23	55	1	270	12	360	99	26	5.9	2.0
Formosa	871620	26.20 S	58.23 W	196	14.596	8293	40	44	30	26	23	28	67	26	64	6	270	17	360	104	34	1.8	3.2
Marcos Juarez	874670	32.70 S	62.15 W	374	14.502	8293	29	32	28	24	22	27	57	24	55	3	50	11	360	101	24	3.4	3.8
Mendoza	874180	32.83 S	68.78 W	2309	13.514	8293	30	33	22	18	15	19	56	15	53	3	230	11	50	102	26	2.7	1.8
Paso De Los Libres	872890	29.68 S	57.15 W	229	14.578	8293	36	39	31	26	24	31	53	27	57	1	180	13	360	101	33	2.0	2.3
Posadas	871780	27.37 S	55.97 W	410	14.483	8293	39	43	22	19	17	23	71	21	66	6	180	11	360	101	34	2.2	4.7
Reconquista	872700	29.18 S	59.67 W	173	14.608	8293	37	40	27	21	20	27	60	21	58	2	200	12	50	103	32	4.9	2.5
Resistencia	871550	27.45 S	59.05 W	170	14.610	8293	35	39	20	17	15	19	69	16	64	2	50	11	20	103	31	2.3	3.4
Rio Gallegos	879250	51.62 S	69.28 W	62	14.667	8293	17	22	52	46	40	40	37	33	39	6	270	20	320	90	12	12.4	7.9
Rosario	874800	32.92 S	60.78 W	82	14.656	8293	30	33	30	26	23	28	53	25	53	2	180	13	360	98	26	2.9	2.7
Salta Airport	870470	24.85 N	65.48 E	3990	12.701	8293	30	33	19	16	14	22	78	17	63	2	270	10	70	97	25	3.1	2.5
San Juan	873110	31.57 S	68.42 W	1961	13.687	8293	28	31	32	28	24	28	52	24	52	1	360	11	180	107	23	2.5	2.9
San Miguel De Tucuman	871210	26.85 S	65.10 W	1476	13.932	8293	37	40	22	18	14	18	58	14	56	5	360	12	90	102	33	3.4	2.7
ARMENIA																							
Yerevan	377890	40.13 N	44.47 E	2919	13.214	8293	7	11	22	16	14	14	41	10	32	1	180	6	210	101	3	6.3	6.8
ASCENSION ISLAND																							
Georgetown	619020	7.97 S	14.40 W	259	14.563	8293	69	70	26	24	23	25	76	24	76	16	90	19	120	87	65	1.8	2.9
AUSTRALIA																							
Adelaide	946720	34.93 S	138.52 E	13	14.693	8293	39	41	27	24	22	26	53	23	54	2	50	13	310	104	35	3.1	2.2
Alice Springs	943260	23.80 S	133.90 E	1774	13.781	8293	34	36	20	17	16	17	64	15	62	2	270	8	100	108	29	1.8	3.6
Brisbane	945780	27.38 S	153.10 E	16	14.691	8293	44	46	22	19	17	22	60	19	61	4	220	11	20	95	39	3.8	2.0
Cairns	942870	16.88 S	145.75 E	22	14.688	8293	56	59	19	17	15	18	73	16	72	8	170	8	120	97	50	2.5	8.8
Canberra	949260	35.30 S	149.18 E	1893	13.722	8293	26	29	24	21	18	24	46	22	48	0	310	12	310	97	20	4.0	8.1
Darwin	941200	12.40 S	130.87 E	98	14.648	8293	64	66	19	17	15	18	80	17	80	7	140	12	290	98	60	2.9	2.5
Kalgoorlie/Boulder	946370	30.77 S	121.45 E	1181	14.083	8293	36	38	22	19	17	24	59	21	58	1	220	9	320	108	31	3.4	1.3
Learmouth	943020	22.23 S	114.08 E	19	14.690	8293	49	51	25	23	21	22	68	20	67	5	210	14	210	112	45	2.7	2.9
Perth	946100	31.93 S	115.95 E	95	14.650	8293	41	43	24	21	19	23	58	19	58	1	50	10	270	107	36	3.4	2.2
Port Hedland	943120	20.37 S	118.62 E	19	14.690	8293	51	54	23	20	19	24	68	22	70	5	160	12	120	111	46	2.5	2.2
Sydney Intl Airport	947670	33.95 S	151.18 E	10	14.695	8293	42	44	25	22	20	25	58	20	56	2	320	12	300	103	38	5.2	3.4
Townsville	942940	19.25 S	146.75 E	19	14.690	8293	48	52	21	19	17	20	72	18	72	0	190	9	50	101	43	3.8	2.7
AUSTRIA																							
Aigen/Ennstal (Mil)	111570	47.53 N	14.13 E	2129	13.604	8293	2	7	19	16	14	22	38	19	34	1	60	7	60	89	−6	2.7	6.3
Graz	112400	47.00 N	15.43 E	1138	14.105	8293	5	12	17	14	11	16	35	11	33	1	180	7	140	90	1	3.4	8.1
Innsbruck	111200	47.27 N	11.35 E	1945	13.695	8293	10	14	19	15	12	18	42	15	39	1	260	8	70	91	5	2.7	6.8
Klagenfurt	112310	46.65 N	14.33 E	1482	13.929	8293	4	9	13	11	9	10	31	7	27	3	310	6	100	90	−1	3.8	7.0
Linz	110100	48.23 N	14.20 E	1026	14.163	8293	5	12	24	20	17	28	39	25	38	6	90	7	110	91	1	3.8	9.9
Salzburg	111500	47.80 N	13.00 E	1476	13.932	8293	7	12	17	15	13	20	42	17	38	3	130	7	330	92	2	4.1	8.3
Vienna, Hohe Warte	110350	48.25 N	16.37 E	656	14.355	8293	12	17	23	19	17	27	43	23	42	6	240	9	140	91	9	3.2	6.7
Vienna, Schwechat	110360	48.12 N	16.57 E	623	14.372	8293	9	14	27	24	21	31	44	27	38	6	320	12	150	92	5	3.1	7.2
Zeltweg	111650	47.20 N	14.75 E	2237	13.550	8293	0	5	18	15	13	19	37	16	36	1	250	7	190	88	−6	3.4	7.6
AZORES																							
Lajes	85090	38.77 N	27.10 W	180	14.604	8293	46	48	28	23	21	30	56	26	57	3	300	8	250	83	41	1.8	3.8
BAHAMAS																							
Nassau	780730	25.05 N	77.47 W	22	14.688	8293	57	60	21	19	17	21	72	19	71	3	300	10	130	93	52	1.3	3.1
BAHRAIN																							
Al-Manamah	411500	26.27 N	50.65 E	6	14.697	8293	52	54	25	23	21	26	56	24	58	13	290	11	340	109	47	2.5	5.6
BELARUS																							
Babruysk (Bobruysk)	269610	53.12 N	29.25 E	541	14.415	8293	−9	−2	20	18	16	20	27	18	30	4	210	8	200	87	−16	4.0	21.1
Homyel (Gomel')	330410	52.45 N	31.00 E	416	14.480	8293	−6	0	18	15	14	17	28	15	28	4	330	7	150	87	−8	4.5	7.9
Hrodna (Grodno)	268250	53.68 N	23.83 E	442	14.466	8293	−5	1	26	22	19	25	33	22	32	5	270	9	180	87	−5	3.8	9.2
Mahilyow (Mogilev)	268630	53.90 N	30.32 E	633	14.367	8293	−9	−3	23	21	18	23	31	21	28	7	30	9	200	85	−11	3.4	7.6
Minsk	268500	53.87 N	27.53 E	767	14.297	8293	−5	0	17	14	13	17	23	15	24	5	300	9	70	85	−8	3.8	7.2
Vitsyebsk (Vitebsk)	266660	55.17 N	30.13 E	577	14.396	8293	−8	−2	19	17	15	21	26	17	28	3	30	8	210	84	−11	3.4	6.3
BELGIUM/LUXEMBOURG																							
Antwerp	64500	51.20 N	4.47 E	45	14.676	8293	16	21	24	21	18	28	47	24	44	7	50	7	90	89	15	4.1	7.2

WMO# = World Meteorological Organization number
Lat = latitude Long. = longitude
Elev = elevation, ft
StdP = standard pressure at station elevation, psia
DB = dry-bulb temperature, °F
WS = wind speed, mph

Table 3B Cooling and Dehumidification Design Conditions—World Locations

	Cooling DB/MWB						WB/MDB						DP/MDB and HR									
	0.4%		1%		2%		0.4%		1%		2%		0.4%			1%			2%			Range
Station	DB	MWB	DB	MWB	DB	MWB	WB	MDB	WB	MDB	WB	MDB	DP	HR	MDB	DP	HR	MDB	DP	HR	MCDB	of DB
ALGERIA																						
Algiers	95	71	92	71	89	72	77	87	76	85	75	83	74	129	82	73	125	81	72	118	80	20.9
Annaba	94	71	90	72	87	73	78	84	77	83	76	83	76	134	82	75	130	81	73	125	80	17.1
Biskra	108	69	106	69	103	69	72	97	71	97	71	96	65	92	83	63	86	82	61	82	83	20.0
Constantine	100	66	96	67	93	66	71	88	70	87	69	86	66	106	76	65	100	76	63	94	75	27.0
El Golea	109	69	107	68	104	68	72	101	71	101	69	99	62	87	82	60	80	84	58	76	85	25.7
Oran	92	69	89	69	86	69	75	84	74	83	73	81	73	123	80	72	119	79	71	115	79	18.4
Tebessa	99	65	97	64	94	64	69	89	68	87	67	86	64	97	73	62	92	72	61	88	72	27.2
ARGENTINA																						
Buenos Aires	93	73	90	72	87	71	76	86	75	84	74	83	74	126	80	72	120	79	71	115	78	21.6
Comodoro Rivadavia	87	61	84	60	80	58	63	82	62	78	60	76	57	70	66	55	65	64	53	60	65	18.7
Cordoba	94	73	91	72	89	71	77	88	76	86	74	84	74	136	84	72	127	82	71	120	80	21.1
Junin Airport	92	72	89	71	87	71	76	85	74	84	73	83	73	125	80	72	118	78	70	111	77	21.6
Formosa	98	76	96	76	94	76	80	91	80	90	79	89	78	145	85	77	141	84	76	136	83	18.5
Marcos Juarez	95	74	92	74	90	73	78	88	77	87	75	85	76	136	85	74	129	82	73	123	80	22.1
Mendoza	96	68	93	67	91	67	73	89	71	87	70	85	68	112	81	66	106	80	65	100	79	22.1
Paso De Los Libres	96	74	94	75	92	73	79	89	78	88	76	86	76	137	84	75	132	83	74	126	82	20.2
Posadas	97	76	95	76	93	75	80	91	79	90	78	89	77	142	87	76	136	85	75	132	84	19.1
Reconquista	96	78	94	77	91	76	81	91	80	89	79	87	79	150	87	77	144	86	76	138	84	17.8
Resistencia	98	76	95	76	93	76	80	91	79	90	78	88	77	144	86	77	140	85	76	134	83	20.2
Rio Gallegos	76	57	72	55	70	54	59	73	57	70	55	67	53	60	63	51	56	61	49	52	60	19.4
Rosario	93	74	90	73	88	72	78	87	76	85	75	83	75	132	83	74	125	81	72	119	79	20.5
Salta Airport	89	65	87	66	84	66	72	82	71	80	70	79	69	123	76	68	119	76	67	115	74	19.1
San Juan	100	69	97	68	94	68	73	93	72	91	71	89	68	110	82	66	102	82	64	97	81	23.8
San Miguel De Tucuman	96	74	93	74	91	74	79	90	78	88	76	87	76	144	85	75	138	84	74	132	83	17.5
ARMENIA																						
Yerevan	96	69	94	69	91	67	72	91	70	90	69	88	65	104	85	63	95	82	61	90	81	24.5
ASCENSION ISLAND																						
Georgetown	86	75	85	75	85	75	77	84	76	83	76	82	75	132	81	74	130	80	74	126	80	7.7
AUSTRALIA																						
Adelaide	95	64	92	64	88	63	70	83	68	81	66	80	66	97	74	64	89	73	61	81	72	19.4
Alice Springs	104	65	102	64	100	64	73	83	72	83	71	83	71	123	77	69	115	77	67	104	77	24.7
Brisbane	88	72	86	72	84	72	77	84	76	82	75	81	75	133	81	74	126	80	73	121	79	13.7
Cairns	91	78	90	77	88	77	81	87	80	87	79	86	79	150	85	78	146	84	77	141	83	13.1
Canberra	90	63	87	63	83	62	67	79	66	78	65	77	64	96	71	62	90	70	60	84	68	23.9
Darwin	93	75	92	76	91	76	82	89	81	88	81	87	81	160	86	79	152	85	79	151	85	13.0
Kalgoorlie/Boulder	102	65	99	64	95	63	70	87	68	85	67	86	65	98	74	63	90	71	61	84	71	24.5
Learmouth	105	70	102	70	99	70	79	88	78	87	77	87	77	140	84	76	134	83	74	127	83	23.6
Perth	99	67	95	66	91	65	72	87	70	85	69	83	67	101	75	66	95	74	64	91	74	22.5
Port Hedland	105	71	102	71	100	71	82	92	81	91	81	90	80	156	87	79	151	86	78	147	86	19.3
Sydney Intl Airport	90	68	85	67	82	68	73	82	72	79	71	78	71	115	77	70	111	76	69	107	75	12.1
Townsville	92	76	91	76	89	76	81	88	80	87	79	85	79	150	84	78	147	84	77	141	83	11.7
AUSTRIA																						
Aigen/Ennstal (Mil)	82	66	79	64	76	63	67	79	66	76	64	74	63	93	74	61	88	72	60	83	69	20.0
Graz	85	69	82	68	79	65	71	81	69	78	67	77	67	105	75	66	99	73	64	94	71	20.5
Innsbruck	85	65	82	64	79	62	66	81	65	78	64	74	62	91	69	61	86	67	59	81	67	20.5
Klagenfurt	85	66	82	65	79	63	68	81	66	79	65	76	64	92	72	62	88	71	61	84	68	23.0
Linz	85	66	82	65	79	63	69	81	67	77	65	74	65	95	70	63	90	69	62	88	68	19.6
Salzburg	86	67	82	65	79	64	68	82	66	80	65	77	63	91	72	61	85	71	60	83	70	18.9
Vienna, Hohe Warte	86	68	83	67	80	66	70	83	69	80	67	77	66	98	74	65	94	73	64	90	72	16.7
Vienna, Schwechat	86	67	83	66	80	65	69	83	67	79	66	78	64	92	73	63	88	72	62	83	71	19.1
Zeltweg	82	66	79	65	76	63	67	80	66	76	64	74	63	94	74	62	89	72	60	85	69	21.1
AZORES																						
Lajes	80	71	78	71	77	69	72	78	72	77	71	75	71	115	75	70	111	75	69	106	74	11.3
BAHAMAS																						
Nassau	91	79	90	78	89	78	81	87	80	87	80	87	79	151	83	79	150	83	78	144	83	12.4
BAHRAIN																						
Al-Manamah	103	77	101	78	99	79	87	94	86	94	85	93	86	189	93	84	181	92	84	176	92	12.4
BELARUS																						
Babruysk (Bobruysk)	82	65	79	65	76	63	68	78	67	76	65	74	65	94	73	63	88	71	61	83	69	20.7
Homyel (Gomel')	83	66	80	65	77	64	69	78	67	76	65	74	65	95	72	64	89	71	62	84	70	16.2
Hrodna (Grodno)	82	66	78	64	75	63	68	79	66	75	64	73	65	92	72	63	87	71	61	81	69	19.4
Mahilyow (Mogilev)	80	66	77	64	75	63	68	77	66	75	64	72	65	94	72	63	88	70	61	83	68	16.9
Minsk	81	65	78	64	75	62	67	78	66	75	64	73	64	91	71	62	85	69	60	80	69	17.3
Vitsyebsk (Vitebsk)	80	65	77	64	74	63	68	77	66	74	64	72	65	94	72	63	88	70	61	83	68	15.1
BELGIUM/LUXEMBOURG																						
Antwerp	82	68	79	66	76	65	70	80	68	77	66	74	66	97	75	65	91	73	63	85	71	16.2

MDB = mean coincident dry-bulb temperature, °F
MWB = mean coincident wet-bulb temperature, °F
MWS = mean coincident wind speed, mph
MWD = mean coincident wind direction
StdD = standard deviation
A = airport
HR = humidity ratio
DP = dew-point temperature, °F

Table 3A Heating and Wind Design Conditions—World Locations

Station	WMO#	Lat.	Long.	Elev. ft	StdP psia	Dates	Heating DB 99.6%	Heating DB 99%	Extreme Wind Speed 1%	Extreme Wind Speed 2.5%	Extreme Wind Speed 5%	Coldest Month WS/MDB 0.4% WS	0.4% MDB	1% WS	1% MDB	MWS/MWD to DB 99.6% MWS	99.6% PWD	0.4% MWS	0.4% PWD	Annual Extreme Daily Mean DB Max	Mean DB Min	StdD DB Max	StdD DB Min
Brussels	64510	50.90 N	4.53 E	190	14.599	8293	15	21	27	23	21	31	46	26	45	7	50	8	60	89	15	3.8	8.6
Charleroi	64490	50.47 N	4.45 E	629	14.368	8293	15	21	25	22	20	29	43	24	43	10	50	8	50	89	14	3.8	7.9
Florennes	64560	50.23 N	4.65 E	980	14.186	8293	13	19	24	21	19	29	44	24	41	9	70	8	170	87	13	3.6	7.9
Koksijde	64000	51.08 N	2.65 E	29	14.684	8293	17	22	30	26	23	33	46	29	45	10	90	10	100	87	14	3.2	7.7
Luxembourg	65900	49.62 N	6.22 E	1243	14.051	8293	13	18	24	21	18	27	41	23	34	11	50	9	80	88	12	3.6	7.2
Oostende	64070	51.20 N	2.87 E	16	14.691	8293	18	22	34	30	26	40	46	33	45	11	70	10	100	86	15	3.2	6.5
Saint Hubert	64760	50.03 N	5.40 E	1827	13.755	8293	11	16	22	20	18	26	35	22	31	11	90	8	30	82	10	2.9	8.1
BENIN																							
Cotonou	653440	6.35 N	2.38 E	29	14.684	8293	71	72	19	17	16	20	80	18	80	5	20	12	200	98	65	4.0	6.7
Parakou	653300	9.35 N	2.62 E	1289	14.028	8293	65	67	14	12	11	12	76	11	76	4	40	5	40	103	55	5.0	9.2
BERMUDA																							
Hamilton, Bermuda NAS	780160	32.37 N	64.68 W	9	14.695	8293	55	57	29	26	23	30	65	28	64	16	310	11	190	90	47	1.3	7.6
BOLIVIA																							
Cochabamba	852230	17.45 S	66.10 W	8303	10.793	8293	35	37	22	18	13	22	66	17	66	0	180	6	360	89	30	2.0	2.3
La Paz	852010	16.52 S	68.18 W	13169	8.926	8293	25	27	19	17	14	22	52	19	52	2	330	8	60	69	21	4.7	2.0
BOSNIA-HERZEGOVINA																							
Banja Luka	132420	44.78 N	17.22 E	511	14.430	8293	10	16	13	10	8	14	52	10	50	2	320	4	360	98	6	4.0	7.6
BRAZIL																							
Archipelago De Fernando	824000	3.85 S	32.42 W	183	14.603	8293	73	74	18	16	15	19	79	17	79	7	130	10	150	98	63	6.5	15.5
Belem	821930	1.38 S	48.48 W	52	14.672	8293	72	73	17	14	12	15	83	13	83	2	90	6	90	96	63	3.2	14.0
Brasilia	833780	15.87 S	47.93 W	3480	12.943	8293	48	51	17	14	12	18	71	15	72	0	90	7	90	94	44	3.2	5.8
Campinas	837210	23.00 S	47.13 W	2168	13.584	8293	47	50	24	22	21	24	65	23	63	8	150	7	330	96	42	2.9	4.5
Campo Grande	836120	20.47 S	54.67 W	1824	13.756	8293	46	50	25	22	21	27	60	24	62	16	180	11	360	98	41	3.2	3.8
Caravelas	834970	17.63 S	39.25 W	13	14.693	8293	61	63	23	20	18	19	75	17	75	1	240	12	60	95	57	4.0	3.2
Curitiba	838400	25.52 S	49.17 W	2979	13.185	8293	36	40	18	16	13	20	68	18	65	2	270	9	300	92	26	3.4	11.5
Fortaleza	823980	3.78 S	38.53 W	82	14.656	8293	72	73	20	18	17	19	84	18	84	5	180	13	120	97	65	4.0	9.7
Goiania	834240	16.63 S	49.22 W	2450	13.444	8293	53	55	18	15	12	16	77	13	78	1	180	4	360	98	42	1.6	10.8
Maceio	829930	9.52 S	35.78 W	377	14.501	8293	67	68	18	15	13	17	78	14	78	0	180	10	100	95	58	4.0	10.6
Manaus, Eduardo Gomes I	821110	3.03 S	60.05 W	6	14.697	8293	70	71	12	11	10	12	85	11	85	1	120	5	60	100	59	1.6	14.0
Manaus, Ponta Pelada	823320	3.15 S	59.98 W	275	14.554	8293	71	73	15	13	12	15	83	13	83	4	30	10	90	98	64	2.0	13.1
Natal	825990	5.92 S	35.25 W	170	14.610	8293	70	71	21	19	18	22	83	20	82	6	150	11	100	95	62	4.7	9.7
Porto Alegre	839710	30.00 S	51.18 W	9	14.695	8293	40	43	19	17	14	20	53	17	55	2	240	7	290	100	37	3.1	5.2
Recife	828990	8.07 S	34.85 W	62	14.667	8293	70	71	17	15	13	19	79	17	79	5	240	12	120	97	66	3.1	4.0
Rio De Janeiro, Galeao	837460	22.82 S	43.25 W	19	14.690	8293	59	61	19	17	15	17	74	15	73	3	320	7	50	107	50	2.5	9.4
Salvador	832480	12.90 S	38.33 W	19	14.690	8293	68	70	21	19	17	21	76	20	76	2	180	12	90	95	59	3.1	12.8
Santarem	822440	2.43 S	54.72 W	236	14.575	8293	72	73	19	18	16	19	81	17	81	6	210	12	90	96	64	3.2	16.0
Sao Paulo	837800	23.62 S	46.65 W	2634	13.353	8293	48	50	15	13	12	14	66	12	64	4	160	6	330	94	43	2.7	3.8
Vitoria	836490	20.27 S	40.28 W	13	14.693	8293	61	63	22	19	17	18	77	16	76	1	210	11	30	98	54	2.0	10.3
BULGARIA																							
Botevgrad	156270	42.67 N	24.83 E	7837	10.987	8293	−7	−2	76	62	53	77	15	64	11	30	320	9	270	66	−8	3.1	6.1
Burgas	156550	42.48 N	27.48 E	91	14.651	8293	17	21	31	24	21	31	37	26	33	9	270	11	110	94	13	3.1	5.2
Lom	155110	43.82 N	23.25 E	108	14.643	8293	14	18	28	23	18	31	35	27	40	2	270	4	50	96	9	4.7	7.7
Musala	156150	42.18 N	23.58 E	9603	10.266	8293	−10	−5	63	45	36	77	19	63	16	17	20	7	320	65	−13	6.8	5.8
Plovdiv	156250	42.13 N	24.75 E	606	14.380	8293	14	19	30	26	23	32	41	27	41	2	40	6	90	99	6	4.1	7.2
Ruse	155350	43.85 N	25.95 E	147	14.622	8293	12	16	32	27	21	32	39	27	33	5	50	8	270	101	7	5.4	4.9
Sofia	156140	42.65 N	23.38 E	1952	13.692	8293	10	14	21	17	14	22	33	18	32	2	360	5	110	94	3	4.9	8.5
Varna	155520	43.20 N	27.92 E	141	14.625	8293	17	20	32	26	21	40	29	30	33	15	360	9	90	92	13	4.0	4.9
BRUNEI																							
Brunei Intl Airport	963150	4.93 N	114.93 E	49	14.674	8293	71	72	16	14	12	18	82	16	82	0	220	8	320	96	67	2.5	4.7
CANARY ISLANDS																							
Las Palmas	600300	27.93 N	15.38 W	82	14.656	8293	55	57	32	30	28	27	65	24	65	4	320	19	30	94	51	4.3	0.9
Santa Cruz De Tenerife	600250	28.05 N	16.57 W	236	14.575	8293	57	57	29	27	24	28	68	25	68	8	360	19	60	101	51	5.4	6.5
CAPE VERDE																							
Sal Island	85940	16.73 N	22.95 W	180	14.604	8293	63	64	27	25	24	28	72	26	72	10	30	15	60	91	58	1.8	4.3
CHILE																							
Antofagasta	854420	23.43 S	70.43 W	393	14.492	8293	50	52	22	20	19	23	60	20	59	6	90	15	190	80	44	2.7	3.6
Arica	854060	18.33 S	70.33 W	193	14.597	8293	52	55	20	18	16	17	64	16	64	4	90	15	210	85	46	2.9	4.3
Concepcion	856820	36.77 S	73.05 W	52	14.672	8293	35	37	30	26	23	36	54	29	53	4	140	19	240	92	30	15.5	1.4
Iquique	854180	20.53 S	70.18 W	170	14.610	8293	54	55	22	20	18	20	64	18	63	6	160	14	210	84	50	2.5	2.3
La Serena	854880	29.92 S	71.20 W	479	14.447	8293	42	44	18	16	15	19	55	15	56	6	120	13	270	81	35	11.5	6.3
Puerto Montt	857990	41.42 S	73.08 W	282	14.551	8293	28	30	27	23	20	29	49	26	48	2	280	10	150	80	24	3.8	2.3
Punta Arenas	859340	53.00 S	70.85 W	121	14.636	8293	23	25	45	39	32	41	43	30	38	6	290	18	270	76	17	6.5	4.3
Santiago	855740	33.38 S	70.78 W	1561	13.889	8293	29	32	19	16	14	16	49	12	50	2	20	12	210	101	25	9.2	2.3
Temuco	857430	38.75 S	72.63 W	393	14.492	8293	29	31	22	19	15	29	51	23	52	2	70	10	240	91	25	8.5	1.8
CHINA																							
Anda	508540	46.38 N	125.32 E	492	14.440	8293	−19	−16	24	21	17	18	7	15	6	3	250	12	250	92	−23	3.1	4.3
Andirlangar	518480	37.93 N	83.65 E	4146	12.627	8293	−1	2	15	12	11	10	32	8	29	1	200	6	40	103	−5	2.3	3.2

WMO# = World Meteorological Organization number
Lat = latitude Long. = longitude
Elev = elevation, ft
StdP = standard pressure at station elevation, psia
DB = dry-bulb temperature, °F
WS = wind speed, mph

Table 3B Cooling and Dehumidification Design Conditions—World Locations

	Cooling DB/MWB						WB/MDB						DP/MDB and HR									
	0.4%		1%		2%		0.4%		1%		2%		0.4%			1%			2%			Range
Station	DB	MWB	DB	MWB	DB	MWB	WB	MDB	WB	MDB	WB	MDB	DP	HR	MDB	DP	HR	MDB	DP	HR	MCDB	of DB
Brussels	82	67	79	66	76	65	69	80	68	77	66	74	66	95	74	64	90	72	62	85	71	16.9
Charleroi	82	67	79	65	76	64	69	79	67	76	65	73	65	94	74	63	89	71	62	85	69	16.7
Florennes	80	66	77	65	74	63	68	78	66	75	65	72	65	97	73	63	90	70	62	85	68	16.2
Koksijde	79	67	75	65	72	64	68	76	66	73	65	71	65	93	72	64	88	70	62	83	68	15.1
Luxembourg	82	65	79	63	76	62	67	78	65	76	64	72	64	92	71	62	87	68	60	82	67	17.1
Oostende	77	66	73	65	70	63	67	74	66	72	64	70	65	92	71	64	88	69	62	83	67	14.0
Saint Hubert	77	64	74	62	72	61	66	74	64	72	62	69	63	91	69	61	85	67	59	80	65	14.6
BENIN																						
Cotonou	90	79	89	80	89	80	82	88	81	88	81	87	80	156	85	80	153	85	79	151	85	8.6
Parakou	98	71	97	71	95	71	78	91	78	90	77	89	75	138	84	74	135	84	74	132	83	20.7
BERMUDA																						
Hamilton, Bermuda NAS	88	78	88	78	86	77	80	85	80	85	79	84	79	150	84	78	144	83	77	141	83	8.3
BOLIVIA																						
Cochabamba	84	56	83	56	81	56	61	76	60	76	59	74	57	97	64	56	91	62	55	90	62	27.0
La Paz	63	44	62	44	61	43	49	58	48	57	47	56	45	73	51	44	71	50	43	68	49	22.7
BOSNIA-HERZEGOVINA																						
Banja Luka	92	69	88	69	85	67	72	85	71	83	69	81	68	106	79	66	99	76	65	94	73	22.9
BRAZIL																						
Archipelago De Fernando	88	78	87	78	86	77	80	85	79	85	79	85	78	147	84	77	144	83	77	141	83	8.1
Belem	92	78	90	78	90	78	81	87	80	87	80	87	79	150	84	79	149	84	77	143	83	14.8
Brasilia	89	65	88	65	86	65	71	80	70	79	70	79	69	120	73	68	118	72	68	116	72	23.4
Campinas	92	73	90	71	88	72	76	85	75	84	74	83	74	136	79	73	134	78	72	128	77	17.8
Campo Grande	95	71	93	71	92	71	77	87	76	86	76	86	75	141	80	74	134	80	73	132	79	18.4
Caravelas	89	77	88	77	87	77	79	87	78	85	78	85	77	142	81	77	140	81	76	137	80	13.7
Curitiba	87	69	85	68	83	68	73	81	72	80	71	78	71	128	75	70	123	74	69	119	74	17.5
Fortaleza	90	78	90	78	89	78	80	87	80	87	79	86	79	151	81	79	149	81	77	143	81	11.2
Goiania	93	67	92	68	91	68	75	85	74	84	73	84	72	129	78	71	125	77	70	123	76	23.0
Maceio	90	75	89	75	88	75	78	85	78	84	77	84	77	141	81	76	136	80	75	134	80	14.2
Manaus, Eduardo Gomes I	97	78	95	78	94	78	83	90	82	89	81	88	81	161	86	80	158	85	79	151	83	20.3
Manaus, Ponta Pelada	94	77	93	78	92	78	81	89	80	88	80	88	79	153	84	79	151	84	78	146	83	14.2
Natal	90	78	90	77	89	77	80	87	79	86	79	86	78	148	82	77	142	82	77	141	82	12.2
Porto Alegre	95	76	92	75	90	74	79	89	77	87	76	86	76	136	83	75	132	82	74	126	80	17.1
Recife	92	78	91	78	90	78	80	89	79	88	79	88	77	143	83	77	142	83	77	140	83	11.3
Rio De Janeiro, Galeao	102	79	99	77	97	77	82	95	81	94	80	91	79	150	86	77	142	84	77	141	84	19.3
Salvador	90	78	88	78	88	78	80	87	80	86	79	85	79	150	85	77	142	84	77	141	84	10.8
Santarem	93	77	92	77	91	78	80	89	80	88	79	88	78	146	83	77	144	82	77	143	82	14.0
Sao Paulo	89	69	88	69	86	69	73	81	72	81	71	80	71	127	77	70	122	75	69	120	75	14.9
Vitoria	93	77	92	77	90	77	80	88	79	87	78	86	78	148	83	77	141	81	76	137	82	14.6
BULGARIA																						
Botevgrad	60	49	57	48	55	47	52	57	50	54	49	53	50	72	53	49	68	52	47	64	50	7.7
Burgas	87	72	84	71	82	69	74	82	73	81	71	79	72	118	79	70	111	77	68	104	76	20.0
Lom	90	74	87	72	85	71	75	88	73	85	72	83	71	114	83	69	108	81	68	102	79	19.1
Musala	56	44	53	43	51	42	47	53	45	51	44	49	44	62	48	43	58	47	41	55	46	9.2
Plovdiv	93	70	90	70	87	69	73	89	72	87	70	84	68	105	83	66	99	80	65	94	78	21.4
Ruse	94	73	91	71	88	70	74	90	72	88	71	86	69	107	85	67	100	80	66	95	78	20.7
Sofia	88	66	85	65	82	64	68	83	67	82	65	80	63	92	74	61	87	72	60	83	72	21.8
Varna	85	72	83	71	81	70	74	82	73	80	71	79	72	120	80	70	112	78	69	106	77	17.3
BRUNEI																						
Brunei Intl Airport	92	79	92	79	91	79	82	89	82	88	81	88	80	158	86	80	155	85	79	152	85	14.0
CANARY ISLANDS																						
Las Palmas	86	68	84	68	81	69	76	80	75	78	73	78	75	132	79	73	125	77	72	118	76	9.9
Santa Cruz De Tenerife	91	68	87	68	84	69	75	82	74	81	73	80	72	119	80	71	117	79	70	111	78	12.4
CAPE VERDE																						
Sal Island	86	75	85	75	84	75	78	82	77	82	77	81	77	141	81	76	136	80	75	133	80	8.8
CHILE																						
Antofagasta	77	68	75	67	74	65	69	75	68	74	67	73	67	100	73	66	97	73	65	92	71	10.4
Arica	83	71	81	69	79	68	73	79	72	78	70	78	71	114	78	69	107	78	67	100	76	11.9
Concepcion	76	63	74	62	72	61	65	72	64	71	63	70	63	85	67	61	80	66	60	78	65	20.2
Iquique	80	68	79	67	78	67	69	78	68	78	67	77	66	95	75	65	92	75	64	89	74	11.7
La Serena	72	63	71	63	70	62	64	71	64	70	63	69	62	84	67	61	81	67	60	78	66	11.9
Puerto Montt	73	62	70	61	68	60	64	71	62	68	61	67	61	81	66	60	77	65	58	72	64	17.6
Punta Arenas	64	54	61	53	59	51	56	62	54	60	53	58	52	59	57	51	55	56	49	52	55	13.0
Santiago	89	65	88	65	86	64	68	85	67	84	66	82	61	85	76	60	81	75	59	78	74	31.5
Temuco	81	64	78	63	75	61	66	77	64	76	62	73	61	82	71	59	76	70	58	72	67	24.7
CHINA																						
Anda	87	69	85	68	82	67	75	82	73	80	71	78	72	122	79	71	115	77	69	108	75	14.9
Andirlangar	98	63	96	62	93	61	66	89	65	88	63	86	60	90	71	58	83	71	55	76	72	26.6

MDB = mean coincident dry-bulb temperature, °F
MWB = mean coincident wet-bulb temperature, °F
MWS = mean coincident wind speed, mph
MWD = mean coincident wind direction
StdD = standard deviation
A = airport
HR = humidity ratio
DP = dew-point temperature, °F

Table 3A Heating and Wind Design Conditions—World Locations

Station	WMO#	Lat.	Long.	Elev. ft	StdP psia	Dates	Heating DB 99.6%	Heating DB 99%	Extreme Wind Speed 1%	Extreme Wind Speed 2.5%	Extreme Wind Speed 5%	Coldest Month 0.4% WS	Coldest Month 0.4% MDB	Coldest Month 1% WS	Coldest Month 1% MDB	MWS/MWD to DB 99.6% MWS	MWS/MWD to DB 99.6% PWD	MWS/MWD to DB 0.4% MWS	MWS/MWD to DB 0.4% PWD	Annual Extreme Daily Mean DB Max	Mean DB Min	StdD DB Max	StdD DB Min
Anyang	538980	36.12 N	114.37 E	249	14.568	8293	18	21	16	14	12	14	36	12	35	1	290	7	180	100	13	2.2	3.6
Baoding	546020	38.85 N	115.57 E	62	14.667	8293	13	16	15	12	11	14	32	11	28	2	220	7	200	100	9	3.4	4.3
Bayan Mod	524950	40.75 N	104.50 E	4360	12.527	8293	−5	−1	28	23	20	26	20	21	22	6	270	8	180	96	−10	2.0	4.1
Beijing	545110	39.93 N	116.28 E	180	14.604	8293	13	15	21	17	14	21	30	18	28	4	340	7	200	101	8	3.4	3.1
Bengbu	582210	32.95 N	117.37 E	72	14.662	8293	22	24	16	14	12	15	38	13	37	3	20	8	180	99	16	2.3	4.1
Changchun	541610	43.90 N	125.22 E	780	14.290	8293	−12	−9	29	23	19	26	18	21	19	6	270	9	250	90	−17	2.5	4.9
Changsha (576790)	576870	28.23 N	112.87 E	223	14.582	8293	29	31	17	15	12	17	42	14	40	3	320	7	200	101	25	3.8	6.1
Chengdu	562940	30.67 N	104.02 E	1666	13.836	8293	32	34	11	9	8	9	44	7	44	0	20	4	200	93	27	2.7	2.5
Dalian	546620	38.90 N	121.63 E	318	14.532	8293	10	13	28	24	22	29	14	26	19	15	360	9	180	91	6	4.0	4.1
Dandong	544970	40.05 N	124.33 E	45	14.676	8293	1	5	21	17	15	23	16	19	18	5	340	6	220	89	−3	2.2	5.2
Datong	534870	40.10 N	113.33 E	3507	12.930	8293	−6	−3	23	19	17	20	15	18	17	5	360	9	160	92	−11	1.3	3.4
Deqen	564440	28.50 N	98.90 E	11443	9.556	8293	18	20	26	22	19	32	29	26	31	2	320	12	180	75	14	5.4	4.7
Dinghai	584770	30.03 N	122.12 E	121	14.636	8293	31	33	21	17	15	20	39	17	41	7	340	6	140	95	27	1.6	2.5
Erenhot	530680	43.65 N	112.00 E	3169	13.093	8293	−20	−16	29	25	21	25	8	21	9	5	70	12	140	96	−26	2.0	4.9
Fuzhou	588470	26.08 N	119.28 E	278	14.552	8293	40	42	18	16	14	17	54	14	52	7	320	10	140	100	35	1.6	3.1
Golmud	528180	36.42 N	94.90 E	9215	10.421	8293	0	4	20	17	15	17	20	14	19	3	250	6	70	86	−6	2.2	4.3
Guangzhou	592870	23.13 N	113.32 E	26	14.686	8293	42	44	15	13	11	15	53	13	53	6	360	5	270	98	37	1.4	2.7
Guilin	579570	25.33 N	110.30 E	544	14.413	8293	34	36	21	18	15	22	43	19	43	11	20	6	20	97	30	1.6	4.5
Guiyang	578160	26.58 N	106.72 E	3523	12.922	8293	28	30	15	12	11	13	48	11	45	4	40	8	160	91	26	3.4	1.6
Hami	522030	42.82 N	93.52 E	2424	13.457	8293	0	3	16	13	10	9	20	8	17	3	40	2	250	102	−4	3.1	4.1
Hangzhou	584570	30.23 N	120.17 E	141	14.625	8293	28	30	17	14	12	17	40	14	40	4	340	8	160	100	22	1.8	3.8
Harbin	509530	45.75 N	126.77 E	469	14.452	8293	−21	−16	23	19	16	18	13	16	8	2	200	11	180	91	−27	2.3	4.1
Hefei	583210	31.87 N	117.23 E	118	14.637	8293	24	27	19	16	14	17	39	15	39	4	340	8	180	98	19	2.3	5.4
Hohhot	534630	40.82 N	111.68 E	3494	12.936	8293	−4	−1	20	17	14	17	11	15	11	2	360	8	180	92	−8	2.5	3.1
Jinan	548230	36.68 N	116.98 E	190	14.599	8293	18	20	20	17	15	18	33	16	36	5	70	10	200	98	13	2.0	4.1
Jingdezhen	585270	29.30 N	117.20 E	196	14.596	8293	28	31	14	12	10	14	43	12	42	4	40	7	250	101	22	3.6	11.0
Jinzhou	543370	41.13 N	121.12 E	229	14.578	8293	2	5	25	21	18	23	18	20	20	6	360	9	200	94	−1	3.1	4.5
Jixi	509780	45.28 N	130.95 E	767	14.297	8293	−14	−10	23	20	17	23	3	21	5	6	250	8	270	91	−17	2.9	4.1
Kashi	517090	39.47 N	75.98 E	4235	12.585	8293	8	12	15	11	9	8	29	6	27	1	320	4	180	96	5	3.2	6.3
Korla	516560	41.75 N	86.13 E	3061	13.145	8293	6	10	21	17	14	14	24	12	24	1	70	4	40	100	2	2.5	7.6
Kowloon	450070	22.33 N	114.18 E	78	14.658	8293	48	51	22	20	18	20	62	19	63	8	330	10	250	96	45	2.2	2.5
Kunming	567780	25.02 N	102.68 E	6207	11.688	8293	32	34	19	16	14	19	60	17	59	2	140	11	250	86	27	5.9	4.3
Lanzhou	528890	36.05 N	103.88 E	4980	12.240	8293	10	13	10	8	7	6	28	5	26	0	90	5	70	94	6	2.0	3.4
Lhasa	555910	29.67 N	91.13 E	11975	9.358	8293	13	16	16	13	11	16	39	13	40	4	90	6	290	85	2	7.7	19.4
Liuzhou	590460	24.35 N	109.40 E	318	14.532	8293	38	40	14	12	10	12	50	11	48	6	340	6	200	99	35	1.8	1.8
Longzhou	594170	22.37 N	106.75 E	423	14.477	8293	43	45	10	9	7	10	61	8	59	1	90	5	200	100	38	2.0	3.4
Macau	450110	22.20 N	113.53 E	193	14.597	8293	44	47	19	17	15	19	50	17	51	13	360	7	200	94	42	1.6	2.7
Mudanjiang	540940	44.57 N	129.60 E	242	14.571	8293	−15	−12	22	18	15	20	5	17	7	3	180	7	200	91	−20	2.5	4.3
Nanchang	586060	28.60 N	115.92 E	164	14.613	8293	30	32	16	14	12	17	42	14	41	6	20	6	220	100	26	2.2	4.0
Nanjing	582380	32.00 N	118.80 E	39	14.679	8293	23	26	18	15	13	17	38	15	37	2	340	9	220	98	18	2.5	3.6
Nanning	594310	22.82 N	108.35 E	239	14.573	8293	42	44	12	10	8	10	62	9	61	2	90	4	160	100	38	2.2	2.2
Nenjiang	505570	49.17 N	125.23 E	797	14.281	8293	−32	−27	27	23	19	20	−1	16	−1	2	20	13	160	91	−36	4.3	5.2
Qingdao	548570	36.07 N	120.33 E	252	14.566	8293	19	21	31	27	23	34	22	29	25	17	340	11	160	90	15	2.2	3.1
Qiqihar	507450	47.38 N	123.92 E	485	14.444	8293	−18	−14	23	19	17	18	4	15	4	4	290	10	180	92	−22	3.8	4.9
Shanghai	583670	31.17 N	121.43 E	22	14.688	8293	26	29	19	17	15	19	38	17	39	6	290	8	200	98	21	2.0	2.7
Shantou	593160	23.40 N	116.68 E	9	14.695	8293	44	47	18	15	13	16	57	14	57	7	20	8	200	95	40	1.8	3.2
Shaoguan	590820	24.80 N	113.58 E	223	14.582	8293	36	39	14	12	10	13	52	11	52	2	360	5	180	101	33	2.2	2.3
Shenyang	543420	41.77 N	123.43 E	141	14.625	8293	−6	−2	22	18	15	19	22	16	20	3	70	9	200	92	−12	2.3	5.0
Shijiazhuang	536980	38.03 N	114.42 E	265	14.559	8293	14	17	16	13	11	16	29	11	34	2	160	6	160	101	10	2.5	4.3
Taiyuan	537720	37.78 N	112.55 E	2555	13.392	8293	4	8	21	17	14	20	27	16	29	2	90	6	180	94	−4	2.0	4.3
Tangshan	545340	39.67 N	118.15 E	95	14.650	8293	7	10	20	17	14	20	27	17	28	2	290	7	180	94	2	1.8	5.0
Tianjin	545270	39.10 N	117.17 E	16	14.691	8293	14	17	18	14	12	19	25	16	25	4	340	6	250	98	8	2.3	4.1
Urumqi	514630	43.78 N	87.62 E	3015	13.167	8293	−10	−5	18	13	11	10	19	8	15	2	180	8	320	98	−11	3.1	6.3
Weifang	548430	36.70 N	119.08 E	167	14.611	8293	12	15	24	21	18	23	30	20	29	6	290	10	200	99	8	2.7	2.5
Wenzhou	586590	28.02 N	120.67 E	22	14.688	8293	33	36	15	12	11	14	46	12	45	4	290	7	90	97	29	2.3	1.8
Wuhan	574940	30.62 N	114.13 E	75	14.660	8293	27	29	15	12	11	13	40	12	39	2	20	8	220	98	21	2.5	5.8
Xi'an	570360	34.30 N	108.93 E	1305	14.019	8293	20	23	14	12	10	12	36	10	35	1	40	5	40	99	14	2.2	4.5
Xiamen	591340	24.48 N	118.08 E	456	14.459	8293	43	45	21	18	16	19	52	18	52	10	70	8	160	95	40	1.3	2.5
Xining	528660	36.62 N	101.77 E	7421	11.163	8293	4	7	15	13	11	14	25	12	25	1	290	6	140	85	−1	2.0	3.1
Xuzhou	580270	34.28 N	117.15 E	137	14.627	8293	19	22	16	14	12	15	36	13	36	2	270	8	160	97	14	2.0	4.5
Yaxian	599480	18.23 N	109.52 E	22	14.688	8293	59	62	17	14	12	18	70	15	71	6	20	8	180	96	56	3.8	3.2
Yichang	574610	30.70 N	111.30 E	439	14.468	8293	30	32	11	9	8	10	46	8	45	2	110	7	140	100	28	2.2	2.3
Yichun	507740	47.72 N	128.90 E	761	14.300	8293	−28	−24	19	16	13	16	1	13	1	1	220	7	220	90	−33	3.1	4.0
Yinchuan	536140	38.48 N	106.22 E	3648	12.863	8293	−1	4	19	15	12	19	27	11	26	2	20	6	20	92	−5	2.0	5.0
Yingkou	544710	40.67 N	122.20 E	13	14.693	8293	1	4	26	22	19	24	26	20	23	6	20	10	200	90	−4	1.6	6.1
Yining	514310	43.95 N	81.33 E	2175	13.581	8293	−10	−3	16	13	10	12	26	9	24	1	70	3	270	97	−13	2.9	6.7
Yueyang	575840	29.38 N	113.08 E	170	14.610	8293	30	32	17	14	13	16	39	14	39	7	20	8	250	98	27	2.2	3.4

WMO# = World Meteorological Organization number
Lat = latitude
Long. = longitude
Elev = elevation, ft
StdP = standard pressure at station elevation, psia
DB = dry-bulb temperature, °F
WS = wind speed, mph

Table 3B Cooling and Dehumidification Design Conditions—World Locations

	Cooling DB/MWB						WB/MDB						DP/MDB and HR									
	0.4%		1%		2%		0.4%		1%		2%		0.4%			1%			2%			Range
Station	DB	MWB	DB	MWB	DB	MWB	WB	MDB	WB	MDB	WB	MDB	DP	HR	MDB	DP	HR	MDB	DP	HR	MCDB	of DB
Anyang	94	73	92	74	90	74	81	88	79	86	78	84	79	151	86	78	144	84	76	138	83	14.4
Baoding	94	72	91	72	89	73	80	87	78	85	77	84	78	146	84	76	139	83	75	132	82	14.8
Bayan Mod	91	59	89	58	86	57	64	80	62	79	60	78	59	88	69	56	80	68	54	74	69	20.7
Beijing	94	71	91	71	89	71	79	87	78	84	76	83	77	142	83	76	134	82	74	130	81	15.7
Bengbu	95	80	93	79	90	77	82	92	82	90	80	88	80	156	87	79	152	87	78	147	86	11.9
Changchun	86	70	84	70	82	69	76	83	74	80	72	78	74	130	81	72	123	78	71	116	76	13.5
Changsha (576790)	96	80	94	80	92	79	82	91	81	90	81	90	80	158	87	79	153	86	78	148	85	12.6
Chengdu	89	76	87	76	85	75	80	86	78	84	77	83	78	155	84	77	148	82	76	144	81	12.2
Dalian	86	74	84	72	82	72	77	82	76	80	75	78	76	135	80	75	132	78	73	126	78	9.9
Dandong	85	75	82	73	80	72	77	82	76	80	75	78	76	135	80	75	131	78	74	126	77	12.2
Datong	87	63	85	62	82	62	69	79	68	77	66	76	67	111	73	65	104	72	63	98	72	19.3
Deqen	67	52	65	52	63	52	55	62	54	61	54	60	53	93	58	52	90	57	51	87	56	12.1
Dinghai	90	81	88	80	86	79	82	89	81	87	80	85	80	156	87	79	151	85	78	148	84	9.7
Erenhot	91	61	88	60	85	59	66	79	64	78	63	77	62	94	70	60	88	69	58	81	69	21.1
Fuzhou	96	81	94	80	92	80	82	93	81	92	80	90	79	154	89	79	150	88	78	148	86	13.9
Golmud	80	52	78	51	75	50	55	75	53	72	51	71	47	69	60	45	62	60	43	57	58	18.9
Guangzhou	94	80	93	79	91	79	82	89	81	89	81	88	80	156	85	80	153	84	79	150	84	12.4
Guilin	94	78	92	78	91	78	81	88	80	87	80	87	79	155	84	78	150	83	78	147	83	13.3
Guiyang	87	70	85	70	83	70	73	82	72	81	72	80	71	130	77	70	126	77	69	123	76	12.6
Hami	97	66	94	65	92	64	70	90	68	89	66	87	63	95	80	61	87	77	58	80	75	23.9
Hangzhou	96	81	94	80	92	80	83	92	82	91	81	89	81	160	89	79	152	86	79	149	85	13.5
Harbin	87	69	84	68	82	68	75	82	74	79	72	78	73	127	80	72	119	78	70	111	76	15.3
Hefei	95	81	92	80	90	79	82	92	82	90	81	89	80	157	88	79	152	87	78	148	86	11.0
Hohhot	88	64	85	63	82	62	70	80	68	77	66	76	67	113	75	65	106	73	63	99	72	18.9
Jinan	95	73	92	74	90	73	80	89	79	87	78	85	78	145	85	76	139	84	75	134	83	13.1
Jingdezhen	97	80	94	79	93	79	82	92	81	91	80	89	79	153	85	78	149	85	78	145	85	14.9
Jinzhou	88	72	85	71	83	70	77	83	76	81	75	79	76	136	81	74	130	79	73	124	78	13.5
Jixi	87	70	83	69	81	68	74	83	72	79	71	77	72	121	79	70	113	77	68	106	76	15.1
Kashi	92	66	90	65	88	64	69	85	68	84	66	83	65	108	78	62	99	76	60	92	75	22.1
Korla	95	67	93	65	90	65	70	90	68	87	66	86	63	97	81	61	89	78	59	83	76	19.1
Kowloon	92	79	91	79	90	79	82	88	81	87	81	87	81	160	85	79	152	85	79	150	85	8.1
Kunming	80	63	78	63	77	63	68	76	67	75	67	74	66	120	72	65	118	71	65	116	71	15.3
Lanzhou	89	64	86	63	84	62	67	81	66	80	65	78	64	106	74	62	99	73	60	92	72	20.0
Lhasa	77	51	74	51	72	51	56	69	55	68	54	67	51	89	59	50	85	58	50	83	58	21.1
Liuzhou	95	78	94	78	92	78	81	90	80	89	80	89	79	151	85	78	148	85	78	145	84	12.6
Longzhou	96	80	95	80	93	79	82	92	82	91	81	90	80	158	86	79	155	86	79	152	85	13.7
Macau	92	81	90	81	89	80	83	90	82	88	81	88	81	161	87	80	157	86	80	156	86	7.6
Mudanjiang	87	71	84	69	81	68	74	84	72	80	71	78	72	120	80	70	113	78	68	106	76	16.0
Nanchang	96	80	94	80	92	80	83	91	82	90	81	89	81	160	87	80	155	87	79	151	86	12.4
Nanjing	94	81	92	80	89	79	83	90	82	89	81	88	81	160	87	80	154	87	79	150	86	11.2
Nanning	95	79	93	79	92	79	82	90	81	89	81	88	80	158	86	80	155	85	79	152	85	11.9
Nenjiang	86	67	82	66	80	66	72	79	70	77	69	75	70	113	76	68	106	74	66	100	73	17.1
Qingdao	85	74	83	74	81	74	78	82	78	80	77	80	78	145	81	77	140	80	76	136	79	8.1
Qiqihar	87	69	84	69	82	68	75	82	73	80	71	78	72	123	80	70	113	77	69	106	75	14.2
Shanghai	94	81	92	81	89	80	83	91	82	89	81	88	81	161	88	80	157	87	79	150	85	11.5
Shantou	91	80	90	80	89	80	82	89	81	87	81	87	80	156	85	79	153	85	79	150	84	9.4
Shaoguan	96	78	95	78	93	78	81	91	80	90	80	89	78	149	85	78	146	84	77	144	84	13.7
Shenyang	88	74	86	73	84	72	78	85	76	83	75	81	75	133	82	74	128	81	73	122	79	14.4
Shijiazhuang	95	72	92	72	90	73	80	87	79	85	77	84	78	149	85	77	141	83	75	134	82	15.5
Taiyuan	89	69	87	68	85	68	75	83	73	81	72	80	72	132	79	71	125	78	69	117	77	19.6
Tangshan	90	74	88	73	86	72	79	85	78	83	76	82	77	142	83	76	136	82	74	130	80	13.9
Tianjin	92	74	90	73	88	73	80	86	78	85	77	84	78	145	84	77	139	83	75	133	82	13.0
Urumqi	91	61	89	60	86	60	64	83	63	82	62	81	58	79	66	56	74	67	54	70	68	18.4
Weifang	93	73	90	73	88	73	81	88	79	85	78	83	79	150	85	77	142	83	76	136	82	14.9
Wenzhou	93	82	91	81	89	80	83	91	82	89	81	88	81	160	88	80	156	87	79	153	85	11.7
Wuhan	95	81	93	80	91	80	83	91	82	90	81	89	81	161	87	80	156	87	79	152	87	11.0
Xi'an	95	73	92	73	90	72	78	89	77	87	76	85	76	141	84	74	134	83	73	128	81	16.0
Xiamen	91	79	90	79	88	79	81	89	80	87	80	86	79	153	85	78	150	84	78	148	83	10.6
Xining	81	57	78	56	75	55	61	72	59	72	58	70	57	91	65	55	86	63	54	81	62	19.6
Xuzhou	94	77	91	76	89	75	82	90	80	88	79	86	80	155	86	78	148	85	77	143	84	12.2
Yaxian	91	80	90	80	89	80	83	89	82	88	82	87	81	160	87	80	157	86	80	155	86	9.5
Yichang	96	79	93	79	91	78	82	92	81	90	80	88	80	158	88	79	152	87	78	148	85	13.0
Yichun	86	69	83	67	80	67	73	81	71	79	69	76	70	113	78	68	106	75	67	101	73	18.2
Yinchuan	88	67	86	66	84	65	71	82	70	81	68	79	68	118	78	66	111	76	64	104	75	19.1
Yingkou	86	76	84	74	83	73	78	84	76	82	75	81	76	136	82	75	130	80	74	125	80	12.1
Yining	91	67	89	66	86	65	69	87	68	85	66	83	63	94	76	62	89	74	60	85	73	23.0
Yueyang	94	81	92	80	91	80	83	91	82	90	81	89	81	161	89	80	155	88	79	150	86	9.0

MDB = mean coincident dry-bulb temperature, °F
MWB = mean coincident wet-bulb temperature, °F
MWS = mean coincident wind speed, mph
MWD = mean coincident wind direction
StdD = standard deviation
A = airport
HR = humidity ratio
DP = dew-point temperature, °F

Table 3A Heating and Wind Design Conditions—World Locations

Station	WMO#	Lat.	Long.	Elev. ft	StdP psia	Dates	Heating DB 99.6%	Heating DB 99%	Extreme Wind Speed 1%	Extreme Wind Speed 2.5%	Extreme Wind Speed 5%	Coldest Month WS/MDB 0.4% WS	Coldest Month WS/MDB 0.4% MDB	Coldest Month WS/MDB 1% WS	Coldest Month WS/MDB 1% MDB	MWS/MWD to DB 99.6% MWS	MWS/MWD to DB 99.6% PWD	MWS/MWD to DB 0.4% MWS	MWS/MWD to DB 0.4% PWD	Annual Extreme Daily Mean DB Max	Annual Extreme Daily Mean DB Min	Annual Extreme Daily StdD DB Max	Annual Extreme Daily StdD DB Min
Zhangjiakou	544010	40.78 N	114.88 E	2381	13.478	8293	1	4	17	15	13	16	15	14	16	7	340	6	140	95	−3	1.1	3.6
Zhanjiang	596580	21.22 N	110.40 E	91	14.651	8293	46	48	16	13	12	14	54	12	56	8	340	8	250	97	42	3.1	4.0
Zhengzhou	570830	34.72 N	113.65 E	364	14.508	8293	19	21	21	17	14	24	40	19	38	3	180	8	160	100	13	2.5	4.1
COLOMBIA																							
Bogota	802220	4.70 N	74.13 W	8359	10.770	8293	36	39	21	18	14	23	64	19	63	0	320	10	90	82	30	8.5	2.7
COOK ISLANDS																							
Rarotonga Island	918430	21.20 S	159.82 W	22	14.688	8293	62	64	25	22	20	26	71	23	72	1	150	11	80	89	58	3.1	2.5
CROATIA																							
Pula	132090	44.90 N	13.92 E	206	14.591	8293	25	27	26	21	17	26	35	22	38	7	20	6	270	92	21	2.0	3.6
Split	133330	43.53 N	16.30 E	68	14.663	8293	29	32	24	19	16	23	41	19	44	9	340	8	230	94	19	7.0	16.7
Zagreb	131310	45.73 N	16.07 E	351	14.514	8293	8	14	19	16	13	17	39	14	39	2	240	6	230	92	2	5.8	8.3
CUBA																							
Guantanamo	783670	19.90 N	75.15 W	55	14.670	8293	67	68	22	20	18	21	85	19	84	8	360	12	130	100	61	4.7	8.3
CYPRUS																							
Akrotiri	176010	34.58 N	32.98 E	75	14.660	8293	40	43	25	22	20	29	53	26	54	5	350	10	260	95	36	3.1	4.3
Larnaca	176090	34.88 N	33.63 E	6	14.697	8293	37	40	27	23	20	28	54	24	54	7	310	12	200	98	33	2.2	3.4
Paphos	176000	34.72 N	32.48 E	26	14.686	8293	39	42	24	21	18	30	55	25	55	9	30	9	280	92	36	3.2	3.4
CZECH REPUBLIC																							
Brno	117230	49.15 N	16.70 E	807	14.276	8293	6	12	24	21	18	26	30	21	31	8	60	10	180	91	4	2.9	7.2
Cheb	114060	50.08 N	12.40 E	1545	13.897	8293	4	10	16	14	12	17	37	14	36	2	40	5	220	90	1	3.8	6.3
Ostrava	117820	49.68 N	18.12 E	839	14.259	8293	1	9	23	20	19	26	32	23	33	5	20	10	190	90	−3	3.1	9.9
Plzen	114480	49.65 N	13.27 E	1194	14.077	8293	2	9	21	19	17	24	41	20	38	2	20	8	120	92	−1	4.0	9.2
Praded Mountain	117350	50.07 N	17.23 E	4895	12.279	8293	−2	2	47	41	36	51	20	43	22	19	20	12	180	72	−4	3.1	7.7
Prague	115180	50.10 N	14.28 E	1200	14.073	8293	3	10	28	23	20	31	39	27	36	4	10	8	160	91	0	3.6	8.8
Pribyslav	116590	49.58 N	15.77 E	1758	13.790	8293	3	9	29	25	22	30	34	27	31	5	360	9	130	87	−2	4.9	7.2
DENMARK																							
Alborg	60300	57.10 N	9.87 E	9	14.695	8293	8	15	29	26	23	32	45	28	42	6	220	11	100	82	7	4.0	12.4
Copenhagen	61800	55.63 N	12.67 E	16	14.691	8293	12	18	29	26	24	30	40	27	38	11	360	11	160	82	13	3.2	8.1
Hammerodde	61930	55.30 N	14.78 E	36	14.681	8293	20	22	44	37	34	45	34	41	34	20	70	12	230	80	22	3.4	5.9
Mon Island	61790	54.95 N	12.55 E	49	14.674	8293	18	22	43	35	32	46	37	41	35	14	320	9	70	78	19	4.0	7.7
Odense	61200	55.47 N	10.33 E	55	14.670	8293	14	18	29	26	23	30	42	27	40	8	40	11	120	84	9	4.1	9.5
Skagen	60410	57.77 N	10.65 E	22	14.688	8293	15	20	41	36	32	41	36	36	38	17	40	10	360	76	16	3.4	7.9
Tirstrup	60700	56.30 N	10.62 E	82	14.656	8293	9	16	27	24	21	28	40	24	39	5	20	11	280	82	7	3.4	11.0
ECUADOR																							
Guayaquil	842030	2.15 S	79.88 W	29	14.684	8293	67	68	16	15	13	17	74	16	74	8	210	7	40	95	51	2.3	11.3
Quito	840710	0.15 S	78.48 W	9225	10.417	8293	45	46	17	15	13	15	64	13	64	1	350	9	150	84	40	7.7	3.2
EGYPT																							
Alexandria	623180	31.20 N	29.95 E	22	14.688	8293	44	46	24	21	18	29	56	25	58	5	190	10	340	102	37	3.2	3.8
Cairo	623660	30.13 N	31.40 E	242	14.571	8293	45	46	21	19	16	23	58	19	62	6	210	13	350	108	38	2.9	4.9
Luxor	624050	25.67 N	32.70 E	288	14.547	8293	40	42	16	14	12	15	64	13	64	2	180	6	330	115	34	3.1	1.8
ESTONIA																							
Kopu/Cape Ristna	261150	58.92 N	22.07 E	29	14.684	8293	5	11	30	25	21	29	38	24	37	5	80	6	70	80	7	4.1	12.1
Tallinn	260380	59.35 N	24.80 E	144	14.623	8293	−4	3	21	18	16	22	34	19	32	6	140	8	40	82	−3	4.3	8.6
FAEROE ISLANDS																							
Torshavn	60110	62.02 N	6.77 W	127	14.632	8293	26	28	41	34	31	48	42	43	43	13	320	11	210	65	22	3.4	2.5
FIJI																							
Nadi	916800	17.75 S	177.45 E	59	14.669	8293	61	63	20	17	16	19	78	17	79	4	120	13	350	95	56	3.6	5.9
Nausori	916830	18.05 S	178.57 E	22	14.688	8293	62	64	20	18	16	20	75	18	74	1	320	11	60	91	59	2.0	1.8
FINLAND																							
Helsinki	29740	60.32 N	24.97 E	183	14.603	8293	−11	−3	22	20	18	24	35	22	32	5	340	11	210	83	−12	3.1	9.5
Jyvaskyla	29350	62.40 N	25.68 E	475	14.449	8293	−21	−13	20	17	15	23	28	19	26	2	330	9	180	83	−22	4.3	7.6
Kauhava	29130	63.10 N	23.03 E	144	14.623	8293	−20	−14	21	19	17	23	31	21	31	2	80	9	230	82	−21	2.3	8.1
Kuopio	29170	63.02 N	27.80 E	334	14.523	8293	−21	−14	19	17	15	21	30	19	28	1	140	7	170	82	−21	2.5	7.9
Lahti	29650	60.97 N	25.63 E	275	14.554	8293	−15	−7	14	12	11	15	33	13	31	1	350	6	150	83	−19	2.7	7.2
Pello	28440	66.80 N	24.00 E	275	14.554	8293	−25	−20	14	13	11	14	25	12	24	1	300	7	340	82	−30	4.9	5.4
Pori	29520	61.47 N	21.80 E	55	14.670	8293	−12	−4	25	22	19	30	37	25	35	5	90	11	140	82	−13	2.5	7.9
Suomussalmi	28790	64.90 N	29.02 E	734	14.314	8293	−21	−17	17	14	13	18	30	15	27	1	360	7	270	83	−26	10.1	6.7
Tampere	29440	61.42 N	23.58 E	367	14.506	8293	−15	−8	19	17	15	21	34	19	32	2	10	9	10	84	−17	2.5	8.6
Turku	29720	60.52 N	22.27 E	193	14.597	8293	−10	−3	21	19	17	25	33	21	32	6	40	9	230	83	−11	2.2	9.7
FRANCE																							
Bordeaux	75100	44.83 N	0.70 W	200	14.594	8293	22	27	22	19	16	24	51	20	51	4	40	7	80	97	19	2.7	7.4
Clermont-Ferrand	74600	45.78 N	3.17 E	1082	14.134	8293	16	20	24	20	17	26	49	22	49	3	360	8	20	97	11	2.9	7.0
Dijon	72800	47.27 N	5.08 E	744	14.309	8293	14	20	23	19	17	25	45	22	43	7	20	10	170	92	12	3.4	8.1
Brest	71100	48.45 N	4.42 W	337	14.521	8293	27	30	26	23	21	28	48	26	46	8	120	9	40	85	24	4.0	4.3
Lyon	74810	45.73 N	5.08 E	787	14.287	8293	17	23	26	22	18	26	44	22	45	3	20	12	180	95	15	2.7	7.4
Marseille	76500	43.45 N	5.23 E	118	14.637	8293	25	28	38	32	28	38	44	32	44	9	360	13	280	95	22	3.2	5.6
Montpellier	76430	43.58 N	3.97 E	19	14.690	8293	25	29	28	25	22	27	49	24	50	7	340	13	180	96	21	2.5	5.4

WMO# = World Meteorological Organization number
Lat = latitude Long. = longitude
Elev = elevation, ft
StdP = standard pressure at station elevation, psia
DB = dry-bulb temperature, °F
WS = wind speed, mph

Table 3B Cooling and Dehumidification Design Conditions—World Locations

	Cooling DB/MWB						WB/MDB						DP/MDB and HR									
	0.4%		1%		2%		0.4%		1%		2%		0.4%			1%			2%			Range
Station	DB	MWB	DB	MWB	DB	MWB	WB	MDB	WB	MDB	WB	MDB	DP	HR	MDB	DP	HR	MDB	DP	HR	MCDB	of DB
Zhangjiakou	89	65	87	65	84	64	72	81	71	80	69	79	70	119	78	67	111	76	66	104	75	17.1
Zhanjiang	93	80	92	80	90	80	82	89	82	88	81	87	81	162	86	80	158	85	80	155	85	8.5
Zhengzhou	94	74	92	74	90	74	81	89	80	87	78	85	79	154	86	78	147	84	77	141	83	14.6
COLOMBIA																						
Bogota	70	56	68	56	68	56	60	66	59	65	58	65	57	97	64	57	94	63	56	90	62	20.7
COOK ISLANDS																						
Rarotonga Island	85	78	85	78	84	77	80	84	79	83	79	83	79	148	82	78	146	82	77	142	82	8.3
CROATIA																						
Pula	89	71	86	69	84	68	74	82	72	82	71	81	71	117	79	69	109	77	68	103	75	19.1
Split	91	70	89	69	86	68	72	86	71	85	70	84	68	104	79	66	97	78	65	92	76	18.5
Zagreb	88	70	85	70	83	68	72	85	71	83	69	80	68	106	78	67	99	77	65	94	74	22.1
CUBA																						
Guantanamo	94	78	93	78	92	78	82	91	81	91	80	90	79	150	89	78	146	88	77	141	87	15.3
CYPRUS																						
Akrotiri	91	71	89	72	87	73	78	85	77	84	76	83	76	137	82	75	132	82	74	127	81	13.0
Larnaca	93	71	91	72	89	72	78	86	77	85	76	84	76	134	83	75	131	83	74	125	82	17.8
Paphos	88	76	86	76	85	76	79	86	78	85	77	83	77	141	85	76	137	84	75	132	83	15.5
CZECH REPUBLIC																						
Brno	85	66	82	65	79	64	68	82	67	79	65	76	64	92	74	63	88	71	61	83	70	19.4
Cheb	83	65	80	64	77	62	67	80	65	76	63	74	62	89	71	61	84	68	59	80	67	19.4
Ostrava	85	67	82	65	78	64	69	82	67	79	65	76	64	93	73	63	88	71	61	84	70	20.7
Plzen	84	67	81	65	78	64	69	82	67	78	65	75	64	94	74	63	89	73	61	84	69	20.3
Praded Mountain	65	55	63	54	60	53	57	63	56	60	54	58	56	79	59	54	74	57	52	70	56	9.7
Prague	84	65	80	64	77	63	67	79	66	76	64	74	63	90	72	62	85	69	60	82	69	20.0
Pribyslav	81	65	77	64	74	62	66	78	65	74	63	72	62	90	71	61	85	69	59	80	67	18.5
DENMARK																						
Alborg	77	63	74	61	71	60	65	75	63	71	61	69	61	80	68	59	75	67	58	71	65	15.1
Copenhagen	77	63	74	62	71	60	65	74	63	71	62	69	61	80	68	60	77	67	59	74	66	14.6
Hammerodde	73	64	70	63	68	62	66	70	64	69	63	67	64	90	68	63	85	67	61	80	66	7.0
Mon Island	74	65	71	63	69	62	66	71	65	70	63	67	64	90	69	62	84	68	61	80	66	10.1
Odense	78	64	75	63	72	61	66	75	64	73	62	70	62	85	70	61	79	68	59	75	66	17.3
Skagen	72	66	69	64	67	63	67	70	65	69	64	66	66	94	69	64	88	67	62	84	65	9.5
Tirstrup	77	64	75	62	72	61	65	74	64	71	62	69	63	85	68	61	80	66	59	75	65	17.8
ECUADOR																						
Guayaquil	92	76	91	76	90	76	80	88	79	87	79	85	78	148	85	77	143	83	77	141	82	13.3
Quito	72	54	70	54	69	54	58	67	57	66	57	65	55	93	61	54	88	59	54	88	59	18.4
EGYPT																						
Alexandria	90	71	88	73	86	74	77	85	76	84	75	83	74	128	82	73	125	82	72	120	81	11.3
Cairo	100	69	97	69	95	69	75	89	74	87	74	86	72	119	79	71	117	78	70	112	78	23.9
Luxor	110	72	108	71	106	71	75	103	74	102	73	101	66	97	92	64	90	91	62	85	90	30.6
ESTONIA																						
Kopu/Cape Ristna	73	64	70	63	68	61	66	71	64	68	62	67	64	88	69	62	83	67	60	78	65	8.6
Tallinn	77	64	74	62	71	61	66	73	64	71	62	69	63	88	69	61	82	68	59	76	66	14.8
FAEROE ISLANDS																						
Torshavn	58	54	56	53	55	53	55	57	54	56	53	54	55	64	56	54	62	55	53	60	54	5.4
FIJI																						
Nadi	90	77	89	77	88	77	80	87	79	86	79	85	78	146	83	77	143	83	77	140	82	14.2
Nausori	88	78	87	78	86	78	80	86	79	85	79	84	79	148	83	78	146	82	77	142	82	11.0
FINLAND																						
Helsinki	79	64	75	61	73	61	66	74	64	72	62	69	63	85	67	61	80	67	59	75	65	17.6
Jyvaskyla	78	62	75	61	71	60	65	73	63	72	61	69	62	84	69	60	78	65	58	72	64	17.6
Kauhava	77	62	74	61	70	59	65	73	63	70	61	68	62	83	68	60	76	65	58	71	64	18.0
Kuopio	78	62	75	61	71	60	65	74	63	71	62	69	62	84	69	60	79	66	58	74	65	12.8
Lahti	79	64	76	63	73	61	66	76	64	74	62	71	63	86	70	60	79	68	59	75	66	19.1
Pello	76	61	72	60	69	58	63	72	61	69	59	66	60	78	66	58	72	64	56	67	62	15.8
Pori	77	62	74	61	71	59	65	73	63	71	61	68	62	83	67	60	76	65	58	72	64	16.2
Suomussalmi	76	61	72	60	69	58	63	73	61	70	59	66	60	78	66	58	73	64	56	69	62	15.1
Tampere	79	62	76	61	73	60	65	75	63	72	61	70	62	83	67	59	76	66	58	71	64	18.7
Turku	79	63	75	61	72	60	65	74	63	72	61	69	62	84	69	60	77	66	58	73	65	16.4
FRANCE																						
Bordeaux	90	70	86	69	83	68	73	85	71	82	70	79	70	111	79	68	104	75	66	98	74	20.3
Clermont-Ferrand	89	70	86	68	82	67	72	85	70	83	68	80	67	104	77	65	98	76	64	93	74	22.7
Dijon	87	68	84	68	81	67	71	82	70	80	68	78	68	106	77	66	99	74	64	93	72	20.9
Brest	78	65	74	65	71	63	67	76	65	72	64	68	65	93	69	64	89	67	62	85	66	13.7
Lyon	90	69	86	69	83	67	72	84	70	82	68	80	68	105	78	66	99	75	64	93	73	21.4
Marseille	90	71	87	70	85	69	75	84	73	82	71	81	72	118	79	70	111	78	68	104	77	18.7
Montpellier	90	71	87	70	84	69	76	82	75	81	73	80	75	131	79	73	123	78	71	116	76	18.5

MDB = mean coincident dry-bulb temperature, °F
MWB = mean coincident wet-bulb temperature, °F
MWS = mean coincident wind speed, mph
MWD = mean coincident wind direction
StdD = standard deviation
A = airport
HR = humidity ratio
DP = dew-point temperature, °F

Table 3A Heating and Wind Design Conditions—World Locations

Station	WMO#	Lat.	Long.	Elev. ft	StdP psia	Dates	Heating DB 99.6%	Heating DB 99%	Extreme Wind Speed 1%	Extreme Wind Speed 2.5%	Extreme Wind Speed 5%	Coldest Month WS/MDB 0.4% WS	Coldest Month WS/MDB 0.4% MDB	Coldest Month WS/MDB 1% WS	Coldest Month WS/MDB 1% MDB	MWS/MWD to DB 99.6% MWS	MWS/MWD to DB 99.6% PWD	MWS/MWD to DB 0.4% MWS	MWS/MWD to DB 0.4% PWD	Annual Extreme Daily Mean DB Max	Annual Extreme Daily Mean DB Min	Annual Extreme Daily StdD DB Max	Annual Extreme Daily StdD DB Min
Nancy	71800	48.68 N	6.22 E	711	14.326	8293	14	17	21	18	16	23	44	21	39	7	60	8	220	92	10	4.0	6.7
Nantes	72220	47.17 N	1.60 W	88	14.653	8293	23	27	24	21	18	27	52	24	50	7	60	9	60	93	20	3.2	6.7
Nice	76900	43.65 N	7.20 E	32	14.683	8293	35	37	25	21	17	24	53	19	52	11	340	8	160	90	31	2.7	4.9
Nimes	76450	43.87 N	4.40 E	203	14.592	8293	26	30	23	20	18	23	39	20	43	10	20	9	40	97	24	2.0	6.3
Orleans	72490	47.98 N	1.75 E	410	14.483	8293	17	22	26	23	20	30	50	26	48	8	60	9	80	93	10	3.1	20.3
Paris, Charles De Gaulle	71570	49.02 N	2.53 E	357	14.511	8293	18	23	27	23	20	32	49	27	46	10	60	9	60	92	16	4.0	8.3
Paris, Orly	71490	48.73 N	2.40 E	314	14.533	8293	19	23	25	22	19	29	48	24	47	8	20	8	100	92	17	3.6	7.6
St.-Quentin	70610	49.82 N	3.20 E	331	14.525	8293	17	22	27	23	20	32	46	28	46	11	60	9	120	88	13	4.1	7.7
Strasbourg	71900	48.55 N	7.63 E	505	14.434	8293	12	17	22	19	16	26	48	21	41	6	340	8	20	93	10	2.5	8.1
Toulouse	76300	43.63 N	1.37 E	501	14.435	8293	22	27	22	19	17	22	48	19	48	5	280	7	140	99	19	3.6	8.1
FRENCH POLYNESIA																							
Moruroa Island	919520	21.82 S	138.80 W	9	14.695	8293	67	68	29	26	24	31	72	28	71	16	140	11	60	90	65	3.8	0.9
Papeete, Tahiti	919380	17.55 S	149.62 W	6	14.697	8293	68	69	22	19	16	22	78	19	79	3	120	7	260	91	65	0.7	2.0
GERMANY																							
Aachen	105010	50.78 N	6.10 E	672	14.346	8293	14	19	23	20	18	26	46	23	44	4	50	6	210	90	14	3.4	8.5
Ahlhorn (Ger-AFB)	102180	52.88 N	8.23 E	184	14.603	8293	11	16	25	22	19	30	45	25	42	7	90	9	10	90	9	4.0	9.5
Berlin	103840	52.47 N	13.40 E	160	14.615	8293	11	15	23	20	18	26	44	21	41	8	80	8	150	93	10	3.8	8.8
Bitburg	106100	49.95 N	6.57 E	1227	14.060	8293	12	18	23	20	18	27	42	23	38	11	60	7	10	90	11	3.6	6.8
Bremen	102240	53.05 N	8.80 E	16	14.691	8293	12	16	25	22	20	28	44	24	42	8	70	10	100	90	9	6.1	9.5
Bremerhaven	101290	53.53 N	8.58 E	36	14.681	8293	15	19	31	27	24	34	43	30	43	8	60	10	130	88	16	4.0	7.4
Dresden	104880	51.13 N	13.78 E	741	14.310	8293	8	13	22	19	16	23	42	20	41	4	320	7	260	92	6	2.9	11.2
Dusseldorf	104000	51.28 N	6.78 E	144	14.623	8293	14	20	23	21	18	26	45	23	44	6	60	9	130	92	13	2.7	8.8
Eggebek (Ger-Navy)	100340	54.63 N	9.35 E	72	14.662	8293	11	17	28	25	22	32	40	28	38	7	30	11	90	86	7	3.1	9.5
Ehrenberg	105440	50.50 N	9.95 E	3034	13.158	8293	5	10	34	30	27	37	27	33	25	15	100	11	190	83	5	3.1	8.5
Frankfurt Am Main	106370	50.05 N	8.60 E	370	14.504	8293	12	17	23	20	17	25	45	21	42	7	30	9	40	93	10	3.1	7.7
Grafenwohr	106870	49.70 N	11.95 E	1361	13.991	8293	−2	5	15	12	11	16	39	13	36	1	10	5	10	92	−7	4.1	9.5
Greifswald	101840	54.10 N	13.40 E	19	14.690	8293	9	15	23	20	18	25	41	21	40	4	250	8	50	89	8	4.1	11.0
Hamburg	101470	53.63 N	10.00 E	52	14.672	8293	11	16	23	20	18	24	42	21	40	6	60	11	90	89	10	4.5	9.0
Hannover	103380	52.47 N	9.70 E	177	14.606	8293	9	14	23	20	18	25	43	22	41	6	80	9	110	90	8	4.0	10.1
Heidelberg	107340	49.40 N	8.65 E	357	14.511	8293	14	19	17	14	12	18	44	15	43	4	170	6	10	96	12	2.9	8.6
Hof	106850	50.32 N	11.88 E	1863	13.737	8293	3	9	22	19	17	23	36	21	34	6	140	7	150	87	0	2.9	7.0
Husum (Ger-AFB)	100260	54.52 N	9.15 E	121	14.636	8293	12	17	29	26	23	32	43	29	42	9	50	9	90	86	8	4.9	9.0
Kap Arkona	100910	54.68 N	13.43 E	134	14.629	8293	17	21	43	38	34	46	37	42	37	15	360	12	70	81	17	3.6	8.3
Kiel/Holtenau (Ger-Navy)	100460	54.38 N	10.15 E	102	14.646	8293	14	19	25	22	19	28	41	24	38	9	40	9	160	85	10	4.9	10.6
Koln	105130	50.87 N	7.17 E	324	14.528	8293	11	17	21	18	16	25	45	21	43	4	110	8	130	92	8	3.1	10.6
Lahr	108050	48.37 N	7.83 E	511	14.430	8293	11	17	19	16	14	22	47	20	45	4	20	5	120	94	8	2.7	10.1
Landsberg (Ger-AFB)	108570	48.07 N	10.90 E	2060	13.638	8293	5	10	26	22	19	30	41	26	37	3	70	6	260	90	0	3.2	7.7
Leck (Ger-AFB)	100220	54.80 N	8.95 E	43	14.677	8293	11	17	28	25	22	32	42	28	40	5	80	10	110	85	4	3.8	11.0
Leipzig	104690	51.42 N	12.23 E	436	14.470	8293	8	13	28	24	21	30	42	25	41	6	70	9	190	92	6	3.2	12.2
Memmingen (Ger-AFB)	109470	47.98 N	10.23 E	2113	13.612	8293	5	10	25	21	18	29	38	26	38	5	50	9	220	90	−1	4.3	8.1
Munich	108660	48.13 N	11.70 E	1735	13.801	8293	4	10	27	22	18	29	42	24	40	4	80	8	30	90	−1	3.6	8.1
Neuburg (Ger-AFB)	108530	48.72 N	11.22 E	1270	14.038	8293	3	10	21	18	15	23	40	20	40	4	60	4	200	92	−1	4.1	11.9
Nordholz (Ger-Navy)	101360	53.77 N	8.67 E	102	14.646	8293	12	17	30	26	23	34	42	29	41	9	80	11	120	88	12	4.5	8.3
Ramstein (US-AFB)	106140	49.43 N	7.60 E	780	14.290	8293	11	16	19	17	14	21	44	18	43	2	10	5	240	93	7	3.8	7.9
Sollingen (Can-AFB)	107220	48.77 N	8.10 E	419	14.478	8293	13	17	22	19	17	26	45	23	44	6	30	6	10	94	10	3.2	8.1
Stuttgart	107380	48.68 N	9.22 E	1374	13.984	8293	9	14	21	18	15	23	41	20	40	4	90	7	90	92	4	4.0	10.3
GEORGIA																							
Batumi	374840	41.65 N	41.63 E	19	14.690	8293	29	32	30	27	24	31	49	28	50	13	130	9	300	91	21	9.0	12.2
K'ut'aisi (Kutaisi)	373950	42.27 N	42.63 E	380	14.499	8293	28	30	49	41	36	49	45	40	47	9	90	22	90	97	23	3.2	4.1
Sokhumi (Sukhumi)	372600	42.87 N	41.13 E	42	14.677	8293	29	31	18	15	13	19	44	16	42	6	50	9	320	90	25	3.2	4.0
Tbilisi	375490	41.68 N	44.95 E	1532	13.904	8293	21	24	49	42	37	51	36	45	36	6	320	10	180	97	16	3.1	4.0
GIBRALTAR																							
North Front	84950	36.15 N	5.35 W	16	14.691	8293	46	48	33	28	25	36	57	32	57	9	270	14	200	97	40	4.7	4.3
GREECE																							
Andravida	166820	37.92 N	21.28 E	39	14.679	8293	32	34	22	19	17	27	56	22	55	2	130	12	350	97	27	2.7	2.7
Athens	167160	37.90 N	23.73 E	49	14.674	8293	34	37	23	21	19	25	47	22	47	8	360	14	30	99	31	4.3	3.1
Elefsis (Hel-AFB)	167180	38.07 N	23.55 E	102	14.646	8293	33	36	23	21	19	24	49	22	46	4	360	12	10	104	30	5.2	3.2
Iraklion	167540	35.33 N	25.18 E	127	14.632	8293	41	44	32	29	25	40	53	33	51	14	340	11	320	97	37	3.6	3.6
Larisa	166480	39.63 N	22.42 E	242	14.571	8293	23	26	19	16	13	20	46	17	46	1	360	7	270	105	16	4.5	5.4
Preveza	166430	38.95 N	20.77 E	13	14.693	8293	36	38	26	22	19	31	51	24	50	10	40	10	250	95	33	4.1	2.7
Rodhos	167490	36.40 N	28.08 E	36	14.681	8293	41	44	24	22	20	28	49	23	49	13	360	14	270	96	38	3.6	3.6
Soudha	167460	35.48 N	24.12 E	495	14.439	8293	39	41	26	22	20	30	52	26	49	10	20	10	300	100	36	2.7	3.2
Thessaloniki	166220	40.52 N	22.97 E	13	14.693	8293	25	28	29	24	20	32	42	28	43	8	110	10	180	99	19	4.5	3.2
GREENLAND																							
Dundas, Thule Ab	42020	76.53 N	68.50 W	193	14.597	8293	−35	−33	30	24	19	30	−2	24	0	7	80	9	310	63	−37	7.6	3.2
Godthab	42500	64.17 N	51.75 W	229	14.578	8293	−12	−8	45	36	30	41	19	33	19	13	350	10	20	64	−8	5.2	6.1
Kangerlussuaq	42310	67.00 N	50.80 W	173	14.608	8293	−42	−38	22	19	16	23	22	19	20	9	60	11	70	69	−41	2.5	8.1

WMO# = World Meteorological Organization number
Lat = latitude Long. = longitude
Elev = elevation, ft
StdP = standard pressure at station elevation, psia
DB = dry-bulb temperature, °F
WS = wind speed, mph

Table 3B Cooling and Dehumidification Design Conditions—World Locations

Station	Cooling DB/MWB 0.4% DB	0.4% MWB	1% DB	1% MWB	2% DB	2% MWB	WB/MDB 0.4% WB	0.4% MDB	1% WB	1% MDB	2% WB	2% MDB	DP/MDB and HR 0.4% DP	0.4% HR	0.4% MDB	1% DP	1% HR	1% MDB	2% DP	2% HR	2% MCDB	Range of DB
Nancy	86	69	82	67	79	65	71	82	69	79	67	76	67	100	76	65	94	73	63	90	71	20.9
Nantes	86	68	83	68	79	66	72	82	70	79	68	76	68	104	76	66	97	73	65	92	70	18.2
Nice	84	74	83	73	81	72	78	82	76	80	75	79	76	136	81	75	131	80	73	123	79	12.2
Nimes	92	69	89	69	87	68	73	84	72	83	71	81	71	113	77	69	107	76	67	100	75	19.6
Orleans	86	68	83	66	80	65	71	82	69	79	67	76	67	102	78	65	93	73	63	88	71	21.2
Paris, Charles De Gaulle	86	69	82	68	79	66	71	81	69	79	67	76	68	104	76	66	97	74	64	91	72	18.7
Paris, Orly	86	69	82	67	79	65	71	82	69	79	67	76	67	99	75	65	93	73	63	88	71	18.4
St.-Quentin	82	68	79	67	76	65	70	79	68	77	66	74	67	102	75	65	94	72	63	88	70	18.9
Strasbourg	87	70	84	68	81	67	71	82	70	80	68	78	68	104	76	66	98	75	64	92	72	20.7
Toulouse	91	70	88	69	84	68	73	86	71	82	70	81	70	111	79	68	104	77	66	98	75	21.4
FRENCH POLYNESIA																						
Moruroa Island	87	79	86	78	85	78	80	85	80	84	79	83	79	151	83	78	147	82	78	144	82	6.8
Papeete, Tahiti	89	79	89	79	88	78	81	88	80	87	79	86	78	148	86	78	145	85	77	141	85	11.0
GERMANY																						
Aachen	83	66	80	65	77	64	68	80	67	78	65	75	64	93	73	63	88	71	61	83	70	15.5
Ahlhorn (Ger-AFB)	84	65	80	65	77	64	68	80	66	78	64	74	64	90	71	62	85	70	61	80	69	18.2
Berlin	86	66	82	65	79	64	68	81	67	79	65	75	64	90	72	62	85	70	61	80	69	16.7
Bitburg	84	66	80	65	77	64	68	80	67	78	65	75	64	94	72	63	89	71	61	83	70	18.5
Bremen	82	66	79	65	76	63	68	78	66	76	65	73	65	91	71	63	85	71	61	80	68	18.0
Bremerhaven	81	66	77	64	73	63	68	77	66	74	65	72	65	94	71	64	88	70	62	83	69	11.7
Dresden	85	66	82	65	78	63	68	81	66	78	65	76	64	91	72	62	86	71	61	81	69	17.6
Dusseldorf	85	67	82	65	79	64	69	81	67	79	66	76	65	92	73	63	88	72	62	83	71	17.5
Eggebek (Ger-Navy)	80	65	77	63	73	62	66	77	65	74	63	71	63	85	70	61	80	68	59	75	67	17.3
Ehrenberg	75	62	72	60	69	59	64	72	62	70	60	67	60	87	68	58	81	65	57	77	63	13.5
Frankfurt Am Main	87	67	83	66	80	64	69	82	67	80	66	77	65	93	73	63	88	71	62	83	70	19.8
Grafenwohr	85	66	82	66	79	64	68	81	66	79	65	76	63	90	69	62	88	71	61	83	69	25.0
Greifswald	81	66	77	65	74	63	68	78	66	75	64	72	65	91	72	63	85	70	61	80	68	16.4
Hamburg	82	66	79	64	75	63	68	78	66	76	64	73	64	90	72	62	85	70	61	80	68	16.7
Hannover	84	67	80	65	77	64	69	80	67	78	65	74	65	93	72	63	87	71	61	82	69	18.7
Heidelberg	90	69	86	67	83	66	71	86	69	84	67	80	66	98	77	64	92	75	63	87	71	20.0
Hof	81	64	77	62	74	61	66	77	64	74	62	71	62	89	70	60	83	67	59	79	66	18.5
Husum (Ger-AFB)	79	64	76	64	72	62	67	76	65	73	63	70	64	90	70	62	84	68	61	79	66	15.5
Kap Arkona	74	65	71	64	69	62	66	72	65	70	63	68	64	90	70	62	85	68	61	80	67	9.2
Kiel/Holtenau (Ger-Navy)	78	64	75	63	72	62	66	76	64	73	62	70	62	84	71	61	79	68	59	75	67	15.5
Koln	85	67	82	65	79	64	69	81	67	79	65	76	65	92	73	63	87	71	62	83	69	19.8
Lahr	86	69	84	68	80	66	71	83	69	81	68	78	67	99	77	65	94	74	64	91	73	20.7
Landsberg (Ger-AFB)	83	66	79	64	77	63	67	81	65	77	64	74	62	90	72	61	85	70	59	80	69	20.2
Leck (Ger-AFB)	79	65	76	63	72	62	67	77	64	75	63	71	63	85	71	61	80	68	59	75	67	16.4
Leipzig	85	66	82	65	78	64	68	81	67	78	65	76	64	91	73	63	86	71	61	81	70	18.5
Memmingen (Ger-AFB)	83	66	80	65	77	63	67	81	65	78	64	74	61	88	73	59	82	70	59	80	69	20.0
Munich	84	66	81	64	78	63	67	80	66	78	65	76	63	91	72	62	87	71	60	83	69	20.2
Neuburg (Ger-AFB)	85	66	81	65	79	64	68	82	66	80	65	76	63	90	73	61	85	71	59	79	69	22.7
Nordholz (Ger-Navy)	81	65	77	64	74	62	67	77	65	74	64	71	64	88	70	62	83	69	60	79	67	14.8
Ramstein (US-AFB)	86	68	83	66	80	65	70	83	67	79	66	78	65	94	72	63	88	72	61	83	71	22.3
Sollingen (Can-AFB)	87	69	84	68	81	66	71	83	69	81	68	78	67	99	75	66	97	74	64	92	72	19.8
Stuttgart	84	66	81	65	78	63	68	81	66	78	65	76	63	91	74	62	87	71	60	83	70	19.4
GEORGIA																						
Batumi	82	73	80	72	79	71	75	80	74	79	72	77	73	123	79	72	118	78	71	113	76	10.4
K'ut'aisi (Kutaisi)	90	71	87	70	84	70	76	83	74	81	73	79	73	125	79	72	120	78	71	116	77	14.9
Sokhumi (Sukhumi)	84	73	82	73	80	72	76	81	75	80	74	79	74	127	79	73	122	78	72	118	77	13.1
Tbilisi	92	70	89	70	87	69	73	88	72	86	70	84	68	111	81	67	105	79	65	99	78	18.4
GIBRALTAR																						
North Front	88	69	85	68	82	68	74	79	73	78	72	77	72	120	76	72	117	76	71	113	75	12.6
GREECE																						
Andravida	91	70	89	71	87	71	75	84	74	83	73	82	73	124	80	72	117	79	70	111	78	21.2
Athens	93	69	91	68	89	68	75	85	73	85	72	83	71	116	83	69	108	82	68	102	80	16.9
Elefsis (Hel-AFB)	97	70	95	68	92	68	74	88	73	88	71	86	70	110	83	68	103	82	66	96	80	18.2
Iraklion	88	66	86	67	84	68	74	82	73	81	72	80	71	116	80	70	110	79	68	104	79	10.6
Larisa	97	69	93	68	91	68	71	91	70	88	69	86	66	98	74	65	92	74	64	89	74	25.2
Preveza	88	71	86	71	84	72	76	83	75	82	74	81	74	126	80	73	122	79	72	118	79	14.4
Rodhos	90	71	87	71	86	70	76	82	75	82	74	81	74	126	79	73	122	79	72	117	79	10.1
Soudha	93	66	90	67	88	66	72	82	71	81	70	80	70	111	75	68	105	75	67	99	75	15.3
Thessaloniki	92	70	90	69	88	69	73	87	72	84	71	84	69	108	81	68	102	79	66	97	78	20.9
GREENLAND																						
Dundas, Thule Ab	54	44	51	43	48	41	45	53	43	50	42	48	39	36	47	37	33	45	36	31	44	8.3
Godthab	57	49	54	47	51	46	50	55	48	53	47	50	48	49	50	46	46	49	44	43	47	10.4
Kangerlussuaq	65	51	63	50	60	49	52	64	51	62	49	59	45	45	52	44	43	52	43	41	52	18.2

MDB = mean coincident dry-bulb temperature, °F MWS = mean coincident wind speed, mph StdD = standard deviation HR = humidity ratio
MWB = mean coincident wet-bulb temperature, °F MWD = mean coincident wind direction A = airport DP = dew-point temperature, °F

Table 3A Heating and Wind Design Conditions—World Locations

Station	WMO#	Lat.	Long.	Elev. ft	StdP psia	Dates	Heating DB 99.6%	Heating DB 99%	Extreme Wind Speed 1%	Extreme Wind Speed 2.5%	Extreme Wind Speed 5%	Coldest Month 0.4% WS	Coldest Month 0.4% MDB	Coldest Month 1% WS	Coldest Month 1% MDB	MWS/MWD to DB 99.6% MWS	99.6% PWD	0.4% MWS	0.4% PWD	Annual Extreme Daily Mean DB Max	Mean DB Min	StdD DB Max	StdD DB Min
Narsarsuaq	42700	61.18 N	45.42 W	85	14.655	8293	−18	−12	47	39	31	53	34	46	35	2	60	17	70	68	−15	2.5	10.8
GUAM																							
Andersen AFB (Guam)	912180	13.58 N	144.93 E	606	14.380	8293	74	75	20	18	16	19	79	18	79	9	70	9	90	91	72	2.9	1.1
HUNGARY																							
Budapest	128390	47.43 N	19.27 E	606	14.380	8293	8	14	36	29	24	35	40	27	40	2	170	10	200	94	2	2.2	7.9
Debrecen	128820	47.48 N	21.63 E	367	14.506	8293	6	11	22	18	15	22	35	19	35	4	50	6	90	92	2	2.7	6.7
Nagykanizsa	129250	46.45 N	16.98 E	462	14.456	8293	8	14	19	16	14	19	33	16	35	5	360	6	230	91	1	2.7	8.3
Pecs	129420	46.00 N	18.23 E	666	14.350	8293	12	16	22	18	15	23	36	20	33	6	320	7	270	93	8	2.3	5.8
Siofok	129350	46.92 N	18.03 E	354	14.513	8293	12	17	30	26	21	30	31	25	39	4	320	5	270	91	7	3.1	8.3
Szombathely	128120	47.27 N	16.63 E	725	14.319	8293	10	15	29	24	20	27	25	21	32	8	270	7	180	91	7	3.1	7.0
ICELAND																							
Akureyri	40630	65.68 N	18.08 W	88	14.653	8293	8	11	30	26	22	34	33	30	37	7	160	11	180	72	2	3.1	2.2
Keflavik	40180	63.97 N	22.60 W	177	14.606	8293	17	20	41	34	30	47	33	41	34	14	20	11	350	66	12	6.5	2.0
Raufarhofn	40770	66.45 N	15.95 W	32	14.683	8293	10	13	37	32	28	43	31	38	31	14	230	14	320	68	4	3.8	3.1
Reykjavik	40300	64.13 N	21.90 W	200	14.594	8293	14	17	41	34	31	47	36	42	36	10	90	12	360	65	10	3.4	2.5
INDIA																							
Ahmadabad	426470	23.07 N	72.63 E	180	14.604	8293	52	55	16	13	11	14	74	12	74	2	360	7	270	111	43	3.1	6.8
Bangalore	432950	12.97 N	77.58 E	3021	13.164	8293	59	60	13	12	10	11	71	10	72	3	90	4	90	99	54	2.0	3.4
Bombay	430030	19.12 N	72.85 E	45	14.676	8293	62	64	15	13	12	12	79	11	80	0	360	7	320	101	56	2.3	2.9
Calcutta	428090	22.65 N	88.45 E	19	14.690	8293	54	56	13	11	9	7	73	7	73	0	360	4	180	103	50	2.0	1.4
Cuddalore	433290	11.77 N	79.77 E	39	14.679	8293	68	69	14	12	11	13	79	12	80	2	320	6	250	105	63	2.7	3.1
Goa/Panaji	431920	15.48 N	73.82 E	196	14.596	8293	67	69	17	14	12	12	83	10	83	5	50	6	320	99	62	2.7	5.4
Hyderabad	431280	17.45 N	78.47 E	1788	13.775	8293	58	60	21	19	17	13	76	11	77	1	50	8	320	107	53	2.0	3.6
Jaipur	423480	26.82 N	75.80 E	1279	14.033	8293	44	47	16	13	11	12	64	10	64	0	90	9	320	110	39	2.2	3.1
Madras	432790	13.00 N	80.18 E	52	14.672	8293	68	69	17	14	13	13	80	11	80	2	290	8	270	106	65	2.2	1.8
Nagpur	428670	21.10 N	79.05 E	1017	14.168	8293	53	55	17	14	12	11	75	8	74	2	360	6	320	113	49	2.3	3.2
New Delhi	421820	28.58 N	77.20 E	708	14.327	8293	44	46	17	14	12	14	66	13	66	2	270	7	320	110	41	2.2	2.0
Poona	430630	18.53 N	73.85 E	1833	13.752	8293	50	52	12	10	8	8	79	6	78	0	70	3	270	105	45	1.3	2.7
Sholapur	431170	17.67 N	75.90 E	1571	13.884	8293	61	63	8	7	6	6	74	5	75	1	90	2	320	108	56	3.1	3.1
Trivandrum	433710	8.48 N	76.95 E	209	14.589	8293	72	73	14	13	11	17	83	15	83	2	360	6	320	99	65	3.1	4.7
INDIAN OCEAN ISLANDS																							
Diego Garcia Isl.	619670	7.30 S	72.40 E	9	14.695	8293	73	75	21	19	17	21	80	20	80	11	110	7	90	95	68	4.0	10.8
IRELAND																							
Belmullet	39760	54.23 N	10.00 W	32	14.683	8293	30	32	39	34	30	45	48	40	47	8	90	11	180	76	26	4.1	3.6
Birr	39650	53.08 N	7.88 W	236	14.575	8293	24	28	23	20	18	28	45	25	45	1	90	7	150	77	20	8.1	5.0
Claremorris	39700	53.72 N	8.98 W	226	14.580	8293	26	28	29	26	22	34	46	30	45	6	70	8	90	77	21	4.3	4.0
Clones	39740	54.18 N	7.23 W	291	14.546	8293	25	28	28	24	21	30	45	28	46	4	60	7	120	78	20	4.1	4.7
Cork	39550	51.85 N	8.48 W	531	14.420	8293	29	32	34	30	27	40	44	34	45	13	40	9	330	75	26	3.6	3.4
Dublin	39690	53.43 N	6.25 W	278	14.552	8293	29	31	31	28	24	35	44	30	44	9	250	11	230	77	26	3.2	3.2
Kilkenny	39600	52.67 N	7.27 W	209	14.589	8293	25	28	27	22	19	29	47	26	47	2	360	7	180	79	20	4.0	4.1
Malin	39800	55.37 N	7.33 W	82	14.656	8293	31	33	45	41	36	50	43	45	43	14	170	16	200	72	28	2.2	3.1
Mullingar	39710	53.53 N	7.37 W	341	14.520	8293	25	28	26	22	20	29	43	26	46	3	70	9	100	77	20	3.8	3.4
Rosslare	39570	52.25 N	6.33 W	82	14.656	8293	32	34	32	29	26	36	41	31	43	15	90	11	220	73	30	3.4	2.3
Shannon	39620	52.70 N	8.92 W	65	14.665	8293	28	31	31	27	23	37	45	31	46	6	70	9	110	78	24	4.3	3.1
Valentia Observatory	39530	51.93 N	10.25 W	45	14.676	8293	31	33	33	30	26	38	47	34	48	7	60	10	270	77	27	4.0	2.7
ISRAEL																							
Jerusalem	401840	31.78 N	35.22 E	2473	13.433	8293	33	35	24	21	19	28	41	24	42	6	270	10	290	98	31	5.4	2.3
Lod	401800	32.00 N	34.90 E	160	14.615	8293	40	42	23	19	17	26	56	22	54	4	150	11	320	103	36	2.9	2.9
Ovda (Isr-AFB/Civ)	401980	30.00 N	34.83 E	1417	13.963	8293	36	39	23	19	17	25	49	21	54	5	210	10	40	105	32	1.3	2.3
Tel Aviv-Yafo	401760	32.10 N	34.78 E	13	14.693	8293	44	46	28	22	19	30	56	24	56	7	120	9	310	99	41	5.4	2.5
ITALY																							
Bologna/Borgo (AFB)	161400	44.53 N	11.30 E	138	14.627	8293	22	25	16	13	11	14	41	11	37	1	220	5	80	97	18	3.4	5.9
Brindisi	163200	40.65 N	17.95 E	32	14.683	8293	36	39	26	22	19	30	51	26	51	9	360	10	180	99	31	4.3	2.3
Catania	164600	37.47 N	15.05 E	55	14.670	8293	35	37	23	19	16	26	54	22	55	5	230	10	90	103	31	5.4	2.2
Genova	161200	44.42 N	8.85 E	9	14.695	8293	32	36	27	24	22	28	43	26	44	15	40	7	50	92	29	2.3	4.1
Messina	164200	38.20 N	15.55 E	167	14.611	8293	43	45	19	16	14	20	56	17	57	4	310	6	60	97	38	4.9	4.0
Milan, Linate	160800	45.43 N	9.28 E	337	14.521	8293	21	25	16	12	10	20	46	14	46	1	90	5	220	95	18	5.9	5.0
Milan, Malpensa	160660	45.62 N	8.73 E	692	14.336	8293	15	18	18	12	9	19	43	11	40	0	360	4	210	94	9	3.1	4.7
Naples	162890	40.85 N	14.30 E	236	14.575	8293	32	34	23	18	15	26	49	21	49	4	340	8	200	98	28	3.8	2.9
Palermo	164050	38.18 N	13.10 E	111	14.641	8293	44	46	30	26	22	32	57	28	55	11	210	11	250	100	38	5.2	4.7
Perugia	161810	43.08 N	12.50 E	672	14.346	8293	24	27	21	18	16	22	46	19	45	1	360	5	270	96	17	2.9	7.2
Pisa	161580	43.68 N	10.38 E	3	14.698	8293	26	29	22	19	16	22	49	19	47	4	90	9	270	96	21	2.0	5.0
Rome	162420	41.80 N	12.23 E	9	14.695	8293	30	32	28	24	21	28	48	23	50	9	60	12	270	94	26	5.2	3.1
Ronchi Legionari Ab	161080	45.82 N	13.48 E	39	14.679	8293	21	25	22	18	14	28	38	20	38	2	10	5	220	96	17	2.3	4.1
Torino	160590	45.22 N	7.65 E	941	14.207	8293	20	23	13	9	7	16	48	10	41	0	260	2	70	92	16	2.9	4.0
Venice	161050	45.50 N	12.33 E	19	14.690	8293	23	26	22	17	13	28	36	21	33	4	60	6	160	92	19	4.3	4.0
JAMAICA																							

WMO# = World Meteorological Organization number
Lat = latitude Long. = longitude
Elev = elevation, ft
StdP = standard pressure at station elevation, psia
DB = dry-bulb temperature, °F
WS = wind speed, mph

Table 3B Cooling and Dehumidification Design Conditions—World Locations

	Cooling DB/MWB						WB/MDB						DP/MDB and HR									
	0.4%		1%		2%		0.4%		1%		2%		0.4%			1%			2%			Range
Station	DB	MWB	DB	MWB	DB	MWB	WB	MDB	WB	MDB	WB	MDB	DP	HR	MDB	DP	HR	MDB	DP	HR	MCDB	of DB
Narsarsuaq	65	50	62	49	60	47	52	60	51	58	50	57	49	52	51	48	49	50	46	46	49	12.2
GUAM																						
Andersen AFB (Guam)	88	79	87	79	87	78	81	86	80	85	80	85	80	158	84	79	154	83	78	150	83	7.6
HUNGARY																						
Budapest	90	69	86	68	84	66	71	87	69	85	67	80	65	94	71	64	92	74	63	87	74	22.0
Debrecen	88	71	85	70	82	68	72	86	71	83	69	81	67	102	80	66	97	78	64	92	76	20.3
Nagykanizsa	87	70	84	69	81	68	72	84	70	82	69	80	68	104	78	66	97	75	65	92	73	22.7
Pecs	88	70	86	69	83	68	72	86	70	84	69	81	67	102	80	65	96	79	64	90	76	18.9
Siofok	86	71	83	70	80	69	73	83	71	81	70	79	69	110	81	68	103	79	66	97	76	14.6
Szombathely	86	69	83	68	80	67	71	83	69	80	68	78	67	103	76	65	97	75	64	92	72	19.8
ICELAND																						
Akureyri	66	56	63	54	61	53	57	65	55	62	53	60	54	62	61	51	57	58	49	52	56	9.5
Keflavik	59	52	57	51	55	51	54	56	52	55	51	54	52	58	54	51	57	53	50	54	52	7.6
Raufarhofn	60	53	56	51	54	50	54	58	52	56	50	53	51	55	57	49	52	54	48	50	52	7.4
Reykjavik	60	53	58	52	56	51	54	58	53	56	52	55	53	59	56	51	56	54	50	54	53	8.5
INDIA																						
Ahmadabad	108	74	106	74	103	75	84	94	83	92	82	91	82	166	88	81	161	87	80	157	86	22.9
Bangalore	94	67	92	67	91	67	74	84	73	82	72	81	72	132	77	71	127	76	70	125	76	19.3
Bombay	95	73	93	74	92	75	82	89	81	88	81	88	80	156	86	80	153	86	79	150	85	9.4
Calcutta	99	78	97	79	95	79	85	94	84	92	83	91	83	171	90	82	167	89	82	164	88	18.0
Cuddalore	99	78	98	78	96	78	84	91	83	90	82	90	82	166	88	81	162	88	81	159	87	14.8
Goa/Panaji	93	77	92	77	91	77	83	88	82	88	81	87	81	163	87	80	158	86	79	153	85	10.4
Hyderabad	105	71	103	71	101	71	77	90	76	88	76	87	75	139	81	74	135	80	73	133	80	18.9
Jaipur	108	69	105	69	103	69	80	88	80	87	79	87	79	157	84	78	153	83	77	149	82	22.3
Madras	101	77	99	77	97	77	83	91	82	90	82	89	81	162	87	81	159	86	80	155	86	14.6
Nagpur	110	71	108	71	106	70	80	90	79	88	79	87	78	151	83	77	148	83	77	145	82	22.9
New Delhi	107	72	105	72	103	73	82	92	82	91	81	90	80	162	87	80	158	86	79	155	86	21.6
Poona	100	67	99	67	97	67	76	86	75	84	74	83	74	134	80	73	131	79	72	127	78	29.0
Sholapur	105	71	104	71	102	71	80	93	78	92	77	90	77	148	86	75	141	83	74	136	82	21.1
Trivandrum	92	78	91	78	90	78	81	88	81	88	80	87	79	153	85	79	150	84	78	148	84	11.7
INDIAN OCEAN ISLANDS																						
Diego Garcia Isl.	90	80	89	79	88	79	82	87	81	86	80	86	81	161	85	80	157	84	79	151	84	9.4
IRELAND																						
Belmullet	70	62	66	61	64	60	64	68	62	65	61	63	62	83	65	61	79	64	59	76	62	8.8
Birr	76	64	72	63	69	61	65	72	64	70	62	67	63	87	69	61	81	66	60	78	65	14.9
Claremorris	73	64	70	62	67	61	65	71	63	68	62	65	63	86	67	61	82	65	60	78	64	13.9
Clones	74	64	71	62	68	61	65	72	63	69	62	67	62	85	67	61	80	66	59	76	65	13.7
Cork	71	62	68	61	66	60	64	69	63	67	61	65	62	86	65	61	82	64	60	78	63	12.1
Dublin	72	63	69	61	67	60	64	69	63	67	61	66	62	85	67	61	80	65	59	76	64	12.6
Kilkenny	76	64	72	62	70	61	65	72	64	70	62	68	63	87	68	61	82	66	60	78	65	15.8
Malin	67	61	65	60	63	59	62	65	61	64	59	62	60	79	64	59	76	62	58	72	61	7.6
Mullingar	74	64	70	62	67	61	65	71	63	69	62	66	63	86	67	61	81	66	60	78	64	14.4
Rosslare	68	62	66	61	64	60	63	66	62	65	61	64	62	83	65	61	79	64	59	76	62	8.8
Shannon	75	64	71	63	68	61	65	72	64	70	62	67	63	85	68	61	81	67	60	78	65	12.1
Valentia Observatory	71	63	68	62	66	61	65	69	64	67	62	65	63	88	66	62	83	65	61	80	64	9.4
ISRAEL																						
Jerusalem	89	65	86	64	84	63	70	81	69	79	68	78	67	110	74	66	104	72	65	100	71	18.4
Lod	94	69	90	72	88	72	78	87	76	85	76	85	75	132	83	74	126	82	73	122	82	17.1
Ovda (Isr-AFB/Civ)	100	65	97	65	95	65	73	89	71	87	69	85	68	109	80	66	102	78	64	95	76	25.0
Tel Aviv-Yafo	88	69	86	74	85	74	78	84	77	83	76	83	76	137	83	75	132	82	74	127	82	9.9
ITALY																						
Bologna/Borgo (AFB)	93	75	90	73	88	72	77	89	75	87	74	85	73	125	83	72	118	81	70	111	81	20.3
Brindisi	90	73	86	74	84	75	80	84	79	83	77	82	79	149	83	77	141	82	75	133	81	13.0
Catania	95	72	91	73	89	72	79	85	78	84	76	83	77	141	82	75	133	81	74	126	80	20.9
Genova	86	72	84	72	82	72	76	83	75	81	74	80	74	129	82	73	123	81	72	117	79	10.4
Messina	89	72	88	73	86	74	79	84	78	83	77	83	77	142	83	76	136	83	75	132	82	9.4
Milan, Linate	89	73	87	72	85	71	76	85	74	84	73	82	73	123	81	71	117	80	70	111	78	18.2
Milan, Malpensa	90	74	87	73	85	72	77	86	75	85	74	83	74	131	83	72	121	81	70	115	79	23.0
Naples	92	73	89	73	87	73	79	85	77	84	76	83	77	142	83	75	133	82	73	125	80	19.8
Palermo	92	71	88	73	86	75	80	85	79	84	78	83	79	149	85	77	142	83	76	137	82	9.5
Perugia	92	70	90	69	86	69	73	87	72	84	70	83	69	110	79	68	104	77	66	99	76	25.0
Pisa	89	72	87	71	85	71	76	84	75	83	73	82	74	125	80	72	118	79	71	113	78	21.2
Rome	87	74	86	74	84	74	79	83	78	82	76	81	77	142	83	76	136	81	75	131	80	17.8
Ronchi Legionari Ab	91	72	88	71	86	71	76	83	74	83	72	82	74	125	80	71	116	79	70	109	78	21.2
Torino	87	72	85	71	83	70	75	84	74	82	72	80	72	125	78	71	120	79	70	113	77	18.9
Venice	87	74	85	73	83	71	77	83	75	82	74	81	75	132	81	73	123	80	71	116	78	16.4
JAMAICA																						

MDB = mean coincident dry-bulb temperature, °F
MWB = mean coincident wet-bulb temperature, °F
MWS = mean coincident wind speed, mph
MWD = mean coincident wind direction
StdD = standard deviation
A = airport
HR = humidity ratio
DP = dew-point temperature, °F

Table 3A Heating and Wind Design Conditions—World Locations

									Extreme Wind			Coldest Month WS/MDB				MWS/MWD to DB				Annual Extreme Daily			
				Elev.	StdP		Heating DB		Speed			0.4%		1%		99.6%		0.4%		Mean DB		StdD DB	
Station	WMO#	Lat.	Long.	ft	psia	Dates	99.6%	99%	1%	2.5%	5%	WS	MDB	WS	MDB	MWS	PWD	MWS	PWD	Max	Min	Max	Min
Kingston	783970	17.93 N	76.78 W	29	14.684	8293	71	72	33	31	29	33	83	31	84	6	330	25	110	96	68	2.3	1.4
Montego Bay	783880	18.50 N	77.92 W	9	14.695	8293	70	71	29	26	23	28	80	26	80	5	140	21	70	95	61	3.6	15.1
JAPAN																							
Aomori	475750	40.82 N	140.77 E	9	14.695	8293	18	20	21	19	17	22	31	19	31	8	230	10	220	92	15	2.0	3.1
Asahikawa	474070	43.77 N	142.37 E	380	14.499	8293	−2	2	13	11	10	12	27	10	25	2	80	6	270	90	−8	2.5	4.9
Atsugi	476790	35.45 N	139.45 E	213	14.587	8293	29	30	23	20	18	21	47	19	46	4	360	10	180	95	24	2.7	2.0
Fukuoka	478080	33.58 N	130.45 E	39	14.679	8293	30	32	21	18	16	21	40	19	41	8	10	11	10	96	26	2.0	2.3
Hakodate	474300	41.82 N	140.75 E	118	14.637	8293	13	16	20	18	15	21	32	18	30	5	290	7	220	86	9	2.2	2.7
Hamamatsu	476810	34.75 N	137.70 E	157	14.617	8293	30	32	22	20	18	22	44	20	44	8	10	13	270	94	26	2.7	1.8
Hiroshima	477650	34.40 N	132.47 E	173	14.608	8293	30	31	21	18	16	20	45	17	43	6	20	10	220	95	27	3.2	2.7
Hyakuri (Jasdf)	477150	36.18 N	140.42 E	114	14.639	8293	19	23	22	19	16	19	43	17	42	2	250	11	130	92	14	2.0	2.2
Kadena	479310	26.35 N	127.77 E	147	14.622	8293	50	52	24	21	18	22	61	20	62	10	70	12	240	94	45	2.0	2.5
Kagoshima	478270	31.57 N	130.55 E	16	14.691	8293	33	35	17	14	13	15	49	13	49	6	300	9	270	94	29	2.0	2.0
Kumamoto	478190	32.82 N	130.72 E	127	14.632	8293	28	30	15	13	11	14	42	12	44	3	360	8	240	96	24	2.2	2.0
Maebashi	476240	36.40 N	139.07 E	370	14.504	8293	26	28	18	15	13	18	43	16	44	7	330	8	110	98	22	2.9	3.1
Maizuru	477500	35.45 N	135.32 E	72	14.662	8293	28	30	19	16	13	18	42	15	41	4	240	7	20	95	25	1.8	2.7
Matsumoto	476180	36.25 N	137.97 E	2004	13.666	8293	16	18	18	16	14	17	40	16	40	3	30	9	170	93	11	2.5	2.5
Matsuyama	478870	33.83 N	132.78 E	111	14.641	8293	31	33	14	12	10	14	42	12	43	4	110	8	270	93	28	1.6	2.2
Miho (Civ/Jasdf)	477430	35.48 N	133.25 E	29	14.684	8293	30	31	24	21	19	26	35	23	37	14	260	13	260	93	26	2.9	2.2
Miyako Jima Island	479270	24.78 N	125.28 E	134	14.629	8293	54	56	26	22	20	22	64	21	64	13	360	13	210	92	51	1.1	2.2
Morioka	475840	39.70 N	141.17 E	515	14.428	8293	14	18	19	17	15	18	31	16	33	4	140	9	190	92	10	1.8	3.2
Nagasaki	478170	32.73 N	129.87 E	114	14.639	8293	33	35	17	15	13	16	49	14	49	5	300	6	230	94	30	1.8	1.6
Nagoya	476350	35.25 N	136.93 E	55	14.670	8293	27	29	22	18	16	22	45	19	44	3	350	9	10	96	23	2.7	2.3
Naha	479300	26.18 N	127.65 E	26	14.686	8293	53	55	29	26	23	30	58	27	59	16	10	13	200	91	51	1.1	2.0
Naze	479090	28.38 N	129.50 E	22	14.688	8293	49	50	16	14	12	16	56	14	56	7	190	7	210	94	45	3.8	1.3
New Tokyo Intl Arpt	476860	35.77 N	140.38 E	144	14.623	8293	23	25	22	19	16	21	46	18	43	5	330	11	10	93	17	2.3	4.0
Niigata	476040	37.92 N	139.05 E	22	14.688	8293	27	29	24	21	18	27	38	23	36	11	200	10	140	97	24	4.0	3.8
Nyutabaru (Jasdf)	478540	32.08 N	131.45 E	269	14.558	8293	28	31	21	18	16	22	46	19	48	7	270	12	230	95	24	2.2	2.0
Oita	478150	33.23 N	131.62 E	42	14.677	8293	30	32	17	14	13	16	44	15	43	6	170	9	10	94	26	1.6	3.2
Osaka	477710	34.78 N	135.45 E	49	14.674	8293	28	30	19	17	15	18	44	16	44	5	10	9	10	96	26	2.0	2.2
Owase	476630	34.07 N	136.20 E	88	14.653	8293	30	32	17	14	12	17	46	15	47	3	280	9	70	95	27	2.5	2.7
Sapporo	474120	43.05 N	141.33 E	62	14.667	8293	12	15	16	14	12	15	32	13	29	3	130	7	150	89	8	2.7	3.6
Sendai	475900	38.27 N	140.90 E	141	14.625	8293	24	26	23	20	17	24	37	21	37	6	350	9	130	91	21	3.6	2.3
Shimonoseki	477620	33.95 N	130.93 E	62	14.667	8293	33	35	24	20	17	21	44	19	44	10	330	9	100	91	30	1.4	2.3
Shizuhama (Jasdf)	476580	34.82 N	138.30 E	32	14.683	8293	30	32	24	22	19	25	47	23	47	9	10	13	260	95	27	2.7	2.0
Tokyo, Intl Airport	476710	35.55 N	139.78 E	26	14.686	8293	31	32	28	24	21	27	50	24	45	7	280	14	10	94	27	2.5	3.1
Tosashimizu	478980	32.72 N	133.02 E	108	14.643	8293	34	36	23	19	17	21	55	17	52	8	350	8	250	88	31	1.8	2.2
Wakkanai	474010	45.42 N	141.68 E	36	14.681	8293	11	13	27	23	21	29	26	25	25	9	190	10	240	81	9	3.6	4.0
JORDAN																							
Amman	402700	31.98 N	35.98 E	2536	13.402	8293	33	35	23	20	17	27	43	22	43	6	90	8	290	101	26	2.9	12.1
KAZAKHSTAN																							
Almaty (Alma Ata)	368700	43.23 N	76.93 E	2778	13.283	8293	−3	3	10	8	6	8	29	6	26	1	360	3	330	98	−4	3.1	5.8
Aqmola (Tselinograd)	351880	51.13 N	71.37 E	1141	14.104	8293	−21	−17	24	21	18	26	20	22	15	5	270	6	230	98	−26	5.8	5.8
Aqtobe (Aktyubinsk)	352290	50.30 N	57.23 E	744	14.309	8293	−19	−14	25	22	19	27	19	23	20	2	190	8	60	99	−26	4.0	3.6
Atyrau (Gur'yev)	357000	47.02 N	51.85 E	−49	14.726	8293	−7	−3	30	26	22	35	18	28	22	4	90	10	140	103	−12	2.5	5.0
Oral (Ural'sk)	351080	51.25 N	51.40 E	118	14.637	8293	−18	−13	28	23	21	31	17	28	16	6	360	11	140	100	−24	4.1	4.9
Pavlodar	360030	52.28 N	76.95 E	403	14.487	8293	−24	−19	22	19	17	23	21	20	20	5	60	7	230	97	−27	3.1	6.5
Qaraghandy (Karaganda)	353940	49.80 N	73.13 E	1820	13.758	8293	−18	−13	22	19	17	26	20	22	19	5	120	9	20	96	−22	3.2	6.8
Qostanay (Kustanay)	289520	53.22 N	63.62 E	511	14.430	8293	−21	−17	25	21	19	27	14	23	19	6	180	9	90	97	−26	3.1	5.2
Semey (Semipalatinsk)	361770	50.35 N	80.25 E	643	14.362	8293	−25	−20	20	17	14	22	27	18	23	1	90	6	90	99	−30	3.4	7.0
Zhambyl (Dzhambul)	383410	42.85 N	71.38 E	2142	13.597	8293	−5	1	27	22	16	27	36	22	36	2	180	9	20	102	−7	3.1	7.0
KENYA																							
Arissa	637230	0.47 S	39.63 E	482	14.446	8293	70	71	29	26	22	30	83	28	83	7	180	8	180	105	59	7.6	8.3
Kisumu	637080	0.10 S	34.75 E	3759	12.810	8293	60	62	22	19	17	19	76	16	79	4	90	13	230	100	52	11.2	7.0
Lodwar	636120	3.12 N	35.62 E	1689	13.824	8293	69	71	23	20	19	21	83	19	83	4	270	13	90	106	58	7.9	9.7
Nairobi	637400	1.32 S	36.92 E	5328	12.082	8293	49	51	23	21	19	17	70	15	69	6	240	14	60	90	41	4.5	4.9
Nakuru	637140	0.27 S	36.10 E	6236	11.675	8293	47	48	19	16	13	17	70	14	70	2	350	9	360	95	40	9.9	5.2
KOREA, NORTH																							
Anju	470500	39.62 N	125.65 E	88	14.653	8293	−1	4	19	16	14	17	20	14	19	3	140	5	230	90	−7	2.7	7.4
Ch'ongjin	470080	41.78 N	129.82 E	141	14.625	8293	7	10	15	12	10	15	17	13	17	6	320	3	90	88	3	3.1	5.0
Changjin	470310	40.37 N	127.25 E	3546	12.911	8293	−19	−15	20	18	16	20	4	18	4	1	320	6	320	85	−25	4.7	3.4
Haeju	470690	38.03 N	125.70 E	265	14.559	8293	10	13	21	18	15	19	22	17	23	7	320	7	180	91	7	3.6	4.5
Hamhung	470410	39.93 N	127.55 E	124	14.634	8293	7	10	18	15	13	21	20	18	22	7	360	9	230	93	2	2.9	4.3
Namp'o	470600	38.72 N	125.37 E	154	14.618	8293	8	11	23	19	17	21	21	18	21	7	320	6	270	91	6	5.8	5.0
P'yongyang	470580	39.03 N	125.78 E	124	14.634	8293	3	7	14	12	10	15	19	13	19	2	110	4	270	91	−1	2.3	6.5
Sinuiju	470350	40.10 N	124.38 E	22	14.688	8293	3	7	17	15	13	19	19	16	18	6	50	5	230	92	−4	3.8	5.0
Wonsan	470550	39.18 N	127.43 E	118	14.637	8293	12	15	16	13	11	16	27	14	25	6	270	4	250	93	7	2.9	4.7

WMO# = World Meteorological Organization number
Lat = latitude Long. = longitude
Elev = elevation, ft
StdP = standard pressure at station elevation, psia
DB = dry-bulb temperature, °F
WS = wind speed, mph

Table 3B Cooling and Dehumidification Design Conditions—World Locations

Station	Cooling DB/MWB						WB/MDB						DP/MDB and HR									Range of DB
	0.4%		1%		2%		0.4%		1%		2%		0.4%			1%			2%			
	DB	MWB	DB	MWB	DB	MWB	WB	MDB	WB	MDB	WB	MDB	DP	HR	MDB	DP	HR	MDB	DP	HR	MCDB	
Kingston	92	78	91	78	90	78	81	89	80	89	79	88	79	150	85	77	143	85	77	141	85	11.7
Montego Bay	90	79	90	79	89	78	80	88	80	88	79	87	78	148	86	77	142	85	77	141	86	11.3
JAPAN																						
Aomori	86	74	83	73	80	71	76	84	74	81	73	79	74	125	80	72	120	80	71	115	78	13.0
Asahikawa	86	73	82	70	80	68	74	84	73	81	71	77	72	118	82	70	112	78	68	106	76	15.7
Atsugi	90	77	88	76	86	76	79	87	78	84	77	84	77	141	82	76	135	83	75	132	81	11.7
Fukuoka	93	77	91	78	88	77	79	89	79	87	78	86	77	141	83	76	134	82	75	133	82	13.1
Hakodate	82	73	80	72	77	71	75	81	73	78	72	76	73	123	79	72	117	77	70	111	76	11.0
Hamamatsu	90	76	88	76	86	76	79	86	79	84	78	82	78	148	81	77	142	81	77	140	81	11.9
Hiroshima	91	78	89	77	87	77	79	88	78	87	78	85	77	142	84	76	138	83	76	134	82	11.7
Hyakuri (Jasdf)	89	78	86	77	84	76	79	87	78	84	77	83	77	142	83	77	139	82	75	134	81	13.3
Kadena	92	81	91	81	90	80	83	90	82	89	82	88	81	162	87	81	161	87	80	159	86	9.7
Kagoshima	91	78	90	78	88	77	80	87	79	86	79	85	78	146	84	77	142	83	77	140	83	11.2
Kumamoto	92	78	91	77	89	77	80	88	79	87	78	85	78	146	83	77	142	82	76	139	82	14.0
Maebashi	92	77	90	76	87	75	78	89	77	87	76	85	75	134	82	74	130	83	73	125	82	13.5
Maizuru	91	77	89	77	87	76	79	88	78	87	77	85	76	137	83	75	132	82	74	130	81	13.7
Matsumoto	89	72	87	72	84	71	74	85	73	83	72	82	71	123	79	70	118	78	69	113	77	16.9
Matsuyama	91	77	89	76	87	76	78	87	77	86	77	85	76	136	82	75	133	82	74	129	82	12.1
Miho (Civ/Jasdf)	90	78	88	77	86	76	79	87	78	85	77	84	77	141	84	76	134	82	75	132	81	11.5
Miyako Jima Island	90	80	89	80	88	80	82	87	81	87	81	86	81	160	85	80	156	84	79	152	84	8.5
Morioka	87	75	84	73	82	71	76	84	75	82	74	79	74	130	80	73	125	79	72	120	78	13.9
Nagasaki	90	77	88	78	86	77	80	85	79	84	78	84	79	150	84	78	146	83	77	141	82	10.1
Nagoya	93	77	90	76	88	75	79	88	78	86	77	85	77	141	82	75	133	81	75	131	82	14.0
Naha	90	80	88	80	88	80	82	87	81	87	81	86	81	159	86	79	151	85	79	151	85	6.8
Naze	90	79	89	79	88	79	81	88	80	87	80	87	79	150	85	78	147	85	78	145	84	9.7
New Tokyo Intl Arpt	89	78	87	78	85	76	79	87	78	85	77	83	77	143	82	77	141	82	75	134	81	13.5
Niigata	90	77	88	76	85	75	78	87	77	85	76	84	76	135	83	75	130	82	74	125	82	10.8
Nyutabaru (Jasdf)	90	78	88	78	86	77	79	86	79	85	78	83	78	149	82	77	143	82	77	141	82	11.0
Oita	91	78	89	77	87	77	79	87	78	86	78	85	77	141	83	76	136	82	75	133	82	12.2
Osaka	93	77	91	76	89	75	79	88	78	87	78	86	77	141	83	76	134	81	75	132	82	14.4
Owase	90	76	87	76	85	76	78	86	78	85	77	84	76	139	83	76	134	82	75	130	81	11.0
Sapporo	84	73	81	71	78	69	74	82	73	80	71	77	72	118	80	70	111	78	69	105	76	11.7
Sendai	86	75	84	74	81	73	77	83	76	82	75	80	75	134	81	74	129	80	73	124	78	9.5
Shimonoseki	88	77	87	77	85	76	79	86	78	85	77	84	77	139	83	76	136	82	75	133	82	7.9
Shizuhama (Jasdf)	91	79	89	78	86	77	80	88	79	86	79	84	79	150	83	77	143	83	77	140	82	11.9
Tokyo, Intl Airport	91	78	88	77	86	77	80	88	79	86	78	84	77	143	83	77	141	83	76	134	82	11.2
Tosashimizu	86	78	85	78	83	78	80	84	80	83	79	83	79	151	83	78	147	83	78	145	82	5.9
Wakkanai	77	71	74	69	72	68	72	76	70	74	69	72	71	113	75	69	106	73	67	100	72	7.6
JORDAN																						
Amman	95	65	92	65	89	64	71	83	70	82	68	81	68	115	76	66	106	74	65	100	72	20.3
KAZAKHSTAN																						
Almaty (Alma Ata)	91	65	89	64	86	63	67	85	66	84	65	83	62	92	76	60	84	73	58	79	73	19.8
Aqmola (Tselinograd)	89	64	85	62	82	62	67	81	65	79	64	78	63	89	71	61	83	69	59	77	69	19.4
Aqtobe (Aktyubinsk)	93	67	90	65	86	64	69	87	67	85	66	82	64	90	73	62	85	72	60	80	72	23.0
Atyrau (Gur'yev)	97	68	94	67	91	66	72	87	70	85	69	84	68	104	78	66	96	76	64	89	76	20.2
Oral (Ural'sk)	93	67	89	66	86	65	70	86	68	84	67	82	65	94	75	64	88	74	62	83	72	22.5
Pavlodar	90	65	87	65	84	63	68	82	67	80	66	79	65	92	72	63	86	71	61	80	71	20.2
Qaraghandy (Karaganda)	89	61	85	61	82	60	65	79	63	78	62	77	61	85	68	59	78	67	57	74	66	20.3
Qostanay (Kustanay)	90	66	86	65	83	64	69	83	67	81	66	79	65	93	73	63	87	72	61	83	71	18.7
Semey (Semipalatinsk)	91	66	87	64	84	64	68	84	67	82	66	80	64	90	72	62	86	72	61	81	71	22.5
Zhambyl (Dzhambul)	96	64	93	64	90	63	67	89	65	88	64	86	60	83	72	58	78	72	56	73	72	25.0
KENYA																						
Arissa	99	74	98	74	97	74	79	90	78	89	77	88	76	139	83	75	135	81	75	132	80	19.1
Kisumu	90	66	89	66	87	67	72	82	71	82	71	81	69	123	76	68	119	76	67	116	75	19.8
Lodwar	100	69	99	69	98	69	75	87	74	86	74	87	73	129	79	72	125	79	71	120	79	19.8
Nairobi	84	60	83	60	81	60	66	75	65	74	64	73	63	106	67	63	104	66	62	102	66	24.3
Nakuru	84	57	83	57	82	57	64	75	63	73	62	72	61	101	66	60	97	65	59	95	65	27.5
KOREA, NORTH																						
Anju	86	75	83	74	82	73	77	83	76	81	75	80	76	136	80	75	130	79	73	125	78	13.5
Ch'ongjin	81	71	78	70	76	69	74	79	73	76	71	74	73	123	77	71	116	75	70	110	74	9.4
Changjin	77	65	74	63	72	62	69	74	67	71	65	70	67	115	71	66	108	70	64	102	68	16.2
Haeju	86	75	84	73	82	72	78	84	76	81	75	79	76	136	80	75	132	79	74	128	78	10.8
Hamhung	87	73	84	72	81	71	77	84	75	82	74	78	74	130	81	73	124	79	72	119	77	12.1
Namp'o	85	76	83	75	82	73	78	83	77	81	76	80	76	139	80	75	134	79	74	130	78	11.3
P'yongyang	87	75	85	74	83	72	78	84	76	82	75	80	76	135	81	75	130	80	74	125	79	13.7
Sinuiju	87	75	84	73	82	72	77	83	76	81	75	79	76	135	80	75	130	78	74	125	78	12.8
Wonsan	88	75	85	72	82	71	78	85	76	82	74	80	75	134	82	74	126	80	72	120	79	9.4

MDB = mean coincident dry-bulb temperature, °F
MWB = mean coincident wet-bulb temperature, °F
MWS = mean coincident wind speed, mph
MWD = mean coincident wind direction
StdD = standard deviation
A = airport
HR = humidity ratio
DP = dew-point temperature, °F

Table 3A Heating and Wind Design Conditions—World Locations

							Heating DB		Extreme Wind Speed			Coldest Month WS/MDB				MWS/MWD to DB				Annual Extreme Daily			
												0.4%		1%		99.6%		0.4%		Mean DB		StdD DB	
Station	WMO#	Lat.	Long.	Elev. ft	StdP psia	Dates	99.6%	99%	1%	2.5%	5%	WS	MDB	WS	MDB	MWS	PWD	MWS	PWD	Max	Min	Max	Min
KOREA, SOUTH																							
Cheju	471820	33.50 N	126.55 E	87	14.653	8293	30	32	27	24	22	28	37	25	40	14	40	14	230	93	26	2.2	2.5
Inch'on	471120	37.48 N	126.63 E	229	14.578	8293	12	15	22	19	16	23	24	20	22	11	320	7	230	92	9	6.8	4.7
Kangnung	471050	37.75 N	128.90 E	88	14.653	8293	16	20	18	16	14	19	29	17	29	12	250	6	90	95	12	3.1	4.1
Kwangju	471560	35.13 N	126.92 E	236	14.575	8293	19	22	17	15	13	17	32	15	33	4	20	7	250	93	15	2.7	3.2
Osan	471220	37.08 N	127.03 E	39	14.679	8293	7	11	17	14	12	16	28	14	29	3	10	6	10	95	2	2.2	5.4
Seoul	471100	37.55 N	126.80 E	62	14.667	8293	7	10	19	17	15	19	24	16	25	3	10	9	160	92	2	1.6	6.1
Taegu	471430	35.88 N	128.62 E	200	14.594	8293	17	20	20	18	16	22	31	19	32	8	290	9	270	96	12	2.5	3.1
Taejon	471330	36.30 N	127.40 E	255	14.565	8293	12	15	15	13	11	12	37	11	34	1	110	6	270	95	8	3.4	2.7
Ulsan	471520	35.55 N	129.32 E	108	14.643	8293	20	22	16	14	12	18	31	16	31	7	320	8	140	95	15	3.1	3.4
KUWAIT																							
Kuwait	405820	29.22 N	47.98 E	180	14.604	8293	38	41	26	23	21	24	61	21	60	4	300	14	340	121	33	2.3	2.3
KYRGYZSTAN																							
Bishkek (Frunze)	383530	42.85 N	74.53 E	2083	13.627	8293	−8	−2	21	17	15	19	32	15	33	3	150	8	220	101	−11	2.2	7.6
Tianshan (Tjan-San) (Mtn	369820	41.92 N	78.23 E	11856	9.402	8293	−27	−23	22	19	17	20	4	17	1	1	360	11	210	68	−32	7.7	4.0
LATVIA																							
Liepaja	264060	56.55 N	21.02 E	26	14.686	8293	1	9	28	24	21	27	39	23	38	7	30	9	120	83	3	2.9	11.0
Riga	264220	56.97 N	24.07 E	9	14.695	8293	−3	4	24	21	18	23	37	21	36	4	40	9	150	85	−3	3.6	13.1
LIBYA																							
Banghazi	620530	32.08 N	20.27 E	433	14.471	8293	44	46	30	27	23	29	55	23	57	5	90	15	350	106	39	1.8	2.9
Tripoli	620100	32.67 N	13.15 E	265	14.559	8293	39	41	23	21	19	22	59	19	58	4	240	13	60	114	35	3.1	2.0
LIECHTENSTEIN																							
Vaduz	69900	47.13 N	9.53 E	1519	13.911	8293	12	17	22	17	13	22	50	18	48	3	180	10	320	89	8	2.0	6.7
LITHUANIA																							
Kaunas	266290	54.88 N	23.88 E	246	14.570	8293	−4	3	23	20	18	23	31	21	32	6	70	8	180	86	−2	3.6	8.5
Klaipeda	265090	55.70 N	21.15 E	32	14.683	8293	1	8	31	26	22	29	40	24	39	8	70	8	140	83	4	3.1	10.3
Vilnius	267300	54.63 N	25.28 E	511	14.430	8293	−5	2	25	23	20	25	29	22	29	5	70	11	140	86	−5	2.9	7.7
MACEDONIA																							
Skopje	135860	41.97 N	21.65 E	784	14.288	8293	10	15	20	17	14	19	36	15	34	1	50	4	270	100	4	4.5	9.4
MADEIRA ISLANDS																							
Funchal	85210	32.68 N	16.77 W	180	14.604	8293	53	55	30	27	23	34	61	29	62	8	310	11	30	87	50	5.2	1.8
MALAYSIA																							
George Town	486010	5.30 N	100.27 E	13	14.693	8293	73	73	15	13	11	13	82	12	83	2	350	8	270	96	70	4.0	1.3
Kota Baharu	486150	6.17 N	102.28 E	16	14.691	8293	71	72	17	15	14	18	81	17	81	1	190	9	90	95	68	2.9	1.4
Kuala Lumpur	486470	3.12 N	101.55 E	72	14.662	8293	71	72	16	14	12	13	85	11	85	1	340	8	270	98	68	3.1	3.4
Kuantan	486570	3.78 N	103.22 E	52	14.672	8293	70	71	16	14	12	16	83	15	82	5	350	8	230	99	58	5.2	23.4
Malacca	486650	2.27 N	102.25 E	29	14.684	8293	72	72	16	13	12	17	84	15	84	3	10	8	20	97	66	3.4	5.6
Sitiawan	486200	4.22 N	100.70 E	26	14.686	8293	71	72	13	12	10	11	84	10	85	1	60	7	180	99	66	5.9	4.9
Kuching	964130	1.48 N	110.33 E	88	14.653	8293	71	72	12	11	9	13	82	11	82	2	260	5	360	99	67	4.1	7.4
Miri	964490	4.33 N	113.98 E	59	14.669	8293	72	73	18	15	13	18	82	16	83	2	120	9	270	99	66	8.3	10.3
MALI																							
Bamako	612910	12.53 N	7.95 W	1250	14.048	8293	59	62	20	17	15	18	77	16	77	7	40	9	80	110	50	6.1	6.7
MALTA																							
Luqa	165970	35.85 N	14.48 E	298	14.542	8293	44	46	26	23	20	29	56	26	56	6	270	9	310	99	38	4.1	3.1
MARSHALL ISLANDS																							
Kwajalein Atoll	913660	8.73 N	167.73 E	26	14.686	8293	76	77	25	23	22	28	81	26	82	12	70	11	70	95	60	7.0	23.6
MAURITANIA																							
Nouadhibou	614150	20.93 N	17.03 W	9	14.695	8293	55	57	32	30	28	30	63	28	63	14	360	14	20	101	48	2.9	6.3
Nouakchott	614420	18.10 N	15.95 W	9	14.695	8293	55	57	23	21	19	26	75	24	75	9	60	14	80	113	44	1.3	6.7
MEXICO																							
Acapulco	768056	16.77 N	99.75 W	16	14.691	8293	68	70	23	19	17	17	84	14	84	2	320	17	200	97	60	2.7	8.6
Merida	766440	20.98 N	89.65 W	32	14.683	8293	57	60	34	23	19	22	77	19	77	4	90	15	140	103	47	2.2	2.2
Mexico City	766790	19.43 N	99.08 W	7329	11.202	8293	39	42	51	22	18	51	52	22	66	5	90	11	360	88	32	2.2	4.1
Puerto Vallarta (766010)	766014	20.68 N	105.25 W	19	14.690	8293	59	60	18	14	12	12	79	12	78	0	10	17	330	94	54	1.6	1.4
Tampico (765491)	765494	22.28 N	97.87 W	78	14.658	8293	50	53	32	24	21	34	59	28	62	9	270	11	90	97	43	4.3	6.5
Veracruz	766910	19.20 N	96.13 W	45	14.676	8293	57	59	46	34	29	47	70	35	68	4	330	22	90	101	50	4.0	4.5
MICRONESIA																							
Truk Intl/Moen Isl	913340	7.47 N	151.85 E	7	14.697	8293	75	76	21	18	17	21	81	19	82	9	100	9	40	102	56	7.9	25.7
MIDWAY ISLAND																							
Midway Island Naf	910660	28.22 N	177.37 W	13	14.693	8293	59	60	24	22	20	29	67	26	67	10	360	9	110	89	46	1.8	20.2
MOLDOVA																							
Chisinau (Kishinev)	338150	47.02 N	28.87 E	590	14.389	8293	6	10	15	13	12	17	31	14	29	5	300	6	200	91	4	3.8	5.8
MONGOLIA																							
Ulaanbataar	442920	47.93 N	106.98 E	4317	12.547	8293	−23	−19	23	21	17	19	0	15	1	2	320	8	270	89	−27	5.2	4.7
Ulaangom	442120	49.97 N	92.08 E	3070	13.140	8293	−40	−37	18	13	11	9	−29	7	−29	1	180	5	50	89	−43	5.0	4.0
MOROCCO																							
Al Hoceima	601070	35.18 N	3.85 W	45	14.676	8293	44	46	24	21	18	24	58	19	58	3	180	12	360	98	39	9.0	4.1

WMO# = World Meteorological Organization number
Lat = latitude
Long. = longitude
Elev = elevation, ft
StdP = standard pressure at station elevation, psia
DB = dry-bulb temperature, °F
WS = wind speed, mph

Table 3B Cooling and Dehumidification Design Conditions—World Locations

	Cooling DB/MWB						WB/MDB						DP/MDB and HR									
	0.4%		1%		2%		0.4%		1%		2%		0.4%			1%			2%			Range
Station	DB	MWB	DB	MWB	DB	MWB	WB	MDB	WB	MDB	WB	MDB	DP	HR	MDB	DP	HR	MDB	DP	HR	MCDB	of DB
KOREA, SOUTH																						
Cheju	89	78	86	79	85	78	82	86	80	85	79	84	81	160	84	79	151	83	78	148	83	9.7
Inch'on	87	76	84	74	82	73	77	84	76	82	75	80	76	136	81	75	131	80	74	127	79	10.4
Kangnung	90	75	87	74	85	72	78	87	76	85	75	82	75	132	82	74	126	82	73	122	80	10.4
Kwangju	90	78	88	76	85	75	80	87	78	85	77	83	78	144	83	76	139	82	76	135	81	12.6
Osan	90	78	88	76	86	75	80	88	78	85	77	83	78	148	86	77	139	83	75	132	82	14.4
Seoul	89	77	86	75	84	74	80	86	78	83	77	80	79	150	82	77	142	81	76	134	79	14.4
Taegu	92	78	90	76	87	74	79	89	78	87	77	85	77	141	85	76	135	83	74	130	82	13.1
Taejon	90	76	88	75	86	74	79	85	77	84	76	83	77	141	82	76	136	81	75	131	80	14.4
Ulsan	91	78	89	77	86	76	80	89	78	86	78	84	77	142	84	76	138	83	76	134	82	11.5
KUWAIT																						
Kuwait	117	69	115	69	113	68	82	94	78	91	75	92	79	153	92	75	132	87	70	112	85	27.7
KYRGYZSTAN																						
Bishkek (Frunze)	95	67	93	65	90	65	69	90	68	87	66	85	63	92	77	61	87	74	60	83	74	25.6
Tianshan (Tjan-San) (Mtn	57	42	54	41	51	39	44	54	42	51	41	49	40	57	46	38	53	44	36	50	43	20.9
LATVIA																						
Liepaja	76	64	73	62	70	62	67	72	65	70	63	68	64	90	70	62	85	68	61	80	66	10.3
Riga	79	65	76	64	73	62	67	75	66	73	64	71	65	91	71	63	85	69	61	80	68	14.2
LIBYA																						
Banghazi	99	72	95	71	92	70	78	89	76	86	75	85	75	134	83	74	127	81	72	123	80	16.7
Tripoli	107	76	103	74	100	73	81	99	78	94	76	90	76	139	88	74	130	84	73	123	83	24.8
LIECHTENSTEIN																						
Vaduz	83	67	80	65	78	64	68	80	67	78	65	75	64	94	74	63	90	72	61	85	71	16.6
LITHUANIA																						
Kaunas	80	67	77	65	74	63	69	78	67	75	65	72	65	94	73	63	88	71	62	83	68	16.6
Klaipeda	77	65	73	64	70	62	67	74	65	71	64	69	65	91	71	63	85	69	61	80	67	9.7
Vilnius	81	65	78	64	75	62	68	78	66	74	64	72	64	91	71	62	85	70	61	80	68	16.2
MACEDONIA																						
Skopje	95	68	92	68	89	67	71	90	70	88	68	86	65	94	78	63	88	76	62	86	75	27.4
MADEIRA ISLANDS																						
Funchal	81	69	79	69	77	68	72	78	71	76	70	76	70	111	76	68	105	75	68	102	74	8.5
MALAYSIA																						
George Town	91	79	90	78	90	78	82	88	81	87	81	87	80	158	85	79	151	84	79	150	84	13.3
Kota Baharu	91	79	90	79	90	79	81	88	80	88	80	87	79	150	85	78	147	84	78	145	84	12.8
Kuala Lumpur	94	78	93	78	92	78	81	90	80	89	80	89	79	152	85	79	149	84	78	146	84	16.2
Kuantan	92	79	91	79	90	79	81	89	80	88	80	88	79	150	85	78	147	84	78	146	84	15.3
Malacca	92	78	91	78	90	78	81	88	81	88	80	87	79	151	84	79	150	84	78	147	83	15.3
Sitiawan	92	79	91	79	90	79	81	90	81	89	80	88	79	151	85	79	150	85	78	147	85	14.8
Kuching	93	79	92	78	91	78	81	90	80	89	80	88	79	151	86	78	148	85	77	143	83	15.8
Miri	90	79	89	79	89	79	82	88	81	87	81	87	80	155	86	79	152	85	79	150	84	11.9
MALI																						
Bamako	104	69	103	69	101	69	79	91	78	89	78	88	77	147	83	76	140	82	75	139	82	22.1
MALTA																						
Luqa	92	71	88	72	86	72	77	84	76	82	75	82	75	134	80	74	127	80	73	125	79	14.4
MARSHALL ISLANDS																						
Kwajalein Atoll	89	79	88	79	88	79	81	87	80	87	80	86	79	150	86	78	147	85	78	146	85	7.6
MAURITANIA																						
Nouadhibou	92	69	88	69	86	69	76	83	74	81	73	81	74	126	79	72	118	78	70	112	77	15.8
Nouakchott	107	70	103	69	100	69	81	88	80	87	79	86	79	150	84	78	148	84	77	141	83	23.0
MEXICO																						
Acapulco	92	80	92	80	91	80	82	90	81	89	81	89	79	151	87	79	150	86	79	150	86	13.0
Merida	100	76	98	76	95	76	81	91	80	90	79	89	78	147	86	77	143	84	77	141	83	22.5
Mexico City	84	57	82	57	80	56	62	74	61	73	60	72	59	98	65	57	92	64	57	92	63	24.8
Puerto Vallarta (766010)	92	81	91	81	90	80	83	90	82	90	81	89	81	160	88	79	153	87	79	150	86	14.2
Tampico (765491)	92	80	90	80	90	80	83	89	82	88	81	87	82	168	88	80	158	86	79	152	84	11.3
Veracruz	94	80	92	80	91	80	82	91	81	90	80	89	79	152	85	79	150	85	78	148	85	14.9
MICRONESIA																						
Truk Intl/Moen Isl	88	80	88	80	87	79	81	87	81	87	80	86	79	151	86	79	148	85	78	147	85	7.4
MIDWAY ISLAND																						
Midway Island Naf	87	75	86	75	86	75	77	84	76	84	76	84	75	132	82	74	129	82	74	126	82	8.1
MOLDOVA																						
Chisinau (Kishinev)	86	67	84	66	81	65	70	81	68	80	67	78	66	98	76	64	92	74	63	87	72	16.4
MONGOLIA																						
Ulaanbataar	82	60	78	59	75	58	63	75	61	73	59	71	58	86	67	56	80	66	55	75	64	17.6
Ulaangom	82	61	79	60	77	59	63	78	62	76	60	73	58	80	68	56	75	68	54	70	66	19.3
MOROCCO																						
Al Hoceima	87	73	84	72	82	72	77	82	76	81	75	80	76	134	80	74	130	79	73	125	78	11.2

MDB = mean coincident dry-bulb temperature, °F
MWB = mean coincident wet-bulb temperature, °F
MWS = mean coincident wind speed, mph
MWD = mean coincident wind direction
StdD = standard deviation
A = airport
HR = humidity ratio
DP = dew-point temperature, °F

Table 3A Heating and Wind Design Conditions—World Locations

Station	WMO#	Lat.	Long.	Elev. ft	StdP psia	Dates	Heating DB 99.6%	Heating DB 99%	Extreme Wind Speed 1%	2.5%	5%	Coldest Month WS/MDB 0.4% WS	0.4% MDB	1% WS	1% MDB	MWS/MWD to DB 99.6% MWS	99.6% PWD	0.4% MWS	0.4% PWD	Annual Extreme Daily Mean DB Max	Mean DB Min	StdD DB Max	StdD DB Min
Casablanca	601550	33.57 N	7.67 W	203	14.592	8293	42	44	21	18	16	23	58	19	58	5	180	8	360	95	37	5.9	2.3
Casablanca/Nouasser	601560	33.37 N	7.58 W	675	14.345	8293	38	40	23	20	18	26	57	20	57	2	160	13	340	107	34	4.0	1.3
Midelt	601950	32.68 N	4.73 W	4970	12.245	8293	29	31	32	27	23	41	46	28	45	6	260	9	200	96	25	1.3	2.2
Ouarzazate	602650	30.93 N	6.90 W	3740	12.819	8293	33	35	32	27	22	29	54	22	51	2	320	12	240	102	29	1.1	1.6
Oujda	601150	34.78 N	1.93 W	1541	13.899	8293	34	36	30	26	22	31	54	27	55	4	240	15	360	106	29	2.7	2.2
Safi	601850	32.28 N	9.23 W	147	14.622	8293	42	44	21	19	17	20	59	18	58	7	60	12	20	106	37	5.2	2.3
Tanger	601010	35.73 N	5.90 W	68	14.663	8293	41	43	43	37	32	41	56	34	57	4	100	24	80	99	36	3.8	3.2
NETHERLANDS																							
Amsterdam	62400	52.30 N	4.77 E	−6	14.703	8293	17	21	31	27	24	35	47	31	44	11	70	11	70	86	16	3.4	8.3
Beek	63800	50.92 N	5.78 E	380	14.499	8293	14	19	27	24	21	30	45	26	43	10	60	8	40	90	13	3.8	8.5
De Bilt	62600	52.10 N	5.18 E	13	14.693	8293	16	20	19	17	15	21	44	19	42	7	50	8	70	87	14	3.6	8.1
Eindhoven	63700	51.45 N	5.42 E	72	14.662	8293	16	21	25	21	19	27	44	23	44	7	40	9	50	89	13	3.6	7.9
Gilze/Rijen	63500	51.57 N	4.93 E	42	14.677	8293	15	20	23	20	18	27	46	23	42	9	20	10	70	89	13	3.2	7.4
Groningen	62800	53.13 N	6.58 E	13	14.693	8293	14	18	28	24	22	31	45	27	44	7	50	9	100	87	11	3.2	7.7
Leeuwarden	62700	53.22 N	5.75 E	6	14.697	8293	16	20	29	26	22	32	43	29	43	8	80	11	80	85	13	2.3	7.4
Rotterdam	63440	51.95 N	4.45 E	−13	14.707	8293	17	21	30	27	24	33	44	29	44	9	50	10	90	86	15	3.4	6.7
NETHERLANDS ANTILLES																							
Willemstad	789880	12.20 N	68.97 W	219	14.584	8293	74	75	23	22	21	23	82	22	81	10	100	18	80	96	71	3.2	1.3
NEW CALEDONIA																							
Noumea	915920	22.27 S	166.45 E	236	14.575	8293	61	62	27	24	22	26	68	23	68	6	60	12	80	93	58	2.9	2.0
NEW ZEALAND																							
Auckland	931190	37.02 S	174.80 E	19	14.690	8293	35	37	31	28	25	32	53	28	53	10	240	13	20	85	35	13.7	2.2
Christchurch	937800	43.48 S	172.55 E	111	14.641	8293	28	30	27	23	21	24	48	21	48	1	280	16	300	92	25	10.8	1.3
Taiaroa Head	938960	45.77 S	170.73 E	249	14.568	8293	38	39	52	46	40	52	44	46	44	19	240	17	320	78	35	3.2	1.4
Wellington (934340)	934360	41.33 S	174.80 E	22	14.688	8293	35	36	42	38	33	40	50	34	50	14	10	17	360	84	35	14.4	2.2
NIGER																							
Agadez	610240	16.97 N	7.98 E	1646	13.846	8293	51	53	32	27	23	35	71	32	71	7	100	11	120	113	40	4.3	7.9
Niamey	610520	13.48 N	2.17 E	744	14.309	8293	60	62	23	19	17	24	73	22	74	6	40	8	40	112	52	3.1	4.5
NORWAY																							
Bergen	13110	60.30 N	5.22 E	164	14.613	8293	16	20	26	23	20	30	41	27	40	3	60	8	240	78	12	2.7	7.2
Bodo	11520	67.27 N	14.37 E	42	14.677	8293	9	13	38	33	29	44	32	39	28	19	80	11	100	76	8	4.0	4.7
Oslo/Fornebu	14880	59.90 N	10.62 E	55	14.670	8293	0	5	19	17	15	22	40	19	39	2	360	7	180	85	−1	5.6	8.6
Oslo/Gardermoen	13840	60.20 N	11.08 E	669	14.348	8293	−8	−2	20	17	15	22	36	19	34	2	30	7	180	82	−10	4.5	10.3
Stavanger	14150	58.88 N	5.63 E	29	14.684	8293	13	18	30	26	23	30	38	27	39	3	150	12	320	79	11	3.4	7.4
Svinoy (Lgt-H)	12050	62.33 N	5.27 E	134	14.629	8293	28	29	53	47	42	58	46	50	42	14	140	13	150	70	24	4.1	4.9
Tromso	10250	69.68 N	18.92 E	32	14.683	8293	6	10	30	27	23	34	36	30	35	2	170	8	180	75	4	2.5	3.6
Trondheim	12710	63.47 N	10.93 E	55	14.670	8293	−1	6	28	23	20	32	36	27	38	9	120	10	260	83	−2	3.2	8.1
Utsira	14030	59.30 N	4.88 E	183	14.603	8293	23	27	48	43	39	49	39	45	37	14	90	11	120	72	17	3.4	18.2
OMAN																							
Masqat	412560	23.58 N	58.28 E	49	14.674	8293	61	63	20	17	15	18	74	16	73	5	200	11	340	116	52	2.3	8.3
Salalah	413160	17.03 N	54.08 E	65	14.665	8293	63	65	21	19	16	28	70	23	73	10	20	12	200	101	52	5.0	10.1
Thamarit	413140	17.67 N	54.03 E	1459	13.941	8293	48	51	32	30	27	21	72	19	71	7	160	11	340	111	42	2.7	2.7
Tur'at Masirah	412880	20.67 N	58.90 E	62	14.667	8293	63	65	27	24	23	25	66	22	68	14	300	13	210	106	53	2.7	9.5
PANAMA																							
Panama	788060	8.92 N	79.60 W	52	14.672	8293	73	73	17	15	13	15	81	12	82	2	10	12	10	99	67	4.5	9.7
Tocumen	787920	9.05 N	79.37 W	36	14.681	8293	68	68	16	14	12	13	80	12	81	0	300	10	30	96	58	3.2	11.3
PARAGUAY																							
Asuncion	862180	25.27 S	57.63 W	331	14.525	8293	41	44	23	20	18	25	71	22	69	2	180	14	360	103	35	4.3	2.7
PERU																							
Arequipa	847520	16.32 S	71.55 W	8267	10.808	8293	42	43	26	21	18	31	54	26	54	6	30	14	240	79	34	3.4	3.4
Cuzco	846860	13.55 S	71.98 W	10659	9.853	8293	32	34	24	20	15	22	62	17	63	0	90	4	330	77	28	3.6	2.5
Iquitos	843770	3.75 S	73.25 W	413	14.482	8293	66	68	19	13	11	16	77	12	78	2	170	4	330	98	53	3.2	18.2
Lima	846280	12.00 S	77.12 W	42	14.677	8293	57	58	24	20	18	21	62	18	63	4	170	13	170	87	50	2.2	6.1
Pisco	846910	13.75 S	76.28 W	22	14.688	8293	53	55	25	21	19	23	65	19	65	1	90	11	210	89	47	4.0	6.3
Talara	843900	4.57 S	81.25 W	295	14.544	8293	60	61	46	42	33	46	64	42	65	22	150	15	190	93	55	3.2	7.7
PHILLIPPINES																							
Angeles, Clark AFB	983270	15.18 N	120.55 E	643	14.362	8293	68	69	15	12	11	13	83	12	82	5	10	7	120	99	65	2.0	2.2
Baguio	983280	16.42 N	120.60 E	4924	12.266	8293	52	54	22	15	12	14	65	12	65	3	90	4	140	93	49	5.4	1.6
Cebu/Mandaue	986460	10.30 N	123.97 E	78	14.658	8293	73	74	18	15	14	19	82	17	82	5	40	9	40	98	67	3.2	10.3
Olongapo	984260	14.80 N	120.27 E	56	14.670	8293	70	71	21	18	16	21	85	19	84	4	70	11	70	100	67	4.3	2.5
Manila, Ninoy Aquino Int'l	984290	14.52 N	121.00 E	68	14.663	8293	69	71	41	36	32	40	83	36	83	2	90	21	90	99	43	1.8	14.2
POLAND																							
Bialystok	122950	53.10 N	23.17 E	495	14.439	8293	−4	3	18	16	14	18	28	16	30	2	310	6	180	87	−5	3.6	11.2
Gdansk	121500	54.38 N	18.47 E	452	14.461	8293	1	9	31	26	23	32	41	27	36	2	130	11	10	86	−2	5.2	11.0
Katowice	125600	50.23 N	19.03 E	931	14.212	8293	3	9	19	17	15	22	40	19	38	2	20	8	250	89	−2	2.9	9.5
Kielce	125700	50.82 N	20.70 E	856	14.251	8293	−1	6	20	18	16	21	31	19	33	4	60	7	190	88	−5	2.9	10.6
Kolobrzeg	121000	54.18 N	15.58 E	16	14.691	8293	10	16	18	16	14	19	38	17	36	5	110	6	140	90	11	4.1	8.1

WMO# = World Meteorological Organization number
Lat = latitude
Long. = longitude
Elev = elevation, ft
StdP = standard pressure at station elevation, psia
DB = dry-bulb temperature, °F
WS = wind speed, mph

Table 3B Cooling and Dehumidification Design Conditions—World Locations

	Cooling DB/MWB						WB/MDB						DP/MDB and HR									
	0.4%		1%		2%		0.4%		1%		2%		0.4%			1%			2%			Range
Station	DB	MWB	DB	MWB	DB	MWB	WB	MDB	WB	MDB	WB	MDB	DP	HR	MDB	DP	HR	MDB	DP	HR	MCDB	of DB
Casablanca	85	72	81	72	79	72	75	80	74	79	73	77	74	126	78	73	122	77	72	118	76	9.2
Casablanca/Nouasser	96	72	91	71	87	70	75	90	73	86	72	84	70	114	80	69	108	77	68	106	77	19.8
Midelt	92	58	90	58	89	58	62	83	61	81	60	80	56	81	66	55	76	67	53	72	67	24.5
Ouarzazate	100	62	98	61	97	61	66	90	64	89	63	89	58	84	72	56	77	71	54	71	71	24.7
Oujda	98	70	94	69	91	69	74	90	72	87	71	85	70	116	80	68	111	78	67	106	77	24.7
Safi	94	71	89	70	85	70	74	86	73	83	72	81	71	116	78	70	111	77	69	107	75	14.8
Tanger	92	71	89	71	86	70	74	86	73	84	72	81	71	114	79	70	110	78	68	104	77	16.7
NETHERLANDS																						
Amsterdam	80	66	77	65	74	64	69	77	67	74	65	72	66	94	72	64	90	69	63	85	68	14.8
Beek	83	67	79	65	76	64	69	79	67	76	66	74	66	97	74	64	92	71	63	87	70	16.4
De Bilt	82	66	79	65	75	64	69	79	67	75	65	73	65	92	73	64	88	70	62	83	69	16.0
Eindhoven	83	67	80	65	77	64	69	80	67	77	65	74	65	92	73	63	87	70	62	83	69	17.8
Gilze/Rijen	82	66	79	65	76	63	68	79	67	76	65	73	65	92	72	63	87	69	62	83	68	17.3
Groningen	81	67	77	65	74	64	69	77	67	74	65	71	66	95	74	64	90	70	63	85	68	17.5
Leeuwarden	79	66	75	64	71	63	67	76	65	72	64	70	64	90	70	63	85	68	61	81	67	13.7
Rotterdam	80	67	77	65	74	64	69	78	67	75	65	72	66	96	73	64	90	71	63	85	68	14.6
NETHERLANDS ANTILLES																						
Willemstad	91	80	90	80	90	79	82	89	81	88	81	87	79	153	86	79	153	86	79	150	86	9.5
NEW CALEDONIA																						
Noumea	88	76	86	76	85	75	79	85	78	84	77	82	77	143	83	76	139	81	75	134	80	9.4
NEW ZEALAND																						
Auckland	77	66	76	66	74	66	70	75	69	73	67	72	68	104	72	67	99	71	66	94	70	11.3
Christchurch	83	62	79	61	76	60	65	77	64	74	62	71	62	83	67	60	78	67	59	74	65	17.5
Taiaroa Head	69	57	66	57	64	56	61	65	60	63	59	62	59	76	62	58	74	61	57	71	60	8.6
Wellington (934340)	74	64	71	63	70	62	66	71	65	69	64	68	64	90	69	63	86	67	62	83	67	9.7
NIGER																						
Agadez	108	67	107	67	105	66	75	92	74	92	73	91	71	122	81	70	116	82	68	111	82	22.5
Niamey	108	71	106	71	105	70	80	95	79	94	78	93	76	142	85	76	138	84	75	135	84	23.8
NORWAY																						
Bergen	73	59	68	57	65	55	61	68	59	64	58	63	59	74	61	57	70	60	56	66	59	11.5
Bodo	70	59	66	57	63	55	60	67	58	64	56	62	57	69	61	55	65	59	54	62	58	9.0
Oslo/Fornebu	80	63	77	62	73	60	65	75	63	71	62	69	62	83	67	60	78	66	59	74	65	15.8
Oslo/Gardermoen	78	60	75	58	71	57	62	73	60	70	59	67	59	76	63	57	71	62	55	66	61	18.0
Stavanger	73	59	70	58	66	57	62	69	60	66	59	64	60	77	64	58	74	62	57	69	61	11.3
Svinoy (Lgt-H)	64	57	61	56	60	56	59	62	58	60	56	59	58	71	60	56	68	59	55	65	58	4.1
Tromso	68	57	64	55	61	54	59	65	56	63	55	60	56	66	61	54	62	59	52	57	57	10.8
Trondheim	75	60	71	59	68	58	63	70	61	68	60	65	61	80	66	59	75	64	57	70	61	12.4
Utsira	67	58	64	58	61	57	60	64	59	62	58	60	59	76	62	58	73	60	57	70	59	5.2
OMAN																						
Masqat	109	73	107	73	105	73	86	93	85	93	84	92	84	181	91	83	174	91	82	169	90	14.9
Salalah	92	71	91	76	90	76	82	88	82	87	81	87	81	160	86	80	158	86	79	152	85	9.7
Thamarit	108	69	106	68	104	68	79	94	78	91	76	90	76	142	86	74	134	84	73	130	82	25.2
Tur'at Masirah	99	74	96	76	94	77	84	90	82	89	82	88	82	168	88	81	160	86	80	156	86	15.5
PANAMA																						
Panama	95	76	93	77	92	77	82	89	81	89	81	88	80	158	86	79	152	85	79	150	85	15.8
Tocumen	93	78	92	77	91	77	81	89	80	88	80	88	79	150	85	78	148	85	77	143	84	17.5
PARAGUAY																						
Asuncion	98	75	95	75	94	75	80	91	79	90	78	89	77	144	86	76	136	83	75	134	83	18.5
PERU																						
Arequipa	75	55	74	54	73	53	59	71	58	69	57	69	55	90	64	54	85	62	53	83	61	23.4
Cuzco	72	52	71	52	70	51	55	67	54	67	53	66	50	80	61	49	76	60	48	74	59	23.9
Iquitos	93	80	92	80	91	80	82	91	81	90	81	90	79	153	87	79	151	87	78	147	87	17.1
Lima	86	75	84	74	82	73	76	83	75	81	74	80	74	126	80	73	124	80	72	118	79	11.5
Pisco	86	75	83	73	82	72	76	84	74	82	73	80	73	123	83	72	117	81	70	111	79	12.4
Talara	90	76	88	75	87	74	79	86	78	84	77	83	77	143	82	76	139	82	75	134	81	14.2
PHILLIPPINES																						
Angeles, Clark AFB	97	78	95	77	93	77	82	89	82	89	81	87	81	164	86	80	161	86	79	154	85	17.6
Baguio	82	71	79	70	77	69	74	78	72	77	71	76	72	145	77	71	135	75	69	130	74	14.8
Cebu/Mandaue	93	81	92	81	91	80	82	90	82	90	81	89	80	153	87	79	152	87	79	150	87	12.4
Olongapo	98	77	96	77	95	78	83	91	82	90	81	89	81	160	88	79	152	86	79	150	86	17.1
Manila, Ninoy Aquino Int'l	95	81	93	80	92	79	83	91	82	90	82	89	81	161	89	80	158	88	79	152	87	15.8
POLAND																						
Bialystok	81	66	78	65	75	64	69	78	67	75	65	73	66	97	74	64	90	71	62	84	69	19.1
Gdansk	80	65	77	63	73	62	67	77	65	73	63	71	64	91	70	61	82	68	59	77	67	17.5
Katowice	83	67	80	65	77	64	68	80	67	77	65	74	64	94	72	63	88	71	62	85	69	18.4
Kielce	83	67	80	65	76	64	68	79	67	76	65	74	65	94	73	63	90	71	62	84	69	20.2
Kolobrzeg	80	65	75	63	71	63	67	74	65	72	64	70	64	90	70	63	85	68	61	80	67	12.1

MDB = mean coincident dry-bulb temperature, °F
MWB = mean coincident wet-bulb temperature, °F
MWS = mean coincident wind speed, mph
MWD = mean coincident wind direction
StdD = standard deviation
A = airport
HR = humidity ratio
DP = dew-point temperature, °F

Table 3A Heating and Wind Design Conditions—World Locations

Station	WMO#	Lat.	Long.	Elev. ft	StdP psia	Dates	Heating DB 99.6%	Heating DB 99%	Extreme Wind Speed 1%	Extreme Wind Speed 2.5%	Extreme Wind Speed 5%	Coldest Month WS/MDB 0.4% WS	Coldest Month WS/MDB 0.4% MDB	Coldest Month WS/MDB 1% WS	Coldest Month WS/MDB 1% MDB	MWS/MWD to DB 99.6% MWS	MWS/MWD to DB 99.6% PWD	MWS/MWD to DB 0.4% MWS	MWS/MWD to DB 0.4% PWD	Annual Extreme Daily Mean DB Max	Annual Extreme Daily Mean DB Min	Annual Extreme Daily StdD DB Max	Annual Extreme Daily StdD DB Min
Krakow	125660	50.08 N	19.80 E	777	14.292	8293	−1	6	20	18	16	24	41	20	36	3	60	6	240	89	−3	3.2	9.4
Lodz	124650	51.73 N	19.40 E	616	14.375	8293	2	9	22	20	17	24	38	20	34	5	90	8	130	92	0	4.0	11.2
Lublin	124950	51.22 N	22.40 E	787	14.287	8293	−2	5	21	18	16	21	29	19	30	5	180	5	220	88	−2	2.5	10.6
Poznan	123300	52.42 N	16.83 E	301	14.540	8293	3	11	22	19	17	23	35	20	35	3	90	8	220	92	3	3.8	11.2
Przemysl	126950	49.80 N	22.77 E	918	14.219	8293	1	7	23	21	19	27	34	23	33	5	270	6	250	86	−2	1.8	9.9
Snezka	125100	50.73 N	15.73 E	5291	12.098	8293	−3	2	80	68	62	85	20	78	20	36	340	13	200	70	−3	3.4	9.5
Suwalki	121950	54.13 N	22.95 E	610	14.379	8293	−5	2	26	21	18	27	36	22	30	3	20	7	300	87	−5	4.0	8.8
Szczecin	122050	53.40 N	14.62 E	9	14.695	8293	7	13	22	19	17	23	42	20	38	4	40	9	220	90	7	4.3	11.0
Torun	122500	53.03 N	18.58 E	236	14.575	8293	1	9	17	15	13	16	36	15	34	4	30	7	110	91	0	2.9	11.5
Warsaw	123750	52.17 N	20.97 E	351	14.514	8293	0	8	24	21	19	25	31	22	34	5	90	9	150	91	0	3.8	11.2
Wroclaw	124240	51.10 N	16.88 E	396	14.490	8293	2	10	21	18	16	22	42	19	39	4	110	8	170	92	0	3.4	10.8
PORTUGAL																							
Beja	85620	38.02 N	7.87 W	810	14.275	8293	36	38	22	19	17	23	55	19	53	7	90	10	180	104	32	2.2	3.1
Braganca	85750	41.80 N	6.73 W	2270	13.533	8293	26	28	22	19	16	24	38	20	43	2	180	8	240	97	22	2.5	3.6
Coimbra	85490	40.20 N	8.42 W	459	14.458	8293	35	38	22	17	14	23	54	19	53	4	180	6	310	101	32	2.2	2.2
Evora	85570	38.57 N	7.90 W	1053	14.149	8293	37	39	23	20	18	23	50	20	48	11	320	8	300	100	33	2.5	3.4
Faro	85540	37.02 N	7.97 W	13	14.693	8293	41	43	24	21	19	25	57	22	57	5	20	11	110	97	36	2.7	3.8
Lisbon	85360	38.78 N	9.13 W	403	14.487	8293	39	41	23	20	18	22	55	19	55	4	50	11	330	101	35	3.1	2.5
Portalegre	85710	39.28 N	7.42 W	1935	13.700	8293	34	37	24	20	18	24	48	22	47	11	290	8	240	98	30	2.7	3.2
Porto	85450	41.23 N	8.68 W	239	14.573	8293	35	37	24	21	18	26	54	22	54	7	90	9	330	95	30	2.3	2.5
Viana Do Castelo	85430	41.70 N	8.80 W	59	14.669	8293	33	35	19	16	13	20	55	17	56	1	50	6	160	97	29	2.3	1.8
PUERTO RICO																							
Cieba, Roosevelt Roads	785350	18.25 N	65.63 W	39	14.679	8293	68	70	17	16	14	18	79	17	79	2	330	11	80	94	66	4.1	1.1
San Juan	785260	18.43 N	66.00 W	62	14.667	8293	69	69	19	17	16	19	81	17	81	3	190	12	170	94	56	2.2	22.3
QATAR																							
Ad Dawhah	411700	25.25 N	51.57 E	32	14.683	8293	51	53	25	22	20	21	65	19	64	7	290	15	350	115	43	1.8	7.0
ROMANIA																							
Bucharest	154200	44.50 N	26.13 E	298	14.542	8293	8	14	20	17	15	20	28	18	29	3	250	5	230	97	2	3.8	6.5
Cluj-Napoca	151200	46.78 N	23.57 E	1354	13.994	8293	4	8	20	17	14	19	32	15	30	2	270	6	140	90	−4	3.4	5.9
Constanta	154800	44.22 N	28.63 E	45	14.676	8293	15	19	31	26	23	35	29	31	34	12	360	8	180	91	11	5.4	6.1
Craiova	154500	44.23 N	23.87 E	639	14.363	8293	10	15	32	23	19	31	30	23	38	5	270	4	180	97	5	3.8	8.5
Galati	153100	45.50 N	28.02 E	236	14.575	8293	7	12	27	21	19	28	25	22	27	9	20	8	230	94	4	3.6	5.4
Omul Mountain	152800	45.45 N	25.45 E	8231	10.823	8293	−13	−7	89	75	54	90	8	89	8	33	230	6	230	73	−18	13.7	10.4
Satu Mare	150100	47.78 N	22.88 E	406	14.485	8293	0	6	22	19	16	23	33	19	33	2	90	5	230	93	−4	4.0	7.7
Timisoara	152470	45.77 N	21.25 E	288	14.547	8293	9	14	19	17	14	18	33	15	34	4	360	6	200	97	2	2.9	7.7
RUSSIA																							
Abakan	298650	53.75 N	91.40 E	803	14.278	8293	−29	−24	23	20	17	22	18	19	18	1	350	5	50	91	−32	3.6	6.3
Aldan	310040	58.62 N	125.37 E	2237	13.550	8293	−41	−37	14	12	11	14	−2	12	−1	2	200	5	180	86	−45	1.8	5.9
Aleksandrovsk-Sahal	320610	50.90 N	142.17 E	102	14.646	8293	−17	−13	31	27	22	35	6	29	18	7	130	12	220	80	−22	4.5	4.9
Anadyr'	255630	64.78 N	177.57 E	203	14.592	8293	−37	−34	48	40	34	52	10	47	14	14	320	9	140	73	−38	5.4	4.9
Apuka	259560	60.45 N	169.58 E	26	14.686	8293	−18	−14	39	33	30	44	12	39	10	15	60	12	270	68	−24	3.8	4.9
Arkhangel'sk	225500	64.53 N	40.47 E	42	14.677	8293	−29	−23	17	15	13	18	22	16	23	2	130	6	140	85	−32	4.9	7.9
Armavir	370310	44.98 N	41.12 E	524	14.423	8293	4	10	23	19	15	24	31	21	32	2	140	6	90	97	−5	4.5	5.8
Astrakhan'	348800	46.27 N	48.03 E	59	14.669	8293	−1	4	26	21	19	27	22	22	22	7	270	11	90	100	−7	4.5	9.2
Barnaul	298380	53.40 N	83.70 E	826	14.266	8293	−21	−16	29	23	20	32	17	26	19	7	170	9	60	91	−25	7.0	9.2
Blagoveshchensk	315100	50.25 N	127.50 E	449	14.463	8293	−27	−23	19	17	14	18	4	14	1	2	310	7	180	91	−31	3.8	5.0
Borzya	309650	50.38 N	116.52 E	2244	13.546	8293	−37	−32	26	22	19	19	−1	16	−6	3	80	10	150	89	−42	3.2	4.9
Bratsk	303090	56.07 N	101.83 E	1604	13.868	8293	−32	−27	22	17	15	19	7	15	7	3	280	7	130	87	−36	4.0	7.9
Bryansk	268980	53.33 N	34.23 E	711	14.326	8293	−8	−3	23	20	18	25	29	21	24	7	110	10	220	86	−12	3.1	6.8
Chelyabinsk	286420	55.30 N	61.53 E	744	14.309	8293	−19	−15	29	24	21	29	17	25	15	7	340	12	190	90	−23	4.3	5.9
Cherepovets	271130	59.12 N	37.93 E	429	14.473	8293	−25	−17	22	18	15	23	26	20	23	2	20	6	170	84	−32	3.1	9.2
Chita	307580	52.02 N	113.33 E	2247	13.545	8293	−33	−29	24	21	19	21	10	17	9	0	310	8	210	90	−36	3.4	3.4
Dudinka	230740	69.40 N	86.17 E	62	14.667	8293	−50	−44	32	27	23	29	−1	24	−2	6	100	10	60	82	−55	4.9	4.5
Egvekinot	253780	66.35 N	179.12 W	85	14.655	8293	−34	−29	35	30	26	41	1	36	1	2	190	8	160	75	−36	9.4	6.7
Groznyy	372350	43.35 N	45.68 E	531	14.420	8293	5	10	24	21	18	25	28	22	29	3	270	9	90	96	−2	4.1	4.7
Habarovsk/Novy	317350	48.52 N	135.17 E	236	14.575	8293	−22	−18	23	20	18	21	2	19	−1	4	200	9	250	90	−26	3.1	2.9
Irkutsk	307100	52.27 N	104.35 E	1683	13.828	8293	−29	−24	23	21	18	21	2	18	2	4	80	8	190	87	−35	2.7	6.8
Izhevsk	284110	56.82 N	53.27 E	518	14.427	8293	−21	−16	24	21	18	26	18	23	19	5	100	11	160	89	−27	4.7	4.3
Juzno-Kurilsk	321650	44.02 N	145.87 E	131	14.630	8293	10	13	34	29	25	36	21	32	22	13	320	8	320	76	9	4.7	4.9
Juzno-Sahalinsk	321500	46.92 N	142.73 E	102	14.646	8293	−11	−8	19	16	14	21	11	18	13	3	360	7	180	83	−16	2.5	4.1
Kaliningrad	267020	54.70 N	20.62 E	88	14.653	8293	−3	6	16	14	12	16	36	14	34	4	350	5	120	88	1	3.8	9.7
Kaluga	277030	54.57 N	36.37 E	659	14.353	8293	−12	−7	21	18	16	23	20	18	21	3	340	9	130	85	−23	2.5	17.6
Kazan'	275950	55.78 N	49.18 E	380	14.499	8293	−18	−12	28	24	21	30	14	27	14	9	330	10	170	88	−23	3.4	5.4
Kirov	271960	58.65 N	49.62 E	482	14.446	8293	−27	−18	22	19	17	23	22	20	18	7	250	9	90	87	−31	3.4	6.8
Kolpashevo	292310	58.30 N	82.88 E	249	14.568	8293	−37	−30	20	17	15	19	14	16	13	3	290	7	160	86	−41	3.1	8.3
Krasnodar	349290	45.03 N	39.15 E	108	14.643	8293	3	9	23	20	18	25	41	21	30	7	50	8	90	94	−4	2.3	8.3
Krasnoyarsk	295740	56.00 N	92.88 E	908	14.224	8293	−29	−24	23	19	16	26	14	22	18	3	290	6	50	90	−31	6.3	7.0

WMO# = World Meteorological Organization number
Lat = latitude Long. = longitude
Elev = elevation, ft
StdP = standard pressure at station elevation, psia
DB = dry-bulb temperature, °F
WS = wind speed, mph

Table 3B Cooling and Dehumidification Design Conditions—World Locations

Station	Cooling DB/MWB						WB/MDB						DP/MDB and HR									Range of DB
	0.4%		1%		2%		0.4%		1%		2%		0.4%			1%			2%			
	DB	MWB	DB	MWB	DB	MWB	WB	MDB	WB	MDB	WB	MDB	DP	HR	MDB	DP	HR	MDB	DP	HR	MCDB	
Krakow	85	69	81	67	77	65	70	82	68	79	67	76	66	99	76	64	93	73	63	88	71	19.6
Lodz	84	66	80	65	77	63	68	80	66	77	65	74	64	92	72	62	86	71	61	82	70	18.7
Lublin	82	67	79	66	76	64	69	80	67	77	65	74	65	96	74	64	90	72	62	85	70	18.0
Poznan	85	66	81	64	78	63	68	80	67	78	65	75	64	92	72	63	86	71	61	80	69	19.6
Przemysl	82	67	79	66	76	64	69	79	67	76	66	74	66	98	74	64	92	72	63	88	70	14.9
Snezka	64	55	60	54	58	52	57	61	55	58	53	57	55	79	58	53	74	57	52	69	55	7.9
Suwalki	80	66	77	65	74	63	68	77	66	74	64	72	65	94	73	63	88	70	61	83	68	18.5
Szczecin	83	67	80	66	77	64	69	80	67	77	66	74	66	96	74	64	90	72	62	84	70	16.9
Torun	84	67	80	65	77	64	69	80	67	77	65	75	65	92	72	63	86	71	61	80	70	18.4
Warsaw	84	67	81	66	77	64	70	82	68	78	66	76	66	97	76	64	91	72	63	86	71	19.8
Wroclaw	84	67	81	66	78	64	69	81	67	78	66	76	65	94	73	63	88	71	62	83	71	19.1
PORTUGAL																						
Beja	99	70	95	68	92	67	71	94	70	92	69	89	65	94	74	63	89	74	62	85	73	29.5
Braganca	92	65	88	64	85	63	67	88	66	85	64	83	60	85	72	59	81	70	58	78	69	24.3
Coimbra	93	70	89	69	85	68	72	88	71	85	69	83	67	100	79	65	96	76	64	92	74	21.4
Evora	96	68	93	66	89	66	69	90	68	88	67	85	64	93	70	63	89	71	62	85	69	23.6
Faro	89	69	86	68	84	69	73	82	72	80	71	80	71	113	77	69	108	77	68	103	76	17.1
Lisbon	93	69	90	68	86	68	73	87	71	83	70	81	68	106	76	68	103	76	66	97	75	18.9
Portalegre	94	66	91	65	88	64	68	89	67	87	66	85	62	90	69	61	85	69	59	81	69	19.4
Porto	86	67	82	66	79	65	69	81	68	78	67	75	66	98	72	65	93	70	64	92	69	17.3
Viana Do Castelo	90	70	86	69	82	67	72	87	70	83	69	80	67	100	76	66	96	74	65	92	72	18.7
PUERTO RICO																						
Cieba, Roosevelt Roads	90	78	89	78	88	77	80	87	80	87	79	86	79	150	85	77	143	84	77	141	84	10.1
San Juan	92	77	90	78	89	78	81	88	80	87	79	87	78	147	85	78	146	84	77	143	84	12.2
QATAR																						
Ad Dawhah	109	71	107	72	105	72	87	94	86	93	85	93	85	184	92	84	180	91	83	171	92	19.4
ROMANIA																						
Bucharest	91	72	88	70	86	69	74	87	73	85	71	83	71	118	78	69	109	77	68	104	75	23.9
Cluj-Napoca	85	68	82	67	79	65	71	80	69	78	67	76	68	107	75	65	99	73	64	93	71	20.5
Constanta	83	72	81	72	79	71	75	80	74	79	72	78	74	126	78	72	118	77	71	113	76	12.2
Craiova	92	74	89	73	86	71	77	89	74	86	73	83	73	127	83	71	118	80	69	110	79	22.0
Galati	89	72	86	70	84	69	74	84	72	83	71	80	71	116	80	69	108	77	67	102	75	20.2
Omul Mountain	58	50	54	48	52	47	52	56	50	53	48	51	50	74	54	49	69	52	47	64	49	11.2
Satu Mare	88	71	85	69	82	68	72	85	71	83	69	80	68	106	78	67	100	76	65	94	74	23.2
Timisoara	92	70	88	69	85	67	71	87	70	85	69	82	67	100	75	66	96	74	64	91	73	23.0
RUSSIA																						
Abakan	85	64	82	63	78	62	67	81	66	77	64	75	63	90	71	62	85	70	60	80	69	18.9
Aldan	81	61	77	60	74	59	64	77	62	74	60	71	60	83	67	58	77	66	56	72	65	18.4
Aleksandrovsk-Sahal	74	64	71	63	68	61	67	72	65	69	63	67	65	92	69	63	85	67	61	80	65	11.3
Anadyr'	65	56	62	55	59	53	57	64	55	61	53	58	54	62	60	52	58	58	50	55	56	9.7
Apuka	61	55	58	53	56	52	56	60	54	57	53	55	54	62	57	53	60	56	51	56	54	8.5
Arkhangel'sk	79	65	75	63	72	62	67	76	65	73	63	70	64	90	72	62	83	69	60	77	67	16.9
Armavir	90	69	87	69	84	68	73	83	71	82	70	81	70	111	79	68	104	77	66	98	75	21.8
Astrakhan'	94	71	91	69	88	68	74	87	72	85	71	83	71	113	79	69	105	78	67	99	76	19.4
Barnaul	85	66	81	64	78	63	68	80	67	78	65	75	64	93	73	63	88	72	61	83	70	16.9
Blagoveshchensk	86	69	83	69	80	67	73	81	72	78	70	77	71	116	77	69	109	76	68	104	74	16.6
Borzya	83	63	80	62	77	62	67	76	66	75	64	73	64	97	72	62	92	70	61	87	69	19.1
Bratsk	81	63	77	62	74	61	66	76	64	74	63	71	62	90	70	61	84	69	59	78	67	16.6
Bryansk	81	66	78	65	75	64	68	76	67	75	65	73	65	97	72	64	90	71	62	85	69	14.6
Chelyabinsk	85	66	82	65	79	64	69	80	67	78	66	76	65	94	74	63	90	72	62	85	70	16.6
Cherepovets	79	67	76	65	73	63	69	77	66	74	65	72	66	97	74	64	90	70	62	83	68	18.2
Chita	84	65	80	63	77	61	68	79	66	76	64	73	64	98	74	62	91	70	60	85	68	20.5
Dudinka	76	62	72	60	67	58	64	73	61	70	59	67	60	77	69	57	69	66	55	64	63	14.4
Egvekinot	64	53	60	52	57	50	55	62	53	60	51	56	51	56	57	50	53	54	48	50	52	9.2
Groznyy	91	71	88	69	86	68	73	87	71	84	70	82	69	108	79	67	102	77	66	99	75	18.5
Habarovsk/Novy	86	70	83	69	80	68	73	81	72	79	70	78	71	115	76	69	108	76	67	102	73	16.2
Irkutsk	81	63	78	62	74	61	66	76	65	74	63	72	63	92	70	61	86	68	59	80	67	20.0
Izhevsk	85	66	81	65	78	64	69	80	67	78	65	75	65	94	74	63	89	72	62	84	70	17.3
Juzno-Kurilsk	68	65	66	64	65	63	66	67	65	66	63	64	65	94	67	64	90	65	63	86	64	5.6
Juzno-Sahalinsk	78	68	75	67	72	65	70	75	68	72	66	70	69	106	73	67	99	71	64	91	69	14.0
Kaliningrad	80	66	77	64	74	62	68	78	66	74	64	72	64	91	72	62	85	70	61	79	68	15.3
Kaluga	80	67	77	66	74	63	69	78	67	75	65	73	65	97	74	64	91	72	62	85	69	16.6
Kazan'	85	67	81	66	78	64	70	81	68	77	66	75	66	98	75	64	91	73	62	85	71	16.6
Kirov	82	67	79	65	75	64	69	79	67	76	65	73	66	97	75	64	90	72	62	85	70	17.3
Kolpashevo	82	66	79	65	76	63	69	78	67	76	65	73	66	97	73	64	90	72	62	84	69	18.0
Krasnodar	90	71	87	70	84	69	74	85	72	83	71	81	70	111	79	69	106	78	67	100	77	21.1
Krasnoyarsk	83	65	80	63	77	62	68	79	66	76	64	73	64	93	72	62	88	70	61	82	69	19.1

MDB = mean coincident dry-bulb temperature, °F
MWB = mean coincident wet-bulb temperature, °F
MWS = mean coincident wind speed, mph
MWD = mean coincident wind direction
StdD = standard deviation
A = airport
HR = humidity ratio
DP = dew-point temperature, °F

Table 3A Heating and Wind Design Conditions—World Locations

Station	WMO#	Lat.	Long.	Elev. ft	StdP psia	Dates	Heating DB 99.6%	Heating DB 99%	Extreme Wind Speed 1%	Extreme Wind Speed 2.5%	Extreme Wind Speed 5%	Coldest Month WS/MDB 0.4% WS	Coldest Month WS/MDB 0.4% MDB	Coldest Month WS/MDB 1% WS	Coldest Month WS/MDB 1% MDB	MWS/MWD to DB 99.6% MWS	MWS/MWD to DB 99.6% PWD	MWS/MWD to DB 0.4% MWS	MWS/MWD to DB 0.4% PWD	Annual Extreme Daily Mean DB Max	Annual Extreme Daily Mean DB Min	Annual Extreme Daily StdD DB Max	Annual Extreme Daily StdD DB Min
Kurgan	286610	55.47 N	65.40 E	259	14.563	8293	−25	−19	28	23	19	30	19	27	20	5	30	10	200	92	−31	3.8	7.6
Kursk	340090	51.73 N	36.27 E	688	14.338	8293	−9	−3	27	22	19	29	18	25	23	5	360	9	90	88	−14	2.3	10.1
Kyakhta	309250	50.37 N	106.45 E	2627	13.357	8293	−21	−18	19	14	12	12	6	9	6	0	150	6	180	91	−25	3.2	3.4
Magadan	259130	59.58 N	150.78 E	387	14.496	8293	−20	−15	22	20	17	24	18	21	15	1	50	9	270	73	−21	5.9	8.1
Magnitogorsk	288380	53.35 N	59.08 E	1253	14.046	8293	−19	−15	23	21	17	25	6	23	10	1	30	9	170	93	−25	5.4	3.6
Markovo	255510	64.68 N	170.42 E	108	14.643	8293	−54	−49	18	15	12	18	10	15	8	0	200	5	180	81	−59	2.9	6.3
Moscow	276120	55.75 N	37.63 E	511	14.430	8293	−10	−4	17	14	13	19	26	16	21	3	20	4	210	86	−14	3.2	5.4
Moscow, Vnukovo	275185	55.65 N	37.27 E	666	14.350	8293	−11	−6	23	21	18	27	27	23	27	6	10	9	240	86	−16	3.4	5.6
Murmansk	221130	68.97 N	33.05 E	167	14.611	8293	−20	−12	27	24	22	30	27	27	26	10	210	10	150	80	−19	3.6	10.3
Nikolayevsk	313690	53.15 N	140.70 E	223	14.582	8293	−28	−23	22	19	17	27	17	19	2	4	290	11	120	85	−31	3.8	4.5
Nikolskoe/Beringa	326180	55.20 N	165.98 E	20	14.690	8293	12	15	40	34	31	43	27	39	28	14	360	10	170	63	9	4.5	4.0
Nizhniy Novgorod (Gor'kiy)	275530	56.22 N	43.82 E	269	14.558	8293	−17	−11	21	19	16	22	18	19	24	4	350	9	150	89	−23	3.2	5.8
Nizhniy Tagil	282400	57.88 N	60.07 E	846	14.256	8293	−26	−20	17	15	13	19	3	16	1	5	20	8	250	88	−30	3.4	7.6
Novokuznetsk	298460	53.73 N	87.18 E	1010	14.171	8293	−23	−17	29	24	21	33	20	28	18	4	340	7	150	88	−26	3.2	7.7
Novosibirsk	296340	55.03 N	82.90 E	580	14.394	8293	−25	−19	27	23	19	28	19	25	18	7	220	10	180	89	−29	4.5	8.6
Nyurba	246390	63.28 N	118.33 E	423	14.477	8293	−63	−58	17	14	12	13	−9	12	−13	2	20	6	20	92	−66	9.0	5.9
Olekminsk	249440	60.40 N	120.42 E	741	14.310	8293	−54	−48	17	14	12	15	−3	12	−2	2	90	6	90	91	−56	4.7	6.7
Omsk	286980	54.93 N	73.40 E	403	14.487	8293	−24	−18	25	21	18	26	17	22	16	7	210	10	120	93	−28	4.3	6.7
Orel	279060	53.00 N	36.03 E	666	14.350	8293	−10	−4	26	22	20	28	23	23	25	7	360	10	280	87	−14	2.7	7.4
Orenburg	351210	51.78 N	55.22 E	357	14.511	8293	−18	−13	27	23	21	29	19	26	19	5	260	11	160	98	−24	4.3	4.5
Ozernaja	325940	51.48 N	156.48 E	95	14.650	8293	4	7	44	36	31	45	28	38	26	11	80	16	90	67	−2	5.0	3.1
Penza	279620	53.13 N	45.02 E	570	14.399	8293	−14	−9	27	23	20	30	18	26	23	6	220	11	130	92	−21	4.9	5.6
Perm'	282250	58.02 N	56.30 E	564	14.403	8293	−24	−18	21	18	16	22	16	18	17	6	280	10	200	90	−28	4.1	6.7
Petropavlovsk-Kamca	325400	52.97 N	158.75 E	79	14.658	8293	5	8	30	25	21	30	24	27	22	8	360	8	330	76	0	3.6	3.1
Petrozavodsk	228200	61.82 N	34.27 E	367	14.506	8293	−19	−11	17	15	13	18	27	16	24	4	90	6	290	83	−19	2.5	10.4
Pskov	262580	57.80 N	28.42 E	137	14.627	8293	−13	−4	20	17	15	22	31	19	30	2	150	6	130	85	−14	4.0	9.0
Rostov-Na-Donu	347310	47.25 N	39.82 E	252	14.566	8293	2	5	31	27	23	35	28	31	20	11	80	11	110	93	−3	3.4	4.9
Rubtsovsk	360340	51.50 N	81.22 E	705	14.329	8293	−26	−20	31	27	23	33	20	30	23	4	360	8	30	96	−29	3.6	7.7
Ryazan'	277310	54.62 N	39.72 E	557	14.406	8293	−10	−6	22	18	15	23	24	20	24	6	310	7	140	88	−15	2.5	6.1
Rybinsk	272250	58.00 N	38.83 E	374	14.502	8293	−19	−12	20	18	16	21	20	19	23	5	320	8	210	85	−24	2.9	7.9
Samara (Kuybyshev)	289000	53.25 N	50.45 E	144	14.623	8293	−17	−12	25	22	18	24	21	22	20	2	320	8	110	93	−23	4.1	4.1
Saratov	341720	51.57 N	46.03 E	511	14.430	8293	−8	−5	27	23	20	29	28	25	21	13	300	11	70	91	−13	3.8	3.6
Smolensk	267810	54.75 N	32.07 E	790	14.285	8293	−9	−4	18	15	13	17	33	15	29	6	280	6	230	84	−12	3.2	6.3
Sochi	371710	43.45 N	39.90 E	52	14.672	8293	28	30	19	16	14	23	43	19	44	8	70	8	270	88	22	2.7	3.8
St Petersburg	260630	59.97 N	30.30 E	13	14.693	8293	−9	−2	19	16	14	23	32	19	31	3	40	6	180	84	−11	3.6	8.5
Svobodnyy	314450	51.45 N	128.12 E	646	14.360	8293	−35	−30	17	15	13	15	−1	14	−3	1	290	7	180	90	−41	3.6	5.0
Syktyvkar	238040	61.72 N	50.83 E	390	14.494	8293	−32	−25	19	17	15	20	19	18	18	4	10	7	140	87	−36	3.2	7.2
Tambov	279470	52.73 N	41.47 E	456	14.459	8293	−13	−8	26	23	20	29	25	23	27	5	360	9	140	92	−20	5.0	7.0
Tayshet	295940	55.95 N	98.00 E	990	14.181	8293	−34	−29	19	16	14	18	16	15	14	2	330	7	40	89	−39	3.1	6.8
Ufa	287220	54.75 N	56.00 E	344	14.518	8293	−24	−18	23	20	18	26	23	22	19	3	140	8	50	93	−30	6.5	6.5
Ulan Ude	308230	51.80 N	107.43 E	1673	13.833	8293	−33	−29	32	27	23	23	4	17	6	0	50	8	20	91	−37	4.0	4.1
Urup Island	321860	46.20 N	150.50 E	230	14.578	8293	12	14	46	40	35	50	25	45	24	24	320	11	180	71	11	3.8	3.2
Ust'ilimsk	301170	58.03 N	102.73 E	1318	14.013	8293	−40	−35	20	17	15	19	10	17	10	9	300	7	170	87	−46	2.5	11.0
Ust-Kamcatsk	324080	56.22 N	162.47 E	89	14.653	8293	−23	−18	29	25	22	34	27	29	25	6	360	9	180	76	−30	8.1	4.5
Vladimir	275320	56.13 N	40.38 E	557	14.406	8293	−16	−10	22	19	17	23	21	21	21	7	10	9	210	86	−21	3.1	7.6
Vladivostok	319600	43.12 N	131.90 E	604	14.382	8293	−8	−4	35	31	27	33	4	30	4	21	360	10	240	86	−12	2.3	5.6
Volgograd	345600	48.68 N	44.35 E	475	14.449	8293	−6	−2	32	28	25	34	21	29	19	9	340	15	110	96	−11	3.8	5.0
Vologda	270370	59.23 N	39.87 E	429	14.473	8293	−27	−18	20	17	15	22	17	18	21	4	330	7	190	84	−32	3.4	7.9
Voronezh	341220	51.70 N	39.17 E	505	14.434	8293	−9	−5	25	21	18	28	24	24	25	6	20	10	160	90	−14	3.4	7.6
Yakutsk	249590	62.08 N	129.75 E	337	14.521	8293	−61	−58	15	13	12	11	−35	9	−37	2	340	7	110	90	−62	4.3	4.7
Yekaterinburg (Sverdlovsk)	284400	56.80 N	60.63 E	777	14.292	8293	−22	−17	20	17	15	19	17	17	14	5	280	10	300	89	−27	4.0	6.5
Yelets	279280	52.63 N	38.52 E	551	14.410	8293	−11	−6	17	15	13	18	27	16	25	4	10	6	140	91	−16	6.8	7.7
Zyryanka	254000	65.73 N	150.90 E	141	14.625	8293	−56	−52	18	15	13	14	−19	12	−21	1	150	6	160	89	−59	3.1	4.1
SAMOA																							
Pago Pago	917650	14.33 S	170.72 W	9	14.695	8293	72	73	25	23	22	26	78	24	78	5	310	11	80	92	58	3.2	16.0
SAUDI ARABIA																							
Abha	411120	18.23 N	42.65 E	6837	11.413	8293	41	44	23	21	19	25	59	22	60	2	180	13	20	92	35	5.6	2.9
Al Jawf	403610	29.78 N	40.10 E	2244	13.546	8293	32	35	26	22	19	26	49	22	52	5	50	9	320	111	28	4.5	3.6
Al Madinah	404300	24.55 N	39.70 E	2070	13.633	8293	48	50	20	18	16	19	65	17	67	7	60	9	300	114	42	2.2	4.0
Al Wajh	404000	26.20 N	36.47 E	52	14.672	8293	53	55	28	25	23	27	68	24	69	7	20	13	270	105	47	3.4	4.7
Ar'ar	403570	30.90 N	41.13 E	1811	13.763	8293	32	34	22	19	17	21	51	18	52	4	270	7	240	111	28	2.0	3.1
At Ta'if	410360	21.48 N	40.55 E	4753	12.344	8293	42	45	23	21	19	24	65	22	66	5	90	10	50	101	35	3.2	2.9
Az Zahran	404160	26.27 N	50.15 E	55	14.670	8293	45	47	26	23	20	22	61	19	61	9	290	13	360	116	40	2.3	3.8
Ha'il	403940	27.43 N	41.68 E	3323	13.018	8293	31	34	23	20	18	23	57	20	60	4	180	8	180	108	25	3.6	4.9
Hafar Al Batin	403730	28.33 N	46.17 E	1165	14.092	8293	36	39	27	23	21	24	56	22	56	6	270	10	240	117	31	2.3	1.8
Jiddah	410240	21.67 N	39.15 E	39	14.679	8293	59	61	23	21	19	23	78	21	77	6	30	13	330	114	54	4.5	3.4
Jizan	411400	16.90 N	42.58 E	9	14.695	8293	68	70	20	18	16	17	82	16	82	5	100	10	230	107	58	4.5	10.3

WMO# = World Meteorological Organization number
Lat = latitude
Long. = longitude
Elev = elevation, ft
StdP = standard pressure at station elevation, psia
DB = dry-bulb temperature, °F
WS = wind speed, mph

Table 3B Cooling and Dehumidification Design Conditions—World Locations

	Cooling DB/MWB						WB/MDB						DP/MDB and HR									
	0.4%		1%		2%		0.4%		1%		2%		0.4%			1%			2%			
Station	**DB**	**MWB**	**DB**	**MWB**	**DB**	**MWB**	**WB**	**MDB**	**WB**	**MDB**	**WB**	**MDB**	**DP**	**HR**	**MDB**	**DP**	**HR**	**MDB**	**DP**	**HR**	**MCDB**	**Range of DB**
Kurgan	87	67	83	66	80	65	70	82	68	80	67	77	66	96	75	64	90	73	63	85	72	18.7
Kursk	83	67	80	66	78	65	69	78	68	77	66	75	66	99	74	64	92	72	62	87	71	17.1
Kyakhta	83	62	80	61	76	60	65	78	64	76	62	73	61	89	69	59	83	68	57	78	67	18.0
Magadan	65	55	61	53	59	52	57	62	55	59	54	57	54	64	58	53	61	56	52	58	55	9.7
Magnitogorsk	85	65	82	64	79	62	68	80	66	78	64	76	64	92	73	62	86	72	60	80	70	18.9
Markovo	76	61	72	59	69	57	62	73	60	70	58	67	58	71	67	55	66	63	54	62	62	18.4
Moscow	82	67	79	65	76	64	69	78	67	76	65	73	65	96	73	64	90	72	62	85	70	14.8
Moscow, Vnukovo	81	66	77	65	75	63	68	77	67	76	65	73	65	94	74	63	88	69	61	83	68	16.4
Murmansk	74	59	70	57	66	56	61	71	59	68	57	65	57	70	64	55	64	63	53	60	60	12.2
Nikolayevsk	78	67	75	65	73	64	69	76	67	73	65	71	66	95	73	64	90	71	62	85	68	15.5
Nikolskoe/Beringa	57	54	55	53	54	52	55	56	53	55	52	54	54	62	56	53	59	54	51	56	53	4.5
Nizhniy Novgorod (Gor'kiy)	83	67	80	66	77	64	70	80	68	77	66	74	66	98	75	64	92	72	63	85	71	17.5
Nizhniy Tagil	83	65	80	64	76	63	68	78	66	76	64	74	64	93	72	63	88	71	61	83	68	18.4
Novokuznetsk	83	64	80	64	77	63	68	79	66	77	65	74	64	93	72	62	88	71	61	83	70	17.8
Novosibirsk	83	66	80	64	77	63	69	79	67	77	65	74	66	97	72	64	91	71	62	85	70	16.9
Nyurba	84	66	80	64	76	61	68	80	66	77	63	73	64	92	73	62	84	71	59	76	68	22.9
Olekminsk	85	65	81	64	77	62	67	81	65	78	64	74	63	88	71	61	82	72	59	76	68	20.2
Omsk	87	66	84	64	81	64	69	82	67	80	65	78	64	90	74	62	85	72	61	80	70	19.3
Orel	82	66	80	66	77	64	69	79	67	76	65	74	65	96	73	64	90	72	62	85	70	16.6
Orenburg	92	66	88	65	85	64	69	85	67	83	66	80	64	90	72	62	85	72	61	80	71	20.9
Ozernaja	60	56	58	54	56	53	57	60	55	57	54	55	56	66	58	54	62	56	53	59	55	5.9
Penza	85	65	82	64	79	63	69	81	67	78	65	75	65	94	74	64	90	72	62	85	70	18.7
Perm'	84	67	81	65	77	64	69	80	67	78	65	75	65	95	75	63	89	72	62	84	70	16.0
Petropavlovsk-Kamca	69	60	66	58	63	56	61	68	58	65	57	62	58	71	64	55	65	61	54	62	59	9.5
Petrozavodsk	77	64	73	62	70	60	66	74	64	71	62	69	64	90	70	61	82	67	59	76	65	13.7
Pskov	80	66	77	64	74	63	69	76	66	74	65	72	65	94	73	63	88	70	61	81	69	16.4
Rostov-Na-Donu	89	69	86	68	83	67	72	85	71	82	69	80	68	104	78	67	99	76	65	95	75	18.5
Rubtsovsk	87	67	84	66	81	64	69	81	68	79	66	77	66	98	73	64	92	71	62	87	71	20.2
Ryazan'	83	67	80	66	77	64	69	80	68	77	66	74	66	97	74	64	92	72	63	87	71	14.9
Rybinsk	80	68	77	65	74	64	69	78	67	74	65	72	66	96	73	64	90	72	63	86	69	13.0
Samara (Kuybyshev)	88	68	85	67	82	66	72	84	70	81	68	78	67	101	77	65	94	76	63	88	73	20.5
Saratov	87	66	84	65	81	64	69	82	68	80	66	78	65	94	74	63	89	73	62	84	72	15.1
Smolensk	79	66	76	65	74	63	68	76	66	74	65	72	65	95	72	63	90	71	62	84	69	15.1
Sochi	83	73	81	73	80	72	75	81	74	80	73	78	74	126	80	72	120	78	71	116	77	13.5
St Petersburg	79	65	76	64	73	62	67	76	66	74	64	71	64	90	72	62	84	69	61	79	68	13.5
Svobodnyy	85	67	82	67	79	66	72	80	70	77	68	76	69	109	75	67	103	73	66	97	72	18.7
Syktyvkar	83	66	79	65	75	62	69	80	67	76	65	73	65	94	74	63	88	71	61	82	68	17.1
Tambov	86	67	83	66	80	64	70	82	68	78	66	77	66	97	75	64	92	73	63	86	71	18.5
Tayshet	83	64	79	63	76	62	67	78	66	76	64	73	64	92	72	62	86	70	60	81	68	19.3
Ufa	87	67	83	66	80	65	70	82	68	80	67	78	66	98	76	64	91	73	63	86	72	19.1
Ulan Ude	85	63	82	62	78	61	67	78	65	76	63	74	62	90	72	61	84	70	59	79	68	20.7
Urup Island	63	59	60	57	57	56	60	62	58	59	56	57	60	78	61	57	71	58	55	65	56	9.2
Ust'ilimsk	82	63	78	62	75	61	66	76	64	74	62	73	63	90	70	61	83	69	58	77	67	20.2
Ust-Kamcatsk	67	57	63	56	60	54	59	64	57	62	55	59	56	66	60	54	63	59	53	60	57	10.6
Vladimir	82	67	78	66	75	64	69	79	67	76	65	73	66	97	74	64	91	71	62	85	70	15.3
Vladivostok	78	69	75	67	72	66	72	75	70	73	69	71	71	116	73	69	110	72	68	104	70	8.6
Volgograd	90	65	88	65	85	64	69	84	68	82	66	80	65	93	73	63	89	72	62	84	72	18.9
Vologda	80	66	76	64	74	63	69	78	67	73	65	71	65	96	74	64	90	71	62	83	68	16.9
Voronezh	85	66	82	65	79	63	69	80	67	78	66	75	65	94	73	64	90	72	62	85	71	17.5
Yakutsk	85	66	81	64	78	62	68	82	66	78	64	75	64	89	74	61	82	71	59	76	72	21.2
Yekaterinburg (Sverdlovsk)	84	66	81	65	77	64	69	80	67	77	65	74	65	96	75	64	90	71	62	85	70	17.1
Yelets	84	67	81	66	78	64	69	80	68	77	66	76	66	97	74	64	92	72	63	88	70	17.1
Zyryanka	83	63	79	62	74	60	66	79	64	76	61	72	60	79	72	58	73	70	56	66	68	17.3
SAMOA																						
Pago Pago	88	80	88	80	87	80	82	87	81	86	81	86	80	155	86	79	153	85	79	150	85	9.4
SAUDI ARABIA																						
Abha	87	56	86	56	84	56	67	76	66	75	65	74	65	119	72	63	111	71	62	108	71	21.4
Al Jawf	105	63	103	62	101	61	65	98	64	95	63	96	57	76	67	55	70	66	52	63	67	26.3
Al Madinah	113	65	110	65	109	64	69	97	68	97	67	98	62	91	74	60	82	75	57	76	76	23.8
Al Wajh	95	72	93	76	91	78	83	90	82	89	82	89	82	167	88	81	159	88	79	152	87	13.1
Ar'ar	107	68	105	68	103	67	73	103	71	100	69	98	62	90	92	60	81	90	57	75	86	25.6
At Ta'if	97	65	95	65	94	65	72	89	70	87	69	87	67	117	80	65	109	80	63	102	79	20.9
Az Zahran	111	71	109	71	107	72	86	94	84	94	83	92	84	180	91	82	168	90	80	158	90	23.9
Ha'il	105	65	104	64	102	64	68	98	67	97	65	97	59	85	71	57	80	72	55	74	71	28.3
Hafar Al Batin	113	67	112	66	110	66	71	99	69	100	67	100	64	94	74	61	85	72	59	78	71	27.5
Jiddah	104	72	102	73	100	74	83	94	82	94	81	92	81	159	89	79	151	89	79	150	88	22.0
Jizan	102	83	100	83	99	83	87	98	86	98	85	97	84	178	96	83	171	96	82	168	95	12.6

MDB = mean coincident dry-bulb temperature, °F
MWB = mean coincident wet-bulb temperature, °F
MWS = mean coincident wind speed, mph
MWD = mean coincident wind direction
StdD = standard deviation
A = airport
HR = humidity ratio
DP = dew-point temperature, °F

Table 3A Heating and Wind Design Conditions—World Locations

									Extreme Wind			Coldest Month WS/MDB				MWS/MWD to DB				Annual Extreme Daily			
				Elev.	StdP		Heating DB		Speed			0.4%		1%		99.6%		0.4%		Mean DB		StdD DB	
Station	WMO#	Lat.	Long.	ft	psia	Dates	99.6%	99%	1%	2.5%	5%	WS	MDB	WS	MDB	MWS	PWD	MWS	PWD	Max	Min	Max	Min
Khamis Mushayt	411140	18.30 N	42.80 E	6738	11.456	8293	40	43	21	18	16	22	61	20	61	2	150	10	30	97	35	6.8	4.1
Makkah	410300	21.48 N	39.83 E	1017	14.168	8293	59	62	14	12	11	15	77	13	78	4	20	8	300	118	53	2.0	4.3
Qasim	404050	26.30 N	43.77 E	2132	13.602	8293	37	39	21	18	16	20	60	17	57	2	30	8	90	115	32	3.8	3.2
Rafha	403620	29.63 N	43.48 E	1466	13.937	8293	33	35	25	22	20	24	54	22	56	4	270	9	300	115	27	2.7	2.7
Riyadh	404380	24.72 N	46.72 E	2007	13.664	8293	41	44	22	19	17	21	60	18	60	4	320	11	360	115	35	1.4	2.7
Tabuk	403750	28.37 N	36.63 E	2526	13.407	8293	34	37	25	20	17	25	60	20	60	2	110	10	270	107	30	2.2	2.3
Turayf	403560	31.68 N	38.67 E	2667	13.337	8293	29	32	25	22	20	26	46	22	46	6	270	9	270	106	25	2.3	3.4
Yanbu'al Bahr	404390	24.15 N	38.07 E	3	14.698	8293	52	54	26	23	21	25	72	22	72	3	10	17	270	114	47	1.8	2.0
SENEGAL																							
Dakar	616410	14.73 N	17.50 W	78	14.658	8293	61	62	23	21	19	23	69	21	69	10	360	10	360	100	54	4.0	7.7
Saint Louis	616000	16.05 N	16.45 W	13	14.693	8293	60	61	23	20	18	23	76	20	76	7	40	11	80	108	53	2.9	3.6
Tambacounda	616870	13.77 N	13.68 W	164	14.613	8293	63	65	17	15	13	17	82	16	81	3	80	6	100	110	53	3.2	6.8
Ziguinchor	616950	12.55 N	16.27 W	75	14.660	8293	61	63	14	12	10	15	82	12	81	1	40	6	60	106	54	0.9	7.4
SINGAPORE																							
Singapore	486980	1.37 N	103.98 E	52	14.672	8293	73	74	18	16	14	18	83	17	84	4	330	11	30	93	65	2.0	12.1
SLOVAKIA																							
Bratislava	118160	48.20 N	17.20 E	427	14.475	8293	9	14	21	18	16	23	35	19	36	3	50	8	160	94	5	2.9	8.8
Chopok Mountain	119160	48.93 N	19.58 E	6601	11.516	8293	−6	−3	52	46	42	61	6	55	6	29	330	11	180	63	−8	2.0	6.5
Kosice	119680	48.70 N	21.27 E	761	14.300	8293	8	12	29	25	21	30	21	26	25	9	350	8	180	89	4	2.5	6.7
Lomnicky Stit(Peak)	119300	49.20 N	20.22 E	8645	10.652	8293	−12	−8	52	44	38	59	−1	51	2	23	310	6	180	58	−13	2.5	5.9
Zilina	118410	49.23 N	18.62 E	1033	14.159	8293	2	8	18	15	13	19	37	15	31	4	70	7	250	89	−3	2.2	7.0
SOLVENIA																							
Ljubljana	130140	46.22 N	14.48 E	1263	14.041	8293	9	13	14	11	9	12	34	10	35	1	290	7	130	93	3	4.5	5.8
SOUTH AFRICA																							
Bloemfontein	684420	29.10 S	26.30 E	4422	12.498	8293	26	28	24	21	19	21	55	18	58	1	220	12	270	98	22	2.5	3.1
Cape Town	688160	33.98 S	18.60 E	137	14.627	8293	38	41	32	29	26	31	57	28	58	0	40	12	170	94	34	2.9	1.4
Durban	685880	29.97 S	30.95 E	26	14.686	8293	50	52	27	23	21	24	70	21	69	1	340	14	30	93	46	2.2	2.0
Johannesburg	683680	26.13 S	28.23 E	5577	11.969	8293	34	37	22	19	17	19	55	18	54	9	210	9	300	89	29	1.8	3.1
Marion Island	689940	46.88 S	37.87 E	72	14.662	8293	30	32	60	53	47	59	38	52	43	17	200	22	290	68	24	9.0	4.3
Port Elizabeth	688420	33.98 S	25.60 E	196	14.596	8293	43	46	33	29	26	31	58	28	60	2	270	10	290	97	38	3.8	2.2
Pretoria	682620	25.73 S	28.18 E	4337	12.538	8293	39	41	14	12	11	13	61	11	59	1	220	4	270	94	35	2.9	2.0
SPAIN																							
Barcelona	81810	41.28 N	2.07 E	19	14.690	8293	32	35	21	17	15	21	50	18	48	8	350	9	210	90	28	2.9	3.4
Granada	84190	37.18 N	3.78 W	1833	13.752	8293	25	28	21	18	16	20	49	17	49	0	230	12	180	103	19	1.6	5.0
La Coruna	80010	43.37 N	8.42 W	219	14.584	8293	39	41	27	23	20	29	53	26	53	6	140	7	60	86	35	3.1	2.7
Madrid	82210	40.45 N	3.55 W	1909	13.714	8293	24	26	22	19	17	23	47	19	47	0	360	8	240	102	20	1.8	3.1
Malaga	84820	36.67 N	4.48 W	22	14.688	8293	39	41	27	23	20	32	55	28	56	10	320	13	320	103	33	3.2	3.1
Palma	83060	39.55 N	2.73 E	26	14.686	8293	31	33	23	20	18	24	54	21	54	0	60	10	60	99	26	3.2	2.2
Salamanca	82020	40.95 N	5.50 W	2608	13.366	8293	23	25	27	22	19	28	46	24	45	1	80	7	300	98	18	2.2	4.5
Santander	80230	43.47 N	3.82 W	213	14.587	8293	36	39	24	19	16	28	51	23	51	5	110	7	40	91	34	4.0	2.5
Santiago De Compostela	80420	42.90 N	8.43 W	1204	14.072	8293	30	32	22	19	16	23	51	21	49	3	90	6	280	97	24	4.1	4.0
Sevilla	83910	37.42 N	5.90 W	101	14.646	8293	34	37	20	18	15	20	55	18	55	2	30	8	240	109	30	2.7	3.1
Valencia	82840	39.50 N	0.47 W	203	14.592	8293	34	36	27	23	19	33	57	27	56	4	280	12	120	100	29	4.0	2.7
Zaragoza	81605	41.67 N	1.05 W	862	14.247	8293	28	30	28	24	22	29	46	26	49	5	10	7	90	101	25	10.3	4.7
SWEDEN																							
Goteborg, Landvetter	25260	57.67 N	12.30 E	554	14.408	8293	3	10	26	23	20	27	38	24	37	9	40	9	310	83	2	4.1	9.7
Goteborg, Save	25120	57.78 N	11.88 E	173	14.608	8293	3	10	27	24	21	28	41	25	39	5	50	9	290	82	3	3.1	9.2
Jonkoping	25500	57.77 N	14.08 E	761	14.300	8293	−4	5	25	22	20	27	41	24	38	7	30	10	50	83	−7	4.7	10.3
Kalmar	26720	56.73 N	16.30 E	52	14.672	8293	5	10	27	23	21	28	41	26	41	6	270	11	270	84	3	4.5	7.0
Karlsborg	25440	58.52 N	14.53 E	334	14.523	8293	2	9	27	23	20	30	38	26	38	11	50	6	190	81	3	3.8	10.1
Karlstad	24180	59.37 N	13.47 E	180	14.604	8293	−5	1	22	20	18	26	38	23	38	4	350	9	200	81	−5	3.6	9.9
Kiruna	20440	67.82 N	20.33 E	1482	13.929	8293	−22	−17	26	23	20	30	29	26	28	4	210	10	190	76	−26	2.3	5.4
Malmo	26360	55.55 N	13.37 E	347	14.516	8293	7	14	30	27	24	31	37	28	36	9	340	12	140	82	8	3.4	9.7
Ostersund/Froso	22260	63.18 N	14.50 E	1213	14.066	8293	−14	−7	27	23	20	35	34	28	32	3	320	7	280	80	−17	3.1	9.4
Soderhamn	23760	61.27 N	17.10 E	118	14.637	8293	−7	0	22	19	17	24	34	21	29	6	290	10	130	84	−7	2.9	7.6
Stockholm, Arlanda	24600	59.65 N	17.95 E	200	14.594	8293	−2	5	24	21	18	28	37	24	36	4	350	8	180	84	−1	3.4	10.3
Stockholm, Bromma	24640	59.35 N	17.95 E	36	14.681	8293	−1	5	21	19	17	20	37	18	36	4	320	9	200	84	−1	3.8	9.7
Sundsvall	23660	62.53 N	17.45 E	32	14.683	8293	−14	−8	24	20	17	28	39	23	32	3	310	10	140	82	−14	3.1	9.4
Ungskar	26660	56.03 N	15.80 E	9	14.695	8293	11	16	41	37	34	43	34	39	37	11	20	12	250	74	13	4.3	9.4
Uppsala	24580	59.88 N	17.60 E	134	14.629	8293	−4	3	24	21	19	28	35	25	37	6	330	9	230	82	−4	2.7	11.9
Visby	25900	57.67 N	18.35 E	154	14.618	8293	12	16	31	28	25	33	33	30	34	12	20	12	210	81	8	3.8	8.5
SWITZERLAND																							
Geneva	67000	46.25 N	6.13 E	1364	13.989	8293	18	23	20	17	15	21	38	18	38	6	230	8	210	92	14	2.0	6.8
Interlaken	67340	46.67 N	7.88 E	1902	13.717	8293	15	19	16	13	11	15	35	12	32	6	190	8	280	88	11	3.4	7.4
Jungfrau Mountain	67300	46.55 N	7.98 E	11732	9.448	8293	−15	−11	48	41	33	48	3	43	6	16	310	11	140	48	−19	6.5	6.8
La Chaux-De-Fonds	66120	47.08 N	6.80 E	3343	13.009	8293	6	12	20	17	15	22	35	19	33	3	230	7	250	86	−1	8.6	11.0
Locarno	67620	46.17 N	8.88 E	649	14.358	8293	21	24	16	13	11	16	43	13	42	4	90	6	240	89	16	3.6	5.2

WMO# = World Meteorological Organization number; Lat = latitude; Long. = longitude
Elev = elevation, ft; StdP = standard pressure at station elevation, psia
DB = dry-bulb temperature, °F; WS = wind speed, mph

Table 3B Cooling and Dehumidification Design Conditions—World Locations

	Cooling DB/MWB						WB/MDB						DP/MDB and HR									
	0.4%		1%		2%		0.4%		1%		2%		0.4%			1%			2%			Range
Station	DB	MWB	DB	MWB	DB	MWB	WB	MDB	WB	MDB	WB	MDB	DP	HR	MDB	DP	HR	MDB	DP	HR	MCDB	of DB
Khamis Mushayt	88	57	87	57	86	56	66	75	65	74	64	74	63	111	71	62	106	71	61	102	70	22.1
Makkah	113	76	111	76	109	75	82	104	81	102	80	101	77	147	93	76	139	94	74	133	94	27.2
Qasim	110	67	109	65	107	64	74	97	70	96	68	98	68	111	84	63	92	80	59	81	75	29.3
Rafha	111	69	109	68	107	67	72	104	70	104	69	103	63	90	75	61	83	72	58	75	75	29.7
Riyadh	111	64	110	64	108	64	69	96	67	97	66	96	62	91	73	60	83	72	57	76	72	25.2
Tabuk	104	64	102	63	100	63	68	95	67	95	65	94	59	80	77	56	74	77	55	71	76	26.6
Turayf	102	64	99	63	97	62	68	92	66	91	65	90	60	87	78	58	78	75	56	73	74	27.4
Yanbu'al Bahr	109	76	106	75	104	76	83	96	82	95	81	94	81	159	90	79	150	89	77	142	88	25.7
SENEGAL																						
Dakar	89	74	88	77	86	77	81	86	80	85	79	84	79	152	84	79	150	83	78	146	83	9.7
Saint Louis	101	69	97	69	94	69	82	88	82	87	81	86	81	160	85	80	156	85	79	153	84	16.2
Tambacounda	106	70	104	70	103	70	81	90	80	89	79	88	79	150	85	78	146	84	77	142	83	22.7
Ziguinchor	101	72	98	71	96	72	83	90	82	90	82	89	81	161	88	80	156	87	80	153	86	27.7
SINGAPORE																						
Singapore	91	79	90	79	90	79	81	88	81	87	80	87	79	152	84	79	152	84	79	150	84	11.3
SLOVAKIA																						
Bratislava	89	69	86	68	83	67	71	85	69	83	68	80	66	98	77	65	92	75	63	88	74	22.1
Chopok Mountain	59	52	56	51	54	50	54	57	52	55	51	53	52	75	54	51	71	53	49	67	52	7.9
Kosice	85	68	82	67	79	65	70	83	68	80	67	77	65	95	76	64	91	74	62	86	72	19.3
Lomnicky Stit(Peak)	53	47	51	45	49	43	49	52	47	50	45	48	47	66	50	45	62	48	43	57	46	8.5
Zilina	85	67	82	65	78	64	68	82	67	79	65	76	64	92	73	62	88	71	61	84	70	21.8
SOLVENIA																						
Ljubljana	86	68	83	67	80	65	70	83	68	80	67	78	65	96	74	64	92	72	62	88	72	22.3
SOUTH AFRICA																						
Bloemfontein	93	60	91	60	89	60	67	79	66	79	65	78	64	104	69	62	99	69	61	94	68	26.3
Cape Town	87	67	83	67	81	65	70	82	69	80	68	77	67	99	73	65	94	72	65	92	72	15.8
Durban	87	75	85	75	83	74	78	84	77	83	76	81	76	136	81	75	132	81	74	128	80	9.9
Johannesburg	84	60	82	60	80	60	65	77	65	76	64	75	62	104	69	61	99	68	60	96	67	18.7
Marion Island	57	54	55	52	53	51	55	56	53	54	52	53	54	62	55	53	59	54	51	56	52	8.1
Port Elizabeth	85	66	81	68	79	68	73	78	72	77	71	76	71	117	76	70	112	74	69	108	74	12.1
Pretoria	89	64	88	63	86	63	69	80	68	80	67	79	66	111	73	64	106	72	63	102	72	17.6
SPAIN																						
Barcelona	85	74	84	74	82	73	77	83	76	82	74	80	75	133	82	74	126	80	73	121	79	15.1
Granada	99	67	96	67	93	66	70	91	69	89	68	88	64	97	76	63	92	75	61	85	74	33.7
La Coruna	77	65	74	65	72	64	67	75	66	72	65	71	65	92	70	64	90	69	63	87	68	9.4
Madrid	97	69	94	68	92	67	71	94	70	91	68	89	64	96	82	62	90	80	60	84	79	29.2
Malaga	93	68	90	68	86	68	75	82	74	81	73	80	73	123	79	72	118	78	70	112	77	16.4
Palma	91	74	89	73	86	73	78	85	77	84	76	83	77	139	83	75	132	82	73	125	81	22.3
Salamanca	93	65	90	64	86	63	67	88	65	86	64	83	61	87	73	59	82	69	57	78	68	28.6
Santander	80	67	76	67	75	66	71	76	69	74	68	73	69	107	73	67	102	72	66	97	71	9.4
Santiago De Compostela	88	69	84	67	80	66	71	85	69	82	67	77	66	101	76	65	95	73	63	90	70	21.2
Sevilla	104	75	100	72	97	71	76	98	74	96	72	92	70	110	84	68	104	80	66	97	78	30.1
Valencia	90	71	88	72	86	72	76	85	75	83	74	82	74	127	81	73	125	80	72	118	80	16.6
Zaragoza	97	69	93	69	90	68	72	89	71	89	70	85	67	101	77	66	99	76	64	93	78	24.1
SWEDEN																						
Goteborg, Landvetter	78	62	75	60	72	59	64	74	62	71	60	68	61	80	66	59	76	64	57	71	62	14.9
Goteborg, Save	78	62	74	61	71	60	65	73	64	70	62	68	63	85	68	61	80	66	59	76	65	13.7
Jonkoping	79	61	75	60	72	58	64	74	62	71	60	68	61	81	65	59	76	64	57	72	62	19.6
Kalmar	79	63	75	62	72	60	66	75	64	72	62	70	63	85	69	61	79	67	59	74	65	18.7
Karlsborg	76	63	73	61	70	60	65	74	63	70	61	68	61	82	68	60	77	66	58	73	65	14.2
Karlstad	77	63	74	62	71	60	65	73	64	70	62	68	63	85	67	61	80	65	59	76	64	15.7
Kiruna	70	56	66	54	63	52	58	66	56	63	54	61	54	66	60	52	62	59	50	57	57	13.7
Malmo	77	62	74	61	71	60	66	72	64	70	62	68	64	90	68	62	85	66	60	80	65	14.2
Ostersund/Froso	74	58	70	57	67	55	61	71	58	67	57	65	56	70	63	54	66	61	53	62	60	13.5
Soderhamn	77	62	73	60	70	59	64	73	62	71	60	68	61	80	67	58	74	65	57	69	64	16.2
Stockholm, Arlanda	80	63	77	61	73	59	65	74	63	71	62	69	63	85	67	61	80	65	59	75	64	16.2
Stockholm, Bromma	79	63	76	61	72	60	66	74	64	72	62	69	63	86	68	61	80	66	59	75	65	15.8
Sundsvall	75	62	72	59	69	58	64	71	62	68	60	66	62	83	66	59	76	64	58	71	62	15.8
Ungskar	71	65	68	63	67	62	66	69	64	68	63	66	64	90	69	63	85	67	61	80	65	6.8
Uppsala	78	62	75	61	71	59	65	73	63	71	61	69	62	82	69	59	76	67	58	71	65	16.7
Visby	75	63	72	62	70	61	66	71	64	70	63	68	64	90	68	62	84	66	61	80	65	13.1
SWITZERLAND																						
Geneva	86	66	83	65	80	64	68	82	67	80	65	77	64	93	72	62	89	71	61	84	70	22.1
Interlaken	82	65	79	65	76	63	67	80	65	77	64	74	63	91	72	61	87	70	60	83	69	17.8
Jungfrau Mountain	43	33	41	32	39	32	37	40	35	38	34	37	35	47	38	33	43	36	32	41	35	6.8
La Chaux-De-Fonds	78	62	74	61	71	59	63	75	62	72	60	70	59	85	67	58	80	66	56	76	65	17.6
Locarno	84	70	82	69	80	67	72	82	71	80	69	78	69	108	78	67	102	76	66	97	74	17.8

MDB = mean coincident dry-bulb temperature, °F
MWB = mean coincident wet-bulb temperature, °F
MWS = mean coincident wind speed, mph
MWD = mean coincident wind direction
StdD = standard deviation
A = airport
HR = humidity ratio
DP = dew-point temperature, °F

Table 3A Heating and Wind Design Conditions—World Locations

												Coldest Month WS/MDB				MWS/MWD to DB				Annual Extreme Daily			
				Elev.	StdP		Heating DB		Extreme Wind Speed			0.4%		1%		99.6%		0.4%		Mean DB		StdD DB	
Station	WMO#	Lat.	Long.	ft	psia	Dates	99.6%	99%	1%	2.5%	5%	WS	MDB	WS	MDB	MWS	PWD	MWS	PWD	Max	Min	Max	Min
Lugano	67700	46.00 N	8.97 E	905	14.225	8293	25	28	18	14	11	17	45	13	45	3	360	6	190	90	17	3.1	9.9
Payerne	66100	46.82 N	6.95 E	1610	13.864	8293	14	19	18	15	13	20	47	18	42	4	40	7	230	90	11	3.1	7.6
Saentis (Aut)	66800	47.25 N	9.35 E	8202	10.835	8293	−3	1	41	36	32	44	16	40	17	14	260	6	210	63	−8	3.2	8.3
San Bernardino	67830	46.47 N	9.18 E	5374	12.061	8293	6	10	23	21	19	26	22	23	22	4	310	7	120	73	1	2.5	8.5
Zurich	66600	47.38 N	8.57 E	1866	13.735	8293	13	18	20	16	13	23	44	20	42	6	60	5	230	89	11	2.7	7.9
SYRIA																							
Damascus	400800	33.42 N	36.52 E	1984	13.676	8293	25	28	26	23	20	25	50	22	49	3	30	7	210	105	19	2.9	3.8
TAIWAN																							
Hsinchu	467570	24.83 N	120.00 E	88	14.653	8293	47	50	22	18	16	19	57	17	55	9	40	9	270	99	44	4.3	3.1
Hualien	466990	23.98 N	121.60 E	62	14.667	8293	53	55	19	15	13	19	63	16	62	6	250	8	140	96	48	6.1	2.5
Kaohsiung	467400	22.58 N	120.35 E	29	14.684	8293	52	55	21	17	15	19	66	17	65	8	360	13	280	97	48	3.6	2.5
T'aichung	467510	24.18 N	120.65 E	367	14.506	8293	46	48	20	17	16	20	62	18	61	5	30	10	240	98	37	1.6	4.1
T'ainan (593580)	467410	23.00 N	120.22 E	45	14.676	8293	51	53	20	17	15	19	62	18	61	11	20	11	200	98	47	5.2	2.9
Taipei	466960	25.07 N	121.55 E	19	14.690	8293	48	50	20	18	16	19	63	18	63	4	110	11	290	98	42	2.7	3.8
Taipei Intl A	466860	25.08 N	121.22 E	108	14.643	8293	48	50	29	27	25	28	56	26	57	16	40	16	260	97	43	1.4	3.2
TAJIKISTAN																							
Dushanbe	388360	38.55 N	68.78 E	2634	13.353	8293	19	23	13	10	8	13	41	9	42	2	60	3	270	105	13	6.3	5.8
Khujand (Leninabad)	385990	40.22 N	69.73 E	1404	13.969	8293	17	21	29	26	23	30	31	27	33	14	240	11	240	105	10	5.8	8.6
THAILAND																							
Bangkok	484560	13.92 N	100.60 E	39	14.679	8293	65	68	19	16	14	14	82	12	81	4	40	11	180	102	61	3.1	2.7
Chiang Mai	483270	18.78 N	98.98 E	1030	14.161	8293	53	56	17	13	11	13	72	10	75	1	360	7	190	103	48	1.6	1.8
Chiang Rai	483030	19.92 N	99.83 E	1295	14.025	8293	49	52	10	8	7	9	69	7	71	0	20	5	180	102	44	1.6	2.5
Chumphon	485170	10.48 N	99.18 E	16	14.691	8293	66	68	18	15	13	19	83	17	83	0	30	9	120	100	63	2.0	2.3
Hat Yai	485690	6.92 N	100.43 E	114	14.639	8293	70	71	18	16	14	19	83	16	83	0	330	7	240	100	67	2.0	1.4
Phetchabun	483790	16.43 N	101.15 E	380	14.499	8293	56	59	10	9	8	10	79	9	79	0	360	6	180	104	46	1.3	19.1
Phrae	483300	18.17 N	100.17 E	531	14.420	8293	55	58	10	8	7	7	76	7	79	0	30	5	240	105	48	2.0	5.8
Tak	483760	16.88 N	99.15 E	406	14.485	8293	57	60	18	14	12	8	80	7	80	0	270	5	270	106	52	1.3	2.9
TRINIDAD																							
Port Of Spain	789700	10.62 N	61.35 W	49	14.674	8293	68	70	19	17	16	20	84	18	84	0	90	12	90	94	60	2.0	11.7
TUNISIA																							
Bizerte	607140	37.25 N	9.80 E	9	14.695	8293	38	40	29	25	22	32	54	27	54	1	320	12	100	106	33	4.9	0.9
Gabes	607650	33.88 N	10.10 E	16	14.691	8293	42	44	20	17	15	22	59	19	57	4	230	10	250	107	36	5.0	3.4
Gafsa	607450	34.42 N	8.82 E	1030	14.161	8293	36	38	26	23	20	26	52	21	53	5	60	9	240	109	31	2.7	1.6
Kelibia	607200	36.85 N	11.08 E	98	14.648	8293	42	44	23	19	17	25	55	22	55	5	300	9	300	97	38	6.3	2.0
Qairouan (Kairouan)	607350	35.67 N	10.10 E	223	14.582	8293	40	42	17	14	12	16	54	13	57	2	240	7	180	111	35	3.1	2.9
Tunis	607150	36.83 N	10.23 E	13	14.693	8293	41	43	26	23	21	28	55	24	56	6	240	11	180	107	35	5.2	1.3
TURKEY																							
Adana	173500	37.00 N	35.42 E	216	14.585	8293	32	34	19	17	14	21	48	19	49	6	30	8	210	103	26	1.8	3.4
Ankara	171280	40.12 N	32.98 E	3113	13.120	8293	2	8	21	18	15	19	33	16	33	1	20	8	270	95	−2	4.9	9.5
Erzurum	170960	39.92 N	41.27 E	5767	11.884	8293	−23	−17	24	22	20	26	24	22	22	0	310	10	90	88	−28	3.2	7.0
Eskisehir	171240	39.78 N	30.57 E	2575	13.382	8293	12	16	19	17	15	19	32	16	31	3	120	9	320	96	6	2.5	6.3
Istanbul	170600	40.97 N	28.82 E	121	14.636	8293	26	29	23	21	20	26	32	23	37	14	360	13	60	95	21	4.1	6.3
Izmir/Cigli(Cv/AFB)	172180	38.50 N	27.02 E	16	14.691	8293	28	31	26	23	21	31	57	26	53	6	360	14	350	101	24	4.0	1.8
Malatya	172000	38.43 N	38.08 E	2785	13.279	8293	10	16	23	21	18	25	31	21	32	4	210	7	60	102	3	2.5	6.5
Van	171700	38.45 N	43.32 E	5449	12.027	8293	6	9	16	13	11	15	32	12	31	4	90	3	300	90	2	4.7	5.8
TURKMENISTAN																							
Ashgabat (Ashkhabad)	388800	37.97 N	58.33 E	688	14.338	8293	20	23	22	19	17	19	42	16	39	4	110	10	90	110	15	2.5	4.0
Dashhowuz (Tashauz)	383920	41.83 N	59.98 E	288	14.547	8293	5	10	22	19	18	21	33	19	33	6	200	10	360	108	−1	1.8	5.9
UNITED KINGDOM & NORTHERN IRELAND																							
Aberdeen/Dyce	30910	57.20 N	2.22 W	213	14.587	8293	22	27	29	25	22	33	41	29	42	3	360	11	170	77	13	2.9	8.6
Aberporth	35020	52.13 N	4.57 W	439	14.468	8293	26	29	41	36	32	47	44	42	46	14	90	13	130	80	24	4.9	4.1
Aughton	33220	53.55 N	2.92 W	183	14.603	8293	26	28	26	23	21	29	44	25	44	8	130	9	130	82	24	4.0	4.1
Aviemore	30630	57.20 N	3.83 W	721	14.321	8293	15	21	29	25	22	33	40	30	41	1	360	9	200	77	9	9.5	8.6
Belfast	39170	54.65 N	6.22 W	265	14.559	8293	27	29	28	25	22	32	43	29	44	4	180	9	110	77	22	3.8	3.4
Birmingham	35340	52.45 N	1.73 W	324	14.528	8293	21	24	23	20	18	26	44	23	43	4	70	9	100	84	15	4.5	7.7
Bournemouth	38620	50.78 N	1.83 W	36	14.681	8293	22	25	27	23	21	30	49	26	47	4	20	10	90	84	18	4.1	4.0
Bristol	37260	51.47 N	2.60 W	36	14.681	8293	26	29	24	21	18	27	47	23	47	8	70	9	90	85	23	4.0	5.4
Camborne	38080	50.37 N	5.53 W	288	14.547	8293	30	32	34	30	28	37	44	34	45	13	50	13	100	76	27	3.8	5.4
Cardiff	37150	51.40 N	3.35 W	219	14.584	8293	25	28	31	28	24	38	44	33	44	14	60	9	60	83	22	4.0	4.5
Edinburgh	31600	55.95 N	3.35 W	134	14.629	8293	21	25	28	25	22	33	47	30	44	2	250	9	250	79	16	3.8	5.9
Exeter	38390	50.73 N	3.42 W	98	14.648	8293	24	27	27	24	21	33	47	28	46	7	40	10	150	82	22	4.5	3.4
Finningley	33600	53.48 N	1.00 W	55	14.670	8293	23	26	28	24	22	31	45	27	47	4	170	10	150	84	18	4.3	3.8
Glasgow	31400	55.87 N	4.43 W	26	14.686	8293	21	24	30	26	23	39	46	34	45	2	270	9	230	80	15	3.4	5.6
Hemsby	34960	52.68 N	1.68 E	45	14.676	8293	27	29	30	26	23	34	42	31	40	15	70	12	230	81	23	3.4	5.2
Herstmonceux	38840	50.87 N	0.33 E	55	14.670	8293	24	27	31	27	23	34	47	30	47	9	40	9	130	80	18	9.7	9.0
Jersey/Channel Islands	38950	49.22 N	2.20 W	275	14.554	8293	28	31	33	29	26	39	47	33	46	13	100	10	70	83	24	4.0	5.9
Kirkwall	30170	58.95 N	2.90 W	68	14.663	8293	29	31	41	34	30	41	41	36	39	11	270	11	120	70	26	2.7	2.0

WMO# = World Meteorological Organization number Lat = latitude Long. = longitude
Elev = elevation, ft StdP = standard pressure at station elevation, psia
DB = dry-bulb temperature, °F WS = wind speed, mph

Table 3B Cooling and Dehumidification Design Conditions—World Locations

	Cooling DB/MWB						WB/MDB						DP/MDB and HR									
	0.4%		1%		2%		0.4%		1%		2%		0.4%			1%			2%			Range
Station	DB	MWB	DB	MWB	DB	MWB	WB	MDB	WB	MDB	WB	MDB	DP	HR	MDB	DP	HR	MDB	DP	HR	MCDB	of DB
Lugano	85	71	83	69	81	68	73	82	71	80	70	79	69	112	79	68	106	77	66	100	76	17.6
Payerne	84	67	81	66	78	64	69	81	67	78	65	76	64	95	74	63	91	73	61	86	70	19.8
Saentis (Aut)	57	47	55	46	53	44	50	54	48	52	47	51	48	67	51	46	63	49	45	60	48	7.7
San Bernardino	69	55	67	54	64	53	58	65	57	64	55	62	56	81	62	54	76	60	52	71	58	14.9
Zurich	83	66	80	65	77	63	67	80	66	77	65	74	64	94	71	62	90	70	61	86	69	16.0
SYRIA																						
Damascus	101	64	98	64	96	64	69	85	68	85	67	84	66	103	72	64	97	72	62	91	71	33.8
TAIWAN																						
Hsinchu	93	81	92	81	91	81	83	91	82	90	81	90	80	158	89	80	154	88	79	150	87	12.4
Hualien	90	80	89	80	88	80	82	88	81	87	81	87	80	156	87	79	153	86	79	149	86	9.7
Kaohsiung	92	79	90	79	90	79	82	88	81	87	80	87	80	158	85	79	151	84	79	150	84	11.5
T'aichung	94	82	93	82	91	81	84	92	83	91	82	90	82	169	90	81	162	89	80	159	89	15.1
T'ainan (593580)	92	81	91	81	90	80	83	89	82	88	81	88	81	162	86	80	158	85	80	155	85	9.9
Taipei	94	80	93	80	92	80	82	91	81	90	81	90	80	153	86	79	150	86	79	148	86	13.3
Taipei Intl A	93	80	92	80	91	80	82	90	82	90	81	89	80	159	88	79	152	86	79	149	86	13.1
TAJIKISTAN																						
Dushanbe	99	67	97	67	95	66	71	93	69	91	68	90	64	98	83	62	91	81	60	85	80	25.6
Khujand (Leninabad)	99	67	96	66	94	66	70	92	69	91	67	90	63	90	79	61	84	79	59	79	78	23.0
THAILAND																						
Bangkok	99	80	97	79	95	78	84	94	83	91	82	89	82	164	88	81	160	87	80	156	87	16.7
Chiang Mai	100	72	98	72	96	73	79	89	78	88	78	87	77	146	83	76	139	82	75	138	81	24.5
Chiang Rai	98	72	96	72	94	73	79	89	79	88	78	87	77	147	84	76	144	83	76	141	82	25.0
Chumphon	95	79	94	79	92	79	82	92	81	91	80	90	79	150	87	78	147	87	78	145	86	16.7
Hat Yai	95	77	94	77	93	77	80	89	80	89	79	88	79	149	83	78	144	82	77	143	82	18.0
Phetchabun	101	78	99	78	97	78	82	92	81	91	81	90	79	153	87	79	150	87	78	148	86	20.9
Phrae	101	76	99	77	97	76	81	92	81	90	80	90	79	153	86	78	149	85	78	146	85	22.1
Tak	102	74	101	74	99	74	80	90	79	89	79	88	78	146	83	77	144	83	76	140	83	18.7
TRINIDAD																						
Port Of Spain	91	77	90	77	90	77	80	87	79	87	79	86	78	147	83	77	143	82	77	141	82	14.2
TUNISIA																						
Bizerte	97	72	92	72	89	72	77	86	76	84	75	83	74	129	81	73	123	81	72	118	80	18.2
Gabes	96	71	92	73	89	73	80	87	78	86	77	85	77	142	85	76	136	85	75	130	84	11.7
Gafsa	105	68	101	69	98	68	73	92	72	90	71	89	69	110	79	67	104	78	66	99	78	23.8
Kelibia	89	73	86	73	85	73	78	83	77	82	76	81	76	138	81	75	133	81	74	127	80	13.1
Qairouan (Kairouan)	105	71	100	71	97	71	76	90	75	88	74	88	73	125	81	72	118	80	70	112	80	25.4
Tunis	98	73	94	73	91	73	78	88	77	86	76	85	76	134	82	74	128	82	73	123	81	21.8
TURKEY																						
Adana	97	71	94	71	92	72	79	89	78	87	77	86	76	137	82	75	133	82	74	128	82	19.8
Ankara	90	63	86	63	84	62	65	84	64	83	63	81	59	83	73	57	78	72	55	74	70	28.4
Erzurum	84	61	82	60	79	59	64	80	62	78	61	76	58	88	74	56	82	72	54	76	70	29.9
Eskisehir	90	68	87	67	85	66	71	85	69	83	67	83	66	105	80	64	97	78	62	92	75	25.9
Istanbul	86	70	84	69	83	69	74	82	72	80	71	78	72	117	78	70	111	76	68	105	76	15.3
Izmir/Cigli(Cv/AFB)	96	72	93	71	91	70	74	92	73	90	72	89	68	104	83	67	98	82	66	95	81	23.0
Malatya	97	68	95	67	93	66	70	94	68	93	67	91	61	88	87	59	82	85	57	77	83	27.4
Van	84	66	82	66	80	65	70	81	68	80	66	78	66	116	80	64	108	78	62	101	77	19.4
TURKMENISTAN																						
Ashgabat (Ashkhabad)	104	67	102	67	99	67	73	94	72	92	70	91	66	99	85	64	92	85	63	87	84	24.1
Dashhowuz (Tashauz)	103	74	99	73	96	71	77	98	75	95	74	93	70	112	92	68	105	90	66	98	88	24.3
UNITED KINGDOM & NORTHERN IRELAND																						
Aberdeen/Dyce	71	62	68	60	65	58	64	70	61	67	60	64	60	79	67	59	75	64	57	70	62	13.0
Aberporth	72	62	68	61	65	60	64	69	62	66	61	64	62	84	65	61	80	64	59	77	62	9.4
Aughton	75	64	72	62	69	60	65	73	63	69	62	67	62	84	68	61	80	66	59	76	64	10.8
Aviemore	75	61	71	60	67	58	63	72	61	69	59	65	59	76	65	57	71	65	56	67	62	15.5
Belfast	72	62	69	61	67	60	64	70	62	67	61	65	61	82	66	60	78	65	58	74	64	12.8
Birmingham	78	64	75	62	72	61	65	75	64	72	62	70	62	84	68	61	80	67	59	76	65	16.9
Bournemouth	78	65	75	63	72	62	66	76	65	72	63	69	63	86	69	62	83	67	61	80	66	18.0
Bristol	79	65	76	63	73	62	67	76	65	73	63	70	63	85	71	62	82	68	60	78	66	12.8
Camborne	71	62	68	61	66	60	64	67	63	66	62	65	63	86	65	62	83	64	60	79	63	8.8
Cardiff	77	64	74	63	71	62	66	74	64	71	63	68	63	87	69	62	83	66	60	79	65	14.8
Edinburgh	72	61	69	60	66	59	63	69	61	67	60	65	60	78	66	59	74	64	57	71	63	14.6
Exeter	78	65	75	64	72	62	67	76	65	73	64	70	64	88	70	62	84	68	61	81	67	15.8
Finningley	78	64	75	63	72	61	65	75	64	72	62	70	62	83	69	60	78	67	59	75	65	17.3
Glasgow	75	63	71	61	67	59	64	72	62	69	60	66	61	79	67	59	75	65	58	71	63	14.6
Hemsby	74	65	71	63	69	62	66	72	64	70	63	67	63	87	69	62	83	67	61	79	65	13.9
Herstmonceux	76	65	74	63	71	62	66	75	65	72	63	69	63	88	70	62	83	67	61	80	66	15.3
Jersey/Channel Islands	76	65	73	63	70	62	66	74	64	71	63	68	63	87	67	62	83	65	61	81	65	11.0
Kirkwall	64	59	62	57	60	56	60	63	58	61	57	59	58	73	62	57	69	59	56	66	58	9.2

MDB = mean coincident dry-bulb temperature, °F
MWB = mean coincident wet-bulb temperature, °F
MWS = mean coincident wind speed, mph
MWD = mean coincident wind direction
StdD = standard deviation
A = airport
HR = humidity ratio
DP = dew-point temperature, °F

Table 3A Heating and Wind Design Conditions—World Locations

Station	WMO#	Lat.	Long.	Elev. ft	StdP psia	Dates	Heating DB 99.6%	Heating DB 99%	Extreme Wind Speed 1%	Extreme Wind Speed 2.5%	Extreme Wind Speed 5%	Coldest Month WS/MDB 0.4% WS	0.4% MDB	1% WS	1% MDB	MWS/MWD to DB 99.6% MWS	99.6% PWD	0.4% MWS	0.4% PWD	Annual Extreme Daily Mean DB Max	Mean DB Min	StdD DB Max	StdD DB Min
Lerwick	30050	60.13 N	1.18 W	275	14.554	8293	28	30	43	38	34	45	42	41	42	14	350	12	180	66	24	3.4	2.0
Leuchars	31710	56.38 N	2.87 W	39	14.679	8293	24	27	32	28	24	38	44	33	43	5	250	13	240	78	19	4.7	4.9
London, Gatwick	37760	51.15 N	0.18 W	203	14.592	8293	22	25	23	20	18	26	45	23	43	3	80	8	70	85	16	4.5	6.1
London, Heathrow	37720	51.48 N	0.45 W	78	14.658	8293	25	28	22	20	18	26	47	22	44	6	20	10	90	87	21	4.1	4.1
Lyneham	37400	51.50 N	1.98 W	511	14.430	8293	22	26	26	23	20	29	43	26	41	11	30	10	70	83	18	4.1	6.3
Lynemouth	32620	55.02 N	1.42 W	98	14.648	8293	28	31	45	38	33	47	42	44	41	16	190	15	260	76	24	5.0	4.3
Manchester	33340	53.35 N	2.27 W	255	14.565	8293	24	27	25	22	20	28	42	25	43	6	90	9	130	83	20	4.3	3.8
Nottingham	33540	53.00 N	1.25 W	383	14.497	8293	23	26	24	21	19	28	41	25	43	8	20	8	210	84	19	4.7	5.8
Oban	31140	56.42 N	5.47 W	13	14.693	8293	26	29	30	24	21	34	42	30	44	3	180	6	180	78	22	3.2	4.0
Plymouth	38270	50.35 N	4.12 W	88	14.653	8293	29	31	34	30	26	39	49	33	48	8	80	10	80	81	26	4.0	3.8
Stansted Airport	36830	51.88 N	0.23 E	347	14.516	8293	23	26	25	22	19	28	44	24	43	7	30	10	130	83	18	4.1	6.3
Stornoway	30260	58.22 N	6.32 W	42	14.677	8293	29	31	37	32	29	42	45	37	43	6	300	10	160	70	24	3.2	3.1
Valley	33020	53.25 N	4.53 W	36	14.681	8293	27	30	40	34	31	42	46	37	47	10	80	9	50	80	25	4.1	3.8
Wyton Raf	35660	52.35 N	0.12 W	134	14.629	8293	22	26	27	23	21	30	44	26	44	8	40	10	100	85	18	4.3	7.4
UKRANIAN																							
Chernihiv (Chernigov)	331350	51.48 N	31.28 E	449	14.463	8293	−7	−1	22	19	17	22	25	20	23	5	110	9	160	89	−10	4.3	9.4
Chernivtsi (Chernovtsky)	336580	48.27 N	25.97 E	787	14.287	8293	2	7	27	22	19	28	26	24	26	8	320	8	110	89	−2	3.4	6.1
Dnipropetrovs'k (Dneprop	345040	48.37 N	35.08 E	465	14.454	8293	0	4	27	23	21	29	27	25	28	9	50	11	90	93	−3	2.5	6.3
Donets'k	345190	48.07 N	37.77 E	741	14.310	8293	−1	3	30	26	22	37	21	32	23	6	70	11	100	91	−5	3.8	5.4
Kerch	339830	45.40 N	36.42 E	160	14.615	8293	11	15	28	24	21	34	26	28	26	11	30	11	40	90	9	3.8	5.6
Kharkiv (Khar'kov)	343000	49.87 N	36.13 E	498	14.437	8293	−3	2	23	20	18	27	28	23	22	6	30	11	110	90	−5	2.7	7.7
Kherson	339020	46.67 N	32.62 E	157	14.617	8293	4	8	23	20	18	28	27	24	26	7	270	8	80	93	2	2.2	5.2
Kirovohrad (Kirovograd)	337110	48.48 N	32.25 E	564	14.403	8293	−2	2	22	19	17	23	28	21	29	5	310	11	100	90	−5	2.5	5.2
Kryvyy Rih (Krivoy Rog)	337910	47.93 N	33.33 E	410	14.483	8293	0	4	27	23	20	32	24	26	26	9	50	10	90	92	−3	2.3	6.3
Kyyiv (Kiev)	333450	50.40 N	30.45 E	551	14.410	8293	−2	3	22	18	16	22	17	18	19	7	270	6	180	87	−3	3.4	9.2
Luhans'k (Voroshilovgrad)	345230	48.60 N	39.27 E	203	14.592	8293	−4	0	30	25	21	32	17	28	22	6	90	10	90	93	−9	3.4	5.8
Mariupol' (Zdanov)	347120	47.07 N	37.50 E	229	14.578	8293	4	8	35	31	27	38	26	36	23	12	70	10	90	89	1	2.3	4.3
Odesa	338370	46.48 N	30.63 E	114	14.639	8293	7	12	27	23	21	30	27	26	27	11	360	10	180	92	4	3.1	6.3
Poltava	335060	49.60 N	34.55 E	521	14.425	8293	−3	2	24	21	18	28	20	23	19	6	360	7	70	89	−6	2.9	8.3
Rivne (Rovno)	333010	50.58 N	26.13 E	767	14.297	8293	−3	3	27	23	19	27	28	24	29	6	270	9	130	87	−5	2.7	10.1
Simferopol'	339460	45.02 N	33.98 E	593	14.387	8293	8	13	27	23	20	28	30	25	28	9	50	11	50	92	5	2.9	5.8
Sumy	332750	50.88 N	34.78 E	570	14.399	8293	−7	−1	23	21	18	23	29	21	28	6	330	8	130	89	−10	4.0	9.4
Uzhhorod (Uzhgorod)	336310	48.63 N	22.27 E	387	14.496	8293	6	10	19	16	14	21	30	16	32	3	100	7	170	90	0	3.2	6.8
Vinnytsya (Vinnitsa)	335620	49.23 N	28.47 E	977	14.188	8293	−2	3	28	23	20	28	25	25	24	8	340	9	180	87	−5	2.9	8.1
Zaporizhzhya (Zaparozh'ye	346010	47.80 N	35.25 E	282	14.551	8293	0	5	24	21	17	28	28	24	29	8	360	9	220	92	−2	2.2	5.4
Zhytomyr (Zhitomir)	333250	50.27 N	28.63 E	744	14.309	8293	−4	2	24	21	18	24	33	21	30	6	90	9	190	87	−5	3.1	8.6
UNITED ARAB EMIRATES																							
Abu Dhabi	412170	24.43 N	54.65 E	88	14.653	8293	52	54	22	19	17	21	69	18	69	4	200	9	320	116	46	0.9	2.5
Dubai	411940	25.25 N	55.33 E	16	14.691	8293	54	55	21	19	17	22	66	19	68	4	170	11	270	114	49	2.5	1.8
Ra's Al Khaymah	411840	25.62 N	55.93 E	101	14.646	8293	49	52	18	15	14	17	69	15	71	2	210	10	320	115	43	1.4	2.3
Sharjah	411960	25.33 N	55.52 E	108	14.643	8293	49	51	19	17	15	20	68	17	69	5	120	10	270	115	42	2.2	4.1
URUGUAY																							
Colonia Del Sacramento	865600	34.45 S	57.83 W	75	14.660	8293	39	41	32	27	23	30	51	26	49	9	50	9	360	95	34	5.9	3.1
Montevideo	865800	34.83 S	56.00 W	104	14.644	8293	35	38	32	28	24	30	53	27	55	8	330	14	360	97	31	3.1	2.2
Paso De Los Toros	864600	32.80 S	56.52 W	246	14.570	8293	34	36	26	23	19	25	55	22	55	2	280	10	330	100	30	2.3	2.2
Rocha	865650	34.48 S	54.30 W	59	14.669	8293	34	36	24	20	18	23	53	20	53	1	310	9	360	96	29	4.3	2.5
Salto	863600	31.38 S	57.95 W	111	14.641	8293	34	37	19	17	14	19	57	17	59	1	120	9	20	103	30	2.5	2.7
Treinta Y Tres	865000	33.22 N	54.38 E	151	14.620	8293	33	35	19	15	12	22	52	17	53	2	270	6	290	99	29	2.3	1.8
UZBEKISTAN																							
Samarqamd (Samarkand)	386960	39.70 N	67.00 E	2375	13.481	8293	12	17	26	22	19	22	38	19	36	8	140	10	50	101	9	2.3	6.3
Tashkent	384570	41.27 N	69.27 E	1604	13.868	8293	13	18	13	11	10	14	45	11	42	2	90	4	300	105	10	2.2	5.4
VANUATU																							
Luganville	915540	15.52 S	167.22 E	144	14.623	8293	66	68	19	17	14	18	77	16	77	2	290	9	100	89	62	1.6	3.1
VENEZUELA																							
Caracas	804150	10.60 N	66.98 W	157	14.617	8293	70	71	12	11	9	12	81	11	81	1	140	6	340	97	59	2.2	14.0
VIETNAM																							
Ho Chi Minh City (Saigon)	489000	10.82 N	106.67 E	62	14.667	8293	68	70	39	26	17	25	83	16	82	4	360	9	160	101	56	5.0	20.5
WAKE ISLAND																							
Wake Island	912450	19.28 N	166.65 E	13	14.693	8293	71	72	29	26	24	30	74	28	75	15	40	14	80	94	67	3.1	2.5
WALLIS & FUTUNA ISLAND																							
Wallis Islands	917530	13.23 S	176.17 W	88	14.653	8293	72	73	21	18	17	21	80	19	80	2	160	9	100	91	68	3.1	2.3
YUGOSLAVIA																							
Belgrade	132720	44.82 N	20.28 E	324	14.528	8293	11	16	25	20	17	23	31	20	32	6	10	6	120	97	6	4.0	8.3
Palic	130670	46.10 N	19.77 E	344	14.518	8293	10	15	17	15	12	17	38	15	36	4	50	5	180	95	4	5.2	7.6
Podgorica (Titograd)	134620	42.37 N	19.25 E	108	14.643	8293	25	27	24	21	17	24	43	21	42	7	360	7	180	99	20	2.2	4.3
ZIMBABWE																							
Harare	677750	17.92 S	31.13 E	4931	12.263	8293	45	46	20	18	16	19	60	17	60	6	120	10	60	90	40	3.2	3.1

WMO# = World Meteorological Organization number
Lat = latitude Long. = longitude
Elev = elevation, ft
StdP = standard pressure at station elevation, psia
DB = dry-bulb temperature, °F
WS = wind speed, mph

Table 3B Cooling and Dehumidification Design Conditions—World Locations

	Cooling DB/MWB						WB/MDB						DP/MDB and HR									
	0.4%		1%		2%		0.4%		1%		2%		0.4%			1%			2%			Range
Station	DB	MWB	DB	MWB	DB	MWB	WB	MDB	WB	MDB	WB	MDB	DP	HR	MDB	DP	HR	MDB	DP	HR	MCDB	of DB
Lerwick	60	56	58	55	57	54	57	59	56	58	55	56	56	69	58	55	66	57	54	64	56	6.8
Leuchars	72	61	69	59	66	58	62	69	61	67	59	64	60	76	65	58	73	63	57	69	62	14.0
London, Gatwick	80	65	76	63	74	62	67	77	65	73	64	71	63	88	70	62	83	68	60	79	67	17.6
London, Heathrow	81	66	78	64	75	63	67	79	66	75	64	72	63	88	70	62	83	69	61	80	68	16.6
Lyneham	78	64	75	62	72	61	66	75	64	72	62	69	62	85	68	61	81	66	59	77	65	15.8
Lynemouth	69	61	67	60	65	58	62	67	61	65	60	64	61	80	64	59	76	63	58	73	62	8.8
Manchester	77	63	74	62	71	60	65	74	63	71	62	69	62	83	68	60	79	67	59	74	65	13.7
Nottingham	78	64	74	63	71	61	66	75	64	72	62	69	63	87	70	61	81	68	60	77	65	16.0
Oban	73	61	69	60	66	58	63	71	61	67	60	64	60	77	65	58	73	64	57	70	62	10.4
Plymouth	75	63	72	62	69	61	65	72	64	69	63	67	63	85	67	62	83	66	61	80	64	11.0
Stansted Airport	79	64	76	63	73	62	66	76	64	72	63	70	63	87	70	61	82	67	60	78	66	16.7
Stornoway	65	59	62	58	60	56	60	64	58	61	57	60	58	74	62	57	71	60	56	67	59	8.6
Valley	74	63	70	61	67	60	64	72	62	68	61	65	62	82	66	60	78	64	59	75	63	10.6
Wyton Raf	79	64	76	63	73	62	67	76	65	73	63	71	63	87	70	62	82	68	60	78	66	16.7
UKRANIAN																						
Chernihiv (Chernigov)	84	68	81	67	78	65	70	80	68	78	67	75	66	99	75	65	93	73	63	89	71	18.2
Chernivtsi (Chernovtsky)	83	68	80	66	78	65	69	80	68	78	66	76	66	98	74	64	92	73	63	88	72	16.2
Dnipropetrovs'k (Dneprop	87	67	85	66	82	65	70	82	69	80	67	78	67	99	75	65	94	74	63	89	72	18.7
Donets'k	86	66	83	66	80	65	70	82	68	80	67	76	65	97	74	64	92	73	63	88	72	19.6
Kerch	85	69	83	69	81	68	73	81	71	80	70	78	70	111	78	68	105	77	67	99	75	14.6
Kharkiv (Khar'kov)	85	67	83	65	80	65	70	80	68	78	67	76	66	98	74	65	93	73	63	88	71	16.4
Kherson	89	68	86	68	83	66	71	83	70	81	68	79	68	102	75	66	97	75	65	92	73	21.2
Kirovohrad (Kirovograd)	86	65	83	65	80	64	69	79	67	78	66	76	65	96	73	64	90	72	62	85	70	20.5
Kryvyy Rih (Krivoy Rog)	87	67	85	66	82	65	70	82	68	80	67	78	66	98	75	64	92	73	63	87	72	21.1
Kyyiv (Kiev)	83	67	80	66	77	65	69	79	68	77	66	75	66	99	74	65	93	73	63	88	71	16.6
Luhans'k (Voroshilovgrad)	88	67	85	66	82	64	70	83	68	81	67	78	66	96	76	64	90	72	63	86	72	19.1
Mariupol' (Zdanov)	84	71	82	70	80	69	74	81	72	79	71	78	72	118	78	69	109	77	68	103	75	15.1
Odesa	86	67	84	67	81	66	72	79	70	79	69	77	70	109	76	67	102	74	66	96	73	18.4
Poltava	85	66	82	66	80	65	69	79	68	78	66	76	66	98	75	64	92	73	63	87	71	17.6
Rivne (Rovno)	82	67	79	66	76	64	69	79	67	77	66	74	66	97	75	64	92	72	62	86	70	18.5
Simferopol'	87	67	85	66	82	65	70	81	69	79	67	78	67	100	74	65	94	73	63	89	72	20.3
Sumy	84	67	81	66	78	64	70	79	68	77	66	75	66	99	74	64	92	72	63	88	71	17.1
Uzhhorod (Uzhgorod)	86	69	83	68	81	66	71	83	69	81	67	79	66	97	77	64	92	75	63	88	73	18.7
Vinnytsya (Vinnitsa)	82	66	79	65	77	64	68	78	67	76	65	74	65	97	73	64	91	71	62	86	70	18.2
Zaporizhzhya (Zaparozh'ye	88	67	85	66	82	65	71	82	69	80	67	78	67	100	76	65	94	74	64	89	72	20.2
Zhytomyr (Zhitomir)	82	67	79	65	77	64	69	79	67	77	65	74	65	96	74	64	91	71	62	86	70	18.9
UNITED ARAB EMIRATES																						
Abu Dhabi	111	74	108	74	106	75	87	95	85	94	85	93	85	182	91	84	178	91	82	169	90	23.0
Dubai	107	75	105	75	103	76	86	94	85	94	85	93	85	181	91	84	177	91	82	169	91	17.5
Ra's Al Khaymah	111	76	109	77	107	77	86	100	85	99	84	98	83	174	94	82	169	93	81	160	92	22.7
Sharjah	110	77	107	77	106	77	86	99	85	97	84	96	84	178	92	82	169	91	81	160	91	23.9
URUGUAY																						
Colonia Del Sacramento	88	74	86	73	84	72	76	85	75	83	74	81	74	127	82	73	122	80	72	117	79	14.9
Montevideo	89	72	86	71	83	70	76	83	74	81	73	79	74	125	79	72	118	77	71	116	76	16.7
Paso De Los Toros	95	72	91	72	88	71	77	87	75	85	74	83	74	127	82	72	122	80	71	117	78	20.3
Rocha	89	73	86	72	83	71	76	84	75	82	74	80	74	130	80	73	123	78	72	118	77	19.1
Salto	98	74	95	74	92	74	79	91	77	89	76	87	76	134	85	74	129	83	73	123	81	22.0
Treinta Y Tres	92	73	89	72	86	71	77	86	75	84	74	81	74	130	80	73	123	79	72	118	78	20.9
UZBEKISTAN																						
Samarqamd (Samarkand)	96	67	94	66	92	66	70	92	68	89	67	89	62	91	81	60	85	77	58	80	76	24.8
Tashkent	100	71	98	68	95	68	75	95	72	92	70	90	68	111	89	65	99	84	63	91	82	26.8
VANUATU																						
Luganville	87	78	86	78	85	77	80	85	79	84	79	84	78	148	83	78	146	82	77	143	82	10.4
VENEZUELA																						
Caracas	92	84	91	83	90	83	86	90	85	89	84	88	86	190	89	84	181	88	84	177	87	12.6
VIETNAM																						
Ho Chi Minh City (Saigon)	95	77	94	77	93	77	81	90	80	89	80	89	79	150	86	78	147	85	77	143	84	14.8
WAKE ISLAND																						
Wake Island	89	79	89	79	88	78	81	86	80	86	80	85	80	155	85	79	150	84	78	146	84	8.1
WALLIS & FUTUNA ISLAND																						
Wallis Islands	88	80	87	80	87	80	82	87	81	86	81	86	80	157	85	79	153	85	79	151	84	8.5
YUGOSLAVIA																						
Belgrade	92	71	89	70	86	69	73	87	71	85	70	83	68	105	80	67	99	78	65	94	76	22.1
Palic	90	70	87	69	84	67	71	87	70	84	69	82	67	99	77	65	94	76	64	90	75	20.3
Podgorica (Titograd)	95	71	93	71	90	70	73	90	72	89	71	87	68	104	79	67	99	80	66	96	78	21.1
ZIMBABWE																						
Harare	86	62	84	61	83	61	68	76	67	76	67	74	66	116	70	65	112	69	65	110	69	21.1

MDB = mean coincident dry-bulb temperature, °F
MWB = mean coincident wet-bulb temperature, °F
MWS = mean coincident wind speed, mph
MWD = mean coincident wind direction
StdD = standard deviation
A = airport
HR = humidity ratio
DP = dew-point temperature, °F

CHAPTER 27

RESIDENTIAL COOLING AND HEATING LOAD CALCULATIONS

THIS chapter covers the engineering basis of modified residential load calculation procedures for the nonengineer. The procedures described in Chapter 28 may be used to calculate a heating or cooling load for residential buildings.

RESIDENTIAL FEATURES

With respect to heating and cooling load calculation and equipment sizing, the unique features distinguishing residences from other types of buildings are the following:

- Unlike many other structures, residences are usually occupied and conditioned 24 h per day, virtually every day of the cooling and heating seasons.
- Residential system loads are primarily imposed by heat loss or gain through structural components and by air leakage or ventilation. Internal loads, particularly those from occupants and lights, are small in comparison to those in commercial or industrial structures.
- Most residences are conditioned as a single zone. Unit capacity cannot be redistributed from one area to another as loads change from hour to hour; however, exceptions do occur.
- Most residential cooling systems use units of relatively small capacity (about 18,000 to 60,000 Btu/h cooling, 60,000 to 110,000 Btu/h heating). Because loads are largely affected by outside conditions, and few days each season are design days, the unit operates at only partial load during most of the season; thus, an oversized unit is detrimental to good system performance, especially for cooling in areas of high wet-bulb temperature.
- Dehumidification occurs during cooling unit operation only, and space condition control is usually limited to use of room thermostats (sensible heat-actuated devices).
- Multifamily living units are similar to single-family detached houses, but the living units may not all have surfaces exposed in all directions. This affects load calculation.

Categories of Residences

Single-Family Detached. A house in this category usually has exposed walls in four directions, often more than one story, and a roof. The cooling system is a single-zone, unitary system with a single thermostat. Two-story houses may have a separate cooling system for each floor. The rooms are reasonably open and generally have a centralized air return. In this configuration, both air and load from rooms are mixed, and a load-leveling effect, which requires a distribution of air to each room that is different from a pure commercial system, results. Because the amount of air supplied to each room is based on the load for that room, proper load calculation procedures must be used.

Multifamily Buildings. Unlike single-family detached units, multifamily units by definition do not have exposed surfaces facing in all directions. Rather, each unit has only one or two exposed surfaces and possibly a roof. Two exposed walls will be at right angles, and both east and west walls will not be exposed in a given living unit. Each living unit has a single unitary cooling system or a single fan-coil unit, and the rooms are relatively open to one another. This configuration does not have the same load-leveling effect as a single-family detached house, but it is not a commercial building. Therefore, a specific load calculation procedure is required.

Other Categories. Many buildings do not fall into either of the above categories. Critical to the designation of a single-family detached building is the exposure of both east and west walls. Therefore, some multifamily structures should be treated as single-family detached when the exposed surfaces are oriented in a particular way. Examples include duplexes or apartments with either exposed east, west, and south walls or exposed east, west, and north walls, with or without a roof; and apartments, town houses, or condominiums with only east and west or north and south exposed walls.

COOLING LOAD

LOAD COMPONENTS

A cooling load calculation determines total sensible cooling load due to heat gain (1) through structural components (walls, floors, and ceilings); (2) through windows; (3) caused by infiltration and ventilation; and (4) due to occupancy. The latent portion of the cooling load is evaluated separately. While the entire structure may be considered a single zone, equipment selection and system design should be based on a room-by-room calculation. For proper design of the distribution system, the amount of conditioned air required by each room must be known.

Peak Load Computation

To select a properly sized cooling unit, the peak or maximum load (**block load**) for each zone must be computed. Because this procedure may vary considerably for different types of buildings, each building type has to be considered; the block load for a single-family detached house with one central system is the sum of all the room loads. If the house has a separate system for each zone, each zone block load (i.e., the sum of the loads for all rooms in each zone) is required. When a house is zoned with one central cooling system, the block load must be computed for the complete house as if it were one zone. In multifamily structures, each living unit has a zone load that equals the sum of the room loads. For apartments with separate systems, the block load for each unit establishes the system size. Apartment buildings with a central cooling system (i.e., a hydronic system with fan-coils in each apartment) require a block load calculation for the complete structure to size the central system; each unit

The preparation of this chapter is assigned to TC 4.1, Load Calculation Data and Procedures.

load establishes the size of the fan-coil and air distribution system for each apartment. One of the methods discussed in Chapter 28 may be used to calculate the block load.

Indoor Temperature Swing

For hour-by-hour load calculations, allowing for a swing in indoor temperature results in lower peak loads. Because the indoor temperature does swing, such an allowance gives a more reasonable equipment capacity. The tables in this section are based on an assumed indoor temperature swing of no more than 3°F on a design day, when the residence is conditioned 24 h per day and the thermostat is set at 75°F.

Cooling Load Due to Heat Gain Through Structure

The sensible cooling load due to heat gains through the walls, floor, and ceiling of each room is calculated using appropriate **cooling load temperature differences** (CLTDs) (Tables 1 and 2) and U-factors for summer conditions. For ceilings under naturally vented attics or beneath vented flat roofs, the combined U-factor for the roof, vented space, and ceiling should be used. The mass of the walls is a variable in Table 2 and is important in calculating energy use, but it is not used in Table 1 because of the averaging technique required to develop the CLTDs. Values in Tables 1 and 2 assume a dark color because color is an unpredictable variable in any residence.

Daily range (outdoor temperature swing on a design day) significantly affects the equivalent temperature difference. Tables 1 and 2 list daily temperature ranges classified as high, medium, and low. Tables 1, 2, and 3 in Chapter 26 list outdoor daily ranges of dry-bulb temperature for different locations.

Cooling Load Due to Heat Gain Through Windows

Direct application of procedures for calculating cooling load due to heat gain for flat glass (discussed in Chapters 28 and 29) results in unrealistically high cooling loads for residential installations. Window **glass load factors** (GLFs), modified for single- and multifamily residential cooling load calculations and including solar heat load plus air-to-air conduction, are given in Tables 3 and 4. Table 5 lists the **shading coefficients** (SCs) and U-factors used to compile Tables 3 and 4.

Table 1 CLTD Values for Single-Family Detached Residences[a]

Daily Temperature Range[b]	Design Temperature, °F 85 L	85 M	90 L	90 M	90 H	95 L	95 M	95 H	100 M	100 H	105 M	110 H
All walls and doors												
North	8	3	13	8	3	18	13	8	18	13	18	23
NE and NW	14	9	19	14	9	24	19	14	24	19	24	29
East and West	18	13	23	18	13	28	23	18	28	23	28	33
SE and SW	16	11	21	16	11	26	21	16	26	21	26	31
South	11	6	16	11	6	21	16	11	21	16	21	26
Roofs and ceilings												
Attic or flat built-up	42	37	47	42	37	51	47	42	51	47	51	56
Floors and ceilings												
Under conditioned space, over unconditioned room, over crawl space	9	4	12	9	4	14	12	9	14	12	14	19
Partitions												
Inside or shaded	9	4	12	9	4	14	12	9	14	12	14	19

[a]Cooling load temperature differences (CLTDs) for single-family detached houses, duplexes, or multifamily, with both east and west exposed walls or only north and south exposed walls, °F.
[b]L denotes low daily range, less than 16°F; M denotes medium daily range, 16 to 25°F; and H denotes high daily range, greater than 25°F.

In application, the area of each window is multiplied by the appropriate GLF. The effects of permanent outside shading devices should be considered separately in determining the cooling load. Shaded glass is considered the same as north-facing glass. The **shade line factor** (SLF) is the ratio of the distance a shadow falls beneath the edge of an overhang to the width of the overhang (Table 6). Therefore, assuming the overhang is at the top of the window, the shade line equals the SLF times the overhang width. The shaded and sunlit glass areas may then be computed separately. The tabulated values are the average of the shade line values for 5 h of maximum solar intensity on each wall orientation shown. Northeast- and northwest-facing windows are not effectively protected by roof overhangs; in most cases, they should not be considered shaded.

Infiltration

Natural air leakage in residential structures is less in summer than in winter, largely because wind velocities are lower in most localities. The data in Tables 7 and 8 showing space air changes per

Table 2 CLTD Values for Multifamily Residences[a]

Daily Temperature Range[b]		Design Temperature, °F 85 L	85 M	90 L	90 M	90 H	95 L	95 M	95 H	100 M	100 H	105 M	110 H
Walls and doors[c]													
N	Light	14	11	19	16	12	24	21	17	26	22	27	32
	Medium	13	10	18	15	11	23	20	16	25	21	26	31
	Heavy	9	6	15	11	7	20	16	12	21	17	22	27
NE	Light	23	17	28	22	17	33	27	22	32	26	31	36
	Medium	20	15	25	20	16	30	25	21	29	25	29	34
	Heavy	16	12	21	17	13	26	22	18	26	22	26	31
E	Light	32	27	37	32	27	43	38	32	42	37	42	47
	Medium	30	24	34	29	24	40	34	29	39	33	39	44
	Heavy	23	18	28	23	18	34	29	23	33	28	33	38
SE	Light	31	27	35	31	26	41	37	31	42	37	42	47
	Medium	28	22	32	27	22	37	32	27	37	33	38	43
	Heavy	21	16	26	22	17	32	27	22	31	27	32	37
S	Light	25	22	29	26	22	35	31	26	36	32	37	43
	Medium	22	18	26	22	18	31	26	22	31	27	32	38
	Heavy	16	11	20	16	12	26	21	17	26	21	27	33
SW	Light	39	36	44	40	35	50	46	40	51	47	52	58
	Medium	33	29	37	34	29	44	40	35	45	40	46	52
	Heavy	23	18	28	24	19	36	31	25	35	30	36	42
W	Light	44	41	48	45	40	54	51	46	56	52	57	63
	Medium	37	33	41	38	33	46	42	38	48	43	49	55
	Heavy	26	22	31	27	23	37	32	27	37	32	38	44
NW	Light	33	30	37	34	30	43	39	34	44	40	45	50
	Medium	28	25	32	29	24	37	33	29	39	35	40	45
	Heavy	20	16	25	20	16	31	26	21	31	26	32	37
Roof and ceiling													
Attic or flat built-up	Light	58	53	65	60	55	70	65	60	70	65	72	77
Flat built-up	Medium or heavy	21	18	23	21	18	25	23	21	25	23	25	28
Floors and ceiling													
Under or over unconditioned space, crawl space		9	4	12	9	4	14	12	9	14	12	14	19
Partitions													
Inside or shaded		9	4	12	9	4	14	12	9	14	12	14	19

[a]Cooling load temperature differences (CLTDs) for multifamily low-rise or single-family detached if zoned with separate temperature control for each zone, °F.
[b]L denotes low daily range, less than 16°F; M denotes medium daily range, 16 to 25°F; and H denotes high daily range, greater than 25°F.
[c]Light denotes lightweight; medium denotes medium-weight; and heavy denotes heavyweight construction.

Table 3 Window Glass Load Factors (GLFs) for Single-Family Detached Residences[a]

Design Temperature, °F	Regular Single Glass						Regular Double Glass						Heat-Absorbing Double Glass						Clear Triple Glass		
	85	90	95	100	105	110	85	90	95	100	105	110	85	90	95	100	105	110	85	90	95
No inside shading																					
North	34	36	41	47	48	50	30	30	34	37	38	41	20	20	23	25	26	28	27	27	30
NE and NW	63	65	70	75	77	83	55	56	59	62	63	66	36	37	39	42	44	44	50	50	53
East and West	88	90	95	100	102	107	77	78	81	84	85	88	51	51	54	56	59	59	70	70	73
SE and SW[b]	79	81	86	91	92	98	69	70	73	76	77	80	45	46	49	51	54	54	62	63	65
South[b]	53	55	60	65	67	72	46	47	50	53	54	57	31	31	34	36	39	39	42	42	45
Horizontal skylight	156	156	161	166	167	171	137	138	140	143	144	147	90	91	93	95	96	98	124	125	127
Draperies, venetian blinds, translucent roller shades, fully drawn																					
North	18	19	23	27	29	33	16	16	19	22	23	26	13	14	16	18	19	21	15	16	18
NE and NW	32	33	38	42	43	47	29	30	32	35	36	39	24	24	27	29	29	32	28	28	30
East and West	45	46	50	54	55	59	40	41	44	46	47	50	33	33	36	38	38	41	39	39	41
SE and SW[b]	40	41	46	49	51	55	36	37	39	42	43	46	29	30	32	34	35	37	35	36	38
South[b]	27	28	33	37	38	42	24	25	28	31	31	34	20	21	23	25	26	28	23	24	26
Horizontal skylight	78	79	83	86	87	90	71	71	74	76	77	79	58	59	61	63	63	65	69	69	71
Opaque roller shades, fully drawn																					
North	14	15	20	23	25	29	13	14	17	19	20	23	12	12	15	17	17	20	13	13	15
NE and NW	25	26	31	34	36	40	23	24	27	30	30	33	21	22	24	26	27	29	23	23	26
East and West	34	36	40	44	45	49	32	33	36	38	39	42	29	30	32	34	35	37	32	32	35
SE and SW[b]	31	32	36	40	42	46	29	30	33	35	36	39	26	27	29	31	32	34	29	29	31
South[b]	21	22	27	30	32	36	20	20	23	26	27	30	18	19	21	23	24	26	19	20	22
Horizontal skylight	60	61	64	68	69	72	57	57	60	62	63	65	52	52	55	57	57	59	56	57	59

[a]Glass load factors (GLFs) for single-family detached houses, duplexes, or multifamily residences, with both east and west exposed walls or only north and south exposed walls, Btu/h·ft^2.

[b]Correct by +30% for latitude of 48° and by −30% for latitude of 32°. Use linear interpolation for latitude from 40 to 48 and from 40 to 32°.

To obtain GLF for other combinations of glass and/or inside shading: $GLF_a = (SC_a/SC_t)(GLF_t - U_tD_t) + U_aD_t$, where the subscripts a and t refer to the alternate and table values, respectively. SC_t and U_t are given in Table 5. $D_t = (t_a - 75)$, where $t_a = t_o - (DR/2)$; t_o is the outdoor design temperature and DR is the daily range.

Table 4 Window Glass Load Factors (GLFs) for Multifamily Residences[a]

Design Temperature, °F	Regular Single Glass						Regular Double Glass						Heat-Absorbing Double Glass						Clear Triple Glass		
	85	90	95	100	105	110	85	90	95	100	105	110	85	90	95	100	105	110	85	90	95
No inside shading																					
North	40	44	49	54	58	64	34	36	39	42	44	47	23	24	26	29	30	33	30	32	34
NE	88	89	91	95	97	100	78	79	80	83	84	85	52	52	53	55	55	57	71	71	73
East	136	137	139	142	144	147	120	121	122	125	126	127	79	79	81	83	83	84	109	109	111
SE	129	130	134	139	141	144	109	113	116	119	120	122	72	75	77	79	79	81	99	103	105
South[b]	88	91	96	101	105	110	76	78	81	84	86	89	50	52	54	56	58	60	68	70	72
SW	154	159	164	169	174	179	134	137	140	143	145	148	89	91	93	95	97	99	121	123	125
West	174	178	183	188	192	197	151	154	157	160	162	165	100	102	104	106	108	110	137	139	141
NW	123	127	132	137	141	147	107	109	112	115	117	121	71	72	75	77	79	81	96	98	100
Horizontal	249	252	256	261	264	268	218	220	223	226	228	230	144	146	148	150	152	154	198	200	202
Draperies, venetian blinds, translucent roller shades, fully drawn																					
North	21	25	29	33	36	40	18	21	23	26	28	31	15	17	19	21	23	25	17	19	21
NE	43	44	46	50	51	52	39	40	41	44	45	46	33	33	34	36	36	37	39	39	40
East	67	68	70	74	75	76	61	62	63	65	66	67	50	50	51	54	54	55	60	60	61
SE	64	65	69	73	74	77	58	59	61	63	64	66	48	48	50	52	52	54	57	57	59
South[b]	45	48	52	56	59	63	40	42	44	47	49	52	33	34	36	39	40	42	38	40	42
SW	79	83	87	91	94	98	70	72	75	78	80	83	57	59	62	64	66	68	68	69	71
West	89	92	96	100	103	107	79	81	84	86	88	91	65	66	69	71	72	75	76	78	80
NW	63	66	70	74	77	81	56	58	61	63	66	68	46	48	50	52	54	56	54	55	57
Horizontal	126	128	132	135	137	141	113	115	117	120	121	124	93	94	96	98	100	102	110	111	113
Opaque roller shades, fully drawn																					
North	17	21	25	29	32	36	15	17	20	23	25	28	14	15	18	20	22	24	15	16	18
NE	33	34	35	39	40	42	31	32	33	36	35	37	29	28	30	32	32	34	32	31	33
East	51	52	53	57	61	65	48	49	50	53	52	55	45	45	46	48	48	49	49	49	50
SE	49	50	53	57	58	61	46	47	49	52	52	55	42	43	45	47	47	49	46	46	48
South[b]	35	38	42	46	49	53	32	34	37	40	42	42	29	31	33	35	37	39	32	33	35
SW	61	65	69	73	77	81	57	59	62	65	67	70	52	54	56	58	60	62	56	58	60
West	68	71	75	80	83	87	64	66	68	71	73	76	58	60	62	64	66	68	63	64	66
NW	49	52	56	60	63	67	45	47	50	53	55	58	41	43	45	47	49	51	45	46	48
Horizontal	97	99	102	106	108	111	91	93	95	97	99	102	83	85	87	89	90	92	90	92	93

[a]Glass load factors (GLFs) for multifamily low-rise or single-family detached residences if zoned with separate temperature control for each zone, Btu/h·ft^2.

[b]Correct by +30% for latitude of 48° and by −30% for latitude of 32°. Use linear interpolation for latitude from 40 to 48 and from 40 to 32°.

To obtain GLF for other combinations of glass and/or inside shading: $GLF_a = (SC_a/SC_t)(GLF_t - U_tD_t) + U_aD_t$, where the subscripts a and t refer to the alternate and table values, respectively. SC_t and U_t are given in Table 5. $D_t = (t_a - 75)$, where $t_a = t_o - (DR/2)$; t_o is the outdoor design temperature and DR is the daily range.

hour (ACH) apply to both single- and multifamily housing, although most of the raw data were for single-family structures (McQuiston 1984). Construction may be defined as follows:

Tight. Good multifamily construction with close-fitting doors, windows, and framing is considered tight. New houses with full vapor retarder, no fireplace, well-fitted windows, weather-stripped doors, one story, and less than 1500 ft^2 floor area fall into this category.

Medium. Medium structures include new, two-story frame houses or one-story houses more than 10 years old with average maintenance, a floor area greater than 1500 ft^2, average fit windows and doors, and a fireplace with damper and glass closure. Below-average multifamily construction falls in this category.

Loose. Loose structures are poorly constructed single- and multifamily residences with poorly fitted windows and doors. Examples include houses more than 20 years old, of average maintenance, having a fireplace without damper or glass closure, or having more than an average number of vented appliances. Average manufactured homes are in this category.

Ventilation

Residential air-conditioning systems may introduce outdoor air, although it is not a code requirement in most localities. Positive ventilation should be considered, however, if the anticipated infiltration is less than about 0.5 ACH. When positive means of introducing outdoor air are used, controls, either manual or automatic, should be provided, and an energy recovery device should be considered.

Occupancy

Even though occupant density is low, occupancy loads should be estimated. Sensible heat gain per sedentary occupant is assumed to be 230 Btu/h. To prevent gross oversizing, the number of occupants should not be overestimated. Recent census studies recommend that the total number of occupants be based on two persons for the first bedroom, plus one person for each additional bedroom. The occupancy load should then be distributed equally among the living areas because the maximum load occurs when most of the residents occupy these areas.

Table 5 Shading Coefficients and U-Factors for Residential Windows

	Inside Shade					
	None		Drapery, Venetian Blind, or Translucent Roller Shade		Opaque Roller Shade	
Glass Type	SC	*U*	SC	*U*	SC	*U*
Single	1.00	1.04	0.50	0.81	0.38	0.81
Double	0.88	0.61	0.45	0.55	0.36	0.55
Heat-absorbing	0.58	0.45	0.37	0.44	0.33	0.44
Triple	0.80	0.44	0.44	0.40	0.36	0.40

Note: *U* is in Btu/h·ft^2·°F.

Table 6 Shade Line Factors (SLFs)

Direction Window Faces	Latitude, Degrees N						
	24	32	36	40	44	48	52
East	0.8	0.8	0.8	0.8	0.8	0.8	0.8
SE	1.8	1.6	1.4	1.3	1.1	1.0	0.9
South	9.2	5.0	3.4	2.6	2.1	1.8	1.5
SW	1.8	1.6	1.4	1.3	1.1	1.0	0.9
West	0.8	0.8	0.8	0.8	0.8	0.8	0.8

Note: Shadow length below the overhang equals the shade line factor times the overhang width. Values are averages for the 5 h of greatest solar intensity on August 1.

Household Appliances

Appliance loads are concentrated mainly in the kitchen and laundry areas. Based on contemporary living conditions in single-family houses, a sensible load of 1600 Btu/h should be divided between the kitchen and/or laundry and the adjoining room or rooms. For multifamily units, the sensible heat gain values should be about 1200 Btu/h. These values assume that the cooking range and clothes dryer are vented. Further allowances should be considered when unusual lighting intensities, computers, or other equipment is present.

Air Distribution System—Heat Loss/Gain

Whenever the air distribution system is outside the conditioned space (i.e., in attics, crawl spaces, or other unconditioned spaces) heat loss or gains to the ducts or pipes must be included in the calculated load and should be considered in equipment selection.

Latent Heat Sources

The latent cooling load has three main sources: outdoor air, occupants, and miscellaneous sources, such as cooking, laundry, and bathing. The miscellaneous latent loads are largely covered by outdoor air

Table 7 Winter Air Exchange Rates (ACH) as Function of Airtightness

	Outdoor Design Temperature, °F									
Class	50	40	30	20	10	0	−10	−20	−30	−40
Tight	0.41	0.43	0.45	0.47	0.49	0.51	0.53	0.55	0.57	0.59
Medium	0.69	0.73	0.77	0.81	0.85	0.89	0.93	0.97	1.00	1.05
Loose	1.11	1.15	1.20	1.23	1.27	1.30	1.35	1.40	1.43	1.47

Note: Values are for 15 mph wind and indoor temperature of 68°F.

Table 8 Summer Air Exchange Rates (ACH) as Function of Airtightness

	Outdoor Design Temperature, °F					
Class	85	90	95	100	105	110
Tight	0.33	0.34	0.35	0.36	0.37	0.38
Medium	0.46	0.48	0.50	0.52	0.54	0.56
Loose	0.68	0.70	0.72	0.74	0.76	0.78

Note: Values are for 7.5 mph wind and indoor temperature of 75°F.

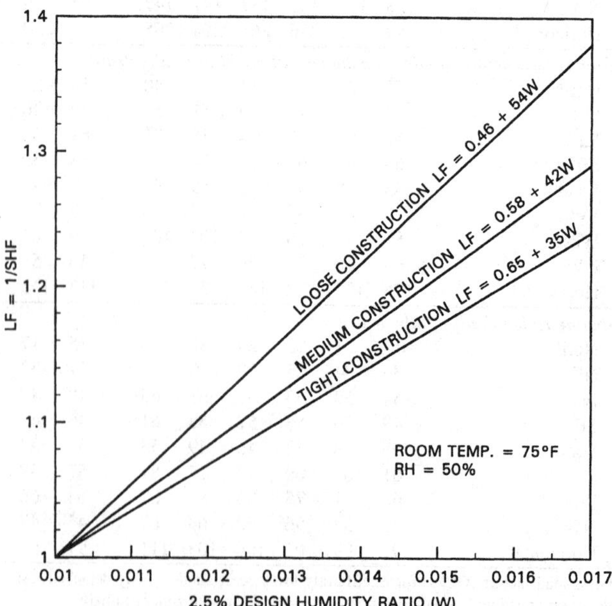

Fig. 1 Effect of Infiltration on Latent Load Factor

because most residences have exhaust fans and clothes dryers that vent most of the moisture from these sources. This vent air is accounted for in the infiltration calculation. McQuiston (1984) estimated latent load factors for typical houses located in geographic regions ranging from very dry to very wet using the transfer function method (Figure 1). A **latent factor** LF (LF = 1/SHF) of 1.3 or a **sensible heat factor** SHF (SHF = sensible load/total load) of 0.77 matches the performance of typical residential vapor compression cooling systems. Homes in almost all other regions of North America have cooling loads with an SHF greater than 0.77 and latent factors less than 1.3. Figure 1 may be used to estimate the total cooling load by reading LF as a function of the design humidity ratio and airtightness. Then q_{total} = (LF)$q_{sensible}$. If the humidity ratio is less than 0.01, set LF = 1.0.

LOAD CALCULATION

The cooling load calculation procedures are summarized in Table 9.

Example 1. A single-family detached house (Figure 2) is located in the south central United States at 36°N latitude.

Roof construction. Conventional roof-attic-ceiling combination, vented to remove moisture with 6 in. of fibrous batt insulation and vapor retarder (U = 0.05 Btu/h·ft²·°F).

Wall construction. Frame with 4 in. face brick, 3.5 in. fibrous batt insulation, 0.75 in. polystyrene sheathing, and 0.5 in. gypsum wallboard (U = 0.06 Btu/h·ft²·°F). Ceiling height is 8 ft throughout.

Floor construction. 4 in. concrete slab on grade.

Fenestration. Clear double glass, 0.125 in. thick, in and out. Assume closed, medium-color venetian blinds. The window glass has a 2 ft overhang at the top.

Doors. Solid core flush with all-glass storm doors (U = 0.32 Btu/h·ft²·°F).

Outdoor design conditions. Temperature of 96°F dry bulb with a 24°F daily range and a humidity ratio of 0.0136 lb vapor/lb dry air (74.6°F wet bulb).

U-factors for all external surfaces are based on a 7.5 mph wind velocity.

Indoor design conditions. Temperature of 75°F dry bulb and 50% rh.

Occupancy. Four persons, based on two for the master bedroom and one for each additional bedroom. Assign to the living room.

Appliances and lights. Assume 1600 Btu/h for the kitchen, and assign 50% to the living room. Assume 1600 Btu/h for the utility room, and assign 25% to the kitchen and 25% to the storage room.

The conditioning equipment is located in the garage, and the construction of the house is considered medium.

Find the sensible, latent, and total cooling load; size the cooling unit; and compute the air quantity for each room.

Solution: The cooling load must be made on a room-by-room basis to determine the proper distribution of air. The calculations follow the procedure outlined in the section on Load Components.

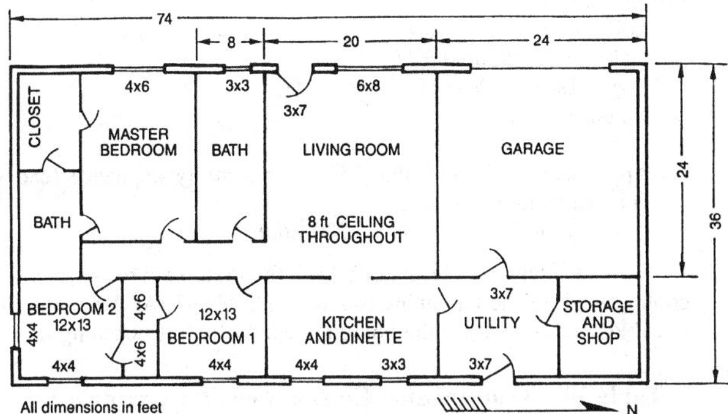

Fig. 2 Floor Plan of Single-Family Detached House

Table 9 Summary of Procedures for Residential Cooling Load Calculations

Load Source	Equation	Tables and Notes
Glass and window areas	q = (GLF)A	Glass load factors may be found in Tables 3 and 4 according to window orientation, type of glass, type of interior shading, and outdoor design temperature. The GLF includes effects of both transmission and solar radiation. Glass shaded by overhangs is treated as north glass. Table 6 gives shade line factors.
Doors	$q = U_d A$(CLTD)	Door CLTD values are in Tables 1 and 2 according to orientation, outdoor design temperature, and design daily temperature range.
Above-grade exterior walls	$q = U_w A$(CLTD)	Wall CLTD values are in Tables 1 and 2 based on the outdoor design temperature, daily range, and orientation.
Partitions to unconditioned space	$q = U_p A \Delta t$	Where Δt is the temperature difference across the partition.
Ceilings and roofs	$q = U_r A$(CLTD)	Tables 1 and 2 for CLTD, based on outdoor design temperature and daily range.
Exposed floors	$q = U_f A$(CLTD)	Tables 1 and 2 for CLTD, based on outdoor design temperature and daily range.
Infiltration	$q = 1.1 Q \Delta t$ Q = ACH × (room volume)/60	Air exchange rates are given in Tables 7 and 8.
Internal loads— People, appliances, lights	Plan 230 Btu/h per person.	Divide occupants evenly among rooms not used as bedrooms. If number of occupants is not known, assume two people for first bedroom and one person for each additional bedroom. The appliance and light load of 1600 Btu/h is divided between the kitchen and adjoining room and the laundry and adjoining room. Use 1200 Btu/h for multifamily units.
Total loads	Total cooling load = LF × (Sum of individual sensible cooling load components)	Load factors are from Figure 1 according to outdoor design humidity ratio and airtightness classification.

q = sensible cooling load, Btu/h
Δt = design temperature difference between outside and inside air, °F
A = area of applicable surface, ft²
U = U-factors for appropriate construction, Btu/h·ft²·°F
Q = volumetric airflow rate, cfm
ACH = air changes per hour, 1/h
GLF = glass load factor, Btu/h·ft²
CLTD = cooling load temperature difference, °F
LF = latent load multiplier

Walls, roof, windows, and doors. The calculations for the living room and the kitchen, where $q = UA(\text{CLTD})$ for the walls, roof, and door and $q = A(\text{GLF})$ for the windows, are outlined in Table 10. The glass shaded by the overhang is treated as north-facing glass, with the shaded area computed using Table 6.

Internal and infiltration sensible cooling loads. Compute as follows.

For the living room:

Infiltration. Using Table 8,

Q = ACH (room volume)/60

$Q = 0.5 \times 3840/60 = 32$

$q = 1.1Q(t_o - t_i) = 1.1Q\Delta t$

$q = 1.1 \times 32(96 - 75) = 740$ Btu/h

Occupants. Assuming 230 Btu/h per person,

$q = 230 \times$ (persons)

$q = 230 \times 4 = 920$ Btu/h

Appliances. Assuming that 50% of the kitchen appliance load is picked up in the living room,

$q = 0.5 \times$ (kitchen appliance load)

$q = 0.5 \times 1600 = 800$ Btu/h

For the kitchen:

Infiltration.

$Q = 0.5 \times 1920/60 = 16$

$q = 1.1 \times 16(96 - 75) = 370$ Btu/h

No occupants.

$q = 0$

Appliances. Assuming that 25% of the utility appliance load is picked up in the kitchen,

$q = (1600/2) + (1600/4) = 1200$ Btu/h

For the total sensible cooling load for these two rooms and the cooling load for the remaining rooms, see Table 11. At this point, the sensible cooling load for the house is 19,679 Btu/h. Depending on the design of the air distribution system, heat losses from the supply and return ducts may add to the cooling load. These may be more accurately estimated after designing the system; however, to size the cooling unit, duct losses should be included initially. If all ducts are in the attic space, a duct loss of 10% of the space sensible cooling load is reasonable. For a counterflow system, with ducts below the slab, a 5% loss is more reasonable.

Table 10 Transmission Cooling Load for Example 1

Item	Net Area, ft²	GLF, Btu/h·ft²	U-Factor, Btu/h·ft²·°F	CLTD, °F	Cooling Load, Btu/h	Reference
Living Room						
West wall	91		0.06	24	131	Table 1
Partition (garage)	192		0.07	12	161	Table 1
Roof	480		0.05	48	1152	Table 1
West door	21		0.32	24	161	Table 1
West glass	35	44			1540	Table 3
Shaded glass	13	19			247	Table 3
Kitchen						
East wall	135		0.06	24	194	Table 1
Roof	240		0.05	48	576	Table 1
East glass	14	44			616	Table 3
Shaded glass	11	19			209	Table 3

Table 11 Summary of Sensible Cooling Load Estimate for Example 1

Room	Roof, Walls, and Doors	Glass	People	Appliances	Infiltration	Total Room Btu/h (q_{rm})	Total Room cfm (Q_{rm})
Living room	1,605	1,787	920	800	740	5,852	296
Kitchen	770	825		1,200	370	3,165	160
Utility and storage	1,404			1,200	443	3,047	154
Bedroom No.1	559	544			278	1,381	70
Bedroom No.2	686	848			278	1,812	91
Master bedroom and bath	1,682	816			813	3,311	167
Bath	540	276			295	1,111	56
Total	7,246	5,096	920	3,200	3,217	19,679	994
Duct loss (10%)						1,968	
Outdoor ventilation air						1,600	
					Total	23,247 Btu/h	

An infiltration rate of 0.5 ACH may not be adequate for good indoor air quality, so some outdoor air should be introduced. This additional cooling load may be estimated in the same way as the infiltration load.

Assume that the entire duct system is in the attic; that is, the total sensible cooling load with a 10% duct loss is $1.1 \times 19{,}679 = 21{,}647$ Btu/h. Also, assume that additional outdoor air is needed to assure good indoor air quality, so the total infiltration and outdoor ventilation air is 0.75 ACH. This increases the infiltration rate by 50%, or about 1600 Btu/h. The total sensible cooling load is then increased to 23,247 Btu/h (Table 11).

The total cooling load (sensible plus latent) may be estimated by applying the latent factor (LF) from Figure 1. For a design humidity ratio of 0.0136 lb vapor per lb dry air, LF = 1.15 for a house of medium construction. Hence, the total cooling load equals $1.15 \times 23{,}247 = 26{,}734$ Btu/h.

The load raises the temperature of the cooling air 18 to 21°F as it leaves the rooms. The total design flow from the air conditioner can be estimated by the following equation:

$$Q_{tot} = \frac{q}{1.1\Delta t} \tag{1}$$

where

Q_{tot} = total airflow, cfm

q = total sensible load, Btu/h

1.1 = density times specific heat of cooling air times 60 min/h

Δt = temperature difference of air entering and leaving room, °F

For a temperature difference of 18°F, the total airflow is estimated from Equation (1) as

$$Q_{tot} = \frac{19{,}679}{1.1 \times 18} = 994 \text{ cfm}$$

The exact design flow can be determined only after the cooling unit has been selected. Then, the supply air quantities can be computed. Air should be supplied to each room on the basis of the room sensible cooling load:

$$Q_{rm} = Q_{tot}(q_{rm}/q)$$

where

Q_{rm} = airflow to each room, cfm

q_{rm} = room sensible cooling load, Btu/h

Thus, for the example,

$$Q_{rm} = (994/19{,}679)q_{rm}$$

If the living space in Example 1 were a multifamily unit (assume that the north, south, and east walls are not exposed surfaces), the calculation procedure would be the same, except that Table 2 would have been used for the CLTDs and Table 4 for the GLFs. Assumptions regarding infiltration, ventilation, and appliance loads are different for smaller multifamily units.

HEATING LOAD

Calculating a residential heating load involves estimating the maximum (block) heat loss of each room or space to be heated and the simultaneous maximum (block) heat loss for the building, while maintaining a selected indoor air temperature during periods of design outdoor weather conditions. Heat losses are mainly

- Transmission losses or heat transferred through the confining walls, glass, ceiling, floor, or other surfaces
- Infiltration losses or energy required to warm outdoor air leaking in through cracks and crevices around doors and windows, through open doors and windows, and through porous building materials

GENERAL PROCEDURE

To calculate a design heating load, prepare the following information about building design and weather data at design conditions.

1. Select outdoor design weather conditions: temperature, wind direction, and wind speed. Winter climatic data can be found in Chapter 26, or selected weather conditions and temperatures appropriate for the application may be used. Weather station data may differ significantly from values in Chapter 26.
2. Select the indoor air temperature to be maintained in each space during design weather conditions.
3. Temperatures in adjacent unheated spaces, attached garages, and attics can be estimated at the outdoor ambient temperature.
4. Select or compute heat transfer coefficients for outside walls and glass; for inside walls, nonbasement floors, and ceilings if these are next to unheated spaces; and for the roof if it is next to heated spaces.
5. Determine the net area of outside wall, glass, and roof next to heated spaces, as well as any cold walls, floors, or ceilings next to unheated spaces. These determinations can be made from building plans or from the actual building, using inside dimensions.
6. Compute transmission heat losses for each kind of wall, glass, floor, ceiling, and roof in the building by multiplying the heat transfer coefficient in each case by the area of the surface and the temperature difference between indoor air and outdoor air or adjacent lower temperature spaces.
7. Compute heat losses from basement or grade-level slab floors using the methods in this chapter.
8. Select unit values, and compute the energy associated with infiltration of cold air around outside doors, windows, porous building materials, and other openings. These unit values depend on the kind or width of crack, wind speed, and the temperature difference between indoor and outdoor air. An alternative method is to use air changes (see Chapter 25).
9. When positive ventilation using outdoor air is provided by an air-heating or air-conditioning unit, the energy required to warm the outdoor air to the space temperature must be provided by the unit. The principle for calculation of this load component is identical to that for infiltration. If mechanical exhaust from the space is provided in an amount equal to the outdoor air drawn in by the unit, the unit must also provide for natural infiltration losses. If no mechanical exhaust is used and the outdoor air supply equals or exceeds the amount of natural infiltration that can occur without ventilation, some reduction in infiltration may occur.
10. The sum of the coincidental transmission losses or heat transmitted through the confining walls, floor, ceiling, glass, and other surfaces, plus the energy associated with cold air entering by infiltration or the ventilation air required to replace mechanical exhaust, represents the total heating load.
11. Include the pickup loads that may be required in intermittently heated buildings using night thermostat setback. Pickup loads frequently require an increase in heating equipment capacity to bring the temperature of structure, air, and material contents to the specified temperature. See Figure 9.
12. Use materials and data in Chapters 24, 25, 26, and others as appropriate to the calculations. See Table 12.

SELECTING HEATING DESIGN CONDITIONS

The ideal solution to a basic heating system design is a plant with a maximum output capacity equal to the heating load that develops with the most severe local weather conditions. However, this solution is usually uneconomical. Weather records show that severe weather conditions do not repeat annually. If heating systems were designed for maximum weather conditions, excess capacity would exist during most of the system's operating life. In many cases, an occasional failure of a heating plant to maintain a preselected indoor design temperature during brief periods of severe weather is not critical.

Table 12 Summary of Loads, Equations, and References for Calculating Design Heating Loads

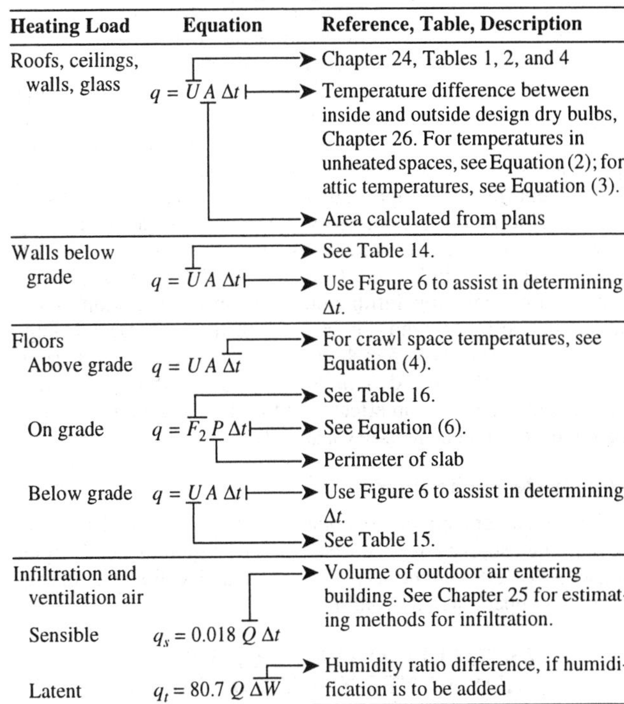

Heating Load	Equation	Reference, Table, Description
Roofs, ceilings, walls, glass	$q = U A \Delta t$	Chapter 24, Tables 1, 2, and 4; Temperature difference between inside and outside design dry bulbs, Chapter 26. For temperatures in unheated spaces, see Equation (2); for attic temperatures, see Equation (3).; Area calculated from plans
Walls below grade	$q = U A \Delta t$	See Table 14.; Use Figure 6 to assist in determining Δt.
Floors		
Above grade	$q = U A \Delta t$	For crawl space temperatures, see Equation (4).
On grade	$q = F_2 P \Delta t$	See Table 16.; See Equation (6).; Perimeter of slab
Below grade	$q = U A \Delta t$	Use Figure 6 to assist in determining Δt.; See Table 15.
Infiltration and ventilation air		
Sensible	$q_s = 0.018\, Q\, \Delta t$	Volume of outdoor air entering building. See Chapter 25 for estimating methods for infiltration.
Latent	$q_l = 80.7\, Q\, \Delta W$	Humidity ratio difference, if humidification is to be added

Outdoor Design Temperature

Before selecting an outdoor design temperature from Chapter 26, the designer should consider the following for residential buildings:

- Is the structure heavy, medium, or light?
- Is the structure insulated?
- Is the structure exposed to high wind?
- Is the load from infiltration or ventilation high?
- Is there more glass area than normal?
- During what part of the day will the structure be used?
- What is the nature of occupancy?
- Will there be long periods of operation at reduced indoor temperature?
- What is the amplitude between local maximum and minimum daily temperatures?
- Are there local conditions that cause significant variation from temperatures reported by the weather service?
- What auxiliary heating devices will be in the building?

Before selecting an outdoor design temperature, the designer must keep in mind that, if the outdoor to indoor design temperature difference is exceeded, the indoor temperature may fall, depending on (1) the thermal mass of the structure and its contents, (2) whether the internal load was included in calculations, (3) the duration of the cold period, and (4) internal heat generated by appliances, etc.

The effect of wind on the heating requirements of any building should be considered because

- Wind movement increases the heat transmission of walls, glass, and roof, affecting poorly insulated walls to a much greater extent than well-insulated walls.
- Wind materially increases the infiltration of cold air through cracks around doors and windows and even through building materials themselves (see Chapter 25).

Theoretically, on a design basis, the most unfavorable combination of temperature and wind speed should be chosen. A building may require more heat on a windy day with a moderately low outdoor temperature than on a quiet day with a much lower outdoor temperature. The worst combination of wind and temperature varies by building because wind speed has a greater effect on buildings with relatively high infiltration rates. The building heating load may be calculated for several combinations of temperature and wind speed on record, and the worst combination may be selected; however, except for critical applications, designers generally find such a degree of refinement unnecessary. No correlation has been shown between the design temperatures in Chapter 26 and the simultaneous maximum wind speed. If a designer prefers the air change method for computing infiltration rates, such correlation is not important. Designers who use the crack method can use a leakage rate at a wind speed of 15 mph, unless local experience has established that another speed is more appropriate. Abnormally high wind speeds may have an effect on infiltration and the U-factor of the building components (see Chapter 22).

Indoor Design Temperature

The indoor temperature for comfort heating may vary depending on building use, type of occupancy, or code requirements. Chapter 8 and ASHRAE *Standards* 55 and 55a define the relationship between temperature and comfort.

ESTIMATING TEMPERATURES IN ADJACENT UNHEATED SPACES

Heat loss from heated rooms to unheated rooms or spaces must be based on the estimated or assumed temperature in such unheated spaces. This temperature will be in between the indoor and outdoor temperatures. If the surface area adjacent to the heated room and that exposed to the outdoors are equal and if the heat transfer coefficients are equal, the temperature in the unheated space may be assumed equal to the mean of the indoor and outdoor design temperatures. If, however, the surface areas and coefficients are unequal, the temperature in the unheated space should be estimated by

$$\begin{aligned} t_u = {} & [t_i(A_1U_1 + A_2U_2 + A_3U_3 + \text{etc.}) \\ & + t_o(60\rho c_p Q_o + A_aU_a + A_bU_b + A_cU_c + \text{etc.})] \\ & \div (A_1U_1 + A_2U_2 + A_3U_3 + \text{etc.} \\ & + 60\rho c_p Q_o + A_aU_a + A_bU_b + A_cU_c + \text{etc.}) \end{aligned} \quad (2)$$

where

ρc_p = density times specific heat of air = 0.018 Btu/ft^3·°F for standard air
t_u = temperature in unheated space, °F
t_i = indoor design temperature of heated room, °F
t_o = outdoor design temperature, °F
A_1, A_2, A_3, etc. = areas of surfaces of unheated space adjacent to heated spaces, ft^2
A_a, A_b, A_c, etc. = areas of surfaces of unheated space exposed to outdoors, ft^2
U_1, U_2, U_3, etc. = heat transfer coefficients of surfaces of A_1, A_2, A_3, etc., Btu/h·ft^2·°F
U_a, U_b, U_c, etc. = heat transfer coefficients of surfaces of A_a, A_b, A_c, etc., Btu/h·ft^2·°F
Q_o = rate of introduction of outside air into unheated space by infiltration and/or ventilation, cfm

Example 2. Calculate the temperature in an unheated space adjacent to a heated room with surface areas (A_1, A_2, and A_3) of 100, 120, and 140 ft^2 and overall heat transfer coefficients (U_1, U_2, and U_3) of 0.15, 0.20, and 0.25 Btu/h·ft^2·°F, respectively. The surface areas of the unheated space exposed to the outdoors (A_a and A_b) are 100 and 140 ft^2, respectively, and the corresponding overall heat transfer coefficients are 0.10 and 0.30 Btu/h·ft^2·°F. The sixth surface is on the ground and can be neglected for this example, as can the effect of introduction of outdoor air into the unheated space. Assume t_i = 70°F and t_o = −10°F.

Solution: Substituting into Equation (2),

$$\begin{aligned} t_u = {} & [70(100 \times 0.15 + 120 \times 0.20 + 140 \times 0.25) \\ & + (-10)(100 \times 0.10 + 140 \times 0.30)] \\ & \div (100 \times 0.15 + 120 \times 0.20 + 140 \times 0.25 \\ & + 100 \times 0.10 + 140 \times 0.30) \\ t_u = {} & 4660/126 = 37°\text{F} \end{aligned}$$

Temperatures in unheated spaces with large glass areas and two or more surfaces exposed to the outdoors (e.g., sleeping porches and sun parlors) are generally assumed to be the same as that of the outdoors.

Attic Temperature

An attic is a space having an average distance of 1 ft or more between a ceiling and the underside of the roof. Estimating attic temperature is a special case of estimating temperature in an adjacent unheated space and can be done using

$$t_a = \frac{A_cU_ct_c + t_o(60\rho c_pA_cV_c + A_rU_r + A_wU_w + A_gU_g)}{A_c(U_c + 60\rho c_pV_c) + A_rU_r + A_wU_w + A_gU_g} \quad (3)$$

where

ρc_p = air density times specific heat = 0.018 Btu/ft^3·°F for standard air
t_a = attic temperature, °F
t_c = indoor temperature near top floor ceiling, °F
t_o = outdoor temperature, °F
A_c = area of ceiling, ft^2
A_r = area of roof, ft^2
A_w = area of net vertical attic wall surface, ft^2
A_g = area of attic glass, ft^2
U_c = heat transfer coefficient of ceiling, Btu/h·ft^2·°F, based on surface conductance of 2.2 Btu/h·ft^2·°F (upper surface, see Table 2 in Chapter 24); 2.2 = reciprocal of one-half the air space resistance
U_r = heat transfer coefficient of roof, Btu/h·ft^2·°F, based on surface conductance of 2.2 Btu/h·ft^2·°F (upper surface, see Table 2 in Chapter 24); 2.2 = reciprocal of one-half the air space resistance
U_w = heat transfer coefficient of vertical wall surface, Btu/h·ft^2·°F
U_g = heat transfer coefficient of glass, Btu/h·ft^2·°F
V_c = rate of introduction of outside air into the attic space by ventilation per square foot of ceiling area, cfm/ft^2

Example 3. Calculate the temperature in an unheated attic assuming t_c = 70°F; t_o = 10°F; A_c = 1000 ft^2; A_r = 1200 ft^2; A_w = 100 ft^2; A_g = 10 ft^2; U_r = 0.50 Btu/h·ft^2·°F; U_c = 0.40 Btu/h·ft^2·°F; U_w = 0.30 Btu/h·ft^2·°F; U_g = 1.13 Btu/h·ft^2·°F; and V_c = 0.5 cfm/ft^2.

Solution: Substituting these values into Equation (3),

$$\begin{aligned} t_a = {} & [(1000 \times 0.40 \times 70) + 10(60 \times 0.018 \times 1000 \times 0.5 + 1200 \times 0.50 \\ & + 100 \times 0.30 + 10 \times 1.13)] \\ & \div [1000(0.40 + 60 \times 0.018 \times 0.5) + 1200 \times 0.50 + 100 \times 0.30 \\ & + 10 \times 1.13] \\ t_a = {} & 39{,}813/1581 = 25.2°\text{F} \end{aligned}$$

Equation (3) includes the effect of air interchange that would take place through attic vents or louvers intended to preclude attic condensation. Test data from Joy et al. (1956), Joy (1958), and Rowley et al. (1940) indicate that a reduction in the temperature difference between attic air and outside air is linear as attic ventilation rates increase from 0 to 0.5 cfm/ft^2 of the ceiling area. When attic ventilation meets the requirements in Chapter 23, 0.5 cfm/ft^2 is the approximate ventilation rate for design conditions. This reduction in temperature difference affects the overall heat loss of a residence with an insulated ceiling by only 1 or 2%.

Equation (3) does not consider factors such as heat exchange between chimney and attic or solar radiation to and from the roof. Because of these effects, attic temperatures are frequently higher than values calculated using Equation (3). However, Equation (3) can be used to calculate attic temperature because the resulting error is generally less than that introduced by neglecting the roof and assuming that the attic temperature is equal to the outdoor air temperature.

When relatively large louvers are installed (customary in southern regions of the United States), the attic temperature is often assumed to be the average of the indoor and outdoor air temperatures.

For an approximate method of calculating heat losses through attics, the combined ceiling and roof coefficient may be used (see Table 5 in Chapter 24).

CALCULATING HEAT LOSS FROM CRAWL SPACES

A crawl space can be considered a half basement. To prevent ground moisture from evaporating and causing a condensation problem, sheets of vapor retarder (e.g., polyethylene film) are used to cover the ground surface (see Chapter 23). Most codes require crawl spaces to be adequately vented all year round. However, venting the crawl space in the heating season causes substantial heat loss through the floor.

The space may be insulated in several ways: the crawl space ceiling (floor above the crawl space) can be insulated, or the perimeter wall can be insulated either on the outside or on the inside. If the floor above is insulated, the crawl space vents should be kept open because the temperature of the crawl space is likely to be below the dew point of the indoor space. If the perimeter wall is insulated, the vents should be kept closed in the heating season and open the remainder of the year.

Crawl Space Temperature

The crawl space temperature depends on such factors as venting, heating ducts, and the heating plant. When the crawl space is well ventilated, its temperature is close to that of the ambient air temperature. When the crawl space vent is closed for the heating season, or if the space is used as a plenum (i.e., part of the forced-air heating system), the crawl space temperature approaches that of the indoor conditioned space. In the former case, the floor above the crawl space, the heating ductwork, and the utility pipes should be insulated similarly to the walls and ceiling of a house.

The following steady-state equation can be used to estimate the temperature of a crawl space.

$$q_f = q_p + q_g + q_a$$

where

q_f = heat loss through floor into crawl space, Btu/h
q_p = heat loss from crawl space through foundation walls and sill box, Btu/h
q_g = heat loss into ground, Btu/h
q_a = heat loss due to ventilation of crawl space, Btu/h

Latta and Boileau (1969) estimated the air exchange rate for an uninsulated basement at 0.67 ACH under winter conditions. In more detail, the above equation can be repeated as

$$U_f A_f (t_i - t_c) = U_p A_p (t_c - t_o) + U_g A_g (t_c - t_g) + 0.67 \rho c_p V_c (t_c - t_o) \quad (4)$$

where

t_i = indoor air temperature (i.e., air above ceiling of crawl space), °F
t_o = outdoor air temperature, °F
t_g = ground temperature (constant), °F
t_c = crawl space temperature, °F
A_f = area of floor above, ft²
A_p = area of perimeter, exposed foundation wall plus sill box, ft²
A_g = area of ground below ($A_f = A_g$), ft²
U_f = average heat transfer coefficient through floor, Btu/h·ft²·°F
U_g = average heat transfer coefficient through ground (horizontal air film and 10 ft of soil), Btu/h·ft²·°F
U_p = combined heat transfer coefficient of sill box and foundation wall (both above and below grade), Btu/h·ft²·°F
V_c = volume of crawl space, ft³
ρc_p = volumetric heat capacity of air = 0.018 Btu/ft³·°F
0.67 = assumed air exchange rate, volumes/hour

Example 4. A crawl space of 1200 ft² with a 140 ft perimeter is considered. The construction of the perimeter wall is shown in Figure 3. The indoor, outdoor, and the deep-down ground temperatures are 70, 10, and 50°F, respectively. Estimate the heat loss and crawl space temperature with and without insulation. The heat transmission coefficient (U-factor) for each component is indicated in Table 13.

Table 13 Estimated U-Factors for Insulated and Uninsulated Crawl Spaces

Component	Uninsulated Btu/h·°F per ft of Perimeter	Insulated[a] Btu/h·°F per ft of Perimeter
16 in. exposed concrete blocks	0.7	0.18
7.5 in. sill box	0.188	0.071
1st 12 in. block wall below grade	0.355	0.127
2nd 12 in. block wall below grade	0.22	0.14
3rd 12 in. block wall below grade	0.133	0.1
Total for perimeter wall	1.6	0.62
	Btu/h·ft²·°F	**Btu/h·ft²·°F**
Ground	0.077	0.077
Floor above crawl space	0.25	0.076[a]

[a]Perimeter walls are insulated with R-5.4; the floor is insulated with R-11 blanket or batts.

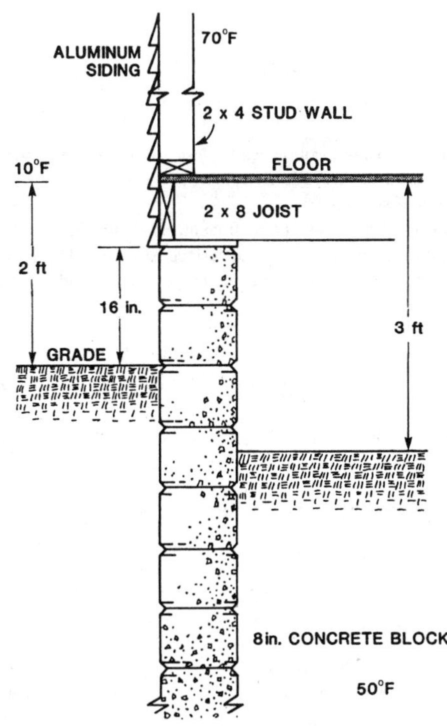

Fig. 3 Uninsulated Crawl Space

Solution: Three cases are examined.

Case A. This base case is a vented and uninsulated crawl space. The crawl space temperature approaches that of the outdoors, 10°F, and the heat loss is 0.25 × 1200(70 − 10) = 18,000 Btu/h.

Case B. The crawl space is vented. The floor above is insulated with an R-11 blanket; no insulation on the perimeter. The temperature of the crawl space approaches that of the outdoors, 10°F. The heat loss is calculated as

$$q_f = 1200 \times 0.076(70 - 10) = 5470 \text{ Btu/h}$$

Case C. The crawl space is not vented during the heating season. The floor above is not insulated, but the perimeter wall is insulated with R-5.4 down to 3 ft below grade.

$$q_f = 1200 \times 0.25(70 - t_c)$$

$$q_p = 140 \times 0.62(t_c - 10)$$

$$q_g = 1200 \times 0.077(t_c - 50)$$

$$q_a = 1200 \times 3 \times 0.67 \times 0.018(t_c - 10)$$

The crawl space temperature is solved using Equation (4): t_c = 51.5°F. The heat loss is 5550 Btu/h.

The results show that base case A can potentially lose the most heat. However, when the floor above is insulated, the crawl space must be vented to eliminate any condensation potential, and the heating ductwork and utility pipeline in the crawl space must be adequately insulated. When the perimeter is insulated, the vents must be closed during the heating season and opened for the rest of the year; the heating ductwork and utility pipeline do not need insulation.

Case	Venting	Insulation	Heat Loss Through Floor Above, Btu/h	Temperature of Crawl Space, °F
A	Yes	None	18,000	10
B	Yes	R-11 on floor above	5,470	10
C	No	R-5.4 on perimeter wall	5,550	51.5

CALCULATING TRANSMISSION HEAT LOSS

Steady-state heat loss by conduction and convection heat transfer through any surface is

$$q = UA(t_i - t_o) \qquad (5)$$

where

q = heat transfer through wall, glass, roof, ceiling, floor, or other exposed surface, Btu/h
A = area of surface, ft^2
U = air-to-air heat transfer coefficient, Btu/h·ft^2·°F
t_i = indoor air temperature near surface involved, °F
t_o = outdoor air temperature or temperature of adjacent unheated space, °F

Example 5. Calculate the transmission loss through an 8 in. brick wall having an area of 150 ft^2, if the indoor temperature t_i is 70°F, and the outdoor temperature t_o is −10°F.

Solution: The overall heat transfer coefficient U of a plain 8 in. brick wall is 0.41 Btu/h·ft^2·°F. Substituting into Equation (5),

$$q = 150 \times 0.41[70 - (-10)] = 4920 \text{ Btu/h}$$

Through Ceiling and Roof

Transmission heat loss through top floor ceilings, attics, and roofs may be estimated by either of two methods:

1. Substitute in Equation (5) the ceiling area A, the indoor/outdoor temperature difference $(t_i - t_o)$, and the proper U-factor:

   Flat roofs. Use appropriate coefficients in Equation (3) if side walls extend appreciably above the ceiling or the floor below.

   Pitched roofs. Calculate the combined roof and ceiling coefficient as outlined in Chapter 24.

2. For *pitched roofs*, estimate the attic temperature (based on the indoor and outdoor design temperatures) using Equation (3), and substitute for t_o in Equation (5), obtaining the value of t_a, together with the ceiling area A and the ceiling U-factor. Attic temperatures do not need to be calculated for *flat roofs*, as the ceiling-roof heat loss can be determined as suggested in Method 1 above.

From the Basement

The basement interior is considered conditioned space if a minimum temperature of 10°F below indoor design air temperature is maintained over the heating season. In many instances, the house heating plant, water heater, and heating ducts are in the basement, so it remains at or above 50°F.

Heat transmission from the below-grade portion of the basement wall to the ambient air cannot be estimated by simple, one-dimensional heat conduction. In fact, field measurement of an uninsulated basement by Latta and Boileau (1969) showed that the isotherms near the wall are not parallel lines but closer to radial lines centered at the intersection of the grade line and the wall. Therefore, heat flow paths approximately follow a concentric circular pattern (Figure 4).

Such heat flow paths are altered when insulation is added to the wall or floor. An extreme case would be no heat loss from the basement wall and floor (i.e., infinite insulation applied to the wall and floor). In this case, the isotherms would be horizontal lines parallel to the grade line, and the heat flow would be vertical. When finite insulation or partial insulation is applied to the wall and floor, the heat flow paths take shapes somewhere between the circular and vertical lines (Figure 5).

Ground Temperature. Ground temperatures assumed for estimating basement heat losses will differ for basement floors and walls. The temperatures under floors are generally higher than those adjacent to walls. This is discussed further in the section on Basement Design Temperatures.

Through Basement Walls

Houghten et al. (1942) observed nonuniform heat flux across the basement wall with respect to the depth of the wall because each heat flow path contains a different thermal resistance. For a basement wall that has its top portion exposed to ambient air, heat may be conducted vertically through the concrete wall and dissipated to the ambient from the top portion of the wall (Wang 1979, Bligh et al. 1978). Under certain conditions, this vertical heat flux becomes significant and should not be ignored.

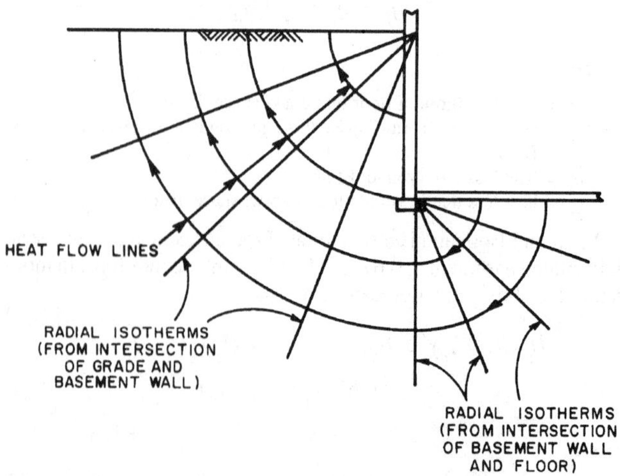

Fig. 4 Heat Flow from Basement

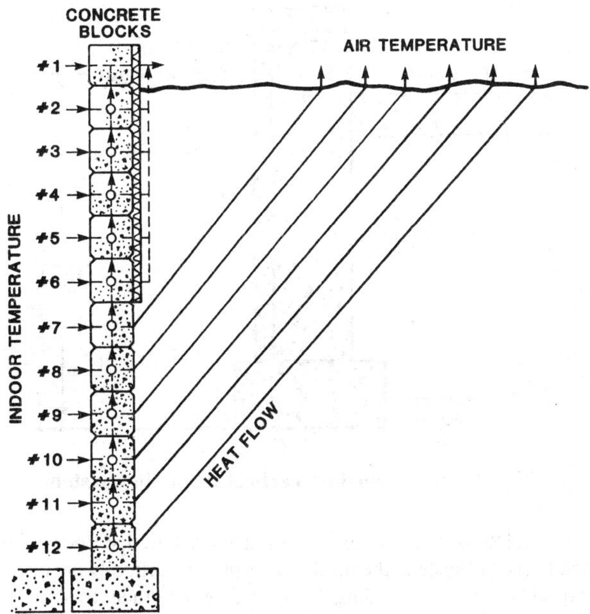

Fig. 5 Heat Flow Path for Partially Insulated Basement Wall

Table 14 Heat Loss Below Grade in Basement Walls

Depth, ft	Path Length Through Soil, ft	Heat Loss Coefficient, Btu/h·ft²·°F[a]							
		Uninsulated		R-4.17		R-8.34		R-12.5	
0 to 1	0.68	0.410	Σ[b]	0.152	Σ[b]	0.093	Σ[b]	0.067	Σ[b]
1 to 2	2.27	0.222	0.632	0.116	0.268	0.079	0.172	0.059	0.126
2 to 3	3.88	0.155	0.787	0.094	0.362	0.068	0.240	0.053	0.179
3 to 4	5.52	0.119	0.906	0.079	0.441	0.060	0.300	0.048	0.227
4 to 5	7.05	0.096	1.002	0.069	0.510	0.053	0.353	0.044	0.271
5 to 6	8.65	0.079	1.081	0.060	0.570	0.048	0.401	0.040	0.311
6 to 7	10.28	0.069	1.150	0.054	0.624	0.044	0.445	0.037	0.348

Source: Latta and Boileau (1969).
[a]Soil conductivity was assumed to be 9.6 Btu·in/h·ft²·°F.
[b]Σ = heat loss to current depth.

Once the heat paths are known or assumed, a steady-state analysis can calculate the overall heat transmission coefficient for each segment of the basement wall. Referring to Figures 4 and 5, the total thermal resistance for each depth increment of the basement wall can be found by summing the thermal resistances along each heat flow path. Based on these resistances, the heat loss at each depth increment can be estimated for a unit temperature difference between the basement and the average mean winter temperature. Table 14 lists such heat loss values at different depths for an uninsulated and an insulated concrete wall (Latta and Boileau 1969). Also listed are the lengths of the heat flow path through the soil (circular path).

Through Basement Floors

The same steady-state design used for the basement wall can be applied to the basement floor, except that the length of the heat flow path is longer (see Figure 4). Thus, the heat loss through the basement floor is much smaller than that through the wall. An average value for the heat loss through the basement floor can be multiplied by the floor area to give total heat loss from the floor. Table 15 lists typical values.

Table 15 Heat Loss Through Basement Floors

Depth of Foundation Wall below Grade, ft	Heat Loss Coefficient, Btu/h·ft²·°F			
	Shortest Width of House, ft			
	20	24	28	32
5	0.032	0.029	0.026	0.023
6	0.030	0.027	0.025	0.022
7	0.029	0.026	0.023	0.021

Note: $\Delta t = (t_a - A)$

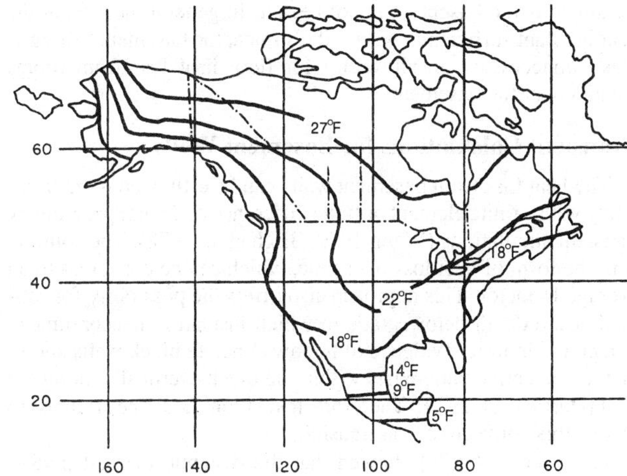

Fig. 6 Lines of Constant Amplitude

Basement Design Temperatures

Although internal design temperature is given by basement air temperature, none of the usual external design air temperatures apply because of the heat capacity of the soil. However, ground surface temperature fluctuates about a mean value by an amplitude A, which varies with geographic location and surface cover. Therefore, suitable external design temperatures can be obtained by subtracting A for the location from the mean winter air temperature t_a. Values for t_a can be obtained from meteorological records, and A can be estimated from the map in Figure 6. This map is part of one prepared by Chang (1958) giving annual ranges in ground temperature at a depth of 4 in.

Example 6. Consider a basement 28 ft wide by 30 ft long sunk 6 ft below grade, with R-8.34 insulation applied to the top 2 ft of the wall below grade. Assume an internal air temperature of 70°F and an external design temperature ($t_a - A$) of 20°F.

Solution:

Wall (using Table 14)

First foot below grade .... 0.093 Btu/h·ft·°F
Second foot below grade .... 0.079 Btu/h·ft·°F
Third foot below grade .... 0.155 Btu/h·ft·°F
Fourth foot below grade .... 0.119 Btu/h·ft·°F
Fifth foot below grade .... 0.096 Btu/h·ft·°F
Sixth foot below grade .... 0.079 Btu/h·ft·°F
Total per foot length of wall .... 0.621 Btu/h·ft·°F

Basement perimeter .... 2(28 + 30) = 116 ft
Total wall heat loss .... 0.62 × 116 = 72 Btu/h·°F

Floor (using Table 15)

Average heat loss per ft² .... 0.025 Btu/h·ft²·°F
Floor area 28 × 30 .... 840 ft²
Total floor heat loss .... 0.025 × 840 = 21 Btu/h·°F

Total

Total basement heat loss below grade...72 + 21 = 93 Btu/h·°F
Design temperature difference .... 70 − 20 = 50°F
Maximum rate of heat loss from
below-grade basement........................ 93 × 50 = 4650 Btu/h

If a basement is completely below grade and unheated, its temperature ranges between that in the rooms above and that of the ground. Basement windows lower the basement temperature when it is cold outdoors, and heat given off by the heating plant increases the basement temperature. The exact basement temperature is indeterminate if the basement is not heated. In general, heat from the heating plant sufficiently warms the air near the basement ceiling to make unnecessary an allowance for floor heat loss from rooms located over the basement.

Transient Calculations for Basement Walls

The heat loss from basement walls can be estimated more accurately with a finite element or finite difference computer program by transient simulations (Wang 1979, Bligh et al. 1978). The solution is in the form of heat loss over time, which can be converted to an average U-factor. This approach also offers the possibility for estimating the depth below grade to which insulation is economical. Direct and indirect evidence of hollow concrete block walls shows that a convective path exists within the blocks vertically along the wall (Harrje et al. 1979). Therefore, insulation should be arranged to reduce this convective heat transfer.

Peony et al. (1979) showed that the dynamic thermal performance of a masonry wall is better when insulation is placed on the exterior. Moreover, transient simulation showed that insulation is more effective when it is placed on the exterior side of the basement wall. Depending on the exposed portion of the block wall and the temperature difference between indoor and outdoor air, exterior application can be 10 to 20% more efficient than a corresponding interior application. However, such exterior insulation must be installed properly to maintain its integrity.

Calculating Transmission Heat Loss from Floor Slabs

Concrete slab floors may be (1) unheated, relying for warmth on heat delivered above floor level by the heating system, or (2) heated, containing heated pipes or ducts that constitute a radiant slab or portion of it for complete or partial heating of the house.

The perimeter insulation of a slab-on-grade floor is quite important for comfort and energy conservation. In unheated slab floors, the floor edge must be insulated in order to keep the floor warm. Downdrafts from windows or exposed walls can create pools of chilly air over considerable areas of the floor. In heated slab floors, the floor edge must be insulated to prevent excessive heat loss from the heating pipe or duct embedded in the floor or from the baseboard heater.

Wang (1979) and Bligh et al. (1978) found that heat loss from an unheated concrete slab floor is mostly through the perimeter rather than through the floor and into the ground. Total heat loss is more nearly proportional to the length of the perimeter than to the area of the floor, and it can be estimated by the following equation for both unheated and heated slab floors:

$$q = F_2 P(t_i - t_o) \quad (6)$$

where

q = heat loss through perimeter, Btu/h
F_2 = heat loss coefficient per foot of perimeter (see Table 16), Btu/h·ft·°F
P = perimeter or exposed edge of floor, ft
t_i = indoor temperature, °F (For the heated slab, t_i is the weighted average heating duct or pipe temperature.)
t_o = outdoor temperature, °F

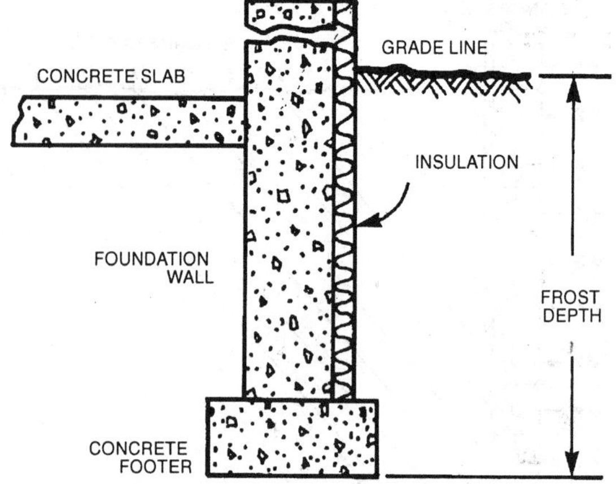

Fig. 7 "I"-Shaped or Vertical Insulation System

Vertical "I"-shaped systems are used to insulate slab floor perimeters. In the "I" system, the insulation is placed vertically next to the exposed slab edge, extending downward below grade, as shown in Figure 7.

Breaks or joints must be avoided when the insulation is installed; otherwise, local thermal bridges can be formed, and the overall efficiency of the insulation is reduced.

Transient Calculations for Floor Slabs

Figure 8 shows four basic slab-on-grade constructions analyzed with a finite element computer program by Wang (1979). Figures 8A-C represent unheated slabs; Figure 8D can be considered a heated slab. Each was investigated with and without insulation of R-5.4 under three climatic conditions (7433, 5350, and 2950 degree-days). Table 16 lists the results in terms of heat loss coefficient F_2, based on degree-days.

Table 16 shows that the heat loss coefficient F_2 is sensitive to both construction and insulation. The reverse loss, or heat loss into the ground and outward through the edges of the slab and foundation wall, is significant when heating pipes, heating ducts, or baseboard heaters are placed near the slab perimeters. To prevent reverse loss, the designer may find it advantageous to use perimeter insulation even in warmer climates. For severe winter regions (above 6000 degree-days), the insulation value should be increased to R >10 ft²·°F·h/Btu.

Table 16 Heat Loss Coefficient F_2 of Slab Floor Construction, Btu/h·°F per ft of Perimeter

		Degree-Days (65°F Base)		
Construction	Insulation	7433	5350	2950
8 in. block wall, brick facing	Uninsulated	0.62	0.68	0.72
	R-5.4 from edge to footer	0.48	0.50	0.56
4 in. block wall, brick facing	Uninsulated	0.80	0.84	0.93
	R-5.4 from edge to footer	0.47	0.49	0.54
Metal stud wall, stucco	Uninsulated	1.15	1.20	1.34
	R-5.4 from edge to footer	0.51	0.53	0.58
Poured concrete wall with duct near perimeter[a]	Uninsulated	1.84	2.12	2.73
	R-5.4 from edge to footer, 3 ft under floor	0.64	0.72	0.90

[a]Weighted average temperature of the heating duct was assumed at 110°F during the heating season (outdoor air temperature less than 65°F).

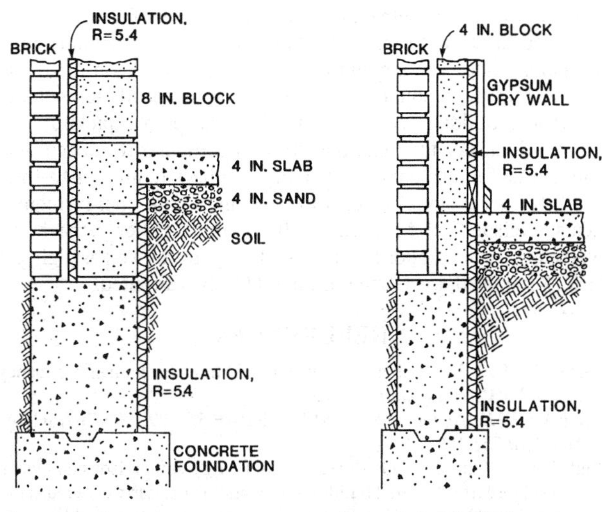

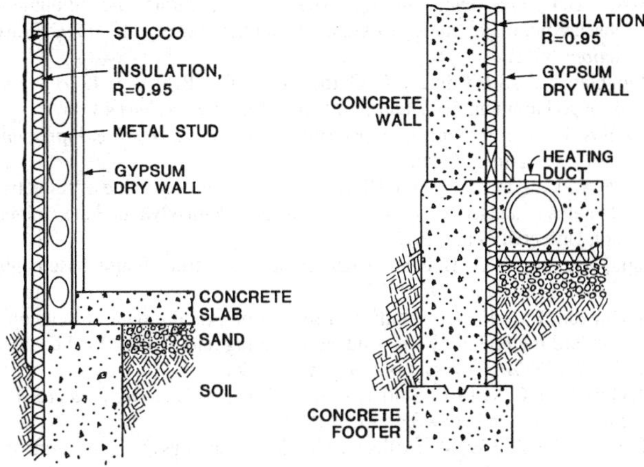

Fig. 8 Slab-on-Grade Foundation Insulation

Figure 8A shows that this construction benefits from the wall insulation between block and brick; the insulation is extended roughly 16 in. below the slab floor. Without this wall insulation, the heat loss coefficient F_2 would be close to that of the 4 in. block wall construction (Figure 8B). Table 16 can be used to estimate F_2 under different degree-days of heating season weather.

CALCULATING INFILTRATION HEAT LOSS

Infiltration of outside air causes both *sensible* and *latent* heat loss. The energy required to raise the temperature of outdoor infiltrating air to indoor air temperature is the sensible component. The energy associated with net loss of moisture from the space is the latent component. Infiltration is discussed in detail in Chapter 25.

Sensible Heat Loss

The energy required to warm outdoor air entering by infiltration to the temperature of the room is given by

$$q_s = c_p Q\rho(t_i - t_o) \tag{7}$$

where

q_s = heat flow required to raise temperature of air leaking into building from t_o to t_i, Btu/h
c_p = specific heat of air, Btu/lb·°F
Q = volumetric flow of outdoor air entering building, ft³/h
ρ = density of air at temperature t_o, lb/ft³

Using standard air (ρ = 0.075 lb/ft³ and c_p = 0.24 Btu/lb·°F), Equation (7) reduces to

$$q_s = 0.018Q(t_i - t_o) \tag{8}$$

The volumetric flow Q of outdoor air entering depends on wind speed and direction, width of cracks or size of openings, type of openings, and other factors explained in Chapter 25. Two methods used to obtain the quantity of infiltration air are the *crack length* and the *air change*. Louvers and doors and the direction they face, as well as any other factors affecting infiltration, may need to be considered.

Latent Heat Loss

When moisture must be added to the indoor air to maintain winter comfort conditions, the energy needed to evaporate an amount of water equivalent to what is lost by infiltration (latent component of infiltration heat loss) must be determined. This energy may be calculated by

$$q_l = Q\rho(W_i - W_o)h_{fg} \tag{9}$$

where

q_l = heat flow required to increase moisture content of air leakage into building from W_o to W_i, Btu/h
Q = volumetric flow of outdoor air entering building, ft³/h
ρ = density of air at temperature t_i, lb/ft³
W_i = humidity ratio of indoor air, lb/lb$_{dry\ air}$
W_o = humidity ratio of outdoor air, lb/lb$_{dry\ air}$
h_{fg} = latent heat of vapor at t_i, Btu/lb

If the latent heat of vapor h_{fg} is 1076 Btu/lb, and the air density is 0.075 lb/ft³, Equation (7) reduces to

$$q_l = 80.7Q(W_i - W_o) \tag{10}$$

Crack Length Method

The basis of calculation for the crack method is that the amount of crack used for computing the infiltration heat loss should not be less than one-half the total length of crack in the outside walls of the room. In a building without partitions, air entering through cracks on the windward side must leave through cracks on the leeward side. Therefore, one-half the total crack for each side and end of the building is used for calculation. In a room with one exposed wall, all the crack is used. With two, three, or four exposed walls, either the wall with the crack that will result in the greatest air leakage or at least one-half the total crack is used, whichever is greater.

In residences, total infiltration loss of the house is generally considered equal to the sum of infiltration losses of the various rooms. But, at any given time, infiltration takes place only on the windward side or sides and not on the leeward. Therefore, for determining total heat requirements of larger buildings, it is more accurate to base total infiltration loss on the wall with the most total crack or on at least half the total crack in the building, whichever is greater. When the crack method rather than Equations (8) and (10) is used for estimating leakage, the heat loss in terms of the crack length may be expressed as

$$q_s = 0.018BL(t_i - t_o) \tag{11}$$

and

$$q_l = 80.7BL(W_i - W_o) \tag{12}$$

where

B = air leakage for wind velocity and type of window or door crack involved, ft³/h per foot of crack
L = length of window or door crack to be considered, ft

Air Change Method

Some designers base infiltration on an estimated number of air changes rather than the length of window cracks. The number of air changes given in Chapter 25 should be considered only as a guide. When calculating infiltration losses by the air change method, Equations (8) and (10) can be used by substituting for Q the volume of the room multiplied by the number of air changes.

Exposure Factors

Some designers use empirical exposure factors to increase calculated heat loss of rooms or spaces on the side(s) of the building exposed to prevailing winds. However, exposure factors are not needed with the method of calculating heat loss described in this chapter. Instead, they may be (1) regarded as safety factors to allow for additional capacity for rooms or spaces exposed to prevailing winds or (2) used to account for the effects of radiation loss, particularly in the case of multistory buildings. Tall buildings may have severe infiltration heat losses induced by stack effect that require special analysis. Although a 15% exposure allowance is often assumed, the actual allowance, if any, is largely a matter of experience and judgment; no test data are available from which to develop rules for the many conditions encountered.

PICKUP LOAD

For intermittently heated buildings and night thermostat setback, additional heat is required to raise the temperature of air, building materials, and material contents of a building to the specified temperature. The pickup load, which is the rate at which this additional heat must be supplied, depends on the heat capacity of the structure, its material contents, and the time in which these are to be heated.

Relatively little information on pickup load exists; however, some early work by Smith (1941, 1942) addressed pickup loads for buildings heated only occasionally, such as auditoriums and churches. Nelson and MacArthur (1978) studied the relationship between thermostat setback, furnace capacity, and recovery time. Based on this limited information, the following design guidelines are offered.

Because design outdoor temperatures generally provide a substantial margin for outdoor temperatures typically experienced during operating hours, many engineers make no allowance for this additional heat in most buildings. However, if a minimum safety factor is to be used, the additional heat should be computed and allowed for, as conditions require. In the case of intermittently heated buildings, an additional 10% capacity should be provided.

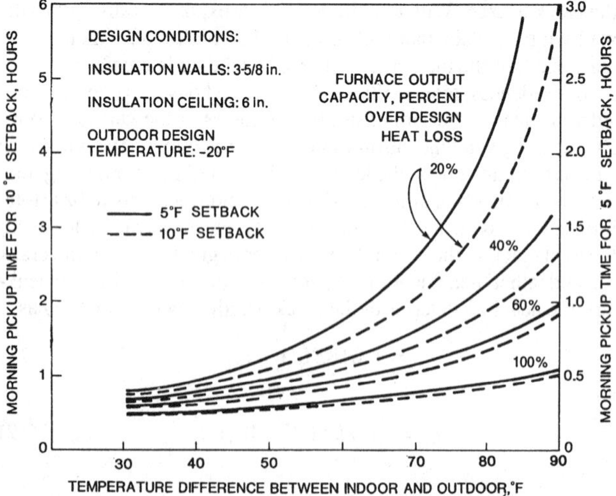

Fig. 9 Furnace Operating Times Required to Pick Up Space Temperature Following 5 and 10°F Night Setback

In buildings with setback-type thermostats, the furnace must be oversized to allow for reestablishing the space temperature in an acceptable time. The amount of oversizing depends on many factors, such as the amount of setback, inside-to-outside temperature difference, building construction, and acceptable pickup time. Figure 9 indicates this relationship for a particular residence. As a general rule for residences, a 10°F night setback requires 40% oversizing for acceptable pickup time and minimum energy requirements (Nelson and MacArthur 1978). For smaller setback, the oversizing can be proportionally less. If daytime as well as night setback is practiced, oversizing of up to 60% is warranted.

REFERENCES

ASHRAE. 1992. Thermal environmental conditions for human occupancy. ANSI/ASHRAE *Standard* 55-1992.

ASHRAE. 1995. Addendum to ANSI/ASHRAE 55-1992. ANSI/ASHRAE *Standard* 55a-1995.

Bligh, T.P., P. Shipp, and G. Meixel. 1978. Energy comparisons and where to insulate earth sheltered buildings and basements. Earth covered settlements, U.S. Department of Energy Conference, Fort Worth, TX.

Chang, J.H. 1958. *Ground temperature*. Bluehill Meteorological Observatory, Harvard University, Cambridge, MA.

Harrje, D.T., G.S. Dutt, and J. Beyea. 1979. Locating and eliminating obscure but major energy losses in residential housing. *ASHRAE Transactions* 85(2).

Houghten, F.C., S.I. Taimuty, C. Gutberlet, and C.J. Brown. 1942. Heat loss through basement walls and floors. *ASHVE Transactions* 48:369.

Joy, F.A. 1958. Improving attic space insulating values. *Heating, Piping and Air Conditioning* 30(1):223.

Joy, F.A., J.J. Zabrony, and S. Bhaduri. 1956. Insulating value of reflective elements in an attic under winter conditions. Pennsylvania State University, University Park, PA.

Latta, J.K. and G.G. Boileau. 1969. Heat losses from house basements. *Canadian Building* 19(10):39.

McQuiston, F.C. 1984. A study and review of existing data to develop a standard methodology for residential heating and cooling load calculations. *ASHRAE Transactions* 90(2A):102-36.

McQuiston, F.C. and J.D. Spitler. 1992. *Cooling and heating load calculation manual*, 2nd ed. ASHRAE, Atlanta.

Nelson, L.W. and J.W. MacArthur. 1978. Energy savings through thermostat setback. *ASHRAE Transactions* 84(2):319-34.

Peony, B.A., F.J. Powell, and D.M. Burch. 1979. Dynamic thermal performance of an experimental masonry building. NBS *Report* 10 664, National Institute of Standards and Technology, Gaithersburg, MD.

Rowley, F.B., A.B. Algren, and C.E. Lund. 1940. Methods of moisture control and their application to building construction. *Bulletin* No. 17 XLIII(4):28. University of Minnesota Engineering Experiment Station.

Smith, E.G. 1941. Heat requirement of intermittently heated buildings. Texas A&M *Engineering Experiment Station Series* No. 62 (November). College Station, TX.

Smith, E.G. 1942. A method of compiling tables for intermittent heating. *Heating, Piping, and Air Conditioning* 14(6):386.

Wang, F.S. 1979. Mathematical modeling and computer simulation of insulation systems in below grade applications. ASHRAE/DOE Conference on Thermal Performance of the Exterior Envelopes of Buildings, Orlando, FL.

BIBLIOGRAPHY

Ackridge, J.M. and J.F. Poulos. 1983. The decremented average ground temperature method for predicting the thermal performance of underground walls. *ASHRAE Transactions* 89(2A).

Burch, D.M., D.F. Krintz, and R.S. Spain. 1984. The effect of wall mass on winter heating loads and indoor comfort—An experimental study. *ASHRAE Transactions* 90(1B).

Kusuda, T. and J.W. Bean. 1984. Simplified methods for determining seasonal heat loss from uninsulated slab-on-grade floors. *ASHRAE Transactions* 90(1B).

Mitalas, G.P. 1983. Calculation of basement heat loss. *ASHRAE Transactions* 89(1B).

Yard, D.C., M. Morton-Gibson, andJ.W. Mitchell. 1984. Simplified dimensionless relations for heat loss from basements. *ASHRAE Transactions* 90(1B).

CHAPTER 28

NONRESIDENTIAL COOLING AND HEATING LOAD CALCULATIONS

THIS chapter presents three methods of calculating air-conditioning cooling load for sizing cooling equipment and a general procedure for calculating heating load, for nonresidential applications. In addition, the fundamental principles for calculating heating loads are presented as a counterpart to cooling load calculation. For residential applications, consult Chapter 27. For information on cooling and/or heating equipment energy use, consult Chapter 30.

The heat balance approach is a fundamental concept in calculating cooling loads. While generally cumbersome for widespread or routine use, this underlying concept is the basis for each of the three simplified procedures outlined for varying purposes.

The cooling calculation procedure most closely approximating the heat balance concept is the transfer function method (TFM), first introduced in the 1972 *ASHRAE Handbook of Fundamentals*. This computer-based procedure takes place in two steps, first establishing the heat gain from all sources and then determining the conversion of such heat gain into cooling load. Developed as an hour-by-hour calculation procedure oriented to simulate annual energy use, its normalizing characteristics make it particularly appropriate for that application.

A simplified version of the TFM, which can be used with certain types of buildings for which application data are available, was presented in the 1977 *ASHRAE Handbook of Fundamentals*. This one-step procedure uses cooling load temperature differences (CLTD), solar cooling load factors (SCL), and internal cooling load factors (CLF), to calculate cooling loads as an approximation of the TFM. Where applicable, this method may be suitable for hand calculation use.

An alternative simplification of the heat balance technique uses total equivalent temperature differential values and a system of time-averaging (TETD/TA) to calculate cooling loads. Also a computer-based, two-step procedure (heat gain, then cooling load), first introduced in the 1967 *ASHRAE Handbook of Fundamentals*, this method gives valid broad-range results to experienced users.

COOLING LOAD PRINCIPLES

The variables affecting cooling load calculations are numerous, often difficult to define precisely, and always intricately interrelated. Many cooling load components vary in magnitude over a wide range during a 24-h period. Since these cyclic changes in load components are often not in phase with each other, each must be analyzed to establish the resultant maximum cooling load for a building or zone. A zoned system (a system of conditioning equipment serving several independent areas, each with its own temperature control) need recognize no greater total cooling load capacity than the largest hourly summary of simultaneous zone loads throughout a design day; however, it must handle the peak cooling load for each zone at its individual peak hour. At certain times of the day during the heating or intermediate seasons, some zones may require heating while others require cooling.

Calculation accuracy. The concept of determining the cooling load for a given building must be kept in perspective. A proper cooling load calculation gives values adequate for proper performance. Variation in the heat transmission coefficients of typical building materials and composite assemblies, the differing motivations and skills of those who physically construct the building, and the manner in which the building is actually operated are some of the variables that make a numerically precise calculation impossible. While the designer uses reasonable procedures to account for these factors, the calculation can never be more than a good estimate of the actual cooling load.

Heat flow rates. In air-conditioning design, four related heat flow rates, each of which varies with time, must be differentiated: (1) space heat gain, (2) space cooling load, (3) space heat extraction rate, and (4) cooling coil load.

Space heat gain. This instantaneous rate of heat gain is the rate at which heat enters into and/or is generated within a space at a given instant. Heat gain is classified by (1) the mode in which it enters the space and (2) whether it is a sensible or latent gain.

Mode of entry. The modes of heat gain may be as (1) solar radiation through transparent surfaces; (2) heat conduction through exterior walls and roofs; (3) heat conduction through interior partitions, ceilings, and floors; (4) heat generated within the space by occupants, lights, and appliances; (5) energy transfer as a result of ventilation and infiltration of outdoor air; or (6) miscellaneous heat gains.

Sensible or latent heat. Sensible heat gain is directly added to the conditioned space by conduction, convection, and/or radiation. Latent heat gain occurs when moisture is added to the space (e.g., from vapor emitted by occupants and equipment). To maintain a constant humidity ratio, water vapor must condense on cooling apparatus at a rate equal to its rate of addition into the space. The amount of energy required to offset the latent heat gain essentially equals the product of the rate of condensation and the latent heat of condensation. In selecting cooling apparatus, it is necessary to distinguish

The preparation of this chapter is assigned to TC 4.1, Load Calculation Data and Procedures.

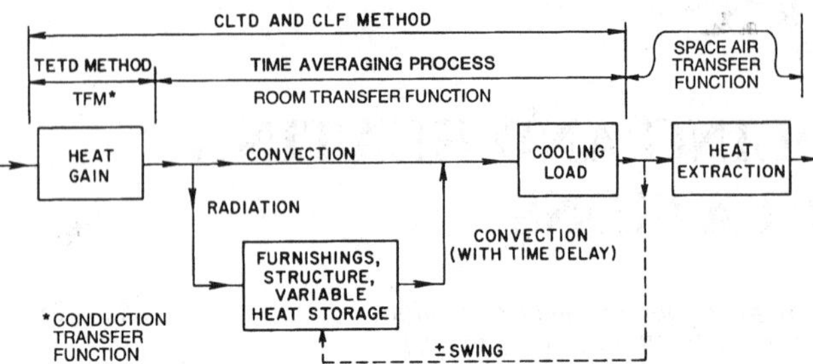

Fig. 1 Origin of Difference Between Magnitude of Instantaneous Heat Gain and Instantaneous Cooling Load

between sensible and latent heat gain. Every cooling apparatus has a maximum sensible heat removal capacity and a maximum latent heat removal capacity for particular operating conditions.

Space cooling load. This is the rate at which heat must be removed from the space to maintain a constant space air temperature. The sum of all space instantaneous heat gains at any given time does not necessarily (or even frequently) equal the cooling load for the space at that same time.

Radiant heat gain. Space heat gain by radiation is not immediately converted into cooling load. Radiant energy must first be absorbed by the surfaces that enclose the space (walls, floor, and ceiling) and the objects in the space (furniture, etc.). As soon as these surfaces and objects become warmer than the space air, some of their heat is transferred to the air in the space by convection. The composite heat storage capacity of these surfaces and objects determines the rate at which their respective surface temperatures increase for a given radiant input, and thus governs the relationship between the radiant portion of heat gain and its corresponding part of the space cooling load (Figure 1). The thermal storage effect is critically important in differentiating between instantaneous heat gain for a given space and its cooling load for that moment. Predicting the nature and magnitude of this elusive phenomenon in order to estimate a realistic cooling load for a particular combination of circumstances has long been a subject of major interest to design engineers. The bibliography lists some of the early work on the subject.

Space Heat Extraction Rate

The rate at which heat is removed from the conditioned space equals the space cooling load only to the degree that room air temperature is held constant. In conjunction with intermittent operation of the cooling equipment, the control system characteristics usually permit a minor cyclic variation or swing in room temperature. Therefore, a proper simulation of the control system gives a more realistic value of energy removal over a fixed time period than using the values of the space cooling load. This concept is primarily important for estimating energy use over time (see Chapter 30); however, it is not needed to calculate design peak cooling load for equipment selection. Space heat extraction rate calculation is discussed later in this chapter; see also Mitalas (1972).

Cooling Coil Load

The rate at which energy is removed at the cooling coil that serves one or more conditioned spaces equals the sum of the instantaneous space cooling loads (or space heat extraction rate if it is assumed that the space temperature does not vary) for all the spaces served by the coil, plus any external loads. Such external loads include heat gain by the distribution system between the individual spaces and the cooling equipment, and outdoor air heat and moisture introduced into the distribution system through the cooling equipment.

SPACE COOLING LOAD CALCULATION TECHNIQUES

Heat Balance Fundamentals

The estimation of cooling load for a space involves calculating a surface-by-surface conductive, convective, and radiative heat balance for each room surface and a convective heat balance for the room air. Sometimes called "the exact solution," these principles form the foundation for all other methods described in this chapter.

To calculate space cooling load directly by heat balance procedures requires a laborious solution of energy balance equations involving the space air, surrounding walls and windows, infiltration and ventilation air, and internal energy sources. To demonstrate the calculation principle, consider a sample room enclosed by four walls, a ceiling, and a floor, with infiltration air, ventilation air, and normal internal energy sources. The calculations that govern energy exchange at each inside surface at a given time are:

$$q_{i,\theta} = \left[h_{ci}(t_{a,\theta} - t_{i,\theta}) + \sum_{j=1,\, j \neq i}^{m} g_{ij}(t_{j,\theta} - t_{i,\theta}) \right] A_i + RS_{i,\theta} + RL_{i,\theta} + RE_{i,\theta} \quad \text{for } i = 1, 2, 3, 4, 5, 6 \tag{1}$$

where

m = number of surfaces in room (6 in this case)
$q_{i,\theta}$ = rate of heat conducted into surface i at inside surface at time θ
A_i = area of surface i
h_{ci} = convective heat transfer coefficient at interior surface i
g_{ij} = radiation heat transfer factor between interior surface i and interior surface j
$t_{a,\theta}$ = inside air temperature at time θ
$t_{i,\theta}$ = average temperature of interior surface i at time θ
$t_{j,\theta}$ = average temperature of interior surface j at time θ
$RS_{i,\theta}$ = rate of solar energy coming through windows and absorbed by surface i at time θ
$RL_{i,\theta}$ = rate of heat radiated from lights and absorbed by surface i at time θ
$RE_{i,\theta}$ = rate of heat radiated from equipment and occupants and absorbed by surface i at time θ

Conduction transfer functions. The equations governing conduction within the six surfaces cannot be solved independently of Equation (1), since the energy exchanges occurring within the room affect the inside surface conditions, in turn affecting the internal conduction. Consequently, the above mentioned six formulations of Equation (1) must be solved simultaneously with the governing equations of conduction within the six surfaces in order to calculate the space cooling load. Typically, these equations are formulated as conduction transfer functions in the form

$$q_{in,\theta} = \sum_{m=1}^{M} Y_{k,m} t_{o,\theta-m+1} - \sum_{m=1}^{M} Z_{k,m} t_{in,\theta-m+1} + \sum_{m=1}^{M} F_m q_{in,\theta-m} \tag{2}$$

where

q = rate of heat conducted into a specific surface at a specific hour
in = inside surface subscript
k = order of CTF
m = time index variable
M = number of nonzero CTF values
o = outside surface subscript
t = temperature
θ = time
x = exterior CTF values
Y = cross CTF values
Z = interior CTF values
F_m = flux (heat flow rate) history coefficients

Space air energy balance. Note that the interior surface temperature, $t_{i,\theta}$ in Equation (1) and $t_{in,\theta}$ in Equation (2), requires simultaneous solution. In addition, Equation (3) representing an energy balance on the space air must also be solved simultaneously

$$Q_{L,\theta} = \left[\sum_{i=1}^{m} h_{ci}(t_{i,\theta} - t_{a,\theta})\right] A_i + \rho C V_{L,\theta}(t_{o,\theta} - t_{a,\theta}) + \rho C V_{v,\theta}(t_{v,\theta} - t_{a,\theta}) + RS_{a,\theta} + RL_{a,\theta} + RE_{a,\theta} \quad (3)$$

where

ρ = air density
C = air specific heat
$V_{L,\theta}$ = volume flow rate of outdoor air infiltrating into room at time θ
$t_{o,\theta}$ = outdoor air temperature at time θ
$V_{v,\theta}$ = volume rate of flow of ventilation air at time θ
$t_{v,\theta}$ = ventilation air temperature at time θ
$RS_{a,\theta}$ = rate of solar heat coming through windows and convected into room air at time θ
$RL_{a,\theta}$ = rate of heat from lights convected into room air at time θ
$RE_{a,\theta}$ = rate of heat from equipment and occupants and convected into room air at time θ

Note that the ventilation air component in Equation (3) is assumed to enter the space directly, rather than through any associated cooling apparatus. Note also that the space air temperature is allowed to float. By fixing the space air temperature, the cooling load need not be determined simultaneously.

This rigorous approach to calculating space cooling load is impractical without the speed at which some computations can be done by modern digital computers. Computer programs in use where instantaneous space cooling loads are calculated in this exact manner are primarily oriented to energy use calculations over extended periods because hourly outdoor temperatures are normalized increments rather than peak design temperature profiles (Mitalas and Stephenson 1967, Buchberg 1958, Walton 1982).

The transfer function concept is a simplification to the strict heat balance calculation procedure. In the transfer function concept, Mitalas and Stephenson (1967) used room thermal response factors. In their procedure, room surface temperatures and cooling load were first calculated by the rigorous method just described, for several typical constructions representing offices, schools, and dwellings of heavy, medium, and light construction. In these calculations, components such as solar heat gain, conduction heat gain, or heat gain from the lighting, equipment, and occupants were simulated by pulses of unit strength. The transfer functions were then calculated as numerical constants representing the cooling load corresponding to the input excitation pulses. Once these transfer functions were determined for typical constructions they were assumed independent of input pulses, thus permitting cooling loads to be determined without the more rigorous calculation. Instead, the calculation requires simple multiplication of the transfer functions by a time-series representation of heat gain and subsequent summation of these products, which can be carried out on a small computer. The same transfer function concept can be applied to calculating heat gain components themselves, as explained later.

Total Equivalent Temperature Differential Method

In the total equivalent temperature differential (TETD) method, the response factor technique is used with a number of representative wall and roof assemblies from which data are derived to calculate TETD values as functions of sol-air temperature and maintained room temperature. Various components of space heat gain are calculated using associated TETD values, and the results are added to internal heat gain elements to get an instantaneous total rate of space heat gain. This gain is converted to an instantaneous space cooling load by the time-averaging (TA) technique of averaging the radiant portions of the heat gain load components for the current hour with related values from an appropriate period of immediately preceding hours. This technique provides a rational means to deal quantitatively with the thermal storage phenomenon, but it is best solved by computer because of its complexity. Its fundamental weakness is that simple averaging of radiant load components is a poor approximation of the actual physics involved, and choosing an appropriate averaging period is subjective and depends on user experience.

Transfer Function Method

Although similar in principle to TETD/TA, the transfer function method (TFM) (Mitalas 1972) applies a series of weighting factors, or conduction transfer function (CTF) coefficients to the various exterior opaque surfaces and to differences between sol-air temperature and inside space temperature to determine heat gain with appropriate reflection of thermal inertia of such surfaces. Solar heat gain through glass and various forms of internal heat gain are calculated directly for the load hour of interest. The TFM next applies a second series of weighting factors, or coefficients of room transfer functions (RTF), to heat gain and cooling load values from all load elements having radiant components, to account for the thermal storage effect in converting heat gain to cooling load. Both evaluation series consider data from several previous hours as well as the current hour. RTF coefficients relate specifically to the spatial geometry, configuration, mass, and other characteristics of the space so as to reflect weighted variations in thermal storage effect on a time basis rather than a straight-line average.

Transfer Functions. These coefficients relate an output function at a given time to the value of one or more driving functions at a given time and at a set period immediately preceding. The CTF described in this chapter is no different from the thermal response factor used for calculating wall or roof heat conduction, while the RTF is the weighting factor for obtaining cooling load components (ASHRAE 1975). The bibliography lists reports of various experimental work that has validated the predictive accuracy of the TFM. While the TFM is scientifically appropriate and technically sound for a specific cooling load analysis, several immediately previous 24-h periods are assumed to be the same as the load hour of interest. Also, a computer is required for effective application in a commercial design environment.

CLTD/SCL/CLF Method

Rudoy and Duran (1975) compared the TETD/TA and TFM. As part of this work, data obtained by using the TFM on a group of applications considered representative were then used to generate cooling load temperature differential (CLTD) data, for direct one-step calculation of cooling load from conduction heat gain through sunlit walls and roofs and conduction through glass exposures (see Bibliography). Cooling load factors (CLF) for similar one-step calculation of solar load through glass and for loads from internal sources were also developed. More recent research (McQuiston 1992) developed an improved factor for solar load through glass, the solar cooling load (SCL) factor, which allows additional influencing parameters to be considered for greater accuracy. CLTDs, SCLs, and CLFs all include the effect of (1) time lag in conductive heat gain through opaque exterior surfaces and (2) time delay by thermal storage in converting radiant heat gain to cooling load. This simplification allows cooling loads to be calculated manually; thus, when data are available and are appropriately used, the results are consistent with those from the TFM, thus making the method popular for instruction.

Application Experience

The CLTD and CLF tables published in previous editions of the Fundamentals volume and in the original *Cooling and Heating Load Calculation Manual* (ASHRAE 1979) are normalized data, based on applications of the original TFM data presented in the 1972 Fundamentals volume. Subsequent studies investigating the effects of 1981 to 1985 RTF data indicated results generally less conservative than those computed with the 1972 data. More recent research, however, suggests otherwise (McQuiston 1992), and the revised values for 1993, including the new SCLs, are currently considered more realistic for design load purposes.

CLTD Data. The originally developed CLTD data were so voluminous that they were first limited to 13 representative flat roof assemblies (with and without ceilings, for 26 total cases) and 7 wall groups (into which 41 different wall assemblies can be categorized). Twenty-four hourly CLTD values were tabulated for each of the 26 roof cases and each of the 7 wall groups, broken down for walls into 8 primary orientations. Adjustments were then required for specific north latitude and month of calculation. Reliability of adjustments was reasonably consistent during summer months but became much less realistic for early and late hours during traditionally noncooling load months.

Solar Heat Gain Data. Solar heat gain through glass required similar data compression to present a corresponding range of conditions. Tables of maximum solar heat gain factors (SHGF) were listed for every 4° of north latitude between 0 and 64°, for each month and by 16 compass directions and horizontal. Cooling load factors (CLF), decimal multipliers for SHGF data, were tabulated for unshaded glass in spaces having carpeted or uncarpeted floors and for inside-shaded glass with any room construction. Unshaded CLFs were presented for each of 24 hours by 8 compass directions plus horizontal, further categorized by light, medium, or heavy room construction. Inside-shaded CLFs disregarded construction mass but included 16 orientations plus horizontal. The product of the selected CLTD and CLF values represented cooling load per unit area as a single process. CLF values published in the *Handbook* were derived for the period May through September as normally the hottest months for load calculation purposes. As with CLTDs, the reliability of CLF data deteriorated rapidly for applications during early and late hours of months considered "noncooling load" periods.

ASHRAE Sponsored Research. For some space geometries and building constructions, the tabulated CLTD and CLF data published through 1989 were found also to be too restrictive or limited. The weighting factors used to generate these data, based on representative spaces in schools, offices, and dwellings at the time of the original research, did not reflect current design and construction practices. ASHRAE research investigated the sensitivity of the weighting factors to variations in space construction, size, exposure, and related conditions to update the tabular data. However, the investigators discovered that the range and amplitude of this sensitivity was much broader than previously thought, rendering even more impractical the generation of enough tabular material to cover the majority of normal applications. Accordingly, two significant changes in direction have occurred:

1. The section describing the CLTD/CLF in the 1985 and 1989 editions of the Fundamentals volume recommended caution in application of this procedure for general practice, and this cautionary notice was also added as an insert to the *Cooling and Heating Load Calculation Manual* (McQuiston and Spitler 1992).
2. The system itself was modified for more specific tabulation of data, abandoning the maximum SHGF concept and incorporating solar cooling load (SCL) factors for estimating cooling load from glass.

The main thrust of ASHRAE sponsored research between 1989 and 1993 was to update the *Cooling and Heating Load Calculation Manual*, published in revised form in 1993. Information from earlier research was used to revise the original factors by incorporating additional parameters, including separating solar load through glass from the CLF category and creating more appropriate SCL factors for that component. Still faced with too much tabular data, information was tabulated only for limited use and representative examples, but it was accompanied by instructions for customizing similar data for specific application; a microcomputer database was also provided to facilitate such calculations. Certain limitations resulting from normalization of data remain, for which anticipated error ranges are listed to aid in evaluating results. The section in this chapter describing the CLTD/SCL/CLF method has incorporated this latest 1993 research, but it does not provide the microcomputer program.

Dissatisfaction with the limitations of CLTD/SCL/CLF led to a reappraisal of prospects for improvement. Because adding flexibility mandated massive extrapolation of tabular material and/or the computational equivalent, the ASHRAE technical committee for load calculations (TC 4.1) decided to leave this method at its present level of development and to direct future research effort toward more promising goals.

TFM Method. Like the CLTD/SCL/CLF method, the TFM method represents, compared to fundamental heat balance principles, a significant compromise with several important physical concepts. Also, the complex computations required of the heat balance method can now be handled by today's desktop computers. For these reasons, ASHRAE is supporting research to clarify heat balance procedures for more general use. Results of this research will appear in the next edition of this Handbook.

TETD/TA Method. Prior to introduction of the CLTD/CLF, most users had turned to computer-based versions of the time-averaging technique, proven successful and practical in ten years of heavy use. Most users, however, recognized the subjectivity of determining the relative percentages of radiant heat in the various heat gain components and selecting the number of hours over which to average such loads—both of which must rely on the individual experience of the user rather than on research or support in the scientific literature. Harris and McQuiston (1988) developed decrement factors and time lag values. In this chapter, these factors have been keyed to typical walls and roofs. All other tabular data pertaining to this method has been deleted, so that since 1989, information has been confined to basic algorithms intended for continued computer applications.

The lack of scientific validation of the time-averaging process led to suspension of further development of TETD/TA. But the need to retain a more simplified computation than heat balance alone led to a study of Radiant Time Series (RTS) coefficients to convert radiant heat gain components to cooling load. Some preliminary results of the relative percentages of various kinds and types of radiant heat gain as compared to convective are included in this chapter.

Alternative Procedures. TFM, CLTD/SCL/CLF and TETD/TA procedures, tables, and related data will continue to be appropriate and dependable when applied within the limits discussed in this chapter. Users will likely incorporate heat balance relationships when developing custom CLTD/SCL/CLF or TETD/TA tabular data for specific projects.

INITIAL DESIGN CONSIDERATIONS

To calculate a space cooling load, detailed building design information and weather data at selected design conditions are required. Generally, the following steps should be followed:

Data Assembly

1. **Building characteristics.** Obtain characteristics of the building. Building materials, component size, external surface colors and shape are usually determined from building plans and specifications.

2. **Configuration.** Determine building location, orientation and external shading from building plans and specifications. Shading from adjacent buildings can be determined by a site plan or by visiting the proposed site, but should be carefully evaluated as to its probable permanence before it is included in the calculation. The possibility of abnormally high ground-reflected solar radiation (i.e., from adjacent water, sand, or parking lots), or solar load from adjacent reflective buildings should not be overlooked.
3. **Outdoor design conditions.** Obtain appropriate weather data and select outdoor design conditions. Weather data can be obtained from local weather stations or from the National Climatic Center, Asheville, NC 28801. For outdoor design conditions for a large number of weather stations, see Chapter 26. Note, however, that the scheduled values for the design dry-bulb and mean coincident wet-bulb temperatures can vary considerably from data traditionally used in various areas. Use judgment to ensure that results are consistent with expectations. Also, consider prevailing wind velocity and the relationship of a project site to the selected weather station.
4. **Indoor design conditions.** Select indoor design conditions, such as indoor dry-bulb temperature, indoor wet-bulb temperature, and ventilation rate. Include permissible variations and control limits.
5. **Operating schedules.** Obtain a proposed schedule of lighting, occupants, internal equipment, appliances, and processes that contribute to the internal thermal load. Determine the probability that the cooling equipment will be operated continuously or shut off during unoccupied periods (e.g., nights and/or weekends).
6. **Date and time.** Select the time of day and month to do the cooling load calculation. Frequently, several different times of day and several different months must be analyzed to determine the peak load time. The particular day and month are often dictated by peak solar conditions, as tabulated in Tables 15 through 21 in Chapter 29. For southern exposures in north latitudes above 32 having large fenestration areas, the peak space cooling load usually occurs in December or January. To calculate a space cooling load under these conditions, the warmest temperature for the winter months must be known. These data can be found in the National Climatic Center's Climatic Atlas of the United States.

Use of Data. Once the data are assembled, the space cooling load at design conditions may be calculated as outlined in the following sections of this chapter.

Additional Considerations

The proper design and sizing of all-air or air-and-water central air-conditioning systems require more than calculation of the cooling load in the space to be conditioned. The type of air-conditioning system, fan energy, fan location, duct heat loss and gain, duct leakage, heat extraction lighting systems, and type of return air system all affect system load and component sizing. Adequate system design and component sizing require that system performance be analyzed as a series of psychrometric processes. Chapter 3 of the 1992 *ASHRAE Handbook—Systems and Equipment* describes some elements of this technique in detail, while others are delineated in this chapter.

HEAT GAIN CALCULATION CONCEPTS

Heat Gain through Fenestration Areas

The primary weather-related variable influencing the cooling load for a building is solar radiation. The effect of solar radiation is more pronounced and immediate in its impact on exposed non-opaque surfaces. The calculation of solar heat gain and conductive heat transfer through various glazing materials and associated mounting frames, with or without interior and/or exterior shading devices, is discussed in Chapter 29. This chapter covers the application of such data to the overall heat gain evaluation and the conversion of the calculated heat gain into a composite cooling load for the conditioned space.

Heat Gain through Exterior Surfaces

Heat gain through exterior opaque surfaces is derived from the same elements of solar radiation and thermal gradient as that for fenestration areas. It differs primarily as a function of the mass and nature of the wall or roof construction, since those elements affect the rate of conductive heat transfer through the composite assembly to the interior surface.

Sol-Air Temperature

Sol-air temperature is the temperature of the outdoor air that, in the absence of all radiation changes, gives the same rate of heat entry into the surface as would the combination of incident solar radiation, radiant energy exchange with the sky and other outdoor surroundings, and convective heat exchange with the outdoor air.

Heat flux into exterior sunlit surfaces. The heat balance at a sunlit surface gives the heat flux into the surface q/A as

$$q/A = \alpha I_t + h_o(t_o - t_s) - \varepsilon \Delta R \qquad (4)$$

where

α = absorptance of surface for solar radiation
I_t = total solar radiation incident on surface, Btu/(h·ft^2)
h_o = coefficient of heat transfer by long-wave radiation and convection at outer surface, Btu/(h·ft^2·°F)
t_o = outdoor air temperature, °F
t_s = surface temperature, °F
ε = hemispherical emittance of surface
ΔR = difference between long-wave radiation incident on surface from sky and surroundings and radiation emitted by blackbody at outdoor air temperature, Btu/(h·ft^2)

Assuming the rate of heat transfer can be expressed in terms of the sol-air temperature t_e

$$q/A = h_o(t_e - t_s) \qquad (5)$$

and from Equations (4) and (5)

$$t_e = t_o + \alpha I_t / h_o - \varepsilon \Delta R / h_o \qquad (6)$$

Horizontal Surfaces. For horizontal surfaces that receive long-wave radiation from the sky only, an appropriate value of ΔR is about 20 Btu/(h·ft^2), so that if $\varepsilon = 1$ and $h_o = 3.0$ Btu/(h·ft^2·°F), the long-wave correction term is about −7°F (Bliss 1961).

Vertical surfaces. Because vertical surfaces receive long-wave radiation from the ground and surrounding buildings as well as from the sky, accurate ΔR values are difficult to determine. When solar radiation intensity is high, surfaces of terrestrial objects usually have a higher temperature than the outdoor air; thus, their long-wave radiation compensates to some extent for the sky's low emittance. Therefore, it is common practice to assume $\Delta R = 0$ for vertical surfaces.

Tabulated Temperature Values. The sol-air temperatures in Table 1 have been calculated based on $\varepsilon \Delta R / h_o$ being 7°F for horizontal surfaces and 0°F for vertical surfaces; total solar intensity values used for the calculations were the same as those used to evaluate the solar heat gain factors (SHGF) for July 21 at 40°N latitude (Chapter 29). These values of I_t incorporate diffuse radiation from a clear sky and ground reflection, but make no allowance for reflection from adjacent walls.

Surface Colors. Sol-air temperature values are given for two values of the parameter α / h_o (Table 1); the value of 0.15 is appropriate for a light-colored surface, while 0.30 represents the usual maximum

Table 1 Sol-Air Temperatures for July 21, 40°N Latitude

$t_e = t_o + \alpha I_t/h_o - \varepsilon \Delta R/h_o$

Time	Air Temp. t_o, °F	Light Colored Surface, $\alpha/h_o = 0.15$								
		N	NE	E	SE	S	SW	W	NW	HOR
1	76	76	76	76	76	76	76	76	76	69
2	76	76	76	76	76	76	76	76	76	69
3	75	75	75	75	75	75	75	75	75	68
4	74	74	74	74	74	74	74	74	74	67
5	74	74	74	74	74	74	74	74	74	67
6	74	80	93	95	84	76	76	76	76	72
7	75	80	99	106	94	78	78	78	78	81
8	77	81	99	109	101	82	81	81	81	92
9	80	85	96	109	106	88	85	85	85	102
10	83	88	91	105	107	95	88	88	88	111
11	87	93	93	99	106	102	93	93	93	118
12	90	96	96	96	102	106	102	96	96	122
13	93	99	99	99	99	108	112	105	99	124
14	94	99	99	99	99	106	118	116	102	122
15	95	100	100	100	100	103	121	124	111	117
16	94	98	98	98	98	99	118	126	116	109
17	93	98	96	96	96	96	112	124	117	99
18	91	97	93	93	93	93	101	112	110	89
19	87	87	87	87	87	87	87	87	87	80
20	85	85	85	85	85	85	85	85	85	78
21	83	83	83	83	83	83	83	83	83	76
22	81	81	81	81	81	81	81	81	81	74
23	79	79	79	79	79	79	79	79	79	72
24	77	77	77	77	77	77	77	77	77	70
Avg.	**83**	**86**	**88**	**90**	**90**	**87**	**90**	**90**	**88**	**90**

Time	Air Temp. t_o, °F	Dark Colored Surface, $\alpha/h_o = 0.30$								
		N	NE	E	SE	S	SW	W	NW	HOR
1	76	76	76	76	76	76	76	76	76	69
2	76	76	76	76	76	76	76	76	76	69
3	75	75	75	75	75	75	75	75	75	68
4	74	74	74	74	74	74	74	74	74	67
5	74	74	75	75	74	74	74	74	74	67
6	74	85	112	115	94	77	77	77	77	77
7	75	84	124	136	113	81	81	81	81	94
8	77	85	121	142	125	86	85	85	85	114
9	80	90	112	138	131	96	89	89	89	131
10	83	94	100	127	131	107	94	94	94	145
11	87	98	99	111	125	118	100	98	98	156
12	90	101	101	102	114	123	114	102	101	162
13	93	104	104	104	106	124	131	117	105	162
14	94	105	105	105	105	118	142	138	111	156
15	95	105	104	104	104	111	146	153	127	146
16	94	102	102	102	102	103	142	159	138	131
17	93	102	99	99	99	99	131	154	142	112
18	91	102	94	94	94	94	111	132	129	94
19	87	87	87	87	87	87	87	88	88	80
20	85	85	85	85	85	85	85	85	85	78
21	83	83	83	83	83	83	83	83	83	76
22	81	81	81	81	81	81	81	81	81	74
23	79	79	79	79	79	79	79	79	79	72
24	77	77	77	77	77	77	77	77	77	70
Avg.	**83**	**89**	**94**	**99**	**97**	**93**	**97**	**99**	**94**	**104**

Note: Sol-air temperatures are calculated based on $\varepsilon \Delta R/h_o$ = 7°F for horizontal surfaces and 0°F for vertical surfaces.

value for this parameter (i.e., for a dark-colored surface, or any surface for which the permanent lightness can not reliably be anticipated).

Air Temperature Cycle. The air temperature cycle used to calculate the sol-air temperatures is given in Column 2, Table 1. Sol-air temperatures can be adjusted to any other air temperature cycle simply by adding or subtracting the difference between the desired air temperature and the air temperature value given in Column 2.

Adjustments. Sol-air temperature cycles can be estimated for other dates and latitudes by using the data in Tables 15 through 21, Chapter 29. For any of the times, dates, and wall orientations listed in those tables, the value of I_t is approximately 1.15 × SHGF. However, the 1.15 factor is approximate and only accounts for the solar energy excluded by a single sheet of ordinary window glass. For surfaces with other orientations or slope angles of other than 0°, and for more accurate estimates at incident angles above 50° (particularly critical for southern exposures), the solar intensity can be found by the method outlined in Chapter 29.

Average Sol-Air Temperature. The average daily sol-air temperature t_{ea} can be calculated for any of the situations covered by Tables 15 through 21 of Chapter 29:

$$t_{ea} = t_{oa} + \frac{\alpha}{h_o}\left(\frac{I_{DT}}{24}\right) - \frac{\varepsilon \Delta R}{h_o} \tag{7}$$

where I_{DT} is the sum of two appropriate half-day totals of solar heat gain in Btu/(h·ft²). For example, the average sol-air temperature for a wall facing southeast at 40°N latitude on August 21 would be

$$t_{ea} = t_{oa} + \frac{\alpha}{h_o}\left[\frac{1.15(956 + 205)}{24}\right]$$

The daily solar heat gain of double-strength sheet glass is 956 + 205 Btu/(h·ft²) in a southeast facade at this latitude and date (Table 18, Chapter 29); and $\varepsilon \Delta R/h_o$ is assumed to be zero for this vertical surface.

Table 2 Percentage of Daily Range

Time, h	%	Time, h	%	Time, h	%
1	87	9	71	17	10
2	92	10	56	18	21
3	96	11	39	19	34
4	99	12	23	20	47
5	100	13	11	21	58
6	98	14	3	22	68
7	93	15	0	23	76
8	84	16	3	24	82

Hourly Air Temperatures. The hourly air temperatures in Column 2, Table 1 are for a location with a design temperature of 95°F and a range of 21°F. To compute corresponding temperatures for other locations, select a suitable design temperature from Table 1 of Chapter 26 and note the outdoor daily range. For each hour, take the percentage of the daily range indicated in Table 2 of this chapter and subtract from the design temperature.

Example 1. Air temperature calculation. Calculate the summer dry-bulb temperature at 1200 h for Reno, Nevada.

Solution: From Table 1, Chapter 26, the daily range is 37.3°F and the 1% design dry-bulb temperature is 95°F. From Table 2, the percentage of the daily range at 1200 hours is 23%. Thus, the dry-bulb temperature at 1200 is Design dry-bulb – (Percentage fraction × Daily range) = 95 – (0.23 × 37.3) = 86.4°F.

Data limitations. The outdoor daily range is the difference between the average daily maximum and average daily minimum temperatures during the warmest month. More reliable results could be obtained by determining or estimating the shape of the temperature curve for typical hot days at the building site and considering each month separately.

Peak cooling load is often determined by solar heat gain through fenestration; this peak may occur in winter months and/or at a time of day when outside air temperature is not at its peak.

Heat Gain through Fenestration

The sections that include Equations (39) through (47) in Chapter 29 describe one method used to calculate space cooling load resulting from heat transfer through fenestration. The solar heat gain profiles listed in Chapter 29 are for fenestration areas with no external shading. The equations for calculating shade angles (Chapter 29) can be used to determine the shape and area of moving shadow falling across a given window from external shading elements during the course of a design day. Thus, a subprofile of heat gain for that window can be created by separating its sunlit and shaded areas for each hour; modifying multipliers for inside shading devices can also be included.

Exterior Shading. Nonuniform exterior shading, caused by roof overhangs, side fins, or building projections, require separate hourly calculations for the externally shaded and unshaded areas of the window in question, with the SC still used to account for any internal shading devices. The areas, shaded and unshaded, depend on the location of the shadow line on a surface in the plane of the glass. Sun (1968) developed fundamental algorithms for analysis of shade patterns. McQuiston and Spitler (1992) provide graphical data to facilitate shadow line calculation, and the north exposure SHGF may be taken for shaded glass (with some loss of accuracy at latitudes less than 24° north).

An alternate, more accurate, method suggested by Todorovic and Curcija (1984) first calculates cooling loads as if the external shading were absent, then adjusts (reduces) the result to account for the shading effect. This correction applies a "negative cooling load factor," calculated in much the same way as a conventional cooling load but using the time-varying area of the shaded portion of the glass as the heat gain element. Todorovic (1987) describes the solution of the moving shade line problem in the context of consequent cooling load.

Temperature Considerations. To estimate the conduction of heat through fenestration at any time, applicable values of the outdoor and indoor dry-bulb temperatures must be used. Chapter 26 gives design values of summer outdoor dry-bulb temperatures for many locations. These are generally mid-afternoon temperatures; for other times, local weather stations or NOAA can supply temperature data. Winter design temperatures should not be used in Equation (15), since such data are for heating design rather than coincident conduction heat gain with sunlit glass during the heating season.

Heat Gain through Interior Surfaces

Whenever a conditioned space is adjacent to a space with a different temperature, transfer of heat through the separating physical section must be considered. The heat transfer rate is given by

$$q = UA(t_b - t_i) \tag{8}$$

where

q = heat transfer rate, Btu/h
U = coefficient of overall heat transfer between adjacent and conditioned space, Btu/(h·ft²·°F)
A = area of separating section concerned, ft²
t_b = average air temperature in adjacent space, °F
t_i = air temperature in conditioned space, °F

Values of U can be obtained from Chapter 24. Temperature t_b may range widely from that in the conditioned space. The temperature in a kitchen or boiler room, for example, may be as much as 15 to 50°F above the outdoor air temperature. Actual temperatures in adjoining spaces should be measured when possible. Where nothing is known, except that the adjacent space is of conventional construction, contains no heat sources, and itself receives no significant solar heat gain, $t_b - t_i$ may be considered the difference between the outdoor air and conditioned space design dry-bulb temperatures minus 5°F. In some cases, the air temperature in the adjacent space will correspond to the outdoor air temperature or higher.

Floors. For floors directly in contact with the ground, or over an underground basement that is neither ventilated nor conditioned, heat transfer may be neglected for cooling load estimates.

HEAT SOURCES IN CONDITIONED SPACES

People

Table 3 gives representative rates at which heat and moisture are given off by human beings in different states of activity. Often these sensible and latent heat gains constitute a large fraction of the total load. Even for short-term occupancy, the extra heat and moisture brought in by people may be significant. Chapter 8 should be referred to for detailed information; however, Table 3 summarizes design data representing conditions commonly encountered.

The conversion of sensible heat gain from people to space cooling load is affected by the thermal storage characteristics of that space and is thus subject to application of appropriate room transfer functions (RTF). Latent heat gains are considered instantaneous.

Lighting

Since lighting is often the major space load component, an accurate estimate of the space heat gain it imposes is needed. Calculation of this load component is not straightforward; the rate of heat gain at any given moment can be quite different from the heat equivalent of power supplied instantaneously to those lights.

Only part of the energy from lights is in the form of convective heat, which is picked up instantaneously by the air-conditioning apparatus. The remaining portion is in the form of radiation, which affects the conditioned space only after having been absorbed and rereleased by walls, floors, furniture, etc. This absorbed energy contributes to space cooling load only after a time lag, with some part of such energy still present and reradiating after the lights have been switched off (Figure 2).

There is always significant delay between the time of switching lights on and a point of equilibrium where reradiated light energy equals that being instantaneously stored. Time lag effect must be considered when calculating cooling load, since load felt by the space can be considerably lower than the instantaneous heat gain being generated, and peak load for the space may be affected significantly.

Instantaneous Heat Gain from Lighting. The primary source of heat from lighting comes from light-emitting elements, or lamps, although significant additional heat may be generated from associated appurtenances in the light fixtures that house such lamps. Generally, the instantaneous rate of heat gain from electric lighting may be calculated from

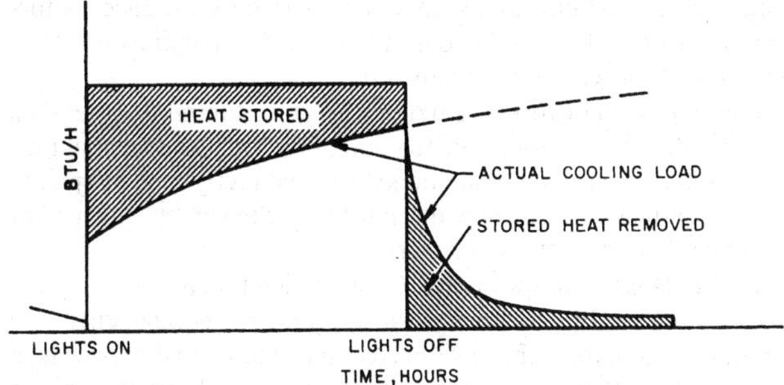

Fig. 2 Thermal Storage Effect in Cooling Load from Lights

Table 3 Rates of Heat Gain from Occupants of Conditioned Spaces

Degree of Activity		Total Heat, Btu/h		Sensible Heat, Btu/h	Latent Heat, Btu/h	% Sensible Heat that is Radiant[b]	
		Adult Male	Adjusted, M/F[a]			Low V	High V
Seated at theater	Theater, matinee	390	330	225	105		
Seated at theater, night	Theater, night	390	350	245	105	60	27
Seated, very light work	Offices, hotels, apartments	450	400	245	155		
Moderately active office work	Offices, hotels, apartments	475	450	250	200		
Standing, light work; walking	Department store; retail store	550	450	250	200	58	38
Walking, standing	Drug store, bank	550	500	250	250		
Sedentary work	Restaurant[c]	490	550	275	275		
Light bench work	Factory	800	750	275	475		
Moderate dancing	Dance hall	900	850	305	545	49	35
Walking 3 mph; light machine work	Factory	1000	1000	375	625		
Bowling[d]	Bowling alley	1500	1450	580	870		
Heavy work	Factory	1500	1450	580	870	54	19
Heavy machine work; lifting	Factory	1600	1600	635	965		
Athletics	Gymnasium	2000	1800	710	1090		

Notes:

1. Tabulated values are based on 75°F room dry-bulb temperature. For 80°F room dry bulb, the total heat remains the same, but the sensible heat values should be decreased by approximately 20%, and the latent heat values increased accordingly.
2. Also refer to Table 4, Chapter 8, for additional rates of metabolic heat generation.
3. All values are rounded to nearest 5 Btu/h.

[a]Adjusted heat gain is based on normal percentage of men, women, and children for the application listed, with the postulate that the gain from an adult female is 85% of that for an adult male, and that the gain from a child is 75% of that for an adult male.

[b]Values approximated from data in Table 6, Chapter 8, where is air velocity with limits shown in that table.

[c]Adjusted heat gain includes 60 Btu/h for food per individual (30 Btu/h sensible and 30 Btu/h latent).

[d]Figure one person per alley actually bowling, and all others as sitting (400 Btu/h) or standing or walking slowly (550 Btu/h).

$$q_{el} = 3.41\,W F_{ul} F_{sa} \tag{9}$$

where

q_{el} = heat gain, Btu/h
W = total light wattage
F_{ul} = lighting use factor
F_{sa} = lighting special allowance factor

The *total light wattage* is obtained from the ratings of all lamps installed, both for general illumination and for display use.

The *special allowance* is the ratio of the wattage in use, for the conditions under which the load estimate is being made, to the total installed wattage. For commercial applications such as stores, the use factor would generally be unity.

The *special allowance factor* is for fluorescent fixtures and/or fixtures that are either ventilated or installed so that only part of their heat goes to the conditioned space. For fluorescent fixtures, the special allowance factor accounts primarily for ballast losses, and can be as high as 2.19 for 32 W single lamp high-output fixtures on 277 V circuits. Rapid-start, 40 W lamp fixtures have special allowance factors that vary from a low of 1.18 for two lamps at 277 V to a high of 1.30 for one lamp at 118 V, with a recommended value of 1.20 for general applications. Industrial fixtures other than fluorescent, such as sodium lamps, may have special allowance factors varying from 1.04 to 1.37, depending on the manufacturer, and should be dealt with individually.

For ventilated or recessed fixtures, manufacturers' or other data must be sought to establish the fraction of the total wattage that may be expected to enter the conditioned space directly (and subject to time lag effect), versus that which must be picked up by return air or in some other appropriate manner.

Light Heat Components. Cooling load caused by lights recessed into ceiling cavities is made up of two components: one part comes from the light heat directly contributing to the space heat gain, and the other is the light heat released into the above-ceiling cavity, which (if used as a return air plenum) is mostly picked up by the return air that passes over or through the light fixtures. In such a ceiling return air plenum, this second part of the load (sometimes referred to as heat-to-return) never enters the conditioned space. It does, however, add to the overall load and significantly influences the load calculation.

Even though the total cooling load imposed on the cooling coil from these two components remains the same, the larger the fraction of heat output picked up by the return air, the more the space cooling load is reduced. The minimum required airflow rate for the conditioned space is decreased as the space cooling load becomes less. Supply fan power reduces accordingly, which ultimately results in reduced energy consumption for the system, and, possibly reduced equipment size as well.

For ordinary design load estimation, the heat gain for each component may simply be calculated as a fraction of the total lighting load by using judgment to estimate heat-to-space and heat-to-return percentages (Mitalas and Kimura 1971).

Return Air Light Fixtures. Two generic types of return air light fixture are available—those that allow and those that do not allow return air to flow through the lamp chamber. The first type is sometimes called a heat-of-light fixture. The percentage of light heat released through the plenum side of various ventilated fixtures can be obtained from lighting fixture manufacturers. For representative data, see Nevens et al. (1971). Even unventilated fixtures lose some heat to plenum spaces; however, most of the heat ultimately enters the conditioned space from a dead-air plenum or is picked up by return air via ceiling return air openings. The percentage of heat to return air ranges from 40 to 60% for heat-to-return ventilated fixtures or 15 to 25% for unventilated fixtures.

Plenum Temperatures. As heat from lighting is picked up by the return air, the temperature differential between the ceiling space and the conditioned space causes part of that heat to flow from the ceiling back to the conditioned space. Return air from the conditioned space can be ducted to capture light heat without passing through a ceiling plenum as such, or the ceiling space can be used as a return air plenum, causing the distribution of light heat to be handled in distinctly different ways. Most plenum temperatures do not rise more than 1 to 3°F above space temperature, thus generating only a relatively small thermal gradient for heat transfer through plenum surfaces but a relatively large percentage reduction in space cooling load. (Many engineers believe that a major reason for plenum temperatures not becoming more elevated is due to leakage into the plenum from supply air duct work normally concealed there, but consideration of this elusive factor is beyond the scope of this chapter.)

Energy Balance. Where the ceiling space is used as a return air plenum, an energy balance requires that the heat picked up from the lights into the return air (1) becomes a part of the cooling load to the return air (represented by a temperature rise of the return air as it passes through the ceiling space), (2) is partially transferred back into the conditioned space through the ceiling material below, and/or (3) may be partially "lost" (from the space) through the floor surfaces above the plenum. In a multistory building, the conditioned space frequently gains heat through its floor from a similar plenum below, offsetting the loss just mentioned. The radiant component of heat leaving the ceiling or floor surface of a plenum is normally so small that all such heat transfer is considered convective for calculation purposes.

Figure 3 shows a schematic diagram of a typical return air plenum. Equations (10) through (14), using the sign convention as shown in Figure 3, represent the heat balance of a return air plenum design for a typical interior room in a multifloor building, as

$$q_1 = U_c A_c (t_p - t_r) \quad (10)$$

$$q_2 = U_f A_f (t_p - t_{fa}) \quad (11)$$

$$q_3 = 1.1Q(t_p - t_r) \quad (12)$$

$$q_{lp} - q_2 - q_1 - q_3 = 0 \quad (13)$$

$$Q = \frac{q_r + q_1}{1.1(t_r - t_s)} \quad (14)$$

where

q_1 = heat gain to space from plenum through ceiling, Btu/h
q_2 = heat loss from plenum through floor above, Btu/h
q_3 = heat gain "pickup" by return air, Btu/h
Q = return airflow, cfm
q_{lp} = light heat gain to plenum via return air, Btu/h
q_{lr} = light heat gain to space, Btu/h
q_f = heat gain from plenum below, through floor, Btu/h
q_w = heat gain from exterior wall, Btu/h
q_r = space cooling load, Btu/h, including appropriate treatment of q_{lr}, q_f, and/or q_w
t_p = plenum temperature, °F
t_r = space temperature, °F
t_{fa} = space temperature of floor above, °F
t_s = supply air temperature, °F

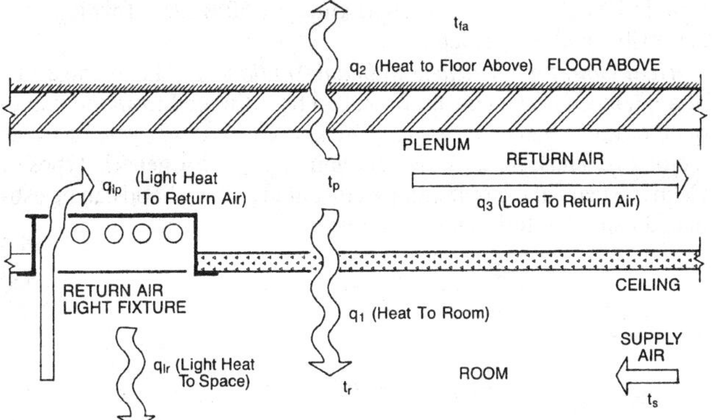

Fig. 3 Heat Balance of Typical Ceiling Return Plenum

By substituting Equations (10), (11), (12), and (14) into heat balance Equation (13), t_p can be found as the resultant return air temperature or plenum temperature, by means of a quadratic equation. The results, although rigorous and best solved by computer, are important in determining the cooling load, which affects equipment size selection, future energy consumption, and other factors.

Equations (10) through (14) are simplified to illustrate the heat balance relationship. Heat gain into a return air plenum is not limited to the heat of lights alone. Exterior walls directly exposed to the ceiling space will transfer heat directly to or from the return air. For single-story buildings or the top floor of a multistory building, the roof heat gain or loss enters or leaves the ceiling plenum rather than entering or leaving the conditioned space directly. The supply air quantity calculated by Equation (14) is for the conditioned space under consideration only, and is assumed equal to the return air quantity.

The amount of airflow through a return plenum above a conditioned space may not be limited to that supplied into the space under consideration; it will, however, have no noticeable effect on plenum temperature if the surplus comes from an adjacent plenum operating under similar conditions. Where special conditions exist, heat balance Equations (10) through (14) must be modified appropriately. Finally, even though the building's thermal storage has some effect, the amount of heat entering the return air is small and may be considered as convective for calculation purposes.

Power

Instantaneous heat gain from equipment operated by electric motors within a conditioned space is calculated as

$$q_{em} = 2545(P/E_M)F_{UM}F_{LM} \quad (15)$$

where

q_{em} = heat equivalent of equipment operation, Btu/h
P = motor power rating, horsepower
E_M = motor efficiency, as decimal fraction < 1.0
F_{UM} = motor use factor, 1.0 or decimal fraction < 1.0
F_{LM} = motor load factor, 1.0 or decimal fraction < 1.0

The motor use factor may be applied when motor use is known to be intermittent with significant nonuse during all hours of operation (e.g., overhead door operator). For conventional applications, its value would be 1.0.

The motor load factor is the fraction of the rated load being delivered under the conditions of the cooling load estimate. In Equation (15), it is assumed that both the motor and the driven equipment are within the conditioned space. If the motor is outside the space or airstream

$$q_{em} = 2545PF_{UM}F_{LM} \quad (16)$$

When the motor is inside the conditioned space or airstream but the driven machine is outside

$$q_{em} = 2545P\left(\frac{1.0 - E_M}{E_m}\right)F_{UM}F_{LM} \quad (17)$$

Equation (17) also applies to a fan or pump in the conditioned space that exhausts air or pumps fluid outside that space.

Average efficiencies, and related data representative of typical electric motors, generally derived from the lower efficiencies reported by several manufacturers of open, drip-proof motors, are given in Tables 4 and 5. These reports indicate that TEFC (totally enclosed fan-cooled) are slightly more efficient. For speeds lower or higher than those listed, efficiencies may be 1 to 3% lower or higher, depending on the manufacturer. Should actual voltages at motors be appreciably higher or lower than rated nameplate voltage, efficiencies in either case will be lower. If electric motor load is an appreciable portion of cooling load, the motor efficiency should be

Table 4 Heat Gain from Typical Electric Motors

				Location of Motor and Driven Equipment with Respect to Conditioned Space or Airstream		
				A	B	C
Motor Name-plate or Rated Horse-power	Motor Type	Nominal rpm	Full Load Motor Effi-ciency, %	Motor in, Driven Equip-ment in, Btu/h	Motor out, Driven Equip-ment in, Btu/h	Motor in, Driven Equip-ment out, Btu/h
0.05	Shaded pole	1500	35	360	130	240
0.08	Shaded pole	1500	35	580	200	380
0.125	Shaded pole	1500	35	900	320	590
0.16	Shaded pole	1500	35	1160	400	760
0.25	Split phase	1750	54	1180	640	540
0.33	Split phase	1750	56	1500	840	660
0.50	Split phase	1750	60	2120	1270	850
0.75	3-Phase	1750	72	2650	1900	740
1	3-Phase	1750	75	3390	2550	850
1.5	3-Phase	1750	77	4960	3820	1140
2	3-Phase	1750	79	6440	5090	1350
3	3-Phase	1750	81	9430	7640	1790
5	3-Phase	1750	82	15,500	12,700	2790
7.5	3-Phase	1750	84	22,700	19,100	3640
10	3-Phase	1750	85	29,900	24,500	4490
15	3-Phase	1750	86	44,400	38,200	6210
20	3-Phase	1750	87	58,500	50,900	7610
25	3-Phase	1750	88	72,300	63,600	8680
30	3-Phase	1750	89	85,700	76,300	9440
40	3-Phase	1750	89	114,000	102,000	12,600
50	3-Phase	1750	89	143,000	127,000	15,700
60	3-Phase	1750	89	172,000	153,000	18,900
75	3-Phase	1750	90	212,000	191,000	21,200
100	3-Phase	1750	90	283,000	255,000	28,300
125	3-Phase	1750	90	353,000	318,000	35,300
150	3-Phase	1750	91	420,000	382,000	37,800
200	3-Phase	1750	91	569,000	509,000	50,300
250	3-Phase	1750	91	699,000	636,000	62,900

Table 5 Typical Overload Limits with Standard Motors

Horsepower	0.05-0.25	0.16-0.33	0.67-0.75	1 and up
AC open	1.4	1.35	1.25	1.15
AC TEFC[a] and DC	—	1.0	1.0	1.0

Note: Some shaded pole, capacitor start, and special purpose motors have a service factor varying from 1.0 up to 1.75.

[a]Some totally enclosed fan-cooled (TEFC) motors have a service factor above 1.0.

obtained from the manufacturer. Also, depending on design, the maximum efficiency might occur anywhere between 75 to 110% of full load; if underloaded or overloaded, the efficiency could vary from the manufacturer's listing.

Overloading or Underloading. Heat output of a motor is generally proportional to the motor load, within the overload limits. Because of typically high no-load motor current, fixed losses, and other reasons, FLM is generally assumed to be unity, and no adjustment should be made for underloading or overloading unless the situation is fixed, can be accurately established, and the reduced load efficiency data can be obtained from the motor manufacturer.

Radiation and Convection. Unless the manufacturer's technical literature indicates otherwise, the heat gain normally should be equally divided between radiant and convective components for the subsequent cooling load calculations.

Appliances

In a cooling load estimate, heat gain from all appliances—electrical, gas, or steam—should be taken into account. Because of the variety of appliances, applications, schedules, use, and installations, estimates can be very subjective. Often, the only information available about heat gain from equipment is that on its nameplate. For electric office equipment in particular, Wilkins (1994) found nameplate data to be very misleading, overstating actual normal usage as much as 400%.

Cooking Appliances. These appliances include common heat-producing cooking equipment found in conditioned commercial kitchens. Marn (1962) concluded that appliance surfaces contributed most of the heat to commercial kitchens and that when installed under an effective hood, the cooling load was independent of the fuel or energy used for similar equipment performing the same operations.

Gordon et al. (1994) and Smith et al. (1995) found that gas appliances may exhibit slightly higher heat gains than their electric counterparts under wall-canopy hoods operated at typical ventilation rates. This is due to the fact that the heat contained in the combustion products exhausted from a gas appliance may increase temperatures of the appliance and surrounding surfaces as well as the hood above the appliance more than that of its electric counterpart. These higher-temperature surfaces radiate heat to the kitchen, adding moderately to the radiant gain directly associated with the appliance cooking surface.

Marn (1962) confirmed that where the appliances are installed under an effective hood, only radiant gain adds to the cooling load; convected and latent heat from the cooking process and combustion products are exhausted and do not enter the kitchen. Gordon et al. (1994) and Smith et al. (1995) substantiated these findings.

Calculating Sensible Heat Gain for Hooded Cooking Appliances. To establish a heat gain value, actual nameplate energy input ratings may should be used with appropriate usage and radiation factors. Where specific rating data are not available (nameplate missing, equipment not yet purchased, etc.) or as an alternative approach, recommended heat gains tabulated in this chapter for a wide variety of commonly encountered equipment items may be used. In estimating the appliance load, probabilities of simultaneous use and operation for different appliances located in the same space must be considered.

The radiant heat gain from hooded cooking equipment can range from 15 to 45% of the actual appliance energy consumption (Talbert et al. 1973, Gordon et al. 1994, Smith et al. 1995). This ratio of heat gain to appliance energy consumption may be expressed as a radiation factor. It is a function of both appliance type and fuel source. The radiant factor, F_R, is applied to the average rate of appliance energy consumption, determined by applying F_U to the nameplate or rated energy input. Marn (1962) found that radiant heat temperature rise can be substantially reduced by shielding the fronts of cooking appliances. Although this approach may not always be practical in a commercial kitchen, radiant gains can aso be reduced by adding side panels or partial enclosures that are integrated with the exhaust hood.

Heat Gain from Meals. For each meal served, the heat transferred to the dining space is approximately 50 Btu/h, of which 75% is sensible and 25% is latent.

Heat Gain for Electric and Steam Appliances. The average rate of appliance energy consumption can be estimated from the nameplate or rated energy input q_{input} by applying a duty cycle or usage factor F_U. Thus the sensible heat gain $q_{sensible}$ for generic types of electric, steam and gas appliances installed under a hood can be estimated using the following equation.

$$q_{sensible} = q_{input} F_U F_R \quad (18)$$

or

$$q_{sensible} = q_{input} F_L \quad (19)$$

where F_L is defined as the ratio of sensible heat gain to the manufacturers rated energy input.

Table 6 Heat Gain Factors of Typical Electric Appliances Under Hood

Appliance	Usage Factor F_U	Radiation Factor F_R	Load Factor $F_L = F_U F_R$ Elec/Steam
Griddle	0.16	0.45	0.07
Fryer	0.06	0.43	0.03
Convection oven	0.42	0.17	0.07
Charbroiler	0.83	0.29	0.24
Open-top range			
without oven	0.34	0.46	0.16
Hot-top range			
without oven	0.79	0.47	0.37
with oven	0.59	0.48	0.28
Steam cooker	0.13	0.30	0.04

Table 7 Heat Gain Factors of Typical Gas Appliances Under Hood

Appliance	Usage Factor F_U	Radiation Factor F_R	Load Factor $F_L = F_U F_R$ Gas
Griddle	0.25	0.25	0.06
Fryer	0.07	0.35	0.02
Convection oven	0.42	0.20	0.08
Charbroiler	0.62	0.18	0.11
Open-top range			
without oven	0.34	0.17	0.06

Table 6 lists usage factors, radiation factors, and load factors based on appliance energy consumption rate for typical electrical and steam appliances under standby or idle conditions (Alereza and Breen 1984, Fisher 1996); Table 7 lists usage factors, radiation factors, and load factors for comparable gas equipment (Fisher 1996).

Unhooded Equipment. For all cooking appliances not installed under an exhaust hood or directly vent-connected and located in the conditioned area, the heat gain may be estimated as 50% (usage factor = 0.50) or the rated hourly input, regardless of the type of energy or fuel used. On average, 34% of the heat may be assumed to be latent and the remaining 66% sensible heat. Note that cooking appliances ventilated by "ductless" hoods should be treated as unhooded appliances from the perspective of estimating heat gain. In other words, all energy consumed by the appliance and all moisture produced by the cooking process is introduced to the kitchen as a sensible or latent cooling load.

Recommended Heat Gain Values. As an alternative procedure, Table 8 lists recommended rates of heat gain from typical commercial cooking appliances (Alereza and Breen 1984, Fisher 1996). The data in the "with hood" columns assume installation under a properly designed exhaust hood connected to a mechanical fan exhaust system.

Hospital and Laboratory Equipment. Hospital and laboratory equipment items are major sources of heat gain in conditioned spaces. Care must be taken in evaluating the probability and duration of simultaneous usage when many components are concentrated in one area, such as laboratory, operating room, etc. Commonly, heat gain from equipment in a laboratory ranges from 15 to 70 Btu/(h·ft^2) or, in laboratories with outdoor exposure, as much as four times the heat gain from all other sources combined. Table 1 in Chapter 13 of the 1995 *ASHRAE Handbook—Applications* lists heat gain values for various hospital and laboratory equipment.

Office Appliances. Electric typewriters, calculators, checkwriters, teletype units, posting machines, etc., can generate 3 to 4 Btu/(h·ft^2) for general offices or 6 to 7 Btu/(h·ft^2) for purchasing and accounting departments. However, in offices having computer display terminals at most desks, heat gains range up to 15 Btu/(h·ft^2) (Table 9A).

A commonly encountered office environment includes a desktop computer and monitor at each workstation, along with a variety of shared devices such as printers, scanners, copy machines, and facsimile (FAX) machines. Nameplate ratings of such equipment should be significantly discounted. Diversity in usage patterns must be considered separately, as peak use of all equipment is unlikely to occur simultaneously. Table 9B summarizes average test results for 270,000 ft^2 of office space measured in five different buildings, in which the average electrical appliance heat gain was approximately 1 W/ft^2 (Wilkins 1994).

Diversity of usage will vary significantly with various types of equipment and the work habits of the occupants. Additionally, earlier models of microcomputers and monitors show little difference in heat generation whether in use or idle, while contemporary units normally have "Power Saver" features that significantly reduce power consumption when idle.

Environmental variations can also have an effect on the heat gain from appliances, primarily with regard to the percent of heat gain that is radiative versus convective. Table 9C illustrates typical measured data for several appliances (Wilkins 1994).

Computer rooms housing mainframe or minicomputer equipment must be considered individually. Computer manufacturers have data pertaining to various individual components. Additional insight should be sought from data processing managers as to schedules, near-term future planning, etc. Heat gain rates from digital computer equipment range from 75 to 175 Btu/(h·ft^2). While the trend in hardware development is toward less heat release on a component basis, the associated miniaturization tends to offset such unitary reduction by a higher concentration of equipment. Chapter 16 of the 1995 *ASHRAE Handbook—Applications* gives further information on the air conditioning of data processing areas.

INFILTRATION AND VENTILATION HEAT GAIN

Ventilation

Outdoor air must be introduced to ventilate conditioned spaces. Chapter 25 suggests minimum outdoor air requirements for representative applications, but the minimum levels are not necessarily adequate for all psychological attitudes and physiological responses. Where maximum economy in space and load is essential, as in submarines or other restricted spaces, as little as 1 cfm of outdoor air per person can be sufficient, provided that recirculated air is adequately decontaminated (Consolazio and Pecora 1947).

Local codes and ordinances frequently specify ventilation requirements for public places and for industrial installations. For example, minimum requirements for safe practice in hospital operating rooms are given in NFPA *Standard* 99. Although 100% outdoor air is sometimes used in operating rooms, this standard does not require it, and limiting the outdoor air to 6 to 8 changes per hour is finding increasing acceptance.

ASHRAE Standard 62 recommends minimum ventilation rates for most common applications. For general applications, such as offices, 20 cfm per person is suggested.

Ventilation air is normally introduced at the air-conditioning apparatus rather than directly into the conditioned space, and thus becomes a cooling coil load component instead of a space load component. Calculations for estimating this heat gain are discussed later.

Reducing heat gain from outdoor air by using filtered recirculated air in combination with outdoor air should be considered. Recirculated air can also be treated to control odor (see Chapter 13 in this volume and Chapter 41 in the 1995 *ASHRAE Handbook—Applications*).

Infiltration

The principles of estimating infiltration in buildings, with emphasis on the heating season, are discussed in Chapter 25. For the cooling

Table 8 Recommended Rate of Heat Gain from Restaurant Equipment Located in Air-Conditioned Areas

		Energy Rate, Btu/h		Recommended Rate of Heat Gain,[a] Btu/h			
				Without Hood			With Hood
Appliance	**Size**	**Rated**	**Standby**	**Sensible**	**Latent**	**Total**	**Sensible**
Electric, No Hood Required							
Barbeque (pit), per pound of food capacity	80 to 300 lb	136	—	86	50	136	42
Barbeque (pressurized), per pound of food capacity	44 lb	327	—	109	54	163	50
Blender, per quart of capacity	1 to 4 qt	1550	—	1000	520	1520	480
Braising pan, per quart of capacity	108 to 140 qt	360	—	180	95	275	132
Cabinet (large hot holding)	16.2 to 17.3 ft^3	7100	—	610	340	960	290
Cabinet (large hot serving)	37.4 to 406 ft^3	6820	—	610	310	920	280
Cabinet (large proofing)	16 to 17 ft^3	693	—	610	310	920	280
Cabinet (small hot holding)	3.2 to 6.4 ft^3	3070	—	270	140	410	130
Cabinet (very hot holding)	17.3 ft^3	21000	—	1880	960	2830	850
Can opener		580	—	580	—	580	0
Coffee brewer	12 cup/2 brnrs	5660	—	3750	1910	5660	1810
Coffee heater, per boiling burner	1 to 2 brnrs	2290	—	1500	790	2290	720
Coffee heater, per warming burner	1 to 2 brnrs	340	—	230	110	340	110
Coffee/hot water boiling urn, per quart of capacity	11.6 qt	390	—	256	132	388	123
Coffee brewing urn (large), per quart of capacity	23 to 40 qt	2130	—	1420	710	2130	680
Coffee brewing urn (small), per quart of capacity	10.6 qt	1350	—	908	445	1353	416
Cutter (large)	18 in. bowl	2560	—	2560	—	2560	0
Cutter (small)	14 in. bowl	1260	—	1260	—	1260	0
Cutter and mixer (large)	30 to 48 qt	12730	—	12730	—	12730	0
Dishwasher (hood type, chemical sanitizing), per 100 dishes/h	950 to 2000 dishes/h	1300	—	170	370	540	170
Dishwasher (hood type, water sanitizing), per 100 dishes/h	950 to 2000 dishes/h	1300	—	190	420	610	190
Dishwasher (conveyor type, chemical sanitizing), per 100 dishes/h	5000 to 9000 dishes/h	1160	—	140	330	470	150
Dishwasher (conveyor type, water sanitizing), per 100 dishes/h	5000 to 9000 dishes/h	1160	—	150	370	520	170
Display case (refrigerated), per 10 ft^3 of interior	6 to 67 ft^3	1540	—	617	0	617	0
Dough roller (large)	2 rollers	5490	—	5490	—	5490	0
Dough roller (small)	1 roller	1570	—	140	—	140	0
Egg cooker	12 eggs	6140	—	2900	1940	4850	1570
Food processor	2.4 qt	1770	—	1770	—	1770	0
Food warmer (infrared bulb), per lamp	1 to 6 bulbs	850	—	850	—	850	850
Food warmer (shelf type), per square foot of surface	3 to 9 ft^2	930	—	740	190	930	260
Food warmer (infrared tube), per foot of length	39 to 53 in.	990	—	990	—	990	990
Food warmer (well type), per cubic foot of well	0.7 to 2.5 ft^3	3620	—	1200	610	1810	580
Freezer (large)	73	4570	—	1840	—	1840	0
Freezer (small)	18	2760	—	1090	—	1090	0
Griddle/grill (large), per square foot of cooking surface	4.6 to 11.8 ft^2	9200	—	615	343	958	343
Griddle/grill (small), per square foot of cooking surface	2.2 to 4.5 ft^2	8300	—	545	308	853	298
Hot dog broiler	48 to 56 hot dogs	3960	—	340	170	510	160
Hot plate (double burner, high speed)		16720	—	7810	5430	13240	6240
Hot plate (double burner, stockpot)		13650	—	6380	4440	10820	5080
Hot plate (single burner, high speed)		9550	—	4470	3110	7580	3550
Hot water urn (large), per quart of capacity	56 qt	416	—	161	52	213	68
Hot water urn (small), per quart of capacity	8 qt	738	—	285	95	380	123
Ice maker (large)	220 lb/day	3720	—	9320	—	9320	0
Ice maker (small)	110 lb/day	2560	—	6410	—	6410	0
Microwave oven (heavy duty, commercial)	0.7 ft^3	8970	—	8970	—	8970	0
Microwave oven (residential type)	1 ft^3	2050 to 4780	—	2050 to 4780	—	2050 to 4780	0
Mixer (large), per quart of capacity	81 qt	94	—	94	—	94	0
Mixer (small), per quart of capacity	12 to 76 qt	48	—	48	—	48	0
Press cooker (hamburger)	300 patties/h	7510	—	4950	2560	7510	2390
Refrigerator (large), per 10 ft^3 of interior space	25 to 74 ft^3	753	—	300	—	300	0
Refrigerator (small), per 10 ft^3 of interior space	6 to 25 ft^3	1670	—	665	—	665	0
Rotisserie	300 hamburgers/h	10920	—	7200	3720	10920	3480
Serving cart (hot), per cubic foot of well	1.8 to 3.2 ft^3	2050	—	680	340	1020	328
Serving drawer (large)	252 to 336 dinner rolls	3750	—	480	34	510	150
Serving drawer (small)	84 to 168 dinner rolls	2730	—	340	34	380	110
Skillet (tilting), per quart of capacity	48 to 132 qt	580	—	293	161	454	218
Slicer, per square foot of slicing carriage	0.65 to 0.97 ft^2	680	—	682	—	682	216
Soup cooker, per quart of well	7.4 to 11.6 qt	416	—	142	78	220	68
Steam cooker, per cubic foot of compartment	32 to 64 qt	20700	—	1640	1050	2690	784
Steam kettle (large), per quart of capacity	80 to 320 qt	300	—	23	16	39	13
Steam kettle (small), per quart of capacity	24 to 48 qt	840	—	68	45	113	32
Syrup warmer, per quart of capacity	11.6 qt	284	—	94	52	146	45

Table 8 Recommended Rate of Heat Gain from Restaurant Equipment Located in Air-Conditioned Areas (*Concluded*)

		Energy Rate, Btu/h		Recommended Rate of Heat Gain,[a] Btu/h			
				Without Hood			With Hood
Appliance	Size	Rated	Standby	Sensible	Latent	Total	Sensible
Toaster (bun toasts on one side only)	1400 buns/h	5120	—	2730	2420	5150	1640
Toaster (large conveyor)	720 slices/h	10920	—	2900	2560	5460	1740
Toaster (small conveyor)	360 slices/h	7170	—	1910	1670	3580	1160
Toaster (large pop-up)	10 slice	18080	—	9590	8500	18080	5800
Toaster (small pop-up)	4 slice	8430	—	4470	3960	8430	2700
Waffle iron	75 in^2	5600	—	2390	3210	5600	1770
Electric, Exhaust Hood Required							
Broiler (conveyor infrared), per square foot of cooking area/minute	2 to 102 ft^2	19230	—	—	—	—	3840
Broiler (single deck infrared), per square foot of broiling area	2.6 to 9.8 ft^2	10870	—	—	—	—	2150
Charbroiler, per linear foot of cooking surface	2 to 8 linear ft	11,000	9300	—	—	—	2800
Fryer (deep fat)	35 - 50 lb oil	48,000	2900	—	—	—	1200
Fryer (pressurized), per pound of fat capacity	13 to 33 lb	1565	—	—	—	—	59
Oven (full-size convection)		41,000	4600	—	—	—	2900
Oven (large deck baking with 537 ft^3 decks), per cubic foot of oven space	15 to 46 ft^3	1670	—	—	—	—	69
Oven (roasting), per cubic foot of oven space	7.8 to 23 ft^3	27350	—	—	—	—	113
Oven (small convection), per cubic foot of oven space	1.4 to 5.3 ft^3	10340	—	—	—	—	147
Oven (small deck baking with 272 ft^3 decks), per cubic foot of oven space	7.8 to 23 ft^3	2760	—	—	—	—	113
Open range top, per 2 element section	2 to 6 elements	14,000	4600	—	—	—	2100
Range (hot top/fry top), per square foot of cooking surface	4 to 8 ft^2	7260	—	—	—	—	2690
Range (oven section), per cubic foot of oven space	4.2 to 11.3 ft^3	3940	—	—	—	—	160
Griddle, per linear foot of cooking surface	2 to 8 linear feet	19,500	3100	—	—	—	1400
Gas, No Hood Required							
Broiler, per square foot of broiling area	2.7 ft^2	14800	660[b]	5310	2860	8170	1220
Cheese melter, per square foot of cooking surface	2.5 to 5.1 ft^2	10300	660[b]	3690	1980	5670	850
Dishwasher (hood type, chemical sanitizing), per 100 dishes/h	950 to 2000 dishes/h	1740	660[b]	510	200	710	230
Dishwasher (hood type, water sanitizing), per 100 dishes/h	950 to 2000 dishes/h	1740	660[b]	570	220	790	250
Dishwasher (conveyor type, chemical sanitizing), per 100 dishes/h	5000 to 9000 dishes/h	1370	660[b]	330	70	400	130
Dishwasher (conveyor type, water sanitizing), per 100 dishes/h	5000 to 9000 dishes/h	1370	660[b]	370	80	450	140
Griddle/grill (large), per square foot of cooking surface	4.6 to 11.8 ft^2	17000	330	1140	610	1750	460
Griddle/grill (small), per square foot of cooking surface	2.5 to 4.5 ft^2	14400	330	970	510	1480	400
Hot plate	2 burners	19200	1325[b]	11700	3470	15200	3410
Oven (pizza), per square foot of hearth	6.4 to 12.9 ft^2	4740	660[b]	623	220	843	85
Gas, Exhaust Hood Required							
Braising pan, per quart of capacity	105 to 140 qt	9840	660[b]	—	—	—	2430
Broiler, per square foot of broiling area	3.7 to 3.9 ft^2	21800	530	—	—	—	1800
Broiler (large conveyor, infrared), per square foot of cooking area/minute	2 to 102 ft^2	51300	1990	—	—	—	5340
Broiler (standard infrared), per square foot of broiling area	2.4 to 9.4 ft^2	1940	530	—	—	—	1600
Charbroiler (large), per linear foot of cooking area	2 to 8 linear feet	36,000	22,000	—	—	—	3800
Fryer (deep fat)	35 to 50 oil cap.	80,000	5600	—	—	—	1900
Oven (bake deck), per cubic foot of oven space	5.3 to 16.2 ft^3	7670	660[b]	—	—	—	140
Oven (convection), full size		70,000	29,400	—	—	—	5700
Oven (pizza), per square foot of oven hearth	9.3 to 25.8 ft^2	7240	660[b]	—	—	—	130
Oven (roasting), per cubic foot of oven space	9 to 28 ft^3	4300	660[b]	—	—	—	77
Oven (twin bake deck), per cubic foot of oven space	11 to 22 ft^3	4390	660[b]	—	—	—	78
Range (burners), per 2 burner section	2 to 10 brnrs	33600	1325	—	—	—	6590
Range (hot top or fry top), per square foot of cooking surface	3 to 8 ft^2	11800	330	—	—	—	3390
Range (large stock pot)	3 burners	100000	1990	—	—	—	19600
Range (small stock pot)	2 burners	40000	1330	—	—	—	7830
Griddle, per linear foot of cooking surface	2 to 8 linear feet	25,000	6300				1600
Range top, open burner (per 2 burner section)	2 to 6 elements	40,000	13,600				2200
Steam							
Compartment steamer, per pound of food capacity/h	46 to 450 lb	280	—	22	14	36	11
Dishwasher (hood type, chemical sanitizing), per 100 dishes/h	950 to 2000 dishes/h	3150	—	880	380	1260	410
Dishwasher (hood type, water sanitizing), per 100 dishes/h	950 to 2000 dishes/h	3150	—	980	420	1400	450
Dishwasher (conveyor, chemical sanitizing), per 100 dishes/h	5000 to 9000 dishes/h	1180	—	140	330	470	150
Dishwasher (conveyor, water sanitizing), per 100 dishes/h	5000 to 9000 dishes/h	1180	—	150	370	520	170
Steam kettle, per quart of capacity	13 to 32 qt	500	—	39	25	64	19

[a]In some cases, heat gain data are given per unit of capacity. In those cases, the heat gain is calculated by: q = (recommended heat gain per unit of capacity) * (capacity)

[b]Standby input rating is given for entire appliance regardless of size.

Table 9A Rate of Heat Gain from Selected Office Equipment

Appliance	Size	Maximum Input Rating, Btu/h	Standby Input Rating, Btu/h	Recommended Rate of Heat Gain, Btu/h
Check processing workstation	12 pockets	16400	8410	8410
Computer devices				
Card puncher	—	2730 to 6140	2200 to 4800	2200 to 4800
Card reader	—	7510	5200	5200
Communication/transmission	—	6140 to 15700	5600 to 9600	5600 to 9600
Disk drives/mass storage	—	3410 to 34100	3412 to 22420	3412 to 22420
Magnetic ink reader	—	3280 to 16000	2600 to 14400	2600 to 14400
Microcomputer	16 to 640 Kbyte[a]	340 to 2050	300 to 1800	300 to 1800
Minicomputer	—	7500 to 15000	7500 to 15000	7500 to 15000
Optical reader	—	10240 to 20470	8000 to 17000	8000 to 17000
Plotters	—	256	128	214
Printers				
Letter quality	30 to 45 char/min	1200	600	1000
Line, high speed	5000 or more lines/min	4300 to 18100	2160 to 9040	2500 to 13000
Line, low speed	300 to 600 lines/min	1540	770	1280
Tape drives	—	4090 to 22200	3500 to 15000	3500 to 15000
Terminal	—	310 to 680	270 to 600	270 to 600
Copiers/Duplicators				
Blue print	—	3930 to 42700	1710 to 17100	3930 to 42700
Copiers (large)	30 to 67[a] copies/min	5800 to 22500	3070	5800 to 22500
Copiers (small)	6 to 30[a] copies/min	1570 to 5800	1020 to 3070	1570 to 5800
Feeder	—	100	—	100
Microfilm printer	—	1540	—	1540
Sorter/collator	—	200 to 2050	—	200 to 2050
Electronic equipment				
Cassette recorders/players	—	200	—	200
Receiver/tuner	—	340	—	340
Signal analyzer	—	90 to 2220	—	90 to 2220
Mailprocessing				
Folding machine	—	430	—	270
Inserting machine	3600 to 6800 pieces/h	2050 to 11300	—	1330 to 7340
Labeling machine	1500 to 30000 pieces/h	2050 to 22500	—	1330 to 14700
Postage meter	—	780	—	510
Wordprocessors/Typewriters				
Letter quality printer	30 to 45 char/min	1200	600	1000
Phototypesetter	—	5890	—	5180
Typewriter	—	270	—	230
Wordprocessor	—	340 to 2050	—	300 to 1800
Vending machines				
Cigarette	—	250	51 to 85	250
Cold food/beverage	—	3920 to 6550	—	1960 to 3280
Hot beverage	—	5890	—	2940
Snack	—	820 to 940	—	820 to 940
Miscellaneous				
Barcode printer	—	1500	—	1260
Cash registers	—	200	—	160
Coffee maker	10 cups	5120	—	3580 sens., 1540 latent
Microfiche reader	—	290	—	290
Microfilm reader	—	1770	—	1770
Microfilm reader/printer	—	3920	—	3920
Microwave oven	1 ft^3	2050	—	1360
Paper shredder	—	850 to 10240	—	680 to 8250
Water cooler	32 qt/h	2390	—	5970

[a]Input is not proportional to capacity.

Table 9B Heat Gain Versus Nameplate Rating From Electrical Office Equipment

Equipment Tested	Nameplate Rating, W	Measured Total Power Consumption, W	Radiant Power, W	Radiant Power, %	Convective Power, %
15 in. monitor energy saver (white screen)	220	78	28.8	37.1	62.9
Laser printer	836	248	26.6	10.7	89.3
Desktop copier	1320	181	25.9	14.3	85.7
Personal computer (Brand 1) and 17 in. monitor (white screen)	575	133	29.7	22.3	77.7
Personal computer (Brand 2) and 17 in. monitor (white screen)	420	125	35.7	28.6	71.4

Table 9C Environmental Effects on Radiant-Convective Split of 15 in. Energy Saver Moniter
(Nominal measured total power consumption = 80 W, Nameplate rating = 220 W)

I. Room Air Velocity and Temperature Effects			
	Radiative Power, %		
	60°F room air	70°F room air	80°F room air
Normal air velocity	35.6	40.7	45
Reduced air velocity	36	37.1	45.7
II. Room Air Temperature Fluctuation Effects (±6°F about mean room of 70°F)			
	Radiative Power, %		
	With air temp. fluctuation	Without air temperature fluctuation	
Normal air velocity	37.7	40.7	
Reduced air velocity	38.2	37.1	
III. Room Wall Temperature Effects (Room air at 70°F)			
	Radiative Power, %		
	With uniform wall temperature	With nonuniform wall temperature (one wall heated)	
Reduced air velocity	37.1	27	

season, infiltration calculations are usually limited to doors and windows. Air leakage through doors can be estimated using the information in Chapter 25. Table 3, Chapter 25, adjusted for the average wind velocity in the locality, may be used to compute infiltration for windows. In calculating window infiltration for an entire structure, the total window area on all sides of the building is not involved, since wind does not act on all sides simultaneously. In any case, infiltration from all windows in any two adjacent wall exposures should be included. A knowledge of the prevailing wind direction and velocity is helpful in selecting exposures.

When economically feasible, sufficient outdoor air should be introduced as ventilation air through the air-conditioning equipment to maintain a constant outward escape of air, and thus eliminate the infiltration portion of the gain. The pressure maintained must overcome wind pressure through cracks and door openings. When the quantity of outside air introduced through the cooling equipment is not sufficient to maintain the required pressure to eliminate infiltration, the entire infiltration load should be included in the space heat gain calculations.

Standard Air Defined

Because the specific volume of air varies appreciably, calculations will be more accurate when made on the basis of air mass instead of volume. However, volume values are often required for selection of coils, fans, ducts, etc., in which cases volume values based on measurement at standard conditions may be used for accurate results. One standard value is 0.075 lb (dry air)/ft^3 (13.33 ft^3/lb). This density corresponds to about 60°F at saturation, and 69°F dry air (at 14.7 psia). Because air usually passes through the coils, fans, ducts, etc. at a density close to standard, the accuracy desired normally requires no correction. When airflow is to be measured at a particular condition or point, such as at a coil entrance or exit, the corresponding specific volume can be read from the psychrometric chart.

Example 2. Standard air calculations. Assume outdoor air at standard conditions is flowing at 1000 cfm. What is the flow rate when the outdoor air is at 95°F dry-bulb and 75°F wet-bulb (14.3 ft^3/lb)? The measured rate at that condition should be 1000(14.3/13.33) = 1070 cfm.

Solution: Air-conditioning design often requires calculation of:

1. Total heat

Total heat gain corresponding to the change of a given standard flow rate Q_s through an enthalpy difference Δh

$$\text{Total heat change} = 60 \times 0.075 Q_s \Delta h = 4.5 Q_s \Delta h \tag{20}$$

where 60 = min/h and 0.075 = lb (dry air)/ft^3.

2. Sensible heat

Sensible heat gain corresponding to the change of dry-bulb temperature Δt for given airflow (standard conditions) Q_s, or sensible heat change q_s, in Btu/h, is

$$q_s = 60 \times 0.075(0.24 + 0.45W)Q_s \Delta t \tag{21}$$

where

0.24 = specific heat of dry air, Btu/(lb·°F)
W = humidity ratio, lb (water)/lb (dry air)
0.45 = specific heat of water vapor, Btu/(lb·°F)

The specific heats are for a range from about −100 to 200°F. When $W = 0$, the value of $60 \times 0.075(0.24 + 0.45W) = 1.08$; when $W = 0.01$, the value is 1.10; when $W = 0.02$, the value is 1.12; and when $W = 0.03$, the value is 1.14. Thus, because a value of $W = 0.01$ approximates conditions found in many air-conditioning problems, the sensible heat change (in Btu/h) can normally be found as

$$q_s = 1.10 Q_s \Delta t \tag{22}$$

3. Latent heat

Latent heat gain corresponding to the change of humidity ratio (ΔW) for given air flow (standard conditions) Q_s is

$$q_l = 60 \times 0.075 \times 1076 Q_s \Delta W = 4840 Q_s \Delta W \tag{23}$$

where 1076 is the approximate heat content of 50% rh vapor at 75°F, less the heat content of water at 50°F. The 50% rh at 75°F is a common design condition for the space, and 50°F is normal condensate temperature from cooling and dehumidifying coils.

The constants 4.5, 1.10, and 4840 are useful in air-conditioning calculations at sea level (14.7 psia) and for normal temperatures and moisture ratios. For other conditions, more precise values should be used. For an altitude of 5000 ft (12.2 psia), appropriate values are 3.73, 0.92, and 4020.

Latent Heat Gain from Moisture through Permeable Building Materials

The diffusion of moisture through all common building materials is a natural phenomenon that is always present. Chapters 22 and 23 cover the principles and specific methods used to control moisture. Moisture transfer through walls is often neglected in the usual comfort air-conditioning application, because the actual rate is quite small, and the corresponding latent heat gain is insignificant. The permeability and permeance values for various building materials are given in Table 9, Chapter 24. Vapor retarders are frequently installed to keep moisture transfer to a minimum.

Special Conditions. Certain industrial applications call for a low moisture content to be maintained in a conditioned space. In such cases, the latent heat gain accompanying moisture transfer through walls may be greater than any other latent heat gain. This gain is computed by

$$q_m = (M/7000) A \Delta p_v (h_g - h_f) \tag{24}$$

where

q_m = latent heat gain, Btu/h
M = permeance of wall assembly in perms, or grains/(ft^2·h·in Hg)
7000 = grains/lb
A = area of wall surface, ft^2
Δp_v = vapor pressure difference, in. Hg
h_g = enthalpy at room conditions, Btu/lb
h_f = enthalpy of water condensed at cooling coil, Btu/lb
= 1076 Btu/lb when room temperature is 75°F and condensate off coil is 50°F

Heat Gain from Miscellaneous Sources

The calculation of the cooling load is affected by such factors as (1) type of HVAC system, (2) effectiveness of heat exchange surfaces, (3) fan location, (4) duct heat gain or loss, (5) duct leakage, (6) heat-extraction lighting systems, (7) type of return air system, and (8) sequence of controls. System performance needs to be analyzed as a sequence of individual psychrometric processes. The most straightforward method first defines all known (or desired) state points on a psychrometric chart. Next, the actual entering and leaving dry- and wet-bulb conditions are calculated for such components as the cooling and/or heating coils (based on zone or space load), the amount of outside air introduced into the system through the equipment, and the amount of heat gain or loss at various points.

This overall process must verify that the space conditions originally sought can actually be met by the designed system by considering all sensible and latent heat changes to the air as it travels from the space conditions through the return air system and equipment back to the conditioned space. If the design is successful (i.e., within the degree of correctness of the various design assumptions), appropriate equipment components can safely be selected. If not, the designer must judge if the results will be "close enough" to satisfy the needs of the project, or if one or more assumptions and/or design criteria must first be modified and the calculations rerun.

Heat Gain from Fans. Fans that circulate air through HVAC systems add energy to the system by one or all of the following processes:

- Temperature rise in the airstream from fan inefficiency. Depending on the equipment, fan efficiencies generally range between 50 and 70%, with an average value of 65%. Thus, some 35% of the energy required by the fan appears as instantaneous heat gain to the air being transported.
- Temperature rise in the airstream as a consequence of air static and velocity pressure. The "useful" 65% of the total fan energy that creates pressure to move air spreads out throughout the entire air transport system in the process of conversion to sensible heat. Designers commonly assume that the temperature change equivalent of this heat occurs at a single point in the system, depending on fan location as noted below.
- Temperature rise from heat generated by motor and drive inefficiencies. The relatively small gains from fan motors and drives are normally disregarded unless the motor and/or drive are physically located within the conditioned airstream. Equations (15), (16), and (17) may be used to estimate heat gains from typical motors. Belt drive losses are often estimated as 3% of the motor power rating. Conversion to temperature rise is calculated by Equation (22).

The location of each fan relative to other elements (primarily the cooling coil), and the type of system (e.g., single zone, multizone, double-duct, terminal reheat, VAV) along with the concept of equipment control (space temperature alone, space temperature and relative humidity, etc.) must be known before the analysis can be completed. A fan located upstream of the cooling coil (blowthrough supply fan, return air fan, outside air fan) adds the heat equivalent of its inefficiency to the airstream at that point; thus, a slightly elevated entering dry-bulb temperature to the cooling coil results. A fan located downstream of the cooling coil raises the dry-bulb temperature of air leaving the cooling coil. This rise can be offset by reducing the cooling coil temperature, or alternatively, by increasing airflow across the cooling coil as long as its impact on space conditions is considered.

Duct Heat Gain and Leakage. Unless return air duct systems are extensive or subjected to rigorous conditions, only the heat gained or lost by supply duct systems is significant; it is normally estimated as a percentage of space sensible cooling load (usually about 1%) and applied to the dry-bulb temperature of the air leaving the coil in the form of an equivalent temperature reduction.

Duct Leakage. Air leakage out of (or into) duct work can have much greater impact than conventional duct heat gain or loss, but it is normally about the same or less. Outward leakage from supply ducts is a direct loss of cooling and/or dehumidifying capacity and must be offset by increased airflow (sometimes reduced supply air temperatures) unless it enters the conditioned space directly. Inward leakage to return ducts causes temperature and/or humidity variations, but these are often ignored under ordinary circumstances due to the low temperature and pressure differentials involved. Chapter 32 has further details on duct sealing and leakage.

A well-designed and installed duct system should not leak more than 1 to 3% of the total system airflow. All HVAC equipment and volume control units connected into a duct system are usually delivered from manufacturers with allowable leakage not exceeding 1 or 2% of maximum airflow rating. Where duct systems are specified to be sealed and leak tested, both low and medium pressure types can be constructed and required to fall within this range, and designers normally assume this loss to approximate 1% of the space load, handled in a similar manner to that for duct heat gain. Latent heat considerations are frequently ignored.

Poorly designed or installed duct systems can have leakage rates of 10 to 30%. Leakage from low-pressure lighting troffer connections lacking proper taping and sealing can be 35% or more of the terminal air supply. Improperly sealed high-pressure systems can leak as much as 10% or more from the high-pressure side alone, before considering the corresponding low-pressure side of such systems. Such extremes destroy the validity of any load calculation procedures. Although not always affecting overall system loads enough to cause problems, they will, however, always adversely impact required supply air quantities for most air-conditioning systems. Also, using uninsulated supply duct work running through return air plenums results in high "thermal leakage," thus loss of space cooling capability by the supply air, and potential condensation difficulties during a warm startup.

HEATING LOAD PRINCIPLES

Techniques for estimating design heating load for commercial, institutional, and industrial applications are essentially the same as for those estimating design cooling loads for such uses, except that (1) temperatures outside the conditioned spaces are generally lower than the space temperatures maintained; (2) credit for solar heat gains or for internal heat gains is not included; and (3) the thermal storage effect of building structure or content is ignored. Heat losses (negative heat gains) are thus considered to be instantaneous, heat transfer essentially conductive, and latent heat treated only as a function of replacing space humidity lost to the exterior environment.

Justification of this simplified approach derives from the purpose of a heating load estimate, as identification of "worst case" conditions that can reasonably be anticipated during a heating season. Traditionally this is considered as the load that must be met under design interior and exterior conditions, including infiltration and/or ventilation, but in the absence of solar effect (at night or cloudy winter days) and before the periodic presence of people, lights, and appliances can begin to have an offsetting effect. The primary orientation is thus toward identification of adequately sized heating equipment to handle the normal worst-case condition.

Safety Factors and Load Allowances. Before mechanical cooling of buildings became a usual procedure, buildings included much less insulation, large operable windows, and generally more infiltration-prone assemblies than the energy-efficient and much tighter buildings typical of post-1975 design. Allowances of 10 to 20% of the net calculated heating load for piping losses to unheated spaces, and 10 to 20% more for a warm-up load were common practice, along with occasional other safety factors reflecting the experience and/or concern of the individual designer. Such measures are infrequently used in estimating heating loads for contemporary buildings, with the uncompensated net heating load normally considered as having an adequate margin for error.

Cooling Needs During Noncooling Months. Perimeter spaces exposed to high solar heat gain often justify mechanical cooling during sunlit portions of traditional heating months, as will completely interior spaces with significant internal heat gain. These conditions require special attention by the system designer for proper accommodation as needed, but such spaces can also represent significant heating loads during nonsunlit hours or after periods of nonoccupancy when adjacent spaces have been allowed to cool below interior design temperatures. The loads involved can be estimated conventionally for the purpose of designing the means to offset or to compensate for them and prevent overheating, but they have no direct relationship to design heating loads for the spaces in question.

Other Considerations. Calculation of design heating load estimates for this general category of applications has essentially become a subset of the more involved and complex estimation of cooling loads for such spaces. Chapter 30 discusses using the heating load estimate to predict or analyze energy consumption over time. Special provisions to deal with atypical problems are relegated to appropriate chapters in the Systems, Equipment, and Applications volumes.

TRANSFER FUNCTION METHOD CALCULATION PROCEDURE

BASIC COOLING LOAD ANALYSIS

The basic procedures for estimating the maximum design cooling load for a conditioned space were developed when all design calculations were performed manually. For this reason, extensive design analysis was not part of the primary load estimate. Today, with computers used for routine design calculations, the individual load elements may be evaluated more thoroughly and a comprehensive design analysis can be included with the results. The TFM method makes it possible to estimate the cooling load for a conditioned space on an hour-by-hour basis and to predict resultant conditions that can be expected in that space for various system types, control strategies, and operating schedules. The equations and sequence of the TFM Procedure in general are summarized in Table 10.

HEAT GAIN BY CONDUCTION THROUGH EXTERIOR WALLS AND ROOFS

Sensible Heat Gain

The transfer function method (TFM) is particularly well suited for use with a computer. This method is a special case of the calculation of heat flow through building components outlined in Chapter 22. This approach uses (1) sol-air temperature to represent outdoor conditions, and (2) an assumed constant indoor air temperature. Furthermore, both indoor and outdoor surface heat transfer coefficients are assumed constant (Mitalas 1968). Thus, the heat gain through a wall or roof is given by

$$q_{e,\theta} = A\left[\sum_{n=0} b_n(t_{e,\theta-n\delta}) - \sum_{n=1} \frac{d_n(q_{e,\theta-n\delta})}{A} - t_{rc}\sum_{n=0} c_n\right] \quad (25)$$

where

$q_{e,\theta}$ = heat gain through wall or roof, at calculation hour θ
A = indoor surface area of a wall or roof
θ = time
δ = time interval
n = summation index (each summation has as many terms as there are non-negligible values of coefficients)
$t_{e,\theta-n\delta}$ = sol-air temperature at time $\theta - n\delta$
t_{rc} = constant indoor room temperature
b_n, c_n, d_n = conduction transfer function coefficients

Conduction Transfer Function Coefficients. Conduction transfer function (CTF) coefficients are usually calculated using combined outdoor heat transfer coefficient h_o = 3.0 Btu/(h·ft²·°F), indoor coefficient h_i = 1.46 Btu/(h·ft²·°F), and the wall or roof constructions, as may be appropriate. The use of h_o = 3.0 limits the application of these coefficients to cases with similarly calculated sol-air temperature values. Specific CTF coefficients for different constructions can be calculated using the procedure and computer program outlined in Mitalas and Arseneault (1970) or as discussed by McQuiston and Spitler (1992) and with the microcomputer software issued with that publication.

Representative Walls and Roofs. Harris and McQuiston (1988) investigated the thermal behavior of approximately 2600 walls and 500 roofs as they influenced transmission of heat gain to conditioned spaces. This work identified 41 representative wall assemblies and 42 roof assemblies with widely varying components, insulating values, and mass, and with the predominant mass concentrated near the inside surface (mass in), outside surface (mass out), or essentially homogeneous (mass integral) with the overall construction. These prototypical assemblies can be used to reflect the overall range of conditions. The CTF and associated data pertaining to these conditions are listed in Tables 11 through 19.

Approximate values of CTF coefficients can be obtained by selecting a set of data from Tables 13 and 14 for a roof construction or Tables 18 and 19 for a wall that is nearly the same as the roof or wall under consideration, and multiplying the s and s by the ratio of the U-factor of the roof or wall under consideration over the U-factor of the selected representative roof or wall.

The physical and thermal properties of the various layers that make up roof and wall assemblies are listed in Table 11. Group numbers for various arrangements of layers with differing insulation R value and placement for roofs are listed in Table 12 and those for walls are listed in Tables 15, 16 and 17. Data from these tables identify prototypical roof or wall CTFs and associated data tabulated in Tables 13, 14, 18, and 19.

Example 3. Heat gain through wall. A light-colored wall is constructed of 4 in. heavy concrete, 2 in. insulation (2.0 lb/ft³, R = 6.667 h·ft²·°F/Btu), 3/4 in. indoor plaster, and with outdoor and indoor surface resistances of 0.333 and 0.68 h·ft²·°F/Btu, respectively. There is an air space between the plaster and the insulation. Find the heat gain through 1 ft² of the wall area (i.e., A = 1.0 ft²) with sol-air temperature as listed in Table 1 for July 21, 40°North latitude, West, α/h_0 = 0.15, a room temperature of 75°F, and assuming that the daily sol-air temperature cycle is repeated on several consecutive days.

Solution: The calculation of heat gain for a particular time requires sol-air temperature values at that and preceding times, as well as the heat flow at preceding times. Heat flow is assumed as zero to start the calculations. The effect of this assumption becomes negligible as the calculation is repeated for successive 24-h cycles.

Sol-Air Temperatures (from Table 1)

Time, h	t_e, °F	Time, h	t_e, °F
1	76	13	110
2	76	14	121
3	75	15	129
4	74	16	131
5	74	17	127
6	75	18	114
7	78	19	87
8	81	20	85
9	85	21	83
10	89	22	81
11	93	23	79
12	97	24	77

For $\theta > 24$, $t_{e,\theta} = t_{e,\theta-24}$.

Table 10 Summary of TFM Load Calculation Procedures

External Heat Gain

$$t_e = t_o + \alpha I_t/h_o - \varepsilon\Delta R/h_o \quad (6)$$

$$t_{ea} = t_{oa} + \alpha/h_o(I_{DT}/24) - \varepsilon\Delta R/h_o \quad (7)$$

where

t_e = sol-air temperature
t_o = current hour dry-bulb temperature, from design db (Chapter 26) adjusted by Table 2 daily range % values

α = absorptance of surface for solar radiation
α/h_o = surface color factor = 0.026 for light colors, 0.052 for dark
I_t = total incident solar load = 1.15 (SHGF), with SHGF per Chapter 29, Tables 15 through 21
$\varepsilon\Delta R/h_o$ = long-wave radiation factor = −7°F for horizontal surfaces, 0°F for vertical
t_e = 24-h average sol-air temperature
t_{oa} = 24-h average dry-bulb temperature
I_{DT} = total daily solar heat gain (Chapter 29, Tables 15 through 21)

Roofs and Walls

$$q_{e,\theta} = A\left[\sum_{n=0} b_n(t_{e,\theta-n\delta}) - \sum_{n=1} d_n[(q_{e,\theta-n\delta})/A] - t_{rc} - \sum_{n=0} c_n\right] \quad (28)$$

where

b and d = conduction transfer coefficients—roof, Table 13; wall, Table 18
c and U_{table} = conduction transfer coefficients—roof, Table 14; wall, Table 19
U_{actual} = design heat transfer coefficient for roof or wall, from Chapter 24, Table 4

Adjust b and c coefficients by ratio U_{actual}/U_{table}.

Roofs

Identify layers of roof construction from Table 11. With R-value of dominant layer, identify R-value Range number R and Roof Group number from Table 12. Proceed to Table 13.

Walls

Identify layers of wall construction from Table 11. With R-value of dominant layer, identify R-value Range number and Wall Group number from Table 15, 16, or 17. Proceed to Table 14.

θ = hour for which calculation is made
δ = time interval (1 h)
n = number of hours for which and values are significant
e = element under analysis, roof or wall assembly
A = area of element under analysis

Glass

Convective $q = UA(t_o - t_i)$
Solar $q = A$(SC)(SHGF)

U = design heat transfer coefficients, glass—Chapter 29
SC = shading coefficient—Chapter 29
SHGF = solar heat gain factor by orientation, north latitude, hour, and month—Chapter 29, Tables 15 to 21.

Partitions, Ceilings, Floors

$$q = UA\,(t_b - t_i) \quad (8)$$

t_b = temperature in adjacent space
t_i = inside design temperature in conditioned space

Internal Heat Gain

People

$q_{sensible} = N$(Sensible heat gain)
$q_{latent} = N$(Latent heat gain)

N = number of people in space, from best available source. Sensible and latent heat gain from occupancy—Table 3, or Chapter 8; adjust as required.

Lights

$$q_{el} = 3.41\,WF_{ul}F_{sa} \quad (9)$$

where

W = watts input from electrical plans or lighting fixture data
F_{ul} = lighting use factor, from the first section, as appropriate
F_{sa} = special allowance factor, from first section, as appropriate

Power

$$q_p = 2545\,PE_F \quad (15), (16), (17)$$

where

P = power rating from electrical plans or manufacturer's data
E_F = efficiency factors and arrangements to suit circumstances

Appliances

where

$$q_{sensible} = q_{input}F_U F_R \quad (18)$$

or

$$q_{sensible} = q_{input}F_L \quad (19)$$

q_{input} = rated energy input from appliances—Tables 5 to 9, or manufacturer's data (set latent heat = 0, if appliance is under exhaust hood)

F_U, F_R, F_L = usage factors, radiation factors, and load factors

Ventilation and Infiltration Air

$$q_{sensible} = 1.10Q(t_o - t_i) \quad (22)$$

$$q_{latent} = 4840Q(W_o - W_i) \quad (23)$$

$$q_{total} = 4.5Q(H_o - H_i) \quad (20)$$

Q = ventilation airflow—ASHRAE *Standard* 62; infiltration cfm—Chapter 25
t_o, t_i = outside, inside air temperature, °F
W_o, W_i = outside, inside air humidity ratio, lb (water)/lb (da)
H_o, H_i = outside, inside air enthalpy, Btu/lb (dry air)

Cooling Load

Sensible $Q\theta = Q_{rf} + Q_{sc}$

$$Q_{rf} = \sum_{i=1}(v_0 q_{\theta,i} + v_1 q_{\theta,i-\delta} + v_2 q_{\theta,i-2\delta} + \ldots) - (w_1 Q_{\theta-\delta} + w_2 Q_{\theta-2\delta} + \ldots) \quad (28)$$

$$Q_{sc} = \sum_{j=1}(q_{c,j}) \quad (30)$$

Q_{rf} = sensible cooling load from heat gain elements having convective and radiant components
v and w = room transfer function coefficients, Tables 24 and 25; select per element type, circulation rate, mass, and/or fixture type
q_θ = each of i heat gain elements having a radiant component; select appropriate fractions for processing, per Tables 24, 25, and 42
δ = time interval (1 h)
Q_{sc} = sensible cooling load from heat gain elements having only convective components
q_c = each of j heat gain elements having only convective component

Latent $Q_l = \sum_{n=1}(q_{c,n})$

q_c = each of n latent heat gain elements

Table 11 Thermal Properties and Code Numbers of Layers Used in Wall and Roof Descriptions for Tables 12 and 13

Code Number	Description	Thickness and Thermal Properties					
		L	k	ρ	c_p	R	Mass
A0	Outside surface resistance	0.0	0.0	0.0	0.0	0.33	0.0
A1	1 in. Stucco	0.0833	0.4	116.0	0.20	0.21	9.7
A2	4 in. Face brick	0.333	0.77	125.0	0.22	0.43	41.7
A3	Steel siding	0.005	26.0	480.0	0.10	0.00	2.4
A4	1/2 in. Slag	0.0417	0.11	70.0	0.40	0.38	2.2
A5	Outside surface resistance	0.0	0.0	0.0	0.0	0.33	0.0
A6	Finish	0.0417	0.24	78.0	0.26	0.17	3.3
A7	4 in. Face brick	0.333	0.77	125.0	0.22	0.43	41.7
B1	Air space resistance	0.0	0.0	0.0	0.0	0.91	0.0
B2	1 in. Insulation	0.083	0.025	2.0	0.2	3.33	0.2
B3	2 in. Insulation	0.167	0.025	2.0	0.2	6.67	0.3
B4	3 in. Insulation	0.25	0.025	2.0	0.2	1.19	0.5
B5	1 in. Insulation	0.0833	0.025	5.7	0.2	3.33	0.5
B6	2 in. Insulation	0.167	0.025	5.7	0.2	6.67	1.0
B7	1 in. Wood	0.0833	0.07	37.0	0.6	1.19	3.1
B8	2.5 in. Wood	0.2083	0.07	37.0	0.6	2.98	7.7
B9	4 in. Wood	0.333	0.07	37.0	0.6	4.76	12.3
B10	2 in. Wood	0.167	0.07	37.0	0.6	2.39	6.2
B11	3 in. Wood	0.25	0.07	37.0	0.6	3.57	9.3
B12	3 in. Insulation	0.25	0.025	5.7	0.2	10.00	1.4
B13	4 in. Insulation	0.333	0.025	5.7	0.2	13.33	1.9
B14	5 in. Insulation	0.417	0.025	5.7	0.2	16.67	2.4
B15	6 in. Insulation	0.500	0.025	5.7	0.2	20.00	2.9
B16	0.15 in. Insulation	0.0126	0.025	5.7	0.2	0.50	0.1
B17	0.3 in. Insulation	0.0252	0.025	5.7	0.2	1.00	0.1
B18	0.45 in. Insulation	0.0379	0.025	5.7	0.2	1.50	0.2
B19	0.61 in. Insulation	0.0505	0.025	5.7	0.2	2.00	0.3
B20	0.76 in. Insulation	0.0631	0.025	5.7	0.2	2.50	0.4
B21	1.36 in. Insulation	0.1136	0.025	5.7	0.2	4.50	0.6
B22	1.67 in. Insulation	0.1388	0.025	5.7	0.2	5.50	0.8
B23	2.42 in. Insulation	0.2019	0.025	5.7	0.2	8.00	1.2
B24	2.73 in. Insulation	0.2272	0.025	5.7	0.2	9.00	1.3
B25	3.33 in. Insulation	0.2777	0.025	5.7	0.2	11.00	1.6
B26	3.64 in. Insulation	0.3029	0.025	5.7	0.2	12.00	1.7
B27	4.54 in. Insulation	0.3786	0.025	5.7	0.2	15.00	2.2
C1	4 in. Clay tile	0.333	0.33	70.0	0.2	1.01	23.3
C2	4 in. Lightweight concrete block	0.333	0.22	38.0	0.2	1.51	12.7
C3	4 in. Heavyweight concrete block	0.333	0.47	61.0	0.2	0.71	20.3
C4	4 in. Common brick	0.333	0.42	120.0	0.2	0.79	40.0
C5	4 in. Heavyweight concrete	0.333	1.0	140.0	0.2	0.33	46.7
C6	8 in. Clay tile	0.667	0.33	70.0	0.2	2.00	46.7
C7	8 in. Lightweight concrete block	0.667	0.33	38.0	0.2	2.00	25.3
C8	8 in. Heavyweight concrete block	0.667	0.6	61.0	0.2	1.11	40.7
C9	8 in. Common brick	0.667	0.42	120.0	0.2	1.59	80.0
C10	8 in. Heavyweight concrete	0.667	1.0	140.0	0.2	0.67	93.4
C11	12 in. Heavyweight concrete	1.0	1.0	140.0	0.2	1.00	140.0
C12	2 in. Heavyweight concrete	0.167	1.0	140.0	0.2	0.17	23.3
C13	6 in. Heavyweight concrete	0.5	1.0	140.0	0.2	0.50	70.0
C14	4 in. Lightweight concrete	0.333	0.1	40.0	0.2	3.33	13.3
C15	6 in. Lightweight concrete	0.5	0.1	40.0	0.2	5.00	20.0
C16	8 in. Lightweight concrete	0.667	0.1	40.0	0.2	6.67	26.7
C17	8 in. Lightweight concrete block (filled)	0.667	0.08	18.0	0.2	8.34	12.0
C18	8 in. Heavyweight concrete block (filled)	0.667	0.34	53.0	0.2	1.96	35.4
C19	12 in. Lightweight concrete block (filled)	1.000	0.08	19.0	0.2	12.50	19.0
C20	12 in. Heavyweight concrete block (filled)	1.000	0.39	56.0	0.2	2.56	56.0
E0	Inside surface resistance	0.0	0.0	0.0	0.0	0.69	0.0
E1	3/4 in. Plaster or gypsum	0.0625	0.42	100.0	0.2	0.15	6.3
E2	1/2 in. Slag or stone	0.0417	0.83	55.0	0.40	0.05	2.3
E3	3/8 in. Felt and membrane	0.0313	0.11	70.0	0.40	0.29	2.2
E4	Ceiling air space	0.0	0.0	0.0	0.0	1.00	0.0
E5	Acoustic tile	0.0625	0.035	30.0	0.2	1.79	1.9

L = thickness, ft
k = thermal conductivity, Btu/h·ft·°F
ρ = density, lb/ft^3
c_p = specific heat, Btu/lb·°F
R = thermal resistance, °F·ft^2·h/Btu
Mass = unit mass, lb/ft^2

Table 12 Roof Group Numbers

Roofs without Suspended Ceilings

Roof Materials[a]		Mass In						Integral Mass						Mass Out					
		R-Value Range Numbers[b]						R-Value Range Numbers[b]						R-Value Range Numbers[b]					
No.	Codes	1	2	3	4	5	6	1	2	3	4	5	6	1	2	3	4	5	6
1	B7							1	2	2	4	4							
2	B8							4	5	9	10	18							
3	B9							19	21	27	27	28							
4	C5	6	7	7	10	10		3						6	7	7	10	11	
5	C12	2	2	4	4	5		2						2	3	4	5	5	
6	C13	7	12	13	13	20		5						7	12	13	13	20	
7	C14		4	5	9	9		2	2						4	5	9	9	
8	C15		5	10	18	18	18		4						5	10	10	18	18
9	C16		9	19	20	27	27		9						9	18	20	27	27
10	A3							1	1	1	2	2							
11	Attic							1	2	2	2	4							
Roof Terrace Systems																			
12	C12-C12							4	5	9	9	9		5	5	7	9	9	
13	C12-C5							6	11	12	18	18		7	12	12	12	20	
14	C12-C13							11	20	20	21	27		12	13	21	21	21	
15	C5-C12							5	10	10	17	17		5	10	11	11	18	
16	C5-C5							10	20	20	26	26		10	13	21	21	21	
17	C5-C13							20	27	28	28	35		20	22	22	22	28	
18	C13-C12							10	18	20	20	26		10	13	20	29	21	
19	C13-C5							18	27	27	28	35		20	22	22	28	28	
20	C13-C13							21	29	30	36	36		21	29	30	31	36	

Roofs with Suspended Ceilings

Roof Materials[a]		Mass In						Integral Mass						Mass Out					
		R-Value Range Numbers[b]						R-Value Range Numbers[b]						R-Value Range Numbers[b]					
No.	Codes	1	2	3	4	5	6	1	2	3	4	5	6	1	2	3	4	5	6
1	B7								4	5	9	10	10						
2	B8								9	20	21	22	28						
3	B9								20	28	30	37	38						
4	C5	8	15	18	18	23		6						7	7	7	10	10	
5	C12	5	8	13	13	14		3						3	3	4	5		
6	C13		18	24	25	25		11	11						12	13	13	20	
7	C14		4	10	11	18	20		4						4	5	9	9	17
8	C15		10	18	21	21	28		9						9	10	18	19	26
9	C16			20	28	29	36			18						18	26	27	27
10	A3							1	1	2	2	4							
11	Attic																		
Roof Terrace Systems																			
12	C12-C12							6	13	22	22	22		5	5	7	9	9	
13	C12-C5							10	21	23	24	31			12	12	18	20	
14	C12-C13							13	23	24	33	33			13	21	21	21	
15	C5-C12							10	20	22	28	29			10	12	18	18	
16	C5-C5							13	23	32	32	33			20	21	21	21	
17	C5-C13							21	32	34	40				22	22	28	28	
18	C13-C12							12	28	30	31	37			13	20	20	21	
19	C13-C5							21	31	39	40	40			22	22	28	28	
20	C13-C13								39	41	41	42	42		29	30	31	36	37

[a]Blank spaces denote a roof that is not possible with the chosen combinations of parameters. Numbers 12 through 20 are roof terrace systems. First material is outer layer, second material is inner layer. Massive material numbers are: 4, 5, 6, 7, 8, 9, 12, 13, 14, 15, 16, 17, 18, 19, and 20. Nonmassive material numbers are: 1, 2, 3, 10, and 11.

[b]R-Value ranges in h·ft²·°F/Btu are:

No.	Range	No.	Range	No.	Range
1	0 to 5	3	10 to 15	5	20 to 25
2	5 to 10	4	15 to 20	6	25 to 30

Table 13 Roof Conduction Transfer Function Coefficients (*b* and *d* Factors)

Roof Group	(Layer Sequence Left to Right = Inside to Outside)		$n = 0$	$n = 1$	$n = 2$	$n = 3$	$n = 4$	$n = 5$	$n = 6$
1	Layers E0 A3 B25 E3 E2 A0	b_n	0.00487	0.03474	0.01365	0.00036	0.00000	0.00000	0.00000
	Steel deck with 3.33 in. insulation	d_n	1.00000	−0.35451	0.02267	−0.00005	0.00000	0.00000	0.00000
2	Layers E0 A3 B14 E3 E2 A0	b_n	0.00056	0.01202	0.01282	0.00143	0.00001	0.00000	0.00000
	Steel deck with 5 in. insulation	d_n	1.00000	−0.60064	0.08602	−0.00135	0.00000	0.00000	0.00000
3	Layers EO E5 E4 C12 E3 E2 A0	b_n	0.00613	0.03983	0.01375	0.00025	0.00000	0.00000	0.00000
	2 in. h.w. concrete deck with suspended ceiling	d_n	1.00000	−0.75615	0.01439	−0.00006	0.00000	0.00000	0.00000
4	Layers E0 E1 B15 E4 B7 A0	b_n	0.00000	0.00065	0.00339	0.00240	0.00029	0.00000	0.00000
	Attic roof with 6 in. insulation	d_n	1.00000	−1.34658	0.59384	−0.09295	0.00296	−0.00001	0.00000
5	Layers E0 B14 C12 E3 E2 A0	b_n	0.00006	0.00256	0.00477	0.00100	0.00002	0.00000	0.00000
	5 in. insulation with 2 in. h.w. concrete deck	d_n	1.00000	−1.10395	0.26169	−0.00475	0.00002	0.00000	0.00000
6	Layers E0 C5 B17 E3 E2 A0	b_n	0.00290	0.03143	0.02114	0.00120	0.00000	0.00000	0.00000
	4 in. h.w. concrete deck with 0.3 in. insulation	d_n	1.00000	−0.97905	0.13444	−0.00272	0.00000	0.00000	0.00000
7	Layers E0 B22 C12 E3 E2 C12 A0	b_n	0.00059	0.00867	0.00688	0.00037	0.00000	0.00000	0.00000
	1.67 in. insulation with 2 in. h.w. concrete RTS	d_n	1.00000	−1.11766	0.23731	−0.00008	0.00000	0.00000	0.00000
8	Layers E0 B16 C13 E3 E2 A0	b_n	0.00098	0.01938	0.02083	0.00219	0.00001	0.00000	0.00000
	0.15 in. insul. with 6 in. h.w. concrete deck	d_n	1.00000	−1.10235	0.20750	−0.00287	0.00000	0.00000	0.00000
9	Layers E0 E5 E4 B12 C14 E3 E2 A0	b_n	0.00000	0.00024	0.00217	0.00251	0.00055	0.00002	0.00000
	3 in. insul. w/4 in. l.w. conc. deck and susp. clg.	d_n	1.00000	−1.40605	0.58814	−0.09034	0.00444	−0.00006	0.00000
10	Layers E0 E5 E4 C15 B16 E3 E2 A0	b_n	0.00000	0.00025	0.00241	0.00303	0.00074	0.00004	0.00000
	6 in. l.w. conc. dk w/0.15 in. ins. and susp. clg.	d_n	1.00000	−1.55701	0.73120	−0.11774	0.00600	−0.00008	0.00000
11	Layers E0 C5 B15 E3 E2 A0	b_n	0.00000	0.00013	0.00097	0.00102	0.00020	0.00001	0.00000
	4 in. h.w. concrete deck with 6 in. insulation	d_n	1.00000	−1.61467	0.79142	−0.13243	0.00611	−0.00008	0.00000
12	Layers E0 C13 B16 E3 E2 C12 A0	b_n	0.00005	0.00356	0.01058	0.00404	0.00019	0.00000	0.00000
	6 in. h.w. deck w/0.15 in. ins. and 2 in. h.w. RTS	d_n	1.00000	−1.59267	0.72160	−0.08275	0.00029	0.00000	0.00000
13	Layers E0 C13 B6 E3 E2 A0	b_n	0.00002	0.00136	0.00373	0.00129	0.00006	0.00000	0.00000
	6 in. h.w. concrete deck with 2 in. insulation	d_n	1.00000	−1.34451	0.44285	−0.04344	0.00016	0.00000	0.00000
14	Layers E0 E5 E4 C12 B13 E3 E2 A0	b_n	0.00000	0.00046	0.00143	0.00057	0.00003	0.00000	0.00000
	2 in. l.w. conc. deck w/4 in. ins. and susp. clg.	d_n	1.00000	−1.33741	0.41454	−0.03346	0.00031	0.00000	0.00000
15	Layers E0 E5 E4 C5 B6 E3 E2 A0	b_n	0.00001	0.00066	0.00163	0.00049	0.00002	0.00000	0.00000
	1 in. insul. w/4 in. h.w. conc. deck and susp. clg.	d_n	1.00000	−1.24348	0.28742	−0.01274	0.00009	0.00000	0.00000
16	Layers E0 E5 E4 C13 B20 E3 E2 A0	b_n	0.00001	0.00060	0.00197	0.00086	0.00005	0.00000	0.00000
	6 in. h.w. deck w/0.76 in. insul. and susp. clg.	d_n	1.00000	−1.39181	0.46337	−0.04714	0.00058	0.00000	0.00000
17	Layers E0 E5 E4 B15 C14 E3 E2 A0	b_n	0.00000	0.00001	0.00021	0.00074	0.00053	0.00010	0.00000
	6 in. insul. w/4 in. l.w. conc. deck and susp. clg.	d_n	1.00000	−1.87317	1.20950	−0.32904	0.03799	−0.00169	0.00002
18	Layers E0 C12 B15 E3 E2 C5 A0	b_n	0.00000	0.00002	0.00027	0.00052	0.00019	0.00002	0.00000
	2 in. h.w. conc. dk w/6 in. ins. and 2 in. h.w. RTS	d_n	1.00000	−2.10928	1.50843	−0.40880	0.03249	−0.00068	0.00000
19	Layers E0 C5 B27 E3 E2 C12 A0	b_n	0.00000	0.00009	0.00073	0.00078	0.00015	0.00000	0.00000
	4 in. h.w. deck w/4.54 in. ins. and 2 in. h.w. RTS	d_n	1.00000	−1.82851	1.02856	−0.17574	0.00556	−0.00003	0.00000
20	Layers E0 B21 C16 E3 E2 A0	b_n	0.00000	0.00002	0.00044	0.00103	0.00049	0.00005	0.00000
	1.36 in. insulation with 8 in. l.w. concrete deck	d_n	1.00000	−1.91999	1.21970	−0.30000	0.02630	−0.00061	0.00000
21	Layers E0 C13 B12 E3 E2 C12 A0	b_n	0.00000	0.00009	0.00072	0.00077	0.00015	0.00000	0.00000
	6 in. h.w. deck w/3 in. insul. and 2 in. h.w. RTS	d_n	1.00000	−1.84585	1.03238	−0.17182	0.00617	−0.00003	0.00000
22	Layers E0 B22 C5 E3 E2 C13 A0	b_n	0.00000	0.00014	0.00100	0.00094	0.00015	0.00000	0.00000
	1.67 in. ins. w/4 in. h.w. deck and 6 in. h.w. RTS	d_n	1.00000	−1.79981	0.94786	−0.13444	0.00360	−0.00001	0.00000
23	Layers E0 E5 E4 C12 B14 E3 E2 C12 A0	b_n	0.00000	0.00002	0.00022	0.00031	0.00008	0.00000	0.00000
	Susp. clg, 2 in. h.w. dk, 5 in. ins, 2 in. h.w. RTS	d_n	1.00000	−1.89903	1.13575	−0.23586	0.01276	−0.00015	0.00000
24	Layers E0 E5 E4 C5 E3 E2 B6 B1 C12 A0	b_n	0.00000	0.00008	0.00047	0.00039	0.00006	0.00000	0.00000
	Susp. clg, 4 in. h.w. dk, 2 in. ins, 2 in. h.w. RTS	d_n	1.00000	−1.73082	0.85681	−0.11614	0.00239	−0.00001	0.00000
25	Layers E0 E5 E4 C13 B13 E3 E2 A0	b_n	0.00000	0.00002	0.00021	0.00031	0.00009	0.00001	0.00000
	6 in. h.w. conc. deck w/4 in. ins. and susp. clg.	d_n	1.00000	−1.63446	0.78078	−0.14422	0.00940	−0.00011	0.00000
26	Layers E0 E5 E4 B15 C15 E3 E2 A0	b_n	0.00000	0.00000	0.00002	0.00014	0.00024	0.00011	0.00002
	6 in. insul. w/6 in. l.w. conc. deck and susp. clg.	d_n	1.00000	−2.29459	1.93694	−0.75741	0.14252	−0.01251	0.00046
27	Layers E0 C13 B15 E3 E2 C12 A0	b_n	0.00000	0.00000	0.00007	0.00024	0.00016	0.00003	0.00000
	6 in. h.w. deck w/6 in. ins. and 2 in. h.w. RTS	d_n	1.00000	−2.27813	1.82162	−0.60696	0.07696	−0.00246	0.00001
28	Layers E0 B9 B14 E3 E2 A0	b_n	0.00000	0.00000	0.00001	0.00010	0.00017	0.00009	0.00001
	4 in. wood deck with 5 in. insulation	d_n	1.00000	−2.41915	2.17932	−0.93062	0.19840	−0.02012	0.00081
29	Layers E0 E5 E4 C12 B13 E3 E2 C5 A0	b_n	0.00000	0.00001	0.00018	0.00026	0.00007	0.00000	0.00000
	Susp. clg, 2 in. h.w. dk, 4 in. ins, 4 in. h.w. RTS	d_n	1.00000	−1.99413	1.20218	−0.20898	0.01058	−0.00010	0.00000
30	Layers E0 E5 E4 B9 B6 E3 E2 A0	b_n	0.00000	0.00000	0.00003	0.00016	0.00018	0.00005	0.00000
	4 in. wood deck w/2 in. insul. and susp. ceiling	d_n	1.00000	−2.29665	1.86386	−0.65738	0.10295	−0.00631	0.00012
31	Layers E0 B27 C13 E3 E2 C13 A0	b_n	0.00000	0.00000	0.00003	0.00014	0.00014	0.00003	0.00000
	4.54 in. ins. w/6 in. h.w. deck and 6 in. h.w. RTS	d_n	1.00000	−2.29881	1.85733	−0.64691	0.10024	−0.00593	0.00006
32	Layers E0 E5 E4 C5 B20 E3 E2 C13 A0	b_n	0.00000	0.00002	0.00024	0.00037	0.00011	0.00001	0.00000
	Susp. clg, 4 in. h.w. dk, 0.76 in. ins, 4 in. h.w. RTS	d_n	1.00000	−2.09344	1.35118	−0.26478	0.01281	−0.00018	0.00000
33	Layers E0 E5 E4 C5 B13 E3 E2 C5 A0	b_n	0.00000	0.00000	0.00005	0.00013	0.00007	0.00001	0.00000
	Susp. clg, 4 in. h.w. dk, 4 in. ins, 4 in. h.w. RTS	d_n	1.00000	−2.07856	1.33963	−0.27670	0.02089	−0.00058	0.00000
34	Layers E0 E5 E4 C13 B23 E3 E2 C5 A0	b_n	0.00000	0.00000	0.00005	0.00013	0.00007	0.00001	0.00000
	Susp. clg, 6 in. h.w. dk, 2.42 in. ins, 4 in. h.w. RTS	d_n	1.00000	−2.13236	1.43448	−0.32023	0.02188	−0.00038	0.00000

Table 13 Roof Conduction Transfer Function Coefficients (*b* and *d* Factors) (*Concluded*)

Roof Group	(Layer Sequence Left to Right = Inside to Outside)		$n = 0$	$n = 1$	$n = 2$	$n = 3$	$n = 4$	$n = 5$	$n = 6$
35	Layers E0 C5 B15 E3 E2 C13 A0	b_n	0.00000	0.00000	0.00002	0.00010	0.00011	0.00003	0.00000
	4 in. h.w. deck w/6 in. ins. and 6 in. h.w. RTS	d_n	1.00000	−2.51234	2.25816	−0.87306	0.14066	−0.00785	0.00016
36	Layers E0 C13 B27 E3 E2 C13 A0	b_n	0.00000	0.00000	0.00002	0.00009	0.00011	0.00003	0.00000
	6 in. h.w. deck w/4.54 in. ins. and 6 in. h.w. RTS	d_n	1.00000	−2.50269	2.23944	−0.88012	0.15928	−0.01176	0.00018
37	Layers E0 E5 E4 B15 C13 E3 E2 C13 A0	b_n	0.00000	0.00000	0.00000	0.00002	0.00005	0.00004	0.00001
	Susp. clg, 6 in. ins, 6 in. h.w. dk, 6 in. h.w. RTS	d_n	1.00000	−2.75535	2.88190	−1.44618	0.36631	−0.04636	0.00269
38	Layers E0 E5 E4 B9 B15 E3 E2 A0	b_n	0.00000	0.00000	0.00000	0.00001	0.00003	0.00003	0.00001
	4 in. wood deck with 6 in. insul. and susp. ceiling	d_n	1.00000	−2.81433	3.05064	−1.62771	0.45499	−0.06569	0.00455
39	Layers E0 E5 E4 C13 B20 E3 E2 C13 A0	b_n	0.00000	0.00000	0.00007	0.00019	0.00011	0.00001	0.00000
	Susp. clg, 6 in. h.w. dk, 0.76 in. ins, 6 in. h.w. RTS	d_n	1.00000	−2.30711	1.77588	−0.52057	0.05597	−0.00118	0.00001
40	Layers E0 E5 E4 C5 B26 E3 E2 C13 A0	b_n	0.00000	0.00000	0.00002	0.00007	0.00006	0.00001	0.00000
	Susp. clg, 4 in. h.w. dk, 3.64 in. ins, 6 in. h.w. RTS	d_n	1.00000	−2.26975	1.68337	−0.45628	0.04712	−0.00180	0.00002
41	Layers E0 E5 E4 C13 B6 E3 E2 C13 A0	b_n	0.00000	0.00000	0.00002	0.00007	0.00006	0.00001	0.00000
	Susp. clg, 6 in. h.w. deck, 2 in. ins, 6 in. h.w. RTS	d_n	1.00000	−2.35843	1.86626	−0.56900	0.06466	−0.00157	0.00001
42	Layers E0 E5 E4 C13 B14 E3 E2 C13 A0	b_n	0.00000	0.00000	0.00000	0.00001	0.00002	0.00001	0.00000
	Susp. clg, 6 in. h.w. deck, 5 in. ins, 6 in. h.w. RTS	d_n	1.00000	−2.68628	2.63091	−1.16847	0.24692	−0.02269	0.00062

Table 14 Roof Conduction Transfer Function Coefficients Σc_n, Time Lag, U-Factors, and Decrement Factors

Roof Group		Σc_n	TL, h	U	DF
1	Layers E0 A3 B25 E3 E2 A0	0.05362	1.63	0.080	0.97
2	Layers E0 A3 B14 E3 E2 A0	0.02684	2.43	0.055	0.94
3	Layers EO E5 E4 C12 E3 E2 A0	0.05997	3.39	0.232	0.75
4	Layers E0 E1 B15 E4 B7 A0	0.00673	4.85	0.043	0.82
5	Layers E0 B14 C12 E3 E2 A0	0.00841	4.82	0.055	0.68
6	Layers E0 C5 B17 E3 E2 A0	0.05668	4.57	0.371	0.60
7	Layers E0 B22 C12 E3 E2 C12 A0	0.01652	5.00	0.138	0.56
8	Layers E0 B16 C13 E3 E2 A0	0.04340	5.45	0.424	0.47
9	Layers E0 E5 E4 B12 C14 E3 E2 A0	0.00550	6.32	0.057	0.60
10	Layers E0 E5 E4 C15 B16 E3 E2 A0	0.00647	7.14	0.104	0.49
11	Layers E0 C5 B15 E3 E2 A0	0.00232	7.39	0.046	0.43
12	Layers E0 C13 B16 E3 E2 C12 A0	0.01841	7.08	0.396	0.40
13	Layers E0 C13 B6 E3 E2 A0	0.00645	6.73	0.117	0.33
14	Layers E0 E5 E4 C12 B13 E3 E2 A0	0.00250	7.06	0.057	0.26
15	Layers E0 E5 E4 C5 B6 E3 E2 A0	0.01477	7.16	0.090	0.16
16	Layers E0 E5 E4 C13 B20 E3 E2 A0	0.00349	7.54	0.140	0.15
17	Layers E0 E5 E4 B15 C14 E3 E2 A0	0.00159	8.23	0.036	0.50
18	Layers E0 C12 B15 E3 E2 C5 A0	0.00101	9.21	0.046	0.41
19	Layers E0 C5 B27 E3 E2 C12 A0	0.00176	8.42	0.059	0.37
20	Layers E0 B21 C16 E3 E2 A0	0.00202	8.93	0.080	0.32
21	Layers E0 C13 B12 E3 E2 C12 A0	0.00174	8.93	0.083	0.26
22	Layers E0 B22 C5 E3 E2 C13 A0	0.00222	8.99	0.129	0.20
23	Layers E0 E5 E4 C12 B14 E3 E2 C12 A0	0.00064	9.26	0.047	0.16
24	Layers E0 E5 E4 C5 E3 E2 B6 B1 C12 A0	0.00100	8.84	0.082	0.12
25	Layers E0 E5 E4 C13 B13 E3 E2 A0	0.00063	8.77	0.056	0.09
26	Layers E0 E5 E4 B15 C15 E3 E2 A0	0.00053	10.44	0.034	0.30
27	Layers E0 C13 B15 E3 E2 C12 A0	0.00050	10.48	0.045	0.24
28	Layers E0 B9 B14 E3 E2 A0	0.00038	11.18	0.044	0.19
29	Layers E0 E5 E4 C12 B13 E3 E2 C5 A0	0.00053	10.57	0.056	0.16
30	Layers E0 E5 E4 B9 B6 E3 E2 A0	0.00042	11.22	0.064	0.13
31	Layers E0 B27 C13 E3 E2 C13 A0	0.00034	11.27	0.057	0.12
32	Layers E0 E5 E4 C5 B20 E3 E2 C13 A0	0.00075	11.31	0.133	0.10
33	Layers E0 E5 E4 C5 B13 E3 E2 C5 A0	0.00026	11.47	0.055	0.08
34	Layers E0 E5 E4 C13 B23 E3 E2 C5 A0	0.00026	11.63	0.077	0.06
35	Layers E0 C5 B15 E3 E2 C13 A0	0.00026	12.29	0.045	0.18
36	Layers E0 C13 B27 E3 E2 C13 A0	0.00025	12.67	0.057	0.13
37	Layers E0 E5 E4 B15 C13 E3 E2 C13 A0	0.00012	13.02	0.040	0.11
38	Layers E0 E5 E4 B9 B15 E3 E2 A0	0.00008	13.33	0.035	0.09
39	Layers E0 E5 E4 C13 B20 E3 E2 C13 A0	0.00039	12.23	0.131	0.07
40	Layers E0 E5 E4 C5 B26 E3 E2 C13 A0	0.00016	12.68	0.059	0.06
41	Layers E0 E5 E4 C13 B6 E3 E2 C13 A0	0.00016	12.85	0.085	0.05
42	Layers E0 E5 E4 C13 B14 E3 E2 C13 A0	0.00005	14.17	0.046	0.03

Table 15 Wall Group Numbers, Walls for Mass-In Case—Dominant Wall Material

Combined with Wall Material A1, E1, or Both

R	1	2	3	4	5	6	7	8	9	10	11	12	13	14	15	16	17	18	19	20	21	22	23	24	25
1	*	*	*	*	*	*	*	*	*	*	*	*	*	*	*	*	2	*	*	*	*	*	*	*	*
2	*	5	*	*	*	*	*	*	*	5	*	*	*	*	11	*	2	6	*	*	*	*	*	*	*
3	*	5	*	*	*	3	*	2	5	6	*	*	5	*	12	18	2	6	*	*	*	*	*	*	*
4	*	5	*	*	*	4	2	2	5	6	*	*	6	12	12	19	2	7	*	*	*	*	*	*	*
5	*	5	*	*	*	4	2	3	6	6	10	4	6	17	12	19	2	7	*	*	*	*	5	*	*
6	*	6	*	*	*	5	2	4	6	6	11	5	10	17	13	19	2	11	*	*	*	*	10	*	16
7	*	6	*	*	*	5	2	4	6	6	11	5	10	18	13	20	2	11	2	*	*	*	10	*	16
8	*	6	*	*	*	5	2	5	10	7	12	5	11	18	13	26	2	12	2	*	*	*	10	*	17
9	*	6	*	*	*	5	4	5	11	7	16	10	11	18	13	20	3	12	4	5	*	*	11	*	18
10	*	6	*	*	*	5	4	5	11	7	17	10	11	18	13	20	3	12	4	9	10	*	11	*	18
11	*	6	*	*	*	5	4	5	11	7	17	10	11	19	13	27	3	12	4	10	15	4	11	*	18
12	*	6	*	*	*	5	4	5	11	11	17	10	11	19	19	27	3	12	4	10	16	4	11	*	24
13	*	10	*	*	*	10	4	5	11	11	17	10	11	19	18	27	4	12	5	11	17	9	12	15	25
14	*	10	*	*	*	10	5	5	11	11	18	11	12	25	19	27	4	12	5	11	17	10	16	16	25
15	*	11	*	*	*	10	5	9	11	11	18	15	16	26	19	28	4	12	5	11	17	10	16	22	25
16	*	11	*	*	*	10	9	9	16	11	18	15	16	26	19	34	4	17	9	16	23	10	16	23	25
17	*	*	*	*	*	*	*	*	*	*	24	16	*	*	*	*	*	*	9	16	24	15	17	24	25

Combined with Wall Material A3 or A6

R	1	2	3	4	5	6	7	8	9	10	11	12	13	14	15	16	17	18	19	20	21	22	23	24	25
1	*	*	*	*	*	*	*	*	*	*	*	*	*	*	*	*	1	*	*	*	*	*	*	*	*
2	*	3	*	*	*	*	*	2	3	5	*	*	*	*	11	*	2	6	*	*	*	*	*	*	*
3	*	5	*	*	*	2	*	2	5	3	*	*	5	*	12	18	2	6	*	*	*	*	*	*	*
4	*	5	*	*	*	3	1	2	5	5	*	*	5	11	12	19	2	7	*	*	*	*	*	*	*
5	*	5	*	*	*	3	2	2	5	5	6	3	5	12	12	19	2	7	*	*	*	*	5	*	*
6	*	6	*	*	*	4	2	2	5	5	10	4	6	12	12	19	2	7	*	*	*	*	5	*	11
7	*	6	*	*	*	5	2	2	6	6	11	5	6	17	13	20	2	7	2	*	*	*	6	*	12
8	*	6	*	*	*	5	2	3	6	6	11	5	6	18	13	20	2	7	2	*	*	*	6	*	17
9	*	6	*	*	*	5	2	3	6	6	11	5	6	18	13	20	2	8	2	5	*	*	10	*	17
10	*	6	*	*	*	5	2	3	6	6	12	5	6	18	14	21	2	12	2	5	10	*	11	*	17
11	*	6	*	*	*	5	2	3	6	6	12	5	6	18	14	21	3	12	4	5	11	4	11	*	18
12	*	6	*	*	*	5	2	3	6	7	12	6	11	19	14	21	3	12	4	10	16	4	11	*	18
13	*	6	*	*	*	5	2	4	6	7	12	10	11	19	14	27	3	12	5	10	17	5	11	10	18
14	*	10	*	*	*	6	4	4	10	7	17	10	11	19	18	27	4	12	5	11	17	9	11	16	18
15	*	10	*	*	*	10	4	4	10	11	17	10	11	25	18	28	4	12	5	11	17	10	11	16	18
16	*	11	*	*	*	10	4	5	11	11	17	10	11	25	18	28	4	12	9	11	18	10	16	17	24
17	*	*	*	*	*	*	*	*	*	*	17	10	*	*	*	*	*	*	9	16	24	11	16	23	25

Combined with Wall Material A2 or A7

R	1	2	3	4	5	6	7	8	9	10	11	12	13	14	15	16	17	18	19	20	21	22	23	24	25
1	*	*	*	*	*	*	*	*	*	*	*	*	*	*	*	*	*	*	*	*	*	*	*	*	*
2	3	*	*	*	*	*	*	*	*	11	*	*	*	*	*	*	6	*	*	*	*	*	*	*	*
3	5	11	*	*	*	*	*	6	11	12	*	*	*	*	18	*	6	12	*	*	*	*	*	*	*
4	5	12	5	*	*	11	*	11	12	12	*	*	12	*	19	26	7	13	*	*	*	*	*	*	*
5	5	12	6	*	*	12	6	12	12	13	*	*	12	24	19	27	7	14	*	*	*	*	*	*	*
6	6	13	6	10	*	13	10	12	12	13	17	11	17	25	20	27	7	18	*	*	*	*	16	*	24
7	6	13	6	11	*	18	11	12	13	13	18	16	17	26	20	28	7	19	11	*	*	*	17	*	25
8	6	13	6	11	*	18	11	12	13	13	24	17	18	26	20	28	12	19	11	*	*	*	17	*	25
9	6	13	6	11	24	18	11	13	18	13	25	17	18	27	20	29	12	19	11	16	*	*	18	*	26
10	6	13	10	16	25	19	11	13	18	13	25	17	18	27	26	35	12	19	11	17	23	*	18	*	26
11	6	14	10	16	32	19	11	13	18	14	25	17	18	33	21	35	12	19	16	23	24	16	18	*	33
12	6	14	10	16	32	19	11	13	18	14	26	18	18	34	27	35	12	19	16	24	31	16	19	*	33
13	6	18	11	16	33	19	12	13	18	18	26	18	18	34	27	36	12	20	17	24	32	17	25	30	33
14	10	18	11	17	33	19	12	13	18	18	26	18	18	34	27	36	12	20	17	24	32	23	25	31	34
15	10	18	11	17	34	19	16	18	18	18	26	24	25	34	27	36	12	20	17	25	33	24	25	32	34
16	11	19	15	23	39	26	16	18	24	19	32	24	25	34	27	36	17	26	23	31	33	24	25	32	34
17	*	*	*	23	39	*	16	*	*	*	33	24	*	35	*	*	*	*	23	32	38	24	25	38	39

R-Value Ranges, h·ft^2·°F/Btu

R	Range
1	0.0 - 2.0
2	2.0 - 2.5
3	2.5 - 3.0
4	3.0 - 3.5
5	3.5 - 4.0
6	4.0 - 4.75
7	4.75 - 5.5
8	5.5 - 6.5
9	6.5 - 7.75
10	7.75 - 9.0
11	9.0 - 10.75
12	10.75 - 12.75
13	12.75 - 15.0
14	15.0 - 17.5
15	17.5 - 20.0
16	20.0 - 23.0
17	23.0 - 27.0

Wall Materials Layers (Table 11)

No.	Layer
1	A1,A3,A6, or E1
2	A2 or A7
3	B7
4	B10
5	B9
6	C1
7	C2
8	C3
9	C4
10	C5
11	C6
12	C7
13	C8
14	C9
15	C10
16	C11
17	C12
18	C13
19	C14
20	C15
21	C16
22	C17
23	C18
24	C19
25	C20

*Denotes a wall not possible with chosen combination of parameters.

Table 16 Walls for Integral Mass Case—Dominant Wall Material

Combined with Wall Material A1, E1, or Both

R	1	2	3	4	5	6	7	8	9	10	11	12	13	14	15	16	17	18	19	20	21	22	23	24	25
1	1	3	*	*	*	*	*	1	3	3	*	*	*	*	11	*	2	5	*	*	*	*	*	*	*
2	1	3	1	*	*	2	*	2	4	4	*	*	5	*	11	17	2	5	*	*	*	*	*	*	*
3	1	4	1	*	*	2	2	2	4	4	*	*	5	10	12	17	4	5	*	*	*	*	*	*	*
4	1	*	1	*	*	2	2	*	*	*	10	4	5	10	*	17	*	*	*	*	*	*	4	*	*
5	1	*	1	2	*	*	4	*	*	*	10	4	*	10	*	*	*	*	*	*	*	*	4	*	10
6	1	*	1	2	*	*	*	*	*	*	10	4	*	*	*	*	*	*	2	*	*	*	4	*	10
7	1	*	1	2	*	*	*	*	*	*	*	*	*	*	*	*	*	*	2	*	*	*	*	*	10
8	1	*	2	4	10	*	*	*	*	*	*	*	*	*	*	*	*	*	4	4	*	*	*	*	*
9	1	*	2	4	11	*	*	*	*	*	*	*	*	*	*	*	*	*	*	4	*	*	*	*	*
10	1	*	2	4	16	*	*	*	*	*	*	*	*	*	*	*	*	*	*	*	9	*	*	*	*
11	1	*	2	4	16	*	*	*	*	*	*	*	*	*	*	*	*	*	*	*	9	4	*	*	*
12	1	*	2	5	17	*	*	*	*	*	*	*	*	*	*	*	*	*	*	*	*	4	*	*	*
13	2	*	2	5	17	*	*	*	*	*	*	*	*	*	*	*	*	*	*	*	*	*	*	15	*
14	2	*	2	5	17	*	*	*	*	*	*	*	*	*	*	*	*	*	*	*	*	*	*	15	*
15	2	*	2	9	24	*	*	*	*	*	*	*	*	*	*	*	*	*	*	*	*	*	*	*	*
16	2	*	4	9	24	*	*	*	*	*	*	*	*	*	*	*	*	*	*	*	*	*	*	*	*
17	*	*	*	9	24	*	*	*	*	*	*	*	*	*	*	*	*	*	*	*	*	*	*	*	*

Combined with Wall Material A3 or A6

R	1	2	3	4	5	6	7	8	9	10	11	12	13	14	15	16	17	18	19	20	21	22	23	24	25
1	1	3	*	*	*	*	*	1	3	2	*	*	*	*	6	*	1	5	*	*	*	*	*	*	*
2	1	3	1	*	*	2	*	1	3	2	*	*	3	*	6	12	1	5	*	*	*	*	*	*	*
3	1	4	1	*	*	2	1	2	4	4	*	*	3	10	11	12	2	5	*	*	*	*	*	*	*
4	1	*	1	*	*	4	1	*	*	*	5	2	4	10	*	12	*	*	*	*	*	*	4	*	*
5	1	*	1	2	*	*	2	*	*	*	5	2	*	10	*	*	*	*	*	*	*	*	4	*	10
6	1	*	1	2	*	*	*	*	*	*	10	4	*	*	*	*	*	*	2	*	*	*	4	*	10
7	1	*	1	2	*	*	*	*	*	*	*	*	*	*	*	*	*	*	2	*	*	*	*	*	10
8	1	*	1	2	10	*	*	*	*	*	*	*	*	*	*	*	*	*	4	4	*	*	*	*	*
9	1	*	1	4	11	*	*	*	*	*	*	*	*	*	*	*	*	*	*	4	*	*	*	*	*
10	1	*	2	4	16	*	*	*	*	*	*	*	*	*	*	*	*	*	*	*	9	*	*	*	*
11	1	*	2	4	16	*	*	*	*	*	*	*	*	*	*	*	*	*	*	*	9	2	*	*	*
12	1	*	2	4	17	*	*	*	*	*	*	*	*	*	*	*	*	*	*	*	*	4	*	*	*
13	1	*	2	5	17	*	*	*	*	*	*	*	*	*	*	*	*	*	*	*	*	*	*	10	*
14	1	*	2	5	17	*	*	*	*	*	*	*	*	*	*	*	*	*	*	*	*	*	*	15	*
15	1	*	2	5	18	*	*	*	*	*	*	*	*	*	*	*	*	*	*	*	*	*	*	*	*
16	2	*	4	9	24	*	*	*	*	*	*	*	*	*	*	*	*	*	*	*	*	*	*	*	*
17	*	*	*	9	24	*	*	*	*	*	*	*	*	*	*	*	*	*	*	*	*	*	*	*	*

Combined with Wall Material A2 or A7

R	1	2	3	4	5	6	7	8	9	10	11	12	13	14	15	16	17	18	19	20	21	22	23	24	25
1	3	6	*	*	*	*	*	*	*	6	*	*	*	*	*	*	3	11	*	*	*	*	*	*	*
2	3	10	*	*	*	*	*	5	10	10	*	*	*	*	17	24	5	11	*	*	*	*	*	*	*
3	4	10	5	*	*	5	*	5	10	11	*	*	10	*	17	25	5	16	*	*	*	*	*	*	*
4	*	11	5	*	*	10	5	5	11	11	15	10	10	17	18	26	5	17	*	*	*	*	10	*	*
5	*	11	5	10	*	10	5	5	11	11	16	10	16	23	18	26	5	17	*	*	*	*	10	*	*
6	*	11	*	11	*	10	5	5	16	11	17	10	16	24	18	33	5	17	*	*	*	*	16	*	23
7	*	11	*	11	*	10	5	10	16	16	17	10	16	25	25	33	5	17	5	*	*	*	16	*	23
8	*	16	*	*	22	10	9	10	16	11	17	11	16	25	25	34	10	18	9	*	*	*	17	*	24
9	*	16	*	*	23	11	9	10	16	16	24	16	16	26	25	34	10	18	10	15	*	*	17	*	25
10	*	16	*	*	*	15	9	10	16	17	24	15	16	26	26	34	10	18	10	15	22	*	17	*	25
11	*	16	*	*	*	15	10	10	17	16	24	16	17	33	26	35	10	18	10	16	23	10	23	*	25
12	*	16	*	*	*	16	10	10	17	17	24	16	17	33	26	35	10	18	10	16	23	15	23	*	32
13	*	16	*	*	*	16	10	10	17	16	25	17	17	33	26	35	10	24	15	23	24	15	24	23	32
14	*	17	*	*	*	16	10	15	23	17	31	23	24	33	26	40	10	24	15	23	31	16	23	30	32
15	*	17	*	*	*	16	15	15	23	23	31	23	24	38	33	40	10	24	15	24	31	16	23	30	32
16	*	23	*	*	*	22	15	16	24	24	32	23	24	38	33	41	15	25	15	23	32	22	23	31	32
17	*	*	*	*	*	*	15	*	*	*	32	23	*	39	*	*	*	*	22	30	32	23	24	32	38

R-Value Ranges,	h·ft²·°F/Btu
1	0.0 - 2.0
2	2.0 - 2.5
3	2.5 - 3.0
4	3.0 - 3.5
5	3.5 - 4.0
6	4.0 - 4.75
7	4.75 - 5.5
8	5.5 - 6.5
9	6.5 - 7.75
10	7.75 - 9.0
11	9.0 - 10.75
12	10.75 - 12.75
13	12.75 - 15.0
14	15.0 - 17.5
15	17.5 - 20.0
16	20.0 - 23.0
17	23.0 - 27.0

Wall Materials	Layers (Table 11)
1	A1,A3,A6, or E1
2	A2 or A7
3	B7
4	B10
5	B9
6	C1
7	C2
8	C3
9	C4
10	C5
11	C6
12	C7
13	C8
14	C9
15	C10
16	C11
17	C12
18	C13
19	C14
20	C15
21	C16
22	C17
23	C18
24	C19
25	C20

*Denotes a wall not possible with chosen combination of parameters.

Table 17 Walls for Mass-Out Case—Dominant Wall Material

Combined with Wall Material A1, E1, or Both

R	1	2	3	4	5	6	7	8	9	10	11	12	13	14	15	16	17	18	19	20	21	22	23	24	25
1	*	*	*	*	*	*	*	*	*	*	*	*	*	*	*	*	1	*	*	*	*	*	*	*	*
2	*	3	*	*	*	*	*	2	3	5	*	*	*	*	6	*	1	5	*	*	*	*	*	*	*
3	*	3	*	*	*	2	*	2	4	5	*	*	5	*	11	18	2	5	*	*	*	*	*	*	*
4	*	3	*	*	*	2	2	2	5	5	*	*	5	16	11	18	2	5	*	*	*	*	*	*	*
5	*	3	*	*	*	2	2	2	5	5	10	4	6	17	11	19	2	6	*	*	*	*	5	*	*
6	*	4	*	*	*	4	2	2	5	5	10	4	6	17	11	19	2	6	*	*	*	*	9	*	16
7	*	4	*	*	*	4	2	2	5	6	11	5	10	17	12	19	2	6	2	*	*	*	10	*	16
8	*	5	*	*	*	4	2	2	5	6	11	5	10	18	11	20	2	6	4	*	*	*	10	*	16
9	*	5	*	*	*	4	2	2	5	6	11	5	10	18	11	26	2	6	4	9	*	*	10	*	17
10	*	5	*	*	*	5	2	4	5	6	16	10	10	18	12	26	2	6	4	9	15	*	10	*	17
11	*	5	*	*	*	5	4	4	5	6	16	10	10	18	12	26	2	6	4	10	15	4	11	*	18
12	*	5	*	*	*	5	4	4	10	6	16	10	10	18	12	26	2	10	5	10	16	9	11	*	18
13	*	5	*	*	*	5	4	4	10	10	17	10	11	18	12	26	2	10	5	11	17	9	11	15	24
14	*	5	*	*	*	5	4	4	10	10	17	10	11	24	18	26	2	10	9	15	23	10	16	16	24
15	*	5	*	*	*	9	4	4	10	10	17	10	15	25	18	26	2	10	9	15	23	10	16	22	24
16	*	9	*	*	*	9	9	9	15	10	17	10	15	25	18	33	4	11	9	16	24	15	16	23	24
17	*	*	*	*	*	*	*	*	*	*	23	15	*	*	*	*	*	*	9	22	24	15	16	24	25

Combined with Wall Material A3 or A6

R	1	2	3	4	5	6	7	8	9	10	11	12	13	14	15	16	17	18	19	20	21	22	23	24	25
1	*	*	*	*	*	*	*	*	*	*	*	*	*	*	*	*	1	*	*	*	*	*	*	*	*
2	*	3	*	*	*	*	*	2	3	2	*	*	*	*	6	*	1	5	*	*	*	*	*	*	*
3	*	3	*	*	*	2	*	2	3	2	*	*	*	*	10	17	1	5	*	*	*	*	*	*	*
4	*	3	*	*	*	2	1	2	4	3	*	*	4	11	11	17	1	5	*	*	*	*	*	*	*
5	*	3	*	*	*	2	2	2	4	3	5	2	5	11	11	18	1	6	*	*	*	*	4	*	*
6	*	3	*	*	*	2	2	2	4	3	10	3	5	12	11	18	2	6	*	*	*	*	5	*	10
7	*	3	*	*	*	2	2	2	5	3	10	4	5	12	11	18	2	6	2	*	*	*	5	*	11
8	*	4	*	*	*	2	2	2	5	3	10	4	5	12	11	18	2	6	2	*	*	*	5	*	12
9	*	4	*	*	*	2	2	2	5	4	11	5	5	17	11	18	2	6	2	5	*	*	6	*	16
10	*	5	*	*	*	2	2	2	5	4	11	5	5	17	11	19	2	6	2	5	10	*	6	*	17
11	*	5	*	*	*	2	2	2	5	4	11	5	5	17	12	19	2	6	4	5	11	4	10	*	17
12	*	5	*	*	*	4	2	2	5	5	11	5	5	17	12	19	2	6	4	10	15	4	10	*	17
13	*	5	*	*	*	4	2	2	5	5	11	5	10	18	12	19	2	10	4	10	16	5	10	10	17
14	*	5	*	*	*	4	2	4	5	5	16	9	10	18	12	25	2	10	4	10	17	9	10	16	17
15	*	5	*	*	*	4	4	4	9	5	16	9	10	18	16	25	2	10	5	11	17	10	10	16	18
16	*	9	*	*	*	4	4	4	9	9	16	10	10	24	17	25	4	10	5	11	17	10	11	17	18
17	*	*	*	*	*	*	*	*	*	*	16	10	*	*	*	*	*	*	9	16	23	10	15	23	24

Combined with Wall Material A2 or A7

R	1	2	3	4	5	6	7	8	9	10	11	12	13	14	15	16	17	18	19	20	21	22	23	24	25
1	*	*	*	*	*	*	*	*	*	*	*	*	*	*	*	*	*	*	*	*	*	*	*	*	*
2	3	*	*	*	*	*	*	*	*	11	*	*	*	*	*	*	5	*	*	*	*	*	*	*	*
3	3	10	*	*	*	*	*	5	10	11	*	*	*	*	17	*	5	12	*	*	*	*	*	*	*
4	3	11	5	*	*	10	*	5	11	11	*	*	11	*	18	26	6	12	*	*	*	*	*	*	*
5	3	11	5	*	*	10	5	6	11	11	*	*	11	24	18	26	6	13	*	*	*	*	*	*	*
6	3	11	5	10	*	10	5	10	11	11	17	10	11	24	18	26	6	13	*	*	*	*	16	*	23
7	3	12	5	10	*	10	9	10	11	12	17	11	16	25	19	27	6	17	9	*	*	*	16	*	23
8	4	12	5	10	*	10	10	10	12	12	17	15	16	25	19	27	6	17	10	*	*	*	16	*	24
9	4	12	5	10	23	11	10	10	12	12	23	16	17	26	19	27	10	18	10	15	*	*	16	*	25
10	5	12	5	15	24	11	10	10	16	12	24	16	17	26	19	34	10	18	10	16	22	*	17	*	25
11	5	12	9	15	30	11	10	10	16	12	24	16	17	26	19	34	10	18	10	16	23	15	17	*	25
12	5	12	10	15	31	11	10	10	17	12	24	16	17	26	25	34	10	18	10	22	24	15	17	*	32
13	5	17	10	16	32	11	10	11	17	17	24	16	17	26	25	34	11	18	15	23	30	15	23	23	32
14	5	17	10	16	32	15	10	11	17	17	25	16	17	33	25	34	11	18	15	23	31	22	23	30	32
15	5	17	10	16	32	16	15	15	17	17	25	22	23	33	26	35	11	18	15	23	31	22	23	30	32
16	9	17	15	16	32	16	15	15	23	17	31	22	23	33	26	40	15	24	15	23	32	23	24	31	32
17	*	*	*	22	38	*	15	*	*	*	31	23	*	33	*	*	*	*	22	30	37	23	24	37	38

R-Value Ranges,	h·ft²·°F/Btu
1	0.0 - 2.0
2	2.0 - 2.5
3	2.5 - 3.0
4	3.0 - 3.5
5	3.5 - 4.0
6	4.0 - 4.75
7	4.75 - 5.5
8	5.5 - 6.5
9	6.5 - 7.75
10	7.75 - 9.0
11	9.0 - 10.75
12	10.75 - 12.75
13	12.75 - 15.0
14	15.0 - 17.5
15	17.5 - 20.0
16	20.0 - 23.0
17	23.0 - 27.0

Wall Materials	Layers (Table 11)
1	A1,A3,A6, or E1
2	A2 or A7
3	B7
4	B10
5	B9
6	C1
7	C2
8	C3
9	C4
10	C5
11	C6
12	C7
13	C8
14	C9
15	C10
16	C11
17	C12
18	C13
19	C14
20	C15
21	C16
22	C17
23	C18
24	C19
25	C20

*Denotes a wall not possible with chosen combination of parameters.

Table 18 Wall Conduction Transfer Function Coefficients (b and d Factors)

Roof Group	(Layer Sequence Left to Right = Inside to Outside)		n =0	n =1	n =2	n =3	n =4	n =5	n =6
1	Layers E0 A3 B1 B13 A3 A0	b_n	0.00768	0.03498	0.00719	0.00006	0.00000	0.00000	0.00000
	Steel siding with 4 in. insulation	d_n	1.00000	−0.24072	0.00168	0.00000	0.00000	0.00000	0.00000
2	Layers E0 E1 B14 A1 A0 A0	b_n	0.00016	0.00545	0.00961	0.00215	0.00005	0.00000	0.00000
	Frame wall with 5 in. insulation	d_n	1.00000	−0.93389	0.27396	−0.02561	0.00014	0.00000	0.00000
3	Layers E0 C3 B5 A6 A0 A0	b_n	0.00411	0.03230	0.01474	0.00047	0.00000	0.00000	0.00000
	4 in. h.w. concrete block with 1 in. insulation	d_n	1.00000	−0.76963	0.04014	−0.00042	0.00000	0.00000	0.00000
4	Layers E0 E1 B6 C12 A0 A0	b_n	0.00001	0.00108	0.00384	0.00187	0.00013	0.00000	0.00000
	2 in. insulation with 2 in. h.w. concrete	d_n	1.00000	−1.37579	0.61544	−0.09389	0.00221	0.00000	0.00000
5	Layers E0 A6 B21 C7 A0 A0	b_n	0.00008	0.00444	0.01018	0.00296	0.00010	0.00000	0.00000
	1.36 in. insulation with 8 in. l.w. concrete block	d_n	1.00000	−1.16043	0.32547	−0.02746	0.00021	0.00000	0.00000
6	Layers E0 E1 B2 C5 A1 A0	b_n	0.00051	0.00938	0.01057	0.00127	0.00001	0.00000	0.00000
	1 in. insulation with 4 in. h.w. concrete	d_n	1.00000	−1.17580	0.30071	−0.01561	0.00001	0.00000	0.00000
7	Layers E0 A6 C5 B3 A3 A0	b_n	0.00099	0.00836	0.00361	0.00007	0.00000	0.00000	0.00000
	4 in. h.w. concrete with 2 in. insulation	d_n	1.00000	−0.93970	0.04664	0.00000	0.00000	0.00000	0.00000
8	Layers E0 A2 C12 B5 A6 A0	b_n	0.00014	0.00460	0.00733	0.00135	0.00002	0.00000	0.00000
	Face brick and 2 in. h.w. concrete with 1 in. insul.	d_n	1.00000	−1.20012	0.27937	−0.01039	0.00005	0.00000	0.00000
9	Layers E0 A6 B15 B10 A0 A0	b_n	0.00000	0.00006	0.00086	0.00146	0.00051	0.00004	0.00000
	6 in. insulation with 2 in. wood	d_n	1.00000	−1.63352	0.86971	−0.18121	0.01445	−0.00031	0.00000
10	Layers E0 E1 C2 B5 A2 A0	b_n	0.00001	0.00102	0.00441	0.00260	0.00024	0.00000	0.00000
	4 in. l.w. conc. block w/1 in. insul. and face brick	d_n	1.00000	−1.66358	0.82440	−0.11098	0.00351	0.00000	0.00000
11	Layers E0 E1 C8 B6 A1 A0	b_n	0.00000	0.00061	0.00289	0.00183	0.00018	0.00000	0.00000
	8 in. h.w. concrete block with 2 in. insulation	d_n	1.00000	−1.52480	0.67146	−0.09844	0.00239	0.00000	0.00000
12	Layers E0 E1 B1 C10 A1 A0	b_n	0.00002	0.00198	0.00816	0.00467	0.00044	0.00001	0.00000
	8 in. h.w. concrete	d_n	1.00000	−1.51658	0.64261	−0.08382	0.00289	−0.00001	0.00000
13	Layers E0 A2 C5 B19 A6 A0	b_n	0.00003	0.00203	0.00601	0.00233	0.00013	0.00000	0.00000
	Face brick and 4 in. h.w. concrete with 0.61 in. ins.	d_n	1.00000	−1.41349	0.48697	−0.03218	0.00057	0.00000	0.00000
14	Layers E0 A2 A2 B6 A6 A0	b_n	0.00000	0.00030	0.00167	0.00123	0.00016	0.00000	0.00000
	Face brick and face brick with 2 in. insulation	d_n	1.00000	−1.52986	0.62059	−0.06329	0.00196	−0.00001	0.00000
15	Layers E0 A6 C17 B1 A7 A0	b_n	0.00000	0.00003	0.00060	0.00145	0.00074	0.00009	0.00000
	8 in. l.w. concrete block (filled) and face brick	d_n	1.00000	−1.99996	1.36804	−0.37388	0.03885	−0.00140	0.00002
16	Layers E0 A6 C18 B1 A7 A0	b_n	0.00000	0.00014	0.00169	0.00270	0.00086	0.00006	0.00000
	8 in. h.w. concrete block (filled) and face brick	d_n	1.00000	−2.00258	1.32887	−0.32486	0.02361	−0.00052	0.00000
17	Layers E0 A2 C2 B15 A0 A0	b_n	0.00000	0.00000	0.00013	0.00044	0.00030	0.00005	0.00000
	Face brick and 4 in. l.w. conc. block with 6 in. ins.	d_n	1.00000	−2.00875	1.37120	−0.37897	0.03962	−0.00165	0.00002
18	Layers E0 A6 B25 C9 A0 A0	b_n	0.00000	0.00001	0.00026	0.00071	0.00040	0.00005	0.00000
	3.33 in. insulation with 8 in. common brick	d_n	1.00000	−1.92906	1.24412	−0.33029	0.03663	−0.00147	0.00002
19	Layers E0 C9 B6 A6 A0 A0	b_n	0.00000	0.00005	0.00064	0.00099	0.00030	0.00002	0.00000
	8 in. common brick with 2 in. insulation	d_n	1.00000	−1.78165	0.96017	−0.16904	0.00958	−0.00016	0.00000
20	Layers E0 C11 B19 A6 A0 A0	b_n	0.00000	0.00012	0.00119	0.00154	0.00038	0.00002	0.00000
	12 in. h.w. concrete with 0.61 in. insulation	d_n	1.00000	−1.86032	1.05927	−0.19508	0.01002	−0.00016	0.00000
21	Layers E0 C11 B6 A1 A0 A0	b_n	0.00000	0.00001	0.00019	0.00045	0.00022	0.00002	0.00000
	12 in. h.w. concrete with 2 in. insulation	d_n	1.00000	−2.12812	1.53974	−0.45512	0.05298	−0.00158	0.00002
22	Layers E0 C14 B15 A2 A0 A0	b_n	0.00000	0.00000	0.00006	0.00026	0.00025	0.00006	0.00000
	4 in. l.w. concrete with 6 in. insul. and face brick	d_n	1.00000	−2.28714	1.85457	−0.63564	0.08859	−0.00463	0.00009
23	Layers E0 E1 B15 C7 A2 A0	b_n	0.00000	0.00000	0.00002	0.00012	0.00019	0.00008	0.00001
	6 in. insulation with 8 in. l.w. concrete block	d_n	1.00000	−2.54231	2.43767	−1.10744	0.24599	−0.02510	0.00101
24	Layers E0 A6 C20 B1 A7 A0	b_n	0.00000	0.00000	0.00015	0.00066	0.00062	0.00015	0.00001
	12 in. h.w. concrete block (filled) and face brick	d_n	1.00000	−2.47997	2.22597	−0.87231	0.14275	−0.00850	0.00018
25	Layers E0 A2 C15 B12 A6 A0	b_n	0.00000	0.00000	0.00004	0.00019	0.00021	0.00006	0.00001
	Face brick and 6 in. l.w. conc. blk. w/3 in. insul.	d_n	1.00000	−2.28573	1.80756	−0.58999	0.08155	−0.00500	0.00013
26	Layers E0 A2 C6 B6 A6 A0	b_n	0.00000	0.00000	0.00010	0.00036	0.00027	0.00005	0.00000
	Face brick and 8 in. clay tile with 2 in. insulation	d_n	1.00000	−2.18780	1.60930	−0.46185	0.05051	−0.00218	0.00003
27	Layers E0 E1 B14 C11 A1 A0	b_n	0.00000	0.00000	0.00001	0.00006	0.00011	0.00005	0.00001
	5 in. insulation with 12 in. h.w. concrete	d_n	1.00000	−2.55944	2.45942	−1.12551	0.25621	−0.02721	0.00107
28	Layers E0 E1 C11 B13 A1 A0	b_n	0.00000	0.00000	0.00002	0.00010	0.00012	0.00004	0.00000
	12 in. h.w. concrete with 4 in. insulation	d_n	1.00000	−2.37671	2.04312	−0.79860	0.14868	−0.01231	0.00037
29	Layers E0 A2 C11 B5 A6 A0	b_n	0.00000	0.00000	0.00004	0.00021	0.00021	0.00006	0.00000
	Face brick and 12 in. h.w. concrete with 1 in. insul.	d_n	1.00000	−2.42903	2.08179	−0.75768	0.11461	−0.00674	0.00015
30	Layers E0 E1 B19 C19 A2 A0	b_n	0.00000	0.00000	0.00001	0.00006	0.00015	0.00010	0.00002
	0.61 in. ins. w/12 in. l.w. blk. (fld.) and face brick	d_n	1.00000	−2.83632	3.10377	−1.65731	0.45360	−0.06212	0.00393
31	Layers E0 E1 B15 C15 A2 A0	b_n	0.00000	0.00000	0.00000	0.00002	0.00007	0.00006	0.00002
	6 in. insul. with 6 in. l.w. conc. and face brick	d_n	1.00000	−2.90291	3.28970	−1.85454	0.55033	−0.08384	0.00599
32	Layers E0 E1 B23 B9 A2 A0	b_n	0.00000	0.00000	0.00000	0.00005	0.00011	0.00007	0.00001
	2.42 in. insulation with face brick	d_n	1.00000	−2.82266	3.04536	−1.58410	0.41423	−0.05186	0.00273
33	Layers E0 A2 C6 B15 A6 A0	b_n	0.00000	0.00000	0.00000	0.00002	0.00006	0.00005	0.00001
	Face brick and 8 in. clay tile with 6 in. insulation	d_n	1.00000	−2.68945	2.71279	−1.28873	0.30051	−0.03338	0.00175

Table 18 Wall Conduction Transfer Function Coefficients (*b* and *d* Factors) (*Concluded*)

Roof Group	(Layer Sequence Left to Right = Inside to Outside)		n =0	n =1	n =2	n =3	n =4	n =5	n =6
34	Layers E0 C11 B21 A2 A0 A0	b_n	0.00000	0.00000	0.00003	0.00015	0.00014	0.00003	0.00000
	12 in. h.w. conc. with 1.36 in. insul. and face brick	d_n	1.00000	−2.67076	2.58089	−1.07967	0.18237	−0.01057	0.00021
35	Layers E0 E1 B14 C11 A2 A0	b_n	0.00000	0.00000	0.00000	0.00001	0.00003	0.00003	0.00001
	5 in. insul. with 12 in. h.w. conc. and face brick	d_n	1.00000	−2.96850	3.45612	−2.02882	0.64302	−0.10884	0.00906
36	Layers E0 A2 C11 B25 A6 A0	b_n	0.00000	0.00000	0.00000	0.00004	0.00007	0.00004	0.00001
	Face brick and 12 in. h.w. conc. with 3.33 in. insul.	d_n	1.00000	−2.55127	2.36600	−0.99023	0.19505	−0.01814	0.00075
37	Layers E0 E1 B25 C19 A2 A0	b_n	0.00000	0.00000	0.00000	0.00001	0.00003	0.00003	0.00002
	3.33 in. ins. w/12 in. l.w. blk. (fld.) and face brick	d_n	1.00000	−3.17762	4.00458	−2.56328	0.89048	−0.16764	0.01638
38	Layers E0 E1 B15 C20 A2 A0	b_n	0.00000	0.00000	0.00000	0.00001	0.00002	0.00003	0.00001
	6 in. ins. w/12 in. h.w. block (fld.) and face brick	d_n	1.00000	−3.14989	3.95116	−2.53790	0.89438	−0.17209	0.01706
39	Layers E0 A2 C16 B14 A6 A0	b_n	0.00000	0.00000	0.00000	0.00001	0.00002	0.00003	0.00001
	Face brick and 8 in. l.w. concrete with 5 in. insul.	d_n	1.00000	−2.99386	3.45884	−1.95834	0.57704	−0.08844	0.00687
40	Layers E0 A2 C20 B15 A6 A0	b_n	0.00000	0.00000	0.00000	0.00001	0.00002	0.00003	0.00001
	Face brick, 12 in. h.w. block (fld.), 6 in. insul.	d_n	1.00000	−2.97582	3.42244	−1.93318	0.56765	−0.08568	0.00652
41	Layers E0 E1 C11 B14 A2 A0	b_n	0.00000	0.00000	0.00000	0.00001	0.00002	0.00002	0.00001
	12 in. h.w. conc. with 5 in. insul. and face brick	d_n	1.00000	−3.08296	3.66615	−2.11991	0.62142	−0.08917	0.00561

Table 19 Wall Conduction Transfer Function Coefficients Σc_n, Time Lag, U-Factors, and Decrement Factors

Roof Group		Σc_n	TL, h	*U*	DF
1	Layers E0 A3 B1 B13 A3 A0	0.04990	1.30	0.066	0.98
2	Layers E0 E1 B14 A1 A0 A0	0.01743	3.21	0.055	0.91
3	Layers E0 C3 B5 A6 A0 A0	0.05162	3.33	0.191	0.78
4	Layers E0 E1 B6 C12 A0 A0	0.00694	4.76	0.047	0.81
5	Layers E0 A6 B21 C7 A0 A0	0.01776	5.11	0.129	0.64
6	Layers E0 E1 B2 C5 A1 A0	0.02174	5.28	0.199	0.54
7	Layers E0 A6 C5 B3 A3 A0	0.01303	5.14	0.122	0.41
8	Layers E0 A2 C12 B5 A6 A0	0.01345	6.21	0.195	0.35
9	Layers E0 A6 B15 B10 A0 A0	0.00293	7.02	0.042	0.58
10	Layers E0 E1 C2 B5 A2 A0	0.00828	7.05	0.155	0.53
11	Layers E0 E1 C8 B6 A1 A0	0.00552	7.11	0.109	0.37
12	Layers E0 E1 B1 C10 A1 A0	0.01528	7.25	0.339	0.33
13	Layers E0 A2 C5 B19 A6 A0	0.01053	7.17	0.251	0.28
14	Layers E0 A2 A2 B6 A6 A0	0.00337	7.90	0.114	0.22
15	Layers E0 A6 C17 B1 A7 A0	0.00291	8.64	0.092	0.47
16	Layers E0 A6 C18 B1 A7 A0	0.00545	8.91	0.222	0.38
17	Layers E0 A2 C2 B15 A0 A0	0.00093	9.36	0.043	0.30
18	Layers E0 A6 B25 C9 A0 A0	0.00144	9.23	0.072	0.24
19	Layers E0 C9 B6 A6 A0 A0	0.00200	8.97	0.106	0.20
20	Layers E0 C11 B19 A6 A0 A0	0.00326	9.27	0.237	0.16
21	Layers E0 C11 B6 A1 A0 A0	0.00089	10.20	0.112	0.13
22	Layers E0 C14 B15 A2 A0 A0	0.00064	10.36	0.040	0.36
23	Layers E0 E1 B15 C7 A2 A0	0.00042	11.17	0.042	0.28
24	Layers E0 A6 C20 B1 A7 A0	0.00159	11.29	0.196	0.23
25	Layers E0 A2 C15 B12 A6 A0	0.00051	11.44	0.060	0.19
26	Layers E0 A2 C6 B6 A6 A0	0.00078	10.99	0.097	0.15
27	Layers E0 E1 B14 C11 A1 A0	0.00024	11.82	0.052	0.12
28	Layers E0 E1 C11 B13 A1 A0	0.00029	11.40	0.064	0.10
29	Layers E0 A2 C11 B5 A6 A0	0.00052	12.06	0.168	0.08
30	Layers E0 E1 B19 C19 A2 A0	0.00034	12.65	0.062	0.24
31	Layers E0 E1 B15 C15 A2 A0	0.00017	12.97	0.038	0.21
32	Layers E0 E1 B23 B9 A2 A0	0.00025	13.05	0.069	0.16
33	Layers E0 A2 C6 B15 A6 A0	0.00015	12.96	0.042	0.12
34	Layers E0 C11 B21 A2 A0 A0	0.00035	12.85	0.143	0.09
35	Layers E0 E1 B14 C11 A2 A0	0.00009	13.69	0.052	0.08
36	Layers E0 A2 C11 B25 A6 A0	0.00016	12.82	0.073	0.06
37	Layers E0 E1 B25 C19 A2 A0	0.00008	14.70	0.040	0.14
38	Layers E0 E1 B15 C20 A2 A0	0.00008	14.39	0.041	0.12
39	Layers E0 A2 C16 B14 A6 A0	0.00007	14.64	0.040	0.10
40	Layers E0 A2 C20 B15 A6 A0	0.00007	14.38	0.041	0.08
41	Layers E0 E1 C11 B14 A2 A0	0.00005	14.87	0.052	0.06

CTF Coefficients (Tables 11 and 15 through 19)

Outside surface resistance = A0
4 in. high density concrete = C5
2 in. insulation = B3
Air space resistance = B1
3/4 in. plaster = E1
Inside surface resistance = E0

The appropriate arrangement of layers in the wall can be found in Table 17. The dominant wall layer C5 is at the outside surface ("mass out"), and has a Wall Material column number of 10; combined with an E1 layer, this dictates use of the upper array of code numbers for wall assembly groups. Entering this array with an R-value range of 9 (R = 6.667), column 10 indicates that Wall Group 6 most nearly represents the wall under consideration.

The CTF coefficients of Wall Group 6 as listed in Table 18 are:

b_0 = 0.00051	d_0 = 1.0000
b_1 = 0.00938	d_1 = −1.17580
b_2 = 0.01057	d_2 = 0.30071
b_3 = 0.00127	d_3 = −0.01561
b_4 = 0.00001	d_4 = 0.00001
b_5 = 0.00000	d_5 = 0.00000
b_6 = 0.00000	d_6 = 0.00000

From Table 19, the U-factor of the wall is 0.199 and $\sum_{n=0} c_n = 0.02174$.

Heat Flow Calculations. The following format of Equation (25) demonstrates heat flow calculations through the wall:

$$q_{e,\theta}/A = \begin{bmatrix} b_0(t_{e,\theta}) \\ +b_1(t_{e,\theta-\delta}) \\ +b_2(t_{e,\theta-2\delta}) \\ \cdot \\ \cdot \\ \cdot \end{bmatrix} - \begin{bmatrix} d_1[(q_{e,\theta-\delta})/A] \\ +d_2[(q_{e,\theta-2\delta})/A] \\ \cdot \\ \cdot \\ \cdot \\ \cdot \end{bmatrix} - \left[t_{rc}\sum_{n=0} c_n\right]$$

This arrangement indicates that the heat gain through the wall is the sum of three parts:

1. Sum of the products of b coefficients and sol-air temperature values. The current value of this temperature is multiplied by b_0, the sol-air temperature of one step in time earlier is multiplied by b_1, etc.
2. Sum of the products of d coefficients and the previous values of heat gain. Note that the first d used is d_1. Again, the order of values is the same as in the first term, i.e., d_1 is multiplied by the heat gain value that was calculated for the previous step in time, d_2 is multiplied by the value calculated for two steps back in time, etc.
3. A constant, since room air temperature is constant and needs to be calculated only once.

The sequence of calculation using numerical values of this example are then as follows (starting at time θ = 1, expressing heat flux in Btu/(h·ft²), setting A = 1.0, and dropping b and d coefficients 4 through 6 as insignificant):

$$q_{e,1} = \begin{bmatrix} +0.00051(76) \\ +0.00938(77) \\ +0.01057(79) \\ +0.00127(81) \end{bmatrix} - \begin{bmatrix} -1.17580(0) \\ +0.30071(0) \\ -0.01561(0) \end{bmatrix} - [0.02174(75)]$$

$$= 0.068$$

$$q_{e,2} = \begin{bmatrix} +0.00051(76) \\ +0.00938(76) \\ +0.01057(77) \\ +0.00127(79) \end{bmatrix} - \begin{bmatrix} -1.17580(0.068) \\ +0.30071(0) \\ -0.01561(0) \end{bmatrix} - [+1.6305]$$

$$= 0.117$$

The values for q_e for this example are given in the summary table.

The convergence of the heat gain values to a periodic steady-state condition is indicated by comparing the average of the last 24 values with the average heat flow. The latter is given by the product of the U-factor and the difference between the average sol-air and room temperature. Thus

$$q_{avg} = 0.199(91.54 - 75.00) = 3.291 \text{ Btu/(h}\cdot\text{ft}^2)$$

The average of the last 24 values of heat gain tabulated in the summary table is given by:

$$q_{e,avg} = \left(\sum_{i=73}^{96} q_{e,i}\right)/24 = \frac{78.954}{24} = 3.29 \text{Btu/h}\cdot\text{ft}^2$$

Summary of Calculations for Example 3

n	$q_{e,n}$	n	$q_{e,n}$	n	$q_{e,n}$	n	$q_{e,n}$
1	0.068	25	3.688	49	3.744	73	3.745
2	0.117	26	3.174	50	3.221	74	3.221
3	0.139	27	2.712	51	2.751	75	2.751
4	0.141	28	2.303	52	2.335	76	2.336
5	0.117	29	1.933	53	1.961	77	1.961
6	0.077	30	1.603	54	1.626	78	1.626
7	0.048	31	1.329	55	1.348	79	1.349
8	0.065	32	1.141	56	1.157	80	1.157
9	0.156	33	1.060	57	1.073	81	1.073
10	0.333	34	1.092	58	1.103	82	1.103
11	0.599	35	1.237	59	1.246	83	1.246
12	0.948	36	1.483	60	1.491	84	1.491
13	1.372	37	1.821	61	1.828	85	1.828
14	1.945	38	2.322	62	2.328	86	2.328
15	2.746	39	3.063	63	3.068	87	3.068
16	3.731	40	3.997	64	4.002	88	4.002
17	4.773	41	4.997	65	5.000	89	5.000
18	5.702	42	5.890	66	5.892	90	5.892
19	6.320	43	6.478	67	6.480	91	6.480
20	6.388	44	6.520	68	6.522	92	6.522
21	5.974	45	6.085	69	6.087	93	6.087
22	5.401	46	5.495	70	5.496	94	5.496
23	4.810	47	4.888	71	4.889	95	4.889
24	4.236	48	4.302	72	4.303	96	4.303

Note: n is in hours and $q_{e,n}$ is in Btu/(h·ft²).

Heat Gain through Interior Partitions, Floors, and Ceilings

Whenever a conditioned space is adjacent to other spaces at different temperatures, the transfer of heat through the partition can be calculated by:

$$q_{e,\theta}/A = \sum_{n=0} b_n(t_{e,\theta-n\delta}) - \sum_{n=1} \frac{d_n(q_{e,\theta-n\delta})}{A} - t_{rc}\sum_{n=0} c_n \tag{26}$$

where

A = area, ft²
t_b = air temperature of adjacent space, °F
b, c, d = CTF coefficients derived from Tables 11 through 19, considering partitions as walls and floors or ceilings as roofs

Heat Gain from Adjacent Spaces. When t_b is constant or at least the variations of t_b are small compared to the difference ($t_b - t_{rc}$), $q_{p,\theta}$ is given by the simple steady-state expression

$$q_{p,\theta} = UA(t_b - t_{rc}) \tag{27}$$

where U = coefficient of overall heat transfer between the adjacent and the conditioned spaces (see Tables 14 or 19 or Chapter 24).

The same expression gives the mean values for $q_{p,\theta}$, when a mean value of t_b is used even though t_b varies. When $q_{p,\theta}$, is relatively small compared to the other room heat gain components, it may be considered constant at its mean value. If this component of heat gain is large, the temperature in the adjacent space should be calculated.

Note the common values $q_{p,\theta}$, A, t_b, and t_{rc} in Equations (26) and (30), illustrating the general functional equivalency of CTF coefficients b, c, and d in dynamic heat transfer over time to the steady-state heat transfer coefficient U, thus setting the rationale for adjustment of tabular CTF values by ratio of U_{actual}/U_{table}.

Conversion of Cooling Load from Heat Gain

The cooling load of a space depends on the magnitude and the nature of the sensible heat gain (i.e., heat conduction through walls, direct and diffuse solar radiation, energy input to lights, etc.) and on the location and mass of room objects that absorb the radiant heat. For example, the cooling load profile resulting from a unit pulse of solar radiation absorbed by window glass is quite different from that absorbed by a floor surface. Thus, each component of the room heat gain gives rise to a distinct component of cooling load, and the sum of these various components at any time is the total cooling load at that time.

Unlike other components, the latent heat gain component of the cooling load may or may not be part of room load depending on the type of air-conditioning system, i.e., ventilation air may be dehumidified at a central location rather than in each room.

Cooling Load by Room Transfer Function. Stephenson and Mitalas (1967), Mitalas and Stephenson (1967), and Kimura and Stephenson (1968) related heat gain to the corresponding cooling load by a room transfer function (RTF), which depends on the nature of the heat gain and on the heat storage characteristics of the space (i.e., of the walls, floor, etc., that enclose the space, and of the contents of that space). Where the heat gain q_θ is given at equal time intervals, the corresponding cooling load Q_θ at time θ can be related to the current value of q_θ and the preceding values of cooling load and heat gain by:

$$Q_\theta = \sum_{i=1}(v_o q_\theta + v_1 q_{\theta-d} + v_2 q_{\theta-2\delta} + \ldots) - (w_1 Q_{\theta-\delta} + w_2 Q_{\theta-2\delta} + \ldots) \tag{28}$$

where i is taken from 1 to the number of heat gain components and δ = time interval. The terms v_0, v_1 ..., w_1, w_2 ... are the coefficients of the RTF

$$K_{(z)} = \frac{v_0 + v_1 z^{-1} + v_2 z^{-2} + \ldots}{1 + w_1 z^{-1} + w_2 z^{-2} + \ldots} \tag{29}$$

which relates the transform of the corresponding parts of the cooling load and of the heat gain. These coefficients depend on (1) the size of the time interval δ between successive values of heat gain and cooling load, (2) the nature of the heat gain (how much is in the form of radiation and where it is absorbed), and (3) on the heat storage capacity of the room and its contents. Therefore, different RTFs are used to convert each distinct heat gain component to cooling load.

While the basic form of Equation (31) anticipates a series of v_n and w_n coefficients, the effect of past v_1 and w_1 is negligible, and data tabulated may generally be used with confidence. A slight inaccuracy does occur in the calculation for the first hour that internal loads begin; up through the second before the hour for which the calculation is made, such load does not exist, and the value generated by the transfer functions is not reached until the end of that hour. The convective component of such load is instantaneous, and the growth of the radiant component (combined with the convective element by the transfer coefficients) as it is absorbed and released by the building mass and contents is realistic throughout the rest of the load period.

Sensitivity of Parameters—Nontypical Applications

The concept of evaluating the thermal storage performance of a given space by means of RTF coefficients is based on the essential similarity of enclosing surfaces, spacial geometry, and related characteristics of that space to corresponding parameters of the space for which the data were calculated. ASHRAE research projects 359-RP [(Chiles and Sowell 1984), (Sowell and Chiles 1984a), (Sowell and Chiles 1984b)], 472-RP [(Harris and McQuiston 1988), (Sowell 1988a), (Sowell 1988b), (Sowell 1988c)], and 626-RP [(Falconer et al. 1993), (Spitler and McQuiston 1993), (Spitler et al. 1993)] investigated the unexpected sensitivity of such attributes and other counterintuitive phenomena regarding apparent responsiveness of relative masses in the storage and rejection of heat, and identified 14 discrete screening parameters with two to five levels of characterization each (Tables 20 through 23) by which to select representative data and to modify factors

Table 20 Zone Parametric Level Definitions

No.	Parameter	Meaning	Levels (in normal order)
1	ZG	Zone geometry	100 ft × 20 ft, 15 ft × 15 ft, 100 ft × 100 ft
2	ZH	Zone height	8 ft, 10 ft, 20 ft
3	NW	No. exterior walls	1, 2, 3, 4, 0
4	IS	Interior shade	100, 50, 0%
5	FN	Furniture	With, Without
6	EC	Exterior wall construction	1, 2, 3, 4 (Table 21)
7	PT	Partition type	5/8 in. gypsum board-air space 5/8 in. gypsum board, 8 in. concrete block
8	ZL	Zone location	Single-story, top floor, bottom floor, mid-floor
9	MF	Mid-floor type	8 in. concrete, 2.5 in. concrete, 1 in. wood
10	ST	Slab type	Mid-floor type, 4 in. slab on 12 in. soil
11	CT	Ceiling type	3/4 in. acoustic tile and air space, w/o ceiling
12	RT	Roof type	1, 2, 3, 4 (Table 23)
13	FC	Floor covering	Carpet with rubber pad, vinyl tile
14	GL	Glass percent	10, 50, 90

Table 21 Exterior Wall Construction Types

Type	Description
1	Outside surface resistance, 1 in. stucco, 1 in. insulation, 3/4 in. plaster or gypsum, inside surface resistance (A0, A1, B1, E1, E0)*
2	Outside surface resistance, 1 in. stucco, 8 in. HW concrete, 3/4 in. plaster or gypsum, inside surface resistance (A0, A1, C10, E1, E0)
3	Outside surface resistance, steel siding, 3 in. insulation, steel siding, inside surface (A0, A3, B12, A3, E0)*
4	Outside surface resistance, 4 in. face brick, 2 in. insulation. 12 in. HW concrete, 3/4 in. plaster or gypsum, inside surface resistance (A0, A2, B3, C11, E1, E0)*

Note: Code letters are defined in Table 11.

Table 22 Floor and Ceiling Types Specified by Zone Location Parameter

Zone Location	Floor	Ceiling
Single story	Slab-on-grade	Roof
Top floor	Mid-floor	Roof
Bottom floor	Slab-on-grade	Mid-floor
Mid-floor	Mid-floor	Mid-floor

Table 23 Roof Construction Types

Type	Description
1	Outside surface resistance, 1/2 in. slag or stone, 3/8 in. felt membrane, 1 in. insulation, steel siding, inside surface resistance (A0, E2, E3, B4, A3, E0)*
2	Outside surface resistance, 1/2 in. slag or stone, 3/8 in. felt membrane, 6 in. LW concrete, inside surface resistance (A0, E2, E3, C15, E0)*
3	Outside surface resistance, 1/2 in. slag or stone, 3/8 in. felt membrane, 2 in. insulation, steel siding, ceiling air space, acoustic tile, inside surface resistance (A0, E2, E3, B6, A3, E4, E5, E0)*
4	Outside surface resistance, 1/2 in. slag or stone, 3/8 in. felt membrane, 8 in. LW concrete, ceiling air space, acoustic tile, inside surface resistance (A0, E2, E3, C16, E4, E5, E0)*

Note: Code letters are defined in Table 11.

Table 24 Room Transfer Functions: Coefficient

Room Air Circulation[a] and S/R Type	Room Envelope Construction[b] 2-in. Wood Floor	3-in. Concrete Floor	6-in. Concrete Floor	8-in. Concrete Floor	12-in. Concrete Floor
	Specific Mass per Unit Floor Area, lb/ft²				
	10	40	75	120	160
Low	−0.88	−0.92	−0.95	−0.97	−0.98
Medium	−0.84	−0.90	−0.94	−0.96	−0.97
High	−0.81	−0.88	−0.93	−0.95	−0.97
Very High	−0.77	−0.85	−0.92	−0.95	−0.97
	−0.73	−0.83	−0.91	−0.94	−0.96

[a]Circulation rate—

Low: Minimum required to cope with cooling load from lights and occupants in interior zone. Supply through floor, wall, or ceiling diffuser. Ceiling space not used for return air, and h = 0.4 Btu/h·ft²·°F (where h = inside surface convection coefficient used in calculation of w_1 value).

Medium: Supply through floor, wall, or ceiling diffuser. Ceiling space not used for return air, and 0.6 Btu/h·ft²·°F.

High: Room air circulation induced by primary air of induction unit or by room fan and coil unit. Ceiling space used for return air, and 0.8 Btu/h·ft²·°F.

Very high: High room circulation used to minimize temperature gradients in a room. Ceiling space used for return air, and 0.8 Btu/h·ft²·°F.

[b]Floor covered with carpet and rubber pad; for a bare floor or if covered with floor tile, take next w_1 value down the column.

appropriately for specific applications. While these selection parameters are arranged so that errors due to deviations are minimal and conservative, careful use is required in situations differing significantly from one or more specific parameters.

Peak Heat Gain Versus Peak Cooling Load. The RTF procedure distributes all heat gained during a 24-h period throughout that period in the conversion to cooling load. Thus, individual heat gain components rarely appear at full value as part of the cooling load unless representing a constant 24-h input (such as a continuously burning light fixture), or in very low mass construction that releases stored radiant heat relatively quickly. This concept is further complicated by the premise of "constant interior space temperature" (i.e., operation of an HVAC system 24 h a day, seven days a week with fixed control settings), which practice is far less prevalent today than in the past. The effect of intermittent system operation is seen primarily during the first hours of operation for a subsequent day, as discussed in the section Heat Extraction Rate, and can impact equipment size selection significantly.

Superposition of Load Components. Finally, a presupposition of the TFM is that total cooling load for a space can be calculated by simple addition of the individual components. For example, radiation heat transfer from individual walls or roofs is assumed to be independent of the other surfaces, which is slightly incorrect in a theoretical sense. However, means for compensation for these

Table 25 Room Transfer Functions: v_0 and v_1 Coefficients

Heat Gain Component	Room Envelope Construction[b]	v_0 (Dimensionless)	v_1 (Dimensionless)
Solar heat gain through glass[c] with no interior shade; radiant heat from equipment and people	Light	0.224	$1 + w_1 - v_0$
	Medium	0.197	$1 + w_1 - v_0$
	Heavy	0.187	$1 + w_1 - v_0$
Conduction heat gain through exterior walls, roofs, partitions, doors, windows with blinds or drapes	Light	0.703	$1 + w_1 - v_0$
	Medium	0.681	$1 + w_1 - v_0$
	Heavy	0.676	$1 + w_1 - v_0$
Convective heat generated by equipment and people, and from ventilation and infiltration air	Light	1.000	0.0
	Medium	1.000	0.0
	Heavy	1.000	0.0

Heat Gain from Lights[d]

Furnishings	Air Supply and Return	Type of Light Fixture	v_0	v_1
Heavyweight simple furnishings, no carpet	Low rate; supply and return below ceiling ($V \le 0.5$)[e]	Recessed, not vented	0.450	$1 + w_1 - v_0$
Ordinary furnishings, no carpet	Medium to high rate, supply and return below or through ceiling ($V \ge 0.5$)[e]	Recessed, not vented	0.550	$1 + w_1 - v_0$
Ordinary furnishings, with or without carpet on floor	Medium to high rate, or induction unit or fan and coil, supply and return below, or through ceiling, return air plenum ($V \ge 0.5$)[e]	Vented	0.650	$1 + w_1 - v_0$
Any type of furniture, with or without carpet	Ducted returns through light fixtures	Vented or free-hanging in air-stream with ducted returns	0.750	$1 + w_1 - v_0$

[a]The transfer functions in this table were calculated by procedures outlined in Mitalas and Stephenson (1967) and are acceptable for cases where all heat gain energy eventually appears as cooling load. The computer program used was developed at the National Research Council of Canada, Division of Building Research.

[b]The construction designations denote the following:

Light construction: such as frame exterior wall, 2-in. concrete floor slab, approximately 30 lb of material per square foot of floor area.

Medium construction: such as 4-in. concrete exterior wall, 4-in. concrete floor slab, approximately 70 lb of building material per square foot of floor area.

Heavy construction: such as 6-in. concrete exterior wall, 6-in. concrete floor slab, approximately 130 lb of building material per square foot of floor area.

[c]The coefficients of the transfer function that relate room cooling load to solar heat gain through glass depend on where the solar energy is absorbed. If the window is shaded by an inside blind or curtain, most of the solar energy is absorbed by the shade, and is transferred to the room by convection and long-wave radiation in about the same proportion as the heat gain through walls and roofs; thus the same transfer coefficients apply.

[d]If room supply air is exhausted through the space above the ceiling and lights are recessed, such air removes some heat from the lights that would otherwise have entered the room. This removed light heat is still a load on the cooling plant if the air is recirculated, even though it is not a part of the room heat gain as such. The percent of heat gain appearing in the room depends on the type of lighting fixture, its mounting, and the exhaust airflow.

[e]V is room air supply rate in cfm/ft² of floor area.

limitations fall within the range of acceptable error that must be expected in any estimate of cooling load.

The previously mentioned research calculated RTF values for all possible combinations of screening parameter levels for a total of 200,640 individual cases. Access to these data is available electronically by techniques outlined in the *Cooling and Heating Load Calculation Manual* (McQuiston and Spitler 1992). A simplified method of RTF selection is presented in this chapter with RTF coefficients for various types and configurations of room construction and room air circulation rates given in Tables 24 and 25.

Use of Room Transfer Functions

To obtain appropriate room transfer function data for use in Equation (28), (1) select the value of w_1 from Table 24 for the approximate space envelope construction and range of air circulation, and (2) select and/or calculate the values of v_0 and v_1 from Table 25 for the appropriate heat gain component and range of space construction mass.

Example 4. Cooling load due to solar radiation through glass. Consider a room having a 1/2 in. air space double-glazed window (shading coefficient = 0.83) in a multistory office building of heavyweight construction (approximately 120 lb/ft² floor area). The building is located at 40°N latitude, the date is June 21, and the window orientation is NW. The U-factor for the window is 0.56 Btu/(h·ft²·°F). Assume the floor to be carpeted, the air circulation rate "medium" (h_i = 0.6 Btu/(h·ft²·°F), and the ceiling space not ventilated. Calculate the cooling load due to solar radiation through glass. Solar heat gain (SHG) to the room through the window is given as SHG = SHGF × Shading Coefficient = SHGF × 0.83.

Time, h	SHGF	SHG	Time, h	SHGF	SHG
0100	0	0	1300	40	33
0200	0	0	1400	63	52
0300	0	0	1500	114	95
0400	0	0	1600	156	129
0500	1	1	1700	172	143
0600	13	11	1800	143	119
0700	21	17	1900	21	17
0800	27	22	2000	0	0
0900	32	27	2100	0	0
1000	35	29	2200	0	0
1100	38	32	2300	0	0
1200	38	32	2400	0	0
Daily total				917	753

Note: SHGF from Table 18, Chapter 29. Units are Btu/h·ft²

Solution: The room transfer function coefficients for 120 lb/ft² construction, solar radiation input, medium air circulation rate, and the condition of "no heat loss for the room" are (see Tables 24 and 25):

$$v_0 = 0.187 \qquad w_0 = 1.000 \text{ (in all cases)}$$
$$v_1 = -0.147 \qquad w_1 = -0.960$$

The cooling load component due to solar radiation through glass at any time θ is given by Equation (28). The calculations can be set up as follows:

$$Q_\theta = \begin{bmatrix} v_0(\text{SHG}_\theta) \\ + v_1(\text{SHG}_{\theta-\delta}) \\ - w_1(Q_{\theta-\delta}) \end{bmatrix}$$

As in the earlier heat gain calculation example, the calculation is started by assuming that the previous *Q*s are zero. Furthermore, in this example, SHG = 0 for θ = 1, 2, 3, and 4; therefore, *Q*s in Btu/(h·ft²) are:

Q_5	*v*, *w*	Hour	SHG	Prev. *Q*s	Factor
	0.187	5	1		0.187
	−0.147	4	0		0.000
	0.96	4		0	0.000
				Q_5 =	0.187

Q_6	*v*, *w*	Hour	SHG	Prev. *Q*s	Factor
	0.187	6	11		0.187
	−0.147	5	1		−0.147
	0.96	5		0.187	0.180
				Q_6 =	2.090

Values of Q_θ for the remainder of the calculations are listed in the following table. The calculations of Q_θ are terminated at θ = 96 h, because by that time, the effect of the assumed zero initial conditions has decreased to negligible proportions.

Values of Q_θ for Example 4

θ	Q_θ	θ	Q_θ	θ	Q_θ	θ	Q_θ
1 =	0.000	25 =	17.041	49 =	23.440	73 =	25.842
2 =	0.000	26 =	16.359	50 =	22.502	74 =	24.808
3 =	0.000	27 =	15.705	51 =	21.602	75 =	23.816
4 =	0.000	28 =	15.077	52 =	20.738	76 =	22.863
5 =	0.187	29 =	14.661	53 =	20.095	77 =	22.135
6 =	2.090	30 =	15.985	54 =	21.201	78 =	23.160
7 =	3.568	31 =	16.908	55 =	21.915	79 =	23.796
8 =	5.040	32 =	17.847	56 =	22.653	80 =	24.459
9 =	6.653	33 =	18.948	57 =	23.562	81 =	25.296
10 =	7.841	34 =	19.644	58 =	24.074	82 =	25.738
11 =	9.248	35 =	20.579	59 =	24.832	83 =	26.429
12 =	10.158	36 =	21.036	60 =	25.119	84 =	26.652
13 =	11.219	37 =	21.662	61 =	25.581	85 =	27.053
14 =	15.643	38 =	25.669	62 =	29.431	86 =	30.844
15 =	25.138	39 =	34.763	63 =	38.375	87 =	39.731
16 =	34.290	40 =	43.530	64 =	46.998	88 =	48.300
17 =	40.696	41 =	49.567	65 =	52.896	89 =	54.146
18 =	40.300	42 =	48.816	66 =	52.012	90 =	53.212
19 =	24.374	43 =	32.549	67 =	35.618	91 =	36.770
20 =	20.900	44 =	28.748	68 =	31.694	92 =	32.800
21 =	20.064	45 =	27.598	69 =	30.426	93 =	31.488
22 =	19.261	46 =	26.494	70 =	29.209	94 =	30.228
23 =	18.491	47 =	25.434	71 =	28.041	95 =	29.019
24 =	17.751	48 =	24.417	72 =	26.919	96 =	27.858

Note: Values carried to 3 decimals to illustrate degree of convergence.

Cooling Load from Nonradiant Heat Gain

Sensible cooling load from strictly convective heat gain elements is instantaneous, added directly to the results of those gains processed by CTF and RTF coefficients, per the following equation.

$$Q_{sc} = \sum_{j=1} (q_{c,j}) \tag{30}$$

where

Q_{sc} = sensible cooling load from heat gain elements having only convective components

q_c = each of *j* heat gain elements having only such convective component

Heat Extraction Rate and Room Temperature

Discussion to this point has concentrated on estimating design cooling load for a conditioned space, assuming the maintenance of a constant interior temperature and the hourly total removal of all cooling load entering the space; and allowing the delaying action of building mass and contents to run its course. Certain minor factors have been ignored, such as the relatively indeterminate radiant heat loss to the outside of the building.

The basic principles of the TFM are also useful in estimating dynamic cooling load requirements over an extended period (see Chapter 30). In such cases, however, the goal is no longer to seek the peak load for equipment selection purposes, and the ebb and flow of heat into and out of the building assume much greater importance; thus, any loss back to the environment must be considered. This concept is also critical in predicting temperature swings in the space and the ability of cooling equipment to extract heat when operated in a building with extended off cycles (nights and weekends).

The cooling loads determined by the TFM serve as input data for estimating the resultant room air temperature and the heat extraction rate with a particular type and size of cooling unit, or set of operating conditions, or both. In addition, the characteristics of the cooling unit (i.e., heat extraction rate versus room air temperature), the schedule of operation, and a space air transfer function (SATF) for the room that relates room air temperature and heat extraction rate must also be included to run these calculations.

The heat extraction characteristics of the cooling unit can be approximated by a linear expression of the form

$$ER_\theta = W_\theta + St_{r\theta} \tag{31}$$

where

ER_θ = rate of heat removal from space at time θ
$t_{r\theta}$ = the air temperature in space at time θ
W, S = parameters characterizing performance of specific types of cooling equipment

This linear relationship only holds when $t_{r\theta}$ is within the throttling range of the control system. When $t_{r\theta}$ lies outside of this range, ER_θ has the value of either ER_{max} or ER_{min}, depending on whether the temperature $t_{r\theta}$ is above or below the throttling range. The value of S is the difference $ER_{max} - ER_{min}$ divided by the width of the throttling range, and W_θ is the value ER_θ would have if the straight-line relationship between it and $t_{r\theta}$ held at t_{r1} equals zero. This intercept depends on the set point temperature of the control system, which may be taken as the temperature at the middle of the throttling range. Thus,

$$W_\theta = \frac{ER_{max} + ER_{min}}{2} - St^*_{r\theta} \tag{32}$$

where $St^*_{r\theta}$ is the thermostat set point temperature at time θ.

Space Air Transfer Function

The heat extraction rate and the room air temperature are related by the space air transfer function (SATF):

$$\sum_{i=0}^{1} p_i(ER_{\theta-\delta} - Q_{\theta-i\delta}) = \sum_{i=0}^{2} g_i(t_{rc} - t_{r,\theta-i\delta}) \tag{33}$$

where g_1 and p_1 are the SATF coefficients, and Q is the calculated cooling load for the room at time θ, based on an assumed constant room temperature of t_{rc}. Normalized values of g and p are given in Table 26 for light, medium, and heavy construction.

Thermal Conductance to Surroundings. In calculating the design cooling load components previously described, it was assumed that all energy transferred into the space eventually appears as space cooling load. However, this is not quite true over an extended period, because a fraction of the input energy can instead be lost back to the surroundings. This fraction F_c depends on the thermal conductance between the space air and the surroundings and can be estimated as

$$F_c = 1 - 0.02K_\theta \tag{34}$$

Table 26 Normalized Coefficients of Space Air Transfer Functions[a]

Room Envelope Construction	g_0^*	g_1^*	g_2^*	p_0	p_1
	Btu/h·ft·°F			Dimensionless	
Light	1.68	−1.73	0.05	1.0	−0.82
Medium	1.81	−1.89	0.08	1.0	−0.87
Heavy	1.85	−1.95	0.10	1.0	−0.93

[a]For simplified procedure for calculating space air transfer function coefficients, see ASHRAE (1975).

[b]The designations Light, Medium, and Heavy denote the same meanings as those footnoted for Table 25.

where K_θ is the unit length conductance between the space air and surroundings given by

$$K_\theta = (1/L_F)(U_R A_R + U_W A_W + U_{OW} A_{OW} + U_P A_P) \tag{35}$$

where

L_F = length of space exterior wall, ft
U = U-factor of space enclosure element (subscript R for roof, W for window, OW for outside wall, and P for partition, should such be adjacent to an unconditioned area), Btu/(h·ft²·°F)
A = area of space enclosure element, ft²

The units of K_θ are Btu/(h·ft·°F). Therefore, if F_c is to be dimensionless, the multiplier is 0.02 h·ft·°F/Btu.

Adjustment of Load Components. To adjust the space cooling loads calculated in the previous sections, multiply the value of the following components by the factor F_c from Equation (34):

- Sensible cooling load from heat gain by conduction through exterior roofs and walls
- Sensible cooling load from conduction and solar heat gain through fenestration areas
- Sensible cooling load from heat gain through interior partitions, ceilings, and floors
- Sensible cooling load from radiant portion of heat gain from lights, people, and equipment

Adjustments to g* Coefficients. To obtain the SATF coefficients for Equation (36), first select the values of p_0, p_1, g_0^*, g_1^*, and g_2^* from Table 26 for the appropriate space envelope construction. Since the * coefficients in Table 26 are for a space with zero heat conductance to surrounding spaces and are normalized to a unit floor area, it is necessary to adjust the 0 and 1 values. To get the g_0 and g_1 coefficients for a space with a floor area A, total conductance K_θ [by Equation (35)] between space air surroundings, ventilation rate, and infiltration rate, the relationships are:

$$g_{0,\theta} = g_0^* A + p_0[K_\theta + 1.10(V_\theta + VI_\theta)] \tag{36}$$

$$g_{1\theta} = g_1^* A + p_1[K_\theta + 1.10(V_{\theta-1\delta} + VI_{\theta-1\delta})] \tag{37}$$

Note that Equation (37) has no second term when calculating $g_{2,\theta}$, since p_2 has no value.

Heat Extraction Rate. For either condition (heat loss to surroundings or not, and using the appropriate values of g), Equations (31) and (32) can be solved simultaneously for ER_θ

$$ER_\theta = \frac{W_\theta g_0}{S + g_0} + \frac{I_\theta S}{S + g_0} \tag{38}$$

where

$$I_\theta = t_{rc}\sum_{i=0}^{2} g_{i,\theta} - \sum_{i=1}^{2} g_{i,\theta}(t_{r,\theta-i\delta}) + \sum_{i=0}^{1} p_i(Q_{\theta-i\delta}) - \sum_{i=1}^{1} p_i(ER_{\theta-i\delta}) \tag{39}$$

If the value of ER_θ calculated by Equation (38) is greater than ER_{max}, it is made equal to ER_{max}; if it is less than ER_{min}, it is made equal to ER_{min}. Then $t_{r\theta}$ is calculated from the expression

$$t_{r\theta} = (1/g_{0,\theta})(I_\theta - ER_\theta) \tag{40}$$

Example 5. Calculation of room air temperature and heat extraction rate. A room is of heavy construction with a floor area of 400 ft². The total room cooling load calculated on the basis of t_{rc} = 70°F is given as:

θ, h	$Q_θ$, Btu/h	θ, h	$Q_θ$, Btu/h	θ, h	$Q_θ$, Btu/h
1	2200	9	2180	17	7630
2	2030	10	2330	18	6880
3	1850	11	2650	19	5530
4	1730	12	3580	20	4380
5	1680	13	4880	21	3630
6	1750	14	6180	22	3130
7	1880	15	7150	23	2730
8	2030	16	7680	24	2450

The cooling unit has a maximum heat extraction capability of 7500 Btu/h and a minimum of zero. The throttling range is 3°F wide. Assume no ventilation and no infiltration, and heat loss to the exterior surroundings at the rate of 100 Btu/(h·°F). Calculate room air temperature and heat extraction rate for:

Schedule A. The control thermostat is set at 77°F from 0700 to 1800 h; during the rest of the time, it is set up to 85°F.

Schedule B. The control thermostat is set at 77°F all the time.

Solution:

(a) Space Air Transfer Functions.

The SATF coefficients for a 400 ft² room of heavy construction are [from Table 26 and Equations (36) and (37) with V and V_l dropping out]:

$$g_{0,θ} \quad 400(+1.85) + 100(+1.0) = 840.00$$
$$g_{1,θ} \quad 400(-1.95) + 100(-0.93) = -873.00$$
$$g_{2,θ} \quad 400(+0.10) = 40.00$$
$$\sum_{i=0}^{2} g_i = 7.00$$

(b) Cooling Unit Characteristics.

$$ER_{max} = 7500 \text{ Btu/h}$$
$$ER_{min} = 0$$
$$t_{tr} = 3°\text{F throttling range}$$
$$S = (7500-0)/3 = 2500 \text{ Btu/(h}\cdot°\text{F)}$$

when $t^*_{rθ}$ = 77.0°F,

$$W_θ = [(7500-0)/2] - 2500(77.0) = -188{,}750 \text{ Btu/h}$$

and when $t^*_{rθ}$ = 85.0°F,

$$W_θ = [(7500-0)/2] - 2500(85.0) = -208{,}750 \text{ Btu/h}$$

(c) Calculation of $ER_θ$ and $t_{rθ}$.

Some prior values for $ER_θ$ and $t_{rθ}$ must be assumed to begin the computation process. The computation is repeated until the results for successive days are the same. At that time, the results are independent of the values assumed initially.

To get the calculation started, assume all previous values of ER = 0 and t_r = 80°F. Thus:

$$I_1 = 70.0(7.0) - \begin{bmatrix} -873(80) \\ +40(80) \end{bmatrix} + \begin{bmatrix} +1.0(2200) \\ -0.93(2450) \end{bmatrix} - [-0.93(0.0)]$$
$$= 67{,}052 \text{ Btu/h}$$
$$ER_1 = \frac{-208{,}750 \times 840}{2500+840} + \frac{67{,}052 \times 2500}{2500+840}$$
$$= -52{,}500 + 50{,}189 = -2311 \text{ Btu/h}$$

As this is less than ER_{min}, $ER_1 = ER_{min} = 0$ and t_{r1} = (1/840)(67,052 − 0.0) = 79.8°F

$$I_2 = 70.0(7.0) - \begin{bmatrix} -873(79.8) \\ +40(80) \end{bmatrix} + \begin{bmatrix} +1.0(2030) \\ -0.93(2200) \end{bmatrix} - [-0.93(0.0)]$$
$$= 66{,}939 \text{ Btu/h}$$
$$ER_2 = \frac{-208{,}750 \times 840}{2500+840} + \frac{66{,}939 \times 2500}{2500+840}$$
$$= -52{,}500 + 50{,}104 = -2396 \text{ Btu/h}$$

As this also is less than ER_{min}, $Er_2 = ER_{min}$ = 0 and t_{r2} = (1/840) (66,939 − 0) =79.7°F, and so on.

The effect of the assumed initial $ER_θ$ and $t_{rθ}$ values has decreased to negligible proportions by the time θ = 145, i.e., $t_{r145} = t_{r169}$ = 82.6°F. The complete set of results for operating schedules A and B is given in Table 27.

Table 27 Room Air Temperature and Heat Extraction Rates for Example 6

	Schedule A (Control thermostat set at 77°F from 0800 to 1800, and at 85°F at all other times)		Schedule B (Control thermostat set at 77°F at all times)	
Time, h	Room Air Temperature t_r, °F	Heat Extraction ER, Btu/h	Room Air Temperature t_r, °F	Heat Extraction ER, Btu/h
0100	82.0	0	76.3	1956
0200	81.9	0	76.2	1806
0300	81.8	0	76.2	1649
0400	81.7	0	76.1	1535
0500	81.7	0	76.1	1474
0600	81.8	0	76.1	1505
0700	82.0	0	76.1	1584
0800	77.2	4235	76.2	1680
0900	77.1	4051	76.2	1780
1000	77.1	4025	76.3	1881
1100	77.2	4138	76.3	2113
1200	77.4	4720	76.6	2807
1300	77.7	5602	77.0	3794
1400	78.1	6507	77.4	4799
1500	78.4	7184	77.7	5571
1600	78.6	7500	77.9	6019
1700	78.5	7477	77.9	6032
1800	78.3	6876	77.7	5513
1900	83.8	786	77.3	4528
2000	83.6	268	77.0	3669
2100	83.3	0	76.7	3099
2200	82.8	0	76.6	2710
2300	82.5	0	76.5	2393
2400	82.3	0	76.4	2164
	Totals	63369		72061

EXAMPLE COOLING LOAD CALCULATION

Example 6. Cooling load calculation of small office building. A one-story small commercial building (Figure 4) is located in the eastern United States near 40°N latitude. The adjoining buildings on the north and west are not conditioned, and the air temperature within them is approximately equal to the outdoor air temperature at any time of day.

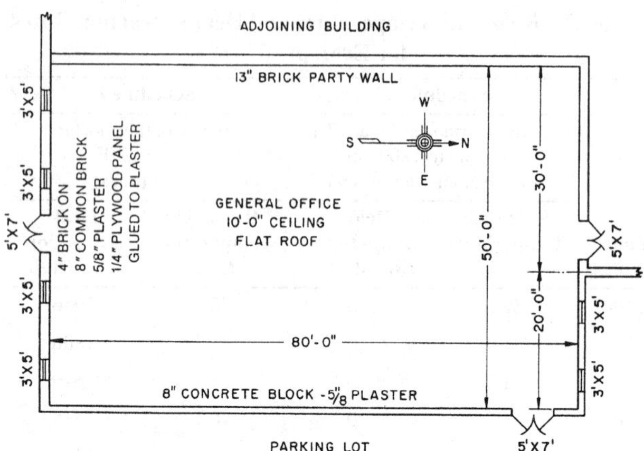

Note: The small commercial building shown in this figure has been in the ASHRAE literature for several decades to illustrate cooling load procedures. In this example, some materials have been updated to reflect currently available products and associated U-factors; the calculation month has been changed to July for better comparison with newer data. Otherwise, all other characteristics of this example remain unchanged.

Fig. 4 Plan of One-Story Office Building

Building Data:

South wall construction. 4-in. light-colored face brick, 8-in. common brick, 0.625-in. plaster, 0.25-in. plywood panel glued on plaster (Summer U = 0.24 Btu/h·ft²·°F, or R = 4.14).

East wall and outside north wall construction. 8-in. light-colored heavy concrete block, 5/8-in. plaster on walls (Summer U = 0.48 Btu/h·ft²·°F, or R = 2.083).

West wall and adjoining north party wall construction. 13-in. solid brick (color n/a), no plaster: with U for a 12-in. brick interior wall = 0.26, R for that wall = 1/0.26 = 3.84; subtracting two still air film coefficients with R_{fc} = 0.68 each leaves R_b = 2.486; thus for this wall:

$$R_w = 0.68 + (2.486 \times 13/12) + 0.68 = 4.053$$

and $$U_w = 1/4.053 = 0.247,\ \text{say}\ U = 0.25\text{Btu}/(\text{h} \cdot \text{ft}^2 \cdot {}^\circ\text{F})$$

Roof construction. 4-1/2-in. (nominal) flat roof of 2-in. gypsum slab on metal roof deck, 2-in. rigid roof insulation, surfaced with two layers of mopped 15-lb felt vapor-seal built-up roofing having dark-colored gravel surface, and with no false ceiling below underside of roof deck; (Summer U = 0.09 Btu/h·ft²·°F, or R = 11.11).

Floor construction. 4-in. concrete on ground.

Fenestration. 3 ft by 5 ftnonoperable windows of regular plate glass with light colored venetian blinds [Summer U = 0.81 Btu/h·ft²·°F].

Door construction. Light-colored 1.75-in. steel door with solid urethane core and thermal break (Summer U = 0.19 Btu/h·ft²·°F or R = 5.26 for exterior doors, and U = 0.18 Btu/h·ft²·°F or R = 5.56 for interior doors).

Front doors. Two 30 in. by 7 ft

Side doors. Two 30 in. by 7 ft

Rear doors. Two 30 in. by 7 ft (interior)

Note: U-factors for all exterior surfaces assume a summer wind velocity of 7.5 mph. Those for party walls and other interior surfaces assume still air.

Summer outdoor design conditions. Dry bulb = 94°F, daily range = 20°F, wet bulb = 77°F, W_o = 0.0161 lb (water)/lb (dry air)0.0159 kg (water)/kg (dry air), h_o = 40.3 Btu/lb (dry air)

Winter outdoor design conditions. Dry bulb = 10°F

Summer indoor design conditions. Dry bulb = 75°F, wet bulb = 62.5°F, W_i = 0.0092 lb (water)/lb (dry air), h_i = 28.07 Btu/lb (dry air)

Winter indoor design conditions. Dry bulb = 75°F

Occupancy. 85 office workers from 0800 to 1700 h

Lights. 17,500 W, fluorescent, operating from 0800 to 1700 hours daily; along with 4000 W, tungsten, operated continuously. Lighting fixtures are non-ventilated type.

Power equipment and appliances. For this example, none are assumed.

Ventilation. A ventilation rate of 15 cfm/person is selected as representative of a drugstore or hardware store. With 85 people, the total ventilation air quantity is thus 1275 cfm. Floor area of 4000 ft² with a 10 ft ceiling height gives a space volume of 40,000 ft³, corresponding to (1275 cfm × 60)/40,000 = 1.91 air changes per hour . In practice, ventilation air is normally conditioned to some extent by the air conditioning equipment before being admitted to the conditioned space. However, the variety of such arrangements and the varying impact felt by the load calculation process are not covered by this chapter and should be evaluated as part of a system analysis procedure. For this example, assume the ventilation air is introduced directly into the space and included as part of the space cooling load, but only during scheduled operating hours of the cooling equipment.

Infiltration. Window infiltration is considered zero, since the windows are sealed. Infiltration through wall surfaces is also neglected as insignificant, particularly with plastered interior surfaces. Calculation of door infiltration however, requires some judgement. The pressure of 1.91 air changes/h in the form of positive ventilation could be sufficient to prevent door infiltration, depending on the degree of simultaneous door openings and the wind direction and velocity. For this example, assume that outside and inside doors are frequently opened simultaneously, and that door infiltration should be included as part of the cooling load, estimating 100 ft³ per person per door passage. Further estimating outside door use at 10 persons hour, and inside doors (to unconditioned space, previously estimated to be at ambient temperature and humidity) at 30 persons per hour, generates the following infiltration rate:

$$Q_{inf} = 40 \times 100/60 = 67 \text{ cfm}$$

Thermal responsiveness of building and contents. For this example, mass of building construction and contents is "medium."

Conditioning equipment location. Conditioning equipment is in an adjoining structure to the north, thus having no direct impact on heat gain.

Find:

1. Sensible cooling load.
2. Latent cooling load.
3. Total cooling load.
4. Capacity of system to maintain:
   (a) Fixed temperature: 75°F indoor temperature, 24-hour "on" period.
   (b) 2°F throttling range: Indoor temperature in the range 75 to 77°F, 24-hour "on" period.
   (c) 4°F throttling range: Indoor temperature in the range 75 to 79°F, 24-hour "on" period.
   (d) 2°F throttling range: Indoor temperature in the range 75 to 77°F, 10-hour "on" period, 0800 to 1700.
   (e) 4°F throttling range: Indoor temperature in the range 75 to 79°F, 12-hour "on" period, 0600 to 1700.

Solution by Transfer Function Method

1. Daily load cycle: Estimated thermal loads are calculated by the TFM once per hour for a 24-h daily cycle.
2. Hourly heat gain components: The methodology using CTF coefficients is used to calculate heat gain components through walls and roof.
3. Thermal storage: The heat storage effect of the building and contents is accounted for by RTF coefficients.
4. Room temperature and heat extraction: TFM approximates resultant room air temperature and heat extraction rates for a specified schedule of thermostat set-points and/or cooling unit operating periods, by applying SATF coefficients to sensible cooling loads, including consideration for heat loss to surroundings. This process can be used to predict the capability of a particular size and type of cooling equipment, its control, and its operating schedule to maintain room air temperature within a specified range.
5. Summary: The data and summary of results using TFM are tabulated in Table 28. The following describes the calculation procedure used to determine the values for this table:

1. **Sensible cooling load**

(a) *General*

Line 1, Time of day in hours: Various temperatures and heat flow rates were calculated for every hour on the hour, assuming that hourly values are sufficient to define the daily profile.

Line 2, Outside air temperatures: Hourly values derived by the abovementioned procedure, using the specified maximum dry bulb temperature of 94°F and daily range of 20°F.

(b) *Solar Heat Gain Factors*

Lines 3, 4, 5, and 6, Solar heat gain through opaque surfaces: SHGF values from Table 18, Chapter 29 for July 21 at 40°N latitude. These values are used to calculate sol-air temperatures of various outside surfaces, and solar heat gain through windows.

Values for June might have been used, since the solar irradiation of horizontal surface (e.g., a roof) is maximum at that time of year and since the heat gain through the roof appears to be the major component of exterior heat gain in this problem. The difference between June and August values is relatively small however, compared to the large percentage increase in solar heat gain through south glass in August versus June at this latitude, thus indicating that August might be the better choice. For this example, data for July were selected as reasonable, and to provide better comparison with the results from other techniques for which tabular data are limited. To determine the month when the maximum building load will occur, the relative loads of various surfaces should first be evaluated and compared for several months.

(c) *Sol-Air Temperatures*

Lines 7, 8, 9, and 10, Sol-air temperatures at opaque surfaces: Sol-air temperatures, calculated by Equation (6), of the various opaque surfaces. These values are used in calculations of heat gain through the roof and outside walls.

(d) *Instantaneous Sensible Heat Gain*

Line 11, Roof heat gain: Instantaneous heat gain through the roof, calculated by CTF coefficients.

From Table 11, the major element of the roof (that layer with the most mass) is the gypsum slab (code number C14). Other elements are the metal deck (A3), rigid insulation (B3), built-up roofing (E3), and gravel surface (E2). Entering Table 12 with these code values, the C14 roof slab designates column 7, and the R-value 11.11 calls for R = 3. From the "mass-in" part of the table and the condition of being "w/o ceiling," the table identifies Roof Group 5 as that whose CTF coefficients will best represent the roof in question.

The CTF coefficients (b, d, and $\sum_{n=0} c_n$) are then obtained from Tables 13 and 14, by selecting roof group 5 and adjusting the tabulated b_n and $\sum_{n=0} c_n$ by the $U_{example}/U_{table}$ = 0.09/0.055 = 1.636.

The adjusted b_n and $\sum_{n=0} c_n$ are:

$$b_0 = 0.00006\,(1.636) = 0.00010$$
$$b_1 = 0.00256\,(1.636) = 0.00419$$
$$b_2 = 0.00477\,(1.636) = 0.00780$$
$$b_3 = 0.00100\,(1.636) = 0.00164$$
$$b_4 = 0.00002 = \text{n/a}$$
$$b_5 = 0.00000 = \text{n/a}$$
$$b_6 = 0.00000 = \text{n/a}$$
$$\sum_{n=0} c_n = 0.00841\,(1.636) = 0.01376$$

The d values (used without modification) are:

$$d_0 = 1.00000$$
$$d_1 = -1.10395$$
$$d_2 = 0.26169$$
$$d_3 = -0.00475$$
$$d_4 = 0.00002$$
$$d_5 = 0.00000$$
$$d_6 = 0.00000$$

The heat gain through the roof is calculated by Equation (25), using the sol-air temperature cycle given in line 7 and t_{rc} = 75°F. The calculations are extended for five daily cycles at which time the daily periodic steady state is effectively reached. The last daily cycle is used as the heat gain through the roof. (*Note*: Three daily cycles are sufficiently accurate in this case, but since calculations do not converge for the more massive wall components before the 93rd hour, all calculations are run to hour 120.)

Lines 12, 13, 14, and 15, Wall heat gain. The instantaneous heat gains through the various walls are calculated by the same approach as that used for the roof. The CTF coefficients selected from Tables 11 and 15 to 19 are:

North and East Exterior Walls
- Dominant element C8, or col. 13 in Integral Mass table (Table 16);
- Interior finish E1;
- R-value indicating R of 2 in Table 16;
- Select Wall Group 5 in Tables 18 and 19 for representative factors.

South Wall
- Dominant element C9, or col. 14 in Table 16;
- Exterior layer A2 or A7;
- Interior layer E1 (plywood panel ignored as trivial);
- R-value indicating R of 6;
- Select Wall Group 24 for representative factors.

North and West Party Walls
- With no specific data for a 13 in. brick wall, use a layer of 8 in. common brick (C9) and a layer of 4 in. face brick (A2 or A7) as an approximation; thus:
- Dominant element C9, or column 14 in Table 16;
- Exterior layer A2 or A7;
- R-value indicating R of 6;
- Select Wall Group 24 in Table 16 for representative factors.

The b_n and $\sum_{n=0} c_n$ require multiplication by the U-factor ratio to account for the difference in U-factors. The heat gain is then calculated by Equation (25), using corresponding wall CTF coefficients and sol-air temperatures for south, east, and north walls, and the outside air temperature cycle for north and west party walls.

Lines 16, 17, and 18, Door heat gain: Heat storage of the doors could be assumed negligible, in which case the heat gain would be calculated by Equation (16) as

$$q_{D\theta} = U_D A_D (t_{D\theta} - t_i)$$

where

U_D = 0.19 Btu/h·ft²·°F, U-factor of doors (0.18 for interior doors)
A_D = 35 ft², area of a door
t_i = 75°F, inside temperature
$t_{D\theta}$ = outside temperature at door, at time θ

For the door in the north party wall, $t_{D\theta}$ equals outside air temperature. For the doors in east and south walls $t_{D\theta}$ equals the east and south wall sol-air temperatures, respectively.

The foregoing would be a reasonable approach for estimating the minor loads involved. For the purpose of this example however, the relatively brief storage effect of the solid core doors has been considered by use of Equation (25), in accordance with:

- Dominant element B7, or column 3 in Table 16;
- Interior finish A6;
- R value indicating *R* of 8;
- Select Wall Group 1 for representative factors.

Lines 19, 20, and 21, Window heat gain: The air to air heat gain (line 19):

$$q_a = U_w A_w (t_{o\theta} - t_i)$$

where

U_w = 0.81, U-factor of window
A_w = 90ft², area of windows
$t_{o\theta}$ = outside air temperature at time θ

The solar radiation heat gain (lines 20 and 21) through south and north windows:

$$Q_r = A_w \times SC \times SHGF_\theta$$

where

$SHGF_\theta$ = Solar heat gain factors given in line 5 for south and line 4 for north.
SC = 0.55; shading coefficient for clear window with light colored curtain or blind.

Table 28 Tabulation of Data for Example 6

1 Time, hour	0100	0200	0300	0400	0500	0600	0700	0800	0900	1000	1100	1200
2 Outside air temperature, °F	76	75	74	74	74	74	75	77	79	82	86	89
3 SHGF, Btu/h·ft², Horizontal	0	0	0	0	0	32	88	145	194	231	254	262
4 East	0	0	0	0	1	37	30	28	32	35	37	38
5 West	0	0	0	0	0	11	21	30	52	81	102	109
6 South	0	0	0	0	2	137	204	216	193	146	81	41
7 Sol-air temperature, °F, Horizontal	69	68	67	67	67	77	94	114	130	144	155	161
8 East	76	75	74	74	74	80	80	81	84	87	92	95
9 West	76	75	74	74	74	76	78	82	87	94	101	105
10 South	76	75	74	74	74	95	106	109	108	104	98	95
Instantaneous Sensible Heat Gain, Btu/h												
11 Roof	5377	3844	2565	1495	599	−140	−555	−208	1218	3715	6989	10682
12 East wall	934	794	664	544	434	337	272	266	292	338	417	532
13 West wall	1466	1454	1424	1379	1321	1255	1181	1103	1024	947	879	824
14 South wall	4390	3723	3107	2542	2029	1579	1471	2235	3679	5277	6646	7551
15 North and east party wall	2498	2535	2534	2497	2429	2335	2218	2086	1943	1797	1656	1529
16 East door (to adjacent building)	19	12	6	1	−3	−5	−4	2	12	26	44	64
17 West door	20	13	6	1	−4	−4	6	20	43	76	118	159
18 South door	20	13	6	1	−4	11	98	176	212	218	200	167
19 Windows, air to air heat gain	117	44	−15	−58	−73	−44	29	160	350	569	816	1050
20 East windows, solar heat gain	0	0	0	0	17	611	495	462	528	578	611	627
21 West windows, solar heat gain	0	0	0	0	0	363	693	990	1716	2673	3366	3597
22 Lights, tungsten (always on)	13640	13640	13640	13640	13640	13640	13640	13640	13640	13640	13640	13640
23 Lights, fluorescent (on-off)	0	0	0	0	0	0	0	71610	71610	71610	71610	71610
24 People	0	0	0	0	0	0	0	21250	21250	21250	21250	21250
25 Infiltration	0	0	0	0	0	0	0	162	354	575	825	1061
26 Ventilation	2244	841	−281	−1122	−1403	−841	561	3086	6732	10940	15708	20196
27 Total instant sensible heat gain	30725	26913	23656	20920	18982	19097	20105	117040	124603	134229	144775	154539
Latent Heat Gain/Cooling Load, Btu/h												
28 People	0	0	0	0	0	0	0	17000	17000	17000	17000	17000
29 Infiltration	0	0	0	0	0	0	0	2205	2205	2205	2205	2205
30 Ventilation	41963	41963	41963	41963	41963	41963	41963	41963	41963	41963	41963	41963
31 Total latent heat gain/cooling load	41963	41963	41963	41963	41963	41963	41963	61168	61168	61168	61168	61168
32 Sum: sens. + latent heat gain, Btu/h	72688	68876	65619	62883	60945	61060	62068	178208	185771	195397	205943	215707
Sensible Cooling Load from Convective Heat Gain, Btu/h												
33 Windows, air to air heat gain	117	44	−15	−58	−73	−44	29	160	350	569	816	1050
34 Lights, tungsten (20% convective)	2728	2728	2728	2728	2728	2728	2728	2728	2728	2728	2728	2728
35 Lights, fluorescent (50% conv.)	0	0	0	0	0	0	0	35805	35805	35805	35805	35805
36 People (67% convective)	0	0	0	0	0	0	0	14237	14237	14237	14237	14237
37 Infiltration (100% convective)	0	0	0	0	0	0	0	162	354	575	825	1061
38 Ventilation (100% convective)	2244	841	−281	−1122	−1403	−841	561	3086	6732	10940	15708	20196
Sensible Cooling Load from Radiant Heat Gain, Btu/h												
39 Lights, tungsten (80% radiant)	10912	10912	10912	10912	10912	10912	10912	10912	10912	10912	10912	10912
40 Lights, fluorescent (50% radiant)	12120	11271	10482	9748	9066	8431	7841	10873	12618	14241	15751	17154
41 People (33% radiant)	2118	1970	1832	1704	1584	1473	1370	2656	2961	3245	3508	3754
Sensible Cooling Load from Convective and Radiant Heat Gain, Btu/h												
42 From SHG through east windows	85	79	73	68	75	476	406	390	440	480	509	527
43 From SHG through west windows	267	249	231	215	200	433	653	858	1362	2038	2555	2769
44 From roof heat gain	7577	6379	5331	4409	3594	2882	2387	2418	3205	4766	6922	9442
45 From east wall heat gain	963	866	773	683	599	521	464	446	452	472	516	587
46 From west wall heat gain	1362	1361	1347	1322	1287	1244	1194	1140	1084	1027	975	931
47 From south wall heat gain	4984	4488	4015	3567	3146	2761	2605	3046	3973	5040	5989	6651
48 From N. and E. party wall heat gain	2304	2343	2356	2343	2308	2252	2178	2091	1993	1890	1788	1692
49 From east door heat gain	32	26	21	16	13	10	10	13	19	28	40	54
50 From west door heat gain	43	36	30	24	20	18	23	32	46	69	98	127
51 From south door heat gain	46	39	33	27	22	30	89	142	169	176	167	147
52 Total sensible cooling load, Btu/h	47902	43632	39868	36586	34078	33286	33450	91195	99440	109238	119849	129824
53 Sum: sens. + lat. cooling load, Btu/h	89865	85595	81831	78549	76041	75249	75413	152363	160608	170406	181017	190992
Air Temperature, °F, and Heat Extraction, Btu/h												
54 Total sensible cooling load, Btu/h (LTS)	45482	41375	37758	34612	32232	31516	31749	89212	97208	106702	116999	126680
55 2°F throttling range: temperature	74.0	74.0	74.0	74.0	74.0	74.0	74.0	75.0	75.0	75.0	75.0	75.0
56 Equipment run 1-24; heat extraction	49487	45493	41936	38803	36358	35399	35314	85674	93257	102010	111520	120541
57 4°F throttling range: temperature	74.0	74.0	74.0	74.0	74.0	74.0	74.0	75.0	75.0	75.0	75.0	76.0
58 Equipment run 1-24; heat extraction	52695	48849	45392	42318	39866	38753	38445	83238	90364	98444	107245	115651
59 2°F throttling range: temperature	90	90	91	91	91	91	91	80	79	79	80	80
60 Equipment run 8-17; heat extraction	0	0	0	0	0	0	0	146600	146600	146600	146600	146600
61 4°F throttling range: temperature	88	88	88	88	88	75	75	76	76	76	76	77
62 Equipment run 6-17; heat extraction	0	0	0	0	0	106261	95365	133756	135291	138399	142778	146600

Table 28 Tabulation of Data for Example 6 (*Concluded*)

1	1300	1400	1500	1600	1700	1800	1900	2000	2100	2200	2300	2400	24 h Total	Heat Loss, Btu/h
2	91	93	94	93	92	89	87	84	82	80	78	77		
3	254	231	194	145	88	32	0	0	0	0	0	0	2150	
4	37	35	32	28	30	37	1	0	0	0	0	0	438	
5	102	81	52	30	21	11	0	0	0	0	0	0	703	
6	37	35	31	26	20	11	0	0	0	0	0	0	1180	
7	160	155	145	130	111	92	80	77	75	73	71	70		
8	97	98	99	97	97	95	87	84	82	80	78	77		
9	106	105	102	98	95	91	87	84	82	80	78	77		
10	97	98	99	97	95	91	87	84	82	80	78	77		
Instantaneous Sensible Heat Gain, Btu/h														
11	14436	17882	20649	22467	23154	22592	20788	18036	14908	11963	9395	7212	239063	23400
12	685	862	1045	1214	1356	1461	1531	1549	1486	1368	1228	1081	20690	5304
13	789	780	801	851	927	1020	1120	1218	1306	1378	1429	1459	27335	6318
14	7939	8025	8075	8159	8246	8247	8099	7757	7214	6548	5829	5100	133467	23868
15	1427	1358	1331	1351	1417	1525	1667	1831	2001	2165	2308	2421	46859	17306
16	84	100	111	116	115	108	96	81	65	51	38	28	1167	410
17	189	203	200	183	162	139	117	91	70	54	40	29	1931	432
18	147	148	153	156	149	136	116	90	70	54	40	29	2406	432
19	1224	1341	1385	1341	1239	1079	889	700	540	393	277	189	13542	4739
20	611	578	528	462	495	611	17	0	0	0	0	0	7231	*
21	3366	2673	1716	990	693	363	0	0	0	0	0	0	23199	*
22	13640	13640	13640	13640	13640	13640	13640	13640	13640	13640	13640	13640	327360	-13640
23	71610	71610	71610	71610	71610	0	0	0	0	0	0	0	716100	-71610
24	21250	21250	21250	21250	21250	0	0	0	0	0	0	0	212500	-21250
25	1238	1356	1400	1356	1253	0	0	0	0	0	0	0	9580	4791
26	23562	25806	26648	25806	23843	20757	17111	13464	10379	7574	5330	3646	260587	91163
27	162197	167612	170542	170952	169549	71678	65191	58457	51679	45188	39554	34834	2043017	
Latent Heat Gain/Cooling Load, Btu/h														
28	17000	17000	17000	17000	17000	0	0	0	0	0	0	0	170000	
29	2205	2205	2205	2205	2205	0	0	0	0	0	0	0	22050	
30	41963	41963	41963	41963	41963	41963	41963	41963	41963	41963	41963	41963	1007112	
31	61168	61168	61168	61168	61168	41963	41963	41963	41963	41963	41963	41963	1199162	
32	223365	228780	231710	232120	230717	113641	107154	100420	93642	87151	81517	76797	3242179	
Sensible Cooling Load from Convective Heat Gain, Btu/h														
33	1224	1341	1385	1341	1239	1079	889	700	540	393	277	189	13542	
34	2728	2728	2728	2728	2728	2728	2728	2728	2728	2728	2728	2728	65472	
35	35805	35805	35805	35805	35805	0	0	0	0	0	0	0	358050	
36	14237	14237	14237	14237	14237	0	0	0	0	0	0	0	142370	
37	1238	1356	1400	1356	1253	0	0	0	0	0	0	0	9580	
38	23562	25806	26648	25806	23843	20757	17111	13464	10379	7574	5330	3646	260587	
Sensible Cooling Load from Radiant Heat Gain, Btu/h														
39	10912	10912	10912	10912	10912	10912	10912	10912	10912	10912	10912	10912	261888	
40	18460	19674	20803	21853	22830	20158	18747	17434	16214	15079	14024	13042	357914	
41	3982	4194	4391	4575	4746	3523	3276	3047	2834	2635	2451	2279	70108	
Sensible Cooling Load from Convective and Radiant Heat Gain, Btu/h														
42	523	507	478	437	461	542	143	122	114	106	98	91	7230	
43	2669	2246	1624	1136	924	683	413	384	357	332	309	288	23195	
44	12086	14596	16711	18225	18990	18898	17928	16254	14249	12289	10518	8953	239009	
45	687	808	937	1059	1167	1251	1314	1341	1313	1245	1158	1063	20685	
46	900	886	893	921	967	1028	1095	1164	1227	1282	1324	1351	27312	
47	6979	7104	7203	7321	7439	7496	7448	7261	6926	6492	6007	5498	133439	
48	1611	1551	1520	1520	1553	1617	1708	1816	1933	2049	2155	2243	46814	
49	68	80	89	94	95	92	85	75	65	55	46	38	1164	
50	150	162	163	154	141	128	113	96	81	70	59	50	1933	
51	134	136	140	143	140	131	118	100	86	74	63	54	2406	
52	137955	144129	148067	149623	149470	91023	84028	76898	69958	63315	57459	52425	2042698	178163
53	199123	205297	209235	210791	210638	132986	125991	118861	111921	105278	99422	94388	3241860	71663
Air Temperature, °F, and Heat Extraction, Btu/h														
54	134552	140509	144275	145693	145432	87215	80409	73476	66754	60326	54673	49830	1974669	
55	75.0	75.0	75.0	75.0	75.0	75.0	75.0	75.0	75.0	74.0	74.0	74.0		
56	128005	133792	137639	139386	139602	88759	82327	75937	69698	63689	58341	53696	1966663	
57	76.0	76.0	76.0	76.0	76.0	75.0	75.0	75.0	75.0	74.0	74.0	74.0		
58	122700	128264	132093	134030	134566	89480	83460	77573	71790	66186	61150	56731	1959283	
59	80	81	81	81	80	89	90	90	90	90	90	90		
60	146600	146600	146600	146600	146600	0	0	0	0	0	0	0	1466000	
61	77	77	78	78	78	86	87	87	87	87	88	88		
62	146600	146600	146600	146600	146600	0	0	0	0	0	0	0	1631450	

Lines 22 and 23, Heat gain from tungsten and fluorescent lights: For the gain from lighting, Equation (9) is used with a use factor of unity and special allowance factors of 1.20 for fluorescent lamps and of unity for tungsten lamps. Thus:

$$q_{el\ tung} = 4000 \times 1 \times 1 \times 3.41 = 13{,}640 \text{ Btu/h},$$

and

$$q_{el\ fluor} = 17{,}500 \times 1 \times 1.20 \times 3.41 = 71{,}610$$

Line 24, Heat gain from people: Sensible heat gain from occupants, for moderately active office work (Table 3):

$$q_{sp} = \text{(number of people)(sensible heat generated per person)}$$
$$= 85 \times 250 = 21{,}250 \text{ Btu/h}$$

Lines 25 and 26, Sensible heat gain from infiltration and ventilation: As developed in Building Data, the value used for infiltration is 67 cfm, and that for ventilation, 1275 cfm.

Heat gain from infiltration air is part of the space load, while that from ventilation air normally is not. In this example however, since ventilation is delivered directly to the space rather than through the cooling equipment first, its gain is also included as a direct space load.

Note: Had the ventilation air instead been mixed with return air leaving the occupied space and before entering the cooling equipment, only (4) that portion which passed through the cooling coil untreated due to coil inefficiency (or "Bypass Factor," normally 3 to 5% for a chilled water coil of six or more rows and close fin spacing up to 15% or more for refrigerant coils in packaged air-conditioning units), and/or (5) that quantity deliberately bypassed around the coil in response to a "face and bypass" or "conventional multizone" space dry-bulb temperature control scheme, would become a part of the space heat gain as such rather than a part of the cooling coil load directly.

The sensible loads are determined from Equation (22). At 1600 hours for example, when $t_o = 94°F$ and $t_i = 75°F$, this generates:

$$q_{si} = 1.1\text{(Infiltration rate)}(t_o - t_i)$$
$$= 1.1 \times 67(94 - 75) = 1400 \text{ Btu/h},$$

and

$$q_{sv} = 1.1\text{(Ventilation rate)}(t_o - t_i)$$
$$= 1.1 \times 1275(94 - 75) = 26{,}600 \text{ Btu/h},$$

Line 27, Total instantaneous sensible heat gain: The sum of instantaneous heat gain values listed in lines 11 through 26. All such values take into account the delaying effects of insulation and mass of the elements enclosing the conditioned space on the heat that ultimately enters that space, but before considering the thermal inertia of the overall mass and configuration of the building and contents in delaying conversion of radiant heat gain to space cooling load.

(e) *Instantaneous Latent Heat Gain*

Line 28, People: The latent heat gain due to people, using Table 3 data:

$$q_{lp} = \text{(number of persons)(latent heat generated per person)}$$
$$= 85 \times 200 = 17{,}500 \text{ Btu/h during the occupied period.}$$

Lines 29 and 30, Latent heat gain from infiltration and ventilation: The latent loads are determined from Equation (23). At 1600 hours for example, when $W_o = 0.0161$ and $W_s = 0.0093$, this generates

$$q_{si} = 4840\text{(Infiltration rate)}(W_o - W_i)$$
$$= 4840 \times 67(0.01661 - 0.0093) = 2{,}205 \text{ Btu/h},$$

and

$$q_{sv} = 4840\text{(Ventilation rate)}(W_o - W_i)$$
$$= 4840 \times 1275(0.01661 - 0.0093) = 41{,}963 \text{ Btu/h},$$

Line 31, Total latent heat gain: The total latent heat gain, i.e., the sum of lines 28, 29, and 30.

Line 32, Sum of instantaneous sensible and latent heat gain: The sum of heat gain values from lines 27 and 31.

(f) *Cooling Load from Convective Sensible Heat Gain Components*

Lines 33 through 38: Direct inclusion of the convective portions of instantaneous heat gain components listed in lines 19, 25, and 26, and 20%, 50%, and 67% of lines 22, 23, and 24 respectively. These room sensible heat gain components (i.e., loads due to air-to-air heat gain through windows, tungsten lights, fluorescent lights, infiltration, ventilation, and heat gain due to people by convection, all appear as cooling load without delay. Percentages of heat gain considered corrective are listed in Table 3 and Table 44 under the section describing TETD/TA procedures. Selection of 33% of sensible gain for people as radiant is an approximation for purposes of this example.

(g) *Cooling Load from Radiant Sensible Heat Gain Components*

Lines 39 through 41: Heat gain data from lights and people (lines 22 through 24) are processed by Equation (28) using RTF coefficients from Tables 24 and 25:

> From Table 24, assuming "medium" mass of building and contents, the 75 lb/ft^2 specific mass classification can be considered representative. Assuming a conventional supply diffuser and nonplenum return air arrangement with inside surface coefficient h = 0.6 Btu/(h·ft^2·°F), or "medium" type indicates a w_1 value of −0.94; except with an uncarpeted floor the next w_1 value down the column is used, or 0.93.
>
> From the lower part of Table 25, assuming ordinary furnishings, no carpet, medium air circulation, supply and return below ceiling, and unvented light fixtures, the v_0 value for lighting is 0.55 and v_1 = 1 + (−0.93) −0.55 = −0.48.
>
> For people, the upper part of Table 25 calls for a v_0 of 1.0 and v_1 of 0 to be applied to convective heat gain (instantaneous conversion to cooling load), and for radiant heat gain a v_0 of 0.197 and v_1 = 1 + (−0.93) − 0.197 = −0.127.
>
> Note that the TFM treatment of lighting heat gain is "generic," without individual regard to the differences in radiant/convective percentages of heat gain from incandescent, fluorescent, or other type lamps, and the RTF coefficients are applied to the combined sensible heat gain values. For the purposes of this example, to facilitate comparison with other calculation methods, the values in lines 39 and 40 represent the hourly results of Equation (28) less the amounts of instantaneous cooling load included and indicated on lines 34 and 35.

(h) *Cooling Load from Convective and Radiant Sensible Heat Gain Components*

Lines 42 through 51: Elements of instantaneous heat gain from solar radiation through windows, walls, doors and roof, i.e., sum of values listed in lines 11 to 21, delayed in being felt as cooling load by the space. Data listed in lines 42 through 51 are the results of applying Equation (28) and appropriate RTF coefficients to the heat gain values from lines 11 through 21, without separately considering radiant or convective components. RTF coefficients are taken from Tables 24 and 25 in the manner above described for lighting loads, producing:

> From Table 24, w = −0.93 in all cases.
>
> From the upper part of Table 25, all cases fall within the second category described, which for "medium" building and contents mass indicates v_0 = 0.681 and v_1 = 1 + (−0.93) − 0.681 = −0.611.
>
> The heat gain by solar radiation transmitted through windows is included with heat gain through walls and roof because the venetian blind intercepts solar radiation and releases it to the room in a similar way as the heat gain through walls and roof.

Note: If the glass had no internal shading, the solar radiation through windows would have to be treated by a different set of RTF coefficients to account otherwise for thermal storage (see Tables 24 and 25). Translucent draperies fall somewhere between these limits, with assumed linear relationship in the absence of specific research on the subject (see Chapter 29).

Line 52, Total room sensible cooling load: Total sensible cooling load felt by the room, and the design sensible load used as the basis for sizing cooling equipment. This total load is the sum of the values listed in lines 33 through 51. The tiny difference between the 24 hour total of 2,042,698 Btu/h on line 52 and the sum of the 24 hour totals for lines 11 through 26 reflects rounding of values during intermediate computation.

2. **Latent Cooling Load**

Line 31—The sum of lines 28, 29, and 30: Total Latent Heat Gain is also the Total Latent Cooling Load, as all components occur instantaneously.

3. **Total Cooling Load**

Line 53—The sum of lines 52 and 31: Note that the Total Cooling Load for this example problem is the theoretical total for the conditions as defined, and may or may not represent the actual total cooling load imposed upon a system of cooling equipment. An appropriate psychrometric analysis should be performed of supply air, space air, return air, and mixed air (where ventilation air is mixed with return air en route back to the cooling equipment), considering the type of cooling equipment and characteristics of the preferred control scheme. Only an analysis of this type can verify that the design will meet the requirements, and determine whether the actual sensible, latent, and total cooling loads are greater or less than the theoretical values calculated.

4. **Capacity of System to Maintain Conditions**

(a) *Fixed temperature*: 75°F indoor temperature, 24 hour "on" period: The basic calculation procedure assumes a fixed indoor temperature, in this case 75°F; thus the results tabulated in lines 1 through 42 are for this condition.

(b) 2°F throttling range: Indoor temperature in the range 75 to 77°F, 24 hour "on" period.

(c) 4°F throttling range: Indoor temperature in the range 75 to 79°F, 24 hour "on" period.

(d) 2°F throttling range: Indoor temperature in the range 75 to 77°F, 10 hour "on" period, 0800 through 1700.

(e) 4°F throttling range: Indoor temperature in the range 75 to 79°F, 12 hour "on" period, 0600 through 1700.

Line 54, Sensible cooling load with loss to surroundings: To be consistent with the concept of heat extraction and resultant space temperatures, certain cooling load elements must be modified to account for heat loss to surroundings. The multiplier F_c = 0.94362 was calculated by the process noted for each of the envelope element areas times the respective U-factors, dividing the sum by the building perimeter to develop K_θ, and generating F_c by Equations (34) and (35); then using F_c to reduce the appropriate load elements. The sum of all modified and unmodified load elements is listed on line 53 as the basis for the various heat extraction/ space temperature evaluations.

Lines 55 through 62, Air temperatures and heat extraction rates: Heat extraction and indoor air temperatures are based on the normalized SATF coefficients for medium weight construction listed in Table 26 and calculated by use of Equations (36) through (40) in the procedure previously described. The SATF coefficients for this example are thus for hour θ (0800 – 1700):

$$\begin{aligned} g_{0\theta} &= \overset{*}{g}_0(\text{Floor area}) + p_0[K_\theta(\text{Perimeter length})] \\ &\quad + 1.1(\text{Ventilation and Infiltration}) \\ &= (1.81 \times 4000) + 1.0[(4.86 \times 260) + 1.1(1275 + 67)] \\ &= 7240 + (1263.4 + 1476.2) = 9980 \end{aligned}$$

$$\begin{aligned} g_{1\theta} &= \overset{*}{g}_1(\text{Floor area}) + p_1[K_\theta(\text{Perimeter length})] \\ &\quad + 1.1(\text{Ventilation and Infiltration}) \\ &= (-1.89 \times 4000)(-0.87)[(4.86 \times 260) + 1.1(1275 + 67)] \\ &= -7560 - 0.87(1263.4 + 1476.2) = -9944 \end{aligned}$$

$$\begin{aligned} g_{2\theta} &= \overset{*}{g}_2(\text{Floor area}) = 0.08 \times 4000 = 320 \\ p_0 &= 1.0000 \\ p_1 &= -0.87 \end{aligned}$$

The heat extraction rates and room air temperatures listed in lines 55 through 62 are calculated using these SATF coefficients, the modified total sensible cooling load values listed in line 54, and the specified throttling ranges and "on" and "off" periods.

The maximum sensible heat extraction capacity required to maintain the space temperature at a constant 75°F can be taken as the design peak value on line 52, or 149,470 Btu/h at 1600 hours.

The maximum sensible heat extraction capacity required to maintain interior temperature within a 75 to 77°F range is 145,693 Btu/h (hour 1600, line 54), and within a 75 to 79°F range is 139,602 Btu/h (hour 1700, line 56), assuming continuous operation of cooling equipment.

Comparable maintenance of space temperature ranges during equipment operation hours (limited to 10 hours and 12 hours respectively) requires heat extraction rates of 146,600 Btu/h (hours 0800 to 1700, line 60) and 146,600 Btu/h (hours 1200 to 1700, line 62) respectively. Here ER_{max} needs to be increased if the heat accumulated overnight is to be overcome; but the total daily heat extraction still will be significantly less than for continuous operation.

5. **Heating Load**

Lines 11 through 19, Heat loss by conduction: The heat loss column lists for each of the building envelope components a single value representing the product of exposed area, U-factor, and the temperature difference between inside design dry bulb and outside design dry bulb temperatures for winter conditions, in an adaptive use of Equation (8). Often, a lower inside design dry bulb temperature is selected for winter conditions than for summer, and, where appropriate, the U-factors are adjusted to reflect different average exterior wind velocities. For this example, the same inside temperatures and U-factors are used year-round.

These results are design heat loss values, which are used to establish a "design heating load" with which to design heating systems and to select properly sized equipment components. When the load calculation is used to analyze energy performance, hourly calculations of heat loss that reflect the profile of outside weather conditions must be run.

Lines 20 through 21, Solar heat gain: For *design* heating loss calculations, offsetting values of solar heat gain are routinely ignored at night or during periods of extended cloud cover, and thus not consistently available to assist the installed heating equipment. Designers must, however, consider the higher solar heat gain values that occur during winter months due to low solar angles that often cause peak cooling loads through large areas of exposed glass. Hourly calculations are required for energy use evaluation.

Lines 22, 23, and 24, Internal heat gains: Like solar heat gain, the heat from internal sources requires year-round cooling for completely interior spaces and contributes to unseasonable cooling requirements in conjunction with glass loads on sunny days. For conventional heating load purposes, however, these loads are normally ignored because of their uncertainty during all hours of need and since their full effect does not occur until some number of hours after occupancy begins during intermittent schedules. Heat gain values in this example are given as "negative heat loss" figures, and not routinely included in design heating load summaries.

Lines 25 and 26, Infiltration and ventilation: Values listed for these variables are calculated on the basis of a single "worst case" hour under winter design temperature conditions, adapting Equation (22) in a similar manner to that noted for conduction heat losses.

Humidification: For this example, the issue of maintaining interior humidity levels during winter months has been ignored. While this represents routine practice for most applications in latitudes 35°N and lower, humidity levels are of major concern in colder climates.

Line 52, Total sensible heat loss: The sum of heat loss values from lines 11 through 19, 25, and 26, and which conventionally represents the design heating load for the building. Internal heat gain figure from lines 22, 23, and 24 are not included in this total.

Line 53, Net sensible heat loss, considering internal heat gains: The heat loss summary value if internal heat gains were to be included in the total, illustrated here only to emphasize the potential significance of such elements and the importance of providing an appropriate means of temperature control for differently affected building areas.

CLTD/SCL/CLF CALCULATION PROCEDURE

To calculate a space cooling load using the CLTD/SCL/CLF convention, the same general procedures outlined for the TFM relative to data assembly and use of data apply. Similarly, the basic heat gain calculation concepts of solar radiation, total heat gain through exterior walls and roofs, heat gain through interior surfaces, and heat gain through infiltration and ventilation are handled in an identical manner.

The CLTD/SCL/CLF method is a one-step, hand calculation procedure, based on the transfer function method (TFM). It may be used to approximate the cooling load corresponding to the first three modes of heat gain (conductive heat gain through surfaces such as windows, walls, and roofs; solar heat gain through fenestrations; and internal heat gain from lights, people, and equipment) and the cooling load from infiltration and ventilation. The acronyms are defined as follows:

CLTD—Cooling Load Temperature Difference
SCL—Solar Cooling Load
CLF—Cooling Load Factor

The following sections give details of how the CLTD/SCL/CLF technique relates to and differs from the TFM. The sources of the space cooling load, forms of equations to use in the calculations, appropriate references, tables, are summarized in Table 29.

SYNTHESIS OF HEAT GAIN AND COOLING LOAD CONVERSION PROCEDURES

Exterior Roofs and Walls

This method was developed by using the TFM to compute one-dimensional transient heat flow through various sunlit roofs and walls. Heat gain was converted to cooling load using the room transfer functions for rooms with light, medium, and heavy thermal characteristics. Variations in the results due to such varying room constructions and other influencing parameters discussed in the TFM description are so large that only one set of factors is presented here for illustration. All calculations for data tabulated were based on the sol-air temperatures in Table 1. The inside air temperature was assumed to be constant at 78°F (cooling system in operation 24 h/day, seven days a week). The mass of building and contents was "light to medium." For application of CLTD/SCL/CLF techniques, refer to McQuiston and Spitler (1992).

Table 29 Procedure for Calculating Space Design Cooling Load by CLTD/SCL/CLF Method

External Cooling Load

Roofs, walls, and conduction through glass

$$q = UA(\text{CLTD}) \qquad (41)$$

U = design heat transfer coefficient for roof or wall from Chapter 24, Table 4; or for glass, Table 5, Chapter 29
A = area of roof, wall, or glass, calculated from building plans
CLTD = cooling load temperature difference, roof, wall, or glass

Solar load through glass

$$q = A(\text{SC})(\text{SCL}) \qquad (43)$$

SC = shading coefficient: Chapter 29
SCL = solar cooling load factor with no interior shade or with shade, Table 36.

Cooling load from partitions, ceilings, floors

$$q = UA(t_o - t_{rc}) \qquad (8)$$

U = design heat transfer coefficient for partition, ceiling, or floor, from Chapter 24, Table 4
A = area of partition, ceiling, or floor, calculated from building plans
t_b = temperature in adjacent space
t_{rc} = inside design temperature (constant) in conditioned space

Internal Cooling Load

People

$$q_{sensible} = N(\text{Sensible heat gain})\text{CLF} \qquad (44)$$

$$q_{latent} = N(\text{Latent heat gain}) \qquad (45)$$

N = number of people in space, from best available source. Sensible and latent heat gain from occupancy—Table 3, or Chapter 8; adjust as required
CLF = cooling load factor, by hour of occupancy, Table 37

Note: CLF 1.0 with high density or 24-h occupancy and/or if cooling off at night or during weekends.

Lights

$$q_{el} = 3.41 W F_{ul} F_{sa}\,(\text{CLF}) \qquad (9)(46)$$

W = watts input from electrical plans or lighting fixture data
F_{ul} = lighting use factor, as appropriate
F_{sa} = special allowance factor, as appropriate
CLF = cooling load factor, by hour of occupancy, Table 38

Note: CLF = 1.0 with 24-h light usage and/or if cooling off at night or during weekends.

Power

$$q_p = 2545\, P E_F\, \text{CLF} \qquad (15)(16)(17)(50)$$

P = horsepower rating from electrical plans or manufacturer's data
E_F = efficiency factors and arrangements to suit circumstances
CLF = cooling load factor, by hour of occupancy, Table 37

Note: CLF = 1.0 with 24-h power operation and/or if cooling off at night or during weekends.

Appliances

$$q_{sensible} = q_{input} F_U F_R(\text{CLF}) \qquad (18)(46)$$

or

$$q_{sensible} = q_{input} F_L(\text{CLF}) \qquad (19)(46)$$

q_{input} = rated energy input from appliances—Tables 5 through 9, or manufacturer's data
F_U, F_R, F_L = usage factors, radiation factors,and load factors from the General Principles section
CLF = cooling load factor, by scheduled hours and hooded or not; Tables 37 and 39

Note 1: CLF = 1.0 with 24-h appliance operation and/or if cooling off at night or during weekends.
Note 2: Set latent load = 0 if appliance under exhaust hood.

Ventilation and Infiltration Air

$$q_{sensible} = 1.10\, Q\, (t_o - t_i) \qquad (25)$$

$$q_{latent} = 4840\, Q\, (W_o - W_i) \qquad (23)$$

$$q_{total} = 4.5\, Q\, (h_o - h_i) \qquad (20)$$

Q = ventilation cfm from ASHRAE *Standard* 62; infiltration from Chapter 25
t_o, t_i = outside, inside air temperature, °F
W_o, W_i = outside, inside air humidity ratio, lb (water)/lb (dry air)
H_o, H_i = outside, inside air enthalpy, Btu/lb (dry air)

Basic CLTD cooling load for exterior surfaces. The results were generalized to some extent by dividing the cooling load by the U-factor for each roof or wall and are in units of total equivalent cooling load temperature difference (CLTD). This establishes the basic cooling load equation for exterior surfaces as:

$$q = UA(\text{CLTD}) \tag{41}$$

where

q = cooling load, Btu/h
U = coefficient of heat transfer, Btu/(h·ft^2·°F)
A = area of surface, ft^2
CLTD = cooling load temperature difference

In developing the method, it was assumed that the heat flow through a similar roof or wall (similar in thermal mass as well as U-factor) can be obtained by multiplying the total CLTDs listed in Tables 30 or 32 by the U-factor of the roof or wall at hand, respectively. The errors introduced by this approach depend on the extent of the differences between the construction in question (components, size, configuration, and general mass of building and contents) and the one used for calculating the CLTDs.

The sol-air temperature value depends on outdoor air temperature as well as the intensity of solar radiation. Consequently, a change in either outdoor air temperature or geographic location changes the sol-air temperature. The CLTD values in the tables were computed for an indoor air temperature of 78°F, an outdoor maximum temperature of 95°F, and an outdoor mean temperature of 85°F, with an outdoor daily range of 21°F and a solar radiation variation typical of 40°N latitude on July 21.

The notes associated with Tables 30 and 32 provide descriptions of the conditions under which the CLTD values were calculated. While variations in exterior color and/or outside and inside surface film resistances do have some effect, their impact on roofs or walls of contemporary construction is relatively minor and can be ignored with data that is already normalized for convenience. Variations in inside space temperature or the mean outdoor temperature are of much more significance, and the means of appropriate adjustment are thus outlined. Additional guidance for specific application may be found along with tables for a broad range of latitudes in McQuiston and Spitler (1992).

Space Cooling Load from Fenestration

The basic principles of calculating heat gain from conduction and solar radiation through fenestration are as previously discussed for the TFM.

CLTD Cooling load from conduction. For conduction heat gain, the overall heat transfer coefficient accounts for the heat transfer processes of (1) convection and long-wave radiation exchange outside and inside the conditioned space, and (2) conduction through the fenestration material. To calculate cooling load for this component, the conduction heat gain is treated in a manner similar to that through walls and roofs. The RTF coefficients used to convert the heat gain to cooling load are thus the same as those for walls and roofs. The resulting CLTDs are given in Table 34, again presenting only a single set of factors for all room construction types, neglecting the effects of mass and latitude due to the generally low density and the small magnitude of these components. The CLTDs from Table 34 can also be used for doors with reasonable accuracy. The cooling load from conduction and convection heat gain is calculated by:

$$q_{cond} = UA(\text{CLTD}) \tag{42}$$

where A is the net glass area of the fenestration in square feet. [Note that the equation is identical to Equation (41).]

Solar Heat Gain. The basic principles of evaluating heat gain from transmitted and absorbed solar energy through fenestration, including the primary terms SHGF and SC, are the same for the CLTD/CLF procedure as previously described for the TFM.

Previous ASHRAE Handbooks tabulated values of maximum solar heat gain factors for sunlit or externally shaded double-strength sheet glass, used as the heat gain input for calculating cooling load factors (CLFs), employing appropriate RTF coefficients as in the TFM discussion. This process, however, introduced new variables into the calculations: (1) the presence or absence of interior shading devices, which is pivotal, and (2) the construction, furnishings, floor coverings, and relative amounts of fenestration, which are critical when interior shading is absent. Results obtained with this method do not recognize the significant variation of solar cooling load profiles due to different latitudes, months, and other factors. A new term, solar cooling load (SCL), is introduced to more closely approximate cooling loads due to solar radiation transmitted through fenestration.

Cooling load caused by solar radiation through fenestration is calculated by:

$$q_{rad} = A(\text{SC})(\text{SCL}) \tag{43}$$

where

q_{rad} = cooling load caused by solar radiation, Btu/h
A = net glass area of fenestration, ft^2
SC = shading coefficient, for combination of fenestration and shading device, obtained from Chapter 29
SCL = solar cooling load from Table 36, Btu/h·ft^2

Total Cooling Load from Fenestration. The total cooling load due to fenestration is the sum of the conductive and radiant components q_{cond} and q_{rad}.

Zone Influencing Parameters. For purposes of estimating a cooling load, a zone is a particular combination of conditions defining the space under consideration, and which govern the absorption and release of radiant energy. The SCL for a particular zone depends on latitude, direction, nature, and quantity of enclosing surfaces, as well as various internal parameters that influence the SHGF for each glass exposure in that zone.

To determine the most appropriate SCL table for a zone, refer to Tables 35A and 35B, where zone types (A, B, C, or D) are given as functions of some of the more dominant of the 16 zone parameters defined in the TFM discussion. The SCLs for sunlit glass at 40°N latitude and one month, July, are tabulated in Table 36 for each zone type. SCLs for externally shaded glass may be taken from these tables as those for North exposure, although with some loss of accuracy at latitudes lower than 24°N. Interpolation between latitudes can be performed with some loss of accuracy. McQuiston and Spitler (1992) include additional data for multistory buildings and for other latitudes, months, and zone types.

Shading Coefficient

Interior Shading. The cooling load from solar radiation must be analyzed for one of two cases: (1) presence of interior shading or (2) absence of interior shading. Blinds (venetian or roller shades) or drapes absorb the solar energy before it can strike the floor or other interior surfaces of the space, which leads to a rapid response in the cooling load due to the low mass of the shading device.

When interior shading is absent, the solar energy is absorbed by the more massive elements of the space, which results in increased delay in such heat gain being converted to cooling load. Many variables, of which the more important are the presence or absence of carpet on the floor, mass of the floor and other surfaces, mass of the contents of the space, amount of glass in the exposed surfaces, presence or absence of a ceiling, the relative size of the space, etc., have influence on this phenomenon.

Table 30 July Cooling Load Temperature Differences for Calculating Cooling Load from Flat Roofs at 40°North Latitude

Roof No.	Hour 1	2	3	4	5	6	7	8	9	10	11	12	13	14	15	16	17	18	19	20	21	22	23	24
1	0	−2	−4	−5	−6	−6	0	13	29	45	60	73	83	88	88	83	73	60	43	26	15	9	5	2
2	2	0	−2	−4	−5	−6	−4	4	17	32	48	62	74	82	86	85	80	70	56	39	25	15	9	5
3	12	8	5	2	0	−2	0	5	13	24	35	47	57	66	72	74	73	67	59	48	38	30	23	17
4	17	11	7	3	1	−1	−3	−3	0	7	17	29	42	54	65	73	77	78	74	67	56	45	34	24
5	21	16	12	8	5	3	1	2	6	12	21	31	41	51	60	66	69	69	65	59	51	42	34	27
8	28	24	21	17	14	12	10	10	12	16	21	28	35	42	48	53	56	57	56	52	48	43	38	33
9	32	26	21	16	13	9	6	4	4	7	12	19	27	36	45	53	59	63	64	63	58	52	45	38
10	37	32	27	23	19	15	12	10	9	10	12	17	23	30	37	44	50	55	57	58	56	52	47	42
13	34	31	28	25	22	20	18	16	16	17	20	24	28	33	38	42	46	48	49	48	46	44	40	37
14	35	32	30	27	25	23	21	20	19	20	22	24	28	32	36	39	42	44	45	45	44	42	40	37

Note: 1. Direct application of data

- Dark surface
- Indoor temperature of 78°F
- Outdoor maximum temperature of 95°F with mean temperature of 85°F and daily range of 21°F
- Solar radiation typical of clear day on 21st day of month
- Outside surface film resistance of 0.333 (h·ft²·°F)/Btu
- With or without suspended ceiling but no ceiling plenum air return systems
- Inside surface resistance of 0.685 (h·ft²·°F)/Btu

Note: 2. Adjustments to table data

- Design temperatures : Corr. CLTD = CLTD + $(78 - t_r) + (t_m - 85)$

  where

  t_r = inside temperature and t_m = mean outdoor temperature

  t_m = maximum outdoor temperature − (daily range)/2
- No adjustment recommended for color
- No adjustment recommended for ventilation of air space above a ceiling

Table 31 Roof Numbers Used in Table 30

Mass Location**	Suspended Ceiling	R-Value, h·ft²·°F/Btu	B7, Wood 1 in.	C12, HW Concrete 2 in.	A3, Steel Deck	Attic-Ceiling Combination
Mass inside the insulation	Without	0 to 5	*	2	*	*
		5 to 10	*	2	*	*
		10 to 15	*	4	*	*
		15 to 20	*	4	*	*
		20 to 25	*	5	*	*
		25 to 30	*	*	*	*
	With	0 to 5	*	5	*	*
		5 to 10	*	8	*	*
		10 to 15	*	13	*	*
		15 to 20	*	13	*	*
		20 to 25	*	14	*	*
		25 to 30	*	*	*	*
Mass evenly placed	Without	0 to 5	1	2	1	1
		5 to 10	2	*	1	2
		10 to 15	2	*	1	2
		15 to 20	4	*	2	2
		20 to 25	4	*	2	4
		25 to 30	*	*	*	*
	With	0 to 5	*	3	1	*
		5 to 10	4	*	1	*
		10 to 15	5	*	2	*
		15 to 20	9	*	2	*
		20 to 25	10	*	4	*
		25 to 30	10	*	*	*
Mass outside the insulation	Without	0 to 5	*	2	*	*
		5 to 10	*	3	*	*
		10 to 15	*	4	*	*
		15 to 20	*	5	*	*
		20 to 25	*	5	*	*
		25 to 30	*	*	*	*
	With	0 to 5	*	3	*	*
		5 to 10	*	3	*	*
		10 to 15	*	4	*	*
		15 to 20	*	5	*	*
		20 to 25	*	*	*	*
		25 to 30	*	*	*	*

*Denotes a roof that is not possible with the chosen parameters.

**The 2 in. concrete is considered massive and the others nonmassive.

Table 32 July Cooling Load Temperature Differences for Calculating Cooling Load from Sunlit Walls 40°North Latitude

Wall Number 1

Wall Face	Hour 1	2	3	4	5	6	7	8	9	10	11	12	13	14	15	16	17	18	19	20	21	22	23	24
N	1	0	−1	−2	−3	−1	7	11	11	13	17	21	25	27	29	29	28	29	27	17	11	7	5	3
NE	1	0	−1	−2	−3	2	24	42	47	43	35	28	27	28	29	29	27	24	20	14	10	7	5	3
E	1	0	−1	−2	−2	2	28	51	62	64	59	48	36	31	30	30	28	25	20	14	10	7	5	3
SE	1	0	−1	−2	−3	0	15	32	46	55	58	56	49	39	33	31	28	25	20	14	10	7	5	3
S	1	0	−1	−2	−3	−2	0	4	11	21	33	43	50	52	50	44	34	27	20	14	10	7	5	3
SW	2	0	−1	−2	−2	−2	0	4	8	13	17	25	39	53	64	70	69	61	45	24	13	8	5	3
W	2	1	−1	−2	−2	−2	1	4	8	13	17	21	27	42	59	73	80	79	62	32	16	9	6	3
NW	2	0	−1	−2	−2	−2	0	4	8	13	17	21	25	29	38	50	61	64	55	29	15	9	5	3

Wall Number 2

Wall Face	Hour 1	2	3	4	5	6	7	8	9	10	11	12	13	14	15	16	17	18	19	20	21	22	23	24
N	5	3	2	0	−1	−2	−1	3	7	9	11	14	18	21	24	26	27	28	28	27	22	17	12	8
NE	5	3	2	0	−1	−2	2	13	26	36	39	37	33	31	29	29	29	28	26	23	18	14	10	7
E	5	3	2	0	−1	−1	2	15	32	47	55	57	52	44	38	34	32	30	27	23	19	14	11	8
SE	5	3	2	0	−1	−2	0	8	20	33	43	50	53	51	45	39	35	31	28	24	19	14	11	8
S	5	3	2	0	−1	−2	−2	−1	2	7	14	24	33	42	47	48	46	40	33	27	21	15	11	8
SW	7	4	2	1	0	−1	−2	0	2	5	9	13	20	30	41	53	61	65	62	53	39	27	17	11
W	8	5	3	1	0	−1	−2	0	2	5	9	13	17	23	33	46	59	69	73	66	50	34	22	14
NW	8	4	2	1	−1	−2	−2	−1	2	5	9	13	17	21	25	32	41	51	57	54	42	29	19	12

Wall Number 3

Wall Face	Hour 1	2	3	4	5	6	7	8	9	10	11	12	13	14	15	16	17	18	19	20	21	22	23	24
N	7	5	3	2	1	0	2	5	7	8	11	14	17	20	23	24	25	26	27	24	20	16	13	10
NE	7	5	3	2	0	0	7	17	26	31	33	31	30	29	29	29	29	28	25	22	18	15	12	9
E	7	5	4	2	1	1	8	21	33	42	47	47	44	40	37	35	33	31	28	24	20	16	13	10
SE	8	5	4	2	1	0	4	12	22	32	39	44	46	44	41	38	35	32	29	24	20	16	13	10
S	8	6	4	2	1	0	0	1	4	9	16	24	31	38	41	42	40	36	31	26	22	17	14	11
SW	12	9	6	4	2	1	1	2	4	6	9	14	21	30	40	49	55	57	54	45	36	28	21	16
W	14	10	7	5	3	1	1	2	4	6	9	13	17	24	34	45	56	63	63	54	43	33	25	19
NW	12	8	6	4	2	1	0	2	3	6	9	13	16	20	25	32	40	48	50	44	35	27	21	16

Wall Number 4

Wall Face	Hour 1	2	3	4	5	6	7	8	9	10	11	12	13	14	15	16	17	18	19	20	21	22	23	24
N	11	8	6	4	2	0	0	1	3	5	7	10	13	16	19	22	24	26	27	27	26	22	19	15
NE	10	7	5	3	2	0	0	4	12	21	29	32	33	32	31	30	30	29	28	26	23	20	16	13
E	10	8	5	4	2	1	1	5	15	27	38	45	49	47	44	40	37	34	32	29	25	21	17	14
SE	11	8	6	4	2	1	0	2	8	17	27	36	43	46	46	44	41	37	34	30	26	22	18	14
S	11	8	6	4	2	1	0	−1	0	2	6	13	20	28	35	41	43	42	39	35	30	24	19	15
SW	18	13	9	6	3	2	0	0	0	2	5	8	12	18	27	36	46	53	57	57	51	42	33	25
W	21	15	10	7	4	2	1	0	1	2	5	8	11	15	21	30	40	51	60	64	60	50	40	30
NW	18	13	9	6	3	1	0	0	0	2	4	8	11	15	19	23	30	37	45	49	48	41	33	25

Wall Number 5

Wall Face	Hour 1	2	3	4	5	6	7	8	9	10	11	12	13	14	15	16	17	18	19	20	21	22	23	24
N	13	10	8	6	5	3	2	3	5	6	8	9	12	14	17	19	21	23	24	24	23	21	18	15
NE	13	10	8	6	5	3	3	7	14	20	25	27	28	28	28	28	28	28	27	26	23	21	18	15
E	14	11	9	7	5	4	4	8	17	26	33	39	40	40	38	37	35	34	32	29	26	23	20	17
SE	14	12	9	7	5	4	3	6	11	18	25	32	37	39	39	38	37	35	33	30	27	24	20	17
S	15	12	9	7	5	4	3	2	3	4	8	13	19	25	31	35	36	36	34	32	28	24	21	18
SW	22	18	14	11	8	6	5	4	4	5	6	9	12	17	25	33	40	46	49	48	44	38	32	26
W	25	20	16	13	10	7	5	4	4	5	7	9	11	15	20	28	37	45	52	54	50	44	37	30
NW	21	17	13	10	8	6	4	3	4	4	6	8	11	14	17	21	27	34	40	42	40	35	30	25

Wall Number 6

Wall Face	Hour 1	2	3	4	5	6	7	8	9	10	11	12	13	14	15	16	17	18	19	20	21	22	23	24
N	13	11	9	8	6	5	4	5	6	7	8	10	12	14	16	18	20	21	22	23	21	20	17	15
NE	14	12	10	8	6	5	6	10	15	20	23	25	25	26	26	27	27	27	26	25	23	21	18	16
E	16	13	11	9	7	6	7	11	18	25	31	35	36	36	35	35	34	33	31	29	26	24	21	18
SE	16	14	11	9	8	6	6	8	13	18	24	29	33	35	36	35	34	33	32	29	27	24	21	18
S	16	13	11	9	7	6	5	4	4	6	9	13	18	24	28	31	33	33	31	29	27	24	21	18
SW	23	19	16	14	11	9	7	6	6	7	8	10	13	18	24	31	37	42	44	43	40	35	31	27
W	26	22	18	15	13	10	8	7	7	7	8	10	12	15	20	27	35	42	47	48	45	40	35	30
NW	21	18	15	12	10	8	7	6	6	6	8	9	11	14	16	21	26	32	36	38	36	32	28	25

Table 32 July Cooling Load Temperature Differences for Calculating Cooling Load from Sunlit Walls 40°North Latitude (*Continued*)

Wall Number 7

Wall Face	1	2	3	4	5	6	7	8	9	10	11	12	13	14	15	16	17	18	19	20	21	22	23	24
N	13	12	10	9	7	6	6	7	8	8	9	11	12	14	16	17	18	19	20	20	19	18	16	15
NE	15	13	11	10	9	8	9	13	17	20	22	23	23	24	24	25	25	25	24	23	22	20	18	16
E	17	15	13	12	10	9	11	16	21	26	30	32	32	32	32	32	31	30	29	27	25	23	21	19
SE	17	15	13	12	10	9	9	12	16	21	25	28	31	32	32	32	31	30	29	27	25	23	21	19
S	16	14	13	11	10	8	7	7	7	9	12	15	19	23	26	28	29	29	28	26	24	22	20	18
SW	23	20	18	16	13	12	10	10	10	10	11	12	15	20	25	30	35	38	39	37	34	31	28	25
W	25	22	20	17	15	13	12	11	11	11	12	13	14	17	22	28	34	39	42	41	38	34	31	28
NW	20	18	16	14	12	10	9	9	9	9	10	11	13	15	17	21	26	30	33	33	30	28	25	23

Wall Number 9

Wall Face	1	2	3	4	5	6	7	8	9	10	11	12	13	14	15	16	17	18	19	20	21	22	23	24
N	17	15	13	11	9	7	5	4	4	4	5	7	8	10	12	15	17	19	21	22	23	23	22	20
NE	18	15	13	11	9	7	5	5	6	10	16	20	23	25	26	27	27	28	28	27	26	25	23	20
E	20	17	14	12	10	8	6	5	7	12	19	26	32	36	37	37	37	36	34	33	31	29	26	23
SE	20	17	15	12	10	8	6	5	6	9	13	19	25	31	34	36	37	36	35	34	32	29	26	23
S	21	18	15	12	10	8	6	5	4	3	4	6	10	14	20	25	29	33	34	34	32	30	27	24
SW	31	26	22	18	15	12	9	7	6	5	5	6	8	10	14	19	26	33	39	43	45	44	40	36
W	35	30	25	21	17	14	11	8	7	6	6	7	8	10	12	16	22	30	37	44	48	48	45	41
NW	29	25	21	17	14	11	9	7	5	5	5	6	7	9	11	14	18	22	28	34	37	38	36	33

Wall Number 10

Wall Face	1	2	3	4	5	6	7	8	9	10	11	12	13	14	15	16	17	18	19	20	21	22	23	24
N	17	15	13	11	9	7	6	5	5	5	6	7	8	10	12	14	17	18	20	22	22	22	21	19
NE	18	16	13	11	9	7	6	6	8	12	16	20	22	24	25	26	27	27	27	27	26	24	22	20
E	20	17	15	12	10	8	7	7	10	14	20	26	31	34	35	36	36	35	34	33	31	28	26	23
SE	21	18	15	13	10	8	7	6	7	10	15	20	25	30	33	34	35	35	34	33	31	29	26	23
S	21	18	15	13	11	9	7	5	4	4	5	7	11	15	20	24	28	31	32	32	31	29	26	24
SW	31	27	23	19	16	13	10	8	7	6	6	7	8	11	15	20	26	32	38	41	42	41	38	35
W	34	30	26	22	18	15	12	9	8	7	7	7	8	10	13	17	23	30	37	42	45	45	42	39
NW	28	24	21	18	15	12	10	8	6	6	6	6	8	10	12	14	18	23	28	33	35	36	34	31

Wall Number 11

Wall Face	1	2	3	4	5	6	7	8	9	10	11	12	13	14	15	16	17	18	19	20	21	22	23	24
N	16	14	13	12	10	9	8	7	7	7	8	9	10	11	12	14	15	17	18	19	20	19	18	17
NE	18	17	15	13	12	10	9	9	11	14	17	20	21	22	23	23	24	24	25	25	24	23	21	20
E	21	19	17	16	14	12	11	11	13	17	22	26	29	30	31	31	31	31	31	30	29	27	25	23
SE	21	19	17	16	14	12	11	10	11	14	17	21	24	27	29	30	31	31	30	30	29	27	25	23
S	20	18	16	15	13	11	10	9	8	8	8	10	13	16	19	23	25	27	28	28	27	25	24	22
SW	28	25	23	20	18	16	14	12	11	11	10	11	12	14	17	21	25	30	33	36	36	35	33	30
W	31	28	25	22	20	18	16	14	12	12	11	12	12	13	15	19	23	28	33	37	39	38	36	33
NW	25	23	20	18	16	14	12	11	10	9	9	10	11	12	13	15	18	22	26	29	31	31	29	27

Wall Number 12

Wall Face	1	2	3	4	5	6	7	8	9	10	11	12	13	14	15	16	17	18	19	20	21	22	23	24
N	16	14	13	12	11	10	8	8	8	8	8	9	10	11	12	14	15	16	17	18	19	19	18	17
NE	18	17	15	14	13	11	10	10	12	14	17	19	21	21	22	23	23	24	24	24	23	22	21	20
E	22	20	18	17	15	13	12	12	14	17	21	25	28	29	30	30	30	30	30	29	28	27	25	24
SE	22	20	18	16	15	13	12	11	12	14	17	21	24	26	28	29	30	30	30	29	28	27	25	23
S	20	19	17	15	14	12	11	10	9	9	9	11	13	16	19	22	24	26	26	26	26	25	23	22
SW	27	25	23	21	19	17	15	14	12	12	12	12	12	14	17	20	24	28	32	34	34	34	32	30
W	30	28	25	23	21	19	17	15	14	13	13	13	13	14	16	19	23	27	32	35	37	36	35	33
NW	24	22	20	19	17	15	13	12	11	10	10	11	11	12	13	15	18	21	25	28	29	29	28	26

Columns 1–24: Hour

Table 32 July Cooling Load Temperature Differences for Calculating Cooling Load from Sunlit Walls 40°North Latitude (*Concluded*)

Wall Number 13

Wall Face	Hour 1	2	3	4	5	6	7	8	9	10	11	12	13	14	15	16	17	18	19	20	21	22	23	24
N	15	14	13	12	11	10	9	9	9	9	9	10	10	11	12	14	15	16	17	18	18	18	17	16
NE	18	17	16	15	13	12	11	12	13	16	18	19	20	21	21	22	23	23	23	23	23	22	21	20
E	22	20	19	17	16	15	14	14	16	19	22	25	27	28	29	29	29	29	29	28	27	26	25	23
SE	22	20	19	17	16	14	13	13	14	16	18	21	24	26	27	28	28	28	28	28	27	26	24	23
S	20	18	17	16	14	13	12	11	10	10	11	12	14	16	19	21	23	24	25	25	24	23	22	21
SW	26	25	23	21	19	18	16	15	14	13	13	13	14	15	18	21	24	28	30	32	32	31	30	28
W	29	27	25	23	21	19	18	16	15	15	14	14	15	15	17	20	23	27	31	34	34	34	32	31
NW	23	22	20	18	17	15	14	13	12	12	12	12	12	13	14	16	18	21	24	26	27	27	26	25

Wall Number 14

Wall Face	Hour 1	2	3	4	5	6	7	8	9	10	11	12	13	14	15	16	17	18	19	20	21	22	23	24
N	15	15	14	13	12	11	10	10	10	10	10	10	10	11	12	13	14	15	15	16	17	17	16	16
NE	19	18	17	16	15	14	13	13	14	15	17	18	19	20	20	21	21	22	22	22	22	22	21	20
E	23	22	21	19	18	17	16	15	16	18	21	23	25	26	27	27	28	28	28	28	27	26	25	24
SE	23	21	20	19	18	16	15	15	15	16	18	20	22	24	25	26	27	27	27	27	26	26	25	24
S	20	19	18	17	16	15	14	13	12	12	12	12	14	15	17	19	21	22	23	23	23	23	22	21
SW	26	25	24	22	21	19	18	17	16	15	15	15	15	16	17	19	22	25	27	29	30	30	29	28
W	29	27	26	24	23	21	20	18	17	16	16	16	16	16	17	19	21	24	27	30	32	32	31	30
NW	23	22	21	19	18	17	16	15	14	13	13	13	13	14	14	15	17	19	21	24	25	25	25	24

Wall Number 15

Wall Face	Hour 1	2	3	4	5	6	7	8	9	10	11	12	13	14	15	16	17	18	19	20	21	22	23	24
N	19	18	16	14	12	10	9	7	6	6	6	6	7	8	9	11	13	15	17	19	20	21	21	20
NE	21	19	17	15	13	11	9	8	7	9	11	14	18	20	22	23	25	25	26	26	26	26	25	23
E	25	22	20	17	15	12	10	9	9	10	14	18	23	27	30	32	34	34	34	33	32	31	29	27
SE	25	22	20	17	15	13	11	9	8	8	10	14	18	22	26	30	32	33	34	33	33	31	30	27
S	25	22	20	17	15	13	11	9	7	6	6	6	7	10	13	17	21	25	28	30	30	30	29	27
SW	35	32	28	25	22	18	16	13	11	9	8	8	8	9	11	14	18	23	28	33	37	39	39	37
W	39	35	32	28	24	21	18	15	12	10	9	8	8	9	10	13	16	21	26	32	38	41	42	41
NW	31	28	26	23	20	17	14	12	10	8	7	7	7	8	9	11	13	16	20	25	29	32	33	33

Wall Number 16

Wall Face	Hour 1	2	3	4	5	6	7	8	9	10	11	12	13	14	15	16	17	18	19	20	21	22	23	24
N	18	17	16	14	13	11	10	9	8	7	7	7	8	9	10	11	13	14	16	17	18	19	19	19
NE	21	20	18	16	14	13	11	10	10	11	13	15	17	19	21	22	23	24	24	25	25	24	24	23
E	25	23	21	19	17	15	13	11	11	12	15	19	22	26	28	30	31	31	32	32	31	30	29	27
SE	25	23	21	19	17	15	13	11	10	11	12	15	18	21	25	27	29	30	31	31	31	30	29	27
S	24	22	20	18	16	14	12	11	9	8	8	8	9	11	14	17	20	23	25	27	27	27	27	25
SW	33	30	28	25	23	20	18	15	13	12	11	10	10	11	12	15	18	22	27	30	33	35	35	34
W	36	33	31	28	25	22	20	17	15	13	12	11	11	11	12	14	17	20	25	30	34	37	38	37
NW	29	27	25	23	20	18	16	14	12	11	10	9	9	10	11	12	14	16	19	23	27	29	30	30

Note 1. Direct application of data

- Dark surface
- Indoor temperature of 78°F
- Outdoor maximum temperature of 95°F with mean temperature of 85°F and daily range of 21°F
- Solar radiation typical of clear day on 21st day of month
- Outside surface film resistance of 0.333 (h·ft²·°F)/Btu
- Inside surface resistance of 0.685 (h·ft²·°F)/Btu

Note 2. Adjustments to table data

- Design temperatures

$$\text{Corr. CLTD} = \text{CLTD} + (78 - t_r) + (t_m - 85)$$

where

t_r = inside temperature and

t_m = maximum outdoor temperature − (daily range)/2

- No adjustment recommended for color

Table 33A Wall Types, Mass Located Inside Insulation, for Use with Table 32

Secondary Material	R-Value, $ft^2 \cdot °F \cdot h/Btu$	Principal Wall Material**														
		A1	A2	B7	B10	B9	C1	C2	C3	C4	C5	C6	C7	C8	C17	C18
Stucco and/or plaster	0.0 to 2.0	*	*	*	*	*	*	*	*	*	*	*	*	*	*	*
	2.0 to 2.5	*	5	*	*	*	*	*	*	*	5	*	*	*	*	*
	2.5 to 3.0	*	5	*	*	*	3	*	2	5	6	*	*	5	*	*
	3.0 to 3.5	*	5	*	*	*	4	2	2	5	6	*	*	6	*	*
	3.5 to 4.0	*	5	*	*	*	4	2	3	6	6	10	4	6	*	5
	4.0 to 4.75	*	6	*	*	*	5	2	4	6	6	11	5	10	*	10
	4.75 to 5.5	*	6	*	*	*	5	2	4	6	6	11	5	10	*	10
	5.5 to 6.5	*	6	*	*	*	5	2	5	10	7	12	5	11	*	10
	6.5 to 7.75	*	6	*	*	*	5	4	5	11	7	16	10	11	*	11
	7.75 to 9.0	*	6	*	*	*	5	4	5	11	7	*	10	11	*	11
	9.0 to 10.75	*	6	*	*	*	5	4	5	11	7	*	10	11	4	11
	10.75 to 12.75	*	6	*	*	*	5	4	5	11	11	*	10	11	4	11
	12.75 to 15.0	*	10	*	*	*	10	4	5	11	11	*	10	11	9	12
	15.0 to 17.5	*	10	*	*	*	10	5	5	11	11	*	11	12	10	16
	17.5 to 20.0	*	11	*	*	*	10	5	9	11	11	*	15	16	10	16
	20.0 to 23.0	*	11	*	*	*	10	9	9	16	11	*	15	16	10	16
	23.0 to 27.0	*	*	*	*	*	*	*	*	*	*	*	16	*	15	*
Steel or other light-weight siding	0.0 to 2.0	*	*	*	*	*	*	*	*	*	*	*	*	*	*	*
	2.0 to 2.5	*	3	*	*	*	*	*	2	3	5	*	*	*	*	*
	2.5 to 3.0	*	5	*	*	*	2	*	2	5	3	*	*	5	*	*
	3.0 to 3.5	*	5	*	*	*	3	1	2	5	5	*	*	5	*	*
	3.5 to 4.0	*	5	*	*	*	3	2	2	5	5	6	3	5	*	5
	4.0 to 4.75	*	6	*	*	*	4	2	2	5	5	10	4	6	*	5
	4.75 to 5.5	*	6	*	*	*	5	2	2	6	6	11	5	6	*	6
	5.5 to 6.5	*	6	*	*	*	5	2	3	6	6	11	5	6	*	6
	6.5 to 7.75	*	6	*	*	*	5	2	3	6	6	11	5	6	*	10
	7.75 to 9.0	*	6	*	*	*	5	2	3	6	6	12	5	6	*	11
	9.0 to 10.75	*	6	*	*	*	5	2	3	6	6	12	5	6	4	11
	10.75 to 12.75	*	6	*	*	*	5	2	3	6	7	12	6	11	4	11
	12.75 to 15.0	*	6	*	*	*	5	2	4	6	7	12	10	11	5	11
	15.0 to 17.5	*	10	*	*	*	6	4	4	10	7	*	10	11	9	11
	17.5 to 20.0	*	10	*	*	*	10	4	4	10	11	*	10	11	10	11
	20.0 to 23.0	*	11	*	*	*	10	4	5	11	11	*	10	11	10	16
	23.0 to 27.0	*	*	*	*	*	*	*	*	*	*	*	10	*	11	16
Face brick	0.0 to 2.0	*	*	*	*	*	*	*	*	*	*	*	*	*	*	*
	2.0 to 2.5	3	*	*	*	*	*	*	*	*	11	*	*	*	*	*
	2.5 to 3.0	5	11	*	*	*	*	*	6	11	12	*	*	*	*	*
	3.0 to 3.5	5	12	5	*	*	11	*	11	12	12	*	*	12	*	*
	3.5 to 4.0	5	12	6	*	*	12	6	12	12	13	*	*	12	*	*
	4.0 to 4.75	6	13	6	10	*	13	10	12	12	13	*	11	*	*	16
	4.75 to 5.5	6	13	6	11	*	*	11	12	13	13	*	16	*	*	*
	5.5 to 6.5	6	13	6	11	*	*	11	12	13	13	*	*	*	*	*
	6.5 to 7.75	6	13	6	11	*	*	11	13	*	13	*	*	*	*	*
	7.75 to 9.0	6	13	10	16	*	*	11	13	*	13	*	*	*	*	*
	9.0 to 10.75	6	14	10	16	*	*	11	13	*	14	*	*	*	16	*
	10.75 to 12.75	6	14	10	16	*	*	11	13	*	14	*	*	*	16	*
	12.75 to 15.0	6	*	11	16	*	*	12	13	*	*	*	*	*	*	*
	15.0 to 17.5	10	*	11	*	*	*	12	13	*	*	*	*	*	*	*
	17.5 to 20.0	10	*	11	*	*	*	16	*	*	*	*	*	*	*	*
	20.0 to 23.0	11	*	15	*	*	*	16	*	*	*	*	*	*	*	*
	23.0 to 27.0	*	*	*	*	*	*	16	*	*	*	*	*	*	*	*

*Denotes a wall that is not possible with the chosen set of parameters.
**See Table 11 for definition of Code letters

Table 33B Wall Types, Mass Located Inside Insulation, for Use with Table 32

Secondary Material	R-Value, $ft^2 \cdot °F \cdot h/Btu$	Principal Wall Material**														
		A1	A2	B7	B10	B9	C1	C2	C3	C4	C5	C6	C7	C8	C17	C18
Stucco and/or plaster	0.0 to 2.0	1	3	*	*	*	*	*	1	3	3	*	*	*	*	*
	2.0 to 2.5	1	3	1	*	*	2	*	2	4	4	*	*	5	*	*
	2.5 to 3.0	1	4	1	*	*	2	2	2	4	4	*	*	5	*	*
	3.0 to 3.5	1	*	1	*	*	2	2	*	*	*	10	4	5	*	4
	3.5 to 4.0	1	*	1	2	*	*	4	*	*	*	10	4	*	*	4
	4.0 to 4.75	1	*	1	2	*	*	*	*	*	*	10	4	*	*	4
	4.75 to 5.5	1	*	1	2	*	*	*	*	*	*	*	*	*	*	*
	5.5 to 6.5	1	*	2	4	10	*	*	*	*	*	*	*	*	*	*
	6.5 to 7.75	1	*	2	4	11	*	*	*	*	*	*	*	*	*	*
	7.75 to 9.0	1	*	2	4	16	*	*	*	*	*	*	*	*	*	*
	9.0 to 10.75	1	*	2	4	16	*	*	*	*	*	*	*	*	4	*
	10.75 to 12.75	1	*	2	5	*	*	*	*	*	*	*	*	*	4	*
	12.75 to 15.0	2	*	2	5	*	*	*	*	*	*	*	*	*	*	*
	15.0 to 17.5	2	*	2	5	*	*	*	*	*	*	*	*	*	*	*
	17.5 to 20.0	2	*	2	9	*	*	*	*	*	*	*	*	*	*	*
	20.0 to 23.0	2	*	4	9	*	*	*	*	*	*	*	*	*	*	*
	23.0 to 27.0	*	*	*	9	*	*	*	*	*	*	*	*	*	*	*
Steel or other lightweight siding	0.0 to 2.0	1	3	*	*	*	*	*	1	3	2	*	*	*	*	*
	2.0 to 2.5	1	3	1	*	*	2	*	1	3	2	*	*	3	*	*
	2.5 to 3.0	1	4	1	*	*	2	1	2	4	4	*	*	3	*	*
	3.0 to 3.5	1	*	1	*	*	4	1	*	*	*	5	2	4	*	4
	3.5 to 4.0	1	*	1	2	*	*	2	*	*	*	5	2	*	*	4
	4.0 to 4.75	1	*	1	2	*	*	*	*	*	*	10	4	*	*	4
	4.75 to 5.5	1	*	1	2	*	*	*	*	*	*	*	*	*	*	*
	5.5 to 6.5	1	*	1	2	10	*	*	*	*	*	*	*	*	*	*
	6.5 to 7.75	1	*	1	4	11	*	*	*	*	*	*	*	*	*	*
	7.75 to 9.0	1	*	2	4	16	*	*	*	*	*	*	*	*	*	*
	9.0 to 10.75	1	*	2	4	16	*	*	*	*	*	*	*	*	2	*
	10.75 to 12.75	1	*	2	4	*	*	*	*	*	*	*	*	*	4	*
	12.75 to 15.0	1	*	2	5	*	*	*	*	*	*	*	*	*	*	*
	15.0 to 17.5	1	*	2	5	*	*	*	*	*	*	*	*	*	*	*
	17.5 to 20.0	1	*	2	5	*	*	*	*	*	*	*	*	*	*	*
	20.0 to 23.0	2	*	4	9	*	*	*	*	*	*	*	*	*	*	*
	23.0 to 27.0	*	*	*	9	*	*	*	*	*	*	*	*	*	*	*
Face brick	0.0 to 2.0	3	6	*	*	*	*	*	*	*	6	*	*	*	*	*
	2.0 to 2.5	3	10	*	*	*	*	*	5	10	10	*	*	*	*	*
	2.5 to 3.0	4	10	5	*	*	5	*	5	10	11	*	*	10	*	*
	3.0 to 3.5	*	11	5	*	*	10	5	5	11	11	15	10	10	*	10
	3.5 to 4.0	*	11	5	10	*	10	5	5	11	11	16	10	16	*	10
	4.0 to 4.75	*	11	*	11	*	10	5	5	16	11	*	10	16	*	16
	4.75 to 5.5	*	11	*	11	*	10	5	10	16	16	*	10	16	*	16
	5.5 to 6.5	*	16	*	*	*	10	9	10	16	11	*	11	16	*	16
	6.5 to 7.75	*	16	*	*	*	11	9	10	16	16	*	16	16	*	*
	7.75 to 9.0	*	16	*	*	*	15	9	10	16	*	*	15	16	*	*
	9.0 to 10.75	*	16	*	*	*	15	10	10	*	16	*	16	*	10	*
	10.75 to 12.75	*	16	*	*	*	16	10	10	*	*	*	16	*	15	*
	12.75 to 15.0	*	16	*	*	*	16	10	10	*	16	*	*	*	15	*
	15.0 to 17.5	*	*	*	*	*	16	10	15	*	*	*	*	*	16	*
	17.5 to 20.0	*	*	*	*	*	16	15	15	*	*	*	*	*	16	*
	20.0 to 23.0	*	*	*	*	*	*	15	16	*	*	*	*	*	*	*
	23.0 to 27.0	*	*	*	*	*	*	15	*	*	*	*	*	*	*	*

*Denotes a wall that is not possible with the chosen set of parameters.
**See Table 11 for definition of Code letters

Table 33C Wall Types, Mass Located Inside Insulation, for Use with Table 32

Secondary Material	R-Value, $ft^2 \cdot °F \cdot h/Btu$	Principal Wall Material**														
		A1	A2	B7	B10	B9	C1	C2	C3	C4	C5	C6	C7	C8	C17	C18
Stucco and/or plaster	0.0 to 2.0	1	3	*	*	*	*	*	1	3	3	*	*	*	*	*
	2.0 to 2.5	1	3	1	*	*	2	*	2	4	4	*	*	5	*	*
	2.5 to 3.0	1	4	1	*	*	2	2	2	4	4	*	*	5	*	*
	3.0 to 3.5	1	*	1	*	*	2	2	*	*	*	10	4	5	*	4
	3.5 to 4.0	1	*	1	2	*	*	4	*	*	*	10	4	*	*	4
	4.0 to 4.75	1	*	1	2	*	*	*	*	*	*	10	4	*	*	4
	4.75 to 5.5	1	*	1	2	*	*	*	*	*	*	*	*	*	*	*
	5.5 to 6.5	1	*	2	4	10	*	*	*	*	*	*	*	*	*	*
	6.5 to 7.75	1	*	2	4	11	*	*	*	*	*	*	*	*	*	*
	7.75 to 9.0	1	*	2	4	16	*	*	*	*	*	*	*	*	*	*
	9.0 to 10.75	1	*	2	4	16	*	*	*	*	*	*	*	*	4	*
	10.75 to 12.75	1	*	2	5	*	*	*	*	*	*	*	*	*	4	*
	12.75 to 15.0	2	*	2	5	*	*	*	*	*	*	*	*	*	*	*
	15.0 to 17.5	2	*	2	5	*	*	*	*	*	*	*	*	*	*	*
	17.5 to 20.0	2	*	2	9	*	*	*	*	*	*	*	*	*	*	*
	20.0 to 23.0	2	*	4	9	*	*	*	*	*	*	*	*	*	*	*
	23.0 to 27.0	*	*	*	9	*	*	*	*	*	*	*	*	*	*	*
Steel or other light-weight siding	0.0 to 2.0	1	3	*	*	*	*	*	1	3	2	*	*	*	*	*
	2.0 to 2.5	1	3	1	*	*	2	*	1	3	2	*	*	3	*	*
	2.5 to 3.0	1	4	1	*	*	2	1	2	4	4	*	*	3	*	*
	3.0 to 3.5	1	*	1	*	*	4	1	*	*	*	5	2	4	*	4
	3.5 to 4.0	1	*	1	2	*	*	2	*	*	*	5	2	*	*	4
	4.0 to 4.75	1	*	1	2	*	*	*	*	*	*	10	4	*	*	4
	4.75 to 5.5	1	*	1	2	*	*	*	*	*	*	*	*	*	*	*
	5.5 to 6.5	1	*	1	2	10	*	*	*	*	*	*	*	*	*	*
	6.5 to 7.75	1	*	1	4	11	*	*	*	*	*	*	*	*	*	*
	7.75 to 9.0	1	*	2	4	16	*	*	*	*	*	*	*	*	*	*
	9.0 to 10.75	1	*	2	4	16	*	*	*	*	*	*	*	*	2	*
	10.75 to 12.75	1	*	2	4	*	*	*	*	*	*	*	*	*	4	*
	12.75 to 15.0	1	*	2	5	*	*	*	*	*	*	*	*	*	*	*
	15.0 to 17.5	1	*	2	5	*	*	*	*	*	*	*	*	*	*	*
	17.5 to 20.0	1	*	2	5	*	*	*	*	*	*	*	*	*	*	*
	20.0 to 23.0	2	*	4	9	*	*	*	*	*	*	*	*	*	*	*
	23.0 to 27.0	*	*	*	9	*	*	*	*	*	*	*	*	*	*	*
Face brick	0.0 to 2.0	3	6	*	*	*	*	*	*	*	6	*	*	*	*	*
	2.0 to 2.5	3	10	*	*	*	*	*	5	10	10	*	*	*	*	*
	2.5 to 3.0	4	10	5	*	*	5	*	5	10	11	*	*	10	*	*
	3.0 to 3.5	*	11	5	*	*	10	5	5	11	11	15	10	10	*	10
	3.5 to 4.0	*	11	5	10	*	10	5	5	11	11	16	10	16	*	10
	4.0 to 4.75	*	11	*	11	*	10	5	5	16	11	*	10	16	*	16
	4.75 to 5.5	*	11	*	11	*	10	5	10	16	16	*	10	16	*	16
	5.5 to 6.5	*	16	*	*	*	10	9	10	16	11	*	11	16	*	16
	6.5 to 7.75	*	16	*	*	*	11	9	10	16	16	*	16	16	*	*
	7.75 to 9.0	*	16	*	*	*	15	9	10	16	*	*	15	16	*	*
	9.0 to 10.75	*	16	*	*	*	15	10	10	*	16	*	16	*	10	*
	10.75 to 12.75	*	16	*	*	*	16	10	10	*	*	*	16	*	15	*
	12.75 to 15.0	*	16	*	*	*	16	10	10	*	16	*	*	*	15	*
	15.0 to 17.5	*	*	*	*	*	16	10	15	*	*	*	*	*	16	*
	17.5 to 20.0	*	*	*	*	*	16	15	15	*	*	*	*	*	16	*
	20.0 to 23.0	*	*	*	*	*	*	15	16	*	*	*	*	*	*	*
	23.0 to 27.0	*	*	*	*	*	*	15	*	*	*	*	*	*	*	*

*Denotes a wall that is not possible with the chosen set of parameters.

**See Table 11 for definition of Code letters

The composite effect of the various forms of interior shading on solar radiation from glass, relative to unshaded clear double-strength glass, is represented by a shading coefficient (SC) or decimal multiplier, tabulated in Chapter 29 for a wide variety of conditions.

Exterior Shading. Where glass is shaded by exterior means of a permanent nature, the hourly mitigating effect of such shading may be estimated by separate evaluations of shaded areas relative to unshaded areas for each situation as previously noted.

Example 7. Cooling load from south and west glass. Determine the cooling load caused by glass on the south and west walls of a building at 1200, 1400, and 1600 h in July. The building is located at 40°N latitude with outside design conditions of 90°F dry-bulb temperature and a 20°F daily range. The inside design dry bulb temperature is 78°F. Assume the room configuration includes two exposed walls, vinyl floor covering, and gypsum partitions, and that the building is a single story. The south glass is insulating type (0.25 in. air space) with an area of 100 ft² and no interior shading. The west glass is 7/32 in. single grey-tinted glass with an area of 100 ft² and with light-colored venetian blinds.

Table 34 Cooling Load Temperature Differences (CLTD) for Conduction through Glass

Solar Time, h	CLTD, °F	Solar Time, h	CLTD, °F
0100	1	1300	12
0200	0	1400	13
0300	−1	1500	14
0400	−2	1600	14
0500	−2	1700	13
0600	−2	1800	12
0700	−2	1900	10
0800	0	2000	8
0900	2	2100	6
1000	4	2200	4
1100	7	2300	3
1200	9	2400	2

Corrections: The values in the table were calculated for an inside temperature of 78°F and an outdoor maximum temperature of 95°F with an outdoor daily range of 21°F. The table remains approximately correct for other outdoor maximums 93 to 102°F and other outdoor daily ranges 16 to 34°F, provided the outdoor daily average temperature remains approximately 85°F. If the room air temperature is different from 78°F and/or the outdoor daily average temperature is different from 85°F see note 2, Table 32.

Solution: By the room configuration described and with inside shading for half the exposed glass area, Table 35B indicates the SCL factors should be selected for a Zone C condition.

Data required for the calculation are as follows:

Variable		South Glass	West Glass
U, Btu/(h·ft²·°F)*		0.61	0.81
Area A, ft²		100	100
SC (Chapter 29)		0.82	0.53
SCL (Table 36, Zone C)	1200	79	37
	1400	80	98
	1600	40	153

*U-factors based on previous edition of this Handbook. See Table 5, Chapter 29 for current values.

Table 35A Zone Types for Use with CLF Tables, Interior Rooms

Zone Parameters[a]				Zone Type	
Room Location	Middle Floor	Ceiling Type	Floor Covering	People and Equipment	Lights
Single story	N/A	N/A	Carpet	C	B
	N/A	N/A	Vinyl	D	C
Top floor	2.5 in. Concrete	With	Carpet	D	C
	2.5 in. Concrete	With	Vinyl	D	D
	2.5 in. Concrete	Without	b	D	B
	1 in. Wood	b	b	D	B
Bottom floor	2.5 in. Concrete	With	Carpet	D	C
	2.5 in. Concrete	b	Vinyl	D	D
	2.5 in. Concrete	Without	Carpet	D	D
	1 in. Wood	b	Carpet	D	C
	1 in. Wood	b	Vinyl	D	D
Mid-floor	2.5 in. Concrete	N/A	Carpet	D	C
	2.5 in. Concrete	N/A	Vinyl	D	D
	1 in. Wood	N/A	b	C	B

[a]A total of 14 zone parameters is fully defined in Table 20. Those not shown in this table were selected to achieve an error band of approximately 10%.
[b]The effect of this parameter is negligible in this case.

Table 35B Zone Types for Use with SCL and CLF Tables, Single-Story Building

Zone Parameters[a]				Zone Type			Error Band	
No. Walls	Floor Covering	Partition Type	Inside Shade	Glass Solar	People and Equipment	Lights	Plus	Minus
1 or 2	Carpet	Gypsum	b	A	B	B	9	2
1 or 2	Carpet	Concrete block	b	B	C	C	9	0
1 or 2	Vinyl	Gypsum	Full	B	C	C	9	0
1 or 2	Vinyl	Gypsum	Half to None	C	C	C	16	0
1 or 2	Vinyl	Concrete block	Full	C	D	D	8	0
1 or 2	Vinyl	Concrete block	Half to None	D	D	D	10	6
3	Carpet	Gypsum	b	A	B	B	9	2
3	Carpet	Concrete block	Full	A	B	B	9	2
3	Carpet	Concrete block	Half to None	B	B	B	9	0
3	Vinyl	Gypsum	Full	B	C	C	9	0
3	Vinyl	Gypsum	Half to None	C	C	C	16	0
3	Vinyl	Concrete block	Full	B	C	C	9	0
3	Vinyl	Concrete block	Half to None	C	C	C	16	0
4	Carpet	Gypsum	b	A	B	B	6	3
4	Vinyl	Gypsum	Full	B	C	C	11	6
4	Vinyl	Gypsum	Half to None	C	C	C	19	−1

[a]A total of 14 zone parameters is fully defined in Table 20. Those not shown in this table were selected to achieve the minimum error band shown in the righthand column for Solar Cooling Load (SCL). The error band for Lights and People and Equipment is approximately 10%.
[b]The effect of inside shade is negligible in this case.

Table 36 July Solar Cooling Load For Sunlit Glass 40°North Latitude

Zone Type A

Glass Face	Hour 1	2	3	4	5	6	7	8	9	10	11	12	13	14	15	16	17	18	19	20	21	22	23	24
												Solar Time												
N	0	0	0	0	1	25	27	28	32	35	38	40	40	39	36	31	31	36	12	6	3	1	1	0
NE	0	0	0	0	2	85	129	134	112	75	55	48	44	40	37	32	26	18	7	3	2	1	0	0
E	0	0	0	0	2	93	157	185	183	154	106	67	53	45	39	33	26	18	7	3	2	1	0	0
SE	0	0	0	0	1	47	95	131	150	150	131	97	63	49	41	34	27	18	7	3	2	1	0	0
S	0	0	0	0	0	9	17	25	41	64	85	97	96	84	63	42	31	20	8	4	2	1	0	0
SW	0	0	0	0	0	9	17	24	30	35	39	64	101	133	151	152	133	93	35	17	8	4	2	1
W	1	0	0	0	0	9	17	24	30	35	38	40	65	114	158	187	192	156	57	27	13	6	3	2
NW	1	0	0	0	0	9	17	24	30	35	38	40	40	50	84	121	143	130	46	22	11	5	3	1
Hor	0	0	0	0	0	24	69	120	169	211	241	257	259	245	217	176	125	70	29	14	7	3	2	1

Zone Type B

Glass Face	Hour 1	2	3	4	5	6	7	8	9	10	11	12	13	14	15	16	17	18	19	20	21	22	23	24
												Solar Time												
N	2	2	1	1	1	22	23	24	28	32	35	37	38	37	35	32	31	35	16	10	7	5	4	3
NE	2	1	1	1	2	73	109	116	101	73	58	52	48	45	41	36	30	23	13	9	6	5	3	3
E	2	2	1	1	2	80	133	159	162	143	105	74	63	55	48	41	34	25	15	10	7	5	4	3
SE	2	2	1	1	1	40	81	112	131	134	122	96	69	58	49	42	35	26	15	10	8	6	4	3
S	2	2	1	1	1	8	15	21	36	56	74	86	87	79	63	46	37	27	16	11	8	6	4	3
SW	6	5	4	3	2	9	16	22	27	31	36	58	89	117	135	138	126	94	46	31	21	15	11	8
W	8	6	5	4	3	9	16	22	27	31	35	37	59	101	139	166	173	147	66	43	30	21	15	11
NW	6	5	4	3	2	9	16	22	27	31	34	37	37	46	76	108	128	119	51	33	22	16	11	8
Hor	8	6	5	4	3	22	60	104	147	185	214	233	239	232	212	180	137	90	53	37	27	19	14	11

Zone Type C

Glass Face	Hour 1	2	3	4	5	6	7	8	9	10	11	12	13	14	15	16	17	18	19	20	21	22	23	24
												Solar Time												
N	5	5	4	4	4	24	23	24	27	30	33	34	35	34	32	29	29	34	14	10	8	7	6	6
NE	7	6	6	5	6	75	106	107	88	61	49	47	45	43	40	36	31	25	16	13	11	10	9	8
E	9	8	8	7	8	83	130	148	145	124	89	62	56	52	47	43	37	30	20	17	15	13	12	11
SE	9	8	7	6	6	45	82	107	121	121	107	82	59	51	47	42	36	29	19	16	14	13	11	10
S	7	7	6	5	5	12	18	23	36	54	70	79	79	70	54	40	33	26	16	13	12	10	9	8
SW	14	12	11	10	9	15	21	26	29	33	36	57	86	110	124	125	111	80	37	28	23	20	17	15
W	17	15	13	12	11	17	22	27	31	34	36	37	59	98	132	153	156	128	50	35	28	24	21	19
NW	12	11	10	9	8	14	20	25	29	32	34	36	36	44	73	102	118	107	39	26	21	17	15	13
Hor	24	21	19	17	16	34	68	107	144	175	199	212	215	207	189	160	123	83	53	44	38	34	30	27

Zone Type D

Glass Face	Hour 1	2	3	4	5	6	7	8	9	10	11	12	13	14	15	16	17	18	19	20	21	22	23	24
												Solar Time												
N	8	7	6	6	6	21	21	21	24	27	29	31	32	31	30	28	29	32	17	14	12	11	10	9
NE	11	10	9	8	9	63	87	90	77	58	49	48	46	44	42	39	35	29	22	19	17	15	14	12
E	15	13	12	11	11	70	107	123	124	110	85	65	60	57	53	48	43	37	29	25	22	20	18	16
SE	14	13	11	10	10	39	68	90	102	104	95	78	60	55	51	47	42	35	27	24	21	19	17	16
S	11	10	9	8	7	12	17	21	32	46	59	67	69	63	52	41	36	30	22	19	17	15	14	12
SW	21	19	17	15	14	18	22	25	28	31	34	51	74	94	106	109	100	78	45	37	33	29	26	23
W	25	23	20	18	17	21	24	28	30	33	34	35	53	84	112	130	135	116	57	46	39	35	31	28
NW	18	16	15	13	12	17	21	24	27	30	32	33	34	41	64	87	101	94	42	33	29	25	22	20
Hor	37	33	30	27	24	38	64	95	124	150	171	185	191	188	176	156	128	96	72	63	56	50	45	41

Notes:

1. Values are in Btu/h·ft².
2. Apply data directly to standard double strength glass with no inside shade.
3. Data applies to 21st day of July.
4. For other types of glass and internal shade, use shading coefficients as multiplier. See text. For externally shaded glass, use north orientation. See text.

The conduction heat gain component of cooling load by Equation (42) is:

Time	CLTD (Table 34)	CLTD Corrected	South Glass, Btu/h	West Glass, Btu/h
1200	9	4	244	324
1400	13	8	488	648
1600	14	9	549	729

The correction factor applied to the above CLTDs was −5°F, computed from the notes of Table 34. Heat gain values are rounded.

The solar heat gain component of cooling load by Equation (43) is:

	South Glass			West Glass		
Time	SC	SCL	SHG, Btu/h	SC	SCL	SHG, Btu/h
1200	0.82	79	6478	0.53	37	1961
1400	0.82	70	5740	0.53	98	5194
1600	0.82	40	3280	0.53	153	8109

The total cooling load due to heat gain through the glass is, therefore:

Time	South Glass, Btu/h	West Glass, Btu/h
1200	6722	2285
1400	6198	5842
1600	3829	8838

HEAT SOURCES WITHIN CONDITIONED SPACE

People

The basic principles of evaluating heat gain and moisture generation from people are the same as those previously described for the TFM. Latent heat gains are considered instantaneous cooling loads.

The total sensible heat gain from people is not converted directly to cooling load. The radiant portion is first absorbed by the surroundings (floor, ceiling, partitions, furniture) then convected to the space at a later time, depending on the thermal characteristics of the room. The radiant portion of the sensible heat gain from people varies widely depending on the circumstances, as indicated by Table 3 and in more detail by Chapter 8. A 70% value was used to generate CLFs for Table 37, which considers the storage effect on this radiant load in its results, plus the 30% convective portion. The instantaneous sensible cooling load is thus:

$$q_s = N(\text{SHG}_p)(\text{CLF}_p) \quad (44)$$

and the latent cooling load is:

$$q_l = N(\text{LHG}_p) \quad (45)$$

Table 37 Cooling Load Factors for People and Unhooded Equipment

Hours in Space	Number of Hours after Entry into Space or Equipment Turned On																							
	1	2	3	4	5	6	7	8	9	10	11	12	13	14	15	16	17	18	19	20	21	22	23	24
Zone Type A																								
2	0.75	0.88	0.18	0.08	0.04	0.02	0.01	0.01	0.01	0.01	0.00	0.00	0.00	0.00	0.00	0.00	0.00	0.00	0.00	0.00	0.00	0.00	0.00	0.00
4	0.75	0.88	0.93	0.95	0.22	0.10	0.05	0.03	0.02	0.02	0.01	0.01	0.01	0.01	0.00	0.00	0.00	0.00	0.00	0.00	0.00	0.00	0.00	0.00
6	0.75	0.88	0.93	0.95	0.97	0.97	0.23	0.11	0.06	0.04	0.03	0.02	0.02	0.01	0.01	0.01	0.01	0.00	0.00	0.00	0.00	0.00	0.00	0.00
8	0.75	0.88	0.93	0.95	0.97	0.97	0.98	0.98	0.24	0.11	0.06	0.04	0.03	0.02	0.02	0.01	0.01	0.01	0.01	0.01	0.00	0.00	0.00	0.00
10	0.75	0.88	0.93	0.95	0.97	0.97	0.98	0.98	0.99	0.99	0.24	0.12	0.07	0.04	0.03	0.02	0.02	0.01	0.01	0.01	0.01	0.01	0.00	0.00
12	0.75	0.88	0.93	0.96	0.97	0.98	0.98	0.98	0.99	0.99	0.99	0.99	0.25	0.12	0.07	0.04	0.03	0.02	0.02	0.02	0.01	0.01	0.01	0.01
14	0.76	0.88	0.93	0.96	0.97	0.98	0.98	0.99	0.99	0.99	0.99	0.99	1.00	1.00	0.25	0.12	0.07	0.05	0.03	0.03	0.02	0.02	0.01	0.01
16	0.76	0.89	0.94	0.96	0.97	0.98	0.98	0.99	0.99	0.99	0.99	0.99	1.00	1.00	1.00	1.00	0.25	0.12	0.07	0.05	0.03	0.03	0.02	0.02
18	0.77	0.89	0.94	0.96	0.97	0.98	0.98	0.99	0.99	0.99	0.99	1.00	1.00	1.00	1.00	1.00	1.00	1.00	0.25	0.12	0.07	0.05	0.03	0.03
Zone Type B																								
2	0.65	0.74	0.16	0.11	0.08	0.06	0.05	0.04	0.03	0.02	0.02	0.01	0.01	0.01	0.01	0.00	0.00	0.00	0.00	0.00	0.00	0.00	0.00	0.00
4	0.65	0.75	0.81	0.85	0.24	0.17	0.13	0.10	0.07	0.06	0.04	0.03	0.03	0.02	0.02	0.01	0.01	0.01	0.01	0.00	0.00	0.00	0.00	0.00
6	0.65	0.75	0.81	0.85	0.89	0.91	0.29	0.20	0.15	0.12	0.09	0.07	0.05	0.04	0.03	0.02	0.02	0.01	0.01	0.01	0.01	0.01	0.00	0.00
8	0.65	0.75	0.81	0.85	0.89	0.91	0.93	0.95	0.31	0.22	0.17	0.13	0.10	0.08	0.06	0.05	0.04	0.03	0.02	0.02	0.01	0.01	0.01	0.01
10	0.65	0.75	0.81	0.85	0.89	0.91	0.93	0.95	0.96	0.97	0.33	0.24	0.18	0.14	0.11	0.08	0.06	0.05	0.04	0.03	0.02	0.02	0.01	0.01
12	0.66	0.76	0.81	0.86	0.89	0.92	0.94	0.95	0.96	0.97	0.98	0.98	0.34	0.24	0.19	0.14	0.11	0.08	0.06	0.05	0.04	0.03	0.02	0.02
14	0.67	0.76	0.82	0.86	0.89	0.92	0.94	0.95	0.96	0.97	0.98	0.98	0.99	0.99	0.35	0.25	0.19	0.15	0.11	0.09	0.07	0.05	0.04	0.03
16	0.69	0.78	0.83	0.87	0.90	0.92	0.94	0.95	0.96	0.97	0.98	0.98	0.99	0.99	0.99	0.99	0.35	0.25	0.19	0.15	0.11	0.09	0.07	0.05
18	0.71	0.80	0.85	0.88	0.91	0.93	0.95	0.96	0.97	0.98	0.98	0.99	0.99	0.99	0.99	0.99	1.00	1.00	0.35	0.25	0.19	0.15	0.11	0.09
Zone Type C																								
2	0.60	0.68	0.14	0.11	0.09	0.07	0.06	0.05	0.04	0.03	0.03	0.02	0.02	0.01	0.01	0.01	0.01	0.01	0.01	0.00	0.00	0.00	0.00	0.00
4	0.60	0.68	0.74	0.79	0.23	0.18	0.14	0.12	0.10	0.08	0.06	0.05	0.04	0.04	0.03	0.02	0.02	0.02	0.01	0.01	0.01	0.01	0.01	0.01
6	0.61	0.69	0.74	0.79	0.83	0.86	0.28	0.22	0.18	0.15	0.12	0.10	0.08	0.07	0.06	0.05	0.04	0.03	0.03	0.02	0.02	0.01	0.01	0.01
8	0.61	0.69	0.75	0.79	0.83	0.86	0.89	0.91	0.32	0.26	0.21	0.17	0.14	0.11	0.09	0.08	0.06	0.05	0.04	0.04	0.03	0.02	0.02	0.02
10	0.62	0.70	0.75	0.80	0.83	0.86	0.89	0.91	0.92	0.94	0.35	0.28	0.23	0.18	0.15	0.12	0.10	0.08	0.07	0.06	0.05	0.04	0.03	0.03
12	0.63	0.71	0.76	0.81	0.84	0.87	0.89	0.91	0.93	0.94	0.95	0.96	0.37	0.29	0.24	0.19	0.16	0.13	0.11	0.09	0.07	0.06	0.05	0.04
14	0.65	0.72	0.77	0.82	0.85	0.88	0.90	0.92	0.93	0.94	0.95	0.96	0.97	0.97	0.38	0.30	0.25	0.20	0.17	0.14	0.11	0.09	0.08	0.06
16	0.68	0.74	0.79	0.83	0.86	0.89	0.91	0.92	0.94	0.95	0.96	0.96	0.97	0.98	0.98	0.98	0.39	0.31	0.25	0.21	0.17	0.14	0.11	0.09
18	0.72	0.78	0.82	0.85	0.88	0.90	0.92	0.93	0.94	0.95	0.96	0.97	0.97	0.98	0.98	0.99	0.99	0.99	0.39	0.31	0.26	0.21	0.17	0.14
Zone Type D																								
2	0.59	0.67	0.13	0.09	0.08	0.06	0.05	0.05	0.04	0.04	0.03	0.03	0.02	0.02	0.02	0.01	0.01	0.01	0.01	0.01	0.01	0.01	0.01	0.00
4	0.60	0.67	0.72	0.76	0.20	0.16	0.13	0.11	0.10	0.08	0.07	0.06	0.05	0.05	0.04	0.03	0.03	0.03	0.02	0.02	0.02	0.01	0.01	0.01
6	0.61	0.68	0.73	0.77	0.80	0.83	0.26	0.20	0.17	0.15	0.13	0.11	0.09	0.08	0.07	0.06	0.05	0.05	0.04	0.03	0.03	0.03	0.02	0.02
8	0.62	0.69	0.74	0.77	0.80	0.83	0.85	0.87	0.30	0.24	0.20	0.17	0.15	0.13	0.11	0.10	0.08	0.07	0.06	0.05	0.05	0.04	0.04	0.03
10	0.63	0.70	0.75	0.78	0.81	0.84	0.86	0.88	0.89	0.91	0.33	0.27	0.22	0.19	0.17	0.14	0.12	0.11	0.09	0.08	0.07	0.06	0.05	0.05
12	0.65	0.71	0.76	0.79	0.82	0.84	0.87	0.88	0.90	0.91	0.92	0.93	0.35	0.29	0.24	0.21	0.18	0.16	0.13	0.12	0.10	0.09	0.08	0.07
14	0.67	0.73	0.78	0.81	0.83	0.86	0.88	0.89	0.91	0.92	0.93	0.94	0.95	0.95	0.37	0.30	0.25	0.22	0.19	0.16	0.14	0.12	0.11	0.09
16	0.70	0.76	0.80	0.83	0.85	0.87	0.89	0.90	0.92	0.93	0.94	0.95	0.95	0.96	0.96	0.97	0.38	0.31	0.26	0.23	0.20	0.17	0.15	0.13
18	0.74	0.80	0.83	0.85	0.87	0.89	0.91	0.92	0.93	0.94	0.95	0.95	0.96	0.97	0.97	0.97	0.98	0.98	0.39	0.32	0.27	0.23	0.20	0.17

Note: See Table 35 for zone type. Data based on a radiative/convective fraction of 0.70/0.30.

where

q_s = sensible cooling load due to people
N = number of people
SHG_p = sensible heat gain per person (Table 3)
CLF_p = cooling load factor for people (Table 37)
q_l = latent cooling load due to people
LHG_p = latent heat gain per person (Table 3)

The CLF for people load is a function of the time such people spend in the conditioned space and the time elapsed since first entering. As defined for estimating cooling load from fenestration, the space under consideration is categorized as a zone, identified in Table 35. The appropriate CLF is selected from Table 37 by zone type, occupancy period, and number of hours after entry.

CLF Usage Exceptions. If the space temperature is not maintained constant during the 24-h period, for example, if the cooling system is shut down during the night (night shutdown), a "pulldown load" results because a major part of the stored sensible heat in the structure has not been removed, thus reappearing as cooling load when the system is started the next day. In this case, a CLF of 1.0 should be used.

When there is a *high occupant density*, as in theaters and auditoriums, the quantity of radiation to the walls and room furnishings is proportionately reduced. In these situations, a CLF of 1.0 should also be used.

Example 8. Cooling load from occupants. Estimate the cooling load in a building at 1200, 1400, and 1600 h from four moderately active people occupying an office from 0900 to 1700 h. The office temperature is 78°F, and the cooling system operates continuously. Assume the conditions of the space as applied to Table 33A; define it as Type D.

Solution: The sensible cooling load is calculated by Equation (44), and the latent cooling load is calculated by Equation (45). The period of occupancy is 8 h. Therefore,

Time	No. of People	Hours in Space	Hours after Entry	Btu/h Each (Table 3) Sen.	Btu/h Each (Table 3) Lat.	CLFp (Table 37) Zone D	Cooling Load Sen., Btu/h	Cooling Load Lat., Btu/h
1200	4	8	3	255	255	0.74	755	1020
1400	4	8	5	255	255	0.80	816	1020
1600	4	8	7	255	255	0.85	867	1020

Lighting

As discussed for the TFM, the cooling load from lighting does not immediately reflect the full energy output of the lights. Kimura and Stephenson (1968), Mitalas and Kimura (1971), and Mitalas (1973) indicated the effect on cooling load of light fixture type, type of air supply and return, space furnishings, and the thermal characteristics of the space. The effect of these influencing parameters have been incorporated in the *Cooling and Heating Load Calculation Manual* (McQuiston and Spitler 1992) into the CLF values for lighting listed in Table 38, and for which selection zones are identified as appropriate by Tables 35. At any time, the space cooling load from lighting can be estimated as:

$$q_{el} = HG_{el}(CLF_{el}) \tag{46}$$

where

q_{el} = cooling load from lighting, Btu/h
HG_{el} = heat gain from lighting, Btu/h, as $WF_{ul}F_{sa}$ [Equation (9)]
W = total lamp watts
F_{ul} = lighting use factor
F_{sa} = lighting special allowance factor
CLF_{el} = lighting cooling load factor (Table 38)

CLF_{el} data in Table 38 are based on the assumptions that (1) the conditioned space temperature is continuously maintained at a constant value, and (2) the cooling load and power input to the lights eventually become equal if the lights are on for long enough.

Operational Exceptions. If the cooling system operates only during occupied hours, the CLF_{el} should be considered 1.0 in lieu of

Table 38 Cooling Load Factors for Lights

Lights On For	1	2	3	4	5	6	7	8	9	10	11	12	13	14	15	16	17	18	19	20	21	22	23	24
Zone Type A																								
8	0.85	0.92	0.95	0.96	0.97	0.97	0.97	0.98	0.13	0.06	0.04	0.03	0.02	0.02	0.02	0.01	0.01	0.01	0.01	0.01	0.01	0.01	0.01	0.01
10	0.85	0.93	0.95	0.97	0.97	0.97	0.98	0.98	0.98	0.98	0.14	0.07	0.04	0.03	0.02	0.02	0.02	0.02	0.02	0.02	0.01	0.01	0.01	0.01
12	0.86	0.93	0.96	0.97	0.97	0.98	0.98	0.98	0.98	0.98	0.98	0.98	0.14	0.07	0.04	0.03	0.03	0.02	0.02	0.02	0.02	0.02	0.02	0.02
14	0.86	0.93	0.96	0.97	0.98	0.98	0.98	0.98	0.98	0.98	0.99	0.99	0.99	0.99	0.15	0.07	0.05	0.03	0.03	0.03	0.02	0.02	0.02	0.02
16	0.87	0.94	0.96	0.97	0.98	0.98	0.98	0.99	0.99	0.99	0.99	0.99	0.99	0.99	0.99	0.99	0.15	0.08	0.05	0.04	0.03	0.03	0.03	0.02
Zone Type B																								
8	0.75	0.85	0.90	0.93	0.94	0.95	0.95	0.96	0.23	0.12	0.08	0.05	0.04	0.04	0.03	0.03	0.03	0.02	0.02	0.02	0.02	0.02	0.02	0.01
10	0.75	0.86	0.91	0.93	0.94	0.95	0.95	0.96	0.96	0.97	0.24	0.13	0.08	0.06	0.05	0.04	0.04	0.03	0.03	0.03	0.03	0.02	0.02	0.02
12	0.76	0.86	0.91	0.93	0.95	0.95	0.96	0.96	0.97	0.97	0.97	0.97	0.24	0.14	0.09	0.07	0.05	0.05	0.04	0.04	0.03	0.03	0.03	0.03
14	0.76	0.87	0.92	0.94	0.95	0.96	0.96	0.97	0.97	0.97	0.97	0.98	0.98	0.98	0.25	0.14	0.09	0.07	0.06	0.05	0.05	0.04	0.04	0.03
16	0.77	0.88	0.92	0.95	0.96	0.96	0.97	0.97	0.97	0.98	0.98	0.98	0.98	0.98	0.98	0.99	0.25	0.15	0.10	0.07	0.06	0.05	0.05	0.04
Zone Type C																								
8	0.72	0.80	0.84	0.87	0.88	0.89	0.90	0.91	0.23	0.15	0.11	0.09	0.08	0.07	0.07	0.06	0.05	0.05	0.05	0.04	0.04	0.03	0.03	0.03
10	0.73	0.81	0.85	0.87	0.89	0.90	0.91	0.92	0.92	0.93	0.25	0.16	0.13	0.11	0.09	0.08	0.08	0.07	0.06	0.06	0.05	0.05	0.04	0.04
12	0.74	0.82	0.86	0.88	0.90	0.91	0.92	0.92	0.93	0.94	0.94	0.95	0.26	0.18	0.14	0.12	0.10	0.09	0.08	0.08	0.07	0.06	0.06	0.05
14	0.75	0.84	0.87	0.89	0.91	0.92	0.92	0.93	0.94	0.94	0.95	0.95	0.96	0.96	0.27	0.19	0.15	0.13	0.11	0.10	0.09	0.08	0.08	0.07
16	0.77	0.85	0.89	0.91	0.92	0.93	0.93	0.94	0.95	0.95	0.95	0.96	0.96	0.97	0.97	0.97	0.28	0.20	0.16	0.13	0.12	0.11	0.10	0.09
Zone Type D																								
8	0.66	0.72	0.76	0.79	0.81	0.83	0.85	0.86	0.25	0.20	0.17	0.15	0.13	0.12	0.11	0.10	0.09	0.08	0.07	0.06	0.06	0.05	0.04	0.04
10	0.68	0.74	0.77	0.80	0.82	0.84	0.86	0.87	0.88	0.90	0.28	0.23	0.19	0.17	0.15	0.14	0.12	0.11	0.10	0.09	0.08	0.07	0.06	0.06
12	0.70	0.75	0.79	0.81	0.83	0.85	0.87	0.88	0.89	0.90	0.91	0.92	0.30	0.25	0.21	0.19	0.17	0.15	0.13	0.12	0.11	0.10	0.09	0.08
14	0.72	0.77	0.81	0.83	0.85	0.86	0.88	0.89	0.90	0.91	0.92	0.93	0.94	0.94	0.32	0.26	0.23	0.20	0.18	0.16	0.14	0.13	0.12	0.10
16	0.75	0.80	0.83	0.85	0.87	0.88	0.89	0.90	0.91	0.92	0.93	0.94	0.94	0.95	0.96	0.96	0.34	0.28	0.24	0.21	0.19	0.17	0.15	0.14

Note: See Table 35 for zone type. Data based on a radiative/convective fraction of 0.59/0.41.

the Table 38 values. Where one portion of the lights serving the space is on one schedule of operation and another portion is on a different schedule, each should be treated separately. Where lights are left on for 24 h a day, the CLF_{el} is 1.0.

Example 9. Cooling load from lighting. Estimate the cooling load in a building at 1200, 1400, and 1600 h from recessed fluorescent lights, turned on at 0800 h and turned off at 1800 h. Lamp wattage is 800 W. The use factor is 1.0, and the special allowance factor is 1.25. The room is an interior type in a one-story building, has tile flooring over a 3-in. concrete floor, and a suspended ceiling. The cooling system runs 24 h/day, including weekends.

Solution: From Table 35B, the room is categorized as Type C for lighting load purposes. Therefore,

Time	Hours in Space	Hours after Entry	Lamp Watts	Heat Gain, Btu/h [Eq. (9)]	CLF_{el} (Table 38) Zone C	Cooling Load, Btu/h [Eq. (39)]
1200	8	3	800	3410	0.85	2489
1400	8	5	800	3410	0.89	2660
1600	8	7	800	3410	0.91	2796

Power and Appliances

Heat gain of power-driven equipment can be estimated by means of Equations (15), (16), or (17) as applicable, or taken directly from Tables 4 and/or 5.

Equations (18) and (19) can be used to estimate heat gain values under various circumstances, and Tables 6 through 9 provide representative data for direct use or as input to the equations.

The radiant component of sensible heat gain from power-driven equipment or appliances is delayed in becoming cooling load in the same manner as that of other load categories already discussed. For power-driven equipment, the CLF values tabulated for unhooded equipment (Table 37) are considered appropriate. Tables 37 and 39 tabulate cooling load factors (CLF_a) for appliances. Multiplying the sensible portion of heat gain by the appropriate CLF_a will produce the following approximate cooling load values:

$$q = \text{SHG (CLF)} \tag{47}$$

Example 10. Appliance cooling load. Determine the cooling load in a building at 1200, 1400, and 1600 h caused by an electric coffee brewer with one brewer and one warmer. The brewer operates continuously

Table 39 Cooling Load Factors for Hooded Equipment

Hours in Operation	Number of Hours after Equipment Turned On 1	2	3	4	5	6	7	8	9	10	11	12	13	14	15	16	17	18	19	20	21	22	23	24
Zone Type A																								
2	0.64	0.83	0.26	0.11	0.06	0.03	0.01	0.01	0.01	0.01	0.00	0.00	0.00	0.00	0.00	0.00	0.00	0.00	0.00	0.00	0.00	0.00	0.00	0.00
4	0.64	0.83	0.90	0.93	0.31	0.14	0.07	0.04	0.03	0.03	0.01	0.01	0.01	0.01	0.00	0.00	0.00	0.00	0.00	0.00	0.00	0.00	0.00	0.00
6	0.64	0.83	0.90	0.93	0.96	0.96	0.33	0.16	0.09	0.06	0.04	0.03	0.03	0.01	0.01	0.01	0.01	0.00	0.00	0.00	0.00	0.00	0.00	0.00
8	0.64	0.83	0.90	0.93	0.96	0.96	0.97	0.97	0.34	0.16	0.09	0.06	0.04	0.03	0.03	0.01	0.01	0.01	0.01	0.01	0.00	0.00	0.00	0.00
10	0.64	0.83	0.90	0.93	0.96	0.96	0.97	0.97	0.99	0.99	0.34	0.17	0.10	0.06	0.04	0.03	0.03	0.01	0.01	0.01	0.01	0.01	0.01	0.00
12	0.64	0.83	0.90	0.94	0.96	0.97	0.97	0.97	0.99	0.99	0.99	0.99	0.36	0.17	0.10	0.06	0.04	0.03	0.03	0.03	0.01	0.01	0.01	0.01
14	0.66	0.83	0.90	0.94	0.96	0.97	0.97	0.99	0.99	0.99	0.99	0.99	1.00	1.00	0.36	0.17	0.10	0.07	0.04	0.04	0.03	0.03	0.03	0.01
16	0.66	0.84	0.91	0.94	0.96	0.97	0.97	0.99	0.99	0.99	0.99	0.99	1.00	1.00	1.00	1.00	0.36	0.17	0.10	0.07	0.04	0.04	0.04	0.03
18	0.67	0.84	0.91	0.94	0.96	0.97	0.97	0.99	0.99	0.99	0.99	1.00	1.00	1.00	1.00	1.00	1.00	1.00	0.36	0.17	0.10	0.08	0.07	0.04
Zone Type B																								
2	0.50	0.63	0.23	0.16	0.11	0.09	0.07	0.06	0.04	0.03	0.03	0.01	0.01	0.01	0.01	0.00	0.00	0.00	0.00	0.00	0.00	0.00	0.00	0.00
4	0.50	0.64	0.73	0.79	0.34	0.24	0.19	0.14	0.10	0.09	0.06	0.04	0.04	0.03	0.03	0.01	0.01	0.01	0.01	0.00	0.00	0.00	0.00	0.00
6	0.50	0.64	0.73	0.79	0.84	0.87	0.41	0.29	0.21	0.17	0.13	0.10	0.07	0.06	0.04	0.03	0.03	0.01	0.01	0.01	0.01	0.01	0.01	0.00
8	0.50	0.64	0.73	0.79	0.84	0.87	0.90	0.93	0.44	0.31	0.24	0.19	0.14	0.11	0.09	0.07	0.06	0.04	0.03	0.03	0.01	0.01	0.01	0.01
10	0.50	0.64	0.73	0.79	0.84	0.87	0.90	0.93	0.94	0.96	0.47	0.34	0.26	0.20	0.16	0.11	0.09	0.07	0.06	0.04	0.03	0.03	0.03	0.01
12	0.51	0.66	0.73	0.80	0.84	0.89	0.91	0.93	0.94	0.96	0.97	0.97	0.49	0.34	0.27	0.20	0.16	0.11	0.09	0.07	0.06	0.05	0.04	0.03
14	0.53	0.66	0.74	0.80	0.84	0.89	0.91	0.93	0.94	0.96	0.97	0.97	0.99	0.99	0.50	0.36	0.27	0.21	0.16	0.13	0.10	0.08	0.07	0.06
16	0.56	0.69	0.76	0.81	0.86	0.89	0.91	0.93	0.94	0.96	0.97	0.97	0.99	0.99	0.99	0.99	0.50	0.36	0.27	0.21	0.16	0.14	0.13	0.10
18	0.59	0.71	0.79	0.83	0.87	0.90	0.93	0.94	0.96	0.97	0.97	0.99	0.99	0.99	0.99	0.99	1.00	1.00	0.50	0.36	0.27	0.23	0.21	0.16
Zone Type C																								
2	0.43	0.54	0.20	0.16	0.13	0.10	0.09	0.07	0.06	0.04	0.04	0.03	0.03	0.01	0.01	0.01	0.01	0.01	0.01	0.00	0.00	0.00	0.00	0.00
4	0.43	0.54	0.63	0.70	0.33	0.26	0.20	0.17	0.14	0.11	0.09	0.07	0.06	0.06	0.04	0.03	0.03	0.03	0.01	0.01	0.01	0.01	0.01	0.01
6	0.44	0.56	0.63	0.70	0.76	0.80	0.40	0.31	0.26	0.21	0.17	0.14	0.11	0.10	0.09	0.07	0.06	0.04	0.04	0.03	0.03	0.02	0.01	0.01
8	0.44	0.56	0.64	0.70	0.76	0.80	0.84	0.87	0.46	0.37	0.30	0.24	0.20	0.16	0.13	0.11	0.09	0.07	0.06	0.06	0.04	0.03	0.03	0.03
10	0.46	0.57	0.64	0.71	0.76	0.80	0.84	0.87	0.89	0.91	0.50	0.40	0.33	0.26	0.21	0.17	0.14	0.11	0.10	0.09	0.07	0.06	0.06	0.04
12	0.47	0.59	0.66	0.73	0.77	0.81	0.84	0.87	0.90	0.91	0.93	0.94	0.53	0.41	0.34	0.27	0.23	0.19	0.16	0.13	0.10	0.09	0.09	0.07
14	0.50	0.60	0.67	0.74	0.79	0.83	0.86	0.89	0.90	0.91	0.93	0.94	0.96	0.96	0.54	0.43	0.36	0.29	0.24	0.20	0.16	0.14	0.13	0.11
16	0.54	0.63	0.70	0.76	0.80	0.84	0.87	0.89	0.91	0.93	0.94	0.94	0.96	0.97	0.97	0.97	0.56	0.44	0.36	0.30	0.24	0.22	0.20	0.16
18	0.60	0.69	0.74	0.79	0.83	0.86	0.89	0.90	0.91	0.93	0.94	0.96	0.96	0.97	0.97	0.99	0.99	0.99	0.56	0.44	0.37	0.33	0.30	0.24
Zone Type D																								
2	0.41	0.53	0.19	0.13	0.11	0.09	0.07	0.07	0.06	0.06	0.04	0.04	0.03	0.03	0.03	0.01	0.01	0.01	0.01	0.01	0.01	0.01	0.01	0.01
4	0.43	0.53	0.60	0.66	0.29	0.23	0.19	0.16	0.14	0.11	0.10	0.09	0.07	0.07	0.06	0.04	0.04	0.04	0.03	0.03	0.03	0.02	0.01	0.01
6	0.44	0.54	0.61	0.67	0.71	0.76	0.37	0.29	0.24	0.21	0.19	0.16	0.13	0.11	0.10	0.09	0.07	0.07	0.06	0.04	0.04	0.04	0.04	0.03
8	0.46	0.56	0.63	0.67	0.71	0.76	0.79	0.81	0.43	0.34	0.29	0.24	0.21	0.19	0.16	0.14	0.11	0.10	0.09	0.07	0.07	0.06	0.06	0.06
10	0.47	0.57	0.64	0.69	0.73	0.77	0.80	0.83	0.84	0.87	0.47	0.39	0.31	0.27	0.24	0.20	0.17	0.16	0.13	0.11	0.10	0.09	0.09	0.07
12	0.50	0.59	0.66	0.70	0.74	0.77	0.81	0.83	0.86	0.87	0.89	0.90	0.50	0.41	0.34	0.30	0.26	0.23	0.19	0.17	0.14	0.13	0.13	0.11
14	0.53	0.61	0.69	0.73	0.76	0.80	0.83	0.84	0.87	0.89	0.90	0.91	0.93	0.93	0.53	0.43	0.36	0.31	0.27	0.23	0.20	0.18	0.17	0.16
16	0.57	0.66	0.71	0.76	0.79	0.81	0.84	0.86	0.89	0.90	0.91	0.93	0.93	0.94	0.94	0.96	0.54	0.44	0.37	0.33	0.29	0.26	0.24	0.21
18	0.63	0.71	0.76	0.79	0.81	0.84	0.87	0.89	0.90	0.91	0.93	0.93	0.94	0.96	0.96	0.96	0.97	0.97	0.56	0.46	0.39	0.35	0.33	0.29

Note: See Table 35 for zone type. Data based on a radiative/convective fraction of 1.0/0.

from 0900 to 1500 h and does not have an exhaust hood. The room is a "midfloor" type in a multistory building, has carpet over a 3-in. concrete floor, and a suspended ceiling. The cooling system runs 24 h/day, including weekends.

Solution: From Table 7, q_s and q_l for an unhooded, two-burner coffee brewer is 3750 and 1910 Btu/h, respectively (thus 1875 and 955 Btu/h each burner), and for a coffee heater (per warming burner) is 230 and 11032 Btu/h, respectively. The brewer is on for 6 h, and 1200 h is 3 h after the brewer is turned on. From Table 35B, the room is categorized as Type D for equipment load purposes. Therefore,

Time	q_s, Btu/h	Hours in Use	Hours after Start	CLF_a (Table 37)	Cooling Load: Sensible, Btu/h	Latent, Btu/h	Total, Btu/h
1200	2105	6	3	0.73	1537	1065	2602
1400	2105	6	5	0.80	1684	1065	2749
1600	2105	6	7	0.26	547	0	547

Total Space Cooling Load

The estimated total space cooling load for a given application is determined by summing the individual components for each hour of interest.

EXAMPLE COOLING LOAD CALCULATION USING CLTD/CLF METHOD

Example 11. For this example, the one-story commercial building in Example 6 will be the basis for calculating a cooling load by the CLTD/ SCL/CLF method. Refer to Example 6 for the statement of conditions.

Find (for stated design conditions):

1. Sensible cooling load
2. Latent cooling load
3. Total cooling load

Solution: By inspection, the cooling load from the roof can be expected to be the variable making the greatest contribution to the overall cooling load for the building. Therefore, the time of maximum cooling load occurrence will probably be close to the time of maximum CLTD for the roof. The maximum cooling load for the building as a whole can be expected to occur in one of the summer months—June, July, or August. From Table 31, a "mass inside" roof with no ceiling and an R-factor of 11.11 is classified as Type 4. From Table 30, the CLTD for Roof No. 4 at 40°N latitude has a maximum tabulated value of 78°F at 1800 h, but is only somewhat less (73) at 1600 h.

South-facing glass can also be expected to wield considerable influence on the cooling load for this particular building. From Table 35A, a one-story building with three exposed walls, uncarpeted floor, masonry partitions, and fully inside-shaded windows is classified B for solar loads, and C for loads from people, equipment, or lights. Cross-checking the variation of SCLs for glass facing south, in Table 36, the maximum cooling load from these windows would be only slightly more at noon or 1300 h than at 1400 for 40°N latitude. Sometime in the early afternoon seems obvious, but it is necessary to make a quick estimate to establish the peak load hour:

Roof [Equation (41)]

$U = 0.09$ Btu/(h·ft²·°F) Area = 4000 ft²

Time	1300	1400	1500	1600	1700
CLTD	42	54	65	73	77
C_1	3	3	3	3	3
C_2	−1	−1	−1	−1	−1
Corr. CLTD	44	56	67	75	79
Btu/h	15840	20160	24120	27000	28440

$C_1 = (78 - 75)$ = indoor design temperature correction

$C_2 = (94 - 74)/2 - 85$ = daily average temperature correction

South Glass, Solar [Equation (43)]:

SC = 0.55 Area = 60 ft²

Time		1300	1400	1500	1600	1700
SCL at 40°,		87	79	63	46	37
Btu/h	=	2871	2607	2079	1518	1221
Roof and south glass, Btu/h	=	18711	22767	26199	28518	29661

Evaluation of the foregoing indicates that 1600 h will be the probable hour of maximum cooling load for this building, considering that although the roof and south glass loads increase another 1143 Btu/h for 1700 h, the trend has slowed and most other load components of significance can be expected to be leveling off or moving toward lower values at that time. In some cases there would be no such clear cut indication, and it would be necessary to estimate the total load for a number of hours, including the potential impact of other significant variables which could exert a determining influence at a different time (such as a major load from appliances, known to occur only in the morning, etc.), before selecting the peak load hour for the overall calculation.

Cooling Load from Heat Gain through Roof, Exposed Walls, and Doors

Such loads are estimated using Equation (41), where the CLTDs are taken from Table 30 after determining appropriate type numbers from Table 31 whose insulation placement, U-factors and general construction are as close to the actual components as possible. Corrections to CLTD values are made in accordance with footnote instructions to Table 30 similar to the above preliminary evaluation. (Note that there are no corrections to CLTD values for building mass variations, per the foregoing discussion, as considered of only limited significance to the overall results. Tabulated data for roofs, walls and doors assume Room Transfer Functions for "light to medium" construction.)

Cooling Load from Heat Gain through Fenestration Areas

The load component from conduction heat gain is calculated using Equation (42), where the CLTD value is taken from Table 34, corrected by −1°F because of a 1°F lower average daily temperature than that for which the table was generated, and +3°F to recognize the 75°F design space temperature. The U-factor of the glass is taken as 0.81 Btu/(h·ft²·°F) for single sheet plate glass, under summer conditions.

The load component from solar heat gain is calculated using Equation (43) as indicated above. A shading coefficient (SC) of 0.55 is used for clear glass with light-colored venetian blinds. SCL values are taken from Table 36 in this chapter, after first identifying the appropriate zone type as B for solar load from Table 35A. Results are tabulated in Table 40.

Table 40 Solar Cooling Load for Windows, Example 11

Section	Net Square Feet	SC	CLF Table 36	Cooling Load, Btu/h
South windows	60	0.55	46	1518
North windows	30	0.55	32	528

Cooling Load from Heat Gain through Party Walls

For the north and west party walls, cooling load is calculated using Equation (16) for wall and door areas, using appropriate U-factors from Chapter 24 and the temperature differential existing at 1600 h, 18.4°F. Results are tabulated in Table 41.

Cooling Load from Internal Heat Sources

For the cooling load component from lights, Equation (9) is first used to obtain the heat gain. Assuming a use factor of 1.0, and a special allowance factor of 1.0 for tungsten lamps and 1.20 for fluorescent lamps, these gains are:

$$q_{tung} = 4000 \times 1.0 \times 1.0 \times 3.41 = 13{,}640 \text{ Btu/h}$$

$$q_{fluor} = 17{,}500 \times 1.0 \times 1.2 \times 3.41 = 71{,}610 \text{ Btu/h}$$

Since the tungsten lamps are operated continuously, the previously stored radiant heat from this source currently being reconvected to the space equals the rate of new radiant heat from this source being stored, thus the cooling load from this source equals heat gain. The fluorescent lamps however are operated only 10 hours per day, 0800 through hour 1700, and thus contribute radiant heat to cooling load in a cyclic and

Table 41 Conduction Cooling Load Summary for Enclosing Surfaces, Example 11

Section	Net ft²	U-Factor, Btu/(h·ft²·°F)	Δt, °F	CLTD, °F	Ref. for CLTD	1600 h Cooling Load, Btu/h
Roof	4000	0.09		73	Table 30 Roof 4	27000
South wall	405	0.24		19	Table 32 Wall 16	1847
East wall	765	0.48		38	Table 32 Wall 10	13954
North exposed wall	170	0.48		16	Table 32 Wall 10	1306
W. and N. party wall	1065	0.25	18.4			4899
Doors in S. Wall	35	0.19		50	Table 32 Wall 2	333
Doors in N. Wall	35	0.18	18.4			116
Doors in E. Wall	35	0.19		36	Table 32 Wall 2	239
South windows	60	0.81		16	Table 34	778
North windows	30	0.81		16	Table 34	389

Summary of Calculations for Example 11

	Dry Bulb, °F	Wet Bulb, °F	Humidity Ratio
Outdoor conditions	93.4	76.8	0.0161
Indoor conditions	75	62.5	0.0093
Difference	19		0.0068

Sensible Cooling Load at 1600 h	Btu/h
Roof and Exposed Walls	
Roof	27,000
South wall	1,847
East wall	13,954
North wall	1,306
South wall doors	333
East wall doors	239
Fenestration Areas	
South windows	2,296
North windows	917
Party Walls	
West and North Walls	4,899
North Wall doors	116
Internal Sources	
People	19,550
Tungsten lights	13,640
Fluorescent lights	65,881
Outside Air	
Infiltration	1,356
Ventilation	25,806
Total	179,140
Latent Cooling Load at 1600 h	**Btu/h**
People	17,000
Infiltration	2,205
Ventilation	41,963
Total	63,208
Grand Total Load	242,348

somewhat delayed manner. From Table 35A, the zone type is identified as C for lighting loads, and from Table 38 a CLF value of 0.92 is obtained for lights which are operated for 10 hours, for a calculation hour 9 hours after the lights have been turned on. The cooling load from fluorescent lights for this estimate is thus:

$$q_{cl\ fluor} = 71{,}610 \times 0.92 = 65{,}881 \text{ Btu/h}$$

For people, Table 3 is used to select heat gains for seated occupants doing light office work, as 250 Btu/h per person, sensible, and 200 Btu/h, latent for 75°F space temperature. The CLF for the sensible component is taken from Table 37 as 0.92, for a condition of 10 total hours in a type C space and a load calculation taken 9 h after entry. Cooling load from people is thus estimated at:

$$q_{ps} = 85 \text{ people} \times 250 \times 0.92 = 19{,}550 \text{ Btu/h}$$

$$q_{pl} = 85 \text{ people} \times 200 = 17{,}000 \text{ Btu/h}$$

Cooling Load from Power Equipment and Appliances

For this example, none are assumed.

Cooling Load from Infiltration and Ventilation Air

As determined in Example 6, ventilation for this building is established at 15 cfm/person, or 1275 cfm, and infiltration (through doors) at 67 cfm. For this example, ventilation is assumed to enter directly into the space (as opposed to first passing through the cooling equipment), and thus is included as part of the space cooling load.

The sensible and latent portions of each load component are calculated using Equations (22) and (23), respectively, where at 1600 h: t_o = 93.4°F; t_i = 75°F; W_o = 0.0161; and W_i = 0.0093; thus:

For ventilation:

Q_s	Factor	Δt	ΔW	q, Btu/h
1275	1.1	18.4		25,806 Sensible
1275	4840		0.0068	41,963 Latent

For infiltration:

Q_s	Factor	Δt	ΔW	q, Btu/h
67	1.1	18.4		1,356 Sensible
67	4840		0.0068	2,205 Latent

Limitations of CLTD/SCL/CLF Methods

The results obtained from using CLTD/CLF data depend on the characteristics of the space and how they vary from those used to generate the weighting factors. Variations can appear in the amplitude and when radiant heat gain components are felt as cooling loads, which affect the hourly cooling loads for the space. Two types of error are possible:

1. The computer software that generated CLTD/SCL/CLF tables uses the TFM to determine cooling loads based on various types of heat gain. The cooling loads for each type of heat gain are normalized appropriately to obtain CLTDs, SCLs, or CLFs. Except, as discussed next, use of the CLTD/SCL/CLF method in conjunction with these tables will yield the same results as the TFM, but only when the same 14 zone parameters are specified.

   Three inherent errors in the TFM are carried through to the CLTD/SCL/CLF data:

   a. Each set of weighting factors or conduction transfer function coefficients are used for a group of walls, roofs, or zones with similar thermal response characteristics. Groups were chosen so that error would be minimal and conservative (Harris and McQuiston 1988, Sowell 1988).

   b. The scheme used for calculating weighting factors is based on 14 discrete parameters applied to a rectangular room. Rarely does a room fit exactly into these parameters. Therefore, engineering judgment must be used to choose the values of the 14 parameters that most closely represent the room for which load calculations are being performed. Deviations of

Table 42 Potential Errors for Roof and Wall CLTDs in Tables 30 and 32

Roof No.	Error, % Plus	Error, % Minus	Wall No.	Error, % Plus	Error, % Minus
1	13	5	1	18	7
2	13	5	2	17	8
3	12	5	3	17	7
4	13	5	4	16	7
5	11	4	5	13	8
6	—	—	6	14	6
7	—	—	7	12	6
8	10	4	—	—	—
9	10	4	9	13	6
10	9	3	10	10	6
11	—	—	11	8	3
12	—	—	12	4	7
13	7	4	13	4	4
14	5	4	14	5	8
15	—	—	15	11	6
16	—	—	16	8	7

Note: Percent error = [(Table Value – TFM Value)/TFM Value] × 100

the room from the available levels of the 14 parameters may result in errors that are not easily quantifiable.

c. A fundamental presupposition of the TFM is that total cooling load for a zone can be calculated by simple addition of the individual components. For example, radiation heat transfer from individual walls and roofs is assumed to be independent of the other surfaces. O'Brien (1985) has shown this assumption can cause some error.

2. The printed tables for CLTDs, SCLs, and CLFs have undergone a further grouping procedure. The maximum potential errors due to the second grouping procedure have been analyzed and are tabulated in Tables 35 and 42. These errors are in addition to those inherent in the TFM. However, for usual construction, these errors are modest.

In summary, the CLTD/SCL/CLF method, as with any method, requires engineering judgment in its application. When the method is used in conjunction with custom tables generated by appropriate computer software (McQuiston and Spitler 1992) and for buildings where external shading is not significant, it can be expected to produce results very close to those produced by the TFM. When the printed tables are used, some additional error is introduced. In many cases, the accuracy should be sufficient.

TETD/TA CALCULATION PROCEDURE

To calculate a space cooling load using the TETD/TA convention, the same general procedures for data assembly and precalculation analysis apply as for the TFM. Similarly, the following factors are handled in an identical manner and are not repeated here.

- Basic heat gain calculation concepts of solar radiation (solar and conductive heat gain through fenestration areas, conversion to cooling load)
- Total heat gain through exterior walls and roofs (sol-air temperature, heat gain through exterior surfaces, tabulated temperature values, surface colors, air temperature cycle and adjustments, average sol-air temperature, hourly air temperatures, and data limitations)
- Heat gain through interior surfaces (adjacent spaces, floors)
- Heat gain through infiltration and ventilation

This section describes how the TETD/TA technique differs from the TFM. For sources of the space cooling load, equations, appropriate references, tables, and sources of other information for an overall analysis, see Table 43.

Treatment of Heat Gain and Cooling Load Conversion Procedures

The TETD/TA method was oriented primarily as a manual procedure. Tables of precalculated time-lags, decrement factors, and total equivalent temperature differential values listed a number of representative wall and roof assemblies for use in the appropriate heat gain equations. These data were based on a Fourier series solution to the one-dimensional unsteady-state conduction equation for a multiple-component slab, as used to calculate the heat flow through each of the walls and roofs selected for that purpose. All calculations were based on an inside air temperature of 75°F and a sol-air temperature at the outside equal to those given in Table 1 for horizontal and vertical surfaces of various orientations, at 2-h increments throughout a typical design day, as outlined by Stewart (1948) and Stephenson (1962). Basic equations were also presented to facilitate a computer solution.

Heat gain through walls and roofs. The results of the foregoing calculations were generalized by dividing the derived hourly heat gain values by the U-factor for each typical wall and roof. The quantity obtained from this generalization is called the total equivalent temperature differential (TETD). This establishes the basic heat gain equation for exterior surfaces as:

$$q = UA(\text{TETD}) \qquad (48)$$

where

q = heat gain, Btu/h
U = coefficient of heat transfer, Btu/(h·ft²·°F)
A = area of surface, ft²
TETD = total equivalent temperature differential (as above)

Heat flow through a similar wall or roof (similar in thermal mass as well as U-factor) can be obtained by multiplying the TETDs listed in the appropriate table by the U-factor of the wall or roof of interest. Any errors introduced by this approach depend on the extent of the differences between the construction in question (components, size, color, and configuration) and the one used for calculating the TETDs.

TETD as Function of Decrement and Time Lag Factors. The heat gain results for representative walls and roofs were also generalized in another way. Effective decrement factors λ and time lags δ were determined for each assembly, such that the equivalent temperature differentials and the corresponding sol-air temperatures are related by:

$$\text{TETD} = t_{ea} - t_i + \lambda(t_{e\delta} - t_{ea}) \qquad (49)$$

where

t_{ea} = daily average sol-air temperature, including consideration for surface color
t_i = indoor air temperature
λ = effective decrement factor
$t_{e,\delta}$ = sol-air temperature δ hours before the calculation hour for which TETD is intended

This relationship permits the approximate calculation of the heat gain through any of the walls or roofs tabulated, or their near equivalents, for any sol-air temperature cycle.

Manual Versus Automated Calculation. Manual application of the TETD/TA procedure, especially the time-averaging calculation itself, is tedious in practice. This fact, plus growing interest in the TFM, led to ASHRAE research with the objective of comparing the differences and similarities of the TETD and TFMs.

Table 43 Summary of TETD/TA Load Calculation Procedures

External Heat Gain

$$t_e = t_o + \alpha I_t / h_o - \varepsilon \Delta R / h_o \quad (6)$$

$$t_{ea} = t_{oa} + \alpha / h_o (I_{DT}/24) - \varepsilon \Delta R / h_o \quad (10)$$

t_e = sol-air temperature
t_o = current hour dry-bulb temperature, from design db (Chapter 26, Table 1) adjusted by Table 2, percentage at daily range values
α = absorptance of surface for solar radiation
α/h_o = surface color factor = 0.15 for light colors, 0.30 for dark
I_t = total incident solar load = 1.15 (SHGF), with SHGF per Chapter 29, Tables 15 through 21
$\varepsilon\Delta R/h_o$ = long-wave radiation factor = −7°F for horizontal surfaces, 0°F for vertical
t_e = 24-h average sol-air temperature
t_{oa} = 24-h average dry-bulb temperature
I_{DT} = total daily solar heat gain (Chapter 29, Tables 15 through 21)

Roofs and Walls

$$q = UA\ (\text{TETD}) \quad (48)$$

$$\text{TETD} = t_{ea} - t_i + 1(t_{e\delta} - t_{ea}) \quad (49)$$

U = design heat transfer coefficient for roof or wall, from Chapter 24, Table 4
A = area of roof or wall, calculated from building plans
TETD = total equivalent temperature difference, roof or wall
t_i = interior design dry-bulb temperature
λ = decrement factor, from Table 14 or 19
$t_{e\delta}$ = sol-air temperature at time lag δ hours (Table 14 or 19) previous to calculation hour

Roofs

Identify layers of roof construction from Table 11. With R-value of dominant layer, identify R-value Range number R and Roof Group number from Table 12. From Table 14 obtain decrement factor and time lag data with which to calculate TETD values for each sol-air temperature value by Equation (52). Calculate hourly heat gain with Equation (48).

Walls

Identify layers of wall construction from Table 11. With R-value of dominant layer, identify R-value Range number and Wall Group number from Table 15, 16, or 17.

Glass

Convective $q = UA(t_o - t_i)$

Solar $q = A(\text{SC})(\text{SHGF})$

U = design heat transfer coefficients, glass—Chapter 29
SC = shading coefficient—Chapter 29
SHGF = solar heat gain factor by orientation, north latitude, hour, and month—Chapter 29, Tables 15 to 21.

Partitions, Ceilings, Floors

$$q = UA\ (t_b - t_i) \quad (8)$$

t_b = temperature in adjacent space
t_i = inside design temperature in conditioned space

Internal Heat Gain

People

$$q_{sensible} = N \times \text{Sensible heat gain}$$

$$q_{latent} = N \times \text{Latent heat gain}$$

N = number of people in space, from best available source. Sensible and latent heat gain from occupancy—Table 3, or Chapter 8; adjust as required.

Lights

$$q_{el} = 3.41\, W F_{ul} F_{sa} \quad (9)$$

W = watts input from electrical plans or lighting fixture data
F_{ul} = lighting use factor, from the first section, as appropriate
F_{sa} = special allowance factor, from from section, as approp.

Power

$$q_p = 2545\, P E_F \quad (15)(16)(17)$$

P = horsepower rating from electrical plans or manufacturer's data
E_F = efficiency factors and arrangements to suit circumstances

Appliances

$$q_{sensible} = q_{input} F_U F_R \quad (18)$$

or

$$q_{sensible} = q_{input} F_L \quad (19)$$

q_{input} = rated energy input from appliances from Tables 5 to 9 or manufacturer's data (Set latent heat = 0, if appliance is under exhaust hood.)
F_U, F_R, F_L = usage factors, radiation factors, flue loss factors

Ventilation and Infiltration Air

$$q_{sensible} = 1.10 Q(t_o - t_i) \quad (22)$$

$$q_{latent} = 4840 Q(W_o - W_i) \quad (23)$$

$$q_{total} = 4.5 Q(H_o - H_i) \quad (20)$$

Q = ventilation airflow—ASHRAE *Standard* 62; infiltration cfm—Chapter 25
t_o, t_i = outside, inside air temperature, °F
W_o, W_i = outside, inside air humidity ratio, lb (water)/lb (da)
H_o, H_i = outside, inside air enthalpy, Btu/lb (dry air)

Cooling Load

Sensible

$$q_{sensible} = q_{cf} + q_{arf} + q_c$$

$$q_{cf} = q_{s,1}(1 - rf_1) + q_{s,2}(1 - rf_2) + \ldots + rf_n$$

$$q_{arf} = \sum_{\gamma = h_{a+1-\theta}}^{\theta} \frac{(q_{s,1} \times rf_1 + q_{s,2} \times rf_2 + \ldots + rf_n)_\gamma}{\theta}$$

$$q_c = (q_{sc,1} + q_{sc,2} + q_{sc,\beta})$$

$q_{sensible}$ = sensible cooling load
q_{cf} = convective fraction of hourly sensible heat gain (current hour) for n load elements
$q_{s,1}$ = sensible hourly heat gain for load element 1, … n
rf_1 = radiation fraction (Table 44) of sensible hourly heat gain for load element 1, … n
q_{arf} = average of radiant fractions of hourly sensible heat gain for n load element 1, … n
θ = number of hour over which to average radiant fractions of sensible heat gain
h_a = current hour, 1 to 24, for which cooling load is to be calculated
γ = one of calculations hours, from $h_{a+1-\theta}$ to h_a, for which the radiant fraction of sensible heat gain is to be averaged for each of n load elements
q_c = convective hourly sensible heat gain (current hour) for β load elements having no radiant component

Latent

$$q_{latent} = (q_{l,1} + q_{l,2} + q_{l,\beta})$$

q_{latent} = latent cooling load
q_l = hourly latent heat gain (current hour) for β load elements

Later research completed the circle of relationships between the TFM, its subsystem CLTD/CLF, and the TETD/TA techniques for dealing with the conversion of heat gain to cooling load. It also confirmed the logic of maintaining these various approaches to solving the problem, depending on the orientation and needs of the individual user and the means available. Finally, the research showed no further need to continue developing manual TETD/TA procedures. Thus, the tabulated values of TETDs have been eliminated from this Handbook in favor of calculation of TETD values by use of the material in the previous section that discusses the TFM.

U-Factors. The values for TETD, originally tabulated in the 1967 *Handbook of Fundamentals*, were calculated using an outside surface conductance of 3.0 Btu/(h·ft²·°F) and an inside surface conductance of 1.2 Btu/(h·ft²·°F), and thus should most appropriately be used with U-factors based on the same surface conductances. TETD data tabulated in the 1972 *Handbook of Fundamentals* and all data listed in this chapter are based on outside and inside surface conductances of 3.0 and 1.46, respectively. U-factors listed in Tables 14 for roofs and 19 for walls can, however, be used with the 1972 TETD data with negligible error, while calculated TETD values are directly compatible with the Table 14 and 19 U-factors.

Example 12. Wall heat gain by TETD. A wall is constructed of 4 in. heavyweight concrete, 2 in. insulation (2.0 lb/ft³, R = 6.667 h·ft²·°F/Btu), 3/4 in. indoor plaster, and with outdoor and indoor surface resistances of 0.333 and 0.685 h·ft²·°F/Btu, respectively. There is an air space between the plaster and the insulation. The wall faces west, the outside design temperature is 95°F, the outdoor daily range is 21°F, the indoor temperature is 75°F, and the color of the exterior surface is light (α/h_o = 0.15). The time is 1400 h on a July day in the central part of the United States (40°N latitude).

Find the heat gain per unit area of wall area.

Solution: Turning first to Table 11, the code numbers for the various layers of the wall described above are:

Outside surface resistance	=	A0
4 in. heavyweight concrete	=	C5
2 in. insulation	=	B3
Air space resistance	=	B1
3/4 in. plaster	=	E1
Inside surface resistance	=	E0

Construction of the wall being "mass out" (as defined in TFM section), Table 17 represents the appropriate arrangement of layers. The dominant wall layer C5 is indicated to have a Wall Material column number of 10, which, combined with an E1 layer, dictates use of the upper array of code numbers for wall assembly "groups." Entering this array with an R value range of 9 (R = 6.667) indicates under column 10 that Wall Group 6 is that most nearly representative of the wall under consideration. The appropriate data from Wall Group 6 as listed in Table 19 are:

$$h = 5.28 \text{ h} = \text{time lag}$$

$$\lambda = 0.54 = \text{effective decrement factor}$$

$$U = 0.199 \text{ Btu/(h·ft}^2\text{·°F)} = \text{heat transfer coefficient}$$

For this example, the sol-air temperature value for 1400 h t_e, as listed in Table 1, is 121°F, that for 0900 h (5 h earlier) is 85°F, and the daily average is 91°F. Thus, from Equation (49):

$$\text{TETD} = 91 - 75 + [0.54(85 - 91)] = 12.76\text{°F}$$

and from Equation (48)

$$q = 0.122 \times 1 \times 12.76 = 2.54 \text{ Btu/(h·ft}^2\text{)}$$

Roof Heat Gain by TETD. The procedure for estimating heat gain from an exposed roof assembly is similar to that described for a wall—first identifying the code letters for the various layers from Table 11; identifying the appropriate roof group number from Table 12; reading the time lag, effective decrement factor, and U-factor for the selected roof group from Table 14; calculating the TETD for the hour of interest from these data and reference to Table 1; and then calculating the heat gain by means of Equation (48).

Heat Gain from Adjacent Unconditioned Spaces. In a manner similar to that described for the TFM, heat gain from adjacent unconditioned spaces can be estimated in two ways, depending on the thermal storage characteristics of the intervening surface. When storage effect is minor, sufficient accuracy can be obtained by use of Equation (8); otherwise, the appropriate TETD value should be calculated by the manner described for an exposed wall surface and the heat gain calculated by Equation (48).

Instantaneous Heat Gain from All Other Sources. Conductive and solar heat gain through fenestration, heat gain from the various internal sources (e.g., people, lighting, power, appliances, etc.), heat gain due to infiltration and ventilation, and latent heat gain from moisture through permeable building surfaces are each calculated in the same manner as described in the TFM section. The basic differences in calculation techniques between TFM and TETD/TA lie in the manner in which the heat gain data are converted to cooling load, as described later.

COOLING LOAD BY TIME AVERAGING

The time-averaging technique for relating instantaneous heat gain to instantaneous cooling load is an approximation of the TFM two-step conversion concept. It recognizes thermal storage by building mass and contents of the radiant portions of heat gain entering a space at any time, with subsequent release of stored heat to the space at some later time. It further recognizes that the cooling load for a space at a given hour is the sum of all convective heat gain and the nonradiant portion of conductive heat gain to that space, plus the amount of previously stored radiant heat gain released back to the space during that same hour.

The effect of room transfer coefficients on hourly heat gain is to generate a load profile that tracks the instantaneous heat gain in amplitude (greater or lower) and delay (negligible for very light structures with a predominance of glass, up to several hours for very massive, monumental construction). Being functions of the mass and configuration of the building and its contents, such coefficients place major emphasis on the immediately preceding hour, and rapidly lessening emphasis on each hour previous to that.

Such TFM-generated cooling load profiles can be closely approximated by averaging the hourly radiant components of heat gain for the previous one to seven or eight hours with those for the current hour, and adding the result to the total convective heat gain for the current hour. As long as results are consistent with results from the more rigorous TFM analysis, those from TETD/TA can be obtained with far less computational effort. The convenient ability to vary the averaging period independently (in recognition of previous experience of the probable thermal performance of an individual building) is also a valuable means of exerting professional judgment on the results.

Success of this approach depends on the accuracy with which the heat gain components are broken down into convective and radiant percentages, as well as on the number of hours used for the averaging period. Weakness of this approach lies in the absence of verified data in the technical literature regarding either determining factor, and the corresponding necessity for experienced judgment by the user.

Heat gain values for either the TFM or TETD/TA method are essentially identical for all load components. Derived cooling load values from properly applied averaging techniques closely track those from the TFM for external heat gain sources. Cooling loads from internal heat gains, however, averaged over the same period as for the external components, normally have peaks that occur more quickly and with greater amplitudes (up to full value of the source heat gain) than those generated by the TFM. This difference is due primarily to the almost constant level of radiant heat input during

Table 44 Convective and Radiant Percentages of Total Sensible Heat Gain for Hour Averaging Purposes

Heat Gain Source	Radiant Heat, %	Convective Heat, %
Window solar, no inside shade	100	—
Window solar, with inside shade	58	42
Fluorescent lights	50	50
Incandescent lights	80	20
People	33	67
Transmission, external roof and walls	60	40
Infiltration and ventilation	—	100
Machinery and appliances	20 to 80	80 to 20

[a]The load from machinery or appliances varies, depending on the temperature of the surface. The higher the surface temperature, the greater the percentage of heat gain that is radiant.

the occupied periods and the resultant "flattening" of the cooling load curves by the TFM as discussed in the TFM section.

The conservative results obtained from the time-averaging method compared with those of the TFM should be viewed in proper perspective. In the CLTD technique, for example, CLF values profile internal loads as a function of time in the space (up to the hour of interest) versus total time to be in effect during the day, and the tabulated fractional values are used *only* when HVAC equipment is operated 24 h a day and space temperature is not allowed to rise during unoccupied periods; otherwise, internal heat gains are considered instantaneous cooling loads at *full* value. On the other hand, the TFM, while not dealing directly with individual load components when space temperatures are permitted to rise overnight, applies space air transfer functions to estimate resultant increased rates of total sensible heat extraction from that space during periods of equipment operation. Regardless of methodology, good engineering judgment must be applied to predict realistic cooling loads from internal heat gains.

Time-averaging data in this chapter are empirical and offered only as information found dependable in practice by users of the technique. Basic assumptions regarding the percentages of radiant heat gain from various sources are used as default values by the TFM in establishing envelope transfer coefficients and room transfer coefficients. The TETD/TA method requires a specific breakdown by the user to determine what values are to be averaged over time. Table 44 suggests representative percentages for this purpose.

The convective portion of heat gain is treated as instantaneous cooling load. The radiant portion of instantaneous heat gain is considered as reduced or averaged over time by the thermal storage of the building and its contents. For lightweight construction, the instantaneous cooling load may be considered as an average of the radiant instantaneous heat gain over a 1 to 3-h period up to and including the hour of calculation interest, plus the nonradiant component of that hour's heat gain. For very heavy construction, the averaging period for hourly values of radiant instantaneous heat gain may be as long as 6 to 8 h, including the hour of calculation interest. Most users of this technique rarely consider application of an averaging period longer than 5 h, with a general norm of 3 h for contemporary commercial construction.

The load from machinery or appliances varies, depending on the temperature of the surface. The higher the surface temperature, the greater the percentage of heat gain that is radiant.

The two-step nature of the TETD/TA procedure offers a unique convenience in calculating cooling load through externally shaded fenestration. As described in the beginning of this section, the hourly history of fenestration heat gain as modified by external shading devices is directly usable for averaging purposes. Thus, the engineer has excellent control and can readily use the effect of external shading on cooling load in the conditioned space. Sun (1968) identified convenient algorithms for analysis of moving shade lines on glass from external projections.

EXAMPLE COOLING LOAD CALCULATION USING TETD/TA

Example 13. Cooling load calculation of small office building. For this example, the one-story building used to illustrate the TFM in Example 6 (and indicated in Figure 4) is also used for calculating a cooling load by the TETD/TA method. Refer to Example 6 for the statement of conditions.

Find (for stated design conditions):
1. Sensible cooling load
2. Latent cooling load
3. Total cooling load

Solution: By TETD/Time-averaging method.

1. Daily Load Cycle

The cooling loads are calculated once per hour for a period of time necessary to cover the hour of anticipated peak design load. For the purposes of this example, the full range of loads over a typical 24-h cycle are presented.

2. Hourly Heat Gain Components

Hourly heat gain values for each load component must be calculated for the same range as those for the cooling load, plus as many preceding hours as will be needed for the purposes of time-averaging (in the case of this example, all 24-h values have been calculated). The methodology involving use of time lag, effective decrement factor, and calculated TETD values is used to calculate heat gain components through walls and roof.

3. Thermal Storage

The heat storage effect of the room is accounted for by averaging the radiant elements of heat gain components for the hour in question with those of the immediately previous hours making up the selected averaging period, and combining the result with the convective heat gain elements for the current hour.

4. Summary

The data and summary of results using TETD/TA are tabulated in Table 45. Following the table is a step by step description of the calculation procedure used to determine the values listed.

1. Sensible Cooling Load

(a) *General*

Line 1, Time of day in hours. Various temperatures and heat flow rates were calculated for every hour on the hour, assuming that hourly values are sufficient to define the daily profile.

Line 2, Outside air temperatures. Hourly values were derived by the procedure given previously, using the specified maximum dry-bulb temperature of 94°F and daily range of 20°F.

(b) *Solar Heat Gain Factors*

Lines 3, 4, 5, and 6, Solar heat gain through opaque surfaces. The values in these columns are copies of the SHGF values listed in Table 4 for July 21 and 40°N latitude. These SHGF values are used to calculate sol-air temperatures of various outside surfaces, and solar heat gain through windows.

The June values might have been used, since the solar irradiation of horizontal surface (i.e., roof) is maximum at that time of year and since the heat gain through the roof appears to be the major component of exterior heat gain in this example problem. The difference between June and August values is relatively small however, compared to the large percentage increase in solar heat gain through south glass in August versus June at this latitude, thus indicating that August might be the better choice. For this example, data for July were selected as reasonable, and to provide better comparison with the results from other techniques for which tabular data are limited. For better assurance of accuracy it is preferable to evaluate and compare the relative loads of various surfaces for several months, before making a final determination as to that in which the maximum load will occur.

(c) *Sol-Air Temperatures*

Lines 7, 8, 9, and 10, Sol-air temperatures at opaque surfaces. Sol-air temperatures were calculated by Equation (6).

(d) *Total Equivalent Temperature Differentials*

Lines 10a through 10h, Calculated TETD values. Hourly TETD values for each of the expose surfaces, are calculated by Equation

Table 45 Tabulation of Data for Example 13—TETD/TA Method

1	Time, hour	0100	0200	0300	0400	0500	0600	0700	0800	0900	1000	1100	1200
2	Outside air temperature, °F	76	75	74	74	74	74	75	77	79	82	86	89
3	SHGF, Btu/h·ft², Horizontal	0	0	0	0	0	32	88	145	194	231	254	262
4	North	0	0	0	0	1	37	30	28	32	35	37	38
5	South	0	0	0	0	0	11	21	30	52	81	102	109
6	East	0	0	0	0	2	137	204	216	193	146	81	41
7	Sol-air temperature, °F, Horizontal	69	68	67	67	67	77	94	114	130	144	155	161
8	North	76	75	74	74	74	80	80	81	84	87	92	95
9	South	76	75	74	74	74	76	78	82	87	94	101	105
10	East	76	75	74	74	74	95	106	109	108	104	98	95
10a	Calculated TETD, °F, Roof	10	8	7	6	5	5	4	3	3	4	12	24
10b	North wall	10	8	7	6	5	5	4	3	3	3	6	7
10c	South wall	16	15	14	14	13	12	11	11	10	10	10	9
10d	East wall	11	10	9	8	7	7	5	5	4	4	17	24
10e	North and west party wall	10	10	10	9	9	8	8	7	7	6	6	6
10f	North door (to adjacent building)	3	2	1	0	0	0	0	0	1	4	6	10
10g	South door	3	2	1	0	0	0	0	2	6	11	17	24
10h	East door	3	2	1	0	0	0	14	27	33	33	30	24
	Instant Sensible Heat Gain, Btu/h												
11	Roof	3722	3193	2718	2347	2106	1814	1580	1393	1264	1681	4518	8802
12	North wall	827	710	605	518	454	446	348	305	273	259	556	592
13	South wall	1599	1536	1457	1379	1306	1213	1141	1090	1044	1005	976	968
14	East wall	4326	3801	3323	2938	2648	2614	2170	1979	1832	1774	6246	9099
15	North and west party wall	2686	2740	2724	2652	2532	2380	2221	2079	1954	1848	1768	1747
16	North door (to adjacent building)	19	17	7	1	−3	−5	−3	2	11	26	44	64
17	South door	21	19	8	2	−2	−5	6	20	41	74	117	160
18	East door	21	19	8	2	−2	−4	93	185	219	223	204	166
19	Windows, air to air heat gain	117	44	−15	−58	−73	−44	29	160	350	569	816	1050
20	North windows, solar heat gain	0	0	0	0	17	611	495	462	528	578	611	627
21	South windows, solar heat gain	0	0	0	0	0	363	693	990	1716	2673	3366	3597
22	Lights, tungsten (always on)	13640	13640	13640	13640	13640	13640	13640	13640	13640	13640	13640	13640
23	Lights, fluorescent (on-off)	0	0	0	0	0	0	0	71610	71610	71610	71610	71610
24	People	0	0	0	0	0	0	0	21250	21250	21250	21250	21250
25	Infiltration	0	0	0	0	0	0	0	162	354	575	825	1061
26	Ventilation	2244	841	−281	−1122	−1403	−841	561	3086	6732	10940	15708	20196
27	Total instant sensible heat gain	29222	26560	24194	22299	21220	22182	22974	118413	122818	128725	14 2255	154629
	Latent Heat Gain/Cooling Load, Btu/h												
28	People	0	0	0	0	0	0	0	17000	17000	17000	17000	17000
29	Infiltration	0	0	0	0	0	0	0	2205	2205	2205	2205	2205
30	Ventilation	41963	41963	41963	41963	41963	41963	41963	41963	41963	419 63	41963	41963
31	Total latent heat gain/cooling load	41963	41963	41963	41963	41963	41963	41963	61168	61168	61168	61168	61168
32	Sum: sensible + latent heat gain, Btu/h	71185	68523	66157	64262	63183	64145	64937	179581	183986	189893	2 03423	215797
	Sensible Cooling Load from Convective Heat Gain, Btu/h												
33	Windows, air to air heat gain	117	44	−15	−58	−73	−44	29	160	350	569	816	1050
34	Lights, tungsten (20% convective)	2728	2728	2728	2728	2728	2728	2728	2728	2728	2728	2728	272 8
35	Lights, fluorescent (50% conv.)	0	0	0	0	0	0	0	35805	35805	35805	35805	35805
36	People (67% convective)	0	0	0	0	0	0	0	14238	14238	14238	14238	14238
37	Infiltration (100% convective)	0	0	0	0	0	0	0	162	354	575	825	1061
38	Ventilation (100% convective)	2244	841	−281	−1122	−1403	−841	561	3086	6732	10940	15708	20 196
	Sensible Cooling Load from Radiant Heat Gain, Btu/h												
39	Lights, tungsten (80% radiant)	10912	10912	10912	10912	10912	10912	10912	10912	10912	10912	10 912	10912
40	Lights, fluorescent (50% radiant)	0	0	0	0	0	0	0	7161	14322	21483	28644	35805
41	People (33% Radiant)	0	0	0	0	0	0	0	1403	2805	4208	5610	7013
	Sensible Cooling Load from Exposed Surfaces; From Convective Heat Gain, Btu/h												
42a	North windows, SHG (42% convective)	0	0	0	0	7	257	208	194	222	243	257	263
43a	South windows, SHG (42% convective)	0	0	0	0	0	152	291	416	721	1123	1414	1511
44a	Roof (40% convective)	1489	1277	1087	939	842	726	632	557	506	672	1807	3521
45a	North wall (40% convective)	331	284	242	207	182	178	139	122	109	104	222	237
46a	South wall (40% convective)	640	614	583	552	522	485	456	436	418	402	390	387
47a	East wall (40% convective)	1730	1520	1329	1175	1059	1046	868	792	733	710	2498	3640
48a	N. and W. party wall (40% conv.)	1074	1096	1090	1061	1013	952	888	832	782	739	707	699
49a	N. door to adj. bldg. (40% conv.)	8	7	3	0	−1	−2	−1	1	4	10	18	26
50a	South door (40% convective)	8	8	3	1	−1	−2	2	8	16	30	47	64
51a	East door (40% convective)	8	8	3	1	−1	−2	37	74	88	89	82	66
	Sensible Cooling Load from Exposed Surfaces; From Radiant Heat Gain, Btu/h												
42b	SHG at north windows (58% radiant)	0	0	0	0	2	73	130	184	245	310	310	325
43b	SHG at south windows (58% radiant)	0	0	0	0	0	42	122	237	436	746	1095	1432
44b	Roof heat gain (60% radiant)	5056	3532	2515	1959	1691	1461	1268	1109	978	928	1253	2119
45b	North wall heat gain (60% radiant)	742	648	545	439	373	328	285	249	219	196	209	238
46b	South wall heat gain (60% radiant)	904	932	932	910	874	827	780	735	695	659	631	610
47b	East wall heat gain (60% radiant)	3532	3110	2697	2325	2044	1839	1643	1482	1349	1244	1681	2511
48b	N. and W. party wall HG (60% rad.)	1428	1515	1576	1605	1600	1563	1501	1423	1340	1258	1185	1127
49b	North door heat gain (60% radiant)	24	18	13	9	5	2	−1	−1	1	4	9	17
50b	South door heat gain (60% radiant)	26	19	14	9	6	3	1	2	8	16	31	49
51b	East door heat gain (60% radiant)	26	19	14	10	6	3	12	33	59	86	111	120
52	Total sensible cooling load, Btu/h	33027	29132	25990	23662	22387	22686	23491	84540	97175	111027	129 243	147770
53	Sum: sens. + lat. cooling load, Btu/h	74990	71095	67953	65625	64350	64649	65454	145708	158343	172195	1 90411	208938

Table 45 Tabulation of Data for Example 13—TETD/TA Method (*Concluded*)

1	1300	1400	1500	1600	1700	1800	1900	2000	2100	2200	2300	2400	24 h Total	Heat Loss, Btu/h
2	91	93	94	93	92	89	87	84	82	80	78	77		
3	254	231	194	145	88	32	0	0	0	0	0	0	2150	
4	37	35	32	28	30	37	1	0	0	0	0	0	438	
5	102	81	52	30	21	11	0	0	0	0	0	0	703	
6	37	35	31	26	20	11	0	0	0	0	0	0	1180	
7	160	155	145	130	111	92	80	77	75	73	71	70		
8	97	98	99	97	97	95	87	84	82	80	78	77		
9	106	105	102	98	95	91	87	84	82	80	78	77		
10	97	98	99	97	95	91	87	84	82	80	78	77		
10a	37	48	57	64	67	66	62	54	44	31	19	12		
10b	7	9	11	14	16	18	18	19	18	17	17	12		
10c	9	9	9	9	9	10	10	11	13	14	16	16		
10d	27	27	24	20	18	19	20	20	19	18	16	13		
10e	6	5	5	5	5	5	6	6	7	8	9	9		
10f	13	15	17	18	18	17	15	12	10	8	6	4		
10g	28	31	30	27	24	20	17	13	10	8	6	4		
10h	21	21	23	23	22	20	17	13	10	8	6	4		
Instantaneous Sensible Heat Gain, Btu/h														
11	13511	17460	20729	23087	24307	23947	22406	19674	15890	11192	6984	4342	238667	23400
12	630	802	962	1184	1358	1488	1529	1561	1490	1462	1404	997	20760	5304
13	930	911	896	889	926	973	1047	1160	1307	1454	1563	1612	28382	6318
14	9969	9958	9110	7627	6870	7300	7487	7630	7311	6767	6033	4990	133802	23868
15	1640	1590	1550	1528	1544	1595	1691	1834	2010	2213	2412	2572	49510	17306
16	84	100	112	117	116	109	96	81	65	51	38	27	1176	410
17	192	206	202	184	161	139	117	90	69	54	41	30	1946	432
18	140	146	153	157	151	137	118	91	70	55	41	30	2423	432
19	1224	1341	1385	1341	1239	1079	889	700	540	393	277	189	13542	4739
20	611	578	528	462	495	611	17	0	0	0	0	0	7231	
21	3366	2673	1716	990	693	363	0	0	0	0	0	0	23199	
22	13640	13640	13640	13640	13640	13640	13640	13640	13640	13640	13640	13640	327360	–13640
23	71610	71610	71610	71610	71610	0	0	0	0	0	0	0	716100	–71610
24	21250	21250	21250	21250	21250	0	0	0	0	0	0	0	212500	–21250
25	1238	1356	1400	1356	1253	0	0	0	0	0	0	0	9580	4791
26	23562	25806	26648	25806	23843	20757	17111	13464	10379	7574	5330	3646	260587	91163
27	163597	169427	171891	171228	169456	72138	66148	59925	52771	44855	37763	32075	2046765	
Latent Heat Gain/Cooling Load, Btu/h														
28	17000	17000	17000	17000	17000	0	0	0	0	0	0	0	170000	
29	2205	2205	2205	2205	2205	0	0	0	0	0	0	0	22050	
30	41963	41963	41963	41963	41963	41963	41963	41963	41963	41963	41963	41963	1007112	
31	61168	61168	61168	61168	61168	41963	41963	41963	41963	41963	41963	41963	1199162	
32	224765	230595	233059	232396	230624	114101	108111	101888	94734	86818	79726	74038	3245927	
Sensible Cooling Load from Convective Heat Gain, Btu/h														
33	1224	1341	1385	1341	1239	1079	889	700	540	393	277	189	13542	
34	2728	2728	2728	2728	2728	2728	2728	2728	2728	2728	2728	2728	65472	
35	35805	35805	35805	35805	35805	0	0	0	0	0	0	0	358050	
36	14238	14238	14238	14238	14238	0	0	0	0	0	0	0	142380	
37	1238	1356	1400	1356	1253	0	0	0	0	0	0	0	9580	
38	23562	25806	26648	25806	23843	20757	17111	13464	10379	7574	5330	3646	260587	
Sensible Cooling Load from Radiant Heat Gain, Btu/h														
39	10912	10912	10912	10912	10912	10912	10912	10912	10912	10912	10912	10912	261888	
40	35805	35805	35805	35805	35805	28644	21483	14322	7161	0	0	0	358050	
41	7013	7013	7013	7013	7013	5610	4208	2805	1403	0	0	0	70130	
Sensible Cooling Load from Exposed Surfaces; From Convective Heat Gain, Btu/h														
42a	257	243	222	194	208	257	7	0	0	0	0	0	3039	
43a	1414	1123	721	416	291	152	0	0	0	0	0	0	9745	
44a	5404	6984	8292	9235	9723	9579	8962	7870	6356	4477	2794	1737	95468	
45a	252	321	385	474	543	595	612	624	596	585	562	399	8305	
46a	372	364	358	356	370	389	419	464	523	582	625	645	11352	
47a	3988	3983	3644	3051	2748	2920	2995	3052	2924	2707	2413	1996	53521	
48a	656	636	620	611	618	638	676	734	804	885	965	1029	19805	
49a	34	40	45	47	46	44	38	32	26	20	15	11	471	
50a	77	82	81	74	64	56	47	36	28	22	16	12	779	
51a	56	58	61	63	60	55	47	36	28	22	16	12	967	
Sensible Cooling Load from Exposed Surfaces; From Radiant Heat Gain, Btu/h														
42b	343	349	343	325	310	310	245	184	130	73	2	0	4193	
43b	1707	1818	1707	1432	1095	746	436	237	122	42	0	0	13452	
44b	3573	5517	7802	10030	11891	13143	13738	13610	12747	11173	9137	6970	143200	
45b	277	341	425	500	592	695	782	855	891	904	893	830	12456	
46b	591	575	562	551	547	552	568	599	650	713	784	851	17032	
47b	3470	4446	5326	5491	5224	4904	4607	4430	4392	4380	4228	3928	80283	
48b	1075	1031	995	967	942	937	949	983	1041	1121	1219	1325	29706	
49b	27	38	48	57	64	66	67	63	56	49	40	31	706	
50b	70	90	105	113	114	107	96	83	69	56	45	34	1166	
51b	115	106	98	91	90	89	86	79	68	56	45	34	1456	
52	156283	163149	167774	169082	168376	105964	92708	78902	64574	49474	43046	37319	204678	178163
53	217451	224317	228942	230250	229544	147927	134671	120865	106537	91437	85009	79282	3245943	71663

(49) to incorporate individual thermal characteristics and orientation.

Line 10a, Roof TETD. Referring to Table 11, the major element of the roof (that with the most mass) is the gypsum slab with code number C14. Other elements are the metal deck (A3), rigid insulation (B3), built-up roofing (E3), and gravel surface (E2). Entering Table 12 with these data, the C14 roof slab with no ceiling and R-values of 11.11calls for an R range of 3. From the "mass-in" part of the upper table these pointers indicate roof group 5 as that whose thermal characteristics will best represent the roof in question.

The time lag and effective decrement factors are then obtained from Table 14, as tabulated for roof group 5. These values are:

$$\text{Time lag } (\delta) = 4.82 \text{ h}$$

$$\text{Effective decrement factor } (\lambda) = 0.68$$

The TETD values were then calculated for the roof surface with Equation (49), using the sol-air temperature cycle given in line 7 and $t_i = 75°F$.

Lines 10b through 10e, Wall TETD. The TETD values for the various walls were calculated by the same approach as that described for the roof. Time lag and effective decrement factors were selected from Table 19, as:

North and East Exterior Walls

Dominant element C8, from Table 11;
Interior finish E1 from Table 11;
R-value indicating R range of 2 from Table 16 (integral mass);
C8 dominant layer indicating Material Layer 13 from Table 16;
From Table 16 select Wall Group 5 from the upper section (combining Layer 13 with E1 finish) as the most representative, and from Table 19 obtain $\delta = 5.11$ h and $\lambda = 0.64$.

South Wall

Dominant element C9, or Layer 14 in Integral mass table;
Exterior layer A2 or A7;
Interior layer E1 (plywood panel ignored as trivial);
R-value indicating R of 6;
Select wall group 24 for representative performance factors of $\delta = 11.29$ h and $\lambda = 0.23$.

North and West Party Walls

With no specific data for a 13 in. brick wall, use a layer of 8 in. common brick (C9) and a layer of 4 in. face brick (A2 or A7) as an approximation;
Dominant element C9, or Layer 14 in Integral Mass table;
Exterior layer A2 or A7;
R-value indicating R of 6;
Select wall group 24 for representative factors $\delta = 11.29$ h and $\lambda = 0.23$. Calculate TETD values as above.

Lines 10f, 10g, and 10h, Door TETD values. Heat storage of the doors may be assumed negligible, and the heat gain, therefore, is calculated with Equation (8) as:

$$q_{DT} = U_D A_D (t_{DT} - t_i)$$

where

U_D = 0.19 Btu/h·ft²·°F, U-factor of doors (0.18 for interior doors)
A_D = 35 ft², area of a door
t_i = 75°F, inside temperature
t_{DT} = outside temperature. For the door in the north party wall, t_{DT} equals outside air temperature. For the doors in east and south walls t_{DT} equals the east and south wall sol-air temperatures, respectively.

While the foregoing calculation would be reasonable in estimating the minor loads involved, for this example, the relatively brief storage effect of the solid core doors has been considered as:

Dominant element B7, or Layer 3 in Integral Mass table;
Interior finish A6;
R-factor indicating R range of 8;

Select wall group 1 for representative time lag δ of 1.30 h and $\lambda = 0.98$, and follow the above procedures to calculate the associated TETD values.

(e) *Instantaneous Sensible Heat Gain*

Line 11, Roof heat gain. Instantaneous heat gain through the roof, calculated with Equation (48) and the TETD values on line 10a.

Lines 12 through 18, Wall and door heat gain. The instantaneous heat gains through the various walls and doors were calculated the same way as heat gain through the roof was calculated. TETD values from lines 10b through 10h were used in Equation (48).

Lines 19, 20, and 21, Window heat gain. The air to air heat gain (line 19) is

$$q_{a-a} = U_w A_w (t_{o\theta} - t_i)$$

where

U_w = 0.81 Btu/h·ft²·°F, U-factor of window
A_w = , area of windows
$t_{o\theta}$ = outside air temperature at hour θ

The solar radiation heat gain (Lines 20 and 21) through south and north windows is:

$$q_r = A_w SC(SHGF)_\theta$$

where

$(SHGF)_\theta$ = Solar heat gain factors given in line 5 for south and line 6 for north
SC = 0.55; shading coefficient for clear window with light-colored curtain or blind

Lines 22 and 23, Heat gain from tungsten and fluorescent lights. For the gain from lighting, Equation (9) was used with a use factor of unity, and special allowance factors of 1.20 for the fluorescent lamps and of unity for the tungsten lamps. Thus:

$$q_{el\ tung} = 4000 \times 1 \times 1 \times 3.41 = 13{,}640 \text{ Btu/h}$$

and

$$q_{el\ fluor} = 17{,}500 \times 1 \times 1.20 \times 3.41 = 71{,}610 \text{ Btu/h}$$

Line 24, People. Sensible heat gain due to people. For the occupants, the data of Table 3 was used for moderately active office work. Thus:

$$q_p = (\text{Number of people})(\text{Sensible heat generated per person})$$
$$= 85 \times 250 = 21{,}250 \text{ Btu/h}$$

Lines 25 and 26, Sensible heat gain from infiltration and ventilation. As developed previously, the value to be used for infiltration was established as 67 cfm, and that for ventilation as 1275 cfm. Heat gain from all air entering as infiltration is routinely part of the space load. In this example (because ventilation is delivered directly to the space, rather than first through the cooling equipment), its gain is also included as a direct space load.

Note: Had the ventilation air been mixed with return air after leaving the occupied space and before entering the cooling equipment, only that portion which passed through the cooling coil without being treated by it—as a function of the coil inefficiency or "Bypass Factor," which is normally 3 to 5% for a chilled water coil of six or more rows and close fin spacing to 15% or more for refrigerant coils in packaged air-conditioning units—and/or that quantity deliberately bypassed around the coil in response to a "face and bypass" or "conventional multizone" space dry-bulb temperature control scheme, would become a part of the space heat gain rather than a part of the cooling coil load. While of potential significance to the design of a cooling system, the details of this concept are not considered in this chapter.

The sensible loads are determined from Equation (22). At 1500 hours for example, when $t_o = 94°F$ and $t_i = 75°F$, this generates:

$$q_{si} = 1.1(\text{Infiltration rate})(t_o - t_i)$$
$$= 1.1 \times 67(94 - 75) = 1400 \text{ Btu/h}$$

and

$$q_{sv} = 1.1(\text{Ventilation rate})(t_o - t_i)$$
$$= 1.1 \times 1275(94 - 75) = 26{,}600 \text{ Btu/h}$$

Line 27, Total instantaneous sensible heat gain. The sum of sensible heat gain values on lines 11 through 26 for each calculation hour. This represents the total amount of such gain that actually enters the building during each hour, including any delaying effects of the individual surfaces on the passage of heat, but before any consideration of the storage and subsequent release of the radiant components of such heat.

(f) Instantaneous Latent Heat Gain

Line 28, People. The latent heat gain due to people, using Table 3 data

= (number of persons)(latent heat generated by one person)

= 85 × 200 = 17,000 Btu/h during the occupied period

Lines 29 and 30, Latent heat gain from infiltration and ventilation. The latent loads are determined from Equation (23). At 1500 h for example, when $W_o = 0.0161$ and $W_s = 0.0093$, this generates:

$$q_{li} = 4840(\text{Infiltration rate})(W_o - W_i)$$
$$= 4840 \times 67(0.0161 - 0.0093) = 2205 \text{ Btu/h}$$

and

$$q_{lv} = 4840(\text{Ventilation rate})(W_o - W_i)$$
$$= 4840 \times 1275(0.0161 - 0.0093) = 41{,}963 \text{ Btu/h}$$

Line 31, Total latent heat gain. The total latent heat gain is the sum of lines 27, 28, and 29.

(g) *Total Instantaneous Heat Gain*

Line 32, Total instantaneous heat gain. The sum of total instantaneous values on lines 27 and 31, sensible and latent heat gain, respectively. The hourly profile of such a total will normally reach a higher level at an earlier time of day than that of the building total cooling load, although the 24-h totals will be identical.

(h) *Cooling Load from Convective Sensible Heat Gain Components*

Lines 33 through 38. Direct inclusion of the instantaneous heat gain components listed in Lines 19, 25, and 26, and 20%, 50%, and 67% of lines 22, 23, and 24, respectively. These room sensible heat gain components (i.e., loads due to air-to-air heat gain through windows, tungsten lights, fluorescent lights, people, infiltration, and ventilation) all appear as cooling load without delay. Percentages of sensible heat gain considered convective are taken from Tables 3 and 42. Selection of 67% of sensible gain from people as convective is an approximation for purposes of this example.

(i) *Cooling Load Involving Time-Averaging*

Radiant elements of instantaneous heat gain will be felt as cooling load in the space only after having first been absorbed by the mass of building and contents, and later released back into the space as convective heat. This delaying action is approximated by time-averaging, or taking the average of such a heat gain value for the current hour with those from some number of immediately previous hours. An averaging period of about 5 h is used for this example, in which, for example, the value of cooling load for hour 1200 is derived as the average of the radiant fractions of hourly sensible heat gain for hours 1200, 1100, 1000, 0900, and 0800; thus delaying the full impact of such heat gain becoming cooling load for 5 h, and extending the period after the heat gain has ended for some amount of cooling load to be felt by the space.

Line 39, Cooling load from tungsten light sensible heat gain. Although 80% of the sensible heat gain from tungsten lights is radiant heat and subject to the storage/re-release phenomenon, data on line 39 appears as a constant value for every hour. This is due to the constant heat input to the room (line 22), from lights switched on all the time and thus with the radiant heat gain component from prior hours being released as cooling load at the same rate as the absorption by the room of the current hour's radiant component.

Line 40, Cooling load from lighting cycled on and off. Fifty percent (the radiant component) of the fluorescent lighting heat gain from line 23, showing the effect of such gain being processed by time-averaging, as indicated above.

Line 41, Cooling load due to radiant heat gain from people. Of sensible heat generated by people, 33% is dissipated by radiation and felt by the space as cooling load only after having been absorbed by the mass of the building and its contents.

(j) *Sensible Cooling Load from Exposed Surfaces*

Elements of instantaneous heat gain from solar radiation through windows, walls, doors and roof, i.e., the sum of values listed in lines 11 through 18, 20, and 21, are also delayed in being felt as cooling. The radiant heat gain by solar radiation transmitted through windows is treated the same way as the radiant portion of heat gain through walls and roof surfaces. However, since the windows have inside shading devices, solar radiation is considered reduced to approximately 58% of the solar heat gain through glass because the venetian blind intercepts about 42% of such solar radiation and releases it to the room in a convective form, similar to the treatment of heat gain through walls and roof (see Table 44).

Note: Had there been *no internal shading* of the glass, the solar radiation through windows would have to be treated as 100% radiant, all subject to time-averaging. Translucent draperies fall between these limits, in a linear relationship. Chapter 29 has more specific information on internal shading.

Lines 42a through 51a, Sensible cooling load from convective heat gain through enclosing surfaces. Data on lines 42b and 43b represent 58% of heat gain values for north and south windows, respectively, form lines 20 and 21, but time-averaged. Data for opaque enclosing surfaces on lines 44b through 51b represent 60% of the corresponding heat gain values on lines 11 through 18, but also time-averaged.

Cooling Load from power equipment and appliances. For this example, none are assumed. Had such loads been involved, with starting or ending periods within the time before the hour of calculation interest that can affect the averaging period, 20 to 80% of the sensible heat gain would have been considered as radiant and subject to time-averaging.

Line 52, Total room sensible cooling load. Total sensible cooling load felt by the room, and the design sensible load which is used as the basis for sizing cooling equipment. It is the sum of the values listed in lines 33 through 51b. The almost exact match between the 24-h total of 2,046,781 Btu/h on line 52 and the sum of the 24-h gain totals on line 27 (differing only by 16 Btu/h) does verify completeness of the computation.

2. **Latent Cooling Load**

Line 31—The sum of lines 28, 29, and 30. Total latent heat gain is also the total latent cooling load, as all components occur instantaneously.

3. **Total Cooling Load**

Line 52, The sum of lines 52 and 31. The total cooling load for this example problem is the theoretical total for the conditions as defined, and may or may not represent the actual total cooling load imposed upon a system of cooling equipment attempting to maintain the specified space conditions. An appropriate psychrometric analysis of supply air, space air, return air, and mixed air [when ventilation air is mixed with return air enroute back to the cooling equipment] should be performed, in conjunction with proper consideration of the type of cooling equipment and characteristics of the preferred control scheme, in order to verify the ability of the design to meet the requirements, and to determine whether the actual sensible, latent, and total cooling loads are greater or less than the theoretical values calculated.

Comparison of Results

Each of the calculation procedures outlined in this chapter, TFM, CLTD/SCL/CLF, and TETD/TA have used the same building in Examples 6, 11, and 13, respectively. Although widely different in purpose, approach, and mathematical processes, the results have many similarities as illustrated by Figure 5.

Tabular data for hourly total instantaneous sensible heat gain and total sensible cooling load values from Tables 28 and 45 are plotted to compare the two computer-based techniques, TFM and TETD/TA. The curves for heat gain are almost identical. Those for cooling load, however, happen to peak at the same hour, 1600, but with different magnitudes. The TETD/TA cooling load peak has reached almost the peak of its companion heat gain curve, but one hour later. The TFM heat gain curve reaches a peak at 1600 with a value only 0.5% different from that for TETD/TA, but the TFM

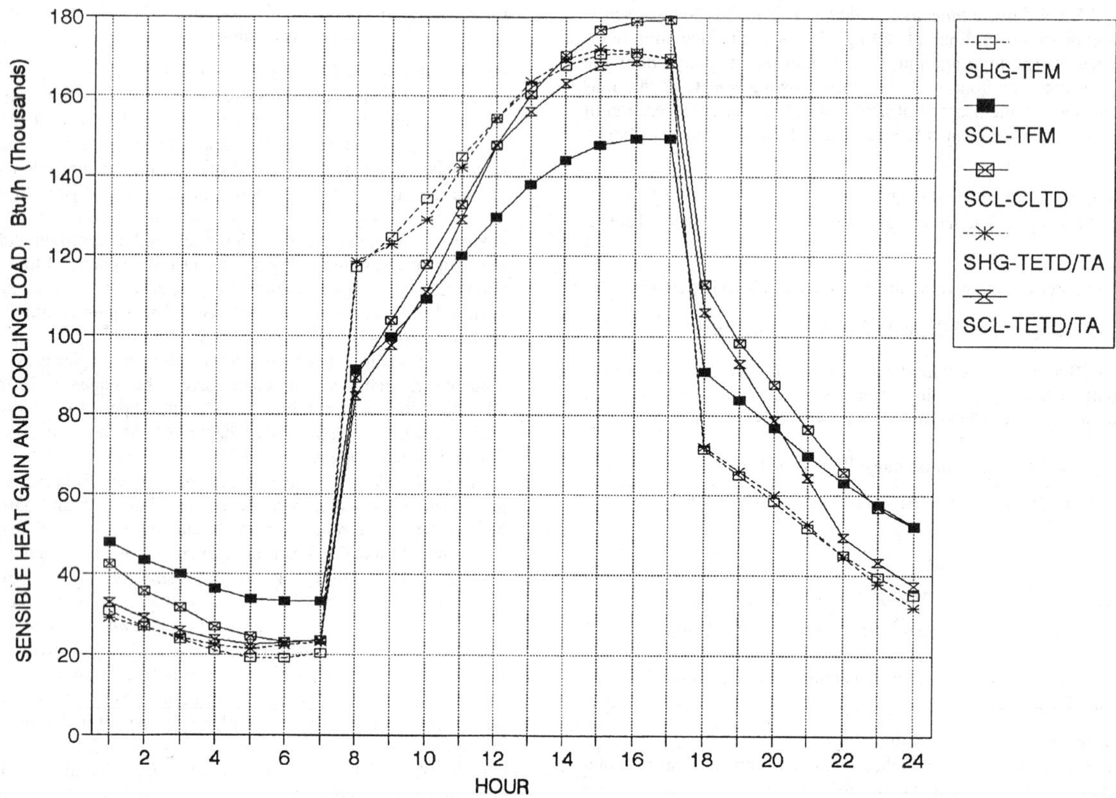

Fig. 5 TFM versus CLTD/SCF/CLF Versus TETD/TA

cooling load curve peaks at only 87.5% of its heat gain curve. All unoccupied hours show substantially greater TFM cooling loads than for TETD/TA, while 24-h totals vary only by 0.15%.

As a manual procedure, Example 11, illustrating the use of CLTD/SCL/CLF, was carried through for hour 1600 only, in the manner that it would primarily be applied by users. For comparison purposes, it was also calculated for each of the daily 24 h and that cooling load profile plotted on Figure 5. There is no comparable heat gain profile, as this method does not produce such values directly. The curve peaks at 1700 hours, one hour later than the others, but with a total value 19.8% greater than TFM. The profile is somewhat different from and between those for TFM and TETD/TA during unoccupied hours.

Note: The small building used in these examples is more massive than typical for a similar function in post-1990 construction, and it would probably not meet ASHRAE *Standard* 90.1-1989 energy requirements. Calculating the entire building as a single simultaneous load could certainly be questioned, particularly in any larger configuration; thus, it is used here purely to illustrate the techniques discussed.

REFERENCES

Alereza, T. and J.P. Breen, III. 1984. Estimates of recommended heat gain due to commercial appliances and equipment. *ASHRAE Transactions* 90(2A): 25-58.

American Gas Association. 1948. A comparison of gas and electric use for commercial cooking. Cleveland, OH.

American Gas Association. 1950. Gas and electric consumption in two college cafeterias. Cleveland, OH.

ASHRAE. 1975. Procedure for determining heating and cooling loads for computerized energy calculations, algorithms for building heat transfer subroutines.

ASHRAE. 1979. *Cooling and heating load calculation manual.*

ASHRAE. 1989. Ventilation for acceptable indoor air quality. ASHRAE *Standard* 62-1989.

Bliss, R.J,V. 1961. Atmospheric radiation near the surface of the ground. Solar Energy 5(3):103.

Buchberg, H. 1958. Cooling load from thermal network solutions ASHAE *Standard* 64:111.

Chiles, D.C. and E.F. Sowell. 1984. A counter-intuitive effect of mass on zone cooling load response. *ASHRAE Transactions* 91(2A):201-208.

Consolazio, W. and L.J. Pecora. 1947. Minimal replenishment air required for living spaces. ASHVE *Standard* 53:127.

Falconer, D.R., E.F. Sowell, J.D. Spitler, and B.B. Todorovic. 1993. Electronic tables for the ASHRAE Load Calculation Manual. *ASHRAE Transactions* 99(1).

Fisher, D.R. 1996. New recommended heat gains for commercial cooking equipment. *Research report*. PG&E Food Service Technology Center, San Ramon, CA.

Gordon, E.B., D.J. Horton, and F.A. Parvin. 1994. Development and application of a standard test method for the performance of exhaust hoods with commercial cooking appliances. *ASHRAE Transactions* 100(2): 988-999.

Harris, S.M. and F.C. McQuiston. 1988. A study to categorize walls and roofs on the basis of thermal response. *ASHRAE Transactions* 94(2): 688-715.

Kimura and Stephenson. 1968. Theoretical study of cooling loads caused by lights. *ASHRAE Transactions* 74(2):189-97.

Marn, W.L. 1962. Commercial gas kitchen ventilation studies. Research Bulletin No. 90 (March). Gas Association Laboratories, Cleveland, Ohio.

McQuiston, F.C. and J.D. Spitler. 1992. *Cooling and heating load calculation manual*, 2nd ed. ASHRAE.

Mitalas, G.P. 1972. Transfer function method of calculating cooling loads, heat extraction rate, and space temperature. *ASHRAE Transactions* 14(12):52.

Mitalas, G.P. 1973. Calculating cooling load caused by lights. *ASHRAE Transactions* 15(6):7.

Mitalas, G.P. and J.G. Arsenault. 1971. Fortran IV program to calculate Z-transfer functions for the calculation of transient heat transfer through walls and roofs. Proceedings of the conference, Use of Computers for Environmental Engineering Related to Buildings. NBS *Building Science Series* 39 (October). Gaithersburg, MD.

Mitalas, G.P. and K. Kimura. 1971. A calorimeter to determine cooling load caused by lights. *ASHRAE Transactions* 77(2)65.

Mitalas, G.P. and D.G. Stephenson. 1967. Room thermal response factors. *ASHRAE Transactions* 73(2):III.2.1.

Nevins, R.G., H.E. Straub, and H.D. Ball. 1971. Thermal analysis of heat removal troffers. *ASHRAE Transactions* 77(2):58-72.

NFPA. 1996. Standard for health care facilities. *Standard* 99-96. National Fire Protection Association, Quincy, MA.

Rudoy, W. and F. Duran. 1975. Development of an improved cooling load calculation method. *ASHRAE Transactions* 81(2):19-69.

Smith, V.A., R.T. Swierczyna, C.N. Claar. 1995. Application and enhancement of the standard test method for the performance of commercial kitchen ventilation systems. *ASHRAE Transactions* 101(2).

Sowell, E.F. 1988a. Classification of 200,640 parametric zones for cooling load calculations. *ASHRAE Transactions* 94(2):754-77.

Sowell, E.F. 1988b. Cross-check and modification of the DOE program for calculation of zone weighting factors. *ASHRAE Transactions* 94(2): 737-53.

Sowell, E.F. 1988c. Load calculations for 200,640 zones. *ASHRAE Transactions* 94(2):71 6-36.

Sowell, E.F. and D.C. Chiles. 1984a. Characterization of zone dynamic response for CLF/CLTD tables. *ASHRAE Transactions* 91(2A):162-78.

Sowell, E.F. and D.C. Chiles. 1984b. Zone descriptions and response characterization for CLF/CLTD calculations. *ASHRAE Transactions* 91(2A):179-200.

Spitler, J.D., F.C. McQuiston, and K.L. Lindsey. 1993. The CLTD/SCL/CLF cooling load calculation method. *ASHRAE Transactions* 99(1).

Spitler, J.D. and F.C. McQuiston. 1993. Development of a revised cooling and heating calculation manual. *ASHRAE Transactions* 99(1).

Stephenson, D.G. 1962. Method of determining non-steady-state heat flow through walls and roofs at buildings. *The Journal of the Institution of Heating and Ventilating Engineers* 30:5.

Stephenson, D.G. and G.P. Mitalas. 1967. Cooling load calculation by thermal response factor method. *ASHRAE Transactions* 73(2):III.1.1.

Stewart, J.P. 1948. Solar heat gain through walls and roofs for cooling load calculations. ASHVE *Transactions* 54:361.

Talbert, S.G., L.J. Canigan, and J.A. Eibling. 1973. An experimental study of ventilation requirements of commercial electric kitchens. *ASHRAE Transactions* 79(1):34.

Todorovic, B. 1987. The effect of the changing shade line on the cooling load calculations. ASHRAE videotape "Practical applications for cooling load calculations."

Todorovic, B. and D. Curcija. 1984. Calculative procedure for estimating cooling loads influenced by window shading, using negative cooling load method. *ASHRAE Transactions* 2:662.

Wilkins, C.K. and N. McGaffin 1994. Measuring computer equipment loads in office buildings. *ASHRAE Journal* 36(8):21-24.

BIBLIOGRAPHY

Historical

Alford, J.S., J.E. Ryan, and F.O. Urban. 1939. Effect of heat storage and variation in outdoor temperature and solar intensity on heat transfer through walls. ASHVE *Transactions* 45:387.

Brisken, W.R. and G.E. Reque. 1956. Thermal circuit and analog computer methods, thermal response. ASHAE *Transactions* 62:391.

Buchberg, H. 1955. Electric analog prediction of the thermal behavior of an inhabitable enclosure. ASHAE *Transactions* 61:339-386.

Buffington, D.E. 1975. Heat gain by conduction through exterior walls and roofs—transmission matrix method. *ASHRAE Transactions* 81(2):89.

Headrick, J.B. and D.P. Jordan. 1969. Analog computer simulation of heat gain through a flat composite roof section. *ASHRAE Transactions* 75(2):21.

Houghton, D.G., C. Gutherlet, and A.J. Wahl. 1935. ASHVE Research Report No. 1001—Cooling requirements of single rooms in a modern office building. ASHVE *Transactions* 41:53.

Leopold, C.S. 1947. The mechanism of heat transfer, panel cooling, heat storage. *Refrigerating Engineering* 7:33.

Leopold, C.S. 1948. Hydraulic analogue for the solution of problems of thermal storage, radiation, convection, and conduction. ASHVE *Transactions* 54:3-9.

Livermore, J.N. 1943. Study of actual vs predicted cooling load on an air conditioning system. ASHVE *Transactions* 49:287.

Mackey, C.O. and N.R. Gay, 1949. Heat gains are not cooling loads. ASHVE *Transactions* 55:413.

Mackey, C.O. and N.R. Gay. 1952. Cooling load from sunlit glass. ASHVE *Transactions* 58:321.

Mackey, C.O. and N.R. Gay. 1954. Cooling load from sunlit glass and wall. ASHVE *Transactions* 60:469.

Mackey, C.O. and L.T. Wright, Jr. 1944. Periodic heat flow—homogeneous walls or roofs. ASHVE *Transactions* 50:293.

Mackey, C.O. and L.T. Wright, Jr. 1946. Periodic heat flow—composite walls or roofs. ASHVE *Transactions* 52:283.

Nottage, H.B. and G.V. Parmelee. 1954. Circuit analysis applied to load estimating. ASHVE *Transactions* 60:59.

Nottage, H.B. and G.V. Parmelee. 1955. Circuit analysis applied to load estimating. ASHAE *Transactions* 61(2):125.

Parmelee, G.V., P. Vance, and A.N. Cherny. 1957. Analysis of an airconditioning thermal circuit by an electronic differential analyzer. ASHAE *Transactions* 63:129.

Paschkis, V. 1942. Periodic heat flow in building walls determined by electric analog method. ASHVE *Transactions* 48:75.

Romine, T.B., Jr. 1992. Cooling load calculation, Art or science? *ASHRAE Journal*, 34(1), p. 14.

Sun, T.-Y. 1968. Computer evaluation of the shadow area on a window cast by the adjacent building, *ASHRAE Journal*, September 1968.

Sun, T.-Y. 1968. Shadow area equations for window overhangs and side-fins and their application in computer calculation. *ASHRAE Transactions* 74(1): I-1.1 to I-1.9.

Vild, D.J. 1964. Solar heat gain factors and shading coefficients. *ASHRAE Journal* 6(10):47.

Transfer Function Method

Burch, D.M., B.A. Peavy, and F.J. Powell. 1974. Experimental validation of the NBS load and indoor temperature prediction model. *ASHRAE Transactions* 80(2):291.

Mast, W.D. 1972. Comparison between measured and calculated hour heating and cooling loads for an instrumented building. ASHRAE Symposium Bulletin No. 72-2.

McBridge, M.F., C.D. Jones, W.D. Mast, and C.F. Sepsey. 1975. Field validation test of the hourly load program developed from the ASHRAE algorithms. *ASHRAE Transactions* 1(1):291.

Mitalas, G.P. 1969. An experimental check on the weighting factor method of calculating room cooling load. *ASHRAE Transactions* 75(2):22.

Peavy, B.A., F.J. Powell, and D.M. Burch. 1975. Dynamic thermal performance of an experimental masonry building. NBS *Building Science Series* 45 (July).

CLTD and CLF Data

DeAlbuquerque, A.J. 1972. Equipment loads in laboratories. *ASHRAE Journal* 14(10):59.

Kusuda, T. 1969. Thermal response factors for multilayer structures of various heat conduction systems. *ASHRAE Transactions* 75(1):246

Mitalas, G.P. 1968. Calculation of transient heat flow through walls and roofs. *ASHRAE Transactions* 74(2):182.

Rudoy, W. 1979. Don't turn the tables. *ASHRAE Journal* 21(7):62.

Stephenson, D.G. and G.P. Mitalas. 1971. Calculation of heat conduction transfer functions for multilayer slabs. *ASHRAE Transactions* 77(2):1.17.

Todorovic B. 1982. Cooling load from solar radiation through partially shaded windows, taking heat storage effect into account. *ASHRAE Transactions* 88(2): 924-937.

Todorovic, B. 1984. Distribution of solar energy following its transmittal through window panes. *ASHRAE Transactions* 90(1B): 806-15.

Todorovic, B. 1989. Heat storage in building structure and its effect on cooling load; Heat and mass transfer in building materials and structure. Hemisphere publishing, New York, 603-14.

CHAPTER 29

FENESTRATION

FENESTRATION refers to any aperture in a building envelope. Fenestration components include glazing material, either glass or plastic; framing, mullions, muntins, dividers, opaque door slabs; external shading devices; internal shading devices; integral (between-glass) shading systems. Fenestration can serve as a physical and/or visual connection to the outdoors, as well as a means to admit solar radiation. The solar radiation provides natural lighting, referred to as daylighting, and heat gain to a space. Fenestration can be fixed or operable, and operable units can allow natural ventilation to a space and egress in low-rise buildings.

Fenestration affects building energy use through four basic mechanisms—thermal heat transfer, solar heat gain, air leakage, and daylighting. The energy impacts of fenestration can be minimized by (1) using daylight to offset lighting requirements, (2) using appropriate glazings and shading strategies to control solar heat gain to supplement heating through passive solar gain and minimize cooling requirements, (3) using appropriate glazing to minimize conductive heat loss, and (4) specifying low air leakage fenestration products.

A designer should consider architectural requirements, thermal performance, economic criteria, and human comfort when selecting fenestration. Typically, a wide range of fenestration products are available that meet the specifications for a project. Refining the specifications to improve the energy performance and enhance a living or work space can result in lower energy costs, increased productivity, and improved thermal and visual comfort. Carmody et al. (1996) and CEA (1995) provide guidance for carrying out these requirements.

Fenestration Components

Fenestration consists of glazing, framing, and in some cases shading devices and insect screens. The glazing unit may have single glazing or multiple glazing. The most common glazing material is glass, although plastic is also used. The glass or plastic may be clear, tinted, obscured, or coated.

Insulating Glass Units. Insulating glazing units (IGUs), which are a sealed assembly with a minimum of two panes of clear or coated glass, are held apart by a spacer bar(s) typically containing a desiccant material and by a sealant. Figure 1 shows the construction of a typical IGU. The most common type of glass is clear. However, low-emittance glazing, which improves the thermal performance of an IGU by reducing energy transfer, has become common. In addition, argon and krypton gas are used in lieu of air in the gap between the panes in combination with low-emittance (low-e) glazing to

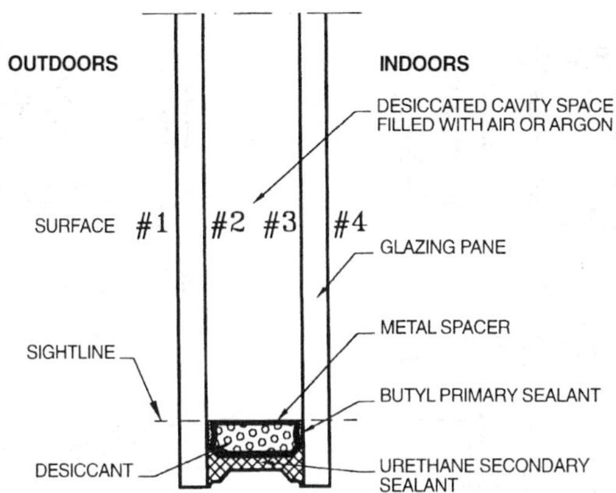

Fig. 1 Insulating Glass Unit (IGU) Construction Detail

further reduce energy transfer. Tinted and reflective glazing are often used to reduce solar heat gain. Some manufacturers construct IGUs with one or more low-e coated plastic films suspended between the glass panes and with a spacer that has better insulating properties and a dual sealant that improves the seal around the gas spaces.

The spacer separates the panes and provides a surface for the primary and secondary sealant adhesion. Traditional spacers are often made of aluminum. Fusing or bending the corners of the spacer minimizes moisture and hydrocarbon vapor transmission into the airspace through the corners. Some spacer designs reduce edge heat transfer by using materials that have lower thermal conductivity than aluminum, (e.g., stainless steel, galvanized steel, polymers, or foamed silicone). Desiccants such as molecular sieves or silica gel are also used to absorb moisture that was initially trapped in the IGU during assembly or gradually diffuses through the seals after construction.

In dual-seal construction, the primary seal minimizes moisture and hydrocarbon transmission and allows the formulation of the secondary seal to provide structural integrity between the lites of the insulating glass unit. A secondary seal ensures long-term adhesion and greater resistance to solvents, oils, and short-term water immersion. In a typical dual-seal a primary seal is of compressed polyisobutylene (PIB) and a secondary seal is of silicone, polysulphide, or polyurethane.

Framing. The three main categories of window framing materials are wood, metal, and polymers. Wood has good structural integrity and insulating value, but low resistance to weather, moisture, warpage, and organic degradation (from mold and insects). Metal is durable, and has excellent structural characteristics, but it has very poor thermal performance. The metal of choice in windows is

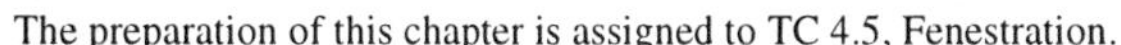

The preparation of this chapter is assigned to TC 4.5, Fenestration.

almost exclusively aluminum, due to ease of manufacture, low cost, and light weight—but aluminum has a thermal conductivity roughly 1000 times that of wood or polymers. The poor thermal performance of metal-frame windows can be improved with a thermal break (a non-metal component that separates the metal frame exposed to the outside from the surfaces exposed to the inside).

Polymer frames are made of extruded vinyl or poltruded fiberglass (glass-reinforced polyester). Their thermal and structural performance is similar to that of wood, although vinyl frames for large windows must be reinforced. Manufacturers sometimes combine these materials as clad units (e.g., vinyl-clad aluminum, aluminum-clad wood, vinyl-clad wood) to increase durability, improve thermal performance, or improve aesthetics. In addition, curtain wall systems for commercial buildings may be structurally glazed and the exterior "framing" is simply rubber gaskets or silicone.

Residential windows can be categorized as shown in Figure 2. The glazing can either be mounted directly in the frame (a direct-glazed or direct-set window, which is not operable) or in a sash that moves in the frame (for an operating window). In operable windows a weather-sealing system between the frame and sash reduces air and water leakage.

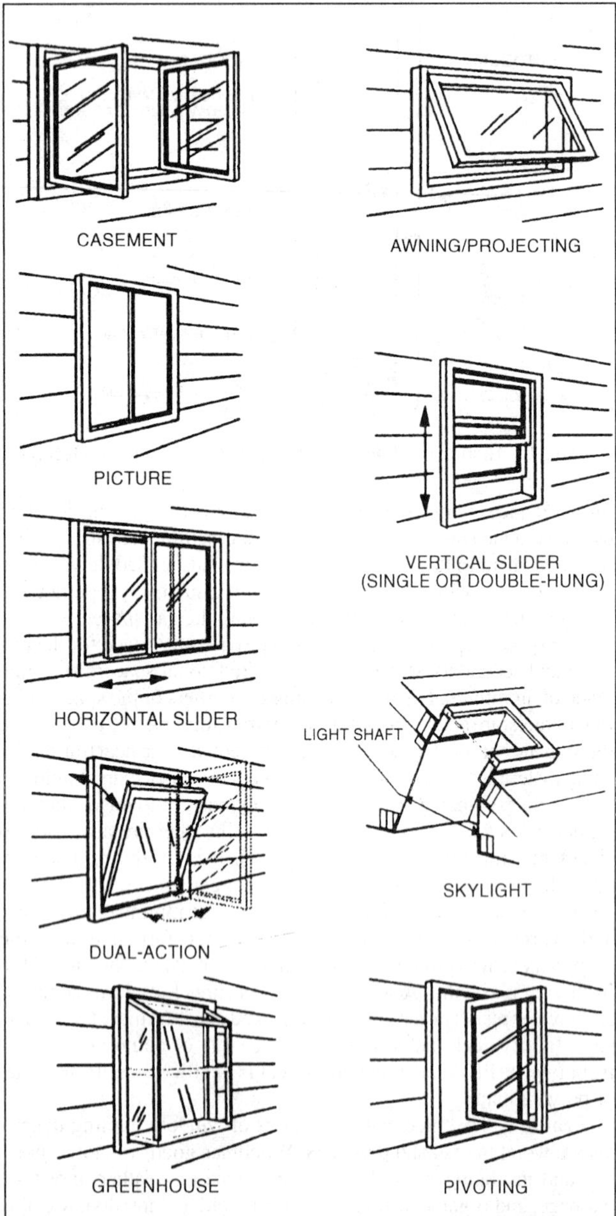

Fig. 2 Types of Residential Windows

Shading. Shading devices are available in a wide range of products that differ greatly in their appearance and energy performance. Shading devices include interior and exterior blinds, integral blinds, interior and exterior screens, shutters, draperies, and roller shades. Shading devices on the exterior of the glazing reduce solar heat gain more effectively than interior devices. However, interior devices are easier to operate and adjust. Some products help insulate the indoors from the outdoors, while others redirect incoming solar radiation to minimize visual and thermal discomfort. Overhangs and vegetation can provide effective shading too.

DETERMINING FENESTRATION ENERGY FLOW

Energy flows through fenestration via (1) conductive and convective heat transfer caused by the temperature difference between outdoor and indoor air; (2) net long-wave (above 2500 nm) radiative exchange between the fenestration and its surrounding and between glazing layers; and (3) short-wave (below 2500 nm) solar radiation incident on the fenestration product, either directly from the sun or reflected from the ground or adjacent objects. Simplified calculations are based on the observation that the temperatures of the sky, ground, and surrounding objects (and hence their radiant emission) correlate with the exterior air temperature. The radiative interchanges are then approximated by assuming that all the radiating surfaces (including the sky) are at the same temperature as the outdoor air. With this assumption, the basic equation for the instantaneous energy flow through a fenestration is

$$q = U_o A_{pf}(t_{out} - t_{in}) + \text{SHGC}_o A_{pf} E_t \qquad (1)$$

where

q = instantaneous energy flow, Btu/h
U_o = overall coefficient of heat transfer (U-factor), Btu/h·ft²·°F
t_{in} = interior air temperature, °F
t_{out} = exterior air temperature, °F
A_{pf} = total projected area of fenestration, ft²
SHGC_o = Overall solar heat gain coefficient, non-dimensional
E_t = incident total irradiance, Btu/h·ft²

The quantities U_o and SHGC_o are instantaneous performance indices. The principal justification for Equation (1) is its simplicity, which is achieved by collecting all the linked radiative, conductive, and convective energy transfer processes into U and SHGC. These quantities vary because (1) convective heat transfer varies as fractional powers of temperature differences or free-stream speeds, (2) variations in temperature due to the weather or climate are small on the absolute temperature scale that controls radiative heat transfer rates, (3) fenestration systems always involve at least two thermal resistances in series, and (4) solar heat gain coefficients depend on solar incident angle and spectral distribution.

U-FACTOR (OVERALL COEFFICIENT OF HEAT TRANSFER)

The first term in Equation (1) represents the rate of thermal heat transfer through fenestration in the absence of sunlight, air infiltration, and moisture condensation. Most fenestration products consist of transparent multi-pane glazing units and opaque elements comprising the sash and frame (hereafter called frame). The glazing unit's heat transfer paths include a one-dimensional center-of-glass contribution and a two-dimensional edge contribution. The frame contribution is primarily two-dimensional.

Consequently, the total heat transfer can be determined by calculating the separate heat transfer contributions of the center glass, edge glass, and frame. (When present, glazing dividers, such as decorative grilles and muntins, also affect heat transfer, and their contribution must be considered.) The overall U-factor may be estimated by adding the area-weighted U-factors for each contribution; or:

$$U_o = \frac{U_{cg}A_{cg} + U_{eg}A_{eg} + U_f A_f}{A_{pf}} \qquad (2)$$

where the subscripts *cg*, *eg*, and *f* refer to the center-of-glass, edge-of-glass, and frame, respectively. A_{pf} is the area of the fenestration product's rough opening in the wall or roof less installation clearances. Where a fenestration product has glazed surfaces in only one direction (typical windows), the sum of the areas equal the projected area. Windows that extend beyond the plane of the wall/roof such as skylights, greenhouse/garden windows, and bay/bow windows have greater a surface area for heat loss than flat windows with a similar glazing option and frame material; consequently, U-factors for such products are expected to be greater.

Center-of-Glass U-Factor

Both convective and radiative heat transfer in the gas space must be considered when calculating heat flow across the central glazed portion of a multi-pane unit. Convective heat transfer may be estimated by using high aspect ratio, natural convection correlations for vertical and inclined air layers (ElSherbiny et al. 1982, Shewen 1986, Wright 1996). Radiative heat transfer (ignoring gas absorption) is quantified using a more fundamental approach. Rubin (1982a,b) and Hollands and Wright (1982) devised computational methods for solving the combined heat transfer problem.

Especially for single glass, U-factors depend strongly on indoor and outdoor film coefficients. The U-factor for single glass is:

$$U = \frac{1}{(1/h_o) + (1/h_i) + (L/k)} \qquad (3)$$

where

h_o, h_i = outdoor and indoor respective glass surface heat transfer coefficients, Btu/h·ft²·°F
L = glass thickness, in.
k = thermal conductivity, Btu·in/h·ft²·°F

Values for U_{cg} at standard indoor and outdoor conditions depend on such glazing construction features as the number of glazing lites, the gas-space dimensions, the orientation relative to vertical, the emittance of each surface, and the composition of the fill gas. Several computer programs have been developed to estimate glazing unit heat transfer for a wide range of glazing construction (Arasteh et al. 1994, Finlayson et al.1993, Wright 1995c). The National Fenestration Rating Council (NFRC) calls for WINDOW 4.1 (LBL 1994) as a standard calculation method for the center glazing. In Canada, the VISION program (Wright 1995b) is used to determine center glazing properties (CSA *Standard* A440.2-93).

Figure 3 shows the effect of gas space width on U_{cg} for vertical double and triple paned glazing units. U-factors are plotted for air, argon, and krypton fill-gases and for high (uncoated) and low (coated) values of surface emittance. Gas space widths greater than 0.5 in. have no significant effect on U_{cg}, but greater glazing unit thicknesses decrease U_o because the length of the shortest heat flow path through the frame increases. A low-emittance coating combined with krypton gas fill offers significant potential for reducing heat transfer in glazing units with narrow gaps.

Edge-of-Glass U-Factor

Insulating glass units usually have continuous spacers around the glass perimeter to separate the glazing and provide an edge seal. Aluminum spacers greatly increase conductive heat transfer between the contacted inner and outer glazing, thereby degrading the thermal performance of the glazing unit. The edge of glass area is typically taken to be a band 2.5 in. wide around the sightline. The width of this area is determined from the extent of two-dimensional heat transfer effects in current computer models, which are based on conduction-only analysis. In reality, due to convective and radiative effects this area may extend beyond 2.5 in. (Beck et al. 1995, Curcija and Goss 1994, Wright and Sullivan 1995b).

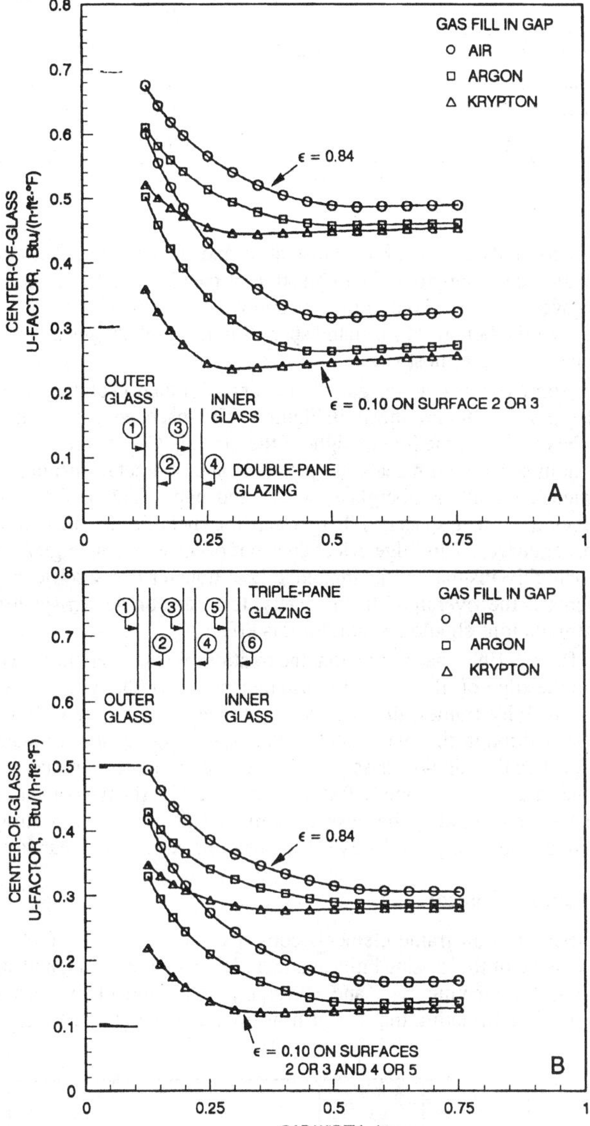

Fig. 3 Center of Glass U-Factor for Vertical, Double- and Triple-Pane Glazing Units

Edge-of-glass heat transfer is two-dimensional and requires detailed modeling for accurate determination. Based on detailed two-dimensional modeling, Arasteh (1989) developed the following correlation to calculate the edge-of-glass U-factor as a function of spacer type and center-of-glass U-factor:

$$U_{eg} = A + BU_{cg} + CU_{cg}^2 \qquad (4)$$

where *A*, *B*, and *C* are correlation coefficients, which are listed in Table 1 for metal, insulating (including wood) and fused-glass

Table 1 Equation (4) Coefficients for Edge-of-Glass U-Factor

	A	*B*	*C*
Metal	0.223	0.842	−0.153
Insulating	0.120	0.682	0.244
Glass	0.158	0.774	0.057
Metal + insulation	0.135	0.706	0.187

Note: A, B and C have units of [Btu/(h·ft^2·°F)]^n, where n = 1, 0, and −1, respectively.

spacers, and a combination of insulating and metal spacers. The correlation constants for the combination of metal and insulated spacers were derived from computer simulations, which showed that 85% of the benefit of insulated spacers in triple-glazing is attributable to the outermost spacer.

Approximate edge-of-glass U-factors as a function of the center-of-glass U-factor are shown in Figure 4. The spacer edge is assumed to be even with the line of sight of the glazing. Curves are for aluminum spacers with sealants (metallic) and nonmetal (insulating) spacers, including fiberglass, wood, and butyl. Values for glass edges and steel spacers fall between the metallic and insulating spacer curves. This edge effect does not occur with single glazing. For highly insulating glazing, edge heat transfer can significantly increase the overall U-factor. Thus, test data or design-specific computations should account for this effect.

Reilly (1994) has shown that the relationship between the spacer and the edge-of-glass U-factor also depends on frame type. In low-conductivity frames, the heat flow at the edge-of-glass and frame area is through the spacer and so the type of spacer has a greater impact on the edge-of-glass and frame U-factor. In metal frames, the edge-of-glass and frame U-factor varies little with the type of spacer (metal or insulating) because a significant amount of heat flows through the highly conductive frame near the edge-of-glass area.

Frame U-Factor

Fenestration frame elements consist of all structural members exclusive of the glazing units and include sash, jamb, head, and sill members; meeting rails and stiles; mullions; and other glazing dividers. Estimating the rate of heat transfer through the frame is

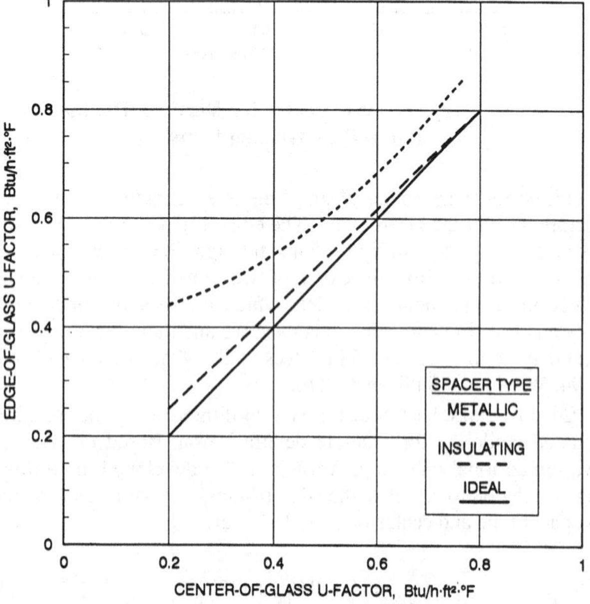

Fig. 4 Relationship Between Edge-of-Glass U-Factor and Center-of-Glass U-Factor for Various Spacers

complicated by (1) the variety of fenestration products and frame configurations, (2) the different combinations of materials used for frames, (3) the different sizes available, and to a lesser extent, (4) the glazing unit width and spacer type. Internal dividers or grilles have little impact on the fenestration U-factor, provided there is at least a 1/8-in. gap between the divider and each panel of glass.

Computer simulations found that frame heat loss in most fenestration is controlled by a single component or controlling resistance, and only changes in this component significantly affect frame heat loss (EEL 1990). For example, the frame U-factor for thermally broken aluminum fenestration products is largely controlled by the depth of the thermal break material in the heat flow direction. For aluminum frames without a thermal break, the inside film coefficient provides most of the resistance to heat flow. For vinyl- and wood-framed fenestrations, the controlling resistance is the shortest distance between the inside and outside surfaces, which usually depends on the thickness of the sealed glazing unit.

Carpenter and McGowan (1993) experimentally validated frame U-factors for a variety of fixed and operable fenestration types, sizes, and materials using computer modeling techniques. Table 2 lists frame U-factors for a variety of frame and spacer materials and glazing unit thicknesses. Frame and edge U-factors are normally determined by 2-D computer simulation. The National Fenestration Rating Council and the Canadian Standards Association require that frame and edge U-factors be determined using the FRAME computer program (EEL 1995). THERM (Arasteh et al. 1995, LBL 1996) is also a computer program developed to determine frame and edge U-factors.

Indoor and Outdoor Surface Heat Transfer Coefficients

Part of the overall thermal resistance of a fenestration system is due to the convective and radiative heat transfer between the exposed surfaces and the environment. Surface heat transfer coefficients at the outer and inner glazing surfaces, h_o and h_i, respectively, combine the effects of radiation and convection.

The wind speed and orientation of the building are important in determining h_o. Many correlations have been proposed for h_o as a function of wind speed. However, no universal relationship has been accepted, and limited field measurements at low air speeds by Klems (1989) show significant difference with values used by others. A value of h_o = 5.1 Btu/h·ft^2·°F corresponding to a 15 mph wind is often used to represent winter design conditions. At near-zero wind speed, h_o varies with outside air and surface temperature, orientation to vertical, and air moisture content. At low wind speeds, the overall surface heat transfer coefficient can be as low as 1.2 Btu/h·ft^2·°F (Yazdanian and Klems 1993).

For natural convection at the inner surface of vertical fenestration, the inner surface coefficient depends on the indoor air and glass surface temperatures and on the emittance of the glass inner surface. Table 3 shows the variation of h_i for winter (t_i = 70°F) and summer (t_i = 75°F) design conditions for a range of glass types and heights. Designers often use h_i = 1.46 Btu/h·ft^2·°F, which corresponds to t_i = 70°F, glass temperature = 15°F, and uncoated glass with e_g = 0.84. For summer conditions, the conventional h_i = 1.46 Btu/h·ft^2·°F corresponds approximately to glass temperature = 95°F and t_i = 75°F. For winter conditions, this most closely approximates single, clear, 2 ft tall glass; but it overestimates the value as the glazing unit conductance decreases and height increases. For summer conditions, this value approximates all types of glass that are 2 ft tall, but again is less accurate as the glass height increases. If the room surface of the glass has a low-emissivity coating, the values are about one-half at both winter and summer conditions.

Heat transfer between the glazing surface and its environment is driven not only by the local air temperatures but also by the radiant temperatures to which the surface is exposed. The radiant temperature of the indoor environment is generally assumed to be equal to the indoor air temperature. While this is a safe assumption where a

Table 2 Representative Fenestration Frame U-Factors in Btu/h·ft²·°F—Vertical Orientation

Frame Material	Type of Spacer	Operable			Fixed			Garden Window		Plant-Assembled Skylight			Curtainwall[e]			Sloped/Overhead Glazing[e]		
		Single[b]	Double[c]	Triple[d]	Single[b]	Double[c]	Triple[d]	Single[b]	Double[c]	Single[b]	Double[c]	Triple[d]	Single[f]	Double[g]	Triple[h]	Single[f]	Double[g]	Triple[h]
Aluminum without thermal break	All	2.38	2.27	2.20	1.92	1.80	1.74	1.88	1.83	7.85	7.02	6.87	3.01	2.96	2.83	3.05	3.00	2.87
Aluminum with thermal break[a]	Metal	1.20	0.92	0.83	1.32	1.13	1.11			6.95	5.05	4.58	1.80	1.75	1.65	1.82	1.76	1.66
	Insulated	n/a	0.88	0.77	n/a	1.04	1.02			n/a	4.75	4.12	n/a	1.63	1.51	n/a	1.64	1.52
Aluminum-clad wood/ reinforced vinyl	Metal	0.60	0.58	0.51	0.55	0.51	0.48			4.86	3.93	3.66						
	Insulated	n/a	0.55	0.48	n/a	0.48	0.44			n/a	3.75	3.43						
Wood/vinyl	Metal	0.55	0.51	0.48	0.55	0.48	0.42	0.90	0.85	2.50	2.08	1.78						
	Insulated	n/a	0.49	0.40	n/a	0.42	0.35	n/a	0.83	n/a	2.02	1.71						
Insulated fiberglass/ vinyl	Metal	0.37	0.33	0.32	0.37	0.33	0.32											
	Insulated	n/a	0.32	0.26	n/a	0.32	0.26											
Structural glazing	Metal												1.80	1.27	1.04	1.82	1.28	1.05
	Insulated												n/a	1.02	0.75	n/a	1.02	0.75

Note: This table should only be used as an estimating tool for the early phases of design.
[a]Depends strongly on width of thermal break. Value given is for 3/8 in.
[b]Single glazing corresponds to individual glazing unit thickness of 1/8 in. (nominal).
[c]Double glazing corresponds to individual glazing unit thickness of 3/4 in. (nominal).
[d]Triple glazing corresponds to individual glazing unit thickness of 1-3/8 in. (nominal).
[e]Glass thickness in curtainwall and sloped/overhead glazing is 1/4 in.
[f]Single glazing corresponds to individual glazing unit thickness of 1/4 in. (nominal).
[g]Double glazing corresponds to individual glazing unit thickness of 1 in. (nominal).
[h]Triple glazing corresponds to individual glazing unit thickness of 1-3/4 in. (nominal).
n/a Not applicable

Table 3 Indoor Surface Heat Transfer Coefficient h_i in Btu/h·ft²·°F—Vertical Orientation (Still Air Conditions)

Glazing ID	Glazing Type	Glazing Height ft	Winter: Glass Temp. °F	Winter: Temp. Diff. °F	Winter: h_i Btu/h·ft²·°F	Summer: Glass Temp. °F	Summer: Temp. Diff. °F	Summer: h_i Btu/h·ft²·°F
1	Single glazing	2	17	53	1.41	89	14	1.41
		4	17	53	1.31	89	14	1.33
		6	17	53	1.25	89	14	1.29
5	Double glazing with 1/2 in. airspace	2	45	25	1.36	89	14	1.41
		4	45	25	1.27	89	14	1.33
		6	45	25	1.22	89	14	1.29
23	Double glazing with e = 0.1 on surface 2 and 1/2 in. argon space	2	56	14	1.31	87	12	1.38
		4	56	14	1.23	87	12	1.31
		6	56	14	1.19	87	12	1.27
43	Triple Glazing with e = 0.1 on surfaces 2 and 5 and 1/2 in. argon spaces	2	63	7	1.25	93	18	1.45
		4	63	7	1.18	93	18	1.36
		6	63	7	1.15	93	18	1.32

Notes: Glazing ID refers to fenestration assemblies in Table 5.
Winter conditions: room air temperature t_i = 70°F, outdoor air temperature t_o = 0°F, no solar radiation
Summer conditions: room air temperature t_i = 75°F, outdoor air temperature t_o = 89°F, direct solar irradiance E_D = 248 Btu/h·ft²

small fenestration product is exposed to a large room with surface temperatures equal to the air temperature, it is not valid in rooms where the fenestration product is exposed to other large areas of glazing surfaces (e.g., greenhouse, atrium) or to other cooled or heated surfaces (Parmelee and Huebscher 1947).

The radiant temperature of the outdoor environment is frequently assumed to be equal to the outdoor air temperature. This assumption may be in error, because additional radiative heat loss occurs between a fenestration and the clear sky. Therefore, for clear-sky conditions, some effective outdoor temperature $t_{o,e}$ should replace t_o, in Equation (1). AGSL (1992) discusses methods for determining $t_{o,e}$. Note that a fully cloudy sky is assumed in ASHRAE design conditions.

The air space in an insulating glass panel made up of glass with no reflective coating on the air space surfaces has a coefficient h_S of 1.3 Btu/h·ft²·°F. When a reflective coating is applied to an air space surface, h_S can be selected from Table 4 by first calculating the effective air space emittance E by Equation (5).

$$E = \frac{1}{(1/e_o) + (1/e_i) - 1} \qquad (5)$$

where e_o and e_i are the hemispherical emittances of the two air space surfaces. Hemispherical emittance of ordinary uncoated glass is 0.84 over a wavelength range of 0.4 to 40 μm.

Representative U-Factors

Table 5 lists computed U-factors for a variety of generic fenestration products, *which should only be used as an estimating tool for the early phases of design*. The table is based on ASHRAE-sponsored research involving laboratory testing and computer simulation of various fenestration products. In the past, test data were used to provide more accurate results for specific products. However, a wide range of measured U-factors for similar products has been reported (Hogan 1988). Also, different test methods sometimes give different U-factors (McCabe et al. 1986). Consequently, computer

Table 4 Air Space Coefficients for Horizontal Heat Flow

Air Space Thickness, in.	Air Space Temp., °F	Air Temp. Diff., °F	Air Space Coefficient h_s, Btu/h·ft²·°F Effective Emittance E 0.82	0.72	0.40	0.20	0.10	0.05
0.5	5	10	0.88	0.82	0.60	0.46	0.39	0.35
		25	0.90	0.83	0.61	0.48	0.41	0.37
		55	1.00	0.93	0.71	0.57	0.50	0.47
		70	1.05	0.98	0.76	0.62	0.55	0.51
		90	1.10	1.03	0.81	0.67	0.60	0.57
	32	10	1.00	0.92	0.66	0.50	0.42	0.38
		25	1.01	0.93	0.67	0.51	0.43	0.39
		55	1.08	1.00	0.74	0.57	0.49	0.45
		70	1.12	1.04	0.78	0.62	0.53	0.49
		90	1.17	1.09	0.83	0.67	0.58	0.54
	50	10	1.09	1.00	0.71	0.53	0.44	0.39
		25	1.10	1.01	0.72	0.54	0.44	0.40
		55	1.14	1.05	0.76	0.58	0.49	0.44
		70	1.18	1.09	0.80	0.62	0.53	0.48
		90	1.23	1.14	0.85	0.67	0.57	0.53
	85	10	1.28	1.16	0.81	0.59	0.48	0.42
		25	1.28	1.17	0.81	0.59	0.48	0.43
		55	1.30	1.19	0.84	0.62	0.51	0.45
		70	1.33	1.21	0.86	0.64	0.53	0.47
		90	1.36	1.25	0.90	0.67	0.56	0.51
	120	10	1.48	1.35	0.92	0.66	0.52	0.46
		25	1.49	1.35	0.92	0.66	0.52	0.46
		55	1.50	1.37	0.94	0.67	0.54	0.47
		70	1.51	1.38	0.95	0.68	0.55	0.48
		90	1.53	1.40	0.97	0.70	0.57	0.50
0.4	5	10	0.96	0.89	0.67	0.54	0.47	0.43
		55	1.00	0.93	0.71	0.57	0.50	0.47
		90	1.07	1.01	0.78	0.64	0.58	0.54
	32	10	1.09	1.00	0.74	0.58	0.50	0.46
		55	1.11	1.03	0.76	0.60	0.52	0.48
		90	1.15	1.07	0.81	0.64	0.56	0.52
	50	10	1.18	1.09	0.79	0.61	0.52	0.48
		55	1.19	1.10	0.81	0.63	0.54	0.49
		90	1.22	1.13	0.84	0.66	0.57	0.52
	85	10	1.37	1.26	0.90	0.68	0.57	0.51
		55	1.38	1.26	0.91	0.69	0.58	0.52
		90	1.40	1.26	0.93	0.70	0.59	0.54
	120	10	1.58	1.45	1.02	0.75	0.62	0.55
		55	1.59	1.45	1.02	0.76	0.62	0.56
		90	1.60	1.46	1.03	0.77	0.63	0.57
0.3	5	<90	1.10	1.03	0.81	0.68	0.61	0.57
	32	<90	1.23	1.15	0.89	0.72	0.64	0.60
	50	<90	1.32	1.23	0.94	0.76	0.67	0.62
	85	<90	1.52	1.41	1.06	0.84	0.72	0.67
	120	<90	1.74	1.61	1.18	0.92	0.78	0.72
0.25	5	<90	1.20	1.13	0.91	0.77	0.70	0.67
	32	<90	1.34	1.26	0.99	0.83	0.75	0.71
	50	<90	1.43	1.34	1.05	0.87	0.78	0.74
	85	<90	1.64	1.53	1.18	0.96	0.84	0.79
	120	<90	1.87	1.74	1.31	1.04	0.91	0.84
0.2	5	<90	1.36	1.29	1.07	0.93	0.86	0.83
	32	<90	1.50	1.42	1.16	1.00	0.92	0.88
	50	<90	1.61	1.52	1.23	1.05	0.95	0.91
	85	<90	1.83	1.71	1.36	1.14	1.03	0.97
	120	<90	2.07	1.93	1.51	1.24	1.10	1.04

simulations (with high/low validation by testing) are now accepted as a standard method for determining accurate, product-specific U-factors. The simulation methodologies are specified in NFRC 100-91 or CSA A440.2-93.

Values are listed at winter design conditions for vertical installation and for skylights and other sloped installations with glazing surfaces that are sloped 20° from the horizontal. Data are based on center-of-glass and edge-of-glass component U-factors and assume that there are no dividers. However, they apply only to the specific design conditions described in the footnotes in the table, and they are typically used only to determine peak load conditions for sizing heating equipment. While these U-factors have been determined for winter conditions, they can also be used to estimate heat gain during peak cooling conditions, because conductive gain is usually a small portion of the total heat gain for fenestration in direct sunlight. Glazing designs and framing materials may be compared in choosing a product that needs a specific winter design U-factor.

Table 5 lists 48 types of glazing. (A subset of these types is included in Table 11 which lists Solar Heat Gain Coefficient and Visible Light Transmittance.) The multiple glazing categories are appropriate for sealed glass units and the combinations of storm sash and other glazing units. No distinction is made between flat and domed units such as skylights. For acrylic domes, an average gas-space width should be used to determine the U-factor. Note that garden window and skylight U-factors are approximately twice those of other similar products. While this difference is partially due to the difference in slope in the skylights, it is largely because these products project from the surface of the wall or roof. For instance, the skylight surface area, which includes the curb, can vary from 13 to 240% greater than the rough opening area, depending on the size and mounting method. Unless otherwise noted, all multiple-glazed units are filled with dry air. Argon units are assumed to be filled with 90% argon (Elmahdy and Yusuf 1995). U-factors for CO_2-filled units are similar to argon fills. For spaces up to 0.5 in., argon/SF_6 mixtures up to 70% SF_6 are generally the same as argon fills. The use of krypton gas can provide U-factors lower than those for argon for glazing spaces less than 0.5 in.

Table 5 provides data for six values of hemispherical emittance and for 0.25 and 0.5 in. gas space width. The emittance of various low-emittance glasses varies considerably between manufacturers and processes. When the emittance is between the listed values, interpolation may be used. When manufacturers' data are not available, assume that glass with a pyrolytic (hard) coating has an emittance of 0.40, and that glass with a sputtered (soft) coating has an emittance of 0.10. Tinted glass does not change the winter U-factor. Also, some reflective glass may have an emittance less than 0.84. Values listed are for insulating glass units using aluminum edge spacers. If an insulated or non-metallic spacer is used, the U-factors are approximately 0.03 Btu/h·ft²·°F lower.

Fenestration products are subdivided by vertical versus sloped installation and then into two general categories—plant assembled and site-assembled. Plant assembled products are intended to be delivered as a complete unit to the site. These products are typically installed in low-rise residential and small commercial/institutional/industrial buildings. For vertical sliders, horizontal sliders, casement, awning, pivoted, and dual-action windows, and for sliding and swinging glass doors, use the operable category in the table. For picture windows, use the fixed category. For products that project from the wall, use the garden window category. For skylights, use the sloped skylight category.

Site-assembled units represent products in which frame extrusions are assembled on site into a fenestration product and the glazing is added on site. These products are typically installed in high-rise residential and large commercial/institutional/industrial buildings. Curtain walls are typically made up of vision (transparent portion) and spandrel (opaque portion) panels. Table 5 contains representative U-factors for the vision panel (including mullions) for these assemblies. The spandrel portion of curtain walls usually consists of a metal pan filled with insulation and covered with a sheet of glass or other weatherproof covering. Although the U-factor in the center of the spandrel panel can be quite low, the metal pan is a thermal bridge and significantly increases the U-factor of the assembly. Two-dimensional simulation validated by testing of a curtain wall having an aluminum frame with a thermal break found that the U-factor for the edge of the spandrel panel (the 2-1/2 in. band around the perimeter adjacent to the frame) was 40% of the way towards the U-factor of the frame. The U-factor was 0.06 Btu/h·ft^2·°F for the center of the spandrel, 0.45 for the edge of the spandrel, and 1.06 for the frame (Carpenter and Elmahdy 1994). Two-dimensional heat transfer analysis or physical testing should be done to determine the U-factor of spandrel panels. The sloped/overhead glazing category for sloped glazing panels comparable to curtain walls should be used.

Physical testing to evaluate the performance of commercial skylights including pyramid and barrel vault geometry showed U-factors of 1.0 Btu/h·ft^2·°F for a thermally broken aluminum pyramidal skylight, and 1.3 Btu/h·ft^2·°F for an aluminum-frame half-round barrel vault (both normalized to a rough opening of 8 ft by 8 ft). Until more conclusive results are available, U-factors for these systems can be estimated by multiplying the "site-assembled sloped/overhead glazing" values in Table 5 by the ratio of total-product surface area (including curbs) to rough opening area. These ratios range from 1.2 to 2.0 for low slope skylights, 1.4 to 2.1 for pyramid assemblies sloped at 45°, and 1.7 to 2.9 for semicircular barrel-vault assemblies. An example calculation is provided in Example 4.

The U-factors in Table 5 are based on the definitions of the types of product, frame sizes, and proportion of frame to glass area as shown in Figure 5. The categories are as follows:

- Operable glazing units are 15 ft^2 in area and the overall size corresponds to a 3 ft by 5 ft fenestration product.
- Fixed (non-operable) units are about 16 ft^2 in area, and the overall size corresponds to a 4 ft by 4 ft window.
- Garden Windows are 15 ft^2 in projected area (35 ft^2 in surface area) and 5 ft wide by 3 ft high by 15 in. deep.
- Plant Assembled Skylights are nominal 8 ft^2 in area corresponding to a 2 ft by 4 ft skylight. The nominal dimensions of a roof-mounted skylight correspond to centerline spacings of roof framing members; consequently, the rough opening dimensions are 22.5 in. by 46.5 in.
- Curtainwall and Sloped/Overhead Glazing are a nominal 16 ft^2 in area representing repeating 4 ft. by 4 ft. panels. The nominal dimensions correspond to centerline spacings of the head and sill and vertical mullions.

Six frame types are listed (though not all for any one category), in order of improving thermal performance. The most conservative frame to assume is the **aluminum frame without thermal break** (although some products on the market have higher U-factors). The **aluminum frame with thermal break** has at least a 3/8-in. thermal break between the inside and outside for all members including both the frame and the operable sash, if applicable. (Products are available with significantly wider thermal breaks, which reduce heat flow considerably.)

The **reinforced vinyl/aluminum clad wood** category represents vinyl-frame products, such as sliding glass doors or large windows. These units have extensive metal reinforcing in the frame and wood products with extensive metal, usually on the exterior surface of the frame. The metal, of course, degrades the thermal performance of the frame material.

The **wood/vinyl frame** represents improved thermal performance over reinforced vinyl/aluminum clad wood. **Insulated fiberglass/vinyl frames** do not have metal reinforcing and the frame cavities are filled with insulation. For several site-assembled products, a **structural glazing frame** category represents products in which sheets of

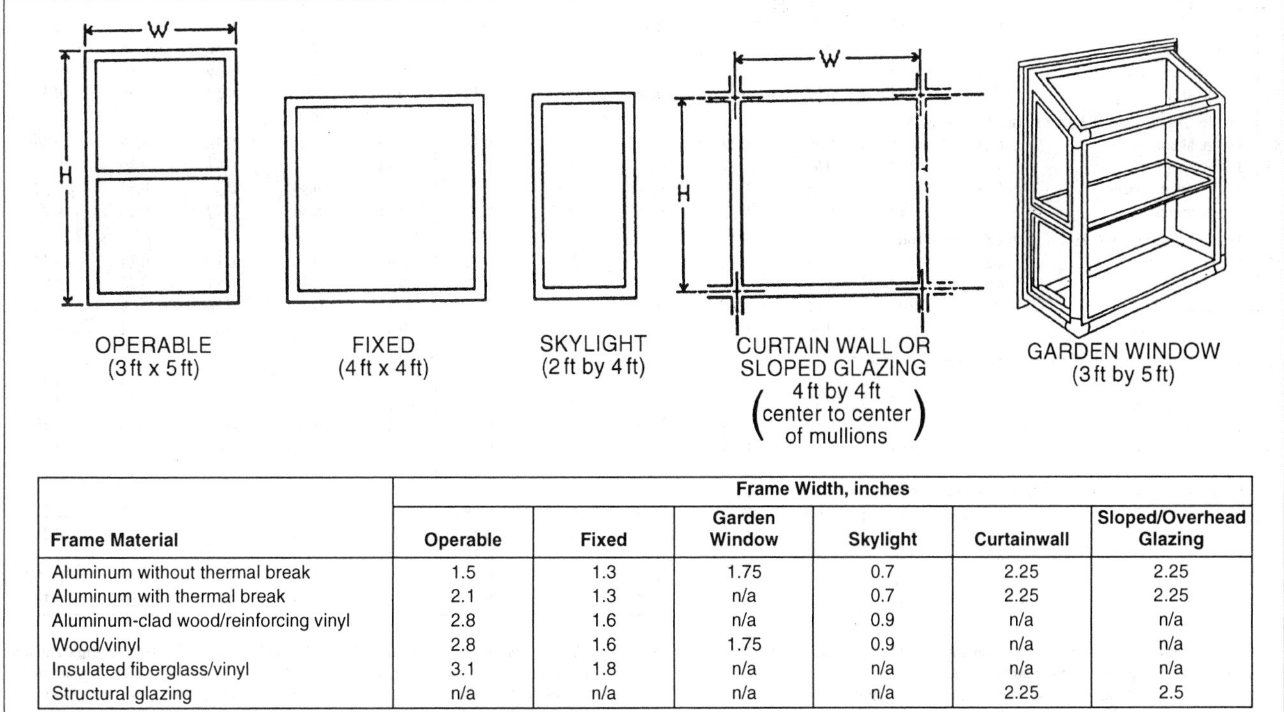

	Frame Width, inches					
Frame Material	**Operable**	**Fixed**	**Garden Window**	**Skylight**	**Curtainwall**	**Sloped/Overhead Glazing**
Aluminum without thermal break	1.5	1.3	1.75	0.7	2.25	2.25
Aluminum with thermal break	2.1	1.3	n/a	0.7	2.25	2.25
Aluminum-clad wood/reinforcing vinyl	2.8	1.6	n/a	0.9	n/a	n/a
Wood/vinyl	2.8	1.6	1.75	0.9	n/a	n/a
Insulated fiberglass/vinyl	3.1	1.8	n/a	n/a	n/a	n/a
Structural glazing	n/a	n/a	n/a	n/a	2.25	2.5

Fig. 5 Standard Fenestration Units

Table 5 U-Factors for Various Fenestration Products in Btu/h·ft²·°F

ID	Product Type / Frame Type / Glazing Type	Glass Only: Center of Glass	Glass Only: Edge of Glass	Vertical Installation, Operable (including sliding and swinging glass doors): Aluminum without Thermal Break	Operable: Aluminum with Thermal Break	Operable: Reinforced Vinyl/ Aluminum Clad Wood	Operable: Wood/ Vinyl	Operable: Insulated Fiberglass/ Vinyl	Vertical Installation, Fixed: Aluminum without Thermal Break	Fixed: Aluminum with Thermal Break	Fixed: Reinforced Vinyl/ Aluminum Clad Wood	Fixed: Wood/ Vinyl	Fixed: Insulated Fiberglass/ Vinyl
	Single Glazing												
1	1/8 in. glass	*1.04*	*1.04*	1.27	1.08	0.90	0.89	0.81	1.13	1.07	0.98	0.98	0.94
2	1/4 in. acrylic/polycarb	*0.88*	*0.88*	1.14	0.96	0.79	0.78	0.71	0.99	0.92	0.84	0.84	0.81
3	1/8 in. acrylic/polycarb	*0.96*	*0.96*	1.21	1.02	0.85	0.83	0.76	1.06	1.00	0.91	0.91	0.87
	Double Glazing												
4	1/4 in. airspace	*0.55*	*0.64*	0.87	0.65	0.57	0.55	0.49	0.69	0.63	0.56	0.56	0.53
5	1/2 in. airspace	*0.48*	*0.59*	0.81	0.60	0.53	0.51	0.44	0.64	0.57	0.50	0.50	0.48
6	1/4 in. argon space	*0.51*	*0.61*	0.84	0.62	0.55	0.53	0.46	0.66	0.59	0.53	0.52	0.50
7	1/2 in. argon space	*0.45*	*0.57*	0.79	0.58	0.51	0.49	0.43	0.61	0.54	0.48	0.48	0.45
	Double Glazing, e = 0.60 on surface 2 or 3												
8	1/4 in. airspace	*0.52*	*0.62*	0.84	0.63	0.55	0.53	0.47	0.67	0.60	0.54	0.53	0.51
9	1/2 in. airspace	*0.44*	*0.56*	0.78	0.57	0.50	0.48	0.42	0.60	0.53	0.47	0.47	0.45
10	1/4 in. argon space	*0.47*	*0.58*	0.81	0.59	0.52	0.50	0.44	0.63	0.56	0.50	0.49	0.47
11	1/2 in. argon space	*0.41*	*0.54*	0.76	0.55	0.48	0.46	0.40	0.58	0.51	0.45	0.44	0.42
	Double Glazing, e = 0.40 on surface 2 or 3												
12	1/4 in. airspace	*0.49*	*0.60*	0.82	0.61	0.53	0.51	0.45	0.64	0.58	0.51	0.51	0.49
13	1/2 in. airspace	*0.40*	*0.54*	0.75	0.54	0.48	0.45	0.40	0.57	0.50	0.44	0.44	0.41
14	1/4 in. argon space	*0.43*	*0.56*	0.78	0.57	0.50	0.47	0.41	0.59	0.53	0.46	0.46	0.44
15	1/2 in. argon space	*0.36*	*0.51*	0.72	0.52	0.45	0.43	0.37	0.53	0.47	0.41	0.40	0.38
	Double Glazing, e = 0.20 on surface 2 or 3												
16	1/4 in. airspace	*0.45*	*0.57*	0.79	0.58	0.51	0.49	0.43	0.61	0.54	0.48	0.48	0.45
17	1/2 in. airspace	*0.35*	*0.50*	0.71	0.51	0.44	0.42	0.36	0.53	0.46	0.40	0.39	0.37
18	1/4 in. argon space	*0.38*	*0.52*	0.74	0.53	0.46	0.44	0.38	0.55	0.48	0.42	0.42	0.40
19	1/2 in. argon space	*0.30*	*0.46*	0.67	0.47	0.41	0.39	0.33	0.48	0.41	0.36	0.35	0.33
	Double Glazing, e = 0.10 on surface 2 or 3												
20	1/4 in. airspace	*0.42*	*0.55*	0.77	0.56	0.49	0.47	0.41	0.59	0.52	0.46	0.45	0.43
21	1/2 in. airspace	*0.32*	*0.48*	0.69	0.49	0.42	0.40	0.35	0.50	0.43	0.37	0.37	0.35
22	1/4 in. argon space	*0.35*	*0.50*	0.71	0.51	0.44	0.42	0.36	0.53	0.46	0.40	0.39	0.37
23	1/2 in. argon space	*0.27*	*0.44*	0.65	0.45	0.39	0.37	0.31	0.46	0.39	0.33	0.33	0.31
	Double Glazing, e = 0.05 on surface 2 or 3												
24	1/4 in. airspace	*0.41*	*0.54*	0.76	0.55	0.48	0.46	0.40	0.58	0.51	0.45	0.44	0.42
25	1/2 in. airspace	*0.30*	*0.46*	0.67	0.47	0.41	0.39	0.33	0.48	0.41	0.36	0.35	0.33
26	1/4 in. argon space	*0.33*	*0.48*	0.70	0.49	0.43	0.41	0.35	0.51	0.44	0.38	0.38	0.36
27	1/2 in. argon space	*0.25*	*0.42*	0.63	0.44	0.38	0.36	0.30	0.44	0.37	0.32	0.31	0.29
	Triple Glazing												
28	1/4 in. airspaces	*0.38*	*0.52*	0.72	0.51	0.44	0.43	0.38	0.55	0.48	0.42	0.41	0.40
29	1/2 in. airspaces	*0.31*	*0.47*	0.67	0.46	0.40	0.39	0.34	0.49	0.42	0.36	0.35	0.34
30	1/4 in. argon spaces	*0.34*	*0.49*	0.69	0.48	0.42	0.41	0.35	0.51	0.45	0.39	0.38	0.36
31	1/2 in. argon spaces	*0.29*	*0.45*	0.65	0.44	0.38	0.37	0.32	0.47	0.40	0.34	0.34	0.32
	Triple Glazing, e = 0.20 on surface 2,3,4, or 5												
32	1/4 in. airspaces	*0.33*	*0.48*	0.69	0.47	0.41	0.40	0.35	0.50	0.44	0.38	0.37	0.36
33	1/2 in. airspaces	*0.25*	*0.42*	0.62	0.41	0.36	0.35	0.30	0.43	0.37	0.31	0.30	0.29
34	1/4 in. argon spaces	*0.28*	*0.45*	0.65	0.44	0.38	0.37	0.32	0.46	0.40	0.34	0.33	0.32
35	1/2 in. argon spaces	*0.22*	*0.40*	0.60	0.39	0.34	0.33	0.28	0.41	0.34	0.29	0.28	0.27
	Triple Glazing, e = 0.20 on surfaces 2 or 3 and 4 or 5												
36	1/4 in. airspaces	*0.29*	*0.45*	0.65	0.44	0.38	0.37	0.32	0.47	0.40	0.34	0.34	0.32
37	1/2 in. airspaces	*0.20*	*0.39*	0.58	0.38	0.32	0.31	0.27	0.39	0.33	0.27	0.26	0.25
38	1/4 in. argon spaces	*0.23*	*0.41*	0.61	0.40	0.34	0.33	0.29	0.42	0.35	0.30	0.29	0.28
39	1/2 in. argon spaces	*0.17*	*0.36*	0.56	0.36	0.30	0.29	0.25	0.37	0.30	0.25	0.24	0.23
	Triple Glazing, e = 0.10 on surfaces 2 or 3 and 4 or 5												
40	1/4 in. airspaces	*0.27*	*0.44*	0.64	0.43	0.37	0.36	0.31	0.45	0.39	0.33	0.32	0.31
41	1/2 in. airspaces	*0.18*	*0.37*	0.57	0.36	0.31	0.30	0.25	0.37	0.31	0.25	0.25	0.23
42	1/4 in. argon spaces	*0.21*	*0.39*	0.59	0.39	0.33	0.32	0.27	0.40	0.34	0.28	0.27	0.26
43	1/2 in. argon spaces	*0.14*	*0.34*	0.54	0.33	0.28	0.27	0.23	0.34	0.28	0.22	0.21	0.20
	Quadruple Glazing, e = 0.10 on surfaces 2 or 3 and 4 or 5												
44	1/4 in. airspaces	*0.22*	*0.40*	0.60	0.39	0.34	0.33	0.28	0.41	0.34	0.29	0.28	0.27
45	1/2 in. airspaces	*0.15*	*0.35*	0.54	0.34	0.29	0.28	0.24	0.35	0.28	0.23	0.22	0.21
46	1/4 in. argon spaces	*0.17*	*0.36*	0.56	0.36	0.30	0.29	0.25	0.37	0.30	0.25	0.24	0.23
47	1/2 in. argon spaces	*0.12*	*0.32*	0.52	0.32	0.27	0.26	0.22	0.32	0.26	0.20	0.20	0.19
48	1/4 in. krypton spaces	*0.12*	*0.32*	0.52	0.32	0.27	0.26	0.22	0.32	0.26	0.20	0.20	0.19

Notes:

1. All heat transmission coefficients in this table include film resistances and are based on winter conditions of 0 °F outdoor air temperature and 70 °F indoor air temperature, with 15 mph outdoor air velocity and zero solar flux. With the exception of single glazing, small changes in the indoor and outdoor temperatures will not significantly affect overall U-factors. The coefficients are for vertical position except skylight and sloped glazing values, which are for 20° from horizontal with heat flow up.

2. Glazing layer surfaces are numbered from the outdoor to the indoor. Double, triple and quadruple refer to the number of glazing panels. All data are based on 1/8 inch glass, unless otherwise noted. Thermal conductivities are: 0.53 Btu/(h·ft·°F) for glass, and 0.11 Btu/(h·ft·°F) for acrylic and polycarbonate.

3. Standard spacers are metal. Edge-of-glass effects assumed to extend over the 2 1/2 inch band around perimeter of each glazing unit as in Figure 3.

Table 5 U-Factors for Various Fenestration Products in Btu/h·ft²·°F (*Concluded*)

Vertical Installation					Sloped Installation									
Garden Windows		Curtainwall			Glass Only (Skylights)		Manufactured Skylight				Site-Assembled Sloped/Overhead Glazing			
Aluminum without Thermal Break	Wood/ Vinyl	Aluminum without Thermal Break	Aluminum with Thermal Break	Structural Glazing	Center of Glass	Edge of Glass	Aluminum without Thermal Break	Aluminum with Thermal Break	Reinforced Vinyl/ Aluminum Clad Wood	Wood/ Vinyl	Aluminum without Thermal Break	Aluminum with Thermal Break	Structural Glazing	ID
2.60	2.31	1.22	1.11	1.11	*1.19*	*1.19*	1.98	1.89	1.75	1.47	1.36	1.25	1.25	1
2.33	2.06	1.08	0.96	0.96	*1.03*	*1.03*	1.82	1.73	1.60	1.31	1.21	1.10	1.10	2
2.46	2.19	1.15	1.04	1.04	*1.11*	*1.11*	1.90	1.81	1.68	1.39	1.29	1.18	1.18	3
1.81	1.61	0.79	0.68	0.63	*0.58*	*0.66*	1.31	1.11	1.05	0.84	0.82	0.70	0.66	4
1.71	1.53	0.73	0.62	0.57	*0.57*	*0.65*	1.30	1.10	1.04	0.84	0.81	0.69	0.65	5
1.76	1.56	0.75	0.64	0.60	*0.53*	*0.63*	1.27	1.07	1.00	0.80	0.77	0.66	0.62	6
1.67	1.49	0.70	0.59	0.55	*0.53*	*0.63*	1.27	1.07	1.00	0.80	0.77	0.66	0.62	7
1.77	1.58	0.76	0.65	0.61	*0.54*	*0.63*	1.27	1.08	1.01	0.81	0.78	0.67	0.63	8
1.65	1.48	0.69	0.58	0.54	*0.53*	*0.63*	1.27	1.07	1.00	0.80	0.77	0.66	0.62	9
1.70	1.52	0.72	0.61	0.56	*0.49*	*0.60*	1.23	1.03	0.97	0.76	0.74	0.63	0.58	10
1.61	1.44	0.67	0.56	0.51	*0.49*	*0.60*	1.23	1.03	0.97	0.76	0.74	0.63	0.58	11
1.73	1.54	0.74	0.63	0.58	*0.51*	*0.61*	1.25	1.05	0.99	0.78	0.76	0.64	0.60	12
1.59	1.43	0.66	0.55	0.51	*0.50*	*0.61*	1.24	1.04	0.98	0.77	0.75	0.64	0.59	13
1.64	1.47	0.69	0.57	0.53	*0.44*	*0.56*	1.18	0.99	0.92	0.72	0.70	0.58	0.54	14
1.53	1.38	0.63	0.51	0.47	*0.46*	*0.58*	1.20	1.00	0.94	0.74	0.71	0.60	0.56	15
1.67	1.49	0.70	0.59	0.55	*0.46*	*0.58*	1.20	1.00	0.94	0.74	0.71	0.60	0.56	16
1.52	1.37	0.62	0.51	0.46	*0.46*	*0.58*	1.20	1.00	0.94	0.74	0.71	0.60	0.56	17
1.56	1.40	0.64	0.53	0.49	*0.39*	*0.53*	1.14	0.94	0.88	0.68	0.65	0.54	0.50	18
1.44	1.30	0.57	0.46	0.42	*0.40*	*0.54*	1.15	0.95	0.89	0.68	0.66	0.55	0.51	19
1.62	1.45	0.68	0.57	0.52	*0.44*	*0.56*	1.18	0.99	0.92	0.72	0.70	0.58	0.54	20
1.47	1.33	0.59	0.48	0.44	*0.44*	*0.56*	1.18	0.99	0.92	0.72	0.70	0.58	0.54	21
1.52	1.37	0.62	0.51	0.46	*0.36*	*0.51*	1.11	0.91	0.85	0.65	0.63	0.52	0.47	22
1.40	1.26	0.55	0.44	0.39	*0.38*	*0.52*	1.13	0.93	0.87	0.67	0.65	0.53	0.49	23
1.61	1.44	0.67	0.56	0.51	*0.42*	*0.55*	1.17	0.97	0.91	0.70	0.68	0.57	0.52	24
1.44	1.30	0.57	0.46	0.42	*0.43*	*0.56*	1.17	0.98	0.91	0.71	0.69	0.58	0.53	25
1.49	1.34	0.60	0.49	0.44	*0.34*	*0.49*	1.09	0.89	0.83	0.63	0.61	0.50	0.45	26
1.37	1.24	0.53	0.42	0.38	*0.36*	*0.51*	1.11	0.91	0.85	0.65	0.63	0.52	0.47	27
see note 7	see note 7	0.63	0.52	0.47	*0.39*	*0.53*	1.12	0.89	0.84	0.64	0.64	0.53	0.48	28
		0.57	0.46	0.41	*0.36*	*0.51*	1.10	0.87	0.81	0.61	0.62	0.51	0.45	29
		0.60	0.49	0.43	*0.35*	*0.50*	1.09	0.86	0.80	0.60	0.61	0.50	0.44	30
		0.55	0.45	0.39	*0.33*	*0.48*	1.07	0.84	0.79	0.59	0.59	0.48	0.42	31
see note 7	see note 7	0.59	0.48	0.42	*0.34*	*0.49*	1.08	0.85	0.79	0.59	0.60	0.49	0.43	32
		0.52	0.41	0.35	*0.31*	*0.47*	1.05	0.82	0.77	0.57	0.57	0.46	0.41	33
		0.54	0.44	0.38	*0.28*	*0.45*	1.02	0.79	0.74	0.54	0.55	0.44	0.38	34
		0.49	0.38	0.33	*0.27*	*0.44*	1.01	0.78	0.73	0.53	0.54	0.43	0.37	35
see note 7	see note 7	0.55	0.45	0.39	*0.29*	*0.45*	1.03	0.80	0.75	0.55	0.56	0.45	0.39	36
		0.48	0.37	0.31	*0.27*	*0.44*	1.01	0.78	0.73	0.53	0.54	0.43	0.37	37
		0.50	0.39	0.34	*0.24*	*0.42*	0.99	0.75	0.70	0.50	0.51	0.40	0.35	38
		0.45	0.34	0.29	*0.22*	*0.40*	0.97	0.74	0.69	0.49	0.50	0.39	0.33	39
see note 7	see note 7	0.54	0.43	0.37	*0.27*	*0.44*	1.01	0.78	0.73	0.53	0.54	0.43	0.37	40
		0.46	0.35	0.29	*0.25*	*0.42*	0.99	0.76	0.71	0.51	0.52	0.41	0.36	41
		0.48	0.38	0.32	*0.21*	*0.39*	0.96	0.73	0.68	0.48	0.49	0.38	0.32	42
		0.42	0.32	0.26	*0.20*	*0.39*	0.95	0.72	0.67	0.47	0.48	0.37	0.31	43
		0.49	0.38	0.33	*0.22*	*0.40*	0.97	0.74	0.69	0.49	0.50	0.39	0.33	44
see note 7	see note 7	0.43	0.32	0.27	*0.19*	*0.38*	0.94	0.71	0.66	0.46	0.47	0.36	0.30	45
		0.45	0.34	0.29	*0.18*	*0.37*	0.93	0.70	0.65	0.45	0.46	0.35	0.30	46
		0.41	0.30	0.24	*0.16*	*0.35*	0.91	0.68	0.63	0.43	0.44	0.33	0.28	47
		0.41	0.30	0.24	*0.13*	*0.33*	0.88	0.65	0.60	0.40	0.42	0.31	0.25	48

4. Product sizes are described in Figure 3 and frame U-factors are from Table 2.
5. Use $U = 0.60$ Btu/h·ft²·°F for glass block with mortar but without reinforcing or framing.
6. The use of this table should be limited to that of an estimating tool for the early phases of design.
7. Values for triple- and quadruple-glazed garden windows are not listed as these are not common products.
8. Minor differences exist between the data in Table 5 and U-factors determined using NFRC 100-91 because the data in Table 5 are generated using modified heat transfer correlations for glazing cavities (Wright 1996) and indoor fenestration surfaces (Curcija and Goss 1995b)

Table 6 Glazing U-Factor for Various Wind Speeds

Wind speed, mph		
15	7.5	0
U-Factor, Btu/h·ft²·°F		
0.10	0.10	0.10
0.20	0.20	0.19
0.30	0.29	0.28
0.40	0.38	0.37
0.50	0.47	0.45
0.60	0.56	0.53
0.70	0.65	0.61
0.80	0.74	0.69
0.90	0.83	0.78
1.00	0.92	0.86
1.10	1.01	0.94
1.20	1.10	1.02
1.30	1.19	1.10

glass are butt glazed with a sealant only and no framing members are exposed to the exterior.

For glazing with a **steel frame**, use aluminum frame values. For an aluminum window with wood trim or vinyl cladding, use the values for aluminum. Frame type refers to the primary unit. Thus, when storm sash is added over another fenestration product, use the values given for the non-storm product.

To estimate the overall U-factor of a fenestration product that differs significantly from the assumptions given in Table 5 and/or Figure 5, first determine the area that is frame/sash, center-of-glass, and edge-of-glass (based on a 2.5 in. band around the perimeter of each glazing unit). Next, determine the appropriate component U-factors. These can be taken either from (1) the standard values listed in italics in Table 5 for glass, (2) the values in Table 2 for frames or from some other source such as test data, or (3) computed factors. Finally, multiply the area and the component U-factors, sum these products, and divide by the rough opening in the building envelope where this product will fit to obtain U_o.

Table 6 provides approximate data to convert the overall U-factor at one wind condition to a U-factor at another.

Representative U-factors for Doors. Doors are often an overlooked component in the thermal integrity of the building envelope. Although swinging and revolving doors represent a small portion of the shell in residential, commercial, and institutional buildings, their U-factor is usually many times higher than the walls or ceilings. In some storage and industrial buildings, loading bay doors (overhead doors) represent a significant area of high heat loss. Table 7 contains representative U-factors for swinging, overhead and revolving doors determined through computer simulation (Carpenter and Hogan 1996). These are generic values and product-specific values determined in accordance with standards should be used whenever available. NFRC 100-91: Section B and CSA A453 provide procedures for evaluating the performance of swinging doors. Overhead doors are often evaluated in accordance with NAGDM 105-1992.

Swinging doors fall into two categories: slab and stile and rail. A stile-and-rail door is a swinging door with a full-glass insert supported by horizontal rails and vertical stiles. The stiles and rails are typically either solid wood members or extruded aluminum or vinyl, as shown in Figure 6. Most residential doors are slab doors with either solid wood, steel, or fiberglass skin over foam insulation in a wood frame with aluminum sill. The edges of the steel skin door are normally wood to provide a thermal break. In commercial construction, doors are either steel skin over foam insulation in a steel frame (i.e., utility doors) or a full glass door with aluminum stiles, rails, and frame (i.e., entrance doors).

The most important factors affecting door U-factor are the construction material and the size and type of glass. Frame depth, slab

Table 7 U-Factors of Doors in Btu/h·ft²·°F

Door Type	No Glazing	Single Glazing	Double Glazing with 1/2 in. Airspace	Double Glazing with $e = 0.10$, 1/2 in. Argon
SWINGING DOORS (Rough Opening—38 in. × 82 in.)				
Slab Doors				
Wood slab in wood frame[a]	0.46			
6% glazing (22 in. × 8 in. lite)	—	0.48	0.46	0.44
25% glazing (22 in. × 36 in. lite)	—	0.58	0.46	0.42
45% glazing (22 in. × 64 in. lite)	—	0.69	0.46	0.39
More than 50% glazing		Use Table 5 (operable)		
Insulated steel slab with wood edge in wood frame[a]	0.16			
6% glazing (22 in. × 8 in. lite)	—	0.21	0.19	0.18
25% glazing (22 in. × 36 in. lite)	—	0.39	0.26	0.23
45% glazing (22 in. × 64 in. lite)	—	0.58	0.35	0.26
More than 50% glazing		Use Table 5 (operable)		
Foam insulated steel slab with metal edge in steel frame[b]	0.37			
6% glazing (22 in. × 8 in. lite)	—	0.44	0.41	0.39
25% glazing (22 in. × 36 in. lite)	—	0.55	0.48	0.44
45% glazing (22 in. × 64 in. lite)	—	0.71	0.56	0.48
More than 50% glazing		Use Table 5 (operable)		
Cardboard honeycomb slab with metal edge in steel frame	0.61			
Style and Rail Doors				
Sliding glass doors/ French doors		Use Table 5 (operable)		
Site-Assembled Style and Rail Doors				
Aluminum in aluminum frame	—	1.32	0.93	0.79
Aluminum in aluminum frame with thermal break	—	1.13	0.74	0.63
REVOLVING DOORS (Rough Opening—82 in. × 84 in.)				
Aluminum in aluminum frame				
Open	—	1.32	—	—
Closed	—	0.65	—	—
SECTIONAL OVERHEAD DOORS (Nominal—10 ft × 10 ft)				
Uninsulated steel (nominal $U = 1.15$)	1.15	—	—	—
Insulated steel (nominal $U = 0.11$)	0.24	—	—	—
Insulated steel with thermal break (nominal $U = 0.08$)	0.13	—	—	—

[a] thermally broken sill (add 0.03 Btu/h·ft²·°F for non-thermally broken sill)
[b] non-thermally broken sill
[c] Nominal U-factors are through the center of the insulated panel before consideration of thermal bridges around the edges of the door sections and due to the frame.

width, and number of panels have a minor effect on door performance. Sidelites and double doors have similar U-factors to a single door of the same construction. The glazing area does not have much effect on the U-factor of wood slab doors in a wood frame. Glazing area has a strong effect, however, on the U-factor of the insulated steel slab in a wood frame. The U-factor of a typical commercial insulated slab door is approximately twice that of a residential insulated door, primarily because of the thermal bridging of the slab edge and the steel frame. Stile and rail doors, even if thermally broken, have U-factors 50% higher than a full-glass commercial steel slab door.

Three generic types of overhead doors are available: roll-up, uninsulated sectional, and insulated sectional. Metal roll-up doors consist of small metal plates approximately 2.5 in. in height that roll up around a metal rod to open. Sectional doors consist of a series of 2 ft high sections that travel in a track to open. Insulated overhead doors are available in a wide range of designs. Factors affecting heat transfer include width of insulation, design of the thermal break (if

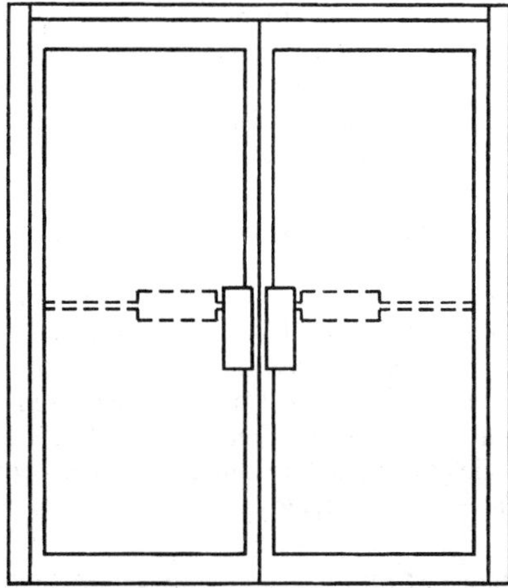

Fig. 6 Details of Stile and Rail Door

any), and design of the interior skin. The uninsulated sectional door has nearly the same U-factor as single glazing. The center of the insulated doors have low U-factors, but thermal bridging at the door and section edges significantly increases the total U-factor. For doors without thermal broken edges, the total value is 2.5 to 3.3 times greater than the center value. The addition of a good thermal break reduces this multiplier to 1.6.

Many commercial buildings have revolving entrance doors. Most of these doors are of similar design: single glazing in an aluminum frame without thermal break. The door, however, can be in two positions: closed or × shape (as viewed from above) or open or + shaped. At night, these doors are locked in the × position, effectively creating a double-glazed system. During the daytime, the door revolves and is often left positioned so that there is only one glazing between the inside and outside (+ position). Table 7 lists U-factors for both positions.

Example 1. Estimate the U-factor for a fixed fenestration product with a reinforced vinyl frame and double-glazing with a sputter-type low-e coating ($e = 0.10$). The argon-filled gap is 0.5 in. wide and the spacer is metal.

Solution: Locate the glazing system type in the first column of Table 5 (ID = 23). Then, find the appropriate product type (fixed) and frame type (reinforced vinyl). The U-factor listed (in the tenth column of U-factors) is 0.33 Btu/h·ft²·°F.

Example 2. Estimate a representative U-factor for a wood-framed, 38 in. by 82 in. swinging French door with eight 11-in. by 16-in. panes (true divided panels), each consisting of clear double-glazing with a 0.25 in. air space and a metal spacer.

Solution: Without more detailed information, assume that the dividers have the same U-factor as the frame, and that the divider edge has the same U-factor as the edge-of-glass. Calculate the center-of-glass, edge-of-glass, and frame areas.

$$A_{cg} = 8(11 - 5)(16 - 5) = 528 \text{ in}^2$$

$$A_{eg} = 8(11 \times 16) - 528 = 880 \text{ in}^2$$

$$A_f = (38 \times 82) - 8(11 \times 16) = 1708 \text{ in}^2$$

Select the center-of-glass, edge-of-glass, and frame U-factors. These component U-factors are 0.55 and 0.64 Btu/h·ft²·°F (from Table 5, glazing ID = 4, U-factor columns 1 and 2) and 0.51 Btu/h·ft²·°F (from Table 2, wood frame, metal spacer, operable, double-glazing), respectively.

$$U_0 = \frac{(0.55 \times 528) + (0.64 \times 880) + (0.51 \times 1708)}{38 \times 82}$$

$$= 0.55 \text{ Btu/h} \cdot \text{ft}^2 \cdot {}^\circ\text{F}$$

Example 3. Estimate the overall average U-factor for a multi-floor curtainwall assembly that is part vision glass and part opaque spandrel. The typical floor to floor height is 12 ft and the building module is 4 ft as reflected in the spacing of the mullions both horizontally and vertically. For a representative section, 4 ft wide and 12 ft tall, one of the modules is glazed and the other two are opaque. The mullions are aluminum frame with a thermal break, 3 in. wide, and centered on the module. The glass is double glazing with a pyrolytic low-emissivity coating ($\varepsilon = 0.40$), has a 1/2 inch gap filled with air and a metal spacer. The spandrel panel has a metal pan backed by R-20 insulation and no intermediate reinforcing members.

Solution: First, calculate the overall U-factor for the glazed module. Calculate the center-of-glass, edge-of-glass, and frame areas. The glass area is 45 in. by 45 in. in a 48 in. square module.

$$A_{cg} = (45 - 5)(45 - 5) = 1600 \text{ in}^2$$

$$A_{eg} = (45 \times 45) - 1600 = 425 \text{ in}^2$$

$$A_f = (48 \times 48) - (45 \times 45) = 279 \text{ in}^2$$

Select the center-of-glass, edge-of-glass, and frame U-factors. These component U-factors are 0.40 and 0.54 Btu/h·ft²·°F (from Table 5, ID = 13, columns 1 and 2) and 1.75 Btu/h·ft²·°F (from Table 2, curtainwall, aluminum frame with a thermal break, double glazing, metal spacer), respectively.

$$U_{glazing\ module} = \frac{(0.40 \times 1600) + (0.54 \times 425) + (1.75 \times 279)}{48 \times 48}$$

$$= 0.59 \text{ Btu/h} \cdot \text{ft}^2 \cdot {}^\circ\text{F}$$

Then, calculate the overall U-factor for the two opaque spandrel modules. The center-of-spandrel, edge-of spandrel, and frame areas are the same as the glazed module. The frame U-factor is the same. Calculate the center-of-spandrel U-factor. In this particular case, the R-value of the insulation does not need to be derated as there are no intermediate framing members penetrating it and providing thermal short-circuits. When the resistance of the R-20 insulation is added to the exterior air film resistance of 0.17 and the interior air film resistance of 0.68 h·ft²·°F/Btu (from Table 1, Chapter 24), the total resistance is 20.85 h·ft²·°F/Btu and the U-factor is 1/20.85 = 0.05 Btu/h·ft²·°F. The edge-of-spandrel U-factor is 40% of the way to the frame U-factor and is 0.05 + [0.40(1.75 – 0.05)] = 0.73 Btu/h·ft²·°F.

$$U_{opaque\ spandrel} = \frac{(0.05 \times 1600) + (0.73 \times 425) + (1.75 \times 279)}{48 \times 48}$$

$$= 0.38 \text{ Btu/h} \cdot \text{ft}^2 \cdot {}^\circ\text{F}$$

Finally, calculate the U-factor for the overall average U-factor for the curtainwall assembly, including the one module of vision glass and the two modules of opaque spandrel.

$$U_{curtainwall} = \frac{(0.59 \times 48 \times 48) + 2 \times 0.38(48 \times 48)}{3(48 \times 48)}$$

$$= 0.45 \text{ Btu/h} \cdot \text{ft}^2 \cdot {}^\circ\text{F}$$

Note that even with double glazing having a low-emissivity coating and with R-20 insulation in the opaque areas, this curtainwall with metal pans only has an overall R-value of approximately 2.

Example 4. Estimate the U-factor for a semicircular barrel vault that is 18 ft wide (9 ft tall) and 30 ft long mounted on a 6 in. curb. The barrel vault has an aluminum frame without a thermal break. The glazing is double with a 1/2 in. gap width filled with air and a low emissivity coating ($e = 0.20$).

Solution: An approximation can be made by multiplying the U-factor for a site assembled sloped/overhead glazing product having the same frame and glazing features by the ratio of the surface area (including the curb) of the barrel vault to the rough opening area in the roof that the barrel vault fits over.

1. Determine surface area (including the curb) of barrel vault.

Area of curved portion of barrel vault

$$= (\pi \times \text{diameter}/2) \times \text{length} = (\pi \times 18/2) \times 30 = 848 \text{ ft}^2$$

Area of two ends of barrel vault

$= 2\pi(\text{radius}^2)/2 = \pi r^2 = \pi 9^2 = 254 \text{ ft}^2$

Area of curb

$= \text{perimeter} \times \text{curb height} = (18 + 30 + 18 + 30)6/12 = 48 \text{ ft}^2$

Total surface area of barrel vault

$= 848 + 254 + 48 = 1150 \text{ ft}^2$

2. Determine rough opening area in roof that barrel vault fits over

$= \text{length} \times \text{width} = 18 \times 30 = 540 \text{ ft}^2$

3. Determine ratio of surface area to rough opening area

$= 1150/540 = 2.13$

4. Determine the U-factor from Table 5 of a site assembled sloped/overhead glazing product having the same frame and glazing features. The U-factor is 0.71 Btu/h·ft²·°F (ID = 17, 12th column on the second page of Table 5).

5. Determine the estimated U-factor of the barrel vault.

$$U_{barrel\ vault} = U_{sloped\ overhead\ glazing} \frac{\text{Surface area}}{\text{Rough opening area}}$$

$= 0.71 \times 2.13 = 1.51 \text{ Btu/h} \cdot \text{ft}^2 \cdot {}^\circ\text{F}$

AIR MOVEMENT

Infiltration Through Fenestration

Infiltration is the uncontrolled inward leakage of air caused by pressure effects of wind or differences in air density, such as the stack effect. While fenestration products can be operated to intentionally provide natural ventilation and increase comfort, infiltration should be reasonably reduced to avoid unpleasant conditions. If additional air is required, controlled ventilation is preferable to infiltration. Mechanical ventilation provides air in a comfortable manner and when desired. For infiltration, however, the peak supply is more likely to occur as an uncomfortable draft and when least desired, such as during a storm or the coldest weather.

Air leakage is determined in accordance with NFRC *Standard* 400-95 and ASTM *Standard* E 283-91. These standards allow all fenestration products (i.e. operable and fixed, windows and doors) to be compared directly. Most manufactured fenestration products have a maximum air leakage of 0.5 cfm/ft²of gross fenestration product area. However, products that do not completely seal, such as jalousie windows or doors, are likely to allow greater infiltration, and are most appropriate for installation in unconditioned spaces.

For products achieving this infiltration level, the energy consumption due to infiltration is likely to be significantly less than the energy consumption due to conduction and solar heat gain. Also, while overall air infiltration is a significant component of a building's heating and cooling loads, infiltration of 0.5 cfm/ft² or less through fenestration generally accounts for a small portion of the total load.

Indoor Air Movement

Because supply air grilles are frequently located directly below fenestration products, air sweeps the interior glass surface. Heated supply air should be directed away from the glass to prevent large temperature differences between the center and edges of the glass. These thermal effects must be considered, particularly when annealed glass is used and when air is forced over the glass surface during the heating season. Direct flow of heated air over the glass surface can increase the heat transfer coefficient and the temperature difference, causing a substantial increase in heat loss; it may also cause glass to break due to thermally induced stress.

An HVAC system that is designed primarily for cooling lowers the glass temperature and rapidly picks up the cooling load. Both conditions tend to improve comfort in the space. However, the space has an increased net heat gain because (1) a larger portion of the absorbed solar heat is delivered to the indoor space; (2) the greater convection at the indoor surface increases the fenestration U-factor; and (3) supply air rather than room air is in contact with the glass surface, which increases the air-to-air temperature difference. The principal increase in heat gain with clear glass is because the glass has a higher U-factor and the temperature difference between the conditioned space and the outside air is greater.

CONDENSATION RESISTANCE

Water vapor condenses in a film on fenestration surfaces that are at temperatures below the dew point of the inside air. If the surface temperature is below freezing, frost forms. Sometimes, condensation occurs first and ice from the condensed water forms when temperatures drop below freezing. Condensation frequently occurs on single glazing and on aluminum frames without a thermal break. The edge-seal creates a thermal bridge at the perimeter of the IGU.

The circulation of fill gas due to temperature differences in the IGU cavity contributes to the condensation problem at the bottom of the indoor glazing (Wright and Sullivan 1995a, 1995b; Curcija and Goss 1994, 1995a). In winter, fill gas near the indoor glazing is warmed and flows up while gas near the outdoor glazing is cooled and flows down. The descending gas becomes progressively colder until it reaches the bottom of the cavity. There the gas turns and flows to the indoor glazing, resulting in higher heat transfer rates at the bottom. Thus, the bottom edge of the indoor glazing is cooled both by edge-seal conduction and by fill-gas convection. The combined effect of these two heat transfer mechanisms is shown in Figure 7. The surface isotherms show a wider band of cold glass at the bottom of the window. Typical condensation patterns match these isotherms. The vertical indoor surface temperature profile also shows the effect of edge-seal conduction and that the minimum indoor surface temperature is near the bottom edge of the glass.

Damage caused by condensation to the fenestration and surrounding structures can be extensive and can cause structural, aesthetic, and health problems. Specific examples include peeling of paint, rotting of wood, saturation of insulation, and mold growth.

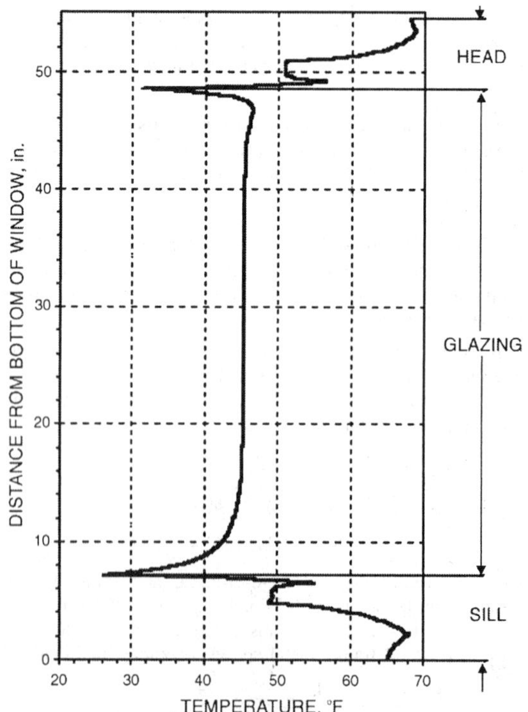

Fig. 7 Temperature Distribution on Indoor Surfaces of Insulated Glazing Unit

Ice can render doors and windows inoperable and prevent egress during an emergency.

Energy efficient housing has been accompanied by reduced ventilation. The resulting increase in indoor humidity has contributed to the condensation problem. However, the solution does not lie in reducing humidity levels to a minimum. Relative humidity below 20% and above 70% can increase health risks and reduce comfort. Generally, a minimum of 30% rh should be maintained, and 40% to 50% is more desirable (Sterling et al. 1985).

Minimum indoor surface temperatures can be quantified in a variety of ways. Sullivan et al. (1996), Griffith et al. (1996), Elmahdy (1996), Zhao et al. (1996), and deAbreu et al. (1996) demonstrated good agreement between detailed two-dimensional numerical simulation and surface temperature measurements using thermographs. Wright and Sullivan (1995c) and Curcija et al. (1996) developed simplified simulation models to predict condensation resistance. Estimates of center-glass and bottom-edge surface temperatures that can be expected for two glazing systems exposed to a range of outdoor temperature are shown in Figure 8. Both glazing systems include insulating foam edge-seals. High performance glazing systems (e.g., low-e/argon and insulated spacers) permits significantly higher indoor humidity levels.

Currently, the measure of condensation resistance of a fenestration system is the Condensation Resistance Factor (CRF) as defined by AAMA (1988) or the Temperature Index (I) as defined in CSA *Standard* A440-90 and A440.1-M90-91). Both standards define the values by a single dimensionless number as

$$\text{CRF or } I = \frac{t - t_c}{t_h - t_c} \qquad (6)$$

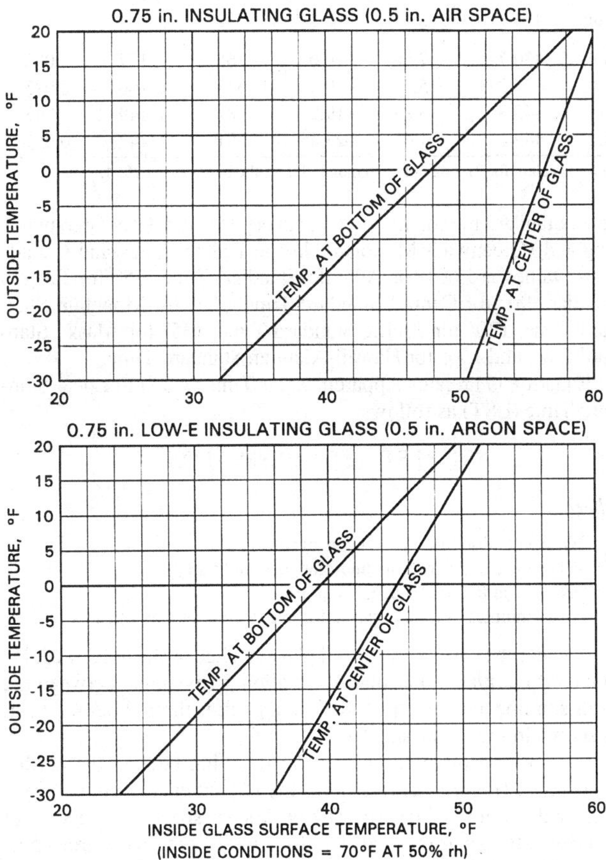

Fig. 8 Minimum Indoor Surface Temperatures Before Condensation Occurs

where t_h and t_c are the warm and cold side temperatures, respectively. Figure 9 can be used to determine the acceptable range of CRF or I for a specific climatic zone.

The two standards differ in the methods used to determine temperature. The CSA test procedure is based on thermocouple measurements at the coldest location on the frame plus three locations on the glass, each 2 in. above the bottom sight line. The AAMA procedure specifies two separate factors: one for the frame (CRF_F), which uses weighted frame temperature obtained from surface temperature measurements at predetermined and roving locations on the frame; and one for the insulated glazing unit or IGU (CRF_G), which uses the average of six temperatures measured at predetermined locations. near the top, middle, and bottom of the glazed area.

Test results for frames must be interpreted carefully because judgment is required in locating test thermocouples. Glazing area results are also difficult to interpret because the influence of the center-glass temperature has been included either because the measured temperatures are either averaged (ASMA 1988) or thermocouples are located away from the bottom sight line (CSA 1990) where condensation is most likely to occur. Because of the limitations in these test methods, and particularly when condensation resistance performance for demanding high-humidity buildings such as hospitals, museums, and art galleries must be specified, many designers specify the location of the test thermocouples and glazing sightlines.

Inside details can significantly alter the potential for condensation on window surfaces. Items such as venetian blinds, roll blinds, insect screens, and drapes increase the thermal resistance between the indoor space and the window and lower the temperature of the window surface. These window treatments do not prevent migration of moisture, so they can cause increased condensation. Figure 10 shows different situations that affect the potential for condensation. Note that window reveal plays an important role. If the window is placed near the outside of the wall, the increase in the outdoor film

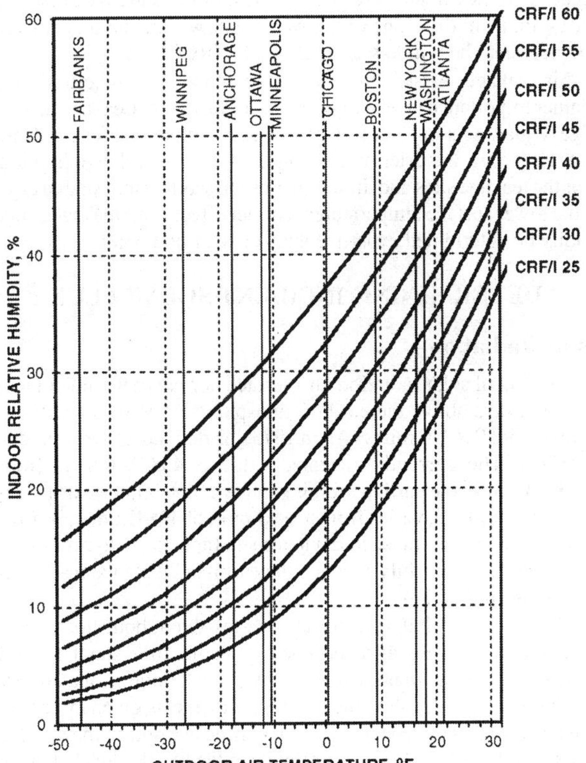

Fig. 9 Minimum Condensation Resistance Requirements (t_h = 68°F)

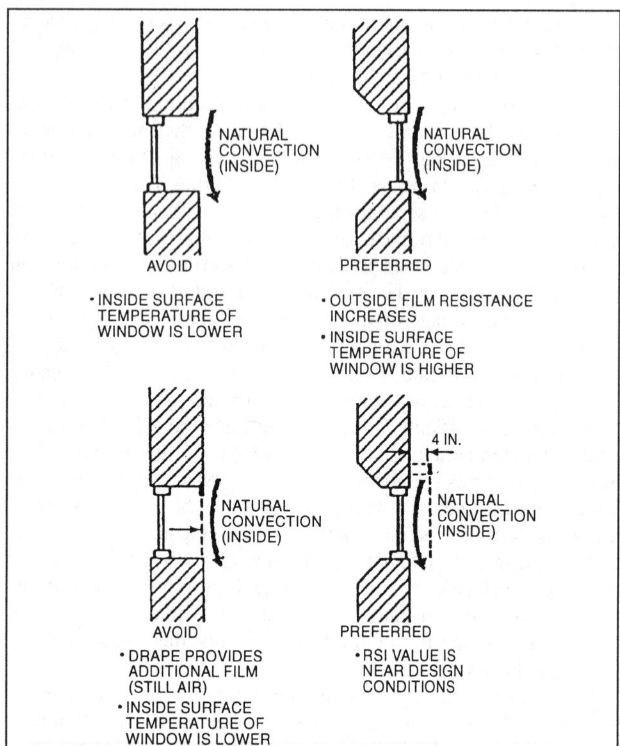

Fig. 10 Location of Fenestration Product Reveals and Blinds/Drapes and Their Effect on Condensation Resistance

coefficient and decrease in indoor film coefficient causes colder window surfaces. This effect is more pronounced near the corners of the recess where the indoor film coefficient is locally suppressed because air movement is restricted. Also, blinds should be placed at least 4 in. from the plane of the wall to allow some natural convection between the window and the blind (NRCC).

Air leakage, especially in operable sections of fenestration, is another important cause of low surface temperature. Leakage near the edge of glass sections can further increase the potential for condensation. However, the drier outdoor air decreases the relative humidity near the leakage sites and, in some cases, offsets the undesirable effect of the lower surface temperatures. The net effect of air leakage cannot readily be determined experimentally or with simulation.

DETERMINING INCIDENT SOLAR FLUX

Solar Radiation

The flux of solar radiation on a surface normal (perpendicular) to the sun's rays above the earth's atmosphere at the mean earth-sun distance [92.9×10^6 miles (Allen 1973)] is defined as the solar constant E_{sc}. The currently accepted value is 433 Btu/h·ft^2 (Iqbal 1983). Because the earth's orbit is slightly elliptical, the extraterrestrial radiant flux E_o varies from a maximum of 448 Btu/h·ft^2 on January 3, when the earth is closest to the sun (aphelion), to a minimum of 419 Btu/h·ft^2 on July 4, when the earth-sun distance reaches its maximum (perihelion).

The earth's orbital velocity also varies throughout the year, so **apparent solar time**, as determined by a solar time sundial, varies somewhat from the **mean time** kept by a clock running at a uniform rate. This variation, called the **equation of time**, is given in Table 8. The conversion between local standard time and solar time involves two steps. First the equation of time is added to the local standard time, then a longitude correction is added. This longitude correction is four minutes of time per degree difference between the local (site) longitude and the longitude of **local standard meridian** for that time

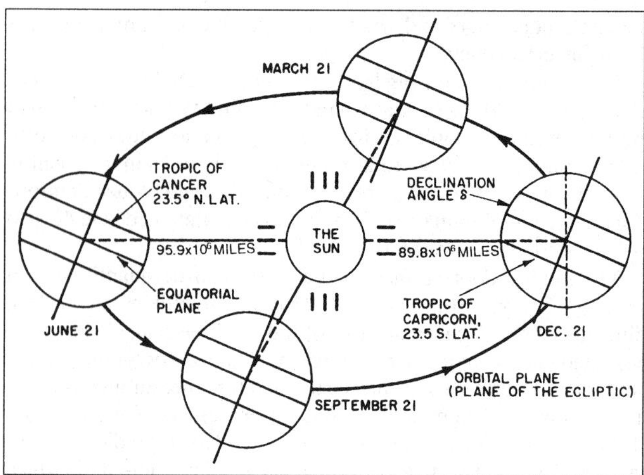

Fig. 11 Motion of Earth around Sun

Table 8 Extraterrestrial Solar Irradiance and Related Data

	I_o Btu/(h·ft^2)	Equation of Time, min.	Declination δ, degrees	*A* Btu/(h·ft^2)	*B* (Dimensionless Ratios)	*C*
Jan	448.8	−11.2	−20.0	390	0.142	0.058
Feb	444.2	−13.9	−10.8	385	0.144	0.060
Mar	437.7	−7.5	0.0	376	0.156	0.071
Apr	429.9	1.1	11.6	360	0.180	0.097
May	423.6	3.3	20.0	350	0.196	0.121
June	420.2	−1.4	23.45	345	0.205	0.134
July	420.3	−6.2	20.6	344	0.207	0.136
Aug	424.1	−2.4	12.3	351	0.201	0.122
Sep	430.7	7.5	0.0	365	0.177	0.092
Oct	437.3	15.4	−10.5	378	0.160	0.073
Nov	445.3	13.8	−19.8	387	0.149	0.063
Dec	449.1	1.6	−23.45	391	0.142	0.057

Note: Data are for 21st day of each month during the base year of 1964.

zone. Standard meridians are found every 15° from 0° at Greenwich, England (Greenwich Meridian). In the United States and Canada these values are 60° for Atlantic Standard Time; 75° for Eastern Standard 90° for Central Standard Time, 105° for Mountain Standard Time, 120° for Pacific Standard Time, 135° for Alaska Standard time, and 150° for Hawaii-Aleutian Standard Time.

Equation (7) relates Apparent Solar Time (AST) to Local Standard Time (LST) as follows:

$$AST = LST + ET + 4(LSM - LON) \quad (7)$$

where

ET = equation of time, minutes of time
LSM = local standard time meridian, degree of arc
LON = local longitude, degrees of arc
4 = minutes of time required for one degree rotation of earth

Because the earth's equatorial plane is tilted at an angle of 23.45° to the orbital plane, the solar declination *L* (the angle between the earth-sun line and the equatorial plane) varies throughout the year, as shown in Figure 11 and Table 8.

The spectral distribution of solar radiation beyond the earth's atmosphere (Figure 12) resembles the radiant energy emitted by a blackbody at about 11000°R.The peak solar spectral irradiance of 675 Btu/h·ft^2·μm is reached at 0.451 μm (451 nm) in the green portion of the visible spectrum.

In passing through the earth's atmosphere, the sun's radiation is reflected, scattered, and absorbed by dust, gas molecules, ozone, water vapor, and water droplets (clouds). The extent of this depletion at any given time is determined by atmospheric composition and length of the atmospheric path traversed by the sun's rays. This length is expressed in terms of the air mass m, the ratio of the mass of atmosphere in the actual earth-sun path to the mass which would exist if the sun were directly overhead at sea level (m = 1.0). For most purposes, the air mass at any time equals the cosecant of the solar altitude, multiplied by the ratio of the existing barometric pressure to standard pressure. Beyond the atmosphere, m = 0.

Most ultraviolet solar radiation is absorbed by the ozone in the upper atmosphere, while part of the radiation in the shortwave portion of the spectrum is scattered by air molecules, imparting the blue color to the sky. The strength of this absorption varies with wavelength and the terrestrial solar spectrum exhibits definite "dips" in regions of strong absorption, called **absorption bands**. The most prominent atmospheric gases contributing to this effect are

- *Ozone*. Strongest absorption in the ultraviolet, some in the visible. Concentration variable.
- *Water vapor* (H_2O). Strongest absorption in near and far infrared. Highly variable.

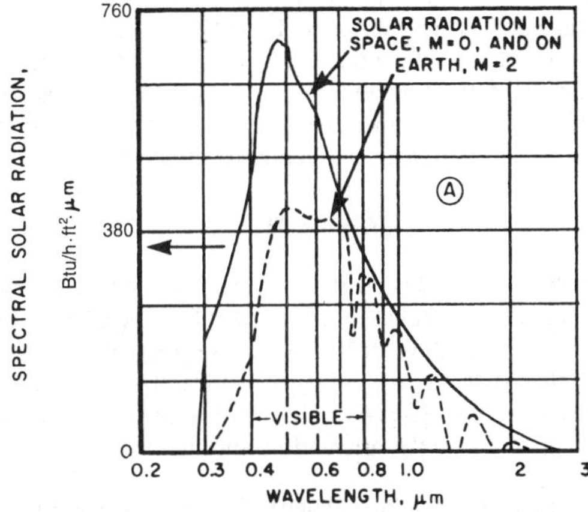

Fig. 12 Terrestrial and Extraterrestrial Solar Spectral Irradiances

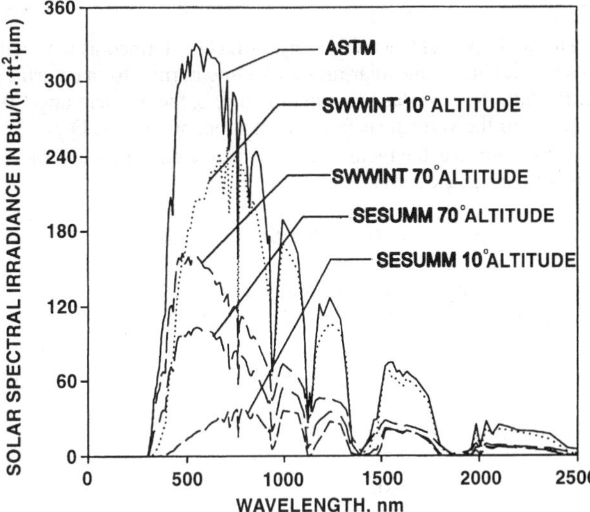

Fig. 13 Comparison of Standard Air Mass m = 1.5 Solar Spectrum with Direct Beam Spectra Through Atmospheres
Characteristic of southwest in winter (SWWINT) and southeastern U.S. in summer (SESUMM) for two solar altitude angles (McCluney 1996)

- *Carbon dioxide* (CO_2). Strongest absorption in near and far infrared. Slightly variable.
- *Oxygen* (O_2), *methane* (CH_4), (N_2O), *clorofluorocarbons* (CFC). Strongest absorption mostly in the infrared. Concentration almost constant.
- *Nitrogen dioxide* (NO_2). Strongest absorption in the visible. Highly variable in polluted areas.

More information on atmospheric optics can be found in Chapter 44 of the *Handbook of Optics* (Optical Society of America 1995) and in Iqbal (1983).

Water vapor in the lower atmosphere causes the characteristic absorption bands observed in the solar spectrum at sea level (Figure 12). For a solar altitude of 41.8° (m = 1.5), the spectrum of the sun's direct radiation on a clear day at sea level shows less than 3% of the total energy in the ultraviolet, 47% in the visible region, and the remaining 50% in the infrared (ASTM E 891-87). The maximum irradiance occurs at 0.61 µm, and little solar energy (less than 5% of the spectrum) exists at wavelengths beyond 2.1 µm.

Because the limits of the visible portion vary from observer to observer (and because the eye is not very sensitive to radiation at the spectral limits of vision) the fractions of total irradiance and illuminance found between different spectral limits at the edge of the visible portion of the spectrum can be calculated. The results are shown in Table 9 for m = 1.5 terrestrial spectrum shown in Figure 13.

The solar spectral distribution shown in Figure 12 for m = 0 is the World Radiation Center's 1985 standard extraterrestrial spectrum, and for a solar constant of 433 Btu/h·ft² (Wehrli 1985) The spectral distribution for m = 1.5 in Figure 13 is from ASTM E 891-87. The ASTM spectral distribution takes no account of monthly variations

Table 9 Portions of Total Solar Spectral Irradiance Contained in Portions of Visible Spectrum

Wavelength, nm		Percent	Percent
Start	End	Irradiance	Illuminance
370	770	54.4	100.0
380	760	52.2	100.0
390	750	50.2	99.9
400	740	47.4	99.9
410	730	44.9	99.8
420	720	41.9	99.8
430	710	39.5	99.8
440	700	36.7	99.8
450	690	35.3	99.5
460	680	31.1	99.1

Note: The integrated total irradiance = 950 W/m² and illuminance = 100 klx.

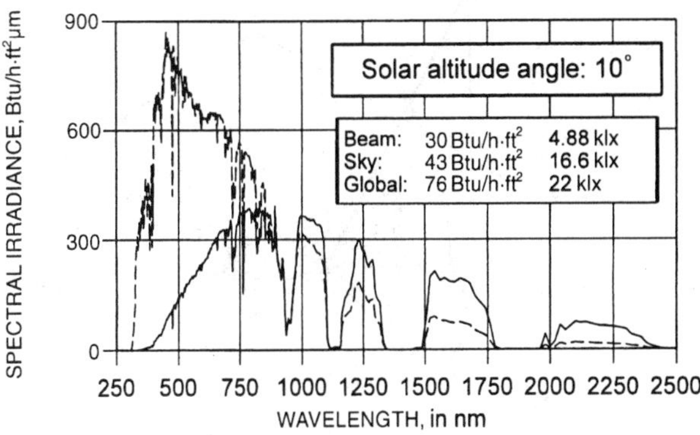

Fig. 14 Comparison of Direct and Diffuse Solar Spectra for Low Solar Altitude Angle

in irradiance caused by changes in the earth-sun distance and by variations in the atmosphere's constituent particulates and gases.

When variations in atmospheric constituents and air mass are considered, the solar spectral distribution is seen to vary, as illustrated in Figure 13 for two different atmospheric conditions and for two solar altitude angles, and in Figure 14 for both direct and diffuse radiation components and a low sun angle. The spectral distribution for low sun angle beam radiation is significantly shifted toward higher wavelengths. This shift can be seen as a reddening of the sun near the horizon. Clear sky diffuse radiation is generally shifted toward the blue end of the spectrum.

Diffuse Sky Radiation

Some short wavelength radiation is scattered by 30 different atmospheric molecules and by larger particles, including aerosols of water, dust, smoke, and other particulates in the atmosphere and reaches the earth in the form of diffuse sky radiation E_d. Because this diffuse radiation comes from all parts of the sky, its irradiance is difficult to predict and varies with the sun angle and as the moisture and particulate content change throughout the day. In completely overcast conditions the diffuse component accounts for all solar radiant heat gain from fenestration.

The total short-wave irradiance E_t reaching a terrestrial surface is the sum of the direct solar radiation E_D, the diffuse sky radiation E_d, and the solar radiation reflected from surrounding surfaces E_r. The irradiance of the direct component is the product of the direct normal irradiation E_{DN} and the cosine of the angle of incidence θ between the incoming solar rays and a line normal (perpendicular) to the surface:

$$E_t = E_{DN} \cos \theta_v + E_d + E_r \quad (8)$$

A method for computing all the factors on the right side of Equation (8) is presented in the section, "Computer Solar Gain Calculations." Perez et al. (1986), Gueymard (1987), and Gueymard (1993) describe more detailed models that separate the diffuse sky radiation into components. Gueymard (1995) provides a comprehensive, spectrally based model for calculating the spectral and broadband totals of all three terms in Equation (8) for cloudless sky conditions. The Gueymard model allows user input of the concentrations of a variety of atmospheric constituents.

The importance of the diffuse component is illustrated in Figure 14, which shows that at low sun angles (1) the diffuse component contains more radiant flux than the direct beam component, even on a clear day, and (2) the spectral distributions of the two components are quite different. Although the total irradiances are relatively modest for both components, they are significant in annual energy performance calculations. Also, the diffuse component that vertical windows receive over a year is an important part of solar radiant heat gain.

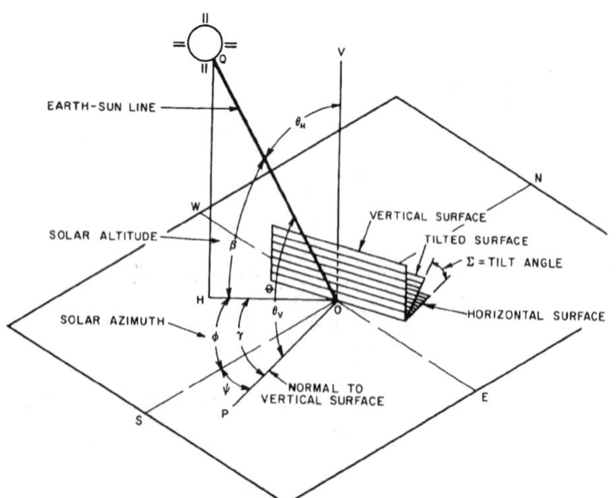

Fig. 15 Solar Angles for Vertical and Horizontal Surfaces

Determining Solar Angle

The sun's position in the sky is expressed in terms of the solar altitude β above the horizontal and the solar azimuth ϕ measured from the south (Figure 15). These angles, in turn, depend on the local latitude L; the solar declination δ, which is a function of the date (Table 6); and the apparent solar time, expressed as the hour angle H, where $H = 0.25$ (number of minutes from local solar noon), in degrees.

Equations (9) and (10) relate β and ϕ to the three angles just mentioned:

$$\sin\beta = \cos L \cos\delta \cos H + \sin L \sin\delta \quad (9)$$

$$\cos\phi = \frac{\sin\beta \sin L - \sin\delta}{\cos\beta \cos L} \quad (10)$$

Figure 15 shows the solar position angles and incident angles for horizontal and vertical surfaces. Line OQ leads to the sun, the north-south line is NOS, and the east-west line is EOW. Line OV is perpendicular to the horizontal plane in which the solar azimuth, angle HOS, and the surface azimuth angle POS (ψ) are located. Angle HOP is the surface solar azimuth defined as:

$$\gamma = \phi - \psi \quad (11)$$

The solar azimuth ϕ is positive for afternoon hours and negative for morning hours. Likewise, surfaces that face west have a positive surface azimuth ψ; those facing east have a negative surface azimuth (Table 10). If γ is greater than 90° or less than 270°, the surface is in the shade. Table 10 gives values in degrees for the surface azimuth ψ, applicable to the orientations of interest.

Table 10 Surface Orientations and Azimuths Measured from South

Orientation	N	NE	E	SE	S	SW	W	NW
Surface azimuth, ψ	180°	−135°	−90°	−45°	0	45°	90°	135°

The angle of incidence θ for any surface is defined as the angle between the incoming solar rays and a line normal to that surface. For the horizontal surface shown in Figure 6, the incident angle θ_H is QOV; for the vertical surface, the incident angle θ_v is QOP.

For any surface, the incident angle θ is related to β, γ, and the tilt angle of the surface Σ by:

$$\cos\theta = \cos\beta \cos\gamma \sin\Sigma + \sin\beta \cos\Sigma \quad (12)$$

where Σ = tilt angle of surface from horizontal.

When the surface is horizontal, $\Sigma = 0°$, and

$$\cos\theta_H = \sin\beta \quad (13)$$

For a vertical surface, $\Sigma = 90°$, and

$$\cos\theta_V = \cos\beta \cos\gamma \quad (14)$$

Example 5. Find the solar azimuth and altitude at 0830 central time on October 21 at 32° north latitude and 95° west longitude.

Solution: Local time is 0830 + 4 (90 - 95) = 0810. The equation of time (Table 6) is +15 min, so apparent solar time (AST) = 0810 + 15 = 0825, or 215 min. before noon, and $H = 0.25 \times 215 = 53.8°$. Table 8 gives the solar declination on October 21 as −10.5°. Thus, by Equation (9):

$$\sin\beta = \cos(32)\cos(-10.5)\cos(53.8) + \sin(32)\sin(-10.5) = 0.396$$
$$\beta = 23.3°$$

Using Equation (10):

$$\cos\phi = \frac{\sin(23.3)\sin(32) - \sin(-10.5)}{\cos(23.3)\cos(32)} = 0.503$$
$$\phi = 59.8°$$

Example 6. For the conditions of Example 5, find the incident angle at a window facing southeast.

Solution: Since the surface azimuth ϕ is to the east (AST < 1200) and the surface azimuth ψ is to the east (Table 10), they are both negative or

$$\phi = -59.8,\ \psi = -45.0.$$

$$\gamma = -59.8 - (-45.0) = -14.8°$$

A negative surface-solar azimuth γ indicates that the sun is east of the normal to the surface. Thus, using Equation (14)

$$\cos\theta_V = \cos(23.3)\cos(-14.8) = 0.888$$
$$\theta_V = 27.4°$$

Direct Normal Irradiance

At the earth's surface on a clear day, direct normal irradiation, or solar irradiance E_{DN} is represented by

$$E_{DN} = \frac{A}{\exp(B/\sin\beta)} \tag{15}$$

where

A = apparent solar irradiation at air mass $m = 0$ (Table 8)
B = atmospheric extinction coefficient (Table 8)

Values of E_{DN} based on these data are given in Tables 15 through 21 for the daylight hours of the 21st day of each month. Values of A and B vary during the year because of seasonal changes in the dust and water vapor content of the atmosphere and because of the changing earth-sun distance. Equation (15) does not give the maximum value of E_{DN} that can occur in each month, but yields values that are representative of conditions on cloudless days for a relatively dry and clear atmosphere. For very clear atmospheres, E_{DN} can be 15% higher than indicated by Equation (15), using values of A and B in Table 8.

For locations where clear, dry skies predominate (e.g., at high elevations), or, conversely, where hazy and humid conditions are frequent, values found by using Equation (15) and Table 8 should be multiplied by the Clearness Numbers in Threlkeld and Jordan (1958), reproduced here as Figure 6 in Chapter 30 of the 1995 *ASHRAE Handbook—Applications*. This broadband model should only be used for determining fenestration solar gain when the glazing system is not strongly spectrally selective.

Example 7. Find the direct component of the solar irradiation on a horizontal roof for the conditions of Example 5.

Solution: From Example 5, sin $\beta = 0.396$, and from Table 8, $A = 378$ Btu/ h·ft² and $B = 0.160$. Therefore,

$$E_{DN} = \frac{378}{\exp(0.160/0.396)} = 252 \text{ Btu/h}\cdot\text{ft}^2$$

$$E_{DH} = E_{DN}\sin\beta = 252 \times 0.396 = 100 \text{ Btu/h}\cdot\text{ft}^2$$

Thermal Infrared Radiation

The rate of electromagnetic radiation emitted from a material can be expressed by the Stefan-Boltzmann equation as follows:

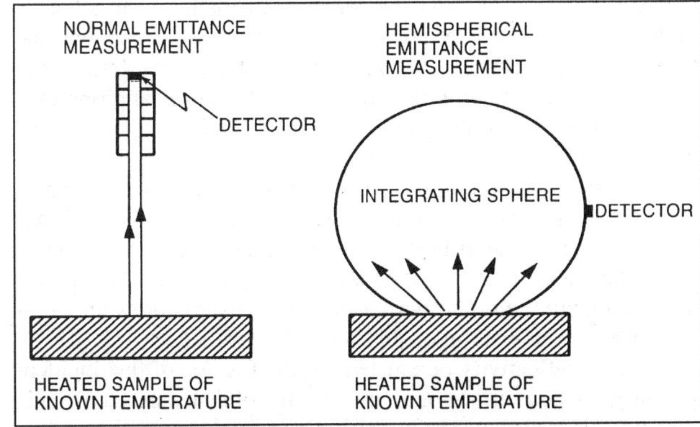

Fig. 16 Illustration of Difference Between Normal and Hemispherical Emittance

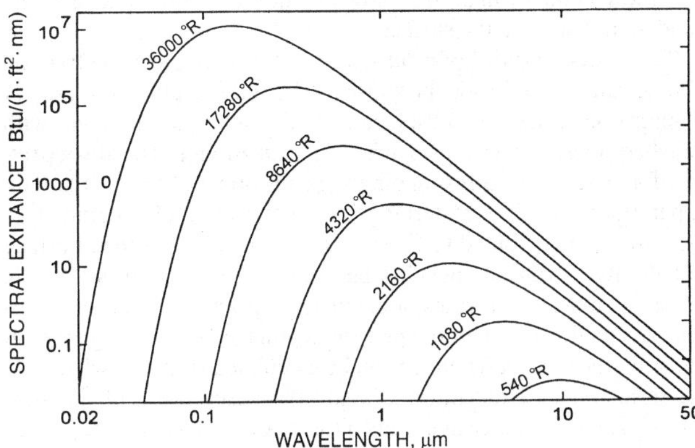

Fig. 17 Spectral Distributions of Blackbody Radiation at Different Source Temperatures

$$M = \varepsilon\sigma T^4 \tag{16}$$

where

M = exitance of material, Btu/h·ft²
T = temperature, °R
σ = Stefan-Boltzmann constant, 0.1714×10^{-8} Btu/h·ft²·°R⁴
ε = hemispherical emittance of material.

The theoretical maximum value of **hemispherical emittance** for any material is 1.0. In this case the radiation emitted by the surface is called **blackbody radiation**. Occasionally, a related quantity, called **normal emittance**, is used. The relationship between hemispherical and normal emittance is illustrated in Figure 16. The spectral distribution of blackbody radiation is illustrated in Figure 17 for temperatures ranging from 540°R (room temperature) to 36,000°R.

OPTICAL PROPERTIES

Solar radiation (including both direct rays from the sun and diffuse rays from the sky, clouds, and surrounding objects) incident on fenestration is partly transmitted and partly reflected by the glazings of that system. An additional fraction is absorbed in the glazings and/or the coatings on their surfaces. The fraction of incident flux that is reflected is called the **reflectance** ρ, the fraction absorbed is called the **absorptance** α, and the fraction transmitted is the **transmittance** τ. The sum of these three terms is unity, or

$$\tau + \alpha + \rho = 1 \tag{17}$$

However, this relationship is complicated by the fact that radiation incident on a surface can have non-constant distributions over the directions of incidence and over the wavelength (or frequency) scale. Thus, when measuring one or more of the optical properties, the wavelength distribution and direction of incident radiation must be specified.

The optical properties transmittance, reflectance, and absorptance are distinguished from the related quantities **transmissivity**, **reflectivity**, and **absorptivity**, as follows. The "*-ivity*" ending refers to the inherent properties of a bulk sample of material. The "*-ance*" ending refers to the property of a specific thickness or sample of a substance or combination of substances (Figure 18).

Thus, the **reflectivity** of a material is the fraction of flux incident upon the polished surface of an infinite slab of that substance that is reflected. If the substance is formed into a parallel plate with polished surfaces a fixed distance apart, the **reflectance** of the particular plate may be determined. Often only the portion of the reflectivity of a polished surface due only to reflection from the interface is considered, as if the volume of material below the interface returns none of the incident radiation to the surface.

The **transmissivity** of a substance can be defined as the fraction of flux transmitted inside the substance per unit length along a ray of propagation. Once the substance is made into a parallel plate with polished sides, it can have an overall **transmittance**. The **absorptivity** of a material is correspondingly the absorptance per unit length. A parallel plate of the material can have an overall **absorptance**.

A formula attributed to Fresnel is used to calculate the interface reflectivity of transparent substances such as glass and plastic, if the refractive index of the transparent material is known. When the substance is made into a parallel plate with polished surfaces the overall transmittance and reflectance can be calculated from knowledge of the reflectivity of the interfaces and the absorptivity of the substance. The reflectance and transmittance of such a plate are plotted in Figure 19.

The **emittance** of the surface ε is the ratio of the emission of thermal radiant flux from a surface to the flux that would be emitted by a blackbody emitter at the same temperature. Given the temperature and spectral emittance of a surface, the emitted irradiance spectrum can be computed.

Angular Dependence

The concept of solid angle is needed to understand angle dependence. A solid angle is a surface formed by all rays joining a point to a closed curve (Figure 20). For a sphere of radius R, the solid angle is the ratio of the projected area A on the sphere to the square of R. A sphere has a solid angle of 4π steradians (4π sr); a hemisphere has a 2π sr solid angle.

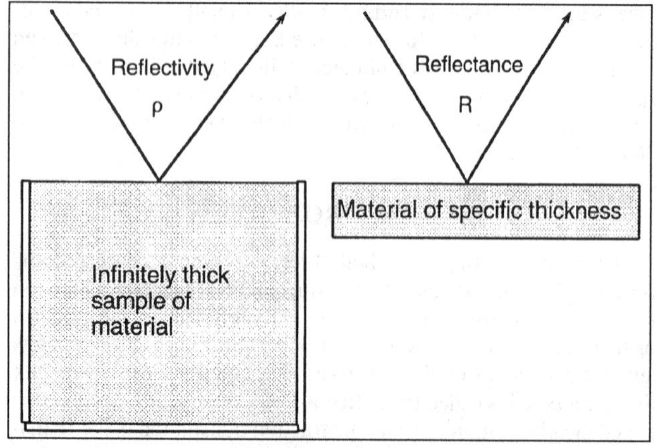

Fig. 18 Illustration of Difference Between "*-ivity*" and "*-ance*" Quantities

Radiation incident on a point in a surface comes to that point from many directions in a conical solid angle. For a cone of half angle α, the solid angle defined by the circular top and point bottom of that cone is given by

$$\Omega = 2\pi(1 - \cos\alpha) \tag{18}$$

In measuring transmittance or reflectance, a sample is illuminated over a specified solid angle. The reflected or transmitted flux is then collected within another solid angle. The size of a conical solid angle and the direction of its axis need to be specified to obtain meaningful results. A conical solid angle is bounded by right circular cone.

ASTM *Standards* E 903-82, E 1084-86, E 971-88, and E 972-88, as well as NFRC 300-94, which refers to the E 971 standard, refer to **conical-hemispherical** measurements of optical properties. For most thermal or HVAC design calculations only the total flux transmitted into a *hemispherical* solid angle, due to the direct solar beam incident in a small *conical* solid angle, is of interest. All transmitted solar irradiance is considered heat gain in these applications, regardless of the directional distribution of the transmitted radiation.

For complex fenestrations (those with nonspecular components), for many daylighting applications, and for some passive solar space heating applications, the directional distribution of transmitted radiation is of interest. In such cases, the **biconical transmittance and**

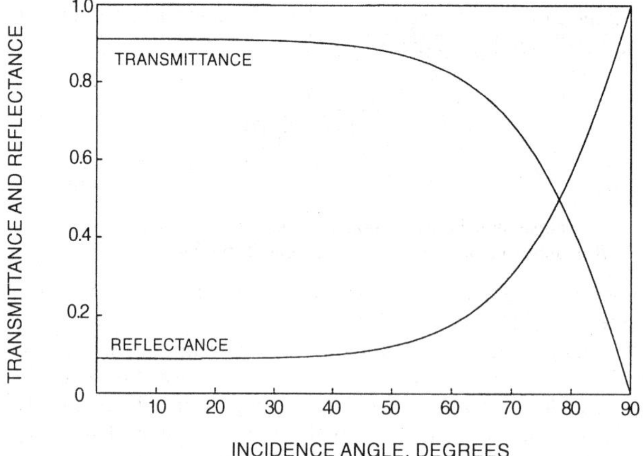

Fig. 19 Fresnel Transmittance and Reflectance of Plane, Parallel, Glass Plate

refractive index n = 1.55, thickness t = 1/8 in., absorptivity a = 0.01/m

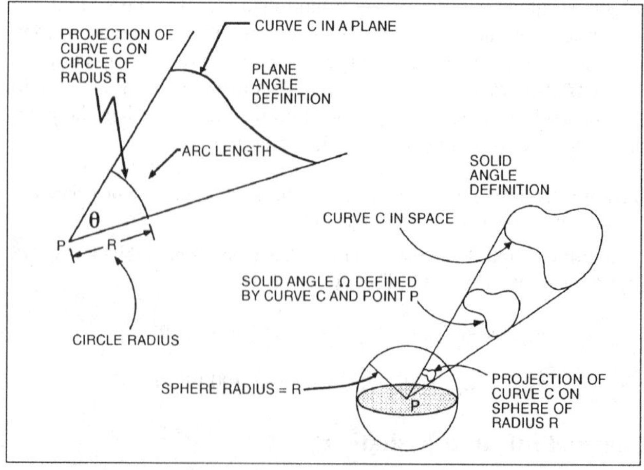

Fig. 20 Geometry for Definition of Plane and Solid Angles

reflectance of fenestration systems, which is described in a subsequent section, is important. The geometry of biconical transmittance and reflectance is illustrated in Figure 21. Biconical optical properties are needed: (1) to treat diffuse sky as well as direct beam radiation, (2) to handle the directional distribution of the flux entering a room through a window, and (3) to calculate the angle-dependent optical and solar gain properties of multiple pane window systems and complex glazing systems, including those with integral or attached shading devices.

Spectral Dependence

Frequency and wavelength are related through the equation

$$\lambda f = c \tag{19}$$

where

λ = wavelength, m
f = frequency, Hz
c = speed of light = 6.7×10^8 mph in air at atmospheric pressure

The wavelength dependence of radiometric quantities is denoted with a subscript λ attached to the optical quantity, thus ϕ_λ, E_λ, and L_λ. The wavelength dependency of optical properties is denoted by the functional notation, thus $\alpha(\lambda)$, $\rho(\lambda)$, $\tau(\lambda)$, and $\varepsilon(\lambda)$.

Kirchhoff's Law (McCluney 1994a) states

$$A(\lambda,\theta,\phi) = \varepsilon(\lambda,\theta,\phi) \tag{20}$$

where θ and ϕ are angles defining the directional dependence of the spectral absorptance $A(\lambda)$ and the spectral emittance $\varepsilon(\lambda)$. A consequence of Equation (17) and Equation (20) is that for opaque materials a good absorber is a good emitter and a poor reflector, and vice versa. But this statement is true only on a wavelength-by-wavelength basis or over a defined wavelength interval. A surface appearing to be an excellent reflector in the visible portion of the spectrum may have a high emittance over most of the infrared spectrum, or vice versa.

Source Spectra. Radiation incident on a surface has a distribution not only over direction within some solid angle but also over a range of wavelengths. The latter distribution is called a **spectrum**. For terrestrial applications, it is only after the extraterrestrial solar spectrum has been modified by passage through the atmosphere that it is of interest (Figure 13 and Figure 14).

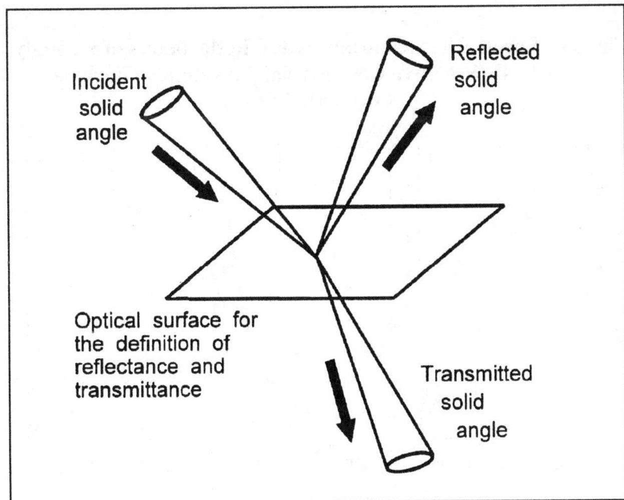

Fig. 21 Geometry for Definition of Biconical Transmittance and Reflectance

SOLAR-OPTICAL PROPERTIES OF GLAZING

The solar-optical properties of a glazing are the integrated or total transmittance, reflectance, and absorptance of the glazing to incident solar radiation. If the spectral optical properties [$\tau(\lambda)$, $\rho(\lambda)$, $\alpha(\lambda)$] of the glazing and the spectral irradiance E_λ incident on the glazing are known, the solar optical properties can be calculated from the equation

$$q = \frac{\int_0^\infty q(\lambda) E_\lambda d\lambda}{\int_0^\infty E_\lambda d\lambda} \tag{21}$$

where q stands for τ, ρ, or α.

If E_λ in Equation (21) is a standard solar spectral irradiance distribution, the optical property resulting from this equation is called the **solar optical property** and it is given the subscript s. If the spectral properties $q(\lambda)$ are available only at a set of discrete wavelengths, λ_k (e.g., measured data), the spectral average of Equation (21) can be calculated from

$$q = \frac{\sum_{k=1}^{M} q(\lambda_k) E(\lambda_k)}{\sum_{k=1}^{M} E(\lambda_k)} \tag{22}$$

Many window glazings do not have strong spectral selectivity over the solar spectrum, so their spectral optical properties can be considered constant, even if the source spectrum changes substantially. In these cases the transmitted irradiance can be determined by multiplying the incident irradiance by the solar transmittance.

Figure 22 shows the normal incidence spectral transmittance of typical architectural glasses. The approximate transmittance of total incident solar radiation through clear float glass at an incident angle of 0° ranges from 86% for 3/32 in. thick glass to 84% for 1/8 in. thick glass to 78% for 1/4 in. thick glass. Actual transmittance varies with the amount of iron or other absorbers in the glass. Low iron content glass has a relatively constant spectral transmittance over the entire solar spectrum.

Glazing that exhibits strong spectral selectivity (strong changes in optical properties over the solar spectrum) pass more or less radiation in different parts of the spectrum. This effect can change the heat gain of the glazing when the shape of the solar spectrum shifts. This selectivity can, in turn, cause errors in predicting solar heat gain when the spectrum of the solar radiation differs from the standard spectrum used to determine the solar heat gain coefficient of

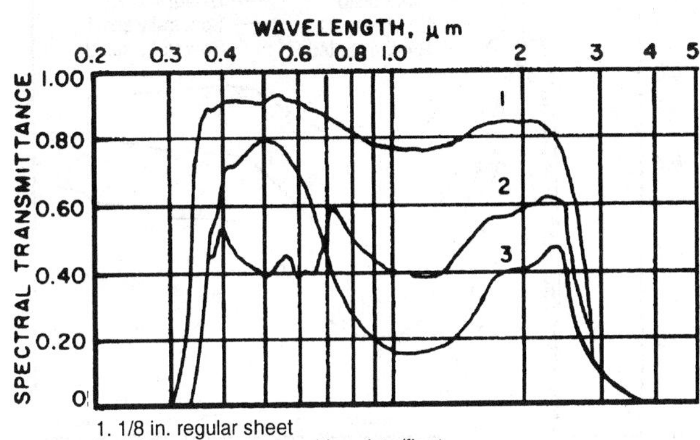

Fig. 22 Spectral Transmittance for Typical Architectural Glass

the fenestration system (McCluney 1996). These errors are typically 5 to 10%, but can be substantially greater in special cases.

Fenestration containing glazing that exhibits strong spectral and/or angular selectivity is available on the market. As a result, broadband methods of determining fenestration solar gain (such as the Shading Coefficient method described in this chapter) are proving to be inadequate. The Solar Heat Gain Coefficient (SHGC) method, which is an angular and spectrally based method, is being developed to replace broadband methods. In the SHGC method the solar gain of a glazing is determined on a wavelength-by-wavelength basis. Computer programs that perform the numerous, repetitive calculations required are available (Arasteh et. al. 1994, AGSL 1992); but an interface between these computer programs and the software used to predict the solar irradiance incident on a window (McCluney and Gueymard 1992, Gueymard 1995, TSC 1996) is required to take full advantage of the SHGC methodology.

Until the SHGC method is more fully developed, the solar heat gain coefficient values presented in this chapter are based on a standard spectral irradiance distribution for air mass m = 1.5 (ASTM E 891). The NFRC recommends this spectrum for rating fenestrations for instantaneous energy performance using defined environmental and incident irradiance conditions. This spectrum differs from Moon's (1940) air mass spectrum (m = 2) used in the past by the glazing industry. Both differ from solar spectral distributions incident upon fenestrations with different atmospheric conditions and for different sun angles.

These differences have little impact on the SHGC of glazing with relatively flat spectral transmittances, i.e., glazings that are not strongly spectrally selective. However, glazings that do exhibit strong spectral selectivity (such as those shown in Figure 23 and Figure 24) can have different SHGC values. The visible transmittances, however, are less sensitive to solar spectral changes. The visible transmittance also can be sensitive to the shape of the incident spectrum for glazings with very strong spectral selectivity in the visible spectrum, such as those exhibiting strong color.

Glazing Angular Selectivity

In addition to spectral selectivity, glazings exhibit angular selectivity—their optical properties change with incidence angle. In most cases this selectivity does not depend strongly on wavelength. Figure 25 shows how solar-optical properties vary with incident angle for typical uncoated glazing materials. As the incident angle increases from zero, the transmittance diminishes, reflectance increases, and absorptance first increases because of the lengthened optical path and then decreases as more incident radiation is reflected (McCluney 1987).

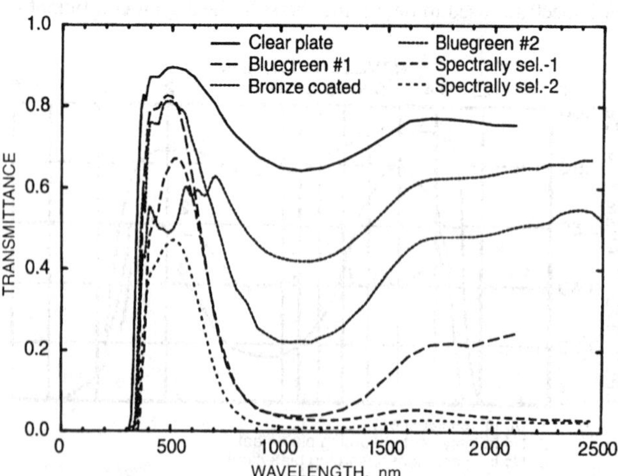

Fig. 23 Spectral Transmittances of Commercially Available Glazings (McCluney 1993)

In North America, peak summer solar gains occur with east- and west-facing vertical windows at angles of incidence ranging from about 25° to 55°. The peak solar gain for horizontal glazings occurs typically at small angles of incidence. For north- and south-facing vertical glazings, however, peak summertime solar gains occur at angles of incidence greater than about 40° (McCluney 1994b).

Angles of incidence important for annual energy performance calculations range from 5° to over 80° for east- and west-facing vertical windows and for horizontal glazings. This range is only slightly diminished for south-facing windows. For north-facing windows, the direct beam solar gains are small and their angles of incidence range from 62° to 86° (McCluney 1994b).

Spectrally Selective Glazing

The spectral range from 350 nm to over 50 μm contains radiation from both the sun and sky incident on the fenestration as well as the longer wavelength **thermal radiation**. Thermal radiation is emitted by warm bodies both outside and inside the building. Figure 26 shows the human eye spectral response, the solar spectrum for an air mass m = 1.5, and a room temperature blackbody radiation spectrum. The blackbody radiation is scaled to compare with the solar spectrum.

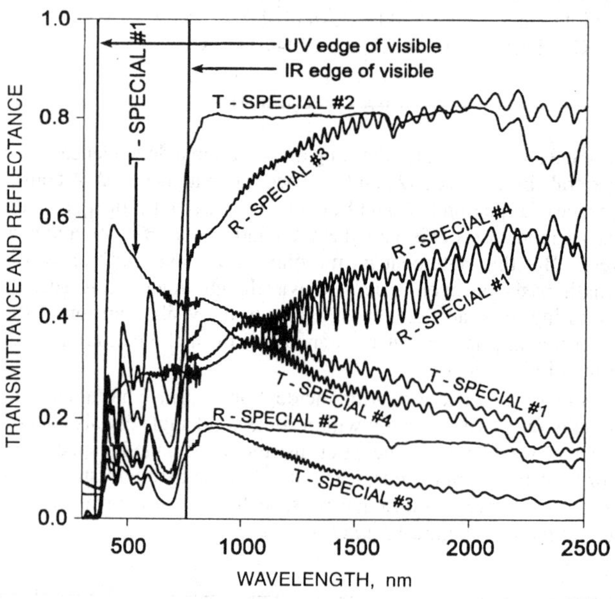

Fig. 24 Spectral Transmittances and Reflectances of Strongly Spectrally Selective Commercially Available Glazings (McCluney 1996)

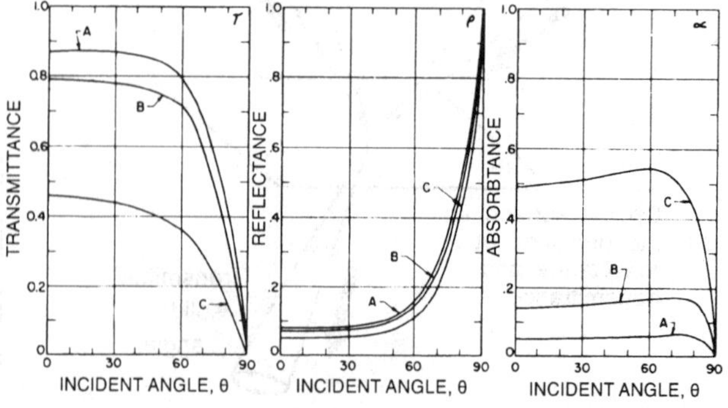

Fig. 25 Variations with Incident Angle of Solar-Optical Properties for (A) Double-Strength Sheet Glass, (B) Clear Plate Glass, and (C) Heat-Absorbing Plate Glass

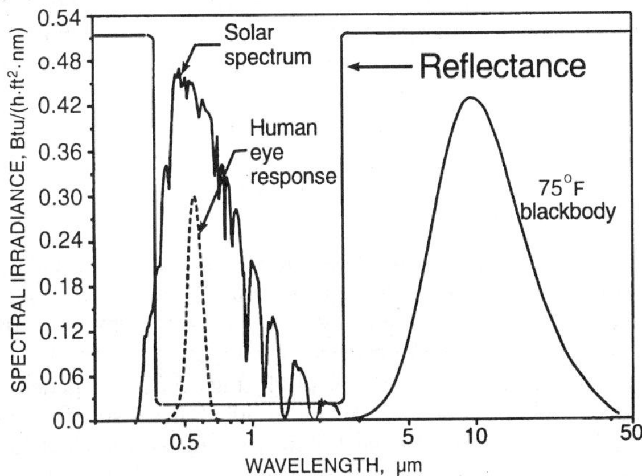

Fig. 26 Solar Spectrum, Human Eye Response Spectrum, Scaled Blackbody Radiation Spectrum, and Idealized Glazing Reflectance Spectrum

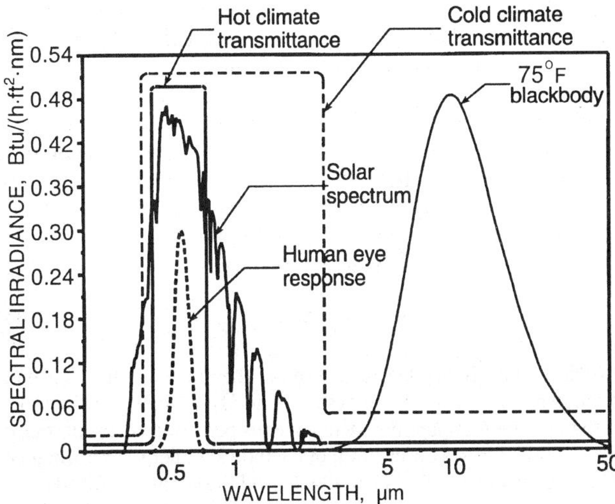

Fig. 27 Demonstration of Two Spectrally Selective Glazing Concepts, Showing Ideal Spectral Transmittances for Glazings Intended for Hot and Cold Climates

Almost all architectural glass is opaque to the long-wave radiation emitted by surfaces at temperatures below about 2200°F. This characteristic produces the **greenhouse effect**, by which radiation absorbed by surfaces in the room is emitted as long-wavelength radiation and cannot escape directly through the glass because it is opaque to all radiation beyond 4.5 μm. Instead, the radiation from the room surfaces is absorbed by the glass and reemitted to both sides as determined by such glazing properties as the inside and outside film heat transfer coefficients and the surface emittances.

A good reflector in the long-wavelength infrared portion of the spectrum can be a poor reflector and a good transmitter in the solar portion. A high reflectance in the long-wavelength infrared portion of the spectrum, because of the conservation of energy ($\tau + \rho + \alpha = 1.0$) means a low transmittance and absorptance and, because of Kirchhoff's Law [Equation (20)], therefore, a low emittance as well. This is the principle of operation of the **low-e coating** on window glass. Such a coating has high transmittance over the entire solar spectrum, which produces high solar heat gain while reflecting the long-wavelength infrared radiation emitted by the interior surfaces back to the room. The term, low-e, refers to a low emittance over the long-wavelength portion of the spectrum.

Another characteristic of glazing is that at a given wavelength, or over a defined range of wavelengths, the transmittance is the same in both directions. A glazing may have one solar transmittance value, but it generally has two reflectance values—one for radiation approaching each side of the system.

Figure 27 shows a hypothetical glazing system with improved performance for hot climates. In this case the sharp **reflectance edge** that the ideal low-e coating exhibits just past the end of the solar spectrum in Figure 26 is shifted closer to the visible portion of the spectrum. The reflectance edge is seen in Figure 27 as a drop in the hot climate transmittance at the right side of the visible portion of the spectrum. The effect is to reflect the near infrared portion of the solar spectrum outside, which reduces solar gain while still admitting visible light below about 800 nm.

Suitable spectrally selective window glazings can be chosen either to trap solar heat inside (low-e coating) or to block its entry into a building (*near* infrared reflective coating). The coating in Figure 27 for hot climate glazings has a low emittance over *both* the near (solar) infrared and far (thermal) infrared portions of the spectrum.

The reduced infrared transmittance of the hot climate glazing in Figure 27 is ideally achieved by high reflectance and low absorptance (meaning low emittance). It can also be achieved with high infrared absorptance, if the flow of the absorbed solar radiation to the interior of the building can be reduced.

One way to reduce solar gain from a spectrally *absorbing* window pane is to *insulate* the interior of the building from this hot, solar-absorbing outer glazing with an insulating dead-air space and an interior clear pane of glass. If a low-e coating is applied to the inside of the hot outer glazing, radiation, conduction, and convection of heat from it to the interior is reduced because of both the insulating gas space and the second pane. Such a glazing system for hot climates is insulated *not* to protect the building from conduction heat losses in winter but to protect the interior from the solar radiant heat absorbed by the hot outer lite in summer. Several manufacturers offer this kind of non-reflecting spectrally selective glazing for commercial buildings having large cooling loads.

Figure 27 shows that glazings for hot climates should have high transmittance over the visible portion of the spectrum to let daylight in for both illumination and view, and low transmittance over all other portions of the spectrum to reduce solar heat gain. In contrast, glazings intended for very cold climates should have high transmittance over the whole solar spectrum, from 380 nm to over 3500 nm, for maximum admission of solar radiant heat gain and light. In addition, glazings for cold climates should have low transmittance over the long-wavelength portion of the spectrum, in order to block the radiant heat emitted by the relatively warm interior surfaces of buildings.

Extreme spectral selectivity in glazing systems in the visible portion of the spectrum can produce an unwanted color shift in the transmitted light. The color of the transmitted light and its color-rendering properties should be considered in the design.

Absorbed Solar Radiation

Absorbed solar radiation, including ultraviolet, visible, and infrared radiation from the sun and sky, becomes heat inside the absorbing material. In a window, the temperature rises to some equilibrium in which the absorbed radiation is dissipated by conduction, convection, and radiation. Some of this heat goes back outside the building and the remainder goes inside, adding to the directly transmitted solar radiation. The magnitude of the **inwardly flowing fraction** N_i of the absorbed radiation depends on the nature of the air boundary layers adjacent to both sides of the glazing, including any gas between the panes of multiple pane glazing. The total solar gain (heat flow per unit area) q_i through the glass is

$$q_i = E_i(\tau_s + N_i\alpha_s) \tag{23}$$

where E_i is the solar irradiance, τ_s the solar transmittance, and α_s the solar absorptance of a single pane of glass; and N_i is the inward flowing fraction of the absorbed radiation.

Solar Heat Gain Coefficient

The quantity in parenthesis in Equation (23) is the **solar heat gain coefficient** or **SHGC**. It is the fraction of incident irradiance that enters the glazing and becomes heat gain. It includes both the directly transmitted portion τ_s and the absorbed and re-emitted portion $N_i\ \alpha_s$. Equation (23) may be rewritten as

$$q_i = E_i \cdot \text{SHGC} \qquad (24)$$

The SHGC is needed to determine the solar radiant heat gain through a window's glazing system. The SHGC should be included with the U-factor and other instantaneous performance properties in any manufacturer's description of a window's energy performance. (Previous versions of this handbook used the symbol F for SHGC.)

Because the optical properties α_s and τ_s vary with angle of incidence θ (defined as the angle between the rays incident on the glazing and the normal (perpendicular) to the glazing), the solar heat gain coefficient is also a function of angle of incidence. Once the incident irradiance and SHGC are known for the angle of incidence, the solar gain can be computed with Equation (24).

The direct beam solar heat gain coefficient SHGC is a characteristic of each type of fenestration and varies with the incident angle, because transmittance and absorptance of the glazing material depend on the angle of incidence θ.

For single glazing

$$\text{SHGC} = \tau + U\alpha/h_o \qquad (25)$$

For double glazing

$$\text{SHGC} = \tau + U\alpha/h_o + [(U/h_o) + (U/h_s)]\alpha_i \qquad (26)$$

For shaded glazing materials and other more complex fenestration, SHGC can be calculated for standard conditions, but such values should be verified by solar calorimeter tests.

Diffuse Radiation. The diffuse solar heat gain coefficient <SHGC> is given by

$$< \text{SHGC} > = \frac{\int_0^{2\pi}\int_0^{\pi/2} L(\theta,\phi)\text{SHGC}(\theta)\cos\theta\sin\theta d\theta d\phi}{\int_0^{2\pi}\int_0^{\pi/2} L(\theta,\phi)\cos\theta\sin\theta d\theta d\phi} \qquad (27)$$

where

$L(\theta,\phi)$ = directional distribution of radiance from the sky, Btu/(h·ft²·sr)
ϕ = azimuth angle (angle of rotation of plane of incidence around the normal to the glazing)
$\text{SHGC}(\theta)$ = solar heat gain coefficient as function of angle of incidence θ

Equation (27) covers a 2π sr (hemispherical) solid angle. Assuming that both sky and ground reflected diffuse radiation have constant radiance, Equation (27) reduces to

$$< \text{SHGC} > = \int_0^{\pi/2} \text{SHGC}(\theta)\cos\theta d\theta \qquad (28)$$

While diffuse radiation from the sky does not have a constant radiance distribution, it is assumed so for many applications. Reflected radiation, which is diffuse, also generally does not have a constant radiance distribution. Thus, Equation (28) is a simplification, which is useful for estimating solar gain. For many glazings, SHGC(θ) is relatively constant from 0 to about 50° of incident angle, so the equation does not generally introduce large errors.

For multiple pane glazings, each glazing has its own inward flowing fraction of the absorbed radiation for that layer. For a system with m layers, the inward flowing fraction of absorbed radiation for the jth layer is $(N_i)_j$ and the solar heat gain coefficient is

$$\text{SHGC} = T_s + \sum_{j=1}^{M}(N_i)_j A_{fj} \qquad (29)$$

where T_s is the solar transmittance of the glazing system and A_{fj} is the effective front solar absorptance of the jth layer in that system. (Although the individual glazing layers may be symmetric from front to back, the total adsorptance may not be, due to reflections.)

Because the optical properties of glazings depend in general on the spectral distribution of the incident radiation, the solar heat gain coefficient varies with the spectral distribution as well (McCluney 1996). In principle, the system properties T_s and A_{fj} in Equation (29) should be calculated from the spectral properties of the individual glazings, and the result spectrally averaged using Equation (21) or (22). In practice, spectrally averaged glazing properties are often more readily available and are used to calculate T_s and A_{fj}.

For most common glazing systems, reversing the order in which the system properties are calculated and spectrally averaged does not have a large effect on the result; but if high accuracy is desired (better than 5% if there is a selective glazing in the system), or if the system contains more than one moderately selective layer, the system spectral properties should be calculated and then averaged.

Solar Heat Gain Calculations

Equations (30) and (31) indicate the preferred way of determining the solar gain of glazings. In addition, Equation (32) is the preferred way to calculate the solar heat gain coefficient. At least two computer programs are available to assist in the calculation (Arasteh et al. 1994, AGSL 1992). This approach has been adopted in NFRC 200-95 and CSA A440.2-93. The method is valid for strongly spectrally selective glazing systems as well as for nonselective ones. For the most accurate calculations, the standard ASTM solar spectral irradiance distribution used in the computational programs should be replaced by a solar spectral distribution more appropriate for the intended application.

$$q_i = \int E_{i\lambda}(\tau_{s\lambda} + N_i\alpha_{s\lambda})d\lambda \qquad (30)$$

$$q_i = \int E_{i\lambda}\, \text{SHGC}_\lambda d\lambda \qquad (31)$$

$$\text{SHGC} = \frac{\int_0^{\infty}[W(\lambda)T(\lambda) + N_i W(\lambda)A(\lambda)]d\lambda}{\int_0^{\infty} W(\lambda)d\lambda} \qquad (32)$$

Gueymard (1995) developed a simplified model for engineering calculations of the atmospheric radiative transfer processes involved in determining terrestrial solar direct and diffuse radiation (see also Gueymard 1987, 1993b). McCluney (1992) developed a computer program, based on Gueymard's model to facilitate these calculations. The program can be used to generate solar spectral data files for use in solar heat gain calculation programs.

Although the spectrally based method is preferred, if a glazing system being considered is not very spectrally selective, or if it can be shown that the SHGC is essentially independent of source spectrum, then the previously described and subsequently detailed shading coefficient (broadband) method can be used.

Solar Gain Through Frame and Other Opaque Elements

Figure 28 illustrates the mechanisms by which a window provides solar gain. All the directly transmitted solar radiation is assumed to be absorbed at indoor surfaces where it is converted to heat. Solar gain also enters a building through opaque elements such as the frame and any mullion or dividers that are part of the fenestration system because a portion of the solar energy absorbed at the surfaces of these elements is redirected to the indoor side by heat transfer.

The solar heat gain coefficient of the fenestration system can be calculated while accounting for solar gain through the opaque elements by area-weighting the separate solar heat gain coefficients of the glazing, frame, and M divider elements. Thus,

$$\mathrm{SHGC} = \frac{\mathrm{SHGC}_g A_g + \mathrm{SHGC}_f A_f + \sum_{i=1}^{M} A_i \mathrm{SHGC}_i}{A_g + A_f + \sum_{i=1}^{M} A_i} \quad (33)$$

where SHGC_g, SHGC_f and SHGC_i are the solar heat gain coefficients of the glazed area, frame and ith divider, respectively. A_g, A_f and A_i are the corresponding projected areas.

SHGC_f can be estimated (Wright 1995a) using:

$$\mathrm{SHGC}_f = \alpha_f^s \left(\frac{U_f}{h_f}\right)\left(\frac{A_f}{A_{surf}}\right) \quad (34)$$

where α_f^s is the solar absorptivity of the outdoor surface of the frame, U_f is the frame U-factor and h_f is the heat transfer coefficient (radiative plus convective) between the frame and the outdoor environment. The projected-to-surface area ratio (A_f/A_{surf}) corrects for the fact that U_f is based on projected area A_f, and h_f is based on the total exposed outdoor frame surface area A_{surf}. SHGC_i can be calculated in the same way.

$$\mathrm{SHGC}_i = \alpha_i^s \left(\frac{U_i}{h_i}\right)\left(\frac{A_i}{A_{surf,i}}\right) \quad (35)$$

The outdoor heat transfer coefficients, h_f and h_i, can be estimated using (ASHRAE 1996):

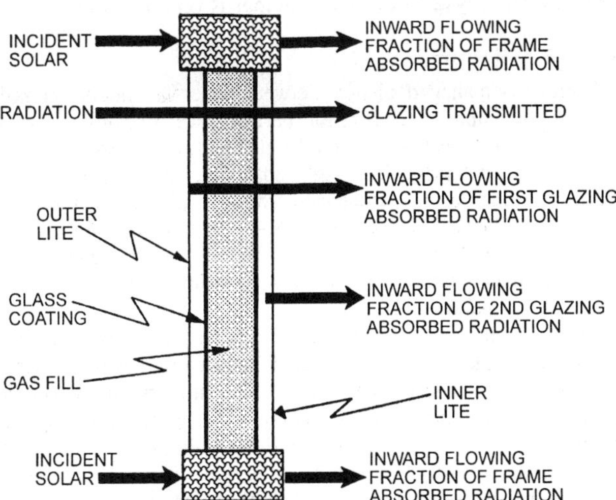

Fig. 28 Components of Solar Radiant Heat Gain with Double Pane Window, Including Both Frame and Glazing Contributions

$$h_f \text{ or } h_i = h_{co} + 4\sigma\varepsilon_f T_{out}^3 \quad (36)$$

where h_{co} is the convective heat transfer coefficient between the frame (or divider) surface and the outdoor environment, ε_f is the longwave emissivity of the outdoor frame (or divider) surface, T_{out} is the outdoor absolute temperature and σ is the Stefan-Boltzmann constant.

Shading Coefficient

The shading coefficient is a multiplier that adjusts the solar gain values for clear glass to a value for tinted glass. The shading coefficient works well for single pane, tinted glass whose angular and spectral depencies are the same as those of single pane, clear glass. But when multiple glazed windows and windows with coatings and other complexities are involved, the shading coefficient method is inadequate and should be replaced with the spectrally based method described previously.

The shading coefficient is defined as the ratio of the solar heat gain coefficient of a glazing system for a particular angle of incidence and incident solar spectrum to that for clear, single pane glass (**standard reference glazing** in which $\tau = 0.86$, $\rho = 0.08$, and $\alpha = 0.06$ at normal incidence) with the same angle and spectral distribution; or:

$$\mathrm{SC} = \frac{\mathrm{SHGC}(\theta)_{test}}{\mathrm{SHGC}(\theta)_{ref}} \quad (37)$$

This ratio remains constant as the solar spectral shape varies and as the angle of incidence varies for clear, single and double pane and many tinted, single pane glazings. Thus a single number can be used to convert from the reference SHGC to the SHGC for tinted glazing at the angle of incidence selected. Because it is not sensitive to angle of incidence, a single SC value works for beam radiation at any angle of incidence, as well as for diffuse radiation.

The value of the SC for standard reference glass is 1.0, but the SHGC for this glass is 0.87 at normal incidence and for the standard ASTM solar spectrum. The relationship between SC and SHGC for tinted, single pane clear glass is

$$\mathrm{SC} = \frac{\mathrm{SHGC}}{0.87} \quad (38)$$

This equation applies to the glazing portions of a fenestration and does not include frame effects. It may be used to determine the SC of commercially available glazing products from the solar heat gain coefficient published by the manufacturer. Equation (38) can be extended to double pane windows and spectrally selective glazings, but the SC then becomes a variable rather than a constant. For the standard solar spectral distribution a SC value can be calculated as a ratio of the appropriate SHGC values, *but only for a specified angle of incidence and for a specified solar spectral distribution.* For more precise calculations, the spectrally based method embodied in Equation (32) should be used to determine solar radiant heat gain (McCluney 1991).

When spectrally and angularly selective *multiple* pane and coated glazings are introduced, the shading coefficient again becomes a variable rather than a constant. Thus the ratio of solar heat gain coefficients that defines the shading coefficient in Equation (37) is no longer constant.

Reilly et al. (1992) compared the spectral based method with the broadband shading coefficient method and concluded:

> "...the shading coefficient method gives incorrect hourly results in two main cases: (1) when the transmittance angular distribution of the actual glazing differs substantially from that of the reference glazing; this is the case for *all* multi-pane glazing at

angles of incidence above 60°; (2) when the solar gain is primarily due to absorption (i.e., the solar transmittance is less than 40%, which is true for half of the glazings in the DOE-2 window library) and the wind speed is not close to the wind speed at which the shading coefficient was determined (7.5 mph). Despite these limitations, which can seriously affect the accuracy of calculating peak cooling loads and, therefore, peak electric demand, the shading coefficient is still commonly used in simplified energy analysis programs."

These authors further state that

> "The shading coefficient approach can over-predict the solar heat gain through a window at a given hour by as much as 35%. The implications of this on load calculations are significant, and we recommend use of the detailed method for such simulations."

Solar Heat Gain Coefficient, Shading Coefficient, and Visible Transmittance Values

Table 11 lists the visible transmittance, shading coefficient, and solar heat gain coefficients for common glazing and window systems. The window systems include windows with aluminum or metal frames and windows with other frames that have a lower conductivity (e.g. thermally-broken aluminum, wood, vinyl, and fiberglass). As can be seen in Table 11, the total window solar heat gain coefficient varies with the type of operator, size of the fenestration product, and type of frame.

The glazing SHGC values have been calculated using manufacturers' spectral data following methods described in this chapter (Finlayson et al. 1994, Wright 1995a). The glazing values are given for 1/8 in. and 1/4 in. glass, and vary with glass thickness and glass manufacturer. The values shown are average values and may vary by ±0.05. While Table 11 can be used for preliminary estimates, actual values should be obtained from the NFRC spectral data disk (NFRC 100-91), or the methods described in this chapter should be used.

The visible transmittances and shading coefficients are center-of-glazing values at normal incidence. A general rule for solar gain reduction is to select a glazing unit whose visible transmittance is greater than its shading coefficient, especially if daylighting will be used in the building.

The solar heat gain coefficients are combined center-glazing values and total window values. The center-glazing solar heat gain coefficients are given at normal incidence (0°), and at 40°, 50°, 60°, and 70° incidence angles. For angles other than those listed, straight-line interpolation can be used.

The total window solar heat gain coefficients in Table 11 assume normal incidence. The operable window and the fixed window sizes in Table 5 were used. To calculate the frame area, the frame heights shown in Figure 5 for aluminum and aluminum clad wood, wood/vinyl were used. The frame area for the aluminum windows is 15% for the operable size and 11% for the fixed size. The frame area for other frames is 27% for the operable size and 13% for the fixed size. The ratio of projected frame area to frame surface area is assumed to be 1.0, based on Wright (1995).

The frame solar heat gain coefficients used to determine the total window solar heat gain coefficients are calculated according to the section on solar heat gain coefficients for frames and other non-glazing elements in this chapter. The frame U-factors are taken from Table 2. The frame absorptance is assumed to be 0.5. The outside film coefficient is 3.9 Btu/h·ft^2·°F. For the aluminum window, the frame solar heat gain coefficient is 0.14 for the operable window and 0.11 for the fixed window. For the other frames, the frame solar heat gain coefficient was found to vary between 0.02 and 0.07 for the various lower conductivity frames. A frame solar heat gain coefficient of 0.04 is used for the operable window, and 0.03 is used for the fixed window. These values correspond directly to the aluminum clad wood/reinforced vinyl frames.

For energy calculations on a daylit building, the visible transmittance for the entire window should be used. The visible transmittance of a window is simply product of the fractional glazing area and the center-glazing visible transmittance (see Example 8).

The U-factor of a window listed in Table 11 can be found in Table 5. The ID for each entry in Table 11 is the ID in Table 5. Remember that while the gap width and gas fill have a negligible impact on the solar heat gain coefficient and other optical properties, they are important factors when determining U-factors.

Example 8. Estimate the overall visible light transmittance for an operable wood casement window with clear, uncoated (1/4 in.) double glazing.

Solution: From Table 11, ID 5b, the center glazing visible light transmittance is 0.78. The operable window has 27% frame area with a wood frame. The overall visible light transmittance is

$$VT = 0.27(0) + 0.73(0.78) = 0.57$$

Passive Solar Gain

Energy analysis of a fenestration product should include the value of passive solar gain through the product in winter. As described in Chapter 30 of the 1995 *ASHRAE Handbook—Applications*, the magnitude of this gain depends on variables such as latitude and orientation. In some cases, properly designed and operated fenestration allows more energy into the building over a heating season than it loses, thus making it energy contributing rather than energy consuming. Excessive solar gain must be controlled during the cooling season, however.

Direct beam admission to occupied spaces can often produce severe local glare and overheating. Judicious use of shades and other fenestration controls, as well as placement and orientation of work stations and furniture can alleviate these problems in most cases.

Solar Gain Rejection and Internal Load Dominated Buildings

For some buildings in certain climates, preventing solar gain is more important than improved thermal insulation using multiple panes of glazing. For example, internal load-dominated buildings even in cool, clear climates can have substantial daytime solar and internal heat gains. These gains can be rejected by controlled conduction through the building envelope and/or forced ventilation through the HVAC system. Preventing excessive solar gain through the fenestration systems of such buildings is very important.

Airflow Windows

If properly managed, airflow between panes of a double-glazed window can improve fenestration performance. In normal use, a venetian blind is located between the glazing layers. Ventilation air from the room enters the double-glazed cavity, flows over the blind, and is, in some designs, exhausted from the building or returned through the ducts to the central HVAC system. In cold, sunny weather, the window acts as a heat exchanger, so that the temperature of the inner glass nearly equals the room air temperature, which improves thermal comfort.

The apparent conductance across the inner glazing is very low, but this is misleading because additional heat carried by the moving airstream in the window cavity is lost to the outdoors. During sunny winter days, the blind acts as a solar collector; heat removed by the moving air can be used elsewhere in the building. If the blinds are appropriately placed in the summer, the window can have a very low shading coefficient because most of the solar gain is removed at the window. These systems can control window heat transfer under many different operating conditions. Sodergren and Bostrom (1971) and Brandle and Boehm (1982) give details on airflow or exhaust windows.

Table 11 Visible Transmission (VT), Shading Coefficient (SC), and Solar Heat Gain Coefficient (SHGC) at Normal Incidence for Single Pane Glass and Insulating Glass

Glazing System				Glazing SHGC at Specified Incidence Angles						Total Window SHGC at Normal Incidence				Total Window VT at Normal Incidence	
				Normal						Aluminum		Other Frames		All Frames	
ID	Glass Thick, in.	Center Glazing VT	Center Glazing SC	0°	40°	50°	60°	70°	Hemis. (Diffuse)	Operable	Fixed	Operable	Fixed	Operable	Fixed
Uncoated Single Glazing															
1a	1/8 Clear	0.90	1.00	0.86	0.85	0.83	0.78	0.67	0.78	0.75	0.78	0.63	0.75	0.65	0.78
1b	1/4 Clear	0.89	0.94	0.81	0.80	0.77	0.73	0.62	0.73	0.71	0.74	0.60	0.71	0.65	0.78
1c	1/8 Bronze	0.68	0.85	0.73	0.71	0.69	0.64	0.55	0.65	0.64	0.67	0.54	0.64	0.49	0.59
1d	1/4 Bronze	0.55	0.73	0.62	0.60	0.58	0.54	0.46	0.55	0.55	0.57	0.46	0.54	0.40	0.48
1e	1/8 Green	0.82	0.82	0.71	0.68	0.66	0.62	0.53	0.63	0.62	0.65	0.53	0.62	0.60	0.71
1f	1/4 Green	0.74	0.68	0.58	0.56	0.54	0.51	0.44	0.52	0.51	0.53	0.43	0.51	0.54	0.64
1g	1/8 Gray	0.62	0.82	0.70	0.68	0.66	0.61	0.53	0.63	0.61	0.64	0.52	0.61	0.45	0.54
1h	1/4 Gray	0.43	0.65	0.56	0.53	0.51	0.48	0.41	0.49	0.50	0.51	0.42	0.49	0.31	0.37
1i	1/4 Bluegreen	0.75	0.72	0.62	0.59	0.57	0.54	0.46	0.55	0.55	0.57	0.46	0.54	0.54	0.65
Reflective Single Glazing															
1j	1/4 SS on CLR 8%	0.08	0.22	0.19	0.19	0.18	0.17	0.15	0.17	0.18	0.18	0.15	0.17	0.06	0.07
1k	1/4 SS on CLR 14%	0.14	0.29	0.25	0.25	0.24	0.23	0.20	0.23	0.23	0.24	0.19	0.22	0.10	0.12
1l	1/4 SS on CLR 20%	0.20	0.36	0.31	0.30	0.30	0.28	0.24	0.28	0.28	0.29	0.24	0.27	0.15	0.17
1m	1/4 SS on GRN 14%	0.12	0.29	0.25	0.25	0.24	0.23	0.20	0.23	0.23	0.24	0.19	0.22	0.09	0.10
1n	1/4 TI on CLR 20%	0.20	0.34	0.29	0.29	0.28	0.26	0.23	0.27	0.27	0.27	0.22	0.26	0.15	0.17
1o	1/4 TI on CLR 30%	0.30	0.45	0.39	0.38	0.37	0.35	0.30	0.35	0.35	0.36	0.29	0.34	0.22	0.26
Uncoated Double Glazing															
5a	1/8 CLR CLR	0.81	0.87	0.75	0.73	0.70	0.63	0.49	0.65	0.66	0.68	0.55	0.66	0.59	0.71
5b	1/4 CLR CLR	0.78	0.81	0.70	0.68	0.65	0.58	0.45	0.60	0.61	0.64	0.52	0.61	0.57	0.68
5c	1/8 BRZ CLR	0.62	0.72	0.62	0.59	0.57	0.51	0.39	0.53	0.55	0.57	0.46	0.54	0.45	0.54
5d	1/4 BRZ CLR	0.48	0.59	0.50	0.47	0.45	0.40	0.31	0.42	0.45	0.46	0.37	0.44	0.35	0.42
5e	1/8 GRN CLR	0.74	0.70	0.60	0.57	0.55	0.49	0.38	0.51	0.53	0.55	0.45	0.53	0.54	0.64
5f	1/4 GRN CLR	0.66	0.54	0.47	0.44	0.42	0.38	0.30	0.40	0.42	0.43	0.35	0.41	0.48	0.57
5g	1/8 GRY CLR	0.56	0.69	0.59	0.57	0.54	0.48	0.37	0.50	0.52	0.54	0.44	0.52	0.41	0.49
5h	1/4 GRY CLR	0.40	0.51	0.44	0.42	0.40	0.35	0.28	0.38	0.39	0.41	0.33	0.39	0.29	0.35
5i	1/4 BLUGRN CLR	0.67	0.58	0.50	0.47	0.45	0.40	0.32	0.43	0.45	0.46	0.37	0.44	0.49	0.58
5j	1/4 HI-P GRN CLR	0.59	0.46	0.39	0.37	0.35	0.31	0.25	0.33	0.35	0.36	0.29	0.34	0.43	0.51
Reflective Double Glazing															
5k	1/4 SS on CLR 8%,CLR	0.07	0.15	0.13	0.13	0.12	0.12	0.10	0.12	0.13	0.13	0.10	0.12	0.05	0.06
5l	1/4 SS on CLR 14%, CLR	0.13	0.20	0.17	0.17	0.16	0.15	0.12	0.15	0.17	0.16	0.13	0.15	0.09	0.11
5m	1/4 SS on CLR 20%, CLR	0.18	0.26	0.22	0.21	0.21	0.19	0.16	0.19	0.21	0.21	0.17	0.20	0.13	0.16
5n	1/4 SS on GRN 14%, CLR	0.11	0.18	0.16	0.16	0.15	0.14	0.12	0.14	0.16	0.16	0.13	0.14	0.08	0.10
5o	1/4 TI on CLR 20%, CLR	0.18	0.24	0.21	0.20	0.20	0.18	0.15	0.19	0.20	0.20	0.16	0.19	0.13	0.16
5p	1/4 TI on CLR 30%, CLR	0.27	0.33	0.29	0.28	0.27	0.25	0.20	0.25	0.27	0.27	0.22	0.26	0.20	0.24
Low-e Double Glazing, e = 0.2 on Surface 2															
17a	1/8 LE CLR	0.76	0.76	0.65	0.63	0.61	0.55	0.43	0.57	0.57	0.59	0.48	0.57	0.55	0.66
17b	1/4 LE CLR	0.73	0.70	0.60	0.58	0.56	0.51	0.40	0.52	0.53	0.55	0.45	0.53	0.53	0.64
Low-e Double Glazing, e = 0.2 on Surface 3															
17c	1/8 CLR LE	0.76	0.81	0.70	0.68	0.65	0.59	0.46	0.61	0.61	0.64	0.52	0.61	0.55	0.66
17d	1/4 CLR LE	0.73	0.75	0.65	0.63	0.60	0.54	0.42	0.56	0.57	0.59	0.48	0.57	0.53	0.64
17e	1/8 BRZ LE	0.58	0.66	0.57	0.54	0.52	0.46	0.36	0.48	0.50	0.52	0.42	0.50	0.42	0.51
17f	1/4 BRZ LE	0.45	0.52	0.45	0.42	0.40	0.35	0.27	0.37	0.40	0.41	0.34	0.40	0.33	0.39
17g	1/8 GRN LE	0.70	0.63	0.55	0.52	0.50	0.44	0.34	0.46	0.49	0.50	0.41	0.48	0.51	0.61
17h	1/4 GRN LE	0.61	0.48	0.42	0.39	0.37	0.33	0.25	0.35	0.38	0.39	0.32	0.37	0.44	0.53
17i	1/8 GRY LE	0.53	0.63	0.54	0.51	0.49	0.43	0.33	0.46	0.48	0.50	0.40	0.47	0.38	0.46
17j	1/4 GRY LE	0.37	0.46	0.39	0.36	0.34	0.31	0.24	0.33	0.35	0.36	0.29	0.34	0.27	0.32
17k	1/4 BLUGRN LE	0.62	0.52	0.45	0.42	0.40	0.35	0.27	0.37	0.40	0.41	0.34	0.40	0.45	0.54
17l	1/4 HI-P GRN LE	0.55	0.40	0.34	0.31	0.29	0.26	0.20	0.28	0.31	0.32	0.26	0.30	0.40	0.48
Low-e Double Glazing, e = 0.1 on Surface 2															
21a	1/8 LE CLR	0.75	0.62	0.54	0.52	0.49	0.44	0.34	0.46	0.48	0.50	0.40	0.47	0.54	0.65
21b	1/4 LE CLR	0.72	0.59	0.51	0.49	0.47	0.42	0.32	0.44	0.45	0.47	0.38	0.45	0.52	0.63
21l	1/4 HI-P GRN W/LE CLR	0.57	0.36	0.31	0.30	0.29	0.26	0.21	0.27	0.28	0.29	0.24	0.27	0.41	0.50
Low-e Double Glazing, e = 0.1 on Surface 3															
21c	1/8 CLR LE	0.75	0.69	0.60	0.58	0.56	0.51	0.41	0.53	0.53	0.55	0.45	0.53	0.54	0.65
21d	1/4 CLR LE	0.72	0.66	0.56	0.54	0.52	0.47	0.38	0.49	0.50	0.51	0.42	0.49	0.52	0.63
21e	1/8 BRZ LE	0.57	0.56	0.48	0.46	0.43	0.39	0.31	0.41	0.43	0.44	0.36	0.42	0.41	0.50
21f	1/4 BRZ LE	0.45	0.45	0.39	0.37	0.34	0.31	0.24	0.33	0.35	0.36	0.29	0.34	0.33	0.39
21g	1/8 GRN LE	0.68	0.57	0.49	0.47	0.44	0.40	0.31	0.42	0.44	0.45	0.37	0.43	0.49	0.59
21h	1/4 GRN LE	0.61	0.45	0.39	0.36	0.34	0.30	0.24	0.33	0.35	0.36	0.29	0.34	0.44	0.53
21i	1/8 GRY LE	0.52	0.53	0.46	0.44	0.41	0.37	0.29	0.39	0.41	0.42	0.34	0.41	0.38	0.45
21j	1/4 GRY LE	0.37	0.40	0.35	0.33	0.31	0.28	0.22	0.29	0.32	0.33	0.26	0.31	0.27	0.32
21k	1/4 BLUGRN LE	0.62	0.48	0.42	0.39	0.37	0.33	0.26	0.35	0.38	0.39	0.32	0.37	0.45	0.54

Table 11 Visible Transmission (VT), Shading Coefficient (SC), and Solar Heat Gain Coefficient (SHGC) at Normal Incidence for Single Pane Glass and Insulating Glass *(Continued)*

Glazing System					Glazing SHGC at Specified Incidence Angles						Total Window SHGC at Normal Incidence				Total Window VT at Normal Incidence	
			Center Glazing	Center Glazing	Normal						Aluminum		Other Frames		All Frames	
ID	Glass Thick, in.		VT	SC	0°	40°	50°	60°	70°	Hemis. (Diffuse)	Operable	Fixed	Operable	Fixed	Operable	Fixed
Low-e Double Glazing, e = 0.05 on Surface 2																
25a	1/8	LE CLR	0.72	0.48	0.41	0.38	0.34	0.26	0.14	0.35	0.37	0.38	0.31	0.36	0.52	0.63
25b	1/4	LE CLR	0.70	0.43	0.37	0.34	0.31	0.24	0.13	0.32	0.33	0.34	0.28	0.33	0.51	0.61
25c	1/4	BRZ W/LE CLR	0.42	0.30	0.26	0.24	0.22	0.18	0.10	0.23	0.24	0.24	0.20	0.23	0.31	0.37
25d	1/4	GRN W/LE CLR	0.60	0.35	0.30	0.28	0.25	0.20	0.11	0.26	0.28	0.28	0.23	0.27	0.44	0.52
25e	1/4	GRY W/LE CLR	0.35	0.27	0.24	0.22	0.20	0.16	0.10	0.20	0.22	0.23	0.18	0.21	0.25	0.30
25f	1/4	BLUE W/LE CLR	0.45	0.32	0.27	0.25	0.23	0.18	0.10	0.23	0.25	0.25	0.21	0.24	0.33	0.39
25g	1/4	HI-P GRN W/LE CLR	0.53	0.31	0.27	0.26	0.25	0.23	0.18	0.24	0.25	0.25	0.22	0.25	0.38	0.46
Triple Glazing																
29a	1/8	CLR CLR CLR	0.74	0.78	0.67	0.65	0.61	0.53	0.39	0.57	0.59	0.61	0.50	0.59	0.54	0.64
29b	1/4	CLR CLR CLR	0.70	0.71	0.61	0.58	0.55	0.48	0.35	0.51	0.54	0.56	0.45	0.54	0.51	0.61
29c	1/4	HI-P GRN CLR CLR	0.53	0.39	0.34	0.31	0.29	0.25	0.19	0.27	0.31	0.32	0.26	0.30	0.38	0.46
Triple Glazing, e = 0.2 on Surface 2																
32a	1/8	LE CLR CLR	0.68	0.69	0.60	0.58	0.55	0.48	0.35	0.51	0.53	0.55	0.45	0.53	0.49	0.59
32b	1/4	LE CLR CLR	0.64	0.62	0.53	0.50	0.47	0.41	0.30	0.44	0.47	0.49	0.39	0.47	0.46	0.56
Triple Glazing, e = 0.2 on Surface 5														0.00	0.00	
32c	1/8	CLR CLR LE	0.68	0.72	0.62	0.60	0.56	0.49	0.36	0.52	0.55	0.57	0.46	0.54	0.49	0.59
32d	1/4	CLR CLR LE	0.64	0.65	0.56	0.53	0.50	0.44	0.32	0.47	0.50	0.51	0.42	0.49	0.46	0.56
Triple Glazing, e = 0.1 on Surface 2 and 5																
40a	1/8	LE CLR LE	0.62	0.52	0.45	0.43	0.40	0.36	0.26	0.38	0.40	0.41	0.34	0.40	0.45	0.54
40b	1/4	LE CLR LE	0.59	0.47	0.41	0.39	0.37	0.32	0.24	0.34	0.37	0.38	0.31	0.36	0.43	0.51
Triple Glazing, e = 0.05 on Surface 2 and 4								0.00	0.00							
40c	1/8	LE LE CLR	0.58	0.37	0.32	0.30	0.29	0.26	0.19	0.27	0.29	0.30	0.24	0.28	0.42	0.51
40d	1/4	LE LE CLR	0.55	0.36	0.31	0.29	0.28	0.25	0.19	0.26	0.28	0.29	0.24	0.27	0.40	0.48

KEY:

CLR = clear, GRN = green, GRY = gray, SS = stainless steel reflective coating, TI = titanium reflective coating

Reflective coating decsriptors include percent visible transmittance as x%.

Hi-P GRN = high performance green tinted glass, LE = glass with a low-emissivity coating with an emittance of e = 0.xx

Low-e coating with an emittance of 0.2 is a pyrolytic coating. Other low-e coatings are sputtered coatings.

VT is Visible Transmittance, SC is Shading Coefficient, SHGC, is Solar Heat Gain Coefficient, and HEMIS is the hemispherical SHGC

ID numberss refer to U-factors in Table 5

SHGC at 90° is 0.

Domed Skylights

Solar and total heat gains for domed skylights can be determined by the same procedure used for windows. The SHGF values for such calculations should be consistent with the dome orientations. For horizontal roofs, Tables 15 through 21 give approximate SHGF values. For sloping roofs, an approximate SHGF can be found from SHGF = E_t/1.15, where E_t = total solar irradiation on the sloping surface. Table 12 lists shading coefficients for plastic domed skylights. Manufacturers' literature has further details.

Glass Block Walls

Glass block can be used for light transmission through exterior walls when the optical clarity of a view is not needed or wanted. Table 13 describes the glass block patterns discussed in the following text and gives solar heat gain coefficients to be applied to the solar irradiances from Tables 15 through 21 so that approximate instantaneous solar heat gains can be calculated.

Convection and low-temperature radiation heat gains for all hollow glass block panels fall in a narrow range. Differences in SHGC values are largely the result of differences in the transmittance of the glass blocks to solar radiation. Solar heat gain coefficients for any particular glass block pattern vary depending on orientation and time of day. The SHGC for western exposures in the morning (in the shade) is depressed because of the heat storage within the block,

Table 12 Shading Coefficients for Domed Horizontal Skylights

TRANSLUCENT DOME
CURB
CURB
HEIGHT
LIGHT DIFFUSER

Dome	Light Diffuser (Translucent)	Curb Height, in.	Curb Width to Height Ratio	Shading Coefficient
Clear	Yes	0	∞	0.61
τ = 0.86	τ = 0.58	9	5	0.58
		18	2.5	0.50
Clear		0	∞	0.99
τ = 0.86	None	9	5	0.88
		18	2.5	0.80
Translucent	None	0	∞	0.57
τ = 0.52		18	2.5	0.46
Translucent	None	0	∞	0.34
τ = 0.27		9	5	0.30
		18	2.5	0.28

Table 13 Shading Coefficients and U-Factors for Standard Hollow Glass Block Wall Panels

A OUTDOOR SURFACE; B; C; D INDOOR SURFACE; E CAVITY PARTITION

Type of Glass Block[a]	Description of Glass Block	Shading Coefficient[b]		U-Factor,[e] Btu/h·ft²·°F
		Panels[c] in Sun	Panels[d] in Shade (N, NW, W, SW)	
Type I	Glass colorless or aqua A, D: Smooth B, C: Smooth or wide ribs, or flutes horizontal or vertical, or shallow configuration E: None	0.65	0.40	0.51
Type IA	Same as Type I except ceramic enamel on A	0.27	0.20	0.51
Type II	Same as Type I except glass fiber screen partition E	0.44	0.34	0.48
Type III	Glass colorless or aqua A, D: Narrow vertical ribs or flutes. B, C: Horizontal light-diffusing prisms, or horizontal light-directing prisms E: Glass fiber screen	0.33	0.27	0.48
Type IIIA	Same as Type III except E: Glass fiber screen with green ceramic spray coating or glass fiber screen and gray glass or glass fiber screen with light-selecting prisms	0.25	0.18	0.48
Type IV	Same as Type I except reflective oxide coating on A	0.16	0.12	0.51

[a]All values are for 7 3/4 by 7 3/4 by 3 7/8 in. block, set in light-colored mortar. For 11 3/4 by 11 3/4 by 3 7/8 in. block, increase coefficients by 15%, and for 5 3/4 by 5 3/4 by 3 7/8 in. block reduce coefficents by 15%.

[b]Shading coefficients are applied to heat gain factors for one hour earlier than the load calculation time to allow for heat storage in the panel.

[c]Shading coefficients are for peak load condition, but provide a close approximation for other conditions.

[d]For NE, E, and SE panels in shade, add 50% to the values listed for panels in the shade.

[e]Values shown are the same for all size block.

whereas the SHGC for eastern exposures in the afternoon (in the shade) is elevated as the stored heat is dissipated. Time lag effects from heat storage are estimated by using solar gains and air-to-air temperature differences for one hour earlier than the time for which the load calculation is made.

Calorimeter tests of Type 1A glass block showed little difference in solar heat gains between glass block with either black or white ceramic enamel on the exterior of the block. Because white and black ceramic enamel surfaces represent the two extremes for reflecting or absorbing solar energy, glass block with enamel surfaces of other colors should have solar heat gain coefficients between these values. Since glass blocks are good examples of strongly angularly selective fenestrations, all the above cautions for such glazings apply here.

Table 14 Solar Optical Properties and Shading Coefficients of Transparent Plastic

Type of Plastic	Transmittance		SC
	Visible	Solar	
Acrylic			
Clear	0.92	0.85	0.98
Gray tint	0.16	0.27	0.52
Gray tint	0.33	0.41	0.63
Gray tint	0.45	0.55	0.74
Gray tint	0.59	0.62	0.80
Gray tint	0.76	0.74	0.89
Bronze tint	0.10	0.20	0.46
Bronze tint	0.27	0.35	0.58
Bronze tint	0.49	0.56	0.75
Bronze tint	0.61	0.62	0.80
Bronze tint	0.75	0.75	0.90
Reflective[a]	0.14	0.12	0.21
Polycarbonate			
Clear, 1/8-in.	0.88	0.82	0.98
Gray, 1/8-in.	0.50	0.57	0.74
Bronze, 1/8-in.	0.50	0.57	0.74

Source: Burkhardt (1975, 1976)

[a]Aluminum metallized polyester film on plastic.

The SC method should be used with glass block walls only for very approximate peak load solar gain estimates.

Plastic Materials For Glazing

Generally, the optical factors outlined for glass apply also to acrylic, polycarbonate, polystyrene, or other plastic panels. If the solar transmittance, absorptance, reflectance, and geometry are known, a SHGC and a shading coefficient can be calculated in the same way as for glass. These properties can be obtained from the manufacturer or determined by laboratory tests such as those described in NFRC *Standards* 301-93 and 300-94.

Table 14 lists solar optical properties of typical plastic materials. Plastic panels are available with transmittance values from 10% to 92% and reflectance values from 4% to over 60%. Emittances for acrylic, polystyrene, and polycarbonate glazing all range between 0.88 and 0.92. Plastics are available in translucent and transparent form; some are corrugated. Many tints (colors) are also available.

In selecting plastic panels for glazing, possible deterioration from the sun, expansion and contraction because of temperature extremes, and possible damage from abrasion are concerns.

SIMPLIFIED METHODS FOR PREDICTING HEAT TRANSFER THROUGH FENESTRATION

Fenestration Heat Balance-Solar Heat Gain

At any instant, the heat balance between a unit area of sunlit single-glazing material and its thermal environment, as shown in Figure 29, is

$$E_t + U(t_o - t_i) = q_R + q_s + q_T + q_{RCo} + q_{RCi} \qquad (39)$$

where q_R, q_s, and q_T represent the heat reflected, stored in the glass, and transmitted, respectively. In general, q_s is relatively small in magnitude, so it is disregarded. The terms q_{RCo} and q_{RCi} are the rates of heat flux outward and inward, respectively, by radiation and convection. The rate of heat rejection to the atmosphere is the sum of the reflected heat q_R and outward radiation/convection heat flux q_{RCo}.

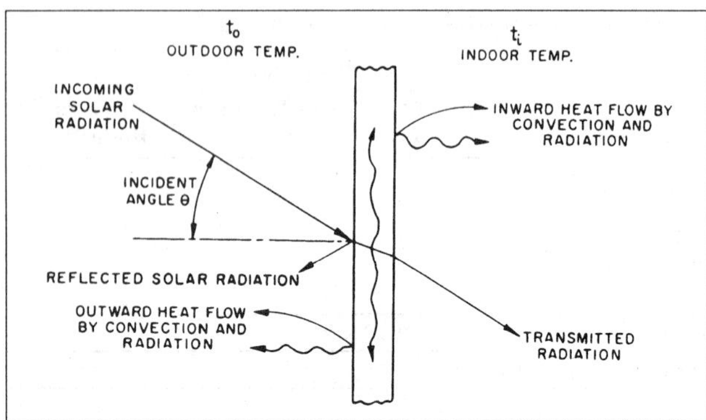

Fig. 29 Instantaneous Heat Balance for Sunlit Glazing Material

The total instantaneous rate of heat gain through the glazing material may be expressed:

Total heat admission through glass	=	Radiation transmitted through glass	+	Heat flow inward by radiation-convection from inner glass surface

$$q_A = E_{D\tau D} + E_{d\tau d} + q_{RCi} \quad (40)$$

Subscript D denotes direct radiation; d denotes diffuse radiation, and τ indicates transmitted radiation. The solar-optical properties for diffuse radiation can be taken as those for direct radiation where $\theta = 60°$. The following equations consider total solar energy values for simplicity.

Equation (40) may also be written as:

Total heat admission through glass	=	Radiation transmitted through glass	+	Inward flow of absorbed solar radiation	+	Heat flow due to outdoor-indoor temperature diff.

For single glazing,

$$q_A = E_t\tau + N_i(\alpha E_t) + U(t_o - t_i) \quad (41)$$

For double glazing, the transmittance through both glasses can be calculated by

$$\bar{\tau} = \frac{(\tau_o\tau_i)}{(1 - \rho_2\rho_3)} \quad (42)$$

The first two terms of Equation (41) are related to the incident solar radiation, while the third occurs whether or not the sun is shining. As a result Equations (41) and (42) may be combined as

$$q_A = E_t\bar{\tau} + N_{jo}(\alpha E_o) + N_i(\alpha E_o) + U(t_o - t_i) \quad (43)$$

Total heat admission through glass	=	Solar heat gain	+	Conduction heat gain

These may both be written as

$$q_A = \text{SHGC } E_t + U(t_o - t_i) \quad (44)$$

where

q_A = instantaneous rate of heat admission through fenestration, Btu/h·ft²
SHGC = solar heat gain coefficient (ratio of solar heat gains to incident solar radiation)

For single glazing, the inward flowing fraction of absorbed radiation N_i may be approximated with the equation

$$N_i \approx U/h_o \quad (45)$$

where U is the U-factor of the glazing system (center-of-glass) and h_o is the outside film coefficient.

For more accurate calculations, especially if the fenestration contains glazings that are strongly angular dependent or spectrally selective, the SHGC should be determined using the wavelength-by-wavelength, spectral approach used in the Solar Heat Gain Coefficient method. However, if clear or lightly tinted glazings are to be used, the shading coefficient method described in the following section can be used.

Simplified Solar Heat Gain Calculations

One way to check the glazing selectivity is to compare the spectral transmittance of the glazing system to a standard solar spectrum. If the spectral transmittance does not vary strongly in regions where the solar spectrum varies greatly, the SHGC will be reasonably constant over modest changes in the incident spectral shape as well. If this condition is not satisfied, or if accurate calculations are needed, then the wavelength-by-wavelength or *spectral method* described previously, should be used.

If the glazing is not particularly selective, the following broad-band shading coefficient method can be used to calculate solar gain.

1. Find the correct latitude, glazing orientation, and time of day from Tables 15 through 21.
2. Write down the combined beam and diffuse solar heat gain factor SHGF for this angle of incidence.
3. Calculate the solar gain using the following equation

$$q_i = \text{SC} \times \text{SHGF} \quad (46)$$

The total instantaneous heat gain is

$$q_A = \text{SC} \times \text{SHGF} + U(t_o - t_i) \quad (47)$$

Equation (47) applies at any time of the year. In winter, when outdoor temperature is lower than indoor, the conduction heat flow occurring in the absence of sunshine is negative or outward. In summer, conduction heat flow is inward whenever the outside temperature exceeds indoor temperature.

Tables in Chapter 24 list design values of summer outdoor dry-bulb temperatures for many locations. These are generally mid-afternoon temperatures; for other times, local weather stations can supply appropriate temperature data. Winter design temperatures should not be used in Equation (47), because they are usually given for early morning hours before sunrise. The indoor air temperature for air-conditioned spaces is likely to be closer to 75°F than to 80°F.

Tables 15 through 21 list solar heat gain through standard reference glazing (clear, single pane glass). The solar heat gain factor (SHGF) has units of heat flow per unit area in contrast to the solar heat gain coefficient (SHGC), which is dimensionless. The tables can also be used to determine the half day and whole day totals of clear sky solar heat gain. To determine the instantaneous hourly SHGFs and the half day total SHGFs for times before apparent solar noon, the orientation legend on the top of the table and the time legend on the left side of each table should be used.

For example, to calculate the entire day clear sky SHGF for an east facing surface, add the half day total for the month from the top column labeled E (7th column from the left) to the half day total from the column labeled E on the bottom (15th column). That is, the tables are read from the top down for morning hours and from the bottom up for afternoon hours.

Table 15 Solar Irradiance (E_{DN}) and Solar Heat Gain Factors (SHGF) for 16° North Latitude

Date	Solar Time	Direct Normal Btu/(h·ft²)	N	NNE	NE	ENE	E	ESE	SE	SSE	S	SSW	SW	WSW	W	WNW	NW	NNW	Hor.	Solar Time
			Solar Heat Gain Factors, Btu/(h·ft²)																	
Jan	0700	141	5	6	44	92	124	134	126	96	49	6	5	5	5	5	5	5	14	1700
	0800	262	14	15	55	147	210	240	233	189	114	25	14	14	14	14	14	14	79	1600
	0900	300	21	21	32	122	200	244	251	219	152	58	22	21	21	21	21	21	150	1500
	1000	317	26	26	27	66	150	209	233	223	178	102	31	26	26	26	26	26	203	1400
	1100	325	29	29	29	31	77	148	195	210	194	146	75	31	29	29	29	29	236	1300
	1200	327	30	30	30	30	32	72	139	184	199	184	138	72	32	30	30	30	248	1200
	HALF DAY TOTALS		110	112	196	461	760	1000	1096	1020	781	426	211	127	111	110	110	110	805	
Feb	0700	182	8	17	84	138	169	172	150	103	36	8	8	8	8	8	8	8	25	1700
	0800	273	17	19	96	180	231	247	224	166	77	18	17	17	17	17	17	17	101	1600
	0900	305	23	24	64	153	214	242	233	188	110	30	23	23	23	23	23	23	174	1500
	1000	319	28	29	33	92	161	202	211	188	134	61	30	28	28	28	28	28	229	1400
	1100	326	32	32	32	37	83	136	167	172	149	102	49	33	32	32	32	32	263	1300
	1200	328	33	33	33	33	34	60	107	142	154	142	106	60	34	33	33	33	275	1200
	HALF DAY TOTALS		124	137	321	609	865	1023	1034	885	582	287	174	132	124	124	124	124	930	124
Mar	0700	201	11	53	124	172	192	183	145	82	15	10	10	10	10	10	10	10	40	1700
	0800	272	20	50	140	205	239	235	195	123	35	19	19	19	19	19	19	19	120	1600
	0900	299	26	35	109	179	218	225	197	138	57	27	26	26	26	26	26	26	192	1500
	1000	312	31	33	61	120	165	182	172	134	76	34	32	31	31	31	31	31	247	1400
	1100	318	34	35	36	53	87	114	125	116	89	55	36	35	34	34	34	34	280	1300
	1200	320	35	35	36	36	37	47	69	87	93	86	68	47	37	36	36	35	291	1200
	HALF DAY TOTALS		141	226	494	755	928	975	879	643	319	187	153	142	139	139	139	139	1025	
Apr	0600	14	2	8	12	14	14	12	8	2	1	1	1	1	1	1	1	1	1	1800
	0700	197	24	94	153	187	191	167	117	45	14	13	13	13	13	13	13	13	53	1700
	0800	256	27	99	172	216	227	204	150	69	24	22	22	22	22	22	22	22	131	1600
	0900	280	31	79	149	193	208	193	147	77	31	29	29	29	29	29	29	29	197	1500
	1000	293	35	54	102	141	158	151	120	73	37	34	33	33	33	33	33	33	249	1400
	1100	299	38	40	54	72	86	88	78	60	43	38	38	36	36	36	36	37	279	1300
	1200	301	39	39	39	40	40	41	43	45	45	45	43	41	40	39	39	39	289	1200
	HALF DAY TOTALS		179	403	674	859	922	851	653	352	174	159	157	156	155	155	155	156	1057	
May	0600	44	14	30	41	45	43	34	19	4	3	3	3	3	3	3	3	3	5	1800
	0700	193	50	120	168	191	185	150	92	24	16	16	16	16	16	16	16	17	62	1700
	0800	244	52	132	189	218	215	179	115	38	25	24	24	24	24	24	24	25	135	1600
	0900	268	49	116	171	198	197	167	109	45	32	30	30	30	30	30	30	32	197	1500
	1000	280	47	89	130	151	150	126	84	44	37	35	35	35	35	35	35	37	245	1400
	1100	286	47	63	79	87	83	70	52	41	40	39	38	38	38	39	39	41	273	1300
	1200	288	46	46	44	43	42	41	41	41	41	41	41	41	42	43	44	46	282	1200
	HALF DAY TOTALS		283	575	804	916	897	748	493	217	172	167	167	167	167	168	169	176	1058	
Jun	0600	53	20	39	52	55	51	39	20	4	4	4	4	4	4	4	4	4	7	1800
	0700	188	62	128	172	190	179	141	80	20	16	16	16	16	16	16	16	18	64	1700
	0800	238	66	142	194	217	207	167	99	31	25	25	25	25	25	25	25	27	135	1600
	0900	261	63	130	178	198	190	154	93	37	31	31	31	31	31	31	31	33	194	1500
	1000	273	59	104	140	154	145	115	70	39	37	36	36	36	36	36	36	38	241	1400
	1100	279	57	76	90	92	82	63	46	41	40	39	39	39	39	40	41	43	268	1300
	1200	281	57	55	50	45	43	42	41	41	41	41	41	42	42	45	50	55	277	1200
	HALF DAY TOTALS		356	648	850	929	876	700	430	194	174	171	171	171	172	173	176	190	1049	
Jul	0600	41	14	29	39	42	40	31	18	4	3	3	3	3	3	3	3	3	6	1800
	0700	184	51	118	164	185	179	145	88	23	16	16	16	16	16	16	16	17	62	1700
	0800	236	55	132	187	214	210	174	111	37	25	25	25	25	25	25	25	26	133	1600
	0900	259	52	117	170	196	193	163	106	44	32	31	31	31	31	31	31	33	194	1500
	1000	272	50	92	131	151	148	123	81	44	38	36	36	36	36	36	36	38	241	1400
	1100	278	49	66	81	88	83	69	52	42	41	40	39	39	39	40	40	42	269	1300
	1200	279	49	48	46	44	43	42	42	42	42	42	42	42	43	44	46	48	277	1200
	HALF DAY TOTALS		296	580	799	903	878	729	478	215	176	172	171	171	171	172	173	182	1043	
Aug	0600	11	2	7	10	12	12	10	6	2	1	1	1	1	1	1	1	1	1	1800
	0700	180	26	92	145	176	180	156	109	42	15	14	14	14	14	14	14	14	53	1700
	0800	240	30	100	168	209	219	196	143	65	25	23	23	23	23	23	23	23	128	1600
	0900	266	33	82	148	190	203	187	142	74	33	30	30	30	30	30	30	30	193	1500
	1000	279	37	58	104	140	155	147	117	71	39	36	35	35	35	35	35	35	243	1400
	1100	285	40	43	57	75	86	87	76	59	44	40	39	38	38	38	38	39	273	1300
	1200	287	41	41	41	42	42	43	44	45	46	45	44	43	42	41	41	41	282	1200
	HALF DAY TOTALS		191	410	666	837	891	817	624	339	180	167	165	164	163	163	163	164	1033	
Sep	0700	179	12	50	114	158	176	168	133	76	15	11	11	11	11	11	11	11	39	1700
	0800	253	21	49	134	196	227	224	186	119	36	20	20	20	20	20	20	20	116	1600
	0900	281	28	36	106	173	211	217	191	134	57	28	27	27	27	27	27	27	185	1500
	1000	295	32	34	61	118	161	178	168	132	76	35	33	32	32	32	32	32	238	1400
	1100	302	35	36	37	54	86	113	123	114	88	56	38	36	35	35	35	35	271	1300
	1200	304	36	36	37	38	39	49	69	86	93	86	69	48	39	38	37	36	282	1200
	HALF DAY TOTALS		146	226	475	722	885	931	842	622	319	192	159	148	145	144	144	144	991	
Oct	0700	166	8	18	79	128	156	159	139	95	33	9	8	8	8	8	8	8	25	1700
	0800	259	17	20	95	174	223	237	215	159	74	19	17	17	17	17	17	17	99	1600
	0900	292	24	25	65	150	209	235	225	182	106	31	24	24	24	24	24	24	170	1500
	1000	307	29	30	34	92	158	197	205	183	130	60	31	29	29	29	29	29	224	1400
	1100	314	32	32	33	39	83	133	163	167	145	100	49	34	32	32	32	32	258	1300
	1200	316	33	33	33	34	35	60	105	139	150	138	104	60	35	34	33	33	270	1200
	HALF DAY TOTALS		127	141	318	592	836	986	996	852	563	283	175	136	128	127	127	127	911	
Nov	0700	134	5	6	43	89	119	129	120	92	47	6	5	5	5	5	5	5	14	1700
	0800	255	15	15	55	145	206	235	228	185	111	25	15	15	15	15	15	15	78	1600
	0900	295	21	21	33	121	197	241	247	215	150	57	22	21	21	21	21	21	149	1500
	1000	312	26	26	28	67	147	206	230	220	176	100	31	26	26	26	26	26	201	1400
	1100	320	29	29	29	31	77	146	192	207	191	144	74	31	29	29	29	29	234	1300
	1200	322	30	30	30	30	32	72	137	181	196	181	137	72	32	30	30	30	246	1200
	HALF DAY TOTALS		112	113	197	456	749	983	1077	1001	767	420	210	128	112	112	112	112	799	
Dec	0700	118	4	5	30	72	101	112	107	85	48	7	4	4	4	4	4	4	10	1700
	0800	255	13	14	41	132	198	233	231	193	124	33	13	13	13	13	13	13	69	1600
	0900	297	20	20	25	108	191	241	254	227	165	72	21	20	20	20	20	20	138	1500
	1000	315	25	25	26	56	144	208	239	233	192	117	35	25	25	25	25	25	191	1400
	1100	323	28	28	28	29	73	150	202	221	207	161	86	30	28	28	28	28	223	1300
	1200	325	29	29	29	29	30	77	149	197	212	196	149	76	30	29	29	29	234	1200
	HALF DAY TOTALS		104	105	159	402	710	975	1099	1050	836	484	228	125	105	104	104	104	748	
			N	NNW	NW	WNW	W	WSW	SW	SSW	S	SSE	SE	ESE	E	ENE	NE	NNE	Hor.	PM

Notes: 1. Clearness number = 1.00; Ground reflectance = 0.20. 2. Figures shown are for 21st day of each month.

Table 16 Solar Irradiance (E_{DN}) and Solar Heat Gain Factors (SHGF) for 24° North Latitude

Date	Solar Time	Direct Normal Btu/(h·ft²)	Solar Heat Gain Factors, Btu/(h·ft²) N	NNE	NE	ENE	E	ESE	SE	SSE	S	SSW	SW	WSW	W	WNW	NW	NNW	Hor.	Solar Time
Jan	0700	71	2	3	21	45	62	67	63	49	25	3	2	2	2	2	2	2	5	1700
	0800	239	12	12	41	128	190	221	218	181	114	28	12	12	12	12	12	12	55	1600
	0900	288	18	18	23	106	190	240	253	227	166	73	19	18	18	18	18	18	121	1500
	1000	308	23	23	24	53	144	211	245	241	200	125	38	24	23	23	23	23	172	1400
	1100	317	26	26	26	27	73	156	211	234	220	173	95	29	26	26	26	26	204	1300
	1200	320	27	27	27	27	29	82	160	210	227	210	160	81	29	27	27	27	214	1200
	HALF DAY TOTALS		95	96	148	372	671	942	1076	1039	840	505	241	120	96	95	95	95	664	
Feb	0700	153	6	12	67	114	141	145	128	90	33	6	6	6	6	6	6	6	17	1700
	0800	262	15	16	80	165	220	240	224	172	89	17	15	15	15	15	15	15	83	1600
	0900	297	21	22	46	138	208	244	243	205	133	42	22	21	21	21	21	21	153	1500
	1000	314	26	26	28	76	157	209	228	213	165	87	28	26	26	26	26	26	205	1400
	1100	321	29	29	29	31	80	148	191	203	185	137	68	31	29	29	29	29	238	1300
	1200	323	30	30	30	30	32	70	134	177	192	177	133	70	32	30	30	30	249	1200
	HALF DAY TOTALS		113	119	257	527	806	1011	1072	965	699	374	200	127	113	113	113	113	820	
Mar	0700	194	11	45	115	164	186	180	145	86	17	10	10	10	10	10	10	10	36	1700
	0800	267	18	35	124	195	234	237	204	138	48	19	18	18	18	18	18	18	112	1600
	0900	295	25	27	85	165	215	232	214	163	82	27	25	25	25	25	25	25	180	1500
	1000	309	30	31	41	103	162	194	195	168	112	47	31	30	30	30	30	30	232	1400
	1100	315	33	33	34	42	85	129	154	155	139	86	43	34	33	33	33	33	264	1300
	1200	317	34	34	34	34	35	56	96	126	137	126	95	56	35	34	34	34	275	1200
	HALF DAY TOTALS		133	189	422	693	906	1011	970	778	458	249	169	139	133	133	133	133	962	
Apr	0600	40	6	21	33	39	39	33	22	7	2	2	2	2	2	2	2	2	4	1800
	0700	203	20	88	151	189	197	176	127	55	15	14	14	14	14	14	14	14	58	1700
	0800	256	24	80	159	209	228	212	164	88	24	22	22	22	22	22	22	22	132	1600
	0900	280	30	54	126	181	208	203	169	105	39	29	28	28	28	28	28	28	195	1500
	1000	292	34	37	75	125	157	165	148	107	56	35	33	33	33	33	33	33	244	1400
	1100	298	36	37	40	59	85	103	106	94	70	45	38	37	36	36	36	36	274	1300
	1200	299	37	37	38	38	39	46	59	70	75	70	58	45	39	38	38	37	283	1200
	HALF DAY TOTALS		168	339	607	826	940	924	773	494	244	180	163	157	155	155	154	154	1048	
May	0600	86	25	57	79	87	84	66	38	8	6	6	6	6	6	6	6	6	13	1800
	0700	203	43	117	171	199	196	163	105	32	17	17	17	17	17	17	17	18	73	1700
	0800	248	38	114	178	214	218	190	132	54	26	25	25	25	25	25	25	26	142	1600
	0900	269	35	88	150	188	198	179	132	66	33	31	31	31	31	31	31	31	201	1500
	1000	280	38	59	103	137	150	141	111	67	39	36	35	35	35	35	35	36	247	1400
	1100	286	40	43	55	72	83	84	75	58	44	40	39	38	38	38	38	39	274	1300
	1200	288	41	41	41	41	42	43	44	46	46	46	44	43	42	41	41	41	282	1200
	HALF DAY TOTALS		238	492	749	909	943	840	614	308	187	176	174	173	172	172	172	175	1089	
Jun	0600	97	36	70	93	101	94	73	39	8	7	7	7	7	7	7	7	8	17	1800
	0700	201	55	127	177	199	192	155	94	26	18	18	18	18	18	18	18	20	77	1700
	0800	242	50	126	184	214	212	179	117	43	27	26	26	26	26	26	26	27	145	1600
	0900	263	43	102	158	189	192	168	116	53	34	32	32	32	32	32	32	33	201	1500
	1000	274	41	72	113	140	146	131	96	55	39	36	36	36	36	36	36	38	245	1400
	1100	279	42	50	65	77	82	77	64	49	42	41	40	39	39	39	40	41	271	1300
	1200	281	43	43	43	43	43	43	43	43	43	43	43	43	43	43	43	43	279	1200
	HALF DAY TOTALS		284	562	802	933	932	797	544	255	187	181	180	179	179	179	180	187	1096	
Jul	0600	81	26	56	76	84	80	63	36	8	6	6	6	6	6	6	6	7	13	1800
	0700	195	45	116	168	194	190	158	101	31	18	18	18	18	18	18	18	19	73	1700
	0800	239	41	115	176	210	213	185	128	52	27	26	26	26	26	26	26	26	141	1600
	0900	261	37	90	150	186	195	175	129	64	34	32	32	32	32	32	32	32	198	1500
	1000	272	39	62	104	137	149	139	108	65	39	37	36	36	36	36	36	37	243	1400
	1100	278	41	44	58	73	83	83	73	57	44	41	40	39	39	39	39	40	270	1300
	1200	280	42	42	42	43	43	44	45	46	46	46	45	43	43	42	42	42	278	1200
	HALF DAY TOTALS		247	498	746	897	925	820	595	300	191	181	178	177	177	177	177	181	1076	
Aug	0600	35	6	20	30	35	35	30	19	6	2	2	2	2	2	2	2	2	4	1800
	0700	186	22	87	144	179	186	165	119	51	16	15	15	15	15	15	15	15	58	1700
	0800	241	26	82	156	203	220	204	157	84	26	24	24	24	24	24	24	24	130	1600
	0900	265	32	57	126	178	202	197	162	101	39	31	30	30	30	30	30	30	191	1500
	1000	278	36	40	78	125	155	161	143	103	55	37	35	35	35	35	35	35	239	1400
	1100	284	38	39	42	61	85	101	104	91	68	46	40	38	37	37	37	37	268	1300
	1200	286	38	39	40	40	41	47	58	69	72	68	58	47	41	40	40	39	277	1200
	HALF DAY TOTALS		179	347	601	806	910	889	740	473	243	186	171	165	164	163	163	162	1028	
Sep	0700	172	11	42	106	149	169	164	132	79	17	10	10	10	10	10	10	10	35	1700
	0800	248	19	36	119	185	222	225	194	132	48	20	19	19	19	19	19	19	108	1600
	0900	278	26	28	84	160	207	223	206	158	81	28	26	26	26	26	26	26	174	1500
	1000	292	31	32	42	101	158	188	190	163	110	48	32	31	31	31	31	31	224	1400
	1100	299	34	34	35	43	84	127	151	151	128	86	44	35	34	34	34	34	256	1300
	1200	301	35	35	35	36	37	57	95	124	134	124	94	57	37	36	35	35	266	1200
	HALF DAY TOTALS		139	190	406	661	863	964	927	749	451	251	174	145	139	138	138	138	930	
Oct	0700	138	6	12	62	104	129	133	117	82	31	7	6	6	6	6	6	6	17	1700
	0800	247	16	17	79	159	211	230	214	164	85	17	16	16	16	16	16	16	82	1600
	0900	284	22	23	47	135	202	237	235	198	128	41	23	22	22	22	22	22	150	1500
	1000	301	27	27	29	77	154	204	222	207	160	85	29	27	27	27	27	27	201	1400
	1100	309	30	30	30	33	80	145	186	198	180	133	67	32	30	30	30	30	233	1300
	1200	311	31	31	31	31	33	70	131	173	187	172	130	69	33	31	31	31	244	1200
	HALF DAY TOTALS		116	123	255	512	778	974	1032	929	675	367	200	131	117	116	116	116	804	
Nov	0700	67	2	3	20	43	59	64	60	46	24	3	2	2	2	2	2	2	5	1700
	0800	232	12	13	42	126	186	216	213	177	111	28	12	12	12	12	12	12	55	1600
	0900	282	19	19	23	106	187	236	249	223	163	71	20	19	19	19	19	19	120	1500
	1000	303	23	23	24	53	143	209	241	237	197	123	37	24	23	23	23	23	171	1400
	1100	312	26	26	26	28	73	154	209	230	217	171	93	29	26	26	26	26	202	1300
	1200	315	27	27	27	27	29	81	158	207	224	207	158	80	29	27	27	27	213	1200
	HALF DAY TOTALS		97	97	149	368	661	926	1056	1020	825	497	239	121	98	97	97	97	659	
Dec	0700	30	1	1	7	18	25	28	27	21	12	2	1	1	1	1	1	1	2	1700
	0800	225	10	10	29	112	174	208	209	178	118	35	11	10	10	10	10	10	44	1600
	0900	281	17	17	19	93	180	234	252	231	174	84	18	17	17	17	17	17	107	1500
	1000	304	22	22	22	44	137	209	247	247	209	137	44	22	22	22	22	22	157	1400
	1100	314	25	25	25	26	69	156	216	241	230	183	104	29	25	25	25	26	188	1300
	1200	317	26	26	26	26	27	85	167	219	237	219	167	84	27	26	26	26	199	1200
	HALF DAY TOTALS		88	88	118	313	611	899	1054	1042	868	550	257	117	89	88	88	88	598	
			N	NNW	NW	WNW	W	WSW	SW	SSW	S	SSE	SE	ESE	E	ENE	NE	NNE	Hor.	PM

Notes: 1. Clearness number = 1.00; Ground reflectance = 0.20. 2. Figures shown are for 21st day of each month.

Table 17 Solar Irradiance (E_{DN}) and Solar Heat Gain Factors (SHGF) for 32° North Latitude

Date	Solar Time	Direct Normal Btu/(h·ft²)	Solar Heat Gain Factors, Btu/(h·ft²) N	NNE	NE	ENE	E	ESE	SE	SSE	S	SSW	SW	WSW	W	WNW	NW	NNW	Hor.	Solar Time
Jan	0700	1	0	0	0	1	1	1	1	1	0	0	0	0	0	0	0	0	0	1700
	0800	203	9	9	29	105	160	189	189	159	103	28	9	9	9	9	9	9	32	1600
	0900	269	15	15	17	91	175	229	246	225	169	82	17	15	15	15	15	15	88	1500
	1000	295	20	20	20	41	135	209	249	250	212	141	46	20	20	20	20	20	136	1400
	1100	306	23	23	23	24	68	159	221	249	238	191	110	29	23	23	23	23	166	1300
	1200	310	24	24	24	24	25	88	174	228	246	228	174	88	25	24	24	24	176	1200
	HALF DAY TOTALS		79	79	107	284	570	856	1015	1014	853	553	264	112	80	79	79	79	512	
Feb	0700	112	4	7	47	82	102	106	95	67	26	4	4	4	4	4	4	4	9	1700
	0800	245	13	14	65	149	205	228	216	170	95	17	13	13	13	13	13	13	64	1600
	0900	287	19	19	32	122	199	242	248	216	149	55	20	19	19	19	19	19	127	1500
	1000	305	24	24	25	62	151	213	241	232	189	112	31	24	24	24	24	24	176	1400
	1100	314	26	26	26	28	76	156	208	227	212	165	87	28	26	26	26	26	207	1300
	1200	316	27	27	27	27	29	79	155	204	221	204	155	79	29	27	27	27	217	1200
	HALF DAY TOTALS		100	103	201	445	735	978	1080	1010	780	452	228	122	100	100	100	100	691	
Mar	0700	185	10	37	105	153	176	173	142	88	20	9	9	9	9	9	9	9	32	1700
	0800	260	17	25	107	183	227	237	209	150	62	18	17	17	17	17	17	17	100	1600
	0900	290	23	25	64	151	210	237	227	183	107	30	23	23	23	23	23	23	164	1500
	1000	304	28	28	30	87	158	202	215	195	144	70	29	28	28	28	28	28	211	1400
	1100	311	31	31	31	34	82	142	179	188	168	120	59	32	31	31	31	31	242	1300
	1200	313	32	32	32	32	33	66	122	162	176	162	122	66	33	32	32	32	252	1200
	HALF DAY TOTALS		124	162	359	629	875	1033	1041	888	589	326	193	136	125	124	124	124	874	
Apr	0600	66	9	35	54	65	66	56	38	12	4	3	3	3	3	3	3	3	7	1800
	0700	206	17	80	146	188	200	182	136	65	16	14	14	14	14	14	14	14	61	1700
	0800	255	23	61	144	200	227	219	177	107	30	22	22	22	22	22	22	22	129	1600
	0900	278	28	36	103	168	206	212	187	133	58	29	28	28	28	28	28	28	188	1500
	1000	290	32	34	52	108	155	177	172	141	87	39	33	32	32	32	32	32	233	1400
	1100	295	35	35	36	47	83	118	135	132	108	70	40	36	35	35	35	35	262	1300
	1200	297	36	36	36	37	38	53	82	106	115	106	82	53	38	37	36	36	271	1200
	HALF DAY TOTALS		161	296	550	792	952	992	889	645	360	228	177	157	153	152	152	152	1015	
May	0600	119	33	77	108	121	116	94	56	13	8	8	8	8	8	8	8	9	21	1800
	0700	211	36	111	170	202	204	174	118	42	19	18	18	18	18	18	18	19	81	1700
	0800	250	29	94	165	208	220	199	149	73	27	25	25	25	25	25	25	25	146	1600
	0900	269	33	61	128	177	198	190	155	93	37	32	31	31	31	31	31	31	201	1500
	1000	280	36	40	76	121	150	156	138	99	54	37	35	35	35	35	35	35	243	1400
	1100	285	38	39	42	59	83	99	102	90	68	47	40	39	37	37	37	37	269	1300
	1200	286	38	39	40	40	41	47	59	70	74	70	59	47	41	40	40	39	277	1200
	HALF DAY TOTALS		222	438	702	900	985	933	747	447	250	199	183	177	175	174	174	175	1098	
Jun	0600	131	44	92	123	135	127	99	55	12	10	10	10	10	10	10	10	11	28	1800
	0700	210	47	122	176	204	201	168	108	35	20	20	20	20	20	20	20	21	88	1700
	0800	245	36	106	171	208	214	189	135	60	28	27	27	27	27	27	27	27	151	1600
	0900	264	35	74	137	178	193	180	139	77	35	32	32	32	32	32	32	32	204	1500
	1000	274	38	47	86	125	146	145	123	83	45	38	36	36	36	36	36	36	244	1400
	1100	279	40	41	47	64	82	91	89	75	56	43	41	40	39	39	39	39	269	1300
	1200	280	41	41	41	42	42	46	52	58	60	58	52	46	42	42	41	41	276	1200
	HALF DAY TOTALS		261	504	762	935	985	897	678	372	225	197	189	185	184	184	183	186	1122	
Jul	0600	113	34	76	105	117	113	90	53	12	9	9	9	9	9	9	9	9	22	1800
	0700	203	38	111	167	198	198	169	114	41	20	19	19	19	19	19	19	19	81	1700
	0800	241	31	95	163	204	215	194	145	70	28	26	26	26	26	26	26	26	145	1600
	0900	261	34	64	129	175	195	186	150	90	37	32	32	32	32	32	32	32	198	1500
	1000	271	37	42	78	121	148	153	134	96	53	38	36	36	36	36	36	36	240	1400
	1100	277	39	40	43	60	83	98	99	88	66	47	41	40	38	38	38	38	265	1300
	1200	279	40	40	41	41	42	48	58	68	72	68	58	48	42	41	41	40	273	1200
	HALF DAY TOTALS		231	444	701	890	967	912	726	433	248	202	187	182	180	179	179	180	1088	
Aug	0600	59	10	33	50	60	60	51	34	11	4	4	4	4	4	4	4	4	8	1800
	0700	190	19	79	141	179	190	172	128	61	17	15	15	15	15	15	15	15	61	1700
	0800	240	25	63	141	195	219	210	170	102	31	23	23	23	23	23	23	23	128	1600
	0900	263	30	39	104	166	200	206	181	127	57	31	29	29	29	29	29	29	185	1500
	1000	276	34	36	55	109	153	173	167	136	84	40	35	34	34	34	34	34	229	1400
	1100	282	36	37	39	50	84	116	131	127	104	69	41	38	36	36	36	36	256	1300
	1200	284	37	37	37	39	40	54	81	103	111	103	81	54	40	39	37	37	265	1200
	HALF DAY TOTALS		171	303	546	774	922	955	854	618	352	231	184	166	162	161	160	160	999	
Sep	0700	163	10	35	96	139	159	156	128	80	20	10	10	10	10	10	10	10	31	1700
	0800	240	18	26	103	173	215	224	198	143	60	19	18	18	18	18	18	18	96	1600
	0900	272	24	26	64	146	202	227	218	177	105	31	24	24	24	24	24	24	158	1500
	1000	287	29	29	32	86	154	196	208	189	141	70	31	29	29	29	29	29	204	1400
	1100	294	32	32	32	36	81	139	174	182	163	118	59	34	32	32	32	32	234	1300
	1200	296	33	33	33	33	35	66	120	158	171	158	120	66	35	33	33	33	244	1200
	HALF DAY TOTALS		130	164	345	598	831	982	993	852	574	325	197	142	130	129	129	129	845	
Oct	0700	99	4	7	43	74	92	96	85	60	24	5	4	4	4	4	4	4	10	1700
	0800	229	13	15	63	143	195	217	206	162	90	17	13	13	13	13	13	13	63	1600
	0900	273	20	20	33	120	193	234	239	208	144	54	21	20	20	20	20	20	125	1500
	1000	293	24	24	26	62	147	207	234	225	183	109	32	24	24	24	24	24	173	1400
	1100	302	27	27	27	29	76	152	203	221	207	160	85	29	27	27	27	27	203	1300
	1200	304	28	28	28	28	30	78	151	199	215	199	151	78	30	28	28	28	213	1200
	HALF DAY TOTALS		103	106	200	433	708	941	1038	972	753	441	226	125	104	103	103	103	679	
Nov	0700	2	0	0	0	1	1	1	1	1	1	0	0	0	0	0	0	0	0	1700
	0800	196	9	9	29	103	156	184	184	155	100	27	9	9	9	9	9	9	32	1600
	0900	263	16	16	17	90	173	225	241	221	166	80	17	16	16	16	16	16	88	1500
	1000	289	20	20	21	41	134	206	245	246	209	138	45	21	20	20	20	20	136	1400
	1100	301	23	23	23	24	67	157	218	245	234	188	109	29	23	23	23	23	165	1300
	1200	304	24	24	24	24	25	87	171	224	243	224	171	87	25	24	24	24	175	1200
	HALF DAY TOTALS		80	81	108	282	561	841	996	995	838	544	261	113	81	80	80	80	509	
Dec	0800	176	7	7	19	84	135	163	166	143	97	31	7	7	7	7	7	7	22	1600
	0900	257	14	14	15	77	162	218	238	222	171	89	15	14	14	14	14	14	72	1500
	1000	288	18	18	18	34	127	204	246	251	216	148	52	19	18	18	18	18	119	1400
	1100	301	21	21	21	22	63	157	222	252	243	197	116	29	21	21	21	21	148	1300
	1200	304	22	22	22	22	23	89	177	232	252	232	177	89	23	22	22	22	158	1200
	HALF DAY TOTALS		71	71	84	227	500	792	965	986	852	578	275	107	71	71	71	71	440	
			N	NNW	NW	WNW	W	WSW	SW	SSW	S	SSE	SE	ESE	E	ENE	NE	NNE	Hor.	PM

Notes: 1. Clearness number = 1.00; Ground reflectance = 0.20. 2. Figures shown are for 21st day of each month.

Table 18 Solar Irradiance (E_{DN}) and Solar Heat Gain Factors (SHGF) for 40° North Latitude

Date	Solar Time	Direct Normal Btu/(h·ft²)	Solar Heat Gain Factors, Btu/(h·ft²)																	Solar Time
			N	NNE	NE	ENE	E	ESE	SE	SSE	S	SSW	SW	WSW	W	WNW	NW	NNW	Hor.	
Jan	0800	142	5	5	17	71	111	132	133	114	75	22	6	5	5	5	5	5	14	1600
	0900	239	12	12	13	74	154	205	224	209	160	82	13	12	12	12	12	12	55	1500
	1000	274	16	16	16	31	124	199	241	246	213	146	51	17	16	16	16	16	96	1400
	1100	289	19	19	19	20	61	156	222	252	244	198	118	28	19	19	19	19	124	1300
	1200	294	20	20	20	20	21	90	179	234	254	234	179	90	21	20	20	20	133	1200
	HALF DAY TOTALS		61	61	73	199	452	734	904	932	813	561	273	101	62	61	61	61	354	
Feb	0700	55	2	3	23	40	51	53	47	34	14	2	2	2	2	2	2	2	4	1700
	0800	219	10	11	50	129	183	206	199	160	94	18	10	10	10	10	10	10	43	1600
	0900	271	16	16	22	107	186	234	245	218	157	66	17	16	16	16	16	16	98	1500
	1000	294	21	21	21	49	143	211	246	243	203	129	38	21	21	21	21	21	143	1400
	1100	304	23	23	23	24	71	160	219	244	231	184	103	27	23	23	23	23	171	1300
	1200	307	24	24	24	24	25	86	170	222	241	222	170	86	25	24	24	24	180	1200
	HALF DAY TOTALS		84	86	152	361	648	916	1049	1015	821	508	250	114	85	84	84	84	548	
Mar	0700	171	9	29	93	140	163	161	135	86	22	8	8	8	8	8	8	8	26	1700
	0800	250	16	18	91	169	218	232	211	157	74	17	16	16	16	16	16	16	85	1600
	0900	282	21	22	47	136	203	238	236	198	128	40	22	21	21	21	21	21	143	1500
	1000	297	25	25	27	72	153	207	229	216	171	95	29	25	25	25	25	25	186	1400
	1100	305	28	28	28	30	78	151	198	213	197	150	77	30	28	28	28	28	213	1300
	1200	307	29	29	29	29	31	75	145	191	206	191	145	75	31	29	29	29	223	1200
	HALF DAY TOTALS		114	139	302	563	832	1035	1087	968	694	403	220	132	114	113	113	113	764	
Apr	0600	89	11	46	72	87	88	76	52	18	5	5	5	5	5	5	5	5	11	1800
	0700	206	16	71	140	185	201	186	143	75	16	14	14	14	14	14	14	14	61	1700
	0800	252	22	44	128	190	224	223	188	124	41	22	21	21	21	21	21	21	123	1600
	0900	274	27	29	80	155	202	219	203	156	83	29	27	27	27	27	27	27	177	1500
	1000	286	31	31	37	92	152	187	193	170	121	56	32	31	31	31	31	41	217	1400
	1100	292	33	33	34	39	81	130	160	166	146	102	52	35	33	33	33	33	243	1300
	1200	293	34	34	34	34	36	62	108	142	154	142	108	62	36	34	34	34	252	1200
	HALF DAY TOTALS		154	265	501	758	957	1051	994	782	488	296	199	157	148	147	147	147	957	
May	0500	1	0	1	1	1	1	1	0	0	0	0	0	0	0	0	0	0	0	1900
	0600	144	36	90	128	145	141	115	71	18	10	10	10	10	10	10	10	11	31	1800
	0700	216	28	102	165	202	209	184	131	54	20	19	19	19	19	19	19	19	87	1700
	0800	250	27	73	149	199	220	208	164	93	29	25	25	25	25	25	25	25	146	1600
	0900	267	31	42	105	164	197	200	175	121	53	32	30	30	30	30	30	30	195	1500
	1000	277	34	36	54	105	148	168	163	133	83	40	35	34	34	34	34	34	234	1400
	1100	283	36	36	38	48	81	113	130	127	105	70	42	38	36	36	36	36	257	1300
	1200	284	37	37	37	38	40	54	82	104	113	104	82	54	40	38	37	37	265	1200
	HALF DAY TOTALS		215	404	666	893	1024	1025	881	601	358	247	200	180	176	175	174	175	1083	
Jun	0500	22	10	17	21	22	20	14	6	2	1	1	1	1	1	1	1	2	3	1900
	0600	155	48	104	143	159	151	121	70	17	13	13	13	13	13	13	13	14	40	1800
	0700	216	37	113	172	205	207	178	122	46	22	21	21	21	21	21	21	21	97	1700
	0800	246	30	85	156	201	216	199	152	80	29	27	27	27	27	27	27	27	153	1600
	0900	263	33	51	114	166	192	190	161	105	45	33	32	32	32	32	32	32	201	1500
	1000	272	35	38	63	109	145	158	148	116	69	39	36	35	35	35	35	35	238	1400
	1100	277	38	39	40	52	81	105	116	110	88	60	41	39	38	38	38	38	260	1300
	1200	279	38	38	38	40	41	52	72	89	95	89	72	52	41	40	38	38	267	1200
	HALF DAY TOTALS		253	470	734	941	1038	999	818	523	315	236	204	191	188	187	186	188	1126	
Jul	0500	2	1	2	2	2	2	1	1	0	0	0	0	0	0	0	0	0	0	1900
	0600	138	37	89	125	142	137	112	68	18	11	11	11	11	11	11	11	12	32	1800
	0700	208	30	102	163	198	204	179	127	53	21	20	20	20	20	20	20	20	88	1700
	0800	241	28	75	148	196	216	203	160	90	30	26	26	26	26	26	26	26	145	1600
	0900	259	32	44	106	163	193	196	170	118	52	33	31	31	31	31	31	31	194	1500
	1000	269	35	37	56	106	146	165	159	129	81	41	36	35	35	35	35	35	231	1400
	1100	275	37	38	40	50	81	111	127	123	102	69	43	39	37	37	37	37	254	1300
	1200	276	38	38	38	40	41	55	80	101	109	101	80	55	41	40	38	38	262	1200
	HALF DAY TOTALS		223	411	666	885	1008	1003	858	584	352	248	204	186	181	180	180	181	1076	
Aug	0600	81	12	44	68	81	82	71	48	17	6	5	5	5	5	5	5	5	12	1800
	0700	191	17	71	135	177	191	177	135	70	17	16	16	16	16	16	16	16	62	1700
	0800	237	24	47	126	185	216	214	180	118	41	23	23	23	23	23	23	23	122	1600
	0900	260	28	31	82	153	197	212	196	151	80	31	28	28	28	28	28	28	174	1500
	1000	272	32	33	40	93	150	182	187	165	116	56	34	32	32	32	32	32	214	1400
	1100	278	35	35	36	41	81	128	156	160	141	99	52	37	35	35	35	35	239	1300
	1200	280	35	35	35	36	38	63	106	138	149	138	106	63	38	36	35	35	247	1200
	HALF DAY TOTALS		164	273	498	741	928	1013	956	751	474	296	205	166	157	156	156	156	946	
Sep	0700	149	9	27	84	125	146	144	121	77	21	9	9	9	9	9	9	9	25	1700
	0800	230	17	19	87	160	205	218	199	148	71	18	17	17	17	17	17	17	82	1600
	0900	263	22	23	47	131	194	227	226	190	124	41	23	22	22	22	22	22	138	1500
	1000	280	27	27	28	71	148	200	221	209	165	93	30	27	27	27	27	27	180	1400
	1100	287	29	29	29	31	78	147	192	207	191	146	77	31	29	29	29	29	206	1300
	1200	290	30	30	30	30	32	75	142	185	200	185	142	75	32	30	30	30	215	1200
	HALF DAY TOTALS		119	142	291	534	787	980	1033	925	672	396	222	137	119	118	118	118	738	
Oct	0700	48	2	3	20	36	45	47	42	30	12	2	2	2	2	2	2	2	4	1700
	0800	204	11	12	49	123	173	195	188	151	89	18	11	11	11	11	11	11	43	1600
	0900	257	17	17	23	104	180	225	235	209	151	64	18	17	17	17	17	17	97	1500
	1000	280	21	21	22	50	139	205	238	235	196	125	38	22	21	21	21	21	140	1400
	1100	291	24	24	24	25	71	156	212	236	224	178	101	28	24	24	24	24	168	1300
	1200	294	25	25	25	25	27	85	165	216	234	216	165	85	27	25	25	25	177	1200
	HALF DAY TOTALS		88	89	152	351	623	878	1006	974	791	493	247	117	89	88	88	88	540	
Nov	0800	136	5	5	18	69	108	128	129	110	72	21	6	5	5	5	5	5	14	1600
	0900	232	12	12	13	73	151	201	219	204	156	80	13	12	12	12	12	12	55	1500
	1000	268	16	16	16	31	122	196	237	242	209	143	50	17	16	16	16	16	96	1400
	1100	283	19	19	19	20	61	154	218	248	240	194	116	28	19	19	19	19	123	1300
	1200	288	20	20	20	20	21	89	176	231	250	231	176	89	21	20	20	20	132	1200
	HALF DAY TOTALS		63	63	75	198	445	721	887	914	798	551	269	101	63	63	63	63	354	
Dec	0800	89	3	3	8	41	67	82	84	73	50	17	3	3	3	3	3	3	6	1600
	0900	217	10	10	11	60	135	185	205	194	151	83	13	10	10	10	10	10	39	1500
	1000	261	14	14	14	25	113	188	232	239	210	146	55	15	14	14	14	14	77	1400
	1100	280	17	17	17	17	56	151	217	249	242	198	120	28	17	17	17	17	104	1300
	1200	285	18	18	18	18	19	89	178	233	253	233	178	89	19	18	18	18	113	1200
	HALF DAY TOTALS		52	52	56	146	374	649	822	867	775	557	276	94	53	52	52	52	282	
			N	NNW	NW	WNW	W	WSW	SW	SSW	S	SSE	SE	ESE	E	ENE	NE	NNE	Hor.	PM

Notes: 1. Clearness number = 1.00; Ground reflectance = 0.20. 2. Figures shown are for 21st day of each month.

Table 19 Solar Irradiance (E_{DN}) and Solar Heat Gain Factors (SHGF) for 48° North Latitude

Date	Solar Time	Direct Normal Btu/(h·ft²)	Solar Heat Gain Factors, Btu/(h·ft²)																	Solar Time
			N	NNE	NE	ENE	E	ESE	SE	SSE	S	SSW	SW	WSW	W	WNW	NW	NNW	Hor.	
Jan	0800	37	1	1	4	18	29	34	35	30	20	6	1	1	1	1	1	1	2	1600
	0900	185	8	8	8	53	118	160	176	166	129	69	10	8	8	8	8	8	25	1500
	1000	239	12	12	12	22	106	175	216	223	195	136	50	12	12	12	12	12	55	1400
	1100	261	14	14	14	15	53	144	208	239	233	190	116	26	14	14	14	14	77	1300
	1200	267	15	15	15	15	16	86	171	226	245	226	171	86	16	15	15	15	85	1200
	HALF DAY TOTALS		43	43	46	117	316	567	729	776	701	512	259	85	43	43	43	43	203	
Feb	0700	4	0	0	1	3	3	3	3	2	1	0	0	0	0	0	0	0	0	1700
	0800	180	8	8	36	103	149	170	166	136	82	17	8	8	8	8	8	8	25	1600
	0900	247	13	13	16	90	168	216	230	209	155	71	14	13	13	13	13	13	66	1500
	1000	275	17	17	17	38	131	203	242	244	207	138	44	18	17	17	17	17	105	1400
	1100	288	19	19	19	20	65	158	221	249	239	192	113	27	19	19	19	19	130	1300
	1200	292	20	20	20	20	22	89	176	231	250	231	176	89	22	20	20	20	138	1200
	HALF DAY TOTALS		68	68	107	274	541	816	968	967	813	531	261	104	68	68	68	68	395	
Mar	0700	153	7	22	80	123	145	145	123	80	23	7	7	7	7	7	7	7	20	1700
	0800	236	14	15	76	154	204	222	206	158	82	15	14	14	14	14	14	14	68	1600
	0900	270	19	19	3	121	193	234	239	207	142	52	20	19	19	19	19	19	118	1500
	1000	287	23	23	24	58	146	208	237	231	189	115	33	23	23	23	23	23	156	1400
	1100	295	25	25	25	26	74	156	210	232	218	172	94	28	25	25	25	25	180	1300
	1200	298	26	26	26	26	27	83	161	211	228	211	161	83	27	26	26	26	188	1200
	HALF DAY TOTALS		100	118	250	494	775	1012	1100	1014	767	465	244	126	101	100	100	100	636	
Apr	0600	108	12	53	86	105	107	93	64	23	6	6	6	6	6	6	6	6	15	1800
	0700	205	15	61	132	180	199	189	148	84	18	14	14	14	14	14	14	14	60	1700
	0800	247	20	32	111	179	219	225	196	138	55	21	20	20	20	20	20	20	114	1600
	0900	268	25	26	60	141	197	223	215	176	106	33	25	25	25	25	25	25	161	1500
	1000	280	28	28	31	77	148	193	209	194	150	80	31	28	28	28	28	28	196	1400
	1100	286	31	31	31	33	78	140	181	193	177	133	69	33	31	31	31	31	218	1300
	1200	288	31	31	31	31	34	71	131	172	186	172	131	71	34	31	31	31	226	1200
	HALF DAY TOTALS		147	242	461	724	957	1098	1081	895	605	370	226	156	141	140	140	140	875	
May	0500	41	17	31	40	42	39	29	14	3	3	3	3	3	3	3	3	3	5	1900
	0600	162	35	97	141	162	160	133	85	24	12	12	12	12	12	12	12	13	40	1800
	0700	219	23	90	158	200	212	191	142	68	21	19	19	19	19	19	19	19	91	1700
	0800	248	26	54	132	190	218	214	178	113	38	25	25	25	25	25	25	25	142	1600
	0900	264	29	32	82	151	194	208	192	147	77	32	29	29	29	29	29	29	185	1500
	1000	274	33	34	39	90	145	178	184	163	116	57	35	33	33	33	33	33	219	1400
	1100	279	35	35	36	40	79	126	155	160	142	101	54	37	35	35	35	35	240	1300
	1200	280	35	35	35	36	38	63	107	139	150	139	107	63	38	36	35	35	247	1200
	HALF DAY TOTALS		215	388	645	893	1065	1114	1007	749	483	316	225	184	174	173	173	174	1045	
Jun	0500	77	35	61	76	80	72	53	24	6	5	5	5	5	5	5	5	8	12	1900
	0600	172	46	110	155	175	169	138	84	22	14	14	14	14	14	14	14	16	51	1800
	0700	220	29	101	165	204	211	187	135	60	23	21	21	21	21	21	21	21	103	1700
	0800	246	29	64	139	191	215	206	168	101	34	27	27	27	27	27	27	27	152	1600
	0900	261	31	36	91	153	190	199	180	133	66	33	31	31	31	31	31	31	193	1500
	1000	269	34	36	45	94	143	169	171	148	101	50	36	34	34	34	34	34	225	1400
	1100	274	36	36	38	44	79	118	142	145	126	88	49	38	36	36	36	36	246	1300
	1200	275	37	37	37	38	40	60	96	124	134	124	96	60	40	38	37	37	252	1200
	HALF DAY TOTALS		257	459	722	955	1095	1102	955	678	436	299	228	197	189	188	188	191	1108	
Jul	0500	43	18	33	42	45	41	30	15	3	3	3	3	3	3	3	3	4	6	1900
	0600	156	37	96	138	159	156	129	82	24	13	13	13	13	13	13	13	14	41	1800
	0700	211	25	90	156	196	207	186	138	66	22	20	20	20	20	20	20	20	92	1700
	0800	240	27	56	132	187	214	209	174	110	38	26	26	26	26	26	26	26	142	1600
	0900	256	30	34	83	149	191	204	187	143	75	33	30	30	30	30	30	30	184	1500
	1000	266	34	35	41	90	143	174	180	158	113	56	36	34	34	34	34	34	217	1400
	1100	271	36	36	37	42	79	124	151	156	138	99	54	38	36	36	36	36	237	1300
	1200	272	36	36	36	37	39	63	104	136	146	136	104	63	39	37	36	36	244	1200
	HALF DAY TOTALS		223	395	646	886	1050	1092	983	730	474	315	229	190	181	179	179	180	1042	
Aug	0600	99	13	51	81	98	100	87	60	22	7	7	7	7	7	7	7	7	16	1800
	0700	190	17	61	128	172	190	179	141	79	19	15	15	15	15	15	15	15	61	1700
	0800	232	22	34	110	174	211	216	188	132	53	23	22	22	22	22	22	22	114	1600
	0900	154	27	28	63	139	192	216	108	169	102	34	27	27	27	27	27	27	159	1500
	1000	266	30	30	33	78	145	188	203	188	144	78	33	30	30	30	30	30	193	1400
	1100	272	32	32	32	36	78	137	175	187	171	129	68	35	32	32	32	32	215	1300
	1200	274	33	33	33	33	36	71	128	167	189	167	128	71	36	33	33	33	223	1200
	HALF DAY TOTALS		157	251	459	709	929	1060	1040	862	587	366	231	165	151	149	149	149	869	
Sep	0700	131	8	21	71	108	128	128	108	71	21	8	7	7	7	7	7	7	20	1700
	0800	215	15	16	72	144	191	207	193	148	77	16	15	15	15	15	15	15	65	1600
	0900	251	20	20	34	116	184	223	227	197	136	52	21	20	20	20	20	20	114	1500
	1000	269	24	24	25	58	141	200	228	221	182	112	34	24	24	24	24	24	151	1400
	1100	278	26	26	26	28	73	151	203	223	210	166	92	29	26	26	26	26	174	1300
	1200	280	27	27	27	27	29	82	156	204	220	204	156	82	29	27	27	27	182	1200
	HALF DAY TOTALS		105	121	240	465	729	953	1040	963	737	453	243	131	106	105	105	105	614	
Oct	0700	4	0	0	2	3	4	4	3	2	1	0	0	0	0	0	0	0	0	1700
	0800	165	8	9	35	96	139	159	155	126	77	16	8	8	8	8	8	8	25	1600
	0900	233	14	14	16	88	161	207	220	199	148	68	15	14	14	14	14	14	66	1500
	1000	262	18	18	18	39	128	196	233	234	199	133	43	18	18	18	18	18	104	1400
	1100	274	20	20	20	21	64	153	213	241	231	186	109	27	20	20	20	20	128	1300
	1200	278	21	21	21	21	23	87	171	223	242	223	171	87	23	21	21	21	136	1200
	HALF DAY TOTALS		71	71	108	266	519	780	925	925	779	513	256	106	72	71	71	71	391	
Nov	0800	36	1	1	4	18	29	34	35	30	20	6	1	1	1	1	1	1	2	1600
	0900	179	8	8	9	52	115	156	171	161	125	67	10	8	8	8	8	8	26	1500
	1000	233	12	12	12	22	104	172	212	218	191	133	49	13	12	12	12	12	55	1400
	1100	255	15	15	15	15	52	142	204	234	228	186	114	26	15	15	15	15	77	1300
	1200	261	15	15	15	15	17	85	168	222	240	222	168	85	17	15	15	15	85	1200
	HALF DAY TOTALS		44	44	47	117	310	555	713	760	686	502	255	85	44	44	44	44	204	
Dec	0900	140	5	5	6	36	86	120	133	127	100	56	8	5	5	5	5	5	13	1500
	1000	214	10	10	10	16	91	156	194	201	179	126	49	10	10	10	10	10	38	1400
	1100	242	12	12	12	13	46	134	195	225	220	180	111	25	12	12	12	12	57	1300
	1200	250	13	13	13	13	14	81	163	215	233	215	168	81	14	13	13	13	65	1200
	HALF DAY TOTALS		33	33	34	73	233	458	610	665	616	468	247	76	34	33	33	33	141	
			N	NNW	NW	WNW	W	WSW	SW	SSW	S	SSE	SE	ESE	E	ENE	NE	NNE	Hor.	PM

Notes: 1. Clearness number = 1.00; Ground reflectance = 0.20. 2. Figures shown are for 21st day of each month.

Table 20 Solar Irradiance (E_{DN}) and Solar Heat Gain Factors (SHGF) for 56° North Latitude

Date	Solar Time	Direct Normal Btu/(h·ft²)	Solar Heat Gain Factors, Btu/(h·ft²) N	NNE	NE	ENE	E	ESE	SE	SSE	S	SSW	SW	WSW	W	WNW	NW	NNW	Hor.	Solar Time
Jan	0900	78	3	3	3	21	49	67	74	70	55	30	4	3	3	3	3	3	5	1500
	1000	170	7	7	7	13	74	126	156	162	143	100	38	7	7	7	7	7	21	1400
	1100	207	9	9	9	10	40	116	169	194	190	156	96	21	9	9	9	9	34	1300
	1200	217	10	10	10	10	11	71	144	190	205	190	144	71	11	10	10	10	40	1200
	HALF DAY TOTALS		23	23	24	46	163	343	468	517	487	378	206	61	24	23	23	23	80	
Feb	0800	115	4	4	21	64	95	109	107	88	55	12	4	4	4	4	4	4	10	1600
	0900	203	10	10	11	71	139	183	197	182	136	66	10	10	10	10	10	10	36	1500
	1000	246	13	13	13	28	115	184	223	227	196	133	45	14	13	13	13	13	65	1400
	1100	262	15	15	15	16	57	148	210	239	232	188	112	25	15	15	15	15	84	1300
	1200	267	16	16	16	16	17	86	171	225	244	225	171	86	17	16	16	16	91	1200
	HALF DAY TOTALS		49	50	66	182	409	666	821	846	737	509	253	89	50	49	49	49	241	
Mar	0700	128	6	16	65	101	121	122	105	70	21	6	6	6	6	6	6	6	14	1700
	0800	215	12	13	61	136	185	205	194	152	84	15	12	12	12	12	12	12	49	1600
	0900	253	16	16	23	105	179	224	233	207	148	61	17	16	16	16	16	16	89	1500
	1000	272	19	19	20	46	136	203	238	236	198	128	39	20	19	19	19	19	122	1400
	1100	282	21	21	21	22	68	156	215	241	230	184	106	27	21	21	21	21	142	1300
	1200	284	22	22	22	22	24	86	170	222	241	222	170	86	24	22	22	22	149	1200
	HALF DAY TOTALS		85	97	200	419	699	956	1071	1016	800	502	258	118	86	85	85	85	491	
Apr	0600	122	13	58	95	118	121	107	75	29	7	7	7	7	7	7	7	7	18	1800
	0700	201	15	51	123	173	195	188	152	91	21	14	14	14	14	14	14	14	56	1700
	0800	239	19	23	95	167	211	223	201	148	68	20	19	19	19	19	19	19	101	1600
	0900	260	23	24	44	126	190	223	223	189	126	44	24	23	23	23	23	23	140	1500
	1000	272	26	26	27	63	142	196	220	212	171	102	33	26	26	26	26	26	170	1400
	1100	278	28	28	28	30	74	147	195	213	200	156	86	31	28	28	28	28	189	1300
	1200	280	28	28	28	28	31	79	149	194	210	194	149	79	31	28	28	28	195	1200
	HALF DAY TOTALS		139	226	430	694	951	1132	1147	982	699	437	252	154	132	131	131	131	772	
May	0500	93	36	68	89	95	88	66	33	7	6	6	6	6	6	6	6	7	14	1900
	0600	175	33	99	148	174	173	147	97	31	14	14	14	14	14	14	14	14	48	1800
	0700	219	21	77	149	195	212	197	152	81	22	19	19	19	19	19	19	19	92	1700
	0800	244	25	38	115	179	215	218	189	131	52	25	24	24	24	24	24	24	135	1600
	0900	259	28	30	62	136	189	213	206	168	102	36	28	28	28	28	28	28	171	1500
	1000	268	31	31	33	75	141	185	200	187	145	80	33	31	31	31	31	31	199	1400
	1100	273	32	32	32	35	76	135	174	187	172	131	71	35	32	32	32	32	216	1300
	1200	275	33	33	33	33	36	71	129	168	181	168	129	71	36	33	33	33	222	1200
	HALF DAY TOTALS		222	391	644	906	1112	1202	1120	878	604	392	256	187	172	170	170	173	986	
Jun	0400	21	13	19	22	21	18	11	3	1	1	1	1	1	1	1	2	5	3	2000
	0500	122	53	94	119	126	115	85	40	10	9	9	9	9	9	9	9	12	25	1900
	0600	185	42	111	160	185	182	152	97	30	16	16	16	16	16	16	16	17	62	1800
	0700	222	25	86	156	199	213	195	147	74	24	22	22	22	22	22	22	22	105	1700
	0800	243	27	46	122	181	213	213	181	122	46	27	26	26	26	26	26	26	146	1600
	0900	257	30	32	69	139	187	206	196	156	91	34	30	30	30	30	30	30	181	1500
	1000	265	33	33	36	79	139	178	190	174	132	71	35	33	33	33	33	33	208	1400
	1100	269	34	34	35	38	76	129	164	174	159	119	65	37	34	34	34	34	225	1300
	1200	271	35	35	35	35	38	68	119	155	168	155	119	68	38	35	35	35	231	1200
	HALF DAY TOTALS		275	473	738	989	1162	1207	1082	822	562	376	260	203	190	189	189	196	1070	
Jul	0500	91	37	69	89	95	88	66	33	8	7	7	7	7	7	7	7	8	16	1900
	0600	169	34	98	145	170	170	143	95	31	15	14	14	14	14	14	14	15	50	1800
	0700	212	23	77	147	192	208	193	148	79	23	20	20	20	20	20	20	20	93	1700
	0800	237	26	40	115	177	211	214	185	128	51	26	25	25	25	25	25	25	135	1600
	0900	252	29	31	63	135	186	209	201	164	99	36	29	29	29	29	29	29	171	1500
	1000	261	32	32	34	76	139	181	196	182	142	78	35	32	32	32	32	32	198	1400
	1100	265	33	33	33	37	76	133	171	183	168	128	70	36	33	33	33	33	215	1300
	1200	267	34	34	34	34	37	71	126	164	177	164	126	71	37	34	34	34	221	1200
	HALF DAY TOTALS		231	398	646	901	1097	1180	1096	859	593	390	259	193	179	177	177	180	987	
Aug	0500	1	0	1	1	1	1	1	0	0	0	0	0	0	0	0	0	0	0	1900
	0600	112	14	56	91	111	114	101	71	28	8	8	8	8	8	8	8	8	20	1800
	0700	187	16	51	119	165	186	179	144	86	22	15	15	15	15	15	15	15	58	1700
	0800	225	20	25	94	162	203	214	192	142	66	22	20	20	20	20	20	20	101	1600
	0900	246	25	26	46	124	184	216	215	182	121	44	26	25	25	25	25	25	140	1500
	1000	258	28	28	30	65	139	191	213	204	165	99	34	28	28	28	28	28	169	1400
	1100	264	30	30	30	32	74	143	189	206	193	152	84	33	30	30	30	30	187	1300
	1200	266	30	30	30	30	30	78	145	188	203	188	145	78	33	30	30	30	198	1200
	HALF DAY TOTALS		149	235	429	680	923	1092	1104	946	678	431	256	163	142	140	140	141	771	
Sep	0700	107	6	15	56	87	104	105	90	60	19	6	6	6	6	6	6	6	14	1700
	0800	194	12	14	58	126	171	189	179	140	78	16	12	12	12	12	12	12	48	1600
	0900	233	17	17	24	100	170	211	220	195	140	59	18	17	17	17	17	17	86	1500
	1000	253	20	20	21	46	131	194	227	225	189	123	39	21	20	20	20	20	118	1400
	1100	263	22	22	22	24	67	150	206	230	220	176	103	28	22	22	22	22	137	1300
	1200	266	23	23	23	23	25	85	163	213	231	213	163	85	25	23	23	23	144	1200
	HALF DAY TOTALS		89	99	191	391	652	893	1004	958	761	484	255	121	90	89	89	89	474	
Oct	0800	104	4	5	20	59	87	100	98	81	50	11	4	4	4	4	4	4	10	1600
	0900	193	10	10	11	68	132	173	186	171	129	63	11	10	10	10	10	10	37	1500
	1000	231	14	14	14	28	111	176	213	216	186	127	44	14	14	14	14	14	64	1400
	1100	248	16	16	16	17	56	142	202	229	222	180	108	25	16	16	16	16	84	1300
	1200	253	16	16	16	16	18	83	164	216	234	216	164	83	18	16	16	16	91	1200
	HALF DAY TOTALS		52	52	68	177	390	633	779	804	702	487	246	90	53	52	52	52	240	
Nov	0900	76	3	3	3	21	48	66	72	69	54	29	4	3	3	3	3	3	6	1500
	1000	165	7	7	7	13	72	122	152	157	139	98	37	7	7	7	7	7	21	1400
	1100	201	9	9	9	10	39	113	165	190	186	152	94	21	9	9	9	9	35	1300
	1200	211	10	10	10	10	11	70	140	186	200	186	140	70	11	10	10	10	40	1200
	HALF DAY TOTALS		24	24	24	47	161	336	457	505	475	369	202	61	24	24	24	24	81	
Dec	0900	5	0	0	0	1	3	4	5	5	4	2	0	0	0	0	0	0	0	1500
	1000	113	4	4	4	7	47	82	103	107	96	68	27	4	4	4	4	4	9	1400
	1100	166	6	6	6	7	30	92	135	156	154	127	78	17	6	6	6	6	19	1300
	1200	180	7	7	7	7	8	59	120	159	171	159	120	59	8	7	7	7	23	1200
	HALF DAY TOTALS		14	14	14	20	88	217	311	354	343	277	163	47	15	14	14	14	40	
			N	NNW	NW	WNW	W	WSW	SW	SSW	S	SSE	SE	ESE	E	ENE	NE	NNE	Hor.	PM

Notes: 1. Clearness number = 1.00; Ground reflectance = 0.20. 2. Figures shown are for 21st day of each month.

Table 21 Solar Irradiance (E_{DN}) and Solar Heat Gain Factors (SHGF) for 64° North Latitude

Date	Solar Time	Direct Normal Btu/(h·ft²)	N	NNE	NE	ENE	E	ESE	SE	SSE	S	SSW	SW	WSW	W	WNW	NW	NNW	Hor.	Solar Time
			Solar Heat Gain Factors, Btu/(h·ft²)																	
Jan	1000	22	1	1	1	1	9	16	20	21	19	13	5	1	1	1	1	1	1	1400
	1100	81	3	3	3	3	15	45	67	77	75	62	38	8	3	3	3	3	6	1300
	1200	100	3	3	3	3	4	33	67	89	96	89	67	33	4	3	3	3	8	1200
	HALF DAY TOTALS		5	5	5	6	25	79	121	142	141	119	75	23		5	5	5	5	11
Feb	0800	18	1	1	3	10	15	17	17	14	9	2	1	1	1	1	1	1	1	1600
	0900	134	5	5	6	43	89	118	128	119	90	45	6	5	5	5	5	5	13	1500
	1000	190	8	8	8	18	87	144	176	180	157	108	38	9	8	8	8	8	28	1400
	1100	215	10	10	10	11	44	122	177	202	197	160	97	20	10	10	10	10	41	1300
	1200	222	11	11	11	11	12	73	147	194	210	194	147	73	12	11	11	11	45	1200
	HALF DAY TOTALS		29	30	33	89	244	446	578	617	560	411	212	66	30	29	29	29	106	
Mar	0700	95	4	11	47	74	90	91	79	53	17	4	4	4	4	4	4	4	9	1700
	0800	185	9	10	46	113	158	177	170	135	78	14	9	9	9	9	9	9	32	1600
	0900	227	13	13	16	88	159	203	215	194	143	64	14	13	13	13	13	13	59	1500
	1000	249	16	16	16	35	122	190	226	228	194	130	42	16	16	16	16	16	84	1400
	1100	260	17	17	17	18	60	148	209	236	228	184	109	25	17	17	17	17	99	1300
	1200	263	18	18	18	18	19	85	168	221	239	221	168	85	19	18	18	18	105	1200
	HALF DAY TOTALS		68	74	150	334	596	854	984	958	779	504	257	104	68	68	68	68	335	
Apr	0500	27	8	18	24	27	26	20	12	2	1	1	1	1	1	1	1	1	2	1900
	0600	133	12	59	102	127	132	118	84	35	8	8	8	8	8	8	8	8	21	1800
	0700	194	14	41	113	163	189	185	153	96	25	13	13	13	13	13	13	13	51	1700
	0800	228	17	19	79	153	201	217	201	153	79	19	17	17	17	17	17	17	85	1600
	0900	248	21	21	32	111	180	219	225	197	138	55	22	21	21	21	21	21	116	1500
	1000	260	23	23	24	51	134	194	225	221	185	118	38	24	23	23	23	23	140	1400
	1100	266	24	24	24	26	68	148	202	225	214	171	99	29	24	24	24	24	155	1300
	1200	268	25	25	25	25	27	83	159	208	224	208	159	83	27	25	25	25	160	1200
	HALF DAY TOTALS		131	218	410	671	943	1150	1186	1036	763	487	273	149	121	120	120	120	651	
May	0400	51	30	44	51	51	43	28	8	3	3	3	3	3	3	3	3	10	6	2000
	0500	132	48	95	125	135	125	96	50	11	9	9	9	9	9	9	9	11	26	1900
	0600	185	28	97	150	181	183	158	109	40	15	15	15	15	15	15	15	15	55	1800
	0700	218	21	63	138	189	211	201	161	94	24	19	19	19	19	19	19	19	90	1700
	0800	239	23	28	97	167	209	220	198	146	68	25	23	23	23	23	23	23	124	1600
	0900	252	26	27	45	122	183	215	215	184	123	46	27	26	26	26	26	26	152	1500
	1000	261	28	28	30	61	135	188	212	205	167	102	36	28	28	28	28	28	174	1400
	1100	265	30	30	30	32	72	141	188	207	195	154	87	33	30	30	30	30	188	1300
	1200	267	30	30	30	30	33	78	146	189	204	189	146	78	33	30	30	30	192	1200
	HALF DAY TOTALS		247	425	680	950	1177	1291	1218	985	708	465	288	191	169	168	168	176	911	
Jun	0300	21	17	21	22	20	14	6	2	1	1	1	1	1	1	1	2	10	3	2100
	0400	93	53	83	96	94	78	50	14	7	7	7	7	7	7	7	7	21	16	2000
	0500	154	62	114	148	158	145	110	55	14	12	12	12	12	12	12	12	14	39	1900
	0600	194	36	107	162	191	192	163	110	39	18	17	17	17	17	17	17	18	71	1800
	0700	221	24	71	145	193	213	200	158	89	25	22	22	22	22	22	22	22	105	1700
	0800	239	25	33	104	170	208	216	192	139	62	27	25	25	25	25	25	25	137	1600
	0900	251	28	29	51	124	181	210	208	175	115	43	29	28	28	28	28	28	165	1500
	1000	258	30	30	32	65	134	183	204	195	157	94	36	30	30	30	30	30	186	1400
	1100	262	32	32	32	34	72	137	180	196	184	144	82	35	32	32	32	32	199	1300
	1200	263	32	32	32	32	35	76	138	179	193	179	138	76	35	32	32	32	203	1200
	HALF DAY TOTALS		326	538	806	1066	1256	1318	1195	947	679	455	297	212	192	191	192	216	1021	
Jul	0400	53	32	47	55	54	46	29	9	4	4	4	4	4	4	4	4	11	8	2000
	0500	128	49	94	123	133	124	95	50	11	10	10	10	10	10	10	10	11	28	1900
	0600	179	30	96	148	177	180	155	106	39	16	15	15	15	15	15	15	15	57	1800
	0700	211	22	64	137	186	207	197	157	92	25	20	20	20	20	20	20	20	92	1700
	0800	231	24	30	97	165	205	215	193	142	67	26	24	24	24	24	24	24	124	1600
	0900	245	27	28	47	121	180	211	211	179	120	46	28	27	27	27	27	27	152	1500
	1000	253	29	29	31	62	134	185	208	200	164	100	37	29	29	29	29	29	174	1400
	1100	257	31	31	31	33	72	139	185	202	191	151	86	34	31	31	31	31	187	1300
	1200	259	31	31	31	31	34	78	143	185	200	185	143	78	34	31	31	31	192	1200
	HALF DAY TOTALS		258	434	684	946	1163	1269	1193	965	697	462	292	198	177	175	175	185	918	
Aug	0500	29	9	20	27	30	28	22	13	2	2	2	2	2	2	2	2	2	3	1900
	0600	123	13	58	97	121	125	111	80	34	9	9	9	9	9	9	9	9	23	1800
	0700	181	15	42	109	157	180	176	145	92	26	14	14	14	14	14	14	14	53	1700
	0800	214	19	21	78	148	193	208	192	147	76	21	19	19	19	19	19	19	87	1600
	0900	234	22	22	34	109	174	211	217	189	133	55	23	22	22	22	22	22	117	1500
	1000	246	25	25	26	52	131	188	217	214	178	114	39	25	25	25	25	25	140	1400
	1100	252	26	26	26	28	69	144	196	217	207	166	97	31	26	26	26	26	154	1300
	1200	254	27	27	27	27	29	82	155	201	217	201	155	82	29	27	27	27	159	1200
	HALF DAY TOTALS		142	226	410	657	914	1109	1141	997	740	478	275	158	131	130	130	130	656	
Sep	0700	77	4	10	39	62	74	75	65	44	15	4	4	4	4	4	4	4	8	1700
	0800	163	10	10	43	103	143	160	154	123	71	14	10	10	10	10	10	10	31	1600
	0900	206	14	14	17	83	148	189	200	181	133	61	15	14	14	14	14	14	57	1500
	1000	229	16	16	17	35	116	179	213	214	183	123	41	17	16	16	16	16	81	1400
	1100	240	18	18	18	19	59	141	198	224	216	174	104	26	18	18	18	18	96	1300
	1200	244	19	19	19	19	21	82	160	209	227	209	160	82	21	19	19	19	101	1200
	HALF DAY TOTALS		71	77	142	307	547	787	910	891	731	480	249	106	72	71	71	71	324	
Oct	0800	17	1	1	3	10	14	16	16	13	8	2	1	1	1	1	1	1	1	1600
	0900	122	5	5	6	40	82	109	118	110	83	42	6	5	5	5	5	5	13	1500
	1000	176	9	9	9	18	83	135	165	169	147	102	36	9	9	9	9	9	29	1400
	1100	201	11	11	11	11	43	116	167	191	186	152	92	20	11	11	11	11	41	1300
	1200	208	11	11	11	11	13	70	140	184	199	184	140	70	13	11	11	11	46	1200
	HALF DAY TOTALS		31	31	34	86	231	420	542	580	527	388	202	66	32	31	31	31	108	
Nov	1000	23	1	1	1	1	10	17	21	22	20	14	5	1	1	1	1	1	1	1400
	1100	79	3	3	3	3	15	44	65	75	74	61	37	8	3	3	3	3	6	1300
	1200	97	4	4	4	4	4	32	66	87	93	87	66	32	4	4	4	4	8	1200
	HALF DAY TOTALS		5	5	5	6	26	79	120	141	140	117	74	23	6	5	5	5	11	
Dec	1100	4	0	0	0	0	1	2	3	4	4	3	2	0	0	0	0	0	0	1300
	1200	16	0	0	0	0	1	5	11	14	15	14	11	5	1	0	0	0	1	1200
	HALF DAY TOTALS		0	0	0	0	1	5	9	11	11	10	7	3	0	0	0	0	1	
			N	NNW	NW	WNW	W	WSW	SW	SSW	S	SSE	SE	ESE	E	ENE	NE	NNE	Hor.	PM

Notes: 1. Clearness number = 1.00; Ground reflectance = 0.20. 2. Figures shown are for 21st day of each month.

Example 9. Find the total heat gain through the center of a double-glazed window consisting of an outdoor light of 0.25-in. clear glass, 0.5-in. air space, and an indoor light of 0.25-in. clear glass, with a reflective film on the No. 3 surface in a west wall at 40° North latitude, at 1600 h on July 21. Outdoor air temperature is 105°F, and indoor air is 75°F. Solar irradiance is 248.3 Btu/h·ft². The spectrally averaged data for the glass are:

$$\text{Outdoor } \tau = 0.80;\ \rho_1 = 0.07;\ \rho_2 = 0.07;\ \alpha_1 = 0.13;\ \alpha_2 = 0.13;$$
$$e_1 = 0.84;\ e_2 = 0.84$$
$$\text{Indoor } \tau_i = 0.12;\ \rho_3 = 0.70;\ \rho_4 = 0.45;\ \alpha_3 = 0.18;\ \alpha_4 = 0.81;$$
$$e_3 = 0.10;\ e_4 = 0.84$$

Solution:
Absorption for the glasses

$$\alpha_o = 0.13 + (0.13)\frac{0.80 \times 0.70}{1 - (0.70 \times 0.70)}$$
$$\alpha_i = (0.18)\frac{0.80 \times 0.70}{1 - (0.70 \times 0.70)} = 0.151$$

The solar radiation absorbed

$$\alpha E_o = 248.3 \times 0.207 = 51.4\ \text{Btu/h} \cdot \text{ft}^2$$
$$\alpha E_i = 248.3 \times 0.151 = 37.5\ \text{Btu/h} \cdot \text{ft}^2$$

Next, the center-of-glazing U-factor for the fenestration must be found. The effective emittance of the air space is

$$e = \frac{1}{(1/e_2) + (1/e_3) - 1} = \frac{1}{(1/0.84) + (1/0.1) - 1} = 0.098$$

Air space and indoor coefficients are determined by trial and error.

Assume the outdoor and indoor glass temperatures are 115°F and 105°F, respectively. The mean temperature is 110°F and the temperature difference 10°F. From Table 4, $h_s = 0.51$. The indoor coefficient from Table 3 is $h_i = 1.59$ for $e_g = 0.84$. The thermal resistance of 0.25-in. thick glass is 0.035°F·ft²·h/Btu. Thus, for an outdoor coefficient of $h_o = 4.0$ Btu/(h·ft²·°F).

$$U = \frac{1}{(1/4.0) + (0.035) + (1/0.51) + 0.035 + (1/1.59)}$$
$$= 0.344\ \text{Btu/h} \cdot \text{ft}^2 \cdot {}^\circ\text{F}$$

The inward radiation and convection gain is

$$q_{RCi} = 0.344\left[\frac{51.4}{4.0} + 36.5\left(\frac{1}{4.0} + \frac{1}{0.51}\right) + (105 - 75)\right]$$
$$= 43.3\ \text{Btu/(h} \cdot \text{ft}^2)$$

To find the glass temperatures, check on first assumed values of $t_{go} = 115$°F and $t_{gi} = 105$°F (inside and outside temperatures, respectively).

$$t_{go} = t_o + (\alpha E_o + E_i - q_{RCi})\left(\frac{1}{h_o} + \frac{R_{go}}{2}\right)$$
$$= 105 + (51.4 + 37.5 - 43.3)\left(\frac{1}{4.0} + \frac{0.035}{2}\right) = 117.2^\circ\text{F}$$

and

$$t_{gi} = t_i + g_{RCi}\left(\frac{1}{h_i} + \frac{R_{gi}}{2}\right)$$
$$t_{gi} = 75 + 43.3\left(\frac{1}{1.59} + \frac{0.035}{2}\right) = 103.0^\circ\text{F}$$

The glass temperatures approximate the assumed temperatures, but repeating the previous calculations for a mean temperature of 110.1°F and a new temperature difference of 14.2°F (by interpolation) results in an $h_s = 0.515$ and an $h_i = 1.57$ Btu/h·ft²·°F and

$$U = 0.345\ \text{Btu/h} \cdot \text{ft}^2 \cdot {}^\circ\text{F}$$

where

$$q_{RCi} = 43.1\ \text{Btu/h} \cdot \text{ft}^2$$
$$t_{go} = 116.3^\circ\text{F}$$
$$t_{gi} = 103.0^\circ\text{F}$$

The transmittance through both panels of glass is

$$\tau = \frac{0.80 \times 0.12}{1 - (0.07 \times 0.70)} = 0.101$$

The SHGC for the fenestration is

$$\text{SHGC} = 0.101 + \frac{0.345}{4.0}(0.207) + \left(\frac{0.345}{4.0} + \frac{0.345}{0.515}\right)(0.151) = 0.233$$

The shading coefficient is

$$\text{SC} = 0.233/0.87 = 0.268$$

The total heat gain is

$$q_A = 0.268(216) + 0.345(105 - 75) = 68.2\ \text{Btu/(h} \cdot \text{ft}^2)$$

Solar Heat Gain Factors for Ground Reflectances Other Than 0.20

The combined direct and diffuse solar heat gain factors in Tables 15 through 21 have been computed using a ground reflectance of 0.20. For other ground reflectance values, computer calculations can be made. For hand computations, the SHGF can be adjusted for the different SHGF of the new foreground reflectances (see Table 22).

Equation (48) can be used to estimate the ground-reflected diffuse radiation falling on a vertical surface. However, the SHGF for a horizontal fenestration is approximately 87% of the total solar radiation falling on a horizontal surface. By using this SHGF in Equation (48) instead of the solar radiation incident on the ground, the ground-reflected contribution to the vertical SHGF will be in the same units. The incident ground-reflected diffuse radiation for the reference reflectance of 0.20 in terms of the SHGF is

$$0.87\ E_{dg} = \text{SHGF} \times 0.20 \times 0.5 \qquad (48)$$

where 0.87 is the conversion factor from incident solar radiation to SHGF, SHGF is the value for the horizontal surface, 0.20 is the reference ground reflectance, and 0.5 is the angle factor for a vertical surface.

The ground component for the new reflectance may be calculated in the same manner, or by means of a ratio of the reflectances and the difference applied to the SHGF for the vertical surface.

Table 22 Solar Reflectances of Foreground Surfaces

	Incident Angle					
Foreground Surface	**20°**	**30°**	**40°**	**50°**	**60°**	**70°**
New concrete	0.31	0.31	0.32	0.32	0.33	0.34
Old concrete	0.22	0.22	0.22	0.23	0.23	0.25
Bright green grass	0.21	0.22	0.23	0.25	0.28	0.31
Crushed rock	0.20	0.20	0.20	0.20	0.20	0.20
Bitumen and gravel roof	0.14	0.14	0.14	0.14	0.14	0.14
Bituminous parking lot	0.09	0.09	0.10	0.10	0.11	0.12

Adapted from Threlkeld (1962).

Example 10. Find the adjusted SHGF for a foreground having a reflectance of 0.32, for an east-facing window at 1000, July 21, 40° North latitude.

Solution: The SHGF for the horizontal surface, from Table 18, is 231 Btu/h·ft². The ground-reflected contribution to the SHGF for the east window, by Equation (48), is

$$0.87\ E_{dg} = 231 \times 0.20 \times 0.5 = 23.1\ \text{Btu/h} \cdot \text{ft}^2$$

For a foreground reflectance of 0.32, it becomes

$$0.87\ E_{dg} = 231 \times 0.32 \times 0.5 = 37.0\ \text{Btu/h} \cdot \text{ft}^2$$

The increase in the SHGF for the higher ground reflectance is 37.0 − 23.1 = 13.9 Btu/h·ft². The SHGF, from Table 18 for the east window, is 146 Btu/h·ft²; and, for the 0.32 ground reflectance, the adjusted SHGF becomes 146 + 13.9 = 159.9 Btu/h·ft².

Computer Calculation of Solar Heat Gain Factors

The following equations can be used to generate SHGF, where all angles are in degrees. The solar azimuth ϕ and the surface azimuth ψ are measured in degrees from south; angles to the east of south are negative, and angles to the west of south are positive.

Variables

- H = hour angle
- L = latitude
- δ = declination
- β = solar altitude
- ϕ = solar azimuth
- ψ = surface azimuth
- γ = surface-solar azimuth
- Σ = surface tilt
- θ = incident angle
- A = apparent solar constant
- B = atmospheric extinction coefficient
- C = sky diffuse factor
- E_{DN} = direct normal irradiance
- E_D = direct irradiance
- Y = ratio of vertical/horizontal sky diffuse
- E_{ds} = diffuse sky irradiance
- ρ_g = ground reflectance
- E_{dg} = diffuse ground reflected irradiance
- E_d = diffuse irradiance
- t_j = transmission coefficients for glass
- a_j = absorption coefficients for glass
- N_i = heat transfer factor, inward flowing fraction
- τ_D = transmittance of DSA glass
- α_D = absorptance of DSA glass

Values of A, B, and C are given in Table 8 for the 21st day of each month. Values for other dates can be obtained by interpolation. The transmission and absorption coefficients for DSA glass are given in Table 23. The absorption or transmission of direct solar radiation incident at an angle θ is

$$\alpha_D = \sum_{j=0}^{5} \alpha_j \cos_j\theta \qquad (49)$$

Table 23 Coefficients for DSA Glass for Calculation of Transmittance and Absorbtance

i	a_j	t_j
0	0.01154	−0.00885
1	0.77674	2.71235
2	−3.94657	−0.62062
3	8.57881	7.07329
4	−8.38135	9.75995
5	3.01188	−3.89922

$$\tau_D = \sum_{j=0}^{5} \tau_j \cos_j\theta \qquad (50)$$

Note that this calculation procedure, using the coefficients found in Table 23, gives a normal transmittance for DSA glass of 0.88, which is slightly higher than values used elsewhere in this chapter.

Tables 15 through 21 are based on a ground reflectance ρ_g of 0.2 and a heat transfer factor, inward flow fraction N_i of 0.267.

Hour angle, degrees	H = 0.25 (minutes of time from local solar noon)
Solar altitude β	$\sin\beta = \cos L \cos\delta \cos H + \sin L \sin\delta$
Solar azimuth ϕ	$\cos\phi = \dfrac{\sin\beta \sin L - \sin\beta}{\cos\beta \cos L}$
Surface-solar azimuth γ	$\gamma = \phi - \psi$
Incident angle θ	$\cos\theta = \cos\beta \cos\gamma \sin\Sigma + \sin\beta \cos\Sigma$
Direct normal irradiance E_{DN}	$E_{DN} = A \exp(-B/\sin\beta)$
Direct irradiance E_D	$E_D = E_{DN}\cos\theta$ if $\cos\theta > 0$ $E_D = 0$ otherwise
Ratio of sky diffuse E on vertical surface to that on horizontal surface Y	If $\cos\theta > -0.2$ $Y = 0.55 + 0.437\cos\theta + 0.313\cos^2\theta$ otherwise, $Y = 0.45$
Diffuse irradiance E_d	$E_d = E_{ds} + E_{dg}$

where

Vertical surfaces	$Ed_s = CY E_{DN}$
Surfaces other than vertical	$E_{ds} = CE_{DN}\left(\dfrac{1+\cos\Sigma}{2}\right)$

and

$$E_{dg} = E_{DN}(C + \sin\beta)\rho_g\left(\frac{1-\cos\Sigma}{2}\right)$$

Solar heat gain

Transmitted component $= E_D\sum_{j=0}^{+5} t_j\cos_j\theta + 2E_d\sum_{j=0}^{5}\dfrac{t_j}{j+2}$

Absorbed component $= E_D\sum_{j=0}^{+5} a_j\cos_j\theta + 2E_d\sum_{j=0}^{5}\dfrac{a_j}{j+2}$

Solar heat gain factor SHGF = Energy transmitted + N_i (Energy absorbed)

COMPLEX FENESTRATION SYSTEMS

A complex fenestration system contains one or more non-specular optical elements in the glazed area of the window. A non-specular optical element is one for which light (or shortwave infrared radiation) incident on the element from a single spatial direction does not emerge traveling in a single transmitted direction and/or a single reflected direction. Examples of non-specular elements are shades, drapes, blinds, honeycombs, figured glass, ground glass and other diffusers, lenses, prisms, and holographic glazings. For these systems the solar heat gain coefficient is extended to a direction-dependent quantity

$$\text{SHGC}(\theta,\phi) = T_{fH}(\theta,\phi) + \sum_{i=0}^{M} N_i A_{fi}(\theta,\phi) \qquad (51)$$

Table 24 Shading Coefficients for Louvered Sun Screens

Profile Angle	Group 1		Group 2		Group 3		Group 4		Group 5		Group 6	
	Transmittance	SC	Transmittance	SC	Transmittance	SC	Transmittance	SC	Transmittance	SC	Transmittance	SC
10°	0.23	0.35	0.25	0.33	0.40	0.51	0.48	0.59	0.15	0.27	0.26	0.45
20°	0.06	0.17	0.14	0.23	0.32	0.42	0.39	0.50	0.04	0.11	0.20	0.35
30°	0.04	0.15	0.12	0.21	0.21	0.31	0.28	0.38	0.03	0.10	0.13	0.26
≥ 40°	0.04	0.15	0.11	0.20	0.07	0.18	0.20	0.30	0.03	0.10	0.04	0.13

Group 1. Black, width over spacing ratio 1.15/1; 23 louvers/in. Group 2. Light color; high reflectance, otherwise same as Group 1. Group 3. Black or dark color; w/s ratio 0.85/1; 17 louvers/in. Group 4. Light color or unpainted aluminum; high reflectance; otherwise same as Group 3. Group 5. Same as Group 1, except two lights of 0.25 in. clear glass with 0.5 in. air space. Group 6. Same as Group 3, except two lights of 0.25 in. clear glass with 0.5 in. air space. U-value = 0.85 Btu/h·ft²·°F for all groups when used with single glazing.

Table 25 Shading Coefficients for Single Glass with Indoor Shading by Venetian Blinds or Roller Shades

Type of Glass	Nominal Thickness,[a] in.	Solar Transmittance[b]	Venetian Blinds: Medium	Venetian Blinds: Light	Roller Shade, Opaque: Dark	Roller Shade, Opaque: White	Roller Shade, Translucent: Light
Clear	3/32[c]	0.87 to 0.80	0.74[d] (0.63)[e]	0.67[d] (0.58)[e]	0.81	0.39	0.44
Clear	1/4 to 1/2	0.80 to 0.71					
Clear pattern	1/8 to 1/2	0.87 to 0.79					
Heat-absorbing pattern	1/8	—					
Tinted	3/16, 7/32	0.74, 0.71					
Heat-absorbing[f]	3/16, 1/4	0.46					
Heat-absorbing pattern	3/16, 1/4	—	0.57	0.53	0.45	0.30	0.36
Tinted	1/8, 7/32	0.59, 0.45					
Heat-absorbing or pattern	—	0.44 to 0.30	0.54	0.52	0.40	0.28	0.32
Heat-absorbing[f]	3/8	0.34					
Heat-absorbing or pattern		0.29 to 0.15					
	—	0.24	0.42	0.40	0.36	0.28	0.31
Reflective coated glass	S.C. = 0.30[g]		0.25	0.23			
	= 0.40		0.33	0.29			
	= 0.50		0.42	0.38			
	= 0.60		0.50	0.44			

[a]Refer to manufacturers' literature for values.
[b]For vertical blinds with opaque white and beige louvers in the tightly closed position, SC is 0.25 and 0.29 when used with glass of 0.71 to 0.80 transmittance.
[c]Typical residential glass thickness.
[d]From Van Dyck and Konen (1982), for 45° open venetian blinds, 35° solar incidence, and 35° profile angle.
[e]Values for closed venetian blinds. Use these values only when operation is automated for solar gain reduction (as opposed to daylight use).
[f]Refers to gray, bronze, and green tinted heat-absorbing glass.
[g]SC for glass with no shading device.

Table 26 Shading Coefficients for Insulating Glass with Indoor Shading by Venetian Blinds or Roller Shades

Type of Glass	Nominal Thickness, Each Light	Solar Transmittance[a]: Outer Pane	Solar Transmittance[a]: Inner Pane	Venetian Blinds[b]: Medium	Venetian Blinds[b]: Light	Roller Shade, Opaque: Dark	Roller Shade, Opaque: White	Roller Shade, Translucent: Light
Clear out	3/32, 1/8 in.	0.87	0.87	0.62[c]	0.58[c]	0.71	0.35	0.40
Clear in				(0.63)[d]	(0.58)[d]			
Clear in	1/4 in.	0.80	0.80					
Heat-absorbing[e] out	1/4 in.	0.46	0.80	0.39	0.36	0.40	0.22	0.30
Clear in								
Reflective coated glass	SC = 0.20[f]			0.19	0.18			
	= 0.30			0.27	0.26			
	= 0.40			0.34	0.33			

Table refers to factory-fabricated units with 3/16, 1/4, or 1/2 in. air space, or to prime windows plus storm windows.

[a]Refer to manufacturers' literature for exact values.
[b]For vertical blinds with opaque white or beige louvers, tightly closed, SC is approximately the same as for opaque white roller shades.
[c]From Van Dyck and Konen (1982), for 45° open venetian blinds, 35° solar incidence, and 35° profile angle.
[d]Values for closed venetian blinds. Use these values only when operation is automated for solar gain reduction (as opposed to daylight use).
[e]Refers to bronze, or green tinted, heat-absorbing glass.
[f]SC for glass with no shading device.

Table 27 Shading Coefficients for Double Glazing with Between-Glass Shading

Type of Glass	Nominal Thickness, Each Pane	Solar Transmittance[a] Outer Pane	Solar Transmittance[a] Inner Pane	Description of Air Space	Type of Shading: Venetian Blinds Light	Venetian Blinds Medium	Louvered Sun Screen
Clear out, Clear in	3/32, 1/8 in.	0.87	0.87	Shade in contact with glass or shade separated from glass by air space.	0.33	0.36	0.43
Clear out, Clear in	1/4 in.	0.80	0.80	Shade in contact with glass-voids filled with plastic.	—	—	0.49
Heat-absorbing[b] out, Clear in				Shade in contact with glass or shade separated from glass by air space.	0.28	0.30	0.37
	1/4 in.	0.46	0.80	Shade in contact with glass-voids filled with plastic.	—	—	0.41

[a]Refer to manufacturers' literature for exact values.
[b]Refers to grey, bronze and green tinted heat-absorbing glass.

Table 28 Properties of Representative Indoor Shading Devices Shown in Tables 25, 26, and 27

Indoor Shade	Solar-Optical Properties (Normal Incidence) Trans.	Reflect.	Absorp.
Venetian blinds[a] (ratio of slat width to slat spacing 1.2, slat angle 45°)			
Light colored slat	0.05	0.55	0.40
Medium colored slat	0.05	0.35	0.60
Vertical blinds			
White louvers	0.00	0.77	0.23
Roller shades			
Light shades (translucent)	0.25	0.60	0.15
White shade (opaque)	0.00	0.65	0.35
Dark colored shade (opaque)	0.00	0.20	0.80

[a]Values in this table and preceding tables are based on horizontal venetian blinds. However, tests show that these values can be used for vertical blinds with good accuracy.

where

θ = incident angle relative to normal to layer
ϕ = azimuth angle (in plane of layer, about normal)
$T_{fH}(\theta,\phi)$ = directional-hemispherical front transmittance of system
$A_{fi}(\theta,\phi)$ = directional absorbance of ith layer
N_i = inward flowing fraction of absorbed energy for ith layer

Usually, complex fenestration solar heat gain is determined either by direct measurement or by calculating SHGC(θ,ϕ) at a specified incident direction, typically either at normal incidence or $\theta = 30°$, $\phi = 0°$. Such values, with the SHGC re-expressed as shading coefficient, are given for a variety of systems in Tables 24–27 and 29. These tables apply only to incident directions of 30° or less, and may not include important azimuth angle dependence. The tables provide an approximate value in the absence of better information. Values at the incident direction of interest should be measured or calculated for accurate and reliable results, unless there is a reason to consider incident angle or azimuthal dependencies unimportant.

Solar-Thermal Separation

Equation (51) may be used to calculate the solar heat gain coefficient by separately determining transmittance $T_{fH}(\theta,\phi)$, layer absorptances $A_{fi}(\theta,\phi)$, and inward-flowing fractions N_i. This method is called **solar-thermal separation** because the processes that determine N_i are thermal in nature, while those that determine $T_{fH}(\theta,\phi)$ and $A_{fi}(\theta,\phi)$ are solar-optical. Table 30 lists calorimetrically determined values of N_i for several generic glazing/shading systems (Klems and Kelley 1995). These values are independent of the solar-optical properties of the particular system.

The corresponding transmittances and layer absorptances may be calculated or measured. Calculation has an advantage, for example, when comparing the performance of different glazings or shading colors in the same general configuration. Optical data is sometimes more available or more economically obtained than overall calorimetric measurements of SHGC. For example, $T_{fH}(\theta,\phi)$ may be measured simply by an optical technique using an integrating sphere. The layer absorptances $A_{fi}(\theta,\phi)$ are more difficult to determine because readily available data is likely to be the directional absorptance of an *isolated* layer $\alpha_{fi}(\theta,\phi)$; whereas $A_{fi}(\theta,\phi)$ is the *in-system* layer absorptance, i.e., it includes the contributions of absorbed radiation multiply reflected back to the ith layer from all other layers in the system. Isolated layer absorptances $[\alpha_{fi}(\theta,\phi)]$ must be corrected for this effect.

Calculating System Transmittance and Absorptances from Layer Properties

The key feature of non-specular elements is that they distribute outgoing radiation (in the solar-optical spectral region) in the transmitted and/or reflected hemisphere, even for incident radiation from a single direction. This means that they are characterized by bidirectional transmittance distribution functions (BTDF) and reflectance distribution functions (BRDF) that give the outgoing radiance (energy flux per unit area per unit solid angle) as a fraction of the incident irradiance (energy flux per unit area):

$$I(\theta_o, \phi_o;\theta_i, \phi_i) = \text{BTDF}(\theta_o, \phi_o;\theta_i, \phi_i) \cdot E(\theta_i, \phi_i)$$
$$J(\theta_o, \phi_o;\theta_i, \phi_i) = \text{BRDF}(\theta_o, \phi_o;\theta_i, \phi_i) \cdot E(\theta_i, \phi_i) \quad (52)$$

where

θ_o, ϕ_o = angles specifying outgoing direction
θ_i, ϕ_i = angles specifying incident direction
E = incident irradiance
I, J = transmitted, reflected radiance

In a real, non-specular element these quantities are also functions of where the radiation falls, as on a venetian blind. However, this information is only useful for finding outgoing radiation patterns. For determining the solar heat gain the process may be considered as spatially averaged over the non-specular device, so that it can be considered as a thin uniform layer with only angular dependence (Klems 1994a). By dividing the (transmission or reflection) hemisphere into a grid of solid angle sections, the bidirectional property functions can be approximated as matrices (Klems 1994B).

Commercial computer programs are available to perform the matrix calculations, including many spreadsheet programs. In the following calculation, use of a biangular grid to characterize a non-specular layer with azimuthal dependence requires handling matrices with a large number of elements. For this level of complexity a

Table 29 Shading Coefficients for Single and Insulating Glass with Draperies

Glazing	Glass Trans-mission	Glass Alone SC (No Drapes)	A	B	C	D	E	F	G	H	I	J
Single glass												
1/8 in. Clear	0.86	1.00	0.87	0.82	0.74	0.69	0.64	0.59	0.53	0.48	0.42	0.37
1/4 in. Clear	0.80	0.95	0.80	0.75	0.70	0.65	0.60	0.55	0.50	0.45	0.40	0.35
1/2 in. Clear	0.71	0.88	0.74	0.70	0.66	0.61	0.56	0.52	0.48	0.43	0.39	0.35
1/4 in. Heat absorbing	0.46	0.67	0.57	0.54	0.52	0.49	0.46	0.44	0.41	0.38	0.36	0.33
1/2 in. Heat absorbing	0.24	0.50	0.43	0.42	0.40	0.39	0.38	0.36	0.34	0.33	0.32	0.30
Reflective coated	—	0.60	0.57	0.54	0.51	0.49	0.46	0.43	0.41	0.38	0.36	0.33
(see manufacturers' literature	—	0.50	0.46	0.44	0.42	0.41	0.39	0.38	0.36	0.34	0.33	0.31
for exact values)	—	0.40	0.36	0.35	0.34	0.33	0.32	0.30	0.29	0.28	0.27	0.26
	—	0.30	0.25	0.24	0.24	0.23	0.23	0.23	0.22	0.21	0.21	0.20
Insulating glass, 1/4-in. air space												
(1/8 in. out and 1/8 in. in)	0.76	0.89	0.75	0.71	0.65	0.63	0.57	0.53	0.48	0.45	0.38	0.36
Insulating glass 1/2-in. air space												
clear out and clear in	0.64	0.83	0.66	0.62	0.58	0.56	0.52	0.48	0.45	0.42	0.37	0.35
Heat absorbing out and clear in	0.37	0.55	0.49	0.47	0.45	0.43	0.41	0.39	0.37	0.35	0.33	0.32
Reflective coated	—	0.40	0.38	0.37	0.37	0.36	0.34	0.32	0.31	0.29	0.28	0.28
(see manufacturers' literature for	—	0.30	0.29	0.28	0.27	0.27	0.26	0.26	0.25	0.25	0.24	0.24
exact values)	—	0.20	0.19	0.19	0.18	0.18	0.17	0.17	0.16	0.16	0.15	0.15

Shading Coefficient Index Letter

Notes:

1. Shading coefficients are for draped fabrics.
2. Other properties are for fabrics in flat orientation.
3. Use fabric reflectance and transmittance to obtain accurate shading coefficients.
4. Use openness and yarn reflectance or openness and fabric reflectance to obtain the various environmental characteristics, or to obtain approximate shading coefficients.

Classification of Fabrics

I = Open weave
II = Semiopen weave
III = Closed weave

D = Dark color
M = Medium color
L = Light color

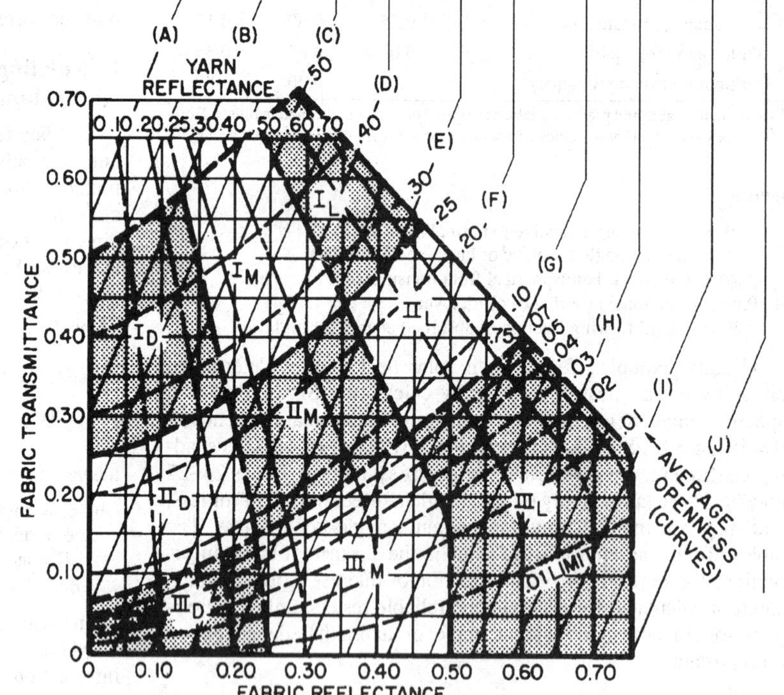

To obtain fabric designator (III_L, I_M, etc.). Using either (1) Fabric Transmittance and Fabric Reflectance coordinates, or (2) Openness and Yarn Reflectance coordinates, find a point on the chart and note the designator for that area. If properties are not known, the classification may be approximated by eye as described in the note in Figure 34. Table 31 lists appropriate classifications for various applications.

To obtain shading coefficient (SC). (1) Locate drapery fabric as a point using its known properties, or approximate using its fabric classification designator. For accuracy, use fabric transmittance and fabric reflectance; (2) follow diagonal SC lines to lettered columns in the table. Find SC on line with glazing used. For example, SC is 0.45 for 0.25-in. clear single glass with III_L drapery (Column H).

Note: Shading coefficients are for 45° incident angle. For 30° or less, add 5% to the number found in the table.

Table 30 Measured Layer Inward-Flowing Fractions N_i for Typical Fenestration

System	Bind Angle Below Horizontal	Inner Shading Layer	Inner Glass	Between Pane Shading	Outer Glass	Exterior Shading Layer
Single glazing with interior shade		0.80±0.08			0.08±0.06	
Single glazing with interior venetian blind	–45°	0.69±0.05			0.24±0.09	
	30°	0.83±0.08			0.21±0.07	
	Closed	0.72±0.07			0.14±0.05	
Single glazing with exterior venetian blind	45°				0.46±0.12	0.04±0.01
Double glazing with interior shade		0.85±0.10	0.52±0.12		0.28±0.06	
Double glazing with interior venetian blind	45°	0.86±0.06	0.69±0.14		0.21±0.09	
Double glazing with between-pane blind	45°		0.69±0.14	0.45±0.06	0.34±0.10	
	–45°		0.76±0.10	0.40±0.07	0.27±0.14	
Low-e double glazing with between-pane blind	35°		0.46±0.12	0.38±0.05	0.32±0.11	
Double glazing with exterior venetian blind	45°		0.73±0.13		0.28±0.12	0.03±0.02

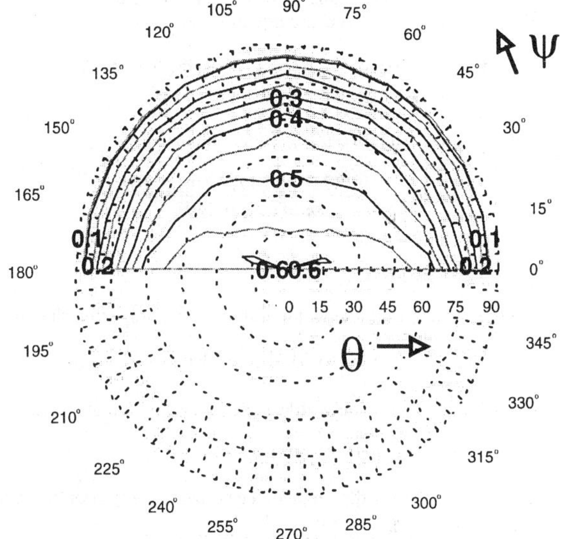

Fig. 30 Contour Plot of Beam SHGC for Double Glazed Window with Interior Venetian Blind with Slats Tilted at 45°

special-purpose computer program for handling the large amount of data involved is probably desirable.

Figure 30 shows the results of a calculation for an interior buff-colored blind (slat reflectance 62%) in combination with sealed double glazing (1/4 in. glass panes). The calculation used bi-directional transmittance and reflectance measurements averaged over a 8-in. square section of the blind, with 1-in.slats. These measurements were used to construct layer property matrices with the clear glazing properties taken from Rubin (1985). The SHGC was calculated using Equation (51), solar-optical transmittances and absorptances calculated by the above method, and inward-flowing fractions from Table 30 (Klems and Kelley 1995, Klems and Warner 1995).

SHADING

EXTERIOR SHADING

The most effective way to reduce the solar load on fenestration is to intercept direct radiation from the sun before it reaches the glass. Fenestration products fully shaded from the outside reduce solar heat gain as much as 80%. In one way or another, fenestration can be shaded by roof overhangs, vertical and horizontal architectural projections, awnings, heavily proportioned exterior louvers, insect or shading screens, patterned screens having a weave designed for sunlight interception, or sun screens of narrow fixed louvers. In all

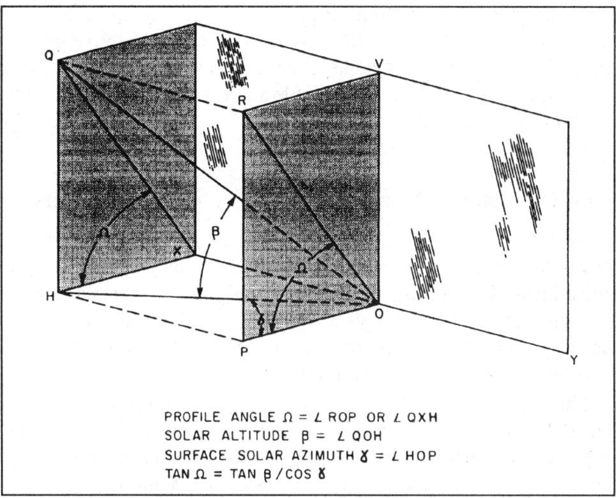

Fig. 31 Profile Angle for South-Facing Slat-Type Sunshades

exterior shading structures, the air must move freely to carry away heat absorbed by the shading and glazing materials. Manufacturer's should include instructions for proper installation that provides suitable free convection ventilation between shading and glazing. Also, the geometry of the structures relative to changing sun position should be considered to determine the times and quantities of direct sunlight penetration. Pennington (1968), Yellott (1972), and Ewing and Yellott (1976) provide details of the effectiveness of various outside shading devices.

Louvers and Sunshades

The ability of horizontal panels or louvers to intercept the direct component of solar radiation depends on their geometry and the profile or shadow-line angle Ω (Figure 31), defined as the angular difference between a horizontal plane and a plane tilted about a horizontal axis in the plane of the fenestration until it includes the sun. The profile angle can be calculated by

$$\tan\Omega = \tan\beta/\cos\gamma \qquad (53)$$

For slat-type sunshades, the transmitted solar radiation consists of straight-through and transmitted-through components. When the profile angle Ω is above the cutoff angle (see Figure 32), straight-through transmission of direct radiation is completely eliminated, but the transmitted diffuse and the reflected-through components remain. Their magnitude depends largely on the reflectance of the sunshade surfaces and of exterior objects.

Narrow horizontal louvers fabricated in conventional width-spacing ratios and framed as window screens, retain their shading

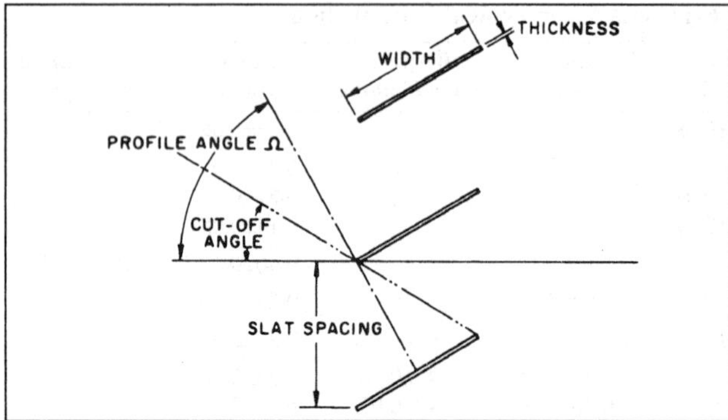

Fig. 32 Geometry of Slat-Type Sunshades

characteristics, while gaining in effective transparency (view) by eliminating the coarse striation pattern of wide louvers. Table 24 gives shading coefficients (SC) for several types of louvered sun screens. Commercially available sun screens completely exclude direct solar radiation when the profile angle exceeds approximately 26° (Groups 1, 2, and 5) or 40° (Groups 3, 4, and 6). Group designations are defined in the table footnote.

Roof Overhangs; Horizontal and Vertical Projections

In the northern hemisphere, horizontal projections can considerably reduce solar heat gain on south, southeast, and southwest exposures during late spring, summer, and early fall. On east and west exposures during the entire year, and on southerly exposures in winter, the solar altitude is generally so low that to be effective horizontal projections must be excessively long.

The shadow width S_W and shadow height S_H (Figure 33), produced by the vertical and horizontal projections (P_V and P_W), respectively, can be calculated using the solar surface azimuth γ and the horizontal profile angle Ω determined by Equation (53).

$$S_W = P_V |\tan \gamma| \quad (54)$$

$$S_H = P_H \tan \Omega \quad (55)$$

Note: When the solar surface azimuth γ is greater than 90° and less than 270°, the fenestration product is completely in the shade; thus, $S_w = W + R_w$ and $A_{SL} = 0$.

The sunlit A_{SL} and shaded A_{SH} area of the fenestration product is variable during the day and can be calculated for each moment using the following relations (see Figure 33)

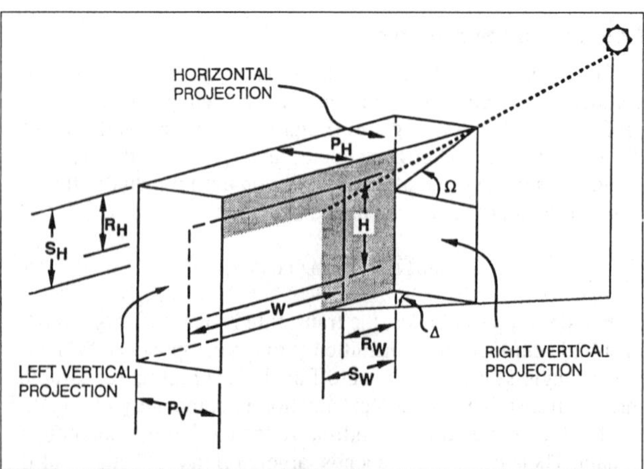

Fig. 33 Vertical and Horizontal Projections and Related Profile Angles for Vertical Surface Containing Fenestration

$$A_{SL} = [W - (S_W - R_W)][H - S_H - R_H] \quad (56)$$

$$A_{SH} = A - A_{SL} \quad (57)$$

where A is total fenestration product area.

Example 11. A window in the southwest wall of a building at 40°N latitude is 34.5 in. wide and 58.5 in. high. The depth of the horizontal and vertical projections are 6 in. and the they are located 3 in. beyond the edges of the window.

Part A. Find the sunlit and shaded area of the window at 1500 on July 21.

Part B. Find the depth of the projections necessary to fully shade the window just described.

Solution:

Part A. The wall azimuth ψ for a southwest wall is +45° (Table 10). The solar azimuth ϕ can be calculated using Equation (10). At 1500, $H = 0.25 \times 180 = 45°$; from Table 8, for July 21, $\delta = 20.6°$.

Find the solar altitude β using Equation (9)

$$\sin \beta = \cos(40)\cos(20.6)\cos(45) + \sin(40)\sin(20.6)$$
$$\beta = 47.2°$$

Find the solar azimuth ϕ using Equation (10)

$$\cos\phi = \frac{\sin(47.2)\sin(40) - \sin(20.6)}{\cos(47.2)\cos(40)}$$
$$\phi = 76.7°$$

Thus, $\gamma = 76.7 - 45 = 31.7°$

Using Equation (54), the width of the vertical projection shadow is

$$S_W = 6|\tan 31.7| = 3.71 \text{ in.}$$

Using Equation (53), the profile angle for the horizontal projection is

$$\tan\Omega = \tan(47.2)/\cos(31.7) = 51.8°$$

Using Equation (55), the height of the horizontal projection shadow is

$$S_H = 6\tan(51.8) = 7.62 \text{ in.}$$

Using Equations (56) and (57), the sunlit and shaded area of the window are now

$$A_{SL} = \frac{[34.5 - (3.71 - 3)][58.5 - (7.62 - 3)]}{144} = 12.64 \text{ ft}^2$$

Part B. The shadow length necessary to fully shade the given window $S_{H(fs)}$ and $S_{W(fs)}$ from the horizontal and vertical projection are given by (see Figure 33)

$$S_{H(fs)} = 58.5 + 3 = 61.5 \text{ in.}$$
$$S_{W(fs)} = 34.5 + 3 = 37.5 \text{ in.}$$

Thus, using Equations (54) and (55)

$$P_{H(fs)} = 61.5\cot(51.8) = 48.4 \text{ in.}$$
$$P_{W(fs)} = 37.5|\cot(31.7)| = 60.7 \text{ in.}$$

For this example, because both horizontal and vertical projections do not need to fully shade the window, a horizontal projection of 48.4 in. is satisfactory. Also, to accurately analyze the influence of external projections, an hour-by-hour calculation must be performed over the periods of the year for which shadowing is desired.

Partially Shaded Fenestration Products

All solar heat gain data in this section are based on the sunlit area of the glass itself. For actual fenestration products, the sunlit glass area is likely to be significantly less than the total fenestration product area, and the effect of external shadowing as well as the effect of the opaque areas must be considered.

Besides the horizontal and vertical projections, external shade can be produced by the mullions and the transom. The shaded area varies continuously throughout the day, but it can be estimated readily by treating the mullions and transom as vertical and horizontal projections and applying Equation (55).

The solar heat gain factors (defined in the section Simplified Methods for Predicting Heat Transfer through Fenestration) through a partially shaded fenestration product, can be estimated using

$$\mathrm{SHGF} = E_D\left(\frac{A_{SL}}{A}\right) + E_d \tag{58}$$

The diffuse component E_d can be found using the SHGF values (Tables 15 through 21) for the nearest facade not receiving direct sunlight. The direct component of solar radiation can be estimated by subtracting the diffuse component E_d from the SHGF values for an unshaded fenestration product.

The solar heat gain for the entire fenestration product is then

$$\text{Solar Heat Gain} = A \times \mathrm{SC} \times \mathrm{SHGF} \tag{59}$$

Example 12. Find the solar heat gain through the window given in Example 11 assuming that the window consists of 0.25 in. clear plate glass.

Solution: At 1500, the SHGF for an unshaded window in a southwest wall from Table 18 is 170 Btu/h·ft^2 of which the diffuse component E_D is 31 Btu/h·ft^2.

$$\mathrm{SHGF} = (170-31)\left(\frac{12.64}{14.02}\right) + 31 = 156 \text{ Btu/h}\cdot\text{ft}^2$$

The SC for 0.25 in. plate glass (assuming in this example that one-half the absorbed energy is dissipated to the room air) is 0.97; thus

$$\text{Solar heat gain} = 0.97 \times 156 \times 14.02 = 2120 \text{ Btu/h}$$

Equations for Computer Calculations of External Shadowing of Inclined Surfaces

Incidence angle: $\theta = \cos^{-1}(\cos\beta\cos\gamma\sin\Sigma + \sin\beta\cos\Sigma)$

Vertical surface $\theta_V = \cos^{-1}(\cos\beta\cos\gamma)$

Horizontal surface $\theta_H = \cos^{-1}(\sin\beta)$

Vertical projection profile angle:

$$\Delta = \tan^{-1}\left(\frac{\sin\gamma\cos\beta}{\cos\theta}\right);\ |2| < 90°$$

For $\theta > 90°$ $A_{SL} = 0$ and $A_{SH} = A$

Vertical surface $\Delta_V = \tan^{-1}(\gamma);\ |\gamma| < 90°$ and $|\gamma| > 270°$

Horizontal surface $\Delta_H = \tan^{-1}\left(\frac{\sin\gamma}{\tan\beta}\right)$; for all γ

Horizontal projection profile:

$$\text{Angle: } \Omega = \tan^{-1}\left(\frac{\sin\beta\sin\Sigma - \cos\beta\cos\gamma\cos\Sigma}{\cos\theta}\right);\ |\theta| < 90°$$

Vertical surface $\Omega_V = \tan^{-1}\left(\frac{\tan\beta}{\cos\gamma}\right);\ |\gamma| < 90°$ and $|\gamma| > 270°$

Horizontal surface $\Omega_H = \tan^{-1}\left(-\frac{\cos\gamma}{\tan\beta}\right);\ 90° < |\gamma| < 270°$

Length of shadow from vertical projection:

$$S_W = P_V|\tan\Delta|$$

Length of shadow from horizontal projection:

$$S_H = P_H|\tan\Omega|$$

Sunlit area of fenestration product:

$$A_{SL} = [W - (S_W - R_W)][H - (S_H - R_H)]$$

Shaded area of fenestration product:

$$A_{SH} = A - A_{SL}$$

where

ϕ = solar azimuth
β = solar altitude
γ = solar surface azimuth
Σ = surface tilt angle
P_V = vertical projection depth
P_H = horizontal projection depth
W = fenestration product width
H = fenestration product height
R_W = width of opaque surface between fenestration product and vertical projection
R_H = height of opaque surface between fenestration product and horizontal projection
A = total projected area of the fenestration product
θ = angle of incidence
Ω = horizontal projection profile angle
Δ = vertical projection profile angle

INDOOR SHADING DEVICES

Venetian Blinds and Roller Shades

Most fenestration has some type of internal shading to provide privacy and aesthetic effects, as well as to give varying degrees of sun control (Ozisik and Schutrum 1960). Approximate SC values for typical internal shading are given in Tables 25, 26, and 27. The effectiveness of any internal shading device depends on its ability to reflect incoming solar radiation back through the fenestration before it can be absorbed and converted into heat within the building. Table 28 lists approximate values of solar-optical properties for the typical indoor shading devices described in Tables 25, 26, and 27.

The values in Table 25 apply both to sunlit fenestration and fenestration on the shaded side of the building. The values are similar in both cases because shades generally are open on shaded exposures. The tabulated values apply specifically to horizontal venetian blinds but are usable for vertical blinds when adjusted so that no direct solar radiation can enter through them.

Table 26 gives SC values for venetian blinds and roller shades used with insulating glass. The first row applies to fenestration products in which both lights are high-transmittance glass; the second applies when the outer light is heat-absorbing and the inner light is clear glass.

Because of the wide variety of glass available, the manufacturer should be consulted for specific data. The SC with no interior shading is included as a reference point for each classification. Note that the energy benefit of a shade decreases as the SC of the unshaded glass decreases. Similarly, the flexibility of the fenestration system decreases as the SC decreases, due to the low transmittances and the inability of the occupant to change this factor.

Draperies

Draperies reduce heating and cooling loads, depending on the type and the use by the occupant. Rudoy and Duran (1975) found annual reductions in load of between 5% and 20%. An approximate model for determining the SC or SHGC of free-standing vertical interior shades was developed by McCluney and Mills (1993). Shading coefficients for representative interior drapes are provided in Table 29.

The solar optical properties of drapery fabrics can be determined accurately by laboratory tests (Yellott 1963), and manufacturers can usually supply solar transmittance and reflectance values of their products. In addition to these properties, the openness factor (ratio of the open area between the fibers to the total area of the fabric) is a useful property that can be measured exactly (Keyes 1967, Moore and Pennington 1967). It can also be estimated by inspection, since the human eye can readily distinguish between tightly woven fabrics that permit little direct radiation to pass between the fibers and loosely woven fabrics which allow the sun's rays to pass freely.

Drapery fabrics can be classified in terms of their solar-optical properties as having specific values of fabric transmittance and reflectance. Fabric reflectance is the major factor in determining the ability of a fabric to reduce solar heat gain. Based on their appearance, draperies can also be classified by yarn color as dark, medium, and light; and by weave as closed, semiopen, and open. The apparent color of a fabric is determined by the reflectance of the yarn itself. The figure in Table 29 shows yarn reflectance. Figure 34 classifies drapery fabrics into nine types, rated by openness and yarn reflectances.

Figure 34, with the aid of Table 29, guides in estimating the probable SC for a fabric-glass combination when the solar-optical properties are unknown. Whenever possible, fabric reflectance and transmittance values should be obtained from the manufacturer, because this data permits more accurate SC estimates to be made. Visual estimations of openness and yarn reflectance, interpreted through Table 29, are valuable in judging the effectiveness of drapes for (1) protection from excessive radiant energy from either sunlight or sun-heated glass, (2) brightness control, (3) providing either outward view or privacy, and (4) sound control.

Table 29 applies to glass and a single drape hung with 100% fullness (drapery width is twice the width of the draped area). If the drapery is hung flat, like a fenestration product shade, a different SC applies; with a low transmittance and high reflectance, the SC is appreciably lower. As an extreme example, a flat opaque drapery having an aluminized or similar coating with reflectance of 0.80, in combination with 0.25-in. clear glass, has an SC of 0.18, as compared with 0.32 that can be extrapolated from Table 29 for this material in draped form. Moore and Pennington (1967) explain the effect of folding drapery materials to provide 100% fullness, and describe a method for calculating SC when materials are used flat.

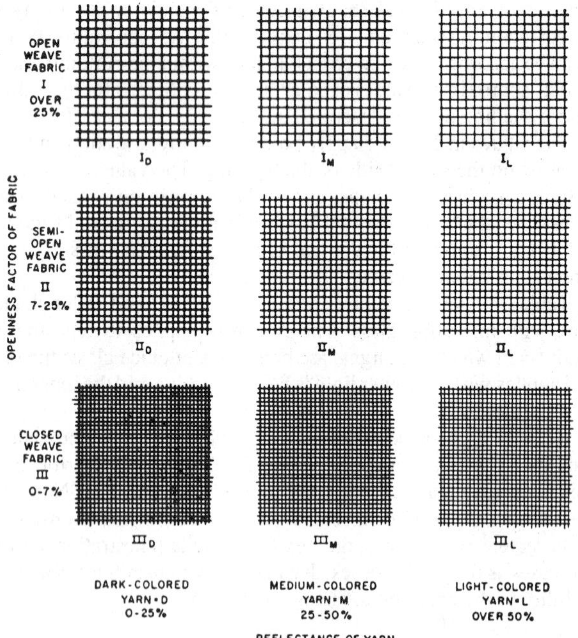

Note: Classes may be approximated by eye. With closed fabrics, no objects are visible through the material, but large light or dark areas may show. Semiopen fabrics do not permit details to be seen, and large objects are clearly defined. Open fabrics allow details to be seen, and the general view is relatively clear with no confusion of vision. The yarn color or shade of light or dark may be observed to determine whether the fabric is light, medium, or dark.

Fig. 34 Classification of Drapery Fabrics

Example 13. A drape with 100% fullness, having a fabric transmittance of 0.20 and a fabric reflectance of 0.40, is used with 0.25-in. glass. What SC should be used?

Solution: From the figure in Table 29, the 0.20 and 0.40 intersection is nearest line F. Table 29 assigns SC 0.55 to column F for 0.25-in. clear single glass. Interpolate if necessary; see notes for other uses.

Example 14. For the same drapery as in Example 13, the incident angle for which the SC is desired is 30°. What SC should be used?

Solution: Add 5% to the value found in Example 12. Thus,

$$SC = (1 + 0.05)\ 0.55 = 0.58$$

Example 15. Determine the Fabric Designator for a fabric having an openness factor of 0.10 and a yarn reflectance of 0.60.

Solution: On the figure in Table 29, these lines intersect in the area of Designator II_L. Refer also to Figure 34. Fabric is semi-open and light in color. Additional information: probable fabric reflectance is 0.50, and fabric transmittance is 0.35.

VISUAL AND THERMAL CONTROLS

The ideal fenestration system permits optimum light, heat, ventilation, and visibility; minimizes moisture and sound transfer between the exterior and the interior; and produces a satisfactory physiological and psychological environment. The controls of an optimum system react to varying climatological and occupant demands. Fixed controls may have operations or cost advantages or both but do not react to physical and psychological variations. Variable controls are, therefore, more effective in energy conservation and environmental satisfaction.

Operational Effectiveness of Shading Devices

Shading devices vary in their operational effectiveness. Some devices such as overhangs, light shelves, and tinted glazings do not require operation, have long life expectancies, and do not degrade significantly over their effective life. Other shading devices, especially operable interior shades, may have reduced effectiveness due to less than optimal operation and degradation of effectiveness over time.

The performance of shading devices for the reduction of peak cooling loads and annual energy use should account for operational effectiveness or reliability in actual operation. Passive devices, such as architectural elements and glazing tinting, are considered 100% effective in operation. Glazing coatings and adherent films may degrade over time. Shade screens are removable and are assumed to be operated seasonally, but some users will keep screens in place all year long and some not install or remove them at optimum times. Automated shading devices controlled for optimum thermal operation are considered more effective than manual devices, but controls require ongoing maintenance. Automated shading devices may also be operated for non-thermal purposes such as glare and daylighting optimization, and this may reduce thermal effectiveness. Manually operated devices are subject to a wide variation in effective use, which should be considered when evaluating performance.

Indoor Shading Devices

Thermal comfort of occupants within the glazed space is paramount. The following factors are also of concern (see Table 29 and Figure 34).

Table 31 Summary of Environmental Control Capabilities of Draperies

Item	Designator (Table 29 and Figure 34)								
	I_D	I_M	I_L	II_D	II_M	II_L	III_D	III_M	III_L
1. Protection from direct solar radiation and long-wave radiation to or from window areas	Fair	Fair	Fair	Fair	Good	Good	Fair	Good	Good
2. Effectiveness in allowing outward vision through fenestration	Good	Good	Fair	Fair	Fair	Some	None	None	None
3. Effectiveness in attaining privacy (limiting inward vision from outside)	None	None	Poor[a] Good[a]	Poor	Fair	Fair[a] Good[a]	Good[b]	Good[b]	Good[b]
4. Protection against excessive brightness and glare from sunshine and external objects	Mild	Mild	Mild[c] Poor[c]	Good	Good	Good[c] Poor[c]	Good	Good	Good[c] Poor[c]
5. Effectiveness in modifying unattractive or distracting view out of window	Little	Little	Some	Some	Good	Good	Blocks	Blocks	Blocks

[a]Good when bright illumination is on the viewing side.
[b]To obscure view completely, material must be completely opaque.
[c]Poor rating applies to white fabric in direct sunlight. Use off-white color to avoid excessive transmitted light.

Radiant Energy Protection. Unshaded fenestration products become sources of radiant heat by transmitting short-wave solar radiation and by emitting long-wave radiation to dissipate some of the absorbed solar energy. In winter, glass temperatures usually fall below room air temperature, which may produce thermal discomfort to occupants near the fenestration. In summer, individuals seated near the unshaded fenestration product may experience discomfort from both direct solar rays and long-wave radiation emitted by sun-heated glass. In winter, loss of heat by radiation to cold glass can also cause discomfort. Tightly woven, highly reflective drapes minimize such discomfort; drapes with high openness factors are less effective because they permit short-wave and long-wave radiation to pass more freely. Light-colored shading devices with maximum total surface usually provide the best protection, because they absorb less heat and tend to lose heat readily by convection to the conditioned air.

Outward Vision. Outward vision is normally desirable in both business and living spaces. Open-weave, dark-colored fabrics of uniform pattern permit maximum outward vision, while uneven pattern weaves reduce the ability to see out. A semi-open weave modifies the view without completely obscuring the outdoors. Tightly woven fabrics block off outward vision completely.

Privacy. Venetian blinds, either vertical or horizontal, can be adjusted and, when completely closed, afford full privacy. When draperies are closed, the degree of privacy is determined by their color and tightness of weave, and the source of the principal illumination. To obscure the view so completely that not even shadows or silhouettes can be detected, fully opaque materials are used.

Brightness Control. Visual comfort is essential in many occupied areas, and freedom from glare is an important factor in performing tasks. Direct sunlight must not strike the eye, and reflected sunlight from bright or shiny surfaces is equally disturbing and even disabling. A tightly woven white fabric with high solar transmittance attains such brilliance when illuminated by direct sunshine that, by contrast with its surroundings, it creates excessive glare. Off-white colors should be used so their surface brightness is not too great. Venetian blinds permit considerable light to enter by inter-reflection between slats. When two shading devices are used, the one on the inside (away from the fenestration product) should be darker and more open. With this arrangement the inside can be used to control brightness for the other shading devices and, when used alone, reduce brightness while still permitting some view of the outside.

View Modification. When the view is unattractive or distracting, draperies modify the view to some degree, depending on the fabric weave and color (summarized in Table 31). Thus, the fenestration product remains an effective connection to the outside.

Sound Control. Indoor shading devices, particularly draperies, can absorb some of the sounds originating within the room, but have little or no effect in preventing outdoor sounds from entering. For

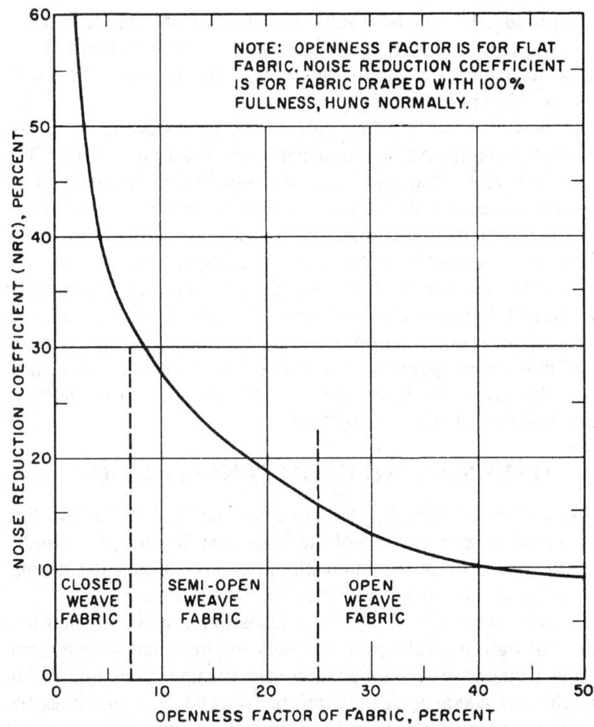

Fig. 35 Noise Reduction Coefficient Versus Openness Factor

excessive internally generated sound, the usual remedy is to apply acoustical treatment to the ceiling and other room surfaces. While these materials can be effective in controlling sound, they are often located on the two horizontal surfaces (ceiling and floor) and leave the opposing vertical surfaces of glass and bare wall to reflect sound. The noise reduction coefficient (NRC = average absorptance coefficient at four frequencies) for venetian blinds is about 0.10, as compared to 0.02 for glass and 0.03 for plaster. For drapery fabrics at 100% fullness, NRC ranges from 0.10 to 0.65, depending on the tightness of weave. Class III (tightly woven) fabrics have NRC values of 0.35 to 0.65. Figure 35 shows the relation between NRC and openness factor for fabrics of normal weight.

Example 16. To select a drapery fabric, consider the five environmental factors listed in Table 31. Choose a fabric designator which has suitable performance for all the factors important to the case being considered. If this is not possible, make compromises resulting in an acceptable designator. Determine from Table 29 if the SC for the chosen designator is satisfactory. Specific cases follow:

1. Where modification of a distracting view is necessary, but a degree of outward vision is needed and an SC of 0.50 with 0.25-in. gray glass is satisfactory (see Table 31, Item 5), select II_M or II_L; Item 2, select II_M. The SC for II_M on Table 29 is approximately 0.46, therefore satisfactory.
2. Where protection from radiation is paramount and minimum SC is necessary (see Table 31, Item 1), select a closed weave, III_M or III_L. Since the SC for III_L is lowest (see Table 29), choose III_L.
3. When good outward vision is desired, together with some reduction in brightness, choose I_M or I_D (Table 31).

Double Drapery

Double draperies (two sets of draperies covering the same area) have a light, open weave on the fenestration product side of the fenestration for outward vision and daylight when desired, and a heavy, close weave or opaque drapery on the room side to block out sunlight and provide privacy when desired. When properly selected and used, double draperies provide a reduced U-factor and a lowered SC.

The reduced U-factor results principally from adding a semi-closed air space to the barrier. A U-factor of about 0.57 Btu/h·ft²·°F is achieved using double draperies with single glass, and about 0.37 Btu/h·ft²·°F with insulating glass.

To most effectively reduce solar heat gain, the drapery exposed to sunlight should have high reflectance and low transmittance. The light, open-weave drapery should be opened when the heavy drapery is closed to prevent entry of sunlight. The open weave drapery works better on the room side, since this arrangement improves the SC and increases the U-factor when both draperies are closed.

Properly used double draperies give (1) extreme flexibility of vision and light intensity, (2) a lowered U-factor and SC, and (3) an improved comfort condition, since the room-side drapery is more nearly at room temperature. Table 31 gives characteristics of individual draperies. For large areas, the SC should be calculated in detail to determine the cooling load.

DAYLIGHTING FROM FENESTRATION

In the United States, a general rule has been that the fenestration area should be at least 20% of the floor area. In Europe, a similar rule was based on a minimum illumination value on the normal work plane from a standard overcast sky condition.

The secondary visual benefit of fenestration is the amount and quality of light it produces in the work environment. One general rule determined the need for auxiliary electric light by assuming that daylight was adequate for a depth of two and one-half times the height of the fenestration product into the room based on a normal sill height. To prevent excessive glare, all fenestration should have sun controls. Variable and removable controls are often more effective in daylight than fixed controls.

IES (1979) describes a method for more accurate evaluation of daylight distribution in a space. This practice shows a simple way of calculating the daylight distribution on the work plane from windows and skylights with and without controls. Many other daylight prediction tools calculate illuminance from radiant flux transfer or ray tracing.

The various daylight prediction tools can be used to compare the relative value of daylight distribution from alternative fenestration systems, but ultimately the designer must evaluate costs and benefits to choose between alternative designs. A simplified prediction tool (BEEM) is available for such analysis (Rundquist 1991). Also, total loss of productivity from an electric brown-out in a space with no natural ventilation or daylight may be as important as the benefits of many energy-saving schemes.

ANNUAL ENERGY PERFORMANCE

Instantaneous energy performance indices (U-factor, solar heat gain coefficient, air leakage, etc.) are typically used to compare fenestration systems under a fixed set of conditions. However, these indices alone are not good indicators of the annual energy performance (energy savings/costs) attributable to the fenestration. Furthermore, fenestration annual energy performance is difficult to quantify because of the dynamic response that occurs between the fenestration and the environment in which it is installed. The four basic mechanisms of fenestration energy performance that were addressed in the chapter—thermal heat transfer, solar heat gains, air leakage and daylighting—should all be considered, but the annual energy performance of fenestration systems can only be accurately determined when a large number of variables are considered. Building type and orientation, climate (weather, temperature, wind speed), microclimate (shading from adjacent buildings, trees, terrain), occupant usage patterns, and certain HVAC parameters can significantly affect the annual energy impact of fenestration.

A detailed dynamic, hourly computer simulation for the specific building and climate of interest is the most effective way to estimate fenestration annual energy performance. These analysis techniques do not consider issues of performance durability for the various instantaneous indices and should only be used as an initial annual energy performance indicator (Mathis and Garries 1995).

Simplified Techniques for Characterizing Fenestration Annual Energy Performance. While dynamic hourly modeling is the most accurate technique for determining fenestration annual energy performance, it is not readily available to many decision makers and end users of fenestration products. Under these circumstances, it may be useful to assess the relative importance of, or balance the tradeoff between, the known instantaneous performance indices of U-factor, SHGC, air leakage, and light transmittance for any given fenestration system when considering heating, cooling, and lighting loads. LBL (1995) and Sullivan et al. (1992) describe personal computer programs that are being developed to run this simplified analysis for residential windows.

Broad generalizations can be made for some types of building and climates. For instance, large commercial buildings often require substantial cooling because they have high internal loads, large thermal mass, and unfavorable orientations. As a result, the primary objective may be to install fenestration with a low SHGC to reduce the cooling load. Also, an evaluation of commercial fenestration annual energy use can determine the tradeoff between artificial lighting and natural daylighting.

In contrast to commercial buildings, the electric lighting loads in low-rise, detached residential buildings are small in comparison to the heating and cooling loads and the energy influence of daylighting may be neglected.

While several simplified annual energy performance indices for fenestration have been developed, at present they only apply to fenestration for detached residential buildings and are not appropriate for multi-family residential or commercial building fenestration.

Simplified Residential Annual Energy Performance Ratings. Annual energy performance ratings can provide a simple product comparison. Such ratings have been derived with many assumptions, usually to suit local climate conditions.

CAN/CSA *Standard* A440.2 describes a simplified Energy Rating (ER) applicable to residential heating in the Canadian climate. The standard also provides for specific energy ratings (ERS) to compare products by orientation and climate. NFRC *Standard* 900 includes an energy rating system to cover a range of U.S. climates that provides both simple heating (FHR) and cooling (FCR) performance indices (Crooks et al. 1995). FHR and FCR can be weighted according to that proportion of residential energy use or cost that is attributable to both heating and cooling, and establish the relative energy efficiency of residential windows.

CHOOSING GLASS FOR BUILDINGS

The choice of glass as a glazing material, should be based on (1) light transmittance and daylight use, (2) occupant comfort, (3) sound reduction, (4) strength (deflection under load), (5) resistance to thermal stress, (6)safety (security), (7) life cycle costs, and (8) aesthetics. In many respects, these factors and the thermal performance of glass are related.

LIGHT TRANSMITTANCE AND DAYLIGHT USE

When daylight is to be the primary lighting system, the minimum expected daylight in the building must be calculated for the building performance cycle and integrated into the lighting calculations. IES (1979) gives daylight design and calculation procedures. In some glazing applications, such as artists' studios and showrooms, maximum transmittance may be required for adequate daylighting of the interior. Regular clear glass, produced by float, plate, or sheet process, may be the logical choice.

When daylight is a supplementary light source, the electric lighting can be designed independently of the daylight system. But adequate switching must be included to substitute available daylight for electric lighting by automatic or prescribed manual control. Photosensitive controls automatically adjust shading devices to provide uniform illumination and reduce energy consumption. Manual control is less effective.

Buildings with large areas of glass usually have insulating glass units with clear, tinted, or reflective coatings. The tinted and reflecting units reduce the brightness contrast between fenestration products and other room surfaces and provide a relatively glare-free environment for most daylight conditions.

Table 11 and Table 32 list typical solar energy transmittances and daylight transmittances for various types of glass. Manufacturers' literature have more appropriate values.

The color of glass chosen for a building depends largely on where and how it is used. For commercial building lobbies, showroom fenestration products, and other areas where maximum visibility from exterior to interior is required, regular clear glass is generally best. For other glass areas, a tinted glass may best complement the interior colors. Bronze, gray, and reflective-film glasses also give some privacy to building occupants during daylight hours. Patterned, etched, or sandblasted glass that diffuses lighting is available.

The primary purpose of a fenestration product to provide a view of the exterior and not to save energy. It is also installed to illuminate the interior. Some buildings are designed especially to use the daylight coming through fenestration to reduce the amount of electric lighting and its attendant energy costs. Daylighting used to displace electric lighting provides a *direct* benefit because it reduces the amount of electrical energy consumed by the lighting, and an *indirect* benefit because it reduces the amount of electrically produced heat that may have to be removed by the air-conditioning.

Table 32 Daylight Transmittance for Various Types of Glass

Type of Glass	Visible Transmittance
1/8 in. regular sheet or float glass	0.86 to 0.91
1/8 in. gray sheet	0.31 to 0.71
3/16 in. gray sheet	0.61
7/32 in. gray sheet	0.14 to 0.56
1/4 in. gray sheet	0.52
1/4 in. green/float glass	0.75
1/4 in. gray plate glass	0.44
1/4 in. bronze plate glass	0.49
1/2 in. gray plate glass	0.21
1/2 in. bronze plate glass	0.25
Coated glasses (single, laminated, insulating)	0.07 to 0.50

The light transmitting or **visible transmittance** (VT) property of a fenestration product is similar to solar transmittance, except that an additional weighting function is included to account for the spectral response of the human eye. In most applications a high visible transmittance is desirable. In northern climates, a good solar heat gain is also important to offset heating costs. In southern climates a low solar heat gain is good for offsetting cooling costs. In the latter situation, it is difficult to have both a high visible transmittance and a low solar heat gain coefficient. Figure 36 and Figure 37 show a plot of visible transmittance versus SHGC for a number of glazing systems covering a range of spectral selectivities. The data is for normal incidence and a single, ASTM standard solar spectral distribution.

Three different zones are shown in Figure 36. In the **neutral zone**, colorless glazing, which has an almost uniform transmittance over the visible spectrum, may be installed. Glazings in this zone may be colored, but it is not necessary.

In the **color zone** the only way to achieve a high level of visible transmittance for a given level of solar heat gain coefficient is to block some of the red and blue wavelengths at the edges of the V-lambda function (human eye spectral response function) with a

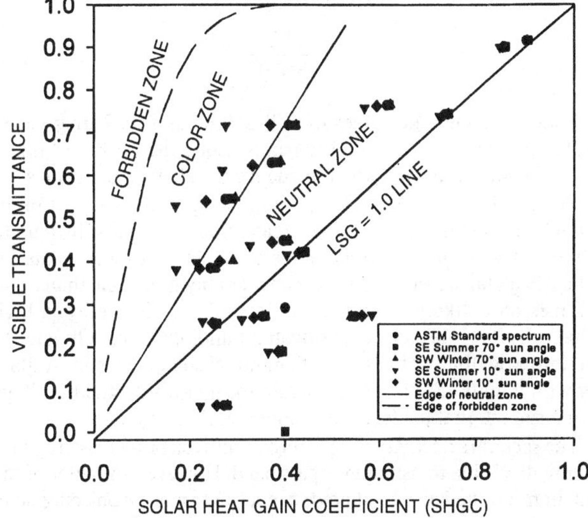

Fig. 36 Visible Transmittance Versus SHGC at Various Spectral Selectivities (McCluney 1996)

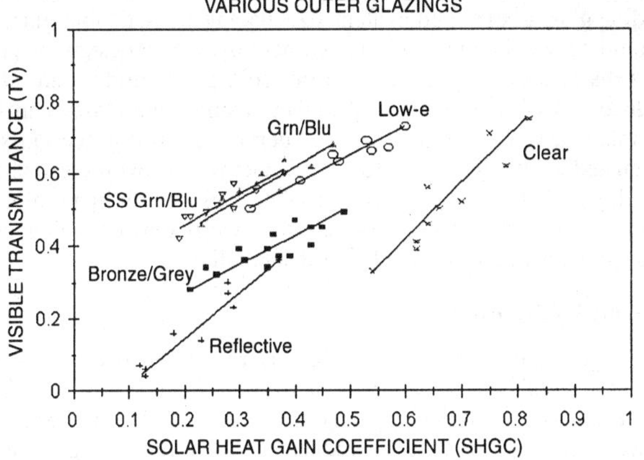

Fig. 37 Visible Transmittance Versus SHGC for Several Glazings with Different Spectral Selectivities

Table 33 Spectral Selectivity of Several Glazings

Glazing	VT	SHGC	LSG
Reflective blue-green	0.33	0.38	0.87
Film on clear glass	0.19	0.22	0.86
Green tinted, medium	0.75	0.69	1.09
Green low-e	0.71	0.49	1.45
Sun-control low e + green	0.36	0.23	1.56
Super low-e + clear	0.71	0.40	1.77
Super low-e + green	0.60	0.30	2.00

spectrally selective glazing. This selective glazing imparts color to the transmitted radiation (or otherwise alters the spectral transmittance and hence color over the visible portion of the spectrum).

In the **forbidden zone**, no combination of visible transmittance and solar heat gain coefficient will work for normal incidence and for the solar spectral distribution used. (Changing the solar spectral distribution used to calculate VT and SHGC shifts the transition curves somewhat. A low solar altitude angle, direct-beam spectrum moves the curves to the left in Figure 36.) It can be seen that the glazings which transmit more solar radiant heat than light cluster on the lower portion of the plot.

The VT versus SHGC chart can be a used to illustrate an index of spectral selectivity attained by a glazing. It is called the **light-to-solar-gain ratio** or LSG, and it is defined as

$$LSG = \frac{VT}{SHGC} \qquad (60)$$

Some characteristic values for VT, SHGC, and LSG are given in Table 33 for several different glazings, using the ASTM standard spectral distribution at normal incidence to calculate the values.

The LSG can be useful in spotting errors in the calculation of SHGC. In general, an LSG value greater than 2.0 is very difficult to achieve. A value near zero should be particularly suspect because it indicates a glazing that transmits considerably more heat than light, and it is an unlikely candidate for general use. Generally, a high value of LSG is desired for residential buildings in hot climates to maximize daylight admission with minimal solar heat gain. A glazing with a high LSG is also desirable for non-residential buildings with large internal loads, even in cool or cold climates.

The spectral selectivity of a glazing system can be increased by narrowing its visible transmittance pass band. However, the color of the radiation admitted may be altered so much as to make it objectionable.

OCCUPANT COMFORT AND ACCEPTANCE

ASHRAE *Standard* 55 recommends that variations in radiant temperature be limited for occupant comfort. For buildings with sliding glass doors or equivalent-size glazing areas, double-glazing should be installed in all climates where the winter design temperature is less than 45°F. Drafts from cold glass surfaces are also reduced in cold weather, and higher relative humidities can be maintained without condensation. Such high-performance glasses often reduce cooling load 15 to 30% below that with clear glass (Rudoy and Duran 1975). Clear insulating glass with light-colored continuous interior shading, or especially with exterior shading, can reduce fenestration cooling load substantially.

Sound Reduction

Proper acoustical treatment of exterior walls can decrease noise levels in certain areas. The air-tightness of a wall is the primary factor to consider in reducing sound transmission from the exterior. Once walls and fenestration products are tight, the choice of glass and draperies becomes important. Draperies do not prevent sound from coming through the fenestration; they act as an absorber for sound that does penetrate. Table 34 lists average sound transmission losses for various types of glass. These averages apply for the frequency range of 125 to 4000 Hz and were determined by tests based on ASTM *Standard* E 90. Chapter 43 of the 1995 *ASHRAE Handbook—Applications* has further details.

Table 34 Sound Transmittance Loss for Various Types of Glass

Type of Glass	Sound Transmittance Loss, dB
1/8 in. double-strength sheet glass	24
1/4 in. plate or float glass	27
1/2 in. plate glass	32
3/4 in. plate glass	35
1 in. plate glass	36
1/4 in. laminated glass (9/20 in. plastic interlayer)	30
1 in. insulating glass	32
1/2 in. laminated glass (9/20 in. plastic interlayer)	34
Insulating glass, 6 in. air space, 1/4 in. plate or float glass	40

Strength and Safety

Wind loads are specified in most building codes, and these requirements may be adequate for many structures. However, detailed wind tunnel tests should be run for tall or unusually shaped buildings and for buildings where the surroundings create unusual wind patterns. The strength of glass for annealed, heat strengthened, tempered, laminated and insulated, is given in ASTM *Standard* E 1300.

Thermal expansion and contraction can break ordinary annealed glass. This expansion and contraction can be caused by solar radiation onto partly shaded glass, by heat traps from drop ceilings and tight fitting drapes, or by HVAC ducts incorrectly directed toward the glazing. High performance tinted and reflective glasses, with low-e coatings, are usually more vulnerable to thermal stress breakage than clear glass.

Heat treating (heat strengthening or fully tempering) the glass reduces thermal stress breakage. Heat strengthened glass, though not a safety glass, is usually preferred to tempered (safety) glass because it typically has less distortion and is much less likely to break spontaneously. Spontaneous breakage can occur on very rare occasions in tempered glass. The glass manufacturer or fabricator should be consulted for information on thermal stress performance.

Building codes may require glass in certain positions to perform with certain breakage characteristics, which can be satisfied by tempered, laminated, or wired glass. In this case, glass should meet *Federal Standard* 16 CRF 1201 or other appropriate breakage performance requirements.

Life-Cycle Costs. Alternative building shells should be compared to assure satisfactory energy use and total energy budget compliance, if required. ASHRAE *Standard* 90.1 and 90.2 should be used as a starting point. A life-cycle cost model should be developed for each system considered (see Chapter 33 of the 1995 *ASHRAE Handbook—Applications*).

DURABILITY

Representative samples of insulating glass (IG) units are usually tested (for the seal durability) according to test methods to ensure the integrity of the seal. Failure of IG units is usually indicated by loss of adhesion of sealant to the glass. As a result, fogging occurs inside the glazing cavity. Elmahdy et al. (1995) studied the durability of IG units filled with argon gas. The results indicated that, under normal conditions, argon loss due to diffusion through the sealant is very small. However, when cracks or pinholes exist in the sealant most of the argon gas escapes from the unit, which implies that the implementation of stringent quality control procedures is essential for the production of durable IG units.

The degradation of organic materials and other chemical components in the IG units, as a result of exposure to ultraviolet radiation, is also among the factors affecting the durability and service life of fenestration. The use of low-emissivity coating on glass tends to enhance the appearance of chemical deposits on the glass surface. Also, muntin bars inserted in the glazing cavities may cause excessive unit failure during the ultraviolet volatile (fogging) test unless strict quality assurance processes are implemented.

The Sealed Insulating Glass Manufacturers Association (SIGMA), in a 15-year correlation study, found that long-term performance and durability of insulating glass correlates well with the test level to which such unit's construction had been manufactured according to the ASTM E 773- test method and ASTM E 774 specification for sealed insulating glass. The units showing the greatest resistance to seal failure were those tested in conformance with the ASTM E 774 Class CBA standard. Units that did not qualify to the A level had a higher percentage of failure. During the field correlation studies, units glazed in compliance with SIGMA recommendations and previously referenced guidelines performed for longer periods. This is compared to units not constructed properly, having deficiencies in the glazing system, or not meeting ASTM standards.

The durability of fenestration also depends on the durability of other components such as weather-stripping, gaskets, glazing tapes, air seals, and hardware. The wear on these elements with time and use may result in excessive air and water leakage, which affects overall performance and the service life. Excess water leakage may damage the fenestration product, especially the IG edge seal and wall section where the product is mounted. Excessive air leaks may cause frost build up and condensation on the fenestration surfaces.

Elmahdy (1995) and Patenaude (1995) showed that when windows are tested at high pressure and temperature differentials, they experience air leakage rates that exceed those determined at 75 Pa and no temperature differential (these conditions are used in rating the window air leakage in U.S. and Canadian standards). In other studies (CANMET 1991, 1993), the effect of pressure and motion cycling on windows caused excessive degradation in almost all window performance factors, particularly condensation resistance, ease of operation, air leakage, and water leakage.

In addition to the factors mentioned previously, durability of glazing may affect long-term energy performance as well. Construction of insulating glass, in order to predict long term performance, should be subjected to a test and certification program such as ASTM E 774 class CBA level and the requirements of SIGMA or the CGSB *Standard* 12.8 certified by the Insulating Glass Manufacturers Association of Canada (IGMAC) or equivalent.

CODES AND STANDARDS

AAMA. 1988. Voluntary test method for thermal transmittance and condensation resistance of windows, doors and glazed wall sections. *Publ.* AAMA 1503.1-88. American Architectural Manufacturers Association.

ASHRAE. 1988. Method of measuring solar-optical properties of materials. ANSI/ASHRAE Standard 74-1988.

ASHRAE. 1989. Energy efficient design of new buildings except low-rise residential buildings. *Standard* 90.1-1989.

ASTM. 1981. Recommended practice for laboratory measurements of airborne sound transmission loss of building partitions. *Standard* E 90-81. American Society for Testing and Materials, West Conshohocken, PA

ASTM. 1991. Standard test method for determining rate of air leakage through exterior windows, curtain walls, and doors under specified pressure differences across the specimen. *Standard* E 283-91.

ASTM. 1982. Standard test method for solar absorptance, reflectance, and transmittance of materials using integrating spheres. *Standard* E 903-82.

ASTM. 1986. Standard test method for solar transmittance (terrestrial) of sheet materials using sunlight. *Standard* E 1084-86.

ASTM. 1987a. Standard tables for terrestrial direct normal solar spectral irradiance for air mass 1.5. *Standard* E 891-87.

ASTM. 1987b. Standard tables for terrestrial solar spectral irradiance at air mass 1.5 for a 37° tilted surface. *Standard* E 892-87.

ASTM. 1988a. Standard practice for calculation of photometric transmittance and reflectance of materials to solar radiation. *Standard* E 971-88.

ASTM. 1988b. Standard test method for solar photometric transmittance of sheet materials using sunlight. *Standard* E972-88.

ASTM. 1987. Standard tables for terrestrial direct normal solar spectral irradiance for air mass 1.5. *Standard* E 891-87.

ASTM. 1988. Standard test method for seal durability of sealed insulated glass units. *Standard* E 773-88.

ASTM. 1992. Standard specification for sealed insulated glass units. *Standard* E 774-92.

ASTM. 1994. Standard practice for determining the minimum thickness and type of glass required to resist a specific load. *Standard* E 1300-94.

CABO. 1995. Model Energy Code 1995 Edition. Council of American Building Officials.

CAN/CSA. 1990. Windows. *Standard* A440. Canadian Standards Association, Rexdale, ON.

CAN/CSA. Windows/User selection guide to CSA standards. A440-M90/ A440.1-M90.

CAN/CSA.1993. Energy performance evaluation of windows and sliding glass doors. *Standard* A440.2-93.

CSA.1995. Energy performance evaluation of swinging doors. *Standard* A453-95. Canadian Standards Association, Rexdale, ON.

Federal Standard 16 CRF 1201. Safety standard for architectural glazing materials.

NAGDM. 1992. Test method for thermal transmittance and air infiltration of garage doors. *Standard* 105-1992. National Association of Garage Door Manufacturers, Chicago.

NFRC. 1991a. Procedure for determining fenestration product thermal properties (currently limited to U-values). *Standard* 100-91. National Fenestration Rating Council, Silver Spring, Maryland.

NFRC. 1991b. Interim standard test method for measuring the steady-state thermal transmittance of fenestration systems using hot box methods. Attachment A, *Standard* 100-91.

NFRC. 1991c. Procedure for determining door system product thermal properties (currently limited to U-values). Section B, *Standard* 100-91.

NFRC. 1993. Standard test method for emittance of specular surfaces using spectrometric measurements. *Standard* 301-93.

NFRC. 1994. Procedures for determining solar optical properties of simple fenestration products. *Standard* 300-94.

NFRC. 1995a. Procedure for determining fenestration product solar heat gain coefficients at normal incidence. *Standard* 200-95.

NFRC. 1995b. Procedure for determining fenestration product air leakage. *Standard* 400-95. National Fenestration Rating Council.

NFRC. 1995c. Procedure for determining the annual heating and cooling energy ratings of fenestration products used in residential dwellings. *Standard* 900-95. National Fenestration Rating Council.

REFERENCES

AGSL. 1992. Vision3, glazing system thermal analysis—User manual. Dept. Of Mechanical Engineering, University of Waterloo, Waterloo, Ontario, Canada.

Arasteh, D. 1989. An analysis of edge heat transfer in residential windows. *Proceedings*, ASHRAE/DOE/BTECC Conference, Thermal Performance of the Exterior Envelopes of Buildings IV, 376-87.

Arasteh, D. et. al. 1994. WINDOW 4.1: A PC program for analyzing window thermal performance in accordance with standard NFRC procedures. *Publ.* LBL-35298, Lawrence Berkeley Laboratory, Energy & Environment Division, Berkeley, CA.

Arasteh, D. et al. 1995. Recent technical improvements to the WINDOW computer program. *Proceedings,* Window Innovations Conference '95. CANMET, Ottawa.

Beck, F.A., B.T. Griffith, D. Turler, D. Arasteh. 1995. Using infrared thermography for the creation of a window surface temperature database to validate computer heat transfer models. *Proceedings*, Window Innovations Conference '95. CANMET, Ottawa.

Brandle, K. and R.F. Boehm. 1982. Air flow windows: Performance and applications. Proceedings, Thermal Performance of the Exterior Envelopes of Buildings II. ASHRAE/DOE Conference.

Burkhardt, W.C. 1975. Solar optical properties of gray and bronze solar control series transparent acrylic sheet. *ASHRAE Transactions* 81(2):384.

Burkhardt, W.C. 1976. Acrylic plastic glazing properties, characteristics and engineering data. *ASHRAE Transactions* 82(1):683-702.

CANMET. 1991. A study of the long term performance of operating and fixed windows subjected to pressure cycling. M91-7/214-1993E. Efficiency and Alternative Energy Technology Branch, CANMET, Ottawa.

CANMET. 1993. Long term performance of operating windows subjected to motion cycling. M91-7/235-1993E. Efficiency and Alternative Energy Technology Branch, CANMET, Ottawa.

Carmody, J., S. Selkowitz, and L. Heschong. 1996. *Residential windows—A guide to new technologies and energy performance.* W.W. Norton.

Carpenter, S.C. and A. McGowan. 1993. Effect of framing systems on the thermal performance of windows. *ASHRAE Transactions* 99(1).

Carpenter, S.C. and Elmahdy, A.H. 1994. Thermal performance of complex fenestration systems. *ASHRAE Transactions* 100(2):1179-86.

Carpenter, S.C. and Hogan, J., 1996. Recommended U-factors for swinging, overhead and revolving doors. *ASHRAE Transactions* 102(1):955-59.

CEA. 1995. *Energy-efficient residential and commercial windows reference guide*. Canadian Electricity Association, Montreal, PQ.

Crooks, B.P., et al. 1995. NFRC efforts to develop a residential fenestration annual energy rating methodology. *Proceedings*, Window Innovations Conference '95. CANMET, Ottawa.

Curcija, D. and W.P. Goss. 1994 Two-dimensional finite element model of heat transfer in complete fenestration systems. *ASHRAE Transactions* 100(2):1207-21.

Curcija, D. and W.P. Goss. 1995a. Three-dimensional finite element model of heat transfer in complete fenestration systems. *Proceedings*, Window Innovations Conference '95. CANMET, Ottawa.

Curcija, D. and W.P. Goss. 1995b. New correlations for convective heat transfer coefficient on indoor fenestration surfaces—Compilation of more recent work. *ASHRAE/DOE/BTECC Conference, Thermal Performance of the Exterior Envelopes of Buildings VI.*

Curcija, D., W.P. Goss, J.P. Power, and Y. Zhao. 1996. "Variable-h" model for improved prediction of surface temperatures in fenestration systems. *Technical Report*, University of Massachusetts at Amherst.

deAbreu, P., R.A. Fraser, H.F. Sullivan, and J.L. Wright. 1996. A study of insulated glazing unit surface temperature profiles using two-dimensional computer simulation. *ASHRAE Transactions* 102(2).

Elmahdy, H. 1996. Surface temperature measurement of insulating glass units using infrared thermography. *ASHRAE Transactions* 102(2).

Elmahdy, A.H. and S.A. Yusuf. 1995. Determination of argon concentration and assessment of the durability of high performance insulating glass units filled with argon gas. *ASHRAE Transactions* 101(2).

Elmahdy, A.H. 1995. Air leakage characteristics of windows subjected to simultaneous temperature and pressure differentials. Window Innovation Conference'95, Proceedings, CANMET.

ElSherbiny, S.M., et al. 1982. Heat transfer by natural convection across vertical and inclined air layers. *Journal of Heat Transfer* 104:96-102.

EEL. 1990. FRAME/VISION Window performance modelling and sensitivity analysis. Institute for Research in Construction, National Research Council of Canada, Ottawa.

EEL. 1995. The FRAMEplus toolkit for heat transfer assessment of building components. Enermodal Engineering. Kitchener, ON and Denver, CO.

Ewing, W.B. and J.I. Yellott. 1976. Energy conservation through the use of exterior shading of fenestration. ASHRAE *Transactions* 82(1):703-33.

Finlayson, E.U. and D. Arasteh. 1993. Window 4.0: Documentation of calculation procedures. LBL-33943/UC-350. Lawrence Berkeley Laboratory, Energy & Environment Division, Berkeley, CA.

Gates, D.M. 1966. Spectral distribution of solar radiation at the earth's surface. *Science* 151(2):3710.

Griffith, B.T., D. Turler, and D. Arasteh. 1996. Surface temperature of insulated glazing units: infrared thermography laboratory measurements, *ASHRAE Transactions* 102(2).

Gueymard, C.A. 1987. An anisotropic solar irradiance model for tilted surfaces and its comparison with selected engineering algorithms. *Solar Energy* 38:367-86. Erratum, *Solar Energy* 40:175 (1988).

Gueymard, C.A. 1993a. Critical analysis and performance assessment of clear sky solar irradiance models using theoretical and measured data. *Solar Energy*.

Gueymard, C.A. 1993b. Development and performance assessment of a clear sky spectral radiation model," *Proceedings*, Solar '93. American Solar Energy Society, 433-438.

Gueymard, C.A. 1995. A simple model of the atmospheric radiative transfer of sunshine: Algorithms and performance assessment. *Report* FSEC-PF-270-95, Florida Solar Energy Center, Cocoa, FL.

Hogan, J.F. 1988. A summary of tested glazing U-values and the case for an industry wide testing program. *ASHRAE Transactions* 94(2).

Hollands, K.G.T. and J.L. Wright. 1982. Heat loss coefficients and effective $\tau\alpha$ products for flat plate collectors with diathermous covers. *Solar Energy* 30:211-16.

IES. 1979. *Recommended practice of daylighting*. IES RP-5. Illuminating Engineering Society of North America, New York.

Iqbal, M. 1983. *An introduction to solar radiation.* Academic Press, Toronto.

Keyes, M.W. 1967. Analysis and rating of drapery materials used for indoor shading. *ASHRAE Transactions* 73(1):VIII.4.1-15.

Klems, J.H. 1989. U-values, solar heat gain, and thermal performance: Recent studies using the MoWitt. *ASHRAE Transactions* 95(1).

Klems, J.H. 1994a. A new method for predicting the solar heat gain of complex fenestration systems—I. Overview and derivation of the matrix layer calculation. *ASHRAE Transactions* 100(1):1065-72.

Klems, J.H. 1994b. A new method for predicting the solar heat gain of complex fenestration systems—II. Detailed description of the matrix layer calculation. *ASHRAE Transactions* 100(1):1073-86.

Klems, J.H. and G.O. Kelley. 1996. Calorimetric measurements of inward-flowing fraction for complex glazing and shading systems. *ASHRAE Transactions* 102(1):947-54.

Klems, J.H. and J.L. Warner. 1995. Measurement of bidirectional optical properties of complex shading devices. *ASHRAE Transactions* 101(1): 791-801.

LBL. 1994. WINDOW 4.1—A PC program for analyzing window thermal performance of fenestration products. LBL-35298. Windows and Daylighting Group, Lawrence Berkeley Laboratory, Berkeley, CA.

LBL. 1995. RESFEN 2.4—A prototype Windows-based PC program for calculating residential fenestration heating and cooling energy use and cost. Windows and Daylighting Group, Lawrence Berkeley Laboratory, Berkeley, CA.

LBL. 1996. THERM 1.0: A PC program for analyzing the two-dimensional heat transfer through building products. LBL-37371. Windows and Daylighting Group, Lawrence Berkeley Laboratory, Berkeley, CA.

Mathis, R.C. and R. Garries. 1995. Instant, annual life: A discussion on the current practice and evolution of fenestration energy performance rating. *Proceedings*, Window Innovations Conference '95. CANMET, Ottawa.

McCabe, M.E., et al. 1986. U-value measurements for windows and movable insulations from hot box tests in two commercial laboratories. *ASHRAE Transactions* 92(1).

McCluney, R. 1987. Determining solar radiant heat gain of fenestration systems. *Passive Solar Journal* 4(4):439-87.

McCluney, R. 1991. The death of the shading coefficient? *ASHRAE Journal* 33(3):36-45.

McCluney, R. and C. Gueymard. 1992. SUNSPEC 1.0 operating manual. FSEC-SW-3-92. Florida Solar Energy Center, Cocoa, FL.

McCluney, R. 1993. Sensitivity of optical properties and solar gain of spectrally selective glazing systems to changes in solar spectrum. *Solar* '93. American Solar Energy Society.

McCluney, R. and L. R. Mills. 1993. Effect of interior shade on window solar gain. *ASHRAE Transactions* 99(2).

McCluney, R. 1994a. *Introduction to Radiometry and Photometry*. Artech House, Boston.

McCluney, R. 1994b. Angle of incidence and diffuse radiation influences on glazing system solar gain. *Proceedings,* Solar '94. American Solar Energy Society.

McCluney, R. 1996. Sensitivity of fenestration solar gain to source spectrum and angle of incidence. *ASHRAE Transactions* 102(2).

Moon, P. 1940. Proposed standard solar radiation curves for engineering use. *Journal of the Franklin Institute* 11:583.

Moore, G.L. and C.W. Pennington. 1967. Measurement and application of solar Properties of drapery shading materials. *ASHRAE Transactions* 73(1):VIII.3.1-15.

Optical Society of America, M. Bass, ed. 1995. *Handbook of optics, Vol. I —Fundamentals, techniques, and design*, 2nd ed. McGraw-Hill, New York.

Ozisik, N. and L.F. Schutrum. 1960. Solar heat gain factors for windows with drapes. *ASHRAE Transactions* 66:228.

Parmelee, G.V. and R.G. Huebscher. 1947. Forced convection heat transfer from flat surfaces. *ASHVE Transactions* 245-84.

Patenaude, A. 1995. Air infiltration rate of windows under temperature and pressure differentials. *Proceedings*, Window Innovation Conference'95. CANMET, Ottawa.

Pennington, C.W. 1968. How louvered sun screens cut cooling, heating loads. *Heating, Piping, Air Conditioning,* December.

Pennington, C.W., et al. 1964. Experimental analysis of solar heat gain through insulating glass with indoor shading. *ASHRAE Journal* 2:27.

Perez, R., et al. 1986. An anisotropic hourly diffuse radiation model for sloping surfaces—Description, performance validation, and site dependency evaluation. *Solar Energy* 36:481-98.

Reilly, S. 1994. Spacer effects on edge-of glass and frame heat transfer. *ASHRAE Transactions* 100(1):1718-23.

Reilly, M.S., F.C. Winkelmann, D.K. Arasteh, and W.L. Carroll, 1992. Modeling windows in DOE-2.1E. *Energy and Buildings* 22:59-66.

Rubin, M. 1982a. Solar optical properties of windows. *Energy Research* 6:122-33.

Rubin, M. 1982b. Calculating heat transfer through windows. *Energy Research* 6:341-49.

Rubin, M. 1985. Optical properties of soda lime silica glasses. *Solar Energy Materials* 12: 275-288.

Rudoy, W. and F. Duran. 1975. Effect of building envelope parameters on annual heating/cooling load. *ASHRAE Journal* 7:19.

Rundquist, R.A. 1991. Calculation procedure for daylighting and fenestration effects on energy and peak demand. *ASHRAE Transactions* 97(2).

Shewen, E.C. 1986. A Peltier-effect technique for natural convection heat flux measurement applied to the rectangular open cavity. Ph.D. thesis, Dept. of Mech. Engineering, Univ. of Waterloo, Ontario, Canada.

Sodergren, D. and T. Bostrom. 1971. Ventilating with the exhaust air window. *ASHRAE Journal* 13(4):51.

Sterling, E.M., A. Arundel, and T.D. Sterling. 1985. Criteria for human exposure in occupied buildings. *ASHRAE Transactions* 91(1).

Sullivan, H.F., J.L. Wright, and R.A. Fraser. 1996. Overview of a project to determine the surface temperatures of insulated glazing units: thermographic measurement and 2-D simulation. *ASHRAE Transactions* 102(2).

Sullivan, H.F. and J.L. Wright. 1987. Recent improvements and sensitivity of the VISION glazing system thermal analysis program. *Proceedings*, 12th Passive Solar Conference, ASES/SESCI, 145-49.

Sullivan, R., et al. 1992. RESFEN: A residential fenestration performance design tool. *ASHRAE Transactions* 98(1).

TSC. 1996. WinSARC: Solar angles and radiation calculation for MS windows. User's Manual. Tait Solar Co., Tempe, AZ.

Threlkeld, J.L. 1970. *Thermal environmental engineering*, 2nd ed. Prentice-Hall, Englewood Cliffs, NJ.

Threlkeld, J.L. and R.C. Jordan. 1958. Direct solar radiation available on clear days. *ASHRAE Transactions* 64:45.

Van Dyke, R.L. and T.P. Konen. 1982. Energy conservation through interior shading of windows: An analysis, test and evaluation of reflective venetian blinds. LBL 14369. Lawrence Berkeley Laboratory, Berkeley, CA.

Vild, D.J. 1964. Solar heat gain factors and shading coefficients. *ASHRAE Journal* 10:47.

Wehrli. 1985. Extraterrestrial solar spectrum. *Publ. No.* 615. Physikalisch Metrologisches Observatorium and World Radiation Data Center, Davos, Switzerland.

Wright, J.L. 1995a. Summary and comparison of methods to calculate solar heat gain. *ASHRAE Transactions* 101(1).

Wright, J.L. 1995b. VISION4 glazing system thermal analysis: User manual. Advanced Glazing System Laboratory, University of Waterloo, Ontario, Canada.

Wright, J.L. 1995c. "VISION4 glazing system thermal analysis: Reference manual. Advanced Glazing System Laboratory, University of Waterloo.

Wright, J.L. 1996. A correlation to quantify convective heat transfer between window glazings. *ASHRAE Transactions* 102(2)

Wright, J.L. and H.F. Sullivan. 1995a. A 2-D numerical model for natural convection in a vertical, rectangular window cavity. *ASHRAE Transactions* 100(2)1193-1206.

Wright, J.L. and H.F. Sullivan. 1995b. A 2-D numerical model for glazing system thermal analysis. *ASHRAE Transactions* 101(1).

Wright, J.L. and H.F. Sullivan. 1995c. A simplified method for the numerical condensation resistance analysis of windows. *Proceedings*, Window Innovations '95. CANMET, Ottawa.

Yazdanian, M. and J.H. Klems. 1994. Measurement of the exterior convective film coefficient for windows in low-rise buildings. *ASHRAE Transactions* 100(1):1087-96.

Yellott, J.I. 1963. Selective reflectance—A new approach to solar heat control. *ASHRAE Transactions* 69:418.

Yellott, J.I. 1966. Shading coefficients and sun-control capability of single glazing. *ASHRAE Transactions* 72(1):72.

Yellott, J.I. 1972. Effect of louvered sun screens upon fenestration heat loss. *ASHRAE Transactions* 78(l):199-204.

Zhao, Y., D. Curcija, and W.P. Goss. 1996. Condensation resistance validation project—Detailed computer simulations using finite element methods. *ASHRAE Transactions* 102(2).

BIBLIOGRAPHY

Allen, C.W. 1973. *Astrophysical quantities*. The Athlone Press, University of London, London.

Brambley, M.R. and S.S. Penner. 1979. Fenestration devices for energy conservation I. Energy savings during the cooling season. *Energy*, February.

Burkhardt, W.C. 1975. Solar optical properties of gray and brown solar control series transparent acrylic sheet. *ASHRAE Transactions* 81(1): 384-97.

Burkhardt, W.C. 1976. Acrylic plastic glazing; properties, characteristics and engineering data. *ASHRAE Transactions* 82(1):683.

Collins, B.L. 1975. Windows and people: A literature survey, psychological reaction with and without windows. *Building Science Series* 70. National Bureau of Standards (now NIST, Gaithersburg, MD).

Duffie, J.A. and W.A. Beckman. 1980. *Solar engineering of thermal processes*. John Wiley & Sons, New York.

Energy, Mines and Resources Canada. 1987. The effect of frame design on window heat loss phase 1. Report prepared by Enermodal Engineering Ltd. for Renewable Energy Branch, Ottawa.

Johnson, B. 1985. Heat transfer through windows. Swedish Council for Building Research, Stockholm.

Iqbal, M. 1983. *An introduction to solar radiation*. Academic Press, New York.

McCluney, W.R. 1968. Radiometry and photometry. *American Journal of Physics* 36:977-79.

Meyer-Arendt, J.R. 1968. Radiometry and photometry: Units and conversion Factors. *Applied Optics* 7:2081-84.

Nicodemus, F.E. 1963. Radiance. *American Journal of Physics* 31:368.

Nicodemus, F.E., et al. 1976-84. Self-study manual on optical radiation measurements, Part I--Concepts. *NBS Technical Notes* 910-1, 910-2, 910-3, and 910-4. National Bureau of Standards (now NIST, Gaithersburg, MD).

Nicodemus, F.E., et al. 1977. Geometrical considerations and nomenclature for reflectance. *NBS Monograph* 160, National Bureau of Standards (now NIST, Gaithersburg, MD).

Pennington, C.W. and D.E. McDuffie, Jr. 1970. Effect of inner surface air velocity and temperature upon heat gain and loss through glass fenestration. *ASHRAE Transactions* 76:190.

Rubin, M. 1984. Optical constants and bulk optical properties of soda lime silica glasses for windows. *Report* 13572. Applied Sciences Division, Lawrence Berkeley Laboratory, Berkeley, CA.

Selkowitz, S.E. 1979. Thermal performance of insulating windows. *ASHRAE Transactions* 85(2):669-85.

Van Dyke, R.L. and T.P. Konen. 1982. Energy conservation through interior shading of windows: An analysis, test and evaluation of reflective venetian blinds. LBL-14369, Lawrence Berkeley Laboratory, Berkeley, CA.

Yellott, J.I. 1965. Drapery fabrics and their effectiveness in sun control. *ASHRAE Transactions* 71(l):260-72.

CHAPTER 30

ENERGY ESTIMATING AND MODELING METHODS

THE energy requirements and fuel consumption of HVAC systems have a direct impact on the cost of operating a building and an indirect impact on the environment. This chapter discusses methods for estimating energy use as a guide in design, for standards compliance, and for economic optimization. These energy estimating methods can provide quantitative energy and cost comparisons among design alternatives. A large number of uncontrolled and unknown factors generally preclude, except in cases where extraordinarily detailed and careful measurements and observations have been made, the use of such methods for the precise calculation of absolute energy consumption. Thus, a careful measurement of energy use is the standard against which energy calculation results should be referenced. In no case should these methods be used to predict future utility bills.

GENERAL CONSIDERATIONS

PURPOSES

A primary objective of building energy analysis is economic—to determine which of the available options has the lowest total cost. Chapter 33 of the 1995 *ASHRAE Handbook—Applications* outlines different analysis techniques that may be used to select an optimum system. Many areas of engineering use life-cycle cost (LCC) optimization techniques in design. Unfortunately, LCC optimization has been restricted by the complexity of buildings, which makes an analysis complex and costly, and by the uniqueness of each building, which makes it difficult to spread the cost of the analysis over several units.

Many factors related to building energy design are not considered in a life cycle cost analysis of a building. One example is the thermal comfort of the occupants, which affects productivity that could be accounted for if the relationship between the two were accurately known. A small loss in productivity due to reduced comfort can quickly offset an energy savings. Therefore, the designer must treat comfort conditions as constraints on a design rather than as variables to be included in the optimization.

The design of the energy system for the building also has an impact on the environment. However, it is difficult to quantify in an life cycle cost optimization unless some authority provides a constraint or price signal.

A few building energy codes and standards allow the use of an energy analysis to demonstrate compliance with the energy performance goals in the code. This use of energy analysis programs may be more prevalent than actual comparative energy studies.

The preparation of this chapter is assigned to TC 4.7, Energy Calculations.

COMMON FACTORS

Although the procedures for estimating energy requirements vary considerably in their degree of complexity, they all have three common elements: the calculation of (1) space load, (2) secondary equipment load, and (3) primary equipment energy requirements. Here, secondary refers to equipment that distributes the heating, cooling, or ventilating medium to conditioned spaces, while primary refers to central plant equipment that converts fuel or electric energy to heating or cooling effect. A major distinction is made between steady-state methods (based on degree days or temperature bins) and dynamic methods (e.g., based on transfer functions).

The first step in calculating energy requirements is to determine the **space load,** which is the amount of energy that must be added to or extracted from a space to maintain thermal comfort. The simplest procedures assume that the energy required to maintain comfort is only a function of the outdoor dry-bulb temperature. More detailed methods consider solar effects, internal gains, heat storage in the walls and interiors, and the effects of wind on both building envelope heat transfer and infiltration. Chapters 27 and 28 discuss load calculation in detail.

While energy calculations are similar to the heating and cooling load calculations used to size equipment, they are not the same. Energy calculations are based on average use and typical weather conditions rather than maximum use and worst case weather. Currently, the most sophisticated procedures are based on hourly profiles for climatic conditions and operational characteristics for a number of typical days of the year or on 8760 h of operation per year.

The second step translates the space load to a **load on the secondary equipment**. This can be a simple estimate of duct or piping losses or gains, or a complex hour-by-hour simulation of an air system, such as variable air volume with outdoor air cooling. This step must include the calculation of all forms of energy required by the secondary system, i.e., electrical energy to operate fans and/or pumps, as well as energy in the form of heated or chilled water.

The third step calculates the fuel and **energy required by the primary equipment** to meet these loads and peak demand on the utility system. It considers equipment efficiencies and part-load characteristics. It is often necessary to keep track of the different forms of energy, such as electrical, natural gas, or oil. In some cases, where calculations are required to assure compliance with codes or standards, these energies must be converted to source energy or resource consumed, as opposed to energy delivered to the building boundary.

Often energy calculations lead to an economic analysis to establish the cost-effectiveness of conservation measures (ASHRAE 1989). Thus, thorough energy analysis provides intermediate data, such as time of energy usage and maximum demand, so that utility charges can be accurately estimated. Although not part of the energy calculations, estimated capital equipment costs should be included in such an analysis.

Complex and often unexpected interactions can occur between the systems or between various modes of heat transfer. For example, radiant heating panels impact the space loads by raising the mean radiant temperature in the space (Howell 1990). As a result, the air temperature can be lowered while maintaining comfort. Compared to a conventional heated air system, radiant panels create a greater temperature difference from the inside surface to the outside air. Thus, conduction losses through the walls and roof increase because the inside surface temperatures are greater. At the same time, the heating load due to infiltration or ventilation decreases because of the reduced indoor air to outdoor air temperature difference and a reduced stack effect. The infiltration rate may also decrease because the reduced air temperature difference reduces the stack effect.

CHOOSING AN ANALYSIS METHOD

The most important step in selecting an energy analysis method is to match the method capabilities with project requirements. The method must be capable of evaluating all design options with sufficient accuracy to make correct choices. The following factors apply generally (Sonderegger 1985):

- **Accuracy.** The method should be sufficiently accurate to allow correct choices. Because of the many parameters involved in energy estimation, absolutely accurate energy prediction is not possible (Waltz 1992).
- **Sensitivity.** The method should be sensitive to the design options being considered. The difference in energy use between two choices should be accurate.
- **Versatility.** The method should allow the analysis of all options under consideration. When different methods must be used to consider different options, an accurate estimate of the differential energy use cannot be made.
- **Speed and cost.** The total time (gathering data, preparing input, calculations, and analysis of output) to make an analysis should be appropriate to the potential benefits gained. With greater speed, more options can be considered in a given time. The cost of analysis is largely determined by the total time of analysis.
- **Reproducibility.** The method should not allow so many vaguely defined choices that different analysts would get completely different results (Corson 1992).
- **Ease of use.** This impacts both on the economics of analysis (speed) and the reproducibility of results.

Selecting Energy Analysis Computer Programs

The selection of a building energy analysis program depends on its application, the number of times it will be used, the experience of the user, and the hardware available to run it. The first criteria is the capability of the program to deal with the application. For example, if the effect of a shading device is to be analyzed on a building that will also be shaded by other buildings part of the time, the capability of analyzing detached shading is an absolute requirement, regardless of any other factors.

Because almost all manual methods have been implemented on a computer, the selection of an energy analysis method is the selection of a computer program. Today, all well-known programs run on microcomputers. However, the cost of the computer facilities and the software itself are typically a small part of running a building energy analysis. The major costs are the cost of learning to use the program and the cost of using it. Major issues that influence the cost of learning a program include: (1) complexity of the input procedures, (2) quality of the user's manual, and (3) availability of a good support system to answer questions. As the user becomes more experienced, the cost of learning becomes less important. However, the requirement for a complex set of input data, will continue to consume the time of even an experienced user.

The **complexity of input** is largely influenced by the availability of default values for the input variables. Default values can be used as a simple set of input data when detail is not needed or when the building design is very conventional; but additional complexity can be supplied when needed. Secondary defaults, which can be supplied by the user, are also useful in the same way. Some programs allow the user to specify a level of detail. Then the program requests only the information appropriate to that level of detail, using default values for all others.

The **quality of the output** is another factor to consider. Reports should be easy to read and uncluttered. The titles and headings should be unambiguous. Units should be stated explicitly. The users' manual should explain the meanings of the data presented. Graphic output can be very helpful. In most cases, simple summaries of overall results are the most useful, but very detailed output is needed for certain studies and also for debugging program input during the early stages of an analysis.

Before making a final decision, manuals for the most suitable programs should be obtained and reviewed; and, if possible, demonstration versions of the programs should be obtained and run. During this last part of the selection process, support from the software supplier should be tested. The availability of training should be considered when choosing a more complex program.

The **availability of weather data** and/or the availability of a weather data processing subroutine or program is a major feature of a program. Some programs include subroutine or supplementary programs that allow the user to create a weather file for any site for which weather data is available. Programs that do not have this capability must have weather files for various sites created by the program supplier. In that case, the available weather data and the terms on which the supplier will create new weather data files must be checked.

Auxiliary capabilities, such as economic analysis and design calculations, are a final concern in selecting a program. An economic analysis may include only the ability to calculate annual energy bills from utility rates, or it might extend to calculations, or even to LCC optimization. An integrated program may save time, because some input data will have been entered already for other purposes.

The results of computer calculations should be accepted with caution as the software vendor does not accept responsibility for the correctness of calculations or the use of the program. A manual calculation should be run to develop a good understanding of the underlying physical processes and building behavior. In addition, the user should (1) review the computer program documentation to determine what calculation procedures are used, (2) compare the results with manual calculations and measured data, and (3) conduct sample tests to confirm that the program delivers acceptable results.

Tools for Energy Analysis

Table 1 lists various tools for energy calculations that are discussed in the first part of the chapter. These techniques move from building design through a series of calculations to ultimately find energy consumption. Other methods discussed in the final section of the chapter move in the inverse direction, i.e., from measured energy consumption to ascertaining key characteristics of the building.

The Need for Modeling. Because systems that consume energy in buildings are nonlinear, dynamic, and very complex, few methods other than computer modeling are available for accurately calculating energy consumption. The most accurate methods for calculating building energy consumption are the most costly because of their intense computational requirements and their needed analyst or designer levels of expertise. In particular, simulation programs that assemble component models into system models and then exercise

Table 1 Energy Estimating Models

Method	Comments
	Steady State Models
Modified degree day method	Based on fixed reference temperature of 65°F
Variable base degree day method	Variable reference temperatures
ASHRAE bin method	Hours in temperature bin times load for that bin
ASHRAE modified bin method	Modified bin method with cooling load factors
	Dynamic Models
Thermal network (Sonderegger 1977)	Uses equivalent thermal parameters (inverse mode)
Response factors (Stephenson & Mitalas 1967)	Tabulated or as used in simulation programs
Fourier Analysis (Shurcliff 1984, Dhar 1995)	Frequency domain analysis convertible to time domain
BEVA, PSTAR (Subbarao 1986)	Combination of ARMA and Fourier series, includes loads in time domain
Modal analysis (Bacot et al. 1984)	Bldg. described by diagonalized differential equation using nodes
Computer simulation (DOE-2, BLAST)	Hourly simulation programs with system models
Computer emulation (HVACSIM+, TRNSYS)	Sub-hourly simulation programs

those models with weather and occupancy data are preferred by experts for determining energy use in buildings.

But often, energy consumption must be estimated quickly to study trends, compare systems, or study building effects such as envelope characteristics. For these purposes, simpler methods may be used. The modified bin and degree-day method are two simpler methods described later in the chapter. Fortunately, complex methods do not need to be understood in order to use the simplified ones.

COMPONENT MODELING AND LOADS

While the equipment and engineering fundamentals discussed here often appear in other *Handbook* chapters, subtle difference often arise when applying these models for energy calculations. *Appropriate models for calculating the load aspects of energy estimating are often different from models used to design or select equipment.* The primary and secondary system models, however, are the same whether used for hourly, bin, or typical day analyses.

CALCULATING INSTANTANEOUS SPACE SENSIBLE LOAD

Calculating instantaneous space sensible load is a key step in any building energy simulation. The **heat balance method** and the **weighting factor method** are two methods used for these calculations. A third method, the **thermal network method**, while not widely used, shows promise.

The *instantaneous space sensible load* is the rate of heat flow into the space air mass. This quantity, sometimes called the cooling load, differs from heat gain in that heat gain usually contains a radiative component that passes through the air and is absorbed by other bounding surfaces. Instantaneous space sensible load is entirely convective; even loads from internal equipment, lights, and occupants enter the air by convection from the surface of such objects or by convection from room surfaces that have absorbed the radiant component of energy emitted from these sources. However, some adjustment must be made when radiant cooling and heating systems are evaluated because some of the space load is offset directly by radiant transfer without convective transfer to the air mass.

For equilibrium, the instantaneous space sensible load must match the heat removal rate of the conditioning equipment. Any imbalance in these rates changes the energy stored in the air mass. Customarily, however, the thermal mass (heat capacity) of the air itself is ignored in an analysis, so that the air is always assumed to be in thermal equilibrium. Under these assumptions, the instantaneous space sensible load and the rate of heat removal are equal in magnitude and opposite in sign.

In sensible load calculations, cooling is generally the reference frame. If there is heat loss, only the algebraic signs change. Therefore, mathematical models that work equally well for either situation.

Weighting factors, also called **room response factors**, are used in the weighting factor method to calculate space instantaneous sensible load components from corresponding heat gain components. **Response factor** and **conduction transfer functions**, on the other hand, refer to conductive heat transfer (e.g., through walls and floors) and are used to calculate the transmission component of heat gain (see Chapter 28 for a discussion about the heat transfer method).

The weighting factor method and the heat balance method use conduction transfer functions (or their equivalents) to calculate transmission heat gain or loss. The principal difference is in the methods used to calculate the subsequent internal heat transfers to the room. Experience with both methods has indicated largely the same results, provided the weighting factors are determined for the specific building under analysis.

Heat Balance Method

The heat balance method for calculating net space sensible loads is based on the first law of thermodynamics (conservation of energy). It is more fundamental and requires fewer assumptions than the weighting factor method. However, the heat balance method requires more calculations at each point in the simulation process, which uses more computer time. The weighting factors used in the weighting factor method are determined by the heat balance equations in a form similar to the one presented here. And, although not necessary, linearization is commonly used to simplify the radiative transfer formulation.

The heat balance method allows the net instantaneous sensible heating and/or cooling load to be calculated on the space air mass. Generally, a heat balance equation is written for each enclosing surface, plus one for room air. This set of equations can then be solved for the unknown surface and air temperatures. Once these temperatures are known, they can be used to calculate the convective heat flow to or from the space air mass.

To write heat balance equations, a space in the building under analysis is considered as an enclosure bounded by a number of discrete surfaces (walls, floors, windows, and ceiling). At any time θ, each of these surfaces is assumed to be at some uniform temperature $t_{i,\theta}$. Also, the space air mass is assumed to be of uniform temperature $t_{a,\theta}$. At any plane boundary, the flux entering the boundary must equal the flux leaving the boundary. Thus, at the inside surface of any room wall, the heat flow into the surface—because of convection from room air, radiation from interior sources such as lights and people, and the net radiant interchange between the wall and all other surfaces in the room—is balanced by the conductive flux leaving the surface to penetrate the solid.

When radiant cooling and heating systems are evaluated, the radiant source should be identified as a room surface. The calculation procedure considers the radiant source in the heat balance analysis. Therefore, this method is preferred over the weighting factor method for evaluating radiant systems.

Similarly, for the outside surface of any exterior wall, the conductive flux leaving the surface to penetrate the solid toward the room is balanced by absorbed solar radiation, net longwave radiant flux from the surroundings, and convective flux from the outdoor air. A heat balance on the room air volume requires the air-conditioning system to remove the net heat added to the volume by *convection* from interior surfaces and interior heat sources (lights, people) and by mass transfer due to infiltration.

Thus, at the ith surface at time q, the energy (per unit area) balance, is

$$q_{i,\theta} = h_i(t_{a,\theta} - t_{i,\theta}) + \sum_{k=1}^{n} g_{i,k}(t_{k,\theta} - t_{i,\theta}) + R_i \tag{1}$$

where the first term represents heat gain due to convection from the room air, the summation represents radiative interchange (long-wave), and the last term R_i represents all other radiant energy absorbed by the surface. The last term includes shortwave radiation from lights, window solar gain, and the radiant component of occupant and equipment loads.

The left side of Equation (1) $q_{i,\theta}$ represents the net transfer to the surface from the air and surroundings. It must match the heat conducted into the solid surface. The latter can be represented in terms of historical values of surface temperatures and heat flux using conduction transfer functions. Response factor and conduction transfer functions are closely related solutions to the transient one-dimensional, multilayered-slab heated conduction problem. Mitalas and Stephenson (1967), Mitalas (1968), and Kusuda (1969) give several theoretically equivalent solutions, and at least two forms appear in Chapters 22 and 26. Equation (2) is a first-order conduction transfer function because it includes only one flux history term CR_i. Higher order conduction transfer functions contain multiple flux history terms—the first flux history term multiplies the previous hours heat flux; the second term multiplies the heat flux two hours ago, etc.

$$\begin{aligned} q_{i,t} &= \sum_{j=1}^{m} X_j t_{i,\theta-j} - \sum_{j=0}^{m} Y_j t_{oi,\theta-j} + CR_i q_{i,\theta-i} \\ &= X_0 t_{i,\theta} + \sum_{j=0}^{m} X_j t_{i,\theta-j} - \sum_{j=0}^{m} Y_j t_{i,\theta-j} + CR_i q_{i,\theta-i} \end{aligned} \tag{2}$$

By substituting Equation (2) into Equation (1), $q_{i,\theta}$ can be eliminated, and after rearrangement

$$\begin{aligned} &\left(X_0 + h_i + \sum_{k=1}^{m} g_{i,k}\right) t_{i,\theta} - \sum_{k=1}^{m} g_{i,k} t_{k,\theta} - h_i t_{a,\theta} \\ &= -\sum_{j=1}^{m} X_j t_{i,\theta-j} + \sum_{j=0}^{m} Y_j t_{oi,\theta-j} + CR_i q_{i,\theta-i} + R_i \end{aligned} \tag{3}$$

In addition to the n equation forms of Equation (3), a heat balance equation also can be written for the space air. Assuming negligible heat capacity of this air relative to the more massive building elements, the sum of all heat flow to it must be zero, yielding

$$\sum_{j=1}^{n} S_j h_j(t_{j,\theta} - t_{a,\theta}) + m_i c_p(t_{o,\theta} - t_{a,\theta}) + QI_\theta + QS_\theta = 0 \tag{4}$$

Here, the summation represents convective transfer from the enclosing surfaces (s_j is the area of the jth surface), the second-term infiltration, QI_θ convective transfer from internal objects, and QS_θ, the rate of energy addition and/or removal by the space-conditioning equipment. This equation can be rearranged to give

$$\sum_{j=1}^{n} S_j h_j(t_{j,\theta} - t_{a,\theta}) + m_i c_p t_{a,\theta} = m_i c_p t_{oa,\theta} - QI_\theta - QS_\theta \tag{5}$$

Equations (3) and (5) together represent a set of $n + 1$ equations in $n + 1$ unknown temperatures. The equations are arranged so that all unknowns are on the left, and all known quantities are on the right. This facilitates rewriting the problem to be solved in vector-matrix notation as

$$\boldsymbol{A} t_\theta = \boldsymbol{B} \tag{6}$$

where $\boldsymbol{A}$ is an $n + 1$ by $n + 1$ matrix, and $\boldsymbol{B}$ is a $(n + 1)$ vector. By convention, the first n elements of the t_θ vector are the surface temperatures, while the $n + 1$ element is the air temperature.

The elements of $\boldsymbol{A}$ and $\boldsymbol{B}$ can be determined from the expanded equations given previously. The first n diagonal elements of $\boldsymbol{A}$ are

$$a_{ii} = X_0 + h_i + \sum_{k=1}^{n} q_{i,k}; \qquad i = 1, n \tag{7A}$$

while the last diagonal element is

$$a_{n+1,n+1} = -\left(\sum_{k=1}^{n} S_k h_k + m_i c_p\right) \tag{7B}$$

and off-diagonal elements are

$$a_{i,j} = -g_{i,j}; \qquad i,j = 1, n \tag{8A}$$

$$a_{i,n+1} = -h_i; \qquad i = 1, n \tag{8B}$$

$$a_{n+1,j} = S_j h_j; \qquad j = 1, n \tag{8C}$$

The elements of $\boldsymbol{B}$ are the terms on the right of Equations (3) and (5).

Through these equations, the heat transfer at any surface is coupled to heat transfer at all other surfaces and to heat transfer to or from the room air. The heat transfer at an inside wall surface, for example, is coupled to the heat transfer at the outside surface through conduction transfer functions for the wall. It is coupled to the heat transfer at all other inside surfaces through equations governing intersurface radiant exchange, and it is coupled to the room air volume through the inside surface convection coefficient. The radiant and convective fluxes from solar effects and the effects of interior energy sources can be calculated independently, so that an equation for the flux at any surface in terms of the surface temperature, the temperature of all other surfaces, and the room air temperature can be written (assume that all *past* temperatures and fluxes are known).

The heat balance method requires that Equation (3) be solved for surface and air temperatures at each point, allowing subsequent calculation of sensible load. The solution can be derived by various mathematical methods, some more efficient than others. As discussed by Sowell and Walton (1980), computational speed can be increased by using the special properties of the coefficient matrix $\boldsymbol{A}$.

Regardless of the solution method, the procedure is to compute updated values of the right vector $\boldsymbol{B}$ and then solve for t_θ. Note that $\boldsymbol{B}$ involves (1) historical values of all surface temperatures, (2) current values of the outside surface temperatures, (3) the previous value of heat flux at each surface, (4) outside air temperature, (5) the absorbed shortwave radiant flux at each surface, (6) the internal load, and (7) the supply heating and/or cooling effect.

Historical data are developed and retained as the solution progresses, starting with some assumed initial conditions and past performance. Internal loads are usually calculated from schedules of occupancy, lighting, and equipment usage. The radiant absorption at each surface is usually approximated by uniformly distributing window solar load and known radiant components of lighting, equipment, and occupants to each unit area of walls, ceiling, and floor.

Outside surface temperatures can be found by writing a heat balance at each outside surface. This produces an equation similar to Equation (1), equating heat flux into the solid material to the algebraic sum of convection from outdoor air, long-wave radiant absorption from the sky and surroundings, and short-wave solar absorption. The resulting equations, one for each outside surface, can be rearranged to give an expression for the outside surface temperatures, shown in Equation (9).

$$t_{oi,\theta} = \left(q_{si,\theta} + q_{li,\theta} + h_o t_{oa,\theta} + \sum_{j=0}^{m} Y_j t_{i,\theta-j} - \sum_{j=1}^{m} Z_j t_{oi,\theta-j} + CR q_{oi,\theta-j} \right) / (h_o - Z_o) \quad (9)$$

Here, $q_{si,\theta}$ and $q_{li,\theta}$ represent the net short- and long-wave radiant flux at the *i*th outside surface at time θ, while h_o is the outside film coefficient. Note that the right side of Equation (9) contains the *inside* surface temperature at time θ, thus this set of equations must be solved simultaneously with Equation (6). However, since this would double the size of the overall heat balance problem, significantly increasing computer memory requirements and solution time, a simplified assumption is often made. For example, some algorithms simply use the outside surface temperatures from the *previous* time step when evaluating the right side of Equation (2). Resulting errors are thought to be small for thermally massive elements. For lightweight walls, however, the equations should probably be solved simultaneously.

The supply heating and/or cooling term in Equation (6) QS_θ presents a similar dilemma. Because of the thermostat and the nature of the air-conditioning process, QS_θ depends on room temperature $t_{a\theta}$, which is one of the unknowns in Equation (6). This requires some type of simultaneous solution procedure for Equation (6) and the space control characteristic. Iteration is usually employed to find a space temperature and supply heating and/or cooling amount that simultaneously satisfies the heat balance equations and the control characteristics.

In principle, the heat balance method extends directly to multiple spaces. An equation such as Equation (6) can be written for each space; this recognizes that the outside of partitions are in fact the inside surfaces of adjacent spaces. In practice, however, the size of the coefficient array required for solving the simultaneous equations becomes prohibitively large, and the solution time excessive. For this reason, many programs solve only one space at a time and assume that the adjacent space temperatures are either the same as the space in question or some assigned, constant value. Other approaches may remove this limitation (Walton 1980).

Example of Heat Balance Method

This example illustrates the steps involved in an hourly load calculation. Weather data are used as the forcing functions for the load calculation, with building envelope and occupancy being other key factors in the final result. Algorithms described in several Handbook chapters are used to obtain intermediate results. The length of calculation is due to the dynamic nature of the thermal response of massive elements in buildings.

Problem Statement

Given the following information, determine the zone loads for a single zone of an office building with the simplified floor plan shown in Figure 1. For this example, calculate the loads for a single day with weather data detailed below. Note that a number of phenomena are not included in this example, e.g. infiltration, interzone airflow, and interzone heat transfer. The building data and weather data are as follows:

1. Building type: Office, single-story
2. Location: Washington, DC (38.9°N latitude).
3. The zone is served by a variable air volume system. The full flow rate is 2924 cfm; the minimum fraction is 30%. The supply air temperature is 55.4°F. Heating is provided by baseboard heat, which has a capacity of 51,200 Btu/h. The system is controlled such that above 78.8°F, the VAV box is fully open; between 78.8°F and 75.2°F the VAV box is modulated (assumed linear) between fully open and the minimum fraction; between 69.8°F and 60.8°F, the baseboard heat is modulated (assumed linear) between 0% and 100%.

   Furthermore, the system is setback so that the above description only applies from 7 a.m. to 7 p.m. From 7 p.m. to 7 a.m., the baseboard heat only comes on if the temperature falls below 53.6°F.
4. Physical data-zone dimensions: 124 ft long, 11.8 ft deep, 12.4 ft high
5. Surface data-zone

Number	Direction	Surface	Opaque Surface Area, ft²	Glass Area, ft²
1	S	Exterior wall	740.3	0
2	S	Window	0	797.3
3	E	Interior partition	146.3	0
4	N	Interior partition	1537.6	0
5	W	Interior partition	146.3	0
6	Down	Floor	1464.4	0
7	Up	Roof	1464.4	0

Ext. wall: From outside in, constructed of the following layers: A2,C2,B15 (face brick, 4 in. low-density concrete block with 6 in. insulation)
Roof: From outside in, constructed of the following layers: A3,B14,E3,E2 (Steel Deck with 5 in. insulation)
Windows: Double glazed, with 0.5 in. airspace, wood/vinyl frame, fixed, overall U = 0.5 Btu/h·ft²·°F; this is a fixed value adopted for this example—it could be improved by accounting for the wind speed and orientation.
Floor: 4 in. thick concrete slab, assumed to be over 12 in. dirt, below which the soil temperature is 63°F.
Partition: 8 in. low-density concrete block

6. Internal loads and schedules

Lights:	8.6 Btu/h·ft², 0700-1900 0.86 Btu/h·ft², 1900-0700
Equipment:	3.5 Btu/h·ft², 0700-1900 0 Btu/h·ft², 1900-0700
People:	1 person/ 100 ft², 0700-1900 0 person/ft², 1900-0700
	$q_{sensible}$ = 256 Btu/h per person q_{latent} = 256 Btu/h per person

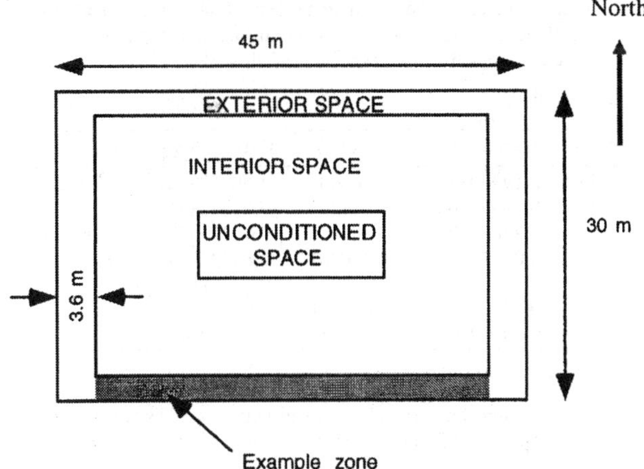

Fig. 1 Floor Plan of Building for Example

7. Weather data for the example day, October 21 are shown in the following table. Typically, weather data for energy analysis are made from a year of typical data, e.g., Typical Meteorological Year (TMY) or Weather Year for Energy Calculations (WYEC) data.

Weather Data for Example Day (October 21)

Hour	Dry-Bulb, °F	Wet-Bulb, °F	Sky Temp., °F	Beam Solar Radiation, Btu/ft²	Diffuse Solar Radiation, Btu/ft²	Ground Reflected Solar Radiation, Btu/ft²	Total Horizontal Solar Radiation, Btu/ft²
1	44.5	38.6	33.7	0	0	0	0
2	43.2	37.9	32.4	0	0	0	0
3	42.1	37.3	31.3	0	0	0	0
4	41.3	36.8	30.5	0	0	0	0
5	41.0	36.7	30.2	0	0	0	0
6	41.5	37.0	30.7	0	0	0	0
7	42.9	37.7	32.1	0	0	0	0
8	45.3	39.0	34.5	170.8	12.5	9.4	46.8
9	48.8	40.9	38.0	245.8	18.0	21.8	109.2
10	52.9	42.9	42.1	276.0	20.2	32.1	160.4
11	57.5	45.2	46.7	290.0	21.2	39.2	195.9
12	61.8	47.2	51.0	295.4	21.6	42.6	213.1
13	65.0	48.7	54.2	294.7	21.5	42.2	210.8
14	67.2	49.6	56.4	287.6	21.0	37.8	189.1
15	68.0	50.0	57.2	270.8	19.8	29.9	149.6
16	67.2	49.6	56.4	234.2	17.1	19.0	95.2
17	65.3	48.8	54.5	137.1	10.0	6.3	31.6
18	62.3	47.4	51.5	0	0	0	0
19	58.8	45.8	48.0	0	0	0	0
20	55.3	44.1	44.5	0	0	0	0
21	52.3	42.7	41.5	0	0	0	0
22	49.6	41.3	38.8	0	0	0	0
23	47.5	40.2	36.7	0	0	0	0
24	45.9	39.3	35.1	0	0	0	0

Solution

The heat balance method is used to determine the hourly zone loads which will be passed to the system simulation. Not every step of the solution is shown because the problem requires a simultaneous solution of 14 surface heat balance equations and one zone air heat balance equation for each hour. Furthermore, each of the 14 surface heat balance equations require knowledge of past values of surface temperatures and heat fluxes. For purposes of simulating a single day, it is assumed that the past weather history consists of a series of days equivalent to the current day.

1. Determine conduction transfer function coefficients for each surface. A variety of techniques may be used to determine the coefficients. In this case, the method described by Hittle and Bishop (1983) was used. The order of the conduction transfer function and the number of coefficients varies depending on the surface construction. The coefficients for each surface type are as follows:

Exterior wall

Time	Internal (Z)	Cross (Y)	External(X)	Flux (CR)
1	0.19000882	0.00000005	4.60562372	1.49882722
2	−0.39597896	0.00011249	−9.71842861	−0.69257158
3	0.28095791	0.00139126	6.80290937	0.09939408
4	−0.07730190	0.00210797	−1.83997715	
5	0.00658080	0.00065834	0.15357184	
6	0.00006232	0.00005821	0.00062279	
7	0.00000192	0.00000257	0.00000947	

Window

The window has effectively no thermal mass; hence the CTF is zeroth order, i.e. no flux history term.

Time	Internal (Z)	Cross (Y)	External(X)
1	1.04471374	1.04471374	1.04471374

Interior Partition

Time	Internal (Z)	Cross (Y)	External(X)	Flux (CR)
1	1.78698206	0.03649400	1.78698206	0.40285680
2	−1.76435661	0.20645089	−1.76435661	−0.00809625
3	0.28009480	0.05590425	0.28009480	
4	−0.00327635	0.00059449	−0.00327635	

Floor++

Time	Internal (Z)	Cross (Y)	External(X)	Flux (CR)
1	5.93112421	0.00000000	1.28655028	2.93992686
2	−21.28324509	0.00000000	−4.53600740	−3.31913614
3	30.12093735	0.00000012	6.36190033	1.79222369
4	−21.24926376	0.00000624	−4.50688505	−0.46112528
5	7.74331141	0.00004866	1.68008733	0.04502894
6	−1.34490633	0.00010723	−0.30481589	
7	0.08200093	0.00008965	0.01922911	
8	0.00031174	0.00003580	0.00022060	
9	0.00002430	0.00000867	0.00001728	
10	0.00000286	0.00000159	0.00000190	
11	0.00000038	0.00000025	0.00000024	
12	0.00000005	0.00000004	0.00000003	

Roof

Time	Internal (Z)	Cross (Y)	External(X)	Flux (CR)
1	0.50617242	0.00433286	1.91402411	0.30613610
2	−0.58148319	0.02842982	−2.40293932	−0.00304688
3	0.11778102	0.00849122	0.53446358	
4	−0.00109058	0.00012561	−0.00416714	

2. Compute the beam and diffuse solar radiation incident on each exterior surface. In this case, only three surfaces are exposed to solar radiation—the roof, the south wall, and the south window. The beam radiation is estimated by determining the incidence angle for each surface for each hour. (See the section that discusses incident angle in Chapter 30 of the 1995 *ASHRAE Handbook—Applications*.) The incident beam radiation is just the normal beam radiation multiplied by the cosine of the incidence angle. The diffuse radiation calculation assumes an isotropic sky, and view factors for the vertical surface as 0.5 from the surface-to-sky and 0.5 from the surface-to-ground. See the section on Computer Calculation of Solar Heat Gain Factors in Chapter 29 for the algorithm used to estimate the transmitted and absorbed components.

	Beam Radiation, Wh/m²			Diffuse Radiation, Wh/m²		
Hour	South Wall	South Window	Roof	South Wall	South Window	Roof
1	0.0	0.0	0.0	0.0	0.0	0.0
2	0.0	0.0	0.0	0.0	0.0	0.0
3	0.0	0.0	0.0	0.0	0.0	0.0
4	0.0	0.0	0.0	0.0	0.0	0.0
5	0.0	0.0	0.0	0.0	0.0	0.0
6	0.0	0.0	0.0	0.0	0.0	0.0
7	0.0	0.0	0.0	0.0	0.0	0.0
8	64.6	64.6	30.5	10.9	10.9	12.5
9	127.2	127.2	86.4	19.9	19.9	18.0
10	174.3	174.3	136.0	26.1	26.1	20.2
11	206.5	206.5	171.8	30.2	30.2	21.2
12	222.8	222.8	190.3	32.1	32.1	21.6
13	222.2	222.2	189.9	31.9	31.9	21.5
14	204.8	204.8	170.4	29.4	29.4	21.0
15	171.1	171.1	133.4	24.9	24.9	19.8
16	121.2	121.2	82.3	18.1	18.1	17.1
17	51.8	51.8	24.5	8.2	8.2	10.0
18	0.0	0.0	0.0	0.0	0.0	0.0
19	0.0	0.0	0.0	0.0	0.0	0.0
20	0.0	0.0	0.0	0.0	0.0	0.0
21	0.0	0.0	0.0	0.0	0.0	0.0
22	0.0	0.0	0.0	0.0	0.0	0.0
23	0.0	0.0	0.0	0.0	0.0	0.0
24	0.0	0.0	0.0	0.0	0.0	0.0

3. For the south window, determine the solar radiation transmitted through the window and the amount absorbed by the window.

Hour	Incident Beam, Btu/h·ft²	Incident Diffuse, Btu/h·ft²	θ	Transmitted Beam, Btu/h	Transmitted Diffuse, Btu/h	Absorbed Beam, Btu/h	Absorbed Diffuse, Btu/h	Total Absorbed, Btu/h	τ
1	0.0	0.0		0.0	0.0	0.0	0.0	0.0	0.00
2	0.0	0.0		0.0	0.0	0.0	0.0	0.0	0.00
3	0.0	0.0		0.0	0.0	0.0	0.0	0.0	0.00
4	0.0	0.0		0.0	0.0	0.0	0.0	0.0	0.00
5	0.0	0.0		0.0	0.0	0.0	0.0	0.0	0.00
6	0.0	0.0		0.0	0.0	0.0	0.0	0.0	0.00
7	0.0	0.0		0.0	0.0	0.0	0.0	0.0	0.00
8	64.6	10.9	67.8	33134.3	3131.3	2628.4	213.0	2841.4	0.71
9	127.2	19.9	58.8	73371.9	5703.6	5131.0	388.0	5519.0	0.80
10	174.3	26.1	50.8	104561.9	7488.7	7282.5	509.5	7791.9	0.84
11	206.5	30.2	44.6	125892.2	8653.9	8524.6	588.8	9113.3	0.85
12	222.8	32.1	41.1	136773.8	9205.1	8947.1	626.3	9573.3	0.86
13	222.2	31.9	41.1	136452.1	9131.5	8926.0	621.2	9547.3	0.86
14	204.8	29.4	44.6	124865.9	8434.3	8455.1	573.8	9028.9	0.85
15	171.1	24.9	50.8	102608.6	7125.8	7146.4	484.8	7631.2	0.84
16	121.2	18.1	58.8	69910.6	5182.9	4888.9	352.6	5241.5	0.80
17	51.8	8.2	67.8	26591.4	2341.8	2109.4	159.3	2268.7	0.71
18	0.0	0.0		0.0	0.0	0.0	0.0	0.0	0.00
19	0.0	0.0		0.0	0.0	0.0	0.0	0.0	0.00
20	0.0	0.0		0.0	0.0	0.0	0.0	0.0	0.00
21	0.0	0.0		0.0	0.0	0.0	0.0	0.0	0.00
22	0.0	0.0		0.0	0.0	0.0	0.0	0.0	0.00
23	0.0	0.0		0.0	0.0	0.0	0.0	0.0	0.00
24	0.0	0.0		0.0	0.0	0.0	0.0	0.0	0.00

4. Once the amounts of beam and diffuse radiation transmitted through the window is determined, an estimate must be made for the distribution on each surface. For convenience, assume the transmitted beam solar radiation is all intercepted by the floor, and that the transmitted diffuse solar radiation is distributed uniformly on all interior surfaces.

Hour	South Wall	South Window	East Wall	North Wall	West Wall	Floor	Roof
1	0	0	0	0	0	0	0
2	0	0	0	0	0	0	0
3	0	0	0	0	0	0	0
4	0	0	0	0	0	0	0
5	0	0	0	0	0	0	0
6	0	0	0	0	0	0	0
7	0	0	0	0	0	0	0
8	758	816	150	1574	150	31320	1499
9	1533	1651	303	3184	303	69068	3033
10	2110	2272	417	4382	417	98279	4173
11	2497	2690	494	5187	494	118243	4941
12	2690	2897	532	5588	532	128418	5322
13	2678	2884	529	5562	529	128104	5297
14	2460	2649	486	5109	486	117245	4866
15	2044	2202	404	4246	404	96391	4044
16	1431	1542	283	2973	283	65751	2831
17	588	633	116	1221	116	25095	1163
18	0	0	0	0	0	0	0
19	0	0	0	0	0	0	0
20	0	0	0	0	0	0	0
21	0	0	0	0	0	0	0
22	0	0	0	0	0	0	0
23	0	0	0	0	0	0	0
24	0	0	0	0	0	0	0

Note: Units are in Btu/h.

5. Based on the scheduled interior heat gains, and their radiative/convective splits, estimate the long-wave radiation incident on each of the interior surfaces.

Hour	South Wall	South Window	East Wall	North Wall	West Wall	Floor	Roof
1	88	95	17	184	17	175	175
2	88	95	17	184	17	175	175
3	88	95	17	184	17	175	175
4	88	95	17	184	17	175	175
5	88	95	17	184	17	175	175
6	88	95	17	184	17	175	175
7	88	95	17	184	17	175	175
8	1614	1738	319	3352	319	3193	3193
9	1614	1738	319	3352	319	3193	3193
10	1614	1738	319	3352	319	3193	3193
11	1614	1738	319	3352	319	3193	3193
12	1614	1738	319	3352	319	3193	3193
13	1614	1738	319	3352	319	3193	3193
14	1614	1738	319	3352	319	3193	3193
15	1614	1738	319	3352	319	3193	3193
16	1614	1738	319	3352	319	3193	3193
17	1614	1738	319	3352	319	3193	3193
18	1614	1738	319	3352	319	3193	3193
19	1614	1738	319	3352	319	3193	3193
20	88	95	17	184	17	175	175
21	88	95	17	184	17	175	175
22	88	95	17	184	17	175	175
23	88	95	17	184	17	175	175
24	88	95	17	184	17	175	175

Note: Units are in Btu/h.

6. Set up the exterior heat balance equations. For convenience, choose the simplest possible exterior convection and radiation models—fixed exterior convection coefficients for each surface, and fixed exterior radiation coefficients. The general form of the exterior heat balance equation is given as Equation (9). If the term for net long-wave radiation is replaced with a linearized expression using fixed exterior radiation coefficients, it can be rewritten as

$$t_{oi,\theta} = \left(q_{si,\theta} + h_{rs}t_{sky,\theta} + h_{rg}t_{ground,\theta} + h_o t_{oa,\theta} + \sum_{j=0}^{m} Y_j t_{i,\theta-j} \right. \qquad (10)$$

$$\left. - \sum_{j=0}^{m} Z_j t_{oi,\theta-j} + CR q_{oi,\theta-j}\right) / (h_o + h_{rs} + h_{rg} + Z_0)$$

For this example, somewhat arbitrarily values have been chosen for the exterior convection coefficients and radiation coefficients as follows.

Surface	South Wall	South Window	Roof
h_o	3.14	3.14	3.14
h_{rs}	0.45	0.45	0.91
h_{rg}	0.45	0.45	0.00

Note: Units are in Btu/h·ft²·°F

More sophisticated convection models (Cooper and Tree 1973, Fracastoro et al. 1982, Melo and Hammond 1991, Walton 1983) that incorporate the effects of wind speed, wind direction, surface orientation, etc. are commonly used instead. Likewise, more detailed models of long-wave radiation transfer, e.g. Cole (1976) and Walton (1983), are implemented in detailed building simulation programs. Once the coefficients have been determined, the exterior heat balance equations can be solved simultaneously with the interior heat balance equations and zone air heat balance equation.

The "exterior" side of the interior partitions are assumed to exchange heat with a zone with a constant temperature of 68°F. Detailed building simulation programs may allow the other zone temperature to be solved simultaneously. The "exterior" temperature for the floor is assumed to be the ground temperature of 62.6°F.

7. Set up the interior surface heat balance equations. Again, simple models for interior convective and radiative heat transfer are used. Convection is modeled using fixed coefficients; radiation is modeled using a mean radiant temperature (MRT) model.

For this example, interior convection coefficients are chosen to be 0.6 Btu/h·ft²·°F for the vertical surfaces, 0.77 Btu/h·ft²·°F for the floor, and 0.22 Btu/h·ft²·°F for the roof. More detailed convection correlations for use in buildings have been published by Alamdari and

Hammond (1982, 1983) Altmayer et al. (1983), Bauman et al. (1983), Bohn et al. (1984), Chandra and Kerestecioglu (1984), Khalifa and Marashall (1990), Spitler et al. (1991), and Walton (1983).

Use of the MRT model involves a simplification of Equation (1). Equation (1) is modified:

$$q_{i,\theta} = h_i(t_{a,\theta} - t_{i,\theta}) + h_{ri}(\text{MRT}_{i,\theta} - t_{i,\theta}) + R_i \qquad (11)$$

where h_{ri} is the radiation coefficient between the *ith* surface, and a fictitious surface, which is at the mean radiant temperature, $\text{MRT}_{i,\theta}$. The MRT is an area-emissivity weighted average of all of the other surface temperatures. See Carroll (1980), Davies (1988), Kamal and Novak (1991), Steinman, et al. (1989) and Walton (1980) for further discussion of interior radiation heat transfer models.

The R_i term would be the sum of the solar radiation absorbed on the interior surface, estimated in step 4, and the long-wave radiation absorbed on each surface, determined in step 5.

8. Set up the zone air heat balance equation. For this example, the energy addition/removal by the space-conditioning equipment will be represented by a piece-wise linear function of zone temperature, sometimes called a control profile.

$$QS_\theta = a + bt_{a,\theta} \qquad (12)$$

This expression for QS_θ is substituted into Equation (5). The control profile is developed from the system description given as part 3 of the problem statement above. The heat added to the zone by the system can be represented with the following coefficients.

$T_{a,\theta}$	*a*	*b*
$64.4 < T_{a,\theta} < 69.8$	451036	−6662.4
$69.8 < T_{a,\theta} < 75.2$	53874	−972.4
$75.2 < T_{a,\theta} < 78.8$	1162983	−15721.2
$78.8 < T_{a,\theta} < 86$	179579	−3241.5

During the setback period, the coefficients are set to zero, except when the zone temperature falls below 53.6°F. The QI_θ term is the sum of the convective portion of each of the internal heat gains.

9. Solve the exterior surface heat balance equations, interior surface heat balance equations, and zone air heat balance equation simultaneously for each hour. Because the temperature and history is initially unknown, some assumption is made for the temperature history, then the first day of the simulation is repeated iteratively until a steady periodic convergence is reached.

Once the steady periodic convergence is reached for this problem, the results are as follows:

Hour	$T_{a,\theta}$	QS_θ	Hour	$T_{a,\theta}$	QS_θ
1	59.7	0	13	72.0	−16062
2	58.6	0	14	73.4	−17455
3	57.6	0	15	73.9	−17985
4	56.8	0	16	73.4	−17473
5	55.9	0	17	71.6	−15742
6	55.2	0	18	69.6	−13110
7	54.5	0	19	69.1	−9196
8	65.1	16733	20	67.3	0
9	66.4	8930	21	65.5	0
10	67.6	866	22	63.9	0
11	68.7	−6996	23	62.4	0
12	69.8	−14014	24	61.0	0

The exterior surface temperatures for each surface, based on Equation (13), are:

Hour	South Wall	South Window	East Wall	North Wall	West Wall	Floor	Roof
1	43.2	45.7	67.1	67.1	67.1	55.4	42.8
2	41.7	44.4	66.7	66.6	66.7	55.4	41.4
3	41.0	43.3	66.4	66.2	66.4	55.4	40.3
4	40.3	42.4	66.0	65.8	66.0	55.4	39.4
5	39.7	42.1	65.7	65.5	65.7	55.4	38.8
6	39.9	42.4	65.3	65.3	65.3	55.4	39.0
7	40.8	43.5	65.1	64.9	65.1	55.4	40.1
8	50.7	47.1	64.9	64.8	64.9	55.4	53.6
9	70.3	51.1	65.1	64.9	65.1	55.4	72.1
10	91.9	55.4	65.7	65.7	65.7	55.4	89.2
11	109.0	60.1	66.4	66.2	66.4	55.4	103.1
12	119.1	64.2	67.1	67.1	67.1	55.4	112.8
13	122.5	67.3	67.8	67.8	67.8	55.4	117.7
14	120.5	69.4	68.5	68.5	68.5	55.4	117.3
15	114.4	70.0	69.1	69.1	69.1	55.4	111.4
16	102.7	68.7	69.6	69.6	69.6	55.4	100.0
17	85.3	66.2	70.0	70.0	70.0	55.4	82.8
18	65.8	62.8	70.0	70.0	70.0	55.4	65.7
19	53.6	59.5	69.8	69.8	69.8	55.4	57.9
20	50.2	56.1	69.4	69.4	69.4	55.4	54.0
21	50.5	53.2	69.1	69.1	69.1	55.4	50.9
22	50.4	50.7	68.5	68.5	68.5	55.4	48.2
23	48.0	48.7	68.0	68.0	68.0	55.4	46.0
24	45.3	47.1	67.6	67.5	67.6	55.4	44.2

The interior surface temperatures for each surface, based on Equation (4), are:

Hour	South Wall	South Window	East Wall	North Wall	West Wall	Floor	Roof
1	59.7	55.0	62.4	62.1	62.4	69.3	59.0
2	58.5	53.8	61.3	61.0	61.3	67.6	57.9
3	57.4	52.7	60.4	60.1	60.4	66.2	56.8
4	43.9	42.8	53.4	53.8	55.0	64.9	55.9
5	55.6	51.3	58.8	58.5	58.8	63.7	55.0
6	54.9	50.9	58.1	57.7	58.1	62.4	54.3
7	54.0	50.9	57.6	57.0	57.6	61.3	53.6
8	61.5	57.4	61.2	60.8	61.2	65.1	58.8
9	64.8	61.3	63.7	63.5	63.7	70.5	63.1
10	67.8	65.3	66.2	65.8	66.2	76.6	67.3
11	70.7	69.3	68.4	68.4	68.4	82.8	71.2
12	73.0	72.5	70.5	70.5	70.5	88.2	74.7
13	75.7	75.4	72.5	72.5	72.5	92.5	77.7
14	77.4	77.2	74.1	74.3	74.1	95.4	79.9
15	78.1	77.4	75.0	75.0	75.0	96.1	80.6
16	77.4	75.9	75.0	75.0	75.0	94.5	79.9
17	75.2	72.7	73.9	73.9	73.9	90.3	77.2
18	72.5	69.3	72.3	72.1	72.3	85.6	73.9
19	71.2	67.1	71.2	71.1	71.2	82.6	71.6
20	68.0	64.0	69.1	68.9	69.1	79.7	67.6
21	66.0	61.7	67.5	67.1	67.5	77.2	65.3
22	64.2	59.7	66.0	65.7	66.0	74.8	63.3
23	62.6	57.9	64.8	64.4	64.8	72.9	61.7
24	61.2	56.5	63.5	63.1	63.5	71.1	60.3

Weighting Factor Method

The weighting factor method of calculating instantaneous space sensible load represents a compromise between simpler methods, such as a steady-state calculation that ignore the ability of building mass to store energy, and more complex methods, such as complete energy balance calculations. With this method, space heat gains at constant space temperature are determined from a physical description of the building, ambient weather conditions, and internal load profiles. Along with the characteristics and availability of heating and cooling systems for the building, space heat gains are used to calculate air temperatures and heat extraction rates. This discussion is in terms of heat gains, cooling loads, and heat extraction rates. Heat losses, heating loads, and heat addition rates are merely different terms for the same quantities when the direction of heat flow is reversed.

The weighting factors represent Z-transfer functions (York and Cappiello 1981, Kerrisk et al. 1981). The Z-transform is a method for solving differential equations with discrete data. Two groups of weighting factors are used: heat gain and air temperature. Heat gain weighting factors represent transfer functions that relate space cooling load to instantaneous heat gains. A set of weighting factors is calculated for each group of heat sources that differ significantly in (1) the relative amounts of energy appearing as convection to the air versus radiation, and (2) in the distribution of radiant energy intensities on different surfaces.

Air temperature weighting factors represent a transfer function that relates room air temperature to the net energy load of the room. The weighting factors for a particular heat source are determined by introducing a unit pulse of energy from that source into the room's network. The network is a set of equations (described in the previous section) that represents a heat balance for the room. At each time step (one hour intervals), including the initial introduction, the energy flow to the room air represents the amount of the pulse that becomes a cooling load. Thus a long sequence of cooling loads can be generated from which the weighting factors are calculated. Similarly, a unit pulse change in room air temperature can be used to produce a sequence of cooling loads.

A two-step process is used to determine the air temperature and heat extraction rate of a room or building zone for a given set of conditions. First, the room's air temperature is assumed to be fixed at some reference value. This reference temperature is usually chosen as the average air temperature expected for the room over the simulation period. Instantaneous heat gains are calculated based on this constant air temperature. Various types of heat gains are considered. Some, such as solar energy entering through windows or energy from lighting, people, or equipment, are independent of the reference temperature. Others, such as conduction through walls, depend directly on the reference temperature.

A space sensible cooling load for the room, defined as the rate at which energy must be removed from the room to maintain the reference value of the air temperature, is calculated for each type of instantaneous heat gain. The cooling load generally differs from the instantaneous heat gain because some energy from the heat gain is absorbed by walls or furniture and stored for later release to the air. At an hour θ, the calculation uses present and past values of the instantaneous heat gain (q_θ, $q_{\theta-1}$), past values of the cooling load ($Q_{\theta-1}$, $Q_{\theta-2}$, ...) and the heat gain weighting factors (v_0, v_1, v_2, ... w_1, w_2, ...) for the type of heat gain under consideration. Thus, for each type of heat gain q_θ, the cooling load Q_θ is calculated as

$$Q_\theta = v_0 q_\theta + v_1 q_{\theta-1} + \ldots - w_1 Q_{\theta-1} - w_2 Q_{\theta-2} \ldots \tag{13}$$

The heat gain weighting factors are a set of parameters that quantitatively determine how much of the energy entering a room is stored and how rapidly the stored energy is released during later hours. Mathematically, the weighting factors are parameters in a Z-transfer function relating the heat gain to the cooling load.

These weighting factors differ for different heat gain sources because the relative amounts of convective and radiative energy leaving various sources differ, and because the distribution of radiative energy can differ. The heat gain weighting factors also differ for different rooms because room construction influences the amount of incoming energy stored by walls or furniture and the rate at which it is released. Sowell (1988) showed the effects of 14 zone design parameters on zone dynamic response. After the first step, the cooling loads from various heat gains are added to give a total cooling load for the room.

In the second step, the total cooling load is used—along with information on the heating, ventilating, and air-conditioning system attached to the room, and a set of **air temperature weighting factors**—to calculate the actual heat extraction rate and air temperature. The actual heat extraction rate differs from the cooling load because, in practice, the air temperature can vary from the reference value used to calculate the cooling load or because of HVAC system characteristics. The deviation of the air temperature t_θ from the reference value at hour θ, is calculated as

$$\begin{aligned} t_\theta = {} & 1/g_o[(Q_\theta - ER_\theta) + P_1(Q_{\theta-1} - ER_{\theta-1}) \\ & + P_2(Q_{\theta-2} - ER_{\theta-2}) + \ldots - g_1 t_{\theta-1} - g_2 t_{\theta-2} - \ldots] \end{aligned} \tag{14}$$

where ER_θ is the energy removal rate of the HVAC system at hour θ, and $g_0, g_1, g_2, \ldots, P_1, P_2, \ldots$ are the air temperature weighting factors, which incorporate information about the room, particularly the thermal coupling between the air and the storage capacity of the massive elements.

Tables of values of weighting factors for typical building rooms are presented in Chapter 28. One of the three groups of weighting factors, for light, medium, and heavy construction rooms, can be used to approximate the behavior of any room. Some automated simulation techniques allow weighting factors to be calculated specifically for the building under consideration. This option improves the accuracy of the calculated results, particularly for a building with an unconventional design. Alternatively, electronic tables of weighting factors for a large number of parametrically defined zones are available on a computer disk that accompanies the *Load Calculation Manual* (McQuiston and Spitler 1992).

Two assumptions are made in the weighting factor method. First, the processes modeled are linear. This assumption is necessary because heat gains from various sources are calculated independently and summed to obtain the overall result (i.e., the superposition principle is used). Therefore, nonlinear processes, such as radiation or natural convection, must be approximated linearly. This assumption does not represent a significant limitation, because these processes can be linearly approximated with sufficient accuracy for most calculations. The second assumption is that system properties influencing the weighting factors are constant, i.e., they are not functions of time. This assumption is necessary because only one set of weighting factors is used during the entire simulation period. This assumption can limit the use of weighting factors in situations where important room properties vary during the calculation. Two examples are the distribution of solar radiation incident on the interior walls of a room that can vary hourly, and inside surface heat transfer coefficients.

When the weighting factor method is used, a combined radiative-convective heat transfer coefficient is used as the inside surface heat transfer coefficient. This value is assumed constant even though in a real room the radiant heat transferred from a surface depends on the temperature of other room surfaces (not on the room air temperature) and the combined heat transfer coefficient is not constant. Under these circumstances, an average value of the property must be used to determine the weighting factors. Cumali et al. (1979) have investigated extensions to the weighting factor method to eliminate this limitation.

Thermal Network Methods

While implementations of the thermal network method vary, they all have in common the discretization of the building into a network of nodes, with interconnecting paths, through which energy flows. In many respects, thermal network models may be considered a refinement of the heat balance method. Where the heat balance model generally uses one node for zone air, the thermal network method might use multiple nodes. For each heat transfer element (wall, roof, floor, etc.) the heat balance model generally has one interior surface node and one exterior surface node; the thermal network model may include additional nodes. Heat balance models generally use simple methods for distributing radiation from lights; thermal network models may model the lamp, ballast, and luminaire housing separately. Furthermore, thermal network models depend on a heat balance at each node to determine the node temperature and the energy flow between all connected nodes. The energy flows may include conduction, convection, short wave radiation, and long wave radiation.

For any mode of energy flow, a range of techniques may be used to model the energy flow between two nodes. Taking conduction heat transfer as an example, the simplest thermal network model would be a resistance-capacitance network (Sowell 1990). By refining the network discretization, the models become what are commonly thought of as finite difference or finite volume models (Clarke 1985, Lewis and Alexander 1990, Walton 1993).

Thermal network models generally use a set of algebraic and differential equations. In most implementations, the solution procedure is separated from the models so that, in theory, different solvers might be used to perform the simulation. In contrast, most heat balance programs and weighting factor programs interweave the solution technique with the models. Various solution techniques have been used in conjunction with thermal network models. Examples include graph theory combined with Newton-Raphson and predictor-corrector ordinary differential equation integration (Buhl et al. 1990) and the use of Euler explicit integration combined with sparse matrix techniques (Walton 1993).

Of the three zone models discussed, thermal network models are the most flexible, and have the greatest potential for high accuracy. As a tradeoff, they also require the most computation time. In current implementations, they require more user effort to take advantage of the flexibility. As a result, most programs that use thermal network models are not in widely used.

SECONDARY SYSTEM COMPONENTS

Secondary HVAC systems generally include all elements of the overall building energy system between a central heating and cooling plant and the building zones. The precise definition depends heavily on the building design. A secondary system typically includes air handling equipment, air distribution systems with the associated ductwork, dampers, fans, and heating, cooling, and humidity conditioning equipment. Secondary systems also include the liquid distribution systems between the central plant and the zone and air handling equipment, including piping, valves, and pumps.

While the exact design of secondary systems varies dramatically among buildings, they are composed of a relatively small set of generic HVAC components. These components include distribution components (e.g., pumps/fans, pipes/ducts, valves/dampers, headers/plenums, fittings) and heat and mass transfer components (e.g., heating coils, cooling and dehumidifying coils, liquid heat exchangers, air heat exchangers, evaporative coolers, steam injectors). Most secondary systems can be described by simply connecting these components to form the complete system.

Energy estimation through computer simulation often mimics the modular construction of secondary systems by using modular simulation elements; e.g. the *ASHRAE HVAC 2 Toolkit* (Brandemuehl 1993, Brandemuehl and Gabel 1994), the simulation program TRNSYS (Klein et al. 1994), and Annex 10 activities of the International Energy Agency. To the extent that the secondary system consumes energy and transfers energy between the building and central plant, an energy analysis can be performed by characterizing the energy consumption of the individual components and the energy transferred among system components. In fact, few of the secondary components consume energy directly, except fans, pumps, furnaces, direct-expansion air-conditioning package units with gas fired heaters, and inline heaters. In this chapter secondary components are divided into two categories: distribution components and heat and mass transfer components.

Fans, Pumps, and Distribution Systems

The distribution system of an HVAC system affects energy consumption in two ways. First, fans and pumps consume electrical energy directly, based on the flow and pressures under which the device operates. Ducts and dampers, or pipes and valves, and the system control strategies affect the flow and pressures at the fan. Second, thermal energy is often transferred to (or from) the fluid due to heat transfer through pipes and ducts and due to the electrical input to fans and pumps. The analysis of system components should, therefore, account for both direct electrical energy consumption and thermal energy transfer.

Fan and pump performance is discussed in Chapters 18 and 38 of the 1996 *ASHRAE Handbook—Systems and Equipment.* In addition, Chapter 32 of this handbook covers pressure loss calculations for airflow in ducts and duct fittings. Chapter 33 presents a similar discussion for fluid flow in pipes. While these chapters do not specifically focus on energy estimation, energy use is governed by the same performance characteristics and engineering relationships.

Strictly speaking, performance calculations of the fan and air distribution systems in a building require a detailed pressure balance on the entire network. For example, in an air distribution system, the airflow through the fan depends on its physical characteristics, the operating speed, and the pressure differential across the fan. The pressure drop through the duct system depends on the duct design, the position of all dampers, and the airflow through the fan. The interaction between the fan and duct system results in a set of coupled, nonlinear algebraic equations. Models and subroutines for performing these calculations are available in the *ASHRAE HVAC 2 Toolkit* (Brandemuehl 1993).

While a detailed analysis of a distribution system requires flow and pressure balancing among the components, nearly all commercially-available energy analysis methods approximate the effect of the interactions with part-load performance curves. This procedure eliminates the need to calculate pressure drop through the distribution system at off-design conditions. The exact shape of the part-load curve depends on the effect of flow control on the pressure and fan efficiency, and may be calculated using a detailed analysis.

The fraction full load power is related to the part-load ratio (defined as the ratio of part-load flow to design flow) as a function of the specific fan control strategy. Figure 2 shows the relationship for three typical fan control strategies, as represented in a simulation program (LBL 1982). Within the simulation program, the curves are represented by polynomial regression equations. Models and subroutines for performing these calculations are also available in the *ASHRAE HVAC 2 Toolkit.*

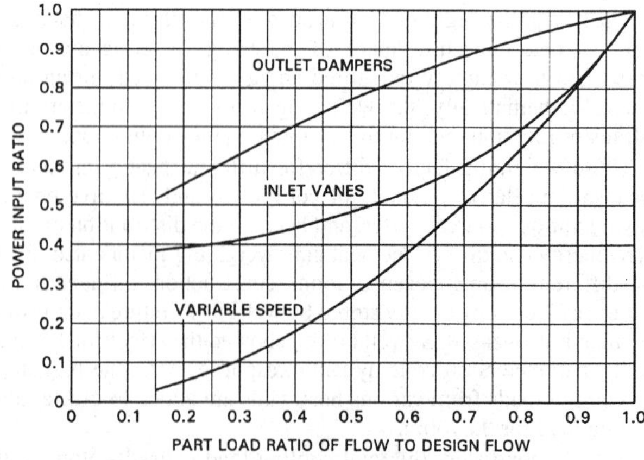

Fig. 2 Part-Load Curves for Typical Fan Operating Strategies

Heat transferred to the airstream due to fan operation, increases the temperature of the air. While the shaft power of the fan has a direct effect on the heat transfer, motor inefficiencies also heat the air if the motor is mounted inside the air stream. (For pumps, this contribution is typically assumed to be zero.)

The following equation provides a convenient and general model to calculate the heat transferred to the fluid.

$$\dot{Q}_{fluid} = [\eta_m + (1 - \eta_m) f_{m,\,loss}]\dot{W} \tag{15}$$

where

$\dot{Q}_{fluid}$ = heat transferred to the fluid, W
$f_{m,loss}$ = fraction of the motor heat loss transferred to the fluid stream, (η = 1 if fan mounted in air stream, η = 0 if fan mounted outside air stream)
$\dot{W}$ = fan motor power, W
η_m = motor efficiency

Heat and Mass Transfer Components

Secondary HVAC systems are comprised of such heat and mass transfer components as steam-based air heating coils, chilled water cooling and dehumidifying coils, shell-and-tube liquid heat exchangers, air-to-air heat exchangers, evaporative coolers, and steam injectors. While these components do not consume energy directly, their thermal performance dictates the interactions between the building loads and the energy-consuming primary components (e.g., chillers, boilers). In particular, the performance of the secondary components determines the entering fluid conditions for primary components, which in turn determine the energy efficiencies of the primary equipment. Accurate energy calculations can not be performed without appropriate models of the system heat and mass transfer components.

For example, the load on a chiller is typically described as the sum of zone sensible and latent loads, plus any heat gain from ducts, plenums, fans, pumps, and piping. However, the energy consumption of the chiller is determined not only by the load, but is also affected by the return chilled water temperature and flow rate. The return water condition is determined by the cooling coil performance and the part-load operating strategy of the air and water distribution system. The cooling coil might typically be controlled to maintain a constant leaving air temperature by modulating the water flow through the coil. In such a scenario, the cooling coil model must be able to calculate the leaving air humidity, leaving water temperature, and leaving water flow rate knowing the cooling coil design characteristics, the entering air temperature and humidity, entering airflow, and entering water temperature.

Virtually all building energy simulation programs include, and require, models of heat and mass transfer components. In general, these models are relatively simple. While a coil designer might use a detailed tube-by-tube analysis of conduction and convection heat transfer and condensation on fin surfaces to develop an optimal combination of fin and tube geometry, an energy analyst is more interested in determining the changes in leaving fluid states as operating conditions vary during the year. In addition, the energy analyst is likely to have limited design data on the equipment and, therefore, requires a model with very few parameters that depend on equipment geometry and detailed design characteristics.

A typical approach to modeling heat and mass transfer components for energy calculations is based on an effectiveness-NTU heat exchanger model (Kays and London 1984). The effectiveness-NTU model is described in most heat transfer textbooks and is briefly discussed in Chapter 3. It is particularly appropriate for describing the leaving fluid conditions when the entering fluid conditions and equipment design characteristics are known. In addition, this model requires only a single parameter to describe the characteristics of the exchanger—the overall transfer coefficient *UA*—which can be determined from limited design performance data.

Because the classical effectiveness methods were developed for sensible heat exchangers, they are used to perform energy calculations for a variety of sensible heat exchangers in HVAC systems. For typical finned-tube air heating coils, the crossflow configuration with both fluid streams unmixed is most appropriate. The same configuration typically applies to air-to-air heat exchangers. For liquid-to-liquid exchangers, tube-in-tube equipment can be modeled as parallel or counter flow, depending on flow directions; and shell-and-tube equipment can be modeled as either counter or crossflow, depending on the extent of baffling and the number of tube passes.

As noted previously, the energy analysis must determine the *UA* to describe the operations of a specific heat exchanger. There are typically two approaches to determine this important parameter: direct calculation and from manufacturers' data. Given detailed information about the materials, geometry, and construction of the heat exchanger, it is possible to apply fundamental heat transfer principles to calculate the overall heat transfer coefficient. An alternative to direct calculation, and the method most appropriate for energy estimation, is to use manufacturers' data. In reporting the design performance of a heat exchanger, a manufacturer typically gives the rate of heat transfer under various operating conditions, with the operating conditions described in terms of entering fluid flow rates and temperatures. From the given the heat transfer rate and entering fluid conditions the effectiveness and *UA* can be calculated.

Application To Cooling and Dehumidifying Coils

The analysis of air cooling and dehumidifying coils requires coupled, non-linear heat and mass transfer relationships. These relationships form the basis for all HVAC components with moisture transfer, including cooling coils, cooling towers, air washers, and evaporative coolers. While the complex heat and mass transfer theory that is presented in many textbooks is often required for cooling coil design, simpler models based on effectiveness concepts are usually more appropriate for energy estimation. For example, the bypass factor is a form of effectiveness in the approach of the leaving air temperature to the apparatus dew-point temperature. Threlkeld (1970), Elmahdy and Mitalas (1977), Braun et al. (1988), and Brandemuehl (1993) present more sophisticated models.

While the effectiveness-NTU method is typically developed and applied in the analysis of sensible heat exchangers, it can also be used to analyze other types of exchangers such as cooling and dehumidifying coils that couple heat and mass transfer. By redefining the state variables, capacity rates, and the overall exchange coefficient of these enthalpy exchangers the effectiveness concept may be used to calculate heat transfer rates and leaving fluid states. For sensible heat exchangers, the state variable is temperature, the capacity is the product of mass flow and fluid specific heat, and the overall transfer coefficient is the conventional overall heat transfer coefficient. For cooling and dehumidifying coils, the state variable becomes moist air enthalpy, the capacity has units of mass flow, and the overall heat transfer coefficient is modified to reflect enthalpy exchange.

The effectiveness model is based on the observation that, for a given set of entering air and liquid conditions, the heat and mass transfer is bounded by thermodynamic maximum values. Figure 3 shows the limits for leaving air states on a psychrometric chart. Specifically, the leaving liquid temperature cannot be warmer than the entering air temperature and the leaving air temperature and humidity cannot be lower than the conditions of saturated moist air at the temperature of the entering liquid.

The effectiveness analysis is accomplished for wet coils by establishing a common state variable for both the moist air and liquid streams. As implied by the lower limit of the entering chilled water temperature, this common state variable is the moist air enthalpy. In other words, all liquid and coil temperatures are transformed to the enthalpy of saturated moist air at the liquid or coil

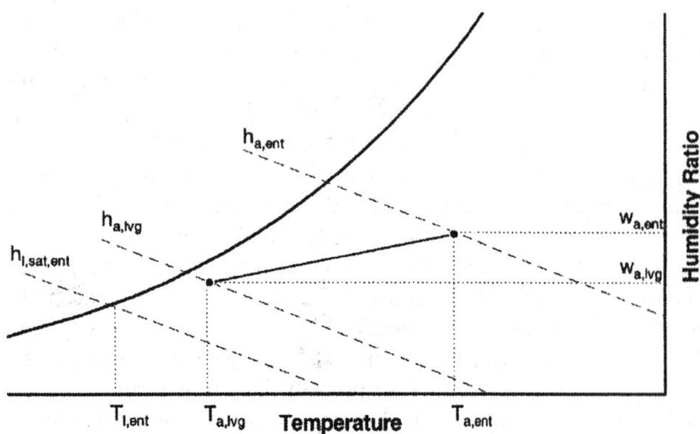

Fig. 3 Psychrometric schematic of cooling coil processes

temperature. Changes in liquid temperature can similarly be expressed in terms of changes in saturated moist air enthalpy through a saturation specific heat $c_{p,sat}$ defined by the following:

$$c_{p,\,sat} = \Delta h_{l,\,sat} / \Delta t_l \tag{16}$$

Using the definition of Equation (16), the basic effectiveness relationships discussed in Chapter 3 can be written as

$$q = c_a(h_{a,\,ent} - h_{a,\,lvg}) = c_l(h_{l,\,sat,\,lvg} - h_{l,\,sat,\,ent}) \tag{17}$$

$$q = \varepsilon c_{min}(h_{a,\,ent} - h_{l,\,sat,\,ent}) \tag{18}$$

$$c_a = m_a \tag{19}$$

$$c_l = \frac{(mc_p)_l}{c_{p,\,sat}} \tag{20}$$

$$c_{min} = \min(c_a, c_l) \tag{21}$$

where

q = heattransfer from air to water, Btu/h
c = fluid heat capacity, Btu/lb·°F
m_a = dry air mass flow rate, lb/h
m_l = liquid mass flow rate, lb/h
$c_{p,l}$ = liquid specific heat, Btu/lb·°F
$c_{p,sat}$ = saturation specific heat, defined by Equation (16), Btu/lb·°F
h_a = enthalpy of moist air, Btu/lb
$h_{l,sat}$ = enthalpy of saturated moist air at the temperature of the liquid, Btu/lb

The cooling coil effectiveness of the above equation is defined, then, as the ratio of moist air enthalpies in Figure 3. As in the case of sensible heat exchangers, the effectiveness is also a function of the physical coil characteristics and can be obtained by modeling the coil as a counterflow heat exchanger. However, since the heat transfer calculations are performed based on enthalpies, the overall transfer coefficient must be based on enthalpy potential rather than temperature potential. The enthalpy-based heat transfer coefficient, UA_h, is related to the conventional temperature-based coefficient by the specific heat.

$$q = UA\Delta t = UA_h \Delta h$$

$$UA_h = \frac{UA\Delta t}{\Delta h} = \frac{UA}{c_p} \tag{22}$$

As with sensible heat exchangers, the overall heat transfer coefficients (UA) can be determined either from direct calculation from coil properties or from manufacturers' performance data. While sensible heat exchangers can be described by a single parameter UA, cooling and dehumidifying coils require at least two parameters. For a fully-wet cooling coil, separate internal and external UAs are required—one describes the heat transfer between the fluid and the surface at its effective temperature and the other between the surface and the moist air. These two parameters can be determined from the sensible and latent capacity of a cooling coil at a single rating condition. A significant advantage of the effectiveness-NTU method is that the component can be described with as little as one measured data point or one manufacturer's design calculation.

PRIMARY SYSTEM COMPONENTS

Primary HVAC systems consume energy and deliver heating and cooling to a building, usually through secondary systems. Primary equipment generally includes chillers, boilers, cooling towers, cogeneration equipment, and plant-level thermal storage equipment. In particular, primary equipment generally represents the major energy-consuming equipment of a building, so accurate characterization of building energy use relies on accurate modeling of primary equipment energy consumption.

Modeling Strategies

The energy consumption characteristics of primary equipment generally depend on equipment design, load conditions, environmental conditions, and equipment control strategies. For example, chiller performance depends on the basic equipment design features (e.g., heat exchange surfaces, compressor design), the temperatures and flow through the condenser and evaporator, and the methods for controlling the chiller at different loads and operating conditions (e.g., inlet guide vane control on centrifugal chillers to maintain leaving chilled water temperature setpoint). In general, these variables that dictate energy consumption vary constantly and require calculations on an hourly basis.

Regression Models. While many secondary components are readily described by fundamental engineering principles (e.g., heat exchangers, valves), the complex nature of most primary equipment has discouraged the use of first-principle models for energy calculations. Instead, the energy consumption characteristics of primary equipment have traditionally been modeled using simple equations developed by running regression analyses on manufacturers' published design data. Because published data are generally only available for full-load design conditions, additional correction functions are used to correct the full-load data to part-load conditions. The functional form of the regression equations and correction functions take many forms, including exponentials, Fourier series, and, most of the time, second or third-order polynomials. The selection of an appropriate functional form depends on the behavior of the equipment. In some cases, energy consumption is calculated using direct interpolation from tables of data. However, this method often requires excessive data input and computer memory.

The typical approach to modeling primary equipment in energy simulation programs is to assume the following functional form for equipment power consumption.

$$P = P_{design} f_1(t_a, t_b, \ldots) f_2(\text{PLR}) \tag{23}$$

where

P = equipment power, kW
P_{design} = equipment power under full-load design conditions, kW
f_1 = function relating full-load power at off-design conditions (t_a, t_b, ...) to full-load power at design conditions
t_a, t_b = various operating temperatures that affect power
f_2 = fraction full-load power function, relating part-load power to full-load power
PLR = part-load ratio

The part-load ratio is defined as the ratio of the load to the available equipment capacity at given off-design operating conditions. Like the power, the available, or full-load, capacity will be a function of operating conditions. The part-load ratio can generally be expressed by the following ratio.

$$PLR = \frac{\text{Load}}{C_{avail}} = \frac{\text{Load}}{C_{nom} f_3(t_a, t_b, \ldots)} \tag{24}$$

where

C_{avail} = available capacity, kW
C_{nom} = nominal capacity, kW
f_3 = function relating available capacity at off-design conditions $(t_a, t_b, \ldots)$ to nominal capacity

The particular forms of the off-design functions f_1 and f_3 depend on the specific type of primary equipment. For example, for fossil-fuel boilers, full-load capacity and power (or fuel use) can be affected by the thermal losses to ambient temperature. However, these off-design functions are typically considered to be unity in most building simulation programs. For chillers, both capacity and power are affected by the condenser and evaporator temperatures. These two temperatures are often characterized in terms of their secondary fluids. For direct expansion air-cooled chillers, the operating temperatures are typically the wet-bulb temperature of the air entering the evaporator and the dry-bulb temperature of the air entering the condenser. For liquid chillers, the temperatures are usually the leaving chilled water temperature and the entering condenser water temperature.

The fraction full-load power function f_2 represents the change in equipment efficiency at part-load conditions and depends heavily on the control strategies used to match load and capacity. Figure 4 shows several possible shapes of these functional relationships. Curve 1 represents equipment with constant efficiency, independent of load. Curve 2 represents equipment that is most efficient in the middle of its operating range, while Curve 3 represents equipment that is most efficient at full-load. Note that these types of curves apply to both boilers and chillers.

First-Principle Models. General models include linear, nonlinear, and neural network constructs. Bourdouxhe et al. (1995), Gordon and Ng (1994, 1995), Gordon et al. (1995), and others have sought to use physical, first-principle models in which unknown model parameters are extracted from measured or published manufacturers' data.

The energy analyst is often faced with choosing the appropriate model for the job. For example, a complex boiler model is not appropriate if the boiler in question operates at virtually constant efficiency. By similar arguments, a regression-based model might be appropriate when the user has a full dataset of reliable in-situ measurements of the plant. However, first-principle physical models generally have several advantages over pure regression models, such as:

- Physical models allow confident extrapolation outside the range of available data.
- Regression is still required to obtain values for unknown physical parameters. However, the values of these parameters usually have physical significance. The engineer can capitalize on this significance to estimate default parameter values, diagnose errors in data analysis though checks for realistic parameter values, and even evaluate potential performance improvements.
- The number of unknown parameters is generally much smaller than the number of unknown coefficients in the typical regression model. For example, the standard ARI compressor model requires as many as 30 coefficients, 10 coefficients each for the regressions of capacity, power, and refrigerant flow. By comparison, a physical compressor model may have as few as 4 or 5 unknown parameters. As a result, the physical models require fewer measured data.
- Data on part-load operation of chillers and boilers are notoriously difficult to obtain. Part-load corrections often represent the greatest uncertainty in the regression models, while causing the greatest effect on annual energy predictions. By comparison, physical models of full-load operation often allow direct extension to part-load operation with little addition required data.

While physical models of primary HVAC equipment are generally based on fundamental engineering analysis and found in many HVAC textbooks (e.g. Kreider and Rabl 1994), the models described here are specifically based on the work of Bourdouxhe et al. (1994a, b, c) in the development of the ASHRAE primary toolkit (Bourdouxhe et al. 1997). The behavior of each elementary component is characterized by a limited number of physical parameters, such as heat exchanger heat transfer area or centrifugal compressor impeller blade angle. Values of these parameters are identified, or tuned, based on regression fits of overall performance compared to measured or published data.

While physical models are based on physical characteristics, the values obtained through a regression analysis of manufacturers' data are not necessarily representative of the actual measured values. Strictly speaking, the parameter values are regression coefficients with somewhat fictitious values, identified to minimize the error in overall system performance. In other words, errors in the fundamental models of the equipment are offset by over or underestimation of the parameter values.

Boiler Model

The thermal model, of a water boiler operating in steady-state regime is shown in Figure 5. It consists of an adiabatic combustion chamber and two heat exchangers. These three components interact through the following fluids: air, fuel, combustion gas, water, and a "fictitious fluid" representing the environment.

The adiabatic combustion chamber and the two heat exchangers can be modeled using classical thermodynamic principles. For example, the boiler heat exchangers can be described using effectiveness-NTU models. That is, knowing the flow rates and entering

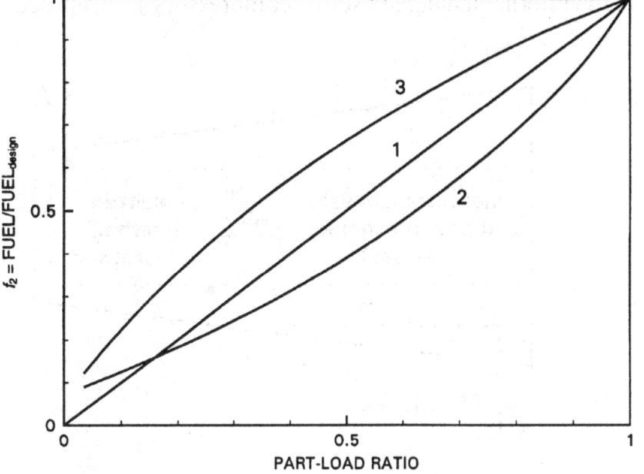

Fig. 4 Possible Part-Load Power Curves

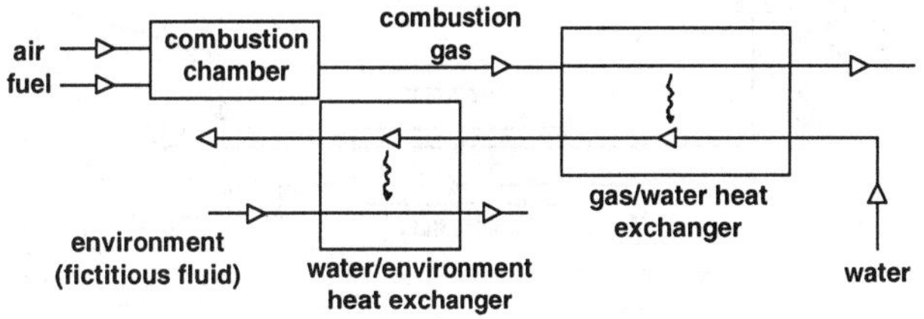

Fig. 5 Boiler Modeled with Elementary Components

fluid temperatures, the heat transfer is calculated using a single overall heat transfer coefficient *UA*. For the water/environment heat exchanger, the environment is considered to be an isothermal reservoir and can be modeled as a fictitious fluid with infinite capacitance. The values of the overall heat transfer coefficients of the two heat exchangers are selected, or tuned to manufacturer's data, to represent the behavior of a particular boiler.

Most boilers are equipped with combustion control systems that vary fuel input to satisfy changing heating loads. The model of Figure 5 can reflect the physical effects of varying fuel, fluid flow, and stand-by losses to realistically show the degradation of performance at part-load operation. In addition, single-stage boilers with on-off control and two-stage boilers with two distinct firing rates can be modeled as "quasi-static" systems, where the cyclic operation is represented by a linear combination of two steady-state regimes.

Vapor Compression Chiller Models

Figure 6 shows a schematic of a vapor compression chiller. In this case, the components include two heat exchangers, an expansion valve, and a compressor with a motor and transmission.

The components of a chiller are linked through the refrigerant. For energy estimating, a simplified approach is sufficient to represent the refrigerant as a "perfect" fluid with fictitious property values. That is, refrigerant liquid is modeled as incompressible and vapor properties are described by ideal gas laws with effective average values of property parameters, such as specific heat. Detailed descriptions of the form and parameters for the perfect fluid models are presented by Bourdouxhe et al. (1997) and associated references.

Condenser and Evaporator Modeling. Both condensers and evaporators are modeled as classical heat exchangers. The two heat exchangers are each assumed to have a constant overall heat transfer coefficient. In addition, the models used in chiller systems suffer from one additional assumption—the refrigerant fluid is assumed to be isothermal for both heat exchangers, which effectively ignores the superheated and subcooled regions of the heat exchanger.

The assumption of an isothermal refrigerant is particularly crude for the condenser, which sees very high refrigerant temperatures from the compressor discharge. The effect of the assumption is to significantly underestimate the mean temperature difference between refrigerant and water in the heat exchanger. Fortunately, this systematic error is compensated by a significant overestimate of the corresponding heat transfer coefficient.

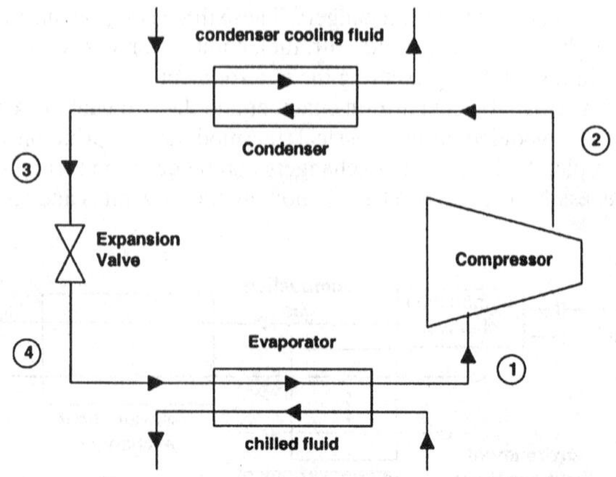

Fig. 6 Chiller Model Using Elementary Components

General Compressor Modeling. The modeling of real compressors requires the description of many thermomechanical losses within the compressor. Such losses could include heat loss, fluid friction, throttling losses in valves, or motor and transmission inefficiencies. While some of these losses can be modeled within the compressor, others are too complex or unknown to describe in a model for energy calculations.

The general approach used here for compressor modeling is described in the Figure 7. The compressor is described by two distinct internal elements: an idealized internal compressor and a motor-transmission element to account for unknown losses. Schematically, the motor-transmission subsystem represents an inefficiency of energy conversion. The losses from these inefficiencies are assumed to heat the fluid prior to compression. Mathematically, it can be modeled by the following linear relationship:

$$W = W_{lo} + (1 + \alpha) W_{in} \tag{25}$$

where

W = electrical power for a hermetic or semi-hermetic compressor, or shaft power for an open compressor
W_{in} = idealized internal compressor power (depends on type of compressor)
W_{lo} = constant electromechanical loss
α = proportional power loss factor

W_{lo} and α are empirical parameters determined by performing a regression analysis on manufacturers' data. Other parameters are also required to model W_{in}, the idealized internal compressor, depending on the type of compressor.

The following sections describe some of the different modeling techniques for reciprocating, screw, and centrifugal compressors. Detailed modeling techniques are available in the *ASHRAE Primary Toolkit* (Bourdouxhe et al. 1997) and associated references.

Modeling the Reciprocating Compressor. The conceptual schematic for a reciprocating compressor, for use with the general model, is shown in Figure 8. The refrigerant enters the compressor at state 1 and is heated to state 1a by the thermomechanical losses of the motor-transmission model in Figure 7. The refrigerant undergoes an isentropic compression process to state 2s, followed by a throttling process to the compressor discharge at state 2. The throttling valve is a simplified approach to model the known losses within the compressor due to pressure drops across the suction and discharge valves. Perhaps a more accurate model would include pressure losses at both the inlet and outlet of the compressor, but analysis of compressor data reveals that this simpler model is adequate for modeling of typical reciprocating compressors. (In fact, many compressors can be adequately modeled with no throttling valve at all.)

Physically, the throttling process is used to describe a pressure loss at the compressor suction and discharge. (It can also be used to model the internal leakage in screw compressors.) This process, can

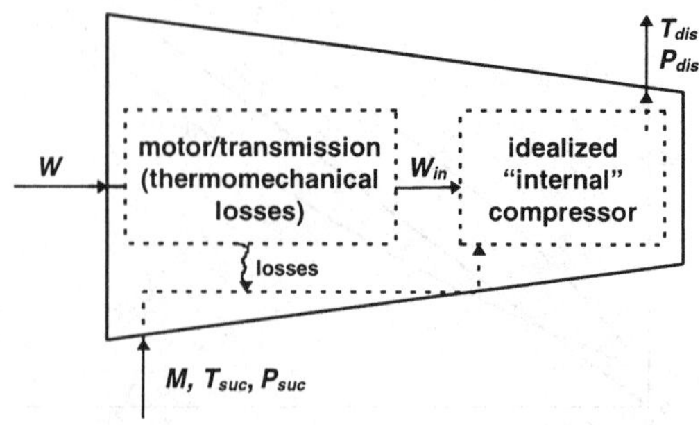

Fig. 7 General Schematic of Compressor

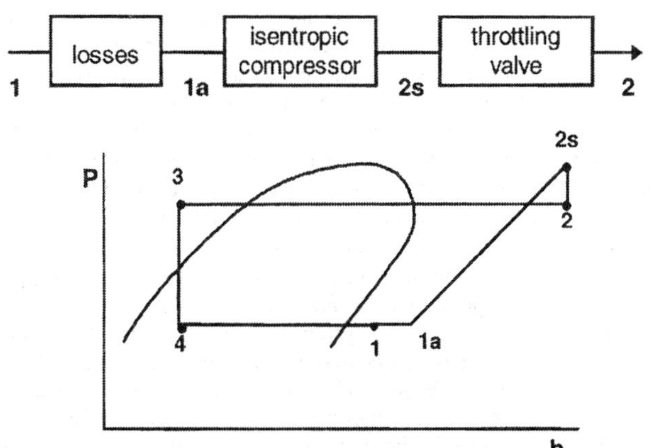

Fig. 8 Schematic of Reciprocating Compressor Model

be represented as an isentropic expansion through an ideal nozzle, followed by an isobaric diffusion. The refrigerant flow rate through the system must be determined to predict chiller and compressor performance. In general, volumetric flow depends on the pressure difference across the compressor. An ideal model of the mechanical cycle of a reciprocating compressor is shown in Figure 9.

The compressor refrigerant flow rate is a decreasing function of the pressure ratio due to the reexpansion of the vapor in the clearance volume. With the refrigerant vapor modeled as an ideal gas, the volumetric flow rate is given by the following:

$$V = V_s\left[1 + C_f - C_f\left(\frac{p_{ex}}{p_{suc}}\right)^{1/\gamma}\right] \tag{26}$$

where

V = volume flow rate
V_s = swept volume flow rate (geometric displacement of the compressor)
C_f = clearance factor = $V_{clearance}/V_{swept}$
p_{ex}/p_{suc} = cylinder pressure ratio
γ = specific heat ratio

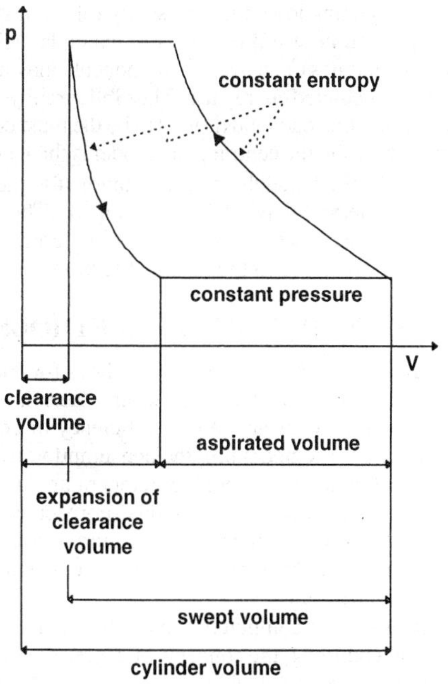

Fig. 9 Ideal Mechanical Cycle

V_s and C_f are two additional parameters that must be identified using data for the actual reciprocating compressor.

While the models discussed apply to full-load operation, Equation (26) is also valid at part-load conditions. However, the internal power use can be different at part-load depending on the particular strategy for capacity modulation, such as on-off cycling, cylinder unloading, hot-gas bypass, or variable speed motor. In most cases, simple physical models can be developed to describe these methods, which generally vary the swept volume rate. Additional thermomechanical losses can also be modeled, but often involve additional parameters. For example, the effect of cylinder unloading can be modeled by the following relationship:

$$W_{in} = W_{in,FL} + \left(1 - \frac{N_c}{N_{c,FL}}\right)W_{pump} \tag{27}$$

where

N_c = number of cylinders in use
$N_{c,FL}$ = number of cylinders in use in full-load regime
W_{pump} = internal power of the compressor when all cylinders are unloaded (pumping power)
$W_{in,FL}$ = full-load power

The variable W_{pump} characterizes the part-load regime of the reciprocating compressor, and it is assumed to be constant throughout the entire part-load range.

In summary, a realistic physical model of a reciprocating compressor, covering both full-load and part-load operations, can be developed based on six parameters: the constant and proportional loss terms of the motor-transmission model W_{lo} and α, the swept volume rate of the compressor cylinders V_s, the cylinder clearance volume factor C_f, the fictitious exhaust valve flow area A_{ex}, and the zero-load pumping power of the unloaded compressor W_{pump}. The entire chiller can then be modeled with two additional parameters for the overall heat transfer coefficients of the condenser and evaporator.

Modeling of Other Compressors and Chillers. From a modeling perspective, the thermodynamic processes of a screw compressor are similar to those of a reciprocating compressor. Physically, the screw compressor transports an initial volume rate of refrigerant vapor to a higher pressure and density by squeezing it into a smaller space. A realistic physical model of a variable-volume-ratio, twin-screw compressor, covering both full-load and part-load operations, can be developed based on five parameters: the constant and proportional loss terms of the motor-transmission model of Equation (25), the swept volume rate of the compressor screw, the internal leakage area, and a pumped pressure differential for diverted flow at part-load (Bourdouxhe et al. 1997). The entire chiller can then be modeled with two additional parameters for the overall heat transfer coefficients of the condenser and evaporator.

An idealized internal model of a centrifugal compressor, to be used in conjunction with Equation (25) and Figure 7, can based on an ideal analysis of a single stage compressor comprised of an isentropic impeller and isentropic diffuser. In addition to the thermomechanical loss parameters of Equation (25), only three additional parameters are required by this centrifugal compressor model: the peripheral speed of the impeller, the inclination of the vanes at the impeller exhaust, and the impeller exhaust area.

The refrigerant cycle of an absorption chiller is the same as for a vapor compression cycle except for the absorption-generation subsystem in place of the compressor. The absorption-generation subsystem includes an absorber, a steam-fired generator, a recovery heat exchanger, a pump and a control valve. All components except the pump and control valve can be modeled as heat exchangers.

Cooling Tower Model

A cooling tower is used in primary systems to reject heat from the chiller condenser. The controls typically control the tower fans and pumps to maintain a desired water temperature entering the condenser. Like cooling and dehumidifying coils in secondary systems, the performance of a cooling tower has a strong influence on the energy consumption of the chiller. In addition, tower fans directly consume electrical energy.

Fundamentally, a cooling tower is a direct-contact heat and mass exchanger. Equations describing the fundamental processes are given in Chapter 5 and in many HVAC textbooks. Chapter 36 of the 1996 *ASHRAE Handbook—Systems and Equipment* describes the specific performance of cooling towers. In addition, cooling tower performance subroutines are available in Bourdouxhe et al. (1997) and Klein et al. (1994).

For energy calculations, cooling tower performance is typically described in terms of the outdoor wet-bulb temperature, the temperature drop of the water flowing through the tower (the range), and the difference between the leaving water temperature and the air wet-bulb temperature (the approach). While simple models assume constant range and approach, more sophisticated models use rating performance data to relate leaving water temperature to the outdoor wet-bulb temperature, water flow, and airflow. Simple cooling tower models, such as those based on a single overall transfer coefficient that can be directly inferred from a single tower rating point, are often appropriate for energy calculations.

SYSTEM MODELING

OVERALL MODELING STRATEGIES

In developing a simulation model for building energy prediction, two basic issues must be considered—modeling of components or subsystems, and the overall modeling strategy. Modeling of components, which was discussed in the previous section, results in sets of equations describing the individual components. Overall modeling strategy refers to the *sequence* and *procedures* used to solve these equations. The accuracy of results and the computer resources required to achieve these results depend on the modeling strategy.

In most building energy programs the load models are executed for every space for every hour of the simulation period. (Practically all models use one hour as the time-step, which excludes any information on phenomena occurring in a shorter time span.) The load model is followed by running models for every secondary system, one at a time, for every hour of the simulation. Finally, the plant simulation model is executed again for the entire period. Each sequential execution processes the *fixed* output of the preceding step.

This procedure is illustrated in Figure 10. The solid lines represent data passed from one model to the next. The dashed lines represent information, usually provided by the user, about one model to the preceding model. For example, the system information consists of a piecewise linear function of zone temperature that gives the system capacity.

Because of this loads-systems-plants sequence, certain phenomena cannot be modeled precisely. For example, if the heat balance method for computing loads is used and some component in the system simulation model cannot meet the load, the program can only report the

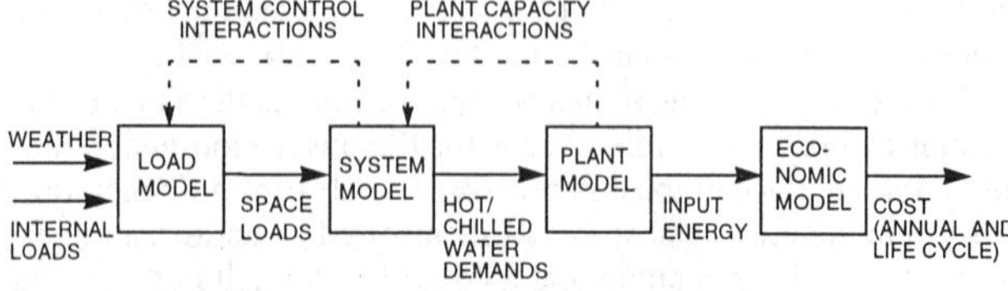

Fig. 10 Overall Modeling Strategy

current load. In actuality, the space temperature should readjust until the load matches the equipment capacity, but this cannot be modeled because the loads have been precalculated and fixed. If the weighting factor method is used for loads, this problem is partially overcome, because loads are continually readjusted during the system simulation. However, the weighting factor technique is based on linear mathematics, and wide departures of room temperatures from those used during execution of the load program can introduce errors.

A similar problem arises in plant simulation. For example, in an actual building, as the load on the central plant varies the supply chilled water temperature also varies. This variation in turn affects the capacity of the secondary system equipment. In an actual building, when the central plant becomes overloaded, space temperatures should rise to reduce the load. However, in most energy estimating programs, this condition cannot occur; thus, only the overload condition can be reported. These are some of the penalties associated with decoupling of the load, system, and plant models.

An alternative strategy, in which all calculations are performed at each time step, is conceivable. Here the load, system, and plant equations are solved simultaneously at each time interval. With this strategy, unmet loads and imbalances cannot occur; conditions at the plant are immediately reflected to the secondary system and then to the load model, forcing them to readjust to the instantaneous conditions throughout the building. The results of this modeling strategy are superior to those currently available, although the magnitude and importance of the improvement are uncertain.

The principal disadvantage of the alternative approach, and the reason that it has not been widely used, is that it demands more computing resources. However, programs that, to one degree or another, implement simultaneous solution of the loads, system, and plant models have been developed by Clarke (1985), Park et al. (1985), Klein et al. (1994) and Metcalfe et al. (1995).

An economic model, as shown in Figure 10, calculates energy costs (and sometimes capital costs) based on the estimated required input energy. Thus, the simulation model calculates energy usage and cost for any given input weather and internal loads. By applying this model (i.e., determining output for given inputs) at each hour (or other suitable interval), the hour-by-hour energy consumption and cost can be determined. Maintaining running sums of these quantities yields monthly or annual energy usage and costs.

These models only compare design alternatives; a large number of uncontrolled and unknown factors usually rules out such models for accurate prediction of utility bills. For example, Miller (1980) found that the dynamics of control of components may have at least minor effects on predicted energy use. The bibliography lists several models. Generally, the load models tend to be the most complex and time-consuming; while the central plant model is the least complex.

Because the detailed models are computationally intensive, several simplified method have been developed. These methods include the degree day method, bin method, and correlation methods, and they are presented in the next two sections.

DEGREE DAY AND BIN METHODS

Degree-day methods are the simplest methods for energy analysis and are appropriate if the building use and the efficiency of the HVAC equipment are constant. Where efficiency or conditions of use vary with outdoor temperature, the consumption can be calculated for different values of the outdoor temperature and multiplied by the corresponding number of hours; this approach is used in various bin methods. When the indoor temperature is allowed to fluctuate or when interior gains vary, models other than simple steady-state models must be used.

Even in an age when computers can easily calculate the energy consumption of a building, the concepts of degree days and balance point temperature remain valuable tools. The severity of a climate can be characterized concisely in terms of degree days. Also, the degree-day method and its generalizations can provide a simple estimate of annual loads, which can be accurate if the indoor temperature and internal gains are relatively constant and if the heating or cooling systems are to operate for a complete season. For these reasons, basic steady-state methods continue to be important.

Balance Point Temperature and Degree Days

The balance point temperature t_{bal} of a building is defined as that value of the outdoor temperature t_o at which, for the specified value of the interior temperature t_i, the total heat loss q_{gain} is equal to the heat gain from sun, occupants, lights, and so forth.

$$q_{gain} = K_{tot}(t_i - t_{bal}) \tag{28}$$

where K_{tot} is the total heat loss coefficient of the building in Btu/h·°F. For any of the steady-state methods described in this section, heat gains must be the average for the period in question, not for the peak values. In particular, solar radiation must be based on averages, not peak values. The balance point temperature is therefore obtained as

$$t_{bal} = t_i - \frac{q_{gain}}{K_{tot}} \tag{29}$$

Heating is needed only when t_o drops below t_{bal}. The rate of energy consumption of the heating system is

$$q_h = \frac{K_{tot}}{\eta_h}[t_{bal} - t_o(\theta)]^+ \tag{30}$$

where η_h is the efficiency of the heating system, also designated on an annual basis as the annual fuel use efficiency (AFUE) and θ is time. If t_{bal}, K_{tot}, and η_h are constant, the annual heating consumption can be written as an integral

$$Q_{h,\,yr} = \frac{K_{tot}}{\eta_h}\int[t_{bal} - t_o(\theta)]^+ d\theta \tag{31}$$

where the plus sign above the bracket indicates that only positive values are to be counted. This integral of the temperature difference conveniently summarizes the effect of outdoor temperatures on a building. In practice, it is approximated by summing averages over short time intervals (daily or hourly); the result is termed *degree days* or *degree hours*.

If daily average values of outdoor temperature are used for evaluating the integral, the degree days for heating $DD_h(t_{bal})$ are obtained as

$$DD_h(t_{bal}) = (1\ \text{day})\sum_{\text{days}}(t_{bal} - t_o)^+ \tag{32}$$

with dimensions of °F·days. Here the summation is to extend over the entire year or over the heating season. It is a function of t_{bal}, reflecting the roles of t_i, heat gain, and loss coefficient. The balance point temperature t_{bal} is also known as the base of the degree days. In terms of degree days, the annual heating consumption is

$$Q_{h,\,yr} = \frac{K_{tot}}{\eta_h}DD_h(t_{bal}) \tag{33}$$

Heating degree days or degree hours for a balance point temperature of 65°F have been widely tabulated based on the observation that this has represented average conditions in typical buildings in the past. The 65°F base is assumed whenever t_{bal} is not indicated explicitly. The extension of degree-day data to different basis discussed later.

Cooling degree days can be calculated using an equation analogous to Equation (32) for heating degree days as

$$DD_c(t_{bal}) = (1\ \text{day})\sum_{\text{days}}(t_o - t_{bal})^+ \tag{34}$$

While the definition of the balance point temperature is the same as that for heating, in a given building its numerical value for cooling is generally different from that for heating because q_i, K_{tot}, and t_i can be different. According to Claridge et al. (1987), t_{bal} can include both solar and internal gains as well as losses to the ground.

In some cases, heating and cooling degree days for shorter periods than a year are desirable, so listings of monthly values are common.

Figure 11A shows how the heating degree-days vary with t_{bal} for a particular site, in this case, New York. The plot is obtained by evaluating Equation (32) with data for the number of hours per year during which t_o is within 5°F temperature intervals centered at 72°F, 67°F, 62°F, ..., 18°F. The data for the number of hours in each interval, or bin, are included as labels in this plot. Analogous curves, without these labels, are shown in Figure 11B for three other sites: Houston, Washington, and Denver.

If the annual average of to is known, the cooling degree days to any base below 72 ±2.5°F can also be found. But calculating cooling energy consumption using degree days is more difficult than heating. For cooling, the equation analogous to Equation (33) is

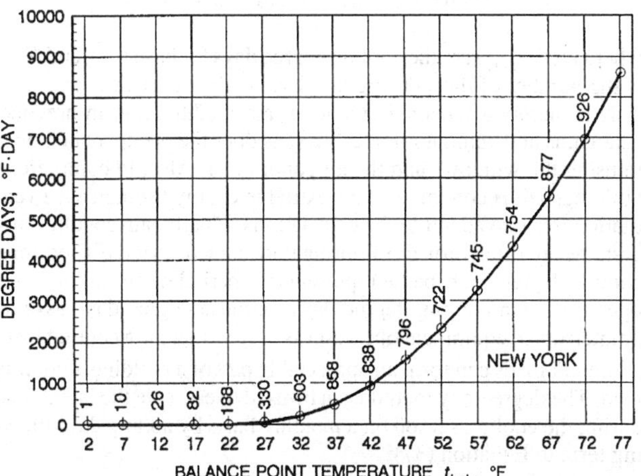

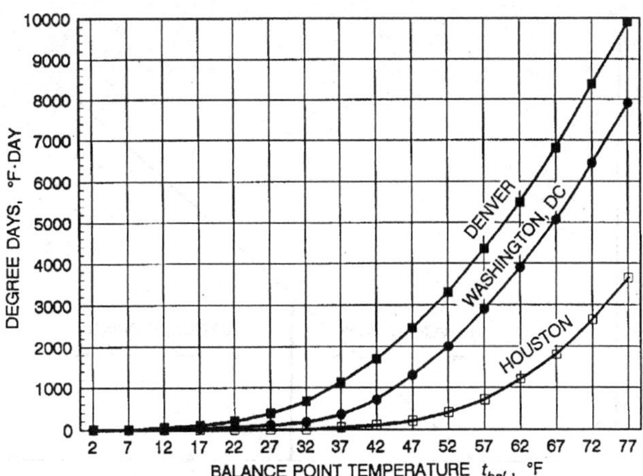

Fig. 11 Annual Heating Days $DD_h(t_{bal})$ as Function of Balance Temperature t_{bal}

$$Q_{c,yr} = \frac{K_{tot}}{\eta_c} DD_c(t_{bal}) \quad (35)$$

for a building whose K_{tot} does not change. That assumption is generally acceptable during the heating season, when windows are closed and the air exchange rate is fairly constant. However, during the intermediate or cooling season, heat gains can be eliminated, and the onset of mechanical cooling can be postponed by opening windows or increasing the ventilation. (In buildings with mechanical ventilation, this is called the *economizer* mode.) Mechanical air conditioning is needed only when the outdoor temperature extends beyond the threshold t_{max}. This threshold is given by an equation analogous to Equation (29), with the replacement of the closed window heat transmission coefficient K_{tot} by its value K_{max} for open windows

$$t_{max} = t_i - \frac{q_{gain}}{K_{max}} \quad (36)$$

K_{max} varies considerably with wind speed; but for simple cases, assume a constant value. The resulting sensible cooling load is shown schematically in Figure 12 as a function of t_o. The solid line is the load with open windows or increased ventilation; the dashed line shows the load if K_{tot} were kept constant. The annual cooling load for this mode can be calculated by breaking the area under the solid line into a rectangle and a triangle, or

$$Q_c = K_{tot}[DD_c(t_{max}) + (t_{max} - t_{bal})N_{max}] \quad (37)$$

where $DD_c\,(t_{max})$ are the cooling degree days for base t_{max}, and N_{max} is the number of days during the season when to rises above t_{max}. This is merely a schematic model of air conditioning. In practice, heat gains and ventilation rates vary, as does the occupant behavior in using the windows and the air conditioner. Also, in commercial buildings with economizers, the extra fan energy for increased ventilation must be added to the calculations. Finally, air-conditioning systems are often turned off during unoccupied periods. Therefore, cooling degree hours better represents the period when equipment is operating than cooling degree days, because degree days assume uninterrupted equipment operation as long as there is a cooling load.

Latent loads can form an appreciable part of a building's cooling load. The degree-day method can be used to estimate the latent load during the cooling season on a monthly basis by adding the following term to Equation (37):

$$Q_{latent} = \dot{m} h_{fg}(W_o - W_i) \quad (38)$$

where

q_{latent} = monthly latent cooling load
$\dot{m}$ = monthly infiltration (total airflow)
h_{fg} = heat of vaporization of water
W_o = outdoor humidity ratio (monthly averaged)
W_i = indoor humidity ratio (monthly averaged)

The degree-day method assumes that t_{bal} is constant which is not well satisfied in practice. Solar gains are zero at night and internal gains tend to be highest during the evening. The pattern for a typical house is shown in Figure 13. As long as to always stays below t_{bal}, the variations average out without changing the consumption. But for the situation in Figure 13, to rises above t_{bal} from shortly after 10:00 A.M. to 10:00 P.M.; the consequences for energy consumption depend on the thermal inertia and on the control of the HVAC system. If this building had low inertia, and if temperature control were critical, heating would be needed at night and cooling during the day. In practice, this effect is reduced by thermal inertia and by the deadband of the thermostat which allows t_i to float.

The closer t_o is to t_{bal}, the greater the uncertainty. If the occupants keep the windows closed during mild weather, t_i will rise above the set point. If they open the windows, the potential benefit of heat gains is reduced. In either case, the true values of t_{bal} become uncertain. Therefore, the degree-day method, like any steady-state method, is unreliable for estimating the consumption during mild weather. In fact, the consumption becomes most sensitive to occupant behavior and cannot be predicted with certainty.

Despite these problems, the degree-day method (using an appropriate base temperature) can give remarkably accurate results for the annual heating energy of single-zone buildings dominated by losses through the walls and roof and/or ventilation. Typical buildings have time constants that are about one day, and a building's thermal inertia essentially averages over the diurnal variations, especially if t_i is allowed to float. Furthermore, the energy consumption in mild weather is small; hence a relatively large error here has only a small effect on the total for the season.

Seasonal Efficiency

The seasonal efficiency of heating equipment η_h depends on such factors as steady-state efficiency, sizing, cycling effects, and energy conservation devices. Sometimes it is much lower than and other times it is comparable to steady-state efficiency. Alereza and Kusuda (1982) developed expressions to estimate the seasonal efficiency for a variety of furnaces, if information on rated input and

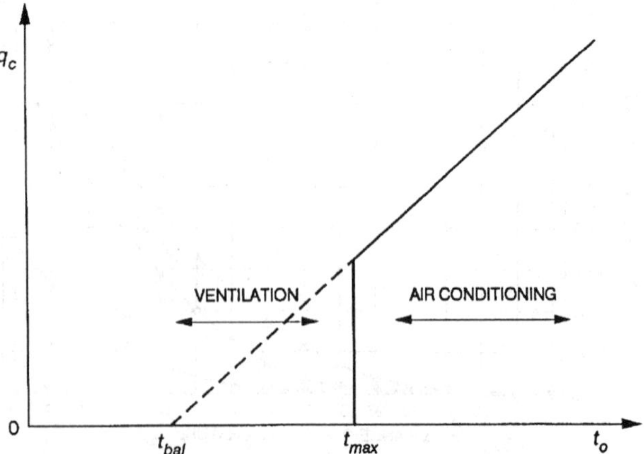

Fig. 12 Cooling Load as Function of Outdoor Temperature t_o
(Below t_{max}, ventilation can be used to avoid cooling.)

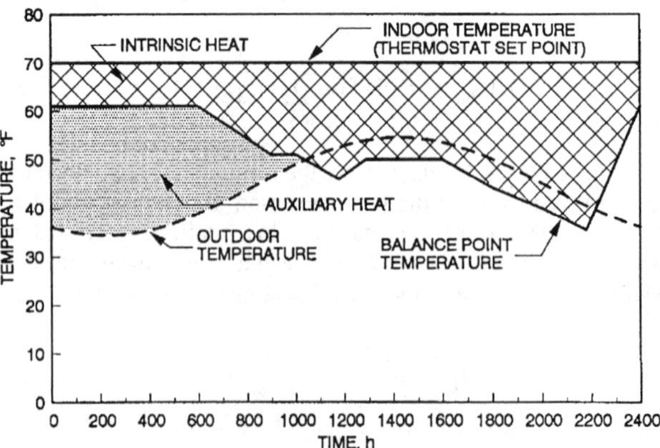

Fig. 13 Variation of Balance Point Temperature and Internal Gains for a Typical House
(Nisson and Dutt 1985)

output is available. These expressions correlate seasonal efficiency with variables determined by using the equipment simulation capabilities of a large hourly simulation program and typical equipment performance curves supplied by the National Institute of Standards and Technology (NIST).

$$\eta = \frac{\eta_{ss} CF_{pl}}{1 + \alpha_D} \tag{39}$$

where

η_{ss} = steady-state efficiency (rated output/input)
α_D = fraction of heat loss from ducts

The term CF_{pl} is a characteristic of the part-load efficiency of the heating equipment, which may be calculated as follows:

Gas-Forced Air Furnaces

With pilot

$$CF_{pl} = 0.6328 + 0.5738\ RLC - 0.3323\ (RLC)^2$$

With intermittent ignition

$$CF_{pl} = 0.7791 + 0.1983\ RLC - 0.0711\ (RLC)^2$$

With intermittent ignition and loose stack damper

$$CF_{pl} = 0.9276 + 0.0732\ RLC - 0.0284\ (RLC)^2$$

Oil furnaces without Stack Damper

$$CF_{pl} = 0.7092 + 0.6515\ RLC - 0.4711\ (RLC)^2$$

Resistance Electric Furnaces

$$CF_{pl} = 1.0$$

These equations are based on many annual simulations for the equipment in which RLC is defined as follows:

$$RLC = \frac{BLC}{CHT}(t_{bal} - t_{od})(1 + a_D)$$

where

t_{od} = outside design temperature
CHT = rated output of equipment

Seasonal efficiency is also discussed by Chi and Kelly (1978), Parker et al. (1980), and Mitchell (1983).

Variable Base Degree Days

The calculation of Q_h from degree days $DD_h(t_{bal})$ depends on the value of t_{bal}. This value varies widely from one building to another because of widely differing personal preferences for thermostat settings and setbacks and because of different building characteristics. In response to the fuel crises of the 1970s, heat transmission coefficients have been reduced, and thermostat setback has become common. At the same time, the energy use by appliances has increased. These trends all reduce t_{bal} (Fels and Goldberg 1986). Hence, in general, degree days with the traditional base 65°F are not to be used.

Many formulas have been proposed for estimating the degree days relative to an arbitrary base when detailed data are not available. The basic idea is to assume a typical probability distribution of temperature data, characterized by its average $\bar{t}_o$ and by its standard deviation σ. Erbs et al. (1983) developed a model that needs as input only the average $\bar{t}_o$ for each month of the year. The standard deviations σ_m for each month are then estimated from the correlation

$$\sigma_m = 3.54 - 0.0290\,\bar{t}_o + 0.0644\,\sigma_{yr} \tag{40a}$$

This is a dimensional equation with t and σ in °F; σ_{yr} is the standard deviation of the monthly average temperatures about the annual average $\bar{t}_{o,yr}$

$$\sigma_{yr} = \sqrt{\frac{1}{12}\sum_{1}^{12}(\bar{t}_o - \bar{t}_{o,yr})^2} \tag{40b}$$

To obtain a simple expression for the degree days, a normalized temperature variable φ is defined as

$$\phi = \frac{\bar{t}_{bal} - \bar{t}_o}{\sigma_m\sqrt{N}} \tag{41}$$

with N = number of days in the month (N has units of day/month and φ has units of $\sqrt{\text{month/day}}$). While temperature distributions can be different from month to month and location to location, most of this variability can be accounted for by the average and the standard deviation of $\bar{t}_o$. Being centered around $\bar{t}_o$ and scaled by σ_m, the quantity φ eliminates these effects. In terms of φ, the monthly heating degree days for any location are well approximated by

$$DD_h(t_{bal}) = \sigma_m N^{1.5}\left[\frac{\phi}{2} + \frac{\ln(e^{-a\theta} + e^{a\theta})}{2a}\right] \tag{42}$$

where $a = 1.698\sqrt{\text{day/mo}}$.

For nine locations spanning most climatic zones of the United States, Erbs et al. (1983) verified that the annual heating degree days can be estimated with a maximum error of 315°F-days if this equation is used for each month. For cooling degree days, the largest error is 270°F-days. Such errors are quite acceptable, representing less than 5% of the total.

Table 2 lists monthly heating degree-days for New York City, using the model of Erbs et al. (1983), given monthly averages of t_o as reproduced in column 2 of Table 2. The degree days are based on a balance temperature of 60°F. Column 2 lists the given values of monthly average outdoor temperature, and N is the number of days in the month. Intermediate quantities are shown in columns 4 and 5, and $t_{o,\,yr}$ and σ_{yr} are shown at the bottom. Column 6 shows the monthly and annual results.

Table 2 Degree-Day Calculation from Monthly Averaged Data

Month	$\bar{t}_o$, °F	N, day/mo.	σ_m, °F	σ, $\sqrt{\text{mo./day}}$	$DD_h(t_{bal})$, °F·day
January	32.2	31	3.65	1.37	864
February	33.4	28	3.62	1.39	746
March	41.1	31	3.40	1.00	592
April	52.1	30	3.08	0.47	265
May	62.3	31	2.79	−0.15	67
June	71.6	30	2.52	−0.84	7
July	76.6	31	2.38	−1.26	2
August	74.9	31	2.41	−1.11	3
September	68.4	30	2.61	−0.59	16
October	58.7	31	2.88	0.08	123
November	46.4	30	3.22	0.72	391
December	35.5	31	3.56	1.24	762
$t_{o,yr}$	54.4			Sum	3837
σ_{yr}	15.8				

Note: Use Equation (42) to calculate $DD_h(t_{bal})$

Table 3 Degree-Day and Monthly Average Temperatures for Various Locations

Site	Variable Base Heating Degree-Day, °F·days[a]					Monthly Average Outdoor Temperature, °F[b]											
	65	60	55	50	45	Jan	Feb	Mar	Apr	May	Jun	Jul	Aug	Sep	Oct	Nov	Dec
Los Angeles, CA	1245	522	158	26	0	54.5	55.6	56.5	58.8	61.9	64.5	68.5	69.6	68.7	65.2	60.5	56.9
Denver, CO	6016	4723	3601	2653	1852	29.9	32.8	37.0	47.5	57.0	66.0	73.0	71.6	62.8	52.0	39.4	32.6
Miami, FL	206	54	8	0	0	67.2	67.8	71.3	75.0	78.0	81.0	82.3	82.9	81.7	77.8	72.2	68.3
Chicago, IL	6127	4952	3912	2998	2219	24.3	27.4	36.8	49.9	60.0	70.5	74.7	73.7	65.9	55.4	40.4	28.5
Albuquerque, NM	4292	3234	2330	1557	963	35.2	40.0	45.8	55.8	65.3	74.6	78.7	76.6	70.1	58.2	44.5	36.2
New York, NY	4909	3787	2806	1980	1311	32.2	33.4	41.1	52.1	62.3	71.6	76.6	74.9	68.4	58.7	47.4	35.5
Bismarck, ND	9044	7656	6425	5326	4374	8.2	13.5	25.1	43.0	54.4	63.8	70.8	69.2	57.5	46.8	28.9	15.6
Nashville, TN	3696	2758	1964	1338	852	38.3	41.0	48.7	60.1	68.5	76.6	79.6	78.5	72.0	60.9	48.4	40.4
Dallas/Ft. Worth, TX	2290	1544	949	526	250	45.4	49.4	55.8	66.4	73.8	81.6	85.7	85.8	78.2	68.0	55.9	48.2
Seattle, WA	4727	3269	2091	1194	602	39.7	43.5	45.5	50.4	56.5	61.3	65.7	64.9	60.6	54.2	45.7	42.0

[a]Source: NOAA (1973) [b]Source: Cinquemani et al. (1978)

Table 4 Sample Annual Bin Data

City	Bin																									
	100/104	95/99	90/94	85/89	80/84	75/79	70/74	65/69	60/64	55/59	50/54	45/49	40/44	35/39	30/34	25/29	20/24	15/19	10/14	5/9	0/4	−5/1	−10/6	−15/11	−20/16	−25/21
Albuquerque	1	54	191	348	511	617	789	785	816	676	637	720	678	676	560	406	180	101	31	3						
Bismarck		11	68	173	252	320	450	590	625	550	583	506	624	539	626	596	424	399	391	306	364	144	131	43	42	3
Chicago			97	222	362	512	805	667	615	622	585	577	636	720	957	511	354	243	125	66	58	6				
Dallas/Ft. Worth	27	210	351	527	804	1100	947	705	826	761	615	615	523	364	289	57	29									
Denver		3	118	235	348	390	472	697	699	762	783	718	665	758	713	565	399	164	106	65	80	22				
Los Angeles	8	8	9	17	53	194	632	1583	2348	2055	1181	394	74	4												
Miami			45	864	1900	2561	1605	871	442	222	105	77	36	12												
Nashville		7	137	407	616	756	1100	866	706	692	650	670	720	582	342	280	107	71	29							
New York City		5	26	170	383	664	820	941	763	699	593	690	765	858	648	377	212	99	20	5						
Seattle				16	62	139	256	450	769	1353	1436	1461	1413	915	358	51	43	51	1							

Table 3 contains degree-day data for several sites and monthly averaged outdoor temperatures needed for the algorithm. More complete tabulations of the latter are contained in Cinquemani et al. (1978) and in local climatological data summaries available from the National Climatic Data Center, Asheville, NC. Monthly degree-day data to the bases of 50°F, 55°F, 60°F, 65°F, and 70°F, as well as other climatic information for 209 U.S. and 14 Canadian cities, may be found in Appendix 3 to Balcomb et al. (1982).

Bin Method

For many applications, the degree-day method should not be used, even with the variable base method, because the heat loss coefficient K_{tot}, the efficiency η_h of the HVAC system, or the balance point temperature may not be sufficiently constant. The efficiency of a heat pump, for example, varies strongly with outdoor temperature; or the efficiency of the HVAC equipment may be affected indirectly by t_o when the efficiency varies with the load, a common situation for boilers and chillers. Furthermore, in most commercial buildings, the occupancy has a pronounced pattern, which affects heat gain, indoor temperature, and ventilation rate.

In such cases, a steady-state calculation can yield good results for the annual energy consumption, if different temperature intervals and time periods are evaluated separately. This approach is known as the bin method, because the consumption is calculated for several values of the outdoor temperature t_o and multiplied by the number of hours N_{bin} in the temperature interval (bin) centered around that temperature

$$Q_{bin} = N_{bin}\frac{K_{tot}}{\eta_h}[t_{bal} - t_o]^+ \tag{43}$$

The superscript plus sign indicates that only positive values are counted; no heating is needed when t_o is above t_{bal}. This equation is evaluated for each bin, and the total consumption is the sum of the Q_{bin} over all bins.

In the United States, the necessary data are available in ASHRAE (1995) and USAF (1978). The bins are usually in 5°F increments and are often collected in three daily 8-h shifts. Mean coincident wet-bulb temperature data (for each dry-bulb bin) are used to calculate latent cooling loads from infiltration and ventilation. The bin method considers both occupied and unoccupied building conditions and gives credit for internal loads by adjusting the balance point. For example, a calculation could be performed for 42°F outdoors (representing all occurrences from 39.5 to 44.5°F) and with building operation during the midnight to 0800 shift. Since there are 23 5°F bins between −10 and 105°F and 3 8-h shifts, 69 separate operating points are calculated. For many applications, the number of calculations can be reduced. A residential heat pump (heating mode), for example, could be calculated for just the bins below 65°F without the three-shift breakdown. The data included in Table 4 are samples of annual totals for a few sites, but ASHRAE (1995) and USAF (1978) include monthly data and data further separated into time intervals during the day.

CORRELATION METHODS

One way to simplify energy analyses is to correlate energy requirements to various inputs. Typically, the results of a correlation is a simple equation that may be used in a calculator or small computer program or to develop a graph provides a quick insight into the energy requirements. Examples of correlation methods are in ASHRAE *Standard* 90.1, which includes several empirical equations that may be used to predict energy consumption by many types of buildings.

The accuracy of correlation methods depends on the size and accuracy of the database and the statistical means used to develop the correlation. A database generated from measured data can lead to accurate correlations (Lachal 1992). The key to the proper use of a correlation is that the case being studied matches the cases used in developing the database. Inputs to the correlation (the independent variables) indicate the factors that are considered to have significant impact on energy consumption. A correlation is invalid when either an input parameter is used beyond its valid range (corresponding to extrapolation rather than interpolation), or when some important feature of the building/system is not included in the available inputs to the correlation.

SIMULATING SECONDARY AND PRIMARY SYSTEMS

Traditionally, most energy analysis programs include a set of preprogrammed models that represent various systems, such as variable air volume, terminal reheat, multizone, etc. In this scheme, the equations for each system are arranged so they can be solved sequentially. If this is not possible, then the smallest number of equations that must be solved simultaneously is solved using an appropriate technique. Furthermore, individual equations may vary from hour to hour in the simulation, depending on controls and operating conditions. For example, a dry coil uses different equations than a wet coil.

The primary disadvantage of this scheme is that it is relatively inflexible—in order to modify a system, the program source code may have to be modified and recompiled. Alternative strategies (Klein et al. 1994, Park et al. 1985) have viewed the system as a series of components (e.g. fan, coil, pump, duct, pipe, damper, thermostat) that may be organized in a component library. Users of the program specify the connections between the components. The program then resolves the specification of components and connections into a set of simultaneous equations.

A refinement of component based modeling is known as equation-based modeling (Sowell and Moshier 1995, Buhl et al. 1993). In this scheme, the models do not follow well defined rules for a solution and input and output variables are not predetermined.

MODELING OF SYSTEM CONTROLS

From a mathematical viewpoint, controls represent equations that must be satisfied at each point during the simulation. For example, the room thermostat can be represented as a function relating heating and cooling delivery to space temperature. Similarly, cooling coil reset controls can be modeled as a relationship between outside or zone temperature and coil discharge temperature. An accurate secondary system model must ensure that all controls are properly represented and that the governing equations are satisfied at each simulation time step. This often creates a need for iteration or, alternately, for use of values from an earlier solution point.

The controls on space temperature affect the interaction between loads calculations and the secondary system simulation. A realistic model might require a dead band in space temperature in which no heating or cooling is called for; within this range, the true space sensible load is zero, and the true space temperature must be adjusted accordingly. If the thermostat has proportional control between zero and full capacity, the space temperature will rise in proportion to the load during cooling and fall similarly during heating. Capacity to heat or cool also varies with space temperature after the control device has reached its maximum because capacity is proportional to the *difference* between supply and space temperatures. Failure to properly model these phenomena results in overestimating required energy.

INTEGRATION OF SYSTEM MODELS

Energy calculations for secondary systems involve construction of the complete system from the set of HVAC components. For example, a VAV system is a single path system that controls zone temperature by modulating the airflow while maintaining a constant supply air temperature. VAV terminal units, located at each zone, adjust the quantity of air reaching each zone depending on its load requirements. Reheat coils may be included to provide required heating for perimeter zones.

This VAV system simulation consists of a central air handling unit and a VAV terminal unit with reheat coil located at each zone, as shown in Figure 14. The central air handling unit includes a fan, cooling coil, preheat coil, and an outside air economizer. The supply air leaving the air handling unit is controlled to a fixed setpoint. The VAV terminal unit at each zone varies the airflow to meet the cooling load. As the zone cooling load decreases, the VAV terminal unit decreases the zone airflow until the unit reaches its minimum posi-

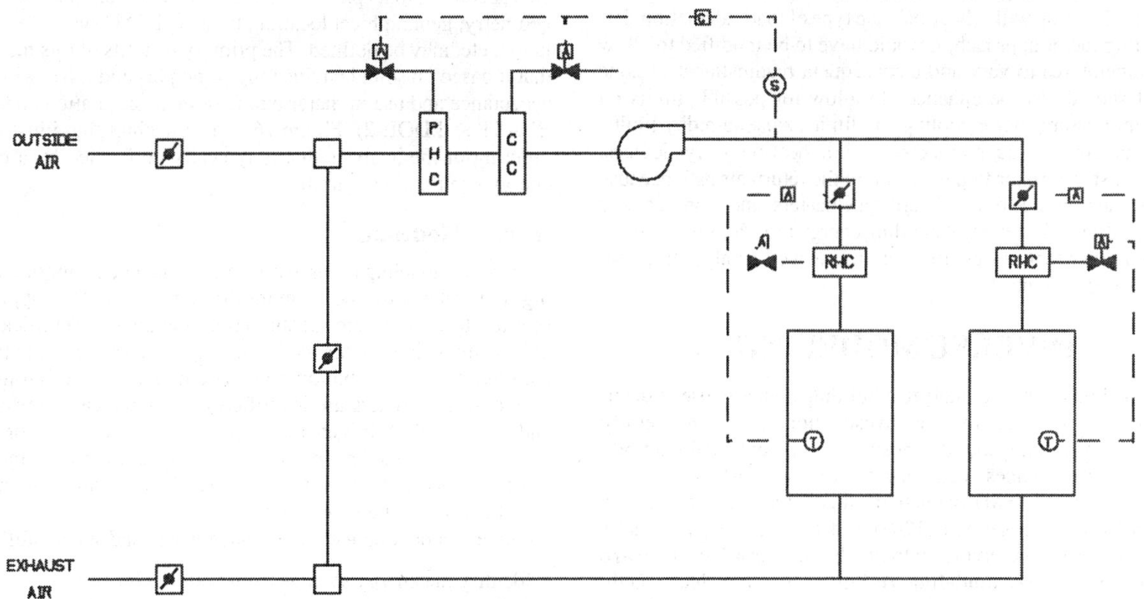

Fig. 14 Schematic of Variable Air Volume System with Reheat

tion. If the cooling load continues to decrease the reheat coil will be activated to meet the zone load. As the supply air volume leaving the unit decreases the fan power consumption will also be reduced. A variable speed drive is used to control the supply fan.

The simulation is based on system characteristics and zone design requirements. For each zone the inputs include the sensible and latent loads, the zone set point temperature, and the minimum zone supply air mass flow. System characteristics include the supply air temperature setpoint, the entering water temperature of the reheat, preheat and cooling coils, the minimum mass flow of outside air, and the economizer temperature/enthalpy set point for minimum airflow.

The algorithm for performing the calculations for this VAV system is shown in Figure 15. The algorithm directs sequential calculations of system performance. Calculations proceed from the zones forward along the return air path to the cooling coil inlet and back through the supply air path to the cooling coil discharge.

Moving back along the supply air path the fan entering air temperature is calculated setting fan outlet air temperature to the system design supply air temperature. The known fan inlet air temperature is then used as both the cooling coil and preheat coil discharge air temperature setpoint. Moving forward along the return air path the cooling coil entering air temperature can be determined by sequentially moving through the economizer cycle and the preheat coil.

Unlike temperature, the humidity ratio at any point in a system cannot be explicitly determined due to the dependence of the cooling coil performance on the mixed air humidity ratio. The latent load defines the difference between zone humidity and supply air humidity. However, the humidity ratio of the supply air depends on the humidity ratio entering the coil, which in turn depends on that of the return air. This calculation must be either performed by solving simultaneous equations or, as in this case, by an iterative process.

Assuming a trial value for the humidity ratio at the cooling coil discharge (e.g., 55°F, 90% RH), the humidity ratio at all other points throughout the system can be calculated. With known cooling coil inlet air conditions and a design discharge air temperature, the inverted cooling coil subroutine iterates on the coil fluid mass flow to converge on the discharge air temperature with the discharge air humidity ratio as an output. The cooling coil discharge air humidity ratio is then compared to the previous discharge humidity ratio. This iterative process continues calculating through the loop several times until the values of the cooling coil discharge air humidity ratio stabilize within a specified tolerance.

This basic algorithm for simulation of a VAV system might be used in conjunction with a heat-balance type of load calculation. For a weighting factor approach, it would have to be modified to allow zone temperatures to vary and a consequent readjustment of zone loads. It should also be enhanced to allow for possible limits on reheat temperature and/or cooling coil limits, zone humidity limits, outside air control (economizers), and/or heat-recovery devices, zone exhaust, return air fan, heat gain in the return air path because of lights, the presence of baseboard-type heaters, and more realistic control profiles. Most current building energy programs incorporate these and other features as user options, as well as algorithms for other types of systems.

```
BEGIN LOOP Calculate zone related design requirements
    • Calculate required supply airflow to meet zone load
    • Sum actual zone mass air flow rate
    • Sum zone latent loads
  IF zone equals last zone THEN Exit Loop
END LOOP
  • Calculate system return air temperature from zone temps
  • Assume an initial cooling coil leaving air humidity ratio

BEGIN LOOPIterate on cooling coil leaving air humidity ratio
    • Calculate return air humidity ratio from latent loads
    • Calculate supply fan power consumption and
      entering fan air temperature
    • Calculate mixed air temperature and humidity
      ratio using an economizer cycle
  IF mixed air temperature is less than design
      supply air temperature THEN
        • Calculate preheat coil load
  ELSE
        • Calculate cooling coil load and leaving air
          humidity ratio
  ENDIF
  IF cooling coil leaving air humidity ratio converged
  THEN Exit Loop
END LOOP

BEGIN LOOP Calculate the zone reheat coil loads
  IF zone supply air temperature is greater than system
  design supply air temperature THEN
        • Calculate reheat coil load
          (Subroutine: COILINV/HCDET)
  ENDIF
    • Sum reheat coil loads for all zones
  IF zone equals last zone THEN Exit Loop
END LOOP
```

Fig. 15 Algorithm for Calculating Performance of VAV with Reheat System

INVERSE MODELING

The method chosen to analyze a building's energy use is determined by the purpose for the investigation, including: retrofit energy savings analysis, diagnosing equipment malfunctions, energy auditing indices, component efficiency testing, Demand Side Management (DSM) evaluation, and others such as defined by MacDonald and Wasserman (1989) and Haberl et al. (1990a,b). These methods can be divided into two basic approaches, **forward modeling** and **inverse modeling**. A third approach includes methods that contain aspects of both definitions, i.e., forward plus inverse (Rabl 1988, Rabl and Riahle 1992).

Forward Modeling

Forward modeling, begins with a description of the building system or component of interest and defines the building being modeled according to its physical description. For example, the building geometry, geographical location, type of HVAC system, wall insulation, etc. may be defined. The primary benefits of this method are that it based on sound engineering principles and it has widespread acceptance and use in major public domain simulation codes (e.g., BLAST and DOE-2). Figure 16 is a flow chart that illustrates the ordering of the analysis that is typically performed by a building energy simulation program.

Inverse Modeling

Inverse modeling is based on the empirical behavior of the building as it relates to one or more driving forces. This approach is referred to as a system identification, parameter identification, or inverse modeling. In this modeling approach a structure or physical configuration of the building or system is assumed first and then important parameters are identified by a statistical analysis (Rabl and Riahle 1992). In general, there are two basic types of inverse models: steady state inverse models and dynamic inverse models. A third category, hybrid models, includes models with characteristics of both forward and inverse models.

Inverse modeling techniques have been used successfully to:

- Identify the energy savings from building retrofits
- Estimate the performance of an existing building under future weather and occupancy conditions

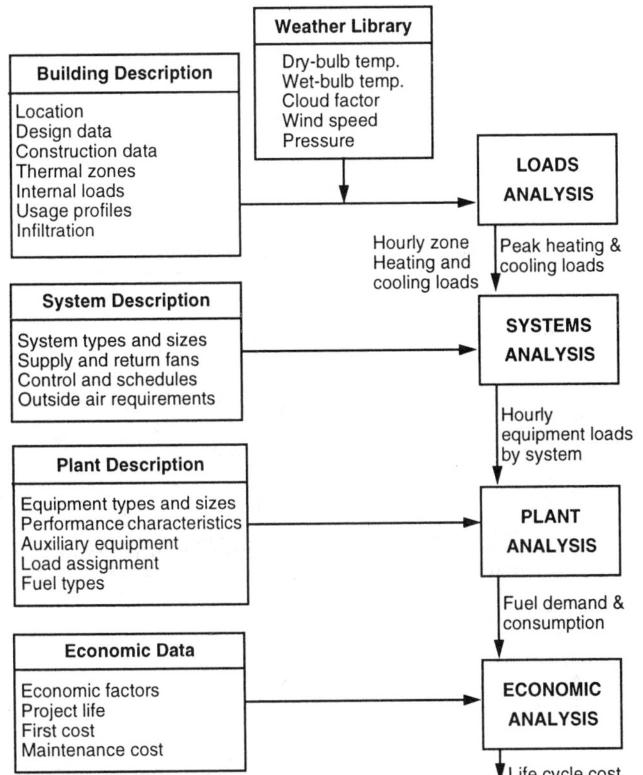

Fig. 16 Flow Chart for Building Energy Simulation Program
(Ayres and Stamper 1995)

- Predict hourly (or subhourly) loads and energy use levels for optimal operation of HVAC systems under demand or real time pricing utility rates
- Construct a model of HVAC subsystems for the optimal, adaptive control of that subsystem
- Fault diagnosis of HVAC systems

Steady-State Inverse Models. The simplest form of an inverse model is a *steady-state inverse model* of a building's energy use. The simplest steady-state inverse model can be developed by performing a regression analysis on monthly utility consumption data against average billing period temperatures. Although simple in concept, the most accurate methods use sophisticated change-point regression procedures that simultaneously solve for several parameters including a weather independent base-level parameter, one or more weather dependent parameters, and the point or points at which the model switches from weather dependent to weather-independent dependent behavior. In its simplest form the 65°F degree day model is a change-point model that has a *fixed* change point at 65°F. Other examples include three and five parameter Princeton Scorekeeping Methods (PRISM) (Fels 1986) and a four parameter model (4P) developed by Ruch and Claridge (1991). The three parameters include: a weather independent base-level use, a change-point, and a temperature dependent parameter or slope of a line that is determined by regression. The four parameters include a change-point, a slope above the change point, a slope below the change point, and the energy use associated with the change point. An inverse bin method has also been proposed to handle more than four change points (Thamilseran and Haberl 1995).

Figure 17 shows several types of steady-state, single variable inverse models. Figure 17A shows a simple one parameter, or constant model, and Equation (44) gives the equivalent notation for calculating the constant energy use using this model. Figure 17B shows a steady-state two parameter (2P) model where β_0 is the y-axis intercept and β_1 is the slope of the regression line for positive values of x, where x represents the ambient air temperature. The 2P model represents cases when either heating or cooling is always required.

Figure 17C shows a three parameter, change-point model. This model is typical of natural gas energy use in a single family residence that uses gas for space heating and domestic water heating. In Equation (46), which is given for the three parameter model, β_0 represents the baseline energy use, β_1 is the slope of the regression line for values of ambient temperature less than the change-point β_2. In this type of notation the exponent (+) indicates that only positive values of the parenthetical expression are considered. Figure 17D shows a three parameter model for cooling energy use and Equation (47) gives the appropriate expression for analyzing cooling energy use with a three parameter model.

Figure 17E and Figure 17F illustrate four parameter models for heating and cooling, respectively. Equations (48) and (49) indicate the appropriate expression for calculating the heating and cooling energy use using a four parameter model. In a four parameter model β_0 represents the baseline energy exactly at the change-point β_3. β_1 and β_2 are the lower and upper region regression slopes for ambient air temperature below and above the change point β_3. Figure 17G illustrates a 5P model (Fels 1986). Such a model is useful for modeling buildings that are electrically heated and cooled. The 5P model has two change points and a base level consumption value as shown in Equation (50).

$$E_{period} = \beta_0 \tag{44}$$

$$E_{period} = \beta_0 + \beta_1 T \tag{45}$$

$$E_{period} = \beta_0 + \beta_1(\beta_2 - T)^+ \tag{46}$$

$$E_{period} = \beta_0 + \beta_1(T - \beta_2)^+ \tag{47}$$

$$E_{period} = \beta_0 + \beta_1(\beta_3 - T)^+ - \beta_2(T - \beta_3)^+ \tag{48}$$

$$E_{period} = \beta_0 - \beta_1(\beta_3 - T)^+ + \beta_2(T - \beta_3)^+ \tag{49}$$

$$E_{period} = \beta_0 + \beta_1(\beta_3 - T)^+ + \beta_2(T - \beta_4)^+ \tag{50}$$

The advantage of these steady-state inverse models is that their use can be easily automated and applied to large numbers of buildings where monthly utility billing data and average daily temperatures for the billing period are available. Steady-state inverse models can also be applied to daily data to compensate for differences in weekday and weekend use (Claridge et al. 1992).

Disadvantages of steady-state inverse models include an insensitivity to dynamic effects (i.e., thermal mass), insensitivity to variables other than temperature (for example humidity and solar gain), and inappropriateness for some building, for example buildings with strong on/off schedule dependent loads or buildings with multiple change-points.

Steady-State Inverse Models Using More than One Independent Variable. Multiple regression techniques allow the analyst to investigate the influence of more than one independent variable (such as outdoor air temperature and humidity, solar radiation, and indicators of scheduling) on a response variable (such as building energy use.) The form of the general linear regression model is

$$Y = \beta_0 + \beta_1 X_1 + \beta_2 X_2 + \ldots + B_p X_p + \varepsilon \tag{51}$$

where Y is the response variable; $X_1, X_2, \ldots X_p$ are the independent variables; $\beta_0, \beta_1, \beta_2, \ldots, \beta_p$ are the p regression parameters; and ε is the error term. When $p = 2$, the response surface is a plane. When $p > 2$, the response surface is a hyperplane.

Interactions between variables can be considered by defining one of the terms as the product of two independent variables or by redefining the independent variables as principle components Reddy and Claridge 1994, Ruch et al. 1993). Curvature in the response surface can be introduced through the use of independent polynomial variables. Equation (52) demonstrates a model with two independent variables, each in quadratic form, with an interaction term.

$$Y = \beta_0 + \beta_1 X_1 + \beta_2 X_1^2 + \beta_3 X_2 + \beta_4 X_2^2 + \ldots \beta_5 X_1 X_2 + \varepsilon \quad (52)$$

The choice of the model should be guided by the analyst's understanding of the physical system and its expected response. Several standard statistical tests exist for evaluating the goodness-of-fit of the model and the degree of influence that each of the independent variables exerts on the response variable (Draper and Smith 1981, Neter et al. 1989).

Dynamic Inverse Models. Examples of dynamic inverse models include equivalent thermal network analysis (Sonderegger 1977), ARMA models (Subbarao 1990, Reddy 1989), Fourier series models (Dhar et al. 1995), machine learning (Miller and Seem 1991), and artificial neural networks (Miller and Seem 1991, Kreider and Wang 1991, Kreider and Haberl 1994). Neural networks seem to hold the most promise.

These models are capable of capturing dynamic effects such as mass dynamics that traditionally have required the solution of a set of differential equations. These models are better suited to handle intercorrelated forcing functions or independent parameters. The advantages of dynamic inverse models include the ability to model complex systems that depend on more than one independent parameter. The disadvantages of dynamic inverse models include their complexity and the need for more detailed measurements to tune the model. Unlike steady-state inverse models, dynamic inverse models usually require a high degree of user interaction and knowledge of the building or system being modeled.

Neural Network Models

Neural Network Construction. An artificial neural network is a massively parallel, dynamic system of interconnected, interacting parts based loosely on some aspects of the brain. Neural networks are considered to be intuitive because they learn by example rather than by following programmed rules. The ability to "learn" is one of the key aspects of neural networks. A neural network consists of several layers of neurons that are connected to each other. A connection is a unique information transport link from one sending neuron to one receiving neuron. The structure of part of a Neural network is schematically shown in Figure 20. The figure shows the input layer, hidden layers, output, and target training values. Hidden and output layers consist of connected neurons; the input layer does not contain neurons. Any number of input, output and hidden layer (only one hidden layer is shown) neurons can be used. One challenge of this technology is to construct a net with sufficient complexity to learn accurately without imposing excessive computational time.

The neuron is the fundamental building block of a neural network. A set of inputs is applied to each. Each element of the input set is multiplied by a weight W and the products are summed at the neuron. The summation of weighted inputs is termed INPUT and must be calculated for each neuron in the network. After INPUT is calculated, an activation function F is applied to modify it, thereby producing the neuron's output.

The weights of a net are initiated with small random numbers. Then the weights are adjusted iteratively or "trained" so that the application of a set of inputs produces the desired set of outputs. Usually a network is trained with a training data set that consists of many input-output pairs. Artificial networks have been trained by a

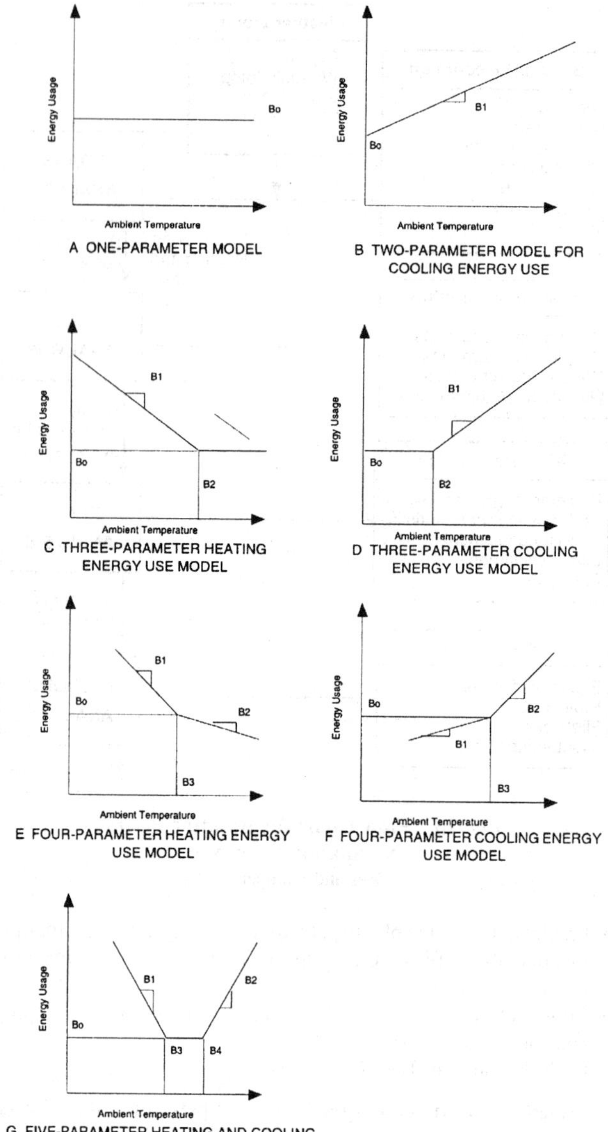

Fig. 17 Steady-State, Single Variable Models Appropriate for Commercial Building Energy Use

wide variety of methods (McClelland and Rumelhart 1988, Wasserman 1989). Training the net using one method called backpropagation requires the following steps:

1. Select a training pair from the training set and apply the input vector to the network input layer.
2. Calculate the output of the network OUT_i.
3. Calculate the error $ERROR_i$ between the network output and the desired output (the target vector from the training pair).
4. Adjust the weights of the network in a way that minimizes the error.
5. Repeat steps 1 through 4 for each vector in the training set until the error for the entire set is lower than the user specified, preset training tolerance.

Neural Networks Applied to Buildings

Commercial Buildings. Neural networks have been useful in predicting energy use in commercial buildings for such reasons as

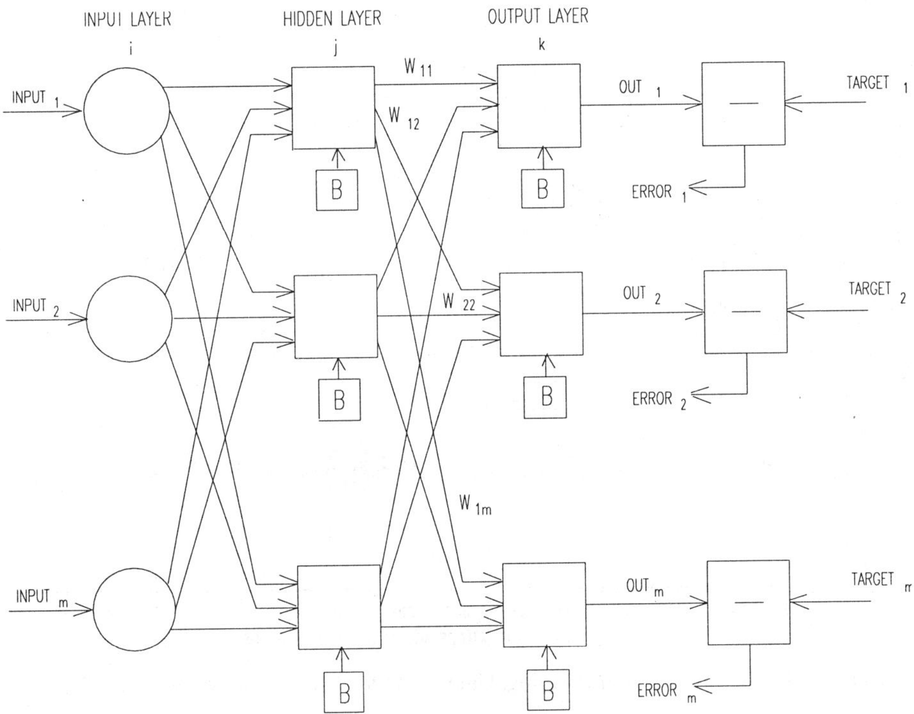

Fig. 18 Schematic of Neural Network

- Prediction of what a properly operating building should be doing compared to actual operation. If there is a difference, it can be used in an expert system to produce early diagnoses of building operation problems
- Prediction of what a building, prior to an energy retrofit, would have consumed under present conditions. When compared to the measured consumption of the retrofitted building, the difference represents a good estimate of the energy savings due to the retrofit. This represents one of the few ways that actual energy savings can be determined after the pre-retrofit building configuration has ceased to exist.

Figure 19 shows results typical of several hundred networks constructed on an academic engineering center located in central Texas. The cooling load is created by solar gains, internal gains, outdoor air sensible heat, and outdoor air humidity loads. The neural network is used to predict the pre-retrofit energy consumption for comparison with measured consumption of the retrofitted building. Six months of pre-retrofit data were available with which to train a network.

The solid lines show the known building consumption data while the dashed lines show the neural network predictions. This figure shows that a neural network trained for one period (here, September of 1989) can predict energy consumption well into the future (here, January of 1990).

The network used for this prediction had two hidden layers. The input layer contained eight neurons that receive eight different types of input data as listed below. The output layer consists of one neuron that gives the output datum (chilled water consumption). Each training fact (i.e., training data set), therefore, contains eight input data (independent variables) and one pattern datum (dependent variable). The eight hourly input data used in each hour's data vector were selected on physical bases (Kreider and Rabl 1994) are as follows:

- Hour number (0 – 2300)
- Ambient dry-bulb temperature
- Horizontal insolation
- Humidity ratio
- Wind speed
- Weekday/weekend binary flag (0,1)
- Past hour's chilled water consumption
- Second past hour's chilled water consumption

These measured independent variables were able to predict the chilled water use to an RMS error of less than 4% [JCEM 1992).

The choice of an optimal network's configuration for a given problem remains an art. The number of hidden neurons and layers must be sufficient to meet the requirement of the given application. However, if too many neurons and layers are used the network tends to memorize data rather than learning, that is, finding the underlying patterns in the data. Further, choosing an excessively large number of hidden layers significantly increases the required training time for certain learning algorithms. Anstett and Kreider (1993), Kreider and Wang (1991), and Wang and Kreider (1992) report on additional case studies on commercial buildings.

Residential Buildings. Neural networks also have been applied to residential buildings. Figure 20 shows a typical prediction for a building in Varese, Italy. Hourly data collected included outdoor temperature and insolation, indoor temperature, and heating energy consumption. The weather data and time information (hour of day, day of year) were used to make the predictions shown.

Application of Inverse Models

The principal applications of inverse models appear to be in the following areas:

- Evaluation of energy conservation programs
- Prescreening indices for energy auditing
- Building energy management
- Optimal control, and
- In-situ characterization of HVAC systems

In each of these applications both steady state and dynamic inverse models have been applied. In general, steady state inverse models are used with monthly and daily data containing one or more independent variables. Dynamic inverse models are usually used

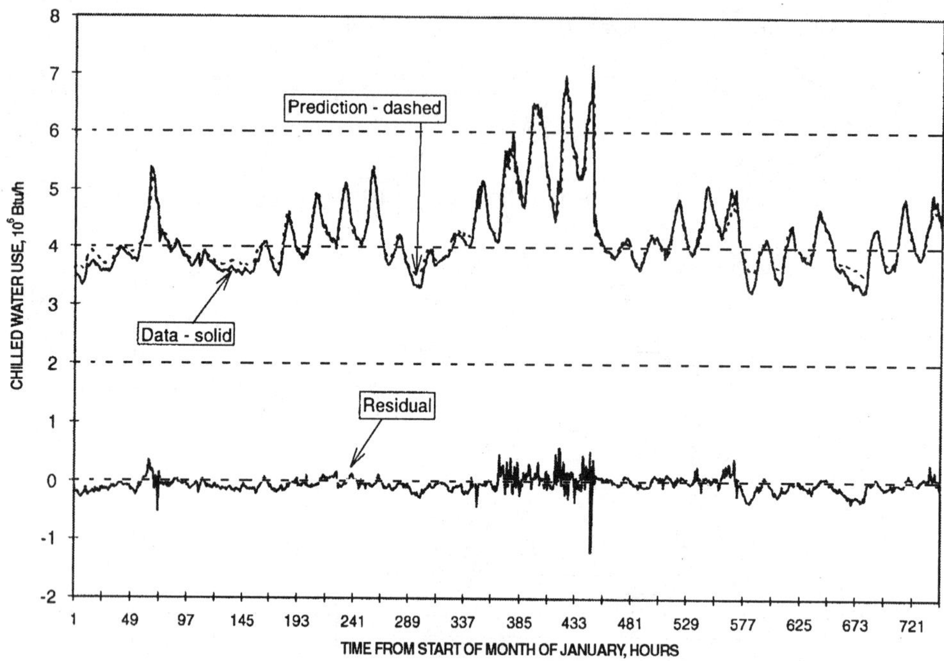

Fig. 19 Neural Network Prediction of Whole Building, Hourly Chilled Water Consumption for Commercial Building

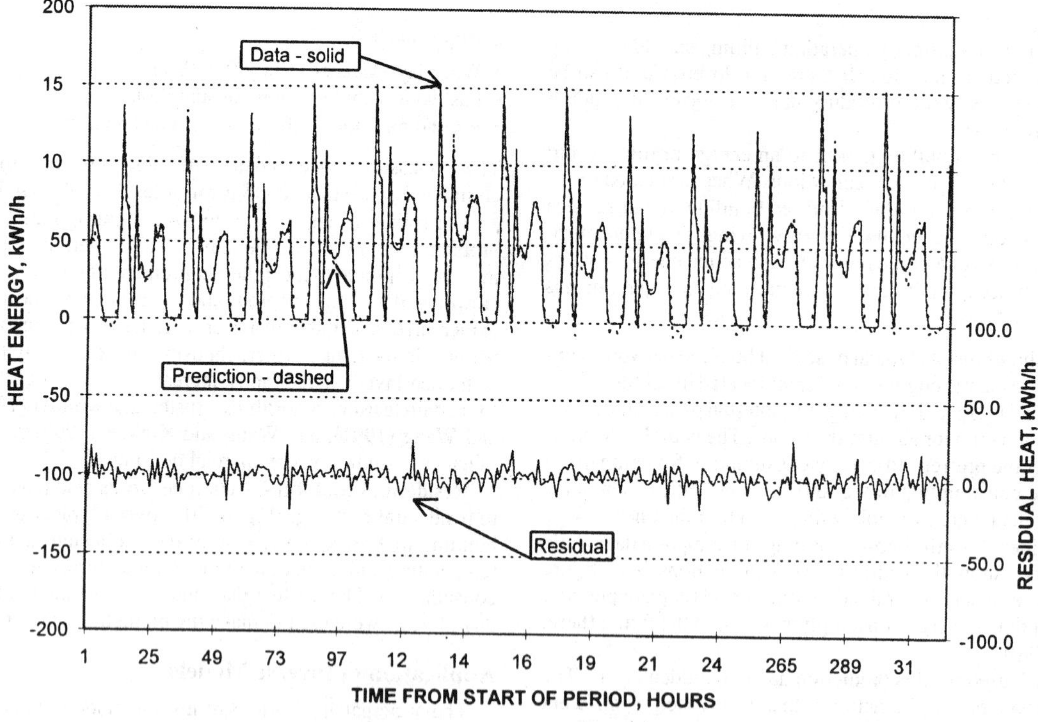

Fig. 20 Neural Network Prediction of Residential Heating
(Test was run January 30 through February 13, 1989 in Varese, Italy.)

with hourly or sub-hourly data in cases where the thermal mass of a building is significant enough to delay the heat gains or losses.

Evaluation of Energy Conservation Programs. Aside from simply running a regression analysis of energy use against temperature, other widely-used steady-state inverse methods for the evaluation of energy conservation retrofits include three, four, and five parameter change points models previously described (Fels 1986, Kissock et al. 1993). Such models are useful for statistically determining average weather-dependent and weather-independent energy use for buildings. Three parameter change-point models can yield (1) baseline energy use, (2) the temperature at which weather-dependent energy use begins to increase energy use above the baseline (i.e., the change-point), and (3) the linear slope of the temperature dependency above (cooling model) or below (heating model) the temperature change point.

Change-point regression models work best with heating data from buildings with systems that have little or no part load nonlinearities (i.e., systems that become less efficient as they begin to cycle on-off with part loads). In general, change-point regression models do not predict cooling loads as well because outdoor humidity has a large influence on latent loads on the cooling coil. Other factors that decrease the accuracy of change point models include: solar effects, thermal lags, and on-off HVAC schedules. Four parameter models exhibit a better statistical fit over three parameter models in buildings with continuous, year-around cooling or heating (i.e., grocery stores and office buildings with high internal loads). However, every model should be checked to ensure that the regression is not falsely indicating an unreasonable relationship).

A major advantage of using a steady-state inverse model to evaluate the effectiveness of energy conservation retrofits lies in its ability to factor out year-to-year weather variations. This variation can be factored out by using a Normalized Annual Consumption or NAC (Fels 1986). Basically, the annual energy conservation savings can be calculated by comparing the difference obtained by multiplying the pre-retrofit and post-retrofit parameters by the weather conditions for the average year. Typically, ten to twenty years of average daily weather data from a nearby weather service site are used to calculate 365 days of average weather conditions which are then used to calculate the average pre-retrofit and post-retrofit conditions.

Utilities and government agencies have found it advantageous to prescreen many buildings against test regression models. Such inverse models can be used to develop comparative figures of merit for buildings in a similar Standard Industrial Code (SIC) classification. In such applications a minimum goodness of fit is usually established that determines whether the monthly utility billing data are well fit by the 1-, 2-, 3-, 4- or 5-parameter model being tested. Comparative figures of merit can then be determined by dividing the parameters by the conditioned floor area to yield average daily energy use per unit area of conditioned space. For example, an area normalized comparison of base-level parameters across residential building would be used to analyze weather independent energy use. Such information can be used by energy auditors to focus their efforts on those systems needing assistance (Haberl and Komor 1990a, b).

Dynamic inverse models can also be used to evaluate energy conservation retrofits, if hourly or more frequent data are available. In this application the same independent variables need to be available after the retrofit to estimate the energy use predicted by the baseline, pre-retrofit model.

Energy Management. Steady state and dynamic inverse models can be used with energy management and control systems to predict energy use (Kreider and Haberl 1994). Hourly or daily comparisons of measured energy use against predicted energy use can be used to determine if systems are being left on unnecessarily or are in need of maintenance. Combinations of predicted energy use and a knowledge-based system can indicate above normal energy use and diagnose the possible cause of the malfunction if sufficient historical information has been previously gathered (Haberl and Claridge 1987). Hourly systems that use artificial neural networks have also been constructed (Kreider and Wang 1991).

Table 5 Classification of Methods for Thermal Analysis of Buildings

Method	Forward	Inverse	Hybrid	Comments	DD[a]
			Steady State Methods		
Simple linear regression		X		One dependent parameter, one independent parameter. May have slope and y-intercept	1
Multiple linear regression		X	X	One dependent parameter, multiple independent parameters	2
Modified degree day method	X			Based on fixed reference temperature of 65°F	1
Variable base degree day method	X			Variable reference temperatures	2
ASHRAE bin method and inverse bin method	X	X	X	Hours in temperature bin times load for that bin	3
Change point models: 3 parameter (PRISM CO, HO), 4 parameter, 5 parameter (PRISM HC).		X	X	Uses daily or monthly utility billing data and average period temperatures	2
Modified bin method	X			Modified bin method with cooling load factors	4
			Dynamic Methods		
Thermal network (Sonderegger 1977)		X	X	Uses equivalent thermal parameters (inverse mode)	3
Response factors (Stephenson and Mitalas 1967)	X		X	Tabulated or as used in simulation programs	4
Fourier Analysis (Shurcliff 1984, Dhar 1995)		X	X	Frequency domain analysis convertible to time domain	4
ARMA Model (Subbarao 1986)		X		Autoregressive Moving Average model	3
ARMA Model (Reddy 1989)		X		Multiple-input autoregressive moving average model	3
BEVA, PSTAR (Subbarao 1986)	X	X	X	Combination of ARMA and Fourier series, includes loads in time domain	4
Modal analysis (Bacot et al. 1984)	X	X	X	Building described by diagonalized diff. equation using nodes	4
Differential equation (Rabl 1988)		X	X	Analytical linear differential equation	4
Computer simulation (DOE-2, BLAST)	X		X	Hourly simulation programs with system models	5
Computer emulation (HVACSIM+, TRNSYS)	X		X	Sub-hourly simulation programs	5
Artificial Neural Networks (Kreider and Wang 1991, Kreider 1992, Kreider and Haberl 1994)		X	X	Connectionist models	4

[a]Degree of difficulty: 1 = simplest to 5 = most time consuming

Table 6 Decision Diagram for Selection of Inverse Model

Method	Usage[a]	Difficulty	Time Scale[b]	Calc. Time	Variables	Accuracy
Simple linear regression	ES	Simple	D,M	Very Fast	T	Low
Multiple linear regression	D,ES	Moderate	D,M	Fast	T,H,S,W,t	Medium
Bin method and inverse bin method	ES,DE	Moderate	H	Fast	T	Medium
Change point models	D,ES	Moderate	H,D,M	Fast	T	Medium
Modified bin method	DE	Moderate	H	Medium	T,S,tm	Medium
Thermal network	D,ES,C	Complex	S,H	Fast	T,S,tm	High
Fourier series analysis	D,ES,C	Complex	S,H	Medium	T,H,S,W,θ,tm	High
ARMA model	D,ES,C	Complex	S,H	Medium	T,H,S,W,θ,tm	High
Modal analysis	D,ES,C	Complex	S,H	Medium	T,H,S,W,θ,tm	High
Differential equation	D,ES,C	Very Complex	S,H	Fast	T,H,S,W,θ,tm	High
Computer simulation (component-based)	D,ES,C,DE	Very Complex	S,H	Slow	T,H,S,W,θ,tm	High
Computer simulation (fixed schematic)	D,ES,DE	Very Complex	H	Slow	T,H,S,W,θ,tm	High
Computer emulation	D,C	Very Complex	S,H	Very Slow	T,H,S,W,θ,tm	High
Artificial neural networks	D,ES,C	Complex	S,H	Fast	T,H,S,W,θ,tm	High

a Usage shown includes diagnostics (D), energy savings calculations (ES), design (DE), and control (C).
b Time scales shown are hourly (H), daily (D), monthly (M), and sub-hourly (S).
c Variables include temperature (T), humidity (H), solar (S), wind (W), time (θ), and thermal mass (tm).

HYBRID MODELING

Forward plus inverse models or hybrid models encompass everything that does not fit into the definition of forward or inverse models. An example is a traditional fixed-schematic simulation program such as DOE-2 or BLAST (or even a component based model) that is used to simulate the energy use of an existing building. In this case a *forward* analysis method is used in an *inverse* application; i.e., the forward simulation model is calibrated or fit to the actual energy consumption data from a building in the same way that energy use is fit by a linear regression model to temperature. Such an application is a *hybrid model.*

Application of Hybrid Models

Several practical difficulties prevent achieving a "calibrated simulation" or a simulation that nearly reflects the actual building performance including (1) the measurement and adaptation of weather data for use by the simulation programs (e.g., converting global horizontal solar into beam and diffuse solar radiation), (2) the choice of methods used to calibrate the model, and (3) the choice of methods used to measure the required input parameters for the simulation (i.e., the weight of the building, infiltration coefficients, and shading coefficients). Truly "calibrated" models have only been achieved in a few applications because they require a very large number of input parameters, a high degree of expertise, and enormous amounts of computing time, patience, and financial resources. However, Bronson, et at. (1992), Haberl et al. (1995), Kaplan et al. (1990), Corson (1992), Bou Saada and Haberl (1995a, 1995b), and Hsieh (1988) provide examples of different methods used to calibrate simulation models.

CLASSIFICATION OF METHODS

In Table 5, different methods of analyzing building energy use are classified using an expanded versions of Rabl's definitions (Rabl 1988). Simple linear regression and multiple linear regression are the most widely used forms of inverse analysis. In the proper application multiple linear regression must adequately address intercorrelations among the independent parameters.

SELECTING AN APPROACH

Table 6 presents a decision diagram for selecting a forward or inverse model where use of the model, degree of difficulty in understanding and applying the model, time scale for the data used by the model, calculation time, and input variables used by the models are the criteria used to choose a particular model.

REFERENCES

Alamdari, F. and G.P. Hammond. 1982. Time-dependent convective heat transfer in warm-air heated rooms. Energy Conservation in the Built Environment: Proceedings CIB W67 Third International Symposium, Dublin, Ireland.209-220.

Alamdari, F. and G.P. Hammond. 1983. Improved data correlations for buoyancy-driven convection in rooms. *Building Services Engineering Research and Technology* 4(3)106-112.

Alereza, T. and T. Kusuda. 1982. Development of equipment seasonal performance models for simplified energy analysis methods. *ASHRAE Transactions* 88(2):249-62.

Altmayer, E.F., A.J. Gadgil, F.S. Bauman, and R.C. Kammerud. 1983. Correlations for convective heat transfer from room surfaces. *ASHRAE Transactions* 89(2A)61-77.

Anstett, M. and J.F. Kreider. 1993. Application of artificial neural networks to commercial building energy use prediction. *ASHRAE Transactions* 99(1):505-517.

ASHRAE. 1995. *Bin and degree hour weather data for simplified energy calculations.*

ASHRAE. 1989. Energy efficient design of new buildings except low-rise residential buildings. *Standard* 90.1-1989.

Ayres, M.J. and E. Stamper. 1995. Historical development of building energy calculations. *ASHRAE Transactions* 101(1).

Bacot, P., A. Neveu, and J. Sicard. 1984. Analyse modale des phenomenes thermiques en regime variable dans le batiment. *Revue Generale de Thermique* 267:189.

Balcomb, J.D., R.W. Jones, R.D. McFarland, and W.O. Wray 1982. Expanding the SLR method. *Passive Solar Journal* 1(2).

Bauman, F., A. Gadgil, R. Kammerud, E. Altmayer, and M. Nansteel. 1983. Convective heat transfer in buildings: Recent research results. *ASHRAE Transactions* 89(1A)215-232.

Bohn, M.S., A.T. Kirkpatrick, and D.A. Olson. 1984. Experimental study of three-dimensional natural convection high-Rayleigh number. *Journal of Heat Transfer* 106:339-345.

Bou Saada, T. and J. Haberl. 1995a. A weather-day typing procedure for disaggregating hourly end-use loads in an electrically heated and cooled building from whole-building hourly data. *Proceedings* 30th IECEC, pp. 349-356.

Bou Saada, T. and J. Haberl. 1995b. An improved procedure for developing calibrated hourly simulation models. *Proceedings* International Building Performance Simulation Association.

Bourdouxhe, J-P., M. Grodent, J. Lebrun, and C. Saavedra. 1994a. A toolkit for primary HVAC system energy calculation—Part 1: Boiler model. ASHRAE *Transaction* 100(2):759-773.

Bourdouxhe, J-P., M. Grodent, J. Lebrun, C. Saavedra, and K. Silva. 1994b. A toolkit for primary HVAC system energy calculation—Part 2: Reciprocating chiller models. ASHRAE *Transactions* 100(2):774-786.

Bourdouxhe, J-P., M. Grodent, and C. Silva. 1994c. Cooling tower model developed in a toolkit for primary HVAC system energy calculation—Part 1: Model description and validation using catalog data. *Proceedings* Fourth International Conference on System Simulation in Buildings.

Bourdouxhe, J-P., M. Grodent, and J. Lebrun. 1997. *HVAC1 Toolkit: Algorithms and subroutines for primary HVAC system energy calculations*. ASHRAE.

Brandemuehl, M.J. 1993. *HVAC2 Toolkit: Algorithms and subroutines for secondary HVAC System energy calculations*. ASHRAE.

Brandemuehl, M.J. and S. Gabel. 1994. Development of a toolkit for secondary HVAC system energy calculations. *ASHRAE Transactions* 100 (1).

Braun, J.E. 1988. Methodologies for the design and control of chilled water systems. Ph.D. Thesis, University of Wisconsin-Madison

Bronson, D., S. Hinchey, J. Haberl, and D. O'Neal. 1992. A procedure for calibrating the DOE-2 simulation program to non-weather dependent loads. *ASHRAE Transactions* 98(1):636-652.

Buhl, W.F., A.E. Erdem, J.M. Nataf, F.C. Winkelmann, M.A. Moshier, and E.F. Sowell. 1990. The US EKS: Advances in the SPANK-based energy kernel system. 3rd International Conference On System Simulation In Buildings, pp. 107-150.

Buhl, W.F., A.E. Erdem, F.C. Winkelmann, and E.F. Sowell. 1993. Recent improvements in SPARK: Strong Component Decomposition, Multivalued Objects and Graphical Interface. *Proceedings* of Building Simulation '93, pp. 283-390.

Carroll, J.A. 1980. An "MRT Method" of computing radiant energy exchange in rooms. Systems Simulation and Economic Analysis, San Diego. 343-348.

Chandra, S. and A. A. Kerestecioglu. 1984. Heat Transfer in Naturally Ventilated Rooms: Data from Full-
scale Measurements. ASHRAE Transactions, Vol. 90, Pt.1b, pp. 211-224.

Chi, J. and G.E. Kelly. 1978. A method for estimating the seasonal performance of residential gas and oil-fired heating systems. *ASHRAE Transactions* 84(1):405.

Cinquemani, V., J.R. Owenby, and R.G. Baldwin. 1978. Input data for solar systems. U.S. Department of Energy Report No. E(49-26)1041.

Clark, D.R. 1985. *HVACSIM+ building systems and equipment simulation program: Reference manual*. NBSIR 84-2996, U.S. Department of Commerce, Washington, DC.

Clarke, J.A. 1985. *Energy simulation in building design*. Adam Hilger Ltd, Boston.

Claridge, D.E., M. Krarti, and M. Bida. 1987. A validation study of variable-base degree-day cooling calculations. *ASHRAE Transactions* 93(2):90-104.

Claridge, D.E., J.S. Haberl, R. Sparks, R. Lopez, and K. Kissock. 1992. Monitored commercial building energy data: Reporting the results. *ASHRAE Transactions* 98(1):636-652.

Cole, R.J. 1976. The longwave radiation incident upon the external surface of buildings. *The Building Services Engineer* 44:195-206.

Cooper, K.W. and D.R. Tree. 1973. A re-evaluation of the average convection coefficient for flow past a wall. *ASHRAE Transactions* 79:48-51.

Corson, G.C. 1992. Input-output sensitivity of building energy simulations. *ASHRAE Transactions* 98(1):618.

Davies, M.G. 1988. Design models to handle radiative and convective exchange in a room. *ASHRAE Transactions* 94(2):173-195.

Dhar, A. 1995. Development of Fourier series and artificial neural network approaches to model hourly energy use in commercial buildings. PhD Thesis, ME Dept., Texas A&M University.

Draper, N. and H. Smith. 1981. *Applied regression analysis*, 2nd ed. John Wiley & Sons, New York.

Elmahdy, A.H. and G.P. Mitalas. 1977. A simple model for cooling and dehumidifying coils for use in calculating energy requirements for buildings. *ASHRAE Transactions* 83(2):103-117.

Erbs, D.G., S.A. Klein, and W.A. Beckman. 1983. Estimation of degree-days and ambient temperature bin data from monthly-average temperatures. *ASHRAE Journal* 25(6):60.

Fels, M. (editor). 1986. Measuring energy savings: The scorekeeping approach. *Energy and Buildings* 9.

Fels, M. and M. Goldberg, 1986. Refraction of PRISM results in components of saved energy. *Energy and Buildings* 9:169.

Fracastoro, G., M. Masoero, and M. Cali. 1982. Surface heat transfer in building components. *Proceedings* Thermal Performance of the Exterior Envelopes of Buildings II, pp. 180-203.

Gordon, J.M. and K.C. Ng. 1994. Thermodynamic modeling of reciprocating chillers. *Journal of Applied Physics* 75(6):2769-74.

Gordon, J.M and K.C. Ng. 1995. Predictive and diagnostic aspects of a universal thermodynamic model for chillers. *International Journal of Heat and Mass Transfer* 38(5):807-818.

Gordon, J.M., K.C. Ng, and H.T. Chua. 1995. Centrifugal chillers: Thermodynamic modeling and a case study. *International Journal of Refrigeration* 18(4):253-257.

Haberl, J. and P. Komor. 1990a. Improving commercial building energy audits: How annual and monthly consumption data can help. *ASHRAE Journal* 32(8):26-33.

Haberl, J. and P. Komor. 1990b. Improving commercial building energy audits: How daily and hourly data can help. *ASHRAE Journal* 32(9):26-36.

Haberl, J., D. Bronson, and D. O'Neal. 1995. An evaluation of the impact of using measured weather data versus TMY weather data in a DOE-2 simulation of an existing building in central Texas. *ASHRAE Transactions* 101(2):558-76.

Haberl, J.S. and D.E. Claridge. 1987. An expert system for building energy consumption analysis: Prototype results. *ASHRAE Transactions* 93(1):979-998.

Haberl, J.S., D.E. Claridge, and D. Harrje. 1990. The design of field experiments and demonstrations. IEA Field Monitoring Workshop, Othenberg, Sweden, pp. 33-58.

Hittle, D.C. and R. Bishop. 1983. An improved root-finding procedure for use in calculating transient heat flow through multilayered slabs. *International Journal of Heat and Mass Transfer* 26:1685-93.

Howell, R.H. and S. Suryanarayana. 1990. Sizing of radiant heating systems: Part I and Part II. *ASHRAE Transactions* 96.

JCEM. 1992. Final report: Artificial neural networks applied to LoanSTAR data. Joint Center for Energy Management *Report* TR/92/15.

Kamal, S. and P. Novak. 1991. Dynamic analysis of heat transfer in buildings with special emphasis on radiation. *Energy and Buildings* 17(3): 231-41.

Kaplan, M., J. McFerran, J. Jansen, and R. Pratt. 1990. Reconciliation of a DOE2.1c model with monitored end-use data from a small office building. *ASHRAE Transactions* 96(1):981.

Kays, W.M. and A.L. London. 1984. *Compact heat exchangers*, 3rd ed. McGraw-Hill Book Company, New York.

Khalifa, A.J.N. and R.H. Marshall. 1990. Validation of heat transfer coefficients on interior building surfaces using a real-sized indoor test cell. *International Journal of Heat and Mass Transfer* 33(10):2219-36.

Kissock, J.K., Reddy, T.A., Haberl, J.S., Claridge, D.E. 1993. E-model: A new tool for analyzing building energy use data. *Proceedings* Intl. Indust. Energy Tech. Conf., Texas A&M University.

Klein, S.A., W.A. Beckman, et al. 1994. *TRNSYS: A transient simulation program*. Engineering Experiment Station *Report* 38-14, University of Wisconsin-Madison.

Kreider, J.F. and X.A. Wang. 1992. Improved artificial neural networks for commercial building energy use prediction. *Journal of Solar Energy Engineering*.

Kreider, J.F. and J. Haberl. 1994. Predicting hourly building energy usage: The great predictor shootout—Overview and discussion of results. *ASHRAE Transactions* 100(.

Kreider, J.F. and A. Rabl. 1994. *Heating and cooling of buildings*. McGraw-Hill Book Company, New York.

Kreider, J.F. and X.A. Wang. 1991. Artificial neural networks demonstration for automated generation of energy use predictors for commercial buildings. *ASHRAE Transactions* 97(1).

Kusuda, T. 1969. Thermal response factors for multi-layer structures of various heat conduction systems. *ASHRAE Transactions* 75(1):246-71.

LBL. 1982. *DOE2 engineers manual*. Lawrence Berkeley Laboratory *Report* LBL-11353 (LA-8520-M, DE83004575). National Technical Information Services, Springfield, VA.

Lewis, P.T. and D.K. Alexander. 1990. HTB2: A flexible model for dynamic building simulation. *Building and Environment*. pp. 7-16.

MacDonald, J.M. and D.M. Wasserman. 1989. Investigation of metered data analysis methods for commercial and related buildings. Oak Ridge National Laboratory *Report* ORNL/CON-279.

McClelland, J.L. and D.E. Rumelhart. 1988. *Exploration in parallel distributed processing*. MIT Press, Cambridge, MA.

Melo, C. and G.P. Hammond. 1991. Modeling and assessing the sensitivity of external convection from building facades in heat and mass transfer in building materials and structures. 683-95. Hemisphere Publishing, New York.

Metcalfe, R.R., R.D. Taylor, C.O. Pederson, R.J. Liesen, and D.E. Fisher. 1995. Incorporating a modular system simulation program into a large energy analysis program: The linking of IBLAST and HVACSIM+. *Proceedings* of Building Simulation'95.

Miller, D.E. 1980. The impact of HVAC process dynamics on energy use.*ASHRAE Transactions* 86(2):535-56.

Miller, R and J. Seem. 1991. Comparison of artificial neural networks with traditional methods of predicting return from night setback. *ASHRAE Transactions* 97(2):500-508.

Mitalas, G.P. 1968. Calculations of transient heat flow through walls and roofs. *ASHRAE Transactions* 74(2):182-88.

Mitalas, G.P. and D.G. Stephenson. 1967. Room thermal response factors. *ASHRAE Transactions* 3(1):2.1-2.10.

Mitchell, J.W. 1983. *Energy engineering*. John Wiley and Sons, Inc., New York.

NOAA. 1973. Degree-days to selected bases. U.S. National Climatic Data Center, Asheville, NC.

Neter, J., W. Wasseran, and M. Kutner. 1989. *Applied linear regression models*, 2nd ed. Richard C. Irwin, Inc., Homewood, IL.

Park, C., D.R. Clark, and G.E. Kelly. 1985. An overview of HVACSIM+, A dynamic building/HVAC control systems simulation program. *Proceedings* First Building Energy Simulation Conference.

Parker, W.H., G.E. Kelly, and D. Didion. 1980. A method for testing, rating, and estimating the heating seasonal performance of heat pumps. National Bureau of Standards, NBSIR 80-2002, April.

Rabl, A. 1988. Parameter estimation in buildings: Methods for dynamic analysis of measured energy use. *Journal of Solar Energy Engineering* 110:52-66.

Rabl, A. and A. Riahle. 1992. Energy signature model for commercial buildings: test with measured data and interpretation. *Energy and Buildings* 19:143-154.

Reddy, T. 1989. Application of dynamic building inverse models to three occupied residences monitored non-intrusively. *Proceedings* Thermal Performance of Exterior Envelopes of Buildings IV, ASHRAE/DOE/BTECC/CIBSE.

Ruch, D., L. Chen, J. Haberl, and D. Claridge. 1993. A change-point principle component analysis (CP/CPA) method for predicting energy use in commercial buildings: The PCA model. *Journal of Solar Energy Engineering* 115:77-84.

Ruch, D. and D. Claridge. 1991. A four parameter change-point model for predicting energy consumption in commercial buildings. *Proceedings ASME-JSES-JSME Int. Solar Energy Conf.* 433-440.

Sonderegger, R.C. 1977. Dynamic models of house heating based on equivalent thermal parameters. Ph.D. Thesis, Center for Energy and Environmental Studies Report No. 57. Princeton University, Princeton, NJ.

Sonderegger, R.C. 1985. Thermal modeling of buildings as a design tool. Proceedings of CHMA 2000, Vol. 1.

Sowell, E.F. and G.N. Walton. 1980. Efficient computation of zone loads. *ASHRAE Transactions* 869(1):49-72.

Sowell, E.F. 1988. Classification of 200,640 parametric zones for cooling load calculations. *ASHRAE Transactions* 94(2).

Sowell, E.F. 1990. Lights: A numerical lighting/HVAC test cell. *ASHRAE Transactions* 96(2):780-786.

Sowell, E.F. and M.A. Moshier. 1995. HVAC component model libraries for equation-based solvers. *ASHRAE Transactions* 101(1).

Sowell, E.F. and G.N. Walton. 1980. Efficient computation of zone loads. *ASHRAE Transactions* 86(1):49-72.

Spitler, J.D., C.O. Pedersen, and D.E. Fisher. 1991. Interior convective heat transfer in buildings with large ventilative flow rates. *ASHRAE Transactions* 97(1):505-515.

Steinman, M., L.N. Kalisperis, and L.H. Summers. 1989. The MRT-correction method: A new method of radiant heat exchange. *ASHRAE Transactions* 95(1):1015-27.

Stephenson, D.G., and G.P. Mitalas. 1967. Cooling load calculations by thermal response factor method. *ASHRAE Transactions* 73.

Subbarao, K., J. Burch, C.E. Hancock. 1990. How to accurately measure the load coefficient of a residential building. *Journal of Solar Energy Engineering.*

Subbarao, K. 1986. Thermal parameters for single and multi-zone buildings and their determination from performance data. SERI *Report* SERI/TR-253-2617. Golden, CO.

Thamilseran, S and J. Haberl. 1995. A bin method for calculating energy conservation retrofit savings in commercial buildings. *Proceedings* 1995 ASME/JSME/JSES Intl. Solar Energy Conference, pp. 111-124.

Threlkeld, J.L. 1970. *Thermal environmental engineering*, 2nd Ed. Prentice-Hall, Englewood Cliffs, NJ.

USAF. 1978. *Engineering weather data*. Dept. of the Air Force *Manual* AFM 88-29. US Government Printing Office, Washington, DC.

Walton, G.N. 1980. A new algorithm for radiant interchange in room loads calculations. *ASHRAE Transactions* 86(2):190-208.

Walton, G.N. 1993. Computer programs for simulation of lighting/HVAC interactions. NISTIR 5322. National Institute of Standards and Technology, Gaithersburg, MD.

Waltz, J.P. 1992. Practical experience in achieving high levels of accuracy in energy simulations of existing buildings. *ASHRAE Transactions* 98.

Wang, X.A. and J.F. Kreider, 1992. Improved Artificial Neural Networks for Commercial Building Energy Use Prediction. *Solar Engineering*. American Society of Mechanical Engineers, New York.

Wasserman, P.D. 1989. *Neural computing, theory and practice*. Van Nostrand Reinhold, New York.

BIBLIOGRAPHY

Knebel, D.E. 1983. Simplified energy analysis using the modified bin method. ASHRAE.

Reddy, T. and D. Claridge. 1994. Using synthetic data to evaluate multiple regression and principle component analyses for statistical modeling of daily building energy consumption. *Energy and Building* 24: 35-44.

Shavit, G. 1995. Short-time-step anaylsis and simulation of homes and buildings during the last 100 years. *ASHRAE Transactions* 101(1).

Sowell, E.F. and D.C. Hittle. 1995. Evolution of building energy simulation methodology. *ASHRAE Transactions* 101(1).

U.S. Army. 1979. *BLAST, The building loads analysis and system thermodynamics program—Users manual*. U.S. Army Construction Engineering Research Laboratory *Report* E-153.

Walton, G.N. 1983. Thermal analysis research program reference manual. NBSIR 83-2655. National Institute of Standards and Technology, Gaithersburg, MD.

Spitler, J.D. 1996. Annotated guide to load calculation models and algorithms. ASHRAE, Atlanta.

Yuill, G.K. 1990. *An annotated guide to models and algorithms for energy calculations relating to HVAC equipment*. ASHRAE.

CHAPTER 31

SPACE AIR DIFFUSION

TERMINOLOGY

Aspect ratio. The ratio of length to width of an opening or core of a grille.

Axial flow jet. A stream of air whose motion is approximately symmetrical along a line, although some spreading and drop or rise can occur from diffusion and buoyancy effects.

Coefficient of discharge. Ratio of area at vena contracta to area of opening.

Cold air. General term used for supply air at 35 to 40°F.

Core area. The total plane area of that portion of a grille, included within lines tangent to the outer edges of the outer openings, through which air can pass.

Damper. A device used to vary the volume of air passing through a confined cross section by varying the cross-sectional area.

Diffuser. An outlet discharging supply air in various directions and planes.

Diffusion. Distribution of air within a space by an outlet discharging supply air in various directions and planes.

Draft. Undesired local cooling of a body caused by low temperature and movement of air.

Drop. Vertical distance that the lower edge of a horizontally projected airstream drops between the outlet and the end of its throw.

Effective area. The net area of an outlet or inlet device through which air can pass; equal to the free area times the coefficient of discharge.

Entrainment. The movement of room air into the jet caused by the airstream discharged from the outlet (secondary air motion).

Entrainment ratio. The total air divided by the air discharged from the outlet.

Envelope. The outer boundary of an airstream moving at a perceptible velocity.

Exhaust opening or inlet. Any opening through which air is removed from a space.

Free area. The total minimum area of the openings in the air outlet or inlet through which air can pass.

Grille. A covering for any area through which air passes.

Induction. See Entrainment.

Isothermal jet. Air jet with the same temperature as the surrounding air.

Lower zone. Room volume below the stratification level created by displacement ventilation.

Nonisothermal jet. Air jet with an initial temperature different from the surrounding air.

Outlet velocity. The average velocity of air emerging from the outlet, measured in the plane of the opening.

Primary air. The air delivered to the outlet by the supply duct.

Radius of diffusion. The horizontal axial distance an airstream travels after leaving an air outlet before the maximum stream velocity is reduced to a specified terminal level (e.g., 50, 100, 150, or 200 fpm).

Register. A grille equipped with a damper or control valve.

Spread. The divergence of the airstream in a horizontal and/or vertical plane after it leaves the outlet.

Stagnant zone. An area characterized by low air motion and stratification. This does not imply poor air quality.

Supply opening or outlet. Any opening through which supply air is delivered into a ventilated space being heated, cooled, humidified, or dehumidified. Supply outlets are classified according to their location in a room as sidewall, ceiling, baseboard, or floor outlets. However, because numerous designs exist, they are more accurately described by their construction features. (See Chapter 17 of the 1996 *ASHRAE Handbook—Systems and Equipment.*)

Temperature differential. Temperature difference between primary and room air.

Terminal velocity. The maximum airstream velocity at the end of the throw.

Throw. The horizontal or vertical axial distance an airstream travels after leaving an air outlet before the maximum stream velocity is reduced to a specified terminal velocity (e.g., 50, 100, 150, or 200 fpm), defined by ASHRAE *Standard* 70.

Total air. The mixture of discharged air and entrained air.

Upper zone. Room volume above the stratification level created by displacement ventilation.

Vane. A thin plate in the opening of a grille.

Vane ratio. The ratio of the depth of a vane to the space between two adjacent vanes.

Vena contracta. Smallest area of a fluid stream leaving an orifice.

METHODS OF ROOM AIR DIFFUSION

Room air diffusion systems can be classified as mixing, displacement, and local systems.

MIXING SYSTEMS

In mixing systems, conditioned air is normally discharged from air outlets at velocities much greater than those acceptable in the occupied zone. Conditioned air temperature may be above, below, or equal to the air temperature in the occupied zone, depending on the heating/cooling load. The diffuser jets mix with the ambient room air by entrainment, which reduces the air velocity and equalizes the air

The preparation of this chapter is assigned to TC 5.3, Room Air Distribution.

temperature. The occupied zone is ventilated either by the decayed air jet directly or by the reverse flow created by the jets.

Mixing air distribution creates relatively uniform air velocity, temperature, humidity, and air quality conditions in the occupied zone.

Outlet Classification and Performance

Straub et al. (1956) and Straub and Chen (1957) classified outlets into five groups:

Group A. Outlets mounted in or near the ceiling that discharge air horizontally.

Group B. Outlets mounted in or near the floor that discharge air vertically in a nonspreading jet.

Group C. Outlets mounted in or near the floor that discharge air vertically in a spreading jet.

Group D. Outlets mounted in or near the floor that discharge air horizontally.

Group E. Outlets mounted in or near the ceiling that project primary air vertically.

Analysis of outlet performance was based on primary air pattern, total air pattern, stagnant air layer, natural convection currents, return air pattern, and room air motion.

Figures 1 through 5 show the room air motion characteristics of the five outlet groups; exterior walls are depicted by heavy lines. The principles of air diffusion emphasized by these figures are as follows:

1. The primary air (shown by clear envelopes in Figures 1 through 5) from the outlet down to a velocity of about 150 fpm can be treated analytically. The heating or cooling load has a strong effect on the characteristics of the primary air.
2. The total air, shown by diagonally lined envelopes in Figures 1 through 5, is influenced by the primary air and is of relatively high velocity (but less than 150 fpm). The total air is also influenced by the environment and drops during cooling or rises during heating; it is not subject to precise analytical treatment.
3. Natural convection currents form a stagnant zone from the ceiling down during cooling, and from the floor up during heating. This zone forms below the terminal point of the total air during heating and above the terminal point during cooling. Because this zone results from natural convection currents, the air velocities within it are usually low (approximately 20 fpm), and the air stratifies in layers of increasing temperatures. The concept of a stagnant zone is important in properly applying and selecting outlets because it considers the natural convection currents from warm and cold surfaces and internal loads.
4. A return inlet affects the room air motion only within its immediate vicinity. The intake should be located in the stagnant zone to return the warmest room air during cooling or the coolest room air during heating. The importance of the location depends on the relative size of the stagnant zone, which depends on the type of outlet.
5. The general room air motion (shown by clear areas in Figures 1 through 5) is a gentle drift toward the total air. Room conditions are maintained by the entrainment of the room air into the total airstream. The room air motion between the stagnant zone and the total air is relatively slow and uniform. The highest air motion occurs in and near the total airstreams.

Group A Outlets. This group includes high sidewall grilles, sidewall diffusers, ceiling diffusers, linear ceiling diffusers, and similar outlets. High sidewall grilles and ceiling diffusers are illustrated in Figure 1.

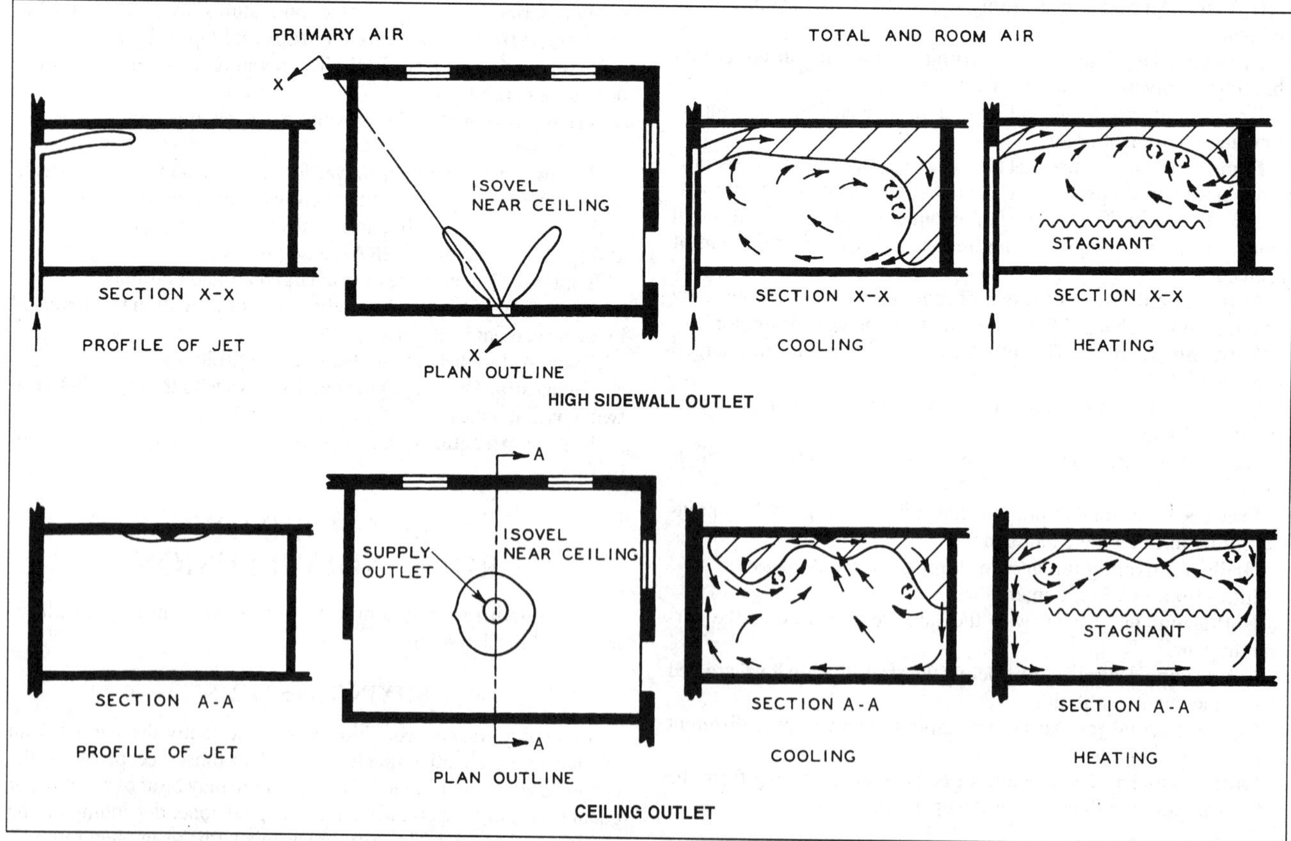

Fig. 1 Air Motion Characteristics of Group A Outlets
(Straub et al. 1956)

The primary air envelopes (isovels) show a horizontal, two-jet pattern for the high sidewall and a 360° diffusion pattern for the ceiling outlet. Although variation of vane settings might cause a discharge in one, two, or three jets in the case of the sidewall outlet, or have a smaller diffusion angle for the ceiling outlet, the general effect in each is the same.

During cooling, the total air drops into the occupied zone at a distance from the outlet that depends on air quantity, supply velocity, temperature differential between supply and room air, deflection setting, ceiling effect, and type of loading within the space. Analytical methods of relating some of these factors are presented in the section on Principles of Jet Behavior.

The cooling diagram for the high sidewall outlet shows an overthrow condition, which causes the total air to drop along the opposite wall and flow slowly for some distance across the floor. Velocities of about 100 to 150 fpm may be found near the wall but will dissipate within about 4 in. of the wall.

The cooling diagram for the ceiling outlet shows that the total air movement is counteracted by the rising natural convection currents on the heated wall and, therefore, drops before reaching the wall. On the other hand, the total air reaches the inside wall and descends for some distance along it. With this type of outlet, temperature variations within the room are minimized, with minimal stagnant volume. The maximum velocity and the maximum temperature variation occur within and near the total air envelope; therefore, the drop region becomes important because it is an area with high effective draft temperature θ. Consequently, it is necessary to know how far the air drops before velocities and temperatures reach acceptable limits.

Because these outlets discharge horizontally near the ceiling, the warmest air in the room is mixed immediately with the cool primary air far above the occupied zone. Therefore, the outlets are capable of handling relatively large quantities of air at large temperature differentials.

During heating, warm supply air introduced at the ceiling can cause stratification in the space if there is insufficient induction of room air at the outlet. Selecting diffusers properly, limiting the room supply temperature differential, and maintaining air supply rates at a level high enough to ensure air mixing by induction provide adequate air diffusion and minimize stratification.

Several building codes and ASHRAE *Standard* 90.1 require sufficient insulation in exterior walls, so most perimeter spaces can be heated effectively by ceiling air diffusion systems. Interior spaces, which generally have only cooling demand conditions, seldom require long-term heating and are seldom a design problem.

Flow rate and velocity for both heating and cooling are the same for outlet types in Figure 1. The heating diagram for the sidewall unit shows that, under these conditions, the total air does not descend along the wall. Consequently, higher velocities might be beneficial in eliminating the stagnant zone, since high velocity causes some warm air to reach floor level and counteract stratification of the stagnant region.

The heating diagram for the ceiling outlet shows the effect of the natural convection currents that produce a larger throw toward the cold exposed wall. The velocity of the total air toward the exposed wall complements the natural convection currents. However, the warm total air loses its downward momentum at its terminal point, and buoyancy forces cause it to rise toward the primary air. Although these forces are complementary, the heating effect of the total air replaces the cool natural convection currents with warm total air.

Group B Outlets. This group includes floor registers, baseboard units, low sidewall units, linear-type grilles in the floor or windowsill, and similar outlets. Figure 2 illustrates a floor outlet adjacent to an inside wall.

Because these outlets have no deflecting vanes, the primary air is discharged in a single vertical jet. When the total air strikes the ceiling, it fans out in all directions from the point of contact and, during cooling, follows the ceiling for some distance before dropping toward the occupied zone. During heating, the total airflow follows the ceiling across the room, then descends partway down the exterior wall.

The cooling diagram shows that a stagnant zone forms outside the total air region above its terminal point. Below the stagnant zone, air temperature is uniform, effecting complete cooling. Also, the space below the terminal point of the total air is cooled satisfactorily. For example, if total airflow is projected upward for 8 ft, the region from this level down to the floor will be cooled satisfactorily. This, however, does not apply to an extremely large space. Judgment to determine the acceptable size of the space outside the total air is needed. A distance of 15 to 20 ft between the drop region and the exposed wall is a conservative design value.

A comparison of Figures 1 and 2 for heating shows that the stagnant region is smaller for Group B outlets than for Group A outlets because the air entrained in the immediate vicinity of the outlet is taken mainly from the stagnant region, which is the coolest air in the room. This results in greater temperature equalization and less buoyancy in the total air than would occur with Group A outlets.

While the temperature gradients for both outlet groups are about the same, the stagnant layer for Group B is lower than that for Group A.

Group C Outlets. This group includes floor diffusers, sidewall diffusers, linear-type diffusers, and other outlets installed in the floor or windowsill (Figure 3).

Although Group C outlets are related to Group B outlets, they are characterized by wide-spreading jets and diffusing action. Total air

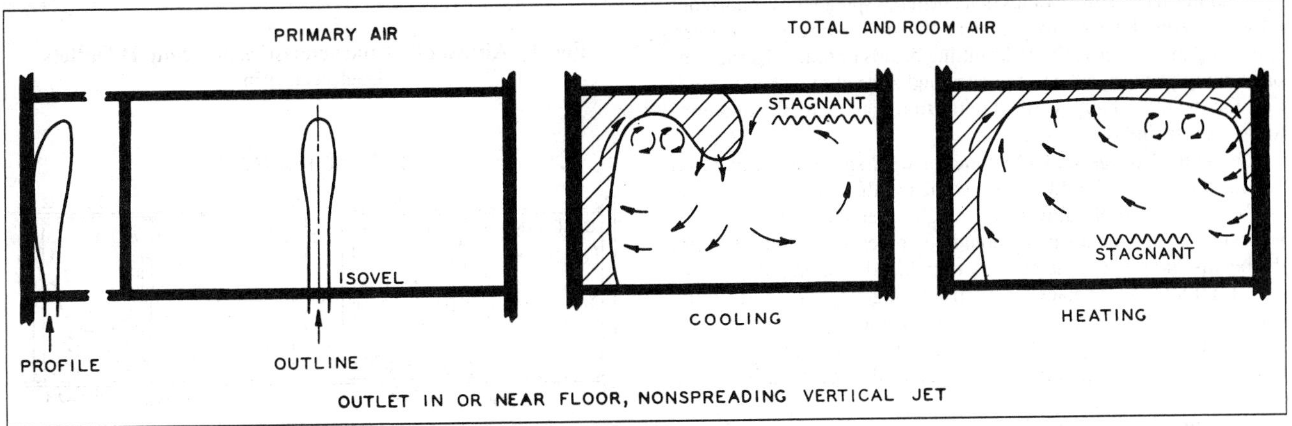

Fig. 2 Air Motion Characteristics of Group B Outlets
(Straub et al. 1956)

and room air characteristics are similar to those of Group B, although the stagnant zone formed is larger during cooling and smaller during heating. Diffusion of the primary air usually causes the total air to fold back on the primary and total air during cooling, instead of following the ceiling. This diffusing action of the outlets makes it more difficult to project the cool air, but it also provides a greater area for induction of room air. This action is beneficial during heating because the induced air comes from the lower regions of the room.

Group D Outlets. This group includes baseboard and low sidewall registers and similar outlets (Figure 4) that discharge the primary air in single or multiple jets. During cooling, because the air is discharged horizontally across the floor, the total air remains near the floor, and a large stagnant zone forms in the entire upper region of the room.

During heating, the total air rises toward the ceiling because of the buoyant effect of warm air. The temperature variations are uniform, except in the total air region.

Group E Outlets. This group includes ceiling diffusers, linear-type grilles, sidewall diffusers and grilles, and similar outlets mounted or designed for vertical downward air projection. Figure 5 shows the heating and cooling diagrams for such a ceiling diffuser.

During cooling, the total air projects to and follows the floor, producing a stagnant region near the ceiling. During heating, the total airflow reaches the floor and folds back toward the ceiling. If projected air does not reach the floor, a stagnant zone results.

Factors Affecting Outlet Performance

Vanes. Vanes affect grille performance if their depth is at least equal to the distance between the vanes (vane ratio ≥ 1). If the vane ratio is less than unity, effective control by the vanes of the airstream discharged from the grille is impossible. Increasing the vane ratio above two has little or no effect, so vane ratios should be between one and two.

A grille discharging air uniformly forward (vanes in straight position) has a spread of 14 to 24°, depending on the type of outlet, the duct approach, and the discharge velocity. Turning the vanes influences the direction and throw of the discharged airstream.

A grille with diverging vanes (vertical vanes with uniformly increasing angular deflection from the centerline to a maximum at each end of 45°) has a spread of about 60° and reduces the throw considerably. With increasing divergence, the quantity of air discharged by a grille for a given upstream total pressure decreases.

A grille with converging vanes (vertical vanes with uniformly decreasing angular deflection from the centerline) has a slightly higher throw than a grille with straight vanes, but the spread is approximately the same for both settings. The airstream converges slightly for a short distance in front of the outlet and then spreads more rapidly than air discharged from a grille with straight vanes.

In addition to vertical vanes that normally spread the air horizontally, horizontal vanes may spread the air vertically. However, spreading the air vertically risks hitting beams or other obstructions or blowing primary air into the occupied zone at excessive velocities. On the other hand, vertical deflection may increase adherence to the ceiling and reduce the drop.

Beamed Ceilings and Obstructions. In spaces with exposed beams, the outlets should be located below the bottom of the lowest beam level, preferably low enough to employ an upward or arched air path. The air path should be arched sufficiently to miss the beams and prevent the primary or induced airstream from striking furniture and obstacles and producing objectionable drafts (Wilson 1970). Obstructions influence airflow patterns and can reduce air distribution efficiency. Obstructions can reduce jet throw, increase air velocities in portions of the occupied zone, and create stagnant zones.

Variable Air Volume (VAV) Systems. The design of air distribution systems is usually based on the full load (heating/cooling). When only a partial load exists, VAV systems reduce the

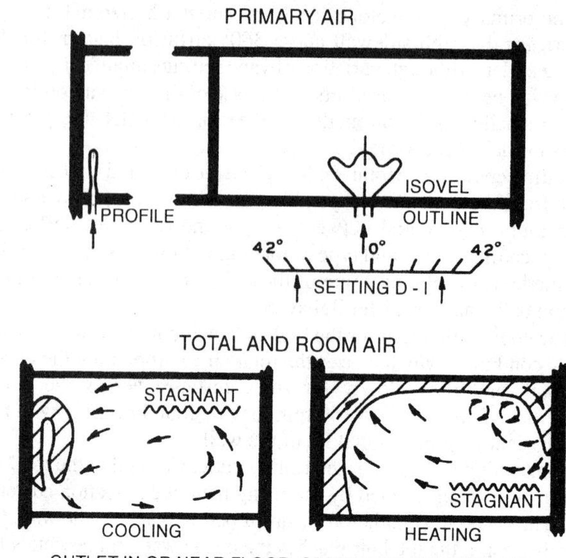

Fig. 3 Air Motion Characteristics of Group C Outlets
(Straub et al. 1956)

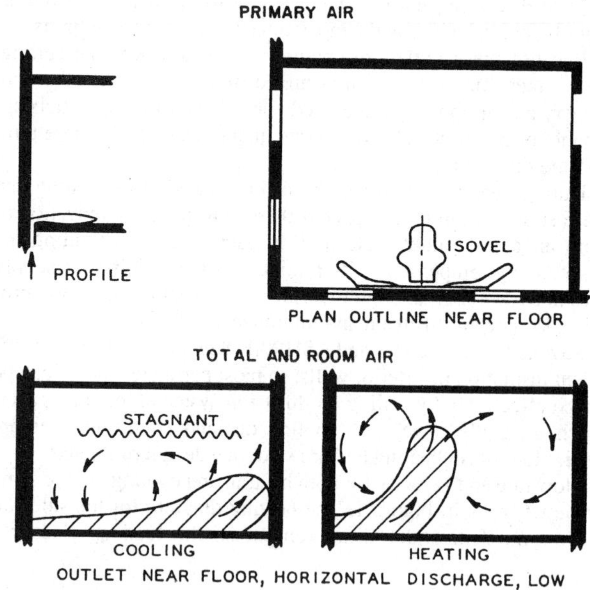

Fig. 4 Air Motion Characteristics of Group D Outlets
(Straub et al. 1956)

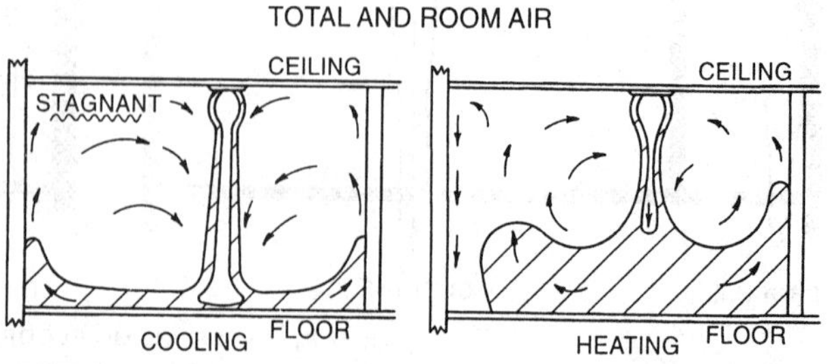

Fig. 5 Air Motion Characteristics of Group E Outlets
(Straub et al. 1956)

supply airflow, which in turn reduces the air velocity at the outlet. Therefore, the different operation modes of the system (airflow and initial temperature difference) should be considered in designing a VAV system air distribution.

DISPLACEMENT VENTILATION

In displacement ventilation, conditioned air with a temperature slightly lower than the desired room air temperature in the occupied zone is supplied from air outlets at low air velocities (100 fpm or less). The outlets are located at or near the floor level, and the supply air is introduced directly to the occupied zone. Returns through which the warm room air is exhausted from the room are located at or close to the ceiling. The supply air is spread over the floor and then rises as it is heated by the heat sources in the occupied zone. Heat sources (e.g., person, computer) in the occupied zone create upward convective flows in the form of thermal plumes. These thermal plumes tend to remove heat and contaminants within the plume from the occupied zone (Figure 6).

The air volume in the plumes increases as they rise because the plumes entrain ambient air. A stratification level exists where the airflow rate in the plumes equals the supply airflow rate. Two distinct zones are thus formed within the room: one lower zone below the stratification level and with no recirculation flow (close to displacement flow), and one upper zone, with recirculation flow (Figure 6). The height of the lower zone depends on the supply airflow rate and the characteristics of heat sources and their distribution across the floor area. In a properly designed displacement ventilation system, the upper boundary of the lower zone is above the occupied zone so that the occupied zone can be ventilated effectively. For this type of system to function properly, a stable vertically stratified temperature field is essential.

In contrast to mixing ventilation, displacement ventilation is designed to minimize mixing of air within the occupied zone. The objective of the displacement ventilation is to create conditions close to supply air conditions in the occupied zone. This type of ventilation was originally used in industrial buildings as an effective method for removing contaminants in the occupied zone. It is now also used for ventilating and cooling office buildings. However, local discomfort due to draft and vertical temperature gradient may be critical (Melikov and Nielsen 1989). Sandberg and Blomqvist (1989) suggest that the maximum convective cooling load in office buildings with displacement ventilation not exceed about 8 Btu/h · ft^2 so that the maximum vertical temperature gradient in the occupied zone will not be larger than 5°F. This is equivalent to 1 cfm/ft^2 at a maximum cooling differential of 7.5°F. Kegel and Schulz (1989) and Svensson (1989) suggest somewhat higher cooling load limits of 10 to 13 Btu/h · ft^2.

One way of increasing the cooling capacity of displacement ventilation systems is to recirculate some of the room air in the occupied zone through an induction circuit; that is, the room air is induced into the supply air and is mixed before discharge through the low-velocity air terminal device into the room. This reduces the room air temperature gradient for a given cooling load, thus allowing a cooling load limit of up to 16 Btu/h · ft^2 (Jackman 1991).

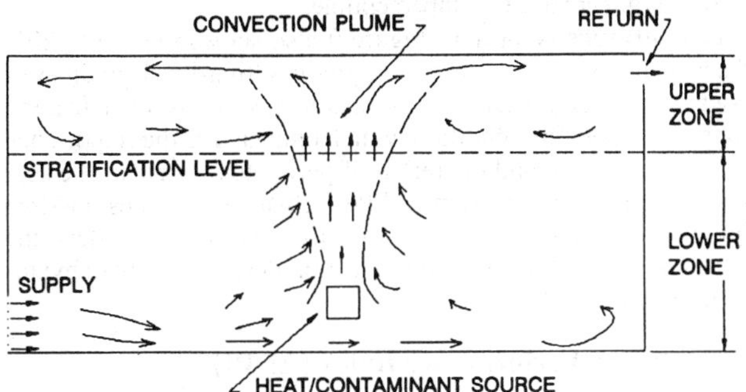

Fig. 6 Schematic of Displacement Ventilation

Air diffusers with a large outlet area are used to supply air at low velocity. Displacement ventilation has been compared with conventional mixing ventilation (Svensson 1989, Seppanen et al. 1989, Stymne et al. 1991). Design guidelines for displacement ventilation can be found in Scaret (1985), Jackman and Appleby (1990), Jackman (1991), and Shilkrot and Zhivov (1992).

UNIDIRECTIONAL AIRFLOW VENTILATION

In this type of ventilation, air is either (1) supplied from the ceiling and exhausted through the floor, or vice versa; or (2) supplied through the wall and exhausted through returns at the opposite wall. The outlets are uniformly distributed over the ceiling, floor, or wall to provide a low-turbulence "plug"-type flow across the entire room. This type of system is mainly used for clean room ventilation, in which the main objective is to remove contaminant particles from the room. Details about clean room ventilation are given in Chapter 15 of the 1995 *ASHRAE Handbook—Applications*. Unidirectional flow ventilation is also used in other areas, such as computer rooms and paint booths.

LOCALIZED VENTILATION

These systems supply conditioned air to localized areas close to the building occupants. In comparison to conventional ceiling-based air diffusion, localized ventilation systems generally have a larger number of supply diffusers directly in the occupied zone of the building (e.g., in floors, desks, workstation partitions, or theater seats). Air is typically returned at or close to ceiling level, so that localized systems benefit from the same overall upward movement of air in the room as displacement ventilation systems. In cooling applications, this allows more efficient removal of heat and contaminant sources from the room. Localized ventilation systems differ from displacement ventilation systems in that (1) they generally use higher supply volumes, which enable higher cooling loads to be met; and (2) they supply air at higher velocities through smaller diffusers. Because air is delivered directly into the occupied zone, supply air temperatures are usually warmer (above 63 to 64°F) than those maintained for conventional ceiling-based systems in order to avoid local draft discomfort for the occupants.

Localized ventilation systems are commonly installed in conjunction with a raised access floor system in which underfloor air distribution is used to deliver supply air through the subfloor plenum to the supply locations (Hanzawa and Nagasawa 1990; McCarry 1995; Shute 1992, 1995; Sodec and Craig 1990; Spoormaker 1990).

One category of localized ventilation systems that is attracting an increasing amount of attention is **task conditioning systems** (Arens et al. 1991; Bauman et al. 1991, 1993, 1995; Bauman and Arens 1996; Faulkner et al. 1993; Fisk et al. 1991; Heinemeier et al. 1990; Matsunawa et al. 1995). Task conditioning systems are unique in their ability to give individuals some control over their local environment. Typically by adjusting the volume and direction of the supply air entering the space, occupants can control the thermal conditions at their work location according to their preferences as well as improve air movement acceptability and indoor air quality.

Figure 7 shows a schematic diagram of an underfloor air distribution system with a task conditioning (personal HVAC) diffuser located in the partition in front of the office worker (Matsunawa et al. 1995). In theaters, air can be supplied through grilles or perforated panels in the chair back (Figure 8A), through the chair legs (Figures 8B and 8D), or through air diffusers installed in the floor under the chairs (Figure 8C) (Scheunemann 1989, Rowlinson and Croom 1987).

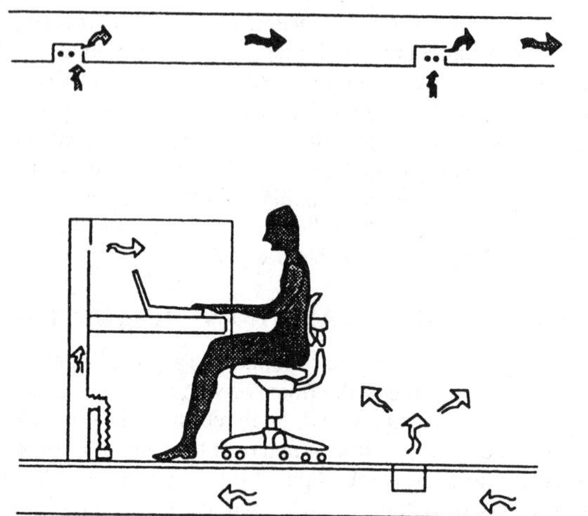

Fig. 7 Underfloor HVAC System and Personal HVAC System (Matsunawa et al. 1995)

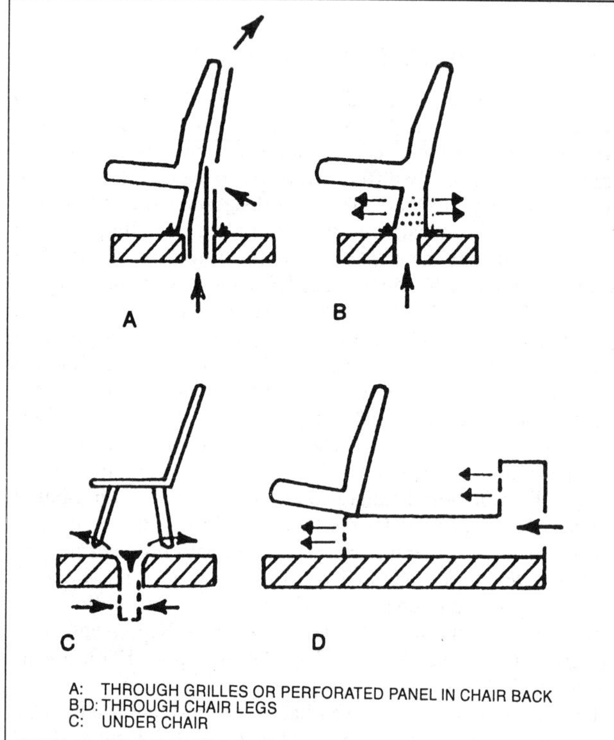

Fig. 8 Local Air Supply in Theaters

METHODS OF EVALUATION

Standards for Satisfactory Conditions

The object of air diffusion in warm-air heating, ventilating, and air-conditioning systems is to create the proper combination of temperature, humidity, and air motion in the occupied zone of the conditioned room—from the floor to 6 ft above floor level (Miller 1989). To obtain comfort conditions within this zone, standard limits have been established for acceptable effective draft temperature, which combines the effects of air temperature, air motion, and relative humidity in terms of their physiological effects on a human body. Variation from accepted standards causes occupant discomfort. Lack of uniform conditions within the space or excessive fluctuation of conditions in the same part of the space also produces discomfort. Discomfort can arise due to any of the following:

- Excessive air motion (draft)
- Excessive room air temperature variations (horizontal, vertical, or both)
- Failure to deliver or distribute air according to the load requirements at different locations
- Overly rapid fluctuation of room temperature

Draft. Koestel and Tuve (1955) and Reinmann et al. (1959) studied the effect of air motion on comfort and defined **draft** as any localized feeling of coolness or warmth of any portion of the body due to both air movement and air temperature, with humidity and radiation considered constant. The warmth or coolness of a draft was measured above or below a controlled room condition of 76°F dry-bulb at the center of the room, 30 in. above the floor, with air moving at about 30 fpm.

To define the **effective draft temperature** θ (the difference in temperature between any point in the occupied zone and the control condition), the investigators used the following equation proposed by Rydberg and Norback (1949) and modified by Straub in discussion of a paper by Koestel and Tuve (1955):

$$\theta = (t_x - t_c) - 0.07(V_x - 30) \tag{1}$$

where

θ = effective draft temperature, °F
t_x = local airstream dry-bulb temperature, °F
t_c = average (control) room dry-bulb temperature, °F
V_x = local airstream centerline velocity, fpm

Equation (1) accounts for the feeling of coolness produced by air motion and is used to establish the neutral line in Figure 9. In summer, the local airstream temperature t_x is below the control temperature t_c. Hence, both temperature and velocity terms are negative when velocity V_x is greater than 30 fpm, and they both add to the feeling of coolness. If, in winter, t_x is above t_c, any air velocity above 30 fpm subtracts from the feeling of warmth produced by t_x. Therefore, it is usually possible to have zero difference in effective temperature between location x and the control point in winter, but not in summer.

Houghten et al. (1938) presented data that make it possible to interpret statistically the percentage of room occupants that will object to a given draft condition. Figure 9 presents the data in the form used by Koestel and Tuve (1955). The data show that a person tolerates higher velocities and lower temperatures at ankle level than at neck level. Because of this, conditions in the zone extending from approximately 30 to 60 in. above the floor are more critical than conditions nearer the floor.

Air Velocity. Room air velocities less than 50 fpm are generally preferred; however, Figure 9 shows that even higher velocities may be acceptable to some occupants. ASHRAE *Standard* 55 recommends elevated air speeds at elevated air temperatures. No minimum air speeds are recommended for comfort, although air speeds below 20 fpm are usually imperceptible.

Temperature Gradient. Figure 9 also shows that up to 20% of occupants will not accept an ankle-to-sitting-level gradient of about 4°F. Poorly designed or operated systems in a heating mode can create this condition, which emphasizes the importance of proper selection and operation of perimeter systems. The section on Outlet Classification and Performance describes possible regions of high room air velocities caused by various outlets; the section on Outlet Location and Selection describes how to evaluate acceptable air diffusion.

Air Diffusion Performance Index (ADPI)

A high percentage of people are comfortable in sedentary (office) occupations where the effective draft temperature θ, as

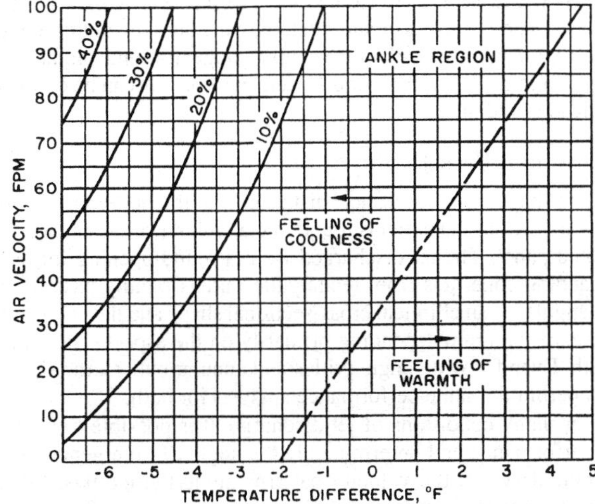

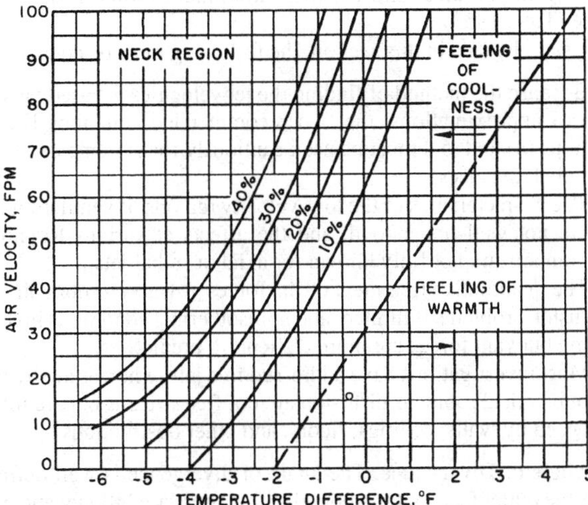

Fig. 9 Percentage of Occupants Objecting to Drafts in Air-Conditioned Rooms

defined in Equation (1), is between −3 and +2°F and the air velocity is less than 70 fpm. If several measurements of air velocity and air temperature are made throughout the occupied zone of an office, the ADPI is the percentage of locations where measurements were taken that meet these specifications for effective draft temperature and air velocity. If the ADPI is maximum (approaching 100%), the most desirable conditions are achieved (Miller and Nevins 1969, 1970, 1972, 1974; Miller 1971; Miller and Nash 1971; Nevins and Ward 1968; Nevins and Miller 1972).

The ADPI is based only on air velocity and effective draft temperature (a combination of local temperature variations from the room average) and is not directly related to the dry-bulb temperature or relative humidity. These and similar effects, such as mean radiant temperature, must be accounted for separately according to ASHRAE *Standard* 55.

The ADPI is for cooling mode conditions; a measurement technique is specified in ASHRAE *Standard* 113. Heating conditions can be evaluated using ASHRAE *Standard* 55 guidelines or ISO *Standard* 7730. The ADPI technique uses isothermal throw data determined under ASHRAE *Standard* 70 (see Table 2) to predict what will happen under isothermal cooling conditions.

Cooling Zone Design Criteria

Jet Throw. The throw of a jet is the distance from the outlet to a point where the maximum velocity in the stream cross section has

Table 1 Characteristic Room Length for Several Diffusers

Diffuser Type	Characteristic Length L
High sidewall grille	Distance to wall perpendicular to jet
Circular ceiling diffuser	Distance to closest wall or intersecting air jet
Sill grille	Length of room in direction of jet flow
Ceiling slot diffuser	Distance to wall or midplane between outlets
Light troffer diffusers	Distance to midplane between outlets plus distance from ceiling to top of occupied zone
Perforated, louvered ceiling diffusers	Distance to wall or midplane between outlets

Table 2 Air Diffusion Performance Index (ADPI) Selection Guide

Terminal Device	Room Load, Btu/h·ft²	T_{50}/L for Maximum ADPI	Maximum ADPI	For ADPI Greater than	Range of T_{50}/L
High sidewall grilles	80	1.8	68	—	—
	60	1.8	72	70	1.5–2.2
	40	1.6	78	70	1.2–2.3
	20	1.5	85	80	1.0–1.9
Circular ceiling diffusers	80	0.8	76	70	0.7–1.3
	60	0.8	83	80	0.7–1.2
	40	0.8	88	80	0.5–1.5
	20	0.8	93	90	0.7–1.3
Sill grille, straight vanes	80	1.7	61	60	1.5–1.7
	60	1.7	72	70	1.4–1.7
	40	1.3	86	80	1.2–1.8
	20	0.9	95	90	0.8–1.3
Sill grille, spread vanes	80	0.7	94	90	0.6–1.5
	60	0.7	94	80	0.6–1.7
	40	0.7	94	—	—
	20	0.7	94	—	—
Ceiling slot diffusers (for T_{100}/L)	80	0.3	85	80	0.3–0.7
	60	0.3	88	80	0.3–0.8
	40	0.3	91	80	0.3–1.1
	20	0.3	92	80	0.3–1.5
Light troffer diffusers	60	2.5	86	80	<3.8
	40	1.0	92	90	<3.0
	20	1.0	95	90	<4.5
Perforated and louvered ceiling diffusers	11–51	2.0	96	90	1.4–2.7
				80	1.0–3.4

been reduced to a selected terminal velocity. To estimate ADPI, **terminal velocity** V_T was selected for all diffusers as 50 fpm, except in the case of ceiling slot diffusers, where it was selected as 100 fpm. Each manufacturer gives data for the throw of a jet from various diffusers for isothermal conditions and without a boundary wall interfering with the jet.

The throw distance of a jet is denoted by T_V, where subscript V indicates the terminal velocity for which the throw is given. The **characteristic room length** L is the distance from the diffuser to the nearest boundary wall in the principle horizontal direction of the airflow. However, where air injected into the room does not impinge on a wall surface but collides with air from a neighboring diffuser, the characteristic length is one-half the distance between diffusers plus the distance the mixed jet travels downward to reach the occupied zone. Table 1 summarizes definitions of characteristic length for various diffusers.

The midplane between diffusers also can be considered the module line when diffusers serve equal modules throughout a space, and a characteristic length consideration can be based on module dimension d.

Load Considerations. The recommendations in Table 2 cover cooling loads of up to 80 Btu/h per square foot of floor surface. The loading is distributed uniformly over the floor up to about 7 Btu/h·ft², lighting contributes about 10 Btu/h·ft², and the remainder is supplied by a concentrated load against one wall that simulates a business machine or a large sun-loaded window. Over this range of data, the maximum ADPI condition is lower for the highest loads; however, the optimum design condition changes only slightly with the load.

Design Conditions. The quantity of air must be known from other design specifications. If it is not known, the solution must be obtained by trial and error.

The devices for which data were obtained are (1) high sidewall grilles, (2) cone-type circular ceiling diffusers, (3) sill grilles, (4) two- and four-slot ceiling diffusers, (5) light troffer diffusers, and (6) square-faced perforated and louvered ceiling diffusers. Table 2 summarizes the results of the recommendations on values of T_V/L by giving the value of T_V/L at which the ADPI is a maximum for various loads, as well as a range of values of T_V/L for which ADPI is above a minimum specified value.

PRINCIPLES OF JET BEHAVIOR

AIR JET CLASSIFICATION

As a rule, air supplied into rooms through the various types of outlets (e.g., grilles, ceiling diffusers, perforated panels) is distributed by turbulent air jets. These air jets are the primary factor affecting room air motion; for further information on the relationship between the air jet and the occupied zone, see Baturin (1972), Christianson (1989), and Murakami (1992). If the air jet is not obstructed by walls, ceiling, or other obstructions, it is considered a **free jet**. If the air jet is attached to a surface, it is an **attached air jet**.

Characteristics of the air jet in a room might be influenced by reverse flows created by the same jet entraining the ambient air. This air jet is called a **confined jet**. If the temperature of the supplied air is equal to the temperature of the ambient room air, the air jet is called an **isothermal jet**. A jet with an initial temperature different from the temperature of the ambient air is called a **nonisothermal jet**. The air temperature differential between supplied and ambient room air generates thermal forces in jets, affecting (1) the trajectory of the jet, (2) the location at which the jet attaches and separates from the ceiling/floor, and (3) the throw of the jet. The significance of such effects depends on the ratio between the thermal buoyancy of the air and inertial forces (characterized by the Archimedes number Ar).

Depending on diffuser type, air jets can be classified as follows:

• **Compact** air jets are formed by cylindrical tubes, nozzles, square or rectangular openings with a small aspect ratio (unshaded or shaded by perforated plates), grilles, etc. Compact air jets are three-dimensional and axisymmetric at least at some distance from the diffuser opening. The maximum velocity in the cross section of the compact jet is on the axis.

• **Linear** air jets are formed by slots or rectangular openings with a large aspect ratio. The jet flows are approximately two-dimensional. Air velocity is symmetric in the plane at which air velocities in the cross section are maximum. At some distance from the diffuser, linear air jets tend to transform into compact jets.

• **Radial** air jets are formed by ceiling cylindrical air diffusers with flat disks or multidiffusers that direct the air horizontally in all directions.

• **Conical** air jets are formed by cone-type or regulated multidiffuser ceiling-mounted air distribution devices. They have an axis of symmetry. The air flows parallel to the conical surface (the angle at the top of the cone is 120°) with the maximum velocities in the cross section perpendicular to the axis.

• **Incomplete radial** air jets are formed by outlets with grilles having diverging vanes and a forced angle of expansion. At a distance, this jet tends to transform into a compact one.

• **Swirling** air jets are formed by diffusers with vortex-forming devices. These devices create rotation, which has, in addition to the axial component of velocity vectors, tangential and radial ones. Depending on the type of air diffuser, swirling jets can be compact, conical, or radial.

Isothermal Free Jets

The shape of jets at a short distance from the outlet face is very similar whether the outlet is round, rectangular, grille-like, or a perforated panel. The jet discharged from a round opening forms an expanding cone; jets from rectangular outlets rapidly pass from a rectangular to an elliptical cross-sectional shape and then to a circular shape, at a rate depending primarily on the aspect ratio and jet width. Even for wide-angle grilles and annular outlets, the similarities permit the same performance analysis for both.

For many conditions of jet discharge, it is possible to analyze jet performance and determine (1) the angle of divergence of the jet boundary, (2) the velocity patterns along the jet axis, (3) the velocity profile at any cross section in the zone of maximum engineering importance, and (4) the entrainment ratios in the same zone (Tuve 1953).

Using the data in this section, the following must be considered:

1. Because the method of finding the jet velocities is based on several approximations, the two recommended equations [Equations (4) and (5)] must be used cautiously for extreme axial and radial distances.
2. The characteristics of the low-velocity regions of ventilating jets are not well understood, and the effects at various Reynolds numbers are not fully known for axial or radial jets.
3. The quantitative treatment of the forces governing room air diffusion problems is limited, and nonisothermal conditions involving buoyant forces are more difficult to predict.
4. Most investigations have addressed free jets, whereas airstreams in practical room air diffusion are not free streams but are influenced by walls, ceilings, floors, and other obstructions.

Angle of Divergence. The angle of divergence is well defined near the outlet face, but the boundary contours are billowy and easily affected by external influences. Near the outlet, as in the room, air movement has local eddies, vortices, and surges. The internal forces governing this air motion are extremely delicate (Nottage et al. 1952b).

Measured angles of divergence (spread) for discharge into large open spaces usually range from 20 to 24° with an average of 22°. Coalescing jets for closely spaced multiple outlets expand at smaller angles, averaging 18°, and jets discharging into relatively small spaces show even smaller angles of expansion (McElroy 1943). In cases where the outlet area is small compared to the dimensions of the space normal to the jet, the jet may be considered free as long as

$$X \le 1.5\sqrt{A_R} \tag{2}$$

where

X = distance from face of outlet, ft
A_R = cross-sectional area of confined space normal to jet, ft^2

Jet Expansion Zones. The full length of an air jet (compact, linear, radial, or conical), in terms of the maximum or centerline velocity and temperature differential at the cross section, can be divided into four zones:

Zone 1. A core zone; a short zone, extending about four diameters or widths from the outlet face, in which the maximum velocity (temperature) of the airstream remains practically unchanged.

Zone 2. A transition zone, the length of which depends on the type of outlet, aspect ratio of the outlet, initial airflow turbulence, and so forth.

Zone 3. A zone of fully established turbulent flow that may be 25 to 100 equivalent air outlet diameters (widths for slot-type air diffusers) long.

Zone 4. A zone of diffuser jet degradation, where the maximum air velocity and temperature decreases rapidly. The distance to this zone and its length depend on the velocities and turbulence characteristics of the ambient air. In a few diameters or widths, the air velocity becomes less than 50 fpm. Although this zone was studied by several researchers (Madison et al. 1946, Weinhold 1969), its characteristics are still not well understood.

Zone 3 is of major engineering importance because, in most cases, the diffuser jet enters the occupied area within this zone.

Centerline Velocities in Zones 1 and 2. In Zone 1, the ratio V_x/V_o is constant and equal to the ratio of the center velocity of the jet at the start of expansion to the average velocity. The ratio V_x/V_o varies from approximately 1.0 for rounded entrance nozzles to about 1.2 for straight pipe discharges; it has much higher values for diverging discharge outlets.

Experimental evidence indicates that in Zone 2,

$$\frac{V_x}{V_o} = \sqrt{\frac{K'H_o}{X}} \tag{3}$$

where

V_x = centerline velocity at distance X from outlet, fpm
V_o = V_c/C_dR_{fa} = average initial velocity at discharge from open-ended duct or across contracted stream at vena contracta of orifice or multiple-opening outlet, fpm
V_c = nominal velocity of discharge based on core area, fpm
C_d = discharge coefficient (usually between 0.65 and 0.90)
R_{fa} = ratio of free area to gross (core) area
H_o = width of jet at outlet or at vena contracta, ft
K' = centerline velocity constant depending on outlet type and discharge pattern (see Table 3)
X = distance from outlet to measurement of centerline velocity V_x, ft

Aspect ratio (Tuve 1953) and turbulence (Nottage et al. 1952b) primarily affect the centerline velocities in Zones 1 and 2. Aspect ratio has little effect on the terminal zone of the jet when H_o is greater than 4 in. This is particularly true of nonisothermal jets. When H_o is very small, it is possible for the induced air to penetrate the core of the jet, thus reducing the centerline velocities. The difference in performance between the radial-type outlet with a small H_o and the axial-type outlet with a large H_o shows the importance of the thickness of the jets.

When air is discharged from relatively large perforated panels, the constant velocity core formed by the coalescence of the individual jets extends a considerable distance from the panel face. In Zone 1, when the ratio $X/\sqrt{A_c}$ (Distance from Panel/$\sqrt{\text{Panel Area}}$) is less than 5, the following equation should be used for estimating centerline velocities (Koestel et al. 1949):

$$V_x = 1.2V_o\sqrt{C_dR_{fa}} \tag{4}$$

Centerline Velocity in Zone 3. In Zone 3, maximum or centerline velocities of straight flow isothermal jets can be determined with accuracy from the following equations:

$$\frac{V_x}{V_o} = \frac{KD_o}{X} = \frac{K'\sqrt{A_o}}{X} \tag{5}$$

$$V_x = \frac{K'V_o\sqrt{A_o}}{X} = \frac{K'Q}{X\sqrt{A_o}} \tag{6}$$

$$V_x = \frac{K'Q}{X\sqrt{A_cC_dR_{fa}}} \tag{7}$$

where

K = proportionality constant, with $K' = 1.13K$
D_o = effective or equivalent diameter of stream at discharge from open-end duct or at contracted section, ft
A_o = $A_cC_dR_{fa}$ = effective area of stream at discharge from open-end duct or at contracted section, ft^2
A_c = measured gross (core) area of outlet, ft^2
Q = discharge from outlet, cfm

Because A_o equals the effective area of the stream, the flow area for commercial registers and diffusers, according to ASHRAE *Standard* 70, can be used in Equation (5) with the appropriate value of K and K'.

Equation (5) is nondimensional and requires only that consistent units be used. Values of K and K' are listed in Table 4 (Tuve 1953, Koestel et al. 1950).

Table 3 Recommended Values of Centerline Velocity Constant K or K' for Commercial Supply Outlets

Outlet Type	Discharge Pattern	Area A_o	K[a]	K'[a]
High sidewall grilles	0° deflection	Flow area	5.0	5.7
	Wide deflection	Flow area	3.7	4.2
High sidewall linear	Core less than 4 in. high	Flow area	3.9	4.4
	Core more than 4 in. high	Flow area	4.4	5.0
Low sidewall	Up and on wall, no spread	Free area	4.4	4.5
	Wide spread	Free area	2.6	3.0
Baseboard	Up and on wall, no spread	Free area	3.9	4.0
	Wide spread	Free area	1.8	2.0
Floor	No spread	Free area	4.1	4.7
	Wide spread	Free area	1.4	1.6
Ceiling circular directional	360° horizontal	Flow area	1.0	1.1
	Four-way—little spread	Flow area	3.3	3.8
Ceiling linear	One-way—horizontal along ceiling	Flow area	4.8	5.5

[a]These values are representative for the commercial type of outlet and discharge pattern given (Straub et al. 1956, Straub and Chen 1957, ADC 1984).

Table 4 Recommended Values of Centerline Velocity Constant for Standard Openings

	K		K'	
Type of Outlet	V_o = 500 to 1000 fpm	V_o = 2000 to 10,000 fpm	V_o = 500 to 1000 fpm	V_o = 2000 to 10,000 fpm
Free openings				
Round or square	5.0	6.2	5.7	7.0
Rectangular, large aspect ratio (<40)	4.3	5.3	4.9	6.0
Annular slots, axial or radial[a]	—	—	3.9	4.8
Grilles and grids				
Free area 40% or more	4.1	5.0	4.7	5.7
Perforated panels				
Free area 3 to 5%	2.7	3.3	3.0	3.7
Free area 10 to 20%	3.5	4.3	4.0	4.9

[a]For radial slots, use X/H instead of $X/\sqrt{A}$. H is the height or width of the slot.
Note: K and K' are indices of loss in axial kinetic energy. Interpolate as required. Departures from maximum value indicate losses in Zones 1 and 2 when compared with the jet from a rounded-entrance, circular nozzle.

In multiple-opening outlets and annular ring outlets, the streams coalesce into a solid jet before actual jet expansion takes place. This coalescence affects the proportionality constants K or K' and accounts for some divergence in reported values for similar outlets.

For perforated panels of relatively large size, the values of K and K' given in Table 4 apply only when the ratio $X/\sqrt{A_c}$ is larger than 5 (see the section on Centerline Velocities in Zones 1 and 2).

Low-velocity test results, in the range $V_x < 150$ fpm, indicate that normal values of K and K' should be reduced about 20% for $V_x = 50$ fpm, as used later in Equation (10) for throw.

Determining Centerline Velocities. To correlate data from all four zones, centerline velocity ratios are plotted against distance from the outlet in Figure 10 according to Equation (3). A nomogram for calculating the parameters $X/\sqrt{A_o}$ and V_x/V_c from $X/\sqrt{A_c}$ and V_x/V_o through R_{fa} and C_d is given in the same illustration.

The variation of the centerline velocity ratio with distance from the outlet or, more properly, from start of jet expansion for Zones 1 and 2 is also shown in Figure 10. V_x/V_o is plotted against X/H_o and, for a range of aspect ratios, against $X/\sqrt{A_o}$ for the single value of $K' = 7.0$. Values of V_x/V_o for other values of K' can be obtained by direct proportioning of $\sqrt{K'}$ to $\sqrt{7.0}$.

The following example illustrates the use of Figure 10.

Example 1. A grille has a core area of 12 in. by 18.75 in., $R_{fa} = 0.90$, $C_d = 0.80$, and $K' = 5.0$. When V_x is 50 fpm for a throw of 50 ft ($X = 50$), find V_c (velocity through core area) and discharge airflow rate.

Solution:

$$A_c = (12 \times 18.75)/144 = 1.56 \text{ ft}^2$$

$$X/\sqrt{A_c} = 50/1.25 = 40$$

$$A_o = 1.56 \times 0.80 \times 0.90 = 1.123 \text{ ft}^2$$

$$X/\sqrt{A_o} = 50/1.06 = 47.2$$

$$V_x/V_o = K'/\sqrt{A_o}/X = 5.0\sqrt{1.123}/50 = 0.106$$

$$V_x/V_c = V_x/V_o C_d R_{fa} = 0.106/(0.80 \times 0.90) = 0.147$$

For $V_x = 50$ fpm, $V_c = 50/0.147 = 340$ fpm. The discharge airflow rate is then

$$Q = V_c A_c = 340 \times 1.56 = 530 \text{ cfm}$$

Throw. Equation (7) can be transposed to determine the throw X of an outlet if the discharge volume and the centerline velocity are known:

$$X = \frac{K'}{V_x} \frac{Q}{\sqrt{A_c C_d R_{fa}}} \tag{8}$$

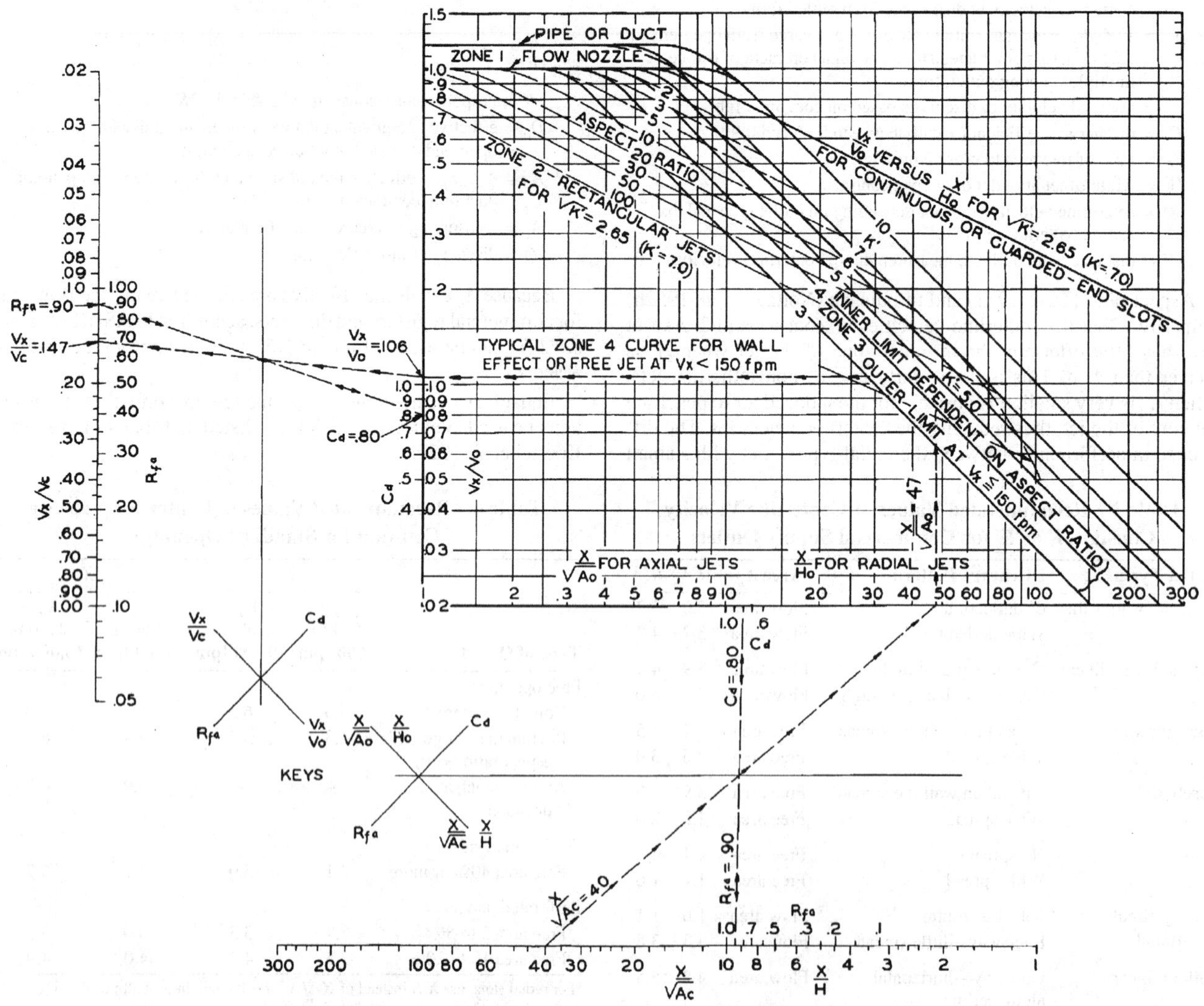

Fig. 10 Chart for Determining Centerline Velocities of Axial and Radial Jets

Or, if $Z = \sqrt{C_d R_{fa}}$, then

$$X = \frac{K'}{V_x} \frac{Q}{Z\sqrt{A_c}} \tag{9}$$

The maximum throw T_V is usually defined as the distance from the outlet face to where the centerline velocity is 50 fpm. Therefore, for V_T = 50 fpm,

$$T_V = X = \frac{K'}{50} \frac{Q}{Z\sqrt{A_c}} \tag{10}$$

Any other terminal centerline velocity could be inserted in Equation (10) for V_T.

Velocity Profiles of Jets. In Zone 3 of both axial and radial jets, the velocity distribution may be expressed by a single curve (Figure 11) in terms of dimensionless coordinates; this same curve can be used as a good approximation for adjacent portions of Zones 2 and 4. Temperature and density differences have little effect on cross-sectional velocity profiles.

Velocity distribution in Zone 3 can be expressed by the Gauss error function or probability curve, which is approximated by the following equation:

$$\left(\frac{r}{r_{0.5V}}\right)^2 = 3.3 \log \frac{V_x}{V} \tag{11}$$

where

r = radial distance of point under consideration from centerline of jet
$r_{0.5V}$ = radial distance in same cross-sectional plane from axis to point where velocity is one-half centerline velocity (i.e., $V = 0.5V_x$)
V_x = centerline velocity in same cross-sectional plane
V = actual velocity at point being considered

Experiments show that the conical angle for $r_{0.5V}$ is approximately one-half the total angle of divergence of a jet. The velocity profile curve for one-half of a straight-flow turbulent jet (the other half being a symmetrical duplicate) is shown in Figure 11. For multiple-opening outlets, such as grilles or perforated panels, the velocity profiles are similar, but the angles of divergence are smaller.

Entrainment Ratios. The following are equations for the entrainment of circular jets and of jets from long slots.

For third-zone expansion of circular jets,

$$\frac{Q_x}{Q_o} = \frac{2X}{K'\sqrt{A_o}} \tag{12}$$

By substituting from Equation (5),

$$\frac{Q_x}{Q_o} = 2\frac{V_o}{V_x} \tag{13}$$

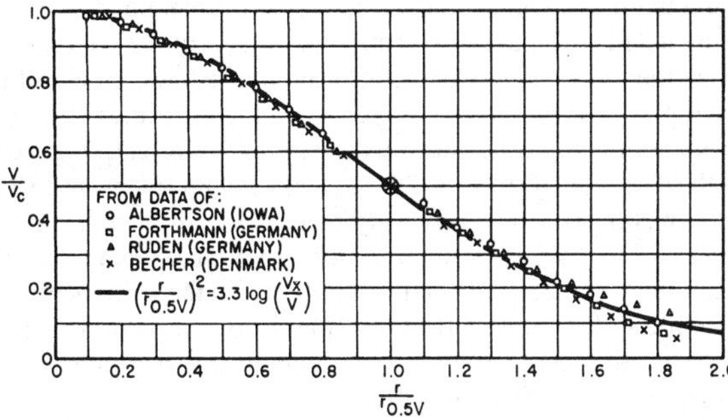

Fig. 11 Cross-Sectional Velocity Profiles for Straight-Flow Turbulent Jets

For a long slot,

$$\frac{Q_x}{Q_o} = \sqrt{\frac{2}{K'}}\sqrt{\frac{X}{H_o}} \tag{14}$$

or, substituting from Equation (3),

$$\frac{Q_x}{Q_o} = \sqrt{2}\frac{V_o}{V_x} \tag{15}$$

where

Q_x = total volumetric flow rate at distance X from face of outlet, cfm
Q_o = discharge from outlet, cfm
X = distance from face of outlet, ft
K' = proportionality constant
A_o = effective area of stream at discharge from open-end duct or at contracted section, ft^2
H_o = width of slot, ft

The entrainment ratio Q_x/Q_o is important in the determination of total air movement at a given distance from an outlet. For a given outlet, the entrainment ratio is proportional to the distance X [Equation (12)] or to the square root of the distance X [Equation (14)] from the outlet. Equations (13) and (15) show that, for a fixed centerline velocity V_x, the entrainment ratio is proportional to the outlet velocity. Equations (13) and (15) also show that, at a given centerline and outlet velocity, a circular jet has greater entrainment and total air movement than a long slot. Comparing Equations (12) and (14), the long slot should have a greater rate of entrainment. The entrainment ratio at a given distance is less with a large K' than with a small K'.

Isothermal Radial Flow Jets

In a radial jet, as with an axial jet, the cross-sectional area at any distance from the outlet varies as the square of this distance. Centerline velocity gradients and cross-sectional velocity profiles are similar to those of Zone 3 of axial jets, and the angles of divergence are about the same.

A jet from a ceiling plaque has the same form as half of a free radial jet. The jet is wider and longer than a free jet, with the maximum velocity close to the surface. Koestel (1957) provides an equation for radial flow outlets.

Nonisothermal Free Jets

When the temperature of introduced air is different from the room air temperature, the behavior of the diffuser air jet is affected by the thermal buoyancy due to air density difference. The trajectory of a nonisothermal jet introduced horizontally is determined by the **Archimedes number** (Baturin 1972):

$$\text{Ar} = \frac{gL_o(t_o - t_s)}{V_o^2 T_s} \tag{16}$$

where

g = gravitational acceleration rate, ft/min^2
L_o = length scale of diffuser outlet equal to hydraulic diameter of outlet, ft
t_o = initial temperature of jet, °F
t_s = temperature of surrounding air, °F
V_o = initial air velocity of jet, fpm
T_s = room air temperature, °R

The paths assumed by horizontally projected heated and chilled jets influenced by buoyant forces are significant in heating and cooling with wall outlets. Koestel's equation (1955) describes the behavior of these jets.

Helander and Jakowatz (1948), Helander et al. (1953, 1954, 1957), Yen et al. (1956), and Knaak (1957) developed equations for outlet characteristics that affect the downthrow of heated air. Koestel (1954, 1955) developed equations for temperatures and velocities in heated and chilled jets. Li et al. (1993, 1995) and Kirkpatrick and Elleson (1996) provide additional information on nonisothermal jets.

Surface Jets (Wall and Ceiling)

Jets discharging parallel to a surface with one edge of the outlet coinciding with the surface take the form of one-half of an axial jet discharging from an outlet twice as large, similar to radial jets from ceiling plaques. Entrainment takes place almost exclusively along the surface of a half cone, and the maximum velocity remains close to the surface (Tuve 1953).

Values of K and K' are approximately those for a free jet multiplied by $\sqrt{2}$; that is, the normal maximum of 7.0 for K' for free jets becomes 9.9 for a similar jet discharged parallel to and adjacent to a surface, and $X/\sqrt{2A_o}$ should replace $X/\sqrt{A_o}$ in Figure 10.

When a jet is discharged parallel to but at some distance from a solid surface (wall, ceiling, or floor), its expansion in the direction of the surface is reduced, and entrained air must be obtained by recirculation from the jet instead of from ambient air (McElroy 1943, Nottage et al. 1952a, Zhang et al. 1990). The restriction to entrainment caused by the solid surface induces the **Coanda effect**, which makes the jet attach to a surface a short distance after it leaves the diffuser outlet. The jet then remains attached to the surface for some distance before separating from the surface again.

In nonisothermal cases, the trajectory of the jet is determined by the balance between the thermal buoyancy and the Coanda effect, which depends on the jet momentum and the distance between the jet exit and the solid surface. The behavior of such nonisothermal surface jets has been studied by Kirkpatrick et al. (1991), Wilson et al. (1970), Oakes (1987), and Zhang et al. (1990), each addressing different factors. A more systematic study of these jets in room ventilation flows is needed to provide reliable guidelines for designing air diffusion systems.

MULTIPLE JETS

Twin parallel air jets act independently until they interfere. The point of interference and its distance from the outlets varies with the distance between the outlets. From the outlets to the point of interference, the maximum velocity, as for a single jet, is on the centerline of each jet. After interference, the velocity on a line midway between and parallel to the two jet centerlines increases until it equals the jet centerline velocity. From this point, a maximum velocity of the combined jet stream is on the midway line, and the profile seems to emanate from a single outlet of twice the area of one of the two outlets.

Koestel and Austin (1956) determined the spacing between outlets for noninterference between the jets. For a K value of 6.5, the outlets should be placed three to eight diameters apart, with V_o values from 500 to 1500 fpm.

AIRFLOW IN OCCUPIED ZONE

Mixing Systems. Laboratory experiments on jets usually involve recirculated air with negligible resistance to flow on the return path of the jet air. Experiments in mine tunnels of small cross-sectional areas, where the return flow of jet air to outlets meets considerable resistance, show that expansion of the jet terminates abruptly at a distance that is independent of discharge velocity and is only slightly affected by the size of the outlet. These distances are determined primarily by the size and length of the return path. In a long tunnel with a cross section of 5 ft by 6 ft, a jet may not travel more than 25 ft; in a tunnel with a relatively large section (25 ft by 60 ft), the jet may travel more than 250 ft. McElroy (1943) provides data on this phase of jet expansion.

Zhang et al. (1990) found that, for a given heat load and room air supply rate, air velocity in the occupied zone increases when the outlet discharge velocity increases. Therefore, the design supply air velocity should be high enough to maintain the jet traveling in the desired direction in order to ensure good mixing before it reaches the occupied zone. Excessively high outlet air velocity would induce high air velocity in the occupied zone and result in thermal discomfort.

Turbulence Production and Transport. The air turbulence within a room is mainly produced at the diffuser jet region by interaction of the supply air with the room air and with the solid surfaces (walls or ceiling) in the vicinity. It is then transported to other parts of the room, including the occupied zone (Zhang et al. 1992). Meanwhile, the turbulence is also damped by viscous effect. Air in the occupied zone usually contains very small amounts of turbulent kinetic energy compared to that in the jet region. Because turbulence may cause thermal discomfort (Fanger et al. 1989), air diffusion systems should be designed so that the primary mixing between the introduced air and the room air occurs away from occupied regions.

SYSTEM DESIGN

DESIGN CONSIDERATIONS

Noise

The noise generated by diffusers transmits to the occupied space directly and cannot be attenuated. Therefore, the diffusion system design should meet the sound level criteria specified in Chapter 43 of the 1995 *ASHRAE Handbook—Applications*.

Duct Approaches to Diffuser Outlets

The manner in which the airstream approaches the diffuser outlet is important. For correct air diffusion, the velocity of the airstream must be as uniform as possible over the entire cross-sectional area of the connecting duct and must be perpendicular to the outlet face. Effects of improper duct approach generally cannot be corrected by the diffuser.

If the system is designed carefully, a wall grille installed at the end of a horizontal duct and a ceiling outlet at the end of a vertical duct receive the air perpendicularly and at uniform velocity over the entire duct cross section. However, few outlets are installed in this way. Most sidewall outlets are installed either at the end of vertical ducts or in the side of horizontal ducts, and most ceiling outlets are attached either directly to the bottom of horizontal ducts or to special vertical takeoff ducts that connect the outlet with the horizontal duct. In all these cases, special devices for directing and equalizing the airflow are necessary for proper direction and diffusion of the air.

The influence of the duct approach on outlet performance has been investigated for vertical stack heads with plain openings (Nelson et al. 1940) or equipped with grilles (Nelson et al. 1942) and side outlets on horizontal ducts (Nelson and Smedberg 1943). In tests conducted with the stack heads, splitters or guide vanes in the elbows at the top of the vertical stacks are needed, regardless of the shape of the elbows (rounded, square, or expanding). Cushion chambers at the top of the stack heads are not beneficial. Figure 12 shows the direction of flow, diffusion, and velocity (measured 12 in. from the opening) of the air for various stack heads tested, expanding from a 14 in. by 6 in. stack to a 14 in. by 9 in. opening, without a grille. The air velocity for each was 500 fpm in the stack below the elbow, but the direction of flow and the diffusion pattern indicate performance obtained with nonexpanding elbows of similar shapes for velocities from 200 to 400 fpm.

In tests conducted with 3 in. by 10 in., 4 in. by 9 in., and 6 in. by 6 in. side outlets in a 6 in. by 20 in. horizontal duct at duct velocities of 200 to 1400 fpm in the horizontal duct section, multiple curved

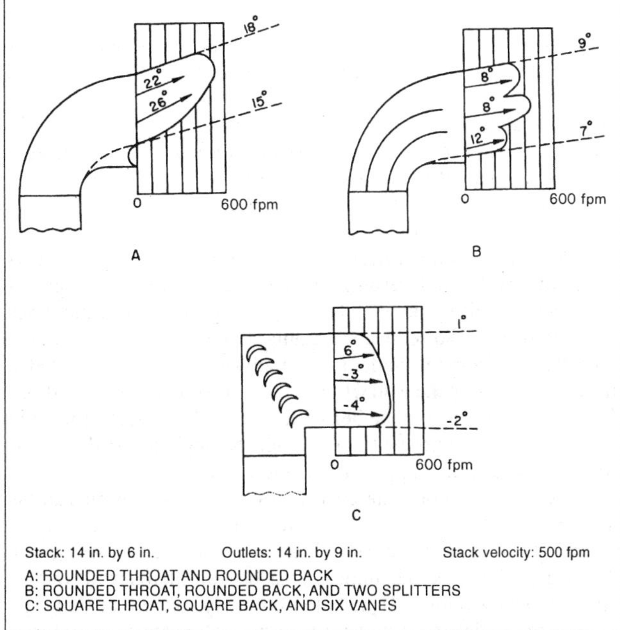

Fig. 12 Outlet Velocity and Air Direction Diagrams for Stack Heads with Expanding Outlets

deflectors produced the best flow characteristics. Vertical guide strips in the outlet were not as effective as curved deflectors. A single scoop-type deflector at the outlet did not improve the flow pattern obtained from a plain outlet and, therefore, was not desirable.

Return and Exhaust Openings

Selection. The selection of return and exhaust openings depends on (1) velocity in the occupied zone near the openings, (2) permissible pressure drop through the openings, and (3) noise.

Velocity. Airflow patterns and room air movement are not influenced by the location of the return and exhaust outlets beyond a distance of one characteristic length of the return or exhaust opening (e.g., square root of the opening area). Air handled by the opening approaches the opening from all directions, and its velocity decreases rapidly as the distance from the opening increases. Therefore, drafty conditions rarely occur near return openings. Table 5 shows recommended return opening face velocities.

Permissible pressure drop. Permissible pressure drop depends on the choice of the designer. Proper pressure drop allowances should be made for control or directive devices.

Noise. The problem of noise in return openings is the same as that in supply outlets. In computing room noise levels resulting from the operation of an air-conditioning system, the return opening must be included as part of the total grille area.

Location. The openings should be located to minimize short-circuiting of supply air. If air is supplied by the jets attached to the ceiling, exhaust openings should be located between the jets or at the side of the room away from the supply air jets. In rooms with vertical temperature stratification, such as foundries, computer rooms, theaters, bars, kitchens, dining rooms, and club rooms, exhaust openings should be located near the ceiling to collect warm air, odors, and fumes.

For industrial rooms with gas release, selection of exhaust opening locations depends on the density of the released gases and their temperature; locations should be specified for each application.

Exhaust outlets located in walls and doors, depending on their elevation, have the characteristics of either floor or ceiling returns. In large buildings with many small rooms, return air may be brought through door grilles or door undercuts into the corridors and then to a common return or exhaust. If the pressure drop through door returns is excessive, the air diffusion to the room may be seriously unbalanced by opening or closing the doors. Outward leakage through doors or windows cannot be counted on for dependable results.

Table 5 Recommended Return Inlet Face Velocities

Inlet Location	Velocity Across Gross Area, fpm
Above occupied zone	> 800
Within occupied zone, not near seats	600 to 800
Within occupied zone, near seats	400 to 600
Door or wall louvers	200 to 300
Through undercut area of doors	200 to 300

System Balancing

Ducts and diffusers in a system should be sized so that the supply of air is distributed properly. However, for flexibility, use standard sizes and allow for future redistribution; the system as designed may not be self-balancing. Chapter 34 of the 1995 *ASHRAE Handbook—Applications* describes the procedures used to balance air distribution systems.

DESIGN PROCEDURE

1. Determine the air volumetric flow requirements based on load and room size. For VAV systems, evaluation should include the range of flow rates from minimum occupied to design load.
2. Select the tentative diffuser type and location within room.
3. Determine the room's characteristic length L (Table 1).
4. Select the recommended T_V/L ratio from Table 2.
5. Calculate the throw distance T_V by multiplying the recommended T_V/L ratio from Table 2 by the room length L.
6. Locate the appropriate outlet size from the manufacturer's catalog.
7. Ensure that this outlet meets other imposed specifications, such as for noise and for static pressure.

Example 2.

Specifications:

Room size	20 ft by 12 ft with 9 ft ceiling
Loading	Uniform, 10 Btu/h·ft^2 or 2400 Btu/h
Air volumetric flow	1 cfm/ft^2 or 240 cfm for the one outlet
Device	High sidewall grille, located at center of 12 ft endwall, 9 in. from ceiling

Calculations:

Characteristic length L	= 20 ft (length of room: Table 1)
Recommended T_V/L	= 1.5 (Table 2)
Throw to 50 fpm T_{50}	= 1.5 × 20 = 30 ft

Refer to the manufacturer's catalog for a size that gives this isothermal throw to 50 fpm. Manufacturer *X* recommends the following sizes, when vanes are straight, discharging 240 cfm: 16 in. by 4 in., 12 in. by 5 in., or 10 in. by 6 in.

OUTLET LOCATION AND SELECTION

No criteria have been established for choosing among the six types of outlets to obtain an optimum ADPI. All outlets tested, when used according to these recommendations, can have ADPI values that are satisfactory (greater than 90% for loads less than 40 Btu/h·ft^2).

The design of an air distribution and air diffusion system is influenced by the same factors that influence the design of an air-conditioning plant—building use, size, and construction type. Location and selection of the supply outlets is further influenced by the interior design of the building, local sources of heat gain or loss, and outlet performance and design.

Local sources of heat gain or loss promote convection currents or cause stratification; they may, therefore, determine both the type and location of the supply outlets. Outlets should be located to neutralize any undesirable convection currents set up by a concentrated load. If a concentrated heat source is located at the occupancy level of the room, the heating effect can be counteracted (1) by directing cool air toward the heat source or (2) by locating an exhaust or return grille adjacent to the heat source. The second method is more economical for cooling applications, since heat is withdrawn at its source rather than dissipated into the conditioned space. Where lighting loads are heavy (5 W/ft^2) and ceilings relatively high (above 15 ft), outlets should be located below the lighting load, and the stratified warm air should be removed by an exhaust or return fan. An exhaust fan is recommended if the wet-bulb temperature of the air is above that of the outdoors; a return fan is recommended if the wet-bulb temperature is below this temperature. These methods reduce the requirements for supply air. Enclosed lights are more economical than exposed lights because a considerable portion of the energy is radiant.

Based on the analysis of the outlet performance tests conducted by Straub et al. (1956) and Straub and Chen (1957), the following are selection considerations for outlets in Groups A through E.

Group A Outlets

Outlets mounted in or near the ceiling with horizontal air discharge should not be used with temperature differentials exceeding 25°F during heating. Researchers have recommended that temperature differentials not exceed 15°F during heating (Hart and Int-Hout 1980, Lorch and Straub 1983). Consequently, such outlets should be used for heating buildings located in regions where winter heating is only a minor problem and, in northern latitudes, solely for interior spaces. However, these outlets are particularly suited for cooling and can be used with high airflow rates and large temperature differentials. They are usually selected for their cooling characteristics.

The performance of these outlets is affected by various factors. Vane deflection settings reduce throw and drop by changing air from a single straight jet to a wide-spreading or fanned-out jet. Accordingly, a sidewall outlet with 0° deflection has a longer throw and a greater drop than a ceiling diffuser with a single 360° angle of deflection. Sidewall grilles and similar outlets with other deflection settings may have performance characteristics between these two extremes.

Wide deflection settings also cause a ceiling effect, which increases the throw and decreases the drop. To prevent smudging, the total air should be directed away from the ceiling, but this is rarely practicable, except for very high ceilings. For optimum air diffusion in areas without high ceilings, total air should scrub the ceiling surface.

Drop increases and throw decreases with larger temperature differentials. For constant temperature differential, airflow rate affects drop more than velocity. Therefore, to avoid drop, several small outlets may be better in a room than one large outlet.

With the data in the section on Principles of Jet Behavior, the throw may be selected for a portion of the distance between the outlet and wall or, preferably, for the entire distance. For outlets in opposite walls, the throw should be one-half the distance between the walls. Following these recommendations, the air drops before striking the opposite wall or the opposing airstream. To counteract specific sources of heat gain or to provide higher air motion in rooms with high ceilings, it may be necessary to select a longer throw. In no case should the drop exceed the distance from the outlet to the 6 ft level.

To maintain maximum ventilation effectiveness with ceiling diffusers, throws should be kept as long as possible. With VAV designs, some overthrow at maximum design volumes will be desirable—the highest induction can be maintained at reduced flows. Adequate induction by a ceiling-mounted diffuser prevents short-circuiting of unmixed supply air between supply outlet and ceiling-mounted returns.

Group B Outlets

In selecting these outlets, it is important to provide enough throw to project the air high enough for proper cooling in the occupied zone. An increase in supply air velocity improves air diffusion during both heating and cooling. Also, a terminal velocity of about 150 fpm is found at the same distance from the floor during both heating and cooling. Therefore, outlets should be selected from the data given in the section on Principles of Jet Behavior, with throw based on a terminal velocity of 150 fpm.

With outlets installed near the exposed wall, the primary air is drawn toward the wall, resulting in a wall effect similar to the ceiling effect for ceiling outlets. This scrubbing of the wall increases heat gain or loss. To reduce scrubbing, outlets should be installed some distance from the wall, or the supply air should be deflected at an angle away from the wall. However, to prevent the air from dropping into the occupied zone before it reaches maximum projection the distance should not be too large nor the angle too wide. A distance of 6 in. and an angle of 15° is satisfactory.

These outlets do not counteract natural convection currents unless sufficient outlets are installed around the perimeter of the space—preferably in locations of greatest heat gain or loss (under windows). The effect of drapes and blinds must be considered with outlets installed near windows. If installed correctly, outlets of this type handle large airflow rates with uniform air motion and temperatures.

Group C Outlets

These outlets can be used for heating, even with severe heat load conditions. Higher supply velocities produce better room air diffusion than lower velocities, but velocity is not critical in selecting these units for heating.

To achieve the required projection for cooling, the outlets should be used with temperature differentials of less than 15°F. With higher temperature differentials, supply air velocity is not sufficient to project the total air up to the desired level.

The outlets have been used successfully for residential heating, but they may also offer a solution for applications where heating requirements are severe and cooling requirements are moderate. For throw, refer to the section on Principles of Jet Behavior.

Group D Outlets

These outlets direct high-velocity total air into the occupied zone, and, therefore, are not recommended for comfort, particularly for summer cooling. For heating, outlet velocities should not be higher than 300 fpm, so that air velocities in the occupied zone will not be excessive. These outlets have been applied successfully to process installations where controlled air velocities are desired.

Group E Outlets

The different throws shown in the heating and cooling diagrams for these outlets become critical in selecting and applying the outlets. Because the total air enters the occupied zone for both cooling and heating, outlets are used for either cooling or heating, but seldom for both.

During cooling, temperature differential, supply air velocity, and airflow rate have considerable influence on projection. Therefore, low values of each should be selected.

During heating, it is important to select the correct supply air velocity to project the warm air into the occupied zone. Temperature differential is also critical because a small temperature differential reduces variation of the throw during the cyclic fluctuation of the supply air temperature. Vane setting for deflection is as important here as it is for Group B and C outlets.

Investigations by Nevins and Ward (1968) and Miller and Nevins (1969) in full-scale interior test rooms indicate that air temperatures and velocities throughout a room cooled by a ventilating ceiling are a linear function of room load (heat load per unit area) and are not affected significantly by variations in ceiling type, total air temperature differential, or air volumetric flow rate. Higher room loading produces wider room air temperature variations and higher velocities, which decrease performance.

These studies also found no appreciable difference in the performance of air-diffusing ceilings and circular ceiling diffusers for lower room loads (20 Btu/h·ft^2). For higher room loads (80 Btu/h·ft^2), an air-diffusing ceiling system has only slightly larger vertical temperature variations and slightly lower room air velocities than a ceiling diffuser system.

When the ventilating ceiling is used at exterior exposures, the additional load at the perimeter must be considered. During heating operation, the designer must provide for the cold wall effect, as with any ceiling supply diffusion system. The sound generated by the air supply device must also be considered in total system analysis to ensure that room sound levels do not exceed the design criteria.

RETURN AIR DESIGN FOR OPTIMUM PERFORMANCE

An HVAC system operating in the cooling mode performs best when generated heat is removed at its source rather than distributed throughout the conditioned space. Heat from solar and miscellaneous loads such as machinery and floor or desk-mounted lamps is difficult to remove at the source. However, return air flowing over ceiling-mounted lighting fixtures keeps most of that heat from being distributed into the conditioned space. In addition to increasing HVAC system efficiency, return air lighting fixtures improve light output and extend the life of the lamps. The manufacturers of fixtures, ceiling grids, and grilles give performance information (airflow rate, pressure drop, and heat removal rate) of their product. Ball et al. (1971) found that the heat removal performance of return air fixtures covers a narrow range.

With a suspended ceiling, low operating static pressure across the ceiling must be maintained. Failure to do so can result in return air being forced around the edges of the ceiling panels or, in some cases, through the ceiling panels. The result is often a soiled ceiling and a mechanical system that is choked for return air. To avoid this, the static pressure difference across the ceiling should be as low as possible. If necessary, slotted tees or grilles can be used with return air fixtures to obtain the specified pressure drop. A maximum pressure drop of 0.02 to 0.03 in. of water is acceptable under most conditions.

At the typical air supply rates found in office interior zone spaces (usually less than 1.5 cfm/ft^2) and with adequate induction at the supply diffusers, the location of the return diffuser has no effect on air patterns in the space. For most office spaces, it is only necessary that sufficient return outlets be provided to maintain inlet velocities within recommendations (see Table 5).

In spaces expected to operate in a cooling mode most of the time, returning the warmest air in the space can effectively reduce energy costs and increase circulation in the space. This is especially true in climates where economizer systems operate for long periods during the year. In spaces having very high ceilings, with atriums, skylights, or large vertical glass surfaces, and where the highest areas are unoccupied, air stratification may be used as an energy-saving measure by locating returns near the occupied zone.

CEILING-MOUNTED AIR DIFFUSER SYSTEMS

For the best thermal comfort conditions and highest ventilation effectiveness in an occupied space (i.e., office or retail store), the entire system performance of air diffusers should be considered. This is particularly true for open spaces, where airstreams from diffusers may interact with each other, and for perimeter spaces, where airstreams from diffusers interact with hot or cold perimeter walls. While throw data for individual diffusers are used in system design, an air diffuser system should maintain a high quality of air diffusion in the occupied space with low temperature variation, good air mixing, and no objectionable drafts in the occupied space (typically 6 in. to 6 ft above the floor).

Adequate ventilation requires that the selected diffusers effectively mix (by entrainment) the total air in the room with the supplied conditioned air, which is assumed to contain adequate ventilation air.

Interior Spaces

An interior space is conditioned exclusively for cooling loads, except after unoccupied periods when the space may have cooled to below a comfortable temperature. Tests by Miller and Nevins (1970), Miller and Nash (1971), Miller (1979), and Hart and Int-Hout (1981) suggest that the air diffusion performance index (ADPI) can be improved by moving diffusers closer together (i.e., specifying more diffusers for a given space and air quantity) and by limiting the value of the supply air/room air temperature difference. In a given system of diffusers, these studies found an optimum operating range of air volumetric flow rates at a given thermal load. The operating load varies with diffuser design, ceiling height, thermal load, and diffuser orientation. This information can be obtained by constructing a mock-up representing the proposed building space, with several alternatives tested for ADPI values, in accordance with ASHRAE *Standard* 113. Usually, the diffuser manufacturer has performed these tests and can provide the best choice of design options for a particular building. For a VAV system, the diffuser spacing selection should not be based on maximum or design air volumes but rather on the air volume range in which the system is expected to operate most of the time. For VAV applications, Miller (1979) recommends that the designer consider the expected variation in the outlet air volume to ensure that ADPI values remain above a specified minimum.

Perimeter Spaces

All-air mechanical systems that handle both heating and cooling thermal loads are commonly used in modern office buildings instead of baseboards for heating and forced air for cooling. State energy codes (most based on ASHRAE *Standard* 90 series) require that commercial buildings have exterior walls that meet minimum thermal performance criteria for a particular location. Typically, walls of new buildings have design heat losses as low as 200 to 300 Btu/h per linear foot of wall.

A successful all-air heating/cooling mechanical system requires the designer to consider several design variables that have been the subject of research by Hart and Int-Hout (1980), Lorch and Straub (1983), and Rousseau (1983). The most important design variables include

- Supply air/room air temperature difference
- Diffuser type and design
- Design heating and cooling loads
- Supply air volumetric flow rates
- Distance between diffusers and perimeter wall
- Direction of air throw (toward wall, away from wall, or both)
- Ceiling height
- Desired air diffusion performance criteria

The diffuser manufacturer is best able to recommend the use of equipment.

For an office environment in cooling mode, the design goal should be an ADPI greater than 80. The ADPI should not be used as a measure of performance for heating conditions. In both cases, ASHRAE *Standard* 55 recommends that the maximum temperature gradient (the difference in temperature between any two points) should not exceed 5°F. Linear diffusers placed parallel to the perimeter wall perform well. For year-round operation, linear diffusers with two-way throw (i.e., both toward and away from the perimeter wall) work best. Lorch and Straub (1983) reported optimum performance with a diffuser that throws warm air toward the perimeter wall under heating load conditions and chilled air in both directions under cooling load conditions. All researchers found less than optimum performance with high discharge temperatures (greater than 15°F above ambient), both with one-way throw of air away from a cold wall and with one-way throw of chilled air toward the perimeter wall. Under heating load conditions, the supply air temperature must be limited to avoid excessive thermal stratification. To resolve any uncertainty about performance, a mock-up should be constructed with provisions for a cold wall; several variations of the design should be tested so that the best diffuser wall spacing and supply air volumes can be selected. The ADPI, room temperature gradients, or both, measured in accordance with ASHRAE *Standard* 113, can help gage system performance.

The following principles provide the best air diffusion quality and minimum energy use:

- For cooling load conditions, return air should exhaust from a location that takes advantage of any thermal stratification design. In many cases, this should be a high point in order to take advantage of rising warm air. Cooling supply air should be introduced as close to the heat sources as possible. Alternately, stratification designs may condition only part of the total space. In these cases, conditioned air is supplied and exhausted as close to the occupants as possible. In either case, comfort zone temperature gradients should be maintained within 5°F.
- For heating load conditions, thermal stratification should be discouraged. Heat should be introduced at points low in the large space. Ceiling-mounted fans may reduce stratification.

REFERENCES

ADC. 1984. Test code for grilles, registers, and diffusers. *Test Code* 1062 GRD-84. Air Diffusion Council, Chicago, IL.

Arens, E.A., F. Bauman, L. Johnston, and H. Zhang. 1991. Testing of localized ventilation systems in a new controlled environment chamber. *Indoor Air* 3:263-81.

ASHRAE. 1990. Method of testing for room air diffusion. ANSI/ASHRAE *Standard* 113-1990.

ASHRAE. 1991. Method of testing for rating the performance of air outlets and inlets. ANSI/ASHRAE *Standard* 70-1991.

ASHRAE. 1992. Thermal environmental conditions for human occupancy. ANSI/ASHRAE *Standard* 55-1992.

Ball, H.D., R.G. Nevins, and H.E. Straub. 1971. Thermal analysis of heat removal troffers. *ASHRAE Transactions* 77(2).

Baturin, V.V. 1972. *Fundamentals of industrial ventilation*, 3rd England ed. Translated by O.M. Blunn. Pergamon Press, New York.

Bauman, F., and E. Arens. 1996. Task/ambient conditioning systems: Engineering and application guidelines. Center for Environmental Design Research, University of California, Berkeley.

Bauman, F.S., E.A. Arens, S. Tanabe, H. Zhang, and A. Baharlo. 1995. Testing and optimizing the performance of a floor-based task conditioning system. *Energy and Buildings* 22(3):173-86.

Bauman, F.S., L. Johnston, H. Zhang, and E. Arens. 1991. Performance testing of a floor-based, occupant-controlled office ventilation system. *ASHRAE Transactions* 97(1).

Bauman, F.S., H. Zhang, E. Arens, and C. Benton. 1993. Localized comfort control with a desktop task conditioning system: Laboratory and field measurements. *ASHRAE Transactions* 99(2).

Christianson, L.L., ed. 1989. *Building systems: Room air and air contaminant distribution*. ASHRAE, Atlanta.

Fanger, P.O., A.K. Melikov, H. Hanzawa, and J. Ring. 1988. Air turbulence and sensation of draft. *Energy and Buildings* 12:21-39.

Faulkner, D., W.J. Fisk, and D.P. Sullivan. 1993. Indoor air flow and pollutant removal in a room with desktop ventilation. *ASHRAE Transactions* 99(2).

Fisk, W.J., D. Faulkner, D. Pih, P. McNeel, F. Bauman, and E. Arens. 1991. Indoor air flow and pollutant removal in a room with task ventilation. *Indoor Air* 3:247-62.

Hanzawa, H., and Y. Nagasawa. 1990. Thermal comfort with underfloor air-conditioning systems. *ASHRAE Transactions* 96(2).

Hart, G.H. and D. Int-Hout. 1980. The performance of a continuous linear diffuser in the perimeter zone of an office environment. *ASHRAE Transactions* 86(2).

Hart, G.H. and D. Int-Hout. 1981. The performance of a continuous linear diffuser in the interior zone of an open office environment. *ASHRAE Transactions* 87(2).

Heinemeier, K.E., G.E. Schiller, and C.C. Benton. 1990. Task conditioning for the workplace: Issues and challenges. *ASHRAE Transactions* 96(2).

Helander, L. and C.V. Jakowatz. 1948. Downward projection of heated air. *ASHVE Transactions* 54:71.

Helander, L., S.M. Yen, and R.E. Crank. 1953. Maximum downward travel of heated jets from standard long radius ASME nozzles. *ASHVE Transactions* 59:241.

Helander, L., S.M. Yen, and L.B. Knee. 1954. Characteristics of downward jets of heated air from a vertical delivery discharge unit heater. *ASHVE Transactions* 60:359.

Helander, L., S.M. Yen, and W. Tripp. 1957. Outlet characteristics that affect the downthrow of heated air jets. *ASHAE Transactions* 63:255.

Houghten, F.C., C. Gutberlet, and E. Witkowski. 1938. Draft temperatures and velocities in relation to skin temperatures and feelings of warmth. *ASHVE Transactions* 44:289.

ISO. 1994. Moderate thermal environments—Determination of the PMV and PPD indices and specification of the conditions for thermal comfort. ISO *Standard* 7730-1994. International Organization for Standardization, Geneva.

Jackman, P.J. 1991. Displacement ventilation. CIBSE National Conference, University of Kent, Canterbury, UK.

Jackman, P.J. and P.A. Appleby. 1990. Displacement flow ventilation. BSRIA Project Report. The Building Services Research and Information Association, Old Bracknell Lane, Berkshire, RG12 4AH, UK.

Kegel, B. and U.W. Schulz. 1989. Displacement ventilation for office buildings. Proceedings of the 10th AIVC Conference, Helsinki. Air Infiltration and Ventilation Center, University of Warwick Science Park, Barclays Venture Centre, Sir William Lyons Road, Coventry CV4 7EZ, UK.

Kirkpatrick, A. and J. Elleson. 1996. *Design guide for cold air distribution systems*. ASHRAE, Atlanta.

Kirkpatrick, A., T. Malmstrom, P. Miller, and V. Hassani. 1991. Use of low temperature air for cooling of buildings. Proceedings Building Simulation. Nice, France.

Knaak, R. 1957. Velocities and temperatures on axis of downward heated jet from 4-inch long-radius ASME nozzle. *ASHAE Transactions* 63:527.

Koestel, A. 1954. Computing temperatures and velocities in vertical jets of hot or cold air. *ASHVE Transactions* 60:385.

Koestel, A. 1955. Paths of horizontally projected heated and chilled air jets. *ASHAE Transactions* 61:213.

Koestel, A. 1957. Jet velocities from radial flow outlets. *ASHAE Transactions* 63:505.

Koestel, A. and J.B. Austin, Jr. 1956. Air velocities in two parallel ventilating jets. *ASHAE Transactions* 62:425.

Koestel, A. and G.L. Tuve. 1955. Performance and evaluation of room air distribution systems. *ASHRAE Transactions* 61:533.

Koestel, A., P. Hermann, and G.L. Tuve. 1949. Air streams from perforated panels. *ASHVE Transactions* 55:283.

Koestel, A., P. Hermann, and G.L. Tuve. 1950. Comparative study of ventilating jets from various types of outlets. *ASHVE Transactions* 56:459.

Li, Z., L.L. Christianson, and J.S. Zhang. 1995. Separation distances of nonisothermal air jets. Research Triangle Park, NC.

Li, Z., J.S. Zhang, A.M. Zhivov and L.L. Christianson. 1993. Characteristics of diffuser air jets and airflow in the occupied regions of mechanically ventilated rooms: A literature review. *ASHRAE Transactions* 99(1): 1119-27.

Lorch, F.A. and H.E. Straub. 1983. Performance of overhead slot diffusers with simulated heating and cooling conditions. *ASHRAE Transactions* 89(1).

Madison, R.D. and W.R. Elliot. 1946. Throw of air from slots and jets. *Heating, Piping, and Air Conditioning* 11:108.

Matsunawa, K., H. Iizuka, and S. Tanabe. 1995. Development and application of an underfloor air-conditioning system with improved outlets for a "smart" building in Tokyo. *ASHRAE Transactions* 101(2):887.

McCarry, B.T. 1995. Underfloor air distribution systems: Benefits and when to use the system in building design. *ASHRAE Transactions* 101(2).

McElroy, G.E. 1943. Air flow at discharge of fan-pipe lines in mines. U.S. Bureau of Mines *Report of Investigations*, 19.

Melikov, A.K. and J.B. Nielsen. 1989. Local thermal discomfort due to draft and vertical temperature difference in rooms with displacement ventilation. *ASHRAE Transactions* 95(2):1050-57.

Miller, P.L. 1971. Room air distribution performance of four selected outlets. *ASHRAE Transactions* 77(2):194.

Miller, P.L. 1979. Design of room air diffusion systems using the air diffusion performance index (ADPI). *ASHRAE Journal* 10:85.

Miller, P.L. 1989. Descriptive methods. In *Building systems: Room air and air contaminant distribution*, L.L. Christianson, ed. ASHRAE, Atlanta.

Miller, P.L. and R.T. Nash. 1971. A further analysis of room air distribution performance. *ASHRAE Transactions* 77(2):205.

Miller, P.L. and R.G. Nevins. 1969. Room air distribution with an air distributing ceiling—Part II. *ASHRAE Transactions* 75:118.

Miller, P.L. and R.G. Nevins. 1970. Room air distribution performance of ventilating ceilings and cone-type circular ceiling diffusers. *ASHRAE Transactions* 76(1):186.

Miller, P.L. and R.G. Nevins. 1972. An analysis of the performance of room air distribution systems. *ASHRAE Transactions* 78(2):191.

Miller, P.L. and R.G. Nevins. 1974. Room air distribution—An ASHRAE engineering practice monograph. *ASHRAE Journal* 1:92.

Murakami, S. 1992. New scales for ventilation efficiency and their application based on numerical simulation of room airflow. International Symposium on Room Air Convection and Ventilation Effectiveness. Tokyo, Japan.

Nelson, D.W. and G.E. Smedberg. 1943. Performance of side outlets on horizontal ducts. *ASHVE Transactions* 49:58.

Nelson, D.W., H. Krans, and A.F. Tuthill. 1940. The performance of stack heads. *ASHVE Transactions* 46:205.

Nelson, D.W., D.H. Lamb, and G.E. Smedberg. 1942. Performance of stack heads equipped with grilles. *ASHVE Transactions* 48:279.

Nevins, R.G. and P.L. Miller. 1972. Analysis, evaluation and comparison of room air distribution performance. *ASHRAE Transactions* 78(2):235.

Nevins, R.G. and E.D. Ward. 1968. Room air distribution with an air distributing ceiling. *ASHRAE Transactions* 74:VI.2.1.

Nottage, H.B., J.G. Slaby, and W.P. Gojsza. 1952a. Isothermal ventilation jet fundamentals. *ASHVE Transactions* 58:107.

Nottage, H.B., J.G. Slaby, and W.P. Gojsza. 1952b. Outlet turbulence intensity as a factor in isothermal-jet flow. *ASHVE Transactions* 58:343.

Oakes, W.C. 1987. Experimental investigation of Coanda jet. M.S. thesis, Michigan State University, East Lansing, MI.

Reinmann, J.J., A. Koestel, and G.L. Tuve. 1959. Evaluation of three room air distribution systems for summer cooling. *ASHRAE Transactions* 65:717.

Rousseau, W.H. 1983. Perimeter air diffusion performance index tests for heating with a ceiling slot diffuser. *ASHRAE Transactions* 89(1).

Rowlinson, D. and D. Croom. 1987. Supply characteristics of the floor mounted diffusers. Air distribution in ventilated spaces. ROOMVENT-'87, Stockholm, Sweden.

Rydberg, J. and P. Norback. 1949. Air distribution and draft. *ASHVE Transactions* 55:225.

Sandberg, M. and C. Blomqvist. 1989. Displacement ventilation in office rooms. *ASHRAE Transactions* 95(2):1041-49.

Scaret, E. 1985. *Ventilation by displacement Characterization and design implications*. Elsevier Science Publishers, New York.

Scheunemann, K.H. 1989. Local ventilating and air conditioning in industrial and culture buildings. The 2nd World Congress on Heating, Ventilating, Refrigerating and Air Conditioning-CLIMA 2000. (Vol. 3, Air Conditioning Components and Systems, pp. 391-97), Sarajevo.

Seppanen, O.A., W.J. Fisk., J. Eto, and D.T. Grimsrud. 1989. Comparison of conventional mixing and displacement air-conditioning and ventilating systems in U.S. commercial buildings. *ASHRAE Transactions* 95(2): 1028-40.

Shilkrot, E. and A. Zhivov. 1992. Room ventilation with designed vertical air temperature stratification. ROOMVENT-'92, Proceedings of the 3rd International Conference on Engineering Aero- and Thermodynamics of Ventilated Rooms, Aalborg, Denmark.

Shute, R.W. 1992. Integrating access floor plenums for HVAC air distribution. *ASHRAE Journal* 34(10).

Shute, R.W. 1995. Integrated access floor HVAC: Lessons learned. *ASHRAE Transactions* 101(2).

Sodec, F., and R. Craig. 1990. The underfloor air supply system—The European experience. *ASHRAE Transactions* 96(2).

Spoormaker, H.J. 1990. Low-pressure underfloor HVAC system. *ASHRAE Transactions* 96(2).

Straub, H.E. and M.M. Chen. 1957. Distribution of air within a room for year-round air conditioning—Part II. University of Illinois Engineering Experiment Station *Bulletin* No. 442.

Straub, H.E., S.F. Gilman, and S. Konzo. 1956. Distribution of air within a room for year-round air conditioning—Part I. University of Illinois Engineering Experiment Station *Bulletin* No. 435.

Stymne, H., M. Sandberg, and M. Mattsson. 1991. Dispersion pattern of contaminants in a displacement ventilation room—Implications for demand control. Proceedings of the 12th Air Movement and Ventilation Control Within Buildings, Ottawa. AIVC Conference.

Svensson, A.G.L. 1989. Nordic experiences of displacement ventilation systems. *ASHRAE Transactions* 95(2):1013-17.

Tuve, G.L. 1953. Air velocities in ventilating jets. *ASHVE Transactions* 59:261.

Weinhold, K., R. Dannecker, U. Schwiegk. 1969. Über Auslegungsverfahren von Lüftungsdecken. *Luft- und Kältetechnik* 2:78-84.

Wilson, J.D., M.L. Esmay, and S. Persson. 1970. Wall-jet velocity and temperature profiles resulting from a ventilation inlet. *ASAE Transactions*.

Yen, S.M., L. Helander, and L.B. Knee. 1956. Characteristics of downward jets from a vertical discharge unit heater. *ASHAE Transactions* 62:123.

Zhang, J.S., L.L. Christianson, and G.L. Riskowski. 1990. Regional airflow characteristics in a mechanically ventilated room under nonisothermal conditions. *ASHRAE Transactions* 96(1):751-59.

Zhang, J.S., L.L. Christianson, G.J. Wu, and G.L. Riskowski. 1992. Detailed measurements of room air distribution for evaluating numerical simulation models. *ASHRAE Transactions* 98(1):58-65.

BIBLIOGRAPHY

Davies, E.L. 1930. The measurement of the flow of air through registers and grilles. *ASHVE Transactions* 36:201-24.

Davies, E.L. 1931. The measurement of the flow of air through registers and grilles—Part II. *ASHVE Transactions* 37:619-31.

Davies, E.L. 1933. Measurement of the flow of air through registers and grilles. *ASHVE Transactions* 39:373.

Elrod, H.G., Jr. 1954. Computation charts and theory for rectangular and circular jets. *ASHVE Transactions* 60:431.

Gilman, S.R., H.E. Straub, A.E. Hershey, and R.B. Engdahl. 1953. Room air distribution research for year-round air conditioning, Part I—Supply outlets at one high sidewall location. *ASHVE Transactions* 59:151.

Greene, A.M., Jr. and M.H. Dean. 1938. The flow of air through exhaust grilles. *ASHVE Transactions* 44:387.

Grimitlyn, M. 1970. Zur Luftverteilung in Räumen. *Luft- und Kältetechnik* 5.

Int-Hout, D. 1981. Measurement of room air diffusion in actual office environments to predict occupant thermal comfort. *ASHRAE Transactions* 87(2).

Koestel, A. and G.L. Tuve. 1948. The discharge of air from a long slot. *ASHVE Transactions* 54:87.

Koestel, A. and C.Y. Young. 1951. The control of airstreams from a long slot. *ASHVE Transactions* 57:407.

Koestel, A., P. Hermann, and G.L. Tuve. 1949. Airstreams from perforated panels. *ASHVE Transactions* 55:283.

Koestel, A., P. Hermann, and G.L. Tuve. 1950. Comparative study of ventilating jets from various types of outlets. *ASHVE Transactions* 56:459.

Kratz, A.P., A.E. Hershey, and R.B. Engdahl. 1940. Development of instruments for the study of air distribution in rooms. *ASHVE Transactions* 46:351.

Nelson, D.W. and G.E. Smedberg. 1943. Performance of side outlets on horizontal ducts. *ASHVE Transactions* 49:58.

Nelson, D.W. and D.J. Stewart. 1938. Air distribution from side wall outlets. *ASHVE Transactions* 44:77.

Nielsen, R.A. 1940. Dirt patterns on walls. *ASHVE Transactions* 46:247.

Nottage, H.B. 1949. Turbulence—A fundamental frontier in air distribution. *ASHVE Transactions* 55:193.

Nottage, H.B. 1950. A simple heated-thermocouple anemometer. *ASHVE Transactions* 56:431.

Nottage, H.B., J.G. Slaby, and W.P. Gojsza. 1952. A smoke-filament technique for experimental research in room air distribution. *ASHVE Transactions* 58:399.

Nottage, H.B., J.G. Slaby, and W.P. Gojsza. 1952. A V-wire direction probe. *ASHVE Transactions* 58:79.

Nottage, H.B., J.G. Slaby, and W.P. Gojsza. 1952. Exploration of a chilled jet. *ASHVE Transactions* 58:357.

Poz, M.Y. 1991. Theoretical investigation and practical applications of nonisothrmal jets for the rooms ventilating. Current East/West HVAC Developments. IEI/CIBSE/ABOK Joint Conference, Dublin, Ireland.

Shepelev, I. 1978. *Airdynamics of the air flows in the premises*. Stroiizdat, Moscow.

Shilkrot, E. 1974. About simulation of radiant-convective heat exchange in the premises with the natural ventilation. Proceedings of the Central Research Institute for Industrial Buildings, p. 37. TsNIIpromzdanii, Moscow.

Shilkrot, E. 1986. Evaluating of the designing loads for the heating and ventilating systems of the premises using the method of the heat balances in its zones. Proceedings of the Central Research Institute for Industrial Buildings. TsNIIpromzdanii, Moscow.

Stewart, D.J. and G.F. Drake. 1937. The noise characteristics of air supply outlets. *ASHVE Transactions* 43:81.

Straub, H.E. and S.F. Gilman. 1954. Room air distribution research for year-round air conditioning, Part II—Supply outlets at three floor locations. *ASHVE Transactions* 60:249.

Tasker, C. 1948. ASHVE research in air distribution and air duct friction. *Heating, Piping, and Air Conditioning* 4:125.

Tuve, G.L. 1953. Air velocities in ventilating jets. *ASHVE Transactions* 59:261.

Tuve, G.L. and G.B. Priester. 1944. Control of airstreams in large spaces. *ASHVE Transactions* 50:153.

Tuve, G.L. and D.K. Wright, Jr. 1940. Air flow measurements at intake and discharge openings and grilles. *ASHVE Transactions* 46:313.

Tuve, G.L., G.B. Priester, and D.K. Wright, Jr. 1942. Entrainment and jet-pump action of airstreams. *ASHVE Transactions* 48:241.

Tuve, G.L., D.K. Wright, Jr., and L.J. Seigel. 1939. The use of air velocity meters. *ASHVE Transactions* 45:645.

Zhivov, A. 1990. Variable-air volume ventilation systems for industrial buildings. *ASHRAE Transactions* 96(2).

CHAPTER 32

DUCT DESIGN

COMMERCIAL, industrial, and residential air duct system design must consider (1) space availability, (2) space air diffusion, (3) noise levels, (4) duct leakage, (5) duct heat gains and losses, (6) balancing, (7) fire and smoke control, (8) initial investment cost, and (9) system operating cost.

Deficiencies in duct design can result in systems that operate incorrectly or are expensive to own and operate. Poor air distribution can cause discomfort; lack of sound attenuators may permit objectionable noise levels. Poorly designed ductwork can result in unbalanced systems. Faulty duct construction or lack of duct sealing produces inadequate airflow rates at the terminals. Proper duct insulation eliminates the problem caused by excessive heat gain or loss.

In this chapter, system design and the calculation of a system's frictional and dynamic resistance to airflow are considered. Chapter 16 of the 1996 *ASHRAE Handbook—Systems and Equipment* examines duct construction and presents construction standards for residential, commercial, and industrial heating, ventilating, air-conditioning, and exhaust systems.

BERNOULLI EQUATION

The Bernoulli equation can be developed by equating the forces on an element of a stream tube in a frictionless fluid flow to the rate of momentum change. On integrating this relationship for steady flow, the following expression (Osborne 1966) results:

$$\frac{v^2}{2g_c} + \int\frac{dP}{\rho} + \frac{gz}{g_c} = \text{constant, ft} \cdot \text{lb}_f/\text{lb}_m \qquad (1)$$

where

v = streamline (local) velocity, fps
g_c = dimensional constant, 32.2 $\text{lb}_m \cdot \text{ft/lb}_f \cdot \text{s}^2$
P = absolute pressure, lb_f/ft^2
ρ = density, lb_m/ft^3
g = acceleration due to gravity, ft/s^2
z = elevation, ft

Assuming constant fluid density within the system, Equation (1) reduces to

$$\frac{v^2}{2g_c} + \frac{P}{\rho} + \frac{gz}{g_c} = \text{constant, ft} \cdot \text{lb}_f/\text{lb}_m \qquad (2)$$

Although Equation (2) was derived for steady, ideal frictionless flow along a stream tube, it can be extended to analyze flow through ducts in real systems. In terms of pressure, the relationship for fluid resistance between two sections is

$$\frac{\rho_1 V_1^2}{2g_c} + P_1 + \frac{g}{g_c}\rho_1 z_1 = \frac{\rho_2 V_2^2}{2g_c} + P_2 + \frac{g}{g_c}\rho_2 z_2 + \Delta p_{t,1\text{-}2} \qquad (3)$$

where

V = average duct velocity, fps
$\Delta p_{t,1\text{-}2}$ = total pressure loss due to friction and dynamic losses between sections 1 and 2, lb_f/ft^2

In Equation (3), V (section average velocity) replaces v (streamline velocity) because experimentally determined loss coefficients allow for errors in calculating $\rho v^2/2g_c$ (velocity pressure) across streamlines.

On the left side of Equation (3), add and subtract p_{z1}; on the right side, add and subtract p_{z2}, where p_{z1} and p_{z2} are the values of atmospheric air at heights z_1 and z_2. Thus,

$$\frac{\rho_1 V_1^2}{2g_c} + P_1 + (p_{z1} - p_{z1}) + \frac{g}{g_c}\rho_1 z_1$$

$$= \frac{\rho_2 V_2^2}{2g_c} + P_2 + (p_{z2} - p_{z2}) + \frac{g}{g_c}\rho_2 z_2 + \Delta p_{t,1\text{-}2} \qquad (4)$$

The atmospheric pressure at any elevation (p_{z1} and p_{z2}) expressed in terms of the atmospheric pressure p_a at the same datum elevation is given by

$$p_{z1} = p_a - \frac{g}{g_c}\rho_a z_1 \qquad (5)$$

$$p_{z2} = p_a - \frac{g}{g_c}\rho_a z_2 \qquad (6)$$

Substituting Equations (5) and (6) into Equation (4) and simplifying yields the total pressure change between sections 1 and 2. Assume no change in temperature between sections 1 and 2 (no heat exchanger within the section); therefore, $\rho_1 = \rho_2$. When a heat exchanger is located within the section, the average of the inlet and outlet temperatures is generally used. Let $\rho = \rho_1 = \rho_2$. $(P_1 - p_{z1})$ and $(P_2 - p_{z2})$ are gage pressures at elevations z_1 and z_2.

$$\Delta p_{t,1\text{-}2} = \left(p_{s,1} + \frac{\rho V_1^2}{2g_c}\right) - \left(p_{s,2} + \frac{\rho V_2^2}{2g_c}\right)$$

$$+ \frac{g}{g_c}(\rho_a - \rho)(z_2 - z_1) \qquad (7a)$$

$$\Delta p_{t,1\text{-}2} = \Delta p_t + \Delta p_{se} \qquad (7b)$$

The preparation of this chapter is assigned to TC 5.2, Duct Design.

$$\Delta p_t = \Delta p_{t,1\text{-}2} - \Delta p_{se} \tag{7c}$$

where

$p_{s,1}$ = static pressure, gage at elevation z_1, lb_f/ft^2
$p_{s,2}$ = static pressure, gage at elevation z_2, lb_f/ft^2
V_1 = average velocity at section 1, fps
V_2 = average velocity at section 2, fps
ρ_a = density of ambient air, lb_m/ft^3
ρ = density of air or gas within duct, lb_m/ft^3
Δp_{se} = thermal gravity effect, lb_f/ft^2
Δp_t = total pressure change between sections 1 and 2, lb_f/ft^2
$\Delta p_{t,1-2}$ = total pressure loss due to friction and dynamic losses between sections 1 and 2, lb_f/ft^2

HEAD AND PRESSURE

The terms **head** and **pressure** are often used interchangeably; however, head is the height of a fluid column supported by fluid flow, while pressure is the normal force per unit area. For liquids, it is convenient to measure the head in terms of the flowing fluid. With a gas or air, however, it is customary to measure pressure on a column of liquid.

Static Pressure

The term $pg_c/\rho g$ is static head; p is static pressure.

Velocity Pressure

The term $V^2/2g$ refers to velocity head, and the term $\rho V^2/2g_c$ refers to velocity pressure. Although velocity head is independent of fluid density, velocity pressure, calculated by Equation (8), is not.

$$p_v = \rho\left(\frac{V}{1097}\right)^2 \tag{8}$$

where

p_v = velocity pressure, in. of water
V = fluid mean velocity, fpm

For air at standard conditions ($0.075\ lb_m/ft^3$), Equation (8) becomes

$$p_v = \left(\frac{V}{4005}\right)^2 \tag{9}$$

Velocity is calculated by Equation (10) or (11).

$$V = 144Q/A \tag{10}$$

where

Q = airflow rate, cfm
A = cross-sectional area of duct, in^2

$$V = Q/A \tag{11}$$

where A = cross-sectional area of duct, ft^2.

Total Pressure

Total pressure is the sum of static pressure and velocity pressure:

$$p_t = p_s + \rho\left(\frac{V}{1097}\right)^2 \tag{12}$$

or

$$p_t = p_s + p_v \tag{13}$$

where

p_t = total pressure, in. of water
p_s = static pressure, in. of water

Pressure Measurement

The range, precision, and limitations of instruments for measuring pressure and velocity are discussed in Chapter 14. The manometer is a simple and useful means for measuring partial vacuum and low pressure. Static, velocity, and total pressures in a duct system relative to atmospheric pressure are measured with a pitot tube connected to a manometer. Pitot tube construction and locations for traversing round and rectangular ducts are presented in Chapter 14.

SYSTEM ANALYSIS

The total pressure change due to friction, fittings, equipment, and net **thermal gravity effect (stack effect)** for each section of a duct system is calculated by the following equation:

$$\Delta p_{t_i} = \Delta p_{f_i} + \sum_{j=1}^{m} \Delta p_{ij} + \sum_{k=1}^{n} \Delta p_{ik} - \sum_{r=1}^{\lambda} \Delta p_{se_{ir}} \tag{14}$$

$$\text{for } i = 1, 2, \ldots, n_{up} + n_{dn}$$

where

Δp_{t_i} = net total pressure change for i-section, in. of water
Δp_{f_i} = pressure loss due to friction for i-section, in. of water
Δp_{ij} = total pressure loss due to j-fittings, including fan system effect (FSE), for i-section, in. of water
Δp_{ik} = pressure loss due to k-equipment for i-section, in. of water
$\Delta p_{se_{ir}}$ = thermal gravity effect due to r-stacks for i-section, in. of water
m = number of fittings within i-section
n = number of equipment within i-section
λ = number of stacks within i-section
n_{up} = number of duct sections upstream of fan (exhaust/return air subsystems)
n_{dn} = number of duct sections downstream of fan (supply air subsystems)

From Equation (7), the thermal gravity effect for each nonhorizontal duct with a density other than that of ambient air is determined by the following equation:

$$\Delta p_{se} = 0.192(\rho_a - \rho)(z_2 - z_1) \tag{15}$$

where

Δp_{se} = thermal gravity effect, in. of water
z_1 and z_2 = elevation from datum in direction of airflow (Figure 1), ft
ρ_a = density of ambient air, lb_m/ft^3
ρ = density of air or gas within duct, lb_m/ft^3

Example 1. For Figure 1, calculate the thermal gravity effect for two cases: (a) air cooled to −30°F, and (b) air heated to 1000°F. The density of air at −30°F and 1000°F is 0.0924 lb_m/ft^3 and 0.0271 lb_m/ft^3, respectively. The density of the ambient air is 0.075 lb_m/ft^3. Stack height is 40 ft.

Solution:

$$\Delta p_{se} = 0.192(\rho_a - \rho)z$$

(a) For $\rho > \rho_a$ (Figure 1A),

$$\Delta p_{se} = 0.192(0.075 - 0.0924)40 = -0.13 \text{ in. of water}$$

(b) For $\rho < \rho_a$ (Figure 1B),

$$\Delta p_{se} = 0.192(0.075 - 0.0271)40 = +0.37 \text{ in. of water}$$

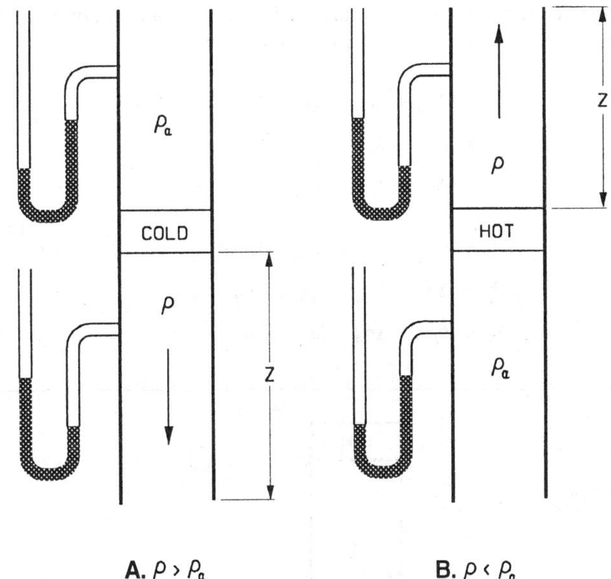

Fig. 1 Thermal Gravity Effect for Example 1

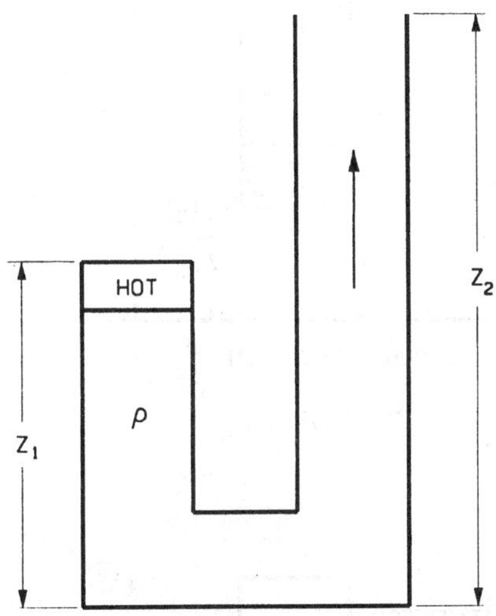

Fig. 2 Multiple Stacks for Example 2

Example 2. Calculate the thermal gravity effect for the two-stack system shown in Figure 2, where the air is 250°F and the stack heights are 50 and 100 ft. The density of 250°F air is 0.0558 lb_m/ft^3; ambient air is 0.075 lb_m/ft^3.

Solution:

$$\Delta p_{se} = 0.192(0.075 - 0.0558)(100 - 50)$$
$$= 0.18 \text{ in. of water}$$

For the system shown in Figure 3, the direction of air movement created by the thermal gravity effect depends on the initiating force. The initiating force could be fans, wind, opening and closing doors, and turning equipment on and off. If for any reason air starts to enter the left stack (Figure 3A), it creates a buoyancy effect in the right stack. On the other hand, if flow starts to enter the right stack (Figure 3B), it creates a buoyancy effect in the left stack. In both cases the produced thermal gravity effect is stable and depends on the stack height and magnitude of heating. The starting direction of flow is important when using natural convection for ventilation.

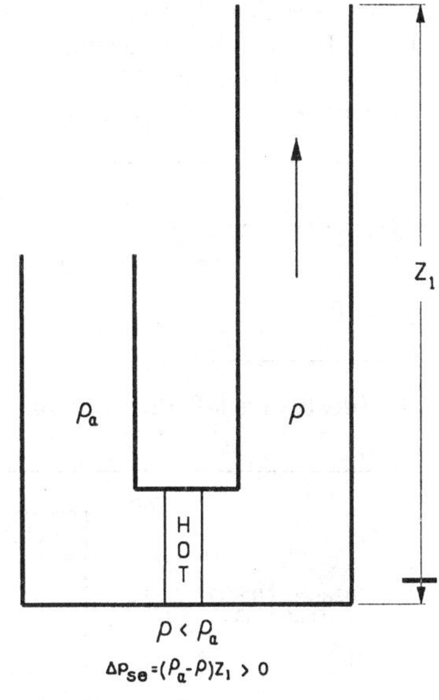

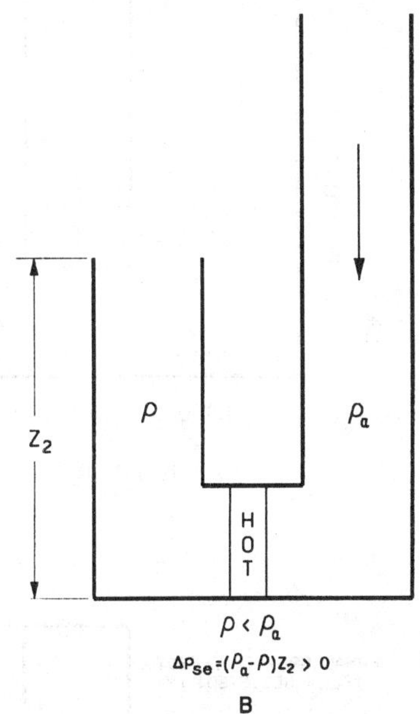

Fig. 3 Multiple Stack Analysis

To determine the fan total pressure requirement for a system, use the following equation:

$$P_t = \sum_{i \varepsilon F_{up}} \Delta p_{t_i} + \sum_{i \varepsilon F_{dn}} \Delta p_{t_i} \qquad \text{for } i = 1, 2, \ldots, n_{up} + n_{dn} \tag{16}$$

where

F_{up} and F_{dn} = sets of duct sections upstream and downstream of a fan
P_t = fan total pressure, in. of water
ε = symbol that ties duct sections into system paths from the exhaust/return air terminals to the supply terminals

Figure 4 illustrates the use of Equation (16). This system has three supply and two return terminals consisting of nine sections connected in six paths: 1-3-4-9-7-5, 1-3-4-9-7-6, 1-3-4-9-8, 2-4-9-7-5, 2-4-9-7-6, and 2-4-9-8. Sections 1 and 3 are unequal area; thus, they are assigned separate numbers in accordance with the rules for identifying sections (see Step 4 in the section on HVAC Duct Design Procedures). To determine the fan pressure requirement, the following six equations, derived from Equation (16), are applied. These equations must be satisfied to attain pressure balancing for design airflow. Relying entirely on dampers is not economical and may create objectionable flow-generated noise.

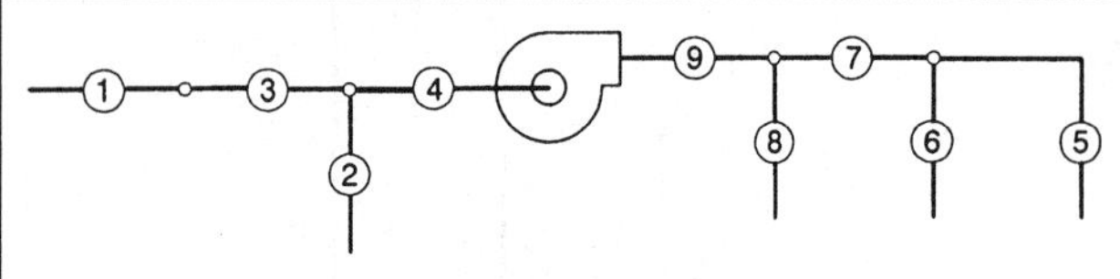

Fig. 4 Illustrative 6-Path, 9-Section System

$$\left\{\begin{array}{l} P_t = \Delta p_1 + \Delta p_3 + \Delta p_4 + \Delta p_9 + \Delta p_7 + \Delta p_5 \\ P_t = \Delta p_1 + \Delta p_3 + \Delta p_4 + \Delta p_9 + \Delta p_7 + \Delta p_6 \\ P_t = \Delta p_1 + \Delta p_3 + \Delta p_4 + \Delta p_9 + \Delta p_8 \\ P_t = \Delta p_2 + \Delta p_4 + \Delta p_9 + \Delta p_7 + \Delta p_5 \\ P_t = \Delta p_2 + \Delta p_4 + \Delta p_9 + \Delta p_7 + \Delta p_6 \\ P_t = \Delta p_2 + \Delta p_4 + \Delta p_9 + \Delta p_8 \end{array}\right. \tag{17}$$

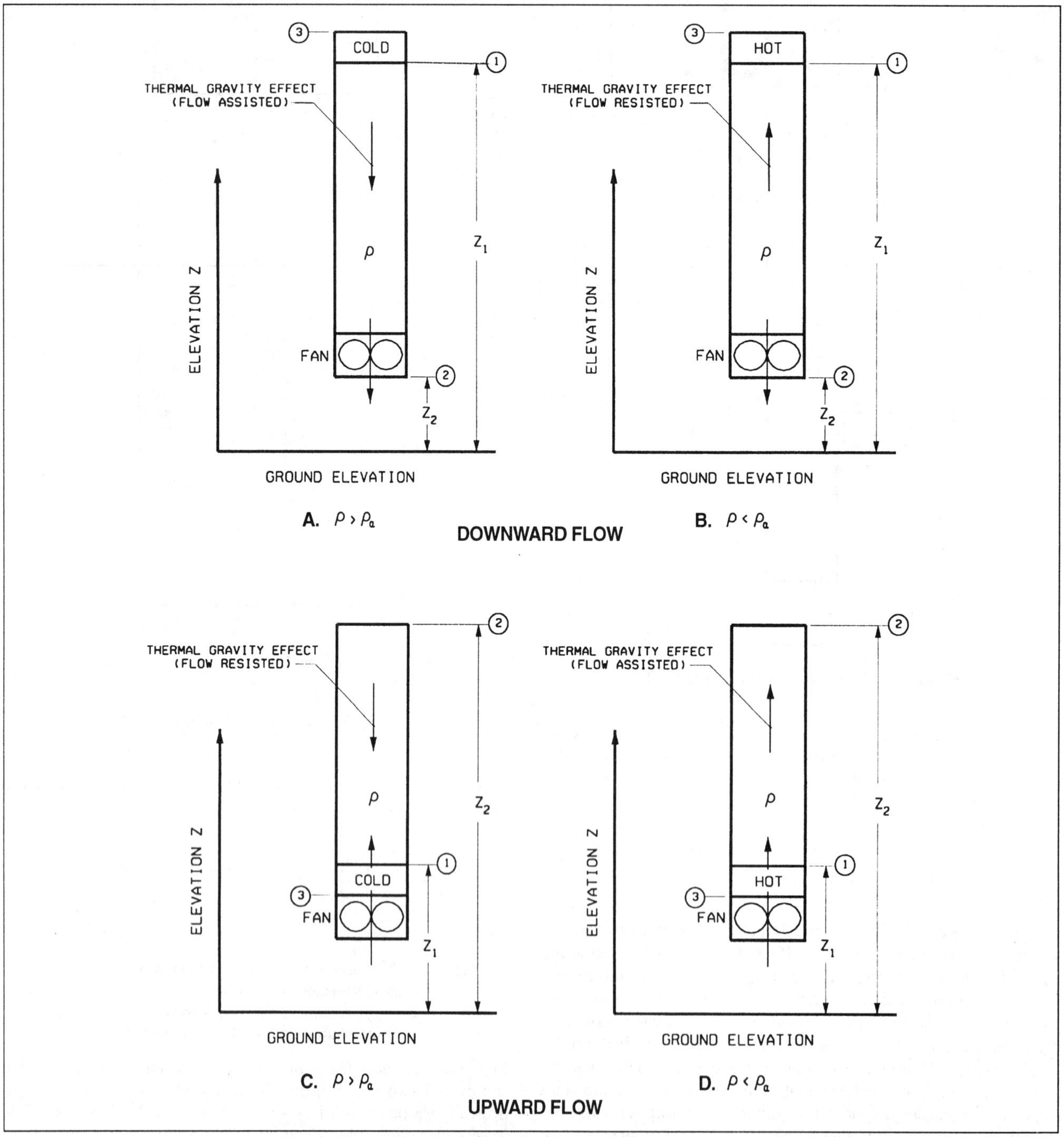

Fig. 5 Single Stack with Fan for Examples 3 and 4

Example 3. For Figures 5A and 5C, calculate the thermal gravity effect and fan total pressure required when the air is cooled to −30°F. The heat exchanger and ductwork (section 1 to 2) total pressure losses are 0.70 and 0.28 in. of water respectively. The density of −30°F air is 0.0924 lb_m/ft^3; ambient air is 0.075 lb_m/ft^3. Elevations are 70 ft and 10 ft as noted in the solutions below.

Solution:

(a) For Figure 5A (downward flow),

$$\begin{aligned} \Delta p_{se} &= 0.192(\rho_a - \rho)(z_2 - z_1) \\ &= 0.192(0.075 - 0.0924)(10 - 70) \\ &= 0.20 \text{ in. of water} \end{aligned}$$

$$\begin{aligned} P_t &= \Delta p_{t,3\text{-}2} - \Delta p_{se} \\ &= (0.70 + 0.28) - (0.20) \\ &= 0.78 \text{ in. of water} \end{aligned}$$

(b) For Figure 5C (upward flow),

$$\begin{aligned} \Delta p_{se} &= 0.192(\rho_a - \rho)(z_2 - z_1) \\ &= 0.192(0.075 - 0.0924)(70 - 10) \\ &= -0.20 \text{ in. of water} \end{aligned}$$

$$\begin{aligned} P_t &= \Delta p_{t,3\text{-}2} - \Delta p_{se} \\ &= (0.70 + 0.28) - (-0.20) \\ &= 1.18 \text{ in. of water} \end{aligned}$$

Example 4. For Figures 5B and 5D, calculate the thermal gravity effect and fan total pressure required when the air is heated to 250°F. The heat exchanger and ductwork (section 1 to 2) total pressure losses are 0.70 and 0.28 in. of water respectively. The density of 250°F air is 0.0558 lb_m/ft^3; ambient air is 0.075 lb_m/ft^3. Elevations are 70 ft and 10 ft as noted in the solutions below.

Solution:

(a) For Figure 5B (downward flow),

$$\begin{aligned} \Delta p_{se} &= 0.192(\rho_a - \rho)(z_2 - z_1) \\ &= 0.192(0.075 - 0.0558)(10 - 70) \\ &= -0.22 \text{ in. of water} \end{aligned}$$

$$\begin{aligned} P_t &= \Delta p_{t,3\text{-}2} - \Delta p_{se} \\ &= (0.70 + 0.28) - (-0.22) \\ &= 1.20 \text{ in. of water} \end{aligned}$$

(b) For Figure 5D (upward flow),

$$\begin{aligned} \Delta p_{se} &= 0.192(\rho_a - \rho)(z_2 - z_1) \\ &= 0.192(0.075 - 0.0558)(70 - 10) \\ &= 0.22 \text{ in. of water} \end{aligned}$$

$$\begin{aligned} P_t &= \Delta p_{t,3\text{-}2} - \Delta p_{se} \\ &= (0.70 + 0.28) - (0.22) \\ &= 0.76 \text{ in. of water} \end{aligned}$$

Example 5. Calculate the thermal gravity effect for each section of the system shown in Figure 6 and the net thermal gravity effect of the system. The density of ambient air is 0.075 lb_m/ft^3, and the lengths are as follows: z_1 = 50 ft, z_2 = 90 ft, z_4 = 95 ft, z_5 = 25 ft, and z_9 = 200 ft. The pressure required at section 3 is −0.1 in. of water. Write the equation to determine the fan total pressure requirement.

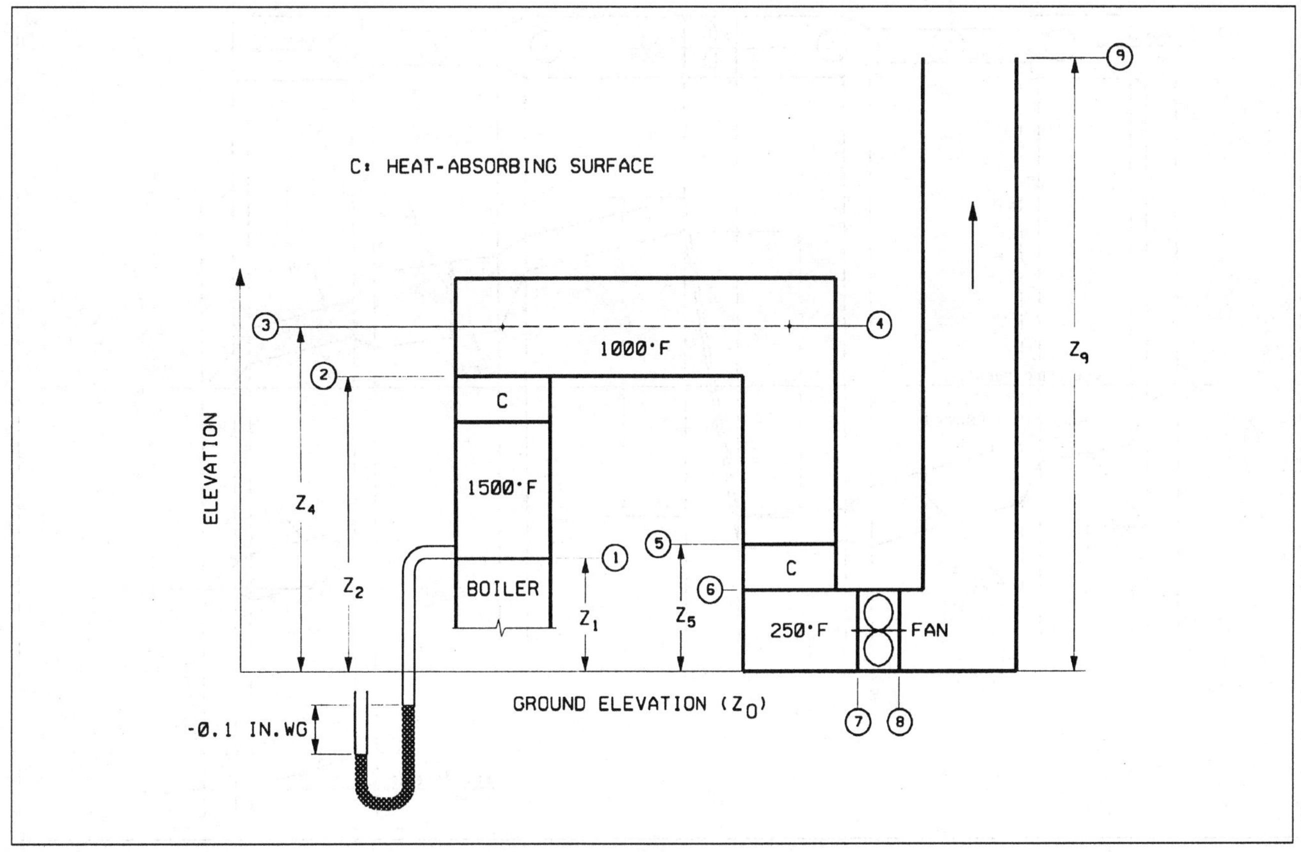

Fig. 6 Triple Stack System for Example 5

Solution: The following table summarizes the thermal gravity effect for each section of the system as calculated by Equation (15). The net thermal gravity effect for the system is 0.52 in. of water. To select a fan, use the following equation:

$$\begin{aligned} P_t &= 0.1 + \Delta p_{t,1\text{-}7} + \Delta p_{t,8\text{-}9} - \Delta p_{se} \\ &= 0.1 + \Delta p_{t,1\text{-}7} + \Delta p_{t,8\text{-}9} - 0.52 \\ &= \Delta p_{t,1\text{-}7} + \Delta p_{t,8\text{-}9} - 0.42 \end{aligned}$$

Path $(x\text{-}x')$	Temp., °F	ρ, lb_m/ft^3	Δz $(z_{x'} - z_x)$, ft	$\Delta\rho$ $(\rho_a - \rho_{x\text{-}x'})$, lb_m/ft^3	Δp_{se}, in. of water [Eq. (15)]
1-2	1500	0.0202	(90 – 50)	+0.0548	+0.42
3-4	1000	0.0271	0	+0.0479	0
4-5	1000	0.0271	(25 – 95)	+0.0479	–0.64
6-7	250	0.0558	0	+0.0192	0
8-9	250	0.0558	(200 – 0)	+0.0192	+0.74
Net Thermal Gravity Effect					0.52

PRESSURE CHANGES IN SYSTEM

Figure 7 shows total and static pressure changes in a fan/duct system consisting of a fan with both supply and return air ductwork. Also shown are the total and static pressure gradients referenced to atmospheric pressure.

For all constant-area sections, the total and static pressure losses are equal. At the diverging transitions, velocity pressure decreases, absolute total pressure decreases, and absolute static pressure can increase. The static pressure increase at these sections is known as **static regain**.

At the converging transitions, velocity pressure increases in the direction of airflow, and the absolute total and absolute static pressures decrease.

At the exit, the total pressure loss depends on the shape of the fitting and the flow characteristics. Exit loss coefficients C_o can be greater than, less than, or equal to one. The total and static pressure grade lines for the various coefficients are shown in Figure 3. Note that for a loss coefficient less than one, static pressure upstream of the exit is less than atmospheric pressure (negative). The static pressure just upstream of the discharge fitting can be calculated by subtracting the upstream velocity pressure from the upstream total pressure.

At section 1, the total pressure loss depends on the shape of the entry. The total pressure immediately downstream of the entrance equals the difference between the upstream pressure, which is zero (atmospheric pressure), and the loss through the fitting. The static pressure of the ambient air is zero; several diameters downstream, static pressure is negative, equal to the sum of the total pressure (negative) and the velocity pressure (always positive).

System resistance to airflow is noted by the total pressure grade line in Figure 7. Sections 3 and 4 include fan system effect pressure losses. To obtain the fan static pressure requirement for fan selection where the fan total pressure is known, use

$$P_s = P_t - p_{v,o} \tag{18}$$

where

P_s = fan static pressure, in. of water
P_t = fan total pressure, in. of water
$p_{v,o}$ = fan outlet velocity pressure, in. of water

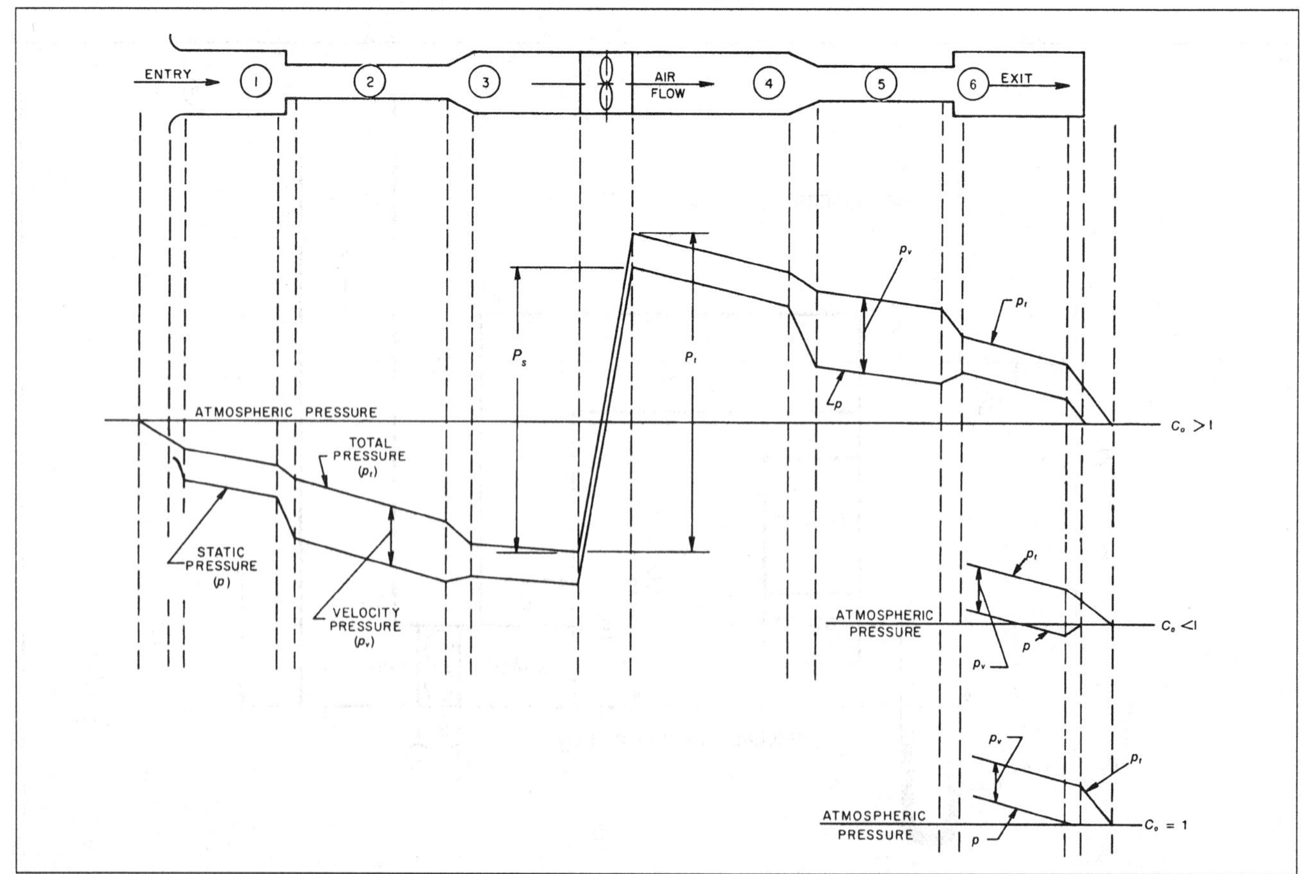

Fig. 7 Pressure Changes During Flow in Ducts

FLUID RESISTANCE

Duct system losses are the irreversible transformation of mechanical energy into heat. The two types of losses are (1) friction losses and (2) dynamic losses.

FRICTION LOSSES

Friction losses are due to fluid viscosity and are a result of momentum exchange between molecules in laminar flow and between individual particles of adjacent fluid layers moving at different velocities in turbulent flow. Friction losses occur along the entire duct length.

Darcy, Colebrook, and Altshul-Tsal Equations

For fluid flow in conduits, friction loss can be calculated by the Darcy equation:

$$\Delta p_f = \frac{12fL}{D_h}\rho\left(\frac{V}{1097}\right)^2 \quad (19)$$

where

Δp_f = friction losses in terms of total pressure, in. of water
f = friction factor, dimensionless
L = duct length, ft
D_h = hydraulic diameter [Equation (24)], in.
V = velocity, fpm
ρ = density, lb_m/ft^3

Within the region of laminar flow (Reynolds numbers less than 2000), the friction factor is a function of Reynolds number only.

For completely turbulent flow, the friction factor depends on Reynolds number, duct surface roughness, and internal protuberances such as joints. Between the bounding limits of hydraulically smooth behavior and fully rough behavior, is a transitional roughness zone where the friction factor depends on both roughness and Reynolds number. In this transitionally rough, turbulent zone the friction factor f is calculated by Colebrook's equation (Colebrook 1938-39). Colebrook's transition curve merges asymptotically into the curves representing laminar and completely turbulent flow. Because Colebrook's equation cannot be solved explicitly for f, use iterative techniques (Behls 1971).

$$\frac{1}{\sqrt{f}} = -2\log\left(\frac{12\varepsilon}{3.7D_h} + \frac{2.51}{\mathrm{Re}\sqrt{f}}\right) \quad (20)$$

where

ε = material absolute roughness factor, ft
Re = Reynolds number

A simplified formula for calculating friction factor, developed by Altshul (Altshul et al. 1975) and modified by Tsal, is

$$f' = 0.11\left(\frac{12\varepsilon}{D_h} + \frac{68}{\mathrm{Re}}\right)^{0.25} \quad (21)$$

If $f' \geq 0.018$: $f = f'$

If $f' < 0.018$: $f = 0.85f' + 0.0028$

Friction factors obtained from the Altshul-Tsal equation are within 1.6% of those obtained by Colebrook's equation.

Reynolds number (Re) may be calculated by using the following equation.

$$\mathrm{Re} = \frac{D_h V}{720\nu} \quad (22)$$

where ν = kinematic viscosity, ft^2/s.

For standard air, Re can be calculated by

$$\mathrm{Re} = 8.56 D_h V \quad (23)$$

Roughness Factors

The roughness factors ε listed in Table 1 are recommended for use with the Colebrook or Altshul-Tsal equation [Equations (20) and (21), respectively]. These values include not only material, but also duct construction, joint type, and joint spacing (Griggs and Khodabakhsh-Sharifabad 1992). Roughness factors for other materials are presented in Idelchik et al. (1986). Idelchik summarizes roughness factors for 80 materials including metal tubes; conduits made from concrete and cement; and wood, plywood, and glass tubes.

Swim (1978) conducted tests on duct liners of varying densities, surface treatments, transverse joints (workmanship), and methods of attachment to sheet metal ducts. As a result of these tests, Swim recommends for design 0.015 ft for spray-coated liners and 0.005 ft for liners with a facing material cemented onto the air side. In both cases, the roughness factor includes the resistance offered by mechanical fasteners and assumes good joints. Liners cut too short result in (1) loss of thermal performance, (2) possible condensation problems, (3) potential damage to the liner (erosion of the blanket or tearing away from the duct surface), and (4) the collection of dirt and debris and the initiation of biological problems. Liner density does not significantly influence flow resistance.

Manufacturers' data indicate that the absolute roughness for fully extended nonmetallic flexible ducts ranges from 0.0035 to 0.015 ft. For fully extended flexible metallic ducts, absolute roughness ranges from 0.0004 to 0.007 ft. This range covers flexible duct with the supporting wire exposed to flow or covered by the material. Figure 8 provides a pressure drop correction factor for straight flexible duct when less than fully extended.

Table 1 Duct Roughness Factors

Duct Material	Roughness Category	Absolute Roughness ε, ft
Uncoated carbon steel, clean (Moody 1944) (0.00015 ft) PVC plastic pipe (Swim 1982) (0.00003 to 0.00015 ft) Aluminum (Hutchinson 1953) (0.000015 to 0.0002 ft)	Smooth	0.0001
Galvanized steel, longitudinal seams, 4 ft joints (Griggs et al. 1987) (0.00016 to 0.00032 ft) Galvanized steel, continuously rolled, spiral seams, 10 ft joints (Jones 1979) (0.0002 to 0.0004 ft) Galvanized steel, spiral seam with 1, 2, and 3 ribs, 12 ft joints (Griggs et al. 1987) (0.00029 to 0.00038 ft)	Medium smooth	0.0003
Galvanized steel, longitudinal seams, 2.5 ft joints (Wright 1945) (0.0005 ft)	Average	0.0005
Fibrous glass duct, rigid Fibrous glass duct liner, air side with facing material (Swim 1978) (0.005 ft)	Medium rough	0.003
Fibrous glass duct liner, air side spray coated (Swim 1978) (0.015 ft) Flexible duct, metallic (0.004 to 0.007 ft when fully extended) Flexible duct, all types of fabric and wire (0.0035 to 0.015 ft when fully extended) Concrete (Moody 1944) (0.001 to 0.01 ft)	Rough	0.01

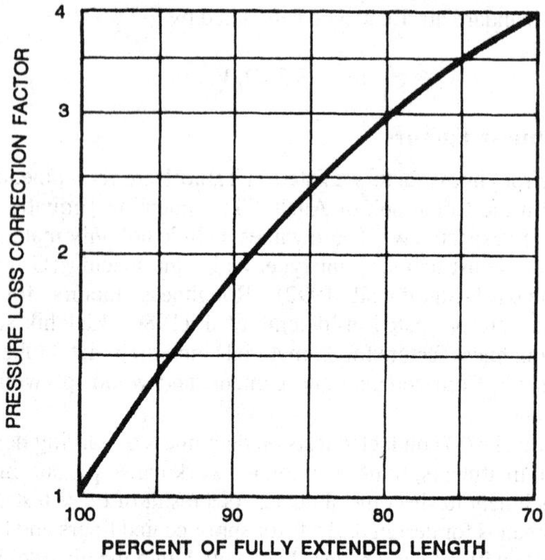

Fig. 8 Correction Factor for Unextended Flexible Duct

Friction Chart

Fluid resistance caused by friction in round ducts can be determined by the friction chart (Figure 9). This chart is based on standard air flowing through round galvanized ducts with beaded slip couplings on 48 in. centers, equivalent to an absolute roughness of 0.0003 ft.

Changes in barometric pressure, temperature, and humidity affect air density, air viscosity, and Reynolds number. No corrections to Figure 9 are needed for (1) duct materials with a medium smooth roughness factor, (2) temperature variations in the order of ±30°F from 70°F, (3) elevations to 1500 ft, and (4) duct pressures from −20 in. of water to +20 in. of water relative to the ambient pressure. These individual variations in temperature, elevation, and duct pressure result in duct losses within ±5% of the standard air friction chart.

For duct materials other than those categorized as medium smooth in Table 1, and for variations in temperature, barometric pressure (elevation), and duct pressures (outside the range listed), calculate the friction loss in a duct by the Altshul-Tsal and Darcy equations [Equations (21) and (19), respectively].

Noncircular Ducts

A momentum analysis can relate average wall shear stress to pressure drop per unit length for fully developed turbulent flow in a passage of arbitrary shape but uniform longitudinal cross-sectional area. This analysis leads to the definition of **hydraulic diameter**:

$$D_h = 4A/P \tag{24}$$

where

D_h = hydraulic diameter, in.
A = duct area, in^2
P = perimeter of cross section, in.

While the hydraulic diameter is often used to correlate noncircular data, exact solutions for laminar flow in noncircular passages show that such practice causes some inconsistencies. No exact solutions exist for turbulent flow. Tests over a limited range of turbulent flow indicated that fluid resistance is the same for equal lengths of duct for equal mean velocities of flow if the ducts have the same ratio of cross-sectional area to perimeter. From a series of experiments using round, square, and rectangular ducts having essentially the same hydraulic diameter, Huebscher (1948) found that each, for most purposes, had the same flow resistance at equal mean velocities. Tests by Griggs and Khodabakhsh-Sharifabad (1992) also indicated that experimental rectangular duct data for airflow over the range typical of HVAC systems can be correlated satisfactorily using Equation (20) together with hydraulic diameter, particularly when a realistic experimental uncertainty is accepted. These tests support using hydraulic diameter to correlate noncircular duct data.

Rectangular Ducts. Huebscher (1948) developed the relationship between rectangular and round ducts that is used to determine size equivalency based on equal flow, resistance, and length. This relationship, Equation (25), is the basis for Table 2.

$$D_e = \frac{1.30(ab)^{0.625}}{(a+b)^{0.250}} \tag{25}$$

where

D_e = circular equivalent of rectangular duct for equal length, fluid resistance, and airflow, in.
a = length of one side of duct, in.
b = length of adjacent side of duct, in.

To determine equivalent round duct diameter, use Table 2. Equations (21) or (20) and (19) must be used to determine pressure loss.

Flat Oval Ducts. To convert round ducts to spiral flat oval sizes, use Table 3. Table 3 is based on Equation (26) (Heyt and Diaz 1975), the circular equivalent of a flat oval duct for equal airflow, resistance, and length. Equations (21) or (20) and (19) must be used to determine friction loss.

$$D_e = \frac{1.55A^{0.625}}{P^{0.250}} \tag{26}$$

where A is the cross-sectional area of flat oval duct defined as

$$A = (\pi b^2/4) + b(a-b) \tag{27}$$

and the perimeter P is calculated by

$$P = \pi b + 2(a-b) \tag{28}$$

where

P = perimeter of flat oval duct, in.
a = major dimension of flat oval duct, in.
b = minor dimension of flat oval duct, in.

DYNAMIC LOSSES

Dynamic losses result from flow disturbances caused by duct-mounted equipment and fittings that change the airflow path's direction and/or area. These fittings include entries, exits, elbows, transitions, and junctions. Idelchik et al. (1986) discuss parameters affecting fluid resistance of fittings and presents local loss coefficients in three forms: tables, curves, and equations.

Local Loss Coefficients

The dimensionless coefficient C is used for fluid resistance, because this coefficient has the same value in dynamically similar streams (i.e., streams with geometrically similar stretches, equal Reynolds numbers, and equal values of other criteria necessary for dynamic similarity). The fluid resistance coefficient represents the ratio of total pressure loss to velocity pressure at the referenced cross section:

$$C = \frac{\Delta p_j}{\rho (V/1097)^2} = \frac{\Delta p_j}{p_v} \tag{29}$$

where

C = local loss coefficient, dimensionless
Δp_j = total pressure loss, in. of water
ρ = density, lb$_m$/ft^3
V = velocity, fpm
p_v = velocity pressure, in. of water

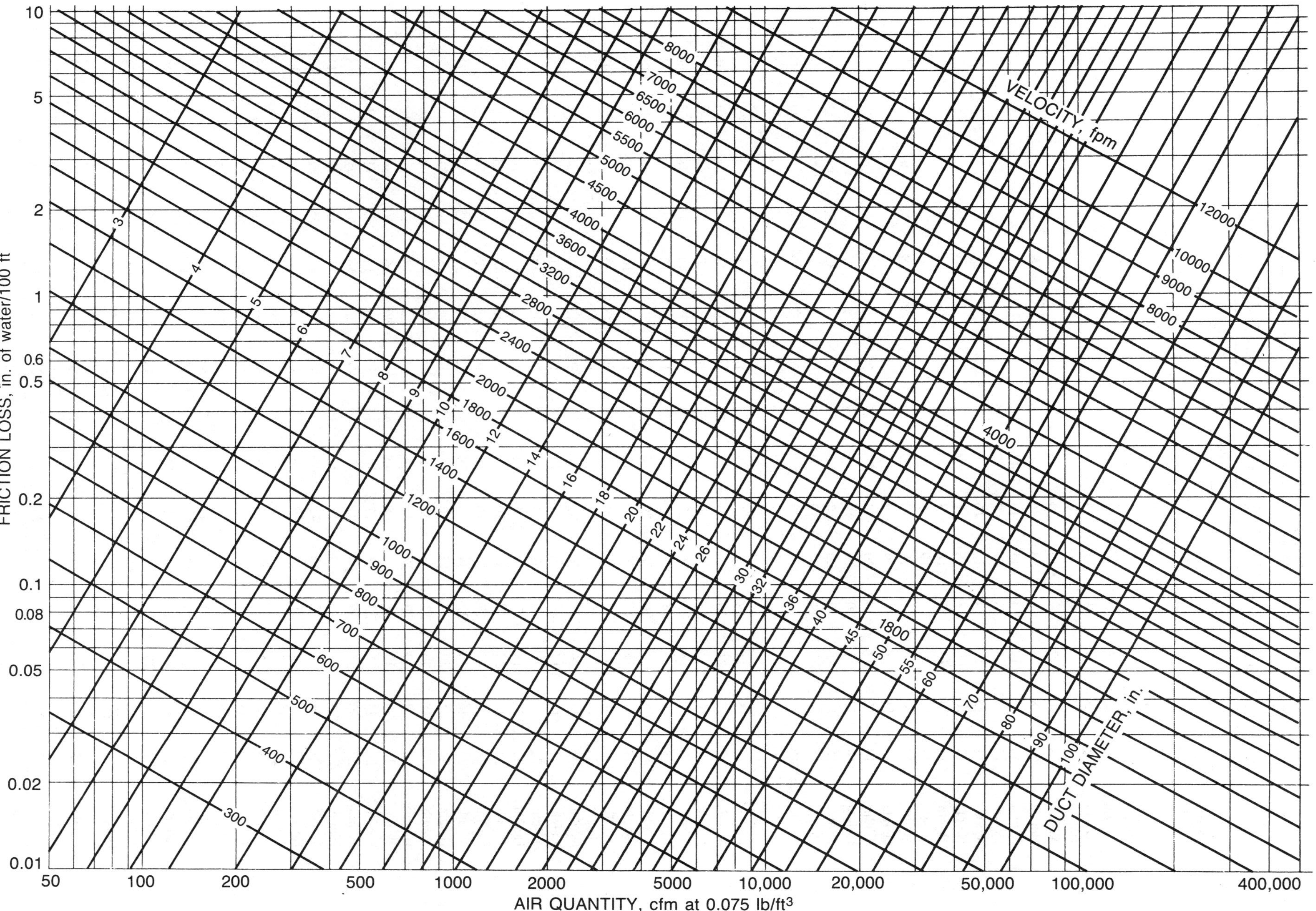

Fig. 9 Friction Chart for Round Duct ($\rho = 0.075\ lb_m/ft^3$ and $\varepsilon = 0.0003$ ft)

Table 2 Equivalent Rectangular Duct Dimension

Duct Diameter, in.		Aspect Ratio 1.00	1.25	1.50	1.75	2.00	2.25	2.50	2.75	3.00	3.50	4.00	5.00	6.00	7.00	8.00
		Rectangular Size, in.														
6	Width	—	6													
	Height	—	5													
7	Width	6	8													
	Height	6	6													
8	Width	7	9	9	11											
	Height	7	7	6	6											
9	Width	8	9	11	11	12	14									
	Height	8	7	7	6	6	6									
10	Width	9	10	12	12	14	14	15	17							
	Height	9	8	8	7	7	6	6	6							
11	Width	10	11	12	14	14	16	18	17	18	21					
	Height	10	9	8	8	7	7	7	6	6	6					
12	Width	11	13	14	14	16	16	18	19	21	21	24				
	Height	11	10	9	8	8	7	7	7	7	6	6				
13	Width	12	14	15	16	18	18	20	19	21	25	24	30			
	Height	12	11	10	9	9	8	8	7	7	7	6	6			
14	Width	13	14	17	18	18	20	20	22	24	25	28	30	36		
	Height	13	11	11	10	9	9	8	8	8	7	7	6	6		
15	Width	14	15	17	18	20	20	23	25	24	28	28	35	36	42	
	Height	14	12	11	10	10	9	9	9	8	8	7	7	6	6	
16	Width	15	16	18	19	20	23	23	25	27	28	32	35	42	42	48
	Height	15	13	12	11	10	10	9	9	9	8	8	7	7	6	6
17	Width	16	18	20	21	22	25	25	28	27	32	32	35	42	49	48
	Height	16	14	13	12	11	11	10	10	9	9	8	7	7	7	6
18	Width	16	19	21	23	24	25	28	28	30	32	36	40	42	49	56
	Height	16	15	14	13	12	11	11	10	10	9	9	8	7	7	7
19	Width	17	20	21	23	24	27	28	30	30	35	36	40	48	49	56
	Height	17	16	14	13	12	12	11	11	10	10	9	8	8	7	7
20	Width	18	20	23	25	26	27	30	30	33	35	40	45	48	56	56
	Height	18	16	15	14	13	12	12	11	11	10	10	9	8	8	7
21	Width	19	21	24	26	28	29	30	33	33	39	40	45	54	56	64
	Height	19	17	16	15	14	13	12	12	11	11	10	9	9	8	8
22	Width	20	23	26	26	28	32	33	36	36	39	44	50	54	56	64
	Height	20	18	17	15	14	14	13	13	12	11	11	10	9	8	8
23	Width	21	24	26	28	30	32	35	36	39	42	44	50	54	63	64
	Height	21	19	17	16	15	14	14	13	13	12	11	10	9	9	8
24	Width	22	25	27	30	32	34	35	39	39	42	48	55	60	63	72
	Height	22	20	18	17	16	15	14	14	13	12	12	11	10	9	9
25	Width	23	25	29	30	32	36	38	39	42	46	48	55	60	70	72
	Height	23	20	19	17	16	16	15	14	14	13	12	11	10	10	9
26	Width	24	26	30	32	34	36	38	41	42	46	52	55	66	70	72
	Height	24	21	20	18	17	16	15	15	14	13	13	11	11	10	9
27	Width	25	28	30	33	36	38	40	41	45	49	52	60	66	70	80
	Height	25	22	20	19	18	17	16	15	15	14	13	12	11	10	10
28	Width	26	29	32	35	36	38	43	44	45	49	56	60	66	77	80
	Height	26	23	21	20	18	17	17	16	15	14	14	12	11	11	10
29	Width	27	30	33	35	38	41	43	44	48	53	56	65	72	77	88
	Height	27	24	22	20	19	18	17	16	16	15	14	13	12	11	11
30	Width	27	31	35	37	40	43	45	47	48	53	60	65	72	77	88
	Height	27	25	23	21	20	19	18	17	16	15	15	13	12	11	11
31	Width	28	31	35	39	40	43	45	50	51	56	60	70	78	84	88
	Height	28	25	23	22	20	19	18	18	17	16	15	14	13	12	11
32	Width	29	33	36	39	42	45	48	50	54	56	60	70	78	84	96
	Height	29	26	24	22	21	20	19	18	18	16	15	14	13	12	12
33	Width	30	34	38	40	44	47	50	52	54	60	64	75	78	91	96
	Height	30	27	25	23	22	21	20	19	18	17	16	15	13	13	12
34	Width	31	35	39	42	44	47	50	52	57	60	64	75	84	91	96
	Height	31	28	26	24	22	21	20	19	19	17	16	15	14	13	12
35	Width	32	36	39	42	46	50	53	55	57	63	68	75	84	91	104
	Height	32	29	26	24	23	22	21	20	19	18	17	15	14	13	13
36	Width	33	36	41	44	48	50	53	55	60	63	68	80	90	98	104
	Height	33	29	27	25	24	22	21	20	20	18	17	16	15	14	13
38	Width	35	39	44	47	50	54	58	61	63	67	72	85	96	105	112
	Height	35	31	29	27	25	24	23	22	21	19	18	17	16	15	14

*Shaded area not recommended.

Table 2 Equivalent Rectangular Duct Dimension (*Continued*)

Duct Diameter, in.		1.00	1.25	1.50	1.75	2.00	2.25	2.50	2.75	3.00	3.50	4.00	5.00	6.00	7.00	8.00
		Aspect Ratio														
		Rectangular Size, in.														
40	Width	37	41	45	49	52	56	60	63	66	70	76	90	96	105	120
	Height	37	33	30	28	26	25	24	23	22	20	19	18	16	15	15
42	Width	38	43	48	51	56	59	63	66	69	74	80	90	102	112	120
	Height	38	34	32	29	28	26	25	24	23	21	20	18	17	16	15
44	Width	40	45	50	54	58	61	65	69	72	81	84	95	108	119	128
	Height	40	36	33	31	29	27	26	25	24	23	21	19	18	17	16
46	Width	42	48	53	56	60	65	68	72	75	84	88	100	114	126	136
	Height	42	38	35	32	30	29	27	26	25	24	22	20	19	18	17
48	Width	44	49	54	60	62	68	70	74	78	88	92	105	120	126	136
	Height	44	39	36	34	31	30	28	27	26	25	23	21	20	18	17
50	Width	46	51	57	61	66	70	75	77	81	91	96	110	120	133	144
	Height	46	41	38	35	33	31	30	28	27	26	24	22	20	19	18
52	Width	48	54	59	63	68	72	78	83	84	95	100	115	126	140	152
	Height	48	43	39	36	34	32	31	30	28	27	25	23	21	20	19
54	Width	49	55	62	67	70	77	80	85	90	98	104	120	132	147	160
	Height	49	44	41	38	35	34	32	31	30	28	26	24	22	21	20
56	Width	51	58	63	68	74	79	83	88	93	102	108	125	138	147	160
	Height	51	46	42	39	37	35	33	32	31	29	27	25	23	21	20
58	Width	53	60	66	70	76	81	85	91	96	105	112	130	144	154	168
	Height	53	48	44	40	38	36	34	33	32	30	28	26	24	22	21
60	Width	55	61	68	74	78	83	90	94	99	109	116	130	144	161	
	Height	55	49	45	42	39	37	36	34	33	31	29	26	24	23	
62	Width	57	64	71	75	82	88	93	96	102	112	120	135	150	168	
	Height	57	51	47	43	41	39	37	35	34	32	30	27	25	24	
64	Width	59	65	72	79	84	90	95	99	105	116	124	140	156		
	Height	59	52	48	45	42	40	38	36	35	33	31	28	26		
66	Width	60	68	75	81	86	92	98	105	108	119	128	145	162		
	Height	60	54	50	46	43	41	39	38	36	34	32	29	27		
68	Width	62	70	77	82	90	95	100	107	111	123	132	150	168		
	Height	62	56	51	47	45	42	40	39	37	35	33	30	28		
70	Width	64	71	80	86	92	99	105	110	114	126	136	155			
	Height	64	57	53	49	46	44	42	40	38	36	34	31			
72	Width	66	74	81	88	94	101	108	113	117	130	140	160			
	Height	66	59	54	50	47	45	43	41	39	37	35	32			
74	Width	68	76	84	91	98	104	110	116	123	133	144	165			
	Height	68	61	56	52	49	46	44	42	41	38	36	33			
76	Width	70	78	86	93	100	106	113	118	126	137	148	165			
	Height	70	62	57	53	50	47	45	43	42	39	37	33			
78	Width	71	80	89	95	102	110	115	121	129	140	152				
	Height	71	64	59	54	51	49	46	44	43	40	38				
80	Width	73	83	90	98	104	113	118	124	132	144	156				
	Height	73	66	60	56	52	50	47	45	44	41	39				
82	Width	75	84	93	100	108	115	123	129	135	147	160				
	Height	75	67	62	57	54	51	49	47	45	42	40				
84	Width	77	86	95	103	110	117	125	132	138	151	164				
	Height	77	69	63	59	55	52	50	48	46	43	41				
86	Width	79	88	98	105	112	119	128	135	141	154	168				
	Height	79	70	65	60	56	53	51	49	47	44	42				
88	Width	80	90	99	107	116	124	130	138	144	158					
	Height	80	72	66	61	58	55	52	50	48	45					
90	Width	82	93	102	110	118	126	133	140	147	161					
	Height	82	74	68	63	59	56	53	51	49	46					
92	Width	84	94	104	112	120	128	138	143	150	165					
	Height	84	75	69	64	60	57	55	52	50	47					
94	Width	86	96	107	116	124	131	140	146	153	168					
	Height	86	77	71	66	62	58	56	53	51	48					
96	Width	88	99	108	117	126	135	143	151	159						
	Height	88	79	72	67	63	60	57	55	53						
98	Width	90	100	111	119	128	137	145	154	162						
	Height	90	80	74	68	64	61	58	56	54						
100	Width	91	103	113	123	132	140	148	157	165						
	Height	91	82	75	70	66	62	59	57	55						
102	Width	93	105	116	124	134	142	153	160	168						
	Height	93	84	77	71	67	63	61	58	56						

*Shaded area not recommended.

Table 2 Equivalent Rectangular Duct Dimension (*Concluded*)

Duct Diameter, in.		Aspect Ratio 1.00	1.25	1.50	1.75	2.00	2.25	2.50	2.75	3.00	3.50	4.00	5.00	6.00	7.00	8.00
		Rectangular Size, in.														
104	Width	95	106	117	128	136	146	155	162							
	Height	95	85	78	73	68	65	62	59							
106	Width	97	109	120	130	140	149	158	165							
	Height	97	87	80	74	70	66	63	60							
108	Width	99	110	122	131	142	151	160	168							
	Height	99	88	81	75	71	67	64	61							
110	Width	101	113	125	135	144	153	163								
	Height	101	90	83	77	72	68	65								
112	Width	102	115	126	137	146	158	165								
	Height	102	92	84	78	73	70	66								
114	Width	104	116	129	140	150	160									
	Height	104	93	86	80	75	71									
116	Width	106	119	131	142	152	162									
	Height	106	95	87	81	76	72									
118	Width	108	121	134	144	154	164									
	Height	108	97	89	82	77	73									
120	Width	110	123	135	147	158										
	Height	110	98	90	84	79										

*Shaded area not recommended.

Table 3 Equivalent Spiral Flat Oval Duct Dimensions

Round Duct Diameter, in.	Minor Dimension (*b*), in. 3	4	5	6	7	8	9	10	11	12	14	16
	Major Dimension (*a*), in.											
5	8											
5.5	9	7										
6	11	9										
6.5	12	10	8									
7	15	12	10	8								
7.5	19	13	—	9								
8	22	15	11	—								
8.5		18	13	11	10							
9		20	14	12	—	10						
9.5		21	18	14	12	—						
10			19	15	13	11						
10.5			21	17	15	13	12					
11				19	16	14	—	12				
11.5				20	18	16	14	—				
12				23	20	17	15	13				
12.5				25	21	—	—	15	14			
13				28	23	19	17	16	—	14		
13.5				30	—	21	18	—	16	—		
14				33	—	22	20	18	17	15		
14.5				36	—	24	22	19	—	17		
15				39	—	27	23	21	19	18		
16				45	—	30	—	24	22	20	17	
17				52	—	35	—	27	24	21	19	
18				59	—	39	—	30	—	25	22	19

Round Duct Diameter, in.	Minor Dimension (*b*), in. 8	9	10	11	12	14	16	18	20	22	24
	Major Dimension (*a*), in.										
19	46	—	34	—	28	23	21				
20	50	—	38	—	31	27	24	21			
21	58	—	43	—	34	28	25	23			
22	65	—	48	—	37	31	29	26			
23	71	—	52	—	42	34	30	27			
24	77	—	57	—	45	38	33	29	26		
25			63	—	50	41	36	32	29		
26			70	—	56	45	38	34	31		
27			76	—	59	49	41	37	34		
28					65	52	46	40	36		
29					72	58	49	43	39	35	
30					78	61	54	46	40	38	
31					81	67	57	49	44	39	37
32						71	60	53	47	42	40
33						77	66	56	51	46	41
34							69	59	55	47	44
35							76	65	58	50	46
36							79	68	61	53	49
37								71	64	57	52
38								78	67	60	55
40									77	69	62
42										75	68
44										82	74

Dynamic losses occur along a duct length and cannot be separated from friction losses. For ease of calculation, dynamic losses are assumed to be concentrated at a section (local) and to exclude friction. Frictional losses must be considered only for relatively long fittings. Generally, fitting friction losses are accounted for by measuring duct lengths from the centerline of one fitting to that of the next fitting. For fittings closely coupled (less than six hydraulic diameters apart), the flow pattern entering subsequent fittings differs from the flow pattern used to determine loss coefficients. Adequate data for these situations are unavailable.

For all fittings, except junctions, calculate the total pressure loss Δp_j at a section by

$$\Delta p_j = C_o p_{v,o} \tag{30}$$

where the subscript o is the cross section at which the velocity pressure is referenced. The dynamic loss is based on the actual velocity in the duct, not the velocity in an equivalent noncircular duct. For the cross section to reference a fitting loss coefficient, refer to Step 4 in the section on HVAC Duct Design Procedures. Where necessary (unequal area fittings), convert a loss coefficient from section o to section i using Equation (31), where V is the velocity at the respective sections.

$$C_i = \frac{C_o}{(V_i/V_o)^2} \tag{31}$$

For converging and diverging flow junctions, total pressure losses through the straight (main) section are calculated as

$$\Delta p_j = C_{c,s} p_{v,c} \tag{32}$$

For total pressure losses through the branch section,

$$\Delta p_j = C_{c,b} p_{v,c} \tag{33}$$

where $p_{v,c}$ is the velocity pressure at the common section c, and $C_{c,s}$ and $C_{c,b}$ are losses for the straight (main) and branch flow paths, respectively, each referenced to the velocity pressure at section c. To convert junction local loss coefficients referenced to straight and branch velocity pressures, use the following equation:

$$C_i = \frac{C_{c,i}}{(V_i/V_c)^2} \tag{34}$$

where

C_i = local loss coefficient referenced to section being calculated (see subscripts), dimensionless
$C_{c,i}$ = straight ($C_{c,s}$) or branch ($C_{c,b}$) local loss coefficient referenced to dynamic pressure at common section, dimensionless
V_i = velocity at section to which C_i is being referenced, fpm
V_c = velocity at common section, fpm

Subscripts:

b = branch
s = straight (main) section
c = common section

The junction of two parallel streams moving at different velocities is characterized by turbulent mixing of the streams, accompanied by pressure losses. In the course of this mixing, an exchange of momentum takes place between the particles moving at different velocities, finally resulting in the equalization of the velocity distributions in the common stream. The jet with higher velocity loses a part of its kinetic energy by transmitting it to the slower moving jet. The loss in total pressure before and after mixing is always large and positive for the higher velocity jet and increases with an increase in the amount of energy transmitted to the lower velocity jet. Consequently, the local loss coefficient, defined by Equation (29), will always be positive. The energy stored in the lower velocity jet increases as a result of mixing. The loss in total pressure and the local loss coefficient can, therefore, also have negative values for the lower velocity jet (Idelchik et al. 1986).

Duct Fitting Database

A duct fitting database, developed by ASHRAE (1994), which includes 228 round and rectangular fittings with the provision to include flat oval fittings, is available from ASHRAE in electronic form with the capability to be linked to duct design programs.

The fittings are numbered (coded) as shown in Table 4. Entries and converging junctions are only in the exhaust/return portion of systems. Exits and diverging junctions are only in supply systems. Equal-area elbows, obstructions, and duct-mounted equipment are common to both supply and exhaust systems. Transitions and unequal-area elbows can be either supply or exhaust fittings. Fitting ED5-1 (see the section on Fitting Loss Coefficients) is an **E**xhaust fitting with a round shape (**D**iameter). The number 5 indicates that the fitting is a junction, and 1 is its sequential number. Fittings SR3-1 and ER3-1 are **S**upply and **E**xhaust fittings, respectively. The R indicates that the fitting is **R**ectangular, and the 3 identifies the fitting as an elbow. Note that the cross-sectional areas at sections 0 and 1 are not equal (see the section on Fitting Loss Coefficients). Otherwise, the elbow would be a **C**ommon fitting such as CR3-6. Additional fittings are reproduced in the section on Fitting Loss Coefficients to support the example design problems (see Table 12 for Example 8; see Table 14 for Example 9).

Table 4 Duct Fitting Codes

Fitting Function	Geometry	Category	Sequential Number
S: Supply	D: round (**D**iameter)	1. Entries 2. Exits	1,2,3...n
E: Exhaust/Return	R: **R**ectangular	3. Elbows 4. Transitions	
C: **C**ommon (supply and return)	O: flat **O**val	5. Junctions 6. Obstructions 7. Fan and system interactions 8. Duct-mounted equipment 9. Dampers 10. Hoods	

DUCTWORK SECTIONAL LOSSES

Darcy-Weisbach Equation

Total pressure loss in a duct section is calculated by combining Equations (19) and (29) in terms of Δp, where ΣC is the summation of local loss coefficients within the duct section. Each fitting loss coefficient must be referenced to that section's velocity pressure.

$$\Delta p = \left(\frac{12fL}{D_h} + \Sigma C\right)\rho\left(\frac{V}{1097}\right)^2 \tag{35}$$

FAN-SYSTEM INTERFACE

Fan Inlet and Outlet Conditions

Fan performance data measured in the field may show lower performance capacity than manufacturers' ratings. The most common causes of deficient performance of the fan/system combination are improper outlet connections, nonuniform inlet flow, and swirl at the fan inlet. These conditions alter the aerodynamic characteristics of the fan so that its full flow potential is not realized. One bad connection can reduce fan performance far below its rating. No data have been published that account for the effects of fan inlet and outlet flexible vibration connectors.

Normally, a fan is tested with open inlets and a section of straight duct attached to the outlet (ASHRAE *Standard* 51). This setup results in uniform flow into the fan and efficient static pressure recovery on the fan outlet. If good inlet and outlet conditions are not provided in the actual installation, the performance of the fan suffers. To select and apply the fan properly, these effects must be considered, and the pressure requirements of the fan, as calculated by standard duct design procedures, must be increased.

Figure 10 illustrates deficient fan/system performance. The system pressure losses have been determined accurately, and a fan has been selected for operation at Point 1. However, no allowance has been made for the effect of system connections to the fan on fan performance. To compensate, a fan system effect must be added to the calculated system pressure losses to determine the actual system curve. The point of intersection between the fan performance curve and the actual system curve is Point 4. The actual flow volume is, therefore, deficient by the difference from 1 to 4. To achieve design flow volume, a fan system effect pressure loss equal to the pressure difference between Points 1 and 2 should be added to the calculated system pressure losses, and the fan should be selected to operate at Point 2.

Fan System Effect Coefficients

The system effect concept was formulated by Farquhar (1973) and Meyer (1973); the magnitudes of the system effect, called

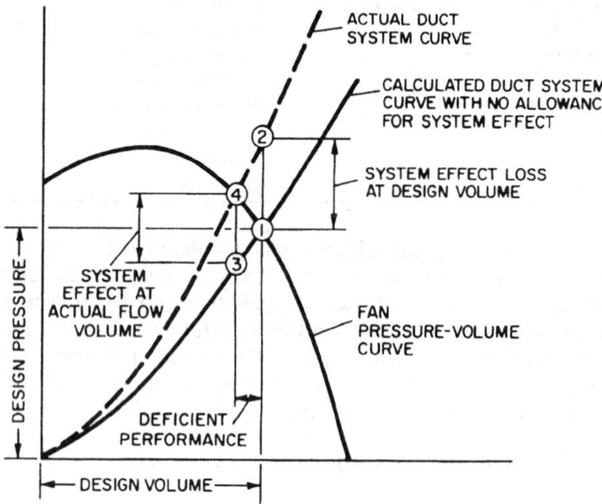

Fig. 10 Deficient System Performance with System Effect Ignored

system effect factors, were determined experimentally in the laboratory of the Air Movement and Control Association (AMCA) (Brown 1973, Clarke et al. 1978) and published in their *Publication* 201 (AMCA 1990a). The system effect factors, converted to local loss coefficients, are in the *Duct Fitting Database* (ASHRAE 1994) for both centrifugal and axial fans. Fan system effect coefficients are only an approximation. Fans of different types and even fans of the same type, but supplied by different manufacturers, do not necessarily react to a system in the same way. Therefore, judgment based on experience must be applied to any design.

Fan Outlet Conditions. Fans intended primarily for duct systems are usually tested with an outlet duct in place (ASHRAE *Standard* 51). Figure 11 shows the changes in velocity profiles at various distances from the fan outlet. For 100% recovery, the duct, including transition, must meet the requirements for 100% effective duct length [L_e (Figure 11)], which is calculated as follows:

For $V_o > 2500$ fpm,

$$L_e = \frac{V_o \sqrt{A_o}}{10{,}600} \qquad (36)$$

For $V_o \le 2500$ fpm,

$$L_e = \frac{\sqrt{A_o}}{4.3} \qquad (37)$$

where

V_o = duct velocity, fpm
L_e = effective duct length, ft
A_o = duct area, in^2

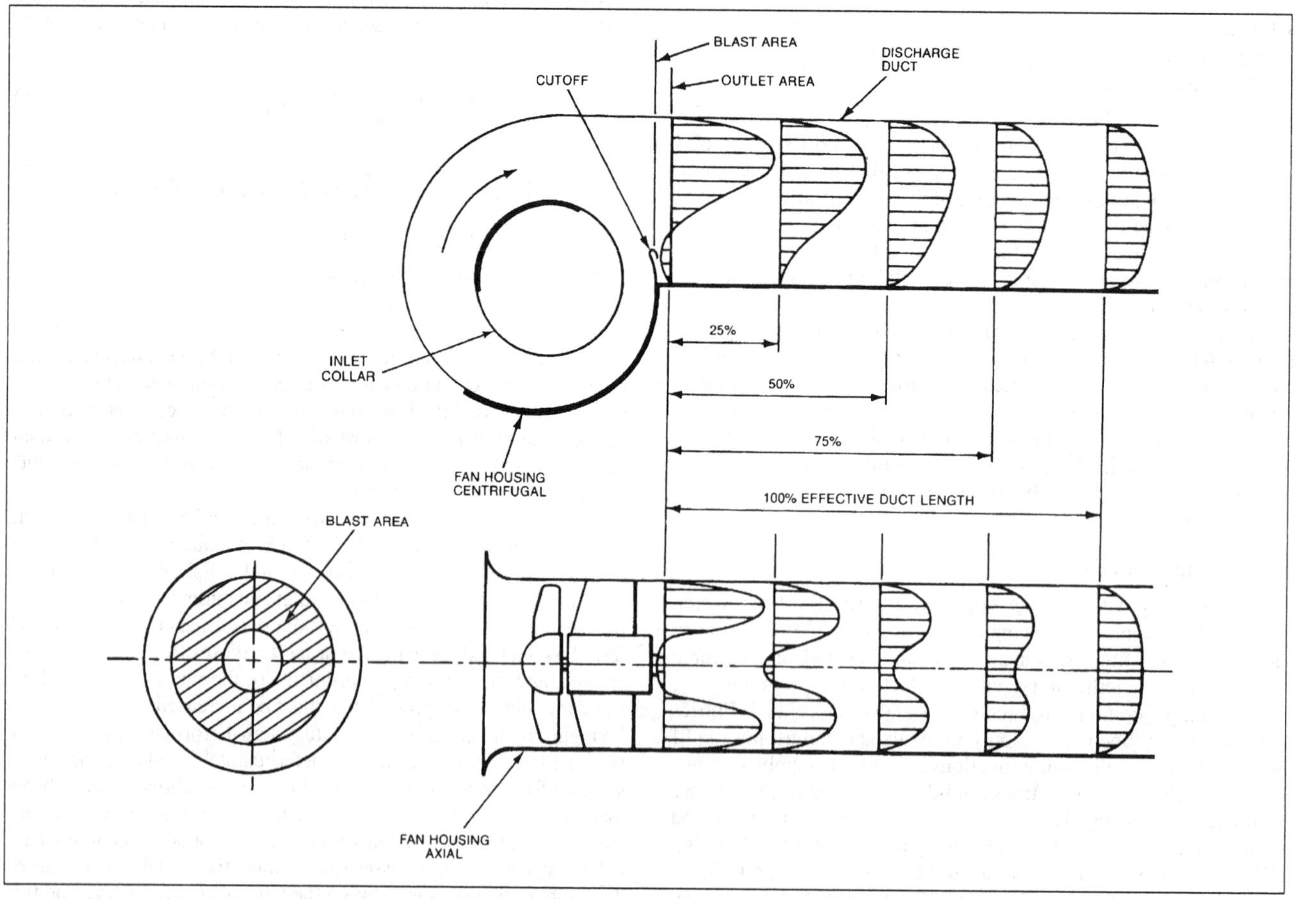

Fig. 11 Establishment of Uniform Velocity Profile in Straight Fan Outlet Duct
(Adapted by permission from AMCA *Publication* 201)

As illustrated by Fitting SR7-1 in the section on Fitting Loss Coefficients, centrifugal fans should not abruptly discharge to the atmosphere. A diffuser design should be selected from Fitting SR7-2 (see the section on Fitting Loss Coefficients) or SR7-3 (see ASHRAE 1994).

Fan Inlet Conditions. For rated performance, air must enter the fan uniformly over the inlet area in an axial direction without prerotation. Nonuniform flow into the inlet is the most common cause of reduced fan performance. Such inlet conditions are not equivalent to a simple increase in the system resistance; therefore, they cannot be treated as a percentage decrease in the flow and pressure from the fan. A poor inlet condition results in an entirely new fan performance. An elbow at the fan inlet, for example Fitting ED7-2 (see the section on Fitting Loss Coefficients), causes turbulence and uneven flow into the fan impeller. The losses due to the fan system effect can be eliminated by including an adequate length of straight duct between the elbow and the fan inlet.

The ideal inlet condition allows air to enter axially and uniformly without spin. A spin in the same direction as the impeller rotation reduces the pressure-volume curve by an amount dependent on the intensity of the vortex. A counterrotating vortex at the inlet slightly increases the pressure-volume curve, but the power is increased substantially.

Inlet spin may arise from a great variety of approach conditions, and sometimes the cause is not obvious. Inlet spin can be avoided by providing an adequate length of straight duct between the elbow and the fan inlet. Figure 12 illustrates some common duct connections that cause inlet spin and includes recommendations for correcting spin.

Fans within plenums and cabinets or next to walls should be located so that air may flow unobstructed into the inlets. Fan performance is reduced if the space between the fan inlet and the enclosure is too restrictive. The system effect coefficients for fans in an enclosure or adjacent to walls are listed under Fitting ED7-1 (see the section on Fitting Loss Coefficients). The manner in which the airstream enters an enclosure in relation to the fan inlets also affects fan performance. Plenum or enclosure inlets or walls that are not symmetrical with the fan inlets cause uneven flow and/or inlet spin.

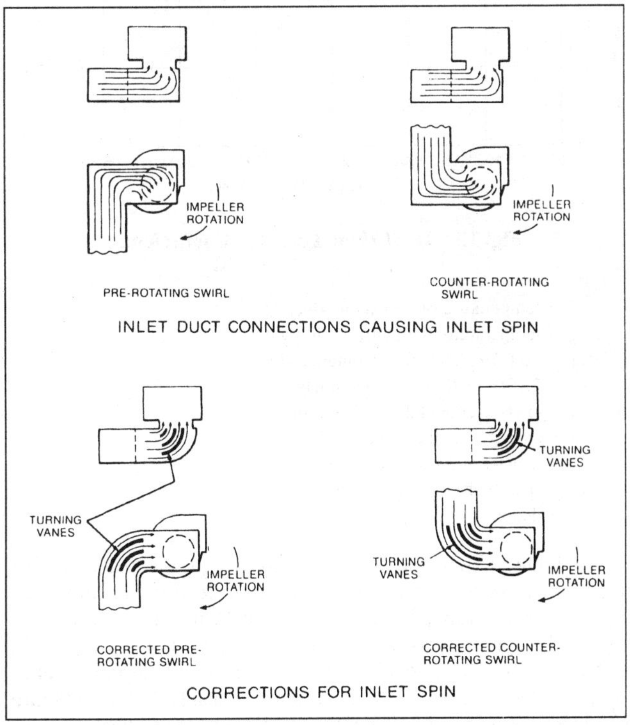

Fig. 12 Inlet Duct Connections Causing Inlet Spin and Corrections for Inlet Spin
(Adapted by permission from AMCA *Publication* 201)

Testing, Adjusting, and Balancing Considerations

Fan system effects (FSEs) are not only to be used in conjunction with the system resistance characteristics in the fan selection process, but are also applied in the calculations of the results of testing, adjusting, and balancing (TAB) field tests to allow direct comparison to design calculations and/or fan performance data. Fan inlet swirl and the effect on system performance of poor fan inlet and outlet ductwork connections cannot be measured directly. Poor inlet flow patterns affect fan performance within the impeller wheel (centrifugal fan) or wheel rotor impeller (axial fan), while the fan outlet system effect is flow instability and turbulence within the fan discharge ductwork.

The static pressure at the fan inlet and the static pressure at the fan outlet may be measured directly in some systems. In most cases, static pressure measurements for use in determining fan total (or static) pressure will not be made directly at the fan inlet and outlet, but at locations a relatively short distance from the fan inlet and downstream from the fan outlet. To calculate fan total pressure for this case from field measurements, use Equation (38), where $\Delta p_{x\text{-}y}$ is the summation of calculated total pressure losses between the fan inlet and outlet sections noted. Plane 3 is used to determine airflow rate. If necessary, use Equation (18) to calculate fan static pressure knowing fan total pressure. For locating measurement planes and calculation procedures, consult AMCA *Publication* 203 (AMCA 1990b).

$$\begin{aligned} P_t &= (p_{s,5} + p_{v,5}) + \Delta p_{2\text{-}5} + \text{FSE}_2 \\ &\quad + (p_{s,4} + p_{v,4}) + \Delta p_{4\text{-}1} + \text{FSE}_1 + \text{FSE}_{1,sw} \end{aligned} \tag{38}$$

where

P_t = fan total pressure, in. of water
p_s = static pressure, in. of water
p_v = velocity pressure, in. of water
FSE = fan system effect, in. of water
$\Delta p_{x\text{-}y}$ = summarization of total pressure losses between planes x and y, in. of water

Subscripts (numerical subscripts same as used by AMCA *Publication* 203):

1 = fan inlet
2 = fan outlet
3 = plane of airflow measurement
4 = plane of static pressure measurement upstream of fan
5 = plane of static pressure measurement downstream of fan
sw = swirl

DUCT SYSTEM DESIGN

DESIGN CONSIDERATIONS

Space Pressure Relationships

Space pressure is determined by fan location and duct system arrangement. For example, a supply fan that pumps air into a space increases space pressure; an exhaust fan reduces space pressure. If both supply and exhaust fans are used, space pressure depends on the relative capacity of the fans. Space pressure is positive if supply exceeds exhaust and negative if exhaust exceeds supply (Osborne 1966). System pressure variations due to wind can be minimized or eliminated by careful selection of intake air and exhaust vent locations (Chapter 15).

Fire and Smoke Management

Because duct systems can convey smoke, hot gases, and fire from one area to another and can accelerate a fire within the system, fire protection is an essential part of air-conditioning and ventilation system design. Generally, fire safety codes require compliance with the standards of national organizations. NFPA *Standard* 90A examines fire safety requirements for (1) ducts, connectors, and appurtenances; (2) plenums and corridors; (3) air outlets, air inlets, and fresh air intakes; (4) air filters; (5) fans; (6) electric wiring and equipment; (7) air-cooling and -heating equipment; (8) building construction, including protection of penetrations; and (9) controls, including smoke control.

Fire safety codes often refer to the testing and labeling practices of nationally recognized laboratories, such as Factory Mutual and Underwriters Laboratories (UL). The *Building Materials Directory* compiled by UL lists fire and smoke dampers that have been tested and meet the requirements of UL *Standards* 555 and 555S. This directory also summarizes maximum allowable sizes for individual dampers and assemblies of these dampers. Fire dampers are 1.5 h or 3 h fire-rated. Smoke dampers are classified by (1) temperature degradation [ambient air or high temperature (250°F minimum)] and (2) leakage at 1 and 4 in. of water pressure difference (8 and 12 in. of water classification optional). Smoke dampers are tested under conditions of maximum airflow. UL's *Fire Resistance Directory* lists the fire resistance of floor/roof and ceiling assemblies with and without ceiling fire dampers.

For a more detailed presentation of fire protection, see Chapter 48 of the 1995 *ASHRAE Handbook—Applications* and the NFPA *Fire Protection Handbook* (NFPA 1991).

Duct Insulation

In all new construction (except low-rise residential buildings), air-handling ducts and plenums installed as part of an HVAC air distribution system should be thermally insulated in accordance with Section 9.4 of ASHRAE *Standard* 90.1. Duct insulation for new low-rise residential buildings should be in compliance with ASHRAE *Standard* 90.2. Existing buildings should meet the requirements of ASHRAE *Standard* 100. The insulation thicknesses in these standards are minimum values. Economic considerations may justify higher insulation levels. Additional insulation, vapor retarders, or both may be required to limit vapor transmission and condensation.

Duct heat gains or losses must be known for the calculation of supply air quantities, supply air temperatures, and coil loads (see Chapter 28 of this volume and Chapter 2 of the 1996 *ASHRAE Handbook—Systems and Equipment*). To estimate duct heat transfer and entering or leaving air temperatures, use the following equations:

$$q_l = \frac{UPL}{12}\left[\left(\frac{t_e + t_l}{2}\right) - t_a\right] \quad (39)$$

$$t_e = \frac{t_l(y+1) - 2t_a}{(y-1)} \quad (40)$$

$$t_l = \frac{t_e(y-1) + 2t_a}{(y+1)} \quad (41)$$

where

q_l = heat loss/gain through duct walls, Btu/h (negative for heat gain)
U = overall heat transfer coefficient of duct wall, Btu/h·ft²·°F
P = perimeter of bare or insulated duct, in.
L = duct length, ft
t_e = temperature of air entering duct, °F

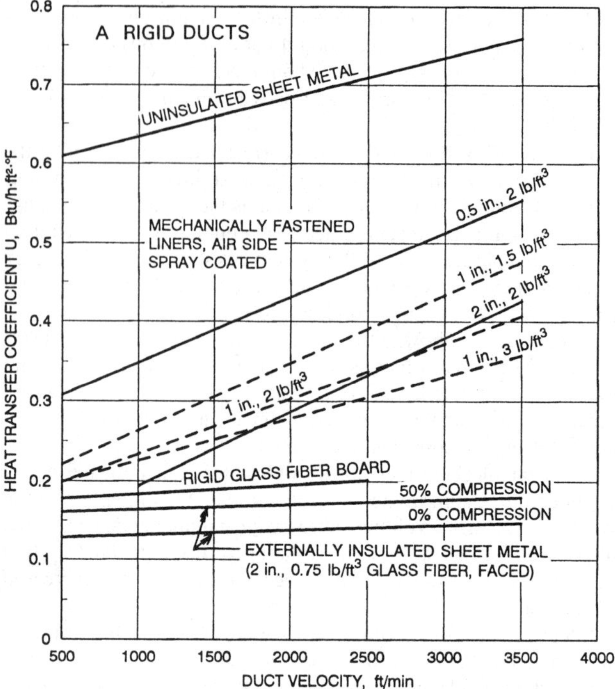

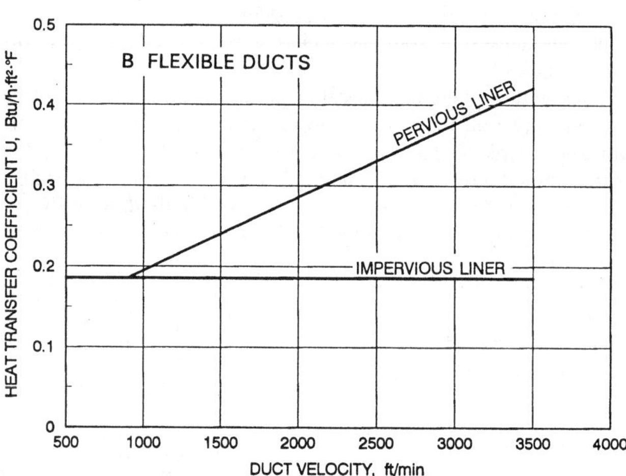

Fig. 13 Duct Heat Transfer Coefficients

t_l = temperature of air leaving duct, °F
t_a = temperature of air surrounding duct, °F
y = $10AV\rho c_p/UPL$ for rectangular ducts
= $2.5DV\rho c_p/UL$ for round ducts
A = cross-sectional area of duct, in²
V = average velocity, fpm
ρ = density of air, lb$_m$/ft³
c_p = specific heat of air, Btu/lb$_m$·°F
D = diameter of duct, in.

Use Figure 13A to determine U-factors for insulated and uninsulated ducts. Lauvray (1978) has shown the effects of (1) compressing insulation wrapped externally on sheet metal ducts and (2) insulated flexible ducts with air-porous liners. For a 2 in. thick, 0.75 lb$_m$/ft³ fibrous glass blanket compressed 50% during installation, the heat transfer rate increases approximately 20% (see Figure 13A). Pervious flexible duct liners also influence heat transfer significantly (see Figure 13B). At 2500 fpm, the pervious liner U-factor is 0.33 Btu/h·ft²·°F; for an impervious liner, U = 0.19 Btu/h·ft²·°F.

Example 6. A 65 ft length of 24 in. by 36 in. uninsulated sheet metal duct, freely suspended, conveys heated air through a space maintained above freezing at 40°F. Based on heat loss calculations for the heated zone, 17,200 cfm of standard air (c_p = 0.24 Btu/lb$_m$·°F) at a supply air temperature of 122°F is required. The duct is connected directly to the heated zone. Determine the temperature of the air entering the duct and the duct heat loss.

Solution: Calculate duct velocity using Equation (10):

$$V = \frac{(144)(17{,}200 \text{ cfm})}{(24 \text{ in.})(36 \text{ in.})} = 2900 \text{ fpm}$$

Calculate entering air temperature using Equation (40):

$$U = 0.73 \text{ Btu/h}\cdot\text{ft}^2\cdot{}^\circ\text{F (from Figure 13A)}$$

$$P = 2(24 \text{ in.} + 36 \text{ in.}) = 120 \text{ in.}$$

$$y = \frac{(10)(24 \text{ in.})(36 \text{ in.})(2900 \text{ fpm})(0.075 \text{ lb}_m/\text{ft}^3)(0.24)}{(0.73 \text{ Btu/h}\cdot\text{ft}^2\cdot{}^\circ\text{F})(120 \text{ in.})(65 \text{ ft})} = 79.2$$

$$t_e = \frac{122^\circ\text{F}(79.2+1)-(2\times 40^\circ\text{F})}{(79.2-1)} = 124.1^\circ\text{F}$$

Calculate duct heat loss using Equation (39):

$$q_l = \frac{(0.73 \text{ Btu/h}\cdot\text{ft}^2\cdot{}^\circ\text{F})(120 \text{ in.})(65 \text{ ft})}{12} \times \left[\frac{124.1^\circ\text{F}+122^\circ\text{F}}{2} - 40^\circ\text{F}\right] = 39{,}400 \text{ Btu/h}$$

Example 7. Same as Example 6, except the duct is insulated externally with 2 in. thick fibrous glass with a density of 0.75 lb$_m$/ft^3. The insulation is wrapped with 0% compression.

Solution: All values except U remain the same as in Example 6. From Figure 13A, U = 0.15 Btu/h·ft^2·°F at 2900 fpm. Therefore,

$$y = 385$$

$$t_e = 122.4^\circ\text{F}$$

$$q_l = 8014 \text{ Btu/h}$$

Insulating this duct reduces heat loss to 20% of the uninsulated value.

Duct System Leakage

Leakage in all unsealed ducts varies considerably with the fabricating machinery used, the methods for assembly, and installation workmanship. For sealed ducts, a wide variety of sealing methods and products exists. Sealed and unsealed duct leakage tests (AISI/SMACNA 1972, ASHRAE/SMACNA/TIMA 1985, Swim and Griggs 1995) have confirmed that longitudinal seam, transverse joint, and assembled duct leakage can be represented by Equation (42) and that for the same construction, leakage is not significantly different in the negative and positive modes. A range of leakage rates for longitudinal seams commonly used in the construction of metal ducts is presented in Table 5. Longitudinal seam leakage for unsealed or unwelded metal ducts is about 10 to 15% of total duct leakage.

Table 5 Unsealed Longitudinal Seam Leakage, Metal Ducts

Type of Duct/Seam	Leakage, cfm per ft Seam Length[a] Range	Average
Rectangular		
Pittsburgh lock	0.01 to 0.56	0.16
Button punch snaplock	0.01 to 0.16	0.08
Round		
Snaplock	0.04 to 0.14	0.11
Grooved	0.11 to 0.18	0.12

[a]Leakage rate is at 1 in. of water static pressure.

$$Q = C\Delta p_s^{\,N} \tag{42}$$

where

Q = duct leakage rate, cfm
C = constant reflecting area characteristics of leakage path
Δp_s = static pressure differential from duct interior to exterior, in. of water
N = exponent relating turbulent or laminar flow in leakage path

Analysis of the AISI/ASHRAE/SMACNA/TIMA data resulted in the categorization of duct systems into **leakage classes** C_L based on Equation (43), where the exponent N is assumed to be 0.65. A selected series of leakage classes based on Equation (43) is shown in Figure 14.

$$C_L = Q/\Delta p_s^{\,0.65} \tag{43}$$

where

Q = leakage rate, cfm/100 ft^2 (surface area)
C_L = leakage class, cfm per 100 ft duct surface at 1 in. of water static pressure

Table 6 is a forecast of the leakage class attainable for commonly used duct construction and sealing practices. Connections of ducts to grilles, diffusers, and registers are not represented in the test data. Leakage classes listed are for a specific duct type, not a system with a variety of duct types, access doors, and other duct-mounted equipment. The designer is responsible for assigning acceptable system leakage rates. It is recommended that this be accomplished by using Table 7 as a guideline to specify a ductwork leakage class or by specifying a duct seal level as recommended by Table 8. The designer should take into account attainable leakage rates by duct type and the fact that casings of volume-controlling air terminal units may leak 2 to 5% of their maximum flow. The effects of such leakage should be anticipated, if allowed, and the ductwork should not be expected to compensate for equipment leakage. When a system leakage class is specified by a designer, it is a performance specification that should not be compromised by prescriptive sealing. A portion of a system may exceed its leakage class if the aggregate

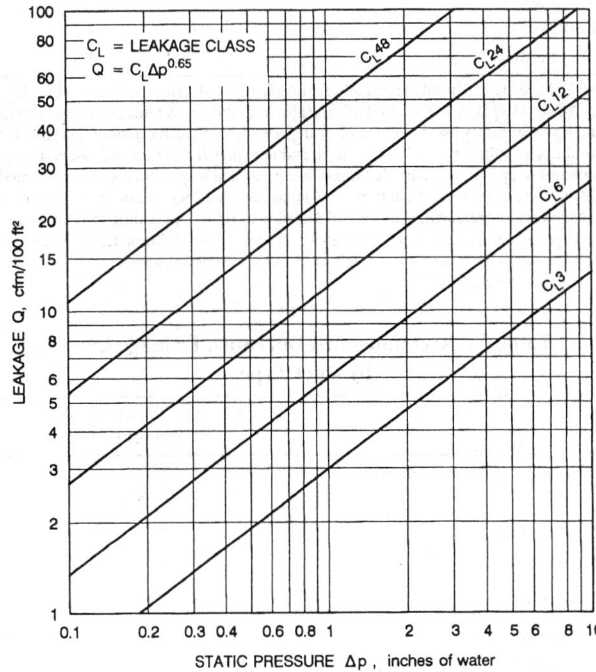

Fig. 14 Duct Leakage Classifications

system leakage meets the allowable rate. Table 9 can be used to estimate the system percent leakage based on the system design leakage class and system duct surface area. Table 9 is predicated on assessment at an average of upstream and downstream pressures because use of the highest pressure alone could indicate an artificially high rate. When several duct pressure classifications occur in a system, ductwork in each pressure class should be evaluated independently to arrive at an aggregate leakage for the system.

Leakage tests should be conducted in compliance with SMACNA's *HVAC Air Duct Leakage Test Manual* (1985) to verify the intent of the designer and the workmanship of the the installing contractor. Leakage tests used to confirm leakage class should be conducted at the pressure class for which the duct is constructed. Leakage testing is also addressed in ASHRAE *Standard* 90.1.

Pressure-sensitive tape should not be used as the primary sealant for sheet metal ducts designed to operate at static pressures in excess of 1 in. of water. UL *Standard* 181A rigid ducts may use UL *Standard* 181A listed pressure-sensitive tapes or other listed closure materials up to the pressure listing of the material. Soldered or

Table 6 Duct Leakage Classification[a]

Duct Type	Predicted Leakage Class C_L [Eq. (43)] Sealed[b,c]	Unsealed[c]
Metal (flexible excluded)		
Round and flat oval	3	30 (6 to 70)
Rectangular		
≤ 2 in. of water (both positive and negative pressures)	12	48 (12 to 110)
> 2 and ≤ 10 in. of water (both positive and negative pressures)	6	48 (12 to 110)
Flexible		
Metal, aluminum	8	30 (12 to 54)
Nonmetal	12	30 (4 to 54)
Fibrous glass		
Round	3	na
Rectangular	6	na

[a]The leakage classes listed in this table are averages based on tests conducted by AISI/SMACNA (1972), ASHRAE/SMACNA/TIMA (1985), and Swim and Griggs (1995).
[b]The leakage classes listed in the sealed category are based on the assumptions that for metal ducts, all transverse joints, seams, and openings in the duct wall are sealed at pressures over 3 in. of water, that transverse joints and longitudinal seams are sealed at 2 and 3 in. of water, and that transverse joints are sealed below 2 in. of water. Lower leakage classes are obtained by careful selection of joints and sealing methods.
[c]Leakage classes assigned anticipate about 25 joints per 100 linear feet of duct. For systems with a high fitting to straight duct ratio, greater leakage occurs in both the sealed and unsealed conditions.

Table 7 Recommended Ductwork Leakage Class by Duct Type

Duct Type	Leakage Class, cfm/100 ft² at 1 in. of water
Metal (flexible excluded)	
Round	3
Flat oval	3
Rectangular	6
Flexible	6
Fibrous glass	
Round	3
Rectangular	6

Table 8A Recommended Duct Seal Levels[a]

Duct Location	Supply ≤ 2 in. of water	Supply > 2 in. of water	Exhaust	Return
Outdoors	A	A	A	A
Unconditioned spaces	B	A	B	B
Conditioned spaces (concealed ductwork)	C	B	B	B
Conditioned spaces (exposed ductwork)				
Office-type spaces	A	A	B	B
Factory-type spaces	C	B	B	B

[a]See Table 8B for definition of seal level.

Table 8B Duct Seal Levels

Seal Level	Sealing Requirements[a]
A	All transverse joints, longitudinal seams, and duct wall penetrations
B	All transverse joints and longitudinal seams
C	Transverse joints only

[a]Transverse joints are connections of two duct or fitting elements oriented perpendicular to flow. Longitudinal seams are joints oriented in the direction of airflow. Duct wall penetrations are openings made by screws, non-self-sealing fasteners, pipe, tubing, rods, and wire. Round and flat oval spiral lock seams need not be sealed prior to assembly but may be coated after assembly to reduce leakage. All other connections are considered transverse joints, including but not limited to spin-ins, taps and other branch connections, access door frames, and duct connection to equipment.

Table 9 Leakage as Percentage of Airflow[a,b]

Leakage Class	System cfm per ft² Duct Surface	Static Pressure, in. of water 0.5	1	2	3	4	6
48	2	15	24	38	49	59	77
	2.5	12	19	30	39	47	62
	3	10	16	25	33	39	51
	4	7.7	12	19	25	30	38
	5	6.1	9.6	15	20	24	31
24	2	7.7	12	19	25	30	38
	2.5	6.1	9.6	15	20	24	31
	3	5.1	8.0	13	16	20	26
	4	3.8	6.0	9.4	12	15	19
	5	3.1	4.8	7.5	9.8	12	15
12	2	3.8	6	9.4	12	15	19
	2.5	3.1	4.8	7.5	9.8	12	15
	3	2.6	4.0	6.3	8.2	9.8	13
	4	1.9	3.0	4.7	6.1	7.4	9.6
	5	1.5	2.4	3.8	4.9	5.9	7.7
6	2	1.9	3	4.7	6.1	7.4	9.6
	2.5	1.5	2.4	3.8	4.9	5.9	7.7
	3	1.3	2.0	3.1	4.1	4.9	6.4
	4	1.0	1.5	2.4	3.1	3.7	4.8
	5	0.8	1.2	1.9	2.4	3.0	3.8
3	2	1.0	1.5	2.4	3.1	3.7	4.8
	2.5	0.8	1.2	1.9	2.4	3.0	3.8
	3	0.6	1.0	1.6	2.0	2.5	3.2
	4	0.5	0.8	1.3	1.6	2.0	2.6
	5	0.4	0.6	0.9	1.2	1.5	1.9

[a]Adapted with permission from HVAC *Air Duct Leakage Test Manual* (SMACNA 1985, Appendix A).
[b]Percentage applies to the airflow entering a section of duct operating at an assumed pressure equal to the average of the upstream and downstream pressures.
[c]The ratios in this column are typical of fan volumetric flow rate divided by total system surface. Portions of the systems may vary from these averages.

welded duct construction is necessary where sealants are not suitable. Sealants used on exterior ducts must be resistant to weather, temperature cycles, sunlight, and ozone.

Shaft and compartment pressure changes affect duct leakage and are important to health and safety in the design and operation of contaminant and smoke control systems. Shafts should not be used for supply, return, and/or exhaust air without accounting for their leakage rates. Airflow around buildings, building component leakage, and the distribution of inside and outside pressures over the height of a building, including shafts, are discussed in Chapters 15 and 24. Smoke management system design is covered in Chapter 48 of the 1995 *ASHRAE Handbook—Applications* and in Klote and Milke (1992).

System Component Design Velocities

Table 10 summarizes face velocities for HVAC components in built-up systems. In most cases, the values are abstracted from pertinent chapters in the 1996 *ASHRAE Handbook—Systems and Equipment*; final selection of the components should be based on data in these chapters or from manufacturers.

Louvers require special treatment since the blade shapes, angles, and spacing cause significant variations in louver-free area and performance (pressure drop and water penetration). Selection and analysis should be based on test data obtained in accordance with AMCA *Standard* 500. This standard presents both pressure drop and water penetration test procedures and a uniform method for calculating louver-free area. Tests are conducted on a 48 in. square louver with the frame mounted flush in the wall. For the water penetration tests, the rainfall is 4 in/h, no wind, and the water flow down the wall is 0.25 gpm per linear foot of louver width.

Use Figure 15 for preliminary sizing of air intake and exhaust louvers. For air quantities greater than 7000 cfm per louver, the air intake gross louver openings are based on 400 fpm; for exhaust louvers, 500 fpm is used for air quantities of 5000 cfm per louver and greater. For air quantities less than these, refer to Figure 15. These criteria are presented on a per louver basis (i.e., each louver in a bank of louvers) to include each louver frame. Representative production-run louvers were used in establishing Figure 15, and all data used in that analysis are based on AMCA standard tests. For louvers larger than 16 ft^2, the free areas are greater than 45%, while for louvers less than 16 ft^2, the free areas are less than 45%. Unless specific louver data are analyzed, no louver should have a face area less than 4 ft^2. If debris collection on the screen of an intake louver is possible, or if louvers are located at grade with adjacent pedestrian traffic, louver face velocity should not exceed 100 fpm.

Table 10 Typical Design Velocities for HVAC Components

Duct Element	Face Velocity, fpm
LOUVERS[a]	
Intake	
7000 cfm and greater	400
Less than 7000 cfm	See Figure 15
Exhaust	
5000 cfm and greater	500
Less than 5000 cfm	See Figure 15
FILTERS[b]	
Panel filters	
Viscous impingement	200 to 800
Dry-type, extended-surface	
Flat (low efficiency)	Duct velocity
Pleated media (intermediate efficiency)	Up to 750
HEPA	250
Renewable media filters	
Moving-curtain viscous impingement	500
Moving-curtain dry media	200
Electronic air cleaners	
Ionizing type	150 to 350
HEATING COILS[c]	
Steam and hot water	500 to 1000 200 min., 1500 max.
Electric	
Open wire	Refer to mfg. data
Finned tubular	Refer to mfg. data
DEHUMIDIFYING COILS[d]	400 to 500
AIR WASHERS[e]	
Spray type	300 to 600
Cell type	Refer to mfg. data
High-velocity spray type	1200 to 1800

[a]Based on assumptions presented in text.
[b]Abstracted from Chapter 24, 1996 *ASHRAE Handbook—Systems and Equipment*.
[c]Abstracted from Chapter 23, 1996 *ASHRAE Handbook—Systems and Equipment*.
[d]Abstracted from Chapter 21, 1996 *ASHRAE Handbook—Systems and Equipment*.
[e]Abstracted from Chapter 19, 1996 *ASHRAE Handbook—Systems and Equipment*.

System and Duct Noise

The major sources of noise from air-conditioning systems are diffusers, grilles, fans, ducts, fittings, and vibrations. Chapter 43 of the 1995 *ASHRAE Handbook—Applications* discusses sound control for each of these sources. Sound control for terminal devices consists of selecting devices that meet the design goal under all operating conditions and installing them properly so that no additional sound is generated. The sound power output of a fan is determined by the type of fan, airflow, and pressure. Sound control in the duct system requires proper duct layout, sizing, and provision for installing duct attenuators, if required. The noise generated by a system increases with both duct velocity and system pressure. Chapter 43 of the 1995 *ASHRAE Handbook—Applications* presents methods for calculating required sound attenuation.

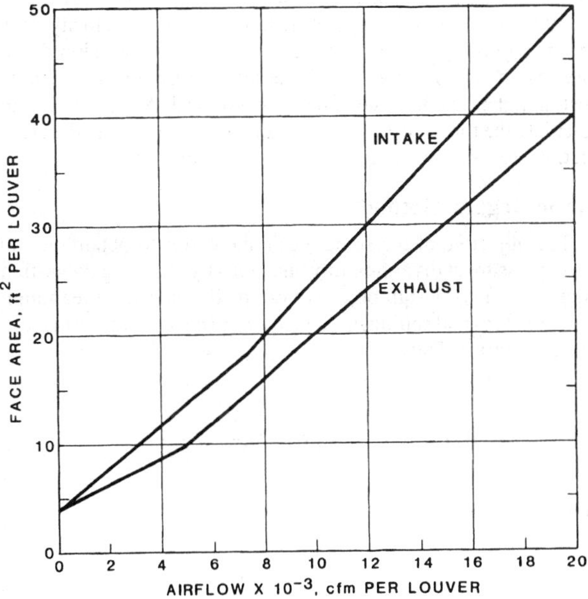

Parameters Used to Establish Figure	Intake Louver	Exhaust Louver
Minimum free area (48 in. square test section), %	45	45
Water penetration, oz/(ft^2·0.25 h)	Negligible (less than 0.2)	na
Maximum static pressure drop, in. of water	0.15	0.25

Fig. 15 Criteria for Louver Sizing

Testing and Balancing

Each air duct system should be tested, adjusted, and balanced. Detailed procedures are given in Chapter 34 of the 1995 *ASHRAE Handbook—Applications*. To properly determine fan total (or static) pressure from field measurements taking into account fan system effect, refer to the section on Fan-System Interface. Equation (38) allows direct comparison of system resistance to design calculations and/or fan performance data. It is important that the system effect magnitudes be known prior to testing. If necessary, use Equation (18) to calculate fan static pressure knowing fan total pressure [Equation (38)]. For TAB calculation procedures of numerous fan/system configurations encountered in the field, refer to AMCA *Publication* 203 (AMCA 1990b).

DUCT DESIGN METHODS

Duct design methods for HVAC systems and for exhaust systems conveying vapors, gases, and smoke are the equal friction method, the static regain method, and the T-method. The section on Industrial Exhaust System Duct Design presents the design criteria and procedures for exhaust systems conveying particulates. Equal friction and static regain are nonoptimizing methods, while the T-method is a practical optimization method introduced by Tsal et al. (1988).

To ensure that system designs are acoustically acceptable, noise generation should be analyzed and sound attenuators and/or acoustically lined duct provided where necessary. Dampers must be installed throughout systems designed by equal friction, static regain, and the T-method because inaccuracies are introduced into these design methods by duct size round-off and the effect of close-coupled fittings on the total pressure loss calculations.

Equal Friction Method

In the equal friction method, ducts are sized for a constant pressure loss per unit length. The shaded area of the friction chart (Figure 9) is the suggested range of friction rate and air velocity. When energy cost is high and installed ductwork cost is low, a low friction rate design is more economical. For low energy cost and high duct cost, a higher friction rate is more economical. After initial sizing, calculate the total pressure loss for all duct sections, and then resize sections to balance pressure losses at each junction.

Static Regain Method

The objective of the static regain method is to obtain the same static pressure at diverging flow junctions by changing downstream duct sizes. This design objective can be developed by rearranging Equation (7a) and setting $p_{s,2}$ equal to $p_{s,1}$ (neglecting thermal gravity effect term). Thus,

$$p_{s,1} - p_{s,2} = \Delta p_{t,1\text{-}2} - \left[\frac{\rho V_1^2}{2g_c} - \frac{\rho V_2^2}{2g_c}\right] \quad (44)$$

and

$$\Delta p_{t,1\text{-}2} = \frac{\rho V_1^2}{2g_c} - \frac{\rho V_2^2}{2g_c} \quad (45)$$

where $\Delta p_{t,1\text{-}2}$ is the total pressure loss from upstream of junction 1 to upstream of junction 2, or the terminal of section 2. The immediate downstream duct size that satisfies Equation (45) is determined by iteration.

To start the design of a system, a maximum velocity is selected for the root section (duct section upstream and/or downstream of a fan). In Figure 17, section 6 is the root for the return air subsystem. Section 19 is the root for the supply air subsystem. The shaded area on the friction chart (Figure 9) is the suggested range of air velocity. When energy cost is high and installed ductwork cost is low, a lower initial velocity is more economical. For low energy cost and high duct cost, a higher velocity is more economical. All other sections, except terminal sections, are sized iteratively by Equation (45). In Figure 17, terminal sections are 1, 2, 4, 7, 8, 11, 12, 15, and 16. Knowing the terminal static pressure requirements, Equation (45) is used to calculate the duct size of terminal sections. If the terminal is an exit fitting rather than a register, diffuser, or terminal box, the static pressure at the exit of the terminal section is zero.

The classical static regain method (Carrier Corporation 1960, Chun-Lun 1983) is based on Equation (46), where R is the static pressure regain factor, and Δp_r is the static pressure regain between junctions.

$$\Delta p_r = R\left(\frac{\rho V_1^2}{2g_c} - \frac{\rho V_2^2}{2g_c}\right) \quad (46)$$

Typically R-values ranging from 0.5 to 0.95 have been used. Tsal and Behls (1988) show that this uncertainty exists because the splitting of mass at junctions and the dynamic (fitting) losses between junctions are ignored. The classical static regain method using an R-value should not be used because R is not predictable.

T-Method Optimization

T-method optimization (Tsal et al. 1988) is a dynamic programming procedure based on the tee-staging idea used by Bellman (1957), except that phase level vector tracing is eliminated by optimizing locally at each stage. This modification reduces the number of calculations but requires iteration.

Optimization Basis. The objective function, Equation (47), includes both initial system cost and the present worth of energy. Hours of operation, annual escalation and interest rates, and amortization period are also required for optimization.

$$E = E_p(\text{PWEF}) + E_s \quad (47)$$

where

E = present worth owning and operating cost
E_p = first year energy cost
E_s = initial cost
PWEF = present worth escalation factor (Smith 1968), dimensionless

Energy cost is determined by

$$E_p = Q_f\left[\frac{1.176 \times 10^4(E_d + E_c T)}{\eta_f \eta_e}\right]P_t \quad (48)$$

where

Q_f = fan airflow rate, cfm
E_c = unit energy cost, cost/kWh
E_d = energy demand cost, cost/kW
T = system operating time, h/year
P_t = fan total pressure, in. of water
η_f = fan total efficiency, decimal
η_e = motor-drive efficiency, decimal

Energy cost depends on both applicable energy rates E_c and demand cost E_d. Since the difference in fan pressure between an optimized and a nonoptimized system is a small part of demand, it is usually neglected. Initial cost includes ducts and HVAC equipment, which is primarily the central handling unit. The cost of duct systems is given by the following equations:

Round $$E_s = S_d \pi DL/12 \tag{49}$$

Rectangular $$E_s = 2S_d(H + W)L/12 \tag{50}$$

where

S_d = unit ductwork cost/ft^2 (including material and labor)
H = duct height, in.
W = duct width, in.
L = duct length, ft

The cost of space required by ducts and equipment is another important factor of duct optimization. Including this cost reduces the size of ducts, thereby increasing energy consumption. Because the space available for ductwork is usually not used for anything else, its cost is ignored.

Both electrical energy rates and ductwork costs vary widely, by a factor of up to eight times for industrial users (DOE). Black iron rectangular ductwork can cost about 3.9 times that of spiral ductwork (Wendes 1989). Combining these ratios yields a factor of 30 to 1 based on locale and type of ductwork. Therefore, a great potential exists for reducing duct system life-cycle cost due to energy and ductwork cost variations.

The following constraints are necessary for duct optimization (Tsal and Adler 1987):

- *Continuity*. For each node, the flow in equals the flow out.
- *Pressure balancing*. The total pressure loss in each path must equal the fan total pressure; or, in effect, at any junction, the total pressure loss for all paths is the same.
- *Nominal duct size*. Ducts are constructed in discrete, nominal sizes. Each diameter of a round duct or height and width of a rectangular duct is rounded to the nearest increment, usually 1 or 2 in. If a lower nominal size is selected, the initial cost decreases, but the pressure loss increases and may exceed the fan pressure. If the higher nominal size is selected, the opposite is true—the initial cost increases, but the section pressure loss decreases. However, this lower pressure at one section may allow smaller ducts to be selected for sections that follow. Therefore, optimization must consider size rounding.
- *Air velocity restriction*. The maximum allowable velocity is an acoustic limitation (ductwork regenerated noise).
- *Construction restriction*. Architectural limits may restrict duct sizes. If air velocity or construction constraints are violated during an iteration, a duct size must be calculated. The pressure loss calculated for this preselected duct size is considered a fixed loss.

Calculation Procedure. The T-method comprises the following major procedures:

- *System condensing*. This procedure condenses a multiple-section duct system into a single imaginary duct section with identical hydraulic characteristics and the same owning cost as the entire system. By Equation (1.41) in Tsal et al. (1988), two or more converging or diverging sections and the common section at a junction can be replaced by one condensed section. By applying this equation from junction to junction in the direction to the root section (fan), the entire supply and return systems can be condensed into one section (a single resistance).
- *Fan selection*. From the condensed system, the ideal optimum fan total pressure P_t^{opt} is calculated and used to select a fan. If a fan with a different pressure is selected, its pressure P^{opt} is considered optimum.
- *System expansion*. The expansion process distributes the available fan pressure P^{opt} throughout the system. Unlike the condensing procedure, the expansion procedure starts at the root section and continues in the direction of the terminals.

Economic Analysis. Tsal et al. (1988) describe the calculation procedure and include an economic analysis of the T-method.

T-Method Simulation

T-method simulation, also developed by Tsal et al. (1990), determines the flow in each duct section of an existing system with a known operating fan performance curve. The simulation version of the T-method converges very efficiently. Usually three iterations are sufficient to obtain a solution with a high degree of accuracy.

Calculation procedure. The simulation version of the T-method includes the following major procedures:

- *System condensing*. This procedure condenses a branched tee system into a single imaginary duct section with identical hydraulic characteristics. Two or more converging or diverging sections and the common section at a junction can be replaced by one condensed section [by Equation (18) in Tsal et al. (1990)]. By applying this equation from junction to junction in the direction to the root section (fan), the entire system, including supply and return subsystems, can be condensed into one imaginary section (a single resistance).
- *Fan operating point*. This step determines the system flow and pressure by locating the intersection of the fan performance and system curves, where the system curve is represented by the imaginary section from the last step.
- *System expansion*. Knowing system flow and pressure, the previously condensed imaginary duct section is expanded into the original system with flow distributed in accordance with the ratio of pressure losses calculated in the system condensing step.

Simulation Applications. The need for duct system simulation appears in many HVAC problems. In addition to the following concerns that can be clarified by simulation, the T-method is an excellent design tool for simulating the flow distribution within a system with various modes of operation.

- Flow distribution in a variable air volume (VAV) system due to terminal box flow diversity
- Airflow redistribution due to HVAC system additions and/or modifications
- System airflow analysis for partially occupied buildings
- Necessity to replace fans and/or motors when retrofitting an air distribution system
- Multiple-fan system operating condition when one or more fans shut down
- Pressure differences between adjacent confined spaces within a nuclear facility when a design basis accident (DBA) occurs (Farajian et al. 1992)
- Smoke management system performance during a fire, when certain fire/smoke dampers close and others remain open

HVAC DUCT DESIGN PROCEDURES

The general procedure for HVAC system duct design is as follows:

1. Study the building plans, and arrange the supply and return outlets to provide proper distribution of air within each space. Adjust calculated air quantities for duct heat gains or losses and duct leakage. Also, adjust the supply, return, and/or exhaust air quantities to meet space pressurization requirements.
2. Select outlet sizes from manufacturers' data (see Chapter 31).
3. Sketch the duct system, connecting supply outlets and return intakes with the air-handling units/air conditioners. Space allocated for supply and return ducts often dictates system layout and ductwork shape. Use round ducts whenever feasible.
4. Divide the system into sections and number each section. A duct system should be divided at all points where flow, size, or shape changes. Assign fittings to the section toward the supply and return (or exhaust) terminals. The following examples are for the fittings identified for Example 6 (Figure 16), and system section numbers assigned (Figure 17). For converging flow fitting 3,

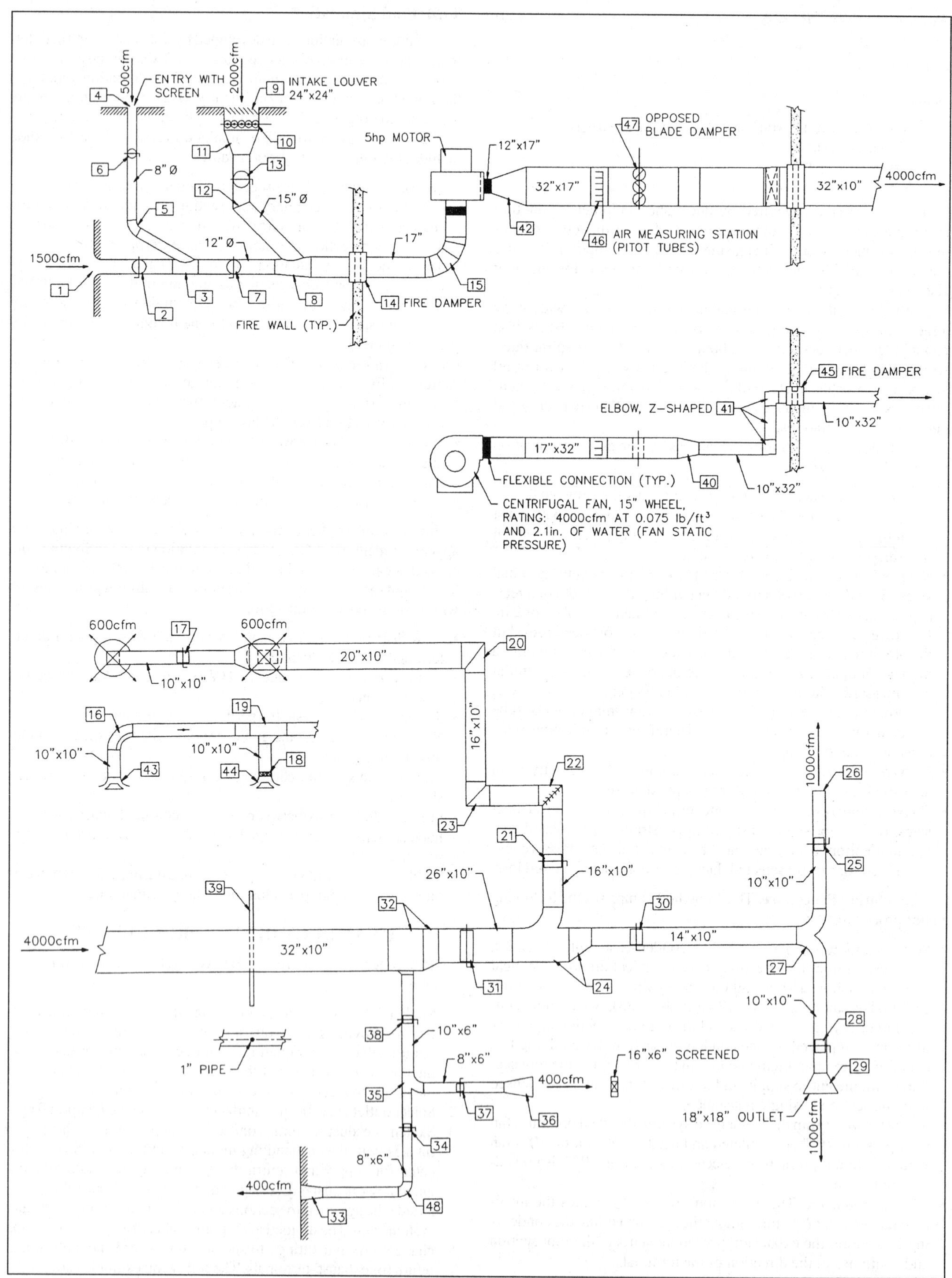

Fig. 16 Schematic for Example 8

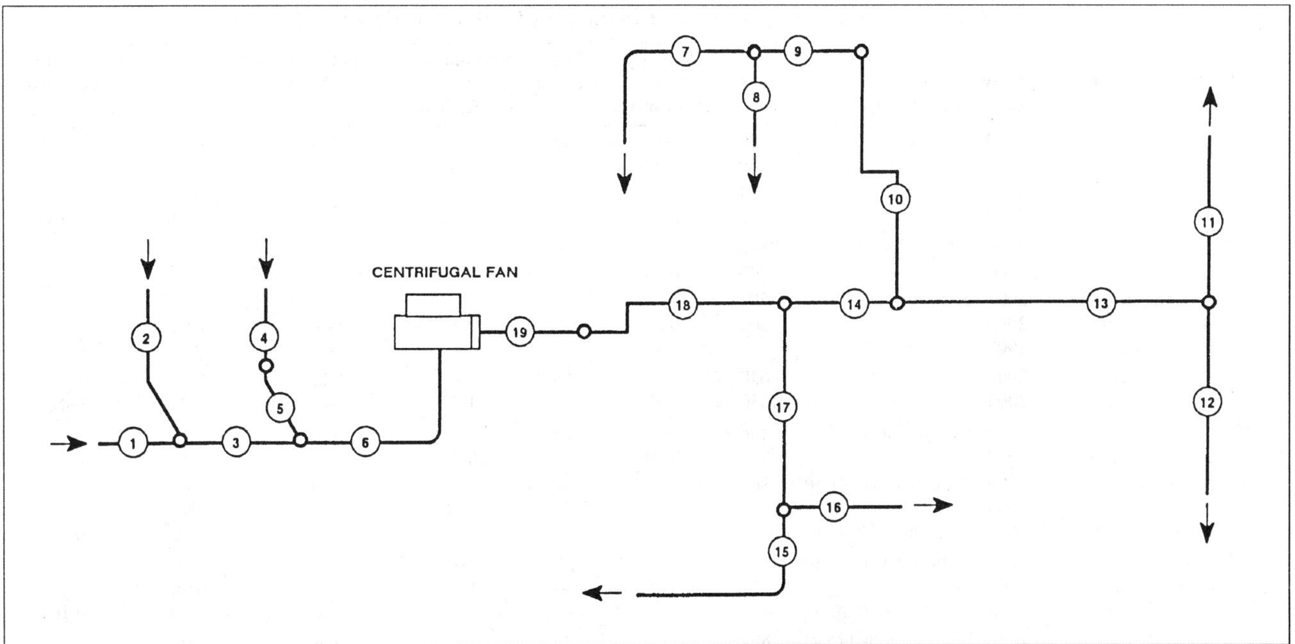

Fig. 17 System Schematic with Section Numbers for Example 8

assign the straight-through flow to section 1 (toward terminal 1), and the branch to section 2 (toward terminal 4). For diverging flow fitting 24, assign the straight-through flow to section 13 (toward terminals 26 and 29) and the branch to section 10 (toward terminals 43 and 44). For transition fitting 11, assign the fitting to upstream section 4 [toward terminal 9 (intake louver)]. For fitting 20, assign the unequal area elbow to downstream section 9 (toward diffusers 43 and 44). The fan outlet diffuser, fitting 42, is assigned to section 19 (again, toward the supply duct terminals).

5. Size ducts by the selected design method. Calculate system total pressure loss; then select the fan (refer to Chapter 18 of the 1996 *ASHRAE Handbook—Systems and Equipment*).
6. Lay out the system in detail. If duct routing and fittings vary significantly from the original design, recalculate the pressure losses. Reselect the fan if necessary.
7. Resize duct sections to approximately balance pressures at each junction.
8. Analyze the design for objectionable noise levels, and specify sound attenuators as necessary. Refer to the section on System and Duct Noise.

Example 8. For the system illustrated by Figures 16 and 17, size the ductwork by the equal friction method, and pressure balance the system by changing duct sizes (use 1 in. increments). Determine the system resistance and total pressure unbalance at the junctions. The airflow quantities are actual values adjusted for heat gains or losses, and ductwork is sealed (assume no leakage), galvanized steel ducts with transverse joints on 4 ft centers (ε = 0.0003 ft). Air is at standard conditions (0.075 lb_m/ft^3 density).

Because the primary purpose of Figure 16 is to illustrate calculation procedures, its duct layout is not typical of any real duct system. The layout includes fittings from the local loss coefficient tables, with emphasis on converging and diverging tees and various types of entries and discharges. The supply system is constructed of rectangular ductwork; the return system, round ductwork.

Solution: See Figure 17 for section numbers assigned to the system. The duct sections are sized within the suggested range of friction rate shown on the friction chart (Figure 9). Tables 11 and 12 give the total pressure loss calculations and the supporting summary of loss coefficients by sections. The straight duct friction factor and pressure loss were calculated by Equations (19) and (20). The fitting loss coefficients are from the *Duct Fitting Database* (ASHRAE 1994). Loss coefficients were calculated automatically by the database program (not by manual interpolation). The pressure loss values in Table 11 for the diffusers (fittings 43 and 44), the louver (fitting 9), and the air-measuring station (fitting 46) are manufacturers' data.

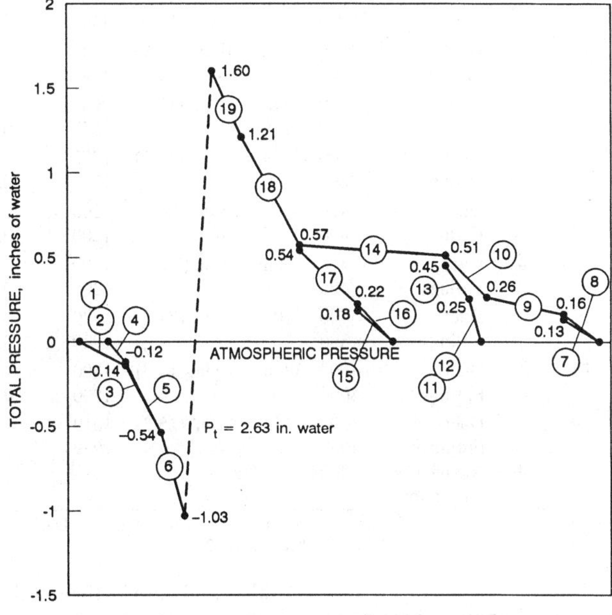

Fig. 18 Total Pressure Grade Line for Example 8

The pressure unbalance at the junctions may be noted by referring to Figure 18, the total pressure grade line for the system. The system resistance P_t is 2.63 in. of water. Noise levels and the need for duct silencers were not evaluated. To calculate the fan static pressure, use Equation (18):

$$P_s = 2.63 - 0.50 = 2.1 \text{ in. of water}$$

where 0.50 in. of water is the fan outlet velocity pressure.

Table 11 Total Pressure Loss Calculations by Sections for Example 8

Duct Section[a]	Fitting No.[b]	Duct Element	Airflow, cfm	Duct Size (Equivalent Round)	Velocity, fpm	Velocity Pressure, in. of water	Duct Length,[c] ft	Summary of Fitting Loss Coefficients[d]	Duct Pressure Loss/100 ft,[e] in. of water	Total Pressure Loss, in. of water	Section Pressure Loss, in. of water
1	—	Duct	1500	12 in. ϕ	1910	—	15	—	0.40	0.06	
	—	Fittings	1500	—	1910	0.23	—	0.33	—	0.08	0.14
2	—	Duct	500	8 in. ϕ	1432	—	60	—	0.39	0.23	
	—	Fittings	500	—	1432	0.13	—	−0.71	—	−0.09	0.14
3	—	Duct	2000	12 in. ϕ	2546	—	20	—	0.67	0.13	
	—	Fittings	2000	—	2546	0.40	—	0.67	—	0.27	0.40
4	—	Duct	2000	24 in. × 24 in. (26.2)	500	—	5	—	0.01	0.00	
	—	Fittings	2000	—	500	0.02	—	1.11	—	0.02	
	9	Louver	2000	24 in. × 24 in.	—	—	—	—	—	0.10[f]	0.12
5	—	Duct	2000	15 in. ϕ	1630	—	55	—	0.23	0.13	
	—	Fittings	2000	—	1630	0.17	—	1.73	—	0.29	0.42
6	—	Duct	4000	17 in. ϕ	2538	—	30	—	0.45	0.14	
	—	Fittings	4000	—	2538	0.40	—	0.87	—	0.35	0.49
7	—	Duct	600	10 in. × 10 in. (10.9)	864	—	14	—	0.11	0.02	
	—	Fittings	600	—	864	0.05	—	0.26	—	0.01	
	43	Diffuser	600	10 in. × 10 in.	—	—	—	—	—	0.10[f]	0.13
8	—	Duct	600	10 in. × 10 in. (10.9)	864	—	4	—	0.11	0.00	
	—	Fittings	600	—	864	0.05	—	1.25	—	0.06	
	44	Diffuser	600	10 in. × 10 in.	—	—	—	—	—	0.10[f]	0.16
9	—	Duct	1200	20 in. × 10 in. (15.2)	864	—	25	—	0.08	0.02	
	—	Fittings	1200	—	864	0.05	—	1.67	—	0.08	0.10
10	—	Duct	1200	16 in. × 10 in. (13.7)	1080	—	45	—	0.13	0.06	
	—	Fittings	1200	—	1080	0.07	—	2.66	—	0.19	0.25
11	—	Duct	1000	10 in. × 10 in. (10.9)	1440	—	10	—	0.29	0.03	
	—	Fittings	1000	—	1440	0.13	—	1.68	—	0.22	0.25
12	—	Duct	1000	10 in. × 10 in. (10.9)	1440	—	22	—	0.29	0.06	
	—	Fittings	1000	—	1440	0.13	—	1.45	—	0.19	0.25
13	—	Duct	2000	14 in. × 10 in. (12.9)	2057	—	35	—	0.47	0.16	
	—	Fittings	2000	—	2057	0.26	—	0.16	—	0.04	0.20
14	—	Duct	3200	26 in. × 10 in. (17.1)	1772	—	15	—	0.27	0.04	
	—	Fittings	3200	—	1772	0.20	—	0.12	—	0.02	0.06
15	—	Duct	400	8 in. × 6 in. (7.6)	1200	—	40	—	0.32	0.13	
	—	Fittings	400	—	1200	0.09	—	0.58	—	0.05	0.18
16	—	Duct	400	8 in. × 6 in. (7.6)	1200	—	20	—	0.32	0.06	
	—	Fittings	400	—	1200	0.09	—	1.74	—	0.16	0.22
17	—	Duct	800	10 in. × 6 in. (8.4)	1920	—	22	—	0.70	0.15	
	—	Fittings	800	—	1920	0.23	—	0.76	—	0.17	0.32
18	—	Duct	4000	32 in. × 10 in. (18.8)	1800	—	23	—	0.25	0.06	
	—	Fittings	4000	—	1800	0.20	—	2.91	—	0.58	0.64
19	—	Duct	4000	32 in. × 17 in. (25.2)	1059	—	12	—	0.06	0.01	
	—	Fittings	4000	—	1059	0.07	—	4.71	—	0.33	
	46	Air-measuring station	4000	—	—	—	—	—	—	0.05[f]	0.39

[a]See Figure 17.
[b]See Figure 16.
[c]Duct lengths are to fitting centerlines.
[d]See Table 12.
[e]Duct pressure based on a 0.0003 ft absolute roughness factor.
[f]Pressure drop based on manufacturers' data.

Table 12 Loss Coefficient Summary by Sections for Example 8

Duct Section	Fitting Number	Type of Fitting	ASHRAE Fitting No.[a]	Parameters	Loss Coefficient
1	1	Entry	ED1-3	$r/D = 0.2$	0.03
	2	Damper	CD9-1	$\theta = 0°$	0.19
	3	Wye (30°), main	ED5-1	$A_s/A_c = 1.0, A_b/A_c = 0.444, Q_s/Q_c = 0.75$	0.11 (C_s)
Summation of Section 1 loss coefficients					0.33
2	4	Entry	ED1-1	$L = 0, t = 0.064$ in. (16 gage)	0.50
	4	Screen	CD6-1	$n = 0.70, A_1/A_o = 1$	0.58
	5	Elbow	CD3-6	60°, $r/D = 1.5$, pleated	0.27
	6	Damper	CD9-1	$\theta = 0°$	0.19
	3	Wye (30°), branch	ED5-1	$A_s/A_c = 1.0, A_b/A_c = 0.444, Q_b/Q_c = 0.25$	−2.25 (C_b)
Summation of Section 2 loss coefficients					−0.71
3	7	Damper	CD9-1	$\theta = 0°$	0.19
	8	Wye (45°), main	ED5-2	$A_s/A_c = 0.498, A_b/A_c = 0.779, Q_s/Q_c = 0.5$	0.48 (C_s)
Summation of Section 3 loss coefficients					0.67

[a]*Duct Fitting Database* (ASHRAE 1994) data for fittings reprinted in the section on Fitting Loss Coefficients.

Table 12 Loss Coefficient Summary by Sections for Example 8 (*Concluded*)

Duct Section	Fitting Number	Type of Fitting	ASHRAE Fitting No.[a]	Parameters	Loss Coefficient
4	10	Damper	CR9-4	$\theta = 0°$, 5 blades (opposed), $L/R = 1.25$	0.52
	11	Transition	ER4-3	$L = 30$ in., $A_o/A_1 = 3.26$, $\theta = 17°$	0.59
Summation of Section 4 loss coefficients					1.11
5	12	Elbow	CD3-17	45°, mitered	0.34
	13	Damper	CD9-1	$\theta = 0°$	0.19
	8	Wye (45°), branch	ED5-2	$Q_b/Q_c = 0.5$, $A_s/A_c = 0.498$, $A_b/A_c = 0.779$	1.20 (C_b)
Summation of Section 5 loss coefficients					1.73
6	14	Fire damper	CD9-3	Curtain type, Type C	0.12
	15	Elbow	CD3-9	90°, 5 gore, $r/D = 1.5$	0.15
	—	Fan and system interaction	ED7-2	90° elbow, 5 gore, $r/D = 1.5$, $L = 34$ in.	0.60
Summation of Section 6 loss coefficients					0.87
7	16	Elbow	CR3-3	90°, $r/W = 0.70$, 1 splitter vane	0.14
	17	Damper	CR9-1	$\theta = 0°$, $H/W = 1.0$	0.08
	19	Tee, main	SR5-13	$Q_s/Q_c = 0.5$, $A_s/A_c = 0.50$	0.04 (C_s)
Summation of Section 7 loss coefficients					0.26
8	19	Tee, branch	SR5-13	$Q_b/Q_c = 0.5$, $A_b/A_c = 0.50$	0.73 (C_b)
	18	Damper	CR9-4	$\theta = 0°$, 3 blades (opposed), $L/R = 0.75$	0.52
Summation of Section 8 loss coefficients					1.25
9	20	Elbow	SR3-1	90°, mitered, $H/W_1 = 0.625$, $W_o/W_1 = 1.25$	1.67
Summation of Section 9 loss coefficients					1.67
10	21	Damper	CR9-1	$\theta = 0°$, $H/W = 0.625$	0.08
	22	Elbow	CR3-10	90°, single-thickness vanes, design 2	0.12
	23	Elbow	CR3-6	$\theta = 90°$, mitered, $H/W = 0.625$	1.25
	24	Tee, branch	SR5-1	$r/W_b = 1.0$, $Q_b/Q_c = 0.375$, $A_s/A_c = 0.538$, $A_b/A_c = 0.615$	1.21 (C_b)
Summation of Section 10 loss coefficients					2.66
11	25	Damper	CR9-1	$\theta = 0°$, $H/W = 1.0$	0.08
	26	Exit	SR2-1	$H/W = 1.0$, Re = 122,500	1.00
	27	Wye, dovetail	SR5-14	$r/W_c = 1.5$, $Q_{b1}/Q_c = 0.5$, $A_{b1}/A_c = 0.714$	0.60 (C_b)
Summation of Section 11 loss coefficients					1.68
12	28	Damper	CR9-1	$\theta = 0°$, $H/W = 1.0$	0.08
	29	Exit	SR2-5	$\theta = 19°$, $A_1/A_o = 3.24$, Re = 130,000	0.77
	27	Wye, dovetail	SR5-14	$r/W_c = 1.5$, $Q_{b2}/Q_c = 0.5$, $A_{b2}/A_c = 0.714$	0.60 (C_b)
Summation of Section 12 loss coefficients					1.45
13	30	Damper	CR9-1	$\theta = 0°$, $H/W = 0.71$	0.08
	24	Tee, main	SR5-1	$r/W_b = 1.0$, $Q_s/Q_c = 0.625$, $A_s/A_c = 0.538$, $A_b/A_c = 0.615$	0.08 (C_s)
Summation of Section 13 loss coefficients					0.16
14	31	Damper	CR9-1	$\theta = 0°$, $H/W = 0.38$	0.08
	32	Tee, main	SR5-13	$Q_s/Q_c = 0.8$, $A_s/A_c = 0.813$	0.04 (C_s)
Summation of Section 14 loss coefficients					0.12
15	48	Elbow	CR3-1	$\theta = 90°$, $r/W = 1.5$, $H/W = 0.75$	0.19
	33	Exit	SR2-6	$L = 18$ in., $D_h = 6.86$	0.28
	34	Damper	CR9-1	$\theta = 0°$, $H/W = 0.75$	0.08
	35	Tee, main	SR5-1	$r/W_b = 1.0$, $Q_s/Q_c = 0.5$, $A_s/A_c = 0.80$, $A_b/A_c = 0.80$	0.03 (C_s)
Summation of Section 15 loss coefficients					0.58
16	36	Exit	SR2-3	$\theta = 20°$, $A_1/A_o = 2.0$, Re = 70,000	0.63
	36	Screen	CR6-1	$n = 0.8$, $A_1/A_o = 2.0$	0.08
	37	Damper	CR9-1	$\theta = 0°$, $H/W = 0.75$	0.08
	35	Tee, branch	SR5-1	$r/W_b = 1.0$, $Q_b/Q_c = 0.5$, $A_s/A_c = 0.80$, $A_b/A_c = 0.80$	0.95 (C_b)
Summation of Section 16 loss coefficients					1.74
17	38	Damper	CR9-1	$\theta = 0°$, $H/W = 0.6$	0.08
	32	Tee, branch	SR5-13	$Q_b/Q_c = 0.2$, $A_b/A_c = 0.187$	0.68 (C_b)
Summation of Section 17 loss coefficients					0.76
18	39	Obstruction, pipe	CR6-4	Re = 15,000, $y = 0$, $d = 1$ in., $S_m/A_o = 0.1$, $y/H = 0$	0.17
	40	Transition	SR4-1	$\theta = 22°$, $A_o/A_1 = 0.588$, $L = 18$ in.	0.04
	41	Elbows, Z-shaped	CR3-17	$L = 42$ in., $L/W = 4.2$, $H/W = 3.2$, Re = 240,000	2.51
	45	Fire damper	CR9-6	Curtain type, Type B	0.19
Summation of Section 18 loss coefficients					2.91
19	42	Diffuser, fan	SR7-17	$\theta_1 = 28°$, $L = 40$ in., $A_o/A_1 = 2.67$, $C_1 = 0.59$	4.19 (C_o)
	47	Damper	CR9-4	$\theta = 0°$, 8 blades (opposed), $L/R = 1.39$	0.52
Summation of Section 19 loss coefficients					4.71

[a]*Duct Fitting Database* (ASHRAE 1994) data for fittings reprinted in the section on Fitting Loss Coefficients.

INDUSTRIAL EXHAUST SYSTEM DUCT DESIGN

Chapter 26 of the 1995 *ASHRAE Handbook—Applications* discusses design criteria, including hood design, for industrial exhaust systems. Exhaust systems conveying vapors, gases, and smoke can be designed by equal friction, static regain, or T-method. Systems conveying particulates are designed by the constant velocity method at duct velocities adequate to convey particles to the system air cleaner. For contaminant transport velocities, see Table 2 in Chapter 26 of the 1995 *ASHRAE Handbook—Applications*.

Two pressure-balancing methods can be considered when designing industrial exhaust systems. One method uses balancing devices (e.g., dampers, blast gates) to obtain design airflow through each hood. The other approach balances systems by adding resistance to ductwork sections (i.e., changing duct size, selecting different fittings, and increasing airflow). This self-balancing method is preferred, especially for systems conveying abrasive materials. Where potentially explosive or radioactive materials are conveyed, the prebalanced system is mandatory because contaminants could accumulate at the balancing devices. To balance systems by increasing airflow, use Equation (51), which assumes that all ductwork has the same diameter and that fitting loss coefficients, including main and branch tee coefficients, are constant.

$$Q_c = Q_d(P_h/P_l)^{0.5} \tag{51}$$

where

Q_c = airflow rate required to increase P_l to P_h, cfm
Q_d = total airflow rate through low-resistance duct run, cfm
P_h = absolute value of pressure loss in high-resistance ductwork section(s), in. of water
P_l = absolute value of pressure loss in low-resistance ductwork section(s), in. of water

For systems conveying particulates, use elbows with a large centerline radius-to-diameter ratio (r/D), greater than 1.5 whenever possible. If r/D is 1.5 or less, abrasion in dust-handling systems can reduce the life of elbows. Elbows are often made of seven or more gores, especially in large diameters. For converging flow fittings, a 30° entry angle is recommended to minimize energy losses and abrasion in dust-handling systems. For the entry loss coefficients of hoods and equipment for specific operations, refer to Chapter 26 of the 1995 *ASHRAE Handbook—Applications* and to ACGIH (1995).

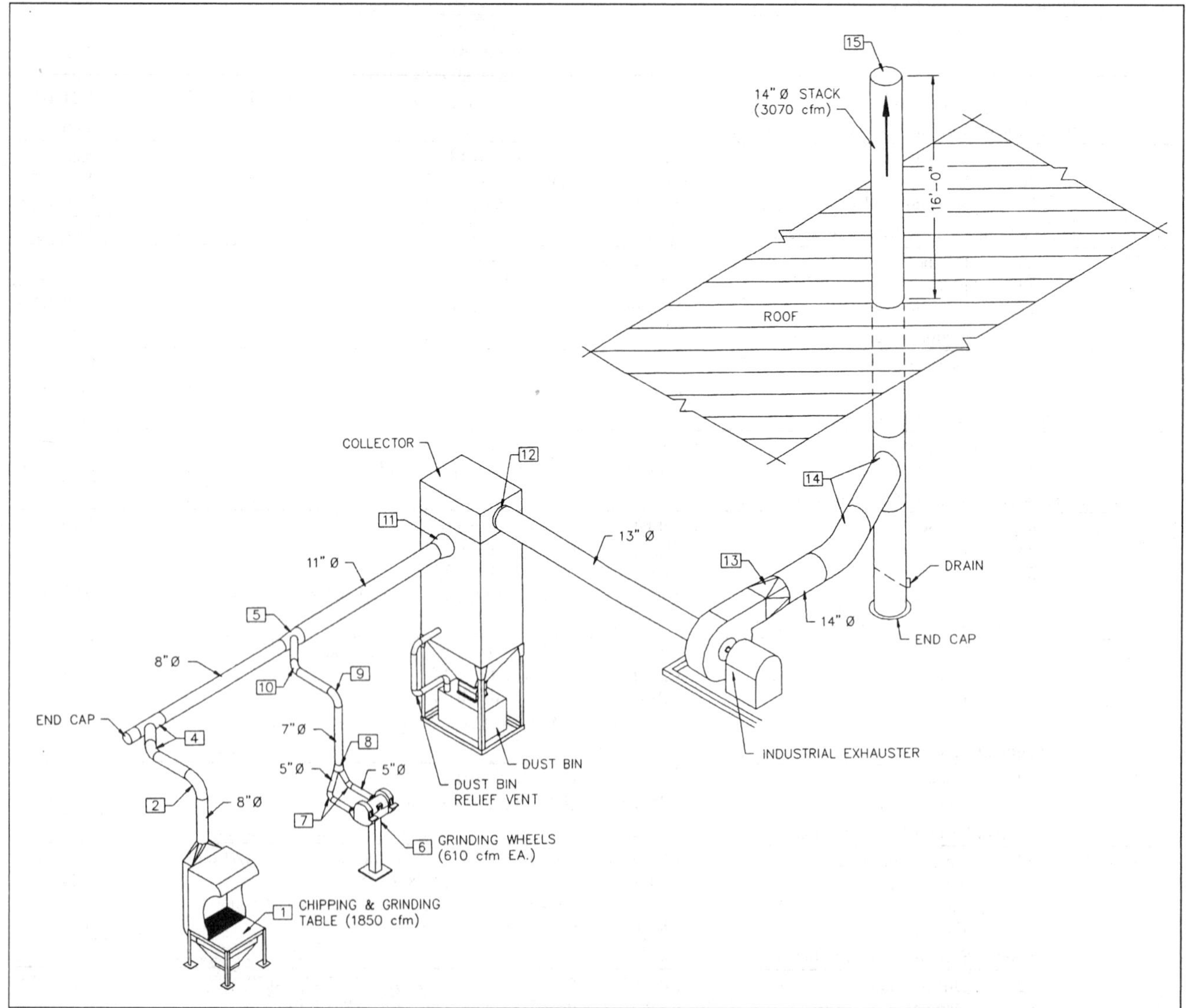

Fig. 19 Metalworking Exhaust System for Example 9

Example 9. For the metalworking exhaust system in Figures 19 and 20, size the ductwork and calculate the fan static pressure requirement for an industrial exhaust designed to convey granular materials. Pressure balance the system by changing duct sizes and adjusting airflow rates. The minimum particulate transport velocity for the chipping and grinding table ducts (sections 1 and 5, Figure 20) is 4000 fpm. For the ducts associated with the grinder wheels (sections 2, 3, 4, and 5), the minimum duct velocity is 4500 fpm. Ductwork is galvanized steel, with the absolute roughness being 0.0003 ft. Assume that air is standard and that duct and fittings are available in the following sizes: 3 in. through 9.5 in. diameters in 0.5 in. increments, 10 in. through 37 in. diameters in 1 in. diameter increments, and 38 in. through 90 in. diameters in 2 in. diameter increments.

The building is one story, and the design wind velocity is 20 mph. For the stack, use Design J shown in Figure 13 in Chapter 15 for complete rain protection. The stack height, determined by calculations from Chapter 15, is 16 ft above the roof. This stack height is based on minimized stack downwash; therefore, the stack discharge velocity must exceed 1.5 times the design wind velocity.

Solution: For the contaminated ducts upstream of the collector, initial duct sizes and transport velocities are summarized below. The 4474 fpm velocity in sections 2 and 3 is acceptable because the transport velocity is not significantly lower than 4500 fpm. For the next available duct size (4.5 in. diameter), the duct velocity is 5523 fpm, significantly higher than 4500 fpm.

Duct Section	Design Airflow, cfm	Transport Velocity, fpm	Duct Diameter, in.	Duct Velocity, fpm
1	1800	4000	9	4074
2,3	610 each	4500	5	4474
4	1220	4500	7	4565
5	3020	4500	11	4576

The following tabulation summarizes design calculations up through the junction after sections 1 and 4.

Design No.	D_1, in.	Δp_1, in. of water	Δp_{2+4}, in. of water	Imbalance, $\Delta p_1 - \Delta p_{2+4}$
1	9	1.46	3.09	−1.63
2	8.5	2.00	3.08	−1.08
3	8	2.79	3.00	−0.21
4	7.5	3.92	2.88	+1.04

Q_1 = 1800 cfm
Q_2 = 610 cfm; D_2 = 5 in. dia.
Q_3 = 610 cfm; D_3 = 5 in. dia.
Q_4 = 1220 cfm; D_4 = 7 in. dia.

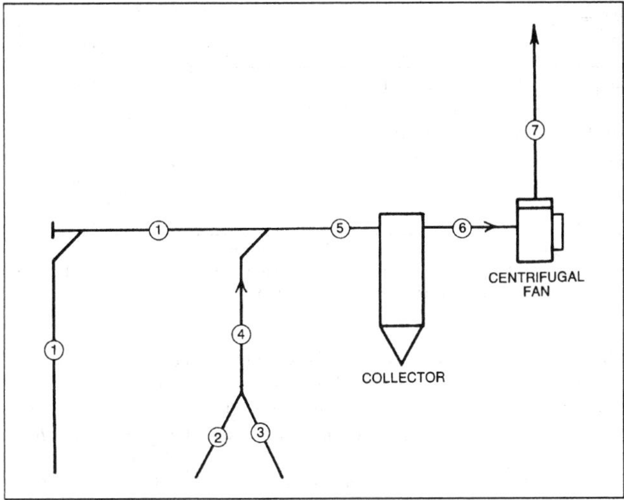

Fig. 20 System Schematic with Section Numbers for Example 9

For the initial design, Design 1, the imbalance between section 1 and section 2 (or 3) is 1.63 in. of water, with section 1 requiring additional resistance. Decreasing section 1 duct diameter by 0.5 in. increments results in the least imbalance, 0.21 in. of water, when the duct diameter is 8 in. (Design 3). Because section 1 requires additional resistance, estimate the new airflow rate using Equation (51):

$$Q_{c,1} = (1800)(3.00/2.79)^{0.5} = 1870 \text{ cfm}$$

At 1870 cfm flow in section 1, 0.13 in. of water imbalance remains at the junction of sections 1 and 4. By trial-and-error solution, balance is attained when the flow in section 1 is 1850 cfm. The duct between the collector and the fan inlet is 13 in. round to match the fan inlet (12.75 in. diameter). To minimize downwash, the stack discharge velocity must exceed 2640 fpm, 1.5 times the design wind velocity (20 mph) as stated in the problem definition. Therefore, the stack is 14 in. round, and the stack discharge velocity is 2872 fpm.

Table 13 summarizes the system losses by sections. The straight duct friction factor and pressure loss were calculated by Equations (19) and (20). Table 14 lists fitting loss coefficients and input parameters necessary to determine the loss coefficients. The fitting loss coefficients are from the *Duct Fitting Database* (ASHRAE 1994). The fitting loss coefficient tables are included in the section on Fitting Loss Coefficients for illustration but can not be obtained exactly by manual interpolation since the coefficients were calculated by the duct fitting database algorithms (more significant figures). For a pressure grade line of the system, see Figure 21. The fan total pressure, calculated by Equation (16), is 7.89 in. of water. To calculate the fan static pressure, use Equation (18):

$$P_s = 7.89 - 0.81 = 7.1 \text{ in. of water}$$

where 0.81 in. of water is the fan outlet velocity pressure. The fan airflow rate is 3070 cfm, and its outlet area is 0.853 ft^3 (10.125 in. by 12.125 in.). Therefore, the fan outlet velocity is 3600 fpm.

The hood suction for the chipping and grinding table hood is 2.2 in. of water, calculated by Equation (19) from Chapter 26 of the 1995 *ASHRAE Handbook—Applications* [$HS = (1 + 0.25)(1.74) = 2.2$ in. of water, where 0.25 is the hood entry loss coefficient C_o, and 1.74 is the duct velocity pressure P_v a few diameters downstream from the hood]. Similarly, the hood suction for each of the grinder wheels is 1.7 in. of water:

$$HS_{2,3} = (1 + 0.4)(1.24) = 1.7 \text{ in. of water}$$

where 0.4 is the hood entry loss coefficient, and 1.24 in. of water is the duct velocity pressure.

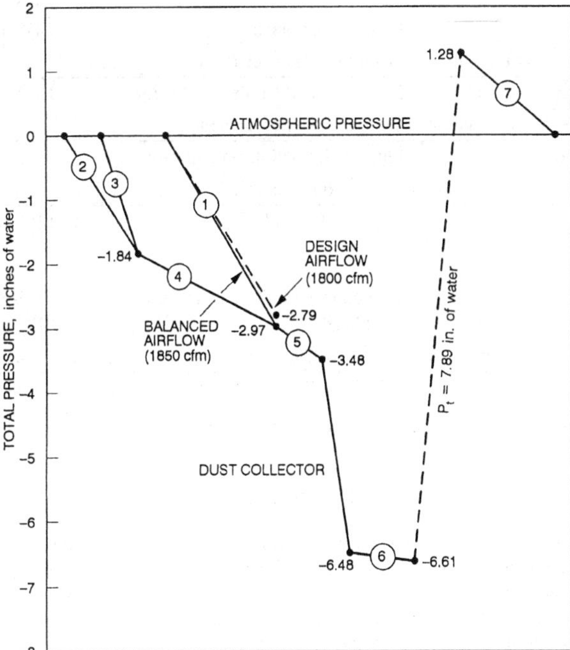

Fig. 21 Total Pressure Grade Line for Example 9

Table 13 Total Pressure Loss Calculations by Sections for Example 9

Duct Section[a]	Duct Element	Airflow, cfm	Duct Size	Velocity, fpm	Velocity Pressure, in. of water	Duct Length,[b] ft	Summary of Fitting Loss Coefficients[c]	Duct Pressure Loss/100 ft, in. of water[d]	Total Pressure Loss, in. of water	Section Pressure Loss, in. of water
1	Duct	1850	8 in. ϕ	5300	—	22.5	—	4.63	1.04	
	Fittings	1850		5300	1.75	—	1.10	—	1.93	2.97
2,3	Duct	610	5 in. ϕ	4474	—	9	—	5.94	0.53	
	Fittings	610		4474	1.24	—	1.06	—	1.31	1.84
4	Duct	1220	7 in. ϕ	4565	—	11.5	—	4.08	0.47	
	Fittings	1220		4565	1.30	—	0.51	—	0.66	1.13
5	Duct	3070	11 in. ϕ	4652	—	8.5	—	2.44	0.21	
	Fittings	3070		4652	1.35	—	0.22	—	0.30	0.51
—	Collector,[e] fabric	3070	—	—	—	—	—	—	3.0	3.0
6	Duct	3070	13 in. ϕ	3331	—	12	—	1.05	0.13	
	Fittings	3070		3331	0.69	—	0.00	—	0.00	0.13
7	Duct	3070	14 in. ϕ	2872	—	29	—	0.72	0.21	
	Fittings	3070		2872	0.51	—	2.09	—	1.07	1.28

[a]See Figure 20.
[b]Duct lengths are to fitting centerlines.
[c]See Table 14.
[d]Duct pressure based on a 0.0003 ft absolute roughness factor.
[e]Collector manufacturers set the fabric bag cleaning mechanism to actuate at a pressure difference of 3.0 in. of water between the inlet and outlet plenums. The pressure difference across the clean media is approximately 1.5 in. of water.

Table 14 Loss Coefficient Summary by Sections for Example 9

Duct Section	Fitting Number	Type of Fitting	ASHRAE Fitting No.[a]	Parameters	Loss Coefficient
1	1	Hood[b]	—	Hood face area: 3 ft by 4 ft	0.25
	2	Elbow	CD3-10	90°, 7 gore, $r/D = 2.5$	0.11
	4	Capped wye (45°), with 45° elbow	ED5-6	$A_b/A_c = 1$	0.64 (C_b)
	5	Wye (30°), main	ED5-1	$Q_s/Q_c = 0.60$, $A_s/A_c = 0.529$, $A_b/A_c = 0.405$	0.10 (C_s)
Summation of Section 1 loss coefficients					1.10
2,3	6	Hood[c]	—	Type hood: For double wheels, dia. = 22 in. each, wheel width = 4 in. each; type takeoff: tapered	0.40
	7	Elbow	CD3-12	90°, 3 gore, $r/D = 1.5$	0.34
	8	Symmetrical wye (60°)	ED5-9	$Q_b/Q_c = 0.5$, $A_b/A_c = 0.51$	0.32 (C_b)
Summation of Sections 2 and 3 loss coefficients					1.06
4	9	Elbow	CD3-10	90°, 7 gore, $r/D = 2.5$	0.11
	10	Elbow	CD3-13	60°, 3 gore, $r/D = 1.5$	0.19
	5	Wye (30°), branch	ED5-1	$Q_b/Q_c = 0.40$, $A_s/A_c = 0.529$, $A_b/A_c = 0.405$	0.21 (C_b)
Summation of Section 4 loss coefficients					0.51
5	11	Exit, conical diffuser to collector	ED2-1	$L = 24$ in., $L/D_o = 2.18$, $A_1/A_o \approx 16$	0.22
Summation of Section 5 loss coefficients					0.22
6	12	Entry, bellmouth from collector	ER2-1	$r/D_1 = 0.20$	0.00 (C_1)
Summation of Section 6 loss coefficients					0.00
7	13	Diffuser, fan outlet[d]	SR7-17	Fan outlet size: 10.125 in. by 12.125 in., $A_o/A_1 = 1.596$ (assume 14 in. by 14 in. outlet rather than 16 in. round), $L = 18$ in.	0.45 (C_o)
	14	Capped wye (45°), with 45° elbow	ED5-6	$A_b/A_c = 1$	0.64 (C_b)
	15	Stackhead	SD2-6	$D_e/D = 1$	1.0
Summation of Section 7 loss coefficients					2.09

[a]*Duct Fitting Database* (ASHRAE 1994) data for fittings reprinted in the section on Fitting Loss Coefficients.
[b]From *Industrial Ventilation* (ACGIH 1995, Figure VS-80-19).
[c]From *Industrial Ventilation* (ACGIH 1995, Figure VS-80-11).
[d]Fan specified: Industrial exhauster for granular materials: 21 in. wheel diameter, 12.75 in. inlet diameter, 10.125 in. by 12.125 in. outlet, 7.5 hp motor.

REFERENCES

ACGIH. 1995. *Industrial ventilation: A manual of recommended practice*, 22nd ed. American Conference of Governmental Industrial Hygienists, Lansing, MI.

AISI/SMACNA. 1972. Measurement and analysis of leakage rates from seams and joints of air handling systems.

Altshul, A.D., L.C. Zhivotovckiy, and L.P. Ivanov. 1987. *Hydraulics and aerodynamics*. Stroisdat Publishing House, Moscow.

AMCA. 1989. Test methods for louvers, dampers and shutters. *Standard* 500. Air Movement and Control Association, Arlington Heights, IL.

AMCA. 1990a. Fans and systems. *Publication* 201.

AMCA. 1990b. Field performance measurement of fan systems. *Publication* 203.

ASHRAE. 1985. Laboratory methods of testing fans for rating. ANSI/ASHRAE *Standard* 51-1985. Also AMCA *Standard* 210-85.

ASHRAE. 1989. Energy efficient design of new buildings except low-rise residential buildings. ASHRAE/IESNA *Standard* 90.1-1989.

ASHRAE. 1993. Energy-efficient design of new low-rise residential buildings. ASHRAE *Standard* 90.2-1993.

ASHRAE. 1994. *Duct fitting database.*

ASHRAE. 1995. Energy conservation in existing buildings. ASHRAE/IESNA *Standard* 100-1995. Addendum 1-1996.

ASHRAE/SMACNA/TIMA. 1985. Investigation of duct leakage. ASHRAE *Research Project* 308.

Behls, H.F. 1971. Computerized calculation of duct friction. *Building Science Series* 39, p. 363. National Institute of Standards and Technology, Gaithersburg, MD.

Bellman, R.E. 1957. *Dynamic programming.* Princeton University Press, New York.

Brown, R.B. 1973. Experimental determinations of fan system effect factors. In *Fans and systems,* ASHRAE Symposium Bulletin LO-73-1, Louisville, KY (June).

Carrier Corporation. 1960. Air duct design. Chapter 2 in *System design manual,* Part 2: Air distribution. pp.17-63. Syracuse, NY.

Chun-Lun, S. 1983. Simplified static-regain duct design procedure. *ASHRAE Transactions* 89(2A):78.

Clarke, M.S., J.T. Barnhart, F.J. Bubsey, and E. Neitzel. 1978. The effects of system connections on fan performance. *ASHRAE Transactions* 84(2): 227-63.

Colebrook, C.F. 1938-39. Turbulent flow in pipes, with particular reference to the transition region between the smooth and rough pipe laws. *Journal of the Institution of Civil Engineers* 11:133. London.

DOE. *Electrical sales and revenue,* latest edition. Department of Energy, Washington, D.C. (To purchase, call Energy Information Administration 202-512-1800.)

Farajian, T., G. Grewal, and R.J. Tsal. 1992. Post-accident air leakage analysis in a nuclear facility via T-method airflow simulation. 22nd DOE/NRC Nuclear Air Cleaning and Treatment Conference, Denver, CO, October.

Farquhar, H.F. 1973. System effect values for fans. In *Fans and systems,* ASHRAE Symposium Bulletin LO-73-1, Louisville, KY (June).

Griggs, E.I. and F. Khodabakhsh-Sharifabad. 1992. Flow characteristics in rectangular ducts. *ASHRAE Transactions* 98(1).

Griggs, E.I., W.B. Swim, and G.H. Henderson. 1987. Resistance to flow of round galvanized ducts. *ASHRAE Transactions* 93(1):3-16.

Heyt, J.W. and M.J. Diaz. 1975. Pressure drop in flat-oval spiral air duct. *ASHRAE Transactions* 81(2):221-32.

Huebscher, R.G. 1948. Friction equivalents for round, square and rectangular ducts. *ASHVE Transactions* 54:101-18.

Hutchinson, F.W. 1953. Friction losses in round aluminum ducts. *ASHVE Transactions* 59:127-38.

Idelchik, I.E., G.R. Malyavskaya, O.G. Martynenko, and E. Fried. 1986. *Handbook of hydraulic resistance,* 2nd ed. Hemisphere Publishing Corp., subsidiary of Harper & Row, New York.

Jones, C.D. 1979. Friction factor and roughness of United Sheet Metal Company spiral duct. United Sheet Metal, Division of United McGill Corp., Westerville, OH (August). Based on data contained in Friction loss tests, United Sheet Metal Company Spiral Duct, Ohio State University Engineering Experiment Station, File No. T-1011, September, 1958.

Klote, J.H. and J. Milke. 1992. *Design of smoke management systems.* ASHRAE, Atlanta.

Lauvray, T.L. 1978. Experimental heat transmission coefficients for operating air duct systems. *ASHRAE Journal* (June):69.

Meyer, M.L. 1973. A new concept: The fan system effect factor. In *Fans and systems,* ASHRAE Symposium Bulletin LO-73-1, Louisville, KY (June).

Moody, L.F. 1944. Friction factors for pipe flow. *ASME Transactions* 66:671.

NFPA. 1991. *Fire protection handbook,* 17th ed. National Fire Protection Association, Quincy, MA.

NFPA. 1993. Installation of air conditioning and ventilating systems. ANSI/NFPA *Standard* 90A-93.

Osborne, W.C. 1966. *Fans.* Pergamon Press Ltd., London.

SMACNA. 1985. *HVAC air duct leakage manual.* Sheet Metal and Air Conditioning Contractors' National Association, Chantilly, VA.

Smith, G.W. 1968. *Engineering economy: Analysis of capital expenditures.* The Iowa State University Press, Ames, IA.

Swim, W.B. 1978. Flow losses in rectangular ducts lined with fiberglass. *ASHRAE Transactions* 84(2):216.

Swim, W.B. 1982. Friction factor and roughness for airflow in plastic pipe. *ASHRAE Transactions* 88(1):269.

Swim, W.B. and E.I. Griggs. 1995. Duct leakage measurement and analysis. *ASHRAE Transactions* 101(1).

Tsal, R.J. and M.S. Adler. 1987. Evaluation of numerical methods for ductwork and pipeline optimization. *ASHRAE Transactions* 93(1):17-34.

Tsal, R.J. and H.F. Behls. 1988. Fallacy of the static regain duct design method. *ASHRAE Transactions* 94(2):76-89.

Tsal, R.J., H.F. Behls, and R. Mangel. 1988. T-method duct design, Part I: Optimization theory; Part II: Calculation procedure and economic analysis. *ASHRAE Transactions* 94(2):90-111.

Tsal, R.J., H.F. Behls, and R. Mangel. 1990. T-method duct design, Part III: Simulation. *ASHRAE Transactions* 96(2).

UL. Published annually. *Building materials directory.* Underwriters Laboratories, Northbrook, IL.

UL. Published annually. *Fire resistance directory.*

UL. 1991. Closure systems for use with rigid air ducts and air connectors. *Standard* 181A-91.

UL. 1995. Fire dampers, 5th ed. *Standard* 555-95.

UL. 1995. Leakage rated dampers for use in smoke control systems, 2nd ed. *Standard* 555S-95.

Wendes, H.C. 1989. *Sheet metal estimating.* Wendes Mechanical Consulting Services, Elk Grove Village, IL.

Wright, D.K., Jr. 1945. A new friction chart for round ducts. *ASHVE Transactions* 51:303-16.

BIBLIOGRAPHY

SMACNA. 1987. Duct research destroys design myths. Videotape (VHS). Sheet Metal and Air Conditioning Contractors' National Association, Chantilly, VA.

FITTING LOSS COEFFICIENTS

Fittings to support Examples 8 and 9 reprinted here. For the complete fitting database see the *Duct Fitting Database* (ASHRAE 1984).

ROUND FITTINGS

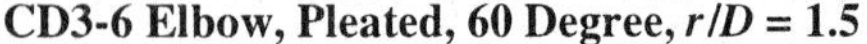

CD3-6 Elbow, Pleated, 60 Degree, r/D = 1.5

D, in.	4	6	8	10	12	14	16
C_o	0.45	0.34	0.27	0.23	0.20	0.19	0.19

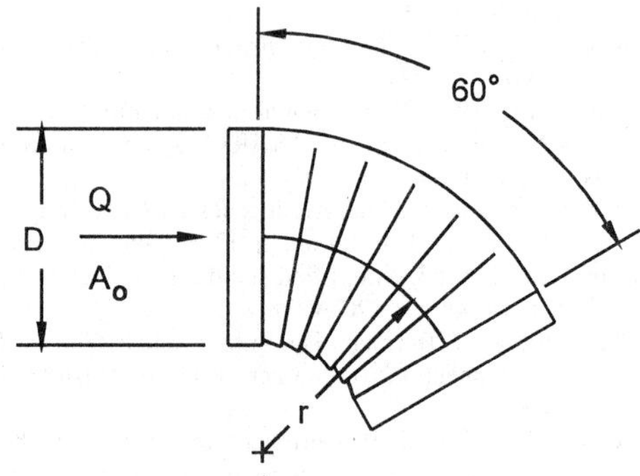

CD3-9 Elbow, 5 Gore, 90 Degree, r/D = 1.5

D, in.	3	6	9	12	15	18	21	24	27	30	60
C_o	0.51	0.28	0.21	0.18	0.16	0.15	0.14	0.13	0.12	0.12	0.12

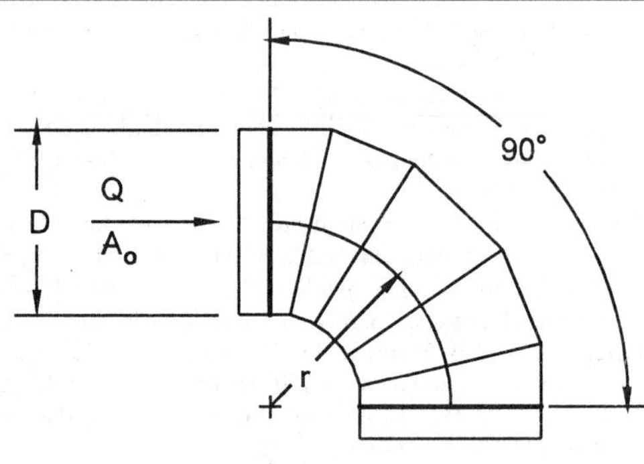

CD3-10 Elbow, 7 Gore, 90 Degree, r/D = 2.5

D, in.	3	6	9	12	15	18	27	60
C_o	0.16	0.12	0.10	0.08	0.07	0.06	0.05	0.03

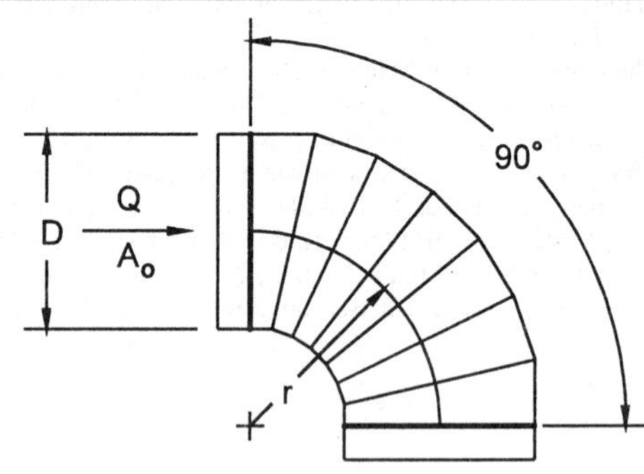

CD3-12 Elbow, 3 Gore, 90 Degree, r/D = 0.75 to 2.0

r/D	0.75	1.00	1.50	2.00
C_o	0.54	0.42	0.34	0.33

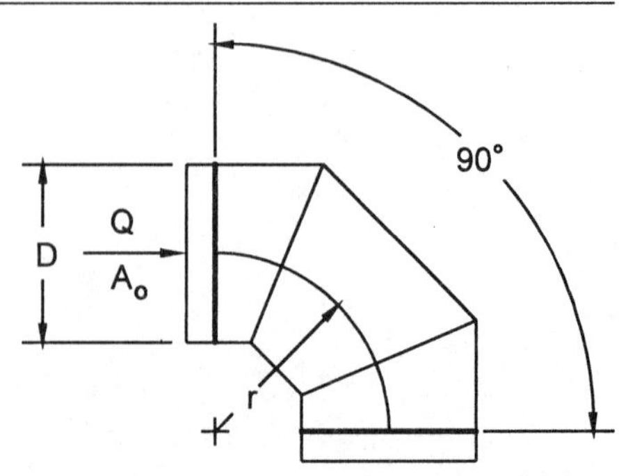

CD3-13 Elbow, 3 Gore, 60 Degree, r/D = 1.5

D, in.	3	6	9	12	15	18	21	24	27	30	60
C_o	0.40	0.21	0.16	0.14	0.12	0.12	0.11	0.10	0.09	0.09	0.09

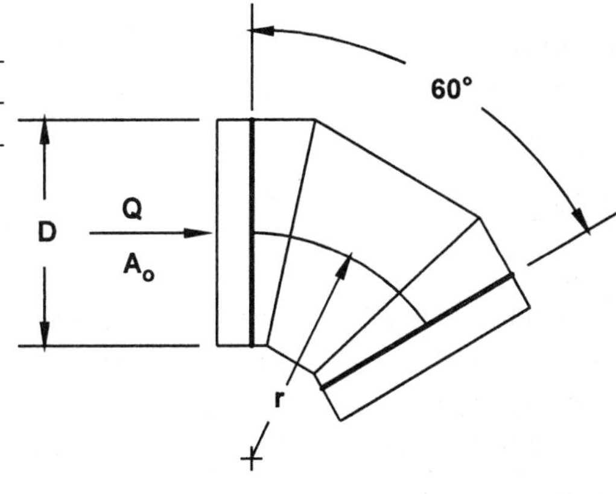

CD3-17 Elbow, Mitered, 45 Degree

D, in.	3	6	9	12	15	18	21	24	27	60
C_o	0.34	0.34	0.34	0.34	0.34	0.34	0.34	0.34	0.34	0.34

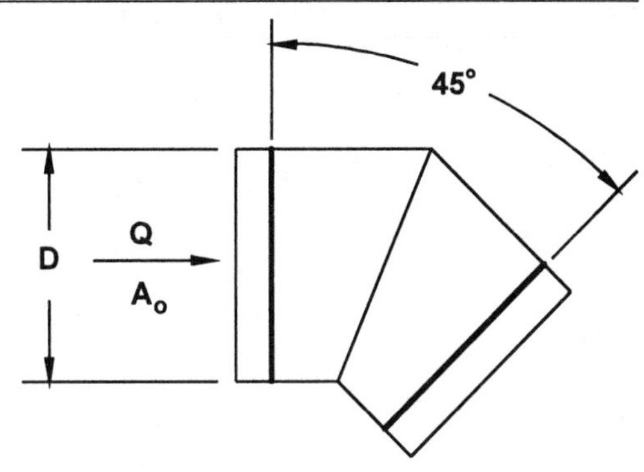

CD6-1 Screen (Only)

	C_o Values												
	n												
A_1/A_o	**0.30**	**0.35**	**0.40**	**0.45**	**0.50**	**0.55**	**0.60**	**0.65**	**0.70**	**0.75**	**0.80**	**0.90**	**1.00**
0.2	155.00	102.50	75.00	55.00	41.25	31.50	24.25	18.75	14.50	11.00	8.00	3.50	0.00
0.3	68.89	45.56	33.33	24.44	18.33	14.00	10.78	8.33	6.44	4.89	3.56	1.56	0.00
0.4	38.75	25.63	18.75	13.75	10.31	7.88	6.06	4.69	3.63	2.75	2.00	0.88	0.00
0.5	24.80	16.40	12.00	8.80	6.60	5.04	3.88	3.00	2.32	1.76	1.28	0.56	0.00
0.6	17.22	11.39	8.33	6.11	4.58	3.50	2.69	2.08	1.61	1.22	0.89	0.39	0.00
0.7	12.65	8.37	6.12	4.49	3.37	2.57	1.98	1.53	1.18	0.90	0.65	0.29	0.00
0.8	9.69	6.40	4.69	3.44	2.58	1.97	1.52	1.17	0.91	0.69	0.50	0.22	0.00
0.9	7.65	5.06	3.70	2.72	2.04	1.56	1.20	0.93	0.72	0.54	0.40	0.17	0.00
1.0	6.20	4.10	3.00	2.20	1.65	1.26	0.97	0.75	0.58	0.44	0.32	0.14	0.00
1.2	4.31	2.85	2.08	1.53	1.15	0.88	0.67	0.52	0.40	0.31	0.22	0.10	0.00
1.4	3.16	2.09	1.53	1.12	0.84	0.64	0.49	0.38	0.30	0.22	0.16	0.07	0.00
1.6	2.42	1.60	1.17	0.86	0.64	0.49	0.38	0.29	0.23	0.17	0.13	0.05	0.00
1.8	1.91	1.27	0.93	0.68	0.51	0.39	0.30	0.23	0.18	0.14	0.10	0.04	0.00
2.0	1.55	1.03	0.75	0.55	0.41	0.32	0.24	0.19	0.15	0.11	0.08	0.04	0.00
2.5	0.99	0.66	0.48	0.35	0.26	0.20	0.16	0.12	0.09	0.07	0.05	0.02	0.00
3.0	0.69	0.46	0.33	0.24	0.18	0.14	0.11	0.08	0.06	0.05	0.04	0.02	0.00
4.0	0.39	0.26	0.19	0.14	0.10	0.08	0.06	0.05	0.04	0.03	0.02	0.01	0.00
6.0	0.17	0.11	0.08	0.06	0.05	0.04	0.03	0.02	0.02	0.01	0.01	0.00	0.00

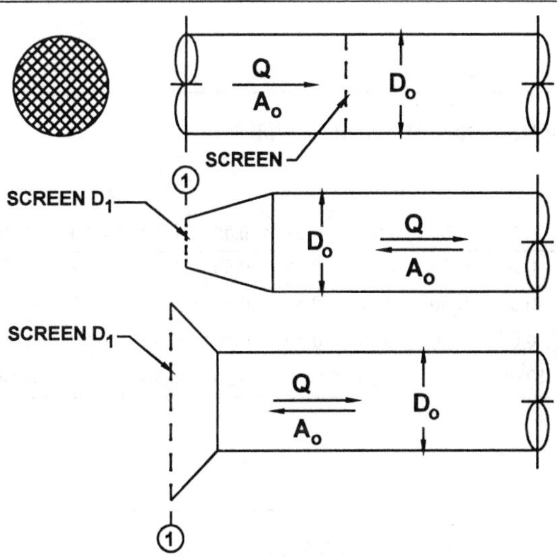

n = free area ratio of screen
A_o = area of duct
A_1 = cross-sectional area of duct or fitting where screen is located

CD9-1 Damper, Butterfly

	C_o Values											
	θ											
D/D_o	0	10	20	30	40	50	60	70	75	80	85	90
0.5	0.19	0.27	0.37	0.49	0.61	0.74	0.86	0.96	0.99	1.02	1.04	1.04
0.6	0.19	0.32	0.48	0.69	0.94	1.21	1.48	1.72	1.82	1.89	1.93	2.00
0.7	0.19	0.37	0.64	1.01	1.51	2.12	2.81	3.46	3.73	3.94	4.08	6.00
0.8	0.19	0.45	0.87	1.55	2.60	4.13	6.14	8.38	9.40	10.30	10.80	15.00
0.9	0.19	0.54	1.22	2.51	4.97	9.57	17.80	30.50	38.00	45.00	50.10	100.00
1.0	0.19	0.67	1.76	4.38	11.20	32.00	113.00	619.00	2010.00	10350.00	99999.00	99999.00

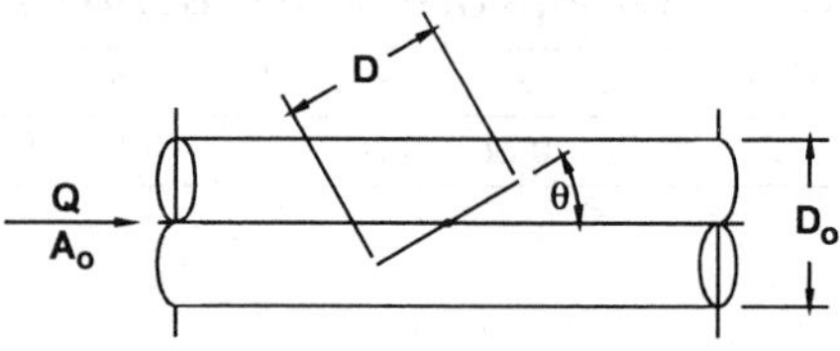

CD9-3 Fire Damper, Curtain Type, Type C

$C_o = 0.12$

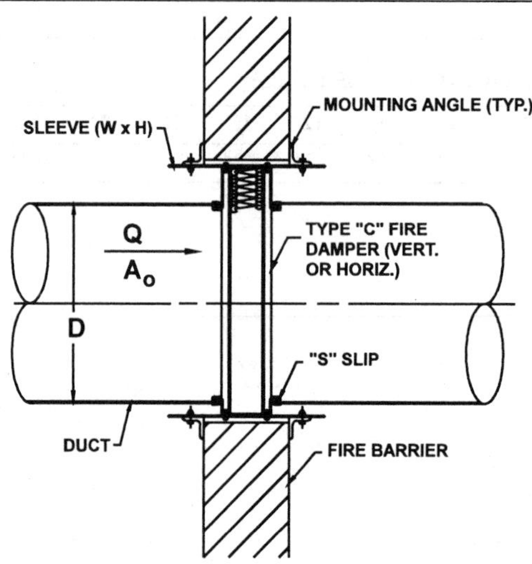

ED1-1 Duct Mounted in Wall

	C_o Values								
	L/D								
t/D	0.00	0.002	0.01	0.05	0.10	0.20	0.30	0.50	10.00
0.00	0.50	0.57	0.68	0.80	0.86	0.92	0.97	1.00	1.00
0.02	0.50	0.51	0.52	0.55	0.60	0.66	0.69	0.72	0.72
0.05	0.50	0.50	0.50	0.50	0.50	0.50	0.50	0.50	0.50
10.00	0.50	0.50	0.50	0.50	0.50	0.50	0.50	0.50	0.50

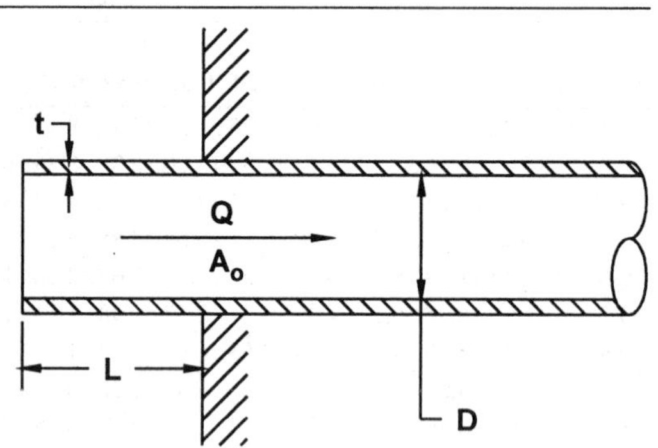

ED1-3 Bellmouth, with Wall

r/D	0.00	0.01	0.02	0.03	0.04	0.05	0.06	0.08	0.10	0.12	0.16	0.20	10.00
C_o	0.50	0.44	0.37	0.31	0.26	0.22	0.20	0.15	0.12	0.09	0.06	0.03	0.03

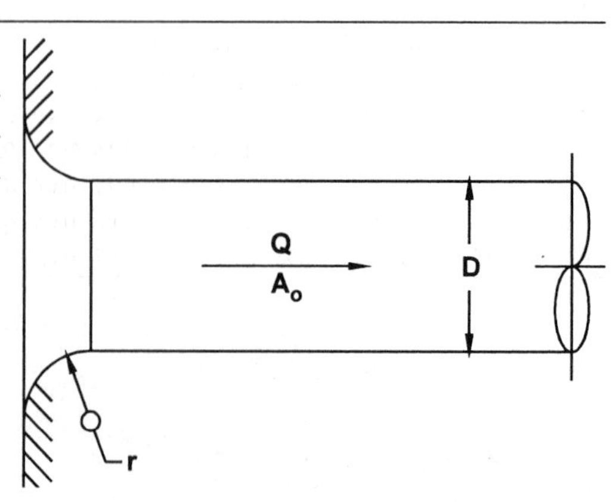

ED2-1 Conical Diffuser, Round to Plenum, Exhaust/Return Systems

	C_o Values										
	L/D_o										
A_1/A_o	0.5	1.0	2.0	3.0	4.0	5.0	6.0	8.0	10.0	12.0	14.0
1.5	0.03	0.02	0.03	0.03	0.04	0.05	0.06	0.08	0.10	0.11	0.13
2.0	0.08	0.06	0.04	0.04	0.04	0.05	0.05	0.06	0.08	0.09	0.10
2.5	0.13	0.09	0.06	0.06	0.06	0.06	0.06	0.06	0.07	0.08	0.09
3.0	0.17	0.12	0.09	0.07	0.07	0.06	0.06	0.07	0.07	0.08	0.08
4.0	0.23	0.17	0.12	0.10	0.09	0.08	0.08	0.08	0.08	0.08	0.08
6.0	0.30	0.22	0.16	0.13	0.12	0.10	0.10	0.09	0.09	0.09	0.08
8.0	0.34	0.26	0.18	0.15	0.13	0.12	0.11	0.10	0.09	0.09	0.09
10.0	0.36	0.28	0.20	0.16	0.14	0.13	0.12	0.11	0.10	0.09	0.09
14.0	0.39	0.30	0.22	0.18	0.16	0.14	0.13	0.12	0.10	0.10	0.10
20.0	0.41	0.32	0.24	0.20	0.17	0.15	0.14	0.12	0.11	0.11	0.10

	Optimum Angle θ										
A_1/A_o	0.5	1.0	2.0	3.0	4.0	5.0	6.0	8.0	10.0	12.0	14.0
1.5	34	20	13	9	7	6	4	3	2	2	2
2.0	42	28	17	12	10	9	8	6	5	4	3
2.5	50	32	20	15	12	11	10	8	7	6	5
3.0	54	34	22	17	14	12	11	10	8	8	6
4.0	58	40	26	20	16	14	13	12	10	10	9
6.0	62	42	28	22	19	16	15	12	11	10	9
8.0	64	44	30	24	20	18	16	13	12	11	10
10.0	66	46	30	24	22	19	17	14	12	11	10
14.0	66	48	32	26	22	19	17	14	13	11	11
20.0	68	48	32	26	22	20	18	15	13	12	11

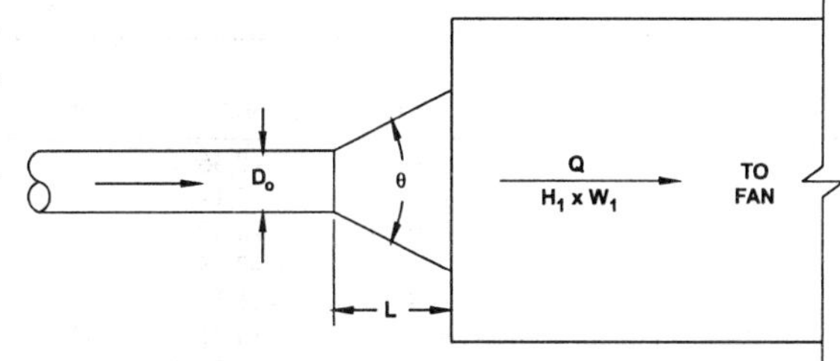

ED5-1 Wye, 30 Degree, Converging

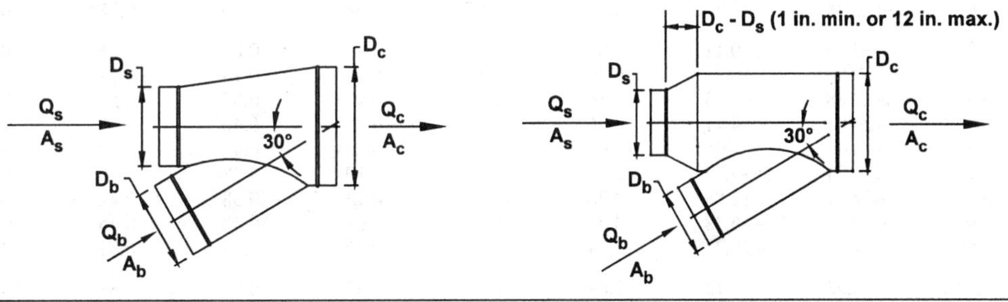

		C_b Values								
		Q_b/Q_c								
A_s/A_c	A_b/A_c	0.1	0.2	0.3	0.4	0.5	0.6	0.7	0.8	0.9
0.2	0.2	−24.17	−3.78	−0.60	0.30	0.64	0.77	0.83	0.88	0.98
	0.3	−55.88	−9.77	−2.57	−0.50	0.25	0.55	0.67	0.70	0.71
	0.4	−99.93	−17.94	−5.13	−1.45	−0.11	0.42	0.62	0.68	0.68
	0.5	−156.51	−28.40	−8.37	−2.62	−0.52	0.30	0.62	0.71	0.69
	0.6	−225.62	−41.13	−12.30	−4.01	−0.99	0.20	0.66	0.78	0.75
	0.7	−307.26	−56.14	−16.90	−5.61	−1.51	0.11	0.73	0.90	0.86
	0.8	−401.44	−73.44	−22.18	−7.44	−2.08	0.04	0.84	1.06	1.01
	0.9	−508.15	−93.02	−28.15	−9.49	−2.71	−0.03	0.99	1.27	1.20
	1.0	−627.39	−114.89	−34.80	−11.77	−3.39	−0.08	1.18	1.52	1.43
0.3	0.2	−13.97	−1.77	0.08	0.59	0.77	0.84	0.88	0.92	1.06
	0.3	−33.06	−5.33	−1.09	0.10	0.51	0.66	0.71	0.72	0.74
	0.4	−59.43	−10.08	−2.52	−0.41	0.32	0.59	0.67	0.68	0.66
	0.5	−93.24	−16.11	−4.30	−1.00	0.14	0.56	0.69	0.70	0.66
	0.6	−134.51	−23.45	−6.44	−1.68	−0.03	0.57	0.76	0.77	0.70
	0.7	−183.25	−32.08	−8.93	−2.45	−0.21	0.61	0.87	0.88	0.79
	0.8	−239.47	−42.01	−11.77	−3.32	−0.38	0.69	1.02	1.03	0.91
	0.9	−303.16	−53.25	−14.97	−4.27	−0.56	0.80	1.21	1.23	1.07
	1.0	−374.32	−65.79	−18.53	−5.32	−0.73	0.94	1.45	1.47	1.27
0.4	0.2	−9.20	−0.85	0.39	0.71	0.82	0.87	0.90	0.94	1.09
	0.3	−22.31	−3.24	−0.38	0.39	0.64	0.73	0.76	0.78	0.85
	0.4	−40.52	−6.48	−1.37	0.02	0.48	0.64	0.67	0.66	0.65
	0.5	−63.71	−10.50	−2.50	−0.33	0.40	0.63	0.69	0.67	0.63
	0.6	−92.00	−15.37	−3.84	−0.71	0.33	0.67	0.75	0.71	0.65
	0.7	−125.40	−21.08	−5.40	−1.13	0.28	0.75	0.85	0.80	0.70
	0.8	−163.90	−27.65	−7.16	−1.59	0.25	0.86	1.00	0.93	0.80
	0.9	−207.52	−35.07	−9.14	−2.09	0.25	1.02	1.18	1.10	0.93
	1.0	−256.25	−43.35	−11.33	−2.63	0.26	1.21	1.42	1.31	1.09

ED5-1 Wye, 30 Degree, Converging (*Continued*)

		C_b Values (*Concluded*)								
		Q_b/Q_c								
A_s/A_c	A_b/A_c	**0.1**	**0.2**	**0.3**	**0.4**	**0.5**	**0.6**	**0.7**	**0.8**	**0.9**
0.5	0.2	−6.62	−0.36	0.54	0.77	0.85	0.88	0.90	0.95	1.11
	0.3	−16.42	−2.11	−0.01	0.54	0.72	0.78	0.80	0.83	0.96
	0.4	−30.26	−4.59	−0.79	0.22	0.54	0.64	0.66	0.64	0.64
	0.5	−47.68	−7.55	−1.61	−0.02	0.48	0.63	0.65	0.62	0.59
	0.6	−68.93	−11.13	−2.56	−0.28	0.45	0.67	0.69	0.65	0.58
	0.7	−94.00	−15.31	−3.65	−0.55	0.44	0.74	0.77	0.71	0.61
	0.8	−122.90	−20.12	−4.88	−0.83	0.46	0.85	0.90	0.81	0.68
	0.9	−155.63	−25.54	−6.25	−1.12	0.51	1.00	1.06	0.94	0.77
	1.0	−192.18	−31.58	−7.77	−1.43	0.59	1.19	1.26	1.12	0.90
0.6	0.2	−5.12	−0.10	0.62	0.79	0.85	0.87	0.90	0.95	1.11
	0.3	−13.00	−1.49	0.18	0.61	0.75	0.79	0.82	0.86	1.02
	0.4	−24.31	−3.55	−0.50	0.30	0.55	0.62	0.63	0.62	0.63
	0.5	−38.41	−5.94	−1.16	0.09	0.48	0.59	0.60	0.57	0.55
	0.6	−55.58	−8.80	−1.92	−0.12	0.45	0.61	0.62	0.57	0.52
	0.7	−75.83	−12.16	−2.79	−0.33	0.44	0.66	0.67	0.60	0.52
	0.8	−99.17	−16.00	−3.76	−0.54	0.46	0.74	0.76	0.67	0.56
	0.9	−125.60	−20.33	−4.83	−0.76	0.51	0.86	0.88	0.77	0.62
	1.0	−155.12	−25.14	−6.02	−0.99	0.58	1.02	1.04	0.90	0.71
0.7	0.2	−4.24	0.05	0.65	0.80	0.85	0.87	0.89	0.94	1.12
	0.3	−11.00	−1.15	0.27	0.63	0.75	0.79	0.82	0.87	1.06
	0.4	−20.82	−3.00	−0.38	0.31	0.52	0.59	0.60	0.59	0.61
	0.5	−32.99	−5.09	−0.98	0.10	0.43	0.53	0.54	0.52	0.51
	0.6	−47.78	−7.58	−1.67	−0.11	0.38	0.52	0.53	0.49	0.45
	0.7	−65.22	−10.50	−2.44	−0.32	0.34	0.53	0.54	0.49	0.43
	0.8	−85.32	−13.83	−3.30	−0.53	0.33	0.58	0.59	0.52	0.43
	0.9	−108.07	−17.58	−4.26	−0.75	0.34	0.66	0.67	0.58	0.46
	1.0	−133.48	−21.76	−5.30	−0.97	0.38	0.76	0.78	0.67	0.51
0.8	0.2	−3.75	0.11	0.65	0.79	0.84	0.86	0.88	0.94	1.12
	0.3	−9.88	−0.99	0.29	0.63	0.74	0.78	0.81	0.87	1.09
	0.4	−18.88	−2.75	−0.36	0.28	0.48	0.55	0.56	0.57	0.61
	0.5	−29.98	−4.71	−0.96	0.04	0.36	0.46	0.47	0.46	0.47
	0.6	−43.46	−7.05	−1.64	−0.20	0.26	0.41	0.43	0.41	0.39
	0.7	−59.34	−9.77	−2.40	−0.44	0.19	0.38	0.41	0.38	0.34
	0.8	−77.64	−12.88	−3.26	−0.69	0.13	0.38	0.42	0.37	0.31
	0.9	−98.35	−16.38	−4.20	−0.95	0.09	0.40	0.45	0.39	0.30
	1.0	−121.48	−20.27	−5.24	−1.23	0.06	0.45	0.51	0.43	0.31
0.9	0.2	−3.52	0.12	0.64	0.78	0.82	0.85	0.88	0.93	1.12
	0.3	−9.34	−0.95	0.28	0.60	0.71	0.76	0.80	0.87	1.10
	0.4	−17.96	−2.70	−0.40	0.22	0.43	0.50	0.53	0.54	0.60
	0.5	−28.58	−4.65	−1.05	−0.07	0.26	0.37	0.40	0.41	0.42
	0.6	−41.45	−6.97	−1.77	−0.35	0.12	0.28	0.32	0.32	0.32
	0.7	−56.61	−9.66	−2.58	−0.65	0.00	0.21	0.27	0.26	0.24
	0.8	−74.08	−12.74	−3.49	−0.97	−0.12	0.16	0.23	0.22	0.18
	0.9	−93.84	−16.21	−4.50	−1.30	−0.23	0.13	0.21	0.19	0.14
	1.0	−115.92	−20.06	−5.61	−1.66	−0.34	0.11	0.21	0.18	0.11
1.0	0.2	−3.48	0.10	0.62	0.76	0.81	0.84	0.87	0.92	1.11
	0.3	−9.22	−1.00	0.23	0.56	0.68	0.74	0.78	0.86	1.11
	0.4	−17.76	−2.79	−0.50	0.14	0.37	0.45	0.49	0.52	0.60
	0.5	−28.31	−4.82	−1.21	−0.20	0.15	0.28	0.33	0.35	0.38
	0.6	−41.06	−7.21	−2.01	−0.55	−0.04	0.15	0.22	0.23	0.25
	0.7	−56.09	−9.99	−2.91	−0.92	−0.23	0.03	0.12	0.14	0.15
	0.8	−73.39	−13.17	−3.92	−1.32	−0.41	−0.07	0.04	0.06	0.06
	0.9	−92.98	−16.75	−5.04	−1.75	−0.60	−0.17	−0.03	−0.01	−0.02
	1.0	−114.85	−20.74	−6.28	−2.21	−0.79	−0.26	−0.09	−0.07	−0.09

		C_s Values								
		Q_s/Q_c								
A_s/A_c	A_b/A_c	**0.1**	**0.2**	**0.3**	**0.4**	**0.5**	**0.6**	**0.7**	**0.8**	**0.9**
0.2	0.2	−16.02	−3.15	−0.80	0.04	0.45	0.69	0.86	0.99	1.10
	0.3	−11.65	−1.94	−0.26	0.32	0.60	0.77	0.90	1.01	1.10
	0.4	−8.56	−1.20	0.05	0.47	0.68	0.82	0.92	1.02	1.11
	0.5	−6.41	−0.71	0.25	0.57	0.73	0.84	0.93	1.02	1.11
	0.6	−4.85	−0.36	0.38	0.63	0.76	0.86	0.94	1.02	1.11
	0.7	−3.68	−0.10	0.48	0.68	0.79	0.87	0.95	1.03	1.11
	0.8	−2.77	0.10	0.56	0.71	0.81	0.88	0.95	1.03	1.11
	0.9	−2.04	0.26	0.62	0.74	0.82	0.89	0.95	1.03	1.11
	1.0	−1.45	0.38	0.66	0.76	0.83	0.89	0.96	1.03	1.11

ED5-1 Wye, 30 Degree, Converging (*Concluded*)

		C_s Values (*Concluded*)								
		Q_s/Q_c								
A_s/A_c	A_b/A_c	0.1	0.2	0.3	0.4	0.5	0.6	0.7	0.8	0.9
0.3	0.2	−36.37	−7.59	−2.48	−0.79	−0.06	0.29	0.47	0.57	0.61
	0.3	−26.79	−5.07	−1.42	−0.27	0.21	0.42	0.53	0.59	0.61
	0.4	−19.94	−3.49	−0.80	0.02	0.35	0.49	0.56	0.60	0.62
	0.5	−15.18	−2.44	−0.41	0.20	0.43	0.54	0.58	0.61	0.62
	0.6	−11.73	−1.70	−0.13	0.32	0.49	0.56	0.60	0.61	0.62
	0.7	−9.13	−1.14	0.07	0.41	0.53	0.58	0.60	0.61	0.62
	0.8	−7.11	−0.72	0.23	0.48	0.57	0.60	0.61	0.62	0.62
	0.9	−5.49	−0.38	0.35	0.53	0.59	0.61	0.62	0.62	0.62
	1.0	−4.17	−0.11	0.45	0.58	0.61	0.62	0.62	0.62	0.62
0.4	0.2	−64.82	−13.76	−4.74	−1.81	−0.59	−0.02	0.24	0.36	0.39
	0.3	−47.92	−9.38	−2.93	−0.94	−0.16	0.19	0.34	0.39	0.40
	0.4	−35.81	−6.62	−1.88	−0.46	0.07	0.30	0.38	0.41	0.40
	0.5	−27.39	−4.78	−1.20	−0.16	0.22	0.36	0.41	0.42	0.41
	0.6	−21.28	−3.48	−0.73	0.04	0.31	0.41	0.43	0.43	0.41
	0.7	−16.68	−2.51	−0.38	0.20	0.38	0.44	0.45	0.43	0.41
	0.8	−13.10	−1.77	−0.12	0.31	0.44	0.46	0.46	0.44	0.41
	0.9	−10.24	−1.18	0.09	0.40	0.48	0.48	0.46	0.44	0.41
	1.0	−7.90	−0.69	0.26	0.47	0.51	0.50	0.47	0.44	0.41
0.5	0.2	−101.39	−21.64	−7.61	−3.07	−1.19	−0.34	0.05	0.22	0.26
	0.3	−75.05	−14.87	−4.83	−1.75	−0.54	−0.03	0.19	0.26	0.27
	0.4	−56.18	−10.59	−3.21	−1.02	−0.20	0.13	0.26	0.29	0.27
	0.5	−43.04	−7.74	−2.16	−0.56	0.02	0.23	0.30	0.30	0.27
	0.6	−33.51	−5.72	−1.43	−0.24	0.16	0.30	0.33	0.31	0.28
	0.7	−26.34	−4.22	−0.90	−0.01	0.27	0.35	0.35	0.32	0.28
	0.8	−20.75	−3.06	−0.49	0.16	0.35	0.39	0.37	0.33	0.28
	0.9	−16.29	−2.14	−0.17	0.30	0.41	0.41	0.38	0.33	0.28
	1.0	−12.64	−1.39	0.10	0.41	0.46	0.44	0.39	0.33	0.28
0.6	0.2	−146.06	−31.26	−11.09	−4.56	−1.89	−0.68	−0.12	0.10	0.16
	0.3	−108.19	−21.55	−7.12	−2.69	−0.97	−0.24	0.07	0.17	0.17
	0.4	−81.04	−15.40	−4.80	−1.65	−0.48	−0.01	0.17	0.20	0.18
	0.5	−62.13	−11.31	−3.30	−0.99	−0.17	0.13	0.22	0.22	0.18
	0.6	−48.43	−8.41	−2.25	−0.54	0.03	0.22	0.26	0.24	0.18
	0.7	−38.10	−6.25	−1.49	−0.22	0.18	0.29	0.29	0.25	0.19
	0.8	−30.07	−4.59	−0.90	0.03	0.30	0.34	0.31	0.25	0.19
	0.9	−23.64	−3.27	−0.44	0.23	0.39	0.38	0.33	0.26	0.19
	1.0	−18.39	−2.20	−0.06	0.39	0.46	0.42	0.34	0.27	0.19
0.7	0.2	−198.85	−42.62	−15.17	−6.31	−2.68	−1.04	−0.29	0.01	0.08
	0.3	−147.33	−29.41	−9.78	−3.77	−1.44	−0.45	−0.04	0.10	0.10
	0.4	−110.40	−21.07	−6.64	−2.36	−0.77	−0.14	0.09	0.15	0.11
	0.5	−84.67	−15.50	−4.60	−1.48	−0.36	0.05	0.17	0.17	0.11
	0.6	−66.02	−11.56	−3.19	−0.86	−0.08	0.18	0.23	0.19	0.12
	0.7	−51.97	−8.63	−2.15	−0.42	0.12	0.27	0.27	0.20	0.12
	0.8	−41.04	−6.37	−1.35	−0.08	0.27	0.34	0.29	0.21	0.12
	0.9	−32.30	−4.58	−0.72	0.19	0.39	0.39	0.32	0.22	0.12
	1.0	−25.16	−3.12	−0.21	0.40	0.49	0.43	0.33	0.23	0.13
0.8	0.2	−259.75	−55.70	−19.86	−8.29	−3.56	−1.43	−0.46	−0.06	0.03
	0.3	−192.48	−38.47	−12.84	−4.99	−1.95	−0.66	−0.12	0.05	0.05
	0.4	−144.25	−27.58	−8.74	−3.16	−1.09	−0.26	0.05	0.11	0.06
	0.5	−110.65	−20.32	−6.08	−2.00	−0.55	−0.01	0.15	0.15	0.07
	0.6	−86.30	−15.17	−4.24	−1.20	−0.19	0.15	0.22	0.17	0.08
	0.7	−67.95	−11.34	−2.88	−0.62	0.08	0.27	0.27	0.19	0.08
	0.8	−53.67	−8.40	−1.84	−0.18	0.28	0.36	0.30	0.20	0.08
	0.9	−42.26	−6.05	−1.02	0.16	0.44	0.43	0.33	0.21	0.08
	1.0	−32.93	−4.15	−0.35	0.44	0.56	0.49	0.36	0.22	0.09
0.9	0.2	−328.76	−70.51	−25.16	−10.53	−4.54	−1.84	−0.62	−0.12	0.00
	0.3	−243.63	−48.72	−16.28	−6.35	−2.50	−0.87	−0.20	0.03	0.03
	0.4	−182.60	−34.94	−11.09	−4.03	−1.41	−0.37	0.02	0.10	0.04
	0.5	−140.07	−25.75	−7.74	−2.57	−0.74	−0.06	0.15	0.14	0.05
	0.6	−109.25	−19.24	−5.40	−1.56	−0.28	0.15	0.23	0.17	0.05
	0.7	−86.04	−14.40	−3.68	−0.83	0.06	0.30	0.30	0.20	0.06
	0.8	−67.96	−10.66	−2.37	−0.27	0.31	0.41	0.34	0.21	0.06
	0.9	−53.52	−7.70	−1.33	0.17	0.51	0.50	0.38	0.22	0.06
	1.0	−41.71	−5.29	−0.49	0.52	0.67	0.57	0.41	0.23	0.07
1.0	0.2	−405.88	−87.06	−31.07	−13.01	−5.62	−2.29	−0.77	−0.16	−0.02
	0.3	−300.78	−60.15	−20.11	−7.85	−3.10	−1.09	−0.26	0.02	0.02
	0.4	−225.44	−43.14	−13.70	−4.99	−1.76	−0.47	0.01	0.11	0.04
	0.5	−172.93	−31.80	−9.56	−3.18	−0.92	−0.09	0.17	0.17	0.05
	0.6	−134.89	−23.76	−6.68	−1.94	−0.35	0.17	0.28	0.20	0.06
	0.7	−106.23	−17.78	−4.56	−1.04	0.06	0.36	0.35	0.23	0.06
	0.8	−83.92	−13.18	−2.93	−0.35	0.37	0.50	0.41	0.25	0.06
	0.9	−66.08	−9.52	−1.65	0.19	0.62	0.61	0.46	0.26	0.07
	1.0	−51.51	−6.54	−0.61	0.63	0.81	0.70	0.49	0.28	0.07

ED5-2 Wye, 45 Degree, Converging

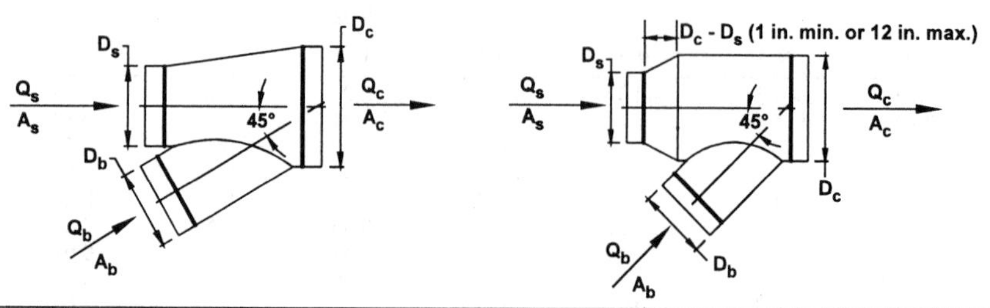

A_s/A_c	A_b/A_c	C_b Values								
		Q_b/Q_c								
		0.1	0.2	0.3	0.4	0.5	0.6	0.7	0.8	0.9
0.2	0.2	−25.19	−3.97	−0.64	0.32	0.67	0.82	0.90	0.96	1.08
	0.3	−58.03	−10.14	−2.63	−0.45	0.36	0.69	0.84	0.93	1.08
	0.4	−104.08	−18.80	−5.40	−1.51	−0.07	0.52	0.77	0.88	1.01
	0.5	−163.36	−29.97	−8.97	−2.87	−0.62	0.29	0.67	0.80	0.84
	0.6	−235.59	−43.47	−13.22	−4.44	−1.20	0.12	0.65	0.83	0.85
	0.7	−320.90	−59.38	−18.21	−6.25	−1.84	−0.04	0.68	0.91	0.93
	0.8	−419.32	−77.73	−23.95	−8.33	−2.56	−0.22	0.72	1.02	1.02
	0.9	−530.86	−98.50	−30.44	−10.66	−3.36	−0.40	0.79	1.16	1.14
	1.0	−655.51	−121.72	−37.68	−13.26	−4.25	−0.59	0.87	1.33	1.28
0.3	0.2	−14.27	−1.77	0.13	0.66	0.85	0.93	0.97	1.03	1.21
	0.3	−33.62	−5.28	−0.95	0.27	0.70	0.87	0.94	1.01	1.19
	0.4	−60.85	−10.26	−2.48	−0.30	0.47	0.77	0.88	0.93	1.04
	0.5	−95.87	−16.64	−4.44	−1.00	0.21	0.66	0.82	0.84	0.84
	0.6	−138.38	−24.26	−6.68	−1.73	0.01	0.66	0.88	0.91	0.88
	0.7	−188.60	−33.25	−9.32	−2.58	−0.20	0.68	0.98	1.02	0.95
	0.8	−246.54	−43.60	−12.34	−3.54	−0.43	0.72	1.11	1.15	1.03
	0.9	−312.21	−55.33	−15.76	−4.61	−0.68	0.78	1.26	1.31	1.13
	1.0	−385.59	−68.43	−19.56	−5.79	−0.94	0.86	1.45	1.49	1.24
0.4	0.2	−8.77	−0.64	0.54	0.85	0.95	0.99	1.03	1.09	1.31
	0.3	−21.41	−2.85	−0.10	0.63	0.87	0.96	1.00	1.06	1.26
	0.4	−39.30	−6.02	−1.05	0.28	0.72	0.87	0.91	0.92	1.00
	0.5	−62.10	−9.96	−2.16	−0.06	0.63	0.85	0.90	0.88	0.86
	0.6	−89.77	−14.65	−3.42	−0.38	0.61	0.93	0.99	0.95	0.90
	0.7	−122.46	−20.19	−4.88	−0.74	0.61	1.04	1.12	1.06	0.95
	0.8	−160.18	−26.56	−6.55	−1.15	0.62	1.18	1.29	1.19	1.01
	0.9	−202.93	−33.77	−8.44	−1.60	0.64	1.36	1.48	1.35	1.07
	1.0	−250.70	−41.83	−10.54	−2.09	0.68	1.56	1.71	1.53	1.15
0.5	0.2	−5.45	0.04	0.79	0.97	1.02	1.04	1.07	1.14	1.39
	0.3	−14.10	−1.39	0.40	0.84	0.97	1.00	1.02	1.07	1.28
	0.4	−26.48	−3.53	−0.24	0.59	0.83	0.89	0.88	0.85	0.86
	0.5	−41.84	−5.96	−0.80	0.51	0.88	0.97	0.95	0.90	0.87
	0.6	−60.61	−8.90	−1.46	0.43	0.97	1.09	1.06	0.97	0.90
	0.7	−82.80	−12.36	−2.22	0.35	1.09	1.25	1.20	1.08	0.93
	0.8	−108.39	−16.35	−3.09	0.27	1.24	1.45	1.38	1.20	0.96
	0.9	−137.41	−20.86	−4.07	0.19	1.42	1.68	1.59	1.35	0.99
	1.0	−169.84	−25.90	−5.15	0.11	1.63	1.95	1.83	1.52	1.02
0.6	0.2	−5.54	−0.08	0.70	0.91	0.98	1.01	1.05	1.14	1.42
	0.3	−14.48	−1.75	0.13	0.64	0.81	0.88	0.92	0.98	1.19
	0.4	−27.10	−4.14	−0.68	0.26	0.57	0.68	0.71	0.72	0.76
	0.5	−42.84	−6.91	−1.50	−0.02	0.47	0.64	0.68	0.69	0.70
	0.6	−62.07	−10.28	−2.48	−0.34	0.37	0.61	0.67	0.66	0.63
	0.7	−84.79	−14.26	−3.62	−0.71	0.27	0.59	0.67	0.63	0.54
	0.8	−111.02	−18.84	−4.92	−1.12	0.16	0.58	0.67	0.61	0.44
	0.9	−140.76	−24.03	−6.40	−1.57	0.04	0.58	0.68	0.59	0.31
	1.0	−174.01	−29.83	−8.04	−2.07	−0.08	0.58	0.70	0.56	0.15
0.7	0.2	−3.96	0.25	0.83	0.97	1.01	1.04	1.08	1.17	1.47
	0.3	−11.07	−1.10	0.34	0.71	0.83	0.87	0.90	0.95	1.13
	0.4	−20.92	−2.92	−0.27	0.43	0.65	0.72	0.73	0.73	0.77
	0.5	−33.20	−5.01	−0.85	0.24	0.59	0.69	0.71	0.69	0.70
	0.6	−48.21	−7.55	−1.55	0.03	0.53	0.68	0.69	0.65	0.61
	0.7	−65.95	−10.56	−2.37	−0.20	0.48	0.68	0.69	0.62	0.49
	0.8	−86.42	−14.01	−3.30	−0.46	0.43	0.68	0.69	0.58	0.35
	0.9	−109.65	−17.93	−4.35	−0.75	0.38	0.70	0.70	0.53	0.18
	1.0	−135.63	−22.32	−5.53	−1.07	0.33	0.72	0.71	0.48	−0.03

ED5-2 Wye, 45 Degree, Converging (*Continued*)

		C_b Values (*Concluded*)								
		Q_b/Q_c								
A_s/A_c	A_b/A_c	0.1	0.2	0.3	0.4	0.5	0.6	0.7	0.8	0.9
0.8	0.2	−2.78	0.50	0.91	1.01	1.03	1.05	1.09	1.18	1.49
	0.3	−8.58	−0.65	0.47	0.74	0.82	0.85	0.86	0.89	1.02
	0.4	−16.29	−2.00	0.05	0.56	0.71	0.75	0.74	0.74	0.78
	0.5	−25.98	−3.59	−0.37	0.44	0.68	0.73	0.72	0.69	0.69
	0.6	−37.82	−5.52	−0.87	0.31	0.65	0.72	0.70	0.64	0.58
	0.7	−51.83	−7.79	−1.44	0.17	0.63	0.73	0.69	0.59	0.43
	0.8	−68.01	−10.42	−2.10	0.01	0.62	0.75	0.69	0.53	0.25
	0.9	−86.37	−13.39	−2.84	−0.16	0.61	0.77	0.68	0.47	0.03
	1.0	−106.91	−16.73	−3.68	−0.35	0.61	0.79	0.68	0.38	−0.25
0.9	0.2	−1.87	0.68	0.98	1.03	1.05	1.06	1.09	1.18	1.49
	0.3	−6.70	−0.33	0.54	0.74	0.79	0.80	0.80	0.81	0.87
	0.4	−12.69	−1.29	0.29	0.66	0.76	0.77	0.75	0.74	0.78
	0.5	−20.37	−2.48	0.00	0.59	0.74	0.75	0.72	0.69	0.67
	0.6	−29.77	−3.94	−0.34	0.52	0.73	0.75	0.70	0.63	0.54
	0.7	−40.89	−5.66	−0.73	0.45	0.74	0.76	0.68	0.56	0.36
	0.8	−53.74	−7.64	−1.18	0.37	0.76	0.78	0.67	0.48	0.13
	0.9	−68.32	−9.89	−1.69	0.28	0.77	0.80	0.65	0.38	−0.15
	1.0	−84.66	−12.42	−2.27	0.18	0.80	0.83	0.62	0.26	−0.49
1.0	0.2	−1.17	0.81	1.02	1.05	1.05	1.06	1.09	1.18	1.48
	0.3	−5.09	−0.02	0.64	0.78	0.81	0.81	0.80	0.80	0.86
	0.4	−9.81	−0.72	0.48	0.74	0.79	0.78	0.76	0.74	0.77
	0.5	−15.89	−1.61	0.29	0.71	0.79	0.77	0.72	0.68	0.65
	0.6	−23.34	−2.69	0.07	0.68	0.80	0.77	0.69	0.60	0.49
	0.7	−32.15	−3.96	−0.18	0.66	0.82	0.78	0.67	0.51	0.27
	0.8	−42.35	−5.44	−0.47	0.64	0.85	0.79	0.63	0.41	0.00
	0.9	−53.94	−7.12	−0.80	0.61	0.88	0.81	0.60	0.28	−0.34
	1.0	−66.93	−9.01	−1.17	0.58	0.92	0.82	0.55	0.13	−0.75

		C_s Values								
		Q_s/Q_c								
A_s/A_c	A_b/A_c	0.1	0.2	0.3	0.4	0.5	0.6	0.7	0.8	0.9
0.2	0.2	−10.16	−2.08	−0.43	0.24	0.62	0.88	1.10	1.29	1.46
	0.3	−7.83	−1.20	0.03	0.50	0.77	0.97	1.14	1.30	1.46
	0.4	−5.62	−0.59	0.30	0.65	0.85	1.01	1.16	1.31	1.46
	0.5	−3.96	−0.18	0.48	0.74	0.90	1.04	1.18	1.32	1.47
	0.6	−2.71	0.12	0.60	0.80	0.94	1.06	1.19	1.32	1.47
	0.7	−1.75	0.34	0.70	0.85	0.96	1.07	1.19	1.32	1.47
	0.8	−0.99	0.52	0.77	0.88	0.98	1.08	1.20	1.32	1.47
	0.9	−0.38	0.66	0.82	0.91	0.99	1.09	1.20	1.33	1.47
	1.0	0.13	0.77	0.87	0.93	1.00	1.10	1.20	1.33	1.47
0.3	0.2	−23.33	−5.14	−1.67	−0.44	0.12	0.42	0.58	0.67	0.72
	0.3	−18.44	−3.44	−0.84	0.00	0.36	0.54	0.64	0.69	0.73
	0.4	−13.64	−2.22	−0.34	0.25	0.49	0.60	0.67	0.70	0.73
	0.5	−10.00	−1.37	0.00	0.41	0.57	0.64	0.69	0.71	0.73
	0.6	−7.26	−0.75	0.24	0.52	0.62	0.67	0.70	0.72	0.73
	0.7	−5.15	−0.29	0.41	0.60	0.66	0.69	0.71	0.72	0.73
	0.8	−3.48	0.07	0.55	0.66	0.69	0.70	0.71	0.72	0.73
	0.9	−2.14	0.36	0.65	0.71	0.72	0.72	0.72	0.72	0.73
	1.0	−1.03	0.60	0.74	0.75	0.73	0.73	0.72	0.72	0.73
0.4	0.2	−42.17	−9.48	−3.34	−1.23	−0.31	0.12	0.33	0.42	0.44
	0.3	−33.68	−6.60	−1.98	−0.53	0.05	0.31	0.41	0.45	0.45
	0.4	−25.24	−4.51	−1.13	−0.13	0.25	0.40	0.46	0.47	0.45
	0.5	−18.83	−3.04	−0.57	0.13	0.37	0.46	0.48	0.48	0.46
	0.6	−13.99	−1.97	−0.17	0.31	0.46	0.50	0.50	0.48	0.46
	0.7	−10.27	−1.17	0.12	0.44	0.52	0.53	0.51	0.49	0.46
	0.8	−7.32	−0.54	0.35	0.54	0.57	0.55	0.52	0.49	0.46
	0.9	−4.94	−0.04	0.53	0.62	0.61	0.57	0.53	0.49	0.46
	1.0	−2.98	0.37	0.68	0.68	0.64	0.58	0.54	0.50	0.46
0.5	0.2	−66.95	−15.18	−5.49	−2.21	−0.81	−0.16	0.14	0.26	0.28
	0.3	−53.80	−10.77	−3.45	−1.17	−0.27	0.11	0.26	0.30	0.29
	0.4	−40.66	−7.54	−2.16	−0.57	0.02	0.25	0.32	0.33	0.30
	0.5	−30.68	−5.27	−1.30	−0.18	0.21	0.33	0.36	0.34	0.30
	0.6	−23.15	−3.62	−0.69	0.09	0.33	0.39	0.38	0.35	0.30
	0.7	−17.34	−2.38	−0.24	0.29	0.42	0.43	0.40	0.35	0.30
	0.8	−12.75	−1.41	0.11	0.44	0.49	0.47	0.41	0.36	0.30
	0.9	−9.04	−0.64	0.39	0.56	0.55	0.49	0.43	0.36	0.30
	1.0	−5.99	0.00	0.61	0.65	0.59	0.51	0.43	0.36	0.30

ED5-2 Wye, 45 Degree, Converging (*Concluded*)

		C_s Values (*Concluded*)								
		Q_s/Q_c								
A_s/A_c	A_b/A_c	0.1	0.2	0.3	0.4	0.5	0.6	0.7	0.8	0.9
0.6	0.2	−97.90	−22.29	−8.18	−3.41	−1.39	−0.46	−0.03	0.13	0.16
	0.3	−79.03	−15.99	−5.28	−1.94	−0.64	−0.09	0.13	0.19	0.17
	0.4	−60.15	−11.37	−3.44	−1.09	−0.23	0.10	0.21	0.22	0.18
	0.5	−45.80	−8.13	−2.22	−0.55	0.03	0.22	0.26	0.24	0.18
	0.6	−34.97	−5.77	−1.35	−0.17	0.20	0.30	0.30	0.25	0.18
	0.7	−26.62	−3.98	−0.71	0.11	0.33	0.36	0.32	0.26	0.19
	0.8	−20.02	−2.59	−0.21	0.33	0.43	0.41	0.34	0.26	0.19
	0.9	−14.68	−1.48	0.18	0.49	0.51	0.44	0.35	0.27	0.19
	1.0	−10.29	−0.57	0.51	0.63	0.57	0.47	0.37	0.27	0.19
0.7	0.2	−135.28	−30.88	−11.42	−4.85	−2.08	−0.80	−0.21	0.02	0.06
	0.3	−109.64	−22.35	−7.50	−2.88	−1.07	−0.31	0.00	0.09	0.07
	0.4	−83.96	−16.08	−5.02	−1.73	−0.52	−0.05	0.11	0.13	0.08
	0.5	−64.44	−11.67	−3.36	−0.99	−0.17	0.11	0.18	0.15	0.09
	0.6	−49.71	−8.47	−2.19	−0.48	0.06	0.22	0.22	0.17	0.09
	0.7	−38.35	−6.04	−1.31	−0.10	0.24	0.30	0.26	0.18	0.09
	0.8	−29.37	−4.16	−0.64	0.18	0.37	0.36	0.28	0.19	0.09
	0.9	−22.12	−2.65	−0.10	0.41	0.47	0.40	0.30	0.19	0.09
	1.0	−16.14	−1.41	0.33	0.60	0.55	0.44	0.32	0.20	0.09
0.8	0.2	−179.32	−41.01	−15.25	−6.55	−2.88	−1.19	−0.41	−0.10	−0.04
	0.3	−145.86	−29.89	−10.14	−3.99	−1.58	−0.55	−0.13	0.00	−0.02
	0.4	−112.34	−21.71	−6.91	−2.50	−0.86	−0.22	0.01	0.05	−0.01
	0.5	−86.85	−15.96	−4.75	−1.54	−0.41	−0.01	0.10	0.08	0.00
	0.6	−67.62	−11.78	−3.22	−0.87	−0.10	0.13	0.16	0.10	0.00
	0.7	−52.79	−8.62	−2.08	−0.38	0.12	0.23	0.20	0.11	0.00
	0.8	−41.06	−6.16	−1.20	0.00	0.29	0.31	0.23	0.12	0.01
	0.9	−31.59	−4.19	−0.51	0.29	0.43	0.37	0.26	0.13	0.01
	1.0	−23.78	−2.58	0.06	0.53	0.54	0.42	0.28	0.14	0.01
0.9	0.2	−230.27	−52.75	−19.69	−8.53	−3.81	−1.63	−0.63	−0.22	−0.13
	0.3	−187.95	−38.69	−13.24	−5.29	−2.16	−0.83	−0.28	−0.10	−0.10
	0.4	−145.53	−28.34	−9.15	−3.41	−1.26	−0.41	−0.10	−0.04	−0.09
	0.5	−113.27	−21.07	−6.42	−2.19	−0.69	−0.15	0.01	0.00	−0.09
	0.6	−88.94	−15.78	−4.48	−1.35	−0.30	0.03	0.09	0.03	−0.08
	0.7	−70.16	−11.78	−3.04	−0.73	−0.02	0.16	0.14	0.04	−0.08
	0.8	−55.33	−8.67	−1.93	−0.25	0.20	0.26	0.18	0.06	−0.07
	0.9	−43.33	−6.18	−1.05	0.12	0.37	0.33	0.21	0.07	−0.07
	1.0	−33.46	−4.14	−0.34	0.42	0.50	0.39	0.24	0.08	−0.07
1.0	0.2	−288.39	−66.15	−24.77	−10.80	−4.88	−2.14	−0.87	−0.35	−0.22
	0.3	−236.14	−48.79	−16.81	−6.80	−2.85	−1.15	−0.44	−0.20	−0.19
	0.4	−183.77	−36.02	−11.76	−4.47	−1.73	−0.63	−0.22	−0.12	−0.18
	0.5	−143.95	−27.05	−8.39	−2.98	−1.03	−0.31	−0.08	−0.08	−0.17
	0.6	−113.91	−20.52	−6.00	−1.93	−0.55	−0.09	0.01	−0.04	−0.16
	0.7	−90.73	−15.58	−4.23	−1.17	−0.20	0.07	0.08	−0.02	−0.16
	0.8	−72.41	−11.74	−2.86	−0.58	0.06	0.19	0.13	−0.01	−0.16
	0.9	−57.61	−8.66	−1.77	−0.12	0.27	0.28	0.16	0.01	−0.15
	1.0	−45.42	−6.15	−0.88	0.25	0.44	0.36	0.20	0.02	−0.15

ED5-6 Capped Wye, Branch with 45-Degree Elbow, Branch 90 Degrees to Main, Converging

A_b/A_c	0.1	0.2	0.3	0.4	0.5	0.6	0.7	0.8	0.9	1.0
C_b	1.26	1.07	0.94	0.86	0.81	0.76	0.71	0.67	0.64	0.64

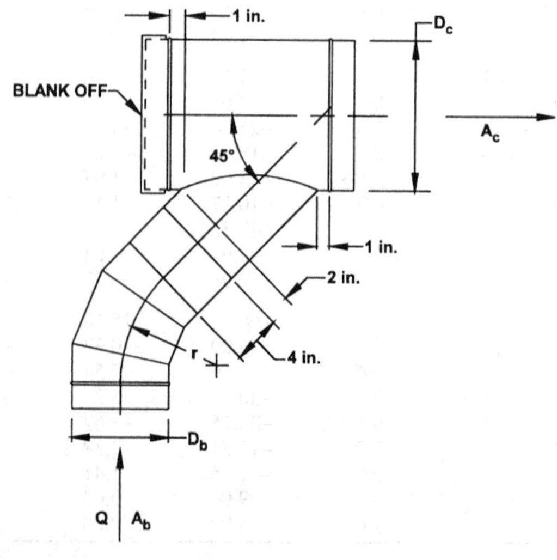

ED5-9 Symmetrical Wye, 60 Degree, $D_{b1} \geq D_{b2}$, Converging

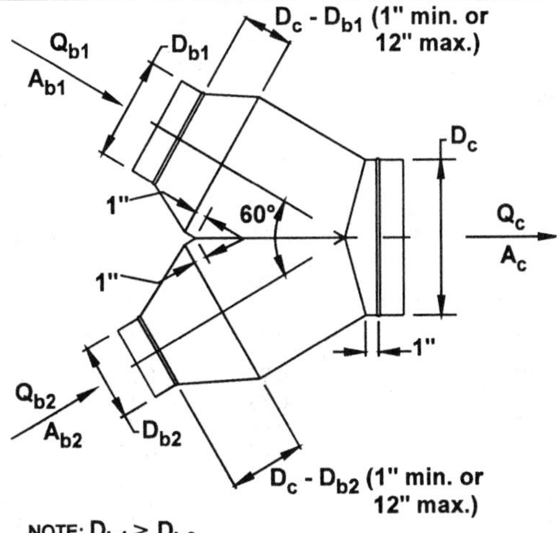

		C_{b1} Values								
		Q_{b1}/Q_c								
A_{b1}/A_c	A_{b2}/A_c	0.1	0.2	0.3	0.4	0.5	0.6	0.7	0.8	0.9
0.2	0.2	−11.95	−1.89	−0.09	0.41	0.62	0.74	0.80	0.80	0.79
	0.3	−11.95	−1.89	−0.09	0.41	0.62	0.74	0.80	0.80	0.79
0.3	0.2	−45.45	−9.39	−2.44	−0.41	0.33	0.68	0.89	1.03	1.13
	0.3	−16.88	−2.92	−0.09	0.59	0.86	1.02	1.09	1.10	1.08
0.4	0.2	−72.04	−14.00	−4.26	−1.24	−0.10	0.33	0.50	0.57	0.63
	0.3	−52.95	−9.91	−2.86	−0.69	0.07	0.30	0.40	0.49	0.62
	0.4	−28.86	−6.22	−2.15	−0.57	0.19	0.55	0.72	0.79	0.85
0.5	0.2	−126.04	−23.80	−7.44	−2.64	−0.85	−0.13	0.16	0.26	0.28
	0.3	−91.07	−16.91	−5.16	−1.73	−0.46	0.04	0.23	0.29	0.28
	0.4	−56.41	−10.07	−2.90	−0.82	−0.07	0.21	0.30	0.31	0.29
	0.5	−30.58	−5.23	−1.06	0.00	0.32	0.43	0.47	0.47	0.41
0.6	0.2	−209.81	−39.31	−12.13	−4.35	−1.54	−0.40	0.06	0.22	0.23
	0.3	−147.43	−27.69	−8.75	−3.20	−1.13	−0.29	0.05	0.17	0.18
	0.4	−85.06	−16.07	−5.38	−2.04	−0.71	−0.17	0.04	0.12	0.13
	0.5	−58.22	−11.03	−3.84	−1.49	−0.50	−0.09	0.07	0.11	0.12
	0.6	−40.57	−7.86	−2.60	−0.99	−0.26	0.00	0.14	0.21	0.25
0.7	0.2	−291.57	−54.52	−17.03	−6.21	−2.27	−0.68	−0.04	0.19	0.21
	0.3	−197.37	−38.02	−12.54	−4.92	−2.01	−0.76	−0.22	0.01	0.08
	0.4	−102.97	−21.41	−8.05	−3.64	−1.75	−0.84	−0.40	−0.17	−0.05
	0.5	−65.15	−14.75	−6.16	−3.07	−1.61	−0.85	−0.44	−0.22	−0.09
	0.6	−48.24	−11.70	−4.97	−2.59	−1.40	−0.76	−0.37	−0.15	−0.03
	0.7	−73.02	−16.68	−6.90	−3.29	−1.61	−0.80	−0.29	0.02	0.22
0.8	0.2	−373.33	−69.73	−21.93	−8.08	−3.00	−0.95	−0.13	0.15	0.20
	0.3	−247.31	−48.35	−16.32	−6.65	−2.89	−1.24	−0.49	−0.15	−0.02
	0.4	−120.88	−26.76	−10.71	−5.24	−2.78	−1.52	−0.84	−0.45	−0.24
	0.5	−72.08	−18.46	−8.48	−4.65	−2.71	−1.61	−0.95	−0.55	−0.31
	0.6	−55.91	−15.54	−7.35	−4.20	−2.54	−1.53	−0.89	−0.51	−0.30
	0.7	−80.68	−20.52	−9.27	−4.90	−2.75	−1.56	−0.80	−0.34	−0.06
	0.8	−105.46	−25.49	−11.19	−5.59	−2.96	−1.60	−0.72	−0.18	0.19
0.9	0.2	−479.24	−89.56	−28.39	−10.59	−4.04	−1.41	−0.36	0.01	0.09
	0.3	−305.31	−61.27	−21.50	−9.28	−4.39	−2.16	−1.07	−0.54	−0.29
	0.4	−131.17	−32.88	−14.60	−7.98	−4.74	−2.91	−1.79	−1.10	−0.68
	0.5	−67.90	−22.76	−12.17	−7.53	−4.89	−3.19	−2.05	−1.30	−0.81
	0.6	−68.95	−23.08	−12.11	−7.45	−4.84	−3.15	−2.01	−1.26	−0.79
	0.7	−90.48	−27.35	−13.58	−7.95	−4.97	−3.16	−1.96	−1.17	−0.65
	0.8	−112.02	−31.63	−15.05	−8.44	−5.11	−3.18	−1.90	−1.07	−0.51
	0.9	−130.32	−35.19	−16.07	−8.70	−5.18	−3.19	−1.88	−1.08	−0.53
1.0	0.2	−585.16	−109.39	−34.85	−13.11	−5.09	−1.86	−0.59	−0.13	−0.01
	0.3	−363.31	−74.20	−26.68	−11.91	−5.90	−3.08	−1.66	−0.94	−0.56
	0.4	−141.46	−39.00	−18.50	−10.71	−6.71	−4.29	−2.74	−1.74	−1.12
	0.5	−63.71	−27.06	−15.85	−10.41	−7.07	−4.77	−3.16	−2.05	−1.31
	0.6	−81.99	−30.62	−16.87	−10.70	−7.13	−4.77	−3.13	−2.02	−1.28
	0.7	−100.28	−34.19	−17.89	−11.00	−7.19	−4.76	−3.11	−1.99	−1.24
	0.8	−118.58	−37.76	−18.91	−11.29	−7.26	−4.76	−3.09	−1.96	−1.20
	0.9	−136.88	−41.32	−19.93	−11.55	−7.32	−4.77	−3.07	−1.98	−1.23
	1.0	−155.18	−44.89	−20.95	−11.80	−7.39	−4.78	−3.05	−1.99	−1.25

ED5-9 Symmetrical Wye, 60 Degree, $D_{b1} \geq D_{b2}$, Converging (*Concluded*)

		C_{b2} Values								
		Q_{b2}/Q_c								
A_{b1}/A_c	A_{b2}/A_c	0.1	0.2	0.3	0.4	0.5	0.6	0.7	0.8	0.9
0.2	0.2	−11.95	−1.89	−0.09	0.41	0.62	0.74	0.80	0.80	0.79
	0.3	−11.95	−1.89	−0.09	0.41	0.62	0.74	0.80	0.80	0.79
0.3	0.2	−8.24	−1.18	0.05	0.42	0.61	0.73	0.78	0.77	0.76
	0.3	−16.88	−2.92	−0.09	0.59	0.86	1.02	1.09	1.10	1.08
0.4	0.2	−6.95	−1.00	0.16	0.53	0.67	0.71	0.72	0.72	0.71
	0.3	−16.21	−2.90	−0.44	0.40	0.79	0.98	1.05	1.06	1.05
	0.4	−28.86	−6.22	−2.15	−0.57	0.19	0.55	0.72	0.79	0.85
0.5	0.2	−4.82	−0.01	0.56	0.71	0.82	0.89	0.92	0.90	0.89
	0.3	−12.27	−1.17	0.44	0.88	1.11	1.25	1.29	1.25	1.23
	0.4	−20.76	−2.93	−0.21	0.48	0.73	0.84	0.88	0.87	0.82
	0.5	−30.58	−5.23	−1.06	0.00	0.32	0.43	0.47	0.47	0.41
0.6	0.2	−3.68	0.07	0.77	0.98	1.06	1.08	1.08	1.06	1.04
	0.3	−9.06	−0.55	0.86	1.27	1.42	1.48	1.49	1.46	1.42
	0.4	−17.62	−2.12	0.06	0.60	0.83	0.95	0.98	0.95	0.91
	0.5	−28.00	−4.26	−0.99	−0.16	0.20	0.39	0.45	0.41	0.38
	0.6	−40.57	−7.86	−2.60	−0.99	−0.26	0.00	0.14	0.21	0.25
0.7	0.2	−5.44	−0.40	0.55	0.86	0.98	1.02	1.04	1.03	1.02
	0.3	−9.36	−0.77	0.73	1.20	1.39	1.47	1.49	1.47	1.44
	0.4	−19.57	−3.09	−0.44	0.36	0.71	0.89	0.97	0.98	0.97
	0.5	−31.88	−6.02	−1.90	−0.63	−0.05	0.26	0.40	0.44	0.46
	0.6	−46.44	−9.82	−3.47	−1.41	−0.48	−0.04	0.21	0.36	0.45
	0.7	−73.02	−16.68	−6.90	−3.29	−1.61	−0.80	−0.29	0.02	0.22
0.8	0.2	−7.21	−0.87	0.33	0.73	0.90	0.97	1.00	1.00	0.99
	0.3	−9.67	−0.99	0.60	1.13	1.36	1.45	1.49	1.48	1.46
	0.4	−21.53	−4.06	−0.93	0.11	0.59	0.83	0.96	1.01	1.03
	0.5	−35.77	−7.77	−2.82	−1.09	−0.29	0.13	0.35	0.48	0.55
	0.6	−52.32	−11.78	−4.34	−1.83	−0.70	−0.09	0.28	0.51	0.65
	0.7	−78.89	−18.64	−7.76	−3.71	−1.83	−0.85	−0.22	0.16	0.42
	0.8	−105.46	−25.49	−11.19	−5.59	−2.96	−1.60	−0.72	−0.18	0.19
0.9	0.2	−4.98	−0.34	0.54	0.85	0.97	1.03	1.04	1.03	1.01
	0.3	−9.97	−1.21	0.48	1.06	1.32	1.44	1.49	1.49	1.48
	0.4	−23.54	−4.98	−1.39	−0.12	0.47	0.78	0.95	1.04	1.09
	0.5	−40.14	−9.57	−3.69	−1.56	−0.55	−0.01	0.31	0.51	0.63
	0.6	−58.25	−14.28	−5.64	−2.53	−1.08	−0.30	0.18	0.49	0.70
	0.7	−84.09	−21.02	−8.91	−4.38	−2.22	−1.04	−0.31	0.15	0.46
	0.8	−109.92	−27.77	−12.18	−6.22	−3.35	−1.79	−0.81	−0.19	0.23
	0.9	−130.32	−35.19	−16.07	−8.70	−5.18	−3.19	−1.88	−1.08	−0.53
1.0	0.2	−2.75	0.19	0.76	0.96	1.05	1.08	1.08	1.06	1.04
	0.3	−10.28	−1.43	0.35	0.99	1.29	1.43	1.49	1.50	1.50
	0.4	−25.56	−5.89	−1.86	−0.36	0.35	0.72	0.93	1.07	1.15
	0.5	−44.52	−11.37	−4.56	−2.02	−0.81	−0.14	0.27	0.54	0.72
	0.6	−64.19	−16.77	−6.94	−3.24	−1.47	−0.50	0.09	0.48	0.74
	0.7	−89.28	−23.41	−10.05	−5.05	−2.61	−1.24	−0.40	0.14	0.50
	0.8	−114.38	−30.04	−13.16	−6.86	−3.75	−1.97	−0.89	−0.20	0.27
	0.9	−134.78	−37.47	−17.06	−9.33	−5.57	−3.38	−1.97	−1.09	−0.49
	1.0	−155.18	−44.89	−20.95	−11.80	−7.39	−4.78	−3.05	−1.99	−1.25

ED7-1 Centrifugal Fan Located in Plenum or Cabinet

L/D_o	0.30	0.40	0.50	0.75
C_o	0.80	0.53	0.40	0.22

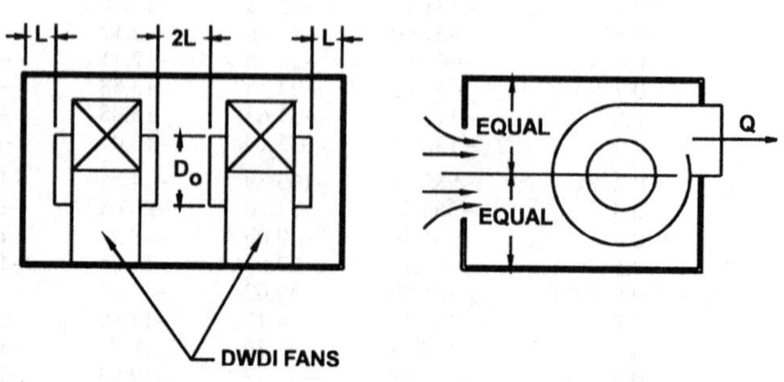

ED7-2 Fan Inlet, Centrifugal, SWSI, with 4 Gore Elbow

	C_o Values			
	L/D_o			
r/D_o	0.0	2.0	5.0	10.0
0.50	1.80	1.00	0.53	0.53
0.75	1.40	0.80	0.40	0.40
1.00	1.20	0.67	0.33	0.33
1.50	1.10	0.60	0.33	0.33
2.00	1.00	0.53	0.33	0.33
3.00	0.67	0.40	0.22	0.22

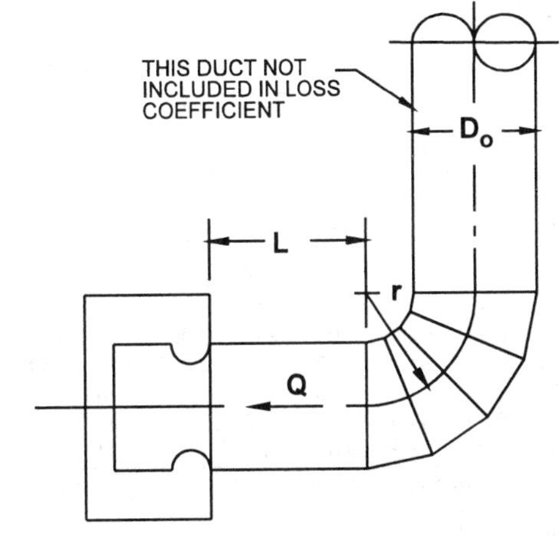

SD2-6 Stackhead

D_e/D	0.3	0.4	0.5	0.6	0.7	0.8	0.9	1.0
C_o	129	41.02	16.80	8.10	4.37	2.56	1.60	1.00

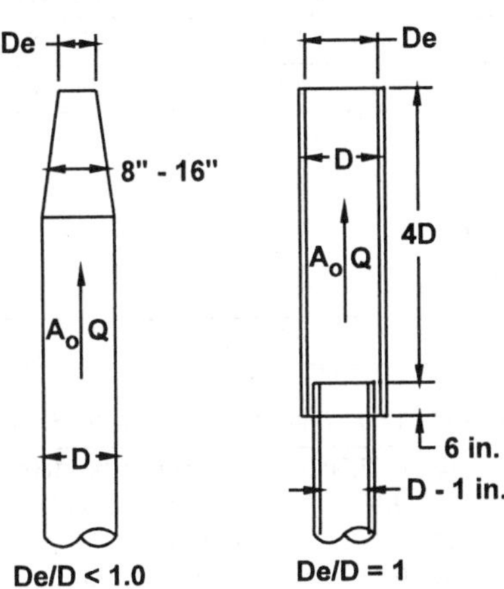

RECTANGULAR FITTINGS

CR3-1 Elbow, Smooth Radius, Without Vanes

	C_p Values										
	H/W										
r/W	0.25	0.50	0.75	1.00	1.50	2.00	3.00	4.00	5.00	6.00	8.00
0.50	1.53	1.38	1.29	1.18	1.06	1.00	1.00	1.06	1.12	1.16	1.18
0.75	0.57	0.52	0.48	0.44	0.40	0.39	0.39	0.40	0.42	0.43	0.44
1.00	0.27	0.25	0.23	0.21	0.19	0.18	0.18	0.19	0.20	0.21	0.21
1.50	0.22	0.20	0.19	0.17	0.15	0.14	0.14	0.15	0.16	0.17	0.17
2.00	0.20	0.18	0.16	0.15	0.14	0.13	0.13	0.14	0.14	0.15	0.15

Angle Factor K

θ	0	20	30	45	60	75	90	110	130	150	180
K	0.00	0.31	0.45	0.60	0.78	0.90	1.00	1.13	1.20	1.28	1.40

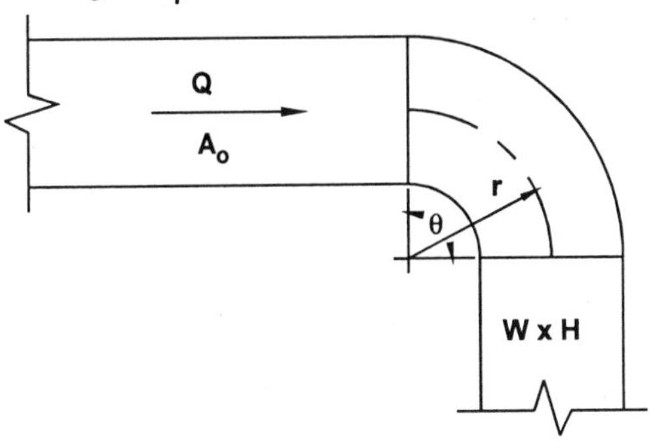

CR3-3 Elbow, Smooth Radius, One Splitter Vane

	C_p Values										
	H/W										
r/W	**0.25**	**0.50**	**1.00**	**1.50**	**2.00**	**3.00**	**4.00**	**5.00**	**6.00**	**7.00**	**8.00**
0.55	0.52	0.40	0.43	0.49	0.55	0.66	0.75	0.84	0.93	1.01	1.09
0.60	0.36	0.27	0.25	0.28	0.30	0.35	0.39	0.42	0.46	0.49	0.52
0.65	0.28	0.21	0.18	0.19	0.20	0.22	0.25	0.26	0.28	0.30	0.32
0.70	0.22	0.16	0.14	0.14	0.15	0.16	0.17	0.18	0.19	0.20	0.21
0.75	0.18	0.13	0.11	0.11	0.11	0.12	0.13	0.14	0.14	0.15	0.15
0.80	0.15	0.11	0.09	0.09	0.09	0.09	0.10	0.10	0.11	0.11	0.12
0.85	0.13	0.09	0.08	0.07	0.07	0.08	0.08	0.08	0.08	0.09	0.09
0.90	0.11	0.08	0.07	0.06	0.06	0.06	0.06	0.07	0.07	0.07	0.07
0.95	0.10	0.07	0.06	0.05	0.05	0.05	0.05	0.05	0.06	0.06	0.06
1.00	0.09	0.06	0.05	0.05	0.04	0.04	0.04	0.05	0.05	0.05	0.05

Angle Factor K

θ	0	30	45	60	90
K	0.00	0.45	0.60	0.78	1.00

Curve Ratio CR

r/W	0.55	0.60	0.65	0.70	0.75	0.80	0.85	0.90	0.95	1.00
CR	0.218	0.302	0.361	0.408	0.447	0.480	0.509	0.535	0.557	0.577

Throat Radius/Width Ratio (R/W)

r/W	0.55	0.60	0.65	0.70	0.75	0.80	0.85	0.90	0.95	1.00
R/W	0.05	0.10	0.15	0.20	0.25	0.30	0.35	0.40	0.45	0.50

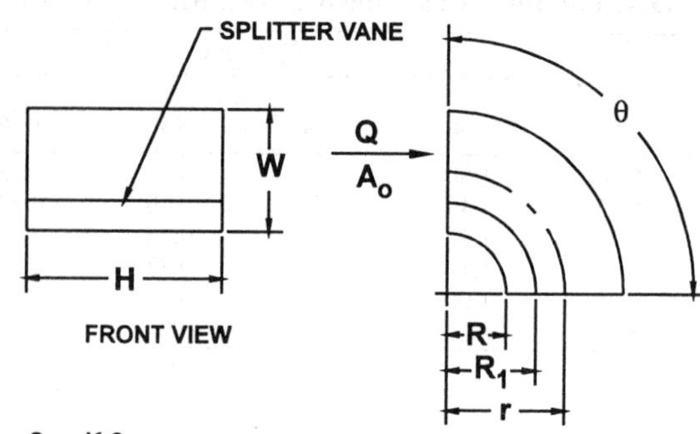

$C_o = K\,C_p$
$R_1 = R/CR$

where
R = throat radius
R_1 = splitter vane radius
CR = curve ratio
K = angle factor

CR3-6 Elbow, Mitered

	C_o Values										
	H/W										
θ	**0.25**	**0.50**	**0.75**	**1.00**	**1.50**	**2.00**	**3.00**	**4.00**	**5.00**	**6.00**	**8.00**
20	0.08	0.08	0.08	0.07	0.07	0.07	0.06	0.06	0.05	0.05	0.05
30	0.18	0.17	0.17	0.16	0.15	0.15	0.13	0.13	0.12	0.12	0.11
45	0.38	0.37	0.36	0.34	0.33	0.31	0.28	0.27	0.26	0.25	0.24
60	0.60	0.59	0.57	0.55	0.52	0.49	0.46	0.43	0.41	0.39	0.38
75	0.89	0.87	0.84	0.81	0.77	0.73	0.67	0.63	0.61	0.58	0.57
90	1.30	1.27	1.23	1.18	1.13	1.07	0.98	0.92	0.89	0.85	0.83

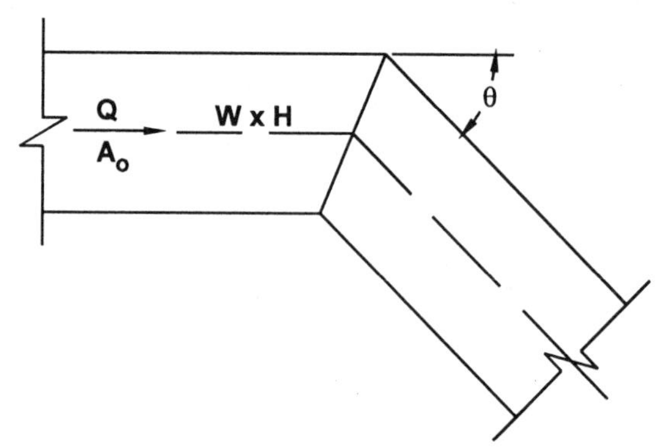

CR3-10 Elbow, Mitered, 90 Degree, Single-Thickness Vanes (Design 2)

$C_o = 0.12$

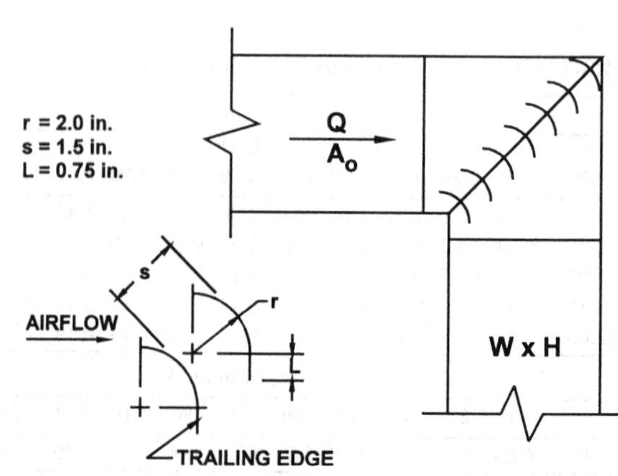

CR3-17 Elbow, Z-Shaped

	C_p Values													
	L/W													
H/W	0.0	0.4	0.6	0.8	1.0	1.2	1.4	1.6	1.8	2.0	4.0	8.0	10.0	100.0
0.25	0.00	0.68	0.99	1.77	2.89	3.97	4.41	4.60	4.64	4.60	3.39	3.03	2.70	2.53
0.50	0.00	0.66	0.96	1.72	2.81	3.86	4.29	4.47	4.52	4.47	3.30	2.94	2.62	2.46
0.75	0.00	0.64	0.94	1.67	2.74	3.75	4.17	4.35	4.39	4.35	3.20	2.86	2.55	2.39
1.00	0.00	0.62	0.90	1.61	2.63	3.61	4.01	4.18	4.22	4.18	3.08	2.75	2.45	2.30
1.50	0.00	0.59	0.86	1.53	2.50	3.43	3.81	3.97	4.01	3.97	2.93	2.61	2.33	2.19
2.00	0.00	0.56	0.81	1.45	2.37	3.25	3.61	3.76	3.80	3.76	2.77	2.48	2.21	2.07
3.00	0.00	0.51	0.75	1.34	2.18	3.00	3.33	3.47	3.50	3.47	2.56	2.28	2.03	1.91
4.00	0.00	0.48	0.70	1.26	2.05	2.82	3.13	3.26	3.29	3.26	2.40	2.15	1.91	1.79
6.00	0.00	0.45	0.65	1.16	1.89	2.60	2.89	3.01	3.04	3.01	2.22	1.98	1.76	1.66
8.00	0.00	0.43	0.63	1.13	1.84	2.53	2.81	2.93	2.95	2.93	2.16	1.93	1.72	1.61

Reynolds No. Correction Factor K_r

Re/1000	10	20	30	40	60	80	100	140	500
K_r	1.40	1.26	1.19	1.14	1.09	1.06	1.04	1.00	1.00

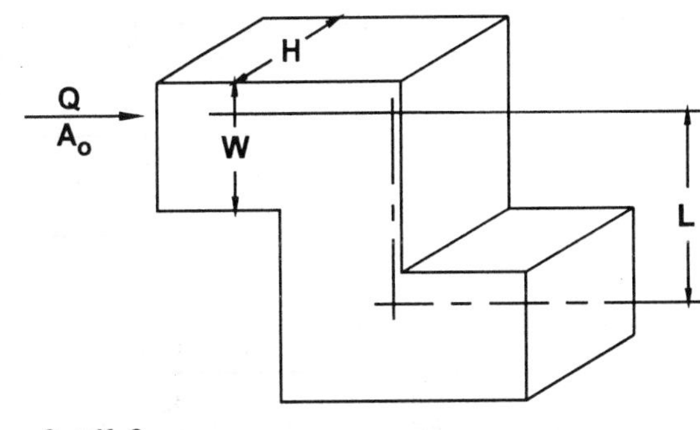

$C_o = K_r C_p$

where K_r = Reynolds no. correction factor

CR6-1 Screen (Only)

	C_o Values												
	n												
A_1/A_o	0.30	0.35	0.40	0.45	0.50	0.55	0.60	0.65	0.70	0.75	0.80	0.90	1.00
0.2	155.00	102.50	75.00	55.00	41.25	31.50	24.25	18.75	14.50	11.00	8.00	3.50	0.00
0.3	68.89	45.56	33.33	24.44	18.33	14.00	10.78	8.33	6.44	4.89	3.56	1.56	0.00
0.4	38.75	25.63	18.75	13.75	10.31	7.88	6.06	4.69	3.63	2.75	2.00	0.88	0.00
0.5	24.80	16.40	12.00	8.80	6.60	5.04	3.88	3.00	2.32	1.76	1.28	0.56	0.00
0.6	17.22	11.39	8.33	6.11	4.58	3.50	2.69	2.08	1.61	1.22	0.89	0.39	0.00
0.7	12.65	8.37	6.12	4.49	3.37	2.57	1.98	1.53	1.18	0.90	0.65	0.29	0.00
0.8	9.69	6.40	4.69	3.44	2.58	1.97	1.52	1.17	0.91	0.69	0.50	0.22	0.00
0.9	7.65	5.06	3.70	2.72	2.04	1.56	1.20	0.93	0.72	0.54	0.40	0.17	0.00
1.0	6.20	4.10	3.00	2.20	1.65	1.26	0.97	0.75	0.58	0.44	0.32	0.14	0.00
1.2	4.31	2.85	2.08	1.53	1.15	0.88	0.67	0.36	0.40	0.31	0.22	0.10	0.00
1.4	3.16	2.09	1.53	1.12	0.84	0.64	0.49	0.38	0.30	0.22	0.16	0.07	0.00
1.6	2.42	1.60	1.17	0.86	0.64	0.49	0.38	0.29	0.23	0.17	0.13	0.05	0.00
1.8	1.91	1.27	0.93	0.68	0.51	0.39	0.30	0.23	0.18	0.14	0.10	0.04	0.00
2.0	1.55	1.03	0.75	0.55	0.41	0.32	0.24	0.19	0.15	0.11	0.08	0.04	0.00
2.5	0.99	0.66	0.48	0.35	0.26	0.20	0.16	0.12	0.09	0.07	0.05	0.02	0.00
3.0	0.69	0.46	0.33	0.24	0.18	0.14	0.11	0.08	0.06	0.05	0.04	0.02	0.00
4.0	0.39	0.26	0.19	0.14	0.10	0.08	0.06	0.05	0.04	0.03	0.02	0.01	0.00
6.0	0.17	0.11	0.08	0.06	0.05	0.04	0.03	0.02	0.02	0.01	0.01	0.00	0.00

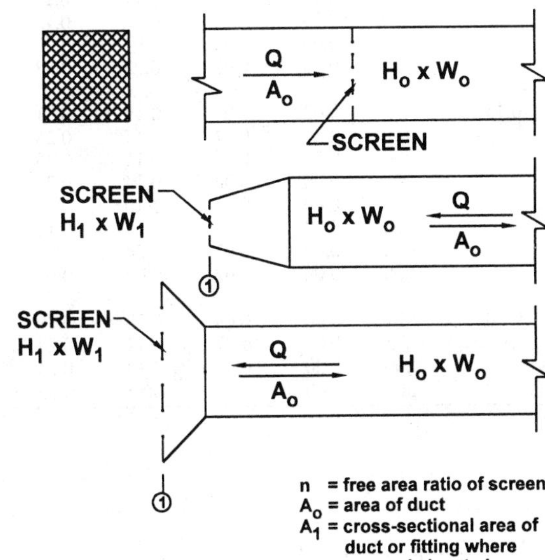

n = free area ratio of screen
A_o = area of duct
A_1 = cross-sectional area of duct or fitting where screen is located

CR6-4 Obstruction, Smooth Cylinder in Rectangular Duct

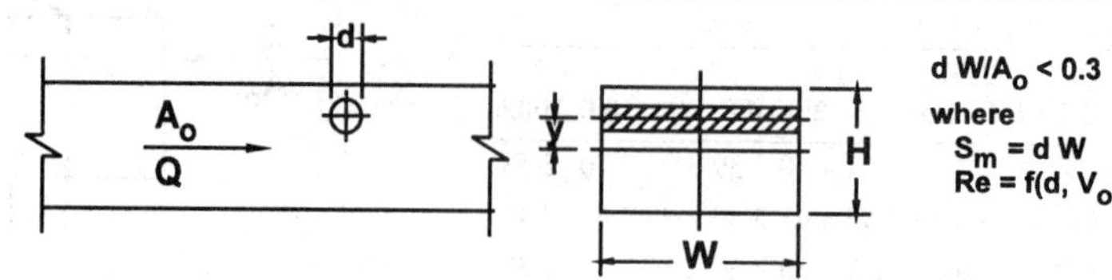

		C_o Values				
		S_m/A_o				
y/H	Re/1000	0.00	0.05	0.10	0.15	0.20
0.00	0.1	0.00	0.10	0.21	0.35	0.47
	0.5	0.00	0.08	0.17	0.28	0.38
	200	0.00	0.08	0.17	0.28	0.38
	300	0.00	0.07	0.16	0.26	0.35
	400	0.00	0.05	0.11	0.19	0.25
	500	0.00	0.04	0.09	0.14	0.19
	600	0.00	0.02	0.05	0.07	0.10
	1000	0.00	0.02	0.05	0.08	0.11
0.05	0.1	0.00	0.10	0.21	0.34	0.46
	0.5	0.00	0.08	0.17	0.27	0.37
	200	0.00	0.08	0.17	0.27	0.37
	300	0.00	0.07	0.15	0.25	0.34
	400	0.00	0.05	0.11	0.18	0.24
	500	0.00	0.04	0.08	0.13	0.18
	600	0.00	0.02	0.04	0.07	0.10
	1000	0.00	0.02	0.05	0.08	0.11
0.10	0.1	0.00	0.09	0.20	0.32	0.44
	0.5	0.00	0.07	0.16	0.26	0.35
	200	0.00	0.07	0.16	0.26	0.35
	300	0.00	0.07	0.15	0.24	0.32
	400	0.00	0.05	0.11	0.17	0.23
	500	0.00	0.04	0.08	0.13	0.18
	600	0.00	0.02	0.04	0.07	0.09
	1000	0.00	0.02	0.05	0.08	0.10
0.15	0.1	0.00	0.09	0.19	0.31	0.42
	0.5	0.00	0.07	0.15	0.25	0.34
	200	0.00	0.07	0.15	0.25	0.34
	300	0.00	0.06	0.14	0.23	0.31
	400	0.00	0.05	0.10	0.17	0.22
	500	0.00	0.04	0.08	0.12	0.17
	600	0.00	0.02	0.04	0.07	0.09
	1000	0.00	0.02	0.04	0.07	0.10
0.20	0.1	0.00	0.08	0.18	0.29	0.40
	0.5	0.00	0.07	0.14	0.24	0.32
	200	0.00	0.07	0.14	0.24	0.32
	300	0.00	0.06	0.13	0.22	0.29
	400	0.00	0.04	0.10	0.16	0.21
	500	0.00	0.03	0.07	0.12	0.16
	600	0.00	0.02	0.04	0.06	0.09
	1000	0.00	0.02	0.04	0.07	0.09
0.25	0.1	0.00	0.08	0.17	0.28	0.38
	0.5	0.00	0.06	0.14	0.22	0.30
	200	0.00	0.06	0.14	0.22	0.30
	300	0.00	0.06	0.12	0.20	0.28
	400	0.00	0.04	0.09	0.15	0.20
	500	0.00	0.03	0.07	0.11	0.15
	600	0.00	0.02	0.04	0.06	0.08
	1000	0.00	0.02	0.04	0.06	0.09
0.30	0.1	0.00	0.07	0.16	0.26	0.35
	0.5	0.00	0.06	0.13	0.21	0.28
	200	0.00	0.06	0.13	0.21	0.28
	300	0.00	0.05	0.12	0.19	0.26
	400	0.00	0.04	0.08	0.14	0.19
	500	0.00	0.03	0.06	0.10	0.14
	600	0.00	0.02	0.03	0.05	0.07
	1000	0.00	0.02	0.04	0.06	0.08
0.35	0.1	0.00	0.07	0.14	0.23	0.32
	0.5	0.00	0.05	0.11	0.19	0.25
	200	0.00	0.05	0.11	0.19	0.25
	300	0.00	0.05	0.11	0.17	0.23
	400	0.00	0.04	0.08	0.12	0.17
	500	0.00	0.03	0.06	0.09	0.13
	600	0.00	0.01	0.03	0.05	0.07
	1000	0.00	0.02	0.03	0.05	0.07
0.40	0.1	0.00	0.06	0.13	0.20	0.28
	0.5	0.00	0.05	0.10	0.16	0.22
	200	0.00	0.05	0.10	0.16	0.22
	300	0.00	0.04	0.09	0.15	0.20
	400	0.00	0.03	0.07	0.11	0.15
	500	0.00	0.02	0.05	0.08	0.11
	600	0.00	0.01	0.03	0.04	0.06
	1000	0.00	0.01	0.03	0.05	0.06

CR9-1 Damper, Butterfly

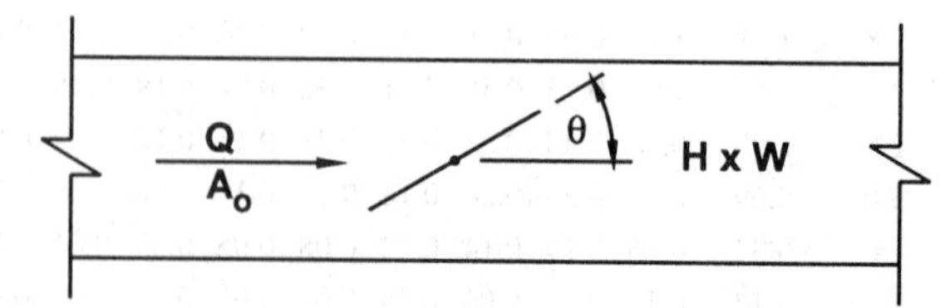

	C_o Values									
	θ									
H/W	0	10	20	30	40	50	60	65	70	90
0.12	0.04	0.30	1.10	3.00	8.00	23.00	60.00	100.00	190.00	99999
0.25	0.08	0.33	1.18	3.30	9.00	26.00	70.00	128.00	210.00	99999
1.00	0.08	0.33	1.18	3.30	9.00	26.00	70.00	128.00	210.00	99999
2.00	0.13	0.35	1.25	3.60	10.00	29.00	80.00	155.00	230.00	99999

CR9-4 Damper, Opposed Blades

	C_o Values								
	θ								
L/R	0	10	20	30	40	50	60	70	80
0.3	0.52	0.79	1.91	3.77	8.55	19.46	70.12	295.21	807.23
0.4	0.52	0.85	2.07	4.61	10.42	26.73	92.90	346.25	926.34
0.5	0.52	0.93	2.25	5.44	12.29	33.99	118.91	393.36	1045.44
0.6	0.52	1.00	2.46	5.99	14.15	41.26	143.69	440.25	1163.09
0.8	0.52	1.08	2.66	6.96	18.18	56.47	193.92	520.27	1324.85
1.0	0.52	1.17	2.91	7.31	20.25	71.68	245.45	576.00	1521.00
1.5	0.52	1.38	3.16	9.51	27.56	107.41	361.00	717.05	1804.40

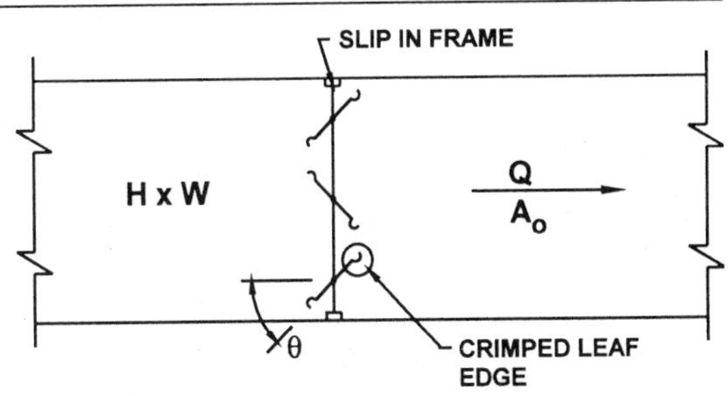

$$L/R = \frac{N\,W}{2(H+W)}$$

where
N = number of damper blades
W = duct dimension parallel to blade axis, in.
H = duct height, in.
L = sum of damper blade lengths, in.
R = perimeter of duct, in.

CR9-6 Fire Damper, Curtain Type, Type B

$C_o = 0.19$

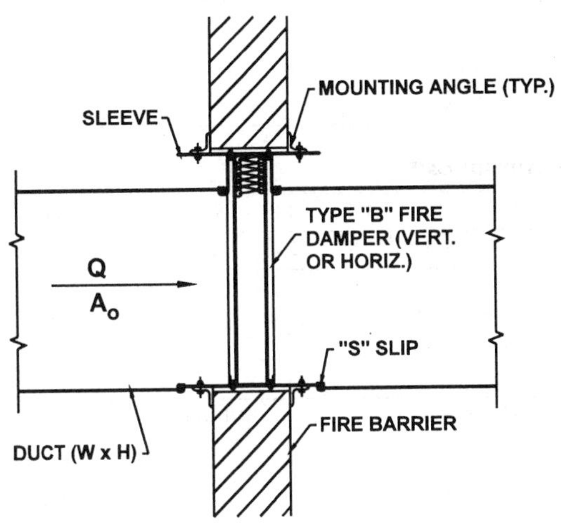

ER2-1 Bellmouth, Plenum to Round, Exhaust/Return Systems

	C_o Values												
	r/D_1												
A_o/A_1	0.00	0.01	0.02	0.03	0.04	0.05	0.06	0.08	0.10	0.12	0.16	0.20	10.00
1.5	0.22	0.20	0.15	0.14	0.12	0.10	0.09	0.07	0.05	0.04	0.03	0.01	0.01
2.0	0.13	0.11	0.08	0.08	0.07	0.06	0.05	0.04	0.03	0.02	0.02	0.01	0.01
2.5	0.08	0.07	0.05	0.05	0.04	0.04	0.03	0.02	0.02	0.01	0.01	0.00	0.00
3.0	0.06	0.05	0.04	0.03	0.03	0.02	0.02	0.02	0.01	0.01	0.01	0.00	0.00
4.0	0.03	0.03	0.02	0.02	0.02	0.01	0.01	0.01	0.01	0.01	0.00	0.00	0.00
8.0	0.01	0.01	0.01	0.00	0.00	0.00	0.00	0.00	0.00	0.00	0.00	0.00	0.00

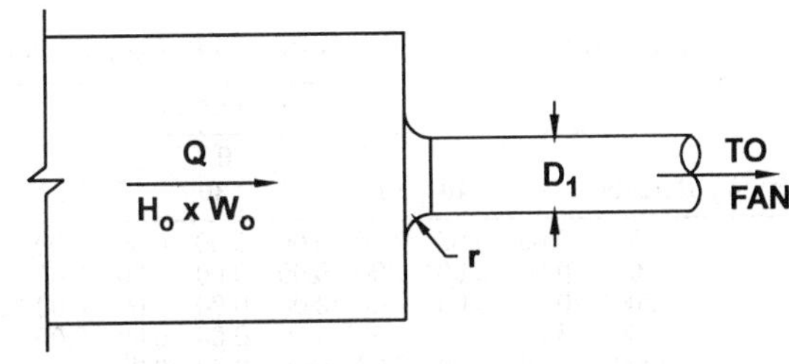

ER3-1 Elbow, 90 Degree, Variable Inlet/Outlet Areas, Exhaust/Return Systems

	C_o Values						
	W_1/W_o						
H/W_o	0.6	0.8	1.0	1.2	1.4	1.6	2.0
0.25	1.76	1.43	1.24	1.14	1.09	1.06	1.06
1.00	1.70	1.36	1.15	1.02	0.95	0.90	0.84
4.00	1.46	1.10	0.90	0.81	0.76	0.72	0.66
100.00	1.50	1.04	0.79	0.69	0.63	0.60	0.55

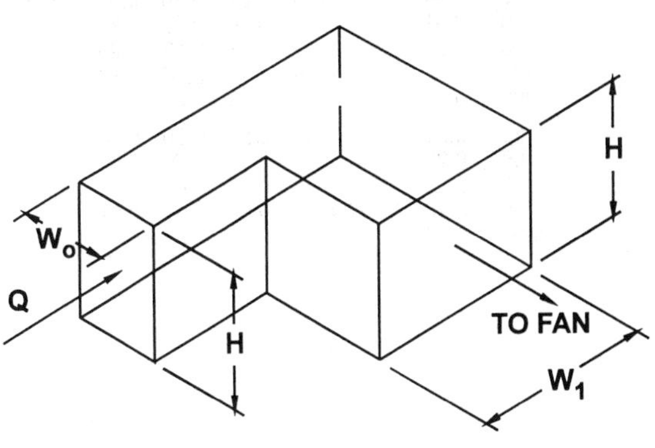

ER4-3 Transition, Rectangular to Round, Exhaust/Return Systems

	C_o Values									
	θ									
A_o/A_1	10	15	20	30	45	60	90	120	150	180
0.06	0.30	0.54	0.53	0.65	0.77	0.88	0.95	0.98	0.98	0.93
0.10	0.30	0.50	0.53	0.64	0.75	0.84	0.89	0.91	0.91	0.88
0.25	0.25	0.36	0.45	0.52	0.58	0.62	0.64	0.64	0.64	0.64
0.50	0.15	0.21	0.25	0.30	0.33	0.33	0.33	0.32	0.31	0.30
1.00	0.00	0.00	0.00	0.00	0.00	0.00	0.00	0.00	0.00	0.00
2.00	0.24	0.28	0.26	0.20	0.22	0.24	0.49	0.73	0.97	1.04
4.00	0.89	0.78	0.79	0.70	0.88	1.12	2.72	4.33	5.62	6.58
6.00	1.89	1.67	1.59	1.49	1.98	2.52	6.51	10.14	13.05	15.14
10.00	5.09	5.32	5.15	5.05	6.50	8.05	19.06	29.07	37.08	43.05

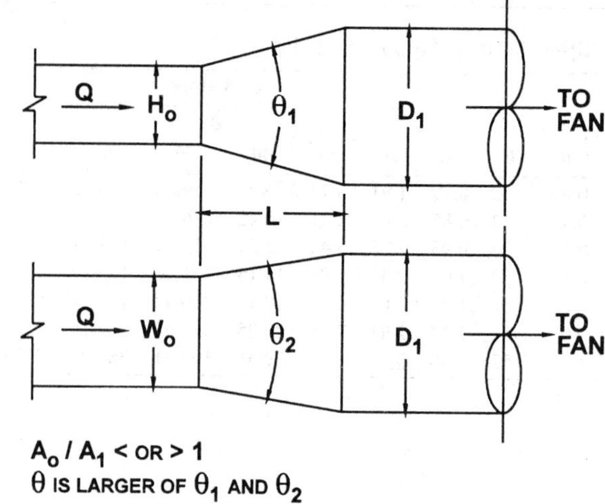

SR2-1 Abrupt Exit

H/W	0.1	0.2	0.9	1.0	1.1	4.0	5.0	10.0
C_o	1.55	1.55	1.55	2.00	1.55	1.55	1.55	1.55

$C_o = 1.0$

Note: Table is LAMINAR flow; $C_o = 1.0$ is TURBULENT flow.

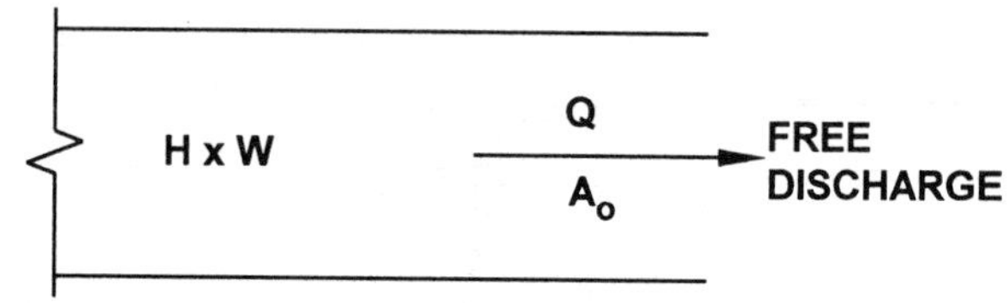

SR2-3 Plain Diffuser (Two Sides Parallel), Free Discharge

		C_o Values								
		θ								
A_1/A_o	Re/1000	8	10	14	20	30	45	60	90	120
1	50	0.00	0.00	0.00	0.00	0.00	0.00	0.00	0.00	0.00
	100	0.00	0.00	0.00	0.00	0.00	0.00	0.00	0.00	0.00
	200	0.00	0.00	0.00	0.00	0.00	0.00	0.00	0.00	0.00
	400	0.00	0.00	0.00	0.00	0.00	0.00	0.00	0.00	0.00
	2000	0.00	0.00	0.00	0.00	0.00	0.00	0.00	0.00	0.00
2	50	0.50	0.51	0.56	0.63	0.80	0.96	1.04	1.09	1.09
	100	0.48	0.50	0.56	0.63	0.80	0.96	1.04	1.09	1.09
	200	0.44	0.47	0.53	0.63	0.74	0.93	1.02	1.08	1.08
	400	0.40	0.42	0.50	0.62	0.74	0.93	1.02	1.08	1.08
	2000	0.40	0.42	0.50	0.62	0.74	0.93	1.02	1.08	1.08
4	50	0.34	0.38	0.48	0.63	0.76	0.91	1.03	1.07	1.07
	100	0.31	0.36	0.45	0.59	0.72	0.88	1.02	1.07	1.07
	200	0.26	0.31	0.41	0.53	0.67	0.83	0.96	1.06	1.06
	400	0.22	0.27	0.39	0.53	0.67	0.83	0.96	1.06	1.06
	2000	0.22	0.27	0.39	0.53	0.67	0.83	0.96	1.06	1.06
6	50	0.32	0.34	0.41	0.56	0.70	0.84	0.96	1.08	1.08
	100	0.27	0.30	0.41	0.56	0.70	0.84	0.96	1.08	1.08
	200	0.24	0.27	0.36	0.52	0.67	0.81	0.94	1.06	1.06
	400	0.20	0.24	0.36	0.52	0.67	0.81	0.94	1.06	1.06
	2000	0.18	0.24	0.34	0.50	0.67	0.81	0.94	1.05	1.05

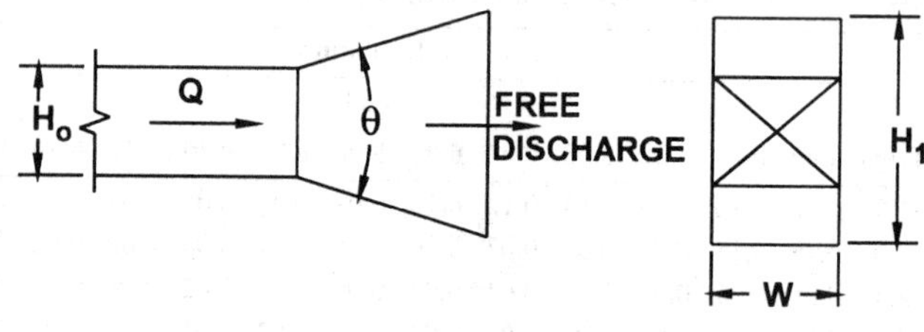

SR2-5 Pyramidal Diffuser, Free Discharge

θ IS LARGER OF θ_1 AND θ_2

		C_o Values								
		θ								
A_1/A_o	Re/1000	8	10	14	20	30	45	60	90	120
1	50	0.00	0.00	0.00	0.00	0.00	0.00	0.00	0.00	0.00
	100	0.00	0.00	0.00	0.00	0.00	0.00	0.00	0.00	0.00
	200	0.00	0.00	0.00	0.00	0.00	0.00	0.00	0.00	0.00
	400	0.00	0.00	0.00	0.00	0.00	0.00	0.00	0.00	0.00
	2000	0.00	0.00	0.00	0.00	0.00	0.00	0.00	0.00	0.00
2	50	0.65	0.68	0.74	0.82	0.92	1.05	1.10	1.08	1.08
	100	0.61	0.66	0.73	0.81	0.90	1.04	1.09	1.08	1.08
	200	0.57	0.61	0.70	0.79	0.89	1.04	1.09	1.08	1.08
	400	0.50	0.56	0.64	0.76	0.88	1.02	1.07	1.08	1.08
	2000	0.50	0.56	0.64	0.76	0.88	1.02	1.07	1.08	1.08
4	50	0.53	0.60	0.69	0.78	0.90	1.02	1.07	1.09	1.09
	100	0.49	0.55	0.66	0.78	0.90	1.02	1.07	1.09	1.09
	200	0.42	0.50	0.62	0.74	0.87	1.00	1.06	1.08	1.08
	400	0.36	0.44	0.56	0.70	0.84	0.99	1.06	1.08	1.08
	2000	0.36	0.44	0.56	0.70	0.84	0.99	1.06	1.08	1.08
6	50	0.50	0.57	0.66	0.77	0.91	1.02	1.07	1.08	1.08
	100	0.47	0.54	0.63	0.76	0.98	1.02	1.07	1.08	1.08
	200	0.42	0.48	0.60	0.73	0.88	1.00	1.06	1.08	1.08
	400	0.34	0.44	0.56	0.73	0.86	0.98	1.06	1.08	1.08
	2000	0.34	0.44	0.56	0.73	0.86	0.98	1.06	1.08	1.08
10	50	0.45	0.53	0.64	0.74	0.85	0.97	1.10	1.12	1.12
	100	0.40	0.48	0.62	0.73	0.85	0.97	1.10	1.12	1.12
	200	0.34	0.44	0.56	0.69	0.82	0.95	1.10	1.11	1.11
	400	0.28	0.40	0.55	0.67	0.80	0.93	1.09	1.11	1.11
	2000	0.28	0.40	0.55	0.67	0.80	0.93	1.09	1.11	1.11

SR2-6 Pyramidal Diffuser, with Wall

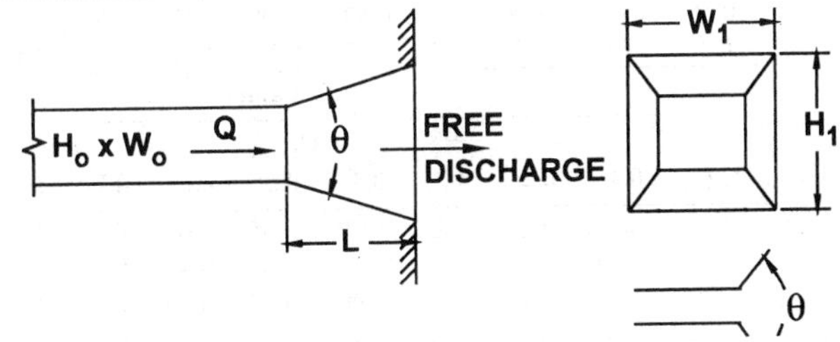

L/D_h	0.5	1.0	2.0	3.0	4.0	5.0	6.0	8.0	10.0	12.0	14.0
C_o	0.49	0.40	0.30	0.26	0.23	0.21	0.19	0.17	0.16	0.15	0.14
θ	26	19	13	11	9	8	7	6	6	5	5

θ is the optimum angle.

SR3-1 Elbow, 90 Degree, Variable Inlet/Outlet Areas, Supply Air Systems

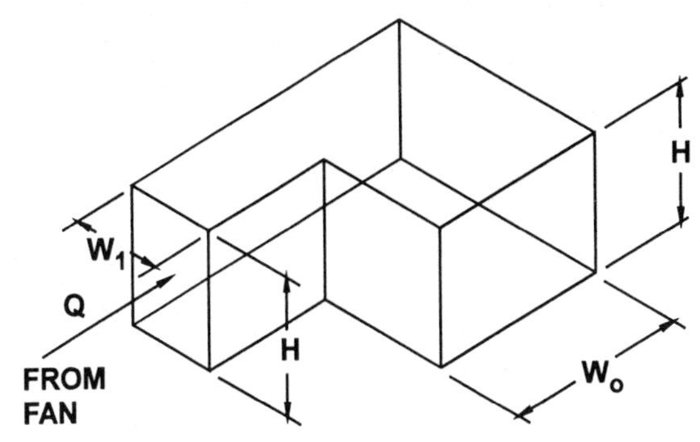

	C_o Values						
	W_o/W_1						
H/W_1	0.6	0.8	1.0	1.2	1.4	1.6	2.0
0.25	0.63	0.92	1.24	1.64	2.14	2.71	4.24
1.00	0.61	0.87	1.15	1.47	1.86	2.30	3.36
4.00	0.53	0.70	0.90	1.17	1.49	1.84	2.64
100.00	0.54	0.67	0.79	0.99	1.23	1.54	2.20

SR4-1 Transition, Rectangular, Two Sides Parallel, Symmetrical, Supply Air Systems

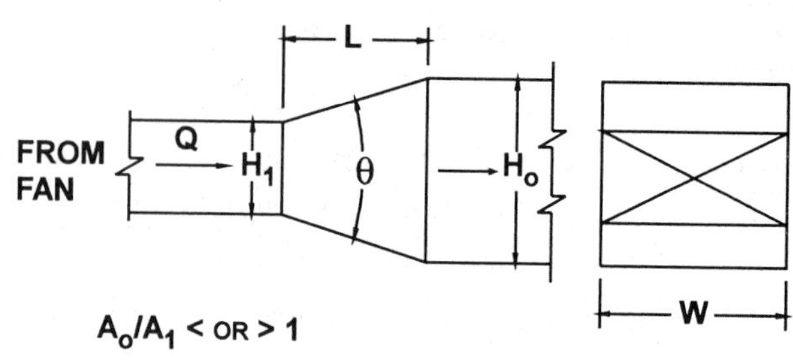

A_o/A_1	C_o Values, θ = 10	15	20	30	45	60	90	120	150	180
0.10	0.05	0.05	0.05	0.05	0.07	0.08	0.19	0.29	0.37	0.43
0.17	0.05	0.04	0.04	0.04	0.05	0.07	0.18	0.28	0.36	0.42
0.25	0.05	0.04	0.04	0.04	0.06	0.07	0.17	0.27	0.35	0.41
0.50	0.06	0.05	0.05	0.05	0.06	0.07	0.14	0.20	0.26	0.27
1.00	0.00	0.00	0.00	0.00	0.00	0.00	0.00	0.00	0.00	1.00
2.00	0.56	0.52	0.60	0.96	1.40	1.48	1.52	1.48	1.44	1.40
4.00	2.72	3.04	3.52	6.72	9.60	10.88	11.20	11.04	10.72	10.56
10.00	24.00	26.00	36.00	53.00	69.00	82.00	93.00	93.00	92.00	91.00
16.00	66.56	69.12	102.40	143.36	181.76	220.16	256.00	253.44	250.88	250.88

SR5-1 Smooth Wye of Type $A_s + A_b \geq A_c$, Branch 90° to Main, Diverging

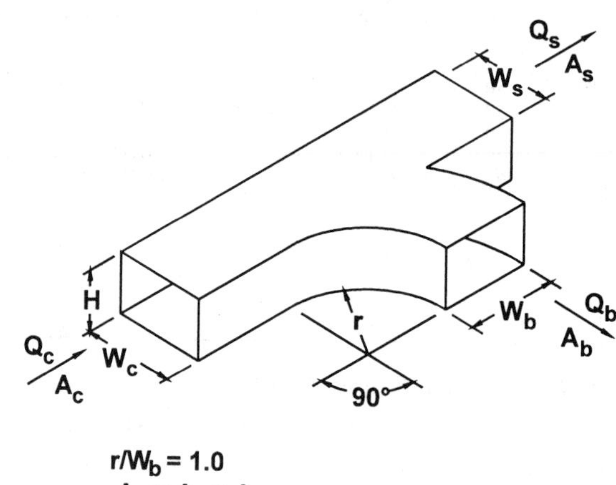

A_s/A_c	A_b/A_c	C_o Values, Q_b/Q_c = 0.1	0.2	0.3	0.4	0.5	0.6	0.7	0.8	0.9
0.50	0.25	3.44	0.70	0.30	0.20	0.17	0.16	0.16	0.17	0.18
	0.50	11.00	2.37	1.06	0.64	0.52	0.47	0.47	0.47	0.48
	1.00	60.00	13.00	4.78	2.06	0.96	0.47	0.31	0.27	0.26
0.75	0.25	2.19	0.55	0.35	0.31	0.33	0.35	0.36	0.37	0.39
	0.50	13.00	2.50	0.89	0.47	0.34	0.31	0.32	0.36	0.43
	1.00	70.00	15.00	5.67	2.62	1.36	0.78	0.53	0.41	0.36
1.00	0.25	3.44	0.78	0.42	0.33	0.30	0.31	0.40	0.42	0.46
	0.50	15.50	3.00	1.11	0.62	0.48	0.42	0.40	0.42	0.46
	1.00	67.00	13.75	5.11	2.31	1.28	0.81	0.59	0.47	0.46

A_s/A_c	A_b/A_c	C_s Values, Q_s/Q_c = 0.1	0.2	0.3	0.4	0.5	0.6	0.7	0.8	0.9
0.50	0.25	8.75	1.62	0.50	0.17	0.05	0.00	−0.02	−0.02	0.00
	0.50	7.50	1.12	0.25	0.06	0.05	0.09	0.14	0.19	0.22
	1.00	5.00	0.62	0.17	0.08	0.08	0.09	0.12	0.15	0.19
0.75	0.25	19.13	3.38	1.00	0.28	0.05	−0.02	−0.02	0.00	0.06
	0.50	20.81	3.23	0.75	0.14	−0.02	−0.05	−0.05	−0.02	0.03
	1.00	16.88	2.81	0.63	0.11	−0.02	−0.05	0.01	0.00	0.07
1.00	0.25	46.00	9.50	3.22	1.31	0.52	0.14	−0.02	−0.05	−0.01
	0.50	35.00	6.75	2.11	0.75	0.24	0.00	−0.10	−0.09	−0.04
	1.00	38.00	7.50	2.44	0.81	0.24	−0.03	−0.08	−0.06	−0.02

SR5-13 Tee, 45 Degree Entry Branch, Diverging

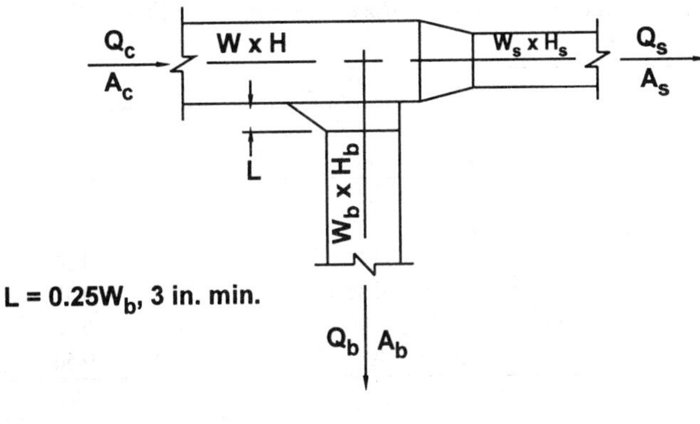

	C_b Values								
	Q_b/Q_c								
A_b/A_c	0.1	0.2	0.3	0.4	0.5	0.6	0.7	0.8	0.9
0.1	0.73	0.34	0.32	0.34	0.35	0.37	0.38	0.39	0.40
0.2	3.10	0.73	0.41	0.34	0.32	0.32	0.33	0.34	0.35
0.3	7.59	1.65	0.73	0.47	0.37	0.34	0.32	0.32	0.32
0.4	14.20	3.10	1.28	0.73	0.51	0.41	0.36	0.34	0.32
0.5	22.92	5.08	2.07	1.12	0.73	0.54	0.44	0.38	0.35
0.6	33.76	7.59	3.10	1.65	1.03	0.73	0.56	0.47	0.41
0.7	46.71	10.63	4.36	2.31	1.42	0.98	0.73	0.58	0.49
0.8	61.79	14.20	5.86	3.10	1.90	1.28	0.94	0.73	0.60
0.9	78.98	18.29	7.59	4.02	2.46	1.65	1.19	0.91	0.73

	C_s Values								
	Q_s/Q_c								
A_s/A_c	0.1	0.2	0.3	0.4	0.5	0.6	0.7	0.8	0.9
0.1	0.04								
0.2	0.98	0.04							
0.3	3.48	0.31	0.04						
0.4	7.55	0.98	0.18	0.04					
0.5	13.18	2.03	0.49	0.13	0.04				
0.6	20.38	3.48	0.98	0.31	0.10	0.04			
0.7	29.15	5.32	1.64	0.60	0.23	0.09	0.04		
0.8	39.48	7.55	2.47	0.98	0.42	0.18	0.08	0.04	
0.9	51.37	10.17	3.48	1.46	0.67	0.31	0.15	0.07	0.04

SR5-14 Wye, Symmetrical, Dovetail, $Q_b/Q_c = 0.5$, Diverging

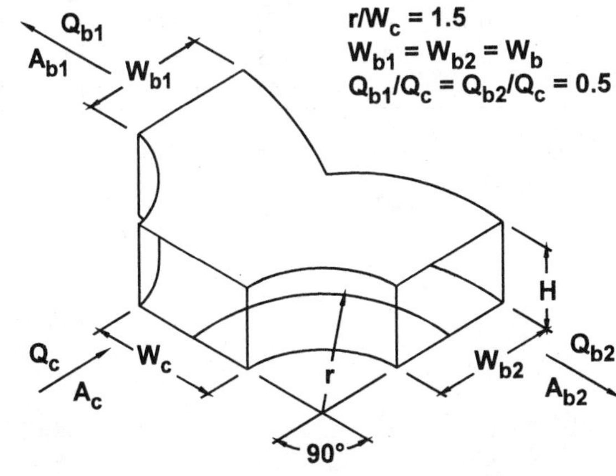

A_b/A_c	0.5	1.0
C_b	0.30	1.00

Branches are identical: $Q_{b1} = Q_{b2} = Q_b$, and $C_{b1} = C_{b2} = C_b$

SR7-1 Fan, Centrifugal, Without Outlet Diffuser, Free Discharge

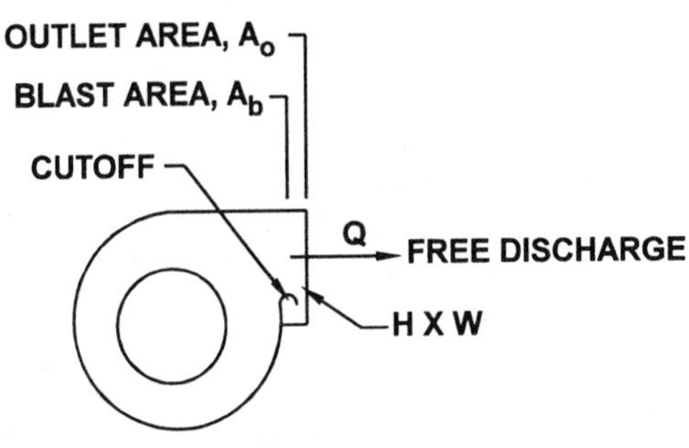

A_b/A_o	0.4	0.5	0.6	0.7	0.8	0.9	1.0
C_o	2.00	2.00	1.00	0.80	0.47	0.22	0.00

SR7-2 Plane Asymmetric Diffuser at Centrifugal Fan Outlet, Free Discharge

	C_o Values					
	A_1/A_o					
θ	1.5	2.0	2.5	3.0	3.5	4.0
10	0.51	0.34	0.25	0.21	0.18	0.17
15	0.54	0.36	0.27	0.24	0.22	0.20
20	0.55	0.38	0.31	0.27	0.25	0.24
25	0.59	0.43	0.37	0.35	0.33	0.33
30	0.63	0.50	0.46	0.44	0.43	0.42
35	0.65	0.56	0.53	0.52	0.51	0.50

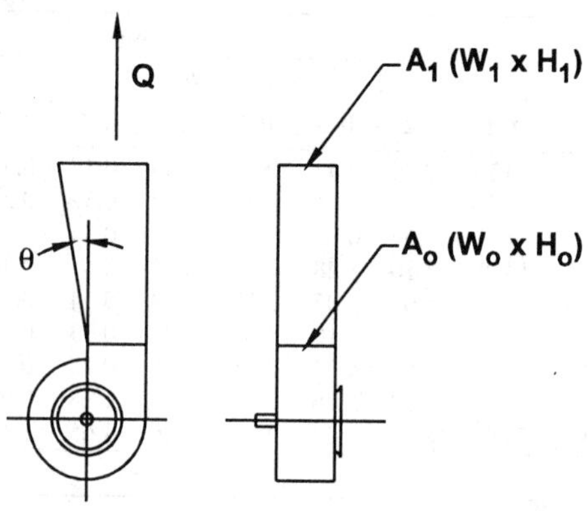

SR7-17 Pyramidal Diffuser at Centrifugal Fan Outlet with Ductwork

	C_1 Values					
	A_o/A_1					
θ	1.5	2.0	2.5	3.0	3.5	4.0
10	0.10	0.18	0.21	0.23	0.24	0.25
15	0.23	0.33	0.38	0.40	0.42	0.44
20	0.31	0.43	0.48	0.53	0.56	0.58
25	0.36	0.49	0.55	0.58	0.62	0.64
30	0.42	0.53	0.59	0.64	0.67	0.69

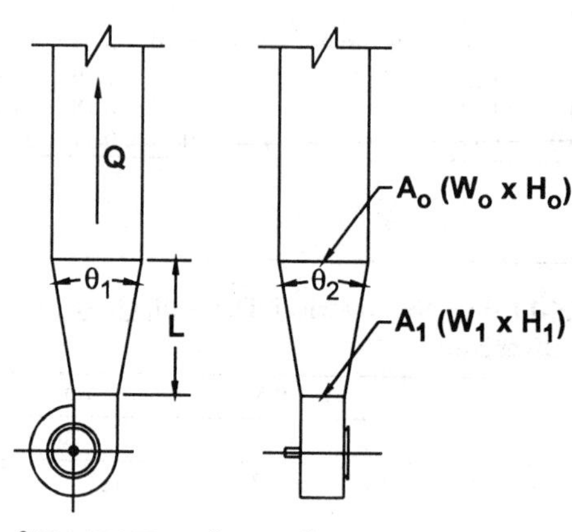

CHAPTER 33

PIPE SIZING

THIS chapter includes tables and charts to size piping for various fluid flow systems. Further details on specific piping systems can be found in appropriate chapters of the ASHRAE Handbook series.

There are two related but distinct concerns when designing a fluid flow system: sizing the pipe and determining the flow-pressure relationship. The two are often confused because they can use the same equations and design tools. Nevertheless, they should be determined separately.

The emphasis in this chapter is on the problem of sizing the pipe, and to this end design charts and tables for specific fluids are presented in addition to the equations that describe the flow of fluids in pipes. Once a system has been sized, it should be analyzed with more detailed methods of calculation to determine the pump head required to achieve the desired flow. Computerized methods are well suited to handling the details of calculating losses around an extensive system.

PRESSURE DROP EQUATIONS

Darcy-Weisbach Equation

Pressure drop caused by fluid friction in fully developed flows of all "well-behaved" (Newtonian) fluids is described by the Darcy-Weisbach equation:

$$\Delta p = f\left(\frac{L}{D}\right)\left(\frac{\rho}{g_c}\right)\left(\frac{V^2}{2}\right) \tag{1}$$

where

Δp = pressure drop, lb_f/ft^2
f = friction factor, dimensionless (from Moody chart, Figure 13 in Chapter 2)
L = length of pipe, ft
D = internal diameter of pipe, ft
ρ = fluid density at mean temperature, lb_m/ft^3
V = average velocity, fps
g_c = units conversion factor, 32.2 $ft \cdot lb_m/lb_f \cdot s^2$

This equation is often presented in head or specific energy form as

$$\Delta h = \left(\frac{\Delta p}{\rho}\right)\left(\frac{g_c}{g}\right) = f\left(\frac{L}{D}\right)\left(\frac{V^2}{2g}\right) \tag{2}$$

where

Δh = head loss, ft
g = acceleration of gravity, ft/s^2

In this form, the density of the fluid does not appear explicitly (although it is in the Reynolds number, which influences f).

The friction factor f is a function of pipe roughness ε, inside diameter D, and parameter Re, the Reynolds number:

$$Re = DV\rho/\mu \tag{3}$$

where

Re = Reynolds number, dimensionless
ε = absolute roughness of pipe wall, ft
μ = dynamic viscosity of fluid, $lb_m/ft \cdot s$

The friction factor is frequently presented on a Moody chart (Figure 13 in Chapter 2) giving f as a function of Re with ε/D as a parameter.

A useful fit of smooth and rough pipe data for the usual turbulent flow regime is the **Colebrook equation**:

$$\frac{1}{\sqrt{f}} = 1.74 - 2\log\left(\frac{2\varepsilon}{D} + \frac{18.7}{Re\sqrt{f}}\right) \tag{4}$$

Another form of Equation (4) appears in Chapter 2, but the two are equivalent. Equation (4) is more useful in showing behavior at limiting cases—as ε/D approaches 0 (smooth limit), the $18.7/Re\sqrt{f}$ term dominates; at high ε/D and Re (fully rough limit), the $2\varepsilon/D$ term dominates.

Equation (4) is implicit in f; that is, f appears on both sides, so a value for f is usually obtained iteratively.

Hazen-Williams Equation

A less widely used alternative to the Darcy-Weisbach formulation for calculating pressure drop is the Hazen-Williams equation, which is expressed as

$$\Delta p = 3.022L\left(\frac{V}{C}\right)^{1.852}\left(\frac{1}{D}\right)^{1.167}\left(\frac{\rho g}{g_c}\right) \tag{5}$$

or

$$\Delta h = 3.022L\left(\frac{V}{C}\right)^{1.852}\left(\frac{1}{D}\right)^{1.167} \tag{6}$$

where C = roughness factor.

Typical values of C are 150 for plastic pipe and copper tubing, 140 for new steel pipe, down to 100 and below for badly corroded or very rough pipe.

Valve and Fitting Losses

Valves and fittings cause pressure losses greater than those caused by the pipe alone. One formulation expresses losses as

$$\Delta p = K\left(\frac{\rho}{g_c}\right)\left(\frac{V^2}{2}\right) \quad \text{or} \quad \Delta h = K\left(\frac{V^2}{2g}\right) \tag{7}$$

where K = geometry- and size-dependent loss coefficient (Tables 1, 2, and 3).

The preparation of this chapter is assigned to TC 6.1, Hydronic and Steam Equipment and Systems.

Table 1 K Factors—Screwed Pipe Fittings

Nominal Pipe Dia., in.	90° Ell Reg.	90° Ell Long	45° Ell	Return Bend	Tee-Line	Tee-Branch	Globe Valve	Gate Valve	Angle Valve	Swing Check Valve	Bell Mouth Inlet	Square Inlet	Projected Inlet
3/8	2.5	—	0.38	2.5	0.90	2.7	20	0.40	—	8.0	0.05	0.5	1.0
1/2	2.1	—	0.37	2.1	0.90	2.4	14	0.33	—	5.5	0.05	0.5	1.0
3/4	1.7	0.92	0.35	1.7	0.90	2.1	10	0.28	6.1	3.7	0.05	0.5	1.0
1	1.5	0.78	0.34	1.5	0.90	1.8	9	0.24	4.6	3.0	0.05	0.5	1.0
1-1/4	1.3	0.65	0.33	1.3	0.90	1.7	8.5	0.22	3.6	2.7	0.05	0.5	1.0
1-1/2	1.2	0.54	0.32	1.2	0.90	1.6	8	0.19	2.9	2.5	0.05	0.5	1.0
2	1.0	0.42	0.31	1.0	0.90	1.4	7	0.17	2.1	2.3	0.05	0.5	1.0
2-1/2	0.85	0.35	0.30	0.85	0.90	1.3	6.5	0.16	1.6	2.2	0.05	0.5	1.0
3	0.80	0.31	0.29	0.80	0.90	1.2	6	0.14	1.3	2.1	0.05	0.5	1.0
4	0.70	0.24	0.28	0.70	0.90	1.1	5.7	0.12	1.0	2.0	0.05	0.5	1.0

Source: *Engineering Data Book* (HI 1979).

Table 2 K Factors—Flanged Welded Pipe Fittings

Nominal Pipe Dia., in.	90° Ell Reg.	90° Ell Long	45° Ell Long	Return Bend Reg.	Return Bend Long	Tee-Line	Tee-Branch	Glove Valve	Gate Valve	Angle Valve	Swing Check Valve
1	0.43	0.41	0.22	0.43	0.43	0.26	1.0	13	—	4.8	2.0
1-1/4	0.41	0.37	0.22	0.41	0.38	0.25	0.95	12	—	3.7	2.0
1-1/2	0.40	0.35	0.21	0.40	0.35	0.23	0.90	10	—	3.0	2.0
2	0.38	0.30	0.20	0.38	0.30	0.20	0.84	9	0.34	2.5	2.0
2-1/2	0.35	0.28	0.19	0.35	0.27	0.18	0.79	8	0.27	2.3	2.0
3	0.34	0.25	0.18	0.34	0.25	0.17	0.76	7	0.22	2.2	2.0
4	0.31	0.22	0.18	0.31	0.22	0.15	0.70	6.5	0.16	2.1	2.0
6	0.29	0.18	0.17	0.29	0.18	0.12	0.62	6	0.10	2.1	2.0
8	0.27	0.16	0.17	0.27	0.15	0.10	0.58	5.7	0.08	2.1	2.0
10	0.25	0.14	0.16	0.25	0.14	0.09	0.53	5.7	0.06	2.1	2.0
12	0.24	0.13	0.16	0.24	0.13	0.08	0.50	5.7	0.05	2.1	2.0

Source: *Engineering Data Book* (HI 1979).

Table 3 Approximate Range of Variation for K Factors

90° Elbow	Regular screwed	±20% above 2 in.	Tee	Screwed, line or branch	±25%
		±40% below 2 in.		Flanged, line or branch	±35%
	Long-radius screwed	±25%	Globe valve	Screwed	±25%
	Regular flanged	±35%		Flanged	±25%
	Long-radius flanged	±30%	Gate valve	Screwed	±25%
45° Elbow	Regular screwed	±10%		Flanged	±50%
	Long-radius flanged	±10%	Angle valve	Screwed	±20%
Return bend (180°)	Regular screwed	±25%		Flanged	±50%
	Regular flanged	±35%	Check valve	Screwed	±50%
	Long-radius flanged	±30%		Flanged	+200% −80%

Source: *Engineering Data Book* (HI 1979).

Example 1. Determine the pressure drop for 60°F water flowing at 4 fps through a nominal 1 in., 90° screwed ell.

Solution: From Table 1, the K for a 1 in., 90° screwed ell is 1.5.

$$\Delta p = 1.5 \times 62.4/32.2 \times 4^2/2 = 23.3 \text{ lb/ft}^2 \text{ or } 0.16 \text{ psi}$$

The loss coefficient for valves appears in another form as C_v, a dimensional coefficient expressing the flow through a valve at a specified pressure drop.

$$Q = C_v\sqrt{\Delta p} \tag{8}$$

where

Q = volumetric flow, gpm
C_v = valve coefficient, gpm at Δp = 1 psi
Δp = pressure drop, psi

See the section on Control Valve Sizing in Chapter 41 of the 1996 *ASHRAE Handbook—Systems and Equipment* for a more complete explanation of C_v.

Example 2. Determine the volumetric flow through a valve with C_v = 10 for an allowable pressure drop of 5 psi.

Solution: $Q = 10\sqrt{5}$ = 22.4 gpm.

Alternative formulations express fitting losses in terms of equivalent lengths of straight pipe (Tables 4 and 5, Figure 4). Pressure loss data for fittings are also presented in Idelchik (1986).

Calculating Pressure Losses

The most common engineering design flow loss calculation selects a pipe size for the desired total flow rate and available or allowable pressure drop.

Because either formulation of fitting losses requires a known diameter, pipe size must be selected before calculating the detailed influence of fittings. A frequently used rule of thumb assumes that the design length of pipe is 50 to 100% longer than actual to account for fitting losses. After a pipe diameter has been selected on this basis, the influence of each fitting can be evaluated.

WATER PIPING

FLOW RATE LIMITATIONS

Stewart and Dona (1987) surveyed the literature relating to water flow rate limitations. This section briefly reviews some of their findings. Noise, erosion, and installation and operating costs all limit the maximum and minimum velocities in piping systems. If piping sizes are too small, noise levels, erosion levels, and pumping costs can be unfavorable; if piping sizes are too large, installation costs are excessive. Therefore, pipe sizes are chosen to minimize initial cost while avoiding the undesirable effects of high velocities.

A variety of upper limits of water velocity and/or pressure drop in piping and piping systems is used. One recommendation places a velocity limit of 4 fps for 2 in. pipe and smaller, and a pressure drop limit of 4 ft of water/100 ft for piping over 2 in. Other guidelines are based on the type of service (Table 4) or the annual operating hours (Table 5). These limitations are imposed either to control the levels of pipe and valve noise, erosion, and water hammer pressure or for economic reasons. Carrier (1960) recommends that the velocity not exceed 15 fps in any case.

Noise Generation

Velocity-dependent noise in piping and piping systems results from any or all of four sources: turbulence, cavitation, release of entrained air, and water hammer. In investigations of flow-related noise, Marseille (1965), Ball and Webster (1976), and Rogers (1953, 1954, 1956) reported that velocities on the order of 10 to 17 fps lie within the range of allowable noise levels for residential and commercial buildings. The experiments showed considerable variation in the noise levels obtained for a specified velocity. Generally, systems with longer pipe and with more numerous fittings and valves were noisier. In addition, sound measurements were taken under widely differing conditions; for example, some tests used plastic-covered pipe, while others did not. Thus, no detailed correlations relating sound level to flow velocity in generalized systems are available.

Table 4 Water Velocities Based on Type of Service

Type of Service	Velocity, fps	Reference
General service	4 to 10	a, b, c
City water	3 to 7	a, b
	2 to 5	c
Boiler feed	6 to 15	a, c
Pump suction and drain lines	4 to 7	a, b

[a]Crane Co. (1976). [b]Carrier (1960). [c]Grinnell Company (1951).

Table 5 Maximum Water Velocity to Minimize Erosion

Normal Operation, h/yr	Water Velocity, fps
1500	15
2000	14
3000	13
4000	12
6000	10

Source: Carrier (1960).

The noise generated by fluid flow in a pipe system increases sharply if cavitation or the release of entrained air occurs. Usually the combination of a high water velocity with a change in flow direction or a decrease in the cross section of a pipe causing a sudden pressure drop is necessary to cause cavitation. Ball and Webster (1976) found that at their maximum velocity of 42 fps, cavitation did not occur in straight 3/8 and 1/2 in. pipe; using the apparatus with two elbows, cold water velocities up to 21 fps caused no cavitation. Cavitation did occur in orifices of 1:8 area ratio (orifice flow area is one-eighth of pipe flow area) at 5 fps and in 1:4 area ratio orifices at 10 fps (Rogers 1954).

Some data are available for predicting hydrodynamic (liquid) noise generated by control valves. The International Society for Measurement and Control compiled prediction correlations in an effort to develop control valves for reduced noise levels (ISA 1985). The correlation to predict hydrodynamic noise from control valves is

$$SL = 10\log C_v + 20\log \Delta p - 30\log t + 5 \tag{9}$$

where

SL = sound level, dB
C_v = valve coefficient, gpm/(psi)$^{0.5}$
Q = flow rate, gpm
Δp = pressure drop across valve, psi
t = downstream pipe wall thickness, in.

Air entrained in water usually has a higher partial pressure than the water. Even when flow rates are small enough to avoid cavitation, the release of entrained air may create noise. Every effort should be made to vent the piping system or otherwise remove entrained air.

Erosion

Erosion in piping systems is caused by water bubbles, sand, or other solid matter impinging on the inner surface of the pipe. Generally, at velocities lower than 100 fps, erosion is not significant as long as there is no cavitation. When solid matter is entrained in the fluid at high velocities, erosion occurs rapidly, especially in bends. Thus, high velocities should not be used in systems where sand or other solids are present or where slurries are transported.

Allowances for Aging

With age, the internal surfaces of pipes become increasingly rough, which reduces the available flow with a fixed pressure supply. However, designing with excessive age allowances may result in oversized piping. Age-related decreases in capacity depend on the type of water, type of pipe material, temperature of water, and type of system (open or closed) and include

- Sliming (biological growth or deposited soil on the pipe walls), which occurs mainly in unchlorinated, raw water systems.
- Caking of calcareous salts, which occurs in hard water (i.e., water bearing calcium salts) and increases with water temperature.
- Corrosion (incrustations of ferrous and ferric hydroxide on the pipe walls), which occurs in metal pipe in soft water. Because oxygen is necessary for corrosion to take place, significantly more corrosion takes place in open systems.

Allowances for expected decreases in capacity are sometimes treated as a specific amount (percentage). Dawson and Bowman (1933) added an allowance of 15% friction loss to new pipe (equivalent to an 8% decrease in capacity). The *HDR Design Guide* (1981) increased the friction loss by 15 to 20% for closed piping systems and 75 to 90% for open systems. Carrier (1960) indicates a factor of approximately 1.75 between friction factors for closed and open systems.

Obrecht and Pourbaix (1967) differentiated between the corrosive potential of different metals in potable water systems and concluded that iron is the most severely attacked, then galvanized steel, lead, copper, and finally copper alloys (i.e., brass). Hunter (1941) and Freeman (1941) showed the same trend. After four years of cold and hot water use, copper pipe had a capacity loss of 25 to 65%. Aged ferrous pipe has a capacity loss of 40 to 80%. Smith (1983) recommended increasing the design discharge by 1.55 for uncoated cast iron, 1.08 for iron and steel, and 1.06 for cement or concrete.

The Plastic Pipe Institute (1971) found that corrosion is not a problem in plastic pipe; the capacity of plastic pipe in Europe and the United States remains essentially the same after 30 years in use.

Extensive age-related flow data are available for use with the Hazen-Williams empirical equation. Difficulties arise in its application, however, because the original Hazen-Williams roughness coefficients are valid only for the specific pipe diameters, water velocities, and water viscosities used in the original experiments. Thus, when the *C*s are extended to different diameters, velocities, and/or water viscosities, errors of up to about 50% in pipe capacity can occur (Williams and Hazen 1933, Sanks 1978).

Water Hammer

When any moving fluid (not just water) is abruptly stopped, as when a valve closes suddenly, large pressures can develop. While detailed analysis requires knowledge of the elastic properties of the pipe and the flow-time history, the limiting case of rigid pipe and instantaneous closure is simple to calculate. Under these conditions,

$$\Delta p_h = \rho c_s V / g_c \tag{10}$$

where

Δp_h = pressure rise caused by water hammer, lb_f/ft^2
ρ = fluid density, lb_m/ft^3
c_s = velocity of sound in fluid, fps
V = fluid flow velocity, fps

The c_s for water is 4720 fps, although the elasticity of the pipe reduces the effective value.

Example 3. What is the maximum pressure rise if water flowing at 10 fps is stopped instantaneously?

Solution: $\Delta p_h = 62.4 \times 4720 \times 10/32.2 = 91{,}468 \text{ lb/ft}^2$
$= 635$ psi

Other Considerations

Not discussed in detail in this chapter, but of potentially great importance, are a number of physical and chemical considerations: pipe and fitting design, materials, and joining methods must be appropriate for working pressures and temperatures encountered, as well as being suitably resistant to chemical attack by the fluid.

Other Piping Materials and Fluids

For fluids not included in this chapter or for piping materials of different dimensions, manufacturers' literature frequently supplies pressure drop charts. The Darcy-Weisbach equation, with the Moody chart or the Colebrook equation, can be used as an alternative to pressure drop charts or tables.

HYDRONIC SYSTEM PIPING

The Darcy-Weisbach equation with friction factors from the Moody chart or Colebrook equation (or, alternatively, the Hazen-Williams equation) is fundamental to calculating pressure drop in hot and chilled water piping; however, charts calculated from these equations (such as Figures 1, 2, and 3) provide easy determination of pressure drops for specific fluids and pipe standards. In addition, tables of pressure drops can be found in HI (1979) and Crane Co. (1976).

The Reynolds numbers represented on the charts in Figures 1, 2, and 3 are all in the turbulent flow regime. For smaller pipes and/or lower velocities, the Reynolds number may fall into the laminar regime, in which the Colebrook friction factors are no longer valid.

Most tables and charts for water are calculated for properties at 60°F. Using these for hot water introduces some error, although the answers are conservative (i.e., cold water calculations overstate the pressure drop for hot water). Using 60°F water charts for 200°F water should not result in errors in Δp exceeding 20%.

Range of Usage of Pressure Drop Charts

General Design Range. The general range of pipe friction loss used for design of hydronic systems is between 1 and 4 ft of water per 100 ft of pipe. A value of 2.5 ft/100 ft represents the mean to which most systems are designed. Wider ranges may be used in specific designs if certain precautions are taken.

Piping Noise. Closed-loop hydronic system piping is generally sized below certain arbitrary upper limits, such as a velocity limit of 4 fps for 2 in. pipe and under, and a pressure drop limit of 4 ft per 100 ft for piping over 2 in. in diameter. Velocities in excess of 4 fps can be used in piping of larger size. This limitation is generally accepted, although it is based on relatively inconclusive experience with noise in piping. Water *velocity noise* is not caused by water but by free air, sharp pressure drops, turbulence, or a combination of these, which in turn cause cavitation or flashing of water into steam. Therefore, higher velocities may be used if proper precautions are taken to eliminate air and turbulence.

Air Separation

Air in hydronic systems is usually undesirable because it causes flow noise, allows oxygen to react with piping materials, and sometimes even prevents flow in parts of a system. Air may enter a system at an air-water interface in an open system or in an expansion tank in a closed system, or it may be brought in dissolved in makeup water. Most hydronic systems use air separation devices to remove air. The solubility of air in water increases with pressure and decreases with temperature; thus, separation of air from water is best achieved at the point of lowest pressure and/or highest temperature in a system. For more information, see Chapter 12, Hydronic Heating and Cooling System Design, of the 1996 *ASHRAE Handbook—Systems and Equipment*.

In the absence of venting, air can be entrained in the water and carried to separation units at flow velocities of 1.5 to 2 fps or more in pipe 2 in. and under. Minimum velocities of 2 fps are therefore recommended. For pipe sizes 2 in. and over, minimum velocities corresponding to a head loss of 0.75 ft/100 ft are normally used. Maintenance of minimum velocities is particularly important in the upper floors of high-rise buildings where the air tends to come out of solution because of reduced pressures. Higher velocities should be used in *downcomer* return mains feeding into air separation units located in the basement.

Example 4. Determine the pipe size for a circuit requiring 20 gpm flow.

Solution: Enter Figure 1 at 20 gpm, read up to pipe size within normal design range (1 to 4 ft/100 ft), and select 1-1/2 in. Velocity is 3.1 fps, which is between 2 and 4. Pressure loss is 2.9 ft/100 ft.

Valve and Fitting Pressure Drop

Valves and fittings can be listed in elbow equivalents, with an elbow being equivalent to a length of straight pipe. Table 6 lists equivalent lengths of 90° elbows; Table 7 lists elbow equivalents for valves and fittings for iron and copper.

Example 5. Determine equivalent feet of pipe for a 4 in. open gate valve at a flow velocity of approximately 4 fps.

Solution: From Table 6, at 4 fps, each elbow is equivalent to 10.6 ft of 4 in. pipe. From Table 7, the gate valve is equivalent to 0.5 elbows. The actual equivalent pipe length (added to measured circuit length for pressure drop determination) will be 10.6 × 0.5, or 5.3 equivalent feet of 4 in. pipe.

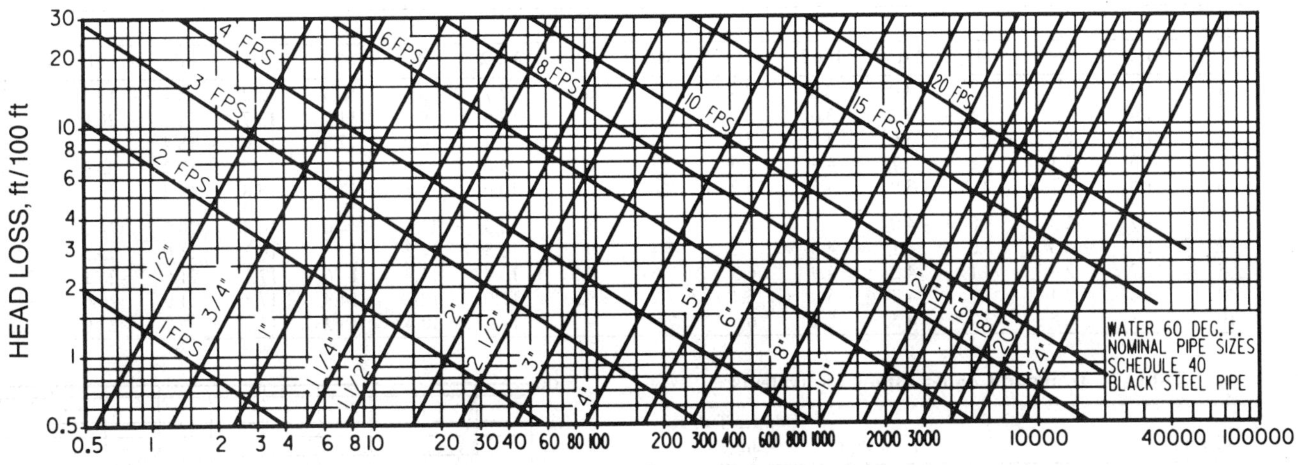

Fig. 1 Friction Loss for Water in Commercial Steel Pipe (Schedule 40)

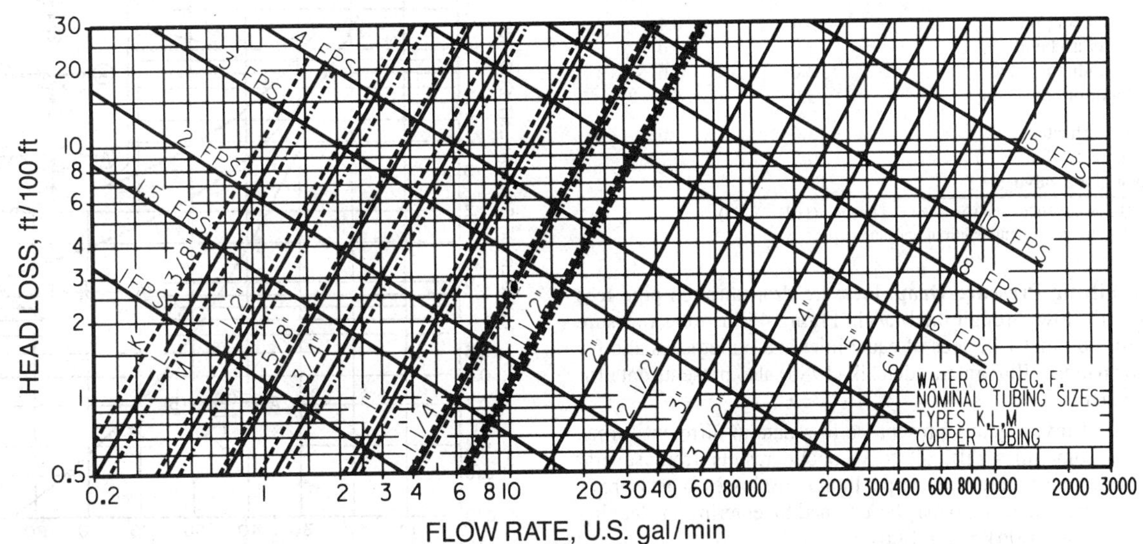

Fig. 2 Friction Loss for Water in Copper Tubing (Types K, L, M)

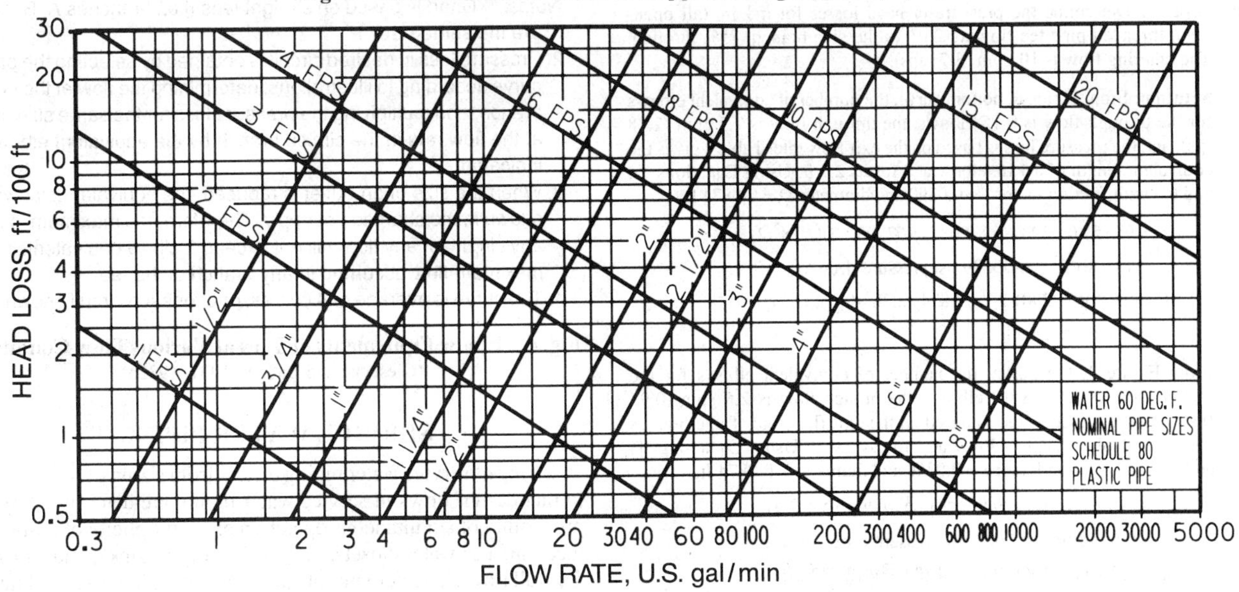

Fig. 3 Friction Loss for Water in Plastic Pipe (Schedule 80)

Table 6 Equivalent Length in Feet of Pipe for 90° Elbows

Veloc-ity, fps	Pipe Size 1/2	3/4	1	1-1/4	1-1/2	2	2-1/2	3	3-1/2	4	5	6	8	10	12
1	1.2	1.7	2.2	3.0	3.5	4.5	5.4	6.7	7.7	8.6	10.5	12.2	15.4	18.7	22.2
2	1.4	1.9	2.5	3.3	3.9	5.1	6.0	7.5	8.6	9.5	11.7	13.7	17.3	20.8	24.8
3	1.5	2.0	2.7	3.6	4.2	5.4	6.4	8.0	9.2	10.2	12.5	14.6	18.4	22.3	26.5
4	1.5	2.1	2.8	3.7	4.4	5.6	6.7	8.3	9.6	10.6	13.1	15.2	19.2	23.2	27.6
5	1.6	2.2	2.9	3.9	4.5	5.9	7.0	8.7	10.0	11.1	13.6	15.8	19.8	24.2	28.8
6	1.7	2.3	3.0	4.0	4.7	6.0	7.2	8.9	10.3	11.4	14.0	16.3	20.5	24.9	29.6
7	1.7	2.3	3.0	4.1	4.8	6.2	7.4	9.1	10.5	11.7	14.3	16.7	21.0	25.5	30.3
8	1.7	2.4	3.1	4.2	4.9	6.3	7.5	9.3	10.8	11.9	14.6	17.1	21.5	26.1	31.0
9	1.8	2.4	3.2	4.3	5.0	6.4	7.7	9.5	11.0	12.2	14.9	17.4	21.9	26.6	31.6
10	1.8	2.5	3.2	4.3	5.1	6.5	7.8	9.7	11.2	12.4	15.2	17.7	22.2	27.0	32.0

Table 7 Iron and Copper Elbow Equivalents[a]

Fitting	Iron Pipe	Copper Tubing
Elbow, 90°	1.0	1.0
Elbow, 45°	0.7	0.7
Elbow, 90° long turn	0.5	0.5
Elbow, welded, 90°	0.5	0.5
Reduced coupling	0.4	0.4
Open return bend	1.0	1.0
Angle radiator valve	2.0	3.0
Radiator or convector	3.0	4.0
Boiler or heater	3.0	4.0
Open gate valve	0.5	0.7
Open globe valve	12.0	17.0

Source: Giesecke (1926) and Giesecke and Badgett (1931, 1932a).

[a]See Table 6 for equivalent length of one elbow.

Tee Fitting Pressure Drop. Pressure drop through pipe tees varies with flow through the branch. Figure 4 illustrates pressure drops for nominal 1 in. tees of equal inlet and outlet sizes and for the flow patterns illustrated. Idelchik (1986) also presents data for threaded tees.

Different investigators present tee loss data in different forms, and it is sometimes difficult to reconcile results from several sources. As an estimate of the upper limit to tee losses, a pressure or head loss coefficient of 1.0 may be assumed for entering and leaving flows (i.e., $\Delta p = 1.0\rho V_{in}^2/2 + 1.0\rho V_{out}^2/2$).

Example 6. Determine the pressure or head losses for a 1 in. (all openings) threaded pipe tee flowing 25% to the side branch, 75% through. The entering flow is 10 gpm (3.71 fps).

Solution: From Figure 4, bottom curve, the number of equivalent elbows for the through-flow is 0.15 elbows; the through-flow is 7.5 gpm (2.78 fps); and the pressure loss is based on the exit flow rate. Table 6 gives the equivalent length of a 1 in. elbow at 3 fps as 2.7 ft. Using Equations (1) and (2) with friction factor $f = 0.0290$ and diameter $D = 0.0874$ ft,

$$\Delta p = (0.15)(0.0290)(2.7/0.0874)(62.4/32.2)(2.78^2/2)$$
$$= 1.01 \text{ lb/ft}^2 = 0.00699 \text{ psi pressure drop, or}$$
$$\Delta h = (0.15)(0.0290)(2.7/0.0874)(2.78^2)/[(2)(32.2)]$$
$$= 0.0161 \text{ ft head loss}$$

From Figure 4, top curve, the number of equivalent elbows for the branch flow of 25% is 13 elbows; the branch flow is 2.5 gpm (0.93 fps); and the pressure loss is based on the exit flow rate. Table 6 gives the equivalent of a 1 in. elbow at 1 fps as 2.2 ft. Using Equations (1) and (2) with friction factor $f = 0.0350$ and diameter = 0.0874 ft,

$$\Delta p = (13)(0.0350)(2.2/0.0874)(62.4/32.2)(0.93^2/2)$$
$$= 9.60 \text{ lb/ft}^2 = 0.0667 \text{ psi pressure drop, or}$$
$$\Delta h = (13)(0.0350)(2.2/0.0874)(0.93^2)/[(2)(32.2)]$$
$$= 0.154 \text{ ft head loss}$$

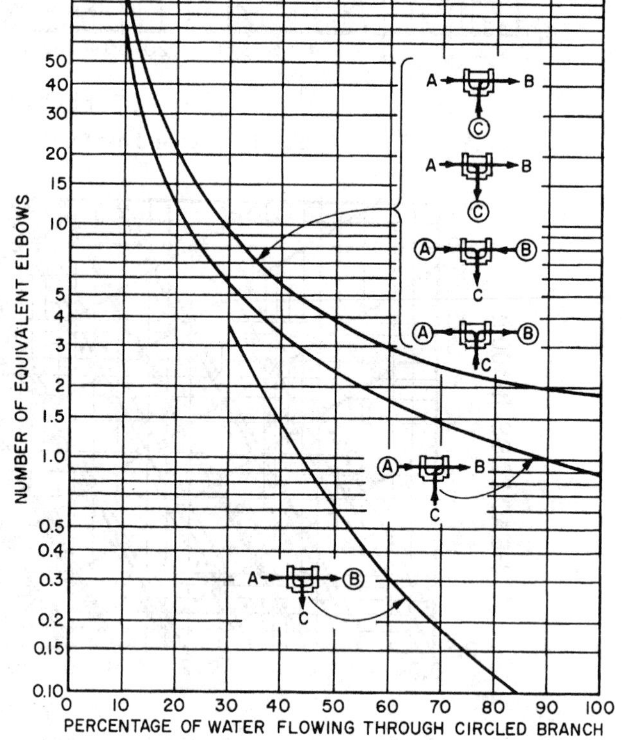

Notes: 1. Chart is based on straight tees (i.e., branches A, B, and C are the same size).
2. Pressure loss in desired circuit is obtained by selecting the proper curve according to illustrations, determining the flow at the circled branch, and multiplying the pressure loss for the same size elbow at the flow rate in the circled branch by the equivalent elbows indicated.
3. When the size of an outlet is reduced, the equivalent elbows shown in the chart do not apply. Therefore, the maximum loss for any circuit for any flow will not exceed 2 elbow equivalents at the maximum flow occurring in any branch of the tee.
4. Top curve is average of 4 curves, one for each circuit shown.

Fig. 4 Elbow Equivalents of Tees at Various Flow Conditions
(Giesecke and Badgett 1931, 1932b)

SERVICE WATER PIPING

Sizing of service water piping differs from sizing of process lines in that design flows in service water piping are determined by the probability of simultaneous operation of a multiplicity of individual loads such as water closets, urinals, lavatories, sinks, and showers. The full flow characteristics of each load device are readily obtained from manufacturers; however, service water piping sized to handle all load devices simultaneously would be seriously oversized. Thus, a major issue in sizing service water piping is to determine the diversity of the loads.

The procedure shown in this chapter uses the work of R.B. Hunter for estimating diversity (Hunter 1940, 1941). The present-day plumbing designer is usually constrained by building or plumbing codes, which specify the individual and collective loads to be used for pipe sizing. Frequently used codes (including the BOCA *National Plumbing Code*, *Standard Plumbing Code*, *Uniform Plumbing Code*, and *National Standard Plumbing Code*) contain procedures quite similar to those shown here. The designer must be aware of the applicable code for the location being considered.

Federal mandates are forcing plumbing fixture manufacturers to reduce design flows to many types of fixtures, but these may not yet be included in locally adopted codes. Also, the designer must be aware of special considerations; for example, toilet usage at sports arenas will probably have much less diversity than the codes allow and thus may require larger supply piping than the minimum specified by the codes.

Table 8 gives the rate of flow desirable for many common fixtures and the average pressure necessary to give this rate of flow. The pressure varies with fixture design.

In estimating the load, the rate of flow is frequently computed in **fixture units**, which are relative indicators of flow. Table 9 gives the demand weights in terms of fixture units for different plumbing fixtures under several conditions of service, and Figure 5 gives the estimated demand in gallons per minute corresponding to any total number of fixture units. Figures 6 and 7 provide more accurate estimates at the lower end of the scale.

The estimated demand load for fixtures used intermittently on any supply pipe can be obtained by multiplying the number of each kind of fixture supplied through that pipe by its weight from Table 9, adding the products, and then referring to the appropriate curve of Figure 5, 6, or 7 to find the demand corresponding to the total fixture units. In using this method, note that the demand for fixture or supply outlets other than those listed in the table of fixture units is not yet included in the estimate. The demands for outlets (e.g., hose connections and air-conditioning apparatus) that are likely to impose continuous demand during heavy use of the weighted fixtures should be estimated separately and added to demand for fixtures used intermittently to estimate total demand.

The Hunter curves in Figures 5, 6, and 7 are based on use patterns in residential buildings and can be erroneous for other usages such as sports arenas. Williams (1976) discusses the Hunter assumptions and presents an analysis using alternative assumptions.

Table 8 Proper Flow and Pressure Required During Flow for Different Fixtures

Fixture	Flow Pressure, psig[a]	Flow, gpm
Ordinary basin faucet	8	3.0
Self-closing basin faucet	12	2.5
Sink faucet—3/8 in.	10	4.5
Sink faucet—1/2 in.	5	4.5
Dishwasher	15-25	—[b]
Bathtub faucet	5	6.0
Laundry tube cock—1/4 in.	5	5.0
Shower	12	3-10
Ball cock for closet	15	3.0
Flush valve for closet	10-20	15-40[c]
Flush valve for urinal	15	15.0
Garden hose, 50 ft, and sill cock	30	5.0

[a]Flow pressure is the pressure in the pipe at the entrance to the particular fixture considered.
[b]Varies; see manufacturers' data.
[c]Wide range due to variation in design and type of flush valve closets.

Table 9 Demand Weights of Fixtures in Fixture Units[a]

Fixture or Group[b]	Occupancy	Type of Supply Control	Weight in Fixture Units[c]
Water closet	Public	Flush valve	10
Water closet	Public	Flush tank	5
Pedestal urinal	Public	Flush valve	10
Stall or wall urinal	Public	Flush valve	5
Stall or wall urinal	Public	Flush tank	3
Lavatory	Public	Faucet	2
Bathtub	Public	Faucet	4
Shower head	Public	Mixing valve	4
Service sink	Office, etc.	Faucet	3
Kitchen sink	Hotel or restaurant	Faucet	4
Water closet	Private	Flush valve	6
Water closet	Private	Flush tank	3
Lavatory	Private	Faucet	1
Bathtub	Private	Faucet	2
Shower head	Private	Mixing valve	2
Bathroom group	Private	Flush valve for closet	8
Bathroom group	Private	Flush tank for closet	6
Separate shower	Private	Mixing valve	2
Kitchen sink	Private	Faucet	2
Laundry trays (1 to 3)	Private	Faucet	3
Combination fixture	Private	Faucet	3

Source: Hunter (1941).
[a]For supply outlets likely to impose continuous demands, estimate continuous supply separately, and add to total demand for fixtures.
[b]For fixtures not listed, weights may be assumed by comparing the fixture to a listed one using water in similar quantities and at similar rates.
[c]The given weights are for total demand. For fixtures with both hot and cold water supplies, the weights for maximum separate demands can be assumed to be 75% of the listed demand for the supply.

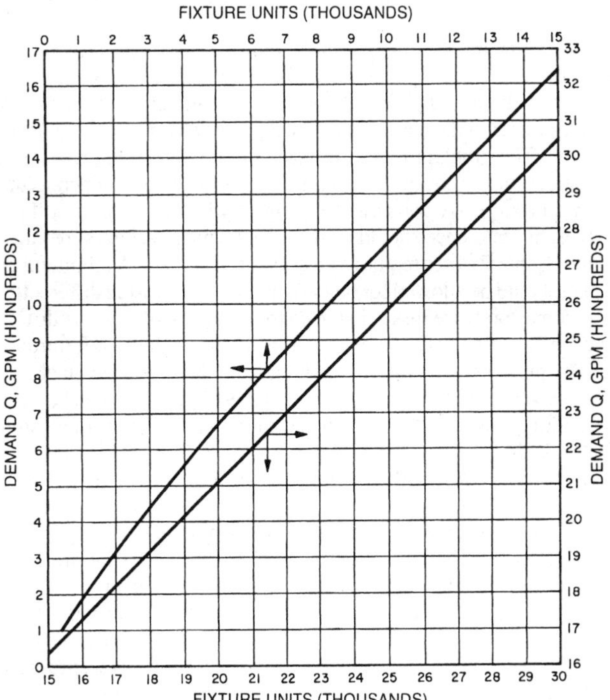

Fig. 5 Demand Versus Fixture Units, Mixed System, High Part of Curve
(Hunter 1941)

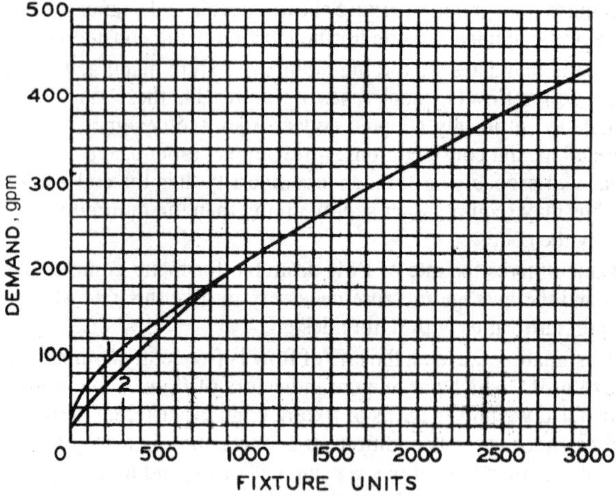

Fig. 6 Estimate Curves for Demand Load
(Hunter 1941)

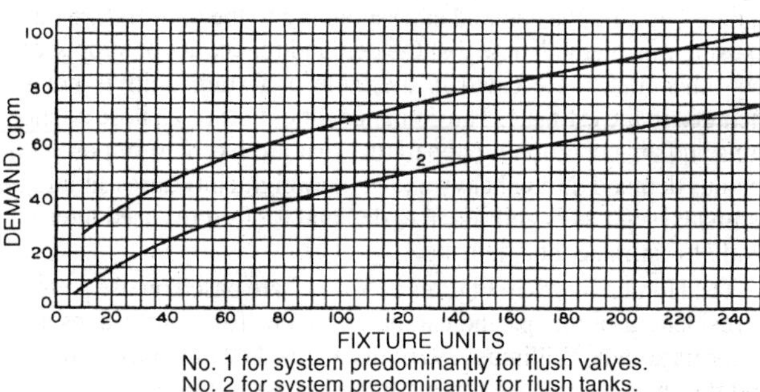

Fig. 7 Section of Figure 6 on Enlarged Scale

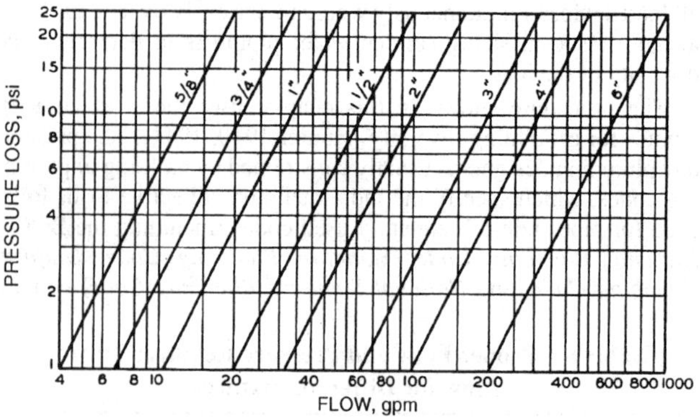

Fig. 8 Pressure Losses in Disk-Type Water Meters

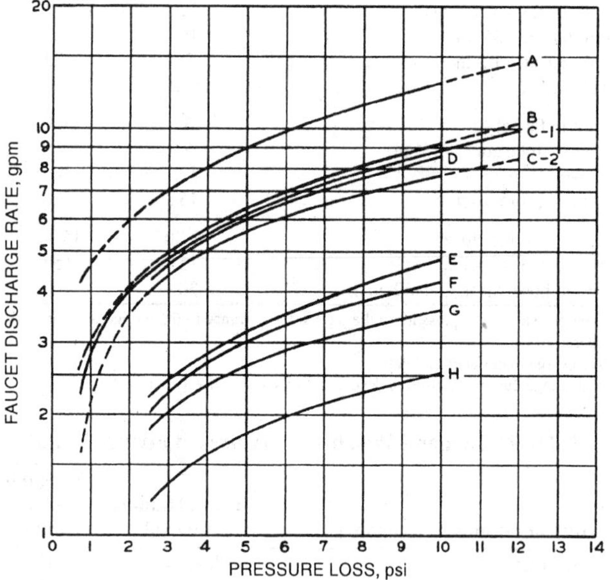

A. 1/2 in. laundry bibb (old style)
B. Laundry compression faucet
C-1. 1/2 in. compression sink faucet (mfr. 1)
C-2. 1/2 in. compression sink faucet (mfr. 2)
D. Combination compression bathtub faucets (both open)
E. Combination compression sink faucet
F. Basin faucet
G. Spring self-closing faucet
H. Slow self-closing faucet
(Dashed lines indicate recommended extrapolation)

Fig. 9 Variation of Pressure Loss with Flow Rate for Various Faucets and Cocks

So far, the information presented shows the *design rate of flow* to be determined in any particular section of piping. The next step is to determine the *size* of piping. As water flows through a pipe, the pressure continually decreases along the pipe due to loss of energy from friction. The problem is then to ascertain the minimum pressure in the street main and the minimum pressure required to operate the topmost fixture. (A pressure of 15 psig may be ample for most flush valves, but reference should be made to the manufacturers' requirements. Some fixtures require a pressure up to 25 psig. A minimum of 8 psig should be allowed for other fixtures.) The pressure differential overcomes pressure losses in the distributing system and the difference in elevation between the water main and the highest fixture.

The pressure loss (in psi) resulting from the difference in elevation between the street main and the highest fixture can be obtained by multiplying the difference in elevation in feet by the conversion factor 0.434.

Pressure losses in the distributing system consist of pressure losses in the piping itself, plus the pressure losses in the pipe fittings, valves, and the water meter, if any. Approximate design pressure losses and flow limits for disk-type meters for various rates of flow are given in Figure 8. Water authorities in many localities require compound meters for greater accuracy with varying flow; consult the local utility. Design data for compound meters differ from the data in Figure 8. Manufacturers give data on exact pressure losses and capacities.

Figure 9 shows the variation of pressure loss with rate of flow for various faucets and cocks. The water demand for hose bibbs or other large-demand fixtures taken off the building main frequently results in inadequate water supply to the upper floor of a building. This condition can be prevented by sizing the distribution system so that the pressure drops from the street main to all fixtures are the same. An ample building main (not less than 1 in. where possible) should be maintained until all branches to hose bibbs have been connected. Where the street main pressure is excessive and a pressure reducing valve is used to prevent water hammer or excessive pressure at the fixtures, the hose bibbs should be connected ahead of the reducing valve.

The principles involved in sizing upfeed and downfeed systems are the same. In the downfeed system, however, the difference in elevation between the overhead supply mains and the fixtures provides the pressure required to overcome pipe friction. Because friction pressure loss and height pressure loss are not additive, as in an upfeed system, smaller pipes may be used with a downfeed system.

Plastic Pipe

The maximum safe water velocity in a thermoplastic piping system under most operating conditions is typically 5 fps; however, higher velocities can be used in cases where the operating characteristics of valves and pumps are known so that sudden changes in flow velocity can be controlled. The total pressure in the system at any time (operating pressure plus surge of water hammer) should not exceed 150% of the pressure rating of the system.

Procedure for Sizing Cold Water Systems

The recommended procedure for sizing piping systems is outlined below.

1. Sketch the main lines, risers, and branches, and indicate the fixtures to be served. Indicate the rate of flow of each fixture.
2. Using Table 9, compute the demand weights of the fixtures in fixture units.
3. Determine the total demand in fixture units and, using Figure 5, 6, or 7, find the expected demand.
4. Determine the equivalent length of pipe in the main lines, risers, and branches. Because the sizes of the pipes are not known, the exact equivalent length of various fittings cannot be determined. Add the equivalent lengths, starting at the street main and proceeding along the service line, the main line of the building, and up the riser to the top fixture of the group served.
5. Determine the average minimum pressure in the street main and the minimum pressure required for the operation of the topmost fixture, which should be 8 to 25 psi.
6. Calculate the approximate design value of the average pressure drop per 100 ft of equivalent length of pipe determined in step 4.

$$\Delta p = (p_s - 0.434H - p_f - p_m)100/L \quad (11)$$

where

Δp = average pressure loss per 100 ft of equivalent length of pipe, psi
p_s = pressure in street main, psig
p_f = minimum pressure required to operate topmost fixture, psig
p_m = pressure drop through water meter, psi
H = height of highest fixture above street main, ft
L = equivalent length determined in step 4, ft

If the system is downfeed supply from a gravity tank, height of water in the tank, converted to psi by multiplying by 0.434, replaces the street main pressure, and the term $0.434H$ is added instead of subtracted in calculating Δp. In this case, H is the vertical distance of the fixture below the bottom of the tank.

7. From the expected rate of flow determined in step 3 and the value of Δp calculated in step 6, choose the sizes of pipe from Figure 1, 2, or 3.

Example 7. Assume a minimum street main pressure of 55 psig; a height of topmost fixture (a urinal with flush valve) above street main of 50 ft; an equivalent pipe length from water main to highest fixture of 100 ft; a total load on the system of 50 fixture units; and that the water closets are flush valve operated. Find the required size of supply main.

Solution: From Figure 7, the estimated peak demand is 51 gpm. From Table 8, the minimum pressure required to operate the topmost fixture is 15 psig. For a trial computation, choose the 1-1/2 in. meter. From Figure 8, the pressure drop through a 1-1/2 in. disk-type meter for a flow of 51 gpm is 6.5 psi.

The pressure drop available for overcoming friction in pipes and fittings is 55 − 0.434 × 50 − 15 − 6.5 = 12 psi.

At this point, estimate the equivalent pipe length of the fittings on the direct line from the street main to the highest fixture. The exact equivalent length of the various fittings cannot be determined since the pipe sizes of the building main, riser, and branch leading to the highest fixture are not yet known, but a first approximation is necessary to tentatively select pipe sizes. If the computed pipe sizes differ from those used in determining the equivalent length of pipe fittings, a recalculation using the computed pipe sizes for the fittings will be necessary. For this example, assume that the total equivalent length of the pipe fittings is 50 ft.

The permissible pressure loss per 100 ft of equivalent pipe is 12 × 100/(100 + 50) = 8 psi or 18 ft/100 ft. A 1-1/2 in. building main is adequate.

The sizing of the branches of the building main, the risers, and the fixture branches follows these principles. For example, assume that one of the branches of the building main carries the cold water supply for 3 water closets, 2 bathtubs, and 3 lavatories. Using the permissible pressure loss of 8 psi per 100 ft, the size of branch (determined from Table 9 and Figures 1 and 7) is found to be 1-1/2 in. Items included in the computation of pipe size are as follows:

Fixtures, No. and Type	Fixture Units (Table 9 and Note c)		Demand (Figure 7)	Pipe Size (Figure 1)
3 flush valves	3 × 6 =	18		
2 bathtubs	0.75 × 2 × 2 =	3		
3 lavatories	0.75 × 3 × 1 =	2.25		
Total	=	23.25	38 gpm	1-1/2 in.

Table 10 is a guide to minimum pipe sizing where flush valves are used.

Table 10 Allowable Number of 1 in. Flush Valves Served by Various Sizes of Water Pipe[a]

Pipe Size, in.	No. of 1 in. Flush Valves
1-1/4	1
1-1/2	2-4
2	5-12
2-1/2	13-25
3	26-40
4	41-100

[a]Two 3/4 in. flush valves are assumed equal to one 1 in. flush valve but can be served by a 1 in. pipe. Water pipe sizing must consider demand factor, available pressure, and length of run.

Velocities exceeding 10 fps cause undesirable noise in the piping system. This usually governs the size of larger pipes in the system, while in small pipe sizes, the friction loss usually governs the selection because the velocity is low compared to friction loss. Velocity is the governing factor in downfeed systems, where friction loss is usually neglected. Velocity in branches leading to pump suctions should not exceed 5 fps.

If the street pressure is too low to adequately supply upper-floor fixtures, the pressure must be increased. Constant or variable speed booster pumps, alone or in conjunction with gravity supply tanks, or hydropneumatic systems may be used.

Flow control valves for individual fixtures under varying pressure conditions automatically adjust the flow at the fixture to a predetermined quantity. These valves allow the designer to (1) limit the flow at the individual outlet to the minimum suitable for the purpose, (2) hold the total demand for the system more closely to the required minimum, and (3) design the piping system as accurately as is practicable for the requirements.

STEAM PIPING

Pressure losses in steam piping for flows of dry or nearly dry steam are governed by Equations (1) through (7) in the section on Pressure Drop Equations. This section incorporates these principles with other information specific to steam systems.

Pipe Sizes

Required pipe sizes for a given load in steam heating depend on the following factors:

- The initial pressure and the total pressure drop that can be allowed between the source of supply and the end of the return system
- The maximum velocity of steam allowable for quiet and dependable operation of the system, taking into consideration the direction of condensate flow
- The equivalent length of the run from the boiler or source of steam supply to the farthest heating unit

Initial Pressure and Pressure Drop. Table 11 lists pressure drops commonly used with corresponding initial steam pressures for sizing steam piping.

Several factors, such as initial pressure and pressure required at the end of the line, should be considered, but it is most important that (1) the total pressure drop does not exceed the initial gage pressure of the system (and in practice it should never exceed one-half the initial gage pressure); (2) the pressure drop is not great enough to cause excessive velocities; (3) a constant initial pressure is maintained, except on systems specially designed for varying initial pressures (e.g., subatmospheric pressure), which normally operate under controlled partial vacuums; and (4) for gravity return systems, the pressure drop to the heating units does not exceed the water column available for removing condensate (i.e., the height above the boiler water line of the lowest point on the steam main, on the heating units, or on the dry return).

Maximum Velocity. For quiet operation, steam velocity should be 8000 to 12,000 fpm, with a maximum of 15,000 fpm. The lower the velocity, the quieter the system. When the condensate must flow against the steam, even in limited quantity, the velocity of the steam must not exceed limits above which the disturbance between the steam and the counterflowing water may (1) produce objectionable sound, such as water hammer, or (2) result in the retention of water in certain parts of the system until the steam flow is reduced sufficiently to permit the water to pass. The velocity at which these disturbances take place is a function of (1) pipe size; (2) the pitch of the pipe if it runs horizontally; (3) the quantity of condensate flowing against the steam; and (4) the freedom of the piping from water pockets that, under certain conditions, act as a restriction in pipe size. Table 12 lists maximum capacities for various size steam lines.

Equivalent Length of Run. All tables for the flow of steam in pipes based on pressure drop must allow for pipe friction, as well as for the resistance of fittings and valves. These resistances are generally stated in terms of straight pipe; that is, a certain fitting produces a drop in pressure equivalent to the stated number of feet of straight run of the same size of pipe. Table 13 gives the number of feet of straight pipe usually allowed for the more common types of fittings and valves. In all pipe sizing tables in this chapter, the *length of run* refers to the *equivalent length of run* as distinguished from the *actual length* of pipe. A common sizing method is to assume the length of run and to check this assumption after pipes are sized. For this purpose, the length of run is usually assumed to be double the actual length of pipe.

Example 8. Using Table 13, determine the equivalent length in feet of pipe for the run illustrated.

132 FT
LAST RISER OR RADIATOR
DRIP
DRIP

Measured length	=	132.0 ft
4 in. gate valve	=	1.9 ft
Four 4 in. elbows	=	36.0 ft
Two 4 in. tees	=	36.0 ft
Equivalent	=	205.9 ft

Sizing Charts

Figure 10 is the basic chart for determining the flow rate and velocity of steam in Schedule 40 pipe for various values of pressure drop per 100 ft, based on 0 psig saturated steam. Using the multiplier chart (Figure 11), Figure 10 can be used at all saturation pressures between 0 and 200 psig (see Example 10).

Figures 10A through 10D present charts for sizing steam piping for systems of 30, 50, 100, and 150 psig at various pressure drops. These charts are based on the Moody friction factor, which considers the Reynolds number and the roughness of the internal pipe surfaces; they contain the same information as the basic chart (Figure 10) but in a more convenient form.

Table 11 Pressure Drops Used for Sizing Steam Pipe[a]

Initial Steam Pressure, psig	Pressure Drop per 100 ft	Total Pressure Drop in Steam Supply Piping
Vacuum return	2 to 4 oz/in^2	1 to 2 psi
0	0.5 oz/in^2	1 oz/in^2
1	2 oz/in^2	1 to 4 oz/in^2
2	2 oz/in^2	8 oz/in^2
5	4 oz/in^2	1.5 psi
10	8 oz/in^2	3 psi
15	1 psi	4 psi
30	2 psi	5 to 10 psi
50	2 to 5 psi	10 to 15 psi
100	2 to 5 psi	15 to 25 psi
150	2 to 10 psi	25 to 30 psi

[a]Equipment, control valves, and so forth must be selected based on delivered pressures.

Table 12 Comparative Capacity of Steam Lines at Various Pitches for Steam and Condensate Flowing in Opposite Directions

Pitch of Pipe, in/10 ft	Nominal Pipe Diameter, in. 3/4 Capacity	3/4 Maximum Velocity	1 Capacity	1 Maximum Velocity	1-1/4 Capacity	1-1/4 Maximum Velocity	1-1/2 Capacity	1-1/2 Maximum Velocity	2 Capacity	2 Maximum Velocity
1/4	3.2	8	6.8	9	11.8	11	19.8	12	42.9	15
1/2	4.1	11	9.0	12	15.9	14	25.9	16	54.0	18
1	5.7	13	11.7	15	19.9	17	33.0	19	68.8	24
1-1/2	6.4	14	12.8	17	24.6	20	37.4	22	83.3	27
2	7.1	16	14.8	19	27.0	22	42.0	24	92.9	30
3	8.3	17	17.3	22	31.3	25	46.8	26	99.6	32
4	9.9	22	19.2	24	33.4	26	50.8	28	102.4	32
5	10.5	22	20.5	25	38.5	31	59.2	33	115.0	33

Source: Laschober et al. (1966). Velocity in fps; capacity in lb/h.

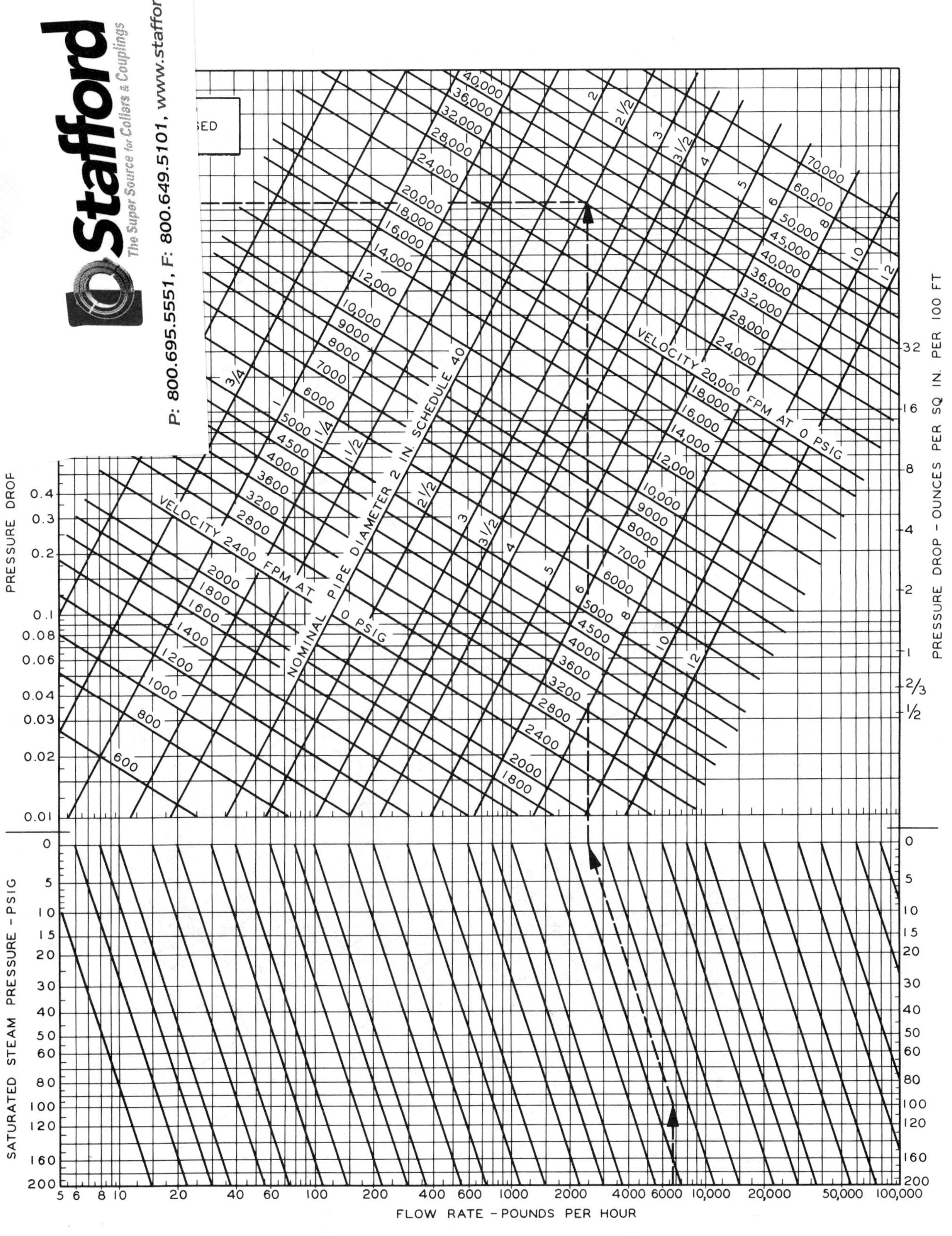

Notes: Based on Moody Friction Factor where flow of condensate does not inhibit the flow of steam.
See Figure 11 for obtaining flow rates and velocities of all saturation pressures between 0 and 200 psig; see also Examples 9 and 10.

Fig. 10 Flow Rate and Velocity of Steam in Schedule 40 Pipe at Saturation Pressure of 0 psig

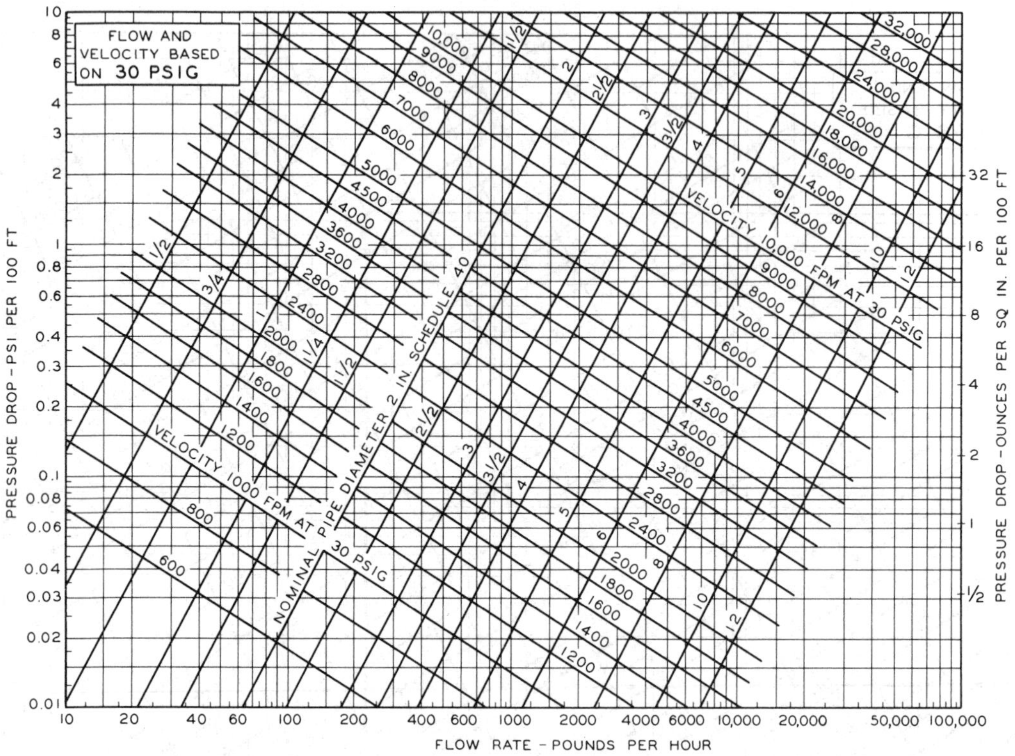

Notes: Based on Moody Friction Factor where flow of condensate does not inhibit the flow of steam.
May be used for steam pressures from 23 to 37 psig with an error not exceeding 9%.

Fig. 10A Flow Rate and Velocity of Steam in Schedule 40 Pipe at Saturation Pressure of 30 psig

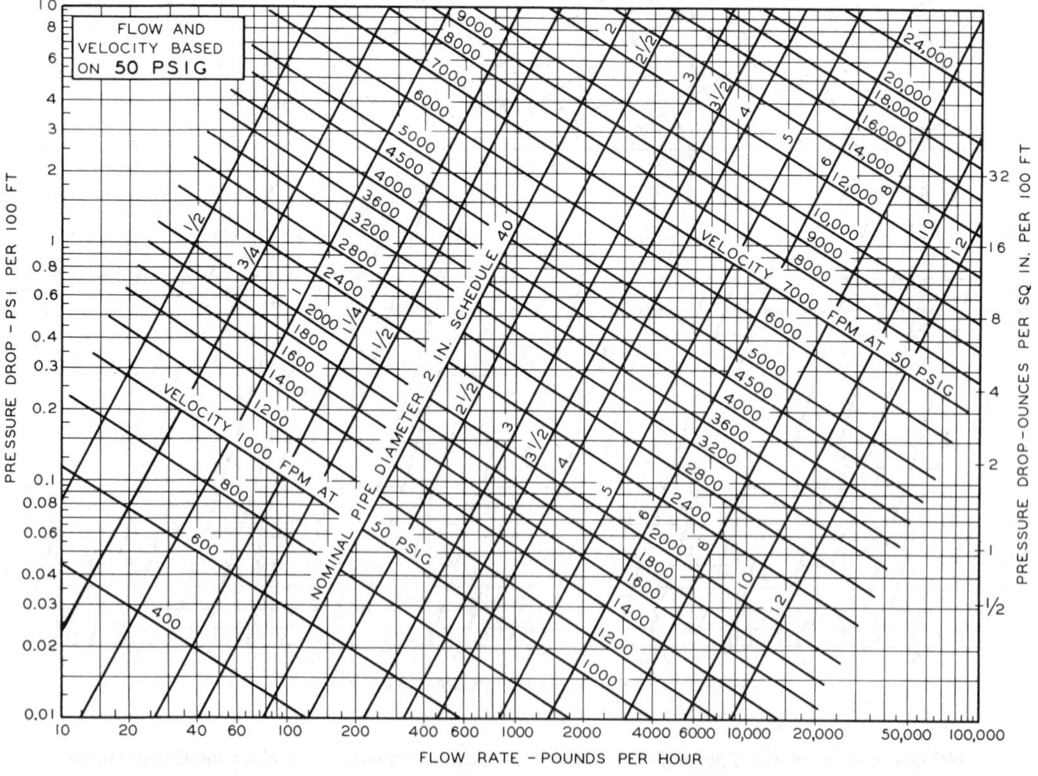

Notes: Based on Moody Friction Factor where flow of condensate does not inhibit the flow of steam.
May be used for steam pressures from 40 to 60 psig with an error not exceeding 8%.

Fig. 10B Flow Rate and Velocity of Steam in Schedule 40 Pipe at Saturation Pressure of 50 psig

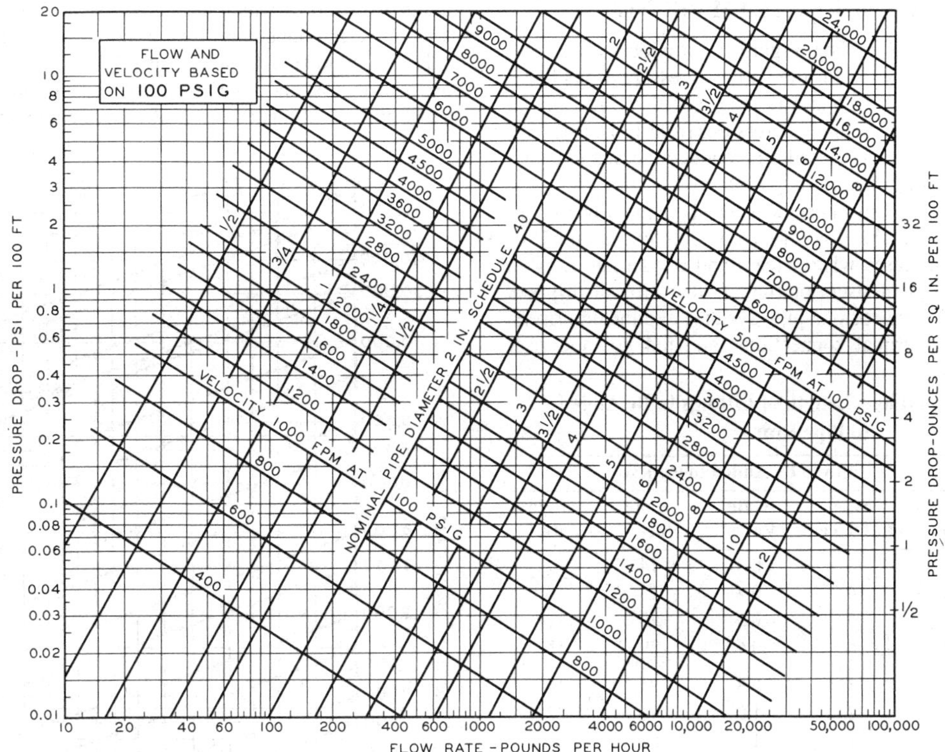

Notes: Based on Moody Friction Factor where flow of condensate does not inhibit the flow of steam.
May be used for steam pressures from 85 to 120 psig with an error not exceeding 8%.

Fig. 10C Flow Rate and Velocity of Steam in Schedule 40 Pipe at Saturation Pressure of 100 psig

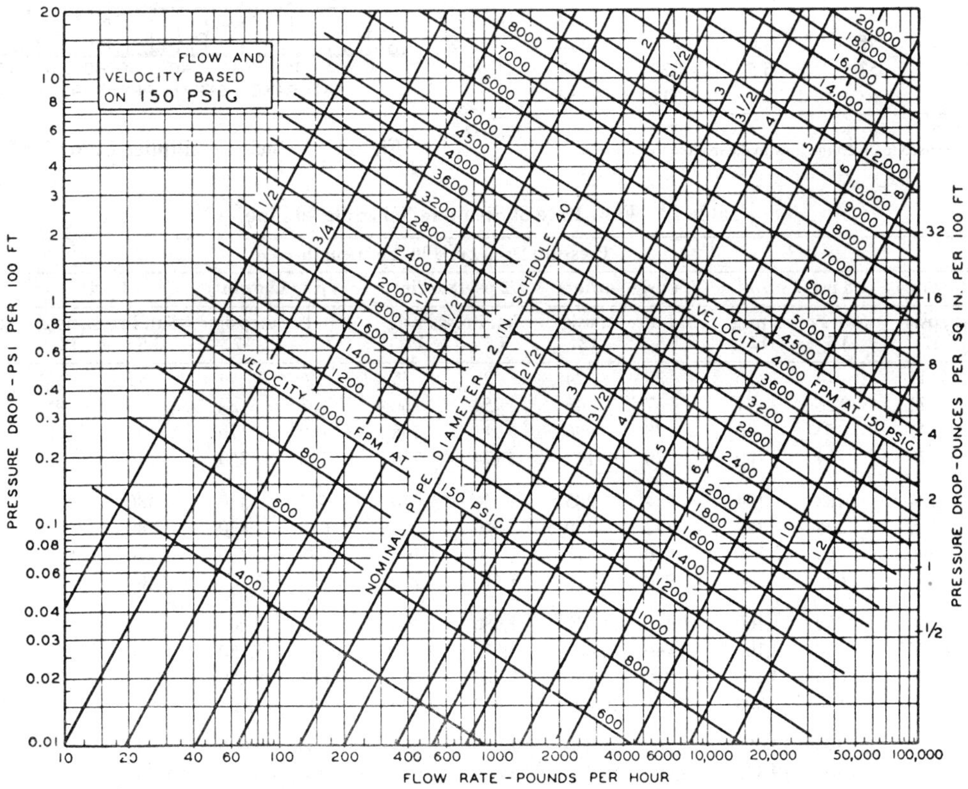

Notes: Based on Moody Friction Factor where flow of condensate does not inhibit the flow of steam.
May be used for steam pressures from 127 to 180 psig with an error not exceeding 8%.

Fig. 10D Flow Rate and Velocity of Steam in Schedule 40 Pipe at Saturation Pressure of 150 psig

LOW-PRESSURE STEAM PIPING

Values in Table 14 (taken from Figure 10) provide a more rapid means of selecting pipe sizes for the various pressure drops listed and for systems operated at 3.5 and 12 psig. The flow rates shown for 3.5 psig can be used for saturated pressures from 1 to 6 psig, and those shown for 12 psig can be used for saturated pressures from 8 to 16 psig with an error not exceeding 8%.

Both Figure 10 and Table 14 can be used where the flow of condensate does not inhibit the flow of steam. Columns B and C of Table 15 are used in cases where steam and condensate flow in opposite directions, as in risers or runouts that are not dripped. Columns D, E, and F are for one-pipe systems and include risers, radiator valves and vertical connections, and radiator and riser runout sizes, all of which are based on the critical velocity of the steam to permit the counterflow of condensate without noise.

Return piping can be sized using Table 16, in which pipe capacities for wet, dry, and vacuum return lines are shown for several values of pressure drop per 100 ft of equivalent length.

Table 13 Equivalent Length of Fittings to Be Added to Pipe Run

Nominal Pipe Diameter, in.	Length to Be Added to Run, ft: Standard Elbow	Side Outlet Tee[b]	Gate Valve[a]	Globe Valve[a]	Angle Valve[a]
1/2	1.3	3	0.3	14	7
3/4	1.8	4	0.4	18	10
1	2.2	5	0.5	23	12
1-1/4	3.0	6	0.6	29	15
1-1/2	3.5	7	0.8	34	18
2	4.3	8	1.0	46	22
2-1/2	5.0	11	1.1	54	27
3	6.5	13	1.4	66	34
3-1/2	8	15	1.6	80	40
4	9	18	1.9	92	45
5	11	22	2.2	112	56
6	13	27	2.8	136	67
8	17	35	3.7	180	92
10	21	45	4.6	230	112
12	27	53	5.5	270	132
14	30	63	6.4	310	152

[a]Valve in full-open position.
[b]Values apply only to a tee used to divert the flow in the main to the last riser.

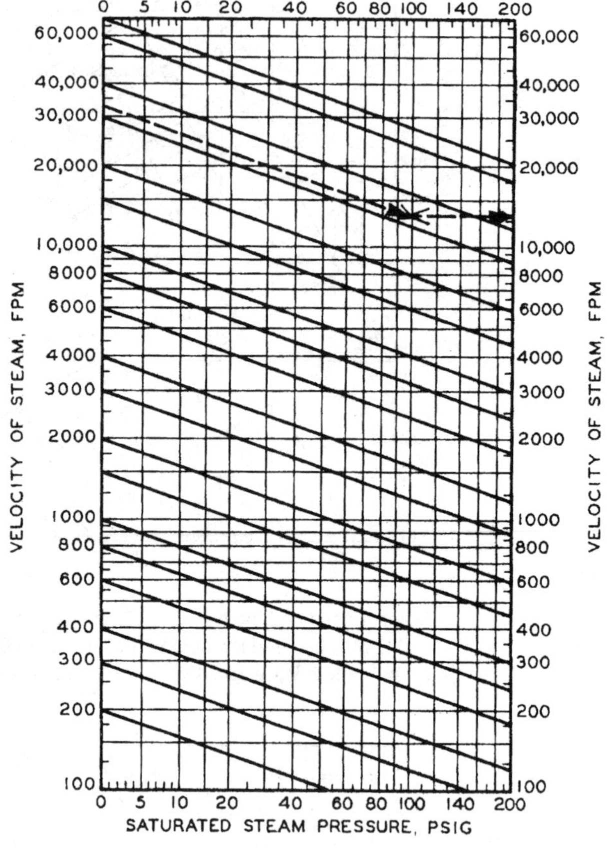

Fig. 11 Velocity Multiplier Chart for Figure 10

Table 14 Flow Rate of Steam in Schedule 40 Pipe

Nominal Pipe Size, in.	1/16 psi (1 oz/in²) 3.5	1/16 psi 12	1/8 psi (2 oz/in²) 3.5	1/8 psi 12	1/4 psi (4 oz/in²) 3.5	1/4 psi 12	1/2 psi (8 oz/in²) 3.5	1/2 psi 12	3/4 psi (12 oz/in²) 3.5	3/4 psi 12	1 psi 3.5	1 psi 12	2 psi 3.5	2 psi 12
	Pressure Drop per 100 ft of Length; Sat. Press., psig													
3/4	9	11	14	16	20	24	29	35	36	43	42	50	60	73
1	17	21	26	31	37	46	54	66	68	82	81	95	114	137
1-1/4	36	45	53	66	78	96	111	138	140	170	162	200	232	280
1-1/2	56	70	84	100	120	147	174	210	218	260	246	304	360	430
2	108	134	162	194	234	285	336	410	420	510	480	590	710	850
2-1/2	174	215	258	310	378	460	540	660	680	820	780	950	1,150	1,370
3	318	380	465	550	660	810	960	1,160	1,190	1,430	1,380	1,670	1,950	2,400
3-1/2	462	550	670	800	990	1,218	1,410	1,700	1,740	2,100	2,000	2,420	2,950	3,450
4	640	800	950	1,160	1,410	1,690	1,980	2,400	2,450	3,000	2,880	3,460	4,200	4,900
5	1,200	1,430	1,680	2,100	2,440	3,000	3,570	4,250	4,380	5,250	5,100	6,100	7,500	8,600
6	1,920	2,300	2,820	3,350	3,960	4,850	5,700	6,800	7,000	8,600	8,400	10,000	11,900	14,200
8	3,900	4,800	5,570	7,000	8,100	10,000	11,400	14,300	14,500	17,700	16,500	20,500	24,000	29,500
10	7,200	8,800	10,200	12,600	15,000	18,200	21,000	26,000	26,200	32,000	30,000	37,000	42,700	52,000
12	11,400	13,700	16,500	19,500	23,400	28,400	33,000	40,000	41,000	49,500	48,000	57,500	67,800	81,000

Notes:
1. Flow rate is in lb/h at initial saturation pressures of 3.5 and 12 psig. Flow is based on Moody friction factor, where the flow of condensate does not inhibit the flow of steam.
2. The flow rates at 3.5 psig cover saturated pressure from 1 to 6 psig, and the rates at 12 psig cover saturated pressure from 8 to 16 psig with an error not exceeding 8%.
3. The steam velocities corresponding to the flow rates given in this table can be found from Figures 10 and 11.

Table 15 Steam Pipe Capacities for Low-Pressure Systems

Nominal Pipe Size, in.	Capacity, lb/h: Two-Pipe System: Condensate Flowing Against Steam: Vertical	Horizontal	One-Pipe Systems: Supply Risers Upfeed	Radiator Valves and Vertical Connections	Radiator and Riser Runouts
A	B[a]	C[b]	D[c]	E	F[b]
3/4	8	7	6	—	7
1	14	14	11	7	7
1-1/4	31	27	20	16	16
1-1/2	48	42	38	23	16
2	97	93	72	42	23
2-1/2	159	132	116	—	42
3	282	200	200	—	65
3-1/2	387	288	286	—	119
4	511	425	380	—	186
5	1,050	788	—	—	278
6	1,800	1,400	—	—	545
8	3,750	3,000	—	—	—
10	7,000	5,700	—	—	—
12	11,500	9,500	—	—	—
16	22,000	19,000	—	—	—

Notes:
1. For one- or two-pipe systems in which condensate flows against the steam flow.
2. Steam at an average pressure of 1 psig is used as a basis of calculating capacities.

[a]Do not use Column B for pressure drops of less than 1/16 psi per 100 ft of equivalent run. Use Figure 10 or Table 13 instead.
[b]Pitch of horizontal runouts to risers and radiators should be not less than 0.5 in/ft. Where this pitch cannot be obtained, runouts over 8 ft in length should be one pipe size larger than that called for in this table.
[c]Do not use Column D for pressure drops of less than 1/24 psi per 100 ft of equivalent run except on sizes 3 in. and over. Use Figure 10 or Table 13 instead.

Example 9. What pressure drop should be used for the steam piping of a system if the measured length of the longest run is 500 ft, and the initial pressure must not exceed 2 psig?

Solution: It is assumed, if the measured length of the longest run is 500 ft, that when the allowance for fittings is added, the equivalent length of run does not exceed 1000 ft. Then, with the pressure drop not over one-half of the initial pressure, the drop could be 1 psi or less. With a pressure drop of 1 psi and a length of run of 1000 ft, the drop per 100 ft would be 0.1 psi; if the total drop were 0.5 psi, the drop per 100 ft would be 0.05 psi. In both cases, the pipe could be sized for a desired capacity according to Figure 10.

On completion of the sizing, the drop could be checked by taking the longest line and actually calculating the equivalent length of run from the pipe sizes determined. If the calculated drop is less than that assumed, the pipe size is adequate; if it is more, an unusual number of fittings is probably involved, and either the lines must be straightened, or the next larger pipe size must be tried.

HIGH-PRESSURE STEAM PIPING

Many heating systems for large industrial buildings use high-pressure steam (15 to 150 psig). These systems usually have unit heaters or large built-up fan units with blast heating coils. Temperatures are controlled by a modulating or throttling thermostatic valve or by face or bypass dampers controlled by the room air temperature, fan inlet, or fan outlet.

Use of Basic and Velocity Multiplier Charts

Example 10. Given a flow rate of 6700 lb/h, an initial steam pressure of 100 psig, and a pressure drop of 11 psi/100 ft, find the size of Schedule 40 pipe required and the velocity of steam in the pipe.

Solution: The following steps are illustrated by the broken line on Figures 10 and 11.

1. Enter Figure 10 at a flow rate of 6700 lb/h, and move vertically to the horizontal line at 100 psig.

Table 16 Return Main and Riser Capacities for Low-Pressure Systems, lb/h

	Pipe Size, in.	1/32 psi (1/2 oz/in^2) Drop per 100 ft: Wet	Dry	Vac.	1/24 psi (2/3 oz/in^2) Drop per 100 ft: Wet	Dry	Vac.	1/16 psi (1 oz/in^2) Drop per 100 ft: Wet	Dry	Vac.	1/8 psi (2 oz/in^2) Drop per 100 ft: Wet	Dry	Vac.	1/4 psi (4 oz/in^2) Drop per 100 ft: Wet	Dry	Vac.	1/2 psi (8 oz/in^2) Drop per 100 ft: Wet	Dry	Vac.
	G	H	I	J	K	L	M	N	O	P	Q	R	S	T	U	V	W	X	Y
Return Main	3/4	—	—	—	—	—	42	—	—	100	—	—	142	—	—	200	—	—	283
	1	125	62	—	145	71	143	175	80	175	250	103	249	350	115	350	—	—	494
	1-1/4	213	130	—	248	149	244	300	168	300	425	217	426	600	241	600	—	—	848
	1-1/2	338	206	—	393	236	388	475	265	475	675	340	674	950	378	950	—	—	1,340
	2	700	470	—	810	535	815	1,000	575	1,000	1,400	740	1,420	2,000	825	2,000	—	—	2,830
	2-1/2	1,180	760	—	1,580	868	1,360	1,680	950	1,680	2,350	1,230	2,380	3,350	1,360	3,350	—	—	4,730
	3	1,880	1,460	—	2,130	1,560	2,180	2,680	1,750	2,680	3,750	2,250	3,800	5,350	2,500	5,350	—	—	7,560
	3-1/2	2,750	1,970	—	3,300	2,200	3,250	4,000	2,500	4,000	5,500	3,230	5,680	8,000	3,580	8,000	—	—	11,300
	4	3,880	2,930	—	4,580	3,350	4,500	5,500	3,750	5,500	7,750	4,830	7,810	11,000	5,380	11,000	—	—	15,500
	5	—	—	—	—	—	7,880	—	—	9680	—	—	13,700	—	—	19,400	—	—	27,300
	6	—	—	—	—	—	12,600	—	—	15,500	—	—	22,000	—	—	31,000	—	—	43,800
Riser	3/4	—	48	—	—	48	143	—	48	175	—	48	249	—	48	350	—	—	494
	1	—	113	—	—	113	244	—	113	300	—	113	426	—	113	600	—	—	848
	1-1/4	—	248	—	—	248	388	—	248	475	—	248	674	—	248	950	—	—	1,340
	1-1/2	—	375	—	—	375	815	—	375	1,000	—	375	1,420	—	375	2,000	—	—	2,830
	2	—	750	—	—	750	1,360	—	750	1,680	—	750	2,380	—	750	3,350	—	—	4,730
	2-1/2	—	—	—	—	—	2,180	—	—	2,680	—	—	3,800	—	—	5,350	—	—	7,560
	3	—	—	—	—	—	3,250	—	—	4,000	—	—	5,680	—	—	8,000	—	—	11,300
	3-1/2	—	—	—	—	—	4,480	—	—	5,500	—	—	7,810	—	—	11,000	—	—	15,500
	4	—	—	—	—	—	7,880	—	—	9680	—	—	13,700	—	—	19,400	—	—	27,300
	5	—	—	—	—	—	12,600	—	—	15,500	—	—	22,000	—	—	31,000	—	—	43,800

2. Follow along inclined multiplier line (upward and to the left) to horizontal 0 psig line. The equivalent mass flow at 0 psig is about 2500 lb/h.
3. Follow the 2500 lb/h line vertically until it intersects the horizontal line at 11 psi per 100 ft pressure drop. Nominal pipe size is 2-1/2 in. The equivalent steam velocity at 0 psig is about 32,700 fpm.
4. To find the steam velocity at 100 psig, locate the value of 32,700 fpm on the ordinate of the velocity multiplier chart (Figure 11) at 0 psig.
5. Move along the inclined multiplier line (downward and to the right) until it intersects the vertical 100 psig pressure line. The velocity as read from the right (or left) scale is about 13,000 fpm.

Note: Steps 1 through 5 would be rearranged or reversed if different data were given.

STEAM CONDENSATE SYSTEMS

The majority of steam systems used in heating applications are two-pipe systems, in which the two pipes are the "steam" pipe and the "condensate" pipe. This discussion is limited to the sizing of the condensate lines in two-pipe systems.

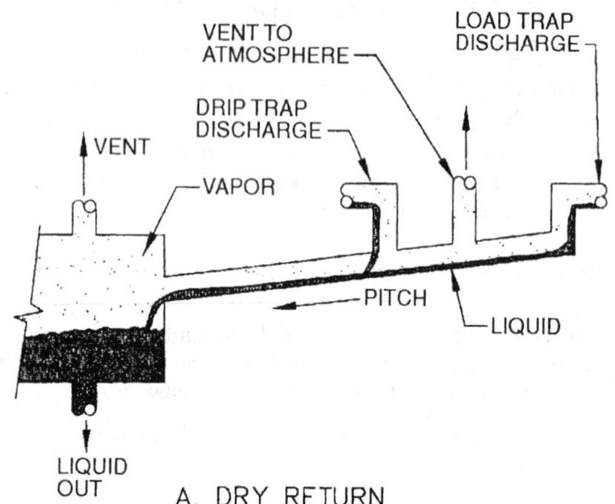

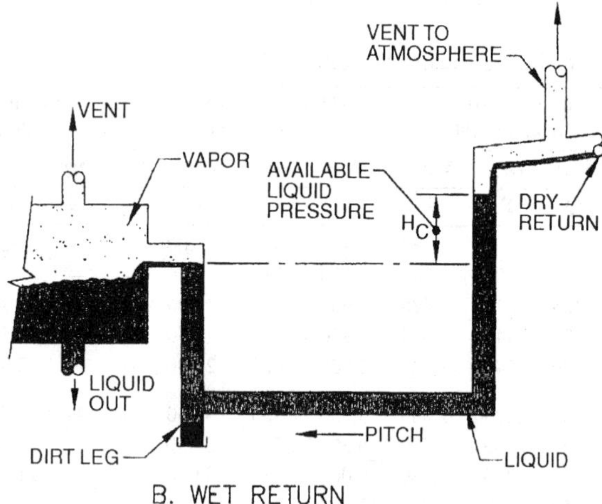

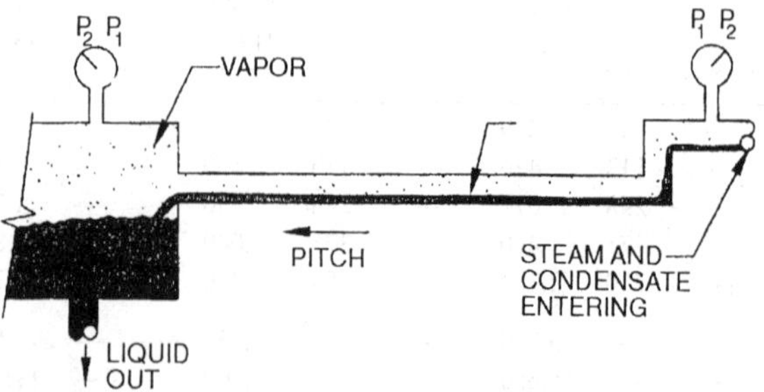

Fig. 12 Types of Condensate Return Systems

Two-Pipe Systems

When steam is used for heating a liquid to 215°F or less (e.g., in domestic water heat exchangers, domestic heating water converters, or air-heating coils), the devices are usually provided with a steam control valve. As the control valve throttles, the absolute pressure in the load device decreases, removing all pressure motivation for flow in the condensate return system. In order to ensure the flow of steam condensate from the load device through the trap and into the return system, it is necessary to provide a vacuum breaker on the device ahead of the trap. This ensures a minimum pressure at the trap inlet of atmospheric pressure plus whatever liquid leg the designer has provided. Then, to ensure flow through the trap, it is necessary to design the condensate system so that it will never have a pressure above atmospheric in the condensate return line.

Vented (Open) Return Systems. To achieve this pressure requirement, the condensate return line is usually vented to the atmosphere (1) near the point of entrance of the flow streams from the load traps, (2) in proximity to all connections from drip traps, and (3) at transfer pumps or feedwater receivers.

With this design, the only motivation for flow in the return system is gravity. Return lines that are below the liquid level in the downstream receiver or boiler and are thus filled with liquid are called wet returns; those above the liquid level have both liquid and gas in the pipes and are called dry returns.

The dry return lines in a vented return system have flowing liquid in the bottom of the line and gas or vapor in the top (Figure 12A). The liquid is the condensate, and the gas may be steam, air, or a mixture of the two. The flow phenomenon for these dry return systems is open channel flow, which is best described by the **Manning equation**:

$$Q = \frac{1.49 A r^{2/3} S^{1/2}}{n} \tag{12}$$

where

Q = volumetric flow rate, cfs
A = cross-sectional area of conduit, ft^2
r = hydraulic radius of conduit, ft
n = coefficient of roughness (usually 0.012)
S = slope of conduit, ft/ft

Table 17 is a solution to Equation (12) that shows pipe size capacities for steel pipes with various pitches. Recommended

Table 17 Vented Dry Condensate Return for Gravity Flow Based on Manning Equation

Nominal Diameter, in. IPS	Condensate Flow, lb/h[a,b] — Condensate Line Slope, in/ft 1/16	1/8	1/4	1/2
1/2	38	54	76	107
3/4	80	114	161	227
1	153	216	306	432
1-1/4	318	449	635	898
1-1/2	479	677	958	1,360
2	932	1,320	1,860	2,640
2-1/2	1,500	2,120	3,000	4,240
3	2,670	3,780	5,350	7,560
4	5,520	7,800	11,000	15,600
5	10,100	14,300	20,200	28,500
6	16,500	23,300	32,900	46,500

[a] Flow is in lb/h of 180°F water for Schedule 40 steel pipes.
[b] Flow was calculated from Equation (12) and rounded.

practice is to size vertical lines by the maximum pitch shown, although they would actually have a capacity far in excess of that shown. As the pitch increases, hydraulic jump that could fill the pipe and other transient effects that could cause water hammer should be avoided. Flow values in Table 17 are calculated for Schedule 40 steel pipe, with a factor of safety of 3.0, and can be used for copper pipes of the same nominal pipe size.

The flow characteristics of **wet return lines** (Figure 12B) are best described by the Darcy-Weisbach equation [Equation (1)]. The motivation for flow is the fluid head difference between the entering section of the flooded line and the leaving section. It is common practice, in addition to providing for the fluid head differential, to slope the return in the direction of flow to a collection point such as a dirt leg in order to clear the line of sediment or solids. Table 18 is a solution to Equation (1) that shows pipe size capacity for steel pipes with various available fluid heads. Table 18 can also be used for copper tubing of equal nominal pipe size.

Nonvented (Closed) Return Systems. For those systems in which there is a continual steam pressure difference between the point where the condensate enters the line and the point where it

Table 18 Vented Wet Condensate Return for Gravity Flow Based on Darcy-Weisbach Equation

Nominal Diameter, in. IPS	Condensate Flow, lb/h[a,b] Condensate Head, ft per 100 ft							
	0.5	1	1.5	2	2.5	3	3.5	4
1/2	105	154	192	224	252	278	302	324
3/4	225	328	408	476	536	590	640	687
1	432	628	779	908	1,020	1,120	1,220	1,310
1-1/4	901	1,310	1,620	1,890	2,120	2,330	2,530	2,710
1-1/2	1,360	1,970	2,440	2,840	3,190	3,510	3,800	4,080
2	2,650	3,830	4,740	5,510	6,180	6,800	7,360	7,890
2-1/2	4,260	6,140	7,580	8,810	9,890	10,900	11,800	12,600
3	7,570	10,900	13,500	15,600	17,500	19,300	20,900	22,300
4	15,500	22,300	27,600	32,000	35,900	39,400	42,600	45,600
5	28,200	40,500	49,900	57,900	64,900	71,300	77,100	82,600
6	45,800	65,600	80,900	93,800	105,000	115,000	125,000	134,000

[a] Flow is in lb/h of 180°F water for Schedule 40 steel pipes.

[b] Flow was calculated from Equation (1) and rounded.

Table 19 Flow Rate for Dry-Closed Returns

Pipe Dia. D, in.	Supply Pressure = 5 psig Return Pressure = 0 psig			Supply Pressure = 15 psig Return Pressure = 0 psig			Supply Pressure = 30 psig Return Pressure = 0 psig			Supply Pressure = 50 psig Return Pressure = 0 psig		
	Δp/L, psi/100 ft											
	1/16	1/4	1	1/16	1/4	1	1/16	1/4	1	1/16	1/4	1
	Flow Rate, lb/h											
1/2	240	520	1,100	95	210	450	60	130	274	42	92	200
3/4	510	1,120	2,400	210	450	950	130	280	590	91	200	420
1	1,000	2,150	4,540	400	860	1,820	250	530	1,120	180	380	800
1-1/4	2,100	4,500	9,500	840	1,800	3,800	520	1,110	2,340	370	800	1,680
1-1/2	3,170	6,780	14,200	1,270	2,720	5,700	780	1,670	3,510	560	1,200	2,520
2	6,240	13,300	a	2,500	5,320	a	1,540	3,270	a	1,110	2,350	a
2-1/2	10,000	21,300	a	4,030	8,520	a	2,480	5,250	a	1,780	3,780	a
3	18,000	38,000	a	7,200	15,200	a	4,440	9,360	a	3,190	6,730	a
4	37,200	78,000	a	14,900	31,300	a	9,180	19,200	a	6,660	13,800	a
6	110,500	a	a	44,300	a	a	27,300	a	a	19,600	a	a
8	228,600	a	a	91,700	a	a	56,400	a	a	40,500	a	a

Pipe Dia. D, in.	Supply Pressure = 100 psig Return Pressure = 0 psig			Supply Pressure = 150 psig Return Pressure = 0 psig			Supply Pressure = 100 psig Return Pressure = 15 psig			Supply Pressure = 150 psig Return Pressure = 15 psig		
	Δp/L, psi/100 ft											
	1/16	1/4	1	1/16	1/4	1	1/16	1/4	1	1/16	1/4	1
	Flow Rate, lb/h											
1/2	28	62	133	23	51	109	56	120	260	43	93	200
3/4	62	134	290	50	110	230	120	260	560	93	200	420
1	120	260	544	100	210	450	240	500	1,060	180	390	800
1-1/4	250	540	1,130	200	440	930	500	1,060	2,200	380	800	1,680
1-1/2	380	810	1,700	310	660	1,400	750	1,600	3,320	570	1,210	2,500
2	750	1,590	a	610	1,300	a	1,470	3,100	6,450	1,120	2,350	4,900
2-1/2	1,200	2,550	a	980	2,100	a	2,370	5,000	10,300	1,800	3,780	7,800
3	2,160	4,550	a	1,760	3,710	a	4,230	8,860	a	3,200	6,710	a
4	4,460	9,340	a	3,640	7,630	a	8,730	18,200	a	6,620	13,800	a
6	13,200	a	a	10,800	a	a	25,900	53,600	a	19,600	40,600	a
8	27,400	a	a	22,400	a	a	53,400	110,300	a	40,500	83,600	a

[a] For these sizes and pressure losses, the velocity is above 7000 fpm. Select another combination of size and pressure loss.

leaves (Figure 12C), Table 16 or Table 19, as applicable, can be used for sizing the condensate lines. Although these tables express condensate capacity without slope, common practice is to slope the lines in the direction of flow to a collection point similar to wet returns to clear the lines of sediment or solids.

When saturated condensate at pressures above the return system pressure enters the return (condensate) mains, some of the liquid flashes to steam. This occurs typically at drip traps into a vented return system or at load traps leaving process load devices that are not valve-controlled and typically have no subcooling. If the return main is vented, the vent lines will relieve any excessive pressure and prevent a back pressure phenomenon that could restrict the flow through traps from valved loads; the pipe sizing would be as described above for vented dry returns. If the return line is not vented, the flash steam results in a pressure rise at that point and the piping could be sized as described above for closed returns, and in accordance with Table 16 or Table 19, as applicable.

The passage of the fluid through the steam trap is a throttling or constant enthalpy process. The resulting fluid on the downstream side of the trap can be a mixture of saturated liquid and vapor. Thus, in nonvented returns, it is important to understand the condition of the fluid when it enters the return line from the trap.

The condition of the condensate downstream of the trap can be expressed by the quality x, defined as

$$x = \frac{m_v}{m_l + m_v} \tag{13}$$

where

m_v = mass of saturated vapor in condensate
m_l = mass of saturated liquid in condensate

Likewise, the volume fraction V_c of the vapor in the condensate is expressed as

$$V_c = \frac{V_v}{V_l + V_v} \tag{14}$$

where

V_v = volume of saturated vapor in condensate
V_l = volume of saturated liquid in condensate

The quality and the volume fraction of the condensate downstream of the trap can be estimated from Equations (13) and (14), respectively.

$$x = \frac{h_1 - h_{f_2}}{h_{g_2} - h_{f_2}} \tag{15}$$

$$V_c = \frac{x v_{g_2}}{v_{f_2}(1-x) + x v_{g_2}} \tag{16}$$

where

h_1 = enthalpy of liquid condensate entering trap evaluated at supply pressure for saturated condensate or at saturation pressure corresponding to temperature of subcooled liquid condensate
h_{f_2} = enthalpy of saturated liquid at return or downstream pressure of trap
h_{g_2} = enthalpy of saturated vapor at return or downstream pressure of trap
v_{f_2} = specific volume of saturated liquid at return or downstream pressure of trap
v_{g_2} = specific volume of saturated vapor at return or downstream pressure of trap

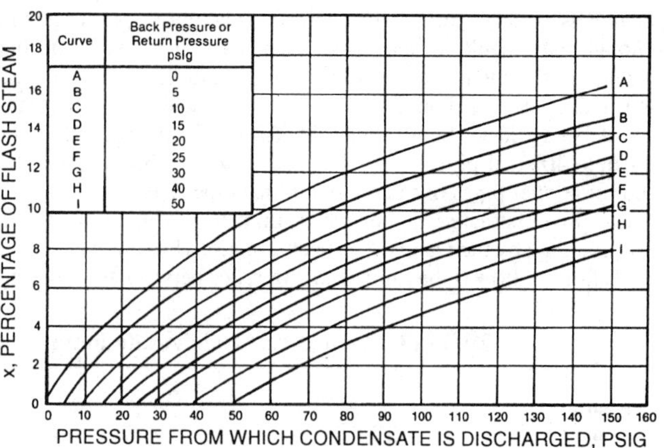

Fig. 13 Working Chart for Determining Percentage of Flash Steam (Quality)

Table 20 Flash Steam from Steam Trap on Pressure Drop

Supply Pressure, psig	Return Pressure, psig	x, Fraction Vapor, Mass Basis	V_c, Fraction Vapor, Volume Basis
5	0	0.016	0.962
15	0	0.040	0.985
30	0	0.065	0.991
50	0	0.090	0.994
100	0	0.133	0.996
150	0	0.164	0.997
100	15	0.096	0.989
150	15	0.128	0.992

Table 21 Estimated Return Line Pressures

Pressure Drop, psi/100 ft	Pressure in Return Line, psig	
	30 psig Supply	150 psig Supply
1/8	0.5	1.25
1/4	1	2.5
1/2	2	5
3/4	3	7.5
1	4	10
2	—	20

Table 20 presents some values for quality and volume fraction for typical supply and return pressures in heating and ventilating systems. Note that the percent of vapor on a mass basis x is small, while the percent of vapor on a volume basis V_c is very large. This indicates that the return pipe cross section is predominantly occupied by vapor. Figure 13 is a working chart to determine the quality of the condensate entering the return line from the trap for various combinations of supply and return pressures. If the liquid is subcooled entering the trap, the saturation pressure corresponding to the liquid temperature should be used for the supply or upstream pressure. Typical pressures in the return line are given in Table 21.

One-Pipe Systems

Gravity one-pipe air vent systems in which steam and condensate flow in the same pipe, frequently in opposite directions, are considered obsolete and are no longer being installed. See Chapter 33 of the 1993 *ASHRAE Handbook—Fundamentals* or earlier ASHRAE Handbooks for descriptions of and design information for one-pipe systems.

Table 22 Maximum Capacity of Gas Pipe in Cubic Feet per Hour

Nominal Iron Pipe Size, in.	Internal Diameter, in.	Length of Pipe, ft													
		10	20	30	40	50	60	70	80	90	100	125	150	175	200
1/4	0.364	32	22	18	15	14	12	11	11	10	9	8	8	7	6
3/8	0.493	72	49	40	34	30	27	25	23	22	21	18	17	15	14
1/2	0.622	132	92	73	63	56	50	46	43	40	38	34	31	28	26
3/4	0.824	278	190	152	130	115	105	96	90	84	79	72	64	59	55
1	1.049	520	350	285	245	215	195	180	170	160	150	130	120	110	100
1-1/4	1.380	1,050	730	590	500	440	400	370	350	320	305	275	250	225	210
1-1/2	1.610	1,600	1,100	890	760	670	610	560	530	490	460	410	380	350	320
2	2.067	3,050	2,100	1,650	1,450	1,270	1,150	1,050	990	930	870	780	710	650	610
2-1/2	2.469	4,800	3,300	2,700	2,300	2,000	1,850	1,700	1,600	1,500	1,400	1,250	1,130	1,050	980
3	3.068	8,500	5,900	4,700	4,100	3,600	3,250	3,000	2,800	2,600	2,500	2,200	2,000	1,850	1,700
4	4.026	17,500	12,000	9,700	8,300	7,400	6,800	6,200	5,800	5,400	5,100	4,500	4,100	3,800	3,500

Note: Capacity is in cubic feet per hour at gas pressures of 0.5 psig or less and a pressure drop of 0.5 in. of water; specific gravity = 0.60.

GAS PIPING

Piping for gas appliances should be of adequate size and installed so that it provides a supply of gas sufficient to meet the maximum demand without undue loss of pressure between the point of supply (the meter) and the appliance. The size of gas pipe required depends on (1) maximum gas consumption to be provided, (2) length of pipe and number of fittings, (3) allowable pressure loss from the outlet of the meter to the appliance, and (4) specific gravity of the gas.

Gas consumption in ft^3/h is obtained by dividing the Btu input rate at which the appliance is operated by the average heating value of the gas in Btu/ft^3. Insufficient gas flow from excessive pressure losses in gas supply lines can cause inefficient operation of gas-fired appliances and sometimes create hazardous operations. Gas-fired appliances are normally equipped with a data plate giving information on maximum gas flow requirements or Btu input as well as inlet gas pressure requirements. The gas utility in the area of installation can give the gas pressure available at the utility's gas meter. Using the information, the required size of gas piping can be calculated for satisfactory operation of the appliance(s).

Table 22 gives pipe capacities for gas flow for up to 200 ft of pipe based on a specific gravity of 0.60. Capacities for pressures less than 1.5 psig may also be determined by the following equation from NFPA/IAS *National Fuel Gas Code*:

$$Q = 2313 d^{2.623} (\Delta p / CL)^{0.541} \tag{17}$$

where

Q = flow rate at 60°F and 30 in. Hg, cfh
d = inside diameter of pipe, in.
Δp = pressure drop, in. of water
C = factor for viscosity, density, and temperature
= $0.00354(t + 460)s^{0.848}\mu^{0.152}$
t = temperature, °F
s = ratio of density of gas to density of air at 60°F and 30 in. Hg
μ = viscosity of gas, centipoise (0.012 for natural gas, 0.008 for propane)
L = pipe length, ft

Gas service in buildings is generally delivered in the "low-pressure" range of 7 in. of water. The maximum pressure drop allowable in piping systems at this pressure is generally 0.5 in. of water but is subject to regulation by local building, plumbing, and gas appliance codes (see also the NFPA/IAS *National Fuel Gas Code*).

Where large quantities of gas are required or where long lengths of pipe are used (e.g., in industrial buildings), low-pressure limitations result in large pipe sizes. Local codes may allow and local gas companies may deliver gas at higher pressures (e.g., 2, 5, or 10 psig). Under these conditions, an allowable pressure drop of 10% of the initial pressure is used, and pipe sizes can be reduced significantly. Gas pressure regulators at the appliance must be specified to accommodate higher inlet pressures. NFPA/IAS (1992) provides information on pipe sizing for various inlet pressures and pressure drops at higher pressures.

More complete information on gas piping can be found in the *Gas Engineers' Handbook* (1970).

FUEL OIL PIPING

The pipe used to convey fuel oil to oil-fired appliances must be large enough to maintain low pump suction pressure and, in the case of circulating loop systems, to prevent overpressure at the burner oil pump inlet. Pipe materials must be compatible with the fuel and must be carefully assembled to eliminate all leaks. Leaks in suction lines cause pumping problems that result in unreliable burner operation. Leaks in pressurized lines create fire hazards. Cast-iron or aluminum fittings and pipe are unacceptable. Pipe joint compounds must be selected carefully.

Oil pump suction lines should be sized so that at maximum suction line flow conditions, the maximum vacuum will not exceed 10 in. Hg for distillate grade fuels and 15 in. Hg for residual oils. Oil supply lines to burner oil pumps should not be pressurized by circulating loop systems or aboveground oil storage tanks to more than 5 psi, or pump shaft seals may fail. A typical oil circulating loop system is shown in Figure 14.

In assembling long fuel pipe lines, care should be taken to avoid air pockets. On overhead circulating loops, the line should vent air at all high points. Oil supply loops for one or more burners should be the continuous circulation type, with excess fuel returned to the storage tank. Dead-ended pressurized loops can be used, but air or vapor venting is more problematic.

Where valves are used, select ball or gate valves. Globe valves are not recommended because of their high pressure drop characteristics.

Oil lines should be tested after installation, particularly if they are buried, enclosed, or otherwise inaccessible. Failure to perform this test is a frequent cause of later operating difficulties. A suction line can be hydrostatically tested at 1.5 times its maximum operating pressure or at a vacuum of not less than 20 in. Hg. Pressure or vacuum tests should continue for at least 60 min. If there is no noticeable drop in the initial test pressure, the lines can be considered tight.

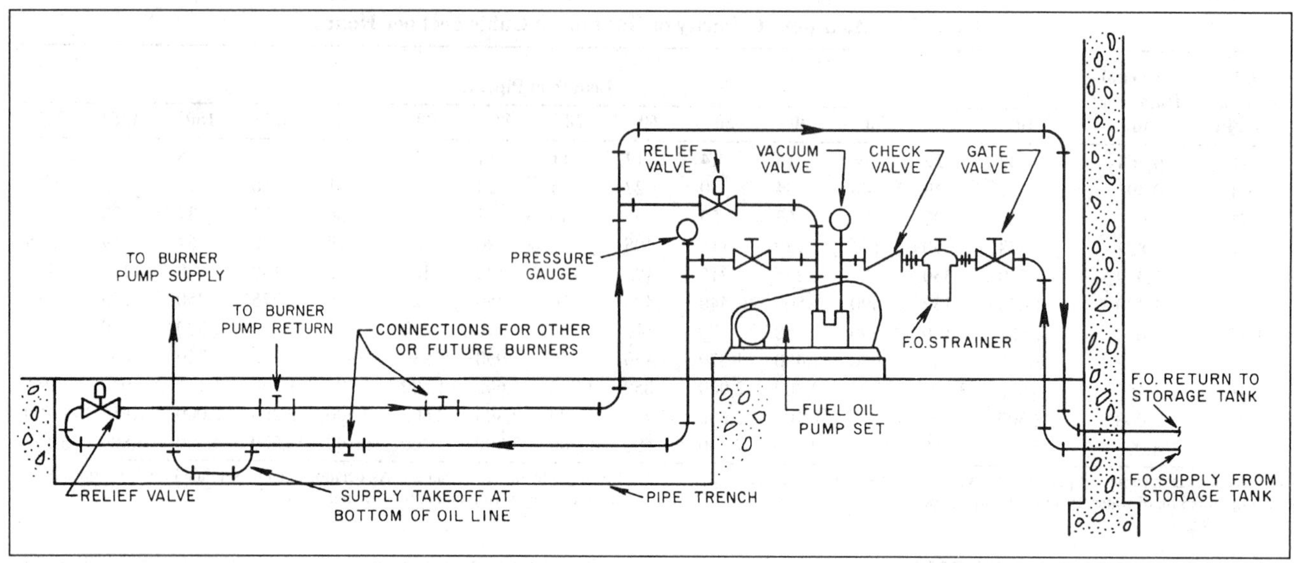

Fig. 14 Typical Oil Circulating Loop

Pipe Sizes for Heavy Oil

Tables 23 and 24 give recommended pipe sizes for handling No. 5 and No. 6 oils (residual grades) and No. 1 and No. 2 oils (distillate grades), respectively.

Table 23 Recommended Nominal Size for Fuel Oil Suction Lines from Tank to Pump (Residual Grades No. 5 and No. 6)

Pumping Rate, gph	Length of Run in Feet at Maximum Suction Lift of 15 ft									
	25	50	75	100	125	150	175	200	250	300
10	1-1/2	1-1/2	1-1/2	1-1/2	1-1/2	1-1/2	2	2	2-1/2	2-1/2
40	1-1/2	1-1/2	1-1/2	2	2	2-1/2	2-1/2	2-1/2	2-1/2	3
70	1-1/2	2	2	2	2	2-1/2	2-1/2	2-1/2	3	3
100	2	2	2	2-1/2	2-1/2	3	3	3	3	3
130	2	2	2-1/2	2-1/2	2-1/2	3	3	3	3	4
160	2	2	2-1/2	2-1/2	2-1/2	3	3	3	4	4
190	2	2-1/2	2-1/2	2-1/2	3	3	3	4	4	4
220	2-1/2	2-1/2	2-1/2	3	3	3	4	4	4	4

Notes:
1. Pipe sizes smaller than 1 in. IPS are not recommended for use with residual grade fuel oils.
2. Lines conveying fuel oil from pump discharge port to burners and tank return may be reduced by one or two sizes, depending on piping length and pressure losses.

Table 24 Recommended Nominal Size for Fuel Oil Suction Lines from Tank to Pump (Distillate Grades No. 1 and No. 2)

Pumping Rate, gph	Length of Run in Feet at Maximum Suction Lift of 10 ft									
	25	50	75	100	125	150	175	200	250	300
10	1/2	1/2	1/2	1/2	1/2	1/2	1/2	3/4	3/4	1
40	1/2	1/2	1/2	1/2	1/2	3/4	3/4	3/4	3/4	1
70	1/2	1/2	3/4	3/4	3/4	3/4	3/4	1	1	1
100	1/2	3/4	3/4	3/4	3/4	1	1	1	1	1-1/4
130	1/2	3/4	3/4	1	1	1	1	1	1-1/4	1-1/4
160	3/4	3/4	3/4	1	1	1	1	1-1/4	1-1/4	1-1/4
190	3/4	3/4	1	1	1	1	1-1/4	1-1/4	1-1/4	2
220	3/4	1	1	1	1	1-1/4	1-1/4	1-1/4	1-1/4	2

Storage tanks and piping and pumping facilities for delivering the oil from the tank to the burner are important considerations in the design of an industrial oil-burning system.

The construction and location of the tank and oil piping are usually subject to local regulations and National Fire Protection Association (NFPA) *Standards* 30 and 31.

REFERENCES

Ball, E.F. and C.J.D. Webster. 1976. Some measurements of water-flow noise in copper and ABS pipes with various flow velocities. *The Building Services Engineer* 44(2):33.

BOCA. 1992. *BOCA National plumbing code*, 9th ed. Building Officials and Code Administrators International, Country Club Hills, IL.

Carrier. 1960. Piping design. In *System design manual*. Carrier Air Conditioning Company, Syracuse, NY.

Crane Co. 1976. Flow of fluids through valves, fittings and pipe. *Technical Paper* No. 410. Crane Company, New York.

Dawson, F.M. and J.S. Bowman. 1933. Interior water supply piping for residential buildings. University of Wisconsin Experiment Station, No. 77.

Freeman, J.R. 1941. *Experiments upon the flow of water in pipes*. American Society of Mechanical Engineers, New York.

Gas engineers' handbook. 1970. The Industrial Press, New York.

Giesecke, F.E. 1926. Friction of water elbows. *ASHVE Transactions* 32:303.

Giesecke, F.E. and W.H. Badgett. 1931. Friction heads in one-inch standard cast-iron tees. *ASHVE Transactions* 37:395.

Giesecke, F.E. and W.H. Badgett. 1932a. Loss of head in copper pipe and fittings. *ASHVE Transactions* 38:529.

Giesecke, F.E. and W.H. Badgett. 1932b. Supplementary friction heads in one-inch cast-iron tees. *ASHVE Transactions* 38:111.

Grinnell Company. 1951. *Piping design and engineering*. Grinnell Company, Cranston, RI.

HDR design guide. 1981. Hennington, Durham and Richardson, Omaha, NE.

HI. 1979. *Engineering data book*. Hydraulic Institute, Parsippany, NJ.

Howell, R.H. 1985. Evaluation of sizing methods for steam condensate systems. *ASHRAE Transactions* 91(1).

Hunter, R.B. 1940. Methods of estimating loads in plumbing systems. NBS *Report* BMS 65. National Institute of Standards and Technology, Gaithersburg, MD.

Hunter, R.B. 1941. Water distributing systems for buildings. NBS *Report* BMS 79. National Institute of Standards and Technology, Gaithersburg, MD.

IAPMO. 1994. *Uniform plumbing code*. International Association of Plumbing and Mechanical Officials, Walnut, CA.

Idelchik, I.E. 1986. *Handbook of hydraulic resistance*. Hemisphere Publishing Corporation, New York.

ISA. 1985. Flow equations for sizing control valves. ANSI/ISA *Standard* S75.01-85. International Society for Measurement and Control, Research Triangle Park, NC.

Laschober, R.R., G.Y. Anderson, and D.G. Barbee. 1966. Counterflow of steam and condensate in slightly pitched pipes. *ASHRAE Transactions* 72(1):157.

Marseille, B. 1965. Noise transmission in piping. *Heating and Ventilating Engineering* (June):674.

NAPHCC. 1996. *National standard plumbing code*. National Association of Plumbing-Heating-Cooling Contractors, Falls Church, VA.

NFPA. 1992. Installation of oil burning equipment. ANSI/NFPA *Standard* 31-92. National Fire Protection Association, Quincy, MA.

NFPA. 1993. Flammable and combustible liquids code. ANSI/NFPA *Standard* 30-93.

NFPA/IAS. 1992. *National fuel gas code*. ANSI/NFPA *Standard* 54-92. National Fire Protection Association, Quincy, MA. ANSI/IAS *Standard* Z223.1-92. American Gas Association, Arlington, VA.

Obrecht, M.F. and M. Pourbaix. 1967. Corrosion of metals in potable water systems. AWWA 59:977. American Water Works Association, Denver, CO.

Plastic Pipe Institute. 1971. *Water flow characteristics of thermoplastic pipe*. Plastic Pipe Institute, New York.

Rogers, W.L. 1953. Experimental approaches to the study of noise and noise transmission in piping systems. *ASHVE Transactions* 59:347-60.

Rogers, W.L. 1954. Sound-pressure levels and frequencies produced by flow of water through pipe and fittings. *ASHRAE Transactions* 60:411-30.

Rogers, W.L. 1956. Noise production and damping in water piping. *ASHAE Transactions* 62:39.

Sanks, R.L. 1978. *Water treatment plant design for the practicing engineer*. Ann Arbor Science Publishers, Ann Arbor, MI.

SBCCI. 1994. *Standard plumbing code*. Southern Building Code Congress International, Birmingham, AL.

Smith, T. 1983. Reducing corrosion in heating plants with special reference to design considerations. *Anti-Corrosion Methods and Materials* 30 (October):4.

Stewart, W.E. and C.L. Dona. 1987. Water flow rate limitations. *ASHRAE Transactions* 93(2):811-25.

Williams, G.J. 1976. The Hunter curves revisited. *Heating/Piping/Air Conditioning* (November):67.

Williams, G.S. and A. Hazen. 1933. *Hydraulic tables*. John Wiley and Sons, New York.

CHAPTER 34

ABBREVIATIONS AND SYMBOLS

THIS chapter contains information about abbreviations and symbols for heating, ventilating, air-conditioning, and refrigerating (HVAC&R) engineers.

Abbreviations are shortened forms of names and expressions used in text, drawings, and computer programs. This chapter discusses conventional English language abbreviations that may be different in other languages. A **letter symbol** represents a quantity or a unit, not its name, and is independent of language. Because of this, use of a letter symbol is preferred over abbreviations for unit or quantity terms. Letter symbols necessary for individual chapters are defined in the chapters where they occur.

Abbreviations are never used for mathematical signs, such as the equality sign (=) or division sign (/), except in computer programming, where the abbreviation functions as a letter symbol. Mathematical operations are performed only with symbols. Abbreviations should be used only where necessary to save time and space; avoid their usage in documents circulated in foreign countries.

Graphical symbols in this chapter are easy to draw and recognize and were selected to save engineering drafting time. Symbols of piping, ductwork, fittings, and in-line accessories can be used on scale drawings and diagrams.

Identifying piping by legend and color promotes greater safety and lessens the chance of error in emergencies. Piping identification is now required throughout the United States by the Occupational Safety and Health Administration (OSHA) for some industries and by many federal, state, and local codes.

ABBREVIATIONS FOR TEXT, DRAWINGS, AND COMPUTER PROGRAMS

Abbreviations for text and drawings have been compiled from Abbreviations for Use on Drawings and in Text, ANSI/ASME *Standard* Y1.1-72. Table 1 gives some of these abbreviations, as well as others commonly found on mechanical drawings and abbreviations (symbols) used in computer programming.

Abbreviations specific to a single subject are defined in the chapters in which they appear. Additional abbreviations used on drawings can be found in the section on Graphical Symbols for Drawings.

The abbreviations (symbols) used for computer programming for the HVAC&R industries have been developed by ASHRAE Technical Committee 1.5, Computer Applications. These symbols identify computer variables, subprograms, subroutines, and functions commonly applied in the industry. Using these symbols enhances comprehension of the program listings and provides a clearly defined nomenclature in applicable computer programs.

Certain programming languages differentiate between *real numbers* (numbers with decimals) and *integers* (numbers without decimals) by reserving certain initial letters of a variable for integer numbers. For instance, in FORTRAN, any variable beginning with the letters H through N is defined by the computer as an integer. Many of the symbols listed in this chapter begin with these letters and, in order to make them real numbers, must be prefixed with a noninteger letter. Thus, HR would become XHR if the programmer wanted to define hours as a decimal value.

Some symbols have two or more options listed. The longest abbreviation is preferred and should be used if possible. However, it is sometimes necessary to shorten the symbol to further identify the variable. For instance, the area of a wall cannot be defined as WALLAREA because some computer languages restrict the number of letters in a variable name. Therefore, a shorter variable symbol is applied, and WALLAREA becomes WALLA or WAREA.

Many advanced computer programming languages such as Basic, C, and C++ do not have the limitations of older computer language compilers. It is good programming practice to include the complete name of each variable and to define any abbreviations in the comments section at the beginning of each module of code. Abbreviations should be used to help clarify the variables in an equation and not to obscure the readability of the code.

In Table 1, the same symbol is sometimes used for different terms. This liberty is taken because it is highly unlikely that the two terms would be used in the same program. If such were the case, one of the terms would require a suffix or prefix to differentiate it from the other.

LETTER SYMBOLS

Letter symbols include symbols for physical quantities (quantity symbols) and symbols for the units in which these quantities are measured (unit symbols). **Quantity symbols**, such as I for electric current, are listed in this chapter and are printed in italic type. A **unit symbol** is a letter or group of letters such as ft for foot or a special sign such as ° for degrees and is printed in Roman type. Subscripts and superscripts are governed by the same principles. Letter symbols are restricted mainly to the English and Greek alphabets.

Quantity symbols may be used in mathematical expressions in any way consistent with good mathematical usage. The product of two quantities, a and b, is indicated by ab. The quotient is a/b, or ab^{-1}. To avoid misinterpretation, parentheses must be used if more than one slash (/) is employed in an algebraic term; for example, $(a/b)/c$ or $a/(b/c)$ is correct, but not $a/b/c$.

Subscripts and superscripts, or several of them separated by commas, may be attached to a single basic letter (kernel), but not to other subscripts or superscripts. A symbol that has been modified by a superscript should be enclosed in parentheses before an exponent is added $(X_a)^3$. Symbols can also have alphanumeric marks such as ′ (prime), + (plus), and * (asterisk).

More detailed information on the general principles of letter symbol standardization can be found in Standard Letter Symbols for Units of Measurement (ANSI/IEEE *Standard* 260.1-93) and Glossary of Terms Concerning Letter Symbols (ANSI/ASME *Standard* Y10.1-72), the source of the above paragraphs.

The letter symbols have, in general, been taken from the following American National Standards Institute (ANSI) standards:

Y10.3M-84 Letter Symbols for Mechanics and Time-Related Phenomena

The preparation of this chapter is assigned to TC 1.6, Terminology.

Table 1 Abbreviations for Text, Drawings, and Computer Programs

Term	Text	Drawings	Program
above finished floor	—	AFF	—
absolute	abs	ABS	ABS
accumulat(e, -or)	acc	ACCUM	ACCUM
air condition(-ing, -ed)	—	AIR COND	—
air-conditioning unit(s)	—	ACU	ACU
air-handling unit	—	AHU	AHU
air horsepower	ahp	AHP	AHP
alteration	altrn	ALTRN	—
alternating current	ac	AC	AC
altitude	alt	ALT	ALT
ambient	amb	AMB	AMB
American National Standards Institute[1]	ANSI	ANSI	—
American wire gage	AWG	AWG	—
ampere (amp, amps)	amp	AMP	AMP, AMPS
angle	—	—	ANG
angle of incidence	—	—	ANGI
apparatus dew point	adp	ADP	ADP
approximate	approx.	APPROX	—
area	—	—	A
atmosphere	atm	ATM	—
average	avg	AVG	AVG
azimuth	az	AZ	AZ
azimuth, solar	—	—	SAZ
azimuth, wall	—	—	WAZ
barometer(-tric)	baro	BARO	—
bill of material	b/m	BOM	—
boiling point	bp	BP	BP
brake horsepower	bhp	BHP	BHP
Brown & Sharpe wire gage	B&S	B&S	—
British thermal unit	Btu	BTU	BTU
Celsius	°C	°C	°C
center to center	c to c	C TO C	—
circuit	ckt	CKT	CKT
clockwise	cw	CW	—
coefficient	coeff.	COEF	COEF
coefficient, valve flow	C_v	C_v	CV
coil	—	—	COIL
compressor	cprsr	CMPR	CMPR
condens(-er, -ing, -ation)	cond	COND	COND
conductance	—	—	C
conductivity	cndct	CNDCT	K
conductors, number of (3)	3/c	3/c	—
contact factor	—	—	CF
cooling load	clg load	CLG LOAD	CLOAD
counterclockwise	ccw	CCW	—
cubic feet	ft^3	CU FT	CUFT, CFT
cubic inch	in^3	CU IN	CUIN, CIN
cubic feet per minute	cfm	CFM	CFM
cfm, standard conditions	scfm	SCFM	SCFM
cubic ft per sec, standard	scfs	SCFS	SCFS
decibel	dB	DB	DB
degree	deg. or °	DEG or °	DEG
density	dens	DENS	RHO
depth or deep	dp	DP	DPTH
dew-point temperature	dpt	DPT	DPT
diameter	dia.	DIA	DIA
diameter, inside	ID	ID	ID
diameter, outside	OD	OD	OD
difference or delta	diff., Δ	DIFF	D, DELTA
diffuse radiation	—		DFRAD
direct current	dc	DC	DC
direct radiation	dir radn	DIR RADN	DIRAD
dry	—		DRY
dry-bulb temperature	dbt	DBT	DB, DBT
effectiveness	—		EFT
effective temperature[2]	ET*	ET*	ET
efficiency	eff	EFF	EFF
efficiency, fin	—		FEFF
efficiency, surface	—		SEFF
electromotive force	emf	EMF	—
elevation	elev.	EL	ELEV
entering	entr	ENT	ENT
entering water temperature	EWT	EWT	EWT
entering air temperature	EAT	EAT	EAT
enthalpy	—	—	H
entropy	—	—	S
equivalent direct radiation	edr	EDR	—
equivalent feet	eqiv ft	EQIV FT	EQFT
equivalent inches	eqiv in	EQIV IN	EQIN
evaporat(-e, -ing, -ed, -or)	evap	EVAP	EVAP
expansion	exp	EXP	XPAN
face area	fa	FA	FA
face to face	f to f	F to F	—
face velocity	fvel	FVEL	FV
factor, correction	—	—	CFAC, CFACT
factor, friction	—	—	FFACT, FF
Fahrenheit	°F	°F	F
fan	—	—	FAN
feet per minute	fpm	FPM	FPM
feet per second	fps	FPS	FPS
film coefficient,[3] inside	—	—	FI, HI
film coefficient,[3] outside	—	—	FO, HO
flow rate, air	—	—	QAR, QAIR
flow rate, fluid	—	—	QFL
flow rate, gas	—	—	QGA, QGAS
foot or feet	ft	FT	FT
foot-pound	ft·lb	FT LB	—
freezing point	fp	FP	FP
frequency	Hz	HZ	—
gage or gauge	ga	GA	GA, GAGE
gallons	gal	GAL	GAL
gallons per hour	gph	GPH	GPH
gph, standard	std gph	SGPH	SGPH
gallons per day	gpd	GPD	GPD
grains	gr	GR	GR
gravitational constant	g	G	G
greatest temp difference	GTD	GTD	GTD
head	hd	HD	HD
heat	—	—	HT
heater	—	—	HTR
heat gain	HG	HG	HG, HEATG
heat gain, latent	LHG	LHG	HGL
heat gain, sensible	SHG	SHG	HGS
heat loss	—	—	HL, HEATL
heat transfer	—	—	Q
heat transfer coefficient	U	U	U
height	hgt	HGT	HGT, HT
high-pressure steam	hps	HPS	HPS
high-temperature hot water	hthw	HTHW	HTHW
horsepower	hp	HP	HP
hour(s)	h	HR	HR
humidity, relative	rh	RH	RH
humidity ratio	W	W	W
incident angle	—	—	INANG
indicated horsepower	ihp	IHP	—
International Pipe Std	IPS	IPS	—
iron pipe size	ips	IPS	—
kelvin	K	K	K
kilowatt	kW	kW	KW
kilowatt hour	kWh	KWH	KWH
latent heat	LH	LH	LH, LHEAT
least mean temp. difference[4]	LMTD	LMTD	LMTD
least temp. difference[4]	LTD	LTD	LTD
leaving air temperature	lat	LAT	LAT
leaving water temperature	lwt	LWT	LWT
length	lg	LG	LG, L
linear feet	lin ft	LF	LF
liquid	liq	LIQ	LIQ
logarithm (natural)	ln	LN	LN
logarithm to base 10	log	LOG	LOG
low-pressure steam	lps	LPS	LPS
low-temp. hot water	lthw	LTHW	LTHW

Term	Text	Drawings	Program
Mach number	Mach	MACH	—
mass flow rate	mfr	MFR	MFR
maximum	max.	MAX	MAX
mean effective temp.	MET	MET	MET
mean temp. difference	MTD	MTD	MTD
medium-pressure steam	mps	MPS	MPS
medium-temp. hot water	mthw	MTHW	MTHW
mercury	Hg	HG	HG
miles per hour	mph	MPH	MPH
minimum	min.	MIN	MIN
noise criteria	NC	NC	—
normally open	n o	N O	—
normally closed	n c	N C	—
not applicable	na	N/A	—
not in contract	n i c	N I C	—
not to scale	—	N T S	—
number	no.	NO	N, NO
number of circuits	—	—	NC
number of tubes	—	—	NT
ounce	oz	OZ	OZ
outside air	oa	OA	OA
parts per million	ppm	PPM	PPM
percent	%	%	PCT
phase (electrical)	ph	PH	—
pipe	—	—	PIPE
pounds	lb	LBS	LBS
pounds per square foot	psf	PSF	PSF
psf absolute	psfa	PSFA	PSFA
psf gage	psfg	PSFG	PSFG
pounds per square inch	psi	PSI	PSI
psi absolute	psia	PSIA	PSIA
psi gage	psig	PSIG	PSIG
pressure	—	PRESS	PRES, P
pressure, barometric	baro pr	BARO PR	BP
critical pressure	—	—	CRIP
pressure, dynamic (velocity)	vp	VP	VP
pressure drop or difference	PD	PD	PD, DELTP
pressure, static	sp	SP	SP
pressure, vapor	vap pr	VAP PR	VAP
primary	pri	PRI	PRIM
quart	qt	QT	QT
radian	—	—	RAD
radiat(-e, -or)	—	RAD	—
radiation	—	RADN	RAD
radius	—	—	R
Rankine	°R	°R	R
receiver	rcvr	RCVR	REC
recirculate	recirc.	RECIRC	RCIR, RECIR
refrigerant (12, 22, etc.)	R-12, R-22	R12, R22	R12, R22
relative humidity	rh	RH	RH
resist(-ance, -ivity, -or)	res	RES	RES, OHMS
return air	ra	RA	RA
revolutions	rev	REV	REV
revolutions per minute	rpm	RPM	RPM
revolutions per second	rps	RPS	RPS
roughness	rgh	RGH	RGH, E
safety factor	sf	SF	SF
saturation	sat.	SAT	SAT
Saybolt seconds Furol	ssf	SSF	SSF
Saybolt seconds Universal	ssu	SSU	SSU
sea level	sl	SL	SE
second	s	s	SEC
sensible heat	SH	SH	SH
sensible heat gain	SHG	SHG	SHG
sensible heat ratio	SHR	SHR	SHR
shading coefficient	—	—	SC
shaft horsepower	sft hp	SFT HP	SHP
solar	—	—	SOL
specification	spec	SPEC	—
specific gravity	SG	SG	—
specific heat	sp ht	SP HT	C
sp ht at constant pressure	c_p	c_p	CP
sp ht at constant volume	c_v	c_v	CV
specific volume	sp vol	SP VOL	V, CVOL
square	sq.	SQ	SQ
standard	std	STD	STD
standard time meridian	—	—	STM
static pressure	SP	SP	SP
suction	suct.	SUCT	SUCT, SUC
summ(-er, -ary, -ation)	—	—	SUM
supply	sply	SPLY	SUP, SPLY
supply air	sa	SA	SA
surface	—	—	SUR, S
surface, dry	—	—	SURD
surface, wet	—	—	SURW
system	—	—	SYS
tabulat(-e, -ion)	tab	TAB	TAB
tee	—	—	TEE
temperature	temp.	TEMP	T, TEMP
temperature difference	TD, Δt	TD	TD, TDIF
temperature entering	TE	TE	TE, TENT
temperature leaving	TL	TL	TL, TLEA
thermal conductivity	k	K	K
thermal expansion coeff.	—	—	TXPC
thermal resistance	R	R	RES, R
thermocouple	tc	TC	TC, TCPL
thermostat	T STAT	T STAT	T STAT
thick(-ness)	thkns	THKNS	THK
thousand circular mils	Mcm	MCM	MCM
thousand cubic feet	Mcf	MCF	MCF
thousand foot-pounds	kip ft	KIP FT	KIPFT
thousand pounds	kip	KIP	KIP
time	—	T	T
ton	—	—	TON
tons of refrigeration	tons	TONS	TONS
total	—	—	TOT
total heat	tot ht	TOT HT	—
transmissivity	—	—	TAU
U-factor	—	—	U
unit	—	—	UNIT
vacuum	vac	VAC	VAC
valve	v	V	VLV
vapor proof	vap prf	VAP PRF	—
variable	var	VAR	VAR
variable air volume	VAV	VAV	VAV
velocity	vel.	VEL	VEL, V
velocity, wind	w vel.	W VEL	W VEL
ventilation, vent	vent	VENT	VENT
vertical	vert.	VERT	VERT
viscosity	visc	VISC	MU, VISC
volt	V	V	E, VOLTS
volt ampere	VA	VA	VA
volume	vol.	VOL	VOL
volumetric flow rate	—	—	VFR
wall	—	—	W, WAL
water	—	—	WTR
watt	W	W	WAT, W
watt-hour	Wh	WH	WHR
weight	wt	WT	WT
wet bulb	wb	WB	WB
wet-bulb temperature	wbt	WBT	WBT
width	—	—	WI
wind	—	—	WD
wind direction	wdir	WDIR	WDIR
wind pressure	wpr	WPR	WP, WPRES
yard	yd	YD	YD
year	yr	YR	YR
zone	z	Z	Z, ZN

[1] Abbreviations of most proper names use capital letters in both text and drawings.
[2] The asterisk (*) is used with ET*, effective temperature, as in Chapter 8 of this volume.
[3] These are surface heat transfer coefficients.
[4] Letter L also used for *Logarithm of* these temperature differences in computer programming.

Y10.4-82 Letter Symbols for Heat and Thermodynamics (Reaffirmed 1988)

260-78 Letter Symbols for Units of Measurement (IEEE)

For standard symbols in related fields, refer to the following ASME standards:

Y10.11-84 Letter Symbols and Abbreviations for Quantities Used in Acoustics

Y10.18-67 Letter Symbols for Illuminating Engineering (Reaffirmed 1987)

Y10.20-75 Mathematical Signs and Symbols for Use in Physical and Technology (Reaffirmed 1988)

Other symbols chosen by an author for a physical magnitude not appearing in any standard list should be ones that do not already have different meanings in the field of the text.

LETTER SYMBOLS

Symbol	Description of Item	Typical Units
a	acoustic velocity	fps or fpm
A	area	ft²
b	breadth or width	ft
B	barometric pressure	psia or in. Hg
c	concentration	lb/ft³, mol/ft³
c	specific heat	Btu/lb·°F
c_p	specific heat at constant pressure	Btu/lb·°F
c_v	specific heat at constant volume	Btu/lb·°F
C	coefficient	—
C	fluid capacity rate	Btu/h·°F
C	thermal conductance	Btu/h·ft²·°F
C_L	loss coefficient	—
C_P	coefficient of performance	—
d	prefix meaning differential	—
d or D	diameter	ft
D_e or D_h	equivalent or hydraulic diameter	ft
D_v	mass diffusivity	ft²/s
e	base of natural logarithms	—
E	energy	Btu
E	electrical potential	V
f	film conductance (alternate for h)	Btu/h·ft²·°F
f	frequency	Hz
f_D	friction factor, Darcy-Weisbach formulation	—
f_F	friction factor, Fanning formulation	—
F	force	lb$_f$
F_{ij}	angle factor (radiation)	—
g	gravitational acceleration	ft/s²
G	mass velocity	lb/h·ft²
h	heat transfer coefficient	Btu/h·ft²·°F
h	hydraulic head	ft
h	specific enthalpy	Btu/lb
h_a	enthalpy of dry air	Btu/lb
h_D	mass transfer coefficient	lb/h·ft²·lb per ft³
h_s	enthalpy of moist air at saturation	Btu/lb
H	total enthalpy	Btu
I	electric current	A
J	mechanical equivalent of heat	ft·lb$_f$/Btu
k	thermal conductivity	Btu/h·ft·°F
k (or γ)	ratio of specific heats, c_p/c_v	—
K	proportionality constant	—
K_D	mass transfer coefficient	lb/h·ft²
l or L	length	ft
L_p	sound pressure	dB
L_w	sound power	dB
m or M	mass	lb
M	molecular weight	lb/lb mol
n or N	number in general	—
N	rate of rotation	rpm
p or P	pressure	psi
p_a	partial pressure of dry air	psi
p_s	partial pressure of water vapor in moist air	psi
p_w	vapor pressure of water in saturated moist air	psi
P	power	hp, watts
q	time rate of heat transfer	Btu/h
Q	total heat transfer	Btu
Q	volumetric flow rate	cfm
r	radius	ft
r or R	thermal resistance	ft²·h·°F/Btu
R	gas constant	ft·lb$_f$/lb$_m$·°R
s	specific entropy	Btu/lb·°R
S	total entropy	Btu/°R
t	temperature	°F
Δt_m or ΔT_m	mean temperature difference	°F
T	absolute temperature	°R
u	specific internal energy	Btu/lb
U	total internal energy	Btu
U	overall heat transfer coefficient	Btu/h·ft²·°F
v	specific volume	ft³/lb
V	total volume	ft³
V	linear velocity	fps
w	mass rate of flow	lb/h
W	weight	lb$_f$
W	humidity ratio of moist air	lb (water)/lb (dry air)
W	work	ft·lb$_f$
W_s	humidity ratio of moist air at saturation	lb (water)/lb (dry air)
x	mole fraction	—
x	quality, mass fraction of vapor	—
x,y,z	lengths along principal coordinate axes	ft
Z	figure of merit	—
α	absolute Seebeck coefficient	V/°C
α	absorptivity, absorptance radiation	—
α	linear coefficient of thermal expansion	per °F
α	thermal diffusivity	ft²/h
β	volume coefficient of thermal expansion	per °F
γ (or k)	ratio of specific heats, c_p/c_v	—
γ	specific weight	lb$_f$/ft³
Δ	difference between values	—
ε	emissivity, emittance (radiation)	—
θ	time	s, h
η	efficiency or effectiveness	—
λ	wavelength	nm
μ	degree of saturation	—
μ	dynamic viscosity	lb/ft·h
ν	kinematic viscosity	ft²/h
ρ	density	lb/ft³
ρ	reflectivity, reflectance (radiation)	—
ρ	volume resistivity	Ω·cm
σ	Stefan-Boltzmann constant	Btu/h·ft²·°R⁴
σ	surface tension	lb$_f$/ft
τ	stress	lb$_f$/ft²
τ	time	s, h
τ	transmissivity, transmittance (radiation)	—
ϕ	relative humidity	—

DIMENSIONLESS NUMBERS

Fo	Fourier number	$\alpha\tau/L^2$
Gr	Grashof number	$L^3\rho^2\beta g(\Delta t)/\mu^2$
Gz	Graetz number	wc_p/kL
j_D	Colburn mass transfer	Sh/ReSc$^{1/3}$
j_H	Colburn heat transfer	Nu/RePr$^{1/3}$
Le	Lewis number	α/D_v
M	Mach number	V/a
Nu	Nusselt number	hD/k
Pe	Peclet number	GDc_p/k
Pr	Prandtl number	$c_p\mu/k$
Re	Reynolds number	$\rho VD/\mu$
Sc	Schmidt number	$\mu/\rho D_v$
Sh	Sherwood number	h_DL/D_v
St	Stanton number	h/Gc_p
Str	Strouhal number	fd/V

MATHEMATICAL SYMBOLS

equal to	$=$
not equal to	$\neq$
approximately equal to	$\approx$
greater than	$>$
less than	$<$
greater than or equal to	$\geq$
less than or equal to	$\leq$
plus	$+$
minus	$-$
plus or minus	$\pm$
a multiplied by *b*	ab, $a \cdot b$, $a \times b$
a divided by *b*	$\frac{a}{b}$, a/b, ab^{-1}
ratio of the circumference of a circle to its diameter	π
a raised to the power *n*	a^n
square root of *a*	$\sqrt{a}$, $a^{0.5}$
infinity	∞
percent	%
summation of	Σ
natural log	ln
logarithm to base 10	log

SUBSCRIPTS

These are to be affixed to the appropriate symbols. Several subscripts may be used together to denote combinations of various states, points, or paths. Often the subscript indicates that a particular property is to be kept constant in a process.

a,b,...	referring to different phases, states or physical conditions of a substance, or to different substances
a	air
a	ambient
b	barometric (pressure)
c	referring to critical state or critical value
c	convection
db	dry bulb
dp	dew point
e	base of natural logarithms
f	referring to saturated liquid
f	film
fg	referring to evaporation or condensation
F	friction
g	referring to saturated vapor
h	referring to change of phase in evaporation
H	water vapor
i	referring to saturated solid
i	internal
if	referring to change of phase in melting
ig	referring to change of phase in sublimation
k	kinetic
L	latent
m	mean value
M	molar basis
o	referring to initial or standard states or conditions
p	referring to constant pressure conditions or processes
p	potential
r	refrigerant
r	radiant or radiation
s	referring to moist air at saturation
s	sensible
s	referring to isentropic conditions or processes
s	static (pressure)
s	surface
t	total (pressure)
T	referring to isothermal conditions or processes
v	referring to constant volume conditions or processes
v	vapor
v	velocity (pressure)
w	wall
w	water
wb	wet bulb
1,2,...	different points in a process, or different instants of time

GRAPHICAL SYMBOLS FOR DRAWINGS

Graphical symbols have been extracted from Graphic Symbols for Pipe Fittings, Valves, and Piping [ANSI/ASME Y32.2.3-1949 (Reaffirmed 1994)]; Graphic Symbols for Heating, Ventilating, and Air Conditioning [ANSI/ASME Y32.2.4-49 (Reaffirmed 1993)]; American Standard Abbreviations for Use on Drawings and in Text (ASME Y1.1-72); and Graphic Symbols for Plumbing Fixtures for Diagrams Used in Architectural and Building Construction [ANSI/ASME Y32.4-77 (Reaffirmed 1994)].

Some of these symbols have been modified, and others have been added to reflect current practice. Symbols and quotations are used with permission of the publisher, the American Society of Mechanical Engineers.

Piping

Heating

High-Pressure Steam	——HPS——
Medium-Pressure Steam	——MPS——
Low-Pressure Steam	——LPS——
High-Pressure Condensate	——HPC——
Medium-Pressure Condensate	——MPC——
Low-Pressure Condensate	——LPC——
Boiler Blowdown	——BBD——
Pumped Condensate	——PC——
Vacuum Pump Discharge	——VPD——
Makeup Water	——MU——
Atmospheric Vent	——ATV——
Fuel Oil Discharge	——FOD——
Fuel Oil Gage	——FOG——
Fuel Oil Suction	——FOS——
Fuel Oil Return	——FOR——
Fuel Oil Tank Vent	——FOV——
Low-Temperature Hot Water Supply	——HWS——
Medium-Temperature Hot Water Supply	——MTWS——
High-Temperature Hot Water Supply	——HTWS——
Low-Temperature Hot Water Return	——HWR——
Medium-Temperature Hot Water Return	——MTWR——
High-Temperature Hot Water Return	——HTWR——
Compressed Air	——A——
Vacuum (Air)	——VAC——
Existing Piping	——(NAME)E——
Pipe to Be Removed	-XX- (NAME) -XX-

Air Conditioning and Refrigeration

Refrigerant Discharge	——RD——
Refrigerant Suction	——RS——
Brine Supply	——B——
Brine Return	——BR——
Condenser Water Supply	——C——
Condenser Water Return	——CR——
Chilled Water Supply	——CWS——
Chilled Water Return	——CWR——
Fill Line	——FILL——
Humidification Line	——H——
Drain	——D——
Hot/Chilled Water Supply	——HCS——
Hot/Chilled Water Return	——HCR——
Refrigerant Liquid	——RL——
Heat Pump Water Supply	——HPWS——
Heat Pump Water Return	——HPWR——

Plumbing

Sanitary Drain above Floor or Grade	——SAN——
Sanitary Drain below Floor or Grade	---- SAN ----
Storm Drain above Floor or Grade	——ST——
Storm Drain below Floor or Grade	---- ST ----

Condensate Drain above Floor or grade	———CD———
Condensate Drain below Floor or grade	———CD— — —
Vent	
Cold Water	
Hot Water	
Hot Water Return	
Gas	—G————G—
Acid Waste	———ACID———
Drinking Water Supply	———DWS———
Drinking Water Return	———DWR———
Vacuum (Air)	———VAC———
Compressed Air	———A———
Chemical Supply Pipes[a]	———(NAME)———
Floor Drain	D
Funnel Drain, open	

Fire Safety Devices[b]

Signal Initiating Detectors

Heat (Thermal)

Gas

Smoke

Flame

Valves

Valves for Selective Actuators

Air Line

Ball

Butterfly

Diaphragm

Gate

Gate, Angle

Globe

Globe, Angle

Plug Valve

Three Way

Valve Actuators

Manual

Non-Rising Stem

Outside Stem & Yoke

Lever

Gear (G)

Electric

Motor (M)

Solenoid (S)

Pneumatic

Motor (M)

Diaphragm (D)

Valves, Special Duty

Check, Swing Gate

Check, Spring

Control, Electric-Pneumatic (EP)

Control, Pneumatic-Electric (PE)

Hose End Drain

Lock Shield

Needle

Pressure Reducing (number and specify) (PRV-1)

Quick Opening

Quick Closing, Fusible Link

Relief (R) or Safety (S)

Solenoid (S)

Square Head Cock

Unclassified (number and specify) (V-1)

[a] See Piping Identification in this chapter.

[b] Refer to *Fire Protection Symbols for Architectural and Engineering Drawing.* (NFPA *Standard* 172 P-86) for additional symbols.

Fittings

The following fittings are shown with screwed connections. The symbol for the body of a fitting is the same for all types of connections, unless otherwise specified. The types of connections are often specified for a range of pipe sizes, but are shown with the fitting symbol where required. For example, an elbow would be:

Flanged Screwed Belt & Spigot

Welded[a] Soldered Solvent Cement

Fitting	*Symbol*
Bushing	
Cap	
Connection, Bottom	
Connection, Top	
Coupling (Joint)	
Cross	
Elbow, 90°	
Elbow, 45°	
Elbow, Turned Up	
Elbow, Turned Down	
Elbow, Reducing, Show Sizes	2 4
Elbow, Base	
Elbow, Long Radius	LR
Elbow, Double Branch	
Elbow, Side Outlet, Outlet Up	
Elbow, Side Outlet, Outlet Down	
Lateral	
Reducer, Concentric	
Reducer, Eccentric Straight Invert	
Reducer, Eccentric Straight Crown	
Tee	
Tee, Outlet Up	
Tee, Outlet Down	
Tee, Reducing (Show Sizes)	2 6 4
Tee, Side Outlet, Outlet Up	
Tee, Side Outlet, Outlet Down	
Tee, Single Sweep	
Union, Screwed	
Union, Flanged	

[a]Includes fusion, specify type.

Piping Specialties

Specialty	Symbol
Air Vent, Automatic	AV
Air Vent, Manual	MV
Air Separator	S
Alignment Guide	
Anchor, Intermediate	
Anchor, Main	
Ball Joint	
Expansion Joint	EJ-1
Expansion Loop	
Flexible Connector	
Flowmeter, Orifice	OFM-1

Flowmeter, Venturi

Flow Switch

Hanger, Rod

Hanger, Spring

Heat Exchanger, Liquid

Heat Transfer Surface (indicate type)

Pitch of Pipe, Rise (R) Drop (D)

Pressure Gauge and Cock

Pressure Switch

Pump (indicate use)

Pump Suction Diffuser

Spool Piece, Flanged

Strainer

Strainer, Blow Off

Strainer, Duplex

Tank (indicate use)

Thermometer

Thermometer Well, only

Thermostat, Electric

Thermostat, Pneumatic

Thermostat, Self-Contained

Traps, Steam (indicate type)

Unit Heater (indicate type)

Air Moving Devices and Components

Fans (indicate use)[a]

Axial Flow

Centrifugal

Propeller

Roof Ventilator, Intake

Roof Ventilator, Exhaust

Roof Ventilator, Louvered

Ductwork[b]

Direction of Flow

Duct Size, first figure is side shown

Duct Section, Positive Pressure, first figure is top

Duct Section, Negative Pressure

Change of Elevation Rise (R) Drop (D)

Access Doors, Vertical or Horizontal

Acoustical Lining (insulation)

Cowl, (Gooseneck) and Flashing

Flexible Connection

Flexible Duct

Sound Attenuator

Terminal Unit, Mixing

Terminal Unit, Reheat

Terminal Unit, Variable Volume

Transition[c]

Turning Vanes

Detectors, Fire and/or Smoke

[a]Units of measurement are not shown herein, but should be shown on drawings. The first of the two dimensions on ducts indicates the side of the duct showing; on duct sections, the top; on grilles and registers, the horizontal edge.

[b]Adapted from SMACNA, Symbols for Ventilation and Air Conditioning Figure 4.2. *HVAC Duct System Design.*

[c]Indicate Flat on Bottom or Top (FOB or FOT) if applicable.

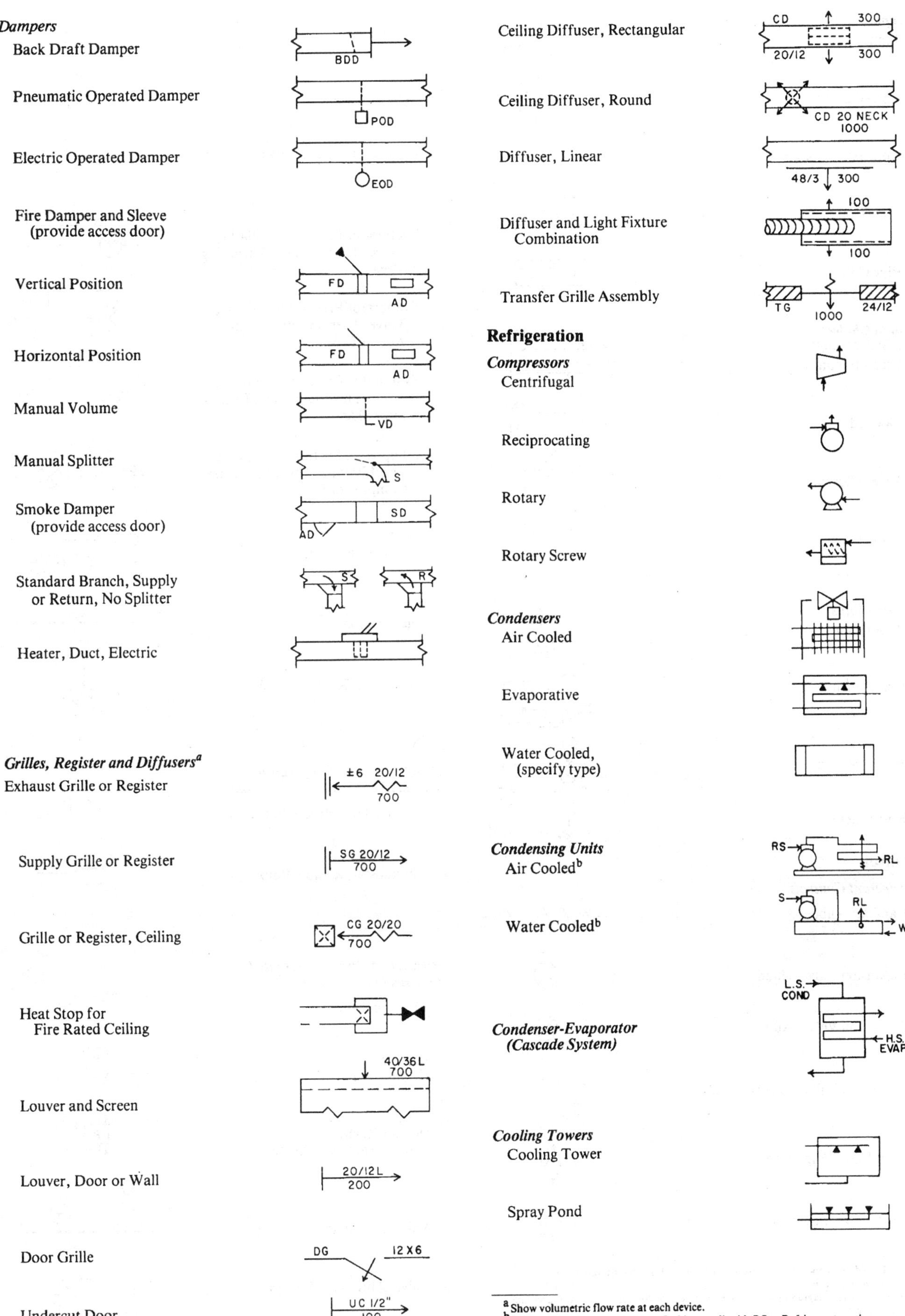

[a]Show volumetric flow rate at each device.
[b]L = Liquid being cooled, RL = Refrigerant liquid, RS = Refrigerant suction.

Evaporators[b]
Finned Coil

Forced Convection

Immersion Cooling Unit

Plate Coil

Pipe Coil[c]

Liquid Chillers
(Chillers only)
Direct Expansion[a]

Flooded[a]

Tank, Closed

Tank, Open

Chilling Units
Absorption

Centrifugal

Reciprocating

Rotary Screw

Controls

Refrigerant Controls
Capillary Tube

Expansion Valve, Hand

Expansion Valve, Automatic

Expansion Valve, Thermostatic

Float Valve, High Side

Float Valve, Low Side

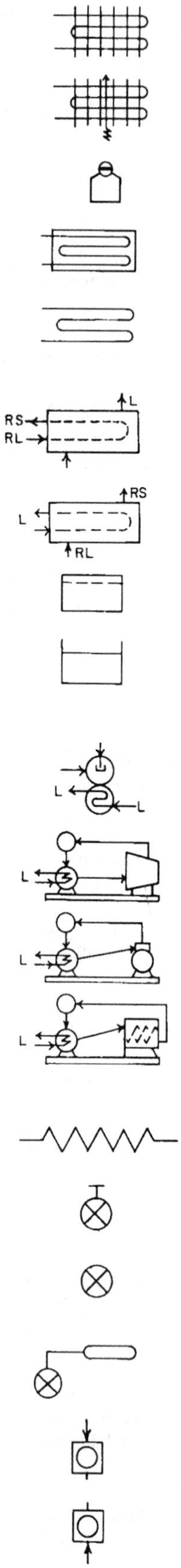

[a]L = Liquid being cooled, RL = Refrigerant liquid, RS = Refrigerant suction.
[b]Specify manifolding.
[c]Frequently used diagrammatically as evaporator and/or condenser with label indicating name and type.

Thermal Bulb

Solenoid Valve

Constant Pressure Valve, Suction

Evaporator Pressure Regulating Valve, Thermostatic, Throttling-Type

Evaporator Pressure Regulating Valve, Thermostatic, Snap-Action Type

Evaporator Pressure-Regulating Valve, Throttling-Type, Evaporator Side

Compressor Suction Valve, Pressure-Limiting, Throttling-Type, Compressor Side

Thermo-Suction Valve

Snap-Action Valve

Refrigerant Reversing Valve

Temperature or Temperature-Actuated Electrical or Flow Controls
Thermostat, Self-Contained

Thermostat, Remote Bulb

Pressure of Pressure-Actuated Electrical or Flow Controls
Pressure Switch

Pressure Switch, Dual (High-Low)

Pressure Switch, Differential Oil Pressure

Valve, Automatic Reducing

Valve, Automatic Bypass

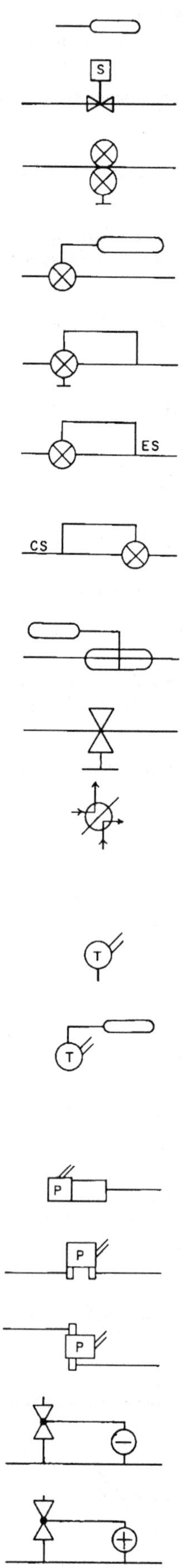

Valve, Pressure-Reducing

Valve, Condenser Water Regulating

Auxiliary Equipment

Refrigerant

Filter

Strainer

Filter and Drier

Scale Trap

Drier

Vibration Absorber

Heat Exchanger

Oil Separator

Sight Glass

Fusible Plug

Rupture Disc

Receiver, High Pressure, Horizontal

Receiver, High Pressure, Vertical

Receiver, Low Pressure

Intercooler

Intercooler/Desuperheater

Energy Recovery Equipment

Condenser, Double Bundle

Air to Air Energy Recovery

Rotary Heat Wheel

Coil Loop

Heat Pipe

Fixed Plate

Plate Fin, Cross Flow

Power Sources

Motor, Electric (number for identification of description in specifications) — M-1

Engine (indicate fuel) — D

Gas Turbine — G

Steam Turbine — S

Steam Turbine, Condensing — S

Electrical Equipment[a]

Symbols for electrical equipment shown on mechanical drawings are usually geometric figures with an appropriate name or abbreviation, with details described in the specifications. The following are some common examples.[b]

Motor Control — MC

Disconnect Switch, Unfused — U DS

Disconnect Switch, Fused — DSF

Time Clock — TC

Automatic Filter Panel — AFP

Lighting Panel — LP

Power Panel — PP

[a] See *Graphic Electrical Symbols for Air-Conditioning and Refrigeration Equipment* (ARI *Standard* 130-82) for preferred symbols of common electrical parts.

[b] Number each symbol if more than one; see *Graphic Symbols for Plumbing Fixtures for Diagrams Used in Architecture and Building Construction* (ANSI Y32.4-77).

PIPING SYSTEM IDENTIFICATION

The material in piping systems is identified to promote greater safety and lessen the chances of error, confusion, or inaction in times of emergency. Primary identification should be by means of a lettered legend naming the material conveyed by the piping. In addition to, but not instead of lettered identification, color can be used to identify the hazards or use of the material.

The data have been extracted from Scheme for the Identification of Piping Systems, ANSI/ASME *Standard* A13.1-81 (Reaffirmed 1993), with the permission of the publisher, the American Society of Mechanical Engineers.

Definitions

Piping Systems. Piping systems include pipes of any kind, fittings, valves, and pipe coverings. Supports, brackets, and other accessories are not included. Pipes are defined as conduits for the transport of gases, liquids, semiliquids, or fine particulate dust.

Materials Inherently Hazardous to Life and Property. There are four categories of hazardous materials:

- Flammable or explosive materials that are easily ignited, including materials known as fire producers or explosives
- Chemically active or toxic materials that are corrosive or are in themselves toxic or productive of poisonous gases
- Materials at extreme temperatures or pressures that, when released from the piping, cause a sudden outburst with the potential for inflicting injury or property damage by burns, impingement, or flashing to vapor state
- Radioactive materials that emit ionizing radiation

Materials of Inherently Low Hazard. All materials that are not hazardous by nature, and are near enough to ambient pressure and temperature that people working on systems carrying these materials run little risk through their release.

Fire Quenching Materials. This classification includes sprinkler systems and other piped fire fighting or fire protection equipment. This includes water (for fire fighting), chemical foam, CO_2, Halon, and so forth.

Method of Identification

Legend. The legend is the primary and explicit identification of content. Positive identification of the content of the piping system is by lettered legend giving the name of the contents, in full or abbreviated form, as shown in Table 2. Arrows should be used to indicate the direction of flow. Use the legend to identify contents exactly and to provide temperature, pressure, and other details necessary to identify the hazard.

Table 2 Examples of Legends

HOT WATER
AIR 100 PSIG
H.P. RETURN
STEAM 100 PSIG

The legend shall be brief, informative, pointed, and simple. Legends should be applied close to valves and adjacent to changes in direction, branches, and where pipes pass through walls or floors, and as frequently as needed along straight runs to provide clear and positive identification. Identification may be applied by stenciling, tape, or markers (see Figure 1). The number and location of identification markers on a particular piping system is based on judgment.

Color. Colors listed in Table 3 are used to identify the characteristic properties of the contents. Color can be shown on or contiguous to the piping by any physical means, but it should be used in combination with a legend. Color can be used in continuous total length coverage or in intermittent displays.

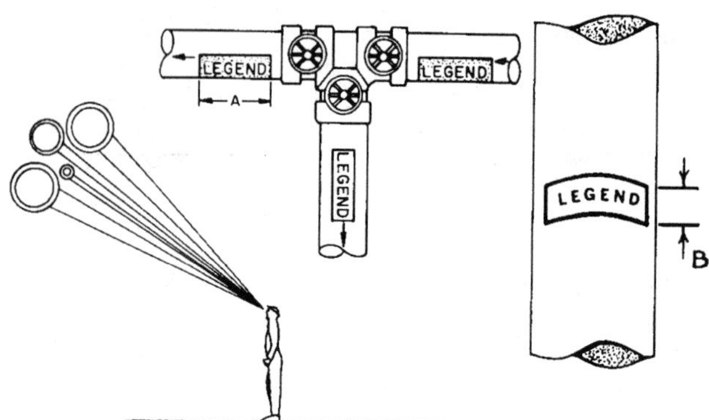

Fig. 1 Visibility of Pipe Markings

Visibility. Pipe markings should be highly visible. If pipe lines are above the normal line of vision, the lettering is placed below the horizontal centerline of the pipe (Figure 1).

Type and Size of Letters. Provide the maximum contrast between color field and legend (Table 3). Table 4 shows the size of letters recommended. Use of standard size letters of 1/2 in. or larger is recommended. For identifying materials in pipes of less than 3/4 in. in diameter and for valve and fitting identification, use a permanently legible tag.

Unusual or Extreme Situations. When the piping layout occurs in or creates an area of limited accessibility or is extremely complex, other identification techniques may be required. While a certain amount of imagination may be needed, the designer should always clearly identify the hazard and use the recommended color and legend guidelines.

Table 3 Classification of Hazardous Materials and Designation of Colors[a]

Classification	Color Field	Colors of Letters for Legend
Materials Inherently Hazardous		
Flammable or explosive	Yellow	Black
Chemically active or toxic	Yellow	Black
Extreme temperatures or pressures	Yellow	Black
Radioactive[b]	Purple	Yellow
Materials of Inherently Low Hazard		
Liquid or liquid admixture[c]	Green	Black
Gas or gaseous admixture	Blue	White
Fire Quenching Materials		
Water, foam, CO_2, Halon, etc.	Red	White

[a]When the color scheme above is used, the colors should be as recommended in the latest revision of Safety Color Code, ANSI/NEMA *Standard* Z535.1.
[b]Previously specified radioactive markers using yellow or purple are acceptable if already installed and/or until existing supplies are depleted, subject to applicable federal regulations.
[c]Markers with black letters on a green color field are acceptable if already installed and/or until existing supplies are depleted.

Table 4 Size of Legend Letters

Outside Diameter of Pipe or Covering, in.	Length of Color Field A, in.	Size of Letters B, in.
3/4 to 1-1/4	8	1/2
1-1/2 to 2	8	3/4
2-1/2 to 6	12	1-1/4
8 to 10	24	2-1/2
over 10	32	3-1/2

CHAPTER 35

UNITS AND CONVERSIONS

Table 1 Conversions to SI Units

Multiply	By	To Obtain
acre	0.4047	ha
atmosphere (standard)	*101.325	kPa
bar	*100	kPa
barrel (42 U.S. gal, petroleum)	159	L
	0.159	m^3
Btu (International Table)	1.055	kJ
Btu/ft^2	11.36	kJ/m^2
Btu/ft^3	37.3	kJ/m^3
Btu/gal	279	kJ/m^3
Btu·ft/h·ft^2·°F	1.731	W/(m·K)
Btu·in/h·ft^2·°F (thermal conductivity, *k*)	0.1442	W/(m·K)
Btu/h	0.2931	W
Btu/h·ft^2	3.155	W/m^2
Btu/h·ft^2·°F (overall heat transfer coefficient, *U*)	5.678	W/(m^2·K)
Btu/lb	*2.326	kJ/kg
Btu/lb·°F (specific heat, c_p)	4.184	kJ/(kg·K)
bushel	0.03524	m^3
calorie, gram	4.184	J
calorie, kilogram (kilocalorie)	4.184	kJ
centipoise (dynamic viscosity, μ)	*1.00	mPa·s
centistokes (kinematic viscosity, ν)	*1.00	mm^2/s
clo	0.155	m^2·K/W
dyne/cm^2	*0.100	Pa
EDR hot water (150 Btu/h)	44.0	W
EDR steam (240 Btu/h)	70.3	W
EER	0.293	COP
ft	*0.3048	m
	*304.8	mm
ft/min, fpm	*0.00508	m/s
ft/s, fps	*0.3048	m/s
ft of water	2.99	kPa
ft of water per 100 ft pipe	0.0981	kPa/m
ft^2	0.09290	m^2
ft^2·h·°F/Btu (thermal resistance, *R*)	0.176	m^2·K/W
ft^2/s (kinematic viscosity, ν)	92,900	mm^2/s
ft^3	28.32	L
	0.02832	m^3
ft^3/min, cfm	0.4719	L/s
ft^3/s, cfs	28.32	L/s
ft·lb_f (torque or moment)	1.356	N·m
ft·lb_f (work)	1.356	J
ft·lb_f/lb (specific energy)	2.99	J/kg
ft·lb_f/min (power)	0.0226	W
footcandle	10.76	lx
gallon (U.S., *231 in^3)	3.7854	L
gph	1.05	mL/s
gpm	0.0631	L/s
gpm/ft^2	0.6791	L/(s·m^2)
gpm/ton refrigeration	0.0179	mL/J
grain (1/7000 lb)	0.0648	g
gr/gal	17.1	g/m^3
gr/lb	0.143	g/kg
horsepower (boiler) (33,470 Btu/h)	9.81	kW
horsepower (550 ft·lb_f/s)	0.746	kW
inch	*25.4	mm
in. of mercury (60°F)	3.377	kPa
in. of water (60°F)	249	Pa
in/100 ft, thermal expansion	0.833	mm/m
in·lb_f (torque or moment)	113	mN·m
in^2	645	mm^2
in^3 (volume)	16.4	mL
in^3/min (SCIM)	0.273	mL/s
in^3 (section modulus)	16,400	mm^3
in^4 (section moment)	416,200	mm^4
km/h	0.278	m/s
kWh	*3.60	MJ
kW/1000 cfm	2.12	kJ/m^3
kilopond (kg force)	9.81	N
kip (1000 lb_f)	4.45	kN
kip/in^2 (ksi)	6.895	MPa
litre	*0.001	m^3
met	58.15	W/m^2
micron (μm) of mercury (60°F)	133	mPa
mile	1.609	km
mile, nautical	*1.852	km
mph	1.609	km/h
	0.447	m/s
millibar	*0.100	kPa
mm of mercury (60°F)	0.133	kPa
mm of water (60°F)	9.80	Pa
ounce (mass, avoirdupois)	28.35	g
ounce (force or thrust)	0.278	N
ounce (liquid, U.S.)	29.6	mL
ounce inch (torque, moment)	7.06	mN·m
ounce (avoirdupois) per gallon	7.49	kg/m^3
perm (permeance)	57.45	ng/(s·m^2·Pa)
perm inch (permeability)	1.46	ng/(s·m·Pa)
pint (liquid, U.S.)	473	mL
pound		
lb (mass)	0.4536	kg
	453.6	g
lb_f (force or thrust)	4.45	N
lb/ft (uniform load)	1.49	kg/m
lb_m/ft·h (dynamic viscosity, μ)	0.413	mPa·s
lb_m/ft·s (dynamic viscosity, μ)	1490	mPa·s
lb_f·s/ft^2 (dynamic viscosity, μ)	47.88	Pa·s
lb/h	0.126	g/s
lb/min	0.00756	kg/s
lb/h [steam at 212°F (100°C)]	0.284	kW
lb_f/ft^2	47.9	Pa
lb/ft^2	4.88	kg/m^2
lb/ft^3 (density, ρ)	16.0	kg/m^3
lb/gallon	120	kg/m^3
ppm (by mass)	*1.00	mg/kg
psi	6.895	kPa
quad (10^{15} Btu)	1.055	EJ
quart (liquid, U.S.)	0.946	L
square (100 ft^2)	9.29	m^2
tablespoon (approximately)	15	mL
teaspoon (approximately)	5	mL
therm (U.S.)	105.5	MJ
ton, long (2240 lb)	1.016	Mg
ton, short (2000 lb)	0.907	Mg; t (tonne)
ton, refrigeration (12,000 Btu/h)	3.517	kW
torr (1 mm Hg at 0°C)	133	Pa
watt per square foot	10.76	W/m^2
yd	*0.9144	m
yd^2	0.836	m^2
yd^3	0.7646	m^3
To Obtain	**By**	**Divide**

*Conversion factor is exact.

Notes: Units are U.S. values unless noted otherwise.

Litre is a special name for the cubic decimetre. 1 L = 1 dm^3 and 1 mL = 1 cm^3.

The preparation of this chapter is assigned to TC 1.6, Terminology.

Table 2 Conversion Factors

Pressure psi	in. of water (60°F)	in. Hg (32°F)	atmosphere	mm Hg (32°F)	bar	kgf/cm²	pascal
1	= 27.708	= 2.0360	= 0.068046	= 51.715	= 0.068948	= 0.07030696	= 6894.8
0.036091	1	0.073483	2.4559×10^{-3}	1.8665	2.4884×10^{-3}	2.537×10^{-3}	248.84
0.491154	13.609	1	0.033421	25.400	0.033864	0.034532	3386.4
14.6960	407.19	29.921	1	760.0	1.01325*	1.03323	1.01325×10^{5}*
0.0193368	0.53578	0.03937	1.31579×10^{-3}	1	1.3332×10^{-3}	1.3595×10^{-3}	133.32
14.5038	401.86	29.530	0.98692	750.062	1	1.01972*	10^{5}*
14.223	394.1	28.959	0.96784	735.559	0.980665*	1	9.80665×10^{4}*
1.45038×10^{-4}	4.0186×10^{-3}	2.953×10^{-4}	9.8692×10^{-6}	7.50×10^{-3}	10^{-5}*	1.01972×10^{-5}*	1

Mass	lb (avoir.)	grain	ounce (avoir.)	kg
	1	= 7000*	= 16*	= 0.45359
	1.4286×10^{-4}	1	2.2857×10^{-3}	6.4800×10^{-5}
	0.06250	437.5*	1	0.028350
	2.20462	1.5432×10^{4}	35.274	1

Volume	cubic inch	cubic foot	gallon	litre	cubic metre (m³)
	1	= 5.787×10^{-4}	= 4.329×10^{-3}	= 0.0163871	= 1.63871×10^{-5}
	1728*	1	7.48055	28.317	0.028317
	231.0*	0.13368	1	3.7854	0.0037854
	61.02374	0.035315	0.264173	1	0.001*
	6.102374×10^{4}	35.315	264.173	1000*	1

Energy	Btu	ft·lb$_f$	calorie (cal)	joule (J) = watt-second (W·s)	watt-hour (W·h)
	1	= 778.17	= 251.9958	= 1055.056	= 0.293071
	1.2851×10^{-3}	1	0.32383	1.355818	3.76616×10^{-4}
	3.9683×10^{-3}	3.08803	1	4.1868*	1.163×10^{-3}*
	9.4782×10^{-4}	0.73756	0.23885	1	2.7778×10^{-4}
	3.41214	2655.22	859.85	3600*	1

Density	lb/ft³	lb/gal	g/cm³	kg/m³
	1	= 0.133680	= 0.016018	= 16.018463
	7.48055	1	0.119827	119.827
	62.4280	8.34538	1	1000*
	0.0624280	0.008345	0.001*	1

Specific Volume	ft³/lb	gal/lb	cm³/g	m³/kg
	1	= 7.48055	= 62.4280	= 0.0624280
	0.133680	1	8.34538	0.008345
	0.016018	0.119827	1	0.001*
	16.018463	119.827	1000*	1

Viscosity (absolute) 1 poise = 1 dyne-sec/cm² = 0.1 Pa·s = 1 g/(cm·s)

Viscosity (absolute)	poise	lb$_f$·s/ft²	lb$_f$·h/ft²	kg/(m·s) = N·s/m²	lb$_m$/ft·s
	1	= 2.0885×10^{-3}	= 5.8014×10^{-7}	= 0.1*	= 0.0671955
	478.8026	1	2.7778×10^{-4}	47.88026	32.17405
	1.72369×10^{6}	3600*	1	1.72369×10^{5}	1.15827×10^{5}
	10*	0.020885	5.8014×10^{-6}	1	0.0671955
	14.8819	0.031081	8.6336×10^{-6}	1.4882	1

Temperature Scale		Temperature K	°C	°R	°F	Temperature Interval	K	°C	°R	°F
Kelvin	x K =	x	$x - 273.15$	$1.8x$	$1.8x - 459.67$	1 K =	1	1	9/5 = 1.8	9/5 = 1.8
Celsius	x°C =	$x + 273.15$	x	$1.8x + 491.67$	$1.8x + 32$	1°C =	1	1	9/5 = 1.8	9/5 = 1.8
Rankine	x°R =	$x/1.8$	$(x - 491.67)/1.8$	x	$x - 459.67$	1°R =	5/9	5/9	1	1
Fahrenheit	x°F =	$(x + 459.67)/1.8$	$(x - 32)/1.8$	$(x + 459.67)/1.8$	$x + 459.67$	1°F =	5/9	5/9	1	1

Notes: Conversions with * are exact.
The Btu and calorie are based on the International Table.

All temperature conversions and factors are exact.
The term centigrade is obsolete and should not be used.

When making conversions, remember that a converted value is no more precise than the original value. For many applications, rounding off the converted value to the same number of significant figures as those in the original value provides sufficient accuracy.

Caution: The conversion values in Table 1 are rounded to three or four significant figures, which is sufficiently accurate for most applications. See ANSI *Standard* SI-10 (available from ASTM or IEEE) for additional conversions with more significant figures.

CHAPTER 36

PHYSICAL PROPERTIES OF MATERIALS

VALUES in the following tables are in consistent units to assist the engineer looking for approximate values. For data on refrigerants, see Chapter 18; for secondary coolants, see Chapter 20. Chapter 24 gives more information on the values for materials used in building construction and insulation. Many properties vary with temperature, material density, and composition. The references document the source of the values and provide more detail or values for materials not listed here. The preparation of this chapter is assigned to TC 1.3, Heat Transfer and Fluid Flow.

Table 1 Properties of Vapor

Material	Molecular Mass	Normal Boiling Point, °F	Critical Temperature, °F	Critical Pressure, psia	Density, lb/ft³	Specific Heat, Btu/lb·°F	Thermal Conductivity, Btu/h·ft·°F	Viscosity, lb/ft·h
Alcohol, Ethyl	46.07[a]	173.3[a]	469.6[b]	927.3[b]		0.362[j]	0.0073[a]	0.0343[j] (60)
Alcohol, Methyl	32.04[a]	148.9[a]	464.0[b]	1157[b]		0.322[j]	0.0174[r]	0.0358[j] (30)
Ammonia	17.03[a]	−28[a]	270.3[b]	1639[b]	0.0482[b]	0.525[aa]	0.0128[b]	0.0225[aa]
Argon	39.948[a]	−302.5*	−188.5*	704.9*	0.1114[b]	0.125[c]	0.0094[a]	0.0507[a]
Acetylene	26.04[a]	−118.5[a]	96.8[b]	911[b]	0.0732[b]	0.377[a]	0.0108[b]	0.0226[a]
Benzene	78.11[a]	176.2[a]	553.1[d]	714.2[d]	0.167[e] (176)	0.31[e] (176)	0.0041[e]	0.017[a]
Bromine	159.82[a]	137.8[a]	591.8[d]	1499[d]	0.38[f] (138)	0.055[f] (212)	0.0035[a]	0.041[a]
Butane	58.12[a]	31.1[a]	305.6[d]	550.7[d]	0.168[g]	0.377[aa]	0.0079[a]	0.017[a]
Carbon dioxide	44.01[a]	−109.3[a]	87.9[d]	1071[d]	0.123[g]	0.20[g]	0.0084[a]	0.033[h]
Carbon disulfide	76.13[h]	115.2[h]	534[h]	1046[h]		0.1431[p] (80)		
Carbon monoxide	28.01[a]	−312.7[a]	−220.4[d]	507[d]	0.078[d]	0.25[f]	0.0133[a]	0.040[a]
Carbon tetrachloride	153.84[g]	169.8[h]	541.8[h]	661[h]		0.206[q] (80)		0.0375[j]
Chlorine	70.91[a]	−30.3[a]	291.2[d]	1118[d]	0.201[d]	0.117[a]	0.0046[a]	0.030[a]
Chloroform	119.39[h]	143.1[h]	506.1[h]	794[h]		0.126[j]	0.0081[r]	0.038[j]
Ethyl chloride	64.52[h]	54.2[h]	369.0[h]	764[h]	0.1793[b]	0.426[r]	0.00504[j]	0.0378[q]
Ethylene	28.03[h]	−154.6[h]	49.9[h]	742[h]	0.0783[b]	0.352[aa]	0.0102[aa]	0.0231[aa]
Ethyl ether	74.12[h]	94.4[h]	378.8[h]	523[h]		0.589[h] (95)		0.0273[q]
Fluorine	38.00[h]	−304.5[h]	−200.5[h]	808[h]	0.1022[b]	0.194[j]	0.0147[j]	0.089[j]
Helium	4.0026[a]	−452.1[i]	−450.2[h]	33.21[i]	0.0111[i]	1.241[aa]	0.0823[aa]	0.0452[aa]
Hydrogen	2.0159[a]	−423.0[i]	−399.9[i]	190.8[i]	0.00562[i]	3.40[j]	0.0972[aa]	0.0203[aa]
Hydrogen chloride	36.461[a]	−120.8[a]	124.5[d]	1198[d]	0.1024[b]	0.191[j]	0.00757[j]	0.0321[j]
Hydrogen sulfide	34.080[a]	−77.3[a]	212.7[d]	1307[d]	0.0961[b]	0.238[j]	0.00751[j]	0.0281[j]
Heptane (m)	100.21[a]	209.2[a]	512.2[b]	394[b]	0.21[k]	0.476[j]	0.0107[j]	0.0168[j]
Hexane (m)	86.18[a]	154[a]	454.5[d]	440[d]	0.21[k]	0.449[j]	0.00971[j]	0.0182[j]
Isobutane	58.12[f]	−11.1*	275.0[j]	529.1[j]	0.154[s] (70)	0.376[aa]	0.0081[aa]	0.0168[aa]
Methane	16.04[a]	−263.2[a]	−115.18[j]	673.1[b]	0.0448[b]	0.520[aa]	0.0178[aa]	0.0250[aa]
Methyl chloride	50.49[a]	−11.6[a]	289.6[j]	968.5[b]	0.1440[b]	0.184[aa]	0.0054[aa]	0.0244[aa]
Naphthalene	128.19[a]	424.4*	876.2[j]	576.1[j]		0.313[q] (77)		
Neon	20.183[a]	−412.6[a]	−379.7[j]	391.3[j]		0.246[aa]	0.0268[aa]	0.0718[aa]
Nitric oxide	30.01[a]	−241.6[a]	−135.2[j]	949.4[j]		0.238[j]		0.0712[j]
Nitrogen	28.01[a]	−320.4[a]	−232.4[j]	492.3[b]		0.248[j]	0.0138[aa]	0.0402[aa]
Nitrous oxide	44.01[a]	−127.3[a]	97.5[j]	1049.3[j]		0.203[j]	0.01001[j] (80.3)	0.0543[j]
Nitrogen tetroxide	92.02[a]		316.8[j]	1469.6[j]		0.201[p] (80)	0.0232[r] (131)	
Oxygen	31.9977*	−297.3*	−181.5*	731.4*		0.218[j]	0.0141[aa]	0.0462[aa]
n-Pentane	72.53[a]	97.0*	385.9[j]	489.5[j]		0.400[a] (80)	0.00877[j] (80.3)	0.0282[j]
Phenol	74.11[b]	358.5[b]	786[b]	889[b]	0.16[k]	0.34[k]	0.0099[k]	0.029[k]
Propane	44.09[g]	−43.76*	206.1*	616.1*	0.126[g]	0.3753[j] (40)	0.0087[j]	0.0179[j]
Propylene	42.08[b]	−53.86[l]	197.2[l]	670.3[l]	0.120[l]	0.349[aa]	0.0081[aa]	0.0195[aa]
Sulfur dioxide	64.06[b]	14.0[b]	315[b]	1142[b]	0.183[b]	0.145[l]	0.0049[j]	0.0281[j]
Water vapor	18.02[b]	212.0[m]	705.18*	3200.0*	0.0373[m]	0.489[aa]	0.0143[m]	0.0293[aa]

*Data source unknown.

Notes: 1. Properties at 14.696 psia and 32°F, or the saturation temperature if higher than 32°F, unless otherwise noted in parentheses.
2. Superscript letters indicate data source from the section on References.

Table 2 Properties of Liquids

Name or Description	Normal Boiling Point, °F at 14.696 psia	Enthalpy of Vaporization, Btu/lb	Specific Heat, c_p Btu/lb·°F	Specific Heat Temp., °F	Viscosity lb/h·ft	Viscosity Temp., °F	Enthalpy of Fusion, Btu/lb	Density lb/ft³	Density Temp., °F	Thermal Conductivity Btu/h·ft·°F	Thermal Conductivity Temp., °F	Vapor Pressure mm of Hg	Vapor Pressure Temp., °F	Freezing Point, °F
Acetic acid	245.3[a]	174.1[b]	0.522[b]	79-203	2.956[f]	68	84.0[b]	65.49[a]	68	0.099[b]	68	400[a]	210	61.9[a]
Acetone	133.2[a]	228.9[b]	0.514[b]	37-73	0.801[f]	68	42.1[b]	49.4[a]	68	0.102[b]	86	400[a]	103	−139.6[a]
Allyl alcohol	206.6[a]	294.1[b]	0.655[b]	70-205	3.298[f]	68		53.31[a]	68	0.104[b]	77-86	400[a]	176	−200.2[a]
n-Amyl alcohol	280.6[i]	216.3[b]			9.686[f]	73.4	48.0[b]	51.06[f]	59	0.094[b]	86	100[a]	186	−110.2[a]
Ammonia	−28[a]	583.2[b]	1.099[b]	32	0.643[f]	−28.3	142.9[b]	43.50[b]	−50	0.29[b]	5-86	400[a]	−49.7	−107.9[a]
Alcohol, Ethyl	173.3[a]	367.5[b]	0.680[b]	32-208	2.889[f]	68	46.4[b]	49.27[a]	68	0.105[b]	68	100[a]	94.8	−179.1[a]
Alcohol, Methyl	148.9[a]	473.0[b]	0.601[b]	59-68	1.434[f]	68	42.7[a]	49.40[a]	68	0.124[b]	68	100[a]	70.2	−144.0[a]
Aniline	363.8[a]	186.6[b]	0.512[b]	46-180	10.806[f]	68	48.8[b]	63.77[a]	68	0.100[b]	32-68	10[a]	156.9	20.84[a]
Benzene	176.2[a]	169.4[b]	0.412[b]	68	1.58[a]	68	54.2[b]	54.9[d]	68	0.085[b]	68	75[d]	68	42[a]
Bromine	137.8[a]	79.4[d]	0.107[f]	68	2.39[a]	68	28.5[d]	194.7[f]	68	0.070[a]	77	165[d]	68	19[a]
n-Butyl alcohol	243.5[a]	254.3[b]	0.563[f]	68	7.13[f]	68	53.9[b]	50.6[a]	68	0.089[b]	68	5[d]	68	−130[a]
n-Butyric acid	326.3[a]	217.0[b]	0.515[f]	68	3.73[a]	68	54.1[a]	60.2[a]	68	0.094[b]	54	0.7[d]	68	20[a]
Calcium chloride brine (20% by mass)			0.744[i]	68	4.8[i]	68		73.8[i]	68	0.332[i]	68			2[i]
Carbon disulfide	115.3[a]	148.8[b]	0.240[i]	68	0.88[a]	68	24.8[d]	78.9[d]	68	0.093[b]	86	295[d]	68	−168[a]
Carbon tetrachloride	170.2[a]	83.7[b]	0.201[f]	68	2.34[a]	68	12.8[d]	99.5[d]	68	0.062[j]	68	87[d]	68	−9[a]
Chloroform	142.3[v]	106[v]	0.234[v]	68	1.36[v]	68		92.96[v]	68	0.075[v]	68	160[v]	68	−81.8[v]
n-Decane	345.2[b]		0.50[b]	68			86.9[b]	45.6[b]	68	0.086[b]	68	1.3[b]	68	−21.5[b]
Ethyl ether	94.06[v]	151[v]	0.541[v]	68	0.56[v]	68	42.4[v]	44.61[v]	68	0.081[b]	68	440[v]	68	−177.3[v]
Ethyl acetate	170.8[v]	183.8[v]	0.468[v]	68	1.09[v]	68	51.2[b]	52.3[v]	68	0.101[b]	68	72[b]	68	−116.3[v]
Ethyl chloride	54.2[j]	165.9[f] (68)	0.368[f]	32			29.68[a]	56.05[a]	68	0.179[f]	33.6	400[y]	53.1	−213.5[a]
Ethyl iodide	162.1[a]	82.1[f] (160)	0.368[f]	32	0.0239[f]	68		120.85[a]	68	0.214[f]	86	100[y]	64.4	−162.4*
Ethylene bromide	268.8[a]	99.2[f] (210)	0.174[f]	68	0.0694[f]	68	24.82[a]	136.05[a]	68			10[y]	65.5	49.2[a]
Ethylene chloride	182.3[a]	153.4[f] (308)	0.301[f]	68	0.0338[f]	68	38.02[a]	77.10[a]	68			60[y]	64.6	−31.64[a]
Ethylene glycol	388.4[a]	344.0[f] (651)					77.86[a]	69.22[a]	68	0.100[f]	68	1[y]	128	12.7[a]
Formic acid	213.3[a]	215.8[f] (420)	0.526[f]	68	0.0719[f]	68	118.89[a]	76.16[a]	68	0.104[a]	33	40[y]	75.2	47.1[a]
Glycerin (glycerol)	359* (20 mm)				43.1[f]	68		78.72[a]	68	0.113[a]	68	1[a]	125.5	68[a]
Heptane	209.2[a]	138[f]	0.532[j]	68	0.990[a]	68	60.4[b]	42.7[a]	68	0.0741[j]	68	35.5[y]	68	−132[a]
Hexane	154[a]	145[f]	0.538[j]	68	0.775[d]	68	65.0[b]	41.1[a]	68	0.0720[j]	68	120.0[y]	68	−139[a]
Hydrogen chloride	−120.8[a]	191[f]					23.6[f]	74.6[d]	b.p.					−174.6[a]
Isobutyl alcohol	226.4[a]	249[f]	0.116[f]	68	9.45[f]	68		50.0[f]	68	0.082[f]	68	9.7[y]	68	−162.4[a]
Kerosene	400-560[b]		0.50[n]	68	6.0[b]	68		51.2[a]	68	0.086[n]	68			
Linseed oil					104[b]	68		58[d]	68					−11†[a]
Methyl acetate	134.6[a]	177[f]	0.468[f]	68	0.940[f]	68		60.6[a]	68	0.093[f]	68	169.8[y]	68	−144.6[a]
Methyl iodide	108.5[a]	82.6[f]			1.21[f]	68		142[a]	68			320[y]	68	−87.7[a]
Naphthalene	411.4[a]	136[f]	0.402[f]	m.p.	2.18[b]	m.p.	64.9[b]	60.9[y]	m.p.			2.18[b]	68	176.4[a]
Nitric acid	186.8[v]	270[v]	0.42[v]	68	2.2[k]	68	71.5[v]	94.45[v]	68	0.16[v]	68	1.77[v]	68	−42.9[v]
Nitrobenzene	411.6[b]	142[b]	0.348[b]	68	5.20[b]	68	40.28[v]	75.2[b]	68	0.96[b]	68	< 0.01[b]	68	42.3[b]
Octane	258.3[b]	131.7[b]	0.51[b]	68	1.36[b]	68	77.70[b]	43.9[b]	68	0.084[b]	68	0.42[b]	68	−69.7[b]
Petroleum		98-165[w]	0.4-0.6[w]	68	19-2900[w]	68		40-66[w]						
n-Pentane	96.8[a]	153.6[b]	0.558[b]	68	0.546[d]	68	50.1[b]	39.1[a]	68	0.066[b]	68	425[d]	68	−201.5[a]
Propionic acid	286.0[a]	177.8[f]	0.473[b]	68	2.666[a]	68		61.9[a]	68	0.100*	54	3[d]	68	−5.4[a]
Sodium chloride brine														
20% by mass	220.8[a]		0.745[x]	68	3.80[x]	68		71.8[x]	68	0.337[x]	68	0.57[x]	68	2.6[x]
10% by mass	215.5[a]		0.865[x]	68	2.85[x]	68		66.9[x]	68	0.343[x]	68	0.65[x]	68	20.6[x]
Sodium hydroxide and water														
15% by mass	215.0[v]		0.864[b]	68				72.4[b]	68					−5.8[b]
Sulfuric acid and water														
100% by mass	550.0[v]		0.335[b]	68	53[b]	68		114.4[v]	68			< 0.01[b]	68	50.9[b]
95% by mass	575.0[v]		0.35[v]	68	52[v]	68		114.6[v]	68			< 0.01[v]	68	−18[v]
90% by mass	500.0[v]		0.39[v]	68	60[v]	68		113.4[v]	68	0.22[b]	68	< 0.01[v]	68	15.0[v]
Toluene ($C_6H_5CH_3$)	231[b]	156[b]	0.404[v]	68	1.42[v]	68	30.9[b]	54.1[b]	68	0.090[b]	68	0.88[b]	68	−139[b]
Turpentine	303[a]	123[v]	0.42[b]	68	1.32[b]	68		53.9[b]	68	0.073[b]	68			
Water	211.9*	970.3[m]	0.999[m]	68	2.39[m]	68	143.5[b]	62.32[m]	68	0.348[m]	68	17.59*	68	32.018[m]
Xylene [$C_6H_4(CH_3)_2$]														
Ortho	291[b]	149[b]	0.411[b]	68	2.01[b]	68	55.1[b]	55.0[b]	68	0.90[b]	68	0.196[b]	68	−13[b]
Meta	283[b]	147[b]	0.400[b]	68	1.52[b]	68	46.9[b]	54.1[b]	68	0.90[b]	68	0.218[b]	68	−53[b]
Para	281[b]	146[b]	0.393[b]	68	1.62[b]	68	69.3[b]	53.8[b]	68			0.227[b]	68	56[b]
Zinc sulfate and water														
10% by mass			0.90[b]	68	3.80[a]	68		69.2[r]	68	0.337[a]	68			29.7[a]
1% by mass			0.80[b]	68	2.54[a]	68		63.0[r]	68	0.346[a]	68			31.7[a]

*Data source unknown.
†Approximate solidification temperature.

Notes: Superscript letters indicate data source from the section on References.
m.p. = melting point b.p. = boiling point

Table 3 Properties of Solids

Material Description	Specific Heat, Btu/lb·°F	Density, lb/ft³	Thermal Conductivity, Btu/h·ft·°F	Emissivity Ratio	Emissivity Surface Condition
Aluminum (alloy 1100)	0.214[b]	171[u]	128[u]	0.09[n]	Commercial sheet
				0.20[n]	Heavily oxidized
Aluminum bronze (76% Cu, 22% Zn, 2% Al)	0.09[u]	517[u]	58[u]		
Asbestos: Fiber	0.25[b]	150[u]	0.097[u]		
Asbestos: Insulation	0.20[t]	36[b]	0.092[b]	0.93[b]	"Paper"
Ashes, wood	0.20[t]	40[b]	0.041[b] (122)		
Asphalt	0.22[b]	132[b]	0.43[b]		
Bakelite	0.35[b]	81[u]	9.7[u]		
Bell metal	0.086[t] (122)				
Bismuth tin	0.040*		37.6*		
Brick, building	0.2[b]	123[u]	0.4[b]	0.93*	
Brass: Red (85% Cu, 15% Zn)	0.09[u]	548[u]	87[u]	0.030[b]	Highly polished
Brass: Yellow (65% Cu, 35% Zn)	0.09[u]	519[u]	69[u]	0.033[b]	Highly polished
Bronze	0.104[t]	530[t]	17[d] (32)		
Cadmium	0.055[a]	540[r]	53.7[b]	0.02[d]	
Carbon (gas retort)	0.17[a]		0.20[b] (2)	0.81[a]	
Cardboard			0.04[b]		
Cellulose	0.32[b]	3.4[t]	0.033[t]		
Cement (portland clinker)	0.16[b]	120[i]	0.017[i]		
Chalk	0.215[t]	143[t]	0.48*	0.34*	About 250°F
Charcoal (wood)	0.20[t]	15[a]	0.03[a] (392)		
Chrome brick	0.17[b]	200[b]	0.67[b]		
Clay	0.22[b]	63[t]			
Coal	0.3[b]	90[t]	0.098[r] (32)		
Coal tars	0.35[b] (104)	75[b]	0.07[b]		
Coke (petroleum, powdered)	0.36[b] (752)	62[b]	0.55[b] (752)		
Concrete (stone)	0.156[b] (392)	144[b]	0.54[b]		
Copper (electrolytic)	0.092[u]	556[u]	227[u]	0.072[n]	Commercial, shiny
Cork (granulated)	0.485[t]	5.4[t]	0.028[t] (23)		
Cotton (fiber)	0.319[u]	95[u]	0.024[u]		
Cryolite ($AlF_3 \cdot 3NaF$)	0.253[b]	181[b]			
Diamond	0.147[b]	151[t]	27[t]		
Earth (dry and packed)		95[t]	0.037*	0.41*	
Felt		20.6[b]	0.03[b]		
Fireclay brick	0.198[b] (212)	112[t]	0.58[b] (392)	0.75[n]	At 1832°F
Fluorspar (CaF_2)	0.21[b]	199[v]	0.63[v]		
German silver (nickel silver)	0.09[u]	545[u]	19[u]	0.135[n]	Polished
Glass: Crown (soda-lime)	0.18[b]	154[u]	0.59[t] (200)	0.94[n]	Smooth
Glass: Flint (lead)	0.117[b]	267[u]	0.79[r]		
Glass: Heat-resistant	0.20[b]	139[t]	0.59[t] (200)		
Glass: "Wool"	0.157[b]	3.25[t]	0.022[t]		
Gold	0.0312[u]	1208[u]	172[t]	0.02[n]	Highly polished
Graphite: Powder	0.165*		0.106*		
Graphite: Impervious	0.16[u]	117[u]	75[u]	0.75[n]	
Gypsum	0.259[b]	78[b]	0.25[b]	0.903[b]	On a smooth plate
Hemp (fiber)	0.323[u]	93[u]			
Ice: 32°F	0.487[t]	57.5[b]	1.3[b]	0.95*	
Ice: −4°F	0.465[t]		1.41*		
Iron: Cast	0.12[v] (212)	450[b]	27.6[b] (129)	0.435[b]	Freshly turned
Iron: Wrought		485[b]	34.9[b]	0.94[b]	Dull, oxidized
Lead	0.0309[u]	707[u]	20.1[u]	0.28[n]	Gray, oxidized
Leather (sole)		62.4[b]	0.092[b]		
Limestone	0.217[b]	103[b]	0.54[b]	0.36* to 0.90	At 145 to 380°F
Linen			0.05[b]		
Litharge (lead monoxide)	0.055[b]	490[b]			
Magnesia: Powdered	0.234[b] (212)	49.7[b]	0.35[b] (117)		
Magnesia: Light carbonate		13[b]	0.034[b]		
Magnesite brick	0.222[b] (212)	158[b]	2.2[b] (400)		
Magnesium	0.241[b]	108[u]	91[u]	0.55[n]	Oxidized
Marble	0.21[b]	162[b]	1.5[b]	0.931[b]	Light gray, polished
Nickel, polished	0.105[u]	555[u]	34.4[u]	0.045[n]	Electroplated
Paints: White lacquer				0.80[n]	
Paints: White enamel				0.91[n]	On rough plate
Paints: Black lacquer				0.80[n]	
Paints: Black shellac		63[u]	0.15[u]	0.91[n]	"Matte" finish
Paints: Flat black lacquer				0.96[n]	
Paints: Aluminum lacquer				0.39[n]	On rough plate

*Data source unknown.

Notes: 1. Values are for room temperature unless otherwise noted in parentheses.

2. Superscript letters indicate data source from the section on References.

Table 3 Properties of Solids (*Concluded*)

Material Description	Specific Heat, Btu/lb·°F	Density, lb/ft³	Thermal Conductivity, Btu/h·ft·°F	Emissivity	
				Ratio	Surface Condition
Paper	0.32*	58[b]	0.075[b]	0.92[b]	Pasted on tinned plate
Paraffin	0.4[bb]	47[bb]	0.14[b] (32)		
Plaster		132[b]	0.43[b] (167)	0.91[b]	Rough
Platinum	0.032[u]	1340[u]	39.9[u]	0.054[b]	Polished
Porcelain	0.18*	162[u]	1.3[u]	0.92[b]	Glazed
Pyrites (copper)	0.131[b]	262[b]			
Pyrites (iron)	0.136[b] (156)	310[v]			
Rock salt	0.219[u]	136[u]			
Rubber, vulcanized: Soft	0.48*	68.6[t]	0.08[t]	0.86[b]	Rough
Hard		74.3[t]	0.092[t]	0.95[b]	Glossy
Sand	0.191[b]	94.6[b]	0.19[b]		
Sawdust		12[b]	0.03[b]		
Silica	0.316[b]	140[v]	0.83[t] (200)		
Silver	0.0560[u]	654[u]	245[u]	0.02[n]	Polished and at 440°F
Snow: Freshly fallen		7[y]	0.34[t]		
At 32°F		31[t]	1.3[t]		
Steel (mild)	0.12[b]	489[b]	26.2[b]	0.12[n]	Cleaned
Stone (quarried)	0.2[b]	95[t]			
Tar: Pitch	0.59[v]	67[u]	0.51[v]		
Bituminous		75[t]	0.41[u]		
Tin	0.0556[u]	455[u]	37.5[u]	0.06[h]	Bright and at 122°F
Tungsten	0.032[u]	1210[u]	116[u]	0.032[n]	Filament at 80°F
Wood: Hardwoods—	0.45/0.65[b]	23/70[z]	0.065/0.148[z]		
Ash, white		43[z]	0.0992[z]		
Elm, American		36[z]	0.0884[z]		
Hickory		50[z]			
Mahogany		34[u]	0.075[u]		
Maple, sugar		45[z]	0.108[z]		
Oak, white	0.570[b]	47[z]	0.102[z]	0.90[n]	Planed
Walnut, black		39[z]			
Softwoods—	See Table 4,	22/46[z]	0.061/0.093[z]		
Fir, white	Chapter 24	27[z]	0.068[z]		
Pine, white		27[z]	0.063[z]		
Spruce		26[z]	0.065[z]		
Wool: Fiber	0.325[u]	82[u]			
Fabric		6.9/20.6[u]	0.021/0.037[u]		
Zinc: Cast	0.092[u]	445[u]	65[u]	0.05[n]	Polished
Hot-rolled	0.094[b]	445[b]	62[b]		
Galvanizing				0.23[n]	Fairly bright

*Data source unknown.

Notes: 1. Values are for room temperature unless otherwise noted in parentheses.

2. Superscript letters indicate data source from References.

REFERENCES

[a]*Handbook of chemistry and physics*, 63rd ed. 1982-83. Chemical Rubber Publishing Co., Cleveland, OH.

[b]Perry, R.H. *Chemical engineers' handbook*, 2nd ed., 1941, 5th ed., 1973. McGraw-Hill, New York.

[c]*Tables of thermodynamic and transport properties of air, argon, carbon dioxide, carbon monoxide, hydrogen, nitrogen, oxygen and steam.* 1960. Pergamon Press, Elmsford, NY.

[d]*American Institute of Physics handbook*, 3rd ed. 1972. McGraw-Hill, New York.

[e]Organick and Studhalter. 1948. *Thermodynamic properties of benzene. Chemical Engineering Progress* (November):847.

[f]Lange. 1972. *Handbook of chemistry*, rev. 12th ed. McGraw-Hill, New York.

[g]ASHRAE. 1969. *Thermodynamic properties of refrigerants.*

[h]Reid and Sherwood. 1969. *The properties of gases and liquids*, 2nd ed. McGraw-Hill, New York.

[i]Chapter 19, 1993 *ASHRAE Handbook—Fundamentals.*

[j]*T.P.R.C. data book.* 1966. Thermophysical Properties Research Center, W. Lafayette, IN.

[k]Estimated.

[l]Canjar, L.N., M. Goldman, and H. Marchman. 1951. Thermodynamic properties of propylene. *Industrial and Engineering Chemistry* (May):1183.

[m]*ASME steam tables.* 1967. American Society of Mechanical Engineers, New York.

[n]McAdams, W.H. 1954. *Heat transmission*, 3rd ed. McGraw-Hill, New York.

[o]Stull, D.R. 1947. Vapor pressure of pure substances (organic compounds). *Industrial and Engineering Chemistry* (April):517.

[p]*JANAF thermochemical tables.* 1965. PB 168 370. National Technical Information Service, Springfield, VA.

[q]*Physical properties of chemical compounds.* 1955-61. American Chemical Society, Washington, D.C.

[r]*International critical tables of numerical data.* 1928. National Research Council of USA, McGraw-Hill, New York.

[s]*Matheson gas data book*, 4th ed. 1966. Matheson Company, Inc., East Rutherford, NJ.

[t]Baumeister and Marks. 1967. *Standard handbook for mechanical engineers.* McGraw-Hill, New York.

[u]Miner and Seastone. *Handbook of engineering materials.* John Wiley and Sons, New York.

[v]Kirk and Othmer. 1966. *Encyclopedia of chemical technology.* Interscience Division, John Wiley and Sons, New York.

[w]Gouse and Stevens. 1960. *Chemical technology of petroleum*, 3rd ed. McGraw-Hill, New York.

[x]*Saline water conversion engineering data book.* 1955. M.W. Kellogg Co. for U.S. Department of Interior.

[y]Timmermans, J. *Physicochemical constants of pure organic compounds*, 2nd ed. American Elsevier, New York.

[z]*Wood handbook.* 1955. Handbook No. 72. Forest Products Laboratory, U.S. Department of Agriculture.

[aa]ASHRAE. 1976. *Thermophysical properties of refrigerants.*

[bb]Lane, G. ed. 1986. *Solar heat storage: Latent heat materials, Vol II—Technology.* CRC Press, Chicago.

CHAPTER 37

FUNDAMENTALS OF CONTROL

AUTOMATIC HVAC control systems are designed to maintain temperature, humidity, pressure, flow, power, lighting levels, and safe levels of indoor contaminants. Automatic control primarily modulates, stages, or sequences mechanical and electrical equipment to meet load requirements and provide safe operation of the equipment. It can use digital, pneumatic, mechanical, electrical, and electric control devises, and implies that human intervention is limited in starting and stopping equipment and adjusting control setpoints.

This chapter focuses on the fundamental concepts and devices normally used by a control system designer. It covers (1) control fundamentals, including terminology; (2) the types of control components; (3) the methods of connecting these components to form various individual control loops or subsystems; and (4) commissioning, operation, and maintenance. Chapter 42 in the 1995 *ASHRAE Handbook—Applications* discusses the design of controls for specific HVAC applications.

TERMINOLOGY

A **closed loop** control, or **feedback** control, measures actual changes in the controlled variable and actuates the control device to bring about a change. The corrective action continues until the variable is brought to a desired value within the design limitations of the controller. This arrangement of sensing the value of the controlled variable by the controller is known as **feedback**.

An **open loop** control does not have a direct link between the value of the controlled variable and the controller. An open loop control anticipates how an external variable will affect the system and adjusts the setpoint to avoid excessive offset. An example is an outdoor thermostat arranged to control heat to a building in proportion to the calculated load caused by changes in outdoor temperature. In essence, the designer presumes a fixed relationship between outside air temperature and the heat requirement of the building and takes control action based on the outdoor air temperature. The actual space temperature has no effect on this controller. Because there is no feedback on the controlled variable (space temperature), in this case, the control is an open loop.

The **control loop**, shown in Figure 1 illustrates the components of the typical control loop.

The **sensor** measures the controlled variable and transmits values to the controller. The **controller** compares the value of the controlled variable with the setpoint and generates a signal to the controlled device for corrective action. A controller can be in the form of hardware or software. Thermostats, humidistats, and pressure controls are examples of hardware controllers. Digital algorithms are examples of software controllers.

The **setpoint** is the desired value of the controlled variable. The controller seeks to maintain this setpoint. The **controlled device** reacts to signals received from the controller to vary the flow of the control agent. The controlled device may be a valve, damper, heating element, or a motor driving a pump or a fan.

The **control agent** is the medium manipulated by the controlled device. It may be air or gas flowing through a damper; gas, steam, or water flowing through a valve; or an electric current.

The **process plant** is the air-conditioning apparatus being controlled. It reacts to the output of the control agent and affects the change in the controlled variable. It may include a coil, fan, or humidifier.

The **controlled variable** is that temperature, humidity, or pressure, etc. being controlled.

A **control loop** can be represented in the form of a block diagram, in which each component of the control loop is modeled and represented in it's own block (Figure 2). The flow of information from one component to the next is shown by lines between the blocks. The figure shows the setpoint being compared to the controlled variable. This difference, or **offset error**, is fed into the controller, which sends a control signal to the controlled device. In this case, the controlled device is a valve. The valve can change the amount of steam flow through the coil of Figure 1. The amount of steam flow is the input to the next block, which represents the process. From the process block comes the controlled variable, which is temperature. The controlled variable is sensed by the sensing element and fed to the controller as feedback, completing the loop.

Each component of Figure 2 can be represented by a transfer function, which is an idealized mathematical representation of the relationship between the input and the output variables of the component. The transfer function must be sufficiently detailed to cover both the dynamic and static characteristics of the device. The dynamics of the component are represented in the time domain by a differential equation. In environmental control, the transfer function of many of the components can be adequately described by a first order differential equation, implying that the dynamic behavior is dominated by a single capacitance factor. For a solution, the differential equation is converted to its LaPlace transform or z-transform.

The **time constant** is defined as the time it takes for the output to reach 63.2% of its final value when a step change in the input is effected. When the time constant of the component is small, its

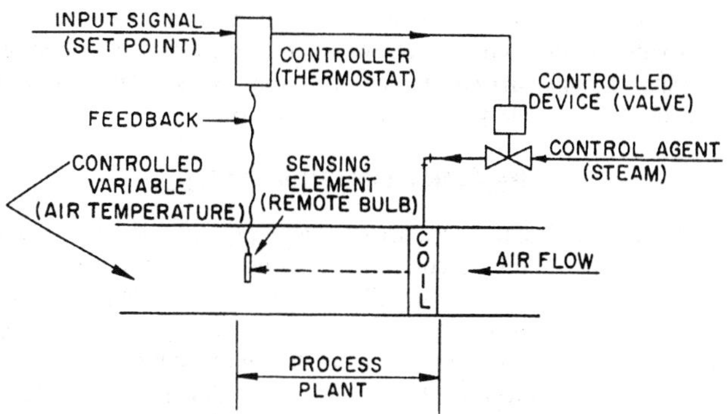

Fig. 1 Discharge Air Temperature Control
(Example of Feedback Control)

The preparation of this chapter is assigned to TC 1.4, Control Theory and Application.

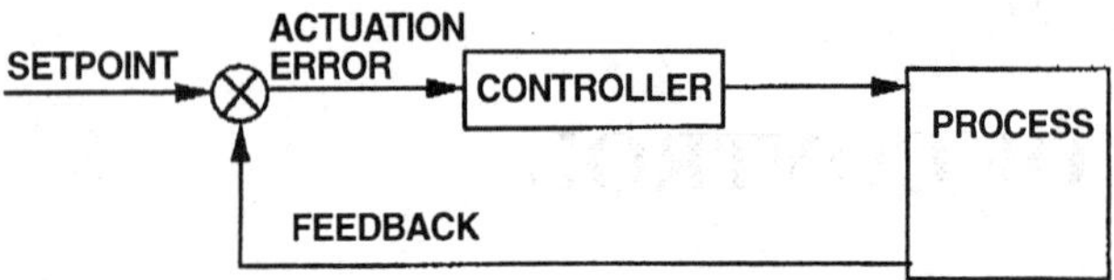

Fig. 2 Block Diagram of Discharge Air Temperature Control

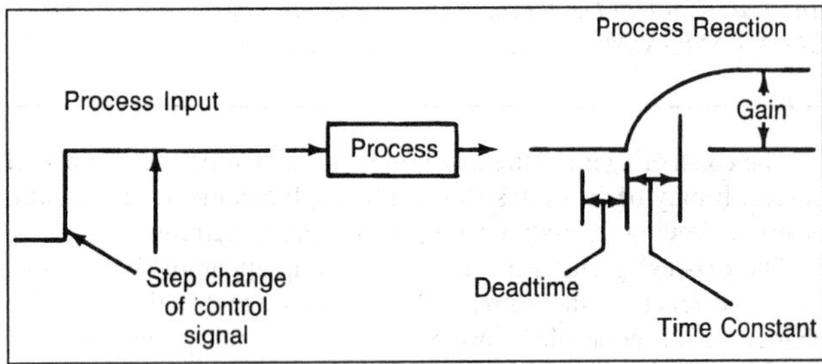

Fig. 3 Process Subjected to a Step Input

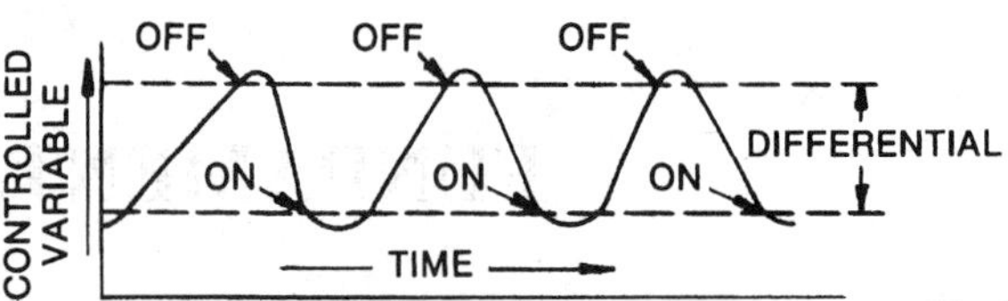

Fig. 4 Two-Position Control

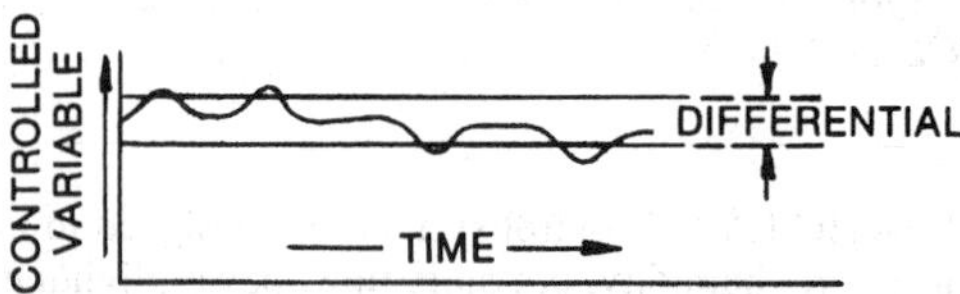

Fig. 5 Floating control Showing Variations in Controlled Variable as Load Changes

output will react rapidly to reflect changes in the input; conversely, components with a larger time constant will be sluggish in responding to changes in the input.

Deadtime is a phase shift that can cause control and modeling problems. Deadtime is the time between a change in the process input and when the change affects the output of the process. Deadtime can occur in the control loop of Figure 1 due to the transportation time of the air from the coil to the space. When a coil temperature is changed there is delay while the affected supply air travels the distribution system and finally affects the sensor in the space. The mass of air within the space further delays the room sensor from detecting the full impact of the coil temperature change. Deadtime can also occur due to a slow sensor, a time lag in the signal from the controller. If the deadtime is small, it could be ignored in the model of the control; if it is significant, it must be considered.

The **gain** of a transfer function is the amount the output of the component changes for a given change of input under steady-state conditions. If the element is linear, its gain remains constant. However, many control components are nonlinear and have varying gains, depending on the operating conditions. Figure 3 shows the response of the first order plus deadtime process to a step change of the input signal. Notice that the process shows no reaction during the deadtime, followed by a response that resembles a first order exponential.

The principle behind a controller is that the sensor sends a signal (pneumatic, electric, or electronic) whose pressure, voltage, or current, is proportional to the value of the variable being measured. The controller compares this signal from the sensor to the desired value, and sends a control signal based on this comparison. The **hardware controller** is an analog device that receives and acts on data continuously. The **software controller** is a digital device that receives and acts on data discretely on a sample-rate basis.

TYPES OF CONTROL ACTION

Closed loop control systems are commonly classified by the type of corrective action the controller is programmed to take when it senses a deviation of the controlled variable from the setpoint. Both hardware and software controller types can be classified according to the following most common types of control action.

Two-Position Action. The control device shown in Figure 4 can be positioned only to a maximum or minimum state, or can be either on or off. A typical home thermostat that starts and stops a furnace is a good example of two-position action.

Controller differential, as it applies to two-position control action, is the difference between a setting at which the controller operates to one position and a setting at which it operates to the other. Thermostat ratings usually refer to the differential (in degrees) that become apparent by raising and lowering the dial setting. This differential is known as the **manual differential** of the thermostat. When the same thermostat is applied to an operating system, the total change in temperature that occurs between a "turn-on" state and a "turn-off" state is usually different from the mechanical differential. The **operating differential** could be greater due to thermostat lag or hysteresis, or it could be due to heating or cooling anticipators built into the thermostat.

Anticipation Applied to Two-Position Action. This common variation of straight two-position action is often used on room thermostats to reduce the operating differential. In heating thermostats, a heater element in the thermostat is energized during "on" periods, thus shortening the on-time because the heater warms the thermostat. This is known as **heat anticipation**. The same anticipation action can be obtained in cooling thermostats by energizing a heater thermostat at "off" periods. In either case, the percentage of on-time is varied in proportion to the load, while the total cycle time remains relatively constant.

Timed Two-Position Action. This action occurs when a heating or cooling element is turned on for a time interval proportional to the deviation from setpoint. For example, an element may by turned on for two minutes and off for one minute when the deviation from setpoint is 3°F. This is similar to incremental action applied to floating control except the time interval is usually shorter for incremental action.

Floating Action. In floating action, the controller can perform only two operations—moving the controlled device to either its open or closed position, usually at a constant rate (Figure 5). Generally, a neutral zone between the two positions allows the controlled device to stop at any position when the controlled variable is within the differential of the controller. When the controlled variable falls outside the differential of the controller, the controller moves the controlled device in the proper direction.

Incremental Action. This action is a variation of floating control. Incremental action varies the pulse action to open or close an actuator depending on how close the controlled variable is to the setpoint. As the controlled variable comes close to the setpoint the pulses become shorter. This action allows closer control using floating motor actuators.

Proportional Action. In proportional action, the controlled device is positioned proportionally in response to changes in the controlled variable (Figure 6). A proportional controller can be described mathematically by:

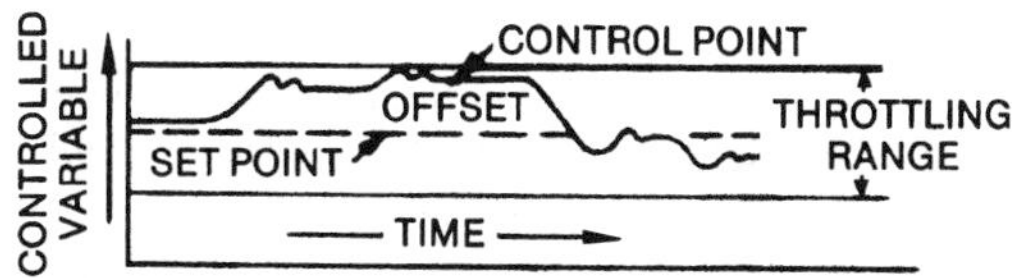

Fig. 6 Proportional Control Showing Variations in Controlled Variable as Load Changes

$$V_p = K_p e + V_o \tag{1}$$

where

V_p = output of proportional controller
K_p = proportional gain (proportional to 1/throttling range)
e = error signal or offset
V_o = offset adjustment parameter

The output of the controller is proportional to the difference between the sensed value, the controller variable, and its setpoint. The controlled device is normally adjusted to be in the middle of its control range at setpoint by using an offset adjustment. This control is similar to that shown in Figure 1.

Throttling range is the amount of change in the controlled variable required to cause the controller to move the controlled device from one extreme to the other. It can be adjusted to meet job requirements. Throttling range is inversely proportional to proportional gain.

Control point is the actual value of the controlled variable at which the instrument is controlling. It varies within the throttling range of the controller and changes with changing load on the system and other variables.

Offset or error signal, is the difference between the setpoint and the actual control point under stable conditions. This is sometimes called drift, deviation, droop, or steady-state error.

Proportional Plus Integral (PI). This type of control improves on simple proportional control by adding another component to the control action that eliminates the offset typical of proportional control (Figure 7). **Reset action** may be described by:

$$V_p = K_p e + K_i \int e d\theta + V_o \tag{2}$$

where

V_p = output of controller
K_i = integral gain
θ = time
e = error
V_o = offset adjustment parameter

The second term in Equation (2) implies that the longer the period during which the error e exists, the more the controller output will change in attempting to eliminate the error. Selecting proportional and integral gain constants is critical to stability. Proper selection eliminates offset, obtaining greater control accuracy. Also energy efficiency can be improved by PI control in applications such as VAV fan control, chiller control, and hot and cold deck control of the air handler.

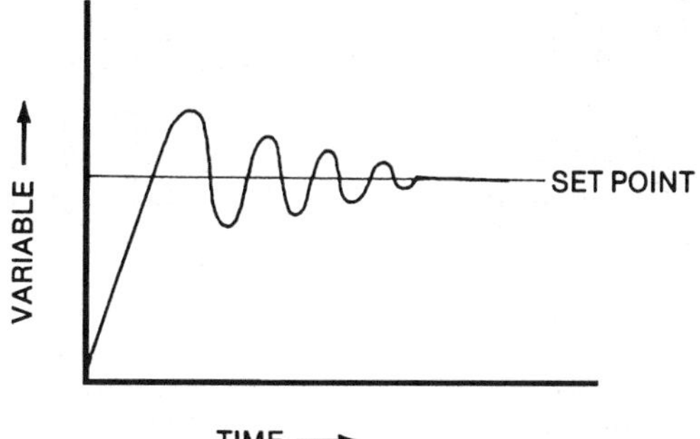

Fig. 7 Proportional Plus Integral (PI) Control

Proportional-Integral-Derivative (PID). This type of control is PI control with a derivative term added to the controller. It varies with the value of the derivative of the error. The equation for PID control is

$$V_p = K_p e + K_i \int e d\theta + K_a \frac{de}{d\theta} + V_o \tag{3}$$

where K_a is the derivative gain of controller $de/d\theta$ is the time derivative of error

Adding the derivative term gives some anticipatory action to the controller, which results in a faster response and greater stability. However, the derivative term also makes the controller more sensitive to noisy signals and harder to tune than a PI controller. Most HVAC control loops perform satisfactorily with PI control alone. Adaptive control, or self tuning, is a form of digital PID control, where the gain factors (K_p, K_i, and K_a) are continuously or periodically modified automatically to compensate for the control loop offset.

Fuzzy logic control is a form of set theory that allows a designer to describe in weighted terms the variables that determine how a process is functioning. The controller calculates the membership class of the process and then determines what control action to take based on a set of control rules created by the designer.

CLASSIFICATION BY ENERGY SOURCE

Control components may be classified according to the primary source of energy as follows:

- **Pneumatic components** use compressed air, usually at a pressure of 15 to 35 psig, as an energy source. The air is generally supplied to the controller, which regulates the pressure supplied to the controlled device.
- **Electric components** use electrical energy, either low or line voltage, as the energy source. The controller regulates electrical energy supplied to the controlled device. Controlled devices in this category includes relays; electro-mechanical, electromagnetic, and hydraulic actuators; and solid state regulating devices. The components that include signal conditioning, modulation, and amplification in their operation are classified as electronic.

  A digital controller receives electronic signals from the sensors, converts the electronic signals to numbers, and performs mathematical operations on these numbers inside a microprocessor. The output from the digital controller takes the form of a number, which is then converted to an electronic signal to operate the actuator. The digital controller must sample its data because the microprocessor requires time for other operation besides reading data. If the sampling interval for the digital controller is properly chosen to avoid second and third order harmonics, no significant degradation in control performance will be seen due to sampling.
- **Self-powered components** apply the power of the measured system to induce the necessary corrective action. The measuring system derives its energy from the process under control, without any auxiliary source of energy. Temperature changes at the sensor result in pressure or volume changes of the enclosed media that are transmitted directly to the operating device of the valve or damper. A component using a thermopile in a pilot flame to generate electrical energy is also self powered.

This method of classification can be extended to individual control loops and to complete control systems. For example, the room temperature control for a particular room that includes a pneumatic room thermostat and a pneumatically actuated reheat coil would be referred to as a pneumatic control loop. Many control systems use a combination of controls and are called **hybrid** systems.

COMPUTERS FOR AUTOMATIC CONTROL

Computers can perform the control described in this chapter. Chapter 36 in the 1995 *ASHRAE Handbook—Applications* covers computer components and some of the ways computers are being used in the HVAC control industry.

CONTROL COMPONENTS

While control components may be classified according to their energy source, this section groups components by their function in a complete control system. The first subsection considers the controlled device or final control element, examples of which are relays, valves, and dampers. Actuators, which are used to drive the valve or damper assembly, are also covered.

The second subsection considers the sensing element the measures changes in the controlled variable. Specific examples of sensor types included are temperature, humidity, flow, and pressure. While many other sensors are available, these types represent the majority of those found in the HVAC control systems.

In the third subsection, various controllers are reviewed. Controllers are classified according to the control action they cause to maintain the desired condition (setpoint)—whether they are two-position, floating, proportional, proportional plus integral (PI) or proportional plus integral plus derivative (PID) control. In addition, this section describes the various techniques available for making the control decision in a modulating control system, such as pneumatic, electronic, and digital controllers. Thermostats (devices that combine a temperature sensor and controller in a single unit) are also described.

Fundamental control systems can be constructed using only the components described in the first three subsections. In practice, however, a fourth group is sometimes necessary. These components are refereed to as auxiliary control components and include transducers, switches, power supplies, and air compressors.

CONTROLLED DEVICES

The controlled device regulates the flow of steam, water, electricity or air in an HVAC system. Water and steam flow regulators are known as **valves**, and airflow control devices are called **dampers**; both devices perform essentially the same function and must be properly sized and selected for the particular application. The control link to the valve or damper is called an operator, or actuator. This device uses electricity, compressed air, or hydraulic fluid to power the motion of the valve stem or damper linkage through its operating range.

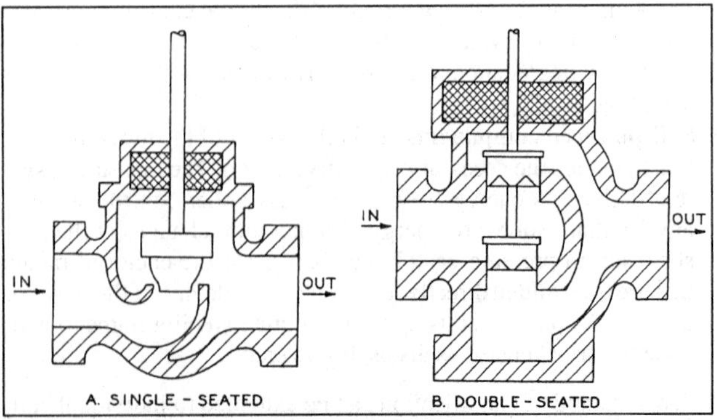

Fig. 8 Typical Single- and Double-Seated Two-Way Valves

Valves

An automatic valve is designed to control the flow of steam, water, gas, or other fluids. It may be considered as a variable orifice positioned by an electric or pneumatic operator in response to impulses, or signals from the controller. It may be equipped with a throttling plug or V-port specially designed to provided a desired flow characteristic.

Renewable composition discs are common. They are made of materials best suited to the media handled by the valve, the operating temperature, and the pressure. For high pressure or for superheated steam, metal discs are often used. Internal parts of valves, such as the seat ring, throttling plug, or V-port skirt, disc holder, and stem, are sometimes made of stainless steel or other hard and corrosion-resistant metal for use in severe service.

Various types of automatic valves include the following:

A **single-seated valve** (Figure 8A) is designed for tight shutoff. Appropriate disc materials for various pressures and media are used.

A **double-sealed** or **balanced valve** (Figure 8B) is designed so that the media pressure acting against the valve disc is essentially balanced, reducing the operator force required. It is widely used where fluid pressure is too high to permit a single-seated valve to close. It cannot be used where a tight shutoff is required.

A **three-way mixing valve** (Figure 9A) has two inlet and one outlet connections and a double-faced disc operating between two seats. It is used to mix two fluids entering through the inlet connections and leaving through the common outlet, according to the position of the valve stem and disc.

A **three-way diverting valve** (Figure 9B) has one inlet and two outlet connections and two separate discs and seats. It is used to divert the flow to either of the outlets or to proportion the flow to both outlets.

A **butterfly valve** consists of a heavy ring enclosing a disc that rotates on an axis at or near its center and is similar to a round single-blade damper. In principle the disc seats against a ring machined within the body or a resilient liner in the body. Two butterfly valves can be used together to act like a three-way valve for mixing or diverting.

Characteristics. The performance of a valve is expressed in terms of its flow characteristics as it operates through its stroke, based on a constant pressure drop. Three common characteristics are shown in Figure 10 and are defined as follows:

- **Quick opening.** Maximum flow is approached rapidly as the device begins to open.
- **Linear.** Opening and flow are related in direct proportion.
- **Equal percentage.** Each equal increment of opening increases the flow by an equal percentage over the previous value.

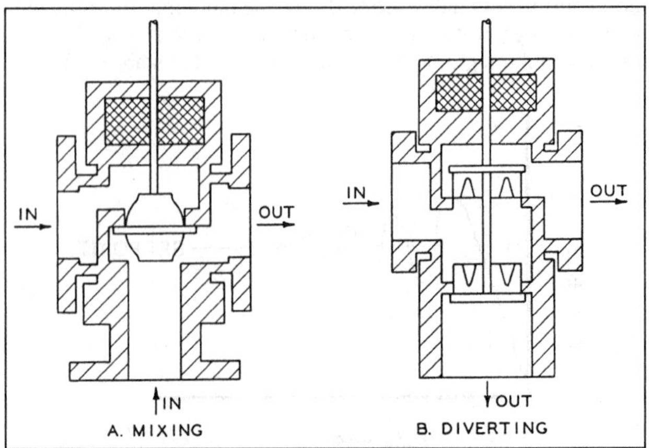

Fig. 9 Typical Three-Way Mixing and Diverting Valves

Because the pressure drop across a valve seldom remains constant as its opening changes, actual performance usually deviates from the published characteristic curve. The magnitude of the deviation is determined by the overall design. For example, in a system arranged so that control valves or dampers can shut off all flow, the pressure drop across a controlled device increases from a minimum at design conditions to the total pressure drop at no flow. Figure 11 shows the extent of the resulting deviations for a valve or damper designed with a linear characteristic, when selection is based on various percentages of total system pressure drop. To allow for adequate control by valve or damper, the design pressure drop should be a reasonably large percentage of the total system pressure drop, or the system should be designed and controlled so that the pressure drop remains relatively constant.

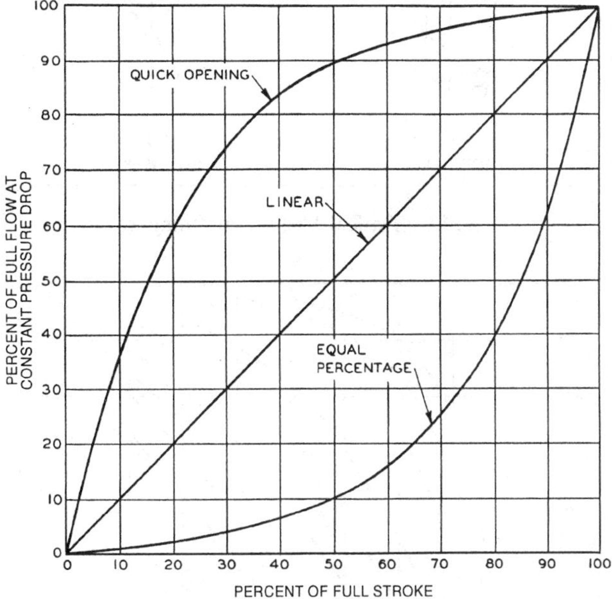

Fig. 10 Typical Flow Characteristics of Valves

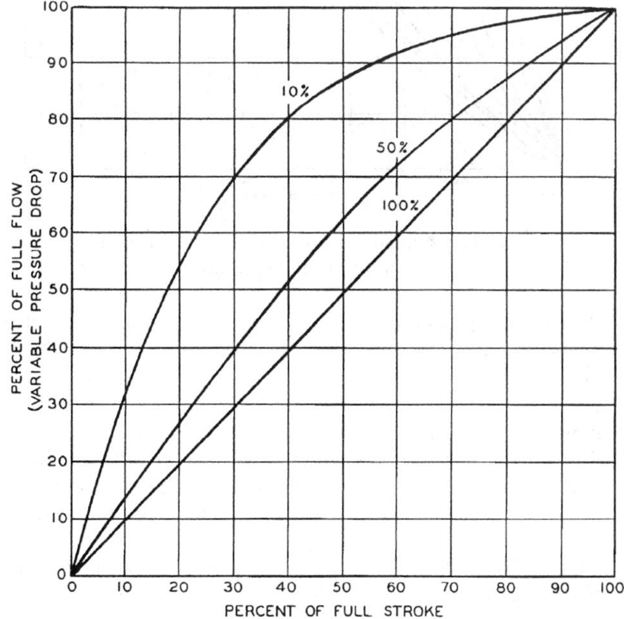

Fig. 11 Typical Performance curves for Linear Devices at Various Percentages of Total System Pressure Drop

Selection and Sizing. Higher pressure drops for controlled devices are obtained by using smaller sizes with a possible increase in size of other equipment in the system. Because sizing techniques are different for steam, water, and air, each is discussed separately.

Steam Valves. Steam-to-water and steam-to-air heat exchanger are typically controlled through regulation of steam flow using a two-way throttling valve. One-pipe steam systems require a line-size, two-position valve for proper condensate drainage and steam flow, while two-pipe steam systems can be controlled by two-position or modulating (throttling) valves.

Water Valves. Valves for water service may be two- or three-way and two-position or proportional. Proportional valves are used most often, but two-position valves are not unusual and are sometimes essential (e.g., on steam preheat coils). While it is possible to design a water system in which the pressure differential from supply to return is kept constant, it is seldom done. It is safer to assume that the pressure drop across the valve increases as it modulates from fully open to fully closed. Figure 12 shows the effect in a simple system with one pump, one two-way control valve, and a heat exchanger. The system curve represents the pressure loss in the piping and heat exchanger at various flow rates. The pump curve is the typical curve for a centrifugal pump. At design flow rates, the valve is selected for a specific pressure drop, A-A'. At part load, the valve must partially close to provide a higher pressure drop, B-B'. The ratio between the design pressure drop A-A' and the zero flow pressure drop C-C' influences the control capability of the valve.

Better control at part load is obtained by using equal percentage valves, particularly in hot water coils where the heat output of the coil is not linearly related to flow. As flow is reduced, a greater amount of heat is transferred from each unit column of water, counteracting the reduction in flow. The use of equal-percentage valves linearizes the heat transfer from the coil with respect to the control signal.

Two-way control valves should be sized to provide from 20 to 60% of the total system pressure drop. The valve operator should be sized to close the valve against the full pump head-pressure to insure complete shut-off during no-flow condition. For additional information on control valve sizing and selection, see chapters 12 and 41 of the 1996 *ASHRAE Handbook—Systems and Equipment.*

Operators. Valve operators include the following general types:

- A **pneumatic operator** consists of a spring-opposed, flexible diaphragm or bellows attached to the valve stem. An increase in air pressure, above the minimum point of the spring range, compresses the spring and simultaneously moves the valve stem. Springs of various pressure ranges, can sequence the operation of two or more devices, if properly selected or adjusted. For example, a chilled water valve operator may modulate the valve from

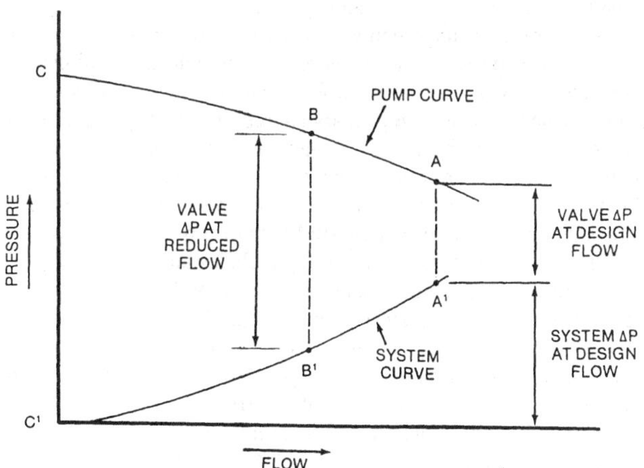

Fig. 12 Pump and System Curves with Valve Control

fully closed to fully open over a spring range of 3 to 8 psig, while a sequenced steam valve may operate from 8 to 13 psig.

Two-position pneumatic control is accomplished using a two-position pneumatic relay to apply either full air pressure or no pressure to the valve operator. Pneumatic valves and valves with spring-return electric operators can be classified as normally open or normally closed.

A **normally open valve** assumes an open position, providing full flow, when all operating force is removed.

A **normally closed valve** assumes a closed position, stopping flow, when all operating force is removed.

- **Springless pneumatic operators**, which use two opposed diaphragms or two sides of a single diaphragm, are generally limited to special applications involving large valves or high fluid pressure.
- An **electric-hydraulic actuator** is similar to a pneumatic one, except that it uses an incompressible fluid circulated by an internal electric pump.
- A **solenoid** consists of a magnetic coil operating a movable plunger. Most are for two-position operation, but modulating solenoid valves are available with a pressure equalization bellows or piston to achieve modulation. Solenoid valves are generally limited to relatively small sizes (up to 4 in.).
- An **electric motor** operates the valve stem through a gear train and linkage. Electric motor operators are classified in the following three types:

  Unidirectional—for two-position operation. The valve opens during one-half revolution of the output shaft and closes during the other one-half revolution. Once started it continues until the half revolution is completed, regardless of subsequent action by the controller. Limit switches in the operator stop the motor at the end of each stroke. If the controller has been satisfied during this interval, the operator continues to the other position.)

  Spring-return—for two-position operation. Electric energy drives the valve to one position and a spring returns the valve to its normal position.

  Reversible—for floating and proportional operation. The motor can run in either direction and can stop in any position. It is sometimes equipped with a return spring. In proportional control applications, a feedback potentiometer for rebalancing the control circuit is also driven by the motor.

Dampers

Types and Characteristics. Automatic dampers are used in air-conditioning and ventilation to control airflow. They may be used (1) for modulating control to maintain a controlled variable such as mixed air temperature or supply air duct static pressure; or (2) for two-position control to initiate operation such as opening minimum outside air dampers when a fan is started.

Two damper arrangements are used for airflow control—parallel-blade and opposed-blade (Figure 13). Parallel-blade dampers are adequate for two-position control and can be used for modulating control when they are the primary source of system pressure drop. However, opposed-blade dampers are preferable, because they normally provide better control (Figures 14 and 15). In these figures, the parameter α is the ratio of the system pressure drop to the drop across the damper at maximum (fully open) flow. Single blades are used for small dampers where multiple blades are not practical.

Damper leakage is important, particularly where tight shutoff is necessary to reduce energy consumption significantly. Also, an outdoor air damper must close tightly to prevent coils and pipes from freezing in cold climates. Low leakage dampers cost more and require larger operators because of the friction of the seals in the closed position; therefore, they should only be used when necessary.

Operators. Either electricity or compressed air is used to operate dampers.

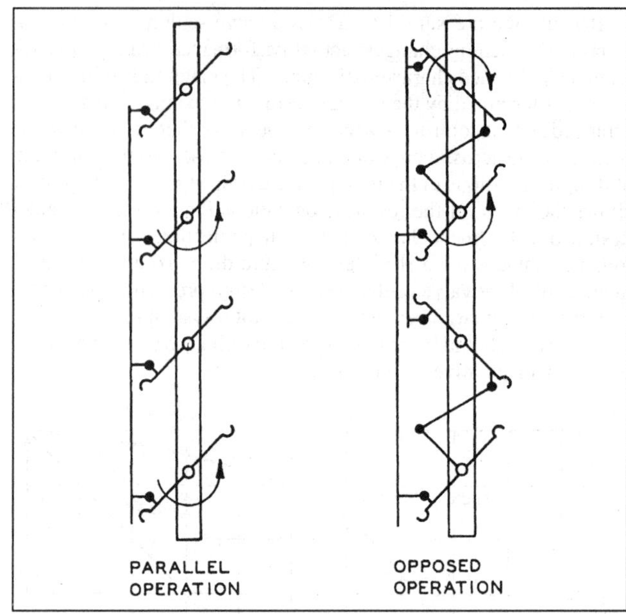

Fig. 13 Typical Multiblade Dampers

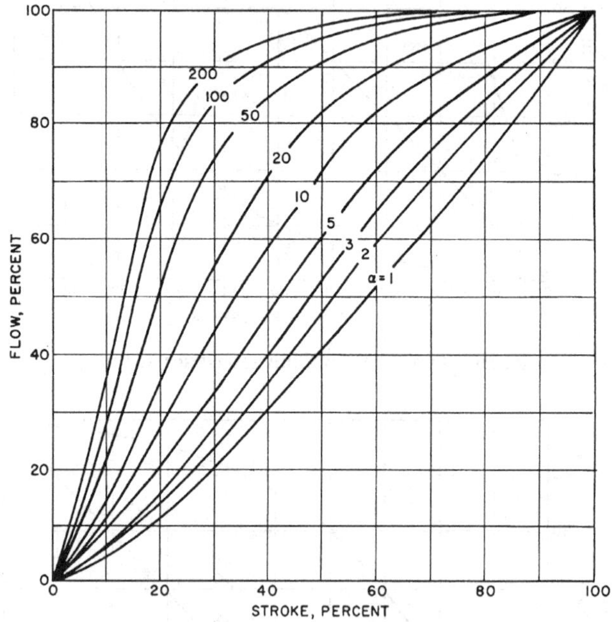

Fig. 14 Characteristic Curves of Installed Parallel Blade Dampers

- **Pneumatic damper operators** are similar to pneumatic valve operators, except that they have a longer stroke or the stroke is increased by a multiplying lever. Increasing the air pressure produces a linear motion of the shaft, which, through a linkage, moves the crank arm to open or close the dampers.
- **Electric damper operators** (actuators) can be either unidirectional, spring-return, or reversible. A reversible operator, which has two sets of motor windings, is frequently used for accurate control in modulating damper applications. Energizing one set of windings turns the actuator output shaft clockwise; energizing the other turns the shaft counter-clockwise.

When neither set of windings is energized, the shaft remains in its last position. The simplest form of control for this actuator is a floating point controller, which causes a contact closure to drive

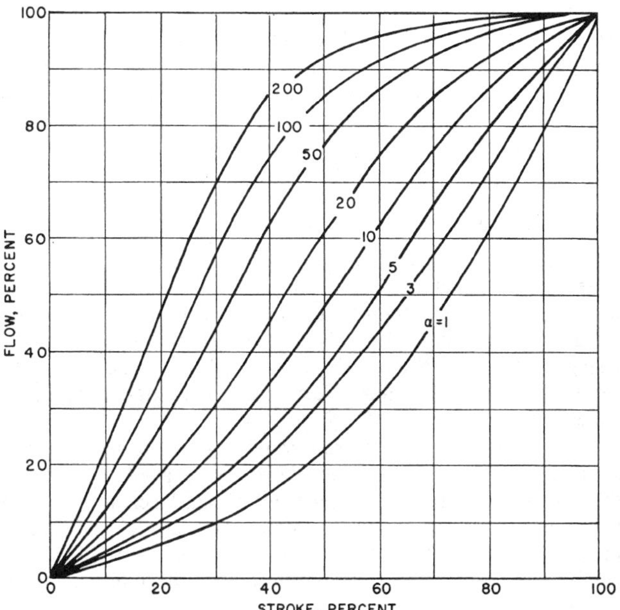

Fig. 15 Characteristic Curves of Installed Opposed Blade Dampers

the motor clockwise and counter-clockwise. This type of actuator is available with a wide range of options for rotational shaft travel (expressed in degrees of rotation) and timing (expressed in the number of seconds to move through the rotational range). In addition, a variety of standard electronics signals from electronic controllers, such as 4–20 mA (dc) or 0–10 V (dc), can be used to control this type of modulating actuator.

A two-position spring return actuator moves in one direction when power is applied to its internal windings; and, when no power is present, the actuator returns (via spring force) to its normal position. Depending on how the actuator is connected to the dampers, this action opens or closes the dampers. A modulating actuator may also have spring return action.

Mounting. Damper operators are mounted in different ways, depending on the damper size, accessibility, and power required to move the dampers. They can be mounted in the airflow on the damper frame and be linked directly to a damper blade; or they can be mounted outside the duct and connected to a crank arm attached to a shaft extension of one of the blades. On large dampers, two or more operators may be needed. In this case, they are usually mounted at separate points on the damper. An alternative is to install the damper in two or more sections, each section being controlled by a single damper operator; however, proper flow control is easier with a single modulating damper. Positive positioners may be required for proper sequencing. A small damper with a two-position, spring return operator may be used for minimum outside flow, with a large damper being independently controlled for economy cycle cooling.

Positive Positioner

A pneumatic operator may not respond quickly or accurately enough to small changes in control pressure due to friction in the actuator or load, or to changing load conditions such as wind acting on a damper blade. Where accurate positioning of a modulating damper or valve in response to load is required, positive positioners should be used. A positive positioner provides up to full main control air pressure to the actuator for any change in position required by the controller.

A positive positioner provides finite and repeatable positioning change and permits adjustment of the control range (spring range of the actuator) to provide a proper sequencing control of two or more controlled devices.

SENSORS

A sensor is a device that responds to a change in the controlled variable (such as the temperature of water flowing in a pipe). The response, which is a change in some physical or electrical property of the primary sensing element, is available for translation or amplification by mechanical or electrical signal. All sensors respond to changes in the control variable to create a signal.

Chapter 42 in the 1995 *ASHRAE Handbook—Applications* and manufacturer's catalogs and tutorials include information on specific applications. In selecting a sensor for a specific application, the following elements should be considered:

- **Operating range of controlled variable.** The sensor must be capable of providing an adequate change in its output signal over the expected output range.
- **Compatibility of controller input.** Electronic and digital controllers accept various ranges and types of electronic signals. The specific controller to be used must be considered in the selection of an electronic sensor; if this is not known, an industry standard signal, such as 4–20 mA (dc) or 0–10 V (dc), should be used.
- **Accuracy and repeatability.** For some control applications, the controlled variable must be maintained within a narrow band around a desired setpoint. Both the accuracy and the sensitivity of the sensor selected must reflect this requirement. However, an accurate sensor alone can not maintain the setpoint if, (1) the controller is unable to resolve the input signal, (2) the controlled device can not be positioned accurately, (3) the controlled device exhibits excessive hysteresis, or (4) disturbances drive its system faster than the controls can regulate it.
- **System response time (or process dynamics).** Associated with a sensor/transducer arrangement is a response curve, which describes the response of the sensor output to change in the controlled variable. If the time constant of the process being controlled is short and stable accurate control is important, the sensor selected must have a fast response time.
- **Control agent properties and characteristics.** The control agent is the medium to which the sensor is exposed, or with which it comes in contact, for measuring a variable such as temperature or pressure. If the agent acts on the sensor so as to corrode or otherwise degrade its performance, different sensor should be selected, or the sensor must be isolated or protected from direct contact with the control agent.
- **Ambient environment characteristics.** Even when the sensor's components are isolated from direct contact with the controlled agent, the ambient environment must be considered. The temperature and humidity range of the ambient environment must not adversely affect the sensor of its accuracy. Likewise the presence of certain gases, chemicals, and electromagnetic interference (EMI) can cause component degradation. In such cases, a special sensor or transducer housing can be used to protect the element, while ensuring a true indication of the controlled variable.

Temperature Sensors

Temperature-sensing elements fall into three general categories: (1) those that use a change in relative dimension due to differences in thermal expansion, (2) those that use a change in state of a vapor or liquid-filled bellows, and (3) those that use a change in some electrical property. Within each category, there are a variety of sensing elements to measure room, duct, water, and surface temperatures. The specific temperature-sensing technologies commonly used in HVAC applications are as followed:

- A **bimetal element** is composed of two thin strips of dissimilar metals fused together. Because the two metals have different coefficients of thermal expansion, the element bends as the temperature varies and produces a change in position. Depending on the space available and the movement required, it may be a straight strip, U-shaped, or wound into a spiral. This element is commonly used in room, insertion, and immersion thermostats.
- A **rod-and-tube element** consists of a high expansion metal tube containing a low expansion rod. One end of the rod is attached to the rear of the tube. The tube changes length with changes in temperature, causing the free end of the rod to move. This element is commonly used in certain insertion and immersion thermostats.
- A **sealed bellows element** is either vapor, gas, or liquid filled. Temperature changes vary the pressure and volume of the gas or liquid, resulting in a change in force or movement.
- A **remote bulb element** is a bulb or capsule that is connected to a sealed bellows or diaphragm by a capillary tube; the entire system is filled with vapor, gas or liquid. Temperature changes at the bulb cause volume or pressure changes that are conveyed to the bellows or diaphragm through the capillary tube. The remote bulb element is useful where the temperature measuring point is remote from the desired thermostat location.
- A **thermistor** is a semiconductor that changes electrical resistance with temperature. It has a negative temperature coefficient; i.e., the resistance decreases as the temperature increases. Its characteristic curve of temperature versus resistance is nonlinear over a wide range. Several techniques are used to convert its response to a linear change over a particular temperature range. With digital control, one technique is to store a computer "look-up table" that maps temperature corresponding to the measured resistance. The table breaks the curve into small segments, and each segment is assumed to be linear over its range. Thermistors are used because of their relatively low cost and the large charge in resistance possible for a small change in temperature.
- A **resistance temperature device (RTD)** is another sensor that changes resistance with temperature. Most metallic materials increase in resistance with increasing temperature. Over limited ranges, this variation is linear for certain metals such as platinum, copper, tungsten, and nickel/iron alloys. Platinum, for example, is linear within ±0.3% from 0 to 300°F. The RTD sensing element is available in several forms for surface or immersion mounting. Flat grid windings are used for measurements of surface temperatures. For direct measurement of fluid temperatures, the windings are encased in a stainless steel bulb to protect them from corrosion.

Humidity Sensors

Humidity sensors or hygrometers are used to measure relative humidity or dew point of ambient or moving air. Materials that respond directly to atmospheric moisture detect relative humidity directly. Two basic types are available for use in automatic control systems—mechanical hygrometers and electronic hygrometers.

A **mechanical hygrometer** operates on the principle that a hygroscopic material, usually a moisture sensitive nylon or bulk polymer material, when exposed to water vapor, retains moisture and expands. The change in size or form is detected by a mechanical linkage and converted to a pneumatic or electronic signal. Mechanical sensors using hair, wood, paper, or cotton do not match the performance of moisture sensitive nylon or bulk polymer sensors and are not widely used.

Electronic hygrometers can use either resistance or capacitance sensing elements. The resistance element is a conductive grid coated with a hygroscopic (water absorbent) substance. The conductivity of the grid varies with the water retained; thus the resistance varies according to the relative humidity. The conductive element is arranged in an alternating current excited Wheatstone bridge and responds rapidly to humidity changes.

The capacitance element is a stretched membrane of non-conductive film. It is coated on both sides with metal electrodes and mounted within a perforated plastic capsule. The change of the sensor's capacity versus relative humidity is nonlinear with respect to rising relative humidity. The signal is linearized and temperature is compensated in the amplifier circuit to provide an output signal as the relative humidity changes from 0 to 100%.

Pressure Transmitters and Transducers

A pneumatic pressure transmitter converts a change in absolute, gage, or differential pressure to a mechanical motion using a bellows, diaphragm, or Bourdon tube mechanism. When corrected through appropriate links, this mechanical motion produces a change in the air pressure to a controller. In some instances, the sensing and control functions are combined in a single component, a pressure controller.

An electronic pressure transducer may use the mechanical actuation of a diaphragm or Bourdon tube to operate a potentiometer or differential transformer. Another type of transducer uses a strain gage bonded to a diaphragm. The strain gage detects the displacement resulting from the force applied to the diaphragm. Electronic circuits provide temperature compensation and amplification to produce a standard output signal.

Flow Rate Sensors

Orifice plate, pitot tube, venturi, turbine, magnetic flow, vortex shedding, and Doppler effect meters are used to sense fluid flow. In general, the pressure differential devices (orifice plates, venturi and pitot tunes) are less expensive and simpler to use but have limited range; thus their accuracy depends on how they are applied and where they are located in a system.

More sophisticated flow devices, such as turbine, magnetic and vortex shedding meters, usually have better range and are more accurate over a wide range. If an existing piping system is being considered for retrofit with a flow device, the expense of shutting down the system and cutting into a pipe must be considered. In this case, a noninvasive meter, such as a Doppler effect meter, can be cost effective.

Indoor Air Quality Sensors

Indoor air quality control can be categorized into two divisions—ventilation control and contamination protection. Ventilation control measures levels of carbon dioxide or other contaminants in a space and controls the amount of outdoor air introduced into the occupied space. The demand control of ventilation helps to maintain proper ventilation rates under all levels of occupancy. Typical control setpoint levels of carbon dioxide are 800 to 1000 ppm. *ASHRAE Standard* 62 provides further information on ventilation for acceptable indoor air quality.

Contamination protection sensors monitor levels of hazardous or toxic substances and either issue warning signals and/or initiate corrective actions through the building automation system. An example of this type of sensor is the carbon monoxide sensor used to control and alarm CO levels in parking garages. Also oxygen depletion sensors used to measure, alarm and initiate ventilation purging in enclosed spaces which house refrigeration equipment to prevent suffocation of occupants upon a refrigeration leak. The application of these sensors determines the type selected, the substances monitored, and the action taken.

Lighting Level Sensors

Analog lighting level transmitters packaged in various configurations allow control of ambient lighting levels using building automation strategies for energy conservation. Some examples include ceiling mounted, indoor light sensors used to measure room lighting levels; outdoor ambient lighting sensors used to control parking, general exterior, security, and sign lighting; and interior skylight sensors used to monitor and control light levels in skylight wells and other atrium spaces.

Power Sensing and Transmission

Passive electronic devices that sense the magnetic field around a conductor carrying current allow low cost instrumentation of power circuits. A wire in the sensor forms an inductive coupling that powers the internal function and senses the level of the power signal. These devices can provide an analog output signal to monitor current flow or operate a switch at a user set level to turn on an alarm or other device.

CONTROLLERS

A controller compares the sensor's signal with a desired setpoint and regulates an output signal to a controlled device. Digital controllers perform the control function using a microprocessor and control algorithm. The controller and sensor can be combined in a single instrument, such as a room thermostat, or they may be two separate devices.

Pneumatic Receiver Controllers

Pneumatic receiver controllers are normally combined with pneumatic elements that use a force or position reaction to the sensed variable to obtain a variable output air pressure. The control mode is usually proportional, but other modes such as proportional-integral can be used. These controllers are generally classified as non-relay, relay direct, or reverse-acting.

The non-relay pneumatic controller uses low volume output. A relay-type pneumatic controller actuates a relay device that amplifies the air volume available for control. The relay provides quicker response to a variable change.

Controllers are further classified by construction as direct or reverse acting. Direct-acting controllers increase the output signal as the controlled variable increases. Reverse-acting controllers increase the output signal as the controlled variable decreases. A reverse acting thermostat increases output pressure when the temperature drops.

Electric/Electronic Controllers

For two-position control, the controller output may be a simple electrical contact that starts a burner or pump, or one that actuates a spring-return valve or damper operator. Single pole, double-throw (SPDT) switching circuits are used to control a three-wire unidirectional motor operator. SPDT circuits are also used for heating and cooling applications. Either single-pole, single-throw (SPST), or SPDT circuits can be modified for timed, two-position action.

Output for floating control is a SPDT switching circuit with a neutral zone where neither contact is made. This control is used with reversible motors. This control action has slow response and a wide throttling range.

Pulse modulation control is an improvement over floating control. It provides closer control by varying the duration of the contact closure. As the actual condition moves closer to the setpoint, the pulse duration shortens for closer control. As the actual condition moves farther from the setpoint, the pulse duration lengthens.

Proportional control gives continuous or incremental changes in output signal to position an electrical actuator or controlled device.

Digital Controllers

A digital controller uses a microprocessor to implement control algorithms on one or multiple control loops. It is fundamentally different from pneumatic or electronic controllers in that the control algorithm is stored as a set of program instructions in memory (software or firmware). the controller itself calculates the proper control signals digitally rather than using an analog circuit or mechanical change.

A digital controller can be either a single- or multi-loop controller. Interface hardware allows the digital computer to process signals from various input devices such as electronic temperature, humidity, or pressure sensors described in the section on sensors. Based on digitized equivalents of the voltage or current signals produced by the inputs, the control software calculates the required state of the output devices, such as valve and damper actuators and fan starters. The output devices are then positioned to the calculated state via interface hardware which converts the digital signal from the computer to an analog voltage or current required to position the actuator or energize a relay.

The operator enters parameters such as setpoints, proportional or integral gains, minimum on and off times, or high and low limits. The control algorithms stored in the computer's memory in conjunction with actual input values, make the control decisions. The computer scans the input devices, executes the control algorithms, and then positions the output device(s), in a stepwise scheme. Digital controllers can be classified with regard to the way control algorithms are stored in memory (such as in firmware and software) and their ability to communicate to higher level devices such as terminals and computers.

Firmware and Software. Preprogrammed control routines, known as firmware, are typically stored in permanent memory such as EEPROM (electronically erasable programmable read only memory). To prevent unauthorized alteration, the operator can modify parameters such as setpoints, limits, and minimum off times within the control routines, but the program logic cannot be changed without replacing the memory chips.

User-programmable controllers allow the algorithms to be changed by the user. The programming language, provided with the controller, can vary from a derivation of a standard language (such as Pascal or Basic) to a custom language developed by the controller's manufacturer to graphically based programming. Preprogrammed routines for proportional, proportional plus integral, Boolean logic, timers, and so forth, are typically included in the language. Standard energy management routines may also be preprogrammed and may interact with other control loops where appropriate.

Digital controllers can be furnished with both preprogrammed firmware and user programmed routines. These routines can automatically modify the parameters of the firmware according to user defined conditions to accomplish the sequence of control designed by the control engineer.

Operator Interface. Some controllers are designed for dedicated purposes and are adjustable only through manual switches and potentiometers mounted on the controller. This type of controller cannot be networked with other controllers. An example of this controller is the programmable room thermostat. A direct digital controller can have manual adjustable features; however, it more typically is adjusted through a built-in LED display, hand held device, or through a terminal or computer. The DDC controller has the ability to communicate digitally, which allows remote connection to other controllers and to higher level computing devices and host operating stations.

A terminal allows the user to communicate with and, where applicable, modify the program in the controller. These terminals can range from hand held units with an LCD display and several buttons to a full size console with a video monitor and keyboard. The terminal can be limited in function to allow only the display of sensor and parameter values, or powerful enough to allow changing or reprogramming the control strategies. In some instances, a terminal can communicate remotely with one or more controllers, thus allowing central displays, alarms, and commands. Usually, hand held terminals are used by technicians for troubleshooting and full consoles located at a fixed location are used to monitor the entire digital control system.

Thermostats

Thermostats combine control and sensing functions in a single device. Microprocessor based thermostats have many of the features described in the following paragraphs.

- The **occupied-unoccupied** or **dual-temperature room thermostat** controls at a reduced temperature at night. It may be indexed (changed from occupied to unoccupied) individually or in groups by a manual switch or time switch from a remote point. Some electric units have an individual clock and switch built into the thermostat.
- The **pneumatic day-night thermostat** uses a two-pressure air supply system--the two pressures often being 13 and 17 psig, or 15 and 20 psig. Changing the pressure at a central point from one value to the other actuates switching devices in the thermostat and indexes it from occupied to unoccupied or vice versa.
- The **heating-cooling** or **summer-winter thermostat** can have its action reversed and its setpoint changed by indexing. It is used to actuate controlled devices, such as valves or dampers, that regulate a heating source at one time and a cooling source at another.
- **Multistage thermostats** are arranged to operate two or more successive steps in sequence.
- A **submaster thermostat** has its setpoint raised or lowered over a predetermined range in accordance with variations in output from a master controller. The master controller can be a thermostat, manual switch, pressure controller, or similar device.

A **dead band thermostat** has a wide differential over which the thermostat remains neutral, requiring neither heating nor cooling. This differential may be adjustable up to 10°F. the thermostat then controls to maximum or minimum output over a small differential at the end of each dead bank (Figure 16).

AUXILIARY CONTROL DEVICES

Auxiliary control devices for electrical systems include:

- **Transformers** to provide current at the required voltage.
- **Occupancy sensors** to automatically adjust controlled variables based on occupancy such as lighting, ventilation rate, and temperature.
- **Signal transducers** change one standard signal into another. The popularity of digital control and other electric based control systems has generated a variety of transducers. The variables usually transformed include voltage [0–10, 0–5, 2–10 V (dc)], milliamperes [4–20 mA], resistance [0–135 Ω], pressure (3–15, 0–30 psig), phase cut voltage [0–20 V (dc)], pulse-width modulation, and time duration pulse. These transducers are useful in a retrofit application to allow the use of an existing control device.

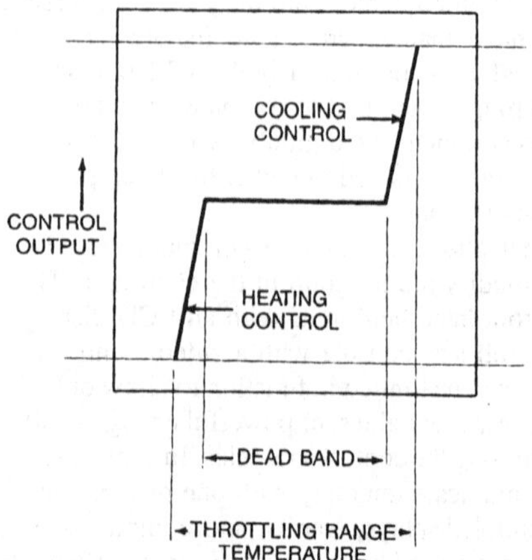

Fig. 16 Dead Band Thermostat

- **Electric relays** to control electric heaters or to start and stop burners, compressors, fans, pumps, or other apparatus for which the electrical load is too large to be handled directly by the controller. Other uses include time-delay and circuit-interlocking safety applications.
- **Potentiometers** for manual positioning of proportional control devices, for remote setpoint adjustment of electronic controllers, and for feedback.
- **Manual switches** for several operations. These can be two-position or multiple-position with single or multiple poles.
- **Auxiliary switches** on valve and damper operators for selecting a sequence of operation.

Auxiliary control devices for pneumatic systems include:

- **Air compressors** and accessories, including dryers and filters, to provide a source of clean, dry air at the required pressure.
- **Electro-pneumatic relays**, electrically actuated air valves for operating pneumatic equipment in accordance with variations in electrical input to the relay.
- **Pneumatic-electric switches** actuated by the pressure from a controller to perform several functions. They may be divided into two groups:

  1. **Two-position relays**, which permit a controller actuating a proportional device to also actuate one or more two-position devices.
  2. **Proportional relays**, which are used to reverse the action of a proportional controller, select the higher or lower of two or more pressures, average two or more pressures, respond to the difference between two pressures, add or subtract pressures, and amplify or retard pressure changes.

- **Positive positioning relays** to ensure accurate positioning of a valve or damper operator in response to changes in pressure from a controller.
- **Switching relays**, which are pneumatically operated air valves used to divert air from one circuit to another or to open and close air circuits.
- **Pneumatic switches**, which are manually operated devices used to divert air from one circuit to another or to open and close air circuits. They can be two-position or multiple- position.
- **Gradual switches** are proportional devices used to manually vary air pressure in a circuit.

Auxiliary control devices common to both electric and pneumatic systems include the following:

- **Step controllers** to operate several switches in sequence by means of a proportional electric or pneumatic operator. They are commonly used to control several steps of refrigeration capacity. They may be arranged to prevent simultaneous starting of compressors and to alternate the sequence to equalize wear. These controllers may also be used for sequence operation of electric heating elements and other equipment.
- **Power controllers** to control electric power input to resistance heating elements. The final controlled device may be a variable auto-transformer, a saturable-core reactor, or a solid state power controller. They are available with various ratings for single or three-phase heater loads, and they are usually arranged to regulate power input to the heater in response to the demands of the proportional electronic or pneumatic controllers. However, solid-state controllers may also be used in two-position control modes.
- **Clocks** or **timers** to turn apparatus on and off at predetermined times, to switch control systems from day to night operation, and for other time sequence functions.
- **Transducers**, which consist of combinations of electric or pneumatic control devices, may be required. For these applications, transducers are used to convert electric signals to pneumatic output

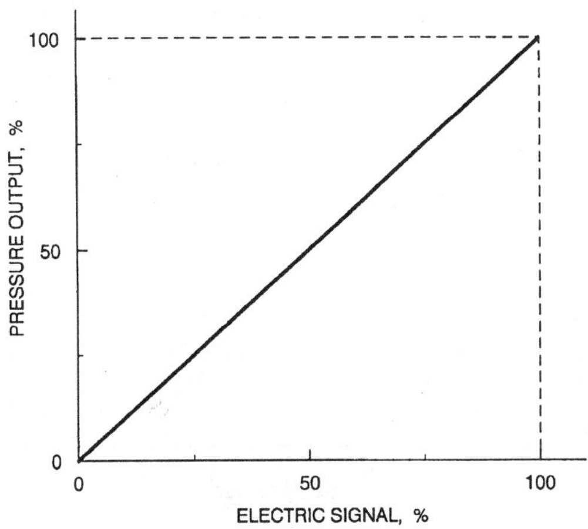

Fig. 17 Response of Electronic-to-Pneumatic (E/P) Transducer

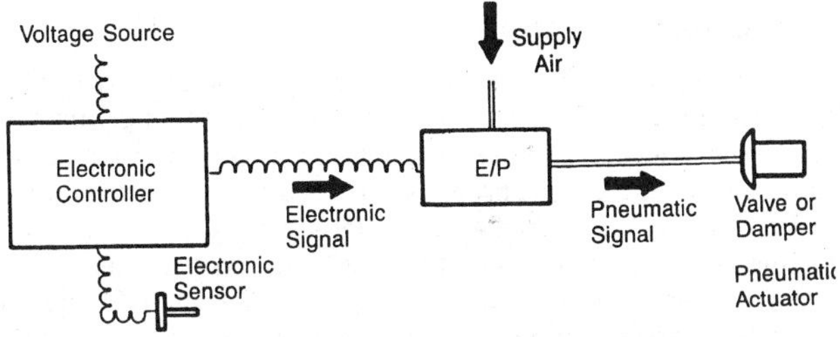

Fig. 18 Example of Electronic and Pneumatic Control Components Combined with Electronic-to-Pneumatic (E/P) Transducer

or vice versa. Transducers may convert proportional input to either proportional or two- position output.

The electronic-to-pneumatic (E/P) transducer is used in many applications. It converts a proportional electronic output signal into a proportional pneumatic signal (as illustrated in Figure 17) and can be used to combine electronic and pneumatic control components to form a control loop, as illustrated in Figure 18. Electronic components are used for sensing and signal conditioning, while pneumatics are used for actuation. The electronic controller can be either analog or digital

The E/P transducer presents a special option for retrofit applications. An existing HVAC system with pneumatic controls can be retrofit with electronic sensors and controllers while retaining the existing pneumatic actuators (Figure 19).

COMMISSIONING

A successful control system requires a proper start-up and testing (commissioning), not merely the adjustment of a few parameters (setpoints and throttling ranges) and a few quick checks. Using the services of an experienced control professional, the typical DDC control system can be used effectively to commission and document the HVAC mechanical system. In general, the increased use of VAV systems and digital controls have increased the importance and need of commissioning.

Design and construction specifications should include specific commissioning procedures. In addition, commissioning should be coordinated with testing and balancing, because each affects the other. The commissioning procedure begins by checking each control device to see that it is installed and connected according to approved drawings. Each electrical and pneumatic connection is

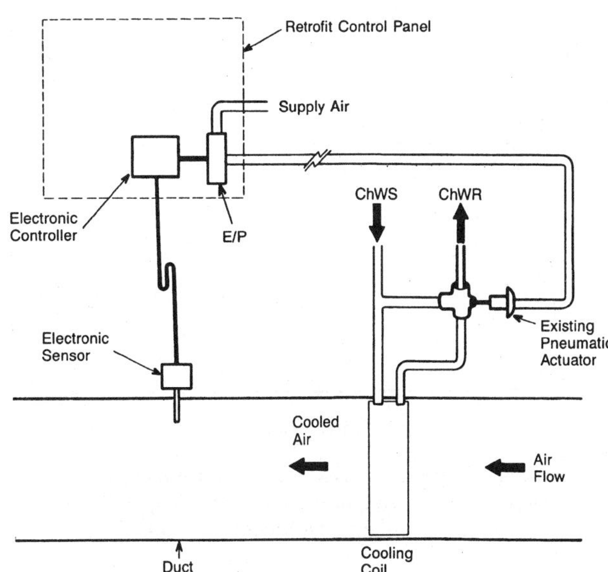

Fig. 19 Retrofit of Existing Pneumatic Control with Electronic Sensors and Controllers

verified, and all interlocks to fan and pump motors and primary heating and cooling equipment are checked. *ASHRAE Guideline* 1-1996, The HVAC Commissioning Process, has further details.

TUNING

The systematic tuning of controllers improves the performance of all controls and is particularly important for digital control. First, the controlled process should be controlled manually between various setpoints to evaluate the following questions:

- Is the process noisy (rapid fluctuations in controlled variable)?
- Is there appreciable hysteresis (backlash) in the actuator?
- How easy (or difficult) is it to maintain and change setpoint?
- In which operating region is the process most sensitive (highest gain)?

If the process cannot be controlled manually, the reason should be identified and corrected before tuning the controller.

Tuning selects control parameters that determine the steady-state and transient characteristics of the control system. HVAC processes are nonlinear, and characteristics change on a seasonal basis. Controllers tuned under one operating condition may become unstable as conditions change. A well- tuned controller will (1) minimize the steady-state error for setpoint, (2) respond quickly to disturbances, and (3) remain stable under all operating conditions. Tuning of proportional controllers is a compromise between minimizing steady-state error and maintaining margins of stability. Proportional plus integral (PI) control minimizes this compromise because the integral action reduces steady-state error, while the proportional term determines the controller's response to disturbances.

Tuning PI Controllers

Popular methods of determining PI controller tuning parameters include closed- and open-loop process identification methods and trial-and-error methods. The closed-loop method increases the gain of the controller in a proportional-only mode until the equipment continuously cycles after a setpoint change (Figure 20, where $K_p = 40$). Proportional and integral terms are then computed from the cycle's period of oscillation and the proportional gain value that caused cycling.

The open-loop method introduces a step change in input into the opened control loop. A graphical technique is used to estimate the

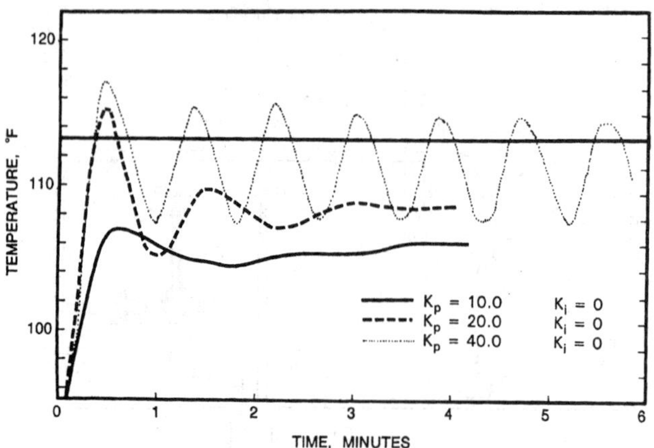

Fig. 20 Response of Discharge Air Temperature to Step Change in Setpoints at Various Proportional Constant and No Integral Action

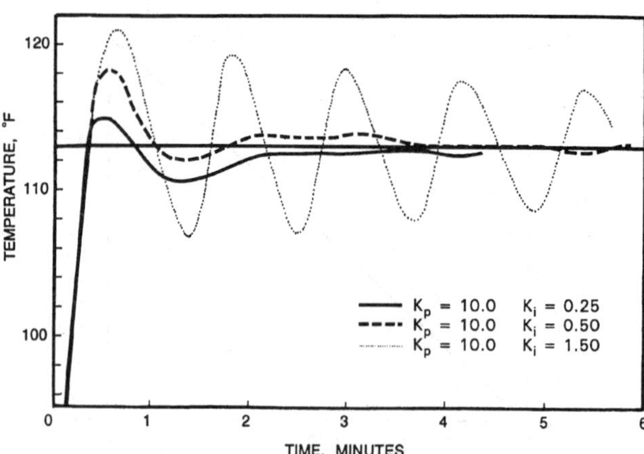

Fig. 21 Response of Discharge Air Temperature to Step Change in Setpoints at Various Integral Constant with Fixed Proportional Constant

process transfer function parameters. Proportional and integral terms are calculated from the estimated process parameters using a series of equations.

The trial-and-error method involves adjusting the gain of the proportion-only controller until the desired response to a setpoint is observed. Conservative tuning dictates that this response should have a small initial overshoot and quickly damp to steady-state conditions. Setpoint changes should be made in the range where controller saturation, or output limit, is avoided. The integral term is then increased until changes in setpoint produce the same dynamic response as the controller under proportional control, but with the response now centered about the setpoint (Figure 21).

Tuning Digital Controls

In tuning digital controllers, additional parameters may need to be specified. The selection of a digital controller sampling interval is a critical selection. The sampling rate can introduce harmonic distortion if it is not selected properly. This is usually set at the factory and may not be adjustable. A controller sampling interval at about one-half of the time constant of the controlled process usually provides adequate control. Many digital control algorithms include an error dead band to eliminate unnecessary control actions when the process is near setpoint. Hysteresis compensation is possible with digital controllers, but it must be carefully applied as over-compensation can cause continuous cycling of the control loop.

CHAPTER 38

CODES AND STANDARDS

THE Codes and Standards listed in Table 1 represent practices, methods, or standards published by the organizations indicated. They are valuable guides for the practicing engineer in determining test methods, ratings, performance requirements, and limits applying to the equipment used in heating, refrigerating, ventilating, and air conditioning. *Copies of the publications can usually be obtained from the organizations listed in the Publisher column.* Addresses of the organizations are given at the end of the chapter.

Table 1 Codes and Standards Published by Various Societies and Associations

Subject	Title	Publisher	Reference
Air Conditioners	Installation Techniques for Perimeter Heating and Cooling, 11th ed.	ACCA	ACCA Manual 4
	Methods of Testing for Rating Ducted Air Terminal Units	ASHRAE	ANSI/ASHRAE 130-1996
	Ducted Air-Conditioners and Air-to-Air Heat Pumps—Testing and Rating for Performance	ISO	ISO 13253:1995
	Non-Ducted Air Conditioners and Heat Pumps—Testing and Rating for Performance	ISO	ISO 5151:1994
	Heating and Cooling Equipment (1995)	UL	UL 1995
		CSA	CAN/CSA-C22.2 No. 236-95
Central	Performance Standard for Split-System Central Air-Conditioners and Heat Pumps	CSA	CAN/CSA-C273.3-M91
	Performance Standard for Single Package Central Air-Conditioners and Heat Pumps	CSA	CAN/CSA-C656-M92
	Performance Standard for Rating Large Air Conditioners and Heat Pumps	CSA	CAN/CSA-C746-93
	Air Conditioners, Central Cooling (1982)	UL	ANSI/UL 465-1984
Gas-Fired	Gas-Fired Absorption Summer Air Conditioning Appliances (with 1982 addenda)	AGA	ANSI Z21.40.1-1994
	Requirements for Gas-Fired, Engine-Driven Air Conditioning Appliances	AGA	4-89
	Requirements for Gas-Fired Desiccant Type Dehumidifiers and Air Conditioners	AGA	9-90
Packaged Terminal	Packaged Terminal Air Conditioners	ARI	ARI 310-93
	Packaged Terminal Heat Pumps	ARI	ANSI/ARI 380-90
	Standards for Packaged Terminal Air-Conditioners and Heat Pumps	CSA	C744-93
Room	Room Air Conditioners	AHAM	ANSI/AHAM RA C-1-1992
	Method of Testing for Rating Room Air Conditioners and Packaged Terminal Air Conditioners	ASHRAE	ANSI/ASHRAE 16-1983 (RA 88)
	Method of Testing for Rating Room Air Conditioner and Packaged Terminal Air Conditioner Heating Capacity	ASHRAE	ANSI/ASHRAE 58-1986 (RA 90)
	Methods of Testing for Rating Room Fan-Coil Air Conditioners	ASHRAE	ANSI/ASHRAE 79-1984 (RA 91)
	Performance Standard for Room Air Conditioners	CSA	CAN/CSA-C368.1-M90
	Room Air Conditioners	CSA	C22.2 No. 117-1970 (R 1992)
	Room Air Conditioners (1993)	UL	ANSI/UL 484
Unitary	Commercial Load Calculation, 4th ed. (1988)	ACCA	ACCA Manual N
	Application of Sound Rated Outdoor Unitary Equipment	ARI	ARI 275-84
	Commercial and Industrial Unitary Air-Conditioning Equipment	ARI	ANSI/ARI 360-93
	Sound Rating of Outdoor Unitary Equipment	ARI	ARI 270-84
	Unitary Air-Conditioning and Air-Source Heat Pump Equipment	ARI	ANSI/ARI 210/240-89
	Method of Rating Computer and Data Processing Room Unitary Air Conditioners	ASHRAE	ANSI/ASHRAE 127-1988
	Method of Rating Unitary Spot Air Conditioners	ASHRAE	ANSI/ASHRAE 128-1989
	Methods of Testing for Rating Heat Operated Unitary Air-Conditioning Equipment for Cooling	ASHRAE	ANSI/ASHRAE 40-1986 (RA 92)
	Methods of Testing for Rating Unitary Air-Conditioning and Heat Pump Equipment	ASHRAE	ANSI/ASHRAE 37-1988
	Methods of Testing for Seasonal Efficiency of Unitary Air Conditioners and Heat Pumps	ASHRAE	ANSI/ASHRAE 116-1995
Air Conditioning	Commercial Low Pressure, Low Velocity Duct System Design (1990)	ACCA	Manual Q
	Residential Duct Systems (1995)	ACCA	ACCA Manual D-R
	Residential Load Calculation, 7th ed. (1986)	ACCA	ACCA Manual J
	Environmental System Technology (1984)	NEBB	NEBB
	Installation of Air Conditioning and Ventilating Systems	NFPA	ANSI/NFPA 90A-1993
	Standard of Purity for Use in Mobile Air-Conditioning Systems	SAE	SAE J 1991-1989
	HVAC Systems—Applications, 1st ed. (1986)	SMACNA	SMACNA
	HVAC Systems—Duct Design (1990)	SMACNA	SMACNA
	Installation Standards for Residential Heating and Air Conditioning Systems (1988)	SMACNA	SMACNA
	Heating and Cooling Equipment (1995)	UL	UL 1995
		CSA	CAN/CSA-22.2 No. 36-95
Aircraft	Air Conditioning of Aircraft Cargo	SAE	SAE AIR 806A-1978 (R 1992)
	Air Conditioning of Subsonic Aircraft at High Altitudes	SAE	SAE AIR 795
	Air Conditioning Systems for Subsonic Airplanes	SAE	ANSI/SAE ARP 85E-1991

Table 1 Codes and Standards Published by Various Societies and Associations (*Continued*)

Subject	Title	Publisher	Reference
Aircraft (*continued*)	Aircraft Fuel Weight Penalty Due to Air Conditioning	SAE	SAE AIR 1168/8-1989
	Aircraft Ground Air Conditioning Service Connection	SAE	SAE AS 4262
	Air Cycle Air Conditioning Systems for Military Air Vehicles	SAE	ANSI/SAE ARP 4073-1993
	Control of Excess Humidity in Avionics Cooling	SAE	SAE ARP 987-1970 (R 1992)
	Engine Bleed Air Systems for Aircraft	SAE	ANSI/SAE ARP 1796-1987
	Guide for Qualification Testing of Aircraft Air Valves	SAE	ANSI/SAE ARP 986B-1980
	Nomenclature, Aircraft Air-Conditioning Equipment	SAE	SAE ARP 147C-1978 (R 1992)
	Testing of Commercial Airplane Environmental Control Systems	SAE	SAE ARP 217B-1973 (R 1992)
Automotive	Automotive Air-Conditioning Hose	SAE	ANSI/SAE J 51-1989
	Design Guidelines for Air Conditioning Systems for Off-Road Operator Enclosures	SAE	SAE J 169-1985
	Extraction and Recycle Equipment for Mobile Automotive Air-Conditioning Systems	SAE	SAE J 1990-1992
	Guide to the Application and Use of Passenger Car Air-Conditioning Compressor Face Seals	SAE	SAE J 1954-1990
	Information Relating to Duty Cycles and Average Power Requirements of Truck and Bus Engine Accessories	SAE	SAE J 1343-1988
	Rating Air Conditioner Evaporator Air Delivery and Cooling Capacities	SAE	ANSI/SAE J 1487-1985
	Service Hose for Automotive Air Conditioning	SAE	SAE J 2196-1992
	Test Method for Measuring Power Consumption of Air Conditioning and Brake Compressors for Trucks and Buses	SAE	ANSI/SAE J 1340-1990
Ships	Mechanical Refrigeration and Air-Conditioning Installations Aboard Ship	ASHRAE	ASHRAE 26-1996
Air Curtains	Air Curtains for Entranceways in Food and Food Service Establishments	NSF	ANSI/NSF-37-1992
	Air Distribution Basics for Residential and Small Commercial Buildings (1989)	ACCA	ACCA Manual T
	Residential Equipment Selection, 2nd ed. (1995)	ACCA	ACCA Manual S
	Test Methods for Air Curtain Units	AMCA	AMCA 220-91
	Air Terminals	ARI	ANSI/ARI 880-94
	Method of Testing for Rating the Performance of Air Outlets and Inlets	ASHRAE	ANSI/ASHRAE 70-1991
	Standard Methods for Laboratory Air Flow Measurement	ASHRAE	ANSI/ASHRAE 41.2-1987 (RA 92)
	Rating the Performance of Residential Mechanical Ventilating Equipment	CSA	CAN/CSA-C260-M90
	Direct Gas-Fired Door Heaters	AGA	ANSI Z83.17-1990; Z83.17a-1991; Z83.17b-1992
Air Diffusion	Test Code for Grilles, Registers and Diffusers	ADC	ADC 1062:GRD-84
	Method of Testing for Rating the Performance of Air Outlets and Inlets	ASHRAE	ANSI/ASHRAE 70-1991
	Method of Testing for Room Air Diffusion	ASHRAE	ANSI/ASHRAE 113-1990
Air Filters	Method for Measuring Performance of Portable Household Electrical Cord Connected Room Air Cleaners	AHAM	ANSI/AHAM AC-1-1988
	Commercial and Industrial Air Filter Equipment	ARI	ARI 850-93
	Residential Air Filter Equipment	ARI	ARI 680-93
	Gravimetric and Dust-Spot Procedures for Testing Air-Cleaning Devices Used in General Ventilation for Removing Particulate Matter	ASHRAE	ANSI/ASHRAE 52.1-1992
	Method for Sodium Flame Test for Air Filters	BSI	BS 3928
	Particulate Air Filters for General Ventilation—Requirements, Testing Marking	BSI	BS EN 779:1993
	Electrostatic Air Cleaners (1995)	UL	UL 867
	High-Efficiency, Particulate, Air Filter Units (1990)	UL	ANSI/UL 586-1990
	Test Performance of Air Filter Units (1994)	UL	UL 900
Air-Handling Units	Commercial Low Pressure, Low Velocity Duct System Design (1990)	ACCA	ACCA Manual Q
	Residential Duct Systems (1995)	ACCA	ACCA Manual D-R
	Central Station Air-Handling Units	ARI	ANSI/ARI 430-89
	Direct Gas-Fired Make-Up Air Heaters	AGA	ANSI Z83.4-1991; Z83.4a-1992
Air Leakage	Air Leakage Performance for Detached Single-Family Residential Buildings	ASHRAE	ANSI/ASHRAE 119-1988 (RA 94)
	A Method of Determining Air Change Rates in Detached Dwellings	ASHRAE	ANSI/ASHRAE 136-1993
	Standard Practices for Air Leakage Site Detection in Building Envelopes	ASTM	ASTM E 1186-87 (R 1992)
	Test Method for Determining Air Change in a Single Zone by Means of a Tracer Gas Dilution	ASTM	ASTM E 741-95
	Test Method for Determining Air Leakage Rate by Fan Pressurization	ASTM	ASTM E 779-87 (R 1992)
	Test Method for Determining the Rate of Air Leakage Through Exterior Windows, Curtain Walls, and Doors Under Specified Pressure and Temperature Differences Across the Specimen	ASTM	ASTM E 1424-91
	Test Method for Determining the Rate of Air Leakage Through Exterior Windows, Curtain Walls, and Doors Under Specified Pressure Differences Across the Specimen	ASTM	ASTM E 283-91
	Test Method for Field Measurement of Air Leakage Through Installed Exterior Window and Doors	ASTM	ASTM E 783-93
Boilers	A Guide to Clean and Efficient Operation of Coal Stoker-Fired Boilers	ABMA	ABMA
	Boiler Water Limits and Steam Purity Recommendations for Watertube Boilers	ABMA	ABMA
	Boiler Water Requirements and Associated Steam Purity—Commercial Boilers	ABMA	ABMA

Table 1 Codes and Standards Published by Various Societies and Associations (*Continued*)

Subject	Title	Publisher	Reference
Boilers *(continued)*	Fluidized Bed Combustion Guidelines	ABMA	ABMA
	Guidelines for Industrial Boiler Performance Improvement	ABMA	ABMA
	Matrix of Recommended Quality Control Requirements	ABMA	ABMA
	Operation and Maintenance Safety Manual	ABMA	ABMA
	Recommended Design Guidelines for Stoker Firing of Bituminous Coals	ABMA	ABMA
	(Selected) Summary of Codes and Standards of the Boiler Industry	ABMA	ABMA
	Thermal Shock Damage to Hot Water Boilers as a Result of Energy Conservation Measures	ABMA	ABMA
	Commercial Applications, Systems and Equipment (1993)	ACCA	ACCA Manual CS
	Methods of Testing for Annual Fuel Utilization Efficiency of Residential Central Furnaces and Boilers	ASHRAE	ANSI/ASHRAE 103-1993
	Boiler and Pressure Vessel Code (11 sections) (1995)	ASME	ASME
	Boiler, Pressure Vessel, and Pressure Piping Code	CSA	B51-95
	Testing and Rating Standard for Heating Boilers (1989)	HYDI	IBR
	Prevention of Furnace Explosions/Implosions in Multiple Burner Boilers	NFPA	ANSI /NFPA 8502-1995
	Heating, Water Supply, and Power Boilers—Electric (1995)	UL	UL 834
Gas or Oil	Gas-Fired Low-Pressure Steam and Hot Water Boilers	AGA	ANSI Z21.13-1991; Z21.13a-1993; Z21.13b-1994
	Gas Utilization Equipment in Large Boilers	AGA	ANSI Z83.3-1971; Z83.3a-1972; Z83.3b-1976 (R 1989)
	Requirements for High Pressure Steam Boilers	AGA	3-89
	Control and Safety Devices for Automatically Fired Boilers	ASME	ANSI/ASME CSD.1-1992
	Industrial and Commercial Gas-Fired Package Boilers	CGA	CAN1-3.1-77 (R 1985)
	Oil-Fired Steam and Hot-Water Boilers for Residential Use	CSA	B140.7.1-1976 (R 1991)
	Oil-Fired Steam and Hot-Water Boilers for Commercial and Industrial Use	CSA	B140.7.2-1967 (R 1991)
	Prevention of Furnace Explosions/Implosions in Multiple Burner Boilers	NFPA	ANSI/NFPA 8502-1995
	Single Burner Boiler Operations	NFPA	ANSI/NFPA 8501-1992
	Commercial-Industrial Gas Heating Equipment (1994)	UL	UL 795
	Oil-Fired Boiler Assemblies (1990)	UL	ANSI/UL 726-1990
	Standards and Typical Specifications for Deaerators, 5th ed. (1992)	HEI	HEI
	Method and Procedure for the Determination of Dissolved Oxygen, 2nd ed. (1963)	HEI	HEI
Building Codes	ASTM Standards Used in Building Codes	ASTM	ASTM
	BOCA National Building Code, 12th ed. (1993)	BOCA	BOCA
	BOCA National Property Maintenance Code, 4th ed. (1993)	BOCA	BOCA
	National Building Code of Canada (1995)	NRCC	NRCC
	One- and Two-Family Dwelling Code (1995)	CABO	CABO
	Model Energy Code (1995)	CABO	BOCA/ICBO/SBCCI
	Uniform Building Code (1994)	ICBO	ICBO
	Uniform Building Code Standards (1994)	ICBO	ICBO
	Directory of Building Codes and Regulations, State and City Volumes (1996 ed.)	NCSBCS	NCSBCS
	Standard Building Code (1994)	SBCCI	SBCCI
Mechanical	Safety Code for Elevators and Escalators (plus two yearly supplements)	ASME	ANSI/ASME A 17.1-1993
	Natural Gas Installation Code	CGA	CAN/CGA-B149.1-M95
	Propane Installation Code	CGA	CAN/CGA-B149.2-M91
	Safety Code for Elevators	CSA	CAN/CSA-B44-94
	BOCA National Mechanical Code, 8th ed. (1993)	BOCA	BOCA
	Uniform Mechanical Code (1994) (with Uniform Mechanical Code Standards)	ICBO	ICBO
		IAPMO	IAPMO
	Standard Gas Code (1994)	SBCCI	SBCCI
	Standard Mechanical Code (1994)	SBCCI	SBCCI
Burners	Guidelines for Burner Adjustments of Commercial Oil-Fired Boilers	ABMA	ABMA
	Domestic Gas Conversion Burners	AGA	ANSI Z21.17-1991; Z21.17a-1993; Z21.17b-1994
	Installation of Domestic Gas Conversion Burners	AGA	ANSI Z21.8-1994
	General Requirements for Oil Burning Equipment	CSA	CAN/CSA-B140.0-M87 (R 1991)
	Installation Code for Oil Burning Equipment	CSA	CAN/CSA-B139-M91
	Oil Burners; Atomizing-Type	CSA	CAN/CSA-B140.2.1-M90
	Pressure Atomizing Oil Burner Nozzles	CSA	B140.2.2-1971 (R 1991)
	Replacement Burners and Replacement Combustion Heads for Residential Oil Burners	CSA	B140.2.3-M1981 (R 1991)
	Vapourizing-Type Oil Burners	CSA	B140.1-1966 (R 1991)
	Commercial/Industrial Gas and/or Oil-Burning Assemblies with Emission Reduction Equipment (1993)	UL	UL 2096
	Commercial-Industrial Gas Heating Equipment (1994)	UL	UL 795
	Oil Burners (1994)	UL	ANSI/UL 296-1995
Chillers	Methods of Testing Liquid Chilling Packages	ASHRAE	ASHRAE 30-1995
	Absorption Water-Chilling and Water Heating Packages	ARI	ARI 560-92
	Centrifugal and Rotary Screw Water-Chilling Packages	ARI	ANSI/ARI 550-92
	Positive Displacement Compressor Water-Chilling Packages	ARI	ANSI/ARI 590-92
	Performance Standard for Rating Packaged Water Chillers	CSA	C743-93

Table 1 Codes and Standards Published by Various Societies and Associations (*Continued*)

Subject	Title	Publisher	Reference
Chimneys	Design and Construction of Masonry Chimneys and Fireplaces	CSA	CAN/CSA-A405-M87
	Chimneys, Fireplaces, Vents, and Solid Fuel-Burning Appliances	NFPA	ANSI/NFPA 211-1992
	Chimneys, Factory-Built, Medium Heat Appliance (1994)	UL	UL 959
	Chimneys, Factory-Built, Residential Type and Building Heating Appliance (1994)	UL	ANSI/UL 103-1995
Clean Rooms	Procedural Standards for Certified Testing of Cleanrooms (1988)	NEBB	NEBB
	Standard Practice for Continuous Sizing and Counting of Airborne Particles in Dust-Controlled Areas and Clean Rooms Using Instruments Capable of Detecting Single Sub-Micrometre and Larger Particles	ASTM	ASTM F 50-92 (R 1996)
Coils	Forced-Circulation Air-Cooling and Air-Heating Coils	ARI	ANSI/ARI 410-91
	Methods of Testing Forced Circulation Air Cooling and Air Heating Coils	ASHRAE	ASHRAE 33-1978
Comfort Conditions	Thermal Environmental Conditions for Human Occupancy	ASHRAE	ANSI/ASHRAE 55-1992 with Addendum 55a-1994
	Ergonomics—Determination of Metabolic Heat Production	ISO	ISO 8996:1990
	Ergonomics of the Thermal Environment—Estimation of the Thermal Insulation and Evaporative Resistance of a Clothing Ensemble	ISO	ISO 9920:1995
	Hot Environments—Estimation of the Heat Stress on Working Man, Based on the WBGT Index (Wet Bulb Globe Temperature)	ISO	ISO 7243:1989
	Moderate Thermal Environments—Determination of the PMV and PPD Indices and Specification of the Conditions for Thermal Comfort	ISO	ISO 7730:1994
Compressors	Compressors and Exhausters (reaffirmed 1986)	ASME	ANSI/ASME PTC 10-1965 (R 1992)
	Displacement Compressors, Vacuum Pumps and Blowers	ASME	ANSI/ASME PTC 9-1974 (R 1992)
	Safety Standard for Air Compressor Systems	ASME	ANSI/ASME B19.1-1990
	Safety Standard for Compressors for Process Industries	ASME	ANSI/ASME B19.3-1991
	Compressed Air and Gas Handbook, 5th ed. (1988)	CAGI	CAGI
Refrigerant	Ammonia Compressor Units	ARI	ANSI/ARI 510-93
	Method for Presentation of Compressor Performance Data	ARI	ARI 540-91
	Positive Displacement Refrigerant Compressors, Compressor Units and Condensing Units	ARI	ANSI/ARI 520-90
	Methods of Testing for Rating Positive Displacement Refrigerant Compressors and Condensing Units	ASHRAE	ANSI/ASHRAE 23-1993
	Safety Code for Mechanical Refrigeration	ASHRAE	ANSI/ASHRAE 15-1994
	Hermetic Refrigerant Motor-Compressors (1991)	UL	UL 984
		CSA	CAN/CSA-C22.2 No.140.2-M91
Computers	Method of Rating Computer and Data Processing Room Unitary Air Conditioners	ASHRAE	ANSI/ASHRAE 127-1988
	Protection of Electronic Computer/Data Processing Equipment	NFPA	ANSI/NFPA 75-1995
Condensers	Commercial Applications, Systems and Equipment (1993) (for equipment selection only)	ACCA	ACCA Manual CS
	Remote Mechanical-Draft Air-Cooled Refrigerant Condensers	ARI	ARI 460-87
	Remote Mechanical Draft Evaporative Refrigerant Condensers	ARI	ANSI/ARI 490-89
	Water-Cooled Refrigerant Condensers, Remote Type	ARI	ARI 450-87
	Methods of Testing for Rating Remote Mechanical-Draft Air-Cooled Refrigerant Condensers	ASHRAE	ASHRAE 20-1970
	Methods of Testing Remote Mechanical-Draft Evaporative Refrigerant Condensers	ASHRAE	ANSI/ASHRAE 64-1995
	Methods of Testing for Rating Water-Cooled Refrigerant Condensers	ASHRAE	ANSI/ASHRAE 22-1992
	Safety Code for Mechanical Refrigeration	ASHRAE	ANSI/ASHRAE 15-1994
	Steam Condensing Apparatus	ASME	ANSI/ASME PTC 12.2-1983 (R 1988)
	Standards for Steam Surface Condensers, 9th ed. (1995)	HEI	HEI
	Standards for Direct Contact Barometric and Low Level Condensers, 6th ed. (1995)	HEI	HEI
Condensing Units	Commercial Applications, Systems and Equipment (1993)	ACCA	ACCA Manual CS
	Residential Equipment Selection, 2nd ed. (1995)	ACCA	ACCA Manual S
	Commercial and Industrial Unitary Air-Conditioning Condensing Units	ARI	ANSI/ARI 365-94
	Methods of Testing for Rating Positive Displacement Refrigerant Compressors and Condensing Units	ASHRAE	ANSI/ASHRAE 23-1993
	Heating and Cooling Equipment (1995)	UL	UL 1995
		CSA	CAN/CSA-C22.2 No. 236-M95
	Refrigeration and Air-Conditioning Condensing and Compressor Units (1987)	UL	ANSI/UL303-1988
Contactors	Definite Purpose Contactors for Limited Duty	ARI	ANSI/ARI 790-86
	Definite Purpose Magnetic Contactors	ARI	ANSI/ARI 780-86
Controls	Quick-Disconnect Devices for Use with Gas Fuel	AGA	ANSI Z21.41-1989; Z21.41a-1990; Z21.41b-1992
	Energy Management Control Systems Instrumentation	ASHRAE	ANSI/ASHRAE 114-1986
	BACnet™—A Data Communication Protocol for Building Automation and Control Networks	ASHRAE	ANSI/ASHRAE 135-1995
	Performance Requirements for Electric Heating Line-Voltage Wall Thermostats	CSA	C273.4-M1978 (R 1992)
	Temperature-Indicating and Regulating Equipment	CSA	C22.2 No. 24-93
	Control Centers for Changing Message Type Electric Signals (1991)	UL	ANSI/UL 1433-1995

Table 1 Codes and Standards Published by Various Societies and Associations (*Continued*)

Subject	Title	Publisher	Reference
Controls *(continued)*	Limit Controls (1994)	UL	ANSI/UL 353-1995
	Primary Safety Controls for Gas- and Oil-Fired Appliances (1994)	UL	ANSI/UL 372-1994
	Solid State Controls for Appliances (1994)	UL	ANSI/UL 244A-1995
	Temperature-Indicating and -Regulating Equipment (1994)	UL	UL 873
	Tests for Safety-Related Controls Employing Solid-State Devices (1995)	UL	UL 991
Commercial and Industrial	Industrial Control and Systems General Requirements	NEMA	ANSI/NEMA ICS 1-1993
	Industrial Control and Systems, Controllers, Contactors, and Overload Relays Rated Not More than 2000 Volts AC or 750 Volts DC	NEMA	ANSI/NEMA ICS 2-1993
	Instructions for the Handling, Installation, Operation and Maintenance of Motor Control Centers	NEMA	NEMA ICS 2.3-1983 (R 1990)
	Preventive Maintenance of Industrial Control and Systems Equipment	NEMA	NEMA ICS 1.3-1986
	Industrial Control Equipment (1993)	UL	UL 508
Residential	Automatic Gas Ignition Systems and Components	AGA	ANSI Z21.20-1993; Z21.20a-1994
	Gas Appliance Pressure Regulators	AGA	ANSI Z21.18-1993; Z21.18a-1994
	Gas Appliance Thermostats	AGA	ANSI Z21.23-1993; Z21.23a-1994
	Manually Operated Gas Valves for Appliances, Appliance Connector Valves and Hose End Valves	AGA	ANSI Z21.15-1992
	Manually-Operated Piezo Electric Spark Gas Ignition Systems and Components	AGA	ANSI Z21.77-1989; Z21.77a-1993
	Hot-Water Immersion Controls	NEMA	NEMA DC-12-1985 (R 1991)
	Line-Voltage Integrally Mounted Thermostats for Electric Heaters	NEMA	NEMA DC 13-1991
	Residential Controls—Quick Connect Terminals	NEMA	NEMA DC 2-1982 (R 1988)
	Residential Controls—Electrical Wall-Mounted Room Thermostats	NEMA	NEMA DC 3-1989
	Residential Controls—Surface Type Controls for Electric Storage Water Heaters	NEMA	NEMA DC 5-1989
	Residential Controls—Temperature Limit Controls for Electric Baseboard Heaters	NEMA	NEMA DC 10-1983 (R 1989)
	Residential Controls—Class 2 Transformers	NEMA	NEMA DC 20-1992
	Safety Guidelines for the Application, Installation, and Maintenance of Solid State Controls	NEMA	NEMA ICS 1.1-1984 (R 1988)
	Electrical Quick-Connect Terminals (1995)	UL	UL 310
Coolers Air	Refrigeration Equipment	CSA	CAN/CSA-C22.2 No. 120-M91
	Unit Coolers for Refrigeration	ARI	ANSI/ARI 420-94
	Methods of Testing Forced Convection and Natural Convection Air Coolers for Refrigeration	ASHRAE	ANSI/ASHRAE 25-1990
	Commercial Bulk Milk Dispensing Equipment	NSF	ANSI/NSF 20-1992
Drinking Water	Self-Contained, Mechanically-Refrigerated Drinking-Water Coolers	ARI	ANSI/ARI 1010-94
	Methods of Testing for Rating Drinking-Water Coolers with Self-Contained Mechanical Refrigeration Systems	ASHRAE	ANSI/ASHRAE 18-1987 (RA 91)
	Drinking-Water Coolers (1993)	UL	ANSI/UL 399-1992
Food and Beverage	Methods of Testing and Rating Bottled and Canned Beverage Vendors and Coolers	ASHRAE	ANSI/ASHRAE 32-1986 (RA 90)
	Refrigerated Vending Machines (1995)	UL	UL 541
	Manual Food and Beverage Dispensing Equipment	NSF	ANSI/NSF 18-1990
Liquid	Refrigerant-Cooled Liquid Coolers, Remote Type	ARI	ANSI/ARI 480-87
	Methods of Testing for Rating Liquid Coolers	ASHRAE	ANSI/ASHRAE 24-1989
	Liquid Cooling Systems	SAE	ANSI/SAE AIR 1811-1985 (R 1992)
Cooling Towers	Commercial Applications, Systems and Equipment (1993)	ACCA	ACCA Manual CS
	Atmospheric Water Cooling Equipment	ASME	ANSI/ASME PTC 23-1986 (R 1992)
	Water-Cooling Towers	NFPA	ANSI/NFPA 214-1992
	Acceptance Test Code for Spray Cooling Systems (1985)	CTI	CTI ATC-133
	Acceptance Test Code for Water Cooling Towers: Mechanical Draft, Natural Draft Fan Assisted Types, Evaluation of Results, and Thermal Testing of Wet/Dry Cooling Towers (1990)	CTI	CTI ATC-105
	Certification Standard for Commercial Water Cooling Towers (1991)	CTI	CTI STD-201
	Code for Measurement of Sound from Water Cooling Towers (1981)	CTI	CTI ATC-128
	Fiberglass-Reinforced Plastic Panels for Application on Industrial Water-Cooling Towers (1986)	CTI	CTI STD-131
	Nomenclature for Industrial Water-Cooling Towers (1983)	CTI	CTI NCL-109
	Recommended Practice for Airflow Testing of Cooling Towers (1994)	CTI	CTI PFM-143
Crop Drying	Density, Specific Gravity, and Mass-Moisture Relationships of Grain for Storage	ASAE	ANSI/ASAE D241.4-1993
	Moisture Measurement—Forages	ASAE	ASAE S358.2-1993
	Moisture Measurement—Unground Grain and Seeds	ASAE	ASAE S352.2-1992
	Moisture Relationships of Grains	ASAE	ASAE D245.4-1994
	Resistance to Airflow of Grains, Seeds, Other Agricultural Products, and Perforated Metal Sheets	ASAE	ASAE D272.2-1994
Dehumidifiers	Commercial Applications, Systems and Equipment (1993)	ACCA	ACCA Manual CS
	Dehumidifiers	AHAM	ANSI/AHAM DH-1-1992
	Dehumidifiers	CSA	C22.2 No. 92-1971 (R 1992)
	Dehumidifiers (1993)	UL	ANSI/UL 474-1992
Desiccants	Method of Testing Desiccants for Refrigerant Drying	ASHRAE	ANSI/ASHRAE 35-1992
Dryers	Method of Testing Liquid Line Refrigerant Driers	ASHRAE	ANSI/ASHRAE 63.1-1995
	Liquid-Line Driers	ARI	ANSI/ARI 710-86

Table 1 Codes and Standards Published by Various Societies and Associations (*Continued*)

Subject	Title	Publisher	Reference
Ducts and Fittings	Fibrous Glass Duct Liner Standards (1994)	NAIMA	NAIMA AH 124
	Hose, Air Duct, Flexible Nonmetallic, Aircraft	SAE	SAE AS 1501C-1994
	Ducted Electric Heat Guide for Air Handling Systems (1971)	SMACNA	SMACNA
	Factory-Made Air Ducts and Air Connectors (1994)	UL	UL 181
	Marine Rigid and Flexible Air Ducting (1986)	UL	ANSI/UL 1136-1986
Construction	Preferred Metric Sizes for Flat Metal Products	ASME	ANSI/ASME B32.3M-1984 (R 1994)
	Sheet Metal Welding Code	AWS	ANSI/AWS D90.1-90
	Pipes, Ducts and Fittings for Residential Type Air Conditioning Systems	CSA	B228.1-1968
	Fibrous Glass Duct Construction Standards, 2nd ed. (1993)	NAIMA	NAIMA AH 116
	Fibrous Glass Duct Construction with 1-1/2" Duct Boards (1994)	NAIMA	NAIMA AH 120
	Fibrous Glass Residential Duct Construction Standards (1993)	NAIMA	NAIMA AH 119
	Fibrous Glass Duct Construction Standards, 6th ed. (1992)	SMACNA	SMACNA
	HVAC Duct Construction Standards—Metal and Flexible, 2nd ed. (1995)	SMACNA	SMACNA
	Rectangular Industrial Duct Construction (1980)	SMACNA	SMACNA
	Round Industrial Duct Construction (1977)	SMACNA	SMACNA
	Thermoplastic Duct (PVC) Construction Manual, 2nd ed. (1994)	SMACNA	SMACNA
Installation	Flexible Duct Performance and Installation Standards	ADC	ADC-91
	Installation of Air Conditioning and Ventilating Systems	NFPA	ANSI/NFPA 90A-1993
	Installation of Warm Air Heating and Air-Conditioning Systems	NFPA	ANSI/NFPA 90B-1993
Materials Specifications	Specification for General Requirements for Flat-Rolled Stainless and Heat-Resisting Steel Plate, Sheet, and Strip	ASTM	ASTM A 480/A 480M-96
	Specification for General Requirements for Steel, Sheet, Carbon, and High-Strength, Low-Alloy, Hot-Rolled and Cold-Rolled	ASTM	ASTM A 568/A 568M-95
	Specification for General Requirements for Steel Sheet, Metallic-Coated by the Hot-Dip Process	ASTM	ASTM A 924/A 924M-95A
	Specification for Steel, Carbon (0.15 Maximum Percent), Hot-Rolled Sheet and Strip Commercial Quality	ASTM	ASTM A 569/A 569M-91A (R 1993)
	Specification for Steel, Sheet, Carbon, Cold-Rolled, Commercial Quality	ASTM	ASTM A 366/A 366M-91 (R 1993)
	Specification for Steel Sheet, Zinc-Coated (Galvanized) or Zinc-Iron Alloy-Coated (Galvannealed) by the Hot-Dip Process	ASTM	ASTM A 653/A 653M-95
System Design	Commercial Low Pressure, Low Velocity Duct System Design (1990)	ACCA	ACCA Manual Q
	Residential Duct Systems (1995)	ACCA	ACCA Manual D-R
	Closure Systems for Use with Rigid Air Ducts and Air Connectors (1994)	UL	UL 181A
Testing	Flexible Air Duct Test Code	ADC	ADC FD-72 (R 1979)
	Test Method for Measuring Acoustical and Airflow Performance of Duct Liner Materials and Prefabricated Silencers	ASTM	ASTM E 477-90
	HVAC Air Duct Leakage Test Manual (1985)	SMACNA	SMACNA
	HVAC Duct Systems Inspection Guide (1989)	SMACNA	SMACNA
Electrical	Voltage Ratings for Electrical Power Systems and Equipment	ANSI	ANSI C84.1-1989
	Canadian Electrical Code, Part I (17th ed.)	CSA	C22.1-1994
	Canadian Electrical Code, Part II—General Requirements	CSA	CAN/CSA-C22.2 No. 0-M91
	Application Guide for Ground Fault Circuit Interrupters	NEMA	NEMA 280-1990
	Application Guide for Ground Fault Protective Devices for Equipment	NEMA	ANSI/NEMA PB 2.2-1988
	Enclosures for Electrical Equipment (1000 Volts Maximum)	NEMA	ANSI/NEMA 250-1991
	Industrial Control and Systems Enclosures	NEMA	ANSI/NEMA ICS 6-1993
	General Requirements for Wiring Devices	NEMA	NEMA WD 1-1983 (R 1989)
	Low Voltage Cartridge Fuses	NEMA	ANSI/NEMA FU 1-1986
	Molded Case Circuit Breakers and Molded Case Switches	NEMA	NEMA AB 1-1993
	Industrial Control and Systems Terminal Blocks	NEMA	NEMA ICS 4-1993
	National Electrical Code	NFPA	ANSI/NFPA 70-1996
	Compatibility of Electrical Connectors and Wiring	SAE	ANSI/SAE AIR 1329A-1988
	Class T Fuses (1988)	UL	ANSI/UL 198H-1987
	Enclosures for Electrical Equipment (1995)	UL	ANSI/UL 50-1995
	Fuseholders (1993)	UL	ANSI/UL 512-1992
	High-Interrupting-Capacity Fuses, Current-Limiting Types (1986)	UL	ANSI/UL 198C-1986
	Molded-Case Circuit Breakers and Circuit-Breaker Enclosures (1991)	UL	ANSI/UL 489-1994
	Terminal Blocks (1993)	UL	UL 1059
	Thermal Cutoffs for Use in Electrical Appliances and Components (1994)	UL	UL 1020
Energy	Air Conditioning and Refrigerating Equipment Nameplate Voltages	ARI	ANSI/ARI 110-90
	Energy Conservation in Existing Buildings (Supersedes *Standards* 100.2, 100.3, 100.4, 100.5, and 100.6)	ASHRAE	ASHRAE/IESNA 100-1995 with Addendum 100a
	Energy-Efficient Design of New Low-Rise Residential Buildings	ASHRAE	ASHRAE/IESNA 90.2-1993
	Energy Efficient Design of New Buildings Except Low-Rise Residential Buildings	ASHRAE	ASHRAE/IESNA 90.1-1989
	Standard Methods of Measuring and Expressing Building Energy Performance	ASHRAE	ANSI/ASHRAE 105-1984 (R 90)
	Model Energy Code (1995)	CABO	BOCA/ICBO/SBCCI
	Uniform Solar Energy Code (1994)	IAPMO	IAPMO
	Model Energy Code, Thermal Envelope Compliance Guide	NAIMA	NAIMA BI407
	Energy Management Guide for Selection and Use of Polyphase Motors	NEMA	NEMA MG 10-1994
	Energy Management Guide for Selection and Use of Single-Phase Motors	NEMA	NEMA MG 11-1977 (R 1992)

Table 1 Codes and Standards Published by Various Societies and Associations (*Continued*)

Subject	Title	Publisher	Reference
Energy (continued)	Building Systems Analysis and Retrofit Manual (1995)	SMACNA	SMACNA
	Energy Conservation Guidelines (1984)	SMACNA	SMACNA
	Energy Recovery Equipment and Systems, Air-to-Air (1991)	SMACNA	SMACNA
	HVAC Commissioning Manual (1994)	SMACNA	SMACNA
	Retrofit of Building Energy Systems and Processes (1982)	SMACNA	SMACNA
	Energy Management Equipment (1994)	UL	ANSI/UL 916-1993
Exhaust Systems	Commercial Low Pressure, Low Velocity Duct System Design (1990)	ACCA	ACCA Manual Q
	Fundamentals Governing the Design and Operation of Local Exhaust Systems	ANSI	ANSI/AIHA Z9.2-1979 (R 1991)
	Laboratory Ventilation	ANSI	ANSI/AIHA Z9.5-1992
	Open-Surface Tanks—Ventilation and Operation	ANSI	ANSI/AIHA Z9.1-1991
	Safety Code for Design, Construction, and Ventilation of Spray Finishing Operations	ANSI	ANSI/AIHA Z9.3-1985
	Abrasive Blasting Operations—Ventilation and Safe Practices	ANSI	ANSI/AIHA Z9.4-1985
	Method of Testing Performance of Laboratory Fume Hoods	ASHRAE	ANSI/ASHRAE 110-1995
	Compressors and Exhausters	ASME	ANSI/ASME PTC 10-1965 (R 1992)
	Mechanical Flue-Gas Exhausters	CSA	CAN 3-B255-M81
	Exhaust Systems for Air Conveying of Materials	NFPA	ANSI/NFPA 91-1995
	Draft Equipment (1993)	UL	UL 378
Expansion Valves	Thermostatic Refrigerant Expansion Valves	ARI	ANSI/ARI 750-94
	Method of Testing for Capacity Rating of Thermostatic Refrigerant Expansion Valves	ASHRAE	ANSI/ASHRAE 17-1986 (RA 90)
Fan-Coil Units	Room Fan-Coil and Unit Ventilators	ARI	ANSI/ARI 440-89
	Methods of Testing for Rating Room Fan-Coil Air Conditioners	ASHRAE	ANSI/ASHRAE 79-1984 (RA 91)
	Fan-Coil Units and Room Fan-Heater Units (1986)	UL	ANSI/UL 883-1986
Fans	Commercial Low Pressure, Low Velocity Duct System Design (1990)	ACCA	ACCA Manual Q
	Residential Duct Systems (1995)	ACCA	ACCA Manual D-R
	Designation of Rotation and Discharge of Centrifugal Fans	AMCA	AMCA 99-2406-83
	Drive Arrangements for Centrifugal Fans	AMCA	AMCA 99-2404-78
	Drive Arrangements for Tubular Centrifugal Fans	AMCA	AMCA 99-2410-82
	Recommended Safety Practices for Users and Installers of Industrial and Commercial Fans	AMCA	AMCA 410-90
	Inlet Box Positions for Centrifugal Fans	AMCA	AMCA 99-2405-83
	Motor Positions for Belt or Chain Drive Centrifugal Fans	AMCA	AMCA 99-2407-66
	Industrial Process/Power Generation Fans: Site Performance Test Standard	AMCA	AMCA 803-94
	Standards Handbook	AMCA	AMCA 99-86
	Fans and Blowers	ARI	ARI 670-90
	Methods for the Measurement of Noise Emitted by Small Air-Moving Devices	ASA	ANSI S12.11-1987 (R 1993)
	Laboratory Methods of Testing Fans for Rating	ASHRAE AMCA	ANSI/ASHRAE 51-1985 ANSI/AMCA 210-85
	Laboratory Method of Testing In-Duct Sound Power Measurement Procedure for Fans	ASHRAE AMCA	ANSI/ASHRAE 68-1986 ANSI/AMCA 330-86
	Methods of Testing Fan Vibration—Blade Vibrations and Critical Speeds	ASHRAE	ANSI/ASHRAE 87.1-1992
	Fans	ASME	ANSI/ASME PTC 11-1984 (R 1990)
	Fans and Ventilators	CSA	C22.2 No. 113-M1984 (R 1993)
	Rating the Performance of Residential Mechanical Ventilating Equipment	CSA	CAN/CSA-C260-M90
	Acoustics—Method for the Measurement of Airborne Noise Emitted by Small Air-Moving Devices	ISO	ISO 10302:1996
	Electric Fans (1994)	UL	UL 507
Ceiling	AC Electric Fans and Regulators	ANSI	ANSI-IEC Pub. 385
Fenestration	Specification for Sealed Insulated Glass Units	ASTM	ASTM E 774-92
	Standard Practice for Calculation of Photometric Transmittance and Reflectance of Materials to Solar Radiation	ASTM	ASTM 971-88 (R 1996)
	Standard Practice for Determining the Minimum Thickness and Type of Glass Required to Resist a Specific Load	ASTM	ASTM E 1300-94
	Standard Tables for Terrestrial Direct Normal Solar Spectral Irradiance for Air Mass 1.5	ASTM	ASTM E 891-87 (R 1992)
	Standard Tables for Terrestrial Solar Spectral Irradiance at Air Mass 1.5 for a 37° Tilted Surface	ASTM	ASTM E 892-87 (R 1992)
	Test Method for Seal Durability of Sealed Insulated Glass Units	ASTM	ASTM E 773-88
	Test Method for Solar Absorptance, Reflectance, and Transmittance of Materials Using Integrating Spheres	ASTM	ASTM E 903-96
	Test Method for Solar Photometric Transmittance of Sheet Materials Using Sunlight	ASTM	ASTM E 972-96
	Test Method for Solar Transmittance (Terrestrial) of Sheet Materials Using Sunlight	ASTM	ASTM E 1084-86 (R 1996)
	Energy Performance Evaluation of Swinging Doors	CSA	CSA-A453-95
	Energy Performance Evaluation of Windows and Sliding Glass Doors	CSA	CAN/CSA-A440.2-93
	Windows	CSA	CAN/CSA-A440-M90

Table 1 Codes and Standards Published by Various Societies and Associations (*Continued*)

Subject	Title	Publisher	Reference
Filters	Flow-Capacity Rating and Application of Suction-Line Filters and Filter Driers	ARI	ANSI/ARI 730-86
	Specification for Octave-Band and Fractional-Octave-Band Analog and Digital Filters	ASA	ANSI S1.11-1986 (R 1993)
	Method of Testing Flow Capacity of Suction Line Filters and Filter Driers	ASHRAE	ANSI/ASHRAE 78-1985 (RA 90)
	Method of Testing Liquid Line Filter-Drier Filtration Capability	ASHRAE	ANSI/ASHRAE 63.2-1996
	Exhaust Hoods for Commercial Cooking Equipment (1990)	UL	ANSI/UL 710-1992
	Grease Filters for Exhaust Ducts (1979)	UL	UL 1046
Fireplaces	Factory-Built Fireplaces (1988)	UL	ANSI/UL 127-1992
	Fireplace Stoves (1995)	UL	ANSI/UL 737-1995
Fire Protection	Standard Method for Fire Tests of Building Construction and Materials	ASTM	ASTM E 119-95A
	Method of Test of Surface Burning Characteristics of Building Materials	ASTM	ASTM E 84-95A
		NFPA	NFPA 255-1996
	BOCA National Fire Prevention Code, 9th ed. (1993)	BOCA	BOCA
	Uniform Fire Code (1994)	IFCI	IFCI
	Uniform Fire Code Standards (1991)	IFCI	IFCI
	Interconnection Circuitry of Noncoded Remote-Station Protective Signalling Systems	NEMA	NEMA SB 3-1969 (R 1989)
	Fire Doors and Fire Windows	NFPA	ANSI/NFPA 80-1995
	Fire Hazard Properties of Flammable Liquids, Gases, and Volatile Solids	NFPA	ANSI/NFPA 325-1994
	Fire Prevention Code	NFPA	ANSI/NFPA 1-1992
	Fire Protection for Laboratories Using Chemicals	NFPA	ANSI/NFPA 45-1996
	Fire Protection Handbook, 18th ed. (1996)	NFPA	NFPA
	Flammable and Combustible Liquids Code	NFPA	ANSI/NFPA 30-1993
	Health Care Facilities	NFPA	ANSI/NFPA 99-1996
	Installation of Sprinkler Systems	NFPA	NFPA 13-1996
	Life Safety Code	NFPA	ANSI/NFPA 101-1994
	National Fire Codes (issued annually)	NFPA	NFPA
	Methods of Fire Tests of Door Assemblies	NFPA	ANSI/NFPA 252-1995
	Standard Fire Prevention Code (1994 ed. with 1995 revisions)	SBCCI	SBCCI
	Fire, Smoke and Radiation Damper Installation Guide for HVAC Systems (1992)	SMACNA	SMACNA
	Fire Dampers (1995)	UL	UL 555
	Fire Tests of Building Construction and Materials (1992)	UL	UL 263
	Fire Tests of Door Assemblies (1993)	UL	UL 10B
	Fire Tests of Through-Penetration Firestops (1994)	UL	UL 1479
	Heat Responsive Links for Fire-Protection Service (1993)	UL	ANSI/UL 33-1995
Smoke Management	Commissioning Smoke Management Systems	ASHRAE	ASHRAE Guideline 5-1994
	Recommended Practice for Smoke Control Systems	NFPA	ANSI/NFPA 92A-1993
	Guide for Smoke Management Systems in Malls, Atria, and Large Areas	NFPA	ANSI/NFPA 92B-1995
	Leakage Rated Dampers for Use in Smoke Control Systems	UL	ANSI/UL 555S-1993
Freezers	Capacity Measurement and Energy Consumption Test Methods for Refrigerators, Combination Refrigerator-Freezers, and Freezers	CSA	CAN/CSA-C300-M91
	Refrigeration Equipment	CSA	CAN/CSA-C22.2 No. 120-M91
Commercial	Dispensing Freezers	NSF	ANSI/NSF 6-1989
	Food Service Refrigerators and Storage Freezers	NSF	NSF 7-1990
	Ice Cream Makers (1993)	UL	ANSI/UL 621-1992
	Commercial Refrigerators and Freezers (1992)	UL	ANSI/UL 471-1991
	Ice Makers (1995)	UL	UL 563
Household	Household Refrigerators, Combination Refrigerator-Freezers and Household Freezers	AHAM	ANSI/AHAM HRF-1-1988
	Household Refrigerators and Freezers (1993)	UL	UL 250
		CSA	C22.2 No. 63-93
Fuels	Standard Classification of Coals by Rank	ASTM	ANSI/ASTM D975-96
	Standard Specification for Diesel Fuel Oils	ASTM	ANSI/ASTM D 975-96
	Standard Specification for Fuel Oils	ASTM	ANSI/ASTM D 396-95
	Standard Specification for Gas Turbine Fuel Oils	ASTM	ANSI/ASTM D 2880-96
Furnaces	Commercial Applications, Systems and Equipment (1993)	ACCA	ACCA Manual CS
	Residential Equipment Selection, 2nd ed. (1995)	ACCA	ACCA Manual S
	Methods of Testing for Annual Fuel Utilization Efficiency of Residential Central Furnaces and Boilers	ASHRAE	ANSI/ASHRAE 103-1993
	BOCA National Mechanical Code, 8th ed. (1993)	BOCA	BOCA
	Prevention of Furnace Explosions/Implosions in Multiple Burner Boilers	NFPA	ANSI /NFPA 8502-1995
	Standard Mechanical Code (1994)	SBCCI	SBCCI
	Heating and Cooling Equipment (1995)	UL	UL 1995
		CSA	CAN/CSA-C22.2 No. 236-95
	Residential Gas Detectors (1991)	UL	UL 1484
	Single and Multiple Station Carbon Monoxide Detectors (1992)	UL	UL 2034
Gas	Direct Vent Central Furnaces	AGA	ANSI Z21.64-1990
	Gas-Fired Central Furnaces	AGA	ANSI Z21.47-1993
		CGA	CAN/CGA-2.3-M93

Table 1 Codes and Standards Published by Various Societies and Associations (*Continued*)

Subject	Title	Publisher	Reference
Gas (*continued*)	Gas-Fired Duct Furnaces	AGA	ANSI Z83.9-1990; Z83.9a-1992
	Gas-Fired Gravity and Fan Type Direct Vent Wall Furnaces	AGA	ANSI Z21.44-1995
	Gas-Fired Gravity and Fan Type Floor Furnaces	AGA	Z21.48-1992
	Gas-Fired Gravity and Fan Type Vented Wall Furnaces	AGA	Z21.49-1992
	Gas-Fired Duct Furnaces	CGA	CAN/CGA-2.8-M86
	Gas-Fired Gravity and Fan Type Direct Vent Wall Furnaces	CGA	CAN1-2.19-M81
	Gas-Fired Gravity and Fan Type Vented Wall Furnaces	CGA	CAN/CGA-2.5-M86
	Industrial and Commercial Gas-Fired Package Furnaces	CGA	CGA 3.2-1976
	Standard Gas Code (1994)	SBCCI	SBCCI
	Commercial-Industrial Gas Heating Equipment (1973)	UL	UL 795
Oil	Standard Specification for Fuel Oils	ASTM	ANSI/ASTM D 396-95
	Test Method for Smoke Density in Flue Gases from Burning Distillate Fuels	ASTM	ANSI/ASTM D 2156-94
	Oil Burning Stoves and Water Heaters	CSA	B140.3-1962 (R 1991)
	Oil-Fired Warm Air Furnaces	CSA	B140.4-1974 (R 1991)
	Installation of Oil-Burning Equipment	NFPA	ANSI/NFPA 31-1992
	Oil-Fired Central Furnaces (1986)	UL	UL 727-1986
	Oil-Fired Floor Furnaces (1987)	UL	ANSI/UL 729-1987
	Oil-Fired Wall Furnaces (1987)	UL	ANSI/UL 730-1986
Solid Fuel	Standard Classification of Coals by Rank	ASTM	ANSI/ASTM D975-96
	Installation Code for Solid-Fuel-Burning Appliances and Equipment	CSA	CAN/CSA-B365-M91
	Solid-Fuel-Fired Central Heating Appliances	CSA	CAN/CSA-B366.1-M91
	Solid-Fuel and Combination-Fuel Central and Supplementary Furnaces (1991)	UL	ANSI/UL 391-1991
Heaters	Gas-Fired Infrared Heaters	AGA	ANSI Z83.6-1990; Z83.6a-1992; Z83.6b-1993
	Requirements for Gas-Fired Infrared Patio	AGA	5-90
	Requirements for Residential Radiant Tube Heaters	AGA	7-89
	Air Heaters	ASME	ANSI/ASME PTC 4.3-1968 (R 1991)
	Direct Gas-Fired Non-Recirculating Make-up Air Heaters	CGA	CAN1-3.7-77
	Gas-Fired Infra-Red Heaters	CGA	CAN1-2.16-M81
	Electric Air Heaters	CSA	C22.2 No.46-M1988
	Electric Duct Heaters	CSA	C22.2 No. 155-M1986 (R 1992)
	Portable Kerosine-Fired Heaters	CSA	CAN 3-B140.9.3 M86
	Standards for Closed Feedwater Heaters, 5th ed. (1992)	HEI	HEI
	Electric Air Heaters (1980)	UL	ANSI/UL 1025-1991
	Electric Dry Bath Heaters (1994)	UL	ANSI/UL 875
	Electric Heating Appliances (1987)	UL	ANSI/UL 499-1987
	Electric Oil Heaters (1991)	UL	ANSI/UL 574-1990
	Oil-Burning Stoves (1993)	UL	UL 896
	Oil-Fired Air Heaters and Direct-Fired Heaters (1993)	UL	UL 733
Combination	Requirements for Gas-Fired Combination Space Heating/Water Heating Appliances	AGA	11-90
Engine	Electric Engine Preheaters and Battery Warmers for Diesel Engines	SAE	SAE J 1310-1993
	Fuel Warmer—Diesel Engines	SAE	ANSI/SAE J 1422-1989
	Selection and Application Guidelines for Diesel, Gasoline, and Propane Fired Liquid Cooled Engine Pre-Heaters	SAE	SAE J 1350-1988
Nonresidential	Direct Gas-Fired Industrial Air Heaters	AGA	ANSI Z83.18-1990; Z83.18a-1991; Z83.18b-1992
	Gas-Fired Construction Heaters	AGA	ANSI Z83.7-1990; Z83.7a-1991; Z83.7b-1993
	Gas-Fired Unvented Commercial and Industrial Heaters	AGA	ANSI Z83.16-1982; Z83.16a-1984; Z83.16b-1989
	Requirements for Direct Gas-Fired Circulating Heaters for Agricultural Buildings	AGA	5-88
	Requirements for High Pressure LP Infrared Poultry and Livestock Heating Systems	AGA	4-87
	Portable Industrial Oil-Fired Heaters	CSA	B140.8-1967 (R 1991)
	Fuel-Fired Heaters—Air Heating—For Construction and Industrial Machinery	SAE	ANSI/SAE J 1024-1989
	Commercial-Industrial Gas Heating Equipment (1994)	UL	UL 795
	Electric Heaters for Use in Hazardous (Classified) Locations (1991)	UL	ANSI/UL 823-1990
Room	Gas-Fired Room Heaters, Vol. I, Vented Room Heaters	AGA	ANSI Z21.11.1-1991; Z21.11.1a-1993; Z21.11.1b-1995
	Gas-Fired Room Heaters, Vol. II, Unvented Room Heaters	AGA	ANSI Z21.11.2-1992 Z21.11.2a-1993; Z21.11.2b-1995
	Gas-Fired Unvented Catalytic Room Heaters for Use with Liquefied Petroleum (LP) Gases	AGA	ANSI Z21.76-1994
	Requirements for Gas-Fired Vented Catalytic Type Room Heaters	AGA	1-81
	Requirements for Unvented Room Heaters Equipped with Oxygen Depletion Safety Shutoff Systems	AGA	2-79
	Fixed and Location-Dedicated Electric Room Heaters (1992)	UL	UL 2021
	Movable and Wall- or Ceiling-Hung Electric Room Heaters (1994)	UL	UL 1278
	Room Heaters, Solid Fuel-Type (1994)	UL	UL 1482
	Unvented Kerosene-Fired Room Heaters and Portable Heaters (1993)	UL	UL 647

Table 1 Codes and Standards Published by Various Societies and Associations (*Continued*)

Subject	Title	Publisher	Reference
Pool	Gas-Fired Pool Heaters	AGA	ANSI Z21.56-1994
	Oil-Fired Service Water Heaters and Swimming Pool Heaters	CSA	B140.12-1976 (R 1991)
Transport	Heater, Aircraft Internal Combustion Heat Exchanger Type	SAE	SAE AS 8040-1988
	Heater, Airplane, Engine Exhaust Gas to Air Heat	SAE	SAE ARP 86A-1952 (R 1992)
	Installation, Heaters, Airplane, Internal Combustion Heater Exchange Type	SAE	SAE ARP 266-1952 (R 1992)
	Motor Vehicle Heater Test Procedure	SAE	SAE J 638-1993
Unit	Gas Unit Heaters	AGA	ANSI Z83.8-1990; Z83.8a-1990; Z83.8b-1992
	Gas Unit Heaters	CGA	CAN/CGA-2.6-M86
	Oil-Fired Unit Heaters (1995)	UL	UL 731
Heat Exchangers	Remote Mechanical-Draft Evaporative Refrigerant Condensers	ARI	ANSI/ARI 490-89
	Method of Testing Air-to-Air Heat Exchangers	ASHRAE	ANSI/ASHRAE 84-1991
	Standard Methods of Test for Rating the Performance of Heat-Recovery Ventilators	CSA	CAN/CSA-C439-88
	Standards for Power Plant Heat Exchangers, 2nd ed. (1990)	HEI	HEI
	Standards of Tubular Exchanger Manufacturers Association, 7th ed. (1988)	TEMA	TEMA
Heating	Commercial Applications, Systems and Equipment (1993)	ACCA	ACCA Manual CS
	Installation Techniques for Perimeter Heating and Cooling, 11th ed.	ACCA	ACCA Manual 4
	Residential Equipment Selection, 2nd ed. (1995)	ACCA	ACCA Manual S
	Determining the Required Capacity of Residential Space Heating and Cooling Appliances	CSA	CAN/CSA-F280-M90
	Automatic Flue-Pipe Dampers for Use with Oil-Fired Appliances	CSA	B140.14-M1979 (R 1991)
	Heater Elements	CSA	C22.2 No.72-M1984 (R 1992)
	Advanced Installation Guide for Hydronic Heating Systems (1991)	HYDI	IBR 250
	Heat Loss Calculation Guide	HYDI	IBR H-21 (1984), IBR H-22 (1989)
	Installation Guide for Residential Hydronic Heating Systems, 6th ed. (1988)	HYDI	IBR 200
	Radiant Floor Heating (1993)	HYDI	IBR 400
	Environmental System Technology (1984)	NEBB	NEBB
	Pulverized Fuel Systems	NFPA	ANSI/NFPA 8503-1992
	Aircraft Electrical Heating Systems	SAE	ANSI/SAE AIR 860-1965 (R 1992)
	Performance Test for Air-Conditioned, Heated, and Ventilated Off-Road Self-Propelled Work Machines	SAE	ANSI/SAE J 1503-1986
	Heating Value of Fuels	SAE	SAE J 1498-1990
	HVAC Systems—Applications, 1st ed. (1986)	SMACNA	SMACNA
	Installation Standards for Residential Heating and Air Conditioning Systems (1988)	SMACNA	SMACNA
	Electric Baseboard Heating Equipment (1994)	UL	ANSI/UL 1042-1995
	Electric Central Air Heating Equipment	UL	ANSI/UL 1096-1986
	Heating and Cooling Equipment (1995)	UL	UL 1995
		CSA	CAN/CSA-C22.2 No. 236-95
Heat Meters	Method of Testing Thermal Energy Meters for Liquid Streams in HVAC Systems	ASHRAE	ANSI/ASHRAE 125-1992
Heat Pumps	Commercial Applications, Systems and Equipment (1993)	ACCA	ACCA Manual CS
	Heat Pump Systems, Principles and Applications (Commercial and Residence), 2nd ed. (1984)	ACCA	ACCA Manual H
	Residential Equipment Selection, 2nd ed. (1995)	ACCA	ACCA Manual S
	Commercial and Industrial Unitary Heat Pump Equipment	ARI	ANSI/ARI 340-86
	Ground Source Closed-Loop Heat Pumps	ARI	ARI 330-93
	Ground Water-Source Heat Pumps	ARI	ANSI/ARI 325-93
	Water-Source Heat Pumps	ARI	ANSI/ARI 320-93
	Methods of Testing for Rating Unitary Air-Conditioning and Heat Pump Equipment	ASHRAE	ANSI/ASHRAE 37-1988
	Methods of Testing for Seasonal of Unitary Air-Conditioners and Heat Pumps	ASHRAE	ANSI/ASHRAE 116-1995
	Installation Requirements for Air-to-Air Heat Pumps	CSA	C273.5-1980 (R 1991)
	Performance Standard for Split-System Central Air-Conditioners and Heat Pumps	CSA	CAN/CSA-C273.3-M91
	Heating and Cooling Equipment (1995)	UL	UL 1995
		CSA	CAN/CSA C22.2 No. 236-95
	Heat Pumps (1985)	UL	ANSI/UL 559-1985
Gas-Fired	Requirements for Gas-Fired, Absorption and Adsorption Heat Pumps	AGA	10-90
Heat Recovery	Gas Turbine Heat Recovery Steam Generators	ASME	ANSI/ASME PTC 4.4-1981 (R 1992)
	Energy Recovery Equipment and Systems (1991)	SMACNA	SMACNA
	Requirements for Heat Reclaimer Devices for Use with Gas-Fired Appliances	AGA	ANSI Z21.40.1-1994
Humidifiers	Method for Measuring Performance of Appliance Humidifiers	AHAM	ANSI/AHAM HU-1-1987
	Central System Humidifiers for Residential Applications	ARI	ANSI/ARI 610-89
	Commercial and Industrial Humidifiers	ARI	ARI 640-90
	Self-Contained Humidifiers for Residential Applications	ARI	ANSI/ARI 620-89
	Humidifiers (1993)	UL	UL 998
		CSA	C22.2 No. 104-93
Ice Makers	Automatic Commercial Ice Makers	ARI	ARI 810-95
	Ice Storage Bins	ARI	ANSI/ARI 820-95

Table 1 Codes and Standards Published by Various Societies and Associations (*Continued*)

Subject	Title	Publisher	Reference
Ice Makers *(continued)*	Methods of Testing Automatic Ice Makers	ASHRAE	ANSI/ASHRAE 29-1988
	Refrigeration Equipment	CSA	CAN/CSA-C22.2 No. 120-M91
	Performance of Automatic Ice-Makers and Ice Storage Bins	CSA	CAN/CSA-C742-94
	Automatic Ice-Making Equipment	NSF	ANSI/NSF 12-1992
	Ice Makers (1995)	UL	UL 563
Incinerators	Large Incinerators	ASME	ANSI/ASME PTC 33-1978 (R 1991)
	Incinerators and Waste and Linen Handling Systems and Equipment	NFPA	ANSI/NFPA 82-1992
	Residential Incinerators (1993)	UL	UL 791
Indoor Air Quality	Standard Practice for Continuous Sizing and Counting of Airborne Particles in Dust-Controlled Areas and Clean Rooms Using Instruments Capable of Detecting Single Sub-Micrometre and Larger Particles	ASTM	ASTM F 50-92 (R 1996)
	Standard Practices for Referencing Suprathreshold Odor Intensity	ASTM	ASTM E 544-75 (R 1993)
Induction Units	Room Air-Induction Units	ARI	ANSI/ARI 445-87
	Frame Assignments for Alternating Current Integral-Horsepower Induction Motors	NEMA	NEMA MG 13-1984 (R 1990)
Industrial Duct	Rectangular Industrial Duct Construction (1980)	SMACNA	SMACNA
	Round Industrial Duct Construction (1977)	SMACNA	SMACNA
Insulation	Classification for Rating Sound Insulation	ASTM	ASTM E 413-87 (R 1994)
	Classification of Potential Health and Safety Concerns Associated with Thermal Insulation Materials and Accessories	ASTM	ASTM C 930-92
	Specification for Adhesives for Duct Thermal Insulation	ASTM	ASTM C 916-85 (R 1990)
	Specification for Thermal and Acoustical Insulation (Glass Fiber, Duct Lining Material)	ASTM	ASTM C 1071-91
	Standard Practice for Determination of Heat Gain or Loss and the Surface Temperature of Insulated Pipe and Equipment Systems by the Use of a Computer Program	ASTM	ASTM C 680-89 (R 1995)
	Standard Practice for Inner and Outer Diameters of Rigid Thermal Insulation for Nominal Sizes of Pipe and Tubing (NPS System)	ASTM	ASTM C 585-90
	Standard Practice for Thermographic Inspection of Insulation Installations in Envelope Cavities of Frame Buildings	ASTM	ASTM C 1060-90
	Standard Terminology Relating to Thermal Insulating Materials	ASTM	ASTM C 168-90
	Test Method for Steady-State Heat Flux Measurements and Thermal Transmission Properties by Means of the Guarded Hot Plate Apparatus	ASTM	ASTM C 177-85 (R 1993)
	Test Method for Steady-State Heat Flux Measurements and Thermal Transmission Properties by Means of the Heat Flow Meter Apparatus	ASTM	ASTM C 518-91
	Test Method for Steady-State Heat Transfer Properties of Horizontal Pipe Insulations	ASTM	ASTM C 335-95
	Test Method for Steady-State and Thermal Performance of Building Assemblies by Means of a Guarded Hot Box	ASTM	ASTM C 236-89 (R 1993)
	Thermal Insulation, Mineral Fibre, for Buildings	CSA	A101-M1983
	National Commercial and Industrial Insulation Standards	MICA	MICA 1993
Louvers	Test Methods for Louvers, Dampers and Shutters	AMCA	AMCA 500-89
Lubricants	Method of Testing the Floc Point of Refrigeration Grade Oils	ASHRAE	ANSI/ASHRAE 86-1994
	Practice for Calculating Viscosity Index from Kinematic Viscosity at 40 and 100°C	ASTM	ASTM D 2270-93
	Practice for Conversion of Kinematic Viscosity to Saybolt Universal Viscosity or to Saybolt Furol Viscosity	ASTM	ASTM D 2161-93
	Test Method for Estimation of Molecular Weight (Relative Molecular Mass) of Petroleum Oils from Viscosity Measurements	ASTM	ANSI/ASTM D 2502-92
	Test Method for Separation of Representative Aromatics and Nonaromatics Fractions of High-Boiling Oils by Elution Chromatography	ASTM	ANSI/ASTM D 2549-91 (R 1995)
	Classification of Industrial Fluid Lubricants by Viscosity System	ASTM	ANSI/ASTM D 2422-86 (R 1993)
	Test Method for Carbon-Type Composition of Insulating Oils of Petroleum Origin	ASTM	ASTM D 2140-91
	Test Method for Dielectric Breakdown Voltage of Insulating Liquids Using Disk Electrodes	ASTM	ASTM D 877-87 (R 1995)
	Test Method for Dielectric Breakdown Voltage of Insulating Oils of Petroleum Origin Using VDE Electrodes	ASTM	ASTM D 1816-84A (R 1990)
	Test Method for Mean Molecular Weight of Mineral Insulating Oils by the Cryoscopic Method	ASTM	ASTM D 2224-78 (R 1983)
	Test Method for Relative Molecular Mass (Molecular Weight) of Hydrocarbons by Thermoelectric Measurement of Vapor Pressure	ASTM	ASTM D 2503-92
	Test Methods for Pour Point of Petroleum Oils	ASTM	ASTM D 97-93
	Semiconductor Graphite	NEMA	NEMA CB 4-1989
Measurements	A Standard Calorimeter Test Method for Flow Measurement of a Volatile Refrigerant	ASHRAE	ANSI/ASHRAE 41.9-1988
	Engineering Analysis of Experimental Data	ASHRAE	ASHRAE Guideline 2-1986 (RA 96)
	Methods of Measuring Solar-Optical Properties of Materials	ASHRAE	ANSI/ASHRAE 74-1988
	Standard Method for Measurement of Proportion of Lubricant in Liquid Refrigerant	ASHRAE	ASHRAE 41.4-1996

Table 1 Codes and Standards Published by Various Societies and Associations (*Continued*)

Subject	Title	Publisher	Reference
Measurements *(continued)*	Standard Method for Measurement of Moist Air Properties	ASHRAE	ANSI/ASHRAE 41.6-1994
	Standard Methods of Measuring and Expressing Building Energy Performance	ASHRAE	ANSI/ASHRAE 105-1984 (RA 90)
	Measurement of Industrial Sound	ASME	ANSI/ASME PTC 36-1985
	Measurement of Rotary Speed	ASME	ANSI/ASME PTC 19.13-1961 (R 1986)
	Measurement Uncertainty	ASME	ANSI/ASME PTC 19.1-1985 (R 1990)
	Method for Establishing Installation Effects on Flowmeter	ASME	ANSI/ASME MFC-10M-1994
	Procedure for Bench Calibration of Tank Level Gaging Tapes and Sounding Rules	ASME	ANSI/ASME MC88.2-1974 (R 1987)
	Specification and Temperature-Electromotive Force (EMF) Tables for Standardized Thermocouples	ASTM	ASTM E 230-96
	Standard Practice for Continuous Sizing and Counting of Airborne Particles in Dust-Controlled Areas and Clean Rooms Using Instruments Capable of Detecting Single Sub-Micrometre and Larger Particles	ASTM	ASTM F 50-92 (R 1996)
	American National Standard for Use of International System of Units (SI)—The Modern Metric System	IEEE/ASTM	ANSI/IEEE/ASTM SI 10-97
	Test Methods for Water Vapor Transmission of Materials	ASTM	ASTM E 96-95
	Ergonomics—Determination of Metabolic Heat Production	ISO	ISO 8996:1990
	Ergonomics of the Thermal Environment—Estimation of the Thermal Insulation and Evaporative Resistance of a Clothing Ensemble	ISO	ISO 9920:1995
	Thermal Environments—Instruments and Methods for Measuring Physical Quantities	ISO	ISO 7726:1985
Fluid Flow	Standard Methods of Measurement of Flow of Liquids in Pipes Using Orifice Flowmeters	ASHRAE	ANSI/ASHRAE 41.8-1989
	Application of Fluid Meters	ASME	ASME 19.5-72
	Fluid Flow in Closed Conduits—Connections for Pressure Signal Transmissions Between Primary and Secondary Devices	ASME	ANSI/ASME MFC-8M-1988
	Glossary of Terms Used in the Measurement of Fluid Flow in Pipes	ASME	ANSI/ASME MFC-1M-1991
	Measurement of Fluid Flow by Means of Coriolis Mass Flowmeters	ASME	ANSI/ASME MFC-11M-1989 (R 1994)
	Measurement of Fluid Flow in Pipes Using Orifice, Nozzle, and Venturi	ASME	ASME MFC-3M-1989
	Measurement of Fluid Flow in Pipes Using Vortex Flow Meters	ASME	ASME/ANSI MFC-6M-1987
	Measurement of Fluid Flow Using Small Bore Precision Orifice Meters	ASME	ANSI/ASME MFC-14M-1995
	Measurement of Liquid Flow in Closed Conduits by Weighting Method	ASME	ANSI/ASME MFC-9M-1988
	Measurement of Liquid Flow in Closed Conduits Using Transit-Time Ultrasonic Flowmeters	ASME	ANSI/ASME MFC-5M-1985 (R 1994)
	Measurement Uncertainty for Fluid Flow in Closed Conduits	ASME	ANSI/ASME MFC-2M-1983 (R 1988)
	Measurement of Fluid Flow in Closed Conduits—Velocity Area Method Using Pitot Static Tubes	ISO	ISO 3966:1977
Gas Flow	Standard Methods for Laboratory Air Flow Measurement	ASHRAE	ANSI/ASHRAE 41.2-1987 (RA 92)
	Standard Method for Measurement of Flow of Gas	ASHRAE	ANSI/ASHRAE 41.7-1984 (RA 91)
	Measurement of Gas Flow by Means of Critical Flow Venturi Nozzles	ASME	ASME/ANSI MFC-7M-1987 (R 1992)
	Measurement of Gas Flow by Turbine Meters	ASME	ANSI/ASME MFC-4M-1986 (R 1990)
Pressure	Standard Method for Pressure Measurement	ASHRAE	ANSI/ASHRAE 41.3-1989
	Gauges—Pressure Indicating Dial Type—Elastic Element	ASME	ANSI/ASME B40.1-91
	Guide for Dynamic Calibration of Pressure Transducers	ASME	ANSI MC88-1-1972 (R 1987)
	Pressure Measurement	ASME	ANSI/ASME PTC 19.2-1987
Temperature	Standard Method for Temperature Measurement	ASHRAE	ANSI/ASHRAE 41.1-1986 (RA 91)
	Temperature Measurement	ASME	ANSI/ASME PTC 19.3-1974 (R 1986)
	Total Temperature Measuring Instruments (Turbine Powered Subsonic Aircraft)	SAE	SAE AS 793-1966 (R 1991)
Thermal Properties	Standard Practice for Determining Thermal Resistance of Building Envelope Components from In-Situ Data	ASTM	ASTM C 1155-95
	Standard Practice for In-Situ Measurement of Heat Flux and Temperature on Building Envelope Components	ASTM	ASTM C 1046-95
	Test Method for Steady-State Heat Flux Measurements and Thermal Transmission Properties by Means of the Guarded Hot Plate Apparatus	ASTM	ASTM C 177-85 (R 1993)
	Test Method for Steady-State Heat Flux Measurements and Thermal Transmission Properties by Means of the Heat Flow Meter Apparatus	ASTM	ASTM C 518-91
	Test Method for Thermal Performance of Building Assemblies by Means of a Calibrated Hot Box	ASTM	ASTM C 976-90 (R 1996)
Mobile Homes and Recreational Vehicles	Residential Load Calculation, 7th ed. (1986)	ACCA	ACCA Manual J
	Recreational Vehicle Cooking Gas Appliances	AGA	ANSI Z21.57-1993
	Mobile Homes	CSA	CAN/CSA-Z240 MH Series-92
	Mobile Home Parks	CSA	Z240.7.1-1972
	Oil-Fired Warm Air Heating Appliances for Mobile Housing and Recreational Vehicles	CSA	B140.10-1974 (R 1991)
	Park Model Trailers	CSA	CAN/CSA-Z41 Series-92
	Recreational Vehicle Parks	CSA	Z240.7.2-1972

Table 1 Codes and Standards Published by Various Societies and Associations (*Continued*)

Subject	Title	Publisher	Reference
Mobile Homes and Recreational Vehicles *(continued)*	Recreational Vehicles	CSA	CAN/CSA-Z240 RV Series-M86 (R 1992)
	Gas Supply Connectors for Manufactured Homes	IAPMO	IAPMO TSC 9-1992
	Manufactured Home Installations	NCSBCS	ANSI A225.1-1994
	Recreational Vehicles	NFPA	NFPA 501C-1993 (ANSI A119.2)
	Plumbing System Components for Manufactured Homes and Recreational Vehicles	NSF	ANSI/NSF 24-1988
	Gas Burning Heating Appliances for Mobile Homes and Recreational Vehicles (1995)	UL	UL 307B
	Gas-Fired Cooking Appliances for Recreational Vehicles (1993)	UL	UL 1075
	Liquid Fuel-Burning Heating Appliances for Manufactured Homes and Recreational Vehicles (1995)	UL	UL 307A
	Low Voltage Lighting Fixtures for Use in Recreational Vehicles (1994)	UL	UL 234
	Roof Jacks for Manufactured Homes and Recreational Vehicles (1994)	UL	ANSI/UL 311-1995
	Roof Trusses for Manufactured Homes (1995)	UL	UL 1298
	Shear Resistance Tests for Ceiling Boards for Manufactured Homes (1992)	UL	UL 1296
Motors and Generators	Steam Generating Units	ASME	ANSI/ASME PTC 4.1-1964 (R 1991)
	Testing of Nuclear Air-Treatment Systems	ASME	ANSI/ASME N510-1989
	Nuclear Power Plant Air Cleaning Units and Components	ASME	ANSI/ASME N509-1989
	Energy Efficiency Test Methods for Three-Phase Induction Motors (Efficiency Quoting Method and Permissible Efficiency Tolerance)	CSA	C390-93
	Motors and Generators	CSA	C22.2 No. 100-95
	Energy Management Guide for Selection and Use of Polyphase Motors	NEMA	NEMA MG 10-1994
	Energy Management Guide for Selection and Use of Single-Phase Motors	NEMA	NEMA MG 11-1977 (R 1992)
	Motion/Position Control Motors, Controls, and Feedback Devices	NEMA	NEMA MG 7-1993
	Motors and Generators	NEMA	NEMA MG 1-1993
	Electric Motors (1994)	UL	UL 1004
	Electric Motors and Generators for Use in Division 1 Hazardous (Classified) Locations (1994)	UL	ANSI/UL 674-1993
	Impedance-Protected Motors (1994)	UL	UL 519
	Thermal Protectors for Motors (1991)	UL	ANSI/UL 547-1991
Operation and Maintenance	Preparation of Operating and Maintenance Documentation for Building Systems	ASHRAE	ASHRAE Guideline 4-1993
Pipe, Tubing, and Fittings	Building Services Piping	ASME	ANSI/ASME B31.9-1988
	Pipe Threads, General Purpose (Inch)	ASME	ANSI/ASME B1.20.1-1983 (R 1992)
	Power Piping	ASME	ANSI/ASME B31.1-1989
	Refrigeration Piping	ASME	ASME/ANSI B31.5-1987
	Scheme for the Identification of Piping Systems	ASME	ANSI/ASME A13.1-1981 (R 1985)
	Welded and Seamless Wrought Steel Pipe	ASME	ASME/ANSI B36.10M-1995
	Specification for Acrylonitrile-Butadiene-Styrene (ABS) Plastic Pipe, Schedules 40 and 80	ASTM	ASTM D 1527-89
	Specification for Chlorinated Polyvinyl Chloride (CPVC) Plastic Pipe, Schedules 40 and 80	ASTM	ASTM F 441/F 441M-96A
	Specification for Pipe, Steel, Black and Hot-Dipped, Zinc-Coated, Welded and Seamless	ASTM	ASTM A 53-96
	Specification for Plastic Insert Fittings for Polybutylene (PB) Tubing	ASTM	ASTM F 845-95
	Specification for Polybutylene (PB) Plastic Hot- and Cold- Water Distribution Systems	ASTM	ASTM D 3309-96
	Specification for Polybutylene (PB) Plastic Pipe (SIDR-PR) Based on Controlled Inside Diameter	ASTM	ASTM D 2662-96
	Specification for Polybutylene (PB) Plastic Pipe (SDR-PR) Based on Outside Diameter	ASTM	ASTM D 3000-95A
	Specification for Polybutylene (PB) Plastic Tubing	ASTM	ASTM D 2666-96
	Specification for Polyethylene (PE) Plastic Pipe, Schedule 40	ASTM	ASTM D 2104-90
	Specification for Polyvinyl Chloride (PVC) Plastic Pipe, Schedules 40, 80, and 120	ASTM	ASTM D 1785-96A
	Specification for Seamless Carbon Steel Pipe for High-Temperature Service	ASTM	ASTMA 106-95
	Specification for Seamless Copper Pipe, Standard Sizes	ASTM	ASTM B 42-89
	Specification for Seamless Copper Tube for Air Conditioning and Refrigeration Field Service	ASTM	ASTM B 280-88
	Specification for Seamless Copper Water Tube	ASTM	ASTM B 88-96
	Specification for Welded Copper and Copper Alloy Tube for Air Conditioning and Refrigeration Service	ASTM	ASTM B 640-90
	Standard Practice for Obtaining Hydrostatic or Pressure Design Basis for "Fiberglass" (Glass-Fiber-Reinforced Thermosetting-Resin) Pipe and Fittings	ASTM	ANSI/ASTM D 2992-96
	Test Method for Obtaining Hydrostatic Design Basis for Thermoplastic Pipe Materials	ASTM	ASTM D 2837-92
	Thickness Design of Ductile-Iron Pipe	AWWA	ANSI/AWWA C150/A21.50-91
	Standards of the Expansion Joint Manufacturers Association, Inc., 6th ed. (1993)	EJMA	EJMA
	Guideline for Quality Piping Installation	MCAA	MCAA
	Pipe Hangers and Supports—Materials, Design and Manufacture	MSS	MSS SP-58-93

Table 1 Codes and Standards Published by Various Societies and Associations (*Continued*)

Subject	Title	Publisher	Reference
Pipe, Tubing, and Fittings (continued)	Pipe Hangers and Supports—Selection and Application	MSS	MSS SP-69-91
	Welding Procedure Specifications	NCPWB	NCPWB
	Corrugated Polyolefin Coilable Plastic Utilities Duct	NEMA	NEMA TC 5-1990
	Corrugated Polyvinyl-Chloride (PVC) Coilable Plastic Utilities Duct	NEMA	NEMA TC 12-1991
	Electrical Nonmetallic Tubing (ENT)	NEMA	NEMA TC 13-1993
	Electrical Plastic Tubing (EPT) and Conduit Schedule EPC-40 and EPC-80	NEMA	NEMA TC 2-1990
	Extra-Strength PVC Plastic Utilities Duct for Underground Installation	NEMA	NEMA TC 8-1990
	Filament-Wound Reinforced Thermosetting Resin Conduit and Fittings	NEMA	NEMA TC 14-1984 (R 1986)
	Fittings, Cast Metal Boxes, and Conduit Bodies for Conduit and Cable Assemblies	NEMA	NEMA FB 1-1993
	Fittings for ABS and PVC Plastic Utilities Duct for Underground Installation	NEMA	NEMA TC 9-1990
	Polyvinyl-Chloride (PVC) Externally Coated Galvanized Rigid Steel Conduit and Intermediate Metal Conduit	NEMA	NEMA RN 1-1989
	PVC and ABS Plastic Utilities Duct for Underground Installation	NEMA	NEMA TC 6-1990
	Smooth-Wall Coilable Polyethylene Electrical Plastic Duct	NEMA	ANSI/NEMA TC 7-1990
	National Fuel Gas Code	NFPA	ANSI/NFPA 54-1992
		AGA	ANSI Z223.1-1992; Z223.1a-1994
	Plastics Piping Components and Related Materials	NSF	ANSI/NSF 14-1990
	Refrigeration Tube Fittings	SAE	ANSI/SAE J 513-1994
	Seismic Restraint Manual Guidelines for Mechanical Systems (1991)	SMACNA	SMACNA
	Rubber Gasketed Fittings for Fire-Protection Service (1993)	UL	UL 213
	Tube Fittings for Flammable and Combustible Fluids, Refrigeration Service, and Marine Use (1993)	UL	UL 109
Plumbing	BOCA National Plumbing Code, 9th ed. (1993)	BOCA	BOCA
	Uniform Plumbing Code (1994) (with IAPMO Installation Standards)	IAPMO	IAPMO
	Safety Requirements for Plumbing	IAPMO/ MCAA/ NAPHCC	ANSI A40-1993 (1996 Pending)
	International Plumbing Code, 1st ed. (1995)	ICC	BOCA/ICBO/SBCCI
	International Private Sewage Disposal Code	ICC	BOCA/ICBO/SBCCI
	National Standard Plumbing Code (NSPC)	NAPHCC	NSPC 1996
	Standard Plumbing Code (1994 ed. with 1995 revisions)	SBCCI	SBCCI
Pumps	Centrifugal Pumps	ASME	ASME PTC 8.2-1990
	Displacement Compressors, Vacuum Pumps and Blowers	ASME	ANSI/ASME PTC 9-1970 (R 1992)
	Liquid Pumps	CSA	CAN/CSA C.22.2 No. 108-M89
	Performance Standard for Liquid Ring Vacuum Pumps, 1st ed. (1987)	HEI	HEI
	Centrifugal Pumps	HI	ANSI/HI 1.1-1.5 (1994)
	Vertical Pumps	HI	ANSI/HI 2.1-2.5 (1994)
	Rotary Pumps	HI	ANSI/HI 3.1-3.5 (1994)
	Sealless Rotary Pumps	HI	ANSI/HI 4.1-4.6 (1994)
	Sealless Centrifugal Pumps	HI	ANSI/HI 5.1-5.6 (1994)
	Reciprocating Power Pumps	HI	ANSI/HI 6.1-6.5 (1994)
	Controlled Volume Pumps	HI	ANSI/HI 7.1-7.5 (1994)
	Direct Acting (Steam) Pumps	HI	ANSI/HI 8.1-8.5 (1994)
	Pumps—General Guidelines	HI	ANSI/HI 9.1-9.6 (1994)
	Engineering Data Book, 2nd ed. (1990)	HI	HI
	Circulation System Components and Related Materials for Swimming Pools, Spas/Hot Tubs	NSF	ANSI/NSF 50-1992
	Swimming Pool Pumps, Filters and Chlorinators (1993)	UL	UL 1081
	Motor-Operated Water Pumps (1991)	UL	ANSI/UL 778-1991
	Pumps for Oil-Burning Appliances (1993)	UL	ANSI/UL 343-1992
Radiators	Testing and Rating Standard for Baseboard Radiation, 6th ed. (1990)	HYDI	IBR
	Testing and Rating Standard for Finned-Tube (Commercial) Radiation (1990)	HYDI	IBR
Receivers	Refrigerant Liquid Receivers	ARI	ARI 495-93
Refrigerant-Containing Components	Refrigerant-Containing Components for Use in Electrical Equipment	CSA	C22.2 No. 140.3-M1987 (R 1993)
	Refrigerant-Containing Components and Accessories, Non-Electrical (1993)	UL	ANSI/UL 207-1994
Refrigerants	Refrigerant Recovery Recycling Equipment	ARI	ARI 740-95
	Specifications for Fluorocarbon and Other Refrigerants	ARI	ARI 700-93
	Format for Information on Refrigerants	ASHRAE	ASHRAE Guideline 6-1996
	Method of Testing Flow Capacity of Refrigerant Capillary Tubes	ASHRAE	ANSI/ASHRAE 28-1996
	Methods of Testing Discharge Line Refrigerant-Oil Separators	ASHRAE	ANSI/ASHRAE 69-1990
	Number Designation and Safety Classification of Refrigerants	ASHRAE	ANSI/ASHRAE 34-1992 with addenda
	Reducing Emission of Halogenated Refrigerants in Refrigeration and Air-Conditioning Equipment and Systems	ASHRAE	ASHRAE Guideline 3-1996
	Refrigeration Oil Description	ASHRAE	ANSI/ASHRAE 99-1981 (RA 87)
	Sealed Glass Tube Method to Test the Chemical Stability of Material for Use Within Refrigerant Systems	ASHRAE	ANSI/ASHRAE 97-1983 (RA 89)
	Field Conversion/Retrofit of Products to Change to an Alternate Refrigerant—Construction and Operation (1993)	UL	ANSI/UL 2170-1995

Table 1 Codes and Standards Published by Various Societies and Associations (*Continued*)

Subject	Title	Publisher	Reference
Refrigerants *(continued)*	Field Conversion/Retrofit of Products to Change to an Alternate Refrigerant—Insulating Material and Refrigerant Compatibility (1993)	UL	ANSI/UL 2171-1995
	Field Conversion/Retrofit of Products to Change to an Alternate Refrigerant—Procedures and Methods (1993)	UL	ANSI/UL 2172-1995
	Refrigerant Recovery/Recycling Equipment (1989)	UL	ANSI/UL 1963-1991
	Refrigerants (1994)	UL	UL 2182
	Recommended Service Procedure for the Containment of HFC-134a	SAE	SAE J 2211-1991
	HFC-134a Recycling Equipment for Mobile Air-Conditioning Systems	SAE	SAE J 2210-1991
	CFC-12 (R-12) Extraction Equipment for Mobile Automotive Air-Conditioning Systems	SAE	SAE J 2209-1992
	HFC-134a (R-134a) Service Hose Fittings for Automotive Air-Conditioning Service Equipment	SAE	SAE J 2197-1992
	Standard of Purity for Recycled HFC-134a for Use in Mobile Air Conditioning Systems	SAE	SAE J 2099-1991
	Recommended Service Procedure for the Containment of R-12	SAE	SAE J 1989-1989
	Procedure for Retrofitting CFC-12 (R12) Mobile Air Conditioning Systems to HFC 134-a (R134a)	SAE	ANSI/SAE J 1661-1993
Refrigeration	Safety Code for Mechanical Refrigeration	ASHRAE	ANSI/ASHRAE 15-1994
	Mechanical Refrigeration Code	CSA	B52-95
	Refrigeration Equipment	CSA	CAN/CSA-C22.2 No. 120-M91
	Equipment, Design and Installation of Ammonia Mechanical Refrigerating Systems	IIAR	ANSI/IIAR 2-1992
	Refrigerated Medical Equipment (1993)	UL	UL 416
Refrigeration Systems			
Steam Jet	Ejectors	ASME	ASME PTC 24-1976 (R 1982)
	Standards for Steam Jet Vacuum Systems, 4th ed. (1988)	HEI	HEI
Transport	Mechanical Transport Refrigeration Units	ARI	ARI 1110-92
	Mechanical Refrigeration and Air-Conditioning Installations Aboard Ship	ASHRAE	ASHRAE 26-1996
	General Requirements for Application of Vapor Cycle Refrigeration Systems for Aircraft	SAE	SAE ARP 731A-1973 (R 1992)
	Safety and Containment of Refrigerant for Mechanical Vapor Compression Systems Used for Mobile Air-Conditioning Systems	SAE	SAE J 639-1994
Refrigerators	Method of Testing Open Refrigerators for Food Stores	ASHRAE	ANSI/ASHRAE 72-1983
	Methods of Testing Closed Refrigerators	ASHRAE	ANSI/ASHRAE 117-1992
Commercial	Energy Performance Standard for Commercial Refrigerated Display Cabinets and Merchandise	CSA	CAN/CSA-C657-95
	Food Carts	NSF	ANSI/NSF 59-1986
	Food Equipment	NSF	ANSI/NSF 2-1992
	Food Service Refrigerators and Storage Freezers	NSF	ANSI/NSF 7-1990
	Soda Fountain and Luncheonette Equipment	NSF	NSF 1-1984
	Commercial Refrigerators and Freezers (1992)	UL	ANSI/UL 471-1991
	Refrigerating Units (1994)	UL	UL 427
	Refrigeration Unit Coolers (1993)	UL	ANSI/UL 412-1992
Household	Refrigerators Using Gas Fuel	AGA	ANSI Z21.19-1990; Z21.19a-1992
	Household Refrigerators, Combination Refrigerator-Freezers and Household Freezers	AHAM	ANSI/AHAM HRF-1-1988
	Capacity Measurement and Energy Consumption Test Methods for Refrigerators, Combination Refrigerator-Freezers, and Freezers	CSA	CAN/CSA C300-M91
	Household Refrigerators and Freezers (1993)	UL CSA	UL 250 CAN/CSA C22.2 No. 63-93
Retrofitting			
Building	Retrofit of Building Energy Systems and Processes (1982)	SMACNA	SMACNA
	Building Systems Analysis and Retrofit Manual (1995)	SMACNA	SMACNA
Refrigerant	Procedure for Retrofitting CFC-12 (R12) Mobile Air Conditioning Systems to HFC-134a (R134a)	SAE	ANSI/SAE J 1661-1993
	Field Conversion/Retrofit of Products to Change to an Alternate Refrigerant—Construction and Operation (1993)	UL	ANSI/UL 2170-1995
	Field Conversion/Retrofit of Products to Change to an Alternate Refrigerant—Insulating Material and Refrigerant Compatibility (1993)	UL	ANSI/UL 2171-1995
	Field Conversion/Retrofit of Products to Change to an Alternate Refrigerant—Procedures and Methods (1993)	UL	ANSI/UL 2172-1995
Roof Ventilators	Commercial Low Pressure, Low Velocity Duct System Design (1990)	ACCA	ACCA Manual Q
	Power Ventilators (1984)	UL	ANSI/UL 705-1984
Solar Equipment	Method of Measuring Solar-Optical Properties of Materials	ASHRAE	ANSI/ASHRAE 74-1988
	Methods of Testing to Determine the Thermal Performance of Flat-Plate Solar Collectors Containing a Boiling Liquid	ASHRAE	ANSI/ASHRAE 109-1986 (RA 90)
	Methods of Testing to Determine the Thermal Performance of Solar Collectors	ASHRAE	ANSI/ASHRAE 93-1986 (RA 96)
	Methods of Testing to Determine the Thermal Performance of Solar Domestic Water Heating Systems	ASHRAE	ASHRAE 95-1981 (RA 87)
	Methods of Testing to Determine the Thermal Performance of Unglazed Flat-Plate Liquid-Type Solar Collectors	ASHRAE	ANSI/ASHRAE 96-1980 (RA 89)

Table 1 Codes and Standards Published by Various Societies and Associations (*Continued*)

Subject	Title	Publisher	Reference
Solar Equipment *(continued)*	Reference Solar Spectral Irradiance at the Ground at Different Receiving Conditions—Part 1: Direct Normal and Hemispherical Solar Irradiance for Air Mass 1.5	ISO	ISO 9845-1:1992
	Solar Energy—Calibration of a Pyranometer Using a Pyrheliometer	ISO	ISO 9846:1993
	Solar Heating—Domestic Water Heating Systems—Part 1: Performance Rating Procedure Using Indoor Test Methods	ISO	ISO 9459-1:1993
	Solar Heating—Domestic Water Heating Systems—Part 2: Outdoor Test Methods for System Performance Characterization and Yearly Performance Prediction of Solar-Only Systems	ISO	ISO 9459-2:1995
	Solar Heating—Swimming Pool Heating Systems—Dimensions, Design and Installation Guidelines	ISO	ISO 12596:1995
	Solar Water Heaters—Elastomeric Materials for Absorbers, Connecting Pipes and Fittings—Method of Assessment	ISO	ISO 9808:1990
	Test Methods for Solar Collectors—Part 2: Qualification Test Procedures	ISO	ISO 9806-2:1995
	Test Methods for Solar Collectors—Part 3: Thermal Performance of Unglazed Liquid Heating Collectors (Sensible Heat Transfer Only) Including Pressure Drop	ISO	ISO 9806-3:1995
Solenoid Valves	Solenoid Valves for Use with Volatile Refrigerants	ARI	ANSI/ARI 760-94
Sound Measurement	Method for the Calibration of Microphones (reaffirmed 1986)	ASA	ANSI S1.10-1966 (R 1986)
	Specification for Sound Level Meters	ASA	ANSI S1.4-1983; ANSI S1.4A-1985 (R 1990)
	Test Method for Laboratory Measurement of Airborne Sound Transmission Loss of Building Partitions	ASTM	ASTM E 90-90
	Test Method for Measuring Acoustical and Airflow Performance of Duct Liner Materials and Prefabricated Silencers	ASTM	ASTM E 477-90
	Sound and Vibration Design and Analysis (1994)	NEBB	NEBB
Fans	Methods for Calculating Fan Sound Ratings from Laboratory Test Data	AMCA	AMCA 301-90
	Reverberant Room Method for Sound Testing of Fans	AMCA	AMCA 300-94
	Methods for the Measurement of Noise Emitted by Small Air-Moving Devices	ASA	ANSI S12.11-1987 (R 1993)
	Laboratory Method of Testing In-Duct Sound Power Measurement Procedure for Fans	ASHRAE AMCA	ANSI/ASHRAE 68-1986 ANSI/AMCA 330-86
	Acoustics—Method for the Measurement of Airborne Noise Emitted by Small Air-Moving Devices	ISO	ISO 10302:1996
Other Equipment	Application of Sound Rated Outdoor Unitary Equipment	ARI	ARI 275-84
	Method of Measuring Machinery Sound Within Equipment Space	ARI	ARI 575-94
	Method of Measuring Sound and Vibration of Refrigerant Compressors	ARI	ANSI/ARI 530-89
	Rating the Sound Levels and Sound Transmission Loss of Packaged Terminal Equipment	ARI	ANSI/ARI 300-88
	Sound Rating of Large Outdoor Refrigerating and Air-Conditioning Equipment	ARI	ARI 370-86
	Sound Rating of Non-Ducted Indoor Air-Conditioning Equipment	ARI	ARI 350-86
	Sound Rating of Outdoor Unitary Equipment	ARI	ARI 270-84
	Statistical Methods for Determining and Verifying Stated Noise Emission Values of Machinery and Equipment	ASA	ANSI S12.3-85 (R 1996)
	Sound Level Prediction for Installed Rotating Electrical Machines	NEMA	NEMA MG 3-1974 (R 1990)
Techniques	Criteria for Evaluating Room Noise	ASA	ANSI S12.2-1995
	Engineering Method for the Determination of Sound Power Levels of Noise Sources Using Sound Intensity	ASA	ANSI S12.12-1992
	Engineering Methods for the Determination of Sound Power Levels of Noise Sources for Essentially Free-Field Conditions over a Reflecting Plane	ASA	ANSI S12.34-1988 (R 1993)
	Guidelines for the Use of Sound Power Standards and for the Preparation of Noise Test Codes	ASA	ANSI S12.30-1990
	Measurement of Sound Pressure Levels in Air	ASA	ANSI S1.13-1995
	Methods for Determination of Insertion Loss of Outdoor Noise Barriers	ASA	ANSI S12.8-1987
	Methods for the Determination of Sound Power Levels of Noise Sources in a Special Reverberation Test Room	ASA	ANSI S12.33-1990
	Precision Methods for the Determination of Sound Power Levels of Broad-Band Noise Sources in Reverberation Rooms	ASA	ANSI S12.31-1990 (R 1996)
	Precision Methods for the Determination of Sound Power Levels of Discrete-Frequency and Narrow-Band Noise Sources in Reverberation Rooms	ASA	ANSI S12.32-1990
	Precision Methods for the Determination of Sound Power Levels of Noise Sources in Anechoic and Hemi-Anechoic Rooms	ASA	ANSI S12.35-1990 (R 1996)
	Preferred Frequencies, Frequency Levels, and Band Numbers for Acoustical Measurements	ASA	ANSI S1.6-1984 (R 1990)
	Procedure for the Computation of Loudness of Noise	ASA	ANSI S3.4-1980 (R 1992)
	Procedures for Outdoor Measurement of Sound Pressure Level	ASA	ANSI S12.18-1994
	Reference Quantities for Acoustical Levels	ASA	ANSI S1.8-1989
	Survey Methods for the Determination of Sound Power Levels of Noise Sources	ASA	ANSI S12.36-1990
	Measurement of Industrial Sound	ASME	ASME/ANSI PTC 36-1985
	Test Method for Evaluating Masking Sound in Open Offices Using A-Weighted and One-Third Octave Band Sound Pressure Levels	ASTM	ASTM E 1573-93
	Test Method for Measurement of Sound in Residential Spaces	ASTM	ASTM E 1574-95
	Acoustics—Determination of Sound Power Levels of Noise Sources Using Sound Intensity—Part 1: Measurement at Discrete Points	ISO	ISO 9614-1:1993

Table 1 Codes and Standards Published by Various Societies and Associations (*Continued*)

Subject	Title	Publisher	Reference
Techniques (*continued*)	Acoustics—Determination of Sound Power Levels of Noise Sources Using Sound Intensity—Part 2: Measurement by Scanning	ISO	ISO 9614-2:1996
	Acoustics—Method for Calculating Loudness Level	ISO	ISO 532:1975
	Procedural Standards for the Measurement and Assessment of Sound and Vibration (1994)	NEBB	NEBB
Terminology	Acoustical Terminology	ASA	ANSI S1.1-1994
	Standard Terminology Relating to Environmental Acoustics	ASTM	ASTM C 634-89
Space Heaters	Method of Testing for Rating Combination Space-Heating and Water-Heating Appliances	ASHRAE	ANSI/ASHRAE 124-1991
	Electric Air Heaters	CSA	C22.2 No. 46-M1988
	Electric Air Heaters (1980)	UL	ANSI/UL 1025-1991
	Fixed and Location-Dedicated Electric Room Heaters (1992)	UL	UL 2021
	Movable and Wall- or Ceiling-Hung Electric Room Heaters (1994)	UL	UL 1278
	Gas-Fired Room Heaters, Vol. I, Vented Room Heaters	AGA	ANSI Z21.11.1-1991; Z21.11.1a-1993; Z21.11.1b-1995
	Gas-Fired Room Heaters, Vol. II, Unvented Room Heaters	AGA	ANSI Z21.11.2-1992 Z21.11.2a-1993; Z21.11.2b-1995
Symbols	Graphic Electrical/Electronic Symbols for Air-Conditioning and Refrigeration Equipment	ARI	ARI 130-88
	Graphic Symbols for Electrical and Electronic Diagrams	IEEE	ANSI/IEEE 315-1975 (R 1994)
	Graphic Symbols for Heating, Ventilating, and Air Conditioning	ASME	ANSI/ASME Y32.2.4-1949 (R 1993)
	Graphic Symbols for Pipe Fittings, Valves and Piping	ASME	ANSI/ASME Y32.2.3-1949 (R 1994)
	Graphic Symbols for Plumbing Fixtures for Diagrams used in Architecture and Building Construction	ASME	ANSI/ASME Y32.4-1977 (R 1994)
	Symbols for Mechanical and Acoustical Elements as used in Schematic Diagrams	ASME	ANSI/ASME Y32.18-1972 (R 1993)
Terminals, Wiring	Residential Controls—Quick Connect Terminals	NEMA	NEMA DC 2-1982 (R 1988)
	Electrical Quick-Connect Terminals (1995)	UL	UL 310
	Equipment Wiring Terminals for Use with Aluminum and/or Copper Conductors (1994)	UL	ANSI/UL 486E-1994
	Splicing Wire Connectors (1991)	UL	ANSI/UL 486C-1990
	Wire Connectors and Soldering Lugs for Use with Copper Conductors (1991)	UL	ANSI/UL 486A-1990
	Wire Connectors for Use with Aluminum Conductors (1991)	UL	ANSI/UL 486B-1990
Testing and Balancing	Industrial Process/Power Generating Fans: Site Performance Test Standard	AMCA	AMCA 803-94
	The HVAC Commissioning Process	ASHRAE	ASHRAE Guideline 1-1996
	Practices for Measurement, Testing, Adjusting, and Balancing of Building Heating, Ventilating, Air-Conditioning, and Refrigeration Systems	ASHRAE	ANSI/ASHRAE 111-1988
	Centrifugal Pump Test	HI	ANSI/HI 1.6-1994
	Vertical Pump Tests	HI	ANSI/HI 2.6-1994
	Rotary Pump Tests	HI	ANSI/HI 3.6-1994
	Reciprocating Pump Tests	HI	ANSI/HI 6.6-1994
	Pumps—General Guidelines (Including "Measurement of Airborne Sound")	HI	HI 9.1-9.6-1994
	Procedural Standards for Certified Testing of Cleanrooms (1988)	NEBB	NEBB
	Procedural Standards for Testing, Adjusting, Balancing of Environmental Systems, 5th ed. (1991)	NEBB	NEBB
	Building Systems Analysis and Retrofit Manual (1995)	SMACNA	SMACNA
	HVAC Systems—Testing, Adjusting and Balancing (1993)	SMACNA	SMACNA
Thermal Storage	Method of Testing Active Sensible Thermal Energy Storage Devices Based on Thermal Performance	ASHRAE	ANSI/ASHRAE 94.3-1986 (RA 96)
	Method of Testing Active Latent Heat Storage Devices Based on Thermal Performance	ASHRAE	ANSI/ASHRAE 94.1-1985 (RA 91)
	Methods of Testing Thermal Storage Devices with Electrical Input and Thermal Output Based on Thermal Performance	ASHRAE	ANSI/ASHRAE 94.2-1981 (RA 96)
	Practices for Measurement, Testing, Adjusting, and Balancing of Building Heating, Ventilation, Air-Conditioning, and Refrigeration Systems	ASHRAE	ANSI/ASHRAE 111-1988
Turbines	Steam Turbines	ASME	ANSI/ASME PTC 6-1996
	Wind Turbines	ASME	ANSI/ASME PTC 42-1988
	Standard Specification for Gas Turbine Fuel Oils	ASTM	ANSI/ASTM D 2880-96
	Land Based Steam Turbine Generator Sets	NEMA	NEMA SM 24-1991
	Steam Turbines for Mechanical Drive Service	NEMA	ANSI/NEMA SM 23-1991
Valves	Methods of Testing Nonelectric, Nonpneumatic Thermostatic Radiator Valves	ASHRAE	ANSI/ASHRAE 102-1983 (RA 89)
	Face-to-Face and End-to-End Dimensions of Valves	ASME	ANSI/ASME B16.10-1992
	Pressure Relief Devices	ASME	ANSI/ASME PTC 25-1994
	Valves—Flanged Threaded, and Welding End	ASME	ANSI/ASME B16.34-1988
	Control Valve Capacity Test Procedure	ISA	ANSI/ISA S75.02-88
	Flow Equations for Sizing Control Valves	ISA	ANSI/ISA S75.01-85 (R 1995)
	High Pressure Oxygen System Filler Valve	SAE	SAE AS 1225
	Electrically Operated Valves (1994)	UL	UL 429
	Pressure Regulating Valves for LP-Gas (1994)	UL	UL 144
	Safety Relief Valves for Anhydrous Ammonia and LP-Gas (1993)	UL	UL 132R

Table 1 Codes and Standards Published by Various Societies and Associations (*Continued*)

Subject	Title	Publisher	Reference
Valves *(continued)* Gas	Valves for Anhydrous Ammonia and LP-Gas (Other than Safety Relief) (1993)	UL	UL 125
	Valves for Flammable Fluids (1993)	UL	ANSI/UL 842-1992
	Automatic Gas Valves for Gas Appliances	AGA	ANSI Z21.21-1993
	Combination Gas Controls for Gas Appliances	AGA	ANSI Z21.78-1992; Z21.78a-1993; Z21.78b-1994
	Manually Operated Gas Valves for Appliances, Appliance Connection Valves, and Hose End Valves	AGA	ANSI Z21.15-1992
	Relief Valves and Automatic Gas Shutoff Devices for Hot Water Supply Systems	AGA	ANSI Z21.22-1986; Z21.22a-1990
	Requirements for Automatic Non-Shutoff Modulating Gas Valves	AGA	1-92
	Requirements for Gas Operated Valves for High Pressure Natural Gas	AGA	3-93
	Requirements for Manually Operated Gas Valves for Use in House Piping Systems	AGA	3-88
	Requirements for Manually Operated Valves for High Pressure Natural Gas	AGA	2-93
	Large Metallic Valves for Gas Distribution (Manually Operated, NPS-2 1/2 to 12, 125 psig Maximum)	ASME	ANSI/ASME B16.38-1985 (R 1994)
	Manually Operated Metallic Gas Valves for Use in Gas Piping Systems up to 125 psig	ASME	ANSI/ASME B16.33-1990
	Manually Operated Thermoplastic Gas Shutoffs and Valves in Gas Distribution Systems	ASME	ANSI/ASME B16.40-1985 (R 1994)
Refrigerant	Refrigerant Access Valves and Hose Connectors	ARI	ANSI/ARI 720-94
	Refrigerant Pressure Regulating Valves	ARI	ARI 770-84
	Solenoid Valves for Use with Volatile Refrigerants	ARI	ANSI/ARI 760-94
	Thermostatic Refrigerant Expansion Valves	ARI	ANSI/ARI 750-94
Vapor Retarders	Standard Practice for Selection of Vapor Retarders for Thermal Insulation	ASTM	ASTM C 755-85 (R 1990)
	Standard Practice for Determining the Properties of Jacketing Material for Thermal Insulation	ASTM	ASTM C 921-89
	Standard Specification for Flexible, Low Permeance Vapor Retarders for Thermal Insulation	ASTM	ASTM C 1136-92
	Test Method for Water Vapor Transmission Rate of Flexible Barrier Materials Using an Infrared Detection Technique	ASTM	ASTM F 372-94
Vending Machines	Methods of Testing Pre-Mix and Post-Mix Soft Drink Vending and Dispensing Equipment	ASHRAE	ANSI/ASHRAE 91-1976 (RA 89)
	Vending Machines	CSA	CAN/CSA-C22.2 No.128-95
	Vending Machines for Food and Beverages	NSF	ANSI/NSF 25-1990
	Vending Machines (1995)	UL	UL 751
	Refrigerated Vending Machines (1995)	UL	UL 541
Vent Dampers	Automatic Vent Damper Devices for Use with Gas-Fired Appliances	AGA	ANSI Z21.66-1994
	Vent or Chimney Connector Dampers for Oil-Fired Appliances (1994)	UL	UL 17-1995
Ventilation	Commercial Low Pressure, Low Velocity Duct System Design (1990)	ACCA	ACCA Manual Q
	Guide for Testing Ventilation Systems	ACGIH	ACGIH
	Industrial Ventilation: A Manual of Recommended Practice, 22nd ed. (1995)	ACGIH	ACGIH
	A Method of Determining Air Change Rates in Detached Dwellings	ASHRAE	ANSI/ASHRAE 136-1993
	Method of Testing for Room Air Diffusion	ASHRAE	ANSI/ASHRAE 113-1990
	Ventilation for Acceptable Indoor Air Quality	ASHRAE	ANSI/ASHRAE 62-1989
	Design of Ventilation Systems for Poultry and Livestock Shelters	ASAE	ASAE D270.5-1991
	Residential Mechanical Ventilation Systems	CSA	CAN/CSA F326-M91
	Installation of Air Conditioning and Ventilating Systems	NFPA	ANSI/NFPA 90A-1993
	Parking Structures	NFPA	ANSI/NFPA 88A-1995
	Repair Garages	NFPA	ANSI/NFPA 88B-1991
	Ventilation Control and Fire Protection of Commercial Cooking Operations	NFPA	ANSI/NFPA 96-1994
	Food Equipment	NSF	ANSI/NSF 2-1992
	Class II (Laminar Flow) Biohazard Cabinetry	NSF	NSF 49-1992
	Test Procedure for Battery Flame Retardant Venting Systems	SAE	SAE J 1495-1992
	Heater, Airplane, Engine Exhaust Gas to Air Heat Exchanger Type	SAE	SAE ARP 86A-1952 (R 1992)
	Aerothermodynamic Systems Engineering and Design	SAE	SAE AIR 1168/3-1990
Venting	Draft Hoods	AGA	ANSI Z21.12-1990; Z21.12a-1993; Z21.12b-1994
	National Fuel Gas Code	AGA	ANSI Z223.1-1992; Z223.1a-1994
		NFPA	ANSI/NFPA 54-1992
	Requirements for Electrically Operated Automatic Combustion and Ventilation Air Control Devices for Use with Gas-Fired Appliances	AGA	1-88
	Requirements for Mechanical Venting Systems	AGA	6-90
	Chimneys, Fireplaces, Vents and Solid Fuel-Burning Appliances	NFPA	ANSI/NFPA 211-1992
	Explosion Prevention Systems	NFPA	ANSI/NFPA 69-1992
	Smoke and Heat Venting	NFPA	ANSI/NFPA 204M-1991
	Guide for Steel Stack Design and Construction (1996)	SMACNA	SMACNA
	Draft Equipment (1993)	UL	UL 378
	Gas Vents (1994)	UL	UL 441
	Low-Temperature Venting Systems, Type L (1994)	UL	ANSI/UL 641-1995

Table 1 Codes and Standards Published by Various Societies and Associations (*Continued*)

Subject	Title	Publisher	Reference
Vibration	Mechanical Vibration of Rotating and Reciprocating Machinery—Requirements for Instruments for Measuring Vibration Severity	ASA	ANSI S2.40-1984 (R 1992)
	Methods for Analysis and Presentation of Shock and Vibration Data	ASA	ANSI S2.10-1971 (R 1990)
	Selection of Calibrations and Tests for Electrical Transducers Used for Measuring Shock and Vibration	ASA	ANSI S2.11-1969 (R 1986)
	Techniques of Machinery Vibration Measurement	ASA	ANSI S2.17-1980 (R 1986)
	Vibrations of Buildings—Guidelines for the Measurement of Vibrations and Evaluation of Their Effects on Buildings	ASA	ANSI S2.47-1990
	Evaluation of Human Exposure to Whole-Body Vibration—Part 2: Continuous and Shock-Induced Vibrations in Buildings (1 to 80 Hz)	ISO	ISO 2631-2:1989
	Guidelines for the Evaluation of the Response of Occupants of Fixed Structures, Especially Buildings and Off-Shore Structures, to Low-Frequency Horizontal Motion (0.063 to 1 Hz)	ISO	ISO 6897:1984
	Procedural Standards for the Measurement and Assessment of Sound and Vibration (1994)	NEBB	NEBB
	Sound and Vibration Design and Analysis (1994)	NEBB	NEBB
Water Heaters	Gas Water Heaters, Vol. I, Storage Water Heaters with Input Ratings of 75,000 Btu per Hour or Less	AGA	ANSI Z21.10.1-1993; Z21.10.1a-1993; Z21.10.1b-1994
	Gas Water Heaters, Vol. III, Storage, with Input Ratings Above 75,000 Btu per Hour, Circulating and Instantaneous Water Heaters	AGA	ANSI Z21.10.3-1993; Z21-10.3a-1993; Z21.10.3b-1994
	Requirements for Non-Metallic Dip Tubes for Use in Gas-Fired Water Heaters	AGA	1-89
	Requirements for Indirect Water Heaters for Use with External Heat Source	AGA	1-91
	Desuperheater/Water Heaters	ARI	ARI 470-87
	Method of Testing for Rating Commercial Gas, Electric, and Oil Water Heaters	ASHRAE	ANSI/ASHRAE 118.1-1993
	Method of Testing for Rating Residential Water Heaters	ASHRAE	ANSI/ASHRAE 118.2-1993
	Methods of Testing for Efficiency of Space-Conditioning/Water-Heating Appliances that Include a Desuperheater Water Heater	ASHRAE	ANSI/ASHRAE 137-1995
	Methods of Testing to Determine the Thermal Performance of Solar Domestic Water Heating Systems	ASHRAE	ANSI/ASHRAE 95-1981 (RA 87)
	Method of Testing for Rating Combination Space-Heating and Water-Heating Appliances	ASHRAE	ANSI/ASHRAE 124-1991
	Construction and Test of Electric Storage-Tank Water Heaters	CSA	CAN/CSA-C22.2 No. 110-M90
	Performance of Electric Storage Tank Water Heaters	CSA	CAN/CSA-C191 Series-M90
	Oil Burning Stoves and Water Heaters	CSA	B140.3-1962 (R 1991)
	Oil-Fired Service Water Heaters and Swimming Pool Heaters	CSA	B140.12-1976 (R 1991)
	Water Heaters, Hot Water Supply Boilers, and Heat Recovery Equipment	NSF	NSF 5-1992
	Commercial-Industrial Gas Heating Equipment (1994)	UL	UL 795
	Electric Booster and Commercial Storage Tank Water Heaters (1995)	UL	ANSI/UL 1453-1987
	Household Electric Storage Tank Water Heaters (1995)	UL	UL 174
	Oil-Fired Storage Tank Water Heaters (1995)	UL	UL 732
Wood-Burning Appliances	Installation Code for Solid Fuel Burning Appliances and Equipment	CSA	CAN/CSA-B365-M91
	Solid-Fuel-Fired Central Heating Appliances	CSA	CAN/CSA-B366.1-M91
	Chimneys, Fireplaces, Vents, and Solid Fuel-Burning Appliances	NFPA	ANSI/NFPA 211-1992
	Commercial Cooking, Rethermalization and Powered Hot Food Holding and Transport Equipment	NSF	ANSI/NSF 4-1992
	Room Heaters, Solid-Fuel Type (1994)	UL	UL 1482

ABBREVIATIONS AND ADDRESSES

ABMA	American Boiler Manufacturers Association, 950 North Glebe Road, Suite 160, Arlington, VA 22203-1824
ACCA	Air Conditioning Contractors of America, 1712 New Hampshire Avenue, NW, Washington, D.C. 20009
ACGIH	American Conference of Governmental Industrial Hygienists, 1330 Kemper Meadow Drive, Cincinnati, OH 45240
ADC	Air Diffusion Council, 11 South Lasalle, Suite 1400, Chicago, IL 60603
AGA	American Gas Association, 1515 Wilson Boulevard, Arlington, VA 22209
	Also available through International Approval Services U.S., Inc., 8501 East Pleasant Valley Road, Cleveland, OH 44131
AHAM	Association of Home Appliance Manufacturers, 20 North Wacker Drive, Suite 1600, Chicago, IL 60606
AIHA	American Industrial Hygiene Association, 2700 Prosperity Avenue, Suite 250, Fairfax, VA 22031
AMCA	Air Movement and Control Association, Inc., 30 West University Drive, Arlington Heights, IL 60004-1893
ANSI	American National Standards Institute, 11 West 42nd Street, 13th floor, New York, NY 10036-8002
ARI	Air-Conditioning and Refrigeration Institute, 4301 North Fairfax Drive, Suite 425, Arlington, VA 22203
ASA	Acoustical Society of America, Standards Secretariat, 120 Wall Street, 32nd floor, New York, NY 10005-3993
	For ordering publications: Standards and Publications Fulfillment Center, P.O. Box 1020, Sewickley, PA 15143-9998
ASAE	American Society of Agricultural Engineers, 2950 Niles Road, St. Joseph, MI 49085-9659
ASHRAE	American Society of Heating, Refrigerating and Air-Conditioning Engineers, Inc., 1791 Tullie Circle, NE, Atlanta, GA 30329
ASME International	The American Society of Mechanical Engineers, 345 East 47 Street, New York, NY 10017-2392
	For ordering publications: ASME Marketing Department, P.O. Box 2350, Fairfield, NJ 07007-2350
ASTM	American Society for Testing and Materials, 100 Barr Harbor Drive, West Conshohocken, PA 19428-2959

ABBREVIATIONS AND ADDRESSES (*Continued*)

AWS	American Welding Society, Inc., 550 N.W. LeJeune Road, Miami, FL 33126
AWWA	American Water Works Association, 6666 W. Quincy Avenue, Denver, CO 80235
BOCA	Building Officials and Code Administrators International, Inc., 4051 West Flossmoor Road, Country Club Hills, IL 60478-5795
BSI	British Standards Institution, 389 Chiswick High Road, London W4 4AL, England
CABO	Council of American Building Officials, 5203 Leesburg Pike, Suite 708, Falls Church, VA 22041
CAGI	Compressed Air and Gas Institute, 1300 Sumner Avenue, Cleveland, OH 44115-2851
CGA	Canadian Gas Association, 55 Scarsdale Road, Toronto, ON M3B 2R3, Canada
CSA	Canadian Standards Association, 178 Rexdale Boulevard, Etobicoke (Toronto), ON M9W 1R3, Canada
CTI	Cooling Tower Institute, P.O. Box 73383, Houston, TX 77273
EJMA	Expansion Joint Manufacturers Association, Inc., 25 North Broadway, Tarrytown, NY 10591-3201
HEI	Heat Exchange Institute, 1300 Sumner Avenue, Cleveland, OH 44115-2851
HI	Hydraulic Institute, 9 Sylvan Way, Parsippany, NJ 07054-3802
HYDI	Hydronics Institute, 35 Russo Place, P.O. Box 218, Berkeley Heights, NJ 07922
IAPMO	International Association of Plumbing and Mechanical Officials, 20001 Walnut Drive South, Walnut, CA 91789-2825
ICBO	International Conference of Building Officials, 5360 Workman Mill Road, Whittier, CA 90601
ICC	International Code Council, 5360 Workman Mill Road, Whittier, CA 90601
IEEE	Institute of Electrical and Electronics Engineers, 445 Hose Lane, P.O. Box 1331 Piscataway, NJ 08855-1331
IESNA	Illuminating Engineering Society of North America, 120 Wall Street, 17th floor, New York, NY 10005-4001
IFCI	International Fire Code Institute, 5360 Workman Mill Road, Whittier, CA 90601-2298
IIAR	International Institute of Ammonia Refrigeration, 1200 19th Street, NW, Suite 300, Washington, DC 20036-2412
ISA	ISA—The International Society for Measurement and Control, P.O. Box 12777, Research Triangle Park, NC 27709
ISO	International Organization for Standardization, 1, rue de Varembé, Case postale 56, CH-1211 Genève 20, Switzerland Publications available in the U.S. from ANSI, 11 West 42nd Street, 13th floor, New York, NY 10036-8002
MCAA	Mechanical Contractors Association of America, 1385 Piccard Drive, Rockville, MD 20850-4329
MICA	Midwest Insulation Contractors Association, 2017 South 139th Circle, Omaha, NE 68144
MSS	Manufacturers Standardization Society of the Valve and Fittings Industry, Inc., 127 Park Street, N.E., Vienna, VA 22180
NAIMA	North American Insulation Manufacturers Association, 44 Canal Center Plaza, Suite 310, Alexandria, VA 22314
NAPHCC	National Association of Plumbing-Heating-Cooling Contractors, P.O. Box 6808, Falls Church, VA 22040
NCPWB	National Certified Pipe Welding Bureau, 1385 Piccard Drive, Rockville, MD 20850
NCSBCS	National Conference of States on Building Codes and Standards, 505 Huntmar Park Drive, Suite 210, Herndon, VA 22070
NEBB	National Environmental Balancing Bureau, 8575 Grovemont Circle, Gaithersburg, MD 20877-4121
NEMA	National Electrical Manufacturers Association, 1300 North 17th Street, Suite 1847, Rosslyn, VA 22209
NFPA	National Fire Protection Association, 1 Batterymarch Park, P.O. Box 9101, Quincy, MA 02269-9101
NRCC	National Research Council of Canada, Client Services, M-20, 1200 Montreal Road, Ottawa, ON K1A 0R6, Canada
NSF	NSF International, P.O. Box 130140, Ann Arbor, MI 48113-0140
SAE	Society of Automotive Engineers, 400 Commonwealth Drive, Warrendale, PA 15096-0001
SBCCI	Southern Building Code Congress International, Inc., 900 Montclair Road, Birmingham, AL 35213-1206
SMACNA	Sheet Metal and Air Conditioning Contractors' National Association, 4201 Lafayette Center Drive, Chantilly, VA 22021-1209
TEMA	Tubular Exchanger Manufacturers Association, Inc., 25 North Broadway, Tarrytown, NY 10591-3201
UL	Underwriters Laboratories Inc., 333 Pfingsten Road, Northbrook, IL 60062-2096

CHAPTER 39

BUILDING ENVELOPES

THIS chapter describes the nature, functions, and performance of building envelopes. It concentrates on the interaction between the components of building envelopes and the subsequent effects on the interior environments of buildings. Understanding this interdependence is essential to executing a reliable building envelope.

The prime functions of the building envelope are to provide:

- shelter for occupants
- strength and rigidity
- stability and durability
- control of heat, air, and moisture vapor flows
- control of liquid water movement
- indoor air quality control
- fire resistance
- cost-effectiveness
- acoustical performance
- aesthetic considerations

In buildings with a large volume to exterior surface area ratio, the internal heat sources may be large compared with gains or losses through the envelope. The effect of the envelope becomes increasingly important as the ratio decreases. In single-family dwellings, heat transfer through the envelope is the dominant factor in determining the overall heating and cooling loads. However, local control (e.g., in business offices) may depend on the ability of glazing to control heat gain or loss.

RESIDENTIAL FOUNDATIONS

Good foundation design entails not only structural soundness and good insulation, but also appropriate moisture, termite, and radon control. In North America, several building energy codes recommend foundation insulation in climates with more than 2500 heating degree days (base 65°F). An uninsulated, conditioned basement may lose up to 50% of the heat in a tightly sealed house that is well insulated above grade.

Although saving energy as a result of insulation is an important incentive for developing good foundation practices, insulating basement foundations also creates a more comfortable, livable space at a relatively low cost. Raising basement temperatures reduces condensation, minimizes problems with mold and mildew, and helps maintain acceptable indoor air quality.

Construction

The three basic types of foundations are basement, crawl space, and slab-on-grade. Each uses specific insulation techniques. The cost of higher insulation levels must be compared with the savings. The comparison can be made in several ways, but a life-cycle cost analysis is the best method (Carmody et al. 1991). It takes into account a number of economic variables, including installation costs, mortgage rates, HVAC efficiencies, and energy costs.

Conditioned Basements. A concrete or masonry basement wall may be insulated by: (1) insulating the exterior upper half, (2) insulating the entire exterior wall, or (3) insulating the entire interior wall.

Unconditioned Foundation Spaces. Unconditioned basements may be insulated in the same way as conditioned basements, or they may be insulated by placing insulation between or below the floor joists in the ceiling above the unconditioned basement. The latter approach separates the basement thermally from the above-grade space and results in lower basement temperatures in winter, usually making it necessary to insulate exposed ducts and pipes.

A crawl space can be designed as a short basement (with a floor). Because the floor is above grade it is less subject to moisture hazards than a slab floor. However, moisture problems are more likely to be noticed in a basement, and it is more accessible than a crawl space.

While most codes require operable louvered vents (1 ft^2 per 1500 ft^2 of area) near each corner of a crawl space, research has shown that a moisture retardant ground cover is more effective and, in most locations, eliminates the need for vents (ASHRAE 1994).

In a vented crawl space, insulation is placed between the floor joists. In an unvented crawl space, either the concrete or masonry walls or the floor above can be insulated. If the walls are insulated, air infiltration paths such as the interface between the crawl space wall and the bottom plate of the floor above must be sealed. Intersections of two hard surfaces are common air infiltration paths.

Walls may be insulated by (1) insulating the entire exterior wall, (2) insulating the entire interior wall, or (3) insulating the entire interior wall and 2 ft of the perimeter of the crawl space floor. When the floor is insulated, an appropriate ground cover must be installed to control moisture in the crawl space and to prevent a high moisture level from degrading the thermal performance of the insulation. In pressure treated wood construction, insulation is placed in the cavities between the wood studs.

The disadvantages of a vented crawl space are that (1) pipes and ducts must be insulated; (2) a larger area must be insulated (the floor area is greater than the wall area), which may increase the cost; and (3) in some climates, circulation of warm, humid air in the crawl space can cause the wood structure to have an excessive moisture content. However, vented crawl spaces are often provided with operable vents that can be closed to reduce winter heat losses (closing them may, however, increase radon levels). Also, these vents can be closed in summer to keep out moist, exterior air that may have a dew point above the crawl space temperature. However, unless steps are taken to control air infiltration, closing vents does not ensure that the crawl space will have no air exchange with the outside; any exchange with outside air reduces the effectiveness of any wall insulation.

There are several advantages to designing crawl spaces as unvented, semi-conditioned zones. Duct and pipe insulation can be reduced; and insulating the foundation at the perimeter instead of the floor above requires less insulation, simplifies installation, and minimizes condensation. On the other hand, venting can complement other moisture and radon control measures such as ground covers and proper drainage. However, while increased airflow can dilute ground source moisture and radon, it may not

The preparation of this chapter is assigned to TC 4.9, Building Envelope Systems.

eliminate them. Venting crawl spaces may be desirable in areas of high radon hazard, but it should not be considered a reliable radon mitigation strategy. Pressurizing the crawl space is an effective method of minimizing soil gas uptake; however, the crawl space walls and ceiling must be tightly constructed for this approach to be effective.

Slab-on-Grade. For slab-on-grade foundations with concrete or masonry walls, insulation is usually placed on the entire exterior surface of the foundation stem wall, on the interior surface of the foundation wall, or horizontally under the slab perimeter. When the insulation is placed on the interior, it is important to place it in the joint between the slab edge and the foundation wall. The insulation placed in this joint does not need to have an R-value greater than 5 °F·ft^2·h/Btu. Exterior vertical insulation with an R-value of 5 to 10 °F·ft^2·h/Btu is justified in most climate zones (Labs et al. 1988).

Another method is to place insulation horizontally on the building exterior extending 2 to 4 ft into the surrounding soil. Shallow footings (less than 2 ft below grade) or slabs with a thickened edge and no foundation wall are common in some regions. In these cases, additional insulation may be placed horizontally along the exterior edge.

Thermal Performance

A key question in foundation design is whether to place insulation inside or outside the basement or crawl space wall. Placing rigid insulation on the exterior of a concrete or masonry basement or crawl space wall has some advantages in that it provides continuity without thermal bridges, maintains the waterproofing and structural wall at moderate temperatures, minimizes moisture condensation problems, does not reduce interior basement floor area, and includes the foundation in the thermal mass in the building. Exterior insulation at the rim joist leaves joists and sills open for inspection from the interior for termites and decay.

However, exterior insulation can provide a path for termites and can prevent inspection of the wall from the outside. A termite shield may need to be installed through the insulation where the sill plate rests on the foundation wall. Vertical, exterior insulation can extend as deep as the top of the footing and can be supplemented by extending the insulation horizontally from the face of the foundation wall.

Interior insulation is an effective alternative, and it is generally less expensive if the costs of the interior finish materials, including a flame spread cover, are not included. However, interior wall insulation increases the exposure of the wall to thermal stress and freezing and may increase the likelihood of condensation on sill plates, band joists, and joist ends.

Placing insulation in the basement ceiling is another acceptable alternative, but it should be used with caution in colder climates where pipes may freeze and structural damage may result from lowering the frost depth in the soil next to the foundation.

In a wood foundation, insulation is placed between studs. Insulation placed around the crawl space floor perimeter can provide additional thermal protection; however, it may also create additional paths for termite entry. A 2 in. air space between the lower end of the insulation in a crawl space or wall cavity and the bottom plate of the foundation wall allows inspection for insect infestation. Batt or tight-fitting rigid foam board insulation is commonly placed inside the rim joist.

A common, low-cost approach to insulating crawl space walls is to drape batts with a vapor retarder facing the inside of the wall. (Most building codes require that the vapor retarder be flame resistant.) The batts can be laid loosely on the ground at the perimeter to reduce heat loss through the footing. However, it is difficult to maintain the continuity of the vapor retarder around the joist ends and to seal the vapor retarders where they join the batts. Good installation is difficult because of cramped working conditions, and vapor-proof installation prevents easy inspection for termites. In a foundation system of wood pressure-treated with preservative, insulation is placed in the stud cavities similarly to above-grade insulation in a wood frame wall. Crawl space walls may also be insulated by spraying an insulating foam.

Insulation may also be incorporated in the concrete or masonry walls. Examples include rigid foam plastic insulation cast within the concrete wall; polystyrene beads or granular insulation poured into the cavities of conventional block walls, or foam insulation field-injected into the cores; concrete blocks with insulating foam inserts; formed interlocking rigid foam units that serve as a permanent, insulating form for cast-in-place concrete; and masonry blocks made with polystyrene beads or a low-density aggregate/regular-density concrete mixture. However, the effectiveness of construction that insulates only a portion of the wall area should be evaluated closely because thermal bridges can reduce the insulation's effectiveness by as much as 80%.

For slab-on-grade systems, the greatest heat loss is through the small area of the foundation wall above grade. Heat is also lost from the slab to the soil. Heat losses to the soil are greatest at the slab edge and diminish rapidly with depth. Both components of the slab heat loss at the edge and through the soil must be considered in designing the insulation system.

Insulation can be placed vertically outside the foundation wall or grade beam. This approach effectively insulates the exposed slab edge above grade and extends downward to reduce heat flow from the floor slab to the ground surface outside the building.

Vertical exterior insulation is the only method of reducing heat loss at the edge of an integral grade beam and slab foundation. A major advantage is that the interior joint between the slab and foundation wall need not be insulated, thus simplifying construction. However, there are several drawbacks to this method. The rigid insulation needs to be covered above grade with a protective board, coating, or flashing material; and brick facings that contact the covering can create a thermal short that bypasses both the foundation and the above-grade insulation. A further limitation is that the depth of the exterior insulation is controlled by the footing depth.

Thermal insulation also can be placed vertically on the interior of the foundation wall or horizontally under the slab. In both cases, heat loss is reduced and the difficulty of placing and protecting exterior insulation is avoided. Interior vertical insulation is limited to the depth of the footing, but under-slab insulation is not limited. Usually, the outer 2 to 4 ft of the slab perimeter is insulated. The joint between the slab and the foundation wall must contain insulation whenever insulation is placed inside the foundation wall or under the slab. Otherwise, a significant amount of heat transfers through the thermal bridge at the slab edge. The insulation is generally limited to a thickness of 1 in. at this point.

Another option for insulating a slab-on-grade foundation is to place insulation above the floor slab. A wood floor deck can be placed on sleepers, leaving cavities that can be filled with rigid board or batt insulation; or a wood floor deck can be placed directly on rigid insulation above the slab. This approach avoids some of the construction detail problems inherent in the more conventional approaches, but it may cause a greater frost depth near the slab edge.

Moisture Control

High moisture levels in a foundation can rapidly degrade the thermal performance of most types of foundation insulation. Properly constructed foundations with good drainage and damp-proofing should keep moisture loads from foundations below 6.3 pints/day; however, Christian (1994) reported on foundations that contribute 40 to 150 pt/day. Elevated moisture levels increase the potential for mold and mildew growth, odors, pathogens, and serious health effects to occupants. Mold and mildew will grow where the relative humidity of a surface is greater than 70%. This level could relate to 50% relative humidity in the building. Because the surface temperature is colder during winter than the crawl space air temperature, the relative humidity of the crawl space air can be much lower than 70% and still result in 70% relative humidity on the surface. Frequently, methods used to solve the problem in existing buildings are energy intensive (e.g., year-around operation of dehumidifiers).

Basement. Rain is carried away by gutters, downspouts, grading away from the building, and a cap of low-permeability backfill material. Subgrade drainage by a drain screen (gravel and footer drain connected to daylight, sump, or storm sewer) prevents water from reaching the foundation wall.

Damp-proof or waterproof materials on the exterior of the foundation wall and over the top of the footing control moisture transport. Capillary moisture movement into the slab is controlled by a 4 in. thick granular layer.

All air leakage openings (i.e., floor slab/wall intersection, rim joist area) are sealed with durable caulking materials, closed with gasket systems, taped, or covered with moisture-vapor-permeable house-wrap.

Damp-proofing on the wall and polyethylene under the slab control vapor diffusion. During heating periods, vapor diffusion from the interior of the basement may transport moisture into the rim joist framing, where it accumulates. Rigid insulation installed on the exterior of the rim joist limits periods of potential condensation.

Slab-On-Grade. The ground should be graded to direct water away from the building. Granular layer under the concrete slab provides a capillary break between the soil and slab. This layer can also be integrated into a subslab ventilation system to reduce radon levels. Extending the vapor retarder over the top of the foundation wall and placing appropriate flashing for any brick facing provide a capillary break to protect the above-grade wall from ground moisture penetration.

A vapor retarder placed under the slab restricts both radon gas entry and water vapor diffusion through the slab. If ducts are placed in the slab, moisture may enter the conditioned space because groundwater and soil gas will likely seep into the ducts.

Air Movement/Radon Control

Construction techniques to minimize radon infiltration through the foundation are appropriate where radon is present. General approaches include sealing joints, cracks, and penetrations in the slab and walls and evacuating soil gas under the foundation.

A tight foundation and a discharge system reduce the chimey or stack effect that draws air from the foundation or basement and into the house. If radon levels are excessive, a passive discharge system can be connected. If further relief is needed, an inline duct fan can be installed in the discharge system (Figure 1).

Subslab depressurization effectively reduces radon concentrations to acceptable levels, even in houses with high concentrations (Dudney et al. 1988). This technique lowers the pressure around the foundation envelope, causing the soil gas to be routed into a collection system and discharged to the outdoors.

A continuously operating fan should not be relied on as a discharge method. Ideally, a passive depressurization system should be installed, radon levels tested, and—if necessary—the system activated by adding a fan. The fan should be located in an accessible section of the stack so that any gas leaking from the positive pressure side of the fan does not enter the living space.

Active subslab depressurization also raises some long term concerns. If the radon barrier techniques are not fully used, indoor air could be discharged, resulting in a larger than expected energy cost. Durability is another concern, particularly with motor driven components. Also, the system is susceptible to owner interference.

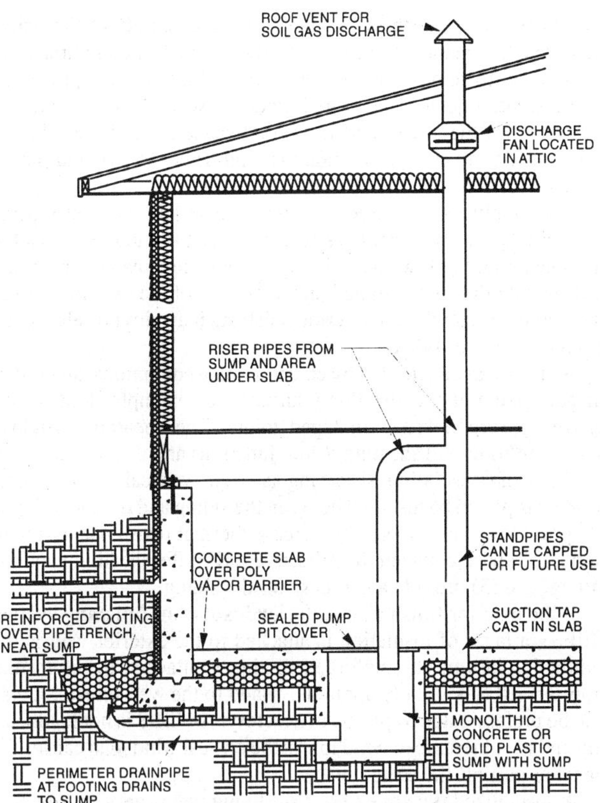

Fig. 1 Soil Gas Collection and Discharge Techniques

RESIDENTIAL WALLS

Lightweight Construction

The most common type of lightweight wall construction is the **wood frame** wall. Most houses built in the United States have exterior walls framed with wood studs, although steel studs offer several advantages; for example, they are more dimensionally stable and are truer than wood studs. Additionally, steel studs can be fabricated from recycled material and are themselves recyclable. However, these studs can create a significant thermal bridge across the cavity insulation, reducing its thermal resistance by up to 50%. Exterior foam sheathing insulation can be used to augment the R-value of the wall (Christian and Kosny 1996).

The interior side of the framing is typically sheathed with gypsum wallboard. In wood construction, to enhance the energy efficiency of the wall, foam insulation sheathing can be applied between the gypsum board and the framing. An advantage of this practice is that the low-permeance sheathing can take the place of a vapor retarder. The disadvantage is that it complicates the application of the gypsum board.

The cavities between framing members are filled with thermal insulation including batt, loose fill, stabilized loose fill, site-injected foam, and rigid foam boards. With all products, the level of performance is related to the quality of the installation of the insulation. Batts should be installed to completely fill the cavities along the back corners and the depth of the cavity. Gaps created by framing members or voids in an insulation caused by material settling or by improper installation can significantly reduce thermal performance. Dry loose fill should be installed at the manufacturer's recommended minimum density so that settling will not occur nor voids form. Wet loose fills should be installed with the proper amount of moisture, as recommended by the manufacturer. Stabilized loose fill insulations are applied with adhesive and water to prevent settling and movement within the cavity. As these products dry, some shrinkage has been observed; the shrinkage can affect the wall's thermal performance. Before the cavity is sealed, enough time must be allowed to remove any excess water added by the application process. Site-injected foams are introduced with a hose and allowed to expand in the cavity and interstitial spaces. Rigid foam board, sometimes used as an insulation material, is cut to size and placed tightly between studs.

Commonly used exterior sheathing materials include gypsum sheathing, plywood, oriented strand board (OSB), and foam insulating sheathing. For low-rise buildings, usually less than three stories, airflow retarders are installed on the outside of the exterior sheathing and covered with an exterior finish such as vinyl or aluminum siding, stucco, or brick.

Airflow can be retarded by constructing a continuous, air-impermeable plane at the sheathing surface; for example, tongue and grove gypsum board with taped joints. The sheathing/window, sheathing/roof, and sheathing/floor junctions are also taped.

When an insulating sheathing is used, vertical or horizontal spacer strips can be installed between the siding and the sheathing to (1) provide a rain screen, (2) increase thermal resistance by maintain an airspace between the reflective foil on the sheathing and the siding, and (3) provide a nail base for the siding.

If the exterior finish is an **exterior insulation finishing system** (EIFS), a layer of insulation is attached to the exterior wall and a synthetic glass-mesh–reinforced coating is attached to the insulation. An EIFS is usually applied by hand to the wall on site; it also can be prefabricated as panels in a factory. An EIFS is nonloadbearing and is vulnerable to wind loads; the critical wind load causes the panel to pull off the structure.

A specialized type of exterior sheathing that is very energy efficient is the **nonstructural truss** (Canadian Home Builders 1989, Nisson and Dutt 1985, Wilson 1988). These trusses are installed on the exterior of a conventionally framed and sheathed wall. The trusses are typically constructed with nominal 2 by 2 wood flanges with plywood, oriented strand board (OSB), or 2 by x webs. The primary advantages are that the trusses are easily installed and they allow high levels of insulation to be installed, although the insulation near the truss must be installed carefully. The additional wall thickness complicates the framing required around wall openings such as windows and doors.

Panelized walls are composed of an insulation layer faced on both sides with a sheathing material. The insulation layer ranges in thickness from 2 to 12 in.. OSB is the most widely used sheathing material, but gypsum sheathing and plywood are also available. Because of evolving techniques for energy-efficient splines, panelized walls can have significantly fewer thermal bridges than wood frame construction.

Interior and exterior finishes for panelized walls are similar to those used in wood frame construction. Some panelized walls may already have suitably finished surfaces. Advantages of using panelized walls include (1) higher energy efficiency than typical wood frame wall systems; (2) little or no thermal bridging in the clear wall area, (3) tighter than average construction due to the reduction in jointing if large panels are used, and (4) integration of the insulation into the panel system, making insulation installation unnecessary. Potential problems with panelized walls are (1) facer delamination in improperly built assemblies, (2) fire safety concerns with the use of cellular plastics (although this is an issue in any wall with insulating sheathing), and (3) insect control (carpenter ants have been reported to nest in hollows carved in the panels).

Masonry Construction

A common masonry wall is made with concrete masonry units (CMUs), blocks of concrete cast in various shapes, and joined with mortar to form a wall. The concrete used to fabricate CMUs is produced with a variety of aggregates and typically varies in density from 85 to 135 lb/ft^3 (Andrews 1989). A CMU wall provides a finished, durable interior and exterior surface. Holes can be drilled through the wall to feedthrough utilities. Many CMUs are available with patterns cast into the exterior face, thus providing diverse exterior finishes. Interior or exterior sheathing, and/or an air retarder, can be applied to improve energy efficiency or to obtain a desired finish.

If an EIFS is attached to the exterior of the wall, the insulation separates the wall from the fluctuating outdoor temperatures. This method of insulation can be used to advantage in energy-efficient buildings in which the thermal energy stored in the wall mass is used to moderate indoor temperature swings. External placement also allows the wall cavity to be used for utilities that might be susceptible to freezing or heat loss. It is important that the edges of EIFSs be tightly abutted so that thermal bridges are minimized.

Thermal Performance

James and Goss (1993) provide a summary of the heat transmission coefficient measurements and calculations for various wall, roof, ceiling, and floor systems. These data are only for clear wall areas and do not include subsystems or the effects of the joints created at intersections with another portion of the building envelope.

To test the validity of applying clear wall data to the wall system, In a series of three-dimensional heat conduction simulations using a finite difference program on a single family, detached house, the average area-weighted R-value for the entire wall was 88% of the R-value of the clear wall. A similar two-dimensional analyses by Tuluca et al. (1997) showed the R-value for the entire wall system to be 40 to 50% of the clear wall R-value. Significant thermal bridging can occur through framing (wood or even more important, steel), metal ties through exterior sheathing, and exposed slab edges.

The thermal bridges created by the webs of concrete masonry units dictate the maximum thermal efficiency that a CMU can attain. To reduce thermal bridging, blocks containing only two webs instead of the usual three have been used, and web thickness has been reduced by up to 40%.

The exterior surface of an EIFS expands and contracts significantly under the influence of changing outdoor temperatures and solar radiation. The thin, highly expansive coating has virtually no thermal storage capacity, but it is reinforced with glass mesh that helps moderate thermal movement. Because the coating is bonded to a layer of plastic insulation, which also is highly thermally expansive, the temperature of the coating fluctuates rapidly. This fluctuation places considerable stress on the coatings themselves and on adjacent sealing joints. This problem is exacerbated when dark colors are used and can become acute, because the maximum service temperature of expanded polystyrene insulation, the type most typically used, is about 170°F.

Moisture

Historically, the primary moisture control strategy for walls is to restrict moisture entry. Another approach to envelope design is to provide walls that balance entry with removal and that control accumulation and pressure differential.

Recognizing that lightweight wall systems can get wet or occasionally start out wet, the designer can promote weather protection and drying. A rain screen (Figure 2) can be placed immediately behind the exterior cladding to control rain penetration by providing a pressure-equalized air space; or a housewrap, which is vapor permeable and water impermeable, can be installed over the exterior sheathing to protect the wall from weather yet allow it to dry if necessary.

For heating-dominated climates and for mixed climates with both significant heating and significant cooling seasons, a vapor retarder at the interior surface is advisable. Moisture originates from the interior of the building and migrates into the wall until it reaches the interior vapor retarder, which is kept warm because it is inside the thermal envelope. In cooling-dominated climates, exterior sheathing of impermeable rigid insulation can act as a vapor diffusion

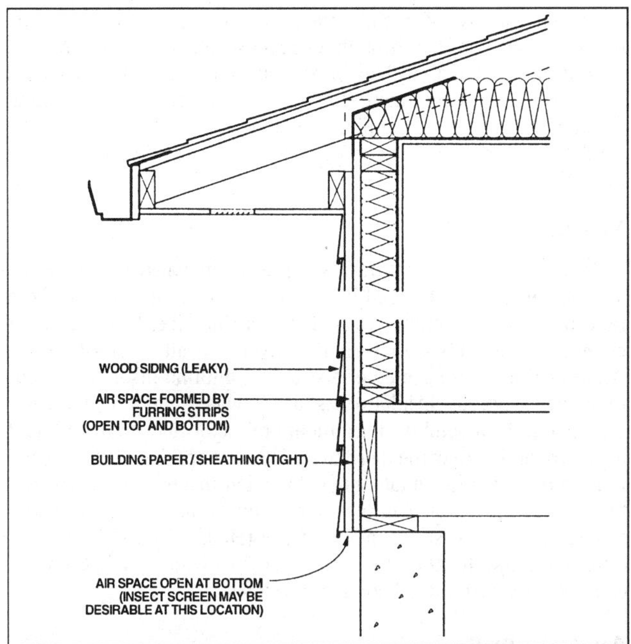

Fig. 2 Example of Rain Screen

retarder. Moisture outside the building moves into the wall until it reaches the exterior vapor retarder, which is kept warm because it is outside of the thermal envelope. As a general rule, to keep moisture condensation accumulation to a minimum, this first condensing surface must be kept above the dew point.

The exterior coating and insulation system of an EIFS is designed to be breathable, allowing moisture to pass through the wall under normal conditions. This design is important, because a vapor retarder could be formed on the wrong side of the wall by the EIFS coatings. Several conditions that can occur should be considered during the design of an EIFS wall system. If a low-vapor-permeance coating is applied on the outside, condensation can occur in the wall. This can happen, for example, if a highly vapor-resistant coating is applied to the outside of the wall for maintenance purposes, or if the wall is refinished with a new EIFS coating. If the ambient temperature and humidity are high in the summer, moisture can flow in and become trapped behind an interior finish that is highly vapor-resistant, such as a heavy vinyl wall covering or polyethylene vapor retarder. This problem is of particular concern in climates with extended hot, humid periods if the space is air-conditioned or refrigerated.

Air Movement

Uncontrolled air movement across walls increases space conditioning costs and may cause moisture to condense. **Air infiltration** is air moving from outside into the conditioned space. **Air exfiltration** is air moving from the conditioned space to outside. **Air intrusion** is air penetrating the building envelope without entirely crossing it. For example, outside air can penetrate a wall at sheathing joints, move around or through the insulation and return to the outside. Similarly, inside air can penetrate a wall at electric outlets, move around or through the insulation and return to the conditioned space. The term **air leakage** designates any type or combination of air movement in and across the envelope (see the "Basic Concepts" and "Residential Infiltration" sections in Chapter 25).

In general, an **air retarder** should be installed in walls. The air retarder reduces both infiltration and exfiltration. It also prevents intrusion by exterior or interior air, depending on the location of the air seal plane. While the air retarder reduces energy use, it also reduces or eliminate moisture condensation.

In heating-dominated climates, the air retarder can be positioned on either the exterior or the interior of the wall structure (e.g., framing). If placed to the exterior of stud construction that does *not* have insulating sheathing, the air retarder protects the insulation located between framing members from cold air intrusion. To avoid moisture condensation, such an air retarder must be moisture permeable, because moist interior air can still intrude into the wall even though it cannot easily exfiltrate across the wall. Examples of exterior moisture-permeable air retarders are

- Certain thin membranes such as spun bonded polyolefin and other house wraps
- Gypsum board, particle board or plywood sheathing taped and sealed at all joints and at the wall/window, wall/roof, and wall/floor junctions
- Pressed paper and cardboard products taped and sealed
- Thin polystyrene accordion-type wraps, stapled, taped and sealed, and used only on houses built according to Housing and Urban Development Department codes

The air retarder can also be placed to the exterior of stud construction that has insulating sheathing, such as steel stud walls with R-11 fibrous insulation, gypsum board sheathing, rigid insulation sheathing, and brick finish. If the air retarder is located on the winter-cold surface of the gypsum board sheathing, it can be moisture impermeable if the R-value of the rigid insulation is sufficient to maintain the air retarder above the dew point during winter conditions. Impermeable air retarders are usually bituminous, rubber-based, or plastic membranes. However, the air retarder can also be composed of a gypsum board layer with the junctions taped and sealed. Masonry construction usually includes air-impermeable air retarders.

If the air retarder is placed to the interior of a wall located in a heating-dominated climate, it protects the wall from air intrusion from moist interior air. In theory, this type of retarder is more advantageous than an exterior retarder for structures where moisture condensation is likely, such as steel studs without insulating sheathing that define a space with high humidity in winter. In practice, it is difficult to achieve a continuous air seal at the interior surface of a wall, especially in multistory buildings, because of the wall/floor and wall/ceiling junctions. A continuous air seal plane could be achieved by gypsum board with or without a polyethylene vapor retarder if particular attention is paid to details and with special construction supervision, .

In hot, humid climates, the air retarder for air-conditioned buildings should be vapor permeable if it is placed nearest the interior of the wall, because the outside moist air can condense on cold interior surfaces. The exterior air retarder can be either moisture permeable or moisture impermeable. If the retarder is moisture impermeable, the insulated layer near the conditioned space must have a sufficient R-value to keep the membrane above the dew point. This caveat applies to each component of the insulated layer separately.

In climates with significant heating and cooling, the heating-dominated construction is usually indicated, but dew point calculations need to be performed if layers with low vapor permeance are used. In climates with some heating and significant cooling and high humidity, it is usually prudent to avoid vapor-impermeable air retarders and layers with low vapor permeance.

In general, with the exception of prefabricated wall panels, two layers with low vapor permeance (e.g., an exterior vapor-impermeable air retarder and an interior vapor retarder) should be avoided, because moisture can be trapped in the wall during construction or operation.

RESIDENTIAL ROOF AND CEILING ASSEMBLIES

Thermal Performance

In pitched roof assemblies, the attic space is typically insulated with mineral fiber batts or blankets, and/or mineral fiber or cellulose loose fill insulation. The total thermal resistance of the ceiling plus insulation, attic space, and roof is the sum of the conduction resistance of the ceiling plus insulation and the effective attic resistance. The roof resistance is generally negligible.

The thermal performance of the insulation varies with its mean temperature. In winter conditions, the thermal resistance of the insulation increases, and in summer conditions it decreases. In installing loose fill insulation, care must be taken to avoid overblowing the material to densities below the intended design density for a given R-value. At densities below the design, the material may not deliver the desired R-value. To prevent overblowing, the correct amount of material for a given coverage area and R-value should always be installed. Manufacturers of loose fill insulation are required to provide maximum net coverage information on all product packaging. Some high-density, loose fill products may gradually settle to a lower thickness after being installed. The reduced thickness decreases the thermal resistance despite the increased density. For materials subject to settling, initial installed and settled thickness information should be available on package labels or from manufacturers so that the desired settled R-values can be achieved.

Some low-density loose fill materials are porous enough that very low temperatures at the top of the insulation may lead to free convection in the insulation. For example, Wilkes et al. (1991) observed that effective thermal resistance decreased by nearly 50% from nominal values when temperature differences greater than 70°F were imposed. Free convection can be suppressed by placing a layer of batt insulation over the low-density loose fill insulation. Alternately, batts or blankets can be installed first and the low-density loose fill blown in on top of the batts or blankets. Mineral fiber batts or blankets should be installed at the proper width so that there are no air gaps between the batts.

In a roof assembly over a ventilated attic space in cooling-dominated climates, insulation may be supplemented by a radiant barrier of high-reflectance material to reduce radiant heat transfer between the hot roof and the relatively cool ceiling. The radiant barrier can be attached to the roof rafters or fastened directly to the underside of the roof sheathing with the operative surface facing down. Care should be taken to not disturb and damage the existing insulation when retrofitting an attic with a radiant barrier. Also, dust decreases the effective reflectance of the radiant barrier over time; however, if the operative surface faces down, the reduction in thermal performance should be minimal.

During the cooling season, a radiant barrier reduces temperatures affecting the air, insulated surfaces, and HVAC ducts and equipment located in the attic space. A ventilated radiant barrier combined with an equally divided, continuous soffit and ridge vents provides optimal performance. The most cost-effective use of radiant barriers is in the cooling-dominated climates. Use of radiant barriers in other climates may result in a net annual increase in energy use because the beneficial attic heat gain in the noncooling seasons is lost. DOE (1991) compares the present value savings of radiant barriers with those of conventional fiberous insulation for several climates and circumstances.

During the cooling season, a radiant barrier under the roof sheathing increases the efficiency of a duct system located in an attic (Hageman and Modera 1996). The DOE publication does not consider this case.

With or without a radiant barrier, powered ventilation increases the effective attic resistance by reducing the temperature difference between the inside conditioned space and the attic space (see Table 5 in Chapter 24). However, this method is rarely cost-effective because of the energy cost of the fan. Additionally, power ventilation can draw air from the conditioned space into the attic.

In low- or no-slope roofs, the insulation is a layered assembly comprising the deck, insulation, and membrane. While there are relatively small gaps between the components, the high thermal resistance of the insulation keeps the total heat flux low. However, the heat flow through metal penetrations can be more than 1000 times greater than that through adjacent insulation. Therefore, the area of these thermal bridges in proportion to the roof area should be kept small. Minimizing this area may require interrupting potential bridges; for example, staggering fasteners that secure the insulation to the deck and those that attach the membrane to the insulation.

Moisture

In addition to moisture normally present in construction materials, moisture can be trapped in the roof and ceiling assembly from rain during construction, leaks that were not fixed before water entered the assembly, or reroofing over partially torn-off roofs. Additional moisture can enter a structure with no insulation vapor retarder. If porous insulation is used, water vapor can diffuse throughout the assembly. The amount of vapor may be insufficient to significantly affect the thermal conductivity of the insulation, but it may add a significant latent heat load. Diurnal temperature variations in the roof can cause water to evaporate, move as vapor through a porous system, and condense (Hedlin 1988). When conditions are right for this effect, the heat flow from this factor alone can easily exceed normal conduction flows.

Air Movement

Whether the roof is pitched, low-slope, or no-slope, a typical roof and ceiling assembly is impervious to significant air movement through the ceiling itself. With pitched roofs, the attic normally is ventilated by airflow from the edges of the assembly, over the insulation, and out through continuous ridge vents or occasional vents in the roof or gable ends of the attic. The ventilation may be by natural or forced convection, but natural convection by a combination of continuous ridge vents and continuous soffit vents seems most effective. A type of low- or no-slope roof called a "ventilated cold deck system" has vents at the eaves and at intervals over the roof to permit the circulation of dry outside air over the insulation. The effectiveness of such vents is uncertain in regions with a cold or a temperate but humid climate (IEA 1994). An alternative is a nonventilated cold deck system with a vapor retarder on the ceiling consisting of staggered strips of polyethylene on the bottom and top of the vapor retarder and exposed fabric between them. Vapor inside the building effectively "sees" a continuous layer of polyethylene because of the staggering. Condensed water wicks through the exposed fabric on the top to the exposed fabric on the bottom and evaporates to the inside (Pedersen et al. 1992).

FENESTRATION

Conduction/Convection and Radiation Effects

Heat transfer through a window resulting from a temperature differential between the inside and outside (i.e., conduction, convection, and radiation) is a complex and interactive phenomenon. While the glass itself is a poor insulator, it is used effectively in a window to create thin insulating air films on either side of single glazing or dead air/gas spaces between glazing layers. Glass also can decrease the direct transmission of radiant energy coming from the room or ambient sources. Chapter 29 discusses fenestration in much greater detail.)

An examination of the modes of heat transfer in a double-glazed window indicates that approximately 70% of the heat flow is through radiation from one glazing layer to another (Selkowitz 1979, Arasteh et al. 1985). Although the glass blocks the direct transmission of radiant energy, it is still absorbed and reemitted. Low-emittance coatings that are transparent to the eye significantly reduce the amount of heat transfer through a glazing cavity.

With the radiation mode of heat transfer miminized, the conductive and convective modes dominate. To reduce these modes, gases

WALL/ROOF INTERFACE

Control of Air Leakage

The roofing membrane forms a plane of continuous airtightness and therefore can and should be connected to the wall airflow retarder. However, the wall airflow retarder could also be connected to other airtight roof components, such as the roof deck or the roof vapor retarder.

Many building systems provide for differential movement between the wall and roof assemblies; in those systems, a flexible airflow retarder connection is required at the location of the movement control joint. In all systems, the airflow retarder must withstand wind-induced pressures and be fully adhered to a rigid, structurally adequate substrate.

Figure 3 shows an assembly that uses the concrete roof deck to achieve a continuous wall/roof airflow retarder. The retarder of the wall assembly is connected to the concrete slab using a sheet airflow retarder material. The connection between the wall and slab edge retarder sheets is designed to accommodate differential movement of the wall and roof.

Figure 4 shows a protected membrane roof assembly in which the roof membrane is an air flow retarder that extends over the edge of the roof perimeter to connect to the wall air flow retarder. A sheet metal backing is used around the roof deck perimeter for attachment and to provide for differential movement at the interface.

In some types of curtain walls, such as the wall in Figure 5, the air seal must be made continuous over the parapet wall. This allows connection of the wall air seal (the metal back pan of the spandrel panel) to the membrane flashing of the roof assembly or to the air/vapor retarder on the parapet, as in asphalt and roofing felt systems.

Control of Moisture Diffusion

The vapor retarder can also act as an airflow retarder, as in the curtain wall shown in Figure 5 in which the metal back pan of the wall spandrel connects to the air/vapor flow retarder membrane on the parapet upstand wall. Because the parapet is encased in vapor retarder sheets, it must be kept dry before it is enclosed because any moisture left in the parapet cannot disperse. A vapor retarder with wicking properties could provide a better solution.

Control of Condensation

Condensation at the connection between the roof and wall can occur in interstitial spaces because of a lack of continuity in either system. Condensation also occurs if the temperature of the vapor retarder falls below the dew point of the air on the high vapor pressure side. Thus thermal insulation applied to the exterior (winter cold) of the vapor retarder must be sufficient to ensure that the temperature at the retarder plane remains higher than the dew point of the air. The temperature of the interior space between the exterior wall system and the parapet upstand wall often falls below the temperature of the interior inhabited space because of poor air circulation above the ceiling. Therefore, the insulation applied over the parapet cap and upstand wall may need to have greater thermal resistance than the insulation for the wall or roof.

In exterior insulated wall systems, such as curtain walls and EIFSs, thermal bridging must be minimized to control condensation between the parapet cap and the parapet upstand wall.

THERMAL BRIDGES IN BUILDINGS

A thermal bridge is an envelope area with a significantly higher rate of heat transfer than the contiguous enclosure. An example of thermal bridging is a steel truss in an attic with glass fiber insulation placed between trusses. Steel conductivity is about 1000 times greater than glass fiber insulation conductivity, so strong thermal bridging results along the trusses. In most construction details, the ratio between the high and low conductivity areas is

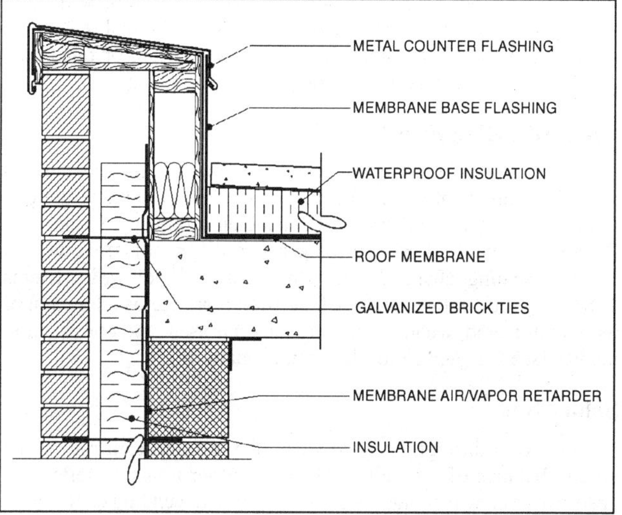

Fig. 3 Detail of Concrete Roof Deck with Continuous Wall/Roof Airflow Retarder

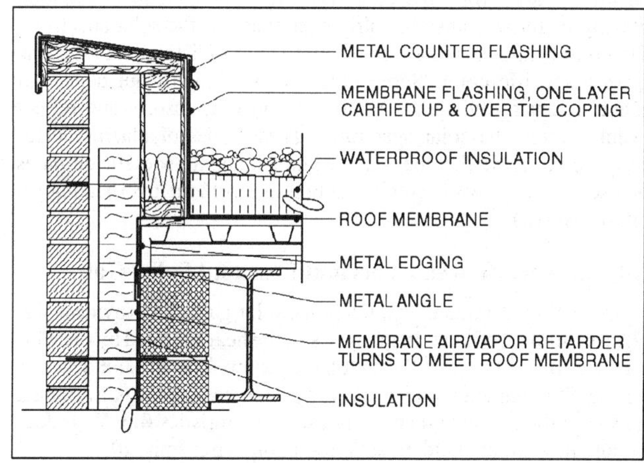

Fig. 4 Protected Membrane Roof Assembly

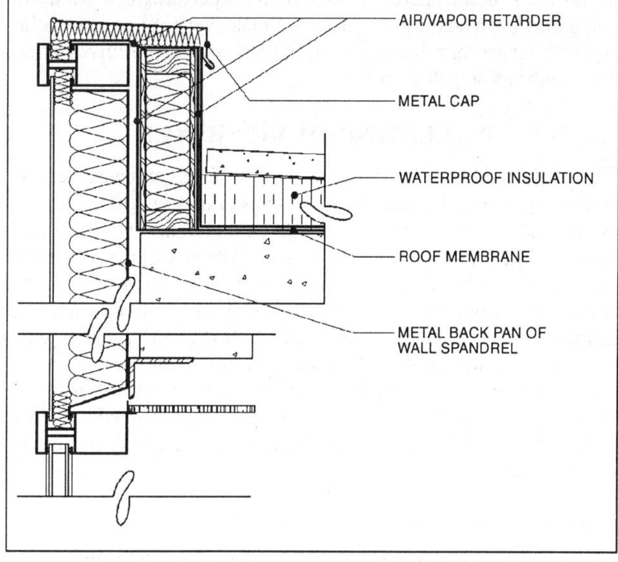

Fig. 5 Continuous Air Seal in Curtain Wall

with low conductivities (e.g., argon or krypton) are used in low-emittance-coated, multiply-glazed windows. Additional reductions in heat transfer are possible through creating triple or quadruple glazing layers (see Chapter 29 for design information).

Air Infiltration Effects

Air Infiltration is the uncontrolled movement of air from the outside to the inside of a building. It is a function of the pressure differential between the inside and outside environments (which is a function of wind speed and temperature differentials) as well as window sealing characteristics (Weidt and Weidt 1980, Klems 1983). The infiltration rate of a fenestration product is a function of its operator type, weatherstripping material used, and construction quality (see Chapter 25 for design information).

Solar Gain

Solar gain through a window can play a significant role in the energy balance of a building. Glazings either transmit, reflect, or absorb a given wavelength of solar radiation, depending on the glazing characteristics. Transmitted solar radiation will contribute heat to a space. Absorbed solar radiation is reemitted and/or conducted either to the inside or outside (depending on the window's glazing system configuration). Reflected solar radiation does not contribute heat to a space (Arasteh et al. 1989). Clear glass, the most common glazing material, transmits fairly evenly across the solar spectrum. Tinted or heat-absorbing glass absorbs solar radiation and gives the glass a specific color. Some tints exhibit a significant degree of spectral sensitivity (i.e., they do not transmit, absorb, and reflect evenly across the solar spectrum). These types of glazings offer great flexibility in that they can be tailor-made for specific climates or uses (e.g., provide ample daylighting without overheating an interior space).

Interactions Between Thermal Loss and Solar Gain

In heating-dominated applications, solar gain can provide a significant amount of heat. In many cases, the heat supplied by the window can offset that lost through the window. The amount depends on the site characteristics (i.e., how much solar gain is available, how cold the climate is) and window characteristics (i.e., its U-factor and how much incident solar radiation is transmitted).

Typical passive solar applications try to maximize the amount of solar heat gain by installing significant amounts of south-facing glass, which receives the most solar radiation during the winter (in the northern hemisphere). However, high-performance windows facing north in a heating-dominated climate can provide more solar gain to a space than heat loss (Arasteh et al 1989, Sullivan et al. 1992, Dubrois and Wilson 1992).

WALL/WINDOW INTERFACE

Air infiltration at the wall/window interface can cause serious damage to surrounding building materials and even remote materials, depending on the leakage path. In cold climates, warm, humid inside air can condense in the wall cavity. The condensate can damage the interior finish, the seals of glazing units, the insulation, the exterior cladding, and possibly the structural elements. Cold air infiltration can affect the health and comfort of the occupants by creating a dry indoor environment, cold drafts, and condensation and mildew on inside window surfaces. Air infiltration also can allow rain to enter the wall. Many materials are installed by many trades at the wall/window interface, making it difficult to inspect and prone to leak.

Control of Air Leakage

The airflow retarder must be continuous across the entire building envelope to provide effective control. To ensure continuity at the window, the components that retard airflow must be unified. Examples include the following:

- In a metal and glass curtainwall, the glazing and sheet metal are the elements that retard airflow in the window and the wall, respectively. Gaskets and caulking make the airtightness plane continuous from the glazing to the metal mullion.
- In a masonry wall, a self-adhesive asphalt membrane mechanically fastened to the masonry makes the wall airtight. A bead of sealant preserves the airtightness between the glazing unit and the inner part of the metal window frame. Thus the asphalt membrane and the inner part of the metal frame are the two materials joined to withstand large air pressure loads. The membrane is extended and mechanically clamped to the inner part of the frame.

Control of Moisture Diffusion

Vapor diffusion across a small gap at the window/wall junction causes little building damage under average environmental conditions. Moreover, typical interior finishes covering the joint, made of metal or plastic or painted drywall, offer some resistance to moisture diffusion.

Control of Heat Flow

Continuity of the plane of thermal insulation is important primarily to reduce the potential for condensation on interior surfaces of the window and the surrounding interior finish. Insulation can be inserted in the joint between the wall and the window frame to compensate for the expected differential movement between the frame and the wall rough opening.

Control of Surface Condensation

To reduce the potential for condensation on the glazing and the frame of windows, as well as the surrounding interior finish of the wall, the inside surface temperature can be controlled in the following ways:

- Seal the wall/window interface and between the sash and frame of operable windows to keep air leakage to a minimum.
- Make the area of window frame exposed to the interior larger than the area exposed to the outside. Metal frame extensions on the inside have a higher resistance to condensation, but contribute to heat loss.

Control of Rain Entry

Applying the rain screen principle at the wall/window interface requires the same features as applying it to the wall—an airflow retarder, a rain deflector on the outside of the interface, much more venting on the outside face of the interface than on the inside, and a drainage path toward the outside.

The line of airtightness is on the inside of the assembly, so the assembly is protected from water, ultraviolet rays, and extremes of temperature. The rain deflector on the outside acts as a rain deterrent only, not as a watertight/airtight seal. A non-airtight rain deflector does not threaten the weathertightness of the system, because the pressure differences across the rain deflector are small. The key is to maintain airtightness on the inside of the joint. With little pressure across the rain deflector and with good detailing for outward drainage of the cavity, rain entry in the wall should be minimal.

Face Seal Approach

The face seal approach calls for perfect sealing of the outer face of the wall/window junction. This sealed surface protects against rain and air infiltration. Such a seal must remain perfect over time; its maintenance schedule is quite demanding. It is also complicated because water leaks can follow indirect paths for entry indoors and, in a large building, it may be difficult to trace these paths back to the entry points.

smaller. An uninsulated CMU wall with brick facing could have a U-factor of 0.33 Btu/h·ft²·°F if the CMU is made with sand and gravel aggregate. A concrete column located within that wall has a U-factor about 50% higher. This 50% higher conductivity may justify classifying the column as a thermal bridge, but in typical construction this classification is relevant only if the concrete column proves to be detrimental for reasons other than energy use (e.g., moisture condensation).

Thermal bridges can be created by conduction, as in the previous examples; by radiation; by convection; or by a combination of those three modes of heat transfer (Silvers et al. 1985). Strong radiation effects occur, for instance, in an uninsulated brick wall that is retrofitted with metal cladding. The cladding is placed at some distance from the exterior of the brick wythe. The warmer brick wall radiates to the cold metal sheet. The different temperatures of the two surfaces also engender air convection.

Detrimental Effects of Thermal Bridges

Thermal bridges increase energy use, promote moisture condensation in and on the envelope, and create nonuniform temperatures

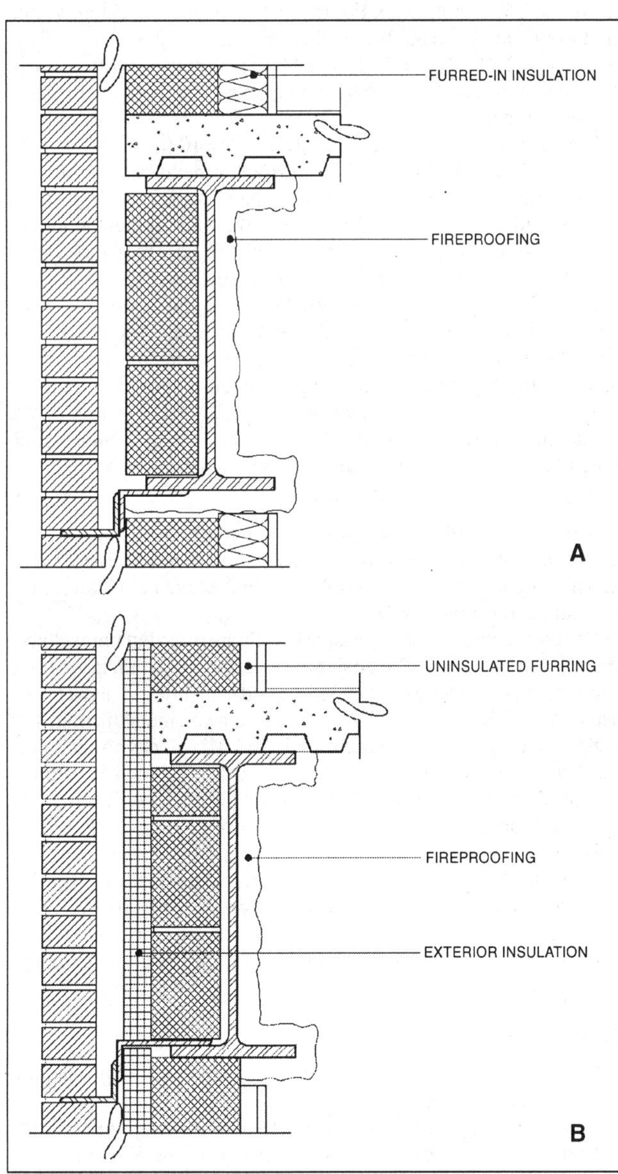

Fig. 6 Detail of Uninsulated (A) and Insulated (B) Slab Edge and Metal Shelf Angle

in conditioned spaces. A comparison between the uninsulated slab edge detail of Figure 6A and the version with slab edge insulation with an R-value of 10 h·ft²·°F/Btu in Figure 6B illustrates the importance of design to reduce thermal bridging. (Refer to Steven Winter Associates (1988) for numerical examples summarized in this section.) This example, and the others in this section, use the term "*total R-value*," where $R_t = 1/U$. This means that R_t includes the effect of exterior and interior air films.

- **Increased energy use.** The portion of the wall that extends from the top of the slab edge to the shelf angle (Figure 6B) has $R_t = 2.3$ h·ft²·°F/Btu when the wall insulation is furred in. Moving the insulation into the wall cavity (Figure 6B) increases R_t by about 2.2 times, to 5.3 h·ft²·°F/Btu.
- **Damage to structural components.** On cold days, the uninsulated wall area collects moisture on the interior surface of the steel beam and on the slab underside. At 20°F outside air temperature, 70°F indoor air temperature, and 50% indoor relative humidity, moisture condenses on about 0.66 ft² of beam and slab for every linear foot of wall. This condensation can rust the metal beam and, if it penetrates into the concrete block, it can rust the brick ties. The insulated detail (Figure 6B) generates moisture condensation only on the bottom flange of the beam.
- **Damage to interior finishes.** Moisture can drip from the slab underside and from the steel beam, staining the ceiling tiles and resulting in higher maintenance costs and loss of rental value. These effects are specific for the heating season and are magnified when thermal bridging is accompanied by air leakage, since air transports and deposits moisture on thermal bridges.

In cold and temperate climates, thermal bridges may slightly decrease the peak cooling load and the cooling energy use. In hot climates, this effect could be reversed. When the climate is hot and humid, moisture condensation can create problems if the building is air conditioned or refrigerated.

Thermal Bridge Mitigation

The primary causes for conduction thermal bridging are (1) high thermal conductivity and (2) geometries that create zones where large exterior surfaces connect to much smaller interior surfaces. Thermal bridges can sometimes be eliminated, but usually they can be mitigated to create systems with significantly lower U-factors. Several examples of conduction thermal bridges and of mitigating details follow.

- **Factory-produced components with high conductivity**

Sandwich panel with insulation fully encased in concrete. (Figure 7A). One possible solution: Connect the exterior and interior wythes of the panel with metal or plastic tie rods (Figure 7B).

This 6 ft high sandwich panel has a 3-in. extruded polystyrene core completely encased in concrete. The concrete has a conductivity over 50 times higher than that of polystyrene, resulting in $R_t =$ 5.4 h·ft²·°F/Btu. Moisture can collect on the top and bottom concrete edges of the panel. For T_{OA} = 20°F, T_{IA} = 70°F and 50% rh, about 20% of the panel area can be covered with moisture.

When the interior and exterior concrete wythes are attached with metal or plastic tie rods, R_t of the panel increases by almost a factor of three to 15.1 h·ft²·°F/Btu or 16 h·ft²·°F/Btu, respectively. No moisture condensation on interior surfaces would be likely.

- **Assemblies that contain high conductance components**

Steel studs in insulated wall with ceramic finish. One possible solution: Provide insulating sheathing (Figure 8).

This steel stud wall is used in low-cost multifamily, commercial, and industrial construction. The exterior gypsum board is provided for fire protection. If the wall has R-11 h·ft²·°F/Btu insulation, the studs reduce the wall R_t to 7.75 h·ft²·°F/Btu. For T_{OA} = 20°F, T_{IA} = 70°F and RH = 50%, every linear foot of steel

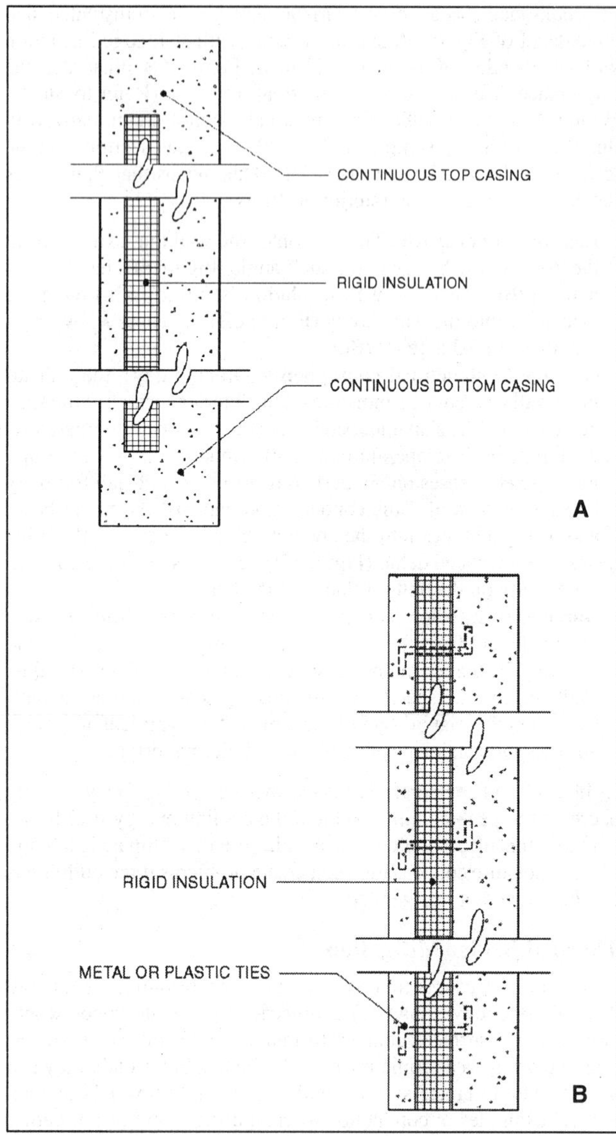

Fig. 7 Sandwich Panel with Insulation Encased in Concrete

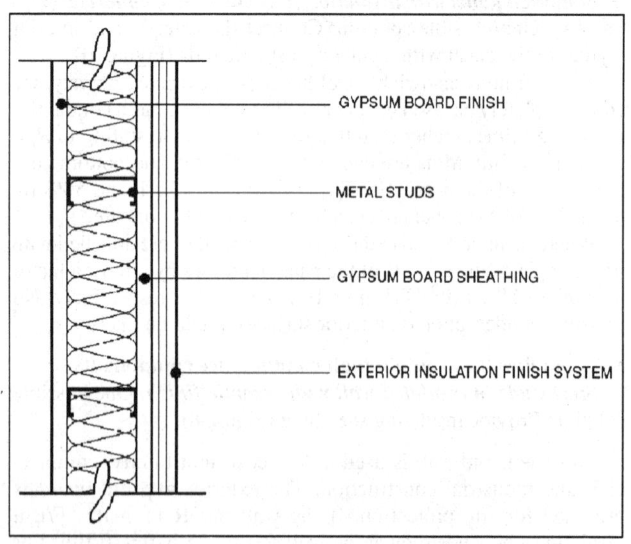

Fig. 8 Steel Studs in Insulated Wall with Ceramic Finish

stud can generate 0.15 ft^2 of moisture condensation on the interior surface of the gypsum wallboard.

Extruded polystyrene sheathing 1 in. thick placed on the winter-cold surface of the gypsum board sheathing would reduce the thermal bridging, resulting in R_t = 14.5 h·ft^2·°F/Btu. For T_{OA} = 20°F, T_{IA} =70°F and 50% rh. Moisture would be less likely to condense on interior surfaces. If polystyrene were placed on the steel studs directly, the increase in R-value would be even higher because the lateral heat flow would be reduced.

If all insulation were placed on the exterior surface of the steel studs, the studs would be kept at a temperature close to that of the conditioned space. Moisture condensation would be practically eliminated. This solution merits special consideration in spaces with high relative humidity such as kitchens, pools, and computer rooms.

- **Junction between envelope systems, achieved with high conductance components**

Uninsulated slab edge and metal shelf angle, at junction between CMU wall and concrete floor (Figure 6A). One possible solution: Place insulation in the wall cavity (Figure 6B).

This detail shows an uninsulated slab edge and perimeter metal beam in a CMU/brick wall. For the wall segment that extends from the top of the slab edge to the shelf angle, R_t = 2.3 h·ft^2·°F/Btu. For T_{OA} = 20°F, T_{IA} = 70°F and 50% rh, about 0.66 ft^2 of slab underside and steel beam could collect moisture along every linear foot of slab edge.

The improved detail of Figure 7B uses R= 10 h·ft^2·°F/Btu rigid insulation in the wall cavity. The perimeter metal beam would be insulated and the shelf angle would remain as the only significant thermal bridge. The new total R-value of the wall segment would be about 2.2 times higher (i.e., 5.3 h·ft^2·°F/Btu). Less moisture would be expected to condense along the slab edge.

Insulation is sometimes specified on the interior surface of the perimeter beam in an effort to increase the thermal resistance of this wall segment. Site conditions rarely allow the installation to be thorough, so the top and bottom flanges remain exposed, creating a strong thermal bypass and marginalizing the effectiveness of the insulation. Moreover, the insulation makes vapor condensation more likely because it decreases the temperature of the beam chord but does not stop water vapor migration.

- **Combinations of geometry with high conductivity**

Column in masonry wall with insulation located in steel stud furring (Figure 9A). One possible solution: Move the insulation in the wall cavity (Figure 9B).

Concrete columns within masonry walls are often left uninsulated. In this example, a 12 × 12 in. column is located at the junction of two exterior walls composed of 6-in. CMU and 4-in. brick. The walls are insulated with R = 10 h·ft^2·°F/Btu polystyrene on metal furring. The corner, which includes the column and two CMU blocks on each side, has R_t = 6 h·ft^2·°F/Btu. This is 38% lower than the total R-value of the wall without the column (R_t = 9.7 h·ft^2·°F/Btu). For T_{OA} = 20°F, T_{IA} = 70°F and 50% rh. Every linear foot of column could collect moisture on about 0.8 ft^2 of interior surfaces.

If the insulation were placed in the wall cavity, the total R-value of the corner would more than double, from R_t = 6 h·ft^2·°F/Btu to R_t = 12.3 h·ft^2·°F/Btu. No moisture condensation would be expected on interior surfaces of the column for the given temperature and RH conditions.

Design Recommendations

Basic design principles, which apply to any thermal bridge, are listed in the following paragraphs.

Consider the combined effects of thermal bridging, air leakage, moisture condensation, and water intrusion. If one of these mechanisms of heat and mass transfer is involved, another one is commonly combined with it in a destructive synergy. In devising a solution for thermal bridging, ensure that moisture condensation is

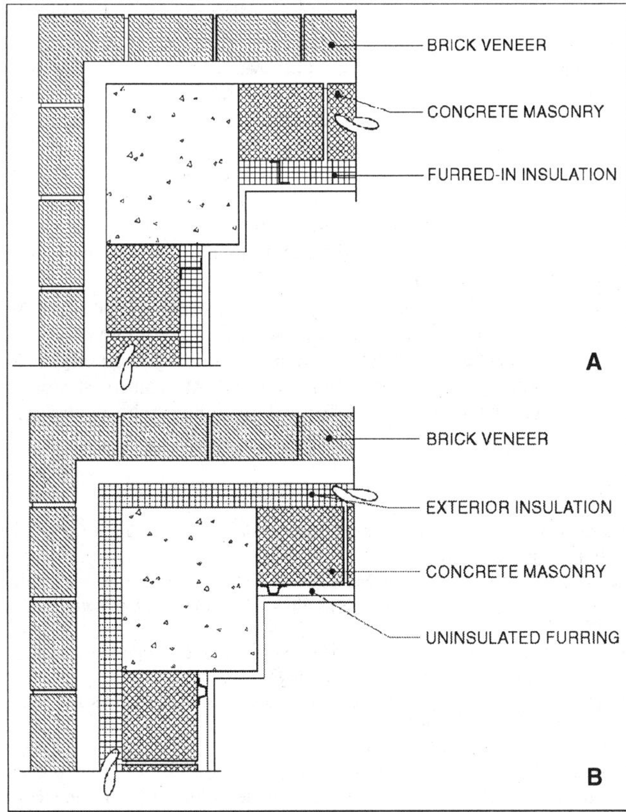

Fig. 9 Details of Insulation Around Column in Masonry Wall

not increased. In climates with significant heating, this is of specialL concern for insulation placed on the interior (winter-warm) surface of materials with low vapor permeance, such as CMU or steel beam. In hot, humid climates, the concern relates to insulation placed on the summer-hot surface of materials with low vapor permeance.

In addition, because many thermal bridges are discontinuities in the envelope system, the chance that air will enter the envelope along these paths should be assessed. To cause damage, air does not need to infiltrate into a space from outdoors, or to exfiltrate from a space to the outside; it is sufficient that air intrude in the envelope. For example, outside air can enter a stud wall cavity, short-circuiting the insulation and intensifying the thermal bridging effect, after which it exits again to the outside. In another example, indoor air can seep from the conditioned space into a furred-in CMU wall, behind the insulation and vapor retarder, increasing the chance for moisture condensation on the CMU surface. Of all forms of air intrusion, exfiltration of moist air along thermal bridge surfaces can be the most damaging. Note that the entry and exit points of air in the envelope can be distant.

It is particularly important to reduce thermal bridging, air intrusion, and moisture condensation if the following conditions are true:

- The building or spaces in it have uses that are likely to generate a relatively high amount of moisture during winter; examples are residences, hospitals, nursing homes, restaurants, kitchens, natatoriums, computer rooms, and bathrooms.
- The building is located in an area that experiences cold winters, or sustained periods of cold during moderate winters. Climates with humid winters with subfreezing periods can also create problems.
- The building is located in a warm but not dry, arid climate and has uses that require low temperatures (e.g., refrigeration), or it is located in a hot, humid area and has uses that require air conditioning.

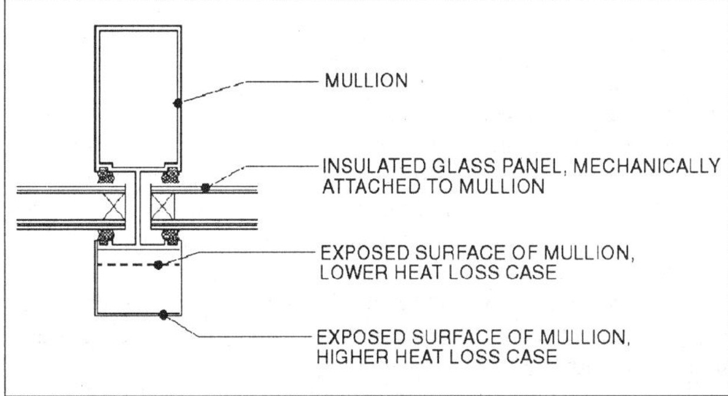

Fig. 10 Mullion with Deep Exterior Projection

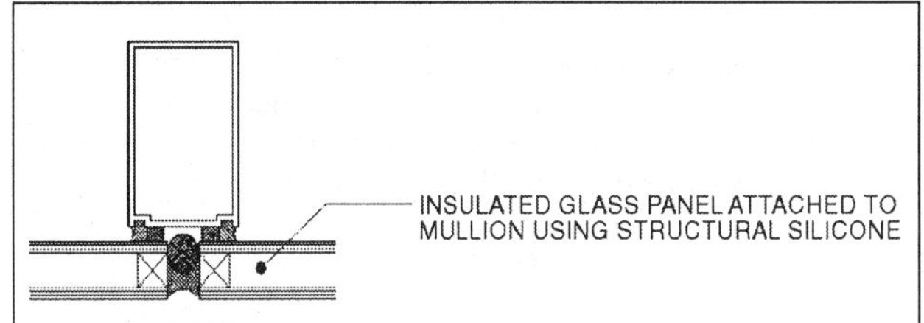

Fig. 11 Mullion without Exterior Projection

Consider the potential for thermal bridging and air intrusion during schematic design when decisions about the envelope system are made. Some systems are inherently more difficult to insulate and air proof, and changes during design development may create budget problems.

Use reasonably accurate methods to determine effect of thermal bridging. Accurate formulas are available for most steel stud configurations (Kosny and Christian 1995), but they are not sufficiently accurate for many other construction types. Two- and three-dimensional models are available and provide better results.

In mitigating a thermal bridge, several techniques can be used, often in combination. Such techniques include:

- Change the thermal bridge material to one with lower conductivity. Plastic ties nearly eliminate thermal bridging in concrete sandwich panels.
- Decouple the thermal bridging elements from the rest of the construction. If the steel stud furring is detached from the CMU wall, the pathway between two materials with high conductivity is broken. The best method to achieve the decoupling is to insert a thin, low-conductivity material, such as foam tape, at each stud. If an air space is created between the insulated steel stud layer and the CMU wythe, air convection can short-circuit the insulation.
- Change the geometry that creates thermal bridging. Figure 10 shows a mullion with a deep exterior projection. This projection acts like a fin during winter, enhancing the heat loss. One methodof reducing the thermal bridging effect would be to reduce the exterior projection (Figure 10, dotted line). If this were not acceptable for aesthetic reasons, a better thermal break system should be implemented.
- Insulate the thermal bridge. For example, use insulating sheathing for steel stud walls.
- Change the insulation system. For example, a furred-in masonry wall with all the insulation located in the furring could be changed to a masonry wall with rigid insulation on the exterior (winter-cold) surface of the CMU.
- Change the construction system. In a natatorium, for instance, structural silicone glazing without an exterior projection (Figure 11) might be a solution to avoid moisture condensation on mullions.

THERMAL MASS

The term thermal mass is commonly used to signify the ability of materials to store significant amounts of thermal energy and delay heat transfer through a building component. This delay leads to three important results. First, the slower response time tends to moderate indoor temperature fluctuations under outdoor temperature swings (Brandemuehl et al. 1990). Second, in hot or cold climates, energy consumption is reduced over that for a similar low mass building (Wilcox et al. 1985, Newell and Snyder 1990). Third, building energy demand can be moved to off-peak periods because energy storage is controlled through correct sizing of the mass and interaction with the HVAC system.

Thermal mass can be characterized by the thermal diffusivity of the building material α, which is defined as

$$\alpha = k/\rho c_p$$

where k = thermal conductivity, ρ = density, and c_p = specific heat. Table 1 lists these properties for some basic building materials.

Heat capacity (HC) is defined as the amount of heat necessary to raise the temperature of a given mass by one degree; or

$$\mathrm{HC} = mc_p$$

where m = component mass. The heat capacity of a building element is the sum of the heat capacities of each of its components.

Heat transfer through a material with high thermal diffusivity is fast, the amount of heat stored in it is relatively small, and the material responds quickly to changes in temperature. The effect of thermal mass on building behavior varies primarily with the climate at the building site and the position of the wall insulation relative to the building mass.

The ideal climate for taking advantage of thermal mass is one that has large daily temperature fluctuations. The mass can be cooled by natural ventilation at night and be allowed to "float" during the warmer day. When outdoor temperatures are at their peak, the inside of the building remains cool because the heat has not yet penetrated the mass. Often, the benefits are greater during spring and fall, when some climates closely approximate this ideal case. In heating-dominated climates, thermal mass can be used effectively to collect and store solar gains or to store heat provided by the mechanical system, allowing the heating system to operate during off-peak hours.

The effectiveness of thermal mass also increases with the allowable temperature swing in the conditioned space. If, for instance, interior temperatures can fluctuate between 68 and 78°F without the intervention of HVAC systems, as is the case in many schools, the mass has the opportunity to charge during warm hours and discharge during cooler periods. Such opportunity is limited in a laboratory that controls its temperatures within a 2°F range.

In nonresidential buildings, thermal mass is often more effective in reducing cooling loads than heating loads. In some climates, buildings with high thermal mass have better thermal performance than those with low mass, regardless of the level of insulation in the low mass building (Wilcox et al. 1985).

Table 1 Typical Densities, Thermal Diffusivities, and Specific Heats of Common Building Materials

Description	Density, lb/ft³	Thermal Diffusivity, ft²/h	Specific Heat, Btu/lb·°F
Concrete	140	0.27-0.054	0.22
Steel	484	0.038	0.12
Wood	22 to 44	0.005 to 0.006	0.40
Insulation	0.6 to 2.0	0.22 to 0.027]	0.2 to 0.38

For maximum effectiveness in moderating indoor temperatures, the thermal mass should be exposed to the interior, conditioned air and insulated from outdoor temperature variations. Studies to quantify the thermal mass effect have concentrated on three wall insulation strategies: interior, integral, and exterior. The latter is the most effective way to insulate the envelope of a thermal mass building.

Calibrated hot box tests (Van Geem 1986) have shown that in walls with identical heat capacities, those with higher R-values cause larger reductions in temperature swings. Insulation is especially significant in very hot climates with large daily temperature variations (Wilcox et al. 1985).

To simulate the complex interactions of all envelope components, computer simulations are necessary. These programs account for material properties of the components, building geometry, orientation, solar gains, internal gains, and HVAC control strategy. The calculations are usually performed on an hourly basis, using a full year of weather data.

REFERENCES

Andrews, S. 1989. *Foam core panels and building systems*. Cutter Information Corp., Arlington, MA.

Arasteh, D.K., M.S. Reilly, and M.D. Rubin. 1989. A versatile procedure for calculating heat transfer through windows. *ASHRAE Transactions* 95(2): 755-65.

Arasteh, D.K., S. Selkowitz, and J. Hartmann. 1985. Detailed thermal performance data on conventional and highly insulating window systems. Thermal Performance of the Exterior Envelopes of Buildings III, 830-845. Available from ASHRAE.

ASHRAE. 1994. Recommended practices for controlling moisture in crawl spaces. *ASHRAE Technical Data Bulletin* 10(3).

Brandemuehl, M.J., J.L. Lepore, and J.F. Kreider. 1990. Modeling and testing the interaction of conditioned air with building thermal mass. *ASHRAE Transactions* 96(2):871-75.

Canadian Home Builders. 1989. *Association builder's manual*. Ottawa, Ontario.

Carmody, J., J. Christian, and K. Labs 1991. *Builder's foundation handbook*. ORNL/CON-295. Oak Ridge National Laboratory.

Christian, J.E. 1994. *Moisture sources*. ASTM *Manual* 18. American Society for Testing and Materials, W. Conshohoken, PA.

Christian, J. and J. Kosny 1996. Thermal performance and wall ratings. *ASHRAE Journal* 38(3):56-65.

DOE. 1991. Radiant barrier fact sheet. DOE/CE-0335P. U.S. Dept. of Energy.

Dubrois, F. and A.G. Wilson. 1992. A simple method for computing energy performance for different locations and orientations. *ASHRAE Transactions* 98(1).

Dudney, C.S., L.M. Hubbard, T.G. Matthews, R.H. Scolow, A.R. Hawthorne, K.J. Gadsby, D.T. Harrje, D.L. Bohac, and D.L. Wilson. 1988. *Investigation of radon entry and effectiveness of mitigation measures in seven houses in New Jersey*. ORNL-6487 (draft). Oak Ridge National Laboratory.

Hageman, R. and M.P. Modera. 1996. Energy savings and HVAC capacity implications of a low-emissivity interior surface for roof sheathing. *ACEEE Summer Study in Energy Efficiency in Building*.

Hedlin, C.P. 1988. Heat flow through a roof insulation having moisture contents between 0 and 1 percent by volume in summer. *ASHRAE Transactions* 94( )1579–94.

IEA. 1994. *A guidebook for insulated low-slope roof systems*, Annex 19: Low-slope roof systems. International Energy Agency. Prepared by Oak Ridge National Laboratory.

James, T.B. and W.P. Goss. 1993. *Manual of heat transmission coefficients for walls, roofs, ceilings, and floors. ASHRAE Manual*.

Klems, J. 1983. Methods of estimating air infiltration through windows. *Energy and Buildings* 5:243–252.

Kosny, J. and J. Christian. 1995. Reducing the uncertainties associated with using the ASHRAE-zone method for R-value calculations of metal frame walls. *ASHRAE Transactions* 101(2):779-88.

Labs, K., J. Carmody, R. Sterling, L. Shen, J. Hwang, and D. Parker. 1988. *Building foundation design handbook*. ORNL/Sub/86-72143/1. Oak Ridge National Laboratory.

Newell, T.A. and M.E. Snyder 1990. Cooling cost minimization using building mass for thermal storage. *ASHRAE Transactions* 96(2).

Nisson, J.D. and G. Dutt 1985. *The superinsulated home book*. J. Wiley and Sons, New York.

Pedersen, C. et al. 1992. Moisture effects in low-slope roofs: Drying rates after water addition with various vapor retarders. ORNL/CON-308, Oak Ridge National Laboratory.

Selkowitz, S. 1979. Thermal performance of insulating window systems. *ASHRAE Transactions* 85(2).

Silvers, J.P., R.P. Tye, D.L. Brownell, and S.E. Smith 1985. A survey of building envelope thermal anomalies and assessment of thermal break materials for anomaly correction. ORNL/Sub/83-70376/1. Dynatec R/C Company for Oak Ridge National Laboratory.

Steven Winter Associates. 1988. *Catalog of thermal bridges in commercial and multi-family residential constructions*. Report 88-SA407/1 for Oak Ridge National Laboratory.

Sullivan, R., et al. 1992. A residential fenestration performance design tool. *ASHRAE Transactions* 98(1).

Tuluca, A., D. Lahiri, and J. Zaidi. 1997. Calculation methods and insulation techniques for steel stud walls in low-rise multifamily housing. *ASHRAE Transactions* 103(1).

Van Geem, M.G. 1986. Summary of calibrated hot box test results for twenty-one wall assemblies. *ASHRAE Transactions* 92(2).

Weidt, J.L. and J. Weidt. 1980. Field air leakage of newly installed residential windows. *LBL Report* 11111. Lawrence Berkeley Laboratory, CA.

Wilcox, B., et al. 1985. The effects of thermal mass exterior walls on heating and cooling loads in commercial buildings. *Proceedings ASHRAE Building Envelopes Conference*.

Wilkes, K.E., et al. 1991. Thermal performance of one loose-fill fiberglass attic insulation. *Insulation Materials: Testing and Applications 2*. ASTM STP 1116:275–291. American Society for Testing And Materials, W. Conshohoken, PA.

Wilson, A. 1988. High performance wall systems. *Journal of Light Construction*, April.

COMPOSITE INDEX

ASHRAE HANDBOOK SERIES

This index covers the current Handbook series published by ASHRAE. The four volumes in the series are identified as follows:

F = 1997 Fundamentals

S = 1996 Systems and Equipment

A = 1995 Applications

R = 1994 Refrigeration

The index is alphabetized in a letter-by-letter system; for example, the entry **Airborne** is before **Air cleaners**, and the entry **Heaters** is before **Heat exchangers.**

The page reference for an index entry includes the book letter and the chapter number, which may be followed by a decimal point and the beginning page or page range within the chapter. For example, the page number F31.4 means the information may be found in the 1997 Fundamentals volume, Chapter 31, beginning on page 4.

Each Handbook is revised and updated on a four-year cycle. Because technology and the interests of ASHRAE members change, the following topics are not included in the current Handbook series but may be found in the Handbooks cited.

- Degree-days
  Average Monthly and Yearly Degree Days for Cities in the United States and Canada
  1981 Fundamentals, Chapter 24, Table 4, pp. 23-28
- One-pipe steam systems
  1993 Fundamentals, Chapter 33, pp. 18-19
- Pipe coils
  1988 Equipment, Chapter 28, Table 5, p. 3
- Steam-jet refrigeration
  1983 Equipment, Chapter 13, Steam-Jet Refrigeration Equipment
- Survival shelters
  1991 Applications, Chapter 11, Environmental Control for Survival
- Three-pipe air-and-water systems
  1976 Systems, Chapter 4, pp. 8-9, 13
- Thermoelectric cooling
  1981 Fundamentals, Chapter 1, pp. 27-33